W9-BEL-812

The Cytokine Handbook
Third Edition

For Robyn, Andrew, Natalie and Emma

The Cytokine Handbook

Third Edition

Edited by
Angus W. Thomson

Departments of Surgery and
Molecular Genetics and Biochemistry,
School of Medicine,
University of Pittsburgh,
USA

Academic Press

San Diego London Boston
New York Sydney Tokyo Toronto

This book is printed on acid-free paper

First edition printed 1991
Second edition printed 1994

Academic Press
525 B Street, Suite 1900, San Diego, California 92101-4495, USA
http://www.apnet.com

Academic Press Limited
24–28 Oval Road, London NW1 7DX, UK
http://www.hbuk.co.uk/ap/

ISBN 0-12-689662-3

Library of Congress Cataloging-in-Publication Data

The cytokine handbook / edited by Angus Thomson.—3rd ed.
p. cm.
ISBN 0-12-689662-3 (alk. paper)
1. Cytokines—Handbooks, manuals, etc. I. Thomson, Angus W.
QR185.8.09509813 1998
616.07′9—dc21 97-44813
CIP

A catalogue record for this book is available from the British Library

Typeset by Paston Press Ltd, Norfolk, UK
Printed in Great Britain by WBC Book Manufacturers,
Bridgend, Mid Glamorgan, UK

98 99 00 01 02 03 WBC 9 8 7 6 5 4 3 2 1

Contents

Contributors

Mary A. Antonysamy Thomas E. Starzl Transplantation Institute and Department of Surgery, University of Pittsburgh Medical Center, 200 Lothrop Street, Pittsburgh, PA 15261, USA

Kevin B. Bacon Neuroscience Bioscience, La Jolla, CA, USA

Jacques Banchereau Baylor Institute for Immunology Research, 3535 Worth Street, Sammons Cancer Center, Suite 4800, Dallas, TX 75246, USA

Pascale Chomarat Baylor Institute for Immunology Research, 3535 Worth Street, Sammons Cancer Center, Suite 4000, Dallas, TX 75246, USA

Lisa Choy Departments of Growth and Development, and Anatomy, University of California at San Francisco, San Francisco, CA 94143-0640, USA

Daniel J. Dairaghi Molecular Medicine Research Institute, 325 East Middlefield Road, Mountain View, CA 94043, USA

Ian D. Davies Department of Surgical Oncology, University of Pittsburgh Medical Center, 200 Lothrop Street, Pittsburgh, PA 15261, USA

Rik Derynck Departments of Growth and Development, and Anatomy, University of California at San Francisco, San Francisco, CA 94143-0640, USA

Charles A. Dinarello Department of Medicine, University of Colorado Health Science Center, Division of Infectious Diseases, Box B168, 4200 East Ninth Avenue, Denver, CO 80262, USA

Gordon W. Duff Division of Molecular and Genetic Medicine, University of Sheffield, Royal Hallamshire Hospital, Glossop Road, Sheffield S10 2JF, UK

Howard D. Edington Department of Surgery, University of Pittsburgh Medical Center, 200 Lothrop Street, Pittsburgh, PA 15261, USA

Marc Feldmann Kennedy Institute of Rheumatology, 1 Aspenlea Road, London W6 8LW, UK

Sarah L. Gaffen Gladstone Institute of Virology and Immunology, PO Box 419100, San Francisco, CA 94141-9100, USA

Alison Glass Division of Molecular Immunology, La Jolla Institute for Allergy and Immunology, 10355 Science Center Drive, San Diego, CA 92121, USA

Mark A. Goldsmith Gladstone Institute of Virology and Immunology, PO Box 419100, San Francisco, CA 94141-9100, USA, and Department of Medicine, University of California at San Francisco, San Francisco, CA, USA

David R. Greaves Sir William Dunn School of Pathology, University of Oxford, Oxford, UK

Warner C. Greene Gladstone Institute of Virology and Immunology, PO Box 419100, San Francisco, CA 94141-9100, USA, and Departments of Medicine and Microbiology and Immunology, University of California at San Francisco, CA, USA

John A. Hamilton Inflammation Research Centre, Department of Medicine, University of Melbourne, Royal Melbourne Hospital, Parkville, Victoria 3050, Australia

Catherine Haworth Department of Haematology, Leicester Royal Infirmary, Leicester LE1 5WW, UK

Toshio Hirano Division of Molecular Oncology, Biomedical Research Centre, Osaka University Medical School, 2-2 Yamada-oka, Suita, Osaka 565, Japan

Sten Eirik Jacobsen Stem Cell Laboratory, Department of Internal Medicine, University Hospital of Lund, S-221 85 Lund, Sweden

Thomas C. Jones Clinical Research Consultants, Webergasse 24, CH-4058 Basel, Switzerland

Mary K. Kennedy Immunex Corporation, 51 University Street, Seattle, WA 98101-2936, USA

Michael T. Lotze Departments of Surgery and Molecular Genetics and Biochemistry, University of Pittsburgh Medical Center and University of Pittsburgh Cancer Institute, 200 Lothrop Street, Pittsburgh, PA 15261, USA

Stewart D. Lyman Immunex Corporation, 51 University Street, Seattle, WA 98101-2935, USA

Markus J. Maeurer Department of Medical Microbiology, University of Mainz, Hochhaus am Augustusplatz, D-55101 Mainz, Germany

Edward De Maeyer CNRS-UMR 177, Institut Curie, Université Paris-Sud, 91405 Orsay, France

Jaqueline De Maeyer-Guignard CNRS-UMR 177, Institut Curie, Université Paris-Sud, 91405 Orsay, France

Ravindir Nath Maini Kennedy Institute of Rheumatology, 1 Aspenlea Road, London W6 8LW, UK

Eugene Maraskovsky Immunex Corporation, 51 University Street, Seattle, WA 98101-2935, USA

Hilary J. McKenna Immunex Corporation, 51 University Street, Seattle, WA 98101-2935, USA

Ian K. McNiece University Hospital, Campus Box B-190, 4200 East Ninth Avenue, Denver, CO 80262, USA

Graham Molineux Amgen Inc., Developmental Hematology Group, 1840 Dehavilland Drive, Thousand Oaks, CA 91320-1789, USA

Kevin W. Moore Department of Molecular Biology, DNAX Research Institute of Molecular and Cellular Biology, 901 California Avenue, Palo Alto, CA 94304-1104, USA

Hiroshi Murakami Osaka Bioscience Institute, Suita, Osaka 565, Japan

Shigekazu Nagata Department of Genetics, Osaka University Medical School, 2-2 Yamada-oka, Suita, Osaka 565, Japan

Robert S. Negrin Bone Marrow Transplant Program, Department of Medicine (Hematology), Room H1353, Stanford University Hospital, Stanford, CA 94305, USA

Joost J. Oppenheim Laboratory of Molecular Immunoregulation, National Cancer Institute, Building 560, Room 21-89A, Frederick, MD 21702-1201, USA

Linda S. Park Immunex Corporation, 51 University Street, Seattle, WA 98101-2936, USA

Raymond J. Paxton Immunex Corporation, 51 University Street, Seattle, WA 98101-2936, USA

Paul Proost Rega Institute for Medical Research, Laboratory of Molecular Immunology, University of Leuven, Minderbroedersstraat 10, B-3000 Leuven, Belgium

R. Geoffrey P. Pugh-Humphreys Cell and Immunobiology Unit, Department of Zoology, University of Aberdeen, Tillydrone Avenue, Aberdeen AB9 2TN, UK

Valerie F.J. Quesniaux Preclinical Research, Novartis Pharma Ltd, S-386.1.55, CH-4058 Basel, Switzerland

Jean-Christophe Renauld Ludwig Institute for Cancer Research, Brussels Branch and Experimental Medicine Unit, Catholic University of Louvain, 74 Avenue Hippocrate, B-1200 Brussels, Belgium

Mary Ellen Rybak Schering-Plough Research Institute, Kenilworth, NJ, USA

Colin J. Sanderson TVW Telethon Institute for Child Health Research, GPO Box 855, West Perth, WA 6872, Australia

Sybil Santee Division of Molecular Immunology, La Jolla Institute for Allergy and Immunology, 10355 Science Center Drive, San Diego, CA 92121, USA

Thomas J. Schall Molecular Medicine Research Institute, 325 East Middlefield Road, Mountain View, CA 94043, USA

John W. Schrader The Biomedical Research Centre, University of British Columbia, Vancouver, British Columbia, Canada V6T 1Z3

Christopher J. Secombes Department of Zoology, University of Aberdeen, Tillydrone Avenue, Aberdeen AB9 2TN, UK

Walter J. Storkus Department of Surgery, University of Pittsburgh School of Medicine, 200 Lothrop Street, Pittsburgh, PA 15261, USA

Hideaki Tahara Departments of Surgery and Molecular Genetics and Biochemistry, University of Pittsburgh School of Medicine, 200 Lothrop Street, Pittsburgh, PA 15261, USA

Angus W. Thomson Thomas E. Starzl Transplantation Institute and Departments of Surgery and Molecular Genetics and Biochemistry, University of Pittsburgh Medical Center, 200 Lothrop Street, Pittsburgh, PA 15261, USA

Robin Thorpe Division of Immunobiology, National Institute for Biological Standards and Control, Blanche Lane, Potters Bar EN6 3QG, UK

Kevin J. Tracey Laboratory of Biomedical Science, North Shore University Hospital and the Picower Institute for Medical Research, 350 Community Drive, Manhasset, NY 11030, USA

Jo Van Damme Rega Institute for Medical Research, Laboratory of Molecular Immunology, University of Leuven, Minderbroedersstraat 10, B-3000 Leuven, Belgium

Jacques Van Snick Ludwig Institute for Cancer Research, Brussels Branch and Experimental Medicine Unit, Catholic University of Louvain, 74 Avenue Hippocrate, B-1200 Brussels, Belgium

Jan Vilček Department of Microbiology, New York University Medical Center, 550 First Avenue, New York, NY 10016-6402, USA

Jan E. de Vries Novartis Research Institute, Brunnerstrasse 59, A-1235 Vienna, Austria

Rene de Waal Malefyt Department of Immunobiology, DNAX Research Institute of Molecular and Cellular Biology, 901 California Avenue, Palo Alto, CA 94304-1104, USA

Meenu Wadhwa Division of Immunobiology, National Institute for Biological Standards and Control, Blanche Lane, Potters Bar EN6 3QG, UK

Carl F. Ware Division of Molecular Immunology, La Jolla Institute for Allergy and Immunology, 10355 Science Center Drive, San Diego, CA 92121, USA

Anja Wuyts Rega Institute for Medical Research, Laboratory of Molecular Immunology, University of Leuven, Minderbroedersstraat 10, B-3000 Leuven, Belgium

Minghuang Zhang Laboratory of Biomedical Science, North Shore University Hospital and the Picower Institute for Medical Research, 350 Community Drive, Manhasset, NY 11030, USA

Preface to the First Edition

Cytokines feature at the forefront of biomedical research. An understanding of their properties is now essential for the immunology student, researcher and teacher and for today's medical practitioner who needs to understand immunologic disease and immunological approaches to therapy. The pace with which this ever-expanding field has developed has been rapid enough to exceed the most optimistic expectations and to bewilder the most assiduous student. Cytokine research is expected to provide the key to pharmacological manipulation of the immune response and commands the attention of a massive and highly focussed biotechnology industry. The chapters in this book are a good representation of the areas to which molecular biology has been most successfully applied. Biotechnology companies provide most of the pure, well characterized cell growth regulatory and effector molecules used in academic and industrial laboratories or in clinical medicine as diagnostic tools or therapeutic agents.

Cytokines represent a sought-after symposium theme in immunology, molecular biology and molecular genetics. The cytokine literature ranges from the most basic to the applied. Unfortunately, technical advances, rapid expansion and diversification have prompted narrower specialization and reduced the ease of communication. The aim of this book is to inform and to provide detailed information and reference material on the many aspects of pure and applied cytokine science. These include the molecular characteristics of cytokines, their genes and receptors, the cellular sources and targets of cytokines, their biological activities and as best can presently be defined, their mechanisms of action. Confronted with such a vast amount of new information, up-to-date coverage is an almost unattainable goal. The scope of cytokine research could only be effectively covered in a multi-authored volume and it is indeed fortunate that each chapter is written by a leading authority(ies). Although many chapters focus on individual cytokines, it is also apparent that aspects such as cell sources, molecular structure, purification and bioassay have many features in common. A certain amount of duplication is, therefore, inevitable. The cytokine network, cytokine interactions, the roles of cytokines in disease pathogenesis and the therapeutic applications of cytokine research are dealt with in detail. In attempting to provide comprehensive coverage, a chapter on phylogeny has been included. The last chapter was commissioned to provide both perspective and a somewhat sobering view of the future.

I am indebted to the many authors around the world who have so generously devoted their knowledge, energy and time to the creation of this book. I also wish to acknowledge the support of Dr Susan King, and her staff at Academic Press in London, whose skill and energy were essential in the genesis of *The Cytokine Handbook*.

Angus W. Thomson
University of Pittsburgh

Preface to the Second Edition

In the Preface to the First Edition, it was stated that to produce a book that provided up-to-date coverage of all aspects of the cytokine field was an 'almost unattainable goal'. Since the First Edition went to press in the spring of 1991, advancement both in our knowledge and understanding of the cytokine network has been predictably rapid, fully justifying the publishers' faith in a Second Edition of *The Cytokine Handbook* within 3 years. The original chapters have been revised and updated, and, with few exceptions, this has been undertaken by the original authors. Synthesis of information in the context of the cytokine network has again been a key objective. Every effort has been made to maintain currency and it is in promptly fulfilling this goal that the contributing authors deserve great credit. 'New' cytokines have, of course, emerged and continue to 'appear'. Thus, this new edition features individual chapters on IL-9, IL-10 and IL-12, with a 'stop-press' overview of IL-13 and even newer molecules that are candidates for designation as interleukins IL-14 and IL-15.

What constitutes a new interleukin? The assignation of a new designation depends on several clearly defined criteria, recently established by a sub-committee of the nomenclature committee of the International Union of Immunological Societies (Paul *et al.*, 1992). The criteria laid down include molecular cloning and expression, a unique nucleotide and inferred amino acid sequence, and the availability of a neutralizing monoclonal antibody. Furthermore, the granting of a new interleukin designation requires that the candidate molecule be a natural product of cells of the immune system (defined loosely as lymphocytes, monocytes and other leukocytes). The new interleukin must also mediate a potentially important function in immune responses and exhibit an additional function(s) so that a simple, functional name might not be adequate. Finally, these characteristic features should have been described in a peer-reviewed publication.

This new edition incorporates separate chapters on G-CSF, M-CSF and GM-CSF, whereas in the first edition, only one chapter – on 'colony stimulating factors' – covered the properties of these molecules. A chapter is also now afforded to TGF-β, which exhibits both inhibitory and stimulatory effects on a variety of cell types and is a potent immunosuppressant. Another significant and substantial new chapter concerns 'chemokines'. IL-8 (the 'highest' interleukin designation afforded a separate chapter in the first edition) was the first member of the chemokine family to be identified. More recently, other low molecular weight chemotactic polypeptides have been discovered that play a key role in cell activation and chemotaxis during the inflammatory response. These include RANTES (*R*egulated upon *A*ctivation, *N*ormal *T* *E*xpressed and *S*ecreted!), macrophage chemotactic and activating factor (MCAF), macrophage inflammatory protein (MIP)-1α and MIP-1β. All share a conserved, four cysteine motif and can be further subclassified on the spacing of the conserved cysteine residues.

Evaluation of the clinical potential of cytokines is one of the most exciting challenges of contemporary medicine. In addition, and in contradistinction to cytokine or cytokine gene therapy, the emergence of a new class of therapeutic agents, comprising soluble cytokine receptors (e.g. soluble (s) IL-1R, sTNFR, sIFN-γR), receptor antagonists (e.g. IL-1ra) and counter-regulatory cytokines, represents one of the most important developments in the cytokine field in recent years. Successes in the sequencing, cloning and expression of cytokine receptors has facilitated production and evaluation of their therapeutic utility in several acute and chronic cytokine-mediated diseases. This exciting and challenging development is one of the themes covered under the several chapters devoted to therapeutic aspects of cytokine biology.

The last few years have seen the advent of new approaches to interrogating the roles of cytokines *in vivo*. Thus cytokine gene 'knockout' mice and mice transgenic for cytokine gene reporter constructs are likely to provide new knowledge about the *in vivo* role(s) of cytokines, especially in experimental autoimmune disorders, cancer or infectious diseases that may be difficult to mimic satisfactorily *in vivo*. The molecular genetics of cytokines (reviewed in chapter 2) is yet another new and exciting aspect of this key field of contemporary molecular biology and medicine that has provided the impetus for the second edition of *The Cytokine Handbook*. Those who acquire knowledge by reading this book, or who are stimulated by the implications of the recent developments described herein, owe thanks to the many experts around the world who have so generously given of their valuable time, energy and expertise. The cordial and enthusiastic support of these scientists and clinicians has been an impelling influence that would be difficult to overrate. I am indebted to my colleague Dr Mike Lotze for constructive suggestions, Ms Shelly Conklin for valuable secretarial help and to Dr Tessa Picknett and her colleagues at Academic Press in London for their resolute support and guidance in ensuring that the notion of a second edition became a tangible reality.

Angus W. Thomson

Reference

Paul, W.E., Kishimoto, T., Melchers, F., Metcalf, D., Mosmann, T., Oppenheim, J., Ruddle, N. and Van Snick, J. (1992). *Clin. Exp. Immunol.* **88**, 367.

Preface to the Third Edition

As the Third Edition goes to press, the stream of cytokine discovery flows unabated. We are confronted by description after description of novel cytokine-controlled systems of cell differentiation and proliferation and of the role of cytokines in immune regulation. In order to keep pace with these developments, this new edition of *The Cytokine Handbook* has been completely revised and updated. The contributors are largely the same body of devoted international authorities, with several notable new additions. Amongst the newcomers to the extensive panoply of cytokines covered in this edition are the recently designated interleukins (IL)-16 (formerly lymphocyte chemotactic factor), IL-17 (identified originally as cytotoxic T lymphocyte antigen-8) and IL-18 (interferon-γ inducing factor). The daunting expansion of new information is perhaps better illustrated by the 'explosion' of chemokines. About sixty distinct human chemokine gene products have been identified to date. Regarded historically as regulators of leukocyte trafficking, chemokines have recently made a major impact in the area of infectious disease, including dramatic new understanding that chemokine receptors are at the center of HIV pathogenesis. Just two examples of chemokines that have recently come to prominence are eotaxin (the CCR-3 receptor-specific, eosinophil-selective chemokine) and the structurally unique fractalkine, with intrinsic adhesion and chemotactic activities. Two chapters are now devoted to the vastly expanding area of chemokines and their receptors.

The Third Edition sees individual chapters afforded to IL-1 through IL-15, with the exception of IL-14, reflecting the current uncertainty about the identity of the IL-14 molecule. The TNF-related cytokine-receptor superfamily of secreted and membrane-bound ligands that includes TNF/lymphotoxin α/β, CD40, Fas and 4-1BB systems, now appears to have important functional roles in the immune response. Each member is paired with a specific cell surface receptor(s) that, together, form a corresponding family of receptors. Recent studies have revealed important and unique roles of the TNF-related ligands and receptors that are covered in an additional new chapter.

In addition to coverage of the colony-stimulating factors G-, M- and GM-CSF, two new chapters in this edition are afforded to individual hematopoietic cytokines – stem cell factor (SCF) (c-kit ligand) and flt-3 ligand – that have many biological functions in common. SCF is a potent costimulatory or synergistic factor in cytokine cocktails for manipulation of hematopoietic cells. It appears to be of value in the combination of *ex vivo* 'expansion' technologies with gene transfer methods for the correction of genetic disease or the support of multiple chemotherapy cancer patients. The recently cloned cytokine flt-3 ligand has been shown to dramatically increase the numbers of functional dendritic cells and also to increase natural killer cells *in vivo*. Moreover it exerts anti-tumor activity, raising expectation of a future role of this molecule as a human immunotherapeutic agent.

Therapeutic applications of cytokines are now covered in detail in several chapters, including a new chapter on cytokine gene therapy. The first therapeutic cytokine gene transfer study began several years ago in patients with advanced malignancy; now numerous trials have been approved for the transfer of cytokine genes to tumor cells, tumor-infiltrating lymphocytes, blood lymphocytes, or fibroblasts. A few studies involve the use of cytokine genes for the treatment of non-malignant conditions, such as rheumatoid arthritis. A new chapter in this edition reviews progress in cytokine gene therapy.

As in the two previous editions, valuable up-to-date practical information is included in the extensive account of cytokine assays. The phylogeny of cytokines has become such a growth area that the information can barely be contained in a single chapter.

I am again immensely indebted to the many experts around the world who have made this new edition possible. On the home front, my colleague Dr Mike Lotze has been a constant source of creative suggestions and lively discussion. My thanks are due once again to Dr Tessa Picknett, senior editor, and also to Mr Duncan Fatz and Ms Emma White at Academic Press in London, and to Ms Shelly Conklin for invaluable secretarial assistance.

Angus W. Thomson

Foreword

Joost J. Oppenheim

Laboratory of Molecular Immunoregulation
National Cancer Institute
Frederick, Maryland, USA

What Are Cytokines?

As so aptly discussed by Vilček in the introductory chapter of *The Cytokine Handbook*, cytokines are regulatory peptides that can be produced by virtually every nucleated cell type in the body. Cytokines have pleiotropic regulatory effects on hematopoietic and many other cell types that participate in host defense and repair processes. Cytokines therefore include lymphocyte-derived factors known as 'lymphokines', monocyte-derived factors called 'monokines', hematopoietic 'colony stimulating factors', connective tissue 'growth factors' and chemotactic chemokines.

Why Do Cytokines Exist?

The evolution of large multicellular organisms required the development of intercellular messengers such as hormones, neuropeptides and cytokines to permit marshaling of coordinated cellular responses. It has been proposed that the structural homology between adhesion proteins that mediate cell contact-dependent interactions and cytokine ligands suggests that soluble cytokines evolved from cell-associated signals (Grumet *et al*., 1991). In fact, a number of cytokines and their receptors exist and are active in both soluble (shed) and cell-associated form.

How Are Cytokines Different from Hormones?

Endocrine hormones, which are generally produced by specialized glands, are present in the circulation and serve to maintain homeostasis. In contrast, most cytokines usually act over short distances as autocrine or paracrine intercellular signals in local tissues, and (with the exception of macrophage colony stimulating factor (M-CSF), stem cell factor, erythropoietin and transforming growth factor-β (TGF-β) only occasionally spill over into the circulation and initiate systemic reactions. Except for the above, cytokines generally are not produced constitutively, but are generated in emergencies to contend with challenges to the integrity of the host. The functions of cytokines are distinct from those of hormones, as they serve to maintain homeostasis by regulating the immune system, through damage control and by promotion of reparative processes.

How Did Lymphokines Come to be Discovered?

The possibility that cell-derived factors mediate biologic activities was first suggested by experiments of Rich and Lewis (1932). They observed that migration of neutrophils and macrophages in cultures of tuberculin-sensitized tissues was inhibited by antigen.

Waksman and Matoltsy (1958) observed that macrophages in monolayer cultures were actually stimulated rather than damaged by exposure to tuberculin antigens. George and Vaughan (1962) improved the technique of evaluating migration of mononuclear cells using capillary tubes. The study of 'lymphokines' was initiated concurrently by David *et al.* (1964) and Bloom and Bennet (1965), who used this technique to show that antigens could stimulate sensitized lymphocytes in cultures to produce macrophage migration inhibitory factors (MIFs). At about the same time, supernatants of mixed leukocyte cultures were found by Kasakura and Lowenstein (1965) to be 'blastogenic' for lymphocytes. This 'blastogenic factor' (BF) was subsequently called 'lymphocyte mitogenic factor' (LMF). This was followed by the discovery of a variety of lymphocyte-derived biologic activities in culture supernatants. Ruddle and Waksman as well as Kolb and Granger described a cytotoxic lymphocyte-derived mediator called 'lymphotoxin' (LT) in 1968.

In vitro monocyte migration in response to supernatants of antigen-activated lymphocytes was attributed to a lymphocyte-derived chemotactic factor (LCF) (Ward *et al.*, 1969) and was correlated with the recruitment of mononuclear cells to *in vivo* inflammatory sites. MIF and the macrophage aggregation factor (MAgF) (Lolekha *et al.*, 1970) presumably served to retain cells at inflammatory sites. The necrotic centers of some granulomas could be attributed to the cytodestructive activity of LT (Kolb and Granger, 1968; Ruddle and Waksman, 1968), and the presence of lymphoblasts and frequent mitotic figures was mediated by LMF (Kasakura and Lowenstein, 1965). Moreover, identification of macrophage activating factor (MAF) (Nathan *et al.*, 1971) and lymphocyte-derived immune interferon (IFN-γ) (Green *et al.*, 1969) provided a biologic basis for acquired resistance to infectious organisms. Consequently, the various biologic activities secreted by cultured antigen-stimulated lymphocytes provided *in vitro* models for the pathogenesis of *in vivo* delayed hypersensitivity reactions. These biochemically undefined, lymphocyte-derived activities were termed 'lymphokines' in 1969 by Dumonde *et al.* Discovery of the lymphokines revolutionized the conceptual basis of cell-mediated immunity and these biologic activities were considered '*in vitro* correlates' of cell-mediated immunity.

What Led to the Recognition that Lymphokines Are Members of the Family of Cytokines?

Gery and coworkers in 1971–1974 showed that the lymphocyte activating factor (LAF) was produced by adherent monocytes and macrophages (Gershon and Kondo, 1971; Gery *et al.*, 1971; Gershon *et al.*, 1974). This was the first demonstration of the existence of non-lymphocyte-derived 'monokines'. Based on this information and on his own observations that some replicating non-lymphoid cell lines, as well as virally infected non-lymphoid cells, could also produce lymphokine-like MIF and chemotactic factors, Cohen *et al.*, in 1974, proposed that all these mediators including lymphokines should be called 'cytokines'. This resulted in the conceptual transformation of lymphokines from subjects of interest to a minor subset of immunologists to cytokines that function as bidirectional intercellular signals between somatic and myeloid as well as lymphoid cells, with potential impact in a great number of biologic disciplines.

What Developments Have Galvanized the Study of Cytokines?

The development of tissue culture techniques enabled immunologists in the 1960s to detect the presence of factors in tissue culture supernatants and to perform *in vitro* studies of the mobility, proliferation, differentiation and functional capabilities of lymphocytes and other leukocytes. Cytokines are very potent and active at picomolar to nanomolar concentrations. Thus, they are active at only trace levels. This makes it particularly difficult to isolate and identify the biochemical structure of these peptides. These factors, which were originally disparagingly termed 'lymphodrek', could be purified only with the development of improved chromatography and microsequencing techniques in the late 1970s. The fortuitous development of molecular biology and monoclonal antibody technologies accelerated the identification of cytokines in the 1980s and has made abundant quantities of recombinant cytokines available during the past decade. This has resulted in an information explosion that is reflected by *The Cytokine Handbook*.

What Was the Origin of the Interleukin Terminology?

By 1978, the confusing plethora of eponyms in existence for monocyte- and lymphocyte-derived activities motivated investigators at the Second International Lymphokine Workshop held near Interlaken, Switzerland, to propose more inclusive 'neutral' terms for these biologic activities. The researchers recognized that the numerous monokine and lymphokine activities they were detecting with a variety of bioassays actually had numerous properties in common. This gave the erroneous impression that these cytokine activities, each with its own name, could all be attributable to one or two molecular entities. Paetkau proposed that LAF/BAF/MCF be renamed interleukin-1 (IL-1), while LMF/BF/T-cell growth factor (TCGF) should be called IL-2 (Mizel and Farrar, 1979). The interleukin terminology symbolized the broader roles of these cytokines, and with the progressive increase in the number of interleukins, now up to 18, has led to an explosive increase in the interest of investigators from a variety of disciplines in these mediators of inflammation and immunity.

Why is the Cytokine Nomenclature so Chaotic?

Of the more than 150 cytokines that have been cloned to date, most retain their initial names, which usually denote only the functional activity that led to their discovery. Chaotic as this may be, such names are easier to recall than an interminable number of interleukins. A minority of investigators request an interleukin designation to focus greater attention on their discovery. To qualify as an interleukin, the cytokine must be documented to have a unique amino acid sequence and functional activity involving leukocytes. The evidence is evaluated by the Nomenclature and Standardization Committee of the International Cytokine Society and the Union of Immunological Societies, which make a recommendation to the World Health Organization. Despite this straightforward process, committees have had to resolve conflicts concerning simultaneous claims to the IL-4 number and even a fraudulent claim (e.g. IL-4α). More recently an error in the nucleotide sequence of the precursor region of IL-16 led to a re-evaluation of the IL-16 designation. Fortunately, the error did not involve the region of the gene coding for the mature protein and the translated amino acid sequence proved to be correct and exhibited the proposed functions of IL-16. In contrast, the nucleotide and consequent amino acid sequences of IL-14 have proven to be incorrect. Unfortunately, the correct protein sequence of IL-14, if any, remains unknown. It is therefore unclear whether IL-14 exists; hence its omission from this new edition of *The Cytokine*

Handbook. Perhaps such errors can be prevented by requiring that novel cytokine sequences be confirmed independently to be eligible for an interleukin designation.

Is There Any Order to the Cytokine Chaos?

Despite the identification of many structurally distinct cytokines, they can be organized into groups that exhibit functional similarities based on shared receptor utilization. This holds for IL-1α and IL-1β as well as tumor necrosis factor-α (TNF-α) and LT, which share receptors. Receptor chains are shared by a number of the cytokine groups. The IL-2γ receptor chain is shared by interleukins 2, 4, 7, 9, 13 and 15. The gp130 chain of the IL-6 receptor is shared by IL-6, IL-11, leukemia inhibitory factor (LIF), oncostatin-M, ciliary neurotrophic factor (CNTF) and cardiotrophin-1. Members of the TNF family of ligands share receptors with homology to the TNF receptors. This includes TNF-α, LT-α, LT-β complex, Fas ligand, CD70, CD40 ligand, CD30 ligand and nerve growth factor. A receptor β chain is shared by IL-3, IL-5 and granulocyte–macrophage colony stimulating factor (GM-CSF). Homologous seven-transmembrane receptors are used by IL-8 and the chemokine family. The receptors for IFN-α, β, ω, γ and IL-10 also show homology, as do those for the TGF-β family of cytokines. Since IL-12 and IL-18 have overlapping activities, their receptors also may be related. These observations permit a very necessary organization of cytokines into families.

The advent of additional information concerning the relationship and location of cytokine genes, shared signal transduction pathways and the tertiary structure of cytokines may enable us further to classify newly discovered cytokines. Although we have always presumed that the most important cytokines and receptors have been discovered, we are repeatedly surprised by the identification, not only of novel cytokines and receptors, but of whole families. Undoubtedly many more cytokines will be identified with the progressive mapping of the genome.

Why Are Cytokines Important?

Recombinant cytokines provide useful laboratory probes for studying the cell biology of immunity and inflammation. Cytokines are the major orchestrators of host defense processes and, as such, are involved in responses to exogenous and endogenous insults, repair and restoration of homeostasis. Microbial pathogens have operated on these principles far longer than immunologists and have been shown to produce variants of proinflammatory cytokine receptors and chemokine antagonists that subvert and suppress the host immune and inflammatory defenses. Deletion of these products reduces the pathogenicity of these viruses. In addition to their role in host defense, cytokines appear to play a major role in development and some of them may account for as yet unidentified embryonic inductive factors. The study of cytokines is also elucidating the mechanisms underlying pathophysiologic processes.

Cytokines mediate not only host responses to invading organisms, tumors and trauma, but also maintain our capacity for daily survival in our germ-laden environment. In fact, the development of more sensitive methods of detection is revealing the presence of local levels of cytokines associated with a variety of binding proteins in the serum. This probably reflects the production of cytokines in response to the many non-pathogenic stimulants present in our conventional environment. Detection of cytokines in disease states may provide useful diagnostic tools. The therapeutic administration of pharmacologic doses of cytokines, or at times gene therapy, is being used for a wide variety of infectious diseases and in immunocompromised patients with acquired immune deficiency syndrome, autoimmune diseases and neoplasia. This has led to the development of numerous biotechnology firms which are also evaluating antagonists of cytokines for anti-inflammatory effects.

Why a Cytokine Handbook?

Studies of cytokines have drawn scientists from a multiplicity of fields including immunologists, hematologists, molecular biologists, neurobiologists, cell biologists, biochemists, physiologists and others. Consequently, the burgeoning field of cytokine research is unique and interdisciplinary. The chapters in this handbook cover the structure and functions of cytokines, their genes, receptors, mechanisms of signal transduction and clinical applications. This third edition of *The Cytokine Handbook* is required, at this relatively early date, to keep up with rapid developments in these dynamic disciplines. All the chapters, and even this foreword, have been updated and a number of new chapters by internationally renowned experts has been added to cover interleukins 13, 15, 16–18, the hematopoietic stem cell stimulating factors (SCFs), and fms-like tyrosine kinase (Flt) Flt3 and cytokine gene therapy. This handbook provides the opportunity to keep up with the rapidly evolving studies of cytokines.

Acknowledgements

I am grateful for the critical input of Drs R. Neta, M. Grimm and S. Durum, and the secretarial assistance of Ms Cheryl Fogle.

References

Bloom, B.R. and Bennet, B. (1965). Science **153**, 80–82.
Cohen, S., Bigazzi, P.E. and Yoshida, T. (1974). *Cell. Immunol.* **12**, 150–159.
David, J.R., Al-Askari, S., Lawrence, H.S. and Thomas, L. (1964). *J. Immunol.* **93**, 264–273.
Dumonde, D.C., Wolstencroft, R.A., Panayi, G.S., Matthew, M., Morley, J. and Howson, W.T. (1969). *Nature* **224**, 38–42.
George, M. and Vaughan, J.H. (1962). *Proc. Soc. Exp. Biol. Med.* **111**, 514–521.
Gershon, R.K. and Kondo, K. (1971). *Immunology* **21**, 903–914.
Gershon, R.K., Gery, I. and Waksman, B.H. (1974). *J. Immunol.* **112**, 215–221.
Gery, I., Gershon, R.K. and Waksman, B.H. (1971). *J. Immunol.* **107**, 1778–1780.
Green, J.A., Cooperland, S.R. and Kibnick, S. (1969). *Science* **164**, 1415–1417.
Grumet, M., Mauro, V., Burgoon, M.P., Edelman, G.M. and Cunningham, B.A. (1991). *J. Cell Biol.* **113**, 1399–1412.
Kasakura, S. and Lowenstein, L. (1965). *Nature* **205**, 794–798.
Kolb, W.P. and Granger, G.A. (1968). *Proc. Natl Acad. Sci. U.S.A.* **61**, 1250–1255.
Lolekha, S., Dray, S. and Gotoff, S.P. (1970). *J. Immunol.* **104**, 296–304.
Mizel, S.B. and Farrar, J.F. (1979). *Cell Immunol.* **48**, 433–436.
Nathan, C.F., Karnovsky, M.L. and David, J.R. (1971). *J. Exp. Med.* **133**, 1356–1376.
Rich, A.R. and Lewis, M.R. (1932). *Bull. Johns Hopkins Hosp.* **50**, 115–131.
Ruddle, N.H. and Waksman, B.H. (1968). *J. Exp. Med.* **128**, 1267–1279.
Waksman, B.H. and Matoltsy, M. (1958). *J. Immunol.* **81**, 220–234.
Ward, P.A., Remold, H.G. and David, J.R. (1969). *Science* **163**, 1079–1081.

Chapter 1

The Cytokines: An Overview

Jan Vilček

Department of Microbiology, New York University Medical Center, New York, NY, USA

GENERAL FEATURES OF CYTOKINES

Origins of Cytokine Research

The field of cytokine research as it exists today has evolved from four originally independent sources. The first and probably most significant source is immunology and, more specifically, the field of lymphokine research. The origins of lymphokine research can be traced to the mid-1960s when it was demonstrated that lymphocyte-derived secreted protein mediators regulate the growth and function of a variety of white blood cells. It soon became apparent that monocytes too are the source of important proteins (monokines) that can modulate leukocyte function.

The second source of cytokine research derives from the study of the interferons. Originally described in the 1950s as selective antiviral agents, interferons gradually became recognized as proteins exerting a broad range of actions on cell growth and differentiation, both within and without the immune system. As a result, the dividing line between lymphokines–monokines and the interferons began to dwindle and today it is clear that interferons in fact are cytokines.

The third source of cytokine research is the field of hematopoietic growth factors, or colony stimulating factors (CSFs). In addition to promoting the growth and differentiation of hematopoietic stem cells, CSFs have been shown to regulate some functions of fully differentiated hematopoietic cells, thus blurring the dividing line between these agents and lymphokines–monokines.

The fourth source of cytokine research derives from the study of growth factors acting on non-hematopoietic cells. One might be reluctant to count 'classic' growth factors, such as platelet-derived growth factor (PDGF), epidermal growth factor (EGF), fibroblast growth factor (FGF) or nerve growth factor (NGF), among the cytokines. Nevertheless, it is clear that, in addition to promoting cell growth, many of these agents exert other effects, which might be referred to as 'cytokine-like' actions. Moreover, at least one of the offspring of growth factor research, i.e. transforming growth factor-β (TGF-β), is now considered a true cytokine.

The Cytokine Handbook, 3rd ed.
ISBN 0-12-689662-3

A Brief Outline of the History of Cytokine Research

The beginnings of lymphokine research are usually traced to the demonstration that migration of normal macrophages is inhibited by material released from sensitized lymphocytes upon exposure to antigen (Bloom and Bennett, 1966; David, 1966). The putative factor responsible for this action was termed macrophage migration inhibitory factor (MIF). The description of MIF activity was followed by the discovery of 'lymphotoxin' activity (i.e. selective cytotoxicity for some target cells) in the supernatant of activated lymphocyte cultures (Ruddle and Waksman, 1968; Williams and Granger, 1968). Dumonde *et al.* (1969) coined the term 'lymphokine' to designate 'cell-free soluble factors (responsible for cell-mediated immunologic reactions), which are generated during interaction of sensitized lymphocytes with specific antigen'.

Among the lymphokines a central role in the regulation of T-cell growth and function is played by interleukin-2 (IL-2). It was known since the early 1970s that lymphocytes can produce one or more factor(s) mitogenic for other lymphocytes (reviewed by Oppenheim *et al.*, 1979). Morgan *et al.* (1976) reported that supernatants of mitogen-activated human mononuclear cells could support the continuous growth of human bone marrow-derived T cells. The responsible mitogenic factor is IL-2, then designated T-cell growth factor (TCGF) and also known under a variety of other names that by now are largely forgotten (Aarden *et al.*, 1979).

The first monocyte–macrophage-derived cytokine described was tumor necrosis factor (TNF), originally identified as a cytotoxic protein present in the serum of animals sensitized with Bacillus Calmette–Guérin and challenged with lipopolysaccharide (LPS) (Carswell *et al.*, 1975). In addition to its direct cytotoxicity for some tumor cells *in vitro*, TNF was identified as the mediator of LPS-induced hemorrhagic necrosis of Meth A sarcoma in mice.

Among the first monocyte-derived cytokines described is also lymphocyte activating factor (LAF), now known as interleukin-1 (IL-1). LAF activity, defined as a mitogenic signal for thymocytes, was originally detected in supernatants of adherent cells isolated from human peripheral blood (Gery *et al.*, 1971). Other investigators described activities that are now known to be mediated by IL-1 under a variety of other names, for example mitogenic protein, leukocytic pyrogen, endogenous pyrogen, B cell-activating factor, leukocyte endogenous mediator (reviewed by Aarden *et al.*, 1979; Oppenheim *et al.*, 1979).

While the early studies of lymphokines and monokines were largely the domain of immunologists who sought a better understanding of delayed-type hypersensitivity and other cell-mediated immune reactions, interferons were the brainchildren of virologists. Interferon was first described by Isaacs and Lindenmann (1957) as a factor produced by a variety of virus-infected cells capable of inducing cellular resistance to infection with homologous or heterologous viruses. That interferons would affect immune reactions was initially not even suspected. However, several years later, Wheelock (1965) described a functionally related virus inhibitory protein (today known as interferon-γ (IFN-γ)) produced by mitogen-activated T lymphocytes. It is now known that T-cell and natural killer (NK) cell-derived IFN-γ is structurally completely distinct from the large family of IFN-α/β proteins that are produced by a variety of cell types, including monocytes, NK cells and B cells in addition to sundry non-hematopoietic cells (reviewed by De Maeyer and De Maeyer-Guignard, 1988; Vilček and Sen, 1996).

CSFs are proteins whose major function is to support the proliferation and differentiation of hematopoietic cells. Their name reflects the early observation that CSFs promote the formation of granulocyte or monocyte colonies in semisolid medium (Bradley and Metcalf, 1966; Sachs, 1987). Years of effort by many groups of investigators have led to the isolation and characterization of several distinct proteins, to be described in greater detail elsewhere in this volume (see chapters 23–26).

Many proteins that can stimulate the growth of non-hematopoietic cells have been identified; best known among these are EGF, PDGF, FGF and vascular endothelial growth factor (VEGF), to name but a few. Although these and other growth factors are generally not included among the cytokines, some cytokine-like actions of classic growth factors on immunocytes and other cells have been described. de Larco and Todaro (1978) described a growth factor, originally termed sarcoma growth factor, whose most interesting property was that it promoted the growth of normal rat fibroblasts in soft agar. Since then, two families of 'transforming growth factors' have been identified—TGF-α and TGF-β; these are distinct peptides with very different spectra of biological activity. TGF-α is closely related to EGF whereas the family of TGF-β proteins plays an important role not only in cell growth control and neoplasia but also in inflammation and immunoregulation. Among the important actions of TGF-β proteins are the recruitment and activation of mononuclear cells, promotion of wound healing, fibrosis and angiogenesis, and a potent immunosuppressive action on numerous functions of T lymphocytes (Roberts and Sporn, 1990). Based on what is known about the actions of TGF-β proteins today, these polypeptides undoubtedly qualify for inclusion among the cytokines.

Cytokine Nomenclature

Inasmuch as the cytokine field evolved from several separate sources, a unifying concept of what cytokines are has been slow to emerge. The term 'lymphokine', which originally denoted the product of sensitized lymphocytes exposed to specific antigen (Dumonde *et al.*, 1969), has been often used less discriminately for secreted proteins from a variety of cell sources, affecting the growth or functions of many types of cells. To dispel the wrong notion that such proteins could be produced by lymphocytes alone, Cohen *et al.* (1974) proposed the term 'cytokines'. After a long-standing reluctance, 'cytokine' has become the generally accepted name for this group of proteins. To designate individual cytokines, a group of participants at the Second International Lymphokine Workshop held in 1979 proposed the term 'interleukin' in order to develop 'a system of nomenclature . . . based on (the proteins') ability to act as communication signals between different populations of leukocytes' (Aarden *et al.*, 1979). As a first step, the group introduced the names 'IL-1' and 'IL-2' for two important cytokines, which until then had been described under a variety of different names. As of this writing, the interleukin series has reached 18 (Ushio *et al.*, 1996). Although the name 'interleukin' implies that these agents function as communication signals among leukocytes, Aarden *et al.* (1979) suggested that the term should not be reserved for factors that can act only on leukocytes. Indeed, a number of the proteins that have been labeled as interleukins not only are produced by a variety of non-hematopoietic cells but also affect the functions of many diverse somatic cells (e.g. IL-1 or IL-6). Whereas many cytokines are now termed interleukins, others remain to be known by their older names (e.g. IFN-α/β,

IFN-γ, TNF, lymphotoxins, TGF-β, leukocyte inhibitory factor (LIF), most CSFs, chemokines and many others). Although these older names are easier to remember, they suggest only one (i.e. the earliest recognized) function of these pleiotropic agents.

In summary, the state of cytokine nomenclature is less than ideal. A look at the history helps to understand how the present situation has arisen, but this does not make it any easier, especially for outsiders, to remember the features and designations of individual cytokines. With new cytokines being characterized (and named) at a rapid rate, no relief is in sight.

Cytokines, Hormones and Growth Factors

Having outlined their history and nomenclature, it is appropriate to try to define what cytokines are. The following general definition is proposed: 'Cytokines are regulatory proteins secreted by white blood cells and a variety of other cells in the body; the pleiotropic actions of cytokines include numerous effects on cells of the immune system and modulation of inflammatory responses'. As no short definition can encompass all essential properties, cytokines are best defined by a set of characteristic features, as listed in Table 1.

In reviewing the characteristic features of cytokines outlined in Table 1, it is evident that many of these properties are shared by two other groups of protein mediators, namely growth factors and hormones. The relationship between cytokines and growth factors was mentioned earlier in this chapter. Although the dividing line is rather tenuous, one difference is that the production of growth factors (e.g. PDGF, EGF or TGF-α) tends to be constitutive and not as tightly regulated as that of cytokines. Another difference is that, unlike cytokine actions, the major actions of growth factors are targeted at non-hematopoietic cells.

It is also not easy to distinguish clearly between cytokines and classic polypeptide hormones (Table 2). One of the major distinguishing features is that classic hormones are produced by specialized cells; for example, insulin is produced by β cells of the pancreas, growth hormone by the anterior pituitary, and parathormone by the

Table 1. Characteristic features of cytokines.

Most cytokines are simple polypeptides or glycoproteins with a molecular weight of 30 kDa or less (but many cytokines form higher molecular weight oligomers and one cytokine (IL-12) is a heterodimer).
Constitutive production of cytokines is usually low or absent; production is regulated by various inducing stimuli at the level of transcription or translation.
Cytokine production is transient and the action radius is usually short (typical action is autocrine or paracrine, not endocrine).
Cytokines produce their actions by binding to specific high-affinity cell surface receptors (K_d in the range 10^{-9} to 10^{-12} M).
Most cytokine actions can be attributed to an altered pattern of gene expression in the target cells. Phenotypically, cytokine actions lead to an increase (or decrease) in the rate of cell proliferation, a change in cell differentiation state and/or a change in the expression of some differentiated functions.
Although the range of actions displayed by individual cytokines can be broad and diverse, at least some action(s) of each cytokine is (are) targeted at hematopoietic cells.

Table 2. Distinguishing features between polypeptide hormones and cytokines.

Hormones		Cytokines	
Characteristic features	Exceptions	Characteristic features	Exceptions
Secreted by one type of specialized cells		Made by more than one type of cell	IL-2, IL-3, IL-4, IL-5 and LT are made only by lymphoid cells
Each hormone is unique in its actions		Structurally dissimilar cytokines have an overlapping spectrum of actions ('redundancy')	
Restricted target cell specificity and a limited spectrum of actions	Insulin	Multiple target cells and multiple actions ('pleiotropy')	
Act at a distant site (endocrine mode of action)		Usually have short action radius (autocrine or paracrine mode of action)	Many (e.g. TNF, IL-1 or IL-6 in septic shock)

parathyroid. In contrast, cytokines tend to be produced by less specialized cells and, more often than not, several unrelated cell types can produce the same cytokine (e.g. IL-1 is produced by monocytes–macrophages, mesangial cells, NK cells, B cells, T cells, neutrophils, endothelial cells, smooth muscle cells, fibroblasts, astrocytes and microglial cells). However, there are exceptions; for example, IL-2, IL-3, IL-4, IL-5 and lymphotoxin are produced essentially only by lymphoid cells, especially T cells. Perhaps the most characteristic features of cytokines, those that distinguish them from hormones, are the redundancy and pleiotropy of cytokine actions, i.e. the fact that structurally dissimilar cytokines (e.g. TNF and IL-1) can be remarkably similar in their actions (Le and Vilček, 1987), and that individual cytokines tend to exert a multitude of actions on different cells and tissues.

Despite some differences, it is apparent that cytokines, growth factors and polypeptide hormones all function as extracellular signaling molecules featuring fundamentally similar mechanisms of action. This conclusion is supported by the finding that receptors for several cytokines and hormones (i.e. IL-2, IL-3, IL-4, IL-5, IL-6, IL-7, granulocyte–macrophage colony stimulating factor (GM-CSF), G-CSF, erythropoietin, prolactin and growth hormone) show several common structural features (Bazan, 1989; D'Andrea *et al.*, 1989; Gearing *et al.*, 1989; Kishimoto *et al.*, 1994; Ihle, 1995; Taga and Kishimoto, 1995; Hirano, 1997). Earlier, Roberts *et al.* (1988) described regions of structural homology in the receptors for PDGF and M-CSF. (Unlike other cytokine receptors, the M-CSF receptor comprises a functional intrinsic tyrosine kinase.) In addition, similar molecular pathways transmit signals from the growth factor, polypeptide hormone or cytokine receptors to the nucleus, and several components in the signal transduction pathways are shared by cytokines, growth factors and polypeptide hormones.

CYTOKINE RECEPTORS AND SIGNAL TRANSDUCTION

Structural Features of Cytokines

Structural analysis has made it possible to group many cytokines within 'families' (Table 3). Some of these families include proteins whose primary sequences show a high degree of homology to one another; for example, all members of the IFN-α/β family (further subdivided into subfamilies IFN-α, IFN-β, IFN-ω and IFN-τ) show at least 30% homology in their amino acid sequences (De Maeyer and De Maeyer-Guignard, 1988; Vilček and Sen, 1996). In other instances, the structural relationship is more distant and is based mainly on a common proposed tertiary structure and spatial organization. An example of the latter type of structural relationship is the IL-2/IL-4 family (Bazan, 1992; Boulay and Paul, 1992). Members of this family contain four α-helical regions in a

Table 3. Structural features of some cytokines permit their grouping into families.

Family	Representative members
Interleukin-2/interleukin-4	IL-2 IL-4 IL-5 GM-CSF
Interleukin-6/interleukin-12	IL-6 IL-12*
Interferons-α/β	IFN-α (many subtypes) IFN-β IFN-ω IFN-τ
Tumor necrosis factors	TNF-α TNF-β (LT-α) LT-β Fas ligand CD40 ligand TNF-related apoptosis-inducing ligand (TRAIL)
Interleukin-1	IL-1α IL-1β IL-1 receptor antagonist IL-18
Transforming growth factor-β	TGF-β Bone morphogenetic proteins Inhibins Activins
Chemokines	C–X–C subfamily (IL-8, many others) C–C subfamily (MIP-1α, many others) C subfamily (lymphotactin)

*IL-12 is a heterodimer in which one subunit is structurally related to IL-6 and the other subunit shows partial homology to the extracellular domain of the IL-6 receptor (α chain).

spatially similar arrangement. Other cytokines that bind to the hematopoietic family of cytokine receptors (see below) might also belong to this family.

Cytokine Receptor Families

Progress in the characterization of cytokine receptors and the cloning of genes encoding cytokine receptors has led to the recognition that many cytokine receptors can be grouped into families based on common structural features (Table 4). The most extensive family is represented by the class I cytokine receptors that encompass receptors for many important cytokines, including IL-2, IL-4, IL-6, IL-12, the hematopoietic growth factor G-CSF, GM-CSF and erythropoietin, as well as growth hormone and prolactin (Bazan, 1989; Kishimoto *et al.*, 1994; Taga and Kishimoto, 1995). Most of these receptors form heterodimers, some are homodimers (e.g. G-CSF and erythropoietin receptors) and some (e.g. IL-2, IL-15 and ciliary neurotrophic factor (CNTF) receptors) are heterotrimers (Heim, 1996). Interestingly, many of these receptors form subfamilies in which one of the chains comprising the heterodimeric or heterotrimeric receptor is common to all members (Taga and Kishimoto, 1995). The latter receptors characteristically contain one or two unique subunits that act as specific binding components for a single cytokine, linked to a signal transducing chain that is shared with other members of the same subfamily. Three such receptor subfamilies, the IL-6, GM-CSF and IL-2 group of receptors, utilize signaling components common to each of these subfamilies (the gp130 chain, β chain and γ chain, respectively). Some members of the class I cytokine receptor family also share sequences in the intracellular domains, termed box1 and box2 regions. These sequences, found in the receptors for IL-2, IL-3, IL-4, IL-6, IL-7, erythropoietin, G-CSF and in gp130 (Ihle, 1995; Thèze *et al.*, 1996), are important for the activation of signaling pathways.

Structural features form the basis for three other cytokine receptor families. One of these is the interferon family receptors, also called class II cytokine receptors because they have features related to the large family of class I cytokine receptors (Bazan, 1990). This family includes the heterodimeric receptors for IFN-α/β and IFN-γ (Bach *et al.*, 1996; Domanski and Colamonici, 1996; Vilček and Sen, 1996; Pestka *et al.*, 1997) and, interestingly, also the IL-10 receptor (Ho *et al.*, 1993). Another large (and still growing) family is that of TNF receptors (Smith *et al.*, 1994; Beutler and van Huffel, 1994; Bazzoni and Beutler, 1996; Wallach, 1996). This family comprises two separate TNF receptors (p55 or TNF-RI, and p75 or TNF-RII), each of which binds both TNF-α and TNF-β. The other known members of the TNF receptor family all function as specific receptors for a single ligand. All members of the TNF receptor family are single-chain receptors (some might form homodimers) that become activated when cross-linked by their trimeric ligands.

A less well defined family is formed by the three known members of the IL-1 receptor family (Sims *et al.*, 1993; Greenfeder *et al.*, 1995; Dinarello, 1996; Vigers *et al.*, 1997). The functional IL-1 receptor is thought to be a heterodimer of the type I receptor and the 'IL-1 receptor accessory protein' (Greenfeder *et al.*, 1995), but the details of IL-1 receptor structure and function have not yet been elucidated.

A number of distinctive features is shared by receptors for the TGF-β family proteins. A unique feature of these receptors is that they contain intracellular serine–threonine kinase domains (Lui *et al.*, 1995; Attisano and Wrana, 1996). Structural features are the

Table 4. Structural features of some cytokine receptors permit their grouping into families.

Receptor family	Common features	Receptor subfamilies	Representative ligands
Class I cytokine receptors ('hematopoietin' family receptors)	Conserved cysteines, WSXWS motif in extracellular domains; some of these receptors also share conserved sequences in intracellular domains (Box1 and Box2)	IL-6 (sharing gp130)	IL-6 IL-11 CNTF* LIF* Oncostatin M* Cardiotrophin 1*
		GM-CSF (sharing β chain)	GM-CSF IL-3 IL-5
		IL-2 (sharing γ chain)	IL-2† IL-4 IL-7 IL-9 IL-15†
		IL-13 (sharing α chain)	IL-13 IL-14
		None	IL-12 G-CSF Erythropoietin Growth hormone Prolactin
Class II cytokine receptors (interferon family receptors)	Conserved cysteines in extracellular domains	None	IFN-α/β family IFN-γ IL-10
TNF receptor family	Partial homology in extracellular domains; conserved 'death domains' in intracellular portions of TNF-RI, Fas, TRAIL and NGF receptors	None	TNF-α TNF-β (LT-α) LT-α/LT-β heteromer NGF Fas ligand CD40 ligand TRAIL
IL-1 receptor family	Immunoglobulin superfamily structure in extracellular domains	None	IL-1α IL-1β IL-1 receptor antagonist
TGF-β receptors	Cysteine-rich extracellular domains, kinase domains, GS domains (type I receptors), serine–threonine-rich tail (type II receptors)	None	TGF-β Bone morphogenetic proteins
Chemokine receptors	Seven-transmembrane domains	C–X–C chemokine receptors	C–X–C chemokines
		C–C chemokine receptors	C–C chemokines

* CNTF, LIF, oncostatin M and cardiotrophin 1 receptors share another common chain, the LIF receptor chain.
† IL-2 and IL-15 receptors also have a common β chain.

basis for the division of TGF-β receptors into type I and type II receptors. TGF-β signaling requires both a type II (the ligand binding component) and a type I (the signaling component) receptor. Finally, there is the family of chemokine receptors, structurally quite dissimilar from the other cytokine receptors (Murphy, 1996). Chemokine receptors are seven-transmembrane-domain, G protein-coupled receptors, related to rhodopsin-like receptors that mediate neurotransmission, light perception and responses to other sensory stimuli.

Cytokine Signal Transduction

Significant progress has been made in recent years in our understanding of the mechanisms and pathways whereby signals are generated and transmitted from the cell-surface cytokine receptors to the inside of cells. Only a very general and superficial overview of this information can be conveyed here. It is not surprising that the patterns of signal transduction are determined largely by the structural characteristics of the cytokine receptors. Thus each of the cytokine receptor families listed in Table 4 exhibits a distinct, characteristic pattern of signal transduction.

The recent revolution in the understanding of cytokine signaling began with the identification of JAK tyrosine kinases and STAT proteins as essential elements in interferon actions (Pellegrini and Schindler, 1993; Darnell *et al.*, 1994). The JAK–STAT signal transduction pathway represents an elegant, efficient, rapid and simple mechanism, utilized by both IFN-α/β and IFN-γ receptors. All essential components pre-exist in unstimulated cells. Although the interferon receptors lack intrinsic kinase activity, their intracellular domains bind cytoplasmic JAK tyrosine kinases. (JAK1 and Tyk2 associate with the two chains of IFN-α/β receptor, JAK1 and JAK2 with the IFN-γ receptor chains.) Binding of the interferons to their receptors results in activation of the receptor-associated JAK kinases. The most important consequence of JAK kinase activation is tyrosine phosphorylation and resulting activation of the SH-2 domain-containing STAT proteins. (IFN-α/β activates STAT1 and STAT2, IFN-γ only STAT1.) Activated STAT proteins then form homo- or hetero-complexes that migrate to the nucleus, bind to recognition sequences in the promoter regions of interferon-inducible genes and thereby activate their transcription. The STAT protein complexes formed upon IFN-α/β receptor activation are partly different from the STAT protein complexes formed on stimulation with IFN-γ, which explains why the actions triggered by the two interferon receptors are both overlapping and unique (Bluyssen *et al.*, 1996).

It appears that all major actions generated by the interferons, including the induction of IRF-1 (Taniguchi *et al.*, 1997), can be traced back to the activation of the JAK–STAT pathway. In contrast, signaling by the class I cytokine receptors relies on elements of the JAK–STAT pathway as well as on additional signaling mechanisms. A case in point is signaling by the IL-2 receptor (Taniguchi *et al.*, 1995; Thèze *et al.*, 1996). IL-2 binds to a receptor composed of three chains of which two, the β and γ chain, generate signals. In a manner similar to the interferon receptors, two JAK kinases, JAK1 and JAK3, associate with the signaling chains of the IL-2 receptor and produce activation of two STAT proteins (STAT3 and STAT5). However, additional signaling pathways are rapidly activated by the IL-2 receptor, including the tyrosine kinases Syk and $p56^{lck}$. IL-2 receptor activation also results in activation of the phosphatidylinositol-3 (PI-3) kinase pathway. Finally, there is evidence that IL-2 receptor activation also leads to the activation of the Ras-dependent mitogen-activated protein (MAP) kinase pathway. Presently, it is not clear which of these signaling pathways is essential for IL-2 actions. The erythropoietin (EPO) receptor, the IL-6 receptor and other members of the class I cytokine receptor family also utilize signaling mechanisms that include the JAK–STAT pathway, the Ras–Raf–MAP kinase pathway, PI-3 kinase and possibly other elements (Ihle, 1995; Taga and Kishimoto, 1995; Hibi *et al.*, 1996).

Signaling by members of the TNF receptor family proceeds through pathways that are distinct from those utilized by the class I and II cytokine receptors. There is no evidence for the involvement of components of the JAK–STAT pathway. Instead, the primary mechanism for signal transduction is the recruitment or activation of a variety of other cytoplasmic proteins to specific regions in the intracellular domains of these receptors (Hsu *et al.*, 1996a,b; Wallach, 1996; Nagata, 1997). The signaling proteins include those that specifically bind to the 'death domains' within some of these receptors (e.g. TNF receptor-associated death domain protein (TRADD)), Fas-associating protein with death domain (FADD)/MORT1 and receptor interacting protein (RIP)) as well as others that bind to different domains (e.g. the TNF receptor-associated factors (TRAFs)). The death domain binding proteins initiate events that lead to apoptosis, but some can also mediate nuclear factor (NF)-κB activation. One of the TRAF proteins (TRAF-2) also induces NF-κB activation.

IL-1 receptor signaling is still poorly understood (Dinarello, 1996). Although it is well known that IL-1 exhibits many activities in common with TNF (Le and Vilček, 1987; Neta *et al.*, 1992), none of the intracellular proteins known to be involved in signaling by the TNF receptors and other members of this receptor family appears to play a role in IL-1 signaling. Recently, however, Malinin *et al.* (1997) identified a new protein kinase termed NF-κB inducing kinase (NIK) which appears to be required for NF-κB activation by the type I IL-1 receptor as well as by both TNF receptors. Thus NIK appears to be at the point of convergence of the TNF and IL-1 pathways. Interestingly, NIK shows sequence similarity with upstream activators of the MAP kinase pathway, suggesting a role for members of the MAP kinase family in IL-1 and TNF signaling. Further support for the relatedness of the TNF- and IL-1-initiated signaling pathways comes from the demonstration that TRAF-6, a recently identified member of the TRAF protein family, is important in IL-1-induced NF-κB activation (Cao *et al.*, 1996).

A unique pattern of signaling is used by the TGF-β receptor family. The ligand binds to the type II receptor (comprising a constitutive kinase) which then recruits the type I receptor into the complex (Liu *et al.*, 1995; Attisano and Wrana, 1996). A phosphorylation event mediated by the type II receptor kinase leads to the activation of the kinase present in the type I receptor; the latter then signals to downstream targets. A number of diverse potential downstream signaling components has been identified; all are different from the components known to function as signal transducers in other cytokine receptor families (Attisano and Wrana, 1996). Strongest evidence for a role in TGF-β signaling is available for the MADR protein family. MADR proteins exist in inactive form in the cytoplasm. TGF-β receptor activation leads to their translocation to the nucleus, possibly as a result of phosphorylation by the TGF-β type I receptor serine–threonine kinase. Once inside the nucleus, MADR proteins are thought to act as transcription factors for specific target genes.

Signaling pathways activated by the G protein-coupled chemokine receptors are still at an early stage of investigation (Murphy, 1996). It is thought that conversion of the receptor-associated heterotrimeric G protein to the activated state brings about activation of a phosphatidylinositol-specific phospholipase C (PI-PLC), leading to the generation of IP-3 and diacylglycerol, and resulting increased Ca^{2+} influx with activation of protein kinase C (Wu *et al.*, 1993). How these events might lead to the many biologic actions associated with chemokine receptor activation is not known.

CYTOKINE NETWORKS

Synergistic and Antagonistic Interactions

Most of the recent studies of cytokine actions are being carried out with homogeneous cytokine preparations produced by recombinant DNA techniques. These studies have led to the assembly of an enormous body of information on the spectrum of actions displayed by individual cytokines. A useful catalog of the repertoire of actions exerted by some of the major cytokines was compiled by Burke *et al.* (1993). Most of this information has been derived from the analysis of recombinant cytokine actions in various *in vitro* systems.

Although this information is important, it may not provide a realistic picture of the functions of cytokines in the intact organism. One reason is the already mentioned pleiotropy and redundancy in cytokine actions. Another reason is that actions of cytokines can be influenced profoundly by the milieu in which they act and especially by the presence or absence of other biologically active agents (i.e. other cytokines, as well as hormones, growth factors, prostaglandins and microbial components). Cytokine action is contextual (Sporn and Roberts, 1988). Under natural conditions a cell rarely, if ever, encounters only one cytokine at a time. Rather, a cell is likely to be exposed to a cocktail of several cytokines and other biologically active agents, with the resulting biologic action reflecting various synergistic and antagonistic interactions of the agents present. A certain pattern of characteristic features of cytokine actions has emerged which, somewhat tongue-in-cheek, might be referred to as the 'molecular philosophy of cytokine actions' (Table 5).

To understand better the actions of cytokines under natural conditions, investigators have begun to analyze mixtures of two or more cytokines. Many examples of synergistic

Table 5. Molecular philosophy of cytokine actions.

Pleiotropy	A cytokine tends to have multiple target cells and multiple actions.
Redundancy	Different cytokines may have similar actions.
Synergism/antagonism	Exposure of cells to two or more cytokines at a time may lead to qualitatively different responses.
Cytokine cascade	A cytokine may increase (or decrease) the production of another cytokine.
Receptor transmodulation	A cytokine may increase (or decrease) the expression of receptors for another cytokine or growth factor.
Receptor transsignaling	A cytokine may increase (or decrease) signaling by receptors for another cytokine or growth factor.

actions have been documented. It is impossible to include in this chapter a comprehensive survey of the myriad of synergistic or antagonistic interactions that have been reported, but some typical examples will be mentioned.

Synergistic interactions are more likely to occur between cytokines that exert related but not identical actions than between cytokines that are functionally closely related. This conclusion is supported by a comparison of the interactions of TNF-α/TNF-β with IL-1α/IL-1β and IFN-γ. As will be explained in other chapters in this volume, TNF-α and TNF-β (the latter is also known as lymphotoxin or LT-α) bind to the same receptors; IL-1α and IL-1β also share common receptors that are distinct from the TNF receptors. A third distinct receptor exists for IFN-γ. (The IFN-γ receptor is also distinct from the IFN-α/β receptor.) Despite the fact that TNF and IL-1 bind to different receptors, they exert many similar actions both *in vitro* and in the intact organism (Le and Vilček, 1987; Neta *et al.*, 1992). However, relatively few examples of a synergistic action between TNF and IL-1 have been described (Neta *et al.*, 1992; Lee *et al.*, 1993). The reason for this infrequent synergy is perhaps the fact that TNF and IL-1 often lead to the activation of the same transcription factors. For example, both TNF and IL-1 are potent activators of the transcription factor NF-κB (Lowenthal *et al.*, 1989; Osborn *et al.*, 1989; Shirakawa *et al.*, 1989; Beg and Baltimore, 1996) and this action is responsible for the activation of expression of several genes by TNF or IL-1. As mentioned above, TNF and IL-1 signaling pathways leading to NF-κB activation have been shown to converge at the level of the NIK kinase (Malinin *et al.*, 1997).

In contrast, a large number of publications described synergistic actions of TNF and IFN-γ. For example, IFN-γ was found to potentiate the cytotoxic action of TNF on tumor cells (Lee *et al.*, 1984; Fransen *et al.*, 1986). Other examples of synergistic actions of TNF and IFN-γ include enhancement of CSF-1 and G-CSF production by monocytes or lymphocytes (Lu *et al.*, 1988), induction of differentiation of human myeloid cell lines (Trinchieri *et al.*, 1986), antiviral activity (Wong and Goeddel, 1986), and the induction of nitric oxide production in murine macrophages (Ding *et al.*, 1988). It is significant that, whereas TNF and IFN-γ act on similar target cells and partly overlap in their ability to activate genes in the target cells (Beresini *et al.*, 1988; Lee *et al.*, 1990), as briefly explained earlier in this chapter they exert their actions through intracellular pathways that are quite distinct. Interestingly, TNF and IFN-γ share the capacity to induce the transcription factor interferon regulatory factor-1 (IRF-1), which is involved in the regulation of expression of IFN-β, the inducible nitric oxide synthase (iNOS) gene and some other IFN-induced genes (Fujita *et al.*, 1989; Reis *et al.*, 1992; Kamijo *et al.*, 1994). However, the pathways whereby TNF and IFN-γ induce IRF-1 synthesis are distinct and therefore synergistic (Pine, 1995).

Not only can a mixture of two cytokines produce an action that represents more than the sum of the separate actions of the individual cytokines (classic definition of a synergistic effect), but a cytokine mixture can result in actions that are qualitatively different from those seen with the individual cytokines. For example, in the HT29 colonic carcinoma cell line, TNF-α alone or IFN-γ alone even at high concentrations does not exert a marked effect on cell viability. However, when the two cytokines are applied together, there is a rapid and marked cytotoxic action resulting from the induction of apoptosis (Feinman *et al.*, 1987; Abreu-Martin *et al.*, 1995). Another example of synergy is the induction of immunoglobulin M secretion by IL-2 and IL-5 (Matsui *et al.*, 1989).

Although not as frequently documented, there are also many examples of antagonistic interactions among cytokines. Since in many types of cells IFNs tend to be growth inhibitory whereas some other cytokines are growth stimulatory, the presence of IFN (either IFN-α/β or IFN-γ) together with a growth-stimulatory cytokine (or growth factor) will result in a mutually antagonistic relationship (De Maeyer and De Maeyer-Guignard, 1988). Other examples of an antagonistic interaction include the actions of IL-4 and IFN-γ on the synthesis of immunoglobulin subclasses in B cells (Snapper *et al.*, 1988), or the inhibitory action of IFN-α/β on IFN-γ-induced enhancement of class II human leukocyte antigen (HLA) antigen expression (Ling *et al.*, 1985; Kamijo *et al.*, 1993b). Analysis of the synergistic or antagonistic interactions involving pairs of cytokines helps to appreciate the complexities of cytokine actions in the intact organism. Nevertheless, it seems that the experimental systems employed still greatly underestimate the variables influencing the actions of cytokines in their natural setting.

Stimulatory and Inhibitory Actions of Cytokines on Cytokine Production

Another characteristic feature of cytokines is their ability to stimulate or inhibit the production of other cytokines. As a result, many cytokine actions are indirect (i.e. they may cause an increase or decrease in the level of production of other cytokines, which then results in an altered biologic response). Among the earliest discovered examples of such an indirect action was the demonstration that the mitogenic action of IL-1 in murine thymocytes involves the stimulation of IL-2 production, and that IL-2 is the actual effector molecule responsible for stimulation of thymocyte proliferation (Smith *et al.*, 1980).

The stimulatory effect of IL-1 on IL-2 production and the role of this interaction in T-cell proliferation has become a paradigm for the actions of many other cytokines. In addition to IL-2, IL-1 was found to stimulate the production of IL-6 (Content *et al.*, 1985), GM-CSF (Zucali *et al.*, 1986) and several chemokines (Matsushima *et al.*, 1988; Larsen *et al.*, 1989) in various types of cells. All of these cytokines are also induced by TNF, in accord with the many other similarities seen between the actions of IL-1 and TNF (Le and Vilček, 1987; Neta *et al.*, 1992). In monocytes both TNF and IL-1 are also autostimulatory and, in addition, they stimulate each other's production (Neta *et al.*, 1992). Other known examples of stimulatory interactions involve the ability of IL-2 and IFN-γ to augment IL-1, TNF-α, IL-6 and TNF-β production (Svedersky *et al.*, 1985; Collart *et al.*, 1986; Kamijo *et al.*, 1993a), and the stimulation of IFN-γ production by IL-2 (Torres *et al.*, 1982). IL-12 (NK cell stimulatory factor) induces IFN-γ production in T and NK cells and appears to be a major regulator of IFN-γ production in the intact organism (Trinchieri, 1995, 1997). Similarly, the recently identified IL-18 (IFN-γ inducing factor) is an inducer of IFN-γ in NK cells and T cells, and in the intact organism controls LPS-induced IFN-γ production (Ghayur *et al.*, 1997). Conversely, IFN-γ, together with TNF-α, stimulates IL-12 production in macrophages, and in mice infected with *Mycobacterium bovis* IL-12 production requires both IFN-γ and TNF (Flesch *et al.*, 1995).

Although not as numerous as the reports of stimulatory interactions, there is increasing evidence of inhibitory actions of cytokines on cytokine production. One example is the inhibitory action of IL-4 (Hart *et al.*, 1989) or IL-6 (Aderka *et al.*, 1989) on the production of TNF or IL-1 by monocytic cells. IL-10 is a cytokine whose major biologic function appears to be inhibition of cytokine production by T helper 1 (TH1) cells and by monocytes–macrophages (Fiorentino *et al.*, 1991). Many of the immunosuppressive and anti-inflammatory actions of TGF-β appear also to be due to its ability to suppress cytokine production in T cells and mononuclear phagocytes (Roberts and Sporn, 1990). Another example is a strong inhibitory activity of IL-13 on inflammatory cytokine production (IL-6, IL-1β, TNF-α, IL-8) in LPS-stimulated monocytes (Minty *et al.*, 1993).

Although the mechanisms of these stimulatory and inhibitory interactions are still not completely elucidated, it is clear that both transcriptional and post-transcriptional events are involved. A case in point is the action of TGF-β on TNF and IL-1 production in monocytes (Chantry *et al.*, 1989). TGF-β was found to increase transcription of IL-1α, IL-1β and TNF-α mRNAs. However, the actual release of the corresponding proteins tended to be inhibited, apparently due to inhibition of translation. IFN-β was found to inhibit TNF-stimulated IL-8 synthesis in some cells at the transcriptional level, but inhibition of IL-8 production by IFN at the post-transcriptional level has also been reported (Aman *et al.*, 1993; Oliveira *et al.*, 1994).

Transmodulation of Cytokine Receptors

Another mechanism important in the network of cytokine actions is the modulation of the level of cytokine receptor expression. One of the earliest studied models involves induction of the high-affinity IL-2 receptor on T cells by IL-1 (Kaye *et al.*, 1984; Lowenthal *et al.*, 1986). Appearance of high-affinity receptors is the consequence of induced expression of the IL-2 receptor complex, mainly due to the regulation of its p55 or α chain (Smith, 1988; Hatakeyama *et al.*, 1989). Other cytokines, including TNF and IL-6 (Noma *et al.*, 1987; Lowenthal *et al.*, 1989), also can affect IL-2 receptor expression, although perhaps not as efficiently as IL-1. Other examples of the modulation of cytokine receptors include the stimulatory action of the interferons, especially IFN-γ, on the expression of TNF receptors on many different cell lines (Aggarwal *et al.*, 1985). This action might contribute to the widely documented synergism between IFN-γ and TNF. Conversely, TNF was also shown to upregulate IFN-γ binding (Raitano and Korc, 1990).

In some instances receptor transmodulation by cytokines results in a reduced level of receptor expression. One example is the downregulation of TNF receptors by IL-1 (Holtmann and Wallach, 1987). There are also examples of downregulation of receptor function that do not result from the decreased expression of cell surface receptors but fall within the general category of 'receptor transsignaling' (Castellino and Chao, 1996). Thus, TNF inhibits insulin signaling by decreasing tyrosine phosphorylation of the insulin receptor and its substrate, IRS-1 (Hotamisligil *et al.*, 1994; Skolnik and Marcusohn, 1996).

How Essential Are Cytokines?

The important roles of cytokines in the regulation of immune and inflammatory responses are now clearly recognized. Availability of purified, potent cytokine preparations and the use of transgenic mouse models has helped to define the major actions of cytokines and their *in vivo* functions. However, the extensive redundancy and pleiotropy in cytokine actions makes it difficult to predict how unique and essential individual cytokine actions are in the intact organism. The quest to learn how cytokines function in their natural environment of the intact host has been aided greatly by development of the technique of targeted gene disruption which utilizes the introduction of suitable constructs into the genome of murine embryonic stem cells by homologous recombination. Genetically altered embryonic stem cells are then introduced into the blastocyst *in utero*, eventually leading to transmission of the disrupted gene through the germline.

Gene targeting has been used for the disruption of several cytokine and cytokine receptor genes in the mouse. The first cytokine gene 'knockout' mice were generated in the early 1990s (Kuhn *et al.*, 1991; Schorle *et al.*, 1991; Shull *et al.*, 1992; Stewart *et al.*, 1992). By 1994, the year in which the previous edition of *The Cytokine Handbook* was printed, all known examples of mice with targeted disruptions of cytokine and cytokine receptor genes could be fitted into one small table. However, the intervening years have witnessed a huge number of publications describing mouse models in which various cytokine genes, cytokine receptor genes and genes encoding proteins important in cytokine signaling have been rendered non-functional by gene targeting. It is no longer possible to include this wealth of information in a brief overview chapter. The reader is referred to a recent separate volume devoted to cytokine gene knockouts (Durum and Muegge, 1997).

Are gene knockout studies providing realistic information about cytokine functions in the intact organism? There is no doubt that the technique represents a very powerful tool for functional analysis, and a wealth of information about the roles of cytokines is emerging from these studies. However, complete deletion of gene function may promote the development of compensatory mechanisms that are not normally operative if the deleted gene is functional. Hence, it is possible that gene knockout studies tend to underestimate the value of specific cytokines or cytokine receptors in a fully functional organism.

REFERENCES

Aarden, L.A., Brunner, T.K., Cerottini, J.-C., Dayer, J.-M., de Weck, A.L., Dinarello, C.A., Di Sabato, G., Farrar, J.J., Gery, I., Gillis, S., Handschumacher, R.E., Henney, C.S., Hoffmann, M.K., Koopman, W.J., Krane, S.M., Lachman, L.B., Lefkowits, I., Mishell, R.I., Mizel, S.B. and Oppenheim, J.J. (1979). Revised nomenclature for antigen-nonspecific T cell proliferation and helper factors. *J. Immunol*. **123**, 2928–2929.

Abreu-Martin, M.T., Vidrich, A., Lynch, D.H. and Targan, S.R. (1995). Divergent induction of apoptosis and IL-8 secretion in HT-29 cells in response to TNF-α and ligation of Fas antigen. *J. Immunol*. **155**, 4147–4154.

Aderka, D., Le, J. and Vilček, J. (1989). IL-6 inhibits lipopolysaccharide-induced tumor necrosis factor production in cultured human monocytes, U937 cells, and in mice. *J. Immunol*. **143**, 3517–3523.

Aggarwal, B.B., Eessalu, T.E. and Hass, P.E. (1985). Characterization of receptors for human tumour necrosis factor and their regulation by γ-interferon. *Nature* **318**, 665–667.

Aman, M.J., Rudolf, G., Goldschmitt, J., Aulitzky, W.E., Lam, C., Huber, C. and Peschel, C. (1993). Type-I interferons are potent inhibitors of interleukin-8 production in hematopoietic and bone marrow stromal cells. *Blood* **82**, 2371–2378.

Attisano, L. and Wrana, J.L. (1996). Signal transduction by members of the transforming growth factor-β superfamily. *Cytokine Growth Factor Rev*. **7**, 327–339.

Bach, E.A., Tanner, J.W., Marsters, S., Ashkenazi, A., Aguet, M., Shaw, A.S. and Schreiber, R.D. (1996). Ligand-induced assembly and activation of the gamma interferon receptor in intact cells. *Mol. Cell. Biol*. **16**, 3214–3221.

Bazan, J.F. (1989). A novel family of growth factor receptors: a common binding domain in the growth hormone, prolactin, erythropoietin and IL-6 receptors, and the p75 IL-2 receptor β-chain. *Biochem. Biophys. Res. Commun*. **164**, 788–795.

Bazan, J.F. (1990). Structural design and molecular evolution of a cytokine receptor superfamily. *Proc. Natl. Acad. Sci. U.S.A.* **87**, 6934–6938.

Bazan, J.F. (1992). Unraveling the structure of IL-2. *Science* **257**, 410–413.

Bazzoni, F. and Beutler, B. (1996). The tumor necrosis factor ligand and receptor families. *N. Engl. J. Med*. **334**, 1717–1725.

Beg, A.A. and Baltimore, D. (1996). An essential role for NK-κB in preventing TNF-α-induced cell death. *Science* **274**, 782–784.

Beresini, M.H., Lempert, M.J. and Epstein, L.B. (1988). Overlapping polypeptide induction in human fibroblasts in response to treatment with interferon-α, interferon-γ, interleukin 1α, and interleukin 1β, and tumor necrosis factor. *J. Immunol*. **140**, 485–493.

Beutler, B. and van Huffel, C. (1994). Unraveling function in the TNF ligand and receptor families. *Science* **264**, 667–668.

Bloom, B.R. and Bennett, B. (1966). Mechanism of a reaction *in vitro* associated with delayed-type hypersensitivity. *Science* **153**, 80–82.

Bluyssen, H.A.R., Durbin, J.E. and Levy, D.E. (1996). ISGF3γ p48, a specificity switch for interferon activated transcription factors. *Cytokine Growth Factor Rev*. **7**, 11–17.

Boulay, J.L. and Paul, W.E. (1992). The interleukin-4 family of lymphokines. *Curr. Opin. Immunol*. **4**, 294–298.

Bradley, T.R. and Metcalf, D. (1966). The growth of mouse bone marrow cells *in vitro*. *Aust. J. Exp. Biol. Med. Sci*. **44**, 287–299.

Burke, F., Naylor, M.S., Davies, B. and Balkwill, F. (1993). The cytokine wall chart. *Immunol. Today* **14**, 165–170.

Cao, Z., Xiong, J., Takeuchi, M., Kurama, T. and Goeddel, D.V. (1996). TRAF6 is a signal transducer for interleukin-1. *Nature* **383**, 443–446.

Carswell, E.A., Old, L.J., Kassel, R.L., Green, S., Fiore, N. and Williamson, B. (1975). An endotoxin-induced serum factor that causes necrosis of tumors. *Proc. Natl. Acad. Sci. U.S.A.* **72**, 3666–3670.

Castellino, A.M. and Chao, M.V. (1996). Trans-signaling by cytokine and growth factor receptors. *Cytokine Growth Factor Rev*. **7**, 297–302.

Chantry, D., Turner, M., Abney, E. and Feldmann, M. (1989). Modulation of cytokine production by transforming growth factor-beta. *J. Immunol*. **142**, 4295–4300.

Cohen, S., Bigazzi, P.E. and Yoshida, T. (1974). Commentary. Similarities of T cell function in cell-mediated immunity and antibody production. *Cell. Immunol*. **12**, 150–159.

Collart, M.A., Belin, D., Vassalli, J.-D., de Kossodo, S. and Vassalli, P. (1986). γ Interferon enhances macrophage transcription of the tumor necrosis factor/cachectin, interleukin 1, and urokinase genes, which are controlled by short-lived repressors. *J. Exp. Med*. **164**, 2113–2118.

Content, J., De Wit, L., Poupart, P., Opdenakker, G., Van Damme, J. and Billiau, A. (1985). Induction of a 26-kDa-protein mRNA in human cells treated with an interleukin-1-related, leukocyte-derived factor. *Eur. J. Biochem*. **152**, 253–257.

D'Andrea, A.D., Fasman, G.D. and Lodish, H.F. (1989). Erythropoietin receptor and interleukin-2 receptor beta chain: a new receptor family. *Cell* **58**, 1023–1024.

Darnell, J.E., Jr., Kerr, I.M. and Stark, G.R. (1994). Jak–STAT pathways and transcriptional activation in response to IFNs and other extracellular signaling proteins. *Science* **264**, 1415–1421.

David, J.R. (1966). Delayed hypersensitivity *in vitro*: its mediation by cell-free substances formed by lymphoid cell–antigen interaction. *Proc. Natl. Acad. Sci. U.S.A.* **56**, 72–77.

de Larco, J.E. and Todaro, G.J. (1978). Growth factors from murine sarcoma virus-transformed cells. *Proc. Natl. Acad. Sci. U.S.A.* **75**, 4001–4005.

De Maeyer, E. and De Maeyer-Guignard, J. (1988). *Interferons and Other Regulatory Cytokines.* John Wiley, New York.

Dinarello, C.A. (1996). Biologic basis for interleukin-1 in disease. *Blood* **87**, 2095–2147.

Ding, A.H., Nathan, C.F. and Stuehr, D.J. (1988). Release of reactive nitrogen intermediates and reactive oxygen intermediates from mouse peritoneal macrophages. Comparison of activating cytokines and evidence for independent production. *J. Immunol*. **141**, 2407–2412.

Domanski, P. and Colamonici, O.R. (1996). The type-I interferon receptor. The long and short of it. *Cytokine Growth Factor Rev*. **7**, 143–151.

Dumonde, D.C., Wolstencroft, R.A., Panayi, G.S., Matthew, M., Morley, J. and Howson, W.T. (1969). 'Lymphokines': non-antibody mediators of cellular immunity generated by lymphocyte activation. *Nature* **224**, 38–42.

Durum, S.K. and Muegge, K. (1997). *Contemporary Immunology: Cytokine Knockouts*. Humana Press, Totowa, NJ.

Feinman, R., Henriksen-DeStefano, D., Tsujimoto, M. and Vilček, J. (1987). Tumor necrosis factor is an important mediator of tumor cell killing by human monocytes. *J. Immunol*. **138**, 635–640.

Fiorentino, D.F., Zlotnik, A., Vieira, P., Mosmann, T.R., Howard, M., Moore, K.W. and O'Garra, A. (1991). IL-10 acts on the antigen-presenting cell to inhibit cytokine production by Th1 cells. *J. Immunol*. **146**, 3444–3451.

Flesch, I.E.A., Hess, J.H., Huang, S., Aguet, M., Rothe, J., Bluethmann, H. and Kaufmann, S.H.E. (1995). Early interleukin 12 production by macrophages in response to mycobacterial infection depends on interferon γ and tumor necrosis factor α. *J. Exp. Med*. **181**, 1615–1621.

Fransen, L., Van der Heyden, J., Ruysschaert, R. and Fiers, W. (1986). Recombinant tumor necrosis factor: its effect and its synergism with interferon-γ on a variety of normal and transformed human cell lines. *Eur. J. Cancer Clin. Oncol*. **22**, 419–426.

Fujita, T., Reis, L.F.L., Watanabe, N., Kimura, Y., Taniguchi, T. and Vilček, J. (1989). Induction of the transcription factor IRF-1 and interferon-β mRNAs by cytokines and activators of second-messenger pathways. *Proc. Natl. Acad. Sci. U.S.A.* **86**, 9936–9940.

Gearing, D.P., King, J.A., Gough, N.M. and Nicola, N.A. (1989). Expression cloning of a receptor for human granulocyte–macrophage colony-stimulating factor. *EMBO J*. **8**, 3667–3676.

Gery, I., Gershon, R.K. and Waksman, B.H. (1971). Potentiation of cultured mouse thymocyte responses by factors released by peripheral leucocytes. *J. Immunol*. **107**, 1778–1780.

Ghayur, T., Banerjee, S., Hugunin, M., Butler, D., Herzog, L., Carter, A., Quintal, L., Sekut, L., Talanian, R., Paskind, M., Wong, W., Kamen, R., Tracey, D. and Allen, H. (1997). Caspase-1 processes IFN-γ-inducing factor and regulates LPS-induced IFN-γ production. *Nature* **386**, 619–623.

Greenfeder, S.A., Nunes, P., Kwee, L., Labow, M., Chizzonite, R.A. and Ju, G. (1995). Molecular cloning and characterization of a second subunit of the interleukin 1 receptor complex. *J. Biol. Chem*. **270**, 13757–13765.

Hart, P.H., Vitti, G.F., Burgess, D.R., Whitty, G.A., Piccoli, D.S. and Hamilton, J.A. (1989). Potential anti-inflammatory effects of interleukin 4: suppression of human monocyte tumor necrosis factor alpha, interleukin 1, and prostaglandin E_2. *Proc. Natl. Acad. Sci. U.S.A.* **86**, 3803–3807.

Hatakeyama, M., Tsudo, M., Minamoto, S., Kono, T., Doi, T., Miyata, T., Miyasaka, M. and Taniguchi, T. (1989). Interleukin-2 receptor β chain gene: generation of three receptor forms by cloned human α and β chain cDNA's. *Science* **244**, 551–556.

Heim, M.H. (1996). The Jak–STAT pathway: specific signal transduction from the cell membrane to the nucleus. *Eur. J. Clin. Invest*. **26**, 1–12.

Hibi, M., Nakajima, K. and Hirano, T. (1996). IL-6 cytokine family and signal transduction: a model of the cytokine system. *J. Mol. Med.* **74**, 1–12.

Ho, A.S.Y., Liu, Y., Khan, T.A., Hsu, D.-H., Bazan, J.F. and Moore, K.W. (1993). A receptor for interleukin 10 is related to interferon receptors. *Proc. Natl. Acad. Sci. U.S.A.* **90**, 11267–11271.

Holtmann, H. and Wallach, D. (1987). Down regulation of the receptors for tumor necrosis factor by interleukin 1 and 4 β-phorbol-12-myristate-13-acetate. *J. Immunol*. **139**, 1161–1167.

Hotamisligil, G.S., Murray, D.L., Choy, L.N. and Spiegelman, B.M. (1994). Tumor necrosis factor alpha inhibits signaling from the insulin receptor. *Proc. Natl. Acad. Sci. U.S.A.* **91**, 4854–4858.

Hsu, H., Shu, H.-B., Pan, M.-G. and Goeddel, D.V. (1996a). TRADD–TRAF2 and TRADD–FADD interactions define two distinct TNF receptor 1 signal transduction pathways. *Cell* **84**, 299–308.

Hsu, H., Huang, J., Shu, H.-B., Baichwal, V. and Goeddel, D.V. (1996b). TNF-dependent recruitment of the protein kinase RIP to the TNF receptor-1 signaling complex. *Immunity* **4**, 387–396.

Ihle, J.N. (1995). Cytokine receptor signalling. *Nature* **377**, 591–594.

Isaacs, A. and Lindenmann, J. (1957). Virus interference. 1. The interferon. *Proc. R. Soc. Lond. [Biol.]* **147**, 258–267.

Kamijo, R., Le, J., Shapiro, D., Havell, E.A., Huang, S., Aguet, M., Bosland, M. and Vilček, J. (1993a). Mice that lack the interferon-γ receptor have profoundly altered responses to infection with Bacillus Calmette–Guerin and subsequent challenge with lipopolysaccharide. *J. Exp. Med*. **178**, 1435–1440.

Kamijo, R., Shapiro, D., Le, J., Huang, S., Aguet, M. and Vilček, J. (1993b). Generation of nitric oxide and induction of major histocompatibility complex class II antigen in macrophages from mice lacking the interferon γ receptor. *Proc. Natl. Acad. Sci. U.S.A.* **90**, 6626–6630.

Kamijo, R., Harada, H., Matsuyama, T., Bosland, M., Gerecitano, J., Shapiro, D., Le, J., Koh, S.I., Kimura, T., Green, S.J., Mak, T.W., Taniguchi, T. and Vilček, J. (1994). Requirement for transcription factor IRF-1 in NO synthase induction in macrophages. *Science* **263**, 1612–1615.

Kaye, J., Gillis, S., Mizel, S.B., Shevach, E.M., Malek, T.R., Dinarello, C.A., Lachman, L.B. and Janeway, C.A., Jr. (1984). Growth of a cloned helper T cell line induced by a monoclonal antibody specific for the antigen receptor: interleukin 1 is required for the expression of receptors for interleukin 2. *J. Immunol*. **133**, 1339–1345.

Kishimoto, T., Taga, T. and Akira, S. (1994). Cytokine signal transduction. *Cell* **76**, 253–262.

Kuhn, R., Rajewsky, K. and Muller, W. (1991). Generation and analysis of interleukin-4 deficient mice. *Science* **254**, 707–710.

Larsen, C.G., Zachariae, C.O.C., Oppenheim, J.J. and Matsushima, K. (1989). Production of monocyte chemotactic and activating factor (MCAF) by human dermal fibroblasts in response to interleukin 1 or tumor necrosis factor. *Biochem. Biophys. Res. Commun*. **160**, 1403–1408.

Le, J. and Vilček, J. (1987). Tumor necrosis factor and interleukin 1: cytokines with multiple overlapping biological activities. *Lab. Invest*. **56**, 234–248.

Lee, S.H., Aggarwal, B.B., Rinderknecht, E., Assisi, F. and Chiu, H. (1984). The synergistic anti-proliferative effect of gamma-interferon and human lymphotoxin. *J. Immunol*. **133**, 1083–1086.

Lee, T.H., Lee, G.W., Ziff, E.B. and Vilček, J. (1990). Isolation and characterization of eight tumor necrosis factor-induced gene sequences from human fibroblasts. *Mol. Cell. Biol*. **10**, 1982–1988.

Lee, T.H., Klampfer, L., Shows, T.B. and Vilček, J. (1993). Transcriptional regulation of TSG6, a tumor necrosis factor- and interleukin-1-inducible primary response gene coding for a secreted hyaluronan-binding protein. *J. Biol. Chem*. **268**, 6154–6160.

Ling, P.D., Warren, M.K. and Vogel, S.N. (1985). Antagonistic effect of interferon-beta on the interferon-gamma-induced expression of la antigen in murine macrophages. *J. Immunol*. **135**, 1857–1863.

Liu, F., Ventura, F., Doody, J. and Massagué, J. (1995). Human type II receptor for bone morphogenic proteins (BMPs): extension of the two-kinase receptor model to the BMPs. *Mol. Cell. Biol*. **15**, 3479–3486.

Lowenthal, J.W., Cerottini, J.-C. and MacDonald, H.R. (1986). Interleukin 1-dependent induction of both interleukin 2 secretion and interleukin 2 receptor expression by thymoma cells. *J. Immunol*. **137**, 1226–1231.

Lowenthal, J.W., Ballard, D.W., Bogerd, H., Bohnlein, E. and Greene, W.C. (1989). Tumor necrosis factor-α activation of the IL-2 receptor-α gene involves the induction of κB-specific DNA binding proteins. *J. Immunol*. **142**, 3121–3128.

Lu, L., Walker, D., Graham, C.D., Waheed, A., Shadduck, R.K. and Broxmeyer, H.E. (1988). Enhancement of release from MHC class II antigen-positive monocytes of hematopoietic colony stimulating factors CSF-1 and G-CSF by recombinant human tumor necrosis factor-α: synergism with recombinant human interferon-γ. *Blood* **72**, 34–41.

Malinin, N.L., Boldin, M.P., Kovalenko, A.V. and Wallach, D. (1997). MAP3K-related kinase involved in NF-κB induction by TNF, CD95 and IL-1. *Nature* **385**, 540–544.

Matsui, K., Nakanishi, K., Cohen, D.I., Hada, T., Furuyama, J., Hamaoka, T. and Higashino, K. (1989). B cell response pathways regulated by IL-5 and IL-2. Secretory microH chain-mRNA and J chain mRNA expression are separately controlled events. *J. Immunol*. **142**, 2918–2923.

Matsushima, K., Morishita, K., Yoshimura, T., Lavu, S., Kobayashi, Y., Lew, W., Appella, E., Kung, H.F., Leonard, E.J. and Oppenheim, J.J. (1988). Molecular cloning of a human monocyte-derived neutrophil chemotactic factor (MDNCF) and the induction of MDNCF mRNA by interleukin 1 and tumor necrosis factor. *J. Exp. Med*. **167**, 1883–1893.

Minty, A., Chalon, P., Derocq, J.-M., Dumont, X., Guillemot, J.-C., Kaghad, M., Labit, C., Leplatois, P., Liauzun, P., Miloux, B., Minty, C., Casellas, P., Loison, G., Lupker, J., Shire, D., Ferrara, P. and Caput, D. (1993). Interleukin-13 is a new human lymphokine regulating inflammatory and immune responses. *Nature* **362**, 248–250.

Morgan, D.A., Ruscetti, F.W. and Gallo, R. (1976). Selective *in vitro* growth of T lymphocytes from normal human bone marrows. *Science* **193**, 1007–1008.

Murphy, P.M. (1996). Chemokine receptors: structure, function and role in microbial pathogenesis. *Cytokine Growth Factor Rev*. **7**, 47–64.

Nagata, S. (1997). Apoptosis by death factor. *Cell* **88**, 355–365.

Neta, R., Sayers, T.J. and Oppenheim, J.J. (1992). Relationship of TNF to interleukins. In *Tumor Necrosis Factors: Structure, Function, and Mechanism of Action* (eds B.B. Aggarwal and J. Vilček), Marcel Dekker, New York, pp. 499–566.

Noma, T., Mizuta, T., Rosen, A., Hirano, T., Kishimoto, T. and Honjo, T. (1987). Enhancement of the interleukin 2 receptor expression on T cells by multiple B-lymphotropic lymphokines. *Immunol. Lett*. **15**, 249–253.

Oliveira, I.C., Mukaida, N., Matsushima, K. and Vilček, J. (1994). Transcriptional inhibition of the interleukin-8 gene by interferon is mediated by the NF-κB site. *Mol. Cell. Biol*. **14**, 5300–5308.

Oppenheim, J.J., Mizel, S.B. and Meltzer, M.S. (1979). Biological effects of lymphocyte and macrophage-derived mitogenic 'amplication' factors. In *Biology of the Lymphokines* (eds S. Cohen, E. Pick and J.J. Oppenheim), Academic Press, New York, pp. 291–323.

Osborn, L., Kunkel, S. and Nabel, G.J. (1989). Tumor necrosis factor α and interleukin 1 stimulate the human immunodeficiency virus enhancer by activation of the nuclear factor κB. *Proc. Natl. Acad. Sci. U.S.A.* **86**, 2336–2340.

Pellegrini, S. and Schindler, C. (1993). Early events in signalling by interferons. *Trends Biochem. Sci*. **18**, 338–342.

Pestka, S., Kotenko, S.V., Muthukumaran, G., Izotova, L.S., Cook, J.R. and Garotta, G. (1997). The interferon gamma (IFN-γ) receptor: a paradigm for the multichain cytokine receptor. *Cytokine Growth Factor Rev*. **8**, 189–206.

Pine, R. (1995). Differential utilization of novel and consensus NFκB recognition sequences in the ISGF-2/IRF-1 promoter for induction by TNFα and synergism of TNFα with IFNγ. *J. Interferon Cytokine Res*. **15**, S154.

Raitano, A.B. and Korc, M. (1990). Tumor necrosis factor up-regulates γ-interferon binding in a human carcinoma cell line. *J. Biol. Chem*. **265**, 10466–10472.

Reis, L.F.L., Harada, H., Wolchok, J.D., Taniguchi, T. and Vilček, J. (1992). Critical role of a common transcription factor, IRF-1, in the regulation of IFN-β and IFN-inducible genes. *EMBO J*. **11**, 185–193.

Roberts, A.B. and Sporn, M.B. (1990). The transforming growth factor-βs. In *Peptide Growth Factors and Their Receptors I* (eds M.B. Sporn and A.B. Roberts), Springer, Berlin, pp. 419–472.

Roberts, W.M., Look, A.T., Roussel, M.F. and Sherr, C.J. (1988). Tandem linkage of human CSF-1 receptor (c-*fms*) and PDGF receptor genes. *Cell* **55**, 655–661.

Ruddle, N.H. and Waksman, B.H. (1968). Cytotoxicity mediated by soluble antigen and lymphocytes in delayed hypersensitivity. 3. Analysis of mechanism. *J. Exp. Med*. **128**, 1267–1279.

Sachs, L. (1987). The molecular control of blood cell development. *Science* **238**, 1374–1379.

Schorle, H., Holtschke, T., Hünig, T., Schimpl, A. and Horak, I. (1991). Development and function of T cells in mice rendered interleukin-2 deficient by gene targeting. *Nature* **352**, 621–624.

Shirakawa, F., Chedid, M., Suttles, J., Pollok, B.A. and Mizel, S.B. (1989). Interleukin 1 and cyclic AMP induce kappa immunoglobulin light-chain expression via activation of an NF-kappa B-like DNA-binding protein. *Mol. Cell. Biol*. **9**, 959–964.

Shull, M.M., Ormsby, I., Kier, A.B., Pawlowski, S., Diebold, R.J., Yin, M., Allen, R., Sidman, C., Proetzel, G., Calvin, D., Annunziata, N. and Doetschman, T. (1992). Targeted disruption of the mouse transforming growth factor-β1 gene results in multifocal inflammatory disease. *Nature* **359**, 693–699.

Sims, J.E., Gayle, M.A., Slack, J.L., Alderson, M.R., Bird, T.A., Giri, J.G., Colotta, F., Re, F., Mantovani, A., Shanebeck, K., Grabstein, K.H. and Dower, S.K. (1993). Interleukin 1 signaling occurs exclusively via the type I receptor. *Proc. Natl. Acad. Sci. U.S.A.* **90**, 6155–6159.

Skolnik, E.Y. and Marcusohn, J. (1996). Inhibition of insulin receptor signaling by TNF: potential role in obesity and non-insulin-dependent diabetes mellitus. *Cytokine Growth Factor Rev*. **7**, 161–173.

Smith, C.A., Farrah, T. and Goodwin, R.G. (1994). The TNF receptor superfamily of cellular and viral proteins: activation, costimulation, and death. *Cell* **76**, 959–962.

Smith, K.A. (1988). Interleukin-2: inception, impact, and implications. *Science* **240**, 1169–1176.

Smith, K.A., Lachman, L.B., Oppenheim, J.J. and Favata, M.F. (1980). The functional relationship of the interleukins. *J. Exp. Med*. **151**, 1551–1556.

Snapper, C.M., Finkelman, F.D. and Paul, W.E. (1988). Regulation of IgG_1 and IgE production of interleukin 4. *Immunol. Rev*. **102**, 51–75.

Sporn, M.B. and Roberts, A.B. (1988). Peptide growth factors are multifunctional. *Nature* **332**, 217–219.

Stewart, C.L., Kaspar, P., Brunet, L.J., Bhatt, H., Gadi, I., Köntgen, F. and Abbondanzo, S.J. (1992). Blastocyst implantation depends on maternal expression of leukaemia inhibitory factor. *Nature* **359**, 76–79.

Svedersky, L.P., Nedwin, G.E., Goeddel, D.V. and Palladino, M.A., Jr. (1985). Interferon-γ enhances induction of lymphotoxin in recombinant interleukin 2-stimulated peripheral blood mononuclear cells. *J. Immunol*. **134**, 1604–1608.

Taga, T. and Kishimoto, T. (1995). Signaling mechanisms through cytokine receptors that share signal transducing receptor components. *Curr. Opin. Immunol*. **7**, 17–23.

Taniguchi, T., Miyazaki, T., Minami, Y., Kawahara, A., Fujii, H., Nakagawa, Y., Hatakeyama, M. and Liu, Z.J.

(1995). IL-2 signaling involves recruitment and activation of multiple protein tyrosine kinases by the IL-2 receptor. *Ann. N.Y. Acad. Sci*. **766**, 235–244.

Taniguchi, T., Lamphier, M.S. and Tanaka, N. (1997). IRF-1: the transcription factor linking the interferon response and oncogenesis. *Biochim. Biophys. Acta* **1333**, M9–M17.

Thèze, J., Alzari, P.M. and Bertoglio, J. (1996). Interleukin 2 and its receptors: recent advances and new immunological functions. *Immunol. Today* **17**, 481–486.

Torres, B.A., Farrar, W.L. and Johnson, H.M. (1982). Interleukin 2 regulates immune interferon (IFN-γ) production by normal and suppressor cell cultures. *J. Immunol*. **128**, 2217–2219.

Trinchieri, G. (1995). Interleukin-12: a proinflammatory cytokine with immunoregulatory functions that bridge innate resistance and antigen-specific adaptive immunity. *Annu. Rev. Immunol*. **13**, 251–276.

Trinchieri, G. (1997). Function and clinical use of interleukin-12. *Curr. Opin. Hematol*. **4**, 59–66.

Trinchieri, G., Kobayashi, M., Rosen, M., Loudon, R., Murphy, M. and Perussia, B. (1986). Tumor necrosis factor and lymphotoxin induce differentiation of human myeloid cell lines in synergy with immune interferon. *J. Exp. Med*. **164**, 1206–1225.

Ushio, S., Namba, M., Okura, T., Hattori, K., Nukada, Y., Akita, K., Tanabe, F., Konishi, K., Micallef, M., Fujii, M., Torigoe, K., Tanimoto, T., Fukuda, S., Ikeda, M., Okamura, H. and Kurimoto, M. (1996). Cloning of the cDNA for human IFN-gamma-inducing factor, expression in *Escherichia coli*, and studies on the biologic activities of the protein. *J. Immunol*. **156**, 4274–4279.

Vigers, G.P., Anderson, L.J., Caffes, P. and Brandhuber, B.J. (1997). Crystal structure of the type-I interleukin-1 receptor complexed with interleukin-1β. *Nature* **386**, 190–194.

Vilček, J. and Sen, G.C. (1996). Interferons and other cytokines. In *Fields Virology* (eds B.N. Fields, D.M. Knipe, P.M. Howley *et al*.), Lippincott–Raven, Philadelphia, PA, pp. 375–399.

Wallach, D. (1996). Suicide by order: some open questions about the cell-killing activities of the TNF ligand and receptor families. *Cytokine Growth Factor Rev*. **7**, 211–221.

Wheelock, E.F. (1965). Interferon-like virus-inhibitor induced in human leukocytes by phytohemagglutinin. *Science* **149**, 310–311.

Williams, T.W. and Granger, G.A. (1968). Lymphocyte *in vitro* cytotoxicity: lymphotoxins of several mammalian species. *Nature* **219**, 1076–1077.

Wong, G.H.W. and Goeddel, D.V. (1986). Tumour necrosis factors α and β inhibit virus replication and synergize with interferons. *Nature* **323**, 819–822.

Wu, D., LaRosa, G.J. and Simon, M.I. (1993). G protein-coupled signal transduction pathways for interleukin-8. *Science* **261**, 101–103.

Zucali, J.R., Dinarello, C.A., Oblon, D.J., Gross, M.A., Anderson, L. and Weiner, R.S. (1986). Interleukin 1 stimulates fibroblasts to produce granulocyte–macrophage colony-stimulating activity and prostaglandin E_2. *J. Clin. Invest*. **77**, 1857–1863.

Chapter 2

Molecular Genetics of Cytokines

Gordon W. Duff
Division of Molecular and Genetic Medicine, University of Sheffield, Sheffield, UK

CYTOKINES IN CHRONIC INFLAMMATORY DISEASE

As extracellular signalling molecules that coordinate the inflammatory and immune responses, cytokines have many roles in the normal processes of host defence against infection and injury. They also appear to be involved, at many stages, in the mechanisms of autoimmunity and the pathogenesis of chronic inflammatory diseases (CIDs). Diseases that fall into this category include rheumatoid arthritis and other chronic inflammatory arthritides, diabetes and its complications, connective tissue diseases such as systemic lupus erythematosus (SLE) and scleroderma, inflammatory bowel diseases and inflammatory skin diseases, and even arteriosclerosis could be classified in this way. Common pathological features of these cytokine-related conditions include inflammatory cell infiltration of target organs, loss of normal cellular components and tissue damage. Such diseases also have in common their chronicity, being more or less life-long once established, and a tendency to occur in families.

The genetic basis for the familial tendencies of common CIDs has not been defined fully in any particular case. However, it seems likely, on epidemiological grounds and from studies of monozygotic and dizygotic twins, that genetic factors contribute to disease susceptibility and severity, although environmental factors are also important or even critical for the development of clinical disease. Thus, common familial diseases appear to be multifactorial with the involvement of an unknown number of genes and unknown environmental factors (Table 1).

Table 1. Some major multifactorial diseases that show familial clustering.

Inflammatory joint disease
Diabetes
Arteriosclerosis
Inflammatory skin diseases
Inflammatory bowel diseases
Connective tissue diseases
Osteoporosis
Osteoarthritis
Periodontitis

The Cytokine Handbook, 3rd ed.
ISBN 0–12–689662–3

In many cases, associations between CIDs and alleles or haplotypes of the major histocompatibility complex (MHC) have been established. These associations have mostly been interpreted as 'immune response' gene effects and more recently in terms of the genetically defined ability of MHC molecules to accommodate a particular linear peptide and interact with a T-cell antigen receptor. This mechanism could explain the immunopathogenic component of familial CIDs.

Within the MHC, diseases are often associated with quite extensive haplotypes, making it difficult to distinguish which particular allele(s) may contribute to pathogenesis and which may be associated with disease through physical linkage to the true disease-related alleles. Not only can it be difficult, for practical reasons, to assess the contribution of an individual allele within an MHC haplotype, but it seems certain that susceptibility to or severity of many CIDs is determined also by other unknown genes located outside the MHC. Progress in defining these non-MHC genes has been, until recently, somewhat slow.

CYTOKINES AS CANDIDATE GENES

There are several approaches to the identification of disease-associated genes. When the mechanisms of disease are not known, linkage analysis of the disease phenotype within families may identify first the chromosome and then the chromosomal region where the gene is located. In the human genome, there are now so many closely spaced polymorphic markers for each chromosome that linkage analysis in families can relatively easily locate disease genes in monogenic diseases with a high degree of success. Clearly, linkage analysis becomes more difficult in multigenic and multifactorial diseases unless major single-gene effects are present.

Another approach is to propose putative genes that may be important. This is often possible when there is some understanding of the disease process. Thus, a 'candidate' gene in this sense is a hypothesis to be tested. The macrophage-derived cytokines that control the inflammatory response would seem to be reasonable 'candidate genes' in CIDs. Not only are they active in the pathogenesis of many CIDs (Duff, 1989) but they also show stable interindividual differences in rates of production (Molvig *et al.*, 1988).

To test the hypothesis that a specified gene may contribute to CID, linkage analysis, population association studies and transmission disequilibrium of the associated allele between a heterozygous parent and affected offspring (TDT) can all be used depending on the clinical resource available.

Family collections of DNA are needed for linkage studies and TDTs, whereas DNA from unrelated affected individuals and an appropriate control population are needed for association studies.

It is, of course, necessary to seek polymorphic markers within or around the gene of interest, whichever method of analysis is used. In an association study an attempt is made to ascertain whether there are differences in allelic frequency between populations of unrelated individuals with a disease and a relevant 'healthy' population. If a disease-associated allele emerges from such a population analysis, a further question is raised: does the identified gene, itself, contribute to the disease process or is it a chromosomal 'marker' for a more important contributory gene with which it is physically linked? Clearly, if a candidate gene has been proposed on the basis of a

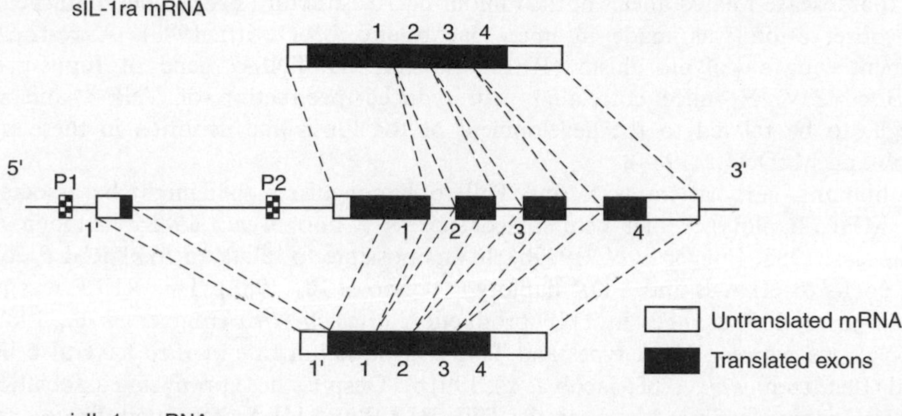

Fig. 1. Structure of the IL-1 receptor antagonist (IL-1ra) gene. The genomic structure is shown between the two alternative mRNA sequences. The genomic sequence is divided into exons (numbered boxes) and introns (intervening sequences between exons). Upstream of exons 1′ and 1, the alternative promoters are indicated by P1 and P2. Cells, such as macrophages, use the downstream promoter (P2) and make an mRNA that encodes a leader sequence. This is translated into a secreted form of the IL-1 receptor antagonist (sIL-1ra). Other cell types, such as skin keratinocytes, use the upstream promoter (P1) and generate an alternative mRNA that lacks the coding region for a leader sequence. This is translated as an intracellular form of the IL-1 receptor antagonist (icIL-1ra).

biological role in the disease process and is then found to possess a disease-associated allele, its likelihood of being a contributory gene may be greater. This would be particularly true if the polymorphism resulted in an altered protein or in altered regulation of the gene. The latter might result in a quantitative difference in gene regulation (e.g. increased rate of gene transcription) or a qualitative difference in gene regulation (e.g. a change in gene transcription in response to a specific signalling pathway or in a particular cell type). The DNA that comprises eukaryotic genes is divided into stretches that contain information for coding RNA which may be translated into protein (exons), broken up by intervening, non-coding sequences (introns). The DNA 'upstream' of the first exon (5′ flanking region) contains short nucleotide sequences that bind transcription factors and thereby controls the process of gene transcription (the promoter region) (Fig. 1). Not all the RNA is translated into protein. Whether it is translated or not, it can affect the production of protein; for example, untranslated sequences at the 3′ end influence the stability of the mRNA of many cytokine genes. Short DNA sequences in the 5′ or 3′ flanking regions or within the gene itself can act to increase gene transcription (enhancer sequences) or to reduce it (repressor sequences).

TUMOUR NECROSIS FACTOR AND THE MHC

The search for disease-related polymorphisms in cytokine genes is relatively recent as specific probes for their identification have been available for only a decade. Because of its biological properties and its location within the MHC, there was early interest in potential disease-related alleles of the tumour necrosis factor (TNF) locus, and the first such observation was made in mice (Jacob and McDevitt, 1988). A restriction fragment length polymorphism (RFLP) within the TNF-α gene of lupus-prone (NZB × NZW) F_1 mice correlated with reduced production of TNF-α and was thought to be related to the development of the lupus-like nephritis in these mice (Jacob and McDevitt, 1988).

In humans, early attempts to find TNF polymorphisms that might be associated with MHC haplotypes met with mixed success (Choo *et al.*, 1988; Partanen and Koskimies, 1988; Fugger *et al.*, 1989b). It was possible to relate an *Nco*I RFLP at the TNF locus to HLA-B and a DR haplotype (Choo *et al.*, 1988). This RFLP was also related to modest changes in TNF production rates *in vitro* (Fugger *et al.*, 1989a). Associations between HLA types and TNF-α production rate *in vitro* have also been noted (Bendtzen *et al.*, 1988; Jacob *et al.*, 1991b). Despite these promising associations, no relationship was found between the TNF RFLP and HLA-associated diseases such as multiple sclerosis and optic neuritis (Fugger *et al.*, 1990). In fact, the *Nco*I RFLP at the TNF locus was, when it was characterized, found to be located within the first intron of the TNF-β gene (Webb and Chaplin, 1990; Messer *et al.*, 1991a). This polymorphism was shown also to be correlated with an amino acid variation at position 26 of TNF-β (lymphotoxin) and with a reduced rate of TNF-β production (Messer *et al.*, 1991a).

The search for further and possibly more informative polymorphisms at the TNF locus continued with the identification of four dinucleotide repeats of variable length (microsatellites). Of these, two were 3.5 kb upstream of the TNF-β gene, one was some 10 kb downstream of the TNF-α gene and one TC repeat was located within the first intron of the TNF-β gene (Nedospasov *et al.*, 1991) (Fig. 2). While the previously discussed polymorphisms are within the TNF-β gene, or at some distance from the tandem TNF genes, polymorphisms have also been described in the promoter region of the murine TNF-α gene (Jongeneel *et al.*, 1990) and at position −308 within the promoter region of the human TNF-α gene (Wilson *et al.*, 1992) (Fig. 3).

TNF POLYMORPHISM AND DISEASES

Several studies have tested whether there may be disease-associated or functionally different alleles of the TNF locus. No association was found between TNF-β polymorphism and the low TNF-β production rates seen in patients with primary biliary cirrhosis (Messer *et al.*, 1991b) and, likewise, there was no relation between TNF-β polymorphism and ankylosing spondylitis (Verjans *et al.*, 1991). Associations were found between TNF microsatellites and other loci within the MHC including class I, II and III alleles (Jongeneel *et al.*, 1991). More recently, the single-base transition polymorphism within the human TNF-α promoter has been highly associated with the 'autoimmune haplotype' HLA-A1, B8, DR3 (Wilson *et al.*, 1993). The TNF-β gene *Nco*I RFLP also correlates with MHC ancestral haplotypes including HLA-A1, B8 and DR3 (Dawkins *et al.*, 1989; Abraham *et al.*, 1993). This HLA haplotype has also been associated with increased TNF-α production by lymphoid cell lines *in vitro* (Jacob *et al.*, 1990; Abraham *et al.*, 1993). In the light of these associations it is not surprising that TNF alleles have been found at raised frequencies in diseases associated with the HLA-A1, B8, DR3 haplotype. For example, the TNF-β RFLP has been associated with

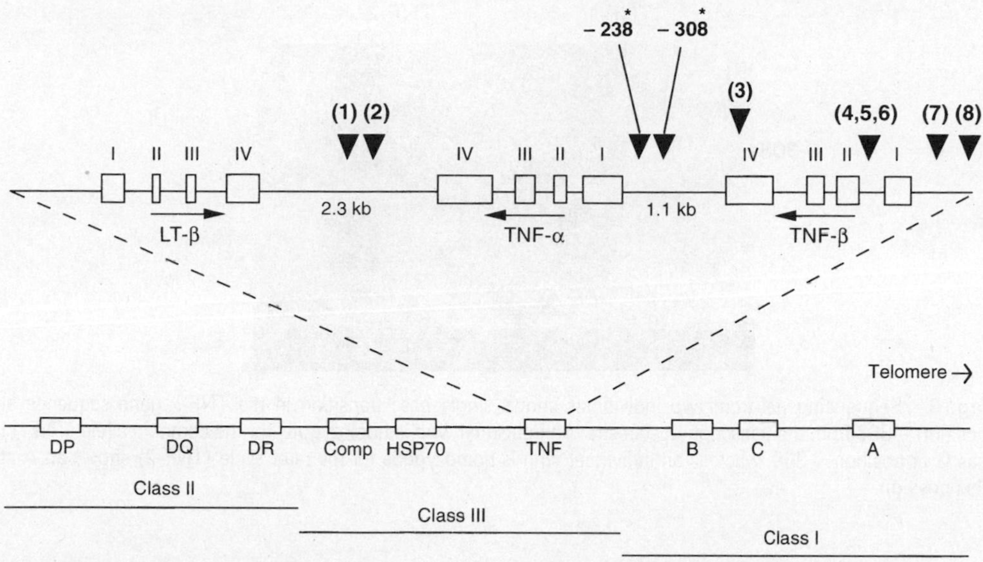

	Nature of gene variant	Reference
(1)	Dinucleotide repeat	Udalova *et al*. (1993)
(2)	Dinucleotide repeat	Udalova *et al*. (1993)
(3)	Single base variation	Partanen and Koskimies (1988)
(4)	Dinucleotide repeat	Jongeneel and Cambon-Thomsen (1991)
(5)	Single base variation	Messer *et al*. (1991)
(6)	Single base variation	Ferencik *et al*. (1991)
(7)	Dinucleotide repeat	Jongeneel and Cambon-Thomsen (1991)
(8)	Dinucleotide repeat	Jongeneel and Cambon-Thomsen (1991)
−238	Single base variation	D'Alfonso and Richiardi (1994)
−308	Single base variation	Wilson *et al*. (1992)

Fig. 2. Structure of the human major histocompatibility complex. The human MHC spans 4–5 Mb on the short arm of chromosome 6 (6q21). It contains genes that encode proteins associated with immune recognition and immune responses, including the TNF family. Polymorphic sites are indicated by arrowheads. The two asterisks indicate promoter region single-base variations in the TNF-α gene.

Graves' disease of the thyroid (Badenhoop *et al*., 1992) and also with SLE (Bettinotti *et al*., 1993). An association has also been found between SLE and the TNF-α base-transition polymorphism in the 5′ region (Wilson *et al*., 1994). With this polymorphism, the association was even stronger between the TNF-α rare allele and the presence of anti-Ro and anti-La autoantibodies (Wilson *et al*., 1994).

In all these studies, however, the association between disease and DR3 has been stronger than that between disease and TNF, suggesting that the TNF association results from linkage disequilibrium between the TNF locus and the DR3 locus. A possible exception is described in a recent study of patients with coeliac disease (McManus *et al*., 1996). It is only in the lupus-prone mouse strains in which disease susceptibility, TNF-α production rate and TNF-α genotype have been associated in a

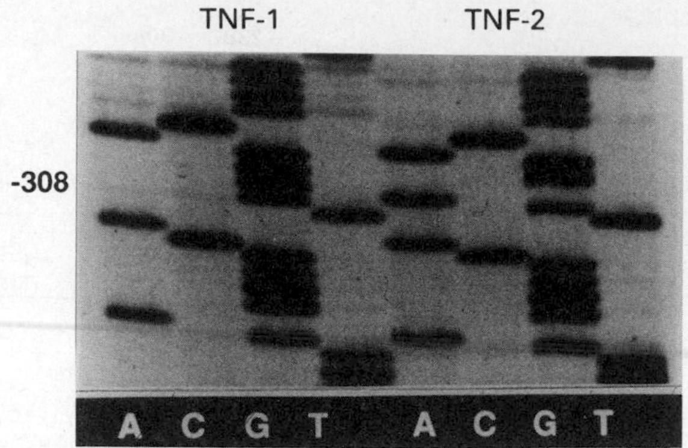

Fig. 3. Sequencing gel from two individuals shows single-base transition in the TNF-α gene sequence at position -308 from the transcription start site. An individual who is homozygous for the common allele (TNF-1) has G at position -308, whereas an individual who is homozygous for the rarer allele (TNF-2) shows an A at this position.

way that suggests a direct involvement of the TNF gene product in disease pathogenesis (Jacob *et al.*, 1991a). These mice have a promoter region polymorphism within the TNF-α gene and it remains to be seen whether polymorphism in the human TNF-α gene is also associated with the production-rate phenotype. In reporter gene assays of the human TNF-α promoter performed in a B-cell line, the rarer allele at -308 was several-fold more efficient than the common promoter-region variant (Wilson *et al.*, 1997). It has, however, not yet been shown convincingly that production of TNF-α protein is correlated with variations in the TNF-α gene.

Although these issues remain outstanding, it is now clear that polymorphisms at the TNF locus form part of extended MHC haplotypes such as the A1, B8, DR3 ancestral haplotype that has been associated with a range of autoimmune human diseases. In infectious diseases, the most striking association with the TNF-α gene was the finding, in The Gambia, that children who were homozygous for the rarer allele of the -308 polymorphism were eightfold more likely to have a severe clinical outcome of cerebral malaria (McGuire *et al.*, 1994).

Observations of this level of importance in the clinical domain underline the need for a fuller understanding of the functional significance of genetic variants of the TNF gene and their haplotypic relation to other genes of the MHC.

THE INTERLEUKIN-1 GENE CLUSTER

The interleukin-1 (IL-1) gene cluster, comprising IL-1α, IL-1β and IL-1 receptor antagonist (IL-1ra), is located on the long arm of human chromosome 2. The IL-1 receptor type I (and possibly IL-1 receptor type II) is also located on 2q (Table 2). Three IL-1 ligand genes have all been mapped to a 430-kb stretch of DNA (Nicklin *et al.*, 1994). The gene products of these loci have been implicated extensively in CIDs of all types and in chronic inflammatory arthritis such as rheumatoid arthritis in particular (Eastgate *et al.*, 1988).

Table 2. Chromosomal locations on the long arm of human chromosome 2 of the IL-1 gene cluster and related genes.

IL-1α	2q13
IL-1β	2q13–21
IL-1ra	2q13–14.1
IL-1 receptor type I	2q12
IL-1 receptor type II	2q12–22

Several different polymorphisms have been found and characterized within the IL-1 gene cluster (Fig. 4). They include single base changes in 5′ flanking DNA of IL-1β (di Giovine *et al.*, 1992) and IL-1α (McDowell *et al.*, 1994) as well as polymorphic variable number tandem repeats (VNTR polymorphism) in IL-1α and IL-1ra. In IL-1α, the VNTR is made up of a 46-bp stretch of DNA that is repeated 5 to 19 times within intron 6. There are six alleles, of which the most common (62%) has nine repeats and the second most frequent (23%) has 18 repeats (Bailly *et al.*, 1993). Each repeat stretch

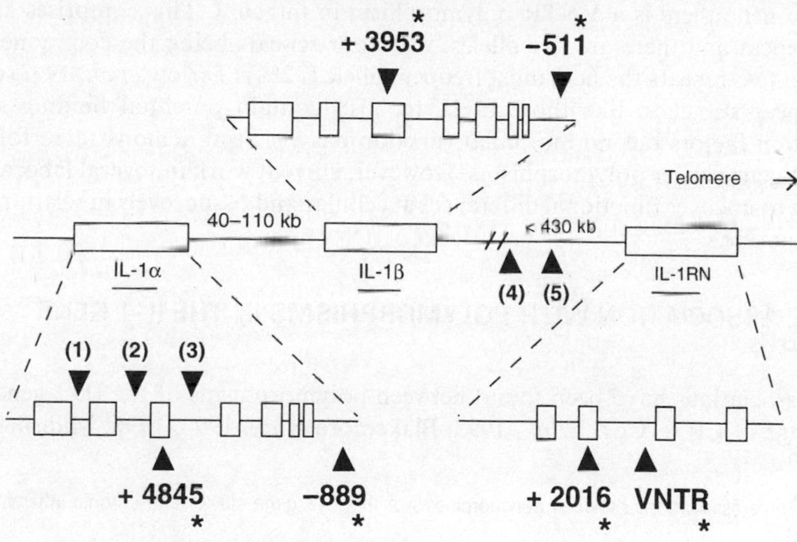

	Nature of gene variant	Reference
(1)	VNTR, intron 6 (86 bp repeats)	Bailly (1993)
(2)	Dinucleotide repeat, intron 5	Todd and Naylor (1991)
(3)	Trinucleotide repeat, intron 4	Zuliani and Hobbs (1990)
(4)	Tetranucleotide repeat	Sunden *et al*. (1996)
(5)	Dinucleotide repeat	Spurr *et al*. (1996)

Fig. 4. Structure of the interleukin-1 locus. The IL-1 locus is on the long arm of human chromosome 2 (2q13). IL-1α, IL-1β and IL-1RN lie within a 430-kb region. Polymorphic sites are indicated by arrowheads. The position with respect to the transcription start site of the gene is shown for polymorphisms that have been associated with CIDs.

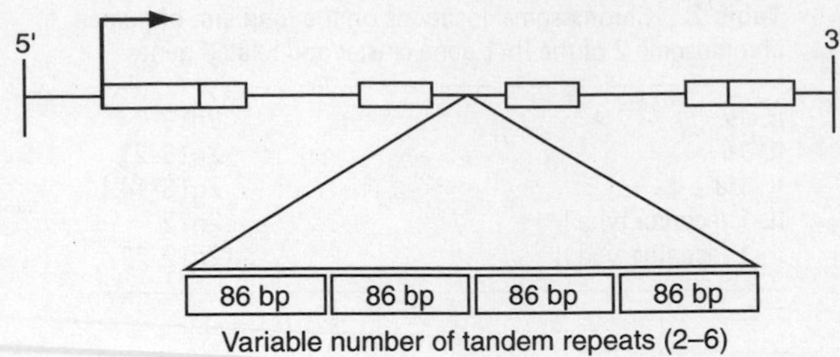

Fig. 5. Diagram to show the nature of the VNTR polymorphism in intron 2 of the human IL-1ra gene. An 86-bp stretch is tandemly repeated up to six times. Four repeats is the commonest allele (74%), and two repeats is the next commonest (22%).

contains sites of potential significance in gene regulation, in particular a glucocorticoid response element and a potential binding site for the transcription factor SP-1. There is also an uncharacterized *Taq* I RFLP in the IL-1 gene cluster which is detected with IL-1β probes (Pociot *et al.*, 1992).

The IL-1ra is a very powerful anti-inflammatory agent *in vivo* (Lennard, 1995). Within the IL-1ra gene there is a VNTR polymorphism in intron 2. This comprises an 86-bp tandem repeat and there are five alleles, with four repeats being the commonest allele (74%) and two repeats the next most frequent allele (22%) (Tarlow *et al.*, 1993) (Fig. 5). These repeat stretches, like those in IL-1α, also contain potential binding sites for transcription factors but no molecular function has yet been demonstrated for any of these IL-1 gene cluster polymorphisms. However, current work in several laboratories is beginning to uncover functional differences at cellular and tissue levels in relation to IL-1 cluster genotypes.

DISEASE ASSOCIATION WITH POLYMORPHISMS IN THE IL-1 GENE CLUSTER

Several associations have been found between polymorphisms of the IL-1 gene cluster and a range of CIDs (Cork *et al.*, 1993; Blakemore *et al.*, 1994, 1996; Mansfield *et al.*,

Table 3. Some associations between polymorphisms of the IL-1 gene cluster and chronic inflammatory diseases (CIDs).

CID	Comment	Reference
Systemic lupus erythematosus (SLE)	Especially discoid rash and photosensitivity (IL-1ra)	Blakemore *et al*. (1994)
Juvenile chronic arthritis	Especially with iridocyclitis (IL-1α)	McDowell *et al*. (1995)
Ulcerative colitis	Especially total colonic involvement (IL-1ra)	Mansfield *et al*. (1994)
Psoriasis	Especially hospital populations (IL-1ra)	Cork *et al*. (1993)
Diabetes	Diabetic complications (IL-1ra)	Blakemore *et al*. (1996)
Diabetes	Type 1	Pociot *et al*. (1992)
Alopecia areata	Especially alopecia universalis	Tarlow *et al*. (1994)
Lichen sclerosus	Especially with extragenital involvement	Clay *et al*. (1994)
Periodontal disease	Severe disease	Kornman *et al*. (1997)

1994; McDowell *et al.*, 1995) (Table 3). Recently a composite genotype involving IL-1β +3953 was found to be significantly over-represented in patients with severe periodontal disease (Kornman *et al.*, 1997). This is of interest because IL-1β has been strongly implicated as a mediator of tissue reabsorption in periodontitis and the +3953 allele that is associated with periodontitis severity is also associated with higher levels of IL-1β production *in vitro* (Pociot *et al.*, 1992).

These associations represent statistically significant differences in allele frequencies between disease populations and matched control populations. What the significance may be in terms of disease pathogenesis is not yet known, but studies to relate cytokine production rates to the genetic markers of the IL-1 gene cluster and direct testing *in vitro* of the functional significance of these polymorphisms in gene regulation are currently being investigated. There has been one report associating an IL-1β RFLP with IL-1β secretion *in vitro* (Pociot *et al.*, 1992). It is notable that in the clinical diseases that show association with the IL-1 gene cluster there have been many reports implicating the products of these loci in their pathogenesis (Dinarello and Wolff, 1993).

MARKERS OF SUSCEPTIBILITY OR SEVERITY?

The disease associations that have been established with IL-1 gene cluster polymorphisms have been detected by comparing allele frequencies in disease and in normal populations. In this way it is possible to generate hypotheses about susceptibility genes, which may be normal variants present in a large proportion of the population. This is in contrast to a mutation in a single gene leading to a disease. For example, mutations in the gene that encodes IL-2 receptor γ chain 'cause' X-linked severe combined immunodeficiency disease (SCID) (Noguchi *et al.*, 1993). Whether genetic linkage analysis will show that other diseases map to single genes encoding cytokines or their receptors remains to be seen.

The data collected to date seem to indicate that IL-1 gene polymorphisms represent markers of disease severity. For example, in ulcerative colitis the carriage rate of IL-1ra allele 2 increases in populations with increasing extent of colonic involvement (Mansfield *et al.*, 1994) (Fig. 6).

CONCLUSION

The discovery in recent years of polymorphisms in cytokine and related genes has made it possible to test allelic associations with CIDs. In the case of the pro-inflammatory cytokine TNF-α, the picture is complicated because of haplotypic association with other genes of the MHC. An allele of TNF-α is associated with the MHC haplotype A1, B8, DR3 which is itself associated with several autoimmune diseases such as SLE, type I diabetes and Graves' disease. TNF-α, therefore, is also associated genetically with these diseases but it might be difficult to isolate the effect of the single gene from that of the extended haplotype.

The IL-1 gene cluster on chromosome 2 also has several polymorphic markers and it has been possible to find disease-associated alleles in CIDS such as SLE, ulcerative colitis, psoriasis and juvenile chronic arthritis. IL-1α, IL-1β and IL-1ra have all been implicated in the pathogenesis of these disorders but it is still possible that the

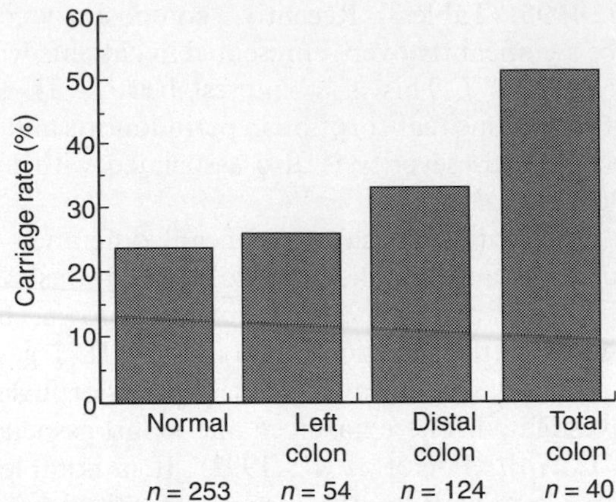

Fig. 6. Allele carriage rate of IL-1ra allele 2 in patients with ulcerative colitis. The carriage rate in a normal population is shown. The carriage rate in patient groups increases with increasing colonic involvement. The number of patients in each group is shown beneath columns.

polymorphisms in this gene cluster are merely markers for disease-related genes elsewhere on chromosome 2. Studies are in progress to resolve this and also to test directly the effect of these polymorphisms on gene function and the production rate of the cytokine.

For many years, the genes of the MHC were the focus of attention for investigators interested in the genetic basis of CIDs. It would now appear that genetic analysis of the cytokine system will also contribute to the understanding of the multifactorial nature of these conditions and also infectious diseases.

Since the days of Garrod, clinical genetics has regarded 'genetic diseases' as those in which a mutation in a critical gene has led directly to the production of disease. With the progress in molecular genetics in the last decade or so, we can now extend the idea of 'genetic diseases' to those common conditions in which DNA variation in the normal human population contributes to susceptibility or to the severity of disease and its clinical outcome. This chapter has focused on sterile inflammatory conditions but we can be confident that similar insights will be forthcoming in lethal infections (such as septic shock, meningococcal and human immunodeficiency virus disease) and also in cancers. The clinical potential for selection of specific therapeutic approaches seems clear.

REFERENCES

Abraham, L.J., French, M.A.H. and Dawkins, R.L. (1993). Polymorphic MHC ancestral haplotypes affect the activity of tumour necrosis factor-alpha. *Clin. Exp. Immunol.* **92**, 14–18.

Badenhoop, K., Schwarz, G., Schleusener, J. *et al.* (1992). Tumor necrosis factor beta gene polymorphisms in Graves' disease. *J. Clin. Endocrinol. Metab.* **74**, 287–291.

Bailly, S. (1993). Genetic polymorphism of human interleukin-1 alpha. *Eur. J. Immunol.* **23**, 1240–1245.

Bailly, S., di Giovine, F.S. and Duff, G.W. (1993). Polymorphic tandem repeat region in interleukin-1 and alpha intron 6. *Hum. Genet.* **91**, 85–86.

Bendtzen, K., Morling, N., Fomsgaard, A. *et al.* (1988). Association between HLA-DR2 and production of tumour necrosis factor alpha and interleukin-1 by mononuclear cells activated by lipopolysaccharide. *Scand. J. Immunol.* **28**, 599–606.

Bettinotti, M.P., Hartung, K., Deicher, H. *et al.* (1993). Polymorphism of the tumor necrosis factor beta gene in systemic lupus erythematosus: TNFB-MHC haplotypes. *Immunogenetics* **37**, 449–454.

Blakemore, A.I.F., Tarlow, J.K., Cork, M.J., Gordon, C., Emery, P. and Duff G.W. (1994). Interleukin-1 receptor antagonist gene polymorphism as a disease severity factor in systemic lupus erythematosus. *Arthritis Rheum.* **37**, 1380–1385.

Blakemore, A.I.F., Cox, A., Gonzalez, A.M., Maskill, J.K., Hughes, M.E., Wilson, R.M., Ward, J.D. and Duff, G.W. (1996). Interleukin-1 receptor antagonist allele (IL1RN*2) associated with nephropathy in diabetes mellitus. *Hum. Genet.* **97**, 369–374.

Choo, S.Y., Speis, T., Strominger, J.L. and Hansen, J. (1988). Polymorphism in the tumor necrosis factor gene: association with HLA-B and DR haplotypes. *Hum. Immunol.* **23**, 86.

Clay, F.E., Cork, M.J., Tarlow, J.K., Blakemore, A.I., Harrington, C.I., Lewis, F. and Duff, G.W. (1994). Interleukin-1 receptor antagonist gene polymorphism association with lichen sclerosis. *Hum. Genet.*, **94**, 407–410.

Cork, M.J., Tarlow, J.K., Blakemore, A.I.F., McDonagh, A.J.G., Messenger, A.G., Bleehan, S.S. and Duff, G.W. (1993). Genetics of interleukin-1 receptor antagonist in inflammatory skin diseases. *J. Invest. Dermatol.* **100**, 736.

D'Alfonso, S. and Richiardi, P.M. (1994). A polymorphic variation in a putative regulation box of the TNFA promoter region. *Immunogenetics* **39**, 150–154.

Dawkins, R.L., Leaver, A., Cameron, P.U. *et al.* (1989). Some disease associated ancestral haplotypes carry a polymorphism of TNF. *Hum. Immunol.* **26**, 91–97.

di Giovine, F.S., Takhsh, E., Blakemore, A.I.F. and Duff, G.W. (1992). Single base polymorphism in the human tumour necrosis factor alpha (TNF-α) gene detectable by NcoI restriction of PCR product. *Hum. Mol. Genet.* **1**, 353.

Dinarello, C.A. and Wolff, S.M. (1993). The role of interleukin-1 in disease. *N. Engl. J. Med.* **328**, 106–113.

Duff, G.W. (1989). Peptide regulatory factors in non-malignant disease. *Lancet* **i**, 1432–1435.

Eastgate, J.A., Symonds, J.A., Wood, N.C., Grinlington, F.M., di Giovine, F.S. and Duff, G.W. (1988). Correlation of plasma interleukin-1 levels with disease activity in rheumatoid arthritis. *Lancet* **ii**, 706–709.

Ferencik, S., Lindemann, M., Horsthemke, B. and Grosse-Wilde, H. (1992). A new restriction fragment length polymorphism on the human TNF-B gene detected by AspHI digest. *Eur. J. Immunogenet.* **19**, 425–430.

Fugger, L., Bendtzen, K., Morling, N. *et al.* (1989a). Possible correlation of TNF alpha-production with RFLP in humans [letter]. *Eur. J. Haematol.* **43**, 255–256.

Fugger, L., Morling, N., Ryder, L.P. *et al.* (1989b). NcoI restriction fragment length polymorphism (RFLP) of the tumor necrosis factor (TNF alpha) region in primary biliary cirrhosis and in healthy Danes. *Scand. J. Immunol.* **30**, 185–189.

Fugger, L., Morling, N., Sandberg-Wollheim, M. *et al.* (1990). Tumor necrosis factor alpha gene polymorphism in multiple sclerosis and optic neuritis. *J. Neuroimmunol.* **27**, 85–88.

Jacob, C.O. and McDevitt, H.O. (1988). Tumour necrosis factor-alpha in murine autoimmune 'lupus' nephritis. *Nature* **331**, 356–358.

Jacob, C.O., Fronek, Z., Lewis, G.D., Koo, M., Hansen, J.A. and McDevitt, H.O. (1990). Heritable major histocompatability complex II-associated differences in production of tumor necrosis factor alpha: relevance to the genetic predisposition to systemic lupus erythematosus. *Proc. Natl Acad. Sci. U.S.A.* **87**, 1233–1237.

Jacob, C.O., Hwang, F., Lewis, G.D. and Stall, A.M. (1991a). Tumor necrosis factor alpha in murine systemic lupus erythematosus disease models: implications for genetic predisposition and immune regulation. *Cytokine* **3**, 551–561.

Jacob, C.O., Lewis, G.D. and McDevitt, H.O. (1991b). MHC class II-associated variation in the production of tumor necrosis factor in mice and humans: relevance to the pathogenesis of autoimmune diseases. *Immunol. Res.* **10**, 156–168.

Jongeneel, C.V. and Cambon-Thomsen, A. (1991). *J. Exp. Med.* **173**, 209–219.

Jongeneel, C.V., Acha-Orbea, H. and Blankenstein, T. (1990). A polymorphic microsatellite in the tumor necrosis factor alpha promoter identifies an allele unique to the NZW mouse strain. *J. Exp. Med.* **171**, 2141–2146.

Jongeneel, C.V., Briant, L., Udalova, I.A. *et al.* (1991). Extensive genetics polymorphism in the human tumor necrosis factor region and relation to extended HLA haplotypes. *Proc. Natl Acad. Sci. U.S.A.* **88**, 9717–9721.

Kornman, D.S., Crane, A., Wang, H.Y., de Giovine, F.S., Newman, M.G., Pirk, F.W., Wilson, T.G., Higginbottom, F.L. and Duff, G.W. (1997). The interleukin-1 genotype as a severity factor in adult periodontal disease. *J. Clin. Periodontol.* **24**, 72–77.

McDowell, T.L., Symons, J.A., Pliski, R., Forre, O. and Duff, G.W. (1993). A polymorphism in the 5′ region of the interleukin-1 alpha gene is associated with juvenile chronic arthritis. *Br. J. Rheumatol.* **32** (Suppl.1), 162.

McDowell, T., Symons, J.A., Ploski, R., Forre, O. and Duff, G.W. (1995). A genetic association between juvenile rheumatoid arthritis and a novel interleukin-1 alpha polymorphism. *Arthritis Rheum.* **38**, 221–228.

McGuire, W., Hill, V.S., Allsopp, C.E.M., Greenwood, B.M. and Kwiatkowski, D. (1994). Variation in the TNF-alpha promoter region associated with susceptibility to cerebral malaria. *Nature* **371**, 508–511.

McManus, R., Wilson, A.G., Mansfield, J., Weir, D.G., Duff, G.W. and Lelleher, D. (1996). TNF2, a polymorphism of the tumor necrosis-α gene promoter, is a component of the celiac disease major histocompatability complex haplotype. *Eur. J. Immunol.* **26**, 2113–2118.

Mansfield, J.C., Holden, H., Tarlow, J.K., di Giovine, F.S., McDowell, T.L., Wilson, A.G., Holdsworth, C.D. and Duff, G.W. (1994). Novel genetic association between ulcerative colitis and anti-inflammatory cytokine interleukin-1 receptor antagonist. *Gastroenterol.* **106**, 637–642.

Messer, G., Spengler, U., Jung, M.C. *et al.* (1991a). Polymorphic structure of the tumor necrosis factor (TNF) locus: an NcoI polymorphism in the first intron of the human TNF-beta gene correlates with a variant amino acid in position 26 and a reduced level of TNF-beta production. *J. Exp. Med.* **173**, 209–219.

Messer, G., Spengler, U. and Jung, M.C. *et al.* (1991b). Allelic variation in the TNF-beta gene does not explain the low TNF-beta response in patients with primary biliary cirrhosis. *Scand. J. Immunol.* **34**, 735–740.

Molvig, J., Back, L., Cristensen, P. *et al.* (1988). Endotoxin-stimulated human monocyte secretion of interleukin-1, tumor necrosis factor alpha, and prostaglandin E2 shows stable interindividual differences. *Scand. J. Immunol.* **27**, 705.

Nedospavov, S.A., Udalova, I.A., Kuprash, D.V. and Turetskaya, R.L. (1991). DNA sequence polymorphism at the human tumor necrosis factor (TNF) locus. Numerous TNF/lymphotoxin alleles tagged by two closely linked microsatellites in the upstream region of the lymphotoxin (TNF-beta) gene. *J. Immunol.* **147**, 1053–1059.

Nicklin, M.J., Weith, A. and Duff, G.W. (1994). A physical map of the region encompassing the human interleukin-1 alpha, interleukin-1 beta, and interleukin-1 receptor antagonist genes. *Genomics* **19**, 382–384.

Noguchi, M., Yi, H. and Rosenblatt, H.M. (1993). Interleukin-2 receptor gamma chain mutation results in x-linked severe combined immunodeficiency in humans. *Cell* **73**, 147–157.

Partanen, J. and Koskimies, S. (1988). Low degree of DNA polymorphism in the HLA-linked lymphotoxin (tumour necrosis factor beta) gene. *Scand. J. Immunol.* **28**, 313–316.

Pociot, F., Molvig, J., Wogensen, L. *et al.* (1992) A Taq1 polymorphism in the human interleukin-1 beta (IL-1 beta) gene correlates with IL-1 beta secretion *in vitro*. *Eur. J. Clin. Invest.* **22**, 396–402.

Spurr, N.K., Hill, N. and Rocchi, M. (1996). Report of the Fourth International Workshop on Human Chromosome 2 Mapping 1996. *Cytogenet. Cell Genet.* **73**, 256–268.

Sunden, S.L.F., Yandave, C.N. and Buetow, K.H. (1996). Chromosomal assignment of 2900 tri- and tetranucleotide repeat markers using NIGMS somatic cell hybrid panel 2. *Genomics* **32**, 15–20.

Tarlow, J.K, Blakemore, A.I.F., Lennard, A., Solari, R., Hughes, H., Steinkasserer, A. and Duff, G.W. (1993). Polymorphism in human IL-1 receptor antagonist gene intron 2 is caused by variable numbers of an 86-bp tandem repeat. *Hum. Genet.* **91**, 403–404.

Tarlow, J.K., Clay, F.E., Cork, M.J., Blakemore, A.I., McDonagh, A.J., Messinger, A.G. and Duff, G.W. (1994). Severity of alopecia areata is associated with a polymorphism in the interleukin-1 receptor antagonist gene. *J. Invest. Dermatol.* **103**, 387–390.

Todd, S. and Naylor, S.L. (1991). Dinucleotide repeat polymorphism in the human surfactant-associated protein 3 gene (SFTP3). *Nucleic Acids Res.* **19**, 3756–3760.

Udalova, I.A., Nedospasov, S.A., Webb, G.C., Chaplin, D.D. and Turetskaya, R.L. (1993). Highly informative typing of the human TNF locus using six adjacent polymorphic markers. *Genomics* **16**, 180–186.

Verjans, G.M., van der Linden, S.M., van Eys, G.J.J.M. *et al.* (1991). Restriction fragment length polymorphism of the tumor necrosis factor region in patients with ankylosing spondylitis. *Arthritis Rheum.* **34**, 486–489.

Webb, G.C. and Chaplin, D.D. (1990). Genetic variability at the human tumor necrosis factor loci. *J. Immunol.* **145**, 1278–1285.

Wilson, A.G., di Giovine, F.S., Blakemore, A.I.F. and Duff, G.W. (1992). Single base polymorphism in the human tumour necrosis factor alpha (TNF alpha) gene detectable by NcoI restriction of PCR product. *Hum. Mol. Genet.* **1**, 353.

Wilson, A.G., de Vries, N., Pociot, F., di Giovine, F.S., van der Putte, L.B. and Duff, G.W. (1993). An allelic polymorphism within the human tumor necrosis factor alpha promoter region is strongly associated with HLA A1, B8, and DR3 alleles. *J. Exp. Med.* **177**, 557–560.

Wilson, A.G., Gordon, C., di Giovine, F.S., de Bries, N., van der Putte, L.B.A., Emery, P. and Duff, G.W. (1994). A genetic association between systemic lupus erythematosus and tumor necrosis factor alpha. *Eur. J. Immunol.* **24**, 191–195.

Wilson, A.G., Symons, J.A., McDowell, T.L., McDevitt, H.O. and Duff, G.W. (1997). Effects of a polymorphism in the human tumor necrosis factor alpha promoter on transcriptional activation. *Proc. Natl Acad. Sci. U.S.A.* **94**, 3195–3199.

Zuliani, G. and Hobbs, H.H. (1990). A high frequency of length polymorphisms in repeated sequences adjacent to Alu sequences. *Am. J. Hum. Genet.* **46**, 963–969.

Chapter 3

Interleukin-1

Charles A. Dinarello

Department of Medicine, University of Colorado Health Science Center, Division of Infectious Diseases, Denver, CO, USA

INTRODUCTION

Interleukin-1 (IL-1) is the prototypic 'multifunctional' cytokine. There are two forms of IL-1, IL-1α and IL-1β, and in most studies their effects in terms of biologic activity are indistinguishable. Unlike the lymphocyte and colony stimulating growth factors, IL-1 affects nearly every cell type, often in concert with other cytokines or small mediator molecules. Although some lymphocyte and colony stimulating growth factors may be therapeutically useful, IL-1 is a highly inflammatory cytokine and the margin between clinical benefit and unacceptable toxicity in humans is exceedingly narrow. In contrast, agents that reduce the production and/or activity of IL-1 are likely to have an impact on clinical medicine. In support of this concept there is growing evidence that the production and activity of IL-1, particularly IL-1β, are tightly regulated events and appear destined to reduce the response to IL-1 during disease. In addition to controlling gene expression, synthesis and secretion, this regulation extends to surface receptors, soluble receptors and a receptor antagonist.

Investigators have studied how production of the different members of the IL-1 family is controlled, the various biologic activities of IL-1, the distinct and various functions of the IL-1 receptor (IL-1R) family and the complexity of intracellular signaling. Mice deficient in IL-1β, IL-1β converting enzyme (ICE) and IL-1R type I have also been studied. Humans have been injected with IL-1 (either IL-1α or IL-1β) to enhance bone marrow recovery and for cancer treatment. The IL-1-specific receptor antagonist (IL-1Ra) has also been tested in clinical trials.

There are three members of the IL-1 gene family: IL-1α, IL-1β and IL-1Ra. IL-1α and IL-1β are agonists and IL-1Ra is a specific receptor antagonist. The naturally occurring IL-1Ra appears to be a unique situation in cytokine biology. The intron–exon organization of the three IL-1 genes suggests duplication of a common gene some 350 million years ago. Before this common IL-1 gene, there may have been an ancestral gene from which fibroblast growth factor (FGF) evolved, as IL-1 and FGF share significant amino acid homologies, lack a signal peptide and form an all β-pleated sheet tertiary structure. IL-1α and β are synthesized as precursors without leader sequences. The molecular weight of each precursor is 31 kDa. Processing of IL-1α or IL-1β to 'mature' forms of

The Cytokine Handbook, 3rd ed.
ISBN 0–12–689662–3

17 kDa requires specific cellular proteases. In contrast, IL-1Ra evolved with a signal peptide and is readily transported out of the cells and termed secreted IL-1Ra (sIL-1Ra).

There are two primary cell surface binding proteins (IL-1 receptors) for IL-1. IL-1 type I receptor (IL-1RI) transduces a signal whereas the type II receptor (IL-1RII) binds IL-1 but does not transduce a signal. In fact, IL-1RII acts as a sink for IL-1β and has been termed a 'decoy' receptor which is somewhat unique to cytokine biology (Colotta *et al.*, 1994). When IL-1 binds to IL-1RI, a complex is formed which then binds to the IL-1R accessory protein (IL-1R-AcP) resulting in high-affinity binding (Greenfeder *et al.*, 1995a). The IL-1R-AcP itself does not bind IL-1. It is likely that the heterodimerization of the cytosolic domains of IL-1RI and IL-1R-AcP triggers IL-1 signal transduction.

As with other cytokines, any importance in health and disease has been revealed using gene deletion in mice. Gene expression and synthesis of IL-1α, IL-1β or IL-1Ra has been shown in ovarian granulosa and theca cells as well as in the dividing embryo. In granulosa cells from *in vitro* fertilization, there are 2000 IL-1 binding sites per cell (Polan *et al.*, 1994). Although IL-1 is found in placental trophoblasts and appears to play a role in embryonic development, studies of implantation, birth and neonatal development of mice deficient in IL-1β, ICE or IL-1RI suggests that ovulation, fertilization, implantation and parturition either do not require IL-1 receptor signaling or that compensatory cytokines are used by these mice. In addition, mice deficient in IL-1β, ICE or IL-1RI are not susceptible to infection in the standard animal facility, in sharp contrast to mice deficient in IL-10 or transforming growth factor-β.

INTERLEUKIN-1α

The 31-kDa IL-1α precursor (proIL-1α) is synthesized in association with cytoskeletal structures (microtubules), unlike most proteins translated in the endoplasmic reticulum (Stevenson *et al.*, 1992). ProIL-1α is fully active as a precursor (Mosley *et al.*, 1987) and remains intracellular (Fig. 1). The opposite is the case with the IL-1β precursor (proIL-1β) which is not fully active and a considerable amount is secreted following cleavage by a specific intracellular protease (see below). When cells die, proIL-1α is released and can be cleaved by extracellular proteases (Kobayashi *et al.*, 1991). ProIL-1α can also be cleaved by activation of the calcium-dependent membrane-associated cysteine proteases called calpains (Kobayashi *et al.*, 1990; Miller *et al.*, 1994). In transformed cell lines constitutively synthesizing proIL-1α, the addition of a calcium ionophore stimulates calpain which cleaves the precursor. Hence, release of the 17-kDa IL-1α can take place in the absence of cell death (Watanabe and Kobayashi, 1994).

Because of the lack of a leader peptide, proIL-1α remains in the cytosol soon after translation and there is no appreciable accumulation of IL-1 in any specific organelle. Immunohistochemical studies of IL-1α in endotoxin-stimulated human blood monocytes reveals a diffuse staining pattern but, by comparison in the same cell, IL-1Ra is localized to the Golgi (Andersson *et al.*, 1992). In experimental inflammatory bowel disease, there is a better correlation of disease severity with colonic tissue levels of IL-1α compared with those of IL-1β (Cominelli *et al.*, 1990), presumably due to the cell-associated nature of IL-1α. IL-1α is not commonly found in the circulation or in body fluids except during severe disease in which case the cytokine may be released from dying

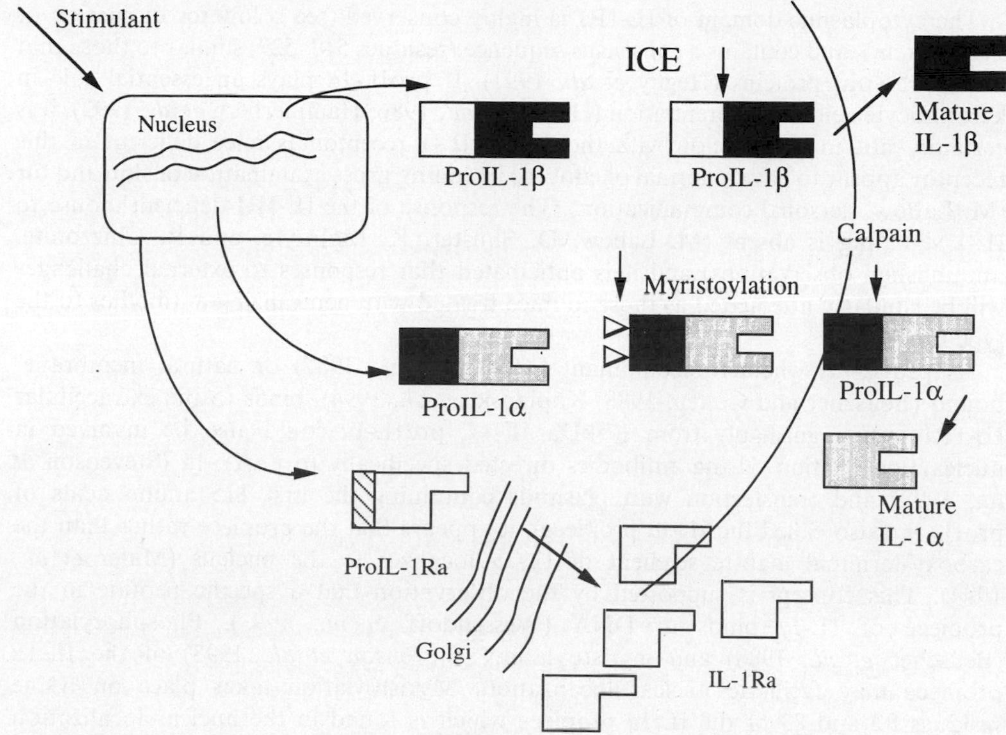

Fig. 1. Scheme for the synthesis and secretion of IL-1α, IL-1β and IL-1Ra. See text for details.

cells (Wakabayashi *et al.*, 1991) or by proteolysis after calpain-mediated cleavage (Watanabe and Kobayashi, 1994).

The concept that IL-1α can be an autocrine growth factor takes into account three distinct mechanisms: first, that proIL-1α is synthesized and remains inside the cell where it exerts a direct effect by binding to the nucleus; second, that intracellular proIL-1α complexes to an intracellular pool of IL-1RI before exerting an effect as a ligand–receptor complex; and third, that either proIL-1α or mature IL-1α bound to surface IL-1RI is internalized with subsequent translocation to the nucleus (similar to steroid receptors). Each mechanism has supporting experimental data.

Initially, Mizel and coworkers (1987) reported that radiolabeled 17-kDa recombinant IL-1α bound to the cell surface receptors was rapidly internalized and after 2–3 h was found associated with the nucleus. It was unclear whether the nuclear binding was comprised of the IL-1α–IL-1R complex or just the ligand. Curtis *et al.* (1990) reported that internalized IL-1α was still bound to its receptor and that internalized IL-1R correlated with increased signal transduction. Using truncated mutants of the cytoplasmic domain of IL-1RI, rapid internalization and nuclear localization of IL-1β was observed with several mutants not capable of transducing a biologic signal (Heguy *et al.*, 1991). It was later shown that the IL-1α–IL-1R complex, but not 17-kDa IL-1α, bound to immobilized DNA and could be eluted under the same salt conditions as that of the estrogen receptor (Weizmann and Savage, 1992).

The cytoplasmic domain of IL-1RI is highly conserved (see below for discussion of *Toll* protein), and contains a consensus sequence (residues 517–529) similar to those that transport viral proteins (Heguy *et al.*, 1991). If proIL-1α plays an essential role in keratinocyte cellular differentiation (Hauser *et al.*, 1986; Hammerberg *et al.*, 1992), it is certainly not in conjunction with the type I IL-1 receptor as mice deficient in this receptor appear to have a normal phenotype including gross examination of skin and fur (M. Labow, personal communication). The response of the IL-1RI-deficient mouse to IL-1 signaling is absent (M. Labow, D. Shuster, K. McIntyre, and R. Chizzonite, unpublished observations) and it is anticipated that responses to external challenges will be similarly attenuated as those in mice treated with neutralizing antibodies to the type I receptor.

As proIL-1α, whether recombinant (Mosley *et al.*, 1987) or natural membrane-bound (Beuscher and Colten, 1988; Kaplanski *et al.*, 1994), binds to the extracellular IL-1RI indistinguishably from 17-kDa IL-1α, proIL-1α could also be involved in nuclear localization. Using antibodies directed specifically to proIL-1α (Stevenson *et al.*, 1992) and transfection with plasmids containing the first 115 amino acids of proIL-1α (also called the IL-1α propiece), it appears that the propiece rather than the carboxy-terminal mature segment of IL-1α localizes to the nucleus (Maier *et al.*, 1994). This concept is supported by the observation that a specific peptide in the propiece of IL-1α binds to DNA (Wessendorf *et al.*, 1993). Phosphorylation (Beuscher *et al.*, 1988) and myristoylation (Stevenson *et al.*, 1993) of the IL-1α propiece may facilitate nuclear localization. Myristoylation takes place on lysine residues 82 and 83 of the IL-1α propiece which is found in the nuclear localization sequence KVLKKRR (Wessendorf *et al.*, 1993). Transfecting endothelial cells with a plasmid containing this sequence revealed nuclear localization (Maier *et al.*, 1994). Transfecting cells with the propiece of IL-1α results in slower rate of proliferation (Maier *et al.*, 1994) consistent with a role for IL-1α in early endothelial senescence (Maier *et al.*, 1990). Transfection of an intracellular IL-1-producing plasmid increases IL-2 production in thymoma cells and this biologic effect is prevented by antisense IL-1 (Falk and Hofmeister, 1994), suggesting that IL-1α without its receptor is functional as an intracellular molecule.

ProIL-1α can be found on the surface of several cells, particularly on monocytes and B lympocytes after stimulation *in vitro*. Approximately 10–15% of the IL-1α is myristoylated (Stevenson *et al.*, 1993) and this form is thought to be transported to the cell surface where it is called 'membrane' IL-1 (Kurt-Jones *et al.*, 1985). Myristoylation on specific lysines facilitates passage to the cell membrane (Stevenson *et al.*, 1993). This 'membrane' IL-1α is biologically active, its biologic activities are neutralized by anti-IL-1α and not anti-IL-1β antibodies and appears to be anchored via a lectin interaction involving mannose residues (Brody and Durum, 1989). Using high concentrations of IL-1Ra to prevent IL-1α binding to the cell surface IL-1R during fixation, the biologic activity of membrane IL-1α was unaffected. In contrast, a mannose-like receptor appears to bind membrane IL-1α (Kaplanski *et al.*, 1994). Although IL-1α has glycosylation sites, recombinant forms of mature IL-1 are biologically active when expressed in *Escherichia coli* which lacks the ability to glycosylate proteins. As membrane IL-1α is probably a glycosylated or myristoylated form of the cytokine, it accounts for no more than 5% of the total proIL-1α synthesized by the cell. There has been some dispute whether membrane IL-1α represents 'leak' of intracellular IL-1 (Minnich-Carruth *et al.*, 1989);

with prolonged fixation, leakage does not account for the activity of membrane IL-1 (Bailly *et al.*, 1990; Kaplanski *et al.*, 1994).

INTERLEUKIN-1β

Nearly all microbes and microbial products induce production of the three IL-1 proteins, but stimulants of non-microbial origin can also stimulate transcription and in many cases synthesis of the IL-1 family. Depending on the stimulant, IL-1β mRNA levels rise rapidly within 15 min but start to fall after 4 h. This decrease is thought to be due to the synthesis of a transcriptional repressor and/or a decrease in mRNA half-life (Fenton *et al.*, 1988; Jarrous and Kaempfer, 1994). Using IL-1 itself as a stimulant of its own gene expression, IL-1β mRNA levels were sustained for over 24 h (Schindler *et al.*, 1990c; Serkkola and Hurme, 1993). Raising cAMP levels in these same cells with histamine enhances IL-1α-induced IL-1β gene expression and protein synthesis (Vannier and Dinarello, 1993). In human peripheral blood mononuclear cells (PBMCs), retinoic acid induces IL-1β gene expression but the primary precursor transcripts fail to yield mature mRNA (Jarrous and Kaempfer, 1994). Inhibition of translation by cyclohex-imide results in enhanced splicing of exons, excision of introns and increased levels of mature mRNA (superinduction) by two orders of magnitude. Thus, synthesis of mature IL-1β mRNA requires an activation step to overcome an apparently intrinsic inhibition to process precursor mRNA.

Stimulants such as the complement component C5a (Schindler *et al.*, 1990b), hypoxia (Ghezzi *et al.*, 1991), adherence to surfaces (Schindler *et al.*, 1990a) or clotting of blood (Mileno *et al.*, 1995) induce the synthesis of large amounts of IL-1β mRNA in monocytic cells without significant translation into the IL-1β protein. This dissociation between transcription and translation is characteristic of IL-1β but also of tumor necrosis factor-α (TNF-α) (Schindler *et al.*, 1990a). It appears that the above stimuli are not sufficient to provide a signal for translation despite a vigorous signal for transcription. Without translation, most of the IL-1β mRNA is degraded and this has been observed in humans undergoing hemodialysis with complement-activating membranes (Schindler *et al.*, 1993). Although the IL-1β mRNA assembles into large polyribosomes, there is little significant elongation of the peptide (Kaspar and Gehrke, 1994). However, adding bacterial endotoxin or IL-1 itself to cells with high levels of steady state IL-1β mRNA results in augmented translation (Schindler *et al.*, 1990a,b) in somewhat the same manner as the removal of cycloheximide following superinduction. One explanation is that stabilization of the AU-rich 3′ untranslated region takes place in cells stimulated with lipopolysaccharide (LPS). These AU-rich sequences are known to suppress normal haemoglobin synthesis. The stabilization of mRNA by microbial products may explain why low concentrations of LPS or a few bacteria or *Borrelia* organisms per cell induce the translation of large amounts of IL-1β (Miller *et al.*, 1992).

Another explanation is that IL-1 stabilizes its own mRNA (Schindler *et al.*, 1990c) by preventing deadenylation as it does for the chemokine gro-α (Stoeckle and Guan, 1993). Removal of IL-1 from cells after 2 h increases the shortening of poly(A) and IL-1 apparently is an important regulator of *gro* synthesis because it prevents deadenylation. In fact, of the several cytokines induced by IL-1, large amounts of the chemokine family are produced in response to low concentrations of IL-1. For example, 1 pM of IL-1 stimulates fibroblasts to synthesize 10 nM of IL-8 (Shapiro *et al.*, 1994).

Following synthesis, proIL-1β remains primarily cytosolic until it is cleaved and transported out of the cell (Fig. 1). The IL-1β propiece (amino acids 1–116) is also myristoylated on lysine residues (Stevenson *et al.*, 1993) but, unlike IL-1α, proIL-1β has no known membrane form and is only marginally active (Jobling *et al.*, 1988). Some IL-1β is found in lysosomes (Bakouche *et al.*, 1987) or associated with microtubules (Rubartelli *et al.*, 1990; Stevenson *et al.*, 1992); either localization may play a role in the secretion of IL-1β. In mononuclear phagocytes, a small amount of proIL-1β is secreted from intact cells (Auron *et al.*, 1987; Beuscher *et al.*, 1990) but the pathway for this secretion remains unknown. On the other hand, release of mature IL-1β appears to be linked to processing at the aspartic acid–alanine (116–117) peptide cleavage by ICE (Black *et al.*, 1988) (see below).

Although well controlled in the setting of laboratory cell culture, death and rupture of inflammatory cells is not an unusual occurrence *in vivo*. There are several sites in the *N*-terminal 16-kDa part of proIL-1β that are vulnerable to cleavage by enzymes in the vicinity of alanine 117. These are trypsin (Kobayashi *et al.*, 1991), elastase (Dinarello *et al.*, 1986), chymotrypsin (Mizutani *et al.*, 1991a), a mast cell chymase (Mizutani *et al.*, 1991b) and a variety of proteases (Hazuda *et al.*, 1990, 1991) which are commonly found in inflammatory fluids. The extent that these proteases play in the *in vivo* conversion of proIL-1β to mature forms is uncertain but, in each case, a biologically active IL-1β species is produced. In the discussion on the soluble IL-1 receptor type II (below), the affinity of proIL-1β for this constitutively produced soluble receptor is high and may prevent haphazard cleavage of the precursor by these enzymes in inflammatory fluids.

INTERLEUKIN-1β CONVERTING ENZYME

As depicted in Fig. 1, proIL-1β requires cleavage before the mature form is secreted. The cDNA encoding ICE has been reported (Cerretti *et al.*, 1992; Thornberry *et al.*, 1992). The 45-kDa precursor of ICE requires two internal cleavages before becoming the enzymatically active heterodimer comprised of a 10- and 20-kDa chain. The active site cysteine is located on the 20-kDa chain. ICE itself contributes to autoprocessing of the ICE precursor by undergoing oligomerization with itself or homologs of ICE (Wilson *et al.*, 1994; Gu *et al.*, 1995b). ICE is the first member of a family of intracellular cysteine proteases called caspases. The term caspase is used to denote the activity of the cysteine proteases cleaving after an aspartic acid residue. ICE is caspase-1.

The tertiary structure of the active site has been reported (Walker *et al.*, 1994; Wilson *et al.*, 1994). Two molecules of the ICE heterodimer form a tetramer with two molecules of proIL-1β for cleavage (Walker *et al.*, 1994; Wilson *et al.*, 1994). The aspartic acid at position 116 of the proIL-1β is the recognition amino acid for ICE cleavage. ICE does not cleave the IL-1α precursor. Enzymes such as elastase (Dinarello *et al.*, 1986) and granzyme A (Irmler *et al.*, 1995) cleave proIL-1β at amino acids 112 and 120 respectively yielding biologically active IL-1β. The propiece of IL-1β can be found both inside and outside the cell (Higgins *et al.*, 1994). In addition, the propiece exhibits biologic activity as a chemoattractant for fibroblasts via an IL-1R-mediated event (Higgins *et al.*, 1993).

In the presence of a tetrapeptide competitive substrate inhibitor of ICE, the generation and secretion of mature IL-1β is reduced and proIL-1β accumulates mostly inside but also outside the cell (Thornberry *et al.*, 1992). This latter finding supports the concept that proIL-1β can be released from a cell independent of processing by ICE. Similar to that of thioredoxin (Rubartelli *et al.*, 1992) and basic FGF (Mignatti and Rifkin, 1991), exocytosis has been proposed as a possible mechanism of proIL-1β release. A putative membrane 'channel' where active ICE is localized has also been proposed. In this model, mature IL-1β is released through this channel (Singer *et al.*, 1993). When ICE activity is blocked by a reversible competitive substrate inhibitor, greater amounts of proIL-1β are found in the supernatant (Thornberry *et al.*, 1992; Singer *et al.*, 1993) and thus the putative channel may provide a passive secretory pathway for both proIL-1β and mature IL-1β. Macrophages from ICE-deficient mice do not release mature IL-1β upon stimulation *in vitro* (Kuida *et al.*, 1995; Li *et al.*, 1995). Although neutrophil enzymes such as elastase and granzyme A (Irmler *et al.*, 1995) can cleave proIL-1β at sites close to alanine 117, proIL-1β accumulates in cells from ICE-deficient mice (Kuida *et al.*, 1995; Li *et al.*, 1995). Interestingly, IL-1α production in macrophages from ICE-deficient mice is reduced, a finding that is consistent with self-induction of IL-1 gene expression and synthesis (Dinarello *et al.*, 1987). ICE-deficient mice can be either resistant (Li *et al.*, 1995) or susceptible (Kuida *et al.*, 1995) to lethal endotoxemia. Recent studies suggest that mice deficient in ICE fail to develop collagen-induced arthritis.

ICE AND INTERFERON-γ-INDUCING FACTOR (IL-18)

In 1989, an endotoxin-induced serum activity that induced interferon-γ (IFN-γ) from mouse spleen cells was described (Nakamura *et al.*, 1989). This serum activity functioned not as a direct inducer of IFN-γ but rather as a costimulant together with IL-2 or mitogens. An attempt to purify the activity from postendotoxin mouse serum revealed an apparently homogeneous 50–55-kDa protein (Nakamura *et al.*, 1993). As other cytokines can act as costimulants for IFN-γ production, the failure of neutralizing antibodies to IL-1, IL-4, IL-5, IL-6 or TNF to neutralize the serum activity suggested it was a distinct factor. In 1995, the third report was published from the same group of scientists demonstrating that the endotoxin-induced costimulant for IFN-γ production was present in extracts of livers from mice preconditioned with *Propionibacterium acnes* (Okamura *et al.*, 1995a). In this model, the hepatic macrophage population (Kupffer cells) expand, and in these mice a low dose of bacterial LPS, which in non-preconditioned mice is not lethal, becomes lethal. The factor, named IFN-γ-inducing factor (IGIF), was purified to homogeneity from 1200 g of *P. acnes*-treated mouse livers. Its molecular weight was 18–19 kDa and an *N*-terminal amino acid sequence was reported (Okamura *et al.*, 1995a). Similar to the endotoxin-induced serum activity, IGIF did not induce IFN-γ by itself but functioned primarily as a costimulant with mitogens or IL-2. Degenerate oligonucleotides derived from amino acid sequences of purified IGIF were used to clone a murine IGIF cDNA (Okamura *et al.*, 1995b). Recombinant IGIF did not induce IFN-γ by itself only in the presence of a mitogen or IL-2. However, the coinduction of IFN-γ was independent of IL-12 induction of IFN-γ.

Neutralizing antibodies to mouse IGIF were shown to prevent the lethality of low-dose LPS in *P. acnes*-preconditioned mice (Okamura *et al.*, 1995b). Other had reported the importance of IFN-γ as a mediator of LPS lethality in preconditioned mice. For example, neutralizing anti-IFN-γ antibodies protected mice against Schwartzmann-like shock (Heremans *et al.*, 1990) and galactosamine-treated mice deficient in the IFN-γ receptor were resistant to LPS-induced death (Car *et al.*, 1994). Hence, it was not unexpected that neutralizing antibodies to murine IGIF protected *P. acnes*-preconditioned mice against lethal LPS (Okamura *et al.*, 1995b). Anti-murine IGIF treatment also protected surviving mice against severe hepatic cytotoxicity. After the murine form was cloned (Okamura *et al.*, 1995b), the human cDNA sequence for IGIF was reported in 1996 (Ushio *et al.*, 1996). Recombinant human IGIF exhibited natural IGIF activity (Ushio *et al.*, 1996). Human recombinant IGIF was without direct IFN-γ-inducing activity on human T cells but acted as a costimulant for production of IFN-γ and other T helper cell 1 (T_{H1}) cytokines (Ushio *et al.*, 1996). IGIF induced T-cell and natural killer (NK)-cell IFN-γ production independently of IL-12 (and *vice versa*) (Okamura *et al.*, 1995b). To data, IGIF is thought of primarily as a costimulant for T_{H1} cytokine production (IFN-γ, IL-2 and granulocyte–macrophage colony stimulating factor (GM-CSF)) (Kohno *et al.*, 1997) and also as costimulant for Fas ligand-mediated cytotoxicity of murine NK cell clones (Tsutsui *et al.*, 1996). *In vivo*, endogenous IGIF activity appears to account for IFN-γ production in *P. acnes*- and LPS-mediated lethality (Okamura *et al.*, 1995b).

Scientists working on other IFN-γ-inducing cytokines analyzed the computer-generated protein folding pattern of murine IGIF and compared its pattern with those of others in the data bank. Using a validated compatibility relatedness program, the mature murine IGIF had the highest score with mature human IL-1β; furthermore, the IGIF amino acid sequence matched best with amino acids that form the all-β-pleated sheet folding pattern of human IL-1β (Bazan *et al.*, 1996). A high degree of alignment was present in the sequences that comprise the 12 β-sheets of the mature IL-1β structure. Using this alignment of conserved amino acids, there is a 19% positional identity of mature murine IGIF with mature human IL-1β and a 12% identity with human IL-1α. Using this same positional alignment, the identity of IL-1β with IL-1α is 23%. It was suggested that the name IGIF be changed to interleukin-1γ (IL-1γ) (Bazan *et al.*, 1996). Does IGIF bind to IL-1 type I receptors? This would be an essential criterion for assigning the name IL-1γ as the type I IL-1 receptor is the signaling receptor for the biologic activity of IL-1. In the absence of evidence that IGIF binds to the IL-1 receptor type I (unpublished data), IL-18 rather than IL-1γ is a more appropriate name (Ushio *et al.*, 1996). Very little is presently known about the spectrum of its activities.

Similar to precursor IL-1β (proIL-1β), precursor IL-18 (proIL-18) does not contain a signal peptide required for the removal of the precursor amino acids with subsequent secretion. The *N*-terminal amino acid sequence of the secreted form of murine IL-18 (Okamura *et al.*, 1995a) is consistent with that following cleavage after an aspartic acid residue, a typical cleavage site for ICE. In fact, this analysis indicated that cleavage of proIL-18 at the aspartic acid site would probably require ICE (Bazan *et al.*, 1996). Therefore, it was not surprising that ICE cleaved proIL-18 (after the aspartic acid 19) and resulted in the mature and active protein (Gu *et al.*, 1997).

ICE AND FAS-MEDIATED CELL DEATH

The gene *ced-3* in the nematode *Caenorhabditis elegans* codes for a protein homologous to human ICE (Yuan *et al.*, 1993). During embryonic development of the worm, this gene is expressed in specialized cells and is thought to be responsible for programmed cell death (apoptosis). In the worm, *ced-9* protects against apoptosis; in the human, the homologous death protecting gene is *bcl-2*. There is a remarkably conserved homology of the five amino acids required for *ced-3* and human ICE activity. Other homologs of ICE have been discovered and each has a similar aspartic acid substrate specificity. *Nedd2* is a mouse gene which is also expressed in cells undergoing apoptosis during development and has recently been shown to be homologous to *ced-3* and ICE (Kumar *et al.*, 1994). Overexpression of ICE or any of its homologs in transiently transfected cells is associated with increased apoptosis (Miura *et al.*, 1993; Fernandes-Alnemri *et al.*, 1994; Kumar *et al.*, 1994; Faucheu *et al.*, 1995). Cell death induced by ICE or its homologs can also be reduced by cotransfection with *crm-A*, a cow pox viral gene coding for an inhibitor of proteases including ICE (Ray *et al.*, 1992). For example, transfection with *crm-A* in neurons prevents programmed cell death due to removal of nerve growth factor (Gagliardini *et al.*, 1994). *Crm-A* is, however, not specific for inhibiting ICE. One highly consistent finding is that cotransfection with *crm-A* prevents cell death associated with cotransfection of several members of the ICE family. At present, no clear proteolytic substrate cascade has been identified that accounts for the initiation of apoptosis by ICE or its homologs.

Recent studies in ICE-deficient mice have shed light on the relationship of ICE to programmed cell death. In these mice, the thymus develops normally; furthermore, stressed-induced apoptosis (corticosteroids or radiation) of thymocytes and macrophades *in vitro* is also normal in these mice (Kuida *et al.*, 1995; Li *et al.*, 1995). However, apoptosis in thymocytes by an activating antibody to Fas, a TNF-related receptor, is diminished in ICE-deficient mice (Kuida *et al.*, 1995). Blocking ICE activity with a specific substrate inhibitor also reduces Fas-induced as well as TNF-induced apoptosis (Enari *et al.*, 1995). Apoptosis triggered by activating Fas or TNF receptors is associated with a 'death domain' on the cytoplasmic segment of these receptors. However, recent studies suggest that ICE (caspase-1) is not required for Fas-mediated cell death (Smith *et al.*, 1997).

Several reports demonstrate that overexpression of ICE leads to programmed cell death. If there is a link between ICE, IL-1β and cell death, it is most likely due to an IL-1R-mediated event rather than an intracellular mechanism. In any cell or tissue in which IL-1β is produced under conditions of disease, agent inhibiting ICE will probably reduce IL-1β-mediated nitric oxide (NO) synthesis and any NO-mediated cell death. Under these conditions, inhibiting ICE should reduce cell death to the same extent as a neutralizing antibody to IL-1β or receptor blockade. Overall, the data support the concept that there are other substrates for caspases and that overexpressing the ICE family of proteases cleaves non-IL-1β intracellular proteins which trigger cell death. Since ICE-deficient mice have normal neuronal and immune cell development, this enzyme does not participate in cell death required for embryonic development.

There is concern that ICE inhibitors which reduce inflammation by inhibiting processing and secretion of IL-1β may inadvertently prolong the life of malignant cells. Is there a risk of aggravating autoimmune or malignant disease using ICE inhibitors? Data support the opposite view; ICE inhibition reduces inflammation without a change in cell death. One explanation may be that cleavage of molecules such as poly ADP-ribose polymerase (PARP) require 50–100-fold more ICE compared with cleavage of proIL-1β (Gu *et al.*, 1995a). Incubation of leukemic blasts from patients with acute myelogenous leukemia (AML) in the presence of ICE inhibitors reduced IL-1β secretion (Estrov *et al.*, 1994); Margolis and Dinarello, 1994) and spontaneous proliferation (Estrov *et al.*, 1994) but did not increase cell survival. Antisense ICE also reduced spontaneous proliferation of AML cells without increasing survival (Stosic-Grujicic *et al.*, 1995). These were anticipated results as IL-1β is a growth factor for AML cells (Cozzolino *et al.*, 1989). Hence, inhibition of ICE in these models does not worsen but improves disease outcome. Nevertheless, the clinical benefit of reducing the release of mature IL-1β will require an inhibitor with a high degree of specificity for the cleavage sites of proIL-1β without affecting the cleavage of substrates of other members of the ICE family.

INTERLEUKIN-1 RECEPTOR ANTAGONIST

As shown in Fig. 1, proIL-1Ra, which possesses a leader sequence, is synthesized, processed and secreted from the cell. Upon stimulation with LPS, human blood monocytes initially express the gene for sIL-1Ra (Arend, 1993). During the first 4–6 h, sIL-1Ra protein can be visualized in the Golgi (Andersson *et al.*, 1992). After 24 h, the primary transcript in these cells is icIL-1Ra which, lacking a leader peptide, stains diffusely in the cytosol and remains intracellular (Andersson *et al.*, 1992). It has been proposed that intracellular (ic) IL-1Ra constitutively produced in keratinocytes and epithelial cells may block the binding of IL-1α to nuclear DNA (Haskill *et al.*, 1991; Arend, 1993).

The primary amino acid homology of mature human IL-1β to IL-1Ra is 26%, greater than that between IL-1α and IL-1β. Each member of the human IL-1 family is comprised of an all-β strand molecule which forms an open barrel-like structure (Preistle *et al.*, 1989; Graves *et al.*, 1990; Vigers *et al.*, 1994) closely related to the structure of FGF (Murzin *et al.*, 1992). As each member of the IL-1 family binds to the IL-1RI, it is not surprising that IL-1α, IL-1β and IL-1Ra share structural similarities. How does IL-1Ra bind to IL-1RI with nearly the same affinity as IL-1α or IL-1β and yet not trigger a response? Crystal structural analysis of the IL-1R–IL-1Ra complex reveals that IL-1Ra contacts all three domains of the IL-1RI (Schreuder, 1995).

IL-1β has two sites of binding to IL-1RI. There is a primary binding site located at the open top of its barrel shape (Gruetter *et al.*, 1994) which is similar but not identical to that of IL-1α (Lambriola-Tomkins *et al.*, 1993). There is a second site in the back side of the IL-1β molecule (Gruetter *et al.*, 1994). IL-1Ra also has two binding sites similar to those of IL-1β (Evans *et al.*, 1994; Vigers *et al.*, 1994). However, the back side site in IL-1Ra is more homologous to that of IL-1β than the primary binding site (Evans *et al.*, 1994). Thus, the present interpretation is that the back side site of IL-1Ra binds to IL-1RI and occupies the receptor. Lacking the second binding site, IL-1Ra does not trigger a signal (see below for discussion of IL-1 receptor dimerization). After IL-1Ra binds to IL-1RI-bearing cells, there is no phosphorylation of the epidermal growth factor receptor (Dripps *et al.*, 1991), a well established and sensitive assessment of IL-1 signal transduction (Bird and Saklatvala, 1989). Moreover, when injected intravenously into humans at doses 1 000 000-fold greater than that of IL-1α or IL-1β (Crown *et al.*, 1991; Smith *et al.*, 1991), IL-1Ra has no agonist activity (Granowitz *et al.*, 1992a,b).

The formation of the heterodimer consisting of the IL-1RI and IL-1R accessory protein (IL-1R-AcP) (Greenfeder *et al.*, 1995a) probably explains the failure of IL-1Ra to trigger a signal. From the structural differences described above between IL-1β and IL-1Ra, one can propose that the second binding site missing from the IL-1Ra is, in fact, the site that binds the accessory protein. The cross-linked complex of radiolabeled

IL-1Ra to the type I receptor was not precipitated by a specific antibody to the accessory protein (Greenfeder *et al.*, 1995a). As shown in Fig. 2, IL-1Ra binds to the type I receptor with the same affinity as IL-1 but, lacking the second binding site, the IL-1R-AcP does not dock to the IL-1Ra and the heterodimer is not formed. The binding of IL-1Ra to the type I receptor probably prevents or disrupts the complex between IL-1 and the type I receptor. This model implies that signal transduction takes place only when the heterodimer is formed. A triple mutation in IL-1Ra (Greenfeder *et al.*, 1995b) may have partially reconstituted the second binding domain so that a degree of dimerization takes place between the cytosolic domains of IL-1RI and IL-1R-AcP, resulting in increased agonist activity of the mutated IL-1Ra (Greedfeder *et al.*, 1995b). Single point mutations or deletions in IL-1β have resulted in molecules with greater than a 100-fold loss in biologic activity but only a small decrease in IL-1RI binding (Gehrke *et al.*, 1990; Simoncsits *et al.*, 1994).

Healthy humans are the most sensitive indicators of IL-1 agonist activity; 1 ng/kg of intravenous IL-1β produces symptoms (Tewari *et al.*, 1990). In contrast, the intravenous infusion of 10 mg/kg of IL-1Ra in healthy humans (a 10 million-fold molar excess) is

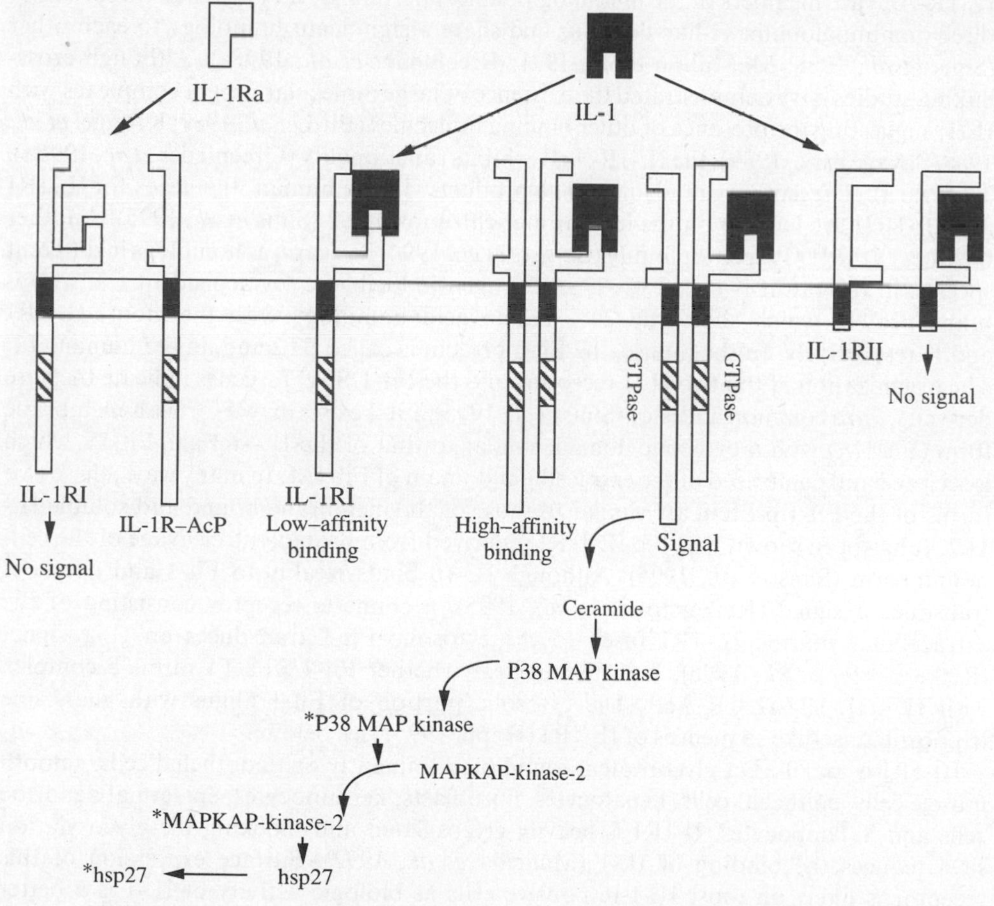

Fig. 2. IL-1 receptors and cell signaling. IL-1β is used in this example. See text for details (* phosphorylated).

without effect (Granowitz *et al.*, 1992a,b). What are the structural requirements of the respective molecules that account for this dramatic difference? The ability of IL-1β to trigger cell signaling optimally requires stability of the overall tertiary structure of the cytokine so that mutations in one amino acid may unfold the molecule resulting in a several 100-fold loss in activity but without a loss in receptor binding. This suggests that biologic activity requires binding of IL-1β to a relatively broad area on the receptor. The tertiary structure of IL-1Ra, which is closely related to that of IL-1β, allows for tight binding to the IL-1RI, but IL-1Ra clearly lacks the second binding site which allows docking of the IL-1R-AcP to form the heterodimer. Without dimerization, no signal is transduced but occupancy of the IL-1RI by IL-1Ra results in effective prevention of IL-1 signal transduction. Small molecules may mimic the near perfect receptor antagonism of IL-1Ra but, to date, none has been reported.

IL-1 RECEPTORS

Two primary IL-1 binding proteins (receptors) have been identified and one receptor accessory protein (IL-1R-AcP). The extracellular domains of the two receptors and the IL-1R-AcP are members of the immunoglobulin superfamily; they are each comprised of three immunoglobulin G-like domains and share a significant homology to each other (Sims *et al.*, 1988; McMahon *et al.*, 1991; Greenfeder *et al.*, 1995a). Although cross-linking studies have demonstrated the existence of large molecular weight complexes with IL-1, suggesting the presence of other binding molecules (Bird *et al.*, 1987; Kroggel *et al.*, 1988; Savage *et al.*, 1989), the IL-1R-AcP exhibits functionality (Greenfeder *et al.*, 1995a). The two IL-1 receptors are distinct gene products. In the human, the genes for IL-1RI and IL-1RII are located on the long arm of chromosome 2 (Sims *et al.*, 1995). Another member of the IL-1 receptor family (Bergers *et al.*, 1994) has various homologs in different species: in the rat, it is called Fit-1, an estrogen-inducible, c-*fos*-dependent transmembrane protein which shares 26–29% amino acid homology with the mouse IL-1RI and II respectively. In the mouse, the Fit-1 protein is called ST2 and, in the human, T1. The organization of the two IL-1 receptors and the Fit-1/ST2/T1 genes indicate they are derived from a common ancestor (Sims *et al.*, 1995). Fit-1 exists in two forms: a membrane form (Fit-1M) with a cytosolic domain similar to that of the IL-1RI and Fit-1S, which is secreted and comprised of the extracellular domain of Fit-1M. In many ways, these two forms of the Fit-1 protein are similar to those of the membrane-bound and soluble IL-1RI. It has been shown that the IL-1sRI is derived from proteolytic cleavage of the cell-bound form (Sims *et al.*, 1995). Although IL-1β binds weakly to Fit-1 and does not transduce a signal (Reikerstorfer *et al.*, 1995), a chimeric receptor consisting of the extracellular murine IL-1RI fused to the cytosolic Fit-1 transduces an IL-1 signal (Reikerstorfer *et al.*, 1995). It is also unclear whether Fit-1/ST2/T1 forms a complex with IL-1/IL-1RI/IL-1R-AcP. The cytosolic portion of Fit-1 aligns with guanosine triphosphatase-like sequences of IL-1RI (Hopp, 1995) (see below).

IL-1RI is an 80-kDa glycoprotein found predominantly on endothelial cells, smooth muscle cells, epithelial cells, hepatocytes, fibroblasts, keratinocytes, epidermal dendritic cells and T lympocytes. Il-1RI is heavily glycosylated and blocking the glycosylation sites reduces the binding of IL-1 (Mancilla *et al.*, 1992). Surface expression of this receptor is likely on most IL-1-responsive cells as biologic activity of IL-1 is a better assessment of receptor expression than ligand binding to cell surfaces (Rosoff *et al.*,

1988). Failure to show specific and saturable IL-1 binding is often due to the low numbers of surface IL-1RI on primary cells. In cell lines, the number of IL-1RI can reach 5000 per cell but primary cells usually express fewer than 200 receptors per cell. In some primary cells there are fewer than 50 per cell (Shirakawa *et al.*, 1987) and IL-1 signal transduction has been observed in cells expressing fewer than ten type I receptors per cell (Stylianou *et al.*, 1992). The low number of IL-1RIs on cells and the discrepancy between binding affinities and biologic activities can be explained by the increased binding affinity of IL-1 in the complex with the IL-1R-AcP (Greenfeder *et al.*, 1995a).

As shown in Fig. 2, IL-1RI has a single transmembrane segment and a cytoplasmic domain. Using specific neutralizing antibodies, IL-1RI but not IL-1RII is the primary signal transducing receptor (Stylianou *et al.*, 1992; Sims *et al.*, 1993, 1994; Dower *et al.*, 1994). Antisense oligonucleotides directed against IL-1RI block IL-1 activities *in vitro* and *in vivo* (Burch and Mahan, 1991). The cytoplasmic domain of IL-1RI has no apparent intrinsic tyrosine kinase activity but when IL-1 binds to only a few receptors the remaining unoccupied receptors appear to undergo phosphorylation (Gallis *et al.*, 1989), probably by a member of the mitogen-activated protein (MAP) kinase family. Interestingly, the cytosolic domain of IL-1RI has 45% amino acid homology with the cystolic domain of the *Drosophila Toll* gene (Gay and Keith, 1991). Toll is a transmembrane protein acting like a receptor, although the ligand for the Toll protein is unknown. Gene organization and amino acid homology suggests that the IL-1RI and the cytosolic Toll are derived from a common ancestor and trigger similar signals (Guida *et al.*, 1992; Heguy *et al.*, 1992).

As shown in Fig. 2, like other models of two-chain receptors, IL-1 binds first to the IL-1RI with a low affinity. Although there is no direct evidence, a structural change may take place in IL-1 allowing for docking of IL-1R-AcP to the IL-1RI–IL-1 complex. Once IL-1RI–IL-1 binds to IL-1R-AcP, a high-affinity binding is observed. Antibodies to the type I receptor and to the IL-1R-AcP block IL-1 binding and activity (Greenfeder *et al.*, 1995a). IL-1R-AcP is essential to IL-1 signaling; in cells deficient of IL-1R-AcP, no IL-1-inducing activation of the stress kinases takes place but this response is restored on transfection with a construct expressing the IL-1R-AcP (Wesche *et al.*, 1997). Therefore, IL-1 may bind to the type I receptor with a low affinity producing a structural change in the ligand followed by recognition by the IL-1R-AcP. Alternatively, cells express IL-1RI–IL-1R-AcP already complexed and high-affinity binding takes place on the preformed complexes.

Similar to IL-1RI and IL-1RII, a soluble form of the IL-1R-AcP exists but this form appears to result from an RNA splice donor/acceptor site resulting in a truncated protein ending before the transmembrane region. Unlike the soluble forms of the IL-1RI and IL-1RII, the soluble IL-1R-AcP is not formed by proteolytic cleavage of the full-length accessory protein. It is unclear how soluble IL-1R-AcP mRNA is expressed compared with the cell-bound protein. Furthermore, as the IL-1R-AcP does not bind IL-1 itself (Greenfeder *et al.*, 1995a), the effect of the soluble IL-1R-AcP on the binding of IL-1 remains unclear. As discussed above and shown in Fig. 2, IL-1Ra does not form a complex with the IL-1RI which probably explains how the IL-1Ra can bind so tightly to the IL-1RI and yet not exhibit any agonist activity. One thus concludes that the IL-1RI–IL-1–IL-1R-AcP complex triggers the cell and that without IL-1R-AcP participation the IL-1 signal via the IL-1RI is weak or non-existent. It is unlikely that the complex IL-1RII–IL-1–IL-1R-AcP exists.

Cell-bound IL-1RII does not appear to form a complex with the type I receptor (Slack *et al.*, 1993; Dower *et al.*, 1994), nor does it transduce a signal (Sims *et al.*, 1993, 1994). In the human and mouse, IL-1RII has a short cytosolic domain consisting of 29 amino acids. The type II receptor appears to act as 'decoy' molecule, particularly for IL-1β. The receptor binds IL-β tightly thus preventing binding to the signal transducing type I receptor (Colotta *et al.*, 1993). It is the lack of a signal transducing cytosolic domain which makes the type II receptor a functionally negative receptor. For example, when the extracellular portion of the type II receptor is fused to the cytoplasmic domain of the type I receptor, a biologic signal occurs (Heguy *et al.*, 1993). The extracellular portion of the type II receptor is found in body fluids where it is termed IL-1-soluble receptor type II (IL-1sRII). It is assumed that proteolytic cleavage of the extracellular domain of IL-1RII from the cell surface is the source of IL-1sRII.

It is likely that, as cell-bound IL-1RII increases, there is a comparable increase in soluble forms (Giri *et al.*, 1990). Similar to soluble receptors for TNF, the extracellular domains of type I and type II IL-1R are found as 'soluble' molecules in the circulation and urine of healthy subjects and in inflammatory synovial and other pathologic body fluids (Symons *et al.*, 1991, 1993; Arend *et al.*, 1994; Orencole *et al.*, 1994; Sims *et al.*, 1994). In healthy humans, the circulating levels of IL-1sRII are 100–200 pM (Arend *et al.*, 1994; Giri *et al.*, 1994; Orencole *et al.*, 1994, 1995), whereas levels of IL-1sRI are ten-fold lower (Arend *et al.*, 1994; Sims *et al.*, 1994; Orencole *et al.*, 1995). The rank of affinities for the two soluble receptors is remarkably different for each of the three IL-1 molecules. The rank for the three IL-1 ligands binding to IL-1sRI is IL-1Ra > IL-1α > IL-1β, whereas for IL-1sRII the rank is IL-1β > IL-1α > IL-1Ra. Raised levels of IL-1sRII are found in the circulation of patients with sepsis (Giri *et al.*, 1994) and in the synovial fluid of patients with active rheumatoid arthritis (Arend *et al.*, 1994), whereas the increased concentration of soluble type I receptor in these fluids is ten-fold lower (Arend *et al.*, 1994). High-dose IL-2 therapy induces IL-1sRI and IL-1sRII (Orencole *et al.*, 1995).

Unlike other cytokine receptors, in cells expressing both IL-1 type I and type II receptors, there is competition to bind IL-1 first. This competition between signaling and non-signaling receptors for the same ligand appears unique to cytokine receptors, although it exists for atrial natriuretic factor receptors (Leitman *et al.*, 1986). As the type II receptor is more likely to bind to IL-1β than IL-1α, this can result in a diminished response to IL-1β. The soluble form of IL-1sRII circulates in healthy humans at molar concentrations that are ten-fold greater than those of IL-1β measured in patients with sepsis and 100-fold greater than the concentration of IL-1β following intravenous administration (Crown *et al.*, 1991). Why do humans have a systemic response to an infusion of IL-1β? One concludes that binding of IL-1β to the soluble form of IL-1R type II exhibits a slow 'on' rate compared with the cell IL-1RI.

In addition to naturally occurring conditions that reduce the biologic response to IL-1β, neutralizing antibodies to IL-1α are present in many subjects and probably reduce the activity of IL-1α. Vaccinia and cowpox virus genes encode for a protein with high amino acid homology to the type II receptor, and this protein binds IL-1β (Alcami and Smith, 1992; Spriggs *et al.*, 1992). Despite the portfolio of soluble receptors and naturally occurring antibodies, IL-1 produced during disease does, in fact, trigger the type I receptor: in animals and humans, blocking receptors or neutralizing IL-1 ameliorates disease. These findings underscore the high functional level of only a few

IL-1 type I receptors. They also imply that the postreceptor triggering events are greatly amplified. It seems reasonable to conclude that treating disease based on blocking IL-1R needs to take into account the efficiency of so few type I receptors initiating a biologic event.

SIGNAL TRANSDUCTION

Early Events in IL-1 Signal Transduction

Within a few minutes following binding to cells, IL-1 induces several biochemical events (Rossi, 1993; Kuno and Matsushima, 1994; Mizel, 1994; O'Neill, 1995). It remains unclear which is the most 'upstream' triggering event or whether several occur at the same time. No sequential order or cascade has been identified but several signaling events appear to be taking place during the first 2–5 min. Some of the biochemical changes associated with signal transduction are likely cell specific. Within 2 min, hydrolysis of GTP (O'Neill *et al.*, 1990), phosphotidylcholine, phosphotidylserine or phosphotidylethanolamine (Rosoff *et al.*, 1988; Rosoff, 1989) and release of ceramide by neutral (Schutze *et al.*, 1994), not acidic, sphingomyelinase (Andrieu *et al.*, 1994) have been reported. In general, multiple protein phosphorylations and activation of phosphatases can be observed within 5 min (Bomalaski *et al.*, 1992) and some are thought to be initiated by the release of lipid mediators. The release of ceramide has attracted attention as a possible early signaling event (Kolesnick and Golde, 1994). Phosphorylation of phospholipase A_2 (PLA_2) activating protein also occurs in the first few minutes (Gronich *et al.*, 1994), leading to a rapid release of arachidonic acid. Multiple and similar signaling events have also been reported for TNF.

Of special consideration to IL-1 signal transduction is the unusual discrepancy between the low number of receptors (fewer than ten in some cells) and the low concentrations of IL-1 that can induce a biologic response (Orencole and Dinarello, 1989). This latter observation, however, may be clarified in studies on high-affinity binding with the IL-1R-AcP complex (Greenfeder *et al.*, 1995a). A rather extensive 'amplification' step(s) takes place following the initial postreceptor binding event. The most likely mechanism for signal amplification is multiple and sequential phosphorylations (or dephosphorylations) of kinases which result in nuclear translocation of transcription factors and activation of proteins participating in translation of mRNA. IL-1RI is phosphorylated following IL-1 binding (Gallis *et al.*, 1989). It is unknown whether the IL-1R-AcP is phosphorylated during receptor complex formation. In primary cells, the number of IL-1RI type I is very low (fewer than 100 per cell) and a biologic response occurs when only as few as 2–3% of IL-1RI receptors are occupied (Gallis *et al.*, 1989; Ye *et al.*, 1992). In IL-1-responsive cells, one assumes that there is constitutive expression of the IL-1R-AcP.

With few exceptions, there is general agreement that IL-1 does not stimulate hydrolysis of phosphatidylinositol nor an increase in intracellular calcium levels. Without a clear increase in intracellular calcium concentration, early postreceptor binding events nevertheless include hydrolysis of GTP with no associated increase in adenyl cyclase concentration (O'Neill *et al.*, 1990; O'Neill, 1995), activation of adenyl cyclase (Mizel, 1990; Munoz *et al.*, 1990), hydrolysis of phospholipids (Rosoff *et al.*,

1988; Kester *et al.*, 1989), release of ceramide (Mathias *et al.*, 1993) and release of arachidonic acid from phospholipids via cytosolic PLA_2 following its activation by PLA_2 activating protein (Clark *et al.*, 1991; Gronich *et al.*, 1994). In addition, tyrosine phosphorylations have been reported (Rzymkiewicz *et al.*, 1995). Each of the above mechanisms occurs within a few minutes of the addition of IL-1 to cultured cells.

Although few comparative studies have been reported, it appears that some IL-1 signaling events are prominent in different cells. Postreceptor signaling mechanisms may therefore provide cellular specificity. For example, in some cells, IL-1 is a growth factor and signaling is associated with serine/threonine phosphorylation of the MAP kinase p42/44 in mesangial cells (Huwiler and Pfeilschifter, 1994). The MAP p38 kinase, another member of the MAP kinase family, is phosphorylated in fibroblasts (Freshney *et al.*, 1994), as is the p54α MAP kinase in hepatocytes (Kracht *et al.*, 1994). These somewhat different phosphorylations may distinguish the phenotypic response in various cells stimulated with IL-1.

Characteristics of the Cytoplasmic Domain of the IL-1RI

The cytoplasmic domain of the IL-1RI does not contain a consensus sequence for intrinsic tyrosine phosphorylation but deletion mutants of the receptor reveal specific functions of some domains. There are four nuclear localization sequences which share homology with the glucocorticoid receptor. Three amino acids (Arg-518, Lys-515 and Arg-518), also found in the *Toll* protein, are essential for IL-1-induced IL-2 production (Heguy *et al.*, 1992). However, deletion of a segment containing these amino acids does not affect IL-1-induced IL-8 (Kuno *et al.*, 1993). If this discrepancy is due to different signaling pathways for IL-1-induced IL-2 or for induction of IL-8, then the cytoplasmic domain of IL-1RI can induce more than one biochemical event. There are also two cytoplasmic domains in the IL-1RI which share homology with the IL-6-signaling gp130 receptor. When these regions are deleted, there is a loss of IL-1-induced IL-8 production (Kuno *et al.*, 1993).

The *C*-terminal 30 amino acids of the IL-1RI can be deleted without affecting biologic activity (Croston *et al.*, 1995). Two independent studies have focused on the area between amino acids 513–529. Amino acids 508–521 contain sites required for the activation of nuclear factor (NF)-κB. In one study, deletion of this segment abolished IL-1-induced IL-8 expression (Kuno *et al.*, 1993), and in another, specific mutations of amino acids 513 and 520 to alanine prevented IL-1-driven E-selectin promoter activity (Croston *et al.*, 1995). This area is also present in the *Toll* protein domain associated with NF-κB translocation and previously shown to be part of the IL-1 signaling mechanism (Hegey *et al.*, 1992). This area (513–520) is also responsible for activating a kinase which associates with the receptor. This kinase, termed 'IL-1RI-associated kinase' phosphorylates a 100-kDa substrate (Croston *et al.*, 1995). Others have reported a serine/threonine kinase which coprecipitates with IL-1RI (Martin *et al.*, 1994). Amino acid sequence comparisons of the cytosolic domain of the IL-1RI have revealed similarities with a protein kinase C (PKC) acceptor site. Because PKC activators usually do not mimic IL-1-induced responses, the significance of this observation is unclear.

GTPase

Hopp (1995) reported a detailed sequence and structural comparison of the cytosolic segment of IL-1RI with the *ras* family of GTPases. In this analysis, the known amino acids residues for GTP binding and hydrolysis by the GTPase family were found to align with residues in the cytoplasmic domain of the IL-1RI. In addition, Rac, a member of the Rho family of GTPases, was also present in the binding and hydrolytic domains of the IL-1RI cytosolic domains. These observations are consistent with the observations that GTP analogs undergo rapid hydrolysis when membrane preparations of IL-1RI are incubated with IL-1 (reviewed by Mizel, 1994; O'Neill, 1995). Amino acid sequences in the cytosolic domain of the IL-1R-AcP also align with the same binding and hydrolytic regions of the GTPases (T.P. Hopp, personal communication). A protein similar to G protein-activating protein has been identified which associates with the cytosolic domain of the IL-1RI (Mitchum and Sims, 1995). This finding is consistent with the hypothesis that an early event in IL-1R signaling involves dimerization of the two cytosolic domains, activation of putative GTP binding sites on the cytosolic domains, binding of a G protein, hydrolysis of GTP and activation of a phospholipase. It then follows that hydrolysis of phospholipids generates diacylglycerol or phosphatidic acids.

Activation of MAP Kinases Following IL-1 Receptor Binding

Multiple phosphorylations take place during the first 15 min following IL-1 receptor binding. A comparison of IL-1-induced phosphorylations with those induced by phorbol esters via PKC reveals some similarities. However, unique, non-PKC-activated kinases are also stimulated by IL-1, possibly through release of diacylglycerol (Rosoff *et al.*, 1988) with activation on non-PKC kinases (Schutze *et al.*, 1994). Most consistently, IL-1 activates protein kinases which phosphorylate serine and threonine residues, which are the targets of the MAP kinase family. An early study reported an IL-1-induced serine/threonine phosphorylation of a 65-kDa protein clearly unrelated to those phosphorylated via PKC (Matsushima *et al.*, 1987). Other studies have implicated a role for PKC (Munoz *et al.*, 1990). As reviewed by O'Neill (1995), before IL-1 activation of serine/threonine kinases, IL-1 receptor binding results in the phosphorylation of tyrosine residues (Freshney *et al.*, 1994; Kracht *et al.*, 1994). Tyrosine phosphorylation induced by IL-1 is probably due to activation of MAP kinase, which then phosphorylates tyrosine and threonine on MAP kinases.

Following activation of MAP kinases, there are phosphorylations on serine and threonine residues of the epidermal growth factor receptor, heat shock protein (hsp) p27, myelin basic protein, and serine 56 and 156 of β-casein, each of which has been observed in IL-1-stimulated cells (Bird *et al.*, 1991, Guesdon *et al.*, 1993). TNF also activates these kinases. There are at least three families of MAP kinases. The p42/44 MAP kinase family is associated with signal transduction by growth factors including ras-raf-1 signal pathways. In rat mesangial cells, IL-1 activates the p42/44 MAP kinase within 10 min and also increases *de novo* synthesis of p42 (Huwiler and Pfeilschifter, 1994).

In addition to p42/44, two members of the MAP kinase family (p38 and p54) have been identified as part of an IL-1 phosphorylation pathway and are responsible for phosphorylating hsp27 (Freshney *et al.*, 1994; Kracht *et al.*, 1994). These MAP kinases are highly conserved proteins homologous to the *HOG-1* stress gene in yeasts. In fact, when *HOG-1* is deleted, yeasts fail to grow in hyperosmotic conditions; however, the mammalian gene coding for the IL-1-inducible p38 MAP kinase (Kracht *et al.*, 1994) can reconstitute the ability of the yeast to grow in hyperosmotic conditions (Galcheva-Gargova *et al.*, 1994). In cells stimulated with hyperosmolar sodium chloride, LPS, IL-1 or TNF, indistinguishable phosphorylation of the p38 MAP kinase takes place (Han *et al.*, 1994). In human monocytes exposed to hyperosmolar sodium chloride (375–425 mOsm/l), IL-8 gene expression and synthesis takes place which is indistinguishable from that induced by LPS or IL-1 (Shapiro and Dinarello, 1995). Thus, the MAP p38 kinase pathways involved in IL-1, TNF and LPS signal transductions share certain elements that are related to the primitive stress-induced pathway. The dependency of Rho members of the GTPase family (see above) for IL-1-induced activation of p38 MAP kinases has been demonstrated (Zhang *et al.*, 1995). This latter observation links the intrinsic GTPase domains of IL-1RI and IL-1R-AcP with activation of the p38 MAP kinase.

Drugs of the pyridinyl imidazole class primarily inhibit the translation rather than the transcription of LPS- and IL-1-induced cytokines (Lee *et al.*, 1994). The target for these drugs has been identified as a homolog of the *HOG-1* family (Lee *et al.*, 1994); its sequence is identical to that of the p38 MAP kinase activating protein-2 (Lee *et al.*, 1994; Han *et al.*, 1995). As expected, this class of imidazoles also prevents the downstream phosphorylation of hsp27 (Cuenda *et al.*, 1995). Compounds of this class appear to be highly specific for inhibition of the p38 MAP kinase in that there was no inhibition of 12 other kinases (Cuenda *et al.*, 1995). Using one of these compounds, both hyperosmotic sodium chloride- and IL-1α-induced IL-8 synthesis were inhibited (Shapiro and Dinarello, 1995). It has been proposed that MAP kinase activating protein-2 is one of the substrates for the p38 MAP kinases and that it is the kinase that phosphorylates hsp27 (Cuenda *et al.*, 1995) (Fig. 2).

Transcription Factors

The cytosolic domain of IL-1RI shares significant homology to the receptor-like *Toll* gene, suggesting that both molecules signal similar events. Following IL-1 stimulation, phosphorylation of inhibitory κB (I-κB) takes place and this is rapidly degraded within the proteosome (DiDonato *et al.*, 1995). Translocation of NF-κB to the nucleus is then observed (Shirakawa and Mizel, 1989; Stylianou *et al.*, 1992). A substrate for the β-casein kinase in IL-1- and TNF-activated cells (Guesdon *et al.*, 1994) has been identified as the p65 subunit of NF-κB (Bird *et al.*, 1995). Most of the biologic effects of IL-1 take place in cells following nuclear translocation of NF-κB and activating protein-1 (AP-1), two nuclear factors common to many IL-1-induced genes. In T lymphocytes and cultured hepatocytes, the addition of IL-1 increases nuclear binding of c-jun and c-fos, the two components of AP-1 (Muegge *et al.*, 1989). Similar to NF-κB, AP-1 sites are present in the promoter regions of many IL-1-inducible genes. IL-1 also increases the transcription of c-*jun* by activating two novel nuclear factors (jun-1 and jun-2) which bind to the promoter of the c-*jun* gene and stimulate c-*jun* transcription (Muegge *et al.*, 1993).

Because so few IL-1RI receptors are expressed on primary cells, and because so few of these receptors need to be triggered to initiate a biologic response to IL-1, one concludes that the signaling mechanism is highly efficient and greatly amplified. Dimerization of the cytosolic domains of type I receptor with the IL-1R-AcP probably initiates the signal. The best explanation for the potency of IL-1-induced signaling is postreceptor amplification through multiple phosphorylations of protein kinases. Phosphorylation and dephosphorylation of transcription factors enables the cell to transcribe genes controlled by IL-1 activation of these transcription factors. The multiple postreceptor phosphorylations may explain why IL-1 induces several genes in the same cell at the same time. LPS and TNF share with IL-1 many of these same MAP kinase pathways, most of them related to the *HOG-1* 'stress' gene family of MAP kinases. These observations support the concept that IL-1 signal transducing events mimic the evolutionary benefit of cell stress.

IL-1 ADMINISTERED TO HUMANS

IL-1α or IL-1β has been injected in patients with various solid tumors. In general, the acute toxicity of either isoform of IL-1 was greater following intravenous compared with subcutaneous injection. Subcutaneous injection was associated with significant local pain, erythema and swelling (Kitamura and Takaku, 1989; Laughlin *et al.*, 1993). Most patients have been given a 15–30 min infusion with doses from 1 ng/kg to 1.0 μg/kg. Chills and fever were observed in nearly all patients, and even those receiving 1 ng/kg experienced fever (Tewari *et al.*, 1990). The febrile response increased in magnitude with increasing dose (Crown *et al.*, 1991; Smith *et al.*, 1992; Nemunaitis *et al.*, 1994). Fever began within 30 min of starting the infusion and reached a peak 1.5–3 h later. When given subcutaneously, IL-1β-induced chills and fever were abated with indomethacin treatment (Iizumi *et al.*, 1991). In several studies, a transient increase in blood pressure and heart rate was observed soon after initiation of the infusion and at the onset of the chill (Smith *et al.*, 1992). Following this transient increase in vascular tone, a progressive dose-dependent fall in systolic blood pressure was observed.

In a trial of 28 patients receiving IL-1α (Smith *et al.*, 1992) and in trials of IL-1β in 19 (Crown *et al.*, 1991) and 17 (Nemunaitis *et al.*, 1994) patients, the acute toxicity of intravenously administered IL-1 was studied in detail. Nearly all patients receiving intravenous IL-1 at a dose of 100 ng/kg or greater experienced significant hypotension. Systolic blood pressure fell steadily and reached a nadir of 90 mmHg or less 3–5 h after infusion of IL-1 (Crown *et al.*, 1991; Smith *et al.*, 1992, 1993). At a dose of 300 ng/kg, most patients required intravenous pressors. Indomethacin treatment did not reduce the hypotensive effect of IL-1, even when increased to 50 mg per day (Smith *et al.*, 1991). By comparison, in a trial of 16 patients given IL-1β, 4–32 ng/kg subcutaneously, there was only one episode of hypotension at the highest dosage (Laughlin *et al.*, 1993).

These results suggest that the hypotension is probably due to induction of NO and raised levels of serum nitrate have been measured in patients with IL-1-induced hypotension (Smith *et al.*, 1992). Despite seven daily infusions of IL-1 and fluids for support of hypotension, there was no dramatic increase in body weight (Smith *et al.*, 1992). This is to be contrasted with IL-2 therapy in humans in which a significant capillary leak syndrome and weight gain is common. In addition, unlike patients on high-dose IL-2 therapy, there was no increase in the number of catheter-related infections. In patients receiving 100 ng/kg or greater of IL-1, transient dyspnea was also noted. Administration of IL-1 was associated with generalized fatigue, headache, nausea, vomiting, myalgia, arthralgia and somnolence (Tewari *et al.*, 1990; Crown *et al.*, 1991; Smith *et al.*, 1992, 1993). Myalgia (lower back pain) and headache were ameliorated by indomethacin treatment. These symptoms are nearly the same as those reported by healthy volunteers receiving intravenous endotoxin, although recombinant IL-1 is essentially free of endotoxins.

Humans injected intravenously with 30–100 ng/kg of IL-1β had a sharp increase in cortisol levels 2–3 h after injection (Crown *et al.*, 1991; Starnes, 1991). Similar increases were noted in patients given IL-1α (Smith *et al.*, 1992). In 13 of 17 patients given IL-1β, there was a fall in serum glucose concentration within the first hour of administration and in 11 patients the glucose level fell to 70 mg/100 ml or lower (Crown *et al.*, 1991). In addition, there were increases in the levels of adrenocorticotrophic hormone and thyroid stimulating hormone but a decrease in testosterone concentration (Smith *et al.*, 1992). No changes were observed in coagulation parameters such as prothrombin time, partial thromboplastin time or fibrinogen degradation products (Smith *et al.*, 1992). This latter finding is to be contrasted to TNF-α infusion into healthy humans, which results in a distinct coagulopathy syndrome (van der Poll *et al.*, 1990).

Not unexpectedly, IL-1 infusion into humans significantly increased circulating IL-6 levels in a dose-dependent fashion (Smith *et al.*, 1992). At a dose of 30 ng/kg, mean IL-6 levels were 500 pg/ml 4 h after IL-1 (baseline <50 pg/ml) and 8000 pg/ml after a dose of 300 ng/kg. In another study, infusion of 30 ng/kg of IL-1α induced increased IL-6 levels within 2 h (Tilg *et al.*, 1994a). These increases in IL-6 levels are associated with a rise in the concentration of C-relative protein and a decrease in that of albumin (Smith *et al.*, 1992). Serum GM-CSF concentration was less than 50 pg/ml. In two studies, one with IL-1α (Tilg *et al.*, 1994b) and one with IL-1β (Bargetzi *et al.*, 1993), a rapid increase in the circulating levels of IL-1Ra and TNF-soluble receptors (p55 and p75) was observed following a 30-min intravenous infusion. The rise in the circulating levels of both naturally occurring antagonists is greater than that measured in human volunteers injected with LPS (Granowitz *et al.*, 1991; Shapiro *et al.*, 1993).

Injecting humans with low doses of either IL-1α or IL-1β confirms the impressive pyrogenic and hypotension-inducing properties of the molecules. The human studies also confirm the effects of IL-1 on stimulating the hypothalamic–pituitary–adrenal axis and on increased cytokine production, particularly IL-6. In many ways, the signs and symptoms following IL-1 injection into humans are indistinguishable from those of low doses of endotoxin (Wolff, 1973). Similar to endotoxin, IL-1 induces a general enhancement of hematopoiesis, particularly in increased neutrophil, monocyte and platelet counts. In patients given marrow-suppressing chemotherapy, cotreatment with IL-1 decreases the nadir and duration of marrow suppression. However, the benefits of IL-1 therapy in these patients are clouded by its formidable toxicity. Low doses of IL-1 may be useful in combination with other hematopoietic growth factors for reducing myelosuppression during chemotherapy or bone marrow transplantation.

IL-1 RECEPTOR ANTAGONIST AND AMELIORATION OF DISEASE

Blocking IL-1 receptors with IL-1Ra has increased our understanding of IL-1 as a mediator of disease. Table 1 lists several studies in which administration of IL-1Ra has reduced the severity of disease. Rat synoviocytes have been transfected with IL-1Ra and expressed *in vivo*; a decrease in arthritis was reported (Makarov *et al.*, 1995) and forms the basis for gene therapy with IL-1Ra. In mice with the functional IL-1Ra gene deletion, there is increased death to fatal endotoxemia (Hirsch *et al.*, 1996). In mice

Table 1. Effects of IL-1Ra.

Models of Infection
Improved survival in endotoxin shock in mice, rats and rabbits
Improved survival in *Klebsiella pneumoniae* infection in newborn rats
Reduction in shock and mortality in rabbits and baboons from *Escherichia coli* or *Staphylococcus epidermidis* bacteremia
Amelioration of shock and reduction in death after cecal ligation and puncture
Attenuation of LPS-induced lung nitric oxide activity
Decreased hypoglycemia, production of colony stimulating factor (CSF) and early tolerance in mice after administration of endotoxin
Reduction in LPS-induced hyperalgesia
Protection against TNF-induced lethality in D-galactosamine-treated mice
Reduction in nematode-induced intestinal nerve dysfunction
Decreased circulating or cellular TNF production in models of sepsis
Decreased IL-6 production after LPS
Protection from *Bacillus anthracis* toxin-induced lethality in mice
Decreased intestinal inflammation and bacterial invasion in shigellosis

Models of Local Inflammation
Decreased neutrophil accumulation in inflammatory peritonitis in mice
Reduction in immune complex-induced neutrophil infiltration, eicosanoid production and tissue necrosis in rabbit colitis
Reduction in acid-induced neutrophil infiltration and enterocolitis in rats
Decreased endotoxin-induced intestinal secretory diarrhea in mice
Reduction in ischemia and excitotoxin-induced brain damage in rats
Decrease in number of necrotic neurons in cerebral artery occlusion
Inhibition of permanganate-induced granulomas in rats
Inhibition of LPS-induced intra-articular neutrophil infiltration
Decreased IL-1-induced synovitis and loss of cartilage proteoglycan
Reduced myocardial neutrophil accumulation after coronary occlusions in dogs
Reduced inflammation and mortality in acute pancreatitis
Decreased hepatic inflammation following haemorrhagic shock

Models of Acute or Chronic Lung Injury
Decreased local LPS-induced neutrophil infiltration in rats
Inhibition of antigen-induced pulmonary eosinophil accumulation and airway hyperactivity in guinea pigs
Prevention of bleomycin or silica-induced pulmonary fibrosis
Reduction in hypoxia-induced pulmonary hypertension
Reduction in carrageenan-induced pleurisy in rats
Decrease intratracheal IL-1-induced fluid leak (systemic administration)
Decreased albumin leak after systemic LPS
Inhibition of antigen-induced eosinophil accumulation in guinea pigs

Models of Metabolic Dysfunction
Reduction in hepatocellular damage following ischemia–reperfusion
Improved survival after hemorrhagic shock in mice
Inhibition of serum amyloid A (SAA) gene expression and synthesis in high-dose IL-2 toxicity
Decreased muscle protein breakdown in rats with peritonitis due to cecal ligation
Reduced muscle protein breakdown in rats with chronic septic peritonitis
Inhibition of weight loss following muscle tissue injury
Decrease in bone loss in ovariectomized rats
Reversal of LPS-induced CRF gene expression in the hypothalamus
Prevention of LPS-induced adrenocorticotrophin release

(*continued*)

Table 1. *Continued*

Models of Autoimmune Disease
Diminution of *Streptococcus* wall-induced arthritis in rats
Reduction in collagen arthritis in mice
Suppression of antibasement membrane glomerulonephritis
Delayed hyperglycemia in the diabetic BB rat
Reduction in streptozotocin-induced diabetes
Models of Immune-Mediated Disease
Prevention of graft *versus* host disease in mice
Prolongation of islet allograft survival
Reduction in autoimmune encephalomyelitis
Reduction in skin contact hypersensitivity
Decrease in coronary artery fibronectin deposition in heterotopic cardiac transplant
Models of Malignant Disease
Reduction in the number and size of metastatic melanoma
Reduction in growth of subcutaneous melanoma tumors
Reduced LPS-induced augmentation of metastatic melanoma
Reduction in tumor-mediated cachexia (intratumoral injection)
Miscellaneous
Inhibition of TNF-induced social depression in mice
Prevention of stress-induced hypothalamic monoamine release
Reduction in LPS-induced sickness behavior in rats
Suppression of crescentic glomerulonephritis in rats
Attenuation of muramyl dipeptide-induced sleep in rabbits
Impairment of Host Responses
Decreased sciatic nerve regeneration in mice
Increased mortality to *K. pneumoniae* in newborn rats (high dose)
Increased mortality to *Listeria* infection
Enhanced growth of *Mycobacterium avium* in organs
Worsening of infectious arthritis (late administration)
Increased vascular leak in mice given high dose IL-2
Studies Without an Effect of IL-1Ra
Antigen-induced arthritis in rabbits
LPS- and *S. epidermidis* bacteremia-induced fever in rabbits
Fever after LPS injection into the brain
Leukopenia in rats after LPS
Hypertriglyceridemia after LPS in mice
LPS-induced increase in skin blood flow

overproducing IL-1Ra, increased protection against fatal endotoxemia has been reported (Hirsch *et al.*, 1996). Other methods such as neutralizing anti-IL-1 antibodies, antibodies to IL-1RI and soluble IL-1 receptors are equally effective, although limited by their animal specificity. In most disease models, other cytokines are produced in addition to IL-1. Therefore, the data depicted in Table 1 reveal that IL-1 plays an important role in the pathogenesis of inflammatory and immunologically mediated disease. In these studies, a reduction of at least 50% is observed, but in many the amelioration of pathologic changes can be complete. One consistent observation is the reduction in the number of infiltrating neutrophils associated with local inflammation; this effect of IL-1Ra may be due to preventing IL-1-induced synthesis of IL-8 and related chemokines (Porat *et al.*, 1992).

There is, however, one important caveat: in the majority of these studies, IL-1Ra has been administered just before the challenging event. This is particularly the case in models of infection where injecting IL-1Ra before a lethal challenge has significantly reduced the mortality rate, but when injected shortly after the challenge IL-1Ra has little or no effect on reducing death. On the other hand, in acute pancreatitis, a dose-dependent administration of IL-1Ra later in the disease reduced the severity of tissue damage (Norman *et al.*, 1995a,b). In some models of chronic disease, administration of IL-1Ra after the onset of disease can still dramatically reduce severity (Dayer-Metroz *et al.*, 1992; Vidal-Vanaclocha *et al.*, 1994).

It is not uncommon to inject a high dose of IL-1Ra (10 mg/kg) in order to observe an ameliorative effect in acute models of infection or inflammation despite the relatively low concentrations of circulating IL-1 in these animals. Why is so much IL-1Ra required? First, it should be pointed out that the plasma half-life of IL-1Ra is short (6 min) and that these models are usually 'severe'. For example, in the bacteremic model, the number of organisms injected intravenously ranges from 10^9 to 10^{11} per kg. Although systemic levels of IL-1 can be low, tissue production of IL-1 can be high (Cominelli *et al.*, 1990), due to membrane IL-1α (Kaplanski *et al.*, 1994) and IL-1 from activated platelets (Kaplanski *et al.*, 1993). In addition to these sources of IL-1, there is rapid excretion of IL-1Ra, a slow receptor 'on rate', increased IL-1RI expression, IL-1Ra binding to the soluble type I receptor and poor tissue penetration of IL-1Ra. Each can also contribute to a requirement for high doses of IL-1Ra. It is also possible that the amount of IL-1β produced during disease has been underestimated because IL-1β binding to the soluble (and cell-bound) type II receptor has prevented accurate measurement (Arend *et al.*, 1994).

As triggering so few IL-1 receptors results in a biologic response, it is necessary to sustain a high level of IL-1Ra to block unoccupied receptors. When exogenous IL-1 is injected intravenously into animals, pretreatment with a 100-fold molar excess of IL-1Ra prevents the response to IL-1. For example, injecting rabbits with 100 ng/kg of IL-1β produces fever; preinjection of 10 μg/kg of IL-1Ra prevents the fever. However, under natural conditions where endogenous IL-1 and other cytokines are released, an IL-1Ra plasma level of 20–30 μg/ml is needed before a reduction in disease is observed (Aiura *et al.*, 1993). In humans, similar levels of IL-1Ra are needed to block the hematologic response to LPS. *In vitro*, considerably lower concentrations of IL-1Ra are needed (Arend *et al.*, 1990). For example, a one to one molar ratio of IL-1Ra to IL-1 blocks 50% of the IL-1-induced response in blood monocytes (Granowitz *et al.*, 1992a,b) and a concentration of 100 ng/ml of IL-1Ra reduces the spontaneous proliferation, colony formation and cytokine production of AML or chronic myeloid leukemia (CML) cells (Estrov *et al.*, 1991; Rambaldi *et al.*, 1991; Schirò *et al.*, 1993).

The molar 'ratio' of endogenous IL-1Ra to IL-1β levels in body fluids from patients with infectious, inflammatory or autoimmune disease is often 10–100-fold more IL-1Ra than IL-1β. In some selected clinical conditions, the ratio is far lower. If the molar ratio of endogenous IL-1Ra to IL-1 falls, does this affect disease outcome? Some data provide important findings regarding this question. In AML cells where IL-1β is expressed spontaneously, IL-1Ra gene expression is suppressed even when stimulated with

GM-CSF (Rambaldi *et al.*, 1991). In 81 patients with CML, cell lysates contained more IL-1β than cells from healthy subjects, whereas the levels of IL-1Ra were the same for both groups (Wetzler *et al.*, 1994). In addition, the survival rate of 44 patients with raised IL-1β levels was lower than that of patients with low IL-1β levels. During accelerated blast crisis, IL-1Ra levels were lower compared with those in patients in a chronic phase (Wetzler *et al.*, 1994). Stromal cultures established from bone marrow of patients with aplastic anemia produced fewer spontaneous as well as induced IL-1Ra compared with stromal cells established from normal bone marrow (Holmberg *et al.*, 1994). Recently we have measured high levels of IL-1 soluble receptor type II (IL-1sRII) (Orencole *et al.*, 1994) in the circulation of 25 patients with hairy cell leukemia which correlate with high levels of IL-1β (Barak *et al.*, 1994); however, there was no increase in IL-1Ra levels in these patients.

In patients with acute Lyme arthritis, the duration of joint inflammation is shortest in those with the highest joint fluid levels of IL-1Ra, whereas it is prolonged in patients with low levels of IL-1Ra (Miller *et al.*, 1993). The reciprocal relationship was found for synovial fluid levels of IL-1β in the same patients. Similar findings were found in the relative production of IL-1Ra and IL-1β in synovial tissue explants of patients with rheumatoid or osteoarthritis (Firestein *et al.*, 1992; Roux-Lombard *et al.*, 1992). In normal skin, icIL-1Ra is present in higher concentrations than IL-1α (Hammerberg *et al.*, 1992), but in psoriatic lesions the balance is in favour of IL-1α (Hammerberg *et al.*, 1992; Kristensen *et al.*, 1992). Alveolar macrophages from smokers produce less IL-1Ra than those from non-smokers (Janson *et al.*, 1993). Under experimental conditions, humans pretreated with corticosteroids before an injection of LPS produce lower circulating levels of TNF, IL-6 and IL-8, but IL-1Ra levels are unaffected by the steroids (Santos *et al.*, 1993).

Does endogenous production of IL-1Ra affect disease outcome? There is little question that increased production of IL-1Ra is an excellent marker of disease, and certainly a better indicator than IL-1 itself. In some clinical conditions, the increase in IL-1Ra concentration rather than that of IL-1 may indicate the presence of a pathologic condition. For example, spontaneous and inducible IL-1Ra production by PBMCs is higher in patients undergoing chronic hemodialysis than in age-matched patients with renal failure (Pereira *et al.*, 1992). Detecting raised IL-1Ra production could indicate a natural compensatory mechanism to counter the activity of IL-1, for example in rheumatoid arthritis (Chomarat *et al.*, 1995) or human immunodeficiency virus-1-infected persons (Thea *et al.*, 1996). Is the amount of IL-1Ra produced in disease sufficient to dampen the response to IL-1? Using specific neutralizing antibodies to mouse IL-1Ra, an increase in the formation of schistosome egg granulomata was observed when endogenous IL-1Ra was neutralized (Chensue *et al.*, 1993). In rabbits with immune complex colitis, infusion of a neutralizing antibody to rabbit IL-1Ra resulted in exacerbation and prolongation of the colitis (Ferretti *et al.*, 1994). As of this writing, the phenotype of an IL-1Ra-deficient mouse is associated with greater susceptibility to LPS.

IL-1Ra ADMINISTERED TO HEALTHY HUMANS

IL-1Ra given intravenously to healthy volunteers is without side-effects or changes in biochemical, hematologic or endocrinologic parameters, even when peak blood levels reach 30 μg/ml and are sustained above 10 μg/ml for several hours (Granowitz *et al.*, 1992a,b). These studies support the concept that there is no role for IL-1 in the regulation of body temperature, blood pressure or hematopoiesis in health. Interestingly, PBMCs taken from these volunteers after receiving IL-1Ra failed to produce IL-6 when stimulated *ex vivo* with LPS (Granowitz *et al.*, 1992a,b). A role for IL-1R on PBMCs in LPS-stimulated IL-6 *in vitro* has been reported (Granowitz *et al.*, 1992a,b).

To evaluate the effect of IL-1 receptor blockade in clinical disease under controlled experimental conditions, healthy volunteers were challenged with intravenous endotoxin and administered an infusion of IL-1Ra at the same time. Even at 10 mg/kg IL-1Ra, there was no effect on endotoxin-induced fever, although blood levels of IL-1Ra were not significantly raised until 1 h after the bolus injection of endotoxin. Humans injected with antibodies to TNF before endotoxin also did not have a reduction in fever (H. Michie, personal communication). In animal studies, peripheral endotoxin induces fever by triggering IL-1 induction of IL-6 synthesis in the central nervous system (LeMay *et al.*, 1990). Because IL-1Ra does not cross the blood–brain barrier, this may account for the inability of IL-1Ra to diminish endotoxin fever (Dinarello *et al.*, 1992). However, IL-1R blockade was accomplished as there was a 50% reduction in the endotoxin-induced neutrophilia and a reduction in the circulating levels of G-CSF compared with those in subjects injected with endotoxin plus saline (Granowitz *et al.*, 1993).

Endotoxin injection suppresses the mitogen-induced proliferative response of PBMCs *in vitro*. However, in volunteers injected with endotoxin plus IL-1Ra, there was no suppression of the response (Granowitz *et al.*, 1993). Mitogen- and antigen-induced proliferation is a well established parameter of immunocompetence and is associated with decreased production of IL-2. Similar to experimental endotoxin injection, this suppression is observed in patients with multiple trauma, sepsis and cardiopulmonary bypass. In experimental endotoxemia and the above clinical conditions, treatment with cyclo-oxygenase inhibitors restores these cell-mediated immune responses (Markewitz *et al.*, 1993). This effect of cyclo-oxygenase inhibitors is consistent with the well known suppressive effects of prostaglandin (PG)E_2 on IL-2 production and T-cell proliferation. As IL-1 is a potent inducer of cyclo-oxygenase-2 (COX-2), it is not surprising that blocking IL-1 receptors during endotoxemia will reduce IL-1-induced PGE_2 production. Thus, these studies establish that, under conditions of low-dose endotoxemia, it is possible to block IL-1 mediated responses with IL-1Ra. Host response parameters that were unaffected by IL-1Ra are probably due to other cytokines such as TNF or IL-6, or the combination of these cytokines with IL-1.

IL-1Ra has been given to patients with septic shock, rheumatoid arthritis, steroid-resistant graft *versus* host disease, and AML and CML. The initial (phase II) trial was a randomized placebo-controlled open-label study in 99 patients. Patients received either placebo or a loading bolus of 100 mg followed by a 3-day infusion of 17, 67 or 133 mg/h IL-1Ra (Fisher *et al.*, 1994b). A dose-dependent improvement in the 28-day mortality rate was observed: 44% in the placebo group *versus* 16% in the group receiving the highest dose of IL-1Ra ($P = 0.015$). In that study, there was a dose-related fall in the circulating levels of IL-6 24 h after initiation of the IL-1Ra infusion. This fall in IL-6 level is consistent with the well established control of circulating IL-6 levels by IL-1 (Gershenwald *et al.*, 1990; Fischer *et al.*, 1991) and the correlation of disease severity and outcome with IL-6 concentration (Casey *et al.*, 1993). A large phase III trial in 893 patients revealed a trend but without a statistically significant reduction in the 28-day mortality rate (Fisher *et al.*, 1994a). However, a retrospective analysis of 563 patients with a predicted risk of mortality of 24% or greater (Knaus *et al.*, 1993) revealed a significant reduction in the 28-day mortality rate (45% in the placebo group *versus* 35% in patients receiving 2 mg/kg per h for 72 h, $P = 0.005$) (Fisher *et al.*, 1994a). A second phase III trial using 10 g of IL-1Ra infused over 3 days was undertaken but terminated following interim analysis because there was no statistical evidence of a reduction in the overall 28-day mortality rate. Patient heterogeneity is thought to contribute to the failure to bridge the gap between animal and clinical data in sepsis.

IL-1Ra was initially tested in a trial in 25 patients with rheumatoid arthritis. In the group receiving a single subcutaneous dose of 6 mg/kg, there was a fall in the mean number of tender joints ($P < 0.05$) (Lebsack *et al.*, 1991). In patients receiving 4 mg/kg per day for 7 days, there was a reduction in the number of tender joints from 24 to 10; the erythrocyte sedimentation rate fell from 48 to 31 mm/m and C-reactive protein concentration decreased from 2.9 to 1.9 μg/ml. In this group the mean plasma concentration of IL-1Ra was 660 $\pm$ 240 ng/ml (Lebsack *et al.*, 1991). In an expanded trial, IL-1Ra was given to 175 patients (Lebsack *et al.*, 1993). Patients were enrolled into the study with active disease and taking non-steroidal anti-inflammatory drugs and/or up to 10 mg/day prednisone. There was an initial phase of 3 weeks of either 20, 70 or 200 mg one, three or seven times per week. Thereafter, patients received the same dose once weekly for 4 weeks. Placebo was given to patients once weekly for the entire 7-week study period. Four measurements of efficacy were used: number of swollen joints, number of painful joints, and patient and physician assessment of disease severity. A reduction of 50% or greater in these scores from baseline was considered significant in the analysis. A statistically significant reduction in the total number of parameters was observed, with the optimal improvement in patients receiving 70 mg per day.

A large double-blind placebo-controlled trial of IL-1Ra in 472 patients with rheumatoid arthritis was recently completed. There were three doses: 30, 75 and 150 mg/day for 24 weeks. There was a dose-dependent reduction in the number of swollen joints and the overall assessment of patient scores ($P = 0.048$) (Bresnihan *et al.*, 1996). In addition, there was a fall in C-reactive protein level and sedimentation rate. In addition, in this trial there was a reduction in new bone erosions (Watt and Cobby, 1996).

A phase I/II trial of escalating doses of IL-1Ra in 17 patients with steroid-resistant graft *versus* host disease has been completed (Antin *et al.*, 1994). IL-1Ra (400–3400 mg/day) was given as a continuous intravenous infusion every 24 h for 7 days. Using an organ-specific acute disease scale, there was improvement in 16 of the 17 patients. Moreover, a decrease in the steady-state mRNA for TNF-α in PBMCs correlated with improvement ($P = 0.001$) (Antin *et al.*, 1994). These studies in humans are similar to the use of IL-1Ra in animal models of graft *versus* host disease (McCarthy *et al.*, 1991).

For clinical efficacy, IL-1Ra in patients with rheumatoid arthritis exhibits a dose-dependent response. Even the reduction of endotoxin-induced neutrophilia in healthy subjects is dose dependent. Animal studies support these clinical observations. The requirement for such high plasma levels of IL-1Ra is not completely understood because IL-1Ra levels are already several logs higher than measurable IL-1 levels in the most severe cases of septic shock (Casey *et al.*, 1993). Rapid renal clearance, binding to the soluble form of the type I receptor and increased type I receptor expression may explain a need for these high levels. In addition, paracrine effects and IL-1 activity on platelets and monocytes may be binding to type I receptors on cells outside the vasculature before tissue levels IL-1Ra match those in the circulation.

In the two phase III trials of IL-1Ra in septic shock, retrospective analysis revealed decreased mortality in patient subgroups, particularly during the first 7 days after entry into the trials. These results suggest that not all patients with life-threatening septic shock benefit from IL-1Ra and that factors other than IL-1 contribute to the cause of death 28 days later. Similar conclusions have been made using antibodies to TNF in clinical trials of septic shock.

ACKNOWLEDGEMENTS

This work was supported by NIH grant no. AI-15614.

REFERENCES

Aiura, K., Gelfand, J.A., Wakabayashi, G., Burke, J.F., Thompson, R.C. and Dinarello, C.A. (1993). Interleukin-1 (IL-1) receptor antagonist prevents *Staphylococcus epidermidis*-induced hypotension and reduces circulating levels of tumor necrosis factor and IL-1β in rabbits. *Infect Immun.* **61**, 3342–3350.

Alcami, A. and Smith, G.L. (1992). A soluble receptor for interleukin-1β encoded by vaccinia virus: a novel mechanism of virus modulation of the host response to infection. *Cell* **71**, 153–167.

Andersson, J., Björk, L., Dinarello, C.A., Towbin, H. and Andersson, U. (1992). Lipopolysaccharide induces human interleukin-1 receptor antagonist and interleukin-1 production in the same cell. *Eur. J. Immunol.* **22**, 2617–2623.

Andrieu, N., Salvayre, R. and Levade, T. (1994). Evidence against involvement of the acid lysosomal shingomyelinase in the tumor necrosis factor and interleukin-1-induced sphingomyelin cycle and cell proliferation in human fibroblasts. *Biochem. J.* **303**, 341–345.

Antin, J.H., Weinstein, H.J., Guinan, E.C., McCarthy, P., Bierer, B.E., Gilliland, D.G., Parsons, S.K., Ballen, K.K., Rimm, I.J., Falzarano, G. and Ferrara, J.L. (1994). Recombinant human interleukin-1 receptor antagonist in the treatment of steroid-resistant graft-*versus*-host disease. *Blood* **84**, 1342–1348.

Arend, W.P. (1993). Interleukin-1 receptor antagonist. *Adv. Immunol.* **54**, 167–227.

Arend, W.P., Welgus, H.G., Thompson, R.C. and Eisenberg, S.P. (1990). Biological properties of recombinant human monocyte-derived interleukin-1 receptor antagonist. *J. Clin. Invest.* **85**, 1694–1697.

Arend, W.P., Malyak, M., Smith, M.F., Whisenand, T.D., Slack, J.L., Sims, J.E., Giri, J.G. and Dower, S.K. (1994). Binding of IL-1α, IL-1β, and IL-1 receptor antagonist by soluble IL-1 receptors and levels of soluble IL-1 receptors in synovial fluids. *J. Immunol.* **153**, 4766–4774.

Auron, P.E., Warner, S.J., Webb, A.C., Cannon, J.G., Bernheim, H.A., McAdam, K.J., Rosenwasser, L.J., LoPreste, G., Mucci, S.F. and Dinarello, C.A. (1987). Studies on the molecular nature of human interleukin 1. *J. Immunol.* **138**, 1447–1456.

Bailly, S., Ferrua, B., Fay, M. and Gougerot-Pocidalo, M.-A. (1990). Paraformaldehyde fixation of LPS-stimulated human monocytes: technical parameters permitting the study of membrane IL-1 activity. *Eur. Cytokine Netw.* **1**, 47–51.

Bakouche, O., Brown, D.C. and Lachman, L.B. (1987). Subcellular localization of human monocyte interleukin 1: evidence for an inactive precursor molecule and a possible mechanism for IL 1 release. *J. Immunol.* **138**, 4249–4255.

Barak, V., Nisman, B., Dann, E.J., Kalickman, I., Ruchlemer, R., Bennett, M.A. and Pollack, A. (1994). Serum interleukin-1β levels as a marker in hairy cell leukemia: correlation with disease status and sIL-2R levels. *Leuk. Lymphom.* **14**, 33–39.

Bargetzi, M.J., Lantz, M., Smith, C.G., Torti, F.M., Olsson, I., Eisenberg, S.P. and Starnes, H.F. (1993). Interleukin-1 beta induces interleukin-1 receptor antagonist and tumor necrosis factor binding proteins. *Cancer Res.* **53**, 4010–4013.

Bazan, J.F., Timans, J.C. and Kaselein, R.A. (1996). A newly defined interleukin-1? *Nature* **379**, 591.

Bergers, G., Reikerstorfer, A., Braselmann, S., Graninger, P. and Busslinger, M. (1994). Alternative promoter usage of the Fos-responsive gene *Fit-1* generates mRNA isoforms coding for either secreted or membrane-bound proteins related to the IL-1 receptor. *EMBO J.* **13**, 1176–1188.

Beuscher, H.U. and Colten, H.R. (1988). Structure and function of membrane IL-1. *Mol. Immunol.* **25**, 1189–1195.

Beuscher, H.U., Nickells, M.W. and Colten, H.R. (1988). The precursor of interleukin-1α is phosphorylated at residue serine 90. *J. Biol. Chem.* **263**, 4023–4028.

Beuscher, H.U., Guenther, C. and Roellinghoff, M. (1990). IL-1β is secreted by activated murine macrophages as biologically inactive precursor. *J. Immunol.* **144**, 2179–2183.

Bird, T.A. and Saklatvala, J. (1989). IL-1 and TNF transmodulate epidermal growth factor receptors by a protein kinase C-independent mechanism. *J. Immunol.* **142**, 126–133.

Bird, T.A. Gearing, A.J. and Saklatvala, J. (1987). Murine interleukin-1 receptor: differences in binding properties between fibroblastic and thymoma cells and evidence for a two-chain receptor model. *FEBS Lett.* **225**, 21–26.

Bird, T.A., Sleath, P.R., de Roos, P.C., Dower, S.K. and Virca, G.D. (1991). Interleukin-1 represents a new modality for the activation of extracellular signal-related kinases/microtubule-associated protein-2 kinases. *J. Biol. Chem.* **266**, 22 661–22 670.

Bird, T.A., Downey, H. and Virca, G.D. (1995). Interleukin-1 regulates casein kinase II-mediated phosphorylation of the p65 subunit of NFκB. *Cytokine* **7**, 603 (abstract).

Black, R.A., Kronheim, S.R., Cantrell, M., Deeley, M.C., March, C.J., Prickett, K.S., Wignall, J., Conlon, P.J., Cosman, D. and Hopp, T.P. (1988). Generation of biologically active interleukin-1 beta by proteolytic cleavage of the inactive precursor. *J. Biol. Chem.* **263**, 9437–9442.

Bomalaski, J.S., Steiner, M.R., Simon, P.L. and Clark, M.A. (1992). IL-1 increases phospholipase A_2 activity, expression of phospholipase A_2-activating protein, and release of linoleic acid from the murine T helper cell line EL-4. *J. Immunol.* **148**, 155–160.

Bresnihan, B., Lookabaugh, J., Witt, K. and Musikic, P. (1996). Treatment with recombinant human interleukin-1 receptor antagonist in rheumatoid arthritis: results of a randomized, double-blind, placebo-controlled multicenter trial. *Arthritis Rheum.* **39**, S73 (abstract).

Brody, D.T. and Durum, S.K. (1989). Membrane IL-1: IL-1α precursor binds to the plasma membrane via a lectin-like interaction. *J. Immunol.* **143**, 1183.

Burch, R.M. and Mahan, L.C. (1991). Oligonucleotides antisense to the interleukin-1 receptor mRNA block the effects of interleukin-1 in cultured murine and human fibroblasts and in mice. *J. Clin. Invest.* **88**, 1190–1196.

Car, B.D., Eng, V.M., Schnyder, B., Ozmen, L., Huang, S., Gallay, P., Heumann, D., Aguet, M. and Ryffel, B. (1994). Interferon γ receptor deficient mice are resistant to endotoxic shock. *J. Exp. Med.* **179**, 1437–1444.

Casey, L.C., Balk, R.A. and Bone, R.C. (1993). Plasma cytokines and endotoxin levels correlate with survival in patients with the sepsis syndrome. *Ann. Intern. Med.* **119**, 771–778.

Cerretti, D.P., Kozlosky, C.J., Mosley, B., Nelson, N., Van Ness, K., Greenstreet, T.A., March, C.J., Kronheim, S.R., Druck, T., Cannizzaro, L.A., Huebner, K. and Black, R.A. (1992). Molecular cloning of the IL-1β processing enzyme. *Science* **256**, 97–100.

Chensue, S.W., Bienkowski, M., Eessalu, T.E., Warmington, K.S., Hershey, S.D., Lukas, N.W. and Kunkel, S.L. (1993). Endogenous IL-1 receptor antagonist protein (IRAP) regulates schistosome egg granuloma formation and the regional lymphoid response. *J. Immunol.* **151**, 3654–3662.

Chomarat, P., Vannier, E., Dechanet, J., Rissoan, M.C., Banchereau, J., Dinarello, C.A. and Miossec, P. (1995). The balance of IL-1 receptor antagonist/IL-1β in rheumatoid synovium and its regulation by IL-4 and IL-10. *J. Immunol.* **154**, 1432–1439.

Clark, M.A., Özgür, L.E., Conway, T.M., Dispoto, J., Crooke, S.T. and Bomalaski, J.S. (1991). Cloning of a phospholipase A_2-activating protein. *Proc. Natl Acad. Sci. U.S.A.* **88**, 5418–5422.

Colotta, F., Re, F., Muzio, M., Bertini, R., Polentarutti, N., Sironi, M., Giri, J., Dower, S.K., Sims, J.E. and Mantovani, A. (1993). Interleukin-1 type II receptor: a decoy target for IL-1 that is regulated by IL-4. *Science* **261**, 472–475.

Colotta, F., Dower, S.K., Sims, J.E. and Mantovani, A. (1994). The type II 'decoy' receptor: a novel regulatory pathway for interleukin-1. *Immunol. Today* **15**, 562–566.

Cominelli, F., Nast, C.C., Clark, B.D., Schindler, R., Llerena, R., Eysselein, V.E., Thompson, R.C. and Dinarello, C.A. (1990). Interleukin-1 gene expression, synthesis and effect of specific IL-1 receptor blockade in rabbit immune complex colitis. *J. Clin. Invest.* **86**, 972–980.

Cozzolino, F., Rubartelli, A., Aldinucci, D., Sitia, R., Torcia, M., Shaw, A. and Di Guglielmo, R. (1989). Interleukin 1 as an autocrine growth factor for acute myeloid leukemia cells. *Proc. Natl Acad. Sci. U.S.A.* **86**, 2369–2373.

Croston, G.E., Cao, Z. and Goeddel, D.V. (1995). NFκB activation by interleukin-1 requires an IL-1 receptor-associated protein kinase activity. *J. Biol. Chem.* **270**, 16514–16517.

Crown, J., Jakubowski, A., Kemeny, N., Gordon, M., Gasparetto, C., Wong, G., Toner, G., Meisenberg, B., Botet, J., Applewhite, J., Sinha, S., Moore, M., Kelsen, D., Buhles, W. and Gabrilove, J. (1991). A phase I trial of recombinant human interleukin-1β alone and in combination with myelosuppressive doses of 5-fluoruracil in patients with gastrointestinal cancer. *Blood* **78**, 1420–1427.

Cuenda, A., Rouse, J., Doza, Y.N., Meier, R., Cohen, P., Gallagher, T.F., Young, P.R. and Lee, J.C. (1995). SB 203580 is a specific inhibitor of a MAP kinase homologue which is stimulated by stresses and interleukin-1. *FEBS Lett.* **364**, 229–233.

Curtis, B.M., Widmer, M.B., de Roos, P. and Quarnstrom, E.E. (1990). IL-1 and its receptor are translocated to the nucleus. *J. Immunol.* **144**, 1295–1303.

Dayer-Metroz, M.D., Duhamel, D., Rufer, N., Izui, S., Carmichaels, D., Wollheim, C.B. and Dayer, J.-M. (1992). IL-1ra delays the spontaneous autoimmune diabetes in the BB rat. *Eur. J. Clin. Invest.* **22**, A50 (abstract).

DiDonato, J.A., Mercurio, F. and Karin, M. (1995). Phosphorylation of IκBα precedes but is not sufficient for its dissociation from NFκB. *Mol. Cell. Biol.* **15**, 1302–1311.

Dinarello, C.A., Cannon, J.G., Mier, J.W., Bernheim, H.A., LoPreste, G., Lynn, D.L., Love, R.N., Webb, A.C., Auron, P.E., Reuben, R.C., Rich, A., Wolff, S.M. and Putney, S.D. (1986). Multiple biological activities of human recombinant interleukin 1. *J. Clin. Invest.* **77**, 1734–1739.

Dinarello, C.A., Ikejima, T., Warner, S.J., Orencole, S.F., Lonnemann, G., Cannon, J.G. and Libby, P. (1987). Interleukin 1 induces interleukin 1. I. Induction of circulating interleukin 1 in rabbits *in vivo* and in human mononuclear cells *in vitro*. *J. Immunol.* **139**, 1902–1910.

Dinarello, C.A., Zhang, X.X., Wen, H.D., Wolff, S.M. and Ikejima, T. (1992). The effect of interleukin-1 receptor antagonist on IL-1, LPS, *Staphylococcus epidermidis* and tumor necrosis factor fever. In *Neuro-Immunology of Fever* (eds T. Bartfai and D. Ottoson), Pergamon Press, Oxford, pp. 11–18.

Dower, S.K., Fanslow, W., Jacobs, C., Waugh, S., Sims, J.E. and Widmer, M.B. (1994). Interleukin-1 antagonists. *Ther. Immunol.* **1**, 113–122.

Dripps, D.J., Brandhuber, B.J., Thompson, R.C. and Eisenberg, S.P. (1991). Effect of IL-1ra on IL-1 signal transduction. *J. Biol. Chem.* **266**, 10331–10336.

Enari, M., Hug, H. and Nagata, S. (1995). Involvement of an ICE-like protease in Fas-mediated apoptosis. *Nature* **375**, 78–81.

Estrov, Z., Kurzrock, R., Wetzler, M., Kantarjian, H., Blake, M., Harris, D., Gutterman, J.U. and Talpaz, M. (1991). Suppression of chronic myelogenous leukemia colony growth by IL-1 receptor antagonist and soluble IL-1 receptors: a novel application for inhibitors of IL-1 activity. *Blood* **78**, 1476–1484.

Estrov, Z., Black, R.A., Kurzrock, R., Wetzler, M., Sleath, P.R., Estey, E.H., Harris, D., Van, Q. and Talpaz, M. (1994). IL-1β converting enzyme (ICE) inhibitor suppresses AML blast proliferation. *Blood* **84**, 380A.

Evans, R.J., Bray, J., Childs, J.D., Vigers, G.P.A., Brandhuber, B.J., Skalicky, J.J., Thompson, R.C. and Eisenberg, S.P. (1994). Mapping receptor binding sites in the IL-1 receptor antagonist and IL-1β by site-directed mutagenesis: identification of a single site in IL-1ra and two sites in IL-1β. *J. Biol. Chem.* **270**, 11477–11483.

Falk, W. and Hofmeister, R. (1994). Intracellular IL-1 replaces signaling by the membrane IL-1 type I receptor. *Cytokine* **6**, 558.

Faucheu, C., Diu, A., Chan, A.W.E., Blanchet, A.-M., Miossec, C., Hervé, F., Collard-Dutilleul, V., Gu, Y., Aldape, R.A., Lippke, J.A., Rocher, C., Su, M.S.-S., Livingston, D.L., Hercend, T. and Lalanne, J.-L. (1995). A novel human protease similar to the interleukin-1β converting enzyme induces apoptosis in transfected cells. *EMBO J.* **14**, 1914–1922.

Fenton, M.J., Vermeulen, M.W., Clark, B.D., Webb, A.C. and Auron, P.E. (1988). Human pro-IL-1 beta gene expression in monocytic cells is regulated by two distinct pathways. *J. Immunol.* **140**, 2267–2273.

Fernandes-Alnemri, T., Litwack, G. and Alnemri, E.S. (1994). CPP32, a novel human apoptotic protein with homology to *Caenorhabditis elegans* cell death protein *Ced-3* and mammalian interleukin-1β converting enzyme. *J. Biol. Chem.* **269**, 30761–30764.

Ferretti, M., Casini-Raggi, V., Pizarro, T.T., Eisenberg, S.P., Nast, C.C. and Cominelli, F. (1994). Neutralization of endogenous IL-1 receptor antagonist exacerbates and prolongs inflammation in rabbit immune colitis. *J. Clin. Invest.* **94**, 449–453.

Firestein, G.S., Berger, A.E, Tracey, D.E., Chosay, J.G., Chapman, D.L., Paine, M.M., Yu, C. and Zvaifler, N.J. (1992). IL-1 receptor antagonist protein production and gene expression in rheumatoid arthritis and osteoarthritis synovium. *J. Immunol.* **149**, 1054–1062.

Firestein, G.S., Boyle, D.L., Yu, C., Paine, M.M., Whisenand, T.D., Zvaifler, N.J. and Arend, W.P. (1994). Synovial IL-1 receptor antagonist and interleukin-1 balance in rheumatoid arthritis. *Arthritis Rheum.* **37**, 644–652.

Fischer, E., Marano, M.A., Barber, A.E., Hudson, A.A., Lee, K., Rock, C.S., Hawes, A.S., Thompson, R.C., Hayes, T.V., Benjamin, W.R., Lowry, S.F. and Moldawer, L.L. (1991). A comparison between the effects of interleukin-1α administration and sublethal endotoxemia in primates. *Am. J. Physiol.* **261**, R442–R449.

Fisher, C.J.J., Dhainaut, J.F., Opal, S.M., Pribble, J.P., Balk, R.A., Slotman, G.J., Iberti, T.J., Rackow, E.C., Shapiro, M.J. and Greenman, R.L. (1994a). Recomblnant human interleukin-1 receptor antagonist in the treatment of patients with sepsis syndrome. Results from a randomized, double blind, placebo-controlled trial. *J.A.M.A.* **271**, 1836–1843.

Fisher, C.J.J., Slotman, G.J., Opal, S.M., Pribble, J., Bone, R.C., Emmanuel, G., Ng, D., Bloedow, D.C. and Catalano, M.A. (1994b). Initial evaluation of human recombinant interleukin-1 receptor antagonist in the treatment of sepsis syndrome: a randomized, open-label, placebo-controlled multicenter trial. *Crit. Care Med.* **22**, 12–21.

Freshney, N.W., Rawlinson, L., Guesdon, F., Jones, E., Cowley, S., Hsuan, J. and Saklatvala, J. (1994). Interleukin-1 activates a novel protein cascade that results in the phosphorylation of hsp27. *Cell* **78**, 1039–1049.

Gagliardini, V., Fernandez, P.-A., Lee, R.K.K., Drexler, H.C.A., Rotello, R.J., Fishman, M.C. and Yuan, J. (1994). Prevention of vertebrate death by the *crmA* gene. *Science* **263**, 826–828.

Galcheva-Gargova, Z., Dérijard, B., Wu, I.-H. and Davis, R.J. (1994). An osmosensing signal transduction pathway in mammalian cells. *Science* **265**, 806–809.

Gallis, B., Prickett, K.S., Jackson, J., Slack, J., Schooley, K., Sims, J.E. and Dower, S.K. (1989). IL-1 induces rapid phosphorylation of the IL-1 receptor. *J. Immunol.* **143**, 3235–3240.

Gay, N.J. and Keith, F.J. (1991). Drosophila *Toll* and IL-1 receptor. *Nature* **351**, 355–356.

Gehrke, L., Jobling, S.A., Paik, L.S., McDonald, B., Rosenwasser, L.J. and Auron, P.E. (1990). A point mutation uncouples human interleukin-1β biological activity and receptor binding. *J. Biol. Chem.* **265**, 5922–5925.

Gershenwald, J.E., Fong, Y.M., Fahey, T.J., Calvano, S.E., Chizzonite, R., Kilian, P.L., Lowry, S.F. and Moldawer, L.L. (1990). Interleukin-1 receptor blockade attenuates the host inflammatory response. *Proc. Natl Acad. Sci. U.S.A.* **87**, 4966–4970.

Ghezzi, P., Dinarello, C.A., Bianchi, M., Rosandich, M.E., Repine, J.E. and White, C.W. (1991). Hypoxia increases production of interleukin-1 and tumor necrosis factor by human mononuclear cells. *Cytokine* **3**, 189–194.

Giri, J., Newton, R.C. and Horuk, R. (1990). Identification of soluble interleukin-1 binding protein in cell-free supernatants. *J. Biol. Chem.* **265**, 17416–17419.

Giri, J.G., Wells, J., Dower, S.K., McCall, C.E., Guzman, R.N., Slack, J., Bird, T.A., Shanebeck, K., Grabstein, K.H., Sims, J.E. and Alderson, M.R. (1994). Elevated levels of shed type II IL-1 receptor in sepsis. *J. Immunol.* **153**, 5802–5813.

Granowitz, E.V., Santos, A., Poutsiaka, D.D., Cannon, J.G., Wilmore, D.A., Wolff, S.M. and Dinarello, C.A. (1991). Circulating interleukin-1 receptor antagonist levels during experimental endotoxemia in humans. *Lancet* **338**, 1423–1424.

Granowitz, E.V., Clark, B.D., Vannier, E., Callahan, M.V. and Dinarello, C.A. (1992a). Effect of interleukin-1 (IL-1) blockade on cytokine synthesis: I. IL-1 receptor antagonist inhibits IL-1-induced cytokine synthesis and blocks the binding of IL-1 to its type II receptor on human monocytes. *Blood* **79**, 2356–2363.

Granowitz, E.V., Porat, R., Mier, J.W., Pribble, J.P., Stiles, D.M., Bloedow, D.C., Catalano, M.A., Wolff, S.M. and Dinarello, C.A. (1992b). Pharmacokinetics, safety, and immunomodulatory effects of human recombinant interleukin-1 receptor antagonist in healthy humans. *Cytokine* **4**, 353–360.

Granowitz, E.V., Porat, R., Mier, J.W., Orencole, S.F., Callahan, M.V., Cannon, J.G., Lynch, E.A., Ye, K., Poutsiaka, D.D., Vannier, E., Shapiro, L., Pribble, J.P., Stiles, D.M., Catalano, M.A., Wolff, S.M. and Dinarello, C.A. (1993). Hematological and immunomodulatory effects of an interleukin-1 receptor antagonist coinfusion during low-dose endotoxemia in healthy humans. *Blood* **82**, 2985–2990.

Graves, B.J., Hatada, M.H., Hendrickson, W.A., Miller, J.K., Madison, V.S. and Satow, Y. (1990). Structure of interleukin-1α at 2.7 Å resolution. *Biochemistry* **29**, 2679–2684.

Greenfeder, S.A., Nunes, P., Kwee, L., Labow, M., Chizzonite, R.A. and Ju, G. (1995a). Molecular cloning and characterization of a second subunit of the interleukin-1 receptor complex. *J. Biol. Chem.* **270**, 13757–13765.

Greenfeder, S.A., Varnell, T., Powers, G., Lombard-Gillooly, K., Shuster, D., McIntyre, K.W., Ryan, D.E., Leven, W., Madison, V. and Ju, G. (1995b). Insertion of a structural domain of interleukin-1β confers agonist activity to the IL-1 receptor antagonist. *J. Biol. Chem.* **270**, 22460–22465.

Gronich, J., Konieczkowski, M., Gelb, M.H., Nemenoff, R.A. and Sedor, J.R. (1994). Interleukin-1α causes a rapid activation of cytosolic phospholipase A_2 by phosphorylation in rat mesangial cells. *J. Clin. Invest.* **93**, 1224–1233.

Gruetter, M.G., van Oostrum, J., Priestle, J.P., Edelmann, E., Joss, U., Feige, U., Vosbeck, K. and Schmitz, A. (1994). A mutational analysis of receptor binding sites of interleukin-1β: differences in binding of human interleukin-1β muteins to human and mouse receptors. *Protein Eng.* **7**, 663–671.

Gu, Y., Sarnecki, C., Aldape, R.A., Livingston, D.L. and Su, M.S. (1995a). Cleavage of poly (ADP-ribose) polymerase by interleukin-1 beta converting enzyme and its homologs TX and Nedd-2. *J. Biol. Chem.* **270**, 18715–18718.

Gu, Y., Wu, J., Faucheu, C., Lalanne, J.-L., Diu, A., Livingston, D.L. and Su, M.S.-S. (1995b). Interleukin-1β converting enzyme requires oligomerization for activity of processed forms *in vivo*. *EMBO J.* **14**, 1923–1931.

Gu, Y., Kuida, K., Tsutsui, H., Ku, G., Hsiao, K., Fleming, M.A., Hayashi, N., Higashino, K., Okamura, H., Nakanishi, K., Kurimoto, M., Tanimoto, T., Flavell, R.A., Sato, V., Harding, M.W., Livingston, D.L. and Su, M.S.-S. (1997). Activation of interferon-γ inducing factor mediated by interleukin-1β converting enzyme. *Science* **275**, 206–209.

Guesdon, F., Freshney, N., Waller, R.J., Rawlinson, L. and Saklatvala, J. (1993). Interleukin 1 and tumor necrosis factor stimulate two novel protein kinases that phosphorylate the heat shock protein hsp27 and beta-casein. *J. Biol. Chem.* **268**, 4236–4243.

Guesdon, F., Waller, R.J. and Saklatvala, J. (1994). Specific activation of β-casein kinase by the inflammatory cytokines interleukin-1 and tumour necrosis factor. *Biochem. J.* **304**, 761–768.

Guida, S., Heguy, A. and Melli, M. (1992). The chicken IL-1 receptor: differential evolution of the cytoplasmic and extracellular domains. *Gene* **111**, 239–243.

Hammerberg, C., Arend, W.P., Fisher, G.J., Chan, L.S., Berger, A.E., Haskill, J.S., Voorhees, J.J. and Cooper, K.D. (1992). Interleukin-1 receptor antagonist in normal and psoriatic epidermis. *J. Clin. Invest.* **90**, 571–583.

Han, J., Lee, J.-D., Bibbs, L. and Ulevitch, R.J. (1994). A MAP kinase targeted by endotoxin and hyperosmolarity in mammalian cells. *Science* **265**, 808–811.

Han, J., Richter, B., Li. Z., Kravchenko, V.V. and Ulevitch, R.J. (1995). Molecular cloning of human p38 MAP kinase. *Biochem. Biophys. Acta* **1265**, 224–227.

Haskill, S., Martin, M., VanLe, L., Morris, J., Peace, A., Bigler, C.F., Jaffe, G.J., Sporn, S.A., Fong, S., Arend, W.P. and Ralph, P. (1991). cDNA cloning of a novel form of the interleukin-1 receptor antagonist associated with epithelium. *Proc. Natl Acad. Sci. U.S.A.* **88**, 3681–3685.

Hauser, C., Saurat, J.-H., Schmitt, A., Jaunin, F. and Dayer, J.-M. (1986). Interleukin-1 is present in normal epidermis. *J. Immunol.* **136**, 3317–3222.

Hazuda, D.J., Strickler, J., Kueppers, F., Simon, P.L. and Young, P.R. (1990). Processing of precursor interleukin-1 beta and inflammatory disease. *J. Biol. Chem.* **265**, 6318–6322.

Hazuda, D.J., Strickler, J., Simon, P. and Young, P.R. (1991). Structure–function mapping of interleukin 1 precursors. Cleavage leads to a conformational change in the mature protein. *J. Biol. Chem.* **266**, 7081–7086.

Heguy, A., Baldari, C., Bush, K., Nagele, R., Newton, R.C., Robb, R.J., Horuk, R., Telford, J.L. and Melli, M. (1991). Internalization and nuclear localization of interleukin 1 are not sufficient for function. *Cell Growth Differ.* **2**, 311–315.

Heguy, A., Baldari, C.T., Macchia, G., Telford, J.L. and Melli, M. (1992). Amino acids conserved in interleukin-1 receptors and the *Drosophila Toll* protein are essential for IL-1R signal transduction. *J. Biol. Chem.* **267**, 2605–2609.

Heguy, A., Baldari, C.T., Censini, S., Ghiara, P. and Telford, J.L. (1993). A chimeric type II/I interleukin-1 receptor can mediate interleukin-1 induction of gene expression in T cells. *J. Biol. Chem.* **268**, 10490–10494.

Heremans, H., van Damme, J., Dillen, C., Dikman, R. and Billiau, A. (1990). Interferon-γ, a mediator of lethal lipopolysaccharide-induced Schwartzman-like shick in mice. *J. Exp. Med.* **171**, 1853–1861.

Higgins, G.C., Foster, J.L. and Postlethwaite, A.E. (1993). Synthesis and biological activity of human interleukin-1β propiece *in vitro. Arthritis Rheum.* **39**, S153.

Higgins, G.C., Foster, J.L. and Postlethwaite, A.E. (1994). Interleukin-1 beta polypeptide is detected intracellularly and extracellularly when human monocytes are stimulated with LPS *in vitro*. *J. Exp. Med.* **180**, 607–614.

Hirsch, E., Irkura, V.M., Paul, S.M. and Hirsch, D. (1996). Functions of interleukin-1 receptor antagonist in gene knockout and overproducing mice. *Proc. Natl Acad. Sci. U.S.A.* **93**, 11 008–11 013.

Holmberg, L.A., Seidel, K., Leisenring, W. and Torok-Storb, B. (1994). Aplastic anemia analysis of stromal cell-function in long-term marrow cultures. *Blood* **84**, 3685–3690.

Hopp, T.P. (1995). Evidence from sequence information that the interleukin-1 receptor is a transmembrane GTPase. *Protein Sci.* **4**, 1851–1859.

Huwiler, A. and Pfeilschifter, J. (1994). Interleukin-1 stimulates *de novo* synthesis of mitogen-activated protein kinase in glomerular mesangial cells. *FEBS Lett.* **350**, 135–138.

Iizumi, T., Sato, S., Iiyama, T., Hata, R., Amemiya, H., Tomomasa, H., Yazaki, T. and Umeda, T. (1991). Recombinant human interleukin-1 beta analogue as a regulator of hematopoiesis in patients receiving chemotherapy for urogenital cancers. *Cancer* **68**, 1520–1523.

Irmler, M., Hertig, S., MacDonald, H.R., Sadoul, R., Becherer, J.D., Proudfoot, A., Solari, R. and Tschopp, J. (1995). Granzyme A is an interleukin-1β-converting enzyme. *J. Exp. Med.* **181**, 1917–1922.

Janson, R.W., King, J.T.E., Hance, K.R. and Arend, W.P. (1993). Enhanced production of IL-1 receptor antagonist by alveolar macrophages from patients with interstitial lung disease. *Am. Rev. Respir. Dis.* **148**, 495–503.

Jarrous, N. and Kaempfer, R. (1994). Induction of human interleukin-1 gene expression by retinoic acid and its regulation at processing of precursor transcripts. *J. Biol. Chem.* **269**, 23141–23149.

Jobling, S.A., Auron, P.E., Gurka, G., Webb, A.C., McDonald, B., Rosenwasser, L.J. and Gehrke, L. (1988). Biological activity and receptor binding of human prointerleukin-1β and subpeptides. *J. Biol. Chem.* **263**, 16372.

Kaplanski, G., Porat, R., Aiura, K., Erban, J.K., Gelfand, J.A. and Dinarello, C.A. (1993). Activated platelets induce endothelial secretion of interleukin-8 *in vitro* via an interleukin-1-mediated event. *Blood* **81**, 2492–2495.

Kaplanski, G., Farnarier, C., Kaplanski, S., Porat, R., Shapiro, L., Bongrand, P. and Dinarello, C.A. (1994). Interleukin-1 induces interleukin-8 from endothelial cells by a juxacrine mechanism. *Blood* **84**, 4242–4248.

Kaspar, R.L. and Gehrke, L. (1994). Peripheral blood mononuclear cells stimulated with C5a or lipopolysaccharide to synthesize equivalent levels of IL-1β mRNA show unequal IL-1β protein accumulation but similar polyribosome profiles. *J. Immunol.* **153**, 277–286.

Kester, M., Siomonson, M.S., Mene, P. and Sedor, J.R. (1989). Interleukin-1 generate transmembrane signals from phospholipids through novel pathways in cultured rat mesangial cells. *J. Clin. Invest.* **83**, 718–723.

Kitamura, T. and Takaku, F. (1989). A preclinical and phase I clinical trial of IL-1. *Exp. Med.* **7**, 170–177.

Knaus, W.A., Harrell, F.E., Fisher, C.J., Wagner, D.P., Opal, S.M., Sadoff, J.C., Draper, E.A., Walawander, C.A., Conboy, K. and Grasela, T.H. (1993). The clinical evaluation of new drugs for sepsis. *J.A.M.A.* **270**, 1233–1241.

Kobayashi, Y., Yamamoto, K., Saido, T., Kawasaki, H., Oppenheim, J.J. and Matsushima, K. (1990). Identification of calcium-activated neutral protease as a processing enzyme of human interleukin-1 alpha. *Proc. Natl Acad. Sci. U.S.A.* **87**, 5548–5552.

Kobayashi, Y., Oppenheim, J.J. and Matsushima, K. (1991). Human pre-interleukin-1α and β: structural features revealed by limited proteolysis. *Chem. Pharm. Bull. (Tokyo)* **39**, 1513–1517.

Kohno, K., Kataoka, J., Ohtsuki, T., Suemoto, Y., Okamoto, I., Usui, M., Ikeda, M. and Kurimoto, M. (1997). IFN-γ-inducing factor (IGIF) is a co-stimulatory factor on the activation of Th1 but not Th2 cells and exerts its effect independently of IL-12. *J. Immunol.* **158**, 1541–1550.

Kolesnick, R. and Golde, D.W. (1994). The sphingomyelin pathway in tumor necrosis factor and interleukin-1 signalling. *Cell* **77**, 325–328.

Kracht, M., Truong, O., Totty, N.F., Shiroo, M. and Saklatvala, J. (1994). Interleukin-1α activates two forms of p54α mitogen-activated proetin kinase in rabbit liver. *J. Exp. Med.* **180**, 2017–2027.

Kristensen, M., Deleuran, B., Eedy, D.J., Feldmann, M., Breathnach, S.M. and Brennan, F.M. (1992). Distribution of interleukin-1 receptor antagonist protein (IRAP), interleukin-1 receptor, and interleukin-1 alpha in normal and psoriatic skin. *Br. J. Dermatol.* **127**, 305–311.

Kroggel, R., Martin, M., Pingoud, V., Dayer, J.-M. and Resch, K. (1988). Two-chain structure of the interleukin-1 receptor. *FEBS Lett.* **229**, 59–62.

Kuida, K., Lippke, J.A., Ku, G., Harding, M.W., Livingston, D.J., Su, M.S.-S. and Flavell, R.A. (1995). Altered cytokine export and apoptosis in mice deficient in interleukin-1β converting enzyme. *Science* **267**, 2000–2003.

Kumar, S., Kinoshita, M., Noda, M., Copeland, N.G. and Jenkins, N.A. (1994). Induction of apoptosis by the mouse *Nedd2* gene, which encodes a protein similar to the product of *Caenorhabditis elegans* cell death gene *ced-3* and mammalian IL-1β converting enzyme. *Genes Dev.* **8**, 1613–1626.

Kuno, K. and Matsushima, K. (1994). The IL-1 receptor signaling pathway. *J. Leukoc. Biol.* **56**, 542–547.

Kuno, K., Okamoto, S., Hirose, K., Murakami, S. and Matsushima, K. (1993). Structure and function of the intracellular portion of the mouse interleukin-1 receptor (type I). *J. Biol. Chem.* **268**, 13 510–13 518.

Kurt-Jones, E.A., Beller, D.I., Mizel, S.B. and Unanue, E.R. (1985). Identification of a membrane-associated interleukin-1 in macrophages. *Proc. Natl Acad. Sci. U.S.A.* **82**, 1204–1208.

Lambriola-Tomkins, E., Chandran, C., Varnell, T.A., Madison, V.S. and Ju, G. (1993). Structure–function analysis of human IL-1α: identification of residues required for binding to the human type I IL-1 receptor. *Protein Eng.* **6**, 535–539.

Laughlin, M.J., Kirkpatrick, G., Sabiston, N., Peters, W. and Kurtzberg, J. (1993). Hematopoietic recovery following high-dose combined alkylating-agent chemotherapy and autologous bone marrow support in patients in phase I clinical trials of colony stimulating factors: G-CSF, GM-CSF, IL-1, IL-2 and MCSF. *Ann. Hematol.* **67**, 267–276.

Lebsack, M.E., Paul, C.C., Bloedow, D.C., Burch, F.X., Sack, M.A., Chase, W. and Catalano, M.A. (1991). Subcutaneous IL-1 receptor antagonist in patients with rheumatoid arthritis. *Arthritis Rheum.* **34** (supplement), S67.

Lebsack, M.E., Paul, C.C., Martindale, J.J. and Catalano, M.A. (1993). A dose- and regimen-ranging study of IL-1 receptor antagonist in patients with rheumatoid arthritis. *Arthritis Rheum.* **36**, S39.

Lee, J.C., Laydon, J.T., McDonnell, P.C., Gallagher, T.F., Kumar, S., Green, D., McNulty, D., Blumenthal, M.J., Heys, J.R., Landvatter, S.W., Strickler, J.E., McLaughlin, M.M., Slemens, I.R., Fisher, S.M., Livi, G.P., White, J.R., Adams, J.L. and Young, P.R. (1994). A protein kinase involved in the regulation of inflammatory cytokine biosynthesis. *Nature* **372**, 739–747.

Leitman, D.C., Andersen, J.W., Kuno, T., Kamisaki, Y., Chang, J. and Murad, F. (1986). Identification of multiple binding sites for atrial natriuretic factor by affinity crosslinking in cultured endothelial cells. *J. Biol. Chem.* **261**, 11 650–11 656.

LeMay, L.G., Otterness, I.G., Vander, A.J. and Kluger, M.J. (1990). *In vivo* evidence that the rise in plasma IL-6 following injection of a fever-inducing dose of LPS is mediated by IL-1 beta. *Cytokine* **2**, 199–204.

Li, P., Allen, H., Banerjee, S., Franklin, S., Herzog, L., Johnston, C., McDowell, J., Paskind, M., Rodman, L., Salfeld, J., Towne, E., Tracey, D., Wardwell, S., Wei, F.-Y., Wong, W., Kamen, R. and Seshadri, T. (1995). Mice deficient in interleukin-1 converting enzyme (ICE) are defective in production of mature interleukin-1β and resistant to endotoxic shock. *Cell* **80**, 401–411.

Maier, J.A.M., Voulalas, P., Roeder, D. and Maciag, T. (1990). Extension of the life span of human endothelial cells by an interleukin-1α antisense oligomer. *Science* **249**, 1570–1574.

Maier, J.A.M., Statuto, M. and Ragnotti, G. (1994). Endogenous interleukin-1 alpha must be transported to the nucleus to exert its activity in human endothelial cells. *Mol. Cell. Biol.* **14**, 1845–1851.

Makarov, S.S., Olsen, J.C., Johnston, W.N., Anderle, S.K., Brown, R.R., Baldwin, A.S., Haskill, J.S. and Schwab, J.H. (1995). Suppression of experimental arthritis by gene transfer of interleukin-1 receptor antagonist cDNA. *Proc. Natl Acad. Sci. U.S.A.* **92**, 11 301–11 315.

Mancilla, J., Ikejima, I. and Dinarello, C.A. (1992). Glycosylation of the interleukin-1 receptor type I is required for optimal binding of interleukin-1. *Lymphokine Cytokine Res.* **11**, 197–205.

Margolis, N.H. and Dinarello, C.A. (1994). Incorporation of ^{3}H-thymidine by peripheral blood mononuclear cells from normal subjects and acute myelogenous leukemia patients is independent of interleukin-1β converting enzyme. *Cytokine* **6**, 566.

Markewitz, A., Faist, E., Lang, S., Endres, S., Fuchs, B. and Reichart, B. (1993). Successful restoration of cell-mediated immune response after cardiopulmonary bypass by immunomodulation. *J. Thorac. Cardiovasc. Surg.* **105**, 15–24.

Martin, M., Bol, G.F., Eriksson, A., Resch, K. and Brigelius-Flohe, R. (1994). Interleukin-1-induced activation of a protein kinase co-precipitating with the type I interleukin-1 receptor in T-cells. *Eur. J. Immunol.* **35**, 1566–1571.

Mathias, S., Younes, A., Kan, C.-C., Orlow, I., Joseph, C. and Kolesnick, R.N. (1993). Activation of the sphingomyelin signaling pathway in intact EL4 cells and in a cell-free system by IL-1β. *Science* **259**, 519–522.

Matsushima, K., Kobayashi, Y., Copeland, T.D., Akahoshi, T. and Oppenheim, J.J. (1987). Phosphorylation of a cytosolic 65-kDa protein induced by interleukin-1 in glucocorticoid pretreated normal human peripheral blood mononuclear leukocytes. *J. Immunol.* **139**, 3367–3374.

McCarthy, P.L., Abhyankar, S., Neben, S., Sieff, C., Thompson, R.C., Burakoff, S. and Ferrara, J.L.M. (1991). Inhibition of interleukin-1 by interleukin-1 receptor antagonist prevents graft *versus* host disease. *Blood* **78**, 1915–1918.

McMahon, C.J., Slack, J.L., Mosley, B., Cosman, D., Lupton, S.D., Brunton, L.L., Grubin, C.E., Wignall, J.M., Jenkins, N.A., Brannan, C.I., Copeland, N.G., Huebner, K., Croce, C.M., Cannizzaro, L.A., Benjamin, D., Dower, S., Spriggs, M.K. and Sims, J.E. (1991). A novel IL-1 receptor cloned form B cells by mammalian expression is expressed in many cell types. *EMBO J.* **10**, 2821–2832.

Mignatti, P. and Rifkin, D.B. (1991). Release of basic fibroblast growth factor, a angiogenic factor devoid of secretory signal sequence: a trivial phenomenon or a novel secretion mechanism? *J. Cell Biochem.* **47**, 201–217.

Mileno, M.D., Margolis, N.H., Clark, B.D., Dinarello, C.A., Burke, J.F. and Gelfand, J.A. (1995). Coagulation of whole blood stimulates interleukin-1β gene expression: absence of gene transcripts in anticoagulated blood. *J. Infect. Dis.* **172**, 308–311.

Miller, A.C., Schattenberg, D.G., Malkinson, A.M. and Ross, D. (1994). Decreased content of the IL-1α processing enzyme calpain in murine bone marrow-derived macrophages after treatment with the bezene metabolite hydroquinone. *Toxicol. Lett.* **74**, 177–184.

Miller, L.C., Isa, S., Vannier, E., Georgilis, K., Steere, A.C. and Dinarello, C.A. (1992). Live *Borrelia burgdorferi* preferentially activate IL-1β gene expression and protein synthesis over the interleukin-1 receptor antagonist. *J. Clin. Invest.* **90**, 906–912.

Miller, L.C., Lynch E.A., Isa, S., Logan, J.W., Dinarello, C.A. and Steere, A.C. (1993). Balance of synovial fluid IL-1β and IL-1 receptor antagonist and recovery from Lyme arthritis. *Lancet* **341**, 146–148.

Minnich-Carruth, L.L., Suttles, J. and Mizel, S.B. (1989). Evidence against the existence of a membrane form of murine IL-1α. *J. Immunol.* **142**, 526.

Mitchum, J.L. and Sims, J.E. (1995). IIP1: a novel human that interacts with the IL-1 receptor. *Cytokine* **7**, 595 (abstract).

Miura, M., Zhu, H., Rotello, R., Hartwieg, E.A. and Yuan, J. (1993). Induction of apoptosis in fibroblasts by IL-1β converting enzyme, a mammalian homolog of the *C. elegans* cell death gene *ced-3*. *Cell* **75**, 653–660.

Mizel, S.B. (1990). Cyclic AMP and interleukin-1 signal transduction. *Immunol. Today* **11**, 390–391.

Mizel, S.B. (1994). IL-1 signal transduction. *Eur. Cytokine Netw.* **5**, 547–561.

Mizel, S.B., Kilian, P.L., Lewis, J.C., Paganelli, K.A. and Chizzonite, R.A. (1987). The interleukin-1 receptor. Dynamics of interleukin 1 binding and internalization in T cells and fibroblasts. *J. Immunol* **138**, 2906–2912.

Mizutani, H., Black, R.A. and Kupper, T.S. (1991a). Human keratinocytes produce but do not process pro-interleukin-1β. *J. Clin. Invest.* **87**, 1066–1071.

Mizutani, H., Schecter, N., Zazarus, G., Black, R.A. and Kupper, T.S. (1991b). Rapid and specific conversion of precursor interleukin-1β to an active IL-1 species by human mast cell chymase. *J. Exp. Med.* **174**, 821–825.

Mosley, B., Urdal, D.L., Prickett, K.S., Larsen, A., Cosman, D., Conlon, P.J., Gillis, S. and Dower, S.K. (1987). The interleukin-1 receptor binds the human interleukin-1α precursor but not the interleukin-1β precursor. *J. Biol. Chem.* **262**, 2941–2944.

Muegge, K., Williams, T.M., Kant, J., Karin, M., Chiu, R., Schmidt, A., Siebenlist, U., Young, H.A. and Durum, S.K. (1989). Interleukin-1 costimulatory activity on the interleukin-2 promoter via AP-1. *Science* **246**, 249–251.

Muegge, K., Vila, M., Gusella, G.L., Musso, T., Herrlich, P., Stein, B. and Durum, S.K. (1993). IL-1 induction of the c-jun promoter. *Proc. Natl Acad. Sci. U.S.A.* **90**, 7054–7058.

Munoz, E., Beutner, U., Zubiaga, A. and Huber, B.T. (1990). IL-1 activates two separate signal transduction pathways in T helper type II cells. *J. Immunol.* **144**, 964–969.

Murzin, A.G., Lesk, A.M. and Chothia, C. (1992). β-trefoil fold. Patterns of structure and sequence in the Kunitz inhibitors interleukins-1β and 1α and fibroblast growth factors. *J. Mol. Biol.* **223**, 531–543.

Nakamura, K., Okamura, H., Wada, M., Nagata, K. and Tamura, T. (1989). Endotoxin-induced serum factor that stimulates gamma interferon production. *Infect. Immun.* **57**, 590–595.

Nakamura, K., Okamura, H., Nagata, K., Komatsu, T. and Tamura, T. (1993). Purification of a factor which provides a costimulatory signal for gamma interferon production. *Infect. Immun.* **61**, 64–70.

Nemunaitis, J., Appelbaum, F.R., Lilleby, K., Buhles, W.C., Rosenfeld, C., Zeigler, Z.R., Shadduck, R.K., Singer, J.W., Meyer, W. and Buckner, C.D. (1994). Phase I study of recombinant interleukin-1β in patients undergoing autologous bone marrow transplantation for acute myelogenous leukemia. *Blood* **83**, 3473–3479.

Norman, J.G., Franz, M.G., Fink, G.S., Messina, J., Fabri, P.J., Gower, W.R. and Carey, L.C. (1995a). Decreased mortality of severe acute pancreatitis after proximal cytokine blockade. *Ann. Surg.* **221**, 625–634.

Norman, J.G., Franz, M.G., Messina, J., Riker, A., Fabri, P.J., Rosemurgy, A.S. and Gower, W.R. (1995b). Interleukin-1 receptor antagonist decreases severity of experimental acute pancreatitis. *Surgery* **117**, 648–655.

O'Neill, L.A.J. (1995). Towards an understanding of the signal transduction pathways for interleukin-1. *Biochim. Biophys. Acta* **1266**, 31–44.

O'Neill, L.A.J., Bird, T.A. and Saklatvala, J. (1990). Interleukin-1 signal transduction. *Immunol. Today* **11**, 392–394.

Okamura, H., Nagata, K., Komatsu, T., Tanimoto, T., Nukata, Y., Tanabe, F., Akita, K., Torigoe, K., Okura, T., Fukuda, S. and Kurimoto, M. (1995a). A novel costimulatory factor for gamma interferon induction found in the livers of mice causes endotoxic shock. *Infect. Immun.* **63**, 3966–3972.

Okamura, H., Tsutsui, H., Komatsu, T., Yutsudo, M., Hakura, A., Takimoto, T., Torigoe, K., Okura, T., Nukada, Y., Hattori, K., Akita, K., Namba, M., Tanabe, F., Konishi, K., Fukuda, S. and Kurimoto, M. (1995b). Cloning of a new cytokine that induces interferon-γ. *Nature* **378**, 88–91.

Orencole, S.F. and Dinarello, C.A. (1989). Characterization of a subclone (D10S) of the D10.G4.1 helper T-cell line which proliferates to attomolar concentrations of interleukin-1 in the absence of mitogens. *Cytokine* **1**, 14–22.

Orencole, S.F., Vannier, E. and Dinarello, C.A. (1994). Detection of soluble IL-1 receptor type II (IL-1sRII) in sera and plasma from healthy volunteers. *Cytokine* **6**, 554 (abstract).

Orencole, S.F., Fantuzzi, G., Vannier, E. and Dinarello, C.A. (1995). Circulating levels of IL-1 soluble receptors in health and after endotoxin or IL-2. *Cytokine* **7**, 642.

Pereira, B.J.G., Poutsiaka, D.D., King, A.J., Strom, J.A., Narayan, G., Levey, A.S. and Dinarello, C.A. (1992). *In vitro* production of interleukin-1 receptor antagonist in chronic renal failure, continuous peritoneal dialysis and hemodialysis. *Kidney Int.* **42**, 1419–1424.

Polan, M.L., Loukides, J.A. and Honig, J. (1994). Interleukin-1 in human ovarian cells and in peripheral blood monocytes increases during the luteal phase: evidence of a midcycle surge in the human. *Am. J. Obstet. Gynecol.* **170**, 1000–1006.

Porat, R., Poutsiaka, D.D., Miller, L.C., Granowitz, E.V. and Dinarello, C.A. (1992). Interleukin-1 (IL-1) receptor blockade reduces endotoxin and *Borrelia burgdorferi*-stimulated IL-8 synthesis in human mononuclear cells. *FASEB J.* **6**, 2482–2486.

Preistle, J.P., Schar, H.P. and Grutter, M.G. (1989). Crystallographic refinement of interleukin 1 beta at 2.0 A resolution. *Proc. Natl Acad. Sci. U.S.A.* **86**, 9667–9671.

Rambaldi, A., Torcia, M., Bettoni, S., Barbui, T., Vannier, E., Dinarello, C.A. and Cozzolino, F. (1991). Modulation of cell proliferation and cytokine production in acute myeloblastic leukemia by interleukin-1 receptor antagonist and lack of its expression by leukemic cells. *Blood* **78**, 3248–3253.

Ray, C.A., Black, R.A., Kronheim, S.R., Greenstreet, T.A., Sleath, P.R., Salvesen, G.S. and Pickup, D.J. (1992). Viral inhibition of inflammation: cowpox virus encodes an inhibitor of the interleukin-1β converting enzyme. *Cell* **69**, 597–604.

Reikerstorfer, A., Holtz, H., Stuynnenberg, H.G. and Busslinger, M. (1995). Low affinity binding of interleukin-1 beta and intracellular signaling via NFκB identify Fit-1 as a distant member of the interleukin-1 receptor family. *J. Biol. Chem.* **270**, 17 645–17 648.

Rosoff, P.M. (1989). Characterization of the interleukin-1-stimulated phospholipase C activity in human T lymphocytes. *Lymphokine Res.* **8**, 407–413.

Rosoff, P.M., Savage, N. and Dinarello, C.A. (1988). Interleukin-1 stimulates diacylglycerol production in T lymphocytes by a novel mechanism. *Cell* **54**, 73–81.

Rossi, B. (1993). IL-1 transduction signals. *Eur. Cytokine Netw.* **4**, 181–187.

Roux-Lombard, P., Modoux, C., Vischer, T., Grassi, J. and Dayer, J.-M. (1992). Inhibitors of interleukin 1 activity in synovial fluids and in cultured synovial fluid mononuclear cells. *J. Rheumatol.* **19**, 517–523.

Rubartelli, A., Cozzolino, F., Talio, M. and Sitia, R. (1990). A novel secretory pathway for interleukin-1 beta, a protein lacking a signal sequence. *EMBO J.* **9**, 1503–1510.

Rubartelli, A., Bajetto, A., Allavena, G., Wollman, E. and Sitia, R. (1992). Secretion of thioredoxin by normal and neoplastic cells through a leaderless secretory pathway. *J. Biol. Chem.* **267**, 24 161–24 164.

Rzymkiewicz, D.M., DuMaine, J. and Morrison, A.R. (1995). IL-1β regulates rat mesangial cyclooxygenase II gene expression by tyrosine phosphorylation. *Kidney Int.* **47**, 1354–1363.

Santos, A.A., Scheltinga, M.R., Lynch, E., Brown, E.F., Lawton, P., Chambers, E., Browning, J., Dinarello, C.A., Wolff, S.M. and Wilmore, D.W. (1993). Elaboration of interleukin-1 receptor antagonist is not attenuated by glucocorticoids after endotoxemia. *Arch. Surg.* **128**, 138–144.

Savage, N., Puren, A.J., Orencole, S.F., Ikejima, T., Clark, B.D. and Dinarello, C.A. (1989). Studies on IL-1 receptors on D10S T-helper cells: demonstration of two molecularly and antigenically distinct IL-1 binding proteins. *Cytokine* **1**, 23–25.

Schindler, R., Clark, B.D. and Dinarello, C.A. (1990a). Dissociation between interleukin-1β mRNA and protein synthesis in human peripheral blood mononuclear cells. *J. Biol. Chem.* **265**, 10 232–10 237.

Schindler, R., Gelfand, J.A. and Dinarello, C.A. (1990b). Recombinant C5a stimulates transcription rather than translation of IL-1 and TNF; cytokine synthesis induced by LPS, IL-1 or PMA. *Blood* **76**, 1631–1638.

Schindler, R., Ghezzi, P. and Dinarello, C.A. (1990c). IL-1 induces IL-1. IV. IFN-γ suppresses IL-1 but not lipopolysaccharide-induced transcription of IL-1. *J. Immunol.* **144**, 2216–2222.

Schindler, R., Linnenweber, S., Schulze, M., Oppermann, M., Dinarello, C.A., Shaldon, S. and Koch, K.-M. (1993). Gene expression of interleukin-1β during hemodialysis. *Kidney Int.* **43**, 712–721.

Schirò, R., Longoni, D., Rossi, V., Maglia, O., Doni, A., Arsura, M., Carrara, G., Masera, G., Vannier, E., Dinarello, C.A., Rambaldi, A. and Biondi, A. (1993). Suppression of juvenile chronic myelogenous leukemia colony growth by interleukin-1 receptor antagonist. *Blood* **83**, 460–465.

Schreuder, H. (1995). Crystal structure of the interleukin-1 receptor antgonist complex. *Cytokine* **7**, 599.

Schutze, S., Machleidt, T. and Kronke, M. (1994). The role of diacylglyercol and ceramide in tumor necrosis factor and interleukin-1 signal transduction. *J. Leukoc. Biol.* **56**, 533–541.

Serkkola, E. and Hurme, M. (1993). Synergism between protein-kinase C and cAMP-dependent pathways in the expression of the interleukin-1β gene is mediated via the activator-protein-1 (AP-1) enhancer activity. *Eur. J. Biochem.* **213**, 243–249.

Shapiro, L. and Dinarello, C.A. (1995). Osmotic regulation of cytokine synthesis *in vitro*. *Proc. Natl Acad. Sci. U.S.A.* **92**, 12 230–12 234.

Shapiro, L., Clark, B.D., Orencole, S.F., Poutsiaka, D.D., Granowitz, E.V. and Dinarello, C.A. (1993). Detection of tumor necrosis factor soluble receptor p55 in blood samples from healthy and endotoxemic humans. *J. Infect. Dis.* **167**, 1344–1350.

Shapiro, L., Panayotatos, N., Meydani, S.N., Wu, D. and Dinarello, C.A. (1994). Ciliary neurotrophic factor combined with soluble receptor inhibits synthesis of pro-inflammatory cytokines and prostaglandin-E_2 *in vitro*. *Exp. Cell. Res.* **215**, 51–56.

Shirakawa, F. and Mizel, S.B. (1989). *In vitro* activation and nuclear translocation of NF-kappa B catalyzed by cyclic AMP-derived protein kinase and protein kinase C. *Mol. Cell Biol.* **9**, 2424–2430.

Shirakawa, F., Tanaka, Y., Ota, T., Suzuki, H., Eto, S. and Yamashita, U. (1987). Expression of interleukin-1 receptors on human peripheral T-cells. *J. Immunol.* **138**, 4243–4248.

Simoncsits, A., Bristulf, J., Tjornhammar, M.L., Cserzo, M., Pongor, S., Rybakina, E., Gatti, S. and Bartfai, T. (1994). Deletion mutants of human IL-1β with significantly reduced agonist properties: search for the agonist/antagonist switch in ligands to the interleukin-1 receptors. *Cytokine* **6**, 206–214.

Sims, J.E., March, C.J., Cosman, D., Widmer, M.B., MacDonald, H.R., McMahon, C.J., Grubin, C.E., Wignall, J.M., Jackson, J.L. and Call, S M. (1988). cDNA expression cloning of the IL-1 receptor, a member of the immunoglobulin superfamily. *Science* **241**, 585–589.

Sims, J.E., Gayle, M.A., Slack, J.L., Alderson, M.R., Bird, T.A., Giri, J.G., Colotta, F., Re, F., Mantovani, A., Shanebeck, K., Grabstein, K.H. and Dower, S.K. (1993). Interleukin-1 signaling occurs exclusively via the type I receptor. *Proc. Natl Acad. Sci. U.S.A.* **90**, 6155–6159.

Sims, J.E., Giri, J.G. and Dower, S.K. (1994). The two interleukin-1 receptors play different roles in IL-1 activities. *Clin. Immunol. Immunopathol.* **72**, 9–14.

Sims. J.E., Painter, S.L. and Gow, I.R. (1995). Genomic organization of the type I and type II IL-1 receptors. *Cytokine* **7**, 483–490.

Singer, I.I., Scott, S., Chin, J., Kostura, M.J., Miller, D.K., Chapman, K. and Bayne, E.K. (1993). Interleukin-1β converting enzyme is localized on the external cell-surface membranes and in the cytoplasmic ground substance of activated human monocytes by immuno-electronmicroscopy. *Lymphokine Cytokine Res.* **12**, 340 (abstract).

Slack, J., McMahon, C.J., Waugh, S., Schooley, K., Spriggs, M.K., Sims, J.E. and Dower, S.K. (1993). Independent binding of interleukin-1 alpha and interleukin-1 beta to type I and type II interleukin-1 receptors. *J. Biol. Chem.* **268**, 2513–2524.

Smith, J.W., Urba, W.J., Curti, B.D., Elwood, L.J., Steis, R.G., Janik, J.E., Sharfman, W.H., Miller, L.L., Fenton, R.G., Conlon, K.C., Rossio, J., Kopp, W., Shimuzut, M., Oppenheim, J.J. and Longo, D. (1991). Phase II trial of interleukin-1 alpha in combination with indomethacin in melanoma patients. *Proc. Am. Soc. Clin. Oncol.* **10**, 293 (abstract).

Smith, J.W., Urba, W.J., Curti, B.D., Elwood, L.J., Steis, R.G., Janik, J.E., Sharfman, W.H., Miller, L.L., Fenton, R.G., Conlon, K.C., Rossio, J., Kopp, W., Shimuzut, M., Oppenheim, J.J. and Longo, D. (1992). The toxic and hematologic effects of interleukin-1 alpha administered in a phase I trial to patients with advanced malignancies. *J. Clin. Oncol.* **10**, 1141–1152.

Smith, J.W., Longo, D., Alford, W.G., Janik, J.E., Sharfman, W.H., Gause, B.L., Curti, B.D., Creekmore, S.P., Holmlund, J.T., Fenton, R.G., Sznol, M., Miller, L.L., Shimzu, M., Oppenheim, J.J., Fiem, S.J., Hursey, J.C., Powers, G.C. and Urba, W.J. (1993). The effects of treatment with interleukin-1α on platelet recovery after high-dose carboplatin. *N. Engl. J. Med.* **328**, 756–761.

Smith, D.J., McGuire, M.J., Tocci, M.J. and Thiele, D.L. (1997). IL-1β convertase (ICE) does not play a requisite role in apoptosis induced in T lymphoblasts by Fas-dependent or Fas-independent CTL effector mechanisms. *J. Immunol.* **158**, 163–170.

Spriggs, M.K., Hruby, D.E., Maliszewski, C.R., Pickup, D.J., Sims, J.E., Buller, R.M. and VanSlyke, J. (1992). Vaccinia and cowpox viruses encode a novel secreted interleukin-1 binding protein. *Cell* **71**, 145–152.

Starnes, H.F. (1991). Biological effects and possible clinical applications of interleukin-1. *Semin. Hematol.* **28**, 41–43.

Stevenson, F.T., Torrano, F., Locksley, R.M. and Lovett, D.H. (1992). Interleukin-1: the patterns of translation and intracellular distribution support alternative secretory mechanisms. *J. Cell Physiol.* **152**, 223–231.

Stevenson, F.T., Bursten, S.L., Fanton, C., Locksley, R.M. and Lovett, D.H. (1993). The 31-kDa precursor of interleukin-1α is myristoylated on specific lysines within the 16-kDa *N*-terminal propiece. *Proc. Natl Acad. Sci. U.S.A.* **90**, 7245–7249.

Stoeckle, M.Y. and Guan, L. (1993). High-resolution analysis of gro-α RNA poly(A) shortening: regulation by interleukin-1β. *Nucl. Acids Res.* **21**, 1613–1617.

Stosic-Grujicic, S., Basara, N., Milenkovic, P. and Dinarello, C.A. (1995). Modulation of acute myeloblastic leukemia (AML) cell proliferation and blast colony formation by antisense oligomer of IL-1 beta converting enzyme (ICE) and IL-1 receptor antagonist (IL-1ra). *J. Chemother.* **7**, 67–70.

Stylianou, E., O'Neill, L.A.J., Rawlinson, L., Edbrooke, M.R., Woo, P. and Saklatvala, J. (1992). Interleukin-1 induces NFκB through its type I but not type II receptor in lymphocytes. *J. Biol. Chem.* **267**, 15836–15841.

Symons, J.A., Eastgate, J.A. and Duff, G.W. (1991). Purification and characterization of a novel soluble receptor for interleukin-1. *J. Exp. Med.* **174**, 1251–1254.

Symons, J.A., Young, P.A. and Duff, G.W. (1993). The soluble interleukin-1 receptor: ligand binding properties and mechanisms of release. *Lymphokine Cytokine Res.* **12**, 381.

Tewari, A., Buhles, W.C., Jr. and Starnes, H.F., Jr. (1990). Preliminary report: effects of interleukin-1 on platelet counts. *Lancet* **336**, 712–714.

Thea, D.M., Porat, R., Nagimbi, K., Baangi, M., St. Louis, M.E., Kaplan, G., Dinarello, C.A. and Keusch, G.T. (1996). Plasma cytokines, plasma cytokine antagonists, and disease progression in African women infected with HIV-1. *Ann. Intern. Med.* **124**, 757–762.

Thornberry, N.A., Bull, H.G., Calaycay, J.R., Chapman, K.T., Howard, A.D., Kostura, M.J., Miller, D.K., Molineaux, S.M., Weidner, J.R., Aunins, J., Schmidt, J.A. and Tocci, M. (1992). A novel heterodimeric cysteine protease is required for interleukin-1 beta processing in monocytes. *Nature* **356**, 768–774.

Tilg, H., Trehu, E., Atkins, M.B., Dinarello, C.A. and Mier, J.W. (1994a). Interleukin-6 (IL-6) as an anti-inflammatory cytokine: induction of circulating IL-1 receptor antagonist and soluble tumor necrosis factor receptor p55. *Blood* **83**, 113–118.

Tilg, H., Trehu, E., Shapiro, L., Pape, D., Atkins, M.B., Dinarello, C.A. and Mier, J.W. (1994b). Induction of circulating soluble tumour necrosis factor receptor and interleukin 1 receptor antagonist following interleukin-1α infusion in humans. *Cytokine* **6**, 215–219.

Tsutsui, H., Nakanishi, K., Matsui, K., Higashino, K., Okamura, H., Miyazawa, Y. and Kaneda, K. (1996). IFN-γ-inducing factor up-regulates Fas ligand-mediated cytotoxic activity of murine natural killer cell clones. *J. Immunol.* **157**, 3967–3973.

Ushio, S., Namba, M., Okura, T., Hattori, K., Nukada, Y., Akita, K., Tanabe, F., Konishi, K., Micallef, M., Fujii, M., Torigoe, K., Tanimoto, T., Fukuda, S., Ikeda, M., Okamura, H. and Kurimoto, M. (1996). Cloning of the cDNA for human IFN-γ-inducing factor, expression in *Escherichia coli*, and studies on the biologic activities of the protein. *J. Immunol.* **156**, 4274–4279.

van der Poll, T., Bueller, H.R., ten Cate, H., Wortel, C.H., Bauer, K.A., van Deventer, S.J.H., Hack, C.E., Sauerwein, H.P., Rosenberg, R.D. and ten Cate, J.W. (1990). Activation of coagulation after administration of tumor necrosis factor to normal subjects. *N. Engl. J. Med.* **322**, 1622–1627.

Vannier, E. and Dinarello, C.A. (1993). Histamine enhances interleukin (IL)-1-induced IL-1 gene expression and protein synthesis *via* H_2 receptors in peripheral blood mononuclear cells: comparison with IL-1 receptor antagonist. *J. Clin. Invest.* **92**, 281–287.

Vidal-Vanaclocha, F., Amézaga, C., Asumendi, A., Kaplanski, G. and Dinarello, C.A. (1994). Interleukin-1 receptor blockade reduces the number and size of murine B16 melanoma hepatic metastases. *Cancer Res.* **54**, 2667–2672.

Vigers, G.P., Caffes, P., Evans, R.J., Thompson, R.C., Eisenberg, S.P. and Brandhuber, B.J. (1994). X-ray structure of interleukin-1 receptor antagonist at 2.0-A resolution. *J. Biol. Chem.* **269**, 12874–12879.

Wakabayashi, G., Gelfand, J.A., Jung, W.K., Connolly, R.J., Burke, J.F. and Dinarello, C.A. (1991). *Staphylococcus epidermidis* induces complement activation, tumour necrosis factor and interleukin-1, a shock-like state and tissue injury in rabbits without endotoxemia. *J. Clin. Invest.* **87**, 1925–1935.

Walker, N.P., Talanian, R.V., Brady, K.D., Dang, L.C., Bump, N.J., Ferenz, C.R., Franklin, S., Ghayur, T., Hackett, M.C. and Hammill, L.D. (1994). Crystal structure of the cysteine protease interleukin-1 beta-converting enzyme: a (p20/p10)2 homodimer. *Cell* **78**, 343–352.

Watanabe, N. and Kobayashi, Y. (1994). Selective release of a processed form of interleukin-1α. *Cytokine* **6**, 597–601.

Watt, I. and Cobby, M. (1996). Recombinant human interleukin-1 receptor antagonist reduces the rate of joint erosion in rheumatoid arthritis. *Arthritis Rheum.* **39**, S123.

Weizmann, M.N. and Savage, N. (1992). Nuclear internalization and DNA binding activities of interleukin-1, interleukin-1 receptor complexes. *Biochem. Biophys. Res. Commun.* **187**, 1166–1171.

Wesche, H., Korherr, C., Kracht, M., Falk, W., Resch, K. and Martin, M.U. (1997). The interleukin-1 receptor accessory protein is essential for IL-1-induced activation of interleukin-1 receptor-associated kinase (IRAK) and stress-activated protein kinases (SAP kinases). *J. Biol. Chem.* **272**, 7727–7731.

Wessendorf, J.H.M., Garfinkel, S., Zhan, X., Brown, S. and Maciag, T. (1993). Identification of a nuclear localization sequence within the structure of the human interleukin-1α precursor. *J. Biol. Chem.* **268**, 22100–22104.

Wetzler, M., Kurrzock, R., Estrov, Z., Kantarjian, H., Gisslinger, H., Underbrink, M.P. and Talpaz, M. (1994). Altered levels of interleukin-1β and interleukin-1 receptor antagonist in chronic myelogenous leukemia: clinical and prognostic correlates. *Blood* **84**, 3142–3147.

Wilson, K.P., Black, J.A., Thomson, J.A., Kim, E.E., Griffith, J.P., Navia, M.A., Murcko, M.A., Chambers, S.P., Aldape, R.A., Raybuck, S.A. and Livingston, D.J. (1994). Structure and mechanism of interleukin-1β converting enzyme. *Nature* **370**, 270–275.

Wolff, S.M. (1973). Biological effects of bacterial endotoxins in man. *J. Infect. Dis.* **128** (supplement), 733–758.

Ye, K., Koch, K.-C., Clark, B.D. and Dinarello, C.A. (1992). Interleukin-1 down regulates gene and surface expression of interleukin-1 receptor type 1 by destabilizing its mRNA whereas interleukin-2 increases its expression. *Immunology* **75**, 427–434.

Yuan, J., Shaam, S., Ledoux, S., Ellis, H.M. and Horvitz, H.R. (1993). The *C. elegans* cell death gene *ced-3* encodes a protein similar to mammalian interleukin-1β-converting enzyme. *Cell* **75**, 641–652.

Zhang, S., Han, J., Sells, M.A., Chernoff, J., Knaus, U.G., Ulevitch, R.J. and Bokoch, G.M. (1995). Rho family GTPases regulate p38 mitogen-activated protein kinase through the downstream mediator Pak1. *J. Biol. Chem.* **270**, 23934–23936.

Chapter 4

Interleukin-2 and the Interleukin-2 Receptor

Sarah L. Gaffen[1], Mark A. Goldsmith[1,2] and Warner C. Greene[1,2,3]

[1]Gladstone Institute of Virology and Immunology and Departments of [2]Medicine and [3]Microbiology and Immunology, University of California San Francisco, San Francisco, CA, USA

INTRODUCTION

Interleukin-2 (IL-2) was first identified in 1975 as a growth-promoting activity for bone marrow-derived T lymphocytes (Morgan *et al.*, 1976). Since then, the spectrum of recognized biologic activity for IL-2 has expanded to include direct effects on the growth and differentiation not only of T cells but also of B cells, natural killer (NK) cells, lymphokine-activated killer (LAK) cells, monocytes, macrophages and oligodendrocytes. These biologic effects of IL-2 are mediated through specific receptors present on these cellular targets. The functional high-affinity IL-2 receptor (IL-2R) is composed of three distinct membrane-associated subunits: a 55-kDa α chain (IL-2Rα, Tac, p55, CD25), a 70–75-kDa β chain (IL-2Rβ, p70/75, CD122) and a 64-kDa γ chain (IL-2Rγ, γc, p64). Substantial progress has been made in defining the biochemical and molecular properties of IL-2 and its receptor as well as the mechanisms by which the ligand–receptor complex transduces specific signals. The structural and functional properties of the IL-2/IL-2R system are reviewed in the following sections.

STRUCTURE AND MOLECULAR BIOLOGY OF IL-2

IL-2 is a 15.5-kDa glycoprotein produced principally by activated T cells, although activated B cells may also have the ability to produce small amounts of IL-2 (Walker *et al.*, 1988; Gaffen *et al.*, 1996b). The complete primary structure of IL-2 in seven mammalian species has been deduced by the cloning of the human (de Vos *et al.*, 1983; Taniguchi *et al.*, 1983), gibbon ape (Chen *et al.*,1985), murine (Kashima *et al.*, 1985; Yokota *et al.*, 1985), bovine (Cerretti *et al.*, 1986; Reeves *et al.*, 1986), rat (McKnight *et al.*, 1989), sheep (Seow *et al.*, 1990) and porcine IL-2 cDNAs (Goodall *et al.*, 1991). Substantial nucleotide sequence and amino acid similarity is seen across these species, including strict conservation of all three cysteine residues. The 153-amino-acid human IL-2 primary translation product undergoes several post-translational processing events, including cleavage of a 20-residue signal peptide, addition of carbohydrate to the threonine residue at position 3 (Thr-3), and formation of a disulfide bond between

The Cytokine Handbook, 3rd ed.
ISBN 0-12-689662-3

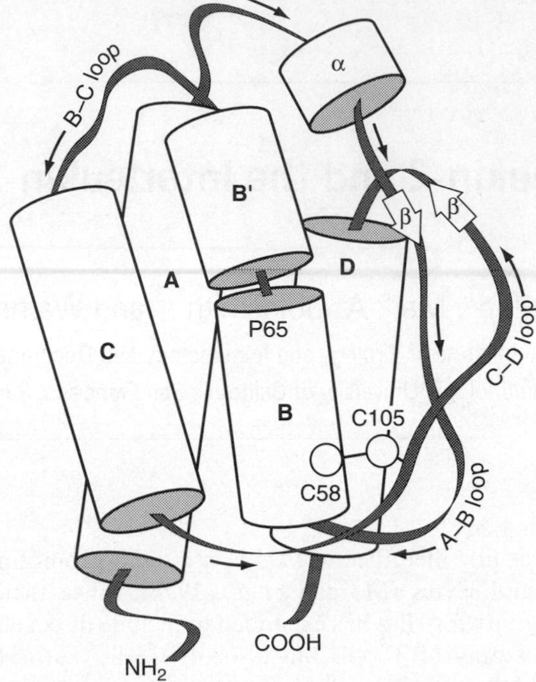

Fig. 1. Model of three-dimensional structure of human IL-2 derived from secondary structure predictions and correlations with other members of the cytokine family. See Bazan (1992) for a detailed discussion. Modified with permission from Bazan (1992). © 1992 American Association for the Advancement of Science.

cysteines 58 and 105. This latter modification is essential for bioactivity, whereas the addition of carbohydrate appears to be dispensable (Robb and Greene, 1987).

The three-dimensional structure of the human IL-2 protein has been described based upon an X-ray crystal structure at 3-Å resolution and secondary structure predictions derived from a comparison of its primary sequence with those of several related cytokines (Fig. 1) (Brandhuber *et al.*, 1987; Bazan, 1992). This model contains four core α helices and two crossover loops containing β strands and appears to accommodate the X-ray scaffold data while incorporating the results of mapping studies of receptor binding epitopes by mutagenesis (Rozwarski *et al.*, 1994).

Site-directed mutagenesis studies of IL-2 have provided significant insights into the subdomains of IL-2 involved in its function. Deletion experiments indicate that the *N*-terminal 20 amino acid residues are critical for its interaction with the receptor (Ju *et al.*, 1987). Similarly, antibodies directed against residues 8–27 block binding of IL-2 to its high-affinity receptor (Kuo and Robb, 1986). Moreover, Asp-20 is conserved across seven species, and substitution of Lys at this position results in the loss of both bioactivity and binding to receptor complexes containing IL-2Rβ (Collins *et al.*, 1988; Zurawski and Zurawski, 1989). These results suggest that residues in the *N*-terminal α-helical segment of IL-2 may interact directly with the ligand-binding domain of IL-2Rβ. Conversely, mutation of Phe-42 or Arg-38 results in markedly diminished binding to IL-2Rα and to high-affinity receptors. However, these mutants retain substantial residual bioactivity (Weigel *et al.*, 1989; Sauve *et al.*, 1991; Grant *et al.*, 1992), consistent with the functional competence of intermediate affinity $\beta\gamma$ receptor complexes. The A–B interloop in the IL-2 structure encompassing these residues probably contains contact residues for IL-2Rα. Other studies indicate that mutation of Gln-141 in murine IL-2 (Zurawski *et al.*, 1990) abrogates ligand binding to $\beta\gamma$ but not $\alpha\beta$ complexes, suggesting that the *C*-terminal D α helix probably contacts IL-2Rγ. Therefore, a structural model of the quaternary IL-2–$\alpha\beta\gamma$ complex places the β and γ chains contacting opposite sides of the bound IL-2 forming a cytoplasmic 'cradle', while the α chain supports the ligand via binding at one end (Bazan, 1993; Rozwarski *et al.*, 1994).

IL-2 RECEPTOR COMPOSITION

The IL-2Rα Subunit

Progress in the characterization of the human high-affinity IL-2R complex was initially facilitated by the development of a sensitive IL-2 bioassay (Gillis *et al.*, 1978) and a radioreceptor binding assay (Robb *et al.*, 1981). A second advance was the generation of an antireceptor monoclonal antibody that did not bind to normal resting T cells but did react strongly with mitogen-activated T cells, hence its name anti-Tac (T activated) (Uchiyama *et al.*, 1981). Anti-Tac was subsequently shown to block the binding of IL-2 to the high-affinity IL-2R, to inhibit IL-2-induced proliferation (Leonard *et al.*, 1982; Miyawaki *et al.*, 1982; Robb and Greene, 1983) and to bind to the 55-kDa IL-2Rα chain (Leonard *et al.*, 1982; Robb and Greene, 1983). The human IL-2Rα gene was cloned (Nikaido *et al.*, 1984; Leonard *et al.*, 1985a) and localized to chromosome 10p band 14–15 (Leonard *et al.*, 1985b). Unlike the other components of the IL-2R complex, the IL-2Rα chain is not a member of the cytokine receptor superfamily (Bazan, 1990). However, it resembles members of the complement receptor family (Medof *et al.*, 1987) and is also structurally related to the newly recognized IL-15Rα chain (Giri *et al.*, 1995). The majority of the IL-2Rα chain is extracellular and contains 11 cysteines, some of which participate in intrachain disulfide bonding. Disruption of any of these cysteines greatly reduces the ability of IL-2Rα to bind IL-2 (Rusk *et al.*, 1988). Although two potential sites of *N*-linked sugar attachment are both utilized in IL-2Rα, neither is required for intracellular transport to the cell membrane or IL-2 binding. In contrast, at least some of the additional *O*-linked sugar attachment sites present in the mid region of the protein appear necessary for cell surface display (Cullen *et al.*, 1988).

Some details have emerged regarding the ligand contact residues and subunit assembly domains within IL-2Rα. Both internal deletion and truncation studies have demonstrated that the *N*-terminal 163 amino acids contain all the information necessary for binding to IL-2 (Cullen *et al.*, 1988; Robb *et al.*, 1988; Rusk *et al.*, 1988). In addition, antibodies that block high-affinity binding sites without impairing low-affinity binding depend on residues 158–160, suggesting that this domain may be involved in subunit interactions with β and/or γ rather than in direct contact with IL-2 (Robb *et al.*, 1988).

The IL-2Rβ Subunit

Several groups observed that certain lymphoid cells respond to IL-2 in the apparent absence of IL-2Rα (Ortaldo *et al.*, 1984; Ralph *et al.*, 1984; Trinchieri *et al.*, 1984). Binding of ^{125}I-labeled IL-2 to these cells followed by chemical cross-linking and immunoprecipitation with anti-IL-2 antibodies revealed a novel IL-2-binding protein of approximately 70–75 kDa (Sharon *et al.*, 1986; Dukovich *et al.*, 1987; Robb *et al.*, 1987; Teshigawara *et al.*, 1987). Using monoclonal antibodies against this protein, a cDNA clone encoding the IL-2Rβ subunit was identified (Hatakeyama *et al.*, 1989b). The human IL-2Rβ gene has been mapped to chromosome 22 (Gnarra *et al.*, 1990). The mature IL-2Rβ protein is composed of 525 amino acids. The extracellular portion of IL-2Rβ contains 214 amino acids with four potential sites of *N*-linked glycosylation and eight cysteine residues, including two canonical pairs with spacing characteristic for members of a growing cytokine receptor superfamily (Bazan, 1990). The *N*-terminal 212 amino acids of the extracellular domain are sufficient for ligand binding in the absence of other receptor subunits, albeit at a very low affinity (Tsudo *et al.*, 1990; Johnson *et al.*, 1994). Although the 286-amino-acid cytoplasmic tail lacks a tyrosine kinase consensus motif, it does contain six tyrosine residues that form potential sites for phosphorylation.

Little is known about regulation of the IL-2Rβ gene. Some normal cell populations, including NK cells, monocytes, and at least a subset of resting T cells constitutively express IL-2Rβ at the cell surface (Dukovich *et al.*, 1987; Nishi *et al.*, 1988; Ben Aribia *et al.*, 1989; Ohashi *et al.*, 1989; Tsudo *et al.*, 1989; Yagita *et al.*, 1989). In unstimulated peripheral blood T cells, IL-2Rβ may be constitutively expressed in CD8$^+$ mature T cells but is inducible in CD4$^+$ T cells (Ohashi *et al.*, 1989; Tsudo *et al.*, 1989). IL-2Rβ expression in B cells may be upregulated by IL-2 in some circumstances (Nakanishi *et al.*, 1992) or by IL-4 in others (Loughnan and Nossal, 1989).

The IL-2Rγ Subunit

Cloning of cDNAs encoding the α and β chains failed to solve all the mysteries surrounding the multiple-affinity forms of the IL-2R. For example, transfection of IL-2Rβ cDNA into non-lymphoid cells failed to generate an IL-2 binding site, whereas transfection into lymphoid cells generated the expected intermediate-affinity binding sites (Hatakeyama *et al.*, 1989b). Coimmunoprecipitation studies performed in the presence of IL-2 led to the identification of a 64-kDa protein that was associated with IL-2Rβ (Takeshita *et al.*, 1990). A cDNA clone encoding this putative receptor component was subsequently obtained, and revealed a 255-amino-acid extracellular domain, a 29-amino-acid hydrophobic transmembrane region, and an 86-amino-acid *C*-terminal cytoplasmic tail (Takeshita *et al.*, 1992). Like the β chain, the extracellular domain of γ contains several features that identify it as a member of the cytokine receptor superfamily. In addition, the γ chain also participates in the IL-4, IL-7, IL-9 and IL-15 receptor complexes (Kondo *et al.*, 1993; Noguchi *et al.*, 1993a; Russell *et al.*, 1993; Giri *et al.*, 1994; Grabstein *et al.*, 1994) and is often referred to as γc for 'common γ'.

THE MULTIMOLECULAR IL-2R COMPLEX

The cloning of human γc helped to clarify the molecular basis of multiple binding sites with distinct binding affinities for IL-2. The IL-2Rα chain alone comprises the low-affinity receptor form (K_d 10–20 nM). The IL-2Rβ chain, which alone binds IL-2 with extremely low affinity (K_d 70–100 nM), can cooperate with IL-2Rα to generate a non-functional 'pseudo-high affinity' binding site (K_d 100–500 pM) or with γc to generate a functional 'intermediate affinity' site (K_d from 500 pM to 2 nM) when expressed on lymphoid cells or multiply transfected non-lymphoid cells. The trimolecular αβγ complex, present on activated T cells, represents the classic 'high affinity' IL-2R (K_d 10–80 pM). Such bimolecular and trimolecular partnerships among these subunits therefore serve to expand dramatically the dynamic range of IL-2 binding.

The distinct binding affinities of the three forms of the receptor derive from the unique binding kinetics characteristic of each of its subunits. Both the association and dissociation of IL-2 to the α chain occur with rapid kinetics ($t_{1/2}$ = 4–10 s), and its association to, and dissociation from, the β chain is significantly slower ($t_{1/2}$ = 40–50 min and 200–400 min, respectively). IL-2 binding to the high-affinity αβγ complex reveals an interesting composite of these binding properties, characterized by a rapid association rate ($t_{1/2}$ = 30–45 s) and a slow dissociation rate ($t_{1/2}$ = 270–300 min) (Robb *et al.*, 1981; Lowenthal and Greene, 1987). A recent study of soluble β and γc chains has confirmed that β binds IL-2 with very low affinity and that weak γc binding is also detectable. Interestingly, combinations of β and γc ectodomains interact to bind IL-2 with higher affinity and greater stability than either chain alone (Johnson *et al.*, 1994). Thus, it appears that the resultant multimolecular IL-2R complexes exhibit the most 'favorable' properties of each of the component subunits.

FUNCTION OF THE IL-2/IL-2R SYSTEM

Role of IL-2/IL-2R in T Cells

The *de novo* synthesis and secretion of IL-2 and the expression of high-affinity IL-2R represent early consequences of antigen- or mitogen-induced activation of mature resting T cells. The subsequent interaction of IL-2 with its high-affinity receptor promotes the rapid clonal expansion of the effector T-cell population originally activated by antigen (reviewed in Smith, 1992). The subsequent decline in both IL-2 synthesis and high-affinity receptor display due to diminished IL-2Rα expression leads to the normal termination of the T-cell immune response. In addition to its growth-promoting function, IL-2 also stimulates T cells to produce other cytokines, including interferon-γ (IFN-γ) (Farrar *et al.*, 1982) and IL-4 (Howard *et al.*, 1983).

While the role of IL-2 in the growth of T cells activated via the T-cell receptor (TCR) by antigen is well established, evidence has emerged that suggests that IL-2 might also function in the thymus during the early stages of T-cell development before the acquisition by thymocytes of a cell surface CD3–TCR complex. The identification of T-cell progenitors within the human thymus that constitutively express the IL-2Rβ chain and secrete low levels of IL-2 has led to the suggestion that an autocrine pathway might promote the growth of these immature thymocytes (De la Hera *et al.*, 1986, 1987;

Toribio *et al.*, 1989). The nature of the activation signals for IL-2 and IL-2Rβ gene expression in these progenitors remains unknown, although interaction with thymic stromal elements and induction by other cytokines represent attractive mechanisms.

In the murine thymus, high levels of IL-2Rα have been detected on fetal and adult immature thymocytes (Ceredig *et al.*, 1985; Habu *et al.*, 1985; Hardt *et al.*, 1985; Raulet, 1985; von Boehmer *et al.*, 1985). However, the importance of IL-2/IL-2R interactions during thymocyte maturation remains uncertain, as mice lacking IL-2 or IL-2Rα synthesis retain normal thymocyte and peripheral T-cell development (Schorle *et al.*, 1991; Willerford *et al.*, 1995). Likewise, mice lacking IL-2Rβ also exhibit normal development of T cells, although these mature cells are largely unresponsive to antigen (Suzuki *et al.*, 1995) (see IL-2/IL-2R Knockout Mice).

Antigen activation of resting T cells is normally followed by induction of the IL-2 gene, cell surface display of high-affinity IL-2R and clonal expansion of the activated cells. However, engagement of the TCR in the absence of costimulation or on self-antigen-specific mature thymocytes and peripheral T cells has been shown to induce a state of clonal anergy, apparently as a result of selective inhibition of IL-2 gene expression (Rammensee *et al.*, 1989; Bonneville *et al.*, 1990). Stimulation of anergic cells with IL-2 reverses the anergic state (Essery *et al.*, 1988; Beverly *et al.*, 1992), probably by forcing the cells to enter the cell cycle. In this model, costimulation by CD28 ligation prevents anergy induction by synergizing with TCR signaling at the IL-2 promoter (see Regulation of Gene Transcription) (Lindsten *et al.*, 1989; Umlauf *et al.*, 1995).

Role of IL-2/IL-2R in Other Cell Types

Other lymphoid cell populations also respond to IL-2. For example, IL-2Rα serves as a marker for the pre-BII stage of B-cell development (Rolink *et al.*, 1994). B cells activated by antigen or mitogen bear high-affinity IL-2R, although at five- to ten-fold lower levels than do activated T cells (Waldmann *et al.*, 1984). Although resting B cells express IL-2Rα, β and γc, expression of β is preferentially upregulated by either IL-2 or IL-4 (Loughnan and Nossal, 1989; Nakanishi *et al.*, 1992), whereas expression of α is preferentially induced by B-cell mitogens (Nakanishi *et al.*, 1992) or IL-5 (Loughnan and Nossal, 1989; Gaffen *et al.*, 1996b). Furthermore, IL-2 can support B-cell growth (Jelinek and Lipsky, 1987; Nakanishi *et al.*, 1992), the induction of immunoglobulin secretion (Jelinek and Lipsky, 1987) and the induction of immunoglobulin J-chain synthesis leading to assembly and secretion of immunoglobulin (Ig)M (Blackman *et al.*, 1986).

NK cells constitutively express IL-2Rβ. These cells not only proliferate in response to high doses of IL-2 but also produce IFN-γ and exhibit enhanced cytolytic activity (Ortaldo *et al.*, 1984; Trinchieri *et al.*, 1984). Similarly, incubation of normal resting T lymphocytes with high concentrations of IL-2 results in the expansion of a population of lymphoid cells that has potent cytolytic activity against fresh tumor target cells. These LAK cells and tumor-infiltrating lymphocytes and their precursors have been characterized extensively, and the generation of these cells *in vitro* and *in vivo* for use in adoptive immunotherapy protocols represents an area of active investigation (Weber *et al.*, 1992; Rosenberg *et al.*, 1993, 1994; Fyfe *et al.*, 1995; Minamoto *et al.*, 1995; Yannelli *et al.*, 1996). Unfortunately, the clinical utility of IL-2 may be limited by its toxicity at high doses, two manifestations of which are a profound capillary leak syndrome and hepatic dysfunction (MacFarlane *et al.*, 1995).

IL-2 responsiveness is not limited to cells of the lymphoid lineage. Myeloid cell populations such as bone marrow-derived macrophage precursors (Baccarini *et al.*, 1989) and primary peripheral blood monocytes (Ohashi *et al.*, 1989) express cell surface IL-2Rβ. Both blood monocytes and alveolar macrophages can be induced to express high- and low-affinity IL-2 binding sites by treatment with IFN-γ and/or lipopolysaccharide (LPS) (Herrmann *et al.*, 1985; Holter *et al.*, 1986, 1987; Hancock *et al.*, 1987). Further, IL-2Rα has also been detected on alveolar macrophages obtained from patients with active pulmonary sarcoidosis (Hancock *et al.*, 1987). Prolonged exposure to high doses of IL-2 results in proliferation and differentiation of macrophage precursors (Baccarini *et al.*, 1989), enhancement of the cytolytic activity of peripheral blood monocytes *in vitro* (Malkovsky *et al.*, 1987) and enhancement of macrophage antibody-dependent tumoricidal activity (Ralph *et al.*, 1988).

Cells outside the immune system may also be sensitive to the biologic effects of IL-2. For instance, isolated oligodendrocytes are reactive with an anti-IL-2Rα monoclonal antibody and express mRNA for IL-2Rα, β and γc (Saneto *et al.*, 1986; Otero and Merrill, 1995). IL-2 has been reported variably to mediate growth stimulation (Benveniste and Merrill, 1986) or growth inhibition in these cells (Saneto *et al.*, 1986).

IL-2/IL-2R Knockout Mice

Experiments with mice bearing targeted deletions in IL-2 or its receptor components have indicated that *in vivo* IL-2 serves to prevent autoimmunity and to maintain homeostasis of the immune system. Somewhat surprisingly, IL-2-deficient mice exhibit normal T-cell development and subset composition (Schorle *et al.*, 1991), in agreement with observations from a reported case of human IL-2 deficiency (Weinberg and Parkman, 1990). However, IL-2$^{-/-}$ mice do display markedly reduced *in vitro* T-cell responses, as well as drastically increased levels of IgG_1 and IgE, implying an effect on B-cell differentiation (Schorle *et al.*, 1991). Although antiviral T-cell responses are impaired *in vitro*, many immune responses *in vivo* proceed relatively normally during the first 4–6 weeks of life, including the T-cell responses to vaccinia and B-cell reactivity to vesicular stomatitis virus. In addition, NK cell activity is reduced but remains inducible, and T helper cell responses are delayed but functional (Kundig *et al.*, 1993). Despite a seemingly normal immune system, approximately 50% of IL-2$^{-/-}$ mice die between 4 and 9 weeks, and those that survive go on to develop an ulcerative colitis-like disease. These findings raise the possibility that IL-2 production normally impedes the emergence of this disease (Sadlack *et al.*, 1993). Thus, while IL-2 clearly plays an important role in the immune response, the ability of T and B cells to develop and respond relatively normally to antigenic assaults indicates that other cytokines may be responsible for these functions or at least are able to compensate for them in the absence of IL-2.

A similar phenotype is observed in IL-2R$\alpha^{-/-}$ and IL-2R$\beta^{-/-}$ animals. Interestingly, these mice also exhibit normal development of T and B cells, despite the presence of IL-2Rα on developing thymocytes. Like IL-2$^{-/-}$ animals, IL-2R$\alpha^{-/-}$ mice develop autoimmune disorders including hemolytic anemia and inflammatory bowel disease. These mice also exhibit a massive enlargement of peripheral lymphoid organs (Willerford *et al.*, 1995). Lymphocyte development is also normal in mice lacking the IL-2Rβ

chain. Spontaneous activation of mature T cells results in deregulated activation of B cells characterized by high serum concentrations of IgG_1 and IgE and hemolytic anemia, as seen in the IL-$2^{-/-}$ animals. The mature T cells present in these animals do not proliferate in response to polyclonal activators or antigen-specific signals. In addition, marked infiltrative granulopoiesis is apparent, and the mice die at about 12 weeks of age (Suzuki *et al.*, 1995). It is noteworthy that these animals are likely to be deficient in signaling by IL-15 as well as IL-2, as these cytokines share the β and γc chains in their receptor complexes (Giri *et al.*, 1994; Grabstein *et al.*, 1994). Thus, the absence of the IL-2, IL-2Rα or IL-2Rβ gene products results in an 'autoimmune' phenotype and disruption in immune homeostasis.

In contrast, mice lacking expression of γc suffer from a form of severe combined immune deficiency (SCID), characterized by a marked ten-fold reduction in the number of circulating mature T and B cells and complete loss of NK cells. IL-2 signaling functions within the remaining cells is impaired, although the cells can respond to mitogenic stimuli (Cao *et al.*, 1995; DiSanto *et al.*, 1995). These results contrasted to observations in humans with mutations in γc (Noguchi *et al.*, 1993b) where B-cell numbers are preserved, although they function poorly. These disparate results raise the possibility that they may be a species-specific difference in the function of γc (Cao *et al.*, 1995).

SIGNAL TRANSDUCTION BY THE IL-2R

The minimal complex competent to transduce signals appears to be the $\beta\gamma$ heterodimer. For example, cells expressing only $\beta\gamma$ rapidly internalize surface-bound IL-2 (Robb and Greene, 1987). In addition, this heterodimer mediates induction by IL-2 of cytolytic activity and of proliferation in NK cells (Seigel *et al.*, 1984; Tsudo *et al.*, 1987; Kehrl *et al.*, 1988) and monocytes (Malkovsky *et al.*, 1987; Baccarini *et al.*, 1989), as well as secretion of immunoglobulin by certain B cells (Ralph *et al.*, 1984). However, the most compelling proof of the functional competence of $\beta\gamma$ heterodimers comes from studies using chimeric receptors, in which the extracellular domains of β and γc are replaced with heterologous receptors. In these experimental systems, dimerization of the extracellular domains by cross-linking antibodies or homodimerizing ligands results in a signaling program that is indistinguishable from that produced by the native IL-2 receptor complex (Nakamura *et al.*, 1994; Nelson *et al.*, 1994; Goldsmith *et al.*, 1995). Using such systems, much has been learned about the contributions of the individual IL-2 receptor subunits to the complete receptor signaling program.

Phosphorylation in Signaling

Early Evidence for Phosphorylation

Like many other receptors, IL-2R undergoes intracellular phosphorylation following stimulation, suggesting a contribution by cellular kinases in the signaling process. The first recognized IL-2 receptor phosphorylation event was serine and threonine phosphorylation of the cytoplasmic domain of IL-2Rα upon exposure to 4β-phorbol 12-myristate 13-acetate (PMA) (Shackelford and Trowbridge, 1991), but site-directed mutagenesis studies proved this event to be non-essential for either receptor down-regulation or the transmission of proliferative signals (Hatakeyama *et al.*, 1986).

Subsequently, inducible tyrosine phosphorylation of the IL-2Rβ and γc chains (Morla *et al.*, 1988; Saltzman *et al.*, 1988: Sharon *et al.*, 1989; Asao *et al.*, 1990; Mills *et al.*, 1990; Sugamura *et al.*, 1990) and of other cellular proteins (Morla *et al.*, 1988; Saltzman *et al.*, 1988; Merida *et al.*, 1990) was observed, and *in vitro* kinase studies demonstrated the association of both a tyrosine kinase and a serine/threonine kinase activity with the IL-2R complex (Fung *et al.*, 1991; Michiel *et al.*, 1991).

The JAK–STAT Signaling Pathway

Studies performed in the IFN-α and IFN-γ signaling systems led to the identification of an important family of tyrosine kinases and their substrates that were subsequently found to be involved in virtually all type I and II cytokine receptor complexes. Somatic mutations in a human fibrosarcoma cell line were generated that rendered the cells unresponsive to either IFN-γ or IFN-α (Pelligrini *et al.*, 1989). The mutations were then complemented with cDNA libraries that rescued the signaling phenotype (Valezquez *et al.*, 1992; Müller *et al.*, 1993; Watling *et al.*, 1993). These studies identified a new family of kinases characterized by both a canonical tyrosine kinase domain and a pseudokinase domain (reviewed in Darnell *et al.*, 1994), termed the Janus kinases or JAKs. The IL-2 receptor is constitutively associated with two JAKs: the β chain with JAK1 and the γc chain with JAK3 (Nelson *et al.*, 1994; Russell *et al.*, 1994). Ligation of the β and γc cytoplasmic tails by IL-2 results in the physical juxtaposition of their associated JAKs; in turn, the JAKs are likely activated by transphosphorylation which triggers a significant increase in their enzymatic activities (Nelson *et al.*, 1994; Russell *et al.*, 1994). Considerable evidence has accumulated emphasizing the importance of these kinases in IL-2 signaling. First, humans and mice deficient in JAK3 exhibit a SCID similar to that observed in those lacking γc itself (Noguchi *et al.*, 1993b; Nosaka *et al.*, 1995; Russell *et al.*, 1995; Thomis *et al.*, 1995). Second, mutations in the domains of the β and γc chains that impair binding of the JAKs also result in loss of signaling (Liu *et al.*, 1995; Russell *et al.*, 1995). Third, overexpression of a dominant negative JAK3 molecule interferes with many IL-2-dependent signaling events (Kawahara *et al.*, 1995). Mutations in IL-2Rβ that lead to loss of JAK function are accompanied by abrogation of tyrosine phosphorylation of the distal tail of IL-2Rβ (Lui *et al.*, 1995), thus rendering the receptor incapable of recruiting phosphotyrosine-associated signaling intermediates. Therefore the Janus kinases appear to play a pivotal role in IL-2 signaling by virtue of both their intrinsic kinase functions and their influence on the assembly of signaling intermediates on the IL-2R signaling platform (Fig. 2).

Activation of the JAKs by cytokine signaling is followed rapidly by the phosphorylation, nuclear import and DNA binding of a family of 'signal transducers and activators of transcription' (STATs) (Fu, 1992; Schindler *et al.*, 1992). STATs contain a key tyrosine residue that serves as a target for phosphorylation by the JAKs. Such phosphorylation allows the STATs to dimerize via a resident SH2-domain and the phosphotyrosine (Shuai *et al.*, 1993a,b). Several STAT factors have been implicated in IL-2 signaling, including STAT1 (Beadling *et al.*, 1994; Lin *et al.*, 1995), STAT3 (Johnston *et al.*, 1995a; Lin *et al.*, 1995), and STAT5 (Fujii *et al.*, 1995; Gaffen *et al.*, 1995; Gilmour *et al.*, 1995; Hou *et al.*, 1995; Johnston *et al.*, 1995a; Lin *et al.*, 1995). Interestingly, STAT1 and STAT3 were observed in phytohemagglutin (PHA)-activated peripheral blood lymphocytes (PBLs) (Lin *et al.*, 1995), but not in most T, proB and myeloid cell lines examined to date, suggesting that the pattern of STAT activation by

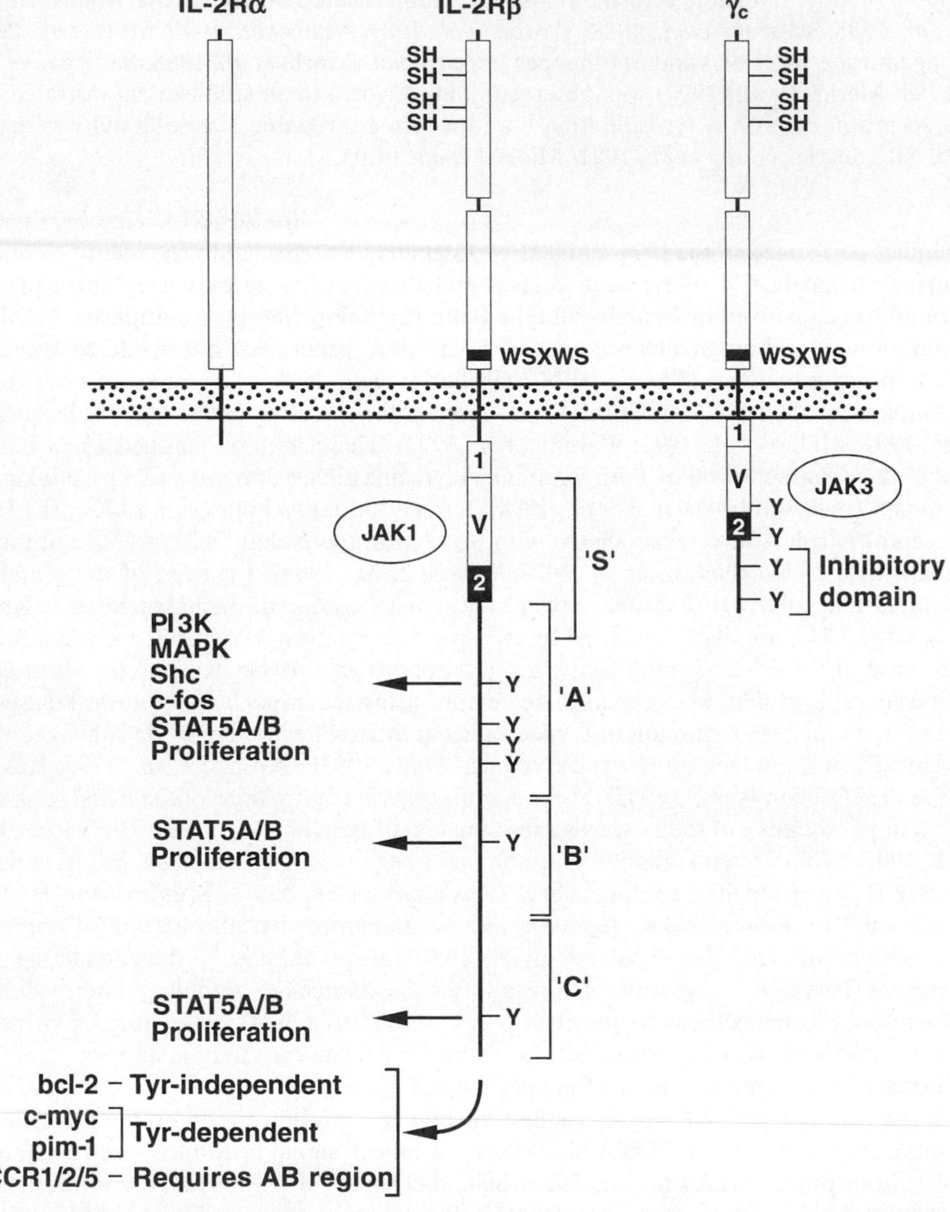

Fig. 2. Model of signal transduction by the IL-2R complex. Characteristic Cys residues (–SH) and WSXWS motifs in extracellular domains of IL-2Rβ and γc are indicated. The conserved 'Box1' and 'Box2' and the variable 'V box' domains are shown as numbered boxes below the transmembrane regions. The boundaries of the 'serine-rich' or 'S' domain, the 'acidic' or 'A' domain, and the 'B' and 'C' domains of IL-2Rβ are indicated. The Janus kinases JAK1 and JAK3 are shown. Cytoplasmic Tyr residues are shown (–Y), with arrows indicating specific signaling responses that have been assigned to the IL-2Rβ Tyr residues. The arrow below the IL-2Rβ chain indicates signaling events driven by IL-2Rβ and the region(s) of the β chain currently known to be responsible. To date, no specific signaling events have been assigned to γc apart from the binding and activation of JAK3. However, recent evidence indicates that the distal tail of γc may negatively regulate IL-2 signaling in the absence of ligand binding (see Structural Features of IL-2R Influencing Signaling Function).

the IL-2R may vary in different cell types or with different states of activation. STAT5 was originally identified in the prolactin signaling system (Wakao *et al.*, 1994) and is also activated by such diverse receptors as the erythropoietin, growth hormone, thrombopoietin, IL-9, IL-5, IL-3, granulocyte–macrophage colony stimulating factor and leptin receptors (reviewed in Darnell, 1996). In humans and mice, two STAT5 genes termed *STAT5A* and *STAT5B* have been identified and shown to be tightly linked on chromosome 11 (Azam *et al.*, 1995; Copeland *et al.*, 1995; Mui *et al.*, 1995; Lin *et al.*, 1996). IL-2 signaling induces the DNA binding activity of both STAT5A and STAT5B and the formation of STAT5A/STAT5B heterodimers in T cells (Gaffen *et al.*, 1996a).

The functional role played by STAT5 activation by IL-2 is likely important but so far undefined. In IL-2-dependent human T-lymphotropic virus-1 (HTLV-1)-transformed cell lines, constitutive activation of STAT5 activity is observed, suggesting a possible role for the JAK–STAT pathway in the transformation process (Migone *et al.*, 1995). In addition, there is a STAT5 binding site in the IL-2Rα promoter (John *et al.*, 1996; Lecine *et al.*, 1996). However, the role of STAT5 in growth signaling is controversial. In the proB-cell line BAF-B03 transfected with human IL-2Rβ, a carboxy-terminal region of the IL-2Rβ cytoplasmic tail is required for IL-2-induced activation of STAT5 DNA binding activity but is dispensable for proliferation signaling, suggesting that these activities can be segregated (Fujii *et al.*, 1995). Similarly, the pharmacologic inhibitor of IL-2-induced proliferation, rapamycin (Dumont *et al.*, 1990), fails to inhibit STAT5 activation in a T-helper cell line under conditions where growth is significantly diminished (Gaffen *et al.*, 1995). In contrast, in the IL-3 receptor system, a dominant-negative STAT5 mutation partially inhibits cell growth in one system (Mui *et al.*, 1996), but has no detectable effect in another (Wang *et al.*, 1996). Thus, the effects of STAT5 on receptor-mediated cell growth are likely to vary depending on the cellular context. The transcriptional targets of STAT5 in the IL-2 signaling pathway have not yet been defined. The two isoforms of murine STAT5 differ dramatically at their *C*-termini, the site of transcriptional activity (Morrigl *et al.*, 1996), which may be important in regulating target genes. Moreover, a naturally occurring variant of STAT5 that lacks transcriptional activity can function as a dominant negative mutant (Morrigl *et al.*, 1996; Wang *et al.*, 1996). Thus, STAT factors potentially play negative as well as positive roles in gene regulation.

Src Family Kinases

Several members of the *src* family of non-receptor-associated tyrosine kinases have been postulated to play a role in IL-2R function. In certain cell lines, IL-2 activates serine/threonine phosphorylation of the lymphoid-specific tyrosine kinase $p56^{lck}$ and alters its activity (Horak *et al.*, 1991). Subsequent experiments demonstrated a physical association between the cytoplasmic tails of IL-2Rβ and $p56^{lck}$ or $p59^{fyn}$ (Hatakeyama *et al.*, 1991; Minami *et al.*, 1993b). Although IL-2-mediated induction of $p56^{lck}$ phosphorylation and augmented kinase activity were observed in certain situations (Hatakeyama *et al.*, 1991; Horak *et al.*, 1991; Minami *et al.*, 1993a), several studies have demonstrated intact IL-2 signaling in cells lacking both $p56^{lck}$ and $p59^{fyn}$ (Mills *et al.*, 1992; Otani *et al.*, 1992), implying either that these or related kinases function in a redundant manner or that they are not essential for the signaling function of the IL-2R.

Other Kinases Involved in IL-2 Signaling

Other types of protein kinases have been implicated in IL-2 signaling. For example, IL-2 activates phosphatidylinositol-3-kinase (PI-3K) (Augustine *et al.*, 1991; Merida *et al.*, 1991, 1993). Recent work has shown that IL-2 stimulates the association of PI-3K with the cytosolic docking protein insulin receptor substrate-1 (IRS-1) and that both IRS-1 and the related IRS-2 protein are phosphorylated on tyrosines upon IL-2 activation. Moreover, JAK1 and JAK3 can associate with and phosphorylate IRS-1 and IRS-2, thus providing a mechanism by which PI-3K activation may be initiated (Johnston *et al.*, 1995b). However, the IL-2-dependent T-cell line HT-2 lacks IRS-1 and IRS-2, yet activates PI-3K, indicating that these adaptor molecules are not always required for this activity (K. Lui, J. Molden and M. Goldsmith, unpublished results).

One report has appeared suggesting that the Syk protein tyrosine kinase associates with the IL-2Rβ chain and contributes to the induction of c-myc (Minami *et al.*, 1995). In addition, IL-2 activates the ribosomal p70S6 kinase, which can be blocked by rapamycin (Calvo *et al.*, 1992; Kuo *et al.*, 1992; Terada *et al.*, 1992). IL-2 signaling also activates cyclin E–Cdk2 complexes to promote entry of cells into the S phase of the cell cycle (Firpo *et al.*, 1994) by triggering the degradation of the Cdk inhibitor protein $p27^{kip1}$. Like p70S6K, both of these events are inhibited by rapamycin (Nourse *et al.*, 1994).

Finally, both the expression and catalytic activity of the non-receptor kinase Raf-1 is regulated by IL-2 (Turner *et al.*, 1991; Zmuidzinas *et al.*, 1991), probably through the association of the IL-2Rβ chain with the adaptor molecule Shc (Evans *et al.*, 1995). This indication may be a proximal event leading to enhancement of both p70S6 kinase activity and the activation of the c-*fos* proto-oncogene (Calvo *et al.*, 1992; Kuo *et al.*, 1992; Terada *et al.*, 1992; Gaffen *et al.*, 1996a) and consequent progression through the cell cycle. However, c-*fos* transcription is not necessary for IL-2-induced proliferation (Gaffen *et al.*, 1996a). It is noteworthy that the GTP-bound (active) form of $p21^{ras}$ accumulates in the presence of IL-2 (Graves *et al.*, 1992), which is blocked by the tyrosine kinase inhibitor herbimycin (Izquierdo *et al.*, 1992; Izquierdo and Cantrell, 1993). IL-2 signaling also induces tyrosine phosphorylation of the signaling molecule *vav* (Evans *et al.*, 1993). The activation of these various kinases and the phosphorylation events they subsequently mediate clearly play a critical role in the signaling program specified by the IL-2–IL-2R complex.

Phosphatases Involved in IL-2 Signaling

The importance of tyrosine phosphorylation in cytokine receptor signaling has been recognized for some time. However, the converse role for tyrosine dephosphorylation in the regulation of signaling is only now emerging. The identification of a subclass of non-transmembrane protein tyrosine phosphatases (PTPs) that contain SH2 domains suggest a role for these phosphatases in signaling from activated receptors (reviewed in Sun and Tonks, 1994). At least one phosphatase, SH-PTP1, associates with the IL-3 and erythropoietin receptors (Yi *et al.*, 1993; Klingmuller *et al.*, 1995), and SH-PTP1 contributes to the termination of erythropoietin (EPO)-induced signaling through the EPO receptor. Of note, mice harboring a SH-PTP1 null mutation (*motheaten*) exhibit a variety of hematopoietic abnormalities (Shultz *et al.*, 1993; Tsui *et al.*, 1993). There is one report that IL-2 signaling induces a sixfold induction in transcription of leukocyte tyrosine phosphatase (LC-PTP) (Adachi *et al.*, 1994). To date, direct evidence for an

IL-2R-associated tyrosine phosphatase(s) has not emerged, but in some cases, IL-2 stimulation is associated with an increase in tyrosine phosphorylation of SHP2, which can function as both a tyrosine phosphatase and an adaptor protein (Barber *et al.*, 1997). It is likely that future work will reveal a role for this emerging class of molecules in IL-2 signaling.

Other Signaling Cascades

Very early studies concluded that several traditional signaling events such as calcium mobilization, G-protein coupling and cyclic AMP generation were not centrally involved in IL-2 signaling. However, several groups have reported the breakdown of glycosylphosphatidylinositol (gly-PI) into potential second messengers as another likely early step in signaling by the IL-2R. These messengers include myristylated diacylglycerol (myr-DAG), inositol phosphate glycan (IPG) and myristylated phosphatidic acid (PA) (Kanoh *et al.*, 1993). For example, in a model B-cell lymphoma line, IL-2 treatment induces expression of the immunoglobulin J chain (Blackman *et al.*, 1986) and stimulates proliferation (Tigges *et al.*, 1989). In these cells, generation of myr-DAG and IPG exhibits the same IL-2 dose–response curve as the growth and differentiative responses (Eardley and Koshland, 1991). Similarly, IL-2-dependent hydrolysis of gly-PI lipids in the T-cell line CTLL-2 correlates with the dose of ligand required for proliferation (Merida *et al.*, 1990). The potential relationship between gly-PI hydrolysis and phosphorylation signaling pathways is intriguing but requires further study.

Interplay of Signaling Cascades in Proliferation Signaling

In recent years, a myriad of growth signaling pathways has been linked to the IL-2R complex. All evidence to date indicates that the JAKs are critically involved in signaling for growth, as mutations in either the IL-2Rβ or γc chain that disrupt JAK1 or JAK3 activation also abrogate proliferation (Russell *et al.*, 1994; Lui *et al.*, 1995), and a JAK3 dominant negative molecule blocks IL-2-dependent proliferation (Kawahara *et al.*, 1995). Mechanistically, the JAKs are required for tyrosine phosphorylation of IL-2Rβ and the subsequent recruitment of cellular signaling intermediates to this chain. Surprisingly, activation of the mitogen-activated protein kinase (MAPK) cascade via Ras and Raf is not absolutely required for IL-2-induced proliferation, as mutations within the β chain that eliminate this activity still support growth (Gaffen *et al.*, 1996a; K. Lui, unpublished results). As described above, the overall importance of the *src* kinases in IL-2 signaling remains unresolved, and the precise role of various kinases such as PI-3K, Syk and p70S6 kinase within the signaling cascades is not yet established. In addition, cellular background may influence which signaling pathways are used. Finally, a recent report examining the interplay among various signaling pathways in the proB-cell line BAF-B03 indicates that activation of any two of three signaling pathways—c-fos/c-jun (activating protein-1 (AP-1)) induction, c-myc induction or bcl-2 induction—is sufficient to promote proliferation, regardless of the activation state of the third pathway (Miyazaki *et al.*, 1995). Future studies of IL-2 signaling need to focus not only on ordering the signaling intermediates within these individual signaling cascades, but also on the integrated network of signaling that ultimately determines the response of a particular cell to IL-2.

Regulation of Gene Transcription

Among the events observed after triggering of the IL-2R is the selective activation of gene transcription. One of the earliest reported IL-2-induced targets is the gene encoding IL-2Rα. As noted, the induction of IL-2Rα plays an important step in the generation of high-affinity IL-2 receptors (Depper *et al.*, 1985; Smith and Cantrell, 1985). Recent work has revealed a composite enhancer element in the IL-2Rα gene that confers IL-2-responsiveness on a heterologous gene. This region contains binding sites for multiple factors, including STAT5, Elf-1, high mobility group protein (HMG-I(Y)) and GATA family proteins (John *et al.*, 1996; Lecine *et al.*, 1996).

A number of other gene targets for IL-2 have been defined. First, a recent search for IL-2-regulated genes has revealed that the chemokine receptors CCR1, CCR2 and CCR5 are induced by IL-2 (Loetscher *et al.*, 1996; W. Xu, S. Lai, W.C. Greene and M.A. Goldsmith, unpublished data). Importantly, CCR5 has also been identified as the human immunodeficiency virus (HIV) coreceptor that determines macrophage tropism of various viral strains (Alkhatib *et al.*, 1996; Choe *et al.*, 1996; Deng *et al.*, 1996; Doranz *et al.*, 1996; Dragic *et al.*, 1996), and a congenital lack of the CCR5 gene appears to confer resistance to HIV infection (Samson *et al.*, 1996). The ability of IL-2 to induce CCR5 may be clinically important, as parenteral IL-2 therapy has been used to boost CD4 counts in HIV disease (Kovacs *et al.*, 1995, 1996). However, such IL-2 therapy produces a transient increase in viral load, perhaps in part resulting from the induction of chemokine receptor expression (Kovacs *et al.*, 1995).

Another target of IL-2 is the immunoglobulin J chain, which functions to close the pentameric ring of secreted IgM multimers (Blackman *et al.*, 1986; Koshland, 1989). Analysis of the J-chain promoter has revealed a repressor molecule, Pax5/B-cell activating protein (BSAP), whose silencing effects are relieved during antigen-driven stages of B-cell differentiation (Rinkenberger *et al.*, 1996). Third, various proto-oncogenes associated with cell cycle regulation are activated upon ligation of the IL-2R. Despite some variability in different cell culture model systems, c-myc, c-myb, pim-1, bcl-2 and the AP-1 factors c-fos/c-jun mRNA levels rise following IL-2 stimulation (Granelli-Piperno *et al.*, 1986; Stern and Smith, 1986; Reed *et al.*, 1987; Dautry *et al.*, 1988; Shibuya *et al.*, 1992). Similarly, IL-2 augments mRNA levels for members of the cyclin family and cdc2 families of cell cycle regulators and for proliferating cell nuclear antigen (Shibuya *et al.*, 1992; Huang *et al.*, 1994). These latter changes are almost certainly involved in the proliferative response induced by IL-2. Finally, a novel SH2 domain-containing protein termed Cis has recently been identified and shown to be regulated by many cytokines, including IL-2 (Yoshimura *et al.*, 1995). Its precise role in the integrated biologic response to IL-2 remains undefined.

Transcriptional regulation of the IL-2 gene itself has been the subject of considerable investigation (reviewed in Jain *et al.*, 1995). IL-2 synthesis involves antigen or mitogen activation via the T-cell antigen receptor (TCR) complex (reviewed in Cantrell, 1996). However, full IL-2 production also requires additional costimulation through engagement of the surface CD28 receptor (Umlauf *et al.*, 1995). CD28 signaling promotes both transcriptional activation of the IL-2 gene and post-transcriptional stabilization of IL-2 mRNA (Lindsten *et al.*, 1989). The IL-2 enhancer encompasses approximately 300 base pairs of sequence located immediately upstream of the transcription initiation site

(Fujita *et al.*, 1983; Holbrook *et al.*, 1984; Sibenlist *et al.*, 1986) and includes binding sites for a variety of known DNA binding proteins, including members of the nuclear factor (NF)-κB, AP-1, Oct-1 and NF-AT families of transcription factors. However, evidence from transgenic mice expressing a reporter gene linked to the IL-2 promoter indicates that additional locus control sequences are required for position-independent, copy number-dependent expression (Brombacher *et al.*, 1994). The interplay among these transcription factors is highly complex, with evidence emerging for cooperative interactions occurring at composite elements and between distal elements. Moreover, these factors are themselves regulated both transcriptionally and by post-translational modifications. Interestingly, the immunosuppressive drugs cyclosporin A and FK506 prevent IL-2 synthesis by blocking translocation of the cytoplasmic component of NF-AT. Normally, NF-AT is dephosphorylated by calcineurin, a calcium-sensitive phosphatase. However, FK506 and cyclosporin bound to immunophilins block calcineurin action (reviewed in Clipstone and Crabtree, 1994; Rao, 1994; Ho *et al.*, 1996). A role for c-rel in the activation of IL-2 gene transcription has also become apparent when c-rel knockout animals were prepared and shown to be deficient in IL-2 production (Kontgen *et al.*, 1995). The IL-2 enhancer has emerged as an attractive experimental model to dissect the transcriptional control mechanisms that mediate the temporal- and lineage-specific expression of this gene (Jain *et al.*, 1995).

STRUCTURAL FEATURES OF IL-2R INFLUENCING SIGNALING FUNCTION

Extracellular Domains of the IL-2Rβ and γc Chains

Unlike the α chain, the IL-2Rβ and γc chains are members of the cytokine receptor superfamily (Bazan, 1990). In the extracellular domain, members of the family are characterized by a series of four periodically spaced cysteine residues ($CX_{9-10}CXWX_{26-32}CX_{10-15}C$) and a W–S–X–W–S sequence located near the cell membrane. In addition, the extracellular domain of γc contains four leucines with heptameric spacing located in a predicted α-helical region, raising the possibility of a leucine zipper-like motif that might be involved in protein–protein interactions (Takeshita *et al.*, 1992). Although a high-resolution crystal structure of the IL-2R has not yet emerged, several homodimeric cytokine receptors have been analyzed (de Vos *et al.*, 1992; Livnah *et al.*, 1996). In the EPO and growth hormone (GH) receptors, the four cysteine residues form disulfide bridges between β sheets located within *N*-terminal fibronectin type III (FBNIII)-like domains. Mutational analysis of the WSXWS motif in the IL-2Rβ chain suggested that the two Trp residues might form the floor of a hormone-binding crevice (Miyazaki *et al.*, 1991). However, the crystal structure of the EPOR and GHR revealed that the WSXWS motif occurs in a β bulge in the membrane-proximal FBNIII domain (Baumgartner *et al.*, 1994; Livnah *et al.*, 1996) and does not appear to play any role in ligand binding or receptor–receptor interactions. Instead, ligand binding occurs in a homophobic crevice between the two *N*-terminal FBNIII domains (Livnah *et al.*, 1996). It will be instructive to compare the structural features of a heterodimeric receptor complex such as the IL-2R with the homodimeric receptors analyzed so far when such structures become available.

Cytoplasmic Domains of the IL-2Rβ and γc Chains

Considerable attention has been paid to the structural requirements of the IL-2Rβ chain in signaling (Fig. 2). Deletion of a membrane-proximal, serine-rich stretch of residues (the 'S domain', residues 267–322) completely abrogated transduction of a proliferative signal by IL-2 through the $\alpha\beta\gamma$ complex, while leaving the ligand-binding and internalization properties of the receptor intact (Hatakeyama *et al.*, 1989a,b). Deletion of this domain also prevented induction of expression of the proto-oncogenes c-*fos* (Hatakeyama *et al.*, 1992) and c-*myc* (Satoh *et al.*, 1992; Merida *et al.*, 1993), coimmunoprecipitation of tyrosine kinase activity with the receptor (Fung *et al.*, 1991), induction of intracellular protein tyrosine kinase activity (Merida *et al.*, 1993), and activation of $p21^{ras}$ (Satoh *et al.*, 1991). The S domain is also essential for the induction of NF-κB by IL-2 (Arima *et al.*, 1992). Moreover, although the S deletion does not prevent physical association of $p56^{lck}$ with IL-2Rβ, induction of its tyrosine kinase activity by IL-2 is abrogated (Minami *et al.*, 1993b). Finally, this region was recently shown to be required for STAT5 activation (Fujii *et al.*, 1995).

Interestingly, this deletion spans a 15-amino-acid region with notable homology to an element termed Box2 found in many other members of the cytokine receptor superfamily, including γc (Murakami *et al.*, 1991). The conserved Box2 and the more *N*-terminal Box1 regions are separated by a variable spacer or 'V box' region. Deletion of either Box1 or Box2 completely abrogates IL-2-induced proliferation; however, alanine-scanning mutagenesis of these regions revealed that only one residue in each box contributed critical side-chains for growth signaling (Goldsmith *et al.*, 1994a). Detailed analysis of the V box demonstrated that several amino acids within this intervening domain are critical for signaling, including proliferation, JAK1 and STAT5 activation (Goldsmith *et al.*, 1994; Liu *et al.*, 1995). The S region overlaps what is so far known to be the JAK1 binding site (Russell *et al.*, 1994), although the precise contact points of IL-2Rβ with JAK1 remain undefined. Notwithstanding, many if not all of the defects in the S deletion and/or Box1/Box2/V box deletions are likely ascribable to the loss of JAK1 association and activation.

The membrane-distal region of β has been divided into three regions. The 'A' region is an acidic region (residues 313–382; Hatakeyama *et al.*, 1989b) that encompasses the first four cytoplasmic tyrosine residues (Tyr-338, -355, -358 and -361). While deletion of the A region does not prevent IL-2-driven proliferation (Hatakeyama *et al.*, 1989a; Goldsmith *et al.*, 1994) or STAT5 activation (Fujii *et al.*, 1995; Gaffen *et al.*, 1995), it does eliminate physical association with $p56^{lck}$ (Hatakeyama *et al.*, 1991), induction of c-*fos* transcription (Hatakeyama *et al.*, 1992), Shc binding (Evans *et al.*, 1995) and activation of $p21^{ras}$ (Satoh *et al.*, 1992). The B region contains the fifth tyrosine residue (Try-392), and deletion of both A and B is still competent to proliferate in the proB-cell line BA/F3 or mature $CD4^+$ HT-2 cells (Goldsmith *et al.*, 1995). Similarly, the C region contains the sixth tyrosine (Tyr-510) and is also unnecessary to support proliferation in this cellular context. Interestingly, a deletion of B and C together in the proB-cell line BA/F3 fails to respond to IL-2 (Goldsmith *et al.*, 1994) but is sufficient for growth in HT-2 cells (Goldsmith *et al.*, 1995). Thus, large-scale deletion analysis of the membrane–distal portion of the IL-2Rβ chain has revealed important structural features of this subunit and different requirements in different cellular backgrounds.

A detailed molecular analysis of the six tyrosine residues within the IL-2Rβ chain has been performed. In the T-cell line HT-2, three redundant tyrosine residues (Tyr-338, -392 and -510) support proliferation signaling (Gaffen *et al.*, 1996a). Similar results were obtained in the myeloid cell line 32D (Friedmann *et al.*, 1996). In contrast, in the BA/F3 cell line, only one of the two most *C*-terminal tyrosines is needed to support proliferation in the context of a full-length receptor (Goldsmith *et al.*, 1995), although a truncated β chain that still retains Tyr-338 has been reported to grow slowly in IL-2 (Fujii *et al.*, 1995). These data reinforce the notion that signaling requirements differ in various cellular backgrounds. The tyrosines of the IL-2Rβ chain have also been examined for their roles in mediating other signaling events. While JAK activation proceeds independently of these tyrosine residues (Goldsmith *et al.*, 1995; Gaffen *et al.*, 1996a), STAT5 activation occurs via the same redundant tyrosine residues that mediate proliferation signaling, namely Tyr-338, -392 and -510 (Gaffen *et al.*, 1995, 1996a; Friedmann *et al.*, 1996). However, it still remains unclear whether STAT5 is required for proliferation signaling (see JAK–STAT Signaling Pathway above). Work in other cytokine receptor systems has suggested that specific STAT responsiveness is contained within short tyrosine-containing sequences present on the activating receptor (Stahl *et al.*, 1995). However, no clear consensus motif for STAT5 binding has emerged. While the sequences flanking the IL-2Rβ, Tyr-392 and Tyr-510 residues are quite similar to each other and to motifs present in other cytokine receptors known to activate STAT5 (D A Y L S/T L X), Tyr-338 (N Q G Y F F F H) differs dramatically in its primary sequence yet still activates STAT5 (Gaffen *et al.*, 1996a). In contrast to the redundant capacities of these three tyrosine residues to activate proliferation and STAT5, association of the IL-2Rβ chain with the adaptor molecule Shc occurs solely through Tyr-338 (Evans *et al.*, 1995; Ravichandran *et al.*, 1996). Consistent with the role of Shc in engaging the Grb2–SOS–Ras–Raf pathway SOS (son of sevenless) that leads to activation of c-*jun* and c-*fos* (AP-1), the proto-oncogene c-*fos* is also uniquely induced via Tyr-338 (Gaffen *et al.*, 1996a). In addition, PI-3K and MAPK activation are also regulated exclusively through Tyr-338, although it is not certain whether their activations are sequentially linked (K. Liu, unpublished results). Combined, these data suggest that Tyr-338 is 'multifunctional', in that it connects the receptor to the JAK–STAT, PI3K–MAPK and Ras–Raf pathways (Fig. 2). Finally, at least some downstream signaling events proceed independently of tyrosine residues. For example, transcription of bcl-2 occurs using a mutant β chain lacking all six cytoplasmic tyrosine residues (Gaffen *et al.*, 1996a). Consistent with the lack of involvement of the STAT pathway, a dominant negative mutation of JAK3 does not inhibit bcl-2 induction (Kawahara *et al.*, 1995). Thus, the IL-2Rβ chain mediates both tyrosine-dependent and -independent forms of signaling.

Like the β chain, γc is a member of the cytokine receptor superfamily and is phosphorylated on tyrosine(s) following receptor stimulation (Sugamura *et al.*, 1990). Initial examination of the primary sequence of γc identified a region with partial homology to an SH2 subdomain, but to date no evidence for phosphotyrosine binding has been ascribed to γc. Although γc is clearly a necessary component of the IL-2 receptor, recent detailed molecular analysis has revealed a surprising degree of asymmetry in the relative signaling functions between β and γc chains. Like β, the conserved Box1 (residues 281–294) and Box2 (residues 321–334) motifs and V box spacer (residues 295–320) within γc appear to be critical for growth signaling (Goldsmith *et al.*, 1995); these regions also encompass the JAK3 binding site (Russell *et al.*, 1994;

Nelson *et al.*, 1996). However, deletion of all of the amino acids *C*-terminal to Box2 has no detectable effect on growth signaling (Goldsmith *et al.*, 1995; Nelson *et al.*, 1996). In addition, mutation of the four cytoplasmic tyrosine residues within γc also fails to affect growth signaling, in contrast to the deleterious effects of an analogous mutation of the six tyrosines in the β cytoplasmic tail (Goldsmith *et al.*, 1995). These observations suggest that γc might play a relatively generic role in the initiation of IL-2 signaling. Evidence in support of this hypothesis came from the analysis of EPO receptor fusion chimeras. Remarkably, these studies revealed that an EPOγ fusion construct could be functionally replaced in the signaling complex with a truncated form of the EPOR, EPO(1–321). The resulting hybrid receptor complex, EPOβ–EPOR(1–321), recapitulated all aspects of IL-2R signaling, except that JAK2 instead of JAK3 was activated (Gaffen *et al.*, 1996a; Lai *et al.*, 1996b). Thus, these studies suggest that γc functions primarily to convey a Janus kinase into the signaling complex to 'trigger' the signaling cascade, whereas the β chain and its distal tyrosine residues 'drive' the full specificity of the signaling program. Hence, the γc chain has been referred to as a 'trigger' chain, and the β chain has been called a 'driver' chain (Lai *et al.*, 1996b). These observations allow predictions to be made about the function of γc in other cytokine receptor complexes. Indeed, γc appears to play a similarly generic role in the IL-7 receptor (Lai *et al.*, 1997). In contrast, γc appears to be entirely dispensable in some forms of the IL-4R complex (Kammer *et al.*, 1996; Lai *et al.*, 1996a). As the IL-15 receptor shares both the β and γc chains with the IL-2R (Giri *et al.*, 1994; Grabstein *et al.*, 1994), the division of labor between these subunits is almost certain to be preserved. Finally, such a separation of receptor subunit functions appears to hold in at least some other heterodimeric receptor complexes, including the IFN-α system (O. Colamonici, personal communication).

Most models of cytokine receptor function view the receptor chains as fairly passive cytoplasmic platforms upon which various cellular signaling intermediates assemble to effect signal transduction. However, recent studies suggest that the distal tail of γc contains an inhibitory domain that directly or indirectly interacts with the Box2 region of IL-2Rβ and blocks JAK–STAT signaling. It is possible that this γc inhibitory domain may safeguard against inadvertent receptor signaling by chains that collide in the membrane in the absence of ligand. Alternatively, this domain may contribute to the termination of IL-2 signaling when ligand concentrations become limiting (S. Lai, S.L. Gaffen, W. Xu, M.A. Goldsmith and W.C. Greene, unpublished data). It remains unknown whether γc plays a similar inhibitory role in other receptor complexes.

CLINICAL IMPLICATIONS

The biologic significance of cytokine–cytokine receptor systems is emphasized by several clinical syndromes or animal models where they appear to play a pathogenic role. For example, leukemic T-cell lines derived from patients with the adult T-cell leukemia/lymphoma syndrome often display extremely high levels of IL-2Rα (Yodoi and Uchiyama, 1992). It has been hypothesized that this deregulated receptor expression is elicited by the Tax gene product derived from the 3′ pX region of the HTLV-I retrovirus. Additionally, recent work has suggested that a second HTLV-I gene product, p12, binds to the cytoplasmic tails of the IL-2Rβ and γc chains (Mulloy *et al.*, 1996). This finding raises the possibility that p12 ligates IL-2Rβ and γc intracellularly, thus activating the JAK–STAT pathway. However, p12 may also prevent IL-2Rβ and γc from reaching the cell surface by sequestering these proteins in the endoplasmic reticulum (Migone *et al.*, 1995). Activation of JAK3 and STAT5 has been observed in malignant T cells derived from patients with cutaneous anaplastic large T-cell lymphoma and Sézary syndrome, suggesting a possible role for IL-2Rβ and γc signaling in the pathogenesis of T-cell malignancy (Zhang *et al.*, 1996). A striking demonstration of the critical role of the IL-2 receptor in human immune function was revealed by studies of X-linked SCID, in which three patients with the condition were shown to have genetic defects in the γc locus that resulted in premature termination of translation of the γc protein (Noguchi *et al.*, 1993b). As described, subsequent studies revealed that the γc chain was shared by the IL-4, IL-7, IL-9 and IL-15 receptor complexes (Kondo *et al.*, 1993; Noguchi *et al.*, 1993a; Russell *et al.*, 1993; Giri *et al.*, 1994; Grabstein *et al.*, 1994). Although these patients would be predicted to be deficient in all of these cytokine functions, studies from knockout mice indicate that the developmental defect in X-linked SCID is likely to be primarily a result of the lack of IL-7 function. This cytokine is required for development of mature B and T cells (Peschon *et al.*, 1994; Lai *et al.*, 1997).

CONCLUSIONS AND FUTURE DIRECTIONS

The IL-2 receptor is now recognized as a heteromeric receptor complex composed of three different chains that mediates many critical cellular responses to IL-2 in several cell types. IL-2 likely lies in a 'cytokine cradle' generated by the N-terminal regions of IL-2Rα, β and γc chains, whereas the more distal domains of these subunits interact in a manner to assemble a high-affinity receptor complex. This configuration, which is at least partly dependent on the binding of IL-2, permits the discharge of signals into the cell that lead to biologic responses including progression through the cell cycle and the transcriptional induction of effector or differentiation genes. Within the IL-2Rβ and γc cytoplasmic regions many domains have been identified that are essential for signaling competence. Some constituents of the complete signal transduction apparatus remain unidentified, although a cadre of tyrosine kinases and other enzymes that participate in the signal transduction cascade has now been identified. In particular, the JAK–STAT pathway appears to play a central role in signal transduction. Finally, the components of the IL-2R appear to play a role in various human lymphoid malignancies and immunodeficiency syndromes. A continued understanding of the molecular basis for IL-2/IL-2R function will be essential to define the agents capable of augmenting or inhibiting the function of this receptor complex *in vivo*.

ACKNOWLEDGEMENTS

The authors acknowledge the valuable comments of K.D. Liu. S.L.G. is supported by the Bank of America–Giannini Foundation. S.L.G., M.A.G. and W.C.G. are supported by the J. David Gladstone Institutes and the National Institutes of Health (AI36452 to W.C.G. and GM54351 to M.A.G.).

REFERENCES

Adachi, M., Sekiya, M., Ishino, M., Sasaki, H., Hinoda, Y., Imai, K. and Yachi, A. (1994). Induction of protein-tyrosine phosphatase LC-PTP by IL-2 in human T cells. *FEBS Lett.* **338**, 47–52.

Alkhatib, G., Combadiere, C., Broder, C.C., Feng, Y., Kennedy, P.E., Murphy, P.M. and Berger, E.A. (1996). CC CKR5: A RANTES, MIP-1α, MIP-1β receptor as a fusion cofactor for macrophage-tropic HIV-1. *Science* **272**, 1955–1958.

Arima, N., Kuziel, W.A., Grdina, T.A. and Greene, W.C. (1992). IL-2 induced signal transduction involves the activation of nuclear NF-κB expression. *J. Immunol.* **149**, 83–91.

Asao, H., Takeshita, T., Nakamura, M., Nagata, K. and Sugamura, K. (1990). Interleukin-2 (IL-2)-induced tyrosine phosphorylation of IL-2 receptor. *J. Exp. Med*. **171**, 637–644.

Augustine, J.A., Sutor, S.L. and Abraham, R.T. (1991). Interleukin 2- and polyomavirus middle T antigen-induced modification of phosphatidylinositol 3-kinase activity in activated T lymphocytes. *Mol. Cell. Biol*. **11**, 4431–4440.

Azam, M., Erdjument-Bromage, H., Kiedler, B.L., Xia, M., Quelle, F., Basu, R., Saris, C., Tempst, P., Ihle, J.N. and Schindler, C. (1995). Interleukin-3 signals through multiple forms of Stat5. *EMBO J*. **14**, 1402–1411.

Baccarini, M., Schwinzer, R. and Lohmann-Matthes, M.L. (1989). Effect of human recombinant IL-2 on murine macrophage precursors. Involvement of a receptor distinct from the p55 (Tac) protein. *J. Immunol*. **142**, 118–125.

Barber, D.L., Corless, C.N., Xia, K., Roberts, T.M. and D'Andrea, A.A. (1997). Erythropoietin activates Raf1 by a Shc-independent pathway in CTLL-EPO-R cells. *Blood* **89**, 55–64.

Baumgartner, J.W., Wells, C.A., Chen, C.-M. and Waters, M.J. (1994). The role of the WSXWS equivalent motif in growth hormone receptor function. *J. Biol. Chem*. **269**, 29 094–29 101.

Bazan, J.F. (1990). Structural design and molecular evolution of a cytokine receptor superfamily. *Proc. Natl Acad. Sci. U.S.A.* **87**, 6934–6938.

Bazan, J.F. (1992). Unraveling the structure of IL-2. *Science* **257**, 410–413.

Bazan, J.F. (1993). Receptor/ligand superfamilies: Emerging families of cytokines and receptors. *Curr. Biol*. **3**, 603–606.

Beadling, C., Gischin, D., Witthuhn, B.A., Ziemiecki, A., Ihle, J.N., Kerr, I.M. and Cantrell, D.A. (1994). Activation of JAK kinases and STAT proteins by interleukin-2 and interferon α, but not the T cell antigen receptor, in human T lymphocyte. *EMBO J*. **13**, 5605–5615.

Ben Aribia, M., Moire, N., Metivier, D., Vaquero, C., Lantz, O., Olive, D., Charpentier, B. and Senik, A. (1989). IL-2 receptors on circulating natural killer cells and T lymphocytes. Similarity in number and affinity but difference in transmission of the proliferation signal. *J. Immunol*. **142**, 490–499.

Benveniste, E.N. and Merrill, J.E. (1986). Stimulation of oligodendroglial proliferation and maturation by interleukin-2. *Nature* **321**, 610–613.

Beverly, B., Kang, S.M., Lenardo, M.J. and Schwartz, R.H. (1992). Reversal of *in vitro* T cell clonal anergy by IL-2 stimulation. *Int. Immunol*. **4**, 661–671.

Blackman, M.A., Tigges, M.A., Minie, M.E. and Koshland, M.E. (1986). A model system for peptide hormone action in differentiation: Interleukin-2 induces a B lymphoma to transcribe the J chain gene. *Cell* **47**, 609–617.

Bonneville, M., Ishida, I. Itohara, S., Verbeek, S., Berns, A., Kanagawa, O., Haas, W. and Tonegawa, S. (1990). Self-tolerance to transgenic γ delta T cells by intrathymic inactivation. *Nature* **344**, 163–165.

Brandhuber, B.J., Boone, T., Kenney, W.C. and McKay, D.B. (1987). Three-dimensional structure of interleukin-2. *Science* **238**, 1707–1709.

Brombacher, F., Schafer, T., Weissenstein, U., Tschopp, C., Andersen, E., Burki, K. and Baumann, G. (1994). IL-2 promoter-driven lacZ expression as a monitoring tool for IL-2 expression in primary T-cells of transgenic mice. *Int. Immunol*. **6**, 189–197.

Calvo, V., Crews, C.M., Vik, T.A. and Bierer, B.E. (1992). Interleukin 2 stimulation of p70 S6 kinase activity is inhibited by the immunosuppressant rapamycin. *Proc. Natl Acad. Sci. U.S.A.* **89**, 7571–7575.

Cantrell, D. (1996). T cell antigen receptor signal transduction pathways. *Annu. Rev. Immunol*. **14**, 259–274.

Cao, X., Shores, E.W., Hu-Li, J., Anver, M.R., Kelsall, B.L., Russell, S.M., Drago, J., Noguchi, M., Grinberg, A., Bloom, E.T., Paul, W.E., Katz, S.I., Love, P.E. and Leonard, W.J. (1995). Defective lymphoid development in mice lacking expression of the common cytokine receptor γ chain. *Immunity* **2**, 223–238.

Ceredig, R., Lowenthal, J.W., Nabholz, M. and MacDonald, H.R. (1985). Expression of interleukin-2 receptors as a differentiation marker on intrathymic stem cells. *Nature* **314**, 98–100.

Cerretti, D.P., McKereghan, K., Larsen, A., Cantrell, M.A., Anderson, D., Gillis, S., Cosman, D. and Baker, P.E. (1986). Cloning, sequence, and expression of bovine interleukin-2. *Proc. Natl Acad. Sci. U.S.A.* **83**, 3223–3227.

Chen, S.J., Holbrook, N.J., Mitchell, K.F., Vallone, C.A., Greengard, J.S., Crabtree, G.R. and Lin, Y. (1985). A viral long terminal repeat in the interleukin 2 gene of a cell line that constitutively produces interleukin 2. *Proc. Natl Acad. Sci. U.S.A.* **82**, 7284–7288.

Choe, H., Farzan, M., Sun, Y., Sullivan, N., Rollins, B., Ponarth, P.D., Wu, L., Mackay, C.R., LaRosa, G., Newman, W., Gerard,, N., Gerard, C. and Sodroski, J. (1996). The β-chemokine receptors CCR3 and CCR5 facilitate infection by primary HIV-1 isolates. *Cell* **85**, 1135–1148.

Clipstone, N.A. and Crabtree, G.R. (1994). Signal transmission between the plasma membrane and nucleus of T lymphocytes. *Annu. Rev. Biochem.* **63**, 1045–1083.

Collins, L., Tsien, W.H., Seals, C., Hakimi, J., Weber, D., Bailon, P., Hoskings, J., Greene, W.C., Toome, V. and Ju, G. (1988). Identification of specific residues of human interleukin-2 that affect binding to the 70-kDa subunit (p70) of the interleukin 2 receptor. *Proc. Natl Acad. Sci. U.S.A.* **85**, 7709–7713.

Copeland, N.G., Gilbert, D.J., Schindler, C., Zhong, Z., Wen, Z., Darnell, J.E., Mui, A.L.-F., Quelle, F.W., Ihle, J.N. and Jenkins, N.A. (1995). Distribution of the mammalian Stat gene family in mouse chromosomes. *Genomics* **29**, 225–228.

Cullen, B.R., Podlaski, F.J., Peffer, N.J., Hosking, J.B. and Greene, W.C. (1988). Sequence requirements for ligand binding and cell surface expression of the Tac antigen, a human interleukin-2 receptor. *J. Biol. Chem.* **263**, 4900–4906.

Darnell, J.E. (1996). Reflections on STAT3, STAT5 and STAT6 as fat STATs. *Proc. Natl Acad. Sci. U.S.A.* **93**, 6221–6224.

Darnell, J.E., Kerr, I.M. and Stark, G.R. (1994). Jak-STAT pathways and transcriptional activation in response to IFNs and other extracellular signaling proteins. *Science* **264**, 1415–1421.

Dautry, F., Weil, D., Yu, J. and Dautry-Varsat, A. (1988). Regulation of pim and myb mRNA accumulation by interleukin 2 and interleukin 3 in murine hematopoietic cell lines. *J. Biol. Chem.* **263**, 17615–17620.

De la Hera, A., Toribio, M.L., Marquez, C., Marcos, M.A., Cabrero, E. and Martinez, A.C. (1986). Differentiation of human mature thymocytes: existence of a T3+4-8-intermediate stage. *Eur. J. Immunol.* **16**, 653–658.

De la Hera, A., Toribio, M.L., Marcos, M.A., Marquez, C. and Martinez, C. (1987). Interleukin-2 pathway is autonomously activated in human T11+3-4-6-8-thymocytes. *Eur. J. Immunol.* **17**, 683–687.

de Vos, R., Plaetinck, G., Cheroutre, H., Simons, G., Degrave, W., Tavernier, J., Remaut, E. and Fiers, W. (1983). Molecular cloning of human interleukin 2 cDNA and its expression in *E. coli*. *Nucleic Acids Res.* **11**, 4307–4323.

de Vos, A.M., Ultsch, M. and Kossiakoff, A.A. (1992). Human growth hormone and extracellular domain of its receptor: Crystal structure of the complex. *Science* **255**, 306–312.

Deng, H., Liu, R., Ellmeier, W., Choe, S., Unutmaz, D., Burkhart, M., DiMarzio, P., Marmon, S., Sutton, R.E., Hill, C.M., Davis, C.B., Peiper, S.C., Schall, T.J., Littman, D.R. and Landau, N.R. (1996). Identification of a major co-receptor for primary isolates of HIV-1. *Nature* **381**, 661–666.

Depper, J.M., Leonard, W.J., Drogula, C., Kronke, M., Waldmann, T.A. and Greene, W.C. (1985). Interleukin 2 (IL-2) augments transcription of the IL-2 receptor gene. *Proc. Natl Acad. Sci. U.S.A.* **82**, 4230–4234.

DiSanto, J.P., Muller, W., Guy-Grand, D., Fischer, A. and Rajewsky, K. (1995). Lymphoid development in mice with a targeted deletion of the interleukin-2 receptor γ chain. *Proc. Natl Acad. Sci. U.S.A.* **92**, 377–381.

Doranz, B.J., Rucker, J., Yi, Y., Smyth, R.J., Samson, M., Peiper, S.C., Parmentier, M., Collman, R.G. and Doms, R.W. (1996). A dual-tropic primary HIV-1 isolate that uses fusin and the β-chemokine receptors CKR-5, CKR-3, and CKR-2b as fusion cofactors. *Cell* **85**, 1149–1158.

Dragic, T., Litwin, V., Allaway, G.P., Martin, S.R., Huang, Y., Nagashima, K.A., Cayanan, C., Maddon, P.J., Koup, R.A., Moore, J.P. and Paxton, W.A. (1996). HIV-1 entry into $CD4^+$ cells is mediated by the chemokine receptor CC-CKR-5. *Nature* **381**, 667–673.

Dukovich, M., Wano, Y., Le, thi, Bich, Thuy, Katz, P., Cullen, B.R., Kehrl, J.H. and Greene, W.C. (1987). A second human interleukin-2 binding protein that may be a component of high-affinity interleukin-2 receptors. *Nature* **327**, 518–522.

Dumont, F.J., Staruch, M.J., Koprak, S.L., Melino, M.R. and Sigal, N.H. (1990). Distinct mechanisms of suppression of murine T cell activation by the related macrolides FK-506 and rapamycin. *J. Immunol.* **144**, 251–258.

Eardley, D.D. and Koshland, M.E. (1991). Glycosylphosphatidylinositol: A candidate system for interleukin-2 signal transduction. *Science* **251**, 78–81.

Essery, G., Feldmann, M. and Lamb, J.R. (1988). Interleukin-2 can prevent and reverse antigen-induced unresponsiveness in cloned human T lymphocytes. *Immunology* **64**, 413–417.

Evans, G.A., Howard, O.M.Z., Erwin, R. and Farrar, W.L. (1993). Interleukin-2 induces tyrosine phosphorylation of the *vav* proto-oncogene product in human T cells: lack of requirement for the tyrosine kinase lck. *Biochem. J.* **294**, 339–342.

Evans, G., Goldsmith, M.A., Johnston, J.A., Xu, W., Weiler, S.R., Erwin, R., Howard, O.M.Z., Abraham, R.T., O'Shea, J.J., Greene, W.C. and Farrar, W.L. (1995). Analysis of interleukin-2-dependent signal transduction through the Shc/Grb2 adapter pathway. *J. Biol Chem*. **270**, 28 858–28 863.

Farrar, J.J., Benjamin, W.R., Hilfiker, M.L., Howard, M., Farrar, W.L. and Fuller-Farrar, J. (1982). The biochemistry, biology, and role of interleukin 2 in the induction of cytotoxic T cell and antibody-forming B cell responses. *Immunol. Rev*. **63**, 129–166.

Firpo, E.J., Koff, A., Solomon, M.J. and Roberts, J.M. (1994). Inactivation of a Cdk2 inhibitor during interleukin 2-induced proliferation of human T lymphocytes. *Mol. Cell. Biol*. **14**, 4889–4901.

Friedmann, M.C., Migone, T.-S., Russell, S.M and Leonard, W.J. (1996). Different interleukin-2 receptor β-chain tyrosines couple to at least two signaling pathways and synergistically mediate interleukin 2-induced proliferation. *Proc. Natl Acad. Sci. U.S.A.* **93**, 2077–2082.

Fu, X.-Y. (1992). A transcription factor with SH2 and SH3 domains is directly activated by an interferon α-induced cytoplasmic protein tyrosine kinase(s). *Cell* **70**, 323–335.

Fujii, H., Nakagawa, Y., Schindler, U., Kawahara, A., Mori, H., Gouilleux, F., Groner, B., Ihle, J.N., Minami, Y., Miyazaki, T. and Taniguchi, T. (1995). Activation of Stat5 by interleukin 2 requires a carboxyl-terminal region of the interleukin-2 receptor β chain but is not essential for the proliferative signal transmission. *Proc. Natl Acad. Sci. U.S.A.* **92**, 5482–5486.

Fujita, T., Takaoka, C., Matsui, H. and Taniguchi, T. (1983). Structure of the human interleukin 2 gene. *Proc. Natl Acad. Sci. U.S.A.* **80**, 7437–7441.

Fung, M.R., Scearce, R.M., Hoffman, J.A., Peffer, N.J., Hammes, S.R., Hosking, J.B., Schmandt, R., Kuziel, W.A., Haynes, B.F., Mills, G.B. and Greene, W.C. (1991). A tyrosine kinase physically associates with the β-subunit of the human IL-2 receptor. *J. Immunol*. **147**, 1253–1260.

Fyfe, G., Fisher, R.I., Rosenberg, S.A., Sznol, M., Parkinson, D.R. and Louie, A.C. (1995). Results of treatment of 255 patients with metastatic renal cell carcinoma who received high-dose recombinant interleukin-2 therapy. *J. Clin. Oncol*. **13**, 688–696.

Gaffen, S.L., Lai, S.Y., Xu, W., Gouilleux, F., Groner, B., Goldsmith, M.A. and Greene, W.C. (1995). Signaling through the IL-2R β chain activates a STAT-5-like DNA binding activity. *Proc. Natl Acad. Sci. U.S.A.* **92**, 7192–7196.

Gaffen, S.L., Lai, S.Y., Ha, M., Liu, X., Hennighausen, L., Greene, W.C. and Goldsmith, M.A. (1996a). Distinct tyrosine residues within the interleukin-2 receptor β chain drive signal transduction specificity, redundancy, and diversity. *J. Biol. Chem*. **271**, 21381–21390.

Gaffen, S.L., Wang, S.L. and Koshland, M.E. (1996b). Expression of the immunoglobulin J chain in a murine B lymphoma is driven by autocrine production of interleukin-2. *Cytokine* **8**, 513–524.

Gillis, S., Ferm, M.M., Ou, W. and Smith, K.A. (1978). T cell growth factor: parameters of production and a quantitative microassay for activity. *J. Immunol*. **120**, 2027–2032.

Gilmour, K.C., Pine, R. and Reich, N.C. (1995). Interleukin 2 activates STAT5 transcription factor (mammary gland factor) and specific gene expression in T lymphocytes. *Proc. Natl Acad. Sci. U.S.A.* **92**, 10 772–10 776.

Giri, J.G., Ahdieh, M., Eisenmann, J., Shanebeck, K., Grabstein, K., Kumaki, S., Namen, A., Park, L.S., Cosman, D. and Anderson, D. (1994). Utilization of the β and γ chains of the IL-2 receptor by the novel cytokine IL-15. *EMBO J*. **13**, 2822–2830.

Giri, J.G., Kumaki, S., Ahdieh, M., Friend, D.J., Loomis, A., Shanebeck, K., DuBose, R., Cosman, D., Park, L.S. and Anderson, D.M. (1995). Identification and cloning of a novel IL-15 binding protein that is structually related to the α chain of the IL-2 receptor. *EMBO J*. **14**, 3654–3663.

Gnarra, J.R., Otani, H., Wang, M.G., McBride, O.W., Sharon, M. and Leonard, W.J. (1990). Human interleukin 2 receptor β-chain gene: chromosomal localization and identification of 5′ regulatory sequences. *Proc. Natl Acad. Sci. U.S.A.* **87**, 3440–3444.

Goldsmith, M.A., Xu, W., Amaral, M.C., Kuczek, E.S. and Greene, W.C. (1994). The cytoplasmic domain of the IL-2 receptor β chain contains both unique and functionally redundant signal transduction elements. *J. Biol. Chem*. **269**, 14 698-14 704.

Goldsmith, M.A., Lai, S.Y., Xu, W., Amaral, M.C., Kuczek, E.S., Parent, L.J., Mills, G.B., Tarr, K.L., Longmore, G.D. and Greene, W.C. (1995). Growth signal transduction by the human IL-2 receptor requires cytoplasmic tyrosines of the β chain and non-tyrosine residues of the γc chain. *J. Biol. Chem*. **270**, 21 729–21 737.

Goodall, J.C., Emery, D.C., Bailey, M., English, L.S. and Hall, L. (1991). cDNA cloning of porcine interleukin 2 by polymerase chain reaction. *Biochim. Biophys. Acta* **1089**, 257–258.

Grabstein, K.H., Eisenman, J., Shanebeck, K., Rauch, C., Srinivasan, S., Fung, V., Beers, C., Richardson, J., Schoenborn, M.A., Ahdieh, M., Johnson, L., Alderson, M.R., Watson, J.D., Anderson, D.M. and Giri, J.G. (1994). Cloning of a T cell growth factor that interacts with the β chain of the interleukin-2 receptor. *Science* **264**, 965–968.

Granelli-Piperno, A., Andrus, L. and Steinman, R.M. (1986). Lymphokine and nonlymphokine mRNA levels in stimulated human T cells. *J. Exp. Med*. **163**, 922–937.

Grant, A.J., Roessler, E., Ju, G., Tsudo, M., Sugamura, K. and Waldmann, T.A. (1992). The interleukin 2 receptor (IL-2R): the IL-2R α subunit alters the function of the IL-2R β subunit to enhance IL-2 binding and signaling by mechanisms that do not require binding of IL-2 to IL-2R α subunit. *Proc. Natl Acad. Sci. U.S.A.* **89**, 2165–2169.

Graves, J.D., Downward, J., Izquierdo-Pastor, M., Rayter, S., Warne, P.H. and Cantrell, D. (1992). The growth factor IL-2 activates $p21^{ras}$ proteins in normal human T lymphocytes. *J. Immunol*. **148**, 2417–2422.

Habu, S., Okumura, K., Diamanstein, T. and Shevach, E.M. (1985). Expression of interleukin 2 receptor on murine fetal thymocytes. *Eur. J. Immunol*. **15**, 456–460.

Hancock, W.W., Muller, W.A. and Cotran, R.S. (1987). Interleukin 2 receptors are expressed by alveolar macrophages during pulmonary sarcoidosis and are inducible by lymphokine treatment of normal human lung macrophages, blood monocytes, and monocyte cell lines. *J. Immunol*. **138**, 185–191.

Hardt, C., Diamanstein, T. and Wagner, H. (1985). Developmentally controlled expression of IL 2 receptors and of sensitivity to IL 2 in a subset of embryonic thymocytes. *J. Immunol*. **134**, 3891–3894.

Hatakeyama, M., Minamoto, S. and Taniguchi, T. (1986). Intracytoplasmic phosphorylation sites of Tac antigen (p55) are not essential for the conformation, function, and regulation of the human interleukin 2 receptor. *Proc. Natl Acad. Sci. U.S.A.* **83**, 9650–9654.

Hatakeyama, M., Mori, H., Doi, T. and Taniguchi, T. (1989a). A restricted cytoplasmic region of the IL-2 receptor β chain is essential for growth signal transduction but not for ligand binding and internalization. *Cell* **59**, 837–845.

Hatakeyama, M., Tsudo, M., Minamoto, S., Kono, T., Doi, T., Miyata, T., Miyasaka, M. and Taniguchi, T. (1989b). Interleukin-2 receptor β chain gene: Generation of three receptor forms by cloned human α and β cDNAs. *Science* **244**, 551–556.

Hatakeyama, M., Kono, T., Kobayashi, N., Kawahara, A., Levin, S.D., Perlmutter, R.M. and Taniguchi, T. (1991). Interaction of the IL-2 receptor with the *src*-family kinase $p56^{lck}$: Identification of novel intermolecular association. *Science* **252**, 1523–1528.

Hatakeyama, M., Kawahara, A., Mori, H., Shibuya, H. and Taniguchi, T. (1992). C-fos gene induction by interleukin-2: Identification of the critical cytoplasmic regions within the interleukin-2 receptor β chain. *Proc. Natl Acad. Sci. U.S.A.* **89**, 2022–2026.

Herrmann, F., Cannistra, S.A., Levine, H. and Griffin, J.D. (1985). Expression of interleukin 2 receptors and binding of interleukin 2 by γ interferon-induced human leukemic and normal monocytic cells. *J. Exp. Med*. **162**, 1111–1116.

Ho, S., Clipstone, N., Timmermann, L., Northrop, J., Graef, I., Fiorentino, D., Nourse, J. and Crabtree, G. (1996). The mechanism of action of cyclosporin A and FK506. *Clin. Immunol. Immunopathol.* **80**, S40–45.

Holbrook, N.J., Lieber, M. and Crabtree, G.R. (1984). DNA sequence of the 5′ flanking region of the human interleukin 2 gene: homologies with adult T-cell leukemia virus. *Nucleic Acids Res*. **12**, 5005–5013.

Holter, W., Grunow, R., Stockinger, H. and Knapp, W. (1986). Recombinant interferon-γ induces interleukin 2 receptors on human peripheral blood monocytes. *J. Immunol*. **136**, 2171–2175.

Holter, W., Goldman, C.K., Casabo, L., Nelson, D.L., Greene, W.C. and Waldmann, T.A. (1987). Expression of functional IL 2 receptors by lipopolysaccharide and interferon-γ stimulated human monocytes. *J. Immunol*. **138**, 2917–2922.

Horak, I.D., Gress, R.E., Lucas, P.J., Horak, E.M., Waldmann, T.A. and Bolen, J.B. (1991). T-lymphocyte interleukin 2-dependent tyrosine protein kinase signal transduction involves the activation of p56lck. *Proc. Natl Acad. Sci. U.S.A.* **88**, 1996–2000.

Hou, J., Schindler, U., Henzel, W.J., Wong, S.C. and McKnight, S.L. (1995). Identification and purification of human Stat proteins activated in response to IL-2. *Immunity* **2**, 325–330.

Howard, M., Matis, L., Malek, T.R., Shevach, E., Kell, W., Cohen, D., Nakanishi, K. and Paul, W.E. (1983). Interleukin 2 induces antigen-reactive T cell lines to secrete BCGF-I. *J. Exp. Med*. **158**, 2024–2039.

Huang, D., Shipman-Appasamy, P.M., Orten, D.J., Hinrichs, S.H. and Prystowsky, M.B. (1994). Promoter activity of the proliferating-cell nuclear antigen gene is associated with inducible CRE-binding proteins in interleukin-2 stimulated T lymphocytes. *Mol. Cell. Biol*. **14**, 4233–4243.

Izquierdo, M. and Cantrell, D.A. (1993). Protein tyrosine kinases couple the interleukin-2 receptor to p21ras. *Eur. J. Immunol*. **23**, 131–135.

Izquierdo, M., Downward, J., Otani, H., Leonard, W.J. and Cantrell, D. (1992). Interleukin (IL)-2 activation of p21-ras in murine myeloid cells transfected with human IL-2 receptor β chain. *Eur. J. Immunol*. **22**, 817–821.

Jain, J., Loh, C. and Rao, A. (1995). Transcriptional regulation of the IL-2 gene. *Curr. Opin. Immunol*. **7**, 333–342.

Jelinek, D.F. and Lipsky, P.E. (1987). Regulation of human B lymphocyte activation, proliferation, and differentiation. *Adv. Immunol*. **40**, 1–59.

John, S., Robbins, C.M. and Leonard, W.J. (1996). An IL-2 response element in the human IL-2 receptor α chain promoter is a composite element that binds Stat5, Elf-1, HMG-1(Y) and a GATA family protein. *EMBO J*. **15**, 5627–5635.

Johnson, K., Choi, Y., Wu, Z., Ciardelli, T., Granzow, R., Whalen, C., Sana, T., Pardee, G., Smith, K. and Creasey, A. (1994). Soluble IL-2 receptor β and γ subunits: ligand binding and cooperativity. *Eur. Cytokine Netw.* **5**, 23–24.

Johnston, J.A., Bacon, C.M., Finbloom, D.S., Rees, R.C., Kaplan, D., Shibuya, K., Ortaldo, J.R., Gupta, S., Chen, Y.Q., Giri, J.D. and O'Shea, J.J. (1995a). Tyrosine phosphorylation and activation of STAT5, STAT3, and Janus kinases by interleukins 2 and 15. *Proc. Natl Acad. Sci. U.S.A.* **92**, 8705–8709.

Johnston, J.A., Wang, L.-M., Hanson, E.P., Sun, X.-J., White, M.F., Oakes, S.A., Pierce, J.H. and O'Shea, J.J. (1995b). Interleukins 2, 4, 7, and 15 stimulate tyrosine phosphorylation of insulin receptor substrates 1 and 2 in T cells. *J. Biol. Chem*. **270**, 28 527–28 530.

Ju, G., Collins, L., Kaffka, K.L., Tsien, W.H., Chizzonite, R., Crowl, R., Bhatt, R. and Kilian, P.L. (1987). Structure-function analysis of human interleukin-2. Identification of amino acid residues required for biological activity. *J. Biol. Chem*. **262**, 5723–5731.

Kammer, W., Lischke, A., Moriggl, R., Groner, B., Ziemieck, A., Gurniak, C.B., Berg, L.J. and Friedrich, K. (1996). Homodimerization of interleukin-4 receptor α chain can induce intracellular signaling. *J. Biol. Chem*. **271**, 23 634–23 637.

Kanoh, H., Sakane, F., Imai, S.-I. and Wada, I. (1993). Diacylglycerol kinase and phosphatidic acid phosphatase – Enzymes metabolizing lipid second messengers. *Cell. Signal.* **5**, 495–503.

Kashima, N., Nishi, T.C., Fujita, T., Taki, S., Yamada, G., Hamuro, J. and Taniguchi, T. (1985). Unique structure of murine interleukin-2 as deduced from cloned cDNAs. *Nature* **313**, 402–404.

Kawahara, A., Minami, Y., Miyazaki, T., Ihle, J.N. and Taniguchi, T. (1995). Critical role of the interleukin 2 (IL-2) receptor γ-chain-associated JAK3 in the IL-2-induced c-fos and c-myc, but not bcl-2, gene induction. *Proc. Natl Acad. Sci. U.S.A.* **92**, 8724–8728.

Kehrl, J.H., Dukovich, M., Whalen, G., Katz, P., Fauci, A.S. and Greene, W.C. (1988). Novel interleukin 2 (IL-2) receptor appears to mediate IL-2-induced activation of natural killer cells. *J. Clin. Invest*. **81**, 200–205.

Klingmuller, U., Lorenz, U., Cantley, L.C., Neel , B.G. and Lodish, H.F. (1995). Specific recruitment of SH-PTP1 to the erythropoietin receptor causes inactivation of JAK2 and termination of proliferative signals. *Cell* **80**, 729–738.

Kondo, M., Takeshita, T., Ishii, N., Nakamura, M., Watanabe, S., Arai, K.-I. and Sugamura, K. (1993). Sharing of the interleukin-2 (IL-2) receptor γ chain between receptors for IL-2 and IL-4. *Science* **262**, 1874–1877.

Kontgen, F., Grumont, R.J., Strasser, A., Metcalf, D., Li, R., Tarlinton, D. and Gerondakis, S. (1995). Mice lacking the c-rel proto-oncogene exhibit defects in lymphocyte proliferation, humoral immunity, and interleukin-2 expression. *Genes Dev*. **9**, 1965–1977.

Koshland, M.E. (1989). The immunoglobulin helper: the J chain. In *Immunoglobulin Genes* (eds T. Honjo, F.W. Alt and T.H. Rabbits), Academic Press, Berkeley, CA, pp. 345–359.

Kovacs, J.A., Baeler, M., Dewar, R.J., Vogel, S., Davey, R.T, Falloon, J., Polis, M.A., Walker, R.E., Stevens, R., Sallzman, N.P., Metcalf, J.A., Masur, H. and Lane, H.C. (1995). Increases in CD4 lymphocytes with intermittent courses of interleukin-2 in patients with human immunodeficiency virus infection. *N. Engl. J. Med*. **332**, 567–575.

Kovacs, J.A., Vogel, S., Albert, J.M., Falloon, J., Davey, R.T., Walker, R.E., Polis, M.A., Spooner, K., Metcalf, J.A., Baseler, M., Fyfe, G. and Lane, H.C. (1996). Controlled trial of interleukin-2 infusions in patients infected with the human immunodeficiency virus. *N. Engl. J. Med*. **335**, 1350–1356.

Kundig, T.M., Schorle, H., Bachmann, M.F., Hengartner, H., Zinkernagel, R.M. and Horak, I. (1993). Immune responses in interleukin-2-deficient mice. *Science* **262**, 1059–1061.

Kuo, C.J., Chung, J., Fiorentino, D.F., Flanagan, W.M., Blenis, J. and Crabtree, G.R. (1992). Rapamycin selectively inhibits interleukin-2 activation of p70 S6 kinase. *Nature* **358**, 70–73.

Kuo, L.M. and Robb, R.J. (1986). Structure–function relationships for the IL 2-receptor system. I. Localization of a receptor binding site on IL 2. *J. Immunol*. **137**, 1538–1543.

Lai, S.Y., Molden, J., Liu, K.D., Puck, J.M., White, M.D. and Goldsmith, M.A. (1996a). Interleukin-4-specific signal transduction events are driven by homotypic interactions of the interleukin-4 receptor α subunit. *EMBO J*. **15**, 4506–4514.

Lai, S.Y., Xu, W., Gaffen, S., Liu, K.D., Greene, W.C. and Goldsmith, M.A. (1996b). The determinants of signal transduction specificity in γc-containing cytokine receptors. *Proc. Natl Acad. Sci. U.S.A.* **93**, 231–235.

Lai, S.Y., Molden, J. and Goldsmith, M.A. (1997). The shared γc subunit within the human interleukin-7 receptor complex: A molecular basis for the pathogenesis of X-linked severe combined immunodeficiency. *J. Clin. Invest*. **99**, 169–177.

Lecine, P., Algate, M., Rameil, P., Beadling, C., Butcher, P., Nabholz, M. and Imbert, J. (1996). Elf-1 and Stat5 bind to a critical element in a new enhancer of the human interleukin-2 receptor α gene. *Mol. Cell. Biol*. **16**, 6829–6840.

Leonard, W.J., Depper, J.M, Uchiyama, T., Smith, K.A., Waldmann, T.A. and Greene, W.C. (1982). A monoclonal antibody that appears to recognize the receptor for human T-cell growth factor; partial characterization of the receptor. *Nature* **300**, 267–269.

Leonard, W.J., Depper, J.M., Kanehisa, M., Kronke, M., Peffer, N.J., Svetlik, P.B., Sullivan, M. and Greene, W.C. (1985a). Structure of the human interleukin-2 receptor gene. *Science* **230**, 633–639.

Leonard, W.J., Donlon, T.A., Lebo, R.V. and Greene, W.C. (1985b). Localization of the gene encoding the human interleukin-2 receptor on chromosome 10. *Science* **228**, 1547–1549.

Lin, J.-X., Migone, T.-S., Tsang, M., Friedmann, M., Weatherbee, J.A., Zhou, L., Yamauchi, A., Bloom, E.T., Mietz, J., John, S. and Leonard, W.J. (1995). The role of shared receptor motifs and common Stat proteins in the generation of cytokine pleiotropy and redundancy by IL-2, IL-4, IL-7, IL-13 and IL-15. *Immunity* **2**, 331–339.

Lin, J.-X., Mietz, J., Modi, W.S., John, S. and Leonard, W.J. (1996). Cloning of human Stat5B. *J. Biol. Chem*. **271**, 10728–10744.

Lindsten, T., June, C.H., Ledbetter, J.A., Stella, G. and Thompson, C.B. (1989). Regulation of lymphokine messenger RNA stability by a surface-mediated T-cell activation pathway. *Science* **244**, 339–343.

Liu, K.D., Lai, S.Y., Goldsmlth, M.A. and Greene, W.C. (1995). Identification of a variable region within the cytoplasmic tail of the IL-2 receptor β chain that is required for growth signal transduction. *J. Biol. Chem*. **270**, 22176–22181.

Livnah, O., Stura, E.A., Johnson, D.L., Middleton, S.A., Mulcahy, L.S., Wrighton, N.C., Dower, W.J., Jolliffe, L.K. and Wilson, I.A. (1996). Functional mimicry of a protein hormone by a peptide agonist: The EPO receptor complex at 2.8 Å. *Science* **273**, 464–471.

Loetscher, P., Seitz, M., Baggliolini, M. and Moser, B. (1996). Interleukin-2 regulates CC chemokine receptor expression and chemotactic responsiveness in T lymphocytes. *J. Exp. Med*. **184**, 569–577.

Loughnan, M.S. and Nossal, G.J.V. (1989). Interleukins 4 and 5 control expression of IL-2 receptor on murine B cells through independent induction of its two chains. *Nature* **340**, 76–79.

Lowenthal, J.W. and Greene, W.C. (1987). Contrasting interleukin 2 binding properties of the α (p55) and β (p70) protein subunits of the human high-affinity interleukin 2 receptor. *J. Exp. Med*. **166**, 1156–1161.

MacFarlane, M.P., Yang, J.C., Guleria, A.S., White, R.L., Seipp, C.A., Einhorn, J.H., White, D.E. and Rosenberg, S.A. (1995). The hematologic toxicity of interleukin-2 in patients with metastatic melanoma and renal cell carcinoma. *Cancer* **75**, 1030–1037.

McKnight, A.J., Mason, D.W. and Barclay, A.N. (1989). Sequence of rat interleukin 2 and anomalous binding of a mouse interleukin 2 cDNA probe to rat MHC class II-associated invariant chain mRNA. *Immunogenetics* **30**, 145–147.

Malkovsky, M., Loveland, B., North, M., Asherson, G.L., Gao, L., Ward, P. and Fiers, W. (1987). Recombinant interleukin-2 directly augments the cytotoxicity of human monocytes. *Nature* **325**, 262–265.

Medof, M.E., Lublin, D.M., Holers, V.M., Ayers, D.J., Getty, R.R., Leykam, J.F., Atkinson, J.P. and Tykocinski, M.L. (1987). Cloning and characterization of cDNAs encoding the complete sequence of decay-accelerating factor of human complement. *Proc. Natl Acad. Sci. U.S.A.* **84**, 2007–2011.

Merida, I., Pratt, J.C. and Gaulton, G.N. (1990). Regulation of interleukin-2 dependent growth responses by glycosylphosphatiylinositol molecules. *Proc. Natl Acad. Sci. U.S.A.* **87**, 9421–9425.

Merida, I., Diez, E. and Gaulton, G.N. (1991). IL-2 binding activates a tyrosine-phosphorylated phosphatidylinositol-3-kinase. *J. Immunol*. **147**, 2202–2207.

Merida, I., Williamson, P., Kuziel, W.A., Greene, W.C. and Gaulton, G.N. (1993). The serine-rich cytoplasmic domain of the interleukin-2 receptor β chain is essential for interleukin-2-dependent tyrosine protein kinase and phosphatidylinositol-3-kinase activation. *J. Biol. Chem*. **268**, 6765–6770.

Michiel, D.F., Garcia, G.G., Evans, G.A. and Farrar, W.L. (1991). Regulation of the interleukin 2 receptor complex tyrosine kinase activity in vitro. *Cytokine* **3**, 428–438.

Migone, T.-S., Lin, J.-X., Cereseto, A., Mulloy, J.C., O'Shea, J.J., Franchini, G. and Leonard, W.J. (1995). Constitutively activated Jak-STAT pathway in T cells transformed with HTLV-1. *Science* **269**, 79–81.

Mills, G.B., May, C., McGill, M., Fung, M., Baker, M., Sutherland, R. and Greene, W.C. (1990). Interleukin 2-induced tyrosine phosphorylation. Interleukin 2 receptor β is tyrosine phosphorylated. *J. Biol. Chem*. **265**, 3561–3567.

Mills, G.B., Arima, N., May, C., Hill, M., Schmandt, R., Li, J., Miyamoto, N.G. and Greene, W.C. (1992). Neither the LCK nor the FYN kinases are obligatory for IL-2-mediated signal transduction in HTLV-1-infected human T cells. *Int. Immunol*. **4**, 1233–1243.

Minami, Y., Kono, T., Miyazaki, T. and Taniguchi, T. (1993a). The IL-2 receptor complex: Its structure, function and target genes. *Annu. Rev. Immunol*. **11**, 245–267.

Minami, Y., Kono, T., Yamada, K., Kobayashi, N., Kawahara, A., Perlmutter, R.M. and Taniguchi, T. (1993b). Association of p56lck with IL-2 receptor β chain is critical for the IL-2-induced activation of p56lck. *EMBO J*. **12**, 759–768.

Minami, Y., Nakagawa, Y., Kawahara, A., Miyazaki, T., Sada, K., Yamamura, H. and Taniguchi, T. (1995). Protein tyrosine kinase syk is associated with and activated by the IL-2 receptor: Possible link with the c-myc induction pathway. *Immunity* **2**, 89–100.

Minamoto, S., Treisman, J., Hawkins, W.D., Sugamura, K. and Rosenberg, S.A. (1995). Acquired erythropoietin responsiveness of interleukin-2-dependent T lymphocytes retrovirally transduced with genes encoding chimeric erythropoietin/interleukin-2 receptors. *Blood* **86**, 2281–2287.

Miyawaki, T., Yachie, A., Uwadana, N., Ohzeki, S., Nagaoki, T. and Taniguchi, T. (1982). Functional significance of Tac antigen expressed on activated human T lymphocytes: Tac antigen interacts with T cell growth factor in cellular proliferation. *J. Immunol*. **129**, 2474–2478.

Miyazaki, T., Maruyama, M., Yamada, G., Hatakeyama, M. and Taniguchi, T. (1991). The integrity of the conserved 'WS motif' common to IL-2 and other cytokine receptors is essential for ligand binding and signal transduction. *EMBO J*. **10**, 3191–3197.

Miyazaki, T., Liu, Z.-J., Kawahara, A., Minami, Y., Yamada, K., Tsujimoto, Y., Barsoumian, E.L., Perlmutter, R.M. and Taniguchi, T. (1995). Three distinct IL-2 signaling pathways mediated by bcl-2, c-myc, and lck cooperate in hematopoietic cell proliferation. *Cell* **81**, 223–231.

Morgan, D.A., Ruscetti, F.W. and Gallo, R. (1976). Selective in vitro growth of T lymphocytes from normal human bone marrows. *Science* **193**, 1007–1008.

Morla, A.O., Schreurs, J., Miyajima, A. and Wang, J.Y. (1988). Hematopoietic growth factors activate the tyrosine phosphorylation of distinct sets of proteins in interleukin-3-dependent murine cell lines. *Mol. Cell. Biol*. **8**, 2214–2218.

Morrigl, R., Gouilleux-Gruart, V., Jahne, R., Berchtold, S., Gartmann, C., Liu, X., Hennighausen, L., Sotiropoulos, A., Groner, B. and Gouilleux, F. (1996). Deletion of the carboxyl-terminal transactivation domain of MGF-Stat5 results in sustained DNA binding and a dominant negative phenotype. *Mol. Cell. Biol*. **16**, 5691–5700.

Mui, A.L.-F., Wakao, H., O'Farrell, A.-M., Harada, N. and Miyajima, A. (1995). Interleukin-3, granulocyte–macrophage colony stimulating factor and interleukin-5 transduce signals through two STAT5 homologs. *EMBO J*. **14**, 1166–1175.

Mui, A.L.-F., Wakao, H., Kinoshita, T., Kitamura, T. and Miyajima, A. (1996). Suppression of interleukin-3-induced gene expression by a C-terminal truncated Stat5: Role of Stat5 in proliferation. *EMBO J*. **15**, 2425–2433.

Müller, M., Briscoe, J., Laxton, C., Guschin, D., Ziemiecki, A., Silvennoinen, O., Harpur, A.G., Berbieri, G., Witthuhn, B.A., Schindler, C., Pellegrini, S., Wilks, A.F., Ihle, J.N., Stark, G.R. and Kerr, I.M. (1993). The protein tyrosine kinase JAK1 complements defects in interferon-α/β and -γ signal transduction. *Nature* **366**, 129–166.

Mulloy, J.C., Crowley, R.W., Fullen, J., Leonard, W.J. and Franchini, G. (1996). The human T-cell leukemia/lymphotrophic virus type 1 $p12^{I}$ protein binds the interleukin-2 receptor β and γc chains and affects their expression on the cell surface. *J. Virol*. **70**, 3599–3605.

Murakami, M., Narazaki, M., Hibi, M., Yawata, H., Yasukawa, K., Hamaguchi, M., Taga, T. and Kishimoto, T. (1991). Critical cytoplasmic region of the interleukin 6 signal transducer gp130 is conserved in the cytokine receptor family. *Proc. Natl Acad. Sci. U.S.A.* **88**, 11 349–11 353.

Nakamura, Y., Russell, S.M., Mess, S.A., Friedmann, M., Erdos, M., Francois, C., Jacques, Y., Adelstein, S. and Leonard, W.J. (1994). Heterodimerization of the IL-2 receptor β- and γ-chain cytoplasmic domains is required for signaling. *Nature* **369**, 330–333.

Nakanishi, K., Hirose, S., Yoshimoto, T., Ishizashi, H., Hiroishi, K., Tanaka, T., Kono, T., Miyasaka, M., Taniguchi, T. and Higashino, K. (1992). Role and regulation of interleukin-2 receptor α and β chains in IL-2 driven growth. *Proc. Natl Acad. Sci. U.S.A.* **89**, 3551–3555.

Nelson, B.H., Lord, J.D. and Greenberg, P.D. (1994). Cytoplasmic domains of the interleukin-2 receptor β and γ chains mediate the signal for T-cell proliferation. *Nature* **369**, 333–336.

Nelson, B.H., Lord, J.D. and Greenberg, P.D. (1996). A membrane-proximal region of the interleukin-2 receptor γc chain sufficient for Jak kinase activation and induction of proliferation in T cells. *Mol. Cell. Biol*. **16**, 309–317.

Nikaido, T. Shimizu, A., Ishida, N., Sabe, H., Teshigawara, K., Maida, M., Uchiyama, T., Yodoi, J. and Honjo, T. (1984). Molecular cloning of cDNA encoding human interleukin-2 receptor. *Nature* **311**, 631–635.

Nishi, M., Ishida, Y. and Honjo, T. (1988). Expression of functional interleukin-2 receptors in human light chain/Tac transgenic mice. *Nature* **331**, 267–269.

Noguchi, M., Nakamura, Y., Russell, S.M., Ziegler, S.F., Tsang, M., Cao, X. and Leonard, W.J. (1993a). Interleukin-2 receptor γ chain: A functional component of the interleukin-7 receptor. *Science* **262**, 1877–1880.

Noguchi, M., Yi, H., Rosenblatt, H.M., Filipovich, A.H., Adelstein, S., Modi, W.S., McBride, O.W. and Leonard, W.J. (1993b). Interleukin-2 receptor γ chain mutation results in X-linked severe combined immunodeficiency in humans. *Cell* **73**, 147–157.

Nosaka, T., van Deursen, J.M.A., Tripp, R.A., Thierfelder, W.E., Witthuhn, B.A., McMickle, A.P., Doherty, P.C., Grosveld, G.C. and Ihle, J.N. (1995). Defective lymphoid development in mice lacking Jak3. *Science* **270**, 800–802.

Nourse, J., Firpo, E., Flanagan, W.M., Coats, S., Polyak, K., Lee, M.-H., Massague, J., Crabtree, G.R. and Roberts, J.M. (1994). Interleukin-2-mediated elimination of the $p27^{Kip1}$ cyclin-dependent kinase inhibitor prevented by rapamycin. *Nature* **372**, 570–573.

Ohashi, Y., Takeshita, T., Nagata, K., Mori, S. and Sugamura, K. (1989). Differential expression of the IL-2 receptor subunits, p55 and p75, on various populations of primary peripheral blood mononuclear cells. *J. Immunol*. **143**, 3548–3555.

Ortaldo, J.R., Mason, A.T., Gerard, J.P., Henderson, L.E., Farrar, W., Hopkins, R.F., 3rd, Herberman, R.B. and Rabin, H. (1984). Effects of natural and recombinant IL 2 on regulation of IFN γ production and natural killer activity: lack of involvement of the Tac antigen for these immunoregulatory effects. *J. Immunol*. **133**, 779–783.

Otani, H., Siegel, J.P., Erdos, M., Gnarra, J.R., Toledano, M.B., Sharon, M., Mostowski, H., Feinberg, M.B., Pierce, J. and Leonard, W.J. (1992). Interleukin(IL)-2 and IL-3 induce distinct but overlapping responses in murine IL-3-dependent 32D cells transduced with human IL-2 receptor β chain: Involvement of tyrosine kinase(s) other than p56-lck. *Proc. Natl Acad. Sci. U.S.A.* **89**, 2789–2793.

Otero, G.C. and Merrill, J.E. (1995). Molecular cloning of IL-2Rα, IL-2Rβ, and IL-2Rγ cDNAs from a human oligodendroglioma cell line: Presence of IL-2R mRNAs in the human central nervous system. *Glia* **14**, 295–302.

Pelligrini, S., John, J., Shearer, M., Kerr, I.M. and Stark, G.R. (1989). Use of a selectable marker regulated by α interferon to obtain mutations in the signaling pathway. *Mol. Cell. Biol*. **9**, 4605–4612.

Peschon, J.J., Morissey, P.J., Grabstein, K.H., Ramsdell, F.J., Maraskovsky, E., Gliniak, B.C., Park, L.S., Ziegler, S.F., Williams, D.E., Ware, C.B., Meyer, J.D. and Davison, B.L. (1994). Early lymphocyte expansion is severely impaired in interleukin-7 receptor-deficient mice. *J. Exp. Med*. **180**, 1955–1960.

Ralph, P., Jeong, G., Welte, K., Mertelsmann, R., Rabin, H., Henderson, L.E., Souza, L.M., Boone, T.C. and Robb, R.J. (1984). Stimulation of immunoglobulin secretion in human B lymphocytes as a direct effect of high concentratons of IL 2. *J. Immunol*. **133**, 2442–2445.

Ralph, P., Nakoinz, I. and Rennick, D. (1988). Role of interleukin 2, interleukin 4, and α, β, and γ interferon in stimulating macrophage antibody-dependent tumoricidal activity. *J. Exp. Med*. **167**, 712–717.

Rammensee, H.G., Kroschewski, R. and Frangoulis, B. (1989). Clonal energy induced in mature V β6+T lymphocytes on immunizing Mls-1b mice with Mls-1a expressing cells. *Nature* **339**, 541–544.

Rao, A. (1994). NFATp: A transcription factor required for the coordinate induction of several cytokine genes. *Immunol. Today* **15**, 274–281.

Raulet, D.H. (1985). Expression and function of interleukin-2 receptors on immature thymocytes. *Nature* **314**, 101–103.

Ravichandran, K.S., Igras, V., Schoelson, S.E., Fesik, S.W. and Burakoff, S.J. (1996). Evidence for a role for the phosphotyrosine-binding domain of Shc in interleukin 2 signaling. *Proc. Natl Acad. Sci. U.S.A.* **93**, 5275–5280.

Reed, J.C., Alpers, J.D., Scherle, P.A., Hoover, R.G., Nowell, P.C. and Prystowsky, M.B. (1987). Proto-oncogene expression in cloned T lymphocytes: mitogens and growth factors induce different patterns of expession. *Oncogene* **1**, 223–228.

Reeves, R., Spies, A.G., Nissen, M.S., Buck, C.D., Weinberg, A.D, Barr, P.J., Magnuson, N.S. and Magnuson, J.A. (1986). Molecular cloning of a functional bovine interleukin 2 cDNA. *Proc. Natl Acad. Sci. U.S.A.* **83**, 3228–3232.

Rinkenberger, J.L., Wallin, J.J., Johnson, K.W. and Koshland, M.E. (1996). An interleukin-2 signal relieves BSAP (Pax5)-mediated repression of the immunoglobulin J chain gene. *Immunity* **5**, 377–386.

Robb, R.J. and Greene, W.C. (1983). Direct demonstration of the identity of T cell growth factor binding protein and the Tac antigen. *J. Exp. Med*. **158**, 1332–1337.

Robb, R.J. and Greene, W.C. (1987). Internalization of interleukin 2 is mediated by the β chain of the high-affinity interleukin 2 receptor. *J. Exp. Med*. **165**, 1201–1206.

Robb, R.J., Munck, A. and Smith, K.A. (1981). T cell growth factor receptors. Quantitation, specificity, and biological relevance. *J. Exp. Med*. **154**, 1455–1474.

Robb, R.J., Rush, C.M., Yodoi, J. and Greene, W.C. (1987). An interleukin 2 binding molecule distinct from the Tac protein: analysis of its role in formation of high affinity receptors. *Proc. Natl Acad. Sci. U.S.A.* **84**, 2002–2006.

Robb, R.J., Rusk, C.M. and Neeper, M.P. (1988). Structure–function relationships for the interleukin 2 receptor: location of ligand and antibody binding sites on the Tac receptor chain by mutational analysis. *Proc. Natl Acad. Sci. U.S.A.* **85**, 5654–5658.

Rolink, A., Grawunder, U., Winkler, T., Karasuyama, H. and Melchers, F. (1994). IL-2 receptor α chain (CD25, TAC) expression defines a crucial stage in pre-B cell development. *Int. Immunol*. **6**, 1257–1264.

Rosenberg, S.A., Lotze, M.T., Yang, J.C., Topelian, S.L., Chang, A.E., Schwartzentruber, D.J., Aebersold, P., Leitman, S., Linehan, W.M. and Seipp, C.A. (1993). Prospective randomized trial of high-dose interleukin-2 alone or in conjunction with lymphokine-activated killer cells for the treatment of patients with advanced cancer. *J. Natl Cancer Inst*. **85**, 622–632.

Rosenberg, S.A., Yannelli, J.R., Yang, J.C., Topalian, S.L., Schwartzentruber, D.J., Weber, J.S., Parkinson, D.R., Seipp, C.A., Einhorn, J.H. and White, D.E. (1994). Treatment of patients with metastatic melanoma with autologous tumor-infiltrating lymphocytes and interleukin 2. *J. Natl Cancer Inst*. **86**, 1159–1166.

Rozwarski, D., Gronenborn, A., Clore, G., Bazan, J., Bohm, A., Wlodawer, A., Hatada, M. and Karplus, P. (1994). Structural comparisons among the short-chain helical cytokines. *Structure* **2**, 159–173.

Rusk, C.M., Neeper, M.P., Kuo, L.M., Kutny, R.M. and Robb, R.J. (1988). A larger number of L chains (Tac) enhance the association rate of interleukin 2 to the high affinity site of the interleukin 2 receptor. *J. Immunol*. **140**, 2249–2259.

Russell, S.M., Keegan, A.D., Harada, N., Nakamura, Y., Noguchi, M., Leland, P., Friedman, M.C., Miyajima, A., Puri, R.K., Paul, W.E. and Leonard, W.J. (1993). Interleukin-2 receptor γ chain: A functional component of the IL-4 receptor. *Science* **262**, 1880–1883.

Russell, S.M., Johnston, J.A., Noguchi, M., Kawamura, M., Bacon, C.M., Friedmann, M., Berg, M., McVicar, D.W., Witthuhn, B.A., Silvennoinen, O., Goldman, A.S., Schmalsteig, F.C., Ihle, J.N., O'Shea, J.J. and Leonard, W.J. (1994). Interaction of the IL-2Rβ and γc chains with Jak1 and Jak3: Implications for XSCID and XCID. *Science* **266**, 1042–1045.

Russell, S.M., Tayebi, N., Nakajima, H., Reidy, M.C., Roberts, J.L, Aman, M.J., Migone, T.-S., Noguchi, M., Markert, M.L., Buckley, R.H., O'Shea, J.J. and Leonard, W.J. (1995). Mutation of Jak3 in a patient with SCID: Essential role of Jak3 in lymphoid development. *Science* **270**, 797–799.

Sadlack, B., Merz, H., Schorle, H., Schimpl, A., Feller, A.C. and Horak, I. (1993). Ulcerative colitis-like disease in mice with a disrupted interleukin-2 gene. *Cell* **75**, 253–261.

Saltzman, E.M., Thom, R.R. and Casnellie, J.E. (1988). Interleukin 2 mediates the inhibition of oligodendrocyte progenitor cell proliferation *in vitro*. *J. Biol. Chem*. **263**, 6956–6959.

Samson, M., Libert, F., Doranz, B.J., Rucker, J., Liesnard, C., Farber, C.-M., Saragosti, S., Lapoumeroulie, C., Cognaux, J., Forceille, C., Muyldermans, G., Verhofstede, C., Burtonboy, G., Georges, M., Imai, T., Rana, S., Yi, Y., Smyth, R.J., Collman, R.G., Doms, R.W., Vassart, G. and Parmentier, M. (1996). Resistance to HIV-1 infection in caucasian individuals bearing mutant alleles of the CCR-5 chemokine receptor gene. *Nature* **382**, 722–725.

Saneto, R.P., Altman, A., Knobler, R.L., Johnson, H.M. and de Vellis, J. (1986). Interleukin-2 mediates the inhibition of oligodendrocyte progenitor cell proliferation in vitro. *Proc. Natl Acad. Sci. U.S.A.* **83**, 9221–9225.

Satoh, T., Nakafuku, M., Miyajima, A. and Kaziro, Y. (1991). Involvement of ras p21 protein in signal-transduction pathways from interleukin-2, interleukin-3, and granulocyte/macrophage colony-stimulating factor, but not from interleukin-4. *Proc. Natl Acad. Sci. U.S.A.* **88**, 3314–3318.

Satoh, T., Minami, Y., Kono, T., Yamada, K., Kawahara, A., Taniguchi, T. and Kaziro, Y. (1992). Interleukin 2-induced activation of Ras requires two domains of interleukin 2 receptor β subunit, the essential region for growth stimulation and Lck-binding domain. *J. Biol. Chem*. **267**, 25423–25427.

Sauve, K., Nachman, M., Spence, C., Bailon, P., Campbell, E., Tsien, W.H., Kondas, J.A., Hakimi, J. and Ju, G. (1991). Localization in human interleukin 2 of the binding site to the α chain (p55) of the interleukin 2 receptor. *Proc. Natl Acad. Sci. U.S.A.* **88**, 4636–4640.

Schindler, C., Shuai, K., Prezioso, V.R. and Darnell, J.E. (1992). Interferon-dependent tyrosine phosphorylation of a latent cytoplasmic transcription factor. *Science* **257**, 809–815.

Schorle, H., Holtschke, T., Hunig, T., Schimpl, A. and Horak, I. (1991). Development and function of T cells in mice rendered interleukin-2 deficient by gene targeting. *Nature* **352**, 621–624.

Seigel, L.J., Harper, M.E., Wong, S.F., Gallo, R.C., Nash, W.G. and O'Brien, S.J. (1984). Gene for T-cell growth factor: location on human chromosome 4q and feline chromosome B1. *Science* **223**, 175–178.

Seow, H.F., Rothel, J.S., Radford, A.J. and Wood, P.R. (1990). Molecular cloning of the ovine interleukin-2 gene by the polymerase chain reaction. *Nucleic Acids Res*. **18**, 7175.

Shackelford, D.A. and Trowbridge, I.S. (1991). Ligand-stimulated tyrosine phosphorylation of the IL-2 receptor β chain and receptor-associated proteins. *Cell Regul*. **2**, 73–85.

Sharon, M., Klausner, R.D., Cullen, B.R., Chizzonite, R. and Leonard, W.J. (1986). A conserved AU sequence from the 3′ untranslated region of GM-CSF mRNA mediates selective mRNA degradation. *Science* **234**, 859–863.

Sharon, M., Gnarra, J.R. and Leonard, W.J. (1989). The β-chain of the IL2 receptor (p70) is tyrosine-phosphorylated on YT and HUT-102B2 cells. *J. Immunol*. **143**, 2530–2533.

Shibuya, H., Yoneyama, M., Ninomiya-Tsuji, J., Matsuomoto, K. and Taniguchi, T. (1992). IL-2 and EGF receptors stimulate the hematopoietic cell cycle via different signaling pathways: Demonstration of a novel role for c-myc. *Cell* **70**, 57–67.

Shuai, K., Stark, G.R., Kerr, I.M. and Darnell, J.E. (1993a). A single phosphotyrosine residue of Stat91 required for gene activation by interferon-γ. *Science* **261**, 1744–1746.

Shuai, K., Ziemiecki, A., Wilks, A.F., Harpur, A.G., Sadowski, H.P., Gilman, M.Z. and Darnell, J.E. (1993b). Polypeptide signaling to the nucleus through tyrosine phosphorylation of Jak and Stat proteins. *Nature* **366**, 580–583.

Shultz, L.D., Schweitzer, P.A., Rajan, T.V., Yi, T., Ihle, J., Matthews, R.J., Thomas, M.L. and Beier, D.R. (1993). Mutations at the murine motheaten locus are within the hematopoietic cell protein-tyrosine phosphatase (Hcph) gene. *Cell* **73**, 1445–1454.

Siebenlist, U., Durand, D.B., Bressler, P., Holbrook, N.J., Norris, C.A., Kamoun, M., Kant, J.A. and Crabtree, G.R. (1986). The IL-2 receptor β chain (p70): role in mediating signals for LAK, NK, and proliferative activities. *Mol. Cell. Biol*. **6**, 3042–3049.

Smith, K.A. (1992). Interleukin-2. *Curr. Opin. Immunol*. **4**, 271–276.

Smith, K.A. and Cantrell, D.A. (1985). Interleukin-2 regulates its own receptors. *Proc. Natl Acad. Sci. U.S.A.* **82**, 864–868.

Stahl, N., Farruggella, T.J., Boulton, T.G., Zhong, Z., Darnell, J.E. and Yancopoulos, G.D. (1995). Choice of STATs and other substrates specified by modular tyrosine-based motifs in cytokine receptors. *Science* **267**, 1349–1353.

Stern, J.B., and Smith, K.A. (1986). Interleukin-2 induction of T-cell G_1 progression and *c-myb* expression. *Science* **233**, 203–206.

Sugamura, K., Takeshita, T., Asao, H., Kumaki, S., Ohbo, K., Ohtani, K. and Nakamura, M. (1990). IL-2-induced signal transduction: Involvement of tyrosine kinase and IL-2 receptor γ chain. *Lymphokine Res*. **9**, 539–543.

Sun, H. and Tonks, N. (1994). The coordinated action of protein tyrosine phosphatases and kinases in cell signaling. *Trends Biochem. Sci*. **19**, 480–485.

Suzuki, H., Kundig, T.M., Furlonger, C., Wakeham, A., Timms, E., Matsuyama, T., Schmits, R., Simard, J.J.L., Ohashi, P.S., Greisser, H., Taniguchi, T., Paige, C.J. and Mak, T.W. (1995). Deregulated T cell activation and autoimmunity in mice lacking interleukin-2 receptor β. *Science* **268**, 1472–1476.

Takeshita, T., Asao, H., Suzuki, J. and Sugamura, K. (1990). An associated molecule, p64, with high-affinity interleukin 2 receptor. *Int. Immunol*. **2**, 477–480.

Takeshita, T., Asao, H., Ohtani, K., Ishii, N., Kumaki, S., Tanaka, N., Munakata, H., Nakamura, M. and Sugamura, K. (1992). Cloning of the γ chain of the human IL-2 receptor. *Science* **257**, 379–382.

Taniguchi, T., Matsui, H., Fujita, T., Takaoka, C., Kashima, N., Yoshimoto, R. and Hamuro, J. (1983). Structure and expression of a cloned cDNA for human interleukin-2. *Nature* **302**, 305–310.

Terada, N., Lucas, J.J., Szepesi, A., Franklin, R.A., Takase, K. and Gelfand, E.W. (1992). Rapamycin inhibits the phosphorylation of p70 S6 kinase in IL-2 and mitogen-activated human T cells. *Biochem. Biophys. Res. Commun*. **186**, 1315–1321.

Teshigawara, K., Wang, H., Kato, K. and Smith, K.A. (1987). Interleukin 2 high-affinity receptor expression requires two distinct binding proteins. *J. Exp. Med*. **165**, 223–238.

Thomis, D.C., Gurniak, C.B., Tivol, E., Sharpe, A.H. and Berg, L.J. (1995). Defects in B lymphocyte maturation and T lymphocyte activation in mice lacking Jak3. *Science* **270**, 794–797.

Tigges, M., Casey, L.S. and Koshland, M.E. (1989). Mechanism of interleukin-2 signaling: mediation of different outcomes by a single receptor and transduction pathway. *Science* **243**, 781–786.

Toribio, M., Gutierrez-Ramos, J., Pezzi, L., Marcos, M. and Martinez, C. (1989). Interleukin-2-dependent autocrine proliferation in T-cell development. *Nature* **342**, 82–85.

Trinchieri, G., Matsumoto-Kobayashi, M., Clark, S.C., Seehra, J., London, L. and Perussia, B. (1984). Response of resting human peripheral blood natural killer cells to interleukin 2. *J. Exp. Med*. **160**, 1147–1169.

Tsudo, M., Goldman, C.K., Bongiovanni, K.F., Chan, W.C., Winton, E.F., Yagita, M., Grimm, E.A. and Waldmann, T.A. (1987). Demonstration of a non-Tac peptide that binds interleukin 2: a potential participant in a multichain interleukin 2 receptor complex. *Proc. Natl Acad. Sci. U.S.A.* **84**, 5394–5398.

Tsudo, M., Kitamura, F. and Miyasaka, M. (1989). Characterization of the interleukin 2 receptor β chain using three distinct monoclonal antibodies. *Proc. Natl Acad. Sci. U.S.A.* **86**, 1982–1986.

Tsudo, M., Karasuyama, H., Kitamura, F., Tanaka, T., Kubo, S., Yamamura, Y., Tamatani, T., Hatakeyama, M., Taniguchi, T. and Miyasaka, M. (1990). The IL-2 receptor β-chain (p70). Ligand binding ability of the cDNA-encoding membrane and secreted forms. *J. Immunol*. **145**, 599–606.

Tsui, H.W., Siminovitch, K.A., deSouza, L. and Tsui, F.W.L. (1993). Motheaten and viable motheaten mice have mutations in the hematopoietic cell phosphatase gene. *Nature Genet*. **4**, 124–129.

Turner, B., Rapp, U., App, H., Greene, M., Dobashi, K. and Reed, J. (1991). Interleukin 2 induces tyrosine phosphorylation and activation of p72–74 Raf-1 kinase in a T-cell line. *Proc. Natl Acad. Sci. U.S.A.* **88**, 1227–1231.

Uchiyama, T., Broder, S. and Waldmann, T.A. (1981). A monoclonal antibody (anti-Tac) reactive with activated and functionally mature human T cells. I. Production of anti-Tac monoclonal antibody and distribution of Tac(+) cells. *J. Immunol*. **126**, 1393–1397.

Umlauf, S.W., Beverly, B., Lantz, O. and Schwartz, R.H. (1995). Regulation of IL-2 gene expression by CD28 costimulation in mouse T-cell clones. Both nuclear and cytoplasmic RNA are regulated with complex kinetics. *Mol. Cell. Biol*. **15**, 3197–3205.

Valezquez, L., Fellous, M., Stark, G.R. and Pellegrini, S. (1992). A protein tyrosine kinase in the interferon α/β signaling pathway. *Cell* **70**, 313–322.

von Boehmer, H., Crisanti, A., Kisielow, P. and Haas, W. (1985). Absence of growth by most receptor-expressing fetal thymocytes in the presence of interleukin-2. *Nature* **314**, 539–540.

Wakao, H., Gouilleux, F. and Groner, B. (1994). Mammary gland factor (MGF) is a novel member of the cytokine regulated transcription factor gene family and confers the prolactin response. *EMBO J*. **13**, 2182–2191.

Waldmann, T.A., Goldman, C.K., Robb, R.J., Depper, J.M., Leonard, W.J., Sharrow, S.O., Bongiovanni, K.F., Korsmeyer, S.J. and Greene, W.C. (1984). Expression of interleukin 2 receptors on activated human B cells. *J. Exp. Med*. **160**, 1450–1466.

Walker, E., Leemhuis, T. and Roeder, W. (1988). Murine B lymphoma cell lines release functionally active interleukin 2 after stimulation with *Staphylococcus aureus*. *J. Immunol.* **140**, 859–865.

Wang, D., Stravopodis, D., Teglund, S., Kitazawa, J. and Ihle, J.N. (1996). Naturally occurring dominant negative variants of Stat5. *Mol. Cell. Biol*. **16**, 6141–6148.

Watling, D., Guschin, D., Müller, M., Silvennoinen, O., Witthuhn, B.A., Quelle, F.W., Rogers, N.C., Schindler, C., Stark, G.R., Ihle, J.N. and Kerr, I.M. (1993). Complementation by the protein tyrosine kinase JAK2 of a mutant cell line defective in the interferon-γ signal transduction pathway. *Nature* **366**, 166–170.

Weber, J.S., Yang, J.C., Topelian, S.L., Schwartzentruber, D.J. White, D.E. and Rosenberg, S.A. (1992). The use of interleukin-2 and lymphokine-activated killer cells for the treatment of patients with non-Hodgkin's lymphoma. *J. Clin. Oncol.* **10**, 33–40.

Weigel, U., Meyer, M. and Sebald, W. (1989). Mutant proteins of human interleukin 2. Renaturation yield, proliferative activity and receptor binding. *Eur. J. Biochem.* **180**, 295–300.

Weinberg, K. and Parkman, R. (1990). Severe combined immunodeficiency due to a specific defect in the production of interleukin-2. *N. Engl. J. Med.* **322**, 1718–1723.

Willerford, D.M., Chen, J., Ferry, J.A., Davidson, L., Ma, A. and Alt, F.W. (1995). Interleukin-2 receptor α chain regulates the size and content of the peripheral lymphoid compartment. *Immunity* **3**, 521–530.

Yagita, H., Nakata, M., Azuma, A., Nitta, T., Takeshita, T., Sugamura, K. and Okumura, K. (1989). Activation of peripheral blood T cells via the p75 interleukin 2 receptor. *J. Exp. Med.* **170**, 1445–1450.

Yannelli, J.R., Hyatt, C., McConnell, S., Hines, K., Jacknin, L., Parker, L., Sanders, M. and Rosenberg, S.A. (1996). Growth of tumor-infiltrating lymphocytes from human solid cancers: summary of a 5-year experience. *Int. J. Cancer* **65**, 413–421.

Yi, T., Mui, A.L.-F., Krystal, G. and Ihle, J.N. (1993). Hematopoietic cell phosphatase associates with the interleukin-3 (IL-3) receptor β chain and down-regulates IL-3-induced tyrosine phosphorylation and mitogenesis. *Mol. Cell. Biol.* **13**, 7577–7586.

Yodoi, J. and Uchiyama, T. (1992). Diseases associated with HTLV-1: virus, IL-2 receptor dysregulation and redox regulation. *Immunol. Today* **13**, 405–411.

Yokota, T., Arai, N., Lee, F., Rennick, D., Mossman, T. and Arai, K. (1985). Use of cDNA expression vector for isolation of mouse interleukin 2 cDNA clones: expression of T-cell growth-factor activity after transfection of monkey cells. *Proc. Natl Acad. Sci. U.S.A.* **82**, 68–72.

Yoshimura, A., Ohkubo, T., Kiguchi, T., Jenkins, N.A., Gilbert, D.J. Copeland, N.G., Hara. T. and Miyajima, A. (1995). A novel cytokine-inducible gene CIS encodes an SH2-containing protein that binds to tyrosine-phosphorylated interleukin 2 and erythropoietin receptors. *EMBO J.* **14**, 2816–2826.

Zhang, Q., Nowak, I., Vonderheid, E., Rook, A.H., Kadin, M.E., Nowell, P.C., Shaw, L.M. and Wasik, M.A. (1996). Activation of Jak/STAT proteins in signal transduction pathway mediated by receptor for interleukin 2 in malignant T lymphocytes derived from cutaneous anaplastic large T-cell lymphoma and Sezary syndrome. *Proc. Natl Acad. Sci. U.S.A.* **93**, 9148–9153.

Zmuidzinas, A., Mamon, H., Roberts, T.M. and Smith, K.A. (1991). Interleukin-2-triggered raf-1 expression, phosphorylation, and associated kinase activity increase through G1 and S in CD3-stimulated primary human T cells. *Mol. Cell. Biol.* **11**, 2794–2803.

Zurawski, S.M. and Zurawski, G. (1989). Mouse interleukin-2 structure–function studies: substitutions in the first α-helix can specifically inactivate p70 receptor binding and mutations in the fifth α-helix can specifically inactivate p55 receptor binding. *EMBO J.* **8**, 2583–2590.

Zurawski, S.M., Imler, J.L. and Zurawski, G. (1990). Partial agonist/antagonist mouse interleukin-2 proteins indicate that a third component of the receptor complex functions in signal transduction. *EMBO J.* **9**, 3899–3905.

Chapter 5

Interleukin-3

John W. Schrader
The Biomedical Research Center, University of British Columbia, Vancouver, Canada

INTRODUCTION

Interleukin-3 (IL-3) acts on numerous target cells within the hemopoietic system, so it is not surprising that it was discovered independently by a number of laboratories studying different biologic activities on a variety of cells. These activities went under a variety of names including persisting cell stimulating factor, histamine-producing cell stimulating factor, multi-colony stimulating factor (CSF), multilineage hemopoietic growth factor, Thy-1 inducing factor, colony forming unit (CFU) stimulating activity, CSF-2α, CSF-2β, hemopoietic cell growth factor, mast cell growth factor, eosinophil CSF, megakaryocyte CSF, erythroid CSF, burst promoting activity, neutrophil–granulocyte CSF, hemopoietin-2 and synergistic activity. It was only with biochemical purification (Ihle *et al.*, 1983; Clark-Lewis *et al.*, 1984), molecular cloning and expression (Fung *et al.*, 1984; Yokota *et al.*, 1984), and chemical synthesis (Clark-Lewis *et al.*, 1986) that it was conclusively established that a single protein mediated all of these bioactivities.

STRUCTURE

IL-3 has broad structural similarities with other interleukins and hemopoietic growth factors. It is a relatively small protein, with a polypeptide chain of 140 amino acids in the mouse (Fung *et al.*, 1984; Yokota *et al.*, 1984) and of 133 in the human (Yang *et al.*, 1986), and is heavily glycosylated. There are no marked amino acid sequence homologies with other cytokines, although the fact that the genes for granulocyte–macrophage colony stimulating factor (GM-CSF) and IL-3 are closely linked on chromosome 5 at 5q23–q31 in the human (Yang *et al.*, 1988) supports the notion of a common evolutionary ancestry. The amino acid sequences of mouse and human IL-3 exhibit only 30% identity, reflecting the lack of cross-species biologic activity (Yang *et al.*, 1986).

Interestingly, IL-3 and a number of other cytokines, including IL-1β, IL-2, GM-CSF and erythropoietin, share a short motif of amino acids at the amino terminus. This is characterized by an *N*-terminal alanine followed in most instances by a proline (Schrader *et al.*, 1986a). The functional significance of this structural feature is obscure; at least in the case of IL-3 and GM-CSF it can be removed without affecting biologic activity *in vitro* (Clark-Lewis *et al.*, 1986, 1988).

The Cytokine Handbook, 3rd ed.
ISBN 0-12-689662-3

Nuclear magnetic resonance studies (Feng *et al.*, 1995) have confirmed that IL-3 has the basic four-helix-bundle three-dimensional structure that is characteristic of members of a family including hemopoietins and many cytokines and hormones. The prototype structure was that of growth hormone, and prolactin, erythropoietin, IL-2, IL-4, IL-5, IL-6 and GM-CSF have all been shown to share this basic pattern of three-dimensional folding. Interestingly several other hemopoietic growth factors, such as CSF-1 and steel locus factor (SLF), which, as discussed below, interact with a family of receptors quite distinct from that utilized by most hemopoietins and cytokines, also share the same overall three-dimensional structure, pointing to a shared evolutionary ancestor.

Analysis of structural analogs of IL-3 and of the effect on biologic activity of antibodies specific for defined parts of the polypeptide chain have yielded information on the structural determinants of IL-3 bioactivity. IL-3 was the first protein of its size to be successfully synthesized by automated chemical methods (Clark-Lewis *et al.*, 1986). This technique, shown by Clark-Lewis to be useful for the synthesis of cytokines in general, allowed a relatively rapid examination of the effects of deleting parts of the IL-3 molecule on bioactivity. In the case of mouse IL-3, these studies showed that the first 16 and the final 22 amino acids could be deleted with very little loss of biologic activity, suggesting that residues 17–118 could form all the structures essential for interaction with the receptor (Clark-Lewis *et al.*, 1986). The notion that the extreme amino-terminus of IL-3 is not involved in interactions with the receptor is supported by the fact that polyclonal antibodies specific for a peptide corresponding to residues 1–29 of IL-3 have relatively weak ability to neutralize IL-3 bioactivity (Ziltener *et al.*, 1987). Moreover antibodies to peptide 1–29 bind to IL-3 molecules that have been allowed to interact with the IL-3 receptor first (Duronio *et al.*, 1991). In contrast, antibodies to peptides corresponding to residues 44–75 (H.J. Ziltener, unpublished data) and 91–112 (Ziltener *et al.*, 1987) strongly neutralize bioactivity, suggesting that these residues are part of or close to the site or sites that interact with the IL-3 receptor.

McKearn and colleagues have mutated multiple amino acid residues in human IL-3 and obtained analogs of IL-3 with increased potency (Thomas *et al.*, 1995). Because the IL-3 receptor is made up of at least two distinct polypeptide chains, both of which associate closely with IL-3 (Duronio *et al.*, 1991), distinct regions of the IL-3 molecule will be involved in binding to the two chains of the receptor termed IL-3Rα and βc. Based on modeling and mutagenesis studies, Bagley and colleagues (1996) have proposed that eight discontinuous residues, Ser-17, Asn-18, Asp-21 and Thr-25 in helix A, and Arg-108, Phe-113, Lys-116 and Glu-119 in helix D, mediate binding of IL-3 to IL-3Rα. Binding to βc involves Glu-22 but, in contrast with the results of studies on the analogous residue in GM-CSF, replacement of Glu-22 with other residues failed to abrogate completely the biologic activity of the mutant IL-3 molecules. Bagley and colleagues (1996) have suggested that βc associates more closely with IL-3 than it does with GM-CSF or, alternatively, that βc interacts with IL-3Rα.

There are other important properties of the IL-3 molecule, apart from its ability to interact with its receptor, that will be regulated by its structure. These include its clearance and half-life in the plasma, and its ability to interact with extracellular matrix. Natural IL-3 occurs in a diversity of glycoforms generated by the addition of carbohydrate groups. Whereas the synthesized polypeptide has a M_r of 14 kDa upon sodium dodecyl sulfate (SDS)–polyacrylamide gel electrophoresis, murine IL-3 released from its natural source, activated T lymphocytes, runs as multiple bands, with major groups of bands with M_r around 22, 28 and 36 kDa (Ziltener *et al.*, 1988). Different T-cell clones appear to produce different proportions of the differently glycosylated forms (H.J. Ziltener, unpublished data). Carbohydrate on murine IL-3 of T-cell origin is exclusively *N*-linked (Ziltener *et al.*, 1988). The glycosylation patterns of IL-3 produced by other physiologic sources, such as the activated mast cells, have not been examined. IL-3 produced by recombinant DNA techniques in unnatural sources such as Chinese hamster ovary cells or COS cells, exhibits on SDS–gel electrophoresis a broad smear of differently glycosylated species, quite different from the pattern of distinct bands seen with IL-3 from the natural source T lymphocytes (Ziltener *et al.*, 1988).

The function of these extensive carbohydrate modifications of the IL-3 polypeptide is unknown. Chemically synthesized IL-3 (Clark-Lewis *et al.*, 1986) or IL-3 produced in *Escherichia coli* (Kindler *et al.*, 1985) exhibit all the biologic activities of naturally glycosylated IL-3. Ziltener and colleagues (1988) purified heavily, lightly and moderately glycosylated forms of IL-3 from T lymphocytes and demonstrated that, at least *in vitro*, they had the same specific activities and target specificities as deglycosylated material. The rate of clearance of naturally glycosylated IL-3 from the blood was approximately half that of non-glycosylated IL-3, but was still rapid with a β half-life of 20 min compared with 10 min for non-glycosylated IL-3 (Ziltener *et al.*, 1994). It is conceivable that the degree or type of glycosylation could regulate interaction with the extracellular matrix and influence diffusion or localization in tissues. One study of *in vitro* interactions of IL-3 with extracellular matrix material found that glycosylated and non-glycosylated IL-3 was bound equally well (Roberts *et al.*, 1988). Ziltener *et al.* (1994) saw no differences in the stimulatory effects on hemopoietic progenitor cells of long-term treatment with glycosylated or non-glycosylated murine IL-3.

ACTIONS ON HEMOPOIETIC STEM AND PROGENITOR CELLS

Although there are reports that IL-3 may affect the growth of certain epithelial cells, for example in carcinomas of the colon (Berdel *et al.*, 1989), the action of IL-3 on normal cells appears to be restricted to derivatives of the pluripotential hemopoietic stem cell. Endothelial cells may be the only exception to this generalization, as IL-3 has been reported to upregulate P-selectin expression (Khew-Goodall *et al.*, 1996).

IL-3 has the broadest target specificity of any of the hemopoietic growth factors. The range of target cells can be summarized as including progenitor cells of every lineage derived from the pluripotential hemopoietic stem cells. Thus IL-3 can stimulate the generation and differentiation of macrophages, neutrophils, eosinophils, basophils, mast cells, megakaryocytes and erythroid cells (summarized in Schrader *et al.*, 1988). In the presence of tumor necrosis factor-α (TNF-α), IL-3 stimulates the generation of dendritic cells from $CD34^+$ cells (Caux *et al.*, 1996). Moreover IL-3 acts on more primitive pluripotential stem cells. IL-3 stimulated the growth *in vitro* of colonies containing mixtures of myeloid and erythroid cells and stimulates both *in vitro* and *in vivo* the division of cells (CFUs) that form splenic colonies in irradiated mice (Iscove *et al.*, 1989; Iscove and Yan, 1990). IL-3 also stimulates the growth of human hemopoietic stem cells with significant capacity for self-renewal (Brugger *et al.*, 1993; Haylock *et al.*, 1992). Stimulation with IL-3, however, may result in a decreased ability of stem cell populations to self-renew as assessed by long-term repopulating capacity (Peters *et al.*, 1996;

Yonemura *et al.*, 1996). IL-3 also directly or indirectly promotes the survival *in vitro* of cells able to repopulate mice with T and B lymphocytes (Schrader *et al.*, 1988).

In vitro, hemopoietic stem and progenitors cells rapidly die if cultured in tissue culture medium alone. Like other hemopoietic growth factors, IL-3 prevents death by apoptosis and promotes survival *in vitro* (G.T. Williams *et al.*, 1990). Populations of mast cells generated by culturing murine bone marrow cells in IL-3 remain dependent on IL-3, not only for their continual proliferation, but also for survival (Schrader, 1981), and when deprived of IL-3, IL-3-dependent cells undergo apoptosis (G.T. Williams *et al.*, 1990). This reaction to the withdrawal of IL-3 may be a control mechanism to ensure that the massive proliferation of cells of the hemopoietic system that can be induced by the release of IL-3 (and other hemopoietic growth factors) during emergency situations like infections is terminated rapidly when the emergency is over and levels of the growth factors drop. Experiments demonstrating the rapid disappearance of IL-3-dependent cells that have been injected into normal animals lacking detectable levels of serum IL-3 (Schrader and Crapper, 1983; Crapper *et al.*, 1984a), are consistent with this notion.

As is common among cytokines, IL-3 shows strong synergistic activities with other cytokines. For example in humans, IL-3 synergizes with CSF-1 in producing macrophages, with G-CSF in producing neutrophils and with IL-11 in producing megakaryocytes. In both humans and mice, IL-3 synergizes with a number of cytokines such as CSF-1, SLF and IL-1 in maximally stimulating the growth of primitive hemopoietic stem cells. In the mouse, the effect of IL-3 in promoting the growth of mast cells can be enhanced by IL-4, IL-9 and IL-10.

EFFECTS OF IL-3 ON MATURE CELLS OF HEMOPOIETIC ORIGIN

In common with other hemopoietic growth factors, IL-3 affects not only immature hemopoietic cells but also the mature members of some lineages. For example, the subset of mast cells associated with mucosal surfaces depends on IL-3 for survival (Crapper *et al.*, 1984a; Schrader *et al.*, 1988). IL-3 also regulates the levels of major histocompatibility antigens on these mast cells, blocking the increased levels of expression induced by interferon-γ (IFN-γ) (Wong *et al.*, 1984).

IL-3 induces limited division of well differentiated murine macrophages and enhances their phagocytosis of yeast (Crapper *et al.*, 1985; Chen *et al.*, 1988). IL-3 stimulation of macrophages results in increased levels of class II major histocompatibility complex antigens and leukocyte functional antigen-1 (LFA-1) (Frendl and Beller, 1990), and increased levels of mRNA encoding IL-1 (Frendl and Beller, 1990), IL-6 and TNF-α (Frendl *et al.*, 1990). IL-3 also blocks the rapid apoptosis of so-called 'plasmacytoid T cells', $CD4^+$, $CD3^-$ and $CD11c^+$ cells present in secondary lymphoid tissue (Grouard *et al.*, 1997). In the presence of IL-3 and CD40 ligand these cells differentiate into dendritic cells.

Murine megakaryocytes differentiate *in vitro* in the presence of IL-3 (Ishibashi and Burstein, 1986). Human basophils are activated by IL-3 (Hirai *et al.*, 1988; Kurimoto *et al.*, 1989) and IL-3 stimulates the survival of human eosinophils (Rothenberg *et al.*, 1988) as well as increasing antibody-dependent cell-mediated cytotoxity, phagocytosis and superoxide anion production in response to stimulation with f-met–leu–phe (Lopez *et al.*, 1987).

LYMPHOID CELLS

The question of whether IL-3 has a key role in regulating the production of T or B lymphocytes has been controversial. Some of this confusion has stemmed from the fact that unfortunate misconceptions or errors in earlier literature have taken some time to be widely recognized.

The early report that first used the term interleukin-3 (Hapel *et al.*, 1981), erroneously ascribed to IL-3 the property of stimulating the growth of helper T lymphocytes. The cells misidentified as helper T lymphocytes were in fact contaminating cells of the myelomonocytic leukemia WEHI-3B, which had been used as a source for purification of the IL-3.

The confusion in this study may in part have been related to the fact that WEHI-3B cells express the Thy-1 antigen. At that time the Thy-1 antigen was thought to be a specific marker for T lymphocytes among lymphohemopoietic cells in the mouse. However, IL-3 induces the expression of high levels of the Thy-1 antigen on hemopoietic progenitor cells, including the precursors of macrophages and neutrophils (Schrader *et al.*, 1982).

Another erroneous link between IL-3 and lymphocytic cells was based on the notion that IL-3 played a critical role in T-lymphocyte development. This arose from the observation that IL-3 induced increased levels of the enzyme 20α-hydroxysteroid dehydrogenase in spleen cells from athymic mice and the postulate that this enzyme was specific for the T-lymphocyte lineage (Ihle *et al.*, 1981). Ihle and colleagues (1983) used the induction of 20α-hydroxysteroid dehydrogenase as the basis of the assay used for the first purification to homogeneity of IL-3. However, the notion that this enzyme was restricted to T lymphocytes was disproved by the demonstration that IL-3 (and GM-CSF) induced this enzyme in cells of a number of myeloid lineages, including mast cells (Hapel and Young, 1988).

T LYMPHOCYTES

As noted above, pluripotential stem cells capable of ultimately giving rise to T and B lymphocytes may be affected directly or indirectly by IL-3 (Schrader *et al.*, 1988). Palacios and Pelkonen (1988) reported the growth of IL-3-responsive lines with characteristics of prothymocytes, but these results have not been reproduced. There is evidence that relatively small subsets of mature T cells may respond to IL-3 in the human (Londei *et al.*, 1989). However, at present there is no evidence that thymic development or the function of the common subsets of T lymphocytes is influenced directly by IL-3.

B LYMPHOCYTES

As noted there is evidence that IL-3 directly or indirectly affects cells that can give rise to B lymphocytes in irradiated animals (Schrader *et al.*, 1988). Palacios and colleagues (1984) reported that IL-3-responsive clones of pre-B lymphocytes could be obtained with high frequency from fetal liver. However, there are no published accounts of the reproduction or extension of this work from other laboratories.

Palacios and Steinmetz (1985) reported the generation of a small number of IL-3-dependent cell lines with the capacity to give rise to B lymphocytes in irradiated animals.

Furthermore there are IL-3-dependent lines that can be induced to undergo B-cell differentiation and immunoglobulin gene rearrangement (Kinashi *et al.*, 1988). However, it is unclear how frequently and how reproducibly such cell lines can be obtained. A proportion of a subset of human acute lymphoblastic leukemias classified as B-cell precursors has also been shown to respond to IL-3 (Uckun *et al.*, 1989).

Ogawa and colleagues (Hirayama *et al.*, 1992; Ball *et al.*, 1995) have reported the *in vitro* growth of primitive lymphohemopoietic stem cells that could give rise to both myeloid cells and B lymphocytes. The growth of these stem cells was optimally supported by combinations of Steel or stem cell factor (SLF) with IL-6, IL-11 or GM-CSF. The generation of B lymphocytes in secondary culture required the presence of IL-7 and SLF. Interestingly, IL-3 was ineffective, either alone or with SLF, in maintaining the potential to generate B lymphocytes in the primary cultures and indeed appeared to inhibit the development of B lymphocytes. This suggests that IL-3 lacked the capacity to maintain or generate cells capable of giving rise to B lymphocytes in response to IL-7 and SLF.

More detailed studies (Ball *et al.*, 1996) suggested that, when given for a restricted period, IL-3 could enhance the generation of pre-B-cell colonies, but that the overall effect of IL-3 was to inhibit B-cell development. Winkler *et al.* (1995) reported that purified, c-Kit$^+$ B220$^+$ precursors of B lymphocytes could survive and proliferate *in vitro* in the presence of an IL-7-deficient stromal cell line and either IL-7 or IL-3. Rennick and colleagues (1989) reported that IL-3 could synergize with a stromal cell factor in stimulating the proliferation of murine pre-B lymphocytes. It is possible that the difficulty in demonstrating the ability of IL-3 to support the growth of precursors of B lymphocytes in many systems relates to the actions of IL-3 on myeloid progenitors. Thus there is *in vivo* and *in vitro* evidence that myeloid cells inhibit B lymphopoiesis and that the absence of myeloid progenitors and macrophages enhances or permits the generation of B cells (Nakano *et al.*, 1994; Rico-Vargas *et al.*, 1994).

In summary, primitive hemopoietic stem cells that are ultimately capable of giving rise to cells contributing to the B- or T-lymphocyte lineages probably respond to IL-3. Such cells or their more committed progeny can give rise to immortal cell lines or leukemic cells. However, overall, there is as yet no compelling evidence that IL-3 has a significant direct influence on B- or T-cell development.

SOURCES OF IL-3

Cells of lymphohemopoietic origin are the only sources of IL-3. The major physiologic source of IL-3 is the activated T lymphocyte (Schrader and Nossal, 1980; Schrader, 1981; Niemeyer *et al.*, 1989). There is as yet no clear understanding of the mechanisms that regulate the spectrum of cytokines produced in response to a given antigen. It is evident that the type or form of antigen and the presence of adjuvants influence the range and quantity of cytokine produced.

Mast cells can also produce IL-3 when immunoglobulin E (IgE) Fc receptors are cross-linked (Burd *et al.*, 1989; Plaut *et al.*, 1989; Wodnar-Filipowicz *et al.*, 1989). The physiologic significance of this phenomenon has yet to be established; it may serve to activate or prime other cells in the vicinity of an allergic response. These could include mast cells themselves, as well as macrophages and other hemopoietic cells. Activation of mast cells and secretion of IL-3 may account for the rapid increase in histamine-producing cell-stimulating activity observed in the serum of parasitized mice, stimulated 6 h before with parasite antigen (Abbud-Filho *et al.*, 1983). Eosinophils also produce IL-3 and other cytokines when activated. Stimuli include cross-linking of FcR and adherence to fibronectin (Moqbel *et al.*, 1994). TNF-α induces the rapid accumulation of IL-3 mRNA in eosinophils, a process that interestingly, is blocked by inhibition of the enzyme p38 mitogen-activated protein (MAP) kinase (T. Tanaka and J.W. Schrader, unpublished data). As discussed below, IL-3 may upregulate the production of cytokines that favor the development of T_{H2} cells.

Interaction of mast cells with fibroblasts *in vitro* can also lead to the accumulation of IL-3 mRNA in mast cells (Razin *et al.*, 1991). The physiologic significance of this is unclear. It may depend on the expression of the SLF on the surface of the fibroblasts, and this may occur only in abnormal situations, for example during inflammation. As discussed below, it is also possible that interaction of hemopoietic stem and progenitor cells expressing the c-kit protein, with stromal cells expressing the kit ligand SLF, could result in a similar phenomenon.

IL-3 IN NORMAL AND IMMUNOLOGICALLY STIMULATED ANIMALS

IL-3 is undetectable in the blood of normal animals (Crapper *et al.*, 1984b). In support of the notion that IL-3 is not present in significant quantities in the blood and extracellular fluids of normal mice, IL-3-dependent cell lines die when injected into normal mice, although they survive if the mice are provided with an artificial source of IL-3 (Schrader and Crapper, 1983; Crapper *et al.*, 1984b).

IL-3 can remain undetectable in the serum of animals undergoing immune responses (Crapper *et al.*, 1984a). However, in these instances, evidence for the local production of IL-3 at sites of immunologic activation can be found (Crapper *et al.*, 1984a). For example, cells from lymph nodes draining the site of injection of an antigen but not from normal lymph nodes produce IL-3 when incubated overnight in tissue culture medium (Crapper *et al.*, 1984a). The local release of IL-3 at sites where T cells are activated results in a characteristic histologic ‘footprint’, namely the local accumulation of mast cells generated by the action of IL-3 on undifferentiated precursors (Crapper and Schrader, 1983).

In cases where there is massive activation of T lymphocytes, for example graft *versus* host disease, small amounts of IL-3 can be detected in the serum (Crapper and Schrader, 1986). Another phenomenon that might have been accounted for by the release of IL-3 into the serum was the transient appearance of histamine-producing cell stimulating factor (HCSF) reported in the serum following the challenge of an immunized animal with a parasite antigen (Abbud-Filho *et al.*, 1983). As noted these observations may reflect the rapid release of IL-3, not from the T lymphocytes, but from mast cells activated by interaction of specific IgE with the injected antigen.

IL-3 IN THE SERUM

The half-life of intravenously injected IL-3 is short, being in the order of only 40 min (Crapper *et al.*, 1984a). A major part of this IL-3 is destroyed in the kidney. IL-3 does not appear to be bound to larger molecules in the serum (Crapper and Schrader, 1986) and enters the glomerular filtrate. Small amounts are detectable in the urine of animals with high serum levels of IL-3, but most of the filtered IL-3 appears to be resorbed and destroyed in the renal tubules (Crapper *et al.*, 1984b).

The release of IL-3 *in vivo* appears to be associated with stimulation of all of the various types of hemopoietic cells predicted from the *in vitro* activities of IL-3. For example, in certain phases of graft *versus* host disease in mice, an increase occurs in the number of mast cells and their precursors and in immature myeloid and erythroid cells in the spleen (Crapper and Schrader, 1986). This coincides with the appearance of small amounts of IL-3 in the serum.

As T-cell activation results in the release of multiple cytokines affecting hemopoiesis, including IL-4, IL-5, IL-6 and GM-CSF, a clearer picture of the effects of the chronic release of IL-3 *in vivo* came from experiments in which mice were inoculated with WEHI-3B, a tumor that produces IL-3 constitutively as a result of insertion of a retroviral DNA into one copy of the IL-3 gene (Ymer *et al.*, 1985). Mice with a localized subcutaneous tumor of WEHI-3B showed dramatic stimulation of hemopoiesis in the spleen, with increased numbers of myeloid cells, mast cells and megakaryocytes (Crapper *et al.*, 1984a). Interestingly, the levels of IL-3 in the serum of these mice were relatively low (fewer than 2 ED_{50} units/ml), suggesting that the chronic maintenance of low concentrations of IL-3 in the serum could achieve marked effects on hemopoiesis.

The effects of IL-3 vary in different tissues depending on the local availability of the different types of target cell. For example, in the gut mucosa, committed mast cell precursors are relatively common, whereas progenitors of other hemopoietic lineages are relatively rare (Crapper and Schrader, 1983). In this tissue the local release of IL-3 induces a mastocytosis. On the other hand, in organs like the murine spleen, where there is a higher frequency of hemopoietic stem and progenitor cells of various lineages, IL-3 stimulates an increase of myeloid and erythroid cells as well as a more modest increase in mast cells and their progenitors (Crapper and Schrader, 1983).

ADMINISTRATION OF IL-3

Administration of IL-3 *in vivo* is complicated by the relatively rapid clearance of IL-3 from the circulation. The subcutaneous administration of 2000 ED_{50} units of chemically synthesized IL-3 three times a day for 3 days resulted in an increase in splenic weight and in the number of mast cells and the progenitors of mast cells, neutrophils and macrophages, and in CFUs (Schrader *et al.*, 1986b). Similar results were obtained using *E. coli*-derived material (Kindler *et al.*, 1985; Metcalf, 1988). The administration of human IL-3 to primates and humans results in effects that are broadly similar to those seen in mice (Donahue *et al.*, 1988; Mayer *et al.*, 1989). In cynomolgus monkeys IL-3 induced extramedullary hematopoiesis at sites of subcutaneous injection (Khan *et al.*, 1996). IL-3 may have particular utility in stimulating platelet production (Ganser *et al.*, 1990a). IL-3 potentiated the mobilization of stem cells into peripheral blood induced by G-CSF (Geissler *et al.*, 1996; Huhn *et al.*, 1996).

ROLE OF IL-3 IN STEADY-STATE LYMPHOHEMOPOIESIS

There is compelling evidence that IL-3 serves as a link between the immune system (which senses intrusion of foreign substances into the body) and the hemopoietic system which generates the phagocytic and granulocytic cells that mediate defense and repair.

There is no evidence, however, that IL-3 is involved in steady-state production of blood cells, despite its potent ability to stimulate almost all phases of hemopoiesis.

The absence of IL-3 in normal serum and the evidence that links its production, whether by T lymphocytes or mast cells to immunologic activation, argue against a role for IL-3 in mediating steady-state hemopoiesis in unperturbed animals. Moreover, the production of a range of hemopoietic cells, including progenitor cells and stem cells capable of generating myeloid, erythroid and lymphoid cells, can occur *in vitro* in long-term bone marrow culture systems in which IL-3 bioactivity is undetectable (Eliason *et al*., 1988). These cultures can support the survival of IL-3-dependent cells unresponsive to other growth factors like GM-CSF, CSF-1 or G-CSF despite the absence of IL-3 (Schrader *et al*., 1984), suggesting the presence of alternative mechanisms.

One mechanism that permits the survival and limited growth of IL-3-dependent mast cells was clarified by characterization of the protein products of the *W* and *Sl* loci as, respectively, a growth factor receptor and its ligand. Both *W* and *Sl* mutant mice exhibited a macrocytic anemia and a deficiency of mast cells that were caused, in the case of *W* mice, by a defect expressed in cells including the hemopoietic stem cells and its derivatives, and in the case of the *Sl* mice by a defect in the microenvironment expressed in tissues including the bone marrow, spleen and skin. Fujita and colleagues (1988a,b) showed that IL-3-dependent mast cells from normal mice but not from *W* mutant mice could survive and proliferate in the absence of IL-3, provided they were allowed to contact fibroblasts from normal mice. Fibroblasts from *Sl* mice could not maintain mast cell survival. The demonstration that the *W* mutations involved the tyrosine kinase receptor encoded by the c-kit gene (Chabot *et al*., 1988; Geissler *et al*., 1988), and that IL-3-dependent mast cells retained expression of this receptor, made mast cells an obvious substrate for assays designed to detect the ligand for this receptor. The kit ligand, or steel locus factor (SLF), was shown to be a homodimer structurally unrelated to IL-3 and encoded, as expected, by the *Sl* locus (Anderson *et al*., 1990; Copeland *et al*., 1990; Huang *et al*., 1990; D.E. Williams *et al*., 1990; Zsebo *et al*., 1990a,b).

Mice that lack functional IL-3 genes show no obvious defects in hemopoiesis (Nishinakamura *et al*., 1996). It is conceivable that IL-3 plays a redundant role in these processes which will be evident only when the IL-3 *null* allele has been bred on to genetic backgrounds where there are defects in other regulators of hemopoiesis.

LINKS BETWEEN STRESS AND STEADY-STATE HEMOPOIESIS

Whether or not small amounts of IL-3 play a subtle role in steady-state hemopoiesis, it is clear that IL-3 is one mediator of the response of the hemopoietic system to stress. The local release of IL-3 from activated T lymphocytes and, in severe immunologic stress, its release into the serum, results in accelerated cycling of stem and progenitor cells and large increases in the production of differentiated cells of multiple lineages (Schrader *et al*., 1988).

IL-3 and the other cytokines released during stress not only increase the blood cell production but also modulate the types of cells produced. The stress response thus involves not only acceleration of the normal mechanisms of hemopoiesis but also overriding of some of the normal processes that regulate the proportions of the different cell types that are produced. Usually this is seen as the result of the positive effects of a lineage-specific factor such as G-CSF or IL-5 that promotes the survival, growth and differentiation of committed progenitor cells expressing the respective receptors. However, another component of this overriding process could be the disengagement of normal regulatory mechanisms.

There is some evidence that IL-3 may be involved in such a disengagement of steady-state mechanisms. Thus exposure to high levels of IL-3 leads to the downregulation of c-kit mRNA and protein in both mast cells and cell lines corresponding to hemopoietic progenitor cells (Welham and Schrader, 1991). GM-CSF and erythropoietin have similar downregulatory effects on expression of c-kit. The downregulation of c-kit by high levels of IL-3 (as well as GM-CSF and erythropoietin) may be part of a mechanism that overrides steady-state regulatory processes and facilitates control of the rate and cellular composition of blood cell production by cytokines released during stress.

One aspect of this overriding process may be simply the facilitation of the exit of stem and early progenitor cells from the bone marrow. The kit ligand exists in a cell-bound form as a transmembrane protein (Anderson *et al.*, 1990; Flanagan and Leder, 1990; Martin *et al.*, 1990) and thus may function as one of the adhesion proteins that retains stem and progenitor cells in the bone marrow microenvironment. Downregulation of c-kit by IL-3 may therefore facilitate the release of stem cells and early committed progenitors from the bone marrow microenvironment. The administration of IL-3 has been shown to result in an increase of stem cells in the circulation (Geissler *et al.*, 1996; Huhn *et al.*, 1996). Seeding of stem and progenitor cells into the blood to sites where cytokine release is occurring would allow local generation of the appropriate effector cells at sites of inflammation.

RECEPTOR FOR IL-3

IL-3, like most other four-helix-bundle cytokines binds to a heterodimeric receptor, both chains of which belong to a large family of hemopoietin or cytokine receptors that have a common evolutionary origin and share distinctive structure features (Bazan, 1990). The family includes receptors for IL-2, IL-4, IL-5, IL-6, IL-7, TSLP, IL-9, IL-11, IL-12, IL-13, IL-15, GM-CSF, G-CSF, ciliary neutrophilic factor, leukemia inhibitory factor, oncostatin M, thrombopoietin, erythropoietin, leptin, cardiotropin, prolactin and growth hormone. In some cases the receptors are homodimers (e.g. those for growth hormone, prolactin, erythropoietin and G-CSF), but most are heterodimers made up of two subunits, each of which is a member of the superfamily. As discussed below, in some cases there is evidence that the functional receptors are more complex oligomers of these basic subunits.

In the human, the IL-3 receptor is made up of two subunits, an α and a β chain, each of which is a member of the hemopoietin receptor superfamily. The smaller, 70-kDa, α chain (IL-3Rα) is a transmembrane protein that binds IL-3 with low affinity. It is homologous with two other α chains, which bind GM-CSF and IL-5 respectively. The larger, 125-kDa, β chain is also a component of the human receptors for IL-5 and GM-CSF. While showing no direct affinity for either human IL-3, IL-5 or GM-CSF, this shared β chain, termed β common (βc), can interact with any of the three distinct complexes of α chains and their respective ligands to generate three specific high-affinity ligand–receptor complexes (Gearing *et al.*, 1989; Kitamura *et al.*, 1991; Tavernier *et al.*, 1991; Kitamura and Miyajima, 1992).

In the mouse the situation is a little more complex. A duplication of the gene encoding βc has occurred. One gene (β common (βc)) encodes a β chain that is functionally equivalent to that in the human, binding neither IL-3, IL-5 nor GM-CSF, but interacting with specific α chains in the presence of the respective ligands to form three specific high-affinity receptors (Kitamura and Miyajima, 1992). The second gene (AIC-2A, now termed β_{IL-3}), in contrast, has a low affinity for IL-3 (Itoh *et al.*, 1990) and interacts only with the IL-3-specific α chain. Thus in the mouse there are two types of high-affinity IL-3 receptor, one corresponding to the human IL-3 receptor and made up of IL-3Rα and βc and the other, consisting of IL-3Rα and β_{IL-3}, peculiar to the mouse. The functional significance of this additional IL-3 receptor in mice is unknown. The intracellular portions of the β_{IL-3} and βc proteins are similar and no differences in the signals they transmit have been detected. Both chains of the IL-3 receptor interact closely with IL-3 and can be detected by cross-linking studies with radiolabelled IL-3 as 70- and 120-K IL-3 binding species (Duronio *et al.*, 1992). In the mouse there is a naturally occurring polymorphism of the IL-3Rα gene which results in abnormally low expression of IL-3Rα and defective responsiveness to IL-3 (Hara *et al.*, 1995; Leslie *et al.*, 1996).

There is also a polymorphism in β_{IL-3} in the mouse, but this does not result in any defects in IL-3 function (Leslie *et al.*, 1996). Mice that lack a functional β_{IL-3} gene exhibit no phenotype, whereas those lacking a functional βc gene exhibit pulmonary alveolar proteinosis (Nishinakamura *et al.*, 1995). This appears to reflect their defective response to GM-CSF, as the same phenotype is seen in mice lacking functional GM-CSF genes. Interestingly, some patients with pulmonary alveolar proteinosis have been shown to have defective expression of βc (Nishinakamura *et al.*, 1995).

IL-3-MEDIATED SIGNAL TRANSDUCTION

Information on the amino sequence of the α and β chains of the IL-3 receptor has provided no clues to the mechanism by which ligand binding activates intracellular signaling. The cytoplasmic domain of the α chain is relatively short and has no homology with known enzymes. However, the cytoplasmic domain of the α chain is essential for the generation of signals required for survival and proliferation (P.C. Orban and J.W. Schrader, unpublished observations). Like other members of this receptor superfamily, the cytoplasmic domain of IL-3Rα exhibits a region of homology termed Box1 which contains a Pro–X–Pro motif. Mutation of these prolines in the IL-5Rα abolishes IL-5-mediated signaling (Takaki *et al.*, 1994). This region may be important for binding the tyrosine kinases that are activated by IL-3 signaling. As discussed below, the JAK2 kinase is one candidate and members of the Src family of kinases, such as Lyn, are others.

The β chain of the IL-3 receptor has a large cytoplasmic domain but, like other receptor subunits in the hemopoietic receptor superfamily, lacks any evidence of catalytic domains. Like the IL-3Rα, βc has a Box1 region. In addition there are multiple tyrosine residues, some of which have been shown to be phosphorylated following binding of IL-3 and to be critical for recruitment of cytoplasmic signaling molecules to the receptor complex, as discussed below.

It is unclear how ligand binding activates intracellular events. Interaction of the complex of IL-3 and the α chain with the β chain results in a stable complex. However, by analogy with GM-CSF signaling, high-affinity binding of IL-3 is not likely to be essential for activation of the receptor. Thus, mutants of murine GM-CSF that bind with only low affinity can nevertheless stimulate the growth of murine factor-dependent cells (Shanafelt and Kastelein, 1992). These experiments suggest that the key signaling event results from interaction of the α and β chains and does not depend directly on high-affinity binding of the ligand and its interaction with the β chain.

One model for receptor activation is that heterodimerization of the cytoplasmic domains of IL-3Rα and βc generates or stabilizes sites for binding of cytoplasmic tyrosine kinases. Another, based on an analogy with the tetrameric structure of the active IL-6 receptor, is that, once formed, the IL-3R$\alpha\beta$c dimer would associate with another IL-3R$\alpha\beta$c dimer (or alternatively a free βc chain) to generate a βc–βc dimer (Ward *et al.*, 1994; Paonessa *et al.*, 1995). Thus, in the case of IL-6 receptor, the key signaling event is dimerization of gp130 (corresponding to βc), and the cytoplasmic domain of IL-6Rα can be deleted without affecting signaling.

In support of a role for dimerization of βc, there is some evidence suggesting that βc already exists as a homodimer in the absence of ligand (Muto *et al.*, 1996). Also supporting the notion that IL-3 signaling involves preformed or induced dimers of βc, is evidence that the generation of dimers of the cytoplasmic domain of $\beta_{\text{IL-3}}$ can induce mitogenic signals. One type of experiment used chimeric receptors in which the cytoplasmic domain of $\beta_{\text{IL-3}}$ was fused with the extracellular domain of the erythropoietin or IL-4 receptors (Satamaki *et al.*, 1993). The addition of erythropoietin or IL-4 to cells expressing these chimeras resulted in mitogenesis. However, these experiments did not exclude the possibility that mitogenesis depended on recruitment of additional receptor subunits by the complexes of the ligands and the extracellular domains of the erythropoietin or IL-4 receptors and their ligands. Indeed, our experiments with a different design of chimeric receptor, in which the extracellular regions of the IL-3Rα or $\beta_{\text{IL-3}}$ were replaced with domains that were not derived from hemopoietin receptors and so should not recruit subunits of the hemopoietin receptor family, have given different results.

Thus expression in an IL-3-dependent cell line of a chimeric molecule made up of the extracellular domain of CD8 (CD8_{ED}) and the cytoplasmic domain of $\beta_{\text{IL-3}}$ ($\beta_{\text{IL-3CD}}$) failed to result in factor-independent growth, despite the fact that CD8_{ED} forms disulfide-linked dimers (P.C. Orban, M.K. Levings and J.W. Schrader, unpublished data). Likewise, expression of a similar chimera of CD8 and the cytoplasmic domain of IL-3Rα did not confer factor-independent growth. However, coexpression of the CD8_{ED}–$\beta_{\text{IL-3CD}}$ and the CD8_{ED}–IL-3Rα_{CD} chimeras resulted in factor-independent growth. This result suggested that heterodimerization of IL-3Rα_{CD} and $\beta_{\text{IL-3CD}}$ was both necessary and sufficient for the initiation of signal transduction. Similar results were obtained by expressing chimeras of the extracellular domain of CD16 and $\beta_{\text{IL-3CD}}$ or IL-3Rα_{CD}. In this instance a monoclonal anti-CD16 antibody was used to induce cross-linking of the chimeric receptors. Once again, only if chimeras of CD16_{ED}–$\beta_{\text{IL-3CD}}$ and CD16_{ED}–IL3Rα_{CD} were coexpressed and subsequently cross-linked with anti-CD16 antibodies, did growth occur.

Patel *et al.* (1996) did observe factor-independent proliferation when leucine zippers were used as extracellular domains to dimerize cytoplasmic domains of βc. In their studies even α_{CD}–α_{CD} dimers of the GM-CSF-Rα had weak but detectable activity in promoting growth and viability. Other evidence that the generation of dimers of $\beta\text{c}_{\text{CD}}$ are sufficient to initiate proliferation has come from experiments involving the expression of chimeras of the intracellular domains of GM-CSF-Rα or IL-5Rα and the cytoplasmic domains of βc in cells that expressed wild-type βc. In both types of experiment, the respective ligands GM-CSF (Muto *et al.*, 1995) or IL-5 (Takaki *et al.*, 1994) induced mitogenesis.

Experiments indicating that expression of mutant βc predisposed to dimerize leads to factor-independent growth also suggest that dimerization of βc can result in mitogenesis (D'Andrea *et al.*, 1994; Hannemann *et al.*, 1995; Jenkins *et al.*, 1995; Shikama *et al.*, 1996). It is noteworthy, however, that β_{CD}–β_{CD} dimers appear to be active in some cell lines but not in others (Jenkins *et al.*, 1995; Shikama *et al.*, 1996).

The simplest explanation for these seemingly contradictory results is to assume that $\alpha_{CD}\alpha_{CD}$, $\beta_{CD}\beta_{CD}$ and $\alpha_{CD}\beta_{CD}$ dimers can all generate signals that suppress apoptosis and promote mitogenesis but differ in their efficacy, the hierarchy being $\alpha_{CD}\alpha_{CD} < \beta_{CD}\beta_{CD} < \alpha_{CD}\beta_{CD}$, with $\alpha_{CD}\alpha_{CD}$ being barely active and $\alpha_{CD}\beta_{CD}$ being very effective. The events triggered by these dimers could differ only quantitatively, but are more likely to differ qualitatively in the number of signal transduction paths activated. One simple model is that both α_{CD} and β_{CD} have affinity for JAK2 kinase. Thus the formation of dimers of α_{CD} and β_{CD} could lead to the juxtaposition of two JAK2 kinase molecules and their transphosphorylation and activation. The greater efficiency of $\beta_{CD}\beta_{CD}$ dimers as compared with $\alpha_{CD}\alpha_{CD}$ dimers could reflect either a greater affinity of JAK2 kinases for β_{CD}, and thus greater efficiency in bringing together two JAK2 kinase molecules, or alternatively a greater capacity of β_{CD} to mediate downstream events, for example by providing multiple tyrosines which when phosphorylated generate docking sites for signal transduction molecules such as Shc, SHP-2, etc.

In contrast to $\alpha_{CD}\alpha_{CD}$ or $\beta_{CD}\beta_{CD}$ dimers, $\alpha_{CD}\beta_{CD}$ dimers function in all cell types tested, consistent with the $\alpha_{CD}\beta_{CD}$ being the most efficient mechanism for initiation of a mitogenic signal. In terms of the above model we would postulate that $\alpha_{CD}\beta_{CD}$ provides the most stable site for docking of two JAK kinase molecules as well as the optimal set of sites for the docking of proteins involved in the activated receptor complex. However, the general arguments are the same if it is postulated that the key event is juxtaposition of other molecules, for example a Src family kinase plus JAK2 kinase.

On the whole it seems likely that although β–β dimers can induce mitogenic signals, α–β dimerization is the physiologic signal for assembly of active receptor complex and is probably sufficient. In support of the notion that ligand binding to receptors of this family induces formation of an only simple α–β dimer, and not more complex structures that include β–β dimers, are the observations of Behrmann *et al.* (1997). These workers coexpressed chimeras of the IL-5Rα_{ED} and βc_{ED} with the cytoplasmic domain of gp130 and showed that IL-5 induced the activation of STAT3 that was expected to follow dimerization of gp130$_{CD}$. They then coexpressed the βc_{ED}–gp130$_{CD}$ chimera and a chimera of IL-5Rα_{ED} and a truncated gp130$_{CD}$ that lacked amino acid residues required for STAT3 activation. IL-5 failed to induce activation of STAT3, as would have been expected had higher-order complexes such as $\alpha\beta\beta$ or $\alpha\beta\beta\alpha$ been generated in the presence of IL-5.

It is clear that one of the earliest detectable changes after binding of IL-3 to its receptor – occurring within seconds – is the phosphorylation of a set of proteins on tyrosine residues (Ferris *et al.*, 1988; Morla *et al.*, 1988). The clue to the likely nature of the tyrosine kinases involved comes from work on interferon-mediated signal transduction. The receptors of the hemopoietin receptor superfamily are close relatives of the receptors for the interferons. The demonstration that signal transduction via the receptors for IFN-α/β or IFN-γ was critically dependent on a new family of tyrosine kinases, the Janus or JAK kinases (Schindler and Darnell, 1995) was rapidly followed by evidence that kinases of this family were also involved in signaling by the hemopoietin receptors. In particular, IL-3 was shown to induce tyrosine phosphorylation of one member of this family, JAK2 kinase (Silvennoinen *et al.*, 1993). In the case of the interferon receptors and some hemopoietin receptors, including the erythropoietin and growth hormone receptors (Argetsinger *et al.*, 1993; Witthuhn *et al.*, 1993), there is evidence for the pre-existing association of JAK family kinases with receptor subunits. In one *in vitro* study JAK2 was shown to bind βc (but not the GM-CSF-Rα) (Quelle *et al.*, 1994), and in another the cytoplasmic domain of βc was shown to interact with Src family kinases and, more weakly, with JAK2 kinase (Rao and Mufson, 1995). Ligand-induced dimerization of the receptors is postulated to bring the receptor-associated JAK kinase into close proximity, leading to cross phosphorylation and association. The precise mechanism of activation of JAK2 kinase by IL-3 is unknown. Other tyrosine kinases may also be involved. IL-3 stimulation results in modest activation of the Src family kinase member Lyn (O'Connor *et al.*, 1992; Torigoe *et al.*, 1992).

Comparison of the tyrosine phosphorylation events stimulated in mast cells by IL-3 with those stimulated by SLF allowed identification of a set of IL-3-specific tyrosine-phosphorylated events (Welham and Schrader, 1992). One of these was tyrosine phosphorylation of the β chains of the IL-3 receptor (Isfort *et al.*, 1988; Duronio *et al.*, 1992). Over a period of 10 min the β chain of the IL-3 receptor increases in apparent molecular weight, shifting from M_r 125 000 in unstimulated cells to M_r 135–150 000. Much of this increase in apparent molecular weight appears to be due to concomitant serine–threonine phosphorylation (Duronio *et al.*, 1992).

There were other tyrosine phosphorylation events that were relatively specific for IL-3 stimulation, in that they were not observed with stimulation of cells with other hemopoietins like SLF, CSF-1 or IL-2. However, in all cases these have involved proteins also involved in signaling by other stimuli. For example, one protein that is prominently phosphorylated on tyrosine in response to IL-3, IL-5 and GM-CSF, but not IL-2 or SLF or CSF-1 has an M_r of 68 kDa (Duronio *et al.*, 1992; Welham and Schrader, 1992). This protein was shown to be the tyrosine phosphatase SHPTP-2, or Syp, now known as SHP-2 (Welham *et al.*, 1994a). Stimulation with IL-3 resulted in an increase in the enzymatic activity of SHP-2 and its association, via its SH-2 domains, with βc, interacting with tyrosines 577 as well as other sites (Itoh *et al.*, 1996; Bone *et al.*, 1997). SHP-2 also became associated with the p85 subunit of phosphatidylinositol-3 (PI-3) kinase and Grb-2 (Welham *et al.*, 1994a). SHP-2 may thus have multiple functions in the receptor complex, by recruiting Grb-2 contributing to the activation of Ras, and by recruiting p85, bringing PI-3 kinase to the membrane and closer to its substrates. Through its tyrosine phosphatase activity, SHP-2 may also downregulate some signaling events. However, SHP-2, while a prominent component of IL-3-stimulated signaling events, is by no means specific for IL-3 signaling, and is involved in signal transduction paths triggered by many growth factors and other stimuli.

Indeed, there is no evidence for any IL-3-specific component of the IL-3 signal transduction process, downstream of the IL-3 receptor itself. Rather IL-3 signaling conforms to the general model, where receptors for extracellular stimuli engage subsets of common intracellular signaling pathways shared by many other receptors. In that

IL-3 exerts effects common to all growth factors, such as stimulation of proliferation or enhanced survival, it would be expected that the intracellular signals triggered by IL-3 would include many that are shared by other growth factors.

One such common event is IL-3-mediated activation of Ras (Sato *et al.*, 1991; Duronio *et al.*, 1992). This is dependent on tyrosine kinase activity (Duronio *et al.*, 1992). The prominent 55-kDa protein that is tyrosine phosphorylated in response to IL-3 (as well as GM-CSF, IL-5, IL-2, SLF and CSF-1) is Shc (Welham *et al.*, 1994b). IL-3-stimulated tyrosine phosphorylation of Shc results in its association with Grb-2, a small adaptor protein that is constitutively associated with mSos-1. The latter is a guanine nucleotide exchange factor and, when it is brought to the membrane in association with Grb-2 with phosphorylated tyrosine residues in active receptors, it stimulates GDP–GTP exchange on Ras and the generation of active GTP-bound Ras (Buday and Downard, 1993). The phosphotyrosine-binding (PTB) domain of Shc binds to the βc when tyrosine phosphorylated, the critical residue being tyrosine 577 (Durstin *et al.*, 1996; Pratt *et al.*, 1996), thus providing a mechanism to recruit Grb-2-m Sos-1 complexes to the receptor. The link between tyrosine phosphorylation of Shc and activation of Ras is strengthened by the observation that IL-4, which fails to activate Ras (Sato *et al.*, 1991; Duronio *et al.*, 1992), also fails to stimulate tyrosine phosphorylation of Shc (Welham *et al.*, 1994b).

Another event common to stimulation by IL-3 and growth factors in general is activation of Erk/MAP kinases. Activation of Erk-1 and Erk-2 involves tyrosine phosphorylation, and the two proteins of M_r 42 and 44 kDa that are phosphorylated on tyrosine in response to the IL-3 were shown to correspond to Erk-2 and Erk-1 MAP kinases (Welham *et al.*, 1992). IL-3 shared this action with GM-CSF, IL-5 and IL-2, as well as SLF and CSF-1 (Welham *et al.*, 1992), although once again IL-4 and IL-13 were notable exceptions (Welham *et al.*, 1994a, 1995). The failure of IL-4 to activate Erk-1/Erk-2 MAP kinases correlates with its inability to activate Ras. The Erk-1/Erk-2 kinases can be activated by a Ras-dependent path that involves Ras-mediated activation of the Raf-1 kinase. Raf-1 phosphorylates and activates MEK-1, a kinase that in turn phosphorylates and activates Erk-1/Erk-2. A minor part of the activation of Erk-1/Erk-2 kinases, however, was inhibited by a specific inhibitor of protein kinase C (Welham *et al.*, 1992). This enzyme can also activate Raf-1 and thus lead to activation of the Erk-1/Erk-2 kinases.

Activation of Erk-1/Erk-2 kinases seems to be an important component of the mitogenic signals stimulated by IL-3. Expression in an IL-3-dependent cell line of a dominant inhibitory mutant of MEK-1 that blocked activation of Erk-MAP kinases resulted in decreased sensitivity to the stimulatory effects of IL-3 on proliferation and the suppression of apoptosis and in inhibition of IL-3-stimulated entry into cell cycle (Perkins *et al.*, 1996).

IL-3 also activates the two other major families of MAP kinases, the stress-activated protein kinases (SAPK) or Jun-*N*-terminal kinases (JNK), and the p38 MAP kinase or hyperosmolarity glycerol response-1 gene (HOG-1) kinases. Both SAPK/JNK and p38 MAP kinase families were initially identified as stress-activated kinases that were activated in response to stresses such as heat, ultraviolet irradiation or hyperosmolarity. It is clear, however, that they are also involved in many physiologic responses to growth factors and other stimuli. Thus p38 MAP kinase is activated by IL-3, both in primary bone marrow-derived mast cells and cell lines (Foltz *et al.*, 1997). The enzyme is also activated in response to other growth factors such as SLF and GM-CSF (Foltz *et al.*, 1997), and to immunologic stimuli like ligation of the antigen receptors on T and B lymphocytes, of the Fc receptors on myeloid cells, or of CD40, or Fas on B and T cells respectively (Salmon *et al.*, 1997). Likewise JNK kinases are activated by stimulation with IL-3 as well as GM-CSF and SLF (Foltz and Schrader, 1997). Interestingly, as was the case with the Erk kinases, IL-4 was an exception and failed to induce activation of either p38 MAP or JNK kinases (Foltz and Schrader, 1997; Foltz *et al.*, 1997). This may reflect the involvement of the Ras pathway in activating p38 MAP or JNK kinases in response to hemopoietins. There is evidence that the activation of JNK kinases by IL-3 is in part mediated via activation of Ras (Terada *et al.*, 1997).

The function of p38 MAP and JNK kinases in IL-3 signaling is unclear. JNK kinases activate c-Jun and Activating Transcription Factor 2 (ATF-2), whereas p38 MAP kinases activate other transcriptional activators such as ATF-2, cAMP Responsive Element Binding protein (CREB) and Elk-1 (Derijard *et al.*, 1994; Kyriakis *et al.*, 1994; Gupta *et al.*, 1995; Raingeaud *et al.*, 1995, 1996; Tan *et al.*, 1996; Iordanov *et al.*, 1997). IL-3-induced activation of MAP-kinase activating protein-2 (MAP-KAP-2) kinase and the subsequent phosphorylation of the small heat shock protein hsp25/27 is completely dependent on activation of p38 MAP kinase (Foltz *et al.*, 1997). Both p38 MAP and JNK kinases have been postulated to play a role in apoptosis, but IL-3, which activates these enzymes, inhibits apoptosis. Moreover, inhibition of p38 MAP kinase activity failed to block apoptosis induced by withdrawal of IL-3 or by ligation of antigen receptors on T and B lymphocytes (I.N. Foltz and J.W. Schrader, unpublished results; Salmon *et al.*, 1997).

Thus, IL-3 activates, via a tyrosine kinase-dependent mechanism, one common path along which lie Shc, Grb-2, mSos-1, Ras, Raf-1 and Erk-1/Erk-2 MAP kinases. Erk-1/Erk-2 MAP kinases are known to activate the $p90^{rsk}$ S6 kinase which phosphorylates Serum Response Factor (SRF), a protein that regulates transcription of the c-*fos* gene. IL-3-mediated activation of the JNK and p38 MAP kinases, which may at least in part depend on Ras activation, may also contribute to regulation of c-*fos* expression and the activity of transcriptional activation complexes like activating protein-1 (AP-1).

Work on interferons not only demonstrated the role of JAK kinases but also identified a new class of transcription factors termed Signal Transducers and Activators of Transcription (STATs). This family of proteins is characterized by an SH-2 domain that recognizes specific phosphotyrosine residues in activated receptors. This SH-2-mediated recognition provides specificity so that different STATs are recruited from the cytoplasm to different receptors. Following binding to the active receptor, STATs are themselves phosphorylated on a tyrosine residue; this phosphotyrosine is within a recognition site for the SH-2 domain of that same STAT. This favors the formation of SH-2-stabilized STAT homodimers, although in certain situations heterodimers, STAT1 and STAT3 are also formed. These STAT dimers are rapidly translocated from the cytoplasm to the nucleus where they bind to characteristic DNA motifs. IL-3 induces activation of STAT5 (Azam *et al.*, 1995; Mui *et al.*, 1995), as do GM-CSF, IL-5, prolactin, erythropoietin and IL-2. There are two closely related STAT5 genes, *STAT5a* and *b* (Mui *et al.*, 1995). Expression of a dominant negative form of STAT5 suppressed induction of *osm*, *cis* and *pim-1* and partially inhibited IL-3-stimulated growth and induction of c-*fos* (Mui *et al.*, 1996).

Another common pathway activated by IL-3 and many other growth factors involves activation of the enzyme PI-3′ kinase (Gold *et al.*, 1994). The function of this enzyme is unknown although it may be involved in suppression of apoptosis (Kauffmann-Zeh *et al.*, 1997) and activation of Rac-1 (Hawkins *et al.*, 1995). In that it is activated by IL-4 (Wang *et al.*, 1992; Gold *et al.*, 1994) it would seem to lie on a common path, distinct from that including $p21^{ras}$ and MAP kinases.

There is evidence that IL-3 results in translocation of protein kinase C from the cytoplasm to the cell membrane (Farrar *et al.*, 1985; Whetton *et al.*, 1988a; Pelech *et al.*, 1990). No increased turnover of phosphatidylinositol has been observed (Whetton *et al.*, 1988b), although there is some evidence that increased turnover of phosphatidylcholine occurs (Duronio *et al.*, 1989), and this may account for the generation of diacylglycerol and activation of protein kinase C.

IL-3 also resembles other growth factors in that it stimulates increases in levels of c-myc RNA (Chang *et al.*, 1991). IL-3-induced increases in c-jun mRNA levels have been reported to be independent of tyrosine kinase activity and to involve protein kinase C, although results of experiments using kinase inhibitors must be interpreted with caution (Mufson *et al.*, 1992). IL-3 stimulates phosphorylation on tyrosine of the adaptor protein Cb1 and its association with Grb-2 and the enzymes Fyn and PI-3 kinase (Anderson *et al.*, 1997). The enzyme phosphatidylinositol-3,4,5-triphosphate 5-phosphatase is constitutively associated with the IL-3 receptor, although its activity does not change following IL-3 stimulation (Liu *et al.*, 1996). Stimulation with IL-3 (and GM-CSF and IL-5) induced expression of a gene *DUB-1* that encodes a deubiquinating enzyme (Zhu *et al.*, 1996).

The path or paths through which IL-3 suppresses apoptosis are not clear. There is some evidence that IL-3 increases expression of bcl-2 and bcl-x_L through activation of the Ras pathway (Kinoshita *et al.*, 1995). There is also evidence that IL-3 can upregulate bcl-2 expression through a protein kinase C-dependent mechanism (Rinaudo *et al.*, 1995). Finally, IL-3 may suppress apoptosis through its activation of PI-3 kinase (Kauffmann-Zeh *et al.*, 1997).

In summary, the binding of IL-3 induces the generation of dimers or possibly higher-order oligomers of the cytoplasmic domains of the IL-3Rα and βc that result in activation of tyrosine kinases including JAK2 kinases and Src kinases. Some of the subsequent events, such as activation of Ras, clearly depend on tyrosine kinase activity. In other cases, such as activation of protein kinase C, it has not been formally shown that activation of an upstream tyrosine kinase is involved.

CLINICAL SIGNIFICANCE OF IL-3

The ability of IL-3 to stimulate early members of hemopoietic differentiation pathways suggests that it may have specific clinical uses. Promising results in accelerating the recovery of bone marrow following bone marrow transplantation or damage by cytotoxic drugs have been obtained with G-CSF and GM-CSF, which appear to be effective in reducing the period of neutropenia (Morstyn *et al.*, 1990). However, there are indications that, unlike G-CSF and GM-CSF, IL-3 may stimulate an increase in platelet levels (Ganser *et al.*, 1990a). Animal experiments suggest that sequential administration of IL-3 and then G-CSF or GM-CSF may provide optimal stimulation of myelopoiesis (Donahue *et al.*, 1988; Mayer *et al.*, 1989). Similar results have been obtained in clinical trials in humans (Lemoli *et al.*, 1996). Other potential uses for IL-3 are in the treatment of conditions such as aplastic anemia (Ganser *et al.*, 1990b) or other anemias (Halperin *et al.*, 1989; Dunbar *et al.*, 1992). In mice the administration of IL-3 (but not erythropoietin) prevents death from acute anemia (Shibata *et al.*, 1990). Because of the likelihood that optimal protocols will involve the use of multiple cytokines, it will be some time before the ultimate clinical potential of IL-3 and combinations of other cytokines is clear.

IL-3 may be useful in managing certain infections. There is some evidence that IL-3 may have a favorable influence on herpes simplex infection in mice (Chan *et al.*, 1991). Modification of lung carcinoma by transfection of IL-3 DNA so that it secreted IL-3 resulted in decreased tumorigenicity and an increase in tumor-infiltrating cytotoxic T lymphocytes (Pulaski *et al.*, 1996). This correlated with an increase in the tumor of macrophage-like cells that presented antigens to T lymphocytes.

Administration of IL-3 to mice infected with *Trichinella spiralis* accelerated the expulsion of worms (Korenaga *et al.*, 1996a) and the development of IgE responses (Korenaga *et al.*, 1996b). These data are consistent with *in vitro* experiments suggesting that IL-3 induces the release of IL-4 and IL-3 from non-T, non-B cells, and thus favors the development of a T_{H2} response (Aoki *et al.*, 1996). IL-3 also induces expression of the T_{H2} cytokines IL-10 and IL-13 by mast cells (Marietta *et al.*, 1996).

The role of IL-3 in promoting T_{H2} responses and its stimulatory effects on mast cells and eosinophils suggest that IL-3 antagonists might prove useful in the management of diseases such as bronchial asthma and allergies. Experimental models in which antibodies that neutralize IL-3 have been used as models of IL-3 antagonists have given encouraging results. Anti-IL-3 antibodies in combination with anti-GM-CSF antibodies block the development of cerebral malaria in mice (Grau *et al.*, 1988). Anti-IL-3 antibodies in combination with anti-IL-4 antibodies block the mastocytosis seen in the parasitized mice (Madden *et al.*, 1991). The administration of IL-3 aggravates leishmaniasis in mice (Feng *et al.*, 1988); *in vitro* anti-IL-3 antibodies have been shown to synergize with anti-IL-4 antibodies in unmasking a macrophage activating activity present in supernatants of cells from *Leishmania*-infected mice (Liew *et al.*, 1989). Analysis of the relative resistance or susceptibility to various diseases of mice in which IL-3 genes have been artificially deleted should yield helpful information on possible therapeutic use of IL-3 agonists and antagonists.

The fact that in humans the receptors for IL-3, IL-5 and GM-CSF share a common β chain, which plays a dominant role in signal transduction, raises the possibility of developing antagonists that will specifically block the activity of this trio of cytokines. Because IL-3, IL-5 and GM-CSF promote the production and survival of eosinophils, basophils and the mast cells that increase in number at sites of allergic reactions, drugs that block these actions could provide a new approach to the treatment of bronchial asthma or allergic diseases. Ideally such an antagonist would leave intact those signal transduction paths required for GM-CSF to block the development of pulmonary alveolar proteinosis.

In the mouse a number of myeloid leukemias have been described in which pathologic activation of an IL-3 gene was a key oncogenic event. The constitutive production of IL-3 results in autostimulation of growth of the myeloid cell (Schrader and Crapper, 1983; Ymer *et al.*, 1985; Leslie and Schrader, 1989). In some instances the growth of such autostimulatory leukemias may be blocked by anti-IL-3 antibodies (J.W. Schrader and

H.J. Ziltener, unpublished results). Autostimulatory production of IL-3, however, does not appear to be an important oncogenic mechanism in human myeloid leukemia, although the leukemic cells usually respond to IL-3 (Budel *et al.*, 1989; Park *et al.*, 1989b) and there has been a report of an acute lymphocytic leukemia in which a translocation joins the IL-3 and immunoglobulin heavy chain genes (Grimaldi and Meeker, 1989).

SUMMARY

IL-3 functions as a link between the T lymphocytes of the immune system, which sense invasion of the body by foreign materials, and the hemopoietic system, which generates the cellular elements that mediate defense and repair responses. IL-3 is also produced rapidly following stimulation of mast cells or eosinophils, for example by cross-linking Fc receptors, and then IL-3, by stimulating production of IL-6, IL-4, IL-13 and IL-10, may promote the development of T_{H2} responses. IL-3 stimulates the broadest range of targets within the hemopoietic system of any of the cytokines and, in addition, has the special ability to stimulate the growth of early stem cells and the progenitors of mast cells and megakaryocytes. IL-3 has no proven role in steady-state hemopoiesis. Evaluation of the clinical utility of IL-3 is still in progress; in the future, antagonists of IL-3 may provide new approaches to the management of allergic and inflammatory diseases.

ACKNOWLEDGEMENTS

Experimental work in the author's laboratory referred to was supported by grants from the MRC of Canada and the National Cancer Institute of Canada.

REFERENCES

Abbud-Filho, M., Dy, M., Lebel, B., Luffau, G. and Hamburger, J. (1983). *In vitro* and *in vivo* histamine-producing cell-stimulating factor (or IL 3) production during *Nippostrongylus brasiliensis* infection: coincidence with self-cure phenomenon. *Eur. J. Immunol*. **13**, 841–845.

Anderson, D.M., Lyman, S.D., Baird, A., Wignall, J.M., Eisenman, J., Rauch, C., March, C.J., Boswell, S., Gimpel, S.D., Cosman, D. and Williams, D.E. (1990). Molecular cloning of mast cell growth factor, a hematopoietin that is active in both membrane bound and soluble forms. *Cell* **63**, 235–243.

Anderson, S.M., Burton, E.A. and Koch, B.L. (1997). Phosphorylation of Cbl following stimulation with interleukin-3 and its association with Grb2, Fyn, and phosphatidyl inositol 3-kinase. *J. Biol. Chem*. **272**, 739–745.

Aoki, I., Tanaka, S., Ishii, N., Minami, M. and Klinman, D.M. (1996). Contribution of interleukin-3 to antigen-induced TH2 cytokine production. *Eur. J. Immunol*. **26**, 1388–1393.

Argetsinger, L.S., Campbell, G.S., Yang, X., Witthuhn, B.A., Silvennoinen, O., Ihle, J. and Carter-Su, C. (1993). Identification of JAK2 as a growth hormone receptor-associated tyrosine kinase. *Cell* **74**, 237–244.

Azam, M., Erdjument-Bromage, H., Kreider, B.L., Xia, M., Quelle, F., Basu, R., Saris, C., Tempst, P., Ihle, J.N. and Schindler, C. (1995). Interleukin-3 signals through multiple isoforms of Stat5. *EMBO J*. **14**, 1402–1411.

Ball, T.C., Hirayama, F. and Ogawa, M. (1995). Lymphohematopoietic progenitors of normal mice. *Blood* **85**, 3086.

Ball, T.C., Hirayama, F. and Ogawa, M. (1996). Modulation of early B lymphopoiesis by interleukin-3. *Exp. Hematol*. **24**, 1225–1231.

Bagley, C.J., Phillips, J., Cambareri, B., Vadas, M.A. and Lopez, A.F. (1996). A discontinuous eight-amino acid epitope in human interleukin-3 binds the alpha-chain of its receptor. *J. Biol. Chem*. **271**, 31 922–31 928.

Bazan, J.F. (1990). Haemopoietic receptors and helical cytokines. *Immunol. Today* **11**, 350–355.

Behrmann, I., Janzen, C., Gerhartz, C., Schmitz-Van de Leur, H., Hermanns, H., Heesel, B., Graeve, L., Horn, F., Tavernier, J. and Heinrich, P. (1997). A single STAT recruitment module in a chimeric cytokine receptor complex is sufficient for STAT activation. *J. Biol. Chem*. **272**, 5269–5274.

Berdel, W.E., Danhauser-Riedl, S., Steinhauser, G. and Winton, E.F. (1989). Various human hematopoietic growth factors (interleukin-3, GM-CSF, G-CSF) stimulate clonal growth of nonhematopoietic tumor cells. *Blood* **73**, 80–83.

Bone, H., Dechert, U., Jirik, F., Schrader, J.W. and Welham, M.J. (1997). SHP1 and SHP2 protein tyrosine phosphatases associate with the IL-3 receptor β subunit after IL-3 induced receptor tyrosine phosphorylation: identification of potential binding sites and substrates. *J. Biol. Chem*. **272**, 14470–14476.

Brugger, W., Mocklin, W., Heimfeld, S., Berenson, R.J., Mertelsmann, R. and Kanz, L. (1993). *Ex vivo* expansion of enriched peripheral blood $CD34^+$ progenitor cells by stem cell factor, interleukin-1β, IL-6, IL-3, interferon-γ and erythropoietin. *Blood* **81**, 2579.

Buday, L. and Downard, J. (1993). Epidermal growth factor regulates p21ras through the formation of a complex of receptor, Grb2 adapter protein, and Sos nucleotide exchange factor. *Cell* **73**, 611.

Budel, L.M., Touw, I.P., Delwel, R., Clark, S.C. and Lowenberg, B. (1989). Interleukin-3 and granulocyte–monocyte colony-stimulating factor receptors on human acute myelocytic leukemia cells and relationship to the proliferative response. *Blood* **74**, 565–571.

Burd, P.R., Rogers, H.W., Gordon, J.R., Martin, C.A., Jayaraman, S., Wilson, S.D., Dvorak, A.M. and Galli, S.J. (1989). Interleukin 3-dependent and -independent mast cells stimulated with IgE and antigen express multiple cytokines. *J. Exp. Med*. **170**, 245–257.

Caux, C., Vanbervliet, B., Massacrier, C., Durand, I. and Banchereau, J. (1996). Interleukin-3 cooperates with tumor necrosis factor alpha for the development of human dendritic/Langerhans cells from cord blood $CD34^+$ hematopoietic progenitor cells. *Blood* **87**, 2376–2385.

Chabot, B., Stephenson, D.A., Chapman, V.M., Besmer, P. and Bernstein, A. (1988). The proto-oncogene c-*kit* encoding a transmembrane tyrosine kinase receptor maps to the mouse *W* locus. *Nature* **335**, 88–89.

Chan, W.-L., Ziltener, H.J. and Liew, F.Y. (1991). Interleukin-3 protects mice from acute herpes simplex virus infection. *Immunology* **71**, 358–363.

Chang, Y., Spicer, D.B. and Sonenshein, G.E. (1991). Effects of IL-3 on promoter usage, attenuation and antisense transcription of the c-*myc* oncogene in the IL-3-dependent Ba/F3 early pre-B cell line. *Oncogene* **6**, 1979–1982.

Chen, B.D., Mueller, M. and Olencki, T. (1988). Interleukin-3 (IL-3) stimulates the clonal growth of pulmonary alveolar macrophage of the mouse: role of IL-3 in the regulation of macrophage production outside the bone marrow. *Blood* **72**, 685–690.

Clark-Lewis, I., Kent, S.B.H. and Schrader, J.W. (1984). Purification to apparent homogeneity of a factor stimulating the growth of multiple lineages of haemopoietic cells. *J. Biol. Chem*. **259**, 7488–7494.

Clark-Lewis, I., Aebersold, R., Ziltener, H., Schrader, J.W., Hood, L.E. and Kent, S.B. (1986). Automated chemical synthesis of a protein growth factor for hemopoietic cells, interleukin-3. *Science* **231**, 134–139.

Clark-Lewis, I., Lopez, A.F., Lo, L.B., Vadas, M., Schrader, J.W., Hood, L.E. and Kent, S.B.H. (1988). Structure–function studies of human granulocyte–macrophage colony-stimulating factor. Identification of residues required for activity. *J. Immunol*. **141**, 881–889.

Copeland, N.G., Gilbert, D.J., Cho, B.C., Donovan, P.J., Jenkins, N.A., Cosman, D., Anderson, D., Lyman, S.D. and Williams, D.E. (1990). Mast cell growth factor maps near the steel locus on mouse chromosome 10 and is deleted in a number of steel alleles. *Cell* **63**, 175–183.

Crapper, R.M. and Schrader, J.W. (1983). Frequency of mast-cell precursors in normal tissues determined by an *in vitro* assay: antigen induces parallel increases in the frequency of P-cell precursors and mast cells. *J. Immunol*. **131**, 923–928.

Crapper, R.M. and Schrader, J.W. (1986). Evidence for the *in vivo* production and release into the serum of a T-cell lymphokine, persisting-cell stimulating factor (PSF), during graft-*versus*-host reactions. *Immunology* **57**, 553–558.

Crapper, R.M., Clark-Lewis, I. and Schrader, J.W. (1984a). The *in vivo* functions and properties of persisting cell-stimulating factor. *Immunology* **53**, 33–42.

Crapper, R.M., Thomas, W.R. and Schrader, J.W. (1984b). *In vivo* transfer of persisting (P) cells; further evidence for their identity with T-dependent mast cells. *J. Immunol*. **133**, 2174–2179.

Crapper, R.M., Vairo, G., Hamilton, J., Clark-Lewis, I. and Schrader, J.W. (1985). P cell stimulating factor stimulates peritoneal and bone marrow derived macrophages. *Blood* **66**, 859–865.

D'Andrea, R.J., Barry, S.C., Moretti, P.A.B., Jones, K., Ellis, S., Vadas, M.A. and Goodall, G.J. (1994). Extracellular truncations of hβ_c, the common signalling subunit for interleukin-3 (IL-3), granulocyte–macrophage colony-stimulating factor (GM-CSF), and IL-5, lead to ligand-dependent activation. *Blood* **87**, 2641–2648.

Derijard, B., Hibi, M., Wu, I-H., Barrett, T., Su, B., Deng, T., Karin, M. and Davis, R.J. (1994). JNK1: a protein kinase stimulated by UV light and Ha-Ras that binds and phosphorylates the c-jun activation domain. *Cell* **76**, 1025–1037.

Donahue, R.E., Seehra, J., Metzger, M., Lefebvre, D., Rock, B., Carbone, S., Nathan, D.G., Garnick, M., Sehgal, P.K., Laston, D., La Vallie, E., McCoy, J., Schendel, P.F., Norton, C., Turner, K., Yang, Y.C. and Clark, S.C. (1988). Human IL-3 and GM-CSF act synergistically in stimulating hematopoiesis in primates. *Science* **241**, 1820–1823.

Dunbar, C.E., Smith, D.A., Kimball, J., Garrison, L., Nienhuis, A.W. and Young, N.S. (1992). Treatment of Diamond–Blackfan anaemia with haematopoietic growth factors, granulocyte–macrophage colony stimulating factor and interleukin 3: sustained remissions following IL-3. *Br. J. Haematol*. **79**, 316–321.

Duronio, V., Nip, L. and Pelech, S.L. (1989). Interleukin 3 stimulates phosphatidylcholine turnover in a mast/megakaryocyte cell line. *Biochem. Biophys. Res. Commun*. **164**, 804–808.

Duronio, V., Granleese, S.R., Clark-Lewis, I., Schrader, J.W. and Ziltener, H.J. (1991). Antibodies to interleukin 3 as probes for the interaction of interleukin 3 with its receptor. *Cytokine* **3**, 414–420.

Duronio, V., Clark-Lewis, I., Federsppiel, B., Wieler, J.S. and Schrader, J.W. (1992). Tyrosine phosphorylation of receptor beta subunits and common substrates in response to interleukin-3 and granulocyte–macrophage colony-stimulating factor. *J. Biol. Chem*. **267**, 21 856–21 863.

Durstin, M., Inhorn, R.C. and Griffin, J.D. (1996). Tyrosine phosphorylation of Shc is not required for proliferation or viability signalling by granulocyte–macrophage colony-stimulating factor in hematopoietic cell lines. *J. Immunol*. **157**, 534–540.

Eliason, J.F., Thorens, B., Kindler, V. and Vassalli, P. (1988). The roles of granulocyte–macrophage colony-stimulating factor and interleukin 3 in stromal cell-mediated hemopoiesis *in vivo*. *Exp. Hematol*. **16**, 307–312.

Farrar, W.L., Thomas, T.P. and Anderson, W.B. (1985). Altered cytosol/membrane enzyme redistribution on interleukin-3 activation of protein kinase C. *Nature* **315**, 235–237.

Feng, Y., Klein, B.K., Vu, L., Aykent, S. and McWherter, C.A. (1995). 1H, 13C and 15N NMR resonance alignments, secondary structure, and backbone topology of a variant of human interleukin-3. *Biochemistry* **34**, 6540.

Feng, Z.Y., Louis, J., Kindler, V., Pedrazzini, T., Eliason, J.F., Behin, R. and Vassalli, P. (1988). Aggravation of experimental cutaneous leishmaniasis in mice by administration of interleukin 3. *Eur. J. Immunol*. **18**, 1245–1251.

Ferris, D.K., Willet-Brown, J., Martensen, T. and Farrar, W.L. (1988). Interleukin 3 stimulation of tyrosine kinase activity in FDC-P1 cells. *Biochem. Biophys. Res. Commun*. **154**, 991–996.

Flanagan, J.G. and Leder, P. (1990). The kit ligand: a cell surface molecule altered in steel mutant fibroblasts. *Cell* **63**, 185–194.

Foltz, I.N. and Schrader, J.W. (1997). Activation of the stress-activated protein kinases by multiple hemopoietic growth factors with the exception of interleukin-4. *Blood* **89**, 3092–3096.

Foltz, I.N., Lee, J.C., Young, P.R. and Schrader, J.W. (1997). Hemopoietic growth factors with the exception of interleukin-4 activate the p38 mitogen-activated protein kinase pathway. *J. Biol. Chem*. **272**, 3296–3301.

Frendl, G. and Beller, D.I. (1990). Regulation of macrophage activation by IL-3. I. IL-3 functions as a macrophage-activating factor with unique properties, inducing Ia and lymphocyte function-associated antigen-1 but not cytotoxicity. *J. Immunol*. **144**, 3392–3399.

Frendl, G., Fenton, M.J. and Beller, D.I. (1990). Regulation of macrophage activation by IL-3. II. IL-3 and lipopolysaccharide act synergistically in the regulation of IL-1 expression. *J. Immunol*. **144**, 3400–3410.

Fujita, J., Nakayama, H., Onoue, H., Ebi, Y., Kanakura, Y., Kuriu, A. and Kitamura, Y. (1988a). Failure of W/Wv mouse-derived cultured mast cells to enter S phase upon contact with NIH/3T3 fibroblasts. *Blood* **72**, 463–468.

Fujita, J., Nakayama, H., Onoue, H., Ebi, Y., Kanakura, Y., Nakano, T., Asai, H., Takeda, S-I., Honjo, T. and Kitamura, Y. (1988b). Fibroblast-dependent growth of mouse mast cells *in vitro*: duplication of mast cell depletion in mutant mice of W/Wv genotype. *J. Cell Physiol*. **134**, 78–84.

Fung, M.C., Hapel, A.J., Ymer, S., Cohen, D.R., Johnson, R.M., Campbell, H.D. and Young, I.G. (1984). Molecular cloning of cDNA for murine interleukin-3. *Nature* **307**, 233–237.

Ganser, A., Lindemann, A., Seipelt, G., Ottmann, O.G., Herrmann, F., Eder, M., Frisch, J., Schulz, G., Mertelsmann, R. and Hoelzer, D. (1990a). Effects of recombinant human interleukin-3 in patients with normal hematopoiesis and in patients with bone marrow failure. *Blood* **76**, 666–676.

Ganser, A., Lindemann, A., Seipelt, G., Ottmann, O.G., Eder, M., Falk, S., Herrmann, F., Kaltwasser, J.P., Meusers, P., Klausmann, M., Frisch, J., Schulz, G., Mertelsmann, R. and Hoelzer, D. (1990b). Effects of recombinant human interleukin-3 in aplastic anemia. *Blood* **76**, 1287–1292.

Gearing, D.P., King, J.A., Gough, N.M. and Nicola, N.A. (1989). Expression cloning of a receptor for human granulocyte–macrophage colony-stimulating factor. *EMBO J*. **8**, 3667–3676.

Geissler, E.N., Ryan, M.A. and Housman, D.E. (1988). The dominant-white spotting (*W*) locus of the mouse encodes the c-*kit* proto-oncogene. *Cell* **55**, 185–192.

Geissler, K., Peschel, C., Niederwieser, D., Strobl, H., Goldschmitt, J., Ohler, L., Bettelheim, P., Kahls, P., Huber, C., Lechner, K., Hocker, P. and Kolbe, K. (1996). Potentiation of granulocyte colony-stimulating factor-induced mobilization of circulating progenitor cells by seven-day pretreatment with interleukin-3. *Blood* **87**, 2732–2739.

Gold, M., Duonio, V., Saxena, S., Schrader, J.W. and Aebersold, R. (1994). Multiple cytokines activate phosphatidylinositol 3-kinase in hemopoietic cells. Association of the enzyme with various tyrosine-phosphorylated proteins. *J. Biol. Chem*. **269**, 5403–5412.

Grau, G.E., Kindler, V., Piguet, E.-E., Lambert, P.-H. and Vassalli, P.J. (1988). Prevention of experimental cerebral malaria by anticytokine antibodies. Interleukin 3 and granulocyte macrophage colony-stimulating factor are intermediates in increased tumor necrosis factor production and macrophage accumulation. *Exp. Med*. **168**, 1499–1504.

Grimaldi, J.C. and Meeker, T.C. (1989). The t(5;14) chromosomal translocation in a case of acute lymphocytic leukemia joins the interleukin-3 gene to the immunoglobulin heavy chain gene. *Blood* **73**, 2081–2085.

Grouard, G., Rissoan, M.C., Filgueira, L., Durand, I., Banchereau, J. and Liu, Y.J. (1997). The enigmatic plasmacytoid T cells develop into dendritic cells with interleukin (IL)-3 and CD40-ligand. *J. Exp. Med*. **185**, 1101.

Gupta, S., Campbell, D., Derijard, B. and Davis, R.J. (1995). Transcription factor ATF2 regulation by the JNK signal transduction pathway. *Science* **276**, 389.

Halperin, D.S., Estrov, Z. and Freedman, M.H. (1989). Diamond–Blackfan anemia: promotion of marrow erythropoiesis *in vitro* by recombinant interleukin-3. *Blood* **73**, 1168–1174.

Hannemann, J., Hara, T., Kawai, M., Miyajima, A., Ostertag, W. and Stocking, C. (1995). Sequential mutations in the interleukin-3 (IL3)/granulocyte–macrophage colony-stimulating factor/IL5 receptor beta-subunit genes are necessary for the complete conversion to growth autonomy mediated by a truncated beta C subunit. *Mol. Cell. Biol*. **15**, 2402–2412.

Hapel, A.J. and Young, I.G. (1988). Lymphokines. In *Molecular Biology of Interleukin 3: A Multilineage Hemopoietic Growth Regulator*, volume 15 (ed. J.W. Schrader), Academic Press, San Diego, CA, pp. 91–126.

Hapel, A.J., Lee, J.C., Farrar, W.L. and Ihle, J.M. (1981). Establishment of continuous cultures of thy1.2+, Lyt1+, 2-T cells with purified interleukin 3. *Cell* **25**, 179–186.

Hara, T., Ichihara, M., Takagi, M. and Miyajima, A. (1995). Interleukin-3 (IL-3) poor-responsive inbred mouse strains carry the identical deletion of a branch point in the IL-3 receptor alpha subunit gene. *Blood* **85**, 2331.

Haylock, D.N., To, L.B., Dowse, T.L., Juttner, C.A. and Simmons, P.J. (1992). *Ex vivo* expansion and maturation of peripheral blood $CD34^+$ cells into the myeloid lineage. *Blood* **80**, 1405.

Hirai, K., Morita, Y., Misaki, Y., Ohta, K., Takashi, T., Suzuki, S., Motoyoshi, K. and Miyamoto, T. (1988). Modulation of human basophil histamine release by hemopoietic growth factors. *J. Immunol*. **141**, 3958–3964.

Hirayama, F., Shih, J.P., Awgulewitsch, A., Warr, G.W., Clark, S.C. and Ogawa, M. (1992). Clonal proliferation of murine lymphohemopoietic progenitors in culture. *Proc. Natl Acad. Sci. U.S.A.* **89**, 5907–5911.

Huang, E., Nocka, K. Beier, D.R., Chu, T.-Y., Buck, J., Lahm, H.-W., Wellner, D., Leder, P. and Besmer, P. (1990). Candidate ligand for the c-kit transmembrane kinase receptor: KL, a fibroblast derived growth factor stimulates mast cells and erythroid progenitors. *Cell* **63**, 225–233.

Huhn, R.D., Yurkow, E.J., Tushinski, R., Clarke, L., Sturgill, M.G., Hoffman, R., Sheay, W., Cody, R., Philipp, C., Resta, D. and George, M. (1996). Recombinant human interleukin-3 (rhIL-3) enhances the mobilization of peripheral blood progenitor cells by recombinant human granulocyte colony-stimulating factor (rhG-CSF) in normal volunteers. *Exp. Hematol*. **24**, 839–847.

Ihle, J.N., Pepersack, L. and Rebar, L. (1981). Regulation of T cell differentiation: *in vitro* induction of 20 alpha-hydroxysteroid dehydrogenase in splenic lymphocytes from athymic mice by a unique lymphokine. *J. Immunol*. **126**, 1284–1289.

Ihle, J.M., Keller, J., Oroszlan, S., Henderson, L.E., Copeland, T.D., Fitch, F., Prystowsky, M.B., Goldwasser, E., Schrader, J.W., Palaszynski, E., Dy, M. and Lebel, B. (1983). Biologic properties of homogeneous interleukin 3. I. Demonstration of WEHI-3 growth factor activity, mast cell growth factor activity, p cell-stimulating factor activity, colony-stimulating factor activity, and histamine-producing cell-stimulating factor activity. *J. Immunol.* **131**, 282–287.

Iordanov, M., Bender, K., Ade, T., Schmid, W., Sachsenmaier, C., Engel, K., Gaestel, M., Rahmsdorf, H.J. and Herrlich, P. (1997). CREB is activated by UVC through a p38/HOG-1-dependent protein kinase. *EMBO J.* **16**, 1009–1022.

Iscove, N.N. and Yan, X.Q. (1990). Precursors (pre-CFCmulti) of multilineage hemopoietic colony-forming cells quantitated *in vitro*. Uniqueness of IL-1 requirement, partial separation from pluripotential colony-forming cells and correlation with long term reconstituting cells *in vivo*. *J. Immunol.* **145**, 190.

Iscove, N.N., Shaw, A.R. and Keller, G. (1989). Net increase of pluripotential hematopoietic precursors in suspension culture in response to IL-1 and IL-3. *J. Immunol.* **142**, 2332–2337.

Isfort, R.J., Stevens, D., May, W.S. and Ihle, J.M. (1988). Interleukin 3 binds to a 140-kDa phosphotyrosine-containing cell surface protein. *Proc. Natl Acad. Sci. U.S.A.* **85**, 7982–7986.

Ishibashi, T. and Burstein, S.A. (1986). Interleukin 3 promotes the differentiation of isolated single megakaryocytes. *Blood* **67**, 1512–1514.

Itoh, N., Yonehara, S., Schreurs, J., Gorman, D.M., Maruyama, K., Ishii, A., Yahara, I., Arai, K.-I. and Miyajima, A. (1990). Interleukin 3 promotes the differentiation of isolated single megakaryocytes. *Science* **247**, 324–327.

Itoh, T., Muto, A., Watanabe, S., Miyajima, A., Yokota, T. and Arai, K.-I. (1996). Granulocyte–macrophage colony-stimulating factor provokes RAS activation and transcription of c-*fos* through different modes of signalling. *J. Biol. Chem.* **271**, 7587–7592.

Jenkins, B.J., D'Andrea, R. and Gonda, T.J. (1995). Activating point mutations in the common β subunit of the human GM-CSF, IL-3 and IL-5 receptors suggest the involvement of β subunit dimerization and cell type-specific molecules in signalling. *EMBO J.* **14**, 4276–4287.

Kauffmann-Zeh, A., Rodriguez-Viciana, P., Ulrich, E., Gilbert, C., Coffer, P., Downard, J. and Evan, G. (1997). Suppression of c-myc-induced apoptosis by Ras signalling through PI(3)K and PKB. *Nature* **385**, 544.

Khan, K.N., Kats, A.A., Fouant, M.M., Snook, S.S., McKearn, J.P., Alden, C.L. and Smith, P.F. (1996). Recombinant human interleukin-3 induces extramedullary hematopoiesis at subcutaneous injection sites in cynomolgus monkeys. *Toxicol. Pathol.* **24**, 391–397.

Khew-Goodall, Y., Butcher, C.M., Litwin, M.S., Newlands, S., Korpelainen, E.I., Noack, L.M., Berndt, M.C., Lopez, A.F., Gamble, J.R. and Vadas, M.A. (1996). Chronic expression of P-selectin on endothelial cells stimulated by the T-cell cytokine, interleukin-3. *Blood* **87**, 1432–1438.

Kinashi, T., Inaba, K., Tsubata, T., Tashiro, K., Palacios, R. and Honjo, T. (1988). Differentiation of an interleukin 3-dependent precursor B-cell clone into immunoglobulin-producing cells *in vitro*. *Proc. Natl Acad. Sci. U.S.A.* **85**, 4473–4477.

Kindler, V., Thorens, S.B., De Kossodo, S., Allet, B., Eliason, J.F., Thatcher, D. and Vassali, P. (1985). Stimulation of hematopoiesis *in vivo* by recombinant bacterial murine interleukin 3. *Proc. Natl Acad. Sci. U.S.A.* **83**, 1001–1005.

Kinoshita, T., Yokota, T., Arai, K. and Miyajima, A. (1995). Regulation of Bcl-2 expression by oncogenic Ras protein in hematopoietic cells. *Oncogene* **10**, 2207–2212.

Kitamura, T. and Miyajima, A. (1992). Functional reconstitution of the human interleukin-3 receptor. *Blood* **80**, 84–90.

Kitamura, T., Sato, N., Arai, K. and Miyajima, A. (1991). Expression cloning of the human IL-3 receptor cDNA reveals a shared beta subunit for the human IL-3 and GM-CSF receptors. *Cell* **66**, 1165–1174.

Korenaga, M., Abe, T. and Hashiguchi, Y. (1996a). Injection of recombinant interleukin-3 hastens worm expulsion in mice infected with *Trichinella spiralis*. *Parasitol. Res.* **82**, 108–113.

Korenaga, M., Watanabe, N., Abe, T. and Hashiguchi, Y. (1996b). Acceleration of IgE responses by treatment with recombinant interleukin-3 prior to infection with *Trichinella spiralis* in mice. *Immunology* **87**, 642–646.

Kurimoto, Y., de Weck, A.L. and Dahinden, C.A. (1989). Interleukin-3 dependent mediator release in basophils triggered by C5a. *J. Exp. Med.* **170**, 467–479.

Kyriakis, J.M., Banerjee, P., Nikolakaki, E., Dai, T., Rubie, E.A., Ahmad, M.F., Avruch, J. and Woodgett, J.R. (1994). The stress-activated protein kinase subfamily of c-Jun kinases. *Nature* **369**, 156.

Lemoli, R.M., Rosti, G., Visani, G., Gherlinzoni, F., Miggiano, M.C., Fortuna, A., Zinzani, P. and Tura, S. (1996). Concomitant and sequential administration of recombinant granulocyte colony-stimulating factor and recombinant human interleukin-3 to accelerate hematopoietic recovery after autologous bone marrow transplantation for malignant lymphoma. *J. Clin. Oncol.* **14**, 3018–3025.

Leslie, K.B. and Schrader, J.W. (1989). Growth factor gene activation and clonal heterogeniety in an autostimulatory myeloid leukemia. *J. Mol. Cell. Biol.* **6**, 2414–2423.

Leslie, K.B., Jalbert, S., Orban, P., Welham, M., Duronio, V. and Schrader, J.W. (1996). Genetic basis of hyporesponsiveness of A/J mice to interleukin-3. *Blood* **87**, 3186–3194.

Liew, F.Y., Millot, S., Li, Y., Lelchuck, R., Chan, W.L. and Ziltener, H.J. (1989). Macrophage activation by interferon-gamma from host-protective T cells is inhibited by interleukin (IL)3 and IL4 produced by disease-promoting T cells in leishmaniasis. *Eur. J. Immunol*. **19**, 1227–1232.

Liu, L., Jefferson, A.B., Zhang, X.L., Norris, F.A., Majerus, P.W. and Krystal, G. (1996). A novel phosphatidylinositol-3,4,5-trisphosphate 5-phosphatase associates with the interleukin-3 receptor. *J. Biol. Chem*. **271**, 29729–29733.

Londei, M., Verhoef, A., De Berardinis, P., Kissonerghis, M., Grubeck-Loebenstein, B. and Feldmann, M. (1989). Definition of a population of CD4-8- T cells that express the alpha beta T-cell receptor and respond to interleukins 2, 3, and 4. *Proc. Natl Acad. Sci. U.S.A.* **86**, 8502–8506.

Lopez, A.F., To, L.B., Yang, Y.-C., Gamble, J.R., Shannon, M.F., Burns, G.F., Dyson, P.G., Juttner, C.A., Clark, S. and Vadas, M.A. (1987). Stimulation of proliferation, differentiation, and function of human cells by primate interleukin 3. *Proc. Natl Acad. Sci. U.S.A.* **84**, 2761–2765.

Madden, K.B., Urban, J.F., Jr., Ziltener, H.J., Schrader, J.W., Finkelman, F.D. and Katona, I.M. (1991). Antibodies to IL-3 and IL-4 suppress helminth-induced intestinal mastocytosis. *J. Immunol*. **147**, 1387–1391.

Marietta, E.V., Chen, Y. and Weis, J.H. (1996). Modulation of expression of the anti-inflammatory cytokines interleukin-13 and interleukin-10 by interleukin-3. *Eur. J. Immunol*. **26**, 49–56.

Martin, F. H., Suggs, S.V., Langley, K.E., Lu, H.S., Ting, J., Okino, K.H., Morris, F., McNiece, I.K., Jacobsen, F.W., Mendiaz, E.A., Birkett, N.C., Smith, D.A., Johnson, M.J., Parker, V.P., Flores, J.C., Patel, A.C., Fisher, E.F., Erjavec, H.O., Herrera, C.J., Wypych, J., Sachdev, R.K., Pope, J.A., Leslie, I., Wen, D., Lin, C.-H., Cupples, R.L. and Zsebo, K.M. (1990). Primary structure and functional expression of rat and human stem cell factor DNAs. *Cell* **63**, 203–211.

Mayer, P., Valent, P., Schmidt, G., Liehl, E. and Bettelheim, P. (1989). The *in vivo* effects of recombinant human interleukin-3: demonstration of basophil differentiation factor, histamine-producing activity, and priming of GM-CSF-responsiveness progenitors in non-human primates. *Blood* **74**, 613–621.

Metcalf, D. (1988). The multipotential colony-stimulating factor, multi-CSF (IL-3). In *Lymphokines 15; Interleukin 3: The Panspecific Hemopoietin* (ed. J.W. Schrader), Academic Press, San Diego, CA, pp. 183–217.

Moqbel, R., Levi-Schaffer, F. and Kay, A.B. (1994). Cytokine generation by eosinophils. *J. Allergy Clin. Immunol*. **94**, 1183.

Morla, A.O., Schreurs, J., Miyajima, A. and Wang, J.Y. (1988). Hematopoietic growth factors activate the tyrosine phosphorylation of distinct sets of proteins in interleukin-3-dependent murine cell lines. *Mol. Cell. Biol*. **8**, 2214–2218.

Morstyn, G., Sheridan, W., Lieschke, G., Cebon, J., Layton, J. and Fox, R. (1990). Towards improved cancer therapy using subcutaneously administered hemopoietic colony stimulated factors. In *Effects of Therapy on Biology and Kinetics of the Residual Tumor, Part B: Clinical Aspects* (eds J. Ragaz, L. Simpson-Herren, M.E. Lippman and B. Fisher), Wiley–Liss, New York, pp. 29–36.

Mufson, R.A., Szabo, J. and Eckert, D. (1992). Human IL-3 induction of c-jun in normal monocytes is independent of tyrosine kinase and involves protein kinase C. *J. Immunol*. **148**, 1129–1135.

Mui, A.L.-F., Wakao, H., O'Farrell, A., Harada, N. and Miyajima, A. (1995). Interleukin-3, granulocyte-macrophage colony stimulating factor and interleukin-5 signals through two STAT5 homologs. *EMBO J*. **14**, 1166–1175.

Mui, A.L.-F., Wakao, H., Kinoshita, T., Kitamura, T. and Miyajima, A. (1996). Suppression of interleukin-3-induced gene expression of a *C*-terminal truncated Stat5: role of Stat5 in proliferation. *EMBO J*. **15**, 2425–2433.

Muto, A., Watanabe, S., Miyajima, A., Yokota, T. and Arai, K. (1995). High affinity chimeric human granulocyte-macrophage colony-stimulating factor receptor carrying the cytoplasmic domain of the beta subunit but not the alpha subunit transduces growth promoting signals in Ba/F3 cells. *Biochem. Biophys. Res. Commun*. **208**, 368.

Muto, A., Watanabe, S., Miyajima, A., Yokota, T. and Arai, K. (1996). The β subunit of human granulocyte-macrophage colony-stimulating factor receptor forms a homodimer and is activated via association with the α subunit. *J. Exp. Med*. **183**, 1911–1916.

Nakano, T., Kodama, H. and Honjo, T. (1994). Generation of lymphohematopoietic cells from embryonic stem cells in culture. *Science* **265**, 1098.

Niemeyer, C.M., Sieff, C.A., Mathey-Prevot, B., Wimperis, J.Z., Bierer, F.E., Clark, S.C. and Nathan, D.G. (1989). Expression of human interleukin-3 (multi-CSF) is restricted to human lymphocytes and T-cell tumor lines. *Blood* **73**, 945–951.

Nishinakamura, R., Nakayama, N., Hirabayashi, Y., Inoue, T., Aud, D., NcNeil, T., Azuma, S., Yoshida, S., Toyoda, Y., Arai, K., Miyajima, A. and Murray, R. (1995). Mice deficient for the IL-3/GM-CSF/IL-5 βc receptor exhibit lung pathology and impaired immune response, while β_{IL3} receptor-deficient mice are normal. *Immunity* **2**, 211.

Nishinakamura, R., Miyajima, A., Mee, P.J., Tybulewicz, V.L.J. and Murray, R. (1996). Hematopoiesis in mice lacking the entire granulocyte–macrophage colony-stimulating factor/interleukin-3/interleukin-5 functions. *Blood* **88**, 2458–2464.

O'Connor, R., Torigoe, T., Reed, J.C. and Santoli, D. (1992). Phenotypic changes induced by interleukin-2 (IL-2) and IL-3 in an immature T-lymphocytic leukemia are associated with regulated expression of IL-2 receptor beta chain and of protein tyrosine kinases LCK and LYN. *Blood* **80**, 1017–1025.

Palacios, R. and Pelkonen, J. (1988). Prethymic and intrathymic mouse T-cell progenitors. Growth requirements and analysis of the expression of genes encoding TCR/T3 components and other T-cell-specific molecules. *Immunol Rev*. **104**, 158.

Palacios, R. and Steinmetz, M. (1985). IL-3-dependent mouse clones that express B-220 surface antigen, contain Ig genes in germ-line configuration, and generate B lymphocytes *in vivo. Cell* **41**, 727–734.

Palacios, R., Henson, G., Steinmetz, M. and McKearn, J.P. (1984). Interleukin-3 supports growth of mouse pre-B-cell clones *in vitro. Nature* **309**, 126–131.

Paonessa, G., Graziani, R., De Serio, A., Savino, R., Ciapponi, L., Lahm, A., Salvati, A.L., Toniatti, C. and Ciliberto, G. (1995). Two distinct and independent sites on IL-6 trigger gp130 dimer formation and signalling. *EMBO J*. **14**, 1942.

Park, L.S., Waldron, P.E., Friend, D. *et al*. (1989b). Interleukin-3, GM-CSF, and G-CSF receptor expression on cell lines and primary leukemia cells: receptor heterogeneity and relationship to growth factor responsiveness. *Blood* **74**, 56–65.

Patel, N., Herrman, J.M., Timans, J.C. and Kastelein, R.A. (1996). Functional replacement of cytokine receptor extracellular domains by leucine zippers. *J. Biol. Chem*. **271**, 30386–30391.

Pelech, S.L., Paddon, H.B., Charest, D.L. and Federsppiel, B.S. (1990). IL-3-induced activation of protein kinases in the mast cell/megakaryocyte R6-XE.4 line. *J. Immunol*. **144**, 1759–1766.

Perkins, G.R., Marshall, C.J. and Collins, M.K. (1996). The role of MAP kinase in interleukin-3 stimulation of proliferation. *Blood* **87**, 3669–3675.

Peters, S.O., Kittler, E.L., Ramshaw, H.S. and Quesenberry, P.J. (1996). *Ex vivo* expansion of murine marrow cells with interleukin-3 (IL-3), IL-6, IL-11, and stem cell factor leads to impaired engraftment in irradiated hosts. *Blood* **87**, 30–37.

Plaut, M., Pierce, J.H., Watson, C.J., Hanley-Hyde, J., Nordan, R.P. and Paul, W.E. (1989). Mast cell lines produce lymphokines in response to crosslinkage of FcERI and calcium ionophores. *Nature* **339**, 64.

Pratt, J.C., Weiss, M., Sieff, C.A., Shoelson, S.E., Burakoff, S.J. and Ravichandran, K.S. (1996). Evidence for a physical association between the Shc-PTB domain and the βc chain of the granulocyte–macrophage colony-stimulating factor receptor. *J. Biol. Chem*. **271**, 12137–12140.

Pulaski, B.A., Yeh, K.Y., Shastri, N., Maltby, K.M., Penney, D.P., Lord, E.M. and Frelinger, J.G. (1996). Interleukin 3 enhances cytotoxic T lymphocyte development and class I major histocompatibility complex 're-presentation' of exogenous antigen by tumor-infiltrating antigen-presenting cells. *Proc. Natl Acad. Sci. U.S.A.* **93**, 3669–3674.

Quelle, F.W., Sato, N., Witthuhn, B.A., Inhorn, R.C., Eder, M., Miyajima, A., Griffin, J.D. and Ihle, J.N. (1994). JAK2 associates with the βc chain of the receptor for granulocyte–macrophage colony-stimulating factor, and its activation requires the membrane-proximal region. *Mol. Cell. Biol*. **14**, 4335.

Raingeaud, J., Gupta, S., Rogers, J.S., Dickens, M., Han, J., Ulevitch, R.J. and Davis, R.J. (1995). Pro-inflammatory cytokines and environmental stress cause p38 mitogen-activated protein kinase activation by dual phosphorylation on tyrosine and threonine. *J. Biol. Chem*. **270**, 7420–7426.

Raingeaud, J., Whitmarsh, A.J., Barrett, T., Derijard, B. and Davis, R.J. (1996). MKK3- and MKK6-regulated gene expression is mediated by the p38 mitogen-activated protein kinase signal transduction pathway. *Mol. Cell. Biol*. **16**, 1247–1255.

Rao, P. and Mufson, R.A. (1995). A membrane proximal domain of the human interleukin-3 receptor βc subunit that signals DNA synthesis in NIH 3T3 cells specifically binds a complex of Src and Janus family tyrosine kinases and phosphatidylinositol 3-kinase. *J. Biol. Chem*. **270**, 6886–6893.

Razin, E., Leslie, K. and Schrader, J. (1991). Connective tissue mast cells in contact with fibroblasts express IL-3 mRNA. Analysis of single cells by polymerase chain reaction. *J. Immunol*. **146**, 981–987.

Rennick, D., Jackson, J., Moulds, C., Lee, F. and Yang, G. (1989). IL-3 and stromal cell-derived factor synergistically stimulate the growth of pre-B cell lines cloned from long-term lymphoid bone marrow cultures. *J. Immunol*. **142**, 161–166.

Rico-Vargas, S.A., Weiskopf, B., Nishikawa, S.-I. and Osmond, D.G. (1994). c-kit expression of B cell genesis by *in vivo* treatment with anti-c-kit antibody. *J. Immunol*. **152**, 2845.

Rinaudo, M.S., Su, K., Falk, L.A., Haldar, S. and Mufson, R.A. (1995). Human interleukin-3 receptor modulates *bcl-2* mRNA and protein levels through protein kinase C in TF-1 cells. *Blood* **86**, 80–88.

Roberts, R., Gallagher, J., Spooncer, E., Allen, T.D., Bloomfield, F. and Dexter, T.M. (1988). Heparan sulphate bound growth factors: a mechanism for stromal cell mediated haemopoiesis. *Nature* **332**, 376–378.

Rothenberg, M.E., Owen, W.F., Jr., Silberstein, D.S., Woods, J., Soberman, R.J., Austen, K.F. and Stevens, R.L. (1988). Human eosinophils have prolonged survival, enhanced functional properties, and become hypodense when exposed to human interleukin 3. *J. Clin. Invest*. **81**, 1986–1992.

Salmon, R.A., Young, P.R. and Schrader, J.W. (1997). The p38 MAP kinase pathway in B lymphocytes is activated by ligation of the B-cell antigen receptor or CD40 but has no role in the regulation of apoptosis (in press).

Satamaki, K., Wang, H.-M., Miyajima, I., Kitamura, T., Todokoro, K., Harada, N. and Miyajima, A. (1993). Ligand-dependent activation of chimeric receptors with the plasmic domain of the interleukin-3 receptor β subunit (β_{IL3}). *J. Biol. Chem*. **268**, 15833–15839.

Sato, T., Nakafuku, M., Miyajima, A. and Kaziro, Y. (1991). Involvement of ras p21 protein in signal-transduction pathways from interleukin 2, interleukin 3, and granulocyte/macrophage colony-stimulating factor, but not from interleukin 4. *Proc. Natl Acad. Sci. U.S.A.* **88**, 3314.

Satoh, T., Minami, Y., Kono, T., Yamada, K., Kawahara, A., Taniguchi, T. and Kaziro, Y. (1992). Interleukin 2-induced activation of Ras requires two domains of interleukin 2 receptor beta subunit, the essential region for growth stimulation and Lck-binding domain. *J. Biol. Chem*. **267**, 25423–25427.

Schindler, C. and Darnell, J.E., Jr. (1995). Transcriptional responses to polypeptide ligands: The JAK–STAT pathway. *Annu. Rev. Biochem*. **64**, 621.

Schrader, J.W. (1981). In *in vitro* production and cloning of the P cell, a bone marrow-derived null cell that expresses H-2 and Ia-antigens, has mast cell-like granules, and is regulated by a factor released by activated T cells. *J. Immunol*. **126**, 452–458.

Schrader, J.W. and Crapper, R.M. (1983). Autogenous production of a hemopoietic growth factor, persisting-cell-stimulating factor, as a mechanism for transformation of bone marrow-derived cells. *Proc. Natl Acad. Sci. U.S.A.* **80**, 6892–6896.

Schrader, J.W. and Nossal, G.J.V. (1980). Strategies for the analysis of accessory-cell function: the *in vitro* cloning and characterization of the P cell. *Immunol. Rev*. **53**, 61–85.

Schrader, J.W., Battye, F. and Scollay, R. (1982). Expression of Thy-1 antigen is not limited to T cells in cultures of mouse hemopoietic cells. *Proc. Natl Acad. Sci. U.S.A.* **79**, 4161–4165.

Schrader, J.W., Schrader, S., Clark-Lewis, I. and Crapper, R.M. (1984). *In vitro* studies on lymphopoiesis, hemopoiesis and oncogenesis. In *Long Term Bone Marrow Culture* (eds D.G. Wright and J.S. Greenberger), Liss, New York, pp. 293–308.

Schrader, J.W., Ziltener, H.J. and Leslie, K.B. (1986a). Structural homologies among the hemopoietins. *Proc. Natl Acad. Sci. U.S.A.* **83**, 2458–2462.

Schrader, J.W., Clark-Lewis, I., Ziltener, H.J., Hood, L.E. and Kent, S.B.H. (1986b). In *Immune Regulation by Characterized Polypeptides* (eds G. Goldstein, J.F. Bach and H. Wigzell), Liss, New York, pp. 475–484.

Schrader, J.W., Clark-Lewis, I., Crapper, R.M., Leslie, K.B., Schrader, S., Varigos, G. and Ziltener, H.J. (1988). The panspecific hemopoietin interleukin-3: Physiology and pathology. In *Lymphokines 15; Interleukin 3: The Panspecific Hemopoietin* (ed. J.W. Schrader), Academic Press, San Diego, CA, pp. 281–311.

Shanafelt, A.B. and Kastelein, R.A. (1992). High affinity ligand binding is not essential for granulocyte–macrophage colony-stimulating factor receptor activation. *J. Biol. Chem*. **267**, 25466–25472.

Shibata, T., Kindler, V., Chicheportiche, Y., Vassalli, P. and Izui, S. (1990). Interleukin 3 perfusion prevents death due to acute anemia induced by monoclonal antierythrocyte autoantibody. *J. Exp. Med*, **171**, 1809–1814.

Shikama, Y., Barber, D.L., D'Andrea, A.D. and Sieff, C.A. (1996). A constitutively activated chimeric cytokine receptor confers factor-independent growth in hematopoietic cell lines. *Blood* **88**, 455–464.

Silvennoinen, O., Witthuhn, B., Quelle, F.W., Cleveland, J.L., Yi, T. and Ihle, J.N. (1993). Structure of the murine Jak2 protein-tyrosine kinase and its role in interleukin 3 signal transduction. *Proc. Natl Acad. Sci. U.S.A.* **90**, 8429–8433.

Takaki, S., Kanazawa, H., Shiiba, M. and Takatsu, K. (1994). A critical cytoplasmic domain of the interleukin-5 (IL-5) receptor α chain and its function in IL-5 mediated growth signal transduction. *Mol. Cell. Biol*. **14**, 7404–7413.

Tan, Y., Rouse, J., Zhang, A., Cariati, S., Cohen, P. and Comb, M.J. (1996). FGF and stress regulate CREB and ATF-1 via a pathway involving a p38 MAP kinase and MAPKAP kinase-2. *EMBO J.* **15**, 4629–4642.

Tavernier, J., Devos, R., Cornelis, S., Tuypens, T., Van der Heyden, J., Fiers, W. and Plaetinck, G. (1991). A human high affinity interleukin-5 receptor (IL5R) is composed of an IL5-specific alpha chain and a beta chain shared with the receptor for GM-CSF. *Cell* **66**, 1175–1184.

Terada, K., Kaziro, Y. and Satoh, T. (1997). Ras-dependent activation of c-Jun *N*-terminal kinase/stress-activated protein kinase in response to interleukin-3 stimulation in hematopoietic BaF3 cells. *J. Biol. Chem*. **272**, 4544–4548.

Thomas, J.W., Baum, C.M., Hood, W.F., Klein, B., Monhan, J.B., Paik, K., Staten, N., Abrams, M. and McKearn, J.P. (1995). Potent interleukin-3 receptor agonist with selectively enhanced hematopoietic activity relative to recombinant human interleukin-3. *Proc. Natl Acad. Sci. U.S.A.* **92**, 3779.

Torigoe, T., O'Connor, R., Santoli, D. and Reed, J.C. (1992). Interleukin-3 regulates the activity of the Lyn protein-tyrosine kinase in myeloid-committed leukemic cell lines. *Blood* **80**, 617–624.

Uckun, F.M., Gesner, T.B., Song, C.W., Myers, D.E. and Mufson, A. (1989). Leukemic B-cell precursors express functional receptors for human interleukin-3. *Blood* **73**, 533–542.

Wang, L.M., Keegan, A.C., Paul, W.E., Heidaran, M.A., Gutkind, J.S. and Pierce, J.H. (1992). IL-4 activates a distinct signal transduction cascade from IL-3 in factor-dependent myeloid cells. *EMBO J*. **11**, 4899–4908.

Ward, L.D., Howlett, G.J., Discolo, G., Yasukawa, K., Hammacher, A., Moritz, R. and Simpson, R.J. (1994). High affinity interleukin-6 receptor is a hexameric complex consisting of two molecules each of interleukin-6, interleukin-6 receptor and gp-130. *J. Biol. Chem*. **269**, 23 286.

Welham, M. and Schrader, J.W. (1991). Modulation of c-kit mRNA and protein by hemopoietic growth factors. *Mol. Cell. Biol*. **11**, 2901–2904.

Welham, M.J. and Schrader, J.W. (1992). Steel factor-induced tyrosine phosphorylation in murine mast cells. Common elements with IL-3-induced signal transduction pathways. *J. Immunol*. **149**, 2772–2783.

Welham, M.J., Duronio, V., Sanghera, J., Pelech, S. and Schrader, J.W. (1992). Multiple hemopoietic growth factors stimulate activation of mitogen-activated protein kinase family members. *J. Immunol*. **149**, 1683–1693.

Welham, M.J., Duronio, V., Leslie, K.B., Bowtell, D. and Schrader, J.W. (1994a). Multiple hemopoietins, with the exception of interleukin-4, induce modification of shc and mSos1, but not their translocation. *J. Biol. Chem*. **269**, 21165–21176.

Welham, M.J., Dechert, U., Leslie, K.B., Jirik, F. and Schrader, J.W. (1994b). IL-3 and GM-CSF, but not IL-4, induce tyrosine phosphorylation, activation and association of SH-PTP2 with grb2 and PI3 kinase. *J. Biol. Chem*. **269**, 23 764.

Welham, M.J., Learmonth, L., Bone, H. and Schrader, J.W. (1995). IL-13 signal transduction in lymphohemopoietic cells: similarities and differences in signal transduction with IL-4 and insulin. *J. Biol. Chem*. **269**, 23 764–23 768.

Whetton, A.D., Vallance, S.J., Monk, P.N., Cragoe, E.J., Dexter, T.M. and Heyworth, C.M. (1988a). Interleukin-3-stimulated haemopoietic stem cell proliferation. Evidence for activation of protein kinase C and Na^+/H^+ exchange without inositol lipid hydrolysis. *Biochem. J*. **256**, 585–592.

Whetton, A.D., Monk, P.N., Consalvey, S.D., Huang, S.J., Dexter, T.M. and Downes, C.P. (1988b). Interleukin 3 stimulates proliferation via protein kinase C activation without increasing inositol lipid turnover. *Proc. Natl Acad. Sci. U.S.A.* **85**, 3284–3288.

Williams, D.E., Eisenman, J., Baird, A., Rauch, C., Van Ness, K., March, C.J., Park, L.S., Martin, U., Mochizuki, D.Y., Boswell, H.S., Burgess, G.S., Cosman, D. and Lyman, S.D. (1990). Identification of a ligand for the c-*kit* proto-oncogene. *Cell* **63**, 167–174.

Williams, G.T., Smith, C.A., Spooncer, E., Dexter, T.M. and Taylor, D.R. (1990). Haemopoietic colony stimulating factors promote cell survival by suppressing apoptosis. *Nature* **343**, 76–79.

Winkler, T.A., Melchers, F. and Rolink, A.G. (1995). Interleukin-3 and Interleukin-7 are alternative growth factors for the same B-cell precursors in the mouse. *Blood* **85**, 2045.

Witthuhn, B.A., Quelle, F.W., Silvennoinen, O., Yi, T., Tang, B., Miura, O. and Ihle, J. (1993). JAK2 associates with the erythropoietin receptor and is tyrosine phosphorylated and activated following stimulation with erythropoietin. *Cell* **74**, 227–236.

Wodnar-Filipowicz, A., Heusser, C.H. and Moroni, C. (1989). Production of the hemopoietic growth factors GM-CSF and interleukin-3 by mast cells in response to IgE receptor-mediated activation. *Nature* **339**, 150–152.

Wong, G.H.W., Clark-Lewis, I., Hamilton, J.A. and Schrader, J.W. (1984). P cell stimulating factor and glucocorticoids oppose the action of interferon-gamma in inducing Ia antigens on T-dependent mast cells (P cells). *J. Immunol.* **133**, 2043–2050.

Yang, Y.-C., Ciarletta, A.B., Temple, P.A., Chung, M.P., Kovacic, S., Witek-Giannotti, J.S., Leary, A.C., Kriz, R., Donahue, R.E., Wong, G.G. and Clark, S.C. (1986). Human IL-3 (multi-CSF): identification of expression cloning of a novel hematopoietic growth factor related to murine IL-3. *Cell* **47**, 3–10.

Yang, Y.-C., Kovacic, S., Kriz, R., Wolf, S., Clark, S., Wellems, T., Nienhuis, A. and Epstein, N. (1988). The human genes for GM-CSF and IL 3 are closely linked in tandem on chromosome 5. *Blood* **71**, 958–961.

Ymer, S., Tucker, W.Q., Sanderson, C.J., Hapel, A.J., Campbell, H.D. and Young, I.G. (1985). Constitutive synthesis of interleukin-3 by leukaemia cell line WEHI-3B is due to retroviral insertion near the gene. *Nature* **317**, 255–258.

Yokota, T., Lee, F., Rennick, D., Hall, C., Arai, N., Mosmann, T., Nabel, G., Cantor, H. and Arai, K.-I. (1984). Isolation and characterization of a mouse cDNA clone that expresses mast-cell growth-factor activity in monkey cells. *Proc. Natl Acad. Sci. U.S.A.* **81**, 1070–1074.

Yonemura, Y., Ku, H., Hirayama, F., Souza, L.M. and Ogawa, M. (1996). Interleukin 3 or interleukin 1 abrogates the reconstituting ability of hematopoietic stem cells. *Proc. Natl Acad. Sci. U.S.A.* **93**, 4040–4044.

Zhu, Y., Pless, M., Inhorn, R., Mathey-Prevot, B. and D'Andrea, A.D. (1996). The murine *DUB-1* gene is specifically induced by the beta-c subunit of interleukin-3 receptor. *Mol. Cell. Biol.* **16**, 4808–4817.

Ziltener, H.J., Clark-Lewis, I., Hood, L.P., Kent, S.B.A. and Schrader, J.W. (1987). Antipeptide antibodies of predetermined specificity recognize and neutralize the bioactivity of the pan-specific hemopoietin interleukin 3. *J. Immunol.* **138**, 1099–1104.

Ziltener, H.J., Fazekas de St. Groth, B., Leslie, K.B. and Schrader, J.W. (1988). Multiple glycosylated forms of T cell-derived interleukin 3 (IL-3). Heterogeneity of IL-3 from physiological and nonphysiological sources. *J. Biol. Chem.* **263**, 14511–14517.

Ziltener, H.J., Clark-Lewis, I., Tomlinson Jones, A. and Dy, M. (1994). Carbohydrate does not modulate the *in vivo* effects of injected interleukin-3. *Exp. Hematol.* **22**, 1070–1075.

Zsebo, K.M., Wypych, J., McNiece, I.K., Lu, H.S., Smith, K.A., Karkare, S.B., Sachdev, R.K., Yuschenkoff, V.N., Birkett, N.C., Williams, L.R., Satyagal, V.N., Tung, W., Bosselman, R.A., Mendiaz, E.A. and Langley, K.E. (1990a). Identification, purification, and biological characterization of hematopoietic stem cell factor from buffalo rat liver-conditioned medium. *Cell* **63**, 195–201.

Zsebo, K.M., Williams, D.A., Geissler, E.N., Broudy, V.C., Martin, F.H., Atkins, H.L., Hsu, R.-Y., Birkett, N.C., Okino, K.H., Murdock, D.C., Jacobsen, F.W., Langley, K.E., Smith, K.A., Takeishi, T., Cattanach, B.M., Galli, S.J. and Suggs, S.V. (1990b). Stem cell factor is encoded at the *Sl* locus of the mouse and is the ligand for the c-kit tyrosine kinase receptor. *Cell* **63**, 213–224.

Chapter 6

Interleukin-4

Pascale Chomarat[1], Mary Ellen Rybak[2] and Jacques Banchereau[1]

[1]Baylor Institute for Immunology Research, Dallas, Texas, USA and
[2]Schering-Plough Research Institute, Kenilworth, New Jersey, USA

INTRODUCTION

Interleukin-4 (IL-4) was identified in 1982 for its ability to induce activated mouse B lymphocytes to proliferate and to secrete immunoglobulin IgG_1. cDNAs coding for both human and murine molecules were isolated in 1986, followed in 1989–1990 by those coding for IL-4 high-affinity glycoprotein receptors. The wide range of cells on which IL-4 induces specific biologic functions highlights the pleiotropic nature of this molecule. Moreover, IL-4 production is restricted to activated T lymphocytes, mast cells and basophils. This implies a precise molecular regulation at the level of both IL-4 gene transcription and IL-4 signaling pathway, which have been the subject of intensive work over the past 3 years. Clinical trials being performed at present aim to determine the therapeutic potential of IL-4 as an antitumor agent. This chapter will refer to studies performed with human or murine IL-4, but additional references can be found in the excellent reviews of O'Garra and Spits (1993) and Brown and Hural (1997).

MOLECULAR ASPECTS OF IL-4

IL-4 Protein

The principal characteristics of IL-4 are summarized in Table 1. Molecular cloning of human IL-4 cDNA revealed a single open reading frame (ORF) of 153 amino acids yielding a secreted glycoprotein of 129 amino acids (Yokota *et al.*, 1986). IL-4 displays 20% homology at the amino acid level with IL-13, another T cell-derived cytokine which shares numerous biologic properties with IL-4 (McKenzie *et al.*, 1993; Minty *et al.*, 1993).

Expression of the recombinant IL-4 protein in mammalian cells demonstrates three variants with apparent M_r values of 15, 18 and 19 kDa, a microheterogeneity related to the nature of the *N*-linked oligosaccharides in the 18- and 19-kDa variants. Human IL-4 contains three disulfide bridges between C3–C127, C4–C65 and C46–C99 (Windsor *et al.*, 1990). X-ray diffraction of IL-4 crystals (Walter *et al.*, 1992), as well as magnetic resonance spectroscopy of IL-4 in solution (Powers *et al.*, 1992) indicate that IL-4 is a

The Cytokine Handbook, 3rd ed.
ISBN 0-12-689662-3

Table 1. Properties of human IL-4 and of the human IL-4 α chain receptor.

	IL-4	IL-4Rα
Protein		
Precursor protein: amino acids	153	825
Mature protein: amino acids	129	800
Molecular weight (kDa)	15–19	140
N-Glycosylation sites	2	6
Disulphide bonds	3 (6 Cys)	3 (6 Cys)
Intracytoplasmic Tyr		5
Gene		
Gene size (kilobase pairs)	10	3.6
Gene introns	3	
Gene location: chromosome	5q23–31	16q12–p11.2
Cell sources		
	T cells (Th2, NK)	Ubiquitous
	Basophils	
	Mast cells	
	Eosinophils	

left-handed four α-helices bundle with short stretches of β sheets. The four α helices are situated between residues 9–21, 45–64, 74–96 and 113–129, and the mini antiparallel β sheets are between residues 32–34 and 110–112 (Garret *et al.*, 1992). This structure bears close resemblance to Granulocyte–Macrophage Colony Stimulating Factor (GM-CSF), M-CSF and growth hormone.

mRNA phenotyping of cytokines has revealed the existence of an additional IL-4 mRNA, which lacks 48 base pairs coding for amino acid residues 22–37 by an alternative splicing of exon 2 (Sorg *et al.*, 1993). Such a transcript would result in a mature protein lacking one of the cysteine bridges and part of the loop connecting helices 1 and 2. More recently, this natural splice variant of IL-4 mRNA, termed IL-4 δ2, has been found to be expressed more strongly in thymocytes and bronchoalveolar lavage cells than IL-4 mRNA, suggesting tissue specificity expression. Unlike IL-4, IL-4 δ2 alone does not act as a costimulator for T-cell proliferation, but inhibits T-cell proliferation induced by full-size IL-4 (Atamas *et al.*, 1996). Moreover, an IL-4 molecule in which the Tyr residue 124 was substituted by an Asp residue was found to act as an antagonist of IL-4 (Kruse *et al.*, 1992).

IL-4 Gene and its Regulation

The human IL-4 gene, composed of four exons and three introns, is localized on the long arm of chromosome 5 on bands q23–31 together with genes of other related cytokines including IL-3, IL-5, IL-9, IL-13 and GM-CSF (Morgan *et al.*, 1992) (Fig. 1). The IL-4 gene displays many potential binding sites for transcriptional factors, suggesting that its regulation is mediated by transcriptional complexes, depending on both cell types and costimulatory signals.

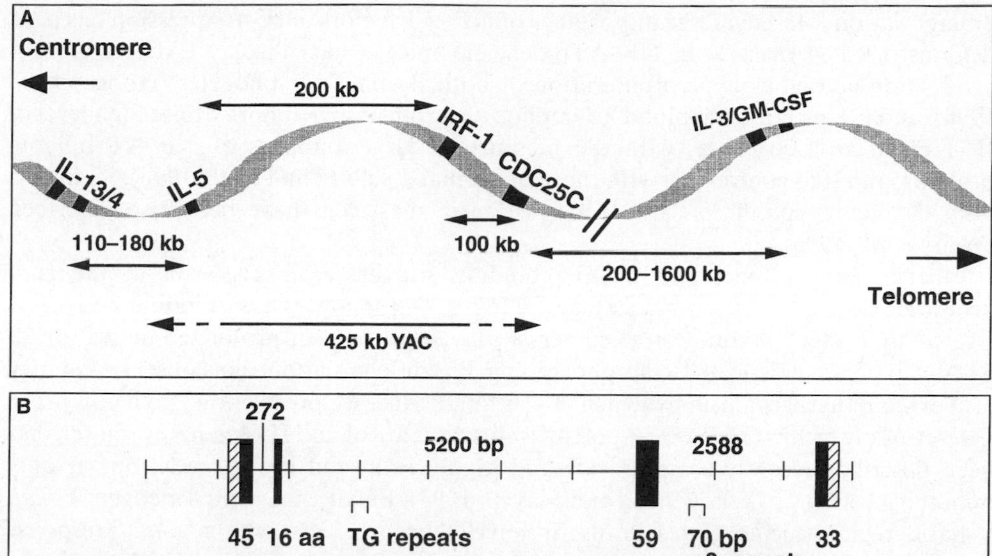

Fig. 1. IL-4 gene. (A) A map showing genes around human chromosome 5q31–1. IRF-1, which encodes a transcription activator of IFNα/β and other IFN-inducible genes. CDC 25C, cell division cycle 25; YAC, yeast artificial chromosome. (B) Schematic representation of human IL-4 gene. Coding regions of exons are indicated by black boxes; untranslated regions are indicated by hatched boxes.

Positive Regulatory Sequences

The PRE-I (positive regulatory enhancer) enhancer element, located between nucleotide −241 and −223, is critical for IL-4 gene expression (Li-Weber *et al.*, 1993), as deletion or mutation in the PRE-I region abolishes basal and PHA/PMA (Phytohemagglutinin/Phorbol Myristate Acetate) induced promoter activity. The PRE-I element interacts with at least two nuclear transcriptional complexes, POS-1 (positive element binding protein) and POS-2 (Li-Weber *et al.*, 1993), constituted upon cell type by transcription factors such as C/EBP-γ (CCAAT/Enhancer Binding Protein) (Davydov *et al.*, 1995a), NF–IL-6 (Nuclear Factor) (Davydov *et al.*, 1995b), NF–IL-6β, jun or NF–AT (Nuclear Factors of Activated T cells) factors (Li-Weber *et al.*, 1997). Of interest, the restricted expression of NF–IL-6 and POS-1 to T helper 2 T_{H2} cells would indicate the importance of PRE-I and the composition of its nuclear transcriptional complexes in the cell type-specific expression of IL-4 (Davydov *et al.*, 1995b; Li-Weber *et al.*, 1997).

The IL-4 gene also contains in its 5′ flanking region, five sites of a purine-rich motif: ARE (Activated Responsive Element) or P sequences that bind the NF–AT transcription factor family, specific of T cells (Abe *et al.*, 1992; Szabo *et al.*, 1993). These P sequences and the IL-4 −741 to +60 base pairs regions located outside the IL-4 promoter have been reported as IL-4 promoter elements conferring T_{H2}-restricted expression (Wenner *et al.*, 1997). Indeed, NF–AT DNA binding is equally induced in both T_{H1} and T_{H2} cells after antigen stimulation, but compared with that of T_{H1} cells only the NF–AT complexes present in T_{H2} cells promote a high level of IL-4 transcription coinciding with an increase in IL-4 production (Rincon and Flavell, 1997). This would suggest that regulation of NF–AT factors appears to be essential in T_{H1}/T_{H2}

differentiation. Moreover, the importance of NF–AT factors in IL-4 expression has been demonstrated *in vivo*: (1) in NF–ATp-deficient mice, which display a defect in IL-4 production as well as hyperproliferation of both B and T lymphocytes (Hodge *et al.*, 1996); and (2) in peripheral blood T lymphocytes from atopic donors whose high level of IL-4 production correlates with the presence of large amounts of NF–AT binding proteins, and thus contrasting with those of normal T cells (Chan *et al.*, 1996). Similarly, NF–AT factors specific of IL-4 transcription in mast cells have been also described (Weiss *et al.*, 1996).

Furthermore, a functional STAT6 binding site has been reported in the IL-4 promoter, located between nucleotides −167 to −135 and whose binding activity is related to T_{H2} cell maturation (Lederer *et al.*, 1996). Indeed, prolonged activation of STAT6 is characteristic of T cells undergoing T_{H2} differentiation. This observation may also explain the mechanism by which IL-4 regulates its own production (see below). The transcription factor GATA-3, a potent transactivator of the IL-4 gene promoter, has been described recently as selectively and highly expressed in T_{H2} cells, but at only minimal levels in T_{H1} cells (Zheng and Flavell, 1997). Finally, the proto-oncogene c-*maf*, a basic region/leucine zipper transcription factor, interacts with a c-*maf* response element in IL-4 promoter and controls also selective expression of IL-4, as shown by its restrictive expression in T_{H2} but not T_{H1} cells (Ho *et al.*, 1996).

The IL-4 promoter contains also two OAP-40 sites that interact with the AP-1 (Activation Protein) transcription factor family (Rooney *et al.*, 1995; Tara *et al.*, 1995). Nucleotide substitution, as a result of nucleotide polymorphism inside these sites, induces overexpression of the IL-4 gene (Song *et al.*, 1996). Another IL-4 promoter site, homologous to the Y box found in all major histocompatibility complex (MHC) class II promoters, binds the ubiquitous nuclear factor NF-Y (Szabo *et al.*, 1993).

Negative Regulatory Sequences

The human IL-4 promoter displays also several negative regulatory elements: (1) a silencer region containing two protein binding sites NRE-I (Negative Regulatory Element) and NRE-II, which interacts with the T cell-specific Neg-1 and the ubiquitous Neg-2 nuclear factors (Li-Weber *et al.*, 1992), and suppresses the activity of the PRE-I enhancer element (Li-Weber *et al.*, 1993); (2) an interferon (IFN)-stimulation response element (ISRE) in which mutations increase by twofold the IL-4 reporter gene activity (Li-Weber *et al.*, 1994), and which binds IRF-2 (Interferon Regulatory Factor), a transcriptional repressor of interferon genes.

Recently, transfection of the nuclear tumor suppressor protein p53 in activated T-cell lines has been described as downregulating IL-4 expression (Pesch *et al.*, 1996). The use of deleted p53 species also demonstrates the importance of the transactivation and oligomerization domains of p53 for IL-4 transcriptional repression. This suggests the presence of regulatory p53 binding sites in the promoter of the IL-4 gene.

Other Mechanisms

IL-4 production is also controlled post-transcriptionally through stabilization of IL-4 mRNA (Dokter *et al.*, 1993a). More recently, it has been also shown that IL-7 increases IL-4 mRNA expression in human anti-CD3/anti-CD28-activated T lymphocytes, via mRNA stabilization (Borger *et al.*, 1996a). In contrast, this anti-CD3/anti-CD28-induced accumulation of IL-4 mRNA is downregulated by prostaglandin (PG)E_2 or

2′-0-dibutyryl cyclic AMP (Borger *et al.*, 1996b), known to activate the protein kinase A pathway, and thus indicating its negative role in IL-4 expression.

Finally, treatment of Jurkat T cells with calcium ionophores (ionomycin), increases the transcriptional activity of IL-4 promoter, suggesting also the involvement of the Ca^{2+} signaling system in IL-4 gene regulation (Paliogianni *et al.*, 1996).

Overall, these findings suggest that the promoter of the IL-4 gene largely contributes to the control of T-cell differentiation, as a key binding site for specific T_{H2} transcription factors.

INTERLEUKIN-4 RECEPTOR COMPLEX

High-affinity (K_d = 40–120 pM) receptors of IL-4 are expressed in low numbers on a wide range of cell types, including T and B lymphocytes, monocytes, granulocytes, fibroblasts, epithelial and endothelial cells (Cabrillat *et al.*, 1987; Park *et al.*, 1987). IL-4 is able to upregulate the expression of its own receptor after inducing its transient downregulation following receptor ligand internalization (Galizzi *et al.*, 1989). Moreover, cross-linking studies show that IL-4 binds to three molecular species of molecular weight 140, 70–75 and 65 kDa (Foxwell *et al.*, 1989; Galizzi *et al.*, 1989), indicating the multimeric structure of IL-4R.

A specific cDNA coding for the human 140-kDa IL-4 binding protein (gp140/IL-4Rα/CDw124) has been isolated (Galizzi *et al.*, 1990; Idzerda *et al.*, 1990) (Table 1). The mature receptor is a glycoprotein composed of 800 amino acids, with an extracellular domain of 207 amino acids containing the two motifs (four conserved Cys and a WSXWS box) characteristic of the cytokine receptor family (Miyajima *et al.*, 1992), a single 24-amino-acid transmembrane domain, and a long 569-amino-acid intracellular portion. The 65-kDa protein represents the IL-2 receptor γc chain, a component common to other cytokine receptors (Noguchi *et al.*, 1993a), which increases by two- to three-fold the affinity of IL-4 for gp140/IL-4Rα in lymphoid cells. Genetic alterations in the γc gene located on the X chromosome, as well as the wide distribution of this chain in cytokine receptors, largely explain the severity of clinical features of patients with X-linked severe combined immune deficiency (X-SCID) (Noguchi *et al.*, 1993b). In addition, this γc chain remains undetectable on human renal cell carcinoma cells, which, however, binds IL-4 efficiently (Obiri *et al.*, 1995), suggesting the participation of another subunit in the IL-4R complex.

Binding experiments on various cell types have also demonstrated the association of the gp140/IL-4Rα with one of the IL-13 receptor chains (Kruse *et al.*, 1992; Obiri *et al.*, 1995; Zurawski *et al.*, 1995), which represents another IL-4R, termed the type II IL-4R. Recently cloned (Aman *et al.*, 1996; Caput *et al.*, 1996), the two IL-13Rα cDNAs both encode glycoproteins of 70–75 kDa with short cytoplasmic domains. Further studies are required to determine the role of these IL-13Rα chains in the IL-4R complex on different cell types (Obiri *et al.*, 1997).

Finally, a soluble form of the extracellular domain of gp140 was found to bind IL-4 with high affinity and to inhibit its biologic effects (Garrone *et al.*, 1991). Similarly, a 40-kDa soluble form of IL-4 binding protein (sIL-4R) has been described as a transport protein preventing IL-4 enzymatic degradation (Fernandez-Botran and Vitetta, 1990, 1991). This results from either an alternative splicing of IL-4R mRNA after IL-4 stimulation or a proteolytic shedding of membrane-bound IL-4R after T-cell receptor (TCR) stimulation (Blum *et al.*, 1996).

SIGNALING CASCADE OF IL-4

After binding of IL-4, dimerization of the IL-4Rα with the γc chain, as well as cytoplasmic regions of both IL-4Rα and the γc chain have been shown to be essential for IL-4-induced growth signal transduction (Harada *et al.*, 1992; Russel *et al.*, 1993; Kawahara *et al.*, 1994). Furthermore, IL-4 induces tyrosine phosphorylation of multiple proteins, including IL-4Rα itself and species of 170, 130, 110–120, 100 and 92 kDa (Izuhara and Harada, 1993). The lack of consensus sequences established for catalytic activity in the cytoplasmic domain of gp140 or that of the γc chain demonstrates the recruitment of non-receptor tyrosine kinases. Thus, IL-4 fails to activate components of the ras (Duronio *et al.*, 1992), MAP (Mitogen Activated Protein) kinase (Foltz *et al.*, 1997) or SAPK (Foltz and Schrader, 1997) pathways, but mobilizes the Janus tyrosine kinase (JAK) family that activates p170/IRS-2 (Insulin Receptor Substrate) substrate (Welham *et al.*, 1995) (Fig. 2). This molecule associates with the 85-kDa subunit of the phosphoinositol-3 kinase (PI-3 kinase) involved in cell proliferation. Another important element of the IL-4 pathway consists in the activation of signal transducers and activators of transcription proteins such as STAT6 (Signal Transducer and Activator of Transcription) (Hou *et al.*, 1994) or STF–IL-4 (Signal Transducing Factor) (Schindler *et al.*, 1994). However, signaling cascades specific to each cell type and distinct from that of the classical IL-4Rα/γc chain, even less clearly defined, have been also described.

B Lymphocytes

IL-4 augments phosphorylation of JAK1, JAK3 and TYK-2 in human normal B lymphocytes, in Epstein–Barr virus-immortalized B lymphocytes (Murata and Puri, 1997) and in malignant B lymphocytes (Tortolani *et al.*, 1995). However, tyrosine phosphorylation of JAK1 by IL-4 in normal B cells is less important than in TF-1 cells (Izuhara *et al.*, 1996b). Moreover, deficient or no tyrosine phosphorylation of JAK3 is detectable in B lymphocytes of patients with X-SCID (Izuhara *et al.*, 1996b; Oakes *et al.*, 1996). Of interest, it has been demonstrated that JAK3 is not expressed constitutively in plasmacytoid cells, which represent the final stage of B-cell differentiation (Tortolani *et al.*, 1995). This suggests that downregulation of JAK3 occurs during B-cell differentiation into plasma cells.

IL-4 induces also phosphorylation of IRS-2 in murine plasmacytoma cell line B9 and splenic B cells (Welham *et al.*, 1995). Moreover, engagement of IL-4 to its receptor on the human B-cell line BL-2 (Burkitt Lymphoma) activates both JAK3 as well as STAT6, which interacts with an IFN-γ activation site located upstream of the I epsilon exon (Fenghao *et al.*, 1995). Generation of STAT6-deficient mice has further demonstrated its requirement in IL-4-mediated maturation and cellular functions of B lymphocytes (Shimoda *et al.*, 1996; Takeda *et al.*, 1996). Indeed, IL-4-induced proliferation and CD23 or MHC class II expression is abolished in STAT6-deficient B lymphocytes. However, activation of STAT6 in B-cell lines derived from patients with X-SCID lacking the functional γc chain or JAK3 remains controversial (Russel *et al.*, 1995; Izuhara *et al.*, 1996b; Oakes *et al.*, 1996).

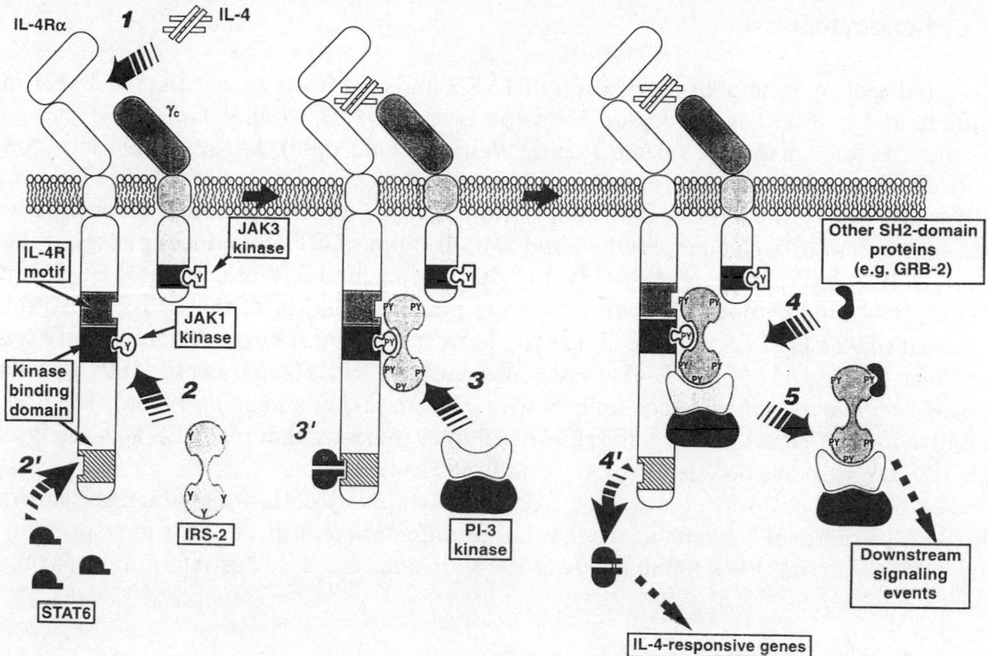

Fig. 2. IL-4 signaling cascade through the gp140/γc IL-4 receptor complex. (1) Binding of IL-4 to the 140 kDa chain induces dimerization of the IL-4Rα chain with the γc chain. After this interaction, tyrosine kinase associated to the receptor becomes activated and tyrosine phosphorylates the Tyr residues of the IL-4Rα. IL-4 binding allows also interactions of JAK-1 with box1 and box2 sequences located near the cellular membrane. (2) The IRS-2 protein interacts with the IL-4Rα and then becomes phosphorylated. (2′) Monomers of STAT-6 bind to the IL-4Rα chain. (3) Phosphorylated IRS-2 displays high affinity binding for SH2 domains of the p85 subunit of the phosphoinositol-3 kinase (PI-3 kinase), inducing its activation. (3′) STAT-6 is tyrosine phosphorylated by JAK-1, allowing its dimerization and its release from the IL-4Rα chain. (4) Phosphorylated IRS-2 proteins may interact also with other SH-2-domain containing proteins. (4′) Translocation of phosphorylated STAT-6 dimers to the nucleus leads to the activation of IL-4-regulated genes. (5) IRS-2–PI-3 kinase complex is released from the IL-4Rα to transduce IL-4 signal.

IL-4-dependent phosphorylation of a B-cell membrane-bound 42–44-kDa protein has been reported early but not further characterized in human B lymphocytes (McGarvie and Cushley, 1989). The contribution of phosphoinositide metabolism to IL-4 signaling is shown by the increase in inositol 1,4,5-triphosphate (IP_3) and intracellular Ca^{2+} levels, followed by intracellular cAMP accumulation via adenylate cyclase (Finney *et al.*, 1990). This increase in cAMP concentration may explain the inhibitory effect of IL-4 on IL-2-induced B-cell proliferation, as cAMP-inducing agents inhibit IL-2-induced B-cell proliferation without altering that of IL-4 (Vasquez *et al.*, 1991; Garrone and Banchereau, 1993). An IL-4-dependent increase in cAMP concentration is also observed in natural killer (NK) cells (Blay *et al.*, 1990). In contrast, IL-4 does not induce the release of IP_3, Ca^{2+} mobilization or protein kinase C (PKC) translocation in murine B cells, but synergizes with non-mitogenic concentrations of anti-immunoglobulin to provoke translocation of PKC from the cytosol to the membrane (Mizuguchi *et al.*, 1986). IL-4 activates two distinct ion channels in B lymphocytes, inducing an inward rectifying K^+ channel and activating a large conductance anion channel (McCann *et al.*, 1991).

T Lymphocytes

IL-4 induces tyrosine phosphorylation of JAK3 and to a lesser extent that of JAK1 in human and murine T lymphocytes (Johnston *et al.*, 1994; Witthuhn *et al.*, 1994), as well as that of IRS-2 in murine splenic T cells (Welham *et al.*, 1995). JAK1 forms complexes with IL-4Rα and the IRS-2 proteins in D10 T cells upon stimulation with IL-4 (Yin *et al.*, 1994). Moreover, the tyrosine phosphorylation of IRS-2 appears to be cell dependent, as IL-4 treatment of T-cell lines HDK-1 and D10, but not of CT.4S, induces phosphorylation of IRS-2 (Welham *et al.*, 1995). In addition, in mouse T-cell lines CTLL-2 and HT-2, IL-4 binding does not induce tyrosine phosphorylation of the IRS-2 substrate, but that of 92-kDa FES tyrosine kinase (Izuhara *et al.*, 1994, 1996a). A domain between residues 353 and 431 of the IL-4Rα associates with p92 FES (Izuhara *et al.*, 1994), which becomes phosphorylated and complexes with PI_3 kinase (Izuhara *et al.*, 1996a). The non-involvement of IRS-2 pathway in these T-cell lines indicates that p92 FES acts solely as an adapter molecule between PI_3 kinase and the IL-4Rα.

The STAT6 pathway is also involved in T-lymphocyte IL-4 signaling, as shown by STAT6-deficient T lymphocytes that fail to differentiate into T_{H2} cells in response to IL-4 (Kaplan *et al.*, 1996; Shimoda *et al.*, 1996). In contrast, T-cell proliferation was only partly blocked.

Myeloid Cells

Tyrosine phosphorylation of JAK3 is activated in response to IL-4 in monocytes (Musso *et al.*, 1995) and in myeloid cells (Witthuhn *et al.*, 1994; Welham *et al.*, 1995). In the human erythroleukemia cell line TF-1, IL-4 stimulates increased tyrosine phosphorylation of JAK1 and TYK-2, even if levels of phosphorylated JAK1 remain very low (Welham *et al.*, 1995). IL-4 induces striking tyrosine phosphorylation of IRS-2 in factor-dependent myeloid cell lines and TF-1 cells (Wang *et al.*, 1992; Welham *et al.*, 1995). Of interest, no tyrosine phosphorylation of IRS-2 is detectable in the myeloid cell line 32D treated with IL-4 (Wang *et al.*, 1993). In contrast, activation of the STAT6 pathway is described in the IL-4-treated cell line 32D, and requires the distal portion of the intracellular domain of IL-4Rα (Quelle *et al.*, 1995). In addition, the IL-4-induced expression of IL-1 receptor antagonist (IL-1Ra) is mediated by STAT6 in the macrophage-like cell line RAW264.7 (Ohmori *et al.*, 1996). STAT6 binds in the IL-1Ra promoter to a region of 50 base pairs containing two inverted repeat elements, termed STAT binding elements (SBEs).

IL-4-induced proliferation of human myeloid cell lines also involves the activation of a tyrosine-specific phosphatase in association with the dephosphorylation of an 80-kDa protein (Mire-Sluis and Thorpe, 1991). It has been also demonstrated that, in monocytes, IL-4 induces the phosphorylation of LsK, a tyrosine kinase with homology to the *C*-terminal Src kinase (Musso *et al.*, 1994) which represents an alternative of the JAK pathway.

Treatment of monocytes with IL-4 induces a significant redistribution of PKC from the cytosol to the nucleus (Arruda and Ho, 1992). IL-4 can increase the levels of cyclic GMP in monocytes by activation of soluble guanylate cyclase (Kolb *et al.*, 1994). This is most clearly seen when phosphodiesterase inhibitors are added because IL-4 itself is able to stimulate the phosphodiesterase activity of monocytes (Li *et al.*, 1989). Interestingly,

IL-4 induces monocytes to release nitric oxide (NO) and this stimulation of NO synthase is likely to explain the accumulation of cyclic GMP (Kolb *et al.*, 1994). Finally, the IL-4-induced inhibition of cytokine production by monocytes may be a consequence of the inhibition of c-*fos* and c-*jun* expression whose complex constitutes the transcription factor AP-1 (Dokter *et al.*, 1993b).

Other Cell Types

In human fibrosarcoma cells, colonic carcinoma cell lines and vascular endothelial cells, IL-4 fails to phosphorylate JAK3 but induces that of JAK1, JAK2 and TYK-2 without implication of the γc chain (Murata *et al.*, 1996; Wang *et al.*, 1997). However, mutagenesis experiments have shown that JAK1 alone is responsible for the subsequent phosphorylation of IRS-2 in human fibrosarcoma cells (Wang *et al.*, 1997). This suggests that the type II IL-4R is found on these cells and that JAK2 may represent the tyrosine kinase associated with the IL-13Rα (Murata *et al.*, 1996).

Of interest, tyrosine phosphorylation of IRS-2 induced by IL-4 in human colorectal carcinoma cell lines is associated to a growth inhibitory signal (Schnyder *et al.*, 1996), suggesting that activation of IRS-2 may act both as a positive and a negative element in the control of cell proliferation. Finally, activation of the STAT6 pathway has also been reported in human vascular endothelial cells (Palmer-Crocker *et al.*, 1996) and in human colonic carcinoma cell lines (Murata *et al.*, 1996).

CELLULAR SOURCES OF IL-4

Unlike IL-1, IL-6 and IL-10, which are produced by many different cell types, IL-4, like IL-2 and IFN-γ, is secreted by restricted cell types.

T Lymphocytes

Studies with murine helper CD4$^+$ T-cell clones have indicated the presence of two cell types on the basis of their pattern of cytokine synthesis: T_{H1} cells secrete IL-2, LT-α (lymphotoxin) and IFN-γ; T_{H2} cells secrete IL-4, IL-5, IL-6 and IL-10; and both types secrete cytokines such as GM-CSF, IL-3 and IL-13 (Romagnani, 1992).

Human CD4$^+$ T-cell clones specific for bacterial antigens, and allergen- or helminth-specific T-cell clones have been found to exhibit T_{H1}- or T_{H2}-like cytokine production profiles (Parronchi *et al.*, 1991; Locksley, 1994). CD4$^+$ T cells with an intermediate cytokine profile T_{H0} have been also described (Rocken *et al.*, 1992). T_{H1} and T_{H2} cells differentiate from a pool of multipotent precursors through differentiation pathways controled either intracellularly by specific transcription factors (see above) or extracellularly by cytokines and costimulatory molecules expressed by environmental lymphocytes or accessory cells (Fig. 3).

Indeed, IFN-γ and IL-12 promote differentiation of T helper precursors into T_{H1} cells (Manetti *et al.*, 1994), whereas IL-4 induces differentiation into T_{H2} cells (Seder *et al.*, 1992). Interestingly, IL-6 has recently been described as inducing the commitment of T_H precursors in T_{H2} cells, as its addition to primary CD4$^+$ T-cell cultures induces an enhancement of IL-4 production with a concomitant decrease in IFN-γ production (Rincon *et al.*, 1997). This differentiation pathway can be mediated by activation of the

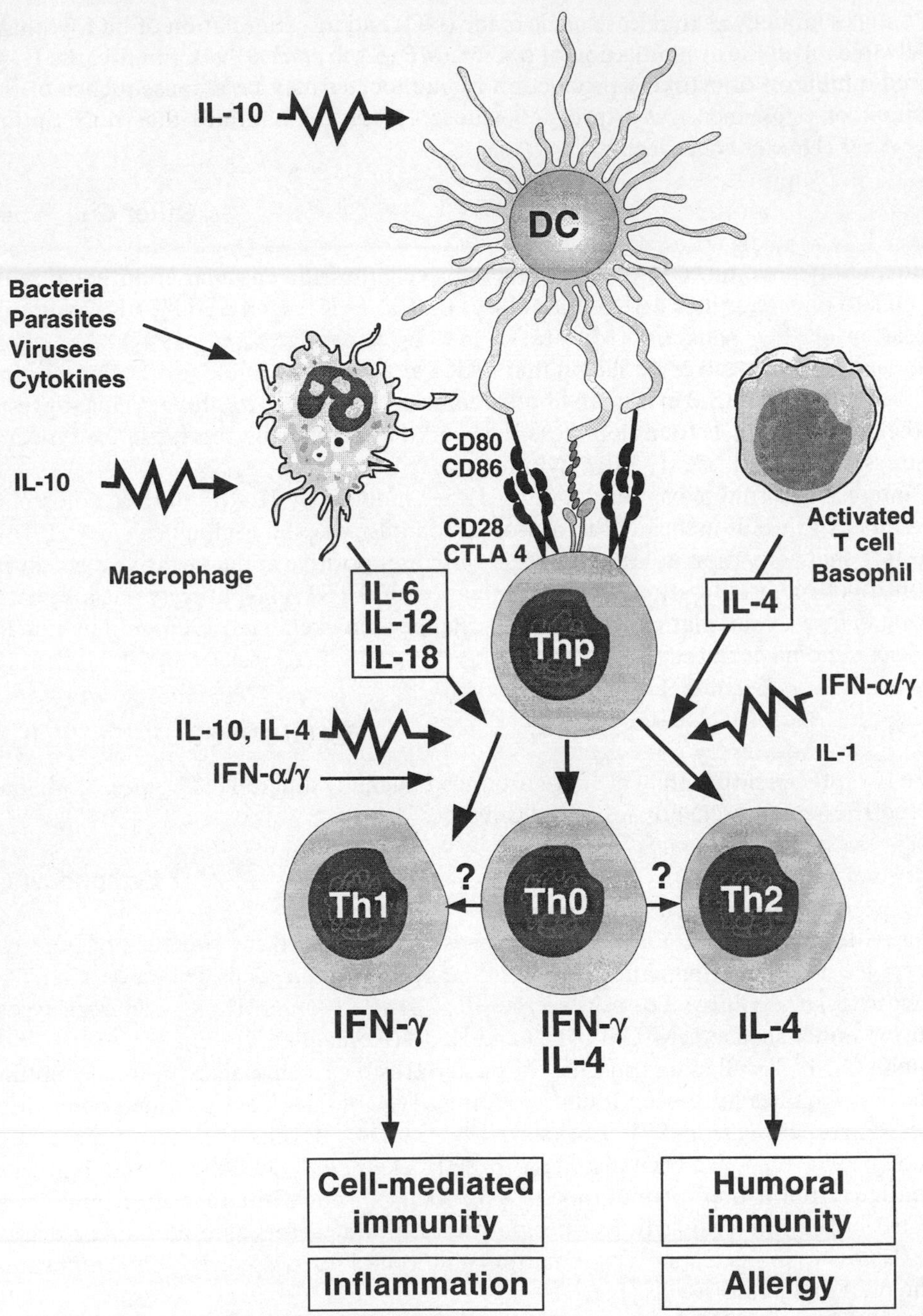

Fig. 3. Functional roles and cytokine control of T_{H1}/T_{H2} development. Protective immunity results from the balanced activation of both subsets. Hyperactivation of either subset leads to immune-mediated diseases. CTLA, cytotoxic T lymphocyte associated antigen.

NF–IL-6 family, whose binding sites are found in the IL-4 gene promoter (Davydov *et al.*, 1995b). Moreover, blockade of IL-6-induced T_{H2} differentiation by anti-IL-4 antibodies indicates that IL-6 acts through induction of the endogenous production of IL-4, which thus promotes maturation of T_{H2} cells. Upregulation of IL-4R expression has recently been reported during T-cell maturation, by a mechanism mediated by IL-4 itself and not suppressed by IL-12 priming (Nakamura *et al.*, 1997). This suggests that endogenous IL-4 autocrine loop favors T_{H2} cell maturation. IL-18, a newly cloned cytokine, has been also described to promote IFN-γ production by T lymphocytes, and thus to favor T_{H1} maturation (Okamura *et al.*, 1995). Studies on the cellular sources of controlling cytokines indicate that IFN-γ and IL-6, IL-12 or IL-18 inducing the maturation of T_{H1} cells originate from NK cells and macrophages respectively. However, the initial source of IL-4 required to promote T_{H2} commitment is still open to debate (Coffman and von der Weid, 1997) and may come from a variety of cell types including $CD4^+$ $NK1.1^+$ T cells (Yoshimoto and Paul, 1994), basophils and mast cells (Moqbel *et al.*, 1995), $CD4^+$ memory T cells or by naive $CD4^+$ T cells themselves (Rincon *et al.*, 1997).

Furthermore, additional evidence demonstrates the role of costimulatory molecules, such as B7-1 and B7-2 and their CD28–CTLA-4 counter-receptors on T lymphocytes, in the development of IL-4-producing T_{H2} cells (Gause *et al.*, 1997). Indeed, numerous studies have observed that priming of T lymphocytes for IL-4 production generally requires CD28–B7 interactions, whereas activation of already differentiated T_{H2} cells occurs in a B7-independent fashion (Wallace *et al.*, 1995; Scholz *et al.*, 1996). In addition, the potential role of CD40–CD40L interactions has been postulated as activation of $CD4^+$ T lymphocytes by CD40L and CD3–CD28 triggering, but not by CD3–CD28 triggering alone, greatly enhances IL-4 synthesis (Blotta *et al.*, 1996). However, *in vivo* administration of anti-CD40L antibodies has no effect on number of secreting IL-4 cells, although it blocks *Heligmosomoides polygyrus*-induced B proliferation and IgG_1 production (Lu *et al.*, 1996).

As expected from the pattern of cytokine production, T_{H1} and T_{H2} cells regulate distinct biologic functions. T_{H1} cells induce the activation of macrophages resulting in delayed-type hypersensitivity responses and the killing of intracellular parasites. In contrast, T_{H2} cells control more particularly humoral responses including the production of IgE and associated eosinophilia. An important feature of T_{H1} and T_{H2} cells is the ability of one subset to regulate the activity of the other. This occurs at the level of the effector cells triggered by these subsets, as indicated by the inhibitory effects of IFN-γ on IL-4-induced B-cell activation or those of IL-4 on IL-2-induced T- and B-lymphocyte proliferation. It also occurs directly at the level of these subsets, as the products of one subset can antagonize the activation of the other: IFN-γ inhibits proliferation of T_{H2} cells (Gajewski *et al.*, 1989), whereas IL-4 inhibits cytokine production by T_{H1} cells (Peleman *et al.*, 1989).

Other Cell Types

Both murine and human IL-4 are also produced by basophils and mast cells activated by cross-linkage of FcεRI and FcγRII (Brown *et al.*, 1987; Brunner *et al.*, 1993). Nasal biopsy specimens from patients with allergic rhinitis and bronchial biopsy specimens from those with allergic asthma display mast cells with intracellular IL-4 (Bradding *et al.*, 1992). Interestingly, anti-IgE induces maximum release of IL-4 within 1 h, suggesting that it is present within the cells in a preformed state.

A minor population of mouse $CD4^+$ $NK1.1^+$ T cells has been shown rapidly to produce large amounts of IL-4 in response to *in vivo* challenge with anti-CD3 (Yoshimoto and Paul, 1994), suggesting that these cells may have the potential to provide initial IL-4 at the onset of T_{H2} maturation. However, *in vivo* depletion of $CD4^+$ $NK1.1^+$ T cells does not abrogate early depletion of IL-4 mRNA in *Leishmania major* antigen-stimulated mice (von der Weid *et al.*, 1996). Finally, freshly isolated and stimulated eosinophils constitutively express IL-4 mRNA and secrete IL-4 protein (Nakajima *et al.*, 1996). Addition of IL-5 to eosinophils enhances the production and intracellular storage of IL-4 proteins.

EFFECTS OF IL-4 ON B LYMPHOCYTES

Ontogeny of B Lymphocytes

IL-4-dependent murine proB-cell lines can be established in the presence of stromal cells (Peschel *et al.*, 1989). In contrast, IL-4 inhibits stromal cell-dependent proliferation of pre-B cells, possibly through the production by stromal cells of inhibitory factors (Peschel *et al.*, 1989). Studies in humans have also shown that IL-4 can inhibit the spontaneous proliferation of progenitor B cells, as well as that induced by IL-7 (Pandrau *et al.*, 1992; Ryan *et al.*, 1994). This inhibition is also observed with human precursor B cells of acute lymphoblastic leukemia and appears to result from the induction of apoptosis in G_0/G_1 phases of cell cycle (Manabe *et al.*, 1994; Renard *et al.*, 1994). IL-4 plays a role in the final maturation step during bone marrow B lymphopoiesis, by enhancing the output of surface IgM^+ cells in murine cultures (Kinashi *et al.*, 1988). Finally, human bone marrow B-cell precursors have been shown to differentiate into immunoglobulin-secreting cells when cultured with IL-4 in association with activated T lymphocytes (Punnonen *et al.*, 1993). The normal B lymphopoiesis observed in IL-4-deficient mice suggests that these observations may represent *in vitro* findings or that alternative cytokines may replace this function of IL-4.

B Lymphocyte Activation

IL-4 increases the number of resting B lymphocytes (Vallé *et al.*, 1989) and induces their homotypic aggregation (Elenström and Severinson, 1989). IL-4 induces hyperexpression of MHC class II antigens on murine B cells (Noelle *et al.*, 1984), but this effect is less pronounced on human resting B lymphocytes, which already express high levels of these surface molecules (Diu *et al.*, 1990). However, IL-4 strongly enhances MHC class II antigen expression on Burkitt lymphoma cell lines and chronic lymphocytic leukemia cells (Rousset *et al.*, 1988).

IL-4 strongly enhances the expression of CD23 (the low-affinity receptor for IgE–FcεRII) on normal and leukemic B lymphocytes (Defrance *et al.*, 1987). The spontaneous expression of CD23 mRNA on B cells isolated from atopic patients suggests a recent encounter with IL-4. Moreover, its association with MHC class II antigens (Bonnefoy *et al.*, 1988a) appears to play an important role in the B-cell presentation of antigen to T lymphocytes (Flores-Romo *et al.*, 1990). IFN-α and γ block the IL-4-dependent increase of CD23 on B cells. IL-4 also induces the release of soluble CD23, which retains the capacity for IgE binding (Bonnefoy *et al.*, 1988b).

IL-4 upregulates the expression of surface IgM (Shields *et al.*, 1989), CD40 (Gordon *et al.*, 1988a), B7-1–CD80 (Vallé *et al.*, 1991) and B7-2–CD86. Mouse IL-4 induces Thy-1 on B cells (Snapper *et al.*, 1988), but decreases that of CDw32–FcγRII, explaining the IL-4-induced reversal of Fc receptor-mediated inhibition of B-lymphocyte activation (O'Garra *et al.*, 1987). In addition, IL-4 induces activated B cells to produce IL-6 and TNF (Smeland *et al.*, 1989), which favor the activation and expansion of activated T lymphocytes, but decreases that of IL-10 (Burdin *et al.*, 1995).

All the above effects of IL-4 on resting B lymphocytes support the role of IL-4 in the enhancement of antigen-presenting capacity of B cells towards T lymphocytes.

B Lymphocyte Proliferation

Antigen Receptor Triggering

IL-4 enhances DNA replication of B cells costimulated or preactivated with insolubilized anti-IgM antibody (Defrance *et al.*, 1987; Clark *et al.*, 1989). IL-4 does not costimulate with particles of *Staphylococcus aureus* strain Cowan (SAC) (Jelinek and Lipsky, 1988), but enhances DNA replication of SAC-preactivated B lymphocytes (Defrance *et al.*, 1987). Addition of PGE_2 and pharmacologic agents inducing intracellular cAMP (cholera toxin, dibutyryl cAMP, forskolin) enhances IL-4-induced DNA synthesis (Vasquez *et al.*, 1991; Garrone and Banchereau, 1993).

Paradoxically, IL-4 antagonizes the IL-2-induced DNA replication of B cells costimulated through their antigen receptor (Defrance *et al.*, 1988). The inhibitory effect is particularly striking in non-Hodgkin's B-cell lymphomas (Defrance *et al.*, 1992), but remains controversial on chronic lymphocytic leukemia B cells (Fluckiger *et al.*, 1994; Chaouchi *et al.*, 1996). This effect may be due to an IL-4-dependent sequestering of the γc chain, which is also a part of IL-2R (Lee *et al.*, 1990). IL-4 also inhibits TNF-α-induced proliferation of phorbol ester activated B-CLL (B-Chronic Lymphocytic Leukemia) cells (van Kooten *et al.*, 1992). However, while blocking DNA synthesis, IL-4 protects the B-CLL cells from death by apoptosis through increased expression of the Bcl-2 protein (Dancescu *et al.*, 1992).

CD40-Dependent Activation

Combinations of soluble anti-CD40 antibodies and either anti-IgM or phorbol esters induce DNA synthesis in B lymphocytes. Addition of IL-4 preferentially boosts the observed proliferation (Gordon *et al.*, 1988b). Although neither of these conditions results in long-term B-cell proliferation, the addition of IL-4 to B lymphocytes cultured in the CD40 system (combining irradiated fibroblastic L cells transfected with human FcγRII–CDw32 and anti-CD40 antibody) or CD40L-transfected L cells results in their sustained proliferation (Banchereau *et al.*, 1991), generating factor-dependent long-term normal B-cell lines. Moreover, agents increasing intracellular cAMP levels strongly enhance IL-4-induced cell proliferation (Garrone and Banchereau, 1993).

B-CLL cells synthesize DNA in the CD40 system, and addition of IL-4 further enhances this (Fluckiger *et al.*, 1992). This results in the expansion of viable leukemic cells, although the extent is lower than that obtained with normal B lymphocytes. Thus, the lack of growth promoting activity of IL-4 on B-CLL cells activated via surface immunoglobulin (sIg) may be due to altered signal transduction through sIg rather than to impaired IL-4R.

B-Lymphocyte Differentiation

Antigen Receptor Triggering

IL-4 induces IgG and IgM but not IgE production by SAC-preactivated B lymphocytes, most likely as a consequence of induced proliferation (Defrance *et al.*, 1988; Jelinek and Lipsky, 1988). Moreover, IL-4 blocks IL-2-induced immunoglobulin secretion by SAC-costimulated B lymphocytes. Inhibitory effects of IL-4 on antigen-specific immunoglobulin production have also been observed. In particular, the secondary response of B lymphocytes to influenza virus, which requires both antigen and IL-2, can be inhibited by IL-4 (Callard *et al.*, 1991). Likewise, the IL-2-dependent primary response to trinitrophenylated polyacrylamide beads is also inhibited by IL-4 (Llorente *et al.*, 1989). However, IL-4 blocks IL-2-dependent B-cell differentiation, whereas it stimulates the antigen-dependent B-cell proliferation (Llorente *et al.*, 1990).

CD40-Dependent Activation

Purified B lymphocytes cultured in the CD40 system produce low amounts of IgM, IgG and IgA. Addition of IL-4 increases the production of IgM and IgG and, more strikingly, the secretion of large amounts of IgE (Jabara *et al.*, 1990; Rousset *et al.*, 1991a). Thus, the IgE secretion dependent on T cell-induced IL-4 production results from CD40 triggering. However, other signals such as Epstein–Barr virus (Thyphronitis *et al.*, 1989) or hydrocortisone (Jabara *et al.*, 1991) can activate IgE production by IL-4-treated B lymphocytes. Addition of IFN-γ or IFN-α to anti-CD40-activated B cells surprisingly fails to inhibit IL-4-induced IgE production, thus contrasting with earlier studies in which B cells were stimulated by T lymphocytes (Pène *et al.*, 1988a,b). Of note, transforming growth factor-β (TGF-β) and PGE_2 were found to inhibit IL-4-induced IgE production of anti-CD40-activated B cells, while TNF-α was found to enhance it (de Waal Malefyt *et al.*, 1992; Garrone *et al.*, 1994). More recently, a role for IL-4 in IgA switching has been reported for lymphoma B cells CH12F3 cultured in IL-4 alone (Nakamura *et al.*, 1996). This effect is enhanced in the presence of CD40L and TGF-β. Similarly, treatment of normal murine B cells with IL-4, CD40L and IL-5 stimulates IgA switching (McIntyre *et al.*, 1995).

IL-4 and Isotype Switching

Many studies performed in both mouse and human have shown that IL-4-induced B lymphocytes produce IgE following isotype switching. Single $sIgD^+$ B lymphocytes cultured for 10 days in the CD40 system with IL-4 yielded B-cell clones whose isolated cells expressed the same VDJ genes coupled to different constant region genes (Galibert *et al.*, 1995). Isotype switching is associated with a DNA recombination event, in which C_H (Heavy Chain) genes originally lying between $S\mu$ and the newly expressed C_H gene are deleted by excision of a circular piece of DNA (Snapper *et al.*, 1997) and is preceded by expression of germline $C\varepsilon$ transcripts that initiate 5′ of the switch region specific for the constant region to be expressed and which encompass the downstream C_H gene. IL-4 is able to induce purified resting B lymphocytes to express 1.8-kilobase germline $C\varepsilon$

transcripts (Gauchat *et al.*, 1990; Shapira *et al.*, 1992). The IL-4-dependent expression of the germline Cε transcripts is enhanced by TNF-α but decreased by TGF-β, thus explaining its inhibitory effects on IgE synthesis (Gauchat *et al.*, 1992). Interestingly, interferons and IL-6, which respectively block and stimulate IL-4 and T cell-dependent IgE synthesis, do not modify the levels of Cε germline transcripts. Molecular studies demonstrate that, in contrast to murine situation, the transcriptional activity of the human IgE germline induced by IL-4 is not mediated by the transcription factor B-cell activator protein (BSAP) (Albrecht *et al.*, 1996), but requires an IFN-γ-activated site in the IgE germline promoter (Ezernieks *et al.*, 1996). Furthermore, treatment of IL-4/CD40-activated B cells by protein tyrosine kinase inhibitors inhibits IgE synthesis as well as Sμ → Sε deletional switch recombination (Loh *et al.*, 1994). This indicates the importance of protein tyrosine kinase, such as JAK3 (Fenghao *et al.*, 1995), in both IL-4 and the CD40 signaling pathway that leads to IgE switching.

Numerous studies in mice with parasitic infections or treated with anti-IgD have confirmed the fundamental role of the IL-4 in the regulation of circulating IgE levels (Finkelman *et al.*, 1990). Anti-IL-4 suppresses the eosinophilia, hyper-IgE and intestinal mastocytosis found in helminth infections, but not IgG_1 and protective immunity to the infection. Moreover, inactivation of the IL-4 gene in mice was associated with normal T- and B-cell development but with a strong reduction of IgE levels after infection with the nematode *Nippostrongylus brasiliensis* (Kühn *et al.*, 1991) or with that of *Brugia malayi* (Lawrence *et al.*, 1995). However, infection of IL-4-deficient mice with a murine retrovirus-induced immunodeficiency syndrome induces raised serum IgE levels comparable to those in control mice (Morawetz *et al.*, 1996), suggesting that an IL-4-independent pathway for IgE switching occurs in mice following retroviral infection. Conversely, IL-4 transgenic mice had increased IgE levels and an allergic-like disease with ocular lesions infiltrated with mast cells and eosinophils (Tepper *et al.*, 1990). As mice that have been induced to a hyper-IgE state display an increased production of IL-4 and a decreased production of IFN-γ through a predominance of a T_{H2} *versus* a T_{H1} response, studies have been performed in humans attempting to correlate IL-4 status with IgE status. Some of these concluded that atopic patient blood mononuclear cells display an increased capacity to produce IL-4 and/or a decreased capacity to produce IFN-γ in response to polyclonal activation (Rousset *et al.*, 1991b).

EFFECTS OF IL-4 ON T LYMPHOCYTES

Effects on Thymocytes

Both murine (Zlotnik *et al.*, 1987) and human (Barcena *et al.*, 1990) thymocytes proliferate in response to IL-4 and phorbol esters. In the mouse, IL-4 induces the proliferation of the most immature $CD4^-$ $CD8^-$ subset, which also has the ability to produce IL-4. The intermediate $CD8^+$ $CD4^+$ subset virtually fails to proliferate under these conditions, whereas the most mature $CD4^+$ $CD8^-$ and $CD4^-$ $CD8^+$ subsets also proliferate in response to IL-4 and PMA. Fetal thymocytes also respond to IL-4, and *in situ* hybridization studies have demonstrated the transcription of the IL-4 gene in fetal thymus (Sideras *et al.*, 1988). In addition, IL-4 can induce the maturation of thymocytes through the induction of mature T-cell antigens (CD3, CD5, TCR) and loss of CD1.

Concomitantly, the $CD4^+$ $CD8^+$ subset disappears and $CD4^-$ $CD8^-$, $CD4^+$ $CD8^-$, $CD4^-$ $CD8^+$ cells are generated (Ueno *et al.*, 1989). These data indicate that IL-4 plays an important role in T-cell ontogeny. Paradoxically, IL-4 transgenic mice were found to have involuted thymuses (Tepper *et al.*, 1990). In accordance with this latter finding, IL-4 inhibits early T-cell development in fetal thymus organ culture, probably during the differentiation of $CD4^-$ $CD8^-$ cells into $CD4^+$ $CD8^+$ thymocytes. The generation of TCR-$\alpha\beta$ thymocytes appears to be more impaired than that of TCR-$\gamma\delta$ thymocytes (Plum *et al.*, 1990). Likewise human IL-4 induces a preferential differentiation of TCR-γ/δ pre-T cells (Barcena *et al.*, 1990).

In addition, murine $CD8^+$ T cells, activated by IL-4, develop into a $CD8^-$ $CD4^-$ population that is not cytolytic, lacks perforin and does not produce IFN-γ, indicating that, like $CD4^+$ T cells, $CD8^+$ T cells may also differentiate into T_{H1} and T_{H2} subsets (Seder and Le Gros, 1995). Furthermore, these cells produce large amounts of IL-4, IL-5 and IL-10, and can induce the growth and differentiation of activated B cells (Erard *et al.*, 1993). IL-4 also appears to play an important role in the development of T_{H2} $CD8^+$ T cells (Sad *et al.*, 1995). In humans, IL-4-producing $CD8^+$ T cells have been observed in diseases such as leprosy (Salgame *et al.*, 1991), acquired immune deficiency syndrome (AIDS) (Paganelli *et al.*, 1995) or leishmaniasis (Uyemura *et al.*, 1993). The emergence of such cells may contribute to the disease, possibly because of reduced overall cytolytic activity against microbe-infected cells.

Effects of IL-4 on Mature T Cells

The T-cell growth-promoting effects of IL-4 were initially discovered in continuous T-cell lines (Mosmann *et al.*, 1986), and then on activated normal $CD4^+$ and $CD8^+$ T cells (Spits *et al.*, 1987) in an IL-2-independent fashion. Studies with antisense oligonucleotides showed that IL-4 and IL-2 are autocrine growth factors of T_{H2} and T_{H1} T-cell clones respectively (Harel-Bellan *et al.*, 1988). However, IL-4 blocks the specific IL-2-induced proliferation of peripheral blood naive $CD4^+$/$CD45RA^+$ T cells (Gaya *et al.*, 1990). IL-4 inhibits the production of IFN-γ by activated T cells (Peleman *et al.*, 1989), and thus favors the generation of T_{H2} cells.

When added to mixed leukocyte cultures, IL-4 increases antigen-specific cytotoxic activity against allogeneic stimulator cells (Spits *et al.*, 1988). It also enhances the development of virus-specific cytotoxic T cells (Horohov *et al.*, 1988). The endogenous production of IL-4 in response to viral challenge is likely to play an important role in the generation of antigen-specific cytotoxic cells. T-cell clones cultured with IL-4 specifically express a lipase, which may be involved in the cytolytic process, and display higher cytolytic ability than those cultured in IL-2 (Grusby *et al.*, 1990). Administration of sIL-4R to mice inhibits an allogenic response *in vivo* and enhances heart allograft survival, thus suggesting an *in vivo* role for IL-4 in the development of cytotoxic T cells (Fanslow *et al.*, 1991).

Effects of IL-4 on NK Cells

$CD3^-$ NK cells proliferate in response to IL-4 (Spits *et al.*, 1987). IL-4 inhibits the IL-2-dependent proliferation of these cells (Kawakami *et al.*, 1989) and accordingly blocks the IL-2-dependent generation of human lymphokine-activated killer (LAK) cells (Nagler *et *al.*, 1988). IL-4 also inhibits the production of TNF-α and serine esterase by NK cells (Blay *et al.*, 1990). The blocking effect of IL-4 on IL-2-induced cytotoxicity appears to be linked to its ability to raise cAMP levels in NK cells.

EFFECTS OF IL-4 ON MYELOMONOCYTIC CELLS

Effects of IL-4 on Hematopoiesis

IL-4 can either inhibit or enhance myelopoiesis from bone marrow progenitor cells. It blocks the development of colonies dependent on M-CSF (Jansen *et al.*, 1989), but stimulates the formation of colonies dependent on G-CSF (Rennick *et al.*, 1987; Broxmeyer *et al.*, 1988). IFN-γ and IL-4 reciprocally regulate the production of monocytes–macrophages and granulocytes (Snoeck *et al.*, 1993). Furthermore, in combination with IL-3, IL-4 induces the generation of basophils–mast cells from human (Favre *et al.*, 1990) and mouse (Rennick *et al.*, 1987) progenitor cells and of eosinophils (Favre *et al.*, 1990) from human progenitor cells. IL-4 also acts on the generation of eosinophils in mice, as administration of plasmacytoma-expressing IL-4 results in eosinophil infiltration of tumors (Tepper *et al.*, 1989, 1992). Furthermore, IL-4 transgenic mice display eye inflammation and conjunctivitis due to large numbers of eosinophils and mast cells (Tepper *et al.*, 1990). IL-4 is an essential growth factor for the *in vitro* growth of connective tissue-type mast cells (Tsuji *et al.*, 1990). Combinations of erythropoietin and IL-4 induce partially purified progenitor cells to generate erythroid colonies (Broxmeyer *et al.*, 1988) but this may be due to an indirect effect of IL-4 on accessory cells. In contrast, IL-4 inhibits the formation of pure and mixed megakaryocyte colonies from enriched human hematopoietic progenitors (Sonoda *et al.*, 1993).

Effects on Monocytes–Macrophages

Activation and Differentiation

IL-4 upregulates the expression of MHC class II antigens, LFA-1, CD13 and CD23 (Vercelli *et al.*, 1988) but downregulates that of CD14 (Vercelli *et al.*, 1989) and of the three FcγRs (Te Velde *et al.*, 1990). Accordingly, it blocks the antibody-dependent cytotoxicity of macrophages without affecting their phagocytic properties. Monocytes cultured in the presence of IL-4 acquire a macrophage-like dendritic cell morphology, as they increase in size and develop extensive processes (Te Velde *et al.*, 1988). Of interest, addition of both GM-CSF and IL-4 to blood monocyte cultures induces their differentiation into immature dendritic cells that can efficiently present soluble antigen to specific T-cell clones (Romani *et al.*, 1994; Sallusto and Lanzavecchia, 1994). Subsequent addition of TNF-α induces differentiation into mature $CD83^+$ dendritic cells that display increased antigen-presenting cell functions (Zhou and Tedder, 1996). Moreover, IL-4 appears to deliver a critical signal for maturation of monocytes into dendritic cells, as dendritic cells generated in the presence of GM-CSF and TNF-α show only low accessory cell capacity (Pickl *et al.*, 1996). Finally, IL-4 induces macrophage mannose receptor expression and thus promotes the formation of giant multinucleated cells (Defife *et al.*, 1997).

Production of Mediators

IL-4 blocks the spontaneous and lipopolysaccharide (LPS)-induced production of IL-1, IL-6, IL-8, IL-10, IL-12 and TNF-α by monocytes (Hart *et al.*, 1989; Standiford *et al.*, 1990). IL-4 also inhibits virus-dependent production of IFN-α/β by monocytes (Gobl and Alm, 1992). This inhibition of cytokine production is not the consequence of a general blockade of protein synthesis as IL-4 stimulates these cells to produce IL-1Ra (Fenton *et al.*, 1992) and the complement protein C2 (Littman *et al.*, 1989). However, the effects of IL-4 on cytokine production appear to be more complex, as pretreatment of monocytes with IL-4 paradoxically induces an enhancement of their production (D'Andrea *et al.*, 1995; Kambayashi *et al.*, 1996). Such a mechanism of cytokine production is controlled by endogenous IL-10, which itself inhibits monokine production: pretreatment by IL-4 before LPS stimulation results in a decrease of IL-10 production, whereas cotreatment markedly increases IL-10 production (Kambayashi *et al.*, 1996). The anti-inflammatory effect of IL-4 appears also to be linked to the differentiation state of monocytes, as only partial inhibition of IL-6 production is observed in monocytes isolated from rheumatoid fluid or synovial membrane (Chomarat *et al.*, 1995; Hart *et al.*, 1995). Moreover, the inhibitory effects of IL-4 on IL-6 release by monocytes are mediated through a reduction of AP-1 and NF–IL-6 binding activity (Dokter *et al.*, 1996).

IL-4 also inhibits the secretion of interstitial collagenase and 92-kDa type IV collagenase (Corcoran *et al.*, 1992; Lacraz *et al.*, 1992) thus reducing the ability of macrophages to degrade extracellular matrix. IL-4 has a profound inhibitory effect on the release of superoxide (Ho *et al.*, 1992), the secretion of PGE_2 (Hart *et al.*, 1989) and the induction of cyclo-oxygenase activity (Endo *et al.*, 1996). IL-4 also blocks transcription of the cellular genes *ISG-54* and *IP-10*, which are upregulated by IFN-α and IFN-γ respectively (Larner *et al.*, 1993). In addition, both surface and soluble p75 TNF receptors are downregulated by IL-4 on monocytes (Hart *et al.*, 1996).

Overall, these properties suggest a powerful and direct anti-inflammatory effect of IL-4. Accordingly *in vivo* administration of IL-4 was found to inhibit the development of an antigen-specific T cell-mediated inflammatory response in a hapten-induced model of contact sensitivity (Gautam *et al.*, 1992). *In vivo* administration of IL-4 to humans also results in increased circulating levels of IL-1Ra (Wong *et al.*, 1993). IL-4 can also exert indirect anti-inflammatory effects, as indicated by: (1) the IL-4-induced increase of expression and activity of aminopeptidase N (CD13), known to inactivate inflammatory peptides (van Hal *et al.*, 1994); and (2) the IL-4-induced activation of the 15-lipo-oxygenase, which induces lipo-oxygenation of arachidonic acid (Nassar *et al.*, 1994). The induction of monocyte apoptosis may contribute to IL-4 anti-inflammatory effects (Mangen *et al.*, 1992).

Finally, IL-4 inhibits the killing by macrophages of various parasites, such as *Leishmania* (Ho *et al.*, 1992) and asexual erythrocytic forms of *Plasmodium falciparum* (Kumaratilake and Ferrante, 1992). Treatment of monocytes with IL-4 resulted in a complete resistance to human immunodeficiency virus-1 infection (Montanier *et al.*, 1993; Denis and Ghadirian, 1994).

Effects on Granulocytes

IL-4 was found to upregulate CD23 expression on the eosinophilic cell lines EoL 1 and 3 (Hosoda *et al.*, 1989), but this property was not confirmed on normal eosinophils (Baskar *et al.*, 1990). Treatment of EoL 3 cells with IL-4 promotes also CD23-mediated cellular aggregation and morphologic changes and enhances their adhesion on human endothelial cells (Yamaoka and Kolb, 1995). Normal eosinophils are nevertheless responsive to IL-4, as they show downregulation of FcγR expression upon IL-4 treatment, correlated with a decrease in secretion of glucuronidase and arylsulphatase in response to IgG-coated beads (Baskar *et al.*, 1990).

IL-4 also acts on neutrophils by enhancing their respiratory burst and their phagocytic properties (Boey *et al.*, 1989). IL-4 inhibits the secretion of IL-8 (Wertheim *et al.*, 1993) but enhances the expression and secretion of membrane and soluble type II IL-1R by neutrophils (Colotta *et al.*, 1993). This latest property may contribute to IL-4 anti-inflammatory effects by blocking IL-1 effects.

Finally, IL-4 promotes proliferation of human mast cells and induces FcεRI expression on their surface. Treatment of mast cells with IL-4 enhances the release of histamine after FcεRI cross-linking (Toru *et al.*, 1996). More recently, IL-4 has been shown to induce leukocyte functional antigen-1 (LFA-1) and intercellular adhesion molecule-1 (ICAM-1) expression, resulting in homotypic aggregation of mast cells (Toru *et al.*, 1997).

EFFECTS OF IL-4 ON OTHER CELL TYPES

IL-4 and Fibroblasts

Fibroblasts are chemoattracted by IL-4 (Postlethwaite and Seyer, 1991). IL-4 also induces dermal fibroblasts to secrete extracellular matrix proteins, such as type I and type III collagen and fibronectin (Fertin *et al.*, 1991; Gillery *et al.*, 1992; Postlethwaite *et al.*, 1992) and stimulates a fibroblast cell line to produce G-GSF and M-CSF (Tushinski *et al.*, 1991). On TNF- or IL-1-stimulated fibroblasts, IL-4 increases in a specific manner the complement protein C3 synthesis, but decreases that of factor B (Katz *et al.*, 1995). Treatment of dermal fibroblasts with IL-4 results in an increased ICAM-1 expression, which allows increased adhesion of LFA-1-bearing T lymphocytes, as well as binding of human rhinovirus (Piela-Smith *et al.*, 1992).

IL-4 blocks cytokine-induced proliferation of synoviocytes (Dechanet *et al.*, 1993). Cell cycle studies have indicated a rapid and persistent blocking effect of IL-4 on the G_1 phase, accompanied by an increased cell volume, which may explain the increase in thymidine uptake. However, viable cell counts clearly demonstrate that IL-4, in fact, inhibits IL-1 and platelet-derived growth factor-induced synoviocyte proliferation. IL-4 also acts on release of mediators by synoviocytes by enhancing IL-6 production and inhibiting spontaneous or cytokine-induced leukocyte inhibitory factor and PGE_2 production (Dechanet *et al.*, 1994). Finally, IL-4 downregulates IFN-γ-induced ICAM-1 expression on synoviocytes and thus has no effect on adhesion of mononuclear cells (Schlaak *et al.*, 1995).

IL-4 and Endothelial Cells

Capillary endothelial cells are induced to proliferate in response to IL-4 (Toi *et al.*, 1992). Treatment of endothelial cells with IL-4 increases their adhesiveness for T cells, eosinophils and basophils but not for neutrophils, owing to increased vascular cell adhesion molecule-1 (VCAM-1) expression (Thornhill *et al.*, 1991; Schleimer *et al.*, 1992). IL-4 pretreatment of vascular constructs, composed of endothelial cells cultivated on extracellular matrix from human fibroblasts, induces adherence and layer penetration of eosinophils but not of neutrophils. For layer penetration, blood eosinophils from non-allergic donors need priming with GM-CSF, IL-3 or IL-5, whereas those from allergic donors transmigrate spontaneously (Moser *et al.*, 1992).

Cytokine-induced upregulation of ICAM-1 and endothelial leukocyte adhesion molecule-1 (ELAM-1) on endothelial cells is inhibited by IL-4. This contrasts with the IL-4-induced increase of ICAM-1 expression observed on dermal fibroblasts, macrophages and mast cells (Thornhill *et al.*, 1991; Valent *et al.*, 1991; Piela-Smith *et al.*, 1992). IL-4 synergizes with TNF or IFN-γ to cause a change of endothelial cell morphology (Thornhill *et al.*, 1990), coinciding with reorganization of the intracellular vimentin matrix from a diffuse pattern to a perinuclear concentration (Klein *et al.*, 1993). Synergy between IL-4 and TNF-α on VCAM-1 expression by endothelial cells is observed *in vitro* (Iademarco *et al.*, 1995) and *in vivo*, following their injection in the skin of baboons (Briscoe *et al.*, 1992).

IL-4 increases IL-6 production by endothelial cells in synergy with IL-1 or IFN-γ (Howells *et al.*, 1991; Colotta *et al.*, 1992), as well as IL-8 production by LPS-stimulated endothelial cells (de Beaux *et al.*, 1995). In contrast, IL-4 decreases that of RANTES (Regulated upon Activation, Normal T Expressed and Secreted) induced by TNF-α and IFN-γ (Marfaing-Koka *et al.*, 1995). Finally, IL-4 counteracts the effect of LPS, IL-6 and TNF-α on the expression of procoagulant activity on endothelial cells and down-regulates the thrombomodulin anticoagulation pathway (Kapiotis *et al.*, 1991). As prothrombotic vascular changes are associated with inflammatory reactions, this illustrates a novel aspect of the anti-inflammatory effect of IL-4 (Mantovani *et al.*, 1992).

IL-4 and Epithelial Cells

IL-4 enhances the production of soluble CD23 by nasopharyngeal carcinoma, one of the human tumors associated with Epstein–Barr virus (Rousselet *et al.*, 1990). IL-4 also enhances expression of the polymeric immunoglobulin receptor by a colonic adenocarcinoma cell line (Phillips *et al.*, 1990). On human thymic epithelial cells, IL-4 increases IL-1-induced IL-6 production and inhibits that induced by GM-CSF (Galy and Spits, 1991).

On tubular epithelial cells, IL-4 upregulates protein expression as well as enzymatic activity of both CD13 and CD26 peptidases (Riemann *et al.*, 1995). Conversely, it inhibits inducible NO synthase expression in lung epithelial cells (Berkman, 1996), RANTES secretion on airway epithelial cells (Berkman *et al.*, 1996), and chloride secretion on intestinal epithelial cells (Zund *et al.*, 1996).

IL-4 and Hepatocytes

IL-4 can also affect human hepatocytes in primary cultures, by decreasing the spontaneous production of haptoglobin and to a lesser extent that of albumin and C-reactive protein while not affecting α_1-antitrypsin and fibrinogen production (Loyer *et al.*, 1993). Furthermore, IL-4 antagonizes the IL-6-enhanced secretion of haptoglobin, thereby demonstrating another level of anti-inflammatory action. IL-4 also inhibits the stimulation of hepatic lipogenesis by TNF-α, IL-1 and IL-6 without altering that induced by IFN-α (Grunfeld *et al.*, 1991). In addition, IL-4 enhances the expression of cytochrome P-450, 2E1 in a specific manner, as levels of cytochromes P-450, 1A2, 2C and 3A are not affected or weakly inhibited (Razzak *et al.*, 1993). Finally, treatment of human hepatocytes in primary cultures by IL-4 upregulates expression of glutathione-*S*-transferases, enzymes involved in cellular defence against lipid peroxidation (Langouet *et al.*, 1995).

ROLE OF IL-4 IN EXPERIMENTAL MODELS OF DISEASE

Direct administration of IL-4 as well as disruption or overexpression of the IL-4 gene have allowed investigations of the *in vivo* impact of IL-4 in etiopathologic processes.

Experimental Model of Autoimmune Disorders

The protective and anti-inflammatory roles of IL-4 have been demonstrated in various animal models of autoimmune disease. In particular, the treatment of experimental allergic encephalomyelitis by retinoids that increase IL-4 production (Racke *et al.*, 1995) or by IL-4-expressing encephalitogenic T cells (Shaw *et al.*, 1997) that allow local delivery of IL-4, markedly improved the time course of the disease. Similarly, the chronic destructive phase of arthritis induced by injection of streptococcal cell wall fragments in rats (Allen *et al.*, 1993) or of collagen in mice (Bessis *et al.*, 1996) can be efficiently counteracted by direct administration of IL-4 or by treatment with IL-4-secreting cells respectively. In addition, the protective effects of IL-4 have also been reported in non-obese diabetic mice (Rapoport *et al.*, 1993) and in a rat model of immune complex-induced lung injury (Mulligan *et al.*, 1993).

However, the role of IL-4 in the murine lupus model remains more elusive and seems to be dependent on the mouse strain. Indeed, the constitutive expression of an IL-4 transgene by B cells completely prevents the spontaneous development of lethal lupus-like glomerulonephritis in the (NZW × C57BL/6) F1 murine model of systemic lupus erythematosus. IL-4 acts by downregulating the appearance of T_{H1}-mediated IgG_3 and IgG_2 autoantibodies, known to be especially nephritogenic (Santiago *et al.*, 1997). Conversely, IL-4 has been shown to promote the autoimmune development of lupus nephritis in NZB/W F1 mice by enhancing the production of IgG anti-double-stranded DNA antibodies. Administration of antibodies against IL-4 prevents the onset of lupus nephritis and decreases the production of autoantibodies (Nakajima *et al.*, 1997).

Experimental Model of Infectious Disorders

Numerous studies suggest that T-cell subsets and their associated cytokine profile critically influence the outcome of infection in mice. However, like autoimmune diseases, IL-4 can elicit both beneficial and detrimental effects on the nature of infectious agents.

During *Borrelia burgdorferi* infection, $CD4^+$ T cell-derived IL-4 plays a critical role in the control of *in vivo* spirochete growth. Administration of IL-4 to mice during the early phase of *B. burgdorferi* infection significantly reduces joint swelling and spirochete numbers (Keane-Myers *et al.*, 1996). This increased resistance is also associated with a decrease in serum levels of specific IgG_{2a} and IgG_3 antibodies and an increase in specific IgG_1 antibodies. In addition, IL-4 exerts protective effects against development of toxoplasmic encephalitis induced by *Toxoplasma gondii*, by preventing formation of *T. gondii* cysts and foci of acute inflammation, and proliferation of tachyzoites in the brain (Suzuki *et al.*, 1996). Control of parasitemia and survival during *Trypanosoma brucei* infection is also related to the ability of the mice to produce IL-4 (Bakhiet *et al.*, 1996). In contrast, the precise role of IL-4 in *Leishmania major* infection in mice remains more controversial, as both disease-promoting and protective effects have been reported (Müller *et al.*, 1991; Leal *et al.*, 1993). Recently, disruption of the IL-4 gene in *L. major*-infected Balb/c mice has also given contradictory results regarding their susceptibility or resistance towards *L. major* infection (Kopf *et al.*, 1996; Noben-Trauth *et al.*, 1996).

Conversely, resistance to *Listeria monocytogenes* requires the production of IFN-γ by NK and T_{H1} cells, whereas IL-4 remains unaffected. However, disruption of the IFN-γ-R gene in mice affected by listeriosis has shown induction of IL-4 production, combined with a decrease in that of TNF-α. Neutralization of IL-4 with anti-IL-4 antibodies in combination with TNF-α reconstitution improves the prognosis of listeriosis (Szalay *et al.*, 1996), and thus demonstrates the pathogenic role of IL-4 in this model.

Finally, besides its protective effects, IL-4 can also exert *in vivo* inflammatory and pathogenic effects, as shown in murine lung in which IL-4 protein is selectively overexpressed (Rankin *et al.*, 1996). Indeed, epithelial cell hypertrophy as well as accumulation of lymphocytes, eosinophils and neutrophils represent the main features of the inflammatory responses induced by IL-4 overexpression in lung.

Experimental Model of Neoplasia

Several independent studies have now demonstrated that IL-4 possesses potent antitumor activity *in vivo* in mice. When a cDNA for IL-4 is introduced into normally tumorigenic lines, the engineered lines fail to form tumors and block tumor proliferation by various other tumor lines transplanted at the same site (Golumbek *et al.*, 1991; Tepper *et al.*, 1989, 1992). The antitumor effect is clearly due to IL-4 as anti-IL-4 antibodies allow re-expression of the tumorigenic potential. Tepper's experiments, carried out in nude mice to exclude any effect of tumor-specific cytotoxic T cells, demonstrate a marked cellular infiltrate composed of eosinophils and activated macrophages at the site of tumor injection. Administration of anti-IL-5 and antigranulocyte antibodies restores the tumorigenicity of the cell lines and results in a marked reduction in the number of tumor-infiltrating eosinophils (Tepper *et al.*, 1992). Thus eosinophil-mediated cytotoxicity appears to be an important mechanism of action for the antitumor activity of IL-4. This finding correlates with earlier clinical observations showing that gastric and colonic malignancies infiltrated with eosinophils have an improved prognosis (Iwasaki *et al.*, 1986).

Golumbek's experiments, carried out in immunocompetent mice, have indicated an important role for cytotoxic $CD8^+$ T cells, as the antitumor effect can also be transferred by T cells from mice administered with renal cancer cells transfected with IL-4 (Golumbek *et al.*, 1991). In this study, after an early influx of macrophages and eosinophils, T cells began infiltrating the tumor site. In addition, this study showed that the renal tumor cells, engineered to secrete IL-4, induce a sufficiently strong systemic immune response to cure animals carrying small amounts of parental tumor. In addition, repeated injections of small amounts of IL-4 into tumor-draining lymph nodes result in an antitumor effect that can be transferred with $CD8^+$ T cells (Bosco *et al.*, 1990).

IL-4 may also display a direct antitumor effect as shown with human colonic, renal, gastric and breast carcinoma (Morisaki *et al.*, 1992; Toi *et al.*, 1992; Obiri *et al.*, 1993), lymphoma (Defrance *et al.*, 1992), acute pre-B-cell leukemia and myelomas (Taylor *et al.*, 1990; Pandrau *et al.*, 1992), chronic lymphocytic and myelogenous leukemias (Karray *et al.*, 1988; Akashi *et al.*, 1991) and some cases of acute myelogenous leukemia (Akashi *et al.*, 1991; Miyauchi *et al.*, 1991; Lahm *et al.*, 1994, 1995). IL-4 may also alter the pattern of metastasis and invasion by colorectal carcinoma (Uchiyama *et al.*, 1996).

CLINICAL STUDIES

Doses and Schedule

Phase I studies have been conducted with both recombinant *Escherichia coli*-produced human IL-4 (Schering-Plough Research Institute) and recombinant yeast-derived human IL-4 (rHuIL-4) (Sterling Winthrop). The phase I studies with rHuIL-4 (*E. coli*) were conducted with daily intravenous administration in doses of 0.25–5 μg/kg per day and daily subcutaneous dosing with 0.25–5 μg/kg in a single dose (Gilleece *et al.*, 1992; Markowitz *et al.*, submitted). Studies were also conducted with rHuIL-4 (yeast), with intravenous bolus injection every 8 h on days 1–5 and 15–19 in doses of 10–15 μg/kg per dose (Atkins *et al.*, 1992; Davies *et al.*, 1997). A series of phase II trials has been conducted with 0.25–4 μg/kg either daily or three times a week, and a dose-ranging study with rHuIL-4 in doses of 1.0–8.0 mg/kg three times weekly to look for safety and efficacy in advanced malignancy.

Toxicity in Clinical Studies

Daily intravenous bolus administration of rHuIL-4, 0.25–5 μg/kg for 10 days, was well tolerated. In ten patients, toxicity consisted primarily of influenza-like symptoms, commonly seen with cytokine treatment, including fever (<39°C in this group), rigor and myalgia. One patient required dose reduction for a grade 3 headache. When given daily by the subcutaneous route, in a dose of 0.25–5 μg/kg daily for 28 days, again the clinical toxicity was that typically seen with cytokine treatment, specifically a flu-like syndrome with fatigue, somnolence, fever and myalgia. Occasional abdominal and back pain, pedal and periorbital edema, and nausea and vomiting were seen. Of note,

Table 2. Maximum clinical laboratory WHO grades recorded for patients treated with SC rhIL-4 SCH 39400.

	Number of patients								
	0.25 μg/kg*			1 or 2 μg/kg*			4 or 5 μg/kg*		
	(*N*=10) Grade			(*N*=15) Grade			(*N*=22) Grade		
Laboratory parameters	*0 or 1*	*2*	*3 or 4*	*0 or 1*	*2*	*3 or 4*	*0 or 1*	*2*	*3 or 4*
Hematology									
Decreased WBC count	10	0	0	12	2	1	21	0	1
Decreased neutrophils	10	0	0	10	4	1	17	0	4
Decreased hemoglobin	9	1	0	11	3	1	15	3	4
Decreased platelets	10	0	0	14	1	0	20	2	0
Hepatic enzyme elevation									
SGOT	8	0	2	13	1	1	17	4	1
SGPT	8	0	1	8	2	0	18	2	1
Alkaline phosphatase	6	2	2	9	2	4	11	8	3
LDH	9	0	0	15	0	0	20	0	2
Increased total bilirubin	6	2	1	14	0	0	21	0	1
Renal function									
Creatinine	10	0	0	15	0	0	21	1	0
BUN	10	0	0	14	1	0	22	0	0

*ECOG Score: Eastern Cooperative Oncology Group Score.
BUN, blood urea nitrogen.

headache was an extremely common complaint in these patients, without any characteristic pattern or localization. The headache, in fact, was a dose-limiting toxicity with this mode of administration. Laboratory abnormalities commonly reported included an increase in the level of liver enzymes, which was seen at all dose levels (Table 2). The severity was generally dose related, reversible and not associated with other evidence of hepatic dysfunction such as hyperbilirubinemia or abnormalities in coagulation profile (one patient with previous hepatitis developed hyperbilirubinemia). Relatively little myeloid toxicity was seen and, in fact, in a number of patients there was an increase in the neutrophil count while on IL-4 therapy (Gilleece *et al.*, 1992).

Glucose values were monitored carefully in these studies because of the appearance of hypoglycemia in some animal studies of rHuIL-4. However, in patients, relatively little impact on serum glucose was noted. One patient did develop hypoglycemia (49 μg/dl) at a daily dose of 5 μg/kg; however, no other patient developed significant hypoglycemia.

In comparison, when rHuIL-4 (yeast) was administered by intravenous bolus at 10–15 μg/kg per dose, more toxicity was seen. In these patients, a similar pattern of fatigue, anorexia, headache and dyspnea was seen (Atkins *et al.*, 1992; Markowitz *et al.*, submitted). However, a greater median weight gain of 6.1% (range 3.4–11.7%) was reported than with the subcutaneous studies. Orthostatic hypotension occurred in two patients at the 15 μg/kg dose level (Atkins *et al.*, 1992). Sustained hypotension sufficient to require pressors was not reported in that study. Decreases in lymphocyte count, sodium, albumin, fibrinogen levels, raised partial thromboplastin time and increases in hematocrit were observed routinely at that dose and mode of administration. This was not seen in the subcutaneous studies. Increases in serum levels of creatinine and hepatic transaminases appeared to occur less frequently than with the subcutaneous route.

Similar to the observations with subcutaneous and intravenous dosing with the *E. coli* rHuIL-4, all side-effects had resolved by the follow-up visits. In the study of Markowitz *et al.*, using daily doses of 20–1280 $\mu g/m^2$ of yeast rHuIL-4, reversible abnormalities in liver function were noted, and grade 1–2 fever was seen 6–8 h after most doses. Frequent grade 1–2 increases in liver function test results, rare grade 3–4 increases in liver function test results, increases in serum creatinine concentration, the characteristic headaches, nasal congestion and a single case (1 of 22 patients) of grade 2 fluid retention were seen. The possibility of an IL-4-associated cardiac toxicity arose due to results in animal toxicologic studies (Leach *et al.*, 1997a,b). However, review of serial cardiac toxicity assessments in over 400 patients and of serial radionuclide cardiac ejection fraction and endomyocardial biopsy data at therapeutic doses of rHuIL-4 (0.25–4 $\mu g/kg$ daily) demonstrated no definitive evidence of dose-related cardiac toxicity. Isolated decreases in cardiac ejection were seen but attribution to IL-4 was not clear. Cardiac effect at much higher doses, as had been suggested by earlier reports with an alternate IL-4 molecule, has not been tested with the currently available material.

Immunologic and Antitumor Activity

Only limited data *in vivo* on immunologic effects of rHuIL-4 are available from these phase I studies. An increase in plasma IL-1Ra and soluble CD23 levels were reported, which were reversed rapidly following the discontinuation of IL-4 (Atkins *et al.*, 1992; Wong *et al.*, 1993). No changes in LAK or NK cell activity were noted. A decrease in measurable serum TNF and IL-1 levels was found in four of four and three of three patients respectively in whom detectable levels were present before treatment (Markowitz *et al.*, submitted). In the subcutaneous study, no consistent trends in changes in lymphocyte counts, owing to wide interpatient variability, were noted. Individual patients, however, did have evidence of an increase in CD4 and CD8 counts. There was no change in CD4/CD8 ratio as the changes tended to occur in parallel. It was notable that, given the known *in vitro* activity of IL-4 in IgE production, there was no evidence of increased IgE levels in the phase I studies.

In phase I studies, limited antitumor activity was observed. Specifically, short-term intravenous bolus administration was not associated with any antitumor response (Atkins *et al.*, 1992; Markowitz *et al.*, submitted). Following subcutaneous administration in the phase I study, antitumor activity was observed, with a response in a patient with advanced refractory Hodgkin's disease, an improvement in a patient with chronic lymphocytic leukemia, and a decrease in M component in a patient with refractory multiple myeloma. These data, although preliminary, suggest that IL-4 does have potential as an antitumor agent, particularly when administered over a more prolonged course, rather than short intravenous bolus. This is consistent with the known pharmacokinetics of this agent, which available data indicate has rapid clearance with a half-life of approximately 45 min (Davies *et al.*, 1997). The duration of immunologic response following discontinuation of rHuIL-4 in patients has yet to be determined, although the data on CD23 levels in patients receiving intravenous bolus therapy suggest a relatively rapid fall off in effect. This is in contrast to some of the animal data (Schering-Plough Research Institute, personal communication) which have shown protracted responses in animals given IL-4, particularly neutrophil activation which lasted as long as 2 weeks following discontinuation of IL-4.

Preliminary results from phase II trials have resulted in individual patients with a long-term response (6 years or more), but with response rates in previously treated patients of 4–10% in those with solid tumors (renal cell carcinoma, melanoma, non-small cell lung cancer) and of 15–30% in non-Hodgkin's lymphoma. No response has been seen in patients with multiple myeloma or Hodgkin's disease.

Despite the low response rate, comparison of survival data with historic controls suggests a potential survival benefit in non-small cell lung cancer (unpublished results). In a randomized dose-ranging trial of 183 patients with advanced, primary, previously treated non-small cell lung cancer, 1.0 μg/kg rHuIL-4 three times daily was associated with a 1-year survival rate greater than 30%, with a median survival of 188 days in the ITT (intention to treat) population. This suggests that a subset of patients may benefit from IL-4 therapy. The correlation between IL-4 receptors on tumor cells and efficacy was not assessed in this study. Predictably for a cytokine, a rapid plateau to dose response effect was seen, with 1 μg/kg three times a day as the optimal dose.

Phase II studies in actual lymphoblastic leukemia, and in combination with chemotherapy for chronic lymphocytic leukemia and gastrointestinal malignancies, are ongoing.

Based on the acceptable tolerability of rHuIL-4 seen in oncology trials in doses of less than 8 μg/kg three times daily, clinical trials have begun in other areas such as rheumatoid arthritis and psoriasis.

Ex vivo application of rHuIL-4 for the generation of dendritic cells for immunotherapeutic vaccines is ongoing in melanoma, gastrointestinal malignancy and prostate carcinoma.

CONCLUSION

In vitro studies have shown that IL-4 can act on many cell types and at various stages of maturation of a given cell type. The biologic effects of IL-4 on a particular cell type depend on the surrounding cells and cytokines. Indeed, IL-4 can play an important biologic role in an indirect fashion, as it modulates cytokine production by T, B and NK cells, monocytes–macrophages, endothelial cells and fibroblasts. *In vivo* studies in animals have demonstrated an important antitumor effect for IL-4. Accordingly, IL-4, when administered to patients with advanced cancer, is well tolerated. This is particularly true for daily subcutaneous administration, but higher doses of intravenous bolus IL-4 are also reasonably well tolerated. There is suggestion of antitumor activity; however, the optimal dose and schedule for this molecule to achieve desired immunomodulatory and antitumor effects *in vivo* remain to be determined. In addition, the tolerability of IL-4, particularly when given subcutaneously, suggest that it may be appropriate for evaluation in a variety of diseases (in addition to treatment of cancer) where its immunomodulatory and anti-inflammatory properties would be of clinical benefit. In particular, the inhibitory effects of IL-4 on the production of cytokines provide a strong rationale for its use in chronic inflammatory diseases, in particular rheumatoid arthritis. Other inflammatory diseases with excess T_{H1} activity may benefit from IL-4 therapy, which may re-equilibrate the balance between T_{H1} and T_{H2} cells. In keeping with this concept, therapy may be designed in the future to counterbalance a

T_{H2} overexpression observed in diseases such as allergy and possibly AIDS (Clerici *et al.*, 1993).

The near future will see many clinical trials with IL-4 itself or with drugs affecting IL-4 production or IL-4 effects, such as interferons. In the long term, structural studies on IL-4, IL-4R and their complex, and studies on the intracellular pathways specifically activated by IL-4 should permit the design of chemical agents that will either inhibit or mimic part or all of the biologic effects of IL-4.

REFERENCES

Akashi, K., Shibuya, T., Harada, M., Takamatsu, Y., Ulke, N., Eto, T. and Niho, Y. (1991). Interleukin-4 suppresses the spontaneous growth of chronic myelomonocytic leukemia cells. *J. Clin. Invest*. **88**, 223–230.

Albrecht, B., Peiritsch, S., Messner, B. and Woisetschlager, M. (1996). The transcription factor B-cell-specific activator protein is not involved in the IL-4-induced activation of the human IgE germline promoter. *J. Immunol.* **157**, 1538–1543.

Allen, J.B., Wong, H.L., Costa, G.L., Bienkowski, M.J. and Wahl, S.M. (1993). Suppression of monocyte function and differential regulation of IL-1 and IL-1ra by IL-4 contribute to resolution of experimental arthritis. *J. Immunol.* **151**, 4344–4351.

Aman, M.J., Tayebi, N., Obiri, N., Puri, R.K., Modi, W.S. and Leonard, W.J. (1996). cDNA cloning and characterization of the human interleukin 13 receptor α chain. *J. Biol. Chem.* **271**, 29 265–29 270.

Arruda, S. and Ho, J.L. (1992). IL-4 receptor signal transduction in human monocytes is associated with protein kinase C translocation. *J. Immunol.* **149**, 1258–1264.

Atamas, S.P., Choi, J., Yurovsky, V.V. and White, B. (1996). An alternative splice variant of human IL-4, IL-4δ2, inhibits IL-4-stimulated T cell proliferation. *J. Immunol.* **156**, 435–441.

Atkins, M.B., Vachino, G., Tllg, H.J., Karp, D.D., Robert, N.J., Kappler, K. and Mier, J.W. (1992). Phase I evaluation of thrice-daily intravenous bolus interleukin-4 in patients wlth refractory malignancy. *J. Clin. Oncol*. **10**, 1802–1809.

Bakhiet, M., Jansson, L., Buscher, P., Holmdahl, R., Kristensson, K. and Olsson, T. (1996). Control of parasitemia and survival during *Trypanosoma brucei brucei* infection is related to strain-dependent ability to produce IL-4. *J. Immunol.* **157**, 3518–3526.

Banchereau, J., de Paoli, P., Vallé, A., Garcia, E. and Rousset, F. (1991). Long term human B cell lines dependent on interleukin-4 and antibody to CD40. *Science* **251**, 70–72.

Barcena, A., Toribio, M.L., Pezzi, L. and Martinez-A.C. (1990). A role for interleukin-4 in the differentiation of mature T cell receptor $\gamma\delta^+$ cells from human intrathymic T cell precursors. *J. Exp. Med.* **172**, 439–446.

Baskar, P., Silberstein, D.S. and Pincus, S.H. (1990). Inhibition of IgG-triggered human eosinophil function by IL-4. *J. Immunol.* **144**, 2321–2326.

Berkman, N. (1996). Inhibition of inducible nitric oxide synthase expression by interleukin-4 and interleukin-13 in human lung epithelial cells. *Immunology* **89**, 363–367.

Berkman, N., Robichaud, A., Krishnan, V.L., Roesems, G., Robbins, R., Jose, P.J., Barnes, P.J. and Chung, K.F. (1996). Expression of RANTES in human airway epithelial cells: effects of corticosteroids and interleukin-4, -10 and -13. *Immunology* **87**, 599–603.

Bessis, N., Boissier, M.C., Ferrara, P., Blankenstein, T., Fradelizi, D. and Fournier, C. (1996). Attenuation of collagen-induced arthritis in mice by treatment with vector cells engineered to secrete interleukin-13. *Eur. J. Immunol.* **26**, 2399–2403.

Blay, J.Y., Branellec, D., Robinet, E., Dugas, B., Gay, F. and Choualb, S. (1990). Involvement of cyclic adenosine monophosphate in the interleukin-4 inhibitory effect on interleukin-2-induced lymphokine-activated killer generation. *J. Clin. Invest.* **85**, 1909–1913.

Blotta, M.H., Marshall, J.D., DeKruyff, R.H. and Umetsu, D.T. (1996). Cross-linking of the CD40 ligand on human $CD4^+$ T lymphocytes generates a costimulatory signal that up-regulates IL-4 synthesis. *J. Immunol.* **156**, 3133–3140.

Blum, H., Wolf, M., Enssle, K., Rollinghoff, M. and Gessner, A. (1996). Two distinct stimulus-dependent pathways lead to production of soluble murine interleukin-4 receptor. *J. Immunol.* **157**, 1846–1853.

Boey, H., Rosembaum, R., Castracane, J. and Borish, L. (1989). Interleukin-4 is a neutrophil activator. *J. Allergy Clin. Immunol.* **83**, 978–984.

Bonnefoy, J.Y., Defrance, T., Péronne, C., Ménétrier, C., Rousset, F., Pène, J., de Vries, J.E. and Banchereau, J. (1988a). Human recombinant interleukin-4 induces normal B cells to produce soluble CD23/IgE-binding factor analogous to that spontaneously released by lymphoblastoid B cell types. *Eur. J. Immunol.* **18**, 117–122.

Bonnefoy, J.Y., Guillot, O., Spits, H., Blanchard, D., Ishizaka, K. and Banchereau, J. (1988b). The low-affinity receptor for IgE (CD23) on B lymphocytes is spatially associated with HLA-DR antigens. *J. Exp. Med.* **167**, 57–72.

Borger, P., Kauffman, H.F., Postma, D.S. and Vellenga, E. (1996a). IL-7 differentially modulates the expression of IFN-γ and IL-4 in activated human T lymphocytes by transcriptional and post-transcriptional mechanisms. *J. Immunol.* **156**, 1333–1338.

Borger, P., Kauffman, H.F., Postma, D.S. and Vellenga, E. (1996b). Interleukin-4 gene expression in activated human T lymphocytes is regulated by the cyclic adenosine monophosphate-dependent signaling pathway. *Blood* **87**, 691–698.

Bosco, M., Giovarelli, M., Forni, M., Modesti, A., Scarpa, S., Masuelli, L. and Forni, G. (1990). Low doses of IL-4 injected perilymphatically in tumor-bearing mice inhibit the growth of poorly and apparently nonimmunogenic tumors and induce a tumor-specific immune memory. *J. Immunol.* **145**, 3136–3143.

Bradding, P., Feather, I.H., Howarth, P.H., Mueller, R., Roberts, J.A., Britten, K., Bews, J.P.A., Hunt, T.G., Okayama, Y. and Heusser, C.H. (1992). Interleukin-4 is localized to and released by human mast cells. *J. Exp. Med.* **176**, 1381–1386.

Briscoe, D.M., Cotran, R.S. and Pober, J.S. (1992). Effects of tumor necrosis factor, lipopolysaccharide, and IL-4 on the expression of vascular cell adhesion molecule-1 *in vivo.* Correlation with $CD3^+$ T cell infiltration. *J. Immunol.* **149**, 2954–2960.

Brown, M.A. and Hural, J. (1997). Functions of IL-4 and control of its expression. *Crit. Rev. Immunol.* **17**, 1–32.

Brown, M.A., Pierce, J.H., Watson, C.J., Falco, J., Ihle, J.N. and Paul, W.E. (1987). B cell stimulatory factor-1/interleukin-4 mRNA is expressed by normal and transformed mast cells. *Cell* **50**, 809–818.

Broxmeyer, H.E., Lu, L., Cooper, S., Tushinski, R., Mochizuki, D., Rubin, B.Y., Gillis, S. and William, D.E. (1988). Synergistic effects of purified recombinant human and murine B cell growth factor-1/IL-4 on colony formation *in vitro* by hematopoietic progenitor cells. Multiple actions. *J. Immunol.* **141**, 3852–3862.

Brunner, T., Heusser, C.H. and Dahinden, C.A. (1993). Human peripheral blood basophils primed by interleukin-3 (IL-3) produce IL-4 in response to immunoglobulin E receptor stimulation. *J. Exp. Med.* **177**, 605–611.

Burdin, N., van Kooten, C., Galibert, L., Abrams, J.S., Wijdenes, J., Banchereau, J. and Rousset, F. (1995). Endogenous IL-6 and IL-10 contribute to the differentiation of CD40-activated human B lymphocytes. *J. Immunol.* **154**, 2533–2544.

Cabrillat, H., Galizzi, J.P., Djossou, O., Arai, N., Yokota, T., Arai, K. and Banchereau, J. (1987). High affinity binding of human interleukin-4 to cell lines. *Biochem. Biophys. Res. Commun.* **149**, 995–1001.

Callard, R.E., Smith, S.H. and Scott, K.E. (1991). The role of interleukin-4 in specific antibody responses by human B cells. *Int. Immunol.* **3**, 157–163.

Caput, D., Laurent, P., Kaghad, M., Lelias, J.M., Lefort, S., Vita, N. and Ferrara, P. (1996). Cloning and characterization of a specific interleukin (IL)-13 binding protein structurally related to the IL-5 receptor α chain. *J. Biol. Chem.* **271**, 16 921–16 926.

Chan, S.C., Brown, M.A., Li, S.H., Stevens, S.R. and Hanifin, J.C. (1996). Abnormal IL-4 gene expression by atopic dermatitis T lymphocytes is reflected in altered nuclear protein interactions with IL-4 transcriptional regulatory element. *J. Invest. Dermatol.* **106**, 1131–1136.

Chaouchi, N., Wallon, C., Goujard, C., Tertian, G., Rudent, A., Caput, D., Ferrera, P., Minty, A., Vazquez, A. and Delfraissy, J.F. (1996). Interleukin-13 inhibits interleukin-2-induced proliferation and protects chronic lymphocytic leukemia B cells from *in vitro* apoptosis. *Blood* **87**, 1022–1029.

Chomarat, P., Banchereau, J. and Miossec, P. (1995). Differential effects of interleukins -10 and -4 on the production of interleukin-6 by blood and synovium monocytes in rheumatoid arthritis. *Arthritis Rheum.* **38**, 1046–1054.

Clark, E.A., Shu, G.L., Luscher, B., Draves, K.E., Banchereau, J., Ledbetter, J.A. and Valentine, M.A. (1989). Activation of human B cells. Comparison of the signal transduced by IL-4 to four different competence signals. *J. Immunol.* **143**, 3873–3880.

Clerici, M. and Shearer, G.M. (1993). A T_{h1}–T_{h2} is a critical step in the etiology of HIV infection. *Immunol. Today* **14**, 107–111.

Coffman, R.L. and von der Weid, T. (1997). Multiple pathways for the initiation of T helper 2 (T_{h2}) responses. *J. Exp. Med.* **185**, 373–375.

Colotta, F., Sironi, M., Borrè, A., Luini, W., Maddalena, F. and Mantovani, A. (1992). Interleukin-4 amplifies monocyte chemotactic protein and interleukin-6 production by endothelial cells. *Cytokine* **4**, 24–28.

Colotta, F., Re, F., Muzio, M., Bertini, R., Polentarutti, N., Sironi, M., Giri, J.G., Dower, S.K., Sims, J.E. and Mantovani, A. (1993). Interleukin-1 Type II receptor: as a decoy target for IL-1 that is regulated by IL-4. *Science* **261**, 472–475.

Corcoran, M.L., Stetler-Stevenson, W.G., Brown, P.D. and Wahl, L.M. (1992). Interleukin-4 inhibition of prostaglandin E_2 synthesis blocks interstitial collagenase and 92-kDa type IV collagenase/gelatinase production by human monocytes. *J. Biol. Chem.* **267**, 515–519.

D'Andrea, A., Ma, X., Aste-Amezaga, M., Paganin, C. and Trinchieri, G. (1995). Stimulatory and inhibitory effects of interleukin (IL)-4 and IL-13 on the production of cytokines by human peripheral blood mononuclear cells: priming for IL-12 and tumor necrosis factor α production. *J. Exp. Med.* **181**, 537–546.

Dancescu, M., Rubio-Trujillo, M., Biron, G., Bron, D., Delespesse, G. and Sarfati, M. (1992). Interleukin-4 protects chronic lymphocytic leukemia B cells from death by apoptosis and upregulates Bcl-2 expression. *J. Exp. Med.* **176**, 1319–1326.

Davydov, I.V., Bohmann, D., Krammer, P.H. and Li-Weber, M. (1995a). Cloning of the cDNA encoding human C/EBP γ, a protein binding to the PRE-1 enhancer element of the human interleukin-4 promoter. *Gene* **161**, 271–275.

Davydov, I.V., Krammer, P.H. and Li-Weber, M. (1995b). Nuclear factor-IL6 activates the human IL-4 promoter in T cells. *J. Immunol.* **155**, 5273–5279.

de Beaux, A.C., Maingay, J.P., Ross, J.A., Fearon, K.C. and Carter, D.C. (1995). Interleukin-4 and interleukin-10 increase endotoxin-stimulated human umbilical vein endothelial cell interleukin-8 release. *J. Interferon Cytokine Res.* **15**, 441–445.

de Waal Malefyt, R., Yssel, H., Roncarolo, M.-G., Spits, H. and de Vries, J.E. (1992). Interleukin-10. *Curr. Opinion Immunol.* **4**, 314–320.

Dechanet, J., Taupin, J.L., Chomarat, P., Rissoan, M.C., Moreau, J.F., Banchereau, J. and Miossec, P. (1994). Interleukin-4 but not interleukin-10 inhibits the production of leukemia inhibitory factor by rheumatoid synovium and synoviocytes. *Eur. J. Immunol.* **24**, 3222–3228.

Dechanet, J., Briolay, J., Rissoan, M.C., Chomarat, P., Galizzi, J.P., Banchereau, J. and Miossec, P. (1993). Interleukin-4 inhibits growth factor-stimulated rheumatoid synoviocyte proliferation by blocking the early phases of the cell cycle. *J. Immunol.* **151**, 4908–4917.

Defife, K.M., Jenney, C.R., McNally, A.K., Colton, E. and Anderson, J.M. (1997). Interleukin-13 induces human monocyte/macrophage fusion and macrophage mannose receptor expression. *J. Immunol.* **158**, 3385–3390.

Defrance, T., Fluckiger, A.C., Rossi, J.F., Magaud, J.P., Sotto, J.J. and Banchereau, J. (1992). Antiproliferative effect of interleukin-4 on freshly isolated non-Hodgkin malignant B-lymphoma cells. *Blood* **79**, 990–996.

Defrance, T., Vanbervliet, B., Aubry, J.P. and Banchereau, J. (1988). Interleukin-4 inhibits the proliferation but not the differentiation of activated human B cells in response to interleukin-2. *J. Exp. Med.* **168**, 1321–1337.

Defrance, T., Vanbervliet, B., Aubry, J.P., Takebe, Y., Arai, N., Miyajima, A., Yokota, T., Lee, T., Arai, K., de Vries, J.E. and Banchereau, J. (1987). B cell growth-promoting activity of recombinant human interleukin-4. *J. Immunol.* **139**, 1135–1141.

Denis, M. and Ghadirian, E. (1994). Interleukin-13 and interleukin-4 protect bronchoalveolar macrophage from productive infection with human immunodeficiency virus-type 1. *AIDS Res. Hum. Retroviruses* **10**, 795–802.

Diu, A., Fevrier, M., Mollier, P., Charron, D., Banchereau, J., Reinherz, E.L. and Theze, J. (1990). Further evidence for a human B cell activating factor distinct from IL-4. *Cell Immunol.* **125**, 14–28.

Dokter, W.H.A., Esselink, M.T., Sierdsema, S.J., Halie, M.R. and Vellenga, E. (1993a). Transcriptional and post-transcriptional regulation of the interleukin-4 and interleukin-3 genes in human T cells. *Blood* **81**, 35–40.

Dokter, W.H.A., Esselink, M.T., Halie, M.R. and Vellenga, E. (1993b). Interleukin-4 inhibits the lipopolysaccharide-induced expression of *c-jun* and *c-fos* messenger RNA and activator protein-1 binding activity in human monocytes. *Blood* **81**, 337–343.

Dokter, W.H.A., Koopmans, S.B. and Vellenga, E. (1996). Effects of IL-10 and IL-4 on LPS-induced transcription factors (AP-1, NF-IL6 and NF-κ B) which are involved in IL-6 regulation. *Leukemia* **10**, 1308–1316.

Duronio, V., Welham, M.J., Abraham, S., Dryden, P. and Schrader, J.W. (1992). p21ras activation via hemopoietin receptors and *c-kit* requires tyrosine kinase activity but not tyrosine phosphorylation of p21ras GTPase-activating protein. *Proc. Natl Acad. Sci. USA* **89**, 1587–1591.

Elenström, C. and Severinson, E. (1989). Interleukin-4 induces cellular adhesion among B lymphocytes. *Growth Factors* **2**, 73–82.

Endo, T., Ogushi, F. and Sone, S. (1996). LPS-dependent cyclooxygenase-2 induction in human monocytes is down-regulated by IL-13, but not by IFN-γ. *J. Immunol.* **156**, 2240–2246.

Erard, F., Wild, M.T., Garcia-Snaz, J.A. and Le Gros, G.G. (1993). Switch of CD8 T cells to noncytolytic $CD8^-$ $CD4^-$ cells that make T_{h2} cytokines and help B cells. *Science* **260**, 1802–1805.

Ezernieks, J., Schnarr, B., Metz, K. and Duschl, A. (1996). The human IgE germline promoter is regulated by interleukin-4, interleukin-13, interferon α and interferon γ via an interferon γ-activated site and its flanking regions. *Eur. J. Biochem.* **240**, 667–673.

Fanslow, W.C., Clifford, K., VandenBos, T., Teel, A., Armitage, R.J. and Beckman, M.P. (1991). A soluble form of the interleukin-4 receptor in biological fluids. *Cytokine* **2**, 398–401.

Favre, C., Saeland, S., Caux, C., Duvert, V. and De Vries, J.E. (1990). Interleukin-4 has basophilic and eosinophilic cell growth-promoting activity on cord blood cells. *Blood* **75**, 67–73.

Fenghao, X., Saxon, A., Nguyen, A., Ke, Z., Diaz-Sanchez, D. and Nel, A. (1995). Interleukin-4 activates a signal transducer and activator of transcription (Stat) protein which interacts with an interferon γ activation site-like sequence upstream of the I ε exon in a human B cell line. Evidence for the involvement of Janus kinase 3 and interleukin-4 Stat. *J. Clin. Invest.* **96**, 907–914.

Fenton, M.J., Buras, J.A. and Donnelly, R.P. (1992). IL-4 reciprocally regulates IL-1 and IL-1 receptor antagonist expression in human monocytes. *J. Immunol.* **149**, 1283–1288.

Fernandez-Botran, R. and Vitetta, E.S. (1990). A soluble, high-affinity, interleukin-4 binding protein is present in the biological fluids of mice. *Proc. Natl Acad. Sci. USA* **87**, 4202–4206.

Fernandez-Botran, R. and Vitetta, E.S. (1991). Evidence that natural murine soluble interleukin-4 receptors may act as transport proteins. *J. Exp. Med.* **174**, 673–681.

Fertin, C., Nicolas, J.F., Gillery, P., Kalis, B., Banchereau, J. and Maquart, F.X. (1991). Interleukin-4 stimulates collagen synthesis by normal and scleroderma fibroblasts in dermal equivalents. *Cell Mol. Biol.* **37**, 823–829.

Finkelman, F.D., Holmes, J., Katona, I.M., Urban, J.F., Beckmann, M.P., Schooley, K.A., Coffman, R.L., Mosmann, T.R. and Paul, W.E. (1990). Lymphokine control of *in vivo* immunoglobulin isotype selection. *Annu. Rev. Immunol.* **8**, 303–333.

Finney, M., Guy, G.R., Michell, R.H., Gordon, J., Dugas, B., Rigley, K.P. and Callard, R.E. (1990). Interleukin-4 activates human B lymphocytes via transient inositol lipid hydrolysis and delayed cyclic adenosine monophosphate generation. *Eur. J. Immunol.* **20**, 151–156.

Flores-Romo, L., Johnson, G.D., Ghaderi, A.A., Stanworth, D.R., Veronesi, A. and Gordon, J. (1990). Functional implication for the topographical relationship between MHC class II and the low-affinity IgE receptor: occupancy of CD23 prevents B lymphocytes from stimulating allogeneic mixed lymphocyte responses. *Eur. J. Immunol.* **20**, 2465–2469.

Fluckiger, A.C., Rossi, J.F., Bussel, A., Bryon, P., Banchereau, J. and Defrance, T. (1992). Responsiveness of chronic lymphocytic leukemia B cells activated via surface Igs or CD40 to B-cell tropic factors. *Blood* **80**, 3173–3181.

Fluckiger, A.C., Briere, F., Zurawski, G., Bridon, J.M. and Banchereau, J. (1994). IL-13 has only a subset of IL-4 like activities on B chronic lymphocytic leukemia cells. *Immunology* **83**, 397–403.

Folz, I.N., Lee, J.C., Young, P.R. and Schrader, J.W. (1997). Hemopoietic growth factors with the exception of interleukin-4 activate the p38 mitogen-activated protein kinase pathway. *J. Biol. Chem.* **272**, 3296–3301.

Folz, I.N. and Schrader, J.W. (1997). Activation of the stress-activated protein kinases by multiple hematopoietic growth factors with the exception of interleukin-4. *Blood* **89**, 3092–3096.

Foxwell, B.M.J., Woerly, G. and Ryffel, B. (1989). Identification of interleukin-4 receptor-associated proteins and expression of both high and low affinity binding on human lymphoid cells. *Eur. J. Immunol.* **19**, 1637–1641.

Gajewski, T.F., Joyce, J. and Fitch, F.W. (1989). Anti-proliferative effect of IFN-γ in immune regulation. III. Differential selection of T_{h1} and T_{h2} murine helper T lymphocyte clones using recombinant IL-2 and recombinant IFN-γ. *J. Immunol.* **143**, 15–22.

Galibert, L., Van Dooren, J., Durand, I., Rousset, F., Jefferis, R., Banchereau, J. and Lebecque, S. (1995). Anti-CD40 plus interleukin-4-activated human naive B cell lines express unmutated immunoglobulin genes with intraclonal heavy chain isotype variability. *Eur. J. Immunol.* **25**, 733–737.

Galizzi, J.P., Zuber, C.E., Cabrillat, H., Djossou, O. and Banchereau, J. (1989). Internalization of human interleukin-4 and transient down-regulation of its receptor in the CD23-inducible Jijoye cells. *J. Biol. Chem.* **264**, 6984–6989.

Galizzi, J.P., Zuber, C.E., Harada, N., Gorman, D.M., Djossou, O., Kastelein, R., Banchereau, J., Howard, M. and

Miyajima, A. (1990). Molecular cloning of a cDNA encoding the human interleukin-4 receptor. *Int. Immunol.* **2**, 669–675.

Galy, A.H. and Spits, H. (1991). IL-1, IL-4, and IFN-γ differentially regulate cytokine production and cell surface molecule expression in cultured human thymic epithelial cells. *J. Immunol.* **147**, 3823–3830.

Garrett, D.S., Powers, R., March, C.J., Frieden, E.A., Clore, G.M. and Gronenborn, A.M. (1992). Determination of the secondary structure and folding topology of human interleukin-4 using three-dimensional heteronuclear magnetic resonance spectroscopy. *Biochemistry* **31**, 4347–4353.

Garrone, P., Galibert, L., Rousset, F., Fu, S.M. and Banchereau, J. (1994). Regulatory effects of prostaglandin E2 on the growth and differentiation of human B lymphocytes activated through their CD40 antigen. *J. Immunol.* **152**, 4282–4290.

Garrone, P. and Banchereau, J. (1993). Agonistic and antagonistic effects of cholera toxin on human B lymphocyte proliferation. *Mol. Immunol.* **30**, 627–635.

Garrone, P., Djossou, O., Galizzi, J.P. and Banchereau, J. (1991). A recombinant extracellular domain of the human interleukin-4 receptor inhibits the biological effects of interleukin-4 on T and B lymphocytes. *Eur. J. Immunol.* **21**, 1365–1369.

Gauchat, J.F., Lebman, D.A., Coffman, R.L., Gascan, H. and de Vries, J.E. (1990). Structure and expression of germline ε transcripts in human B cells induced by interleukin-4 to switch IgE production. *J. Exp. Med.* **172**, 463–473.

Gauchat, J.F., Gascan, H., de Waal Malefyt, R. and de Vries, J.E. (1992). Regulation of germ-line ε transcription and induction of ε switching in cloned EBV-transformed and malignant human B cell lines by cytokines and $CD4^+$ T cells. *J. Immunol.* **148**, 2291–2299.

Gause, W.C., Halvorson, M.J., Lu, P., Greenwald, R., Linsley, P., Urban, J.F. and Finkelman, F.D. (1997). The function of costimulatory molecules and the development of IL-4 producing T cells. *Immunol. Today* **18**, 115–120.

Gautam, S.C., Chikkala, N.F. and Hamilton, T.A. (1992). Anti-inflammatory action of IL-4. *J. Immunol.* **148**, 1411–1415.

Gaya, A., Alsinet, E., Martorell, J., Places, L., De La Calle, O., Yagüe, J. and Vives, J. (1990). Inhibitory effect of IL-4 on the sepharose-CD3-induced proliferation of the CD4CD45RO human T cell subset. *Int. Immunol.* **2**, 685–689.

Gilleece, M.H., Scarffe, J.H., Ghosh, A., Heyworth, C.M., Bonnem, E., Testa, N., Stern, P. and Dexter, T.M. (1992). Recombinant human interleukin-4 (IL-4) given as daily subcutaneous injections—a phase 1 dose toxicity trial. *Br. J. Cancer* **66**, 204–210.

Gillery, P., Fertin, C., Nicolas, J.F., Chastang, F., Kalis, B., Banchereau, J. and Maquart, F.X. (1992). Interleukin-4 stimulates collagen gene expression in human fibroblast monolayer cultures. *FEBS Lett.* **302**, 231–234.

Gobi, A.E. and Alm, G.V. (1992). Interleukin-4 down-regulates Sendai virus-induced production of interferon-α and -β in human peripheral blood monocytes *in vitro*. *Scand. J. Immunol.* **35**, 167–175.

Golumbek, P.T., Lazenby, A.J., Levitsky, H.I., Jaffee, L.M., Karasuyama, H., Baker, M. and Pardoll, D.M. (1991). Treatment of established renal cancer by tumor cells engineered to secrete interleukin-4. *Science* **254**, 713–716.

Gordon, J., Cairns, J.A., Millsum, M.J., Gillis, S. and Guy, G.R. (1988a). Interleukin-4 and soluble CD23 as progression factors for human B lymphocytes: analysis of their interactions with agonists of the phosphoinositide "dual pathway" of signalling. *Eur. J. Immunol.* **18**, 1561–1565.

Gordon, J., Millsum, M.J., Guy, G.R. and Ledbetter, J.A. (1988b). Resting B lymphocytes can be triggered directly through the Cdw40 (Bp50) antigen. A comparison with IL-4-mediated signaling. *J. Immunol.* **140**, 1425–1430.

Grunfeld, C., Soued, M., Adi, S., Moser, A.H., Fiers, W., Dinarello, C.A. and Feingold, K.R. (1991). Interleukin-4 inhibits stimulation of hepatic lipogenesis by tumor necrosis factor, interleukin-1 and interleukin-6 but not interferon-α. *Cancer Res.* **51**, 2803–2807.

Grusby, M.J., Nabavi, N., Wong, H., Dick, R.F., Bluestone, J.A., Schotz, M.C. and Glimcher, L.H. (1990). Cloning of an interleukin-4 inducible gene from cytotoxic T lymphocytes and its identification as a lipase. *Cell* **60**, 451–459.

Harada, N., Yang, G., Miyajima, A. and Howard, M. (1992). Identification of an essential region for growth signal transduction in the cytoplasmic domain of the human interleukin-4 receptor. *J. Biol. Chem.* **267**, 22752–22758.

Harel-Bellan, A., Durum, S., Muegge, K., Abbas, A.K. and Farrar, W.L. (1988). Specific inhibition of lymphokine biosynthesis and autocrine growth using antisense oligonucleotides in T_{h1} and T_{h2} helper T cell clones. *J. Exp. Med.* **168**, 2309–2318.

Hart, P.H., Vitti, G.F., Burgess, D.R., Whitty, G.A., Piccoli, D.S. and Hamilton, J.A. (1989). Potential anti-inflammatory effects of interleukin-4: suppression of human monocyte tumor necrosis factor α, interleukin 1, and prostaglandin E_2. *Proc. Natl Acad. Sci. USA* **86**, 3803–3807.

Hart, P.H., Ahern, M.J., Smith, M.D. and Finlay-Jones, J.J. (1995). Regulatory effects of IL-13 on synovial fluid macrophages and blood monocytes from patients with inflammatory arthritis. *Clin. Exp. Immunol.* **99**, 331–337.

Hart, P.H., Hunt, E.K., Bonder, C.S., Watson, C.J. and Finlay-Jones, J.J. (1996). Regulation of surface and soluble TNF receptor expression on human monocytes and synovial fluid macrophages by IL-4 and IL-10. *J. Immunol.* **157**, 3672–3680.

Ho, I.C., Hodge, M.R., Rooney, J.W. and Glimcher, L.H. (1996). The proto-oncogene c-maf is responsible for tissue-specific expression of interleukin-4. *Cell* **85**, 973–983.

Ho, J.L., He, S.H., Rios, M.J. and Wick, E.A. (1992). Interleukin-4 inhibits human macrophage activation by tumor necrosis factor, granulocyte-monocyte colony-stimulating factor, and interleukin-3 for antileishmanial activity and oxidative burst capacity. *J. Infect. Dis.* **165**, 344–351.

Hodge, M.R., Ranger, A.M., Charles de la Brousse, F., Hoey, T., Grusby, M.J. and Glimcher, L.H. (1996). Hyper-proliferation and dysregulation of IL-4 expression in NF-ATp-deficient mice. *Immunity* **4**, 397–405.

Horohov, D.W., Crim, J.A., Smith, P.L. and Siegel, J.P. (1988). IL-4 (B cell-stimulatory factor 1) regulates multiple aspects of influenza virus-specific cell-mediated immunity. *J. Immunol.* **141**, 4217–4223.

Hosoda, M., Makino, S., Kawabe, T., Maeda, Y., Satoh, S., Takami, M., Mayumi, M., Arai, K.I., Saitoh, H. and Yodoi, J. (1989). Differential regulation of the low affinity Fc receptor for IgE (Fc ε R2/CD23) and the IL-2 receptor (Tac/p55) on eosinophil leukemia cell line (EoL-1 and EoL-3). *J. Immunol.* **143**, 147–152.

Hou, J., Schindler, U., Henzel, W.J., Ho, T.C., Brasseur, M. and McKnight, S.L. (1994). An interleukin-4-induced transcription factor: IL-4 stat. *Science* **265**, 1701–1706.

Howells, G., Pham, P., Taylor, D., Foxwell, B. and Feldman, M. (1991). Interleukin-4 induces interleukin-6 production by endothelial cells: synergy with interferon-γ. *Eur. J. Immunol.* **21**, 97–101.

Iademarco, M.F., Barks, J.L. and Dean, D.C. (1995). Regulation of vascular cell adhesion molecule-1 expression by IL-4 and TNF-α in cultured endothelial cells. *J. Clin. Invest.* **95**, 264–271.

Idzerda, R.L., March, C.J., Mosley, B., Lyman, S.D., Vanden Bos, T., Gimpel, S.D., Din, W.S., Grabstein, K.H., Widmer, M.B., Park, L.S., Cosman, D. and Beckmann, M.P. (1990). Human Interleukin-4 receptor confers biological responsiveness and defines a novel receptor superfamily. *J. Exp. Med.* **171**, 861–873.

Iwasaki, K., Torisu, M. and Fujimura, T. (1986). Malignant tumor and eosinophils. I. Prognostic significance in gastric cancer. *Cancer* **58**, 1321–1327.

Izuhara, K. and Harada, N. (1993). Interleukin-4 (IL-4) induces protein tyrosine phosphorylation of the IL-4 receptor and association of phosphatidylinositol 3-kinase to the IL-4 receptor in a mouse T cell line, HT2. *J. Biol. Chem.* **268**, 13 097–13 102.

Izuhara, K., Feldman, R.A., Greer, P. and Harada, N. (1994). Interaction of the *c-fes* proto-oncogene product with the interleukin-4 receptor. *J. Biol. Chem.* **269**, 18 623–18 629.

Izuhara, K., Heike, T., Otsuka, T., Yamaoka, K., Mayumi, M., Imamura, T., Niho, Y. and Harada, N. (1996a). Signal transduction pathway of interleukin-4 and interleukin-13 in human B cells derived from X-linked severe combined immunodeficiency patients. *J. Biol. Chem.* **271**, 619–622.

Izuhara, K., Feldman, R.A., Greer, P. and Harada, N. (1996b). Interleukin-4 induces association of *c-fes* proto-oncogene product with phosphatidylinositol-3 kinase. *Blood* **88**, 3910–3918.

Jabara, H.H., Fu, S.M., Geha, R.S. and Vercelli, D. (1990). CD40 and IgE: synergism between anti-CD40 monoclonal antibody and interleukin-4 in the induction of IgE synthesis by highly purified human B cells. *J. Exp. Med.* **172**, 1861–1864.

Jabara, H.H., Ahern, D.J., Vercelli, D. and Geha, R.S. (1991). Hydrocortisone and IL-4 induce IgE isotype switching in human B cells. *J. Immunol.* **147**, 1557–1560.

Jansen, J.H., Wientjens, G.J.H.M., Fibbe, W.E., Willemze, R. and Kluin-Nelemans, H.C. (1989). Inhibition of human macrophage colony formation by interleukin-4. *J. Exp. Med.* **170**, 577–582.

Jelinek, D.F. and Lipsky, P.E. (1988). Inhibitory influence of IL-4 on human B cell responsiveness. *J. Immunol.* **141**, 164–173.

Johnston, J.A., Kawamura, M., Kirken, R.A., Chen, Y.Q., Blake, T.B., Shibuya, K., Ortaldo, J.R., McVicar, D.W. and O'Shea, J.J. (1994). Phosphorylation and activation of the JAK-3 Janus kinase in response to interleukin-2. *Nature* **370**, 151–153.

Kambayashi, T., Jacob, C.O. and Strassmann, G. (1996). IL-4 and IL-13 modulate IL-10 release in endotoxin-stimulated murine peritoneal mononuclear phagocytes. *Cell. Immunol.* **171**, 153–158.

Kapiotis, S., Besemer, J., Bevec, D., Valent, P., Bettelheim, P., Lechner, K. and Speiser, W. (1991). Interleukin-4 counteracts pyrogen-induced downregulation of thrombomodulin in cultured human vascular endothelial cells. *Blood* **78**, 410–415.

Kaplan, M.H., Schindler, U., Smiley, S.T. and Grusby, M.J. (1996). Stat6 is required for mediating responses to IL-4 and for the development of T_{h2} cells. *Immunity* **4**, 313–319.

Katz, Y., Stav, D., Barr, J. and Passwell, J.H. (1995). IL-13 results in differential regulation of the complement proteins C3 and factor B in tumor necrosis factor (TNF)-stimulated fibroblasts. *Clin. Exp. Immunol.* **101**, 150–156.

Karray, S., Delfraissy, J.F., Merle-Beral, H., Wallon, C., Debre, P. and Galanaud, P. (1988). Positive effects of interferon-α on B cell-type chronic lymphocytic leukemia proliferative response. *J. Immunol.* **140**, 774–778.

Kawahara, A., Minami, Y. and Taniguchi, T. (1994). Evidence for a critical role for the cytoplasmic region of the interleukin-2 (IL-2) receptor γ chain in IL-2, IL-4, and IL-7 signaling. *Mol. Cell. Biol.* **14**, 5433–5440.

Kawakami, Y., Custer, M.C., Rosenberg, S.A. and Lotze, M.T. (1989). IL-4 regulates IL-2 induction of lymphokine-activated killer activity from human lymphocytes. *J. Immunol.* **142**, 3452–3461.

Keane-Myers, A., Maliszewski, C.R., Finkelman, F.D. and Nickell, S.P. (1996). Recombinant IL-4 treatment augments resistance to *Borrelia burgdorferi* infections in both normal susceptible and antibody-deficient susceptible mice. *J. Immunol.* **156**, 2488–2494.

Kinashi, T., Inaba, K., Tsubata, T., Tashiro, K., Palacios, R. and Honjo, T. (1988). Differentiation of an interleukin 3-dependent precursor B-cell clone into immunoglobulin-producing cells *in vitro*. *Proc. Natl Acad. Sci. USA* **85**, 4473–4477.

Klein, N.J., Rigley, K.P. and Callard, R.E. (1993). IL-4 regulates the morphology, cytoskeleton, and proliferation of human umbilical vein endothelial cells: relationship between vimentin and CD23. *Int. Immunol.* **5**, 293–301.

Kolb, J.P., Paul-Eugène, N., Damais, C., Yamaoka, K., Drapier, J.C. and Dugas, B. (1994). Interleukin-4 stimulates cGMP production by IFN-γ-activated human monocytes. Involvement of the nitric oxide synthase pathway. *J. Biol. Chem.* **269**, 9811–9816.

Kopf, M., Brombacher, F., Kohler, G., Kienzle, G., Widmann, K.H., Lefrang, K., Humborg, C., Ledermann, B. and Soibach, W. (1996). IL-4-deficient Balb/c mice resist infection with *Leishmania major*. *J. Exp. Med.* **184**, 1127–1136.

Kruse, N., Tony, H.P. and Sebald, W. (1992). Conversion of human interleukin-4 into a high affinity antagonist by a single amino acid replacement. *EMBO J.* **11**, 3237–3244.

Kühn, R., Rajewsky, K. and Müller, W. (1991). Generation and analysis of interleukin-4 deficient mice. *Science* **254**, 707–710.

Kumaratilake, L.M. and Ferrante, A. (1992). IL-4 inhibits macrophage-mediated killing of *Plasmodium falciparum in vitro*. *J. Immunol.* **149**, 194–199.

Lacraz, S., Nicod, L., Galve-de-Rochemonteix, B., Baumberger, C., Dayer, J.-M. and Welgus, H.G. (1992). Suppression of metalloproteinase biosynthesis in human alveolar macrophages by interleukin-4. *J. Clin. Invest.* **90**, 382–388.

Lahm, H., Schnyder, B., Wyniger, J., Borbenyi, Z., Yilmaz, A., Car, B.D., Fischer, J.R., Givel, J.C. and Ryffel, B. (1994). Growth inhibition of human colorectal-carcinoma cells by interleukin-4 and expression of functional interleukin-4 receptors. *Int. J. Cancer* **59**, 440–447.

Lahm, H., Arnstad, P., Yilmaz, A., Borbenyi, Z., Wyniger, J., Fischer, J.R., Suardet, L., Givel, J.C. and Odartchenko, N. (1995). Interleukin-4 down-regulates expression of *c-kit* and autocrine stem cell factor in human colorectal carcinoma cells. *Cell Growth Differ.* **6**, 1111–1118.

Langouet, S., Corcos, L., Abdel-Razzak, Z., Loyer, P., Ketterer, B. and Guillouzo, A. (1995). Up-regulation of glutathione S-transferases α by interleukin-4 in human hepatocytes in primary culture. *Biochem. Biophys. Res. Commun.* **216**, 793–800.

Larner, A.C., Petricoin, E.F., Nakagawa, Y. and Finbloom, D.S. (1993). IL-4 attenuates the transcriptional activation of both IFN-α and IFN-γ-induced cellular gene expression in monocytes and monocytic cell lines. *J. Immunol.* **150**, 1944–1950.

Lawrence, R.A., Allen, J.E., Gregory, W.F., Kopf, M. and Maizels, R.M. (1995). Infection of IL-4-deficient mice with the parasitic nematode *Brugia malayi* demonstrates that host resistance is not dependent on a T helper 2-dominated immune response. *J. Immunol.* **154**, 5995–6001.

Leach, M.W., Rybak, M.E. and Rosenblum, I.Y. (1997). Safety evaluation of recombinant human interleukin-4. II. Clinical studies. *Clin. Immunol. Immunopathol.* **83**, 12–14.

Leach, M.W., Snyder, E.A., Sinha, D.P. and Rosenblum, I.Y. (1997). Safety evaluation of recombinant human interleukin-4. I. Preclinical studies. *Clin. Immunol. Immunopathol.* **83**, 8–11.

Leal, L.M., Moss, D.W., Kühn, R., Müller, W. and Liew, F.Y. (1993). Interleukin-4 transgenic mice of resistant background are susceptible to *Leishmania major* infection. *Eur. J. Immunol.* **23**, 566–569.

Lederer, J.A., Perez, V.L, DesRoches, L., Sim, S.M., Abbas, A.K. and Lichtman, A.H. (1996). Cytokine transcriptional events during helper T cell subset differentiation. *J. Exp. Med.* **184**, 397–406.

Lee, H.K., Xia, X. and Choi, Y.S. (1990). IL-4 blocks the up-regulation of IL-2 receptors induced by IL-2 in normal human B cells. *J. Immunol.* **144**, 3431–3436.

Li, Y.S., Kouassi, E. and Revillard, J.P. (1989). Cyclic AMP can enhance mouse B cell activation by regulating progression into the late G1/S phase. *Eur. J. Immunol.* **19**, 1721–1725.

Litman, B.H., Dastvan, F.F., Carlson, P.L. and Sanders, K.M. (1989). Regulation of monocyte/macrophage C2 production and HLA-DR expression by IL-4 (BSF-1) and IFN-γ. *J. Immunol.* **142**, 520–525.

Li-Weber, M., Eder, A., Krafft-Czepa, H. and Krammer, P.H. (1992). T cell-specific negative regulation of transcription of the human cytokine IL-4. *J. Immunol.* **148**, 1913–1918.

Li-Weber, M., Krafft, H. and Krammer, P.H. (1993). A novel enhancer element in the human IL-4 promoter is suppressed by a position-independent silencer. *J. Immunol.* **151**, 1371–1382.

Li-Weber, M., Davydov, I.V., Krafft, H. and Krammer, P.H. (1994). Role of NF-Y and IRF-2 in the regulation of human IL-4 gene expression. *J. Immunol.* **153**, 4122–4133.

Li-Weber, M., Salgame, P., Hu, C., Davydov, I.V. and Krammer, P.H. (1997). Differential interaction of nuclear factors with the PRE-I enhancer element of the human IL-4 promoter in different T cell subsets. *J. Immunol.* **158**, 1194–1200.

Llorente, L., Crevon, M.L., Karray, S., Defrance, T., Banchereau, J. and Galanaud, P. (1989). Interleukin (IL) 4 counteracts the helper effect of IL-2 on antigen-activated human B cells. *Eur. J. Immunol.* **19**, 765–769.

Llorente, L., Mitjavila, F., Crevon, M.C. and Galanaud, P. (1990). Dual effects of interleukin-4 on antigen-activated human B cells: induction of proliferation and inhibition of interleukin-2-dependent differentiation. *Eur. J. Immunol.* **20**, 1887–1892.

Locksley, R.M. (1994). T_{h2} cells: help for helminths. *J. Exp. Med.* **179**, 1405–1407.

Loh, R.K., Jabara, H.H., Ren, C.L., Fu, S.M. and Geha, R.S. (1994). Role of protein tyrosine kinases in CD40/interleukin-4-mediated isotype switching to IgE. *J. Allergy Clin. Immunol.* **94**, 784–792.

Loyer, P., Ilyin, G., Razzak, Z.A., Dézier, J.F., Banchereau, J., Campion, J.P., Guguen-Guillouzo, C. and Guillouzo, A. (1993). IL-4 modulates production of acute phase proteins in adult human hepatocytes. *FEBS Lett.* **336**, 215–220.

Lu, P., Urban, J.F., Zhou, X.D., Chen, S.J., Madden, K., Moorman, M., Nguyen, H., Morris, S.C., Finkelman, F.D. and Gause, W.C. (1996). CD40-mediated stimulation contributes to lymphocyte proliferation, antibody production, eosinophilia and mastocytosis during an *in vivo* type 2 response, but is not required for T cell IL-4 production. *J. Immunol.* **156**, 3327–3333.

Manabe, A., Coustan-Smith, E., Kumagai, M.A., Behm, F.G., Raimondi, S.C., Pui, C.H. and Campana, D. (1994). Interleukin-4 induces programmed cell death (apoptosis) in cases of high-risk acute lymphoblastic leukemia. *Blood* **83**, 1731–1737.

Manetti, R., Gerosa, F., Giudizi, M.G., Biagiotti, R., Parronchi, P., Piccini, M.P., Sampognaro, S., Maggi, E., Romagnani, S. and Trinchieri, G. (1994). Interleukin-12 induces stable priming for interferon γ (IFN γ) production during differentiation of human T helper (T_h) cells and transient IFN-γ production in established T_{h2} cell clones. *J. Exp. Med.* **179**, 1273–1283.

Mangan, D.F., Robertson, B. and Wahl, S.M. (1992). IL-4 enhances programmed cell death (apoptosis) in stimulated human monocytes. *J. Immunol.* **148**, 1812–1816.

Mantovani, A., Bussolino, F. and Dejana, E. (1992). Cytokine regulation of endothelial cell functions. *FASEB J.* **6**, 2591–2599.

Marfaing-Koka, A., Devergne, O., Gorgone, G., Portier, A., Schall, T.J., Galanaud, P. and Emilie, D. (1995). Regulation of the production of the RANTES chemokine by endothelial cells. Synergistic induction by IFN-γ plus TNF-α and inhibition by IL-4 and IL-13. *J. Immunol.* **154**, 1870–1878.

McCann, F.V., McCarthy, D.C. and Noelle, R.J. (1991). Interleukin-4 activates ion channels in B lymphocytes. *Cell Signall.* **3**, 483–490.

McGarvie, G.M. and Cushley, W. (1989). The effect of recombinant interleukin-4 upon protein kinase activities associated with murine and human B lymphocyte plasma membranes. *Cell Signal.* **1**, 447–460.

McIntyre, T.M., Kehry, M.R. and Snapper, C.M. (1995). Novel *in vitro* model for high rate IgA class switching. *J. Immunol.* **154**, 3156–3161.

McKenzie, A.N.J., Culpepper, J.A., de Waal Malefyt, R., Brière, F., Punnonen, J., Aversa, G., Sato, A., Dang, W., Cocks, B.G., Menon, S., de Vries, J.E., Banchereau, J. and Zurawski, G. (1993). Interleukin-13, a novel T cell-derived cytokine that regulates human monocyte and B cell function. *Proc. Natl Acad. Sci. USA* **90**, 3735–3739.

Minty, A., Chalon, P., Derocq, J.M., Dumont, X., Guillemot, J.C., Kaghad, M., Labit, C., Leplatois, P., Liauzun, P., Miloux, B., Minty, C., Casellas, P., Loison, G., Lupker, J., Shire, D., Ferrara, P. and Caput, D. (1993). Interleukin-13 is a new human lymphokine regulating inflammatory and immune responses. *Nature* **362**, 248–250.

Mire-Sluis, A.R. and Thorpe, R. (1991). Interleukin-4 proliferative signal transduction involves the activation of a tyrosine-specific phosphatase and the dephosphorylation of an 80-kDa protein. *J. Biol. Chem.* **266**, 18113–18118.

Miyajima, A., Kitamura, T., Harada, N., Yokota, T. and Arai, K. (1992). Cytokine receptors and signal transduction. *Annu. Rev. Immunol.* **10**, 295–331.

Miyauchi, J., Clark, S.C., Tsunematsu, Y., Shimizu, K., Park, J.W., Ogawa, T. and Toyama, K. (1991). Interleukin-4 as a growth regulator of clonogenic cells in acute myelogenous leukemia in suspension culture. *Leukemia* **5**, 108–115.

Mizuguchi, J., Beaven, M.A., O'Hara, J. and Paul, W.E. (1986). BSF-1 action on resting B cells does not require elevation of inositol phospholipid metabolism or increased [Ca^{2+}]i. *J. Immunol.* **137**, 2215–2219.

Montanier, L.J., Doyle, A.G., Collin, M., Herbein, G., Illei, P., James, W., Minty, A., Caput, D., Ferrara, P. and Gordon, S. (1993). Interleukin-13 inhibits human immunodeficiency virus type 1 production in primary blood-derived human macrophages *in vitro*. *J. Exp. Med.* **178**, 743–747.

Moqbel, R., Ying, S., Barkins, J., Newman, T.M., Kimmit, P., Wakelin, M., Taborda-Barata, L., Meng, Q., Corrigan, C.J., Durham, S.R. and Kay, A.B. (1995). Identification of messenger RNA for IL-4 in human eosinophils with granule localization and release of the translated product. *J. Immunol.* **155**, 4939–4947.

Morawetz, R.A., Gabriele, L., Rizzo, L.V., Noben-Trauth, N., Kühn, R., Rajewsky, K., Muller, W., Doherty, T.M., Finkelman, F., Coffman, R.L. and Morse, H.C. (1996). Interleukin (IL)-4-independent immunoglobulin class switch to immunoglobulin (Ig)E in the mouse. *J. Exp. Med.* **184**, 1651–1661.

Morgan, J.G., Dolganov, G.M., Robbins, S.E., Hinton, L.M. and Lovett, M. (1992). The selective isolation of novel cDNAs encoded by the regions surrounding the human interleukin-4 and -5 genes. *Nucleic Acids Res.* **20**, 5173–5179.

Morisaki, T., Yuzuki, D.H., Lin, R.T., Foshag, L.J., Motron, D.L. and Hoon, D.S. (1992). Interleukin-4 receptor expression and growth inhibition of gastric carcinoma cells by interleukin-4. *Cancer Res.* **52**, 6059–6065.

Moser, R., Fehr, J. and Bruijnzeel, P.L.B. (1992). IL-4 controls the selective endothelium-driven transmigration of eosinophils from allergic individuals. *J. Immunol.* **149**, 1432–1438.

Mosmann, T.R., Bond, M.W., Coffman, R.L., Ohara, J. and Paul, W.E. (1986). T-cell and mast cell lines respond to B-cell stimulatory factor 1. *Proc. Natl Acad. Sci. USA* **83**, 5654–5658.

Müller, I., Pedrazzini, T., Kropf, P., Louis, J. and Milon, G. (1991). Establishment of resistance to *Leishmania major* infection in susceptible BALB/c mice requires parasite-specific $CD8^+$ T cells. *Int. Immunol.* **3**, 587–597.

Mulligan, M.S., Jones, M.L., Vaporciyan, A.A., Howard, M.C. and Ward, P.A. (1993). Protective effects of IL-4 and IL-10 against immune complex-induced lung injury. *J. Immunol.* **151**, 5666–5674.

Murata, T. and Puri, R.K. (1997). Comparison of IL-13 and IL-4-induced signaling in EBV-immortalized human B cells. *Cell Immunol.* **175**, 33–40.

Murata, T., Noguchi, P.D. and Puri, R.K. (1996). IL-13 induces phosphorylation and activation of JAK2 Janus kinase in human colon carcinoma cell lines: similarities between IL-4 and IL-13 signaling. *J. Immunol.* **156**, 2972–2978.

Musso, T., Varesio, L., Zhang, X., Rowe, T.K., Ferrara, P., Ortaldo, J.R., O'Shea, J.J. and McVicar, D.W. (1994). IL-4 and IL-13 induce Lsk, a Csk-like tyrosine kinase, in human monocytes. *J. Exp. Med.* **180**, 2383–2388.

Musso, T., Johnston, J.A., Linnekin, D., Varesio, L., Rowe, T.K., O'Shea, J.J. and McVicar, D.W. (1995). Regulation of JAK3 expression in human monocytes: phosphorylation in response to interleukins-2, -4, and -7. *J. Exp. Med.* **181**, 1425–1431.

Nagler, A., Lanier, L.L. and Phillips, J.H. (1988). The effects of IL-4 on human natural killer cells. A potent regulator of IL-2 activation and proliferation. *J. Immunol.* **141**, 2349–2351.

Nakajima, H., Gleich, G.J. and Kita, H. (1996). Constitutive production of IL-4 and IL-10 and stimulated production of IL-8 by normal peripheral blood eosinophils. *J. Immunol.* **156**, 4859–4866.

Nakajima, A., Hirose, S., Yagita, H. and Okumura, K. (1997). Roles of IL-4 and IL-12 in the development of lupus in NZB/W F1 mice. *J. Immunol.* **158**, 1466–1472.

Nakamura, M., Kondo, S., Sugai, M., Nazarea, M., Imamura, S. and Honjo, T. (1996). High frequency class switching of an IgM$^+$ B lymphoma clone CH12F3 to IgA$^+$ cells. *Int. Immunol.* **8**, 193–201.

Nakamura, T., Kamogawa, Y., Bottomly, K. and Flavell, R.A. (1997). Polarization of IL-4 and IFN-γ-producing CD4$^+$ T cells following activation of naive CD4$^+$ T cells. *J. Immunol.* **158**, 1085–1094.

Nassar, G.M., Morrow, J.D., Roberts, L.J., Lakkis, F.G. and Badr, K.F. (1994). Induction of 15-lipoxygenase by interleukin-13 in human blood monocytes. *J. Biol. Chem.* **269**, 27631–27634.

Noben-Trauth, N., Kropf, P. and Müller, I. (1996). Susceptibility to *Leishmania major* infection in interleukin-4 deficient mice. *Science* **271**, 987–990.

Noelle, R., Krammer, P.H., Ohara, J., Uhr, J.W. and Vitetta, E.S. (1984). Increased expression of Ia antigens on resting B cells: an additional role for B-cell growth factor. *Proc. Natl Acad. Sci. USA* **81**, 6149–6153.

Noguchi, M., Nakamura, Y., Russel, S.M., Ziegler, S.F., Tsang, M., Cao, X. and Leonard, W.J. (1993a). Interleukin-2 receptor γ chain: a functional component of the interleukin-7 receptor. *Science* **262**, 1877–1880.

Noguchi, M., Yi, H., Rosenblatt, H.M., Filipovich, A.H., Adelstein, S., Modi, W.S., McBride, O.W. and Leonard, W.J. (1993b). Interleukin-2 receptor γ chain mutation results in X-linked severe combined immunodeficiency in humans. *Cell* **73**, 147–157.

O'Garra, A. and Spits, H. (1993). The immunobiology of interleukin-4. *Res. Immunol.* **144**, 567–643.

O'Garra, A., Rigley, K.P., Holman, M., McLaughlin, J.B. and Klaus, G.G.B. (1987). B-cell-stimulatory factor 1 reverses Fc receptor mediated inhibition of B lymphocyte activation. *Proc. Natl Acad. Sci. USA.* **84**, 6254–6258.

Oakes, S.A., Candotti, F., Johnston, J.A., Chen, Y.Q., Ryan, J.J., Taylor, N., Liu, X., Hennighausen, L., Notarangelo, L.D., Paul, W.E., Blaese, R.M. and O'Shea, J.J. (1996). Signaling via IL-2 and IL-4 in JAK3-deficient severe combined immunodeficiency lymphocytes: JAK-3 dependent and independent pathways. *Immunity* **5**, 605–615.

Obiri, N.I., Hillman, G.G., Haas, G.P., Sud, S. and Puri, R.K. (1993). Expression of high affinity interleukin-4 receptors on human renal cell carcinoma cells and inhibition of tumor cell growth *in vitro* by interleukin-4. *J. Clin. Invest.* **91**, 88–93.

Obiri, N.I., Debinski, W., Leonard, W.J. and Puri, R.K. (1995). Receptor for interleukin-13. Interaction with interleukin-4 by a mechanism that does not involve the common γ chain shared by receptors for interleukins-2,-4,-7,-9 and -15. *J. Biol. Chem.* **270**, 8797–8804.

Obiri, N.I., Leland, P., Murata, T., Debinski, W. and Puri, R.K. (1997). The IL-13 receptor structure differs on various cell types and may share more than one component with IL-4 receptor. *J. Immunol.* **158**, 756–764.

Ohmori, Y., Smith, M.F. and Hamilton, T.A. (1996). IL-4-induced expression of the IL-1 receptor antagonist gene is mediated by STAT6. *J. Immunol.* **157**, 2058–2065.

Okamura, H., Tsutsui, H., Komatsu, T., Yutsudo, M., Hakura, A., Tanimoto, T., Torigoe, K., Okura, T., Nukada, Y., Hattori, K., Akita, K., Namba, M., Tanabe, F., Konishi, K., Fukuda, S. and Kurimoto, M. (1995). Cloning of a new cytokine that induces IFNγ production by T cells. *Nature* **378**, 88–91.

Paganelli, R., Scala, E., Ansotegui, I.L., Ausiello, C.M., Halapi, E., Fanales-Belasio, E., D'Offizi, G., Mezzaroma, I., Pandolfi, F., Fiorilli, M., Cassone, A. and Aiuti, F. (1995). CD8$^+$ T lymphocytes provide helper activity for IgE synthesis in human immunodeficiency virus-infected patients with hyper-IgE. *J. Exp. Med.* **181**, 423–428.

Paliogianni, F., Hama, N., Mavrothalassitis, G.J., Thyphronitis, G. and Boumpas, D.T. (1996). Signal requirements for interleukin-4 promoter activation in human T cells. *Cell Immunol.* **168**, 33–38.

Palmer-Crocker, R.L., Hughes, C.C. and Pober, J.S. (1996). IL-4 and IL-13 activate the JAK2 tyrosine kinase and Stat6 in cultured human vascular endothelial cells through a common pathway that does not involve the γc chain. *J. Clin. Invest.* **98**, 604–609.

Pandrau, D., Saeland, S., Duvert, V., Durand, I., Manel, A.M., Zabot, M.T., Philippe, N. and Banchereau, J. (1992). Interleukin-4 inhibits *in vitro* proliferation of leukemic and normal human B cell precursors. *J. Clin. Invest.* **90**, 1697–1706.

Park, L.S., Friend, D., Sassenfeld, H.M. and Urdal, D.L. (1987). Characterization of the human B cell stimulatory factor 1 receptor. *J. Exp. Med.* **166**, 476–488.

Parronchi, P., Macchia, D., Piccini, M.P., Biswas, P., Simonelli, C., Maggi, E., Ricci, M., Ansari, A.A. and Romagnani, S. (1991). Allergen- and bacterial antigen-specific T-cell clones established from atopic donors show a different profile of cytokine production. *Proc. Natl Acad. Sci. USA* **88**, 4538–4542.

Peleman, R., Wu, J., Fargeas, C. and Delespesse, G. (1989). Recombinant interleukin-4 suppresses the production of interferon γ by human mononuclear cells. *J. Exp. Med.* **170**, 1751–1756.

Pène, J., Rousset, F., Brière, F., Chrétien, I., Paliard, X., Banchereau, J., Spits, H. and De Vries, J.E. (1988a). IgE production by normal human B cells induced by alloreactive T cell clones is mediated by IL-4 and suppressed by IFN-γ. *J. Immunol.* **141**, 1218–1224.

Pène, J., Rousset, F., Brière, F., Chrétien, I., Wideman, J., Bonnefoy, J.Y. and De Vries, J.E. (1988b). Interleukin-5 enhances interleukin-4-induced IgE production by normal human B cells. The role of soluble CD23 antigen. *Eur. J. Immunol.* **18**, 929–935.

Pesch, J., Brehm, U., Staib, C. and Grummt, F. (1996). Repression of interleukin-2 and interleukin-4 promoters by tumor suppressor protein p53. *J. Interferon Cytokine Res.* **16**, 595–600.

Peschel, C., Green, I. and Paul, W.E. (1989). Interleukin-4 induces a substance in bone marrow stromal cells that reversibly inhibits factor-dependent and factor-independent cell proliferation. *Blood* **73**, 1130–1141.

Phillips, J.O., Everson, M.P., Moldoveanu, Z., Lue, C. and Mestecky, J. (1990). Synergistic effect of IL-4 and IFN-γ on the expression of polymeric Ig receptor (secretory component) and IgA binding by human epithelial cells. *J. Immunol.* **145**, 1740–1741.

Pickl, W.F., Majdic, O., Kohl, P., Stockl, J., Riedl, E., Scheinecker, C., Bello-Fernandez, C. and Knapp, W. (1996). Molecular and functional characteristics of dendritic cells generated from highly purified $CD14^+$ peripheral blood monocytes. *J. Immunol.* **157**, 3850–3859.

Piela-Smith, T.H., Broketa, G., Hand, A. and Korn, J.H. (1992). Regulation of ICAM-1 expression and function in human dermal fibroblasts by IL-4. *J. Immunol.* **148**, 1375–1381.

Plum, J., De Smedt, M., Leclercq, G. and Tison, B. (1990). Inhibitory effect of murine recombinant IL-4 on thymocyte development in fetal thymus organ cultures. *J. Immunol.* **145**, 1066–1073.

Postlethwaite, A.E. and Seyer, J.M. (1991). Fibroblast chemotaxis induction by human recombinant interleukin-4. Identification by synthetic peptide analysis of two chemotactic domains residing in amino acid sequences 70–88 and 89–122. *J. Clin. Invest.* **87**, 2147–2152.

Postlethwaite, A.E., Holness, M.A., Katai, H. and Raghow, R. (1992). Human fibroblasts synthetize elevated levels of extracellular matrix proteins in response to interleukin-4. *J. Clin. Invest.* **90**, 1479–1485.

Powers, R., Garrett, D.S., March, C.J., Frieden, E.A., Gronenborn, A.M. and Clore, G.M. (1992). Three-dimensional solution structure of human interleukin-4 by multidimensional heteronuclear magnetic resonance spectroscopy. *Science* **256**, 1673–1677.

Punnonen, J., Aversa, G. and de Vries, J.E. (1993). Human pre-B cells differentiate into Ig-secreting plasma cells in the presence of Interleukin-4 and activated $CD4^+$ T cells or their membranes. *Blood* **82**, 2781–2789.

Quelle, F.W., Shimoda, K., Thierfelder, W., Fischer, C., Kim, A., Ruben, S.M., Cleveland, J.L., Pierce, J.H., Keegan, A.D., Nelms, K., Paul, W.E. and Ihle, J.N. (1995). Cloning of murine Stat6 and human Stat6, Stat proteins that are tyrosine phosphorylated in responses to IL-4 and IL-3 but are not required for mitogenesis. *Mol. Cell. Biol.* **15**, 3336–3343.

Racke, M.K., Burnett, D., Pak, S.H., Albert, P.S., Cannella, B., Raine, C.S., McFarlin, D.E. and Scott, D.E. (1995). Retinoid treatment of experimental allergic encephalomyelitis. IL-4 production correlates with improved disease course. *J. Immunol.* **154**, 450–458.

Rankin, J.A., Picarella, D.E., Geba, G.P., Temannn, U.A., Prasad, B., DiCosmo, B., Tarallo, A., Stripp, B., Whitsett, J. and Flavell, R.A. (1996). Phenotypic and physiologic characterization of transgenic mice expressing interleukin-4 in the lung: lymphocytic and eosinphilic inflammation without airway hyperreactivity. *Proc. Natl Acad. Sci. USA* **93**, 7821–7825.

Rapoport, M.J., Jaramillo, A., Zipris, D., Lazarus, A.H., Serreze, D.V., Leiter, E.H., Cyopick, P., Dnaska, J.S. and Deloviqtch, T.L. (1993). Interleukin-4 reverses T cell proliferative unresponsiveness and prevents the onset of diabetes in nonobese diabetic mice. *J. Exp. Med.* **178**, 87–99.

Razzak, Z.A., Loyer, P., Fautrel, A., Gautier, J.C., Corcos, L., Turlin, B., Beaune, P. and Guillouzo, A. (1993). Cytokines down-regulate expression of major cytochrome P-450 enzymes in adult human hepatocytes in primary culture. *Mol. Pharmacol.* **44**, 707–715.

Renard, N., Duvert, V., Banchereau, J. and Saeland, S. (1994). Interleukin-13 inhibits the proliferation of normal and leukemic human B-cell precursors. *Blood* **84**, 2253–2260.

Rennick, D., Yang, G., Muller-Sieburg, C., Smith, C., Arai, N., Takabe, Y. and Gemmell, L. (1987). Interleukin-4 (B-cell stimulatory factor 1) can enhance or antagonize the factor-dependent growth of hemopoietic progenitor cells. *Proc. Natl Acad. Sci. USA* **84**, 6889–6893.

Riemann, D., Kehlen, A. and Langner, J. (1995). Stimulation of the expression and the enzyme activity of aminopeptidase N/CD13 and dipeptidylpeptidase IV/CD26 on human renal cell carcinoma cells and renal tubular epithelial cells by T cell-derived cytokines, such as IL-4 and IL-13. *Clin. Exp. Immunol.* **100**, 277–283.

Rincon, M. and Flavell, R.A. (1997). Transcription mediated by NFAT is highly inducible in effector $CD4^+$ T helper 2 (T_{h2}) cells but not in T_{h1} cells. *Mol. Cell Biol.* **17**, 1522–1534.

Rincon, M., Anguita, J., Nakamura, T., Fikrig, E. and Flavell, R.A. (1997). IL-6 directs the differentiation of IL-4-producing CD4$^+$ T cells. *J. Exp. Med.* **185**, 461–469.

Rocken, M., Saurat, J.H. and Hauser, C. (1992). A common precursor for CD4$^+$ T cells producing IL-2 or IL-4. *J. Immunol.* **148**, 1031–1036.

Romagnani, S. (1992). Human T_{h1} and T_{h2} subsets: regulation of differentiation and role in protection and immunopathology. *Int. Arch. Allergy Immunol.* **98**, 279–285.

Romani, N., Gruner, S., Brang, D., Kämpgen, E., Lenz, A., Trockenbacher, B., Konwalinka, G., Fritsch, P.O., Steinman, R.M. and Schuler, G. (1994). Proliferating dendritic cell progenitors in human blood. *J. Exp. Med.* **180**, 83–93.

Rooney, J.W., Hoey, T. and Glimcher, L.H. (1995). Coordinate and cooperative roles for NF-AT and AP-1 in the regulation of the murine IL-4 gene. *Immunity* **2**, 473–483.

Rousselet, G., Busson, P., Billaud, M., Guillon, J.M., Scamps, C., Wakasugi, H., Lenoir, G. and Tursz, T. (1990). Structure and regulation of the Blast-2/CD23 antigen in epithelial cells from nasopharyngeal carcinoma. *Int. Immunol.* **2**, 1159–1166.

Rousset, F., de Waal Malefit, R., Slierendregt, B., Aubry, J.P., Bonnefoy, J.Y., Defrance, T., Banchereau, J. and de Vries, J.E. (1988). Regulation of Fc receptor for IgE (CD23) and class II MHC antigen expression on Burkitt's lymphoma cell lines by human IL-4 and IFN-γ. *J. Immunol.* **140**, 2625–2632.

Rousset, F., Garcia, E. and Banchereau, J. (1991a). Cytokine-induced proliferation and immunoglobulin production of human B lymphocytes triggered through their CD40 antigen. *J. Exp. Med.* **173**, 705–710.

Rousset, F., Robert, J., Andary, M., Bonnin, J.P., Souillet, G., Chrétien, I., Brière, F., Pène, J. and de Vries, J.E. (1991b). Shifts in interleukin-4 and interferon-γ production by T cells of patients with elevated serum IgE levels and the modulatory effects of these lymphokines on spontaneous IgE synthesis. *J. Allergy Clin. Immunol.* **87**, 58–69.

Russell, S.M., Keegan, A.D., Harada, N., Nakamura, Y., Noguchi, M., Leland, P., Friedmann, M.C., Miyajima, A., Puri, R.K., Paul, W.E. and Leonard, W.J. (1993). Interleukin-2 receptor γ chain: a functional component of the interleukin-4 receptor. *Science* **262**, 1880–1883.

Russell, S.M., Tayebi, N., Nakajima, H., Riedy, M.C., Roberts, J.L., Aman, M.J., Migone, T.S., Noguchi, M., Markert, M.L., Buckley, R.H., O'Shea, J.J. and Leonard, W.J. (1995). Mutation of Jak3 in a patient with SCID: essential role in JAK3 in lymphoid development. *Science* **270**, 797–800.

Ryan, D.H., Nuccie, B.L., Ritterman, I., Liesveld, J.L. and Abboud, C.N. (1994). Cytokine regulation of early human lymphopoiesis. *J. Immunol.* **152**, 5250–5258.

Sad, S., Marcotte, R. and Mosmann, T.R. (1995). Cytokine-induced differentiation of precursor mouse CD8$^+$ T cells into cytotoxic CD8$^+$ T cells secreting T_{h1} or T_{h2} cytokines. *Immunity* **2**, 271–279.

Salgame, P., Abrams, J.S., Clayberger, C., Goldstein, H., Convit, J., Modlin, R.L. and Bloom, B.R. (1991). Differing lymphokine profiles of functional subsets of human CD4 and CD8 T cell clones. *Science* **254**, 279–282.

Sallusto, F. and Lanzavecchia, A. (1994). Efficient presentation of soluble antigen by cultured human dendritic cells is maintained by granulocyte/macrophage colony-stimulating factor plus interleukin-4 and downregulated by tumor necrosis factor α. *J. Exp. Med.* **179**, 1109–1118.

Santiago, M.L., Fossati, L., Jacquet, C., Muller, W., Izui, S. and Reininger, L. (1997). Interleukin-4 protects against a genetically linked lupus-like autoimmune syndrome. *J. Exp. Med.* **185**, 65–70.

Schindler, C., Kashleva, H., Pernis, A., Pine, R. and Rothman, P. (1994). STF-IL-4: a novel IL-4-induced signal transducing factor. *EMBO J.* **13**, 1350–1356.

Schlaak, J.F., Schwarting, A., Knolle, P., Meyer zum Buschenfelde, K.H. and Mayet, W. (1995). Effects of T_{h1} and T_{h2} cytokines on cytokine production and ICAM-1 expression on synovial fibroblasts. *Ann. Rheum. Dis.* **54**, 560–565.

Schleimer, R.P., Sterbinsky, S.A., Kaiser, J., Bickel, C.A., Klunk, D.A., Tomioka, K., Newman, W., Luscinskas, F.W., Gimbrone, M.A., McIntyre, B.W. and Bochner, B.S. (1992). IL-4 induces adherence of human eosinophils and basophils but not neutrophils to endothelium. Association with expression of VCAM-1. *J. Immunol.* **148**, 1086–1092.

Schnyder, B., Lahm, H., Woerly, G., Odartchenko, N., Ryffel, B. and Car, B.D. (1996). Growth inhibition signalled through the interleukin-4/interleukin-13 receptor complex is associated with tyrosine phosphorylation of insulin receptor substrate-1. *Biochem. J.* **315**, 767–774.

Scholz, C., Freeman, G.J., Greenfield, E.A., Hafler, D.A. and Hollsberg, P. (1996). Activation of human T cell lymphotropic virus type-1 infected T cells is independent of B7 costimulation. *J. Immunol.* **157**, 2932–2938.

Seder, R.A. and Le Gros, G.G. (1995). The functional role of CD8$^+$ T helper type 2 cells. *J. Exp. Med.* **181**, 5–7.

Seder, R.A., Boulay, J.L., Finkelman, F., Barbier, S., Ben-Sasson, S.Z., Le Gros, G. and Paul, W.E. (1992). $CD8^+$ T cells can be primed *in vitro* to produce IL-4. *J. Immunol.* **148**, 1652–1656.

Shapira, S.K., Vercelli, D., Jabara, H.H., Fu, S.M. and Geha, R.S. (1992). Molecular analysis of the induction of immunoglobulin E synthesis in human B cells by interleukin-4 and engagement of CD40 antigen. *J. Exp. Med.* **175**, 289–292.

Shaw, M.K., Lorens, J.B., Dhawan, A., Dalcanto, R., Tse, H.Y., Tran, A.B., Bonpana, C., Eswaran, S.L., Brocke, S., Sarvetnick, N., Steinman, L., Nolan, G.P. and Fathman, C.G. (1997). Local delivery of interleukin-4 by retrovirus-transduced T lymphocytes ameliorates experimental autoimmune encephalomyelitis. *J. Exp. Med.* **185**, 1711–1714.

Shields, J.G., Armitage, R.J., Jamieson, B.N., Beverley, P.C.L. and Callard, R.E. (1989). Increased expression of surface IgM but not IgD or IgG on human B cells in response to IL-4. *Immunol.* **66**, 224–227.

Shimoda, K., van Deursen, J., Sangster, M.Y., Sarawar, S.R., Carson, R.T., Tripp, R.A., Chu, C., Quelle, F.W., Nosaka, T., Vignali, D.A.A., Doherty, P.C., Grosveld, G., Paul, W.E. and Ihle, J.N. (1996). Lack of IL-4-induced T_{h2} response and IgE class switching in mice with disrupted Stat6 gene. *Nature* **380**, 630–633.

Sideras, P., Funa, K., Zalcberg-Quintana, I., Xanthopoulos, K.G., Kisielow, P. and Palacios, R. (1988). Analysis by *in situ* hybridization of cells expressing mRNA for interleukin-4 in the developing thymus and in peripheral lymphocyte from mice. *Proc. Natl Acad. Sci. USA* **85**, 218–221.

Smeland, E.B., Blomhoff, H.K., Funderud, S., Shalaby, M.R. and Espevik, T. (1989). Interleukin-4 induces selective production of interleukin-6 from normal human B lymphocytes. *J. Exp. Med.* **170**, 1463–1468.

Snapper, C.M., Marcu, K.B. and Zelazowski, P. (1997). The immunoglobulin class switch: beyond "accessibility". *Immunity* **6**, 217–223.

Snapper, C.M., Hornebeck, P.V., Atasoy, U., Pereira, G.M.B. and Paul, W.E. (1988). Interleukin-4 induces membrane Thy-1 expression on normal murine B cells. *Proc. Natl Acad. Sci. USA* **85**, 6107–6111.

Snoeck, H.W., Lardon, F., Lenjou, M., Nys, G., Van Bockstaele, D.R. and Peetermans, M.E. (1993). Interferon-γ and interleukin-4 reciprocally regulate the production of monocytes/macrophages and neutrophils through a direct effect on committed monopotential bone marrow progenitor cells. *Eur. J. Immunol.* **23**, 1072–1077.

Song, Z., Casolaro, V., Chen, R., Georas, S.N., Monos, D. and Ono, S.J. (1996). Polymorphic nucleotides within the human IL-4 promoter that mediate overexpression of the gene. *J. Immunol.* **156**, 424–429.

Sonoda, Y., Kuzuyama, Y., Tanaka, S., Yokota, S., Maekawa, T., Clark, S.C. and Abe, T. (1993). Human interleukin-4 inhibits proliferation of megakaryocyte progenitor cells in culture. *Blood* **81**, 624–630.

Sorg, R.V., Enczmann, J., Sorg, U.R., Schneider, E.M. and Wernet, P. (1993). Identification of an alternatively spliced transcript of human interleukin-4 lacking the sequence encoded by exon 2. *Exp. Hematol.* **21**, 560–563.

Splits, H., Yssel, H., Takebe, Y., Arai, N., Yokota, T., Lee, F., Arai, K., Banchereau, J. and de Vries, J.E. (1987). Recombinant interleukin-4 promotes the growth of human T cells. *J. Immunol.* **135**, 1142–1147.

Spits, H., Yssel, H., Paliard, X., Kastelein, R., Figdor, D. and de Vries, J.E. (1988). Interleukin-4 inhibits interleukin-2 mediated induction of human lymphokine activated killer cells, but not the generation of antigen specific cytotoxic T lymphocytes in mixed leucocyte cultures. *J. Immunol.* **141**, 29–36.

Suzuki, Y., Yang, Q., Yang, S., Nguyen, N., Lim, S., Liesenfeld, O., Kojima, T. and Remington, J.S. (1996). IL-4 is protective against development of toxoplasmic encephalitis. *J. Immunol.* **157**, 2564–2569.

Standiford, T.J., Strieter, R.M., Kasahara, K. and Kunkel, S.L. (1990). Disparate regulation of interleukin-8 gene expression from blood monocytes, endothelial cells, and fibroblasts by interleukin-4. *Biochem. Biophys. Res. Commun.* **171**, 531–536.

Szabo, S.J., Gold, J.S., Murphy, T.L. and Murphy, K.M. (1993). Identification of cis-acting regulatory elements controlling interleukin-4 gene expression in T cells. *Mol. Cell. Biol.* **13**, 4793–4805.

Szalay, G., Ladel, C.H., Blum, C. and Kaufmann, S.H. (1996). IL-4 neutralization or TNF-α treatment ameliorate disease by an intracellular pathogen in IFN-γ receptor-deficient mice. *J. Immunol.* **157**, 4746–4750.

Takeda, K., Tanaka, T., Shi, W., Matsumoto, M. and Minami, M. (1996). Essential role of Stat6 in IL-4 signalling. *Nature* **380**, 627–630.

Tara, D., Weiss, D.L. and Brown, M.A. (1995). Characterization of the constitutive and inducible components of a T cell IL-4 activation responsive element. *J. Immunol.* **154**, 4592–4602.

Taylor, C.W., Grogan, T.M. and Salmon, S.E. (1990). Effects of interleukin-4 on the *in vitro* growth of human lymphoid and plasma cell neoplasms. *Blood* **75**, 1114–1118.

Te Velde, A.A., Klomp, J.P.G., Yard, B.A., de Vries, J.E. and Figdor, C.G. (1988). Modulation of phenotypic and functional properties of human peripheral blood monocytes by IL-4. *J. Immunol.* **140**, 1548–1554.

Te Velde, A.A, Huybens, R.J.F., de Vries, J.E. and Figdor, C.G. (1990). IL-4 decreases Fc γ R membrane expression and Fc γ R-mediated cytotoxic activity of human monocytes. *J. Immunol.* **144**, 3046–3051.

Tepper, R.I., Levinson, D.A. Stanger, B.Z., Campos-Torres, J., Abbas, A.K. and Leder, P. (1990). IL-4 induces allergic-like inflammatory disease and alters T cell development in transgenic mice. *Cell* **62**, 457–467.

Tepper, R.I., Coffman, R.L. and Leder, P. (1992). An eosinophil-dependent mechanism for the antitumor effect of interleukin-4. *Science* **257**, 548–551.

Tepper, R.I., Pattengale, P.K. and Leder, P. (1989). Murine interleukin-4 displays potent anti-tumor activity *in vivo. Cell* **57**, 503–512.

Thornhill, M.H., Kyan-Aung, U. and Haskard, D.O. (1990). IL-4 increases human endothelial cell adhesiveness for T cells but not for neutrophils. *J. Immunol.* **144**, 3060–3065.

Thornhill, M.H., Wellicome, S.M., Mahiouz, D.L., Lanchbury, J.S.S., Kyan-Aung, U. and Haskard, D.O. (1991). Tumor necrosis factor combines with IL-4 or IFN-γ to selectively enhance endothelial cell adhesiveness for T cells. The contribution of vascular cell adhesion molecule-1-dependent and -independent binding mechanisms. *J. Immunol.* **146**, 592–598.

Thyphronitis, G., Tsokos, G.C., June, C.H., Levine, A.D. and Finkelman, F.D. (1989). IgE secretion by Epstein-Barr virus-infected purified human B lymphocytes is stimulated by interleukin-4 and suppressed by interferon γ. *Proc. Natl Acad. Sci. USA* **86**, 5580–5584.

Toi, M., Bicknell, R. and Harris, A.L. (1992). Inhibition of colon and breast carcinoma cell growth by interleukin-4. *Cancer Res.* **52**, 275–279.

Tortolani, P.J., Lal, B.K., Riva, A., Johnston, J.A., Chen, Y.Q., Reaman, G.H., Beckwith, M., Longo, D., Ortaldo, J.R., Bhatia, K., McGrath, I., Kehrl, J., Tuscano, J., McVicar, D.W. and O'Shea, J.J. (1995). Regulation of JAK3 expression and activation in human B cells and B cell malignancies. *J. Immunol.* **155**, 5220–5226.

Toru, H., Ra, C., Nonoyama, S., Suzuki, K., Yata, J.I. and Nakahata, T. (1996). Induction of the high-affinity IgE receptor (Fc ε RI) on human mast cells by IL-4. *Int. Immunol.* **8**, 1367–1373.

Toru, H., Kinashi, T., Ra, C., Nonoyama, S., Yata, J.I. and Nakahata, T. (1997). Interleukin-4 induces homotypic aggregation of human mast cells by promoting LFA-1/ICAM-1 adhesion molecules. *Blood* **89**, 3296–3302.

Tsuji, K., Nakahata, T., Takagi, M., Kobayashi, T., Ishiguro, A., Kikuchi, T., Naganuma, K., Koike, K., Miyajima, A., Arai, K. and Akabane, T. (1990). Effects of interleukin-3 and interleukin-4 on the development of "connective tissue-type" mast cells: interleukin-3 supports their survival and interleukin-4 triggers and supports their proliferation synergistically with interleukin-3. *Blood* **75**, 421–427.

Tushinski, R.J., Larsen, A., Park, L.S. and Spoor, E. (1991). Interleukin-4 alone or in combination with interleukin-1 stimulates 3T3 fibroblasts to produce colony-stimulating factors. *Exp. Hematol.* **19**, 238–244.

Uchiyama, A., Essner, R., Doi, F., Nguyen, T., Ramming, K.P., Nakamura, T., Morton, D.L. and Hoon, D.S. (1996). Interleukin-4 inhibits hepatocyte growth factor-induced invasion and migration of colon carcinomas. *J. Cell. Biochem.* **62**, 443–453.

Ueno, Y., Boone, T. and Uittenbogaart, C.H. (1989). Selective stimulation of human thymocyte subpopulations by recombinant IL-4 and IL-3. *Cell. Immunol.* **118**, 382–393.

Uyemura, K., Pirmez, C., Sieling, P.A., Kiene, K., Paes-Oliveira, M. and Modlin, R.L. (1993). $CD4^+$ type 1 and $CD8^+$ type 2 T cell subsets in human leishmaniasis have distinct T cell receptor repertoires. *J. Immunol.* **151**, 7095–7104.

Valent, P., Bevec, D., Maurer, D., Besemer, J., Di Padova, F., Butterfield, J.H., Speiser, W., Majdic, O., Lechner, K. and Bettelheim, P. (1991). Interleukin-4 promotes expression of mast cell ICAM-1 antigen. *Immunol.* **88**, 3339–3342.

Vallé, A., Garrone, P., Aubry, J.P. and Banchereau, J. (1989). Identification of a 45-60kDa human B-cell activation antigen (Ag 104) and its modulation by IL-4 and IL-2. In *Leukocyte Typing IV. White Cell Differentiation Antigens* (eds W. Knapp, B. Dorken, E.P. Rieber, H. Stein, W.R. Gilks, R.E. Schmidt, and A.E.G.K. van dem Borne), Oxford Univesity Press. Part 3: Activation antigens, p. 510.

Vallé, A., Aubry, J.P., Durand, I. and Banchereau, J. (1991). IL-4 and IL-2 upregulate the expression of antigen B7, the B cell counterstructure to T cell CD28: an amplification mechanism for T-B cell interactions. *Int. Immunol.* **3**, 229–235.

van Hal, P.T.W., Hopstaken-Broos, J.P.M., Prins, A., Favaloro, E.J., Huijbens, R.J.F., Hilvering, C., Figdor, C.G. and Hoogsteden, H.C. (1994). Potential indirect anti-inflammatory effects of IL-4. Stimulation of human monocytes, macrophages, and endothelial cells by IL-4 increases aminopeptidase-N activity (CD13; EC 3.4.11.2). *J. Immunol.* **153**, 2718–2728.

van Kooten, C., Rensink, I., Aarden, L. and van Oers, R. (1992). Interleukin-4 inhibits both paracrine and autocrine tumor necrosis factor-α-induced proliferation of B chronic lymphocytic leukemia cells. *Blood* **80**, 1299–1306.

Vazquez, A., Auffredou, M.T., Chaouchi, N., Taib, J., Sharma, S., Galanaud, P. and Leca, G. (1991). Differential inhibition of interleukin 2- and interleukin 4-mediated human B cell proliferation by ionomycin: a possible regulatory role for apoptosis. *Eur. J. Immunol.* **21**, 2311–2316.

Vercelli, D., Jabara, H.H., Arai, K. and Geha, R.S. (1989). Induction of human IgE synthesis requires interleukin-4 and T/B cell interactions involving the T cell receptor/CD3 complex and MHC class II antigens. *J. Exp. Med.* **169**, 1295–1307.

Vercelli, D., Jabara, H.H., Lee, W., Woodland, N., Geha, R.S. and Leung, D.Y.M. (1988). Human recombinant interleukin-4 induces Fc ε R2/CD23 on normal human monocytes. *J. Exp. Med.* **167**, 1406–1416.

von der Weid, T., Beebe, A.M., Roopenian, D.C. and Coffman, R.L. (1996). Early production of IL-4 and induction of T_{h2} responses in the lymph node originate from an MHC class I-independent $CD4^+$ NK1.1-T cell population. *J. Immunol.* **157**, 4421–4427.

Wallace, P.M., Rodgers, J.N., Leytze, G.M., Johnson, J.S. and Linsley, P.S. (1995). Induction and reversal of long-lived specific unresponsiveness to a T-dependent antigen following CTLA4Ig treatment. *J. Immunol.* **154**, 5885–5895.

Walter, M.R., Cook, W.J., Zhao, B.G., Cameron, R.P., Ealick, S.E., Walter, R.L., Reichert, P., Nagabhushan, T.L., Trotta, P.P. and Bugg, C.E. (1992). Crystal structure of recombinant human interleukin-4. *J. Biol. Chem.* **267**, 20 371–20 376.

Wang, H.Y., Zamorano, J., Yoerkie, J.L., Paul, W.E. and Keegan, A.D. (1997). The IL-4-induced tyrosine phosphorylation of the insulin receptor substrate is dependent on JAK-1 expression in human fibrosarcoma cells. *J. Immunol.* **158**, 1037–1040.

Wang, L.M., Keegan, A.D., Paul, W.E., Heidaran, M.A., Gutkind, J.S. and Pierce, J.H. (1992). IL-4 activates a distinct signal transduction cascade from IL-3 in factor-dependent myeloid cells. *EMBO J.* **11**, 4899–4908.

Wang, L.W., Myers, M.G., Sun, X.J., Aaronson, S.A., White, M. and Pierce, J.H. (1993). IRS-1: essential for insulin- and IL-4-stimulated mitogenesis in hematopoietic cells. *Science* **261**, 1591–1594.

Weiss, D.L., Hural, J., Tara, D., Timmerman, L.A., Henkel, G. and Brown, M.A. (1996). Nuclear factor of activated T cells is associated with a mast cell interleukin-4 transcription complex. *Mol. Cell Biol.* **16**, 228–235.

Welham, M.J., Learmonth, L., Bone, H. and Schrader, J.W. (1995). Interleukin-13 signal transduction in lymphohemopoietic cells. Similarities and differences in signal transduction with interleukin-4 and insulin. *J. Biol. Chem.* **270**, 12 286–12 296.

Wenner, C.A., Szabo, S.J. and Murphy, K.M. (1997). Identification of IL-4 promoter elements conferring T_{h2}-restricted expression during T helper cell subset development. *J. Immunol.* **158**, 765–773.

Wertheim, W.A., Kunkel, S.L., Standiford, T.J., Burdick, M.D., Becker, F.S., Wilke, C.A., Gilbert, A.R. and Strieter, R.M. (1993). Regulation of neutrophil-derived IL-8: the role of prostaglandin E2, dexamethasone, and IL-4. *J. Immunol.* **151**, 2166–2175.

Windsor, W.T., Syto, R., Durkin, J., Das, P., Reichert, P., Pramanik, B., Tindall, S., Le, H.V., Labdon, J., Nagabhushan, T.L. and Trotta, P.P. (1990). Disulfide bond assignment of mammalian cell-derived recombinant human interleukin-4. *Biophys. J.* **57**, 423–436.

Witthuhn, B.A., Silvennoinen, O., Miura, O., Lai, K.S., Cwik, C., Liu, E.T. and Ihle, J.N. (1994). Involvement of the Jak-3 Janus kinase in signalling by interleukin-2 and -4 in lymphoid and myeloid cells. *Nature* **370**, 153–157.

Wong, H.L., Costa, G.L., Lotze, M.T. and Wahl, S.M. (1993). Interleukin (IL) 4 differentially regulates monocyte IL-1 family gene expression and synthesis *in vitro* and *in vivo*. *J. Exp. Med.* **177**, 775–781.

Yamaoka, K.A. and Kolb, J.P. (1995). Involvement of CD23/Fc ε RII in the homotypic and heterotypic cytoadhesion of the human eosinophilic cell line Eol-3. *Eur. Cytokine. Netw.* **6**, 145–155.

Yin, T., Tsang, M.L.S. and Yang, Y.C. (1994). JAK1 kinase forms complexes with interleukin-4 receptor and 4PS/insulin receptor substrate-1-like protein and is activated by interleukin-4 and interleukin-9 in T lymphocytes. *J. Mol. Biol.* **269**, 26 614–26 617.

Yokota, T., Otsuka, T., Mosmann, T., Banchereau, J., Defrance, T., Blanchard, D., de Vries, J.E., Lee, F. and Arai, K. (1986). Isolation and characterization of a human interleukin cDNA clone, homologous to mouse B-cell stimulatory factor 1, that expresses B-cell-stimulatory activities. *Proc. Natl Acad. Sci. USA* **83**, 5894–5898.

Yoshimoto, T. and Paul, W.E. (1994). $CD4^{pos}$, $NK1.1^{pos}$ T cells promptly produce interleukin-4 in response to *in vivo* challenge with anti-CD3. *J. Exp. Med.* **179**, 1285–1295.

Zheng, W. and Flavell, R.A. (1997). The transcription factor GATA-3 is necessary and sufficient for T_{h2} cytokine gene expression in CD4 T cells. *Cell* **89**, 587–596.

Zhou, L.J. and Tedder, T.F. (1996). $CD14^+$ blood monocytes can differentiate into functionally mature $CD83^+$ dendritic cells. *Proc. Natl Acad. Sci. USA* **93**, 2588–2592.

Zlotnik, A., Ramsom, J., Franck, G., Fischer, M. and Howard, M. (1987). Interleukin-4 is a growth factor for activated thymocytes: possible role in T-cell ontogeny. *Proc. Natl Acad. Sci. USA* **84**, 3856–3860.

Zund, G., Madara, J.L., Dzus, A.L., Awtrey, C.S. and Colgan, S.P. (1996). Interleukin-4 and interleukin-13 differentially regulate epithelial chloride secretion. *J. Biol. Chem.* **271**, 7460–7464.

Zurawski, S.M., Chomarat, P., Djossou, O., Bidaud, C., McKenzie, A.N.J., Miossec, P., Banchereau, J. and Zurawski, G. (1995). The primary binding subunit of the human interleukin-4 receptor is also a component of the interleukin-13 receptor. *J. Biol. Chem.* **270**, 13 869–13 878.

Chapter 7

Interleukin-5

Colin J. Sanderson
TVW Telethon Institute for Child Health Research, Perth, Western Australia

INTRODUCTION

Interleukin-5 (IL-5) is produced by T lymphocytes as a glycoprotein with an M_r of 40–45 kDa and is unusual among the T-cell-produced cytokines in being a disulphide-linked homodimer. It is the most highly conserved member of a group of evolutionally related cytokines, including also IL-3, IL-4 and granulocyte–macrophage colony stimulating factor (GM-CSF), which are closely linked on human chromosome 5.

Historically, two lines of research converged when it was demonstrated that two very different biological activities were properties of this molecule. On the one hand, in the early 1970s a number of different factors was emerging which showed activity on mouse B cells *in vitro*. These preparations were mixtures of cytokines, particularly IL-4 and IL-5, and the assays available did not distinguish between the two molecules. It was not until the early 1980s that a group of high-molecular-weight activities emerged that were clearly based on IL-5, although in published reviews in 1984 the identity of the activities was not appreciated (Howard *et al.*, 1984; Vitetta *et al.*, 1984). Three main groups were working with this molecule: Takatsu in Japan, who considered the activity to be various forms of T-cell replacing factor (TRF) (Takatsu *et al.*, 1988), Swain and Dutton in California, who used the term B-cell growth factor-II (BCGF-II) (Swain *et al.*, 1988), and Vitetta in Texas, who called it B-cell differentiation factor-μ (BCDF-μ) (Vitetta *et al.*, 1984). In 1985, Takatsu's group purified TRF and showed that it was identical to BCGF-II (Harada *et al.*, 1985).

On the other hand, and also in the early 1970s, work on the colony stimulating factors (CSFs) by Metcalf's group in Australia had demonstrated the production of eosinophilic colonies from mouse bone marrow in the presence of crude spleen cell-conditioned medium (Metcalf *et al.*, 1974). It is now clear that this medium contained IL-5 because it was shown to produce a selective stimulation of human eosinophil colonies (Metcalf *et al.*, 1983) and IL-5 is the only eosinophil haematopoietic growth factor that cross-reacts between human and mouse. This line of work culminated in the identification of murine eosinophil differentiation factor in the author's laboratory in the mid-1980s. This was done using a liquid assay system which in the mouse is a much more sensitive assay than the colony assay (Sanderson *et al.*, 1985; Warren and Sanderson, 1985). However, as is discussed in detail in this review, IL-5 is a CSF for which the name Eo-CSF would have been appropriate.

The Cytokine Handbook, 3rd ed.
ISBN 0-12-689662-3

These two lines of research came together in 1986, when our group purified eosinophil differentiation factor (EDF) and showed that it was identical to BCGF-II (Sanderson *et al.*, 1986). It is interesting to note that the observations on the identity of TRF, BCGF-II and EDF were made before the cloning of IL-5, which thus confirmed the biochemical data.

There are two intriguing aspects of the dual biological activities of IL-5. First, although there is a well-known association between eosinophilia and immunoglobulin (Ig)E levels, IL-5 does not appear to be involved in the IgE response, where IL-4 is the major controlling cytokine. This raises the possibility of some common features in the control of IL-4 and IL-5 expression. Second, although the activity on murine B cells *in vitro* is well characterized (see below), human IL-5 (hIL-5) is not active in assays on human B cells analogous to those used in the mouse system.

The role of IL-5 in eosinophilia, coupled with a better understanding of the part played by eosinophils in the development of tissue damage in chronic allergy, suggests that IL-5 will be a major target for a new generation of antiallergy drugs.

GENE STRUCTURE AND EXPRESSION

The coding sequence of the IL-5 gene forms four exons (Fig. 1). The introns show areas of similarity between the mouse and human sequences, although the mouse has a considerable number of sequences (including repeat sequences) that are not present in the human gene. The mouse includes a 738-base-pair fragment in the 3′-untranslated region (also known as the Alu-like repeat) which is not present in the human gene, and thus the mouse mRNA is 1.6 kilobases (kb) whereas the human is 0.9 kb (Campbell *et al.*, 1987). Each of the exons contains the codons for an exact number of amino acids.

Coexpression[a] with Other Cytokines

In T helper (T_{H2}) clones IL-4 and IL-5 are often coexpressed (Coffman *et al.*, 1988), which could explain the frequent association of eosinophilia and IgE. However, the molecular mechanism of this remains unclear as alignment of the two promoter regions shows no significant areas of homology. Even the CLE0 element (see below) is only 56% homologous. In fact, many studies have indicated that different signals are required for the induction of these two genes. Anti-CD3 induces the expression of IL-4, IL-5 and GM-CSF messenger RNA (mRNA) in mouse T cells, whereas treatment with IL-2 induced IL-5 mRNA expression but did not induce detectable amounts of IL-4 and GM-CSF messengers (Bohjanen *et al.*, 1990). In humans, a T_{H2}-like pattern of cytokine mRNA expression could be demonstrated in asthmatic patients (Robinson *et al.*, 1993; Okudaira *et al.*, 1995). T cells in the bronchoalveolar lavage of mild atopic asthmatics showed increased expression of IL-4 and IL-5 (Robinson *et al.*, 1993). However, T cells purified from peripheral blood of non-atopic asthmatics secreted raised amounts of IL-5, but not IL-4, compared with normal controls, whereas those from atopic patients secreted increased quantities of both IL-4 and IL-5 (Corrigan and Kay, 1992). These

---

[a] Coexpression is used here to denote the simultaneous expression of genes in an organism, not necessarily in the same cells, and not due to coordinate genetic control. Coordinate expression implies the simultaneous (or sequential) expression of genes in the same cell controlled at the genetic level.

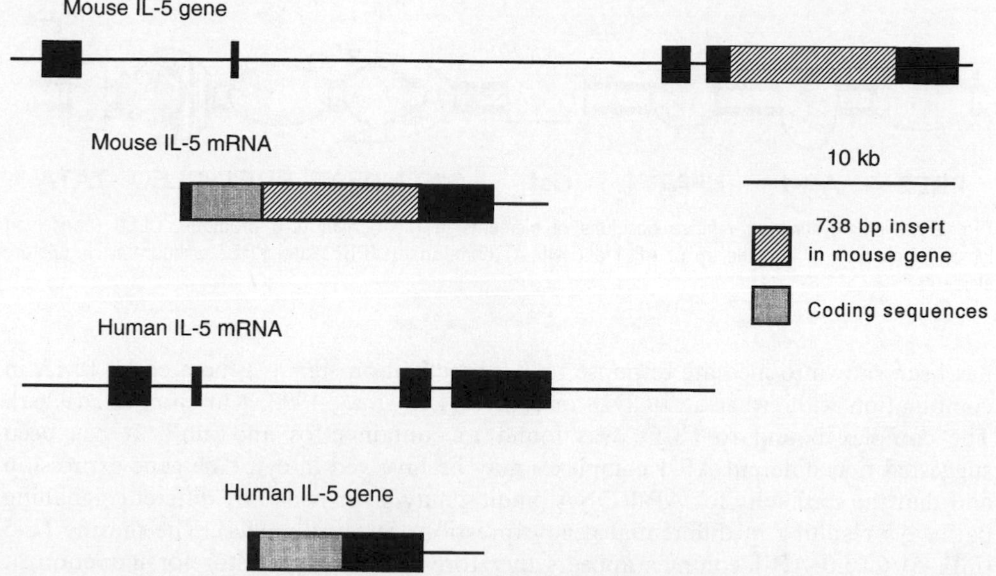

Fig. 1. Maps of the human and mouse IL-5 genomic genes and corresponding mRNA. Exons are shown as boxes. The shaded area indicates the insert in the 3′-untranslated region of the mouse gene.

differences are emphasized by the observation that cyclic AMP and 4*β*-phorbol 12-myristate 13-acetate (PMA) have different effects on the induction of different cytokines in the mouse lymphoma EL4. PMA induces IL-2 and IL-4 but has only a low effect on IL-5 induction. While cAMP markedly enhances the effect of PMA on the induction of IL-5, it has an inhibitory effect on IL-2, IL-3, GM-CSF and IL-10, and no effect on IL-4 (Derig *et al.*, 1990; Lee *et al.*, 1993; Chen and Rothenberg, 1994; Karlen *et al.*, 1996a). Similarly, pertussis toxin induces IL-4 but not IL-5 mRNA synthesis, whereas cyclophosphamide stimulates the transcription of IL-5 but not of IL-4 (Sewell and Mu, 1996). From these observations it appears that IL-4 and IL-5 expression is not regulated coordinately, suggesting that unique control mechanisms for these lymphokines exist.

This raises two questions: (1) What is the basis for the coexpression of IL-4 and IL-5 in certain parasitic and allergic diseases? There is no clear answer at this point. (2) Are there control elements that are specific to these genes? By developing a reporter system in which IL-5 transcription is studied in its genomic context, we have found evidence that regulatory sites exist which may be involved in the specific regulation of the gene.

Promoter Region

The 5′ flanking region of the IL-5 RNA initiation site contains several motifs involved in the transcription of the gene (Fig. 2). There is a short sequence called the conserved lymphoid element 0 (CLE0), between nucleotides −56 and −42 in the hIL-5 promoter region. This element is essential for promoter activity (Naora *et al.*, 1994). CLE0 is conserved among the regulatory regions of several other lymphokine genes such as IL-3, IL-4 and GM-CSF (Masuda *et al.*, 1993). It binds factors of the activating protein-1 (AP-1) family together with nuclear factor (NF)–AT regulatory proteins, and

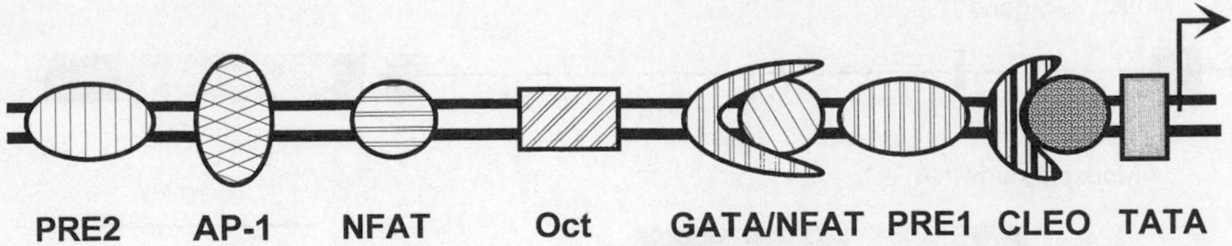

Fig. 2. Diagram showing relative positions of elements in the human IL-5 promoter. CLE0 (conserved lymphoid element 0) is made up of AP-1 and NF–AT components. PRE1 and PRE2, palindromic regulatory elements.

has been shown to mediate response to T-cell activation signals generated by PMA in combination with either anti-CD28 or cAMP (Lee *et al.*, 1993; Karlen *et al.*, 1996a). The complex bound to CLE0 was found to contain c-fos and junB. It has been suggested that different AP-1 complexes may be involved in cytokine gene expression and that the specificity for AP-1 DNA binding may be provided by different signalling pathways resulting in differential gene expression (Su *et al.*, 1994). The murine IL-5 (mIL-5) CLE0–AP-1 complex appears therefore to be a key factor for induction of IL-5 transcription. The participation of Oct factors (Gruart-Gouilleux *et al.*, 1995) and GATA4 binding proteins (Yamagata *et al.*, 1995) at two sites immediately upstream of the CLE0 element (nucleotides −89 to −56) is also critical for regulation of the gene. Binding to the Oct element was found to be dependent on activation signals, whereas the binding of the GATA4 protein was constitutive. The characterization of a transcriptionally active NF–AT binding site around position −110 in the proximal promoter region has been reported recently (Lee *et al.*, 1995; Prieschl *et al.*, 1995). The proximal part of this NF–AT element is flanked by a consensus binding sequence for members of the GATA transcription factor family. NF–AT and GATA cooperate in the induction of the IL-5 gene by IgE and antigen in the mouse mast cell line CPII; however, only NF–AT binding is inducible whereas GATA activity is constitutive (Prieschl *et al.*, 1995).

A series of positive and negative regulatory elements involved in the induction of the mIL-5 gene has been identified using the mouse T-cell clone D10.G4.1 (Stranick *et al.*, 1995). A sequence located between positions −130 and −176 contains at least two protein binding sites which participate in activation of the IL-5 promoter, and two negatively acting elements were mapped between positions −261 and −300 and positions −392 and −431. Finally, a palindromic element is found at position −481 in the hIL-5 regulatory region. In GM-CSF, this element has a strong positive effect on gene expression in response to induction signals (Staynov *et al.*, 1995). Analysis of the hIL-5 promoter sequence also reveals the presence of several sites that can bind factors resembling AP-1, NF–AT and NF–κB (A.D. Singh and V.A. Mordvinov, unpublished data). Most of these elements participate in the regulation of many other lymphokine genes in activated T cells. The fact that IL-5 shares many of these regulatory elements with other lymphokines may explain the coexpression of those genes in T cell-mediated allergic processes.

Two interacting palindromic regulatory elements between PRE1-IL5 and PRE2-IL5 have been identified. The murine mPRE1 is located at position −79 to −90. This element consists of a 3 base pairs (bp) inverted repeat and a 5 bp core sequence, and has been shown to be involved in the positive induction of the mIL-5 gene. The mPRE2-IL5 is located at a distal site upstream in the mIL-5 promoter. This element has identical inverted repeats but a dissimilar core sequence and is also involved in positive induction of the mIL-5 gene. These elements bind proteins of identical electrophoretic mobility from nuclear extracts of the murine thymoma cell line EL4-23 and murine primary T cells, and can cross-compete for DNA–protein complex formation. A model for interaction of these two elements resulting in formation of a higher order nucleoprotein complex has been proposed due to the cooperative effect observed on IL-5 promoter activity when both these elements are mutated. Interestingly, the mPRE-IL5 elements are present in all known IL-5 genes (mouse, rat, human and ovine) but not in any other gene on the database. This suggests the possibility of a role for these elements in the specific regulation of the IL-5 gene (Mordvinov *et al.*, unpublished data; Schwenger *et al.*, unpublished data).

Post-transcriptional Regulation

Deletion of the 3′ untranslated region from murine IL-5 complementary DNA (cDNA) in an expression vector resulted in more than a ten-fold increase in IL-5 production in COS cells. While a small component of this might be attributed to mRNA stability, the major effect was on translation. Thus *in vitro* translation in reticulocyte lysates was much higher from the deleted mRNA than from the full-length messenger. Surprisingly, deletion of the 5′ region corresponding to a predicted stem–loop structure gave a similar effect. This demonstrates that IL-5 translation is blocked by a mechanism requiring both intact 5′ and intact 3′ regions. The control of this block is not yet understood, but as it operates in a variety of cell types and in rabbit reticulocyte lysates it is unlikely to be specific for IL-5 or even T cell-derived cytokines (Karlen *et al.*, unpublished data).

PROTEIN STRUCTURE

IL-5 belongs to a family of structurally related proteins which includes IL-2, IL-4, macrophage colony stimulating factor (M-CSF), GM-CSF and growth hormone. These proteins fold to produce a tertiary structure of four α helices arranged in an up–up, down–down configuration (Milburn *et al.*, 1993). IL-5 is a homodimer which forms two helical bundles, each containing three helices (A–C) from one chain and one from the other (D′). The dimeric protein is held together by two disulphide bridges between residues C44 and C86 of opposing chains (Minamitake *et al.*, 1990; McKenzie *et al.*, 1991a; Proudfoot *et al.*, 1991). In addition, the structure also features two antiparallel β sheets on opposite sides of the molecule (between residues 32–35 of chain 1 and 89–92 of chain 2) (Fig. 3).

Characterization of recombinant human IL-5 (from *Escherichia coli*) by electrospray mass spectrometry gives a molecular mass of 26 kDa (Graber *et al.*, 1993), which is close to the theoretical molecular mass of 24 kDa; however, when analysed by sodium dodecyl sulphate–polyacrylamide gel electrophoresis, recombinant mammalian IL-5 is heterogeneous, with a molecular mass of 45–60 kDa under non-reducing conditions (Sanderson *et al.*, 1985). Thus approximately half of the mass is carbohydrate. Human IL-5 has carbohydrate *N*-linked at position Asn-28 and *O*-linked at Thr-3 (Minamitake *et al.*, 1990), and mIL-5 is glycosylated at analogous sites, plus further *N*-linked at Asn-55

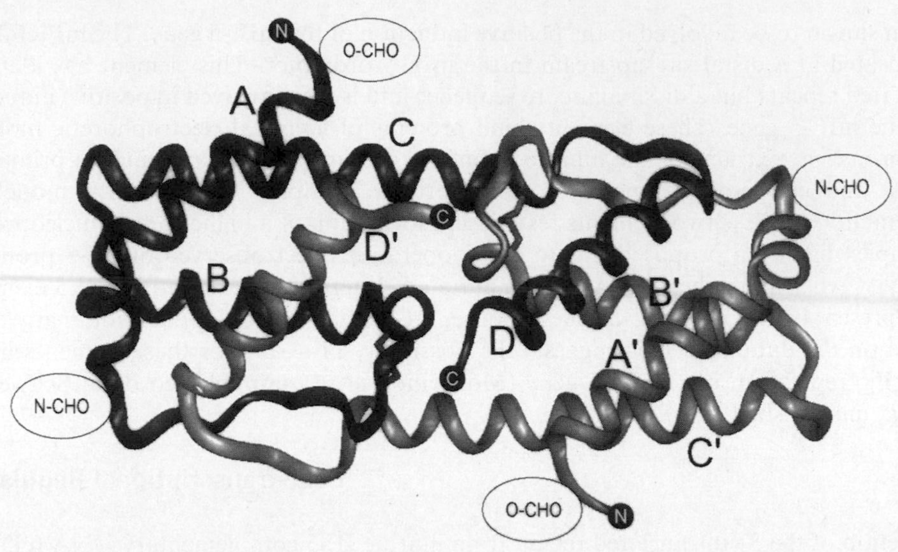

Fig. 3. Diagram based on the crystal structure (Milburn *et al.*, 1993) of hIL-5 showing the main structural features. One monomer is shown in light grey and the other in dark grey. Helices are indicated A–D for one monomer and A′–D′ for the other starting at the *N*-terminus (N). The disulphide bridges connecting cysteines at positions 44 and 86 are shown as rods.

(Kodama *et al.*, 1992). A potential *N*-linked site at Asn-69/71 is not glycosylated in either species. The removal of carbohydrate from IL-5 does not appear to affect its biological activity *in vitro* (Tominaga *et al.*, 1990; Kodama *et al.*, 1993), although this does cause loss of thermostability and may be the reason for the strong hydrophobicity observed for recombinant bacterial protein.

A better understanding of IL-5 structure has come from construction of a biologically active monomer. An extra eight amino acids were inserted in the loop between helices C and D (loop 3), thereby allowing a different folding of the protein and permitting the D helix to align with the other three helices of the same chain (Dickason and Huston, 1996). This contrasts with a previous study in which the dimer-bridging cysteine residues were mutated to threonine, thereby producing monomeric IL-5 which was not biologically active (McKenzie *et al.*, 1991a), and thus confirming that in the native material it is the shorter loop 3 that inhibits monomeric polypeptide folding.

The point of contact between IL-5 and its receptor chains has been investigated by mutation analysis, constructing species hybrid molecules and mapping neutralizing antibodies (McKenzie *et al.*, 1991a; Cornelis *et al.*, 1995; Graber *et al.*, 1995; Morton *et al.*, 1995; Dickason *et al.*, 1996; Li *et al.*, 1996). These studies have indicated that residues around loop 3 (between the C and D helices; centred on Arg-91) are important for binding to the receptor α chain, with other central regions of the molecule also implicated, in particular the *C*-terminus (around Glu-110). Although IL-5 has two potential sites for interaction with the receptor, only a 1:1 complex is formed (Devos *et al.*, 1993). Studies of the contact point of IL-5 with the receptor β chain have highlighted E13 (of helix A) as the critical residue involved. An IL-5 mutant (E13Q) has been shown to have antagonistic properties (Tavernier *et al.*, 1995).

RECEPTORS

The receptors for each of the three cytokines, IL-3, IL-5 and GM-CSF consist of an α chain, which is different for each ligand, and a β chain, which is common to each receptor complex (Lopez *et al.*, 1992b). The α chain forms the low-affinity interaction with its ligand and serves to increase the affinity to give the high-affinity interaction (Tavernier *et al.*, 1991; Tominaga *et al.*, 1991). The β chain appears not to have a measurable interaction with any of the ligands in the absence of the α chain. It seems likely that the cross-inhibition exhibited within the group is due to limiting numbers of β chains (Lopez *et al.*, 1989, 1990; Nicola and Metcalf, 1991).

Both the α and the β subunits share a number of features with other receptors which has led to their classification as members of the cytokine–haematopoietin superfamily. They have a modular structure build up of fibronectin-III-like domains, and a number of conserved amino acids.

Murine eosinophils have been calculated to express approximately 50 high-affinity receptors for IL-5, and approximately 5000 low-affinity receptors (Barry *et al.*, 1991). A number of murine cell lines have been established which require the presence of IL-5 for growth; these are all of B-cell origin and have relatively high numbers of IL-5 receptors. They represent an important tool in the study of the IL-5 receptor in the mouse, but no analogues of these B-cell lines have been described from human tissues. Initial work on the human receptor utilized a clone of the human promyelocytic leukaemia cell line HL-60 (Plaetinck *et al.*, 1990). These cells express only a single population of high-affinity IL-5 binding sites. Similarly, only a single high-affinity receptor population has been identified on human eosinophils (Ingley and Young, 1991; Lopez *et al.*, 1991; Migita *et al.*, 1991).

Receptor Structure

Using a panel of human/mouse IL-5Rα hybrids it has been determined that species specificity resides in the *N*-terminal region (Cornelis *et al.*, 1995). Mutation of amino acids within this region indicated that residues D55, D56, Y57 and E58 are involved in interaction with residue N108 in the D helix of IL-5. The importance of the *N*-terminal region was supported by the abolition of IL-5 binding by covalent coupling of isothiazolone derivatives to residue C66 of IL-5Rα (Devos *et al.*, 1993). As mentioned above, E13 in hIL-5 has been shown to be involved in interacting with the subunit. There is evidence that the critical residues in the β subunit interacting with this IL-5 residue are Y365, H367 and I368 (Woodcock *et al.*, 1994). It is interesting to note that these residues are also important in the GM-CSF interaction, but not for IL-3. This suggests that different regions of the receptor may be involved with each ligand. On the other hand a single tyrosine residue (Y421) on the β chain is critical for the interaction of all three cytokines (Woodcock *et al.*, 1996).

Of the residues on the receptor identified so far, it appears that the membrane distal domain of the α chain and the membrane proximal domain of the β chain are involved in interaction with IL-5.

Receptor Isoforms

The α chains of each receptor exist in a number of isoforms generated by alternative splicing. In each case the membrane-bound isoform results from splicing that retains a transmembrane domain (Tavernier *et al.*, 1991). The other isoforms lack the transmembrane domain, resulting in soluble forms of the receptor. The soluble form of the hIL-5Rα chain is antagonistic *in vitro* (Tavernier *et al.*, 1991) but there is not yet any evidence that it is active *in vivo*.

The cloning of cDNAs for both the human and mouse IL-5Rα chain has revealed different isoforms. Remarkably, in humans the major transcript is a soluble isoform. The structure of the IL-5Rα gene provides an explanation for the isotypes observed (Tavernier *et al.*, 1992; Tuypens *et al.*, 1992). A number of different splicing events can occur (Fig. 4), but the main difference with other receptors appears to lie in the fact that exon 11 encodes four amino acids followed by a stop codon and a polyadenylation site. Normal splicing leads to the inclusion of exon 11, and gives rise to a transcript truncated by the polyA site at the 3′ end of the exon. This soluble isoform (S1) is the most abundant transcript in HL60 cells and cultured eosinophils, and yet can not mediate signalling. The biologically active receptor is produced only when this exon is skipped, to give rise to the functional membrane-bound form of the receptor.

In a study of receptor expression in asthma it was shown that over 90% of the cells expressing IL-5R were eosinophils; furthermore, analysis of the two main isotopes expressed suggested that expression of the soluble form was associated with less severe asthma. Thus it is possible that a switch from the soluble to the membrane-bound

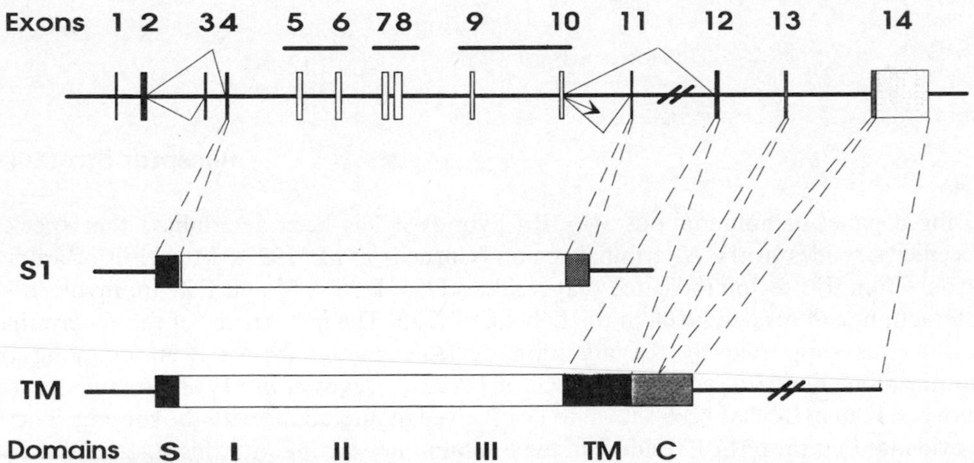

Fig. 4. Diagram showing the exon structure (not to scale) of hIL-5Rα chain (top) and two of the alternatively spliced transcripts. S1 indicates the major transcript in human eosinophilic cells, a soluble form formed by splicing in exon 11 causing transcription to end before the exon encoding the transmembrane domain. TM indicates the transcript of the active form of the receptor formed by splicing out exon 11. Exons numbers are indicated at the top. Alternative splicing events are indicated by fine lines (the arrow indicates no splicing at this donor site, forming the soluble S2 transcript; see text). The structural domains of the protein are indicated at the bottom: S, signal peptide; I, II and III, the three fibronectin type III-like domains derived from exons 5 and 6, 7 and 8, and 9 and 10 respectively; TM, transmembrane domain; C, intracytoplasmic tail. Adapted from Tuypens *et al.* (1992).

isoform might be associated with an increased sensitivity of the eosinophils to IL-5 with a greater tendency to cause tissue damage (Yasruel *et al.*, 1997).

EOSINOPHILIA

T-Cell Dependence

Eosinophilia is T-cell dependent and therefore it is not surprising that the controlling factor is a T-cell-derived cytokine (Sanderson *et al.*, 1985). It is characteristic of a limited number of disease states, most notably parasitic infections and allergy. Clearly, as eosinophilia is not characteristic of all immune responses, it is obvious that the factors controlling eosinophilia are not produced by all T cells. Similarly, as it is now clear that IL-5 is the main controlling cytokine for eosinophilia (see below), then if IL-5 has other biological activities, it is likely that these will coincide with the production of eosinophils.

Biological Specificity

One of the features of eosinophilia that has attracted the curiosity of haematologists for several decades is the apparent independence of eosinophil numbers on the numbers of other leukocytes. Thus eosinophils are present in low numbers in normal individuals but can increase dramatically and independently of the number of neutrophils. Such changes are common during the summer months in individuals with allergic rhinitis (hay fever), or in certain parasitic infections. Clearly, such conditions will result in more broadly based leukocytosis when complicated by other infections. Although this specificity has been known for many years, somewhat surprisingly it is not easy to find clear examples in the early literature. More recently, in experimental infection of volunteers with hookworms (*Necator americanus*), it was noted that an increase in eosinophils was the only significant change (Maxwell *et al.*, 1987), and our own work with *Mesocestoides corti* in the mouse has demonstrated massive increases in eosinophils, independent of changes in neutrophils (Strath and Sanderson, 1986). This biological specificity suggests a mechanism of control that is independent of the control of other leukocytes. This, coupled with the normally low numbers of eosinophils, provides a useful model for the study of the control of haematopoiesis by the immune system.

Eosinophil Production *In Vitro*

In the mouse system, IL-5 induces the production of eosinophils in liquid bone marrow cultures, and this is much more sensitive than the corresponding colony assay in semi-solid medium (Sanderson *et al.*, 1985, 1988; Sanderson, 1990). In contrast, both IL-3 and GM-CSF induce eosinophils as well as other cell types, most notably neutrophils and macrophages in bone marrow cultures (Campbell *et al.*, 1988). The production of eosinophils is considerably higher when the bone marrow is taken from mice infected with *M. corti* than it is from normal marrows. This suggests that marrow from infected mice contains more eosinophil precursors than that from normal mice (Sanderson *et al.*, 1985, 1988; Sanderson, 1993).

In human bone marrow cultures both IL-3 and GM-CSF, but not IL-5, appear to amplify the number of eosinophil colony precursors (Clutterbuck and Sanderson, 1990). This led to the concept of IL-5 as a late acting factor in eosinopoiesis. However, experiments *in vivo* have not substantiated this, and it seems likely that culture systems do not fully mimic the production of eosinophils *in vivo*. One possibility is that the action of IL-5 is uniquely dependent on stromal cells for the production of the progenitor cells. Although there is no direct evidence for this, there are a number of factors suggesting that IL-5 may be at least partly dependent on stromal cells, even in the later stages of eosinophil differentiation. For example, in the mouse system few eosinophil colonies form in semi-solid medium, whereas large numbers of eosinophils are produced in the adherent layer of stromal cells in liquid culture (Sanderson *et al.*, 1985; Warren and Sanderson, 1985). Second, in human liquid bone marrow cultures more eosinophils are produced in round-bottomed vessels than in flat-bottomed vessels, possibly due to better cell–cell interactions (Clutterbuck and Sanderson, 1988). Third, although, in contrast to the mouse, human eosinophil colonies are produced in semi-solid cultures, the number is significantly lower in the presence of IL-5 than with either IL-3 or GM-CSF. However, in liquid cultures the situation is reversed and IL-5 stimulates the production of more eosinophils than either IL-3 or GM-CSF. This is again consistent with a requirement for stromal cells by IL-5.

Recent results from this Institute have shown that IL-5 binds strongly to certain proteoglycans, suggesting that the presentation of IL-5 on the extracellular matrix in the bone marrow could be important in its biological activity (Lipscombe *et al.*, unpublished data).

Eosinophil Production *In Vivo*

An approach to understanding the role of IL-5 *in vivo* is to alter the expression of IL-5 in transgenic mice. As IL-5 is normally a T-cell product and the gene is transcribed for only a relatively short period of time after antigen stimulation, transgenic mice in which IL-5 is constitutively expressed by all T cells have been produced (Dent *et al.*, 1990). These mice have detectable levels of IL-5 in the serum. They show a profound and lifelong eosinophilia, with large numbers of eosinophils in the blood, spleen and bone marrow. This indicates that the expression of IL-5 is sufficient to induce the full pathway of eosinophil differentiation. If other cytokines are required for the development of eosinophilia, then either they must be expressed constitutively, or their expression is secondary to expression of the IL-5 gene. This clear demonstration that expression of the IL-5 gene in transgenic animals is sufficient for the production of eosinophilia provides an explanation for the biological specificity of eosinophilia. It therefore seems likely that, because eosinophilia can occur without a concomitant neutrophilia or monocytosis, a mechanism must exist by which IL-5 is the dominant haemopoietic cytokine produced by the T-cell system in natural eosinophilia.

Another important aspect of these transgenic animals is that, despite their massive long-lasting eosinophilia, the mice remained normal. This illustrates that an increased number of eosinophils is not of itself harmful, and that the tissue damage seen in allergic reactions and other diseases must be due to agents that trigger the eosinophils to degranulate.

The observation that IL-5 in transgenic mice is capable of inducing the full pathway of eosinophil production leaves unresolved the question of why IL-5 appears to be unable to induce the production of eosinophil progenitors *in vitro*.

Another important approach to the understanding of the biological role of IL-5 comes from the administration of neutralizing antibody. Mice infected with *Trichinella spiralis* develop eosinophilia and increased levels of IgE; however, when treated with an anti-IL-5 antibody, no eosinophils are observed (Coffman *et al.*, 1989). Indeed, the number of eosinophils is lower than that seen in control animals. These experiments illustrate the unique role of IL-5 in the control of eosinophilia in this parasite infection. They also show that the apparent redundancy seen *in vitro*, where both IL-3 and GM-CSF are also able to induce eosinophil production, does not operate in these infections. Furthermore, IL-5 plays no role in the development of IgE antibody (this activity is controlled by IL-4), or in the development of the granuloma seen surrounding schistosomes in the tissues (Sher *et al.*, 1990).

The generation of mice with an inactive IL-5 gene (IL-5 knockout mice) has confirmed the key role of IL-5 in the control of eosinophilia (Foster *et al.*, 1996; Kopf *et al.*, 1996). No eosinophils were produced in response to either a parasite infection or aero-allergen sensitization with ovalbumin. In fact, the low background level of eosinophils seen in normal control mice was substantially reduced in non-sensitized knockout mice, to leave a very small number of eosinophils produced in the absence of IL-5. The lack of effect on other cell types or on antibody production confirmed the unique specificity of IL-5 for the eosinophil lineage, which the author had proposed over a number of years (Sanderson, 1992). The knockout mice have provided an important animal model to test the biological role of IL-5 and eosinophils. Following the induction of eosinophils in the lung by the challenge of sensitized mice with an antigen aerosol, normal mice develop a high degree of lung inflammation. However, in knockout mice lung eosinophilia is not observed, and there is very little development of inflammation and lung damage (Foster *et al.*, 1996). This provides an experimental rationale for the role of eosinophils in human asthma, as has been proposed over many years (Gleich, 1990; Corrigan and Kay, 1992).

ACTIVATION OF EOSINOPHILS

The ability of eosinophils to perform in functional assays can be increased markedly by incubation with a number of different agents, including IL-5. The phenomenon of activation is apparently independent of differentiation. It appears to have a counterpart *in vivo*, as eosinophils from different individuals vary in functional activity. It has been demonstrated that the ability of eosinophils to kill schistosomula increases in proportion to the degree of eosinophilia (David *et al.*, 1980; Hagan *et al.*, 1985). This is consistent with a common control mechanism for both the production and activation of eosinophils in these cases.

The first observations on selective activation of human eosinophils by IL-5 showed that the ability of purified peripheral blood eosinophils to lyse antibody-coated tumour cells was increased when IL-5 was included in the assay medium (Lopez *et al.*, 1986). Similarly, the phagocytic ability of these eosinophils towards serum-opsonized yeast particles was increased in the presence of IL-5. There was a 90% increase in surface C3bi complement receptors, as well as an approximately 50% increase in the granulocyte functional antigens GFA-1 and GFA-2. Later studies demonstrated that IL-5 increases 'polarization', including membrane ruffling and pseudopod formation, which appear to reflect changes in the cytoskeletal system. IL-5 also induces a rapid increase in superoxide anion production by eosinophils (Lopez *et al.*, 1988). In addition, IL-5 increases the survival of peripheral blood eosinophils (Begley *et al.*, 1986).

A further interesting observation in this context was the demonstration that IL-5 is a potent inducer of immunoglobulin-induced eosinophil degranulation, as measured by the release of eosinophil-derived neurotoxin (EDN). IL-5 increased EDN release by 48% for secretory IgA and by 136% for IgG. This enhancing effect appeared by 15 min and reached a maximum by 4 h (Fujisawa *et al.*, 1990). The finding that secretory IgA can induce eosinophil degranulation is particularly important because eosinophils are frequently found at mucosal surfaces where IgA is the most abundant immunoglobulin.

TISSUE LOCALIZATION

Another aspect of the pathology of diseases characterized by eosinophilia is the preferential accumulation of eosinophils in tissues. As the blood contains both eosinophils and neutrophils, there must exist a specific mechanism that allows the eosinophils to pass preferentially from the blood vessels to the tissues. A number of factors, including IL-5, are reported to have a specific chemotactic activity for eosinophils (Yamaguchi *et al.*, 1988b; Wang *et al.*, 1989).

The different tissue distribution of eosinophils in the two transgenic mice systems probably results from the different tissue expression of IL-5. Using the metallothionein promoter, transgene expression was demonstrated in the liver and skeletal muscle, and eosinophils were observed in these tissues (Murata *et al.*, 1992). In contrast, the CD2–IL-5 mice with IL-5 expression in T cells did not have eosinophils in the liver or skeletal muscle (Dent *et al.*, 1990). This suggests that eosinophils migrate into tissues where IL-5 is expressed. While IL-5 is reported to be chemotactic for eosinophils (Yamaguchi *et al.*, 1988; Wang *et al.*, 1989), the activity is relatively weak, and it is not clear what role this could play *in vivo*.

An alternative mechanism for extravasation of eosinophils is suggested by experiments in which IL-5 has been shown to upregulate adhesion molecules. Thus, it was demonstrated that IL-5 increased the expression of the integrin CD11b on human eosinophils (Lopez *et al.*, 1986), and this increased expression was accompanied by an increased adhesion to endothelial cells (Walsh *et al.*, 1990). Adhesion was inhibited by antibody to CD11b or CD18, suggesting that the integrins are involved in eosinophil adhesion to endothelial cells (Walsh *et al.*, 1990). More recently it has been shown that eosinophils can use the integrin very late activation antigen-4 (VLA-4) (CD49d/CD29) in adherence to endothelial cells. In this case the ligand is vascular cell adhesion molecule-1 (VCAM-1). In contrast neutrophils do not express VLA-4 and do not use this adherence mechanism (Walsh *et al.*, 1991).

Thus, while IL-5 appears to be involved in eosinophil localization, the recent identification of eotaxin, a member of the chemokine family, provides an important step in understanding the specificity of eosinophil localization. Like IL-5, eotaxin is specific for the eosinophil lineage, but is a powerful chemoattractant and activator of eosinophils. There is good evidence for a biological interaction between IL-5 and eotaxin (Collins *et al.*, 1995).

ACTIVITIES ON OTHER CELL TYPES

Basophils

The most pronounced effect of IL-5 on cells other than eosinophils in humans, is the effect on basophils. While our studies suggested that IL-5 induces only eosinophils, a detailed study by electron microscopy of cells produced in human cord blood cultures has revealed a small number of basophils (Dvorak *et al.*, 1989). Other studies have shown that IL-5 primes basophils for increased histamine production and leukotriene generation (Bischoff *et al.*, 1990; Hirai *et al.*, 1990), and basophils in the blood clearly express the IL-5 receptor (Lopez *et al.*, 1990). Thus, while the effect of IL-5 on the production of basophils may be minor, the priming effect on mature basophils may be of significance in the allergic response.

B Cells

As discussed in the Introduction, the characterization of the activities of IL-5 on mouse B cells developed around several different *in vitro* assay systems. In the TRF assay, IL-5 induces specific antibody production by B cells primed with antigen *in vivo* (Takatsu *et al.*, 1988). The BCGF-II assay was based on the ability of IL-5 to induce DNA synthesis in normal splenic B cells in the presence of dextran sulphate, and later on the ability of IL-5 to increase DNA synthesis in the BCL1 cell (a mouse B-cell tumour) line (Swain *et al.*, 1988). The BCDF-μ assay depends on the ability of IL-5 to induce BCL1 cells to secrete IgM (Vitetta *et al.*, 1984).

IL-5 is a late acting factor in the differentiation of primary B cells, requiring a priming stimulus to make resting B cells responsive. This can be either polyclonal stimulants such as dextran sulphate, bacterial lipopolysaccharide (LPS), anti-immunoglobulin or specific antigen. Large splenic B cells, presumed to have been activated *in vivo*, when cultured with IL-5 for 7 days show markedly enhanced numbers of IgM- and IgG-producing cells (O'Garra *et al.*, 1986). IL-5 in combination with antigen is sufficient to induce growth and differentiation of B cells at the single cell level (Alderson *et al.*, 1987). Combinations of IL-2, IL-4 and IL-5 appear to regulate the amount of IgG_1 isotype secreted by B cells (McHeyzer-Williams, 1989; Purkerson *et al.*, 1992). Neutralizing antibody to IL-5 was found to inhibit the polyclonal antibody response induced by T-cell clones on B cells, suggesting a critical role for IL-5 in this system (Rasmussen *et al.*, 1988).

A possible role for IL-5 in the development of autoimmunity in mice was suggested by the observation that this cytokine stimulates B cells from NZB mice to produce high levels of IgM anti-DNA antibody (Howard *et al.*, 1984). In another study the B cells from autoimmune NZB/W mice were found to be hyper-responsive to IL-5, whereas two other strains of mice which are prone to autoimmunity did not show this response. As NZB/W mice have raised numbers of Ly-1-positive B cells, these were tested and found to show a higher response to IL-5 than the negative cells, suggesting that the increased responsiveness to IL-5 in these mice may be due to the increased numbers of Ly-1 B cells (Umland *et al.*, 1989). In support of this, freshly isolated peritoneal Ly-1 B cells express high levels of IL-5 receptor, and IL-5 increases the frequency of cells that produce autoantibodies (Wetzel, 1989). As these effects concern mainly the production of IgM, whereas autoimmune disease appears to be due mainly to IgG, the significance of these findings for autoimmunity are unclear. However, these experiments point to the possible restriction of IL-5 activity to the Ly-1 subpopulation of B cells.

A potentially interesting observation is the demonstration that IL-5 appears preferentially to enhance IgA production. When added to cultures in the presence of LPS, the highest increase over background occurs with the IgA-producing cells, with significant increases in IgM and IgG_1 as well (Bond *et al.*, 1987; Yokota *et al.*, 1987). The interpretation of these experiments is not straightforward, as the LPS itself induces a large effect, and the activity on IgA and IgG_1 was small in comparison to the total levels of IgM produced. In a study of B cells from gut-associated lymphoid tissue (Peyer's patches), IL-5 increased the production of IgA but maximum enhancement of IgA in these cultures requires IL-4 (Murray *et al.*, 1987; Lebman and Coffman, 1988). This effect of IL-5 was shown to be due to the induction of a high rate of IgA synthesis in cells positive for surface IgA expression. No IgA secretion was induced in the surface IgA-negative cells (Murray *et al.*, 1987; Harriman *et al.*, 1988; Kunimoto *et al.*, 1988). This suggests that IL-5 does not induce switching to IgA production, but acts after switching to enhance the production of IgA.

In contrast to these studies, which suggest a key role for IL-5 in the production of IgA, more recent studies have indicated that its effect is minor compared with the activity of other cytokines, and that it may only augment these activities. For example, IL-5 was shown to enhance IgA secretion from B cells isolated from Peyer's patches, but the effect was small compared with the effect of IL-6 (Beagley *et al.*, 1989). A combination of IL-5 and IL-6 had a greater effect than either cytokine alone (Kunimoto *et al.*, 1989). It has been shown that transforming growth factor has an important activity in the switching to IgA production in LPS-stimulated B cells, and while IL-5 enhances this effect it is less active than IL-2 (Sonoda *et al.*, 1989). The action of IL-5 allows the cells to respond to IL-2 by amplification of J chain mRNA. Thus IL-5 and IL-2 are both necessary for IgM secretion (Matsui *et al.*, 1989). A possible mechanism for the effect of IL-2 on B cells is suggested by the observation that IL-5 increases the expression of the IL-2 receptor (Loughnan *et al.*, 1988).

The significance of these activities remains unclear. While transgenic mice expressing IL-5 on a metallothionein promoter were shown to develop autoimmunity (Tominaga *et al.*, 1991), no effects on B cells or antibody levels were detected in transgenic mice expressing IL-5 under control of the CD2 locus control region (Sanderson *et al.*, 1993). Similarly, treatment of mice with anti-IL-5 antibody completely blocked eosinophil production but had no effect on antibody levels (Finkelman *et al.*, 1990).

Despite this large body of research on IL-5 as a B-cell growth factor *in vitro*, IL-5 knockout mice show a surprising absence of effect on immunoglobulin levels or ability to mount an antibody response. In fact, the only detectable effect of IL-5 gene disruption was a transient decrease in Ly^+ B cells in the first few weeks of life. This is in contrast to the major effects on eosinophil production discussed above.

In view of the well characterized activity of IL-5 on mouse B cells it was surprising that no activity could be demonstrated in a wide range of human B-cell assay systems (Clutterbuck *et al.*, 1987). This lack of activity of human IL-5 has been confirmed in many different systems (Bende *et al.*, 1992). Although two reports of low activity in some human assays has reopened this question (Bertolini *et al.*, 1993; Huston *et al.*, 1996), the failure to detect IL-5 receptor on the surface of human B cells and the absence of significant effects in IL-5 knockout mice probably dismiss the concept of IL-5 as an important cytokine in the generation of an antibody response.

Nervous System

An intriguing observation that both IL-4 and IL-5 regulate nerve growth factor production by astrocytes suggests a possible role in the regulation of the neural system (Awatsuji *et al.*, 1993).

SOURCES OF IL-5

All of the original reports on the characterization, purification and cloning of murine IL-5 utilized T-cell lines or lymphomas as the source of material, suggesting that T cells are an important source of the cytokine. The demonstration that IL-5 as well as other cytokine mRNAs are produced by mast cell lines opens the possibility that these cells may serve to induce or amplify the development of eosinophilia (Burd *et al.*, 1989; Plaut *et al.*, 1989). Similarly, the observation that human Epstein–Barr virus-transformed B cells produce IL-5 raises the possibility that B cells may be an additional source of this cytokine (Paul *et al.*, 1990). Furthermore, eosinophils themselves have been demonstrated to produce IL-5 (Broide *et al.*, 1992), although they do not appear to produce enough to sustain their own survival.

In a careful study of cells producing IL-5 in bronchial biopsies from asthmatic subjects it was concluded that T cells are the major source of IL-5. The apparent dominance of mast cells in some studies was attributed to the fact that mast cells store IL-5 in their granules, whereas T cells rapidly secrete IL-5 as it is synthesized. Thus immunohistological staining for IL-5 under-represents the number of T cells compared with *in situ* hybridization (Ying *et al.*, 1994, 1997).

It is not clear whether the non-T cell-derived IL-5 plays a significant biological role in the development of eosinophilia. Eosinophilia has been observed in a significant proportion of a wide range of human tumours. In many cases the presence of eosinophils has been found to be of positive prognostic significance (reviewed by Sanderson, 1992). Clearly, it is important to understand the mechanism of production of these eosinophils. In a study of Hodgkin's disease with associated eosinophilia, all 16 cases gave a positive signal for IL-5 mRNA by *in situ* hybridization (Samoszuk and Nansen, 1990). This suggests that IL-5 may be responsible for the production of eosinophils in these patients, and raises the possibility that eosinophilia in other tumours may also be due to the production of IL-5 by tumour cells.

REFERENCES

Alderson, M.R., Pike, B.L., Harada, N., Tominaga, A., Takatsu, K. and Nossal, G.J. (1987). Recombinant T cell replacing factor (interleukin 5) acts with antigen to promote the growth and differentiation of single hapten-specific B lymphocytes. *J. Immunol.* **139**, 2656–2660.

Awatsuji, H., Furukawa, Y., Hirota, M., Murakami, Y., Nii, S., Furukawa, S. and Hayashi, K. (1993). Interleukin-4 and -5 as modulators of nerve growth factor synthesis/secretion in astrocytes. *J. Neurosci. Res.* **34**, 539–545.

Barry, S.C., McKenzie, A.N., Strath, M. and Sanderson, C.J. (1991). Analysis of interleukin 5 receptors on murine eosinophils: a comparison with receptors on B13 cells. *Cytokine* **3**, 339–344.

Beagley, K.W., Eldridge, J.H., Lee, F., Kiyono, H., Everson, M.P., Koopman, W.J., Hirano, T., Kishimoto, T. and McGhee, J.R. (1989). Interleukins and IgA synthesis. Human and murine interleukin 6 induce high rate IgA secretion in IgA-committed B cells. *J. Exp. Med.* **169**, 2133–2148.

Begley, C.G., Lopez, A.F., Nicola, N.A., Warren, D.J., Vadas, M.A., Sanderson, C.J. and Metcalf, D. (1986). Purified colony stimulating factors enhance the survival of human neutrophils. *Blood* **68**, 162–166.

Bende, R.J., Jochems, G.J., Frame, T.H., Klein, M.R., Van Eijk, R.V.W., Van Lier, R.A.W. and Zeijlemaker, W.P. (1992). Effects of IL-4, IL-5, and IL-6 on growth and immunoglobulin production of Epstein–Barr. *Cell. Immunol.* **143**, 310–323.

Bertolini, J.W., Sanderson, C.J. and Benson, E.M. (1993). Human interleukin-5 induces staphylococcal A Cowan 1 strain-activated human B cells to secrete IgM. *Eur. J. Immunol.* **23**, 398–402.

Bischoff, S.C., Brunner, T., De Weck, A.L. and Dahinden, C.A. (1990). Interleukin 5 modifies histamine release and leukotriene generation by human basophils in. *J. Exp. Med.* **172**, 1577–1582.

Bohjanen, P.R., Okajima, M. and Hodes, R.J. (1990). Differential regulation of interleukin 4 and interleukin 5 gene expression: a comparison of T cell gene induction by anti-CD3 antibody or by exogenous lymphokines. *Proc. Natl Acad. Sci. U.S.A.* **87**, 5283–5287.

Bond, M.W., Shrader, B., Mosmann, T.R. and Coffman, R.L. (1987). A mouse T cell product that preferentially enhances IgA production. II. Physicochemical characterization. *J. Immunol.* **139**, 3691–3696.

Broide, D.H., Paine, M.M. and Firestein, G.S. (1992). Eosinophils express interleukin 5 and granulocyte macrophage-colony-stimulating factor. *J. Clin. Invest.* **90**, 1414–1424.

Burd, P.R., Rogers, H.W., Gordon, J.R., Martin, C.A., Jayaraman, S., Wilson, S.D., Dvorak, A.M., Galli, S.J. and Dorf, M.E. (1989). Interleukin 3-dependent and -independent mast cells stimulated with IgE and antigen. *J. Exp. Med.* **170**, 245–257.

Campbell, H.D., Tucker, W.Q., Hort, Y., Martinson, M.E., Mayo, G., Clutterbuck, E.J., Sanderson, C.J. and Young, I.G. (1987). Molecular cloning, nucleotide sequence, and expression of the gene encoding human. *Proc. Natl Acad. Sci. U.S.A.* **84**, 6629–6633.

Campbell, H.D., Sanderson, C.J., Wang, Y., Hort, Y., Martinson, M.E., Tucker, W.Q., Stellwagen, A., Strath, M. and Young, I.G. (1988). Isolation, structure and expression of cDNA and genomic clones for murine eosinophil. *Eur. J. Biochem.* **174**, 345–352.

Chen, D. and Rothenberg E.V. (1994). Interleukin 2 transcription factors as molecular targets of cAMP inhibition: delayed inhibition kinetics and combinatorial transcription roles. *J. Exp. Med.* **179**, 931–942.

Clutterbuck, E.J. and Sanderson, C.J. (1988). Human eosinophil hematopoiesis studied *in vitro* by means of murine eosinophil. *Blood* **71**, 646–651.

Clutterbuck, E.J. and Sanderson, C.J. (1990). The regulation of human eosinophil precursor production by cytokines: a comparison of recombinant human interleukin-1 (rhIL-1), rhIL-3, rhIL-6, and rh granulocyte macrophage colony-stimulating factor. *Blood* **75**, 1774–1779.

Clutterbuck, E., Shields, J.G., Gordon, J., Smith, S.H., Boyd, A., Callard, R.E., Campbell, H.D., Young, I.G. and Sanderson, C.J. (1987). Recombinant human interleukin-5 is an eosinophil differentiation factor but has no activity. *Eur. J. Immunol.* **17**, 1743–1750.

Coffman, R.L., Seymour, B.W.P., Lebman, D.A., Hiraki, D.D., Christiansen, J.A., Shrader, B., Cherwinski, H.M., Savelkoul, H.F.J., Finkelman, F.D., Bond, M.W. and Mosmann, T.R. (1988). The role of helper T cell products in mouse B cell differentiation and isotype regulation. *Immunol. Rev.* **102**, 5–28.

Coffman, R. L., Seymour, B.W., Hudak, S., Jackson, J. and Rennick, D. (1989). Antibody to interleukin-5 inhibits helminth-induced eosinophilia in mice. *Science* **245**, 308–310.

Collins, P.D., Weg, V.B., Faccioli, L.H., Watson, M.L., Moqbel, R. and Williams, T.J. (1993). Eosinophil accumulation induced by human interleukin-8 in the guinea-pig *in vivo. Immunology* **79**, 312–318.

Collins, P.D., Marleau, S., Griffiths-Johnson, D.A., Jose, P.J. and Williams, T.J. (1995). Cooperation between interleukin-5 and the chemokine cotaxin to induce eosinophil accumulation *in vivo. J. Exp. Med.* **182**, 1169–1174.

Cornelis, S., Plaetinck, G., Devos, R., Van der Heyden, J., Tavernier, J., Sanderson, C.J., Guisez, Y. and Fiers, W. (1995). Detailed analysis of the IL5/IL5R interaction: characterisation of crucial residues on the ligand and the receptor. *EMBO J.* **14**, 3395–3402.

Corrigan, C.J. and Kay, A.B. (1992). T cells and eosinophils in the pathogenesis of asthma. *Immunol. Today* **13**, 501–505.

David, J.R., Vadas, M.A., Butterworth, A.E., de Brito, P.A., Carvalho, E.M., David, R.A., Bina, J.C. and Andrade, Z.A. (1980). Enhanced helminthotoxic capacity of eosinophils from patients with eosinophilia. *N. Engl. J. Med.* **303**, 1147–1152.

Dent, L.A., Strath, M., Mellor, A.L. and Sanderson, C.J. (1990). Eosinophilia in transgenic mice expressing interleukin 5. *J. Exp. Med.* **172**, 1425–1431.

Derig, H.G., Burgess, G.S., Klinberg, D., Dahreine, T.S., Mochizuki, D.Y., Williams, D.E. and Boswell, H.S. (1990). Role for cyclic AMP in the postreceptor control of cytokine-stimulated stromal cell growth factor production. *Leukemia* **4**, 471–479.

Devos, R., Guisez, Y., Cornelis, S., Verhee, A., Van der Heyden, J., Manneberg, M., Lahm, H.-W., Fiers, W., Tavernier, J. and Plaetinck, G. (1993). Recombinant soluble human interleukin-5 (hIL-5) receptor molecules. Cross linking and stoichiometry of binding to IL-5. *J. Biol. Chem.* **268**, 6581–6587.

Dickason, R.R. and Huston, D.P. (1996). Creation of a biologically active interleukin-5 monomer. *Nature* **379**, 652–655.

Dickason, R.R., Huston, M.M. and Huston, D.P. (1996). Delineation of IL-5 domains predicted to engage the IL-5 receptor complex. *J. Immunol.* **156**, 1030–1037.

Dvorak, A.M., Saito, H., Estrella, P., Kissell, S., Arai, N. and Ishizaka, T. (1989). Ultrastructure of eosinophils and basophils stimulated to develop in human cord blood. *Lab. Invest.* **61**, 116–132.

Finkelman, F.D., Holmes, J., Katona, I.M., Urban, J.F., Beckmann, M.P., Park, L.S., Schooley, K.A., Coffman, R.L., Mosmann, T.R. and Paul, W.E. (1990). Lymphocyte control of *in vivo* immunoglobulin isotype selection. *Ann. Rev. Immunol.* **8**, 303–333.

Foster, P.S., Hogan, S.P., Ramsay, A.J., Matthaei, K.I. and Young, I.G. (1996). Interleukin 5 deficiency abolishes eosinophilia, airways hyperreactivity, and lung damage in a mouse asthma model. *J. Exp. Med.* **183**, 195–201.

Fujisawa, T., Abu-Ghazaleh, R., Kita, H., Sanderson, C.J. and Gleich, G.J. (1990). Regulatory effect of cytokines on eosinophil degranulation. *J. Immunol.* **144**, 642–646.

Gleich, G.J. (1990). The eosinophil and bronchial asthma: current understanding. *J. Allergy Clin. Immunol.* **85**, 422–436.

Graber, P., Bernard, A.R., Hassell, A.M., Milburn, M.V., Jordan, S.R., Proudfoot, A.E., Fattah, D. and Wells, T.N. (1993). Purification, characterisation and crystallisation of selenomethionyl recombinant human interleukin-5 from *Escherichia coli. Eur. J. Biochem.* **212**, 751–755.

Graber, P., Proudfoot A.E., Talabot F., Bernard A., McKinnon, M., Banks, M., Fattah, D., Solari, R., Peitsch, M.C. and Wells, T.N. (1995). Identification of key charged residues of human interleukin-5 in receptor binding and cellular activation. *J. Biol. Chem.* **270**, 15 762–15 769.

Gruart-Gouilleux, V., Engels, P. and Sullivan, M. (1995). Characterization of the human interleukin-5 gene promoter: involvement of octamer binding sites in the gene promoter activity. *Eur. J. Immunol.* **25**, 1431–1435.

Hagan, P., Moore, P.J., Adjukiewicz, A.B., Greenwood, B.M. and Wilkins, H.A. (1985). *In-vitro* antibody-dependent killing of schistosomula of *Schistosoma haematobium* by human eosinophils. *Parasite Immunol.* **7**, 617–624.

Harada, N., Kikuchi, Y., Tominaga, A., Takaki, S. and Takatsu, K. (1985). BCGFII activity on activated B cells or purified T cell replacing factor (TRF) from a T cell hybridoma (B151K12). *J. Immunol.* **134**, 3944–3951.

Harriman, G.R., Kunimoto, D.Y., Elliott, J.F., Paetkau, V. and Strober, W. (1988). The role of IL-5 in IgA B cell differentiation. *J. Immunol.* **140**, 3033–3039.

Hirai, K., Yamaguchi, M., Misaki, Y., Takaishi, T., Ohta, K., Morita, Y., Ito, K. and Miyamoto, T. (1990). Enhancement of human basophil histamine release by interleukin 5. *J. Exp. Med.* **172**, 1525–1528.

Howard, M., Nakanishi, K. and Paul, W.E. (1984). B cell growth and differentiation factors. *Immunol. Rev.* **78**, 185–210.

Huston, M.M., Moore, J.P., Mettes, H.J., Tavana, G. and Huston, D.P. (1996). Human B cells express IL-5 receptor messenger ribonucleic acid and respond to IL-5 with enhanced IgM production after mitogen stimulation with *Moraxella catarrhalis. J. Immunol.* **156**, 1392–1401.

Ingley, E. and Young, I.G. (1991). Characterization of a receptor for interleukin-5 on human eosinophils and the myeloid. *Blood* **78**, 339–344.

Karlen, S., D'Ercole, M. and Sanderson, C.J. (1996a). Two pathways can activate the interleukin-5 gene and induce binding to the conserved lymphokine element 0. *Blood* **88**, 211–221.

Karlen, S., Mordvinov, S. and Sanderson, C.J. (1996b). How is expression of the interleukin-5 gene regulated? *Immunol. Cell Biol.* **74**, 218–223.

Kodama, S., Endo, T., Tsujimoto, M. and Kobata, A. (1992). Characterization of recombinant murine interleukin 5 expressed in Chinese hamster ovary. *Glycobiology* **2**, 419–427.

Kodama, S., Tsujimoto, M., Tsuruoka, N., Sugo, T., Endo, T. and Kobata, A. (1993). Role of sugar chains in the *in-vitro* activity of recombinant human interleukin 5. *Eur. J. Biochem.* **211**, 903–908.

Kopf, M., Brombacher, F., Hodgkin, P.D., Ramsay, A.J., Milbourne, E.A., Dai, W.J., Ovington, K.S., Behm, C.A., Kohler, G., Young, I.G. and Matthaei, K.I. (1996). IL-5-deficient mice have a developmental defect in $CD5^+$ B-1 cells and lack eosinophilia but have normal antibody and cytotoxic T cell responses. *Immunity* **4**, 15–24.

Kunimoto, D.Y., Harriman, G.R. and Strober, W. (1988). Regulation of IgA differentiation in CH12LX B cells by lymphokines. IL-4 induces membrane IgM-positive CH12LX cells to express membrane IgA and IL-5 induces membrane IgA-positive CH12LX cells to secrete IgA. *J. Immunol.* **141**, 713–720.

Kunimoto, D.Y., Nordan, R.P. and Strober, W. (1989). IL-6 is a potent cofactor of IL-1 in IgM synthesis and of IL-5 in IgA synthesis. *J. Immunol.* **143**, 2230–2235.

Lebman, D.A. and Coffman, R.L. (1988). The effects of IL-4 and IL-5 on the IgA response by murine Peyer's patch B cell subpopulations. *J. Immunol.* **141**, 2050–2056.

Lee, H.J., Koyano-Nakagawa, N., Naito, Y., Nishida, J., Arai, N., Arai, K.-I. and Yokota, T. (1993). cAMP activates the IL5 promoter synergistically with phorbol ester through the signalling pathway involving protein kinase. *J. Immunol.* **151**, 6135–6142.

Lee, H.J., Masuda, E.S., Arai, N., Arai, K. and Yokota, T. (1995). Definition of *cis*-regulatory elements of the mouse interleukin-5 gene promoter. Involvement of nuclear factor of activated T cells-related factors in interleukin-5 expression. *J. Biol. Chem.* **270**, 17 541–17 550.

Li, J., Cook, R., Dede, K. and Chaiken, I. (1996). Single chain human interleukin 5 and its asymmetric mutagenesis for mapping receptor binding sites. *J. Biol. Chem.* **271**, 1817–1820.

Lopez, A.F., Begley, C.G., Williamson, D.J., Warren, D.J., Vadas, M.A. and Sanderson, C.J. (1986). Murine eosinophil differentiation factor. An eosinophil-specific colony stimulating factor. *J. Exp. Med.* **163**, 1085–1099.

Lopez, A.F., Sanderson, C.J., Gamble, J.R., Campbell, H.D., Young, I.G. and Vadas, M.A. (1988). Recombinant human interleukin 5 is a selective activator of human eosinophil function. *J. Exp. Med.* **167**, 219–224.

Lopez, A.F., Eglinton, J.M., Gillis, D., Park, L.S., Clark, S. and Vadas, M.A. (1989). Reciprocal inhibition of binding between interleukin 3 and granulocyte–macrophage. *Proc. Natl Acad. Sci. U.S.A.* **86**, 7022–7026.

Lopez, A.F., Eglinton, J.M., Lyons, A.B., Tapley, P.M., To, L.B., Park, L.S., Clark, S.C. and Vadas, M.A. (1990). Human interleukin-3 inhibits the binding of granulocyte–macrophage colony-stimulating factor. *J. Cell Physiol.* **145**, 69–77.

Lopez, A.F., Vadas, M.A., Woodcock, J.M., Milton, S.E., Lewis, A., Elliott, M.J., Gillis, D., Ireland, R., Olwell, E. and Park, L.S. (1991). Interleukin-5, interleukin-3, and granulocyte–macrophage colony-stimulating factor. *J. Biol. Chem.* **267**, 24 741–24 747.

Lopez, A.F., Shannon, M.F., Hercus, T., Nicola, N.A., Camareri, B., Dottore, M., Layton, M.J., Eglinton, L. and Vadas, M.A. (1992). Residue 21 of human granulocyte–macrophage colony-stimulating factor is critical for biological activity and for high but not low affinity binding. *EMBO J.* **11**, 901–916.

Loughnan, M.S., Sanderson, C.J. and Nossal, G.J. (1988). Soluble interleukin 2 receptors are released from the cell surface of normal murine B lymphocytes stimulated with interleukin 5. *Proc. Natl Acad. Sci. U.S.A.* **85**, 3115–3119.

Masuda, E.S., Tokumitsu, H., Tsuboi, A., Schlomai, J., Hung, P., Arai, K.-I. and Arai, N. (1993). The granulocyte–macrophage colony-stimulating factor promoter *cis*-acting element CLE0 mediates induction signals in T cells and is recognised by factors related to API and NFAT. *Mol. Cell. Biol.* **13**, 7399–7407.

Matsui, K., Nakanishi, K., Cohen, D.I., Hada, T., Furuyama, J., Hamaoka, T. and Higashino, K. (1989). B cell response pathways regulated by IL-5 and IL-2. Secretory microH chain-mRNA and J chain mRNA expression are separately controlled events. *J. Immunol.* **142**, 2918–2923.

Maxwell, C., Hussian, R., Nutman, T.B., Poindexter, R.W., Little, M.D., Schad, G.A. and Ottesen, E.A. (1987). The clinical and immunologic responses of normal human volunteers to low dose. *Am. J. Trop. Med. Hyg.* **37**, 126–134.

McHeyzer-Williams, M.G. (1989). Combinations of interleukins 2, 4 and 5 regulate the secretion of murine immunoglobulin isotypes. *Eur. J. Immunol.* **19**, 2025–2030.

McKenzie, A.N.J., Barry, S.C., Strath, M. and Sanderson, C.J. (1991a). Structure–function analysis of interleukin-5 utilizing mouse/human chimeric molecules. *EMBO J.* **10**, 1193–1199.

McKenzie, A.N.J., Ely, B. and Sanderson, C.J. (1991b). Mutated interleukin-5 monomers are biologically inactive. *Mol. Immunol.* **28**, 155–158.

Metcalf, D., Parker, J.W., Chester, H.M. and Kincade, P.W. (1974). Formation of eosinophil-like granulocytic colonies by mouse bone marrow cells *in vitro*. *J. Cell. Physiol.* **84**, 275–290.

Metcalf, D., Cutler, R.L. and Nicola, N.A. (1983). Selective stimulation by mouse spleen conditioned medium of human eosinophil colony. *Blood* **61**, 999–1005.

Migita, M., Yamaguchi, N., Mita, S., Higuchi, S., Hitoshi, Y., Yoshida, Y., Tominaga, M., Matsuda, F., Tominaga, A. and Takatsu, K. (1991). Characterization of the human IL-5 receptors on eosinophils. *Cell. Immunol.* **133**, 484–497.

Milburn, M., Hassell, A.M., Lambert, M.H., Jordan, S.R., Proudfoot, A.E.I., Graber, P. and Wells, T.N.C. (1993). A novel dimer configuration revealed by the crystal structure at 2.4A resolution of human. *Nature* **363**, 172–176.

Minamitake, Y., Kodama, S., Katayama, T., Adachi, H., Tanaka, S. and Tsujimoto, M. (1990). Structure of recombinant human interleukin 5 produced by Chinese hamster ovary cells. *J. Biochem. (Tokyo)* **107**, 292–297.

Morton, T., Li, J., Cook, R. and Chaiken, I. (1995). Mutagenesis in the *C*-terminal region of human interleukin 5 reveals a central patch for receptor alpha chain recognition. *Proc. Natl Acad. Sci. U.S.A.* **92**, 10879–10883.

Murata, Y., Takaki, S., Migita, M., Kikuchi, Y., Tominaga, A. and Takatsu, K. (1992). Molecular cloning and expression of the human interleukin 5 receptor. *J. Exp. Med.* **175**, 341–351.

Murray, P.D., McKenzie, D.T., Swain, S.L. and Kagnoff, M.F. (1987). Interleukin 5 and interleukin 4 produced by Peyer's patch T cells selectively enhance immunoglobulin A expression. *J. Immunol.* **139**, 2669–2674.

Naora, H., Van Leeuwen, B.H., Bourke, P.F. and Young, I.G. (1994). Functional role and signal-induced modulation of proteins recognizing the conserved TCATTT-containing promoter elements in the murine IL-5 and GM-CSF genes in T lymphocytes. *J. Immunol.* **153**, 3466–3475.

Nicola, N.A. and Metcalf, D. (1991). Subunit promiscuity among hemopoietic growth factor receptors. *Cell* **67**, 1–4.

O'Garra, A., Warren, D.J., Holman, W., Popham, A.M., Sanderson, C.J. and Klaus, G.G.B. (1986). Interleukin 4 (B-cell growth factor II/eosinophil differentiation factor) is a mitogen and differentiation factor for preactivated murine B lymphocytes. *Proc. Natl Acad. Sci. U.S.A.* **83**, 5228–5232.

Okudaira, H., Mori, A., Suko, M., Etoh, T., Nakagawa, H. and Ito, K. (1995). Enhanced production and gene expression of interleukin-5 in patients with bronchial asthma: possible management of atopic diseases by down-regulation of interleukin-5 gene transcription. *Int. Arch. Allergy Immunol.* **107**, 255–258.

Paul, C.C., Keller, J.R., Armpriester, J.M. and Baumann, M.A. (1990). Epstein–Barr virus transformed B lymphocytes produce interleukin-5. *Blood* **75**, 1400–1403.

Plaetinck, G., der Heyden, J.V., Tavernier, J., Fache, I., Tuypens, T., Fischkoff, S., Fiers, W. and Devos, R. (1990). Characterization of interleukin 5 receptors on eosinophilic sublines from human promyelocytic leukaemia (HL-60) cells. *J. Exp. Med.* **172**, 683–691.

Plaut, M., Pierce, J.H., Watson, C.J., Hanley-Hyde, J., Nordan, R.P. and Paul, W.E. (1989). Mast cell lines produce lymphokines in response to cross-linkage of Fc epsilon RI or to calcium ionophores. *Nature* **339**, 64–67.

Prieschl, E.E., Gouilleux-Gruart, V., Walker, C., Harrer, N.E. and Baumruker, T. (1995). A nuclear factor of activated T cell-like transcription factor in mast cells is involved in IL-5 gene regulation after IgE plus antigen stimulation. *J. Immunol.* **154**, 6112–6119.

Proudfoot, A.E., Davies, J.G., Turcatti, G. and Wingfield, P.T. (1991). Human interleukin-5 expressed in *Escherichia coli*: assignment of the disulfide bridges. *FEBS Lett.* **283**, 61–64.

Purkerson, J.M. and Isakson, P.C. (1992). Interleukin 5 (IL-5) provides a signal that is required in addition to IL-4 for isotype. *J. Exp. Med.* **175**, 973–982.

Rasmussen, R., Takatsu, K., Harada, N., Takahashi, T. and Bottomly, K. (1988). T cell-dependent hapten-specific and polyclonal B cell responses require release of interleukin 5. *J. Immunol.* **140**, 705–712.

Robinson, D., Hamid, Q., Bentley, A., Ying, S., Kay, A.B. and Durham, S.R. (1993). Activation of $CD4^+$ T cells, increased T_{H2}-type cytokine mRNA expression, and eosinophil recruitment in bronchoalveolar lavage after allergen inhalation challenge in patients with atopic asthma. *J. Allergy Clin. Immunol.* **92**, 313–324.

Samoszuk, M. and Nansen, L. (1990). Detection of interleukin-5 messenger RNA in Reed–Sternberg cells of Hodgkin's disease. *Blood* **75**, 13–16.

Sanderson, C.J. (1990). Eosinophil differentiation factor (interleukin-5). In *Colony Stimulating Factors: Molecular and Cellular Biology* (eds T.M. Dexter, J.M. Garland and N.G. Testa), Marcel Dekker, New York, pp. 231–256.

Sanderson, C.J. (1992). IL5, eosinophils and disease. *Blood* **79**, 3101–3109.

Sanderson, C.J. (1993). Interleukin-5 and the regulation of eosinophil production. In *Immunopharmacology of Eosinophils* (eds H. Smith and R.M. Cook), Academic Press, London, pp. 11–24.

Sanderson, C.J., Warren, D.J. and Strath, M. (1985). Identification of a lymphokine that stimulates eosinophil differentiation *in vitro*: its relationship to IL-3, and functional properties of eosinophils produced in cultures. *J. Exp. Med.* **162**, 60–74.

Sanderson, C.J., O'Garra, A., Warren, D.J. and Klaus, G.G.B. (1986). Eosinophil differentiation factor also has B-cell growth factor activity: proposed name interleukin 4. *Proc. Natl Acad. Sci. U.S.A.* **83**, 437–440.

Sanderson, C.J., Campbell, H.D. and Young, I.G. (1988). Molecular and cellular biology of eosinophil differentiation factor (interleukin-5) and its effects on human and mouse B cells. *Immunol. Rev.* **102**, 29–50.

Sanderson, C.J., Strath, M., Mudway, I. and Dent, L.A. (1994). Transgenic experiments with interleukin-5. In *Eosinophils: Immunological and Clinical Aspects* (eds G.J. Gleich and A.B. Kay), Marcel Dekker, New York, pp. 335–351.

Sewell, W.A. and Mu, H.-H. (1996). Dissociation of production of interleukin-4 and interleukin-5. *Immunol. Cell Biol.* **74**, 274–277.

Sher, A., Coffman, R.L., Hieny, S., Scott, P. and Cheever, A.W. (1990). Interleukin 5 is required for the blood and tissue eosinophilia but not granuloma formation. *Proc. Natl Acad. Sci. U.S.A.* **87**, 61–65.

Sonoda, Y., Arai, N. and Ogawa, M. (1989). Humoral regulation of eosinophilopoiesis *in vitro*: analysis of the targets of interleukin-3. *Leukemia* **3**, 14–18.

Staynov, D.Z., Cousins, D.J. and Lee, T.H. (1995). A regulatory element in the promoter of the human granulocyte–macrophage colony-stimulating factor gene that has related sequences in other T-cell-expressed cytokine genes. *Proc. Natl Acad. Sci. U.S.A.* **92**, 3606–3610.

Stranick, K.S., Payvandi, F., Zambas, D.N., Umland, S.P., Egan, R.W. and Billah, M.M. (1995). Transcription of the murine interleukin 5 gene is regulated by multiple promoter elements. *J. Biol. Chem.* **270**, 20 575–20 582.

Strath, M. and Sanderson, C.J. (1986). Detection of eosinophil differentiation factor and its relationship to eosinophilia in *Mesocestoides corti*-infected mice. *Exp. Hematol.* **14**, 16–20.

Su, B., Jacinto, E., Hibi, M., Kallunki, T., Karin, M. and Ben-Neriah, Y. (1994). JNK is involved in signal integration during costimulation of T lymphocytes. *Cell* **77**, 727–736.

Swain, S.L., McKenzie, D.T., Dutton, R.W., Tonkonogy, S.L. and English, M. (1988). The role of IL4 and IL5: characterization of a distinct helper T cell subset that makes IL4 and IL5 (Th2) and requires priming before induction of lymphokine secretion. *Immunol. Rev.* **102**, 77–105.

Takatsu, K., Tominaga, A., Harada, N., Mita, S., Matsumoto, M., Takashi, T., Kikuchi, Y. and Yamaguchi, N. (1988). T cell-replacing factor (TRF)/interleukin 5 (IL-5): molecular and functional properties. *Immunol. Rev.* **102**, 107–135.

Tavernier, J., Devos, R., Cornelis, S., Tuypens, T., Van der Heyden, J., Fiers, W. and Plaetinck, G. (1991). A human high affinity interleukin-5 receptor (IL5R) is composed of an IL5-specific x chain. *Cell* **66**, 1175–1184.

Tavernier, J., Tuypens, T., Plaetinck, G., Verhee, A., Fiers, W. and Devos, R. (1992). Molecular basis of the membrane-anchored and two soluble isoforms of the human interleukin 5 receptor alpha subunit. *Proc. Natl Acad. Sci. U.S.A.* **89**, 7041–7045.

Tavernier, J., Tuypens, T., Verhec, A., Plaetinck, G., Devos, R., Van der Heyden, J., Guidez, Y. and Oefner, C. (1995). Identification of receptor-binding domains on human interleukin 5 and design of an interleukin 5-derived receptor antagonist. *Proc. Natl Acad. Sci. U.S.A.* **92**, 5194–5198.

Tominaga, A., Takahashi, T., Kikuchi, Y., Mita, S., Noami, S., Harada, N., Yamaguchi, N. and Takatsu, K. (1990). Role of carbohydrate moiety of IL5: effect of tunicamycin on the glycosylation of IL5 and the biologic activity of deglycosylated IL5. *J. Immunol.* **144**, 1345–1352.

Tominaga, A., Takaki, S., Koyama, N., Katoh, S., Matsumoto, R., Migita, M., Hitoshi, Y., Hosoya, Y., Yamauchi, S., Kanai, Y., Miyazaki, J.-I., Usuku, G., Yamamura, K.-I. and Takatsu, K. (1991). Transgenic mice expressing a B cell growth and differentiation factor gene (interleukin 5). *J. Exp. Med.* **173**, 429–437.

Tuypens, T., Plaetinck, G., Baker, E., Sutherland, G., Brusselle, G., Fiers, W., Devos, R. and Tavernier, J. (1992). Organization and chromosomal localization of the human interleukin 5 receptor alpha-chain. *Eur. Cytokine Netw.* **3**, 451–459.

Umland, S.P., Go, N.F., Cupp, J.E. and Howard, M. (1989). Responses of B cells from autoimmune mice to IL-5. *J. Immunol.* **142**, 1528–1535.

Vitetta, E.S., Brooks, K., Chen, Y.-W., Isakson, P., Jones, S., Layton, J., Mishra, G.C., Pure, E., Weiss, E., Word, C., Yuan, D., Tucker, P., Uhr, J.W. and Krammer, P.H. (1984). T cell derived lymphokines that induce IgM and IgG secretion in activated murine B cells. *Immunol. Rev.* **78**, 137–157.

Walsh, G.M., Hartnell, A., Wardlaw, A.J., Kurihara, K., Sanderson, C.J. and Kay, A.B. (1990). IL-5 enhances the *in vitro* adhesion of human eosinophils, but not neutrophils, in a leucocyte integrin (CD11/18)-dependent manner. *Immunology* **71**, 258–265.

Walsh, G.M., Mermod, J.-J., Hartnell, A., Kay, A.B. and Wardlaw, A.J. (1991). Human eosinophil, but not neutrophil, adherence to IL-1-stimulated human umbilical vascular endothelial cells is 41 (very late antigen-4) dependent. *J. Immunol.* **146**, 3419–3423.

Wang, J.M., Rambaldi, A., Biondi, A., Chen, Z.G., Sanderson, C.J. and Mantovani, A. (1989). Recombinant human interleukin 5 is a selective eosinophil chemoattactant. *Eur. J. Immunol.* **19**, 701–705.

Warren, D.J. and Sanderson, C.J. (1985). Production of a T cell hybrid producing a lymphokine stimulating eosinophil differentiation. *Immunology* **54**, 615–623.

Wetzel, G.D. (1989). Interleukin 5 regulation of peritoneal Ly-1 B lymphocyte proliferation, differentiation and autoantibody secretion. *Eur. J. Immunol.* **19**, 1701–1707.

Woodcock, J.M., Bagley, C.J., Qiyu, S., Hercus, T., Plaetinck, G., Tavernier, J. and Lopez, A.F. (1994). Three residues in the common beta chain of the human GM-CSF, IL-3 and IL-5 receptors are essential for GM-CSF and IL-5 but not IL-3 high affinity binding and interact with Glu21 of GM-CSF. *EMBO J.* **13**, 5176–5185.

Woodcock, J.M., Bagley, C.J., Zaccharakis, B. and Lopez, A.F. (1996). A single tyrosine residue in the membrane-proximal domain of the granulocyte–macrophage colony-stimulating factor, interleukin (IL)-3, and IL-5 receptor common beta-chain is necessary and sufficient for high affinity binding and signaling by all three ligands. *J. Biol. Chem.* **271**, 25 999–26 006.

Yamagata, T., Nishida, J., Sakai, R., Tanaka, T., Honda, H., Hirano, N., Mano, H., Yazaki, Y. and Hirai, H. (1995). Of the GATA-binding proteins, only GATA-4 selectively regulates the human interleukin-5 gene promoter in interleukin-5-producing cells which express multiple GATA-binding proteins. *Mol. Cell. Biol.* **15**, 3830–3839.

Yamaguchi, Y., Hayashi, Y., Sugama, Y., Miura, Y., Kasahara, T., Kitamura, S., Torisu, M., Mita, S.,Tominaga, A. and Takatsu, K. (1988). Highly purified murine interleukin 5 (IL-5) stimulates eosinophil function and prolongs *in vitro* survival. IL-5 as an eosinophil chemotactic factor. *J. Exp. Med.* **167**, 1737–1742.

Yasruel, Z., Humbert, M., Kotsimos, A.T.C., Ploysongsang, Y., Minshall, E., Durham, S.R., Pfister, R., Menz, G., Tavernier, J., Kay, A.B. and Hamid, Q. (1997). Membrane-bound and soluble alpha IL-5 receptor mRNA in the bronchial mucosa of atopic and nonatopic asthmatics. *Am. J. Respir. Crit. Care Med.* **155**, 1413–1418.

Ying, S., Durham, S.R., Jacobson, M.R., Rak, S., Masuyama, K., Lowhagen, O., Kay, A.B. and Hamid, Q.A. (1994). T lymphocytes and mast cells express messenger RNA for interleukin-4 in the nasal mucosa in allergen-induced rhinitis. *Immunology* **82**, 200–206.

Ying, S., Durham, S.R., Corrigan, C.J., Hamid, Q. and Kay, A.B. (1995). Phenotype of cells expressing mRNA for T_{H2}-type (IL-4 and IL-5) and T_{H1}-type (IL-2 and interferon γ) cytokines in bronchoalveolar lavage and bronchial biopses from atopic asthmatic and normal control subjects. *Am. J. Cell. Mol. Biol.* **12**, 477–487.

Ying, S., Humbert, M., Barkans, J., Corrigan, C.J., Pfister, T., Menz, G. Larche, M., Robinson, D.S., Durham, S.R. and Kay, A.B. (1997). Expression of IL-4 and IL-5 mRNA and protein product by $CD4^+$ and $CD8^+$ T cells, eosinophils, and mast cells in bronchial biopsies obtained from atopic and nonatopic (intrinsic) asthmatics. *J. Immunol.* **158**, 3539–3544.

Yokota, T., Coffman, R.L., Hagiwara, H., Rennick, D.M., Takebe, Y., Yokota, K., Gemmell, L., Shrader, B., Yang, G., Meyerson, P., Luh, J., Hoy, P., Pene, J., Briere, F., Spits, H., Banchereau, J., de Vries, J., Lee, F.D., Aria, N. and Aria, K. (1987). Isolation and characterization of lymphokine cDNA clones encoding mouse and human IgA-enhancing factor and eosinophil colony-stimulating factor activities: relationship to IL-5. *Proc. Natl Acad. Sci. U.S.A.* **84**, 7388–7392.

Chapter 8

Interleukin-6

Toshio Hirano
Division of Molecular Oncology, Biomedical Research Centre,
Osaka University Medical School, Osaka, Japan

INTRODUCTION

Interleukin-6 (IL-6) is a multifunctional cytokine which is produced by both lymphoid and non-lymphoid cells and regulates immune responses, acute-phase reactions and haematopoiesis (Le and Vilcek, 1989; Sehgal *et al.*, 1989; Heinrich *et al.*, 1990; Hirano and Kishimoto, 1990; Van Snick, 1990; Hirano, 1992a; Hirano and Kishimoto, 1992). IL-6 has been called by a variety of names, such as interferon-β_2 (IFN-β_2) (Weissenbach *et al.*, 1980; May *et al.*, 1986; Zilberstein *et al.*, 1986), T-cell replacing factor (TRF)-like factor (Teranishi *et al.*, 1982), B-cell differentiation factor (BCDF) (Okada *et al.*, 1983), BCDF2 (Hirano *et al.*, 1984a,b), 26-kDa protein (Haegeman *et al.*, 1986), B-cell stimulatory factor-2 (BSF-2) (Hirano *et al.*, 1985, 1986), hybridoma–plasmacytoma growth factor (HPGF or IL-HP1) (Aarden *et al.*, 1985; Nordan and Potter, 1986; Van Damme *et al.*, 1987a; Van Snick *et al.*, 1988), hepatocyte stimulating factor (HSF) (Andus *et al.*, 1987; Gauldie *et al.*, 1987), and monocyte–granulocyte inducer type 2 (MGI-2) (Shabo *et al.*, 1988). However, molecular cloning of IFN-β2 (May *et al.*, 1986; Zilberstein *et al.*, 1986), 26-kDa protein (Haegeman *et al.*, 1986) and BSF-2 (Hirano *et al.*, 1986) revealed that all these molecules are identical (Sehgal *et al.*, 1987a), and it was proposed that this molecule be referred as IL-6 at the end of 1988 (Le and Vilcek, 1989; Sehgal *et al.*, 1989; Van Snick, 1990; Hirano and Kishimoto, 1990; Heinrich *et al.*, 1990; Hirano, 1992a). In the following sections, the structure and function of IL-6 and its receptor, the regulatory mechanisms governing IL-6 gene expression, signal transduction mechanisms and the possible involvement of IL-6 in a variety of diseases are described.

BIOLOGICAL ACTIVITY

IL-6 and Immune Responses

B cells stimulated with antigen proliferate and differentiate to antibody-forming cells under the control of a variety of cytokines produced by T cells and macrophages (Kishimoto and Hirano, 1988). IL-6 was identified as one of the factors acting on B cells

The Cytokine Handbook, 3rd ed.
ISBN 0–12–689662–3

in the culture supernatants of phytohaemagglutinin (PHA)- or antigen-stimulated peripheral mononuclear cells which induce immunoglobulin production in Epstein–Barr virus (EBV)-transformed B-cell lines (Muraguchi *et al.*, 1981; Teranishi *et al.*, 1982). Furthermore, it was demonstrated that IL-6 functions in the late phase of *Staphylococcus aureus* Cowan I (SAC) stimulation of normal B cells (Hirano *et al.*, 1984a; Teranishi *et al.*, 1984) or leukaemic B cells (Yoshizaki *et al.*, 1982), inducing immunoglobulin production when other factors such as IL-2 are available. Further effects of IL-6 on B cells have been demonstrated utilizing recombinant human IL-6. The original observation that IL-6 acts on B-cell lines at the mRNA level and induces biosynthesis of secretory-type immunoglobulin (Kikutani *et al.*, 1985) was confirmed and it was demonstrated that transcriptional activation is the primary mechanism for the quantitative increase of secretory immunoglobulin mRNAs (Raynal *et al.*, 1989). Furthermore, IL-6 was found to activate immunoglobulin heavy chain enhancer (Eμ) in large but not unstimulated small B cells obtained from transgenic mice carrying the Eμ and κ light chain promoter-driving chloramphenicol acetyltransferase (CAT) gene (Miller *et al.*, 1992). IL-6 acts on B cells activated with SAC or pokeweed mitogen (PWM) to induce immunoglobulin (Ig)M, IgG and IgA production, but not on resting B cells (Muraguchi *et al.*, 1988). Anti-IL-6 antibody inhibits PWM-induced immunoglobulin production, indicating that IL-6 is essential for PWM-induced immunoglobulin production (Muraguchi *et al.*, 1988).

An essential role of IL-6 was also demonstrated in IL-4-dependent IgE synthesis (Vercelli *et al.*, 1989) and in polysaccharide-specific antibody production (Ambrosino *et al.*, 1990) in human B cells, and in the influenza A virus-specific primary response in murine B cells (Hilbert *et al.*, 1989). Anti-IL-6 antibody inhibits IL-4-driven IgE production, suggesting that endogenous IL-6 plays an obligatory role in the IL-4-dependent induction of IgE (Vercelli *et al.*, 1989). Indeed, it has been demonstrated that IL-4 induces IL-6 production in normal human B cells (Smeland *et al.*, 1989). An obligatory role for IL-6 in antibody production has also been shown in IL-2-induced immunoglobulin production in SAC-stimulated B cells (Xia *et al.*, 1989). In this case IL-2 does not induce IL-6 production but may induce the IL-6 responsiveness in SAC-activated B cells which produce IL-6 spontaneously. The dependence on IL-2 of the action of IL-6 in B cells was demonstrated previously utilizing partly purified IL-6 (Teranishi *et al.*, 1984) and this has been confirmed with recombinant IL-6 (Splawski *et al.*, 1990), indicating that, as well as antigenic stimulation, additional signals provided by growth factors such as IL-2 are required for B cells to acquire IL-6 responsiveness.

IL-6 has been demonstrated to be required differentially for antigen-specific antibody production by primary and secondary murine B cells. The former response is dependent on IL-6 but the latter is not (Hilbert *et al.*, 1989). IL-6 and IL-1 synergistically stimulate the growth and differentiation of murine B cells activated with anti-immunoglobulin or dextran sulphate (Vink *et al.*, 1988). In addition, IL-6 increases IgA production in murine Peyer's patch B cells (Beagley *et al.*, 1989; Kunimoto *et al.*, 1989) or human appendix B cells that express IL-6 receptor (Fujihashi *et al.*, 1991). This effect of IL-6 is not the result of isotype switching, as membrane-bound IgA-negative B cells were not induced to secrete IgA by IL-6 (Beagley *et al.*, 1989). These facts indicate that IL-6 plays a role in mucosal immune response (Fujihashi *et al.*, 1992). IL-6 is also reported to augment the *in vivo* production of anti sheep red blood cell (SRBC) antibodies in mice (Takasuki *et al.*, 1988). Consistent with this, IL-6 transgenic mice or mice bearing a retrovirus vector expressing IL-6 show massive plasmacytosis, hypergammaglobulinaemia (Suematsu *et al.*, 1989; Brandt *et al.*, 1990). All results showed that IL-6 plays roles in immunoglobulin production *in vivo*. However, IL-6 may not be essential for immunoglobulin production and could be compensated for by other factors *in vivo*. IL-6-deficient mice showed a reduced IgG response, but no reduction in the IgM response to both a soluble protein antigen and vesicular stomatitis virus (VSV) antigen (Kopf *et al.*, 1994). A striking effect was observed in the mucosal IgA antibody response in IL-6-deficient mice; the number of IgA-producing cells was greatly reduced (Ramsay *et al.*, 1994). This reduced IgA response was completely restored after intranasal infection with recombinant vaccinia viruses engineered to express IL-6.

IL-6 is involved in T-cell activation, growth and differentiation (see reviews by Van Snick, 1990; Houssiau and Van Snick, 1992). IL-6 induces IL-2 receptor (Tac antigen) expression in one T-cell line (Noma *et al.*, 1987) and in thymocytes (Le *et al.*, 1988), and functions as a second signal for IL-2 production by T cells (Garman *et al.*, 1987). IL-6 promotes the growth of human T cells stimulated with PHA (Houssiau *et al.*, 1988; Lotz *et al.*, 1988) or mouse peripheral T cells (Uyttenhove *et al.*, 1988). It also acts on murine thymocytes to induce proliferation (Helle *et al.*, 1988; Le *et al.*, 1988; Uyttenhove *et al.*, 1988). The effects of IL-6 are synergistic with IL-1 and tumour necrosis factor (TNF) (Le *et al.*, 1988). IL-6 enhances the proliferative response of thymocytes to IL-4 and phorbol myristate acetate (Hodgkin *et al.*, 1988). As IL-6 stimulates thymocyte proliferation and IL-1 can induce IL-6 production in thymocytes (Helle *et al.*, 1989), the effect of IL-1 on thymocyte proliferation is possibly mediated by induced IL-6. After the removal of thymocytes with low buoyant density which are capable of producing IL-6 following stimulation with IL-1, IL-1 cannot induce cell proliferation but IL-6 or IL-2 is still comitogenic; the IL-1-induced proliferation of thymocytes thus seems to be dependent on endogenous IL-6 production (Helle *et al.*, 1989). A part of the effect of IL-6 on T-cell growth is mediated by endogenously produced IL-2. Anti-IL-2R α chain (Tac) antibody generally inhibits IL-6-induced T-cell proliferation (Garman *et al.*, 1987; Le *et al.*, 1988; Helle *et al.*, 1989; Kawakami *et al.*, 1989; Tosato *et al.*, 1990). IL-1 and IL-6 synergistically induce IL-2 production (Holsti and Raulet, 1989; Houssiau *et al.*, 1989) and IL-2R α chain expression in T cells (Houssiau *et al.*, 1989). IL-6 also induces the differentiation of cytotoxic T lymphocytes (CTLs) in the presence of IL-2 from murine as well as from human thymocytes and splenic T cells (Okada *et al.*, 1988; Takai *et al.*, 1988; Uyttenhove *et al.*, 1988). Utilizing purified murine T cells, both IL-1 and IL-6 were demonstrated to be required for the generation of CTLs and, in this case, induction of the IL-2R α chain and IL-2 production by IL-1 and IL-6 were critical for CTL generation (Renauld *et al.*, 1989). IL-6 also induces serine esterase and perforin, required for mediating target cell lysis in the granules of CTLs (Takai *et al.*, 1988; Liu *et al.*, 1990), suggesting a role in the differentiation and expression of cytotoxic T-cell function. In IL-6-deficient mice, the generation of cytotoxic T cells against vaccinia virus was three- to ten-fold reduced, although CTL function against lymphocytic choriomeningitis virus (LCMV) was not reduced (Kopf *et al.*, 1994).

IL-6 and Haematopoiesis

IL-6 and IL-3 induce synergistically the proliferation of murine pluripotential haematopoietic progenitors *in vitro* (Ikebuchi *et al.*, 1987). The combination of IL-6 and IL-3 acts on blast cell colony forming cells to cause them to leave G_0 earlier. IL-6 appears to trigger the entry into the cell cycle of the dormant progenitor cells whereas IL-3 can support continued proliferation of progenitors after they exit from the G_0 phase (Ogawa, 1992). The colony forming units in spleen (CFU-S) were increased by culturing bone marrow cells in the presence of both IL-6 and IL-3 (Bodine *et al.*, 1989; Okana *et al.*, 1989a). Bone marrow cells cultured with IL-3 and IL-6 for 6 days had a much higher capability to rescue lethally irradiated mice than did cells cultured with IL-3 alone. These data indicate that the combination of IL-6 and IL-3 stimulates haematopoietic stem cells *in vitro* and therefore could be applied in bone marrow transplantation. IL-6 synergizes with macrophage colony stimulating factor (M-CSF) in the stimulation of colony forming unit-macrophage (CFU-M) with respect to both the number and size of macrophage colonies (Bot *et al.*, 1989). IL-6 has also been found to act synergistically with granulocyte–macrophage colony stimulating factor (GM-CSF) (Caracciolo *et al.*, 1989). Colony-forming units in culture (CFU-C) in the spleen and femur of mice that had been exposed to 750 rads and reconstituted with bone marrow cells were increased when IL-6 was injected (Okano *et al.*, 1989b). Furthermore, the survival rate of lethally irradiated mice transplanted with 5×10^4 bone marrow cells was increased by IL-6 treatment from 20% to 75% at day 21. The effect of IL-6 was more pronounced if it was administered as a continuous perfusion via an osmotic minipump (Suzuki *et al.*, 1989). One of the interesting reports on IL-6 and the haematopoeitic system is that the defect in differentiation of the haematopoiesis in Fanconi anaemia may be due to the deficiency in IL-6 production (Rosselli *et al.*, 1992). Consistent with the possible role of IL-6 in haematopoietic stem cells, IL-6-deficient mice showed a decrease in the absolute number of CFU-Sd12 and pre-CFU-S progenitors and reduced functionality of long-term repopulating stem cells (Bernad *et al.*, 1994).

In vitro megakaryopoiesis is supported by several haematopoietic CSFs. IL-6 was found to induce the maturation of megakaryocytes synergistically with IL-3 (Ishibashi *et al.*, 1989a); IL-6 promoted marked increments in megakaryocyte size and acetylcholinesterase activity. Furthermore, IL-6 induced a significant shift towards higher ploidy classes. These effects of IL-6 on megakaryocytes have subsequently been confirmed (Lotem *et al.*, 1989; Williams *et al.*, 1990). The role of IL-6 in megakaryocyte development is further demonstrated by the fact that anti-mouse IL-6 monoclonal antibody inhibits megakaryocyte development in mouse bone marrow cultures in both the absence and presence of IL-3 (Lotem *et al.*, 1989). Human megakaryocytes were demonstrated to express IL-6 receptor and produce IL-6, suggesting that IL-6 may regulate terminal maturation of megakaryocytes by autocrine manner (Hegyi *et al.*, 1990). IL-6 can function *in vivo*. The number of mature megakaryocytes in the bone marrow was increased in IL-6 transgenic mice (Suematsu *et al.*, 1989). Moreover, it was found that administration of IL-6 increased platelet numbers in both mice (Ishibashi *et al.*, 1989b) and monkeys (Asano *et al.*, 1990). The additive or synergistic effect of IL-3 and IL-6 on megakaryocytopoiesis was further demonstrated in mice (Carrington *et al.*, 1991) and monkeys (Geissler *et al.*, 1992). The *in vivo* effect of IL-6 was consistent with the results obtained in IL-6-deficient mice, which showed the reduction of megakaryocyte progenitors (Bernad *et al.*, 1994).

Human and mouse myeloid leukaemic cell lines, such as human histiocytic U937 cells and mouse myeloid M1 cells, can be induced to differentiate into macrophages and granulocytes *in vitro* by several synthetic and natural products. Several factors have been identified that can induce differentiation of leukaemic cells, such as G-CSF (Nicola *et al.*, 1983), macrophage granulocyte inducer type 2 (MGI-2) (Sachs, 1987), which was found to be identical to IL-6 (Shabo *et al.*, 1988), D-factor (Tomida *et al.*, 1984) and leukaemia inhibitory factor (LIF) (Gearing *et al.*, 1987). IL-6 actually induces growth inhibition and macrophage differentiation of several human and murine myeloid leukaemic cell lines, suggesting a role for IL-6 in the final maturation of cells of the granulocyte–monocyte lineage (Miyaura *et al.*, 1988; Shabo *et al.*, 1988; Onozaki *et al.*, 1989; Oritani *et al.*, 1992; Revel, 1992). Consistent with this, the predominant cell type of GM–CFU colonies was monoblast–myeloblast and macrophage in IL-6-deficient mice and wild-type mice respectively (Bernad *et al.*, 1994).

Acute-Phase Reactions

The biosynthesis of acute-phase proteins by hepatocytes is regulated by several factors, including IL-1, TNF and HSF. It was found that recombinant IL-6 can function as HSF (Gauldie *et al.*, 1987) and that the activity of crude HSF can be neutralized by anti-IL-6 (Andus *et al.*, 1987), indicating that HSF activity is exerted by the IL-6 molecule (see reviews by Heinrich *et al.*, 1990; Gauldie *et al.*, 1992). IL-6 can induce a variety of acute-phase proteins, such as fibrinogen, α_1-antichymotrypsin, α_1-acid glycoprotein and haptoglobin, in the human hepatoma cell line HepG2. In addition to these proteins, it induces serum amyloid A, C-reactive protein (CRP) and α_1-antitrypsin in human primary hepatocytes (Castell *et al.*, 1988). The proteins induced in rats by IL-6 are fibrinogen, cysteine proteinase inhibitor, α_2-macroglobulin and α_1-acid glycoprotein (Andus *et al.*, 1987; Gauldie *et al.*, 1987; Heinrich *et al.*, 1990). *In vivo* administration of IL-6 in rats induces typical acute-phase reactions similar to those induced by turpentine, and the IL-6-induced expression of mRNAs for acute-phase proteins is more rapid than that induced by turpentine (Geiger *et al.*, 1988). These results confirm the *in vivo* effect of IL-6 in the acute-phase reaction. It has also been reported that serum levels of IL-6 correlate well with those of CRP and with fever in patients with severe burns (Nijstein *et al.*, 1987), and an increase in serum IL-6 concentration has been observed before an increase in serum CRP levels in patients undergoing surgical operation (Nishimoto *et al.*, 1989; Shenkin *et al.*, 1989), supporting a causal role of IL-6 in the acute-phase response. In fact, IL-6-deficient mice are severely defective in the inflammatory acute-phase response after tissue damage or infection (Kopf *et al.*, 1994). It is also likely that different patterns of cytokines are involved in systemic and localized tissue damage: IL-6 is an essential mediator of the inflammatory response to localized inflammation, such as a turpentine-induced response, but not to a systemic response induced by lipopolysaccharide (Fattori *et al.*, 1994).

Other Activities

IL-1 stimulation of glioblastoma or astrocytoma cells was found to induce the expression of IL-6 mRNA (Yasukawa *et al.*, 1987). Both virus-infected microglial cells and astrocytes were also found to produce IL-6 (Frei *et al.*, 1989), indicating the possible involvement of IL-6 in nerve cell functions. In fact, IL-6 induces neurite outgrowth of PC12 cells into neural cells (Satoh *et al.*, 1988; Ihara *et al.*, 1996; Wu and Bradshaw, 1996). Furthermore, IL-6 can support the survival of cultured cholinergic neurons (Hama *et al.*, 1989).

IL-6 stimulates the secretion of adrenocorticotrophic hormone either through the corticotrophin-releasing hormone (Naitoh *et al.*, 1988) or directly (Fukata *et al.*, 1989). IL-6 also stimulates the release of a variety of anterior pituitary hormones, such as prolactin, growth hormone and luteinizing hormone (Spangelo *et al.*, 1989). Anterior pituitary cells produce IL-6 spontaneously (Spangelo *et al.*, 1990). Trophoblast produces IL-6 *in vivo*, although the biological significance of IL-6 in the placenta is unknown (Kameda *et al.*, 1990). Because IL-6 stimulates hepatic lipogenesis in mice, and IL-6 is induced by TNF, the lipogenic effects of TNF may be in part mediated by IL-6 (Grunfeld *et al.*, 1990). IL-6 is produced by vascular smooth muscle cells (Loppnow and Libby, 1990) and may induce their growth (Nabata *et al.*, 1990), suggesting the possible involvement of IL-6 in arteriosclerosis.

IL-6 may directly or indirectly affect osteoclast development and play a role in postmenopausal osteoporosis, because osteoclast development was enhanced in ovariectomized mice which were released from the oestrogen-induced suppression of IL-6 gene expression. Furthermore, the enhanced osteoclast development in ovariectomized mice was prevented by administration of anti-IL-6 antibody (Jilka *et al.*, 1992) and oestrogen can inhibit the IL-1- and TNF-α-induced production of IL-6 (Girasole *et al.*, 1992). The involvement of IL-6 in ovarectomy-induced osteoporosis is now evident because in IL-6-deficient mice, ovariectomy does not induce any change in either bone mass or bone remodelling rates (Poli *et al.*, 1994).

Intraperitoneal injections of either lipopolysaccharide (LPS) or IL-1β failed to evoke a fever response in IL-6-deficient mice and the fever response was recovered by the intracerebroventricular injection of recombinant human IL-6, but not of IL-1, showing that IL-6 is a necessary component of the fever response to both IL-1 and LPS (Chai *et al.*, 1996). IL-6 is a growth factor for various cells, including plasmacytoma, myeloma, hybridoma, renal cell carcinoma, Kaposi's sarcoma and keratinocyte (see below). IL-6 is essentially involved in the regeneration of hepatocytes because IL-6-deficient mice have impaired liver regeneration characterized by liver necrosis and failure (Cressman *et al.*, 1996). On the other hand, IL-6 acts as a growth inhibitor for a number of carcinoma and leukaemia cell lines, including breast carcinoma, ovarian carcinoma and myeloleukaemic cell lines (Revel, 1992).

An inability to clear *Listeria monocytogenes* is observed in IL-6-deficient mice (Kopf *et al.*, 1994; Dalrymple *et al.*, 1995). This inability is most likely due to the inability of neutrophils to function in IL-6-deficient mice, suggesting that IL-6 plays a critical role in listeriosis by stimulating neutrophils (Dalrymple *et al.*, 1995). IL-6-deficient mice show an increased susceptibility to *Escherichia coli* infection and are unable to induce neutrophilia following challenge with *E. coli* (Dalrymple *et al.*, 1996). Furthermore, IL-6-deficient mice are more susceptible than wild-type mice to virulent *Candida albicans*. Impairment of the macrophage, neutrophil and T helper 1 (T_{H1})-associated protective immunity are observed in IL-6-deficient mice (Romani *et al.*, 1996). These indicate the role of IL-6 in the function of macrophages and neutrophils *in vivo*.

STRUCTURE OF IL-6 AND ITS GENE

IL-6 is a glycoprotein with a molecular mass ranging from 21 to 28 kDa. Post-translational modifications include *N*- and *O*-linked glycosylations and phosphorylations (May *et al.*, 1988, Santhanam *et al.*, 1989). Human IL-6 consists of 212 amino

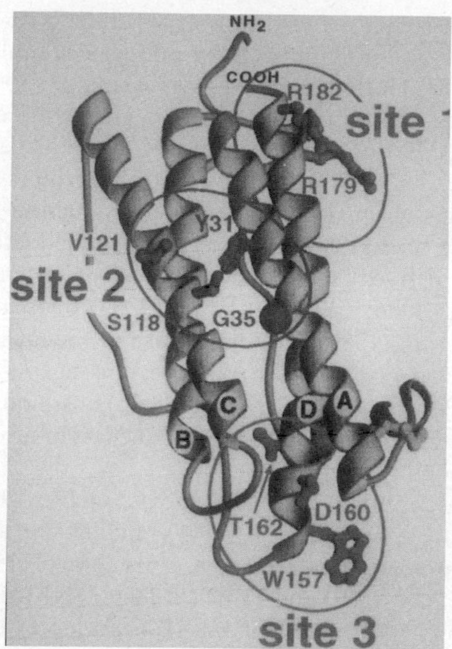

Fig. 1. Three-dimensional model of IL-6. IL-6 is composed of four (A, B, C, D) helical bundles oriented in an 'up–up–down–down' configuration. Three sites are involved in receptor interaction: site 1 is the binding site for IL-6Rα and sites 2 and 3 are those for gp130. From Paonessa *et al.* (1995), with permission.

acids including a hydrophobic signal sequence of 28 amino acids (Hirano *et al.*, 1986). Human IL-6 shows homology with IL-6 from the mouse and rat by 65% and 68% at the DNA level and 42% and 58% at the protein level respectively (Van Snick *et al.*, 1988; Northemann *et al.*, 1989). The mouse and rat protein sequences are 93% identical (Van Snick *et al.*, 1988; Northemann *et al.*, 1989). Both the *C*- and the *N*-terminus play a critical role for its biological functions (Brakenhoff *et al.*, 1989; Ida *et al.*, 1989; Krüttgen *et al.*, 1990). A computer-aided structural analysis predicts that IL-6 consists of four anti-parallel α helices with two long and one short loop connections (Fig. 1), like other cytokines, including growth hormone, prolactin, erythropoietin, IL-2, IL-4, G-CSF and GM-CSF, LIF, oncostatin M (OSM) and ciliary neurotrophic factor (CNTF) (Bazan, 1992). The evidence suggest an evolutionary relationship among these molecules acting in the immune, haematopoietic, endocrine and nerve system. The human and mouse IL-6 genes are approximately 5 and 7 kilobases in length, respectively, and both consist of five exons and four introns (Yasukawa *et al.*, 1987; Tanabe *et al.*, 1988). The genomic genes for human and murine IL-6 are mapped to chromosomes 7p21 and 5 respectively (Sehgal *et al.*, 1986; Bowcock *et al.*, 1988; Mock *et al.*, 1989). The sequence similarity in the coding region of human and mouse IL-6 genes is about 60%, whereas the 3′ untranslated region and the first 300 base pairs of the 5′ flanking region are highly conserved (80%) (Tanabe *et al.*, 1988).

The production of IL-6 is regulated by a variety of stimuli. IL-6 production is induced in T cells or T-cell clones by T-cell mitogens or antigenic stimulation (Van Snick *et al.*, 1987; Hodgkin *et al.*, 1988; Horii *et al.*, 1988; Espevik *et al.*, 1990). LPS enhances IL-6 production in monocytes and fibroblasts, whereas glucocorticoids inhibit it (Helfgott *et al.*, 1987; Sehgal, 1992). Various viruses induce IL-6 production in fibroblasts (Sehgal *et al.*, 1988; Van Damme *et al.*, 1989) or in the central nervous system (Frei *et al.*, 1988). Human immunodeficiency virus also induces IL-6 production (Nakajima *et al.*, 1989; Breen *et al.*, 1990; Emilie *et al.*, 1990). A variety of peptide factors, such as IL-1, TNF, IL-2, IFN-β and platelet-derived growth factor (PDGF) (Content *et al.*, 1985; May *et al.*, 1986; Wong and Goeddel, 1986; Zilberstein *et al.*, 1986; Kohase *et al.*, 1987; Van Damme *et al.*, 1987a,b; Kasid *et al.*, 1989), protein kinase C (Sehgal *et al.*, 1987b), calcium ionophore A23187 (Sehgal *et al.*, 1987b) and various agents causing an increase in intracellular cyclic AMP levels (Zhang *et al.*, 1988a,b), also induce IL-6 production. In contrast to these, IL-4 and IL-13 inhibit IL-6 production in monocytes (Gibbons *et al.*, 1990; Velde *et al.*, 1990; Minty *et al.*, 1993).

Several potential transcriptional control elements, such as glucocorticoid-responsive elements (GRFs), an activating protein-1 (AP-1) binding site, a c-fos serum-responsive element (c-fos SRE) homology, c-fos retinoblastoma control element (RCE) homology, a cAMP-responsive element (CRE) and a nuclear factor (NF)-κB binding site have been identified within the conserved region of the IL-6 promoter (Ray *et al.*, 1988, 1989; Tanabe *et al.*, 1988; Sehgal, 1992), as shown in Fig. 2. Among them, c-fos SRE and AP-1-like elements appear to contain the major *cis*-acting regulatory elements that confer responsiveness to several reagents (including serum, forskolin and phorbol ester) upon the heterologous herpes virus thymidine kinase (TK) promoter (Ray *et al.*, 1989). The 23-base-pair oligonucleotide designated as multiple response element (MRE) within the IL-6 enhancer region (173 to 151), which contains a CGTCA motif, binds nuclear proteins. A single copy of MRE inserted upstream of the herpes virus TK promoter renders this heterologous promoter inducible by IL-1α, TNF and serum, as well as by activators of protein kinase A (forskolin) and protein kinase C (phorbol ester). The IL-1-responsive element was also mapped within the region from 180 to 111 base pairs of the IL-6 gene and a nuclear factor, NF-IL6 CCAAT/enhancer binding protein β (C/EBP-β), was identified which binds specifically to a 14-base-pair palindrome (Akira *et al.*, 1990; Isshiki *et al.*, 1990). Shimizu *et al.* (1990) showed that the NF-κB binding motif located

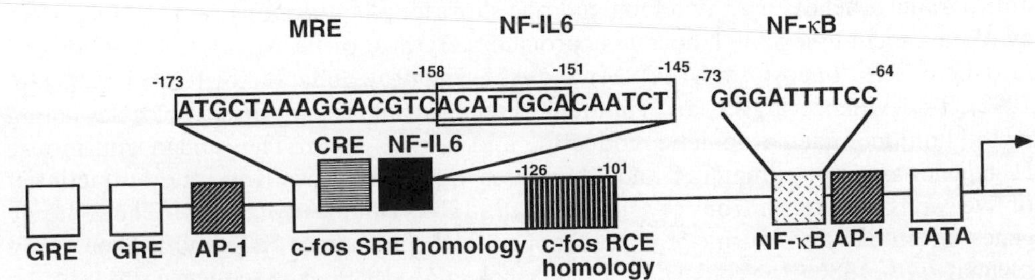

Fig. 2. *Cis*-regulatory elements in the 5′ flanking region of the IL-6 gene. GRE, glucocorticoid-responsive element; CRE, c-AMP-responsive element; SRE, serum-responsive element; RCE, retinoblastoma control element.

between 73 and 63 base pairs relative to the mRNA cap site is required for IL-1/TNF-α-induced expression of the IL-6 gene. Libermann and Baltimore (1990) and Zhang *et al.* (1990) also demonstrated the involvement of an NF-κB-like molecule in IL-6 gene expression. In fact, antisense oligonucleotide of NF-κB inhibits the expression of IL-6 mRNA in tumour cells derived from human T-lymphotropic virus-1 tax transgenic mice (Kitajima *et al.*, 1992) and p40tax induces IL-6 mRNA through the NF-κB binding site, concomitantly inducing NF-κB binding protein (Muraoka *et al.*, 1993).

The involvement of NF-κB is also implicated in IL-6 gene induction by non-structural regulatory protein 1 (NS-1), a non-structural regulatory protein of human parvovirus B19 (Moffatt *et al.*, 1996). In monocytic cell lines, the NF-κB site is crucial for LPS-induced IL-6 gene expression (Dendorfer *et al.*, 1994; Sanceau *et al.*, 1995). Synergistic induction of the IL-6 gene by IFN-γ and TNF-α, in monocytic cells, involves cooperation between the interferon regulatory factor-1 (IRF-1) and NF-κB p65 homodimers with concomitant removal of the negative effect of the retinoblastoma control element (Sanceau *et al.*, 1995). NF-κB is also involved in CD40-mediated IL-6 gene expression (Hess *et al.*, 1995). Although the NF-κB site functions as a potent IL-1/TNF-responsive element in non-lymphoid cells, its activity is repressed in lymphoid cells and NF-κB binding factor containing c-Rel seems to act as a repressor in lymphoid cells (Nakayama *et al.*, 1992). p53 and retinoblastoma (RB) also repress IL-6 gene promoter, although biological significance remains to be evaluated (Sehgal, 1992).

IL-6 RECEPTORS

The Cytokine Receptor Superfamily and Sharing of a Receptor Subunit Among Cytokine Receptors

IL-6 receptor consists of two molecules, one is an 80-kDa IL-6 binding protein (α chain) and the other a 130-kDa signal transducer, gp130 (β chain) (Yamasaki *et al.*, 1988; Taga *et al.*, 1989; Hibi *et al.*, 1990) (Fig. 3). gp130 does not bind IL-6, but IL-6Rα and gp130 form a high-affinity IL-6 binding site. Many cytokine receptors, including the receptors for growth hormone, CNTF, IL-2, erythropoietin, G-CSF and IL-5, are similar in structure to IL-6Rα and constitute the type I cytokine receptor superfamily (Bazan, 1990, 1992).

The most striking features of these receptors are the conservation of four cysteine residues and a tryptophan–serine–X–tryptophan–serine (W–S–X–W–S) motif (WS motif) located just outside the transmembrane domain. The cytoplasmic domain of IL-6Rα is not required for the IL-6-mediated signal transduction (Taga *et al.*, 1989; Hibi *et al.*, 1990). Insertion of the intracisternal A particle gene-long terminal repeat (IAP-LTR) in the cytoplasmic domain of murine IL-6Rα is present in a mouse plasmacytoma. This plasmacytoma abundantly expresses an abnormal but functional IL-6Rα whose cytoplasmic domain is replaced with IAP sequence (Sugita *et al.*, 1990). Furthermore, even the complex of IL-6 and soluble IL-6Rα can generate IL-6-mediated signal transduction (Taga *et al.*, 1989; Hibi *et al.*, 1990).

The binding of IL-6 to IL-6Rα induces the formation of a hexamer composed of two each of IL-6, IL-6Rα and gp130 (Fig. 4) (Paonessa *et al.*, 1995). In fact, IL-6 has three topologically distinct receptor binding sites: site 1 around Arg-179 on helix D is the

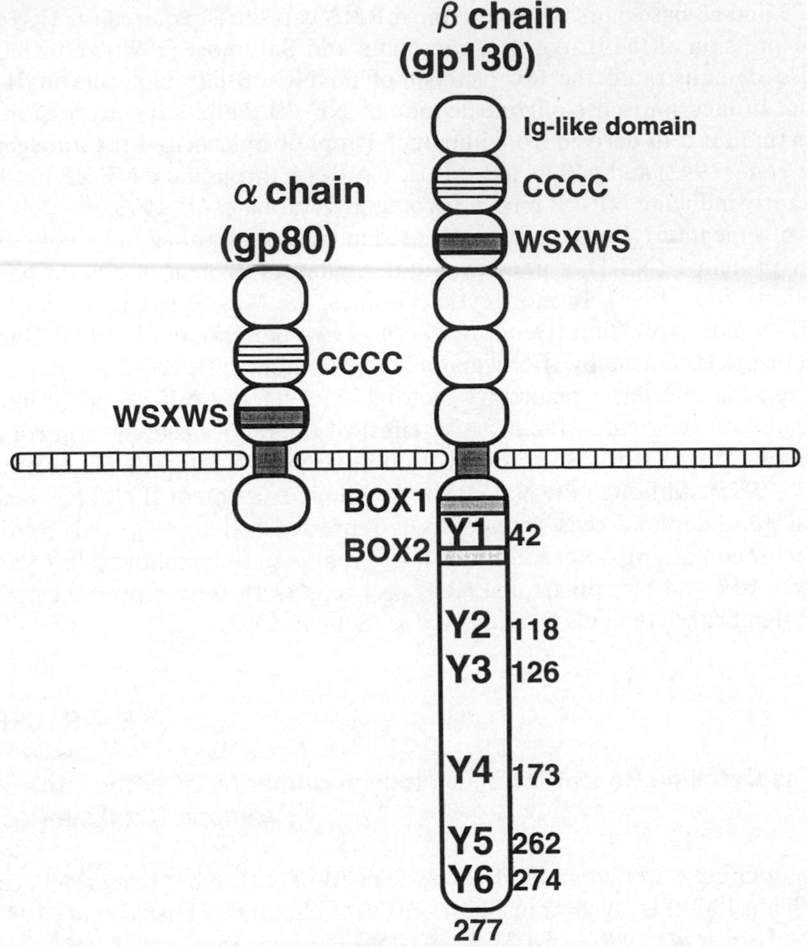

Fig. 3. Schematic representation of the IL-6 receptor. Human IL-6Rα consists of 468 amino acids including a putative signal peptide of 19 amino acids, a transmembrane domain of 28 amino acids and a cytoplasmic domain of 82 amino acids (Yamasaki *et al.*, 1988). Mature IL-6Rα is a glycoprotein with a molecular weight of 80 kDA and binds IL-6 with low affinity. The first domain of the IL-6Rα consisting of 90-amino-acid residues is similar to an immunoglobulin domain. Murine IL-6Rα consists of 460 amino acids with essentially the same structure as human IL-6Rα (Sugita *et al.*, 1990). The overall homology between murine and human IL-6Rα is 69% and 54% at the DNA and protein levels respectively. IL-6Rα belongs to the type 1 cytokine receptor superfamily. Human gp130 consists of an extracellular domain of 597 amino acids, a transmembrane domain of 22 amino acids, and a cytoplasmic domain of 277 amino acids; it encodes a protein belonging to the type 1 cytokine receptor superfamily, having a WS motif and a contactin-like sequence (Hibi *et al.*, 1990). Murine gp130 shows 76.8% homology with human gp130 at the amino acid level (Saito *et al.*, 1992). G-CSFR, LIFR, IL-12R and obesity protein receptor (OB-R) constitute the family of gp130.

binding site for IL-6Rα (Savino *et al.*, 1993); site 2, composed of residues from helix A and C, is the binding site for one gp130; site 3, centred around the *N*-terminal end of helix D, is the binding site for another gp130 (Brakenhoff *et al.*, 1994; Savino *et al.*, 1994a,b) (Figs 1 and 4). Furthermore, residues around Asn-230 and His-280 of IL-6Rα are involved in the interaction between IL-6Rα and gp130 (Yawata *et al.*, 1993; Salvati

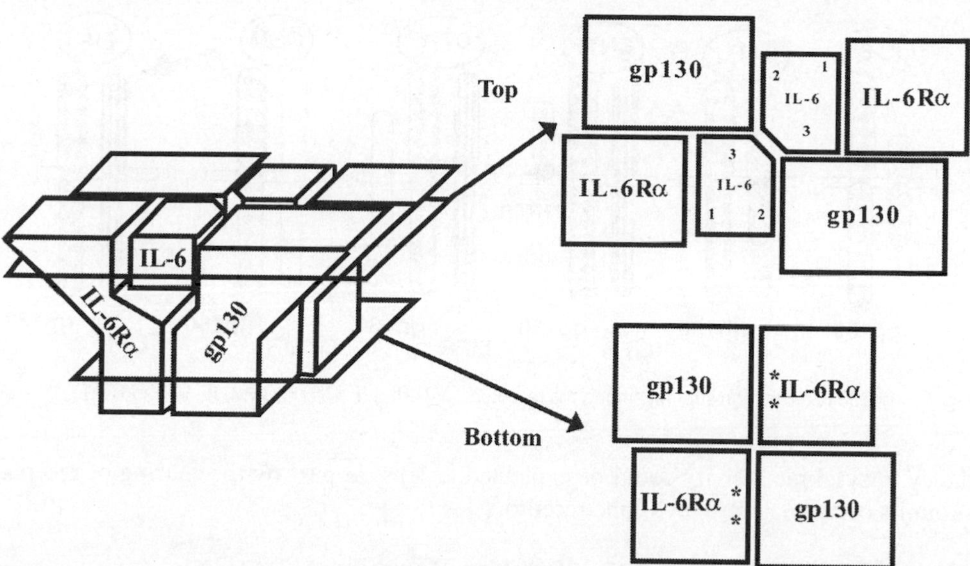

Fig. 4. Model of the hexameric IL-6 receptor complex. Shown are a side view of the IL-6 hexameric receptor complex (left) and two cross-sections through the complex (right): one that includes the bound cytokine (top), and a second (bottom) formed by the second subdomains of the cytokine binding domain (CBDs) of both IL-6Rα and gp130 where only interactions between the receptor molecules are postulated to occur. At this level, two asterisks in the IL-6Rα indicate the hypothetical location of the amino acids H280 and D281 mutagenized in the IL-6Rα variant Mut1. For IL-6, numbers identify the sites of interaction with the various receptor components (sites 1–3). For the receptors, only the part corresponding to their CBDs is shown. From Paonessa *et al.* (1995), with permission.

et al., 1995). Actually, substitutions at Asn-230, His-280 and Asp-281 impaired the capability of soluble IL-6Rα to associate with gp130, without affecting its affinity for IL-6, and antagonized IL-6 bioactivity (Salvati *et al.*, 1995). As both sites 2 and 3 of IL-6 are required for the dimerization of gp130, which is essential for signal transduction, IL-6 variants carrying amino acid substitutions in either site 2 or site 3, or both, function as an inhibitor for IL-6 (Sporeno *et al.*, 1996).

Redundancy of activity is another feature of cytokines. For example, IL-6, LIF and OSM induce macrophage differentiation in the myeloid leukaemia cell line, M1 (Miyaura *et al.*, 1988; Shabo *et al.*, 1988; Metcalf, 1989; Rose *et al.*, 1991; Oritani *et al.*, 1992) and acute-phase protein synthesis in hepatocytes (Andus *et al.*, 1987; Baumann *et al.* 1987; Gauldie *et al.*, 1987; Baumann and Wong, 1989; Richards *et al.*, 1992). One of the important findings on cytokine receptors is that one constituent of a certain cytokine receptor is shared by several other cytokine receptors, as first demonstrated for GM-CSF, IL-3 and IL-5 receptor systems (Miyajima *et al.*, 1992). Another example is gp130, which is shared by the receptors for CNTF, LIF, OSM, IL-11 and CT-1, as illustrated in Fig. 5 (Hirano *et al.*, 1994; Kishimoto *et al.*, 1995; Hibi *et al.*, 1996). Furthermore, the β chain of the IL-2 receptor (IL-2Rβc—c for common) is shared by the IL-15 receptor, and the γ chain of the IL-2 receptor (γc) is shared by IL-4, IL-7, IL-9 and IL-15 receptors (Sugamura *et al.*, 1995; Taniguchi, 1995). Thus, the molecular mechanisms of redun-

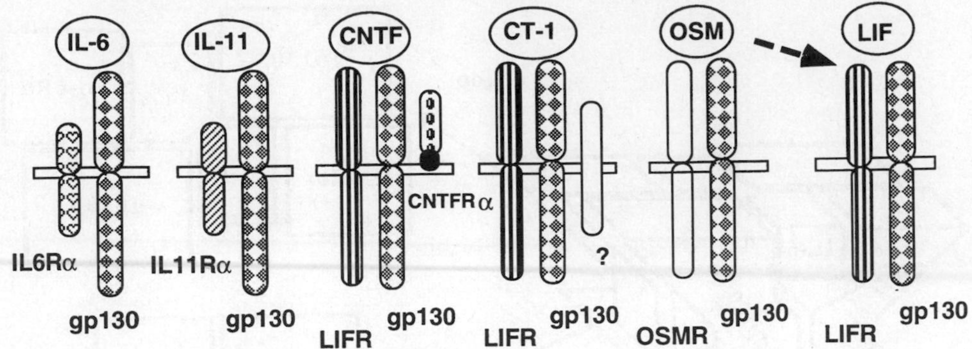

Fig. 5. gp130 is a common subunit among the receptors for IL-6, LIF, CNTF, OSM, IL-11 and CT-1.

dancy in cytokine activity could be explained at least in part by the sharing of receptor subunits between several cytokine receptors.

A Novel Mechanism Generating Cytokine Diversity

Investigations of the IL-6R system have provided evidence that a complex of IL-6 and a soluble form of IL-6Rα could act on the cells that express gp130 but not IL-6Rα (Fig. 6). Another molecule that acts in a similar manner to soluble IL-6Rα–IL-6 complexes is IL-12. IL-12 consists of a disulphide heterodimer of 40-kDa (p40) and 35-kDa (p35) subunits (Wolf *et al.*, 1991). p35 itself is a helical cytokine and p40 shows similarity to the soluble form of IL-6Rα (Gearing *et al.*, 1991). Therefore, IL-12 is a complex of a cytokine and a soluble form of its presumed receptor. Thus, the p40–p35 heterodimer can act through IL-12R, which is most closely related to gp130 (Chua *et al.*, 1994). Another example is a CNTFRα that is anchored to the cell membrane by a glycosyl-

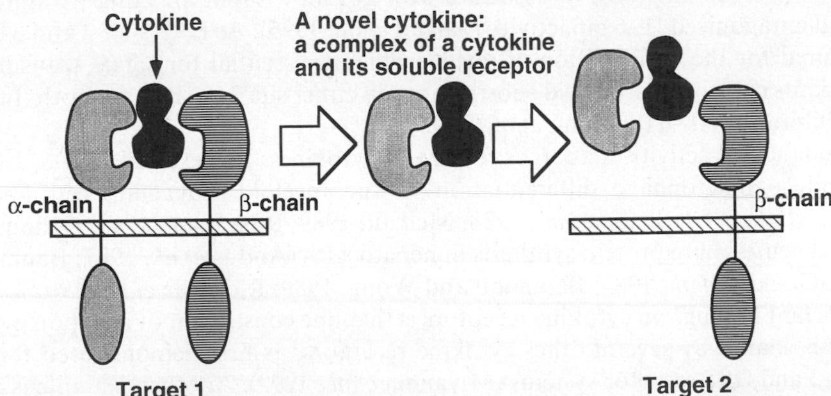

Fig. 6. A novel mechanism generating cytokine diversity. A cytokine acts on cells (target 1) that express a specific receptor. With certain cytokines, such as IL-6 and CNTF, the complex, consisting of the cytokine and its soluble form of receptor subunit, also acts on cells (target 2) that express only a subset of the receptor subunits, on which the original cytokine cannot act. IL-12, identified as a cytokine, is a complex composed of a cytokine and the soluble form of a cytokine receptor.

phosphatidyl inositol (GPI) linkage (Davis *et al.*, 1991). The complex of soluble CNTFRα and CNTF acts on the cells that express LIFRβ and gp130 (Davis *et al.*, 1993). Potential physiological roles for the soluble CNTFRα are suggested by the presence of the soluble form of the α chain in cerebrospinal fluid and its release from skeletal muscle in response to peripheral nerve injury.

Based on these facts, the author originally proposed a novel mechanism by which the cytokine system generates functional diversity (Fig. 6) (Hirano *et al.*, 1994). A complex consisting of a soluble cytokine receptor and its corresponding cytokine acquires a different target specificity from the original cytokine and, therefore, it should express functions distinct from the original cytokine. Actually, double transgenic mice expressing human IL-6 and IL-6Rα showed myocardial hypertrophy (Hirota *et al.*, 1995), indicating that the complex of IL-6 and the soluble form of IL-6Rα acts on heart muscle cells that express gp130, on which IL-6 alone cannot act, leading to the induction of cardiac hypertrophy similar to the effect of cardiotrophin-1 (CT-1).

This model could also be applied to the glial cell line-derived neurotrophic factor (GDNF) receptor system, which consists of a GDNF-specific binding molecule, GDNFRα, a GPI-anchored membrane molecule, and a signal transducing GDNFR, Ret, which is a receptor for tyrosine kinase (Jing *et al.*, 1996; Treanor *et al.*, 1996). This novel mechanism may be applied to a wide range of other receptor systems. The mechanism may contribute to generation of the functional diversity of cytokines and may also play pathological roles in various diseases, as an increase in the serum-soluble form of various cytokine receptors has been reported to occur in a variety of diseases.

SIGNAL TRANSDUCTION THROUGH THE IL-6 RECEPTOR

JAK–STAT Signal Transduction Pathway

As the cytoplasmic domain of most cytokine receptors, including gp130, does not have an intrinsic catalytic domain, one of the hottest issues until 1993 was the identification of catalytic molecules that associate with cytokine receptors. This issue was resolved by the findings that several Janus family tyrosine kinases (JAK1, JAK2, JAK3, Tyk-2) are involved in the signal transduction of cytokines and hormones. Furthermore, the signal transducer and activator of transcription (STAT) plays a central role in a variety of cytokine signal transduction pathways (Darnell *et al.*, 1994; Ihle *et al.*, 1994; Schindler and Darnell, 1995).

JAK1, JAK2 and Tyk-2 associate constitutively with gp130 and are tyrosine phosphorylated in response to IL-6, CNTF, LIF or OSM (Lutticken *et al.*, 1993; Stahl *et al.*, 1993; Matsuda *et al.*, 1994). Furthermore, IL-6 activates STAT3, STAT1 and STAT5 (Akira *et al.*, 1994; Fujitani *et al.*, 1994, 1997; Zhong *et al.*, 1994; Lai *et al.*, 1995; Nakajima *et al.*, 1995). In the absence of JAK1, transcriptional factor STATs are not activated efficiently following stimulation by IL-6, although JAK2 and Tyk-2 are activated, suggesting that there is a hierarchy among gp130-associated JAKs (Guschin *et al.*, 1995).

Two types of IL-6 responsive element (RE) have been identified in the genes encoding acute-phase proteins. Type I IL-6 RE, which is a binding site for NF–IL-6/IL-6DBP/LAP/C/EBPβ (Akira *et al.*, 1990; Poli *et al.*, 1990; Cao *et al.*, 1991; De Groot *et al.*,

1991), is present in the CRP, haemopexin A and haptoglobin genes. The binding activity of NF–IL-6 is probably induced by IL-6 through the increased expression of the NF–IL-6 gene, rather than through its post-translational modification (Baumann *et al.*, 1992, 1993). The type II IL-6 RE is present in the fibrinogen, α_2-macroglobin, α_1-acid glycoprotein and haptoglobin genes. IL-6 triggers the rapid activation of a nuclear factor, termed acute-phase response factor (APRF), which binds to type II IL-6 RE (Wegenka *et al.*, 1993). The purification and molecular cloning of APRF revealed that it is identical with STAT3 (Akira *et al.*, 1994; Zhong *et al.*, 1994).

In parallel with these studies, Nakajima *et al.* (1993) identified the IL-6 responsive element of the *junB* gene (JRE–IL-6), which consists of a putative Ets binding site (JEBS) and CRE-like site. The IL-6-inducible JEBS binding protein mainly contains STAT3, although the JEBS is a low-affinity binding site for STAT3 (Fujitani *et al.*, 1994; Nakajima *et al.*, 1995). IL-6 induces the formation of a complex consisting of STAT3 and p36-CRE-like site binding molecules on the JRE–IL-6 and on the IL-6/IFN-γ RE in the IRF-1 promoter (Kojima *et al.*, 1996). Such binding complex formation seems to be important for STAT to act on a low-affinity binding site, such as the JEBS, and may contribute to generating the diversity of target genes of STAT proteins. In addition to the tyrosine phosphorylation of STAT3 by JAK tyrosine kinase, the H7-sensitive pathway, most likely a serine/threonine kinase, is required for STAT3 to be transcriptionally active on both the JRE-IL-6 and type II IL-6 RE (Nakajima *et al.*, 1993, 1995). In certain cell lines, STAT3 requires phosphorylation on serine to form a STAT3–STAT3 homodimer (Zhang *et al.*, 1995). Maximal activation of transcription by STAT1 and STAT3 requires both tyrosine and serine phosphorylation in line with the involvement of a serine/threonine kinase in the STAT signal pathway (Wen *et al.*, 1995).

Multiple Signal Transduction Pathways Through gp130 Involved in Cell Growth, Differentiation and Survival

Human gp130 has 277-amino-acid residues in its cytoplasmic domain, which contains two motifs conserved among the cytokine receptor family, termed Box1 and Box2 (Fig. 3) (Hibi *et al.*, 1990; Fukunaga *et al.*, 1991; Murakami *et al.*, 1991). The membrane-proximal region containing Box1 and Box2 is sufficient for JAK to be activated through gp130 (Narazaki *et al.*, 1994). Human gp130 has six tyrosine residues in its cytoplasmic domain, and the tyrosine phosphorylation of Src homology 2 protein tyrosine phosphatase-2 (SHP-2) (also called protein tyrosine phosphatase-1D (PTP1-D), SHPTP-2, PTP2C and Syp), a phosphotyrosine phosphatase, and that of STAT3 are dependent on the second tyrosine from the membrane (Y2), and any one of the four tyrosines (Y3, Y4, Y5, Y6) in the carboxy terminus that have a glutamine residue at the third position behind tyrosine (Y–X–X–Q) respectively (Fig. 7) (Stahl *et al.*, 1995; Yamanaka *et al.*, 1996). In accordance with the fact that STAT3 is involved in the activation of type II acute-phase genes, the membrane-proximal region of gp130, containing 133 amino acids and Y3, is necessary for the activation of IL-6 responsive acute-phase genes (Baumann *et al.*, 1994).

IL-6 induces growth arrest and macrophage differentiation in the murine myeloid leukaemic cell lines, M1 and Y6 (Miyaura *et al.*, 1988; Shabo *et al.*, 1988; Oritani *et al.*, 1992). The membrane-proximal region of gp130, consisting of 133 amino acids, is sufficient to generate the signals for growth arrest, macrophage differentiation, down-

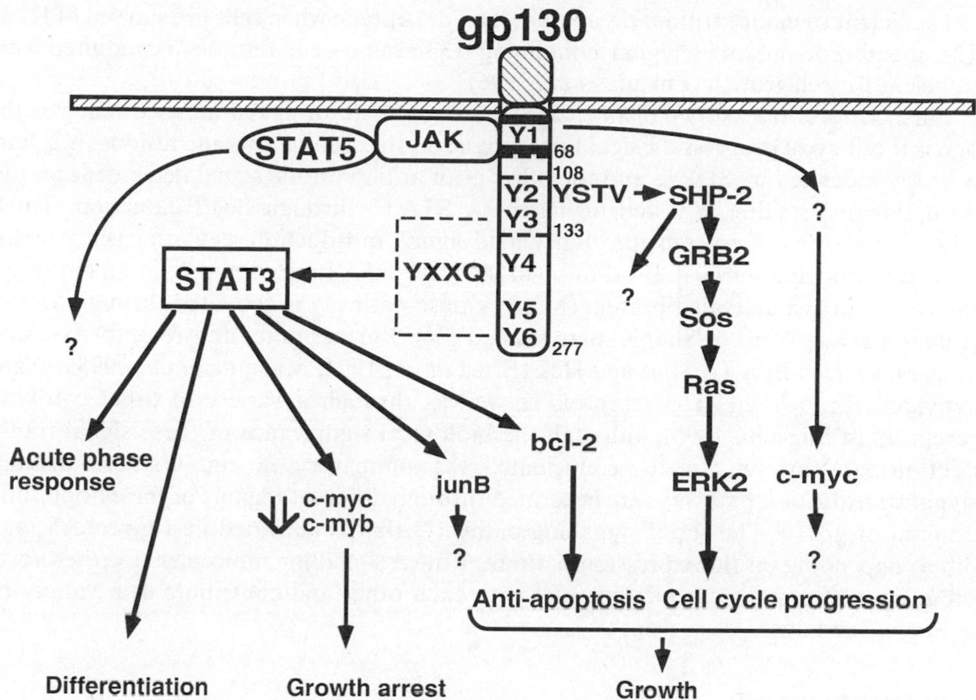

Fig. 7. Multiple signal transduction pathways through gp130.

regulation of c-*myc* and c-*myb*, induction of *junB* and *IRF-1*, and the activation of STAT3 (Yamanaka *et al.*, 1996). The region between amino acids 108 and 133 contains two tyrosine residues (Figs 3 and 7): one (Y3) at amino acid position 126 with the YXXQ motif, and the other one (Y2) without the motif at amino acid position 118. Y2 is essential for gp130-mediated *egr-1* gene induction (Yamanaka *et al.*, 1996), whereas Y3 is critical in generating the signals not only for STAT3 activation but also for growth arrest and differentiation, accompanied by the downregulation of c-*myc* and c-*myb* and the immediate early induction of *junB* and *IRF-1*. These results show the tight correlation of STAT3 activation and the signals for growth arrest and differentiation. In fact, dominant-negative forms of STAT3 inhibit both IL-6-induced growth arrest and macrophage differentiation in the M1 transformants (Nakajima *et al.*, 1996). Blocking the STAT3 activation inhibits IL-6-induced repression of c-*myb* and c-*myc*, but not *egr-1* induction. Furthermore, IL-6 enhances the growth of M1 cells when STAT3 is suppressed. Thus, IL-6 generates both growth-enhancing signals and growth arrest- and differentiation-inducing signals at the same time, but the former is apparent only when STAT3 activation is suppressed. The essential role of STAT3 in the IL-6-induced macrophage differentiation of M1 cells is also shown (Minami *et al.*, 1996).

For the growth signal, it was shown that a 65-amino-acid region proximal to the transmembrane domain is sufficient for the growth response, by using gp130 transfectants of an IL-3-dependent proB-cell line BAF/B03 (Murakami *et al.*, 1991; Kishimoto *et al.*, 1995). However, the membrane-proximal region of 68 amino acids is not sufficient to induce tritium thymidine (^{3}H-Tdr) uptake when cells are starved of IL-3. The membrane-proximal region containing 133-amino-acid residues is required and sufficient for cell growth (Fukada *et al.*, 1996).

Furthermore, at least two distinct signals are required for gp130-induced cell growth: one is a cell cycle progression signal dependent on the second tyrosine residue, Y2, and possibly mediated by SHP-2, and the other is an antiapoptotic signal dependent on the third tyrosine residue, Y3, and mediated by STAT3 through Bcl-2 induction. Thus, STAT3 plays pivotal roles in gp130-mediated signal transduction regulating cell growth, differentiation and survival. In addition to the JAK–STAT signal transduction pathway, the Ras–mitogen-activated protein (MAP) kinase pathway is activated through SHP-2 (Fukada *et al.*, 1996) or Shc (Kumar *et al.*, 1994). Furthermore, non-receptor tyrosine kinases, such as Btk, Tec, Fes and Hck (Ernst *et al.*, 1994; Matsuda *et al.*, 1995a,b) are activated through the IL-6 receptor, as well as through a variety of other cytokine receptors (Taniguchi, 1995), although the biological significance of these signal transduction pathways remains to be elucidated. As summarized in Fig. 7, several distinct signal transduction pathways are generated through different regions of the cytoplasmic domain of gp130. The set of signalling pathways that is activated in a given cell may differ, depending on the expression pattern of these signalling molecules. Furthermore, these signalling pathways may interact with each other and contribute to a variety of biological activities.

IL-6 AND DISEASE

A possible involvement of deregulated expression of the IL-6 gene in polyclonal B-cell abnormalities was first demonstrated in patients with cardiac myxoma (Hirano *et al.*, 1987). Since then, much evidence has been accumulated indicating that deregulated production of IL-6 could be involved in a variety of diseases, including inflammation, autoimmune disease and malignancy. Considering the possible involvement of IL-6 in such diseases, it might be worth noting that IL-6 was identified as virus-induced IFN-β_2 (Weissenbach *et al.*, 1980) or found in the culture supernatants of cells infiltrating the pleural effusion of patients with tuberculous pleurisy (Teranishi *et al.*, 1982), which also contained factor(s) capable of inducing immunoglobulin production in activated human B cells (Hirano *et al.*, 1981).

Interleukin-6 and B-Cell Abnormalities in Chronic Inflammation

Patients with cardiac myxoma show a variety of autoimmune symptoms, such as hypergammaglobulinaemia, presence of autoantibodies and an increase in acute-phase proteins, all of which disappear following resection of the tumour cells. Involvement of IL-6 in B-cell abnormalities was first suggested in patients with cardiac myxoma (Hirano *et al.*, 1987; Jourdan *et al.*, 1990). Before this finding, it was demonstrated that pleural effusion cells of patients with pulmonary tuberculosis, when stimulated with purified protein derivative (PPD), produced a large amount of factors capable of inducing immunoglobulin production in activated normal B cells (Hirano *et al.*, 1981); one of these factors was identified as IL-6 (previously called either thyrotrophin-releasing factor-like factor, BCDF-II or BSF-2) (Teranishi *et al.*, 1982; Hirano *et al.*, 1984a,b, 1985, 1986). It is noteworthy that patients with pulmonary tuberculosis often have a wide range of autoantibodies (Shoenfeld and Isenberg, 1988), and in certain cases a significant diffuse hypergammaglobulinaemia has been reported (Sela *et al.*, 1987). Taken together, all the evidence indicated that overproduction of IL-6 may play a critical role in autoimmune disease (Hirano *et al.*, 1987).

Abnormal IL-6 production was also observed in patients with Castleman's disease (Yoshizaki *et al.*, 1989) and rheumatoid arthritis (Hirano *et al.*, 1988; Houssiau *et al.*, 1988; Bhardwaj *et al.*, 1989). IL-6 production was also seen in type II collagen-induced arthritis in mice (Takai *et al.*, 1989) and MRL/lpr mice (Tang *et al.*, 1991), which develop autoimmune disease with proliferative glomerulonephritis and arthritis. IL-6 was also found to be produced by islet β cells and the thyroid (Bendtzen *et al.*, 1989; Campbell *et al.*, 1989), suggesting that, by enhancing the response of autoreactive T cells, IL-6 may be involved in type I diabetes (Campbell *et al.*, 1990). The evidence suggests that IL-6 plays a role in autoimmune disease, although IL-6 alone may not be sufficient for its generation (Hirano, 1992b). The observation that anti-IL-6 antibody inhibits the development of insulin-dependent diabetes in NOD/Wehi mice may support the role of IL-6 in autoimmune disease (Campbell *et al.*, 1991). Other interesting evidence is that a strikingly increased prevalence of agalactosyl IgG has been observed in a variety of autoimmune and/or IL-6-related diseases, such as pulmonary tuberculosis, rheumatoid arthritis, Crohn's disease, sarcoidosis, leprosy, Castleman's disease, Takayasu's arteritis, multiple myeloma and pristane-induced arthritis (Nakao *et al.*, 1991; Rook *et al.*, 1991; Rook and Stanford, 1992). In accordance with these facts, IL-6 transgenic mice showed an increase in agalactosyl IgG activity (Rook *et al.*, 1991), further strengthening the intimate relationship between IL-6 and certain autoimmune diseases.

Chronic Inflammatory Proliferative Disease

Glomerulonephritis is commonly accompanied by a variety of autoimmune diseases, and several growth factors have been suggested as candidates that induce the pathological growth of mesangial cells. IL-6 is a possible autocrine growth factor for rat mesangial cells (Horii *et al.*, 1989; Ruef *et al.*, 1990). It is produced by renal mesangial cells in patients with mesangial proliferative glomerulonephritis (Horii *et al.*, 1989). IL-6 is detected in urine samples from patients with mesangial proliferative glomerulonephritis, but not from those with other types of glomerulonephritis. There is a correlation between the levels of IL-6 in urine and the progressive stage of mesangial proliferative glomerulonephritis.

Other chronic proliferative diseases that may be related to IL-6 are psoriasis (Grossman *et al.*, 1989) and Kaposi's sarcoma (Miles *et al.*, 1990) in which IL-6 is considered to be one of the growth factors for keratinocytes and Kaposi's sarcoma cells. Because mesangial cell proliferative glomerulonephritis, psoriasis and Kaposi's sarcoma are diseases in which abnormal cell growth and inflammatory and/or immunological reactions are operating, they may be categorized as chronic inflammatory proliferative diseases. From this point of view, rheumatoid arthritis is also considered in this disease category, because one of its prominent features is chronic expansion of synovial cells.

Plasma Cell Neoplasia

IL-6 is a potent growth factor for murine plasmacytoma cells (Aarden *et al.*, 1985; Van Damme *et al.*, 1987a; Van Snick *et al.*, 1988) and human myeloma cells (Kawano *et al.*, 1988), suggesting the possible involvement of IL-6 in the generation of plasmacytoma or myeloma (Hirano, 1991). There is a significant association between the occurrence of plasma cell neoplasias and chronic inflammation (Isobe and Osserman, 1971; Isomaki *et al.*, 1978). Plasmacytomas can be induced in BALB/c mice by mineral oil or pristane, both of which are potent inducers of chronic inflammation and IL-6 biosynthesis (Potter and Boyce, 1962; Nordan and Potter, 1986). The *in vitro* growth of the primary mouse plasmacytoma thus developed was found to be dependent on IL-6 (Namba and Hanaoka, 1972).

IL-6 was also found to be a possible autocrine growth factor for human myeloma cells (Kawano *et al.*, 1988). Freeman *et al.* (1989) demonstrated that myelomas and plasma cell leukaemias express IL-6 mRNA. Cytoplasmic IL-6 was detected in myeloma cells of the bone marrow by light and electron microscopy (Ohtake *et al.*, 1990). It was reported that the growth-inducing activity of IL-1 or TNF on freshly isolated myeloma cells could be due to IL-6-mediated autocrine mechanism (Carter *et al.*, 1990). Rearrangement of the IL-6 gene was reported in certain myeloma cells that expressed the IL-6 gene (Fiedler *et al.*, 1990).

Constitutive IL-6 production in a murine plasmacytoma cell line due to the insertion of an IAP retrotransposon in the IL-6 gene has also been reported (Blankenstein *et al.*, 1990). Expression of the IL-6 gene in an IL-6-dependent murine plasmacytoma cell line caused the cells to proliferate in an autocrine manner (Tohyama *et al.*, 1990; Vink *et al.*, 1990). These cells displayed greatly enhanced tumorigenicity, and monoclonal antibodies capable of blocking the binding of IL-6 to its receptor inhibited their growth *in vivo* (Vink *et al.*, 1990). IL-6 was also demonstrated to be an autocrine growth factor for EBV-transformed B-cell lines (Yokoi *et al.*, 1990) and expression of an exogenous IL-6 gene in these B-cell lines confered growth advantage and *in vivo* tumorigenicity (Scala *et al.*, 1990). However, it has been controversial whether all myeloma cells produce IL-6, because only some myeloma cell lines were found to produce it (Kawano *et al.*, 1988; Klein *et al.*, 1989; Shimizu *et al.*, 1989; Hata *et al.*, 1990) and bone marrow adherent cells rather than bone marrow non-adherent cell populations containing myeloma cells were demonstrated to be major producers of IL-6 (Klein *et al.*, 1989). Evidence indicated that IL-6 plays an important role in the *in vivo* growth of myeloma cells and generation of plasma cell neoplasia in an autocrine or paracrine manner. This was further supported by the following findings. The *in vitro* IL-6 responsiveness of myeloma cells obtained from patients with multiple myeloma correlated directly with the *in vivo* labelling index of these tumours (Zhang *et al.*, 1989) and increased serum IL-6 levels correlated well with disease severity in multiple myeloma and plasma cell leukaemia (Bataille *et al.*, 1989). Finally, administration of anti-IL-6 antibodies suppresses myeloma cell growth (Klein *et al.*, 1991).

However, IL-6 alone is not sufficient for the generation of plasmacytoma. Plasma cells generated in the IL-6 transgenic mice were not transplantable to syngeneic animals, indicating that additional factors may be required for malignant transformation. An interesting finding is that C57BL/6 IL-6 transgenic mice, when backcrossed to BALB/c mice, showed a progression from polyclonal plasmacytosis to fully transformed monoclonal plasmacytoma which contained chromosomal translocation with c-*myc* gene rearrangement (Suematsu *et al.*, 1992). The evidence strongly support the hypothesis that deregulated expression of the IL-6 gene can trigger polyclonal plasmacytosis, resulting in the generation of malignant monoclonal plasmacytoma (Hirano, 1991). In fact, in IL-6-deficient mice pristane cannot induce plasmacytoma (Hilbert *et al.*, 1995).

CONCLUSIONS

IL-6 has been found to play a central role in defence mechanism(s), the immune response, haematopoiesis and acute-phase reactions. On the other hand, deregulated expression of the IL-6 gene has been implicated in the pathogenesis of a variety of diseases, especially autoimmune diseases, plasmacytoma–myeloma and several chronic inflammatory proliferative diseases. Future studies on the regulation of IL-6 gene expression and the mechanisms of IL-6 action through its receptor, and development of inhibitors for IL-6 action, would provide critical information on the molecular mechanisms of a variety of diseases and the development of new therapeutic methods.

REFERENCES

Aarden, L., Lansdorp, P. and De Groot, E. (1985). A growth factor for B cell hybridomas produced by human monocytes. *Lymphokines* **10**, 175–185.

Akira, S., Isshiki, H., Sugita, T., Tanabe, O., Kinoshita, S., Nishio, Y., Nakajima, T., Hirano, T. and Kishimoto, T. (1990). A nuclear factor for IL-6 expression (NF-IL6) is a member of a C/EBP family. *EMBO J.* **9**, 1897–1906.

Akira, S., Nishio, Y., Inoue, M., Wang, X.J., Wei, S., Matsusaka, T., Yoshida, K., Sudo, T., Naruto, M. and Kishimoto, T. (1994). Molecular cloning of ABRF, a novel IFN-stimulated gene factor 3 p91-related transcription factor involved in the gp130-mediated signaling pathway. *Cell* **77**, 63–71.

Ambrosino, D.M., Delaney, N.R. and Shamberger R.C. (1990). Human polysaccharide-specific B cells are responsive to pokeweed mitogen and IL-6. *J. Immunol.* **144**, 1221–1226.

Andus T., Geiger, T., Hirano, T., Northoff, H., Ganter, U., Bauer, J., Kishimoto, T. and Heinrich, P.C. (1987). Recombinant human B cell stimulatory factor 2 (BSF-2/IFN-beta 2) regulates beta-fibrinogen and albumin mRNA levels in Fao-9 cells. *FEBS Lett.* **221**, 18–22.

Asano, S., Okano, A., Ozawa, K., Nakahata, T., Ishibashi, T., Koike, K., Kimura, H., Tanioka, Y., Shibuya, A., Hirano, T., Kishimoto, T., Takaku, F. and Akiyama, T. (1990). *In vivo* effects of recombinant human interleukin-6 in primates: stimulated production in platelets. *Blood* **75**, 1602–1605.

Bataille, R., Jourdan, M., Zhang, X.-G. and Klein, B. (1989). Serum levels of interleukin-6, a potent myeloma cell growth factor, as a reflection of disease severity in plasma cell dyscrasias. *J. Clin. Invest.* **84**, 2008–2011.

Baumann, H. and Wong, G.G. (1989). Hepatocyte-stimulatory factor III shares structural and functional identity with leukemia inhibitory factor. *J. Immunol.* **143**, 1163–1167.

Baumann, H., Onorato, V., Gauldie, J. and Jahreis, G.P. (1987). Distinct sets of acute phase plasma proteins are stimulated by separate human hepatocyte-stimulating factors and monokines in rat hepatoma cells. *J. Biol. Chem.* **262**, 9756–9768.

Baumann, H., Morella, K.K., Campos, S.P., Cao, Z. and Jahreis, G.P. (1992). Role of CAAT-enhancer binding protein isoforms in the cytokine regulation of acute-phase plasma protein genes. *J. Biol. Chem.* **267**, 19744–19751.

Baumann, H., Ziegler, S.F., Mosley, B., Morella, K.K., Pajovie, S. and Gearing, D.P. (1993). Reconstitution of the response to leukemia inhibitory factor, oncostatin M, and ciliary neurotrophic factor in hepatoma cells. *J. Biol. Chem.* **268**, 8414–8417.

Baumann, H., Symes, A.J., Comeau, M.R., Morella, K.K., Wang, Y., Friend, D., Ziegler, S.F., Fink, J.S. and Gearing, D.P. (1994). Multiple regions within the cytoplasmic domains of the leukemia inhibitory factor receptor and gp130 cooperate in signal transduction in hepatic and neuronal cells. *Mol. Cell. Biol.* **14**, 138–146.

Bazan, J.F. (1990). Haemopoietic receptors and helical cytokines. *Immunol. Today* **11**, 350–354.

Bazan, J.F. (1992). Neurotropic cytokines in the hematopoietic fold. *Neuron* **7**, 1–12.

Beagley, K.W., Eldridge, J.H., Lee, F., Kiyono, H., Everson, M.P., Koopman, W.J., Hirano, T., Kishimoto, T. and McGhee, J.R. (1989). Interleukins and IgA synthesis. Human and murine interleukin 6 induce high rate IgA secretion in IgA-committed B cells. *J. Exp. Med.* **169**, 2133–2148.

Bendtzen, K., Buschard, K., Diament, M., Horn, T. and Svenson, M. (1989). Possible role of IL-1, TNF-α, and IL-6 in insulin-dependent diabetes mellitus and autoimmune thyroid disease. *Lymphokine Res.* **8**, 335–340.

Bernad, A., Kopf, M., Kulbacki, R., Weich, N., Koehler, G. and Gutierrez-Ramos, J.C. (1994). Interleukin-6 is required *in vivo* for the regulation of stem cells and committed progenitors of the hematopoietic system. *Immunity* **1**, 725–731.

Bhardwaj, N., Santhanam, U., Lau, L.L., Tatter, S.B., Ghrayeb, J., Rivelis, M., Steinman, R.M., Sehgal, P.B. and May, L.T. (1989). IL-6/IFN-β 2 in synovial effusions of patients with rheumatoid arthritis and other arthritides. Identification of several isoforms and studies of cellular sources. *J. Immunol.* **143**, 2153–2159.

Blankenstein, T., Qin, Z., Li, W. and Diamanstein, T. (1990). DNA rearrangement and constitutive expression of the interleukin 6 gene in a mouse plasmacytoma. *J. Exp. Med.* **171**, 965–970.

Bodine, D.M., Karlsson, S. and Nienhuis, A.W. (1989). Combination of interleukins 3 and 6 preserves stem cell function in culture and enhances retrovirus-mediated gene transfer into hematopoietic stem cells. *Proc. Natl Acad. Sci. U.S.A.* **86**, 8897–8901.

Bot, F.J., Van Eijk, L., Broeders, L., Aarden, L.A. and Lowenber, B. (1989). Interleukin-6 synergizes with M-CSF in the formation of macrophage colonies from purified human marrow progenitor cells. *Blood* **73**, 435–437.

Bowcock, A.M., Kidd, J.R., Lathrop, M., Danshvar, L., May, L.T., Ray, A., Sehgal, P.B., Kidd, K.K. and Cavallisforza, L.L. (1988). The human 'interferon-β 2/hepatocyte stimulating factor/interleukin-6' gene: DNA polymorphism studies and localization to chromosome 7p21. *Genomics* **3**, 8–16.

Brakenhoff, J.P., Hart, M. and Aarden, L.A. (1989). Analysis of human IL-6 mutants expressed in *Escherichia coli.* Biologic activities are not affected by deletion of amino acids 1–28. *J. Immunol.* **143**, 1175–1182.

Brakenhoff, J.P., De Hon, F.D., Fontaine, V., Ten Boekel, E., Schooltink, H., Rose-John, S., Heinrich, P.C., Content, J. and Aarden, L.A. (1994). Development of a human interleukin-6 receptor antagonist. *J. Biol. Chem.* **269**, 86–93.

Brandt, S.J., Bodine, D.M., Dunbar, C.E. and Nienhuis, A. (1990). Dysregulated interleukin 6 expression produces a syndrome resembling Castleman's disease in mice. *J. Clin. Invest.* **86**, 592–599.

Breen, E.C., Rezai, A.R., Nakajima, K., Beall, G.N., Mitsuyasu, R.T., Hirano, T., Kishimoto, T. and Martinez-Maza, O. (1990). Infection with HIV is associated with elevated IL-6 levels and production. *J. Immunol.* **144**, 480–484.

Campbell, I.L. and Harrison, C. (1990). A new view of the β cell as an antigen-presenting cell and immunogenic target. *J. Autoimmun.* **3**, 53–62.

Campbell, I.L., Cutri, A., Wilson, A. and Harrison, L.C. (1989). Evidence for IL-6 production by and effects on the pancreatic β-cell. *J. Immunol.* **143**, 1188–1191.

Campbell, I.L., Kay, T.W., Oxbrow, L. and Harrison, L.C. (1991). Essential role for interferon-γ and interleukin-6 in autoimmune insulin-dependent diabetes in NOD/wehi mice. *J. Clin. Invest.* **87**, 739–742.

Cao, Z., Umkek, R.M. and McKnight, S.L. (1991). Regulated expression of three C/EBP isoforms during adipose conversion of 3T3–L1 cells. *Genes Dev.* **5**, 1538–1552.

Caracciolo, D., Clark, S.C. and Rovera, G. (1989). Human interleukin-6 supports granulocytic differentiation of hematopoietic progenitor cells and acts synergistically with GM-CSF. *Blood* **73**, 666–670.

Carrington, P.A., Hill, R.J., Stenberg, P.E., Levin, J., Corash, L., Schreurs, J., Baker, G. and Levin, F.C. (1991). Multiple *in vivo* effects of interleukin-3 and interleukin-6 on murine megakaryocytopoiesis. *Blood* **77**, 34–41.

Carter, A., Merchav, S., Silvian Draxler, I. and Tatarsky, I. (1990). The role of interleukin-1 and tumour necrosis factor-α in human multiple myeloma. *Br. J. Haematol.* **74**, 424–431.

Castell, J.V., Gomez-Lechon, M.J., David, M., Hirano, T., Kishimoto, T. and Heinrich, P.C. (1988). Recombinant human interleukin-6 (IL-6/BSF-2/HSF) regulates the synthesis of acute phase proteins in human hepatocytes. *FEBS Lett.* **232**, 347–350.

Chai, Z., Gatti, S., Toniatti, C., Poli, V. and Barfai, T. (1996). Interleukin (IL)-6 gene expression in the central nervous system is necessary for fever response to lipopolysaccharide or IL-1β: a study on IL-6-deficient mice. *J. Exp. Med.* **183**, 311–316.

Chua, A.O., Chizzonite, R., Desai, B.B., Truitt, T.P., Nunes, P., Minetti, L.J., Warrier, R.R., Presky, D.H., Levine, J.F., Gately, M.K. and Gubler, U. (1994). Expression cloning of a human IL-12 receptor component. A new member of the cytokine receptor superfamily with strong homology to gp120. *J. Immunol.* **153**, 128–136.

Content, J., De Wit, L., Poupart, P., Opdenakker, G., Van Damme, J. and Billiau, A. (1985). Induction of a 26-kDa-protein mRNA in human cells treated with an interleukin-1-related, leukocyte-derived factor. *Eur. J. Biochem.* **152**, 253–257.

Cressman, D.E., Greenbaum, L.E., DeAngelis, R.A., Ciliberto, G., Furth, E.E., Poli, V. and Taub, R. (1996). Liver failure and defective hepatocyte regeneration in interleukin-6-deficient mice. *Science* **274**, 1379–1383.

Dalrymple, S.A., Lucian, L.A., Slattery, R., McNeil, T., Aud, D.M., Fuchino, S., Lee, F. and Murray, R. (1995). Interleukin-6-deficient mice are highly susceptible to *Listeria monocytogenes* infection: correlation with inefficient neutrophilia. *Infect. Immun.* **63**, 2262–2268.

Dalrymple, S.A., Slattery, R., Aud, D.M., Krishna, M., Lucian, L.A. and Murray, R. (1996). Interleukin-6 is required for a protective immune response to systemic *Escherichia coli* infection. *Infect. Immun.* **64**, 3231–3235.

Darnell, J.E., Kerr, I.M. and Stark, G.M. (1994). Jak–STAT pathways and transcriptional activation in response to IFNs and other extracellular signaling proteins. *Science* **257**, 803–806.

Davis, S., Aldrich, T.H., Valenzula, D.M., Wong, V., Furth, M.E., Squinto, S.P. and Yancopoulos, G.D. (1991). The receptor for ciliary neurotrophic factor. *Science* **253**, 59–63.

Davis, S, Aldrich, T.H., Ip, N.Y., Stahl, N., Scherer, S., Farruggella, T., DiStefano, P.S., Curtis, R., Panayotatos, N., Gascan, H., Chevalier, S. and Yancopoulos, G.D. (1993). Released form of CNTF receptor a components as a soluble mediator of CNTF responses. *Science* **259**, 1736–1739.

De Groot, R.P., Auwerx, J., Karperien, M., Staels, B. and Kruijer, W. (1991). Activation of junB by PKC and PKA signal transduction through a novel *cis*-acting element. *Nucleic Acids Res.* **19**, 775–781.

Dendorfer, U., Oettgen, P. and Libermann, T.A. (1994). Multiple regulatory elements in the interleukin-6 gene mediate induction by prostaglandins, cyclic AMP, and lipopolysaccharide. *Mol. Cell. Biol.* **14**, 4443–4454.

Emilie, D., Peuchmaur, M., Malillot, M.C., Crevon, M.C., Brousse, N., Delfraissy, J.F., Dormont, J. and Galanaud, P. (1990). Production of interleukins in human immunodeficiency virus-1-replicating lymph nodes. *J. Clin. Invest.* **86**, 148–159.

Ernst, M., Gearing, D.P. and Dunn, A.R. (1994). Functional and biochemical association of Hck with the LIF/IL-6 receptor signal transducing subunit gp130 in embryonic stem cells. *EMBO J.* **13**, 1574–1584.

Espevik, T., Waage, A., Faxvaag, A. and Shalaby, M.R. (1990). Regulation of interleukln-2 and interleukin-6 production from T-cells: involvement of interleukin-1β and transforming growth factor-β. *Cell. Immunol.* **126**, 47–56.

Fattori, E., Cappelletti, M., Costa, P., Sellitto, C., Cantoni, L., Carelli, M., Faggioni, R., Fantuzzi, G., Ghezzi, P. and Poli, V. (1994). Defective inflammatory response in interleukin 6-deficient mice. *J. Exp. Med.* **180**, 1243–1250.

Fiedler, W., Weh, H.J., Suciu, E., Wittlief, C., Stocking, C. and Hossfeld, D.K. (1990). The IL-6 gene but not the IL-6 receptor gene is occasionally rearranged in patients with multiple myeloma. *Leukemia* **4**, 462–465.

Freeman, G.J., Freedman, A.S., Rabinowe, S.N., Segil, J.M., Horowitz, J., Rosen, K., Whitman, J.F. and Nadler, L.M. (1989). Interleukin 6 gene expression in normal and neoplastic B cells. *J. Clin. Invest.* **83**, 1512–1518.

Frei, K., Leist, T.P., Meager, A., Gallo, P., Leppert, D., Zinkernagel, R.M. and Fontana, A. (1988). Production of B cell stimulatory factor-2 and interferon-γ in the central nervous system during viral meningitis and encephalitis. Evaluation in a murine model infection and in patients. *J. Exp. Med.* **168**, 449–453.

Frei, K., Malipiero, U.V., Leist, T.P., Zinkernagel, R.M., Schwab, M.E. and Fontana, A. (1989). On the cellular source and function of interleukin 6 produced in the central nervous system in viral diseases. *Eur. J. Immunol.* **19**, 689–694.

Fukada, T., Hibi, M., Yamanaka, Y., Takahashi-Tezuka, M., Fujitani, Y., Yamaguchi, T., Nakajima, K. and Hirano, T. (1996). Two signals are necessary for cell proliferation induced by a cytokine receptor gp130: involvement of STAT3 in anti-apoptosis. *Immunity* **5**, 449–460.

Fukata, J., Usui, T., Naitoh, Y., Nakai, Y. and Imura, H. (1989). Effects of recombinant human interleukin-1α, -1β, 2 and 6 on ACTH synthesis and release in the mouse pituitary tumor cell line AT-20. *J. Endocrinol.* **122**, 33–39.

Fukunaga, R., Ishizaka Ikeda, E., Pan, C.X., Seto, Y. and Nagata, S. (1991). Functional domains of the granulocyte colony-stimulating factor receptor. *EMBO J.* **10**, 2855–2865.

Fujihashi, K., McGhee, J.R., Lue, C., Beagley, K.W., Taga, T., Hirano, T., Kishimoto, T., Mestecky, J. and Kiyono, H. (1991). Human appendix B cells naturally express receptors for and respond to interleukin 6 with selective IgA_1 and IgA_2 synthesis. *J. Clin. Invest.* **88**, 248–252.

Fujihashi, K., Kono, Y. and Kiyono, H. (1992). Effects of IL6 on B cells in mucosal immune response and inflammation. *Res. Immunol.* **143**, 744–749.

Fujitani, Y., Nakajima, K., Kojima, H., Nakae, K., Takeda, T. and Hirano, T. (1994). Transcriptional activation of the IL-6 response element in the junB promoter is mediated by multiple Stat family proteins. *Biochem. Biophys. Res. Commun.* **202**, 1181–1187.

Fujitani, Y., Hibi, M., Fukada, T., Takahashi Tezuka, M., Yoshida, H., Yamaguchi, T., Sugiyama, K., Yamanaka, Y., Nakajima, K. and Hirano, T. (1997). An alternative pathway for STAT activation that is mediated by the direct interaction between JAK and STAT. *Oncogene* **14**, 751–761.

Garman, R.D., Jacobs, K.A., Clark, S.C. and Raulet, D.H. (1987). B-cell-stimulatory factor 2 (β_2 interferon) functions as a second signal for interleukin 2 production by mature murine T cells. *Proc. Natl Acad. Sci. U.S.A.* **84**, 7629–7633.

Gauldie, J., Richards, C., Harnish, D., Lansdorp, P. and Baumann, H. (1987). Interferon b2/B-cell stimulatory factor type 2 shares identity with monocyte-derived hepatocyte-stimulating factor and regulates the major acute phase protein response in liver cells. *Proc. Natl Acad. Sci. U.S.A.* **84**, 7251–7255.

Gauldie, J., Richards, C. and Baumann, H. (1992). IL6 and the acute phase reaction. *Res. Immunol.* **143**, 755–759.

Gearing, D., Gough, N.M., King, J.A., Hilton, D.J., Nicola, N.A., Simpson, R.J., Nice, E.C., Kelso, A. and Metcalf, D. (1987). Molecular cloning and expression of cDNA encoding a murine myeloid leukaemia inhibitory factor (LIF). *EMBO J.* **6**, 3995–4002.

Gearing, D.P., Thut, C.J., VandenBos, T., Gimpel, S.D., Delaney, P.B., King, J., Price, V., Cosman, D. and Beckman, M.P. (1991). Leukemia inhibitory factor receptor is structurally related to the IL-6 signal transducer, gp130. *EMBO J.* **10**, 2839–2848.

Geiger, T., Andus, T., Klapproth, J., Hirano, T., Kishimoto, T. and Heinrich, P.C. (1988). Induction of rat acute-phase proteins by interleukin 6 *in vivo*. *Eur. J. Immunol.* **18**, 717–721.

Geissler, K., Valent, P., Bettelheim, P., Sillaber, C., Wagner, B., Kyrle, P., Hinterberger, W., Lechner, K., Liehl, E. and Mayer, P. (1992). *In vivo* synergism of recombinant human interleukin-3 and recombinant human interleukin-6 on thrombopoiesis in primates. *Blood* **79**, 1155–1160.

Gibbons, R., Martinez, O., Matli, M., Heinzel, F., Bernstein, M. and Warren, R. (1990). Recombinant IL-4 inhibits IL-6 synthesis by adherent peripheral blood cells *in vitro*. *Lymphokine Res.* **9**, 283–293.

Girasole, G., Jilka, R.L., Passeri, G., Boswell, S., Boder, G., Williams, D.C. and Manolagas, S.C. (1992). 17β-Estradiol inhibits interleukin-6 production by bone marrow-derived stromal cells and osteoblasts *in vitro*: a potential mechanism for the antiosteoporotic effect of estrogens. *J. Clin. Invest.* **89**, 883–891.

Grossman, R.M., Krueger, J., Yourish, D., Granelli-Piperno, A., Murphy, D.P., May, L.T., Kupper, T.S., Sehgal, P.B. and Gottlieb, A.B. (1989). Interleukin 6 is expressed in high levels in psoriatic skin and stimulates proliferation of cultured human keratinocytes. *Proc. Natl Acad. Sci. U.S.A.* **86**, 6367–6371.

Grunfeld, C., Adi, S., Soued, M., Moser, A., Fiers, W. and Feingold, K.R. (1990). Search for mediators of the lipogenic effects of tumor necrosis factor: potential role for interleukin 6. *Cancer Res.* **50**, 4233–4238.

Guschin, D., Rogers, N., Briscoe, J., Witthuhn, B.A., Wathing, D., Horn, F., Pellegrini, S., Yasukawa, K., Heinrich, P., Stark, G.R., Ihle, J.N. and Kerr, I.M. (1995). A major role for the protein kinase JAK1 in the JAK/STAT signal transduction pathway in response to the interleukin-6. *EMBO J.* **14**, 1421–1429.

Haegeman, G., Content, J. Volckaert, G., Derynck, R., Tavernier, J. and Fiers, W. (1986). Structural analysis of the sequence encoding for an inducible 26-kDa protein in human fibroblasts. *Eur. J. Biochem.* **159**, 625–632.

Hama, T., Myamoto, M., Tsukui, H., Nishio, C. and Hatanaka, H. (1989). Interleukin-6 as a neurotrophic factor for promoting the survival of cultured basal forebrain cholinergic neurons from postnatal rats. *Neurosci. Lett.* **104**, 340–344.

Hata, H., Matsuzaki, H. and Takatsuki, K. (1990). Autocrine growth by two cytokines, interleukin-6 and tumor necrosis factor α, in the myeloma cell line KHM-1A. *Acta Haematol.* **83**, 133–136.

Hegyi, E., Navarro, S., Debili, N., Mouthon, M.A., Katz, A., Breton-Gorius, J. and Vainchenker, W. (1990). Regulation of human megakaryocytopoiesis: analysis of proliferation, ploidy and maturation in liquid cultures. *Int. J. Cell Cloning* **8**, 236–244.

Heinrich, P.C., Castell, J.V. and Andus, T. (1990). Interleukin-6 and the acute phase response. *Biochem. J.* **265**, 621–636.

Helfgott, D.C., May, L.T., Sthoeger, Z., Tamm, I. and Sehgal, P.B. (1987). Bacterial lipopolysaccharide (endotoxin) enhances expression and secretion of β_2 interferon by human fibroblasts. *J. Exp. Med.* **166**, 1300–1309.

Helle, M., Brakenhoff, J.P.J., De Groot, E.R. and Aarden, L.A. (1988). Interleukin 6 is involved in interleukin 1-induced activities. *Eur. J. Immunol.* **18**, 957–959.

Helle, M., Boeije, L. and Aarden, L.A. (1989). IL-6 is an intermediate in IL-1-induced thymocyte proliferation. *J. Immunol.* **142**, 4335–4338.

Hess, S., Rensing-Ehl, A., Schwabe, R., Bufler, P. and Engelmann, H. (1995). CD40 function in nonhematopoietic cells. Nuclear factor κB mobilization and induction of IL-6 production. *J. Immunol.* **155**, 4588–4595.

Hibi, M., Murakami, M., Saito, M., Hirano, T., Taga, T. and Kishimoto, T. (1990). Molecular cloning and expression of an IL-6 signal transducer, gp130. *Cell* **63**, 1149–1157.

Hibi, M., Nakajima, K. and Hirano, T. (1996). IL-6 cytokine family and signal transduction: a model of the cytokine system. *J. Mol. Med.* **74**, 1–12.

Hilbert, D.M., Cancro, M.P., Scherle, P.A., Nordan, R.P., Van-Snick, J., Gerhard, W. and Rudikoff, S. (1989). T cell derived IL-6 is differentially required for antigen-specific antibody secretion by primary and secondary B cells. *J. Immunol.* **143**, 4019–4024.

Hilbert, D.M., Kopf, M., Mock, B.A., Kohler, G. and Rudikoff, S. (1995). Interleukin 6 is essential for *in vivo* development of B lineage neoplasms. *J. Exp. Med.* **182**, 243–248.

Hirano, T. (1991). Interleukin 6 (IL-6) and its receptor: their role in plasma cell neoplasias. *Int. J. Cell Cloning* **9**, 166–184.

Hirano, T. (1992a). The biology of interleukin-6. *Chem. Immunol.* **51**, 153–180.

Hirano, T. (1992b). Interleukin 6 and autoimmunity and plasma cell neoplasias. *Res. Immunol.* **143**, 759–763.

Hirano, T. and Kishimoto, T. (1990). Interleukin 6. In *Handbook of Experimental Pharmacology, Peptide Growth Factors and Their Receptors* (eds M.B. Sporn and A.B. Roberts), Springer, Berlin, Vol. 95/I, pp. 633–665.

Hirano, T. and Kishimoto, T. (organizers) (1992). In the 46th Forum in Immunology 'Molecular Biology and Immunology of Interleukin-6. *Res. Immunol.* **143**, 723–783.

Hirano, T., Teranishi, T., Toba, T., Sakaguchi, N., Fukukawa, T. and Tsuyuguchi, I. (1981). Human helper T cell factor(s) (ThF). I. Partial purification and characterization. *J. Immunol.* **126**, 517–522.

Hirano, T., Teranishi, T., Lin, B.H. and Onoue, K. (1984a). Human helper T cell factor(s). IV. Demonstration of a human late-acting B cell differentiation factor acting on *Staphylococcus aureus* Cowan I-stimulated B cells. *J. Immunol.* **133**, 798–802.

Hirano, T., Teranishi, T. and Onoue, K. (1984b). Human helper T cell factor(s). III. Characterization of B cell differentiation factor I (BDCF I). *J. Immunol.* **132**, 229–234.

Hirano, T., Taga, T., Nakano, N., Yasukawa, K., Kashiwamura, S., Shimizu, K., Nakajima, K., Pyun, K.H. and Kishimoto, T. (1985). Purification to homogeneity and characterization of human B-cell differentiation factor (BDCF or BSFp-2). *Proc. Natl Acad. Sci. U.S.A.* **82**, 5490–5494.

Hirano, T., Yasukawa, K., Harada, H., Taga, T., Watanabe, Y., Matsuda, T., Kashiwamura, S., Nakajima, K., Koyama, K., Iwamatsu, A., Tsunasawa, S., Sakiyama, F., Matsui, H., Takahara, Y., Taniguchi, T. and Kishimoto, T. (1986). Complementary DNA for a novel human interleukin (BSF-2) that induces B lymphocytes to produce immunoglobulin. *Nature* **324**, 73–76.

Hirano, T., Taga, T., Yasukawa, K., Nakajima, K., Nakano, N., Takatsuki, F., Shimizu, M., Murashima, A., Tsunasawa, S., Sakiyama, F. and Kishimoto, T. (1987). Human B cell differentiation factor defined by an anti-peptide antibody and its possible role in autoantibody production. *Proc. Natl Acad. Sci. U.S.A.* **84**, 228–231.

Hirano, T., Matsuda, T., Turner, M., Miyasaka, N., Buchan, G., Tang, B., Sato, K., Shimizu, M., Maini, R., Feldman, M. and Kishimoto, T. (1988). Excessive production of interleukin 6/B cell stimulatory factor-2 in rheumatoid arthritis. *Eur. J. Immunol.* **18**, 1797–1801.

Hirano, T., Matsuda, T. and Nakajima, K. (1994). Signal transduction through gp130 that is shared among the receptors for the interleukin 6 related cytokine subfamily. *Stem Cells* **12**, 262–277.

Hirota, H., Yoshida, K., Kishimoto, T. and Taga, T. (1995). Continuous activation of gp130, a signal transducing receptor component for interleukin 6-related cytokines, causes myocardial hypertrophy in mice. *Proc. Natl Acad. Sci. U.S.A.* **92**, 4862–4866.

Hodgkin, P.D., Bond, M.W., O'Garra, A., Frank, G., Lee, F., Coffman, R.L., Zlotnik, A. and Howard, M. (1988). Identification of IL-6 as a T cell-derived factor that enhances the proliferative response of thymocytes to IL-4 and phorbol myristate acetate. *J. Immunol.* **141**, 151–157.

Holsti, M.A. and Raulet, D.H. (1989). IL-6 and IL-1 synergize to stimulate IL-2 production and proliferation of peripheral T cells. *J. Immunol.* **143**, 2514–2519.

Horii, Y., Muraguchi, A., Suematu, S., Matsuda, T., Yoshizaki, K., Hirano, T. and Kishimoto, T. (1988). Regulation of BSF-2/IL-6 production by human mononuclear cells. Macrophage-dependent synthesis of BSF-2/IL-6 by T cells. *J. Immunol.* **141**, 1529–1535.

Horii, Y., Muraguchi, A., Iwano, M., Matsuda, T., Hirayama, T., Yamada, H., Fujii, Y., Dohi, K., Ishikawa, H.,

Ohmoto, Y., Yoshizaki, K., Hirano, T. and Kishimoto, T. (1989). Involvement of IL-6 in mesangial proliferative glomerulonephritis. *J. Immunol.* **143**, 3949–3955.

Houssiau, F. and Van Snick, J. (1992). IL6 and the T-cell response. *Res. Immunol.* **143**, 740–743.

Houssiau, F.A., Coulie, P.G., Olive, D. and Van Snick, J. (1988). Synergistic activation of human T cells by interleukin 1 and interleukin 6. *Eur. J. Immunol.* **18**, 653–656.

Houssiau, F.A., Coulie, P.G. and Van Snick, J. (1989). Distinct roles of IL-1 and IL-6 in human T cell activation. *J. Immunol.* **143**, 2520–2524.

Ida, N., Sakurai, S., Hosaka, T., Hosoi, K., Kunitomo, T., Shimazu, T., Maruyama, Y. and Kahase, M. (1989). Establishment of strongly neutralizing monoclonal antibody to human interleukin-6 and its epitope analysis. *Biochem. Biophys. Res. Commun.* **165**, 728–734.

Ihara, S., Iwamatsu, A., Fujiyoshi, T., Komi, A., Yamori, T. and Fukui, Y. (1996). Identification of interleukin-6 as a factor that induces neurite outgrowth by PC12 cells primed with NGF. *J. Biochem.* **120**, 865–868.

Ihle, J.N., Witthuhn, B.A., Quelle, F.W., Yamamoto, K., Thierfelder, W.E., Kreider, B. and Silvennoinen, O. (1994). Signaling by the cytokine receptor superfamily: JAKs and STATs. *Trends Biochem. Sci.* **19**, 222–227.

Ikebuchi, K., Wong, G.G., Clark, S.C., Ihle, J.N., Hirai, Y. and Ogawa, M. (1987). Interleukin 6 enhancement of interleukin 3-dependent proliferation of multipotential hemopoietic progenitors. *Proc. Natl Acad. Sci. U.S.A.* **84**, 9035–9039.

Ishibashi, T., Kimura, H., Uchida, T., Kariyone, S., Friese, P. and Burstein, S.A. (1989a). Human interleukin 6 is a direct promoter of maturation of megakaryocytes *in vitro*. *Proc. Natl Acad. Sci. U.S.A.* **86**, 5953–5957.

Ishibashi, T., Kimura, H., Shikama, Y., Uchida, T., Kariyone, S., Hirano, T., Kishimoto, T., Takasuki, F. and Akiyama, Y. (1989b). Interleukin-6 is a potent thrombopoietic factor *in vivo* in mice. *Blood* **74**, 1241–1244.

Isobe, T. and Osserman, E.F. (1971). Pathologic conditions associated with plasma cell dyscrasias: a study of 806 cases. *Ann. N.Y. Acad. Sci.* **190**, 507–517.

Isomaki, H.A., Hakulinen, T. and Joutsenlahti, U. (1978). Excess risk of lymphomas, leukemia and myeloma in patients with rheumatoid arthritis. *J. Chronic Dis.* **31**, 691–696.

Isshiki, H., Akira, S., Tanabe, O., Nakajima, T., Shimamoto, T., Hirano, T. and Kishimoto, T. (1990). Constitutive and interleukin-1 (IL-1)-inducible factors interact with the IL-1-responsive element in the IL-6 gene. *Mol. Cell. Biol.* **10**, 2757–2764.

Jilka, R.L., Hangoc, G., Girasole, G., Passeri, G., Williams, D.C., Abrams, J.S., Boyce, B., Broxmeyer, H. and Manolagas, S.C. (1992). Increased osteoclast development after estrogen loss: mediation by interleukin-6. *Science* **257**, 88–91.

Jing, S., Wen, D., Yu, Y., Holst, P.L., Luo, Y., Fang, M., Tamir, R., Antonio, L., Hu, Z., Cupples, R., Louis, J.-C., Hu, S., Altrock, W.B. and Fox, G.M. (1996). GDNF-induced activation of the ret protein tyrosine kinase is mediated by GDNFR-α, a novel receptor for GDNF. *Cell* **85**, 1113–1124.

Jourdan, M., Bataille, R., Seguin, J., Zhang, X.G., Chaptal, P.A. and Klein, B. (1990). Constitutive production of interleukin-6 and immunologic features in cardiac myxomas. *Arthritis Rheum.* **33**, 398–402.

Kameda, T., Matsuzaki, N., Sawai, K., Okada, T., Saji, F., Matuda, T., Hirano, T., Kishimoto, T. and Tanizawa, O. (1990). Production of interleukin-6 by normal human trophoblast. *Placenta* **11**, 205–213.

Kasid, A., Director, E.P. and Rosenberg, S.A. (1989). Induction of endogenous cytokine-mRNA in circulating peripheral blood mononuclear cells by IL-2 administration to cancer patients. *J. Immunol.* **143**, 736–739.

Kawakami, K., Kakimoto, K., Shinbori, T. and Onoue, K. (1989). Signal delivery by physical interaction and soluble factors from accessory cells in the induction of receptor-mediated T-cell proliferation. Synergistic effect of BSF-2/IL-6 and IL-1. *Immunology* **67**, 314–320.

Kawano, M., Hirano, T., Matsuda, T., Taga, T., Horii, Y., Iwato, K., Asaoku, H., Tang, B., Tanabe, O., Tanaka, H., Kuramoto, A. and Kishimoto, T. (1988). Autocrine generation and requirement of BSF-2/IL-6 for human multiple myelomas. *Nature* **332**, 83–85.

Kikutani, H., Taga, T., Akira, S., Kishi, H., Miki, Y., Saiki, O., Yamamura, Y. and Kishimoto, T. (1985). Effect of B cell differentiation factor (BCDF) on biosynthesis and secretion of immunoglobulin molecules in human B cell lines. *J Immunol.* **134**, 990–995.

Kishimoto, T. and Hirano, T. (1988). Molecular regulation of B lymphocyte response. *Ann. Rev. Immunol.* **6**, 485–512.

Kishimoto, T., Akira, S., Narazaki, M. and Taga, T. (1995). Interleukin-6 family of cytokines and gp130. *Blood* **86**, 1243–1254.

Kitajima, I., Shinohara, T., Bilakovic, J., Brown, D.A., Xu, X. and Nerenberg, M. (1992). Ablation of transplanted HTLV-I Tax-transformed tumors in mice by antisense inhibition of NF-κ B. *Science* **258**, 1792–1794.

Klein, B., Zhang, X.G., Jourdan, M., Content, J., Houssiau, F., Aarden, L., Piechaczyk, M. and Bataille, R.

(1989). Paracrine rather than autocrine regulation of myeloma-cell growth and differentiation by interleukin-6. *Blood* **73**, 517–526.

Klein, B., Wijdenes, J., Zhang, X.G., Jourdan, M., Boiron, J.M., Brochier, J., Liautard, J., Merlin, M., Clement, C., Morel-Fournier, B., Lu, Z.Y., Mannoni, P., Sany, J. and Bataille, R. (1991). Murine anti-interleukin-6 monoclonal antibody therapy for a patient with plasma cell leukemia. *Blood* **78**, 1198–1204.

Kohase, M., May, L.T., Tamm, I., Vilcek, J. and Sehgal, P.B. (1987). A cytokine network in human diploid fibroblasts: interactions of β-interferons, tumor necrosis factor, platelet-derived growth factor, and interleukin-1. *Mol. Cell. Biol.* **7**, 273–280.

Kojima, H., Nakajima, J. and Hirano, T. (1996). IL-6-inducible complexes on an IL-6 response element of the junB promoter contain Stat3 and 36 kDa CRE-like site binding protein(s). *Oncogene* **12**, 547–554.

Kopf, M., Baumann, H., Freer, G., Freudenberg, M., Lamers, M., Kishimoto, T., Zinkernagel, R., Bluethmann, H. and Kohler, G. (1994). Impaired immune and acute-phase responses in interleukin-6-deficient mice. *Nature* **368**, 339–342.

Krüttgen, A., Rosejohn, S., Môller, C., Wroblowski, B., Wollmer, A., Müllberg, J., Hirano, T., Kishimoto, T. and Heinrich, P.C. (1990). Structure–function analysis of human interleukin-6. Evidence for the involvement of the carboxy-terminus in function. *FEBS Lett.* **262**, 323–326.

Kumar, G., Gupta, S., Wang, S. and Nel, A.E. (1994). Involvement of Janus kinases, p52shc, Raf-1, and MEK-1 in the IL-6-induced mitogen-activated protein kinase cascade of a growth-responsive B cell line. *J. Immunol.* **153**, 4436–4447.

Kunimoto, D.Y., Nordan, R.P. and Stober, W. (1989). IL-6 is a potent cofactor of IL-1 in IgM synthesis and of IL-5 in IgA synthesis. *J. Immunol.* **143**, 2230–2235.

Lai, C.F., Ripperger, J., Morella, K.K., Wang, Y., Gearing, D.P., Horseman, N.D., Campos, S.P., Fey, G.H. and Baumann, H. (1995). STAT3 and STAT5B are targets of two different signal pathways activated by hematopoietin receptors and control transcription via separate cytokine response elements. *J. Biol. Chem.* **270**, 23 254–23 257.

Le, J. and Vilcek, J. (1989). Interleukin 6: a multifunctional cytokine regulating immune reaction and the acute phase protein response. *Lab. Invest.* **61**, 588–602.

Le. J., Fredrickson, G., Reis, L.F.L., Diamanstein, T., Hirano, T., Kishimoto, T. and Vilcek, J. (1988). Interleukin 2-dependent and interleukin 2-independent pathways of regulation of thymocyte function by interleukin 6. *Proc. Natl Acad. Sci. U.S.A.* **85**, 8643–8647.

Libermann, T.A. and Baltimore, D. (1990). Activation of interleukin-6 gene expression through the NF-κ B transcription factor. *Mol. Cell. Biol.* **10**, 2327–2334.

Liu, C.C., Joag, S.V., Kwon, B.S. and Young, J.D. (1990). Induction of perforin and serine esterases in a murine cytotoxic T lymphocyte clone. *J. Immunol.* **144**, 1196–1201.

Loppnow, H. and Libby, P. (1990). Proliferating or interleukin 1-activated human vascular smooth muscle cells secrete copious interleukin 6. *J. Clin. Invest.* **85**, 731–738.

Lotem, J., Shabo, Y. and Sachs, L. (1989). Regulation of megakaryocyte development by interleukin-6. *Blood* **74**, 1545–1551.

Lotz, M., Jirik, F., Kabouridis, R., Tsoukas, C., Hirano, T., Kishimoto, T. and Carson, D.A. (1988). B cell stimulating factor 2/interleukin 6 is a costimulant for human thymocytes and T lymphocytes. *J. Exp. Med.* **167**, 1253–1258.

Lutticken, C., Wegenka, U.M., Yuan, J., Buschmann, J., Schindler, C., Ziemiecki, A., Harpur, A.G., Wilks, A.F., Yasukawa, K., Taga, T., Kishimoto, T., Barbieri, G., Pellegrini, S., Sendtner, M., Heinrich, P.C. and Horn, F. (1993). Association of transcription factor APRF and protein kinase Jak1 with the interleukin-6 signal transducer gp130. *Science* **263**, 89–92.

Matsuda, T., Yamanaka, Y. and Hirano, T. (1994). Interleukin-6-induced tyrosine phosphorylation of multiple proteins in murine hematopoietic lineage cells. *Biochem. Biophys. Res. Commun.* **200**, 821–828.

Matsuda, T., Takahashi-Tezuka, M., Fukada, T., Okuyama, Y., Funitani, Y., Tshukada, S., Mano, H., Hirai, H., Witte, O.N. and Hirano, T. (1995a). Activation of Fes tyrosine kinase by gp130, an interleukin-6 family cytokine signal transducer, and their association. *Blood* **85**, 627–633.

Matsuda, T., Fukada, T., Takahashi-Tezuka, M., Okuyama, Y., Fujitani, Y., Hanazono, Y., Hirai, H. and Hirano, T. (1995b). Association and activation of Btk and Tec tyrosine kinases by gp130, a signal transducer of the interleukin-6 family of cytokines. *J. Biol. Chem.* **270**, 11 037–11 039.

May, L.T., Helfgott, D.C. and Sehgal, P.B. (1986). Anti-β-interferon antibodies inhibit the increased expression of HLA-B7 mRNA in tumor necrosis factor-treated human fibroblasts: structural studies of the β_2 interferon involved. *Proc. Natl Acad. Sci. U.S.A.* **83**, 8957–8961.

May, L.T., Grayeb, J., Santhanam, U., Tatter, S.B., Sthoeger, Z., Helfgott, D.C., Chiorazzi, N., Grieninger, G. and Sehgal, P.B. (1988). Synthesis and secretion of multiple forms of β_2-interferon/B-cell differentiation factor2/ hepatocyte-stimulating factor by human fibroblasts and monocytes. *J. Biol. Chem.* **263**, 7760–7766.

Metcalf, D. (1989). Actions and interactions of G-CSF, LIF, and IL-6 on normal and leukemic murine cells. *Leukemia* **3**, 349–355.

Miles, S.A., Rezai, A.R., Salazar-Gonzalez, J.F., Meyden, M.V., Stevens, R.H., Logan, D.M., Mitsuyasu, R.T., Taga, T., Hirano, T., Kishimoto, T. and Martinez-Maza, O. (1990). AIDS Kaposi sarcoma-derived cells produce and respond to interleukin 6. *Proc. Natl Acad. Sci. U.S.A.* **87**, 4068–4072.

Miller, A.E., Ennist, D.L., Ozatao, K. and Westphal, H. (1992). Activation of immunoglobulin control elements in transgenic mice. *Immunogenetics* **35**, 24–32.

Minami, M., Inoue, M., Wei, S., Takeda, K., Matsumoto, M., Kishimoto, T. and Akira, S. (1996). STAT3 activation is a critical step in gp130-mediated terminal differentiation and growth arrest of a myeloid cell line. *Proc. Natl Acad. Sci. U.S.A.* **93**, 3963–3966.

Minty, P., Chalon, P., Derocq, J.-M., Dumont, X., Guillemont, J.-C., Kaghad, M., Labit, C., Leplatois, P., Liauzum, P., Miloux, B., Minty, C., Casellas, P., Loison, G., Lupker, J., Shire, D., Ferrara, P. and Caput, D. (1993). Interleukin-13 is a new human lymphokine regulating inflammatory and immune responses. *Nature* **362**, 248–250.

Miyajima, A., Kitamura, T., Harada, N., Yokota, T. and Arai, K. (1992). Cytokine receptors and signal transduction. *Annu. Rev. Immunol.* **10**, 295–331.

Miyaura, C., Onozaki, K., Akiyama, Y., Taniyama, T., Hirano, T., Kishimoto, T. and Suda, T. (1988). Recombinant human interleukin 6 (B-cell stimulatory factor 2) is a potent inducer of differentiation of mouse myeloid leukemia cells (M1). *FEBS Lett.* **234**, 17–21.

Mock, B.A., Nordan, R.P., Justine, M.J., Kozak, C., Jenkins, N.A., Copeland, N.G., Clask, S.C., Wong, G.G. and Rudikoff, S. (1989). The murine IL-6 gene maps to the proximal region of chromosome 5. *J. Immunol.* **142**, 1372–1376.

Moffatt, S., Tanaka, N., Tada, K., Nose, M., Nakamura, M., Muraoka, O., Hirano, T. and Sugamura, K. (1996). A cytotoxic nonstructural protein, NS1, of human parvovirus B19 induces activation of interleukin-6 gene expression. *J. Virol.* **70**, 8485–8491.

Muraguchi, A., Kishimoto, T., Miki, Y., Kuritani, T., Kaieda, T., Yoshizaki, K. and Yamamura, Y. (1981). T cell-replacing factor- (TRF) induced IgG secretion in a human B blastoid cell line and demonstration of receptors for TRF. *J. Immunol.* **127**, 412–416.

Muraguchi, A., Hirano, T., Tang, B., Matsuda, T., Horii, Y., Nakajima, K. and Kishimoto, T. (1988). The essential role of B cell stimulatory factor 2 (BSF-2/IL-6) for the terminal differentiation of B cells. *J. Exp. Med.* **67**, 332–344.

Murakami, M., Narazaki, M., Hibi, M., Yawata, H., Yasukawa, K., Hamaguchi, M., Taga, T. and Kishimoto, T. (1991). Critical cytoplasmic region of the interleukin 6 signal transducer gp130 is conserved in the cytokine receptor family. *Proc. Natl Acad. Sci. U.S.A.* **88**, 11 349–11 353.

Muraoka, O., Kaisho, T., Tanabe, M. and Hirano, T. (1993). Transcriptional activation of the interleukin-6 gene by HTLV-1 p40tax through an NFκ B-like binding site. *Immunol. Lett.* **37**, 159–165.

Nabata, T., Morimoto, S., Koh, E., Shiraishi, T. and Ogihara, T. (1990). Interleukin-6 stimulates c-myc expression and proliferation of cultured vascular smooth muscle cells. *Biochem. Int.* **20**, 445–453.

Naitoh, Y., Fukata, J., Tominaga, T., Nakai, Y., Tamai, S., Mori, K. and Imura, H. (1988). Interleukin-6 stimulates the secretion of adrenocorticotropic hormone in conscious, freely-moving rats. *Biochem. Biophys. Res. Commun.* **155**, 1459–1463.

Nakajima, K., Martinez-Maza, O., Hirano, T., Nishanian, P., Salazar-Gonzalez, J.F., Fahey, J.L. and Kishimoto, T. (1989). Induction of IL-6 (B cell stimulatory factor-2/IFN-β_2) production by HIV *J. Immunol.* **142**, 144–147.

Nakajima, K., Kusafuka, T., Takeda, T., Fujitani, Y., Nakae, K. and Hirano, T. (1993). Identification of a novel interleukin-6 response element containing an Ets-binding site and a CRE-like site in the junB promoter. *Mol. Cell. Biol.* **13**, 3027–3041.

Nakajima, K., Matsuda, T., Fujitani, Y., Kojima, H., Yamanaka, Y., Nakae, K., Takeda, T. and Hirano, T. (1995). Signal transduction through IL-6 receptor: involvement of multiple protein kinases, stat factors, and a novel H7-sensitive pathway. *Ann. N.Y. Acad. Sci.* **762**, 55–70.

Nakajima, K., Yamanaka, Y., Nakae, K., Kojima, H., Kiuchi, N., Ichiba, M., Kitaoka, T., Fukada, T., Hibi, M. and Hirano, T. (1996). A central role for Stat3 in IL-6-induced regulation of growth and differentiation in M1 leukemia cells. *EMBO J.* **15**, 3651–3658.

Nakao, H. Nishikawa, A., Nishiura, T., Kanayama, Y., Tarui, S. and Taniguchi, N. (1991). Hypogalactosylation of immunoglobulin G sugar chains and elevated serum interleukin 6. *Clin. Chem. Acta* **197**, 221–228.

Nakayama, K., Shimizu, H., Mitomo, K., Watanabe, T., Okamoto, S. and Yamamoto, K. (1992). A lymphoid cell-specific nuclear factor containing c-Rel-like proteins preferentially interacts with interleukin-6 κB-related motifs whose activities are repressed in lymphoid cells. *Mol. Cell. Biol.* **12**, 1736–1746.

Namba, Y. and Hanaoka, M. (1972). Immunocytology of cultured IgM-forming cells of mouse. I. Requirement of phagocytic cell factor for the growth of IgM-forming tumor cells in tissue culture. *J. Immunol.* **109**, 1193–1200.

Narazaki, M., Witthuhn, B.A., Yoshida, K., Silvennoinen, O., Yasukawa, K., Ihle, J.N., Kishimoto, T. and Taga, T. (1994). Activation of JAK2 kinase mediated by the interleukin 6 signal transducer gp130. *Proc. Natl Acad. Sci. U.S.A.* **91**, 2285–2289.

Nicola, N.A., Metcalf, D., Matsumoto, M. and Johnson, G.R. (1993). Purification of a factor inducing differentiation in murine myelomonocytic leukemia cells. Identification as granulocyte colony-stimulating factor. *J. Biol. Chem.* **258**, 9017–9023.

Nijstein, M.W.N., De Groot, E.R., Ten Duis, H.J., Klasen, H.J., Hack, C.E. and Aarden, L.A. (1987). Serum levels of interleukin-6 and acute phase responses. *Lancet* **ii**, 921–921 (letter).

Nishimoto, N., Yoshizaki, K., Tagoh, H., Monden, M., Kishimoto, S., Hirano, T. and Kishimoto, T. (1989). Elevation of serum interleukin 6 prior to acute phase proteins on the inflammation by surgical operation. *Clin. Immunol. Immunopathol.* **50**, 399–401.

Noma, T., Mizuta, T., Rosen, A., Hirano, T., Kishimoto, T. and Honja, T. (1987). Enhancement of the interleukin 2 receptor expression on T cells by multiple B-lymphotropic lymphokines. *Immunol. Lett.* **15**, 249–253.

Nordan, R.P. and Potter, M. (1986). A macrophage-derived factor required by plasmacytomas for survival and proliferation *in vitro*. *Science* **233**, 566–569.

Northemann, W., Braciak, T.A., Hattori, M., Lee, F. and Fey, G.H. (1989). Structure of the rat interleukin 6 gene and its expression in macrophage-derived cells. *J. Biol. Chem.* **264**, 16 072–16 082.

Ogawa, M. (1992). IL6 and haematopoietic stem cells. *Res. Immunol.* **143**, 749–751.

Ohtake, K., Yano, T., Kameda, K. and Ogawa, T. (1990). Detection of interleukin-6 (IL-6) in human bone marrow myeloma cells by light and electron microscopy. *Am. J. Hematol.* **35**, 84–87.

Okada, M., Sakaguchi, N., Yoshimura, N., Hara, H., Shimizu, K., Yoshida, H., Yoshizaki, K., Kishimoto, S., Yamamura, Y. and Kishimoto, T. (1983). B cell growth factor (BCGF) and B cell differentiation factor from human T hybridomas: two distinct kinds of BCGFs and their synergism in B cell proliferation. *J. Exp. Med.* **157**, 583–590.

Okada, M., Kitahara, M., Kishimoto, S., Matsuda, T., Hirano, T. and Kishimoto, T. (1988). IL-6/BSF-2 functions as a killer helper factor in the *in vitro* induction of cytotoxic T cells. *J. Immunol.* **141**, 1543–1549.

Okada, M., Suzuki, C., Takatsuki, F., Akiyama, Y., Koike, K., Ozawa, K., Hirano., T., Kishimoto, T., Nakahata, T. and Asano, S. (1989a). *In vitro* expansion of the murine pluripotent hemopoietic stem cell population in response to interleukin 3 and interleukin 6. Application to bone marrow transplantation. *Transplantation* **48**, 495–498.

Okano, A., Suzuki, C., Takatsuki, F., Akiyama, Y., Koike, K., Nakahata, T., Hirano, T., Kishimoto, T., Ozawa, K. and Asano, S. (1989b). Effects of interleukin-6 on hematopoiesis in bone marrow-transplanted mice. *Transplantation* **47**, 738–740.

Onozaki, K., Akiyama, Y., Okano, A., Hirano, T., Kishimoto, T., Hashimoto, T., Yoshizawa, K. and Taniyama, T. (1989). Synergistic regulatory effects of interleukin 6 and interleukin 1 on the growth and differentiation of human and mouse myeloid leukemic cell lines. *Cancer Res.* **49**, 3602–3607.

Oritani, K., Kaisho, T., Nakajima, K. and Hirano, T. (1992). Retinoic acid inhibits interleukin-6-induced macrophage differentiation and apoptosis in a murine hematopoietic cell line, Y6. *Blood* **80**, 2298–2305.

Paonessa, G., Graziani, R., Serio, A.D., Svio, R., Ciappori, L., Lahm, A., Salvati, A.L., Tniatti, C. and Ciliberto, G. (1995). Two distinct and independent sites on IL-6 trigger gp130 dimer formation and signalling. *EMBO J.* **14**, 1942–1951.

Poli, V., Mancini, F.P. and Cortese, R. (1990). IL-6DBP, a nuclear protein involved in interleukin-6 signal transduction, defines a new family of leucine zipper proteins related to C/EBP. *Cell* **63**, 643–653.

Poli, V., Balena, R., Fattori, E., Markatos, A., Yamamoto, M., Tanaka, H., Ciliberto, G., Rodan, G.A. and Costantini, F. (1994). Interleukin-6 deficient mice are protected from bone loss caused by estrogen depletion. *EMBO J.* **13**, 1189–1196.

Potter, M. and Boyce, C. (1962). Induction of plasma cell neoplasms in strain Balb/c mice with mineral oil and mineral oil adjuvants. *Nature* **193**, 1086–1087.

Ramsay, A.J., Husband, A.J., Ramshaw, I.A., Bao, S., Matthaei, K.I., Koehler, G. and Kopf, M. (1994). The role of interleukin-6 in mucosal IgA antibody responses *in vivo*. *Science* **264**, 561–563.

Ray, A., Tatter, S.B., May, L.T. and Sehgal, P.B. (1988). Activation of the human 'beta 2-interferon/hepatocyte-stimulating factor/interleukin 6' promoter by cytokines, viruses, and second messenger agonists. *Proc. Natl Acad. Sci. U.S.A.* **85**, 6701–6705.

Ray, A., Sassone Corsi, P. and Sehgal, P.B. (1989). A multiple cytokine- and second messenger-responsive element in the enhancer of the human interleukin-6 gene: similarities with c-*fos* gene regulation. *Mol. Cell. Biol.* **9**, 5537–5547.

Raynal, M.C., Liu, Z.Y., Hirano, T., Mayer, L., Kishimoto, T. and Chen Kiang, S. (1989). Interleukin 6 induces secretion of IgG_1 by coordinated transcriptional activation and differential mRNA accumulation. *Proc. Natl Acad. Sci. U.S.A.* **86**, 8024–8028.

Renauld, J.C., Vink, A. and Van Snick, J. (1989). Accessory signals in murine cytolytic T cell responses. Dual requirement for IL-1 and IL-6. *J. Immunol.* **143**, 1894–1898.

Revel, M. (1992). Growth regulatory functions of IL6 and antitumour effects. *Res. Immunol.* **143**, 769–773.

Richards, C.D., Brown, T.J., Shoyab, M., Baumann, H. and Gauldie, J. (1992). Recombinant oncostatin M stimulates the production of acute phase protein in HepG2 cells and rat primary hepatocytes *in vitro*. *J. Immunol.* **148**, 1731–1736.

Romani, L., Mencacci, A., Cenci, E. Spaccapelo, R., Toniatti, C., Puccetti, P., Bistoni, F. and Poli, V. (1996). Impaired neutrophil response and $CD4^+$ T helper cell 1 development in interleukin 6-deficient mice infected with *Candida albicans*. *J. Exp. Med.* **183**, 1345–1355.

Rook, G.A.W. and Stanford, J.L. (1992). Slow bacterial infections or autoimmunity. *Immunol. Today* **13**, 160–164.

Rook, G.A.W., Thompson, S., Buckley, M., Elson, C., Brealey, R., Lambert, C., White, T. and Rademacher, T. (1991). The role of oil and agalactosyl IgG in the induction of arthritis in rodent models. *Eur. J. Immunol.* **21**, 1027–1032.

Rose, T.M. and Bruce, A.G. (1991). Oncostatin M is a member of a cytokine family which includes leukemia inhibitory factor, granulocyte colony-stimulatory factor and interleukin-6. *Proc. Natl Acad. Sci. U.S.A.* **88**, 8641–8645.

Rosselli, F., Sanceau, J., Wietzerbin, J. and Moustacchi, E. (1992). Abnormal lymphokine production: a novel feature of the genetic disease Fanconi anemia. I. Involvement of interleukin-6. *Hum. Genet.* **89**, 42–48.

Ruef, C., Budde, K., Lacy, J., Northemann, W., Baumann, M., Sterzel, R.B. and Coleman, D.L. (1990). Interleukin 6 is an autocrine growth factor for mesangial cells. *Kidney Int.* **38**, 249–257.

Sachs, L. (1987). The molecular control of blood cell development. *Science* **238**, 1374–1379.

Saito, M., Yoshida, K., Hibi, M., Taga, T. and Kishimoto, T. (1992). Molecular cloning of a murine IL-6 receptor-associated signal transducer, gp130, and its regulated expression *in vivo*. *J. Immunol.* **148**, 4066–4071.

Salvati, A.L., Lahm, A., Paonessa, G., Ciliberto, G. and Toniatti, C. (1995). Interleukin-6 (IL-6) antagonism by soluble IL-6 receptor α mutated in the predicted gp130-binding interface. *J. Biol. Chem.* **270**, 12 242–12 249.

Sanceau, J., Kaisho, T., Hirano, T. and Wietzerbin, J. (1995). Triggering of the human interleukin-6 gene by interferon-γ and tumor necrosis factor-α in monocytic cells involves cooperation between interferon regulatory factor-1, NF κB, and Sp1 transcription factors. *J. Biol. Chem.* **270**, 27 920–27 931.

Santhanam, U., Ghrayeb, J., Sehgal, P.B. and May, L.T. (1989). Post-translational modifications of human interleukin-6. *Arch. Biochem. Biophys.* **274**, 161–170.

Satoh, T., Nakamura, S., Taga, T., Matsuda, T., Hirano, T., Kishimoto, T. and Kaziro, Y. (1988). Induction of neuronal differentiation in PC12 cells by B-cell stimulatory factor 2/interleukin 6. *Mol. Cell. Biol.* **8**, 3546–3549.

Savino, R., Lahm, A., Giorgio, M., Cabibbo, A., Tramontano, A. and Ciliberto, G. (1993). Saturation mutagenesis of the human interleukin 6 receptor-binding site: implications for its three-dimensional structure. *Proc. Natl Acad. Sci. U.S.A.* **90**, 4067–4071.

Savino, R., Lahm, A., Salvati, A.L., Ciapponi, L., Sporeno, E., Altamura, S., Paonessa, G., Toniatti, C. and Ciliberto, G. (1994a). Generation of interleukin-6 receptor antagonists by molecular-modeling guided mutagenesis of residues important for gp130 activation. *EMBO J.* **13**, 1357–1367.

Savino, R., Ciapponi, L., Lahm, A., Demartis, A., Cabibbo, A., Toniatti, C., Delmastro, P., Altamura, S. and Ciliberto, G. (1994b). Rational design of a receptor super-antagonist of human interleukin-6. *EMBO J.* **13**, 5863–5870.

Scala, G., Quinto, I., Ruocco, M.R., Arcucci, A., Mallardo, M., Caretto, P., Forni, G. and Venuta, S. (1990). Expression of an exogenous interleukin 6 gene in human Epstein–Barr virus B cells confers growth advantage and *in vivo* tumorigenicity. *J. Exp. Med.* **172**, 61–68.

Schindler, C. and Darnell, J.E., Jr. (1995). Transcriptional responses to polypeptide ligands: the JAK–STAT pathway. *Annu. Rev. Biochem.* **64**, 621–651.

Sehgal, P.B. (1992). Regulation of IL6 gene expression. *Res. Immunol.* **143**, 724–734.

Sehgal, P.B., Zilberstein, A., Ruggieri, R.M., May, L.T., Ferguson Smith, A., Slate, D.L., Revel, M. and Ruddle, F. (1986). Human chromosome 7 carries the β_2 interferon gene. *Proc. Natl Acad. Sci. U.S.A.* **83**, 5219–5222.

Sehgal, P.B., May, L.T., Tamm, I. and Vilcek, J. (1987a). Human β_2 interferon and B-cell differentiation factor BSF-2 are identical. *Science* **235**, 731–732.

Sehgal, P.B., Walther, Z. and Tamm, I. (1987b). Rapid enhancement of β_2-interferon/B-cell differentiation factor BSF-2 gene expression in human fibroblasts by diacylglycerols and the calcium ionophore A23187. *Proc. Natl Acad. Sci. U.S.A.* **84**, 3663–3667.

Sehgal, P.B., Helfgott, D.C., Santhanam, U., Tatter, S.B., Clarick, R.H., Ghrayeb, J. and May, L.T. (1988). Regulation of the acute phase and immune responses in viral disease. Enhanced expression of the β_2-interferon/hepatocyte-stimulating factor/interleukin 6 gene in virus-infected human fibroblasts. *J. Exp. Med.* **167**, 1951–1956.

Sehgal, P.B., Grienger, G. and Tosata, G. (1989). Regulation of the acute phase and immune responses: interleukin-6. *Ann. N.Y. Acad. Sci.* **557**, 1–583.

Sela, O., El-Roeiy, O., Isenberg, D.A., Kennedy, R.C., Colaco, C.B., Pinkhas, J. and Shoenfeld, Y. (1987). A common anti-DNA idiotype in sera of patients with active pulmonary tuberculosis. *Arthritis Rheum.* **30**, 50–56.

Shabo, Y., Lotem, J., Rubinstein, M., Revel, M., Clark, S.C., Wolf, S.F., Kamen, R. and Sachs, L. (1988). The myeloid blood cell differentiation-inducing protein MGI-2A is interleukin-6. *Blood* **72**, 2070–2073.

Shenkin, A., Fraser, W.D., Series, J., Winstanley, F.P., McCartney, A.C., Burns, H.J. and Van Damme, J. (1989). The serum interleukin 6 response to elective surgery. *Lymphokine Res.* **8**, 123–127.

Shimizu, S., Yoshioka, R., Hirose, Y., Sugai, S., Tachibana, J. and Konda, S. (1989). Establishment of two interleukin 6 (B cell stimulatory factor 2/interferon β_2-dependent human bone marrow-derived myeloma cell lines. *J. Exp. Med.* **169**, 339–344.

Shimizu, H., Mitomo, K., Watanabe, T., Okamoto, S. and Yamamoto, K. (1990). Involvement of a NF-κ B-like transcription factor in the activation of the interleukin-6 gene by inflammatory lymphokines. *Mol. Cell. Biol.* **10**, 561–568.

Shoenfeld, Y. and Isenberg, D.A. (1988). Mycobacteria and autoimmunity. *Immunol. Today* **9**, 178–182.

Smeland, E.B., Blomhoff, H.K., Funderud, S., Shalaby, M.R. and Espevik, T. (1989). Interleukin 4 induces selective production of interleukin 6 from normal human B lymphocytes. *J. Exp. Med.* **170**, 1463–1468.

Spangelo, B.L., Judd, A.M., Isakson, P.C. and MacLeod, R.M. (1989). Interleukin-6 stimulates anterior pituitary hormone release *in vitro*. *Endocrinology* **125**, 575–577.

Spangelo, B.L., MacLeod, R.M. and Isakson, P.C (1990). Production of interleukin-6 by anterior pituitary cells *in vitro*. *Endocrinology* **126**, 582–586.

Splawski, J.B., McAnally, L.M. and Lipsky, P.E. (1990). IL-2 dependence of the promotion of human B cell differentiation by IL-6 (BSF-2). *J. Immunol.* **144**, 562–569.

Sporeno, E., Savino, R., Ciapponi, L., Paonessa, G., Cabibbo, A., Lahm, A., Pulkki, K., Sun, R., Toniatti, C., Klein, B. and Ciliberto, G. (1996). Human interleukin-6 receptor super-antagonists with high potency and wide spectrum on multiple myeloma cells. *Blood* **87**, 4510–4519.

Stahl, N., Boulton, T.G., Farruggella, T., Ip, N.Y., Davis, S., Witthuhn, B.A., Quelle, F.W., Silvernnoinen, O., Barbieri, G., Pellegrini, S., Ihle, J.N. and Yancopoulos, G.D. (1993). Association and activation of Jak–Tyk kinases by CNTF-LIF-OSM-IL-6 β receptor components. *Science* **263**, 92–95.

Stahl, N., Farruggella, T.J., Boulton, T.G., Zhong, Z., Darnell, J.J. and Yancopoulos, G.D. (1995). Choice of STATs and other substrates specified by modular tyrosine-based motifs in cytokine receptors. *Science* **267**, 1349–1353.

Suematsu, S., Matsuda, T., Aozasa, K., Akira, S., Nakano, N., Ohno, S., Miyazaki, J., Yamamura, K., Hirano, T. and Kishimoto, T. (1989). IgG_1 plasmacytosis in interleukin 6 transgenic mice. *Proc. Natl Acad. Sci. U.S.A.* **86**, 7547–7551.

Suematsu, S., Matsusaka, T., Matsuda, T., Ohno, S., Miyazaki, J., Yamamura, K., Hirano, T. and Kishimoto, T. (1992). Generation of plasmacytomas with the chromosomal translocation t(12; 15) in interleukin 6 transgenic mice. *Proc. Natl Acad. Sci. U.S.A.* **89**, 232–235.

Sugamura, K., Asao, H., Kondo, M., Tanaka, N., Ishii, N., Nakamura, M. and Takeshita, T. (1995). The common γ-chain for multiple cytokine receptors. *Adv. Immunol.* **59**, 225–277.

Sugita, T., Totsuka, T., Saito, M., Tamasaki, K., Taga, T., Hirano, T. and Kishimoto, T. (1990). Functional murine interleukin 6 receptor with the intracisternal A particle gene product at its cytoplasmic domain. Its possible role in plasmacytomagenesis. *J. Exp. Med.* **171**, 2001–2009.

Suzuki, C., Okano, A., Takasuki, F., Miyasaka, Y., Hirano, T., Kishimoto, T., Ejima, D. and Akiyama, Y. (1989). Continuous perfusion with interleukin 6 (IL-6) enhances production of hematopoietic stem cells (CFU-S). *Biochem. Biophys. Res. Commun.* **159**, 933–938.

Taga, T., Hibi, M., Hirata, Y., Yamasaki, K., Yasukawa, K., Matsuda, T., Hirano, T. and Kishimoto, T. (1989). Interleukin-6 triggers the association of its receptor with a possible signal transducer, gp130. *Cell* **58**, 573–581.

Takai, Y., Wong, G.G., Clark, S.C., Burakoff, S.J. and Herrmann, S.H. (1988). B cell stimulatory factor-2 is involved in the differentiation of cytotoxic T lymphocytes. *J. Immunol.* **140**, 508–512.

Takai, Y., Seki, N., Senoh, H., Yokota, T., Lee, F., Hamaoka, T. and Fujiwara, H. (1989). Enhanced production of interleukin-6 in mice with type II collagen-induced arthritis. *Arthritis Rheum.* **32**, 594–600.

Takatsuki, F., Okano, A., Suzuki, C., Chieda, R., Takahara, Y., Hirano, T., Kishimoto, T., Hamuro, J. and Akiyama, Y. (1988). Human recombinant IL-6/B cell stimulatory factor 2 augments murine antigen-specific antibody responses *in vitro* and *in vivo*. *J. Immunol.* **141**, 3072–3077.

Tanabe, O., Akira, S., Kamiya, T., Wong, G.G., Hirano, T. and Kishimoto, T. (1988). Genomic structure of the urine IL-6 gene. High degree conservation of potential regulatory sequences between mouse and human. *J. Immunol.* **41**, 3875–3881.

Tang, B., Matsuda, T., Akira, S., Nagata, N., Ikehara, S., Hirano, T. and Kishimoto, T. (1991). Age-associated increase in interleukin 6 in MRL/lpr mice. *Int. Immunol.* **3**, 273–278.

Taniguchi, T. (1995). Cytokine signaling through nonreceptor protein tyrosine kinase. *Science* **268**, 251–255.

Teranishi, T., Hirano, T., Arima, N. and Onoue, K. (1982). Human helper T cell factor(s) (ThF). II. Induction of IgG production in B lymphoblastoid cell lines and identification of T cell replacing factor (TRF)-like factor(s). *J. Immunol.* **128**, 1903–1908.

Teranishi, T., Hirano, T., Lin, B.H. and Onoue, K. (1984). Demonstration of the involvement of interleukin 2 in the differentiation of *Staphylococcus aureus* Cowan I-stimulated B cells. *J. Immunol.* **133**, 3062–3067.

Tohyama, N., Karasuyama, H. and Tada, T. (1990). Growth autonomy and tumorigenicity of interleukin 6-dependent B cells transfected with interleukin 6 cDNA. *J. Exp. Med.* **171**, 389–400.

Tomida, M., Yamamoto-Yamaguchi, Y. and Hozumi, M. (1984). Purification of a factor inducing differentiation of mouse myeloid leukemic M1 cells from conditioned medium of mouse fibroblast L929 cells. *J. Biol. Chem.* **259**, 10978–10982.

Tosato, G., Miller, J., Marti, G. and Pike, S.E. (1990). Accessory function of interleukin-1 and interleukin-6: preferential costimulation of T4 positive lymphocytes. *Blood* **75**, 922–930.

Treanor, J.J.S., Goodman, L., de Sauvage, F., Stone, D.M., Poulsen, K.T., Beck, C.D., Gray, C., Armanini, M.P., Pollock, R.A., Hefti, F., Phillips, H.S., Goddard, A., Moore, M.W., Buj-Bello, A., Davies, A.M., Asai, N., Takahashi, M., Vandlen, R., Henderson, C.E. and Rosenthal, A. (1996). Characterization of a multicomponent receptor for GDNF. *Nature* **382**, 80–83.

Uyttenhove, C., Coulie, P.G. and Van Snick, J. (1988). T cell growth and differentiation induced by interleukin-HP1/IL-6, the murine hybridoma/plasmacytoma growth factor. *J. Exp. Med.* **167**, 1417–1427.

Van Damme, J., Opdenakker, G., Simpson, R.J., Rubira, M.R., Cayphas, S., Vink, A., Biliau, A. and Van Snick, J. (1987a). Identification of the human 26-kD protein, interferon-β_2 (IFN-β_2), as a B cell hybridoma/plasmacytoma growth factor induced by interleukin 1 and tumor necrosis factor. *J. Exp. Med.* **165**, 914–919.

Van Damme, J., Cayphas, S., Opdenakker, G., Billiau, A. and Van Snick, J. (1987b). Interleukin 1 and poly(rl).poly(rC) induce production of a hybridoma growth factor by human fibroblasts. *Eur. J. Immunol.* **17**, 1–7.

Van Damme, J., Schaafsma, M.R., Fibbe, W.E., Falkenburg, J.H., Opdenakker, G. and Billiau, A. (1989). Simultaneous production of interleukin 6, interferon-β and colony-stimulating activity by fibroblasts after viral and bacterial infection. *Eur. J. Immunol.* **19**, 163–168.

Van Snick, J. (1990). Interferon-6: an overview. *Annu. Rev. Immunol.* **8**, 253–278.

Van Snick, J., Vink, A., Cayphas, S. and Uyttenhove, C. (1987). Interleukin-HP1, a T cell-derived hybridoma growth factor that supports the *in vitro* growth of murine plasmacytomas. *J. Exp. Med.* **165**, 641–649.

Van Snick, J., Cayphas, S., Szikora, J.-P., Renauld, J.-C., Van Roost, E., Boon, T. and Simpson, R.J. (1988). cDNA cloning of murine interleukin-HP1: homology with human interleukin 6. *Eur. J. Immunol.* **18**, 193–197.

Velde, A., Huijbens, R.J., Heije, K., Varies, K. and Figdor, C.G. (1990). Interleukin-4 (IL-4) inhibits secretion of IL-1β, tumor necrosis factor α, and IL-6 by human monocytes. *Blood* **76**, 1392–1397.

Vercelli, D., Jabara, H.H., Arai, K., Yokota, T. and Geha, R.S. (1989). Endogenous interleukin 6 plays an obligatory role in interleukin 4-dependent human IgE synthesis. *Eur. J. Immunol.* **19**, 1419–1424.

Vink, A., Coulie, P.G., Wauters, P., Nordan, R.P. and Van Snick, J. (1988). B cell growth and differentiation activity of interleukin-HP1 and related murine plasmacytoma growth factors. Synergy with interleukin 1. *Eur. J. Immunol.* **18**, 607–612.

Vink, A., Coulie, P., Warnier, G., Renauld, J.C., Stevens, M., Donckers, D. and Van Snick, J. (1990). Mouse plasmacytoma growth *in vivo*: enhancement by interleukin 6 (IL-6) and inhibition by antibodies directed against IL-6 or its receptor. *J. Exp. Med.* **172**, 997–1000.

Wegenka, U.M., Buschmann, J., Lutticken, C., Heinrich, P.C. and Horn, F. (1993). Acute-phase response factor, a nuclear factor binding to acute-phase response elements, is rapidly activated by interleukin-6 at the post-translational level. *Mol. Cell. Biol.* **13**, 276–288.

Weissenbach, J., Chernajovsky, Y., Zeevi, M., Shulman, L., Soreq, H., Nir, U., Wallach, D., Perricaudet, M., Tiollais, P. and Revel, M. (1980). Two interferon mRNAs in human fibroblasts: *in vitro* translation and *Escherichia coli* cloning studies. *Proc. Natl Acad. Sci. U.S.A.* **77**, 7152–7156.

Wen, Z., Zhong, Z. and Darnell, J.E. (1995). Maximal activation of transcription by Stat1 and Stat3 requires both tyrosine and serine phosphorylation. *Cell* **82**, 241–250.

Williams, N., De Giorgio, T., Banu, N., Withy, R., Hirano, T. and Kishimoto, T. (1990). Recombinant interleukin 6 stimulates immature murine megakaryocytes. *Exp. Hematol.* **18**, 69–72.

Wolf, S.F., Temple, P.A., Kobayashi, M., Young, D., Dicig, M., Lowe, L., Dzialo, R., Fitz, L., Ferenz, C., Hewick, R.M., Kelleher, K., Herrmann, S.H., Clark, S., Azzoni, L., Chan, S., Trinchieri, G. and Perussia, B. (1991). Cloning of cDNA for natural killer cell stimulatory factor, a heterodimeric cytokine with multiple biologic effects on T and natural killer cells. *J. Immunol.* **146**, 3074–3081.

Wong, G.H.W. and Goeddel, D.V. (1986). Tumour necrosis factors α and β inhibit virus replication and synergize with interferons. *Nature* **323**, 819–822.

Wu, Y.Y. and Bradshaw, R.A. (1996). Induction of neurite outgrowth by interleukin-6 is accompanied by activation of Stat3 signaling pathway in a variant PC12 cell (E2) line. *J. Biol. Chem.* **271**, 13023–13032.

Xia, X., Lee, H.K., Clark, S.C. and Choi, Y.S. (1989). Recombinant interleukin (IL) 2-induced human B cell differentiation is mediated by autocrine IL6. *Eur. J. Immunol.* **19**, 2275–2281.

Yamanaka, Y., Nakajima, K., Fukada, T., Hibi, M. and Hirano, T. (1996). Differentiation and growth arrest signals are generated through the cytoplasmic region of gp130 that is essential for Stat3 activation. *EMBO J.* **15**, 1557–1565.

Yamasaki, K., Taga, T., Hirata, Y., Yawata, H., Kawanishi, Y., Seed, B., Taniguchi, T., Hirano, T. and Kishimoto, T. (1988). Cloning and expression of the human interleukin-6 (BSF-2/IFN β_2) receptor. *Science* **241**, 825–828.

Yasukawa, K., Hirano, T., Watanabe, Y., Muratani, K., Matsuda, T. and Kishimoto, T. (1987). Structure and expression of human B cell stimulatory factor-2 (BSF-2/IL-6) gene. *EMBO J.* **6**, 2939–2945.

Yawata, H., Yasukawa, K., Natsuka, S., Murakami, M., Yamasaki, K., Hibi, M., Taga, T. and Kishimoto, T. (1993). Stucture–function analysis of human IL-6 receptor: dissociation of amino acid residues required for IL-6-binding and for IL-6 signal transduction through gp130. *EMBO J.* **12**, 1705–1712.

Yokoi, T., Miyawaki, T., Yachie, A., Kato, K., Kasahara, Y. and Taniguchi, N. (1990). Epstein–Barr virus-immortalized B cells produce IL-6 as an autocrine growth factor. *Immunology* **70**, 100–105.

Yoshizaki, K., Nakagawa, T., Kaieda, T., Muraguchi, A., Yamamura, Y. and Kishimoto, T. (1982). Induction of proliferation and Ig production in human B leukemic cells by anti-immunoglobulins and T cell factors. *J. Immunol.* **128**, 1296–1301.

Yoshizaki, K., Matsuda, T., Nishimoto, N., Kuritani, T., Taeho, L., Aozasa, K., Nakahata, T., Kawai, H., Tagoh, H., Komori, T., Kishimoto, S., Hirano, T. and Kishimoto, T. (1989). Pathogenic significance of interleukin-6 (IL-6/BSF-2) in Castleman's disease. *Blood* **74**, 1360–1367.

Zhang, Y., Lin, J.-X. and Vilcek, J. (1988a). Synthesis of interleukin 6 (interferon-β_2/B cell stimulatory factor 2) in human fibroblasts is triggered by an increase in intracellular cyclic AMP. *J. Biol. Chem.* **263**, 6177–6182.

Zhang, Y., Lin, J.-X., Yip, Y.K. and Vilcek, J. (1988b). Enhancement of cAMP levels and of protein kinase activity by tumor necrosis factor and interleukin 1 in human fibroblasts: role in the induction of interleukin 6. *Proc. Natl Acad. Sci. U.S.A.* **85**, 6802–6805.

Zhang, X.-G., Klein, B. and Bataille, R. (1989). Interleukin-6 is a potent myeloma-cell growth factor in patients with aggressive multiple myeloma. *Blood* **74**, 11–13.

Zhang, Y., Lin, J.-X. and Vilcek, J. (1990). Interleukin-6 induction by tumor necrosis factor and interleukin-1 in human fibroblasts involves activation of a nuclear factor binding to a κB-like sequence. *Mol. Cell. Biol.* **10**, 3818–3823.

Zhang, X., Blenis, J., Li, H.C., Schindler, C. and Chen-Kiang, S. (1995). Requirement of serine phosphorylation for formation of STAT-promoter complexes. *Science* **267**, 1990–1994.

Zhong, Z., Wen, Z. and Darnell, J.E. (1994). Stat3: a STAT family member activated by tyrosine phosphorylation in response to epidermal growth factor and interleukin-6. *Science* **264**, 95–98.

Zilberstein, A., Ruggieri, R., Korn, J.H. and Revel, M. (1986). Structure and expression of cDNA and genes for human interferon-β_2, a distinct species inducible by growth-stimulatory cytokines. *EMBO J.* **5**, 2529–2537.

Chapter 9

Interleukin-7

Markus J. Maeurer[1], Howard D. Edington[2] and Michael T. Lotze[2,3]

[1]Department of Medical Microbiology, University of Mainz, Mainz, Germany and Departments of [2]Surgery and [3]Biochemistry and Molecular Genetics, University of Pittsburgh Medical School and University of Pittsburgh Cancer Institute, Pittsburgh, PA, USA

INTRODUCTION

Interleukin-7 (IL-7) is an exceptional cytokine, as it mediates lymphopoiesis in mice in a non-redundant fashion. In contrast, targeted gene deletion of other cytokines, including IL-2, IL-4 or IL-10 (Schorle *et al.*, 1991; Kuhn *et al.*, 1991, 1993), revealed that these cytokines are not essential for development and proper function of B or T lymphocytes. IL-7 is secreted by both immune and non-immune cells and appears not only to be involved in the development of an effective immune system, but also in the generation and maintenance of strong and effective cellular immune responses directed against cancer cells, or infectious diseases. IL-7 appears to serve as the major growth and differentiation factor for both thymic and extrathymic development of $\gamma\delta^+$ T lymphocytes. IL-7 promotes immune effector functions in T lymphocytes, natural killer (NK) cells and monocytes–macrophages, and modulates the quantity and quality of immune responses *in vitro* and *in vivo*. The availability of IL-7 targeted gene-deleted mice, or IL-7 transgenic animals, allowed a more detailed study of the physiology and pathophysiology of the paracrine and systemic effects of IL-7. The use of IL-7 in the treatment of different diseases, including immunodeficiency disorders and malignancy, suggests that IL-7 may facilitate a number of therapeutic endeavors including bone marrow and organ transplantation, cancer immunotherapy and the treatment of infectious diseases.

CLONING AND PURIFICATION

Following the development of techniques for studying bone marrow cultures, it was apparent that B-cell maturation occurred in the presence of bone marrow stromal cells, suggesting the existence of a growth and/or maturation enhancing cytokine (Hunt *et al.*, 1987). Namen and coworkers subsequently demonstrated that conditioned medium from stromal cell cultures stimulated the growth of B-cell precursors. They immortalized a stromal cell line by transfecting it with the plasmid pSV3neo (encoding both the large and small T antigens of SV40) and isolated a clone (I × N/A6) which produced a factor initially called lymphopoietin-1 (LP-1) that stimulated the growth of B-cell precursors.

The Cytokine Handbook, 3rd ed.
ISBN 0–12–689662–3

Conditioned medium from the growth of this clone was then purified. High-performance liquid chromatography column fractions containing LP-1 bioactivity were isolated. A single unit of LP-1 activity is that causing half maximal ^{3}H-TdR incorporation in a culture of precursor B cells (LP-1 bioassay). At this stage of purification it was clear that several proteins were present in the fraction that could account for the biologic activity. Additional sodium dodecyl sulfate–polyacrylamide gel electrophoresis analysis under non-reducing conditions associated bioactivity with a protein of 25×10^3 Da, substantiated by ^{125}I-labeled LP-1 binding experiments. The purified protein exhibits a specific activity of approx. 4×10^6 units/μg of protein and is active at a half-maximal concentration of 10^{-13} M (Namen *et al.*, 1988b).

The same murine stromal cell clone provided a cDNA library which was screened for LP-1 activity following expression in COS-7 cells. A clone (1046) was identified that was associated with high biologic activity (Fig. 1). The sequence contains a 548 base pairs (bp) 5′ non-coding region which may be involved in expression regulation, as its removal results in increased COS cell expression of protein. The sequence includes a 462 bp open reading frame and a 579 bp 3′ non-coding region containing a consensus polyadenylation signal and terminating in 15 adenine residues. Purified protein was subjected to *N*-terminal analysis, which suggested that the nucleotide sequence from clone 1046 codes for the same protein identified in the biologic assay; the protein was designated IL-7. The mature protein has a 25-amino-acid leader sequence followed by 129 amino acids with two *N*-linked glycosylation sites and six cysteine residues which may be involved with intramolecular disulfide bond formation. The importance of disulfide bond formation is suggested by loss of activity following treatment with 2-mercaptoethanol, which breaks disulfide bonds.

The calculated molecular weight of IL-7 is 14.9 kDa. The disparity between calculated molecular weight and that predicted by migration of the native protein may be accounted for by glycosylation (Namen *et al.*, 1988a,b). Two such *N*-linked glycosylation sites in murine IL-7 are located at amino acids 69 and 90 (Namen *et al.*, 1988a). IL-7 mRNA has been detected in murine thymus, spleen, kidney and liver by Northern blot analysis. Interestingly, although message was present in thymus and spleen, no biologic activity could be detected in these tissues.

Goodwin and colleagues characterized human IL-7 by nucleic acid hybridization of cDNA prepared from a hepatocarcinoma cell line (SK-HEP-1, ATCC HTB 52) with the murine IL-7 probe. There is considerable homology between the two IL-7 nucleotide sequences (81% in the coding region) and up to 60% amino acid homology, with all six cysteine residues being conserved (Goodwin *et al.*, 1989). The human IL-7 gene contains six exons over 33 kilobases (kb) (Lupton *et al.*, 1990). The human IL-7 cDNA is composed of 534 nucleotides encoding a protein of 177-amino-acid residues with a signal sequence of 25-amino-acid residues and three potential *N*-linked glycosylation sites (Goodwin *et al.*, 1989). There is a 19-amino-acid insert for human IL-7 (coded for by exon 5 in the human genome) which does not exist in murine IL-7 (Fig. 1) and appears not be essential for biologic IL-7 activity using a proliferation assay of progenitor B cells (Goodwin *et al.*, 1989). Additionally, an apparently alternatively spliced human IL-7 mRNA lacking the entire exon 4 (44-amino-acid residues) was isolated from the SK-HEP-1 line, which results in loss of the capability to stimulate proliferation of murine progenitor B cells. Human recombinant IL-7 is active on murine and human B-cell progenitors. In contrast, murine IL-7 acts only on murine, but not on human cells.

```
        -25                    +1        10        20        30        40        50        60        70
hIL-7   MFHVSFRYIFGLPPLILVLLPVASSDCDIEGKDGKQYESVLMVSIDQLLDSMKEIGSNCLNNEFNFFKRHICDANKEGMFLFRAARKLRQFLKMNSTGDF
hIL-7   MFHVSFRYIFGLPPLILVLLPVASSDCDIEGKDGKQYESVLMVSIDQLLDSMKEIGSNCLNNEFNFFKRHICDANK------------------------
mIL-7   MFHVSFRYIFGIPPLILVLLPVTSSECHIKDKEGKAYESVLMISID-ELDKMTGTDSNCPNNEPNFFRKHVCDDTKEAAFLNRAARKLKQFLKMNISEEF
        exon 1 ---------- exon 2 ---------- exon 3 ---------- exon 4 ---- \\

         80        90        100       110       120       130       140       150
IL-7    DLHLLKVSEGTTILLNCTGQVKGRKPAALGEAQPTKSLEENKSLKEQKKLNDLCFLKRLLQEIKTCWNKILMGTKEH
IL-7    -------------------VKGRKPAALGEAQPTKSLEENKSLKEQKKLNDLCFLKRLLQEIKTCWNKILMGTKEH
mIL-7   NVHLLTVSQGTQTLVNCTSK-------------------EEKNVKEQKK-NDACFLKRLLREIKTCWNKILKGSI--
        // ---------- exon 5 ---------- exon 6 ----------
```

Fig. 1. Amino acid sequences of human and murine IL-7. The hIL-7 gene codes for a 173-amino-acid molecule (top sequence). A differentially spliced IL-7 mRNA has initially been identified (middle sequence) by probing a cDNA library derived from a human hepatocellular carcinoma cell line, with the mIL-7 cDNA by nucleic acid hybridization (Goodwin *et al.*, 1989; Lupton *et al.*, 1990). The alternative IL-7 transcript lacks exon 4 coding for 132 bp, thereby reducing the protein by 44 amino acids. Both the entire IL-7 mRNA and the alternatively spliced IL-7 mRNA have been identified reproducibly in chronic lymphatic B-cell leukemia (Frishman *et al.*, 1993), in follicular dendritic cells (Kröncke *et al.*, 1996a,b) and in renal cell cancer (authors' unpublished observations). The mIL-7 cDNA (bottom sequence) lacks a region that codes for 19 amino acids and would correspond to exon 5 of the human IL-7 gene (Namen *et al.*, 1988a). The lack of exon 5 apparently does not impair biologic IL-7 functions in conventional assay systems. Human and murine IL-7 exhibit up to 81% sequence homology with regard to the nucleotide sequence and up to 60% homology in amino acid residues. The leader peptide is shaded.

Human IL-7 is predicted to conform to the 'small hematopoietin' subclass members, forming a four-α-helix bundle structure. The human IL-7 gene appears to be a single copy gene located on chromosome 8q12–13 (Sutherland *et al.*, 1989) in proximity to the $p53/p56^{lyn}$ gene, a member of the Src tyrosine kinases and the *HYRC* gene, which is potentially involved in VDJ recombination located at 8q11 (Corey and Shapiro, 1994; Seckinger and Fougereau, 1994). The murine IL-7 gene lacks transcription regulatory elements that have commonly been identified in eukaryotic promoters (e.g. the TATA box, CAAT sequences, Sp-1 or GC-rich regions). Only one Sp-1 binding site has been identified in the human IL-7 gene.

Additionally, a potential binding site for E12 is conserved in both murine and human IL-7 genes, as well as other sequences conforming to the 'helix–loop–helix' class of DNA binding proteins (Lupton *et al.*, 1990). Of note, several IL-7 mRNA species of 1.5, 1.7, 2.6 and 2.9 kb have been identified in murine tissues. However, all four transcripts appeared to be present in the thymus. In contrast, only the 2.6- and the 2.9-kb IL-7 mRNA transcripts could be detected in kidney (Namen *et al.*, 1988a).

Regulation of murine IL-7 mRNA expression has been addressed by examining IL-7 RNA resolution in murine PAM 212 keratinocytes (Ariizumi *et al.*, 1995). Treatment with interferon-γ (IFN-γ) yields preferential expression of the 1.5- and 2.6-kb mRNA species in addition to the constitutively expressed 2.9- and 1.7-kb mRNAs by the use of alternative transcription initiation sites. The 1.5- and 2.6-kb mRNAs are transcribed within 250 bp from the coding sequence. In contrast, the 1.7- and 2.9-kb mRNA species contained more than 400 bp in the 5′ untranslated region. IFN-γ promotes conversion to 1.5- and 2.5-kb mRNA expression through the interferon-stimulated response element (ISRE) located 270 bp upstream from the coding sequence (GAAACTGAAAGT). This ISRE is immediately followed by a non-TAT-type transcription 'initiator' element (CTTACTCTTG). It appears that IFN-γ-induced transcription of IL-7 may be controlled through the ISRE–control complex, whereas other 'initiator sequences' may be responsible for the baseline IL-7 transcriptional activity in certain cell types (Ariizumi *et al.*, 1995). It has been suggested the IFN-γ-inducible IL-7 transcripts may be more translationally active than the conventional 'baseline' 1.7- and 2.9-kb transcripts, a concept that may impact on the molecular definition of the cellular interaction of keratinocytes, a source of IL-7 secretion, and IFN-γ-producing cells in skin (see below).

In addition to murine and human IL-7 cDNA, IL-7 cDNA from a bovine leukemia virus-induced B-cell line has been characterized. Bovine IL-7 is 176 amino acids long and shows 75% homology with human IL-7 and 65% homology with murine IL-7 (Cludts *et al.*, 1992).

THE IL-7 RECEPTOR

A cell line absolutely dependent on IL-7 for growth (I × N/2b) (Park *et al.*, 1990) was used to characterize the IL-7 receptor (IL-7R) designated as CD127 (Schlossman *et al.*, 1994; Kishimoto *et al.*, 1997). The IL-7R complex, exhibiting both high affinity (approximately K_d 100 pM) and low affinity (K_d 1 nM) binding sites (Page *et al.*, 1993), is composed of at least two subunits: the IL-7Rα chain, identified by a direct expression cloning strategy (Goodwin *et al.*, 1990), maps to the human chromosome 5p13 (Lynch *et al.*, 1992) and the common γ chain (γc) shared with receptors for IL-2, IL-4, IL-9 and IL-15 (Kondo *et al.*, 1993, 1994; Noguchi *et al.*, 1993; Russell *et al.*, 1993, 1994; Giri *et al.*,

1994). The γc chain appears to be expressed constitutively (Taniguchi and Minami, 1993). The IL-7Rα receptor subunit forms a heterodimer with the γc chain which is required for the high-affinity IL-7 binding (Noguchi *et al.*, 1993; Kondo *et al.*, 1994). It has been suggested that IL-7 receptors expressed on some cells may contain an as yet poorly defined subunit, as IL-7 binds (albeit at low affinity) to COS cells in the absence of transfected IL-7Rα and γc chains (Goodwin *et al.*, 1990; Noguchi *et al.*, 1993; Kondo *et al.*, 1994). Alternatively, because COS cells represent primate cells, endogenously expressed IL-7R may provide IL-7 binding sites.

The existence of an as yet undefined additional IL-7R subunit is also substantiated by blocking experiments with antibodies (Abs) directed against the (murine) IL-7Rα (Ab TUGm2) and the murine γc subunit (Ab A7R34). Proliferation of the IL-7-dependent cell line I × N/2b is decreased by IL-7Rα-specific Ab TUGm2 and totally abrogated by mixtures of the Abs TUGm2 (IL-7Rα) and A7R34 (γc-chain) (Kondo *et al.*, 1994). However, TUGm2 (IL-7Rα) reduces the high-affinity IL-7 receptor from 79 to 255 pM, but does not affect the low-affinity IL-7 receptor (Kondo *et al.*, 1994). It has been postulated that the high-affinity IL-7 receptor may be composed of the IL-7Rα chain, the γc chain and an as yet undefined moiety (IL-7 'Rx'). Thus, the decreased IL-7R affinity obtained by blocking with the TUGm2 (anti-IL-7R Ab) from 79 to 255 pM could reflect an 'intermediate' IL-7 receptor complex composed of the unkown IL-7 'Rx' and the γc subunit (Sugamura *et al.*, 1996).

Six extracellular and four intracellular cysteine residues are present in human as well as in the murine IL-7 Rα coding for a 439-amino-acid protein with a calculated molecular weight of 49.5 kDa. The IL-7Rα domain exhibits the characteristic features of the cytokine receptor superfamily: the cytoplasmic IL-7 Rα domain, composed of 195-amino-acid residues, does not exhibit consensus proteine kinase sequences (reviewed in Sugamura *et al.*, 1996). A differential splicing event results in mRNA encoding for the secreted form of the IL-7 receptor capable of binding IL-7 in solution, a form that may be important for binding circulating IL-7 (Goodwin *et al.*, 1990; Park *et al.*, 1990; Mehrotra *et al.*, 1995). IL-7Rα was detected on pre-B cells, thymocytes, some T lineage cells (Park *et al.*, 1990; Rich *et al.*, 1993), on human intestinal cells (Reinecker and Podolsky, 1995), colorectal cancer cells, renal cell cancer cells (authors' unpublished observations), bone marrow-derived macrophages but not mature B cells (Foxwell *et al.*, 1992), cutaneous T-cell lymphomas (Bagot *et al.*, 1996) and thymic NK1.1$^+$ T cells (Miyaji *et al.*, 1996). The total number of IL-7 binding sites on individual cells may range from 1×10^4 up to 5×10^5 per cell (Armitage *et al.*, 1992b). Interestingly, IL-2 and IL-7 reciprocally induce IL-2Rα and IL-7Rα receptor expression on $\gamma\delta^+$ T lymphocytes, which may be important for proliferation and T-cell response to locally produced cytokine of intraepithelial lymphocytes (iIELs). However, there at two different subsets of such iIELs: $\gamma\delta^+$ T-cell receptor (TCR) dim cells ($\gamma\delta^+$ TCR^{dim}) are responsive to IL-2 (presumably provided *in situ* by $\alpha\beta^+$ T cells) and IL-7 (presumably provided *in situ* by epithelial cells or macrophages). In contrast, $\gamma\delta^+$ TCR^{bright} cells do not respond by upregulation of their IL-2 and IL-7 receptors (Fujihashi *et al.*, 1996). Additionally, IL-7 binding to the p90 IL-7R leads to expression of the transferrin receptor and the 4F2 antigen (Costello *et al.*, 1993).

Sequence analysis of the 5′ flanking region of murine IL-7Rα revealed that it contains CAATT, TATA sequences and potential glucocorticoid receptor binding sites, as well as a potential ISRE element (Pleiman *et al.*, 1991). There appears to be an alternate IL-7 receptor structure on T cells according to their state of activation, an observation that may account for differential IL-7-induced signaling events in T cells (Foxwell *et al.*, 1992; Lin *et al.*, 1995). The expression of the 90-kDa IL-7R is stimulated by IL-7, ionomycin and phorbol esters, and inhibited by cyclosporin A and FK506 (Foxwell *et al.*, 1993). Expression of the 90-kDa receptor on freshly isolated human T cells could not be increased with phytohemagglutinin (PHA), concanavalin A or CD3 (Armitage *et al.*, 1992a,b). Interestingly, activation of peripheral blood mononuclear cells (PBMCs) with anti-CD3 results in a fourfold downregulation of IL-7 receptors (high and low affinity) (Foxwell *et al.*, 1992). These findings have recently been substantiated by the observation that IL7Rα–γc chain complexes are detectable in activated, but not in resting, T cells, independent of total cell surface γc chain expression. Thus, stimulation of T cells may lead to assembly of IL7Rα–γc chain complexes, which correlates with JAK3 expression (Page *et al.*, 1997).

However, as IL-2 or IL-4 gene-deleted mice do not exhibit severe defects in T-cell differentiation, such as those observed in either IL-7 or IL-7Rα gene-deleted mice, IL-7 may account for most of the immunologic defects observed in murine models of the X-SCID defect associated with defects of the common γc chain receptor unit (Takeshita *et al.*, 1992; Noguchi *et al.*, 1993; DiSanto *et al.*, 1994; Leonard *et al.*, 1994). The X-SCID defects can also be observed in humans (Lai *et al.*, 1997).

In unstimulated human T cells, the p90 IL-7R is constitutively associated with the Src kinase enzymes $p59^{fyn}$ and $p56^{lck}$ (Page *et al.*, 1995). IL-7 binding the p90 IL-7R leads to both increased $p59^{fyn}$ and $p56^{lck}$ levels in stimulated and unstimulated T cells (Page *et al.*, 1995). Signaling via the p90 IL-7R also leads to increased activity of the Src kinase, suggesting that activation of $p59^{fyn}$ and $p56^{lck}$ is not exclusively responsible for IL-7-driven T-cell proliferation and that other signaling events (e.g. mediated through the γc chain) may be required (Page *et al.*, 1995). However, targeted gene deletion for $p59^{fyn}$ in mice did not show a major impact on lymphopoiesis (Stein *et al.*, 1992; Grabstein *et al.*, 1993; Sudo *et al.*, 1993). In contrast, in $p56^{lck}$ gene-deleted mice, a thymocyte maturation block at the double negative state could be observed (Molina *et al.*, 1992). However, similar effects could not be detected in CD4, CD8, or IL-2 gene-deficient deleted mice. These observations suggest that $p56^{lck}$ is also involved in the signaling pathways (Fung-Leung *et al.*, 1991; Schorle *et al.*, 1991). Thus, the observed effects of $p56^{lck}$ on lymphopoiesis may be attributed to the lack of IL-7-driven $p56^{lck}$-mediated cellular responses. IL-7-mediated phosphatidylinositol-3 (PI-3) kinase activation induced by tyrosine phosphorylation of the PI-3 kinase p85 subunit appears to be to essential to the IL-7 proliferative signal (Sharfe *et al.*, 1995). A different protein tyrosine kinase, termed pim-1, may also be involved in IL-7-mediated signaling, as IL-7-mediated pre-B-cell expansion is decreased in pim-1-deficient mice (Domen *et al.*, 1993).

IL-7 activates members of the Janus (JAK) family of non-receptor tyrosine kinases, JAK1 and JAK3 (Russell *et al.*, 1994; Zeng *et al.*, 1994; Musso *et al.*, 1995), which are both activated by γc chain-sharing cytokines including IL-2, IL-4 and IL-9. These kinases may serve as the signal transduction pathway to the nucleus by phosphorylation and activation of signal transducers and activators of transcription (STATs). IL-7 has been shown to activate STAT1, STAT3 and STAT5 (Zeng *et al.*, 1994; Lin *et al.*, 1995; van der Plas *et al.*, 1996; Perumal *et al.*, 1997) by interacting with an area spanning the tyrosine residue 409 at the *C*-terminal end of the IL-7R (Lin *et al.*, 1995). Thus, at least several alternate signal transduction pathways (e.g. $p56^{lck}$, $p59^{fyn}$, JAKs, STATs) may be operational in IL-7-responsive cells (e.g. T cells, epithelial cells). It is possible that IL-7 may exert its functions in a cell- or tissue-specific manner dependent on differential activation of the IL-7R signalling transduction pathway(s). For instance, recent data suggest that the IL-7 receptor complex delivers signals of different quality to lymphoid progenitor cells during rearrangement of the antigen receptors (reviewed in Candeias *et al.*, 1997a). First, the IL-7Rα mediates a 'trophic' or 'maintenance' effect regarding cell viability during gene rearrangement. Earlier studies showed that immature thymocytes undergo apoptosis when separated from the thymus. IL-7 is capable of sustaining these cells without inducing significant cell proliferation (Watson *et al.*, 1989). These antiapoptotic effects delivered by the IL-7Rα can also be observed in mature lymphoid cells (Komschlies *et al.*, 1994) and may be attributed to the induction of Bcl-2 members (Hernandez-Caselles *et al*, 1995; Lee *et al.*, 1996; Vella *et al.*, 1997). However, other Bcl-2-related proteins, inducing Bcl-x_L or Bcl-w, or other as yet ill-defined antiapoptotic factors, may also be involved, as bcl-2 knockout ($-/-$) mice exhibit a different picture concerning T-cell development compared with alterations identified in IL-7R$\alpha^{-/-}$ mice (Veis *et al.*, 1993; Matsuzaki *et al.*, 1997).

Second, the IL-7Rα may also deliver 'mechanistic' signals required for gene rearrangement. IL-7R$\alpha^{-/-}$ mice exhibit impaired γ gene rearrangement (Maki *et al.*, 1996; Candeias *et al.*, 1997b). The same was found to be true for IL-7Rα-mediated signals, required for immunoglobulin (Ig) heavy chain and TCR β-chain rearrangement (Corcoran *et al.*, 1996; Crompton *et al.*, 1997). However, it appears that alternate strategies concerning the IL-7Rα-mediated function may operate in gene rearrangement: the TCR γ-chain rearrangement appears to be dependent on IL-7-mediated signals. Ig heavy chain and TCR-β chain rearrangement requires IL-7, but not absolutely. The TCR δ rearrangement may not be exclusively IL-7 dependent, as IL-7R$\alpha^{-/-}$ mice exhibit δ-chain rearrangements *in vivo* (Corcoran *et al.*, 1996; Candeias *et al.*, 1997a,b; Oosterwegel *et al.*, 1997; Peschon *et al.*, 1997).

Such effects may derive from several factors. First, IL-7 induces RAG-1 (recombinant activation gene) and RAG-2 expression (Muegge *et al.*, 1993). IL-7R$\alpha^{-/-}$ mice exhibit decreased RAG expression in double-negative, but not in double-positive, cells (Crompton *et al.*, 1997). Therefore, decreased recombinase activity may affect recombinatorial events in distinct thymic cells. Second, IL-7Rα-mediated signals may be required to prevent untimely apoptosis in thymocytes. It has been suggested that IL-7Rα-mediated signals may unmask genes associated with proliferation and antiapoptotic properties (Peschon *et al.*, 1997). This is substantiated by the observation that peripheral T cells in IL-7R$\alpha^{-/-}$ mice undergo apoptosis upon stimulation (Maraskovsky *et al.*, 1996). However, future studies may address in greater detail the antiapoptotic properties and the effects of IL-7Rα-mediated signals on gene rearrangements in immune cells.

IL-7 AND B LYMPHOCYTES

The most compelling evidence that IL-7 represents an important lymphopoietin and possibly one with clinical importance comes from a number of *in vivo* investigations. IL-7 administration to normal mice (5 μg twice daily for 4–7 days) results in a two- to five-fold increase in the number of peripheral and splenic white cells with no significant change in bone marrow cellularity. Analysis of the bone marrow showed an increase in B-cell precursors (B220$^+$, secretory immunoglobulin (sIg)$^-$) with a concurrent decrease in 8C5 and MAC-1 cells (myelomonocytic marker positive) (Damia *et al.*, 1992).

A general scheme for B-cell maturation is outlined in Table 1. For purposes of clarity, the nomenclature of Hardy and coworkers, defining the early stages of differentiation of murine B cells, has been adapted. These cells can be identified in liver or bone marrow and are divided into distinct classes (A–F) based on cell surface marker expression (Hardy *et al.*, 1991; Li *et al.*, 1993). Adult stem cells develop into 'conventional' B2 cells. Fetal liver stem cells are capable of differentiating into B1 cells, which persist in adult animals, reside primarily into the peritoneal cavity and stain positively for the CD5 antigen. The role of B1 and B2 cells in the context of IL-7 is discussed further below in the section entitled 'IL-7 and antimicrobial immune responses'. The early stages of B-cell development will occur in the bone marrow in response to stromal cell contact and cytokines. Hematopoietic stem cells (HSCs) of the adult bone marrow have been characterized by cell surface marker analysis. HSCs can be derived from murine bone marrow using the CD34 (sialomucin) antibody; other cell surface markers include the antigens CD4, major histocompatibility complex class I, ER-MP12 and AA4.1 (Katz *et al.*, 1985; Berenson *et al.*, 1988; Szilvassy *et al.*, 1989; Wineman *et al.*, 1992; Orlic *et al.*, 1993; Slieker *et al.*, 1993; Szilvassy and Cory, 1993).

Additionally, B-cell differentiation may be defined by DJ or VDJ rearrangement (see Table 1; Hardy *et al.*, 1991; Hardy and Hayakawa, 1991). The antigen receptors of B cells (and those of T cells) are encoded in the germline by individual DNA segments, termed V, D and J, which are joined during lymphocyte differentiation. This process (VDJ recombination) is initiated by the RAG-1 and RAG-2 proteins, which act together at the junctions between the coding segments and the recombination signal sequence to produce two types of DNA ends: a signal end (terminating in a blunt double-stranded break) and a coding end, which terminates in a DNA hairpin. The involvement of double-stranded DNA cleavage has suggested that this process is linked to the cell cycle; several lines of evidence indicate that the initiation of VDJ recombination takes place in the G_0–G_1 phase of the cell cycle (Oettinger *et al.*, 1990; Lewis, 1994). IL-7 appears to sustain expression of the *RAG-1* and *RAG-2* genes (Muegge *et al.*, 1993). The precise mechanism of this process is ill defined. However, more recent data suggest that IL-7 does not alter RAG mRNA levels, but rather affects post-transcriptional regulatory mechanisms. Alternatively, other as yet undefined IL-7-responsive gene products may additionally be involved, as IL-7 appears to be required for induction as well as for maintenance of VDJ recombination. In contrast, IL-7 reduces VDJ recombinatorial events in pre-B cells (Dobbeling, 1996).

To discriminate progenitor cells from cells that are already commited to the B-cell lineage, Hardy and coworkers recently investigated bone marrow stromal cells for expression of the B-cell lineage marker B220 and HSA in combination with the CD4 and AA4.1 markers (Li *et al.*, 1996). The latter marker is expressed on HSCs, B-cell–myeloid progenitors and early B-cell lineage cells (McKearn *et al.*, 1985; Loken *et al.*, 1988; Cumano and Paige, 1992). About 50% of the B220$^+$, CD43$^+$ and HSA$^-$ cells (formerly termed A) stained positive for AA4.1 expression (Li *et al.*, 1996). This cell population was capable of proliferating on a stromal cell layer, indicating that it may indeed represent B-cell lineage precursors. Thus, the earlier designation of fraction-A B cells had to be revised. Two AA4.1 fractions (A_1 and A_2) appear to represent the

Table 1. IL-7 in B-cell lineage commitment. Cell surface marker expression, V(D)J rearrangement according to Hardy and Hayakawa (1991), Hardy *et al.* (1991), Kitamura *et al.* (1991, 1992), Peschon *et al.* (1994), von Freeden-Jeffrey *et al.* (1995) and Li *et al.* (1996).

Classification	Characteristic cell surface markers	Expression of antigen receptors	IL-7 responsiveness	Anatomic compartment
Pre-pro-B cells: A_0	$B220^-$, $CD43^+$, $AA41^+$, $CD4^{low+}$	A_0 cells may not yet be lineage committed		Bone marrow
Pre-pro-B cells: A_1	$B220^+$, $CD43^+$, $AA41^+$, $CD4^{low+}$			Bone marrow
Pre-pro-B cells: A_2	$B220^+$, $CD43^+$, $AA41^+$, $CD4^{low-}$	Dependent on stromal contact for growth; immunoglobulin genes in germline configuration	Not IL-7 responsive	Bone marrow
Early pro-B cells: B	$B220^+$, $CD43^+$, $AA41^+$, $CD4^{low-}$, $CD19^+$ upregulation of heat-stable antigen(HSA)	DJ rearrangement	Growth in response to IL-7 and stroma	Bone marrow
Late pro-B cells: C	$B220^+$, $CD43^+$, HSA^+, $CD19^+$	VDJ rearrangement has occurred	IL-7 response in the absence of stroma	Bone marrow, periphery
Pre-B cells: D	$B220^+$, $CD19^+$, downregulation of CD43 expression		IL-7 alone may stimulate the *in vitro* growth of early pre-B cells, but not late pre-B cells. Development of the small resting B cells requires membrane-bound immunoglobulin heavy chain, λ5 and IL-7	Bone marrow, periphery
Immature B cells: E	$B220^+$, $CD43^-$, $CD19^+$	Light chain rearrangement and detection of IgM on the cell surface (sIgM)		Bone marrow, periphery
Mature B cells: F	$B220^+$, $CD43^-$, $CD19^+$	Encounter of antigen in association with T-cell help may lead to proliferation. Somatic hypermutation of immunoglobulin genes. In response to cytokines, mature B cells undergo immunoglobulin class switching		(Bone marrow), periphery

earliest stages of B-cell lineage development. The B220$^-$, AA4.1$^+$, CD4^{low} fraction has been designated as A_0 cells and appears to represent yet uncommited progenitor cells. However, these 'earliest' stages identified in B-cell development will have to be characterized for activity of B-cell differentiation factors, such as IL-7, kit ligand (Flanagan and Leder, 1990; Williams *et al.*, 1990) and flk2/flt3 ligand (FL) (Matthews *et al.*, 1991; Rosnet *et al.*, 1991). More recent studies have in part addressed this issue. IL-7 does not support *in vitro* growth of cells of the granulocytic–monocytic or erythroid lineage, but does stimulate eosinophil colony formation. This activity can be abolished by anti-IL-5 antibody treatment, suggesting that IL-7 acts by stimulating release of IL-5 or that potentially IL-5 represents an obligate cofactor (Vellenga *et al.*, 1992).

The growth factor combination of IL-11 and mast cell growth factor (MGF) supports bipotential progenitor cells to commit either to the B or to the macrophage lineage (Kee and Paige, 1996). Single-cell cloning assays suggest that IL-7 does not act directly to determine whether cells commit to the B-cell or macrophage lineage. However, bipotential cells respond to IL-7 by an increase in number, and IL-7 added to the IL-11–MGF mixture promotes expression of mRNA transcripts coding for B cell-specific genes (Kee and Paige, 1996). Furthermore, the growth factor combination of IL-11–flt3 ligand–IL-7 appears to maintain the potential of bipotential precursors (Ray *et al.*, 1996). Yet, in a different report, uncommited Lin-SCA-1$^+$ (SCA, stem cell antigen) bone marrow progenitor cells were shown to differentiate into B220$^+$, CD43$^+$, HSA$^+$ B cells (without expressing cytoplasmic μ heavy chain or sIgM) using a combination of flt3 ligand and IL-7 which proved to be superior in driving B-cell differentiation compared with the combination of stem cell factor and IL-7; the latter combination leads to the production of mature granulocytes (Veiby *et al.*, 1996a,b).

Concerning already committed B cells, early pro-B cells require a combination of IL-7 and factors provided by stromal cell layers; late pro-B cells are capable of proliferating in IL-7 without stromal cell support. The same has been found for early pre-B cells, but probably not for late pre-B cells (Hardy *et al.*, 1991; Hardy and Hayakawa, 1991). IL-7-mediated effects in B-cell differentiation may in part be mediated by regulation of the G_1–S transition of the cell cycle (Yasunaga *et al.*, 1995).

Rearrangement of κ light chains and sIgM expression correlates with IL-7Rα downregulation and therefore IL-7 unresponsiveness (Cumano *et al.*, 1990; Park *et al.*, 1990; Era *et al.*, 1991; Henderson *et al.*, 1992). In μ-chain transgenic animals, there is a reduction in the IL-7- and stromal cell-dependent cell population. In animals with a κ-chain transgene, an increase in IL-7-dependent cell populations could be observed. Expression of cytosolic μ chain promotes differentiation to an IL-7-dependent stage. The μ chain-positive cells with a functional light chain gene become IL-7 unresponsive. These results imply that B-cell precursors are driven to the next stage of differentiation by functional immunoglobulin molecules provided by the transgene (Era *et al.*, 1991).

The most precise data concerning the role of IL-7 in B-cell development are provided from IL-7$^{-/-}$ or IL-7R$\alpha^{-/-}$ mice. B lymphopoiesis in bone marrow appeared to be blocked at the transition to pre-B cells (see Table 1). IL-7$^{-/-}$ mice were blocked in the transition between the pro-B (fractions B/C B220$^+$/IgM$^-$/S7$^+$/HSA$^+$) to the pre-B-cell population (fraction D, B220$^+$, IgM$^-$, S7$^-$, HSA$^+$). Thus, differentiation and maturation of B–C fraction B cells to fraction D appears to be IL-7 dependent (Von Freeden-

Jeffrey *et al.*, 1995; Moore *et al.*, 1996). However, IL-7 receptor (IL-7Rα) gene-deleted mice show a block in B-cell development at the transition of pre-pro-B cells (formerly fraction A) to pro-B cells (fraction B) (Peschon *et al.*, 1994). This may be due to the action of other growth factors, potentially the thymic stroma-derived lymphopoietin (TSLP) (Friend *et al.*, 1994; Peschon *et al.*, 1994) or flt3 ligand (Namikawa *et al.*, 1996). Application of IL-7 neutralizing monoclonal antibodies of mice resulted in a similar B-cell maturation blockade to that observed in IL-7Rα knockout animals, but not to the B-cell maturation blockade observed in the IL-7 gene-deleted animals (Grabstein *et al.*, 1993; Peschon *et al.*, 1994). One potential explanation is that other cytokines (e.g. TSLP) may utilize the IL-7 receptor as well. Other cytokines, including TLSP, stem cell factor (SCF)/c-kit or flk2/flt3 ligand (Veiby *et al.*, 1996a,b), may synergize with IL-7 to regulate B-cell development. The SCF–kit ligand, which represents a growth factor for myeloid and erythroid progenitor cells, synergizes with IL-7 in stimulating B-cell precursor cells (McNiece *et al.*, 1991; Billips *et al.*, 1992; Funk *et al.*, 1993). However, some cytokines appear to counteract the IL-7-mediated effects. For instance, IL-1α (Suda *et al.*, 1989), IFN-γ (Garvy and Riley, 1994) and transforming growth factor-β (TGF-β) (Lee *et al.*, 1989) are able to inhibit IL-7-mediated B-cell precursor growth.

Additionally, a number of genes involved in B-cell development may be upregulated by IL-7, including n-*myc*, c-*myc* (Morrow *et al.*, 1992), CD19 (Wolf *et al.*, 1993), the precursor lymphocyte-specific regulatory light chain (PLRLC) (Oltz *et al.*, 1992) and the aminopeptidase BP-1/6C3. Incubation with IL-7 is associated with an increase in 6C3Ag expression by pre-B cells, but not mature B cells. The BP-1/6C3 molecule is expressed by early B-lineage cells and some stromal cells, and represents a type II integral membrane glycoprotein that belongs to the zinc family of metallopeptidases (Sherwood and Weissman, 1990).

In humans, IL-7 does not stimulate proliferation of B-cell lineage cells expressing CD24 (heat-stable antigen). Human pro-B cells but not pre-B cells respond to IL-7 (Ryan *et al.*, 1994; Dittel and LeBien, 1995); this is in contrast to the data for murine cells which suggests that species-specific differences in mode of action exist between humans and mice (Tushinski *et al.*, 1991). This human–rodent dichotomy exists for other cytokines—perhaps most notably IL-4.

In general, human HSC commitment and differentiation has not been as extensively characterized as that of the murine system. However, recent data suggest that certain stages of human B-cell development may not necessarily depend on the presence of IL-7. Using a human bone marrow stromal cell culture system, human HSC CD34 cells underwent commitment, differentiation and expansion into the B-cell lineage as defined by loss of CD34, increased CD19 cell surface expression, and appearance of μ/κ or μ/λ cell surface immunoglobulin receptor expressing immature B cells. This was not significantly influenced either by exogenously added IL-7 or by addition of anti-IL-7 neutralizing antibody (Prieyl and Le Bien, 1996). The implementation of the flt3 ligand in combination with IL-7 or IL-3 using human fetal bone marrow-derived CD34 CD19$^+$ pro-B cells in a stromal cell-independent and serum-deprived culture system revealed that flt3 ligand, like IL-3, synergizes with IL-7 in promoting B-cell growth and differentiation of the majority of cells into CD43$^-$, CD19$^+$, c (cytoplasmic) IgM$^+$, sIgM$^-$ pre-B cells; a minority of pro-B cells matured into sIgM$^+$ B cells (Namikawa *et al.*, 1996). However, the precise role of IL-7 in human B-cell commitment and differentiation has to be analyzed further.

IL-7 AND T LYMPHOCYTES

IL-7 added to murine fetal thymic organ cultures (day 13) causes a preferential expansion of immature cells exhibiting the $CD4^-$, $CD8^- CD3^-$, $CD2^-$, $SCA\text{-}1^+$ phenotype. Cells expressing $\gamma\delta^+$ TCR are increased and the number of $\alpha\beta^+$ TCRs is decreased. Neutralizing anti-IL-7 antibody inhibits growth of fetal thymocytes (Leclercq *et al.*, 1992; Plum *et al.*, 1993). *In vitro* culture of human fetal thymocytes in recombinant IL-7 results in the proliferation of $CD4^+$ and $CD8^+$ thymocytes and partial differentiation of thymocytes with preferential expansion of the $CD4^+ CD8^-$ population (Uckun *et al.*, 1991). IL-7 promotes the growth of pre-T cells from fetal liver at day 14 and promotes the expression of TCR-γ, α and β genes (Appasamy, 1992). IL-7 mRNA can be detected in the fetal thymus as early as day 12, peaking at day 15 (Wiles *et al.*, 1992). IL-7 stimulates the generation of $CD3^+$ cells from human bone marrow cultures, with the production of both $CD4^+$ and $CD8^+$ populations (Tushinski *et al.*, 1991). These results suggest that IL-7 may be produced locally in the thymic and bone marrow microenvironments and that it plays a role in the proliferation and potential differentiation of immature T cells (Watanabe *et al.*, 1992). Similar studies have indicated that IL-7 induces the proliferation and maintenance of T-lymphocyte numbers, but not T-cell differentiation. However, with the advent of IL-7, or IL-7Rα gene-deleted mice, several central questions concerning the role of IL-7 in lymphopoiesis could be addressed in more detail.

The macroscopic examination of IL-$7^{-/-}$ mice indicated apparently normal development of both fertile sexes. The lymphatic organs or tissues, including thymus and spleen, were dramatically reduced in size and the peripheral lymph nodes and immune cells within Peyer's patches were not detectable (Von Freeden-Jeffrey *et al.*, 1995). Accordingly, the reduced white blood count in IL-7 gene-deleted mice appeared to be due to an absolute reduction in lymphocytes. However, the normal ratio, as well as the absolute numbers of granulocytes and monocytes, was decreased. Overall, the massive lymphocyte reduction in these animals was due to decreased B- and T-cell numbers (Von Freeden-Jeffrey *et al.*, 1995; Moore *et al.*, 1996), reflecting the inefficient thymic development of IL-7-deficient mice. Only 5% of normal thymocyte numbers and 15% of splenic cell numbers could be detected in IL-7 gene-deleted mice (Von Freeden-Jeffrey *et al.*, 1995; Moore *et al.*, 1996). However, these remaining cells appeared to be similar to those observed in normal mice with regard to function, as defined by testing B cells in response to lipopolysaccharide (LPS), splenic T cells to concanavalin A, or proliferation of thymocytes to a mixture of concanavalin A and IL-2 (Von Freeden-Jeffrey *et al.*, 1995).

Similar T-cell abnormalities to those observed in IL-7 gene-deleted mice have been identified in IL-2Rγ receptor chain knockout mice (Takeshita *et al.*, 1992; Noguchi *et al.*, 1993; DiSanto *et al.*, 1994). As discussed above, the common γc chain is shared by several other cytokines, including IL-2, IL-4, IL-9 and IL-15 (Takeshita *et al.*, 1992; Kondo *et al.*, 1993; Giri *et al.*, 1994). Because IL-2 or IL-4 gene-deleted mice do not exhibit defects in T-cell development, IL-7, but not other cytokines, appears to account for most of the lymphocyte defects observed in murine models of X-SCID associated with abnormalities of the γc chain receptor (Takeshita *et al.*, 1992, 1993; Noguchi *et al.*, 1993; DiSanto *et al.*, 1994).

Thymic T-cell development has been separated into sequential stages based on expression of distinct cell surface markers. Thymic IL-7 is produced primarily during fetal development (Chantry *et al.*, 1989; Conlon *et al.*, 1989; Okazaki *et al.*, 1989). $CD4^- CD8^-$ fetal and adult immature thymocytes proliferate well in response to IL-7. In contrast, $CD4^+ CD8^+$ thymocytes respond rather poorly. The capability to respond to IL-7 correlates with expression of the IL-7 receptor α chain (IL-7Rα) expressed by $CD4^- CD8^-$, $CD4^+ CD8^-$ and $CD4^- CD8^+$, but not by $CD4^+ CD8^+$ thymocytes (Chantry *et al.*, 1989; Conlon *et al.*, 1989; Okazaki *et al.*, 1989; Everson *et al.*, 1990; Suda and Zlotnik, 1991). Additional studies have indicated that IL-7 mediates effects on TCR rearrangement. T-cell precursors from thymus or fetal liver cultured in IL-7 express rearranged β- or γ-chain transcripts (Appasamy, 1992; Appasamy *et al.*, 1993; Muegge *et al.*, 1993). IL-7, sustaining expression of the *RAG* genes (Muegge *et al.*, 1993; see above) induces rearrangement of Vγ2 and Vγ4, but not Vγ3 or Vγ5, TCR chains in mice (Appasamy *et al.*, 1993). Further evaluation of IL-7 gene-deleted mice showed reduced numbers of total T lymphocytes with preservation of the normal CD4/CD8 ratio and an increased percentage of $\alpha\beta^+$ T cells compared with $\gamma\delta^+$ T cells (Von Freeden-Jeffrey *et al.*, 1995; Moore *et al.*, 1996). These data suggest that proliferation, and not T-cell differentiation, may be affected. However, more recent data indicate that IL-7 may also be involved in T-cell differentiation.

Immature thymocytes have been divided into four distinct phenotypes based on differential expression of the cell surface markers CD25, CD44 and CD117 (c-kit). $CD4^{low}$ cells ($CD44^+$, $CD25^-$, $CD117^+$, $CD3^- CD8^-$) and pro-T cells ($CD44^+$, $CD25^+$, $CD117^+$), representing the early stages of thymic differentiation, are present in $IL7^{-/-}$ mice. In contrast, transition of pro-T cells to pre-T cells ($CD44^- CD25^+$, $CD117^-$) and post-pre-T cells ($CD44^-$, $CD25^-$, $CD117^-$) could not be detected in IL-7 gene-deleted mice (Moore *et al.*, 1995, 1996). Interestingly, lack of IL-7 in such animals resulted in decreased expression of the CD117 (c-kit) marker on $CD4^{low}$ and pro-T cells as well, indicating that IL-7 may induce expression of yet undefined cytokine receptors during thymic T-cell maturation. Based on these data, current models of IL-7-mediated effects may have to be revised, because IL-7 may be critically involved in T-cell differentiation and not only in thymocyte proliferation. However, other thymic factors (e.g. TSLP) may also be critical for thymic differentiation. Future studies may address whether IL-7-deficient animals exhibit a qualitatively different TCR repertoire in peripheral $\alpha\beta^+$ T lymphocytes, particularly in variable TCR chain transcripts which have been shown to be influenced by IL-7 (Appasamy, 1992; Muegge *et al.*, 1993).

More recent studies have scrutinized the role of IL-7 in the development of $\gamma\delta^+$ T lymphocytes. $IL\text{-}7^{-/-}$ mice showed a profound reduction of $CD4^- CD8^- \gamma\delta^+$ T cells to approximately 1% of normal levels (Von Freeden-Jeffrey *et al.*, 1995; Moore *et al.*, 1996). A substantial body of evidence supports the notion that IL-7 preferentially promotes development of $\gamma\delta$ TCR^+ thymocytes over $\alpha\beta^+$ thymocytes, as a result of differential IL-7Rα expression on $\gamma\delta^+$ thymocytes compared with $\alpha\beta^+$ TCR thymocytes. This notion is supported by the fact that $\gamma\delta^+$ T cells are absent in thymus, gut, liver and spleen in IL-7R$\alpha^{-/-}$ mice (Peschon *et al.*, 1994).

In contrast, $\alpha\beta$ TCR^+ lymphocytes, and NK cells appear to be reduced in number, but to develop normally (He and Malek, 1996; Maki *et al.*, 1996). However, $NK1^+$ T cells can be detected in thymus, liver and spleen of IL-7R$\alpha^{-/-}$ mice. Recent data suggested that differentiation of these $NK1^+$ cells is dependent of signaling via the γc chain and expansion on IL-7Rα-mediated signals (Boesteanu *et al.*, 1997). These results provide reasonable evidence that signal transduction mediated by the IL-7 receptor is a prerequisite for $\gamma\delta$ T-cell development in both thymic and extrathymic pathways. Of note, thymocyte development in IL-7R$\alpha^{-/-}$ mice can be reconstituted by the introduction of a transgenic TCR, suggesting that one of several functions of the IL-7Rα may be to initiate TCR gene rearrangement. This notion is further consolidated by the observation that expression of the *RAG-1* and *RAG-2* genes is also significantly reduced in the thymus of IL-7R$\alpha^{-/-}$ mice, but restored in double-positive thymocytes observed in TCR transgenic IL-7R$\alpha^{-/-}$ mice (Crompton *et al.*, 1997). Thus, signaling through the IL-7Rα appears be necessary for *RAG* expression and initiation of VDJ rearrangement, as described for VDJ recombinatorial events in B-cell differentiation (see above). VDJ rearrangement may impact on organ-specific immunity. For instance, pulmonary cells with the canonical fetal-type Vγ6 chain are missing in nude mice owing to a preferred thymic pathway of TCR gene rearrangement, and not to thymic selection. These cells can be restored *in vitro* and *in vivo* by administration of IL-7 (Hayes *et al.*, 1996).

In murine fetal development, T-cell production can be detected at day 15 of gestation. T cells at this stage express the invariant TCR complex composed of Vγ3 and Vδ1 chains. Maturation of thymocytes is accompanied by differential expression of CD24 (heat-stable antigen) expression. First, immature Vγ3 cells exhibit a TCR Vγ3^{low} and CD24^{+} phenotype, and progress to mature Vγ3^{high} and CD24^{-} cells. These $\gamma\delta^{+}$ T cells may populate the epidermis, or potentially other epithelial sites, and represent the dendritic epidermal T cells (DETCs). Alternatively, $\gamma\delta^{+}$ T cells may mature extrathymically. Interestingly, IL-7$^{-/-}$ mice characteristically exhibit a block of maturation of Vγ3^{low} CD24^{+} T cells to Vγ3^{high} CD24^{low} T cells (Moore *et al.*, 1996). This observation provides another piece of evidence that IL-7 does not serve exclusively as a 'maintenance' factor for thymocytes, but may also be involved in T-cell maturation and differentiation.

In recent years, characterization of T lymphocytes residing primarily in the intestine (intestinal intraepithelial lymphocytes; iIELs) has revealed a distinct phenotype as well as different functional activity for such immune cells compared with 'conventional' $\alpha\beta$ T cells in the periphery (Van Kerckhove *et al.*, 1992; Boismenu and Havran, 1994; Guy-Grand *et al.*, 1994; Havran and Boismenu, 1994; Rocha *et al.*, 1994). Of note, thymic and intestinal epithelial cells share the same embryologic origin as they are both derived from entoderm and may both be capable of secreting IL-7 *in situ* (Namen *et al.*, 1988a; Heufler *et al.*, 1993; Matsue *et al.*, 1993a,b; Ariizumi *et al.*, 1995; Watanabe *et al.*, 1995; Maeurer *et al.*, 1997). Thus, given the fact that IL-7$^{-/-}$ mice (Moore *et al.*, 1995, 1996; Von Freeden-Jeffrey *et al.*, 1995), γc chain knockout mice (Takeshita *et al.*, 1992; Kondo *et al.*, 1993; Noguchi *et al.*, 1993; DiSanto *et al.*, 1994) as well as JAK3-deficient mice (Nosaka *et al.*, 1995; Park *et al.*, 1995) lack $\gamma\delta^{+}$ T cells, IL-7 appears to represent the major growth/differentiation factor required for thymic and extrathymic development of $\gamma\delta$ T cells. Of note, $\alpha\beta^{+}$ TCR iIELs are detectable in IL-7$^{-/-}$ mice, but not in γc or in JAK3-deficient mice, suggesting that other cytokines may be critical for generation of $\alpha\beta$ TCR^{+} iIELs, but not necessarily for $\gamma\delta^{+}$ TCR iIELs (Takeshita *et al.*, 1992; Noguchi *et al.*, 1993; Di Santo *et al.*, 1994; Moore *et al.*, 1995, 1996; Nosaka *et al.*, 1995; Park *et al.*, 1995; Von Freeden-Jeffrey *et al.*, 1995).

There is strong experimental evidence that TCR^{+} iIELs may develop *in situ*. Such immune cells are present in both congenitally athymic nude mice and in athymic radiation chimeras (for review see Poussier and Julius, 1994; Klein, 1996). Much more controversy surrounds the origin of the various subsets of iIELs which show a limited TCR repertoire (Van Kerckhove *et al.*, 1992; Guy-Grand *et al.*, 1994; Poussier and Julius, 1994). IL-7 gene-deleted mice may help to define the impact of IL-7 in the generation and TCR composition of $\alpha\beta^+$ TCR lymphocytes at different anatomic sites, preferentially in the intestine. Recently, clusters of lymphocytes located in crypt lamina propria (designated cryptopatches) have been characterized within the murine small and large intestinal mucosa (Kanamori *et al.*, 1996). Such lymphoid cells are characterized by cell surface expression of $CD117^+$ (c-kit), IL-7Rα^+ and $Thy1^+$, and by the absence of markers for CD3, $\alpha\beta$ TCR, $\gamma\delta$ TCR, sIgM and B220. It has been proposed that the immune cell population first detected at days 14–17 after birth may represent the lymphohematopoietic progenitors for T and B cells in the intestine. The prominent role of IL-7 in lymphopoietic development is further underscored by the observation that such cryptopatch-associated lymphoid cells are virtually absent in IL-7Rα-deficient mice (Kanamori *et al.*, 1996).

IL-7 IN ANTI-MICROBIAL IMMUNE RESPONSES

The role of cytokines in regulation of the host immune response to intracellular and extracellular pathogens has become increasingly understood. Regarding the infection of mice or humans with obligate intracellular pathogens, the T helper cell-1 (T_{H1})-type response, as defined by secretion of IFN-γ, IL-2 and IL-12, appears to represent the principal mediator directed against intracellular infection. Recently, a number of studies have indicated a central role for IL-7 in infections with intracellular bacteria or parasites. Interestingly, IL-7 has a somewhat 'Janus-faced' role, depending on the infection model studied or on the time of IL-7 application in the course of the disease.

Examples of apparently beneficial effects are to be found in murine models of infections with *Mycobacterium* species, or with the parasite *Toxoplasma gondii*. Female A/J mice treated with IL-7, commencing at the time of infection (2 μg daily for 2 weeks) with *T. gondii*, survived. In contrast, mice treated after infection (or not treated) died. *In vivo* depletion experiments have revealed that asialo $GM1^+$ NK cells as well $CD8^+$ T cells are required for protection against the intracellular parasite. Additionally, the IL-7-mediated effects appear to be predominantly mediated by IFN-γ secretion, as *in vivo* depletion of IFN-γ abolished the IL-7 protective effects (Kasper *et al.*, 1995).

In a different infection model, a combination of IL-7 with IL-1β augments anti-*Listeria monocytogenes*-directed immune responses in mice. The cellular immune response is predominantly mediated by peritoneal $\gamma\delta$ T lymphocytes which specifically react to heat-killed *Listeria* preparations in the presence of macrophages as accessory cells in a non-H2-restricted manner. Additionally, the IL-7 responsiveness of $\gamma\delta$ T cells was enhanced in the presence of accessory cells. This effect could be replaced by exogenous IL-1 (Skeen and Ziegler, 1993).

Similarly, IL-7 appears to be involved in the successful immune response directed against infections with mycobacteria. The infection with *Mycobacterium leprae* represents a particular spectrum of the disease, in which the clinical manifestations correlate with the quantity and quality of the cellular immune response. Increased IL-7 mRNA and IL-7 receptor mRNA expression correlates with the tuberculoid form of the disease, in which the infection is limited. In contrast, the lepromatous form, which shows progressing disease, does not show significant IL-7 mRNA expression (Sieling *et al.*, 1995). Additionally, IL-7 inhibits the intracellular growth of *M. avium complex* (MAC) in human macrophages *in vitro* (Tantawichien *et al.*, 1996). MAC represents a common opportunistic pathogen common in patients with human immunodeficiency virus (HIV) infection. Mononuclear phagocytes represent a major reservoir for MAC in the susceptible host. Such infections are often resistant to standard treatment protocols. Therefore, additional treatment modalities may be required to control MAC infections. For instance, previous studies have shown that treatment of human macrophages with tumor necrosis factor-γ (TNF-α) or granulocyte–macrophage colony stimulating factor (GM-CSF) leads to mycobacteriostatic or mycobactericidal activity (Denis, 1991). Additionally, MAC infection of human macrophages results in the generation of TGF-β, which inhibits the capacity of infected cells to control bacterial growth (Bermudez, 1993). Treatment of human macrophages with IL-7 results in a dose-dependent reduction in the number of intracellular bacteria. When IL-7 was added to cultured macrophages before infection, the anti-MAC activity was diminished compared with that obtained when IL-7 was added to MAC-infected cells (Tantawichien *et al.*, 1996). The authors have obtained similar results with virulent *M. tuberculosis* bacilli.

Treatment of Balb/c mice preinfected with *M. tuberculosis* resulted in up to 100% increased survival compared with that in non-treated mice, or in mice treated with IL-2 or IL-4. These IL-7-mediated effects can be transferred passively, using spleen cells derived from IL-7 treated and *M. tuberculosis*-infected animals, to mice that have been preinfected with mycobacteria. In contrast, transfer of cells from mice treated with IL-7 alone did not result in an increased survival rate compared with control animals, suggesting that priming with *M. tuberculosis* is required to elicit antimicrobial immune responses facilitated by IL-7 treatment (the authors' unpublished observations). In other studies using IL-7 as a treatment, the IL-7-mediated effects could be abolished by antihuman TNF-α antibody. In contrast, IL-7 did not decrease the TGF-β secretion by macrophages upon infection, an observation that was found to be true for IL-7-mediated downregulation of IL-2 or LPS-induced TGF-β mRNA expression in murine macrophages (Dubinett *et al.*, 1993). Therefore, IL-7 may exert some of its effects by inducing or potentiating proinflammatory cytokines, such as IL-1α, IL-1β, IL-6 and TNF-α (Alderson *et al.*, 1991). Additionally, some of the antibactericidal effects of macrophages are mediated by generation of nitric oxide and superoxide radicals, both of which are induced by IL-7 (Alderson *et al.*, 1991; Gessner *et al.*, 1993).

However, IL-7 does not always appear clinically to benefit animals with intracellular infections. For instance, earlier results indicated that IL-7 mediates antimicrobial activity against the intracellular parasite *Leishmania major* in murine macrophages *in vitro* (Gessner *et al.*, 1993). In contrast, treatment of susceptible Balb/c mice with IL-7 at the onset of infection leads to enhanced lesion development and accelerates death of treated animals, correlating with an up to 40-fold increased parasite burden in the spleen and lymph nodes. Analysis of cellular immune responses of such animals has revealed that lymphocytes obtained from IL-7-treated mice produced comparable amounts of the T_{H2} cytokines IL-4 and IL-10, but less IFN-γ in response to antigen (Gessner *et al.*, 1995). This observation suggests that a number of other factors may be involved in the complex interactions of cytokines; for instance, IL-7 upregulates the anti-CD3, or anti-CD3/anti-CD28-induced IFN-γ and IL-4 mRNA expression in (human) T lymphocytes (Borger *et al.*, 1996).

One of the major alterations in the cellular composition of IL-7-treated animals appears to be a rise in total cell numbers in the B-cell compartment. To elucidate the nature of the potentially deleterious B-cell responses, mice with the X-SCID were evaluated. Such animals typically lack B1 cells and exhibit reduced numbers and functions of B2 cells. B1 cells (formerly referred to as Ly-1 B or $CD5^+$ B cells) represent a small subpopulation with a distinct phenotype, and developmental and functional properties. B1 cells express a unique array of cell surface molecules, in addition to expression of the CD5 marker; they are preferentially generated from fetal or neonatal sources of progenitors, and the antibodies derived from B1 cells are predominantly of the IgM class, show minimal hypermutation and a high frequency of low-affinity, poly- or self-reactive specificities (for review see Stall and Wells, 1996). The absence of these cells leads to reduced susceptibility against infections with intracellular parasites (e.g. *Leishmania* species). Of note, following application of a single IL-7 dose concomitant with *Leishmania* infection, the clinical course resembled that in susceptible Balb/mice with an up to 100-fold enhanced parasite load in treated animals. Again, examination of $CD4^+$ *Leishmania*-specific T lymphocytes revealed that IFN-γ secretion is reduced in IL-7-treated Balb/c X-SCID mice compared with that in control animals, and that the population of B2 ($B220^+$, $sIgM^+$, MHC class II^+) cells appeared to be significantly enhanced. However, the nature of the disease-aggravating effects of IL-7 remain to be elucidated. One potential mechanism may be antigen presentation by B cells, an event that may lead to preferential activation and expansion of the T_{H2}-type lymphocytes.

A similar dichotomy of IL-7 has emerged in infection with HIV. Previous studies suggested that exogenous recombinant human IL-7 (rhIL-7) is capable of augmenting the generation of antivirus-directed cytotoxic T-lymphocyte (CTL) responses (Carini *et al.*, 1994). Examination of HIV-infected individuals testing negative for anti-HIV-1-specific CTL reactivity indicated that $CD8^+$ and $CD4^+$ T cells lack IL-7 receptor cell surface expression, which may be attributed to production of insufficient numbers of IL-7R upon retroviral infection or, alternatively, to increased shedding of the IL-7 receptor (Carini and Essex, 1994; Carini *et al.*, 1994). Because HIV infection is accompanied with reduced numbers in the $CD4^+$ cell compartment and associated with loss of cytotoxic $CD8^+$ T-cell activity, several cytokines capable of modulating the immune system have been contemplated for implementation in treatment of HIV-positive individuals. IL-7 is such a candidate (as well as IL-2, IL-12 and IL-15) because it not only enhances anti-HIV-directed $CD8^+$ T-cell responses, but also augments both $CD4^+$ T helper cell-dependent humoral immune responses and $CD8^+$ cytotoxic T-cell reactivity in mice immunized with the HIV envelope protein (Bui *et al.*, 1994).

However, caution must be exercised before implementing these cytokines, including IL-7, into clinical protocols, as exogenous IL-7 induces virus replication and increases proviral DNA levels in PBMC cultures, and increases the levels of doubly spliced HIV-1 TAT RNA (Smithgall *et al.*, 1996). These effects are not inhibited by neutralizing IL-1β, IL-2, IL-6 or TNF-α activity. However, $CD8^+$ T cells inhibit the increase in viral replication induced by IL-7 stimulation, although they do not prevent virus replication following CD3 ligation in the presence of IL-7, an event that can also be mimicked by adding IL-7 to anti-CD3 antibody-stimulated HIV^+ PBMC cultures, resulting in enhanced HIV production (Moran *et al.*, 1993). However, the results obtained from such studies have addressed the role of exogenous IL-7 in HIV replication *in vitro*. They have not addressed the role of endogenous IL-7 on viral load or viremia.

The concentration of IL-7 in plasma from HIV-seronegative individuals as measured by enzyme-linked immunosorbent assay ranges from 1.6 to 9.8 pg/ml (R&D Systems, 1994). These concentrations are 1000-fold lower than the amount of IL-7 implemented for *in vitro* assays. However, there are clinical conditions in which raised IL-7 levels have been determined. For instance, plasma and synovial fluid levels of IL-7 are increased significantly in patients with systemic juvenile rheumatoid arthritis, but not in those with polyarticular or pauciarticular juvenile rheumatoid arthritis, or in patients with other rheumatic diseases (De Benedetti *et al.*, 1995), in patients with untreated Hodgkin's lymphoma (Trumper *et al.*, 1994) and in some with colorectal or renal cell cancer (authors' unpublished observation). However, the precise source and the functional consequence of such increased IL-7 serum levels remain to be determined.

IL-7 mRNA has been observed in a number of different infections, and exogenously added IL-7 (provided either by the recombinant protein or by retroviral infections) has been shown to augment specific cellular immune responses. For instance, IL-7 mRNA has been detected in patients with *Helicobacter pylori*-positive gastritis, but not in *H. pylori*-negative controls (Yamaoka *et al.*, 1995). Others have shown that IL-7 helps the induction of antiviral specific T-cell responses using synthetic peptides and IL-7 as an adjuvant (Kos and Mullbacher, 1992, 1993), or that IL-7 overcomes anergy in parasite-specific cellular immune responses (Sartono *et al.*, 1995), and facilitates expansion of tetanus toxoid (Kim *et al.*, 1994), or dengue virus-specific cytotoxic $CD4^+$ T-cell clones (Berrios *et al.*, 1996).

IL-7 IN 'TISSUE-SPECIFIC' IMMUNITY

Detection of IL-7 mRNA in various tissues has rekindled interest in the role of IL-7 in promoting local immune responses (see Table 2). IL-7 has been reported to be produced by human and murine keratinocytes (Heufler *et al.*, 1993; Matsue *et al.*, 1993a,b; Ariizumi *et al.*, 1995) and serves a major growth factor for dendritic epidermal T cells (DETCs) which express the $\gamma\delta$ TCR (Matsue *et al.*, 1993a,b). The mouse epidermis harbors a T-cell population characterized by expression of CD3, asialo-GM1, CD2, Thy-1, Ly48 and E-cadherin, but not CD4 or CD8 phenotypic markers (Steiner *et al.*, 1988). Such DETCs express the $\gamma\delta$ TCR composed of the Vγ3 and Vδ1 chains without junctional diversity (Steiner *et al.*, 1988; Matsue *et al.*, 1993a,b; Moore *et al.*, 1996). These $\gamma\delta$ T-cell effector cells may monitor stressed keratinocytes, or recognize class Ib antigens (e.g. TIa or Qa) (for review see Hayday, 1995). Keratinocytes express constitutively IL-7 mRNA for IL-7 and secrete *in vitro* biologically meaningful amounts of IL-7 protein. IL-7 appears not only to serve as the principal growth factor for DETCs (Fig. 2) but also prevents apoptosis in DETCs exposed to ultraviolet B radiation or corticosteroid treatment (Takashima *et al.*, 1995). This fits well into earlier observations that IL-7 is superior, when compared with IL-2, in maintaining viability and responsiveness in antigen-specific T-cell lines. Additionally, IFN-γ secreted by $\gamma\delta$ T cells appears to inhibit growth of murine keratinocytes (Takashima and Bergstresser, 1996). IL-7 has been shown to augment expression of leukocyte functional antigen-1 (LFA-1) and very late activation antigen-4 in human phorbol myristate acetate (PMA) and calcium-stimulated peripheral blood lymphocytes (PBLs), enhancing the capacity of these cells to adhere to parenchymal cell monolayers (Fratazzi and Carini, 1996). Additionally, IL-7 appears to induce cell surface expression of the costimulatory

Table 2. IL-7 mRNA or protein detection in cell lines and tissues.

Tissue or cell type	Detection of mRNA	Detection of protein	Reference
Bone marrow stromal cells	+ (h, m)	+ (h, m)	Namen *et al.* (1988a,b) Sudo *et al.* (1989) Witte *et al.* (1993)
Spleen	+ (h, m)	n.d.	Namen *et al.* (1988a,b) Goodwin *et al.* (1989)
Kidney	+ (h, m)	n.d.	Namen *et al.* (1988a,b) Authors' unpublished results
Kidney allograft	+ (h)	n.d.	Strehlau *et al.* (1997)
Renal cell cancer (RCC) tissue sections or RCC cell lines	+ (h)*	+ (h)	Authors' unpublished results
Fetal and adult thymus	+ (h, m)	+ (h, m)	Namen *et al.* (1988a,b) Goodwin *et al.* (1989) Montgomery and Dallman (1991) Sakata *et al.* (1990) Wiles *et al.* (1992) Authors' unpublished results
Thymic stromal cells	+ (m)	+	Sakata *et al.* (1990) Gutierrez and Palacios (1991)
Hassall's corpuscles	+ (h)	n.d.	He *et al.* (1995)
Keratinocytes	+ (h, m)	+ (h, m)	Heufler *et al.* (1993) Matsue *et al.* (1993a,b)
Intestinal epithelial cells, epithelial goblet cells	+ (h)	+ (h)	Watanabe *et al.* (1995)
Colorectal cancer cells	+ (h)*	+ (h)	Maeurer *et al.* (1997) Authors' unpublished results
Uterus	+ (m)	n.d.	Appasamy (1997)
Brain	+ (h)	n.d.	Appasamy (1997)
Adult liver	+ (r)	n.d.	Appasamy (1997)
Hepatocarcinoma	+ (h)*	n.d.	Goodwin *et al.* (1989)
EBV^+ B-cell lines	+ (h)	+ (h)	Benjamin *et al.* (1994)
Burkitt's lymphoma cells	+ (h)	+ (h)	Benjamin *et al.* (1994)
Chronic B-lymphocytic leukemia cells	+ (h)*	+ (h)	Frishman *et al.* (1993) Long *et al.* (1995)
Bladder cancer	+ (h)	n.d.	Kaashoek *et al.* (1991)
Inflammatory malignant fibrous histiocytoma	+ (h)	n.d.	Melhem *et al.* (1993)
Follicular dendritic cells	+ (h)*	+ (h)	Kröncke *et al.* (1996a,b)
Fibroblasts	+ (m)		Aiba *et al.* (1994)
Oral mucosa	+ (h)	+ (h)	Kröncke *et al.* (1996a,b)
Vascular endothelial cells	+ (h)	+ (h)	Kröncke *et al.* (1996a,b)
Hodgkin cell line, nodular sclerosing type	+ (h)	n.d.	Bargou *et al.* (1993)
Sézary lymphoma cells	+ (h)†	n.d.	Foss *et al.* (1994, 1995) Asadullah *et al.* (1996)†
Lesions from tuberculoid lepra	+ (h)	+ (h)	Sieling *et al.* (1995)

n.d., Not determined; h, human; m, mouse; r, rat.

*Cloning and sequence analysis of mRNA exhibits alternatively spliced forms(s) of IL-7; †IL-7 mRNA did not appear to be overexpressed as determined by semiquantitative analysis of skin biopsies from patients with mycosis fungoides or with pleiomorphic T-cell lymphoma compared with biopsies obtained from normal skin, psoriatic lesions or atopic dermatitis.

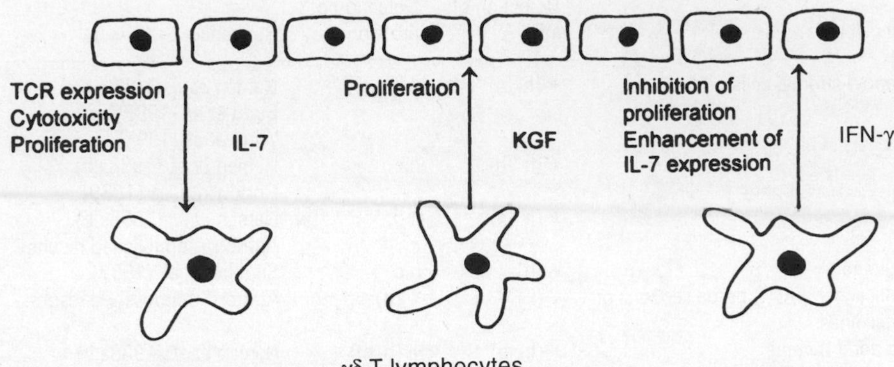

Fig. 2. Tissue-specific immunity: a central role for IL-7 in the interactive milieu of cytokines elaborated by epithelial cells and $\gamma\delta^+$ T lymphocytes. IL-7 secreted by epithelial cells (e.g. keratinocytes) *in situ* is able to maintain viability of $\gamma\delta$ T cells and is critically involved in thymic and extrathymic development of $\gamma\delta$ T cells. IFN-γ produced by $\gamma\delta^+$ T cells alters the IL-7 mRNA transcript pattern in murine keratinocytes (Ariizumi *et al.*, 1995). IFN-γ is able to augment IL-7 secretion by epithelial cells (Ariizumi *et al.*, 1995; Sieling *et al.*, 1995) and inhibits proliferation and immune functions of keratinocytes and downregulates the mitotic capacity of Langerhans cells (Sillevis Smith *et al.*, 1992; Matsue *et al.*, 1993b; Xu *et al.*, 1995). In contrast, keratinocyte growth factor (KGF) secreted by $\gamma\delta^+$ T cells induces proliferation of murine epithelial cells (Boismenu and Havran, 1994).

molecule B7/BB1 as well as intercellular adhesion molecule-1 (ICAM-1) (CD54) on pre-B cells (see Table 1). Acquisition of B7 molecule(s) may be biologically relevant if B cells act as antigen-presenting cells (Dennig and O'Reilly, 1994).

IL-7 as a growth factor for $\gamma\delta$ T cells homing to epidermis represents a critical growth factor in the evolution of contact sensitivity to trinitrochlorbenzene, which can be abrogated by administering monoclonal antibodies to $\gamma\delta$ T cells *in vivo*. $\gamma\delta$ T cells invading the respective antigen challenge site typically exhibit a $CD8\alpha^+$, $CD8\beta^-$, IL-4R$^+$ Vγ3$^+$ phenotype, and proliferate in response to IL-7 but not to IL-2 or IL-4. Moreover, *in vivo* application of IL-7 neutralizing antibody inhibits accumulation of Vγ3$^+$ T cells in the skin, as well as in the regional lymph nodes adjacent to the sensitization site (Dieli *et al.*, 1997). In addition to keratinocytes, other cell types (e.g. fibroblasts) within the epidermis may also provide biologically meaningful IL-7 levels *in vivo* (Aiba *et al.*, 1994). Perhaps the best illustration that IL-7 serves as the major growth factor for IELs is provided by studies from Williams and Kupper, who showed that the epidermal density of DETCs is increased substantially in keratin-14 promoter-driven IL-7 transgenic mice, in which ectopic IL-7 is produced exclusively by keratinocytes (cited as a personal communication in Takashima and Bergstresser, 1996).

The role of IL-7 in skin immune reactions is underscored by the observation that IL-7 mRNA appears to be upregulated in mite allergen patch test reactions in patients with atopic dermatitis (Yamada *et al.*, 1996). The site of positive patch reactions also tested positive for eosinophilic infiltrations. Noteworthy in this context, IL-7 is also capable of upregulating the low-affinity receptor for IgE (CD23) in activated PBLs (Fratazzi and Carini, 1996).

IL-7 may not only be involved in creating an interactive environment of keratinocytes and $\gamma\delta$ T cells; it may also play a role in the germinal center reaction (Kröncke *et al.*, 1996a). IL-7 mRNA and protein have been detected in human follicular dendritic cells (FDCs) obtained from tonsils. However, mature peripheral IgM$^+$ B cells are non-responsive to IL-7, whereas anti-μ-stimulated tonsilar B cells proliferate in response to IL-7 without secreting immunoglobulins, suggesting that IL-7 may be indeed be able to regulate B-cell responses in the periphery. In addition to skin and tonsils, IL-7 mRNA and protein have been detected in human intestinal cells (Reinecker and Podolsky, 1995; Watanabe *et al.*, 1995), in human colorectal tumor cells (Maeurer *et al.*, 1997), in normal kidney (Goodwin *et al.*, 1989) and in human renal cell cancer cell lines (authors' unpublished observation). These cells produce significant amounts of IL-7 protein *in vitro*. Of interest, raised IL-7 mRNA expression appears to represent one of the most sensitive markers of graft rejection in patients after kidney transplantation (Strehlau *et al.*, 1997).

IL-7 promotes the growth of lamina propria lymphocytes and inhibits CD3-dependent proliferation of these cells (Watanabe *et al.*, 1995). IL-7, in comparison with IL-2, promotes the preferential expansion of (short term, day 14) cultured tumor-infiltrating lymphocytes obtained from patients with colorectal cancer (Maeurer *et al.*, 1997). Long-term *in vitro* culture of human iIELs harvested from patients with colorectal cancer with IL-7 results in preferential outgrowth of V$\delta 1^+$ T cells (Maeurer *et al.*, 1995, 1996) which recognize colorectal cancer cells, renal cell cancer and pancreatic cancer cell lines (Maeurer *et al.*, 1996). Such immune effector cells release significant amounts of IFN-γ. Interestingly, human intestinal cells also appear to express the IL-7Rα. Stimulation of such cells with IL-7 leads to rapid tyrosine phosphorylation of proteins (Reinecker and Podolsky, 1995).

The physiologic role of such IL-7-responsive epithelial cell lines remains to be elucidated. Presumably, IL-7 may represent a member of a family of epithelial growth factors, which promote homing, maturation and maintenance of IL-7-responsive immune cells (see Fig. 2). Thus, IL-7 may be an important cytokine involved in creating an interactive environment of epithelial cells and lymphocytes. Murine $\gamma\delta$ T cells secrete keratinocyte growth factor (KGF), which promotes proliferation of epithelial cells (Boismenu and Havran, 1994). Conversely, human epidermal growth factor (EGF) (Reinecker and Podolsky, 1995) increases IL-7Rα mRNA expression in human colorectal cancer cell lines. Human KGF (authors' unpublished observation) stimulates IL-7 mRNA expression and IL-7 protein secretion by human intestinal cells. Therefore, EGF and/or KGF and IL-7 may represent cytokines involved in the homeostasis of epithelial and immune cells *in vivo* (see Fig. 2). The 'nourishing' capacity of intestinal cells is further substantiated by the observation that bone marrow cells develop into phenotypically mature T cells using a co-culture system with the intestinal epithelial cell line MODE-K (Vidal *et al.*, 1993; Maric *et al.*, 1996). Whether IL-7 is one of the principal factors mediating these effects remains to be determined.

More recently, several reports have addressed the role of IL-7 and IL-7Rα mRNA expression in developing tissues. The observation that IL-7 stimulates maturation of embryonic hippocampal progenitor cells in culture suggests that IL-7 may effect proliferation and differentiation of immature cells of non-hematopoietic origin (Mehler *et al.*, 1993). IL-7 and IL-7R mRNA expression can also be observed in the developing brain, and treatment of culture of embryonic brain with exogenous IL-7 leads to increased neuronal survival and greater numbers of cells exhibiting neurite outgrowth.

One of the IL-7-mediated effects may be via phosphorylation of $p59^{fyn}$ (Michaelson *et al.*, 1996).

IL-7 may also be involved in local immune reactions affecting the eye. The neuroectodermis-derived retinal pigment epithelium (RPE) contributes to the blood–retina barrier regulating infiltration of immune cells in retinal diseases. Activation of RPE cells leads to expression of MHC class II antigens and adhesion molecules. Additionally, IL-7 is able to induce monocyte chemotactic protein-1 and IL-8 in such RPE cells (Elner *et al.*, 1996). However, further studies may address whether IL-7 can be detected in retina-associated diseases *in vivo*.

IL-7 AND CANCER

IL-7 may play different roles in cancer-bearing hosts, dependent on the tumor and status of the immune system. First, IL-7 mRNA, IL-7 protein and the IL-7Rα have been demonstrated in some hematologic malignancies, suggesting that IL-7 may serve as a growth factor in an autocrine fashion. Some tumor cells exhibit expression of the IL-7Rα without IL-7 expression, and may be responsive to IL-7 provided by different cell types. Second, IL-7 may be implemented as a treatment for cancer as IL-7 increases immune effector cell functions by T lymphocytes, NK cells and macrophages. IL-7 may be provided by systemic application, or it may be secreted by genetically engineered tumor cells to induce a strong and long-lived immune response. Third, IL-7 may be one of several growth factors to be used for recovery from bone marrow transplantation in the setting of treatment for hematologic malignancy or, alternatively, for bone marrow rescue after high-dose chemotherapy treatment of solid tumors (e.g. breast cancer).

IL-7 and IL-7Rα Expression in Cancer

IL-7 mRNA and protein expression have been identified in solid tumors including colorectal and renal cell cancers (Watanabe *et al.*, 1995; Maeurer *et al.*, 1997; the authors' unpublished observations). Both tumor cell types express the IL-7Rα receptor and the common γc chain. IL-7 mRNA has also been observed in tumor cells of nodular sclerosing and mixed cellularity type of Hodgkin's disease (Bargou *et al.*, 1993; Foss *et al.*, 1995). The prominent immune cell infiltrate observed in most cases of Hodgkin's disease may be attributed to local delivery of IL-7 *in vivo*. IL-7 serum levels may also be increased in patients with Hodgkin's disease (Trumper *et al.*, 1994; Gorschluter *et al.*, 1995). Similarly, Sézary's lymphoma cells express IL-7Rα and proliferate in response to IL-7. However, some of these lymphoma cells obtained from different patients (3/5) also expressed IL-7 mRNA (Foss *et al.*, 1994).

It has been presumed that keratinocyte-secreted IL-7 may serve as a growth factor for cutaneous T-cell lymphomas. This hypothesis is substantiated by examination of IL-7 transgenic mice. In transgenic mice, in which the IL-7 gene is expressed under the control of the mouse MHC class II (Eα) promoter (Mertsching *et al.*, 1995), a lymphoproliferative syndrome characterized by early polyclonal expansion of T lymphocytes followed by development of pro-pre-B and bipotential myeloid/B-cell tumors can be observed in about 25% of C57Bl/6, and in up to 100% of Balb/c mice (Mertsching *et al.*, 1995, 1996). If the IL-7 gene is controlled by the Srα promoter, which is expressed constitutively in many tissues, development of cutaneous ($\gamma\delta^+$ TCR) lymphomas may be observed

(Uehira *et al.*, 1993). A number of leukemia and lymphoma cells isolated from patients have been screened for their growth responses and/or dependence on IL-7: many but not all proliferate when exposed to IL-7 and the cell types include B- and T-cell malignancies (Eder *et al.*, 1990; Touw *et al.*, 1990; Makrynikola *et al.*, 1991; Shand and Betlach, 1991; Skjonsberg *et al.*, 1991; Lu *et al.*, 1992; Yoshioka *et al.*, 1992). Evidence of lymphoid maturation of the tumor cells in response to IL-7 incubation was not observed (Eder *et al.*, 1990).

In a separate study, pre-B cells transformed by a variety of oncogenes were tested for IL-7 production. None produced any IL-7 bioactivity. IL-7 overexpression achieved by removing portions of the 5′ flanking region was not associated with dramatic colony formation in agar, and most clones were not tumorigenic *in vivo* (syngeneic mice) (Young *et al.*, 1991).

It seems, therefore, that production of IL-7 does not represent a final common step in the malignant transformation of lymphoid cells, but that in selected malignancies it may represent a target for therapeutic intervention. For instance, IL-7 may represent an 'antiapoptotic' factor for some hematopoietic malignancies: IL-7 induces in murine T-cell lymphoma cells (CS-21) expression of the Bcl2 protein and suppression the CPP32-like protease (Lee *et al.*, 1996). An additional role for IL-7 in malignant progression has been suggested by the observation that IL-7 upregulates ICAM expression by melanoma cells, a phenotype correlated with metastatic behavior (for review see Möller *et al.*, 1996).

A detailed study addressed IL-7Rα expression in several types of cutaneous and nodal lymphoma. IL-7Rα was not expressed in cutaneous B-cell lymphomas, benign cutaneous lymphoid infiltrates or reactive lymph nodes. In contrast, IL-7Rα was expressed in over 50% of all histologic types of cutaneous T-cell lymphoma (Bagot *et al.*, 1996). IL-7 mRNA and protein are also readily detectable in B-cell chronic lymphocytic leukemia (B-CLL) cells. The coincidence of IL-7 mRNA downregulation and apoptosis in B-CLL cells suggested that IL-7 gene expression may be required for B-CLL viability *in vivo*. Of note, IL-7 downregulation and apoptosis could be prevented by co-culture of B-CLL cells with human umbilical cord endothelial hybrid cells (EA.hy926). Cell–cell contact appears to be a prerequisite, as cell culture supernatant could not reconstitute the effect, indicating that poorly defined integrins expressed on B-CLL cells may affect IL-7 gene expression and apoptosis (Long *et al.*, 1995). Moreover, IL-7 mRNA and protein elaborated by B-CLL cells may account for some of the clinical symptoms: some patients with CLL may experience suppression of immune responses and also autoimmune symptoms (Frishman *et al.*, 1993). Of note, cloning of the IL-7 gene product from B-CLL cells revealed that at least a different, alternatively spliced, IL-7 mRNA is expressed in tumor cells. The alternatively spliced form appears to be identical with an original IL-7 cDNA clone obtained by screening a (hepatocarcinoma) cDNA library for the human IL-7 gene (Goodwin *et al.*, 1989). The alternative transcript lacks the entire exon 4 (132 bp) coding for 44-amino-acid residues, as depicted in Fig. 1 (Goodwin *et al.*, 1989; Frishman *et al.*, 1993). The present authors have also observed that IL-7 mRNA expressing cells derived from renal or colorectal cancer cells contain the 'canonical' IL-7 full-length IL-7 mRNA and additionally differentially spliced IL-7 mRNA (authors' unpublished observations). The biology of these IL-7 mRNA species remains to be determined.

Based on these data, it appears that IL-7 may exert differential effects on tumor cells. IL-7 may exert growth-promoting, but potentially also growth-arresting, activities. For instance, proliferation of some pre-B acute lymphoblastic leukemia (B-ALL) cells can be specifically inhibited by exogenous IL-7. This effect can be abrogated by blocking of the IL-7 receptor (Pandrau *et al.*, 1994). In contrast, other ALL cells appear to be IL-7 responsive (Greil *et al.*, 1994). Additionally, targeting of IL-7 receptor-positive cells implementing a recombinant fusion toxin (DAB_{389}–IL-7) has been suggested as a treatment for lymphoma (Sweeney *et al.*, 1995). Thus, IL-7-induced effects mediated by the IL-7Rα may also be dependent on the actual cell type, as proliferation of early pre-B cells (see Table 1) can be augmented by IL-7.

IL-7 as a Treatment Option for Cancer-Bearing Hosts

Immunotherapy has evolved to become a reasonable treatment option for some patients with cancer. This approach, at least theoretically, assumes that an antigenic difference between malignant and normal cells exists, can be recognized by the host and can be manipulated. In addition, it must be presumed that the tumor-bearing host is functionally immunodeficient in that the tumor somehow blocks or inactivates the patient's own antitumor response. For instance, alterations in expression and functions of signal transduction molecules associated with the TCR are indeed responsible for inefficient immune responsiveness in T lymphocytes in several human malignancies, including renal cell cancer, colorectal cancer, ovarian cancer and melanoma.

Decreased CD3 expression and inefficient CD3-mediated signaling in tumor infiltrating lymphocytes (TILs) and PBLs has been observed primarily in tumor-bearing mice (Mizoguchi *et al.*, 1992; Salvadori *et al.*, 1994; Levey and Srivastava, 1995) and recently in cancer-bearing patients as well (Finke *et al.*, 1993; Nakagomi *et al.*, 1993; Matsuda *et al.*, 1995; Tartour *et al.*, 1995; Zea *et al.*, 1995; Lai *et al.*, 1996; Rabinowich *et al.*, 1996).

One of several mechanisms of CD3 downregulation and inefficient signaling appears to be due to reduced expression of the ζ chain of the TCR, presumably related to hydrogen peroxide secretion elaborated by tumor-derived macrophages (Kono *et al.*, 1996). This defect can be reversed *in vitro* and *in vivo* using exogenous IL-2 or IL-2 transfected into tumor cells to be implemented as a vaccine (Salvadori *et al.*, 1994; Rabinowich *et al.*, 1996). However, IL-7 is also capable of upregulating the TCR (Ono *et al.*, 1996) and can enhance protein expression of molecules associated with TCR expression and signaling functions (e.g. ZAP-70, ζ-chain, $p56l^{ck}$ and $p59^{fyn}$; authors' unpublished data). Additionally, suppressive factors released by tumors may impair antitumor-directed immune responses, such as TGF-β. Macrophage-derived TGF-β mRNA can be downregulated by IL-7 (Dubinett *et al.*, 1993). The same effect of IL-7 has been found to be true for TGF-β downregulation in a murine fibrosarcoma (Dubinett *et al.*, 1995). In contrast, TGF-β is able to reduce stromal IL-7 mRNA expression and protein secretion using a human *in vitro* lymphoid progenitor cell culture system (Tang *et al.*, 1997). Biologic therapy approaches including immunotherapy seek to reverse this apparent state of anergy and to augment antitumor-directed immune responses.

Clinical trials utilizing IL-2 as a cytokine-based immunotherapy have demonstrated that this approach is successful in treating some patients. The challenge for both clinicians and researchers is to increase the efficacy and decrease the non-specific effects of the therapy. IL-7 appears to have a number of 'IL-2-like' properties and preclinical testing suggests a potential role for IL-7-based immunotherapy trials. The IL-7-mediated effects may be segregated into effects due to non-specific, MHC non-restricted, lysis of tumor cell targets (e.g. due to lymphokine-activated killer (LAK) cells), MHC class I- or II-specific recognition of tumor cells by $\alpha\beta^+$ T lymphocytes, and tumor-restricted and presumably classic MHC non-restricted recogniton by $\gamma\delta^+$ T-cell effectors.

Characterization of the LAK phenomenon was reported in 1980 by Yron and associates. The phenomenon describes the *in vitro* lysis of labeled fresh tumor targets by lymphoid cells that have been preincubated in IL-2 or other lymphokines. The effect is not MHC restricted and is relatively non-specific, in that a variety of different fresh tumors are lysed yet most normal cells are spared. IL-7 is able to generate LAK activity from thymocytes and PBMCs.

In comparison to IL-2, IL-7 appears to be a relatively weak LAK inducer. IL-2 stimulates fivefold more LAK precursors than IL-7 (Alderson *et al.*, 1990). Thymocytes from cultures grown in IL-2 are highly cytolytic, whereas those grown in IL-7 exhibit minimal cytolytic activity; however, cultures grown in IL-7 and then switched to IL-2 become cytolytic. The addition of IL-4 does not induce cytolytic activity of the cells grown in IL-7 but rather downregulates IL-2-induced proliferation and cytolytic activity (Widmer *et al.*, 1990). IL-7 can generate human LAK cell activity in the absence of IL-2, and induces or upregulates expression of CD25, CD54 and CD69. LAK cell generation is negatively influenced by TGF-β and IL-4. Anti-IL-4 antibody and anti-IL-4 antisense enhances IL-7-induced LAK cell activity (Stötter and Lotze, 1991; Stötter *et al.*, 1991). IL-7 promotes secretion of TNF but not of IFN-γ (Stötter and Lotze, 1991).

The nature of the LAK cell precursor for IL-7-induced LAK is not totally clear. One study showed that LAK cell activity (comparable to IL-2) could be generated from a population of NK cells ($CD56^+$) whereas no LAK activity was generated in PBMCs (Naume and Espevik, 1991; Naume *et al.*, 1992).

Another study using murine cells compared IL-7-induced LAK with IL-2 LAK. IL-7 LAK peaked at day 6–8. IL-7 was more effective at maintaining cytotoxic activity over longer periods of time than IL-2. IL-7 LAK was induced from secondary lymphoid tissue (spleen and nodes) but not from primary lymphoid tissue (thymus and bone marrow). LAK cell activity was abrogated by anti-CD8 or anti-Thy-1 + C and unaffected by anti-CD4, anti-asialo GM-1 or anti-NK-1.1 + C, suggesting that IL-7 LAK may not necessarily be mediated by NK cells, but rather by T lymphocytes (Lynch and Miller, 1990). In comparison to IL-2 and IL-12, IL-7 stimulates the $CD56^+$ NK cells to secrete significantly lower amounts of soluble TNF receptor compared with IL-2-, or IL-12-mediated stimulation. In comparison to IL-2, IL-7 induced lower levels of GM-CSF, but significantly higher GM-CSF levels when compared with IL-12 (Naume *et al.*, 1993).

If indeed, IL-7-induced LAK cell activity resides within the T-cell population, then it might be possible to create a LAK immunotherapy treatment regimen that lacks some of the deleterious effects of the IL-2 treatment which have been blamed on the NK cell population. Circulating human T cells also proliferate when incubated in IL-7. Both $CD4^+$ and $CD8^+$ subsets respond to a similar degree; however, when T cells are separated on the basis of reactivity with an antibody (anti-CD45) that reacts with a 220-kDa isoform (CD45RA) of the common leukocyte antigen, memory T cells (CD45RO) appear to respond more readily than naive T cells (CD45RA) (Welch *et al.*, 1989). Additionally, a variety of effects of IL-7 on monocytes has been reported. Activation of monocytes with IL-7 can result in the development of a tumor lytic phenotype using melanoma cells as targets (Alderson *et al.*, 1991). Induction of mRNA for both IL-8 and human macrophage inflammatory protein-1 β gene is induced in monocytes by IL-7 (Ziegler *et al.*, 1991; Standiford *et al.*, 1992). Monocytes incubated in IL-7 are stimulated to secrete large quantities of IL-6 as well as IL-1α, IL-1β and TNF-α. This response is abrogated by provision of IL-4.

T cell-based immunotherapy has the advantage of increased specificity compared with LAK cell therapy. T-cell tumor lysis has been shown to be MHC restricted when antimelanoma or renal cell cancer reactive MHC class I- or class II-restricted $\alpha\beta^+$ T lymphocytes are examined. Various cytokines, including IL-2 and IL-4, are active in promoting the clonal expansion *in vitro* of T cells while maintaining their tumor lytic activity. IL-7 appears to have similar properties.

IL-7 alone generates modest CTL activity which is augmented by IL-2, IL-6 or IL-4 (Bertagnolli and Herrmann, 1990; Hickman *et al.*, 1990). Removal of CD8$^+$ cells results in decreased killing, whereas removal of CD4$^+$ cells enhances the CTL response. IL-7 enhances cell proliferation and duration of growth more than IL-2. Allospecific cytotoxicity was maintained for at least 60 days in these cultured cells. Addition of anti-IL-4, anti-IL-2 or anti-IL-6 decreases the proliferation of CTLs in culture (Bertagnolli and Herrmann, 1990; Jicha *et al.*, 1992).

CTLs harvested from draining nodes of tumor-bearing animals and incubated in IL-7 were fourfold more effective than CTLs grown in medium alone in adoptive transfer experiments (Lynch *et al.*, 1991). CTLs incubated in IL-7 and adoptively transferred to mice bearing 3-day pulmonary metastases (MCA tumor) were effective in mediating tumor regression (Jicha *et al.*, 1991).

IL-7 stimulates proliferation of human TILs derived from renal cell carcinoma but only if the TILs are first incubated in either IL-2 alone or in IL-2 plus IL-7. IL-7 stimulates proliferation of CD4$^+$ or CD8$^+$ TIL lines specific for renal cell carcinoma. IL-7 synergizes with anti-CD3 in the induction of IFN-γ from short-term TIL cultures (Sica *et al.*, 1993). Human T cells harvested from peripheral blood and incubated in IL-7 when restimulated with phorbol ester and ionomycin secrete IL-2, IL-4, IL-6 and IFN-γ. This effect was not seen as readily in cultures initiated with IL-2 or IL-4. Both CD4$^+$ and CD8$^+$ subsets responded by cytokine secretion. Almost all the potential to secrete IL-4 and IL-6 in response to IL-7 preincubation resides within the memory subset as opposed to the naive population (Armitage *et al.*, 1992a). The preceding observations suggested that IL-7, either alone or in conjunction with IL-2, acts to stimulate proliferation and tumor lytic activity in sensitized T cells and therefore may be clinically useful in the immunotherapy of malignancy. The most promising data have come from a study demonstrating that anti-tumor-specific T lymphocytes can be grown and expanded *in vitro* without restimulation for extended periods (up to 22 months) compared with T lymphocytes grown in IL-2 (Lynch and Miller, 1994). IL-7 alone (Maeurer *et al.*, 1997), admixture of IL-7 to IL-2 and INF-γ appears also preferentially to expand and maintain tumor-specific and MHC class II-restricted CD4$^+$ T lymphocytes (Cohen *et al.*, 1993) from tumor-bearing patients.

Of note, some of these tumor-reactive and MHC class II-restricted T lymphocytes secrete IFN-γ preferentially in response to autologous tumor cells (Maeurer *et al.*, 1997). This observation is in concordance with an earlier report demonstrating IL-7-mediated effects in adoptive immunotherapy in human colorectal cancer xenografts in SCID mice.

Exclusively the combination of IL-7 treatment and passive transfer of human autologous T cells resulted in enhanced survival of mice engrafted with the respective tumors; treatment with IL-7 alone showed no effect. The antitumor effects are correlated with IFN-γ secretion by the passively transferred T cells and not by their cytolytic capability (Murphy *et al.*, 1993). The ability of IL-7 to generate antitumor-directed immune reactivity may also be dependent on the tumor type and availability of T cells capable of recognizing tumor-associated peptides presented either in the context of MHC class I or II molecules. For instance, application of IL-7 resulted in up to a 75% reduction in pulmonary metastases of the murine renal cell cancer line Renca (Komschlies *et al.*, 1994). However, the pharmacokinetics of IL-7 administered to humans have not yet been evaluated. Some toxic side-effects have been observed in mice treated with IL-7 systemically (Komschlies *et al.*, 1994).

However, IL-7 may also be used to restore the immune system in primary or secondary immunodeficiencies (e.g. induced by viral infections or inherited abnormalities, such as Di George syndrome) or after bone marrow transplantation (BMT). For instance, the successful outcome of autologous BMT is limited by susceptibility to infection. As the effective restitution of an immune system does not only require the quantitative replacement of immune cells (usually achieved within 3–4 months after transplantation), the quality of the immune system is often impaired. Since IL-7 has not only growth-promoting but also differentiation effects on both B- and T-cell lymphopoiesis, it may represent an attractive cytokine, potentially in combination with flt3 ligand, to reconstitute a competent immune system. Several studies have addressed this issue. For instance IL-7 treatment of Balb/c mice after syngeneic BMT leads to increased thymic cellularity, increased RAG-1 expression, and to promotion of Vβ8(D)J gene rearrangement of TCRs. The increased 'quality' of IL-7-treated mice is reflected in better mitogenic responses of thymic cells and in enhanced cytokine production provoked by influenza virus challenge (Abdul-Hai *et al.*, 1996). Additionally, IL-7 accelerates PBL recovery of mice after cyclophosphamide, 5-fluorouracil treatment (Damia *et al.*, 1992) or radiation (Faltynek *et al.*, 1992). Using a metastatic breast cancer model in mice, IL-7 and BMT could significantly prolong survival, presumably due to enhanced immune cell reconstitution after split-dose chemotherapy using cyclophosphamide, cisplatin and nitrosourea (Talmadge *et al.*, 1993).

A more recent study has shown that IL-7 may be able to mobilize long-term reconstituting peripheral CD34$^+$ stem cells (Grzegorzewski *et al.*, 1994). Such cells may be useful for stem cell transplantation, or for therapies using CD34$^+$ cells either for gene transduction or for maturation *in vitro* in order to generate potent antigen-presenting cells capable of initiating potent antitumor-directed cellular immune responses. Additionally, development of tumors in older individuals may reflect not only accumulative genetic alternations, but also a decreased capacity of the humoral and cellular immune system to identify and eradicate transformed cells efficiently. Several studies have demonstrated age-related alterations in T and B lymphocytes (Zharhary, 1994). In murine models, the ability of pro-B cells to proliferate in response to stroma cells decreases with age (Stephan *et al.*, 1996). This functional alteration is due to an impaired response of pro-B cells to IL-7, but not to other stroma-associated cytokines, including stem cell factor or insulin-like growth factor (Stephan *et al.*, 1997). The reduced IL-7 responsiveness appears not to be induced by inefficient IL-7R expression, but by as yet poorly defined intracellular signaling events mediated through the IL-7 receptor complex (Stephan *et al.*, 1997). Thus, IL-7 may not only be implemented for primary or secondary immundeficiency disorders: the functional impairment of the immune system in older individuals may in part be mediated by reduced IL-7-responsive immune cells. Future studies may devise therapeutic strategies to overcome age-related immunodeficiencies which may play a role in decreased immune surveillance.

The effects of locally secreted IL-7 elaborated by genetically engineered tumor cells may be overlapping with some of the effects observed by systemic application. Transfection of cytokine genes into tumor cell lines has been developed as a theoretic strategy to increase the local regional response to the tumor in the hope that a heightened *in situ* response might translate to an enhanced systemic response, not only to the transfected tumor but also to the non-transfected or wild-type tumor. IL-7 transfection experiments have yielded some provocative results. Transfection of IL-7 into the murine tumor line J5581 leads to tumor rejection *in vivo*.

$CD8^+$ cells were required for long-term tumor eradication, but short-term regression was noted in the absence of $CD8^+$ cells. While tumor transfected with IL-2, IL-4, TNF or IFN-γ regressed when placed in nude or SCID mice IL-7-transfected tumor required the presence of $CD4^+$ cells for regression, and no regression was observed in nude mice bearing tumor transfected with IL-7.

In most of the murine studies, tumors were eventually rejected by the animals, while the *in vitro* growth was not affected by IL-7 (Hock *et al.*, 1991, 1993; Aoki *et al.*, 1992; McBride *et al.*, 1992; Miller *et al.*, 1993; Allione *et al.*, 1994; Tepper and Mule, 1994). It appears that $CD8^+$ T cells play a major role in mediating tumor rejection (Hock *et al.*, 1991, 1993; Aoki *et al.*, 1992; McBride *et al.*, 1992; Miller *et al.*, 1993) and that antigen-specific T cells are elicited upon immunization with IL-7-secreting tumor cell lines (Aoki *et al.*, 1992). However, other immune cells may also contribute to antitumor responses, as not just T lymphocytes, but also macrophages, eosinophils and basophils, are present at the site of tumor rejection (Hock *et al.*, 1991; McBride *et al.*, 1992).

A more recent study examined in detail the effects of locally secreted IL-7 and induction of tumor-specific cellular immune responses (Cayeux *et al.*, 1995). In B7-transfected mammary adenocarcinoma cells TS/A, T cells showed predominantly $CD28^+$ and $CD25^-$ marker expression, in IL-7-transduced tumor cells CD28 and $CD25^+$ marker expression, whereas in $B7^+IL7^+$ tumor cells the T-cell infiltrate showed typically $CD28^+CD25^+$ expression. The double-transfected tumor elicited a more strong immunity compared with tumor cells expressing IL-7, or B7 alone, or non-transfected tumor cells admixed with *Corynebacterium parvum* (Cayeux *et al.*, 1995). Human non-small-cell lung cancer cell lines infected with a retroviral construct containing the human IL-7 cDNA show alterations of cell surface expression of molecules (e.g. MHC class I, LFA-3) on co-cultured PBLs favoring antitumor-directed immune responses (Sharma *et al.*, 1996). Thus, IL-7-transfected tumor cells may represent a reasonable vaccine for eliciting a strong antitumor-directed immunity (Möller *et al.*, 1996). IL-7-transfected tumor cells are now actively being scrutinized in the setting of tumor vaccines in Heidelberg, Germany (D. Schadendorf) (Möller *et al.*, 1996) and at the University of California at Los Angeles in the USA (J. Economou).

SUMMARY

IL-7 is an important lymphopoietin and plays a critical role in both B- and T-cell development. It promotes expansion of T lymphocytes exhibiting antigen-specific reactivity. IL-7 may be implemented to promote strong and effective immune responses directed against tumor cells, or against microbial or viral infection. It may also be useful to restore an effective and functional immune system after bone marrow transplantation. Clinical trials of IL-7 will begin at the University of Pittsburgh and other institutions soon (supplied by the National Cancer Institute).

ACKNOWLEDGEMENTS

The authors thank M. O'Malley and Maria Bond for excellent help in preparing the manuscript, and W. Walter and G. Winternheimer for discussion and help in preparing the figures.

REFERENCES

Abdul-Hai, A., Or, R., Slavin, S., Friedman, G., Weiss, L., Matsa, D. and Ben Yehuda, A. (1996). Stimulation of immune reconstitution by interleukin-7 after syngeneic bone marrow transplantation in mice. *Exp. Hematol*. **24**, 1416–1422.

Aiba, S., Nakagawa, S., Hara, M., Tomioka, Y., Deguchi, M. and Tagami, H. (1994). Cultured murine dermal cells can function like thymic nurse cells. *J. Invest. Dermatol*. **103**, 162–167.

Alderson, M.R., Sassenfeld, H.M. and Widmer, M.B. (1990). Interleukin 7 enhances cytolytic T lymphocyte generation and induces lymphokine-activated killer cells from human peripheral blood. *J. Exp. Med*. **172**, 577–587.

Alderson, M.R., Tough, T.W., Ziegler, S.F. and Grabstein, K.H. (1991). Interleukin 7 induces cytokine secretion and tumoricidal activity by human peripheral blood monocytes. *J. Exp. Med*. **173**, 923–930.

Allione, A., Consalvo, M., Nanni, P., Lollini, P.L., Cavallo, F., Giovarelli, M., Forni, M., Gulino, A., Colombo, M.P. and Dellabona, P. (1994). Immunizing and curative potential of replicating and nonreplicating murine mammary adenocarcinoma cells engineered with interleukin (IL)-2, IL-4, IL-6, IL-7, IL-10, tumor necrosis factor α, granulocyte–macrophage colony-stimulating factor, and γ-interferon gene or admixed with conventional adjuvants. *Cancer Res*. **54**, 6022–6026.

Aoki, T., Tashiro, K., Miyatake, S., Kinashi, T., Nakano, T., Oda, Y., Kikuchi, H. and Honjo, T. (1992). Expression of murine interleukin 7 in a murine glioma cell line results in reduced tumorigenicity *in vivo*. *Proc. Natl Acad. Sci. U.S.A*. **89**, 3850–3854.

Appasamy, P.M. (1992). IL 7-induced T cell receptor-γ gene expression by pre-T cells in murine fetal liver cultures. *J. Immunol*. **149**, 1649–1656.

Appasamy, P.M. (1997). In *Cytokines in Health and Disease* (in press).

Appasamy, P.M., Kenniston, T.W., Jr., Weng, Y., Holt, E.C., Kost, J. and Chambers, W.H. (1993). Interleukin 7-induced expression of specific T cell receptor γ variable region genes in murine fetal liver cultures. *J. Exp. Med*. **178**, 2201–2206.

Ariizumi, K., Meng, Y., Bergstresser, P.R. and Takashima, A. (1995). IFN-γ-dependent IL-7 gene regulation in keratinocytes. *J. Immunol*. **154**, 6031–6039.

Armitage, R.J., Macduff, B.M., Ziegler, S.F. and Grabstein, K.H. (1992a). Multiple cytokine secretion by IL-7-stimulated human T cells. *Cytokine* **4**, 461–469.

Armitage, R.J., Ziegler, S.F., Friend, D.J., Park, L.S. and Fanslow, W.C. (1992b). Identification of a novel low-affinity receptor for human interleukin-7. *Blood* **79**, 1738–1745.

Asadullah, K., Haeussler, A., Friedrich, M., Siegling, A., Olaizola Horn, S., Trefzer, U., Volk, H.D. and Sterry, W. (1996). IL-7 mRNA is not overexpressed in mycosis fungoides and pleomorphic T-cell lymphoma and is likely to be an autocrine growth factor *in vivo*. *Arch. Dermatol. Res*. **289**, 9–13.

Bagot, M., Charue, D., Boulland, M.L., Gaulard, P., Revuz, J., Schmitt, C. and Wechsler, J. (1996). Interleukin-7 receptor expression in cutaneous T-cell lymphomas. *Br. J. Dermatol*. **135**, 572–575.

Bargou, R.C., Mapara, M.Y., Zugck, C., Daniel, P.T., Pawlita, M., Dohner, H. and Dorken, B. (1993). Characterization of a novel Hodgkin cell line, HD-MyZ, with myelomonocytic features mimicking Hodgkin's disease in severe combined immunodeficient mice. *J. Exp. Med*. **177**, 1257–1268.

Benjamin, D., Sharma, V., Knobloch, T.J., Armitage, R.J., Dayton, M.A. and Goodwin, R.G. (1994). B cell IL-7. Human B cell lines constitutively secrete IL-7 and express IL-7 receptors. *J. Immunol*. **152**, 4749–4757.

Berenson, R.J., Andrews, R.G., Bensinger, W.I., Kalamasz, D., Knitter, G., Buckner, C.D. and Bernstein, I.D. (1988). Antigen CD34$^+$ marrow cells engraft lethally irradiated baboons. *J. Clin. Invest*. **81**, 951–955.

Bermudez, L.E. (1993). Production of transforming growth factor-β by *Mycobacterium avium*-infected human macrophages is associated with unresponsiveness to IFN-γ. *J. Immunol*. **150**, 1838–1845.

Berrios, V., Kurane, I. and Ennis, F.A. (1996). Immunomodulatory effects of IL-7 on dengue virus-specific cytotoxic CD4$^+$ T cell clones. *Immunol. Invest*. **25**, 231–240.

Bertagnolli, M. and Herrmann, S. (1990). IL-7 supports the generation of cytotoxic T lymphocytes from thymocytes. Multiple lymphokines required for proliferation and cytotoxicity. *J. Immunol*. **145**, 1706–1712.

Billips, L.G., Petitte, D., Dorshkind, K., Narayanan, R., Chiu, C.P. and Landreth, K.S. (1992). Differential roles of stromal cells, interleukin-7, and kit-ligand in the regulation of B lymphopoiesis. *Blood* **79**, 1185–1192.

Boesteanu, A., De Silva, A.D., Nakajima, H., Leonard, W.J., Peschon, J.J. and Joyce, S. (1997). Distinct roles for signals relayed through the common cytokine receptor γ chain and interleukin 7 receptor α chain in natural T cell development. *J. Exp. Med*. **186**, 331–336.

Boismenu, R. and Havran, W.L. (1994). Modulation of epithelial cell growth by intraepithelial gamma δ T cells. *Science* **266**, 1253–1255.

Borger, P., Kauffman, H.F., Postma, D.S. and Vellenga, E. (1996). IL-7 differentially modulates the expression of IFN-γ and IL-4 in activated human T lymphocytes by transcriptional and post-transcriptional mechanisms. *J. Immunol*. **156**, 1333–1338.

Bui, T., Dykers, T., Hu, S.L., Faltynek, C.R. and Ho, R.J. (1994). Effect of MTP-PE liposomes and interleukin-7 on induction of antibody and cell-mediated immune responses to a recombinant HIV-envelope protein. *J. Acquir. Immune Defic. Syndr*. **7**, 799–806.

Candeias, S., Muegge, K. and Durum, S.K. (1997a). IL-7 receptor and VDJ recombination: trophic *versus* mechanistic actions. *Immunity* **6**, 501–508.

Candeias, S., Peschon, J.J., Muegge, K. and Durum, S.K. (1997b). Defective T-cell receptor γ gene rearrangement in interleukin 7 receptor knockout mice. *Immunol. Lett*. **57**, 9–14.

Carini, C. and Essex, M. (1994). Interleukin 2-independent interleukin 7 activity enhances cytotoxic immune response of. *AIDS Res. Hum. Retroviruses* **10**, 121–130.

Carini, C., McLane, M.F., Mayer, K.H. and Essex, M. (1994). Dysregulation of interleukin-7 receptor may generate loss of cytotoxic T cell response in human immunodeficiency virus type 1 infection. *Eur. J. Immunol*. **24**, 2927–2934.

Cayeux, S., Beck, C., Aicher, A., Dorken, B. and Blankenstein, T. (1995). Tumor cells cotransfected with interleukin-7 and B7.1 genes induce CD25 and CD28 on tumor-infiltrating T lymphocytes and are strong vaccines. *Eur. J. Immunol*. **25**, 2325–2331.

Chantry, D., Turner, M. and Feldmann, M. (1989). Interleukin 7 (murine pre-B cell growth factor/lymphopoietin 1) stimulates thymocyte growth: regulation by transforming growth factor β. *Eur. J. Immunol*. **19**, 783–786.

Cludts, I., Droogmans, L., Cleuter, Y., Kettmann, R. and Burny, A. (1992). Sequence of bovine interleukin 7. *DNA Seq*. **3**, 55–59.

Cohen, P.A., Kim, H., Fowler, D.H., Gress, R.E., Jakobsen, M.K., Alexander, R.B., Mule, J.J., Carter, C. and Rosenberg, S.A. (1993). Use of interleukin-7, interleukin-2, and interferon-γ to propagate CD4$^+$ T cells in culture with maintained antigen specificity. *J. Immunother*. **14**, 242–252.

Conlon, P.J., Morrissey, P.J., Nordan R.P., Grabstein, K.H., Prockett, K.S., Reed, S.G., Goodwin, R., Cosman, D. and Namen, A.E. (1989). Murine thymocytes proliferate in direct response to interleukin 7. *Blood* **74**, 1368–1373.

Corcoran, A.E., Smart, F.M., Cowling, R.J., Crompton, T., Owen, M.J. and Venkitaraman, A.R. (1996). The interleukin-7 receptor α chain transmits distinct signals for proliferation and differentiation during B lymphopoiesis. *EMBO J*. **15**, 1924–1932.

Corey, S.J. and Shapiro, D.N. (1994). Localization of the human gene for Src-related protein tyrosine kinase LYN to chromosome 8q11–12: a lymphoid signaling cluster? *Leukemia* **8**, 1914–1917.

Costello, R., Imbert, J. and Olive, D. (1993). Interleukin-7, a major T-lymphocyte cytokine. *Eur. Cytokine Netw*. **4**, 253–262.

Crompton, T., Outram, S.V., Buckland, J. and Owen, M.J. (1997). A transgenic T cell receptor restores thymocyte differentiation in interleukin-7 receptor α chain-deficient mice. *Eur. J. Immunol*. **27**, 100–104.

Cumano, A. and Paige, C.J. (1992). Enrichment and characterization of uncommitted B-cell precursors from fetal liver at day 12 of gestation. *EMBO J*. **11**, 593–601.

Cumano, A., Dorshkind, K., Gillis, S. and Paige, C.J. (1990). The influence of S17 stromal cells and interleukin 7 on B cell development. *Eur. J. Immunol*. **20**, 2183–2189.

Damia, G., Komschlies, K.L., Faltynek, C.R., Ruscetti, F.W. and Wiltrout, R.H. (1992). Administration of recombinant human interleukin-7 alters the frequency and number of myeloid progenitor cells in the bone marrow and spleen of mice. *Blood* **79**, 1121–1129.

De Benedetti, F., Massa, M., Pignatti, P., Kelley, M., Faltynek, C.R. and Martini, A. (1995). Elevated circulating interleukin-7 levels in patients with systemic juvenile rheumatoid arthritis. *J. Rheumatol*. **22**, 1581–1585.

Denis, M. (1991). Tumor necrosis factor and granulocyte macrophage-colony stimulating factor stimulate human macrophages to restrict growth of virulent *Mycobacterium avium* and to kill avirulent *M. avium*: killing effector mechanism depends on the generation of reactive nitrogen intermediates. *J. Leukoc. Biol*. **49**, 380–387.

Dennig, D. and O'Reilly, R.J. (1994). IL-7 induces surface expression of B7/BB1 on pre-B cells and an associated increase in their costimulatory effects on T cell proliferation. *Cell Immunol*. **153**, 227–238.

Dieli, F., Asherson, G.L., Sireci, G., Dominici, R., Gervasi, F., Vendetti, S., Colizzi, V. and Salerno, A. (1997). $\gamma\delta$ cells involved in contact sensitivity preferentially rearrange the Vγ3 region and require interleukin-7. *Eur. J. Immunol*. **27**, 206–214.

DiSanto, J.P., Muller, W., Guy Grand, D., Fischer, A. and Rajewsky, K. (1995). Lymphoid development in mice with a targeted deletion of the interleukin 2 receptor γ chain. *Proc. Natl Acad. Sci. U.S.A*. **92**, 377–381.

Dittel, B.N. and LeBien, T.W. (1995). The growth response to IL-7 during normal human B cell ontogeny is restricted to B-lineage cells expressing CD34. *J. Immunol*. **154**, 58–67.

Dobbeling, U. (1996). The influence of IL-7 V(D)J recombination. *Immunology* **89**, 569–572.

Domen, J., van der Lugt, N.M., Acton, D., Laird, P.W., LInders, K. and Berns, A. (1993). Pim-1 levels determine the size of early B lymphoid compartments in bone marrow. *J. Exp. Med*. **178**, 1665–1673.

Dubinett, S.M., Huang, M., Dhanani, S., Wang, J. and Beroiza, T. (1993). Down-regulation of macrophage transforming growth factor-β messenger RNA expression by IL-7. *J. Immunol*. **151**, 6670–6680.

Dubinett, S.M., Huang, M., Dhanani, S., Economou, J.S., Wang, J., Lee, P., Sharma, S., Dougherty, G.J. and McBride, W.H. (1995). Down-regulation of murine fibrosarcoma transforming growth factor-β_1 expression by interleukin 7. *J. Natl Cancer Inst*. **87**, 593–597.

Eder, M., Ottmann, O.G., Hausen Hagge, T.E., Bartram, C.R., Gillis, S., Hoelzer, D. and Gauser, A. (1990). *Leukemia* **4**, 533–540.

Elner, V.M., Elner, S.G., Standiford, T.J., Lukacs, N.W., Strieter, R.M. and Kunkel, S.L. (1996). Interleukin-7 (IL-7) induces retinal pigment epithelial cell MCP-1 and IL-8. *Exp. Eye Res*. **63**, 297–303.

Era, T., Ogawa, M., Nishikawa, S., Okamoto, M., Honjo, T., Akagi, K., Miyazaki, J. and Yamamura, K. (1991). Differentiation of growth signal requirement of B lymphocyte precursor is directed by expression of immunoglobulin. *EMBO J*. **10**, 337–342.

Everson, M.P., Eldridge, J.H. and Koopman, W.J. (1990). Synergism of interleukin 7 with the thymocyte growth factors interleukin 2, interleukin 6, and tumor necrosis factor α in the induction of thymocyte proliferation. *Cell. Immunol*. **127**, 470–482.

Faltynek, C.R., Wang, S., Miller, D., Young, E., Tiberio, L., Kross, K., Kelley, M. and Kloszewski, E. (1992). Administration of human recombinant IL-7 to normal and irradiated mice increases the numbers of lymphocytes and some immature cells of the myeloid lineage. *J. Immunol*. **149**, 1276–1282.

Finke, J.H., Zea, A.H., Stanley, J., Longo, D.L., Mizoguchi, H., Tubbs, R.R., Wiltrout, R.H., O'Shea, J.J., Kudoh, S., Klein, E., Bukowski, R.M. and Ochoa, A. (1993). Loss of T-cell receptor zeta chain and p56lck in T-cells infiltrating human renal cell carcinoma. *Cancer Res*. **53**, 5613–5616.

Flanagan, J.G. and Leder, P. (1990). The kit ligand: a cell surface molecule altered in Steel mutant fibroblasts. *Cell* **63**, 185–194.

Foss, F.M., Koc, Y., Stetler Stevenson, M.A., Nguyen, D.T., O'Brien, M.C., Turner, R. and Sausville, E.A. (1994). Costimulation of cutaneous T-cell lymphoma cells by interleukin-7 and interleukin-2: potential autocrine or paracrine effectors in the Sézary syndrome. *J. Clin. Oncol*. **12**, 326–335.

Foss, H.D., Hummel, M., Gottstein, S., Ziemann, K., Falini, B., Herbst, H. and Stein, H. (1995). Frequent expression of IL-7 gene transcripts in tumor cells of classical Hodgkin's disease. *Am. J. Pathol*. **146**, 33–39.

Foxwell, B.M., Taylor Fishwick, D.A., Simon, J.L., Page, T.H. and Londei, M. (1992). Activation induced changes in expression and structure of the IL-7 receptor on human T cells. *Int. Immunol*. **4**, 277–282.

Foxwell, B.M., Willcocks, J.L., Taylor Fishwick, D.A., Kulig, K., Ryffel, B. and Londei, M. (1993). *Eur. J. Immunol*. **23**, 85–89.

Fratazzi, C. and Carini, C. (1996). Interleukin-7 modulates intracytoplasmatic CD23 production and induces adhesion molecule expression and adhesiveness in activated $CD4^+CD23^+$ T cell subsets. *Clin. Immunol. Immunopathol*. **81**, 261–270.

Friend, S.L., Hosier, S., Nelson, A., Foxworthe, D., Williams, D.E. and Farr, A. (1994). A thymic stromal cell line supports *in vitro* development of surface IgM^+ B cells and produces a novel growth factor affecting B and T lineage cells. *Exp. Hematol*. **22**, 321–328.

Frishman, J., Long, B., Knospe, W., Gregory, S. and Plate, J. (1993). Genes for interleukin 7 are transcribed in leukemic cell subsets of individuals with chronic lymphocytic leukemia. *J. Exp. Med*. **177**, 955–964.

Fujihashi, K., Kawabata, S., Hiroi, T., Yamamoto, M., McGhee, J.R., Nishikawa, S. and Kiyono, H. (1996). Interleukin 2 (IL-2) and interleukin 7 (IL-7) reciprocally induce IL-7 and IL-2 receptors on $\gamma\delta$ T-cell receptor-positive intraepithelial lymphocytes. *Proc. Natl Acad. Sci. U.S.A*. **93**, 3613–3618.

Fung Leung, W.P., Schilham, M.W., Rahemtulla, A., Kundig, T.M., Vollenweider, M., Potter, J., van Ewijk, W. and Mak, T.W. (1991). CD8 is needed for development of cytotoxic T cells but not helper T cells. *Cell* **65**, 443–449.

Funk, P.E., Varas, A. and Witte, P.L. (1993). Activity of stem cell factor and IL-7 in combination on normal bone marrow B lineage cells. *J. Immunol*. **150**, 748–752.

Garvy, B.A. and Riley, R.L. (1994). IFN-γ abrogates IL-7-dependent proliferation in pre-B cells, coinciding with onset of apoptosis. *Immunology* **81**, 381–388.

Gessner, A., Vieth, M., Will, A., Schroppel, K. and Rollinghoff, M. (1993). Interleukin-7 enhances antimicrobial activity against *Leishmania major* in murine macrophages. *Infect. Immun*. **61**, 4008–4012.

Gessner, A., Will, A., Vieth, M., Schroppel, K. and Rollinghoff, M. (1995). Stimulation of B-cell lymphopoiesis by interleukin-7 leads to aggravation of murine leishmaniasis. *Immunology* **84**, 416–422.

Giri, J.G., Ahdieh, M., Eisenman, J., Shanebeck, K., Grabstein, K., Kumaki, S., Namen, A., Park, L.S., Cosman, D. and Anderson, D. (1994). Utilization of the β and γ chains of the IL-2 receptor by the novel cytokine IL-15. *EMBO J*. **13**, 2822–2830.

Goodwin, R.G., Lupton, S., Schmierer, A., Hjerrild, K.J., Jerzy, R., Clevenger, W., Gillis, S., Cosman, D. and Namen, A.E. (1989). Human interleukin 7: molecular cloning and growth factor activity on human and murine B-lineage cells. *Proc. Natl Acad. Sci. U.S.A*. **86**, 302–306.

Goodwin, R.G., Friend, D., Ziegler, S.F., Jerzy, R., Falk, B.A., Gimpel, S., Cosman, D., Dower, S.K., March, C.J. and Namen, A.E. (1990). Cloning of the human and murine interleukin-7 receptors: demonstration of a soluble form and homology to a new receptor superfamily. *Cell* **60**, 941–951.

Gorschluter, M., Bohlen, H., Hasenclever, D., Diehl, V. and Tesch, H. (1995). Serum cytokine levels correlate with clinical parameters in Hodgkin's disease. *Ann. Oncol*. **6**, 477–482.

Grabstein, K.H., Waldschmidt, T.J., Finkelman, F.D., Hess, B.W., Alpert, A.R., Boiani, N.E., Namen, A.E. and Morrissey, P.J. (1993). Inhibition of murine B and T lymphopoiesis *in vivo* by an anti-interleukin 7 monoclonal antibody. *J. Exp. Med*. **178**, 257–264.

Greil, J., Gramatzki, M., Burger, R., Marschalek, R., Peltner, M., Trautmann, U., Hansen Hagge, T.E., Bartram, C.R., Fey, G.H. and Stehr, K. (1994). The acute lymphoblastic leukaemia cell line SEM with t(4;11). chromosomal rearrangement is biphenotypic and responsive to interleukin-7. *Br. J. Haematol*. **86**, 275–283.

Grzegorzewski, K., Komschlies, K.L., Mori, M., Kaneda, K., Usui, N., Faltynek, C.R., Keller, J.R., Ruscetti, F.W. and Wiltrout, R.H. (1994). Administration of recombinant human interleukin-7 to mice induces the exportation of myeloid progenitor cells from the bone marrow to peripheral sites. *Blood* **83**, 377–385.

Gutierrez, J.C. and Palacios, R. (1991). Heterogeneity of thymic epithelial cells in promoting T-lymphocyte differentiation *in vivo*. *Proc. Natl Acad. Sci. U.S.A*. **88**, 642–646.

Guy Grand, D., Rocha, B., Mintz, P., Malassis Seris, M., Selz, F., Malissen, B. and Vassalli, P. (1994). Different use of T cell receptor transducing modules in two populations of gut intraepithelial lymphocytes are related to distinct pathways of T cell differentiation. *J. Exp. Med*. **180**, 673–679.

Hardy, R.R. and Hayakawa, K. (1991). A developmental switch in B lymphopoiesis. *Proc. Natl Acad. Sci. U.S.A*. **88**, 11 550–11 554.

Hardy, R.R., Carmack, C.E., Shinton, S.A., Kemp, J.D. and Hayakawa, K. (1991). Resolution and characterization of pro-B and pre-pro-B cell stages in normal mouse bone marrow. *J. Exp. Med*. **173**, 1213–1225.

Havran, W.L. and Boismenu, R. (1994). Activation and function of $\gamma\delta$ T cells. *Curr. Opin. Immunol*. **6**, 442–446.

Hayday, A.C. (1995). $\gamma\delta$ T cell specificity and function. In *T Cell Receptors* (eds J.I. Bell, M.J. Owen and E. Simpson), Oxford University Press, Oxford, pp. 70–91.

Hayes, S.M., Sirr, A., Jacob, S., Sim, G.K. and Augustin, A. (1996). Role of IL-7 in the shaping of the pulmonary $\gamma\delta$ T cell repertoire. *J. Immunol*. **156**, 2723–2729.

He, Y.W., Adkins, B., Furse, R.K. and Malek, T.R. (1995). Expression and function of the γc subunit of the IL-2, IL-4, and IL-7 receptors. Distinct interaction of γc in the IL-4 receptor. *J. Immunol*. **154**, 1596–1605.

He, Y.W. and Malek, T.R. (1996). Interleukin-7 receptor α is essential for the development of $\gamma\delta^+$ T cells, but not natural killer cells. *J. Exp. Med*. **184**, 289–293.

Henderson, A.J., Narayanan, R., Collins, L. and Dorshkind, K. (1992). Status of κ L chain gene rearrangements and c-kit and IL-7 receptor expression in stromal cell-dependent pre-B cells. *J. Immunol*. **149**, 1973–1979.

Hernandez Caselles, T., Martinez Esparza, M., Sancho, D., Rubio, G. and Aparicio, P. (1995). Interleukin-7 rescues human activated T lymphocytes from apoptosis induced by glucocorticoesteroids and regulates bcl-2 and CD25 expression. *Hum. Immunol*. **43**, 181–189.

Heufler, C., Topar, G., Grasseger, A., Stanzl, U., Koch, F., Romani, N., Namen, A.E. and Schuler, G. (1993). Interleukin 7 is produced by murine and human keratinocytes. *J. Exp. Med*. **178**, 1109–1114.

Hickman, C.J., Crim, J.A., Mostowski, H.S. and Siegel, J.P. (1990). Regulation of human cytotoxic T lymphocyte development by IL-7. *J. Immunol*. **145**, 2415–2420.

Hock, H., Dorsch, M., Diamantstein, T. and Blankenstein, T. (1991). Interleukin 7 induces CD4$^+$ T cell-dependent tumor rejection. *J. Exp. Med*. **174**, 1291–1298.

Hock, H., Dorsch, M., Kunzendorf, U., Qin, Z., Diamantstein, T. and Blankenstein, T. (1993). Mechanisms of rejection induced by tumor cell-targeted gene transfer of interleukin 2, interleukin 4, interleukin 7, tumor necrosis factor, or interferon γ. *Proc. Natl Acad. Sci. U.S.A*. **90**, 2774–2778.

Hunt, P., Robertson, D., Weiss, D., Rennick, D., Lee, F. and Witte, O.N. (1987). A single bone marrow-derived stromal cell type supports the *in vitro* growth of early lymphoid and myeloid cells. *Cell* **48**, 997–1007.

Jicha, D.L., Mule, J.J. and Rosenberg, S.A. (1991). Interleukin 7 generates antitumor cytotoxic T lymphocytes against murine sarcomas with efficacy in cellular adoptive immunotherapy. *J. Exp. Med*. **174**, 1511–1515.

Jicha, D.L., Schwarz, S., Mule, J.J. and Rosenberg, S.A. (1992). Interleukin-7 mediates the generation and expansion of murine allosensitized and antitumor CTL. *Cell Immunol*. **141**, 71–83.

Kaashoek, J.G., Mout, R., Falkenburg, J.H., Willemze, R., Fibbe, W.E. and Landegent, J.E. (1991). Cytokine production by the bladder carcinoma cell line 5637: rapid analysis of mRNA expression levels using a cDNA–PCR procedure. *Lymphokine Cytokine Res*. **10**, 231–235.

Kanamori, Y., Ishimaru, K., Nanno, M., Maki, K., Ikuta, K., Nariuchi, H. and Ishikawa, H. (1996). Identification of novel lymphoid tissues in murine intestinal mucosa where clusters of c-kit$^+$ IL-7R$^+$ Thy1$^+$ lympho-hemopoietic progenitors develop. *J. Exp. Med*. **184**, 1449–1459.

Kasper, L.H., Matsuura, T. and Khan, I.A. (1995). IL-7 stimulates protective immunity in mice against the intracellular pathogen, *Toxoplasma gondii*. *J. Immunol*. **155**, 4798–4804.

Katz, F.E., Tindle, R., Sutherland, D.R. and Greaves, M.F. (1985). Identification of a membrane glycoprotein associated with haemopoietic progenitor cells. *Leuk. Res*. **9**, 191–198.

Kee, B.L. and Paige, C.J. (1996). *In vitro* tracking of IL-7 responsiveness and gene expression during commitment of bipotent B-cell/macrophage progenitors. *Curr. Biol*. **6**, 1159–1169.

Kim, J.H., Ratto, S., Sitz, K.V., Mosca, J.D., McLinden, R.J., Tencer, K.L., Vahey, M.T., St Louis, D., Birx, D.L. and Redfield, R.R. (1994). Consequences of stable transduction and antigen-inducible expression of the human interleukin-7 gene on tetanus-toxoid-specific T cells. *Hum. Gene Ther*. **5**, 1457–1466.

Kishimoto, T., Goyert, S., Kikutani, H., Mason, D., Miyasaka, M., Moretta, L., Ohno, T., Okumura, K., Shaw, S., Springer, T.A., Sugamura, K., Sugawara, H., von dem Borne, A.E. and Zola, H. (1997). Update of CD antigens, 1996. *J. Immunol*. **158**, 3035–3036.

Kitamura, D., Roes, J., Kuhn, R. and Rajewsky, K. (1991). A B cell-deficient mouse by targeted disruption of the membrane exon of the immunoglobulin μ chain gene. *Nature* **350**, 423–426.

Kitamura, D., Kudo, A., Schaal, S., Muller, W., Melchers, F. and Rajewsky, K. (1992). A critical role of λ5 protein in B cell development. *Cell* **69**, 823–831.

Klein, J.R. (1996). Whence the intestinal intraepithelial lymphocyte? *J. Exp. Med*. **184**, 1203–1206.

Komschlies, K.L., Gregorio, T.A., Gruys, M.E., Back, T.C., Faltynek, C.R. and Wiltrout, R.H. (1994). Administration of recombinant human IL-7 to mice alters the composition of B-lineage cells and T cell subsets, enhances T cell function, and induces regression of established metastases. *J. Immunol*. **152**, 5776–5784.

Kondo, M., Takeshita, T., Ishii, N., Nakamura, M., Watanabe, S., Arai, K. and Sugamura, K. (1993). Sharing of the interleukin-2 (IL-2) receptor γ chain between receptors for IL-2 and IL-4. *Science* **262**, 1874–1877.

Kondo, M., Takeshita, T., Higuchi, M., Nakamura, M., Sudo, T., Nishikawa, S. and Sugamura, K. (1994). Functional participation of the IL-2 receptor γ chain in IL-7 receptor complexes. *Science* **263**, 1453–1454.

Kono, K., Salazar Onfray, F., Petersson, M., Hansson, J., Masucci, G., Wasserman, K., Nakazawa, T., Anderson, P. and Kiessling, R. (1996). Hydrogen peroxide secreted by tumor-derived macrophages down-modulates signal-transducing zeta molecules and inhibits tumor-specific T cell- and natural killer cell-mediated cytotoxicity. *Eur. J. Immunol*. **26**, 1308–1313.

Kos, F.J. and Mullbacher, A. (1992). Induction of primary anti-viral cytotoxic T cells by *in vitro* stimulation with short synthetic peptide and interleukin-7. *Eur. J. Immunol*. **22**, 3183–3185.

Kos, F.J. and Mullbacher, A. (1993). IL-2-independent activity of IL-7 in the generation of secondary antigen-specific cytotoxic T cell responses *in vitro*. *J. Immunol*. **150**, 387–393.

Kröncke, R., Loppnow, H., Flad, H.D. and Gerdes, J. (1996a). Human follicular dendritic cells and vascular cells produce interleukin-7: a potential role for interleukin-7 in the germinal center reaction. *Eur. J. Immunol*. **26**, 2541–2544.

Kröncke, R., Steffen, S., Flad, H.-D. and Gerdes, J. (1996b). IL-7 mRNA exists in different alternatively spliced isoforms. *Immunobiology* **196**, 12 (Abstract A18).

Kuhn, R., Rajewsky, K. and Muller, W. (1991). Generation and analysis of interleukin-4 deficient mice. *Science* **254**, 707–710.

Kuhn, R., Lohler, J., Rennick, D., Rajewsky, K. and Muller, W. (1993). Interleukin-10-deficient mice develop chronic enterocolitis. *Cell* **75**, 263–274.

Lai, P., Rabinowich, H., Crowley-Nowick, P.A., Bell, M.C., Mantovani, G. and Whiteside, T.L. (1996). Alteration in expression and function of signal transducing proteins in tumor-associated T and NK cells in patients with ovarian carcinoma. *Clin. Cancer Res*. **2**, 161–173.

Lai, S.Y., Molden, J. and Goldsmith, M.A. (1997). Shared γ(c) subunit within the human interleukin-7 receptor complex. A molecular basis for the pathogenesis of X-linked severe combined immunodeficiency. *J. Clin. Invest*. **99**, 169–177.

Leclercq, G., De Smedt, M. and Plum, J. (1992). Presence of CD8 α-CD8 β-positive TcR γ/δ thymocytes in the fetal murine thymus and their *in vitro* expansion with interleukin-7. *Eur. J. Immunol*. **22**, 2189–2193.

Lee, G., Namen, A.E., Gillis, S., Ellingsworth, L.R. and Kincade, P.W. (1989). Normal B cell precursors responsive to recombinant murine IL-7 and inhibition of IL-7 activity by transforming growth factor-β. *J. Immunol*. **142**, 3875–3883.

Lee, S.H., Fujita, N., Mashima, T. and Tsuruo, T. (1996). Interleukin-7 inhibits apoptosis of mouse malignant T-lymphoma cells by both suppressing the CPP32-like protease activation and inducing the Bcl-2 expression. *Oncogene* **13**, 2131–2139.

Leonard, W.J., Noguchi, M., Russell, S.M. and McBride, O.W. (1994). The molecular basis of X-linked severe combined immunodeficiency: the role of the interleukin-2 receptor γ chain as a common γ chain, γ c. *Immunol. Rev*. **138**, 61–86.

Levey, D.L. and Srivastava, P.K. (1995). T cells from late tumor-bearing mice express normal levels of p56lck, p59fyn, ZAP-70, and CD3 zeta despite suppressed cytolytic activity. *J. Exp. Med*. **182**, 1029–1036.

Lewis, S.M. (1994). The mechanism of V(D)J joining: lessons from molecular, immunological, and comparative analyses. *Adv. Immunol*. **56**, 27–150.

Li, Y.S., Hayakawa, K. and Hardy, R.R. (1993). The regulated expression of B lineage associated genes during B cell differentiation in bone marrow and fetal liver. *J. Exp. Med*. **178**, 951–960.

Li, Y.S., Wasserman, R., Hayakawa, K. and Hardy, R.R. (1996). Identification of the earliest B lineage stage in mouse bone marrow. *Immunity* **5**, 527–535.

Lin, J.X., Migone, T.S., Tsang, M., Friedmann, M., Weatherbee, J.A., Zhou, L., Yamauchi, A., Bloom, E.T., Mietz, J. and John, S. (1995). The role of shared receptor motifs and common Stat proteins in the generation of cytokine pleiotropy and redundancy by IL-2, IL-4, IL-7, IL-13, and IL-15. *Immunity* **2**, 331–339.

Loken, M.R., Shah, V.O., Hollander, Z. and Civin, C.I. (1988). Flow cytometric analysis of normal B lymphoid development. *Pathol. Immunopathol. Res*. **7**, 357–370.

Long, B.W., Witte, P.L., Abraham, G.N., Gregory, S.A. and Plate, J.M. (1995). Apoptosis and interleukin 7 gene expression in chronic B-lymphocytic leukemia cells. *Proc. Natl Acad. Sci. U.S.A*. **92**, 1416–1420.

Lu, L., Zhou, Z., Wu, B., Xiao, M., Shen, R.N., Williams, D.E., Kim, Y.J., Kwon, B.S., Ruscetti, S. and Broxmeyer, H.E. (1992). Influence of recombinant human interleukin (IL)-7 on disease progression in mice infected with Friend virus complex. *Int. J. Cancer* **52**, 261–265.

Lupton, S.D., Gimpel, S., Jerzy, R., Brunton, L.L., Hjerrild, K.A., Cosman, D. and Goodwin, R.G. (1990). Characterization of the human and murine IL-7 genes. *J. Immunol*. **144**, 3592–3601.

Lynch, D.H. and Miller, R.E. (1990). Induction of murine lymphokine-activated killer cells by recombinant IL-7. *J. Immunol*. **145**, 1983–1990.

Lynch, D.H. and Miller, R.E. (1994). Interleukin 7 promotes long-term *in vitro* growth of antitumor cytotoxic T lymphocytes with immunotherapeutic efficacy *in vivo*. *J. Exp. Med*. **179**, 31–42.

Lynch, D.H., Namen, A.E. and Miller, R.E. (1991). *In vivo* evaluation of the effects of interleukins 2, 4 and 7 on enhancing the immunotherapeutic efficacy of anti-tumor cytotoxic T lymphocytes. *Eur. J. Immunol*. **21**, 2977–2985.

Lynch, M., Baker, E., Park, L.S., Sutherland, G.R. and Goodwin, R.G. (1992). The interleukin-7 receptor gene is at 5p13. *Hum. Genet*. **89**, 566–568.

MacNeil, I.A., Suda, T., Moore, K.W., Mosmann, T.R. and Zlotnik, A. (1990). IL-10, a novel growth cofactor for mature and immature T cells. *J. Immunol*. **145**, 4167–4173.

Maeurer, M., Zitvogel, L., Elder, E., Storkus, W.J. and Lotze, M.T. (1995). Human intestinal $V\delta1^+$ T cells obtained from patients with colon cancer respond exclusively to SEB but not to SEA. *Nat. Immun*. **14**, 188–197.

Maeurer, M.J., Martin, D., Walter, W., Liu, K., Zitvogel, L., Halusczcak, K., Rabinowich, H., Duquesnoy, R., Storkus, W. and Lotze, M.T. (1996). Human intestinal $V\delta1^+$ lymphocytes recognize tumor cells of epithelial origin. *J. Exp. Med*. **183**, 1681–1696.

Maeurer, M.J., Walter, W., Martin, D., Zitvogel, L., Elder, E., Storkus, W. and Lotze, M.T. (1997). Interleukin-7 (IL-7) in colorectal cancer: IL-7 is produced by tissues from colorectal cancer and promotes preferential expansion of tumour infiltrating lymphocytes. *Scand. J. Immunol*. **45**, 182–192.

Maki, K., Sunaga, S., Komagata, Y., Kodaira, Y., Mabuchi, A., Karasuyama, H., Yokomuro, K., Miyazaki, J.I. and Ikuta, K. (1996). Interleukin 7 receptor-deficient mice lack $\gamma\delta$ T cells. *Proc. Natl Acad. Sci. U.S.A*. **93**, 7172–7177.

Makrynikola, V., Kabral, A. and Bradstock, K.F. (1991). Effects of recombinant human cytokines on precursor-B acute lymphoblastic leukemia cells. *Exp. Hematol*. **19**, 674–679.

Maraskovsky, E., Teepe, M., Morrissey, P.J., Braddy, S., Miller, R.E., Lynch, D.H. and Peschon, J.J. (1996). Impaired survival and proliferation in IL-7 receptor-deficient peripheral T cells. *J. Immunol*. **157**, 5315–5323.

Maric, D., Kalserlian, D. and Croitoru, K. (1996). Intestinal epithelial cell line induction of T cell differentiation from bone marrow precursors. *Cell Immunol*. **172**, 172–179.

Matsuda, M., Petersson, M., Lenkei, R., Taupin, J.L., Magnusson, I., Mellstedt, H., Anderson, P. and Kiessling, R. (1995). Alterations in the signal-transducing molecules of T cells and NK cells in colorectal tumor-infiltrating, gut mucosal and peripheral lymphocytes: correlation with the stage of the disease. *Int. J. Cancer* **61**, 765–772.

Matsue, H., Bergstresser, P.R. and Takashima, A. (1993a). Keratinocyte-derived IL-7 serves as a growth factor for dendritic epidermal T cells in mice. *J. Immunol*. **151**, 6012–6019.

Matsue, H., Bergstresser, P.R. and Takashima, A. (1993b). Reciprocal cytokine-mediated cellular interactions in mouse epidermis: promotion of $\gamma\delta$ T-cell growth by IL-7 and TNF α and inhibition of keratinocyte growth by γ IFN. *J. Invest. Dermatol*. **101**, 543–548.

Matsuzaki, Y., Nakayama, K., Tomita, T., Isoda, M., Loh, D.Y. and Nakauchi, H. (1997). Role of bcl-2 in the development of lymphoid cells from the hematopoietic stem cell. *Blood* **89**, 853–862.

Matthews, W., Jordan, C.T., Wiegand, G.W., Pardoll, D. and Lemischka, I.R. (1991). *Cell* **65**, 1143–1152.

McBride, W.H., Thacker, J.D., Comora, S., Economou, J.S., Kelley, D., Hogge, D., Dubinett, S.M. and Dougherty, G.J. (1992). Genetic modification of a murine fibrosarcoma to produce interleukin 7 stimulates host cell infiltration and tumor immunity. *Cancer Res*. **52**, 3931–3937.

McKearn, J.P., McCubrey, J. and Fagg, B. (1985). Enrichment of hematopoietic precursor cells and cloning of multipotential B-lymphocyte precursors. *Proc. Natl Acad. Sci. U.S.A*. **82**, 7414–7418.

McNiece, I.K., Langley, K.E. and Zsebo, K.M. (1991). The role of recombinant stem cell factor in early B cell development. Synergistic interaction with IL-7. *J. Immunol*. **146** 3785–3790.

Mehler, M.F., Rozental, R., Dougherty, M., Spray, D.C. and Kessler, J.A. (1993). Cytokine regulation of neuronal differentiation of hippocampal progenitor cells. *Nature* **362**, 62–65.

Mehrotra, P.M., Grant, A.J. and Siegel, J.P. (1995). Synergistic effects of IL-7 and IL-12 on human T-cell activation. *J. Immunol.* **154**, 5093–5102.

Melhem, M.F., Meisler, A.I., Saito, R., Finley, G.G., Hockman, H.R. and Koski, R.A. (1993). Cytokines in inflammatory malignant fibrous histiocytoma presenting with leukemoid reaction. *Blood* **82**, 2038–2044.

Mertsching, E., Burdet, C. and Ceredig, R. (1995). IL-7 transgenic mice: analysis of the role of IL-7 in the differentiation of thymocytes *in vivo* and *in vitro*. *Int. Immunol*. **7**, 401–414.

Mertsching, E., Grawunder, U., Meyer, V., Rolink, T. and Ceredig, R. (1996). Phenotypic and functional analysis of B lymphopoiesis in interleukin-7-transgenic mice: expansion of pro/pre-B cell number and persistence of B lymphocyte development in lymph nodes and spleen. *Eur. J. Immunol*. **26**, 28–33.

Michaelson, M.D., Mehler, M.F., Xu, H., Gross, R.E. and Kessler, J.A. (1996). Interleukin-7 is trophic for embryonic neurons and is expressed in developing brain. *Dev. Biol*. **179**, 251–263.

Miller, A.R., McBride, W.H., Dubinett, S.M., Dougherty, G.J., Thacker, J.D., Shau, H., Kohn, D.B., Moen, R.C., Walker, M.J. and Chiu, R. (1993). Transduction of human melanoma cell lines with the human interleukin-7 gene using retroviral-mediated gene transfer: comparison of immunologic properties with interleukin-2. *Blood* **82**, 3686–3694.

Miyaji, C., Watanabe, H., Osman, Y., Kuwano, Y. and Abo, T. (1996). A comparison of proliferative response to IL-7 and expression of IL-7 receptors in intermediate TCR cells of the liver, spleen, and thymus. *Cell. Immunol*. **169**, 159–165.

Mizoguchi, H., O'Shea, J.J., Longo, D.L., Loeffler, C.M., McVicar, D.W. and Ochoa, A.C. (1992). Alterations in signal transduction molecules in T lymphocytes from tumor-bearing mice. *Science* **258**, 1795–1798.

Molina, T.J., Kishihara, K., Siderovski, D.P., van Ewijk, W., Narendran, A., Timms, E., Wakeham, A., Paige, C.J., Hartmann, K.U., Veillette, A., Davidson, D. and Mak, T.W. (1992). Profound block in thymocyte development in mice lacking p56lck. *Nature* **357**, 161–164.

Möller, P., Bohm, M., Czarnetszki, B.M. and Schadendorf, D. (1996). Interleukin-7. Biology and implications for dermatology. *Exp. Dermatol*. **5**, 129–137.

Montgomery, R.A. and Dallman, M.J. (1991). Analysis of cytokine gene expression during fetal thymic ontogeny using the polymerase chain reaction. *J. Immunol*. **147**, 554–560.

Moore, T.A., von Freeden-Jeffrey, U., Murray, R. and Zlotnik, A. (1995). *Ninth International Congress on Immunology*, Abstract 441, p. 75.

Moore, T.A., von Freeden-Jeffrey, U., Murray, R. and Zlotnik, A. (1996). Inhibition of $\gamma\delta$ T cell development and early thymocyte maturation in IL-$7^{-/-}$ mice. *J. Immunol*. **157**, 2366–2373.

Moran, P.A., Diegel, M.L., Sias, J.C., Ledbetter, J.A. and Zarling, J.M. (1993). Regulation of HIV production by blood mononuclear cells from HIV-infected donors: I. Lack of correlation between HIV-1 production and T cell activation. *AIDS Res. Hum. Retroviruses* **9**, 455–464.

Morrow, M.A., Lee, G., Gillis, S., Yancopoulos, G.D. and Alt, F.W. (1992). Interleukin-7 induces N-myc and c-myc expression in normal precursor B lymphocytes. *Genes Dev*. **6**, 61–70.

Muegge, K., Vila, M.P. and Durum, S.K. (1994). Interleukin-7: a cofactor for V(D)J rearrangement of the T cell receptor β gene. *Science* **261**, 93–95.

Murphy, W.J., Back, T.C., Conlon, K.C., Komschlies, K.L., Ortaldo, J.R., Sayers, T.J., Wiltrout, R.H. and Longo, D.L. (1993). Antitumor effects of interleukin-7 and adoptive immunotherapy on human colon carcinoma xenografts. *J. Clin. Invest*. **92**, 1918–1924.

Musso, T., Johnston, J.A., Linnekin, D., Varesio, L., Rowe, T.K., O'Shea, J.J. and McVicar, D.W. (1995). Regulation of JAK3 expression in human monocytes: phosphorylation in response to interleukins 2, 4, and 7. *J. Exp. Med*. **181**, 1425–1431.

Nakagomi, H., Petersson, M., Magnusson, I., Juhlin, C., Matsuda, M., Mellstedt, H., Taupin, J.L., Vivier, E., Anderson, P. and Kiessling, R. (1993). Decreased expression of the signal-transducing zeta chains in tumor-infiltrating T-cells and NK cells of patients with colorectal carcinoma. *Cancer Res*. **53**, 5610–5612.

Namen, A.E., Lupton, S., Hjerrild, K., Wignall, J., Mochizuki, D.Y., Schmierer, A., Mosley, B., March, C.J., Urdal, D. and Gillis, S. (1988a). Stimulation of B-cell progenitors by cloned murine interleukin-7. *Nature* **333**, 571–573.

Namen, A.E., Schmierer, A.E., March, C.J., Overell, R.W., Park, L.S., Urdal, D.L. and Mochizuki, D.Y. (1988b). B cell precursor growth-promoting activity. Purification and characterization of a growth factor active on lymphocyte precursors. *J. Exp. Med*. **167**, 988–1002.

Namikawa, R., Muench, M.O., de Vries, J.E. and Roncarolo, M.G. (1996). The FLK2/FLT3 ligand synergizes with interleukin-7 in promoting stromal-cell-independent expansion and differentiation of human fetal pro-B cells *in vitro*. *Blood* **87**, 1881–1890.

Naume, B. and Espevik, T. (1991). Effects of IL-7 and IL-2 on highly enriched CD56$^+$ natural killer cells. A comparative study. *J. Immunol*. **147**, 2208–2214.

Naume, B., Gately, M. and Espevik, T. (1992). A comparative study of IL-12 (cytotoxic lymphocyte maturation factor)-, IL-2-, and IL-7-induced effects on immunomagnetically purified CD56$^+$ NK cells. *J. Immunol*. **148**, 2429–2436.

Naume, B., Johnsen, A.C., Espevik, T. and Sundan, A. (1993). Gene expression and secretion of cytokines and cytokine receptors from highly purified CD56$^+$ natural killer cells stimulated with interleukin-2, interleukin-7 and interleukin-12. *Eur. J. Immunol.* **23**, 1831–1838.

Noguchi, M., Nakamura, Y., Russell, S.M., Ziegler, S.F., Tsang, M., Cao, X. and Leonard, W.J. (1993). Interleukin-2 receptor γ chain: a functional component of the interleukin-7 receptor. *Science* **262**, 1877–1880.

Nosaka, T., van Deursen, J.M., Tripp, R.A., Thierfelder, W.E., Witthuhn, B.A., McMickle, A.P., Doherty, P.C., Grosveld, G.C. and Ihle, J.N. (1995). Defective lymphoid development in mice lacking Jak3. *Science* **270**, 800–802.

Oettinger, M.A., Schatz, D.G., Gorka, C. and Baltimore, D. (1990). *RAG-1* and *RAG-2*, adjacent genes that synergistically activate V(D)J recombination. *Science* **248**, 1517–1523.

Okazaki, H., Ito, M., Sudo, T., Hattori, M., Kano, S., Katsura, Y. and Minato, N. (1989). IL-7 promotes thymocyte proliferation and maintains immunocompetent thymocytes bearing α/β or $\gamma\delta$ T-cell receptors *in vitro*: synergism with IL-2. *J. Immunol.* **143**, 2917–2922.

Oltz, E.M., Yancopoulos, G.D., Morrow, M.A., Rolink, A., Lee, G., Wong, F., Kaplan, K., Gillis, S., Melchers, F. and Alt, F.W. (1992). A novel regulatory myosin light chain gene distinguishes pre-B cell subsets and is IL-7 inducible. *EMBO J.* **11**, 2759–2767.

Ono, M., Ariizumi, K., Bergstresser, P.R. and Takashima, A. (1996). IL-7 upregulates T cell receptor/CD3 expression by cultured dendritic epidermal T cells. *J. Dermatol. Sci.* **11**, 89–96.

Oosterwegel, M.A., Haks, M.C., von Freeden-Jeffrey, U., Murray, R. and Kruisbeek, A.M. (1997). Induction of TCR gene rearrangements in uncommitted stem cells by a subset of IL-7 producing, MHC class-II-expressing thymic stromal cells. *Immunity* **6**, 351–360.

Orlic, D., Fischer, R., Nishikawa, S., Nienhuis, A.W. and Bodine, D.M. (1993). Purification and characterization of heterogeneous pluripotent hematopoietic stem cell populations expressing high levels of c-kit receptor. *Blood* **82**, 762–770.

Page, T.H., Willcocks, J.L., Taylor Fishwick, D.A. and Foxwell, B.M. (1993). Characterization of a novel high affinity human IL-7 receptor. Expression on T cells and association with IL-7 driven proliferation. *J. Immunol.* **151**, 4753–4763.

Page, T.H., Lali, F.V., Groome, N. and Foxwell, B.M.J. (1997). Association of the common γ-chain with the human IL-7 receptor is modulated by T cell activation. *J. Immunol.* **158**, 5727–5735.

Pandrau Garcia, D., de Saint Vis, B., Saeland, S., Renard, N., Ho, S., Moreau, I., Banchereau, J. and Galizzi, J.P. (1994). Growth inhibitory and agonistic signals of interleukin-7 (IL-7). can be mediated through the CDw127 IL-7 receptor. *Blood* **83**, 3613–3619.

Park, L.S., Friend, D.J., Schmierer, A.E., Dower, S.K. and Namen, A.E. (1990). Murine interleukin 7 (IL-7). receptor. Characterization on an IL-7-dependent cell line. *J. Exp. Med.* **171**, 1073–1089.

Park, S.Y., Saijo, K., Takahashi, T., Osawa, M., Arase, H., Hirayama, N., Miyake, K., Nakauchi, H., Shirasawa, T. and Saito, T. (1995). Developmental defects of lymphoid cells in Jak3 kinase-deficient mice. *Immunity* **3**, 771–782.

Perumal, N.B., Kenniston, T.W., Jr., Tweardy, D.J., Dyer, K.F., Hoffman, R., Peschon, J. and Appasamy, P.M. (1997). *J. Immunol.* **158**, 5744–5750.

Peschon, J.J., Morrissey, P.J., Grabstein, K.H., Ramsdell, F.J., Maraskovsky, E., Gliniak, B.C., Park, L.S., Ziegler, S.F., Williams, D.E. and Ware, C.B. (1994). Early lymphocyte expansion is severely impaired in interleukin 7 receptor-deficient mice. *J. Exp. Med.* **180**, 1955–1960.

Peschon, J.J., Gliniak, B.C., Morissey, P. and Maraskowsky, E. (1997). In *Cytokine Knockouts* (eds S.K. Durak and U. Muegge), Humana Press, Totowo, New Jersey.

Pleiman, C.M., Gimpel, S.D., Park, L.S., Harada, H., Taniguchi, T. and Ziegler, S.F. (1991). *Mol. Cell. Biol.* **11**, 3052–3059.

Plum, J., De Smedt, M. and Leclercq, G. (1993). Exogenous IL-7 promotes the growth of CD3$^-$ CD4$^-$ CD8$^-$ CD44$^+$ CD25$^{+/-}$ precursor cells and blocks the differentiation pathway of TCR-$\alpha\beta$ cells in fetal thymus organ culture. *J. Immunol.* **150**, 2706–2716.

Poussier, P. and Julius, M. (1994). Thymus independent T cell development and selection in the intestinal epithelium. *Annu. Rev. Immunol.* **12**, 521–553.

Prieyl, J.A. and LeBien, T.W. (1996). Interleukin 7 independent development of human B cells. *Proc. Natl Acad. Sci. U.S.A.* **93**, 10 348–10 353.

R&D Systems (1994). Levels of IL-1, IL-3, IL-6, IL-7 TNF, and basic FGF in normal blood and urine. In *Cytokine Bulletin*, R&D Systems, Minneapolis, p. 3.

Rabinowich, H., Banks, M., Reichert, T., Logan, T.F., Kirkwood, J.M. and Whiteside, T.L. (1996). Expression and activity of signalling molecules in T lymphocytes obtained from patients with metastatic melanoma before and after interleukin-2 therapy. *Clin. Cancer Res*. **2**, 1263–1274.

Ray, R.J., Paige, C.J., Furlonger, C., Lyman, S.D. and Rottapel, R. (1996). Flt3 ligand supports the differentiation of early B cell progenitors in the presence of interleukin-11 and interleukin-7. *Eur. J. Immunol*. **26**, 1504–1510.

Reinecker, H.C. and Podolsky, D.K. (1995). Human intestinal epithelial cells express functional cytokine receptors sharing the common γc chain of the interleukin 2 receptor. *Proc. Natl Acad. Sci. U.S.A*. **92**, 8353–8357.

Rich, B.E., Campos Torres, J., Tepper, R.I., Moreadith, R.W. and Leder, P. (1993). Cutaneous lymphoproliferation and lymphomas in interleukin 7 transgenic mice. *J. Exp. Med*. **177**, 305–316.

Rocha, B., Vassalli, P. and Guy Grand, D. (1994). Thymic and extrathymic origins of gut intraepithelial lymphocyte populations in mice. *J. Exp. Med*. **180**, 681–686.

Rosnet, O., Marchetto, S., deLapeyriere, O. and Birnbaum, D. (1991). Murine Flt3, a gene encoding a novel tyrosine kinase receptor of the PDGFR/CSF1R family. *Oncogene* **6**, 1641–1650.

Russell, S.M., Keegan, A.D., Harada, N., Nakamura, Y., Noguchi, M., Leland, P., Friedmann, M.C., Miyajima, A., Puri, R.K. and Paul, W.E. (1993). Interleukin-2 receptor γ chain: a functional component of the interleukin-4 receptor. *Science* **262**, 1880–1883.

Russell, S.M., Johnston, J.A., Noguchi, M., Kawamura, M., Bacon, C.M., Friedmann, M., Berg, M., McVicar, D.W., Witthuhn, B.A. and Silvennoinen, O. (1994). Interaction of IL-2R β and γc chains with Jak1 and Jak3: implications for XSCID and XCID. *Science* **266**, 1042–1045.

Ryan, D.H., Nuccie, B.L., Ritterman, I., Liesveld, J.L. and Abboud, C.N. (1994). Cytokine regulation of early human lymphopoiesis. *J. Immunol*. **152**, 5250–5258.

Sakata, T., Iwagami, S., Tsuruta, Y., Teraoka, H., Tatsumi, Y., Kita, Y., Nishikawa, S., Takai, Y. and Fujiwara, H. (1990). Constitutive expression of interleukin-7 mRNA and production of IL-7 by a cloned murine thymic stromal cell line. *J. Leukoc. Biol*. **48**, 205–212.

Salvadori, S., Gansbacher, B., Pizzimenti, A.M. and Zier, K.S. (1994). Abnormal signal transduction by T cells of mice with parental tumors is not seen in mice bearing IL-2-secreting tumors. *J. Immunol*. **153**, 5176–5182.

Sartono, E., Kruize, Y.C., Partono, F., Kurniawan, A., Maizels, R.M. and Yazdanbakhsh, M. (1995). Specific T cell unresponsiveness in human filariasis: diversity in underlying mechanisms. *Parasite Immunol*. **17**, 587–594.

Schlossman, S.F., Boumsell, L., Gilks, W., Harlan, J.M., Kishimoto, T., Morimoto, C., Ritz, J., Shaw, S., Silverstein, R.L., Springer, T.A., Tedder, T.F. and Todd, R.F. (1994). CD antigens 1993. *J. Immunol*. **152**, 1.

Schorle, H., Holtschke, T., Hunig, T., Schimpl, A. and Horak, I. (1991). Development and function of T cells in mice rendered interleukin-2 deficient by gene targeting. *Nature* **352**, 621–624.

Seckinger, P. and Fougereau, M. (1994). Activation of src kinases in human pre-B cells by IL-7. *J. Immunol*. **153**, 97–109.

Shand, R.F. and Betlach, M.C. (1991). Expression of the bop gene cluster of *Halobacterium halobium* is induced by low oxygen tension and by light. *J. Bacteriol*. **173**, 4692–4699.

Sharfe, N., Dadi, H.K. and Roifman, C.M. (1995). JAK3 protein tyrosine kinase mediates interleukin-7-induced activation of phosphatidylinositol-3′ kinase. *Blood* **86**, 2077–2085.

Sharma, S., Wang, J., Huang, M., Paul, R.W., Lee, P., McBride, W.H., Economou, J.S., Roth, M.D., Kiertscher, S.M. and Dubinett, S.M. (1996). Interleukin-7 gene transfer in non-small-cell lung cancer decreases tumor proliferation, modifies cell surface molecule expression, and enhances antitumor reactivity. *Cancer Gene Ther*. **3**, 302–313.

Sherwood, P.J. and Weissman, I.L. (1990). The growth factor IL-7 induces expression of a transformation-associated antigen in normal pre-B cells. *Int. Immunol*. **2**, 399–406.

Sica, D., Rayman, P., Stanley, J., Edinger, M., Tubbs, R.R., Klein, E., Bukowski, R. and Finke, J.H. (1993). Interleukin 7 enhances the proliferation and effector function of tumor-infiltrating lymphocytes from renal-cell carcinoma. *Int. J. Cancer* **53**, 941–947.

Sieling, P.A., Sakimura, L., Uyemura, K., Yamamura, M., Oliveros, J., Nickoloff, B.J., Rea, T.H. and Modlin, R.L. (1995). IL-7 in the cell-mediated immune response to a human pathogen. *J. Immunol*. **154**, 2775–2783.

Sillevis Smitt, J.H., Weening, R.S., Krieg, S.R. and Bos, J.D. (1992). The skin in severe combined immunodeficiency: a case with transient cutaneous presence of γ/δ ($TRC1^+$) T cells. *Br. J. Dermatol*. **127**, 281–285.

Skeen, M.J. and Ziegler, H.K. (1993). Induction of murine peritoneal γ/δ T cells and their role in resistance to bacterial infection. *J. Exp. Med*. **178**, 971–984.

Skjonsberg, C., Erikstein, B.K., Smeland, E.B., Lie, S.O., Funderud, S., Beiske, K. and Blomhoff, H.K. (1991). Interleukin-7 differentiates a subgroup of acute lymphoblastic leukemias. *Blood* **77**, 2445–2450.

Slieker, W.A., van der Loo, J.C., de Rijk de Bruijn, M.F., Godfrey, D.I., Leenen, P.J. and van Ewijk, W. (1993). ER-MP12 antigen, a new cell surface marker on mouse bone marrow cells with thymus-repopulating ability: II. Thymus-homing ability and phenotypic characterization of ER-MP12-positive bone marrow cells. *Int. Immunol*. **5**, 1099–1107.

Smithgall, M.D., Wong, J.G., Critchett, K.E. and Haffar, O.K. (1996). IL-7 up-regulates HIV-1 replication in naturally infected peripheral blood mononuclear cells. *J. Immunol*. **156**, 2324–2330.

Stall, A.M. and Wells, S.M. (1996). Introduction: B-1 cells: origins and functions. *Semin. Immunol*. **8**, 1–2.

Standiford, T.J., Strieter, R.M., Allen, R.M., Burdick, M.D. and Kunkel, S.L. (1992). IL-7 up-regulates the expression of IL-8 from resting and stimulated human blood monocytes. *J. Immunol*. **149**, 2035–2039.

Stein, P.L., Lee, H.M., Rich, S. and Soriano, P. (1992). pp59fyn mutant mice display differential signaling in thymocytes and peripheral T cells. *Cell* **70**, 741–750.

Steiner, G., Koning, F., Elbe, A., Tschachler, E., Yokoyama, W.M., Shevach, E.M., Stingl, G. and Coligan, J.E. (1988). Characterization of T cell receptors on resident murine dendritic epidermal T cells. *Eur. J. Immunol*. **18**, 1323–1328.

Stephan, R.P., Sanders, V.M. and Witte, P.L. (1996). Stage-specific alterations in murine B lymphopoiesis with age. *Int. Immunol*. **8**, 509–518.

Stephan, R.P., Lill Elghanian, D.A. and Witte, P.L. (1997). Development of B cells in aged mice: decline in the ability of pro-B cells to respond to IL-7 but not to other growth factors. *J. Immunol*. **158**, 1598–1609.

Stötter, H. and Lotze, M.T. (1991). Human lymphokine-activated killer cell activity. Role of IL-2, IL-4, and IL-7. *Arch. Surg*. **126**, 1525–1530.

Stötter, H., Custer, M.C., Bolton, E.S., Guedez, L. and Lotze, M.T. (1991). IL-7 induces human lymphokine-activated killer cell activity and is regulated by IL-4. *J. Immunol*. **146**, 150–155.

Strehlau, J., Pavlakis, M., Lipman, M., Shapiro, M., Vasconcellos, L., Harmon, W. and Strom, T.B. (1997). Quantitative detection of immune activation transcripts as a diagnostic tool in kidney transplantation. *Proc. Natl Acad. Sci. U.S.A*. **94**, 695–700.

Suda, T. and Zlotnik, A. (1991). IL-7 maintains the T cell precursor potential of $CD3^-$ $CD4^-$ $CD8^-$ thymocytes. *J. Immunol*. **146**, 3068–3073.

Suda, T., Okada, S., Suda, J., Miura, Y., Ito, M., Sudo, T., Hayashi, S., Nishikawa, S. and Nakauchi, H. (1989). A stimulatory effect of recombinant murine interleukin-7 (IL-7) on B-cell colony formation and an inhibitory effect of IL-1α. *Blood* **74**, 1936–1941.

Sudo, T., Ito, M., Ogawa, Y., Iizuka, M., Kodama, H., Kunisada, T., Hayashi, S., Ogawa, M., Sakai, K. and Nishikawa, S. (1989). Interleukin 7 production and function in stromal cell-dependent B cell development. *J. Exp. Med*. **170**, 333–338.

Sudo, T., Nishikawa, S., Ohno, N., Akiyama, N., Tamakoshi, M. and Yoshida, H. (1993). Expression and function of the interleukin 7 receptor in murine lymphocytes. *Proc. Natl Acad. Sci. U.S.A*. **90**, 9125–9129.

Sugamura, K., Asao, H., Kondo, M., Tanaka, N., Ishii, N., Ohbo, K., Nakamura, M. and Takeshita, T. (1996). The interleukin-2 receptor γ chain: its role in the multiple cytokine receptor complexes and T cell development in XSCID. *Annu. Rev. Immunol*. **14**, 179–205.

Sutherland, G.R., Baker, E., Fernandez, K.E., Callen, D.F., Goodwin, R.G., Lupton, S., Namen, A.E., Shannon, M.F. and Vadas, M.A. (1989). The gene for human interleukin 7 (IL7) is at 8q12–13. *Hum. Genet*. **82**, 371–372.

Sweeney, E., van der Spek, J. and Murphy, J. (1995). IL-7 receptor-specific cell killing by DAB389-IL-7: a novel agent for the elimination of IL-7 receptor-positive malignant cells. *Fourth International Symposium on Immunotoxins*, p. 74 (abstract).

Szilvassy, S.J. and Cory, S. (1993). Phenotypic and functional characterization of competitive long-term repopulating hematopoietic stem cells enriched from 5-fluorouracil-treated murine marrow. *Blood* **81**, 2310–2320.

Szilvassy, S.J., Lansdorp, P.M., Humphries, R.K., Eaves, A.C. and Eaves, C.J. (1989). Isolation in a single step of a highly enriched murine hematopoietic stem cell population with competitive long-term repopulating ability. *Blood* **74**, 930–939.

Takashima, A. and Bergstresser, P.R. (1996). Cytokine-mediated communication by keratinocytes and Langerhans cells with dendritic epidermal T cells. *Semin. Immunol*. **8**, 333–339.

Takashima, A., Matsue, H., Bergstresser, P.R. and Ariizumi, K. (1995). Interleukin-7-dependent interaction of dendritic epidermal T cells with keratinocytes. *J. Invest. Dermatol*. **105**, 50S–53S.

Takeshita, T., Asao, H., Ohtani, K., Ishii, N., Kumaki, S., Tanaka, N., Munakata, H., Nakamura, M. and Sugamura, K. (1992). Cloning of the γ chain of the human IL-2 receptor. *Science* **257**, 379–382.

Talmadge, J.E., Jackson, J.D., Kelsey, L., Borgeson, C.D., Faltynek, C. and Perry, G.A. (1993). T-cell reconstitution by molecular, phenotypic, and functional analysis in the thymus, bone marrow, spleen, and blood following split-dose polychemotherapy and therapeutic activity for metastatic breast cancer in mice. *J. Immunother*. **14**, 258–268.

Tang, J., Nuccie, B.L., Ritterman, I., Liesveld, J.L., Abboud, C.N. and Ryan, D.H. (1997). TGF-β down-regulates stromal IL-7 secretion and inhibits proliferation of human B cell precursors. *J. Immunol*. **159**, 117–125.

Taniguchi, T. and Minami, Y. (1993). The IL-2/IL-2 receptor system: a current overview. *Cell* **73**, 5–8.

Tantawichien, T., Young, L.S. and Bermudez, L.E. (1996). Interleukin-7 induces anti-*Mycobacterium avium* activity in human monocyte-derived macrophages. *J. Infect. Dis*. **174**, 574–582.

Tartour, E., Latour, S., Mathiot, C., Thiounn, N., Mosseri, V., Joyeux, I., D'Enghien, C.D., Lee, R., Debre, B. and Fridman, W.H. (1995). Variable expression of CD3-zeta chain in tumor-infiltrating lymphocytes (TIL) derived from renal-cell carcinoma: relationship with TIL phenotype and function. *Int. J. Cancer* **63**, 205–212.

Tepper, R.I. and Mule, J.J. (1994). Experimental and clinical studies of cytokine gene-modified tumor cells. *Hum. Gene Ther*. **5**, 153–164.

Touw, I., Pouwels, K., van Agthoven, T., van Gurp, R., Budel, L., Hoogerbrugge, H., Delwel, R., Goodwin, R., Namen, A. and Lowenberg, B. (1990). Interleukin-7 is a growth factor of precursor B and T acute lymphoblastic leukemia. *Blood* **75**, 2097–2101.

Trumper, L., Jung, W., Dahl, G., Diehl, V., Gause, A. and Pfreundschuh, M. (1994). Interleukin-7, interleukin-8, soluble TNF receptor, and p53 protein levels are elevated in the serum of patients with Hodgkin's disease. *Ann. Oncol*. **5 (supplement 1)**, 93–96.

Tushinski, R.J., McAlister, I.B., Williams, D.E. and Namen, A.E. (1991). The effects of interleukin 7 (IL-7) on human bone marrow *in vitro*. *Exp. Hematol*. **19**, 749–754.

Uckun, F.M., Tuel Ahlgren, L., Obuz, V., Smith, R., Dibirdik, I., Hanson, M., Langlie, M.C. and Ledbetter, J.A. (1991). Interleukin 7 receptor engagement stimulates tyrosine phosphorylation, inositol phospholipid turnover, proliferation, and selective differentiation to the CD4 lineage by human fetal thymocytes. *Proc. Natl Acad. Sci. U.S.A*. **88**, 6323–6327.

Uehira, M., Matsuda, H., Hikita, I., Sakata, T., Fujiwara, H. and Nishimoto, H. (1993). The development of dermatitis infiltrated by $\gamma\delta$ T cells in IL-7 transgenic mice. *Int. Immunol*. **5**, 1619–1627.

van der Plas, D.C., Smiers, F., Pouwels, K., Hoefsloot, L.H., Lowenberg, B. and Touw, I.P. (1996). Interleukin-7 signaling in human B cell precursor acute lymphoblastic leukemia cells and murine BAF3 cells involves activation of STAT1 and STAT5 mediated via the interleukin-7 receptor α chain. *Leukemia* **10**, 1317–1325.

Van Kerckhove, C., Russell, G.J., Deusch, K., Reich, K., Bhan, A.K., DerSimonian, H. and Brenner, M.B. (1992). Oligoclonality of human intestinal intraepithelial T cells. *J. Exp. Med*. **175**, 57–63.

Veiby, O.P., Lyman, S.D. and Jacobsen, S.E. (1996a). Combined signaling through interleukin-7 receptors and flt3 but not c-kit potently and selectively promotes B-cell commitment and differentiation from uncommitted murine bone marrow progenitor cells. *Blood* **88**, 1256–1265.

Veiby, O.P., Jacobsen, F.W., Cui, L., Lyman, S.D. and Jacobsen, S.E. (1996b). The flt3 ligand promotes the survival of primitive hemopoietic progenitor cells with myeloid as well as B lymphoid potential. Suppression of apoptosis and counteraction by TNF-α and TGF-β. *J. Immunol*. **157**, 2953–2960.

Vellenga, E., Esselink, M.T., Straaten, J., Stulp, B.K., De Wolf, J.T., Brons, R., Giannotti, J., Smit, J.W. and Halie, M.R. (1992). The supportive effects of IL-7 on eosinophil progenitors from human bone marrow cells can be blocked by anti-IL-5. *J. Immunol*. **149**, 2992–2995.

Veis, D.J., Sorenson, C.M., Shutter, J.R. and Korsmeyer, S.J. (1993). Bcl-2-deficient mice demonstrate fulminant lymphoid apoptosis, polycystic kidneys, and hypopigmented hair. *Cell* **75**, 229–240.

Vella, A., Teague, T.K., Ihle, J., Kappler, J. and Marrack, P. (1997). Interleukin 4 (IL-4) or IL-7 prevents the death of resting T-cells: Stat6 is probably not required for the effect of IL-4. *J. Exp. Med*. **186**, 325–330.

Vidal, K., Grosjean, I., Evillard, J.P., Gespach, C. and Kaiserlian, D. (1993). Immortalization of mouse intestinal epithelial cells by the SV40-large T gene. Phenotypic and immune characterization of the MODE-K cell line. *J. Immunol. Methods* **166**, 63–73.

von Freeden-Jeffrey, U., Vieira, P., Lucian, L.A., McNeil, T., Burdach, S.E. and Murray, R. (1995). Lymphopenia in interleukin (IL)-7 gene-deleted mice identifies IL-7 as a nonredundant cytokine. *J. Exp. Med*. **181**, 1519–1526.

Watanabe, M., Ueno, Y., Yajima, T., Iwao, Y., Tsuchiya, M., Ishikawa, H., Aiso, S., Hibi, T. and Ishii, H. (1995). Interleukin 7 is produced by human intestinal epithelial cells and regulates the proliferation of intestinal mucosal lymphocytes. *J. Clin. Invest*. **95**, 2945–2953.

Watanabe, Y., Mazda, O., Aiba, Y., Iwai, K., Gyotoku, J., Ideyama, S., Miyazaki, J. and Katsura, Y. (1992). A murine thymic stromal cell line which may support the differentiation of $CD4^-8^-$ thymocytes into $CD4^+8^-$ $\alpha\beta$ T cell receptor positive T cells. *Cell. Immunol.* **142**, 385–397.

Watson, J.D., Morrissey, P.J., Namen, A.E., Conlon, P.J. and Widmer, M.B. (1989). Effect of IL-7 on the growth of fetal thymocytes in culture. *J. Immunol.* **143**, 1215–1222.

Welch, P.A., Namen, A.E., Goodwin, R.G., Armitage, R. and Cooper, M.D. (1989). Human IL-7: a novel T cell growth factor. *J. Immunol.* **143**, 3562–3567.

Widmer, M.B., Morrissey, P.J., Namen, A.E., Voice, R.F. and Watson, J.D. (1990). Interleukin 7 stimulates growth of fetal thymic precursors of cytolytic cells: induction of effector function by interleukin 2 and inhibition by interleukin 4. *Int. Immunol.* **2**, 1055–1061.

Wiles, M.V., Ruiz, P. and Imhof, B.A. (1992). Interleukin-7 expression during mouse thymus development. *Eur. J. Immunol.* **22**, 1037–1042.

Williams, D.E., Eisenman, J., Baird, A., Rauch, C., Van Ness, K., March, C.J., Park, L.S., Martin, U., Mochizuki, D.Y., Boswell, H.S., Burgess, G.S., Cosman, D. and Lyman, S.D. (1990). Identification of a ligand for the c-*kit* proto-oncogene. *Cell* **63**, 167–174.

Wineman, J.P., Gilmore, G.L., Gritzmacher, C., Torbett, B.E. and Muller Sieburg, C.E. (1992). CD4 is expressed on murine pluripotent hematopoietic stem cells. *Blood* **80**, 1717–1724.

Witte, P.L., Frantsve, L.M., Hergott, M. and Rahbe, S.M. (1993). Cytokine production and heterogeneity of primary stromal cells that support B lymphopoiesis. *Eur. J. Immunol.* **23**, 1809–1817.

Wolf, M.L., Weng, W.K., Stieglbauer, K.T., Shah, N. and LeBien, T.W. (1993). Functional effect of IL-7-enhanced CD19 expression on human B cell precursors. *J. Immunol.* **151**, 138–148.

Xu, S., Ariizumi, K., Edelbaum, D., Bergstresser, P.R. and Takashima, A. (1995). Cytokine-dependent regulation of growth and maturation in murine epidermal dendritic cell lines. *Eur. J. Immunol.* **25**, 1018–1024.

Yamada, N., Wakugawa, M., Kuwata, S., Nakagawa, H. and Tamaki, K. (1996). Changes in eosinophil and leukocyte infiltration and expression of IL-6 and IL-7 messenger RNA in mite allergen patch test reactions in atopic dermatitis. *J. Allergy Clin. Immunol.* **98**, S201–S206.

Yamaoka, Y., Kita, M., Kodama, T., Sawai, N., Kashima, K. and Imanishi, J. (1995). Expression of cytokine mRNA in gastric mucosa with *Helicobacter pylori* infection. *Scand. J. Gastroenterol.* **30**, 1153–1159.

Yasunaga, M., Wang, F., Kunisada, T. and Nishikawa, S. (1995). Cell cycle control of c-kit^+IL-$7R^+$ B precursor cells by two distinct signals derived from IL-7 receptor and c-kit in a fully defined medium. *J. Exp. Med.* **182**, 315–323.

Yoshioka, R., Shimizu, S., Tachibana, J., Hirose, Y., Fukutoku, M., Takeuchi, Y., Sugai, S., Takiguchi, T. and Konda, S. (1992). Interleukin-7 (IL-7)-induced proliferation of $CD8^+$ T-chronic lymphocytic leukemia cells. *J. Clin. Immunol.* **12**, 101–106.

Young, J.C., Gishizky, M.L. and Witte, O.N. (1991). Hyperexpression of interleukin-7 is not necessary or sufficient for transformation of a pre-B lymphoid cell line. *Mol. Cell Biol.* **11**, 854–863.

Yron, I., Wood, T.A. Jr., Spiess, P.J. and Rosenberg, S.A. (1980). *In vitro* growth of murine T cells. V. The isolation and growth of lymphoid cells infiltrating syngeneic solid tumors. *J. Immunol.* **125**, 238–245.

Zea, A.H., Cutrie, B.D., Longo, D.L., Alvord, W.G., Strobl, S.L., Mizoguchi, H., Creekmore, S.P., O'Shea, J.J., Powers, G.C., Urba, W.J. and Ochoa, A.C. (1995). Alterations in T cell receptor and signal transduction molecules in melanoma patients. *Clin. Cancer Res.* **1**, 1327–1335.

Zeng, Y.X., Takahashi, H., Shibata, M. and Hirokawa, K. (1994). JAK3 Janus kinase is involved in interleukin 7 signal pathway. *FEBS Lett.* **353**, 289–293.

Ziegler, S.F., Tough, T.W., Franklin, T.L., Armitage, R.J. and Alderson, M.R. (1991). Induction of macrophage inflammatory protein-1β gene expression in human monocytes by lipopolysaccharide and IL-7. *J. Immunol.* **147**, 2234–2239.

Zharhary, D. (1994). In *Handbook of B and T lymphocytes* (ed. E.C. Snow), Academic Press, San Diego, CA, p. 91.

Chapter 10

Interleukin-8 and Other CXC Chemokines

Anja Wuyts, Paul Proost and Jo Van Damme
Rega Institute for Medical Research, Laboratory of Molecular Immunology, University of Leuven, Leuven, Belgium

INTRODUCTION

During the last decade, a family of structurally related cytokines showing chemotactic activity for specific types of leukocytes has been identified. These chemotactic cytokines are now called chemokines. Molecules such as the anaphylatoxin C5a and the bacterium-derived peptide formyl-methionyl-leucyl-phenylalanine (fMLP) are less specific chemoattractants, active on both neutrophils and mononuclear cells. Other terms used in the past to designate chemokines are intercrines, the scy (small cytokine) family and SIS (small inducible, secreted) cytokines. The molecular hallmark of the chemokine family is the conservation of four cysteine residues which are important for the tertiary structure of the proteins. Disulfide bridges are formed between cysteines 1 and 3 and between cysteines 2 and 4. Depending on whether the first two cysteines are separated by one amino acid or not, chemokines can be divided into a CXC or a CC subfamily, respectively. CXC chemokines mainly attract and activate neutrophils, whereas CC chemokines attract and activate monocytes, lymphocytes, basophils, eosinophils, natural killer (NK) cells and dendritic cells, but not neutrophils. Lymphotactin is the only member of a third subclass (C chemokine subfamily) and lacks two of the four cysteines. Recently, a fourth chemokine type that contains the motif CX_3C has been identified (Bazan *et al.*, 1997).

Chemokines are low molecular weight proteins with a basic nature and with affinity for heparin. They are produced by many different cell types after stimulation with endogenous or exogenous inducers. It is believed that chemokines are preferentially immobilized through low-affinity binding to proteoglycans on the vascular endothelium and to extracellular matrix proteins in the tissues, rather than remaining in solution. Functional binding to leukocytes and subsequent signaling is mediated by high-affinity receptors belonging to the G protein-coupled serpentine-like receptor family.

Some chemokines have been identified by sequence analysis of the protein after purification based on the bioactivity. Other chemokines have been discovered as a result of molecular cloning. The similarity between the CXC chemokines ranges from 23% to 88% identical amino acids. Many CXC chemokines possess the Glu–Leu–Arg (ELR) motif just in front of the first cysteine residue, whereas other CXC chemokines do not contain this motif (Fig. 1). The ELR^+ CXC chemokines all attract neutrophils and share

The Cytokine Handbook, 3rd ed.
ISBN 0-12-689662-3

CXC chemokines		% identity
IL-8	EGAVLPRSAKELRCQCIKTYSKPFHPKFIKELRVIESGPHCANTEIIVKLSD GRELCLDPKENWVQRVVEKFLKRAENS	100
PF-4	EAEEDGDLQCLCVKTTSQ VRPRHITSLEVIKAGPHCPTAQLIATLKN GRKICLDLQAPLYKKIIKKLLES	34
NAP-2	AELRCMCIKTTSG IHPKNIQSLEVIGKGTHCNQVEVIATLKD GRKICLDPDAPRIKKIVQKKLAGDESAD	48
GRO-α	ASVATELRCQCLQTLQG IHPKNIQSVNVKSPGPHCAQTEVIATLKN GRKACLNPASPIVKKIIEKMLNSDKSN	42
GRO-β	APLATELRCQCLQTLQG IHLKNIQSVKVKSPGPHCAQTEVIATLKN GQKACLNPASPMVKKIIEKMLKNGKSN	41
GRO-γ	ASVVTELRCQCLQTLQG IHLKNIQSVNVRSPGPHCAQTEVIATLKN GKKACLNPASPMVQKIIEKILNKGSTN	40
ENA-78	AGPAAAVLRELRCVCLQTTQG VHPKMISNLQVFAIGPQCSKVEVVASLKN GKEICLDPEAPFLKKVIQKILDGGNKEN	34
GCP-2	GPVSAVLTELRCTCLRVTLR VNPKTIGKLQVFPAGPQCSKVEVVASLKN GKQVCLDPEAPFLKKVIQKILDSGNKKN	30
IP-10	VPLSRTVRCTCISISNQPVNPRSLEKLEIIPASQFCPRVEIIATMKKKGEKRCLNPESKAIKNLLKAVSKEMSKRSP	23
Mig	TPVVRKGRCSCISTNQGTIHLQSLKDLKQFAPSPSCEKIEIIATLKN GVQTCLNPDSADVKELIKKWEKQVSQKKK...	28
	...QKNGKKHQKKKVLKVRKSQRSRQKKTT	
SDF-1	DGKPVSLSYRCPCRFFESH VARANVKHLK ILNTPNCAL QIVARLKNNNRQVCIDPKLKWIQEYLEKALNKRFKM	30

CC chemokines		
MCP-1	QPDAINAPVTCCYNFTNRK ISVQRLASYRRITSSK CPKEAVIFKTIVAKEICADPKQKWVQDSMDHLDKQTQTPKT	100
MCP-2	QPDSVSIPITCCFNVINRK IPIQRLESYTRITNIQ CPKEAVIFKTKRGKEVCADPKERWVRDSMKHLDQIFQNLKP	62
MCP-3	QPVGINTSTTCCYRFINKK IPKQRLESYRRTTSSH CPREAVIFKTKLDKEICADPTQKWVQDFMKHLDKKTQTPKL	71
MCP-4	QPDALNVPSTCCFTFSSKK ISLQRLKSY VITTSR CPQKAVIFRTKLGKEICADPKEKWVQNYMKHLGRKAHTLKT	61
Eotaxin	GPASVPTTCCFNLANRK IPLQRLESYRRITSGK CPQKAVIFKTKLAKDICADPKKKWVQDSMKYLDQKSPTPKP	65
RANTES	SPYSSDTTPCCFAYIARPL PRAHIKEY FYTSGK CSNPAVVFVTRKNRQVCANPEKKWVREYINSLEMS	28
MIP-1α	ASLAADTPTACCFSYTSRQ IPQNFIADY FETSSQ CSKPGVIFLTKRSRQVCADPSEEWVQKYVSDLELSA	37
MIP-1β	APMGSDPPTACCFSYTARKL PRNFVVDY YETSSL CSQPAVVFQTKRSKQVCADPSESWVQEYVYDLELN	37
MIP-3	DRFHATSADCCISYTPRS IPCSLLESY FETNSE CSKPGVIFLTKKGRRFCANPSDKQVQVCMRMLKLDTRIKTRKN	34
LARC/MIP-3α/exodus	ASNFD CCLGYTDRILHP KFIVGFTRQLANEGCDINAIIFHTKKKLSVCANPKQTWVKYIVRLLSKKVKNM	28
MIP-3β	GTNDAEDCCLSVTQKP IPGYIVRNFHYLLIKDGCRVPAVVFTTLRGRQLCAPPDQPWVERIIQRLQRTSAKMKRRSS	25
PARC/MIP-4	AQVGTNKELCCLVYTSWQ IPQKFIVDY SETSPQ CPKPGVILLTKRGRQICADPNKKWVQKYISDLKLNA	34
HCC-2/MIP-5	SFHFAADCCTSYISQS IPCSLMKSY FETSSE CSKPGVIFLTKKGRQVCAKPSGPGVQDCMKKLKPYSI	32
HCC-1	TKTESSSRGPYHPSECCFTYTTYK IPRQRIMDY YETNSQ CSKPGIVFITKRGHSVCTNPSDKWVQDYIKDMKEN	32
I-309	KSMQVPFSRCCFSFAEQE IPLRAILCY RNTSSI CSNEGLIFKLKRGKEACALDTVGWVQRHRKMLRHCPSKRK	32
MDC	GPYGANMEDSVCCRDYVRYRL PLR VVKHFYWTSDS CPRPGVVLLTFRDKEICADPRVPWVKMILNKLSQ	28
TARC	ARGTNVGRECCLEYFKGA IPLRKLKTW YQTSED CSRDAIVFVTVQGRAICSDPNNKRVKNAVKYLQSLERS	25

Fig. 1. Amino acid sequence alignment of human CXC and CC chemokines. LARC: liver and activation-regulated chemokine; PARC: pulmonary and activation-regulated chemokine; TARC: thymus and activation-regulated chemokine; HCC-1: hemofiltrate CC chemokine-1; MDC: macrophage-derived chemokine.

the usage of the CXC chemokine receptor-2 (CXCR-2). In general, ELR^+ CXC chemokines are angiogenic, whereas ELR^- CXC chemokines are angiostatic (Strieter *et al.*, 1995). Based on this criterion, the CXC subfamily can thus be divided into two groups: the ELR^+ CXC chemokines—interleukin-8 (IL-8), growth-regulated oncogenes GRO-α, GRO-β, GRO-γ, neutrophil activating protein-2 (NAP-2), epithelial cell-derived, neutrophil attractant-78 (ENA-78) and granulocyte chemotactic protein-2 (GCP-2)—and the ELR^- CXC chemokines—platelet factor-4 (PF-4), interferon-γ (IFN-γ)-inducible protein-10 (IP-10), monokine induced by IFN-γ (Mig) and stromal cell-derived factor-1 (SDF-1).

CXC chemokines are produced as precursor molecules containing a signal sequence of 17–34 amino acids. After cleavage, mature proteins of 70–103 amino acids are secreted. The CXC chemokines contain no *N*-glycosylation sites and are probably not glycosylated. The genes for all known CXC chemokines, except one (SDF-1), are localized on human chromosome 4q12–21. The genes probably arose through gene duplication and subsequent divergence. The 5′ flanking region of the chemokine genes contains regulatory sequences for gene expression and the stability of their mRNA may be influenced by RNA instability elements in the 3′ untranslated region. The genes for IL-8, GRO, ENA-78, IP-10 and SDF-1 consist of four exons and three introns, whereas the genes for PF-4 and NAP-2 contain three exons and two introns (Luster and Ravetch, 1987a; Mukaida *et al.*, 1989; Baker *et al.*, 1990; Eisman *et al.*, 1990; Majumdar *et al.*, 1991; Lee and Farber, 1996). The first and second introns are conserved within this family.

PRODUCTION OF HUMAN CXC CHEMOKINES: CELLULAR SOURCES AND INDUCERS

Several laboratories independently identified IL-8 as a neutrophil activating peptide produced by stimulated peripheral blood monocytes (neutrophil activating factor (NAF), monocyte-derived neutrophil activating peptide (MONAP), monocyte-derived neutrophil chemotactic factor (MDNCF)) (Walz *et al.*, 1987; Schröder *et al.*, 1987; Yoshimura *et al.*, 1987), and as a factor (GCP) isolated from stimulated mononuclear cells which induces granulocytosis and neutrophil infiltration in rabbit skin (Van Damme *et al.*, 1988). IL-8 has also been purified as a T-cell chemotactic peptide produced by peripheral blood mononuclear cells (PBMCs) (Larsen *et al.*, 1989). In addition to mononuclear cells, IL-8 is produced by other leukocyte cell types (myeloid precursors, NK cells, neutrophils, eosinophils, mast cells), various tissue cells (e.g. fibroblasts, endothelial cells, epithelial cells) and tumor (e.g. sarcoma, carcinoma) cells (Table 1). In contrast, other CXC chemokines such as PF-4 and NAP-2 are secreted predominantly by platelets, whereas IP-10, ENA-78 and GCP-2 seem to be produced by a more restricted number of cell types than IL-8.

IL-8 can be induced by a variety of stimuli including cytokines (e.g. IL-1, tumor necrosis factor-α (TNF-α)), bacterial products (e.g. lipopolysaccharide (LPS)), viral products (e.g. double-stranded RNA), plant products (e.g. concanavalin A) and various other inducers (Table 2). In fibroblasts, IL-1, double-stranded RNA and virus are the best inducers of IL-8 and GRO, whereas virus, LPS and mitogen are superior inducers in monocytes. In contrast, IP-10 is inducible in mononuclear cells, fibroblasts and endothelial cells by IFN-γ (Luster *et al.*, 1985), whereas this cytokine

Table 1. Cellular sources of human CXC chemokines.

Cell type	IL-8	PF-4	NAP-2	GRO-$\alpha/\beta/\gamma$	ENA-78	GCP-2	IP-10
Endothelial cells	+			+	+		+
Mesothelial cells	+						
Mesangial cells	+						
Fibroblasts	+			+	+	+	+
Chondrocytes	+						
Epithelial cells	+			+	+		
Keratinocytes	+						+
Smooth muscle cells	+						
Astrocytes	+						
Hepatocytes	+						
Synovial cells	+						+
Amnion cells	+					+	
Endocrine cells	+						
Choriodecidual cells	+						
Stromal cells	+						+
Myeloid precursors	+						
Monocytes/macrophages	+			+	+		+
Lymphocytes	+						+
Natural killer cells	+						
Neutrophils	+			+	+		+
Eosinophils	+						
Mast cells	+						
Platelets		+	+		+		
Tumor cells	+			+	+	+	+

For a detailed list of references, see Table 1 from Van Damme (1994).

inhibits the induction of IL-8 (Oliveira *et al.*, 1992; Cassatella *et al.*, 1993). As chemokines can be produced by a variety of cell types, the immunologic response retains its specificity by the nature of stimulus to trigger or inhibit production by the cells. Indeed, after bacterial infection, IL-8 is efficiently produced by peripheral blood monocytes, whereas fibroblasts seem to respond better during the course of a viral infection. Furthermore, endogenous IL-1 generated from monocytes can efficiently stimulate the release of CXC chemokines. Antibodies against the IL-1 type I receptor inhibit CXC chemokine induction in monocytes. However, antibodies against the IL-1 type II receptor augment chemokine production, indicating that the latter is acting as decoy target for IL-1 (Colotta *et al.*, 1993). Parallel secretion of chemokines in response to IL-1 and other cytokine inducers can be explained through common responsive elements, such as the nuclear factor-κB (NF-κB) consensus motif, as possible sites of gene transactivation. Chemokines coproduced by stimulated cells can be isolated from the conditioned medium and fractionated based on their different biochemical characteristics (size, charge, ligand affinity). As an example, the behavior of the various CXC chemokines during isolation is illustrated in Table 3. This standard procedure allows for the simultaneous purification of natural chemokines (Proost *et al.*, 1993a).

Table 2. Inducers of human CXC chemokine production.

Inducer	IL-8	PF-4	NAP-2	GRO-α	GCP-2, ENA-78	IP-10
Cytokines						
IL-1	+			+	+	+
IL-2	+					
IL-3	+					
IL-7	+					
IL-13	+					
TNF-α	+			+	+	+
IFN-γ						+
GM-CSF	+					
Leukoregulin	+					
Bacterial products						
LPS	+			+	+	+
Staphylococcal enterotoxin A (SEA)	+					
FMLP	+					
Viral products						
Measles virus	+					
Double-stranded RNA	+				+	
Plant products						
Concanavalin A	+					
PHA	+			+		
PMA	+			+	+	
Others						
Thrombin		+	+	+		
Adherence	+			+		
Serum	+		+	+		+

For a detailed list of references, see Table 1 from Van Damme (1994).

Table 3. Isolation procedure and biochemical characteristics of human CXC chemokines.

		Chemokine elution						
Purification step	Elution buffer	IL-8	PF-4	NAP-2	GRO-α	ENA-78	GCP-2	IP-10
1. CPG adsorption*	pH 2.0	+	+	+	+	+	+	+
2. Heparin–Sepharose†	0–2 M NaCl	0.6–0.8 M	2.0 M	0.6–0.8 M	0.6–0.8 M	0.6–0.8 M	0.6–0.8 M	—
3. Cation-exchange‡	0–1 M NaCl	0.85 M	1.0 M	0.7 M	0.6 M	0.55–0.7 M‖	0.7 M	0.8 M
4. RP-HPLC§	0–80% CH_3CN	30%	30%	23.5%	26%	30–37%‖	29–33%‖	26–28%
5. SDS–PAGE¶	kDa	7.0	10	6.5	5.5	6.0	6.0	6.5

*Adsorption to controlled pore glass at neutral pH and elution with 0.3 M glycine–hydrochloric acid, pH 2.0.
†Binding to heparin–Sepharose and elution with a 0–2 M NaCl gradient.
‡Cation-exchange chromatography (mono S fast protein liquid chromatography (FPLC)) at pH 4.0; elution with a 0–1 M NaCl gradient.
§Reversed-phase high-performance liquid chromatography, elution with a 0–80% acetonitrile gradient.
¶Sodium dodecyl sulfate–polyacrylamide gel electrophoresis under reducing conditions.
‖Broad elution pattern due to different truncated forms.

PROTEIN AND GENE STRUCTURE OF HUMAN CXC CHEMOKINES

Interleukin-8

IL-8 is stable at extremes of pH (pH 2.0 and 9.0) and is resistant to mild oxidizing and reducing conditions. Heating to 100°C does not alter its activity. IL-8 resists exposure to detergents and organic solvents as well as to plasma peptidases. It is inactivated only slowly at 37°C by cathepsin G, elastase and proteinase 3 (Schröder *et al.*, 1987; Padrines *et al.*, 1994). These properties are presumably due to the conformation of the chemokine. IL-8 can be inactivated by a protease found in serosal fluid (Ayesh *et al.*, 1993). Reduction of disulfide bonds leads to unfolding and inactivation of IL-8, but its activity is regained after dialysis. IL-8 can be frozen, stored at 4°C in 0.1% trifluoroacetic acid pH 2.0 or lyophilized with only little loss of activity. Storage without additional protein in phosphate-buffered saline at 4°C results in a drastic loss of activity, apparently by binding of IL-8 to plastic material.

The IL-8 cDNA encodes a 99-amino-acid precursor protein with a signal sequence of 22 amino acids which is cleaved to yield the 77-residue mature protein (Schmid and Weissmann, 1987). A 79-residue form of IL-8 has been detected and originates from cleavage within the predicted signal peptide. IL-8 is processed further at the NH_2-terminus yielding different truncation analogs (77-, 72-, 71-, 70-, 69-amino-acid forms). The occurrence of the NH_2-terminal forms depends on the producer cells and culture conditions (Van Damme *et al.*, 1989a,c, 1990). The truncation is due to proteases that are released by the IL-8-secreting cells or by accessory cells (Hébert *et al.*, 1990; Nakagawa *et al.*, 1991; Padrines *et al.*, 1994). The two major forms are the 77- and the 72-amino-acid proteins. Fibroblasts and endothelial cells predominantly produce intact IL-8, whereas leukocytes mainly secrete NH_2-terminally truncated forms. *In vitro*, the 72-amino-acid form of IL-8 is a more potent chemoattractant and activator of neutrophils than the 77-residue form. *In vivo*, both IL-8 forms are equipotent, possibly due to proteolytic cleavage of 77-amino-acid IL-8 (Nourshargh *et al.*, 1992).

The three-dimensional structure of IL-8 has been studied by nuclear magnetic resonance (NMR) spectroscopy and by X-ray crystallography. In a concentrated solution and on crystallization, IL-8 occurs as a dimer of two identical subunits (Clore and Gronenborn, 1991). The monomer contains a disordered NH_2-terminus, followed by a loop region, three antiparallel β strands and a prominent α helix extending from amino acid 57 to the COOH-terminus. The dimer is stabilized by six hydrogen bonds between the first β strands of the partner molecules (residues 23–29) and by other side-chain interactions. The dimer consists of two antiparallel α helices lying on top of a six-stranded antiparallel β sheet. However, monomeric IL-8 is probably the active form. A chemically synthesized IL-8 analog, containing a methylated Leu-25 to block dimerization, is equivalent to IL-8 for neutrophil activation and receptor binding, indicating that the monomer is a functional form (Rajarathnam *et al.*, 1994).

It has also been shown that the association state of IL-8 is characterized by an equilibrium between monomers and dimers. Dimers predominate at concentrations above 100 μM, but at nanomolar concentrations, which are physiologically relevant and induce maximal biologic activity, the chemokine occurs almost exclusively in its monomeric form (Burrows *et al.*, 1994; Paolini *et al.*, 1994).

Based on these findings, the three-dimensional structure of monomeric IL-8_{4-72}, using the chemically synthesized analog with methylated Leu-25, has been determined by NMR spectroscopy. The structure is well defined except for NH_2-terminal residues 4–6 and COOH-terminal residues 67–72. The structure consists of a series of turns and loops at the NH_2-terminal region followed by three β strands and a COOH-terminal α helix. The structure is largely similar to the native dimeric IL-8 structure, but a major difference is that the COOH-terminal residues 67–72 are disordered in the monomeric structure, whereas they are helical in the dimeric structure (Rajarathnam *et al.*, 1995).

Platelet Factor-4

For PF-4, a similar three-dimensional structure has been described as for IL-8. PF-4 aggregates into dimers and tetramers, but at lower concentrations the equilibrium favors the monomeric state (Mayo and Chen, 1989; Zhang *et al.*, 1994; Clark-Lewis *et al.*, 1995). PF-4 is released from the α granules upon platelet stimulation, together with platelet-derived growth factor (PDGF), transforming growth factor-β (TGF-β) and β-thromboglobulin (β-TG). PF-4 binds with high affinity to heparin and promotes blood coagulation. The amino acid sequence of PF-4 was already determined in 1977 (Deuel *et al.*, 1977), 10 years before the discovery of IL-8, and revealed a protein of 70 amino acids containing four cysteines. PF-4 cDNA was isolated only in 1987 from a library derived from a human erythroleukemic (HEL) cell line (Poncz *et al.*, 1987). The cDNA encodes a protein of 101 residues, yielding mature PF-4 after cleavage of the 31-amino-acid signal sequence.

Neutrophil Activating Protein-2

In addition to β-TG (Begg *et al.*, 1978), secreted from α granules of blood platelets, an NH_2-terminally truncated form with chemotactic activity for neutrophils has been isolated from thrombin-stimulated platelets (Van Damme *et al.*, 1989c). This truncated protein is more abundantly generated in the conditioned medium of blood platelets cultured in the presence of monocytes, and is called NAP-2 (Walz and Baggiolini, 1989). β-TG itself is an NH_2-terminally processed form of platelet basic protein (PBP). The cDNA for PBP encodes a protein of 128 amino acids corresponding to mature 94-residue PBP after cleavage of the signal peptide (Wenger *et al.*, 1989). Recently, the precursor protein of PBP has been isolated from conditioned media of neutrophils, macrophages and T lymphocytes as a mitogenic factor for fibroblasts and has been called leukocyte-derived growth factor (LDGF) (Iida *et al.*, 1996).

Processing of PBP at the NH_2-terminus yields at least three distinct molecules with different biologic activities: connective tissue activating protein-III (CTAP-III), also called low-affinity PF-4 (85 residues), β-TG (81 residues) and NAP-2 (70 amino acids). Multiple biologic effects (see below) have been ascribed to these platelet products, but only NAP-2 has neutrophil activating properties. PBP and CTAP-III are both storage proteins present in α granules of platelets. CTAP-III is believed to be the primary species in the α granules, but further processing after degranulation is suspected (Van Damme *et al.*, 1989c; Walz and Baggiolini, 1990).

PBP and its derivatives can form dimers and tetramers. Removal of the NH_2-terminal residues from PBP attenuates dimer and tetramer formation. For all forms, the monomeric form is favored under physiologic conditions and is probably the biologically active state. The tertiary structure of PBP is similar to that of IL-8. An α helix and antiparallel β-sheet structure are conserved among PBP and its derivatives. For PBP, there is additional α-helix formation within the elongated NH_2-terminal segment. Possibly, the NH_2-terminus of PBP, CTAP-III and β-TG folds over the NAP-2 structure important for neutrophil activation, thereby blocking normal activity (Yang *et al.*, 1994).

GRO-α/Melanoma Growth-Stimulatory Activity, GRO-β and GRO-γ

Melanoma growth-stimulatory activity (MGSA) or GRO-α has been isolated as an autocrine growth factor from a human melanoma cell line (Richmond *et al.*, 1988). The protein has been found to be identical to the gene product of a cDNA, called GRO, which is overexpressed in transformed fibroblasts (Anisowicz *et al.*, 1987). The cDNA encodes a precursor protein of 107 residues which yields the mature protein after cleavage of 34 amino acids. Two highly related cDNAs, GRO-β and GRO-γ, have been cloned from a leukocyte library by Haskill *et al.* (1990). The three GRO proteins show 84–88% sequence similarity. GRO-α, β and γ are produced by leukocytes, other tissue cells and tumor cells (Table 1) after stimulation with different inducers. At physiologic concentrations, GRO-α occurs as a monomer (Clark-Lewis *et al.*, 1995). The three-dimensional structure of GRO-α is similar to that of IL-8 (Fairbrother *et al.*, 1994; Kim *et al.*, 1994).

Epithelial Cell-Derived Neutrophil Attractant-78 and Granulocyte Chemotactic Protein-2

ENA-78 is a 78-residue chemokine originally isolated from the supernatant of a lung type II alveolar epithelial cell line stimulated with IL-1 or TNF-α (Walz *et al.*, 1991a). The cDNA encodes a protein of 114 amino acids. Corbett *et al.* (1994) suppose that this precursor protein contains a signal peptide of 31 amino acids, whereas Chang *et al.* (1994) propose a signal peptide of 17 residues. Both groups suggest that ENA-78 is processed proteolytically after secretion to yield the 78-residue protein. In addition to epithelial cells, monocytes, neutrophils, fibroblasts and endothelial cells (Strieter *et al.*, 1992a), as well as platelets (Power *et al.*, 1995) and tumor cells (Froyen *et al.*, 1997), can produce ENA-78 (Table 1).

GCP-2 is a 6-kDa protein (77 residues) isolated from the supernatant of cytokine-stimulated osteosarcoma cells (Proost *et al.*, 1993a,b). The protein shows high structural similarity with ENA-78 (77% identical residues) and contains no *N*-glycosylation sites. Synthetic GCP-2 has the same apparent relative molecular mass as natural GCP-2, indicating that the natural chemokine is not glycosylated (Wuyts *et al.*, 1997). Different forms of GCP-2, missing two, five or eight NH_2-terminal amino acids of the mature protein, have been identified and could be separated from each other by reversed-phase–high-performance liquid chromatography (RP-HPLC). Recently, GCP-2 cDNA has been cloned and expressed in *Escherichia coli* (Froyen *et al.*, 1997). In addition to sarcoma and carcinoma cells, stimulated diploid fibroblasts express GCP-2 (Table 1).

IFN-γ-Inducible Protein-10 and Monokine Induced by IFN-γ (Mig)

IP-10 has been identified as a gene induced by IFN-γ in the histiocytic lymphoma cell line U937 (Luster *et al.*, 1985). The cDNA encodes a protein of 98 amino acids containing a signal peptide of 21 residues. IP-10 is produced by macrophages, T cells, thymocytes, fibroblasts, endothelial cells, keratinocytes, synovial cells and tumor cells, and can be induced by stimuli other than IFN-γ (Tables 1 and 2). Thymic and splenic stromal cells constitutively express high levels of IP-10, which explains the abundant presence of the protein in lymphoid tissue. High levels of IP-10 have also been observed in the liver, possibly due to the presence of activated macrophages (Luster and Ravetch, 1987b; Gattass *et al.*, 1994; Cassatella *et al.*, 1997).

Mig has been cloned from the THP-1 monocytic cell line treated with IFN-γ (Farber, 1993). Mig can also be induced in PBMCs by IFN-γ, but not by IFN-α or LPS. The gene encodes a 125-residue protein containing a signal peptide of 22 amino acids. Monocytes and THP-1 cells stimulated with IFN-γ secrete various COOH-terminally truncated forms of Mig, which is the result of proteolytic processing (Liao *et al.*, 1995).

Stromal Cell-Derived Factor-1α and -1β

In contrast to the other CXC chemokine genes, the gene for SDF-1 is mapped to chromosome 10q11.1. Human SDF-1α and SDF-1β have been cloned from a cDNA library of a pro-B-cell line (Shirozu *et al.*, 1995). SDF-1β differs from SDF-1α by only four additional amino acids at the COOH-terminus. Both proteins are encoded by a single gene and arise by alternative splicing. The SDF-1 gene consists of four exons and three introns. SDF-1α and SDF-1β are encoded by three and four exons respectively. The cDNAs encode proteins of 89 and 93 amino acids respectively, with a signal peptide of 19 amino acids. The SDF-1 gene is expressed in almost all organs except in blood cells. The gene contains no TATA box in the 5′ flanking region but a GC-rich sequence which is suggestive for ubiquitous expression.

Mouse SDF-1α has been cloned from a bone marrow stromal cell line as a factor that supports proliferation of a pre-B-cell clone and that augments the growth of bone marrow B-cell progenitor cells in the presence of IL-7 (Nagasawa *et al.*, 1994). A highly efficient lymphocyte chemoattractant, purified from the supernatant of a murine bone marrow stromal cell line, has been identified as SDF-1α lacking the COOH-terminal lysine and two or seven NH_2-terminal residues (67- and 62-residue form respectively) (Bleul *et al.*, 1996b).

Structural Requirements of CXC Chemokines for Receptor Binding

The existence of two classes of IL-8 receptors has been indicated by binding competition assays and cross-desensitization experiments: one receptor with high affinity for IL-8, GRO-α, GRO-β, GRO-γ, NAP-2 and ENA-78 (CXCR2), and a second receptor with high affinity for IL-8 and low affinity for the other tested chemokines (CXCR1) (see below). Regions of ELR^+ CXC chemokines, important for binding to and activation of neutrophils, have been determined. Removal of the entire COOH-terminal sequence after the fourth cysteine residue decreases, but does not abrogate, the binding and the biologic effect of IL-8 on neutrophils. In contrast, receptor binding and activity of IL-8 are abrogated by removal of the ELR motif in the NH_2-terminal region (Clark-Lewis *et al.*, 1991). The ELR motif is essential for neutrophil activation by NAP-2, but the remainder of the protein structure also participates in the interaction with the receptor

(Yan *et al.*, 1994). For GRO-α it has been shown that the ELR motif, Lys-49 and His-19 are important for binding to CXCR2 (Hesselgesser *et al.*, 1995).

The importance of the ELR motif is also shown by the differences in activity and IL-8 receptor binding between ELR^+ and ELR^- CXC chemokines. ELR^+ CXC chemokines all bind to the IL-8 receptors and attract and activate neutrophils. PF-4 and IP-10, which do not contain the ELR motif, do not bind to IL-8 receptors and are much less potent or inactive on neutrophils. PF-4, modified to contain the ELR motif in the NH_2-terminus, binds to IL-8 receptors and activates neutrophils. Modified IL-8 with ELQ instead of the ELR motif becomes inactive. Modification of the NH_2-terminal region of IP-10 or the CC chemokine monocyte chemotactic protein-1 (MCP-1) to contain the ELR motif does not result in IL-8 receptor binding, indicating a role for other structural parts in receptor binding (Clark-Lewis *et al.*, 1993). Considering the importance of the NH_2-terminal region in receptor binding, IL-8 antagonists have been synthesized. The analogs IL-8_{6-72}, starting with Arg, and Ala–Ala–Il-8_{6-72}, which lack the ELR motif, are potent antagonists: they inhibit both IL-8 binding and activity. These antagonists have a much lower receptor affinity than IL-8, indicating that the ELR motif is both a binding and activating motif (Moser *et al.*, 1993b). Besides the ELR motif, the two disulfide bridges, residues 10–22 and 30–35, are also important for the binding and activity of IL-8 (Clark-Lewis *et al.*, 1994).

The above-mentioned structural requirements for IL-8 receptor binding and activation have been achieved using neutrophils as test cells. Neutrophils express both IL-8 receptors, thus the obtained results show the combined requirements for both CXCR1 and CXCR2. Using cells transfected with either CXCR1 or CXCR2, the important regions of IL-8 for selective receptor binding and activation have been determined separately by testing mutants and chimeric molecules of IL-8 and GRO-α. The ELR motif is important for IL-8 binding to the two IL-8 receptors, but additional regions are necessary for high-affinity binding. The COOH-terminal α helix is important for binding to CXCR2, but does not mediate high-affinity binding to CXCR1. Recently, it has been shown that Tyr-13 and Lys-15 of human IL-8 are important residues for interaction with CXCR1 (Schraufstätter *et al.*, 1993, 1995).

BIOLOGIC PROPERTIES OF HUMAN CXC CHEMOKINES

Interleukin-8

In Vitro *Effects*

IL-8 stimulates neutrophil chemotaxis (Schröder *et al.*, 1987; Yoshimura *et al.*, 1987; Van Damme *et al.*, 1988) and haptotactic migration (i.e. towards a concentration gradient of immobilized chemokines) (Rot, 1993) and induces shape change (Thelen *et al.*, 1988), actin polymerization (Detmers *et al.*, 1991) and degranulation (release of β-glucuronidase, elastase, myeloperoxidase, gelatinase B, vitamin B_{12}-binding protein and lactoferrin) of neutrophils (Schröder *et al.*, 1987; Walz *et al.*, 1987; Peveri *et al.*, 1988; Willems *et al.*, 1989; Masure *et al.*, 1991). IL-8 increases the intracellular calcium concentration ($[Ca^{2+}]_i$) in neutrophils (Thelen *et al.*, 1988) and induces a weak (in comparison to fMLP) respiratory burst (Schröder *et al.*, 1987; Peveri *et al.*, 1988). The chemokine primes neutrophils for an enhanced superoxide anion production in response to fMLP, phorbol 12-myristate 13-acetate (PMA) and platelet activating factor (PAF), and upregulates formylpeptide receptor expression (Wozniak *et al.*, 1993; Metzner *et al.*, 1995; Green *et al.*, 1996). IL-8 enhances intracellular killing of *Mycobacterium fortuitum* by human neutrophils by priming the cells to enhance hydrogen peroxide production upon stimulation with preopsonized bacteria (Nibbering *et al.*, 1993). Thus, IL-8 facilitates the elimination of microorganisms by increasing the bactericidal activity of neutrophils. It stimulates phagocytosis of opsonized particles (Detmers *et al.*, 1991) and enhances the growth inhibitory activity of neutrophils to *Candida albicans* (Djeu *et al.*, 1990). IL-8 activates arachidonate-5-lipoxygenase with release of leukotriene B_4 (LTB_4) and 5-hydroxyeicosatetraenoic acid (5-HETE) in the presence of exogenous arachidonic acid (Schröder, 1989), and induces the synthesis of PAF in neutrophils (Bussolino *et al.*, 1992). IL-8 inhibits nitric oxide induction in peritoneal neutrophils (McCall *et al.*, 1992), whereas it stimulates nitric oxide production in osteoclast-like cells (Sunyer *et al.*, 1996). It induces the release of the IL-1 type II decoy receptor from neutrophils (Colotta *et al.*, 1995). The chemokine stimulates transendothelial migration of neutrophils (Huber *et al.*, 1991; Smith *et al.*, 1991). The protein induces shedding of the adhesion molecule L-selectin and upregulates β_2 integrins (CD11b/CD18 and CD11c/CD18) and complement receptor type 1 (CR1 or CD35) on neutrophils, and alters the avidity of the constitutively expressed integrin molecules (Carveth *et al.*, 1989; Detmers *et al.*, 1990, 1991; Huber *et al.*, 1991). IL-8 promotes adhesion of neutrophils to plastic, extracellular matrix proteins and unstimulated as well as cytokine-stimulated endothelial monolayers. The adhesion is inhibited by antibodies to the β_2 integrin CD11b/CD18 (Carveth *et al.*, 1989; Smith *et al.*, 1991). Binding of IL-8 to heparan sulfate enhances its neutrophil chemotactic activity (Webb *et al.*, 1993).

IL-8 has also effects on other leukocyte cell types including lymphocytes, basophils and eosinophils (Table 4). However, it shows no chemotactic activity for monocytes but induces an increase in $[Ca^{2+}]_i$ and a weak respiratory burst in monocytes (Schröder *et*

Table 4. Target cells of CXC chemokines.

Cell type	IL-8	PF-4	NAP-2	GRO-α	ENA-78	GCP-2	IP-10	Mig	SDF-1
Neutrophils	+	+	+	+	+	+			+
Monocytes	+	+		+			+		+
T lymphocytes	+			+			+	+	+
NK cells	+						+		
Tumor infiltrating lymphocytes							+	+	
B lymphocytes	+			+					
Basophils	+		+	+					
Mast cells		+							
Eosinophils	+	+							
Hematopoietic progenitor cells	+	+					+		+
Fibroblasts	+	+		+					
Melanocytes				+					
Melanoma cells	+			+					
Endothelial cells	+	+		+			+		
Keratinocytes	+								
Smooth muscle cells	+								
Tumor cells		+							

al., 1987; Yoshimura *et al.*, 1987; Walz *et al.*, 1991b). These effects are much less pronounced than in neutrophils. The chemokine enhances the expression of integrins on monocytes as well as their adherence to endothelial cells (Oppenheim *et al.*, 1991). Some controversy exists about the chemotactic activity of IL-8 on T lymphocytes. Larsen *et al.* (1989) isolated IL-8 as a T-cell chemoattractant but this activity could not be confirmed by others. Xu *et al.* (1995) showed that the cell purification procedure can influence the response of T lymphocytes to IL-8. Freshly isolated T cells, both $CD4^+$ and $CD8^+$ cells, migrate in response to IL-8, but incubation of the cells at 37°C reduces the migration response. The chemokine enhances the $[Ca^{2+}]_i$ in a subset of T cells (Bacon *et al.*, 1995). IL-8 suppresses the spontaneous production of IL-4 and stimulates its own production by $CD4^+$ T cells (Gesser *et al.*, 1995, 1996). The protein causes chemokinesis of IL-2-activated NK cells (Sebok *et al.*, 1993). IL-8 is a chemoattractant for B lymphocytes (Jinquan *et al.*, 1997) but selectively inhibits IL-4-induced immunoglobulin E (IgE) production and growth of B cells (Kimata *et al.*, 1992; Kimata and Lindley, 1994). IL-8 is chemotactic for basophils (Geiser *et al.*, 1993) and stimulates the adhesion of these cells to endothelial cells, mediated by β_2 integrins (Bacon *et al.*, 1994). IL-8 also augments the $[Ca^{2+}]_i$ in basophils (Krieger *et al.*, 1992) and induces a release of histamine and leukotrienes from IL-3-pretreated cells (Dahinden *et al.*, 1989). However, lower concentrations of IL-8 than those required to stimulate the release of histamine have been shown to inhibit such release induced by IL-8, MCP-1 and IL-3 (Bischoff *et al.*, 1991; Kuna *et al.*, 1991; Alam *et al.*, 1992). NAP-2 also inhibits, although weakly, the IL-8-induced histamine release, whereas CTAP-III and PF-4 have no effect (Bischoff *et al.*, 1991). IL-8 shows no chemotactic activity for normal eosinophils (Schröder *et al.*, 1987). However, the protein induces an increase in $[Ca^{2+}]_i$, shape change and release of eosinophil peroxidase in eosinophils, isolated from patients with hypereosinophilic syndrome (Kernen *et al.*, 1991). Circulating eosinophils from patients with atopic dermatitis and allergic asthma show an increased migratory response towards IL-8 (Bruijnzeel *et al.*, 1993; Warringa *et al.*, 1993). IL-8 also chemoattracts eosinophils pretreated with IL-3, IL-5 or granulocyte–macrophage colony stimulating factor (GM-CSF). IL-8 suppresses colony formation of immature subsets of myeloid progenitor cells stimulated by GM-CSF plus steel factor (Broxmeyer *et al.*, 1993) and also inhibits megakaryocyte colony formation (Gewirtz *et al.*, 1995).

In addition to the effects on leukocytes, IL-8 has activities on several other cell types (Table 4). The chemokine inhibits collagen expression in rheumatoid synovial fibroblasts (Unemori *et al.*, 1993). IL-8 enhances the replication of cytomegalovirus in fibroblasts, possibly through interaction with low numbers of the IL-8 receptor CXCR1, which are expressed on these cells after infection with the virus (Murayama *et al.*, 1994). IL-8 induces a loss of focal adhesions (structures linking the cytoskeleton to the underlying extracellular matrix) in fibroblasts, and promotes chemotaxis and chemokinesis of these cells (Dunlevy and Couchman, 1995). The protein induces proliferation and haptotactic migration of melanoma cells (Wang *et al.*, 1990; Schadendorf *et al.*, 1993), as well as chemotaxis and proliferation of endothelial cells (Koch *et al.*, 1992). IL-8 is chemotactic for keratinocytes, increases $[Ca^{2+}]_i$ and promotes their proliferation (Michel *et al.*, 1992; Tuschil *et al.*, 1992). IL-8 is a mitogen and chemoattractant for smooth muscle cells and also stimulates these cells to produce prostaglandin E_2 (PGE_2) (Yue *et al.*, 1993, 1994).

In Vivo *Effects*

IL-8 has been isolated as a factor that provokes at high doses (3 μg) an early (3 h) skin reaction in rabbits characterized by swelling, redness and neutrophil infiltration (Van Damme *et al.*, 1988). Upon coinjection with a vasodilator (e.g. PGE_2), IL-8 induces local edema formation and neutrophil accumulation in rabbit skin from 0.2 ng onwards, with a maximum effect after 30 min (Foster *et al.*, 1989; Rampart *et al.*, 1989). The plasma leakage is neutrophil dependent and the effects can be inhibited by antibodies against leukocyte integrins (Forrest *et al.*, 1992). Histology of IL-8-induced lesions has also revealed intravascular neutrophil accumulation, cell aggregate formation and venular wall damage (Colditz *et al.*, 1989). IL-8, injected subcutaneously into the lymphatic drainage areas of rat lymph nodes, causes accelerated emigration only of lymphocytes from high endothelial venules (Larsen *et al.*, 1989). However, injection of IL-8 in rat ear skin results within 3 h in accumulation of both lymphocytes and neutrophils in the connective tissue. Lower doses of IL-8 (20 pg) selectively attract lymphocytes, whereas higher doses (2 ng) predominantly recruit neutrophils (Larsen *et al.*, 1989). Taub *et al.* (1996a) reported that subcutaneous injection of IL-8 (100 ng) in mice with severe combined immunodeficiency (SCID) engrafted with human peripheral blood lymphocytes (huPBL–SCID), induces neutrophil accumulation at 4 h after administration, whereas a T-cell infiltrate is detected only after 72 h. The T-cell accumulation may be mediated by chemoattractants, released by neutrophils after stimulation with IL-8. The chemokine (10 μg, 4 h) induces an influx of neutrophils and lymphocytes into the skin of sheep, with a neutrophil to lymphocyte ratio of 45:1 (Colditz and Watson, 1992). Intradermal injection of IL-8 in human subjects causes perivascular neutrophil infiltration, detectable already after 30 min and further increasing until 3 h after injection; no basophils or eosinophils are found and the number of lymphocytes is not increased (Leonard *et al.*, 1991). At moderate doses (400 ng), no wheal and flare, itching or pain is detected, suggesting that IL-8 does not elicit histamine release from local mast cells. The observed differences in lymphocyte accumulation after intradermal injection may be due to differences in animal species and chemokine dose.

Intravenous injection of IL-8 into animals results in an immediate leukopenia followed by a profound neutrophilia (Van Damme *et al.*, 1988). The neutrophilia is accompanied by the release of non-segmented neutrophils from the bone marrow reservoir (Jagels and Hugli, 1992). Intravenous injection of IL-8 can inhibit emigration of neutrophils to inflammatory sites (Taub, 1996). After intraperitoneal injection of IL-8 in mice, rapid mobilization of progenitor cells and pluripotent stem cells occurs. These mobilized cells are able to rescue lethally irradiated mice and to repopulate host hematopoietic tissues completely and permanently (Laterveer *et al.*, 1995). Intracranial injection of IL-8 in mice provokes neutrophil recruitment in the central nervous system, which is associated with breaching of the blood–brain barrier (Bell *et al.*, 1996). Intracerebroventricular administration of IL-8 in rats induces fever by a prostaglandin-independent mechanism (Zampronio *et al.*, 1994) and suppresses food intake (Plata-Salamàn and Borkoski, 1993). IL-8 is angiogenic when implanted in the rat cornea and induces neovascularization in a rabbit corneal pocket model (Koch *et al.*, 1992; Strieter *et al.*, 1992b, 1995). IL-8 is also a primary mediator of angiogenesis in human bronchogenic carcinoma (Smith *et al.*, 1994). Knockout mice that lack the murine IL-8 receptor homolog look outwardly healthy but show lymphadenopathy (due to an increase in B cells) and splenomegaly (due to an increase in metamyelocytes and mature neutrophils).

This indicates that the receptor may participate in the expansion and development of neutrophils and B cells. As expected, these mice also show less neutrophil migration to sites of inflammation (Cacalano *et al.*, 1994).

Platelet Factor-4

The reports of the neutrophil activating properties of PF-4 are rather controversial. Earlier studies reported chemotactic migration, degranulation and enhanced adhesion of neutrophils to plastic surfaces and endothelial cells, whereas others have failed to demonstrate any neutrophil activating capacity (Petersen *et al.*, 1996). Studies with highly purified PF-4 revealed that only weak degranulation of specific secondary neutrophil granules and no exocytosis of azurophilic granules occurred in response to this chemokine. The degranulation of secondary granules can be significantly enhanced by preincubation or coincubation of the cells with TNF-α. Thus, PF-4 has no neutrophil chemotactic activity and no effect on the $[Ca^{2+}]_i$, nor does it compete for IL-8 binding to neutrophils in the absence or presence of TNF-α (Petersen *et al.*, 1996). However, PF-4 might be weakly chemotactic for monocytes (Deuel *et al.*, 1981; Van Damme *et al.*, 1989b), IgE-stimulated mast cells (Taub *et al.*, 1995a) and normal eosinophils, but the migration response of the latter is potentiated markedly by IL-5. Eosinophils from patients with atopic dermatitis have an increased migratory response to PF-4 (Bruijnzeel *et al.*, 1993). PF-4 suppresses colony formation of immature subsets of myeloid progenitor cells stimulated by GM-CSF plus steel factor (Broxmeyer *et al.*, 1993) and inhibits megakaryocytopoiesis (Gewirtz *et al.*, 1989).

PF-4 is chemotactic for fibroblasts (Senior *et al.*, 1983). The chemokine is a growth inhibitor of various tumor cell lines including human erythroid leukemia and osteosarcoma cells (Han *et al.*, 1992; Tatakis, 1992). PF-4 inhibits endothelial cell proliferation and migration, as well as angiogenesis in the chicken chorioallantoic membrane system. These effects may be due to the ability of PF-4 to block binding of basic fibroblast growth factor (bFGF) to its receptor on endothelial cells (Maione *et al.*, 1990; Sato *et al.*, 1990). A PF-4 derivative originating from cleavage of the peptide bond between Thr-16 and Ser-17 has been isolated from activated human leukocytes. This cleavage leads to structural changes and a 30- to 50-fold greater growth inhibitory activity on endothelial cells than for intact PF-4, suggesting that the CXC motif and the intramolecular disulfide bridges downmodulate the growth inhibitory effect of PF-4 (Gupta *et al.*, 1995). Finally, PF-4 alleviates concanavalin A-induced immunosuppression in mice (Schall, 1994).

Neutrophil Activating Protein-2

The precursor of PBP, called LDGF, is a potent mitogenic factor for connective tissue cells (Iida *et al.*, 1996). So far, no biologic activity has been assigned to PBP (Holt *et al.*, 1986), whereas CTAP-III stimulates the proliferation of connective tissue cells (Mullenbach *et al.*, 1986). β-TG is chemotactic for fibroblasts (Senior *et al.*, 1983), but not for neutrophils or monocytes. NAP-2 induces chemotaxis, a rise in $[Ca^{2+}]_i$, degranulation and a weak respiratory burst in neutrophils as well as actin polymerization and phagocytosis of opsonized particles (Walz *et al.*, 1989, 1991b; Detmers *et al.*, 1991). NAP-2 enhances the expression of the adhesion molecule CD11b/CD18 and of CR1, as well as the binding activity of CD11b/CD18 (Detmers *et al.*, 1991). Priming of neutrophils with a non-stimulatory dose of NAP-2 leads to a reduced degranulation response to higher doses of NAP-2, GRO-α or IL-8. This reduced response is correlated with a downregulation and internalization of NAP-2 and IL-8 high-affinity binding sites (Petersen *et al.*, 1994). NAP-2 also augments $[Ca^{2+}]_i$ in basophils and stimulates histamine release from IL-3-pretreated basophils (Krieger *et al.*, 1992). On monocytes, no effects have been observed. NAP-2 inhibits megakaryocytopoiesis (Gewirtz *et al.*, 1995).

Intradermal injection of NAP-2 together with a vasodilator results in neutrophil accumulation and plasma extravasation in rabbit skin. Intravenous injection provokes rapid granulocytosis in rabbits. Although the kinetics of neutrophilia are similar for NAP-2 and IL-8, the effect with IL-8 is more pronounced. Taken together, NAP-2 has a lower specific activity *in vitro* and *in vivo*, both as neutrophil activator and chemoattractant (Van Damme *et al.*, 1990; Van Osselaer *et al.*, 1991). NAP-2 variants containing three to five additional amino acids at the NH_2-terminus are less potent neutrophil activators than 70-residue NAP-2 (Walz and Baggiolini, 1990). A COOH-terminally truncated variant of NAP-2, lacking the four COOH-terminal residues, has also been isolated. This truncated protein shows a three- to four-fold enhanced potency for neutrophil degranulation as well as three-fold enhanced receptor binding affinity (Ehlert *et al.*, 1995).

GRO-α/Melanoma Growth-Stimulatory Activity, GRO-β and GRO-γ

GRO-α, GRO-β and GRO-γ induce neutrophil chemotaxis, shape change, actin polymerization, degranulation, rise in $[Ca^{2+}]_i$, phagocytosis and a weak respiratory burst in neutrophils. GRO-α upregulates the expression of the β_2 integrin CD11b/CD18 and of CR1, but decreases the expression of L-selectin on neutrophils (Detmers *et al.*, 1991; Geiser *et al.*, 1993). GRO-α primes the superoxide anion production by fMLP-stimulated neutrophils and upregulates the formylpeptide receptor (Metzner *et al.*, 1995; Green *et al.*, 1996). Pretreatment of neutrophils with a non-stimulatory dose of GRO-α reduces the degranulation response to higher doses of GRO-α, NAP-2 or IL-8 (Petersen *et al.*, 1994). GRO-α is not chemotactic for monocytes, but elicits a weak increase in $[Ca^{2+}]_i$ and a weak respiratory burst in these cells (Walz *et al.*, 1991b). The GRO molecules play a role in monocyte adhesion to endothelium, stimulated with minimally modified low density lipoprotein (Schwartz *et al.*, 1994). GRO-α is also chemotactic for T lymphocytes (Jinquan *et al.*, 1995) and B lymphocytes (Jinquan *et al.*, 1997). GRO-α, β and γ are chemotactic for basophils and augment the $[Ca^{2+}]_i$, whereas no chemotactic activity for eosinophils has been detected (Geiser *et al.*, 1993). GRO-β suppresses colony formation of immature subsets of myeloid progenitor cells stimulated by GM-CSF plus steel factor (Broxmeyer *et al.*, 1993). GRO-α, β and γ are also active on cells other than leukocytes. GRO-α has melanoma growth-stimulatory activity (MGSA) and is mitogenic for melanocytes (Richmond *et al.*, 1988; Bordoni *et al.*, 1990). The protein has growth-stimulatory activity for fibroblasts and decreases collagen expression in rheumatoid synovial fibroblasts (Schröder *et al.*, 1990; Unemori *et al.*, 1993). GRO-α and GRO-β inhibit bFGF-stimulated proliferation of capillary endothelial cells, whereas GRO-γ has no inhibitory effect (Cao *et al.*, 1995). GRO-β has angiostatic properties: it inhibits blood vessel formation in the chicken chorioallantoic membrane assay and can suppress bFGF-induced corneal neovascularization after systemic administration in mice (Cao *et al.*, 1995).

Intradermal injection of GRO-α induces neutrophil accumulation in rabbit skin (Proost *et al.*, 1993a). Transgenic mice expressing KC, the murine homolog of GRO-α, in the thymus or epidermis show a marked infiltration of neutrophils to the sites of transgene expression without morphologic evidence of injury. Thus, KC expression results in neutrophil recruitment without inflammatory reaction (Lira *et al.*, 1994). Central nervous system-specific KC expression induces remarkable neutrophil infiltration into perivascular, meningeal and parenchymal sites. Unexpectedly, these mice develop a neurologic syndrome of pronounced postural instability and rigidity. The major neuropathologic alterations are florid microglial activation and blood–brain barrier disruption (Tani *et al.*, 1996). Intracranial injection of macrophage inflammatory protein-2 (MIP-2), the murine counterpart of GRO-β/γ, induces neutrophil recruitment in the central nervous system which is associated with a breaching of the blood–brain barrier (Bell *et al.*, 1996).

Epithelial Cell-Derived Neutrophil Attractant-78 and Granulocyte Chemotactic Protein-2

ENA-78 stimulates neutrophil chemotaxis, rise in $[Ca^{2+}]_i$ and exocytosis (Walz *et al.*, 1991a). Preincubation of neutrophils with ENA-78 enhances their ability to generate superoxide anion in response to fMLP (Green *et al.*, 1996). ENA-78 is capable of desensitizing the neutrophil response to human GCP-2, indicating a common receptor usage (Wuyts *et al.*, 1997).

GCP-2 occurs in several NH_2-terminally truncated forms, which are equally potent in chemoattracting neutrophils. GCP-2 also induces gelatinase B release from neutrophils and augments $[Ca^{2+}]_i$. GCP-2 and GRO-α have a comparable specific activity in these assays, but are about ten-fold weaker than IL-8. Intradermal injection of GCP-2 in rabbit skin results in neutrophil accumulation and plasma extravasation (Proost *et al.*, 1993a,b; Wuyts *et al.*, 1997). The murine counterpart of GCP-2, isolated from stimulated fibroblasts and epithelial cells, occurs in 11 different NH_2-terminally truncated forms. In contrast to human GCP-2, the more truncated forms of mouse GCP-2 have a higher specific activity in neutrophil chemotaxis and activation assays than the longer forms. Mouse GCP-2 is a better chemoattractant than human GCP-2 on human neutrophils and more active than murine KC (GRO-α) and MIP-2 (GRO-β/γ) on mouse neutrophils (Wuyts *et al.*, 1996). Human GCP-2 is not chemotactic for monocytes, lymphocytes and eosinophils (Wuyts *et al.*, 1997).

IFN-γ-Inducible Protein-10 and Monokine Induced by IFN-γ (Mig)

IP-10 is not chemotactic for neutrophils, but attracts monocytes and T lymphocytes. The protein promotes T-cell adhesion to endothelial cells (Taub *et al.*, 1993), to the adhesion molecules intercellular adhesion molecule-1 (ICAM-1) and vascular cell adhesion molecule-1 (VCAM-1), and to extracellular matrix proteins via $\beta 1$ and $\beta 2$ integrins (Lloyd *et al.*, 1996). In contrast to the findings of P. Loetscher *et al.* (1996), Taub *et al.* (1995b) reported that IP-10 induces NK cell migration and degranulation, and that it enhances NK cell-mediated killing of tumor cells. In the presence of growth factors,

IP-10 suppresses *in vitro* colony formation by early human bone marrow progenitor cells (Sarris *et al.*, 1993). IP-10 inhibits endothelial cell proliferation *in vitro* (Luster *et al.*, 1995) and is a potent inhibitor of angiogenesis *in vivo* (Angiolillo *et al.*, 1995). After subcutaneous injection in mice, mainly monocytes, but few or no lymphocytes or neutrophils, are recruited within 4 h. Subcutaneous injection of IP-10 into huPBL–SCID mice induces infiltration of murine monocytes and human $CD3^+$ T cells within 24–48 h (Taub *et al.*, 1996b).

Mig is chemotactic for tumor-infiltrating lymphocytes and activated T cells. Its biologic activity is diminished by COOH-terminal truncation (Liao *et al.*, 1995). In huPBL–SCID mice, Mig induces infiltration of monocytes within 4 h after injection, whereas $CD3^+$ T-cell recruitment is observed between 4 and 48 h (Taub, 1996).

Stromal Cell-Derived Factor-1α and -1β

The 67-residue form of human SDF-1α is chemotactic for monocytes, phytohemagglutinin (PHA)-activated peripheral blood lymphocytes and freshly purified peripheral blood T lymphocytes, but also for neutrophils. The 62-residue form of SDF-1α is inactive. SDF-1 inhibits infection of PBMCs by T-tropic human immunodeficiency virus-1 (HIV-1) strains, but does not affect infection by M-tropic or dual-tropic primary HIV-1 strains (Bleul *et al.*, 1996a; Oberlin *et al.*, 1996).

Subcutaneous injection of mouse SDF-1α in mice results in a mononuclear cell infiltrate (Bleul *et al.*, 1996b). SDF-1 augments $[Ca^{2+}]_i$ in $CD34^+$ cells and chemoattracts these hematopoietic progenitor cells *in vitro* and *in vivo*. This SDF-1-induced chemotaxis is increased by IL-3. $CD34^+$ progenitors from peripheral blood are less responsive than $CD34^+$ progenitors from bone marrow (Aiuti *et al.*, 1997). SDF-1 is responsible for B-cell lymphopoiesis and bone marrow myelopoiesis. Mutant mice with a targeted disruption of the gene encoding SDF-1 die perinatally. The mutants have a cardiac ventricular septal defect (Nagasawa *et al.*, 1996).

HUMAN CXC CHEMOKINE RECEPTORS

High-affinity receptors ($20–90 \times 10^3$ per cell) for IL-8 have been identified on the surface of neutrophils, with a K_d ranging from 0.2 to 4 nM (Samanta *et al.*, 1989; Grob *et al.*, 1990; Moser *et al.*, 1991; Schumacher *et al.*, 1992). The IL-8 receptors do not bind unrelated chemoattractants such as LTB_4, fMLP, C5a or PAF. After binding, IL-8 is rapidly internalized and degraded (Samanta *et al.*, 1990). More than 90% of the ligand-bound receptors are endocytosed within 10 min. The receptors reappear rapidly on the cell surface, probably through recycling. The addition of lysomotropic agents does not inhibit ligand binding or internalization, but partially inhibits reappearance of the receptors and also neutrophil chemotaxis. This indicates that recycling of the receptors may be essential for the chemotactic response of neutrophils (Samanta *et al.*, 1990).

By cross-linking experiments with iodinated IL-8, two membrane proteins (44–59 and 67–70 kDa) can be demonstrated. The 44–59-kDa receptor (CXCR1) corresponds to the receptor with high affinity for IL-8 and low affinity for the other ELR^+ CXC chemokines (Samanta *et al.*, 1989; Schumacher *et al.*, 1992). On basophils, a single class of high-affinity receptors (K_d 0.15 nM, 9600 per cell) for IL-8 is present, which also weakly binds NAP-2 (Krieger *et al.*, 1992). A receptor for GRO-α that does not bind

IL-8 has been reported on a human melanoma cell line (Horuk *et al.*, 1993), as well as on rheumatoid synovial fibroblasts (Unemori *et al.*, 1993).

Four different receptors for human CXC chemokines have been cloned. These chemokine receptors are members of the rhodopsin or serpentine receptor superfamily. They contain seven hydrophobic transmembrane spanning regions and couple to guanine nucleotide binding proteins (G proteins). The NH_2-terminus and COOH-terminus are located extracellularly and intracellularly respectively. Chemokine receptors are small, relative to other rhodopsin-like receptors, due to a short third intracellular loop and a short NH_2- and COOH-terminus. The receptors contain *N*-linked glycosylation sites in the NH_2-terminus and in extracellular regions. The intracellular COOH-terminus contains several serine and threonine residues which may function as phosphorylation sites, important for desensitization of the receptors (Kelvin *et al.*, 1993; Murphy, 1996).

IL-8 Receptors (CXCR1 and CXCR2)

Cloning, Expression and Function

The cDNA for IL-8 receptor A (IL-8RA) or CXCR1 has been identified from human neutrophils and encodes a 350-amino-acid protein (Holmes *et al.*, 1991). The cDNA for IL-8RB or CXCR2 has been cloned from HL-60 cells differentiated into neutrophils and encodes a 355- or 360-residue protein (Murphy and Tiffany, 1991). CXCR1 and CXCR2 show 77% homology in their amino acid sequences. The calculated molecular mass is approximately 40 kDa for both receptors. Glycosylation may explain the difference between the theoretic molecular mass of the cloned receptors and the relative molecular mass of the native receptors on neutrophils. Both receptors have an extracellular NH_2-terminus which is rich in acidic residues (Holmes *et al.*, 1991; Murphy and Tiffany, 1991; Lee *et al.*, 1992). The genes for the two IL-8 receptors as well as for an IL-8RB pseudogene have been assigned to human chromosome 2q34–35 (Morris *et al.*, 1992; Mollereau *et al.*, 1993). The open reading frame of both CXCR1 and CXCR2 corresponds to a single exon. The gene for CXCR1 consists of two exons, whereas for CXCR2 seven distinct neutrophil mRNAs are formed by alternative splicing of 11 exons (Ahuja *et al.*, 1994).

Messenger RNA for CXCR1 has been detected in neutrophils, PHA-activated T-cell blasts, $CD4^+$ T cells, monocytes (Moser *et al.*, 1993a), platelets (Power *et al.*, 1995), megakaryocytes (Gewirtz *et al.*, 1995) and leukocytic cell lines, as well as in melanocytes, fibroblasts, keratinocytes, endothelial cells and melanoma cells (Schönbeck *et al.*, 1995). The gene for CXCR2 is expressed in neutrophils, monocytes (Moser *et al.*, 1993a), platelets, megakaryocytes (Gewirtz *et al.*, 1995), HL-60 cells, AML 193 cells, melanocytes (Norgauer *et al.*, 1996), keratinocytes (Mueller *et al.*, 1994) and melanoma cells (Norgauer *et al.*, 1996). With specific antibodies for CXCR1 or CXCR2, IL-8 receptors can be detected on neutrophils, monocytes, a subset of $CD8^+$ T cells and of $CD56^+$ NK cells. Neutrophils express the highest numbers of both CXCR1 and CXCR2 at an approximately equal ratio, whereas monocytes and IL-$8R^+$ lymphocytes express higher levels of surface protein for CXCR2 than for CXCR1 (Chuntharapai *et al.*, 1994; Morohashi *et al.*, 1995). Recently, Jinquan *et al.* (1997) showed that about 15% of freshly isolated B cells functionally express IL-8 receptors. Also astrocytes, microglia and the human astrocyte cell line HSC2 contain specific RNA and express surface protein for CXCR2 (Lacy *et al.*, 1995).

The expression of CXCR1 and CXCR2 can be regulated by cytokines. Granulocyte colony stimulating factor (G-CSF) augments specific mRNA for both IL-8 receptors in neutrophils, and this correlates with increased binding of IL-8 and subsequent neutrophil chemotaxis. In contrast, a dramatic reduction in IL-8 receptor mRNA is observed after treatment of neutrophils with LPS, which is correlated with decreased IL-8 binding and a reduced chemotactic response. A less marked IL-8 receptor downregulation occurs after treatment with TNF-α. The G-CSF effect is dependent on enhanced transcriptional activity of the IL-8 receptor genes. LPS inhibits IL-8 receptor expression on neutrophils by a combination of transcriptional inhibition and a decrease in mRNA stability (Lloyd *et al.*, 1995). In contrast to these findings, Manna *et al.* (Manna and Samanta, 1995; Manna *et al.*, 1995) reported that LPS and fMLP augment IL-8 receptor expression on neutrophils. IFN-γ upregulates both CXCR1 and CXCR2 on freshly isolated $CD4^+$ and $CD8^+$ T lymphocytes (Jinquan *et al.*, 1995). TNF-α and IL-2 increase both IL-8 receptors on $CD4^+$ T lymphocytes. On B cells, the IL-8 receptors are downregulated by IFN-γ, IL-2 and TNF-α (only CXCR1) and upregulated by IL-4 and IL-13 (Jinquan *et al.*, 1997).

Expression of CXCR1 or CXCR2 in mammalian cells has confirmed that IL-8 binds to both receptors with high affinity to induce chemotaxis and calcium flux. GRO-α, NAP-2 and IL-8 are equally potent as attractants for CXCR2-transfected cells, but IL-8 is 300–1000 times more potent than GRO-α or NAP-2 for chemotaxis of CXCR1 transfectants (P. Loetscher *et al.*, 1994). The ELR^+ CXC chemokines GRO-α, GRO-β, GRO-γ, NAP-2 and ENA-78 are potent agonists to increase $[Ca^{2+}]_i$ in CXCR2-transfected cells, but not in CXCR1 transfectants. These chemokines have high affinity for CXCR2 whereas they compete weakly for the high-affinity IL-8 binding site on CXCR1 (Ahuja and Murphy, 1996). In contrast to the other ELR^+ CXC chemokines, GCP-2 is equipotent in augmenting $[Ca^{2+}]_i$ in both CXCR1- and CXCR2-transfected cells. The calcium rise in neutrophils or IL-8 receptor-transfected cells after stimulation with GCP-2 can be inhibited by prestimulation of the cells with IL-8. Similarly, the IL-8 response can be abolished by pretreatment with GCP-2. These data indicate that both chemokines use the same receptors (Wuyts *et al.*, 1997). The ELR^- CXC chemokines Mig and IP-10 do not induce signal transduction in CXCR1- or CXCR2-tranfected cells (Ahuja and Murphy, 1996).

Both IL-8 receptors are functionally different. GRO-α stimulates neutrophil chemotaxis exclusively through CXCR2, whereas chemotaxis to IL-8 is mediated predominantly by CXCR1 (Hammond *et al.*, 1995). In contrast, Chuntharapai and Kim (1995) postulate that CXCR2 may play an active role in the initiation of neutrophil migration distant from the site of inflammation (where the IL-8 concentration is at the picomolar level), whereas CXCR1 may be important for mediating the IL-8 signal at the site of inflammation (where the concentration of IL-8 is high). Elastase release by neutrophils in response to IL-8 is decreased by anti-CXCR1 and even more substantially by anti-CXCR1 plus anti-CXCR2, whereas the elastase release induced by GRO-α is inhibited only by anti-CXCR2. The superoxide anion production by IL-8 is affected only by anti-CXCR1 and not by anti-CXCR2 (Jones *et al.*, 1996). The priming effect of IL-8 on fMLP-induced superoxide anion production is mediated predominantly by CXCR1, whereas the same effect of GRO-α and ENA-78 is mediated by CXCR2 (Green *et al.*, 1996). T- and B-lymphocyte chemotaxis in response to IL-8 and GRO-α is probably mediated by CXCR2 (Jinquan *et al.*, 1995, 1997). Binding of GRO-α to melanoma cells is partially blocked by anti-CXCR2. The interaction of GRO-α with CXCR2 is required for melanoma cell growth (Norgauer *et al.*, 1996).

Structural Requirements for CXCR1 and CXCR2 to Respond to CXC Chemokines

ELR^+ CXC chemokines show different affinities for CXCR1 and CXCR2. Several studies indicate that the NH_2-terminal segment of CXCR1 is a dominant determinant of receptor subtype selectivity (LaRosa *et al.*, 1992; Gayle *et al.*, 1993; Wu *et al.*, 1996), whereas Ahuja *et al.* (1996) reported that the NH_2-terminal segment of CXCR2 is dominant. Determinants of GRO-α selectivity have been found both in the NH_2-terminal segment before transmembrane domain one beyond the first 15 residues and in the region from transmembrane domain four to the end of the second extracellular loop of CXCR2. The NH_2-terminal amino acid sequences of both IL-8 receptors are rich in acidic residues and have been suggested to play a major role in binding IL-8. The residues 10–15 (WDFDDL) of CXCR1 and the amino acids 6–10 (FEDFW) of CXCR2 play an important role in ligand binding and functionality of the receptors (Wu *et al.*, 1996). The cysteines in the extracellular domains of CXCR1, which may be part of a binding site or may be critical for the overall folding of the receptor, are also important for IL-8 binding and IL-8-mediated signaling (Samanta *et al.*, 1993; Leong *et al.*, 1994). Besides the extracellular domain cysteines, the three residues Arg-199, Arg-203 and Asp-265 of CXCR1 are involved in IL-8 binding and signal transduction. These residues are found in extracellular loops two and three and are conserved in CXCR2.

Substitution of several other residues in the extracellular domains of CXCR1 yields mutants with a reduced affinity for IL-8, but which can still increase $[Ca^{2+}]_i$ in response to IL-8. These amino acids might provide the initial contacts that facilitate formation of successively more stable interactions or possibly influence the tertiary structure of the receptor itself (Hébert *et al.*, 1993; Leong *et al.*, 1994). The finding that an antibody against CXCR1 can block IL-8-induced activity, without inhibiting binding, confirms this model. It is possible that the receptor cannot undergo IL-8-induced conformational changes to elicit biologic responses or that the antibody blocks the access of a signaling domain of IL-8 to the appropriate site of the receptor.

Thus, IL-8 binding to, and signaling through, CXCR1 occur in at least two discrete steps involving distinct domains of the receptor. A conformational change of the receptor or ligand after IL-8 binding may play an important role in signaling (Wu *et al.*, 1996). Amino acids Tyr-136, Leu-137, Ile-139 and Val-140 in the second intracellular loop and Met-241 in the third intracellular loop are essential for binding of the G protein α subunit to CXCR1 and for mediating the calcium signaling in response to IL-8. These residues are conserved in CXCR2 where they are also important for signal transduction (Damaj *et al.*, 1996). Amino acids 317–324 of the COOH-terminus of CXCR2 are also essential for signaling (Ben-Baruch *et al.*, 1995).

IP-10–Mig Receptor (CXCR3)

A seven transmembrane domain receptor selective for IP-10 and Mig has been cloned from a $CD4^+$ T-cell library. The receptor cDNA encodes a protein of 368 amino acids with a molecular mass of 40 kDa. The receptor shows about 40% homology with the two IL-8 receptors. The IP-10–Mig receptor or CXCR3 is highly expressed in IL-2-activated

T lymphocytes, but is not detectable in resting T lymphocytes, B lymphocytes, monocytes or granulocytes. The receptor mediates Ca^{2+} mobilization and chemotaxis in response to IP-10 and Mig, but does not recognize the other CXC or CC chemokines, nor the C chemokine lymphotactin (M. Loetscher *et al.*, 1996).

LESTR/Fusin, the SDF-1 Receptor (CXCR4)

A cDNA clone, called LESTR, encoding a protein of 352 amino acids and corresponding to a seven transmembrane domain, G protein-coupled receptor has been isolated from a human blood monocyte cDNA library and shows about 34% similarity to CXCR1 and CXCR2. There is no binding of the CXC chemokines IL-8, NAP-2, GRO-α, PF-4 and IP-10, or of the CC chemokines MCP-1, MCP-2, MCP-3, MIP-1α, I-309 and RANTES (Regulated on Activation, Normally T cell Expressed and Secreted), nor of C3a or LTB_4 to this receptor. A high level of LESTR mRNA is found in neutrophils, PBLs, PHA-activated T-cell blasts and 1α,25$(OH)_2$-vitamin D_3-differentiated HL-60 cells; moderate levels are found in monocytes and undifferentiated HL-60 cells, but only low levels are detectable in U937 cells, Jurkat cells and dimethyl sulfoxide-differentiated HL-60 cells (M. Loetscher *et al.*, 1994). This receptor has also been cloned from fetal brain, lung, spleen and PBMC libraries (Federsppiel *et al.*, 1993; Jazin *et al.*, 1993; D'Souza and Harden, 1996). Recently, it has been shown that SDF-1 is the ligand for LESTR, which is now called CXCR4. SDF-1 induces calcium and chemotaxis in CXCR4-transfected cells (Bleul *et al.*, 1996a; Oberlin *et al.*, 1996). CXCR4, also designated fusin, has also been identified as the cofactor for T-tropic HIV-1 infection (D'Souza and Harden, 1996; Feng *et al.*, 1996). The gene for CXCR4 is located on chromosome 2 (Premack and Schall, 1996).

Duffy Antigen Receptor for Chemokines (DARC)

A promiscuous seven transmembrane spanning domain chemokine receptor with an estimated molecular weight of approximately 39 kDa is expressed by erythrocytes (Horuk, 1994; Horuk *et al.*, 1996). Its cDNA encodes a protein of 338 amino acids with a theoretic mass of about 36 kDa. The difference between the theoretic mass and the observed molecular mass is due to glycosylation. The gene for DARC is located on chromosome 1q22–23. In contrast to other identified chemokine receptors, which specifically bind CXC or CC chemokines, this receptor binds both CXC and CC chemokines with a K_d of about 5 nM and is present at about 5000 molecules per cell. This receptor is identical to the Duffy antigen, the receptor for the malarial parasite *Plasmodium vivax*, and is now called the Duffy antigen receptor for chemokines (DARC). DARC binds the ELR^+ CXC chemokines IL-8, GRO, NAP-2 and ENA-78, and the CC chemokines RANTES, MCP-1 and MCP-3 with high affinity. The ELR^- CXC chemokines PF-4 and IP-10 have low affinity for DARC, whereas the MIP-1 proteins and lymphotactin do not bind to DARC.

The NH_2-terminal extracellular domain of DARC is the critical region in determining the promiscuous ligand-binding repertoire. This erythrocyte chemokine receptor does not internalize the ligand after binding and is not coupled to G proteins. There is no signal transduction after binding of chemokines to this receptor on erythrocytes but, by binding, chemokines can inhibit infection with *P. vivax*. DARC may represent a mechanism for controlling the levels of proinflammatory chemokines in circulation. This regulatory function could be complemented by naturally occurring antichemokine antibodies. DARC is also expressed on endothelial cells of postcapillary venules, where it could act as a chemokine-presenting molecule (Horuk, 1994; Horuk, *et al.*, 1996).

ANIMAL MODELS AND THE ROLE OF CXC CHEMOKINES IN PATHOLOGY

Chemokines are important for the recruitment of leukocytes to sites of infection, which is essential in host defense and may lead to clearance of inciting factors. However, the accumulation of leukocytes in (non-)infectious disorders can also be deleterious for the host and contributes to the pathogenesis of several disorders (e.g. glomerulonephritis, rheumatoid arthritis and ischemia–reperfusion-induced injury). Chemokines have been detected in body fluids and tissues in a variety of pathologic conditions (Table 5).

CXC Chemokines and Skin Disorders

IL-8 has been demonstrated in, and isolated from, skin lesions of patients with psoriasis (Schröder and Christophers, 1986). Psoriasis is an inflammatory skin disorder which is characterized by hyperproliferation of keratinocytes and by infiltration of neutrophils and T lymphocytes. Besides IL-8, GRO-α (Schröder *et al.*, 1992) and IP-10 (Gottlieb *et al.*, 1988), as well as the IL-8 receptor (CXCR1) (Schulz *et al.*, 1993), have been detected in psoriatic plaques. Keratinocytes are a cellular source of these three chemokines and express the IL-8 receptor. In addition, IL-8 has been detected in the infiltrating neutrophils and GRO-α, present in the dermis, is produced by vessel-associated cells (Gottlieb *et al.*, 1988; Gillitzer *et al.*, 1996). The expression of IL-8 and GRO-α in psoriatic scales is reduced after treatment with the T cell-directed immunosuppressive agent cyclosporin A. However, this agent has no effect on the chemokine production by keratinocytes *in vitro*. The overexpression of IL-8 and GRO-α is probably a keratinocyte response to activated T cells (Elder *et al.*, 1993; Kojima *et al.*, 1993). Another immunosuppressive antipsoriatic drug, FK-506 or tacrolimus, reduces the IL-8 concentration in psoriatic lesions and in plasma (Lemster *et al.*, 1995).

IL-8 plays an important role in the development of a delayed-type hypersensitivity response. In rabbits, infusion of anti-IL-8 suppresses the development of a tuberculin skin reaction, which represents the delayed-type hypersensitivity reaction. Neutrophil and lymphocyte infiltration at the tuberculin injection site, as well as skin swelling, are decreased. In the skin of positive tuberculin reactions, IL-8 is produced by keratinocytes (Larsen *et al.*, 1995). In humans, the presence of IP-10 in epidermal keratinocytes and dermal macrophages and endothelial cells has been detected during a delayed-type hypersensitivity response (Kaplan *et al.*, 1987).

CXC Chemokines and Glomerulonephritis

Glomerulonephritis is a major cause of renal failure and defines a group of disorders that may essentially be considered as immunologically mediated injury to glomeruli. The major pathogenic mechanisms are deposition or *in situ* formation of immune complexes, and deposition of antiglomerular basement membrane antibodies. Glomerulonephritis is characterized by infiltration of leukocytes, mesangial cell proliferation and increased

Table 5. CXC chemokines and human pathology.

Chemokine	System	Pathologic condition
IL-8, GRO-α, IP-10	Skin	Psoriasis
IP-10		Tuberculoid leprosy
IP-10		Cutaneous leishmaniasis
IL-8		Contact dermatitis
IL-8		Atopic dermatitis
IL-8, IP-10		Cutaneous T-cell lymphoma
IL-8, ENA-78, GRO-α	Joints	Rheumatoid arthritis
IL-8		Gout
IL-8, ENA-78, GRO-α	Respiratory	Adult respiratory distress syndrome (ARDS)
IL-8		Risk for ARDS
IL-8		Idiopathic pulmonary fibrosis
IL-8		Cystic fibrosis
IL-8		Asthma
IL-8		Pleural empyema
IL-8		Metal fume fever
IL-8, GRO-α		Bacterial pneumonia
IL-8		Chronic bronchitis
IL-8		Hypersensitivity pneumonia
IL-8		*Mycobacterium tuberculosis* infection
IL-8		Viral respiratory tract infection
IL-8		Allergic rhinitis
IL-8		Sinusitis
IL-8		Bronchogenic carcinoma
IL-8	Cardiovascular	Atherosclerosis
IL-8		Sepsis, septic shock
IL-8		Cardiopulmonary bypass
IL-8		Aortic surgery
IL-8		Heart transplantation
IL-8		Myocardial infarction
IL-8, GRO-α	Central nervous	Bacterial meningitis
IL-8		Viral meningitis
IL-8		Multiple sclerosis
IL-8		Brain tumor (astrocytoma, glioblastoma)
IL-8	Gastrointestinal	Pancreatitis
IL-8		Ulcerative colitis
IL-8		Crohn's disease
IL-8, GRO-α		Alcoholic hepatitis
IL-8		Viral hepatitis
IL-8		*Helicobacter pylori* gastritis
IL-8		Gastric carcinoma
IL-8		Hepatocellular carcinoma
IL-8, GRO-α		Peritonitis
IL-8	Genitourinary	Urinary tract infection
IL-8		Glomerulonephritis
IL-8		*Trichomonas vaginalis* infection
IL-8, GRO-α		Endometriosis
IL-8	Various	Periodontitis
IL-8, GRO-α		HIV infection
IL-8		Chronic lymphatic leukemia

matrix deposition, which will lead to impaired filtration, proteinuria and glomerulosclerosis (Brown *et al.*, 1996). *In vitro* studies have shown that IL-8 is produced by mesangial, epithelial and endothelial renal cells as well as by fibroblasts after stimulation with TNF-α and IL-1β. GRO-α and GRO-β, and their rat counterparts cytokine-induced neutrophil chemoattractant (CINC) and MIP-2 respectively, are induced in mesangial cells and epithelial cells in response to TNF-α, IL-1β and LPS, and ENA-78 production by epithelial cells is stimulated by IL-1β (Schlöndorff *et al.*, 1997). Increased urinary IL-8 levels, which correlate with IL-8 immunostaining and leukocyte recruitment in the glomeruli, have been detected in several types of glomerulonephritis. Urinary IL-8 levels decrease after treatment with glucocorticoids (Wada *et al.*, 1994b).

A rabbit model of immune complex-mediated acute glomerulonephritis has been used to evaluate the role of IL-8 in this disease. Induction of glomerulonephritis causes an influx of neutrophils into the glomeruli and increased urinary protein levels. Anti-IL-8 antibodies decrease the number of neutrophils in the glomeruli and normalize the urinary protein levels, indicating that IL-8 participates in the impairment of renal functions in experimental acute immune complex-mediated glomerulonephritis through attraction and activation of neutrophils (Wada *et al.*, 1994a). In a rat model of antiglomerular basement membrane antibody-induced glomerulonephritis, CINC (GRO-α) and MIP-2 (GRO-β/γ) are induced in the kidney, as well as TNF-α and IL-1β, which may induce the chemokine expression. IP-10 and PF-4 mRNA have also been detected in the glomeruli in this model. Administration of glucocorticoids suppresses glomerular chemokine mRNA expression, as well as leukocyte infiltration. Anti-CINC or anti-MIP-2 antibodies attenuate neutrophil influx into the glomerulus as well as proteinuria (Wu *et al.*, 1994; Feng *et al.*, 1995; Tang *et al.*, 1995).

CXC Chemokines and Ischemia–Reperfusion-Induced Injury

Reperfusion of ischemic organs occurs in myocardial infarction, cerebral infarction, multiple organ failure after hypovolemia, frostbite, etc. Re-establishing the blood flow to ischemic tissues causes greater injury than that induced during the ischemic period, and neutrophils play an important role in this type of injury. Reperfusion is characterized by the generation of reactive oxygen intermediates, which may regulate the production of chemokines and thereby the infiltration of neutrophils.

The role of IL-8 has been investigated in a rabbit model of lung reperfusion injury. Reperfusion of ischemic lung induces neutrophil infiltration in the lung tissue and in bronchoalveolar lavage fluid. The IL-8 levels in bronchoalveolar lavage fluid and lung tissue homogenates are increased. IL-8-producing cells are mainly bronchiolar ciliary cells and alveolar macrophages. Ischemia alone does not induce a significant increase in IL-8 levels. Anti-IL-8 antibodies prevent neutrophil infiltration and tissue injury (Sekido *et al.*, 1993). IL-8 is produced after reperfusion of ischemic myocardium in dogs and probably participates in neutrophil-mediated myocardial injury. Immunohistochemical IL-8 staining has been detected in the inflammatory infiltrate and in the endothelium of veins (Kukielka *et al.*, 1995). In addition to IL-8, ENA-78 has been shown to be involved in ischemia–reperfusion injury. In the rat, hepatic ischemia–reperfusion injury causes synthesis of TNF-α in the liver and production of ENA-78, as well as neutrophil accumulation in the lung. ENA-78 production is reduced by anti-TNF antiserum, and anti-ENA-78 antibodies decrease the neutrophil recruitment as well as the lung injury.

Thus, pulmonary ENA-78, produced in response to hepatic-derived TNF-α, is an important mediator in hepatic ischemia–reperfusion-induced lung injury (Colletti *et al.*, 1995). These animal studies indicate that CXC chemokines indeed play an important role in ischemia–reperfusion-induced injury by the attraction of neutrophils.

CXC Chemokines and Rheumatoid Arthritis

CXC chemokines are important mediators in the pathogenesis of human rheumatoid arthritis, an autoimmune disease characterized by joint inflammation and destruction and by the recruitment and activation of leukocytes in the synovial space and joint tissue. The infiltration of neutrophils into the joints may contribute to tissue damage. Synovial fluid and plasma of patients with rheumatoid arthritis contain IL-8 (Rampart *et al.*, 1992), ENA-78 and GRO-α, and the neutrophil chemotactic activity of the synovial fluid can be reduced by antibodies against IL-8, ENA-78 and GRO-α respectively. The presence of these chemokines in the synovial fluid correlates with the neutrophil influx. Synovial tissue macrophages of patients with rheumatoid arthritis express IL-8 and ENA-78 constitutively. Synovial tissue fibroblasts constitutively produce significant levels of ENA-78 but release only a small amount of IL-8 and GRO-α. The production of the three chemokines is enhanced upon stimulation of the cells with TNF-α or IL-1β, cytokines that are produced locally in the joints of patients with rheumatoid arthritis. Synovial fluid mononuclear cells express IL-8, GRO-α and ENA-78, which are enhanced upon stimulation with LPS, and neutrophils from the synovial fluid have been shown to express IL-8 and GRO-α. After intra-articular injection of glucocorticoids, the production of IL-8 by synovial fluid mononuclear cells is inhibited almost completely (Koch *et al.*, 1991, 1994, 1995; Seitz *et al.*, 1991; Beaulieu and McColl, 1994). Immunohistochemical analysis shows that the macrophages are the predominant cellular source for IL-8 in rheumatoid arthritis. However, the expression of ENA-78 is mainly increased in synovial tissue macrophages and fibroblasts, and that of GRO-α is enhanced in macrophages and synovial lining cells. Chondrocytes present in the joint surface cartilage also stain positively for IL-8 (Koch *et al.*, 1991, 1994, 1995; Deleuran *et al.*, 1994).

Injection of IL-8 into rabbit knee joints induces neutrophil infiltration and provokes neutrophil elastase release, which leads to cartilage destruction (Matsukawa *et al.*, 1995). In acute arthritis, induced in rabbits by injection of LPS or IL-1α into the joints, the recruitment of neutrophils, as well as tissue injury, can be reduced by anti-IL-8 antibodies (Akahoshi *et al.*, 1994). In a mouse model of type II collagen-induced arthritis, MIP-2 (GRO-β/γ) is expressed by infiltrating macrophages in the joint and the protein can also be detected in aqueous joint tissue extracts. Passive immunization with anti-MIP-2 antibodies causes a decrease in the incidence of arthritis as well as in the degree of joint inflammation (Kunkel *et al.*, 1996). These studies show that CXC chemokines contribute to the pathogenesis of rheumatoid arthritis.

CXC Chemokines and Pulmonary Disease

Chemokines have been detected in a variety of lung disorders (Table 5). For example, patients with idiopathic pulmonary fibrosis have increased expression of IL-8 mRNA in alveolar macrophages, which correlates with increased levels of IL-8 protein and the number of neutrophils in bronchoalveolar lavage fluid, as well as with disease severity

(Carré *et al.*, 1991; Lynch *et al.*, 1992). High levels of IL-8 and neutrophils, which correlate with mortality, are found in the bronchoalveolar lavage fluid of patients with the adult respiratory distress syndrome (Jorens *et al.*, 1992; Miller *et al.*, 1992).

Chemokines are important in antibacterial host defense in the lung. Effective host defense against bacterial infections is characterized by recruitment and activation of inflammatory cells. In the setting of a low bacterial burden or exposure of the lungs to less virulent Gram-positive organisms, the alveolar macrophages can effectively phagocytose and kill the microorganisms, but when the bacterial burden is large or when more virulent Gram-negative organisms (e.g. *Pseudomonas aeruginosa* and *Klebsiella pneumoniae*) gain access to the lower airspaces, the recruitment of neutrophils is essential for bacterial clearance. There are several indications that chemokines are involved in the host defense against bacterial infection. Bacteria and cell wall components induce chemokine production by mononuclear phagocytes and stromal cells. IL-8 has been detected in sputum and bronchoalveolar lavage fluid from patients with acute pulmonary infections. The produced chemokines attract and activate leukocytes. *In vitro* studies have shown that IL-8 and MIP-2 augment the ability of neutrophils to phagocytose and kill *E. coli*. IL-8 has also been shown to enhance neutrophil microbicidal activity against *Mycobacterium fortuitum* and *C. albicans* (Djeu *et al.*, 1990; Standiford *et al.*, 1996).

In a rabbit model of endotoxin-induced pleurisy, IL-8 was detected in the pleural liquid and anti-IL-8 antibodies reduced neutrophil influx into the lung (Broaddus *et al.*, 1994). A murine model of *Klebsiella* pneumonia has been used to investigate the role of MIP-2 in antibacterial host defense. Intratracheal inoculation of mice with *K. pneumoniae* causes respiratory distress and accumulation of neutrophils in the lung airspace and interstitium, which is correlated with the presence of MIP-2 in lung homogenates. The main cellular source of MIP-2 is the alveolar macrophage, but the recruited neutrophils also express MIP-2. Passive immunization of mice with anti-MIP-2 antibodies before inoculation with *K. pneumoniae* causes a reduced neutrophil influx in the lung, an increase in the number of viable bacteria in the lung, and dissemination to blood and liver, and an increase in (early, but not late) mortality. This indicates that MIP-2 is an important mediator of neutrophil influx and bacterial clearance in murine *Klebsiella* pneumonia (Standiford *et al.*, 1996).

CXC Chemokines and Other Infections

IL-8 has been detected in the plasma of patients with sepsis, and higher levels of IL-8 are present in patients who die than in those who survive (Friedland *et al.*, 1992; Hack *et al.*, 1992). In lethal and sublethal sepsis, induced by infusing primates with live *E. coli* and LPS respectively, IL-8 is detectable in the circulation. The IL-8 concentrations are higher in animals with lethal bacteremia than in those with sublethal endotoxemia (Van Zee *et al.*, 1991). *E. coli* administration to primates has also been shown to increase neutrophil elastase levels in plasma (Redl *et al.*, 1991). IL-8 is induced in human volunteers or primates following intravenous administration of endotoxin or of the cytokines IL-1β or TNF-α, which may be intermediate factors in the endotoxin-induced IL-8 release (van Deventer *et al.*, 1993).

Recently, chemokines have been shown to be involved in the pathogenesis of acquired immune deficiency syndrome (AIDS). CD4 is not sufficient for HIV to enter mononuclear cells. The required cofactors for infection are chemokine receptors. Whereas

M-tropic strains use CC chemokine receptors as cofactors, T-tropic strains, which occur during later phases of infection and are associated with the progression to AIDS, utilize the CXC chemokine receptor CXCR4. SDF-1, the natural ligand for CXCR4, can inhibit HIV infection of cells by T-tropic strains *in vitro* (Bleul *et al.*, 1996a; D'Souza and Harden, 1996; Oberlin *et al.*, 1996; Doranz *et al.*, 1997).

CXC Chemokines and Tumor Growth

Chemokines can have tumor-promoting or tumor-inhibitory effects. Angiogenesis is important for tumor growth and metastasis. Due to the angiogenic or angiostatic properties of the CXC chemokines, they can promote or inhibit tumor growth respectively. Tumor cells are a good source of various CXC chemokines, including IL-8, GRO, IP-10, GCP-2 and ENA-78 (Proost *et al.*, 1993a). The recruitment of leukocytes by tumor-produced chemokines can also influence tumor growth.

Human bronchogenic carcinoma tissue contains IL-8, which is produced by the tumor cells. The angiogenic activity of tissue homogenates from these tumors is decreased by anti-IL-8 antibodies, indicating that IL-8 is a primary mediator for new blood vessel formation in bronchogenic carcinoma (Smith *et al.*, 1994). In SCID mice, IL-8 promotes tumorigenesis of lung cancer cell lines through its angiogenic properties: passive immunization with anti-IL-8 antibodies results in a reduction of tumor size and a decrease in metastasis, which are associated with a decline in angiogenic activity of the tumor tissue and a decrease in tumor-associated vascular density. However, IL-8 does not stimulate the growth of the tumor cells *in vitro* (Arenberg *et al.*, 1996). IL-8 is an autocrine growth factor for melanoma cells, and the expression of IL-8 by these cells correlates with their metastatic potential. Ultraviolet B irradiation of a melanoma cell line, which expresses negligible levels of IL-8 and is not tumorigenic in nude mice, induces the production of IL-8. These irradiated melanoma cells show enhanced tumorigenicity and metastatic potential after injection in mice. This may be explained by an increase in collagenase activity and angiogenesis attributed to the induction of IL-8 by ultraviolet B radiation (Singh *et al.*, 1995). The cooperative role of proteases and chemokines in tumor growth and metastasis has previously been postulated (Opdenakker and Van Damme, 1992).

PF-4 inhibits the growth of murine melanoma and of human colonic carcinoma cells in nude mice, whereas these cell lines are insensitive to PF-4 *in vitro* (Sharpe *et al.*, 1990). The antitumor effects are due to the angiostatic properties of this chemokine and have led to current clinical trials with PF-4 as an anticancer therapeutic agent in humans. Similarly, GRO-β, which is also angiostatic, has no effect on the growth of Lewis lung carcinoma cells *in vitro*. However, it inhibits the growth of these cells in immunocompetent as well as in immunodeficient mice by suppression of tumor-induced neovascularization (Cao *et al.*, 1995).

Transfection of murine plasmacytoma cells or mammary adenocarcinoma cells with the IP-10 gene has no effect on the growth of these cells *in vitro*, but these transfected cells fail to grow in mice. The antitumor response is T-lymphocyte dependent, is also observed after injection of a 1:1 mixture of non-transfected and transfected tumor cells, and appears to be mediated by the recruitment of an inflammatory infiltrate composed of lymphocytes, monocytes and neutrophils (Luster and Leder, 1993). However, IP-10 has also been shown to be an antitumor agent against human Burkitt's lymphoma in athymic mice. This chemokine, which has angiostatic properties, damages established tumor vasculature and causes tissue necrosis in nude mice, but complete regression of the tumor is not observed. Constitutive expression of IP-10 in Burkitt's lymphoma cells reduces their ability to grow and induces visible tumor necrosis (Sgadari *et al.*, 1996).

ACKNOWLEDGEMENTS

The authors acknowledge the financial support of the InterUniversity Attraction Pole (IUAP) initiative of the Belgian Federal Government, the Concerted Research Actions (GOA) of the Regional Government of Flanders, the Fund for Scientific Research of Flanders (FWO-Vlaanderen), the General Savings and Retirement Fund (ASLK), the Belgian Cancer Association and the European Commission. They thank Inge Aerts and Willy Put for editorial help and Ghislain Opdenakker and Sofie Struyf for critical reading of the manuscript. A. Wuyts is a research assistant of the FWO-Vlaanderen.

REFERENCES

Ahuja, S.K. and Murphy, P.M. (1996). The CXC chemokines growth-regulated oncogene (GRO) α, GROβ, GROγ, neutrophil-activating peptide-2, and epithelial cell-derived neutrophil-activating peptide-78 are potent agonists for the type B, but not the type A, human interleukin-8 receptor. *J. Biol. Chem.* **271**, 20 545–20 550.

Ahuja, S.K., Shetty, A., Tiffany, H.L. and Murphy, P.M. (1994). Comparison of the genomic organization and promoter function for human interleukin-8 receptors A and B. *J. Biol. Chem.* **269**, 26 381–26 389.

Ahuja, S.K., Lee, J.C. and Murphy, P.M. (1996). CXC chemokines bind to unique sets of selectivity determinants that can function independently and are broadly distributed on multiple domains of human interleukin-8 receptor B. *J. Biol. Chem.* **271**, 225–232.

Aiuti, A., Webb, I.J., Bleul, C., Springer, T. and Gutierrez-Ramos, J.C. (1997). The chemokine SDF-1 is a chemoattractant for human $CD34^+$ hematopoietic progenitor cells and provides a new mechanism to explain the mobilization of $CD34^+$ progenitors to peripheral blood. *J. Exp. Med.* **185**, 111–120.

Akahoshi, T., Endo, H., Kondo, H., Kashiwazaki, S., Kasahara, T., Mukaida, N., Harada, A. and Matsushima, K. (1994). Essential involvement of interleukin-8 in neutrophil recruitment in rabbits with acute experimental arthritis induced by lipopolysaccharide and interleukin-1. *Lymphokine Cytokine Res.* **13**, 113–116.

Alam, R., Forsythe, P.A., Lett-Brown, M.A. and Grant, J.A. (1992). Interleukin-8 and RANTES inhibit basophil histamine release induced with monocyte chemotactic and activating factor/monocyte chemoattractant peptide-1 and histamine releasing factor. *Am. J. Respir. Cell Mol. Biol.* **7**, 427–433.

Angiolillo, A.L., Sgadari, C., Taub, D.D., Liao, F., Farber, J.M., Maheshwari, S., Kleinman, H.K., Reaman, G.H. and Tosato, G. (1995). Human interferon-inducible protein 10 is a potent inhibitor of angiogenesis *in vivo*. *J. Exp. Med.* **182**, 155–162.

Anisowicz, A., Bardwell, L. and Sager, R. (1987). Constitutive overexpression of a growth-regulated gene in transformed Chinese hamster and human cells. *Proc. Natl Acad. Sci. U.S.A.* **84**, 7188–7192.

Arenberg, D.A., Kunkel, S.L., Polverini, P.J., Glass, M., Burdick, M.D. and Strieter, R.M. (1996). Inhibition of interleukin-8 reduces tumorigenesis of human non-small cell lung cancer in SCID mice. *J. Clin. Invest.* **97**, 2792–2802.

Ayesh, S.K., Azar, Y., Babior, B.M. and Matzner, Y. (1993). Inactivation of interleukin-8 by the C5a-inactivating protease from serosal fluid. *Blood* **81**, 1424–1427.

Bacon, K.B., Flores-Romo, L., Aubry, J.-P., Wells, T.N.C. and Power, C.A. (1994). Interleukin-8 and RANTES induce the adhesion of the human basophilic cell line KU-812 to human endothelial cell monolayers. *Immunology* **82**, 473–481.

Bacon, K.B., Flores-Romo, L., Life, P.F., Taub, D.D., Premack, B.A., Arkinstall, S.J., Wells, T.N., Schall, T.J. and Power, C.A. (1995). IL-8-induced signal transduction in T lymphocytes involves receptor-mediated activation of phospholipases C and D. *J. Immunol.* **154**, 3654–3666.

Baker, N.E., Kucera, G. and Richmond, A. (1990). Nucleotide sequence of the human melanoma growth stimulatory activity (MGSA) gene. *Nucleic Acids Res.* **18**, 6453.

Bazan, J.F., Bacon, K.B., Hardiman, G., Wang, W., Soo, K., Rossi, D., Greaves, D.R., Zlotnik, A. and Schall, T.J. (1997). A new class of membrane-bound chemokine with a CX_3C motif. *Nature* **385**, 640–644.

Beaulieu, A.D. and McColl, S.R. (1994). Differential expression of two major cytokines produced by neutrophils, interleukin-8 and the interleukin-1 receptor antagonist, in neutrophils isolated from the synovial fluid and peripheral blood of patients with rheumatoid arthritis. *Arthritis Rheum.* **37**, 855–859.

Begg, G.S., Pepper, D.S., Chesterman, C.N. and Morgan, F.J. (1978). Complete covalent structure of human beta-thromboglobulin. *Biochemistry* **17**, 1739–1744.

Bell, M.D., Taub, D.D. and Perry, V.H. (1996). Overriding the brain's intrinsic resistance to leukocyte recruitment with intraparenchymal injections of recombinant chemokines. *Neuroscience* **74**, 283–292.

Ben-Baruch, A., Bengali, K.M., Biragyn, A., Johnston, J.J., Wang, J.-M., Kim, J., Chuntharapai, A., Michiel, D.F., Oppenheim, J.J. and Kelvin, D.J. (1995). Interleukin-8 receptor β. The role of the carboxyl terminus in signal transduction. *J. Biol. Chem.* **270**, 9121–9128.

Bischoff, S.C., Baggiolini, M., de Weck, A.L. and Dahinden, C.A. (1991). Interleukin 8—inhibitor and inducer of histamine and leukotriene release in human basophils. *Biochem. Biophys. Res. Commun.* **179**, 628–633.

Bleul, C.C., Farzan, M., Choe, H., Parolin, C., Clark-Lewis, I., Sodroski, J. and Springer, T.A. (1996a). The lymphocyte chemoattractant SDF-1 is a ligand for LESTR/fusin and blocks HIV-1 entry. *Nature* **382**, 829–833.

Bleul, C.C., Fuhlbrigge, R.C., Casasnovas, J.M., Aiuti, A. and Springer, T.A. (1996b). A highly efficacious lymphocyte chemoattractant, stromal cell-derived factor 1 (SDF-1). *J. Exp. Med.* **184**, 1101–1109.

Bordoni, R., Fine, R., Murray, D. and Richmond, A. (1990). Characterization of the role of melanoma growth stimulatory activity (MGSA) in the growth of normal melanocytes, nevocytes, and malignant melanocytes. *J. Cell. Biochem.* **44**, 207–219.

Broaddus, V.C., Boylan, A.M., Hoeffel, J.M., Kim, K.J., Sadick, M., Chuntharapai, A. and Hébert, C.A. (1994). Neutralization of IL-8 inhibits neutrophil influx in a rabbit model of endotoxin-induced pleurisy. *J. Immunol.* **152**, 2960–2967.

Brown, Z., Robson, R.L. and Westwick, J. (1996). Regulation and expression of chemokines: potential role in glomerulonephritis. *J. Leukocyte Biol.* **59**, 75–80.

Broxmeyer, H.E., Sherry, B., Cooper, S., Lu, L., Maze, R., Beckmann, M.P., Cerami, A. and Ralph, P. (1993). Comparative analysis of the human macrophage inflammatory protein family of cytokines (chemokines) on proliferation of human myeloid progenitor cells. *J. Immunol.* **150**, 3448–3458.

Bruijnzeel, P.L.B., Kuijper, P.H.M., Rihs, S., Betz, S., Warringa, R.A.J. and Koenderman, L. (1993). Eosinophil migration in atopic dermatitis. I: Increased migratory responses to *N*-formyl-methionyl-leucyl-phenylalanine, neutrophil-activating factor, platelet-activating factor, and platelet factor 4. *J. Invest. Dermatol.* **100**, 137–142.

Burrows, S.D., Doyle, M.L., Murphy, K.P., Franklin, S.G., White, J.R., Brooks, I., McNulty, D.E., Scott, M.O., Knutson, J.R., Porter, D., Young, P.R. and Hensley, P. (1994). Determination of the monomer-dimer equilibrium of interleukin-8 reveals it is a monomer at physiological concentrations. *Biochemistry* **33**, 12741–12745.

Bussolino, F., Sironi, M., Bocchietto, E. and Mantovani, A. (1992). Synthesis of platelet-activating factor by polymorphonuclear neutrophils stimulated with interleukin-8. *J. Biol. Chem.* **267**, 14598–14603.

Cacalano, G., Lee, J., Kikly, K., Ryan, A.M., Pitts-Meek, S., Hultgren, B., Wood, W.I. and Moore, M.W. (1994). Neutrophil and B cell expansion in mice that lack the murine IL-8 receptor homolog. *Science* **265**, 682–684.

Cao, Y., Chen, C., Weatherbee, J.A., Tsang, M. and Folkman, J. (1995). Gro-β, a C-X-C chemokine, is an angiogenesis inhibitor that suppresses the growth of Lewis lung carcinoma in mice. *J. Exp. Med.* **182**, 2069–2077.

Carré, P.C., Mortenson, R.L., King, T.E., Noble, P.W., Sable, C.L. and Riches, D.W. (1991). Increased expression of the interleukin-8 gene by alveolar macrophages in idiopathic pulmonary fibrosis. A potential mechanism for the recruitment and activation of neutrophils in lung fibrosis. *J. Clin. Invest.* **88**, 1802–1810.

Carveth, H.J., Bohnsack, J.F., McIntyre, T.M., Baggiolini, M., Prescott, S.M. and Zimmerman, G.A. (1989). Neutrophil activating factor (NAF) induces polymorphonuclear leukocyte adherence to endothelial cells and to subendothelial matrix proteins. *Biochem. Biophys. Res. Commun.* **162**, 387–393.

Cassatella, M.A., Guasparri, I., Ceska, M., Bazzoni, F. and Rossi, F. (1993). Interferon-gamma inhibits interleukin-8 production by human polymorphonuclear leucocytes. *Immunology* **78**, 177–184.

Cassatella, M.A., Gasperini, S., Calzetti, F., Bertagnin, A., Luster, A.D. and McDonald, P.P. (1997). Regulated production of the interferon-γ-inducible protein-10 (IP-10) chemokine by human neutrophils. *Eur. J. Immunol.* **27**, 111–115.

Chang, M.-S., McNinch, J., Basu, R. and Simonet, S. (1994). Cloning and characterization of the human neutrophil-activating peptide (ENA-78) gene. *J. Biol. Chem.* **269**, 25 277–25 282.

Chuntharapai, A. and Kim, K.J. (1995). Regulation of the expression of IL-8 receptor A/B by IL-8: possible functions of each receptor. *J. Immunol.* **155**, 2587–2594.

Chuntharapai, A., Lee, J., Hébert, C.A. and Kim, K.J. (1994). Monoclonal antibodies detect different distribution patterns of IL-8 receptor A and IL-8 receptor B on human peripheral blood leukocytes. *J. Immunol.* **153**, 5682–5688.

Clark-Lewis, I., Schumacher, C., Baggiolini, M. and Moser, B. (1991). Structure–activity relationships of interleukin-8 determined using chemically synthesized analogs. Critical role of the NH_2-terminal residues and evidence for uncoupling of neutrophil chemotaxis, exocytosis, and receptor binding activities. *J. Biol. Chem.* **266**, 23 128–23 134.

Clark-Lewis, I., Dewald, B., Geiser, T., Moser, B. and Baggiolini, M. (1993). Platelet factor 4 binds to interleukin 8 receptors and activates neutrophils when its *N* terminus is modified with Glu–Leu–Arg. *Proc. Natl Acad. Sci. U.S.A.* **90**, 3574–3577.

Clark-Lewis, I., Dewald, B., Loetscher, M., Moser, B. and Baggiolini, M. (1994). Structural requirements for interleukin-8 function identified by design of analogs and CXC chemokine hybrids. *J. Biol. Chem.* **269**, 16 075–16 081.

Clark-Lewis, I., Kim, K.-S., Rajarathnam, K., Gong, J.-H., Dewald, B., Moser, B., Baggiolini, M. and Sykes, B.D. (1995). Structure–activity relationships of chemokines. *J. Leukocyte Biol.* **57**, 703–711.

Clore, G.M. and Gronenborn, A.M. (1991). Comparison of the solution nuclear magnetic resonance and crystal structures of interleukin-8. Possible implications for the mechanism of receptor binding. *J. Mol. Biol.* **217**, 611–620.

Colditz, I. and Watson, D.L. (1992). The effect of cytokines and chemotactic agonists on the migration of T lymphocytes into skin. *Immunology* **76**, 272–278.

Colditz, I., Zwahlen, R., Dewald, B. and Baggiolini, M. (1989). *In vivo* inflammatory activity of neutrophil-activating factor, a novel chemotactic peptide derived from human monocytes. *Am. J. Pathol.* **134**, 755–760.

Colletti, L.M., Kunkel, S.L., Walz, A., Burdick, M.D., Kunkel, R.G., Wilke, C.A. and Strieter, R.M. (1995). Chemokine expression during hepatic ischemia/reperfusion-induced lung injury in the rat. The role of epithelial neutrophil activating protein. *J. Clin. Invest.* **95**, 134–141.

Colotta, F., Re, F., Muzio, M., Bertini, R., Polentarutti, N., Sironi, M., Giri, J.G., Dower, S.K., Sims, J.E. and Mantovani, A. (1993). Interleukin-1 type II receptor: a decoy target for IL-1 that is regulated by IL-4. *Science* **261**, 472–475.

Colotta, F., Orlando, S., Fadlon, E.J., Sozzani, S., Matteucci, C. and Mantovani, A. (1995). Chemoattractants induce rapid release of the interleukin 1 type II decoy receptor in human polymorphonuclear cells. *J. Exp. Med.* **181**, 2181–2188.

Corbett, M.S., Schmitt, I., Riess, O. and Walz, A. (1994). Characterization of the gene for human neutrophil-activating peptide 78 (ENA-78). *Biochem. Biophys. Res. Commun.* **205**, 612–617.

Dahinden, C.A., Kurimoto, Y., de Weck, A.L., Lindley, I., Dewald, B. and Baggiolini, M. (1989). The neutrophil-activating peptide NAF/NAP-1 induces histamine and leukotriene release by interleukin 3-primed basophils. *J. Exp. Med.* **170**, 1787–1792.

Damaj, B.B., McColl, S.R., Neote, K., Songqing, N., Ogborn, K.T., Hébert, C.A. and Naccache, P.H. (1996). Identification of G-protein binding sites of the human interleukin-8 receptors by functional mapping of the intracellular loops. *FASEB J.* **10**, 1426–1434.

Deleuran, B., Lemche, P., Kristensen, M., Chu, C.Q., Field, M., Jensen, J., Matsushima, K. and Stengaard-Pedersen, K. (1994). Localisation of interleukin 8 in the synovial membrane, cartilage–pannus junction and chondrocytes in rheumatoid arthritis. *Scand. J. Rheumatol.* **23**, 2–7.

Detmers, P.A., Lo, S.K., Olsen-Egbert, E., Walz, A., Baggiolini, M. and Cohn, Z.A. (1990). Neutrophil-activating protein 1/interleukin 8 stimulates the binding activity of the leukocyte adhesion receptor CD11b/CD18 on human neutrophils. *J. Exp. Med.* **171**, 1155–1162.

Detmers, P.A., Powell, D.E., Walz, A., Clark-Lewis, I., Baggiolini, M. and Cohn, Z.A. (1991). Differential effects of neutrophil-activating peptide 1/IL-8 and its homologues on leukocyte adhesion and phagocytosis. *J. Immunol.* **147**, 4211–4217.

Deuel, T.F., Keim, P.S., Farmer, M. and Heinrikson, R.L. (1977). Amino acid sequence of human platelet factor 4. *Proc. Natl Acad. Sci. U.S.A.* **74**, 2256–2258.

Deuel, T.F., Senior, R.M., Chang, D., Griffin, G.L., Heinrikson, R.L. and Kaiser, E.T. (1981). Platelet factor 4 is chemotactic for neutrophils and monocytes. *Proc. Natl Acad. Sci. U.S.A.* **78**, 4584–4587.

Djeu, J.Y., Matsushima, K., Oppenheim, J.J., Shiotsuki, K. and Blanchard, D.K. (1990). Functional activation of human neutrophils by recombinant monocyte-derived neutrophil chemotactic factor/IL-8. *J. Immunol.* **144**, 2205–2210.

Doranz, B.J., Berson, J.F., Rucker, J. and Doms, R.W. (1997). Chemokine receptors as fusion cofactors for human immunodeficiency virus type 1 (HIV-1). *Immunol. Res.* **16**, 15–28.

D'Souza, M.P. and Harden, V.A. (1996). Chemokines and HIV-1 second receptors. Confluence of two fields generates optimism in AIDS research. *Nature Medicine* **2**, 1293–1300.

Dunlevy, J.R. and Couchman, J.R. (1995). Interleukin-8 induces motile behavior and loss of focal adhesions in primary fibroblasts. *J. Cell Sci.* **108**, 311–321.

Ehlert, J.E., Petersen, F., Kubbutat, M.H.G., Gerdes, J., Flad, H.-D. and Brandt, E. (1995). Limited and defined truncation at the *C* terminus enhances receptor binding and degranulation activity of the neutrophil-activating peptide 2 (NAP-2). *J. Biol. Chem.* **270**, 6338–6344.

Eisman, R., Surrey, S., Ramachandran, B., Schwartz, E. and Poncz, M. (1990). Structural and functional comparison of the genes for human platelet factor 4 and PF4alt. *Blood* **76**, 336–344.

Elder, J.T., Hammerberg, C., Cooper, K.D., Kojima, T., Nair, R.P., Ellis, C.N. and Voorhees, J.J. (1993). Cyclosporin A rapidly inhibits epidermal cytokine expression in psoriasis lesions, but not in cytokine-stimulated cultured keratinocytes. *J. Invest. Dermatol.* **101**, 761–766.

Fairbrother, W.J., Reilly, D., Colby, T.J., Hesselgesser, J. and Horuk, R. (1994). The solution structure of melanoma growth stimulating activity. *J. Mol. Biol.* **242**, 252–270.

Farber, J.M. (1993). HuMIG: a new human member of the chemokine family of cytokines. *Biochem. Biophys. Res. Commun.* **192**, 223–230.

Federsppiel, B., Melhado, I.G., Duncan, A.M.V., Delaney, A., Schappert, K., Clark-Lewis, I. and Jirik, F.R. (1993). Molecular cloning of the cDNA and chromosomal localization of the gene for a putative seven-transmembrane segment (7-TMS) receptor isolated from human spleen. *Genomics* **16**, 707–712.

Feng, L., Xia, Y., Yoshimura, T. and Wilson, C.B. (1995). Modulation of neutrophil influx in glomerulonephritis in the rat with anti-macrophage inflammatory protein-2 (MIP-2) antibody. *J. Clin. Invest.* **95**, 1009–1017.

Feng, Y., Broder, C.C., Kennedy, P.E. and Berger, E.A. (1996). HIV-1 entry cofactor: functional cDNA cloning of a seven-transmembrane, G protein-coupled receptor. *Science* **272**, 872–877.

Forrest, M.J., Eiermann, G.J., Meurer, R., Walakovits, L.A. and MacIntyre, D.E. (1992). The role of CD18 in IL-8 induced dermal and synovial inflammation. *Br. J. Pharmacol.* **106**, 287–294.

Foster, S.J., Aked, D.M., Schroder, J.M. and Christophers, E. (1989). Acute inflammatory effects of a monocyte-derived neutrophil-activating peptide in rabbit skin. *Immunology* **67**, 181–183.

Friedland, J.S., Suputtamongkol, Y., Remick, D.G., Chaowagul, W., Strieter, R.M., Kunkel, S.L., White, N.J. and Griffin, G.E. (1992). Prolonged elevation of interleukin-8 and interleukin-6 concentrations in plasma and of leukocyte interleukin-8 mRNA levels during septicemic and localized *Pseudomonas pseudomallei* infection. *Infect. Immun.* **60**, 2402–2408.

Froyen, G., Proost, P., Ronsse, I., Mitera, T., Haelens, A., Wuyts, A., Opdenakker, G., Van Damme, J. and Billiau, A. (1997). Cloning, bacterial expression and biological characterization of recombinant human granulocyte chemotactic protein-2 and differential expression of granulocyte chemotactic protein-2 and epithelial cell-derived neutrophil activating peptide-78 mRNAs. *Eur. J. Biochem.* **243**, 762–769.

Gattass, C.R., King, L.B., Luster, A.D. and Ashwell, J.D. (1994). Constitutive expression of interferon γ-inducible protein 10 in lymphoid organs and inducible expression in T cells and thymocytes. *J. Exp. Med.* **179**, 1373–1378.

Gayle, R.B. III, Sleath, P.R., Srinivason, S., Birks, C.W., Weerawarna, K.S., Cerretti, D.P., Kozlosky, C.J., Nelson, N., Vanden Bos, T. and Beckmann, M.P. (1993). Importance of the amino terminus of the interleukin-8 receptor in ligand interactions. *J. Biol. Chem.* **268**, 7283–7289.

Geiser, T., Dewald, B., Ehrengruber, M.U., Clark-Lewis, I. and Baggiolini, M. (1993). The interleukin-8-related chemotactic cytokines GROα, GROβ, and GROγ activate human neutrophil and basophil leukocytes. *J. Biol. Chem.* **268**, 15 419–15 424.

Gesser, B., Deleuran, B., Lund, M., Vestergård, C., Lohse, N., Deleuran, M., Jensen, S.L., Pedersen, S.S., Thestrup-Pedersen, K. and Larsen, C.G. (1995). Interleukin-8 induces its own production in CD4$^+$ T lymphocytes: a process regulated by interleukin 10. *Biochem. Biophys. Res. Commun.* **210**, 660–669.

Gesser, B., Lund, M., Lohse, N., Vestergaard, C., Matsushima, K., Sindet-Pedersen, S., Jensen, S.L., Thestrup-

Pedersen, K. and Larsen, C.G. (1996). IL-8 induces T cell chemotaxis, suppresses IL-4, and up-regulates IL-8 production by $CD4^+$ T cells. *J. Leukocyte Biol.* **59**, 407–411.

Gewirtz, A.M., Calabretta, B., Rucinski, B., Niewiarowski, S. and Xu, W. (1989). Inhibition of human megakaryocytopoiesis *in vitro* by platelet factor 4 (PF4) and a synthetic COOH-terminal PF4 peptide. *J. Clin. Invest.* **83**, 1477–1486.

Gewirtz, A.M., Zhang, J., Ratajczak, J., Ratajczak, M., Park, K.S., Li, C., Yan, Z. and Poncz, M. (1995). Chemokine regulation of human megakaryocytopoiesis. *Blood* **86**, 2559–2567.

Gillitzer, R., Ritter, U., Spandau, U., Goebeler, M. and Bröcker, E.-B. (1996). Differential expression of GRO-α and IL-8 mRNA in psoriasis: a model for neutrophil migration and accumulation *in vivo*. *J. Invest. Dermatol.* **107**, 778–782.

Gottlieb, A.B., Luster, A.D., Posnett, D.N. and Carter, D.M. (1988). Detection of a γ interferon-induced protein IP-10 in psoriatic plaques. *J. Exp. Med.* **168**, 941–948.

Green, S.P., Chuntharapai, A. and Curnutte, J.T. (1996). Interleukin-8 (IL-8), melanoma growth-stimulatory activity, and neutrophil-activating peptide selectively mediate priming of the neutrophil NADPH oxidase through the type A or type B IL-8 receptor. *J. Biol. Chem.* **271**, 25 400–25 405.

Grob, P.M., David, E., Warren, T.C., DeLeon, R.P., Farina, P.R. and Homon, C.A. (1990). Characterization of a receptor for human monocyte-derived neutrophil chemotactic factor/interleukin-8. *J. Biol. Chem.* **265**, 8311–8316.

Gupta, S.K., Hassel, T. and Singh, J.P. (1995). A potent inhibitor of endothelial cell proliferation is generated by proteolytic cleavage of the chemokine platelet factor 4. *Proc. Natl Acad. Sci. U.S.A.* **92**, 7799–7803.

Hack, C.E., Hart, M., van Schijndel, R.J., Eerenberg, A.J., Nuijens, J.H., Thijs, L.G. and Aarden, L.A. (1992). Interleukin-8 in sepsis: relation to shock and inflammatory mediators. *Infect. Immun.* **60**, 2835–2842.

Hammond, M.E.W., Lapointe, G.R., Feucht, P.H., Hilt, S., Gallegos, C.A., Gordon, C.A., Giedlin, M.A., Mullenbach, G. and Tekamp-Olson, P. (1995). IL-8 induces neutrophil chemotaxis predominantly via type I IL-8 receptors. *J. Immunol.* **155**, 1428–1433.

Han, Z.C., Maurer, A.M., Bellucci, S., Wan, H., Kroviarski, Y., Bertrand, O. and Caen, J.P. (1992). Inhibitory effect of platelet factor 4 (PF4) on the growth of human erythroleukemia cells: proposed mechanism of action of PF4. *J. Lab. Clin. Med.* **120**, 645–660.

Haskill, S., Peace, A., Morris, J., Sporn, S.A., Anisowicz, A., Lee, S.W., Smith, T., Martin, G., Ralph, P. and Sager, R. (1990). Identification of three related human GRO genes encoding cytokine functions. *Proc. Natl Acad. Sci. U.S.A.* **87**, 7732–7736.

Hébert, C.A., Luscinskas, F.W., Kiely, J.-M., Luis, E.A., Darbonne, W.C., Bennett, G.L., Liu, C.C., Obin, M.S., Gimbrone, M.A. and Baker, J.B. (1990). Endothelial and leukocyte forms of IL-8. Conversion by thrombin and interactions with neutrophils. *J. Immunol.* **145**, 3033–3040.

Hébert, C.A., Chuntharapai, A., Smith, M., Colby, T., Kim, J. and Horuk, R. (1993). Partial functional mapping of the human interleukin-8 type A receptor. Identification of a major ligand binding domain. *J. Biol. Chem.* **268**, 18 549–18 553.

Hesselgesser, J., Chitnis, C.E., Miller, L.H., Yansura, D.G., Simmons, L.C., Fairbrother, W.J., Kotts, C., Wirth, C., Gillece-Castro, B.L. and Horuk, R. (1995). A mutant of melanoma growth stimulating activity does not activate neutrophils but blocks erythrocyte invasion by malaria. *J. Biol. Chem.* **270**, 11 472–11 476.

Holmes, W.E., Lee, J., Kuang, W.-J., Rice, G.C. and Wood, W.I. (1991). Structure and functional expression of a human interleukin-8 receptor. *Science* **253**, 1278–1280.

Holt, J.C., Harris, M.E., Holt, A.M., Lange, E., Henschen, A. and Niewiarowski, S. (1986). Characterization of human platelet basic protein, a precursor form of low-affinity platelet factor 4 and beta-thromboglobulin. *Biochemistry* **25**, 1988–1996.

Horuk, R. (1994). The interleukin-8-receptor family: from chemokines to malaria. *Immunol. Today* **15**, 169–174.

Horuk, R., Yansura, D.G., Reilly, D., Spencer, S., Bourell, J., Henzel, W., Rice, G. and Unemori, E. (1993). Purification, receptor binding analysis, and biological characterization of human melanoma growth stimulating activity (MGSA). Evidence for a novel MGSA receptor. *J. Biol. Chem.* **268**, 541–546.

Horuk, R., Martin, A., Hesselgesser, J., Hadley, T., Lu, Z.-H., Wang, Z.-X. and Peiper, S.C. (1996). The duffy antigen receptor for chemokines: structural analysis and expression in the brain. *J. Leukocyte Biol.* **59**, 29–38.

Huber, A.R., Kunkel, S.L., Todd, R.F. and Weiss, S.J. (1991). Regulation of transendothelial neutrophil migration by endogenous interleukin-8. *Science* **254**, 99–102.

Iida, N., Haisa, M., Igarashi, A., Pencev, D. and Grotendorst, G.R. (1996). Leukocyte-derived growth factor links the PDGF and CXC chemokine families of peptides. *FASEB J.* **10**, 1336–1345.

Jagels, M.A. and Hugli, T.E. (1992). Neutrophil chemotactic factors promote leukocytosis. A common mechanism for cellular recruitment from bone marrow. *J. Immunol.* **148**, 1119–1128.

Jazin, E.E., Yoo, H., Blomqvist, A.G., Yee, F., Weng, G., Walker, M.W., Salon, J., Larhammar, D. and Wahlestedt, C. (1993). A proposed bovine neuropeptide Y (NPY) receptor cDNA clone, or its human homologue, confers neither NPY binding sites nor NPY responsiveness on transfected cells. *Regul. Pept.* **47**, 247–258.

Jinquan, T., Frydenberg, J., Mukaida, N., Bonde, J., Larsen, C.G., Matsushima, K. and Thestrup-Pedersen, K. (1995). Recombinant human growth-regulated oncogene-α induces T lymphocyte chemotaxis. *J. Immunol.* **155**, 5359–5368.

Jinquan, T., Moller, B., Storgaard, M., Mukaida, N., Bonde, J., Grunnet, N., Black, F.T., Larsen, C.G., Matsushima, K. and Thestrup-Pedersen, K. (1997). Chemotaxis and IL-8 receptor expression in B cells from normal and HIV-infected subjects. *J. Immunol.* **157**, 475–484.

Jones, S.A., Wolf, M., Qin, S., Mackay, C.R. and Baggiolini, M. (1996). Different functions for the interleukin 8 receptors (IL-8R) of human neutrophil leukocytes: NADPH oxidase and phospholipase D are activated through IL-8R1 but not IL-8R2. *Proc. Natl Acad. Sci. U.S.A.* **93**, 6682–6686.

Jorens, G.P., Van Damme, J., De Backer, W., Bossaert, L., De Jongh, R.F., Herman, A.G. and Rampart, M. (1992). Interleukin 8 (IL-8) in the bronchoalveolar lavage fluid from patients with the adult respiratory distress syndrome (ARDS) and patients at risk for ARDS. *Cytokine* **4**, 592–597.

Kaplan, G., Luster, A.D., Hancock, G. and Cohn, Z.A. (1987). The expression of a γ interferon-induced protein (IP-10) in delayed immune responses in human skin. *J. Exp. Med.* **166**, 1098–1108.

Kelvin, D.J., Michiel, D.F., Johnston, J.A., Lloyd, A.R., Sprenger, H., Oppenheim, J.J. and Wang, J.-M. (1993). Chemokines and serpentines: the molecular biology of chemokine receptors. *J. Leukocyte Biol.* **54**, 604–612.

Kernen, P., Wymann, M.P., von Tscharner, V., Deranleau, D.A., Tai, P.C., Spry, C.J., Dahinden, C.A. and Baggiolini, M. (1991). Shape changes, exocytosis, and cytosolic free calcium changes in stimulated human eosinophils. *J. Clin. Invest.* **87**, 2012–2017.

Kim, K.-S., Clark-Lewis, I. and Sykes, B.D. (1994). Solution structure of GRO/melanoma growth stimulatory activity determined by ^{1}H NMR spectroscopy. *J. Biol. Chem.* **269**, 32 909–32 915.

Kimata, H. and Lindley, I. (1994). Interleukin-8 differentially modulates interleukin-4- and interleukin-2-induced human B cell growth. *Eur. J. Immunol.* **24**, 3237–3240.

Kimata, H., Yoshida, A., Ishioka, C., Lindley, I. and Mikawa, H. (1992). Interleukin-8 (IL-8) selectively inhibits immunoglobulin E production induced by IL-4 in human B cells. *J. Exp. Med.* **176**, 1227–1231.

Koch, A.E., Kunkel, S.L., Burrows, J.C., Evanoff, H.L., Haines, G.K., Pope, R.M. and Strieter, R.M. (1991). Synovial tissue macrophage as a source of the chemotactic cytokine IL-8. *J. Immunol.* **147**, 2187–2195.

Koch, A.E., Polverini, P.J., Kunkel, S.L., Harlow, L.A., DiPietro, L.A., Elner, V.M., Elner, S.G. and Strieter, R.M. (1992). Interleukin-8 as a macrophage-derived mediator of angiogenesis. *Science* **258**, 1798–1801.

Koch, A.E., Kunkel, S.L., Harlow, L.A., Mazarakis, D.D., Haines, G.K., Burdick, M.D., Pope, R.M., Walz, A. and Strieter, R.M. (1994). Epithelial neutrophil activating peptide-78: a novel chemotactic cytokine for neutrophils in arthritis. *J. Clin. Invest.* **94**, 1012–1018.

Koch, A.E., Kunkel, S.L., Shah, M.R., Hosaka, S., Halloran, M.M., Haines, G.K., Burdick, M.D., Pope, R.M. and Strieter, R.M. (1995). Growth-related gene product α. A chemotactic cytokine for neutrophils in rheumatoid arthritis. *J. Immunol.* **155**, 3660–3666.

Kojima, T., Cromie, M.A., Fisher, G.J., Voorhees, J.J. and Elder, J.T. (1993). GRO-α mRNA is selectively overexpressed in psoriatic epidermis and is reduced by cyclosporin A *in vivo*, but not in cultured keratinocytes. *J. Invest. Dermatol.* **101**, 767–772.

Krieger, M., Brunner, T., Bischoff, S.C., von Tscharner, V., Walz, A., Moser, B., Baggiolini, M. and Dahinden, C.A. (1992). Activation of human basophils through the IL-8 receptor. *J. Immunol.* **149**, 2662–2667.

Kukielka, G.L., Smith, C.W., LaRosa, G.J., Manning, A.M., Mendoza, L.H., Daly, T.J., Hughes, B.J., Youker, K.A., Hawkins, H.K., Michael, L.H., Rot, A. and Entman, M.L. (1995). Interleukin-8 gene induction in the myocardium after ischemia and reperfusion *in vivo*. *J. Clin. Invest.* **95**, 89–103.

Kuna, P., Reddigari, S.R., Kornfeld, D. and Kaplan, A.P. (1991). IL-8 inhibits histamine release from human basophils induced by histamine-releasing factors, connective tissue activating peptide III, and IL-3. *J. Immunol.* **147**, 1920–1924.

Kunkel, S.L., Lukacks, N., Kasama, T. and Strieter, R.M. (1996). The role of chemokines in inflammatory joint disease. *J. Leukocyte Biol.* **58**, 6–12.

Lacy, M., Jones, J., Whittemore, S.R., Haviland, D.L., Wetsel, R.A. and Barnum, S.R. (1995). Expression of the receptors for the C5a anaphylatoxin, interleukin-8 and fMLP by human astrocytes and microglia. *J. Neuroimmunol.* **61**, 71–78.

LaRosa, G.J., Thomas, K.M., Kaufmann, M.E., Mark, R., White, M., Taylor, L., Gray, G., Witt, D. and Navarro, J. (1992). Amino terminus of the interleukin-8 receptor is a major determinant of receptor subtype specificity. *J. Biol. Chem.* **267**, 25 402–25 406.

Larsen, C.G., Anderson, A.O., Appella, E., Oppenheim, J.J. and Matsushima, K. (1989). The neutrophil-activating protein (NAP-1) is also chemotactic for T lymphocytes. *Science* **243**, 1464–1466.

Larsen, C.G., Thomsen, M.K., Gesser, B., Thomsen, P.D., Deleuran, B.W., Nowak, J., Skodt, V., Thomsen, H.K., Deleuran, M., Thestrup-Pedersen, K., Harada, A., Matsushima, K. and Menné, T. (1995). The delayed-type hypersensitivity reaction is dependent on IL-8. Inhibition of a tuberculin skin reaction by an anti-IL-8 monoclonal antibody. *J. Immunol.* **155**, 2151–2157.

Laterveer, L., Lindley, I.J.D., Hamilton, M.S., Willemze, R. and Fibbe, W.E. (1995). Interleukin-8 induces rapid mobilization of hematopoietic stem cells wlth radioprotective capacity and long-term myelolymphoid repopulating ability. *Blood* **85**, 2269–2275.

Lee, H.-H. and Farber, J.M. (1996). Localization of the gene for the human MIG cytokine on chromosome 4q21 adjacent to INP10 reveals a chemokine 'mini-cluster'. *Cytogenet Cell Genet.* **74**, 255–258.

Lee, J., Horuk, R., Rice, G.C., Bennett, G.L., Camerato, T. and Wood, W.I. (1992). Characterization of two high affinity human interleukin-8 receptors. *J. Biol. Chem.* **267**, 16 283–16 287.

Lemster, B.H., Carroll, P.B., Rilo, H.R., Johnson, N., Nikaein, A. and Thomson, A.W. (1995). IL-8/IL-8 receptor expression in psoriasis and the response to systemic tacrolimus (FK506) therapy. *Clin. Exp. Immunol.* **99**, 148–154.

Leonard, E.J., Yoshimura, T., Tanaka, S. and Raffeld, M. (1991). Neutrophil recruitment by intradermally injected neutrophil attractant/activation protein-1. *J. Invest. Dermatol.* **96**, 690–694.

Leong, S.R., Kabakoff, R.C. and Hébert, C.A. (1994). Complete mutagenesis of the extracellular domain of interleukin-8 (IL-8) type A receptor identifies charged residues mediating IL-8 binding and signal transduction. *J. Biol. Chem.* **269**, 19 343–19 348.

Liao, F., Rabin, R.L., Yannelli, J.R., Koniaris, L.G., Vanguri, P. and Farber, J.M. (1995). Human Mig chemokine: biochemical and functional characterization. *J. Exp. Med.* **182**, 1301–1314.

Lira, S.A., Zalamea, P., Heinrich, J.N., Fuentes, M.E., Carrasco, D., Lewin, A.C., Barton, D.S., Durham, S. and Bravo, R. (1994). Expression of the chemokine N51/KC in the thymus and epidermis of transgenic mice results in marked infiltration of a single class of inflammatory cells. *J. Exp. Med.* **180**, 2039–2048.

Lloyd, A.R., Biragyn, A., Johnston, J.A., Taub, D.D., Xu, L., Michiel, D., Sprenger, H., Oppenheim, J.J. and Kelvin, D.J. (1995). Granulocyte-colony stimulating factor and lipopolysaccharide regulate the expression of interleukin 8 receptors on polymorphonuclear leukocytes. *J. Biol. Chem.* **270**, 28 188–28 192.

Lloyd, A.R., Oppenheim, J.J., Kelvin, D.J. and Taub, D.D. (1996). Chemokines regulate T cell adherence to recombinant adhesion molecules and extracellular matrix proteins. *J. Immunol.* **156**, 932–938.

Loetscher, M., Geiser, T., O'Reilly, T., Zwahlen, R., Baggiolini, M. and Moser, B. (1994). Cloning of a human seven-transmembrane domain receptor, LESTR, that is highly expressed in leukocytes. *J. Biol. Chem.* **269**, 232–237.

Loetscher, M., Gerber, B., Loetscher, P., Jones, S.A., Piali, L., Clark-Lewis, I., Baggiolini, M. and Moser, B. (1996). Chemokine receptor specific for IP10 and Mig: structure, function, and expression in activated T-lymphocytes. *J. Exp. Med.* **184**, 963–969.

Loetscher, P., Seitz, M., Clark-Lewis, I., Baggiolini, M. and Moser, B. (1994). Both interleukin-8 receptors independently mediate chemotaxis. Jurkat cells transfected with IL-8R1 or IL-8R2 migrate in response to IL-8, GROα and NAP-2. *FEBS Lett.* **341**, 187–192.

Loetscher, P., Seitz, M., Clark-Lewis, I., Baggiolini, M. and Moser, B. (1996). Activation of NK cells by CC chemokines. Chemotaxis, Ca^{2+} mobilization, and enzyme release. *J. Immunol.* **156**, 322–327.

Luster, A.D. and Leder, P. (1993). IP-10, a -C-X-C- chemokine, elicits a potent thymus-dependent antitumor response *in vivo*. *J. Exp. Med.* **178**, 1057–1065.

Luster, A.D. and Ravetch, J.V. (1987a). Genomic characterization of a gamma-interferon-inducible gene (IP-10) and identification of an interferon-inducible hypersensitive site. *Mol. Cell. Biol.* **7**, 3723–3731.

Luster, A.D. and Ravetch, J.V. (1987b). Biochemical characterization of a γ interferon-inducible cytokine (IP-10). *J. Exp. Med.* **166**, 1084–1097.

Luster, A.D., Unkeless, J.C. and Ravetch, J.V. (1985). γ-Interferon transcriptionally regulates an early-response gene containing homology to platelet proteins. *Nature* **315**, 672–676.

Luster, A.D., Greenberg, S.M. and Leder, P. (1995). The IP-10 chemokine binds to a specific cell surface heparan sulfate site shared with platelet factor 4 and inhibits endothelial cell proliferation. *J. Exp. Med.* **182**, 219–231.

Lynch, J.P., Standiford, T.J., Rolfe, M.W., Kunkel, S.L. and Strieter, R.M. (1992). Neutrophilic alveolitis in idiopathic pulmonary fibrosis. The role of interleukin-8. *Am. Rev. Respir. Dis.* **145**, 1433–1439.

Maione, T.E., Gray, G.S., Petro, J., Hunt, A.J., Donner, A.L., Bauer, S.I., Carson, H.F. and Sharpe, R.J. (1990). Inhibition of angiogenesis by recombinant human platelet factor-4 and related peptides. *Science* **247**, 77–79.

Majumdar, S., Gonder, D., Koutsis, B. and Poncz, M. (1991). Characterization of the human beta-thromboglobulin gene. Comparison with the gene for platelet factor 4. *J. Biol. Chem.* **266**, 5785–5789.

Manna, S.K. and Samanta, A.K. (1995). Upregulation of interleukin-8 receptor in human polymorphonuclear neutrophils by formyl peptide and lipopolysaccharide. *FEBS Lett.* **367**, 117–121.

Manna, S.K., Bhattacharya, C., Gupta, S.K. and Samanta, A.K. (1995). Regulation of interleukin-8 receptor expression in human polymorphonuclear neutrophils. *Mol. Immunol.* **32**, 883–893.

Masure, S., Proost, P., Van Damme, J. and Opdenakker, G. (1991). Purification and identification of 91-kDa neutrophil gelatinase. Release by the activating peptide interleukin-8. *Eur. J. Biochem.* **198**, 391–398.

Matsukawa, A., Yoshimura, T., Maeda, T., Ohkawara, S., Takagi, K. and Yoshinaga, M. (1995). Neutrophil accumulation and activation by homologous IL-8 in rabbits. IL-8 induces destruction of cartilage and production of IL-1 and IL-1 receptor antagonist *in vivo*. *J. Immunol.* **154**, 5418–5425.

Mayo, K.H. and Chen. M.J. (1989). Human platelet factor 4 monomer–dimer–tetramer equilibria investigated by ^{1}H NMR spectroscopy. *Biochemistry* **28**, 9469–9478.

McCall, T.B., Palmer, R.M. and Moncada, S. (1992). Interleukin-8 inhibits the induction of nitric oxide synthase in rat peritoneal neutrophils. *Biochem. Biophys. Res. Commun.* **186**, 680–685.

Metzner, B., Barbisch, M., Parlow, F., Kownatzki, E., Schraufstatter, I. and Norgauer, J. (1995). Interleukin-8 and GROα prime human neutrophils for superoxide anion production and induce up-regulation of *N*-formyl peptide receptors. *J. Invest. Dermatol.* **104**, 789–791.

Michel, G., Kemény, L., Peter, R.U., Beetz, A., Ried, C., Arenberger, P. and Ruzicka, T. (1992). Interleukin-8 receptor-mediated chemotaxis of normal human epidermal cells. *FEBS Lett.* **305**, 241–243.

Miller, E.J., Cohen, A.B., Nagao, S., Griffith, D., Maunder, R.J., Martin, T.R., Weiner-Kronish, J.P., Sticherling, M., Christophers, E. and Matthay, M.A. (1992). Elevated levels of NAP-1/interleukin-8 are present in the airspaces of patients with the adult respiratory distress syndrome and are associated with increased mortality. *Am. Rev. Respir. Dis.* **146**, 427–432.

Mollereau, C., Muscatelli, F., Mattei, M.-G., Vassart, G. and Parmentier, M. (1993). The high-affinity interleukin 8 receptor gene (IL-8RA) maps to the 2q33–q36 region of the human genome: cloning of a pseudogene (IL8RBP) for the low-affinity receptor. *Genomics* **16**, 248–251.

Morohashi, H., Miyawaki, T., Nomura, H., Kuno, K., Murakami, S., Matsushima, K. and Mukaida, N. (1995). Expression of both types of human interleukin-8 receptors on mature neutrophils, monocytes, and natural killer cells. *J. Leukocyte Biol.* **57**, 180–187.

Morris, S.W., Nelson, N., Valentine, M.B., Shapiro, D.N., Look, A.T., Kozlosky, C.J., Beckmann, M.P. and Cerretti, D.P. (1992). Assignment of the genes encoding human interleukin-8 receptor types 1 and 2 and an interleukin-8 receptor pseudogene to chromosome 2q35. *Genomics* **14**, 685–691.

Moser, B., Schumacher, C., von Tscharner, V., Clark-Lewis, I. and Baggiolini, M. (1991). Neutrophil-activating peptide 2 and gro/melanoma growth-stimulatory activity interact with neutrophil-activating peptide 1/interleukin-8 receptors on human neutrophils. *J. Biol. Chem.* **266**, 10666–10671.

Moser, B., Barella, L., Mattei, S., Schumacher, C., Boulay, F., Colombo, M.P. and Baggiolini, M. (1993a). Expression of transcripts for two interleukin 8 receptors in human phagocytes, lymphocytes and melanoma cells. *Biochem. J.* **294**, 285–292.

Moser, B., Dewald, B., Barella, L., Schumacher, C., Baggiolini, M. and Clark-Lewis, I. (1993b). Interleukin-8 antagonists generated by *N*-terminal modification. *J. Biol. Chem.* **268**, 7125–7128.

Mueller, S.G., Schraw, W.P. and Richmond, A. (1994). Melanoma growth stimulatory activity enhances the phosphorylation of the class II interleukin-8 receptor in non-hematopoietic cells. *J. Biol. Chem.* **269**, 1973–1980.

Mukaida, N., Shiroo, M. and Matsushima, K. (1989). Genomic structure of the human monocyte-derived neutrophil chemotactic factor IL-8. *J. Immunol.* **143**, 1366–1371.

Mullenbach, G.T., Tabrizi, A., Blacher, R.W. and Steimer, K.S. (1986). Chemical synthesis and expression in yeast of a gene encoding connective tissue activating peptide-III. *J. Biol. Chem.* **261**, 719–722.

Murayama, T., Kuno, K., Jisaki, F., Obuchi, M., Sakamuro, D., Furukawa, T., Mukaida, N. and Matsushima, K. (1994). Enhancement of human cytomegalovirus replication in a human lung fibroblast cell line by interleukin-8. *J. Virol.* **68**, 7582–7585.

Murphy, P.M. (1996). Chemokine receptors: structure, function and role in microbial pathogenesis. *Cytokine and Growth Factor Reviews* **7**, 47–64.

Murphy, P.M. and Tiffany, H.L. (1991). Cloning of complementary DNA encoding a functional human interleukin-8 receptor. *Science* **253**, 1280–1283.

Nagasawa, T., Kikutani, H. and Kishimoto, T. (1994). Molecular cloning and structure of a pre-B-cell growth-stimulating factor. *Proc. Natl Acad. Sci. U.S.A.* **91**, 2305–2309.

Nagasawa, T., Hirota, S., Tachibana, K., Takakura, N., Nishikawa, S.-I., Kitamura, Y., Yoshida, N., Kikutani, H. and Kishimoto, T. (1996). Defects of B-cell lymphopoiesis and bone-marrow myelopoiesis in mice lacking the CXC chemokines PBSF/SDF-1. *Nature* **382**, 635–638.

Nakagawa, H., Hatakeyama, S., Ikesue, A. and Miyai, H. (1991). Generation of interleukin-8 by plasmin from AVLPR-interleukin-8, the human fibroblast-derived neutrophil chemotactic factor. *FEBS Lett.* **282**, 412–414.

Nibbering, P.H., Pos. O., Stevenhagen, A. and Van Furth, R. (1993). Interleukin-8 enhances nonoxidative intracellular killing of *Mycobacterium fortuitum* by human granulocytes. *Infect. Immun.* **61**, 3111–3116.

Norgauer, J., Metzner, B. and Schraufstätter, I. (1996). Expression and growth-promoting function of the IL-8 receptor β in human melanoma cells. *J. Immunol.* **156**, 1132–1137.

Nourshargh, S., Perkins, J.A., Showell, H.J., Matsushima, K., Williams, T.J. and Collins, P.D. (1992). A comparative study of the neutrophil stimulatory activity *in vitro* and pro-inflammatory properties *in vivo* of 72 amino acid and 77 amino acid IL-8. *J. Immunol.* **148**, 106–111.

Oberlin, E., Amara, A., Bachelerie, F., Bessia, C., Virelizier, J.-L., Arenzana-Seisdedos, F., Schwartz, O., Heard, J.-M., Clark-Lewis, I., Legler, D.F., Loetscher, M., Baggiolini, M. and Moser, B. (1996). The CXC chemokine SDF-1 is the ligand for LESTR/fusin and prevents infection by T-cell-line-adapted HIV-1. *Nature* **382**, 833–835.

Oliveira, I.C., Sciavolino, P.J., Lee, T.H. and Vilcek, J. (1992). Downregulation of interleukin 8 gene expression in human fibroblasts: unique mechanism of transcriptional inhibition by interferon. *Proc. Natl Acad. Sci. U.S.A.* **89**, 9049–9053.

Opdenakker, G. and Van Damme, J. (1992). Chemotactic factors, passive invasion and metastasis of cancer cells. *Immunol. Today* **13**, 463–464.

Oppenheim, J.J., Zachariae, C.O.C., Mukaida, N. and Matsushima, K. (1991). Properties of the novel proinflammatory supergene 'intercrine' cytokine family. *Annu. Rev. Immunol.* **9**, 617–648.

Padrines, M., Wolf, M., Walz, A. and Baggiolini, M. (1994). Interleukin-8 processing by neutrophil elastase, cathepsin G and proteinase-3. *FEBS Lett.* **352**, 231–235.

Paolini, J.F., Willard, D., Consler, T., Luther, M. and Krangel, M.S. (1994). The chemokines IL-8, monocyte chemoattractant protein-1, and I-309 are monomers at physiologically relevant concentrations. *J. Immunol.* **153**, 2704–2717.

Petersen, F., Flad, H.-D. and Brandt, E. (1994). Neutrophil-activating peptides NAP-2 and IL-8 bind to the same sites on neutrophils but interact in different ways. *J. Immunol.* **152**, 2467–2478.

Petersen, F., Ludwig, A., Flad, H.-D. and Brandt, E. (1996). TNF-α renders human neutrophils responsive to platelet factor 4. Comparison of PF-4 and IL-8 reveals different activity profiles of the two chemokines. *J. Immunol.* **156**, 1954–1962.

Peveri, P., Walz, A., Dewald, B. and Baggiolini, M. (1988). A novel neutrophil-activating factor produced by human mononuclear phagocytes. *J. Exp. Med.* **167**, 1547–1559.

Plata-Salamàn, C.R. and Borkoski, J.P. (1993). Interleukin-8 modulates feeding by direct action in the central nervous system. *Am. J. Physiol.* **265**, R877–R882.

Poncz, M., Surrey, S., LaRocco, P., Weiss, M.J., Rappaport, E.F., Conway, T.M. and Schwartz, E. (1987). Cloning and characterization of platelet factor 4 cDNA derived from a human erythroleukemic cell line. *Blood* **69**, 219–223.

Power, C.A., Clemetson, J.M., Clemetson, K.J. and Wells, T.N.C. (1995). Chemokine and chemokine receptor mRNA expression in human platelets. *Cytokine* **7**, 479–482.

Premack, B.A. and Schall, T.J. (1996). Chemokine receptors: gateways to inflammation and infection. *Nature Medicine* **2**, 1174–1178.

Proost, P., De Wolf-Peeters, C., Conings, R., Opdenakker, G., Billiau, A. and Van Damme, J. (1993a). Identification of a novel granulocyte chemotactic protein (GCP-2) from human tumor cells. *In vitro* and *in vivo* comparison with natural forms of GRO, IP-10 and IL-8. *J. Immunol.* **150**, 1000–1010.

Proost, P., Wuyts, A., Conings, R., Lenaerts, J.-P., Billiau, A., Opdenakker, G. and Van Damme, J. (1993b). Human and bovine granulocyte chemotactic protein-2: complete amino acid sequence and functional characterization as chemokines. *Biochemistry* **32**, 10170–10177.

Rajarathnam, K., Sykes, B.D., Kay, C.M., Dewald, B., Geiser, T., Baggiolini, M. and Clark-Lewis, I. (1994). Neutrophil activation by monomeric interleukin-8. *Science* **264**, 90–92.

Rajarathnam, K., Clark-Lewis, I. and Sykes, B.D. (1995). ^{1}H NMR solution structure of an active monomeric interleukin-8. *Biochemistry* **34**, 12 983–12 990.

Rampart, M., Van Damme, J., Zonnekeyn, L. and Herman, A.G. (1989). Granulocyte chemotactic protein/interleukin-8 induces plasma leakage and neutrophil accumulation in rabbit skin. *Am. J. Pathol.* **135**, 21–25.

Rampart, M., Herman, A.G., Grillet, B., Opdenakker, G. and Van Damme, J. (1992). Development and application of a radioimmunoassay for interleukin-8: detection of interleukin-8 in synovial fluids from patients with inflammatory joint disease. *Lab. Invest.* **66**, 512–518.

Redl, H., Schlag, G., Bahrami, S., Schade, U., Ceska, M. and Stütz, P. (1991). Plasma neutrophil-activating peptide-1/interleukin-8 and neutrophil elastase in a primate bacteremia model. *J. Infect. Dis.* **164**, 383–388.

Richmond, A., Balentien, E., Thomas, E.G., Flaggs, G., Barton, D.E., Spiess, J., Bordoni, R., Francke, U. and Derynck, R. (1988). Molecular characterization and chromosomal mapping of melanoma growth stimulatory activity, a growth factor structurally related to β-thromboglobulin. *EMBO J.* **7**, 2025–2033.

Rot, A. (1993). Neutrophil attractant/activation protein-1 (interleukin-8) induces *in vitro* neutrophil migration by haptotactic mechanism. *Eur. J. Immunol.* **23**, 303–306.

Samanta, A.K., Oppenheim, J.J. and Matsushima, K. (1989). Identification and characterization of specific receptors for monocyte-derived neutrophil chemotactic factor (MDNCF) on human neutrophils. *J. Exp. Med.* **169**, 1185–1189.

Samanta, A.K., Oppenheim, J.J. and Matsushima, K. (1990). Interleukin 8 (monocyte-derived neutrophil chemotactic factor) dynamically regulates its own receptor expression on human neutrophils. *J. Biol. Chem.* **265**, 183–189.

Samanta, A.K., Dutta, S. and Ali, E. (1993). Modification of sulfhydryl groups of interleukin-8 (IL-8) receptor impairs binding of IL-8 and IL-8-mediated chemotactic response of human polymorphonuclear neutrophils. *J. Biol. Chem.* **268**, 6147–6153.

Sarris, A.H., Broxmeyer, H.E., Wirthmueller, U., Karasavvas, N., Cooper, S., Lu, L., Krueger, J. and Ravetch, J.V. (1993). Human interferon-inducible protein 10: expression and purification of recombinant protein demonstrate inhibition of early human hematopoietic progenitors. *J. Exp. Med.* **178**, 1127–1132.

Sato. Y., Abe, M. and Takaki, R. (1990). Platelet factor 4 blocks the binding of basic fibroblast growth factor to the receptor and inhibits the spontaneous migration of vascular endothelial cells. *Biochem. Biophys. Res. Commun.* **172**, 595–600.

Schadendorf, D., Moller, A., Algermissen, B., Worm, M., Sticherling, M. and Czarnetzki, B.M. (1993). IL-8 produced by human malignant melanoma cells *in vitro* is an essential autocrine growth factor. *J. Immunol.* **151**, 2667–2675.

Schall, T.J. (1994). The chemokines. In *The Cytokine Handbook*, 2nd ed. (ed. A.W. Thomson), Academic Press, London, pp. 419–460.

Schlöndorff, D., Nelson, P.J., Luckow, B. and Banas, B. (1997). Chemokines and renal disease. *Kidney Int.* **51**, 610–621.

Schmid, J. and Weissmann, C. (1987). Induction of mRNA for a serine protease and a β-thromboglobulin-like protein in mitogen-stimulated human leukocytes. *J. Immunol.* **139**, 250–256.

Schönbeck, U., Brandt, E., Petersen, F., Flad, H.-D. and Loppnow, H. (1995). IL-8 specifically binds to endothelial but not to smooth muscle cells. *J. Immunol.* **154**, 2375–2383.

Schraufstätter, I.U., Barritt, D.S., Ma, M., Oades, Z.G. and Cochrane, C.G. (1993). Multiple sites on IL-8 responsible for binding to α and β IL-8 receptors. *J. Immunol.* **151**, 6418–6428.

Schraufstätter, I.U., Ma, M., Oades, Z.G., Barritt, D.S. and Cochrane, C.G. (1995). The role of Tyr^{13} and Lys^{15} of interleukin-8 in the high affinity interaction with the interleukin-8 receptor type A. *J. Biol. Chem.* **270**, 10 428–10 431.

Schröder, J.-M. (1989). The monocyte-derived neutrophil activating peptide (NAP/interleukin 8) stimulates human neutrophil arachidonate-5-lipoxygenase, but not the release of cellular arachidonate. *J. Exp. Med.* **170**, 847–863.

Schröder, J.-M. and Christophers, E. (1986). Identification of C5ades arg and an anionic neutrophil-activating peptide (ANAP) in psoriatic scales. *J. Invest. Dermatol.* **87**, 53–58.

Schröder, J.-M., Mrowietz, U., Morita, E. and Christophers, E. (1987). Purification and partial biochemical characterization of a human monocyte-derived, neutrophil-activating peptide that lacks interleukin 1 activity. *J. Immunol.* **139**, 3474–3483.

Schröder, J.-M., Persoon, N.L.M. and Christophers, E. (1990). Lipopolysaccharide-stimulated human monocytes secrete apart from neutrophil-activating peptide 1/interleukin 8, a second neutrophil-activating protein. NH_2-terminal amino acid sequence identity with melanoma growth stimulatory activity. *J. Exp. Med.* **171**, 1091–1100.

Schröder, J.-M., Gregory, H., Young, J. and Christophers, E. (1992). Neutrophil-activating proteins in psoriasis. *J. Invest. Dermatol.* **98**, 241–247.

Schulz, B.S., Michel, G., Wagner, S., Süss, R., Beetz, A., Peter, R.U., Kemény, L. and Ruzicka, T. (1993). Increased expression of epidermal IL-8 receptor in psoriasis. Down-regulation by FK-506 *in vitro*. *J. Immunol.* **151**, 4399–4406.

Schumacher, C., Clark-Lewis, I., Baggiolini, M. and Moser, B. (1992). High- and low-affinity binding of GROα and neutrophil-activating peptide 2 to interleukin 8 receptors on human neutrophils. *Proc. Natl Acad. Sci. U.S.A.* **89**, 10542–10546.

Schwartz, D., Andalibi, A., Chaverri-Almada, L., Berliner, J.A., Kirchgessner, T., Fang, Z.-T., Tekamp-Olson, P., Lusis, A.J., Gallegos, C., Fogelman, A.M. and Territo, M.C. (1994). Role of the GRO family of chemokines in monocyte adhesion to MM-LDL-stimulated endothelium. *J. Clin. Invest.* **94**, 1968–1973.

Sebok, K., Woodside, D., Al-Aoukaty, A., Ho, A.D., Gluck, S. and Maghazachi, A.A. (1993). IL-8 induces the locomotion of human IL-2-activated natural killer cells. Involvement of a guanine nucleotide binding (G_0) protein. *J. Immunol.* **150**, 1524–1534.

Seitz, M., Dewald, B., Gerber, N. and Baggiolini, M. (1991). Enhanced production of neutrophil-activating peptide-1/interleukin-8 in rheumatoid arthritis. *J. Clin. Invest.* **87**, 463–469.

Sekido, N., Mukaida, N., Harada, A., Nakanishi, I., Watanabe, Y. and Matsushima, K. (1993). Prevention of lung reperfusion injury in rabbits by a monoclonal antibody against interleukin-8. *Nature* **365**, 654–657.

Senior, R.M., Griffin, G.L., Huang, J.S., Walz, D.A. and Deuel, T.F. (1983). Chemotactic activity of platelet alpha granule proteins for fibroblasts. *J. Cell. Biol.* **96**, 382–385.

Sgadari, C., Angiolillo, A.L., Cherney, B.W., Pike, S.E., Farber, J.M., Koniaris, L.G., Vanguri, P., Burd, P.R., Sheikh, N., Gupta, G., Teruya-Feldstein, J. and Tosato, G. (1996). Interferon-inducible protein-10 identified as a mediator of tumor necrosis *in vivo*. *Proc. Natl Acad. Sci. U.S.A.* **93**, 13791–13796.

Sharpe, R.J., Byers, H.R., Scott, C.F., Bauer, S.I. and Maione, T.E. (1990). Growth inhibition of murine melanoma and human colon carcinoma by recombinant human platelet factor 4. *J. Natl Cancer Inst.* **82**, 848–853.

Shirozu, M., Nakano, T., Inazawa, J., Tashiro, K., Tada, H., Shinohara, T. and Honjo, T. (1995). Structure and chromosomal localization of the human stromal cell-derived factor 1 (SDF1) gene. *Genomics* **28**, 495–500.

Singh, R.K., Gutman, M., Reich, R. and Bar-Eli, M. (1995). Ultraviolet B irradiation promotes tumorigenic and metastatic properties in primary cutaneous melanoma via the induction of interleukin-8. *Cancer Res.* **55**, 3669–3674.

Smith, D.R., Polverini, P.J., Kunkel, S.L., Orringer, M.B., Whyte, R.I., Burdick, M.D., Wilke, C.A. and Strieter, R.M. (1994). Inhibition of interleukin 8 attenuates angiogenesis in bronchogenic carcinoma. *J. Exp. Med.* **179**, 1409–1415.

Smith, W.B., Gamble, J.R., Clark-Lewis, I. and Vadas, M.A. (1991). Interleukin-8 induces neutrophil transendothelial migration. *Immunology* **72**, 65–72.

Standiford, T.J., Kunkel, S.L., Greenberger, M.J., Laichalk, L.L. and Strieter, R.M. (1996). Expression and regulation of chemokines in bacterial pneumonia. *J. Leukocyte Biol.* **59**, 24–28.

Strieter, R.M., Kunkel, S.L., Burdick, M.D., Lincoln, P.M. and Walz, A. (1992a). The detection of a novel neutrophil-activating peptide (ENA-78) using a sensitive ELISA. *Immunol. Invest.* **21**, 589–596.

Strieter, R.M., Kunkel, S.L., Elner, V.M., Martonyi, C.L., Koch, A.E., Polverini, P.J. and Elner, S.G. (1992b). Interleukin-8. A corneal factor that induces neovascularization. *Am. J. Pathol.* **141**, 1279–1284.

Strieter, R.M., Polverini, P.J., Kunkel, S.L., Arenberg, D.A., Burdick, M.D., Kasper, J., Dzuiba, J., Van Damme, J., Walz, A., Marriott, D., Chan, S.-Y., Roczniak, S. and Shanafelt, A.B. (1995). The functional role of the ELR motif in CXC chemokine-mediated angiogenesis. *J. Biol. Chem.* **270**, 27348–27357.

Sunyer, T., Rothe, L., Jiang, X., Osdoby, P. and Collin-Osdoby, P. (1996). Proinflammatory agents, IL-8 and IL-10, upregulate inducible nitric oxide synthase expression and nitric oxide production in avian osteoclast-like cells. *J. Cell. Biochem.* **60**, 469–483.

Tang, W.W., Yin, S., Wittwer, A.J. and Qi, M. (1995). Chemokine gene expression in anti-glomerular basement membrane antibody glomerulonephritis. *Am. J. Physiol.* **269**, F323–F330.

Tani, M., Fuentes, M.E., Peterson, J.W., Trapp, B.D., Durham, S.K., Loy, J.K., Bravo, R., Ransohoff, R.M. and Lira, S.A. (1996). Neutrophil infiltration, glial reaction, and neurological disease in transgenic mice expressing the chemokine N51/KC in oligodendrocytes. *J. Clin. Invest.* **98**, 529–539.

Tatakis, D.N. (1992). Human platelet factor 4 is a direct inhibitor of human osteoblast-like osteosarcoma cell growth. *Biochem. Biophys. Res. Commun.* **187**, 287–293.

Taub, D.D. (1996). Chemokine–leukocyte interactions. The voodoo that they do so well. *Cytokine and Growth Factor Reviews* **7**, 355–376.

Taub, D.D., Lloyd, A.R., Conlon, K., Wang, J.M., Ortaldo, J.R., Harada, A., Matsushima, K., Kelvin, D.J. and Oppenheim, J.J. (1993). Recombinant human interferon-inducible protein 10 is a chemoattractant for human monocytes and T lymphocytes and promotes T cell adhesion to endothelial cells. *J. Exp. Med.* **177**, 1809–1814.

Taub, D., Dastych, J., Inamura, N., Upton, J., Kelvin, D., Metcalfe, D. and Oppenheim, J. (1995a). Bone marrow-derived murine mast cells migrate, but do not degranulate, in response to chemokines. *J. Immunol.* **154**, 2393–2402.

Taub, D.D., Sayers, T.J., Carter, C.R.D. and Ortaldo, J.R. (1995b). α and β chemokines induce NK cell migration and enhance NK-mediated cytolysis. *J. Immunol.* **155**, 3877–3888.

Taub, D.D., Anver, M., Oppenheim, J.J., Longo, D.L. and Murphy, W.J. (1996a). T lymphocyte recruitment by interleukin-8 (IL-8). IL-8-induced degranulation of neutrophils releases potent chemoattractants for human T lymphocytes both *in vitro* and *in vivo*. *J. Clin. Invest.* **97**, 1931–1941.

Taub, D.D., Longo, D.L. and Murphy, W.J. (1996b). Human interferon-inducible protein-10 induces mononuclear cell infiltration in mice and promotes the migration of human T lymphocytes into the peripheral tissues of human peripheral blood lymphocytes-SCID mice. *Blood* **87**, 1423–1431.

Thelen, M., Peveri, P., Kernen, P., von Tscharner, V., Walz, A. and Baggiolini, M. (1988). Mechanism of neutrophil activation by NAF, a novel monocyte-derived peptide agonist. *FASEB J.* **2**, 2702–2706.

Tuschil, A., Lam, C., Haslberger, A. and Lindley, I. (1992). Interleukin-8 stimulates calcium transients and promotes epidermal cell proliferation. *J. Invest. Dermatol.* **99**, 294–298.

Unemori, E.N., Amento, E.P., Bauer, E.A. and Horuk, R. (1993). Melanoma growth-stimulatory activity/GRO decreases collagen expression by human fibroblasts. Regulation by C-X-C but not C-C cytokines. *J. Biol. Chem.* **268**, 1338–1342.

Van Damme, J. (1994). Interleukin-8 and related chemotactic cytokines. In *The Cytokine Handbook*, 2nd ed. (ed. A.W. Thomson), Academic Press, London, pp. 185–208.

Van Damme, J., Van Beeumen, J., Opdenakker, G. and Billiau, A. (1988). A novel, NH_2-terminal sequence-characterized human monokine possessing neutrophil chemotactic, skin-reactive, and granulocytosis-promoting activity. *J. Exp. Med.* **167**, 1364–1376.

Van Damme, J., Decock, B., Conings, R., Lenaerts, J.-P., Opdenakker, G. and Billiau, A. (1989a). The chemotactic activity for granulocytes produced by virally infected fibroblasts is identical to monocyte-derived interleukin 8. *Eur. J. Immunol.* **19**, 1189–1194.

Van Damme, J., Decock, B., Lenaerts, J.-P., Conings, R., Bertini, R., Mantovani, A. and Billiau, A. (1989b). Identification by sequence analysis of chemotactic factors for monocytes produced by normal and transformed cells stimulated with virus, double-stranded RNA or cytokine. *Eur. J. Immunol.* **19**, 2367–2373.

Van Damme, J., Van Beeumen, J., Conings, R., Decock, B. and Billiau, A. (1989c). Purification of granulocyte chemotactic peptide/interleukin-8 reveals *N*-terminal sequence heterogeneity similar to that of β-thromboglobulin. *Eur. J. Biochem.* **181**, 337–344.

Van Damme, J., Rampart, M., Conings, R., Decock, B., Van Osselaer, N., Willems, J. and Billiau, A. (1990). The neutrophil-activating proteins interleukin 8 and β-thromboglobulin: *in vitro* and *in vivo* comparison of NH_2-terminally processed forms. *Eur. J. Immunol.* **20**, 2113–2118.

van Deventer, S.J.H., Hart, M., van der Poll, T., Hack, C.E. and Aarden, L.A. (1993). Endotoxin and tumor necrosis factor-α-induced interleukin-8 release in humans. *J. Infect. Dis.* **167**, 461–464.

Van Osselaer, N., Van Damme, J., Rampart, M. and Herman, A.G. (1991). Increased microvascular permeability *in vivo* in response to intradermal injection of neutrophil-activating protein (NAP-2) in rabbit skin. *Am. J. Physiol.* **138**, 23–27.

Van Zee, K.J., Deforge, L.E., Fischer, E., Marano, M.A., Kenney, J.S., Remick, D.G., Lowry, S.F. and Moldawer, L.L. (1991). IL-8 in septic shock, endotoxemia, and after IL-1 administration. *J. Immunol.* **146**, 3478–3482.

Wada, T., Tomosugi, N., Naito, T., Yokoyama, H., Kobayashi, K.-I., Harada, A., Mukaida, N. and Matsushima, K. (1994a). Prevention of proteinuria by the administration of anti-interleukin 8 antibody in experimental acute immune complex-induced glomerulonephritis. *J. Exp. Med.* **180**, 1135–1140.

Wada, T., Yokoyama, H., Tomosugi, N., Hisada, Y., Ohta, S., Naito, T., Kobayashi, K.-I., Mukaida, N. and Matsushima, K. (1994b). Detection of urinary interleukin-8 in glomerular diseases. *Kidney Int.* **46**, 455–460.

Walz, A. and Baggiolini, M. (1989). Novel cleavage product of β-thromboglobulin formed in cultures of stimulated mononuclear cells activates human neutrophils. *Biochem. Biophys. Res. Commun.* **159**, 969–975.

Walz, A. and Baggiolini, M. (1990). Generation of the neutrophil-activating peptide NAP-2 from platelet basic protein or connective tissue-activating peptide III through monocyte proteases. *J. Exp. Med.* **171**, 449–454.

Walz, A., Peveri, P., Aschauer, H. and Baggiolini, M. (1987). Purification and amino acid sequencing of NAF, a novel neutrophil-activating factor produced by monocytes. *Biochem. Biophys. Res. Commun.* **149**, 755–761.

Walz, A., Dewald, B., von Tscharner, V. and Baggiolini, M. (1989). Effects of the neutrophil-activating peptide NAP-2, platelet basic protein, connective tissue-activating peptide III, and platelet factor 4 on human neutrophils. *J. Exp. Med.* **170**, 1745–1750.

Walz, A., Burgener, R., Car, B., Baggiolini, M., Kunkel, S.L. and Strieter, R.M. (1991a). Structure and neutrophil-activating properties of a novel inflammatory peptide (ENA-78) with homology to interleukin 8. *J. Exp. Med.* **174**, 1355–1362.

Walz, A., Meloni, F., Clark-Lewis, I., von Tscharner, V. and Baggiolini, M. (1991b). $[Ca^{2+}]_i$ changes and respiratory burst in human neutrophils and monocytes induced by NAP-1/interleukin-8, NAP-2, and gro/MGSA. *J. Leukocyte Biol.* **50**, 279–286.

Wang, J.M., Taraboletti, G., Matsushima, K., Van Damme, J. and Mantovani, A. (1990). Induction of haptotactic migration of melanoma cells by neutrophil activating protein/interleukin-8. *Biochem. Biophys. Res. Commun.* **169**, 165–170.

Warringa, R.A.J., Mengelers, H.J.J., Raaijmakers, J.A.M., Bruijnzeel, P.L.B. and Koenderman, L. (1993). Upregulation of formyl-peptide and interleukin-8-induced eosinophil chemotaxis in patients with allergic asthma. *J. Allergy Clin. Immunol.* **91**, 1198–1205.

Webb, L.M.C., Ehrengruber, M.U., Clark-Lewis, I., Baggiolini, M. and Rot, A. (1993). Binding to heparan sulfate or heparin enhances neutrophil responses to interleukin 8. *Proc. Natl Acad. Sci. U.S.A.* **90**, 7158–7162.

Wenger, R.H., Wicki, A.N., Walz, A., Kieffer, N. and Clemetson, K.J. (1989). Cloning of cDNA coding for connective tissue activating peptide III from a human platelet-derived lambda gt11 expression library. *Blood* **73**, 1498–1503.

Willems, J., Joniau, M., Cinque, S. and Van Damme, J. (1989). Human granulocyte chemotactic peptide (IL-8) as a specific neutrophil degranulator: comparison with other monokines. *Immunology* **67**, 540–542.

Wozniak, A., Betts, W.H., Murphy, G.A. and Rokicinski, M. (1993). Interleukin-8 primes human neutrophils for enhanced superoxide anion production. *Immunology* **79**, 608–615.

Wu, X., Wittwer, A.J., Carr, L.S., Crippes, B.A., DeLarco, J.E. and Lefkowith, J.B. (1994). Cytokine-induced neutrophil chemoattractant mediates neutrophil influx in immune complex glomerulonephritis in rat. *J. Clin. Invest.* **94**, 337–344.

Wu, L., Ruffing, N., Shi, X., Newman, W., Soler, D., Mackay, C.R. and Qin, S. (1996). Discrete steps in binding and signaling of interleukin-8 with its receptor. *J. Biol. Chem.* **271**, 31 202–31 209.

Wuyts, A., Haelens, A., Proost, P., Lenaerts, J.-P., Conings, R., Opdenakker, G. and Van Damme, J. (1996). Identification of mouse granulocyte chemotactic protein-2 from fibroblasts and epithelial cells. Functional comparison with natural KC and macrophage inflammatory protein-2. *J. Immunol.* **157**, 1736–1743.

Wuyts, A., Van Osselaer, N., Haelens, A., Samson, I., Herdewijn, P., Ben-Baruch, A., Oppenheim, J.J., Proost, P. and Van Damme, J. (1997). Characterization of synthetic human granulocyte chemotactic protein 2: usage of chemokine receptors CXCR1 and CXCR2 and *in vivo* inflammatory properties. *Biochemistry* **36**, 2716–2723.

Xu, L., Kelvin, D.J., Ye, G.Q., Taub, D.D., Ben-Baruch, A., Oppenheim, J.J. and Wang, J.M. (1995). Modulation of IL-8 receptor expression on purified human T lymphocytes is associated with changed chemotactic responses to IL-8. *J. Leukocyte Biol.* **57**, 335–342.

Yan, Z., Zhang, J., Holt, J.C., Stewart, G.J., Niewiarowski, S. and Moncz, M. (1994). Structural requirements of platelet chemokines for neutrophil activation. *Blood* **84**, 2329–2339.

Yang, Y., Mayo, K.H., Daly, T.J., Barry, J.K. and La Rosa, G.J. (1994). Subunit association and structural analysis of platelet basic protein and related proteins investigated by ^{1}H NMR spectroscopy and circular dichroism. *J. Biol. Chem.* **269**, 20110–20118.

Yoshimura, T., Matsushima, K., Tanaka, S., Robinson, E.A., Appella, E., Oppenheim, J.J. and Leonard, E.J. (1987). Purification of a human monocyte-derived neutrophil chemotactic factor that has peptide sequence similarity to other host defense cytokines. *Proc. Natl Acad. Sci. U.S.A.* **84**, 9233–9237.

Yue, T.-L., Mckenna, P.J., Gu, J.-L. and Feuerstein, G.Z. (1993). Interleukin-8 is chemotactic for vascular smooth muscle cells. *Eur. J. Pharmacol.* **240**, 81–84.

Yue, T.-L., Wang, X., Sung, C.-P., Olson, B., McKenna, P.J., Gu, J.-L. and Feuerstein, G.Z. (1994). Interleukin-8. A mitogen and chemoattractant for vascular smooth muscle cells. *Circ. Res.* **75**, 1–7.

Zampronio, A.R., Souza, G.E.P., Silva, C.A.A., Cunha, F.Q. and Ferreira, S.H. (1994). Interleukin-8 induces fever by a prostaglandin-independent mechanism. *Am. J. Physiol.* **266**, R1670–R1674.

Zhang, X., Chen, L., Bancroft, D.P., Lai, C.K. and Maione, T.E. (1994). Crystal structure of recombinant human platelet factor 4. *Biochemistry* **33**, 8361–8366.

Chapter 11

Interleukin-9

Jean-Christophe Renauld and Jacques Van Snick

Ludwig Institute for Cancer Research, Brussels Branch and Experimental Medicine Unit, Catholic University of Louvain, Brussels, Belgium

INTRODUCTION

Like many other members of the interleukin family, interleukin-9 (IL-9) is involved in the regulation of multiple functions in the immune system. Produced preferentially by T_{H2} lymphocytes, IL-9 is a 30–40-kDa glycoprotein active on various cell subsets in the immune and hematopoietic systems. Its gene is located on human chromosome 5, in the near vicinity of other cytokine genes such as IL-3, IL-4, IL-5, granulocyte–macrophage colony stimulating factor (GM-CSF) and IL-13, and its receptor is related structurally to the hematopoietic receptor superfamily.

Originally described as a T-cell growth factor present in the supernatant of murine activated T-cell clones, this molecule was characterized by a narrow specificity for some T helper clones and by its ability to sustain long-term growth in the absence of feeder cells and antigen. Taking advantage of such factor-dependent cell lines, the protein was purified to homogeneity, provisionally designated P40, and cloned molecularly (Uyttenhove *et al.*, 1988; Van Snick *et al.*, 1989).

Independently, Hültner and coworkers observed that the proliferation of mast cell lines in response to IL-3 or IL-4 could be enhanced by a factor produced by activated splenocytes (Hültner *et al.*, 1989; Moeller *et al.*, 1989). This activity was partially purified and called mast cell growth-enhancing activity (MEA). High levels of MEA were also found in the supernatant from a murine T-helper cell line derived by Schmitt and colleagues, who had observed that these cells produced a T-cell growth factor distinct from IL-2 and IL-4, which was called TCGF-III (Moeller *et al.*, 1990). The molecular cloning of a murine P40 cDNA and the availability of recombinant protein led to the demonstration that the same factor, now termed IL-9, was responsible for all these biologic activities (Hültner *et al.*, 1990).

Human IL-9 cDNA was identified by expression cloning of a factor stimulating the proliferation of a human megakaryoblastic leukemia (Yang *et al.*, 1989) and by cross-hybridization with a mouse probe (Renauld *et al.*, 1990a). Subsequently, IL-9 activities were described on normal hemopoietic progenitors (Donahue *et al.*, 1990), human T cells (Houssiau *et al.*, 1993), B cells (Dugas *et al.*, 1993; Petit-Frère *et al.*, 1993), fetal thymocytes (Suda *et al.*, 1990a), thymic lymphomas (Vink *et al.*, 1993) and murine

The Cytokine Handbook, 3rd ed.
ISBN 0-12-689662-3

neuronal progenitors (Mehler *et al.*, 1993). This chapter reviews the currently available information on the structure and function of this molecule.

IDENTIFICATION AND CLONING OF MOUSE AND HUMAN IL-9

The purification of mouse IL-9 from the supernatants of activated helper T cells was made possible by the use of stable factor-dependent T-cell lines, derived from normal antigen-dependent clones. The purified protein, originally designated P40 on the basis of its apparent size in gel filtration, was characterized by a high pI (approximately 10) and a high level of glycosylation (Uyttenhove *et al.*, 1988). Partial amino acid sequences obtained after cyanogen bromide treatment allowed the cloning of a full-length cDNA encoding the murine IL-9 protein (Van Snick *et al.*, 1989). In parallel with the cDNA cloning, complete sequencing of the purified protein has been achieved (Simpson *et al.*, 1989), confirming the deduced amino acid sequence of the mature protein. A cDNA encoding the human homolog of mouse IL-9 was cloned independently by expression cloning of a factor stimulating the growth of a human megakaryoblastic leukemia (Yang *et al.*, 1989) and by cross-hybridization with the mouse gene (Renauld *et al.*, 1990a). A comparison of the mouse and human IL-9 sequences is shown in Fig. 1. Both deduced protein sequences contain 144 residues with a typical signal peptide of 18 amino acids. The overall identity reached 69% at the nucleotide level and 55% at the protein level. In accordance with the heavy glycosylation observed with the natural murine protein, four potential *N*-linked glycosylation sites were noted in both sequences. This glycosylation is probably responsible for the discrepancy observed between the predicted relative molecular mass (14 150) and the M_r measured for native IL-9. The sequence is also characterized by the presence of ten cysteines, which are perfectly matched in both

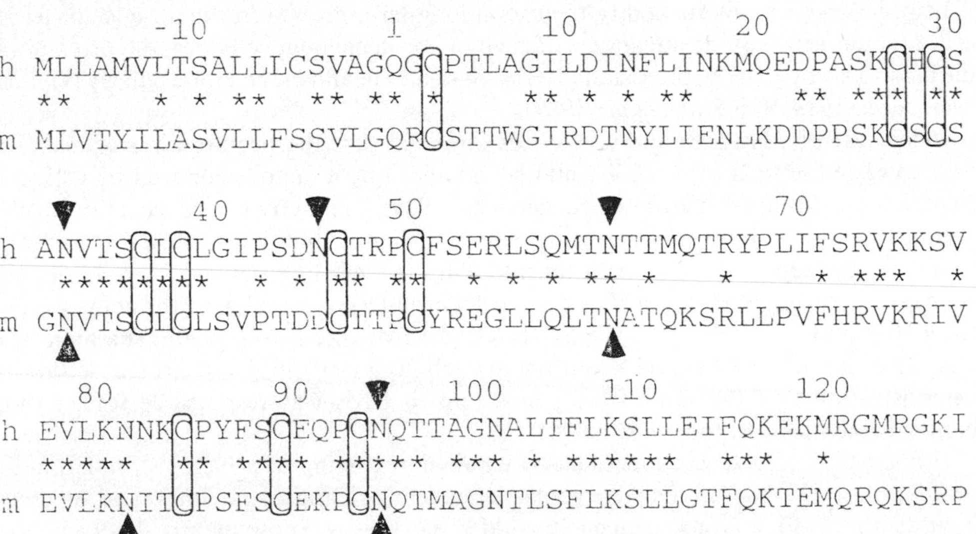

Fig. 1. Alignment of human and mouse IL-9 protein sequences. Amino acids are indicated in one-letter code. The ten cysteine residues of the mature protein are boxed and arrows indicate the potential *N*-linked glycosylation sites. Amino acid number 1 refers to the *N*-terminus of the mature mouse protein.

mature proteins, and a strong predominance of cationic residues, which explains the raised pI measured with purified natural IL-9.

THE IL-9 GENE: STRUCTURE AND EXPRESSION

Genomic Organization

The human IL-9 gene is a single-copy gene and was mapped on chromosome 5, in the 5q31→q35 region (Modi *et al.*, 1991). Interestingly, this region also contains various growth factor and growth factor receptor genes such as IL-3, IL-4, IL-5, CSF-1 and CSF-1R, and has been shown to be deleted in a series of hematologic disorders (Sokal *et al.*, 1975). Radiation hybrid mapping analysis has located the IL-9 gene between the IL-3 gene and the EGR-1 (early growth response-1) gene (Warrington *et al.*, 1992). However, in the mouse, the IL-9 gene does not seem to be linked to the same gene cluster: it has been localized on mouse chromosome 13 (Mock *et al.*, 1990) whereas the IL-3, IL-4, IL-5 and GM-CSF genes are located on chromosome 11.

As shown in Fig. 2, the human and murine IL-9 genes share a similar structure with five exons and four introns stretching over about 4 kilobases (kb) (Renauld *et al.*, 1990b). The five exons are identical in size for both species and show homology levels ranging from 56% to 74%. In contrast, no significant sequence homology was found in the introns (except for intron 2, which is also the smallest one), although their size is roughly conserved. However, 3′ and 5′ untranslated regions show a high level of identity, supporting a possible involvement of these sequences in the transcriptional or post-transcriptional regulation of IL-9 expression. In particular, numerous ATTTA motifs were found in the 3′ untranslated region of both genes. These sequences, frequently noticed in cytokine mRNAs, are thought to be related to the short half-life of these messengers by modulating their stability.

The transcription start has been mapped by S1 nuclease protection 22 to 24 nucleotides downstream from a classic TATA box sequence. The promoter of the IL-9 gene contains potential recognition sites for several tetradecanoyl phorbol acetate (TPA)-inducible transcription factors such as activating protein-1 (AP-1) and AP-2, which could provide a structural basis for the induction of IL-9 expression by phorbol

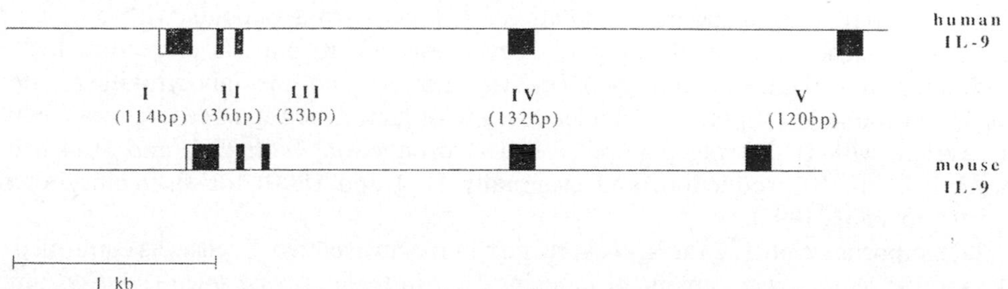

Fig. 2. Map of the human and murine IL-9 genes. Closed boxes represent the coding regions and open boxes correspond to the 5′ untranslated sequence. Exons are numbered and their respective size is indicated; homology levels are 56, 67, 64, 73 and 74% respectively.

esters. A consensus sequence for interferon regulatory factor-1 (IRF-1) was also identified in both promoters but its physiologic relevance remains elusive (Renauld *et al.*, 1990b). Kelleher and coworkers (1991) identified other consensus sequences in the 5′ untranslated region of the human gene (SP-1, NF-κB, Octamer, AP-3, AP-5, glucocorticoid responsive element, cAMP response element) and suggested that the NF-κB site and cAMP response element could be involved in the constitutive expression of IL-9 by human T-lymphotropic virus-1 (HTLV-1)-transformed T cells. The importance of the NF-κB site in this model was further confirmed by Zhu *et al.* (1996), who showed that this region of the promoter was critical for gene expression and bound various proteins including NF-κB, c-jun and an unidentified 35-kDa protein.

Regulation of IL-9 Expression

T cells are the main source of IL-9. In the mouse, its expression is typical of *in vitro* activated T_{H2} clones (Schmitt *et al.*, 1989; Gessner *et al.*, 1993). *In vivo*, potent T_{H2} stimuli such as anti-immunoglobulin D (IgD)-triggered polyclonal activation (Svetic *et al.*, 1991) and helminth infections (Grencis *et al.*, 1991; Else *et al.*, 1992; Svetic *et al.*, 1993) also induce IL-9 expression. Conversely, during *Nippostrongylus brasiliensis* infection, administration of IL-12, which completely supresses T_{H2} activation, abolished IL-9 expression as well (Finkelman *et al.*, 1994). A similar correlation between T_{H2} responses and IL-9 was observed in *Leishmania major* infection, where susceptible T_{H2} responder mice produce IL-9 while resistant T_{H1} strains do not (Gessner *et al.*, 1993).

The regulation of human IL-9 expression has been studied *in vitro* using freshly isolated peripheral blood mononuclear cells (PBMCs). Absent in resting cells, IL-9 message is strongly induced after T-cell mitogenic stimulation with phytohemagglutinin (PHA) or anti-CD3 monoclonal antibodies and can be totally ascribed to the $CD4^+$ population (Renauld *et al.*, 1990b). When anti-CD3 and anti-CD28 antibodies are used as stimulators, IL-9 production is restricted to 'memory' ($CD45RO^+$) $CD4^+$ T cells (Houssiau *et al.*, 1995).

After stimulation of peripheral T cells with lectins or other mitogenic agents, IL-9 mRNA expression is not an early event, as it peaks at 28 h. Moreover it is completely abrogated by cyclohexamide, an inhibitor of protein synthesis, indicating the involvement of secondary signals in this process. IL-2 apparently plays a crucial role, as anti-IL-2 receptor antibodies completely block IL-9 production (Houssiau *et al.*, 1992). The importance of IL-2 in this phenomenon was confirmed by the low IL-9 production observed after antigen-specific stimulation of T cells from patients with a primary immunodeficiency disease affecting IL-2 gene expression, and its correction by the addition of IL-2 (Hauber *et al.*, 1995). IL-2 was, however, not the only cytokine required for IL-9 production. In fact, a complex cascade of factors acting in synergy was finally unraveled, with IL-2 being required for IL-4 production, both IL-2 and IL-4 being needed for IL-10 production, and eventually IL-4 and IL-10 for IL-9 biosynthesis (Houssiau *et al.*, 1995).

The importance of IL-2 for IL-9 expression in freshly isolated T cells was confirmed in the mouse, as IL-2 was capable of inducing IL-9 in fresh murine splenocytes costimulated with phorbol myristate acetate (PMA) (P. Monteyne, personal communication) and IL-9 production is significantly reduced in IL-2-deficient mice (Schmitt *et al.*, 1994). However, other regulatory mechanisms could be involved in IL-9 expression by some

T-cell lines and tumors. It was indeed reported that IL-1, and not IL-2, serves as a secondary signal for IL-9 expression in murine T-cell lines (Schmitt *et al.*, 1991) and that transforming growth factor-β is also a potent inducer of IL-9 production by murine peripheral T cells (Schmitt *et al.*, 1994).

Another characteristic of IL-9 expression is its association with HTLV-1, a retrovirus involved in adult T-cell leukemia, as HTLV-1-transformed T cells produce IL-9 constitutively (Yang *et al.*, 1989; Kelleher *et al.*, 1991). The tax transactivator of HTLV-1 might be implicated in this process through the NF-κB consensus site in the IL-9 promoter (Zhu *et al.*, 1996). Interestingly, in another system of T-cell transformation by murine polytropic retroviruses, viral infection also resulted in IL-9 expression (Flubacher *et al.*, 1994).

BIOASSAYS FOR IL-9

The original bioassay for murine IL-9 is based on the proliferation of a factor-dependent T-cell line called TS1, the half-maximal proliferation being obtained with 15 pg/ml of purified IL-9 (Uyttenhove *et al.*, 1988). These cells are not responsive to IL-2 but proliferate in response to IL-4 (Fig. 3). Most of the IL-9-dependent lines have lost the α chain of the IL-2 receptor as well as the capacity to proliferate in the presence of IL-2, with one exception, the ST2.K9 line, which is responsive to IL-2 and IL-9 but not to IL-4 (Schmitt *et al.*, 1989). Leukemia inhibitory factor (LIF) was also found to stimulate the proliferation of some IL-9-responsive T-cell lines (Van Damme *et al.*, 1992). Murine mast cell lines such as MC9 also respond to IL-9, in addition to IL-3, IL-4 and IL-10.

Human IL-9 was identified as having growth-promoting activity for the Mo7E cell line, isolated from a child with acute megakaryoblastic leukemia (Yang *et al.*, 1989). This cell line is also responsive to Steel factor, GM-CSF and IL-3, which are more potent stimulators of these cells (Fig. 4) (Hendrie *et al.*, 1991), a clear disadvantage when IL-9

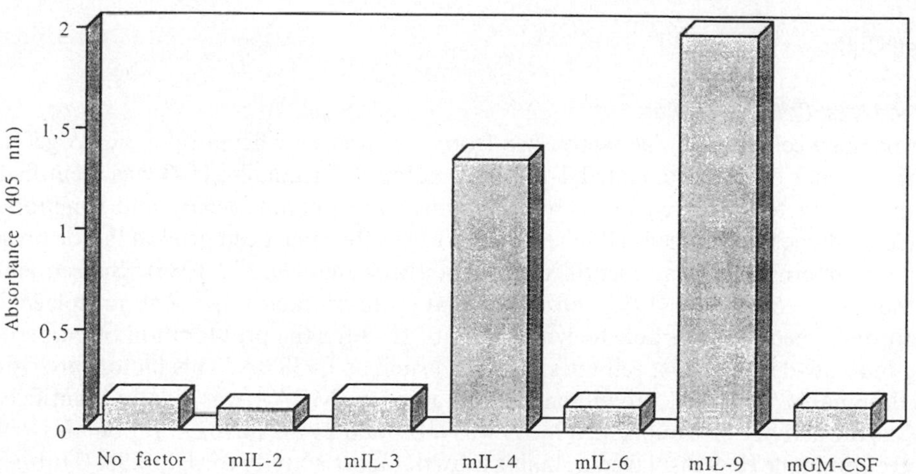

Fig. 3. Factor-dependent proliferation of the TS1 cell line. The TS1 bioassay was performed as described by Uyttenhove *et al.* (1988), in the presence or absence of the indicated cytokines at a concentration of 100 units/ml.

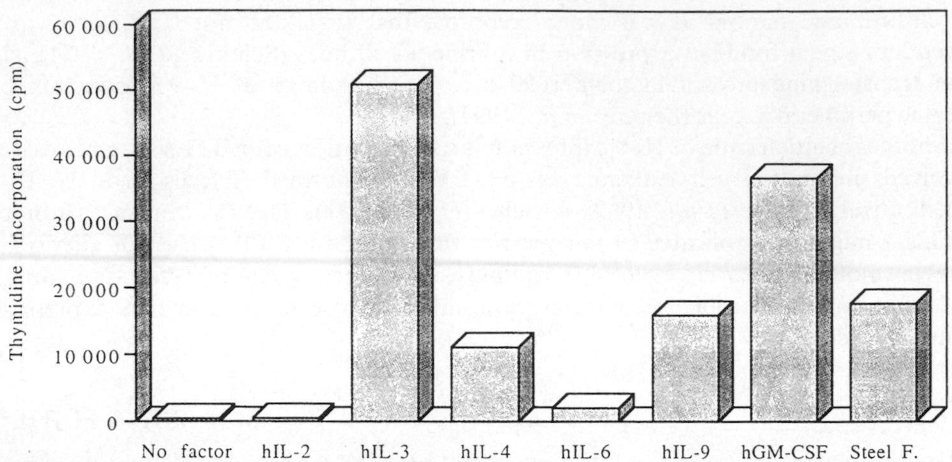

Fig. 4. Factor-dependent proliferation of the Mo7E cell line. The Mo7E bioassay was performed as described by Yang *et al.* (1989). Thymidine incorporation was measured after 24 h in the presence or absence of hIL-2 (20 units/ml), hIL-3 (3 ng/ml), hIL-4 (1/100 of crude baculovirus supernatant), hIL-6 (1000 units/ml), hIL-9 (100 units/ml), hGM-CSF (0.1 ng/ml) and hSteel factor (1/100 of transfected COS cells supernatant).

must be measured in complex cytokine mixtures. While human and mouse IL-9 are equally active in a Mo7E proliferation assay, human IL-9 is not active on murine cells. An easier and more specific bioassay for human IL-9 was obtained when murine TS1 cells were transfected with the human IL-9 receptor cDNA (Renauld *et al.*, 1992). Of the human factors described so far, only LIF–human interleukin for DA cells (HILDA) and insulin have been shown to promote the proliferation of this transfected murine cell line.

BIOLOGIC ACTIVITIES OF IL-9

Mast Cells

IL-9 and Mast Cell Lines In Vitro

Murine mast cell lines can be established from bone marrow hematopoietic progenitors in the presence of lectin-activated T-cell supernatants. Originally, IL-3 was identified as the growth factor required by these cells, which are phenotypically and functionally related to mucosal mast cells (Ihle *et al.*, 1983). Another mast cell growth factor present in these supernatants was identified as IL-4 (Mossman *et al.*, 1986). Subsequently, Hültner and colleagues (1989) observed that a third factor, present in spleen cell conditioned medium, was able to synergize with IL-3 for the proliferation of permanent bone marrow-derived mast cell lines (BMMC) such as L138.8A. This factor, provisionally designated mast cell growth-enhancing activity (MEA), was finally identified as IL-9, as the activity of semipurified MEA was inhibited by an anti-IL-9 rabbit antiserum, and recombinant IL-9 displayed a similar growth factor activity for L138.8A (Hültner *et al.*, 1990). The proliferative activity of IL-9, alone or in synergy with IL-3, was confirmed on other permanent mast cell lines such as MC-6, H7 and MC-9 (Williams *et al.*, 1990 and Renauld *et al.*, unpublished data). In fact, responses of BMMCs to IL-9 are highly dependent on the time spent *in vitro*. When primary BMMCs are derived from hematopoietic progenitors, IL-9 alone is not sufficient to sustain mast cell growth, but synergistically enhances the proliferation induced by IL-3 or the combination of IL-3 and IL-4. Moreover, in the absence of other factors, IL-9 significantly increases the survival of these primary BMMCs. After some more time *in vitro*, when stable mast cell lines are obtained, IL-9 alone becomes capable of inducing proliferation, without the need for additional factors. Finally, autonomous cells, which induce tumors in syngeneic animals, can be derived from the originally factor-dependent cell lines (Hültner and Moeller, 1990).

In addition to its proliferative activity, IL-9 seems to be a potent regulator of mast cell effector molecules. Thus, IL-9 induces IL-6 secretion by mast cell lines (Hültner and Moeller, 1990; Hültner *et al.*, 1990) and the identification of genes specifically induced by IL-9 has shown that protease genes belonging to the granzyme family are expressed and produced efficiently in response to IL-9 (Louahed *et al.*, 1995). Other mast cell-specific proteases of the mouse mast cell protease (mMCP) family are also upregulated upon IL-9 stimulation (Eklund *et al.*, 1993). Moreover, expression of the α chain of the high-affinity IgE receptor is also upregulated by IL-9, indicating that this factor could be a key mediator of mast cell differentiation (Louahed *et al.*, 1995).

IL-9 and Mastocytosis In Vivo

Analysis of IL-9 transgenic mice has shown that in all transgenic animals increased mast cell infiltration is found inside the gastric and intestinal epithelium as well as in the upper airway epithelium and kidneys, but not in other organs, notably in skin. Although IL-9 by itself does not induce mast cell development *in vitro*, it strongly synergizes with Steel factor or stem cell factor (SCF) to promote the growth and differentiation of mast cells from bone marrow progenitors. *In vivo*, antibodies directed against c-kit, the SCF receptor, block the IL-9 transgenic mastocytosis. Because similar constitutive SCF expression was observed in both IL-9 transgenic and control mice, these observations indicate that neither SCF nor IL-9 is sufficient to induce mastocytosis, but that the synergistic activity of these cytokines is responsible for the *in vivo* amplification of this cell population in IL-9 transgenic mice (Godfraind *et al.*, 1998).

Unexpectedly, intestinal IL-9 transgenic mast cells showed a phenotype related to connective tissue type mast cells, usually found in the skin and peritoneal cavity. They were stained by safranin and strong expression of mMCP-4 and -5 proteases was found in this organ. However, they also expressed proteases related to the mucosal mast cells such as mMCP-1 and -2. Interestingly, mast cells derived *in vitro* in the presence of IL-9 and SCF expressed the same protease pattern as observed in IL-9 transgenic mice. These observations suggest that the synergistic activity of IL-9 and SCF induces the proliferation and differentiation of a new mast cell subset expressing an extended pattern of proteases characteristic of both connective tissue type and mucosal mast cells.

Infection with helminth parasites typically induces an immune response characterized by strong IgE production and massive mucosal mastocytosis resulting from T_{H2} cytokine production. During infection with *Trichinella spiralis*, IL-9 production is induced in mesenteric lymph nodes, as well as other T_{H2} cytokines such as IL-3, IL-4 and IL-5 (Grencis *et al.*, 1991). The protective potential of a T_{H1}- or T_{H2}-like response seems to vary depending on the parasites, so that a T_{H2} response may be harmful in some helminth infections but protective for others. For instance, resistance to *Trichuris muris* was found to correlate with the production of IL-5 and IL-9 in mesenteric lymph nodes (Else *et al.*, 1992). Interestingly, the IL-9 transgenic mice were found to be particularly resistant to infection with the intestinal nematode *T. spiralis*. Furthermore, depression of the mast cell response with anti-c-kit antibodies resulted in the inability of these mice to expel the parasite (Faulkner *et al.*, 1997). Quite similar observations were reported with *N. brasiliensis* infections. When *N. brasiliensis* larvae were inoculated subcutaneously, IL-9 mRNA induction was found in the lungs by 24 h, and in mesenteric lymph nodes after 4 days. As the larvae reach the lungs within 1 day and the gut after 3–4 days, IL-9 expression appears to be an early event in the immune response, and actually precedes the production of other cytokines such as IL-4. By contrast, IL-9 expression was substantially decreased from its peak level by day 6 after inoculation, whereas expression of IL-3, IL-4 and IL-5 genes was maintained at least to day 12 (Madden *et al.*, 1991).

The role of IL-3 and IL-4 in *Nippostrongylus*-induced mastocytosis was demonstrated by the observation that either anti-IL-3 or anti-IL-4 antibodies partially (40–50%) inhibited the mucosal mast cell response. As the combination of both anti-IL-3 and anti-IL-4 antibodies failed to block more than 85% of the mastocytosis, a third factor was possibly involved in the mast cell response (Madden *et al.*, 1992). Although anti-IL-9 antiserum by itself did not influence mast cell hyperplasia, the addition of such antibodies to anti-IL-3 and anti-IL-4 antibodies increased the suppression of the mast cell response from 85% to 95%. Moreover, when mice received suboptimal doses of anti-IL-3 plus anti-IL-4 antibodies, the suppression of mastocytosis increased from 60% to 94% after injection of an anti-IL-9 antiserum (Madden *et al.*, unpublished data), demonstrating that IL-9 may contribute to the development of mucosal mast cell hyperplasia induced by worm infections particularly under conditions where IL-3 and IL-4 are limiting.

B Cells

The involvement of IL-9 in mast cell activation and proliferation, as well as IL-9 production during parasite infections and T_{H2} activation, raised the hypothesis of a potential role for this factor in IgE-mediated responses. The ability of IL-9 to modulate IgE production by mouse B cells has been investigated by Petit-Frère *et al.* (1993). Their results indicated that IL-9 synergized with suboptimal doses of IL-4 for the IgE and IgG_1 production by lipopolysaccharide (LPS)-activated semipurified B cells, but did not induce any IgE or IgG_1 production in the absence of IL-4. The IL-9 activity on the IL-4-induced IgG_1 production correlated with an increase in the number of IgG_1 secreting cells. By contrast, IL-9 did not affect IL-4-induced CD23 expression by LPS-activated B cells, indicating that its activity did not consist of a simple upregulation of IL-4 responsiveness by the B cells. Moreover, these experiments have not ruled out the possibility that the effect of IL-9 observed on murine B cells might be mediated by accessory cells.

In the human, very similar observations have been reported by Dugas and coworkers (1993) using semipurified peripheral B cells. In this experimental system, IL-9 synergized with IL-4 for IgE and IgG but not for IgM production. Moreover, IL-9 also potentiated IL-4-induced IgE production by sorted $CD20^+$ cells upon costimulation by irradiated EL4 murine T cells, thereby suggesting a direct activity on B cells.

Contrasting with the rather weak activity of IL-9 on *in vitro* immunoglobulin production, preliminary observations on IL-9 transgenic mice strongly support its implication in humoral responses *in vivo*. Both basal titers of all immunoglobulin classes and antigen-specific antibody responses are indeed increased in the serum of these animals. In addition, these mice are characterized by a dramatic increase in the number of peritoneal B1b cells, but not in conventional B cells, suggesting that IL-9 may be specifically active in B-cell responses involving this particular subset, such as autoimmune processes (J.-C. Renauld *et al.*, unpublished data).

The potential role of IL-9 in humoral autoimmunity is illustrated by preliminary data obtained in non-obese diabetic (NOD) mice. NOD mice are considered as a model of cell-mediated autoimmunity as they spontaneously develop within a few months a T cell-mediated insulitis leading to diabetes. This autoimmune process can be inhibited by T_{H2} cytokines such as IL-4 and IL-10, or accelerated by a T_{H1}-promoting cytokine such as IL-12. In addition, NOD mice are also susceptible to autoimmune processes such as iodide-induced thyroiditis. When a high iodide dose is administered to goitrous NOD mice, the iodide-induced thyroid cell necrosis is followed by diffuse infiltration by macrophages and CD4 and CD8 T cells, leading to follicular destruction similar to Hashimoto's thyroiditis in the human.

A short course of IL-9 treatment completely abrogated T-lymphocyte and macrophage infiltration. In addition, IL-9 induced an increase in germinal center formation in draining lymph nodes, indicating that inhibition of the cell-mediated response is accompanied by B-cell activation. However, no significant change in T_{H1} or T_{H2} cytokine expression was detected in the thyroid gland or lymph nodes of IL-9-treated NOD mice. A similar inhibition of cellular infiltrate was observed in the pancreatic islets of NOD mice, where a 4–6-day treatment with IL-9 significantly suppressed the insulitis (M.C. Many *et al.*, unpublished results). Taken together, these data indicate that IL-9 favors humoral autoimmunity but inhibits cell-mediated autoimmune processes.

T Cells

Although T cells were the first identified targets for IL-9, the physiologic role of IL-9 for T cells remains puzzling. Initial observations in a murine system suggested that the activity of IL-9 was restricted to some T helper cell clones (Uyttenhove *et al.*, 1988). Noticeably, freshly isolated T cells never responded to this cytokine. However, it appeared that sensitivity to the growth-promoting activity of IL-9 is not a characteristic of a particular T-cell subpopulation but can be gradually acquired by long-term *in vitro* culture. Interestingly, murine T cells were found to undergo some phenotypical changes *in vitro* that were reminiscent of—and eventually lead to—tumoral transformation.

After several months of culture with the antigen and antigen-presenting cells, T-cell cultures could be maintained by addition of IL-9, and permanent IL-9-dependent cell lines could be derived. An increase in cell size was noticed, as well as an accelerated growth rate. Progressively, most T-cell markers such as Thy1, CD4, CD3 and TCR expression were lost. When such cells were transfected with an IL-9 cDNA expression vector, and injected into syngeneic mice, the mice died in 3–4 months as a result of widespread lymphoma development (Uyttenhove *et al.*, 1991). In addition, factor-independent cells were sporadically generated during *in vitro* culture and were similarly found to form tumors *in vivo*. Similar observations have been reported in a rat model, using an IL-2-dependent T-cell lymphoma which, after infection with murine polytropic retroviruses, became IL-2 independent by induction of an autocrine loop involving IL-9 and its receptor (Flubacher *et al.*, 1994).

The significance of these observations is supported by the analysis of IL-9 transgenic mice that constitutively express high levels of this cytokine, as 5–10% of these mice spontaneously develop lymphoblastic lymphomas (Renauld *et al.*, 1994). Interestingly, no preneoplastic T-cell hyperplasia has ever been observed in these mice, thereby confirming the lack of activity of IL-9 on normal resting T cells *in vitro*. This contrasts with other transgenic models such as the IL-7 transgenic mice, in which a proliferation of normal T cells in the skin precedes the onset of lymphoma (Rich *et al.*, 1993). Moreover, the IL-9 transgenic mice were highly susceptible to chemical mutagenesis as all transgenic animals developed T-cell lymphomas after injection of doses of a mutagen (*N*-methyl-*N*-nitrosourea) that were totally innocuous in control mice. Similarly, these transgenic mice exhibited a high sensitivity to the tumorigenic effect of γ irradiation.

The growth-promoting activity of IL-9 for T-cell tumors was also investigated *in vitro*, using other models of thymic lymphomas generated in normal mice. In these experiments, IL-9 was found to significantly stimulate the *in vitro* proliferation of primary lymphomas induced either by chemical mutagenesis in DBA/2 mice or by X irradiation in B6 mice (Vink *et al.*, 1993). Moreover, IL-9 was found to protect such tumor cells against dexamethasone-induced apoptosis, even for cell lines whose *in vitro* proliferation is completely independent of any cytokine. With some cell lines, IL-9 turned out to be more potent in protecting cells against apoptosis than in inducing proliferation. By contrast, IL-2 seems to be more efficient as a proliferation inducer (Renauld *et al.*, 1995). This activity could therefore be particularly relevant for the oncogenic potential of IL-9 *in vivo*.

In the human, a link between dysregulated IL-9 production and lymphoid malignancy has been suggested initially by the observation that lymph nodes from patients with Hodgkin and large cell anaplastic lymphomas produce IL-9 constitutively (Merz *et al.*, 1991). Constitutive IL-9 expression was also detected in HTLV-1-transformed T cells (Kelleher *et al.*, 1991) and in Hodgkin cell lines (Merz *et al.*, 1991; Gruss *et al.*, 1992; Trümper *et al.*, 1993). Moreover, an *in vitro* autocrine loop involving IL-9 has been observed for one of these Hodgkin cell lines (Gruss *et al.*, 1992), suggesting potential involvement of IL-9 in the pathology of this disease. Such an autocrine loop may also play a role in HTLV-1 leukemias, as illustrated by the *cis*/*trans*-activation of the IL-9 receptor gene by insertion of the HTLV-1 long terminal repeat (LTR) in one leukemia cell line (Kubota *et al.*, 1996).

Besides this well documented activity on T-cell lymphomas and leukemias, the function of IL-9 in normal T-cell responses remains unclear. In line with previous reports in the murine model, IL-9 did not induce any proliferation of freshly isolated human T cells, either alone or in synergy with other cytokines or T-cell costimuli. By contrast, significant proliferation could be induced by IL-9 when PBMCs were preactivated for only 10 days with PHA, IL-2 and irradiated allogeneic feeder cells, indicating that responses to IL-9 require previous activation (Houssiau *et al.*, 1993). Similar results have been reported for T cells activated with PHA for 3 or 7 days (Lehrnbecher *et al.*, 1994). When human T-cell clones were derived from established PHA-stimulated T-cell lines, most of these clones proliferated in response to IL-9, irrespective of their CD4 or CD8 phenotype (Houssiau *et al.*, 1993). Taken together, these results indicate that at least two different mechanisms may render T cells responsive to IL-9: tumoral transformation or potent activation of normal T cells. It is not yet clear whether these observations only reflect the regulation of the IL-9 receptor or whether they are linked to the specificity of the signal transduced by this receptor.

IL-9 and Hematopoietic Progenitors

The first evidence for an involvement of IL-9 in the hematopoietic system was provided by the identification and cloning of the human protein as a growth factor for the megakaryoblastic leukemia Mo7E, a cell line displaying early markers of differentiation, such as CD33 and CD34, and markers for bipotent erythromegakaryoblastic hematopoietic precursors (Avanzi *et al.*, 1988; Yang *et al.*, 1989; Hendrie *et al.*, 1991). In fact, IL-9 did not seem to be active on normal megakaryoblastic precursors but supported the clonogenic maturation of erythroid progenitors in the presence of erythropoietin (Donahue *et al.*, 1990). This activity was confirmed by several groups and reproducibly observed with highly purified progenitors after sorting for $CD34^+$ cells and T-cell depletion (Birner *et al.*, 1992; Lu *et al.*, 1992; Sonoda *et al.*, 1992; Schaafsma *et al.*, 1993), particularly in synergy with SCF (Lemoli *et al.*, 1994; Sonoda *et al.*, 1994). In the mouse, a similar erythroid burst-promoting activity has been described, but appears to be dependent on the presence of T cells (Williams *et al.*, 1990).

By contrast, granulocyte or macrophage colony formation (CFU-GM, CFU-G or CFU-M) was usually not enhanced by IL-9. An activity on early multipotent progenitors was, however, observed by a two-step liquid culture assay with $CD34^+$ $CD33^-$ DR^- cells (Lemoli *et al.*, 1994). In this assay, the majority of the colonies observed with IL-9 or IL-9 and SCF corresponded to CFU-GM. Noticeably, Schaafsma and colleagues (1993) observed that IL-9 also promoted some granulocytic as well as monocytic colony (CFU-GM) growth from $CD34^+$ $CD2^-$ progenitors from some bone marrow donors. Experiments comparing the effects of IL-9 on fetal and adult progenitors have shown that addition of IL-9 to cultures of fetal progenitors induces maturation of CFU-Mix and CFU-GM, whereas IL-9 is also more effective on fetal cells of the erythroid lineage (Holbrook *et al.*, 1991). In addition, IL-9 was found to increase the *in vitro* proliferation of human myeloid leukemic cells in a clonogenic assay in methylcellulose, suggesting a preferential activity on transformed myeloid cells compared with their normal progenitors (Lemoli *et al.*, 1996). These observations are in line with findings on murine T cells. Thus, murine fetal thymocytes (Suda *et al.*, 1990a) and thymic lymphomas (Vink *et al.*, 1993), but not adult thymocytes (Suda *et al.*, 1990b), respond to IL-9. This raises the hypothesis that the spectrum of activities of IL-9 is greater on fetal and tumoral progenitors.

THE IL-9 RECEPTOR

A single class of high-affinity binding sites for IL-9 ($K_d \sim 100$ pM) has been detected on various IL-9-responsive murine cells. Cross-linking data indicate that IL-9 binds to a single-chain receptor consisting of a 64-kDa glycoprotein (Druez *et al.*, 1990). The murine IL-9 receptor (IL-9R) is a 468-amino-acid polypeptide with an extracellular domain, composed of 233 amino acids, including a WSEWS motif and six cysteine residues, whose position indicates that the IL-9 receptor is a member of the hematopoietin receptor superfamily. The human IL-9 receptor consists of a 522-amino-acid protein with 53% identity with the mouse molecule. The extracellular region is particularly conserved with 67% identity, whereas the cytoplasmic domain is significantly larger in the human receptor (231 *versus* 177 residues) (Renauld *et al.*, 1992). As observed for many members of the hematopoietin receptor superfamily, IL-9R mRNAs have been identified that lack the sequences encoding the transmembrane and cytoplasmic domains, as a result of alternative splicing. However, the frequency of these mRNAs seems quite low and it is not clear yet whether they really encode a soluble IL-9 binding protein. A more frequent alternative splicing of the human gene generates an intriguing heterogeneity in the 5′ untranslated region of the mRNA and introduces some short open reading frames that might represent an additional level in the regulation of IL-9R translation, as suggested for many genes involved in cell growth (Kozak, 1991; Kermouni *et al.*, 1995).

In the mouse, the IL-9R gene is a single-copy gene located on chromosome 11 and composed of nine exons and eight introns, sharing many characteristics with other genes encoding cytokine receptors (Renauld *et al.*, unpublished data; Vermeesch *et al.*, 1997). By contrast, the human genome contains at least four IL-9R pseudogenes with approximately 90% homology with the IL-9R gene, which is located in the subtelomeric region of chromosomes X and Y (Kermouni *et al.*, 1995). *IL-9R* was thus the first gene to be identified in the long arm pseudoautosomal region. Using a polymorphism in the coding region of this gene, Vermeesch and colleagues (1997) have shown that *IL-9R* is expressed both from X and Y chromosomes, and escapes X inactivation. In addition, comparative mapping of *IL-9R* in mouse, great apes and human made it possible to reconstitute the evolution of this pseudoautosomal region (Vermeesch *et al.*, 1997). Further studies will have to determine whether this unusual localization in the vicinity of the telomere may play a role in the regulation of transcription, as reported in various organisms (Biessmann and Mason, 1992).

The IL-9 receptor was found to interact with the γ chain of the IL-2 receptor, which is required for signal transduction but not for IL-9 binding, as an antibody directed against this molecule completely inhibits the activity of IL-9 without affecting the K_d of IL-9 binding (Kimura *et al.*, 1995). The fact that this molecule, now called γc, is shared by the IL-2, IL-4, IL-7, IL-9 and IL-15 receptors could explain the overlapping activities observed for these T-cell growth factors. In addition, IL-9 was recently reported to inhibit IL-2 binding on leukemic cells, raising the possibility of multiple interactions within this subset of cytokines (Schumann *et al.*, 1996).

So far, the main function of γc is to recruit the tyrosine kinase JAK3, while the IL-9R is associated with JAK1. This association of JAK1 with the IL-9R has been ascribed to a 98-residue juxtamembrane region of the receptor (Demoulin *et al.*, 1996). This region contains a Pro–X–Pro sequence preceded by a cluster of hydrophobic residues, which partially fits a recently described consensus sequence shared by many cytokine receptors (e.g. IL-4R, IL-7R, IL-3R, erythropoietin receptor (EPOR), IL-2Rβ, G-CSFR) (Murakami *et al.*, 1991). Downstream from this Pro–X–Pro motif, a striking homology was observed with the β chain of the IL-2 receptor and with the erythropoietin receptor. As a result, for the first 33 amino acids of the cytoplasmic domain, a 40% identity was noticed between the human IL-9R and IL-2Rβ. This homologous region probably explains that IL-9, like other cytokines such as IL-2, induces JAK1 and JAK3 phoshorylation (Russel *et al.*, 1994; Yin *et al.*, 1994; Demoulin *et al.*, 1996).

Upon IL-9 binding, both JAK1 and JAK3 become phosphorylated and catalytically active. These kinases are likely to be responsible for IL-9R phosphorylation on one of its five tyrosine residues. This single phosphorylated residue acts as a docking site for signal transducers and activators of transcription STAT1, STAT3 and STAT5, three transcription factors which, after phosphorylation by the JAK kinases associated with the receptor, form heterodimers or homodimers and migrate to the nucleus. Although several signal transduction studies of other cytokine receptors have shown that activation of STAT transcription factors is often dispensable for cell growth regulation, mutation of a single tyrosine of the IL-9R abolished both STAT activation and control of cell growth by IL-9, including protection against apoptosis and positive as well as negative effects on proliferation (Demoulin *et al.*, 1996). STAT activation may also explain the synergy observed between IL-9 and Steel factor, as observed in Mo7E cells, where Steel factor induces STAT3 phosphorylation on serine and IL-9 induces its tyrosine phosphorylation (Gotoh *et al.*, 1996).

The role of other signal transduction pathways for IL-9 activity remains more elusive. Opposing observations have been reported concerning the involvement of the ras–mitogen-activated protein (MAP) kinase pathway. On the one hand, IL-9 did not induce or enhance phosphorylation of the serine–threonine kinases Raf-1 or MAP in the Mo7E leukemia cell line (Miyazawa *et al.*, 1992). On the other hand, Raf-1 antisense oligonucleotides inhibited 70% of the response to IL-9 for the very same cell line (Brennscheidt *et al.*, 1994).

More clearly established is the activation by IL-9 of an adaptor protein called IL-4-induced phosphorylated substrate-insulin receptor substrate 2 (4PS–IRS2) (Yin *et al.*, 1995; Demoulin *et al.*, 1996), a feature shared with IL-4 signal transduction, where this pathway was shown to be critical for growth regulation (Keegan *et al.*, 1994). Phosphorylation of 4PS–IRS2 is not dependent on phosphorylation of the IL-9 receptor, contrasting with the IL-4 system in which 4PS–IRS2 associates with the IL-4 receptor through a phosphotyrosine residue. Preliminary observations suggest that 4PS–IRS2 and JAK1 activation require the same region of the IL-9 receptor (J.B. Demoulin *et al.*, unpublished results). These two molecules were also shown to associate in response to IL-9 (Yin *et al.*, 1995). Taken together, these observations raise the possibility that, upon IL-9 activation, 4PS–IRS2 becomes phosphorylated by interacting directly with the JAK1 tyrosine kinase. After phosphorylation, 4PS–IRS2 binds the SH2 domain of various signaling proteins including the p85 subunit of phosphatidylinositol-3 kinase (J.B. Demoulin *et al.*, unpublished results). Although the importance of this pathway for IL-9 activity remains to be established, the observation that overexpression of IRS1, a protein closely related to 4PS–IRS2, enhances the sensitivity of a T-cell line that proliferates in response to IL-9 (Yin *et al.*, 1995) suggests that molecules of this family are functionally involved in proliferative responses. A schematic representation of the currently described signal transduction pathways triggered by IL-9 is shown in Fig. 5.

CONCLUSIONS

Like many other cytokines, IL-9 was discovered as a growth factor, originally for mouse T and mast cell lines and subsequently for a human megakaryoblastic leukemia line. Cloning of its receptor showed that IL-9 interacts with a molecule that belongs to the hematopoietic cytokine receptor superfamily, and that, like IL-2, IL-4, IL-7 and IL-15, it also interacts with γc and activates JAK kinases and STAT factors.

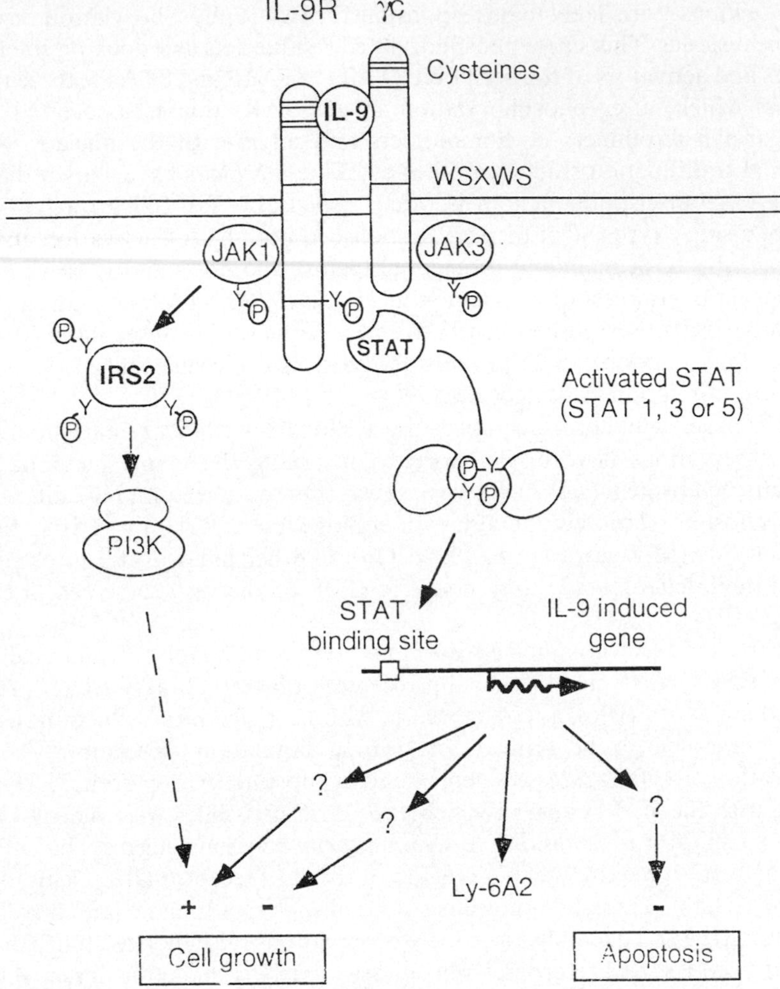

Fig. 5. Schematic representation of the signal transduction pathways triggered by IL-9.

Although these observations provide fairly essential information, they fall short of delivering a clear picture of the biologic significance of this molecule, partly because of the broad range of detected activities. If, for example, one considers only stimulation of cell growth, IL-9 has been reported to act on embryonic thymocytes, T-cell lines, mast cells, hematopoietic precursors and a variety of hematopoietic tumors. As a regulatory factor, it has been shown to induce neuronal differentiation, to stimulate IgE production by B cells, to inhibit apoptosis of T-cell tumors and to activate selective transcription of a variety of genes, including oncogenes in T cells.

Given the wide spectrum of these activities, only educated guesses can be made with regard to the essential biologic activities of this molecule. In the realm of the immune system, our present guess would be that IL-9 plays an important role in stimulating T_{H2} immunity and, conversely, in inhibiting T_{H1} responses. This conclusion derives from observations made in parasitic infections and in the NOD mouse model of autoimmunity. The fact that IL-9 plays a significant role in mast cell activation and proliferation is in line with this idea. In addition, recent genetic data point to a role for IL-9 in the complex pathogenesis of bronchial hyperresponsiveness in a mouse model (Nicolaides *et al.*, 1997). The importance of this activity now needs to be evaluated further using IL-9 antagonists or agonists *in vivo* in systems where T_{H2} responses need to be either stimulated, as in certain parasitic infections, or inhibited, as in allergy.

REFERENCES

Avanzi, G.C., Lista, P., Giovinazzo, B., Mineiro, R., Saglio, G., Benetton, G., Coda, R., Cattoretti, G. and Pegoraro, L. (1988). Selective growth response to IL-3 of a human leukemic cell line with megakaryoblastic features. *Br. J. Haematol*. **69**, 359–366.

Biessmann, H. and Mason, J.M. (1992). Genetics and molecular biology of telomeres. *Adv. Genet*. **30**, 185–249.

Birner, A., Hültner, L., Mergenthaler, H.G., Van Snick, J. and Dörmer, P. (1992). Recombinant murine interleukin 9 enhances the erythropoietin-dependent colony formation of human BFU-E. *Exp. Hematol*. **20**, 541–545.

Brennscheidt, U., Riedel, D., Kölch, W., Bonifer, R., Brach, M.A., Ahlers, A., Mertelsmann, R.H. and Herrmann, F. (1994). Raf-1 is a necessary component of the mitogenic response of the human megakaryoblastic leukemia cell line Mo7 to human stem cell factor, granulocyte-macrophage-colony-stimulating factor, interleukin-3, and interleukin-9. *Cell Growth Differ*. **5**, 367–372.

Demoulin, J.B., Uyttenhove, C., Van Roost, E., de Lestré, B., Donckers, D., Van Snick, J. and Renauld, J.-C. (1996). A single tyrosine of the interleukin-9 (IL-9) receptor is required for STAT activation, antiapoptotic activity and growth regulation by IL-9. *Mol. Cell. Biol*. **16**, 4710–4716.

Donahue, R.E., Yang, Y.C. and Clark, S.C. (1990). Human P40 T-cell growth factor (interleukin-9) supports erythroid colony formation. *Blood* **75**, 2271–2275.

Druez, C., Coulie, P., Uyttenhove, C. and Van Snick, J. (1990). Functional and biochemical characterization of mouse P40/IL-9 receptors. *J. Immunol*. **145**, 2494–2499.

Dugas, B., Renauld, J.-C., Bonnefoy, J.Y., Petit-Frère, C., Braquet, P., Van Snick, J. and Mencia-Huerta, J.M. (1993). Interleukin-9 potentiates the interleukin-4-induced immunoglobulin (IgG, IgM and IgE) production by normal human B lymphocytes. *Eur. J. Immunol*. **23**, 1687–1692.

Eklund, K.K., Ghildyal, N., Austen, K.F. and Stevens, R.L. (1993). Induction by IL-9 and suppression by IL-3 and IL-4 of the levels of chromosome 14-derived transcripts that encode late-expressed mouse mast cell proteases. *J. Immunol*. **151**, 4266–4273.

Else, K., Hültner, L. and Grencis, R. (1992). Cellular immune responses to the murine nematode parasite *Trichuris muris. Immunology* **75**, 232–237.

Faulkner, H., Humphreys, N., Renauld, J.-C., Van Snick, J. and Grencis, R. (1997). IL-9 is involved in host protective immunity to intestinal nematode infection. *Eur. J. Immunol*. (in press).

Finkelman, F.D., Madden, K.B., Cheever, A.W., Katona, I.M., Morris, S.C., Gately, M.K., Hubbard, B.R., Gause, W.C. and Urban, J.F. (1994). Effects of interleukin-12 on immune responses and host protection in mice infected with intestinal parasites. *J. Exp. Med*. **179**, 1563–1572.

Flubacher, M.M., Bear, S.E. and Tsichlis, P.N. (1994). Replacement of interleukin-2-generated mitogenic signals by a mink cell focus-forming (MCF) or xenotropic virus-induced IL-9-dependent autocrine loop: implications for MCF virus-induced leukemogenesis. *J. Virol*. **68**, 7709–7716.

Gessner, A., Blum, H. and Röllinghoff, M. (1993). Differential regulation of IL-9 expression after infection with *Leishmania major* in susceptible and resistant mice. *Immunobiology* **189**, 419–435.

Godfraind, C., Louahed, J., Faulkner, H., Vink, A., Warnier, G., Grencis, R. and Renauld, J.-C. (1998). Intraepithelial infiltration by mast cells with both connective tissue type and mucosal type characteristics in gut, trachea and kidneys of interleukin-9 transgenic mice. *J. Immunol.* (in press).

Gotoh, A., Takahira, H., Mantel, C., Litz-Jeackson, S., Boswell, H.S. and Broxmeyer, H.E. (1996). Steel factor induces serine phosphorylation of Stat3 in human growth factor-dependent myeloid cell lines. *Blood* **88**, 138–145.

Grencis, R.K., Hültner, L. and Else, K.J. (1991). Host protective immunity to *Trichinella spiralis* in mice: activation of Th cell subsets and lymphokine secretion in mice expressing different response phenotypes. *Immunology* **74**, 329–332.

Gruss, H.J., Brach, M.A., Drexler, H.G., Bross, K.J. and Herrman, F. (1992). Interleukin 9 is expressed by primary and cultured Hodgkin and Reed-Sternberg cells. *Cancer Res*. **52**, 1026–1031.

Hauber, I., Fisher, M.B., Maris, M. and Eibl, M.M. (1995). Reduced IL-2 expression upon antigen stimulation is accompanied by deficient IL-9 gene expression in T-cells of patients with CVID. *Scand. J. Immunol*. **41**, 215–219.

Hendrie, P., Miyazawa, K., Yang, Y.C., Langefeld, C. and Broxmeyer, H. (1991). Mast cell growth factor (c-kit ligand) enhances cytokine stimulation of proliferation of the human factor-dependent cell line, Mo7e. *Exp. Hematol.* **19**, 1031–1037.

Holbrook, S.T., Ohls, R.K., Schribler, K.R., Yang, Y.C. and Christensen, R.D. (1991). Effect of interleukin-9 on chronogenic maturation and cell cycle status of fetal and adult hematopoietic progenitors. *Blood* **77**, 2129–2134.

Houssiau, F., Renauld, J-C., Fibbe, W. and Van Snick, J. (1992). IL-2 dependence of IL-9 expression in human T lymphocytes. *J. Immunol*. **148**, 3147–3151.

Houssiau, F., Renauld, J.-C., Stevens, M., Lehmann, F., Coulie, P.G. and Van Snick, J. (1993). Human T cell lines and clones respond to IL-9. *J. Immunol*. **150**, 2634–2640.

Houssiau, F., Schandené, L., Stevens, M., Cambiaso, C., Goldman, M., Van Snick, J. and Renauld, J.-C. (1995). A cascade of cytokines is responsible for IL-9 expression in human T cells: involvement of IL-2, IL-4 and IL-10. *J. Immunol*. **154**, 2624–2630.

Hültner, L. and Moeller, J. (1990). Mast cell growth enhancing activity (MEA) stimulates interleukin 6 production in a mouse bone marrow-derived mast cell line and malignant subline. *Exp. Hematol*. **18**, 873–877.

Hültner L., Moeller J., Schmitt E., Jäger G., Reisbach G., Ring, J. and Dörmer, P. (1989). Thiol-sensitive mast cell lines derived from mouse bone marrow respond to a mast cell growth-enhancing activity different from both IL-3 and IL-4. *J. Immunol*. **142**, 3440–3446.

Hültner, L., Druez, C., Moeller, J., Uyttenhove, C., Schmitt, E., Rüde, E., Dörmer, P. and Van Snick, J. (1990). Mast cell growth enhancing activity (MEA) is structurally related and functionally identical to the novel mouse T cell growth factor P40/TCGFIII (interleukin-9). *Eur. J. Immunol*. **20**, 1413–1416.

Ihle, J., Keller, J., Oroszlan, S., Henderson, L., Copeland, T., Fitch, F., Prystowsky, M., Goldwasser, E., Schrader, J., Palaszynski, E., Dy, M. and Lebel, B. (1983). Biologic properties of homogenous interleukin-3. I. Demonstration of WEHI-3 growth factor activity, mast cell growth factor activity, P cell stimulating factor activity, colony stimulating factor activity, and histamine-producing cell-stimulating activity. *J. Immunol*. **131**, 282–287.

Keegan, A., Nelms, K., Wang, L.M., Pierce, J. and Paul, W. (1994). Interleukin-4 receptor: signaling mechanisms. *Immunol. Today* **15**, 423–432.

Kelleher, K., Bean, K., Clark, S.C., Leung, W.Y., Yang-Feng, T.L., Chen, J.W., Lin, P.F., Luo, W. and Yang, Y.C. (1991). Human interleukin-9: genomic sequence, chromosomal location, and sequences essential for its expression in human T-cell leukemia virus (HTLV)-I-transformed human T cells. *Blood* **77**, 1436–1441.

Kermouni, A., Van Roost, E., Arden, K.C., Vermeesch, J.R., Weiss, S., Godelaine, D., Flint, J., Lurquin, C., Szikora, J.P., Higgs, D.R., Marynen, P. and Renauld, J.-C. (1995). The IL-9 receptor gene: genomic structure, chromosomal localization in the pseudo-autosomal region of the long arm of the sex chromosomes and identification of IL-9R pseudogenes at 9pter, 10pter, 16pter and 18pter. *Genomics* **29**, 371–382.

Kimura, Y., Takeshita, T., Kondo, M., Ishii, N., Nakamura, M., Van Snick, J. and Sugamura, K. (1995). Sharing of the IL-2 receptor γ chain with the functional IL-9 receptor complex. *Int. Immunol*. **7**, 115–120.

Kubota, S., Siomi, H., Hatanaka, M. and Pomerantz, R.J. (1996). Cis/trans-activation of the interleukin-9 receptor gene in an HTLV-1 transformed human lymphocytic cell. *Oncogene* **12**, 1441–1447.

Kozak, M. (1991). An analysis of vertebrate mRNA sequences: intimations of translational control. *J. Cell. Biol*. **115**, 887–903.

Lehrnbecher, T., Poot, M., Orscheschek, K., Sebald, W., Feller, A.C. and Merz, H. (1994). Interleukin-7 and interleukin-9 drives phytohemagglutinin-activated T cells through several cell cycles: no synergism between interleukin-7, interleukin-9 and interleukin-4. *Cytokine* **6**, 279–284.

Lemoli, R.M., Fortiuna, A., Fogli, M., Motta, M.R., Rizzi, S., Benini, C. and Tura, S. (1994). Stem Cell Factor (c-kit ligand) enhances the interleukin-9-dependent proliferation of human CD34+ and CD34+CD33-DR- cells. *Exp. Hematol*. **22**, 919–924.

Lemoli, R.M., Fortuna, A., Tafuri, A., Fogli, M., Amabile, M., Grande, A., Ricciardi, M.R., Petrucci, M.T., Bonsi, L., Bagnara, G., Visani, G., Martinelli, G., Ferrari, S. and Tura, S. (1996). Interleukin-9 stimulates the proliferation of human myeloid leukemic cells. *Blood* **87**, 3852–3859.

Louahed, J., Kermouni, A., Van Snick, J. and Renauld, J.-C. (1995). IL-9 induces expression of granzymes and high affinity IgE receptors in murine T helper clones. *J. Immunol*. **154**, 5061–5070.

Lu, L., Leemhuis, T., Srour, E. and Yang, Y.C. (1992). Human interleukin (IL)-9 specifically stimulates proliferation of CD34+++DR+CD33-erythroid progenitors in normal human bone marrow in the absence of serum. *Exp. Hematol*. **20**, 418–424.

Madden, K., Urban, K., Ziltener, H., Schrader, J., Finkelman, F. and Katona, I. (1991). Antibodies to IL-3 and IL-4 suppress helminth-induced intestinal mastocytosis. *J. Immunol*. **147**, 1387–1391.

Mehler, M.F., Rozental, R., Dougherty, M., Spray, D.C. and Kessler, J.A. (1993). Cytokine regulation of neuronal differentiation of hippocampal progenitor cells. *Nature* **362**, 62–65.

Merz, H., Houssiau, F., Orscheschek, K., Renauld, J.-C., Fliedner, A., Herin, M., Noël, H., Kadin, M., Mueller-Hermelink, K.H., Van Snick, J. and Feller, A.C. (1991). IL-9 expression in human malignant lymphomas: unique association with Hodgkin's disease and large cell anaplastic lymphoma. *Blood* **78**, 1311–1317.

Miyazawa, K., Hendrie, P., Kim, Y.J., Mantel, C., Yang, Y.C., Se Kwon, B. and Broxmeyer, H. (1992). Recombinant human interleukin-9 induces protein tyrosine phosphorylation and synergizes with steel factor to stimulate proliferation of the human factor-dependent cell line Mo7e. *Blood* **80**, 1685–1692.

Mock, B.A., Krall, M., Kozak, C.A., Nesbitt, M.N., McBride, O.W., Renauld, J.-C. and Van Snick, J. (1990). *IL-9* maps to mouse chromosome 13 and human chromosome 5. *Immunogenetics* **31**, 265–270.

Modi, W.S., Pollock, D.D., Mock, B.A., Banner, C., Renauld, J.-C. and Van Snick, J. (1991). Regional localization of the human glutaminase (GLS) and interleukin-9 (IL-9) genes by in situ hybridization. *Cytogenet. Cell. Genet*. **57**, 114–116.

Moeller, J., Hültner, L., Schmitt, E. and Dörmer, P. (1989). Partial purification of a mast cell growth-enhancing activity and its separation from IL-3 and IL-4. *J. Immunol*. **142**, 3447–3451.

Moeller, J., Hültner, L., Schmitt, E., Breuer, M. and Dörmer, P. (1990). Purification of MEA, a mast cell growth-enhancing activity to apparent homogeneity and its partial amino acid sequencing. *J. Immunol*. **144**, 4231–4234.

Mossman, T., Bond, M., Coffman, R., Ohara, J. and Paul, W. (1986). T-cell and mast cell lines respond to B-cell stimulatory factor 1. *Proc. Natl Acad. Sci. U.S.A*. **83**, 5654–5658.

Murakami, M., Narazaki, M., Hibi, M., Yawata, H., Yasukawa, K., Hamaguchi, M., Taga, T. and Kishimoto, T. (1991). Critical cytoplasmic region of the interleukin-6 signal transducer gp130 is conserved in the cytokine receptor family. *Proc. Natl Acad. Sci. U.S.A*. **88**, 11 349–11 353.

Nicolaides, N.C., Holroyd, K.J., Ewart, S.L., Eleff, S.M., Kiser, M.B., Dragwa, C.R., Sullivan, C.D., Grasso, L., Zhang, L.Y., Messler, C.J., Zhou, T., Kleeberger, S.R., Buetow, K.H. and Levitt, R.C. (1997). Interleukin 9: a candidate gene for asthma. *Proc. Natl Acad. Sci. U.S.A*. **94**, 13175–13180.

Petit-Frère, C., Dugas, B., Braquet, P. and Mencia-Huerta, J.M. (1993). Interleukin-9 potentiates the interleukin-4-induced IgE and IgG1 release from murine B lymphocytes. *Immunology* **79**, 146–151.

Renauld, J.-C., Goethals, A., Houssiau, F., Van Roost, E. and Van Snick, J. (1990a). Cloning and expression of a cDNA for the human homology of mouse T-cell and mast cell growth factor P40. *Cytokine* **2**, 9–12.

Renauld, J.-C., Goethals, A., Houssiau, F., Merz, H., Van Roost, E. and Van Snick, J. (1990b). Human P40/IL-9-expression in activated $CD4^+$ T cells, genomic organization, and comparison with the mouse gene. *J. Immunol*. **144**, 4235–4241.

Renauld, J.-C., Druez, C., Kermouni, A., Houssiau, F., Uyttenhove, C., Van Roost, E. and Van Snick, J. (1992). Expression cloning of the murine and human interleukin 9 receptor cDNAs. *Proc. Natl Acad. Sci. U.S.A*. **89**, 5690–5694.

Renauld, J.-C., van der Lugt, N., Vink, A., van Roon, M., Godfraind, C., Warnier, G., Merz, H., Feller, A., Berns, A. and Van Snick, J. (1994). *Oncogene* **9**, 1327–1332.

Renauld, J.-C., Vink, A., Louahed, J. and Van Snick, J. (1995). Thymic lymphomas in interleukin 9 transgenic mice. *Blood* **85**, 1300–1305.

Rich, B., Campos-Torres, J., Tepper, R., Moreadith, R. and Leder, P. (1993). Cutaneous lymphoproliferation and lymphomas in interleukin 7 transgenic mice. *J. Exp. Med*. **177**, 305–317.

Russel, S.M., Johnston, J.A., Noguchi, M., Kawamura, M., Bacon, C.M., Friedman, M., Berg, M., McVicar, D.W., Witthuhn, B.A., Silvennonien, O., Goldmann, A.S., Schmalstieg, F.C., Ihle, J.M., O'Shea, J.J. and Leonard, W.J. (1994). Interaction of IL-2RB and γc chains with Jak1 and Jak3: implications for XSCID and XCID. *Science* **266**, 1042–1045.

Schaafsma, M.R., Falkenburg, J.H., Duinkerken, N., Van Snick, J., Landergent, J.E., Willemze, R. and Fibbe, W.E. (1993). Interleukin-9 stimulates the proliferation of enriched human erythroid progenitor cells: additive effect with GM-CSF. *Ann. Hematol*. **66**, 45–49.

Schmitt, E., van Brandwijk, R., Van Snick, J., Siebold, B. and Rüde, E. (1989). TCGFIII/P40 is produced by naive murine CD4+ T cells but is not a general T cell growth factor. *Eur. J. Immunol*. **19**, 2167–2170.

Schmitt, E., Beuscher, H.U., Huels, C., Monteyne, P., van Brandwijk, R., Van Snick, J. and Rüde, E. (1991). IL-1 serves as a secondary signal for IL-9 expression. *J. Immunol*. **147**, 3848–3854.

Schmitt, E., Germann, T., Goedert, S., Hoehn, P., Huels, C., Koelsch, S., Kuhn, R., Muller, W., Palm, N. and Rüde, E. (1994). IL-9 production of naive CD4+ T cells depends on IL-2, is synergistically enhanced by a combination of TGF-B and IL-4, and is inhibited by IFN-γ. *J. Immunol*. **153**, 3989–3996.

Schumann, R.R., Nakarai, T., Gruss, H.-J., Brach, M.A., von Arnim, U., Kirschning, C., Karawajew, L., Ludwig, W.-D., Renauld, J.-C., Ritz, J. and Herrmann, F. (1996). Transcript synthesis and surface expression of the interleukin-2 receptor (α-, β- and γ-chain) by normal and malignant myeloid cells. *Blood* **87**, 2419–2427.

Simpson, R.J., Moritz, R.L., Gorman, J.J. and Van Snick, J. (1989). Complete amino acid sequence of a new murine T cell growth factor P40. *Eur. J. Biochem*. **183**, 715–722.

Sokal, G., Michaux, J.L., Van Den Berghe, H., Cordier, A., Rodhain, J., Ferrant, A., Moriau, M., De Bruyère, M. and Sonnet, J. (1975). A new hematologic syndrome with a distinct karyotype: the 5q-chromosome. *Blood* **46**, 519–533.

Sonoda, Y., Maekawa, T., Kuzuyama, Y., Clark, S. and Abe, T. (1992). Human interleukin-9 supports formation of a subpopulation of erythroid bursts that are responsive to interleukin-3. *Am. J. Hematol*. **41**, 84–91.

Sonoda, Y., Sakabe, H., Ohmisono, Y., Tanimukai, S., Yokota, S., Nakagawa S., Clark S.C. and Abe, T. (1994). Synergistic actions of stem cell factor and other burst-promoting activities on proliferation of CD34+ highly purified blood progenitors expressing HLA-DR or different levels of c-kit protein. *Blood* **84**, 4099–4106.

Suda, T., Murray, R., Fischer, M., Tokota, T. and Zlotnik, A. (1990a). Tumor necrosis factor-alpha and P40 induce day 15 murine fetal thymocyte proliferation in combination with IL-2. *J. Immunol*. **144**, 1783–1787.

Suda, T., Murray, R., Guidos, C. and Zlotnik, A. (1990b). Growth-promoting activity of IL-1a, IL-6, and tumor necrosis factor-a in combination with IL-2, IL-4, or IL-7 on murine thymocytes. Differential effects on CD4/CD8 subsets and on CD3+/CD3− double negative thymocytes. *J. Immunol*. **144**, 3039–3045.

Svetic, A., Finkelman, F.D., Jian, Y.C., Dieffenbach, C.W., Scott, D.E., McCarthy, K.F., Steinberg, A.D. and Gause, W.C. (1991). Cytokine gene expression after *in vivo* primary immunization with goat antibody to mouse IgD antibody. *J. Immunol*. **147**, 2391–2397.

Svetic, A., Madden, K.B., Zhou, X.D., Lu, P., Katona, I.M., Finkelman, F.D., Urban J.F. and Gause, W.C. (1993). A primary intestinal helminthic infection rapidly induces a gut-associated elevation of the Th2-associated cytokines and IL-3. *J. Immunol*. **150**, 3434–3441.

Trümper, L.H., Brady, G., Bagg, A., Gray, D., Loke, S.L., Griesser, H., Wagman, R., Graziel, R., Gascoyne, R.D., Vicini, S., Iscove, N.N., Cossman, J. and Mak, T.W. (1993). Single-cell analysis of Hodgkin and Reed-Sternberg cells: molecular heterogeneity of gene expression and P53 mutations. *Blood* **81**, 3097–3115.

Uyttenhove, C., Simpson, R.J. and Van Snick, J. (1988). Functional and structural characterization of P40, a mouse glycoprotein with T cell growth factor activity. *Proc. Natl Acad. Sci. U.S.A*. **85**, 6934–6938.

Uyttenhove, C., Druez, C., Renauld, J.-C., Herin, M., Noel, H. and Van Snick, J. (1991). Autonomous growth and tumorigenicity induced by P40/interleukin-9 cDNA transfection of a mouse P40-dependent T-cell line. *J. Exp. Med*. **173**, 519–522.

Van Damme, J., Uyttenhove, C., Houssiau, F., Put, W., Proost, P. and Van Snick, J. (1992). Human growth factor for murine interleukin (IL)-9 responsive T cell lines: co-induction with IL-6 in fibroblasts and identification as LIF-HILDA. *Eur. J. Immunol*. **22**, 2801–2808.

Van Snick, J., Goethals, A., Renauld, J.-C., Van Roost, E., Uyttenhove, C., Rubira, M.R., Moritz, R.L. and Simpson, R.J. (1989). Cloning and characterization of a cDNA for a new mouse T cell growth factor (P40). *J. Exp. Med*. **169**, 363–368.

Vermeesch, J.R., Petit, P., Kermouni, A., Renauld, J.-C., Van Den Berghe, H. and Marynen, P. (1997). The IL-9 receptor gene, located in the Xq/Yq pseudoautosomal region, has an autosomal origin and escapes X inactivation. *Hum. Mol. Genet*. **6**, 1–8.

Vink, A., Renauld, J.-C., Warnier, G. and Van Snick, J. (1993). Interleukin-9 stimulates *in vitro* growth of mouse thymic lymphomas. *Eur. J. Immunol*. **23**, 1134–1138.

Warrington, J., Bailey, S., Armstrong, E., Aprelikova, O., Alitalo, K., Dolganov, G., Wilcox, A., Sikela, J., Wolfe, S., Lovett, M. and Vasmuth, J. (1992). A radiation hybrid map of 18 growth factor, growth factor receptor, hormone receptor, or neurotransmitter receptor genes on the distal region of the long arm of chromosome 5. *Genomics* **13**, 803–808.

Williams, D.E., Morrissey, P.J., Mochizuki, D.Y., de Vries, P., Anderson, D., Cosman, D., Boswell, H.S., Cooper, S., Grabstein, K.H. and Broxmeyer, H.E. (1990). T-cell growth factor P40 promotes the proliferation of myeloid cell lines and enhances erythroid burst formation by normal murine bone marrow cells *in vitro*. *Blood* **76**, 906–911.

Yang, Y., Ricciardi, S., Ciarletta, A., Calvetti, J., Kelleher, K. and Clark, S.C. (1989). Expression cloning of a cDNA encoding a novel human hematopoietic growth factor: human homologue of mouse T-cell growth factor. *Blood* **74**, 1880–1884.

Yin, T., Lik-Shing Tsang, M. and Yang, Y.-C. (1994). JAK1 kinase forms complexes with interleukin-4 receptor and 4ps/insulin receptor substrate-1-like protein and is activated by interleukin-4 and interleukin-9 in T lymphocytes. *J. Biol. Chem*. **269**, 26 614–26 617.

Yin, T., Keller, S.R., Quelle, F.W., Witthuhn, B.A., Lik-Shing Tsang, M., Lienhard, G.E., Ihle, J.N. and Yang, Y.-C. (1995). Interleukin-9 induces tyrosine phosphorylation of insulin receptor substrate-1 via JAK tyrosine kinases. *J. Biol. Chem*. **270**, 20 497–20 502.

Zhu, Y.X., Kang, L.Y., Luo, W., Li, C.C.H., Yang, L. and Yang, Y.C. (1996). Multiple transcription factors are required for activation of human interleukin 9 gene in T cells. *J. Biol. Chem*. **271**, 15 815–15 822.

Chapter 12

Interleukin-10

Rene de Waal Malefyt[1] and Kevin W. Moore[2]

Departments of [1]Immunobiology and [2]Molecular Biology, DNAX Research Institute of Molecular and Cellular Biology, Palo Alto, CA, USA

INTRODUCTION

The cytokine interleukin-10 (IL-10) was discovered as 'cytokine synthesis inhibitory factor' (CSIF) which was produced by mouse T_{H2} clones and inhibited cytokine production by activated T helper 1 (T_{H1}) clones (Fiorentino *et al.*, 1989). cDNA clones for mouse and human IL-10 (mIL-10, hIL-10) were described shortly thereafter, and the relationship of IL-10 with an Epstein–Barr virus (EBV) gene (viral IL-10; vIL-10) was uncovered (Moore *et al.*, 1990; Vieira *et al.*, 1991). The ability of IL-10 to inhibit cytokine synthesis by both T cells (de Waal Malefyt *et al.*, 1991b; Fiorentino *et al.*, 1991b) and natural killer (NK) cells (Hsu *et al.*, 1992) was soon found to be due to its inhibitory effects on the macrophage–monocyte accessory cell. This inhibition of activation of cells of the macrophage–monocyte and dendritic cell lineages led to IL-10 being termed a 'macrophage deactivating factor' (Bogdan *et al.*, 1991; Moore *et al.*, 1993; Ho and Moore, 1994).

Use of recombinant IL-10 and anti-IL-10 antibodies (Mosmann *et al.*, 1990; de Waal Malefyt *et al.*, 1991a) for *in vitro* and *in vivo* experimentation revealed additional activities of the cytokine on mast cells, T cells and B cells (Go *et al.*, 1990; MacNeil *et al.*, 1990; Thompson-Snipes *et al.*, 1991; Rousset *et al.*, 1992; de Waal Malefyt *et al.*, 1993b), and confirmed *in vivo* several of the activities observed *in vitro* (see below). The severe chronic inflammatory bowel disease and other exaggerated inflammatory responses manifested by IL-10-deficient (IL-$10^{-/-}$) mice showed that a crucial role of IL-10 *in vivo* is limitation of inflammatory responses (Kuhn *et al.*, 1993; Berg *et al.*, 1995a,b). The first reports of *in vivo* studies in humans have also appeared (Chernoff *et al.*, 1995; Schreiber *et al.*, 1995; Pajkrt *et al.*, 1997).

Mouse and human receptors for IL-10 have been cloned and characterized (Ho *et al.*, 1993; Tan *et al.*, 1993; Liu *et al.*, 1994), and several mechanisms of signal transduction described (Finbloom and Winestock, 1995; Ho *et al.*, 1995; Weber-Nordt *et al.*, 1996b; Wehinger *et al.*, 1996). The ability of IL-10 to inhibit macrophage activation offers an opportunity to understand signal transduction mechanisms involved in 'cross-talk' between stimulatory and inhibitory cytokines, about which little is known.

A recent database search revealed well over 1000 reports of research involving IL-10. This chapter aims to provide an overview of the cytokine and its functions.

The Cytokine Handbook, 3rd ed.
ISBN 0–12–689662–3

IL-10 BIOCHEMISTRY AND MOLECULAR BIOLOGY

mIL-10, hIL-10 and rat IL-10 amino acid sequences were deduced from cloned cDNAs (Moore *et al.*, 1990; Vieira *et al.*, 1991; Goodman *et al.*, 1992). A recent search of the Genbank database yielded entries for dog, cat, cow, rabbit, gerbil, sheep, pig, red deer, pigtailed macaque, mangabey and rhesus monkey IL-10 cDNA sequences. The predicted 178-amino-acid sequences of mIL-10 and hIL-10 were 73% identical and, together with the X-ray structures, revealed IL-10 as a 'long chain' four α-helical bundle cytokine (Sprang and Bazan, 1993; Walter and Nagabhushan, 1995; Zdanov *et al.*, 1995).

Biochemical evidence indicated that mIL-10 and hIL-10 were acid-sensitive noncovalent homodimers (Fiorentino *et al.*, 1989; Tan *et al.*, 1993; Windsor *et al.*, 1993). The X-ray structures of hIL-10 (Walter and Nagabhushan, 1995; Zdanov *et al.*, 1995) and vIL-10 (Zdanov *et al.*, 1997) supported this notion and revealed that IL-10 is a dimer of two interpenetrating polypeptide chains. It is not certain whether the monomeric cytokine is biologically active or can even form a proper cytokine-like tertiary structure. Both mIL-10 and hIL-10 contain two intrachain disulfide bonds; mIL-10 has a fifth cysteine which remains unpaired. Reduction of the disulfide bonds abolishes the activity of hIL-10 (Windsor *et al.*, 1993).

Recombinant hIL-10 is an 18-kDa polypeptide which is not glycosylated. Both recombinant and T cell-derived mIL-10 are expressed as a mixture of 17-, 19- and 21-kDa polypeptides due to heterogeneous *N*-glycosylation at a site near the *N*-terminus (Moore *et al.*, 1990; Mosmann *et al.*, 1990). Glycosylation of mIL-10 has no known effect on biologic activity. hIL-10 is active on both mouse and human cells, but mIL-10 is not active on human cells. At least one rat monoclonal antibody cross-reacts with mIL-10 and hIL-10 (Liu *et al.*, 1994).

The mIL-10 and hIL-10 genes are composed of five exons and are located on the respective chromosomes 1 (Kim *et al.*, 1992; de Waal Malefyt and de Vries, 1996). Transcription generates predominantly single mRNAs of approximately 1.4 kilobases (kb) (mIL-10) or 2 kb (hIL-10), although a shorter mRNA was reported in a mouse T helper 2 (T_{H2}) clone (Moore *et al.*, 1990). A number of potential transcriptional regulatory sequence elements were identified in the mIL-10 and hIL-10 genes which together somewhat resembled those found in IL-6 genes. Relatively little is known about regulation of IL-10 transcription, except that in most cases it requires activation of the cell. However, mIL-10 production stimulated by lipopolysaccharide (LPS) *in vivo* was not inhibited by cyclosporin, whereas anti-CD3 stimulated IL-10 expression was (Durez *et al.*, 1993; Blancho *et al.*, 1995), suggesting that stimulation of T-cell IL-10 production (anti-CD3) could be inhibited by cyclosporin, whereas monocyte IL-10 expression (LPS) could not. These data suggested different cell type-specific regulatory mechanisms of IL-10 expression.

VIRAL HOMOLOGS OF IL-10

The EBV, equine herpes virus 2 (EHV2), and poxvirus Orf virus genomes encode viral homologs of IL-10 (Vieira *et al.*, 1991; Moore *et al.*, 1993; Rode *et al.*, 1993; Fleming *et al.*, 1997) which, unlike the genomic IL-10 genes, lack introns. Homology to the mammalian cytokines lies principally in the mature protein, with very little similarity found in the leader peptide or flanking untranslated DNA. Excluding the signal sequences, the hIL-10 and EBV (vIL-10) amino acid sequences are 84% identical, with most differences in the *N*-terminal 20 amino acids. Most anti-hIL-10 antibodies cross-react with vIL-10 (de Waal Malefyt *et al.*, 1991a; Ho and Moore, 1994; Liu *et al.*, 1997). The EHV2 and Orf IL-10 homologs also differ from the mouse, rat and human proteins, mostly in the *N*-terminal region.

vIL-10 is a 17-kDa non-glycosylated polypeptide which, as discussed below, expresses some but not all of the activities of IL-10 *in vitro* and *in vivo* (Go *et al.*, 1990; MacNeil *et al.*, 1990; Vieira *et al.*, 1991; Suzuki *et al.*, 1995; Berman *et al.*, 1996; Qin *et al.*, 1996; Liu *et al.*, 1997). Moreover, in *in vitro* systems where vIL-10 lacked activity, it did not detectably antagonize the activity of the cellular cytokine (cIL-10) (Moore *et al.*, 1993), consistent with the observed 1000-fold or greater impaired binding of vIL-10 to the IL-10 receptor (IL-10R) (Liu *et al.*, 1997). These observations suggest the existence of different IL-10R complexes and/or signal transduction mechanisms on vIL-10-responsive and vIL-10-non-responsive cells.

THE IL-10 RECEPTOR

Mouse and human IL-10R are members of the interferon receptor family which bind ligand (mIL-10 or hIL-10) with high affinity (K_d approximately 35–200 pM) (Ho *et al.*, 1993; Tan *et al.*, 1993; Liu *et al.*, 1994, 1997). The affinity for vIL-10 is at least 1000-fold lower (Liu *et al.*, 1997). Cross-linking studies with ^{35}S-Met- or ^{125}I-hIL-10, and more recent immunoprecipitation data, indicate a molecular weight of 90–120 kDa for IL-10R, although a number of both larger and smaller ^{125}I-labeled species have been reported (Tan *et al.*, 1993), suggesting proteolytic degradation or possibly multiple polypeptides cross-linked to ^{125}I-hIL-10.

mIL-10R and hIL-10R cDNAs have been isolated from mouse macrophage and mast cell lines, and from a human B-cell line. The hIL-10R cDNA sequence is 70% homologous to the mIL-10R cDNA. The predicted hIL-10R and mIL-10R amino acid sequences are 578 and 576 amino acids respectively, and are 60% identical. mIL-10R encoded by the mast cell and macrophage-derived cDNA clones are identical, despite differences in the responses of these cells to cIL-10 and vIL-10 (Ho *et al.*, 1993). hIL-10 binds to m,hIL-10R, and mIL-10 binds only to mIL-10R, as expected from the species specificity of IL-10. IL-10R mRNA was detected in all IL-10-responsive cells examined (Ho *et al.*, 1993; Liu *et al.*, 1994). Most hemopoietic cells express IL-10R, although at very low levels of a few hundred IL-10R per cell (Ho *et al.*, 1993; Tan *et al.*, 1993; Liu *et al.*, 1994, 1997; Carson *et al.*, 1995). IL-10R expression is induced in fibroblasts by LPS (Weber-Nordt *et al.*, 1994).

An anti-hIL-10R monoclonal antibody (mAb) was reported, which is an antagonist of the receptor (Liu *et al.*, 1997). This antibody blocked all known responses to hIL-10 and vIL-10, indicating that IL-10R is required for responses to both the cellular and viral cytokines. Recombinant soluble hIL-10R (shIL-10R) has been expressed, and also exhibited antagonist activity *in vitro* (Tan *et al.*, 1995; Liu *et al.*, 1997). Evidence suggested that hIL-10–shIL-10R complexes could be multimeric, consisting of up to two ligand dimers bound to up to four shIL-10R molecules (Tan *et al.*, 1995). So far no evidence has been reported for detection of sIL-10R *in vivo*.

Several lines of evidence suggest that there are additional IL-10R components. A mouse fibroblast cell line transfected with mIL-10R showed no evidence of signaling in response to ligand (Weber-Nordt *et al.*, 1994). Second, the collective observations that (a) some cells respond poorly to vIL-10 but normally to cIL-10, whereas others are comparably responsive (Ho and Moore, 1994; Liu *et al.*, 1997), (b) the affinity of vIL-10 for IL-10R is 1000-fold or more lower than that of cIL-10, and (c) IL-10 activates tyrosine kinases JAK1 and TYK2 (see below), but IL-10R associates only with JAK1 (S. Wei, K. Moore and A. Mui, unpublished results) all point to the existence of at least one, and perhaps more, additional IL-10R subunits. So far, we have been unable to demonstrate that the other known interferon (IFN) receptor family IFN-αR1, IFN-γRβ and class II cytokine receptor family (CRF2–4) members can complement deficiency in responsiveness to vIL-10 (Liu *et al.*, 1997).

SIGNAL TRANSDUCTION IN RESPONSE TO IL-10

Mouse pro-B cell (Ba/F3) and human erythroleukemia cell (TF1) lines transfected with wild-type and mutant IL-10R were used to study IL-10R function (Ho *et al.*, 1995; Liu *et al.*, 1997). These cell lines both exhibited short-term proliferative responses to IL-10. In addition, Ba/F3 transfectants expressed new cell surface antigens when stimulated with IL-10, and TF1 cells exhibited enhanced FcγR expression (C. Parham and K. Moore, unpublished results). Site-directed mutagenesis and deletion analysis enabled delineation of functional domains of mIL-10R, including those mediating proliferation, induction of cell surface antigen expression and a 'negative regulatory domain', which may function in deactivation of activated IL-10R or associated molecules.

The most studied aspect of IL-10 signaling has been the JAK–STAT system. Like IFN-α, IL-10 induces tyrosine phosphorylation of tyrosine kinases JAK1 and TYK2 (Finbloom and Winestock, 1995; Ho *et al.*, 1995). In addition, tyrosine phosphorylated, DNA binding STAT1, STAT3 and, in some cases, STAT5 were detected in cells stimulated with IL-10 (Lai *et al.*, 1996; Weber-Nordt *et al.*, 1996b; Wehinger *et al.*, 1996). Both STAT activation and biologic responses to IL-10 require the presence of at least one of two tyrosine residues (Y427 and Y477) in the mIL-10R cytoplasmic domain. These two tyrosines appear to be substantially functionally redundant with respect to IL-10-induced short-term proliferation and STAT activation, and have been implicated directly in recruitment of STAT3 to IL-10R (Ho *et al.*, 1995; Weber-Nordt *et al.*, 1996b). IL-10 induces transcription and expression of STAT3-responsive and possibly one or more STAT1-responsive genes (FcγR) (te Velde *et al.*, 1992; Ho *et al.*, 1995; Lai *et al.*, 1996).

A few studies suggested involvement of other signaling pathways. IL-10 reportedly inhibits nuclear factor-κB (NF-κB) activation in response to macrophage activation (Wang *et al.*, 1995), but activates activating protein-1 (AP-1) and NF-κB in $CD8^+$ T cells (Hurme *et al.*, 1994). IL-10 induced Bcl-2 expression in $CD34^+$ progenitors and germinal center B cells (Levy and Brouet, 1994; Weber-Nordt *et al.*, 1996a), and also activated c-fos expression in human B cells (Bonig *et al.*, 1996). In monocytes, IL-10 activated p85 phosphatidylinositol-3 (PI-3) and p70 S6 kinases (Crawley *et al.*, 1996). Activation of the raf–ras–mitogen activated kinase (MAP) cascade does not occur in response to IL-10, and may be inhibited in some cases (Geng *et al.*, 1994).

PRODUCTION OF IL-10

Monocytes–macrophages, T cells and B cells express IL-10 following activation. Monocytes–macrophages produce IL-10 after stimulation by LPS, immune complexes or cross-linking of CD23 receptors (Fiorentino *et al.*, 1991a; de Waal Malefyt *et al.*, 1991a; Tripp *et al.*, 1995; Dugas *et al.*, 1996). LPS-induced IL-10 production is enhanced by tumor necrosis factor-α (TNF-α) and agents that increase intracellular cAMP levels such as α-melanocyte-stimulating hormone, prostaglandin E_2, cyclic nucleotide phosphodiesterase type IV inhibitors, adenosine, catecholamines (epinephrine) and chlorpromazine (dopamine receptor type I agonist), and is inhibited by IL-4, IL-10 itself, IL-13, IFN-γ and transforming growth factor-β (TGF-β) (de Waal Malefyt *et al.*, 1991a, 1993a; Chomarat *et al.*, 1993; Wanidworanun and Strober, 1993; Mengozzi *et al.*, 1994; Kambayashi *et al.*, 1995; Reinhold *et al.*, 1995; Tarazona *et al.*, 1995; Bhardwaj *et al.*, 1996; Huang *et al.*, 1996; Le Moine *et al.*, 1996; van der Poll *et al.*, 1996a).

Certain epithelial cells, such as normal human bronchial epithelial cells, produce IL-10 constitutively (Bonfield *et al.*, 1995), whereas keratinocytes or keratinocyte cell lines secrete IL-10 upon exposure to ultraviolet light in mice (Enk and Katz, 1992; Rivas and Ullrich, 1992), but not in humans (Kang *et al.*, 1994; Enk *et al.*, 1995; Teunissen *et al.*, 1997). Human placental cytotrophoblasts produce IL-10 (Roth *et al.*, 1996), and constitutive expression of IL-10 mRNA and protein was shown in decidua of pregnant mice (Lin *et al.*, 1993).

IL-10 is produced predominantly by T_{H2} and T_{H0} T-cell clones in the mouse (Fiorentino *et al.*, 1989; Moore *et al.*, 1993), but T_{H2}, T_{H0} and T_{H1} T-cell subsets are able to produce IL-10 following activation in humans (Yssel *et al.*, 1992; Del Prete *et al.*, 1993). Both memory and naive T cells isolated from adult peripheral blood, as well as cord blood T cells, produce IL-10 (Yssel *et al.*, 1992). IL-10 production by T cells is enhanced by IL-12 and IFN-α (Aman *et al.*, 1996; Daftarian *et al.*, 1996; Jeannin *et al.*, 1996; Meyaard *et al.*, 1996), and can be blocked by cyclosporin A (de Waal Malefyt and de Vries, 1996). IL-12 also primes T cells for high levels of IL-10 production in short-term cultures (Gerosa *et al.*, 1996), and long-term culture of T cells in the presence of IL-10 results in the differentiation of T cells that are able to produce high levels of IL-10 (De Vries *et al.*, 1997). Together with the inhibitory effects of IL-10 on its own production by monocytes, these results demonstrate that IL-10 has a cell type-dependent positive or negative autoregulation.

B cells also produce IL-10 following activation (O'Garra *et al.*, 1990). Ly1 (B-1) B cells are the main IL-10-producing B-cell subset in mice (O'Garra *et al.*, 1992), and EBV-transformed human B cells produce high levels of IL-10 (Benjamin *et al.*, 1992; Burdin *et al.*, 1993). IL-10 may act as an autocrine growth factor for B-1 B cells or their malignancies and EBV transformed cells (Ishida *et al.*, 1992; Peng *et al.*, 1995; Ramachandra *et al.*, 1996). LPS-activated mouse mesanginal cells produce IL-10 (Fouqueray *et al.*, 1995) and IL-10 is produced by certain tumor cell lines including melanomas (Chen *et al.*, 1994; Dummer *et al.*, 1996) and a variety of carcinomas (Gastl *et al.*, 1993; Smith *et al.*, 1994a; Kim *et al.*, 1995).

EFFECTS OF IL-10 ON MONOCYTES–MACROPHAGES AND DENTRITIC CELLS

IL-10 significantly modulates expression of cytokines, soluble mediators and cell surface molecules on cells of myeloid origin. These changes have important consequences for the functions of these cells in activating and maintaining immune and inflammatory responses. IL-10 strongly inhibits the production of IL-1α, IL-1β, IL-6, IL-8, IL-10 itself, IL-12, granulocyte–macrophage colony stimulating factor (GM-CSF), G-CSF, M-CSF, TNF-α, macrophage inflammatory protein-1 (MIP-1α), MIP-1β, MIP-2, RANTES and leukemia inhibitory factor (LIF) by activated monocytes–macrophages (Fiorentino *et al.*, 1991a; de Waal Malefyt *et al.*, 1991a, 1993a; D'Andrea *et al.*, 1993; Gruber *et al.*, 1994; Berkman *et al.*, 1995; Marfaing-Koka *et al.*, 1996). However, IL-10 enhanced production of IL-1 receptor agonist (IL-1RA) (de Waal Malefyt *et al.*, 1991a; Cassatella *et al.*, 1994; Jenkins *et al.*, 1994) and expression of soluble p55 and p75 TNFR (Hart *et al.*, 1996; Joyce and Steer, 1996; Linderholm *et al.*, 1996), indicating that IL-10 induces a shift from production of proinflammatory to anti-inflammatory mediators. IL-10 also inhibited expression of major histocompatibility complex (MHC) class II antigens, CD54 (intercellular adhesion molecule-1), CD80 (B7.1) and CD86 (B7.2) on monocytes, even following induction of these molecules by IL-4 or IFN-γ (de Waal Malefyt *et al.*, 1991b; Ding *et al.*, 1993; Kubin *et al.*, 1994; Willems *et al.*, 1994). Downregulation of these stimulatory or costimulatory molecules significantly affected the T cell-activating capacity of monocytes as antigen-presenting cells (APCs) (de Waal Malefyt *et al.*, 1991b; Fiorentino *et al.*, 1991b; Ding and Shevach, 1992).

In contrast, IL-10 upregulated expression of FcR on monocytes, including CD16 and CD64 (te Velde *et al.*, 1992; de Waal Malefyt *et al.*, 1993a; Calzada-Wack *et al.*, 1996) but downregulated the expression of IL-4-induced CD23 (Morinobu *et al.*, 1996). Upregulation of CD64 correlated with enhanced antibody-dependent cell-mediated cytotoxicity (te Velde *et al.*, 1992). However, in contrast, IL-10 inhibited tumor cytotoxicity (Nabioullin *et al.*, 1994). Upregulation of FcR by IL-10 enhanced the capacity of monocytes–macrophages to phagocytose opsonized particles, bacteria or fungi (Capsoni *et al.*, 1995; Spittler *et al.*, 1995), but IL-10 reduced the ability of the cells to kill the ingested organisms by decreasing the generation of superoxide anion (O_2^-) and nitric oxide (NO) (Bogdan *et al.*, 1991; Cunha *et al.*, 1992; Gazzinelli *et al.*, 1992; Niiro *et al.*, 1992, 1995; Cenci *et al.*, 1993; Wu *et al.*, 1993; Kuga *et al.*, 1996; Roilides *et al.*, 1997). The inhibitory effects of IL-10 on production of NO by mouse macrophages can occur by an indirect mechanism involving inhibition of cytokine synthesis (Oswald *et al.*, 1992a,b; Flesch *et al.*, 1994). IL-10 inhibited production of prostaglandin (PG) E_2, another proinflammatory mediator (Niiro *et al.*, 1994). IL-10 inhibited the ability of monocytes–macrophages to modulate turnover of extracellular matrix through its inhibitory effects on the production of gelatinase and collagenase, and its ability to enhance production of tissue inhibitor of metalloproteinases (Mertz *et al.*, 1994; Lacraz *et al.*, 1995). IL-10 also inhibited C-reactive protein and LPS, but not CD40-induced tissue factor expression by human monocytes, which leads to a reduction in procoagulant activity (Pradier *et al.*, 1993, 1996; Ramani *et al.*, 1993, 1994; Jungi *et al.*, 1994).

The effects of IL-10 on cytokine production and function of human macrophages are generally similar to those on monocytes, although a little less pronounced (Berkman *et al.*, 1995; Nicod *et al.*, 1995; Wilkes *et al.*, 1995; Armstrong *et al.*, 1996; Park and Skerrett, 1996; Thomassen *et al.*, 1996; Zissel *et al.*, 1996). Furthermore, IL-10 inhibited production of IL-12 and expresssion of costimulatory molecules on various types of dendritic cells (Macatonia *et al.*, 1993; Peguet-Navarro *et al.*, 1994; Buelens *et al.*, 1995; Ludewig *et al.*, 1995; Mitra *et al.*, 1995), which correlates with its ability to inhibit primary alloantigen-specific T-cell responses (Bejarano *et al.*, 1992; Caux *et al.*, 1994).

EFFECTS OF IL-10 ON GRANULOCYTES AND EOSINOPHILS

Granulocyte function is also affected by IL-10. LPS-induced cytokine and chemokine production, generation of superoxide anions, and survival by polymorphonuclear leukocytes or neutrophils are inhibited by IL-10, whereas production of IL-1Ra is enhanced (Cassatella *et al.*, 1993, 1994, 1997; Kasama *et al.*, 1994; Wang *et al.*, 1994; Chaves *et al.*, 1996; Cox, 1996; Marie *et al.*, 1996). IL-10 also inhibits LPS-induced survival and cytokine production by purified eosinophils (Takanaski *et al.*, 1994). IL-10 reduces neutrophil migration in immunoglobulin G (IgG) complex-induced lung injury and LPS- or antigen-induced pulmonary inflammation (Shanley *et al.*, 1995; Zuany-Amorim *et al.*, 1995; Cox, 1996), which may be related to its regulation of chemokine expression. On the other hand, neutralization of IL-10 leads to enhanced survival in murine models of *Klebsiella pneumoniae*, *Streptococcus pneumoniae* and *Mycobacterium avium* infections, since killing of phagocytosed bacteria was partly suppressed by endogenous IL-10 (Denis and Ghadirian, 1993; Greenberger *et al.*, 1995; van der Poll *et al.*, 1996b). Inhibition by IL-10 of production of chemokines, proinflammatory cytokines, mediators, and granulocyte survival, no doubt collectively limits the duration of inflammatory responses.

EFFECTS OF IL-10 ON MAST CELLS

IL-10 is a cofactor for the IL-3- or IL-4-induced proliferation of mucosal-type mouse mast cells (Thompson-Snipes *et al.*, 1991; Rennick *et al.*, 1995). In addition, it induces the expression of two mouse mast cell-specific proteases mMCP-1 and mMCP-2 (Ghildyal *et al.*, 1992a,b, 1993). These activities of IL-10 may contribute to immune responses against helminth infections (Behnke *et al.*, 1993). IL-10 also has an inhibitory effect on mast cell TNF-α and GM-CSF production following triggering of the high-affinity IgE receptor (Arock *et al.*, 1996). However, the same activation induces prolonged cyclo-oxygenase (Cox-2) expression which results in increased PGD_2 production (Ashraf *et al.*, 1996). IL-10 is produced constitutively by mucosal-type mast cells but needs to be induced by IL-3 in connective tissue-type mast cells (Marietta *et al.*, 1996).

EFFECTS OF IL-10 ON B CELLS

IL-10 induces enhanced expression of MHC class II antigens and survival of resting mouse B cells (Go *et al.*, 1990). In contrast, IL-10 inhibits motility and IL-5-induced antibody production against thymus-independent type I and II antigens (produced by B-1 B cells), but not against thymus-dependent antigens. This inhibition was reversed by IL-4 (Pecanha *et al.*, 1992, 1993; Clinchy *et al.*, 1994). However, serum immunoglobulin levels and development of B-1 B cells are normal in IL-$10^{-/-}$ mice (Kuhn *et al.*, 1993). Furthermore, B cells do not seem to play a role in the inflammatory bowel disease that develops in these animals (Davidson *et al.*, 1996). Current data suggest that the *in vivo* role of IL-10 in murine B-cell responses may be limited.

More extensive activities of IL-10 have been described on human B cells. IL-10 enhanced the survival of normal human B cells (depending on their activation state), which correlated with increased expression of the antiapoptotic protein bcl-2 (Levy and Brouet, 1994; Itoh and Hirohata, 1995). IL-10 strongly enhanced proliferation of human B-cell precursors and mature B cells activated by anti-IgM mAbs, *Staphylococcus aureus* Cowan (SAC) or cross-linking of CD40 (Rousset *et al.*, 1992; Saeland *et al.*, 1993). Both IL-2 and IL-4 further enhanced B-cell proliferation induced by IL-10 with anti-CD40 activation. The synergistic effects of IL-2 correlated with IL-10-enhanced expression of high-affinity IL-2 receptors on B cells (Fluckiger *et al.*, 1993).

B cell-derived and exogenous IL-10 also affected B-cell differentiation and isotype switching (Burdin *et al.*, 1995). Isotype-committed B cells, activated by SAC or anti-CD40 mAbs, produced large amounts of IgM, IgG_{1-3}, and IgA in the presence of IL-10, and IgG_4 and IgE in the presence of both IL-10 and IL-4 (Rousset *et al.*, 1992; Nonoyama *et al.*, 1993; Punnonen *et al.*, 1993; Garrone *et al.*, 1994; Uejima *et al.*, 1996). Activation (anti-CD40 and IL-10) of surface IgD^+ ($sIgD^+$) naive B cells also resulted in production of IgM, IgG_{1-3} and IgA (Defrance *et al.*, 1992; Briere *et al.*, 1994b). Furthermore, anti-CD40 and IL-10 stimulation of $sIgD^+$ B cells induced isotype switching to IgG_1 and IgG_3 (Malisan *et al.*, 1996) and, in combination with TGF-β, to IgA (Defrance *et al.*, 1992). IL-10 induced production of IgA by anti-CD40-activated B cells of patients suffering from IgA deficiency, although no defects in the production of IL-10 were observed (Briere *et al.*, 1994a). Long-term culture of B cells stimulated by either anti-CD40 or activated T cells with IL-10 resulted in differentiation of B cells into plasma cells (Merville *et al.*, 1995; Rousset *et al.*, 1995).

The effects of IL-10 on B cells and antibody production may play a role in the pathogenesis of systemic lupus erythematosus (SLE) where positive correlations between serum IL-10 levels and severity of disease, and between the production of IL-10 and autoantibodies by the B cells of patients with SLE, were demonstrated (Llorente *et al.*, 1994, 1995; Houssiau *et al.*, 1995; al-Janadi *et al.*, 1996; Hagiwara *et al.*, 1996). A role for IL-10 in development of SLE was suggested by the delayed onset of autoimmunity in lupus-prone NZB mice treated with anti-IL-10 mAbs (Ishida *et al.*, 1994). These results suggested that IL-10 antagonists could be useful therapeutic agents in SLE.

DIRECT EFFECTS OF IL-10 ON T CELLS

IL-10 strongly inhibited cytokine production and proliferation of T cells and T-cell clones activated in the presence of APCs. This effect of IL-10 is due mostly to its downregulatory effects on APC function (de Waal Malefyt *et al.*, 1991b; Fiorentino *et al.*, 1991b). However, T cells do express IL-10R (Liu *et al.*, 1994, 1997) and IL-10 modulates T-cell function. IL-10 inhibits production of IL-2 and TNF-α, but not that of IL-4 and IFN-γ, when T cells are stimulated without APCs (Taga and Tosato, 1992; de Waal Malefyt *et al.*, 1993b). In addition, inhibition of IL-2 production may also indirectly inhibit CD28- and phorbol myristate acetate (PMA)-induced production of IL-5, which is dependent on IL-2 (Schandene *et al.*, 1994). Interestingly, activation of T cells in the presence of IL-10 can induce a long-lasting state of non-responsiveness or anergy, which cannot be reversed by IL-2 or by stimulation with anti-CD3 and anti-

CD28 (Groux *et al.*, 1996). A role for IL-10 in induction and maintenance of non-responsiveness or anergy was suggested by studies of antitumor cell responses, ultraviolet light-induced tolerance, hapten-specific tolerance, parasitic and human immunodeficiency virus (HIV) infection, and superantigen induced hyporesponsiveness (Enk *et al.*, 1993, 1994; Flores *et al.*, 1993; Becker *et al.*, 1994; Flores Villanueva *et al.*, 1994; Suzuki *et al.*, 1995; King *et al.*, 1996; Schols and De Clercq, 1996; Sundstedt *et al.*, 1997). In addition, IL-10 inhibits apoptosis of normal or infectious mononucleosis T cells following growth factor deprivation, and has chemotactic activity on CD8-positive T cells (Jinquan *et al.*, 1993; Taga *et al.*, 1993, 1994). Stimulatory activities of IL-10 on T cells were also described in the mouse, where it acts as a thymocyte growth factor and augments the outgrowth of cytotoxic T-cell precursors (MacNeil *et al.*, 1990; Chen and Zlotnik, 1991).

EFFECTS OF IL-10 ON NK CELLS

IL-10 directly or indirectly affects the function of NK cells in several ways. IL-10 inhibits monocyte-dependent enhancement in production of IFN-γ and TNF-α by IL-2 activated human NK cells. This effect is indirect and mediated via inhibition of production of monokines such as IL-1 and IL-12 (Hsu *et al.*, 1992; D'Andrea *et al.*, 1993). Inhibition by IL-10 of IFN-γ production following infection of mice by intracellular parasites occurs through similar mechanisms (Tripp *et al.*, 1993; Hunter *et al.*, 1994; Cardillo *et al.*, 1996). IL-10 also inhibits production of IL-5 by human NK cell clones stimulated by tumor cells and IL-2 (Warren *et al.*, 1995). In contrast, IL-10 enhances IL-2-induced proliferation of $CD56^{bright}$ NK cells and cytokine production and cytotoxicity by $CD56^{bright}$ and $CD56^{dim}$ subsets (Carson *et al.*, 1995).

Studies with IL-10 also revealed a role for NK cells in antitumor responses. At low concentrations, the inhibition of tumor metastasis by IL-10 in experimental or spontaneous models correlated with localization of high numbers of NK cells at the tumor site. These findings were repeated in T and B cell-deficient mice with severe combined immune deficiency (SCID), but not in NK cell-deficient mice (Zheng *et al.*, 1996). Furthermore, administration of IL-10 or transduction of tumor cells with IL-10 expression plasmids resulted in NK cell-dependent inhibition of tumor growth and tumorigenity, and in the establishment of protective memory responses (Giovarelli *et al.*, 1995; Gerard *et al.*, 1996; Kundu *et al.*, 1996). Interestingly, IL-10 transfection of a B-cell lymphoma resulted in inhibition of MHC class I expression and a change in phenotype from being cytotoxic T lymphocyte (CTL) sensitive and NK resistant to being CTL resistant and NK sensitive (Salazar-Onfray *et al.*, 1995). It is possible that the absence of MHC class I on these cells abrogated the inhibition of target cell lysis mediated by KIRs (killer inhibitory receptors) on NK cells (Lanier, 1997). Downregulation of MHC class I antigens on human melanoma cells and EBV–lymphoblastoid cell line (LCL) following treatment of these targets with IL-10 also increased sensitivity to NK cell cytotoxicity and inhibited CTL activity (Matsuda *et al.*, 1994).

IN VIVO EFFECTS OF IL-10: SYSTEMIC INFLAMMATION

The effects of IL-10 on individual cell types suggested that it could have potent anti-inflammatory or immunosuppresive activities *in vivo*; this has been tested in a variety of experimental models. IL-10 rescued Balb/c and BDF1 mice from LPS- or staphylococcus enterotoxin B (SEB)-induced toxic shock, which correlated with reduced serum levels of TNF-α (Bean *et al.*, 1993; Gerard *et al.*, 1993; Howard *et al.*, 1993). Intratracheal IL-10 gene transfer protected mice from a lethal intraperitoneal endotoxin challenge and furthermore reduced pulmonary TNF-α levels and neutrophil infiltration following LPS challenge (Rogy *et al.*, 1995). When administered at the time of challenge, IL-10 did not prevent lethality in *Corynebacterium parvum*-primed BDF1 or Balb/c mice following endotoxin challenge, but it prevented lethality to endotoxin challenge when administered at priming (Smith *et al.*, 1994b, 1996). Although IL-10 reduced TNF-α and IFN-γ levels in the former situation, synergistic effects of TNF-α and IFN-γ still resulted in mortality in this model. IL-10 also reduced serum levels of TNF-α, IL-6 and IL-8, as well as neutrophil accumulation, elastase production and cortisone levels in human volunteers challenged with a low dose of endotoxin (Pajkrt *et al.*, 1997).

Administration of endotoxin induced IL-10 production in mice, chimpanzees, baboons and humans (Durez *et al.*, 1993; van der Poll *et al.*, 1994; Jansen *et al.*, 1996; Pajkrt *et al.*, 1997). Interestingly, LPS-induced IL-10 expression by monocytes *in vitro* occurs late after activation (de Waal Malefyt *et al.*, 1991a), but IL-10 serum levels *in vivo* were already raised 3–6 h after endotoxin challenge (Durez *et al.*, 1993; van der Poll *et al.*, 1994; Jansen *et al.*, 1996; Pajkrt *et al.*, 1997), indicating that other factors or cells may contribute to IL-10 production in these circumstances. Endogenous IL-10 confers significant protection to endotoxin challenge and reduces TNF-α, IFN-γ and MIP-2 production (Marchant *et al.*, 1994a; Standiford *et al.*, 1995). This is also clearly observed in mice treated from birth with anti-IL-10 mAbs and IL-$10^{-/-}$ mice which are killed by 20-fold lower doses of LPS (Ishida *et al.*, 1993; Berg *et al.*, 1995a). Furthermore, IL-$10^{-/-}$ mice were extremely vulnerable to a generalized Swartzman reaction, in which previous exposure to a small amount of LPS primes the host for a lethal response to a subsequent, otherwise sublethal, dose (Berg *et al.*, 1995a). In a model of septic peritonitis in which mice undergo cecal ligation and puncture and are exposed to live bacteria, endogenous IL-10 also had a protective effect (Kato *et al.*, 1995; van der Poll *et al.*, 1995). IL-10 also protected neonatal mice from lethal streptococcal B infections (Cusumano *et al.*, 1996). Furthermore, endogenous IL-10 prevented lethality from SEB-induced shock, which is dependent on IL-2 and IFN-γ production by T cells (Florquin *et al.*, 1994).

In humans, IL-10 is produced during septicemia and septic shock, and serum levels correlate with intensity of the inflammatory response (Marchant *et al.*, 1994b, 1995; Gomez-Jimenez *et al.*, 1995; Sherry *et al.*, 1996; van der Poll *et al.*, 1997), which was especially evident in patients suffering from septic shock associated with meningococcal infections (Frei *et al.*, 1993; Derkx *et al.*, 1995; Lehmann *et al.*, 1995; van Furth *et al.*, 1995).

The inhibitory effects of IL-10 on the endotoxin-induced production of proinflammatory cytokines, including indirect effects on endothelial cell activation and adhesion (Pugin *et al.*, 1993; Eissner *et al.*, 1996), monocyte tissue factor expression (which contributes to procoagulant activity and ultimately to disseminated intravascular coagulation) (Pradier *et al.*, 1993, 1996; Ramani *et al.*, 1993, 1994; Jungi *et al.*, 1994), as well as the observation that many strategies to intervene in sepsis affected IL-10 production (Mengozzi *et al.*, 1994; Bourrie *et al.*, 1995; Suberville *et al.*, 1996; van der Poll *et al.*, 1996a), indicate an important role for this cytokine in controlling systemic inflammatory responses.

EFFECTS OF IL-10 IN LOCALIZED INFLAMMATORY DISEASES

Similarly, IL-10 exhibited protective effects in a variety of experimental models of local inflammation, such as pancreatitis (Van Laethem *et al.*, 1995), uveitis (Rosenbaum and Angell, 1995; Li *et al.*, 1996), keratitis (Tumpey *et al.*, 1994), hepatitis (Arai *et al.*, 1995), peritonitis (Kato *et al.*, 1995; van der Poll *et al.*, 1995; Cusumano *et al.*, 1996) and lung injury (Mulligan *et al.*, 1993; Shanley *et al.*, 1995). In addition, IL-10 inhibited allergic airway inflammation and infiltration of neutrophils and eosinophils in sensitized mice (Zuany-Amorim *et al.*, 1995). In IL-$10^{-/-}$ mice, a role for IL-10 as a natural suppressor of cytokine production and inflammation in allergic bronchopulmonary aspergillosis was revealed, indicating that IL-10 suppresses not only inflammatory T_{H1}- but also T_{H2}-related responses (Grünig *et al.*, 1997).

IL-10 also inhibited both the sensitization and effector phases of delayed-type hypersensitivity (DTH) responses (Li *et al.*, 1994; Schwarz *et al.*, 1994) as well as the elicitation of contact hypersensitivity reactions in sensitized mice (Ferguson *et al.*, 1994; Kondo *et al.*, 1994). IL-10 synergized with IL-4 to inhibit tuberculin type DTH reactions in Balb/c mice that had recovered from a *Leishmania major* infection (Powrie *et al.*, 1993b). The role of endogenous IL-10 in these responses is critical as reactions to irritants, DTH and contract hypersensitivity (CHS) responses in IL-$10^{-/-}$ mice are exaggerated and prolonged (Berg *et al.*, 1995b). Keratinocyte-derived IL-10 plays a major role in ultraviolet light-induced suppression of DTH, CHS and antitumor responses (Enk and Katz, 1992; Rivas and Ullrich, 1992, 1994; Enk *et al.*, 1994; Beissert *et al.*, 1995; Yagi *et al.*, 1996), because IL-$10^{-/-}$ mice were completely resistant to UV-induced immunosuppression (Beissert *et al.*, 1996).

INFLAMMATORY BOWEL DISEASE

IL-10 is also involved in the development of inflammatory bowel disease (IBD) as revealed by two in *vivo* models. Transfer of naive $CD45RB^{hi}$ T cells (which develop T cells that produce high levels of IFN-γ and TNF-α in response to enteric antigens) from Balb/c into CB-17 SCID mice resulted in the spontaneous development of IBD. The disease could be prevented by coadministration of $CD45RB^{lo}$ cells (which contain regulatory T cells producing IL-10 and IL-4), treatment with anti-IFN-γ or anti-TNF-α mAbs, or administration of IL-10 (Powrie *et al.*, 1993a, 1994; Powrie, 1995). Interestingly, both $CD45RB^{hi}$ and $CD45RB^{lo}$ cells from IL-$10^{-/-}$ mice could induce IBD in $Rag2^{-/-}$ mice, suggesting that IL-10 is involved in the protective effects of $CD45RB^{lo}$ cells in normal mice (Rennick *et al.*, 1997). In addition, IL-$10^{-/-}$ mice spontaneously develop enterocolitis (Kuhn *et al.*, 1993), with the appearance of adenocarcinomas at advanced stages (Berg *et al.*, 1996). Enterocolitis in IL-$10^{-/-}$ mice could be prevented by administration of IL-10 from birth and by treatment with anti-IFN-γ or anti-IL-12 mAbs. IL-10 ameliorated but could not completely cure established disease. $CD4^+$ T cells isolated from diseased colon produced high levels of IFN-γ and TNF-α, and could transfer disease, indicating that the dysregulated interaction between enteric flora and inflammatory cells in the absence of IL-10 leads to uncontrolled T_{H1} responses (Davidson *et al.*, 1996). IL-10 was detected in the serum of patients with IBD (Kucharzik *et al.*, 1995).

RHEUMATOID ARTHRITIS

A possible beneficial role for IL-10 in pathogenesis of rheumatoid arthritis (RA) has been described. Production of IL-10 was demonstrated in serum, synovial fluid and synovial explant cultures from patients with RA (Katsikis *et al.*, 1994; Llorente *et al.*, 1994; Cush *et al.*, 1995; al-Janadi *et al.*, 1996; Bucht *et al.*, 1996). IL-10 inhibited TNF-α, IL-1β and IL-8 production by synovial macrophages and synoviocytes (Deleuran *et al.*, 1994; Chomarat *et al.*, 1995; Hart *et al.*, 1995), and synergistic effects were observed in combination with IL-4 (Sugiyama *et al.*, 1995). Endogenous IL-10 partly suppressed cytokine production from synovial cells, indicating that IL-10 *in vivo* may have a protective role (Katsikis *et al.*, 1994). IL-10-producing cells in the joint include both synovial macrophages and T cells (Katsikis *et al.*, 1994; Cohen *et al.*, 1995). On the other hand IL-10 production in RA has been linked to increased autoantibody production, serum factor and B-cell activation (Cush *et al.*, 1995; Perez *et al.*, 1995; al-Janadi *et al.*, 1996). IL-10 is protective in animal models of RA: it reduced joint swelling, infiltration, cytokine production and cartilage degradation in collagen- and streptococcal cell wall-induced arthritis (Kasama *et al.*, 1995; Persson *et al.*, 1996; Tanaka *et al.*, 1996; van Roon *et al.*, 1996; Walmsley *et al.*, 1996). IL-4 and IL-10 synergistically reduced joint inflammation in models of acute and chronic athritis (Joosten *et al.*, 1997).

ROLE OF IL-10 IN OTHER MODELS OF AUTOIMMUNITY

IL-10 has been tested in mouse and rat models of experimental autoimmune encephalomyelitis (EAE) which are in part representative of the pathology observed in multiple sclerosis in humans. Systemic administration of IL-10 during the initiation phase of the disease suppressed disease induction, and reduced T-cell proliferation to myelin basic protein and central nervous system infiltration (Rott *et al.*, 1994). In addition, IL-10 prevented TNF-α-induced relapses in SJL mice that had partially or completely recovered from acute EAE. Furthermore, neutralization of endogenous IL-10 increased both the incidence and severity of SEB- or TNF-induced relapses, suggesting a protective role for IL-10 in the development of disease (Crisi *et al.*, 1995). Increases in IL-10 mRNA levels were observed in the recovery phase of acute EAE in mice and rats (Kennedy *et al.*, 1992; Issazadeh *et al.*, 1995), and lack of IL-10 production was observed in chronic relapsing EAE (Issazadeh *et al.*, 1996).

IL-10 expression was detected in mononuclear cells from the cerebrospinal fluid of patients with multiple sclerosis (Navikas *et al.*, 1995). IFN-β, which was used successfully to treat patients with multiple sclerosis, induced expression of IL-10 in their peripheral blood mononuclear cells (PBMCs), indicating that IL-10 could potentially be useful for the treatment of multiple sclerosis.

Daily administration of IL-10 also inhibited the spontaneous onset of insulin-dependent diabetes mellitus (IDDM), a chronic T_{H1}-mediated autoimmune disease which results in destruction of pancreatic islet cells, in non-obese diabetic (NOD) mice (Pennline *et al.*, 1994). However, expression of IL-10 under islet-specific promoters accelerated the development IDDM in transgenic NOD mice (Wogensen *et al.*, 1993, 1994; Lee *et al.*, 1994, 1996; Moritani *et al.*, 1994, 1996; Balasa and Sarvetnick, 1996; Mueller *et al.*, 1996). Localized expression of high levels of IL-10 in the pancreas, compared with systemic administration or production at a different site, may activate endothelium and promote chemoattraction of lymphocytes and/or myeloid cells, thus predisposing the mice to development of disease. In contrast, transgenic mice expressing IL-10 under control of the IL-2 promoter (Hagenbaugh *et al.*, 1997) do not become diabetic.

IL-10 IN INFECTIOUS DISEASES

A number of infectious diseases are characterized by a lack of cell-mediated immunity and hyporesponsiveness to the inducing pathogen. IL-10 contributes to immunosuppression observed in lepromatous leprosy, filariasis, visceral leishmaniasis, toxoplasmosis, listerisosis, schistosomiasis, *Legionella pneumophila* infection, *Trypanosoma* infection, malaria and HIV infection. Involvement in human infections is revealed in many studies by enhanced production of IL-10 by patients' PBMCs and the restoration of antigen-specific proliferative responses of PBMCs *in vitro* by neutralizing anti-IL-10 mAbs (Sieling *et al.*, 1993; Peyron *et al.*, 1994; Carvalho, 1995; Mahanty and Nutman, 1995; Sieling and Modlin, 1995; King *et al.*, 1996; Mahanty *et al.*, 1996).

The role of IL-10 in animal models of infection has also been clearly established. Neutralization of IL-10 leads to enhanced antigen-specific IFN-γ production from the CD4$^+$ T cells of mice infected with *Leishmania major* or *Schistosoma mansoni*, and the *in vivo* use of IL-10 or anti-IL-10 mAbs can convert susceptible strains into resistant ones or *vice versa*, respectively (Sher *et al.*, 1991; Powrie and Coffman, 1993; Flores Villanueva *et al.*, 1994, 1996; Reed *et al.*, 1994). IL-10 acts in several ways to modulate immune responses to parasites, including inhibition of production of IFN-γ, which is neccesary to activate macrophages for intracellular killing, and of production of IL-12, which induces protective cellular immune responses.

On the other hand, the complete absence of IL-10 in acute infections also results in pathology. Infection of IL-10$^{-/-}$ mice with *Trypanosoma cruzi* resulted in reduced parasitemia and enhanced IFN-γ production (Abrahamsohn and Coffman, 1996). Furthermore, IL-10$^{-/-}$ mice infected with *Toxoplasma gondii* succumbed to the infection and had high levels of TNF-α, IFN-γ, IL-1-β and IL-12 in serum and lung, implying an important role for endogenous IL-10 synthesis in downregulating monokine and IFN-γ responses to acute intracellular infection, and preventing host immunopathology (Gazzinelli *et al.*, 1996). These data imply a required balance in the production of IL-10, IL-12 and IFN-γ which determines the outcome of infection.

A similar dual role for IL-10 has been described in HIV infection. IL-10 was shown to inhibit viral replication and virus production by infected monocytes (Akridge *et al.*, 1994; Kootstra *et al.*, 1994; Montaner *et al.*, 1994; Saville *et al.*, 1994; Weissman *et al.*, 1994; Ludewig *et al.*, 1996). This effect was attributed to inhibition of production of TNF-α and IL-6 (which activate the HIV long terminal repeat (LTR)), and to inhibitory effects on post-translational processing and viral assembly. In addition, IL-10 inhibited acute HIV infection in SCID-hu mice (Kollmann *et al.*, 1996). On the other hand, HIV or recombinant gp120 induced production of IL-10 by PBMCs (Ameglio *et al.*, 1994a,b; Borghi *et al.*, 1995; Chehimi *et al.*, 1996; Schols and De Clercq, 1996). Raised levels of IL-10 production were also observed during end-stage disease which correlated with diminished cellular responses against viral, mycobacterial and recall antigens. Anti-IL-10 mAbs restored proliferative responses of PBMCs from HIV-infected patients with fewer than 200 CD4 cells per ml, indicating a suppressive role for IL-10 at late stages of infection (Clerici and Shearer, 1993; Clerici *et al.*, 1994, 1996; Landay *et al.*, 1996). The authors suggested that a switch from T_{H1} to T_{H2} responses might be critical in pathology of acquired immune deficiency syndrome (AIDS). Furthermore, B-cell lines derived from patients with AIDS produce large amounts of IL-10 (Benjamin *et al.*, 1992). At present, the results suggest that IL-10 may inhibit HIV replication or production at early stages of infection, but may be detrimental in advanced stages of the disease when immune responses are compromised.

BIOLOGIC ACTIVITIES OF vIL-10 AND RELEVANCE TO THE EBV LIFECYCLE

vIL-10 retains many properties of cIL-10, including CSIF and macrophage deactivating factor activity on mouse and human cells (Hsu *et al.*, 1990, 1992; de Waal Malefyt *et al.*, 1991a,b; Vieira *et al.*, 1991; Niiro *et al.*, 1992, 1994), and the known activities of hIL-10 on human B cells (Defrance *et al.*, 1992; Rousset *et al.*, 1992). Like cIL-10, vIL-10 enhances viability of mouse B cells in culture (Go *et al.*, 1990).

However, vIL-10 lacks several cIL-10 activities on mouse and human cells. vIL-10 neither effectively enhanced class II MHC expression on mouse B cells (Go *et al.*, 1990), nor costimulated mouse thymocyte or mast cell proliferation (MacNeil *et al.*, 1990; Vieira *et al.*, 1991). Similarly, vIL-10 has a greatly reduced ability to inhibit IL-2 production by an activated human $CD4^+$ T-cell clone, which was mirrored by its approximately 1000-fold lower affinity for IL-10R (Liu *et al.*, 1997). These *in vitro* data recently received remarkable support from animal model studies in which immunogenic and allogeneic mouse tumor cells were infected with recombinant retroviruses expressing mIL-10 or vIL-10 (Suzuki *et al.*, 1995; Berman *et al.*, 1996). Rejection of mIL-10-expressing tumors was significantly accelerated compared with controls, as observed for administration of exogenous IL-10 to mouse cardiac allograft recipients (Qian *et al.*, 1996). In contrast, vIL-10-expressing tumor cells were either not rejected at all or the rejection kinetics were substantially slowed. Likewise, mouse heart allograft rejection was markedly inhibited when the grafts expressed vIL-10 (Qin *et al.*, 1996). Taken together, the *in vitro* and *in vivo* studies have: (a) indicated that vIL-10 has conserved only a subset of activities of cIL-10; and (b) shown that targeting of a limited subset of IL-10-responsive cells by vIL-10 *in vivo* can have a profound effect on the outcome of an immune response.

It is not yet certain whether the restricted activities of vIL-10 are merely a consequence of evolutionary drift or of actual benefit to EBV. However, as vIL-10 is expressed during the lytic phase (Hudson *et al.*, 1985), presumably the target cells for which it retains specificity—dendritic cells, macrophages–monocytes and B cells—are those specifically relevant to the EBV lifecycle. The ability of vIL-10 to inhibit dendritic cell and macrophage–monocyte activation probably suppresses an antiviral immune response during the early (lytic) phase of infection, thus allowing the virus to establish latency. This idea was supported by studies which showed that cells infected with a vIL-10-deleted EBV were, unlike wild-type virus-infected cells, unable to block IFN-γ production by autologous human peripheral blood cells (Swaminathan *et al.*, 1993). The B-cell growth and differentiation promoting activity could enhance the numbers and susceptibility to infection of the principal host cell of EBV. However, perhaps due to the unavailability of good animal models of EBV infection, it has proven difficult to confirm such notions. Whether vIL-10 plays any part in EBV-induced B-cell transformation is controversial (Burdin *et al.*, 1993; Miyazaki *et al.*, 1993; Swaminathan *et al.*, 1993). Low IL-10R binding affinity may restrict vIL-10 to local effects in the vicinity of the infected cell, as suggested by recent *in vivo* studies (Suzuki *et al.*, 1995; Qin *et al.*, 1996). In any case, because it has been so highly conserved, vIL-10 probably confers a number of adaptive advantages upon EBV in its interaction with the immune system. Moreover, capture of an IL-10 gene by the ancestor of EBV could have been an important step in the ultimate development of the virus as a sophisticated, and mostly benign, parasite.

CONCLUSIONS

IL-10 exhibits striking activities *in vitro* and *in vivo* suggesting important functions in immune regulation. In particular, studies with IL-$10^{-/-}$ mice have shown that a key 'steady state' role for IL-10 is control of inflammatory responses via inhibition of monocyte–macrophage activation.

Of considerable importance also are the potent deactivating effects of IL-10 on monocytes–macrophages, granulocytes and dendritic cells. These include inhibitory effects on the expression of MHC class II and accessory molecules, monokines, chemokines and release of inflammatory mediators, ultimately leading to inhibition of APC and accessory cell functions, inhibition of antigen-specific T-cell proliferation and cytokine production, and altered cell trafficking and migration. In addition, IL-10 has direct inhibitory effects on T-cell proliferation and IL-2 production, and can induce an anergic state under physiologic conditions. We believe it possible that these latter activities may have profound significance in the regulation of T-cell tolerance.

Finally, the demonstrated activities of IL-10 as a costimulator of B-cell growth and differentiation *in vitro* still await definitive confirmation *in vivo*, although evidence supports a possible role in B cell-mediated autoimmune diseases such as SLE.

Future research on IL-10 may uncover additional activities both *in vivo* and *in vitro*. In particular, the availability of IL-10R cDNA clones and anti-IL-10R mAbs will make possible further elucidation of signaling molecules and pathways activated by IL-10. Of particular importance are: (a) to understand which signaling pathways are responsible for which biologic responses to IL-10; and (b) to explain at the molecular level why IL-10 elicits such different responses from various cell types, for example B cells and macrophages.

IL-10 has also showed possible clinical promise in two phase I trials to test its safety, tolerance, pharmacokinetics and hematologic effects in healthy humans. IL-10 could be administered intravenously at doses of up to 100 μg/kg, with only mild to moderate influenza-like symptoms at the highest doses, while inhibiting production of IL-6, IL-1β and TNF-α in *ex vivo* stimulation of whole blood cells with LPS (Chernoff *et al.*, 1995; Huhn *et al.*, 1996). *In vitro* and *in vivo* studies suggest that IL-10 could be a potentially useful therapeutic agent for acute and chronic, systemic and localized, inflammatory reactions. On the other hand, antagonists of IL-10 production or function may be useful, activating strongly inhibited T_{H1} responses associated with pathology of certain autoimmune and infectious diseases. Studies on the expression and function of IL-10 and its possible role in disease states will continue to be pursued vigorously.

REFERENCES

Abrahamsohn, I.A. and Coffman, R.L. (1996). *Trypanosoma cruzi*: IL-10, TNF, IFN-gamma, and IL-12 regulate innate and acquired immunity to infection. *Exp. Parasitol.* **84**, 231–244.

Akridge, R.E., Oyafuso, L.K. and Reed, S.G. (1994). IL-10 is induced during HIV-1 infection and is capable of decreasing viral replication in human macrophages. *J. Immunol.* **153**, 5782–5789.

al-Janadi, M., al-Dalaan, A., al-Balla, S., al-Humaidi, M. and Raziuddin, S. (1996). Interleukin-10 (IL-10) secretion in systemic lupus erythematosus and rheumatoid arthritis: IL-10-dependent $CD4^+CD45RO^+$ T cell-B cell antibody synthesis. *J. Clin. Immunol.* **16**, 198–207.

Aman, M.J., Tretter, T., Eisenbeis, I., Bug, G., Decker, T., Aulitzky, W.E., Tilg, H., Huber, C. and Peschel, C. (1996). Interferon-alpha stimulates production of interleukin-10 in activated $CD4^+$ T cell and monocytes. *Blood* **87**, 4731–4736.

Ameglio, F., Capobianchi, M.R., Castilletti, C., Cordiali Fei, P., Fais, S., Trento, E. and Dianzani, F. (1994a). Recombinant gp120 induces IL-10 in resting peripheral blood mononuclear cells; correlation with the induction of other cytokines. *Clin. Exp. Immunol.* **95**, 455–458.

Ameglio, F., Cordiali Fei, P., Solmone, M., Bonifati, C., Prignano, G., Giglio, A., Caprilli, F., Gentili, G. and Capobianchi, M.R. (1994b). Serum IL-10 levels in HIV-positive subjects: correlation with CDC stages. *J. Biol. Regul. Homeost. Agents* **8**, 48–52.

Arai, T., Hiromatsu, K., Kobayashi, N., Takano, M., Ishida, H., Nimura, Y. and Yoshikai, Y. (1995). IL-10 is involved in the protective effect of dibutyryl cyclic adenosine monophosphate on endotoxin-induced inflammatory liver injury. *J. Immunol.* **155**, 5743–5749.

Armstrong, L., Jordan, N. and Millar, A. (1996). Interleukin 10 (IL-10) regulation of tumour necrosis factor alpha (TNF-alpha) from human alveolar macrophages and peripheral blood monocytes. *Thorax* **51**, 143–149.

Arock, M., Zuany-Amorim, C., Singer, M., Benhamou, M. and Pretolani, M. (1996). Interleukin-10 inhibits cytokine generation from mast cells. *Eur. J. Immunol.* **26**, 166–170.

Ashraf, M., Murakami, M. and Kudo, I. (1996). Cross-linking of the high-affinity IgE receptor induces the expression of cyclo-oxygenase 2 and attendant prostaglandin generation requiring interleukin 10 and interleukin 1 beta in mouse cultured mast cells. *Biochem. J.* **320**, 965–973.

Balasa, B. and Sarvetnick, N. (1996). The paradoxical effects of interleukin 10 in the immunoregulation of autoimmune diabetes. *J. Autoimmun.* **9**, 283–286.

Bean, A.G., Freiberg, R.A., Andrade, S., Menon, S. and Zlotnik, A. (1993). Interleukin 10 protects mice against staphylococcal enterotoxin B-induced lethal shock. *Infect. Immun.* **61**, 4937–4939.

Becker, J.C., Czerny, C. and Brocker, E.B. (1994). Maintenance of clonal anergy by endogenously produced IL-10. *Int. Immunol.* **6**, 1605–1612.

Behnke, J.M., Wahid, F.N., Grencis, R.K., Else, K.J., Ben, S.A. and Goyal, P.K. (1993). Immunological relationships during primary infection with *Heligmosomoides polygyrus* (*Nematospiroides dubius*): downregulation of specific cytokine secretion (IL-9 and IL-10) correlates with poor mastocytosis and chronic survival of adult worms. *Parasite Immunol.* **15**, 415–421.

Beissert, S., Ullrich, S.E., Hosoi, J. and Granstein, R.D. (1995). Supernatants from UVB radiation-exposed keratinocytes inhibit Langerhans cell presentation of tumor-associated antigens via IL-10 content. *J. Leukoc. Biol.* **58**, 234–240.

Beissert, S., Hosoi, J., Kuhn, R., Rajewsky, K., Muller, W. and Granstein, R.D. (1996). Impaired immunosuppressive response to ultraviolet radiation in interleukin-10-deficient mice. *J. Invest. Dermatol.* **107**, 553–557.

Bejarano, M.T., de Waal Malefyt, R., Abrams, J.S., Bigler, M., Bacchetta, R., de Vries, J. and Roncarolo, M.G. (1992). Interleukin 10 inhibits allogeneic proliferative and cytotoxic T cell responses generated in primary mixed lymphocyte cultures. *Int. Immunol.* **4**, 1389–1397.

Benjamin, D., Knobloch, T.J. and Dayton, M.A. (1992). Human B-cell interleukin-10: B-cell lines derived from patients with acquired immunodeficiency syndrome and Burkitt's lymphoma constitutively secrete large quantities of interleukin-10. *Blood* **80**, 1289–1298.

Berg, D.J., Kuhn, R., Rajewsky, K., Muller, W., Menon, S., Davidson, N., Grunig, G. and Rennick, D. (1995a). Interleukin-10 is a central regulator of the response to LPS in murine models of endotoxic shock and the Shwartzman reaction but not endotoxin tolerance. *J. Clin. Invest.* **96**, 2339–2347.

Berg, D.J., Leach, M.W., Kuhn, R., Rajewsky, K., Muller, W., Davidson, N.J. and Rennick, D. (1995b). Interleukin 10 but not interleukin 4 is a natural suppressant of cutaneous inflammatory responses. *J. Exp. Med.* **182**, 99–108.

Berg, D.J., Davidson, N., Kuhn, R., Muller, W., Menon, S., Holland, G., Thompson-Snipes, L., Leach, M.W. and Rennick, D. (1996). Enterocolitis and colon cancer in interleukin-10-deficient mice are associated with aberrant cytokine production and CD4(+) TH1-like responses. *J. Clin. Invest.* **98**, 1010–1020.

Berkman, N., John, M., Roesems, G., Jose, P.J., Barnes, P.J. and Chung, K.F. (1995). Inhibition of macrophage inflammatory protein-1 alpha expression by IL-10. Differential sensitivities in human blood monocytes and alveolar macrophages. *J. Immunol.* **155**, 4412–4418.

Berman, R.M., Suzuki, T., Tahara, H., Robbins, P.D., Narula, S.K. and Lotze, M.T. (1996). Systemic administration of cellular IL-10 induces an effective, specific, and long-lived immune response against established tumors in mice. *J. Immunol.* **157**, 231–238.

Bhardwaj, R.S., Schwarz, A., Becher, E., Mahnke, K., Aragane, Y., Schwarz, T. and Luger, T.A. (1996). Pro-opiomelanocortin-derived peptides induce IL-10 production in human monocytes. *J. Immunol.* **156**, 2517–2521.

Blancho, G., Gianello, P., Germana, S., Baetscher, M., Sachs, D.H. and LeGuern, C. (1995). Molecular identification of porcine interleukin 10: regulation of expression in a kidney allograft model. *Proc. Natl Acad. Sci. U.S.A.* **92**, 2800–2804.

Bogdan, C., Vodovotz, Y. and Nathan, C. (1991). Macrophage deactivation by interleukin 10. *J. Exp. Med.* **174**, 1549–1555.

Bonfield, T.L., Konstan, M.W., Burfeind, P., Panuska, J.R., Hilliard, J.B. and Berger, M. (1995). Normal bronchial epithelial cells constitutively produce the anti-inflammatory cytokine interleukin-10, which is downregulated in cystic fibrosis. *Am. J. Respir. Cell Mol. Biol.* **13**, 257–261.

Bonig, H., Korholz, D., Pafferath, B., Mauz-Korholz, C. and Burdach, S. (1996). Interleukin 10 induced c-fos expression in human B cells by activation of divergent protein kinases. *Immunol. Invest.* **25**, 115–128.

Borghi, P., Fantuzzi, L., Varano, B., Gessani, S., Puddu, P., Conti, L., Capobianchi, M.R., Ameglio, F. and Belardelli, F. (1995). Induction of interleukin-10 by human immunodeficiency virus type 1 and its gp120 protein in human monocytes/macrophages. *J. Virol.* **69**, 1284–1287.

Bourrie, B., Bouaboula, M., Benoit, J.M., Derocq, J.M., Esclangon, M., Le Fur, G. and Casellas, P. (1995). Enhancement of endotoxin-induced interleukin-10 production by SR 31747A, a sigma ligand. *Eur. J. Immunol.* **25**, 2882–2887.

Briere, F., Bridon, J.M., Chevet, D., Souillet, G., Bienvenu, F., Guret, C., Martinez-Valdez, H. and Banchereau, J. (1994a). Interleukin 10 induces B lymphocytes from IgA-deficient patients to secrete IgA. *J. Clin. Invest.* **94**, 97–104.

Briere, F., Servet-Delprat, C., Bridon, J.M., Saint-Remy, J.M. and Banchereau, J. (1994b). Human interleukin 10 induces naive surface immunoglobulin D+ (sIgD+) B cells to secrete IgG1 and IgG3. *J. Exp. Med.* **179**, 757–762.

Bucht, A., Larsson, P., Weisbrot, L., Thorne, C., Pisa, P., Smedegard, G., Keystone, E.C. and Gronberg, A. (1996). Expression of interferon-gamma (IFN-gamma), IL-10, IL-12 and transforming growth factor-beta (TGF-beta) mRNA in synovial fluid cells from patients in the early and late phases of rheumatoid arthritis (RA). *Clin. Exp. Immunol.* **103**, 357–367.

Buelens, C., Willems, F., Delvaux, A., Pierard, G., Delville, J.P., Velu, T. and Goldman, M. (1995). Interleukin-10 differentially regulates B7-1 (CD80) and B7-2 (CD86) expression on human peripheral blood dendritic cells. *Eur. J. Immunol.* **25**, 2668–2672.

Burdin, N., Peronne, C., Banchereau, J. and Rousset, F. (1993). Epstein–Barr virus transformation induces B lymphocytes to produce human interleukin-10. *J. Exp. Med.* **177**, 295–304.

Burdin, N., Van Kooten, C., Galibert, L., Abrams, J.S., Wijdenes, J., Banchereau, J. and Rousset, F. (1995). Endogenous IL-6 and IL-10 contribute to the differentiation of CD40-activated human B lymphocytes. *J. Immunol.* **154**, 2533–2544.

Calzada-Wack, J.C., Frankenberger, M. and Ziegler-Heitbrock, H.W. (1996). Interleukin-10 drives human monocytes to CD16 positive macrophages. *J. Inflamm.* **46**, 78–85.

Capsoni, F., Minonzio, F., Ongari, A.M., Carbonelli, V., Galli, A. and Zanussi, C. (1995). IL-10 up-regulates human monocyte phagocytosis in the presence of IL-4 and IFN-gamma. *J. Leukoc. Biol.* **58**, 351–358.

Cardillo, F., Voltarelli, J.C., Reed, S.G. and Silva, J.S. (1996). Regulation of *Trypanosoma cruzi* infection in mice by gamma interferon and interleukin 10: role of NK cells. *Infect. Immun.* **64**, 128–134.

Carson, W.E., Lindemann, M.J., Baiocchi, R., Linett, M., Tan, J.C., Chou, C.-C., Narula, S. and Caligiuri, M.A. (1995). The functional characterization of interleukin-10 receptor expression on human natural killer cells. *Blood* **85**, 3577–3585.

Carvalho, E. (1995). IL-10 in human Leishmaniasis. In *Interleukin-10* (eds J.E. de Vries and R. de Waal Malefyt), R.G. Landes Company, Austin, Texas, pp. 79–87.

Cassatella, M.A., Meda, L., Bonora, S., Ceska, M. and Constantin, G. (1993). Interleukin 10 (IL-10) inhibits the release of proinflammatory cytokines from human polymorphonuclear leukocytes. Evidence for an autocrine role of tumor necrosis factor and IL-1 beta in mediating the production of IL-8 triggered by lipopolysaccharide. *J. Exp. Med.* **178**, 2207–2211.

Cassatella, M.A., Meda, L., Gasperini, S., Calzetti, F. and Bonora, S. (1994). Interleukin 10 (IL-10) upregulates IL-1 receptor antagonist production from lipopolysaccharide-stimulated human polymorphonuclear leukocytes by delaying mRNA degradation. *J. Exp. Med.* **179**, 1695–1699.

Cassatella, M.A., Gasperini, S., Calzetti, F., Bertagnin, A., Luster, A.D. and McDonald, P.P. (1997). Regulated production of the interferon-γ-inducible protein-10 (IP-10) chemokine by human neutrophils. *Eur. J. Immunol.* **27**, 111–115.

Caux, C., Massacrier, C., Vanbervliet, B., Barthelemy, C., Liu, Y.J. and Banchereau, J. (1994). Interleukin 10 inhibits T cell alloreaction induced by human dendritic cells. *Int. Immunol.* **6**, 1177–1185.

Cenci, E., Romani, L., Mencacci, A., Spaccapelo, R., Schiaffella, E., Puccetti, P. and Bistoni, F. (1993). Interleukin-4 and interleukin-10 inhibit nitric oxide-dependent macrophage killing of *Candida albicans*. *Eur. J. Immunol.* **23**, 1034–1038.

Chaves, M.M., Silvestrini, A.A., Silva-Teixeira, D.N. and Nogueira-Machado, J.A. (1996). Effect *in vitro* of gamma interferon and interleukin-10 on generation of oxidizing species by human granulocytes. *Inflamm. Res.* **45**, 313–315.

Chehimi, J., Ma, X., Chouaib, S., Zyad, A., Nagashunmugam, T., Wojcik, L., Chehimi, S., Nissim, L. and Frank, I. (1996). Differential production of interleukin 10 during human immunodeficiency virus infection. *AIDS Res. Hum. Retroviruses* **12**, 1141–1149.

Chen, Q., Daniel, V., Maher, D.W. and Hersey, P. (1994). Production of IL-10 by melanoma cells: examination of its role in immunosuppression mediated by melanoma. *Int. J. Cancer* **56**, 755–760.

Chen, W.-F. and Zlotnik, A. (1991). IL-10: A novel cytotoxic T cell differentiation factor. *J. Immunol.* **147**, 528–534.

Chernoff, A.E., Granowitz, E.V., Shapiro, L., Vannier, E., Lonnemann, G., Angel, J.B., Kennedy, J.S., Rabson, A.R., Wolff, S.M. and Dinarello, C.A. (1995). A randomized, controlled trial of IL-10 in humans. Inhibition of inflammatory cytokine production and immune responses. *J. Immunol.* **154**, 5492–5499.

Chomarat, P., Rissoan, M.C., Banchereau, J. and Miossec, P. (1993). Interferon gamma inhibits interleukin 10 production by monocytes. *J. Exp. Med.* **177**, 523–527.

Chomarat, P., Vannier, E., Dechanet, J., Rissoan, M.C., Banchereau, J., Dinarello, C.A. and Miossec, P. (1995). Balance of IL-1 receptor antagonist/IL-1 beta in rheumatoid synovium and its regulation by IL-4 and IL-10. *J. Immunol.* **154**, 1432–1439.

Clerici, M. and Shearer, G.M. (1993). A Th1 to Th2 switch is a critical step in the etiology of HIV infection. *Immunol. Today* **14**, 107–111.

Clerici, M., Wynn, T.A., Berzofsky, J.A., Blatt, S.P., Hendrix, C.W., Sher, A., Coffman, R.L. and Shearer, G.M. (1994). Role of interleukin-10 in T helper cell dysfunction in asymptomatic individuals infected with the human immunodeficiency virus. *J. Clin. Invest.* **93**, 768–775.

Clerici, M., Balotta, C., Salvaggio, A., Riva, C., Trabattoni, D., Papagno, L., Berlusconi, A., Rusconi, S., Villa, M.L., Moroni, M. and Galli, M. (1996). Human immunodeficiency virus (HIV) phenotype and interleukin-2/interleukin-10 ratio are associated markers of protection and progression in HIV infection. *Blood* **88**, 574–579.

Clinchy, B., Bjorck, P., Paulie, S. and Moller, G. (1994). Interleukin-10 inhibits motility in murine and human B lymphocytes. *Immunology* **82**, 376–382.

Cohen, S.B., Katsikis, P.D., Chu, C.Q., Thomssen, H., Webb, L.M., Maini, R.N., Londei, M. and Feldmann, M. (1995). High level of interleukin-10 production by the activated T cell population within the rheumatoid synovial membrane. *Arthritis Rheum.* **38**, 946–952.

Cox, G. (1996). IL-10 enhances resolution of pulmonary inflammation *in vivo* by promoting apoptosis of neutrophils. *Am. J. Physiol.* **271**, L566–L571.

Crawley, J.B., Williams, L.M., Mander, T., Brennan, F.M. and Foxwell, B.M.J. (1996). Interleukin-10 stimulation of phosphatidylinositol 3-kinase and p70 S6 kinase is required for the proliferative but not the antiinflammatory effects of the cytokine. *J. Biol. Chem.* **271**, 16 357–16 362.

Crisi, G.M., Santambrogio, L., Hochwald, G.M., Smith, S.R., Carlino, J.A. and Thorbecke, G.J. (1995). Staphylococcal enterotoxin B and tumor-necrosis factor-alpha-induced relapses of experimental allergic encephalomyelitis: protection by transforming growth factor-beta and interleukin-10. *Eur. J. Immunol.* **25**, 3035–3040.

Cunha, F.Q., Moncada, S. and Liew, F.Y. (1992). Interleukin-10 (IL-10) inhibits the induction of nitric oxide synthase by interferon-gamma in murine macrophages. *Biochem. Biophys. Res. Commun.* **182**, 1155–1159.

Cush, J.J., Splawski, J.B., Thomas, R., McFarlin, J.E., Schulze-Koops, H., Davis, L.S., Fujita, K. and Lipsky, P.E. (1995). Elevated interleukin-10 levels in patients with rheumatoid arthritis. *Arthritis Rheum.* **38**, 96–104.

Cusumano, V., Genovese, F., Mancuso, G., Carbone, M., Fera, M.T. and Teti, G. (1996). Interleukin-10 protects neonatal mice from lethal group B streptococcal infection. *Infect. Immun.* **64**, 2850–2852.

D'Andrea, A., Aste, A.M., Valiante, N.M., Ma, X., Kubin, M. and Trinchieri, G. (1993). Interleukin 10 (IL-10) inhibits human lymphocyte interferon gamma-production by suppressing natural killer cell stimulatory factor/IL-12 synthesis in accessory cells. *J. Exp. Med.* **178**, 1041–1048.

Daftarian, P.M., Kumar, A., Kryworuchko, M. and Diaz-Mitoma, F. (1996). IL-10 production is enhanced in human T cells by IL-12 and IL-6 and in monocytes by tumor necrosis factor-alpha. *J. Immunol.* **157**, 12–20.

Davidson, N.J., Leach, M.W., Fort, M.M., Thompson-Snipes, L., Kuhn, R., Muller, W., Berg, D.J. and Rennick, D.M. (1996). T helper cell 1-type CD4+ T cells, but not B cells, mediate colitis in interleukin 10-deficient mice. *J. Exp. Med.* **184**, 241–251.

de Vries, J.E., Groux, H., de Waal Malefyt, R. and Roncarolo, M.G. (1997). IL-10 induces long term antigen specific T cell tolerance: consequences for inflammatory bowel disease. In *Inflammatory Bowel Diseases From Bench to Bedside* (eds J. Scholmerich, H. Goebbel, T. Andus and P. Layer), Kluwer, Lancaster, UK, pp. 315–322.

de Waal Malefyt, R., Abrams, J., Bennett, B., Figdor, C. and de Vries, J. (1991a). IL-10 inhibits cytokine synthesis by human monocytes: an autoregulatory role of IL-10 produced by monocytes. *J. Exp. Med.* **174**, 1209–1220.

de Waal Malefyt, R., Haanen, J., Spits, H., Roncarolo, M.-G., te Velde, A., Figdor, C., Johnson, K., Kastelein, R., Yssel, H. and de Vries, J.E. (1991b). IL-10 and viral IL-10 strongly reduce antigen-specific human T cell proliferation by diminishing the antigen-presenting capacity of monocytes via downregulation of class II MHC expression. *J. Exp. Med.* **174**, 915–924.

de Waal Malefyt, R., Figdor, C.G., Huijbens, R., Mohan, P.S., Bennett, B., Culpepper, J., Dang, W., Zurawski, G. and de Vries, J. (1993a). Effects of IL-13 on phenotype, cytokine production, and cytotoxic function of human monocytes. Comparison with IL-4 and modulation by IFN-gamma or IL-10. *J. Immunol.* **151**, 6370–6381.

de Waal Malefyt, R., Yssel, H. and de Vries, J.E. (1993b). Direct effects of IL-10 on subsets of human CD4+ T cell clones and resting T cells. Specific inhibition of IL-2 production and proliferation. *J. Immunol.* **150**, 4754–4765.

de Waal Malefyt, R. and de Vries, J. (1996). Interleukin-10. In *Human Cytokines: A Handbook for Basic and Clinical Research* (eds B. Aggarwal and J. Gutterman), Blackwell Science, Cambridge, MA, pp. 19–42.

Defrance, T., Vanbervliet, B., Briere, F., Durand, I., Rousset, F. and Banchereau, J. (1992). Interleukin 10 and transforming growth factor β cooperate to induce anti-CD40-activated naive human B cells to secrete immunoglobulin A. *J. Exp. Med.* **175**, 671–682.

Del Prete, G., De Carli, M., Almerigogna, F., Giudizi, M.G., Biagiotti, R. and Romagnani, S. (1993). Human IL-10 is produced by both type 1 helper (Th1) and type 2 helper (Th2) T cell clones and inhibits their antigen-specific proliferation and cytokine production. *J. Immunol.* **150**, 353–360.

Deleuran, B., Iversen, L., Kristensen, M., Field, M., Kragballe, K., Thestrup-Pedersen, K. and Stengaard-Pedersen, K. (1994). Interleukin-8 secretion and 15-lipoxygenase activity in rheumatoid arthritis: *in vitro* anti-inflammatory effects by interleukin-4 and interleukin-10, but not by interleukin-1 receptor antagonist protein. *Br. J. Rheumatol.* **33**, 520–525.

Denis, M. and Ghadirian, E. (1993). IL-10 neutralization augments mouse resistance to systemic *Mycobacterium avium* infection. *J. Immunol.* **151**, 5425–5430.

Derkx, B., Marchant, A., Goldman, M., Bijlmer, R. and van Deventer, S. (1995). High levels of interleukin-10 during the initial phase of fulminant meningococcal septic shock. *J. Infect. Dis.* **171**, 229–232.

Ding, L. and Shevach, E.M. (1992). IL-10 inhibits mitogen-induced T cell proliferation by selectively inhibiting macrophage costimulatory function. *J. Immunol.* **148**, 3133–3139.

Ding, L., Linsley, P.S., Huang, L.Y., Germain, R.N. and Shevach, E.M. (1993). IL-10 inhibits macrophage costimulatory activity by selectively inhibiting the up-regulation of B7 expression. *J. Immunol.* **151**, 1224–1234.

Dugas, N., Vouldoukis, I., Becherel, P., Arock, M., Debre, P., Tardieu, M., Mossalayi, D.M., Delfraissy, J.F., Kolb, J.P. and Dugas, B. (1996). Triggering of CD23b antigen by anti-CD23 monoclonal antibodies induces interleukin-10 production by human macrophages. *Eur. J. Immunol.* **26**, 1394–1398.

Dummer, W., Bastian, B.C., Ernst, N., Schanzle, C., Schwaaf, A. and Brocker, E.B. (1996). Interleukin-10 production in malignant melanoma: preferential detection of IL-10-secreting tumor cells in metastatic lesions. *Int. J. Cancer* **66**, 607–610.

Durez, P., Abramowicz, D., Gerard, C., Van Mechelen, M., Amraoui, Z., Dubois, C., Leo, O., Velu, T. and Goldman, M. (1993). *In vivo* induction of interleukin-10 by anti-CD3 monoclonal antibody or bacterial lipopolysaccharide: differential modulation by cyclosporin A. *J. Exp. Med.* **177**, 551–555.

Eissner, G., Lindner, H., Behrends, U., Kolch, W., Hieke, A., Klauke, I., Bornkamm, G.W. and Holler, E. (1996). Influence of bacterial endotoxin on radiation-induced activation of human endothelial cells *in vitro* and *in vivo*: protective role of IL-10. *Transplantation* **62**, 819–827.

Enk, A.H. and Katz, S.I. (1992). Identification and induction of keratinocyte-derived IL-10. *J. Immunol.* **149**, 92–95.

Enk, A.H., Angeloni, V.L., Udey, M.C. and Katz, S.I. (1993). Inhibition of Langerhans cell antigen-presenting function by IL-10. A role for IL-10 in induction of tolerance. *J. Immunol.* **151**, 2390–2398.

Enk, A.H., Saloga, J., Becker, D., Madzadeh, M. and Knop, J. (1994). Induction of hapten-specific tolerance by interleukin 10 *in vivo. J. Exp. Med.* **179**, 1397–1402.

Enk, C.D., Sredni, D., Blauvelt, A. and Katz, S.I. (1995). Induction of IL-10 gene expression in human keratinocytes by UVB exposure *in vivo* and *in vitro. J. Immunol.* **154**, 4851–4856.

Ferguson, T.A., Dube, P. and Griffith, T.S. (1994). Regulation of contact hypersensitivity by interleukin 10. *J. Exp. Med.* **179**, 1597–1604.

Finbloom, D.S. and Winestock, K.D. (1995). IL-10 induces the tyrosine phosphorylation of tyk2 and Jak1 and the differential assembly of STAT1 alpha and STAT3 complexes in human T cells and monocytes. *J. Immunol.* **155**, 1079–1090.

Fiorentino, D.F., Bond, M.W. and Mosmann, T.R. (1989). Two types of mouse helper T cell. IV. Th2 clones secrete a factor that inhibits cytokine production by Th1 clones. *J. Exp. Med.* **170**, 2081–2095.

Fiorentino, D.F., Zlotnik, A., Mosmann, T.R., Howard, M. and O'Garra, A. (1991a). IL-10 inhibits cytokine production by activated macrophages. *J. Immunol.* **147**, 3815–3822.

Fiorentino, D.F., Zlotnik, A., Vieira, P., Mosmann, T.R., Howard, M., Moore, K.W. and O'Garra, A. (1991b). IL-10 acts on the antigen-presenting cell to inhibit cytokine production by Th1 cells. *J. Immunol.* **146**, 3444–3451.

Fleming, S.B., McCaughan, C.A., Andrews, A.E., Nash, A.D. and Mercer, A.A. (1997). A homolog of interleukin-10 is encoded by the poxvirus Orf virus. *J. Virol.* **71**, 4857–4861.

Flesch, I.E., Hess, J.H., Oswald, I.P. and Kaufmann, S.H. (1994). Growth inhibition of *Mycobacterium bovis* by IFN-gamma stimulated macrophages: regulation by endogenous tumor necrosis factor-alpha and by IL-10. *Int. Immunol.* **6**, 693–700.

Flores, V.P., Chikunguwo, S.M., Harris, T.S. and Stadecker, M.J. (1993). Role of IL-10 on antigen-presenting cell function for schistosomal egg-specific monoclonal T helper cell responses *in vitro* and *in vivo. J. Immunol.* **151**, 3192–3198.

Flores-Villanueva, P.O., Reiser, H. and Stadecker, M.J. (1994). Regulation of T helper cell responses in experimental murine schistosomiasis by IL-10. Effect on expression of B7 and B7-2 costimulatory molecules by macrophages. *J. Immunol.* **153**, 5190–5199.

Flores-Villanueva, P.O., Zheng, X.X., Strom, T.B. and Stadecker, M.J. (1996). Recombinant IL-10 and IL-10/Fc treatment down-regulate egg antigen-specific delayed hypersensitivity reactions and egg granuloma formation in schistosomiasis. *J. Immunol.* **156**, 3315–3320.

Florquin, S., Amraoui, Z., Abramowicz, D. and Goldman, M. (1994). Systemic release and protective role of IL-10 in staphylococcal enterotoxin B-induced shock in mice. *J. Immunol.* **153**, 2618–2623.

Fluckiger, A.C., Garrone, P., Durand, I., Galizzi, J.P. and Banchereau, J. (1993). Interleukin 10 (IL-10) upregulates functional high affinity IL-2 receptors on normal and leukemic B lymphocytes. *J. Exp. Med.* **178**, 1473–1481.

Fouqueray, B., Boutard, V., Philippe, C., Kornreich, A., Marchant, A., Perez, J., Goldman, M. and Baud, L. (1995). Mesangial cell-derived interleukin-10 modulates mesangial cell response to lipopolysaccharide. *Am. J. Pathol.* **147**, 176–182.

Frei, K., Nadal, D., Pfister, H.W. and Fontana, A. (1993). Listeria meningitis: identification of a cerebrospinal fluid inhibitor of macrophage listericidal function as interleukin 10. *J. Exp. Med.* **178**, 1255–1261.

Garrone, P., Galibert, L., Rousset, F. and Banchereau, J. (1994). Regulatory effects of prostaglandin E2 on the growth and differentiation of human B lymphocytes activated through their CD40 antigen. *J. Immunol.* **152**, 82–90.

Gastl, G.A., Abrams, J.S., Nanus, D.M., Oosterkamp, R., Silver, J., Liu, F., Chen, M., Albino, A.P. and Bander, N.H. (1993). Interleukin-10 production by human carcinoma cell lines and its relationship to interleukin-6 expression. *Int. J. Cancer* **55**, 96–101.

Gazzinelli, R.T., Oswald, I.P., Hieny, S., James, S.L. and Sher, A. (1992). The microbicidal activity of interferon-gamma-treated macrophages against *Trypanosoma cruzi* involves an L-arginine-dependent, nitrogen oxide-mediated mechanism inhibitable by interleukin-10 and transforming growth factor-beta. *Eur. J. Immunol.* **22**, 2501–2506.

Gazzinelli, R.T., Wysocka, M., Hieny, S., Scharton-Kersten, T., Cheever, A., Kuhn, R., Muller, W., Trinchieri, G. and Sher, A. (1996). In the absence of endogenous IL-10, mice acutely infected with *Toxoplasma gondii* succumb to a lethal immune response dependent on CD4+ T cells and accompanied by overproduction of IL-12, IFN-gamma and TNF-alpha. *J. Immunol.* **157**, 798–805.

Geng, Y., Gulbins, E., Altman, A. and Lotz, M. (1994). Monocyte deactivation by interleukin 10 via inhibition of tyrosine kinase activity and the Ras signaling pathway. *Proc. Natl Acad. Sci. U.S.A.* **91**, 8602–8606.

Gerard, C., Bruyns, C., Marchant, A., Abramowicz, D., Vandenabeele, P., Delvaux, A., Fiers, W., Goldman, M. and Velu, T. (1993). Interleukin 10 reduces the release of tumor necrosis factor and prevents lethality in experimental endotoxemia. *J. Exp. Med.* **177**, 547–550.

Gerard, C.M., Bruyns, C., Delvaux, A., Baudson, N., Dargent, J.L., Goldman, M. and Velu, T. (1996). Loss of tumorigenicity and increased immunogenicity induced by interleukin-10 gene transfer in B16 melanoma cells. *Hum. Gene Ther.* **7**, 23–31.

Gerosa, F., Paganin, C., Peritt, D., Paiola, F., Scupoli, M.T., Aste-Amezaga, M., Frank, I. and Trinchieri, G. (1996). Interleukin-12 primes human CD4 and CD8 T cell clones for high production of both interferon-gamma and interleukin-10. *J. Exp. Med.* **183**, 2559–2569.

Ghildyal, N., McNeil, H.P., Gurish, M.F., Austen, K.F. and Stevens, R.L. (1992a). Transcriptional regulation of the mucosal mast cell-specific protease gene, MMCP-2, by interleukin 10 and interleukin 3. *J. Biol. Chem.* **267**, 8473–8477.

Ghildyal, N., McNeil, H.P., Stechschulte, S., Austen, K.F., Silberstein, D., Gurish, M.F., Somerville, L.L. and Stevens, R.L. (1992b). IL-10 induces transcription of the gene for mouse mast cell protease-1, a serine protease preferentially expressed in mucosal mast cells of *Trichinella spiralis*-infected mice. *J. Immunol.* **149**, 2123–2129.

Ghildyal, N., Friend, D.S., Nicodemus, C.F., Austen, K.F. and Stevens, R.L. (1993). Reversible expression of mouse mast cell protease 2 mRNA and protein in cultured mast cells exposed to IL-10. *J. Immunol.* **151**, 3206–3214.

Giovarelli, M., Musiani, P., Modesti, A., Dellabona, P., Casorati, G., Allione, A., Consalvo, M., Cavallo, F., di Pierro, F. and De Giovanni, C. (1995). Local release of IL-10 by transfected mouse mammary adenocarcinoma cells does not suppress but enhances antitumor reaction and elicits a strong cytotoxic lymphocyte and antibody-dependent immune memory. *J. Immunol.* **155**, 3112–3123.

Go, N.F., Castle, B.E., Barrett, R., Kastelein, R., Dang, W., Mosmann, T.R., Moore, K.W. and Howard, M. (1990). Interleukin 10 (IL-10), a novel B cell stimulatory factor: unresponsiveness of X chromosome-linked immunodeficiency B cells. *J. Exp. Med.* **172**, 1625–1631.

Gomez-Jimenez, J., Martin, M.C., Sauri, R., Segura, R.M., Esteban, F., Ruiz, J.C., Nuvials, X., Boveda, J.L., Peracaula, R. and Salgado, A. (1995). Interleukin-10 and the monocyte/macrophage-induced inflammatory response in septic shock. *J. Infect. Dis.* **171**, 472–475.

Goodman, R.E., Oblak, J. and Bell, R.G. (1992). Synthesis and characterization of rat interleukin-10 (IL-10) cDNA clones from the RNA of cultured OX8-OX22-thoracic duct T cells. *Biochem. Biophys. Res. Commun.* **189**, 1–7.

Greenberger, M.J., Strieter, R.M., Kunkel, S.L., Danforth, J.M., Goodman, R.E. and Standiford, T.J. (1995). Neutralization of IL-10 increases survival in a murine model of *Klebsiella* pneumonia. *J. Immunol.* **155**, 722–729.

Groux, H., Bigler, M., de Vries, J.E. and Roncarolo, M.G. (1996). Interleukin-10 induces a long-term antigen-specific anergic state in human CD4+ T cells. *J. Exp. Med.* **184**, 19–29.

Gruber, M.F., Williams, C.C. and Gerrard, T.L. (1994). Macrophage-colony-stimulating factor expression by anti-CD45 stimulated human monocytes is transcriptionally up-regulated by IL-1 beta and inhibited by IL-4 and IL-10. *J. Immunol.* **152**, 1354–1361.

Grünig, G., Corry, D.B., Leach, M.W., Seymour, B.W.P., Kurup, V.P. and Rennick, D.M. (1997). Interleukin-10 is a natural suppressor of cytokine production and inflammation in a murine model of allergic bronchopulmonary aspergillosis. *J. Exp. Med.* **185**, 1089–1099.

Hagenbaugh, A., Sharma, S., Dubinett, S.M., Wei, S.H.-Y., Aranda, R., Cheroutre, H., Fowell, D., Binder, S., Tsao, B., Locksley, R.M., Moore, K.W. and Kronenberg, M. (1997). Altered immune responses in IL-10 transgenic mice. *J. Exp. Med.* **185**, 2101–2110.

Hagiwara, E., Gourley, M.F., Lee, S. and Klinman, D.K. (1996). Disease severity in patients with systemic lupus erythematosus correlates with an increased ratio of interleukin-10:interferon-gamma-secreting cells in the peripheral blood. *Arthritis Rheum.* **39**, 379–385.

Hart, P.H., Ahern, M.J., Smith, M.D. and Finlay-Jones, J.J. (1995). Comparison of the suppressive effects of interleukin-10 and interleukin-4 on synovial fluid macrophages and blood monocytes from patients with inflammatory arthritis. *Immunology* **84**, 536–542.

Hart, P.H., Hunt, E.K., Bonder, C.S., Watson, C.J. and Finlay-Jones, J.J. (1996). Regulation of surface and soluble TNF receptor expression on human monocytes and synovial fluid macrophages by IL-4 and IL-10. *J. Immunol.* **157**, 3672–3680.

Ho, A.S.-Y. and Moore, K.W. (1994). Interleukin-10 and its receptor. *Ther. Immun.* **1**, 173–185.

Ho, A.S.-Y., Liu, Y., Khan, T.A., Hsu, D.-H., Bazan, J.F. and Moore, K.W. (1993). A receptor for interleukin-10 is related to interferon receptors. *Proc. Natl Acad. Sci. U.S.A.* **90**, 11 267–11 271.

Ho, A.S., Wei, S.H., Mui, A.L., Miyajima, A. and Moore, K.W. (1995). Functional regions of the mouse interleukin-10 receptor cytoplasmic domain. *Mol. Cell. Biol.* **15**, 5043–5053.

Houssiau, F.A., Lefebvre, C., Vanden Berghe, M., Lambert, M., Devogelaer, J.P. and Renauld, J.C. (1995). Serum interleukin 10 titers in systemic lupus erythematosus reflect disease activity. *Lupus* **4**, 393–395.

Howard, M., Muchamuel, T., Andrade, S. and Menon, S. (1993). Interleukin 10 protects mice from lethal endotoxemia. *J. Exp. Med.* **177**, 1205–1208.

Hsu, D.-H., de Waal Malefyt, R., Fiorentino, D.F., Dang, M.-N., Vieira, P., de Vries, J., Spits, H., Mosmann, T.R. and Moore, K.W. (1990). Expression of IL-10 activity by Epstein–Barr Virus Protein BCRFI. *Science* **250**, 830–832.

Hsu, D.-H., Moore, K.W. and Spits, H. (1992). Differential effects of interleukin-4 and -10 on interleukin-2-induced interferon-γ synthesis and lymphokine-activated killer activity. *Int. Immunol.* **4**, 563–569.

Huang, M., Sharma, S., Mao, J.T. and Dubinett, S.M. (1996). Non-small cell lung cancer-derived soluble mediators and prostaglandin E2 enhance peripheral blood lymphocyte IL-10 transcription and protein production. *J. Immunol.* **157**, 5512–5520.

Hudson, G.S., Bankier, A.T., Satchwell, S.C. and Barrell, B.G. (1985). The short unique region of the B95-8 Epstein–Barr virus genome. *Virology* **147**, 81–98.

Huhn, R.D., Radwanski, E., O'Connell, S.M., Sturgill, M.G., Clarke, L., Cody, R.P., Affrime, M.B. and Cutler, D.L. (1996). Pharmacokinetics and immunomodulatory properties of intravenously administered recombinant human interleukin-10 in healthy volunteers. *Blood* **87**, 699–705.

Hunter, C.A., Subauste, C.S., Van Cleave, V.H. and Remington, J.S. (1994). Production of gamma interferon by natural killer cells from *Toxoplasma gondii*-infected SCID mice: regulation by interleukin-10, interleukin-12, and tumor necrosis factor alpha. *Infect. Immun.* **62**, 2818–2824.

Hurme, M., Henttinen, T., Karppelin, M., Varkila, K. and Matikainen, S. (1994). Effect of interleukin-10 on NF-κB and AP-1 activities in interleukin-2 dependent CD8 T lymphoblasts. *Immunol. Lett.* **42**, 129–133.

Ishida, H., Hastings, R., Kearney, J. and Howard, M. (1992). Continuous anti-interleukin 10 antibody administration depletes mice of Ly-1 B cells but not conventional B cells. *J. Exp. Med.* **175**, 1213–1220.

Ishida, H., Hastings, R., Thompson, S.L. and Howard, M. (1993). Modified immunological status of anti-IL-10 treated mice. *Cell. Immunol.* **148**, 371–384.

Ishida, H., Muchamuel, T., Sakaguchi, S., Andrade, S., Menon, S. and Howard, M. (1994). Continuous administration of anti-interleukin 10 antibodies delays onset of autoimmunity in NZB/W F1 mice. *J. Exp. Med.* **179**, 305–310.

Issazadeh, S., Ljungdahl, A., Hojeberg, B., Mustafa, M. and Olsson, T. (1995). Cytokine production in the central nervous system of Lewis rats with experimental autoimmune encephalomyelitis: dynamics of mRNA expression for interleukin-10, interleukin-12, cytolysin, tumor necrosis factor alpha and tumor necrosis factor beta. *J. Neuroimmunol.* **61**, 205–212.

Issazadeh, S., Lorentzen, J.C., Mustafa, M.I., Hojeberg, B., Mussener, A. and Olsson, T. (1996). Cytokines in relapsing experimental autoimmune encephalomyelitis in DA rats: persistent mRNA expression of proinflammatory cytokines and absent expression of interleukin-10 and transforming growth factor-beta. *J. Neuroimmunol.* **69**, 103–115.

Itoh, K. and Hirohata, S. (1995). The role of IL-10 in human B cell activation, proliferation, and differentiation. *J. Immunol.* **154**, 4341–4350.

Jansen, P.M., van der Pouw Kraan, T.C., de Jong, I.W., van Mierlo, G., Wijdenes, J., Chang, A.A., Aarden, L.A., Taylor, F.B., Jr. and Hack, C.E. (1996). Release of interleukin-12 in experimental *Escherichia coli* septic shock in baboons: relation to plasma levels of interleukin-10 and interferon-gamma. *Blood* **87**, 5144–5151.

Jeannin, P., Delneste, Y., Seveso, M., Life, P. and Bonnefoy, J.Y. (1996). IL-12 synergizes with IL-2 and other stimuli in inducing IL-10 production by human T cells. *J. Immunol.* **156**, 3159–3165.

Jenkins, J.K., Malyak, M. and Arend, W.P. (1994). The effects of interleukin-10 on interleukin-1 receptor antagonist and interleukin-1 beta production in human monocytes and neutrophils. *Lymphokine Cytokine Res.* **13**, 47–54.

Jinquan, T., Larsen, C.G., Gesser, B., Matsushima, K. and Thestrup, P.K. (1993). Human IL-10 is a chemoattractant for CD8+ T lymphocytes and an inhibitor of IL-8-induced CD4+ T lymphocyte migration. *J. Immunol.* **151**, 4545–4551.

Joosten, L.A., Lubberts, E., Durez, P., Helsen, M.M., Jacobs, M.J., Goldman, M. and van den Berg, W.B. (1997). Role of interleukin-4 and interleukin-10 in murine collagen-induced arthritis. Protective effect of interleukin-4 and interleukin-10 treatment on cartilage destruction. *Arthritis Rheum.* **40**, 249–260.

Joyce, D.A. and Steer, J.H. (1996). IL-4, IL-10 and IFN-gamma have distinct, but interacting, effects on differentiation-induced changes in TNF-alpha and TNF receptor release by cultured human monocytes. *Cytokine* **8**, 49–57.

Jungi, T.W., Brcic, M., Eperon, S. and Albrecht, S. (1994). Transforming growth factor-beta and interleukin-10, but not interleukin-4, down-regulate procoagulant activity and tissue factor expression in human monocyte-derived macrophages. *Thromb. Res.* **76**, 463–474.

Kambayashi, T., Jacob, C.O., Zhou, D., Mazurek, N., Fong, M. and Strassmann, G. (1995). Cyclic nucleotide phosphodiesterase type IV participates in the regulation of IL-10 and in the subsequent inhibition of TNF-alpha and IL-6 release by endotoxin-stimulated macrophages. *J. Immunol.* **155**, 4909–4916.

Kang, K., Hammerberg, C., Meunier, L. and Cooper, K.D. (1994). CD11b+ macrophages that infiltrate human epidermis after *in vivo* ultraviolet exposure potently produce IL-10 and represent the major secretory source of epidermal IL-10 protein. *J. Immunol.* **153**, 5256–5264.

Kasama, T., Strieter, R.M., Lukacs, N.W., Burdick, M.D. and Kunkel, S.L. (1994). Regulation of neutrophil-derived chemokine expression by IL-10. *J. Immunol.* **152**, 3559–3569.

Kasama, T., Strieter, R.M., Lukacs, N.W., Lincoln, P.M., Burdick, M.D. and Kunkel, S.L. (1995). Interleukin-10 expression and chemokine regulation during the evolution of murine type II collagen-induced arthritis. *J. Clin. Invest.* **95**, 2868–2876.

Kato, T., Murata, A., Ishida, H., Toda, H., Tanaka, N., Hayashida, H., Monden, M. and Matsuura, N. (1995). Interleukin 10 reduces mortality from severe peritonitis in mice. *Antimicrob. Agents Chemother.* **39**, 1336–1340.

Katsikis, P.D., Chu, C.Q., Brennan, F.M., Maini, R.N. and Feldmann, M. (1994). Immunoregulatory role of interleukin 10 in rheumatoid arthritis. *J. Exp. Med.* **179**, 1517–1527.

Kennedy, M.K., Torrance, D.S., Picha, K.S. and Mohler, K.M. (1992). Analysis of cytokine mRNA expression in the central nervous system of mice with experimental autoimmune encephalomyelitis reveals that IL-10 mRNA expression correlates with recovery. *J. Immunol.* **149**, 2496–2505.

Kim, J.M., Brannan, C.I., Copeland, N.G., Jenkins, N.A., Khan, T.A. and Moore, K.W. (1992). Structure of the mouse interleukin-10 gene and chromosomal localization of the mouse and human genes. *J. Immunol.* **148**, 3618–3623.

Kim, J., Modlin, R.L., Moy, R.L., Dubinett, S.M., McHugh, T., Nickoloff, B.J. and Uyemura, K. (1995). IL-10 production in cutaneous basal and squamous cell carcinomas. A mechanism for evading the local T cell immune response. *J. Immunol.* **155**, 2240–2247.

King, C.L., Medhat, A., Malhotra, I., Nafeh, M., Helmy, A., Khaudary, J., Ibrahim, S., El-Sherbiny, M., Zaky, S., Stupi, R.J., Brustoski, K., Shehata, M. and Shata, M.T. (1996). Cytokine control of parasite-specific anergy in human urinary schistosomiasis. IL-10 modulates lymphocyte reactivity. *J. Immunol.* **156**, 4715–4721.

Kollmann, T.R., Pettoello-Mantovani, M., Katopodis, N.F., Hachamovitch, M., Rubinstein, A., Kim, A. and Goldstein, H. (1996). Inhibition of acute *in vivo* human immunodeficiency virus infection by human interleukin 10 treatment of SCID mice implanted with human fetal thymus and liver. *Proc. Natl Acad. Sci. U.S.A.* **93**, 3126–3131.

Kondo, S., McKenzie, R.C. and Sauder, D.N. (1994). Interleukin-10 inhibits the elicitation phase of allergic contact hypersensitivity. *J. Invest. Dermatol.* **103**, 811–814.

Kootstra, N.A., van't Wout, A., Huisman, H.G., Miedema, F. and Schuitemaker, H. (1994). Interference of interleukin-10 with human immunodeficiency virus type 1 replication in primary monocyte-derived macrophages. *J. Virol.* **68**, 6967–6975.

Kubin, M., Kamoun, M. and Trinchieri, G. (1994). Interleukin-12 synergizes with B7/CD28 interaction in inducing efficient proliferation and cytokine production by human T cells. *J. Exp. Med.* **180**, 263–274.

Kucharzik, T., Stoll, R., Lugering, N. and Domschke, W. (1995). Circulating antiinflammatory cytokine IL-10 in patients with inflammatory bowel disease (IBD). *Clin. Exp. Immunol.* **100**, 452–456.

Kuga, S., Otsuka, T., Niiro, H., Nunoi, H., Nemoto, Y., Nakano, T., Ogo, T., Umei, T. and Niho, Y. (1996). Suppression of superoxide anion production by interleukin-10 is accompanied by a downregulation of the genes for subunit proteins of NADPH oxidase. *Exp. Hematol.* **24**, 151–157.

Kuhn, R., Lohler, J., Rennick, D., Rajewsky, K. and Muller, W. (1993). Interleukin-10 deficient mice develop chronic enterocholitis. *Cell* **75**, 263–274.

Kundu, N., Beaty, T.L., Jackson, M.J. and Fulton, A.M. (1996). Antimetastatic and antitumor activities of interleukin 10 in a murine model of breast cancer [see comments]. *J. Natl Cancer Inst.* **88**, 536–541.

Lacraz, S., Nicod, L.P., Chicheportiche, R., Welgus, H.G. and Dayer, J.M. (1995). IL-10 inhibits metalloproteinase and stimulates TIMP-1 production in human mononuclear phagocytes. *J. Clin. Invest.* **96**, 2304–2310.

Lai, C.F., Ripperger, J., Morella, K.K., Jurlander, J., Hawley, T.S., Carson, W.E., Kordula, T., Caligiuri, M.A., Hawley, R.G., Fey, G.H. and Baumann, H. (1996). Receptors for interleukin (IL)-10 and IL-6 type cytokines use similar signaling mechanisms for inducing transcription through IL-6 response elements. *J. Biol. Chem.* **271**, 13968–13975.

Landay, A.L., Clerici, M., Hashemi, F., Kessler, H., Berzofsky, J.A. and Shearer, G.M. (1996). *In vitro* restoration of T cell immune function in human immunodeficiency virus-positive persons: effects of interleukin (IL)-12 and anti-IL-10. *J. Infect. Dis.* **173**, 1085–1091.

Lanier, L. (1997). Natural killer cells: from no receptors to too many. *Immunity* **6**, 371–378.

Le Moine, O., Stordeur, P., Schandene, L., Marchant, A., de Groote, D., Goldman, M. and Deviere, J. (1996). Adenosine enhances IL-10 secretion by human monocytes. *J. Immunol.* **156**, 4408–4414.

Lee, M.S., Wogensen, L., Shizuru, J., Oldstone, M.B. and Sarvetnick, N. (1994). Pancreatic islet production of murine interleukin-10 does not inhibit immune-mediated tissue destruction. *J. Clin. Invest.* **93**, 1332–1338.

Lee, M.S., Mueller, R., Wicker, L.S., Peterson, L.B. and Sarvetnick, N. (1996). IL-10 is necessary and sufficient for autoimmune diabetes in conjunction with NOD MHC homozygosity. *J. Exp. Med.* **183**, 2663–2668.

Lehmann, A.K., Halstensen, A., Sornes, S., Rokke, O. and Waage, A. (1995). High levels of interleukin 10 in serum are associated with fatality in meningococcal disease. *Infect. Immun.* **63**, 2109–2112.

Levy, Y. and Brouet, J.C. (1994). Interleukin-10 prevents spontaneous death of germinal center B cells by induction of the bcl-2 protein. *J. Clin. Invest.* **93**, 424–428.

Li, L., Elliott, J.F. and Mosmann, T.R. (1994). IL-10 inhibits cytokine production, vascular leakage, and swelling during T helper 1 cell-induced delayed-type hypersensitivity. *J. Immunol.* **153**, 3967–3978.

Li, Q., Sun, B., Dastgheib, K. and Chan, C.C. (1996). Suppressive effect of transforming growth factor beta1 on the recurrence of experimental melanin protein-induced uveitis: upregulation of ocular interleukin-10. *Clin. Immunol. Immunopathol.* **81**, 55–61.

Lin, H., Mosmann, T.R., Guilbert, L., Tuntipopipat, S. and Wegmann, T.G. (1993). Synthesis of T helper 2-type cytokines at the maternal–fetal interface. *J. Immunol.* **151**, 4562–4573.

Linderholm, M., Ahlm, C., Settergren, B., Waage, A. and Tarnvik, A. (1996). Elevated plasma levels of tumor necrosis factor (TNF)-alpha, soluble TNF receptors, interleukin (IL)-6, and IL-10 in patients with hemorrhagic fever with renal syndrome. *J. Infect. Dis.* **173**, 38–43.

Liu, Y., Wei, S.H.-Y., Ho, A.S.-Y., de Waal Malefyt, R. and Moore, K.W. (1994). Expression cloning and characterization of a human interleukin-10 receptor. *J. Immunol.* **152**, 1821–1829.

Liu, Y., de Waal Malefyt, R., Briere, F., Parham, C., Bridon, J.-M., Banchereau, J., Moore, K.W. and Xu, J. (1997). The Epstein–Barr Virus interleukin-10 (IL-10) homolog is a selective agonist with impaired binding to the IL-10 receptor. *J. Immunol.* **158**, 604–613.

Llorente, L., Richaud-Patin, Y., Fior, R., Alcocer-Varela, J., Wijdenes, J., Fourrier, B.M., Galanaud, P. and Emilie, D. (1994). *In vivo* production of interleukin-10 by non-T cells in rheumatoid arthritis, Sjogren's syndrome, and systemic lupus erythematosus. A potential mechanism of B lymphocyte hyperactivity and autoimmunity. *Arthritis Rheum.* **37**, 1647–1655.

Llorente, L., Zou, W., Levy, Y., Richaud-Patin, Y., Wijdenes, J., Alcocer-Varela, J., Morel-Fourrier, B., Brouet,

J.C., Alarcon-Segovia, D. and Galanaud, P. (1995). Role of interleukin 10 in the B lymphocyte hyperactivity and autoantibody production of human systemic lupus erythematosus. *J. Exp. Med.* **181**, 839–844.

Ludewig, B., Graf, D., Gelderblom, H.R., Becker, Y., Kroczek, R.A. and Pauli, G. (1995). Spontaneous apoptosis of dendritic cells is efficiently inhibited by TRAP (CD40-ligand) and TNF-alpha, but strongly enhanced by interleukin-10. *Eur. J. Immunol.* **25**, 1943–1950.

Ludewig, B., Gelderblom, H.R., Becker, Y., Schafer, A. and Pauli, G. (1996). Transmission of HIV-1 from productively infected mature Langerhans cells to primary CD4+ T lymphocytes results in altered T cell responses with enhanced production of IFN-gamma and IL-10. *Virology* **215**, 51–60.

Macatonia, S.E., Doherty, T.M., Knight, S.C. and O'Garra, A. (1993). Differential effect of IL-10 on dendritic cell-induced T cell proliferation and IFN-gamma production. *J. Immunol.* **150**, 3755–3765.

MacNeil, I., Suda, T., Moore, K.W., Mosmann, T.R. and Zlotnik, A. (1990). IL-10: a novel cytokine growth cofactor for mature and immature T cells. *J. Immunol.* **145**, 4167–4173.

Mahanty, S. and Nutman, T.B. (1995). Immunoregulation in human lymphatic filariasis: the role of interleukin 10. *Parasite Immunol.* **17**, 385–392.

Mahanty, S., Mollis, S.N., Ravichandran, M., Abrams, J.S., Kumaraswami, V., Jayaraman, K., Ottesen, E.A. and Nutman, T.B. (1996). High levels of spontaneous and parasite antigen-driven interleukin-10 production are associated with antigen-specific hyporesponsiveness in human lymphatic filariasis. *J. Infect. Dis.* **173**, 769–773.

Malisan, F., Briere, F., Bridon, J.M., Harindranath, N., Mills, F.C., Max, E.E., Banchereau, J. and Martinez-Valdez, H. (1996). Interleukin-10 induces immunoglobulin G isotype switch recombination in human CD40-activated naive B lymphocytes. *J. Exp. Med.* **183**, 937–947.

Marchant, A., Bruyns, C., Vandenabeele, P., Ducarme, M., Gerard, C., Delvaux, A., De Groote, D., Abramowicz, D., Velu, T. and Goldman, M. (1994a). Interleukin-10 controls interferon-gamma and tumor necrosis factor production during experimental endotoxemia. *Eur. J. Immunol.* **24**, 1167–1171.

Marchant, A., Deviere, J., Byl, B., De Groote, D., Vincent, J.L. and Goldman, M. (1994b). Interleukin-10 production during septicaemia. *Lancet* **343**, 707–708.

Marchant, A., Alegre, M.L., Hakim, A., Pierard, G., Marecaux, G., Friedman, G., De Groote, D., Kahn, R.J., Vincent, J.L. and Goldman, M. (1995). Clinical and biological significance of interleukin-10 plasma levels in patients with septic shock. *J. Clin. Immunol.* **15**, 266–273.

Marfaing-Koka, A., Maravic, M., Humbert, M., Galanaud, P. and Emilie, D. (1996). Contrasting effects of IL-4, IL-10 and corticosteroids on RANTES production by human monocytes. *Int. Immunol.* **8**, 1587–1594.

Marie, C., Pitton, C., Fitting, C. and Cavaillon, J.M. (1996). IL-10 and IL-4 synergize with TNF-alpha to induce IL-1Ra production by human neutrophils. *Cytokine* **8**, 147–151.

Marietta, E.V., Chen, Y. and Weis, J.H. (1996). Modulation of expression of the anti-inflammatory cytokines interleukin-13 and interleukin-10 by interleukin-3. *Eur. J. Immunol.* **26**, 49–56.

Matsuda, M., Salazar, F., Petersson, M., Masucci, G., Hansson, J., Pisa, P., Zhang, Q.J., Masucci, M.G. and Kiessling, R. (1994). Interleukin 10 pretreatment protects target cells from tumor- and allo-specific cytotoxic T cells and downregulates HLA class I expression. *J. Exp. Med.* **180**, 2371–2376.

Mengozzi, M., Fantuzzi, G., Faggioni, R., Marchant, A., Goldman, M., Orencole, S., Clark, B.D., Sironi, M., Benigni, F. and Ghezzi, P. (1994). Chlorpromazine specifically inhibits peripheral and brain TNF production, and up-regulates IL-10 production in mice. *Immunology* **82**, 207–210.

Mertz, P.M., DeWitt, D.L., Stetler-Stevenson, W.G. and Wahl, L.M. (1994). Interleukin 10 suppression of monocyte prostaglandin H synthase. 2. Mechanism of inhibition of prostaglandin-dependent matrix metalloproteinase production. *J. Biol. Chem.* **269**, 21 322–21 329.

Merville, P., Dechanet, J., Grouard, G., Durand, I. and Banchereau, J. (1995). T cell-induced B cell blasts differentiate into plasma cells when cultured on bone marrow stroma with IL-3 and IL-10. *Int. Immunol.* **7**, 635–643.

Meyaard, L., Hovenkamp, E., Otto, S.A. and Miedema, F. (1996). IL-12-induced IL-10 production by human T cells as a negative feedback for IL-12-induced immune responses. *J. Immunol.* **156**, 2776–2782.

Mitra, R.S., Judge, T.A., Nestle, F.O., Turka, L.A. and Nickoloff, B.J. (1995). Psoriatic skin-derived dendritic cell function is inhibited by exogenous IL-10. Differential modulation of B7-1 (CD80) and B7-2 (CD86) expression. *J. Immunol.* **154**, 2668–2677.

Miyazaki, I., Cheung, R.K. and Dosch, H.-M. (1993). Viral interleukin 10 is critical for the induction of B cell growth transformation by Epstein–Barr virus. *J. Exp. Med.* **178**, 439–447.

Montaner, L.J., Griffin, P. and Gordon, S. (1994). Interleukin-10 inhibits initial reverse transcription of human immunodeficiency virus type 1 and mediates a virostatic latent state in primary blood-derived human macrophages *in vitro*. *J. Gen. Virol.* **75**, 3393–3400.

Moore, K.W., Vieira, P., Fiorentino, D.F., Trounstine, M.L., Khan, T.A. and Mosmann, T.R. (1990). Homology of cytokine synthesis inhibitory factor (IL-10) to the Epstein-Barr Virus gene BCRFI. *Science* **248**, 1230–1234.

Moore, K.W., O'Garra, A., de Waal Malefyt, R., Vieira, P. and Mosmann, T.R. (1993). Interleukin-10. *Ann. Rev. Immunol.* **11**, 165–190.

Morinobu, A., Kumagai, S., Yanagida, H., Ota, H., Ishida, H., Matsui, M., Yodoi, J. and Nakao, K. (1996). IL-10 suppresses cell surface CD23/Fc epsilon RII expression, not by enhancing soluble CD23 release, but by reducing CD23 mRNA expression in human monocytes. *J. Clin. Immunol.* **16**, 326–333.

Moritani, M., Yoshimoto, K., Tashiro, F., Hashimoto, C., Miyazaki, J., Ii, S., Kudo, E., Iwahana, H., Hayashi, Y. and Sano, T. (1994). Transgenic expression of IL-10 in pancreatic islet A cells accelerates autoimmune insulitis and diabetes in non-obese diabetic mice. *Int. Immunol.* **6**, 1927–1936.

Morltanl, M., Yoshlmoto, K., Ii, S., Kondo, M., Iwahana, H., Yamaoka, T., Sano, T., Nakano, N., Kikutani, H. and Itakura, M. (1996). Prevention of adoptively transferred diabetes in nonobese diabetic mice with IL-10-transduced islet-specific Th1 lymphocytes. A gene therapy model for autoimmune diabetes. *J. Clin. Invest.* **98**, 1851–1859.

Mosmann, T.R., Schumacher, J., Fiorentino, D.F., Leverah, J., Moore, K.W. and Bond, M.W. (1990). Isolation of monoclonal antibodies specific for IL4, IL5, IL6, and a new Th2-specific cytokine (IL-10), cytokine synthesis inhibitory factor, by using a solid phase radioimmunoadsorbent assay. *J. Immunol.* **145**, 2938–2945.

Mueller, R., Lee, M.S., Sawyer, S.P. and Sarvetnick, N. (1996). Transgenic expression of interleukin 10 in the pancreas renders resistant mice susceptible to low dose streptozotocin-induced diabetes. *J. Autoimmun.* **9**, 151–158.

Mulligan, M.S., Jones, M.L., Vaporciyan, A.A., Howard, M.C. and Ward, P.A. (1993). Protective effects of IL-4 and IL-10 against immune complex-induced lung injury. *J. Immunol.* **151**, 5666–5674.

Nabioullin, R., Sone, S., Mizuno, K., Yano, S., Nishioka, Y., Haku, T. and Ogura, T. (1994). Interleukin-10 is a potent inhibitor of tumor cytotoxicity by human monocytes and alveolar macrophages. *J. Leukoc. Biol.* **55**, 437–442.

Navikas, V., Link, J., Palasik, W., Soderstrom, M., Fredrikson, S., Olsson, T. and Link, H. (1995). Increased mRNA expression of IL-10 in mononuclear cells in multiple sclerosis and optic neuritis. *Scand. J. Immunol.* **41**, 171–178.

Nicod, L.P., el Habre, F., Dayer, J.M. and Boehringer, N. (1995). Interleukin-10 decreases tumor necrosis factor alpha and beta in alloreactions induced by human lung dendritic cells and macrophages. *Am. J. Respir. Cell Mol. Biol.* **13**, 83–90.

Niiro, H., Otsuka, T., Abe, M., Satoh, H., Ogo, T., Nakano, T., Furukawa, Y. and Niho, Y. (1992). Epstein–Barr virus BCRF1 gene product (viral interleukin 10) inhibits superoxide anion production by human monocytes. *Lymphokine Cytokine Res.* **11**, 209–214.

Niiro, H., Otsuka, T., Kuga, S., Nemoto, Y., Abe, M., Hara, N., Nakano, T., Ogo, T. and Niho, Y. (1994). IL-10 inhibits prostaglandin E2 production by lipopolysaccharide-stimulated monocytes. *Int. Immunol.* **6**, 661–664.

Niiro, H., Otsuka, T., Tanabe, T., Hara, S., Kuga, S., Nemoto, Y., Tanaka, Y., Nakashima, H., Kitajima, S. and Abe, M. (1995). Inhibition by interleukin-10 of inducible cyclooxygenase expression in lipopolysaccharide-stimulated monocytes: its underlying mechanism in comparison with interleukin-4. *Blood* **85**, 3736–3745.

Nonoyama, S., Hollenbaugh, D., Aruffo, A., Ledbetter, J.A. and Ochs, H.D. (1993). B cell activation via CD40 is required for specific antibody production by antigen-stimulated human B cells. *J. Exp. Med.* **178**, 1097–1102.

O'Garra, A., Stapleton, G., Dhar, V., Pearce, M., Schumacher, J., Rugo, H., Barbis, D., Stall, A., Cupp, J., Moore, K., Vieira, P., Mosmann, T., Whitmore, A., Arnold, L., Haughton, G. and Howard, M. (1990). Production of cytokines by mouse B cells: B lymphomas and normal B cells produce interleukin 10. *Int. Immunol.* **2**, 821–832.

O'Garra, A., Chang, R., Go, N., Hastings, R., Haughton, G. and Howard, M. (1992). Ly-1 B (B-1) cells are the main source of B cell-derived interleukin 10. *Eur. J. Immunol.* **22**, 711–717.

Oswald, I.P., Gazzinelli, R.T., Sher, A. and James, S.L. (1992a). IL-10 synergizes with IL-4 and transforming growth factor-beta to inhibit macrophage cytotoxic activity. *J. Immunol.* **148**, 3578–3582.

Oswald, I.P., Wynn, T.A., Sher, A. and James, S.L. (1992b). Interleukin 10 inhibits macrophage microbicidal activity by blocking the endogenous production of tumor necrosis factor alpha required as a costimulatory factor for interferon gamma-induced activation. *Proc. Natl Acad. Sci. U.S.A.* **89**, 8676–8680.

Pajkrt, D., Camoglio, L., Tiel-van Buul, M.C.M., de Bruin, K., Cutler, D., Affrime, M.B., Rikken, G., van der Poll, T., ten Cate, J.W. and van Deventer, S.J.H. (1997). Attenuation of proinflammatory response by recombinant human IL-10 in human endotoxemia: Effect of timing of recombinant human IL-10 administration. *J. Immunol.* **158**, 3971–3977.

Park, D.R. and Skerrett, S.J. (1996). IL-10 enhances the growth of *Legionella pneumophila* in human mononuclear phagocytes and reverses the protective effect of IFN-gamma: differential responses of blood monocytes and alveolar macrophages. *J. Immunol.* **157**, 2528–2538.

Pecanha, L.M., Snapper, C.M., Lees, A. and Mond, J.J. (1992). Lymphokine control of type 2 antigen response. IL-10 inhibits IL-5- but not IL-2-induced Ig secretion by T cell-independent antigens. *J. Immunol.* **148**, 3427–3432.

Pecanha, L.M., Snapper, C.M., Lees, A., Yamaguchi, H. and Mond, J.J. (1993). IL-10 inhibits T cell-independent but not T cell-dependent responses *in vitro*. *J. Immunol.* **150**, 3215–3223.

Peguet-Navarro, J., Moulon, C., Caux, C., Dalbiez-Gauthier, C., Banchereau, J. and Schmitt, D. (1994). Interleukin-10 inhibits the primary allogeneic T cell response to human epidermal Langerhans cells. *Eur. J. Immunol.* **24**, 884–891.

Peng, B., Mehta, N.H., Fernandes, H., Chou, C.C. and Raveche, E. (1995). Growth inhibition of malignant CD5+B (B-1) cells by antisense IL-10 oligonucleotide. *Leuk. Res.* **19**, 159–167.

Pennline, K.J., Roque-Gaffney, E. and Monahan, M. (1994). Recombinant human IL-10 prevents the onset of diabetes in the nonobese diabetic mouse. *Clin. Immunol. Immunopathol.* **71**, 169–175.

Perez, L., Orte, J. and Brieva, J.A. (1995). Terminal differentiation of spontaneous rheumatoid factor-secreting B cells from rheumatoid arthritis patients depends on endogenous interleukin-10. *Arthritis Rheum.* **38**, 1771–1776.

Persson, S., Mikulowska, A., Narula, S., O'Garra, A. and Holmdahl, R. (1996). Interleukin-10 suppresses the development of collagen type II-induced arthritis and ameliorates sustained arthritis in rats. *Scand. J. Immunol.* **44**, 607–614.

Peyron, F., Burdin, N., Ringwald, P., Vuillez, J.P., Rousset, F. and Banchereau, J. (1994). High levels of circulating IL-10 in human malaria. *Clin. Exp. Immunol.* **95**, 300–303.

Powrie, F. (1995). T cells in inflammatory bowel disease: protective and pathogenic roles. *Immunity* **3**, 171–174.

Powrie, F. and Coffman, R.L. (1993). Inhibition of cell-mediated immunity by IL4 and IL10. *Res. Immunol.* **144**, 639–643.

Powrie, F., Leach, M.W., Mauze, S. and Coffman, R.L. (1993a). Phenotypically distinct subsets of CD4+ T cells induce or protect from chronic intestinal inflammation in C. B-17 scid mice. *Int. Immunol.* **5**, 1461–1471.

Powrie, F., Menon, S. and Coffman, R.L. (1993b). Interleukin-4 and Interleukin-10 synergize to inhibit cell-mediated immunity *in vivo*. *Eur. J. Immunol.* **23**, 3043–3049.

Powrie, F., Leach, M.W., Mauze, S., Menon, S., Barcomb Caddle, L. and Coffman, R.L. (1994). Inhibition of Th1 responses prevents inflammatory bowel disease in scid mice reconstituted with CD45RBhi CD4+ T cells. *Immunity* **1**, 553–562.

Pradier, O., Gerard, C., Delvaux, A., Lybin, M., Abramowicz, D., Capel, P., Velu, T. and Goldman, M. (1993). Interleukin-10 inhibits the induction of monocyte procoagulant activity by bacterial lipopolysaccharide. *Eur. J. Immunol.* **23**, 2700–2703.

Pradier, O., Willems, F., Abramowicz, D., Schandene, L., de Boer, M., Thielemans, K., Capel, P. and Goldman, M. (1996). CD40 engagement induces monocyte procoagulant activity through an interleukin-10 resistant pathway. *Eur. J. Immunol.* **26**, 3048–3054.

Pugin, J., Ulevitch, R.J. and Tobias, P.S. (1993). A critical role for monocytes and CD14 in endotoxin-induced endothelial cell activation. *J. Exp. Med.* **178**, 2193–2200.

Punnonen, J., de Waal Malefyt, R., van Vlasselaer, P., Gauchat, J.F. and de Vries, J.E. (1993). IL-10 and viral IL-10 prevent IL-4-induced IgE synthesis by inhibiting the accessory cell function of monocytes. *J. Immunol.* **151**, 1280–1289.

Qian, S., Li, W., Li, Y., Fu, F., Lu, L., Fung, J.J. and Thompson, A.W. (1996). Systemic administration of cellular interleukin-10 can exacerbate cardiac allograft rejection in mice. *Transplantation* **62**, 1709–1714.

Qin, L., Chavin, K.D., Ding, Y., Tahara, H., Favaro, J.P., Woodward, J.E., Suzuki, T., Robbins, P.D., Lotze, M.T. and Bromberg, J.S. (1996). Retrovirus-mediated transfer of viral IL-10 gene prolongs murine cardiac allograft survival. *J. Immunol.* **156**, 2316–2323.

Ramachandra, S., Metcalf, R.A., Fredrickson, T., Marti, G.E. and Raveche, E. (1996). Requirement for increased IL-10 in the development of B-1 lymphoproliferative disease in a murine model of CLL. *J. Clin. Invest.* **98**, 1788–1793.

Ramani, M., Ollivier, V., Khechai, F., Vu, T., Ternisien, C., Bridey, F. and de Prost, D. (1993). Interleukin-10 inhibits endotoxin-induced tissue factor mRNA production by human monocytes. *FEBS Lett.* **334**, 114–116.

Ramani, M., Khechai, F., Ollivier, V., Ternisien, C., Bridey, F., Hakim, J. and de Prost, D. (1994). Interleukin-10 and pentoxifylline inhibit C-reactive protein-induced tissue factor gene expression in peripheral human blood monocytes. *FEBS Lett.* **356**, 86–88.

Reed, S.G., Brownell, C.E., Russo, D.M., Silva, J.S., Grabstein, K.H. and Morrissey, P.J. (1994). IL-10 mediates susceptibility to *Trypanosoma cruzi* infection. *J. Immunol.* **153**, 3135–3140.

Reinhold, D., Bank, U., Buhling, F., Lendeckel, U. and Ansorge, S. (1995). Transforming growth factor beta 1 inhibits interleukin-10 mRNA expression and production in pokeweed mitogen-stimulated peripheral blood mononuclear cells and T cells. *J. Interferon Cytokine Res.* **15**, 685–690.

Rennick, D., Hunte, B., Holland, G. and Thompson-Snipes, L. (1995). Cofactors are essential for stem cell factor-dependent growth and maturation of mast cell progenitors: comparative effects of interleukin-3 (IL-3), IL-4, IL-10, and fibroblasts. *Blood* **85**, 57–65.

Rennick, D.M., Fort, M.M. and Davidson, N.J. (1997). Studies with IL-10−/− mice: an overview. *J. Leukoc. Biol.* **61**, 389–396.

Rivas, J.M. and Ullrich, S.E. (1992). Systemic suppression of delayed-type hypersensitivity by supernatants from UV-irradiated keratinocytes. An essential role for keratinocyte-derived IL-10. *J. Immunol.* **149**, 3865–3871.

Rivas, J.M. and Ullrich, S.E. (1994). The role of IL-4, IL-10, and TNF-alpha in the immune suppression induced by ultraviolet radiation. *J. Leukoc. Biol.* **56**, 769–775.

Rode, H.-J., Janssen, W., Rosen-Wolff, A., Bugert, J.J., Thein, P., Becker, Y. and Darai, G. (1993). The genome of equine herpesvirus type 2 harbors an interleukin-10 (IL-10)-like gene. *Virus Genes* **7**, 111–116.

Rogy, M.A., Auffenberg, T., Espat, N.J., Philip, R., Remick, D., Wollenberg, G.K., Copeland, E.M.R. and Moldawer, L.L. (1995). Human tumor necrosis factor receptor (p55) and interleukin 10 gene transfer in the mouse reduces mortality to lethal endotoxemia and also attenuates local inflammatory responses. *J. Exp. Med.* **181**, 2289–2293.

Roilides, E., Dimitriadou, A., Kadiltsoglou, I., Sein, T., Karpouzas, J., Pizzo, P.A. and Walsh, T.J. (1997). IL-10 exerts suppressive and enhancing effects on antifungal activity of mononuclear phagocytes against *Aspergillus fumigatus*. *J. Immunol.* **158**, 322–329.

Rosenbaum, J.T. and Angell, E. (1995). Paradoxical effects of IL-10 in endotoxin-induced uveitis. *J. Immunol.* **155**, 4090–4094.

Roth, I., Corry, D.B., Locksley, R.M., Abrams, J.S., Litton, M.J. and Fisher, S.J. (1996). Human placental cytotrophoblasts produce the immunosuppressive cytokine interleukin 10. *J. Exp. Med.* **184**, 539–548.

Rott, O., Fleischer, B. and Cash, E. (1994). Interleukin-10 prevents experimental allergic encephalomyelitis in rats. *Eur. J. Immunol.* **24**, 1434–1440.

Rousset, F., Garcia, E., Defrance, T., Peronne, C., Hsu, D.-H., Kastelein, R., Moore, K.W. and Banchereau, J. (1992). IL-10 is a potent growth and differentiation factor for activated human B lymphocytes. *Proc. Natl Acad. Sci. U.S.A.* **89**, 1890–1893.

Rousset, F., Peyrol, S., Garcia, E., Vezzio, N., Andujar, M., Grimaud, J.A. and Banchereau, J. (1995). Long-term cultured CD40-activated B lymphocytes differentiate into plasma cells in response to IL-10 but not IL-4. *Int. Immunol.* **7**, 1243–1253.

Saeland, S., Duvert, V., Moreau, I. and Banchereau, J. (1993). Human B cell precursors proliferate and express CD23 after CD40 ligation. *J. Exp. Med.* **178**, 113–120.

Salazar-Onfray, F., Petersson, M., Franksson, L., Matsuda, M., Blankenstein, T., Karre, K. and Kiessling, R. (1995). IL-10 converts mouse lymphoma cells to a CTL-resistant, NK-sensitive phenotype with low but peptide-inducible MHC class I expression. *J. Immunol.* **154**, 6291–6298.

Saville, M.W., Taga, K., Foli, A., Broder, S., Tosato, G. and Yarchoan, R. (1994). Interleukin-10 suppresses human immunodeficiency virus-1 replication *in vitro* in cells of the monocyte/macrophage lineage. *Blood* **83**, 3591–3599.

Schandene, L., Alonso-Vega, C., Willems, F., Gerard, C., Delvaux, A., Velu, T., Devos, R., de Boer, M. and Goldman, M. (1994). B7/CD28-dependent IL-5 production by human resting T cells is inhibited by IL-10. *J. Immunol.* **152**, 4368–4374.

Schols, D. and De Clercq, E. (1996). Human immunodeficiency virus type 1 gp120 induces anergy in human peripheral blood lymphocytes by inducing interleukin-10 production. *J. Virol.* **70**, 4953–4960.

Schreiber, S., Heinig, T., Thiele, H.G. and Raedler, A. (1995). Immunoregulatory role of interleukin 10 in patients with inflammatory bowel disease. *Gastroenterology* **108**, 1434–1444.

Schwarz, A., Grabbe, S., Riemann, H., Aragane, Y., Simon, M., Manon, S., Andrade, S., Luger, T.A., Zlotnik, A.

and Schwarz, T. (1994). *In vivo* effects of interleukin-10 on contact hypersensitivity and delayed-type hypersensitivity reactions. *J. Invest. Dermatol.* **103**, 211–216.

Shanley, T.P., Schmal, H., Friedl, H.P., Jones, M.L. and Ward, P.A. (1995). Regulatory effects of intrinsic IL-10 in IgG immune complex-induced lung injury. *J. Immunol.* **154**, 3454–3460.

Sher, A., Fiorentino, D., Caspar, P., Pearce, E. and Mosmann, T. (1991). Production of IL-10 by CD4+ T lymphocytes correlates with down-regulation of Th1 cytokine synthesis in helminth infection. *J. Immunol.* **147**, 2713–2716.

Sherry, R.M., Cue, J.I., Goddard, J.K., Parramore, J.B. and DiPiro, J.T. (1996). Interleukin-10 is associated with the development of sepsis in trauma patients. *J. Trauma* **40**, 613–616; discussion 616–617.

Sieling, P.A. and Modlin, R.L. (1995). IL-10 in mycobacterial infection. In *Interleukin-10* (eds J.E. de Vries and R. de Waal Malefyt), R. G. Landes Company, Austin, Texas, pp. 79–87.

Sieling, P.A., Abrams, J.S., Yamamura, M., Salgame, P., Bloom, B.R., Rea, T.H. and Modlin, R.L. (1993). Immunosuppressive roles for IL-10 and IL-4 in human infection. *In vitro* modulation of T cell responses in leprosy. *J. Immunol.* **150**, 5501–5510.

Smith, D.R., Kunkel, S.L., Burdick, M.D., Wilke, C.A., Orringer, M.B., Whyte, R.I. and Strieter, R.M. (1994a). Production of interleukin-10 by human bronchogenic carcinoma. *Am. J. Pathol.* **145**, 18–25.

Smith, S.R., Terminelli, C., Kenworthy-Bott, L., Calzetta, A. and Donkin, J. (1994b). The cooperative effects of TNF-alpha and IFN-gamma are determining factors in the ability of IL-10 to protect mice from lethal endotoxemia. *J. Leukoc. Biol.* **55**, 711–718.

Smith, S.R., Terminelli, C., Denhardt, G., Narula, S. and Thorbecke, G.J. (1996). Administration of interleukin-10 at the time of priming protects *Corynebacterium parvum*-primed mice against LPS- and TNF-alpha-induced lethality. *Cell Immunol.* **173**, 207–214.

Spittler, A., Schiller, C., Willheim, M., Tempfer, C., Winkler, S. and Boltz-Nitulescu, G. (1995). IL-10 augments CD23 expression on U937 cells and down-regulates IL-4-driven CD23 expression on cultured human blood monocytes: effects of IL-10 and other cytokines on cell phenotype and phagocytosis. *Immunology* **85**, 311–317.

Sprang, S.R. and Bazan, J.F. (1993). Cytokine structural taxonomy and mechanisms of receptor engagement. *Curr. Opin. Struct. Biol.* **3**, 815–827.

Standiford, T.J., Strieter, R.M., Lukacs, N.W. and Kunkel, S.L. (1995). Neutralization of IL-10 increases lethality in endotoxemia. Cooperative effects of macrophage inflammatory protein-2 and tumor necrosis factor. *J. Immunol.* **155**, 2222–2229.

Suberville, S., Bellocq, A., Fouqueray, B., Philippe, C., Lantz, O., Perez, J. and Baud, L. (1996). Regulation of interleukin-10 production by beta-adrenergic agonists. *Eur. J. Immunol.* **26**, 2601–2605.

Sugiyama, E., Kuroda, A., Taki, H., Ikemoto, M., Hori, T., Yamashita, N., Maruyama, M. and Kobayashi, M. (1995). Interleukin 10 cooperates with interleukin 4 to suppress inflammatory cytokine production by freshly prepared adherent rheumatoid synovial cells. *J. Rheumatol.* **22**, 2020–2026.

Sundstedt, A., Hoiden, I., Rosendahl, A., Kalland, T., van Rooijen, N. and Dohlsten, M. (1997). Immunoregulatory role of IL-10 during superantigen-induced hyporesponsiveness *in vivo*. *J. Immunol.* **158**, 180–186.

Suzuki, T., Tahara, H., Narula, S., Moore, K.W., Robbins, P.D. and Lotze, M.T. (1995). Viral interleukin 10 (IL-10), the human herpes virus 4 cellular IL-10 homologue, induces local anergy to allogeneic and syngeneic tumors. *J. Exp. Med.* **182**, 477–486.

Swaminathan, S., Hesselton, R., Sullivan, J. and Kieff, E. (1993). Epstein–Barr virus recombinants with specifically mutated BCRF1 genes. *J. Virol.* **67**, 7406–7413.

Taga, K. and Tosato, G. (1992). IL-10 inhibits human T cell proliferation and IL-2 production. *J. Immunol.* **148**, 1143–1148.

Taga, K., Cherney, B. and Tosato, G. (1993). IL-10 inhibits apoptotic cell death in human T cells starved of IL-2. *Int. Immunol.* **5**, 1599–1608.

Taga, K., Chretien, J., Cherney, B., Diaz, L., Brown, M. and Tosato, G. (1994). Interleukin-10 inhibits apoptotic cell death in infectious mononucleosis T cells. *J. Clin. Invest.* **94**, 251–260.

Takanaski, S., Nonaka, R., Xing, Z., O'Byrne, P., Dolovich, J. and Jordana, M. (1994). Interleukin 10 inhibits lipopolysaccharide-induced survival and cytokine production by human peripheral blood eosinophils. *J. Exp. Med.* **180**, 711–715.

Tan, J.C., Indelicato, S., Narula, S.K., Zavodny, P.J. and Chou, C.-C. (1993). Characterization of interleukin-10 receptors on human and mouse cells. *J. Biol. Chem.* **268**, 21 053–21 059.

Tan, J.C., Braun, S., Rong, H., DiGiacomo, R., Dolphin, E., Baldwin, S., Narula, S.K., Zavodny, P.J. and Chou, C.-C. (1995). Characterization of recombinant extracellular domain of human interleukin-10 receptor. *J. Biol. Chem.* **270**, 12906–12911.

Tanaka, Y., Otsuka, T., Hotokebuchi, T., Miyahara, H., Nakashima, H., Kuga, S., Nemoto, Y., Niiro, H. and Niho, Y. (1996). Effect of IL-10 on collagen-induced arthritis in mice. *Inflamm. Res.* **45**, 283–288.

Tarazona, R., Gonzalez-Garcia, A., Zamzami, N., Marchetti, P., Frechin, N., Gonzalo, J.A., Ruiz-Gayo, M., van Rooijen, N., Martinez, C. and Kroemer, G. (1995). Chlorpromazine amplifies macrophage-dependent IL-10 production *in vivo*. *J. Immunol.* **154**, 861–870.

te Velde, A.A., de Waal Malefyt, R., Huijbens, R.J.F., de Vries, J.E. and Figdor, C.G. (1992). IL-10 stimulates monocyte FcγR surface expression and cytotoxic activity: distinct regulation of ADCC by IFNγ, IL-4, and IL-10. *J. Immunol.* **149**, 4048–4052.

Teunissen, M.B., Koomen, C.W., Jansen, J., de Waal Malefyt, R., Schmitt, E., Van den Wijngaard, R., Das, P.K. and Bos, J.D. (1997). In contrast to their murine counterparts, normal human keratinocytes and human epidermoid cell lines A431 and HaCaT fail to express IL-10 mRNA and protein. *Clin. Exp. Immunol.* **107**, 213–223.

Thomassen, M.J., Divis, L.T. and Fisher, C.J. (1996). Regulation of human alveolar macrophage inflammatory cytokine production by interleukin-10. *Clin. Immunol. Immunopathol.* **80**, 321–324.

Thompson-Snipes, L., Dhar, V., Bond, M.W., Mosmann, T.R., Moore, K.W. and Rennick, D. (1991). Interleukin-10: a novel stimulatory factor for mast cells and their progenitors. *J. Exp. Med.* **173**, 507–510.

Tripp, C.S., Wolf, S.F. and Unanue, E.R. (1993). Interleukin 12 and tumor necrosis factor alpha are costimulators of interferon gamma production by natural killer cells in severe combined immunodeficiency mice with listeriosis, and interleukin 10 is a physiologic antagonist. *Proc. Natl Acad. Sci. U.S.A.* **90**, 3725–3729.

Tripp, C.S., Beckerman, K.P. and Unanue, E.R. (1995). Immune complexes inhibit antimicrobial responses through interleukin-10 production. Effects in severe combined immunodeficient mice during Listeria infection. *J. Clin. Invest.* **95**, 1628–1634.

Tumpey, T.M., Elner, V.M., Chen, S.H., Oakes, J.E. and Lausch, R.N. (1994). Interleukin-10 treatment can suppress stromal keratitis induced by herpes simplex virus type 1. *J. Immunol.* **153**, 2258–2265.

Uejima, Y., Takahashi, K., Komoriya, K., Kurozumi, S. and Ochs, H.D. (1996). Effect of interleukin-10 on anti-CD40- and interleukin-4-induced immunoglobulin E production by human lymphocytes. *Int. Arch. Allergy Immunol.* **110**, 225–232.

van der Poll, T., Jansen, J., Levi, M., ten Cate, H., ten Cate, J.W. and van Deventer, S.J. (1994). Regulation of interleukin 10 release by tumor necrosis factor in humans and chimpanzees. *J. Exp. Med.* **180**, 1985–1988.

van der Poll, T., Marchant, A., Buurman, W.A., Berman, L., Keogh, C.V., Lazarus, D.D., Nguyen, L., Goldman, M., Moldawer, L.L. and Lowry, S.F. (1995). Endogenous IL-10 protects mice from death during septic peritonitis. *J. Immunol.* **155**, 5397–5401.

van der Poll, T., Coyle, S.M., Barbosa, K., Braxton, C.C. and Lowry, S.F. (1996a). Epinephrine inhibits tumor necrosis factor-alpha and potentiates interleukin 10 production during human endotoxemia. *J. Clin. Invest.* **97**, 713–719.

van der Poll, T., Marchant, A., Keogh, C.V., Goldman, M. and Lowry, S.F. (1996b). Interleukin-10 impairs host defense in murine pneumococcal pneumonia. *J. Infect. Dis.* **174**, 994–1000.

van der Poll, T., de Waal Malefyt, R., Coyle, S.M. and Lowry, S.F. (1997). Antiinflammatory cytokine responses during clinical sepsis and experimental endotoxemia: sequential measurements of plasma soluble interleukin (IL)-1 receptor type II, IL-10, and IL-13. *J. Infect. Dis.* **175**, 118–122.

van Furth, A.M., Seijmonsbergen, E.M., Langermans, J.A., Groeneveld, P.H., de Bel, C.E. and van Furth, R. (1995). High levels of interleukin 10 and tumor necrosis factor alpha in cerebrospinal fluid during the onset of bacterial meningitis [see comments]. *Clin. Infect. Dis.* **21**, 220–222.

Van Laethem, J.L., Marchant, A., Delvaux, A., Goldman, M., Robberecht, P., Velu, T. and Deviere, J. (1995). Interleukin 10 prevents necrosis in murine experimental acute pancreatitis. *Gastroenterology* **108**, 1917–1922.

van Roon, J.A., van Roy, J.L., Gmelig-Meyling, F.H., Lafeber, F.P. and Bijlsma, J.W. (1996). Prevention and reversal of cartilage degradation in rheumatoid arthritis by interleukin-10 and interleukin-4. *Arthritis Rheum.* **39**, 829–835.

Vieira, P., de Waal Malefyt, R., Dang, M.-N., Johnson, K.E., Kastelein, R., Fiorentino, D.F., de Vries, J.E., Roncarolo, M.-G., Mosmann, T.R. and Moore, K.W. (1991). Isolation and expression of human cytokine synthesis inhibitory factor (CSIF/IL-10) cDNA clones: homology to Epstein–Barr virus open reading frame BCRFI. *Proc. Natl Acad. Sci. U.S.A.* **88**, 1172–1176.

Walmsley, M., Katsikis, P.D., Abney, E., Parry, S., Williams, R.O., Maini, R.N. and Feldmann, M. (1996). Interleukin-10 inhibition of the progression of established collagen-induced arthritis. *Arthritis Rheum.* **39**, 495–503.

Walter, M.R. and Nagabhushan, T.L. (1995). Crystal structure of interleukin 10 reveals an interferon γ-like fold. *Biochemistry* **34**, 12118–12125.

Wang, P., Wu, P., Anthes, J.C., Siegel, M.I., Egan, R.W. and Billah, M.M. (1994). Interleukin-10 inhibits interleukin-8 production in human neutrophils. *Blood* **83**, 2678–2683.

Wang, P., Wu, P., Siegel, M.I., Egan, R.W. and Billah, M.M. (1995). Interleukin (IL)-10 inhibits nuclear factor kappa B (NF kappa B) activation in human monocytes. IL-10 and IL-4 suppress cytokine synthesis by different mechanisms. *J. Biol. Chem.* **270**, 9558–9563.

Wanidworanun, C. and Strober, W. (1993). Predominant role of tumor necrosis factor-alpha in human monocyte IL-10 synthesis. *J. Immunol.* **151**, 6853–6861.

Warren, H.S., Kinnear, B.F., Phillips, J.H. and Lanier, L.L. (1995). Production of IL-5 by human NK cells and regulation of IL-5 secretion by IL-4, IL-10, and IL-12. *J. Immunol.* **154**, 5144–5152.

Weber-Nordt, R.M., Meraz, M.A. and Schreiber, R.D. (1994). Lipopolysaccharide-dependent induction of IL-10 receptor expression on murine fibroblasts. *J. Immunol.* **153**, 3734–3744.

Weber-Nordt, R.M., Henschler, R., Schott, E., Wehinger, J., Behringer, D., Mertelsmann, R. and Finke, J. (1996a). Interleukin-10 increases Bcl-2 expression and survival in primary human CD34+ hematopoietic progenitor cells. *Blood* **88**, 2549–2558.

Weber-Nordt, R.M., Riley, J.K., Greenlund, A.C., Moore, K.W., Darnell, J.E. and Schreiber, R.D. (1996b). Stat3 recruitment by two distinct ligand-induced, tyrosine-phosphorylated docking sites in the interleukin-10 receptor intracellular domain. *J. Biol. Chem.* **271**, 27 954–27 961.

Wehinger, J., Gouilleux, F., Groner, B., Finke, J., Mertelsmann, R. and Weber-Nordt, R.M. (1996). IL-10 induces DNA binding activity of three STAT proteins (Stat1, Stat3, and Stat5) and their distinct combinatorial assembly in the promoters of selected genes. *FEBS Lett.* **394**, 365–370.

Weissman, D., Poli, G. and Fauci, A.S. (1994). Interleukin 10 blocks HIV replication in macrophages by inhibiting the autocrine loop of tumor necrosis factor alpha and interleukin 6 induction of virus. *AIDS Res. Hum. Retroviruses* **10**, 1199–1206.

Wilkes, D.S., Neimeier, M., Mathur, P.N., Soliman, D.M., Twigg, H.L.R., Bowen, L.K. and Heidler, K.M. (1995). Effect of human lung allograft alveolar macrophages on IgG production: immunoregulatory role of interleukin-10, transforming growth factor-beta, and interleukin-6. *Am. J. Respir. Cell. Mol. Biol.* **13**, 621–628.

Willems, F., Marchant, A., Delville, J.P., Gerard, C., Delvaux, A., Velu, T., de Boer, M. and Goldman, M. (1994). Interleukin-10 inhibits B7 and intercellular adhesion molecule-1 expression on human monocytes. *Eur. J. Immunol.* **24**, 1007–1009.

Windsor, W.T., Syto, R., Tsarbopoulos, A., Zhang, R., Durkin, J., Baldwin, S., Paliwal, S., Mui, P.W., Pramanik, B., Trotta, P.P. and Tindall, S.H. (1993). Disulfide bond assignments and secondary structure analysis of human and murine interleukin 10. *Biochemistry* **32**, 8807–8815.

Wogensen, L., Huang, X. and Sarvetnick, N. (1993). Leukocyte extravasation into the pancreatic tissue in transgenic mice expressing interleukin 10 in the islets of Langerhans. *J. Exp. Med.* **178**, 175–185.

Wogensen, L., Lee, M.S. and Sarvetnick, N. (1994). Production of interleukin 10 by islet cells accelerates immune-mediated destruction of beta cells in nonobese diabetic mice. *J. Exp. Med.* **179**, 1379–1384.

Wu, J., Cunha, F.Q., Liew, F.Y. and Weiser, W.Y. (1993). IL-10 inhibits the synthesis of migration inhibitory factor and migration inhibitory factor-mediated macrophage activation. *J. Immunol.* **151**, 4325–4332.

Yagi, H., Tokura, Y., Wakita, H., Furukawa, F. and Takigawa, M. (1996). TCRV beta 7+ Th2 cells mediate UVB-induced suppression of murine contact photosensitivity by releasing IL-10. *J. Immunol.* **156**, 1824–1831.

Yssel, H., de Waal Malefyt, R., Roncarolo, M.G., Abrams, J.S., Lahesmaa, R., Spits, H. and de Vries, J.E. (1992). IL-10 is produced by subsets of human CD4+ T cell clones and peripheral blood T cells. *J. Immunol.* **149**, 2378–2384.

Zdanov, A., Schalk-Hihi, C., Gustchina, A., Tsang, M., Weatherbee, J. and Wlodawer, A. (1995). Crystal structure of interleukin-10 reveals the functional dimer with an unexpected topological similarity to interferon γ. *Structure* **3**, 591–601.

Zdanov, A., Schalk-Hihi, C., Menon, S., Moore, K.W. and Wlodawer, A. (1997). Crystal structure of Epstein–Barr virus protein BCRF1, a homolog of cellular interleukin-10. *J. Mol. Biol.* **268**, 460–467.

Zheng, L.M., Ojcius, D.M., Garaud, F., Roth, C., Maxwell, E., Li, Z., Rong, H., Chen, J., Wang, X.Y., Catino, J.J.

and King, I. (1996). Interleukin-10 inhibits tumor metastasis through an NK cell-dependent mechanism. *J. Exp. Med.* **184**, 579–584.

Zissel, G., Schlaak, J., Schlaak, M. and Muller-Quernheim, J. (1996). Regulation of cytokine release by alveolar macrophages treated with interleukin-4, interleukin-10, or transforming growth factor beta. *Eur. Cytokine Netw.* **7**, 59–66.

Zuany-Amorim, C., Haile, S., Leduc, D., Dumarey, C., Huerre, M., Vargaftig, B.B. and Pretolani, M. (1995). Interleukin-10 inhibits antigen-induced cellular recruitment into the airways of sensitized mice. *J. Clin. Invest.* **95**, 2644–2651.

Chapter 13

Interleukin-11: A Cytokine Signaling Through gp130 with Pleiotropic Effects and Potential Clinical Utility Within and Outside the Hematopoietic System

Sten Eirik W. Jacobsen

Stem Cell Laboratory, Department of Internal Medicine, University Hospital of Lund, Lund, Sweden

INTRODUCTION

The large turnover of mature blood cells as well as progenitor and stem cells at various levels of differentiation in the hematopoietic hierarchy is tightly regulated, involving, at least in part, a large number of soluble and membrane-bound cytokines (Moore, 1991; Spangrude *et al.*, 1991; Metcalf, 1993; Ogawa, 1993). More than 20 cytokines with stimulatory effects on hematopoiesis have now been cloned and characterized, of which interleukin-11 (IL-11) is one (Metcalf, 1993; Ogawa, 1993). Although initially identified as a cytokine with hematopoietic activity, it has become increasingly clear that IL-11 has important and pleiotropic effects (and potential utility), outside the hematopoietic system as well. This review summarizes and discusses the discovery and cloning, structure, expression, regulation, effects and potential utility of this cytokine (and its receptor) in different systems, focusing primarily on its hematopoietic activities.

DISCOVERY, CLONING AND STRUCTURE OF THE IL-11 GENE AND PROTEIN

Steady-state hematopoiesis is regulated through interactions between the stromal microenvironment and progenitor–stem cells, and many cytokines have initially been identified as hematopoietic factors produced by stromal cells (Metcalf, 1993; Ogawa, 1993; Muller-Sieburg and Deryugina, 1995). Thus, IL-11 was originally identified as a factor produced by a non-human primate bone marrow stroma cell line, capable of stimulating proliferation of a growth factor-dependent murine plasmacytoma cell line (Paul *et al.*, 1990). Subsequently, an adipogenesis inhibitory factor was found to be identical to IL-11 (Kawashima *et al.*, 1991).

The cDNA encoding the growth-stimulatory activity towards the plasmacytoma cell line PU-34 was isolated by functional expression cloning, and contained a single reading frame of 597 nucleotides, encoding a 199-amino-acid protein, which includes a signal

The Cytokine Handbook, 3rd ed.
ISBN 0-12-689662-3

peptide of 21 amino acids, and has a molecular weight of 23 kDa (Paul *et al.*, 1990; Yang, 1995). Two IL-11 RNA transcripts of approximately 1.5 and 2.5 kb were isolated from simian as well as human cell lines (Paul *et al.*, 1990; Kawashima *et al.*, 1991). The different sizes of the two transcripts result from different polyadenylation signals, and thus both transcripts encode the same 199-amino-acid peptide (Paul *et al.*, 1990; Kawashima *et al.*, 1991). The human and primate coding regions for IL-11 are 97% identical at the nucleotide level (Paul *et al.*, 1990). Recently, the cDNA encoding murine IL-11 has been cloned from a fetal thymic cell line, and demonstrated (as the human cDNA) to encode 199 amino acids (Morris *et al.*, 1996). Human and murine IL-11 have 88% identity at the amino acid level, are equally active on murine cells, and both can be neutralized by a murine antibody raised against human IL-11 (Morris *et al.*, 1996). Although IL-11 shares many biologic activities with IL-6 (see separate sections), there is no sequence homology to IL-6 or other known cytokines at the nucleotide or amino acid level.

At variance with a number of other growth factors, the IL-11 protein lacks cysteine residues (potential sites for disulfide bonds) and *N*-linked glycosylation sites. Like many other cytokines IL-11 contains four α-helical bundles and two β-sheets (Czupryn *et al.*, 1995a,b). Recent mutagenesis studies have identified amino acids critical for receptor binding and biologic activity (Czupryn *et al.*, 1995a,b).

The human IL-11 gene is 7 kb long, contains five exons and four introns, and has been mapped to the long arm of chromosome 19 (McKinley *et al.*, 1992), whereas the murine gene is localized to the proximal part of chromosome 7 (Morris *et al.*, 1996). The 5′ flanking region of the gene contains multiple elements implicated in transcriptional regulation, including sites for activating protein-1 (AP-1) and nuclear factor-κB (Yang, 1995).

EXPRESSION OF IL-11

IL-11 gene expression is observed in a number of different cell types and tissues as determined through studies of cell lines and primary cells or tissues. Among the many cells and tissues demonstrated to express IL-11 transcripts are endothelial cells (Suen *et al.*, 1994), fibroblasts (Paul *et al.*, 1990, 1991; Kawashima *et al.*, 1991; Elias *et al.*, 1994a,b; Suen *et al.*, 1994), chondrocytes (Maier *et al.*, 1993), synoviocytes (Maier *et al.*, 1993), osteoblasts (Romas *et al.*, 1996), epithelial cells (Elias *et al.*, 1994a), neurons (Du *et al.*, 1996), trophoblasts (Paul *et al.*, 1990), spermatids (Du *et al.*, 1996) and thymus (Morris *et al.*, 1996). IL-11 protein is normally expressed at low or undetectable levels and cannot be detected in normal serum.

Depending on the cell type, the expression of IL-11 can be regulated at the transcriptional and/or post-transcriptional level (reviewed by Yang, 1995). IL-11 expression can be induced by the proinflammatory cytokines IL-1α and tumor necrosis factor-α (TNF-α), the anti-inflammatory transforming growth factor-β (TGF-β), phorbol esters, retinoic acid and histamine (Paul *et al.*, 1990; Maier *et al.*, 1993; Elias *et al.*, 1994a,b; Suen *et al.*, 1994; Yang and Yang, 1994, 1995; Zheng *et al.*, 1994; Yang *et al.*, 1996; Morris *et al.*, 1996).

The molecular and structural characteristics and expression pattern of IL-11 are summarized in Table 1.

Table 1. Molecular and structural characteristics and expression pattern of IL-11.

Chromosome	7 (mouse), 19 (human)
Protein size	22–23 kDa, 199 amino acids including a 21-amino-acid leader sequence (mouse and human)
Protein structure	No cysteine residues, no *N*-glycosylation sites Four α-helical bundles and two β-sheets
Homology mouse/human	80% identity at nucleotide level, 88% at amino acid level
Expression pattern	mRNA transcripts expressed widely (endothelium, fibroblasts, chondrocytes, synoviocytes, osteoblasts, osteoclasts, epithelium, neurons, trophoblasts, neurons, spermatids, thymus) Inducers: IL-1, TNF, TGF-β, phorbol esters, retinoic acid, histamine

THE IL-11 RECEPTOR COMPLEX: CLONING, STRUCTURE, EXPRESSION AND SIGNAL TRANSDUCTION

IL-11 is a member of a cytokine family which also includes IL-6, leukemia inhibitory factor (LIF), oncostatin M (OSM), ciliary neurotrophic factor (CNTF) and cardiotrophin-1, which all mediate signal transduction through gp130, a protein originally identified as part of the IL-6 receptor complex (Taga *et al.*, 1989; Gearing *et al.*, 1992; Ip *et al.*, 1992; Liu *et al.*, 1992; Yin *et al.*, 1993; Fourcin *et al.*, 1994; Zhang *et al.*, 1994). The use of a common and essential signal transduction receptor subunit explains the largely overlapping and redundant activities of these different ligands (reviewed by Kishimoto *et al.*, 1994). However, the fact that these ligands in part also promote distinct activities is likely to be explained by the finding that the receptor for each of these ligands consists of at least two subunits (reviewed by Kishimoto *et al.*, 1994; Taga and Kishimoto, 1997). In that regard, IL-11 binds to an α-receptor unit which binds IL-11 exclusively (Hilton *et al.*, 1994; Nandurkar *et al.*, 1996). The complex of IL-11 and IL-11 receptor α (IL-11Rα) has been speculated to induce homodimerization of gp130, as has been demonstrated for the IL-6–IL-6Rα complex (Murakami *et al.*, 1993; Yin *et al.*, 1993; Fourcin *et al.*, 1994; Taga and Kishimoto, 1997). In this regard, IL-6 and IL-11 differ from the remaining known ligands utilizing gp130 for signal transduction, which rather induce heterodimer complexes of gp130 and the LIF receptor (LIFR), or (in the case of OSM) also a gp130-OSMR heterodimer (reviewed by Taga and Kishimoto, 1997). Importantly, a recent study has demonstrated that, similar to IL-6 and CNTF, gp130 can induce dimerization of the IL-11–IL-11Rα complex but, unlike IL-6 and CNTF, IL-11 does not induce gp130 homodimerization or gp130–LIFR heterodimerization, implicating the involvement of a novel (yet to be identified) receptor subunit involved in IL-11 signal transduction (Neddermann *et al.*, 1996).

The murine IL-11Rα was cloned by screening of a cDNA library of adult mouse liver with oligonucleotides corresponding to the conserved WSXWS motif of the hematopoietin receptor superfamily (Bazan, 1990; Hilton *et al.*, 1994). The IL-11 binding receptor consists of an extracellular domain characteristic for the hematopoietin superfamily, with two potential *N*-glycosylation sites, a transmembrane portion and a short cytoplasmic domain (Hilton *et al.*, 1994). The extracellular domain of the IL-11Rα shows sequence and structural homology to the IL-6Rα (24% amino acid identity), CNTFR (22% amino acid identity) and the p40 subunit of IL-12 (16% amino acid identity)

(Hilton *et al.*, 1994). When expressed in the absence of gp130, the IL-11Rα binds IL-11 with low affinity (K_d 10 nM), which is converted to high affinity (K_d 300–800 pM) upon interaction with gp130 (Hilton *et al.*, 1994). Furthermore, coexpression of IL-11Rα and gp130 appears to be required absolutely for IL-11-induced signal transduction (Hilton *et al.*, 1994). The murine IL-11Rα gene is located on chromosome 4.

Recently, the human IL-11Rα has also been cloned and demonstrated to share 85% nucleotide and 84% amino acid identity with the murine receptor (Cherel *et al.*, 1995; Nandurkar *et al.*, 1996). Interestingly, two isoforms of human IL-11Rα have been described, which differ exclusively in the presence and absence of a short cytoplasmic domain (Cherel *et al.*, 1995; Nandurkar *et al.*, 1996). Based on homology with the IL-6-R system, it is not surprising that the isoform lacking the cytoplasmic tail is fully active (Neddermann *et al.*, 1996).

The human IL-11Rα gene has been mapped to chromosome 9; it is 9 kb long and contains 12 exons and 12 introns (Cherel *et al.*, 1996; Van Leuven *et al.*, 1996). Also the murine gene for IL-11Rα has been described along with a second related locus (Robb *et al.*, 1996). The two forms of IL-11Rα are 99% identical at the amino acid level (Bilinski *et al.*, 1996; Robb *et al.*, 1997). Both the murine IL-11Rα genes can be expressed, but the relative expression of the two transcripts varies considerably between different cell types (Bilinski *et al.*, 1996; Robb *et al.*, 1997).

Studies of IL-11Rα mRNA expression on cell lines and primary tissues or cells suggest that the receptor is expressed widely (Hilton *et al.*, 1994; Bilinski *et al.*, 1996; Nandurkar *et al.*, 1996; Robb *et al.*, 1996, 1997; Davidson *et al.*, 1997), although more extensive studies of cell surface expression on purified primary cell populations needs to be performed. The tissues investigated and found to express IL-11Rα mRNA include bone marrow, spleen, thymus, lymph nodes, kidney, liver, brain, epididymis, testis, heart, muscle, salivary gland, lung, placenta and intestine. Of cell lines, several of hematopoietic origin (myeloid, erythroid and megakaryocytic) express IL-11Rα, whereas one investigated B-cell line and one T-cell line were negative (Cherel *et al.*, 1995). Furthermore osteosarcoma, keratinocyte and breast cancer cell lines express IL-11Rα as well (Cherel *et al.*, 1995; Douglas *et al.*, 1997).

Signal transduction through interaction of IL-11 with the IL-11Rα–gp130 complex does not involve intrinsic tyrosine kinase activity, but results in activation of the Janus kinase (JAK2) and ras–mitogen-activated protein kinase (MAPK) pathways, and the signal transducers and activators of transcription (STAT91). For a more detailed overview regarding the involvement of these signal transduction pathways and activation of primary response genes in response to IL-11, other reviews can be recommended (Ihle *et al.*, 1995; Yang and Yin, 1995; Taga and Kishimoto, 1997).

The complex of IL-6 and soluble IL-6R has been demonstrated to activate cells expressing gp130, even in the absence of membrane-bound IL-6R (Taga *et al.*, 1989; Murakami *et al.*, 1993). Although it remains to be established to what degree soluble IL-11Rα is shed into biologic fluids under physiologic and pathologic conditions, as has been demonstrated for soluble IL-6Rα (Rose-John and Heinrich, 1994), it has been demonstrated that a complex of IL-11 and soluble IL-11Rα is biologically active, thus potentially allowing IL-11 to activate gp130$^+$ cells lacking expression of IL-11Rα (Karow *et al.*, 1996; Neddermann *et al.*, 1996; Robb *et al.*, 1996).

Table 2 summarizes the molecular and structural characteristics and expression pattern of the IL-11 receptor complex.

Table 2. Characteristics of the IL-11 receptor complex.

Chromosome (IL-11Rα)	4 (mouse), 9 (human)
Receptor complex	IL-11Rα: ligand binding subunit (two functional isoforms) gp130: signaling and high-affinity converting subunit Possibly third novel subunit
Homology human/mouse	85% identity at nucleotide level, 84% at amino acid level
Homology other receptors (mouse)	IL-6Rα: 24% amino acid identity CNTFR: 22% amino acid identity IL-12, p40 subunit: 16% amino acid identity
Expression pattern	Widely expressed (one isoform in more restricted pattern) Tissue expression: bone marrow, spleen, thymus, lymph nodes, kidney, liver, brain, epididymis, testis, heart, muscle, salivary gland, lung, placenta, intestine Normal hematopoietic progenitor cells: myeloid, megakaryocytic (including mature megakaryocytes), erythroid, stem cells Malignant hematopoietic cell lines: myeloid, erythroid and megakaryocytic leukemia. Myeloma Other malignant cell lines: osteosarcomas, breast cancer

POTENTIAL PHYSIOLOGIC AND NON-REDUNDANT ROLE OF IL-11

Studies of mice deficient in IL-11 or IL-11Rα are yet to be described but will ultimately provide important information regarding the non-redundant and redundant activities of this receptor–ligand pair. Owing to the high degree of overlap in activity between IL-11 and other cytokines signaling through gp130, in particular IL-6, it might be expected that most of the pleiotropic functions mediated by IL-11 could be replaced by other members of this cytokine family. However, a loss of function phenotype is expected if IL-11 has a crucial function that cannot be replaced by the other members. Accordingly, although IL-6-deficient mice show no severe developmental abnormalities, they exhibit an impaired immune response following viral infection as well as abnormal acute-phase responses (Kopf *et al.*, 1994). Furthermore, the finding that the IL-11 receptor complex may consist of an additional and unique subunit (Neddermann *et al.*, 1996), might result in a more severe phenotype than would otherwise be expected. In that regard, inactivation of either gp130 (Yoshida *et al.*, 1996) or the LIFR (Ware *et al.*, 1995) yields severely compromised and lethal phenotypes, in the case of gp130 resulting in death between 12.5 days postcoitum and term, resulting from severe deficiencies in hematopoietic and cardiac development, demonstrating that this family of cytokines collectively has unique and non-redundant activities.

IN VITRO HEMATOPOIETIC EFFECTS OF IL-11

Effects on Candidate Murine and Human Stem Cells

For many decades intensive efforts have been put into the isolation and characterization of the true hematopoietic stem cells, that is the infrequent cells that each have the unique long-term (life-long) ability to reconstitute all blood cell lineages (Spangrude *et al.*, 1991; Watt and Visser, 1992). Although it is arguable and hard to prove whether or not the stage has yet been reached at which a pure stem cell population can be obtained (in particular in the human), it is obvious that the last decade has produced much progress towards this aim (reviewed by Spangrude *et al.*, 1991; Watt and Visser, 1992; Uchida *et al.*, 1993; Orlic and Bodine, 1994).

Murine stem cells have been purified to a high degree by different techniques, and have proved capable of efficiently supporting long-term reconstitution of all blood lineages at very low numbers, down to single stem cells (Till and McCulloch, 1961; Hodgson and Bradley, 1979; Visser *et al.*, 1984; Spangrude *et al.*, 1988; Wolf *et al.*, 1993; Morrison *et al.*, 1995; Jones *et al.*, 1996; Osawa *et al.*, 1996; Randall *et al.*, 1996; Doi *et al.*, 1997). From these studies and others, several characteristics have evolved with regard to hematopoietic stem cells, which are important to consider when evaluating *in vitro* studies of candidate stem cell populations:

(1) Long-term reconstituting murine stem cells are heterogeneous, and likely to be organized in a hierarchy with the most quiescent stem cell at the top of the hierarchy.
(2) Although the long-term reconstituting stem cell(s) have been proposed to be predominantly dormant in steady state (Lemischka *et al.*, 1986), it is clear that a significant fraction of long-term reconstituting stem cells are actively cycling (although slowly) (Goodell *et al.*, 1996; Bradford *et al.*, 1997).
(3) Populations of 'purified' murine stem cells are highly clonogenic *in vitro* when stimulated by defined cytokines (in the absence of stroma), and thus at least some long-term reconstituting stem cells can grow under defined *in vitro* conditions (Heimfeld *et al.*, 1991; Rebel *et al.*, 1994; Li and Johnson, 1995; Trevisan and Iscove, 1995).

Methods have been developed to enrich candidate human stem cells to high purity, although their exact phenotype remains more uncertain owing to the lack of an optimal long-term reconstituting assay for human stem cells (Baum *et al.*, 1992; Watt and Visser, 1992; Huang and Terstappen, 1994; Berardi *et al.*, 1995; Petzer *et al.*, 1996a; Rusten *et al.*, 1996). However, human $CD34^+$ $CD38^-$ cells have recently been demonstrated to be highly enriched in cells capable of reconstituting preimmune fetal sheep and immune-deficient mice (Civin *et al.*, 1996; Kawashima *et al.*, 1996; Larochelle *et al.*, 1996). In addition, and contrary to what was previously believed, it might be possible also to recruit human candidate stem cell populations efficiently into active cell cycle through the action of defined cytokines *in vitro*, although such recruitment appears to occur somewhat slower than in mice (Petzer *et al.*, 1996a; Ramsfjell *et al.*, 1997).

In general, the most primitive hematopoietic progenitor and stem cells must be stimulated simultaneously by multiple cytokines to be recruited into active proliferation (Metcalf, 1993; Ogawa, 1993). Such synergistic recruitment can best be obtained when cytokines of different (functional) types or classes are combined (Table 3). In that regard, IL-11 belongs to a large class of synergistic cytokines, along with IL-6 and LIF, which alone have little or no ability to promote *in vitro* growth of primitive (or commited) progenitor or stem cells, but which efficiently synergize with cytokines from two other classes (Table 3) (Metcalf, 1993; Ogawa, 1993). The first class of synergistic partners are the colony stimulating factors (granulocyte colony stimulating factor (G-CSF), granu-

Table 3. Classification of cytokines with growth-promoting activities on candidate stem cells.

Class	Members
1. Purely synergistic factors	IL-1, IL-4, IL-6, IL-11, IL-12, LIF
2. Colony-stimulating factors	G-CSF, M-CSF, GM-CSF, IL-3, Tpo
3. Stem cell factors	KL, FL

Classification is based on the ability of different cytokines to promote the growth of primitive murine progenitor–stem cell populations *in vitro*. In general, recruitment can be obtained only when a minimum of two to three cytokines is combined. Most efficient recruitment and growth is obtained when cytokines from different classes are combined. A similar pattern is observed for candidate human stem cells.

locyte–macrophage CSF (GM-CSF), macrophage CSF (M-CSF) or CSF-1, IL-3 and thrombopoietin (Tpo)), which individually promote growth of more commited progenitor cells, but which also appear to have an important role in early hematopoiesis in that they can interact with other cytokines (such as IL-11) to promote growth of candidate stem cells (Table 3). Finally, the stem cell factors, which to date include the ligands for two tyrosine kinase receptors, c-kit and flt3, have been demonstrated to be efficient and crucial stimulators of the earliest stages of hematopoiesis, through interaction with cytokines of the two other classes (Table 3) (reviewed by Galli *et al.*, 1994; Lyman, 1995).

The status of IL-11Rα expression on candidate murine and human stem cells remains to be investigated in detail, but preliminary data suggest that human stem cell populations express IL-11Rα mRNA (Turner *et al.*, 1996). In addition, there are ample functional data implicating that the IL-11Rα is expressed at an early stage of hematopoiesis. In particular, extensive studies have been performed on various populations of candidate murine stem cells, and have revealed that IL-11 can enhance multilineage growth synergistically and efficiently in combination with a number of other cytokines. These include the c-kit ligand (KL), flt3 ligand (FL), IL-3, GM-CSF, M-CSF, G-CSF, IL-4, IL-12 and Tpo; in general the effects of IL-11 appear to parallel those of IL-6 (Musashi *et al.*, 1991a,b; Tsuji *et al.*, 1991; Ploemacher *et al.*, 1993; Hirayama *et al.*, 1995; Hudak *et al.*, 1995; F.W. Jacobsen *et al.*, 1995a; S.E.W. Jacobsen *et al.*, 1995; Fujimoto *et al.*, 1996; Ku *et al.*, 1996; Ramsfjell *et al.*, 1996). Importantly, these effects appear to be mediated directly on the progenitor–stem cells, as they can be observed when candidate stem cells are cultured at the single cell level (F.W. Jacobsen *et al.*, 1995a; S.E.W. Jacobsen *et al.*, 1995; Ramsfjell *et al.*, 1996). IL-11 appears most efficient at promoting growth of primitive murine progenitor–stem cells when combined with KL, FL and/or IL-3, cytokines which in general are the most efficient *in vitro* stimulators of candidate stem cells (Metcalf, 1993; Ogawa, 1993; Hirayama *et al.*, 1995; Jacobsen *et al.*, 1995). In fact, the ability of IL-11 to synergize with cytokines such as Tpo and IL-4 might require the presence of one of these cytokines, as a synergistic interaction between IL-11 and each of these cytokines alone appears to be serum-dependent (Tsuji *et al.*, 1991; Ramsfjell *et al.*, 1996). This is supported by the fact that serum contains KL in biologically active amounts (Langley *et al.*, 1993) and that neutralization of KL in serum-supplemented cultures reverses most of the synergy between IL-11 and Tpo on multipotent progenitor cells (Ku *et al.*, 1996).

Enhanced growth in response to IL-11 can be observed as an increased number of progenitor cells being recruited into active proliferation (enhanced cloning frequency) as well as enhanced proliferation within the clones (enhanced size of individual clones). One mechanism for the enhanced recruitment and proliferation seen in response to IL-11 appears to be through promoting entry of quiescent cells into active cell cycle (Musashi *et al.*, 1991a,b).

Whereas IL-11 when acting alone has little or no growth-promoting activity, it can maintain a viable fraction of candidate murine stem cells in the absence of other cytokines, although much less efficiently than KL, IL-3 or Tpo (Bodine *et al.*, 1992; Katayama *et al.*, 1993; Li and Johnson, 1994; F.W. Jacobsen *et al.*, 1995b; Keller *et al.*, 1995; Borge *et al.*, 1996). Whereas KL has been demonstrated to suppress apoptosis of long-term reconstituting stem cells *in vitro* (Bodine *et al.*, 1992; Li and Johnson, 1994; Keller *et al.*, 1995), it remains to be determined whether the fraction of progenitor–stem cells surviving in response to IL-11 includes true stem cells.

Whereas most cytokines investigated have been demonstrated either to synergize or to have no interaction with IL-11 on candidate murine stem cells, TNF-α and TGF-β have been shown to be direct inhibitors of the growth as well as the viability-promoting activities of IL-11 on primitive murine progenitor–stem cells (F.W. Jacobsen *et al.*, 1995b, 1996; S.E.W. Jacobsen *et al.*, 1997), in agreement with their predominantly growth inhibitory effects on candidate murine stem cells (reviewed by Keller *et al.*, 1997).

IL-11 also appears to promote stroma-independent growth of enriched populations of primitive human progenitor cells isolated from bone marrow as well as cord blood and, as in mice, primarily in combination with IL-3 and KL, but also GM-CSF (Leary *et al.*, 1992; Schibler *et al.*, 1992; Lemoli *et al.*, 1993; van de Ven *et al.*, 1995). Because IL-11 alone promotes some colony formation of human progenitors in serum-supplemented but not serum-free cultures (as in mice), it appears likely that also this colony formation is KL dependent. It is noteworthy that IL-11, like other early acting synergistic cytokines, appears to have much less dramatic effects on primitive human than murine progenitor cells, and that in general human stem cells appear to require activation by more cytokines to acquire optimal *in vitro* growth. This might be due to an intrinsic difference between murine and human stem cells, but could also be a consequence of the candidate human stem cells studied being less purified (and thus less prone to synergy), or it might simply reflect that current *in vitro* culture conditions have not been sufficiently optimized to support growth of human stem cells. In fact, recent studies have suggested that the growth of candidate human stem cells might not be supported in semisolid medium, and that serum-depleted liquid cultures might represent the most optimal culture conditions for human stem cells (Petzer *et al.*, 1996a,b). As all clonogenic studies of the effect of IL-11 on primitive human progenitor–stem cells to date have been performed in semisolid cultures, additional studies under more optimal conditions and at the single cell level would be of interest.

Murine stem cell activity is best evaluated through *in vivo* repopulation studies, whereas an equivalent and optimal assay for human stem cells does not yet exist (Watt and Visser, 1992). As a consequence of this, the ability of human candidate stem cells to produce committed progenitor cells on established functional stroma over a prolonged time (minimum 5 weeks) has proved useful as a surrogate human stem cell assay (Gartner and Kaplan, 1980; Eaves *et al.*, 1991; Watt and Visser, 1992). In addition, such stroma-dependent cultures are useful for studies of the regulation of human as well as murine hematopoiesis, as they appear to be more physiologic and can support hematopoietic development in the absence of exogenously added cytokines (Eaves *et al.*, 1991; Spooncer *et al.*, 1993). In such long-term bone marrow cultures (human and murine), IL-11 has been demonstrated to enhance hematopoietic cell production, including various early and multipotent progenitors, and progenitors committed to the granulocyte–macrophage and megakaryocyte lineages (Keller *et al.*, 1993; Du *et al.*, 1995). However, addition of IL-11 rather depletes cultures of long-term repopulating murine stem cells and candidate human stem cells (Du *et al.*, 1995). These findings implicate that IL-11 might deplete the stem cell reserve in long-term cultures, potentially through induction of commitment of stem cells (Du *et al.*, 1995). However, the observed depletion of stem cells in IL-11-supplemented long-term bone marrow cultures might equally well be a consequence of IL-11 inducing cycling of stem cells which subsequently might then be induced to commitment by other soluble or cell–matrix-bound factors produced in the cultures. Alternatively, IL-11 might have an effect on the stromal environment which negatively affects the maintenance of the stem cell pool. Accordingly, this interesting finding must be viewed together with the effects of IL-11 observed in (stroma-independent) stem cell *ex vivo* expansion studies, as well as various *in vivo* studies (reviewed in separate sections).

Effects on *In Vitro* Megakaryocytopoiesis

Like IL-6, IL-11 potently stimulates *in vitro* (and *in vivo*; see separate section) megakaryocytopoiesis (reviewed by Du and Williams, 1994; Turner *et al.*, 1996). Similar to its effects on primitive hematopoietic progenitor–stem cells, IL-11 alone has limited effects, and rather acts as a synergistic factor to increase the growth-promoting activities of other thrombopoietic cytokines, primarily IL-3, KL and the recently cloned Tpo on murine (Paul *et al.*, 1990; Burstein *et al.*, 1992; Yonemura *et al.*, 1992; Broudy *et al.*, 1995) and human (Bruno *et al.*, 1991; Burstein *et al.*, 1992; Teramura *et al.*, 1992; Imai and Nakahata, 1994) megakaryocyte progenitor cells.

IL-11 appears to promote *in vitro* megakaryocytopoiesis at multiple levels, by recruiting more progenitors (primitive burst-forming unit as well as more mature colony-forming unit megakaryocyte (BFU-Mk and CFU-Mk respectively)) into active proliferation, by enhancing the size of the resulting colonies (increased number of megakaryocytes produced per progenitor), and by promoting megakaryocyte maturation and ploidy–endoreduplication (Bruno *et al.*, 1991; Burstein *et al.*, 1992; Teramura *et al.*, 1992; Yonemura *et al.*, 1992; Imai and Nakahata, 1994; Broudy *et al.*, 1995). In agreement with this, highly purified human megakaryocytes express IL-11Rα mRNA (Turner *et al.*, 1996). However, unlike Tpo (Kaushansky, 1995), IL-11 does not appear to have any effects on mature platelets (Novik *et al.*, 1995; Turner *et al.*, 1996). Thus, based on *in vitro* observations it is evident that IL-11 has effects on most stages of megakaryocyte formation and development, from the earliest progenitor–stem cells to the fully mature megakaryocyte, but that it apparently lacks effects on platelet activation. IL-11 has also been demonstrated to act as an autocrine growth factor for human megakaryocytoblastic cell lines (Kobayashi *et al.*, 1993).

The recently cloned thrombopoietin has been identified as a key regulator of steady-state megakaryocyte and platelet production (Gurney *et al.*, 1994; Kaushansky, 1995).

In agreement with this, c-mpl (the receptor for Tpo) is expressed, and Tpo has potent stimulatory effects, at all stages of megakaryocyte development, from stem cells down to platelets (reviewed by Kaushansky, 1995; S.E.W. Jacobsen *et al.*, 1996). Although the observation that c-mpl- or Tpo-deficient mice have an 85% reduction in platelet and megakaryocyte numbers confirms the crucial role of Tpo in regulating steady-state megakaryocytopoiesis, it also shows that other Tpo-independent regulators or mechanisms can promote platelet production (Gurney *et al.*, 1994; Carver-Moore *et al.*, 1996). In one study, neutralization of Tpo *in vitro* was demonstrated to block the ability of IL-11 (as well as IL-6 and KL) to promote megakaryocyte formation, suggesting that endogenous Tpo in the cultures was absolutely required for megakaryocyte development in response to these cytokines (Kaushansky *et al.*, 1995). However, in studies of c-mpl-deficient mice (lacking functional c-mpl), IL-11 (as well as IL-3, KL and IL-6) stimulated megakaryocyte production *in vitro* as well as platelet formation *in vivo* (Carver-Moore *et al.*, 1996), suggesting that IL-11 might play a role in both Tpo-dependent and Tpo-independent platelet formation.

In Vitro Effects of IL-11 on Myeloid (Granulocyte–Macrophage) and Erythroid Progenitor Cells

The most prominent and important hematopoietic effects of IL-11 might be those involving the stem cell compartment and megakaryocyte development. However, IL-11 has also been found to promote the *in vitro* growth of granulocyte–macrophage (GM) and erythroid (E) progenitors. In addition to promoting GM colony formation from primitive progenitor–stem cells, IL-11 can also enhance growth of murine and human progenitor cells already committed to the myeloid lineage, acting in combination with G-CSF, M-CSF, GM-CSF, IL-3 and KL (Musashi *et al.*, 1991b; Lemoli *et al.*, 1993; Jacobsen *et al.*, 1995a).

Functional data suggest also that erythroid progenitor cells might express IL-11Rα, as IL-11 synergizes with erythropoietin (Epo), IL-3, GM-CSF and KL to enhance clonogenic growth of predominantly early erythroid progenitors, BFU-erythroid (BFU-E), but also the more mature CFU-E (Quesniaux *et al.*, 1992; Schibler *et al.*, 1992; Lemoli *et al.*, 1993). The ability of IL-11 to synergistically promote Epo-dependent growth of BFU-E does not appear to require interaction with additional synergistic factors.

Effects of IL-11 on *In Vitro* B-cell Development

Postnatally, B-cell genesis occurs from hematopoietic stem cells residing in the bone marrow (Hardy *et al.*, 1991; Rolink *et al.*, 1995). Whereas a number of cytokines can promote myeloerythroid commitment and development from uncommitted bone marrow progenitors, much less is known about the cytokines governing the earliest stages of B-cell development (Metcalf, 1993; Ogawa, 1993; Rolink *et al.*, 1995). Whereas normal B-cell development is IL-7 dependent, the process of B-cell commitment appears to be IL-7 independent (Peschon *et al.*, 1994; von Freeden-Jeffrey *et al.*, 1995). Both KL and FL have been implicated in early B-cell development and, in combination with IL-7, each of these ligands has been shown to promote B-cell development from uncommitted murine progenitor cells isolated from bone marrow, fetal liver or yolk sac (Hirayama *et*

Table 4. *In vitro* hematopoietic effects of IL-11.

Cell formation	Response
Candidate stem cells	As single factor: promotes viability Synergistically enhances growth in combination with other early acting cytokines (primarily KL, FL, IL-3 and GM-CSF)
Megakaryocytopoiesis	Synergizes with other cytokines (IL-3, GM-CSF, KL, Tpo) to enhance growth of Mk progenitors (BFU-Mk and CFU-Mk) Stimulates maturation and endoreduplication of megakaryocytes
Myelopoiesis	Synergizes with other cytokines (G-CSF, M-CSF, GM-CSF, IL-3, KL) to enhance growth of GM progenitors, leading primarily to macrophage differentiation Inhibits proinflammatory cytokine secretion from macrophages
Erythropoiesis	Synergizes with other cytokines (Epo, IL-3, KL, GM-CSF) to enhance growth of erythroid progenitors (primarily BFU-E)
B lymphopoiesis	Synergizes with KL or FL to promote formation of IL-7-responsive B-cell progenitors from candidate murine stem cells Promotes differentiation–activation of B cells (indirectly through T-cell activation)

al., 1992, 1995; Hirayama and Ogawa, 1994; Ray *et al.*, 1996; Veiby *et al.*, 1996). Although IL-11 cannot replace either IL-7 or FL/KL in these systems, it can further enhance the generation of early B-cell progenitors (Hirayama *et al.*, 1992, 1995; Hirayama and Ogawa, 1994; Ray *et al.*, 1996). IL-11 has not been demonstrated to stimulate the growth of committed B-cell progenitors, but it remains possible that subpopulations of early B-cell progenitors might be responsive to IL-11.

The ability of IL-11 to stimulate differentiation and activation of mature murine and human B cells appears to be indirect, mediated through activation of T cells (Yin *et al.*, 1992; Bitko *et al.*, 1997).

Table 4 summarizes known *in vitro* activities of IL-11 in the hematopoietic system.

IN VIVO HEMATOLOGIC AND HEMATOPOIETIC EFFECTS OF IL-11

The dominating hematologic effects observed following IL-11 administration include those that could be predicted from *in vitro* studies, that is stimulation of megakaryocyte and platelet production, as well as progenitor–stem cell expansion and mobilization (reviewed by Goldman, 1995; Turner *et al.*, 1996). However, at variance with the *in vitro* observations, the corresponding *in vivo* effects can be observed when IL-11 is administered alone. This is frequently observed for other synergistic cytokines as well, and is likely to result from the presence of other endogenously produced synergistic partners, such as KL (Metcalf, 1993; Ogawa, 1993).

When IL-11 is administered to normal mice, a time-dependent increase in platelet numbers is observed, together with an increased frequency of splenic (but not bone marrow) megakaryocytes and increased high-ploidy megakaryocytes (32N) (Neben *et al.*, 1993). Furthermore, megakaryocyte progenitor cells in bone marrow and particularly in spleen are enhanced. Similar findings are made when splenectomized mice are treated with IL-11 (Neben *et al.*, 1993). Comparable increases in platelet and megakaryocyte numbers and ploidy are seen in rats, in which IL-11 and IL-6 are equally efficient at promoting thrombocytopoiesis (Cairo *et al.*, 1993, 1994), as well as in non-human primates (Goldman, 1995; Schlerman *et al.*, 1996). When the treatment or dosage schedule is optimized, platelet numbers in normal mice as well as in non-human primates can be raised as much as 150% above those in control animals (Goldman, 1995; Leonard *et al.*, 1996; Schlerman *et al.*, 1996). However, when compared with Tpo, the quantitative increases in platelet as well as megakaryocyte numbers and ploidy are relatively modest (Hughes and Howells, 1993; Kaushansky *et al.*, 1994; Harker *et al.*, 1996).

A phase I trial of the thrombopoietic effects of IL-11 on a group of women with advanced breast cancer (but no evidence of bone marrow involvement) also demonstrated an increase in platelet numbers, with an accompanying increase in the number, cycling and ploidy of bone marrow megakaryocytes (Gordon, 1996; Gordon *et al.*, 1996; Orazi *et al.*, 1996a). An increase was also seen in the number of megakaryocyte progenitor cells, whereas the number of other types of progenitors (in bone marrow or peripheral blood) was not significantly affected, although an increased fraction of actively cycling progenitors was observed (Orazi *et al.*, 1996a). This is somewhat in contrast to the situation in normal mice treated with IL-11, where a concomitant increase in myeloid (GM), erythroid and multipotent progenitor cells can be seen. These progenitor–stem cells appear to be lodging preferentially in the spleen (Hangoc *et al.*, 1993; Leonard *et al.*, 1996), but are also mobilized to peripheral blood (Mauch *et al.*, 1995). Potentially, similar effects might be observed in humans on administration of higher doses, although studies of non-human primates also demonstrate a more marginal effect than in mice (Goldman, 1995).

IL-11 administration has little or no effect on neutrophil count, but might induce a mild anemia, potentially resulting from expansion of the plasma volume (de Haan *et al.*, 1995; Orazi *et al.*, 1996a; Schlerman *et al.*, 1996). When IL-11 is combined with KL or G-CSF, myelopoiesis is stimulated as well, and in the case of KL also early erythropoiesis (Cairo *et al.*, 1993, 1994; de Haan *et al.*, 1995).

Overexpression of IL-11 in murine bone marrow accelerates recovery of platelet and neutrophil numbers following transplantation, but primarily increases platelet number following reconstitution (Hawley *et al.*, 1993, 1996). Furthermore, stem cells constitutively expressing IL-11 appear to maintain their long-term reconstituting ability following serial transplantations, suggesting that continuous exposure to IL-11 does not deplete the stem cell pool (Hawley *et al.*, 1996).

EFFECTS OF IL-11 OUTSIDE THE HEMATOPOIETIC SYSTEM: PLEIOTROPIC EFFECTS ON BONE REMODELING, CHONDROCYTES, NEURONS, ADIPOCYTES, AND GASTROINTESTINAL AND BRONCHIAL EPITHELIUM

Although not yet as extensively characterized, it is evident that IL-11 has many potent effects outside the hematopoietic system, underscoring its pleiotropic nature. Some of these effects will be briefly summarized here. For a more detailed overview other reviews can be recommended (Du and Williams, 1994; Yang, 1995).

IL-11 enhances the generation of osteoclasts from bone marrow, thereby inhibiting bone formation and promoting bone resorption *in vitro* (Hughes and Howells, 1993; Girasole *et al.*, 1994). As osteoclasts and macrophages derive from the same GM progenitor cells, it is noteworthy that IL-11 also appears preferentially to promote *in vitro* formation of macrophage rather than granulocyte colonies in regular hematopoietic cultures (Musashi *et al.*, 1991b). Normal osteoclasts express IL-11Rα and gp130, and the effects of other stimulators of osteoclasts, including IL-1, TNF-α, parathyroid hormone and 1,25-dihydroxyvitamin D_3, appear to be dependent largely on IL-11 production (Girasole *et al.*, 1994). Of particular interest, IL-11-stimulated osteoclast formation is estrogen independent, at variance with the finding for IL-6 (Girasole *et al.*, 1994). As IL-11Rα and IL-11 expression also can be induced in osteoblasts, it is possible that IL-11 might also be involved in regulation of osteoblast activity (Romas *et al.*, 1996).

IL-11 is also expressed in chondrocytes and synoviocytes, in which it stimulates the production of the tissue inhibitor of metalloproteases, and might thus have a protective effect in connection with inflammatory lesions in the joints (Maier *et al.*, 1993).

Neurons have also been demonstrated to express IL-11, and exogenously added IL-11 can promote growth and differentiation of neuronal progenitor cell lines (Fann and Patterson, 1994; Du *et al.*, 1996).

IL-11 inhibits adipogenesis and efficiently suppresses the formation of adipocytes in long-term bone marrow cultures (Kawashima *et al.*, 1991; Ohsumi *et al.*, 1991; Keller *et al.*, 1993).

IL-11 suppresses the growth–cell cycling of gastrointestinal epithelial stem cells and protects the gastrointestinal mucosa against radiation, chemotherapy and other inflammatory injuries to the intestine (Du *et al.*, 1994; Keith *et al.*, 1994; Booth and Potten, 1995; Peterson *et al.*, 1996).

Human lung fibroblasts and epithelial cells produce IL-11 in connection with airway inflammation, causing airway hyperresponsiveness, and when overexpressed in the airways IL-11 causes obstruction (Elias *et al.*, 1994b; Zheng *et al.*, 1994; Romas *et al.*, 1996; Tang *et al.*, 1996).

As has been demonstrated for IL-6, acute-phase protein synthesis by hepatocytes is also enhanced by IL-11 (Baumann and Schendel, 1991).

The *in vitro* biologic activities of IL-11 on non-hematopoietic cells and tissues are summarized in Table 5.

Table 5. *In vitro* effects of IL-11 on normal tissues and cells outside the hematopoietic system.

Cell type/tissue	Response
Bone	Stimulates formation of osteoclasts from bone marrow Inhibits bone formation and promotes bone resorption
Chondrocytes and synoviocytes	Stimulates production of tissue inhibitor of metalloproteases
Neurons	Promotes growth of neuronal progenitor cell lines
Adipocytes	Inhibits adipogenesis
Gastrointestinal epithelial cells	Inhibits growth
Hepatocytes	Stimulates acute-phase protein synthesis

IL-11 IN DISEASE

Although the clinical significance remains to be determined, IL-11 has, like many other cytokines (reviewed by Hassan and Drexler, 1995), been shown to promote growth of myeloid leukemic cell lines and primary myeloid leukemic cells (Hu *et al.*, 1993; Lemoli *et al.*, 1995). Also leukemic cell lines of the megakaryocytic lineage are growth stimulated by IL-11, whereas those of the lymphoid lineages appear to be unaffected (Hu *et al.*, 1993; Kobayashi *et al.*, 1993). Like normal progenitors, IL-11 alone has little or no effect, and acts predominantly as a synergistic factor in combination with IL-3, GM-CSF or KL (Hu *et al.*, 1993; Lemoli *et al.*, 1995).

IL-11 can be expressed in leukemic cells, and neutralizing anti-IL-11 antibodies or IL-11 antisense oligonucleotides inhibit their growth, implying that IL-11 in certain cases might be involved in autocrine growth stimulation of leukemic cells (Kobayashi *et al.*, 1993; Lemoli *et al.*, 1995). It is also noteworthy that in studies of mice constitutively expressing IL-11 in the bone marrow, one case of development of myeloid leukemia has been observed in two secondary recipients that received bone marrow from the same donor (Hawley *et al.*, 1993).

Whereas IL-11 was originally identified through its ability to promote growth of a murine plasmacytoma cell line (Paul *et al.*, 1990), most human myeloma cell lines and all primary myeloma cells investigated have been found to lack IL-11Rα and IL-11 expression, and to be unresponsive to IL-11 stimulation (Paul *et al.*, 1992; Zhang *et al.*, 1994). However, other studies have found that some human myeloma cell lines express IL-11Rα and are IL-11-responsive, and that IL-10 stimulation can confer IL-11 responsiveness of otherwise unresponsive myeloma cell lines (Lu *et al.*, 1995).

In a recent study, most human breast cancer cell lines as well as primary breast cancer cells were found to express gp130 along with the α chains for IL-6, LIF, IL-11 and CNTF, whereas other cytokine receptors were predominantly not detectable (by polymerase chain reaction or at the cell surface) (Douglas *et al.*, 1997). Interestingly, IL-11 (although to a lesser degree than OSM) was found to inhibit the growth of some receptor-positive cell lines, suggesting a potential role in the growth regulation of normal and malignant breast cells (Douglas *et al.*, 1997).

It is obvious from the non-hematopoietic effects described in the preceding section that IL-11 might also play a role in other diseases, including inflammatory diseases of the gastrointestinal system, joints, and airway/lungs. In that regard, IL-11 has been demonstrated to attenuate inflammatory responses, involving downregulation of proinflammatory cytokines and mediators (Trepicchio *et al.*, 1996). However, the regulatory effects of IL-11 on inflammatory processes are clearly likely to be complex and pleiotropic, as underscored by other studies suggesting that high levels of IL-11 in the airway may promote inflammation (Tang *et al.*, 1996).

PRECLINICAL AND CLINICAL EXPERIENCE AND POTENTIAL UTILITY OF IL-11

Preclinical and clinical studies with IL-11 have been reviewed in detail elsewhere (Du and Williams, 1994; Goldman, 1995; Turner *et al.*, 1996), and will only be summarized here.

Myelosuppression

IL-11 has been studied extensively for its ability to promote recovery following myelosuppression (Du *et al.*, 1993a,b; Hangoc *et al.*, 1993; Hawley *et al.*, 1993, 1996; Leonard *et al.*, 1994; Paul *et al.*, 1994; Schlerman *et al.*, 1996), as it may prove useful for treatment of chemotherapy-induced cytopenia and in bone marrow transplantation. In mice, administration of IL-11 reduces the morbidity and mortality associated with these procedures, promoting primarily accelerated platelet recovery, although neutrophils frequently also recover more quickly (Du *et al.*, 1993a,b; Hangoc *et al.*, 1993; Hawley *et al.*, 1993, 1996; Leonard *et al.*, 1994; Paul *et al.*, 1994). In non-human primates, IL-11 appears to promote only platelet recovery (Goldman, 1995; Schlerman *et al.*, 1996).

In clinical trials, IL-11 has been well tolerated, the main side-effect being mild anemia, likely to result from expansion of the plasma volume (Goldman, 1995; Lemoli *et al.*, 1995; Gordon, 1996; Gordon *et al.*, 1996; Orazi *et al.*, 1996a; Turner *et al.*, 1996). Other side-effects (myalgia, arthralgia, fatigue, edema) are mild at doses of up to 75 μg/kg daily.

IL-11 treatment of patients with cancer following chemotherapy has been found to promote only platelet recovery, with no effect on neutropenia (Goldman, 1995; Lemoli *et al.*, 1995; Gordon, 1996; Gordon *et al.*, 1996; Orazi *et al.*, 1996a; Turner *et al.*, 1996). However, in combination with other cytokines (such as G-CSF, IL-3 or KL), it might prove more efficient at promoting platelet as well as neutrophil recovery (Du *et al.*, 1993a; Cairo *et al.*, 1993, 1994; Galmiche *et al.*, 1996).

Because the recently cloned Tpo is likely to prove more efficient at promoting platelet recovery following myelosuppression than IL-11 (Nichol, 1996), the potential of IL-11 in this setting lies in its ability to act as a synergistic factor (combination therapy) and/or the fact that it has simultaneous and seemingly unique protective effects on non-hematopoietic organs (outlined below) subject to toxicity following treatment with chemotherapy and/or radiation. It is important also to emphasize that in certain clinical settings IL-11 may still prove to have effects on platelet recovery that might be similar or even superior to those of Tpo. In particular, the fact that IL-11 is a cytokine also with potent effects on the stem cell compartment might be important, although recent studies also suggest that Tpo has such an effect (reviewed by S.E.W. Jacobsen *et al.*, 1996).

Whereas both Tpo and IL-11 concentrations are increased in the serum of patients with thrombocytopenia following myeloablative treatment or bone marrow transplantation, the level of IL-6 is not (Chang *et al.*, 1996). In patients with immune thrombocytopenic purpura (decreased platelet survival), serum levels of IL-11 are increased, whereas levels of Tpo are undetectable (Chang *et al.*, 1996).

Mobilization and *Ex Vivo* Expansion of Progenitor–Stem Cells

There is currently extensive interest in optimizing protocols for mobilization of progenitor–stem cells to peripheral blood, and for expanding stem–progenitor cells *in vitro* (Williams, 1993; Emerson, 1996; Lange *et al.*, 1996). Mobilized progenitor cells are currently being used in various transplantation settings, in particular autologous and allogeneic stem cell transplantations of patients with cancer following high-dose chemotherapy. In addition, mobilized and expanded stem cells are likely targets for gene therapy (Bodine *et al.*, 1994; Kohn *et al.*, 1995).

Whereas IL-11 alone has limited mobilizing ability, it synergizes with KL to promote mobilization of long-term repopulating stem cells, similar to what has previously been shown for KL in combination with G-CSF (Mauch *et al.*, 1995). Such an enhanced number of progenitor–stem cells might decrease the number of leukapheresis sessions required to yield sufficient cells for a transplant. More importantly, there are indications that progenitor–stem cells mobilized by combination therapy might be better targets for stem cell gene therapy (Bodine *et al.*, 1994; Dunbar *et al.*, 1996). Moreover, the potential qualitative differences in stem cell populations mobilized by different cytokine combinations require further study.

As it is possible to efficiently increase the number of committed progenitor cells (required for short-term reconstitution), the main focus of most current protocols is how best to maintain or even raise the number of long-term reconstituting stem cells (Williams, 1993; Emerson, 1996; Lange *et al.*, 1996). In that regard, IL-11 has been utilized in a number of *ex vivo* expansion protocols (primarily mouse studies), some of which have demonstrated depletion (Keller *et al.*, 1993; Du *et al.*, 1995; Peters *et al.*, 1995, 1996), while others have shown maintenance or expansion of stem cells (Holyoake *et al.*, 1996). However, the protocols that have demonstrated depletion have either been performed in bone marrow stroma cultures (discussed above) or in combination with IL-3, which recently has been demonstrated to compromise long-term reconstituting stem cells in culture (Yonemura *et al.*, 1996). In contrast, short-term incubation of murine bone marrow with IL-11 in combination with KL augments both short- and long-term repopulating ability, allowing successful reconstitution of quaternary recipients (Holyoake *et al.*, 1996).

As the combination of IL-11 and soluble IL-11Rα has been shown to activate gp130 (discussed earlier), the soluble IL-11Rα–IL-11 complex might prove particularly efficient at expanding stem cells in combination with KL, as has recently been demonstrated for the soluble IL-6R–IL-6 complex (Sui *et al.*, 1995).

Protective Effects on Gastrointestinal Mucosal Cells

One of the most exciting, unique and potentially clinically important effects of IL-11 is its ability to protect gastrointestinal mucosal cells from damage following various types of injury, including irradiation and chemotherapy (Du *et al.*, 1994; Keith *et al.*, 1994; Liu *et al.*, 1996; Orazi *et al.*, 1996b). The mucosal protective effect of IL-11 in this setting appears to correlate with its ability to inhibit the growth of intestinal epithelial cells before the insult (Booth and Potten, 1995; Peterson *et al.*, 1996), as well as enhanced growth and recovery after the damage has taken place (Du *et al.*, 1994). IL-11 also decreases the severity and improves survival in models of oral mucositis (induced by chemotherapy), ischemic intestinal necrosis and inflammatory bowel disease (Keith *et al.*, 1994; Du *et al.*, 1997). Chemotherapy of patients with cancer frequently induces simultaneous damage to the hematopoietic and gastrointestinal systems, and thus IL-11 might promote recovery of both systems, which could be a unique property when compared with other cytokines currently used to accelerate hematopoietic recovery.

Anti-inflammatory Effects of IL-11

As described elsewhere in this review, IL-11 has anti-inflammatory effects that might prove to be of value in the treatment of a number of inflammatory conditions. For instance, the ability of IL-11 to reduce mortality following sepsis, enterotoxin-induced toxic shock syndrome and radiation-induced thoracic injury (Barton *et al.*, 1996; Van Leuven *et al.*, 1996) might (at least in part) result from its ability to downregulate production and release of TNF and other proinflammatory mediators (Trepicchio *et al.*, 1996; Van Leuven *et al.*, 1996). It is also likely that these anti-inflammatory effects might explain some of the protective effects of IL-11 in the gastrointestinal system.

ACKNOWLEDGEMENTS

The assistance of Ole Johan Borge, Li Cui and Veslemøy Ramsjell in carefully reviewing this manuscript is highly appreciated.

REFERENCES

Barton, B.E., Shortall, J. and Jackson, J.V. (1996). Interleukins 6 and 11 protect mice from mortality in a staphylococcal enterotoxin-induced toxic shock model. *Infect. Immun.* **64**, 714–718.

Baum, C.M., Weissman, I.L., Tsukamoto, A.S., Buckle, A.-M. and Peault, B. (1992). Isolation of a candidate human hematopoietic stem-cell population. *Proc. Natl Acad. Sci. U.S.A.* **89**, 2804–2808.

Baumann, H. and Schendel, P. (1991). Interleukin-11 regulates the hepatic expression of the same plasma protein genes as interleukin-6. *J. Biol. Chem.* **266**, 20424–20427.

Bazan, J.F. (1990). Structural design and molecular evolution of a cytokine receptor superfamily. *Proc. Natl Acad. Sci. U.S.A.* **87**, 6934–6938.

Berardi, A.C., Wang, A., Levine, J.D., Lopez, P. and Scadden, D.T. (1995). Functional isolation and characterization of human hematopoietic stem cells. *Science* **267**, 104–108.

Bilinski, P., Hall, M.A., Neuhaus, H., Gissel, C., Heath, J.K. and Gossler, A. (1996). Two differentially expressed interleukin-11 receptor genes in the mouse genome. *Biochem. J.* **320**, 359–363.

Bitko, V., Velazquez, A., Yang, L., Yang, Y.C. and Barik, S. (1997). Transcriptional induction of multiple cytokines by human respiratory syncytial virus requires activation of NF-kappa B and is inhibited by sodium salicylate and aspirin. *Virology* **232**, 369–378.

Bodine, D.M., Orlic, D., Birkett, N.C., Seidel, N.E. and Zsebo, K.M. (1992). Stem cell factor increases colony-forming unit-spleen number *in vitro* in synergy with interleukin-6, and *in vivo* in Sl/Sl^d mice as a single factor. *Blood* **79**, 913–919.

Bodine, D.M., Seidel, N.E., Gale, M.S., Nienhuis, A.W. and Orlic, D. (1994). Efficient retrovirus transduction of mouse pluripotent hematopoietic stem cells mobilized into the peripheral blood by treatment with granulocyte colony-stimulating factor and stem cell factor. *Blood* **84**, 1482–1491.

Booth, C. and Potten, C.S. (1995). Effects of IL-11 on the growth of intestinal epithelial cells *in vitro*. *Cell Prolif.* **28**, 581–594.

Borge, O.J., Ramsfjell, V., Veiby, O.P., Murphy, M.J., Lok, S. and Jacobsen, S.E.W. (1996). Thrombopoietin, but not erythropoietin promotes viability and inhibits apoptosis of multipotent murine hematopoietic progenitor cells *in vitro*. *Blood* **88**, 2859–2870.

Bradford, G.B., Williams, B. and Bertoncello, I. (1997). Quiescence, cycling, and turnover in the primitive hematopoietic stem cell compartment. *Exp. Hematol.* **25**, 445–453.

Broudy, V.C., Lin, N.L. and Kaushansky, K. (1995). Thrombopoietin (c-mpl ligand) acts synergistically with erythropoietin, stem cell factor, and interleukin-11 to enhance murine megakaryocyte ploidy *in vitro*. *Blood* **85**, 1719–1726.

Bruno, E., Briddell, R.A. and Cooper, R.J. (1991). Effects of recombinant interleukin 11 on human megakaryocyte progenitor cells. *Exp. Hematol.* **19**, 378–381.

Burstein, S.A., Mei, R.-L., Henthorn, J., Friese, P. and Turner, K. (1992). Leukemia inhibitory factor and interleukin-11 promote maturation of murine and human megakaryocytes *in vitro. J. Cell. Physiol.* **153**, 305–312.

Cairo, M.S., Plunkett, J.M., Nguyen, A., Schendel, P. and van de Ven, C. (1993). Effect of interleukin-11 with and without granulocyte colony-stimulating factor on *in vivo* neonatal rat hematopoiesis: induction of neonatal thrombocytosis by interleukin-11 and synergistic enhancement of neutrophilia by interleukin-11 + granulocyte colony-stimulating factor. *Pediatr. Res.* **34**, 56–61.

Cairo, M.S., Plunkett, J.M., Schendel, P. and van de Ven, C. (1994). The combined effects of interleukin-11, stem cell factor, and granulocyte colony-stimulating factor on newborn rat hematopoiesis: significant enhancement of the absolute neutrophil count. *Exp. Hematol.* **22**, 1118–1123.

Carver-Moore, K., Broxmeyer, H.E., Luoh, S.-M., Cooper, S., Peng, J., Burstein, S.A., Moore, M.W. and de Sauvage, F.J. (1996). Low levels of erythroid and myeloid progenitors in thrombopoietin and c-mpl-deficient mice. *Blood* **88**, 803–808.

Chang, M., Suen, Y., Meng, G., Buzby, J.S., Bussel, J., Shen, V., van de Ven, C. and Cairo, M.S. (1996). Differential mechanisms in the regulation of endogenous levels of thrombopoietin and interleukin-11 during thrombocytopenia: insight into the regulation of platelet production. *Blood* **88**, 3354–3362.

Cherel, M., Sorel, M., Lebeau, B., Dubois, S., Moreau, J.-F., Bataille, R., Minvielle, S. and Jacques, Y. (1995). Molecular cloning of two isoforms of a receptor for the human hematopoietic cytokine interleukin-11. *Blood* **86**, 2534–2540.

Cherel, M., Sorel, M., Apiou, F., Lebeau, B., Dubois, S., Jacques, Y. and Minvielle, S. (1996). The human interleukin-11 receptor alpha gene (IL-11RA): genomic organization and chromosome mapping. *Genomics* **32**, 49–53.

Civin, C.I., Almeida-Porada, G., Lee, M.-J., Olweus, J., Terstappen, L.W.M.M. and Zanjani, E.D. (1996). Sustained retransplantable, multilineage engraftment of highly purified adult human bone marrow stem cells *in vivo. Blood* **88**, 4102–4109.

Czupryn, M., Bennett, F., Dube, J., Grant, K., Scoble, H., Sookdeo, H. and McCoy, J.M. (1995a). Alanine-scanning mutagenesis of human interleukin-11: identification of regions important for biological activity. *Ann. N.Y. Acad. Sci.* **762**, 152–164.

Czupryn, M.J., McCoy, J.M. and Scooble, H.A. (1995b). Structure–function relationships in human interleukin-11. *J. Biol. Chem.* **270**, 978–985.

Davidson, A.J., Freeman, S.A., Crosier, K.E., Wood, C.R. and Crosier, P.S. (1997). Expression of the murine interleukin 11 and its receptor alpha-chain in adult and embryonic tissues. *Stem Cells* **15**, 119–124.

de Haan, G., Dontje, B., Engel, C., Loeffler, M. and Nijhof, W. (1995). *In vivo* effects of interleukin-11 and stem cell factor in combination with erythropoietin in the regulation of erythropoiesis. *Br. J. Haematol.* **90**, 783–790.

Doi, H., Inaba, M., Yamamoto, Y., Taketani, S., Mori, S.-I., Sugihara, A., Ogata, H., Toki, J., Hisha, H., Inaba, K., Sogo, S., Adachi, M., Matsuda, T., Good, R.A. and Ikehara, R.A. (1997). Pluripotent hemopoietic stem cells are c-kit$^{<low}$. *Proc. Natl Acad. Sci. U.S.A.* **94**, 2513–2517.

Douglas, A.M., Goss, G.A., Sutherland, R.L., Hilton, D.J., Berndt, M.C., Nicola, N.A. and Begley, C.G. (1997). Expression and function of members of the cytokine receptor superfamily on breast cancer cells. *Oncogene* **14**, 661–669.

Du, X.X. and Williams, D.A. (1994). Interleukin-11: a multifunctional growth factor derived from the hematopoietic microenvironment. *Blood* **83**, 2023–2030.

Du, X.X., Keller, D., Maze, R. and Williams, D.A. (1993a). Comparative effects of *in vivo* treatment using interleukin-11 and stem cell factor on reconstitution in mice after bone marrow transplantation. *Blood* **82**, 1016–1022.

Du, X.X., Neben, T., Goldman, S. and Williams, D.A. (1993b). Effects of recombinant human interleukin-11 on hematopoietic reconstitution in transplant mice: acceleration of recovery of peripheral blood neutrophils and platelets. *Blood* **81**, 27–34.

Du, X.X., Doerschuk, C.M., Orazi, A. and Williams, D.A. (1994). A bone marrow stromal-derived growth factor, interleukin-11, stimulates recovery of small intestinal mucosal cells after cytoablative therapy. *Blood* **83**, 33–37.

Du, X.X., Scott, D., Yang, Z.X., Cooper, R., Xiao, X.L. and Williams, D.A. (1995). Interleukin-11 stimulates multilineage progenitors, but not stem cells, in murine and human long-term marrow cultures. *Blood* **86**, 128–134.

Du, X.X., Everett, E.T., Wang, G., Lee, W.-H., Yang, Z. and Williams, D.A. (1996). Murine interleukin-11 (IL-11) is expressed at high levels in the hippocampus and expression is developmentally regulated in the testis. *J. Cell. Physiol.* **168**, 362–372.

Du, X., Liu, Q., Yang, Z., Orazi, A., Rescorla, F.J., Grosfeld, J.L. and Williams, D.A. (1997). Protective effects of interleukin-11 in a murine model of ischemic bowel necrosis. *Am. J. Physiol.* **272**, G545–G552.

Dunbar, C.E., Seidel, N.E., Doren, S., Sellers, S., Cline, A.P., Metzger, M.E., Agricola, B.A., Donahue, R.E. and Bodine, D.M. (1996). Improved retroviral gene transfer into murine and rhesus peripheral blood or bone marrow repopulating cells primed *in vivo* with stem cell factor and granulocyte colony-stimulating factor. *Proc. Natl Acad. Sci. U.S.A.* **93**, 11 871–11 876.

Eaves, C.J., Cashman, J.D. and Eaves, A.C. (1991). Methodology of long-term culture of human hemopoietic cells. *J. Tissue Culture Methods* **13**, 55–62.

Elias, J.A., Zheng, T., Einarsson, O., Landry, M., Trow, T., Rebert, N. and Panuska, J. (1994a). Epithelial interleukin 11. Regulation by cytokines, respiratory syncytial virus, and retinoic acid. *J. Biol. Chem.* **269**, 22 261–22 268.

Elias, J.A., Zheng, T., Whiting, N.L., Trow, T.K., Merrill, W.W., Zitnik, R., Ray, P. and Alderman, B.M. (1994b). IL-1 and transforming growth factor-β regulation of fibroblast-derived IL-11. *J. Immunol.* **152**, 2421–2429.

Emerson, S.G. (1996). *Ex vivo* expansion of hematopoietic precursors, progenitors, and stem cells: the next generation of cellular therapeutics. *Blood* **87**, 3082–3088.

Fann, M.J. and Patterson, P.H. (1994). Neuropoietic cytokines and activin A differentially regulate the phenotype of cultured sympathetic neurons. *Proc. Natl Acad. Sci. U.S.A.* **91**, 43–47.

Fourcin, M., Chevalier, S., Lebrun, J.-J., Kelly, P., Pouplard, A., Wijdenes, J. and Gascan, H. (1994). Involvement of gp130/interleukin-6 receptor transducing component in interleukin-11 receptor. *Eur. J. Immunol.* **24**, 277–280.

Fujimoto, K., Lyman, S.D., Hirayama, F. and Ogawa, M. (1996). Isolation and characterization of primitive hematopoietic progenitors of murine fetal liver. *Exp. Hematol.* **24**, 285–290.

Galli, S.J., Zsebo, K.M. and Geissler, E.N. (1994). The kit ligand, stem cell factor. *Adv. Immunol.* **55**, 1–96.

Galmiche, M.C., Vogel, C.A., Delaloye, A.B., Schmidt, P.M., Healy, F., Mach, J.P. and Buchegger, F. (1996). Combined effects of interleukin-3 and interleukin-11 on hematopoiesis in irradiated mice. *Exp. Hematol.* **24**, 1298–1306.

Gartner, S. and Kaplan, H.S. (1980). Long-term culture of human bone marrow cells. *Proc. Natl Acad. Sci. U.S.A.* **77**, 4756–4759.

Gearing, D.P., Comeau, M.R., Friend, D.J., Gimpel, S.D., Thut, C.J., McGourty, J., Brasher, K.K., King, J.A., Gillis, S., Mosley, B., Ziegler, S.F. and Cosman, D. (1992). The IL-6 signal transducer, gp130: an oncostatin M receptor and affinity converter for the LIF receptor. *Science* **255**, 1434–1437.

Girasole, G., Passeri, G., Jilka, R.L. and Manolagas, S.C. (1994). Interleukin-11: a new cytokine critical for osteoclast development. *J. Clin. Invest.* **93**, 1516–1524.

Goldman, S.J. (1995). Preclinical biology of interleukin 11: a multifunctional hematopoietic cytokine with potent thrombopoietic activity. *Stem Cells* **13**, 462–471.

Goodell, M.A., Brose, K., Paradis, G., Conner, A.S. and Mulligan, R.C. (1996). Isolation and functional properties of murine hematopoietic stem cells that are replicating *in vivo*. *J. Exp. Med.* **183**, 1797–1806.

Gordon, M.S. (1996). Thrombopoietic activity of recombinant human interleukin 11 in cancer patients receiving chemotherapy. *Cancer Chemother. Pharmacol.* **38 (supplement)**, 96–98.

Gordon, M.S., McCaskill-Stevens, W.J., Battiato, L.A., Loewy, J., Loesch, D., Breeden, E., Hoffman, R., Beach, K.J., Kuca, B., Kaye, J. and Sledge, G.W., Jr. (1996). A phase I trial of recombinant human interleukin-11 (neumega rhIL-11 growth factor) in women with breast cancer receiving chemotherapy. *Blood* **87**, 3615–3624.

Gurney, A.L., Carver-Moore, K., de Sauvage, F.J. and Moore, M.W. (1994). Thrombocytopenia in c-mpl-deficient mice. *Science* **265**, 1445–1450.

Hangoc, G., Yin, T., Cooper, S., Schendel, P., Yang, Y.C. and Broxmeyer, H.E. (1993). *In vivo* effects of recombinant interleukin-11 on myelopoiesis in mice. *Blood* **81**, 965–972.

Hardy R.R., Carmack, C.E., Shinton, S.A., Kemp, J.D. and Hayakawa, K. (1991). Resolution and characterization of pro-B and pre-pro-B cell stages in normal mouse bone marrow. *J. Exp. Med.* **173**, 1213–1225.

Harker, L.A., Hunt, P., Marzec, U.M., Kelly, A.B., Tomer, A., Hanson, S.R. and Stead, R.B. (1996). Regulation of platelet production and function by megakaryocyte growth and development factor in nonhuman primates. *Blood* **87**, 1833–1844.

Hassan, H.T. and Drexler, H.G. (1995). Interleukins and colony stimulating factors in human myeloid leukemia cell lines. *Leuk. Lymphoma* **20**, 1–15.

Hawley, R.G., Fong, A.Z., Ngan, B.Y., de Lanux, V.M., Clark, S.C. and Hawley, T.S. (1993). Progenitor cell hyperplasia with rare development of myeloid leukemia in interleukin 11 bone marrow chimeras. *J. Exp. Med.* **178**, 1175–1188.

Hawley, R.G., Hawley, T.S., Fong, A.Z., Quinto, C., Collins, M., Leonard, J.P. and Goldman, S.J. (1996). Thrombopoietic potential and serial repopulating ability of murine hematopoietic stem cells constitutively expressing interleukin 11. *Proc. Natl Acad. Sci. U.S.A.* **93**, 10 297–10 302.

Heimfeld, S., Hudak, S., Weissman, I. and Rennick, D. (1991). The *in vitro* response of phenotypically defined mouse stem cells and myeloerythroid progenitors to single or multiple growth factors. *Proc. Natl Acad. Sci. U.S.A.* **88**, 9902–9906.

Hilton, D.J., Hilton, A.A., Raicevic, A., Rakar, S., Harrison-Smith, M., Gough, N.M., Begley, C.G., Metcalf, D., Nicola, N.A. and Willson, T.A. (1994). Cloning of a murine IL-11 receptor α-chain; requirement for gp130 for high affinity binding and signal transduction. *EMBO J.* **13**, 4765–4775.

Hirayama, F. and Ogawa, M. (1994). Cytokine regulation of early B-lymphopoiesis assessed in culture. *Blood Cells* **20**, 341–346; Discussion 346–347.

Hirayama, F., Shih, J.-P., Awgulewitsch, A., Warr, G.W., Clark, S.C. and Ogawa, M. (1992). Clonal proliferation of murine lymphohemopoietic progenitors in culture. *Proc. Natl Acad. Sci. U.S.A.* **89**, 5907–5911.

Hirayama, F., Lyman, S.D., Clark, S.C. and Ogawa, M. (1995). The flt3 ligand supports proliferation of lymphohematopoietic progenitors and early B-lymphoid progenitors. *Blood* **85**, 1762–1768.

Hodgson, G.S. and Bradley, T.R. (1979). Properties of haemopoietic stem cells surviving 5-fluorouracil treatment: evidence for a pre-CFU-S cell? *Nature* **281**, 381–382.

Holyoake, T.L., Freshney, M.G., McNair, L., Parker, A.N., McKay, P.J., Steward, W.P., Fitzsimons, E., Graham, G.J. and Pragnell, I.B. (1996). *Ex vivo* expansion with stem cell factor and interleukin-11 augments both short-term recovery posttransplant and the ability to serially transplant marrow. *Blood* **87**, 4589–4595.

Hu, J.P., Cesano, A., Santoli, D., Clark, S.C. and Hoang, T. (1993). Effects of interleukin-11 on the proliferation and cell cycle status of myeloid leukemic cells. *Blood* **81**, 1586–1592.

Huang, S. and Terstappen, L.W.M.M. (1994). Lymphoid and myeloid differentiation of single human $CD34^+$, HLA-DR^+, $CD38^-$ hematopoietic stem cells. *Blood* **83**, 1515–1526.

Hudak, S., Hunte, B., Culpepper, J., Menon, S., Hannum, C., Thompson-Snipes, L. and Rennick, D. (1995). Flt3/flk2 ligand promotes the growth of murine stem cells and the expansion of colony-forming cells and spleen colony-forming units. *Blood* **85**, 2747–2755.

Hughes, F.J. and Howells, G.L. (1993). Interleukin-11 inhibits bone formation *in vitro. Calcif. Tissue Int.* **53**, 362–364.

Ihle, J.N., Witthuhn, B.A., Quelle, F.W., Yamamoto, K. and Silvennoinen, O. (1995). Signaling through the hematopoietic cytokine receptors. *Annu. Rev. Immunol.* **13**, 369–398.

Imai, T. and Nakahata, T. (1994). Stem cell factor promotes proliferation of human primitive megakaryocytic progenitors, but not megakaryocytic maturation. *Int. J. Hematol.* **59**, 91–98.

Ip, N.Y., Nye, S.H., Boulton, T.G., Davis, S., Taga, T., Yanping, L., Birren, S.J., Yasukawa, K., Kishimoto, T., Anderson, D.J., Stahl, N. and Yancopoulos, G.D. (1992). CNTF and LIF act on neuronal cells via shared signaling pathways that involve the IL-6 signal transducing receptor component gp130. *Cell* **69**, 1121–1132.

Jacobsen, F.W., Keller, J.R., Ruscetti, F.W., Veiby, O.P. and Jacobsen, S.E.W. (1995a). Direct synergistic effects of IL4 and IL11 on proliferation of primitive hematopoietic progenitor cells. *Exp. Hematol.* **23**, 990–995.

Jacobsen, F.W., Stokke, T. and Jacobsen, S.E.W. (1995b). Transforming growth factor-β potently inhibits the viability-promoting activity of stem cell factor and other cytokines and induces apoptosis of primitive murine hematopoietic progenitor cells. *Blood* **86**, 2957–2966.

Jacobsen, F.W., Veiby, O.P., Stokke, T. and Jacobsen, S.E.W. (1996a). TNF-α bidirectionally modulates the viability of primitive murine hematopoietic progenitor cells *in vitro. J. Immunol.* **157**, 1193–1199.

Jacobsen, S.E.W., Okkenhaug, C., Myklebust, J., Veiby, O.P. and Lyman, S.D. (1995). The flt3 ligand potently and directly stimulates the growth and expansion of primitive murine bone marrow progenitor cells *in vitro*: synergistic interactions with interleukin (IL) 11, IL-12, and other hematopoietic growth factors. *J. Exp. Med.* **181**, 1357–1363.

Jacobsen, S.E.W., Borge, O.J., Ramsfjell, V., Cui, L., Cardier, J.E., Veiby, O.P., Murphy, M.J. and Lok, S. (1996b). Thrombopoietin, a direct stimulator of viability and multilineage growth of primitive bone marrow progenitor cells. *Stem Cells* **14 (supplement 1)**, 173–180.

Jacobsen, S.E.W., Veiby, O.P., Myklebust, J., Okkenhaug, C. and Lyman, S.D. (1996c). Ability of flt3 ligand to stimulate the *in vitro* growth of primitive murine hematopoietic progenitors is potently and directly inhibited by transforming growth factor-β and tumor necrosis factor-α. *Blood* **87**, 5016–5026.

Jones, R.J., Collector, M.I., Barber, J.P., Vala, M.S., Fackler, M.J., May, S.W., Griffin, C.A., Hawkins, A.L., Zehnbauer, B.A., Hilton, J., Colvin, O.M. and Sharkis, S.J. (1996). Characterization of mouse lymphohematopoietic stem cells lacking spleen colony-forming activity. *Blood* **88**, 487–491.

Karow, J., Hudson, K.R., Hall, M.A., Vernallis, A.B., Taylor, J.A., Gossler, A. and Heath, J.K. (1996). Mediation of interleukin-11-dependent biological responses by a soluble form of the interleukin-11 receptor. *Biochem. J.* **318**, 489–495.

Katayama, N., Clark, S.C. and Ogawa, M. (1993). Growth factor requirement for survival in cell-cycle dormacy of primitive murine lymphohematopoietic progenitors. *Blood* **81**, 610–616.

Kaushansky, K. (1995). Thrombopoietin: the primary regulator of platelet production. *Blood* **86**, 419–430.

Kaushansky, K., Lok, S., Holly, R.D., Broudy, V.C., Lin, N., Bailey, M.C., Forstrom, J.W., Buddle, M.M., Oort, P.J. and Hagen, F.S. (1994). Promotion of megakaryocyte progenitor expansion and differentiation by the c-Mpl ligand thrombopoietin. *Nature* **369**, 568–571.

Kaushansky, K., Broudy, V.C., Lin, N., Jorgensen, M.J., McCarty, J., Fox, N., Zucker-Franklin, D. and Lofton-Day, C. (1995). Thrombopoietin, the Mp1 ligand, is essential for full megakaryocyte development. *Proc. Natl Acad. Sci. U.S.A.* **92**, 3234–3238.

Kawashima, I., Ohsumi, J., Mita-Honjo, K., Shimoda-Takano, K., Ishikawa, H., Sakakibara, S., Miyadai, K. and Takiguchi, Y. (1991). Molecular cloning of cDNA encoding adipogenesis inhibitory factor and identity with interleukin-11. *FEBS Lett.* **283**, 199–202.

Kawashima, I., Zanjani, E.D., Almaida-Porada, G., Flake, A.W., Zeng, H. and Ogawa, M. (1996). $CD34^+$ human marrow cells that express low levels of kit protein are enriched for long-term marrow-engrafting cells. *Blood* **87**, 4136–4142.

Keith, J.C., Jr., Albert, L., Sonis, S.T., Pfeiffer, C.J. and Schaub, R.G. (1994). IL-11, a pleiotropic cytokine: exciting new effects of IL-11 on gastrointestinal mucosal biology. *Stem Cells* **12 (supplement 1)**, 79–89; Discussion 89–90.

Keller, D.C., Du, X.X., Srour, E.F., Hoffman, R. and Williams, D.A. (1993). Interleukin-11 inhibits adipogenesis and stimulates myelopoiesis in human long-term marrow cultures. *Blood* **82**, 1428–1435.

Keller, J.R., Ortiz, M. and Ruscetti, F.W. (1995). Steel factor (c-kit ligand) promotes the survival of hematopoietic stem/progenitor cells in the absence of cell division. *Blood* **86**, 1757–1764.

Keller, J.R., Ruscetti, F.W., Grzegorzewski, K.J., Wiltrout, R.H., Bartelmez, S.H., Sitnicka, E. and Jacobsen, S.E.W. (1997). Transforming growth factor-β, tumor necrosis factor-α, and macrophage inflammatory protein-1α are bidirectional regulators of hematopoietic cell growth. In *Colony Stimulating Factors. Molecular and Cellular Biology*, 2nd edn. (eds J.M. Garland, P.J. Quesenberry and D.J. Hilton), Marcel Dekker, New York, pp. 445–466.

Kishimoto, T., Taga, T. and Akira, S. (1994). Cytokine signal transduction. *Cell* **76**, 253–262.

Kobayashi, S., Teramura, M., Sugawara, I., Oshimi, K. and Mizoguchi, H. (1993). Interleukin-11 acts as an autocrine growth factor for human megakaryoblastic cell lines. *Blood* **81**, 889–893.

Kohn, D.B., Weinberg, K.I., Nolta, J.A., Heiss, L.N., Lenarsky, C., Crooks, G.M., Hanley, M.E., Annett, G., Brooks, J.S., El-Khoureiy, A., Lawrence, K., Wells, S., Moen, R.C., Bastian, J., Williams-Herman, D.E., Elder, M., Wara, D., Bowen, T., Hershfield, M.S., Mullen, C.A., Blaese, R.M. and Parkman, R. (1995). Engraftment of gene-modified umbilical cord blood cells in neonates with adenosine deaminase deficiency. *Nature Med.* **1**, 1017–1023.

Kopf, M., Baumann, H., Freer, G., Freudenberg, M., Lamers, M., Kishimoto, T., Zinkernagel, R., Bluethmann, H. and Kohler, G. (1994). Impaired immune and acute-phase responses in interleukin-6-deficient mice. *Nature* **368**, 339–342.

Ku, H., Yonemura, Y., Kaushansky, K. and Ogawa, M. (1996). Thrombopoietin, the ligand for the mpl receptor, synergizes with Steel factor and other early acting cytokines in supporting proliferation of primitive hematopoietic progenitors in mice. *Blood* **87**, 4544–4551.

Lange, W., Henschler, R. and Mertelsmann, R. (1996). Biological and clinical advances in stem cell expansion. *Leukemia* **10**, 943–945.

Langley, K.E., Bennett, L.G., Wypych, J., Yancik, S.A., Liu, X.D., Westcott, K.R., Chang, D.G., Smith, K.A. and Zsebo, K.M. (1993). Soluble stem cell factor in human serum. *Blood* **81**, 656–660.

Larochelle, A., Vormoor, J., Hanenberg, H., Wang, J.C.Y., Bhatia, M., Lapidot, T., Moritz, T., Murdoch, B., Xiao, X.L., Kato, I., Williams, D.A. and Dick, J.E. (1996). Identification of primitive human hematopoietic cells capable of repopulating NOD/SCID mouse bone marrow: implications for gene therapy. *Nature Med.* **2**, 1329–1337.

Leary, A.G., Zeng, H.Q., Clark, S.C. and Ogawa, M. (1992). Growth factor requirements for survival in G0 and entry into the cell cycle of primitive human hemopoietic progenitors. *Proc. Natl Acad. Sci. U.S.A.* **89**, 4013–4017.

Lemischka, I.R., Raulet, D.H. and Mulligan, R.C. (1986). Developmental potential and dynamic behavior of hematopoietic stem cells. *Cell* **45**, 917–927.

Lemoli, R.M., Fogli, M., Fortuna, A., Motta, M.R., Rizzi, S., Benini, C. and Tura, S. (1993). Interleukin-11 stimulates the proliferation of human hematopoietic $CD34^+$ and $CD34^+CD33^-DR^-$ cells and synergizes with stem cell factor, interleukin-3, and granulocyte–macrophage colony-stimulating factor. *Exp. Hematol.* **21**, 1668–1672.

Lemoli, R.M., Fogli, M., Fortuna, A., Amabile, M., Zucchini, P., Grande, A., Martinelli, G., Visani, G., Ferrari, S. and Tura, S. (1995). Interleukin-11 (IL-11) acts as a synergistic factor for the proliferation of human myeloid leukaemic cells. *Br. J. Haematol.* **91**, 319–326.

Leonard, J.P., Quinto, C.M., Kozitza, M.K., Neben, T.Y. and Goldman, S.J. (1994). Recombinant human interleukin-11 stimulates multilineage hematopoietic recovery in mice after a myelosuppressive regimen of sublethal irradiation and carboplatin. *Blood* **83**, 1499–1506.

Leonard, J.P., Neben, T.Y., Kozitza, M.K., Quinto, C.M. and Goldman, S.J. (1996). Constant subcutaneous infusion of rhIL-11 in mice: efficient delivery enhances biological activity. *Exp. Hematol.* **24**, 270–276.

Li, C.L. and Johnson, G.R. (1994). Stem cell factor enhances the survival but not the self-renewal of murine hematopoietic long-term repopulating cells. *Blood* **84**, 408–414.

Li, C.L. and Johnson, G.R. (1995). Murine hematopoietic stem and progenitor cells: I. Enrichment and biologic characterization. *Blood* **85**, 1472–1479.

Liu, J., Modrell, A., Aruffo, A., Marken, J.S., Taga, T., Murakami, M., Yasukawa, K., Kishimoto, T. and Shoyab, M. (1992). Interleukin-6 signal transducer gp130 mediates oncostatin M signaling. *J. Biol. Chem.* **267**, 16763–16766.

Liu, Q., Du, X.X., Schindel, D.T., Yang, Z.X., Rescorla, F.J., Williams, D.A. and Grosfeld, J.L. (1996). Trophic effects of interleukin-11 in rats with experimental short bowel syndrome. *J. Pediatr. Surg.* **31**, 1047–1050.

Lu, Z.Y., Gu, Z.J., Zhang, X.G., Wijdenes, J., Neddermann, P., Rossi, J.F. and Klein, B. (1995). Interleukin-10 induces interleukin-11 responsiveness in human myeloma cell lines. *FEBS Lett.* **377**, 515–518.

Lyman, S.D. (1995). Biology of flt3 ligand and receptor. *Int. J. Hematol.* **62**, 63–73.

Maier, R., Ganu, V. and Lotz, M. (1993). Interleukin-11, an inducible cytokine in human articular chondrocytes and synoviocytes, stimulates the production of the tissue inhibitor of metalloproteinases. *J. Biol. Chem.* **266**, 21527–21532.

Mauch, P., Lamont, C., Neben, T.Y., Quinto, C., Goldman, S.J. and Witsell, A. (1995). Hematopoietic stem cells in the blood after stem cell factor and interleukin-11 administration: evidence for different mechanisms of mobilization. *Blood* **86**, 4674–4680.

McKinley, D., Wu, Q., Yang-Feng, T. and Yang, Y.-C. (1992). Genomic sequence and chromosomal location of human interleukin (IL)-11 gene. *Genomics* **13**, 814–819.

Metcalf, D. (1993). Hematopoietic regulators: redundancy or subtlety. *Blood* **82**, 3515–3523.

Moore, M.A.S. (1991). Clinical implications of positive and negative hematopoietic stem cell regulators. *Blood* **78**, 1–19.

Morris, J.C., Neben, S., Bennett, F., Finnerty, H., Long, A., Beier, D.R., Kovacic, S., McCoy, J.M., DiBlasio-Smith, E., LaVallie, E.R., Caruso, A., Calvetti, J., Morris, G., Weich, N., Paul, S.R., Crosier, P.S., Turner, K.J. and Wood, C.R. (1996). Molecular cloning and characterization of murine interleukin-11. *Exp. Hematol.* **24**, 1369–1376.

Morrison, S.J., Hemmati, H.D., Wandycz, A.M. and Weissman, I.L. (1995). The purification and characterization of fetal liver hematopoietic stem cells. *Proc. Natl Acad. Sci. U.S.A.* **92**, 10302–10306.

Muller-Sieburg, C.M. and Deryugina, E. (1995). The stromal cells' guide to the stem cell universe. *Stem Cells* **13**, 477–486.

Murakami, M., Hibi, M., Nakagawa, N., Nakagawa, T., Yasukawa, K., Yamanishi, K., Taga, T. and Kishimoto, T. (1993). IL-6-induced homodimerization of gp130 and associated activation of a tyrosine kinase. *Science* **260**, 1808–1810.

Musashi, M., Clark, S.C., Sudo, T., Urdal, D.L. and Ogawa, M. (1991a). Synergistic interactions between interleukin-11 and interleukin-4 in support of proliferation of primitive hematopoietic progenitors of mice. *Blood* **78**, 1448–1451.

Musashi, M., Yang, Y.-C., Paul, S.R., Clark, S.C., Sudo, T. and Ogawa, M. (1991b). Direct and synergistic effects of interleukin 11 on murine hemopoiesis in culture. *Proc. Natl Acad. Sci. U.S.A.* **88**, 765–769.

Nandurkar, H.H., Hilton, D.J., Nathan, P., Willson, T., Nicola, N. and Begley, C.G. (1996). The human IL-11 receptor requires gp130 for signalling: demonstration by molecular cloning of the receptor. *Oncogene* **12**, 585–593.

Neben, T.Y., Loebelenz, J., Hayes, L., McCarthy, K., Stoudemire, J., Schaub, R. and Goldman, S.J. (1993). Recombinant human interleukin-11 stimulates megakaryocytopoiesis and increases peripheral platelets in normal and splenectomized mice. *Blood* **81**, 901–908.

Neddermann, P., Graziani, R., Ciliberto, G. and Paonessa, G. (1996). Functional expression of soluble human interleukin-11 (IL-11) receptor α and stochiometry of *in vitro* IL-11 receptor complexes with gp130. *J. Biol. Chem.* **271**, 30 986–30 991.

Nichol, J.L. (1996). Preclinical biology of megakaryocyte growth and development factor: a summary. *Stem Cells* **14 (supplement 1)**, 48–52.

Novik, Y., Dutcher, J.P. and Oleksowicz, L. (1995). Interleukin-3 and interleukin 11: absence of an effect on *in vitro* agonist-induced platelet aggregation and platelet gmp-140 expression. *Blood* **86**, 701 (abstract).

Ogawa, M. (1993). Differentiation and proliferation of hematopoietic stem cells. *Blood* **81**, 2844–2853.

Ohsumi, J., Miyadai, K., Kawashima, I., Ishikawa-Ohsumi, H., Sakakibara, S., Mita-Honjo, K. and Takiguchi, Y. (1991). Adipogenesis inhibitory factor. A novel inhibitory regulator of adipose conversion in bone marrow. *FEBS Lett.* **288**, 13–16.

Orazi, A., Cooper, R.J., Tong, J., Gordon, M.S., Battiato, L., Sledge, G.W., Jr., Kaye, J.A., Kahsai, M. and Hoffman, R. (1996a). Effects of recombinant human interleukin-11 (Neumega rhIL-11 growth factor) on megakaryocytopoiesis in human bone marrow. *Exp. Hematol.* **24**, 1289–1297.

Orazi, A., Du, X., Yang, Z., Kashai, M. and Williams, D.A. (1996b). Interleukin-11 prevents apoptosis and accelerates recovery of small intestinal mucosa in mice treated with combined chemotherapy and radiation. *Lab. Invest.* **75**, 33–42.

Orlic, D. and Bodine, D.M. (1994). What defines a pluripotent hematopoietic stem cell (PHSC)? Will the real PHSC please stand up! *Blood* **84**, 3991–3994.

Osawa, M., Hanada, K.-I., Hamada, H. and Nakauchi, H. (1996). Long-term lymphohematopoietic reconstitution by a single CD34-low/negative hematopoietic stem cell. *Science* **273**, 242–245.

Paul, S.R., Bennett, F., Calvetti, J.A., Kelleher, K., Wood, C.R., O'Hara, R.M.J., Leary, A.C., Sibley, B., Clark, S.C., Williams, D.A. and Yang, Y.-C. (1990). Molecular cloning of a cDNA encoding interleukin 11, a stromal cell-derived lymphopoietic and hematopoietic cytokine. *Proc. Natl Acad. Sci. U.S.A.* **87**, 7512–7516.

Paul, S.R., Yang, Y.-C., Donahue, R.E., Goldring, S. and Williams, D.A. (1991). Stromal cell-associated hematopoiesis immortalization and characterization of a primate bone marrow-derived stromal cell line. *Blood* **77**, 1723–1733.

Paul, S.R., Barut, B.A., Bennett, F., Cochran, M.A. and Anderson, K.C. (1992). Lack of a role of interleukin 11 in the growth of multiple myeloma. *Leuk. Res.* **16**, 247–252.

Paul, S.R., Hayes, L.L., Palmer, R., Morris, G.E., Neben, T.Y., Loebelenz, J., Pedneault, G., Brooks, J., Blue, I. and Moore, M.A. (1994). Interleukin-11 expression in donor bone marrow cells improves hematological reconstitution in lethally irradiated recipient mice. *Exp. Hematol.* **22**, 295–301.

Peschon, J.J., Morrissey, P.J., Grabstein, K.H., Ramsdell, F.J., Maraskovsky, E., Gliniak, B.C., Park, L.S., Ziegler, S.F., Williams, D.E., Ware, C.B., Meyer, J.D. and Davison, B.L. (1994). Early lymphocyte expansion is severely impaired in interleukin 7 receptor-deficient mice. *J. Exp. Med.* **180**, 1955–1960.

Peters, S.O., Kittler, E.L., Ramshaw, H.S. and Quesenberry, P.J. (1995). Murine marrow cells expanded in culture with IL-3, IL-6, IL-11, and SCF acquire an engraftment defect in normal hosts. *Exp. Hematol.* **23**, 461–469.

Peters, S.O., Kittler, E.L., Ramshaw, H.S. and Quesenberry, P.J. (1996). *Ex vivo* expansion of murine marrow cells with interleukin-3 (IL-3), IL- 6, IL-11, and stem cell factor leads to impaired engraftment in irradiated hosts. *Blood* **87**, 30–37.

Peterson, R.L., Bozza, M.M. and Dorner, A.J. (1996). Interleukin-11 induces intestinal epithelial cell growth arrest through effects on retinoblastoma protein phosphorylation. *Am. J. Pathol.* **149**, 895–902.

Petzer, A.L., Hogge, D.E., Lansdorp, P.M., Reid, D.S. and Eaves, C.J. (1996a). Self-renewal of primitive human hematopoietic cells (long-term-culture-initiating cells) *in vitro* and their expansion in defined medium. *Proc. Natl Acad. Sci. U.S.A.* **93**, 1470–1474.

Petzer, A.L., Zandstra, P.W., Piret, J.M. and Eaves, C.J. (1996b). Differential cytokine effects on primitive ($CD34^{+}CD38^{-}$) human hematopoietic cells: novel responses to flt3-ligand and thrombopoietin. *J. Exp. Med.* **183**, 2551–2558.

Ploemacher, R.E., van Soest, P.L., Boudewijn, A. and Neben, S. (1993). Interleukin-12 enhances interleukin-3-

dependent multilineage hematopoietic colony formation stimulated by interleukin-11 or Steel factor. *Leukemia* **7**, 1374–1380.

Quesniaux, V.F., Clark, S.C., Turner, K. and Fagg, B. (1992). Interleukin-11 stimulates multiple phases of erythropoiesis *in vitro*. *Blood* **80**, 1218–1223.

Ramsfjell, V., Borge, O.J., Veiby, O.P., Cardier, J., Murphy, M.J., Lyman, S.D., Lok, S. and Jacobsen, S.E.W. (1996). Thrombopoietin, but not erythropoietin, directly stimulates multilineage growth of primitive murine bone marrow progenitor cells in synergy with early acting cytokines: distinct interactions with the ligands for c-kit and flt3. *Blood* **88**, 4481–4492.

Ramsfjell, V., Borge, O.J., Li, C. and Jacobsen, S.E.W. (1997). Thrombopoietin directly and potently stimulates multilineage growth and progenitor cell expansion from primitive ($CD34^{+}CD38^{-}$) human bone marrow progenitor cells: effects of combined signaling through c-mpl, c-kit and flt3, and inhibitory effects of transforming growth factor-β and tumor necrosis factor-α. *J. Immunol.* **158**, 5169–5177.

Randall, T.D., Lund, F.E., Howard, M.C. and Weissman, I.L. (1996). Expression of murine CD38 defines a population of long-term reconstituting hematopoietic stem cells. *Blood* **87**, 4057–4067.

Ray, R.J., Paige, C.J., Furlonger, C., Lyman, S.D. and Rottapel, R. (1996). Flt3 ligand supports the differentiation of early B cell progenitors in the presence of interleukin-11 and interleukin-7. *Eur. J. Immunol.* **26**, 1504–1510.

Rebel, V.I., Dragowska, W., Eaves, C.J., Humphries, R.K. and Lansdorp, P.M. (1994). Amplification of $Sca\text{-}1^{+}$ Lin^{-} WGA^{+} cells in serum-free cultures containing Steel factor, interleukin-6, and erythropoietin with maintenance of cells with long-term *in vivo* reconstituting potential. *Blood* **83**, 128–136.

Robb, L., Hilton, D.J., Willson, T.A. and Begley, C.G. (1996). Structural analysis of the gene encoding the murine interleukin-11 receptor α-chain and a related locus. *J. Biol. Chem.* **271**, 13754–13761.

Robb, L., Hilton, D.J., Brook-Carter, P.T. and Begley, C.G. (1997). Identification of a second murine interleukin-11 receptor alpha-chain gene (IL11Rα2) with a restricted pattern of expression. *Genomics* **40**, 387–394.

Rolink, A., Ghia, P., Grawunder, U., Haasner, D., Karasuyama, H., Kalberer, C., Winkler, T. and Melchers, F. (1995). *In-vitro* analyses of mechanisms of B-cell development. *Semin. Immunol.* **7**, 155–167.

Romas, E., Udagawa, N., Zhou, H., Tamura, T., Saito, M., Taga, T., Hilton, D.J., Suda, T., Ng, K.W. and Martin, T.J. (1996). The role of gp130-mediated signals in osteoclast development: regulation of interleukin 11 production by osteoblasts and distribution of its receptor in bone marrow cultures. *J. Exp. Med.* **183**, 2581–2591.

Rose-John, S. and Heinrich, P.C. (1994). Soluble receptors for cytokines and growth factors: generation and biological function. *Biochem. J.* **300**, 281–290.

Rusten, L.S., Lyman, S.D., Veiby, O.P. and Jacobsen, S.E.W. (1996). The flt3 ligand is a direct and potent stimulator of the growth of primitive and committed human $CD34^{+}$ bone marrow progenitor cells *in vitro*. *Blood* **87**, 1317–1325.

Schibler, K.R., Yang, Y.C. and Christensen, R.D. (1992). Effect of interleukin-11 on cycling status and clonogenic maturation of fetal and adult hematopoietic progenitors. *Blood* **80**, 900–903.

Schlerman, F.J., Bree, A.G., Kaviani, M.D., Nagle, S.L., Donnelly, L.H., Mason, L.E., Schaub, R.G., Grupp, S.A. and Goldman, S.J. (1996). Thrombopoietic activity of recombinant human interleukin 11 (rHuIL-11) in normal and myelosuppressed nonhuman primates. *Stem Cells* **14**, 517–532.

Spangrude, G.J., Heimfeld, S. and Weissman, I.L. (1988). Purification and characterization of mouse hematopoietic stem cells. *Science* **241**, 58–62.

Spangrude, G.J., Smith, L., Uchida, N., Ikuta, K., Heimfeld, S., Friedman, J. and Weissman, I.L. (1991). Mouse hematopoietic stem cells. *Blood* **78**, 1395–1402.

Spooncer, E., Eliason, J. and Dexter, T.M. (1993). Long-term mouse bone marrow cultures. In *Haemapoiesis. A Practical Approach* (eds N.G. Testa and G. Molineux), IRL Press, Oxford, pp. 55–106.

Suen, Y., Chang, M., Lee, S.M., Buzby, J.S. and Cairo, M.S. (1994). Regulation of IL-11 protein and mRNA expression in neonatal and adult fibroblasts and endothelial cells. *Blood* **84**, 4125–4134.

Sui, X., Tsuji, K., Tanaka, R., Tajima, S., Muraoka, K., Ebihara, Y., Ikebuchi, K., Yasukawa, K., Taga, T. and Kishimoto, T. (1995). gp130 and c-Kit signalings synergize for *ex vivo* expansion of human primitive hemopoietic progenitor cells. *Proc. Natl Acad. Sci. U.S.A.* **92**, 2859–2863.

Taga, T. and Kishimoto, T. (1997). gp130 and the interleukin-6 family of cytokines. *Annu. Rev. Immunol.* **15**, 797–819.

Taga, T., Hibi, M., Hirata, Y., Yamasaki, K., Yasukawa, K., Matsuda, T., Hirano, T. and Kishimoto, T. (1989). Interleukin-6 triggers the association of its receptor with a possible signal transducer, gp130. *Cell* **58**, 573–581.

Tang, W., Geba, G.P., Zheng, T., Ray, P., Homer, R.J., Kuhn, C., III, Flavell, R.A. and Elias, J.A. (1996). Targeted expression of IL-11 in the murine airway causes lymphocytic inflammation, bronchial remodeling, and airway obstruction. *J. Clin. Invest.* **98**, 2845–2853.

Teramura, M., Kobayashi, S., Hoshino, S., Oshimi, K. and Mizoguchi, H. (1992). Interleukin-11 enhances human megakaryocytopoiesis *in vitro*. *Blood* **79**, 327–331.

Till, J.E. and McCulloch, E.A. (1961). A direct measurement of the radiation sensitivity of normal mouse bone marrow cells. *Radiat. Res.* **14**, 213.

Trepicchio, W.L., Bozza, M., Pedneault, G. and Dorner, A.J. (1996). Recombinant human IL-11 attenuates the inflammatory response through down-regulation of proinflammatory cytokine release and nitric oxide production. *J. Immunol.* **157**, 3627–3634.

Trevisan, M. and Iscove, N.N. (1995). Phenotypic analysis of murine long-term hemopoietic reconstituting cells quantified competitively *in vivo* and comparison with more advanced colony-forming progeny. *J. Exp. Med.* **181**, 93–103.

Tsuji, K., Zsebo, K.M. and Ogawa, M. (1991). Enhancement of murine blast cell colony formation in culture by recombinant rat stem cell factor, ligand for c-kit. *Blood* **78**, 1223–1229.

Turner, K.J., Neben, S., Weich, N., Schaub, R.G. and Goldman, S.J. (1996). The role of recombinant interleukin 11 in megakaryocytopoiesis. *Stem Cells* **14 (supplement 1)**, 53–61.

Uchida, N., Fleming, W.H., Alpern, E.J. and Weissman, I.L. (1993). Heterogeneity of hematopoietic stem cells. *Curr. Opin. Immunol.* **5**, 177–184.

van de Ven, C., Ishizawa, L., Law, P. and Cairo, M.S. (1995). IL-11 in combination with SLF and G-CSF or GM-CSF significantly increases expansion of isolated $CD34^+$ cell population from cord blood *vs.* adult bone marrow. *Exp. Hematol.* **23**, 1289–1295.

Van Leuven, F., Stas, L., Hilliker, C., Miyake, Y., Bilinski, P. and Gossler, A. (1996). Molecular cloning and characterization of the human interleukin-11 receptor alpha-chain gene, IL11RA, located on chromosome 9p13. *Genomics* **31**, 65–70.

Veiby, O.P., Lyman, S.D. and Jacobsen, S.E.W. (1996). Combined signaling through interleukin-7 receptors and flt3 but not c-kit potently and selectively promotes B cell commitment and differentiation from uncommitted murine bone marrow progenitor cells. *Blood* **88**, 1256–1265.

Visser, J.W.M., Bauman, J.G.J., Mulder, A.H., Eliason, J.F. and deLeeuw, A.M. (1984). Isolation of murine pluripotent hemopoietic stem cells. *J. Exp. Med.* **59**, 1576–1590.

von Freeden-Jeffrey, U., Vieira, P., Lucian, L.A., McNeil, T., Burdach, S.E.G. and Murray, R. (1995). Lymphopenia in interleukin (IL)-7 gene-deleted mice identifies IL-7 as a nonredundant cytokine. *J. Exp. Med.* **181**, 1519–1526.

Ware, C.B., Horowitz, M.C., Renshaw, B.R., Hunt, J.S., Liggitt, D., Koblar, S.A., Gliniak, B.C., McKenna, H.J., Papayannopoulou, T. and Thoma, B. (1995). Targeted disruption of the low-affinity leukemia inhibitory factor receptor gene causes placental, skeletal, neural and metabolic defects and results in perinatal death. *Development* **121**, 1283–1299.

Watt, S.M. and Visser, J.W.M. (1992). Recent advances in the growth and isolation of primitive human haemopoietic progenitor cells. *Cell Prolif.* **25**, 263–297.

Williams, D.A. (1993). *Ex vivo* expansion of hematopoietic stem and progenitor cells—robbing Peter to pay Paul? *Blood* **81**, 3169–3172 (editorial).

Wolf, N.S., Kone, A., Priestly, G.V. and Bartelmez, S.H. (1993). *In vivo* and *in vitro* characterization of long-term repopulating primitive hematopoietic cells isolated by sequential Hoechst 33342-rhodamine 123 FACS selection. *Exp. Hematol.* **21**, 614–622.

Yang, L. and Yang, Y.-C. (1994). Regulation of interleukin (IL)-11 gene expression in IL-1-induced primate bone marrow stromal cells. *J. Biol. Chem.* **269**, 32732–32739.

Yang, L. and Yang, Y.-C. (1995). Heparin inhibits the expression of interleukin-11 and granulocyte–macrophage colony-stimulating factor in primate bone marrow stromal fibroblasts through mRNA destabilization. *Blood* **86**, 2526–2533.

Yang, L., Steussy, C.N., Fuhrer, D.K., Hamilton, J. and Yang, Y.-C. (1996). Interleukin-11 mRNA stabilization in phorbol ester-stimulated primate bone marrow stromal cells. *Mol. Cell. Biol.* **16**, 3300–3307.

Yang, Y.-C. (1995). Interleukin-11 (IL-11) and its receptor: biology and potential clinical applications in thrombocytopenic states. In *Cytokines: Interleukins and their Receptors* (ed. R. Kurzrock), Kluwer Academic Publishers, pp. 321–340.

Yang, Y.-C. and Yin, T. (1995). Interleukin (IL)-11-mediated signal transduction. *Ann. N.Y. Acad. Sci.* **762**, 31–41.

Yin, T., Schendel, P. and Yang, Y.-C. (1992). Enhancement of *in vitro* and *in vivo* antigen-specific antibody response by interleukin 11. *J. Exp. Med.* **175**, 211–216.

Yin, T., Taga, T., Tsang, M.L.-S., Yasukawa, K., Kishimoto, T. and Yang, Y.-C. (1993). Involvement of IL-6 signal transducer gp130 in IL-11-mediated signal transduction. *J. Immunol.* **151**, 2555–2561.

Yonemura, Y., Kawakita, M., Masuda, T., Fujimoto, K., Kato, K. and Takatsuki, K. (1992). Synergistic effects of interleukin 3 and interleukin 11 on murine megakaryopoiesis in serum-free culture. *Exp. Hematol.* **20**, 1011–1016.

Yonemura, Y., Ku, H., Hirayama, F., Souza, L.M. and Ogawa, M. (1996). Interleukin 3 or interleukin 1 abrogates the reconstituting ability of hematopoietic stem cells. *Proc. Natl Acad. Sci. U.S.A.* **93**, 4040–4044.

Yoshida, K., Taga, T., Saito, M., Suematsu, S., Kumanogoh, A., Tanaka, T., Fujiwara, H., Hirata, M., Yamagami, T., Nakahata, T., Hirabayashi, T., Yoneda, Y., Tanaka, K., Wang, W.Z., Mori, C., Shiota, K., Yoshida, N. and Kishimoto, T. (1996). Targeted disruption of gp130, a common signal transducer for the interleukin 6 family of cytokines, leads to myocardial and hematological disorders. *Proc. Natl Acad. Sci. U.S.A.* **93**, 407–411.

Zhang, X.-G., Gu, J.-J., Lu, Z.-Y., Yasukawa, K., Yancopoulos, G.D., Turner, K., Shoyab, M., Taga, T., Kishimoto, T., Bataille, R. and Klein, B. (1994). Ciliary neurotrophic factor, interleukin 11, leukemia inhibitory factor, and oncostatin M are growth factors for human myeloma cell lines using the interleukin 6 signal transducer gp130. *J. Exp. Med.* **179**, 1337–1342.

Zheng, T., Nathanson, M.H. and Elias, J.A. (1994). Histamine augments cytokine-stimulated IL-11 production by human lung fibroblasts. *J. Immunol.* **153**, 4742–4752.

Chapter 14

Interleukin-12

Walter J. Storkus, Hideaki Tahara and Michael T. Lotze
Departments of Surgery and Molecular Genetics and Biochemistry,
University of Pittsburgh School of Medicine and University of Pittsburgh Cancer Institute,
Pittsburgh, PA, USA

INTRODUCTION

Since its original identification in 1986, interleukin-12 (IL-12) has been the focus of significant scrutiny owing to its pleiotropic immunomodulatory effects (Gately *et al.*, 1986). In particular, the ability of IL-12 to enhance T helper 1 (T_{H1})-type immunity has prompted the use of this cytokine to treat infectious disease and neoplasia. This chapter focuses on the multifaceted roles of IL-12 in promoting both the innate and adaptive arms of the immune response *in vitro* and *in vivo*. The recent results of preclinical and clinical trials and future applications of this important biologic agent will also be discussed.

BIOLOGY, BIOCHEMISTRY AND GENETICS OF IL-12

IL-12 was originally identified based on its potent capacity to synergize with IL-2 in potentiating the secretion of interferon-γ (IFN-γ) by, and the cytolytic activity of, both natural killer (NK) cells and cytotoxic T lymphocytes (CTLs) (Gately *et al.*, 1986; Germann *et al.*, 1987; Wong *et al.*, 1987; Kobayashi *et al.*, 1989; Chan *et al.*, 1991). The active component associated with this biologic activity found in crude Epstein–Barr virus (EBV)-transformed B-cell culture supernatants or IL-2-depleted phytohaemagglutinin (PHA) T-blast supernatants was initially termed NK stimulatory factor (NKSF) (Kobayashi *et al.*, 1989), cytotoxic lymphocyte maturation factor (CLMF) (Wong *et al.*, 1987) or T-cell stimulatory factor (TSF) (Germann *et al.*, 1987). The ability of cultured human EBV-B cell lines, such as NC-37 or RPMI-8866, to secrete IL-12 when appropriately stimulated with mitogens such as phorbol diester provided an excellent source of a secreted 70-kDa (35 kDa + 40 kDa heterodimer) protein exhibiting the biologic characteristics of NKSF or CLMF (Kobayashi *et al.*, 1989). Furthermore, these producer cell lines yielded a ready source of p70 mRNA, ultimately used to expression clone the p35/p40 gene(s). This factor is now known as interleukin-12 (IL-12) based on its pleiotropic effects on multiple lymphoid subsets (Gubler *et al.*, 1991; Wolf *et al.*, 1991; Podlaski *et al.*, 1992). The IL-12 p35 chain bears significant homology to IL-6

The Cytokine Handbook, 3rd ed.
ISBN 0-12-689662-3

and granulocyte colony stimulating factor G-CSF (Merberg *et al.*, 1992), whereas the IL-12 p40 subunit resembles the IL-6 receptor (IL-6R) and the G-CSFR (Gearing and Cosman, 1991). Thus, structurally, the IL-12 p70 heterodimer is homologous to a soluble cytokine–cytokine receptor complex. Murine IL-12 p35 and p40 subunits have similarly been expression cloned (Schoenhaut *et al.*, 1992; Chua *et al.*, 1994; Yoshimoto *et al.*, 1996).

As noted above, unlike most previously defined cytokines, biologically active IL-12 was determined to be a heterodimeric protein, composed of a heavy chain (p40) and a covalently associated light chain (p35) (Stern *et al.*, 1990). mIL-12 and hIL-12 will both potentiate activated human T-cell proliferation, whereas mIL-12 is effective only in promoting murine T-cell expansion, which appears to reflect the species specificity of the IL-12 mp35 chain (Schoenhaut *et al.*, 1992). The IL-12 p35 and p40 chains are expressed independently and the genes encoding these subunits are localized to different chromosomes (Table 1). Most cells analyzed were determined to express p35, either constitutively or after appropriate stimulation (Gubler *et al.*, 1991). In marked contrast, the IL-12 p40 gene was expressed in a far more restricted manner (Gately *et al.*, 1994a; Wolf *et al.*, 1994; Zeh *et al.*, 1994), with IL-12 p40 expressed primarily in antigen-presenting cells (APCs) (Table 2) (Trinchieri, 1995). Biologically active IL-12 can be produced only in cell types that coexpress both the IL-12 p35 and IL-12 p40 chains (Schoenhaut *et al.*, 1992). Of significant note, IL-12 producers typically express far more IL-12 p40 chain than IL-12 p35 chain (Gately *et al.*, 1993; Gately and Mulqueca, 1996), resulting in the formation and secretion of IL-12 p40 homodimers (Wolf *et al.*, 1994; Heinzel *et al.*, 1997). The mIL-12 p40 homodimer binds to the IL-12R, but fails to mediate a signal, thereby serving as a functional antagonist to IL-12 p70 heterodimers (Mattner *et al.*, 1993; Ling *et al.*, 1995). This is less prominent or apparent with the hp40 homodimer. IL-12 p40 homodimers inhibit IL-12-induced murine concanavalin A blast proliferation, splenocyte secretion of IFN-γ and NK cell activation (Gillessen *et al.*, 1995); mIL-12 p40 homodimers block mIL-12, but not hIL-12, from binding their respective isologous high-affinity IL-12 receptors.

While originally isolated from EBV-transformed B-cell lines, normal B cells do not secrete the functional IL-12 heterodimer (Guery *et al.*, 1997). Macrophages and dendritic cells appear instead to be the principal producers of IL-12. Macatonia *et al.* (1993) demonstrated that monocytes treated with heat-inactivated *Listeria monocytogenes* produce IL-12 and direct the promotion of cellular immunity of the T helper (T_{H1}) type *in vitro* and *in vivo*. Similarly murine macrophages expressing the CD40 costimu-

Table 1. Chromosomal location and degree of protein homology between hIL-12 and mIL-12 subunits.

Species	Chain	Protein homology to human (%)	Chromosomal location	Reference
Human	p40	100	5q31–q33	Sieburth *et al.* (1992), Ma *et al.* (1996a,b)
	p35	100	3p12–p13.2	Sieburth *et al.* (1992), Ma *et al.* (1996a,b)
Murine	p40	79	11A5–B2	Noben-Trauth *et al.* (1996), Tone *et al.* (1996), Yoshimoto *et al.* (1996)
	p35	60	3	Schweitzer *et al.* (1996), Tone *et al.* (1996)

Table 2. Cells that produce and secrete IL-12.

IL-12 producer	Stimulus	Reference
Dendritic cells	Anti-CD40, T cells plus antigen, particulates, bacteria, bacterial products	Macatonia *et al.* (1995), Murphy *et al.* (1995), O'Garra *et al.* (1995), Scheicher *et al.* (1995), Cella *et al.* (1996), Heufler *et al.* (1996), Kato *et al.* (1996b), Koch *et al.* (1996)
Macrophages	Anti-CD40, LPS, IFN-γ, poly I:C, HIV gp120, bacteria, bacterial products	D'Andrea *et al.* (1992), Hsieh *et al.* (1993), Manetti *et al.* (1995), Fantuzzi *et al.* (1996), Kennedy *et al.* (1996), Ma *et al.* (1996a), Snidjers *et al.* (1996)
Langerhans	*In vitro* culture	Kang *et al.* (1996), Yawalkar *et al.* (1996a)
EBV-transformed B cells	LPS plus IFN-γ, phorbol esters, SAC	Ma *et al.* (1996b), Trinchieri (1995, 1996)
Neutrophils	LPS	Trinchieri (1995, 1996)
Keratinocytes	Ultraviolet B radiation, phorbol esters, *in vitro* culture	Trinchieri (1995, 1996), Enk *et al.* (1996), Yawalkar *et al.* (1996b)
Microglia	LPS	Becher *et al.* (1996), Constantinescu *et al.* (1996)
Astrocytes	LPS	Constantinescu *et al.* (1996)

LPS, bacterial lipopolysaccharide; SAC, *Staphylococcus aureus* Cowan A strain; poly I:C, polyinosidic acid:polycytidylic acid; HIV, human immunodeficiency virus.

latory molecule and interacting with anti-CD3 activated CD40L$^+$ splenic T cells or CD40L$^+$ T-cell clones are induced to produce and secrete IL-12 (Kennedy *et al.*, 1996). The CD40–CD40L interaction dependence of IL-12 production by APCs was verified by the ability of anti-CD40L antibodies to block IL-12 production and by the inability of activated T cells derived from CD40L knockout mice to elicit IL-12 secretion from splenic monocytes and to mount T_{H1}-type immunity (Koch *et al.*, 1996; Stuber *et al.*, 1996; see also below). In addition, soluble trimeric CD40L promotes the secretion of IL-12 from macrophages, albeit at levels lower than those observed in macrophage-activated T-cell cultures (Kennedy *et al.*, 1996).

Additional stimuli that enhance APC production include microparticulate ingestion (Scheicher *et al.*, 1995), IFN-γ plus lipopolysaccharide (LPS) (Hayes *et al.*, 1995), *Mycobacterium tuberculosis* (Cooper *et al.*, 1995), *Listeria* (Hsieh *et al.*, 1993), *Neisseria meningitidis*-derived LPS (van der Pouw Kraan *et al.*, 1995), polyinosidic acid (Manetti *et al.*, 1995), phagocytosis of *Borrelia burgdorferi* (Filgueira *et al.*, 1996), lipoteichoic acid preparations of Gram-positive bacteria (Cleveland *et al.*, 1996), recombinant human immunodeficiency virus (HIV) gp120 protein (Fantuzzi *et al.*, 1996) and nitric oxide generating compounds (Rothe *et al.*, 1996). CpG motifs present in hypomethylated oligodeoxynucleotides or bacterial DNA also elicit IL-12 production from both human and murine responder APCs (Ballas *et al.*, 1996; Klinman *et al.*, 1996). The production of IL-12 by activated APCs appears to represent a critical early physiologic event in cellular immunity, as dendritic cells appear to represent the principal producers of IL-12 *in situ* in draining lymphoid organs (Muller *et al.*, 1995).

The regulatory regions of the IL-12 p40 promoter include multiple nuclear factor

(NF)-κB sites, IFN-γ response elements (IREs), as well as ETS (ets-2) binding sites (Murphy *et al.*, 1995; Ma *et al.*, 1996b, 1997; Tone *et al.*, 1996; Yoshimoto *et al.*, 1996). The latter appears to be critically important for IFN-γ (with LPS)-induced APC production of IL-12 *in situ* and is directly related to the restricted range of cell types capable of producing IL-12 (Ma *et al.*, 1996b, 1997). Overexpression of ets-2 in T cells drives IL-12 production in this atypical producer cell type (Ma *et al.*, 1997).

There are diverse stimuli that suppress the synthesis and secretion of IL-12. Macrophage production of IL-12 is inhibited by ultraviolet B irradiation (Kremer *et al.*, 1996). Soluble tumor necrosis factor receptor (TNFR)–immunoglobulin inhibits IL-12 production by LPS-stimulated human microglial cells *in vitro* (Becher *et al.*, 1996). IL-4 and IL-10 suppress dendritic cell production of IL-12 when stimulated with anti-CD40 and/or anti-major histocompatibility complex (MHC) class II antibodies (Koch *et al.*, 1996), principally by inhibiting the steady-state levels of IL-12 p40 mRNA (Hino and Nariuchi, 1996). Neuropeptides secreted by peripheral nerves, such as calcitonin gene-related peptide (CGRP), inhibit APC expression of costimulatory molecules (B7.1), IL-12 production and APC function in an IL-10-dependent manner (Fox *et al.*, 1997). The chemokine monocyte chemoattractant protein-1 (MCP-1), the cytokines IL-13, transforming growth factor-β (TGF-β), IFN-α, prostaglandin E_2 and CD46 ligation also inhibit macrophage production of IL-12 (Trinchieri, 1995; Van der Pouw Kraan *et al.*, 1995; Chensue *et al.*, 1996; Cousens *et al.*, 1997).

IL-12 RECEPTOR AND SIGNALING

Based on radiolabeled Scatchard binding assays implementing ^{125}I-labeled IL-12, IL-12 receptors were initially determined to be expressed only on the cell surfaces of NK cells and activated (IL-2, PHA) T lymphocytes (Chizzonite *et al.*, 1992). These results were consistent with the known ability of IL-12 to promote the secretion of IFN-γ from these two cell types. IFN-γ, in turn, leads to enhanced APC function by increasing both the expression of MHC gene products and antigen-processing functions that are requisite for the promotion of primary T-cell immune responsiveness (Robertson, 1991). T lymphocytes stimulated for 4 days with PHA were determined to express approximately 1000–10 000 binding sites per cell with three classes of binding affinities consistent with low, intermediate and high affinity IL-12 receptors displaying K_ds in the range of 2–6 nM, 50–200 pM and 5–20 pM, respectively (Chizzonite *et al.*, 1992). Recombinant IL-12 (rIL-12) binding to its receptor was inhibited only by IL-12 itself, but not by IL-1β, IL-1RA, IL-2, IFN-α_{2A} or IFN-γ (Chizzonite *et al.*, 1992). The low-affinity IL-12 receptor was characterized as a multichain complex with an approximate molecular weight (M_r) of 135–210 kDa. This receptor complex was expressed optimally 2–4 days after PHA stimulation of T cells or 7–8 days after IL-2 activation of T cells or NK cells.

Based on expression cloning studies performed by Presky *et al.* (1996a) and Wu *et al.* (1996), we now know that the IL-12 receptor is composed of a dimeric complex ($\beta_1\beta_2$) in which the β_1 and β_2 chains (designated as β chains because of their homology to the gp130 molecule, serving as a common β chain for a number of alternate cytokines (Gubler and Presky, 1996)), each confer low–intermediate binding affinity of the complex for IL-12 (Table 3). Using DNA cross-hybridization techniques, the murine IL-12Rβ chains have been isolated and cloned (Chua *et al.*, 1995). In the murine system, the IL-12Rβ_1 chain appears to play the dominant role in binding the IL-12 ligand,

Table 3. Characteristics of IL-12 receptor subunits.

IL-12 receptor subunit	K_d for IL-12	Distribution	Reference
β_1	2–5 nM	T (T_{H1}, T_{H2}, $CD8^+$), NK, B	Wu *et al.* (1996) Rogge *et al.* (1997)
β_2	3–5 nM	T_{H1} $CD4^+$ T cells	Rogge *et al.* (1997)
β_1 plus β_2 (COS)	50 pM	T_{H1} $CD4^+$ T cells	Presky *et al.* (1996b)

whereas in human cells expressing the high-affinity IL-12R, both the β_1 and β_2 chains appear to play significant roles in binding hIL-12 (Gubler and Presky, 1996; Presky *et al.*, 1996b). The β_1 chain appears to be occupied principally with binding to the IL-12p40 chain while the IL-12Rβ_2 chain appears to associate with the IL-12 p35 chain (Gillessen *et al.*, 1995; Presky *et al.*, 1996b). This explains the capacity of the IL-12 p40 homodimer to antagonize effectively the binding of the IL-12 heterodimer, largely owing to its capacity to bind to the IL-12β_1 binding site of the multimeric IL-12 receptor complex in mice. Of note, antibodies directed against the IL-12Rβ chains efficiently inhibit IL-12-induced IFN-γ production and activated T-cell proliferation, but have no effect on the ability of other T-cell growth factors (i.e. IL-2, IL-4 or IL-7) to promote T-cell expansion (Wu *et al.*, 1996).

Both the IL-12Rβ_1 and β_2 chains are inducible on naive ($CD45RA^+$) and memory $CD4^+$ T-cell clones within 48 h of activation with antigen, anti-CD3 or PHA treatment (Gollob *et al.*, 1997; Rogge *et al.*, 1997; Wu *et al.*, 1997). This TCR-induced upregulation is further enhanced by the crosslinking of T cell or NK cell-expressed CD28 molecules (Wu *et al.*, 1997). IL-2, IL-7, IL-12 and IL-15 augment the expression level of the IL-12Rβ_1 chain (Naume *et al.*, 1992; Mehrotra *et al.*, 1995; Wu *et al.*, 1997). Treatment of T cells with IL-10 and TGF-β completely blocks anti-CD3 induction of T cell-expressed high-affinity IL-12R by suppressing the expression of IL-12Rβ_2, but not of IL-12Rβ_1 (Pardoux *et al.*, 1997; Wu *et al.*, 1997). T_{H2} clones could be induced transiently to express low levels of IL-12Rβ_2 chains and to secrete low levels of IFN-γ. The transience of IL-12Rβ_2 expression by T_{H2} $CD4^+$ T cells may be due to IL-4 production by T_{H2} clones, which may enhance the extinction of IL-12R expression and IL-12 responsiveness (Rogge *et al.*, 1997). Szabo *et al.* (1997) have also observed that IL-4 and IFN-γ modify expression of the high-affinity IL-12R after T-cell activation. IL-4 inhibits IL-12Rβ_2 expression and, concomitantly, IL-12-mediated signaling. T_{H2} $CD4^+$ T-cell clones (low-affinity IL-12R$^+$) generated from the peripheral blood of allergen-specific atopic individuals are not responsive to IL-12, producing little or no IFN-γ (Hilkens *et al.*, 1996). IFN-γ promoted sustained IL-12Rβ_2 expression by developing T_{H2} cells, but did not promote IFN-γ production by these cells as a result of IL-12 treatment (Szabo *et al.*, 1997). Cumulative data support the ability of factors promoting the expression or maintenance of expression of IL-12Rβ_2 to push the balance of a mixed T_{H1}–T_{H2} immune response towards a T_{H1}-biased response (see below). Inhibitors of IL-12Rβ_2 expression preclude or minimize T_{H1}-associated immunity from occurring. Furthermore, if IL-4 is present during T-cell activation, IFN-γ production may be required for IL-12Rβ_2 to be expressed by $CD4^+$ T cells and for T_{H1}-associated immune reactivity to develop (Hsieh *et al.*, 1993; Macatonia *et al.*, 1993; Szabo *et al.*, 1997).

Interestingly, the IL-12Rβ_1 chain is readily documented on the surface of non-T cell or NK cell types such as resting B lymphocytes (Vogel *et al.*, 1996); however, such complexes do not appear capable of mediating signals when IL-12 is bound (Metzger *et al.*, 1996; Wu *et al.*, 1996). Recent evidence suggests that the IL-12Rβ_2, but not the IL-12β_1, chain mediates signaling (Chua *et al.*, 1996; Presky *et al.*, 1996a,b). This difference in function appears to be due, in part, to the expression of tyrosine residues in the cytoplasmic domain of the β_2, but not β_1, subunits, consistent with signaling potential. Cotransfection of the IL-12Rβ_1 and IL-12Rβ_2 chains into Ba/F3 cells yields transfectants with high-affinity IL-12 receptor sites. IL-12 promotes enhanced proliferation of these cells with an ED_{50} of 1 pM IL-12, making this a sensitive *in vitro* assay system for p70 IL-12 (Presky *et al.*, 1996a,b).

The IL-12R appears to use intracellular signaling apparatus that readily distinguishes IL-12 from most alternate cytokines (Fig. 1). Specifically, functional signaling via the IL-12R results in the phosphorylation of p56^{lck}, the Janus kinases TYK2 and JAK2, and the signal transducer and activator of transcription (STAT) kinases 3 and 4 (Bacon *et al.*, 1995a,b; Pignata *et al.*, 1995; Klein *et al.*, 1996; Yu *et al.*, 1996a,b). STAT4 is similarly used in IFN-αR-mediated signaling (Cho *et al.*, 1996). Fusion of the transmembrane and intracellular domains of the IL-12Rβ_1 and IL-12Rβ_2 subunits to the extracellular domain of the epidermal growth factor (EGF) receptor results in the production of

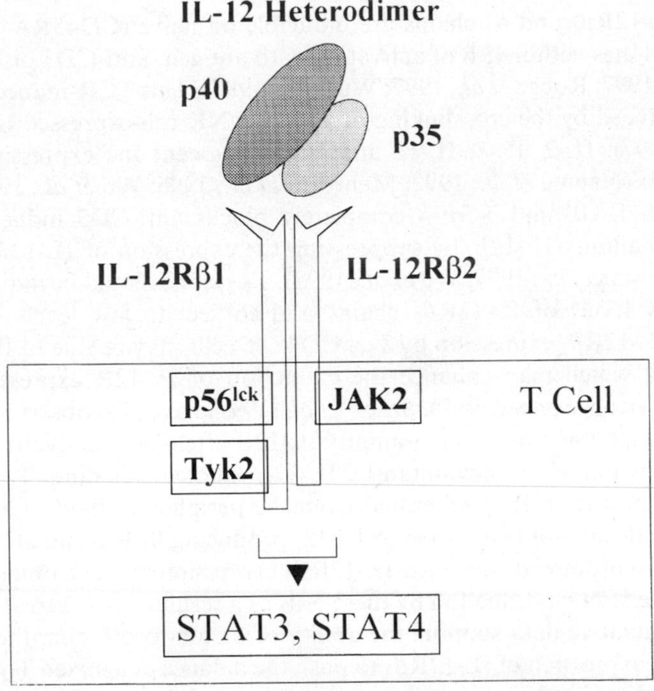

Fig. 1. Schematic of the high-affinity IL-12R. The IL-12 p35–p40 heterodimer binds with high affinity (approximately 50 pM) to the high-affinity IL-12 receptor composed of the IL-12Rβ_1 and IL-12Rβ_2 subunits. IL-12 p40 binds preferentially to IL-12Rβ_1, whereas IL-12 p35 appears to bind IL-12Rβ_2. Intracellular signaling, mediated principally through the IL-12Rβ_2 chain, results in the phosphorylation of the kinases p56^{lck}, TYK2, JAK2, and STATs 3 and 4.

chimeric molecules that allow for the evaluation of the individual and combined signaling capacity mediated by each of the IL-12R subunits (Zou *et al.*, 1997). Only cells transfected with the fusion protein containing the IL-12Rβ_2 subunit intracellular domain were able to mediate signals upon EGF stimulation resulting in proliferation and JAK2 kinase phosphorylation. Interestingly, although unable to promote proliferation by itself, cells transfected with IL-12Rβ_1 alone or with IL-12Rβ_2, but not IL-12Rβ_2 alone, demonstrated association with TYK2. Overall, activation through the chimeric receptor resulted in the phosphorylation of JAK2, TYK2 and STAT3 (in STAT4-deficient BaF3 cells). Addition of IL-12 to T_{H1} $CD4^+$ T cells leads to the rapid phosphorylation of STAT4 kinase (Rogge *et al.*, 1997). T_{H2} $CD4^+$ T-cell clones do not phosphorylate STAT4 in response to IL-12, despite expressing the low-affinity IL-12R (Hilkens *et al.*, 1996). Studies performed in STAT4 knockout mice support a similar biology to that observed in IL-12 p40 knockout animals. T_{H2} immune responses (and IL-4, IL-5 and IL-10 production) predominate in the absence of functional STAT4 kinase activity (Kaplan *et al.*, 1996). In particular, although T_{H1}-associated immune responses do occur, they are generally far weaker than those observed in heterozygous littermates or in wild-type control animals in response to IL-12 or to pathogenic organisms such as *L. monocytogenes* (Kaplan *et al.*, 1996; Thierfelder *et al.*, 1996). IL-12 activation of NK cells does not occur in STAT4 knockout mice (Thierfelder *et al.*, 1996). IFN-γ produced as a consequence of IL-12 administration is minimized and long-term antigen-specific cellular immunity is lessened in these knockout animals.

Interestingly, other T cell surface-expressed molecules may impact the signaling capacity of IL-12. Gollob and Ritz (1996) developed murine antibodies against activated human T cells that blocked IL-12-induced T-cell proliferation. Two antibodies recognizing CD2 (leukocyte functional antigen-2 (LFA-2)) or CD58 (LFA-3) were identified with this ability. Neither antibody blocked the binding of IL-12 to its receptor expressed by T cells, supporting an IL-12R-independent mechanism.

IMMUNOBIOLOGY OF IL-12

IL-12 (alone or in combination with low-dose IL-2) promotes the *in vitro* expansion and survival of preactivated (mitogen, phorbol ester, anti-CD3) T cells, $CD4^+$ and $CD8^+$ T-cell clones, tumor infiltrating lymphocyte (TIL) T lymphocytes, and NK–lymphokine-activated killer (LAK) cells (Trinchieri, 1995). IL-12 enhances the cytolytic reactivity in, and low level proliferation of, both $CD56^{dim^-}$ and $CD56^{bri^+}$ NK–LAK cell subsets and promotes the efficient cloning of NK cells *in vitro*. Based on these characteristics, IL-12 appears well suited to promote the T_{H1}-associated cellular immune response against established tumors, to generate a protective immune response to subsequent tumor challenge, and to potentiate the efficacy of TIL T lymphocytes and activated NK cells *in vivo*.

IL-12 was originally identified as a factor that synergized with IL-2 in the presence of hydrocortisone to generate LAK cell activity from human peripheral blood mononuclear cells (PBMCs) (Gately *et al.*, 1986). IL-12 alone provided only small levels of LAK cell induction compared with IL-2 alone or IL-2 with IL-12 (Gately *et al.*, 1991; Zeh *et al.*, 1994), with optimal IL-12 bioactivity in the 10–50 pM range (Gately *et al.*, 1991). Based on antibody neutralization studies, LAK cell induction by IL-12 could be inhibited effectively by up to 70% by either anti-TNF-α or anti-IFN-γ, with the combination of these antibodies virtually ablating IL-12 efficacy (Gately *et al.*, 1991). Based on this *in vitro* biology, several *in vivo* applications of IL-12 have been demonstrated to exhibit enhanced efficacy when combined with systemic IL-2 (see below).

The synergy that IL-12 displays with other single agents is not restricted to IL-2. IL-12 synergizes with IL-4 to promote the proliferation of NK cells, but not T cells (Naume *et al.*, 1993). IL-12 synergizes with IL-7 to augment the proliferation of anti-CD3-activated $CD4^+$ and $CD8^+$ T cells, with $CD8^+$ T cells generally proving responsive to this combination of cytokines at lower doses than $CD4^+$ T cells (Mehrotra *et al.*, 1995). IL-12 and IL-2 or IL-15 synergize to promote the proliferation of activated 'memory' T cells but not naive T cells (Maruo *et al.*, 1996; Seder, 1996). IL-12 and IL-1 synergize to induce IFN-γ production by, and proliferation of, $TCR\gamma/\delta^+$ T cells (Pignata *et al.*, 1995).

Kubin *et al.* (1994) demonstrated the synergistic induction (proliferation, cytokine secretion) of freshly isolated or mitogen-activated human T cells by IL-12 and the costimulatory interaction of T cell-expressed CD28 with its ligand B7. As PBMC generation of IFN-γ by IL-12 is inhibited by the soluble $CTLA_4$–immunoglobulin molecule, which blocks B7-mediated signaling, it appears that the CD28–B7 interaction is required for optimal IL-12-induced cell activation (Kubin *et al.*, 1994). IL-12–B7 synergy in proliferation and cytokine release assays has also been demonstrated for murine T_{H1} clones (Murphy *et al.*, 1994). These data support a requirement for both IL-12 and CD28–B7 signaling pathways in the activation and differentiation of both naive and memory T cells. B7-1 and IL-12 in conjunction with IL-6 have been shown to be sufficient for the *in vitro* generation of antigen-specific murine antitumor cytolytic T cells (Gajewski *et al.*, 1995). Both IL-10 and prostaglandin E_2 (PGE_2) may antagonize IL-12-associated immunobiology by suppressing IL-12 production by accessory cells and by diminishing the APC expression of B7 costimulatory molecules (Kubin *et al.*, 1994; van der Pouw Kraan *et al.*, 1995).

IL-12 appears to modulate cell-mediated cytoxicity effectively. IL-12 can enable NK–LAK cell activity from the PBMCs of HIV-infected individuals (Chehimi *et al.*, 1992) and facilitates antibody-dependent cell-mediated cytotoxicity (ADCC) function against tumor target cell lines *in vitro* (Lieberman *et al.*, 1991). While IL-12 has been reported to synergize with other cytokines in promoting T and NK cell expansion (Chizzonite *et al.*, 1992; Desai *et al.*, 1992; Naume *et al.*, 1992; Zeh *et al.*, 1994), the proliferative indices found and attributable to IL-12 are generally low. Overall, the available data appear to support the ability of IL-12 to enhance cytolytic effector function (Gately *et al.*, 1986), but not necessarily the number or frequency of effector cells generated in *in vitro* cultures containing IL-12 (Bhardwaj *et al.*, 1996). Using IL-12 to augment the ability of influenza-infected dendritic cells to promote $CD8^+$ CTLs *in vitro*, Bhardwaj *et al.* (1996) noted significant enhancement in antigen-specific CTL reactivity without a notable increase in the viral-specific pCTL frequency *in vitro*. This supports the maturational, but not proliferative, impact of IL-12 on CTL generation. Thus, IL-12 may be directly promoting the cytolytic apparatus of the CTL (i.e. upregulation of perforin, serine proteases and cytolytic granules) (Bloom and Horvath, 1994; Chouaib *et al.*, 1994; Mehrotra *et al.*, 1995). IL-12-induced effector maturation is not restricted to NK cells or $CD8^+$ T cells. In the absence of $CD8^+$ T cells, IL-12 promotes the expansion and differentiation of effector $CD4^+$ T cells (DiIulio *et al.*, 1996).

Some of the perceived proliferative-inducing capacity of IL-12 may be due to its ability to promote enhanced cellular expression of the antiapoptotic transcription factors bcl-2 and bcl_x1, thereby preventing the death of cultured responder cells. This may be the case not only for T and NK cells, but also for APCs such as dendritic cells (W.J. Storkus *et al.*, unpublished data). Human IL-2-dependent T_{H1} T-cell clones treated with IL-12 were protected from apoptosis resulting from T-cell receptor (TCR) ligation (anti-CD3, anti-TCR, rgp120a and anti-gp120 monoclonal antibody (mAb)) or IL-2 deprivation (Radrizzani *et al.*, 1995). The apoptosis of $CD4^+$ (but not $CD8^+$) T cells derived from HIV-positive patients and treated with anti-TCR or anti-Fas (CD95) antibodies may also be prevented by IL-12 treatment (Estaquier *et al.*, 1995). Conversely, *in vivo* application of anti-IL-12 antibodies promotes the apoptosis of OVA-specific T cells in TCR-transgenic mice (Marth *et al.*, 1996). Such antiapoptotic effects of IL-12 may not be universal, however. Armant *et al.* (1995) showed that IL-2 and IL-7, but not IL-12, prevent glucocorticoid-induced programmed cell death in $CD56^+$ NK cells in association with upregulated expression of the proto-oncogene *bcl-2*.

Given the ability of IL-12 to promote the production and release of potent immunomodulatory factors such as IFN-γ and TNF-α from bulk PBMC populations, it is perhaps not surprising that the surface marker phenotype of treated lymphocytes should be modified in concert with IL-12 treatment. Thus, IL-12 upregulates the IL-2Rα (CD25), TNF-Rα (p55), CD2, CD11a, CD54, CD56, CD69, CD71 (transferrin receptor), human leukocyte antigen (HLA) class I and II expression on NK cells (Robertson, 1991; Naume *et al.*, 1992; Rabinowich *et al.*, 1993). IL-12 also promotes the enhanced expression of the MHC class II binding lymphocyte activation gene-3 (LAG-3) antigen by IFN-γ-producing (T_{H0} or T_{H1}), but not IL-4-producing (T_{H2}), $CD4^+$ T-cell clones (Annunziato *et al.*, 1996). These modifications may result in enhanced intercellular communication (CD2, CD4, CD11a, CD54, CD56, MHC class I, MHC class II, LAG-3) or cellular proliferation (CD25, CD69, CD71).

Importance of IFN-γ and IL-10 in IL-12 Immunobiology

In virtually every study performed where the biochemical mechanistic action of IL-12 has been evaluated, a critical dependency on IFN-γ has been uncovered (Ozmen *et al.*, 1995; Trinchieri, 1995). This dependency is perhaps not surprising given the early literature defining the biology of IL-12 and its linkage to the cellular production of IFN-γ. The ability of neutralizing antibodies to IFN-γ to block effectively the *in vitro* and *in vivo* biologic effects of IL-12 is a widespread finding, and IFN-γ-independent effects of IL-12 have been principally demonstrated only in IFN-γ knockout or IFN-γR knockout mice. In these animals, IL-12 is still able to facilitate the promotion of T_{H1}-associated immune reactivity, but the ability of IL-12 to serve as a biologic adjuvant is lost. It is also worth noting that the IFN-γ-dependent toxicity associated with systemic application of IL-12 (see below) is nominally ablated in these knockout animals.

IFN-γ potentiates IL-12-dependent immunity. Thus IFN-γ pretreatment of APCs appears to enhance their subsequent ability to produce heightened levels of IL-12, thereby begetting yet additional IFN-γ production by IL-12-responsive T or NK cells (Hayes *et al.*, 1995) . Clearly such autopotentiation by IFN-γ and IL-12 must be buffered by a negative feedback loop. This may occur in the form of the observed IL-12 induction of IL-10 production by T cells, with IL-10 subsequently inhibiting the production of

IL-12 from APCs (Van de Pouw Kraan *et al.*, 1995) and suppressing alternate IFN-γ-modulated attributes of APCs such as costimulatory molecule expression (i.e. CD86) and MHC class II expression (Jeannin *et al.*, 1995; Gerosa *et al.*, 1996; Ma *et al.*, 1996a,b; Peritt *et al.*, 1996; Windhagen *et al.*, 1996).

Enhancement of T_{H1}-Associated Immunity

The APC-biased production of IL-12 noted above suggested an important role for IL-12 in modulating the efficacy of antigen processing and/or presentation and the activation, expansion and differentiation of T cells and/or B cells. We now know that IL-12 has profound modulatory effects on both cellular and humoral immunity.

It has been appreciated for some time that mature $CD4^+$ T helper cells can be segregated into two principal cytokine secretion categories, T_{H1} and T_{H2} (Mosmann *et al.*, 1986). More recently cytotoxic $CD8^+$ T cells have been similarly discriminated into T_{C1} and T_{C2} categories, based on their ability preferentially to secrete IFN-γ (T_{H1},T_{C1}) or IL-4/IL-5 (T_{H2},T_{C2}) (Mosmann *et al.*, 1997). While this dichotomy in T-cell subsets was originally documented in the mouse, recent reports support a similar segregation of such T-cell subpopulations in humans (Mosmann *et al.*, 1997). In the mouse, T_{H1} cells secreting IFN-γ, lymphotoxin (LT) and IL-2 were demonstrated to be affiliated with enhanced cellular immunity and dominated the natural responses to complex antigens in certain strains of mice (i.e. B6 mice responding to virus). T_{H2} cells producing high levels of IL-4 and IL-5 in response to specific antigens promoted raised humoral immune responses and were again characteristic of the prevalent immune response made by certain mouse strains (i.e. Balb/c antibacteria or allergens). The ability of an organism to mount a T_{H1}-biased immune response has been generally affiliated with enhanced ability to eliminate certain pathogens and tumors. T helper cells specific for the dust mite allergen *Der pI* have a reduced capacity to secrete IL-4 and a concomitantly increased capacity to secrete IFN-γ, with T-cell clones biased towards T_{H0} or T_{H1} phenotypes (Manetti *et al.*, 1995). In contrast, T-cell clones specific for purified protein derivative (PPD) generated in the presence of anti-IL-12 neutralizing antibodies were typically IL-4 producers when stimulated antigenically (T_{H2}).

IL-12 appears to play an important role in determining whether a T_{H1}- or T_{H2}-biased immune response is mounted to antigenic challenge. Functionally, IL-12 has been demonstrated to bias the immune response towards a T_{H1} phenotype *in vitro* and *in vivo* (Bradley *et al.*, 1995), resulting in the augmentation of cellular (i.e. $CD8^+$ T cell) immunity against viral, bacterial and protozoan infections. Such immune augmentation results in reduced pathogen load, hastened lesional clearance, and the prolonged survival of both normal and severe combined immune deficiency (SCID) animals infected with virus. As would be expected, pathogens that promote the production of IL-12 by APC types bias the resulting immune response towards a T_{H1} cellular immune response. Thus, monocytes treated with *L. monocytogenes* promote T_{H1}-associated immunity, which could be mimicked by the application of exogenous IL-12 (Macatonia *et al.*, 1993). In addition to promoting T_{H1} development from naive and memory T-cell populations, IL-12 has also been shown to suppress the development of T_{H2} cells (Murphy *et al.*, 1994; Trinchieri *et al.*, 1996). Conversely, as noted above, the T_{H2}-associated cytokine IL-10 ablates the induction of T_{H1} T cells, presumably by preventing the induction and secretion of IL-12 by monocytes.

As IL-12 produced by APCs drives the preferential induction of T_{H1} immune responses, this suggests, at the most simplistic level, that there is biased reactivity of T_{H1} $CD4^+$ T cells to IL-12. Rogge *et al.* (1997) recently demonstrated that the IL-12Rβ_2 subunit is a differentiation marker expressed on developing antigen-specific T_{H1} cells, is not expressed by T_{H2}-cloned T cells, and mediates the biased response of T_{H1} clones to IL-12. The IL-12Rβ_2 chain is inducible by IFN-α/β, but not IFN-γ, supporting the utility of these early type I interferons (principally secreted by APCs such as dendritic cells) (Ferbas *et al.*, 1994) in enhancing T_{H1} immune responses (Rogge *et al.*, 1997).

Studies performed in IL-12 knockout mice have confirmed the importance of IL-12 in promoting T_{H1}-type immunity. In IL-12 p40 knockout mice, T_{H1} immunity can occur, but is significantly depressed (Magram *et al.*, 1996a,b). The hyporesponsiveness of T_{H1} $CD4^+$ T cells correlated with a reduced ability to produce IFN-γ in response to endotoxin *in vivo*. Mattner *et al.* (1996) evaluated the capacity of mice normally resistant to *Leishmania major* to respond to bacterial infection when the IL-12 p35 or IL-12 p40 genes were knocked out. While the wild-type 129/Sv/Ev strain of mice presents only small lesions that are quickly resolved, both the IL-12-deficient mice displayed large progressive lesions after infection. IL-12-deficient mice failed to develop *Leishmania* antigen-specific delayed-type hypersensitivity (DTH) responses or to produce IFN-γ efficiently, resulting in a predominantly T_{H2}-type immune response. Despite such biasing of the T_{H1}–T_{H2} balance of immunity in these mice, no apparent developmental abnormalities were noted.

Dendritic cells promote the induction of T_{H1} cells from antigen-specific naive lectin cell adhesion model-1 (LECAM-1^{bri+}) $CD4^+$ T cells derived from TCR transgenic mice in the presence of IL-12 and the absence of IL-4 (Macatonia *et al.*, 1995). T cells activated through ligation of their TCR express CD40 ligand (CD40L) that allows for functional crosslinking with CD40 expressed on dendritic cells (Fig. 2). This two-way intercellular interaction results in raised production of APC-derived IL-12 and in T-cell expression of IL-12R and B7–1 (Peng *et al.*, 1996). CD40L-mediated signaling into the T cell, in the presence of IL-12, strongly biases the production of IFN-γ by T_{H1}-type $CD4^+$ T cells (Peng *et al.*, 1996). Humans lacking CD40L exhibit increased susceptibility to opportunistic infections normally regulated by cellular immunity (Kroczek *et al.*, 1994), whereas CD40L knockout mice fail to generate protective immunity against *L. major* (Kennedy *et al.*, 1996). Polyinosidic acid, which enhances macrophage production of IL-12 and IFN-α, promotes the induction of T_{H1}-biased cellular immunity against *Dermatophagoides pteronyssinus* (*Der pI*) (Manetti *et al.*, 1995). Synergy between APC-secreted IFN-α and IL-12 may also prevent herpetic viral infections (Rogerson *et al.*, 1996).

In situ, in draining lymphoid organs, dendritic cells appear to represent the principal producers of IL-12 (Muller *et al.*, 1995). Animals painted with the contact allergen trinitrochlorobenzene (TNCB) or dinitrofluorobenzene (DNFB) yield immunoproliferative responses in the draining lymph nodes and the spleen, with IL-12 mRNA evident within the first 72 h of challenge. Immunodepletion of dendritic cells results in the loss of IL-12 mRNA signal. This study also documented the ability of IL-12 to overcome antigen-specific tolerance associated with ultraviolet irradiation. Thus, the anergic responses of mice to TNCB-treated skin promoted by ultraviolet radiation were ablated by systemic administration of IL-12, resulting in contact hypersensitivity response to topical hapten application to the ear.

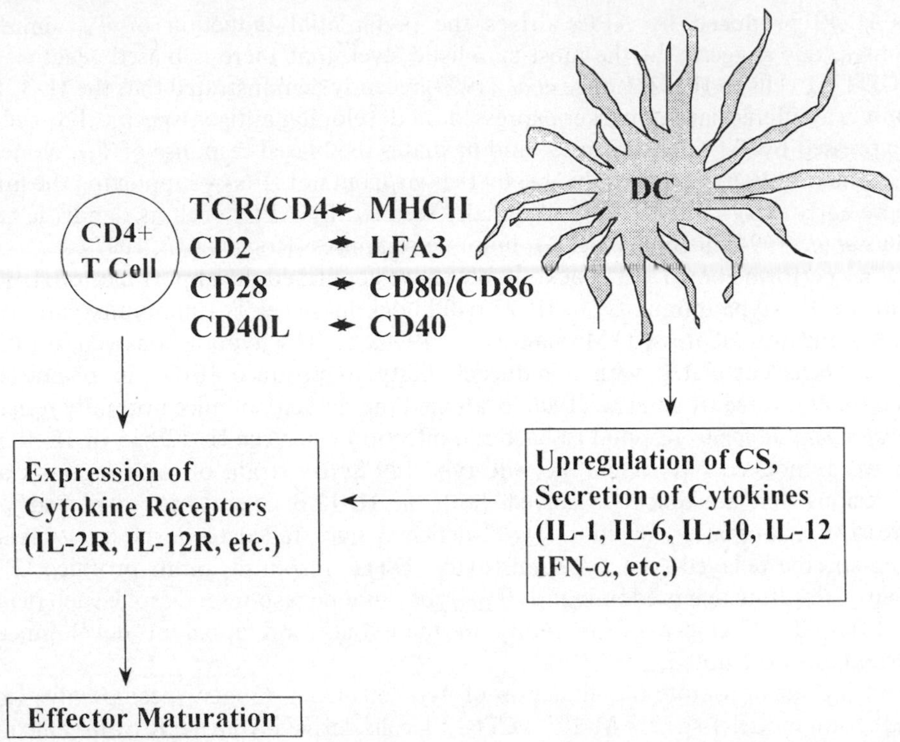

Fig. 2. IL-12 is produced by potent APCs such as dendritic cells upon T-cell interaction and promotes the effector maturation of T cells. APCs expressing antigenic MHC complexes interact with antigen-specific $CD4^+$ T cells through a series of intercellular ligand–receptor complexes. This two-way activation event promotes the expression of cytokine receptors, including the high-affinity IL-12R by T_{H1} $CD4^+$ T cells. Dendritic cells activated through cognate T-cell interaction increase their expression levels of costimulatory (CS) molecules such as CD40, CD54, CD80 and CD86, and are induced to produce cytokines, including IL-1, IL-6, IL-10 and IL-12. These cytokines, produced in the paracrine environment of the lymph node or spleen bias the resulting T_{H1}/T_{H2} balance of antigen-specific immunity and the expansion and differentiation of effector T cells.

Adjuvanticity of IL-12

IL-12 has also been reported to serve as a potent 'adjuvant', promoting T_{H1}-associated cellular (i.e. $CD8^+$ CTL) and humoral (IgG_{2a}) responses to protein or peptide antigens *in vivo* (Bliss *et al.*, 1996). IL-12 is a promising adjuvant for incorporation in vaccines against pathogens that require cell-mediated immunity for resistance, such as most intracellular parasites and viruses (Scott, 1993; Miller *et al.*, 1995). IL-12 substitutes for the bacterial adjuvant *Cornybacterium parvum* in a *L. major* vaccine in promoting T_{H1}-mediated immunity in normally T_{H2}-biased responder Balb/c mice (Afonso *et al.*, 1994). In these vaccinated mice, IL-4 production is minimized whereas IFN-γ production is augmented. Of note, the effective induction of T_{H1}-mediated immunity was dependent on asialo $SGM1^+$ NK cells (and probably NK cell-secreted IFN-γ), as depletion of this subset *in vivo* ablated protective immunity (Afonso *et al.*, 1994). IL-12 similarly serves as an adjuvant in promoting T_{H1}-biased immunity in mice vaccinated with inactivated respiratory syncytial virus (Tang and Graham, 1995), resulting in increased production of IgG_{2a} antiviral antibodies. IL-12 also induces a T_{H1}-associated immune response and protection of mice against an aerosolized challenge with *Bordetella pertussis* when applied as adjuvant with an acellular bacterial vaccine (Mahon *et al.*, 1996). Further, IL-12 provided systemically also potentiates T_{H1}-type immunity in response to orally administered protein antigens and cholera toxin (Marinaro *et al.*, 1996).

IL-12-induced IFN-γ appears crucial to the 'adjuvanticity' of IL-12 in many experimental models. Schijns *et al.* (1995) evaluated the response of wild-type and IFN-γR knockout mice to inactivated pseudorabies virus. Injection of wild-type mice with both IL-12 and inactivated virus resulted in enhanced resistance to a subsequent challenge with virulent virus that was associated with the production of an IFN-γ-dependent IgG_{2a} antiviral antibody response. The identical experiment performed in IFN-γR knockout mice yielded no protection against challenge with a lethal dose of virus nor production of an IgG_{2a} antiviral antibody response, despite the production of IFN-γ induced by IL-12 administration. Lastly, recombinant IFN-γ could substitute for IL-12 as an adjuvant in wild-type mice.

Bertagnolli *et al.* (1991) demonstrated that IL-12 is an effective 'mitogen' for both allospecific and influenza peptide-specific T cells, with optimal T-cell responsiveness occurring 3–7 days after antigenic stimulation. IL-12 applied in combination with low-dose IL-2 appeared to synergize in the activation and proliferation of 'memory' T cells. Thus, unlike IL-2, which is able to promote the proliferation of resting T cells expressing the intermediate-affinity IL-2R, IL-12 appears to promote the limited expansion of preactivated (TCR-triggered, antigen-specific) responder cells.

Bhardwaj *et al.* (1996) showed that human dendritic cells infected with influenza virus effectively promote the *in vitro* expansion of autologous $CD8^+$ CTLs in an apparently IL-12-independent manner (i.e. not blocked with anti-IL-12). Addition of rhIL-12 into primary cultures promotes enhanced antigen-specific T-cell proliferation, cytotoxicity and IFN-γ production. Dendritic cells pulsed with tumor-associated peptides elicit T_{H1}-biased antitumor immunity *in vivo*, which is IL-12 dependent (Bianchi *et al.*, 1996; Zitvogel *et al.*, 1996a). These vaccinations could be enhanced if IL-12 was applied systemically with the dendritic cell-peptide adoptive transfers and required the participation of both $CD4^+$ and $CD8^+$ T cells *in vivo*. Tumor antigen-specific DTH skin responses were also induced by such vaccines. Furthermore, the early production of IL-12 (presumably by transferred APCs) resulted in subsequent IL-12-dependent production of host IL-12, required for optimal vaccine efficacy.

In addition to promoting antigen-specific cellular immune reactivity, IL-12 may also modulate the apparent immunogenicity of a given CTL epitope applied in the context of a multiepitope vaccine (Eberl *et al.*, 1996), thereby altering the comparative immunodominance of the epitope. The exact mechanism for such modulation of immunogenicity remains unknown, but may reflect IL-12- and secondarily produced IFN-γ-mediated effects on APC processing and presentation of specific epitopes.

PRECLINICAL AND CLINICAL APPLICATIONS OF IL-12

IL-12 has been evaluated in both infectious disease and tumor models for its ability to promote the induction of protective and/or therapeutic cellular and humoral immunity. The preferential activation of T_{H1}-type T cells *in vitro* and *in vivo* supports the ability of IL-12 to skew the cumulative immune response. In particular this is observed in both

T_{H1}- and T_{H2}-biased murine strains. In B6 mice, cellular immunity is enhanced, whereas in Balb/c mice, which are typically affiliated with T_{H2}-type immunity, IL-12 converts immunity to T_{H1}-type immune responses. This is also reflected in pronounced IgG_{2a} *versus* IgG_1 humoral immune response biasing, resulting from IL-12-induced production of IFN-γ.

IL-12 Effects on Hematopoiesis

Gately and coworkers (1993, 1994a,b) initially evaluated the *in vivo* impact of microgram quantities of recombinant IL-12 administered to normal mice. Predictably, enhanced NK cell lytic activity was documented from the spleen and liver of IL-12-treated animals. Furthermore, increased prevalence of IFN-γ-producing NK cells, NK1.1$^+$ T cells, and CD8$^+$ T cells was observed in the liver of treated mice. Perhaps the most profound immune deviation resulting from IL-12 treatment was the reproducible splenomegaly noted by numerous groups, which was the result of extramedullary hematopoiesis. Erythroid, myeloid and megakaryocytic lineages were all represented within the impacted splenic regions (Bertolini *et al.*, 1995; Tare *et al.*, 1995). In additional murine studies, IL-12 was found to synergize with stem cell factor (SCF) in supporting the ontogeny of lymphohematopoietic precursors (Jacobsen *et al.*, 1993). IL-12 also synergizes with IL-3 and serum factor to promote granulocyte macrophage colony forming unit (CFU-GM) from hematopoietic progenitors (Hirao *et al.*, 1995) and promotes the expansion of Lin$^-$ Sca-1$^+$ immature blast colonies from bone marrow progenitors when combined with Flt3L (Jacobsen *et al.*, 1995, 1996). Of prime concern for the design of clinical applications of IL-12, no significant toxicities were noted in these murine models (i.e. fur ruffling, mild lethargy). Tare *et al.* (1995) demonstrated that, despite the ability of rIL-12 to synergize with c-kit ligand in driving the proliferation of murine hematopoietic stem cells *in vitro*, chronic administration of IL-12 resulted in a dose- and time-dependent anemia, leukopenia and thrombocytopenia *in vivo*. Profound suppression of hematopoiesis was noted in the bone marrow of mice treated chronically with IL-12, with CFU-GM, erythrocyte colony forming unit (CFU-E) and erythrocyte blast forming unit (BFU-E) colony forming cells affected. This decrease in bone marrow hematopoiesis was countered by the aforementioned extramedullary hematopoiesis noted in the spleen of IL-12-treated animals, presumably due in part to the mobilization of hematopoietic progenitor cells from the bone marrow to the periphery (Gately and Mulqueca, 1996). Similar studies performed in IFN-γR knockout mice suggest that the bone marrow mobilization of hematopoietic progeny by IL-12 is IFN-γ dependent, as bone marrow progenitors increase in knockout mice *versus* the decrease observed in wild-type mice (Eng *et al.*, 1995). Interestingly, splenic hyperplasia (extramedullary hematopoiesis) remained strong in the knockout mice (Eng *et al.*, 1995).

IL-12 in Cancer

Perhaps the single most directed translational use of IL-12 has been in the cancer setting (Lotze *et al.*, 1996). IL-12 appears important in the immunologically intact host to prevent the occurrence of spontaneous tumors. Furthermore, when applied as a vaccine adjuvant or therapeutic agent, IL-12 promotes profound antitumor immunity. In cases where indolent tumors have promoted general immunosuppression, IL-12 may reactivate or induce the antitumor reactivity of hyporesponsive effector cells.

Role of Natural IL-12 in Tumor Immunity

IL-12 appears to play a significant role in the host's natural resistance to tumorigenesis. Noguchi *et al.* (1996) evaluated the impact of IL-12 administration on the induction of tumors *in situ* by the tumor-promoting agent 3-methylcholanthrene (3-MC). Chronic treatment of 3-MC-injected mice with IL-12 for up to 18 weeks resulted in delayed tumor appearance with reduced tumor incidence which was associated with *in vivo* production of IFN-γ, TNF-α, IL-10 and serum nitrates (Noguchi *et al.*, 1996). Similarly, the incidence of primary tumors in c-Ha-ras transgenic mice induced by the chemical carcinogen methylnitrosourea was significantly reduced as a result of IL-12 application (Nishimura *et al.*, 1995, 1996). Fallarino *et al.* (1996) have also noted that the P815 mastocytoma exhibits a reproducible 30% spontaneous regression rate when transplanted into DBA/2 mice that is IL-12 dependent. Thus, all mice inoculated with P815 and anti-IL-12 mAb develop tumors. This supports the role of IL-12 in the natural, inherent, immune response to tumor development.

Recombinant IL-12 in Tumor Models

Studies focusing on the impact of systemic rIL-12 administration on antigen-specific T-cell immune responses subsequently demonstrated that IL-12 enhanced (i.e. 1–2 log-fold) allospecific T-cell responses in vaccinated animals. Daily administration of rIL-12 via intraperitoneal injection was also observed to be associated with increased resistance to the outgrowth of transplanted syngeneic tumors in diverse murine models (Table 4). Highly immunogenic murine tumors were found to respond better to IL-12 therapy in both metastatic and subcutaneous delivery models, often with optimal efficacy observed if treatment began at the time of, or within 14 days of, tumor injection (Brunda *et al.*, 1994; Nastala *et al.*, 1994). In spontaneous metastatic tumor models, the ability of IL-12 to mediate complete regression of cancer was apparent even when IL-12 was applied as late as 30 days after tumor injection (Lotze *et al.*, 1996; Verbik *et al.*, 1996). Poorly immunogenic models similarly displayed some degree of therapeutic efficacy derived from IL-12 treatment, although the window of treatment efficacy was far more restricted (0–5 days post-tumor transplant) than that observed for the more immunogenic tumors.

Based on *in vivo* neutralization and cellular depletion studies, the efficacy of IL-12-dependent therapy required the *in situ* production of IFN-γ and TNF-α, with long-term immunity mediated by both $CD4^+$ and $CD8^+$ T cells. It should be noted, however, that the antitumor efficacy of IL-12 is not strictly IFN-γ dependent (Brunda *et al.*, 1995). While antitumor response was diminished in IL-12-treated mice bearing Renca renal cell carcinoma (RCC) upon injection with anti-IFN-γ antibodies, IFN-γ application did not approximate the efficacy observed for IL-12. Further, IL-12 administration to Renca-bearing nude mice was far less effective despite raised (i.e. tenfold) serum levels of IFN-γ compared with control mice.

Therapies implementing IL-12 have also been combined with additional cytokines in both recombinant protein and cDNA formats. In particular, IL-12 plus IL-2 has been observed by a number of groups to enhance therapeutic efficacy (Leder *et al.*, 1995; Pappo *et al.*, 1995; Rodolfo *et al.*, 1996; Wigginton *et al.*, 1996). In gene-based protocols, IL-12 and the costimulatory molecule B7.1 (CD80) appear to promote heightened

Table 4. Murine tumor models demonstrating anti-tumor efficacy of IL-12.

Tumor model	Strain	E/T	SC/M	IFN-γ	TNF-α	$CD4^+$	$CD8^+$	Reference
A20 ly.[GT]	Balb/c	T	SC	Y	NT	NT	Y	Nishimura *et al.* (1996)
B16F10 mel.[R]	B6	T	SC	Y	NT	N	Y	Brunda *et al.* (1994)
C26 co.ca.[R]	Balb/c	T	SC	Y	NT	N	Y	Colombo *et al.* (1996), Rodolfo *et al.* (1996)
C51 co.ca.[GT]	Balb/c	T	M	Y	NT	Y	Y	Rodolfo *et al.* (1996)
CSA1M fsar.[R]	Balb/c	T	SC	Y	NT	Y	Y	Yu *et al.* (1996a)
CT26 co.ca.[R]	Balb/c	T	SC	Y	NT	NT	Y	Tannenbaum *et al.* (1996)
K1735 mel.[R]	C3H	T	SC	Y	NT	Y	Y	Coughlin *et al.* (1995)
M5076 ret.sar.[R]	B6	T	SC	Y	NT	N	Y	Brunda *et al.* (1994)
MC38 co.ad.[R,GT]	B6	T	M, SC	Y	N	Y	Y	Nastala *et al.* (1994), Zitvogel *et al.* (1996b)
MCA 105 sar.[R]	B6	T	SC	Y	Y	Y	Y	Nastala *et al.* (1994), Pappo *et al.* (1995)
MCA 205 sar.[GT]	B6	E, T	SC	Y	NT	Y	Y	Tahara *et al.* (1995), Zitvogel *et al.* (1996a,c)
MCA 207 fsar.[GT]	B6	E, T	SC	Y	Y	Y	Y	Zitvogel *et al.* (1996b)
Meth A sar.[R,GT]	Balb/c	T	SC	NT	NT	NT	Y	Noguchi *et al.* (1995), Rakhmilevich *et al.* (1996)
OV-HM ov.ca.[R]	B6	T	SC	Y	NT	Y	Y	Yu *et al.* (1996)
P815 mast.[R,GT]	DBA/2	T	SC, M	Y	NT	Y	Y	Fallarino *et al.* (1996), Rakhmilevich *et al.* (1996)
RAW117 ly.[R]	Balb/c	T	M	NT	NT	NT	NT	Verbik *et al.* (1996)
Renca RCC[R]	Balb/c	T	SC	Y	NT	N	Y	Brunda *et al.* (1994), Wigginton *et al.* (1996)
SCK mam.ca.[R]	C3H	T	SC	Y	NT	Y	Y	Coughlin *et al.* (1995)
TS/A mam.ad.[GT]	Balb/c	E, T	SC	Y	Y	Y	Y	Zitvogel *et al.* (1996a,b)
X5563 ly.[R]	C3H	T	SC	Y	NT	N	Y	Brunda *et al.* (1994)

Tumor types include: sarcoma (sar.), fibrosarcoma (fsar.), renal cell carcinoma (RCC), colorectal carcinoma (co.ca.), mammary carcinoma (mam.ca.), mammary adenocarcinoma (mam.ad.), lymphoma (ly.), mastocytoma (mast.), melanoma (mel.), ovarian carcinoma (ov.ca.), colonic adenocarcinoma (co.ad.), and reticulum sarcoma (ret.sar.). [GT], IL-12 gene therapy approach; [R], recombinant IL-12 approach. E, tumor establishment model; T, tumor therapy model; SC, subcutaneous tumor model; M, metastatic tumor model; NT, not tested; Y, yes shown to be involved in IL-12-induced antitumor immunity; N, not shown to be involved in IL-12-induced antitumor immunity.

cellular antitumor immunity (Zitvogel *et al.*, 1996b,c). Balb/c mice bearing subcutaneous or metastatic Renca RCCs were effectively treated with daily rmIL-12 (0.5 μg/day) plus 'pulse' IL-2 (300 000 units given twice daily for 1 day each week), with 60–100% cure rates observed (Wigginton *et al.*, 1996). In these models, IL-12 or IL-2 given alone proved poorly therapeutic (8–17% cure rates). More recent studies support the effective synergy of IL-12 and IL-18 (Osaki *et al.*, 1997) and IL-12 and Flt3L (Peron *et al.*, 1997) in promoting antitumor immunity *in vivo*.

IL-12 Inhibits Tumor Progression via Anti-Angiogenic Activity

The anticancer mechanism of action for IL-12 appears to be complex and multifactorial. IL-12 clearly promotes the elaboration of IFN-γ, leading to enhanced APC function and antigen presentation, the potentiation of T_{H1} $CD4^+$ T-cell responsiveness, and the maturation of both adaptive and non-adaptive immune effector cells. However, the antitumor and antimetastatic effects of IL-12 have also been observed in SCID mice, and in both nude and CD8-depleted mice (Brunda *et al.*, 1994, 1995), suggesting that IL-12 may affect additional non-immune events associated with tumor progression.

It is now known that IL-12 induction of IFN-γ triggers the subsequent synthesis of a cascade of interferon-induced proteins (IPs). One of these proteins, a CXC chemokine termed IP-10, appears to display significant antiangiogenic activity leading to reduced vascularization of transplanted tumors in treated animals (Tannenbaum *et al.*, 1996). Murine IL-12 profoundly inhibits basic fibroblast growth factor (bFGF)-induced Matrigel neovascularization *in vivo*, with these effects neutralized by mAbs specific for either IFN-γ or IP-10 (Sgadari *et al.*, 1996). IL-12 appears to induce IP-10 expression in splenocytes, with human IL-12 promoting the production of IP-10 by human endothelial cells (Sgadari *et al.*, 1996). Additional cell types known to express IP-10 upon IFN-γ stimulation include mononuclear cells, fibroblasts, keratinocytes, endothelial cells and T cells (Luster *et al.*, 1995).

Tumor cell-induced angiogenesis is a requisite component for tumor establishment and progression. Majewski *et al.* (1996) injected a series of human tumor cell lines (mammary carcinoma, vulval carcinoma, bowenoid papulosis) intradermally into X-irradiated Balb/c mice followed by systemic administration of rmIL-12. Tumor-induced angiogenesis was inhibited in an IL-12 dose- and time-dependent manner, with this effect ablated by administration of anti-mIFN-γ neutralizing antibody. Interestingly, hIFN-γ administered to tumor-bearing animals also inhibited tumor angiogenesis, supporting both recipient and tumor species-specific IFN-γ effects. Neither hIL-12 or mIL-12 had any effect on the *in vitro* growth of tumor cells.

The indirect antiangiogenic action of IL-12 coupled with its ability to activate both non-adaptive and adaptive immunity in the tumor microenvironment presumptively leads to the necrotic or apoptotic demise of tumor cells. This in turn promotes the subsequent APC uptake of tumor-associated materials, the processing and presentation of tumor epitopes, and the induction of tumor-reactive T cells, principally within lymphoid organs draining tumor sites.

IL-12 Gene Therapy of Cancer

The efficacy of IL-12 has also been evaluated systematically in approaches designed to target the delivery of IL-12 to the tumor microenvironment using ‘gene therapy’ approaches. NIH3T3 fibroblasts transfected to express both the mIL-12 p35 and mIL-12 p40 chains (4 ng IL-12 per 10^6 cells per 24 h) were injected perilesionally into B6 mice bearing MCA 105 sarcomas (Pappo *et al.*, 1995). As a single modality therapy, this paracrine delivery of IL-12 reduced tumor area by more than 80% on day 32 post-tumor inoculation, but was not fully curative. Systemic administration of IL-2 in conjunction with perilesional injection of 3T3 IL-12-transfected fibroblasts yielded complete tumor regression in all treated animals. Using a polycistronic retroviral vector encoding both the IL-12 p35 and IL-12 p40 subunits, Tahara *et al.* (1995) engineered tumor cells (MCA 207 fibrosarcoma, MCA 102 sarcoma) stably to secrete nanogram quantities of IL-12 *in vitro*. These IL-12-producing tumors, but not control transfected tumors, failed to grow when injected intradermally into B6 mice. Non-adaptive and systemic tumor-specific immunity was induced, protecting animals against contralateral challenge with control tumors, with protection dependent on the production of IFN-γ. In an alternate approach (Nishimura *et al.*, 1995), injection of recombinant vaccinia virus encoding mIL-12 into subcutaneous MCA 105 sarcomas in B6 mice resulted in a 60% cure rate compared with 0% in control vaccinia injected groups. Perhaps most readily, IL-12 cDNA simply injected intradermally results in systemic immune augmentation (NK cell activity, IFN-γ secretion) and enhanced resistance to Renca RCC tumor progression (Tan *et al.*, 1996). Such augmented antitumor immunity occurs in the absence of the profound splenomegaly affiliated with systemic administration of rmIL-12. IL-12 cDNA has also been inserted into the dermis overlying subcutaneous tumors using the biolistic gene gun as a delivery system (Rakhmilevich *et al.*, 1996). Transfected skin produced 200–300 pg IL-12 at the vaccine site and promoted the expansion of antigen-specific $CD8^+$ T cells and the complete regression of mice bearing well-established Renca, Meth A, SA-1, or L5178Y syngeneic tumors. Dermally delivered IL-12 also appeared capable of reducing metastatic spread of the P815 tumor. Colombo *et al.* (1996) demonstrated that paracrine delivery of 30–80 pg IL-12 delivered via IL-12 cDNA transfected C26 colonic carcinoma was equally efficacious as 1 μg IL-12 delivered systemically, with enhanced efficacy mediated by $CD8^+$ T cells and NK cells in CD4-depleted mice.

The previously noted synergy of IL-12 and CD28–B7 signaling in promoting T-cell responses (Trinchieri, 1996) supported the assessment of the efficacy of these modalities in murine tumor models (Zitvogel and Lotze, 1995). Zitvogel *et al.* (1996b) evaluated the antitumor benefit associated with combined IL-12 and B7.1 gene delivery in two poorly immunogenic models (MCA 207 fibrosarcoma in B6 mice and TS/A mammary adenocarcinoma in Balb/c mice). Coinoculation of tumor cells infected with retroviruses encoding IL-12 with the same tumor cells transduced to express B7.1 into mice resulted in disease-free animals with long-term tumor-specific immunity in 80% of cases, compared with 20% cure rates observed for animals injected with tumor cells transduced to express either IL-12 or B7.1 alone. The therapeutic efficacy of these vaccines was dependent on the dose of IL-12 delivered by transfected tumor cells and was ablated by *in situ* administration of $CTLA_4$–immunoglobulin or neutralizing antibodies against mIFN-γ or mTNF-α. Similar synergistic action of IL-12 and B7-1 has been noted in C3H mice bearing mB7-1 transfected K1735 melanoma or SCK mammary carcinomas (Coughlin *et al.*, 1995). Systemic application of rIL-12 resulted in 75–100% cure rates of these B7-1^+ tumors whereas only 0–40% was observed for IL-12 treatment of B7-1^- parental tumors. As noted above, IFN-γ, $CD4^+$ and $CD8^+$ T cells were required for therapeutic efficacy resulting in systemic protective immunity.

IL-12 Clinical Trials for the Treatment of Cancer

Initial phase I multicenter trials (Atkins *et al.*, 1997) evaluated patients with advanced cancer using 3–1000 ng per kg per day doses of rhIL-12 given as a bolus (pretest) injection 2 weeks before chronic administration of IL-12 over the subsequent 3-week period (same dose as the pretest dose daily for 5 days of each week). Mild toxicity was noted, including anemia, thrombocytopenia, leukopenia, lymphopenia, hyperglycemia and hypoalbuminemia. These side-effects, as well as low-grade fever, nausea and headache, were rapidly reversible on discontinuation of IL-12 administration or application of acetominophen and anti-inflammatory drugs. Dose-limiting toxicity for IL-12 was determined to be at the 1000 ng/kg level, with oral stomatitis of unknown etiology and raised levels of transaminases noted in several patients. Of six patients evaluated, four exhibited stable disease, with one stable (for more than 2 years) partial response in a patient with renal carcinoma and one transient complete responder with melanoma noted. Based on the *in vitro* capacity of IL-12 alone or in combination with IFN-α, to suppress the growth of $CD4^+$ cutaneous T (Sézary) cell lymphomas and to counter the predominant T_{H2}-biased immune responses observed in cutaneous lesions, phase I trials implementing IL-12 are also now ongoing for patients with Sézary T-cell lymphoma (Rook *et al.*, 1996).

The application of rhIL-12 in the treatment of cancer has provided equivocal results to date, with severe toxicity noted in early phase II trials (Cohen, 1995). In particular, several patients with RCC displayed pathology in multiple organ systems, with two patients succumbing to toxicity attributable to IL-12. These patients did not receive a pretest dose of IL-12 or a rest period before more chronic administration of IL-12, which was resident in phase I clinical trials. This pretest dose of IL-12 appears to limit the high systemic production of IFN-γ and the associated toxicity in patients with cancer noted in the phase II trials. This approach has been redesigned in ongoing revised phase II trials of rIL-12 carried out by the Genetics Institute.

Based on the success of genetic approaches employing IL-12 cDNA in preclinical murine tumor models, IL-12 'gene therapy' approaches have been designed and are being implemented for the treatment of patients with advanced cancer. We have recently completed a phase I–II clinical trial at the University of Pittsburgh School of Medicine and the University of Pittsburgh Cancer Institute using perilesional injection of IL-12 cDNA transduced autologous fibroblasts for the treatment of cutaneous cancer (basal cell carcinoma, melanoma, squamous cell carcinoma). Patient fibroblasts are infected with amphotropic retrovirus engineered to deliver the IL-12 p35 and p40 chains, as well as the neo^r gene product. After infection and selection in G418, transduced fibroblasts are injected perilesionally in numbers sufficient to deliver up to 7 μg hIL-12 per 24 h at the cumulative injection sites. By provision of this level of IL-12 in the locoregional tumor site, and not systemically, untoward toxicity affiliated with the early phase II clinical trials applying rhIL-12 would theoretically be averted. Three patients at the highest doses applied (7000–9000 ng/day) have exhibited acute necrosis of the tumor within 30–60 min of fibroblast–IL-12 injection, which was associated with severe local pain. Thus far in 32 patients treated to date, the treatment has been otherwise well tolerated, with 28% of the 26 patients treated exhibiting an objective clinical response (tumor regression more than 50%). IL-12 gene therapy approaches using IL-12 transfected autologous melanoma cells as a vaccine have also been initiated in multi-center trials (Schadendorf *et al.*, 1995). Additional vector delivery systems allowing for locoregional production of IL-12 *in situ* include recombinant adenoviral vectors (Bramson *et al.*, 1996; Caruso *et al.*, 1996), vaccinia constructs (Meko *et al.*, 1996) and Semliki Forest virus vectors (Zhang *et al.*, 1997).

IL-12 in Transplantation

The ability of IL-12 to promote cellular immunity against alloantigens (Gately *et al.*, 1986; Oshima and Delespesse, 1997) is clearly counter to the goals of maintenance of tissue allografts in an immunocompetent host. Strategies designed to minimize or neutralize the production of IL-12 in the recipient of an allograft in hope of promoting antigen-specific tolerance are desired and would serve as the conceptual basis for the construction of clinical trials in the transplant setting. Gorczynski *et al.* (1996) have demonstrated that the pretransplant transfer of donor splenocytes or NLDC145$^+$ dendritic cells into the recipient portal vein enhances the durability of allogenic skin transplants (B10.Br > C3H). Graft survival, associated with T_{H2} immune response, can be further augmented by injection of neutralizing anti-IL-12 mAb. Further, *in vivo* production of IL-12 p40 homodimer, which antagonizes the binding of bioactive IL-12 to the IL-12R, can prevent or prolong allograft survival. Kato *et al.* (1996a) showed that murine myoblast cell line C2C12 transfected with the mIL-12 p40 cDNA via retroviral-mediated gene transfer had far greater survival than non-transduced C2C12 cells when transplanted into allogeneic recipients. This enhanced graft retention was associated with depressed splenic IFN-γ production, CTL generation, and DTH responses promoted against C2C12 target cells. These results support the potential clinical benefit of paracrine IL-12 p40 homodimer production in potentiating allografts.

IL-12 in Autoimmunity

The role of IL-12 in autoimmune disease has been evaluated in a series of murine models (reviewed in Seder *et al.*, 1996). In many experimental models of autoimmunity, T_{H1}-type immunity is associated with pathogenesis, whereas T_{H2}-type immunity is associated with host protection against autoimmune disease (Adorini *et al.*, 1996). These data support the role of IL-12 in activating or exacerbating auotimmune disease and support the development of IL-12 antagonists for clinical applications designed to treat patients with at least certain forms of T_{H1}-associated autoimmunity. rIL-12 strongly enhances cellular and humoral autoimmunity in DBA/1 mice injected with chicken type II collagen and incomplete Freund's adjuvant, with more than 80% of animals developing degenerative arthritis (Szeliga *et al.*, 1996).

Segal and Shevach (1996) compared non-MHC-related factors involved in the inherent susceptibility of SJL (H-2^s) or resistance of B10.S (H-2^s) mice to experimental allergic encephalitis (EAE). The resistance of B10.S mice was determined to reflect an inability of APCs to elicit T_{H1} immunity and IFN-γ production in response to myelin basic protein (MBP) challenge. This apparent defect could be overcome by anti-MBP T-cell exposure to IL-12, with the resulting T cells capable of transferring EAE to naive recipient animals. The treatment of animals receiving encephalitogenic T cells with neutralizing anti-IL-12 antibodies attenuates the disease (Leonard *et al.*, 1995). Furthermore, *in vivo* administration of the IL-12 'antagonists' IL-4 and IL-10 results in clinical remission of EAE (Racke *et al.*, 1994).

In situ production of IL-12 has been documented in active lesions in multiple sclerosis in humans (Windhagen *et al.*, 1995) and in EAE in mice (Becher *et al.*, 1996). Autoreactive T cells recovered from the central nervous system of affected animals with EAE exhibit T_{H1}-type immune reactivity that is dependent on IL-12 production. IL-12 p70 is found at increased levels in the serum of patients with chronic progressive multiple sclerosis (Nicoletti *et al.*, 1996).

Germann *et al.* (1995) demonstrated that IL-12, like *Mycobacterium tuberculosis*, when emulsified in oil with type II collagen can induce severe arthritis in DBA/1 mice. Enhanced T_{H1}-associated immunity was evidenced with heightened production of IFN-γ and increased titers of collagen-specific IgG_{2a} and IgG_{2b} antibody titers in arthritic animals. This same group observed that IL-12 administered into CBA/J (T_{H2}-biased strain) mice receiving alum–protein vaccines (KLH, hemocyanin, phospholipase A_2) increase their antigen-specific titers of T_{H1} isotype antibodies (IgG_{2a}, IgG_{2b}, and IgG_3) by 100–1000-fold. No impact was observed for IgG_1 antibody titers, while T_{H2}-associated IgE titers were reduced.

IL-12 in Infectious Disease

The balance of T_{H1}- *versus* T_{H2}-type immunity is important in determining the resulting host resistance or susceptibility to various infectious diseases. The reader is referred to several recent reviews on this topic (Locksley, 1993; Scharton-Kersten *et al.*, 1995; Tripp and Unanue, 1995). Unanue and coworkers showed that T_{H1}-type cellular immunity and IFN-γ production was required for mice effectively to regulate infections with the Gram-positive bacterium *L. monocytogenes* (Bancroft *et al.*, 1987). In SCID mice, deficient in mature B- and T-cell function, *L. monocytogenes* is controlled, but not rejected. This control of the pathogen is IFN-γ dependent, with IL-12 induction of IFN-γ from NK cells in SCID mice appearing to represent the dominant relevant mechanism (Tripp *et al.*, 1993; Tripp and Unanue, 1995). Similar studies by Gazzinelli *et al.* (1994) supported IL-12 induction of TNF-α and IFN-γ in SCID mice infected with the intracellular pathogen *T. gondii* as critical steps in host resistance. Administration of rIL-12 prolongs the life of *T. gondii*-infected SCID mice and has been shown to provide similar therapeutic benefit to mice infected with alternate intracellular pathogens such as *Leishmania* (Heinzel *et al.*, 1993). Outbred CD1 mice pretreated or treated with IL-12 during group A (Gram-positive) streptoccocal infection exhibit enhanced survival over controls, with protection affiliated with augmented natural immunity (Metzger *et al.*, 1995). Mice infected with murine cytomegalovirus (MCMV) exhibit early IL-12 and IFN-α/β production by APCs, leading to NK cell secretion of IFN-γ and NK cell proliferation and cytotoxicity, respectively (Orange and Biron, 1996).

Scharton-Kersten *et al.* (1995) showed that IL-12 was essential for NK cell activation and T_{H1}-type cell development in C3H mice infected with *L. major*. Neutralizing antibodies to IL-12 abrogate T_{H1}, and augment T_{H2}, immune reactivity. Interestingly, *L. major* also promoted early production of IL-12 in the 'T_{H2}-biased' Balb/c strain of mice (Scharton-Kersten *et al.*, 1995), despite the subsequent inability to enhance NK cell activity and IFN-γ production. This refractiveness to T_{H1}-biased immunity appears to be due to the antagonistic action of other early-induced cytokines, such as TGF-β, IL-4 and IL-10 in Balb/c mice.

IL-12 therapy of susceptible A/J mice infected with blood-stage *Plasmodium chabaudi AS* decreased parasitemia and, at 0.1 μg levels, promoted increased survival of the animals (Stevenson *et al.*, 1995). Interestingly, while IL-12 treatment of resistant B6 strain mice infected with *P. chabaudi AS* also yielded a lower parasitic load, 40% of treated animals still died. Therapeutic efficacy required the *in situ* production of IFN-γ, TNF-α and nitric oxide, and involved T_{H1} $CD4^+$ T cells.

Intratracheal infection of CBA/J mice with *Klebsiella pneumoniae* induced *in situ* production of IL-12 in the lung (Greenberger *et al.*, 1996). Neutralization of IL-12 using specific antibody administration or injection of recombinant adenovirus encoding IL-12 p35 and IL-12 p40 resulted in exacerbation of disease pathogenesis or enhanced recovery of treated animals, respectively. IL-12-promoted immune resistance to infection was determined to be dependent on *in vivo* production of TNF-α and IFN-γ.

Early stages in *M. tuberculosis* infection provide *in vitro* and *in vivo* models of pathogen-induced IL-12 production from host macrophages with a subsequent induction of IFN-γ-producing CD4$^+$ T cells (Cooper *et al.*, 1995). *In vivo* antibody neutralization of IL-12 resulted in the reduced ability of mice to regulate bacterial growth. Treatment of mice with rIL-12 enhanced resistance to bacterial infection. Large and numerous granulomas develop in genetically susceptible Balb/c, but not genetically resistant DBA/2, mice infected chronically with *M. avium* (Kobayashi *et al.*, 1996). This inability in Balb/c mice to regulate the progression of chronic mycobacterial infection correlates with a deficiency in *in situ* IL-12 production, which may be overcome by systemic administration of rmIL-12, resulting in a reduced bacterial burden.

Patients with chronic disease (i.e. HIV, cancer) exhibit immune dysfunction that is correlated with decreased capacity of host mononuclear cells to produce IL-12 and increased susceptibility to opportunistic infection (Gately *et al.*, 1996). Similarly, Balb/c and CB6F1 mice are susceptible to chronic infection with *Leishmania* (Li *et al.*, 1996). The CB6F1 mice develop a mixed T_{H1} and T_{H2} immune response during the first month of infection that biases towards the T_{H1} cellular response as the cutaneous lesions are resolved. This response may be facilitated by application of IL-12 and anti-IL-4 antibody, resulting in faster shrinkage in lesional size and reduced parasite numbers. Mice injected with soluble *L. monocytogenes* antigens or heat-killed *Listeria* in the context of intraperitoneally injected rIL-12 evolve strong T_{H1}-associated immunity, enhanced APC function, and protection against virulent *Leishmania*. Similarly, PBMCs drawn from patients with visceral leishmaniasis exhibit anergy to leishmanial antigens assessed in T-cell proliferative and IFN-γ secretion assays *in vitro* (Bacellar *et al.*, 1996). This hyporesponsiveness in the cellular immune response is largely eradicated when rIL-12 is added to the induction assays.

Interestingly, IL-12 has been shown to inhibit HIV replication in infected human monocytes *in vitro* (Akridge *et al.*, 1996). IL-12 blocked viral reverse transcriptase (RT) activity by more than 75% in *in vitro* cocultures of infected macrophages and autologous PBMCs. This effect appears to involve both direct effects on the infected macrophages (suppression of HIV replication) and indirect effects on responder PBMCs (suppression of HIV proliferation). In addition to such clinical ramifications, this suppression of viral transcription has been noted in *in vitro* studies. The ability effectively to transduce human CD34$^+$ myeloid (CFU-GM) and erythroid (BFU-E) progenitor cells with amphotropic retroviral vectors is inhibited by IL-12 in a dose-dependent manner and is correlated with IL-12-induced TNF-α production (Xiao *et al.*, 1996). IL-12 does not appear to impact negatively on the retrovirus receptor density on target cells.

Wild-type mice develop increased titers of IgG_{2a} and IgG_{2b} antibodies directed against schistosomal antigens with repeated vaccination with IL-12 and attenuated cercariae or soluble larval antigens (Mountford *et al.*, 1996; Wynn, 1996). NK1.1$^+$ cells were required for IFN-γ production affiliated with the T_{H1} bias of the vaccine-induced immunity. In many cases, vaccinated animals were completely resistant to challenge with *Schistosoma mansoni* (Wynn, 1996). In marked contrast, Wynn *et al.* (1995) observed that IL-12 exacerbates rather than suppresses T_{H2}-dependent pathology in IFN-γ knockout mice. IFN-γ knockout mice develop enhanced T_{H2}-associated eosinophilia and granuloma formation in response to schistosome eggs, with raised serum IgE titers compared with wild-type mice.

The importance of IL-12 in promoting T_{H1}-mediated immunity prompts the immediate concern that pathogens would have evolved mechanisms that prevent, neutralize or diminish the utility of IL-12 in the evolving, normally effective, immune clearance mechanisms. This may actually occur in the case of certain virus infections (i.e. EBV) that encode a viral homolog of the IL-12 p40 chain called EBI-3 (Devergne *et al.*, 1996). While not yet shown to exhibit antagonistic activity in IL-12 signaling pathways, EBI-3 may dimerize with IL-12 p35 chains, thereby diminishing infected cellular production of bioactive IL-12 heterodimers and promoting enhanced secretion of IL-12 p40 homodimers. We have recently demonstrated that EBI-3 is found in DCs as well as B cells and that EBI-3 overproduction in cells results in diminution of IL-12 production (Kadakia *et al.*, 1997). It is likely that additional pathogen-associated immune debilitating mechanisms targeting IL-12 production will be identified with a more systematic evaluation of the genomes of pathogenic organisms.

IL-12 Clinical Trials in Infectious Diseases

Patients with HIV infection display suppressed cellular cytotoxicity afflicting T and NK cells (Trinchieri, 1995). In a phase I clinical trial of rIL-12 (300–1000 ng/kg) applied to 15 patients with HIV, Kohl *et al.* (1996) discerned a rapid (within 24 h) reduction in the number of PBMC-derived NK cells and of NK cell-mediated cytotoxicity. This reduced presence of effector NK cells in the peripheral blood of treated patients may reflect IL-12-modified trafficking of these cells *in vivo*.

FUTURE DIRECTIONS

Given its central role in the promotion of T_{H1}-type cellular immunity, direct application of IL-12 or modulation of IL-12 levels will remain the prime focus of research for the treatment of cancer, infectious disease, transplantation and autoimmunity. In the cases of cancer and infectious disease therapies, the goal of clinical trials is to enhance the provision of IL-12, at the appropriate time and preferably within sites of active T-cell induction. In the cases of transplantation and autoimmunity, the goal is often to reduce the production level of the proinflammatory cytokine IL-12, at least within the target tissues of interest.

While the initial application of rhIL-12 in phase II clinical trials was associated with significant toxicity owing to the lack of a desensitizing effect observed in precursor protocols employing a preliminary test dose of IL-12, significant antitumor efficacy has been observed in a series of clinical protocols. The ability of paracrine delivery of IL-12 through gene-based therapies to result in objective tumor regression and the efficacy of IL-12 to serve as an effective biologic adjuvant in various vaccine formats similarly reinforces the translational importance of this T_{H1}-biasing cytokine. Given its potent antiangiogenic activity in the cancer setting, IL-12 may be useful in the treatment of angiogenesis-dependent malignancy, particularly in combination with alternate anti-angiogenic agents such as retinoids, IFN-α or vitamin D_3 analogs (Majewski *et al.*, 1996). In addition, the adjuvant qualities of rIL-12 make it an attractive candidate for local subcutaneous administration in vaccines containing soluble (synthetic) tumor antigens. For instance, Balb/c mice injected subcutaneously with a mutant peptide epitope derived from the p53 gene product emulsified with the QS-21 adjuvant failed to provide therapeutic benefit to mice bearing Meth A sarcoma (mutant p53$^+$), unless systemic IL-12 was administered simultaneously with the vaccine (Noguchi *et al.*, 1995).

We are conducting a phase I clinical trial at the University of Pittsburgh and University of Pittsburgh Cancer Institute that evaluates the ability of melanoma peptides (derived from the tyrosinase, gp100 and melanoma antigen recognized by T cells-1 (MART-1) antigens) and IL-12 injected subcutaneously to elicit CTL responses and to provide therapeutic efficacy in HLA-A2$^+$ patients with metastatic melanoma.

The ability of type I interferons to enhance expression of the high-affinity IL-12R, preferentially on T_{H1} clones, supports the combined application of IL-12 and IFN-α/β in trials designed to foster the development of T_{H1}-associated immunity (Wenner *et al.*, 1996). The contrasting capacity of IL-4 or IL-10 to reduce the stability of IL-12R complexes (Rogge *et al.*, 1997; Szabo *et al.*, 1997) supports the use of these factors in protocols designed to diminish the T_{H1} immune response due to its associated clinical pathology, such as that observed in autoimmunity.

Recent evidence suggests that vaccines best designed to elicit effective antiviral or antitumor immunity implement or access the most potent APCs, dendritic cells (Mayordomo *et al.*, 1995; Paglia *et al.*, 1996, Porgador and Gilboa, 1996; Zitvogel *et al.*, 1996a). Further, despite the ability of at least certain subsets of dendritic cells to produce IL-12 upon T-cell interaction or CD40 ligation, the provision of picogram quantities of exogenous IL-12 may significantly enhance the ability of these APCs to promote antigen-specific T-cell responses *in vitro* (Bhardwaj *et al.*, 1996). Similar systemic application of low-dose rIL-12 may promote enhanced T_{H1}-based immunity *in vivo* when conjoined with dendritic cell-based vaccines in individuals with weak immune reactivity or in cases where dendritic cell/T cell ratios are concluded to be rate limiting. Alternatively, gene therapy approaches that target the delivery of IL-12 to vaccine-dependent dendritic cells have been shown to enhance the induction of antitumor immunity *in vivo* and to augment tumor antigen-specific T-cell immune reactivity *in vitro* (Zitvogel *et al.*, 1996c; Tueting *et al.*, 1997).

CONCLUSIONS

IL-12 is an important player in the host's ability to promote protective immunity against infectious disease, foreign grafts and tumorigenesis. Produced principally by the sentinel APCs of the body after antigenic challenge and T-cell interaction, IL-12 promotes the rapid release of IFN-γ from T and NK cells, and facilitates a biased T_{H1}-type, antigen-specific, cellular immune response. This bias towards T_{H1}-type immunity is determined, at least in part, by the preferential expression of the high-affinity IL-12R by T_{H1} CD4$^+$ T cells. IL-12 also drives the preferential switching of antigen-specific immunoglobulins towards IgG_{2a} and IgG_{2b} isotypes, which may enhance phagocytic opsonization, complement fixation and ADCC mechanisms designed to clear disease or foreign tissue. The ability of IL-12 to synergize effectively with a number of alternate factors in the induction of T_{H1}-mediated immunity makes it an attractive candidate for current and future application in multicomponent vaccines and therapies.

ACKNOWLEDGEMENTS

The authors thank Drs Cynthia Brissette-Storkus, Cara Wilson, Lisa Salvucci Kierstead and Thomas Tueting for careful review of this chapter. This work was supported by NIH grant CA 68067 (WJS, HT, MTL).

REFERENCES

Adorini, L., Guery, J.C. and Trembleau, S. (1996). Manipulation of the Th1/Th2 cell balance: an approach to treat human autoimmune disease? *Autoimmunity* **23**, 53–68.

Afonso, L.C., Scharton, T.M., Vieira, L.Q., Wysocka, M., Trinchieri, G. and Scott, P. (1994). The adjuvant effect of interleukin-12 in a vaccine against *Leishmania major*. *Science* **263**, 235–237.

Akridge, R.E. and Reed, S.G. (1996). Interleukin-12 decreases human immunodeficiency virus type 1 replication in human macrophage cultures reconstituted with autologous peripheral blood mononuclear cells. *J. Infect. Dis.* **173**, 559–564.

Annunziato, F., Manetti, R., Tomasevic, I., Guidizi, M.G., Biagiotti, R., Gianno, V., Germano, P., Mavilia, C., Maggi, E. and Romagnani, S. (1996). Expression and release of LAG-3-encoded protein by human $CD4^+$ T cells are associated with IFN-γ production. *FASEB J.* **10**, 769–776.

Armant, M., Delespesse, G. and Sarfati, M. (1995). IL-2 and IL-7 but not IL-12 protect natural killer cells from death by apoptosis and up-regulate bcl-2 expression. *Immunology* **85**, 331–337.

Atkins, M.B., Robertson, M.J., Gordon, M., Lotze, M.T., DeCoste, M., DuBois, J.S., Ritz, J., Sandler, A.B., Edington, H.D., Garzone, P.D., Mier, J.W., Canning, C.M., Battiato, L., Tahara, H. and Sherman, M.L. (1997). Phase I evaluation of intravenous recombinant human interleukin-12 (rhIL-12) in patients with advanced malignancies. *Clin. Cancer Res.* **3**, 409–418.

Bacellar, O., Brodskyn, C., Guerreiro, J., Barral-Netto, M., Costa, C.H., Coffman, R.L., Johnson, W.D. and Carvalho, E.M. (1996). Interleukin-12 restores interferon-γ production and cytotoxic responses in visceral leishmaniasis. *J. Infect. Dis.* **173**, 1515–1518.

Bacon, C.M., McVicar, D.W., Ortaldo, J.R., Rees, R.C., O'Shea, J.J. and Johnston, J.A. (1995a). Interleukin 12 (IL-12) induces tyrosine phosphorylation of JAK2 and TYK2: differential use of Janus family tyrosine kinases by IL-2 and IL-12. *J. Exp. Med.* **181**, 399–404.

Bacon, C.M., Petricoin, E.F. III, Ortaldo, J.R., Rees, R.C., Larner, A.C., Johnston, J.A. and O'Shea, J.J. (1995b). Interleukin 12 induces tyrosinse phosphorylation and activation of STAT4 in human lymphocytes. *Proc. Natl Acad. Sci. U.S.A.* **92**, 7307–7311.

Ballas, Z.K., Rasmussen, W.L. and Kreig, A.M. (1996). Induction of NK activity in murine and human cells by CpG motifs in oligodeoxynucleotides and bacterial DNA. *J. Immunol.* **157**, 1840–1845.

Bancroft, G.J., Schreiber, R.D., Bosma, G.C., Bosma, M.J. and Unanue, E.R. (1987). A T cell-independent mechanism of macrophage activation by interferon-γ. *J. Immunol.* **139**, 1104–1107.

Becher, B., Dodelet, V., Fedorowicz, V. and Antel, J.P. (1996). Soluble tumor necrosis factor receptor inhibits interleukin 12 production by stimulated human adult microglial cells *in vitro*. *J. Clin. Invest.* **98**, 1539–1543.

Bertagnolli, M.M., Herrmann, S.H., Pinto, V.M., Schoof, D.D. and Eberlein, T.J. (1991). Approaches to immunotherapy of cancer: characterization of lymphokines as second signals for cytotoxic T-cell generation. *Surgery* **110**, 459–468.

Bertolini, F., Soligo, D., Lazzari, L., Corsini, C., Servida, F. and Sirchia, G. (1995). The effect of interleukin-12 in *ex-vivo* expansion of human haematopoietic progenitors. *Br. J. Haematol.* **90**, 935–938.

Bhardwaj, N., Seder, R.A., Reddy, A. and Feldman, M.V. (1996). IL-12 in conjunction with dendritic cells enhances antiviral $CD8^+$ CTL responses *in vitro*. *J. Clin. Invest.* **98**, 715–722.

Bianchi, R., Grohmann, U., Belladonna, M.L., Silla, S., Fallarino, F., Ayroldi, E., Fioretti, M.C. and Puccetti, P. (1996). IL-12 is both required and sufficient for initiating T cell reactivity to a class I-restricted tumor peptide (P815AB) following transfer of P815AB-pulsed dendritic cells. *J. Immunol.* **157**, 1589–1597.

Bliss, J., Van Cleave, V., Murray, K., Wiencis, A., Ketchum, M., Maylor, R., Haire, T., Resmini, C., Abbas, A.K. and Wolf, S.F. (1996). IL-12, as an adjuvant, promotes a T helper 1 cell, but does not suppress a T helper 2 cell recall response. *J. Immunol.* **156**, 887–894.

Bloom, E.T. and Horvath, J.A. (1994). Cellular and molecular mechanisms of the IL-12-induced increase in allospecific murine cytolytic T cell activity. Implications for the age-related decline in CTL. *J. Immunol.* **152**, 4242–4254.

Bradley, L.M., Yoshimoto, K. and Swaim, S.L. (1995). The cytokines IL-4, IFN-γ and IL-12 regulate the development of subsets of memory effector helper T cells *in vitro*. *J. Immunol.* **155**, 1713–1724.

Bramson, J., Hitt, M., Gallichan, W.S., Rosenthal, K.L., Gauldie, J. and Graham, F.L. (1996). Construction of a double recombinant adenovirus vector expressing a heterodimeric cytokine: *in vitro* and *in vivo* production of biologically active interleukin-12. *Hum. Gene Ther.* **7**, 333–342.

Brunda, M.J., Luistro, L., Warrier, R.R., Wright, R.B., Hubbard, B.R., Murphy, M., Wolf, S.F. and Gately, M.K.

(1994). Antitumor and antimetastatic activity of interleukin 12 against murine tumors. *J. Exp. Med.* **178**, 1223–1230.

Brunda, M., Luistro, L., Hendrzak, J.A., Fountoulakis, M., Garotta, G. and Gately, M.K. (1995). Role of interferon-γ in mediating the antitumor efficacy of interleukin-12. *J. Immunother.* **17**, 71–77.

Caruso, M., Pham-Nguyen, K., Kwong, Y.L., Xu, B., Kosai, K.I., Finegold, M., Woo, S.L. and Chen, S.H. (1996). Adenovirus-mediated interleukin-12 gene therapy for metastatic colon carcinoma. *Proc. Natl Acad. Sci. U.S.A.* **93**, 11 302–11 306.

Cella, M., Scheidegger, D., Palmer-Lehmann, K., Lane, P., Lanzevecchia, A. and Alber, G. (1996). Ligation of CD40 on dendritic cells triggers production of high levels of interleukin-12 and enhanced T cell stimulatory capacity: T-T help via APC activation. *J. Exp. Med.* **184**, 747–752.

Chan, S.H., Perussia, B., Gupta, J.W., Kobayashi, M., Pospisil, M., Young, H.A., Wolf, S.F., Young, D., Clark, S.C. and Trinchieri, G. (1991). Induction of interferon γ production by natural killer cell stimulatory factor: characterization of the responder cells and synergy with other inducers. *J. Exp. Med.* **173**, 869–879.

Chehimi, J., Starr, S.E., Frank, I., Rengaraju, M., Jackson, S.J., Llanes, C., Kobayashi, M., Perussia, B., Young, D. and Nickbarg, E. (1992). Natural killer (NK) cell stimulatory factor increases the cytotoxic activity of NK cells from both healthy donors and human immunodeficiency virus-infected patients. *J. Exp. Med.* **175**, 789–796.

Chensue, S.W., Warmington, K.S., Rith, J.H., Sanghi, P.S., Lincoln, P. and Kunkel, S.L. (1996). Role of monocyte chemoattractant protein-1 (MCP-1) in Th1 (mycobacterial) and Th2 (schistosomal) antigen-induced granuloma formation: relationship to local inflammation, Th cell expression and IL-12 production. *J. Immunol.* **157**, 4602–4608.

Chizzonite, R., Truitt, T., Desai, B., Nunes, P., Podlaski, F.J., Stern, A. and Gately, M.K. (1992). IL-12 receptor I. Characterization of the receptor on phytohemagglutinin-activated human lymphoblasts. *J. Immunol.* **148**, 3117–3124.

Cho, S.S., Bacon, C.M., Sudarshan, C., Rees, R.C., Finbloom, D., Pine, R. and O'Shea, J.J. (1996). Activation of STAT4 by IL-12 and IFN-α: evidence for the involvement of ligand-induced tyrosine and serine phosphorylation. *J. Immunol.* **157**, 4781–4789.

Chouaib, S., Chehimi, J., Bani, L., Gentet, N., Tursz, T., Gay, F., Trinchieri, G. and Mami-Chouaib, F. (1994). Interleukin 12 induces the differentiation of major histocompatibility complex class I-primed cytotoxic T-lymphocyte precursors into allospecific cytotoxic effectors. *Proc. Natl Acad. Sci. U.S.A.* **91**, 12 659–12 663.

Chua, A.O., Chizzonite, R., Desai, B.B., Truitt, T.P., Nunes, P., Minetti, L.J., Warrier, R.R., Presky, D.H., Levine, J.F. and Gately, M.K. (1994). Expression cloning of a human IL-12 receptor component. A new member of the cytokine receptor superfamily with strong homology to gp130. *J. Immunol.* **153**, 128–136.

Chua, A.O., Wilkinson, V.L., Presky, D.H. and Gubler, U. (1995). Cloning and chracterization of a mouse IL-12 receptor-β component. *J. Immunol.* **155**, 4286–4294.

Cleveland, M.G., Gorham, J.D., Murphy, T.L., Tuomanen, E. and Murphy, K.M. (1996). Lipoteichoic acid preparations of Gram-positive bacteria induce interleukin-12 through a CD14-dependent pathway. *Infect. Immun.* **64**, 1906–1912.

Cohen, J. (1995). IL-12 deaths: explanation and a puzzle. *Science* **270**, 908.

Colombo, M.P., Vagliani, M., Spreafico, F., Parenza, M., Chiodoni, C., Melani, C. and Stoppacciaro, A. (1996). Amount of IL-12 available at the tumor site is critical for tumor regression. *Cancer Res.* **56**, 2531–2534.

Constantinescu, C.S., Frei, K., Wysocka, M., Trinchieri, G., Malipiero, U., Rostami, A. and Fontana, A. (1996). Astrocytes and microglia produce interleukin-12 p40. *Ann. N.Y. Acad. Sci. U.S.A.* **795**, 328–333.

Cooper, A.M., Roberts, A.D., Rhoades, E.R., Callahan, J.E., Getzy, D.M. and Orme, I.M. (1995). The role of interleukin-12 in acquired immunity to *Mycobacterium tuberculosis* infection. *Immunology* **84**, 423–432.

Coughlin, C.M., Wysocka, M., Kurzawa, H.L., Lee, W.M., Trinchieri, G. and Eck, S.L. (1995). B7-1 and interleukin 12 synergistically induce effective anti-tumor immunity. *Cancer Res.* **55**, 4980–4987.

Cousens, L.P., Orange, J.S., Su, H.C. and Biron, C.A. (1997). Interferon-α/β inhibition of interleukin 12 and interferon-γ production *in vitro* and endogenously during viral infection. *Proc. Natl Acad. Sci. U.S.A.* **94**, 634–639.

D'Andrea, A., Rengaraju, M., Valiante, N.M., Chehimi, J., Kubin, M., Aste, M., Chan, S.H., Kobayashi, M., Young, D. and Nickbarg, E. (1992). Production of natural killer cell stimulatory factor (interleukin-12) by peripheral blood mononuclear cells. *J. Exp. Med.* **176**, 1387–1398.

Desai, B.B., Quinn, P.M., Wolitzky, A.G., Mongini, P.K., Chizzonite, R. and Gately, M.K. (1992). IL-12 receptor. II. Distribution and regulation of receptor expression. *J. Immunol.* **148**, 3125–3132.

Devergne, O., Hummel, M., Koeppen, H., Le Beau, M.M., Nathanson, E.C., Kieff, E. and Birkenbach, M. (1996).

A novel interleukin-12 p40-related protein induced by latent Epstein–Barr virus infection in B lymphocytes. *J. Virol.* **70**, 1143–1153.

Dilulio, N.A., Xu, H. and Fairchild, R.L. (1996). Diversion of CD4$^+$ T cell development from regulatory T helper to effector T helper cells alters the contact hypersensitivity response. *Eur. J. Immunol.* **26**, 2606–2612.

Eberl, G., Kessler, B., Eberl, L.P., Brunda, M.J., Valmori, D. and Corradin, G. (1996). Immunodominance of cytotoxic T lymphocyte epitopes co-injected *in vivo* and modulation by interleukin-12. *Eur. J. Immunol.* **26**, 2709–2716.

Eng, V.M., Car, B.D., Schnyder, B., Lorenz, M., Lugli, S., Aguet, M., Anderson, T.D., Ryfdfel, B. and Quesniaux, V.F. (1995). The stimulatory effects of interleukin (IL)-12 on hematopoiesis are antagonzied by IL-12-induced interferon γ *in vivo*. *J. Exp. Med.* **181**, 1893–1898.

Enk, C.D., Mahanty, S., Blauvelt, A. and Katz, S.I. (1996). UVB induces IL-12 transcription in human keratinocytes *in vivo* and *in vitro*. *Photochem. Photobiol.* **63**, 854–859.

Estaquier, J., Idziorek, T., Zou, W., Emilie, D., Farber, C.M., Bourez, J.M. and Ameisen, J.C. (1995). T helper type 1/T helper type 2 cytokines and T cell death: preventive effect of interleukin 12 on activation-induced and CD95 (FAS/APO-1)-mediated apoptosis of CD4$^+$ T cells from human immunodeficiency virus-infected persons. *J. Exp. Med.* **182**, 1759–1767.

Fallarino, F., Uyttenhove, C., Boon, T. and Gajewski, T.F. (1996). Endogenous IL-12 is necessary for rejection of P815 tumor variants *in vivo*. *J. Immunol.* **156**, 1095–1100.

Fantuzzi, L., Gessani, S., Birghi, P., Varano, B., Conti, L., Puddu, P. and Belardelli, F. (1996). Induction of interleukin-12 (IL-12) by recombinant glycoprotein gp120 of human immunodeficiency virus type 1 in human monocytes/macrophages: requirement of γ interferon for IL-12 secretion. *J. Virol.* **70**, 4121–4124.

Ferbas, J.J., Toso, J.F., Logar, A.J., Navratil, J.S. and Rinaldo, C.R. Jr. (1994). CD4$^+$ blood dendritic cells are potent producers of IFN-α in responses to *in vitro* HIV-1 infection. *J. Immunol.* **152**, 4649–4662.

Filgueira, L., Nestle, F.O., Rittig, M., Joller, H.I. and Groscurth, P. (1996). Human dendritic cells phagocytose and process *Borrelia burgdorferi*. *J. Immunol.* **157**, 2998–3005.

Fox, F.E., Kubin, M., Cassin, M., Niu, Z., Hosoi, J., Toril, H., Granstein, R.D., Trinchieri, G. and Rook, A.H. (1997). Calcitonin gene-related peptide inhibits proliferation and antigen presentation by human peripheral blood mononuclear cells: effects on B7, interleukin 10, and interleukin 12. *J. Invest. Dermatol.* **108**, 43–48.

Gajewski, T.F., Renauld, J.C., Van Pel, A. and Boon, T. (1995). Costimulation with B7-1, IL-6 and IL-12 is sufficient for primary generation of murine antitumor cytolytic T lymphocytes *in vitro*. *J. Immunol.* **154**, 5637–5648.

Gately, M.K. (1993). Interleukin-12: a recently discovered cytokine with potential for enhancing cell-mediated immune responses to tumors. *Cancer Invest.* **11**, 500–506.

Gately, M.K. and Mulqueca, M.J. (1996). Interleukin-12: potential clinical applications in the treatment and prevention of infectious diseases. *Drugs* **52**, 18–25.

Gately, M.K., Wilson, D.E. and Wong, H.L. (1986). Synergy between interleukin 2 (rIL2) and IL-2-depleted lymphokine-containing supernatants in facilitating allogeneic human cytolytic T lymphocyte responses *in vitro*. *J. Immunol.* **136**, 1274–1282.

Gately, M.K., Desai, B.B., Wolitzky, A.G., Quinn, P.M., Dwyer, C.M., Podlaski, F.J., Familletti, P.C., Sinigaglia, F., Chizzonite, R., Gubler, U. and Stern, A.S. (1991). Regulation of human lymphocyte proliferation by a heterodimeric cytokine, IL-12 (cytokine lymphocyte maturation factor). *J. Immunol.* **147**, 874–882.

Gately, M.K., Gubler, U., Brunda, M.J., Nadeau, R.R. anderson, T.D., Lipman, J.M. and Sarmiento, U. (1994a). Interleukin-12: a cytokine with therapeutic potential in oncology and infectious diseases. *Ther. Immunol.* **1**, 187–196.

Gately, M.K., Warrier, R.R., Honasoge, S., Carvajal, D.M., Faherty, D.A., Connaughton, S.E., Anderson, T.D., Sarmiento, U., Hubbard, B.R. and Murphy, M. (1994b). Administration of recombinant IL-12 to normal mice enhances cytotolytic lymphocyte activity and induces production of IFN-γ *in vivo*. *Int. Immunol.* **6**, 157–167.

Gately, M.K., Carvajal, D.M., Connaughton, S.E., Gillessen, S., Warrier, R.R., Kolinsky, K.D., Wilkinson, V.L., Dwyer, C.M., Higgins, G.F., Jr., Podlaski, F.J., Faherty, D.A., Familletti, P.C., Stern, A.S. and Presky, D.H. (1996). Interleukin-12 antagonist activity of mouse interleukin-12 p40 homodimer *in vitro* and *in vivo*. *Ann. N.Y. Acad. Sci.* **795**, 1–12.

Gazzinelli, R.T., Wysocka, M., Hayashi, S., Denkers, E.Y., Hieny, S., Casper, P., Trinchieri, G. and Sher, A. (1994). Parasite-induced IL-12 stimulates early IFN-γ synthesis and resistance during acute infection with *Toxoplasma gondii*. *J. Immunol.* **153**, 2533–2543.

Gearing, D.P. and Cosman, D. (1991). Homology of the p40 subunit of natural killer cell stimulatory factor (NKSF) with the extracellular domain of the interleukin-6 receptor. *Cell* **66**, 9–10.

Germann, T., Huhn, H., Zimmerman, F. and Rude, E. (1987). An antigen-independent physiological activation pathway for L3T4$^+$ T lymphocytes. *Eur. J. Immunol.* **17**, 775–781.

Germann, T., Bonggartz, N., Dlugonska, H., Hess, H., Schmitt, E., Kolbe, L., Kolsch, E., Podlaski, F.J., Gately, M.K. and Rude, E. (1995). Interleukin-12 profoundly up-regulates the synthesis of antigen-specific complement-fixing IgG_{2a}, IgG_{2b} and IgG_3 antibody subclasses *in vivo*. *Eur. J. Immunol.* **25**, 823–829.

Gerosa, F., Paganin, C., Peritt, D., Paiola, F., Scupoli, M.T., Aste-Amezaga, M., Frank, I. and Trinchieri, G. (1996). Interleukin-12 primes human CD4 and CD8 T cell clones for high production of both interferon-γ and IL-10. *J. Exp. Med.* **183**, 2559–2569.

Gillessen, S., Carvehal, D., Ling, P., Podlaski, F.J., Stremlo, D.L., Familletti, P.C., Gubler, U., Presky, D.H., Stern, A.S. and Gately, M.K. (1995). Mouse interleukin-12 (IL-12) p40 homodimer: a potent IL-12 antagonist. *Eur. J. Immunol.* **25**, 200–206.

Gollob, J.A. and Ritz, J. (1996). CD2–CD58 interaction and the control of T-cell interleukin-12 responsiveness. Adhesion molecules link innate and acquired immunity. *Ann. N.Y. Acad. Sci.* **795**, 71–81.

Gollob, J.A., Kawasaki, H. and Ritz, J. (1997). Interferon-γ and interleukin-4 regulate T cell interleukin-12 responsiveness through the differential modulation of high-affinity interleukin-12 receptor expression. *Eur. J. Immunol.* **27**, 647–652.

Gorczynski, R.M., Cohen, Z., Fu, X.M., Hua, Z., Sun, Y. and Chen, Z. (1996). Interleukin-13, in combination with anti-interleukin-12, increases graft prolongation after portal venous immunization with cultured allogeneic bone marrow-derived dendritic cells. *Transplantation* **62**, 1592–1600.

Greenberger, M.J., Kunkel, S.L., Streiter, R.M., Lukacs, N.W., Bramson, J., Gauldie, J., Graham, F.L., Hitt, M., Danforth, J.M. and Standiford, T.J. (1996). IL-12 gene therapy protects mice in lethal *Klebsiella* pneumonia. *J. Immunol.* **157**, 3006–3012.

Gubler, U. and Presky, D.H. (1996). Molecular biology of interleukin-12 receptors. *Ann. N.Y. Acad. Sci.* **795**, 36–40.

Gubler, U., Chua, A.O., Schoenhaut, D.S., Dwyer, C.M., McComas, W., Motyka, R., Nabavi, N., Wolitzky, A.G., Quinn, P.M., Familletti, P.C. and Gately, M.K. (1991). Coexpression of two distinct genes is required to generate secreted bioactive cytotoxic lymphocyte maturation factor. *Proc. Natl Acad. Sci. U.S.A.* **88**, 4143–4147.

Guery, J.-C., Ria, F., Galbaiati, F. and Adorini, L. (1997). Normal B cells fail to secrete interleukin-12. *Eur. J. Immunol.* **27**, 1632–1639.

Hayes, M.P., Wang, J. and Norcross, M.A. (1995). Regulation of interleukin-12 expression in human monocytes: selective priming by interferon-γ of lipopolysaccharide-inducible p35 and p40 genes. *Blood* **86**, 646–650.

Heinzel, F.P., Schoenhaut, D.S., Rerko, R.M., Rosser, L.E. and Gately, M.K. (1993). Recombinant interleukin 12 cures mice infected with *Leishmania major*. *J. Exp. Med.* **177**, 1505–1507.

Heinzel, F.P., Hujer, A.M., Ahmed, F.N. and Rerko, R.M. (1997). *In vivo* production and function of IL-12 p40 homodimers. *J. Immunol.* **158**, 4381–4388.

Heufler, C., Koch, F., Stanzl, U., Topar, G., Wysocka, M., Trinchieri, G., Enk, A., Steinman, R.M., Romani, N. and Schuler, G. (1996). Interleukin-12 is produced by dendritic cells and mediates T helper 1 development as well as interferon-γ production by T helper 1 cells. *Eur. J. Immunol.* **26**, 659–668.

Hilkens, C.M., Messer, G., Tesselaar, K., van Rietschoten, A.G., Kapsenberg, M.L. and Wierenga, E.A. (1996). Lack of IL-12 signaling in human allergen-specific Th2 cells. *J. Immunol.* **157**, 4316–4321.

Hino, A. and Nariuchi, H. (1996). Negative feedback mechanism suppresses interleukin-12 production by antigen-presenting cells interacting with T helper 2 cells. *Eur. J. Immunol.* **26**, 623–628.

Hirao, A., Takaue, Y., Kawano, Y., Sato, T., Abe, T., Saito, S., Kawahito, M., Okamoto, Y. and Makimoto, A. (1995). Synergism of interleukin 12, interleukin 3 and serum factor on primitive human hematopoietic progenitor cells. *Stem Cells* **13**, 47–53.

Hsieh, C.S., Macatonia, S.E., Tripp, C.S., Wolf, S.F., O'Garra, A. and Murphy, K.M. (1993). Development of Th1 CD4$^+$ T cells through IL-12 produced by *Listeria*-induced macrophages. *Science* **260**, 547–549.

Jacobsen, S.E., Veiby, O.P. and Smeland, E.B. (1993). Cytotoxic lymphocyte maturation factor (IL-12) is a synergistic growth factor for hematopoietic stem cells. *J. Exp. Med.* **178**, 413–418.

Jacobsen, S.E., Okkenhaug, C., Myklebust, J., Veiby, O.P. and Lyman, S.D. (1995). The FLT3 ligand potently and directly stimulates the growth and expansion of primitive murine bone marrow progenitor cells *in vitro*: synergistic interactions with interleukin (IL)-11, IL-12 and other hematopoietic growth factors. *J. Exp. Med.* **181**, 1357–1363.

Jacobsen, S.E., Veiby, O.P., Myklebust, J., Okkenhaug, C. and Lyman, S.D. (1996). Ability of flt3 ligand to stimulate the *in vitro* growth of primitive murine hematopoietic progenitors is potently and directly inhibited by transforming growth factor-β and tumor necrosis factor-α. *Blood* **87**, 5016–5026.

Jeannin, P., Delneste, Y., Life, P., Gauchat, J.F., Kaiserlian, D. and Bonnefoy, J.Y. (1995). Interleukin-12 increases interleukin-4 production by established human Th0 and Th2-like T cell clones. *Eur. J. Immunol.* **25**, 2247–2252.

Kadakia, M.P., Birkenbach, M., Baar, J., Farhood, H. and Lotze, M.T. (1997). Modulation of interleukin-12 by Epstein-Barr virus induced gene product 3 (EBI-3). *12th Annual Society of Biologic Therapy Meeting*, 22–25 October, Pasadena, CA, USA.

Kang, K., Kubin, M., Cooper, K.D., Lessin, S.R., Trinchieri, G. and Rook, A.H. (1996). IL-12 synthesis by human Langerhans cells. *J. Immunol.* **156**, 1402–1407.

Kaplan, M.H., Sun, Y.L., Hoey, T. and Grusby, M.J. (1996). Impaired IL-12 responses and enhanced development of Th2 cells in Stat4-deficient mice. *Nature* **382**, 174–177.

Kato, K., Shimozato, O., Hoshi, K., Wakimoto, H., Hamada, H., Yagita, H. and Okumura, K. (1996a). Local production of the p40 subunit of interleukin 12 suppresses T-helper 1-mediated immune responses and prevents allogeneic myoblast rejection. *Proc. Natl Acad. Sci. U.S.A.* **93**, 9085–9089.

Kato, T., Hakamada, R., Yamane, H. and Nariuchi, H. (1996b). Induction of IL-12 p40 messenger RNA expression and IL-12 production of macrophages via CD40–CD40 ligand interaction. *J. Immunol.* **156**, 3932–3938.

Kennedy, M.K., Picha, K.S., Fanslow, W.C., Grabstein, K.H., Alderson, M.R., Clifford, K.N., Chin, W.A. and Nohler, K.M. (1996). CD40/CD40 ligand interactions are required for T cell-dependent production of interleukin-12 by mouse macrophages. *Eur. J. Immunol.* **26**, 370–378.

Klein, J.L., Fichenscher, H., Holliday, J.E., Biesinger, B. and Fleckenstein, B. (1996). Herpesvirus saimiri immortalized $\gamma\delta$ T cell line activated by IL-12. *J. Immunol.* **156**, 2754–2760.

Klinman, D.M., Yi, A.K., Beaucage, S.L., Conover, J. and Kreig, A.M. (1996). CpG motifs present in bacterial DNA rapidly induce lymphocytes to secrete interleukin 6, interleukin 12 and interferon γ. *Proc. Natl Acad. Sci. U.S.A.* **93**, 2879–2883.

Kobayashi, M., Fitz, L., Ryan, M., Hewick, R.M., Clark, S.C., Chang, S., Loudon, R., Sherman, F., Perussla, B. and Trinchieri, G. (1989). Identification and purification of natural killer cell stimulatory factor (NKSF), a cytokine with multiple biologic effects on human lymphocytes. *J. Exp. Med.* **170**, 827–845.

Kobayashi, K., Yamazaki, J., Kasama, T., Katsura, T., Kasahara, K., Wolf, S.F. and Shimamura, T. (1996). Interleukin (IL)-12 deficiency in susceptible mice infected with *Mycobacterium avium* and amelioration of established infection by IL-12 replacement therapy. *J. Infect. Dis.* **174**, 564–573.

Koch, F., Stanzl, U., Jennewein, P., Janke, K., Heufler, C., Kampgen, E., Romani, N. and Schuler, G. (1996). High level IL-12 production by murine dendritic cells: upregulation via MHC class II and CD40 molecules and downregulation by IL-4 and IL-10. *J. Exp. Med.* **184**, 741–746.

Kohl, S., Sigaroudinia, M., Charlebois, E.D. and Jacobson, M.A. (1996). Interleukin-12 administered *in vivo* decreases human NK cell cytotoxicity and antibody-dependent cellular cytotoxicity to human immunodeficiency virus-infected cells. *J. Infect. Dis.* **174**, 1105–1108.

Kremer, I.B., Hilkens, C.M., Sylva-Steenland, R.M., Koomen, C.W., Kapsenberg, M.L., Bos, J.D. and Teumissen, M.B. (1996). Reduced IL-12 production by monocytes upon ultraviolet-B irradiation selectively limits activation of T helper-1 cells. *J. Immunol.* **157**, 1913–1918.

Kroczek, R.A., Graf, D., Brugnoni, D., Giliani, S., Korthuer, U., Ugazio, A., Senger, G., Mages, H.W., Villa, A. and Notarangelo, L.D. (1994). Defective expression of CD40 ligand on T cells causes 'X-linked immunodeficiency with hyper-IgM (HIgM1)'. *Immunol. Rev.* **138**, 39–59.

Kubin, M., Kamoun, M. and Trinchieri, G. (1994). Interleukin-12 synergizes with B7/CD28 interaction in inducing efficient proliferation of human T cells. *J. Exp. Med.* **180**, 211–222.

Leder, G.H., Oppenheim, M., Rosenstein, M., Lotze, M.T. and Beger, H.G. (1995). Addition of interleukin 12 to low dose interleukin 2 treatment improves antitumor efficacy *in vivo*. *Z. Gastroenterol.* **33**, 449–502.

Leonard, J.P., Waldburger, K.E. and Goldman, S.J. (1995). Prevention of experimental autoimmune encephalomyelitis by antibodies against interleukin 12. *J. Exp. Med.* **181**, 381–386.

Li, J., Scott, P. and Farrell, J.P. (1996). *In vivo* alterations in cytokine production following interleukin-12 (IL-12) and anti-IL-4 antibody treatment of CB6F1 mice with chronic cutaneous leishmaniasis. *Infect. Immun.* **64**, 5248–5254.

Lieberman, M.D., Sigal, R.K., Williams, N.N., II and Daly, J.M. (1991). Natural killer cell stimulatory factor (NKSF) augments natural killer cell and antibody-dependent tumoricidal response against colon carcinoma cell lines. *J. Surg. Res.* **50**, 410–415.

Ling, P., Gately, M.K., Gubler, U., Stern, A.S., Lin, P., Hollfelder, K., Su, C., Pan, Y.C. and Hakimi, J. (1995). Human IL-12 p40 homodimer binds to the IL-12 receptor but does not mediate biologic activity. *J. Immunol.* **154**, 116–127.

Locksley, R.M. (1993). Interleukin 12 in host defense against microbial pathogens. *Proc. Natl Acad. Sci. U.S.A.* **90**, 5879–5880.

Lotze, M.T., Zitvogel, L., Campbell, R., Robbins, P.D., Elder, E., Haluszczak, C., Martin, D., Whiteside, T.L., Storkus, W.J. and Tahara, H. (1996). Cytokine gene therapy of cancer using interleukin-12: murine and clinical trials. *Ann. N.Y. Acad. Sci.* **795**, 440–454.

Luster, A.D., Greenberg, S.M. and Leder, P. (1995). The IP-10 chemokine binds to a specific cell surface heparin sulfate site shared with platelet factor 4 and inhibits endothelial cell proliferation. *J. Exp. Med.* **182**, 219–231.

Ma, X., Chow, J.M., Gri, G., Carra, G., Gerosa, F., Wolf, S.F., Dzialo, R. and Trinchieri, G. (1996a). The interleukin 12 p40 gene promoter is primed by interferon γ in monocytic cells. *J. Exp. Med.* **183**, 147–157.

Ma, X., Aste-Amezaga, M. and Trinchieri, G. (1996b). Regulation of interleukin-12 production. *Ann. N.Y. Acad. Sci.* **795**, 13–25.

Ma, X., Neurath, M., Gri, G. and Trinchieri, G. (1997). Identification and characterization of a novel ets-2-related nuclear complex implicated in the activation of the human interleukin-12 p40 gene promoter. *J. Biol. Chem.* **272**, 10389–10395.

Macatonia, S.E., Hsieh, C.S., Murphy, K.M. and O'Garra, A. (1993). Dendritic cells and macrophages are required for Th1 development of $CD4^+$ T cells from $TCR\alpha/\beta$ transgenic mice: IL-12 substitution for macrophages to stimulate IFN-γ production is IFN-γ-dependent. *Int. Immunol.* **5**, 1119–1128.

Macatonia, S.E., Hosken, N.A., Litton, M., Vieira, P., Hsieh, C.S., Culpepper, J.A., Wysocka, M., Trinchieri, G., Murphy, K.M. and O'Garra, A. (1995). Dendritic cells produce IL-12 and direct the development of Th1 cells from naive $CD4^+$ T cells. *J. Immunol.* **154**, 5071–5079.

Magram, J., Sfarra, J., Connaughton, S., Faherty, D., Warrier, R., Carvejal, D., Wu, Y., Stewart, C., Sarmiento, U. and Gately, M.K. (1996a). IL-12-deficient mice are defective but not devoid of type 1 cytokine responses. *Ann N.Y. Acad. Sci.* **795**, 60–70.

Magram, J., Connaughton, S.E., Warrier, R.R., Carvejal, D.M., Wu, C.Y., Ferrante, J., Stewart, C., Sarmiento, U., Faherty, D.A. and Gately, M.K. (1996b). IL-12-deficient mice are defective in IFN γ production and type 1 cytokine responses. *Immunity* **4**, 471–481.

Mahon, B.P., Ryan, M.S., Griffin, F. and Mills, K.H. (1996). Interleukin-12 is produced by macrophages to live or killed *Bordatella pertussis* and enhances the efficacy of an acellular pertussis vaccine by promoting induction of Th1 cells. *Infect. Immun.* **64**, 5295–5301.

Majewski, S., Marczak, M., Sznurlo, A., Jablonska, S. and Bollag, W. (1996). Interleukin-12 inhibits angiogenesis induced by human tumor cell lines *in vivo*. *J. Invest. Dermatol.* **106**, 1114–1118.

Manetti, R., Annunziato, F., Tomasevic, L., Gianno, V., Partonchi, P., Romagnani, S. and Maggi, E. (1995). Polyinosinic acid: polycytidylic acid promotes T helper type 1-specific immune responses by stimulating macrophage production of interferon-α and IL-12. *Eur. J. Immunol.* **25**, 2656–2660.

Marinaro, M., Boyaka, P.N., Jackson, R.J., Finkelman, F.D., Kiyono, H. and McGhee, J.R. (1996). Interleukin-12 alters helper T-cell subsets and antibody profiles induced by the mucosal adjuvant cholera toxin. *Ann N.Y. Acad. Sci.* **795**, 361–365.

Marth, T., Strober, W. and Kelsall, B.L. (1996). High dose oral tolerance in ovalbumin TCR-transgenic mice: systemic neutralization of IL-12 augments TGF-β secretion and T cell apoptosis. *J. Immunol.* **157**, 2348–2357.

Maruo, S., Toyo-oka, K., Oh-hora, M., Tai, X.G., Iwata, H., Takenaka, H., Yamada, S., Ono, S., Hamaoka, T., Kobayashi, M., Wysocka, M., Trinchieri, G. and Fujiwara, H. (1996). IL-12 produced by antigen-presenting cells induces IL-2-independent proliferation of T helper cell clones. *J. Immunol.* **156**, 1748–1755.

Mattner, F., Fischer, S., Guckes, S., Jin, S., Kaulen, H., Schmitt, E., Rude, E. and Germann, T. (1993). The interleukin-12 subunit p40 specifically inhibits effects of the interleukin-12 heterodimer. *Eur. J. Immunol.* **23**, 2202–2208.

Mattner, F., Magram, J., Ferrante, J., Launois, P., Di Padova, K., Behin, R., Gately, M.K., Louis, J.A. and Alber, G. (1996). Genetically resistant mice lacking interleukin-12 are susceptible to infection with *Leishmania major* and mount a polarized Th2 cell response. *Eur. J. Immunol.* **26**, 1553–1559.

Mayordomo, J.I., Zorina, T., Storkus, W.J., Zitvogel, L., Celluzzi, C., Falo, L.D., Melief, C.J., Ildstad, S.T., Kast, W.M., DeLeo, A.B. and Lotze, M.T. (1995). Bone marrow-derived dendritic cells pulsed with synthetic tumour peptides elicit protective and therapeutic antitumour immunity. *Nature Med.* **1**, 1297–1302.

Mehrotra, P.T., Grant, A.J. and Seigel, J.P. (1995). Synergistic effects of IL-7 and IL-12 on human T cell activation. *J. Immunol.* **154**, 5093–5102.

Meko, J.B., Tsung, K. and Norton, J.A. (1996). Cytokine production and antitumor effect of a nonreplicating, noncytopathic recombinant vaccinia virus expressing interleukin-12. *Surgery* **120**, 274–282.

Merberg, D.M., Wolf, S.F. and Clark, S.C. (1992). Sequence similarity between NKSF and the IL-6/G-CSF family. *Immunol. Today* **13**, 77–78.

Metzger, D.W. Raeder, R., Van Cleave, V.H. and Boyle, M.D. (1995). Protection of mice from group A streptococcal skin infection by interleukin-12. *J. Infect. Dis.* **171**, 1643–1645.

Metzger, D.W., Buchanan, J.M., Collins, J.T., Lester, T.L., Murray, K.S., Van Cleave, V.H., Vogel, L.A. and Dunnick, W.A. (1996). Enhancement of humoral immunity by interleukin-12. *Ann. N.Y. Acad. Sci.* **795**, 100–115.

Miller, M.A., Skeen, M.J. and Ziegler, H.K. (1995). Nonviable bacterial antigens administered with IL-12 generate antigen-specific T cell responses and protective immunity against *Listeria monocytogenes*. *J. Immunol.* **155**, 4817–4828.

Mosmann, T.R., Cherwinski, H., Bond, M.W., Giedlin, M.A. and Coffman, R.L. (1986). Two types of murine helper T cell clones. I. Definition according to profiles of lymphokine activities and secreted proteins. *J. Immunol.* **136**, 2348–2357.

Mosmann, T.R., Li, L., Hengartner, H., Kagi, D., Fu, W. and Sad, S. (1997). Differentiation and functions of T cell subsets. *Ciba Found. Symp.* **204**, 148–154.

Mountford, A.P., Anderson, S. and Wilson, R.A. (1996). Induction of Th1 cell-mediated protective immunity to *Schistosoma mansoni* by co-administration of larval antigens and IL-12 as an adjuvant. *J. Immunol.* **156**, 4739–4745.

Muller, G., Saloga, J., Germann, T., Schuler, G., Knop, J. and Enk, A.H. (1995). IL-12 as mediator and adjuvant for the induction of contact sensitivity *in vivo*. *J. Immunol.* **155**, 4661–4668.

Murphy, E.E., Terres, G., Macatonia, S.E., Hsieh, C.-S., Mattson, J., Lanier, L., Wysocka, M., Trinchieri, G., Murphy, K. and O'Garra, A. (1994). B7 and interleukin-12 cooperate for proliferation and interferon γ production by mouse T helper clones that are nonresponsive to B7 costimulation. *J. Exp. Med.* **180**, 223–231.

Murphy, T.L., Cleveland, M.G., Kulesza, P., Magram, J. and Murphy, K.M. (1995). Regulation of interleukin 12 p40 expression through an NF-κB half-site. *Mol. Cell. Biol.* **15**, 5258–5267.

Nastala, C.L., Edington, H.D., McKinney, T.G., Tahara, H., Nalesnik, M.A., Brunda, M.J., Gately, M.K., Wolf, S.F., Schreiber, R.D., Storkus, W.J. and Lotze, M.T. (1994). Recombinant IL-12 administration induces tumor regression in association with IFN-γ production. *J. Immunol.* **153**, 1697–1706.

Naume, B., Gately, M. and Espevik, T. (1992). A comparative study of IL-12 (cytotoxic lymphocyte maturation factor)-, IL-2- and IL-7-induced effects on immunomagnetically purified CD56$^+$ NK cells. *J. Immunol.* **148**, 2429–2436.

Naume, B., Gately, M.K., Desai, B.B., Sundan, A. and Espevik, T. (1993). Synergistic effects of interleukin 4 and interleukin 12 on NK cell proliferation. *Cytokine* **5**, 38–46.

Nicoletti, F., Patti, F., Cocuzza, C., Zaccone, P., Nicoletti, A., Di Marco, R. and Reggio, A. (1996). Elevated serum levels of interleukin-12 in chronic progressive multiple sclerosis. *J. Neuroimmunol.* **70**, 87–90.

Nishimura, T., Watanabe, K., Lee, U., Yahata, T., Ando, K., Kimura, M., Hiroyama, Y., Kobayashi, M., Herrmann, S.H. and Habu, S. (1995). Systemic *in vivo* antitumor activity of interleukin-12 against both transplantable and primary tumor. *Immunol. Lett.* **48**, 149–152.

Nishimura, T., Watanabe, K., Yahata, T., Ushaku, L., Ando, K., Kimura, M., Saiki, I., Uede, T. and Habu, S. (1996). Application of interleukin 12 to antitumor and gene therapy. *Cancer Chemother. Pharmacol.* **38**, S27–34.

Noben-Trauth, N., Schweitzer, P.A., Johnson, K.R., Wolf, S.F., Knowles, B.B. and Shultz, L.D. (1996). The interleukin-12β subunit (p40) maps to mouse chromosome 11. *Mammal. Genome* **7**, 392.

Noguchi, Y., Richards, E.C., Chen, Y.T. and Old, L.J. (1995). Influence of interleukin 12 on p53 peptide vaccination against established Meth A sarcoma. *Proc. Natl Acad. Sci. U.S.A.* **92**, 2219–2223.

Noguchi, Y., Jungbluth, A., Richards, E.C. and Old, J. (1996). Effect of interleukin 12 on tumor induction by 3-methylcholanthrene. *Proc. Natl Acad. Sci. U.S.A.* **93**, 11 798–11 801.

O'Garra, A., Hosken, N., Macatonia, S., Wenner, C.A. and Murphy, K. (1995). The role of macrophage- and dendritic cell-derived IL-12 in Th1 phenotype development. *Res. Immunol.* **146**, 466–472.

Orange, J.S. and Biron, C.A. (1996). Characterization of early IL-12, IFN-α/β and TNF effects on antiviral state and NK cell responses during murine cytomegalovirus infection. *J. Immunol.* **156**, 4746–4756.

Oshima, Y. and Delespesse, G. (1997). T cell-derived IL-4 and dendritic cell-derived IL-12 regulate the lymphokine-producing phenotype of alloantigen-primed naive human CD4 T cells. *J. Immunol.* **158**, 629–636.

Ozmen, L., Aguet, M., Trinchieri, G. and Garotta, G. (1995). The *in vivo* antiviral activity of interleukin-12 is mediated by γ interferon. *J. Virol.* **69**, 8147–8150.

Paglia, P., Chiodoni, C., Rodolfo, M. and Colombo, M.P. (1996). Murine dendritic cells loaded *in vitro* with soluble protein prime cytotoxic T lymphocytes against tumor antigens *in vivo. J. Exp. Med.* **183**, 317–322.

Pappo, I., Tahara, H., Robbins, P.D., Gately, M.K., Wolf, S.F., Barnes, A. and Lotze, M.T. (1995). Administration of systemic or local interleukin-2 enhances the anti-tumor effects of interleukin-12 gene therapy. *J. Surg. Res.* **58**, 218–226.

Pardoux, C., Asselin-Paturel, C., Chehimi, J., Gay, F., Mami-Chouaib, F. and Chouaib, S. (1997). Functional interaction between TGF-β and IL-12 in human allogeneic cytotoxicity and proliferation response. *J. Immunol.* **158**, 136–143.

Peng, X., Kasran, A., Warmerdam, P.A., de Boer, M. and Ceuppens, J.L. (1996). Accessory signaling by CD40 for T cell activation: induction of Th1 and Th2 cytokines and synergy with interleukin-12 for interferon-γ production. *Eur. J. Immunol.* **26**, 1621–1627.

Peritt, D., Aste-Amezaga, M., Gerosa, F., Paganin, C. and Trinchieri, G. (1996). Interleukin-10 induction by IL-12: a possible modulatory mechanism? *Ann. N.Y. Acad. Sci.* **795**, 387–389.

Peron, J.M., Esche, C., Hunter, O., Subbotin, V.M., Lotze, M.T. and Shurin, M.R. (1997). Effective treatment of murine liver metastases using FLT3 ligand (FL) and/or IL-12. *12th Annual Society of Biologic Therapy Meeting*, 22–25 October, Pasadena, CA, USA.

Pignata, C., Prasad, K.V., Haliek, M., Druker, B., Rudd, C.E., Robertson, M.J. and Ritz, J. (1995). Phosphorylation of src family lck tyrosine kinase following interleukin-12 activation of human natural killer cells. *Cell. Immunol.* **165**, 211–216.

Podlaski, F.J., Nanduri, V.B., Hulmes, J.D., Pan, Y.E., Levin, W., Danho, W., Cizzonite, R., Gately, M.K. and Stern, A.S. (1992). Molecular characterization of interleukin 12. *Arch. Biochem. Biophys.* **294**, 230–237.

Porgador, A. and Gilboa, E. (1996). Bone marrow-generated dendritic cells pulsed with a class I-restricted peptide are potent inducers of cytotoxic T lymphocytes. *J. Exp. Med.* **182**, 255–260.

Presky, D.H., Yang, H., Minetti, L.J., Chua, A.O., Nabavi, N., Wu, C.Y., Gately, M.K. and Gubler, U. (1996a). A functional interleukin 12 receptor complex is composed of two β-type cytokine receptor subunits. *Proc. Natl Acad. Sci. U.S.A.* **93**, 14 002–14 007.

Presky, D.H., Minetti, L.J., Gillessen, S., Gubler, U., Chizzonite, R., Stern, A.S. and Gately, M.K. (1996b). Evidence for multiple sites of interaction between IL-12 and its receptor. *Ann. N.Y. Acad. Sci.* **795**, 390–393.

Rabinowich, H., Herberman, R.B. and Whiteside, T.L. (1993). Differential effects of IL12 and IL2 on expression and function of cellular adhesion molecules on purified human natural killer cells. *Cell. Immunol.* **152**, 481–498.

Racke, M.K., Bonomo, A., Scott, D.E., Cannella, B., Levine, A., Raine, C.S., Shevach, E.M. and Rocken, M. (1994). Cytokine-induced immune deviation as a therapy for inflammatory autoimmune disease. *J. Exp. Med.* **180**, 1963–1966.

Radrizzani, M., Accornero, P., Anidei, A., Aiello, A., Delia, D., Kurrle, R. and Colombo, M.P. (1995). IL-12 inhibits apoptosis induced in a human Th1 clone by gp120/CD4 crosslinking and CD3/TCR activation or by IL-2 deprivation. *Cell. Immunol.* **161**, 14–21.

Rakhmilevich, A.L., Turner, J., Ford, M.J., McCabe, D., Sun, W.H., Sondel, P.M., Grota, K. and Yang, N.S. (1996). Gene gun-mediated skin transfection with interleukin 12 gene results in regression of established primary and metastatic murine tumors. *Proc. Natl Acad. Sci. U.S.A.* **93**, 6291–6296.

Robertson, M. (1991). Antigen processing: proteasomes in the pathway. *Nature* **353**, 300–309.

Rodolfo, M., Zilocchi, C., Melari, C., Ceppetti, B., Arioli, I., Parmiani, G. and Colombo, M. (1996). Immunotherapy of experimental metastases by vaccination with interleukin gene-transduced adenocarcinoma cells sharing tumor-associated antigens. Comparison between IL-12 and IL-2 gene transduced tumor cell vaccines. *J. Immunol.* **157**, 5536–5542.

Rogerson, J.A., Pole, D.S., Carr, J.A., Roberts, N.A. and Lamont, A.G. (1996). Co-therapy with suboptimal doses of interleukin-12 and interferon-α promotes synergistic protection against herpes virus infection. *Ann. N.Y. Acad. Sci.* **795**, 354–356.

Rogge, L., Barberis-Maino, L., Biffi, M., Passini, N., Presky, D.H., Gubler, U. and Sinigaglia, F. (1997). Selective expression of an interleukin-12 receptor component by human T helper 1 cells. *J. Exp. Med.* **185**, 825–831.

Rook, A.H., Kubin, M., Fox, F.E., Niu, Z., Cassin, M., Vowels, B.R., Gottleib, S.L., Voderheid, E.C., Lessin, S.R.

and Trinchieri, G. (1996). The potential therapeutic role of interleukin-12 in cutaneous T-cell lymphoma. *Ann. N.Y. Acad. Sci.* **795**, 310–318.

Rothe, H., Hartmann, B., Geerlings, P. and Kolb, H. (1996). Interleukin-12 gene-expression of macrophages is regulated by nitric oxide. *Biochem. Biophys. Res. Commun.* **224**, 159–163.

Schadendorf, D., Czarnetzki, B.M. and Wittig, B. (1995). Interleukin-7, interleukin-12 and GM-CSF gene transfer in patients with metastatic melanoma. *J. Mol. Med.* **73**, 473–477.

Scharton-Kersten, T., Afonso, L.C., Wysocka, M., Trinchieri, G. and Scott, P. (1995). IL-12 is required for natural killer cell activation and subsequent T helper 1 cell development in experimental leishmaniasis. *J. Immunol.* **154**, 5320–5330.

Scheicher, C., Mehlig, M., Dienes, H.-P. and Reske, K. (1995). Uptake of microparticulate-absorbed protein antigen by bone-marrow derived dendritic cells results in up-regulation of interleukin-1α and interleukin-12 p40/p35 and triggers prolonged, efficient antigen presentation. *Eur. J. Immunol.* **24**, 1566–1572.

Schijns, V.E., Haagmans, B.L. and Horzinek, M.C. (1995). IL-12 stimulates an antiviral type 1 cytokine response but lacks adjuvant activity in IFN-γ-receptor-deficient mice. *J. Immunol.* **155**, 2525–2532.

Schoenhaut, D.S., Chua, A.O., Wolitzky, A.G., Quinn, P.M., Dwyer, C.M., McComas, W., Familletti, P.C., Gately, M.K. and Gubler, U. (1992). Cloning and expression of murine IL-12. *J. Immunol.* **148**, 3433–3440.

Schweitzer, P.A., Noben-Trauth, N., Pelsue, S.C., Johnson, K.R., Wolf, S.F. and Shultz, L.D. (1996). Genetic mapping of the IL-12 α chain (IL12α) on mouse chromosome 3. *Mammal. Genome* **7**, 394–395.

Scott, P. (1993). IL-12: initiation cytokine for cell-mediated immunity. *Science* **260**, 496–497.

Seder, R.A. (1996). High-dose IL-2 and IL-15 enhance the *in vitro* priming of naive $CD4^+$ T cells for IFN-γ but have differential effects on priming for IL-4. *J. Immunol.* **156**, 2413–2422.

Seder, R.A., Kelsall, B.L. and Jankovic, D. (1996). Differential roles for IL-12 in the maintenance of immune responses *versus* autoimmune disease. *J. Immunol.* **157**, 2745–2748.

Segal, B.M. and Shevach, E.M. (1996). IL-12 unmasks latent autoimmune disease in resistant mice. *J. Exp. Med.* **184**, 771–775.

Sgadari, C., Angiolillo, A.L. and Tosato, G. (1996). Inhibition of angiogenesis by interleukin-12 is mediated by the interferon-inducible protein 10. *Blood* **87**, 3877–3882.

Sieburth, D., Jabs, E.W., Warrington, J.A., Li, X., Lasota, J., LaForgia, S., Kelleher, K., Huebner, K., Wasmuth, J.J. and Wolf, S.F. (1992). Assignment of genes encoding a unique cytokine (IL12) composed of two unrelated subunits to chromosomes 3 and 5. *Genomics* **14**, 59–62.

Snidjers, A., Hilkens, C.M., van der Pouw Kraan, T.C., Engel, M., Aarden, L.A. and Kapsenberg, M.L. (1996). Regulation of bioactive IL-12 production in lipopolysaccharide-stimulated human monocytes is determined by the expression of the p35 subunit. *J. Immunol.* **156**, 1207–1212.

Stern, A.S., Podlaski, F.J., Hulmes, J.D., Pan, Y.C.E., Quinn, P.M., Wolitzky, A.G., Familletti, P.C., Stremlo, D.L., Truitt, T., Chizzonite, R. and Gately, M.K. (1990). Purification to homogeneity and partial characterization of cytotoxic lymphocyte maturation factor from human B-lymphoblastoid cells. *Proc. Natl Acad. Sci. U.S.A.* **87**, 6808–6812.

Stevenson, M.M., Tam, M.F., Wolf, S.F. and Sher, A. (1995). IL-12-induced protection against blood-stage *Plasmodium chabaudi* AS requires IFN-γ and TNF-α and occurs via a nitric oxide-dependent mechanism. *J. Immunol.* **155**, 2545–2556.

Stuber, E., Strober, W. and Neurath, M. (1996). Blocking the CD40L–CD40 interaction *in vivo* specifically prevents priming of T helper 1 cells through the inhibition of interleukin 12 secretion. *J. Exp. Med.* **183**, 693–698.

Szabo, S.J., Dighe, A.S., Gubler, U. and Murphy, K.M. (1997). Regulation of the interleukin (IL)-12R β_2 subunit expression in developing T helper 1 (Th1) and Th2 cells. *J. Exp. Med.* **185**, 817–824.

Szeliga, J., Hess, H., Rude, E., Schmitt, E. and Germann, T. (1996). IL-12 promotes cellular but not humoral type II collagen-specific Th1-type responses in C57BL/6 and B10.Q mice and fails to induce arthritis. *Int. Immunol.* **8**, 1221–1227.

Tahara, H., Zitvogel, L., Storkus, W.J., Zeh, H.J., III, McKinney, T.G., Schreiber, R.D., Gubler, U., Robbins, P.D. and Lotze, M.T. (1995). Effective eradication of established murine tumors with IL-12 gene therapy using a polycistronic retroviral vector. *J. Immunol.* **154**, 6466–6474.

Tan, J., Newton, C.A., Djeu, J.Y., Gutsch, D.E., Chang, A.E., Yang, N.S., Klein, T.W. and Hua, Y. (1996). Injection of complementary DNA encoding interleukin-12 inhibits tumor establishment at a distant site in a murine renal carcinoma model. *Cancer Res.* **56**, 3399–3403.

Tang, Y.W. and Graham, B.S. (1995). Interleukin-12 treatment during immunization elicits a T helper cell type 1-like immune response in mice challenged with respiratory syncytial virus and improves vaccine immunogenicity. *J. Infect. Dis.* **172**, 734–738.

Tannenbaum, C.S., Wicker, N., Armstrong, D., Tubbs, R., Finke, J., Bukowski, R.M. and Hamilton, T.A. (1996). Cytokine and chemokine expression in tumors of mice receiving systemic therapy with IL-12. *J. Immunol.* **156**, 693–699.

Tare, N.S., Bowen, S., Warrier, R.R., Carvejal, D.M., Benjamin, W.R., Riley, J.H., Anderson, T.D. and Gately, M.K. (1995). Administration of recombinant interleukin-12 to mice suppresses hematopoiesis in the bone marrow but enhances hematopoiesis in the spleen. *J. Interferon Cytokine Res.* **15**, 377–383.

Thierfelder, W.E., van Deursen, J.M., Yamamoto, K., Tripp, R.A., Sarawar, S.R., Carson, R.T., Sangster, M.Y., Vignali, D.A., Doherty, P.C., Grosveld, G.C. and Ihle, J.N. (1996). Requirement for Stat4 in interleukin-12-mediated responses of natural killer cells and T cells. *Nature* **382**, 171–174.

Tone, Y., Thompson, S.A., Babik, J.M., Nolan, K.F., Tone, M., Raven, M. and Waldmann, H. (1996). Structure and chromosomal location of the mouse interleukin-12 p35 and p40 subunit genes. *Eur. J. Immunol.* **26**, 1222–1227.

Trinchieri, G. (1995). The two faces of interleukin 12: a pro-inflammatory cytokine and a key immunoregulatory molecule produced by antigen-presenting cells. *Ciba Found. Symp.* **195**, 203–214.

Trinchieri, G. (1996). Role of IL-12 in human Th1 response. *Chem. Immunol.* **63**, 14–29.

Tripp, C.E. and Unanue, E.R. (1995). Macrophage production of IL12 is a critical link between the innate and specific immune responses to *Listeria*. *Res. Immunol.* **146**, 515–520.

Tripp, C.S., Wolf, S.F. and Unanue, E.R. (1993). Interleukin 12 and tumor necrosis factor α are costimulators of interferon γ production by natural killer cells in severe combined immunodeficiency mice with listeriosis and interleukin 10 is a physiologic antagonist. *Proc. Natl Acad. Sci. U.S.A.* **90**, 3725–3729.

Tueting, T., Wilson, C.C., Martin, D.M., Kasamon, Y.L., Rowles, J., Ma, D.I., Slingluff, C.L. Jr., Wagner, S.N., van der Bruggen, P., Baar, J., Lotze, M.T. and Storkus, W.J. (1998). Autologous human monocyte-derived dendritic cells genetically modified to express melanoma antigens elicit primary cytotoxic T cell responses *in vitro*: enhancement by cotransfection of genes encoding the Th1-biasing cytokines IL-12 and IFN-a. *J. Immunol.* **160**, 1139–1147.

Van der Pouw Kraan, T.C., Boeije, L.C., Smeenk, R.J., Wijdened, J. and Aarden, L.A. (1995). Prostaglandin-E_2 is a potent inhibitor of human interleukin 12 production. *J. Exp. Med.* **181**, 775–779.

Verbik, D.J., Stinson, W.W., Brunda, M.J., Kessinger, A. and Joshi, S.S. (1996). *In vivo* therapeutic effects of interleukin-12 against highly metastatic residual lymphoma. *Clin. Exp. Metastasis* **14**, 219–229.

Vogel, L.A., Showe, L.C., Lester, T.L., McNutt, R.M., Van Cleave, V.H. and Metzger, D.W. (1996). Direct binding of IL-12 to human and murine B lymphocytes. *Int. Immunol.* **8**, 1955–1962.

Wenner, C.A., Guler, M.L., Macatonia, S.E., O'Garra, A. and Murphy, K.M. (1996). Roles of IFN-γ and IFN-α in IL-12-induced T helper cell-1 development. *J. Immunol.* **156**, 1442–1447.

Wigginton, J.M., Komschilies, K.L., Back, T.C., Franco, J.L., Brunda, M.J. and Wiltrout, R.H. (1996). Administration of interleukin 12 with pulse interleukin 2 and the rapid and complete eradication of murine renal carcinoma. *J. Natl Cancer Inst.* **88**, 38–43.

Windhagen, A., Newcombe, J., Dangond, F., Strand, C., Woodroofe, M.N., Cuzner, M.L. and Hafler, D.A. (1995). Expression of costimulatory molecules B7-1 (CD80), B7-2 (CD86) and interleukin 12 cytokine in multiple sclerosis lesions. *J. Exp. Med.* **182**, 1985–1996.

Windhagen, A., Anderson, D.E., Carrizosa, A., Williams, R.E. and Hafler, D.A. (1996). IL-12 induces human T cells secreting IL-10 with IFN-γ. *J. Immunol.* **157**, 1127–1131.

Wolf, S.F., Temple, P.A., Kobayashi, M., Young, D., Dicig, M., Lowe, L., Dzialo, R., Fitz, L., Ferenz, C., Hewick, R.M., Kelleher, K., Herrmann, S.H., Clark, S.C., Azzoni, L., Chan, S.H., Trinchieri, G. and Perussia, B. (1991). Cloning of cDNA for natural killer cell stimulatory factor, a heterodimeric cytokine with multiple biologic effects on T and natural killer cells. *J. Immunol.* **146**, 3074–3081.

Wolf, S.F., Sieburth, D. and Sypek, J. (1994). Interleukin 12: a key modulator of immune function. *Stem Cells* **12**, 154–168.

Wong, H.L., Wilson, D.E., Jensen, J., Familletti, P., Stremlo, P. and Gately, M.K. (1987). Characterization of a factor(s) which synergizes with recombinant interleukin 2 in promoting allogeneic human cytolytic T-lymphocyte responses *in vitro*. *Cell. Immunol.* **111**, 39–54.

Wu, C.Y., Warrier, R.R., Carvajal, D.M., Chua, A.O., Minetti, L.J., Chizzonite, R., Mongini, P.K., Stern, A.S., Gubler, U., Presky, D.H. and Gately, M.K. (1996). Biological function and distribution of human interleukin-12 receptor β chain. *Eur. J. Immunol.* **26**, 345–350.

Wu, C., Warrier, R.R., Wang, X., Presky, D.H. and Gately, M.K. (1997). Regulation of interleukin-12 receptor β_1 chain expression and interleukin-12 binding by human peripheral blood mononuclear cells. *Eur. J. Immunol.* **27**, 147–154.

Wynn, T.A. (1996). Development of an antipathology vaccine for schistosomiasis. *Ann. N.Y. Acad. Sci.* **795**, 191–195.

Wynn, T.A., Jankovic, D., Hieny, S., Zioncheck, K., Jardieu, P., Cheever, A.W. and Sher, A. (1995). IL-12 exacerbates rather than suppresses T helper 2-dependent pathology in the absence of endogenous IFN-γ. *J. Immunol.* **154**, 3999–4009.

Xiao, M., Li, Z.H., McMahel, J., Broxmeyer, H.E. and Lu, L. (1996). Inhibitory effects of interleukin 12 on retroviral gene transduction into CD34 cord blood myeloid progenitors mediated by induction of tumor necrosis factor-α. *J. Hematol.* **5**, 171–177.

Yawalkar, N., Brand, C.U. and Braathen, L.R. (1996a). IL-12 gene expression in human skin-derived $CD1a^+$ dendritic lymph cells. *Arch. Dermatol. Res.* **288**, 79–84.

Yawalkar, N., Limat, A., Brand, C.U. and Braathen, L.R. (1996b). Constitutive expression of both subunits of interleukin-12 in human keratinocytes. *J. Inv. Dermatol.* **106**, 80–83.

Yoshimoto, T, Kojima, K., Funakoshi, T., Endo, Y., Fujita, T. and Nariuchi, H. (1996). Molecular cloning and characterization of murine IL-12 genes. *J. Immunol.* **156**, 1082–1088.

Yu, W.G., Yamamoto, N., Takenaka, H., Mu, J., Tai, X.G., Zou, J.P., Ogawa, M., Tautsui, T., Wijesuriya, R., Yoshida, R., Herrmann, S., Fujiwara, H. and Hamaoka, T. (1996a). Molecular mechanisms underlying IFN-γ-mediated tumor growth inhibition induced during tumor immunotherapy with rIL-12. *Int. Immunol.* **8**, 855–865.

Yu, C.R., Lin, J.K., Fink, D.W., Akira, S., Bloom, E.T. and Yamauchi, A. (1996b). Differential utilization of Janus kinase-signal transducer activator of transcription signaling pathways in the stimulation of human natural killer cells by IL-2, IL-12 and interferon-α. *J. Immunol.* **157**, 126–137.

Zeh, H.J. III, Tahara, H. and Lotze, M.T. (1994). Interleukin-12. In *The Cytokine Handbook*, 2nd ed. (ed. A. Thomson), Academic Press, New York, pp. 239–256.

Zhang, J., Asselin-Paturel, C., Bex, F., Bernard, J., Chehimi, J., Willems, F., Caignard, A., Berglund, P., Liljestrom, P., Burny, A. and Chouaib, S. (1997). Cloning of human IL-12 p40 and p35 DNA into the Semliki Forest virus vector: expression of IL-12 in human tumor cells. *Gene Ther.* **4**, 367–374.

Zitvogel, L. and Lotze, M.T. (1995). Role of interleukin-12 as an anti-tumor agent: experimental biology and clinical application. *Res. Immunol.* **146**, 628–638.

Zitvogel, L., Mayordomo, J.I., Tjandrawan, T., DeLeo, A.B., Clarke, M.R., Lotze, M.T. and Storkus, W.J. (1996a). Therapy of murine tumors with tumor peptide-pulsed dendritic cells: dependence on T cells, B7 costimulation and T helper cell 1-associated cytokines. *J. Exp. Med.* **183**, 87–97.

Zitvogel, L., Robbins, P.D., Storkus, W.J., Clarke, M.R., Maeurer, M.J., Campbell, R.L., Davis, C.G., Tahara, H., Schreiber, R.D. and Lotze, M.T. (1996b). Interleukin-12 and B7.1 co-stimulation cooperate in the induction of effective antitumor immunity and therapy of established tumors. *Eur. J. Immunol.* **26**, 1335–1341.

Zitvogel, L., Couderc, B., Mayordomo, J.I., Robbins, P.D., Lotze, M.T. and Storkus, W.J. (1996c). IL-12-engineered dendritic cells serve as effective tumor vaccine adjuvants *in vivo*. *Ann. N.Y. Acad. Sci.* **795**, 284–293.

Zou, J., Presky, D.H., Wu, C.Y. and Gubler, U. (1997). Differential associations between the cytoplasmic regions of the interleukin-12 receptor subunits β_1 and β_2 and JAK kinases. *J. Biol. Chem.* **272**, 6073–6077.

Chapter 15

Interleukin-13

René de Waal Malefyt[1] and Jan E. de Vries[2]

[1]DNAX Research Institute, Palo Alto, CA, USA and [2]Novartis Research Institute, Vienna, Austria

INTRODUCTION

In 1989 Brown *et al.* isolated an unknown cDNA by differential hybridization of cDNA libraries from T helper 1 (T_{H1}) and T_{H2} cells and designated it p600. However, it was not until 1992 that the human homolog of this murine sequence was cloned (McKenzie *et al.*, 1993b; Minty *et al.*, 1993a). Expression of this human cDNA led to the discovery that p600 encoded for a cytokine that had structural features and functional characteristics in common with interleukin-4 (IL-4). Following the demonstration of its immune modulating effects on B cells and monocytes, the name interleukin-13 (IL-13) was proposed and accepted (Zurawski and de Vries, 1994).

IL-13 PROTEIN AND GENE

Human IL-13 is a protein with a molecular mass of 12 kDa, which folds into four α-helical bundles. It contains four potential *N*-glycosylation sites (McKenzie *et al.*, 1993b; Minty *et al.*, 1993a). Alternative splicing results in two functionally similar isoforms of mature hIL-13 consisting of 131 or 132 amino acids, the latter having an additional glutamine residue at position 98 (McKenzie *et al.*, 1993b). The hydrophobic signal sequence contains 20 amino acids. IL-13 contains four cysteine residues, which form two intramolecular disulfide bonds (McKenzie *et al.*, 1993b; Minty *et al.*, 1993a). IL-13 protein is approximately 25% homologous with IL-4, and all residues that contribute to the hydrophobic structural core of IL-4 are conserved, or have conservative replacements in IL-13, suggesting that the overall three-dimensional structures of IL-4 and IL-13 are identical (Zurawski *et al.*, 1993; Bamborough *et al.*, 1994). Both human and mouse IL-13 are biologically active on human and murine cells. The gene encoding human IL-13 is located on chromosome 5q31, in the same 3000-kb cluster of genes encoding IL-3, IL-4, IL-5, IL-9 and granulocyte–macrophage colony stimulating factor (GM-CSF) (Morgan *et al.*, 1992; Smirnov *et al.*, 1995). The IL-13 gene is only 12 kb upstream from the gene encoding IL-4 and lies in the same orientation, indicating that a gene duplication event took place during evolution. Mouse and human IL-13 genes are composed of four exons separated by three relatively short introns, which makes the total gene size approximately 3 kb (GenBank accession no. L13029) (McKenzie *et al.*,

The Cytokine Handbook, 3rd ed.
ISBN 0-12-689662-3

1993a). Two kilobases upstream of the IL-13 gene is a CpG island, but the overall homology between IL-13 and IL-4 5′ flanking sequences is relatively low. An IL-4P site which binds the nuclear factor of activated T cells (NFATp) transcription factor has been identified in the IL-13 promoter, and is active in regulating expression of IL-13 in T cells (Dolganov *et al.*, 1996). IL-13 mRNA is 1280 nucleotides long and contains four copies of the (A/T)ATTTA(A/T) repeat implicated in mRNA instability and present in most cytokine mRNAs (GenBank accession no. L06801).

IL-13R COMPLEXES

IL-13 and IL-4 share many, but not all, biologic properties, which is partially due to the sharing and differential expression of IL-4 and IL-13 receptor components on various cell types (Zurawski *et al.*, 1993). IL-13 receptors are expressed on B cells, monocytes–macrophages, basophils, mast cells, endothelial cells and certain tumor cells, but not on T cells or mouse B cells (Zurawski *et al.*, 1993; de Waal Malefyt *et al.*, 1995). Receptors are usually present at 200 to 3000 sites per cell and bind IL-13 with high affinity (K_d 30 pM). The high-affinity IL-13 receptor complex comprises the 140-kDa IL-4R α chain and an IL-13 binding protein. The IL-4R α chain alone binds IL-4 with a relatively high affinity (K_d approximately 10^{-10} M), but does not bind IL-13. Two different cDNAs encoding IL-13 binding proteins have been recently cloned and termed IL-13Rα1 and IL-13Rα2. IL-13Rα1 was first identified in the mouse (Hilton *et al.*, 1996) and subsequently three groups have reported the cloning of the human homolog (Genbank accession no. Y09328, U62858) (Aman *et al.*, 1996; Gauchat *et al.*, 1997; Miloux *et al.*, 1997). IL-13Rα1 consists of a 427-amino-acid protein that specifically binds IL-13 with low affinity (K_d approximately 4 nM) but not IL-4. It is encoded by a 4-kb mRNA and the gene is located on the X chromosome. IL-13Rα2 is a 380-amino-acid protein which binds IL-13 with high affinity (K_d 50 pM) in the absence of the IL-4Rα chain (Genbank accession no. X95302) (Caput *et al.*, 1996). Human IL-13Rα1 and IL-13Rα2 are 27% homologous, whereas human IL-13Rα1 is 76% homologous to the mouse IL-13Rα1. Both IL-13Rα1 and IL-13Rα2 show homology to the IL-5Rα and the prolactin receptor. They contain two consensus patterns characteristic for the hematopoietic cytokine receptor family. One is the WSXWS motif in the extracellular domain and the other is a consensus binding motif for a signal transducer and activator of transcription (STAT) protein present in their short cytoplasmic tails. The IL-13R α chains are expressed as 65–70-kDa glycosylated molecules.

The IL-13R complexes also function as a second receptor for IL-4, whereas the other IL-4R complex, consisting of the IL-4Rα chain and the common γ chain (γc), a shared component of receptors for IL-2, IL-4, IL-7, IL-9 and IL-15, is specific for IL-4 (Fig. 1).

This model of IL-4–IL-13 receptor complexes is supported by competitive binding studies, which show that IL-4 can always completely compete for IL-13 binding on hematopoietic cells, but that IL-13 can compete only partially for IL-4 binding (Zurawski *et al.*, 1993; Feng *et al.*, 1995; Obiri *et al.*, 1995). In addition, an IL-4 mutant protein in which the tyrosine residue at position 124 was replaced by aspartic acid could antagonize the effects of both IL-4 and IL-13 (Aversa *et al.*, 1993). Furthermore, monoclonal antibodies against the 140-kDa IL-4Rα chain were able to block both IL-4 and IL-13 responses. Finally, some cell types that respond to IL-4 do not respond to IL-13 (Renard *et al.*, 1994; Tony *et al.*, 1994; Lefort *et al.*, 1995; Zurawski *et al.*, 1995).

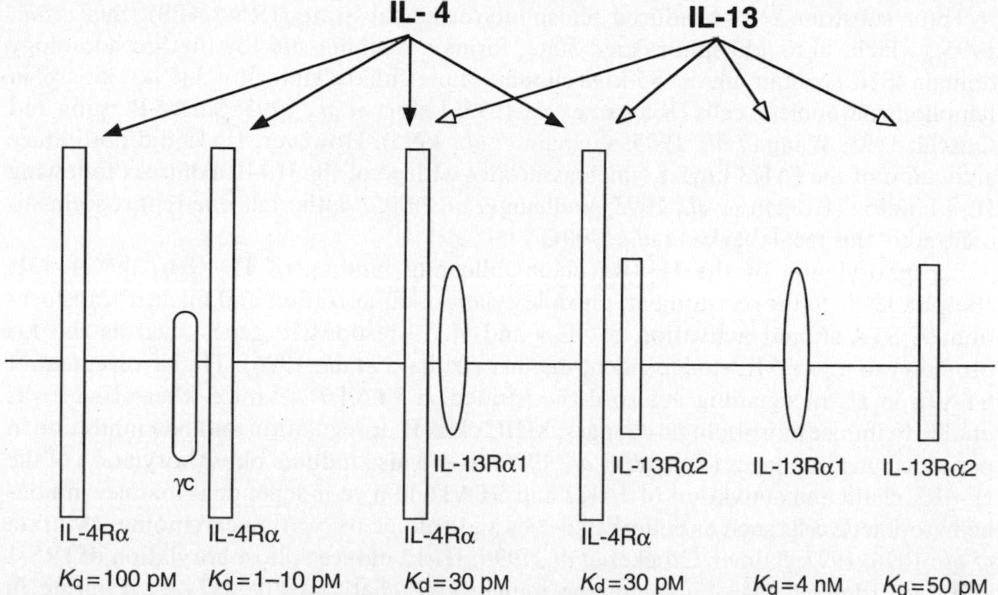

Fig. 1. IL-4 and IL-3 receptor complexes. IL-4 interacts with the 140 K_d IL-4Rα binding protein in combination with either γc, IL-13Rα1 or IL-13Rα2. The K_d given for the interaction of IL-4 with IL-4Rα is from transfected cell lines. IL-13 interacts with IL-13Rα1 or IL-13Rα2 in the absence or presence of IL-4Rα.

Additional insights that γc did not participate in IL-13 receptor complexes were obtained from data with T and B cells from patients with X-linked severe combined immunodeficiency. These patients have mutations in the γc gene and their T cells are unable to respond to IL-2, IL-4, IL-7, IL-9 and IL-15, causing severe abnormalities in both T- and B-cell functions. However, their B cells could proliferate and produce immunoglobulin E (IgE) following stimulation through CD40 in the presence of IL-4 or IL-13, indicating that both IL-4 and IL-13 can mediate their biologic effects through the IL-13R complex in the absence of a functional γc (Matthews *et al.*, 1995, 1997; Izuhara *et al.*, 1996). In addition, macrophages from $\gamma c^{-/-}$ mice respond both to IL-4 and IL-13, which results in upregulation of major histocompatibility proteins (MHC) class II and inhibition of nitric oxide (NO) production (Andersson *et al.*, 1997). IL-4 and IL-13 signaling in the absence of γc has also been observed on renal cell carcinoma, glioblastoma and endothelial cells, which apparently express high levels of IL-13Rα1 and/or IL-13Rα2, but no IL-4Rα chains, as IL-4 did not compete for binding of IL-13 in these cases (Debinski *et al.*, 1995, 1996; He *et al.*, 1995; Obiri *et al.*, 1995, 1997; Schnyder *et al.*, 1996).

IL-13 SIGNAL TRANSDUCTION

Binding of IL-13 to the IL-13Rα chain leads to its dimerization with the IL-4Rα chain and initiation of the signal transduction cascade. A strong similarity is observed in the activation of signal transduction pathways following stimulation of cells with IL-4 and IL-13, as the IL-4Rα chain is the signal transducing receptor component. IL-13, like IL-4, activated the Janus tyrosine kinase 1 (JAK1) and tyrosine kinase 2 (TYK2) kinases and induced tyrosine phosphorylation of the IL-4Rα chain and the 170-kDa insulin receptor substrate-2/IL-4-induced phosphotyrosine substrate (IRS-2/4PS) (Sun *et al.*, 1995), which, in its phosphorylated state, forms a docking site for the Src homology domain (SH2) containing the 85-kDa subunit of phosphatidylinositol-3 (PI-3) kinase, in lymphohematopoietic cells (Keegan *et al.*, 1995; Lefort *et al.*, 1995; Smerz-Bertling and Duschl, 1995; Wang *et al.*, 1995; Welham *et al.*, 1995). However, IL-13 did not induce activation of the JAK3 kinase which associates with γc of the IL-4R complex following IL-4 binding (Keegan *et al.*, 1995; Welham *et al.*, 1995), although one report seems to contradict this (Malabarba *et al.*, 1996).

Phosphorylation of the IL-4Rα chain following binding of IL-13 to the IL-13R complex leads to the recruitment, phosphorylation, dimerization and nuclear translocation of STAT6 and activation of IL-4 and IL-13-responsive genes such as the Igε promoter in a γc–JAK3-independent manner (Izuhara *et al.*, 1996). The involvement of STAT6 in IL-13 signaling is also demonstrated in STAT6$^{-/-}$ mice where IL-13 was unable to induce morphologic changes, MHC class II upregulation and NO inhibition in peritoneal macrophages (Takeda *et al.*, 1996). IL-13 also induces phosphorylation of the IL-4Rα chain and activation of JAK2 and STAT6 in a γc-independent manner in non-hematopoietic cells such as endothelial cells and colonic or ovarian carcinomas (Murata *et al.*, 1996, 1997; Palmer-Crocker *et al.*, 1996) IL-13 induces phosphorylation of IRS-1 and activation of PI-3 kinase in the human epithelial cell line HT-29, resulting in inhibition of inducible nitric oxide synthase expression (Wright *et al.*, 1997). Whether IL-13, like IL-4, leads to activation of c-fes or a c-fes-like protein has not yet been reported, but both cytokines induce the expression of lck in human monocytes (Musso *et al.*, 1994).

IL-13 PRODUCTION

IL-13 is not only produced by CD4$^+$ T_{H2} cells, but also by T_{H0} and CD8$^+$ T cells and, in contrast to IL-4, by naive CD45RA$^+$ T cells and T_{H1} cells, albeit at lower levels (de Waal Malefyt *et al.*, 1995). However, all T cells that produce IL-4 produce IL-13 when studied at the single cell level (Jung *et al.*, 1996). T-cell receptor (TCR) activation of naive CD4$^+$ CD45RA$^+$ T cells from peripheral blood in the absence of IL-4 resulted in T cells producing IL-13, and addition of IL-4 enhanced the percentage of IL-13-expressing cells whereas IL-12 had no inhibitory effects (Brinkmann and Kristofic, 1995; Jung *et al.*, 1996). IL-13 production was also observed in IL-4$^{-/-}$ mice following challenge with *Onchocerca volvulus* antigens (Pearlman *et al.*, 1996). These results indicate that IL-4 is not absolutely required for the induction of IL-13-producing cells.

The kinetics of IL-13 production by activated T cells also differs significantly from that of IL-4. IL-4 mRNA appears 2–4 h after activation and is almost undetectable after 12 h, whereas IL-13 mRNA can still be observed 48 h after T-cell activation, suggesting that IL-13 is produced for significantly longer periods of time after antigen-specific T-cell activation. This was also observed in mice following administration of goat antimouse IgD antibodies (Kricek *et al.*, 1995). The activation requirements for IL-13 induction are also less stringent than those necessary for IL-4 induction, in that activation of protein kinase C alone or a rise in intracellular Ca^{++} concentration is sufficient to allow IL-13 mRNA expression in human T-cell clones (de Waal Malefyt *et al.*, 1995). Interestingly, TCR engagement of anti-CD28 and phorbol myristate acetate (PMA)-activated T cells resulted in a decrease of IL-13 production, which could be reversed by cyclosporin A. However, cyclosporin A did inhibit IL-13 production by T cells following activation by phytohemagglutinin (PHA) or CD3 activation. These results indicate a complex regulation of IL-13 with possible involvement of the NFATp transcription factor, as NFATp$^{-/-}$ mice showed unexpectedly enhanced immune responses and changes in cytokine production, revealing its possible role as a transcription suppressor (Hodge *et al.*, 1996).

IL-13 is also produced by Epstein–Barr virus-transformed B-cell lines (Fior *et al.*, 1994; de Waal Malefyt *et al.*, 1995), B-cell lymphomas (Emilie *et al.*, 1997) , keratinocytes (Michel *et al.*, 1994), mast cells (Burd *et al.*, 1995; Marietta *et al.*, 1996; Pawankar *et al.*, 1997) and basophils (Gibbs *et al.*, 1996; Li *et al.*, 1996). The production of IL-13 by mast cells and basophils following cross-linking of FcεRI could contribute to initiation and maintenance of allergic responses (Gauchat *et al.*, 1993; Pawankar *et al.*, 1995; Marietta *et al.*, 1996).

BIOLOGIC ACTIVITY OF IL-13

Based on the composition of IL-13 receptor complexes and the requirement for the IL-4Rα chain for signal transduction, it is not surprising that IL-13 shares many of its biologic activities with IL-4. However, we were unable to demonstrate binding of ^{125}I-labeled IL-13 to human PHA blasts or T-cell clones and failed to demonstrate any biologic activity of IL-13 on T cells, indicating that IL-13 lacks the T-cell stimulatory or inhibitory activities of IL-4, presumably through a lack of IL-13 binding protein expression (Zurawski *et al.*, 1993; de Waal Malefyt *et al.*, 1995). In contrast to IL-4, IL-13 could not support the proliferation of activated T cells or induce the expression of CD8α on CD4$^+$ T cells (de Waal Malefyt *et al.*, 1995). In addition, IL-13 was unable to drive the differentiation of naive CD4$^+$ cord blood T cells towards a T_{H2} phenotype (Sornasse *et al.*, 1996). However, IL-13 may indirectly affect T-cell functions and differentiation through its downregulatory effects on the production of IL-12 and interferon-α (IFN-α), which direct T_{H1} development. IL-13 also lacked activities on mouse T cells and, furthermore, did not have any of the B-cell stimulatory effects of IL-4 in the mouse. These results indicate that murine T and B cells may also lack expression of the IL-13 receptor binding proteins.

EFFECTS OF IL-13 ON HUMAN B LYMPHOCYTES

The effects of IL-13 on human B cells are largely similar to those of IL-4, but IL-13 is slightly less (two- to fivefold) potent than IL-4. IL-13 does not have additive or synergistic effects with IL-4 when both cytokines are added at optimal concentrations. IL-13 modulates the phenotype of normal human B cells. It upregulates the expression of CD23, CD71, CD72, surface IgM (sIgM) and MHC class II molecules on purified human B cells (McKenzie *et al.*, 1993b; Punnonen *et al.*, 1993; Defrance *et al.*, 1994). The IL-13-induced upregulation of CD23 is observed on only a subpopulation of B cells. IL-13 has growth promoting activity on normal B cells stimulated by anti-IgM or anti-CD40 monoclonal antibodies (mAbs) (Aversa *et al.*, 1993; Cocks *et al.*, 1993; McKenzie *et al.*, 1993b; Minty *et al.*, 1993a), but it inhibits IL-2-induced proliferation of chronic lymphocytic leukemia B cells (Chaouchi *et al.*, 1996) and the proliferation of pre-B cells induced by stromal cells and IL-7 (Renard *et al.*, 1994).

IL-13 enhances the production of IgM, IgG and IgA and induces IgG_4 and IgE synthesis by human B cells cultured in the presence of activated $CD4^+$ T cells, anti-CD40 mAbs or CD40 ligand (L) transfectants (Aversa *et al.*, 1993; Cocks *et al.*, 1993; Punnonen *et al.*, 1993). IL-13 also induces proliferation of IgM, total IgG, IgG4 and IgE synthesis when total fetal bone marrow (BM) cells or highly purified $sIgM^+$, $CD10^+$, $CD19^+$ fetal B cells are cultured in the presence of anti-CD40 mAbs or activated $CD4^+$ T cells (Punnonen and de Vries, 1994). Even highly purified $sIgM^-$, $CD10^+$, $CD19^+$ pre-B cells cocultured in the presence of IL-13, activated cloned $CD4^+$ T cells and IL-7 can be induced to secrete IgG_4 and IgE. However, IL-13 does not enhance CD23, CD40 and human leukocyte antigen (HLA)-DR expression on cultured $sIgM^-$ pre-B cells in the absence of other stimuli, nor can it induce germline ε transcription by itself, suggesting that IL-13 alone, unlike IL-4 alone, does not activate pre-B cells and that IL-13Rs are expressed later during B-cell differentiation than IL-4Rs (Punnonen and de Vries, 1994; Punnonen *et al.*, 1995). IL-13, like IL-4, induces germline ε transcription, a prerequisite for switching to and successful IgE production. IL-13-induced germline ε induction, IgE switching and IgE production is enhanced by CD40 signaling, tumor necrosis factor-α (TNF-α) and IL-10 and inhibited by IFN-γ, IFN-α and transforming growth factor-β (Cocks *et al.*, 1993; Punnonen *et al.*, 1993; Ezernieks *et al.*, 1996).

Although both IL-4 and IL-13 induce IgE synthesis by human B cells *in vitro*, the relative contribution of the two cytokines to IgE synthesis under physiologic conditions remains to be determined. Neutralization of either IL-4 or IL-13 activity in the supernatants of activated allergen-specific T_{H2} cells with anti-IL-4 or anti-IL-13 mAbs indicated that IL-4 is the dominant cytokine in inducing IgE synthesis. However, IgE production induced by supernantants of T_{H1} or $CD8^+$ T-cell clones could be completely blocked by anti-IL-13 mAbs, indicating that IL-13 may be the dominant cytokine inducing IgE synthesis in situations where IL-4 is absent or present at low levels, for example during the initiation of allergic responses (Essayan *et al.*, 1996; Levy *et al.*, 1997; Punnonen *et al.*, 1997). This notion is supported by the observations that IL-13 is also produced by naive T cells.

Several lines of indirect evidence suggest that both IL-4 and IL-13 are required for optimal induction of human IgE synthesis. Both IL-4 and IL-13 are upregulated in the lungs of asthmatic patients after allergen challenge and in atopic dermatitis, indicating that both cytokines play a role in the regulation of allergic inflammatory responses (Huang *et al.*, 1995; Hamid *et al.*, 1996; Kotsimbos *et al.*, 1996; Kroegel *et al.*, 1996; Tengvall Linder *et al.*, 1996; Miadonna *et al.*, 1997). In addition, IL-4-deficient mice produce low levels of IgE *in vivo*, following viral or parasitic infections, indicating that IL-4-independent IgE synthesis occurs in these mice (von der Weid *et al.*, 1994; Noben-Trauth *et al.*, 1996). However, it is not known whether this IgE production is mediated by IL-13.

EFFECTS OF IL-13 ON MONOCYTES

The effects of IL-13 on monocytes–macrophages and endothelial cells are similar to those of IL-4. IL-13 modulates the phenotype of both human and murine monocytes–macrophages. IL-13 upregulates the expression of various adhesion molecules on monocytes, including CD11b, CD11c, CD18, CD29 and CD49e (very late activating antigen-5) (de Waal Malefyt *et al.*, 1993). This may contribute to the changes in morphology induced in monocytes–macrophages such as homotypic aggregation, strong adherence and the development of long cytoplasmic processes. IL-13 will also upregulate the expression of MHC class II molecules on human monocytes as well as that of CD80 and CD86, the ligands of CD28, on T cells, leading to an enhanced capacity to stimulate alloantigen-specific T-cell responses. Long-term culture of monocytes and macrophage precursors in the presence of IL-13 and GM-CSF leads to the differentiation of monocytes into dendritic cells and inhibition of monocyte proliferation (Piemonti *et al.*, 1995; Sakamoto *et al.*, 1995; Romani *et al.*, 1996). IL-13 also acts as chemoattractant for human monocytes (Magazin *et al.*, 1994).

In addition to these immunostimulatory properties, IL-13 has also important immune suppressive and anti-inflammatory activities. IL-13 inhibits synthesis of IL-1α, IL-1β, IL-6, IL-12, TNF-α and the chemokines IL-8, macrophage inflammatory protein-1α (MIP-1α), MIP-1β and monocyte chemotactic protein-3 (MCP-3) by lipopolysaccharide (LPS)-activated monocytes, synovial fluid and alveolar macrophages (Doherty *et al.*, 1993; Minty *et al.*, 1993a,b; Cosentino *et al.*, 1995; Hart *et al.*, 1995; Yanagawa *et al.*, 1995; Berkman *et al.*, 1996; Yano *et al.*, 1996).

Furthermore, IL-13 enhances the production of the IL-1 receptor antagonist and the release of decoy IL-1RII molecules, which both possess anti-inflammatory properties, by antagonizing IL-1 activity (Colotta *et al.*, 1994, 1996; Muzio *et al.*, 1994; Vannier *et al.*, 1996). However, IL-13 decreased the glucocorticoid receptor binding affinity of LPS-activated monocytes, resulting in a decreased sensitivity to the immunosuppressive effects of dexamethasone on IL-6 production (Spahn *et al.*, 1996). Although IL-13 inhibits the production of IL-12 by LPS-activated monocytes, overnight priming of monocytes with IL-13 before activation with LPS or *Staphylococcus aureus* Cowan type 1 (SAC) increased the production of IL-12 (D'Andrea *et al.*, 1995).

IL-13 induces expression of 15-lipoxygenase, which catalyzes the formation of 15-S hydroxyeicosatetraenoic acid (HETE) and lipoxin A4, mediators that antagonize proinflammatory leukotrines (Nassar *et al.*, 1994). In addition, IL-13 inhibits the formation of prostaglandin E_2 (PGE_2) from arachidonic acid through the inhibition of cox-2 induction by LPS-stimulated monocytes (Endo *et al.*, 1996) and osteoclasts, which results in inhibition of IL-1-induced bone resorption (Onoe *et al.*, 1996).

IL-13 inhibits the production of NO by LPS-activated mouse macrophages and mesangial cells, and inhibits superoxide anion production by human monocytes (Doherty *et al.*, 1993; Doyle *et al.*, 1994; Sozzani *et al.*, 1995; Saura *et al.*, 1996). Inhibition of NO production results in an enhanced iron uptake by murine macrophages, which contributes to the downregulation of their effector functions (Weiss *et al.*, 1997).

IL-13 increases the expression of mannose receptors on mouse and human monocytes and macrophages, which results in the elimination of proteins bearing terminal mannosyl ligands, such as lysosomal hydrolases and plasminogen activators (Doyle *et al.*, 1994; DeFife *et al.*, 1997). In addition, enhanced expression of mannose receptors results in human macrophage fusion and the formation of foreign-body giant cells (DeFife *et al.*, 1997).

IL-13, like IL-4, downregulates tissue factor expression on LPS-activated monocytes and thus inhibits procoagulant activity (Herbert *et al.*, 1993; Del Prete *et al.*, 1995; Ernofsson *et al.*, 1996).

Finally, IL-13 downregulates the expression of Fc γ receptors, CD16, CD32 and

CD64 on human monocytes, but induces expression of the low-affinity FcεRII (CD23) (de Waal Malefyt *et al.*, 1993; McKenzie *et al.*, 1993b). Downregulation of CD64 expression correlates with decreased antibody-dependent cell-mediated cytotoxicity by these cells (de Waal Malefyt *et al.*, 1993).

EFFECTS OF IL-13 ON ENDOTHELIAL CELLS AND EOSINOPHILS

IL-13 induces vascular cell adhesion molecule-1 (VCAM-1) expression on human umbilical vein endothelial cells, which results in adhesion of eosinophils to these cells (Sironi *et al.*, 1994; Bochner *et al.*, 1995). Eosinophils accumulate at sites of allergic inflammation, and they are thought to play a crucial role in the pathogenesis of the lung inflammation and in lung epithelial cell destruction in asthmatic patients. Consequently, the ability of IL-13, like that of IL-4, to induce VCAM-1 expression on endothelial cells and the subsequent interactions of VCAM-1 on endothelial cells and $\alpha_4\beta_1$ integrins on eosinophils, leading to eosinophil recruitment, further emphasizes the importance of these cytokines in the pathogenesis of allergic and asthmatic responses (Ying *et al.*, 1997). Furthermore IL-13 induces expression of CD69 on eosinophils and prolongs the survival of these cells (Luttmann *et al.*, 1996). In addition, it upregulates GM-CSF expression by bronchial epithelial cells, which also may contribute to the survival of eosinophils (Nakamura *et al.*, 1996). On the other hand, IL-13 has been shown to inhibit the expression of the eosinophil attractant RANTES by airway smooth muscle and endothelial cells (Marfaing-Koka *et al.*, 1995; John *et al.*, 1997).

EFFECTS OF IL-13 ON OTHER CELL TYPES

As discussed, IL-13 inhibits the production of proinflammatory cytokines, chemokines and NO by human monocytes and alveolar macrophages, but the effects of IL-13 on other cell types can be variable. IL-13 induced the production of IL-1 receptor antagonist (IL-1RA), IL-1RII and IL-8 by human neutrophils (Girard *et al.*, 1996). IL-13 also activated immediate early gene expression (c-fos) and upregulated intercellular adhesion molecule-1 expression on human mast cells (Nilsson and Nilsson, 1995). Furthermore, IL-13 enhanced the production of IL-6 by human microglial cells and keratinocytes (Derocq *et al.*, 1994; Sebire *et al.*, 1996), IL-1-induced production of IL-8 by an epithelial cell line and it induced neutral endopeptidase (CD13) expression on monocytes, renal carcinoma cell lines and tubular renal epithelial cells (de Waal Malefyt *et al.*, 1993; Riemann *et al.*, 1995; Kolios *et al.*, 1996). IL-13 also enhanced TNFα-induced expression of complement C3 and inhibited factor B production by dermal fibroblasts (Katz *et al.*, 1995) but inhibited IL-8 and PGE_2 production by synovial fibroblasts (Seitz *et al.*, 1996). IL-13 inhibited the production of fibrinogen by HepG2 cells, which might contribute to its anti-inflammatory activities (Vasse *et al.*, 1996).

IL-13 also affects hematopoiesis. IL-13 increased proliferation of mouse Lin^- Sca^+ bone marrow cells in combination with stem cell factor and G-CSF or IL-11, leading to the formation of macrophage colonies (Jacobsen *et al.*, 1994). Furthermore, IL-13 promotes megakaryocyte formation from $CD34^+$ human cord blood cells (Xi *et al.*, 1995). These effects of IL-13 were confirmed by continuous infusion of IL-13 in Balb/c mice, which led to splenomegaly and extramedullary hematopiesis due to expansion of erythroid and megakaryocyte colonies as well as monocytosis (Lai *et al.*, 1996). As mentioned above, IL-13 inhibited granylocyte–macrophage and macrophage colony formation from $CD34^+$ precursors from human bone marrow or cord blood (Sakamoto *et al.*, 1995).

CONCLUSIONS

IL-13 is a cytokine with important immunomodulating activities. Most notable are the effects on B cells and the role of IL-13 in IgE production and allergic responses, where it has overlapping, but also different, functions with IL-4. IL-13, in contrast to IL-4, is earlier and more abundantly produced by naive T cells and therefore may play a role in initiating IgE production and IgE-mediated allergic responses. On the other hand, IL-13, in contrast to IL-4, does not drive T_{H2} cell differentiation, which reflects one of the hallmarks of the allergic response. This suggests that IL-4 is the more dominant cytokine for the expansion and maintenance of T_{H2} cell-mediated allergic responses. The other prominent effects of IL-13 are its anti-inflammatory activities through the downregulation of proinflammatory cytokines, chemokines and mediators by activated monocytes–macrophages. This is exemplified *in vivo* by the finding that IL-13 could rescue mice from LPS-induced endotoxemia (Muchamuel *et al.*, 1997). However, whether endogenously produced IL-13 has a protective role in sepsis remains to be determined, since IL-13 was not detected in the serum from septic patients (van der Poll *et al.*, 1997).

The deactivating effects of IL-13 on monocytes could also be beneficial in chronic inflammatory diseases such as rheumatoid arthritis and inflammatory bowel disease. IL-13 inhibited IL-1β and TNF-α production by synovial fluid macrophages from patients with rheumatoid arthritis or from cocultures of synoviocytes and peripheral blood monocytes (Chomarat *et al.*, 1995; Hart *et al.*, 1995; Isomaki *et al.*, 1996). *In vivo*, L-13, delivered through a single injection of CHO cells which were genetically engineered to express it, reduced the severity and incidence of collagen-induced arthritis in DBA/1 mice, which correlated with a reduced production of TNF-α in the spleen (Bessis *et al.*, 1996). This same approach was also beneficial in autoimmune models. IL-13 suppressed the development of experimental autoimmune encephalomyelitis, as assessed by a reduction in mean duration, severity and incidence of the disease. Although this is primarily a T-cell disease, the anti-inflammatory effects of IL-13 on macrophages–monocytes seem to ameliorate its pathology (Cash *et al.*, 1994). IL-13 inhibited the pokeweed mitogen-induced production of IL-1, IL-6 and TNF-α by monocytes isolated from patients with inactive inflammatory bowel disease, but not from those with active disease. However, in combination with IL-10, both IL-13 and IL-4 synergized in inhibiting the production of proinflammatory cytokines, by monocytes from patients with active disease (Kucharzik *et al.*, 1996, 1997).

Some of the immune stimulatory effects of IL-13 could have therapeutic potential as well. IL-13 showed beneficial effects in models of tumor formation. Coinjection of CHO cells expressing IL-13 and HeLa cells prevented tumor cell formation in nude mice. Similar protective effects were observed with tumor cells engineered to express IL-13 in DBA/2 mice, where even immunity to the parental cell line was induced (Lebel-Binay *et al.*, 1995). These results were extended by the observations that IL-13 could reverse tumor-induced immune suppression in a rat model of colonic cancer (Reisser *et al.*, 1996).

Finally, IL-13 may have therapeutic potential in human immunodeficiency virus (HIV) infection, as it has been shown that IL-13 is able to inhibit virus production by primary monocytes, macrophages, chronically infected monocytic cell lines and alveolar macrophages (Montaner *et al.*, 1993; Denis and Ghadirian, 1994; Mikovits *et al.*, 1994; Naif *et al.*, 1997). However, IL-13 did not reduce virus production by T cells and enhanced the production of human cytomegalovirus (hCMV) from alveolar macrophages that were coinfected with HIV and hCMV (Montaner *et al.*, 1993; Hatch *et al.*, 1997).

REFERENCES

Aman, M.J., Tayebi, N., Obiri, N.I., Puri, R.K., Modi, W.S. and Leonard, W.J. (1996). cDNA cloning and characterizaton of the human interleukin 13 receptor α chain. *J. Biol. Chem.* **271**, 29 265–29 270.

Andersson, A., Grunewald, S.M., Duschl, A., Fischer, A. and DiSanto, J.P. (1997). Mouse macrophage development in the absence of the common γ chain: defining receptor complexes responsible for IL-4 and IL-13 signaling. *Eur. J. Immunol.* **27**, 1762–1768.

Aversa, G., Punnonen, J., Cocks, B.G., de Waal Malefyt, R., Vega, F., Jr., Zurawski, S.M., Zurawski, G. and de Vries, J.E. (1993). An interleukin 4 (IL-4) mutant protein inhibits both IL-4 or IL-13-induced human immunoglobulin G_4 (IgG_4) and IgE synthesis and B cell proliferation: support for a common component shared by IL-4 and IL-13 receptors. *J. Exp. Med.* **178**, 2213–2218.

Bamborough, P., Duncan, D. and Richards, W.G. (1994). Predictive modelling of the 3-D structure of interleukin-13. *Protein Eng.* **7**, 1077–1082.

Berkman, N., John, M., Roesems, G., Jose, P., Barnes, P.J. and Chung, K.F. (1996). Interleukin 13 inhibits macrophage inflammatory protein-1α production from human alveolar macrophages and monocytes. *Am. J. Respir. Cell Mol. Biol.* **15**, 382–389.

Bessis, N., Boissier, M.C., Ferrara, P., Blankenstein, T., Fradelizi, D. and Fournier, C. (1996). Attenuation of collagen-induced arthritis in mice by treatment with vector cells engineered to secrete interleukin-13. *Eur. J. Immunol.* **26**, 2399–2403.

Bochner, B.S., Klunk, D.A., Sterbinsky, S.A., Coffman, R.L. and Schleimer, R.P. (1995). IL-13 selectively induces vascular cell adhesion molecule-1 expression in human endothelial cells. *J. Immunol.* **154**, 799–803.

Brinkmann, V. and Kristofic, C. (1995). TCR-stimulated naive human $CD4^+$ $45RO^-$ T cells develop into effector cells that secrete IL-13, IL-5, and IFN-γ, but no IL-4, and help efficient IgE production by B cells. *J. Immunol.* **154**, 3078–3087.

Brown, K.D., Zurawski, S.M., Mosmann, T.R. and Zurawski, G. (1989). A familly of small inducible proteins secreted by leukocytes are members of a new superfamily that includes leukocyte and fibroblast-derived inflammatory agents, growth factors and indicators of various activation processes. *J. Immunol.* **142**, 679–687.

Burd, P.R., Thompson, W.C., Max, E.E. and Mills, F.C. (1995). Activated mast cells produce interleukin 13. *J. Exp. Med.* **181**, 1373–1380.

Caput, D., Laurent, P., Kaghad, M., Lelias, J.M., Lefort, S., Vita, N. and Ferrara, P. (1996). Cloning and characterization of a specific interleukin (IL)-13 binding protein structurally related to the IL-5 receptor α chain. *J. Biol. Chem.* **271**, 16 921–16 926.

Cash, E., Minty, A., Ferrara, P., Caput, D., Fradelizi, D. and Rott, O. (1994). Macrophage-inactivating IL-13 suppresses experimental autoimmune encephalomyelitis in rats. *J. Immunol.* **153**, 4258–4267.

Chaouchi, N., Wallon, C., Goujard, C., Tertian, G., Rudent, A., Caput, D., Ferrera, P., Minty, A., Vazquez, A. and Delfraissy, J. F. (1996). Interleukin-13 inhibits interleukin-2-induced proliferation and protects chronic lymphocytic leukemia B cells from *in vitro* apoptosis. *Blood* **87**, 1022–1029.

Chomarat, P., Rissoan, M.C., Pin, J.J., Banchereau, J. and Miossec, P. (1995). Contribution of IL-1, CD14, and CD13 in the increased IL-6 production induced by *in vitro* monocyte–synoviocyte interactions. *J. Immunol.* **155**, 3645–3652.

Cocks, B.G., de Waal Malefyt, R., Galizzi, J.P., de Vries, J.E. and Aversa, G. (1993). IL-13 induces proliferation an differentiation of human B cells activated by the CD40 ligand. *Int. Immunol.* **5**, 657–663.

Colotta, F., Re, F., Muzio, M., Polentarutti, N., Minty, A., Caput, D., Ferrara, P. and Mantovani, A. (1994). Interleukin-13 induces expression and release of interleukin-1 decoy receptor in human polymorphonuclear cells. *J. Biol. Chem.* **269**, 12 403–12 406.

Colotta, F., Saccani, S., Giri, J.G., Dower, S.K., Sims, J.E., Introna, M. and Mantovani, A. (1996). Regulated expression and release of the IL-1 decoy receptor in human mononuclear phagocytes. *J. Immunol.* **156**, 2534–2541.

Cosentino, G., Soprana, E., Thienes, C.P., Siccardi, A.G., Viale, G. and Vercelli, D. (1995). IL-13 down-regulates CD14 expression and TNF-α secretion in normal human monocytes. *J. Immunol.* **155**, 3145–3151.

D'Andrea, A., Ma, X., Aste-Amezaga, M., Paganin, C. and Trinchieri, G. (1995). Stimulatory and inhibitory effects of interleukin (IL)-4 and IL-13 on the production of cytokines by human peripheral blood mononuclear cells: priming for IL-12 and tumor necrosis factor α production. *J. Exp. Med.* **181**, 537–546.

de Waal Malefyt, R., Figdor, C.G., Huijbens, R., Mohan-Peterson, S., Bennett, B., Culpepper, J., Dang, W., Zurawski, G. and de Vries, J.E. (1993). Effects of IL-13 on phenotype, cytokine production, and cytotoxic function of human monocytes. Comparison with IL-4 and modulation by IFN-γ or IL-10. *J. Immunol.* **151**, 6370–6381.

de Waal Malefyt, R., Abrams, J.S., Zurawski, S.M., Lecron, J.C., Mohan-Peterson, S., Sanjanwala, B., Bennett, B., Silver, J., de Vries, J.E. and Yssel, H. (1995). Differential regulation of IL-13 and IL-4 production by human $CD8^+$ and $CD4^+$ Th0, Th1 and Th2 T cell clones and EBV-transformed B cells. *Int. Immunol.* **7**, 1405–1416.

Debinski, W., Obiri, N.I., Pastan, I. and Puri, R.K. (1995). A novel chimeric protein composed of interleukin 13 and *Pseudomonas* exotoxin is highly cytotoxic to human carcinoma cells expressing receptors for interleukin 13 and interleukin 4. *J. Biol. Chem.* **270**, 16 775–16 780.

Debinski, W., Miner, R., Leland, P., Obiri, N.I. and Puri, R.K. (1996). Receptor for interleukin (IL) 13 does not interact with IL4 but receptor for IL4 interacts with IL13 on human glioma cells. *J. Biol. Chem.* **271**, 22 428–22 433.

DeFife, K.M., Jenney, C.R., McNally, A.K., Colton, E. and Anderson, J.M. (1997). Interleukin-13 induces human monocyte/macrophage fusion and macrophage mannose receptor expression. *J. Immunol.* **158**, 3385–3390.

Defrance, T., Carayon, P., Billian, G., Guillemot, J.C., Minty, A., Caput, D. and Ferrara, P. (1994). Interleukin 13 is a B cell stimulating factor. *J. Exp. Med.* **179**, 135–143.

Del Prete, G., De Carli, M., Lammel, R.M., D'Elios, M.M., Daniel, K.C., Giusti, B., Abbate, R. and Romagnani, S. (1995). Th1 and Th2 T-helper cells exert opposite regulatory effects on procoagulant activity and tissue factor production by human monocytes. *Blood* **86**, 250–257.

Denis, M. and Ghadirian, E. (1994). Interleukin 13 and interleukin 4 protect bronchoalveolar macrophages from productive infection with human immunodeficiency virus type 1. *AIDS Res. Hum. Retroviruses* **10**, 795–802.

Derocq, J.M., Segui, M., Poinot-Chazel, C., Minty, A., Caput, D., Ferrara, P. and Casellas, P. (1994). Interleukin-13 stimulates interleukin-6 production by human keratinocytes. Similarity with interleukin-4. *FEBS Lett.* **343**, 32–36.

Doherty, T.M., Kastelein, R., Menon, S. Andrade, S. and Coffman, R.L. (1993). Modulation of murine macrophage function by IL-13. *J. Immunol.* **151**, 7151–7160.

Dolganov, G., Bort, S., Lovett, M., Burr, J., Schubert, L., Short, D., McGurn, M., Gibson, C. and Lewis, D.B. (1996). Coexpression of the interleukin-13 and interleukin-4 genes correlates with their physical linkage in the cytokine gene cluster on human chromosome 5q23-31. *Blood* **87**, 3316–3326.

Doyle, A.G., Herbein, G., Montaner, L.J., Minty, A.J., Caput, D., Ferrara, P. and Gordon, S. (1994). Interleukin-13 alters the activation state of murine macrophages *in vitro*: comparison with interleukin-4 and interferon-γ. *Eur. J. Immunol.* **24**, 1441–1445.

Emilie, D., Zou, W., Fior, R., Llorente, L., Durandy, A., Crevon, M.C., Maillot, M.C., Durand-Gasselin, I., Raphael, M., Peuchmaur, M. and Galamaud, P. (1997). Production and roles of IL-6, IL-10, and IL-13 in B-lymphocyte malignancies and in B-lymphocyte hyperactivity of HIV infection and autoimmunity. *Methods* **11**, 133–142.

Endo, T., Ogushi, F. and Sone, S. (1996). LPS-dependent cyclooxygenase-2 induction in human monocytes is down-regulated by IL-13, but not by IFN-γ. *J. Immunol.* **156**, 2240–2246.

Ernofsson, M., Tenno, T. and Siegbahn, A. (1996). Inhibition of tissue factor surface expression in human peripheral blood monocytes exposed to cytokines. *Br. J. Haematol.* **95**, 249–257.

Essayan, D.M., Han, W.F., Li, X.M., Xiao, H.Q., Kleine-Tebbe, J. and Huang, S.K. (1996). Clonal diversity of IL-4 and IL-13 expression in human allergen-specific T lymphocytes. *J. Allergy Clin. Immunol.* **98**, 1035–1044.

Ezernieks, J., Schnarr, B., Metz, K. and Duschl, A. (1996). The human IgE germline promoter is regulated by interleukin-4, interleukin-13, interferon-α and interferon-γ via an interferon-γ-activated site and its flanking regions. *Eur. J. Biochem.* **240**, 667–673.

Feng, N., Schnyder, B., Vonderschmitt, D.J., Ryffel, B. and Lutz, R.A. (1995). Characterization of interleukin-13 receptor in carcinoma cell lines and human blood cells and comparison with the interleukin-4 receptor. *J. Recept. Signal Transduct. Res.* **15**, 931–949.

Fior, R., Vita, N., Raphael, M., Minty, A., Maillot, M.C., Crevon, M.C., Caput, D., Biberfeld, P., Ferrara, P. and

Galanaud, P. (1994). Interleukin-13 gene expression by malignant and EBV-transformed human B lymphocytes. *Eur. Cytokine Netw.* **5**, 593–600.

Gauchat, J.F., Henchoz, S., Mazzei, G., Aubry, J.P., Brunner, T., Blasey, H., Life, P., Talabot, D., Flores-Romo, L. and Thompson, J. (1993). Induction of human IgE synthesis in B cells by mast cells and basophils. *Nature* **365**, 340–343.

Gauchat, J.F., Schlagenhauf, E., Feng, N.P., Moser, R., Yamage, M., Jeannin, P., Alouani, S., Elson, G., Notarangelo, L.D., Wells, T., Eugster, H.P. and Bonnefoy, J.Y. (1997). A novel 4-kb interleukin-13 receptor α mRNA expressed in human B, T, and endothelial cells encoding an alternate type-II interleukin-4/interleukin-13 receptor. *Eur. J. Immunol.* **27**, 971–978.

Gibbs, B.F., Haas, H., Falcone, F.H., Albrecht, C., Vollrath, I.B., Noll, T., Wolff, H.H. and Amon, U. (1996). Purified human peripheral blood basophils release interleukin-13 and preformed interleukin-4 following immunological activation. *Eur. J. Immunol.* **26**, 2493–2498.

Girard, D., Paquin, R., Naccache, P.H. and Beaulieu, A.D. (1996). Effects of interleukin-13 on human neutrophil functions. *J. Leukoc. Biol.* **59**, 412–419.

Hamid, Q., Naseer, T., Minshall, E.M., Song, Y.L., Boguniewicz, M. and Leung, D.Y. (1996). *In vivo* expression of IL-12 and IL-13 in atopic dermatitis. *J. Allergy Clin. Immunol.* **98**, 225–231.

Hart, P.H., Ahern, M.J., Smith, M.D. and Finlay-Jones, J.J. (1995). Regulatory effects of IL-13 on synovial fluid macrophages and blood monocytes from patients with inflammatory arthritis. *Clin. Exp. Immunol.* **99**, 331–337.

Hatch, W.C., Freedman, A.R., Boldt-Houle, D.M., Groopman, J.E. and Terwilliger, E.F. (1997). Differential effects of interleukin-13 on cytomegalovirus and human immunodeficiency virus infection in human alveolar macrophages. *Blood* **89**, 3443–3450.

He, Y.W., Adkins, B., Furse, R.K. and Malek, T.R. (1995). Expression and function of the γc subunit of the IL-2, Il-4, and IL-7 receptors. Distinct interaction of γc in the IL-4 receptor. *J. Immunol.* **154**, 1596–1605.

Herbert, J.M., Savi, P., Laplace, M.C., Lale, A., Dol, F., Dumas, A., Labit, C. and Minty, A. (1993). IL-4 and IL-13 exhibit comparable abilities to reduce pyrogen-induced expression of procoagulant activity in endothelial cells and monocytes. *FEBS Lett.* **328**, 268–270.

Hilton, D.J., Zhang, J.G., Metcalf, D., Alexander, W.S., Nicola, N.A. and Willson, T.A. (1996). Cloning and characterization of a binding subunit of the interleukin 13 receptor that is also a component of the interleukin 4 receptor. *Proc. Natl Acad. Sci. U.S.A.* **93**, 497–501.

Hodge, M.R., Ranger, A.M., Charles de la Brousse, F., Hoey, T., Grusby, M.J. and Glimcher, L.H. (1996). Hyperproliferation and dysregulation of IL-4 expression in NF-ATp-deficient mice. *Immunity* **4**, 397–405.

Huang, S.K., Xiao, H.Q., Kleine-Tebbe, J., Paciotti, G., Marsh, D.G., Lichtenstein, L.M. and Liu, M.C. (1995). IL-13 expression at the sites of allergen challenge in patients with asthma. *J. Immunol.* **155**, 2688–2694.

Isomaki, P., Luukkainen, R., Toivanen, P. and Punnonen, J. (1996). The presence of interleukin-13 in rheumatoid synovium and its antiinflammatory effects on synovial fluid macrophages from patients with rheumatoid arthritis. *Arthritis Rheum.* **39**, 1693–1702.

Izuhara, K., Heike, T., Otsuka, T., Yamaoka, K., Mayumi, M., Imamura, T., Niho, Y. and Harada, N. (1996). Signal transduction pathway of interleukin-4 and interleukin-13 in human B cells derived from X-linked severe combined immunodeficiency patients. *J. Biol. Chem.* **271**, 619–622.

Jacobsen, S.E., Okkenhaug, C., Veiby, O.P., Caput, D., Ferrara, P. and Minty, A. (1994). Interleukin 13: novel role in direct regulation of proliferation and differentiation of primitive hematopoietic progenitor cells. *J. Exp. Med.* **180**, 75–82.

John, M., Hirst, S.J., Jose, P.J., Robichaud, A., Berkman, N., Witt, C., Twort, C.H., Barnes, P.J. and Chung, K.F. (1997). Human airway smooth muscle cells express and release RANTES in response to T helper 1 cytokines: regulation by T helper 2 cytokines and corticosteroids. *J. Immunol.* **158**, 1841–1847.

Jung, T., Wijdenes, J., Neumann, C., de Vries, J.E. and Yssel, H. (1996). Interleukin-13 is produced by activated human $CD45RA^+$ and $CD45RO^+$ T cells: modulation by interleukin-4 and interleukin-12. *Eur. J. Immunol.* **26**, 571–577.

Katz, Y., Stav, D., Barr, J. and Passwell, J.H. (1995). IL-13 results in differential regulation of the complement proteins C3 and factor B in tumour necrosis factor (TNF)-stimulated fibroblasts. *Clin. Exp. Immunol.* **101**, 150–156.

Keegan, A.D., Johnston, J.A., Tortolani, P.J., McReynolds, L.J., Kinzer, C., O'Shea, J.J. and Paul, W.E. (1995). Similarities and differences in signal transduction by interleukin 4 and interleukin 13: analysis of Janus kinase activation. *Proc. Natl Acad. Sci. U.S.A.* **92**, 7681–7685.

Kolios, G., Robertson, D.A., Jordan, N.J., Minty, A., Caput, D., Ferrara, P. and Westwick, J. (1996). Interleukin-

8 production by the human colon epithelial cell line HT-29: modulation by interleukin-13. *Br. J. Pharmacol.* **119**, 351–359.

Kotsimbos, T.C., Ernst, P. and Hamid, Q.A. (1996). Interleukin-13 and interleukin-4 are coexpressed in atopic asthma. *Proc. Assoc. Am. Phys.* **108**, 368–373.

Kricek, F., Ruf, C., Zunic, M., De Jong, G., Dukor, P. and Bahr, G.M. (1995). Induction in mice of serum IgE levels after treatment with anti-mouse IgD antibodies is preceded by differential modulation of tissue cytokine gene transcription. *Eur. J. Immunol.* **25**, 936–941.

Kroegel, C., Julius, P., Matthys, H., Virchow, J.C., Jr. and Luttmann, W. (1996). Endobronchial secretion of interleukin-13 following local allergen challenge in atopic asthma: relationship to interleukin-4 and eosinophil counts. *Eur. Respir. J.* **9**, 899–904.

Kucharzik, T., Lugering, N., Weigelt, H., Adolf, M., Domschke, W. and Stoll, R. (1996). Immunoregulatory properties of IL-13 in patients with inflammatory bowel disease; comparison with IL-4 and IL-10. *Clin. Exp. Immunol.* **104**, 483–490.

Kucharzik, T., Lugering, N., Adolf, M., Domschke, W. and Stoll, R. (1997). Synergistic effect of immunoregulatory cytokines on peripheral blood monocytes from patients with inflammatory bowel disease. *Dig. Dis. Sci.* **42**, 805–812.

Lai, Y.H., Heslan, J.M., Poppema, S., Elliot, J.F. and Mosmann, T.R. (1996). Continuous administration of IL-13 to mice induces extramedullary hemopoiesis and monocytosis. *J. Immunol.* **156**, 3166–3173.

Lebel-Binay, S., Laguerre, B., Quintin-Colonna, F., Conjeaud, H., Magazin, M., Miloux, B., Pecceu, F., Caput, D., Ferrara, P. and Fradelizi, D. (1995). Experimental gene therapy of cancer using tumor cells engineered to secrete interleukin-13. *Eur. J. Immunol.* **25**, 2340–2348.

Lefort, S., Vita, N., Reeb, R., Caput, D. and Ferrara, P. (1995). IL-13 and IL-4 share signal transduction elements as well as receptor components in TF-1 cells. *FEBS Lett.* **366**, 122–126.

Levy, F., Kristofic, C., Heusser, C. and Brinkmann, V. (1997). Role of IL-13 in CD4 T cell-dependent IgE production in atopy. *Int. Arch. Allergy Immunol.* **112**, 49–58.

Li, H., Sim, T.C. and Alam, R. (1996). IL-13 released by and localized in human basophils. *J. Immunol.* **156**, 4833–4838.

Luttmann, W., Knoechel, B., Foerster, M., Matthys, H., Virchow, J.C., Jr. and Kroegel, C. (1996). Activation of human eosinophils by IL-13. Induction of CD69 surface antigen, its relationship to messenger RNA expression, and promotion of cellular viability. *J. Immunol.* **157**, 1678–1683.

Magazin, M., Guillemot, J.C., Vita, N. and Ferrara, P. (1994). Interleukin-13 is a monocyte chemoattractant. *Eur. Cytokine Netw.* **5**, 397–400.

Malabarba, M.G., Rui, H., Deutsch, H.H., Chung, J., Kalthoff, F.S., Farrar, W.L. and Kirken, R.A. (1996). Interleukin-13 is a potent activator of JAK3 and STAT6 in cells expressing interleukin-2 receptor-γ and interleukin-4 receptor-α. *Biochem. J.* **319**, 865–872.

Marfaing-Koka, A., Devergne, O., Gorgone, G., Portier, A., Schall, T.J., Galanaud, P. and Emilie, D. (1995). Regulation of the production of the RANTES chemokine by endothelial cells. Synergistic induction by IFN-γ plus TNF-α and inhibition by IL-4 and IL-13. *J. Immunol.* **154**, 1870–1878.

Marietta, E.V., Chen, Y. and Weis, J.H. (1996). Modulation of expression of the anti-inflammatory cytokines interleukin-13 and interleukin-10 by interleukin-3. *Eur. J. Immunol.* **26**, 49–56.

Matthews, D.J., Clark, P.A., Herbert, J., Morgan, G., Armitage, R.J., Kinnon, C., Minty, A., Grabstein, K.H., Caput, D. and Ferrara, P. (1995). Function of the interleukin-2 (IL-2) receptor γ-chain in biologic responses of X-linked severe combined immunodeficient B cells to IL-2, IL-4, IL-13, and IL-15. *Blood* **85**, 38–42.

Matthews, D.J., Hibbert, L., Friedrich, K., Minty, A. and Callard, R.E. (1997). X-SCID B cell responses to interleukin-4 and interleukin-13 are mediated by a receptor complex that includes the interleukin-4 receptor α chain (p140) but not the γc chain. *Eur. J. Immunol.* **27**, 116–121.

McKenzie, A.N., Li, X., Largaespada, D.A., Sato, A., Kaneda, A., Zurawski, S.M., Doyle, E.L., Milatovich, A., Francke, U. and Copeland, N.G. (1993a). Structural comparison and chromosomal localization of the human and mouse IL-13 genes. *J. Immunol.* **150**, 5436–5444.

McKenzie, A.N.J., Culpepper, J.A., de Waal Malefyt, R., de Vries, J.E. and Zurawski, G. (1993b). Interleukin-13, a T cell derived cytokine that regulates human monocyte and B cell function. *Proc. Natl Acad. Sci. U.S.A.* **150**, 5436–5444.

Miadonna, A., Gibelli, S., Lorini, M., Tedeschi, A., Oddera, S., Rossi, G.A. and Crimi, E. (1997). Expression of cytokine mRNA in bronchoalveolar lavage cells from atopic asthmatics before late antigen-induced reaction. *Lung* **175**, 195–209.

Michel, G., Ried, C. and Beetz, A. (1994). IL-6 is a potent inducer of IL-13 mRNA in normal human keratinocytes. *J. Invest. Dermatol.* **103**, 433.

Mikovits, J.A., Meyers, A.M., Ortaldo, J.R., Minty, A., Caput, D., Ferrara, P. and Ruscetti, F.W. (1994). IL-4 and IL-13 have overlapping but distinct effects on HIV production in monocytes. *J. Leukoc. Biol.* **56**, 340–346.

Miloux, B., Laurent, P., Bonnin, O., Lupker, J., Caput, D., Vita, N. and Ferrara, P. (1997). Cloning of the human IL-13R α_1 chain and reconstitution with the IL4R α of a functional IL-4/IL-13 receptor complex. *FEBS Lett.* **401**, 163–166.

Minty, A., Chalon, P., Derocq, J. M., Dumont, X., Guillemot, J. C., Kaghad, M., Labit, C., Leplatois, P., Liauzun, P. and Miloux, B. (1993a). Interleukin-13 is a new human lymphokine regulating inflammatory and immune responses. *Nature* **362**, 248–250.

Minty, A., Chalon, P., Guillemot, J.C., Kaghad, M., Liauzun, P., Magazin, M., Miloux, B., Minty, C., Ramond, P. and Vita, N. (1993b). Molecular cloning of the MCP-3 chemokine gene and regulation of its expression. *Eur. Cytokine Netw.* **4**, 99–110.

Montaner, L.J., Doyle, A.G., Collin, M., Herbein, G., Illei, P., James, W., Minty, A., Caput, D., Ferrara, P. and Gordon, S. (1993). Interleukin 13 inhibits human immunodeficiency virus type 1 production in primary blood-derived human macrophages *in vitro*. *J. Exp. Med.* **178**, 743–747.

Morgan, J.G., Dolganov, G.M., Robbins, S.E. and Paul, W. (1992). The selective isolation of novel cDNAs encoded by the regions surrounding the human interleukin 4 and 5 genes. *Nucl. Acids Res.* **20**, 5173–5179.

Muchamuel, T., Menon, S., Pisacane, P., Howard, M.C. and Cockayne, D.A. (1997). IL-13 protects mice from lipopolysaccharide-induced lethal endotoxemia: correlation with down-modulation of TNF-α, IFN-γ, and IL-12 production. *J. Immunol.* **158**, 2898–2903.

Murata, T., Noguchi, P.D. and Puri, R.K. (1996). IL-13 induces phosphorylation and activation of JAK2 Janus kinase in human colon carcinoma cell lines: similarities between IL-4 and IL-13 signaling. *J. Immunol.* **156**, 2972–2978.

Murata, T., Obiri, N.I. and Puri, R.K. (1997). Human ovarian-carcinoma cell lines express IL-4 and IL-13 receptors: comparison between IL-4 and IL-13-induced signal transduction. *Int. J. Cancer* **70**, 230–240.

Musso, T., Varesio, L., Zhang, X., Rowe, T.K., Ferrara, P., Ortaldo, J.R., O'Shea, J.J. and McVicar, D.W. (1994). IL-4 and IL-13 induce Lsk, a Csk-like tyrosine kinase, in human monocytes. *J. Exp. Med.* **180**, 2383–2388.

Muzio, M., Re, F., Sironi, M., Polentarutti, N., Minty, A., Caput, D., Ferrara, P., Mantovani, A. and Colotta, F. (1994). Interleukin-13 induces the production of interleukin-1 receptor antagonist (IL-1ra) and the expression of the mRNA for the intracellular (keratinocyte) form of IL-1ra in human myelomonocytic cells. *Blood* **83**, 1738–1743.

Naif, H.M., Li, S., Ho-Shon, M., Mathijs, J.M., Williamson, P. and Cunningham, A.L. (1997). The state of maturation of monocytes into macrophages determines the effects of IL-4 and IL-13 on HIV replication. *J. Immunol.* **158**, 501–511.

Nakamura, Y., Azuma, M., Okano, Y., Sano, T., Takahashi, T., Ohmoto, Y. and Sone, S. (1996). Upregulatory effects of interleukin-4 and interleukin-13 but not interleukin-10 on granulocyte/macrophage colony-stimulating factor production by human bronchial epithelial cells. *Am. J. Respir. Cell Mol. Biol.* **15**, 680–687.

Nassar, G.M., Morrow, J.D., Roberts, L.J.D., Lakkis, F.G. and Badr, K.F. (1994). Induction of 15-lipoxygenase by interleukin-13 in human blood monocytes. *J. Biol. Chem.* **269**, 27 631–27 634.

Nilsson, G. and Nilsson, K. (1995). Effects of interleukin (IL)-13 on immediate-early response gene expression, phenotype and differentiation of human mast cells. Comparison with IL-4. *Eur. J. Immunol.* **25**, 870–873.

Noben-Trauth, N., Kropf, P. and Muller, I. (1996). Susceptibility to *Leishmania major* infection in interleukin-4 deficient mice. *Science* **271**, 987–990.

Obiri, N.I., Debinski, W., Leonard, W.J. and Puri, R.K. (1995). Receptor for interleukin 13 interaction with interleukin 4 by a mechanism that does not involve the common γ chain shared by receptors for interleukins 2, 4, 7, 9, and 15. *J. Biol. Chem.* **270**, 8797–8804.

Obiri, N.I., Leland, P., Murata, T., Debinski, W. and Puri, R.K. (1997). The IL-13 receptor structure differs on various cell types and may share more than one component with IL-4 receptor. *J. Immunol.* **158**, 756–764.

Onoe, Y., Miyaura, C., Kaminakayashiki, T., Nagai, Y., Noguchi, K., Chen, Q.R., Seo, H., Ohta, H., Nozawa, S., Kudo, I. and Suda, T. (1996). IL-13 and IL-4 inhibit bone resorption by suppressing cyclooxygenase-2-dependent prostaglandin synthesis in osteoblasts. *J. Immunol.* **156**, 758–764.

Palmer-Crocker, R.L., Hughes, C.C. and Pober, J.S. (1996). IL-4 and IL-13 activate the JAK2 tyrosine kinase and Stat6 in cultured human vascular endothelial cells through a common pathway that does not involve the γc chain. *J. Clin. Invest.* **98**, 604–609.

Pawankar, R.U., Okuda, M., Hasegawa, S., Suzuki, K., Yssel, H., Okubo, K., Okumura, K. and Ra, C. (1995).

Interleukin-13 expression in the nasal mucosa of perennial allergic rhinitis. *Am. J. Respir. Crit. Care Med.* **152**, 2059–2067.

Pawankar, R., Okuda, M., Yssel, H., Okumura, K. and Ra, C. (1997). Nasal mast cells in perennial allergic rhinitics exhibit increased expression of the Fc εRI, CD40L, IL-4, and IL-13, and can induce IgE synthesis in B cells. *J. Clin. Invest.* **99**, 1492–1499.

Pearlman, E., Lass, J.H., Bardenstein, D.S., Diaconu, E., Hazlett, F.E., Jr., Albright, J., Higgins, A.W. and Kazura, J.W. (1996). *Onchocerca volvulus*-mediated keratitis: cytokine production by IL-4-deficient mice. *Exp. Parasitol.* **84**, 274–281.

Piemonti, L., Bernasconi, S., Luini, W., Trobonjaca, Z., Minty, A., Allavena, P. and Mantovani, A. (1995). IL-13 supports differentiation of dendritic cells from circulating precursors in concert with GM-CSF. *Eur. Cytokine Netw.* **6**, 245–252.

Punnonen, J. and de Vries, J.E. (1994). IL-13 induces proliferation, Ig isotype switching, and Ig synthesis by immature human fetal B cells. *J. Immunol.* **152**, 1094–1102.

Punnonen, J., Aversa, G., Cocks, B.G., McKenzie, A.N., Menon, S., Zurawski, G., de Waal Malefyt, R. and de Vries, J.E. (1993). Interleukin 13 induces interleukin 4-independent IgG_4 and IgE synthesis and CD23 expression by human B cells. *Proc. Natl Acad. Sci. U.S.A.* **90**, 3730–3734.

Punnonen, J., Cocks, B.G. and de Vries, J.E. (1995). IL-4 induces germ-line IgE heavy chain gene transcription in human fetal pre-B cells. Evidence for differential expression of functional IL-4 and IL-13 receptors during B cell ontogeny. *J. Immunol.* **155**, 4248–4254.

Punnonen, J., Yssel, H. and de Vries, J.E. (1997). The relative contribution of IL-4 and IL-13 to human IgE synthesis induced by activated $CD4^+$ or $CD8^+$ T cells. *J. Allergy Clin. Immunol.* (in press).

Reisser, D., Pinard, D. and Jeannin, J.F. (1996). IL-13 restores an *in vitro* specific cytolytic response of spleen cells from tumor-bearing rats. *J. Leukoc. Biol.* **59**, 728–732.

Renard, N., Duvert, V., Banchereau, J. and Saeland, S. (1994). Interleukin-13 inhibits the proliferation of normal and leukemic human B-cell precursors. *Blood* **84**, 2253–2260.

Riemann, D., Kehlen, A. and Langner, J. (1995). Stimulation of the expression and the enzyme activity of aminopeptidase N/CD13 and dipeptidylpeptidase IV/CD26 on human renal cell carcinoma cells and renal tubular epithelial cells by T cell-derived cytokines such as IL-4 and IL-13. *Clin. Exp. Immunol.* **100**, 277–283.

Romani, N., Reider, D., Heuer, M., Ebner, S., Kampgen, E., Eibl, B., Niederwieser, D. and Schuler, G. (1996). Generation of mature dendritic cells from human blood. An improved method with special regard to clinical applicability. *J. Immunol. Methods* **196**, 137–151.

Sakamoto, O., Hashiyama, M., Minty, A. Ando, M. and Suda, T. (1995). Interleukin-13 selectively suppresses the growth of human macrophage progenitors at the late stage. *Blood* **85**, 3487–3493.

Saura, M., Martinez-Dalmau, R., Minty, A., Perez-Sala, D. and Lamas, S. (1996). Interleukin-13 inhibits inducible nitric oxide synthase expression in human mesangial cells. *Biochem. J.* **313**, 641–646.

Schnyder, B., Lugli, S., Feng, N., Etter, H., Lutz, R.A., Ryffel, B., Sugamura, K., Wunderli-Allenspach, H. and Moser, R. (1996). Interleukin-4 (IL-4) and IL-13 bind to a shared heterodimeric complex on endothelial cells mediating vascular cell adhesion molecule-1 induction in the absence of the common γ chain. *Blood* **87**, 4286–4295.

Sebire, G., Delfraissy, J.F., Demotes-Mainard, J., Oteifeh, A., Emilie, D. and Tardieu, M. (1996). Interleukin-13 and interleukin-4 act as interleukin-6 inducers in human microglial cells. *Cytokine* **8**, 636–641.

Seitz, M., Loetscher, P., Dewald, B., Towbin, H. and Baggiolini, M. (1996). Opposite effects of interleukin-13 and interleukin-12 on the release of inflammatory cytokines, cytokine inhibitors and prostaglandin E from synovial fibroblasts and blood mononuclear cells. *Eur. J. Immunol.* **26**, 2198–2202.

Sironi, M., Sciacca, F.L., Matteucci, C., Conni, M., Vecchi, A., Bernasconi, S., Minty, A., Caput, D., Ferrara, P. and Colotta, F. (1994). Regulation of endothelial and mesothelial cell function by interleukin-13: selective induction of vascular cell adhesion molecule-1 and amplification of interleukin-6 production. *Blood* **84**, 1913–1921.

Smerz-Bertling, C. and Duschl, A. (1995). Both interleukin 4 and interleukin 13 induce tyrosine phosphorylation of the 140-kDa subunit of the interleukin 4 receptor. *J. Biol. Chem.* **270**, 966–970.

Smirnov, D.V., Smirnova, M.G., Korobko, V.G. and Frolova, E.I. (1995). Tandem arrangement of human genes for interleukin-4 and interleukin-13: resemblance in their organization. *Gene* **155**, 277–281.

Sornasse, T., Larenas, P.V., Davis, K.A., de Vries, J.E. and Yssel, H. (1996). Differentiation and stability of T helper 1 and 2 cells derived from naive human neonatal $CD4^+$ T calls, analyzed at the single-cell level. *J. Exp. Med.* **184**, 473–483.

Sozzani, P., Cambon, C., Vita, N., Seguelas, M.H., Caput, D., Ferrara, P. and Pipy, B. (1995). Interleukin-13 inhibits protein kinase C-triggered respiratory burst in human monocytes. Role of calcium and cyclic AMP. *J. Biol. Chem.* **270**, 5084–5088.

Spahn, J.D., Szefler, S.J., Surs, W., Doherty, D.E., Nimmagadda, S.R. and Leung, D.Y. (1996). A novel action of IL-13: induction of diminished monocyte glucocorticoid receptor-binding affinity. *J. Immunol.* **157**, 2654–2659.

Sun, X.J., Wang, L.M., Zhang, Y., Yenush, L., Myers, M.G., Jr., Glasheen, E., Lane, W.S., Pierce, J.H. and White, M.F. (1995). Role of IRS-2 in insulin and cytokine signalling. *Nature* **377**, 173–177.

Takeda, K., Kamanaka, M., Tanaka, T., Kishimoto, T. and Akira, S. (1996). Impaired IL-13-mediated functions of macrophages in STAT6-deficient mice. *J. Immunol.* **157**, 3220–3222.

Tengvall Linder, M., Johansson, C., Zargari, A., Bengtsson, A., van der Ploeg, I., Jones, I., Harfast, B. and Scheynius, A. (1996). Detection of *Pityrosporum orbiculare* reactive T cells from skin and blood in atopic dermatitis and characterization of their cytokine profiles. *Clin. Exp. Allergy* **26**, 1286–1297.

Tony, H.P., Shen, B.J., Reusch, P. and Sebald, W. (1994). Design of human interleukin-4 antagonists inhibiting interleukin-4-dependent and interleukin-13-dependent responses in T-cells and B-cells with high efficiency. *Eur. J. Biochem.* **225**, 659–665.

van der Poll, T., de Waal Malefyt, R., Coyle, S.M. and Lowry, S.F. (1997). Antiinflammatory cytokine responses during clinical sepsis and experimental endotoxemia: sequential measurements of plasma soluble interleukin (IL)-1 receptor type II, IL-10, and IL-13. *J. Infect. Dis.* **175**, 118–122.

Vannier, E., de Waal Malefyt, R., Salazar-Montes, A., de Vries, J.E. and Dinarello, C.A. (1996). Interleukin-13 (IL-13) induces IL-1 receptor antagonist gene expression and protein synthesis in peripheral blood mononuclear cells: inhibition by an IL-4 mutant protein. *Blood* **87**, 3307–3315.

Vasse, M., Paysant, I., Soria, J., Mirshahi, S.S., Vannier, J.P. and Soria, C. (1996). Down-regulation of fibrinogen biosynthesis by IL-4, IL-10 and IL-13. *Br. J. Haematol.* **93**, 955–961.

von der Weid, T., Kopf, M., Kohler, G. and Langhorne, J. (1994). The immmune response to *Plasmodium chabaudi* malaria in interleukin-4 deficient mice. *Eur. J. Immunol.* **24**, 2285–2293.

Wang, L.M., Michieli, P., Lie, W.R., Liu, F., Lee, C.C., Minty, A., Sun, X.J., Levine, A., White, M.F. and Pierce, J.H. (1995). The insulin receptor substrate-1-related 4PS substrate but not the interleukin-2R γ chain is involved in interleukin-13-mediated signal transduction. *Blood* **86**, 4218–4227.

Weiss, G., Bogdan, C. and Hentze, M.W. (1997). Pathways for the regulation of macrophage iron metabolism by the anti-inflammatory cytokines IL-4 and IL-13. *J. Immunol.* **158**, 420–425.

Welham, M.J., Learmonth, L., Bone, H. and Schrader, J.W. (1995). Interleukin-13 signal transduction in lymphohemopoietic cells. Similarities and differences in signal transduction with interleukin-4 and insulin. *J. Biol. Chem.* **270**, 12 286–12 296.

Wright, K., Ward, S.G., Kolios, G. and Westwick, J. (1997). Activation of phosphatidylinositol 3-kinase by interleukin-13: an inhibitory signal for inducible nitric-oxide synthase expression in epithelial cell line HT-29. *J. Biol. Chem.* **272**, 12 626–12 633.

Xi, X., Schlegel, N., Caen, J.P., Minty, A., Fournier, S., Caput, D., Ferrara, P. and Han, Z.C. (1995). Differential effects of recombinant human interleukin-13 on the *in vitro* growth of human haemopoietic progenitor cells. *Br. J. Haematol.* **90**, 921–927.

Yanagawa, H., Sone, S., Haku, T., Mizuno, K., Yano, S., Ohmoto, Y. and Ogura, T. (1995). Contrasting effect of interleukin-13 on interleukin-1 receptor antagonist and proinflammatory cytokine production by human alveolar macrophages. *Am. J. Respir. Cell Mol. Biol.* **12**, 71–76.

Yano, S., Yanagawa, H., Nishioka, Y., Mukaida, N., Matsushima, K. and Sone, S. (1996). T helper 2 cytokines differently regulate monocyte chemoattractant protein-1 production by human peripheral blood monocytes and alveolar macrophages. *J. Immunol.* **157**, 2660–2665.

Ying, S., Meng, Q., Barata, L. T., Robinson, D.S., Durham, S.R. and Kay, A.B. (1997). Associations between IL-13 and IL-4 (mRNA and protein), vascular cell adhesion molecule-1 expression, and the infiltration of eosinophils, macrophages, and T cells in allergen-induced late-phase cutaneous reactions in atopic subjects. *J. Immunol.* **158**, 5050–5057.

Zurawski, G. and de Vries, J.E. (1994). Interleukin 13, an interleukin 4-like cytokine that acts on monocytes and B cells, but not on T cells. *Immunol. Today* **15**, 19–26.

Zurawski, S.M., Vega, F., Jr., Huyghe, B. and Zurawski, G. (1993). Receptors for interleukin-13 and interleukin-4 are complex and share a novel component that functions in signal transduction. *EMBO J.* **12**, 2663–2670.

Zurawski, S.M., Chomarat, P., Djossou, O., Bidaud, C., McKenzie, A.N., Miossec, P., Banchereau, J. and Zurawski, G. (1995). The primary binding subunit of the human interleukin-4 receptor is also a component of the interleukin-13 receptor. *J. Biol. Chem.* **270**, 13 869–13 878.

Chapter 16

Interleukin-15

Mary K. Kennedy, Linda S. Park and Raymond J. Paxton
Immunex Corporation, Seattle, WA, USA

INTRODUCTION

Interleukin-15 (IL-15) is a cytokine that was discovered because of its 'IL-2-like' activity. The shared activities of IL-2 and IL-15 stem from their use of common receptor subunits. IL-15 interacts with a receptor complex that consists of both the β and γ chains of the interleukin-2 receptor (IL-2R) as well as a specific high-affinity IL-15 binding subunit known as IL-15Rα. Although IL-15 shares many *in vitro* activities with IL-2, it is becoming increasingly evident that these cytokines exert differential effects on a number of cell types. These recent observations, coupled with the observations that the two cytokines and the specific α chains of their receptors are differentially expressed among tissues and cell types, suggest that IL-15 and IL-2 have distinct roles *in vivo*.

THE IL-15 GENE AND PROTEIN

Purification, Cloning and Sequence

IL-15 was first detected in concentrated supernatants of a simian kidney epithelial cell line, CV-1/EBNA, based on its ability to support proliferation of the IL-2-dependent mouse T-cell line, CTLL-2 (Grabstein *et al.*, 1994). A similar activity, designated IL-T, was detected in supernatants from a human adult T-cell leukemia line, HuT-102 (Burton *et al.*, 1994). IL-15 and IL-T are now known to be identical proteins (Bamford *et al.*, 1996). IL-15 was purified from CV-1/EBNA supernatants and subjected to amino-terminal sequence analysis, which led to the isolation of cDNA clones for simian and human IL-15 (Grabstein *et al.*, 1994). cDNA clones for mouse IL-15 (Anderson *et al.*, 1995a) and rat IL-15 (Reinecker *et al.*, 1996) have also been isolated.

The cDNA clones encode 162-amino-acid precursor proteins (Fig. 1). Based on the amino-terminal sequence obtained from native simian IL-15, mature IL-15 begins at residue 49. Mature IL-15 contains four cysteine residues that are predicted to form two disulfide bonds, Cys-35–Cys-85 and Cys-42–Cys-88 and, depending on the animal species, two to four potential asparagine-linked glycosylation sites. The mature amino acid sequences are 96% identical for human *versus* simian IL-15 or rat *versus* mouse IL-15, but only 70% identical for human *versus* mouse IL-15.

The Cytokine Handbook, 3rd ed.
ISBN 0-12-689662-3

```
           -48                                          -1 1                                                           50
Human      MRISKPHL RSISIQCYLC LLLNSHFLTE AGIHVFILGC FSAGLPKTEA NWVNVISDLK KIEDLIQSMH IDATLYTESD VHPSCKVTAM KCFLLELQVI
Simian     MRISKPHL RSISIQCYLC LLLKSHFLTE AGIHVFILGC FSAGLPKTEA NWVNVISDLK KIEDLIQSMH IDATLYTESD VHPSCKVTAM KCFLLELQVI
Rat        MKILKPYM RNTSILYYLC FLLNSHFLTE AGIHVFILGC VSVGLPKTEA NWIDVRYDLE KIESLIQFIH IDTTLYTDSD FHPSCKVTAM NCFLLELQVI
Mouse      MKILKPYM RNTSISCYLC FLLNSHFLTE AGIHVFILGC VSVGLPKTEA NWIDVRYDLE KIESLIQSIH IDTTLYTDSD FHPSCKVTAM NCFLLELQVI
Consensus  M-I-KP-- R--SI--YLC -LL-SHFLTE AGIHVFILGC -S-GLPKTEA NW--V--DL- KIE-LIQ--H ID-TLYT-SD -HPSCKVTAM -CFLLELQVI

           51                                          100 101      114
Human      SLESGDASIH DTVENLIILA NNSLSSNGNV TESGCKECEE LEEKNIKEFL QSFVHIVQMF INTS
Simian     SHESGDTDIH DTVENLIILA NNILSSNGNI TESGCKECEE LEEKNIKEFL QSFVHIVQMF INTS
Rat        LHEYSNMTLN ETVRNVLYLA NSTLSSNKNV IESGCKECEE LEERNFTEFL QSFIHIVQMF INTS
Mouse      LHEYSNMTLN ETVRNVLYLA NSTLSSNKNV AESGCKECEE LEEKTFTEFL QSFIRIVQMF INTS
Consensus  --E------- -TV-N---LA N--LSSN-N- -ESGCKECEE LEE----EFL QSF--IVQMF INTS
```

Fig. 16.1. Amino acid sequence alignment of human, simian, rat and mouse IL-15. The signal peptide is numbered from −48 to −1, and the mature protein is numbered from 1 to 114. The consensus sequence shows residues that are identical in all four sequences, and potential asparagine-linked glycosylation sites are underlined.

Protein Structure

Molecular modeling predicts that IL-15 has a four-helix bundle-like structure and is a member of the helical cytokine family (Grabstein *et al.*, 1994). Although human IL-15 and human IL-2 show little sequence homology, it is possible to align their sequences based on the helical cytokine structure of IL-2 and the predicted structure of IL-15 (Pettit *et al.*, 1997). In this alignment there are 21-amino-acid identities between IL-15 and IL-2, and the Cys-42–Cys-88 disulfide bond in IL-15 corresponds to the lone disulfide bond in IL-2, Cys-58–Cys-105. Furthermore, Asp-8 and Gln-108 in IL-15 correspond to Asp-20 and Gln-126 in IL-2, residues that are critical for IL-2 biologic activity and which interact with the IL-2R β and γ chains respectively (Collins *et al.*, 1988; Zurawski *et al.*, 1990). When Asp-8 and Gln-108 in IL-15 are individually mutated to serine, neither mutant is biologically active, suggesting that these residues are critical for IL-15 activity. However, both mutants inhibit the ability of wild-type IL-15 to bind to cells with high affinity, which suggests that the mutants retain their overall structure and the ability to bind to the IL-15Rα chain. These data support the prediction that IL-15 is a member of the helical cytokine family (Pettit *et al.*, 1997).

Chromosomal Location and Genomic Structure

The mouse *il15* gene maps to the central region of chromosome 8 (Anderson *et al.*, 1995a). The mouse *il15* gene contains eight exons and seven introns, and is at least 34 kb in length (Anderson *et al.*, 1995a). The majority of the protein coding region is contained within exons 5–8, and the intron–exon structure that defines this region is similar to that of the mouse *il2* gene as well as other members of the helical cytokine family. Characterization of partial human *IL15* genomic clones suggests that this genomic structure is also conserved in human *IL15*. The human *IL15* gene maps to chromosome 4q31 (Anderson *et al.*, 1995a), which shares regions of homology with mouse chromosome 8. The human *IL2* gene maps to chromosome 4q26–27 and a number of other cytokine–growth factor genes map in the region 4q13–q31. The proximity of a number of cytokine genes and the similar gene structures for *IL15* and *IL2* suggest a common ancestry within the helical cytokine family.

DISTRIBUTION OF IL-15 mRNA

In contrast to the restricted expression of IL-2 mRNA, which is limited to activated T cells and to tissues that contain T cells, IL-15 mRNA is expressed constitutively in a wide variety of cells and in many tissues (Table 1), but is not expressed in normal T, B or natural killer (NK) cells (Grabstein *et al.*, 1994; Bamford *et al.*, 1996; Doherty *et al.*, 1996). However, a human T-lymphotropic virus-I (HTLV-I)-transformed T-cell line, HuT-102, expresses mRNA for IL-15 (Bamford *et al.*, 1996). The IL-15 message produced by the HuT-102 line is smaller than that produced by non-transformed human cells and is a chimeric mRNA that joins a segment of the R region of the HTLV-I long terminal repeat (LTR) with the 5′ untranslated region (UTR) of IL-15 (Bamford *et al.*, 1996).

Table 16.1. Constitutive expression of IL-15 mRNA in tissues and cells.

Cell type or tissue	References
Skeletal muscle, placenta, heart, kidney, spleen, liver, lung, skin	Grabstein *et al.* (1994), Mohamadzadeh *et al.* (1995), Doherty *et al.* (1996), Sorel *et al.* (1996)
Bone marrow stromal cells and cell lines	Grabstein *et al.* (1994), Mrózek *et al.* (1996)
Epithelial and fibroblast cell lines, thymic and intestinal epithelial cells, keratinocytes and dermal fibroblasts	Grabstein *et al.* (1994), Mohamadzadeh *et al.* (1995), Blauvelt *et al.* (1996), S. Kumaki *et al.* (1996), Leclercq *et al.* (1996), Reinecker *et al.* (1996), Sorel *et al.* (1996)
Monocytes–macrophages, Langerhans cells, blood-derived dendritic cells (DC), astrocytes and microglia	Grabstein *et al.* (1994), Blauvelt *et al.* (1996), Doherty *et al.* (1996), Lee *et al.* (1996)
Umbilical vein endothelial cells	Cosman *et al.* (1994)
Osteosarcomas, rhabdosarcomas and small cell lung cancer cell lines	Meazza *et al.* (1996), Lollini *et al.* (1997)

REGULATORY CONTROLS OF IL-15 TRANSCRIPTION AND SYNTHESIS

Transcriptional Control

The expression of IL-15 mRNA can be enhanced in response to a variety of stimuli. For example, the message for IL-15 is upregulated in spleen, liver and lung tissue obtained from mice infected with bacillus Calmette–Guérin (BCG) (Doherty *et al.*, 1996) and in lung tissue of *Toxoplasma gondii*-infected mice (Hunter *et al.*, 1997). Enhanced IL-15 mRNA expression is also seen in human renal allografts undergoing rejection (Pavlakis *et al.*, 1996), in alveolar macrophages from patients with active sarcoidosis (Agostini *et al.*, 1996) and in skin biopsies from patients with the self-healing (tuberculoid) form of leprosy (Jullien *et al.*, 1997). In addition, *in vitro* studies have shown that IL-15 mRNA expression by various cells is affected by exposure to cytokines or infectious agents and their products. For example, IL-15 mRNA expression by human retinal pigment epithelial cells is upregulated by interferon-γ (IFN-γ) or tumor necrosis factor-α (TNF-α) (N. Kumaki *et al.*, 1996). IFN-γ has also been shown to upregulate IL-15 mRNA expression by a human intestinal epithelial cell line (Reinecker *et al.*, 1996). The expression of IL-15 mRNA in macrophages can be upregulated by exposure to lipopolysaccharide (LPS), BCG, *T. gondii*, *Mycobacterium tuberculosis* or human herpes virus-6 (HHV-6) (Grabstein *et al.*, 1994; Doherty *et al.*, 1996; Flamand *et al.*, 1996). The optimal production of IL-15 mRNA by mouse macrophages requires priming by IFN-γ and triggering by infectious agents or their products (Doherty *et al.*, 1996). Interestingly, the production of IL-15 mRNA by mouse macrophages is upregulated, rather than inhibited, by IL-10 (Doherty *et al.*, 1996). Finally, the effect of ultraviolet B radiation on IL-15 mRNA expression is not clear as it has been reported to enhance (Mohamadzadeh *et al.*, 1995) or inhibit (Blauvelt *et al.*, 1996) IL-15 mRNA expression by human keratinocytes.

Although IL-15 mRNA is upregulated in response to a variety of stimuli, transcription does not appear to be the predominant mechanism of control of IL-15 production as it is difficult to detect IL-15 protein in the supernatant of most cells that express IL-15 mRNA. A notable exception is the HuT-102 cell line which effectively synthesizes and secretes IL-15 into the supernatant (Bamford *et al.*, 1996). The predominant IL-15 message expressed by this cell line represents a fusion message joining a segment of the R region of the long terminal repeat of HTLV-I with the 5′ UTR of human IL-15. The increased production of IL-15 by this transformed cell line appears to involve a marked increase in IL-15 mRNA transcription (resulting from regulatory control exerted by the HTLV-I-R element) as well as a modest increase in translation due to the elimination of a number of the upstream AUGs that are present in the normal IL-15 message.

Translational Control

The discordance between the expression of IL-15 mRNA and protein may be due in part to an impairment of the translation of IL-15 mRNA caused by the presence of multiple AUGs in the 5′ UTR of the message (Bamford *et al.*, 1996). There are ten AUGs upstream of the authentic initiation codon in the human IL-15 message and five in the mouse IL-15 message (Anderson *et al.*, 1995a; Giri and Paxton, 1996). In contrast, no AUGs are found in the 5′ UTRs of IL-2, IL-3, IL-4, IL-5, IL-6, IL-10, IL-13, granulocyte–macrophage colony stimulating factor (GM-CSF), G-CSF and IFN-γ, although they are present in the 5′ UTRs of IL-7 and IL-9 (Kozak, 1991; Bamford *et al.*, 1996). cDNAs with encumbered 5′ non-coding sequences may represent either mRNA precursors that are regulated at a post-transcriptional step which precedes translation or mRNAs that encode critical regulatory proteins and are intended to be translated poorly (Kozak, 1991). An IL-15 expression construct that codes for an mRNA lacking all upstream AUGs yields fourfold more protein than the wild-type IL-15 construct when analyzed in an *in vitro* transcription–translation coupled system (Bamford *et al.*, 1996). However, it does not appear that translational controls alone account for the disparity between the expression of IL-15 mRNA and protein as even the elimination of all of the upstream AUGs does not lead to a high level of protein expression by transfected cells.

Secretory Control

The unusually long leader sequence of IL-15 may be part of a complex processing pathway involved in the release of mature IL-15 (Grabstein *et al.*, 1994). When IL-15 is expressed in cells transfected with a chimeric construct containing the IL-2 leader sequence linked to the mature IL-15 protein-coding sequence, the amount of IL-15 protein produced is increased approximately 20-fold. Conversely, in cells transfected with a construct consisting of the IL-15 leader sequence linked to the mature IL-2 protein coding sequence, the production of IL-2 protein is reduced as much as 50-fold (Tagaya *et al.*, 1996a). In fact, it has been suggested that the low level of the IL-15 detected in certain supernatants might be indicative of a protein that is not a typical secretory protein, but is released from the cytoplasm of dying cells (Ferrini *et al.*, 1996).

THE IL-15R COMPLEX

Subunits Shared with the IL-2R Complex

Early studies demonstrated that the biologic similarities between IL-2 and IL-15 stem from their use of common receptor signaling subunits. In T cells and NK cells, IL-15 and IL-2 interact with receptor complexes that contain both the β and γ chains of the IL-2R. The high-affinity IL-2R complex consists of at least three chains, designated α, β and γ (reviewed by Minami *et al.*, 1993). IL-2Rα is required for high-affinity binding of IL-2 by the IL-2R complex, but only the IL-2R β and γ chains are required for signaling and internalization of IL-2. The γ chain is also a component of the IL-4, IL-7 and IL-9 receptors and is now referred to as γc, for common γ chain (Leonard *et al.*, 1995). Antibodies specific for either IL-2Rβ or γc block IL-15-induced responses *in vitro*, whereas antibodies specific for IL-2Rα have no inhibitory effect (Bamford *et al.*, 1994; Grabstein *et al.*, 1994). In addition, results of binding and receptor transfection studies clearly demonstrate that IL-15 utilizes both IL-2Rβ and γc, but not IL-2Rα, and that IL-2Rβ and γc are required for signaling and efficient internalization of IL-15 in many types of cells (Giri *et al.*, 1994, 1995; Kumaki *et al.*, 1995).

Unique IL-15Rα Chain

The high-affinity IL-15R complex contains an additional IL-15-specific binding subunit that is not shared with the IL-2R. This unique subunit is designated IL-15Rα because of its close structural relationship with IL-2Rα. However, neither IL-15Rα nor IL-2Rα belongs to the cytokine receptor superfamily to which IL-2Rβ and γc belong (Minami *et al.*, 1993). Instead, they are considered to define a new cytokine receptor family that is part of a larger family of proteins that contain protein-binding motifs known as 'sushi domains'. IL-15Rα differs from IL-2Rα in that it has a longer cytoplasmic tail and it contains one sushi domain, whereas IL-2Rα contains two (Giri *et al.*, 1995). In addition, IL-15Rα has a broader distribution of expression in different cell types and shows a much higher affinity for its ligand in the absence of IL-2Rβ and γc (Table 2).

Protein Structure

Mouse IL-15Rα is a type I membrane protein with a predicted 32-amino-acid signal sequence, a 173-amino-acid extracellular domain, a 21-amino-acid transmembrane domain and a 37-amino-acid cytoplasmic domain (Giri *et al.*, 1995). Three alternatively spliced forms of human IL-15Rα have been identified and all three proteins encoded by these alternative clones are capable of binding IL-15 (Anderson *et al.*, 1995b). One of the alternatively spliced forms is homologous to the cDNA encoding mouse IL-15Rα, one has a deleted exon in the extracellular domain, and the third has an alternative cytoplasmic domain. No alternatively spliced mRNA encoding a soluble IL-15Rα (mouse or human) has been detected. Although a soluble, shed form of IL-2Rα is present in biologic fluids and raised levels of soluble IL-2Rα are observed in certain disease states (Rose-John and Heinrich, 1994), it is not yet known whether a soluble, shed form of IL-15Rα exists.

Table 16.2. Comparison of the IL-2R and IL-15R systems.

	IL-2R	IL-15R
Subunits required for signaling	IL-2Rβ and γc	IL-2Rβ and γc
Role of specific α chain	IL-2Rα is an 'affinity converter' which is required, but not sufficient, for high-affinity binding of IL-2	IL-15Rα is required and sufficient for high-affinity binding of IL-15
Affinity of specific α chain for its ligand	$K_a = 10^8$ M^{-1} for IL-2	$K_a = 10^{11}$ M^{-1} for IL-15
Affinity of trimeric receptor complex for ligand	$K_a = 10^{11}$ M^{-1} for IL-2	$K_a = 10^{11}$ M^{-1} for IL-15 (possibly higher; see text)
Ligand-induced regulation of α chain	IL-2 upregulates IL-2Rα expression	IL-15 upregulates IL-2Rα and downregulates IL-15Rα expression in PBTs and B-cell lines
Distribution of α chain mRNA	IL-2Rα mRNA is expressed in monocytes, B and T cells, and in tissues that contain T cells	IL-15Rα mRNA is expressed in a wide variety of cells and tissues, including fibroblasts, epithelial, T and B cells, monocytes, prostate, liver, testis, ovary, small intestine, colon

Chromosomal Location

The genes coding for human and mouse IL-15Rα map to human chromosome 10, bands p14–p15 and the proximal region of mouse chromosome 2 respectively (Anderson *et al.*, 1995b). The *IL15RA* and *IL2RA* genes have similar intron–exon organization and are closely linked in the human and mouse genomes. A few human diseases are potential candidates for *IL15RA* and/or *IL2RA* mutation or translocation. Rare cases of DiGeorge syndrome are associated with deletions at the 10p13–10p14 boundary (Daw *et al.*, 1996) and rare, but recurrent, non-random chromosomal translocations of the distal 10p region are found in acute lymphocytic and acute myeloid leukemias (Mitelman, 1994).

IL-15Rα is Required for High-Affinity Binding, but not Signaling, by IL-15

Similar to the IL-2–IL-2R system, IL-15Rα is required for high-affinity binding, but not signaling, by IL-15 (Anderson *et al.*, 1995b). However, although IL-2Rα is required for high-affinity binding of IL-2 by the IL-2R complex, IL-2Rα alone exhibits only a low affinity for IL-2, whereas IL-15Rα alone binds IL-15 with high affinity (Table 2) (Minami *et al.*, 1993; Anderson *et al.*, 1995b; Giri *et al.*, 1995). Coexpression of IL-15Rα with either or both IL-2Rβ and γc generates an affinity for IL-15 very similar to that observed with IL-15Rα alone. However, it is possible that the heterotrimeric IL-15R complex binds IL-15 with an even higher affinity which cannot be determined accurately because of technical limitations in measuring very high binding affinities.

An important distinction between the IL-2–IL-2R system and the IL-15–IL-15R system is that cells exhibiting high-affinity binding sites for IL-2 must express all three IL-2R subunits and are likely to respond to IL-2, whereas cells that exhibit high-affinity binding sites for IL-15 might express IL-15Rα in the absence of IL-2Rβ and/or γc, and thus may be unresponsive to IL-15. It is possible that IL-15-binding, but non-responsive, cells sequester IL-15 from cells capable of responding to it or 'tether' IL-15 for presentation to cells capable of responding to it. It is also possible that IL-15Rα associates with as yet unidentified receptor subunits to signal in cells that express IL-15Rα but lack IL-2Rβ and/or γc.

In the absence of the appropriate α chains, IL-2Rβ and γc can transduce a signal in response to relatively high concentrations of either IL-2 or IL-15. However, unlike binding of IL-15 to the IL-15Rα subunit or the IL-15R$\alpha\beta\gamma$c complex, which shows little species specificity, binding of IL-15 to the IL-2R$\beta\gamma$c complex in the absence of IL-15Rα shows a high degree of species specificity (see below).

As neither IL-2Rα nor IL-15Rα appears to be directly required for signaling in response to its respective ligand, it is not surprising that IL-2 and IL-15 share signal transduction pathways. In human T cells or phytohemagglutinin (PHA)-activated peripheral blood lymphocytes, both IL-2 and IL-15 induce tyrosine phosphorylation of JAK1 and JAK3 (Janus kinase) and cause rapid induction of DNA binding complexes that contain signal transducers and activators of transcription (STAT3 and STAT5) (Johnston *et al.*, 1995; Lin *et al.*, 1995). However, it cannot be assumed that a given cell type or population will be equally responsive to both cytokines because specific high-affinity binding of IL-2 and IL-15 requires IL-2Rα and IL-15Rα respectively, and these α chains may be differentially expressed on a particular cell type.

Distribution and Regulation of IL-15Rα versus *IL-2Rα*

The expression of IL-2Rα mRNA is more restricted than that of IL-15Rα, which is found in a wide variety of tissues and cells (Table 2). This observation suggests that there is a wider range of cellular targets for IL-15 than for IL-2. However, as both IL-2Rβ and γc are required for responsiveness to IL-15 in many cell types, and the expression of mRNA for these subunits appears to be more limited than that of IL-15Rα, the expression of IL-15Rα may not necessarily correlate with responsiveness to IL-15 (Anderson *et al.*, 1995b; Giri *et al.*, 1995).

IL-15Rα can be upregulated in response to various stimuli. Stimulation of monocyte cell lines with IFN-γ or T cells with phorbol myristate acetate (PMA) or anti-CD3 results in a dramatic increase in the number of high-affinity binding sites for IL-15 on these cells (Giri *et al.*, 1995). It is known that IL-2 upregulates the expression of IL-2Rα (Malek *et al.*, 1986); however, there is no evidence to suggest that IL-15 upregulates the expression of IL-15Rα. In fact, stimulation of PHA-activated human peripheral blood T cells (PBTs) or Epstein–Barr virus-transformed B-cell lines with IL-15 upregulates the expression of IL-2Rα but downregulates that of high-affinity binding sites for IL-15 (S. Kumaki *et al.*, 1996). The majority of these sites are downregulated within 2 h of IL-15 addition and they remain downregulated in the continued presence of IL-15. The ability of IL-15 to downregulate expression of IL-15Rα requires the presence of γc, as IL-15 fails to downregulate high-affinity IL-15 binding sites on a cell line that expresses IL-15Rα and IL-2Rβ, but not γc. It is not known whether downregulation of high-affinity IL-15 binding sites occurs primarily via receptor internalization or via the release of IL-15Rα from the cell surface.

Putative IL-15R-X

IL-15 stimulates proliferation of mast cells that do not respond to IL-2 (Tagaya *et al.*, 1996b). Stimulation of mast cells with IL-15 is independent of the IL-2R subunits and results in recruitment of JAK2 and STAT5 rather than JAK1/3 and STAT3/5. The results suggest that the mast cells respond to IL-15 via the use of a distinct receptor, which has been tentatively designated IL-15R-X. The apparent affinity of this receptor for IL-15 is approximately 100-fold lower than the affinity of IL-15Rα for IL-15.

IN VITRO BIOLOGIC EFFECTS OF EXOGENOUS IL-15

Species Specificity

Binding inhibition studies have indicated that human and mouse IL-15 bind human and mouse IL-15Rα or IL-15R$\alpha\beta\gamma$c with an equal high affinity. In contrast, binding of IL-15 to IL-2R$\beta\gamma$c is of much lower affinity and shows a high degree of species specificity. Thus, mouse cells that bear IL-2Rβ and γc, but either lack or express only low levels of IL-15Rα, are much more responsive to mouse IL-15 than to simian or human IL-15, whereas the reverse is true of human cells that bear only IL-2Rβ and γc (J. Eisenman and L. Park, unpublished results). For example, the mouse CTLL-2 cell line, which expresses a high level of IL-15Rα, responds to low concentrations (10–30 pg/ml or greater) of either mouse or human IL-15. However, other types of mouse cells, such as NK1.1$^+$ cells, fresh bone marrow cells or CD4$^+$ T-cell clones, require higher concentrations of mouse IL-15 in order to respond (1 ng/ml or greater mouse IL-15) and are up to 100-fold more responsive to mouse IL-15 than to simian or human IL-15 (Puzanov *et al.*, 1996; J. Eisenman, K. Brasel and M. Kennedy, unpublished results). In contrast, the dose–response curves for human and mouse IL-2 are virtually identical when using these same cell types. It is likely that these cells lack or express only a low level of IL-15Rα, and are binding and responding to IL-15 primarily via IL-2Rβ and γc.

Effects on NK and T Cells

Induction of Cytotoxic Activity of NK and T Cells

IL-15 can substitute for IL-2 in the induction of alloantigen- or virus-specific cytotoxic T lymphocytes (CTLs) (Grabstein *et al.*, 1994; Kanai *et al.*, 1996) and in the generation of lymphokine-activated killer (LAK) cell activity from cultures of human peripheral blood mononuclear cells (PBMCs), large granular lymphocytes, purified CD56^{dim} NK cells or mouse splenocytes (Burton *et al.*, 1994; Carson *et al.*, 1994; Grabstein *et al.*, 1994; Gamero *et al.*, 1995; Flamand *et al.*, 1996; Ye *et al.*, 1996a). In addition, the ability of human NK cells to mediate antibody-dependent cellular cytotoxicity (ADCC) is augmented by IL-15 (Carson *et al.*, 1994). IL-15, like IL-2, induces the expression of a number of cytolytic mediators that are associated with both perforin-dependent and Fas-dependent cytolytic pathways (Gamero *et al.*, 1995; Ye *et al.*, 1996a).

Induction of Proliferation and Cytokine Production by NK Cells

Both IL-15 and IL-2 induce strong proliferative responses in the $CD56^{bright}$ subset of human NK cells, which represents approximately 10% of human NK cells and constitutively expresses the IL-2R α, β and γc chains. In contrast, the majority of human NK cells, which are $CD56^{dim}$, constitutively express IL-2Rβ and γc, but not IL-2Rα, and proliferate only weakly to either IL-2 or IL-15 alone (Carson *et al.*, 1994). However, these cells will proliferate when cocultured in the presence of certain irradiated stimulator cell lines and IL-2. IL-15 cannot replace the costimulatory activity of IL-2 in this system, but the combination of IL-15 with either IL-12 or IL-10 is costimulatory (Warren *et al.*, 1996). Despite its weak proliferative response to IL-2 or IL-15 alone, the $CD56^{dim}$ subset develops strong cytotoxic activity and produces low levels of certain cytokines in response to either cytokine. In addition, IL-12 synergizes with either IL-2 or IL-15 to induce IFN-γ and TNF-α production and to enhance GM-CSF production by $CD56^{dim}$ NK cells (Carson *et al.*, 1994). Human NK cells also produce moderate levels of macrophage inflammatory protein-1α (MIP-1α) in response to either IL-2 or IL-15, but not in response to IL-10, TNF-α or IL-1β. The production of high levels of MIP-1α by IL-15-stimulated human NK cells requires costimulation with IL-12 (Bluman *et al.*, 1996).

IL-15 costimulates IFN-γ production by severe combined immune deficiency (SCID) splenocytes stimulated with *T. gondii* or IL-12, and induces mouse NK cells to express CD28. Because the production of IFN-γ by mouse NK cells can be costimulated by the B7/CD28 interaction, IL-15 probably enhances the ability of mouse NK cells to produce IFN-γ via direct as well as indirect effects (Hunter *et al.*, 1997).

Induction of Proliferation and Cytokine Production by T Cells

IL-15 is a potent growth factor for activated T cells. Like IL-2, IL-15 stimulates the proliferation of PHA-activated human PBMCs, PBTs or $CD4^+$ and $CD8^+$ PBTs (Grabstein *et al.*, 1994). IL-15 enhances proliferation of human $\gamma\delta$ T cells in cultures of PMBCs depleted of $CD4^+$ cells and stimulated with malarial antigen (Elloso *et al.*, 1996) and supports the outgrowth of tumor-derived T cells in primary human tumor cell cultures (Lewko *et al.*, 1995). IL-15 enhances mitogen- and antigen-specific proliferation of PBMCs from human immunodeficiency-virus-infected patients (Seder *et al.*, 1995; Patki *et al.*, 1996) and the relatively weak *M. leprae*-specific responses of PBMCs from patients with the lepromatous form of leprosy (Jullien *et al.*, 1997). Highly purified human memory/primed ($CD45RO^+$) $CD4^+$ and $CD8^+$ T cells and naive ($CD45RO^-$) $CD8^+$ T cells proliferate in response to either IL-2 or IL-15 in the absence of exogenous mitogen or antigen. In contrast, the phenotypically naive ($CD45RO^-$) $CD4^+$ T cells, which make up the majority of circulating $CD4^+$ T cells in humans, are unresponsive to IL-2 or IL-15 (Kanegane and Tosato, 1996). T cells enriched from inflammatory sites such as the synovium of patients with rheumatoid arthritis (RA) or the lungs of patients with active sarcoidosis are responsive to IL-15 in the absence of exogenous antigen (Agostini *et al.*, 1996; McInnes *et al.*, 1996, 1997). The responsiveness of synovial T cells from patients with RA to IL-15 or IL-2 (McInnes *et al.*, 1997) is consistent with the observation that the majority of T cells from these sites belong to the memory/primed ($CD45RO^+$) subset (Kohem *et al.*, 1996).

Mouse T-cell populations that proliferate in response to IL-15 include $CD4^+$ $\alpha\beta$ T-cell clones (T helper 1 T_{H1}, T_{H2} and T_{H0}) (M. Kennedy and K. Picha, unpublished results), antigen-activated $CD4^+$ cell blasts from T-cell receptor (TCR) transgenic mice (Seder, 1996), concanavalin A-activated, epidermal $\gamma\delta$ T cells (Edelbaum *et al.*, 1995) and $\gamma\delta$ T cells isolated from the peritoneum of mice infected with *Salmonella* (Nishimura *et al.*, 1996). In addition, IL-15 synergizes with IL-12 to induce proliferation of mouse T_{H1} clones (Kennedy *et al.*, 1994) and $CD4^+$ T cell blasts (Seder, 1996).

IL-15 also induces cytokine production by T cells. Human $CD4^+$ T-cell clones that are capable of producing IL-2, IL-4 and IL-5 in response to stimulation with anti-CD3 produce IL-5, but not IL-4, in response to stimulation with either IL-2 or IL-15 (Mori *et al.*, 1996). T cells isolated from the inflamed synovial tissue of patients with RA produce TNF-α in response to stimulation with IL-15. In addition, IL-15-activated synovial T cells or PBTs from patients with RA induce the production of TNF-α by monocytes in a contact-dependent manner. Interestingly, IL-2 is much less effective at inducing TNF-α production in either system, despite the fact that IL-2 and IL-15 induce similar proliferative responses of the responding T cells (McInnes *et al.*, 1997). Mouse $\gamma\delta$ T cells induced in response to *Salmonella* infection produce low levels of IL-4 and IFN-γ in response to stimulation with IL-15 (Nishimura *et al.*, 1996). Mouse T_{H1} clones produce low levels of IFN-γ in response to either IL-12 or IL-15 alone but produce high levels in response to the combination of the two cytokines (M. Kennedy, unpublished results).

Inhibition of T-Cell Apoptosis

IL-2, IL-4, IL-7 and IL-15 inhibit apoptosis of cytokine-deprived, activated T cells (Akbar and Salmon, 1997). These cyokines prevent apoptosis *in vitro* by upregulating Bcl-2 and Bcl-x_L expression relative to Bax and CD95. The authors speculate that stromal cell-derived cytokines such as IL-7 and IL-15 may be important for survival of activated T cells in the periphery.

Induction of Motility and Migration of T Cells

IL-15, like IL-2, induces locomotion and directed migration of human T cells but appears to have no chemokinetic or chemotactic activity on human monocytes, neutrophils or B cells (Wilkinson and Liew, 1995). The formation of uropods by IL-2 or IL-15-treated T cells is associated with a redistribution of cell adhesion markers (most notably intercellular adhesion molecule-3 (ICAM-3)) and is thus likely to modulate the adhesive properties of the cells (Nieto *et al.*, 1996).

Differentiation of NK, T and Granulated Metrial Gland Cells

In vitro studies indicate that IL-15 may be a differentiation factor as well as a growth factor for NK cells. Mrózek *et al.* (1996) have shown that adult bone marrow-derived human $CD34^+$ hematopoietic progenitor cells differentiate into $CD3^-$ $CD56^+$ NK cells in long-term cultures containing IL-15 alone, but not in cultures containing stem cell factor (SCF, c-*kit* ligand) alone. However, the addition of SCF to long-term cultures containing IL-15 results in greatly enhanced expansion of the NK cells. Cavazzana-Calvo *et al.* (1996) have shown that the combination of IL-15 and SCF induces differentiation of human NK cells from cord blood $CD34^+$ hematopoietic progenitors. The combination of IL-15 and SCF causes committed $CD34^+$ $CD7^+$ as well as immature $CD34^+$ $CD7^-$ cord blood cells to proliferate, to express NK cell markers, and to develop ADCC and NK-dependent cytolytic activity. The majority of the $CD56^+$ cells generated from the long-term cultures of $CD34^+$ hematopoietic progenitors derived from either adult bone marrow or cord blood are $CD2^-$ $CD16^-$ $CD56^{bright}$, in contrast to the majority of peripheral blood NK cells, which are $CD2^+$ $CD16^+$ $CD56^{dim}$. Thus additional factors may be involved in the generation of mature circulating NK cells.

Studies in the mouse also support a role for IL-15 in NK cell differentiation. Mouse IL-15 induces non-lytic immature mouse $NK1.1^+$ cells to become lytic and to acquire cell surface markers associated with mature $NK1.1^+$ cells. Although mouse IL-2 can also activate the immature $NK1.1^+$ cells, mouse IL-15 is 10- to 50-fold more effective than mouse IL-2 (Puzanov *et al.*, 1996). Another study has shown differential effects of IL-2 and IL-15 on expansion and differentiation of bipotential T/NK cells in fetal thymic organ cultures (FTOCs) (Leclercq *et al.*, 1996). When highly purified IL-2Rβ^+ $CD8^-$ TCR^- fetal thymus cells are transferred to deoxyguanosine-treated thymic lobes and cultured *in vitro*, they develop into TCR$\alpha\beta$ and TCR$\gamma\delta$ cells. Addition of low concentrations (20 or 100 ng/ml) of simian IL-15 to the FTOC increases the cell yield by 2.5-fold and expands cells at all stages of differentiation, including TCR^-, TCR$\alpha\beta$ and TCR$\gamma\delta$ cells, although the largest expansion occurs in the TCR$\gamma\delta$ compartment. The addition of a high concentration (500 ng/ml) of IL-15 results in a fourfold increase in cell yield, a block in TCR$\alpha\beta$ but not TCR$\gamma\delta$ expansion, and a shift in differentiation toward NK cells. The majority of the cells in these cultures are IL-2Rβ^+ TCR^- and appear to be NK cells as they have lost their T-cell progenitor potential but have acquired the ability to lyse NK-sensitive targets. Interestingly, the addition of low or high concentrations of human IL-2 to the FTOC does not increase the cell yield above that of control cultures and does not expand the TCR$\gamma\delta$ compartment, although high concentrations of IL-2 shift differentiation toward NK cells. The authors speculate that the lack of expansion in the IL-2-treated FTOC is due to the delivery of a negative regulatory signal via IL-2Rα, as the addition of anti-IL-2Rα to the IL-2-treated FTOC results in a 2.5-fold increase in cell yield.

Studies of IL-2Rβ knockout mice indicate a role for IL-15 and/or IL-2 in the development of NK cells and certain populations of intestinal intraepithelial lymphocytes (IELs) (Suzuki *et al.*, 1997). $NK1.1^+$ $CD3^-$ cells in the spleen and peripheral blood are reduced, if not totally absent, in these mice and NK activity is absent. In contrast, NK activity in IL-2 knockout mice is reduced but inducible (Kündig *et al.*, 1993). Together these observations suggest that the development of NK cells is dependent on IL-15 or on the combination of IL-15 and IL-2. In addition, although the IL-2Rβ knockout mice have normal numbers of IELs, the IELs consist primarily of the TCR$\alpha\beta$CD8$\alpha\beta$ IEL subset, whereas the TCR$\gamma\delta$ and TCR$\alpha\beta$CD8$\alpha\alpha$ IEL subsets are reduced dramatically (Suzuki *et al.*, 1997). The authors speculate that their results and the FTOC studies by Leclercq *et al.* (1996) implicate IL-15 as a necessary factor in the gut microenvironment for the development of lymphocyte subsets of extrathymic origin.

IL-15 has also been implicated in the differentiation of granulated metrial gland (GMG) cells, which are bone marrow-derived lymphoid cells which differentiate in the mouse pregnant uterus into NK-like cells (Ye *et al.*, 1996b). The factors that regulate the differentiation of GMG cells are not known. Interestingly, IL-15 and IL-15Rα mRNA are expressed transiently in the pregnant uterus and their expression is restricted to the late–early to the mid-gestational period (days 6–11). The expression of IL-15 and IL-15Rα mRNA coincides with the expression of mRNA for various cytolytic mediators that are expressed in the pregnant uterus predominantly, if not exclusively, by differentiating GMG cells. In contrast, IL-2, IL-2Rα and IL-7 mRNAs are not detectable in the pregnant uterus throughout gestation. *In vitro* studies of uterine tissue explants prepared from pregnant mice indicate that human IL-15 stimulates perforin mRNA expression by cells present in the pregnant uterus. The ability of IL-15 to increase perforin mRNA expression by the explants was restricted to gestational days 8–12.

Effects on Other Cell Types

B Cells

IL-15 costimulates proliferation of PMA- or anti-IgM-activated, purified human B cells and works in combination with soluble CD40 ligand to induce secretion of polyclonal immunoglobulin M (IgM), IgG_1 and IgA, but not IgG_4 or IgE. Although human IL-2 and IL-15 have similar activities in these assays, the effects of IL-15 are weaker and less consistent than those of IL-2 on the generation of primary antigen-specific plaque-forming cells (PFCs) from unfractionated human PBMCs. It is possible that the difference in activity of IL-2 and IL-15 in the latter system stems from differential effects of these cytokines on the non-B cells within the PBMCs (Armitage *et al.*, 1995).

Neutrophils

IL-15, but not IL-2, is an agonist of human neutrophils. The lack of response to IL-2 may be related to the observation that IL-2Rα is absent on neutrophils and appears to be uninducible. IL-15 induces morphologic cell shape changes that are typical of activated neutrophils, increases their ability to phagocytose opsonized sheep red blood cells, and affects their gene and protein expression, but has no effect on respiratory burst. Although IL-15 appears to be a significant neutrophil agonist its effects are more selective than those of other agonists such as GM-CSF and formyl-methionyl-leucyl-phenylalanine (fMLP) (Girard *et al.*, 1996).

Mast Cells

Mouse mast cell lines and bone marrow-derived mouse mast cells proliferate in response to human IL-15 but not IL-2. The mast cell response apparently involves a novel 'IL-15R-X' that does not utilize IL-15Rα or any of the IL-2R subunits (Tagaya *et al.*, 1996b).

Skeletal Muscle Cells

IL-15 stimulates the accumulation of muscle-specific myosin heavy chain protein in cultures of a mouse skeletal myoblast cell line or primary bovine skeletal myoblasts. As IL-15 has no effect on proliferation or the rate of differentiation in either culture system, the results suggest that IL-15 is an anabolic factor that stimulates the expression and/or accumulation of contractile proteins in differentiated skeletal myocytes (Quinn *et al.*, 1995).

Intestinal Epithelial Cells

Relatively high concentrations of either IL-15 or IL-2 induce a modest (1.5–1.8-fold over background) proliferative response of the human intestinal epithelial cell line, Caco-2 (Reinecker *et al.*, 1996).

IN VIVO BIOLOGIC EFFECTS OF EXOGENOUS IL-15

Induction of LAK/NK Cell Activity and Vascular Leak Syndrome

Multiple intraperitoneal injections of simian IL-15 or human IL-2 into mice induces cytolytic splenocytes that are capable of killing YAC (NK and LAK sensitive) target cells (Munger *et al.*, 1995). IL-15 or IL-2 treatment also enhances the generation of alloantigen-specific CTLs in mice that are primed with allogeneic splenocytes (Munger *et al.*, 1995). Interestingly, simian IL-15 is three to four times more potent than human IL-2 in the generation of splenic cytolytic activity, whereas human IL-2 is three times more potent than simian IL-15 in enhancing the generation of allogeneic CTL. In these same studies, it was noted that the induction of pulmonary vascular leak syndrome in response to multiple high-dose injections of IL-2 or IL-15 requires six times more simian IL-15 than human IL-2. However, as noted above, recent *in vitro* studies show that, although some mouse cells exhibit similar responsiveness to mouse and human IL-15, certain mouse T and NK cell populations are 10- to 100-fold more responsive to mouse IL-15 than to human or simian IL-15, but exhibit similar responsiveness to human and mouse IL-2. Thus, the efficacy of human or simian IL-15 in mouse systems might not reflect the activity of mouse IL-15 in the same system.

Antitumor Effects

In mice injected intravenously with a syngeneic fibrosarcoma, intensive high-dose treatment with either simian IL-15 or human IL-2 results in a dramatic decrease in the number of tumor foci present in the lungs 6 days after tumor inoculation (Munger *et al.*, 1995). IL-2 mediates its protective effects in this tumor model primarily through the induction of LAK cell activity (Rosenberg *et al.*, 1985). IL-15 or IL-2 therapy also increases the survival of mice injected with a syngeneic, NK-resistant mastocytoma. Surviving mice are resistant to a secondary challenge with the same tumor (W. Munger, personal communication). In this tumor model, the protective effect of IL-2 is believed to involve the enhanced generation of tumor-specific CTLs, rather than the induction of LAK cell activity (Maas *et al.*, 1992). IL-15 administration also prolongs cyclophosphamide-induced remissions in rhabdomyosarcoma-bearing mice and enhances the efficacy of combination cyclophosphamide–adoptive immunotherapy in this same model. IL-2 has not been tested in this model (Evans *et al.*, 1997).

Treatment of Chemotherapy-Induced Mucositis

In normal rats, IL-15 administration reduces the mortality, weight loss and incidence of diarrhea induced by treatment with 5-fluorouracil (5-FU). Moreover, in a rat bearing a transplanted colonic carcinoma, IL-15 administration not only reduces the mortality and gastrointestinal toxicity associated with 5-FU treatment, but increases the maximum tolerated dose of 5-FU and the frequency of complete tumor regressions (Y. Rustum, personal communication).

Vaccination Against Lethal Infection with *T. gondii*

Treatment of immunocompetent mice with human IL-15 in combination with a soluble *Toxoplasma* lysate induces protective immunity against a lethal infection with *T. gondii* (Khan and Kasper, 1996). Protective immunity induced by vaccination with the *Toxoplasma* lysate and IL-15 is associated with increased serum levels of IFN-γ and enhanced parasite-specific CTL responses, and can be adoptively transferred to naive mice with $CD8^+$, but not $CD4^+$, T cells.

T-Cell Infiltration

Injection of IL-15 into the footpads of mice primed with *Corynebacterium parvum* induces a cellular infiltrate in the hypodermis and muscle layers of the footpad which persists for at least 3 days and consists primarily of T cells. The duration of the response suggests that IL-15 not only serves as a chemoattractant for activated T cells but induces the expression of cytokines and/or adhesion molecules that contribute to the inflammation (McInnes *et al.*, 1996).

DETERMINATION OF IL-15 IN BIOLOGIC SAMPLES

Enzyme-linked immunosorbent assay (ELISA) kits are available commercially for the determination of immunoreactive human IL-15. Bioactive human or mouse IL-15 can be detected based on the ability of either species of cytokine to induce strong proliferation of the mouse CTLL-2 cell line. This bioassay is sensitive to 10–30 pg/ml or greater human or mouse IL-15 and is described in detail elsewhere (Giri and Paxton, 1996; Paxton, 1996). However, it is important to note that the CTLL-2 line also responds to human IL-2, mouse IL-2 and mouse IL-4. Thus, the detection of IL-15 in supernatants and other biologic fluids by the CTLL-2 bioassay requires that saturating amounts of neutralizing antibodies to the interfering cytokines are included in the assay. In addition, neutralizing antibodies to human and mouse IL-15 are now available and should always be used to demonstrate that the proliferative activity induced by a particular sample is due to IL-15.

EVIDENCE FOR IL-15 PRODUCTION *IN VITRO* AND *IN VIVO*

IL-15 in Culture Supernatants

Although IL-15 mRNA expression can be easily detected in a variety of cell types, it has been difficult to detect IL-15 in unconcentrated supernatants from such cells. In fact, simian IL-15 was first discovered when CV-1/EBNA cells were cocultured in the presence of the IL-2 (and IL-15)-responsive cell line, CTLL-2. In cell-free supernatants, strong IL-15 activity was apparent only when the supernatants were concentrated. Low levels of IL-15 (less than 26 pg/ml after correction for concentration; detected by an ELISA) have also been detected in concentrated supernatants from cultures of human bone marrow stromal cells (Mrózek *et al.*, 1996) and human rhabdomyosarcoma, osteosarcoma and keratinocyte cell lines (Blauvelt *et al.*, 1996; Lollini *et al.*, 1997).

Immunoblot analysis has also been used to detect IL-15 in concentrated supernatants from human keratinocytes (Mohamadzadeh *et al.*, 1995). As mentioned previously, low or undetectable levels of IL-15 protein in unconcentrated supernatants may reflect strict regulatory controls over the translation and/or secretion of IL-15. In addition, because IL-15Rα can be expressed by the same cells that produce IL-15, it is possible that the IL-15 is bound to, and/or is being utilized by, the producing cells.

Despite the difficulties associated with detecting IL-15 in culture supernatants, a number of studies have demonstrated that human PBMCs or monocytes produce IL-15 in response to stimulation with bacterial or viral products. In some cases, the presence of IL-15 in unconcentrated supernatants has not been demonstrated by direct methods, but its presence can be inferred by the ability of anti-IL-15 antibodies to inhibit known activities of IL-15. For example, Flamand *et al.* (1996) have shown that infection of PBMCs with HHV-6 upregulates NK cell-dependent cytotoxic activity. The induction of the cytotoxic activity is blocked by antibodies specific for HHV-6, IL-15 or IL-2Rβ, but not by those specific for IL-2, IL-12, IFN-γ, IFN-α, TNF-α or lymphotoxin α (LT-α). Carson *et al.* (1995) have demonstrated that LPS-stimulated human monocytes produce IL-15 within 4 h of stimulation. IL-15 was detected by immunohistochemical staining and by immunoblot analysis. In addition, coculture of purified human monocytes and NK cells in the presence of LPS induces the production of IFN-γ, IL-12 and low levels of IL-15 (200 pg/ml; detected by an ELISA). The production of IFN-γ requires the presence of both NK cells and LPS-activated monocytes and is inhibited by antibodies specific for IL-12, IL-15 or IL-2Rβ, but not by an antibody specific for IL-2. As NK cells are not known to produce IL-12 or IL-15, it is likely that monocyte-derived IL-12 and IL-15 synergize to induce IFN-γ production by the NK cells. Human PBMCs also produce IL-15 in response to stimulation with *M. leprae*, but not in response to *Listeria monocytogenes* (Jullien *et al.*, 1997). Adherent PBMCs stimulated with *M. leprae* produce relatively high levels of IL-15 (1–5 ng/ml) detectable by an ELISA. The supernatants from these cultures also stimulate the proliferation of the mouse CTLL-2 line in the presence of an anti-IL-2 antiserum capable of neutralizing up to 20 ng/ml human IL-2.

IL-15 Protein *In Vivo*

Synovial fluids from patients with RA have been reported to contain high levels of IL-15 (up to 1200 ng/ml; detected by an ELISA) (McInnes *et al.*, 1996). Although the synovial fluids have chemotactic activity that can be attributed in part to IL-15, they do not cause proliferation in the highly sensitive CTLL-2 bioassay. In a follow-up study by the same group (McInnes *et al.*, 1997), a different IL-15-specific ELISA was used and the levels of IL-15 detected in synovial fluids from patients with RA ranged from 50 to 1134 pg/ml. If the much lower IL-15 levels detected in the latter study more accurately reflect the levels of IL-15 present in RA synovial fluids, it is not surprising that bioactive IL-15 is not detectable in the CTLL-2 bioassay of these fluids, as human serum and synovial fluids are inhibitory in this assay when used at dilutions of less than 1:100. However, it is possible that bioactive IL-15 was the 'IL-2-like factor' detected in an earlier study in which RA synovial fluids were found to be capable of stimulating proliferation of the CTLL-2 cell line even though they lacked detectable IL-2 as assessed by an ELISA (Firestein *et al.*, 1988).

Immunohistochemical analysis of tissues and flow cytometric analysis of cells taken from inflammatory sites has also provided evidence that IL-15 is expressed *in vivo*. Immunohistochemical analysis of joint tissues has shown that numerous IL-15-positive cells are present in the synovial tissue from patients with RA, whereas far fewer positive cells are present in the synovial tissue from patients with osteoarthritis (McInnes *et al.*, 1996). The IL-15-positive cells are most abundant in the synovial membrane lining layer and the pattern of staining suggests that they are macrophages, and possibly type A synoviocytes, which are derived from fibroblast-like cells. Immunohistochemical analysis of skin lesions from leprosy patients has shown that numerous IL-15-positive cells are evident in skin biopsies from patients with the self-healing (tuberculoid) form of leprosy, whereas such cells are rare in the lesions from patients with the progressive (lepromatous) form of leprosy (Jullien *et al.*, 1997). In addition, IL-15-positive cells are found in the suprabasal and granular layer of the epidermis as well as in the stratum corneum of normal human skin (Sorel *et al.*, 1996). Flow cytometric analysis has been used to demonstrate that alveolar macrophages isolated from patients with active sarcoidosis, and PBMCs isolated from patients with active ulcerative colitis, express IL-15 in their cytoplasm, whereas the same cell populations isolated from healthy subjects and patients with inactive sarcoidosis or ulcerative colitis do not (Agostini *et al.*, 1996; Kirman and Nielsen, 1996).

ROLE OF IL-15 IN IMMUNE AND INFLAMMATORY FUNCTIONS

The presence of IL-15 mRNA in many normal tissues and the increased production of IL-15 in response to a variety of stimuli suggest that this cytokine plays a role in protective immune responses, allograft rejection and the pathogenesis of autoimmune disorders. Early in infection, IL-15 derived from infected stromal or antigen-presenting cells might contribute to the recruitment and/or activation of inflammatory cells such as neutrophils, NK or T cells. However, in contrast to many other cytokines whose production is controlled primarily at the level of transcription and message stabilization, the production and secretion of IL-15 appears to be regulated at multiple levels, some of which are not well understood (Tagaya *et al.*, 1996a). IL-15 might actually represent a cytokine that is released only on the death of the cell (Ferrini *et al.*, 1996). If this is the case, IL-15 could serve as a warning signal of tissue damage in response to infection or other insults.

It is possible that dysregulation of the normally strict controls over IL-15 production or secretion contributes to inflammatory and autoimmune disorders. For example, overexpression of IL-15 within transplanted tissues could lead to the recruitment and activation of T cells involved in the process of graft rejection. In this regard, increased expression of IL-15 mRNA, but not of IL-2 mRNA, correlates with rejection of human renal allografts (Pavlakis *et al.*, 1996). McInnes *et al.* (1997) speculate that aberrant production of IL-15 by synovial macrophages or fibroblasts contributes to the pathogenesis of RA. In this scenario, IL-15 serves to attract polyclonal memory or primed T cells to the synovial membrane, supports their expansion, and leads to the overproduction of TNF-α by the T cells and by synovial monocytes that come in contact with the T cells.

The presence of IL-15 mRNA in many normal tissues has also led to the suggestion that stromal cell-derived cytokines such as IL-15 serve as survival factors for activated T cells in the periphery (Akbar and Salmon, 1997). In addition, a number of studies suggest that IL-15 is a differentiation factor for NK cells and certain subsets of IELs (Cavazzana-Calvo *et al.*, 1996; Leclercq *et al.*, 1996; Mrózek *et al.*, 1996; Puzanov *et al.*, 1996; Ye *et al.*, 1996b; Suzuki *et al.*, 1997). The use of IL-15 antagonists and the development of IL-15 knockout mice should help to elucidate the role of IL-15 in the development and function of the normal immune response as well as in infectious disease or inflammatory disorders.

ACKNOWLEDGEMENTS

The authors thank Drs Anthony Troutt and David Lynch for critical review of the chapter, Drs Robert Evans, William Munger and Youcef Rustum for sharing unpublished results, and Anne C. Bannister for preparation of the manuscript.

REFERENCES

Agostini, C., Trentin, L., Facco, M., Sancetta, R., Cerutti, A., Tassinari, C., Cimarosto, L., Adami, F., Cipriani, A., Zambello, R. and Semenzato, G. (1996). Role of IL-15, IL-2, and their receptors in the development of T cell alveolitis in pulmonary sarcoidosis. *J. Immunol.* **157**, 910–918.

Akbar, A.N. and Salmon, M. (1997). Cellular environments and apoptosis: tissue microenvironments control activated T-cell death. *Immunol. Today* **18**, 72–76.

Anderson, D.M., Johnson, L., Glaccum, M.B., Copeland, N.G., Gilbert, D.J., Jenkins, N.A., Valentine, V., Kirstein, M.N., Shapiro, D.N., Morris, S.W., Grabstein, K. and Cosman, D. (1995a). Chromosomal assignment and genomic structure of *Il15*. *Genomics* **25**, 701–706.

Anderson, D.M., Kumaki, S., Ahdieh, M., Bertles, J., Tometsko, M., Loomis, A., Giri, J., Copeland, N.G., Gilbert, D.J., Jenkins, N.A., Valentine, V., Shapiro, D.N., Morris, S.W., Park, L.S. and Cosman, D. (1995b). Functional characterization of the human interleukin-15 receptor α chain and close linkage of *IL15RA* and *IL2RA* genes. *J. Biol. Chem.* **270**, 29862–29869.

Armitage, R.J., Macduff, B.M., Eisenman, J., Paxton, R. and Grabstein, K.H. (1995). IL-15 has stimulatory activity for the induction of B cell proliferation and differentiation. *J. Immunol.* **154**, 483–490.

Bamford, R.N., Grant, A.J., Burton, J.D., Peters, C., Kurys, G., Goldman, C.K., Brennan, J., Roessler, E. and Waldmann, T.A. (1994). The interleukin (IL) 2 receptor β chain is shared by IL-2 and a cytokine, provisionally designated IL-T, that stimulates T-cell proliferation and the induction of lymphokine-activated killer cells. *Proc. Natl Acad. Sci. U.S.A.* **91**, 4940–4944.

Bamford, R.N., Battiata, A.P., Burton, J.D., Sharma, H. and Waldmann, T.A. (1996). Interleukin (IL) 15/IL-T production by the adult T-cell leukemia cell line HuT-102 is associated with a human T-cell lymphotrophic virus type IR region/IL-15 fusion message that lacks many upstream AUGs that normally attenuate IL-15 mRNA translation. *Proc. Natl Acad. Sci. U.S.A.* **93**, 2897–2902.

Blauvelt, A., Asada, H., Klaus-Kovtun, V., Altman, D.J., Lucey, D.R. and Katz, S.I. (1996). Interleukin-15 mRNA is expressed by human keratinocytes, Langerhans cells, and blood-derived dendritic cells and is downregulated by ultraviolet B radiation. *J. Invest. Dermatol.* **106**, 1047–1052.

Bluman, E.M., Bartynski, K.J., Avalos, B.R. and Caligiuri, M.A. (1996). Human natural killer cells produce abundant macrophage inflammatory protein-1α in response to monocyte-derived cytokines. *J. Clin. Invest.* **97**, 2722–2727.

Burton, J.D., Bamford, R.N., Peters, C., Grant, A.J., Kurys, G., Goldman, C.K., Brennan, J., Roessler, E. and Waldmann, T.A. (1994). A lymphokine, provisionally designated interleukin T and produced by a human adult T-cell leukemia line, stimulates T-cell proliferation and the induction of lymphokine-activated killer cells. *Proc. Natl Acad. Sci. U.S.A.* **91**, 4935–4939.

Carson, W.E., Giri, J.G., Lindemann, M.J., Linett, M.L., Ahdieh, M., Paxton, R., Anderson, D., Eisenman, J., Grabstein, K. and Caligiuri, M.A. (1994). Interleukin (IL) 15 is a novel cytokine that activates human natural killer cells via components of the IL-2 receptor. *J. Exp. Med.* **180**, 1395–1403.

Carson, W.E., Ross, M.E., Baiocchi, R.A., Marien, M.J., Boiani, N., Grabstein, K. and Caligiuri, M.A. (1995).

Endogenous production of interleukin 15 by activated human monocytes is critical for optimal production of interferon-γ by natural killer cells *in vitro*. *J. Clin. Invest.* **96**, 2578–2582.

Cavazzana-Calvo, M., Hacein-Bey, S., de Saint Basile, S., De Coene, C., Selz, F., Le Deist, F. and Fischer, A. (1996). Role of interleukin-2 (IL-2), IL-7, and IL-15 in natural killer cell differentiation from cord blood hematopoietic progenitor cells and from γc transduced severe combined immunodeficiency X1 bone marrow cells. *Blood* **88**, 3901–3909.

Collins, L., Tsien, W.-H., Seals, C., Hakimi, J., Weber, D., Bailon, P., Hoskings, J., Greene, W.C., Toome, V. and Ju, G. (1988). Identification of specific residues of human interleukin 2 that affect binding to the 70-kDa subunit (p70) of the interleukin 2 receptor. *Proc. Natl Acad. Sci. U.S.A.* **85**, 7709–7713.

Cosman, D., Anderson, D.M., Grabstein, K.H., Shanebeck, K., Kumaki, S., Ahdieh, M. and Giri, J.G. (1994). IL-15, a novel T cell growth factor, that shares components of the receptor for IL-2. In *Cytokines: Basic Principles and Practical Applications* (eds S. Romagnani, G. Del Prete and A.K. Abbas), Ares-Serono Symposia Publications, Rome, pp. 79–86.

Daw, S.C.M., Taylor, C., Kraman, M., Call, K., Mao, J., Schuffenhauer, S., Meitinger, T., Lipson, T., Goodship, J. and Scambler, P. (1996). A common region of 10p deleted in DiGeorge and velocardiofacial syndromes. *Nature Genet.* **13**, 458–460.

Doherty, T.M., Seder, R.A. and Sher, A. (1996). Induction and regulation of IL-15 expression in murine macrophages. *J. Immunol.* **156**, 735–741.

Edelbaum, D., Mohamadzadeh, M., Bergstresser, P.R., Sugamura, K. and Takashima, A. (1995). Interleukin (IL)-15 promotes the growth of murine epidermal γδ T cells by a mechanism involving the β- and γc-chains of the IL-2 receptor. *J. Invest. Dermatol.* **105**, 837–843.

Elloso, M.M., van der Heyde, H.C., Troutt, A., Manning, D.D. and Weidanz, W.P. (1996). Human γδ T cell subset-proliferative response to malarial antigen *in vitro* depends on $CD4^+$ T cells or cytokines that signal through components of the IL-2R. *J. Immunol.* **157**, 2096–2102.

Evans, R., Fuller, J.A., Christianson, G., Krupke, D.M. and Troutt, A.B. (1997). IL-15 mediates anti-tumor effects after cyclophosphamide injection of tumor-bearing mice and enhances adoptive immunotherapy: the potential role of NK cell subpopulations. *Cell. Immunol.* **179**, 66–73.

Ferrini, S., Azzorone, B. and Jasmin, C. (1996). Is IL-15 a suitable candidate for cancer gene therapy? *Gene Therapy* **3**, 656–657.

Firestein, G.S., Xu, W.-D., Townsend, K., Broide, D., Alvaro-Gracia, J., Glasebrook, A. and Zvaifler, N.J. (1988). Cytokines in chronic inflammatory arthritis. I. Failure to detect T cell lymphokines (interleukin 2 and interleukin 3) and presence of macrophage colony-stimulating factor (CSF-1) and a novel mast cell growth factor in rheumatoid synovitis. *J. Exp. Med.* **168**, 1573–1586.

Flamand, L., Stefanescu, I. and Menezes, J. (1996). Human herpesvirus-6 enhances natural killer cell cytotoxicity via IL-15. *J. Clin. Invest.* **97**, 1373–1381.

Gamero, A.M., Ussery, D., Reintgen, D.S., Puleo, C.A. and Djeu, J.Y. (1995). Interleukin 15 induction of lymphokine-activated killer cell function against autologous tumor cells in melanoma patient lymphocytes by a CD18-dependent, perforin-related mechanism. *Cancer Res.* **55**, 4988–4994.

Girard, D., Paquet, M.E., Paquin, R. and Beaulieu, A.D. (1996). Differential effects of interleukin-15 (IL-15) and IL-2 on human neutrophils: modulation of phagocytosis, cytoskeleton rearrangement, gene expression, and apoptosis by IL-15. *Blood* **88**, 3176–3184.

Giri, J.G. and Paxton, R.J. (1996). Interleukin-15. In *Human Cytokines: Handbook for Basic and Clinical Research*, Vol. II (eds B.B. Aggarwal and J.U. Gutterman), Blackwell Science, Cambridge, MA, pp. 135–145.

Giri, J.G., Ahdieh, M., Eisenman, J., Shanebeck, K., Grabstein, K., Kumaki, S., Namen, A., Park, L.S., Cosman, D. and Anderson, D. (1994). Utilization of the β and γ chains of the IL-2 receptor by the novel cytokine IL-15. *EMBO J.* **13**, 2822–2830.

Giri, J.G., Kumaki, S., Ahdieh, M., Friend, D.J., Loomis, A., Shanebeck, K., DuBose, R., Cosman, D., Park, L.S. and Anderson, D.M. (1995). Identification and cloning of a novel IL-15 binding protein that is structurally related to the α chain of the IL-2 receptor. *EMBO J.* **14**, 3654–3663.

Grabstein, K.H., Eisenman, J., Shanebeck, K., Rauch, C., Srinivasan, S., Fung, V., Beers, C., Richardson, J., Schoenborn, M.A., Ahdieh, M., Johnson, L., Alderson, M.R., Watson, J.D., Anderson, D.M. and Giri, J.G. (1994). Cloning of a T cell growth factor that interacts with the β chain of the interleukin-2 receptor. *Science* **264**, 965–968.

Hunter, C.A., Ellis-Neyer, L., Gabriel, K.E., Kennedy, M.K., Grabstein, K.H., Linsley, P.S. and Remington, J.S. (1997). The role of the CD28/B7 interaction in the regulation of NK cell responses during infection with *Toxoplasma gondii*. *J. Immunol.* **158**, 2285–2293.

Johnston, J.A., Bacon, C.M., Finbloom, D.S., Rees, R.C., Kaplan, D., Shibuya, K., Ortaldo, J.R., Gupta, S., Chen, Y.Q., Giri, J.G. and O'Shea, J.J. (1995). Tyrosine phosphorylation and activation of STAT5, STAT3, and Janus kinases by interleukins 2 and 15. *Proc. Natl Acad. Sci. U.S.A.* **92**, 8705–8709.

Jullien, D., Sieling, P.A., Uyemura, K., Mar, N.D., Rea, T.H. and Modlin, R.L. (1997). IL-15, an immunomodulator of T cell responses in intracellular infection. *J. Immunol.* **158**, 800–806.

Kanai, T., Thomas, E.K., Yasutomi, Y. and Letvin, N.L. (1996). IL-15 stimulates the expansion of AIDS virus-specific CTL. *J. Immunol.* **157**, 3681–3687.

Kanegane, H. and Tosato, G. (1996). Activation of naive and memory T cells by interleukin-15. *Blood* **58**, 230–235.

Kennedy, M.K., Picha, K.S., Shanebeck, K.D., Anderson, D.M. and Grabstein, K.H. (1994). Interleukin-12 regulates the proliferation of Th1, but not Th2 or Th0, clones. *Eur. J. Immunol.* **24**, 2271–2278.

Khan, I.A. and Kasper, L.H. (1996). IL-15 augments $CD8^+$ T cell-mediated immunity against *Toxoplasma gondii* infection in mice. *J. Immunol.* **157**, 2103–2108.

Kirman, I. and Nielsen, O.H. (1996). Increased numbers of interleukin-15 expressing cells in active ulcerative colitis. *Am. J. Gastroenterol.* **91**, 1789–1794.

Kohem, C.L., Brezinschek, R.I., Wisbey, H., Tortorella, C., Lipsky, P.E. and Oppenheimer-Marks, N. (1996). Enrichment of differentiated $CD45RB^{dim}$, $CD27^-$ memory T cells in the peripheral blood, synovial fluid, and synovial tissue of patients with rheumatoid arthritis. *Arthritis Rheum.* **39**, 844–854.

Kozak, M. (1991). An analysis of vertebrate mRNA sequences: intimations of translational control. *J. Cell Biol.* **115**, 887–903.

Kumaki, N., Anderson, D.M., Cosman, D. and Kumaki, S. (1996). Expression of interleukin-15 and its receptor by human retinal pigment epithelial cells. *Curr. Eye Res.* **15**, 876–882.

Kumaki, S., Ochs, H.D., Timour, M., Schooley, K., Ahdieh, M., Hill, H., Sugamura, K., Anderson, D., Zhu, Q., Cosman, D. and Giri, J.G. (1995). Characterization of B-cell lines established from two X-linked severe combined immunodeficiency patients: interleukin-15 binds to the B cells but is not internalized efficiently. *Blood* **86**, 1428–1436.

Kumaki, S., Armitage, R., Ahdieh, M., Park, L. and Cosman, D. (1996). Interleukin-15 up-regulates interleukin-2 receptor α chain but down-regulates its own high-affinity binding sites on human T and B cells. *Eur. J. Immunol.* **26**, 1235–1239.

Kündig, T.M., Schorle, H., Bachmann, M.F., Hengartner, H., Zinkernagel, R.M. and Horak, I. (1993). Immune responses in interleukin-2-deficient mice. *Science* **262**, 1059–1061.

Leclercq, G., Debacker, V., De Smedt, M. and Plum, J. (1996). Differential effects of interleukin-15 and interleukin-2 on differentiation of bipotential T/natural killer progenitor cells. *J. Exp. Med.* **184**, 325–336.

Lee, Y.B., Satoh, J., Walker, D.G. and Kim, S.U. (1996). Interleukin-15 gene expression in human astrocytes and microglia in culture. *NeuroReport* **7**, 1062–1066.

Leonard, W.J., Shores, E.W. and Love, P.E. (1995). Role of the common cytokine receptor γ chain in cytokine signaling and lymphoid development. *Immunol. Rev.* **148**, 97–114.

Lewko, W.M., Smith, T.L., Bowman, D.J., Good, R.W. and Oldham, R.K. (1995). Interleukin-15 and the growth of tumor derived activated T-cells. *Cancer Biother.* **10**, 13–20.

Lin, J.-X., Migone, T.-S., Tsang, M., Friedmann, M., Weatherbee, J.A., Zhou, L., Yamauchi, A., Bloom, E.T., Mietz, J., John, S. and Leonard, W.J. (1995). The role of shared receptor motifs and common Stat proteins in the generation of cytokine pleiotropy and redundancy by IL-2, IL-4, IL-7, IL-13, and IL-15. *Immunity* **2**, 331–339.

Lollini, P.-L., Palmieri, G., De Giovanni, C., Landuzzi, L., Nicoletti, G., Rossi, I., Griffoni, C., Frabetti, F., Scotlandi, K., Benini, S., Baldini, N., Santoni, A. and Nanni, P. (1997). Expression of interleukin 15 (IL-15) in human rhabdomyosarcoma, osteosarcoma, and Ewing's sarcoma. *Int. J. Cancer* **71**, 732–736.

Maas, R.A., Roest, P.A., Becker, M.J., Weimar, I.S., Dullens, H.F. and Den Otter, W. (1992). Effector cells of low-dose IL-2 immunotherapy in tumor bearing mice: tumor cell killing by $CD8^+$ cytotoxic T lymphocytes and macrophages. *Immunobiology* **186**, 214–229.

Malek, T.R., Ashwell, J.D., Germain, R.N., Shevach, E.M. and Miller, J. (1986). The murine interleukin-2 receptor: biochemical structure and regulation of expression. *Immunol. Rev.* **92**, 81–101.

McInnes, I.B., Al-Mughales, J., Field, M., Leung, B.P., Huang, F., Dixon, R., Sturrock, R.D., Wilkinson, P.C. and Liew, F.Y. (1996). The role of interleukin-15 in T-cell migration and activation in rheumatoid arthritis. *Nature Med.* **2**, 175–182.

McInnes, I.B., Leung, B.P., Sturrock, R.D., Field, M. and Liew, F.Y. (1997). Interleukin-15 mediates T cell-dependent regulation of tumor necrosis factor-α production in rheumatoid arthritis. *Nature Med.* **3**, 189–195.

Meazza, R., Verdiani, S., Biassoni, R., Coppolecchia, M., Gaggero, A., Orengo, A.M., Columbo, M.P., Azzarone, B. and Ferrini, S. (1996). Identification of a novel interleukin-15 (IL-15) transcript isoform generated by alternative splicing in human small cell lung carcinoma cell lines. *Oncogene* **12**, 2187–2192.

Minami, Y., Kono, T., Miyazaki, T. and Taniguchi, T. (1993). The IL-2 receptor complex: its structure, function, and target genes. *Annu. Rev. Immunol.* **11**, 245–268.

Mitelman, F. (1994). *Catalog of Chromosome Aberrations in Cancer.* Wiley-Liss, New York, pp. 1656–1787.

Mohamadzadeh, M., Takashima, A., Dougherty, I., Knop, J., Bergstresser, P.R. and Cruz, P.D., Jr. (1995). Ultraviolet B radiation up-regulates the expression of IL-15 in human skin. *J. Immunol.* **155**, 4492–4496.

Mori, A., Suko, M., Kaminuma, O., Inoue, S., Ohmura, T., Nishizaki, Y., Nagahori, T., Asakura, Y., Hoshino, A., Okumura, Y., Sato, G., Ito, K. and Okudaira, H. (1996). IL-15 promotes cytokine production of human T helper cells. *J. Immunol.* **156**, 2400–2405.

Mrózek, E., Anderson, P. and Caligiuri, M.A. (1996). Role of interleukin-15 in the development of human $CD56^+$ natural killer cells from $CD34^+$ hematopoietic progenitor cells. *Blood* **87**, 2632–2640.

Munger, W., DeJoy, S.Q., Jeyaseelan, R., Sr., Torley, L.W., Grabstein, K.H., Eisenman, J., Paxton, R., Cox, T., Wick, M.M. and Kerwar, S.S. (1995). Studies evaluating the antitumor activity and toxicity of interleukin-15, a new T cell growth factor: comparison with interleukin-2. *Cell. Immunol.* **165**, 289–293.

Nieto, M., Angel del Pozo, M. and Sánchez-Madrid, F. (1996). Interleukin-15 induces adhesion receptor redistribution in T lymphocytes. *Eur. J. Immunol.* **26**, 1302–1307.

Nishimura, H., Hiromatsu, K., Kobayashi, N., Grabstein, K.H., Paxton, R., Sugamura, K., Bluestone, J.A. and Yoshikai, Y. (1996). IL-15 is a novel growth factor for murine $\gamma\delta$ T cells induced by *Salmonella* infection. *J. Immunol.* **156**, 663–669.

Patki, A.H., Quinoñes-Mateu, M.E., Dorazio, D., Yen-Lieberman, B., Boom, W.H., Thomas, E.K. and Lederman, M.M. (1996). Activation of antigen-induced lymphocyte proliferation by interleukin-15 without the mitogenic effect of interleukin-2 that may induce human immunodeficiency virus-1 expression. *J. Clin. Invest.* **98**, 616–621.

Pavlakis, M., Strehlau, J., Lipman, M., Shapiro, M., Maslinski, W. and Strom, T.B. (1996). Intragraft IL-15 transcripts are increased in human renal allograft rejection. *Transplantation* **62**, 543–545.

Paxton, R.J. (1996). Measurement of interleukin-15. In *Current Protocols in Immunology* (eds J.E. Colligan, A.M. Kruisbeek, D.H. Margulies, E.M. Shevach and W. Strober), John Wiley, New York, pp. 6.22.1–6.22.7.

Pettit, D.K., Bonnert, T.P., Eisenman, J., Srinivasan, S., Paxton, R., Beers, C., Lynch, D., Miller, B., Yost, J., Grabstein, K.H. and Gombotz, W.R. (1997). Structure–function studies of interleukin 15 using site-specific mutagensis, polyethylene glycol conjugation, and homology modeling. *J. Biol. Chem.* **272**, 2312–2318.

Puzanov, I.J., Bennett, M. and Kumar, V. (1996). IL-15 can substitute for the marrow microenvironment in the differentiation of natural killer cells. *J. Immunol.* **157**, 4282–4285.

Quinn, L.S., Haugk, K.L. and Grabstein, K.H. (1995). Interleukin-15: a novel anabolic cytokine for skeletal muscle. *Endocrinology* **136**, 3669–3672.

Reinecker, H.-C., MacDermott, R.P., Mirau, S., Dignass, A. and Podolsky, D.K. (1996). Intestinal epithelial cells both express and respond to interleukin 15. *Gastroenterology* **111**, 1706–1713.

Rose-John, S. and Heinrich, P.C. (1994). Soluble receptors for cytokines and growth factors: generation and biological function. *Biochem. J.* **300**, 281–290.

Rosenberg, S.A., Mulé, J.J., Spiess, P.J., Reichert, C.M. and Schwarz, S.L. (1985). Regression of established pulmonary metastases and subcutaneous tumor mediated by the systemic administration of high-dose recombinant interleukin 2. *J. Exp. Med.* **161**, 1169–1188.

Seder, R.A. (1996). High-dose IL-2 and IL-15 enhance the *in vitro* priming of naive $CD4^+$ T cells for IFN-γ but have differential effects on priming for IL-4. *J. Immunol.* **156**, 2413–2422.

Seder, R.A., Grabstein, K.G., Berzofsky, J.A. and McDyer, J.F. (1995). Cytokine interactions in human immunodeficiency virus-infected individuals: roles of (IL)-2, IL-12, and IL-15. *J. Exp. Med.* **182**, 1067–1078.

Sorel, M., Cherel, M., Dreno, B., Bouyge, I., Guilbert, J., Dubois, S., Lebeau, B., Raher, S., Minvielle, S. and Jacques, Y. (1996). Production of interleukin-15 by human keratinocytes. *Eur. J. Dermatol.* **6**, 209–212.

Suzuki, H., Duncan, G.S., Takimoto, H. and Mak, T.W. (1997). Abnormal development of intestinal intraepithelial lymphocytes and peripheral natural killer cells in mice lacking the IL-2 receptor β chain. *J. Exp. Med.* **185**, 499–505.

Tagaya, Y., Bamford, R.N., DeFilippis, A.P. and Waldmann, T.A. (1996a). IL-15: a pleiotropic cytokine with diverse receptor/signaling pathways whose expression is controlled at multiple levels. *Immunity* **4**, 329–336.

Tagaya, Y., Burton, J.D., Miyamoto, Y. and Waldmann, T.A. (1996b). Identification of a novel receptor/signal transduction pathway for IL-15/T in mast cells. *EMBO J.* **15**, 4928–4939.

Warren, H.S., Kinnear, B.F., Kastelein, R.L. and Lanier, L.L. (1996). Analysis of the costimulatory role of IL-2 and IL-15 in initiating proliferation of resting ($CD56^{dim}$) human NK cells. *J. Immunol.* **156**, 3254–3259.

Wilkinson, P.C. and Liew, F.Y. (1995). Chemoattraction of human blood T lymphocytes by interleukin-15. *J. Exp. Med.* **181**, 1255–1259.

Ye, W., Young, J.D. and Liu, C.-C. (1996a). Interleukin-15 induces the expression of mRNAs of cytolytic mediators and augments cytotoxic activities in primary murine lymphocytes. *Cell. Immunol.* **174**, 54–62.

Ye, W., Zheng, L.-M., Young, J.D. and Liu, C.-C. (1996b). The involvement of interleukin (IL)-15 in regulating the differentiation of granulated metrial gland cells in the mouse pregnant uterus. *J. Exp. Med.* **184**, 2405–2410.

Zurawski, S.M., Imler, J.-L. and Zurawski, G. (1990). Partial agonist/antagonist mouse interleukin-2 proteins indicate that a third component of the receptor complex functions in signal transduction. *EMBO J.* **9**, 3899–3905.

Chapter 17

Interleukins 16, 17 and 18

Mary A. Antonysamy[1,2], Michael T. Lotze[2,3,4], Hideaki Tahara[2,4] and Angus W. Thomson[1,2,3,4]

[1]Thomas E. Starzl Transplantation Institute and Departments of [2]Surgery and [3]Molecular Genetics and Biochemistry, University of Pittsburgh Medical Center, and [4]University of Pittsburgh Cancer Institute, Pittsburgh, PA, USA

INTERLEUKIN-16

Background

Interleukin-16 (IL-16), formerly termed lymphocyte chemoattractant factor (LCF), was initially described by Center and Cruikshank (1982) as a factor that stimulated $CD4^+$ T-cell chemotaxis. IL-16 exhibits unique characteristics with regard to its structure and function. The secreted, biologically active form of IL-16 is a 14–17-kDa sialoprotein that multimerizes into cationic homotetramers. The isoelectric point of the tetramer is 9.1. The tetrameric structure is reported to be absolutely required for the biologic function of IL-16 (Cruikshank and Center, 1982). The only known receptor for IL-16 is the CD4 molecule, through which it transmits its functional signals (Cruikshank *et al.*, 1991).

IL-16 Protein

The IL-16 protein is encoded by a gene located at chromosome 15q26.1, in an area distant from other known chemokine or cytokine genes. Screening of a COS cell expression library of mitogen-activated peripheral blood mononuclear cells (PBMCs) led to the isolation of a 2150 base pairs (bp) IL-16 cDNA encoding a bioactive protein of 130 amino acids (Cruikshank *et al.*, 1994) (Fig. 1). However, Western blot analysis of human and murine cells identified an 80-kDa protein when antibodies were directed against the 14-kDa polypeptide, implying the presence of a larger precursor (Center *et al.*, 1996). The presence of this 80-kDa protein suggests that the open reading frame in the IL-16 cDNA encodes a larger IL-16 precursor, which would then be processed to the 130-amino-acid bioactive form. Recently, Mukhtar *et al.* (1997) reported the isolation of a full-length human IL-16 cDNA of 3070 bp from a human genomic clone. The new open reading frame, with a translation start site in-frame with the previously identified 130-amino-acid coding sequence, encoded a putative IL-16 precursor molecule of 630

The Cytokine Handbook, 3rd ed.
ISBN 0-12-689662-3

```
1     TTCCTCGAGAGCTGTCAACACAGGCTGAGGAATCTCAAGGCCCAGTGCTCAAGATGCCT
60    AGCCAGCGAGCACGGAGCTTCCCCCTGACCAGGTCCCAGTCCTGTGAGACGAAGCTACT
119   TGACGAAAAGACCAGCAAACTCTATTCTATCACCAGCCAGTGTCATCGGCTGTCATGAA
178   ATCCTTGCTGTGCCTTCCATCTTCTATCTCCTGTGCCCAGACTCCCTGCATCCCCAAGG
237   CAGGGGCATCTCCAACATCATCATCCAACGAAGACTCAGCTGCAAATGGTTCTGCTGAA
296   ACATCTGCCTTGGACACGGGGTTCTCGCTCAACCTTTCAGAGCTGAGAGAATATACAGA
355   GGGTCTCACGGAAGCCAAGGAAGACGATGATGGGGACCACAGTTCCTTCAGTCTGGTCA
414   GTCCGTTATCTCCCTGCTGAGCTCAGAAGAATTAAAAAAACTCATCGAGGAGGTGAAGG
473   TTCTGGATGAAGCAACATTAAAGCAATTAGACGGCATCCATGTCACCATCTTACACAAG
532   GAGGAAGGTCGTGGTCTTGGGTTCAGCTTGGCAGGAGGAGCAGATCTAGAAAACAAGGT
591   GATTACGGTTCACAGAGTGTTTCCAAATGGGCTGGCCTCCCAGGAAGGGACTATTCAGA
650   AGGGCAATGAGGTTCTTTCCATCAACGGCAAGTCTCTCAAGGGGACCACGCACCATGAT
709   GCCTTGGCCATCCTCCGCCAAGCTCGAGAGCCCAGGCAAGCTGTGATTGTCACAAGGAA
768   GCTGACTCCAGAGCC ATG CCC GAC CTC AAC TCC TCC ACT GAC TCT GCA
  1                 Met Pro Asp Leu Asn Ser Ser Thr Asp Ser Ala
816   GCC TCA GCC TCT GCA GCC AGT GAT GTT TCT GTA GAA TCT ACA GCA
 12   Ala Ser Ala Ser Ala Ala Ser Asp Val Ser Val Glu Ser Thr Ala
861   GAG GCC ACA GTC TGC ACG GTG ACA CTG GAG AAG ATG TCG GCA GGG
 27   Glu Ala Thr Val Cys Thr Val Thr Leu Glu Lys Met Ser Ala Gly
906   CTG GGC TTC AGC CTG GAA GGA GGG AAG GGC TCC CTA CAC GGA GAC
 42   Leu Gly Phe Ser Leu Glu Gly Gly Lys Gly Ser Leu His Gly Asp
951   AAG CCT CTC ACC ATT AAC AGG ATT TTC AAA GGA GCA GCC TCA GAA
 57   Lys Pro Leu Thr Ile Asn Arg Ile Phe Lys Gly Ala Ala Ser Glu
996   CAA AGT GAG ACA GTC CAG CCT GGA GAT GAA ATC TTG CAG CTG GGT
 72   Gln Ser Glu Thr Val Gln Pro Gly Asp Glu Ile Leu Gln Leu Gly
1041  GGC ACT GCC ATG CAG GGC CTC ACA CGG TTG GAA GCC TGG AAC ATC
 87   Gly Thr Ala Met Gln Gly Leu Thr Arg Phe Glu Ala Trp Asn Ile
1086  ATC AAG GCA CTG CCT GAT GGA CCT GTC ACG ATT GTC ATC AGG AGA
102   Ile Lys Ala Leu Pro Asp Gly Pro Val Thr Ile Val Ile Arg Arg
1131  AAA AGC CTC CAG TCC AAG GAA ACC ACA GCT GCT GGA GAC TCC TAG
117   Lys Ser Leu Gln Ser Lys Glu Thr Thr Ala Ala Gly Asp Ser  -
1176  GCAGGACATGCTGAAGCCAAAGCCAATAACACACAGCTAACACACAGCTCCCATAACCG
1235  CTGATTCTCAGGGTCTCTGCTGCCGCCCCACCCAGATGGGGAAAGCACAGGTGGGCTT
1294  CCCAGTGGCTGCTGCCCAGGCCCAGACCTTCTAGGACGCCACCCAGCAAAAGGTTGTTC
1353  CTAAAATAAGGGCAGAGTCACACTGGGGCAGCTGATACAAATTGCAGACTGTGTAAAAA
1412  GAGAGCTTAATGATAATATTGTGGTGCCACAAATAAAATGGATTTATTAGAATTCCATA
1471  TGACATTCATGCCTGGCTTCGCAAAATGTTTCAAGTACTGTAACTGTGTCATGATTCAC
1530  CCCCAAACAGTGACATTTATTTTTCTCATGAATCTGCAATGTGGGCAGAGATTGGAATG
1589  GGCAGCTCATCTCTGTCCCACTTGGCATCAGCTGGCGTCATGCAAAGTCATGCAAAGGC
1648  TGGGACCACCTGAGATCATTCACTCATACATCTGGCCGTTGATGTTGGCTGGGAACTCA
1707  CCTGGGGCTGCTGGCCTGAATGCTTATAGGTGGCCTCTCCTTGTTGCCTGGGCTCCTCA
1766  CAACATGGTGTCTGGATTCCCAGGATGAGCATCCCAGGATCGCAAGAGCCACGTAGAAG
1825  CTGCATCTTGTTTATACCTTTGCCTTGGAAGTTGCATGGCATCACCTCCACCATACTCC
1884  ATCAGTTAGAGCTGACACAAACCTGCCTGGGTTTAAGGGGAGAGGAAATATTGCTGGGG
1943  TCATTTATGAAAAATACAGTTTGTCACATGAAACATTTGCAAAATTGTTTTTGGTTGGA
2002  TTGGAGAAGTAATCCTAGGGAAGGGTGGTGGAGCCAGTAAATAGAGGAGTACAGTGTAA
2061  GCACCAAGCTCAAAGCGTGGACAGGTGTGCCGACAGAAGGAACCAGCGTGTATATGAGG
2120  GTATCAAATAAAATTGCTACTACTTACCACC
```

Fig. 1. The LCF (IL-16) cDNA nucleotide sequence and the putative amino acid sequence of the encoded protein. The deduced amino acid sequence of the LCF protein is shown beneath the corresponding nucleotide sequence. The sole potential glycosylation location is an asparagine (Asn-5) residue. The three ATTTA sequences potentially associated with message stability are underlined and the two polyadenylylation sequences are doubly underlined. Succeeding the last indicated nucleotide (2150) is the polyA tail, which has been omitted. From Cruikshank *et al.* (1994), with permission. Copyright (1994) National Academy of Sciences, USA.

amino acids (Mukhtar *et al.*, 1997). The precursor IL-16 molecule is expressed in the cytoplasm of both resting $CD4^+$ and $CD8^+$ cells (Chupp *et al.*, 1997). Upon proteolytic cleavage, this larger precursor molecule would yield a 14–17-kDa polypeptide, which would then autoaggregate into the biologically active 50–60-kDa tetrameric protein. The IL-16 mRNA contains three AUUUA sequences and two polyadenylation sites in its 3′ untranslated region (Cruikshank *et al.*, 1994). The predicted sequence lacks a hydrophobic signal peptide and a potential transmembrane domain, evoking the question as to how the molecule is secreted.

The sequence and structure of the secreted form of IL-16 appear to be conserved across species. African green monkey and murine cDNAs are reported to have high sequence homology in the region encoding the secreted peptide (Baier *et al.*, 1995). Mouse (Center *et al.*, 1996) and rat protein homologs (McFadden and Vickers, 1989) of IL-16 have been identified. The molecular sizes of the secreted homologs from the African green monkey, mouse and rat were found to be similar to the human protein. Functional studies describe similar bioactivities for both murine and human IL-16 proteins (Wu *et al.*, 1997). Thus, there is substantial cross-reactivity among species; in fact, the amino acid sequences of the bioactive forms of IL-16 are highly homologous among species (D.M. Center, personal communication). Deletional analysis of recombinant IL-16 (rIL-16) ascertained all biologic activities to reside in the *C*-terminal 114 residues. Synthetic peptides representing the 15 hydrophilic *C*-terminal amino acids and polyclonal antibody raised against the *C*-terminal peptide were both found to inhibit IL-16 functions. These studies indicate that the 15 hydrophilic *C*-terminal amino acids might be important in the binding of IL-16 to its receptor, CD4. The binding region of IL-16 on CD4 is reported to be distant from major histocompatibility complex (MHC) and human immunodeficiency virus (HIV) gp120 binding sites, and is located in the CDR3 region, close to the OKT4 binding domain (Center *et al.*, 1996).

Production of IL-16

IL-16 was originally identified as a product of $CD8^+$ T lymphocytes. $CD8^+$ T cells were subsequently demonstrated to be able constitutively to synthesize and store the protein in its bioactive form (Laberge *et al.*, 1995). The $CD8^+$ lymphocytes release the protein in response to histamine stimulation through H_2 receptors (Center *et al.*, 1983a) or serotonin stimulation through S_2-type receptors (Laberge *et al.*, 1996), as well as stimulation with mitogen or antigen. In contrast to histamine and serotonin stimulation, where IL-16 is released within 1–4 h, stimulation with antigen, mitogen or anti-CD3 antibody releases IL-16 only after 18–24 h. This is because the latter mode of stimulation requires the process of transcription, translation and synthesis of the new protein (Center *et al.*, 1996). At this point in time the mechanism of IL-16 synthesis and release is not clearly understood.

$CD4^+$ cells can synthesize and secrete IL-16 as well. Even though the IL-16 mRNA is constitutively expressed in $CD4^+$ cells, these cells contain no preformed bioactive protein. However, they will release the bioactive IL-16 in 18–24 h, following stimulation with antigen, mitogen or anti-CD3, by a process that requires transcription and translation of the new protein (Center *et al.*, 1995, and personal communication). IL-16 is also produced by eosinophils, an additional mechanism by which eosinophils may contribute to immune responses in allergic inflammation (Lim *et al.*, 1996). Bronchial epithelial cells of asthmatic patients are another source of IL-16 (Bellini *et al.*, 1993). Laberge *et al.* (1997a) have recently demonstrated that the amount of IL-16 present in the epithelium of atopic asthmatics (but not in that of normals or atopic non-asthmatics) correlates with the number of $CD4^+$ cells located within the submucosa. Furthermore, treatment of allergic rhinitics with dexamethasone not only decreases the constitutive expression of IL-16, but also inhibits the upregulation of IL-16 message upon subsequent antigen challenge (Laberge *et al.*, 1997b). Mast cells are also known to express transcripts for IL-16 constitutively, and when induced by the C5a cleavage product as well as other stimuli, they synthesize and secrete the bioactive protein (Rumsaeng *et al.*, 1996). Cruikshank *et al.* (1997) have recently identified IL-16 message and bioactivity in several human fibroblast lines derived from orbital connective tissue, thyroid and skin. Treatment of the fibroblast cultures with proinflammatory cytokines results in the release of IL-16 along with other chemoattractants. Based on their findings, they suggest that exposure of human fibroblasts from several anatomic regions to inflammatory stimuli can induce IL-16 expression and may contribute to their recruitment of $CD4^+$ cells to sites of inflammation (Cruikshank *et al.*, 1997). Northern blot analysis has revealed IL-16 message in the human brain, thymus, spleen, whole blood leukocytes and pancreas (Center *et al.*, 1996).

IL-16 Signaling

The activity of IL-16 is absolutely dependent on the presence of functional CD4 on the cell surface. The evidence that all IL-16-mediated functions are specifically inhibited by Fab fragments of OKT4 monoclonal antibody (mAb), that IL-16 can be purified by means of CD4 affinity chromatography (Center *et al.*, 1995) and that transfection of IL-16-unresponsive murine T-cell hybridomas with normal human CD4 renders them responsive (Cruikshank *et al.*, 1991) confirms the association of IL-16 with the immunoglobulin superfamily member, CD4. IL-16 is therefore the first described soluble ligand for CD4.

In normal $CD4^+$ T cells, IL-16 induces a rise in intracytoplasmic calcium concentration, as well as inositol trisphosphate and the autophosphorylation of $p56^{lck}$ (Ryan *et al.*, 1995). In mutants lacking the intracytoplasmic tail of CD4 essential for $p56^{lck}$ interaction, these effects are not observed. However, cells bearing chimeric CD4-$p56^{lck}$ that lacks the kinase domain of lck display normal chemoattractant function. These studies suggest that, although the enzymatic activity of $p56^{lck}$ may not be required, the absolute need for CD4-linked tyrosine kinase $p56^{lck}$ for the transmission of migratory signals is related to its ability to recruit other signaling molecules via the SH2–SH3 domains (Ryan *et al.*, 1995) (Fig. 2). Several intracellular proteins require recruitment through the SH2–SH3 domains, and some of these proteins could play a role in the generation of a motile response. Phospholipase Cγ (PLCγ) and phosphatidylinositol 3-kinase (PI3-kinase) are two such candidate molecules. Selective inhibition of protein kinase C (PKC) or PI3-kinase, abrogate the motile response in $CD4^+$ cells, thereby supporting an essential role for PLCγ and PI3-kinase in the IL-16 signal transduction pathway (Fig. 2). The evidence for PKC dependence in certain non-IL-16-induced T-cell motile responses led Parada *et al.* (1996) to evaluate the role of PKC in IL-16-induced migratory responses. IL-16 induced rapid translocation of PKC from the cytosol to the membrane in human $CD4^+$ T cells and T-cell lines. Furthermore, addition of PKC inhibitors abolished IL-16-induced lymphocyte migration, suggesting PKC dependence in IL-16-induced motile responses. From these observations it could be inferred that IL-16 interaction with CD4 results in PLC-associated signal transduction events. However, the precise downstream biochemical signals following early PLC signaling events that lead to the induction of migratory responses are as yet unclear.

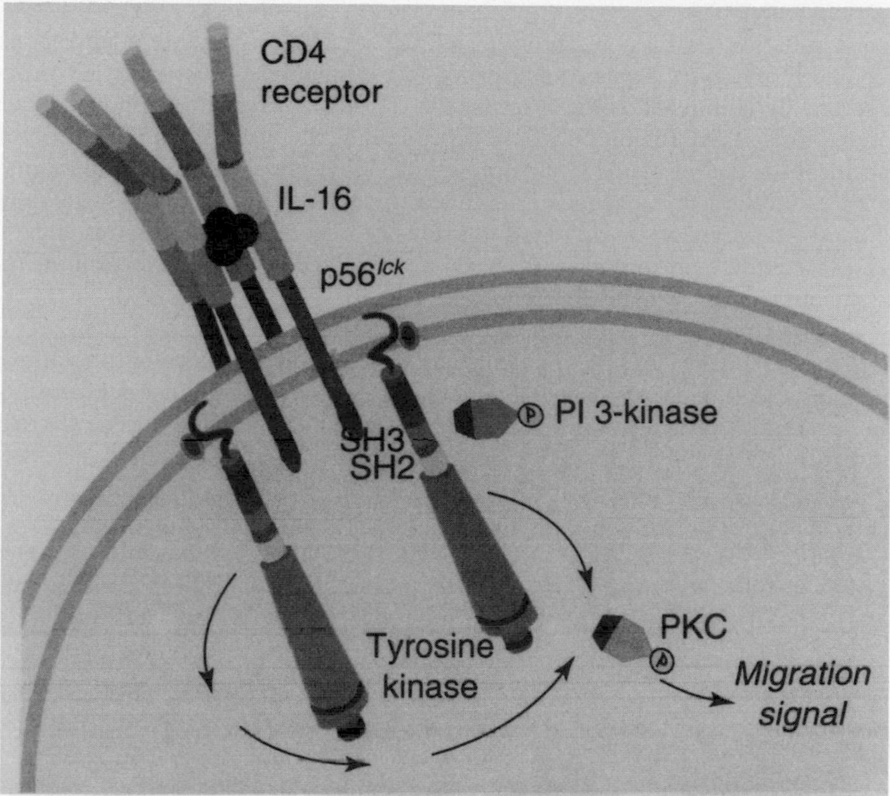

Fig. 2. IL-16–CD4 signaling in T cells. IL-16-induced chemotaxis requires the physical association but not the enzymatic activity of $p56^{lck}$. The migratory response is associated with translocation of PKC and activation of PI3-kinase. Inhibitors of PKC or PI3-kinase abrogate IL-16-induced T-cell migration. IL-16, interleukin-16; PI3-kinase, phosphoinositide 3-kinase; PKC, protein kinase C. From Center *et al.* (1996), with permission. Copyright (1995) Anne Meneghetti.

Biologic Effects of IL-16

The functions of IL-16 are varied as well as interesting. IL-16 is a potent chemotactic factor for $CD4^+$ T cells, with an ED_{50} (half the maximal effective dose) of 10^{-11} M. In addition, it is also a competence growth factor for $CD4^+$ T cells, augmenting their proliferation, a response that can be inhibited by anti-IL-16 mAb (Cruikshank *et al.*, 1991). IL-16 induces cell surface expression of high-affinity IL-2 receptors (IL-2Rs) without eliciting the expression or secretion of IL-2. These cells are subsequently induced to enter the G_1 phase of the cell cycle (Center *et al.*, 1995). T cells responding to IL-16 are rendered unresponsive to subsequent activation by antigen, presumably because CD4 can no longer associate with CD3. Nevertheless, these cells remain potentially reactive to other proinflammatory cytokines (Center *et al.*, 1996). Recently, Center and coworkers (personal communication) have reported that IL-16 primes resting $CD4^+$ T cells to proliferate to IL-2 with the retention of antigen responsiveness as late as 6 weeks following culture.

Binding of CD4 to alternate ligands (HIV gp120, anti-CD4 antibodies) inhibits antigen stimulation via T-cell receptor (TCR)/CD3, leading to functional unresponsiveness or T-cell anergy. Likewise, addition of recombinant IL-16, a soluble ligand for CD4, reversibly inhibits alloantigen responsive T cells in a mixed lymphocyte reaction (Theodore *et al.*, 1996). Inhibition is dose dependent, with maximum inhibition observed at concentrations as low as 10^{-11} M. Although IL-2R expression on responder cells was not altered, addition of exogenous IL-2 failed to restore responsiveness. Unlike other CD4 ligands, suppression by IL-16 did not downmodulate CD4 expression, suggesting that IL-16-induced unresponsiveness may have resulted from an interruption in the IL-2R signaling pathway (Theodore *et al.*, 1996).

IL-16 is an extremely potent chemoattractant for eosinophils, with an ED_{50} of 10^{-12} to 10^{-11} M, two to three logs more potent than other recognized chemoattractants, C5a and platelet-activating factor (Rand *et al.*, 1991a). Although IL-16 increased eosinophil adhesion to matrix proteins, it did not influence degranulation, leukotriene C4 release, respiratory burst activity, superoxide generation, *in vitro* survival, human leukocyte antigen (HLA)-DR expression or CR3/CD11b expression by eosinophils. p55 IL-2R that is detected normally on eosinophils by flow cytometry can be increased by IL-16 (Rand *et al.*, 1991b). In a recent study, Lim *et al.* (1996) reported that IL-16 secreted by eosinophils enabled recruitment of additional populations of $CD4^+$ cells into inflammatory sites. In addition, treatment of monocytes with IL-16 leads to increased chemotaxis and enhanced MHC class II expression, similar in magnitude to that induced by interferon-γ (IFN-γ), but independent of IFN-γ (Center *et al.*, 1995).

IL-16 in Clinical Conditions

IL-16 may have roles in many clinical situations, especially in disease states characterized by increased numbers of activated $CD4^+$ cells, such as asthma, sarcoidosis, rheumatoid arthritis and Graves' disease. For example, IL-16 is present in bronchoalveolar lavage (BAL) fluid from patients with atopic asthma 6 h after antigen challenge. Most of the chemoattractant activity in the BAL fluid was associated with IL-16 (Cruikshank *et al.*, 1995). In a mouse model of allergic asthma, IL-16 is reported to be important for immunoglobulin E (IgE) synthesis and airway hyperresponsiveness (Hessel *et al.*, 1997). Bellini *et al.* (1993) reported IL-16 in the airway epithelial cells of asthmatics. It is possible that histamine released by mast cells, and serotonin released by platelets, induce the release of preformed IL-16 from bronchial epithelial cells and resting $CD8^+$ T cells (Fig. 3). IL-16, a chemotactic factor for $CD4^+$ cells, would then initiate early infiltration of $CD4^+$ T cells, eosinophils and monocytes. This would further augment IL-16 secretion and selective recruitment of more $CD4^+$ T cells and eosinophils into the airway epithelium, culminating in an enhanced inflammatory response in asthmatics.

IL-16 has also been described in the blister fluid of bullous pemphigoid (Center *et al.*, 1983b). This disease is characterized by dermal infiltration of lymphocytes, followed by a striking influx of eosinophils as the lesion progresses to the bullous phase. Atopic dermatitis is an inflammatory skin disease with increased numbers of $CD4^+$ T cells in acute lesions. Laberge *et al.* (1997c) have demonstrated augmentation of IL-16 expression in acute atopic dermatitis and suggest that IL-16 could be initiating inflammation by recruiting more $CD4^+$ T cells in this disease model as well. Another disease where IL-16 has been implicated is sarcoidosis, a granulomatous disease typified by $CD4^+$ T-cell

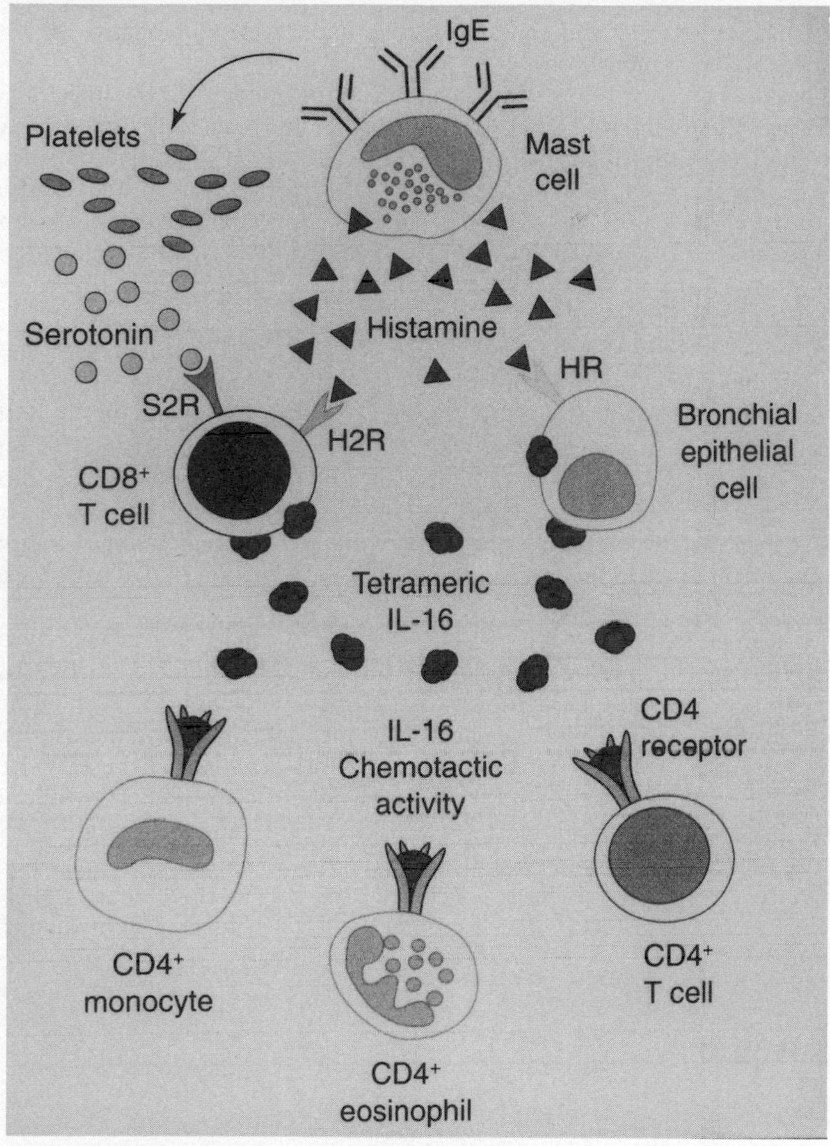

Fig. 3. The suggested role of IL-16 in triggering CD4$^+$ inflammation in allergic asthmatics. In allergic asthmatics, activation of mast cells by the allergen leads to the release of platelet-activity factor (PAF) and granule-associated histamine. Histamine triggers the release of preformed IL-16 from the asthmatic epithelium or CD8$^+$ T cells (resident or recruited), and this IL-16 acts as a migratory factor for CD4$^+$ T cells, monocytes and eosinophils. Likewise, PAF-induced secretion of serotonin from platelets causes the release of preformed IL-16 from CD8$^+$ T cells. However, the effects of serotonin on epithelial cells are unknown. IL-16, interleukin-16; IL-2R, interleukin-2 receptor. From Center *et al.* (1996), with permission. Copyright (1995) Anne Meneghetti.

alveolitis. High levels of IL-16 were detected in the BAL fluid of patients with active disease, but IL-16 was no longer detectable following corticosteroid therapy (Center *et al.*, 1995). Graves' disease is yet another disorder characterized by CD4$^+$ T-cell infiltration in the retro-orbital space. Recently Smith *et al.* have shown fibroblasts isolated from the retro-orbital space to secrete high levels of bioactive IL-16 (D.M. Center, personal communication).

IL-16 has been described to play a crucial role in the biology of HIV infection. $CD8^+$ T-cell-derived IL-16 inhibits HIV-1 and simian immunodeficiency virus replication in PBMCs. Antiviral activity correlates with the health of HIV-positive patients and is maximal during the asymptomatic stages of infection (Baier *et al.*, 1995). Nevertheless, Scala *et al.* (1997) have provided evidence that C–C chemokines, IL-16 and soluble HIV antiviral factor activity are all independently increased in T-cell clones of long-term non-progressors, and therefore postulate that all these factors might together be necessary to prevent HIV disease progression.

Since IL-16 and gp120 both use the CD4 receptor, one could suggest that IL-16 sterically inhibits HIV-1 binding. However this is not so. Maciaszek *et al.* (1997) in their recent report have shown rIL-16 to repress HIV-1 promoter activity. Both CD4 expression and a 24-h pretreatment with IL-16 is required to cause this repressed expression from the HIV-1 long terminal repeat (LTR). Experiments using mutant LTR constructs have revealed that sequences within the core enhancer element are required for IL-16-repressive effects. Repression was not due to downregulation of transcription factors, but rather to the induction of a transcriptional repressor that acted via sequences within or close to the core enhancer.

In addition to its antiviral activity, IL-16 potentiates $CD4^+$ T-cell growth by inducing IL-2R expression. This finding, coupled with the fact that a major predictive factor for the onset of acquired immune deficiency syndrome is depletion of $CD4^+$ T cells, led Parada *et al.* (1997) to determine the effects of sequential IL-16–IL-2 therapy on $CD4^+$ T-cell reconstitution in HIV-1-infected patients. PBMCs isolated from HIV-1-infected patients and normal individuals were treated with either IL-16 alone, IL-2 alone or both cytokines. Six weeks of *in vitro* treatment with both IL-16 and IL-2 resulted in a 3000-fold increase in $CD4^+$ T cells in normal and HIV-infected PBMCs when compared with the individual treatments. It was concluded that IL-16–IL-2 therapy could help reconstitute or maintain $CD4^+$ T-cell numbers in HIV-1-infected individuals (Parada *et al.*, 1997).

INTERLEUKIN-17

Background

The designation interleukin-17 (IL-17) has been assigned recently to a T-cell-derived lymphokine that acts on a wide variety of cells. Murine IL-17 was identified originally by Rouvier *et al.* (1993) as cytotoxic T lymphocyte-associated antigen-8 (CTLA-8). It was generated using a subtractive hybridization approach from an activated T-cell hybridoma in a search for molecules with immune functions. The CTLA-8 gene encodes a putative 150-amino-acid protein with a signal peptide and contains several AU-rich repeats in its 3′ untranslated region associated with mRNA instability, as found in several cytokines, growth factors and proto-oncogenes (Rouvier *et al.*, 1993). This presumably suggests tightly controlled gene expression at the level of transcription in the molecules bearing these repeats. Conserved AU sequences in the 3′ untranslated region of cytokine mRNAs are reported to serve as recognition signals for an mRNA processing pathway that mediates selective mRNA degradation, resulting in transient expression of the gene (Caput *et al.*, 1986; Shaw and Kamen, 1986). The CTLA-8 protein was found to be strikingly homologous to the protein encoded by the open reading frame 13 of herpesvirus saimiri (HVS-13) (56.7% homology on 146 amino acids) (Albrecht *et al.*, 1992). This phenomenon is analogous to the presence of the IL-10 homolog called BamH1 C rightward fragment 1 (BCRF1) in the Epstein–Barr virus (Moore *et al.*, 1990) and that of the IL-6 homolog in the recently identified human herpesvirus-8 (HHV-8) or Kaposi's sarcoma herpesvirus, itself closely related to HVS.

Murine IL-17

Yao *et al.* (1995a) suggested that the cDNA encoding CTLA-8 isolated from a rodent T-cell hybridoma by Rouvier *et al.* (1993) was of rat origin and that it may have been derived from the rat lymphoma rather than the murine cytotoxic T cell. Although they reported the mouse IL-17 message to have been detected in activated T cells and the murine thymoma cell line EL4, the exact origin of this molecule was not identified. In an attempt to address the above issues, and to generate a molecular profile for $\alpha\beta$ TCR$^+$ CD4$^-$ CD8$^-$ (DN) cells, Kennedy *et al.* (1996) used subtraction techniques to isolate genes specifically expressed by activated mouse $\alpha\beta$ TCR$^+$ DN thymocytes. One of the molecules isolated from this cDNA library shared 87.3% homology with the previously described murine CTLA-8 cDNA. Results from polymerase chain reaction (PCR) analysis of activated rat splenocytes, using primers from the murine CTLA-8 sequence, confirmed that murine CTLA-8 was derived from the rat lymphoma fusion partner used to generate the hybridoma described by Rouvier *et al.* (1993). Mouse IL-17 shared 62.5% amino acid identity with the human IL-17 protein and 57.7% identity with the HVS-13 protein (Kennedy *et al.*, 1996) (Fig. 4).

The $\alpha\beta$ TCR$^+$ DN cells, generally found in the thymus, spleen, lymph nodes and bone marrow, when activated, appeared to be the only abundant source of IL-17 (Kennedy *et al.*, 1996). PCR analysis of other activated T-cell subsets revealed IL-17 message, although this was not significantly high to be detected by Northern analysis. These $\alpha\beta$ TCR$^+$ DN cells are also reported to be capable of producing T helper 2 (T_{H2})-type cytokines (IL-4, IL-5, IL-10 and IL-13) and to express genes commonly associated with CD8$^+$ T cells, such as IFN-γ, lymphotactin, RANTES, granzyme B and Fas ligand (FasL) (Kennedy *et al.*, 1996). They also appear to be among the first cells to be activated during an immune response.

mIL-17 and HVS-13: Sequence and Production

The close homology observed between the murine CTLA-8 and the simian HVS-13 sequences indicates that CTLA-8 and HVS-13 are probably derived from the same primordial gene. Herpesvirus saimiri, a T-lymphotropic virus, is a member of the gammaherpesvirinae subgroup (rhadinoviruses). Although squirrel monkeys serve as natural hosts for HVS, the virus causes fulminant T-cell lymphomas in New World primates. HVS is capable of transforming non-saimiri New World primate and human T cells, resulting in continuous growth *in vitro* (Albrecht *et al.*, 1992). HVS also induces IL-2-dependent autocrine growth of human lymphocytes following cross-linking of CD2 (Rouvier *et al.*, 1993). The HVS genome has at least 15 genes homologous to cellular

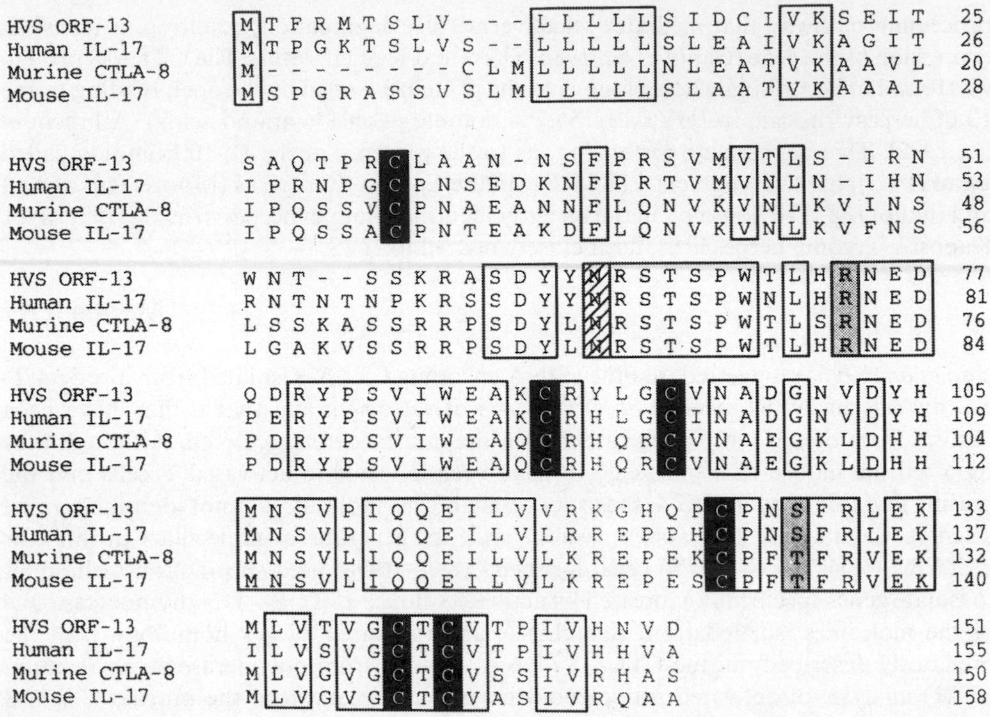

```
HVS ORF-13      M T F R M T S L V - - - L L L L L S I D C I V K S E I T   25
Human IL-17     M T P G K T S L V S - - L L L L L S L E A I V K A G I T   26
Murine CTLA-8   M - - - - - - - - - C L M L L L L L N L E A T V K A A V L   20
Mouse IL-17     M S P G R A S S V S L M L L L L L S L A A T V K A A A I   28

HVS ORF-13      S A Q T P R C L A A N - N S F P R S V M V T L S - I R N   51
Human IL-17     I P R N P G C P N S E D K N F P R T V M V N L N - I H N   53
Murine CTLA-8   I P Q S S V C P N A E A N N F L Q N V K V N L K V I N S   48
Mouse IL-17     I P Q S S A C P N T E A K D F L Q N V K V N L K V F N S   56

HVS ORF-13      W N T - - S S K R A S D Y Y N R S T S P W T L H R N E D   77
Human IL-17     R N T N T N P K R S S D Y Y N R S T S P W N L H R N E D   81
Murine CTLA-8   L S S K A S S R R P S D Y L N R S T S P W T L S R N E D   76
Mouse IL-17     L G A K V S S R R P S D Y L N R S T S P W T L H R N E D   84

HVS ORF-13      Q D R Y P S V I W E A K C R Y L G C V N A D G N V D Y H   105
Human IL-17     P E R Y P S V I W E A K C R H L G C I N A D G N V D Y H   109
Murine CTLA-8   P D R Y P S V I W E A Q C R H Q R C V N A E G K L D H H   104
Mouse IL-17     P D R Y P S V I W E A Q C R H Q R C V N A E G K L D H H   112

HVS ORF-13      M N S V P I Q Q E I L V V R K G H Q P C P N S F R L E K   133
Human IL-17     M N S V P I Q Q E I L V L R R E P P H C P N S F R L E K   137
Murine CTLA-8   M N S V L I Q Q E I L V L K R E P E K C P F T F R V E K   132
Mouse IL-17     M N S V L I Q Q E I L V L K R E P E S C P F T F R V E K   140

HVS ORF-13      M L V T V G C T C V T P I V H N V D   151
Human IL-17     I L V S V G C T C V T P I V H H V A   155
Murine CTLA-8   M L V G V G C T C V S S I V R H A S   150
Mouse IL-17     M L V G V G C T C V A S I V R Q A A   158
```

Fig. 4. Homology between mouse IL-17, HVS open reading frame-13 (ORF-13), human IL-17 and murine CTLA-8 proteins. The homologous residues are within boxes. The conserved cysteines between the homologs are boxed in black. The conserved potential phosphorylation sites are within shaded boxes. The conserved potential *N*-linked glycosylation site is within the hatched box. From Kennedy *et al.* (1996), with permission.

DNA or proteins, possibly acquired by retroposition (as most of the viral counterparts have no introns), suggesting that it perhaps functions as a spontaneous vector for cellular genes (Albrecht *et al.*, 1992). HVS-13 has higher homology with hIL-17 (72%) than with mCTLA-8 (56.7%), and thus may be yet another virus-captured host gene that has provided an as yet unknown evolutionary advantage to the virus.

Northern blot analysis using a CTLA-8 probe on an extended panel of cells revealed that the 1.35-kb transcript could be detected only in two T-cell clones (d10S and d18S which arose from the 10th and 18th subcloning respectively) upon activation with phorbol myristate acetate (PMA) and ionomycin. It could not be identified in resting lymphocytes, the M12.4.1 lymphoma, lipopolysaccharide (LPS)-activated B-cell blasts, concanavalin A-activated T cells, the KB5C20 cytotoxic T-cell line or in several T-cell hybridomas, either constitutively or following activation with PMA and ionomycin (Rouvier *et al.*, 1993). Using *in situ* hybridization, the CTLA-8 gene was mapped to the IA band of the mouse genome, and the human homolog of CTLA-8 to 2q31. Southern blot studies of mouse, rat and human genomic DNA revealed the presence of a single-copy CTLA-8 gene in all three species (Rouvier *et al.*, 1993).

Yao and colleagues (1995a) studied the expression patterns and the biologic activity of both HVS-13 and CTLA-8. When CTLA-8 and HVS-13 were transfected into CV1/EBNA cells, supernatants and lysates from CTLA-8-expressing cells were found to contain two unique proteins with molecular masses of 17 and 20 kDa, and those from HVS-13-expressing cells contained four proteins with apparent masses of 17, 20, 23 and 26 kDa. Treatment of transfectants with tunicamycin, which inhibits the addition of *N*-linked oligosaccharides to proteins, led to augmented expression of the 17-kDa form alone, suggesting that the other proteins contain *N*-linked glycans. Whereas HVS-13 possesses three potential *N*-glycosylation sites, mIL-17 has only a single site.

Human IL-17

A partial hIL-17 sequence was obtained by amplification of fragments of human genomic DNA with degenerate primers from HVS-13 and mIL-17 nucleotide sequences (Yao *et al.*, 1995b). The degenerate PCR-derived partial hIL-17 sequence was then used as a probe to screen a $CD4^+$ T-cell clone λgt10 library, which led to the isolation of a 155-amino-acid peptide bearing an *N*-terminal hydrophobic leader sequence of 19 amino acids. The predicted mass of this protein was 17.5 kDa. Human IL-17 exhibits an overall sequence identity of 72% at the amino acid level with HVS-13 and 63% with mIL-17. Expression in CV1/EBNA cells leads to proteins of 15 and 20 kDa under reducing conditions, with the 20-kDa protein disappearing in the presence of tunicamycin. When immunoprecipitates were analyzed under non-reducing conditions, the expressed proteins had masses of 30 and 38 kDa, suggesting disulfide-linked dimers. Northern analysis of various tissues suggests the constitutive expression of hIL-17 to be quite restricted. Human IL-17 mRNA was detected in PI-activated $CD4^+$ T cells (Fossiez *et al.*, 1996), in a $CD4^+$ T-cell clone (clone 22) and in peripheral blood T cells (Yao *et al.*, 1995b) but not in resting PBMCs, $CD4^+$ T cells, $CD8^+$ T cells, monocytes, B cells, or a variety of cell lines and human tissues tested. The abundant expression of hIL-17 mRNA in activated human peripheral T cells is quite unlike that found with mCTLA-8, which could not be induced by similar stimuli (Rouvier *et al.*, 1993).

Human IL-17, a glycoprotein of 155 amino acids secreted primarily by activated $CD4^+$ T cells (predominantly by the memory cell subset), stimulated stromal cells derived from various tissues to secrete IL-6, IL-8, granulocyte colony stimulating factor (G-CSF) and prostaglandin E_2 (PGE_2), which were blocked by a specific anti-IL-17 mAb. Both tumor necrosis factor-α (TNF-α) and IFN-γ synergized with IL-17 and augmented IL-17-induced secretion of IL-6 by rheumatoid synoviocytes. On the other hand, neither IL-17 nor TNF-α alone had any effect on granulocyte–macrophage colony stimulating factor (GM-CSF) secretion, but together they were capable of inducing GM-CSF production by synovial fibroblasts (Fossiez *et al.*, 1996). In addition to inducing cytokine production, human IL-17.Fc fusion protein and supernatants from cells transfected with hIL-17 enhanced surface expression of intercellular adhesion molecule-1 on human fibroblasts. Addition of hIL-17 to fibroblast cells enhanced their ability to sustain growth of $CD34^+$ hematopoietic progenitors and directed their preferential maturation into neutrophils (Fossiez *et al.*, 1996). The ability of IL-17 to induce IL-6, IL-8 and PGE_2 secretion suggests a proinflammatory role for IL-17 on T cell-dependent inflammatory reactions. The capacity of IL-17 to induce the production of hematopoietic cytokines, such as IL-6, IL-8 and G-CSF, supports an indirect hematopoietic activity for IL-17 as well. Recently Fine *et al.* (1997) examined the mechanism by which IL-17 enhanced G-CSF expression in fibroblasts. Treatment of 3T3 fibroblasts with IL-17 resulted in an increased steady state of mRNA expression and augmented production of G-CSF protein. Addition of LPS and IL-17 to fibroblasts further enhanced the level of G-CSF mRNA and protein. Stability studies revealed that IL-17 stabilizes LPS-induced G-CSF mRNA expression, with a $t_{1/2}$ of 6 h compared with 2 h in cells treated with LPS alone (Fine *et al.*, 1997).

The IL-17 Receptor

Recombinant HVS-13 and mCTLA-8 exhibit cytokine-like activities on a variety of cell types that include: activation of the transcription factor, nuclear factor κB (NF-κB), induction of IL-6 secretion in fibroblasts, and enhancement of T-cell proliferation induced by a suboptimal costimulus (phytohemagglutinin (PHA)). Based on these observations, Yao *et al.* (1995a) proposed the term IL-17 for mCTLA-8 and vIL-17 for HVS-13. The mIL-17R was identified subsequently by searching for proteins that bound to an HVS-13.Fc fusion protein (Yao *et al.*, 1995a). The chimeric protein was constructed by fusing a portion of the Fc region of human IgG_1 with residues 19–151 of HVS-13. Screening of a cDNA expression system for binding proteins led to the isolation of two cDNA clones that specifically bound HVS-13.Fc. The open reading frame of these clones encoded an extremely long type I transmembrane protein of 864 amino acids, which included an *N*-terminal signal peptide with a cleavage site after amino acid 31, followed by a 291-amino-acid extracellular domain, a 21-amino-acid transmembrane domain and a 521-amino-acid cytoplasmic tail. The IL-17R has no sequence similarity with any other known cytokine receptor family (Fig. 5). The protein has a predicted molecular mass of 97.8 kDa and an estimated isoelectric point of 4.85 (Yao *et al.*, 1995a). The receptor has 12 cysteine residues in the extracellular domain with relative positions not characteristic of members belonging to the immunoglobulin superfamily or the TNF receptor family. Despite the presence of relatively large proportions of acidic and proline residues similar to that observed in many growth factor receptors, the large cytoplasmic tail has no sequence homology with any other growth factor receptor known to be a tyrosine kinase. Northern analysis with a probe to mIL-17R revealed expression in virtually all cells and tissues (Yao *et al.*, 1995a). This ubiquitous tissue distribution of IL-17R is in stark contrast with the restricted expression of mIL-17. A soluble IL-17R.Fc fusion protein was found significantly to inhibit IL-2 production and T-cell proliferation induced by concanavalin A, PHA, anti-TCRαβ mAb and anti-CD28 mAb, suggesting a key role for endogenously produced IL-17 in T-cell proliferation (Yao *et al.*, 1995a).

IL-17 in Immunopathologic Conditions

Based on the observations that IL-17 is a proinflammatory cytokine, the present authors along with researchers from the Immunex Corporation (Seattle, WA) hypothesized that blocking the effects of IL-17 using a soluble mIL-17R.Fc fusion protein would modify alloimmune responses. It was found that administration of smIL-17R.Fc fusion protein to murine cardiac allograft recipients led to a significant increase in cardiac allograft survival (Antonysamy *et al.*, 1997). This novel finding that IL-17R.Fc can enhance organ allograft survival indicates a role for IL-17 in alloimmunity and may have potential for therapy of allograft rejection, either alone or in combination with other

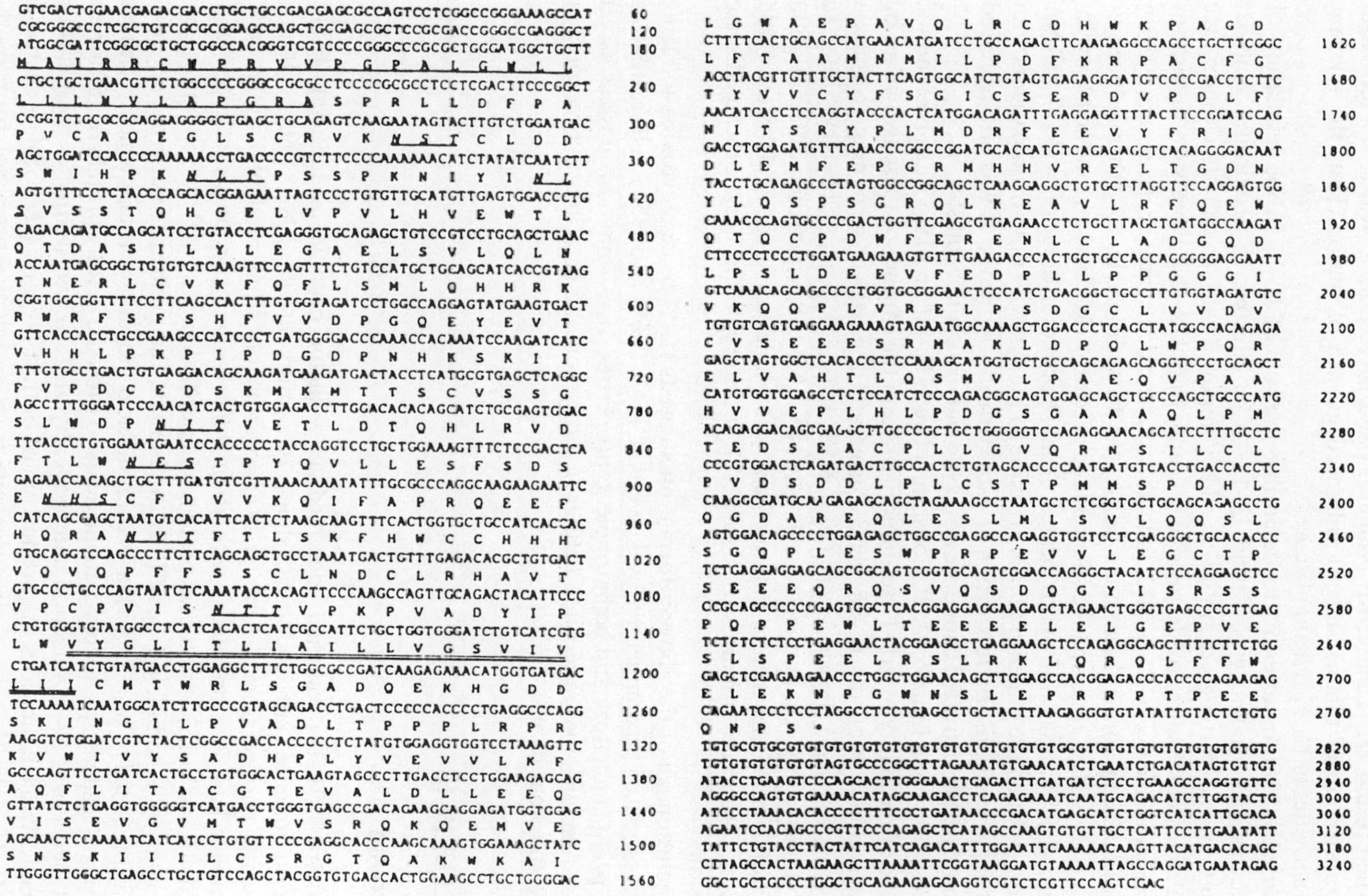

```
GTCGACTGGAACGAGACGACCTGCTGCCGACGAGCGCCAGTCCTCGGCCGGGAAAGCCAT   60
CGCGGGCCCTCGCTGTCGCGCGGAGCCAGCTGCGAGCGCTCCGCGACCGGGCCGAGGGCT   120
ATGGCGATTCGGCGCTGCTGGCCACGGGTCGTCCCCGGGCCCGCGCTGGGATGGCTGCTT   180
M  A  I  R  R  C  W  P  R  V  V  P  G  P  A  L  G  W  L  L
CTGCTGCTGAACGTTCTGGCCCCGGGCCGCGCCTCCCCGCGCCTCCTCGACTTCCCGGCT   240
L  L  L  N  V  L  A  P  G  R  A  S  P  R  L  L  D  F  P  A
CCGGTCTGCGCGCAGGAGGGGCTGAGCTGCAGAGTCAAGAATAGTACTTGTCTGGATGAC   300
P  V  C  A  Q  E  G  L  S  C  R  V  K  N  S  T  C  L  D  D
AGCTGGATCCACCCCAAAAACCTGACCCCGTCTTCCCCAAAAAACATCTATATCAATCTT   360
S  W  I  H  P  K  N  L  T  P  S  S  P  K  N  I  Y  I  N  L
AGTGTTTCCTCTACCCAGCACGGAGAATTAGTCCCTGTGTTGCATGTTGAGTGGACCCTG   420
S  V  S  S  T  Q  H  G  E  L  V  P  V  L  H  V  E  W  T  L
CAGACAGATGCCAGCATCCTGTACCTCGAGGGTGCAGAGCTGTCCGTCCTGCAGCTGAAC   480
Q  T  D  A  S  I  L  Y  L  E  G  A  E  L  S  V  L  Q  L  N
ACCAATGAGCGGCTGTGTGTCAAGTTCCAGTTTCTGTCCATGCTGCAGCATCACCGTAAG   540
T  N  E  R  L  C  V  K  F  Q  F  L  S  M  L  Q  H  H  R  K
CGGTGGCGGTTTTCCTTCAGCCACTTTGTGGTAGATCCTGGCCAGGAGTATGAAGTGACT   600
R  W  R  F  S  F  S  H  F  V  V  D  P  G  Q  E  Y  E  V  T
GTTCACCACCTGCCGAAGCCCATCCCTGATGGGGACCCAAACCACAAATCCAAGATCATC   660
V  H  H  L  P  K  P  I  P  D  G  D  P  N  H  K  S  K  I  I
TTTGTGCCTGACTGTGAGGACAGCAAGATGAAGATGACTACCTCATGCGTGAGCTCAGGC   720
F  V  P  D  C  E  D  S  K  M  K  M  T  T  S  C  V  S  S  G
AGCCTTTGGGATCCCAACATCACTGTGGAGACCTTGGACACACAGCATCTGCGAGTGGAC   780
S  L  W  D  P  N  I  T  V  E  T  L  D  T  Q  H  L  R  V  D
TTCACCCTGTGGAATGAATCCACCCCCTACCAGGTCCTGCTGGAAAGTTTCTCCGACTCA   840
F  T  L  W  N  E  S  T  P  Y  Q  V  L  L  E  S  F  S  D  S
GAGAACCACAGCTGCTTTGATGTCGTTAAACAAATATTTGCGCCCAGGCAAGAAGAATTC   900
E  N  H  S  C  F  D  V  V  K  Q  I  F  A  P  R  Q  E  E  F
CATCAGCGAGCTAATGTCACATTCACTCTAAGCAAGTTTCACTGGTGCTGCCATCACCAC   960
H  Q  R  A  N  V  T  F  T  L  S  K  F  H  W  C  C  H  H  H
GTGCAGGTCCAGCCCTTCTTCAGCAGCTGCCTAAATGACTGTTTGAGACACGCTGTGACT   1020
V  Q  V  Q  P  F  F  S  S  C  L  N  D  C  L  R  H  A  V  T
GTGCCCTGCCCAGTAATCTCAAATACCACAGTTCCCAAGCCAGTTGCAGACTACATTCCC   1080
V  P  C  P  V  I  S  N  T  T  V  P  K  P  V  A  D  Y  I  P
CTGTGGGTGTATGGCCTCATCACACTCATCGCCATTCTGCTGGTGGGATCTGTCATCGTG   1140
L  W  V  Y  G  L  I  T  L  I  A  I  L  L  V  G  S  V  I  V
CTGATCATCTGTATGACCTGGAGGCTTTCTGGCGCCGATCAAGAGAAACATGGTGATGAC   1200
L  I  I  C  M  T  W  R  L  S  G  A  D  Q  E  K  H  G  D  D
TCCAAAATCAATGGCATCTTGCCCGTAGCAGACCTGACTCCCCCACCCCTGAGGCCCAGG   1260
S  K  I  N  G  I  L  P  V  A  D  L  T  P  P  P  L  R  P  R
AAGGTCTGGATCGTCTACTCGGCCGACCACCCCCTCTATGTGGAGGTGGTCCTAAAGTTC   1320
K  V  W  I  V  Y  S  A  D  H  P  L  Y  V  E  V  V  L  K  F
GCCCAGTTCCTGATCACTGCCTGTGGCACTGAAGTAGCCCTTGACCTCCTGGAAGAGCAG   1380
A  Q  F  L  I  T  A  C  G  T  E  V  A  L  D  L  L  E  E  Q
GTTATCTCTGAGGTGGGGGTCATGACCTGGGTGAGCCGACAGAAGCAGGAGATGGTGGAG   1440
V  I  S  E  V  G  V  M  T  W  V  S  R  Q  K  Q  E  M  V  E
AGCAACTCCAAAATCATCATCCTGTGTTCCCGAGGCACCCAAGCAAAGTGGAAAGCTATC   1500
S  N  S  K  I  I  I  L  C  S  R  G  T  Q  A  K  W  K  A  I
TTGGGTTGGGCTGAGCCTGCTGTCCAGCTACGGTGTGACCACTGGAAGCCTGCTGGGGAC   1560
L  G  W  A  E  P  A  V  Q  L  R  C  D  H  W  K  P  A  G  D
CTTTTCACTGCAGCCATGAACATGATCCTGCCAGACTTCAAGAGGCCAGCCTGCTTCGGC   1620
L  F  T  A  A  M  N  M  I  L  P  D  F  K  R  P  A  C  F  G
ACCTACGTTGTTTGCTACTTCAGTGGCATCTGTAGTGAGAGGGATGTCCCCGACCTCTTC   1680
T  Y  V  V  C  Y  F  S  G  I  C  S  E  R  D  V  P  D  L  F
AACATCACCTCCAGGTACCCACTCATGGACAGATTTGAGGAGGTTTACTTCCGGATCCAG   1740
N  I  T  S  R  Y  P  L  M  D  R  F  E  E  V  Y  F  R  I  Q
GACCTGGAGATGTTTGAACCCGGCCGGATGCACCATGTCAGAGAGCTCACAGGGGACAAT   1800
D  L  E  M  F  E  P  G  R  M  H  H  V  R  E  L  T  G  D  N
TACCTGCAGAGCCCTAGTGGCCGGCAGCTCAAGGAGGCTGTGCTTAGGTTCCAGGAGTGG   1860
Y  L  Q  S  P  S  G  R  Q  L  K  E  A  V  L  R  F  Q  E  W
CAAACCCAGTGCCCCGACTGGTTCGAGCGTGAGAACCTCTGCTTAGCTGATGGCCAAGAT   1920
Q  T  Q  C  P  D  W  F  E  R  E  N  L  C  L  A  D  G  Q  D
CTTCCCTCCCTGGATGAAGAAGTGTTTGAAGACCCACTGCTGCCACCAGGGGGAGGAATT   1980
L  P  S  L  D  E  E  V  F  E  D  P  L  L  P  P  G  G  G  I
GTCAAACAGCAGCCCCTGGTGCGGGAACTCCCATCTGACGGCTGCCTTGTGGTAGATGTC   2040
V  K  Q  Q  P  L  V  R  E  L  P  S  D  G  C  L  V  V  D  V
TGTGTCAGTGAGGAAGAAAGTAGAATGGCAAAGCTGGACCCTCAGCTATGGCCACAGAGA   2100
C  V  S  E  E  E  S  R  M  A  K  L  D  P  Q  L  W  P  Q  R
GAGCTAGTGGCTCACACCCTCCAAAGCATGGTGCTGCCAGCAGAGCAGGTCCCTGCAGCT   2160
E  L  V  A  H  T  L  Q  S  M  V  L  P  A  E  Q  V  P  A  A
CATGTGGTGGAGCCTCTCCATCTCCCAGACGGCAGTGGAGCAGCTGCCCAGCTGCCCATG   2220
H  V  V  E  P  L  H  L  P  D  G  S  G  A  A  A  Q  L  P  M
ACAGAGGACAGCGAGGCTTGCCCGCTGCTGGGGGTCCAGAGGAACAGCATCCTTTGCCTC   2280
T  E  D  S  E  A  C  P  L  L  G  V  Q  R  N  S  I  L  C  L
CCCGTGGACTCAGATGACTTGCCACTCTGTAGCACCCCAATGATGTCACCTGACCACCTC   2340
P  V  D  S  D  D  L  P  L  C  S  T  P  M  M  S  P  D  H  L
CAAGGCGATGCAAGAGAGCAGCTAGAAAGCCTAATGCTCTCGGTGCTGCAGCAGAGCCTG   2400
Q  G  D  A  R  E  Q  L  E  S  L  M  L  S  V  L  Q  Q  S  L
AGTGGACAGCCCCTGGAGAGCTGGCCGAGGCCAGAGGTGGTCCTCGAGGGCTGCACACCC   2460
S  G  Q  P  L  E  S  W  P  R  P  E  V  V  L  E  G  C  T  P
TCTGAGGAGGAGCAGCGGCAGTCGGTGCAGTCGGACCAGGGCTACATCTCCAGGAGCTCC   2520
S  E  E  E  Q  R  Q  S  V  Q  S  D  Q  G  Y  I  S  R  S  S
CCGCAGCCCCCCGAGTGGCTCACGGAGGAGGAAGAGCTAGAACTGGGTGAGCCCGTTGAG   2580
P  Q  P  P  E  W  L  T  E  E  E  E  L  E  L  G  E  P  V  E
TCTCTCTCTCCTGAGGAACTACGGAGCCTGAGGAAGCTCCAGAGGCAGCTTTTCTTCTGG   2640
S  L  S  P  E  E  L  R  S  L  R  K  L  Q  R  Q  L  F  F  W
GAGCTCGAGAAGAACCCTGGCTGGAACAGCTTGGAGCCACGGAGACCCACCCCAGAAGAG   2700
E  L  E  K  N  P  G  W  N  S  L  E  P  R  R  P  T  P  E  E
CAGAATCCCTCCTAGGCCTCCTGAGCCTGCTACTTAAGAGGGTGTATATTGTACTCTGTG   2760
Q  N  P  S  •
TGTGCGTGCGTGTGTGTGTGTGTGTGTGTGTGTGTGCGTGTGTGTGTGTGTGTGTG   2820
TGTGTGTGTGTGTAGTGCCCGGCTTAGAAATGTGAACATCTGAATCTGACATAGTGTTGT   2880
ATACCTGAAGTCCCAGCACTTGGGAACTGAGACTTGATGATCTCCTGAAGCCAGGTGTTC   2940
AGGGCCAGTGTGAAAACATAGCAAGACCTCAGAGAAATCAATGCAGACATCTTGGTACTG   3000
ATCCCTAAACACACCCCTTTCCCTGATAACCCGACATGAGCATCTGGTCATCATTGCACA   3060
AGAATCCACAGCCCGTTCCCAGAGCTCATAGCCAAGTGTGTTGCTCATTCCTTGAATATT   3120
TATTCTGTACCTACTATTCATCAGACATTTGGAATTCAAAAACAAGTTACATGACACAGC   3180
CTTAGCCACTAAGAAGCTTAAAATTCGGTAAGGATGTAAAATTAGCCAGGATGAATAGAG   3240
GGCTGCTGCCCTGGCTGCAGAAGAGCAGGTCGTCTCGTTCCAGTCGAC
```

Fig. 5. Nucleotide sequence of mIL-17R cDNA. The nucleotide sequence of a cDNA encoding the IL-17R and the predicted amino acid sequence is illustrated. The signal peptide and the putative transmembrane region are underlined and doubly underlined respectively. The potential *N*-linked glycosylation sites are italicized and underlined. From Yao *et al.* (1995a), with permission.

immunosuppressive agents exhibiting complementary modes of action. Conversely, our unpublished results (H. Tahara, M. Numasaki and M.T. Lotze) with IL-17 retroviral vector transfection of tumors suggest enhanced growth of tumor. Thus the relevant biologic activities of IL-17 remain obscure.

Lotz *et al.* (1996), in an effort to determine the role of IL-17 in the regulation of chondrocytes and synoviocytes, found IL-17 to be effective in promoting synovial inflammation and cartilage degradation. They report IL-17 to inhibit chondrocyte proliferation, induce nitric oxide production and stimulate the expression of various genes (i.e. inducible cyclo-oxygenase, inducible nitric oxide synthase, IL-6, stromelysin and collagenase) in normal human chondrocytes. In addition, IL-17 either induced or augmented IL-6, IL-1, macrophage chemotactic protein-1 (MCP-1) and inducible cyclo-oxygenase gene expression in synoviocytes. Analysis of joint tissues for IL-17 mRNA expression revealed the presence of this cytokine in the synovial fluid from patients with rheumatoid arthritis, psoriatic arthritis or osteoarthritis. However, it was not expressed by unstimulated or activated chondrocytes, synovial or skin fibroblasts.

The possibility of a functional relationship between IL-8R and IL-17 has been hinted at because of the ability of hIL-17 to induce IL-8 and the capture of these two genes by HVS. In addition, both the IL-8R gene cluster and the hIL-17 gene map to human chromosome 2q31–q35 (Fossiez *et al.*, 1996). HVS-13 (vIL-17) secreted by HVS-infected cells may bind to adjacent uninfected cells expressing IL-17R, enhancing differentiation and/or proliferation, and thereby making them susceptible targets for viral infection and replication. Therefore, abnormalities in IL-17 secretory levels may indicate disease progression.

In conclusion, all of the above information suggests IL-17 to be a cytokine-inducing cytokine, produced predominantly by activated $CD4^+$ T cells and DN T cells exhibiting indirect proinflammatory and hematopoietic properties. IL-17 may also possess as yet unknown functions that necessitate further investigation and creation of a receptor or cytokine knockout mouse.

INTERLEUKIN-18

Background

Interleukin-18 (IL-18) was first identified as a factor capable of inducing IFN-γ production in mice with endotoxic shock (Okamura *et al.*, 1982, 1995a). It was subsequently purified from the mouse liver (presumably from resident Kupffer cells) after animals had been pretreated with *Propionibacterium acnes* and challenged with LPS. This 18.3-kDa 157-amino-acid protein was purified and the cDNA from mouse, rat and human (65% homology) cloned. In the human, the molecular mass of the entire protein is predicted to be a 22.3-kDa 192-amino-acid precursor protein which is cleaved of an unusual leader sequence of 35 amino acids to form the mature protein of 157 amino acids. It increases IFN-γ production independently of IL-12 and stimulates proliferation of T_{H1} cells. While it increases the production of both IFN-γ and GM-CSF, it decreases that of IL-10. Like IL-1, it requires cleavage from an inactive precursor form by the enzyme caspase-1 and appears to be a better substrate for it than IL-1. It is likely that IL-18 plays an important role in the initiation of the immune response and may be responsible, in part, for the pathogenesis of septic shock and tissue injury occurring during inflammation. Information available about this cytokine is quite limited and, although cloning of the receptor and production of animals with targeted deletion have been completed, information is not yet available in the published database.

Cloning and Purification

Mice treated with *P. acnes* and then subsequently challenged with LPS develop septic shock and very high serum levels of IFN-γ. The livers of such animals were utilized to extract the protein, IFN-γ-inducing factor (IGIF), from which tryptic digests and partial sequences were obtained. Degenerate oligonucleotide primers were utilized to generate (by reverse transcriptase (RT)–PCR) partial cDNA fragments from mRNA extracted from the liver. Subsequently these fragments were used to probe a liver cDNA library and a previously unrecognized full-length clone was obtained which encoded a sequence of 192 amino acids (Fig. 6B) (Okamura *et al.*, 1995a). This protein was found to have approximately 12% homology with IL-1α and IL-1β, leading to the suggestion that it be termed IL-1γ (Bazan *et al.*, 1996). This also included the presence of a leader of 35 amino acids, consistent with that associated with the precursor form of IL-1β and, when cleaved, predicted the release of a 157-amino-acid active cytokine. Direct identification for a role of caspase-1 in the cleavage of such a precursor was subsequently identified (Ghayur *et al.*, 1997; Gu *et al.*, 1997). The IL-1β-converting enzyme (ICE), which processes the inactive IL-1β precursor to the proinflammatory cytokine, was also shown to cleave the IL-18 precursor at the authentic processing site with high efficiency, thereby activating IL-18 and facilitating its export. Kupffer cells from LPS-activated ICE-deficient (ICE$^{-/-}$) mice synthesized the IL-18 precursor, but were unable to process it into the bioactive form. The serum of ICE$^{-/-}$ mice exposed to *P. acnes* and LPS also had decreased levels of IFN-γ and IL-18. Subsequently, both human (Fig. 6A) (Torigoe *et al.*, 1996; Ushio *et al.*, 1996) and rat (Fig. 6C) (Conti *et al.*, 1997) IL-18 molecules were cloned and expressed.

Production of IL-18

As described above, IL-18 was cloned initially from the liver in association with a response to pathogens. It was presumed that resident Kupffer cells were the source of IL-18. More recently, in the process of cloning of rat IL-18, it was shown that IL-18 was produced by the adrenal gland as well (Conti *et al.*, 1997). Conti *et al.* (1997) examined the adrenal gland of reserpine-treated rats with differential display techniques using RT–PCR, and two transcripts were isolated. They were identified as rat IL-18 on the basis of the high homology with mouse IL-18: 91% at both the nucleotide and the amino acid level. Subsequently, the effects of stress on IL-18 mRNA levels were examined and it was found that acute cold stress strongly induced IL-18 gene expression. *In situ* hybridization analysis showed that IL-18 was synthesized in the adrenal cortex, specifically in the zona reticularis and fasciculata that produce glucocorticoids. IL-18 transcripts were also detected in the neurohypophysis, although induction by stress was not significant.

Recent results have shown IL-18 production by other cell types as well. IL-18 was detected in keratinocytes of the skin, osteoblasts and epithelial cells of small intestine by immunohistochemistry using anti-IL-18 antibodies. Northern blot and RT–PCR

A. Human liver RNA and translation for interferon-γ inducing factor (IGIF)

MAAEPVEDNCINFVAMKFIDNTLYFIAEDDENLESDYFGKLESKLSVIRNLNDQVLFIDQ
GNRPLFEDMTDSDCRDNAPRTIFIISMYKDSQPRGMAVTISVKCEKISTLSCENKIISFKE
MNPPDNIKDTKSDIIFFQRSVPGHDNKMQFESSSYEGYFLACEKERDLFKLILKKEDELG
DRSIMFTVQNED

polyA_signal 1062..1080; polyA_site 1102; BASE COUNT 360 a 228 c 231 g 283 t
ORIGIN

1 gcctggacag tcagcaagga attgtctccc agtgcatttt gccctcctgg ctgccaactc
61 tggctgctaa agcggctgcc acctgctgca gtctacacag cttcgggaag aggaaaggaa
121 cctcagacct tccagatcgc ttcctctcgc aacaaactat ttgtcgcagg aataaagatg
181 gctgctgaac cagtagaaga caattgcatc aactttgtgg caatgaaatt tattgacaat
241 acgctttact ttatagctga agatgatgaa aacctggaat cagattactt tggcaagctt
301 gaatctaaat tatcagtcat aagaaatttg aatgaccaag ttctcttcat tgaccaagga
361 aatcggcctc tatttgaaga tatgactgat tctgactgta gagataatgc accccggacc
421 atatttatta taagtatgta taaagatagc cagcctagag gtatggctgt aactatctct
481 gtgaagtgtg agaaaatttc aactctctcc tgtgagaaca aaattatttc ctttaaggaa
541 atgaatcctc ctgataacat caaggataca aaaagtgaca tcatattctt tcagagaagt
601 gtcccaggac atgataataa gatgcaattt gaatcttcat catacgaagg atactttcta
661 gcttgtgaaa aagagagaga cctttttaaa ctcattttga aaaaagagga tgaattgggg
721 gatagatcta taatgttcac tgttcaaaac gaagactagc tattaaaatt tcatgccggg
781 cgcagtggct cacgcctgta atcccagccc tttgggaggc tgaggcgggc agatcaccag
841 aggtcaggtg ttcaagacca gcctgaccaa catggtgaaa cctcatctct actaaaaata
901 ctaaaaatta gctgagtgta gtgacgcatg ccctcaatcc cagctactca agaggctgag
961 gcaggagaat cacttgcact ccggaggtag aggttgtggt gagccgagat tgcaccattg
1021 cgctctagcc tgggcaacaa cagcaaaact ccatctcaaa aaataaaata aataaataaa
1081 caaataaaaa attcataatg tg

Fig. 6. Amino acid sequences of human (A), mouse (B) and rat (C) IL-18. The initial sequence of murine IL-18 was useful in the subsequent identification of both mouse and rat IL-18. Although significant homology exists at the genetic and protein levels, human IL-18 does not act on murine cells.

analyses have revealed the presence of IL-18 mRNA in the pancreas, kidney, skeletal muscle, liver, lungs and unstimulated PBMCs of humans (H. Okamura, personal communication; Table 1). It is quite possible there are many more cell types that elaborate IL-18 and this needs further investigation. As IL-18 exists as both precursor and mature forms, as described above, and requires post-translational regulation via the ICE mechanism, detection of mRNA does not always indicate that biologically active IL-18 is expressed by the cell types examined. Careful investigation, taking account of the activation cascade of IL-18, should be pursued.

The IL-18 Receptor

It is quite clear that the receptors for IL-18 are distinct from those shared by IL-1 (Akita *et al.*, 1997). However, no published information currently exists about the nature and distribution of the IL-18 receptor. Based on the limited information available, the IL-18 receptor appears to be expressed only on T cells, B cells and natural killer (NK) cells.

B. Mouse liver mRNA and translation for interferon-γ inducing factor (IGIF)

MAAMSEDSCVNFKEMMFIDNTLYFIPEENGDLESDNFGRLHCTTAVIRNINDQVLFVDK
RQPVFEDMTDIDQSASEPQTRLIIYMYKDSEVRGLAVTLSVKDSKMSTLSCKNKIISFEE
MDPPENIDDIQSDLIFFQKRVPGHNKMEFESSLYEGHFLACQKEDDAFKLILKKKDENG
DKSVMFTLTNLHQS

polyA_site 866 BASE COUNT 262 a 187 c 187 g 230 t

ORIGIN
1 ggcacagctg gacctggtgg gggttctctg tggttccatg ctttctggac tcctgcctgc
61 tggctggagc tgctgacagg cctgacatct tctgcaacct ccagcatcag gacaaagaaa
121 gccgcctcaa accttccaaa tcacttcctc ttggcccagg aacaatggct gccatgtcag
181 aagactcttg cgtcaacttc aaggaaatga tgtttattga caacacgctt tactttatac
241 ctgaagaaaa tggagacctg gaatcagaca actttggccg acttcactgt acaaccgcag
301 taatacggaa tataaatgac caagttctct tcgttgacaa aagacagcct gtgttcgagg
361 atatgactga tattgatcaa agtgccagtg aaccccagac cagactgata atatacatgt
421 acaaagacag tgaagtaaga ggactggctg tgaccctctc tgtgaaggat agtaaaatgt
481 ctaccctctc ctgtaagaac aagatcattt cctttgagga aatggatcca cctgaaaata
541 ttgatgatat acaaagtgat ctcatattct ttcagaaacg tgttccagga cacaacaaga
601 tggagtttga atcttcactg tatgaaggac actttcttgc ttgccaaaag gaagatgatg
661 ctttcaaact cattctgaaa aaaaaggatg aaaatgggga taaatctgta atgttcactc
721 tcactaactt acatcaaagt taggtgggga gggtttgtgt tccagaaaga tgattagcac
781 acatgcgcct tgtgatgacc tcgcctgtat ttccataaca gaatacccga ggctgcatga
841 tttatagagt aaacacgttt atttgt

C. Rat adrenal mRNA and translation for interferon-γ inducing factor (IGIF)

MAAMSEEGSCVNFKEMMFIDNTLYLIPEDNGDLESDHFGRLHCTTAVIRSINDQVLFVD
KRNPPVFEDMPDIDRTANESQTRLIIYMYKDSEVRGLAVTLSVKDGRMSTLSCKNKIISF
EEMNPPENIDDIKSDLIFFQKRVPGHNKMEFESSLYEGHFLACQKEDDAFKLVLKRKDE
NGDKSVMFTLTNLHQS

BASE COUNT 252 a 138 c 140 g 192 t
ORIGIN
1 atggctgcca tgtcagaaga aggctcttgt gtcaacttca aagaaatgat gtttattgac
61 aacacacttt accttatacc tgaagataat ggagacttgg aatcagacca ctttggcaga
121 cttcactgta caaccgcagt aatacggagc ataaatgacc aagttctctt cgttgacaaa
181 agaaacccgc ctgtgttcga ggacatgcct gatatcgacc gaacagccaa cgaatcccag
241 accagactga taatatatat gtacaaagat agtgaagtaa gaggactggc tgtgacccta
301 tctgtgaagg atggaaggat gtctaccctc tcctgtaaaa acaaaatcat ttcctttgag
361 gaaatgaatc cacctgaaaa tattgatgat ataaaaagtg atctcatatt ctttcagaaa
421 cgtgtgccag gacacaacaa aatggaattt gaatcttccc tgtatgaagg acactttcta
481 gcttgccaaa aggaagatga tgctttcaaa ctcgttttga aaaggaagga tgaaaatggg
541 gataaatctg taatgttcac tcttactaac ttacatcaaa gttaggtatt aaggtttctg
601 tattccagaa agacgattag tatacacgag ccttatgata acctactctg tatttctatg
661 acaaaatacc tgaggccgca tgatttatag agtaaacaag cttgattgcc caaaaaaaaa
721 aa

Fig. 6. (*continued*).

Table 1. IL-18 mRNA or protein detection in cell lines and tissues.

Tissue or cell type	Detection of mRNA	Detection of protein	References
Macrophage	+ (h, m)	+ (h, m)	Okamura *et al.* (1995a)
Adrenal	+ (rat)	n.d.	Conti *et al.* (1997)
Keratinocytes (skin)	+ (m)	+ (h, m)	Stoll *et al.* (1997) and p.c.
Epithelial cells (small intestine)	n.d.	+ (h)	p.c.
Pancreas	+ (h)	n.d.	p.c.
Skeletal muscle	+ (h)	n.d.	p.c.
Liver	+ (h)	n.d.	p.c.
Lung	+ (h)	n.d.	p.c.
PBMC (unstimulated)	+ (h)	n.d.	p.c.

h, Human; m, mouse; n.d., not done; p.c., personal communication.

Biologic Functions of IL-18

IFN-γ Induction

As described above, IL-18 was initially designated as IGIF, because it was cloned as a cytokine that strongly induced IFN-γ production (Okamura *et al.*, 1995b). Okamura and coworkers initiated the study of this molecule when they noticed that large amounts of IFN-γ were secreted into the blood of mice following sequential treatment with *P. acnes* and LPS (Okamura *et al.*, 1982). Although the molecular weight of this IGIF (70–80 kDa) was found to be similar to that of IL-12, it appeared to be functionally distinct from IL-12 (Trinchieri and Scott, 1994; Gavett *et al.*, 1995) because of its ability to induce much higher (100-fold) levels of IFN-γ.

Recombinant mIL-18 and mIL-12 induce IFN-γ production by T cells and NK cells in a dose-dependent and synergistic manner (Micallef *et al.*, 1996; Kohno *et al.*, 1997). The levels of IFN-γ induced by IL-18 are much higher than those induced by IL-12. Moreover, addition of a neutralizing anti-mIL-18 antibody did not suppress IFN-γ production induced by IL-12 and, likewise, a neutralizing antibody to IL-12 did not inhibit IFN-γ induced by mIL-18. In addition, IL-18 was found to be effective in inducing IFN-γ production in IL-12 knockout mice. These results suggest that these two cytokines induce IFN-γ production via independent pathways (Magram *et al.*, 1996, and personal communication). IL-18 knockout mice have now been created and are being backcrossed with the IL-12 knockout mice.

Induction of IFN-γ is also observed when recombinant (r) mIL-18 is administered to mice *in vivo*. Although the presumed half-life of rmIL-18 is 8–10 h (determined by enzyme-linked immunosorbent assay (ELISA) (H. Okamura, personal communication), which is a bit longer than that of rmIL-12 (6–8 h) in mice, the level of IFN-γ induction by IL-18 is lower than that of IL-12 (Osaki *et al.*, 1998). This relatively weak potency in IFN-γ induction by IL-18 *in vivo* has not been well explained. A synergistic effect of IL-12 and IL-18 has also been observed *in vivo*. Mice that received 7 daily injections of 1.0 μg of IL-18 and 0.1 μg of IL-12 after inoculation of MCA205 tumor cells had raised serum IFN-γ levels about 300 times higher than observed with IL-18 alone (Osaki *et al.*, 1998).

Effects on T Lymphocytes and NK Cells

Mouse IL-18 stimulates the proliferation of T cells and enhances the killing activity of NK cells from mouse spleen, but the synergistic effect with IL-2 is minimal. When anti-CD3 mAb-stimulated human enriched T cells were exposed to IL-18, the cytokine enhanced cell proliferation in a dose-dependent fashion. This could be inhibited completely by a neutralizing antibody against IL-2 at lower concentrations of IL-18. However, at higher concentrations of IL-18 a neutralizing antibody against IL-2 had only insignificant inhibitory effects on T-cell proliferation. ELISAs revealed that, like PBMCs, T cells exposed to IL-18 produced large amounts of IFN-γ; however, changes in the production of IL-4 and IL-10 were minimal (Kohno *et al.*, 1997). Conversly, IL-12 clearly suppresses T_{H2} cytokines (Gavett *et al.*, 1995). IL-18, but not IL-12, significantly enhances IL-2 and GM-CSF production in T-cell cultures, as determined by CTLL-2 bioassay and ELISA respectively; however, both IL-18 and IL-12 enhance IFN-γ production by T cells (Micallef *et al.*, 1996). When T cells were exposed to a combination of IL-18 and IL-12, a synergistic effect was observed on the production of IFN-γ, but not on production of IL-2 and GM-CSF. Together with IL-12, IL-18 may also induce IFN-γ from B cells (Yoshimoto *et al.*, 1997). In conclusion, IL-18 enhances T-cell proliferation apparently through an IL-2-dependent pathway. It enhances T_{H1} cytokine production *in vitro*, and exhibits synergism when combined with IL-12 in terms of enhanced IFN-γ production but not that of IL-2 and GM-CSF (Ushio *et al.*, 1996).

IL-18 also enhances the expression of Fas ligand (FasL) on NK and T_{H1} cells, and induces apoptosis of these cells in mice (Dao *et al.*, 1996; Tsutsui *et al.*, 1996). FasL, expressed on activated T cells, plays a central role in regulating the immune response by inducing apoptosis in activated lymphocytes through binding to its receptor, Fas. IL-18 selectively enhances the FasL-mediated cytotoxicity of cloned murine T_{H1} cells, but not that of T_{H0} or T_{H2} cells. Anti-IFN-γ antibody did not block the IL-18-induced cytotoxicity of T_{H1} cells, and recombinant IFN-α, IFN-γ or TNF-α protein failed to augment the cytotoxic activity of these cells. These results indicate that this enhanced cytotoxicity of T_{H1} cells was mediated by IL-18. In the same study (Dao *et al.*, 1996), IL-12 was also found to enhance the FasL-mediated cytotoxicity of T_{H1} cells. These findings suggest that T_{H1} cells possess receptors for both cytokines, but their FasL expression is induced by these cytokines through different pathways. The results also suggest that IL-18 might play a potential role in immunoregulation or in inflammation by augmenting the functional activity of FasL on T_{H1} cells.

Tsutsui *et al.* (1996) investigated the mechanism regulating the killing apparatus of hepatic NK cells. They established IL-2-dependent NK cell clones from liver lymphocytes of BALB/c nude mice. To generate these NK cell clones, they incubated liver lymphocytes with high doses of IL-2 in the presence of irradiated Kupffer cells. Kupffer cells were added as feeder cells expressing IL-12, originally identified as NK cell stimulatory factor. Unless liver lymphocytes were incubated with both IL-2 and Kupffer cells, no cell growth was observed. The surface phenotype of cloned hepatic NK cells was IL-2Rβ^+ CD16$^+$ CD3$^-$ IgM$^-$ and NK2.1$^-$. These NK cell clones contained dense granules reactive with mAb against perforin, but exerted no conventional cytolytic activity against YAC-1. They constitutively expressed FasL and specifically killed Fas-positive target cells, with subsequent evidence of fragmented DNA. This Fas–FasL-mediated killing activity was enhanced by IL-18, produced by activated Kupffer cells. Other Kupffer cell-produced cytokines, including IL-12, IL-1β

and TNF-α did not enhance Fas–FasL-mediated killing. These findings suggest that the Fas–FasL system is an effector mechanism of hepatic NK cells, which is induced by IL-18. Tsutsui *et al.* (1997) have recently shown that administration of anti-IL-18 to mice with endotoxin-induced liver injury resulted in total abrogation of the disease, whereas administration of anti-TNF-α markedly, although incompletely, suppressed liver injury. Data suggest that IL-18 activates both TNF-α- and FasL-mediated hepatocytotoxic pathways in LPS-induced liver injury.

IL-18 in Antimicrobial Immune Responses

As IFN-γ induction appears to be involved as a consequence of stimulation with microbial products, it is presumed that IL-18 is involved in antimicrobial immune responses. Although there are no reports available to date examining these effects, extensive investigations are now in progress.

Antitumor Effects

As IL-18 has multiple immunoregulatory functions, some of which appear to promote antitumor immune responses, the antitumor effects of IL-18 have been examined. Micallef *et al.* (1997a,b) recently evaluated the antitumor effects of this cytokine. In their study, IL-18 exhibited significant antitumor effects in BALB/c mice challenged intraperitoneally with syngeneic Meth A sarcoma when the cytokine was administered on days 1, 2 and 3 after the tumor challenge. Intravenous administration also induced antitumor effects; however, subcutaneous administration did not. When pretreated twice with 1 mg of IL-18, 3 days and 6 h before tumor challenge, all mice survived whereas control mice died within 3 weeks of challenge. Inhibitory effects on Meth A cell growth *in vitro* were not observed with administration of either IL-18 or IFN-γ. The effects of IL-18 pretreatment were abrogated by abolition of NK cell activity when mice were given anti-asialo GM1 antibody 48 h before as well as 24 and 72 h after tumor challenge. Mice that rejected the tumor with IL-18 pretreatment also rejected a subsequent challenge with the Meth A tumor, but not the Ehrlich ascites carcinoma. Splenocytes were obtained from mice that survived the Meth A challenge following IL-18 treatment, restimulated with mitomycin C-treated Meth A cells for 5 days, and tested for cytotoxic activity against Meth A cells *in vitro*. These effector cell preparations from resistant mice had raised cytotoxic activity when compared with that of naive animals, and appeared to be $CD4^+$ cells, as cytolytic activity was significantly inhibited after depletion of this subset by mAbs and complement. These results suggest that IL-18 is associated with significant antitumor effects in mice challenged with the syngeneic Meth A sarcoma. These antitumor effects appear to be mediated by the immunomodulatory functions of IL-18 primarily through NK cell functions, but expansion of cytotoxic $CD4^+$ cells and generation of immunologic memory were also observed.

Others (Hanaya *et al.*, 1996; Sanchez-Bueno *et al.*, 1996; Tan *et al.*, 1996) have also observed antitumor activity of this cytokine in murine models. We have also independently observed significant antitumor effects of IL-18 in our systems (Osaki *et al.*, 1998). The potency of the IL-18 antitumor effects was somewhat inferior to that of IL-12, but appears to be superior to that of IFN-γ. Administration of IL-12 with IL-18 was associated with more potent antitumor effects than that of IL-12 or IL-18 alone. Whether the antitumor activity of IL-18, like IL-12 (Sgadari *et al.*, 1996), also involves effects on angiogenesis is under investigation, as are other putative mechanisms.

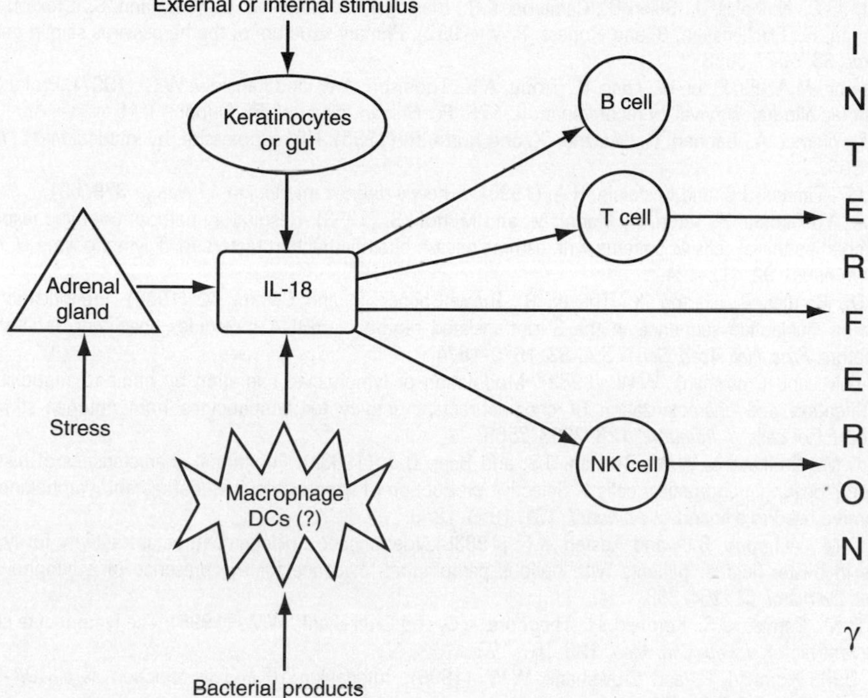

Fig. 7. 'Danger signals': a central role for IL-18 in the initiation of the immune response. A variety of stresses including cold, bacterial challenge and stimuli to the skin and gut provoke IL-18 production. IL-18 is considered to be important in the initial response to a variety of inflammatory stimuli. DCs, dendritic cells.

In conclusion, IL-18 clearly has multiple functions which appear to be important in regulating immune reactions. Factors influencing its production and IFN-γ-inducing capacity are depicted in Fig. 7. Further investigation of its immunologic functions may lead to a better understanding of its role in the cytokine network, in the regulation of immune responses, and its potential application in the regulation of resistance to infection or cancer.

ACKNOWLEDGEMENTS

The authors thank Dr David M. Center, Boston University School of Medicine, for up-to-date information on IL-16, Dr Martin Lotz, The Scripps Research Institute, for sharing his recent observations on IL-17, Dr Muneo Numasaki for helpful discussions on IL-17 and Drs Haruki Okamura and Tadashi Osaki for the helpful discussions on IL-18.

REFERENCES

Akita, K., Ushio, S., Ohtsuki, T., Tanimoto, T., Ikeda, M. and Kurimoto, M. (1997). Comparison between the biological and biochemical aspects of IL-18 (IFN-gamma-inducing factor) and IL-1 beta. *Proceedings of the American Association for Cancer Research* **38**, A2389.

Albrecht, J.-C., Nicholas, J., Biller, D., Cameron, K.R., Biesinger, B., Newman, C., Wittmann, S., Craxton, M.A., Coleman, H., Fleckenstein, B. and Honess, R.W. (1992). Primary structure of the herpesvirus saimiri genome. *J. Virol.* **66**, 5047–5058.

Antonysamy, M.A., Fu, F., Li, W., Qian, S., Troutt, A.B., Thomson, A.W. and Fanslow, W.C. (1997). Prolongation of cardiac allograft survival by recombinant IL-17R:Fc. *Human Immunol.* **55** (suppl. 1), 15.

Baier, M., Werner, A., Bannert, N., Metzner, K. and Kurth, R. (1995). HIV suppression by interleukin-16. *Nature* **378**, 563.

Bazan, J.F., Timans, J.C. and Kastelein, R.A. (1996). A newly defined interleukin-1? *Nature* **379**, 591.

Bellini, A., Yoshimura, H., Vitori, E., Marini, M. and Mattoli, S. (1993). Respiratory pathophysiologic responses. Bronchial epithelial cells of patients with asthma release chemoattractant factors for T lymphocytes. *J. Allergy Clin. Immunol.* **92**, 412–424.

Caput, D., Beutler, B., Hartog, K., Thayer, R., BrownShimer, S. and Cerami, A. (1986). Identification of a common nucleotide sequence in the 3′-untranslated region of mRNA molecules specifying inflammatory mediators. *Proc. Natl Acad. Sci. U.S.A.* **83**, 1670–1674.

Center, D.M. and Cruikshank, W.W. (1982). Modulation of lymphocyte migration by human lymphokines. 1. Identification and characterization of chemoattractant activity for lymphocytes from mitogen-stimulated mononuclear cells. *J. Immunol.* **128**, 2563–2568.

Center, D.M., Cruikshank, W.W., Berman, J.S. and Beer, D.J. (1983a). Functional characteristics of histamine receptor-bearing mononuclear cells 1. Selective production of lymphocyte chemoattractant lymphokines with histamine used as a ligand. *J. Immunol.* **131**, 1854–1859.

Center, D.M., Wintroub, B.U. and Austen, K.F. (1983b). Identification of chemoattractant activity for lymphocytes in blister fluid of patients with bullous pemphigoid: evidence for the presence of a lymphokine. *J. Invest. Dermatol.* **81**, 204–208.

Center, D.M., Berman, J.S., Kornfeld, H., Theodore, A.C. and Cruikshank, W.W. (1995). The lymphocyte chemoattractant factor. *J. Lab. Clin. Med.* **125**, 167–172.

Center, D.M., Kornfeld, H. and Cruikshank, W.W. (1996). Interleukin 16 and its function as a CD4 ligand. *Immunol. Today* **17**, 476–481.

Chupp, G., Zhang, Y., Wright, E., Cruikshank, W.W., Kornfeld, H., Center, D.M. and Berman, J. (1997). Pro-IL-16 is an 80 kDa cytoplasmic protein expressed in blood T-lymphocytes. *J. Allergy Clin. Immunol.* **99**, 224.

Conti, B., Jahng, J.W., Tinti, C., Son, J.H. and Joh, T.H. (1997). Induction of interferon-gamma inducing factor in the adrenal cortex. *J. Biol. Chem.* **272**, 2035–2037.

Cruikshank, W.W. and Center, D.M. (1982). Modulation of lymphocyte migration by human lymphokines. II. Purification of a lymphotactic factor (LCF). *J. Immunol.* **128**, 2569–2571.

Cruikshank, W.W., Greenstein, J.L., Theodore, A.C. and Center, D.M. (1991). Lymphocyte chemoattractant factor induces CD4-dependent intracytoplasmic signaling in lymphocytes. *J. Immunol.* **146**, 2928–2934.

Cruikshank, W.W., Center, D.M., Nisar, N., Natke, B., Theodore, A. and Kornfeld, H. (1994). Molecular and functional analysis of a lymphocyte chemoattractant factor: association of biologic function with CD4 expression. *Proc. Natl Acad. Sci. U.S.A.* **91**, 5109–5113.

Cruikshank, W.W., Long, A., Tarpy, R.E., Kornfeld, H., Carroll, M.P., Teran, L., Holgate, S.T. and Center, D.M. (1995). Early identification of interleukin-16 (lymphocyte chemoattractant factor) and macrophage inflammatory protein 1α (MIP1α) in bronchoalveolar lavage fluid of antigen-challenged asthmatics. *Am. J. Respir. Cell Mol. Biol.* **13**, 738–747.

Cruikshank, W.W., Sciaky, D., Brazer, W.F., Center, D.M. and Smith, T.J. (1997). Primary human fibroblasts express IL-16 mRNA and bioactivity in culture. *J. Allergy Clin. Immunol.* **99**, 221.

Dao, T., Ohashi, K., Kayano, T., Kurimoto, M. and Okamura, H. (1996). Interferon-gamma-inducing factor, a novel cytokine, enhances Fas ligand-mediated cytotoxicity of murine T helper 1 cells. *Cell. Immunol.* **173**, 230–235.

Fine, J.S., Gommoll, C.G., Justice, L. and Cai, X-Y. (1997). Regulation of G-CSF gene expression by IL-17 (CTLA-8). *J. Allergy Clin. Immunol.* **99**, 225.

Fossiez, F., Djossou, O., Chomarat, P., Flores-Romo, L., Ait-Yahia, S., Maat, C., Pin, J.-J., Garrone, P., Garcia, E., Saeland, S., Blanchard, D., Gaillard, C., Mahapatra, B.D., Rouvier, E., Golstein, P., Banchereau, J. and Lebecque, S. (1996). T cell interleukin-17 induces stromal cells to produce proinflammatory and hematopoietic cytokines. *J. Exp. Med.* **183**, 2593–2603.

Gavett, S.H., O'Hearn, D.J., Li, X., Huang, S.K., Finkelman, F.D. and Wills-Karp, M. (1995). Interleukin 12 inhibits antigen-induced airway hyperresponsiveness, inflammation, and Th2 cytokine expression in mice. *J. Exp. Med.* **182**, 1527–1536.

Ghayur, T., Banerjee, S., Hugunin, M., Butler, D., Herzog, L., Carter, A., Quintal, L., Sekut, L., Talanian, R., Paskind, M., Wong, W., Kamen, R., Tracey, D. and Allen, H. (1997). Caspase-1 processes IFN-gamma-inducing factor and regulates LPS-induced IFN-gamma production. *Nature* **386**, 619–623.

Gu, Y., Kuida, K., Tsutsui, H., Ku, G., Hsiao, K., Fleming, M.A., Hayashi, N., Higashino, K., Okamura, H., Nakanishi, K., Kurimoto, M., Tanimoto, T., Flavell, R.A., Sato, V., Harding, M.W., Livingston, D.J. and Su, M.S. (1997). Activation of interferon-γ-inducing factor mediated by interleukin-1 converting enzyme. *Science* **275**, 206–208.

Hanaya, T., Kawai, S., Arai, S., Micallef, M., Tanimoto, T., Ikeda, M., Okamura, H. and Kurimoto, M. (1996). Antitumor effect of a new cytokine, IGIF on the metastasis and growth of murine colon 26 adenocarcinoma. *Proc. Am. Assoc. Cancer Res.* **37**, A3081.

Hessel, E.M., Cruikshank, W.W., Van Ark, I., De Bie, J.J., Van Esch, B., Hofman, G, Nijkamp, F.P., Center, D.M. and Van Oosterhout, A.J.M. (1997). IL-16 is important for IgE synthesis and airway hyperresponsiveness in a mouse model for allergic asthma. *J. Allergy Clin. Immunol.* **99**, 1082.

Kennedy, J., Rossi, D.L., Zurawski, S.M., Vega, F. Jr., Kastelein, R.A., Wagner, J.L., Hannum, C.H. and Zlotnik, A. (1996). Mouse IL-17: a cytokine preferentially expressed by $\alpha\beta TCR^+CD4^-CD8^-$ T cells. *J. Interferon Cytokine Res.* **16**, 611–617.

Kohno, K., Kataoka, J., Ohtsuki, T., Suemoto, Y., Okamoto, I., Usui, M., Ikeda, M. and Kurimoto, M. (1997). IFN-inducing factor (IGIF) is a costimulatory factor on the activation of Th1 but not Th2 cells and exerts its effect independently of IL-12. *J. Immunol.* **158**, 1541–1550.

Laberge, S., Cruikshank, W.W., Kornfeld, H. and Center, D.M. (1995). Histamine-induced secretion of lymphocyte chemoattractant factor from $CD8^+$ T cells is independent of transcription and translation. Evidence for constitutive protein synthesis and storage. *J. Immunol.* **155**, 2902–2910.

Laberge, S., Cruikshank, W.W., Beer, D.J. and Center, D.M. (1996). Secretion of IL-16 (lymphocyte chemoattractant factor) from serotonin-stimulated $CD8^+$ T cells *in vitro*. *J. Immunol.* **156**, 310–315.

Laberge, S., Ernst, P., Ghaffer, O., Cruikshank, W.W., Kornfeld, H., Center, D.M. and Hamid, Q. (1997a). Increased expression of interleukin-16 in bronchial mucosa of subjects with atopic asthma. *Am. J. Respir. Cell Mol. Biol.* **17**, 193–202.

Laberge, S., Durham, S.R., Ghaffar, O., Rak, S., Center, D.M., Jacobson, M. and Hamid, Q. (1997b). Expression of IL-16 in allergen-induced late phase nasal responses and relation to topical glucocorticosteroid treatment. *J. Allergy Clin. Immunol.* **100**, 569–574.

Laberge, S., Leung, D.Y.M., Ghaffar, O, Bogunlewicz, M., Center, D.M. and Hamid, Q. (1997c). Interleukin-16 expression in acute *versus* chronic atopic dermatitis. *J. Allergy Clin. Immunol.* **99**, 1079.

Lim, K.G., Wan, H.-C., Bozza, P.T., Resnick, M.B., Wong, D.T.W., Cruikshank, W.W., Kornfeld, H., Center, D.M. and Weller, P.F. (1996). Human eosinophils elaborate the lymphocyte chemoattractants, IL-16 (lymphocyte chemoattractant factor) and RANTES. *J. Immunol.* **156**, 2566–2570.

Lotz, M., Hashimoto, S., Quach, J., Bober, L., Narula, S., Dudler, J. and Geng, Y. (1996). IL-17 promotes cartilage degradation. *Arthritis Rheum.* **39** (suppl.), 559.

Maclaszek, J.W., Parada, N.A., Cruikshank, W.W., Center, D.M., Kornfeld, H. and Viglianti, G.A. (1997). IL-16 represses HIV-1 promoter activity. *J. Immunol.* **158**, 5–8.

Magram, J., Connaughton, S.E., Warrier, R.R., Carvajal, D.M., Wu, C.Y., Ferrante, J., Stewart, C., Sarmiento, U., Faherty, D.A. and Gately, M.K. (1996). IL-12-deficient mice are defective in IFN gamma production and type 1 cytokine responses. *Immunity* **4**, 471–481.

McFadden, R.G. and Vickers, K.E. (1989). Rat lymphokines control the migration of nonsensitized lymphocytes. *Cell. Immunol.* **118**, 345–357.

Micallef, M.J., Ohtsuki, T., Kohno, K., Tanabe, F., Ushio, S., Namba, M., Tanimoto, T., Torigoe, K., Fuji, M., Ikeda, M., Fukuda, S. and Kurimoto, M. (1996). Interferon-gamma-inducing factor enhances T helper 1 cytokine production by stimulated human T cells: synergism with interleukin-12 for interferon-gamma production. *Eur. J. Immunol.* **26**, 1647–1651.

Micallef, M.J., Yoshida, K., Kawai, S., Hanaya, T., Kohno, K., Arai, S., Tanimoto, T., Torigoe, K., Fuji, M., Ikeda, M. and Kurimoto, M. (1997a). *In vivo* antitumor effects of murine interferon-gamma-inducing factor/interleukin-18 in mice bearing syngeneic meth A sarcoma malignant ascites. *Cancer Immunol. Immunother.* **43**, 361–367.

Micallef, M.J., Janaya, T., Kohno, K., Arai, S., Tanimoto, T., Ikeda, M. and Kurimoto, M. (1997b). Interleukin 18 (interferon gamma-inducing factor) exhibits antitumor effects in mice challenged intraperitoneally with syngeneic Meth A sarcoma. *Proc. Am. Assoc. Cancer Res.* **38**, A2388.

Moore, K.W., Vieira, P., Fiorentino, D.F., Trounstine, M. L., Khan, T.A. and Mosmann, T.R. (1990). Homology of cytokine synthesis inhibitory factor (IL-10) to the Epstein–Barr virus gene BCRF1. *Science* **248**, 1230–1234.

Mukhtar, A.S., Kim, S.J., Nicoll, J.M., Wu, D.M. and Kornfeld, H. (1997). Identification of a full length human IL-16 cDNA. *J. Allergy Clin. Immunol.* **99**, 223.

Okamura, H., Kawaguchi, K., Shoji, K. and Kawade, Y. (1982). High-level induction of gamma interferon with various mitogens in mice pretreated with *Propionibacterium acnes. Infect. Immun.* **38**, 440–443.

Okamura, H., Tsutsui, H., Komatsu, T., Yutsudo, M., Hakura, A., Tanimoto, T., Torigoe, K., Okura, T., Nukada, Y., Hattori, K., Akita, K., Namba, M., Tanabe, F., Konisl, K., Fukada, S. and Kurimoto, M. (1995a). Cloning of a new cytokine that induces IFN-gamma production by T cells. *Nature* **378**, 88–91.

Okamura, H., Nagata, K., Komatsu, T., Tanimoto, T., Nukata, Y., Tanabe, F., Akita, K., Torigoe, K., Okura, T. and Fukuda, S. (1995b). A novel costimulatory factor for gamma interferon induction found in the liver of mice causes endotoxic shock. *Infect. Immun.* **63**, 3966–3972.

Osaki, T., Péron, J.-M., Cai, Q., Okamura, H., Robbins, P.D., Kurimoto, M., Lotze, M.T. and Tahara, H. (1998). IFN-γ-inducing factor/IL-18 administration mediates IFN-γ and IL-12 independent anti-tumor effects. *J. Immunol.* (in press).

Parada, N.A., Cruikshank, W.W., Danis, H.L., Ryan, T.C. and Center, D.M. (1996). IL-16-and other CD4 ligand-induced migration is dependent upon protein kinase C. *Cell. Immunol.* **168**, 100–106.

Parada, N.A., Cruikshank, W.W., Kornfeld, H. and Center, D.M. (1997). Synergistic effects of IL-16 and IL-2 on immune reconstitution of $CD4^+$ T cells. *J. Allergy Clin. Immunol.* **99**, 431.

Rand, T.H., Cruikshank, W.W., Center, D.M. and Weller, P.F. (1991a). CD4-mediated stimulation of human eosinophils: lymphocyte chemoattractant factor and other CD4-binding ligands elicit eosinophil migration. *J. Exp. Med.* **173**, 1521–1528.

Rand, T.H., Silberstein, D.S., Kornfeld, H. and Weller, P.F. (1991b). Human eosinophils express functional interleukin 2 receptors. *J. Clin. Invest.* **88**, 825–832.

Rouvier, E., Luciani, M.-F., Mattei, M.-G., Denizot, F. and Golstein, P. (1993). CTLA-8, cloned from an activated T cell, bearing AU-rich messenger RNA instability sequences, and homologous to a herpesvirus saimiri gene. *J. Immunol.* **150**, 5445–5456.

Rumsaeng, V., Cruikshank, W.W., Foster, B., Kornfeld, H., Center, D.M. and Metcalf, D.D. (1996). Interleukin-16 production by human leukemic mast cells (HMC-1). *FASEB J.* **10**, 1776.

Ryan, T.C., Cruikshank, W.W., Kornfeld, H., Collins, T.L. and Center, D.M. (1995). The CD4-associated tyrosine kinase $p56^{lck}$ is required for lymphocyte chemoattractant factor-induced T lymphocyte migration. *J. Biol. Chem.* **270**, 17 081–17 086.

Sanchez-Bueno, A., Verkhusha, V., Tanaka, Y., Takikawa, O. and Yoshida, R. (1996). Interferon-gamma-dependent expression of inducible nitric oxide synthase, interleukin-12, and interferon-gamma-inducing factor in macrophages elicited by allografted tumor cells. *Biochem. Biophys. Res. Commun.* **224**, 555–563.

Scala, E., D'Offizi, G., Rosso, R., Turriziani, O., Ferrara, R., Mazzone, A. M., Antonelli, G., Aiuti, F. and Paganelli, R. (1997). C–C chemokines, IL-16, and soluble antiviral factor activity are increased in cloned T cells from subjects with long-term nonprogressive HIV infection. *J. Immunol.* **158**, 4485–4492.

Sgadari, C., Angiolillo, A.L. and Tosato, G. (1996). Inhibition of angiogenesis by interleukin-12 is mediated by the interferon-inducible protein 10. *Blood* **87**, 3877–3882.

Shaw, G. and Kamen, R. (1986). A conserved AU sequence from the 3′ untranslated region of GM-CSF mRNA mediates selective mRNA degradation. *Cell* **46**, 659–667.

Stoll, S., Muller, G., Kurimoto, M., Saloga, J., Tanimoto, T., Yamauchi, H., Okamura, H., Knop, J. and Enk, A.H. (1997). Production of IL-18 (IFN-γ-inducing factor) messenger RNA and functional protein by murine keratinocytes. *J. Immunol.* **159**, 298–302.

Tan, J., Crucian, B.E., Chang, A.E., Aruga, A., Dovhey, S.E. and Yu, H. (1996). The novel cytokine IFN-gamma-inducing factor induces antitumor immunity against renal carcinoma. *Cancer Vaccines*, T100.

Theodore, A.C., Center, D.M., Nicoll, J., Fine, G., Kornfeld, H. and Cruikshank, W.W. (1996). CD4 ligand IL-16 inhibits the mixed lymphocyte reaction. *J. Immunol.* **157**, 1958–1964.

Torigoe, K., Tanimoto, T., Nukada, Y., Akita, K., Tanabe, F., Konishi, K., Fujii, M., Itoh, M. and Kurimoto, M. (1996). Structure and function of a novel human interferon-γ inducing factor (HuIGIF). *Proc. Am. Assoc. Cancer Res.* **37**, A3083.

Trinchieri, G. and Scott, P. (1994). The role of interleukin 12 in the immune response, disease and therapy. *Immunol. Today* **15**, 460–463.

Tsutsui, H., Nakanishi, K., Matsui, K., Higashino, K., Okamura, H., Miyazawa, Y. and Kaneda, K. (1996). IFN-gamma-inducing factor up-regulates Fas ligand-mediated cytotoxic activity of murine natural killer cell. *J. Immunol.* **157**, 3967–3973.

Tsutsui, H., Matsui, K., Kawada, N., Hyodo, Y., Hayashi, N., Okamura, H., Higashino, K. and Nakanishi, K. (1997). IL-18 accounts for both TNF-α- and Fas ligand-mediated hepatotoxic pathways in endotoxin-induced liver injury in mice. *J. Immunol.* **159**, 3961–3967.

Ushio, S., Namba, M., Okura, T., Hattori, K., Nukada, Y., Askita, K., Tanabe, R., Konishi, K., Micallef, M., Fujii, M., Torigoe, K., Tanimoto, T., Fukada, S., Ikeda, M., Okamura, H. and Kurimoto, M. (1996). Cloning of the cDNA for human IFN-inducing factor, expression in *Escherichia coli*, and studies on the biologic activities of the protein. *J. Immunol.* **156**, 4274–4279.

Wu, D.M.H., Nicoll, J., Keane, J., Kornfeld, H., Cruikshank, W.W. and Center, D.M. (1997). Cloning and functional characterization of the murine CD4 ligand interleukin-16. *J. Allergy Clin. Immunol.* **99**, 222.

Yao, Z., Fanslow, W.C., Seldin, M.F., Rousseau, A-M., Painter, S.L., Comeau, M.R., Cohen, J.I. and Spriggs, M.K. (1995a). Herpesvirus saimiri encodes a new cytokine, IL-17, which binds to a novel cytokine receptor. *Immunity* **3**, 811–821.

Yao, Z., Painter, S.L., Fanslow, W.C., Ulrich, D., Macduff, B.M., Spriggs, M.K. and Armitage, R.J. (1995b). Human IL-17: a novel cytokine derived from T cells. *J. Immunol.* **155**, 5483–5486.

Yoshimoto, T., Okamura, H., Tagawa, Y., Iwakura, Y. and Nakamishi, K. (1997). Interleukin 18 together with IL-12 inhibits IgE sythesis by induction of IFN production from activated B-cells. *J. Allergy Clin. Immunol.* **99**, S466.

Chapter 18

Interferons

Edward De Maeyer and Jaqueline De Maeyer-Guignard
CNRS-UMR 177, Institut Curie, Université Paris-Sud, Orsay, France

INTRODUCTION

Type I interferons (IFNs) are major contributors to the first line of antiviral defence, which by itself is sufficient to make them of great interest; in addition to inhibiting virus replication, they exert many other important effects on cells. They belong to the network of cytokines that are involved in the control of cellular function and replication and that become actively engaged in host defence during infection. Type I interferons are used in the clinic as antiviral and antitumour agents, and also to treat multiple sclerosis. Type II interferon, also called IFN-γ, in addition to its antiviral activities, is an important modulator of the immune system.

STRUCTURE OF INTERFERONS

Type I IFNs are all derived from the same ancestral gene and have retained sufficient structural homology to act via the same cell surface receptor. They comprise IFN-α, IFN-ω, IFN-β and IFN-τ (trophoblast IFN) (see Table 1). The latter species has been described only in cattle and sheep. Type II IFN or IFN-γ is a lymphokine that displays no molecular homology with type I IFNs, but shares some important biological activities.

IFN-α, IFN-ω and IFN-β

The Human IFN-α, IFN-ω and IFN-β Gene Cluster

Thirteen non-allelic genes code for structurally different forms of human IFN-α. The IFN-α genes encode mature proteins of 165 or 166 amino acids, and the IFN-ω gene encodes a mature protein of 172 amino acids. IFN-ω shares about 60% of its amino acid sequence with the various species of IFN-α, and only about 30% with IFN-β. Unlike the majority of structural genes in the human genome, including the genes for all other cytokines, the IFN-α, ω and β genes lack introns, a feature they share with all other known mammalian IFN-α and β genes. It has been estimated that IFN-α and IFN-ω genes diverged more than 100 million years ago, before the mammalian radiation.

Contrary to the many genes coding for the different IFN-α species, there is only a single gene coding for human IFN-β. The mature peptide contains 166 amino acids and

The Cytokine Handbook, 3rd ed.
ISBN 0-12-689662-3

Table 1. Characteristics of human and murine IFN-α, ω and β genes.

	IFN-α	IFN-ω	IFN-β
Human			
Number of amino acids	165 or 166	172	166
Number of structural genes coding for active proteins	At least13	1	1
Gene designation and chromosomal localization	*IFNA* 9p21-pter	*IFNW* 9p21-pter	*IFNB* 9p21-pter
Murine			
Number of amino acids	166 or 167 (IFN-α4: 161)	No IFN-ω genes described	161
Number of structural genes coding for active proteins	At least 12		1
Gene designation and chromosomal localization	*Ifa* 4		*Ifb* 4

In bovines and sheep, in addition to IFN-α/β and ω, a fourth class, IFN-τ, has been described. The genes coding for these four IFN species belong to the same gene cluster.

the significant degree of homology between human IFN-α and IFN-β—about 30% at the amino acid and 45% at the nucleotide level—suggests that the genes are derived from a common ancestor by gene duplication.

The human IFN-α, IFN-ω and IFN-β genes are clustered in the same chromosomal region, on the short arm of chromosome 9. The IFN-α and IFN-ω genes are interspersed, and the IFN-β gene is situated distal from the IFN-α–ω cluster (Diaz *et al*., 1994; Weissmann and Weber, 1986; De Maeyer and De Maeyer-Guignard, 1988). The reason for the existence of multiple IFN-α subtypes is not clear; the relative biological activities of the different IFN-α subtypes can vary markedly and, for example, on a molar basis, IFN-α8 and IFN-β are more efficient antiviral agents than many other IFN-α subtypes (Foster *et al*., 1996).

The Murine IFN-α/β Gene Cluster

At least 12 non-allelic intronless genes have been identified in the murine IFN-α gene family. Murine IFN-ω genes have not been described so far.

The general structure of the murine IFN-α genes is comparable to that of the corresponding human genes. The mature proteins contain 166 or 167 amino acids (except for murine IFN-α4, which has a five codon deletion between codons 102 and 108), and the maximum divergence for replacement sites is about 13% (Kelley and Pitha, 1985a,b; Ryals *et al*., 1985; Seif and De Maeyer-Guignard, 1986).

The single-copy intronless murine IFN-β gene codes for a mature protein consisting of 161 residues, with three potential *N*-glycosylation sites at position 29, 69 and 76, which explains the difference between the 17-kDa molecular weight calculated from the amino acid sequence and the apparent molecular weight of about 28–35 kDa for natural murine IFN-β (De Maeyer-Guignard *et al*., 1978). The amino acid sequence displays 48% homology with that of human IFN-β.

Like the human IFN genes, the murine IFN-α genes and the IFN-β gene are clustered on murine chromosome 4, with the IFN-β gene distal from the IFN-α cluster (Dandoy *et al*., 1984; De Maeyer and Dandoy, 1987).

IFN-τ or Trophoblast IFN

In sheep and cattle, a novel type I IFN, called trophoblast IFN or IFN-τ, has been identified. These IFNs are the major secretory products of the trophoblast of ruminant ungulates during pregnancy, in the period immediately before attachment and implantation of the fertilized ovum. Trophoblast IFNs share most of the biological activities of other type I IFNs but are poorly responsive to viral induction; their major function is to create the conditions for efficient implantation of the ovum. Bovine trophoblast IFN shows more homology with bovine IFN-ω than with bovine IFN-α (Roberts *et al.*, 1991).

IFN-γ

Human IFN-γ

The single-copy human IFN-γ gene, situated on chromosome 12, contains three introns. The four exons code for 38, 23, 61 and 44 amino acids respectively, resulting in a polypeptide of 166 amino acids, 23 of which constitute the signal peptide. There are two potential *N*-glycosylation sites at positions 25–27 and 97–99, which explains the existence of two species of different molecular weight, 20 and 25 kDa respectively, glycosylated on either one or both sites (Devos *et al.*, 1982; Yip *et al.*, 1982; Rinderknecht *et al.*, 1984). The protein has a very high content of basic residues, more particularly two clusters of four residues each, Lys–Lys–Lys–Arg at positions 86–90 and Lys–Arg–Lys–Arg at positions 128–132. The basic nature of the mature protein probably explains its acid lability.

Murine IFN-γ

Like the human IFN-γ gene, the murine IFN-γ gene contains four exons and three introns. The coding part of the gene displays an overall nucleotide homology with the human gene of about 65%; the overall protein homology is 40%. Mature murine IFN-γ has 136 amino acids (Gray and Goeddel, 1982). The murine IFN-γ structural gene (*Ifg* locus) is on chromosome 10 (Naylor *et al.*, 1984).

INDUCTION OF IFN SYNTHESIS

Induction of IFN-α and IFN-β

The production of IFN-α and IFN-β is not a specialized cell function, and all cells of the organism are probably capable of producing these IFNs. In the absence of viral infection, most cells, whether in the organism or in culture, do not release measurable amounts of IFN-α/β. However, spontaneous IFN production has frequently been observed in cultures of cells derived from the hemopoietic system (as reviewed in De Maeyer and De Maeyer-Guignard, 1988), and there is a low level of constitutive IFN synthesis in murine BALB/c 3T3 cells (Seif *et al.*, 1991). As growth factors and other cytokines can induce IFN-α and IFN-β, it is possible that very low levels of some IFN species are made by many cells during a certain period of the cell cycle. Indeed, low levels of human IFN-α1 and IFN-α2 mRNA are constitutively transcribed in spleen, liver, kidney and peripheral blood lymphocytes of normal individuals, and *in situ* hybridization has shown that IFN-α and IFN-β genes are actively transcribed in normal murine bone marrow and peritoneal cells (Tovey *et al.*, 1987; Proietti *et al.*, 1992).

If cells either do not normally produce or at the most produce a very low level of IFN-α/β, exposure to a variety of agents triggers the production and secretion of IFN-α or IFN-β, or of a mixture of both. IFN was discovered during the study of viral interference (Isaacs and Lindenmann, 1957), and viruses are the most efficient natural inducers of IFN-α/β, but other infectious agents can also induce IFN. Many bacteria, especially those that replicate inside animal cells such as *Listeria monocytogenes*, induce IFN-α/β during systemic infection of the host. Endotoxin, the lipopolysaccharide derived from the cell walls of Gram-negative bacteria, and the M proteins of group A streptococci are also IFN inducers; IFN induction by bacterial endotoxin takes place mainly in macrophages.

Induction by Double-Stranded RNA and Viruses

Both natural and synthetic double-stranded RNAs (dsRNAs) induce IFN-α/β with high efficiency (Field *et al.*, 1967). Of the synthetic polynucleotides, the homopolymer pair polyriboinosinic–ribocytidilic acid (polyI:rC) is the most active and the most widely used for induction studies.

Despite their differences in structure and mode of replication, all animal viruses can induce IFN production under appropriate conditions and a unifying hypothesis would be the formation of dsRNA as a common pathway for induction. For RNA viruses, there are many arguments in favour of dsRNA as an essential intermediate for IFN induction; one of the most compelling is the demonstration with vesicular stomatitis virus particles that viral dsRNA generated within the cell is the actual trigger for IFN production. Theoretical considerations suggest that a single molecule of dsRNA is sufficient for induction (Sekellick and Marcus, 1982) and, indeed, dsRNA is present in the cell with most viruses (Jacobs and Langland, 1996).

Induction by Growth Factors and Other Cytokines

Several growth factors and cytokines have been shown to induce the synthesis of IFN-α or IFN-β. For example, stimulation of murine bone marrow cells with colony stimulating factor-1 (CSF-1) results in the production of murine IFN-α/β (Moore *et al.*, 1984). Two other cytokines, interleukin-1 (IL-1) and tumour necrosis factor (TNF), induce the synthesis of human IFN-β in human diploid fibroblasts, and IL-2 can induce the production of murine IFN-α/β in mouse bone marrow cells (Reis *et al.*, 1989). IFN-γ can sometimes act as an inducer of IFN-α or IFN-β (Gessani *et al.*, 1989; Cantell and Pirhonen, 1996).

Induction of IFN-γ

Production of IFN-γ is a function of T cells and of natural killer NK cells. All IFN-γ inducers activate T cells either in a polyclonal (mitogens or antibodies) or in a clonally restricted, antigen-specific, manner.

Cells of the Cytotoxic/Suppressor Phenotype (Human CD8 or Murine Ly-2 Phenotype)

Both in humans and in mice, IFN-γ synthesis has been observed in T cells of the cytotoxic/suppressor phenotype, bearing either the CD8 or the Ly-2 antigen respectively. T cells expressing the CD8 antigen, isolated from individuals after an infection with influenza virus or after immunization against rabies virus, are stimulated to release IFN-γ when exposed to the corresponding viral antigens *in vitro* (Celis *et al.*, 1986). Similarly, cells from a line of cytotoxic T cells derived from BALB/c mice immunized with influenza virus react specifically against influenza virus-infected target cells in a major histocompatibility complex (MHC)-restricted way and release IFN-γ (Morris *et al.*, 1982; Taylor *et al.*, 1985).

Cells of the T-Helper Phenotype (Human CD4 or Murine L3T4 Phenotype)

IFN-γ is produced during infection, and, for example, antigen-specific IFN-γ-producing circulating T cells of the helper phenotype are found in patients with recurrent herpes labialis, and rabies virus-specific, MHC class II-restricted, IFN-γ-producing T-helper cells are present in rabies vaccine recipients (Cunningham *et al.*, 1985; Celis *et al.*, 1986).

Based on the array of lymphokines secreted, two different subsets of mouse T helper (T_H) cells have been described. T_{H1} cells secrete IL-2, IL-3, TNF-β and IFN-γ, whereas T_{H2} cells mainly produce IL-3, IL-4, IL-5 and IL-10, but little or no IFN-γ (Mosmann *et al.*, 1986). IFN-γ preferentially inhibits the proliferation of T_{H2} but not of T_{H1} cells, indicating that the presence of IFN-γ during an immune response will result in the preferential expansion of T_{H1} cells (Gajewski and Fitch, 1988). Evidence that IFN-γ is one of the natural regulators that limit expansion of T-cell clones has been provided by a comparison of the rate of proliferation of mitogen-stimulated splenocytes derived from normal and knockout mice lacking a functional IFN-γ gene (Dalton *et al.*, 1993).

The Molecular Mechanism of IFN-α/β Induction

The production of IFN-α and IFN-β is controlled both at the transcriptional and the post-transcriptional levels. Upon induction, transcription starts rapidly, reaches a peak after several hours, and is then terminated, despite the continuous presence of the inducer. The cause of the shut-off of IFN synthesis is unknown; it is unlikely that negative feedback by IFN itself is responsible for the arrest of transcription, as treatment of cells, even with very high doses of IFN-α/β, increases rather than decreases polyI–C-induced IFN synthesis (De Maeyer-Guignard *et al.*, 1980).

Although very often IFN-α and IFN-β are induced coordinately in the same cell, selective induction of either IFN-α or IFN-β can occur and the ratio of IFN-α to IFN-β mRNA transcripts, as well as the proportion of individual IFN-α mRNAs, varies significantly with cell type and with the inducer (Hiscott *et al.*, 1984). Comparative analysis of the promoter regions of the IFN-β and the different IFN-α genes and their corresponding positive and negative transcriptional regulators shows that there are significant differences between the various promoters that regulate the expression of IFN genes (MacDonald *et al.*, 1990). The extreme example of this is the promoter of the bovine trophoblast IFNs (IFN-τ), which, although still virus inducible, is functionally quite distinct from other IFN-ω genes and contains a region that directs trophoblast-specific expression (Cross and Roberts, 1991; Hansen *et al.*, 1991).

Regulation of Human IFN-α/β Gene Expression

The regulatory region of the IFN-β gene spans about 200 base pairs (bp) immediately upstream from the transcription start site. A minimal sequence, necessary for virus induction and called IFN response element (IRE) or VRE (virus response element) has been described by Goodbourn and Maniatis (1988). The complete viral induction region spans the region from −125 to −38. This region contains an array of overlapping positive and negative regulatory domains, comprising at least four positive regulatory domains (PRD-I to IV). Each domain contains binding sites for one or several transcriptional regulators, and at least two of the domains are also implicated in post-transcriptional shut-off (Goodbourn and Maniatis, 1988; Fan and Maniatis, 1989; Lenardo *et al.*, 1989). PRD-I contains two copies of the hexamer sequence AAGTG(A/G), and the PRD-II domain contains a recognition site GGGAAATTCC for the NF-κB transcription factor which probably explains why many agents that can activate NF-κB are IFN inducers (Hiscott *et al.*, 1989; Lenardo and Baltimore, 1989; Xanthoudakis *et al.*, 1989). Two negative regulatory domains, NRD-I (−79 to −39) and NRD-II (−168 to −94) are possibly implicated in the normal repression of IFN gene expression (Zinn and Maniatis, 1986).

Several transcriptionally regulatory proteins that interact with the PRD-I have been characterized. Of these, interferon regulatory factor (IRF-1) (also called IFN-stimulated gene factor-2 (ISGF2)) acts as a transcriptional activator, and IRF-2 can function as a repressor of IRF-1-activated expression. IRF-2 is probably involved in preinduction and postinduction silencing of the IFN-β gene. The gene coding for IRF-1 is not only inducible by viruses or dsRNA, but its expression is also stimulated by TNF, IL-1 and by IFN-β itself (Goodbourn *et al.*, 1985; Miyamoto *et al.*, 1988; Fujita *et al.*, 1989a,b,c; Harada *et al.*, 1989; Pine *et al.*, 1990; Palombella and Maniatis, 1992; Sims *et al.*, 1993). Interestingly, constitutive expression of the IRF-1/ISGF-2 gene, obtained in cells transfected with this gene, causes resistance to infection with several different viruses (Pine, 1992). Another protein that interacts with PRD-I, PRD-I–BF-l, has been identified as a postinduction repressor (Keller and Maniatis, 1991).

INTERFERON RECEPTORS

IFN-α, IFN-β and IFN-ω Receptors

There are sufficient structural homologies between IFN-α, IFN-β, IFN-ω and IFN-τ to allow for common receptor binding domains, and nothing is unusual in the fact that they share a common receptor and these IFN species evolved from the same ancestral gene. Competitive receptor binding experiments using several human IFN-α subspecies indicate that they can have different binding affinities (Aguet *et al.*, 1984; Uzé *et al.*, 1990; Flores *et al.*, 1991).

The IFN-α/β Receptor

The cDNA coding for the α chain of the human IFN-α/β receptor (IFNAR1) codes for a 590 amino-acid protein, corresponding to a size of about 66 kDa. The extracellular *N*-terminal part of the molecule contains two distinct 200 amino-acid domains, which suggests that it belongs to the class II cytokine receptor family. Binding studies carried out with different IFN-α subtypes as well as an analysis using antireceptor monoclonal antibodies suggest that, to ensure complete binding activity, accessory proteins have to associate with the receptor chain. The exact position of the IFNAR locus is on the distal part of the long arm of chromosome 21, in the 21q22.1 band; it contains 11 exons and is one of the more polymorphic loci of this chromosomal region (Uzé *et al.*, 1990; Lutfalla *et al.*, 1990, 1992; Benoît *et al.*, 1993; Ling *et al.*, 1995).

Evidence for the existence of a second receptor chain has been provided by the work of several groups (Colamonici and Domanski, 1993; Novick *et al.*, 1994). The binding proteins of the IFN-α/β receptor do not contain a functional enzyme, but recruit the cytoplasmic tyrosine kinases JAK1 and TYK2 (Uzé *et al.*, 1995).

A cDNA sequence coding for the 590-amino-acid murine IFN-α/β receptor α chain has been isolated and expressed. The overall organization of the human and murine IFN receptor protein appears similar, in that the putative extracellular domain of the murine receptor also appears to be organized in two 200-amino-acid domains. As is the case for other receptors of the cytokine receptor family, accessory intracytoplasmic proteins are probably required for full activity of the receptor (Kumar *et al.*, 1989; Uzé *et al.*, 1992).

The IFN-γ Receptor

IFN-γ binds to a species-specific 90-kDa cell surface receptor, the IFN-γ α-receptor chain. One IFN-γ homodimer binds two receptor α chains, which do not interact with one another, but remain separated (Walter *et al.*, 1995). Both the human and the murine receptor α chains have an equally large extracellular domain consisting of 288 amino acids, a single transmembrane domain, and an intracellular domain of 222 amino acids and 220 amino acids for the human and murine chain respectively. The structural gene for the receptor α chain is on chromosome 6 in humans, and on chromosome 10 in mice.

To be functionally active, the human IFN-γ α-receptor chain requires the presence of at least one other component, the β chain, whose structural gene is on human chromosome 21 (Farrar *et al.*, 1991; Soh *et al.*, 1994). In addition to the α chain, the presence of the β chain is required (and sufficient) for the induction of MHC class II antigens by IFN-γ ; it is a transmembrane protein with an overall structural homology to the class 2 cytokine receptor family (Hibino *et al.*, 1991, 1992; Kalina *et al.*, 1993).

The murine IFN-γ receptor β chain, encoded by a gene on chromosome 16, consists of a large extracellular domain and a relatively short cytoplasmic domain, estimated at 66 amino acids (Gray *et al.*, 1989; Hemmi *et al.*, 1989, 1994).

Mice with a disrupted IFN-γ receptor gene that no longer have a functional IFN-γ receptor show no overt anomalies, and their immune system appears to develop normally. However, such animals display decreased resistance when challenged with an intracellular parasite such as *Listeria monocytogenes* and also after infection with vaccinia virus, the latter in spite of an apparently normal development of cell-mediated immunity against this virus (Huang *et al.*, 1993).

In humans, two different mutations in the coding sequence of the IFN-γ receptor α chain have been reported, each one responsible for the synthesis of a truncated non-functional α-receptor chain. Individuals that are homozygous for such a mutation display a greatly enhanced susceptibility to mycobacterial infection (Jouanguy *et al.*, 1996; Newport *et al.*, 1996). IFN-γ receptor-deficient mice are furthermore extremely sensitive to the development of vasculitis as a result of infection with murine γ-herpesvirus 68, suggesting γ-herpesviruses as candidate etiologic agents for vasculitis (Weck *et al.*, 1997).

THE IFN SIGNALLING PATHWAY

IFN-α/β

Many genes are transcriptionally activated in IFN-treated cells, and a short summary of some important IFN-induced genes and of some important biological activities is given in Tables 2 and 3. A *cis*-acting DNA element, called ISRE (IFN-stimulated response element), common to the promotors of IFN-α, IFN-β and of some IFN-γ-stimulated genes, mediates the transcriptional response. The core ISRE is 13 bp long and is highly conserved among the many IFN-induced genes, but there is no obvious homology in the sequences flanking them. Several transcriptional activators involved in the signal transduction from the IFN receptor to the IFN-induced genes have been characterized and cloned; the best characterized of these is ISGF-3. The activated ISGF-3 is a complex of four distinct polypeptides; three of these, with molecular masses of 113, 91 and 84 kDa, are normally already present in the cytoplasm and, after IFN receptor binding, become activated very rapidly, in a matter of minutes, by phosphorylation of tyrosine residues. The activated proteins then form a complex, termed ISGF-3α, that immediately translocates to the nucleus and binds to the fourth component, a 48-kDa protein, called the ISGF-3γ protein. This 48-kDa nuclear protein, which is probably the major subunit that binds to the ISRE, normally already displays some ISRE binding activity, which is boosted 20-fold in the active ISGF-3 complex. The 84- and 91-kDa proteins are products of the same gene; amino acid sequence analysis has shown that the ISGF-3 proteins belong to a new family of tyrosine kinase-activated signal transducers (Levy *et al.*, 1989; Fu *et al.*, 1990, 1992; Imam *et al.*, 1990; Kessler *et al.*, 1990; McKendry *et al.*, 1991; Fu, 1992; Schindler *et al.*, 1992a,b; Veals *et al.*, 1992; Constantinescu *et al.*, 1994). The

Table 2. Some important proteins induced by IFN-α and IFN-β in humans.

Protein	Size (kDa)	Chromosomal localization	Principal activity
2–5A synthetases	100	—	Antiviral and antitumoral, via activation of RNase L
	69	—	
	46	12	
	40	12	
PKR kinase	68	2	Inhibition of translation of viral mRNA
MxA	76	21	Inhibition of orthomyxovirus and of vesicular stomatitis virus replication
MxB	73	21	?
GBP-1	67	—	Guanylate binding protein
15-kDa protein	15	—	IFN-γ induction
17-kDa protein	17	—	Inhibition of cellular replication
9–27 protein	14	—	Antiviral activity
ISG15	15	—	Expansion of NK cells and induction of IFN-γ
IRF-1	—	—	Transcriptional activator; antiviral, antitumour, antibacterial activity
MHC class I	45	6	Antigen processing and immune regulation
β_2 Microglobulin	12	15	Antigen processing and immune regulation
MHC class II	24–33	6	Antigen processing and immune regulation
Metallothionein II	—	16	Heavy metal binding

Table 3. A synopsis of the principal biological activities of IFN-α/β.

Antiviral effect	Broad-spectrum antiviral activity, owing to a variety of mechanisms that, depending on the virus involved, can act at different stages of the infectious cycle. Some viruses have developed more or less efficient ways of escaping the antiviral effect.
	Clinical use: type I IFNs are used to treat Kaposi sarcoma and infection with hepatitis B and C virus.
Effect on cell growth and division	Inhibition of the replication of normal and tumour cells. IFN-α and β cause a prolongation of G1, a reduced rate of entry into the S phase, and a lengthening of S and G2, resulting in a slower replication rate. Cells show highly different sensitivities to this activity.
Modulation of the expression of the MHC I and, to a lesser extent, of MHC II cell surface molecules	This is one of the mechanisms responsible for the immunomodulatory activities of IFN-α/β. Modulation of MHC class II antigens is more specifically a function of IFN-γ.
Stimulation of macrophage activity	Stimulation of receptor- and non-receptor-mediated phagocytosis (the major macrophage activator is IFN-γ).
Stimulation of CTL activity	The lytic activity of cytotoxic T cells is upregulated.
Stimulation of NK cell activity	Contributes to the antitumour activity, but treatment of target cells with IFN-α/β sometimes results in protection against NK cell activity, due to upregulation of MHC I expression.
Up- or downregulation of delayed-type hypersensitivity	DTH is an important immune mechanism which can be up- or downregulated by IFN-α/β, depending on timing of action.
Antitumoral activity	The mechanism of the antitumour activity is complex and only partly understood. Stimulation of macrophages, T cells and NK cells is involved, as is the direct antiproliferative activity.
	Clinical use: for example, renal carcinoma, hairy cell leukaemia, chronic myelogenous leukaemia, squamous carcinoma, haemangioma.
Other	IFN-β is used for the treatment of multiple sclerosis.

tyrosine kinases that activate the ISGF-3α subunits after binding of IFN-α to its receptor have been characterized as JAK1 and TYK2. The kinases are probably in close physical contact with the IFN receptor, as their presence has been demonstrated in cell membrane extracts. The signal transduction pathway thus activated by IFN provides the first example of a very rapid and direct route from the cell surface to the promoter regions of genes transcriptionally activated as a result of ligand–receptor binding; no intervention of second messengers, such as for example cyclic adenosine monophosphate, is needed (Darnell *et al.*, 1994). Although activation of JAK1 and TYK2 is required for signal transduction after receptor binding of most IFN-α subtypes, human IFN-α8 as well as human IFN-β are biologically active in cells that express only JAK1 but not TYK2, indicating that these two subtypes are somehow capable of TYK2-independent activation of IFN-inducible genes (Foster *et al.*, 1996).

IFN-γ

As a result of the binding of IFN-γ to its cell surface receptor, the transcription of many previously quiescent genes is activated. IFN-γ-induced transcriptional activation can be either late, requiring first the synthesis of specific proteins that are then responsible for gene activation, or immediate, and take place via activation of specific signal-transducing proteins that are already present in the cytoplasm and/or the nucleus in an inactive form (McKendry *et al.*, 1991; Velazquez *et al.*, 1992; Watling *et al.*, 1993; Darnell *et al.*, 1994; Shuai *et al.*, 1994).

Receptor binding of IFN-γ activates two tyrosine kinases of the JAK family: JAK1 and JAK2, which catalyse the tyrosine phosphorylation of the transcription factors involved in signal transduction (Harpur *et al.*, 1992; Müller *et al.*, 1993). It seems likely that the kinases are associated physically with the intracellular part of the receptor, as binding of IFN-γ to its receptor results in rapid phosphorylation of the latter (Shuai *et al.*, 1994). A latent cytoplasmic factor, the γ activation factor (GAF), is then activated through tyrosine phosphorylation. GAF is a dimer of a 91-kDa protein, called STAT91 (or STAT84) protein (STAT is an acronym for signal transducer and activator of transcription) (Imam *et al.*, 1990; Decker *et al.*, 1991; Shuai *et al.*, 1992; Müller *et al.*, 1993). The STAT91 and STAT84 proteins are encoded by the same gene, but STAT91 contains 38 carboxy-terminal amino acids that are lacking in STAT84 (Schindler *et al.*, 1992; Zhong *et al.*, 1994). STAT91 is also one of the proteins that make up the ISGF-3 complex that is involved in the immediate transcriptional activation of IFN-α-inducible genes and is made up of four different subunits, consisting of 113-, 91-, 84- and 48-kDa protein (Müller *et al.*, 1993). Inactive STAT91 in the cytoplasm of untreated cells is a monomer, but upon IFN-γ-induced phosphorylation it forms a stable dimer, which then migrates to the nucleus. Only the dimer is capable of binding to the γ activation site (GAS) (Shuai *et al.*, 1994).

GAS is a consensus immediate response element of nine nucleotides (TTNCNNNAA) which is present in genes that are activated immediately after IFN-γ receptor–ligand binding. Although STAT91 is also present in ISGF-3, the transcriptional activator of the IFN-α response, the major DNA binding component of ISGF-3 appears to be the 48-kDa protein, which is normally already present in the nucleus, and which, after activation and inclusion into the ISGF-3 complex, binds to the ISRE (a consensus sequence present in most genes that respond to IFN-α). Several of the STAT proteins characterized in the IFN-α and IFN-γ signal transduction pathways will undoubtedly turn out also to be involved in the control of expression of genes activated by other cytokines, and the knowledge gained from the study of IFN signal transduction has important implications for the understanding of gene expression in general. It has indeed become evident that several cytokines other than interferons also use tyrosine phosphorylation to activate putative transcription factors, and, for example, STAT91 is phosphorylated after the interaction of epidermal growth factor and its receptor, and then mediates activation of the c-*fos* gene promoter (Fu and Zhang, 1993; Larner *et al.*, 1993).

INTERFERONS AS ANTIVIRAL AGENTS

Type I IFNs

IFN-α and IFN-β play an important role in host resistance to virus infections as they occupy the first line of defence, before immune mechanisms come into play. During viral infection, IFN-γ is produced only after T cells have been sensitized to viral antigens, and, although its key function then resides in the activation of antiviral immune reactions, it has also been shown to exert direct antiviral effects that contribute to host defence (see, for example, Huang *et al.*, 1993).

The production of IFN-α/β has been demonstrated in humans during viral disease and in mouse models using a wide variety of different viruses. The kinetics of the appearance of IFN-α/β stand out as an important factor in determining the efficacy of endogenous IFN action: early IFN production is instrumental in limiting infection, whereas late production generally has much less apparent protective effect. This is explained by the time required to mount a specific antiviral immune response; during this interval, IFN-α/β production is the only known active defence mechanism.

Treatment of mice with anti-IFN-α/β globulin markedly enhances the severity of infection with many different viruses (Gresser, 1984). Moreover, mice that are normally resistant to infection with some viruses can become susceptible as a result of treatment with anti-IFN globulin. For example, C3H/He mice resistant to infection with mouse hepatitis virus (MHV-3) become fully susceptible when treated with anti-IFN globulin, and die a few days after inoculation of the virus (Virelizier and Gresser, 1978). It is clear from mouse model studies that the antiviral activity of IFN-α/β *in vivo* results not only from an induction of the antiviral state in cells, but also from a wide variety of other IFN effects on host defence mechanisms, such as stimulation of NK cells and of cell-mediated immunity, and activation of macrophages. Knockout mice lacking the IFN-AR1 subunit of the type I IFN receptor are extremely susceptible to viral infections such as, for example, infection with vesicular stomatitis virus or Semliki Forest virus, despite otherwise normal immune responses (Muller *et al.*, 1994).

Type II IFNs

An analysis of two knockout mouse strains, one lacking a functional IFN-γ structural gene, and the other lacking a functional IFN-γ receptor, shows no gross histological abnormalities in the lymphoid organs and no significant difference in the number of cells in the spleen and thymus of these mice (Dalton *et al.*, 1993; Huang *et al.*, 1993). There are, furthermore, no differences in the expression of CD3, CD4, CD8, cell surface immunoglobulin M (IgM) and MHC class I and II on splenic and thymic cell populations or on peripheral blood cells. However, MHC class II antigen expression on macrophages is reduced. Thus, IFN-γ is not essential for the development of the immune system, and is not required for survival under specific pathogen-free conditions. The importance of IFN-γ in the resistance to some virus infections is confirmed when mutant mice are infected with vaccinia virus. The early defence against vaccinia virus is severely defective: within the first few days after infection, virus replication is about three orders of magnitude above control values, resulting in death. With another virus, vesicular stomatitis virus, the course of infection is identical in wild-type and mutant animals, and the titres of neutralizing antibody that appear as a result of the infection are no different from those of control mice. Knockout mice have decreased total serum $IgG_{2\alpha}$ concentrations and, after immunization with the appropriate antigens, show decreased titres of hapten-specific $IgG_{2\alpha}$ and IgG_3 antibodies. Significantly reduced $IgG_{2\alpha}$ titres are also observed after immunization with pseudorabies virus. When infected with a dose of *L. monocytogenes* that does not kill wild-type animals, knockout mice succumb to the infection and the bacterial titres found in the liver and spleen are up to 100-fold higher than those observed in wild-type mice. Similarly, infection of knockout mice with a dose of *Mycobacterium tuberculosis* (BCG) that is sublethal in normal animals results in a significantly enhanced mortality rate. The generation of nitric oxide is severely decreased in knockout mice, as a result of which the infectivity of intracellular parasites is greatly enhanced in these animals (Kamijo *et al.*, 1993).

Multiple Mechanisms of the Antiviral State

The interaction of IFNs with their specific cell surface receptors is followed by the rapid activation of DNA binding factors that stimulate the transcription of a set of genes containing IFN response sequences (IRSs) homologous to the prototypic sequence GGGAAANNGAAACT (Cohen *et al.*, 1988; Hug *et al.*, 1988; Levy *et al.*, 1988; Shirayoshi *et al.*, 1988; Dale *et al.*, 1989a,b; Reich and Darnell, 1989). Several IFN-stimulated genes code for proteins responsible for the antiviral state, such as the dsRNA-dependent PKR kinase, the (2′–5′)oligoadenylate synthetase and the Mx proteins. In addition to these, many other proteins are induced in IFN-treated cells (see Table 2 and De Maeyer and De Maeyer-Guignard (1988) for an extensive review), but the discussion here will be limited to those whose contribution to the antiviral state has been well established.

Oligoadenylate Synthetase

This enzyme, also called (2′–5′)A_n synthetase or (2′–5′)oligo(A) adenyltransferase, is constitutively present in many cells but at very low levels; its concentration increases by several orders of magnitude after IFN treatment. IFN-γ is a less efficient inducer of (2′–5′)A_n synthetase than are IFN-α or IFN-β (Baglioni and Maroney, 1980; Verhaegen-Lewalle *et al.*, 1982). When activated by dsRNA, (2′–5′)A_n synthetase polymerizes ATP into a series of 2′–5′-linked oligomers (ppp(A2′p)$_n$), of which the trimer is the most abundant. These oligomers, collectively called 2–5A, then activate a latent cellular endoribonuclease, designated RNase L, which is responsible for the antiviral activity. The third enzyme of this system, present in both untreated and IFN-treated cells, is a 2′–5′ phosphodiesterase that catalyses the degradation of 2–5A. The dsRNA required to activate the (2′–5′)A_n synthetases are probably intermediates or side products of viral RNA replication (Gribaudo *et al.*, 1991).

Functional mRNAs of different sizes, 1.65 and 1.85 kb in human and 1.8 and 3.6 kb in murine cells, have been described for this enzyme. The human 1.6- and 1.8-kb mRNAs are transcribed from the same gene but are spliced differently (Merlin *et al.*, 1983; St Laurent *et al.*, 1983; Benech *et al.*, 1985a,b), and the 40- and 46-kDa synthetases derived from them are identical in their first 346 residues but differ at their *C*-terminal ends (Benech *et al.*, 1985b; Wathelet *et al.*, 1986). The structural gene is located on human chromosome 12; it contains six exons, spread over 12 kb of DNA (Williams *et al.*, 1986). Altogether, human cells contain four different (2′–5′)A_n synthetases, all antigenically related, with sizes of 40, 46, 67 and 100 kDa. The latter two forms are in all likelihood encoded by genes different from the one encoding the 40- and 46-kDa forms. These four enzymes differ in their preferential intracellular localization and in their optimal conditions for activity (Chebath *et al.*, 1987; Hovanessian *et al.*, 1987).

In the mouse, 40-, 75- and 100-kDa forms have been described (Hovanessian *et al.*, 1987). The murine 42-kDa (2′–5′)A_n synthetase displays 62% homology at the amino acid level and 73% homology at the nucleotide level with the human 46-kDa enzyme (Ichii *et al.*, 1986). Two different genes, closely linked and each encoding (2′–5′)A_n synthetase have been identified in mice (Rutherford *et al.*, 1991).

The best documented function of the 2–5A oligomers made by the (2′–5′)A_n synthetase is activation of a latent endonuclease, ribonuclease L, resulting in the degradation of viral RNA. *In vitro*, however, degradation is not limited to viral RNA, as cellular mRNA and ribosomal RNA are also degraded by ribonuclease L (Nilsen *et al.*, 1981; Silverman *et al.*, 1983).

PKR Protein Kinase

Activation of the protein kinase PKR (a serine–threonine kinase) is the second pathway of IFN-induced translational control dependent on the presence of dsRNA. Treatment of cells with type I IFN enhances the transcription of the PKR gene. PKR kinase, when activated by low levels of dsRNA, first autophosphorylates and then phosphorylates the α subunit of eIF2, the eukaryotic protein synthesis initiation factor. As a result, recycling of the α subunit is inhibited and initiation of translation cannot take place (Samuel *et al.*, 1984). When expressed in the yeast *Saccharomyces*, the human PKR kinase has growth-suppressing activity, which correlates with phosphorylation of yeast eIF2α (Chong *et al.*, 1992). Activated PKR also phosphorylates the NF-κB, and migration of this transcription factor to the nucleus results in the transcription of certain genes (Kumar *et al.*, 1994).

In addition to its role in the IFN-activated antiviral, and probably also antiproliferative, mechanism, the possibility has been raised that the PKR kinase plays a role in normal cells as a homeostatic regulator, whose aberrant expression can lead to malignant transformation. This is based on the observation that expression of a functionally defective mutant in NIH 3T3 cells, acting as a dominant negative mutation, leads to malignant transformation (Koromilas *et al.*, 1992; Meurs *et al.*, 1993). The possibility has been raised that PKR is involved in apoptosis, resulting from the presence of dsRNA in the cell (Lee and Esteban, 1994; Kibler *et al.*, 1997).

The human PKR kinase is a 68-kDa polypeptide, and the murine kinase a 65-kDa polypeptide; the corresponding cDNAs have been isolated, and the dsRNA binding domains of the protein have been identified. The dsRNA required to activate the enzyme is usually of viral origin; for example, in the case of human immunodeficiency virus (HIV), efficient binding and activation of the kinase is a result of the interaction with the Tat-responsive sequence of the virus (Hovanessian *et al.*, 1988; Roy *et al.*, 1991; Feng *et al.*, 1992).

Several viruses have developed strategies to counter the effects of the PKR kinase. In the case of adenovirus, small viral RNA transcripts bind to the protein kinase in such a way that subsequent interaction with activating dsRNA is prevented (Galabru *et al.*, 1989). In poliovirus-infected cells, the kinase is degraded by a cellular protease that somehow becomes activated (Black *et al.*, 1993). Influenza virus inhibits activation of PKR by activating the cellular protein P58 (Lee *et al.*, 1990).

Mx Protein

Mx proteins constitute a family of IFNα/β-inducible GTPases that display antiviral activity against specific viruses. The murine Mx protein produces an antiviral state that is directed specifically against influenza virus replication. It is a 72-kDa nuclear protein that is induced by IFN-α/β, but not by IFN-γ, in cells from mice with the *Mx*$^+$ genotype. Mice of *Mx1*$^-$ strains that are genetically susceptible to influenza virus infection carry a defective *Mx* gene, with deletions in the coding exons. The *Mx1* gene maps to murine chromosome 16. A corresponding gene, *MxA*, has been found in humans; it maps to chromosome 21. The human IFN-induced MxA protein, however, is a cytoplasmic protein that confers resistance not only to influenza virus but also to vesicular stomatitis virus (for a review of this interesting system, see Staeheli, 1990).

Other Antiviral Mechanisms

There are abundant indications that IFNs can increase the antiviral resistance of cells, for example against retroviral infection, including infection with HIV (Vieillard *et al.*, 1994), by a mechanism other than the above-mentioned, but the molecular basis for these activities remains to be elucidated.

Conclusion

The study of the antiviral state in specific virus–host cell systems has provided examples of interference with virus production at every stage of the infectious cycle in cells treated with IFN-α or IFN-β. The major mechanism is inhibition of translation, with involvement of the 2–5A and protein kinase pathways, but other mechanisms are activated and remain to be resolved.

The molecular mechanisms of the antiviral action of IFN-γ have received less attention. IFN-γ also activates the 2–5A pathway and the dsRNA-dependent protein kinase, but to a lesser extent than IFN-α or IFN-β. However, and in contrast to IFN-α and β, an important exception to the usual antiviral activity of IFN-γ has been observed, in that the expression of HIV can be stimulated after IFN-γ treatment of promonocytic cells chronically infected with this virus (Biswas *et al.*, 1992).

MODULATION OF THE EXPRESSION OF THE MAJOR HISTOCOMPATIBILITY ANTIGENS

Modulation of MHC Class II Antigen Expression by IFN-γ

A critical step in immune responses is the recognition by cells belonging to the immune system of peptide fragments of foreign antigens. Short peptide fragments from proteins degraded in the cytosol are bound to MHC class I molecules and the complex recognized by CD8$^+$ T cells. Peptides from proteins degraded in endosomal cellular vesicles bind to MHC class II molecules and the complex then migrates to the cell surface. The complex of peptide and MHC class II molecules is then recognized by CD4$^+$ T helper cells. This critical step in the immune response is stimulated by IFN-γ, which induces or enhances the expression of MHC class II antigens on macrophages and T cells. The stimulation of MHC class II expression by IFN-γ is not limited to these two classes of cells, but is also observed on B cells and on many different tumour cells. The genes encoding MHC class II molecules belong to the class of genes that need several hours for activation by IFN-γ, as opposed to the 'early' genes that are activated in a matter of minutes (Blanar *et al.*, 1988; Amaldi *et al.*, 1989; Lew *et al.*, 1991).

The importance of endogenous IFN-γ production in augmenting MHC class II expression has been confirmed by the use of knockout mice lacking a functional IFN-γ gene; in such animals, class II expression on macrophages from BCG-infected mice is significantly reduced compared with that in normal mice (Dalton *et al.*, 1993).

Modulation of MHC Class I Antigen Expression by IFN-α/β

Modulation of the expression of the cell surface antigens of the MHC is one of the major mechanisms by which all three IFN species can influence the immune system. MHC modulation is furthermore implicated in the antiviral activity of IFNs, as a successful antiviral cell-mediated immune response depends on the ability of the target cells to present the antiviral antigens in conjunction with class I antigens (Zinkernagel and Doherty, 1974). It has been shown, for example, that by increasing the expression of MHC class I antigens murine IFN-α/β enhances the susceptibility of vaccinia or lymphocytic choriomeningitis virus-infected fibroblasts to lysis by cytotoxic T cells (Bukowski and Welsh, 1985).

Although IFN-α and IFN-β can stimulate the expression of class II antigens, their major effect on MHC expression is induction of class I antigens. These effects have been observed *in vitro* and also *in vivo* after systemic administration of IFN to animals, which means that they can occur as a result of IFN production during viral infection. Increased expression of MHC class I antigens induced by murine IFN-α/β on malignant cells significantly reduces the tumorigenicity of these cells in immunocompetent hosts, which shows the importance of class I antigens for the antitumour effect of IFN (Hayashi *et al.*, 1985).

STIMULATION OF MACROPHAGE ACTIVITY BY INTERFERONS

IFN-α, IFN-β and especially IFN-γ play an important role in macrophage activation, mainly as a result of effects on the expression and activity of cell surface receptors.

Tumoricidal Activity

Lymphokines released by activated T cells are capable of priming macrophages for tumoricidal activity. This activity, called macrophage activating factor (MAF), is at least partly due to IFN-γ, as it is inactivated by highly specific antibodies to IFN-γ and can be reproduced by recombinant murine IFN-γ (Le *et al.*, 1983; Nathan *et al.*, 1983; Svedersky *et al.*, 1984). Activation of cell killing by macrophages is an important function of IFN-γ, as tumour cell lysis by activated macrophages is in all likelihood part of the mechanism of natural resistance to cancer. Although IFN-γ is undoubtedly a major macrophage priming agent, it shares this function with other lymphokines that also prime or activate macrophages to kill tumour cells by releasing reactive oxygen intermediates and by producing other cytotoxic molecules such as TNF-α (Urban *et al.*, 1986). IFN-γ induces the formation and release of TNF by macrophages (Philip and Epstein, 1986).

Destruction of Parasites by IFN-Activated Macrophages

Intracellular parasite killing is activated significantly when infected macrophages are exposed to IFN-γ, for example in IFN-γ-treated suspensions of human monocytes infected with *Leishmania donovani*. The effect can be either preventive or curative, because parasite killing is also enhanced when already infected monocytes are treated subsequently with IFN-γ (Hoover *et al.*, 1985). Production of reactive oxygen intermediates and secretion of hydrogen peroxide are correlated with the capacity of macrophages to kill intracellular parasites (Murray, 1981). For example, in human and murine macrophages, exposure for a few hours to recombinant IFN-γ leads to substantial activation of hydrogen peroxide-releasing capacity that can last for several days, and is accompanied by a stimulation of intracellular killing of *Toxoplasma gondii*. Stimulation of the secretion of reactive oxygen intermediates seems to be a function exclusive to IFN-γ. It has not been found for IFN-α, IFN-β, CSF-1, TNF or IL-2 (Murray *et al.*, 1985; Nathan *et al.*, 1985; Nathan and Tsunawaki, 1986). Moreover, in human monocytes recombinant human IFN-α2 and IFN-β antagonize the hydrogen peroxide-stimulating activity of IFN-α (Garotta *et al.*, 1986).

IFN-γ induces indoleamine 2,3-dioxygenase (IDO), an enzyme of tryptophan catabolism, which is responsible for the conversion of tryptophan to kynurenine. IDO activity has been implicated in the killing of intracellular parasites such as *T. gondii* or *Chlamydia trachomatis* and *Chlamydia psitacci*. It is believed that the inhibition results, at least in part, from tryptophan starvation of the parasites (as reviewed by Taylor and Feng, 1991), and, indeed, in mutant cell lines lacking the capacity to synthesize IDO, the inhibition by IFN-γ of intracellular parasites is reduced (Thomas *et al.*, 1993).

MODULATION OF T, B AND NK CELL ACTIVITY BY IFNs

Effects on T-Cell Function

T Cells with Cytolytic Activity

Cytotoxic T cells react with target cells in the context of class I MHC antigens, and both IFN-α/β and IFN-γ stimulate the expression of these antigens, on most target cells, as a result of which the cytotoxic activity of T cells is boosted. Thus, virus-induced IFNs restrict infection not only by inducing the antiviral state but also by conditioning infected cells for destruction by cytotoxic T cells (Blackman and Morris, 1985; Bukowski and Welsh, 1985). Moreover, IFN-α/β enhances the specific cytotoxicity of sensitized lymphocytes against allogeneic tumour cells (Belardelli and Gresser, 1996).

T Cells with Suppressive Activity

The effects of IFNs on T suppressor cells can result in either boosting or downregulation of suppression. IFN-γ is capable of stimulating accessory cells to induce T suppressor cells, and stimulation by IFN-γ of MHC class II antigen expression on macrophages is followed by an increase in the ability of these macrophages to induce the generation of T suppressor cells (Noma and Dorf, 1985). The effect of IFNs on suppressor cell generation is not always stimulatory, and under certain conditions suppressor cell formation is inhibited. Frequently, T suppressor cells have been found to be particularly sensitive to inhibition by IFNs in delayed hypersensitivity and in other immune reactions. For example, human leukocyte IFN added to mixed lymphocyte cultures causes a marked decrease in suppressor cell activity, as a result of inhibition of the differentiation of presuppressor cells into active suppressor cells (Fradelizi and Gresser, 1982). Similarly, the stimulatory effect of IFN-α/β on the expression of delayed hypersensitivity is due to specific inhibition of either the generation or the expansion of T suppressor cells (Knop *et al.*, 1982, 1987).

Effects on NK Cells

NK cells can be activated without previous sensitization and they are therefore, like macrophages, in the first line of defence against tumour cells and infectious agents. NK cell activity is boosted by IFN-α/β and IFN-γ, with an optimal IFN dosage above which NK cell activity is often decreased instead of being enhanced (Herberman *et al.*, 1982; Edwards *et al.*, 1985).

IFNs of all three species can protect target cells against NK cell activity, as shown for example in HeLa cells, which can be protected by human IFN-α, β or γ (Wallach, 1983).

EFFECTS OF INTERFERONS ON TUMOUR CELLS

The first cytokines to have found large-scale use in the clinic were the IFNs, and human IFN-α and IFN-β are currently used as therapeutic agents in patients with different forms of carcinoma, mestastatic melanoma, myeloma, ovarian cancer, chronic myelogenous leukaemia, haemangioma and hairy cell leukaemia. The effects of IFNs on tumours *in vivo* can result either from a direct action of IFN on the tumour cell or, indirectly, via the activation of several, not completely defined, effector mechanisms. These include stimulation of MHC antigen expression, macrophage activation, and stimulation of T and NK cell activity. Many murine model studies have shown tumour inhibition as a result of IFN treatment. Host genes can upregulate or downregulate the antitumour activity of IFN-α/β in the mouse and, on some genetic backgrounds, IFN treatment has no effect or even enhances tumour development. This may help to explain the apparent discordance between mouse model studies, showing inhibition of tumour formation by IFN, and findings in the clinic, where only a certain proportion of individuals, high or low, depending on the nature of the tumour, show tumour regression (De Maeyer-Guignard *et al.*, 1993). Among the direct effects of IFNs on tumour cells that have been reported are the following.

Antiproliferative Effects

IFNs slow down the growth and proliferation of normal and tumour cells by prolonging the cell cycle. Some tumour cells are extremely sensitive to this effect, whereas others can be totally resistant. Mutants resistant to the antiproliferative effect can be isolated from IFN-sensitive tumour cell lines (Gresser *et al.*, 1974); such mutants can retain their sensitivity to the antiviral activity (Lin *et al.*, 1982). On freshly isolated tumour cells, the antiproliferative activity of human IFNs ranges from complete inhibition of cell replication to total resistance, with no obvious correlation between tumour cell type and degree of inhibition (Ludwig *et al.*, 1983). Evidence for a direct cytostatic effect of

IFNs on the growth and development of tumour cells *in vivo* is provided by the inhibition of human tumour growth in nude mice treated with human IFNs of various origins.

Effects on Oncogene Expression

IFNs have various effects on oncogene expression. They can, for example, downregulate c-*myc* expression in malignant cells. In Burkitt's lymphoma Daudi cells that have been treated with IFN-β, a reduction in c-*myc* mRNA levels occurs as early as 3 h after the addition of IFN and precedes the inhibition of cell growth, suggesting a correlation between inhibition of c-*myc* expression and cessation of cell proliferation (Jonak *et al.*, 1987).

Like the expression of c-*myc*, the expression of the *ras* oncogene, either endogenous or transfected, can be influenced by IFNs. When murine 3T3 cells transformed with the human Ha-*ras*1 gene are cultured in the continuous presence of murine IFN-α/β, revertant colonies arise that no longer give rise to tumours in nude mice. In the revertant cells, there is a significant reduction in c-Ha-*ras*-specific mRNA and of the c-Ha-*ras* p21 protein (Samid *et al.*, 1984).

Effects on Differentiation

IFNs can cooperate with other agents to stimulate differentiation of malignant cells. A good example of this capacity is provided by the effects of murine IFN-α/β on Friend erythroleukaemic cells, which become more responsive to a differentiation-inducing chemical after IFN treatment (Rossi, 1985). Moreover, even in the absence of any other differentiation-inducing agent, IFNs have the potential to redirect tumour cells towards a more differentiated state. A pertinent example of this is the plasmacytoid differentiation and refractoriness to growth factors that occurs in Daudi cells after IFN treatment. This is a result of the capacity of IFN-α to act sometimes as a B-cell differentiation factor, which may also explain the success of IFN-α treatment in hairy cell leukaemia (Quesada *et al.*, 1984; Exley *et al.*, 1987). The combination of human IFN-β and the antileukaemic compound mezerein can result in loss of proliferative capacity and terminal differentiation of human melanoma cells (Lin *et al.*, 1995).

CONCLUSION

In this short overview, we have summarized some important properties and activities of IFNs, a family of cytokines that have a truly impressive broad range of activities, with a much wider spectrum than it has been possible to discuss here. By virtue of their broad-spectrum antiviral activity, IFNs stand out from the other cytokines in that they play a unique role in host response to viral infection. But IFNs also belong to the cytokine network, and many of the other cytokines discussed in this book influence the production and action of IFNs, and have their own production and activity influenced by IFNs. Unravelling the intricacy, complexity and biological significance of these interactions is of prime importance for the clinical use of these substances.

REFERENCES

Aguet, M., Grobke, M. and Dreiding, P. (1984). Various human interferon alpha subclasses cross-react with common receptors: their binding affinities correlate with their specific biological activities. *Virology* **132**, 211–216.

Amaldi, I., Reith, W., Berte, C. and Mach, B. (1989). Induction of HLA class II gene by IFN-γ is transcriptional and requires a trans-acting protein. *J. Immunol.* **142**, 999–1004.

Baglioni, C. and Maroney, P.A. (1980). Mechanisms of action of human interferons: induction of 2′,5′-oligo(A) polymerase. *J. Biol. Chem.* **255**, 8390–8393.

Belardelli, F. and Gresser, I. (1996). *Immunol. Today* **17**, 369–372.

Benech, P., Merlin, G., Revel, M. and Chebath, J. (1985a). 3′ end structure of the human (2′–5′) oligo A synthetase gene: prediction of two distinct proteins with cell type-specific expression. *Nucleic Acids Res.* **13**, 1267–1281.

Benech, P., Mory, Y., Revel, M. and Chebath, J. (1985b). Structure of two forms of the interferon-induced (2′–5′) oligo A synthetase of human cells based on cDNAs and gene sequences. *EMBO J.* **4**, 2249–2256.

Benoît, P., Maguire, D., Plavec, I., Kocher, H., Tovey, M. and Meyer, F. (1993). A monoclonal antibody to recombinant human IFN-alpha receptor inhibits biologic activity of several species of human IFN-alpha, IFN-beta, and IFN-omega. Detection of heterogeity of the cellular type I IFN receptor. *J. Immunol.* **150**, 707–716.

Biswas, P., Poli, G., Kinter, A.L., Justement, J.S., Stanley, S.K., Maury, W.J., Bressler, P., Orenstein, J.M. and Fauci, A.S. (1992). Interferon gamma induces the expression of human immunodeficiency virus in persistently infected promonocytic cells (U1) and redirects the production of virions to intracytoplasmic vacuoles in phorbol myristate acetate-differentiated U1 cells. *J. Exp. Med*. **176**, 739–750.

Black, T.L., Barber, G.N. and Katze, M.G. (1993). Degradation of the interferon-induced 68,000-M_r protein kinase by poliovirus requires RNA. *J. Virol.* **67**, 791–800.

Blackman, M.J. and Morris, A.G. (1985). The effect of interferon treatment of targets on susceptibility to cytotoxic T-lymphocyte killing: augmentation of allogenic killing: augmentation of allogenic killing and virus-specific killing relative to viral antigen expression. *Immunology* **56**, 451–457.

Blanar, M.A., Boettger, E.C. and Flavell, R.A. (1988). Transcriptional activation of *HLA-DR*α by interferon γ requires trans-acting protein. *Proc. Natl Acad. Sci. U.S.A.* **85**, 4672–4676.

Bukowski, J.F. and Welsh, R.M. (1985). Inability of interferon to protect virus-infected cells against lysis by natural killer (NK) cells correlates with NK cell-mediated antiviral effects *in vivo*. *J. Exp. Med.* **161**, 257–262.

Cantell, K. and Pirhonen, J. (1996). IFN-gamma enhances production of IFN-alpha in human macrophages but not in monocytes. *J. Interferon Cytokine Res.* **16**, 461–463.

Celis, E., Miller, R.M., Wiktor, T.J., Dietzschold, B. and Koprowski, H. (1986). *J. Immunol.* **136**, 692–697.

Chebath, J., Benech, P., Hovanessian, A., Galabru, J. and Revel, M. (1987). Four different forms of interferon-induced 2′,5′-oligo(A) synthetase identified by immunoblotting in human cells. *J. Biol. Chem.* **262**, 3852–3857.

Chong, K.L., Feng, L., Schappert, K., Meurs, E., Donahue, T.F., Friesen, J.D., Hovanessian, A.G. and Williams, B.R. (1992). Human p68 kinase exhibits growth suppression in yeast and homology to the translational regulator GCN2. *EMBO J.* **11**, 1553–1562.

Cohen, B., Peretz, D., Vaiman, D., Benech, P. and Chebath, J. (1988). Enhancer-like interferon responsive sequences of the human and murine (2′–5′) oligoadenylate synthetase gene promoters. *EMBO J.* **7**, 1411–1419.

Colamonici, O.R. and Domanski, P. (1993). The identification of a novel subunit of the type I interferon receptor localized to human chromosome 21. *J. Biol. Chem.* **268**, 10 895–10 899.

Constantinescu, S.N., Croze, E., Wang, C., Murti, A., Basu, L., Mullersman, J.E. and Pfeffer, L.M. (1994). Role of interferon alpha/beta receptor chain 1 in the structure and transmembrane signaling of the interferon alpha/beta receptor complex. *Proc. Natl Acad. Sci. U.S.A.* **91**, 9602–9606.

Cross, J.C. and Roberts, M.R. (1991). Constitutive and trophoblast-specific expression of a class of bovine interferon genes. *Proc. Natl Acad. Sci. U.S.A.* **88**, 3817–3821.

Cunningham, A.L., Nelson, P.A., Fathman, C.G. and Merigan, T.C. (1985). Interferon gamma production by herpes simplex virus antigen-specific T cell clones from patients with recurrent herpes labialis. *J. Gen. Virol.* **66**, 249–258.

Dale, T.C., Imam, A.M.A., Kerr, I.M. and Stark, G.R. (1989a). Rapid activation by interferon α of a latent DNA-binding protein present in the cytoplasm of untreated cells. *Proc. Natl Acad. Sci. U.S.A.* **86**, 1203–1207.

Dale, T.C., Rosen, J.M., Guille, M.J., Lewin, A.R., Porter, A.G.C., Kerr, I.M. and Stark, G.R. (1989b). Overlapping sites for constitutive and induced DNA binding factors involved in interferon-stimulated transcription. *EMBO J.* **8**, 831–839.

Dalton, D.K., Pitts-Meek, S., Keshav, S., Figari, I.S., Bradley, A. and Stewart, T.A. (1993). Multiple defects of immune cell function in mice with disrupted interferon-γ genes. *Science* **259**, 1739–1742.

Dandoy, F., Kelley, K.A., De Maeyer-Guignard, J., De Maeyer, E. and Pitha, P.M. (1984). Linkage analysis of the murine interferon-α locus on chromosome 4. *J. Exp. Med.* **160**, 294–302.

Darnell, J.E., Jr., Kerr, I.M. and Stark G.R. (1994). Jak-STAT pathways and transcriptional activation in response to IFNs and other extracellular signalling proteins. *Science* **264**, 1415–1421.

De Maeyer, E. and Dandoy, F. (1987). Linkage analysis of the murine interferon alpha locus (Ifa) on chromosome 4. *Heredity* **78**, 143–146.

De Maeyer, E. and De Maeyer-Guignard, J. (1988). *Interferons and Other Regulatory Cytokines*, John Wiley, New York.

De Maeyer-Guignard, J., Tovey, M.G., Gresser, I. and De Maeyer, E. (1978). Purification of mouse interferon by sequential affinity chromatography on poly(U)—and antibody—agarose columns. *Nature* **271**, 622–625.

De Maeyer-Guignard, J., Cachard, A. and De Maeyer, E. (1980). Electrophoretically pure mouse interferon has priming but no blocking activity in poly (I.C.)-induced cells. *Virology* **102**, 222–225.

De Maeyer-Guignard, J., Lauret, E., Eusebe, L. and De Maeyer, E. (1993). Accelerated tumor development in interferon-treated B6.C-Hyal-1 a mice. *Proc. Natl Acad. Sci. U.S.A.* **90**, 5708–5712.

Decker, T., Lew, D.J., Mirkovitch, J. and Darnell, J.E., Jr. (1991). Cytoplasmic activation of GAF, and IFN-γ regulated DNA-binding factor. *EMBO J.* **10**, 927–932.

Devos, R., Cheroutre, H., Taya, Y., Degrave, W., Van Heuverswyn, H. and Fiers, W. (1982). Molecular cloning of human immune interferon cDNA and its expression in eukaryotic cells. *Nucleic Acids Res.* **10**, 2487–2501.

Diaz, M.O., Pomykala, H.M., Bohlander, S.K., Maltepe, E., Malik, K., Brownstein, B. and Olopade, O.I. (1994). Structure of the human type I interferon gene cluster determined from a YAC clone contig. *Genomics* **22**, 540–552.

Edwards, B.S., Merrin, J.A., Fuhlbridge, R.C. and Borden, E.C. (1985). Low doses of interferon alpha result in more effective clinical natural killer cell activation. *J. Clin. Invest.* **75**, 1908–1913.

Exley, R., Gordon, J., Nathan, P., Walker, L. and Clemens, M.J. (1987). Anti-proliferative effects of interferons on Daudi Burkitt lymphoma cells: induction of cell differentiation and loss of response to autocrine growth factors. *Int. J. Cancer* **40**, 53–57.

Fan, C.-M. and Maniatis, T. (1989). Two different virus-inducible elements are required for human β-interferon gene regulation. *EMBO J.* **8**, 101–110.

Farrar, M.A., Fernandez-Luna, J. and Schreiber, R.D. (1991). Identification of two regions within the cytoplasmic domain of the human interferon-γ receptor required for function. *J. Biol. Chem.* **266**, 19626–19635.

Feng, G.S., Chong, K., Kumar, A., and Williams, B.R.G. (1992). Identification of double-stranded-RNA binding domains in the interferon-induced soluble stranded RNA-activated p68 kinase. *Proc. Natl Acad. Sci. U.S.A.* **89**, 5447–5451.

Field, A.K., Tytell, A.A., Lampson, G.P. and Hilleman, M.R. (1967). Inducers of interferon and host resistance. II. Multistranded synthetic polynucleotide complexes. *Proc. Natl Acad. Sci. U.S.A.* **58**, 1004–1010.

Flores, I., Mariano, T.M. and Pestka, S. (1991). Human interferon omega (omega) binds to the alpha/beta receptor. *J. Biol. Chem.* **266**, 19875–19877.

Foster, G.R., Rodrigues, O., Ghouze, F., Schulte-Frohlinde, E., Testa, D., Liao, M.J., Stark, G.R., Leadbeater, L. and Thomas, H.C. (1996). *J. Interferon Cytokine Res.* **16**, 1027–1033.

Fradelizi, D. and Gresser, I. (1982). Interferon inhibits the regeneration of allospecific suppressor T lymphocytes. *J. Exp. Med.* **155**, 1610–1622.

Fu, X.Y. (1992). A transcription factor with SH2 and SH3 domains is directly activated by an interferon α-induced cytoplasmic protein tyrosine kinase(s). *Cell* **70**, 323–335 .

Fu, X.Y. and Zhang, J.J. (1993). Transcription factor p91 interacts with the epidermal growth factor receptor and mediates activation of the c-fos gene promoter. *Cell* **74**, 1135–1145.

Fu, X.Y., Kessler, D.S., Veals, S.A., Levy, D.E. and Darnell, J.E., Jr. (1990). ISGF3, the transcriptional activator induced by interferon α, consists of multiple interacting polypeptide chains. *Proc. Natl Acad. Sci. U.S.A.* **87**, 8555–8559.

Fu, X.Y., Schindler, C., Improta, T., Aebersold, R. and Darnell, J.E., Jr. (1992). The proteins of ISGF-3, the interferon α-induced transcriptional activator, define a gene family involved in signal transduction. *Proc. Natl Acad. Sci. U.S.A.* **89**, 7840–7843.

Fujita, T., Kimura, Y., Miyamoto, M., Barsoumian, E.L. and Taniguchi, T. (1989a). Induction of endogenous IFN-α and IFN-β genes by a regulatory transcription factor, IRF-1. *Nature* **337**, 270–272.

Fujita, T., Miyamoto, M., Kimura, Y., Hammer, J. and Taniguchi, T. (1989b). Involvement of a cis-element that binds an H2TF-1NFκB like factor(s) in the virus-induced interferon-β gene expression. *Nucleic Acids Res.* **17**, 3335–3346.

Fujita, T., Reis, L.F.L., Watanabe, N., Kimura, Y., Taniguchi, T. and Vilcek, J. (1989c). Induction of the transcription factor IRF-1 and interferon-β mRNAs by cytokines and activators of second-messenger pathways. *Proc. Natl Acad. Sci. U.S.A.* **86**, 9936–9940.

Gajewski, T.F. and Fitch, F.W. (1988). Anti-proliferative effect of IFN-gamma in immune regulation. I. IFN-gamma inhibits the proliferation of Th2 but not Th1 murine helper T lymphocyte clones. *J. Immunol.* **140**, 4245–4252.

Galabru, J., Katze, M.G., Robert, N. and Hovanessian, A.G. (1989). The binding of double-stranded RNA and adenovirus VAI RNA to the interferon induced protein kinase. *Eur. J. Biochem.* **178**, 581–589.

Garotta, G., Talmadge, K.W., Pink, J.R., Dewald, B. and Baggiolini, M. (1986). Functional antagonism between type I and type II interferons on human macrophages. *Biochem. Biophys. Res. Commun.* **140**, 948–954.

Gessani, S., Belardelli, F., Pecorelli, A., Puddu, P. and Baglioni, C. (1989). Bacterial lipopolysaccharide and gamma interferon induce transcription of beta interferon mRNA and interferon secretion in murine macrophages. *J. Virol.* **63**, 2785–2789.

Goodbourn, S. and Maniatis, T. (1988). Overlapping positive and negative regulatory domains of the human beta-interferon gene. *Proc. Natl Acad. Sci. U.S.A.* **85**, 1447–1451.

Goodbourn, S., Zinn, K. and Maniatis, T. (1985). Human β-interferon gene expression is regulated by an inducible enhancer element. *Cell* **41**, 509–520.

Gray, P.W. and Goeddel, D.V. (1982). *Nature* **298**, 859–863.

Gray, P.W., Leung, D.W., Pennica, D., Yelverton, E., Najarian, R., Simonsen, C.C., Derynck, R., Sherwood, P.J., Wallace, D.M., Berger, S.L., Levinson, A.D. and Goeddel, D.V. (1982). Expression of human immune interferon cDNA in *E. coli* and monkey cells. *Nature* **295**, 503–508.

Gray, P.W., Leong, S., Fennie, E.H., Farrar, M.A., Pingel, J.T., Fernandez-Luna, J. and Schreiber, R.D. (1989). Cloning and expression of the cDNA for the murine interferon γ receptor. *Proc. Natl Acad. Sci. U.S.A.* **86**, 8497–8501.

Gresser, I. (1984). Role of interferon in resistance to viral infection *in vivo*. In: *Interferon 2: Interferons and the Immune System* (eds J. Vilcek and E. De Maeyer), Elsevier Science Publishers, Amsterdam, pp. 221–247.

Gresser, I., Bandu, M.T. and Brouty-Boye, D. (1974). Interferon and cell division. IX. Interferon-resistant L1210 cells: characteristics and origin. *J. Natl Cancer Inst.* **52**, 553–559.

Gribaudo, G., Lembo, D., Cavallo, G., Landolfo, S., and Lengyel, P. (1991). Interferon action: binding of viral RNA to the 40-kilodalton 2′-5′-oligoadenylate synthetase in interferon-treated HeLa cells infected with encephalomyocarditis virus. *J. Virol.* **65**, 1748–1757.

Hansen, T.R., Leaman, D.W., Cross, J.C., Mathialagan, N., Bixby, J.A. and Roberts, R.M. (1991). The genes for the trophoblast interferons and the related interferon-α II possess distinct 5′-promoter and 3′-flanking sequences. *J. Biol. Chem.* **266**, 3060–3067.

Harada, H., Fujita, T., Miyamoto, M., Kimura, Y., Maruyama, M., Furia, A., Miyata, T. and Taniguchi, T. (1989). Structurally similar but functionally distinct factors, IRF-1 and IRF-2, bind to the same regulatory elements of IFN and IFN-inducible genes. *Cell* **58**, 729–739.

Harpur, A.G., Andres, A.-C., Ziemiecki, A., Aston, R.R. and Wilks, A.F. (1992). JAK2, a third member of the JAK family of protein tyrosine kinases. *Oncogene* **7**, 1347–1353.

Hayashi, H., Tanaka, K., Jay, F., Khoury, G. and Jay, G. (1985). Modulation of the tumorigenicity of human adenovirus-12-transformed cells by interferon. *Cell* **43**, 263–267.

Hemmi, S., Peghini, P., Metzler, M., Merlin, G., Dembic, Z. and Aguet, M. (1989). Cloning of murine interferon gamma receptor cDNA: expression in human cells mediates high-affinity binding but is not sufficient to confer sensitivity to murine interferon gamma. *Proc. Natl Acad. Sci. U.S.A.* **86**, 9901–9905.

Hemmi, S., Böhni, R., Stark, G., Di Marco, F. and Aguet M. (1994). A novel member of the interferon receptor family complements functionality of the murine interferon γ receptor in human cells. *Cell* **76**, 803–810.

Herberman, R.B., Ortaldo, J.R., Mantovani, A., Hobbs, D.S., Kung, H.F. and Pestka, S. (1982). Effect of human recombinant interferon on cytotoxic activity of natural killer (NK) cells and monocytes. *Cell Immunol.* **67**, 160–167.

Hibino, Y., Mariano, T.M., Kumar, C.S., Kozak, C.A. and Pestka, S. (1991). Expression and reconstitution of a biologically active mouse interferon gamma receptor in hamster cells: chromosomal location of an accessory factor. *J. Biol. Chem.* **266**, 6948–6951.

Hibino, Y., Kumar, C.S., Mariano, T.M., Lai, D.H. and Pestka, S. (1992). Chimeric interferon-gamma receptors demonstrate that an accessory factor required for activity interacts with the extracellular domain. *J. Biol. Chem.* **267**, 3741–3749.

Hiscott, J., Cantell, K. and Weissmann, C. (1984). Differential expression of human interferon genes. *Nucleic Acids Res.* **12**, 3727–3746.

Hiscott, J., Alper, D., Cohen, L., Leblanc, J.F., Sportza, L., Wong, A. and Xanthoudakis, S. (1989). Differential expression of human interferon genes. *J. Virol.* **63**, 2557–2566.

Hoover, D.L., Nacy, C.A. and Meltzer, M.S. (1985). Human monocyte activation for cytotoxicity against intracellular *Leishmania donovani amastigotes*: induction of microbicidal activity by interferon-gamma. *Cell. Immunol.* **94**, 500–511.

Hovanessian, A.G., Laurent, A.G., Chebath, J., Galabru, J., Robert, N. and Svab, J. (1987). Identification of 69-kd and 100-kd forms of 2–5A synthetase in interferon-treated human cells by specific monoclonal antibodies. *EMBO J.* **6**, 1273–1280.

Hovanessian, A.G., Svab, J., Marie, I., Robert, N., Chamaret, S. and Laurent, A.G. (1988). Characterization of 69- and 100-kDa forms of 2–5A synthetase from interferon-treated human cells. *J. Biol. Chem.* **263**, 4945–4949.

Huang, S., Hendriks, W., Althage, A., Hemmi, S., Bluethmann, H., Kamijo, R., Vilcek, J., Zinkernagel, R.M. and Aguet, M. (1993). Immune response in mice that lack the interferon-gamma receptor. *Science* **259**, 1742–1745.

Hug, H., Costa, M., Staeheli, P., Aebi, M. and Weissmann, C. (1988). Organisation of the murine *Mx* gene and characterization of its interferon- and virus-inducible promoter. *Mol. Cell. Biol.* **8**, 3065–3079.

Ichii, Y., Fukunaga, R., Shiojiri, S. and Sokawa, Y. (1986). Mouse 2–5A synthetase cDNA: nucleotide sequence and comparison to human 2–5A synthetase. *Nucleic Acids Res.* **14**, 101–117.

Imam, A.M.A., Ackrill A.M., Dale, T.C., Kerr, I.M. and Stark, G.R. (1990). Transcription factors induced by interferon α and γ. *Nucleic Acids Res.* **18**, 6573–6580.

Isaacs, A. and Lindenmann, J. (1957). Virus interference. I. The interferon. *Proc. R. Soc. (London), Ser. B.* **147**, 258–273.

Jacobs, B.L. and Langland, J.O. (1996). When two strands are better than one: the mediators and modulators of the cellular responses to ds-RNA. *Virology* **219**, 339–349.

Jiang, H., Lin, J., Young, S.M., Goldstein, N.I., Waxman, S., Davila, V., Chellappan, S.P. and Fisher, P.B. (1995). Cell cycle gene expression and E2F transcription factor complexes in human melanoma cells induced to terminally differentiate. *Oncogene* **11**, 1179–1189.

Jonak, G.J., Friedland, B.K., Anton, E.D. and Knight, E., Jr. (1987). Regulation of c-myc RNA and its proteins in Daudi cells by interferon-beta. *J. Interferon Res.* **7**, 41–52.

Jouanguy, E., Altare, F., Lamhamedi, S., Revy, P., Emile, J.F., Newport, M., Levin, M., Blanche, S., Seboun, E., Fischer, A. and Casanova, J.L. (1996). Interferon-γ-receptor deficiency in an infant with fatal bacille Calmette-Guérin infection. *N. Engl. J. Med.* **335**, 1956–1961.

Kalina, U., Ozmen, L., Di Padova, K., Gentz, R. and Garotta, G. (1993). The human gamma interferon receptor accessory factor encoded by chromosome 21 transduces the signal for the induction of 2′,5′-oligoadenylate-synthetase, resistance to virus cytopathic effect, and major histocompatibility complex class I antigens. *J. Virol.* **67**, 1702–1706.

Kamijo, R., Shapiro, D., Le, J., Huang, S., Aguet, M. and Vilcek, J. (1993). Generation of nitric oxide and induction of major histocompatibility complex class II antigen in macrophages from mice lacking the interferon γ receptor. *Proc. Natl Acad. Sci. U.S.A.* **90**, 6626–6630.

Keller, A.D. and Maniatis, T. (1991). Identification and characterization of a novel repressor of beta-interferon gene expression. *Genes Dev.* **5**, 868–879.

Kelley, K.A. and Pitha, P.M. (1985a). Characterization of a mouse interferon gene locus. I. Isolation of a cluster of four α interferon genes. *Nucleic Acids Res.* **13**, 805–823.

Kelley, K.A. and Pitha, P.M. (1985b). Characterization of a mouse interferon gene locus. II. Differential expression of α-interferon genes. *Nucleic Acids Res.* **13**, 825–839.

Kessler, D.S., Veals, S.A., Fu, X.Y. and Levy D.E. (1990). Interferon-α regulates nuclear translocation and DNA-binding affinity of ISGF3. *Genes Dev.* **4**, 1753–1765.

Kibler, K.V., Shors, T., Perkins, K.B., Zeman, C.C., Banaszak, M.P., Biesterfeldt, J., Langland, J.O. and Jacobs, B.L. (1997). Double-stranded RNA is a trigger for apoptosis in vaccinia virus-infected cells. *J. Virol.* **71**, 1992–2003.

Kleinerman, E.S., Zicht, R., Sarin, P.S., Gallo, R.C. and Fidler, I.J. (1984). Constitutive production and release of a lymphokine with macrophage-activating factor activity distinct from gamma-interferon by a human T-cell leukemia virus-positive cell line. *Cancer Res.* **44**, 4470–4475.

Knop, J., Stremmer, R., Neumann, C., De Maeyer, E. and Macher, E. (1982). Interferon inhibits the suppressor T cell response of delayed-type hypersensitivity. *Nature* **296**, 775–776.

Knop, J., Taborksi, B. and De Maeyer-Guignard, J. (1987). Selective inhibition of the generation of T suppressor cells of contact sensitivity *in vitro* by interferon. *J. Immunol.* **138**, 3684–3687.

Koromilas, A.E., Roy, S., Barber G.N., Katze, M.G. and Sonenberg, N. (1992). *Science* **257**, 1685–1689.

Kumar, A., Haque, J., Lacoste, J., Hiscott, J. and Williams, B.R. (1994). Ds-RNA dependent protein kinase activates transcription factor NF-κB by phosphorylating I kappaB. *Proc. Natl Acad. Sci. U.S.A.* **91**, 6288–6292.

Kumar, C.S., Muthukumaran, G., Frost, L.J., Noe, M., Ahn, Y.H., Mariano, T.M. and Pestka, S. (1989). Molecular characterization of the murine interferon gamma receptor cDNA. *J. Biol. Chem.* **264**, 17 939–17 946.

Larner, A.C., David, M., Feldman, G.M., Igarashi, K.-I., Hackett, R.H., Webb, D.S.A., Sweitzer, S.M., Petricoin, E.F. III and Finbloom, D.S. (1993). Tyrosine phosphorylation of DNA binding proteins by multiple cytokines. *Science* **261**, 1730–1733.

Le, J., Prensky, W., Yip, Y.K., Chang, Z., Hoffman, T., Stevenson, H.C., Balazs, I., Sadlik, J.R. and Vilcek, J. (1983). Activation of human monocyte cytotoxicity by natural and recombinant immune interferon. *J. Immunol.* **131**, 2821–2826.

Lee, S.B. and Esteban, M. (1994). The IFN-induced ds-RNA activated PKR-induced apoptosis *Virology* **199**, 491–496.

Lee, T.G., Tomita, J., Hovanessian, A.G. and Katze, M.G (1990). Purification and partial characterization of a cellular inhibitor of the interferon-induced protein kinase from influenza virus-infected cells. *Proc. Natl Acad. Sci. U.S.A.* **87**, 6208–6212.

Lenardo, M.J. and Baltimore, D. (1989). NF-κB: a pleiotropic mediator of inducible and tissue-specific gene control. *Cell* **58**, 227–229.

Lenardo, M.J., Fan, C.-M., Maniatis, T. and Baltimore, D. (1989). The involvement of NF-κB in β-interferon gene regulation reveals its role as widely inducible mediator of signal transduction. *Cell* **57**, 287–294.

Levy, D.E., Kessler, D.S., Pine, R., Reich, N. and Darnell, J.E., Jr. (1988). Interferon-induced nuclear factors that bind a shared promoter element correlate with positive and negative transcriptional control. *Genes Dev.* **2**, 383–393.

Levy, D.E., Kessler, D.S., Pine, R. and Darnell, J.E., Jr. (1989). Cytoplasmic activation of ISGF3, the positive regulator of interferon-α-stimulated transcription, reconstituted *in vitro. Genes Dev.* **3**, 1362–1371.

Lew, D.J., Decker, T., Strehlow, I. and Darnell, J.E. (1991). Overlapping elements in the guanylate-binding protein gene promoter mediate transcriptional induction by alpha and gamma interferons. *Mol. Cell. Biol.* **11**, 182–191.

Lin, S.L., Greene, J.J., Ts'o, P.O. and Carter, W.A. (1982). Sensitivity and resistance of human tumor cells to interferon and rIn.rCn. *Nature* **297**, 417–419.

Ling, L.E., Zafari, M., Reardon, D., Brickelmeier, M., Goelz, S.E. and Benjamin, C.D. (1995). Human type I interferon receptor, IFNAR, is a heavily glycosylated 120–130 kD membrane protein. *J. Interferon Cytokine Res.* **15**, 55–61.

Ludwig, C.U., Durie, B.G., Salmon, S.E. and Moon, T.E. (1983). Tumor growth stimulation *in vitro* by interferons. *Eur. J. Cancer Clin. Oncol.* **19**, 1625–1632.

Lutfalla, G., Roeckel, N., Mogensen, K.E., Mattei, M.-G. and Uzé, G. (1990). Assignment of the human interferon-α receptor gene to chromosome 21q22.1 by *in situ* hybridization. *J. Interferon Res.* **10**, 515–517.

Lutfalla, G., Gardiner, K., Proudhon, D., Vielh, E. and Uzé, G. (1992). The structure of the human interferon α/β receptor gene. *J. Biol. Chem.* **267**, 2802–2809.

MacDonald, N.J., Kuhl, D., Maguire, D., Naf, D., Gallant, P., Goswamy, A., Hug, H., Bueler, H., Chaturvedi, M., de la Fuente, J., Ruffner, H., Meyer, F. and Weissmann, C. (1990). Different pathways mediate virus inducibility of the human IFN-α1 and IFN-β genes. *Cell* **60**, 767–779.

McKendry, R., John, J., Flavell, D., Muller, M., Kerr, I.M. and Stark, G.R. (1991). High-frequency mutagenesis of human cells and characterization of a mutant unresponsive to both alpha and gamma interferons. *Proc. Natl Acad. Sci. U.S.A.* **88**, 11 455–11 459.

Merlin, G., Chebath, J., Benech, P., Metz, R. and Revel, M. (1983). Molecular cloning and sequence of partial cDNA for interferon-Induced (2′–5′)oligo(A) synthetase mRNA from human cells. *Proc. Natl Acad. Sci. U.S.A.* **80**, 4904–4908.

Meurs, D.F., Galabru, J., Barber, G.N., Katze, M.G. and Hovanessian A.G. (1993). Tumor suppressor function of the interferon-induced ds-RNA activated protein kinase. *Proc. Natl Acad. Sci. U.S.A.* **90**, 232–236.

Miyamoto, M., Fujita, T., Kumura, Y., Maruyama, M., Harada, H., Sudo, Y., Miyata, T. and Taniguchi, T. (1988). Regulated expression of a gene encoding a nuclear factor, IRF-1, that specifically binds to IFN-β gene regulatory elements. *Cell* **54**, 903–913.

Moore, R.N., Larsen, H.S., Horohov, D.W. and Rouse, B.T. (1984). Endogenous regulation of macrophage proliferative expansion by CSF-induced interferon. *Science* **223**, 178–181.

Morris, A.G., Lin, Y.L. and Askonas, B.A. (1982). Immune interferon release when a cloned cytotoxic T-cell line meets its correct influenza-infected target cell. *Nature* **295**, 150–152.

Mossman, T.R., Cherwinski, H., Bond, M.W., Giedlin, M.A. and Coffman, R.L. (1986). Two types of murine helper T cell clone. I. Definition according to profiles of lymphokine activities and secreted proteins. *J. Immunol*. **136**, 2348–2357.

Müller, M., Briscoe, J., Laxton, C., Guschin, D., Zlemlecki, A., Silvennoinen, O., Harpur, A.G., Barbleri, G., Witthuhn, B.A., Schindler, C., Pellegrini, S., Wilks, A.F., Ihle, J.N., Stark, G.R. and Kerr, I.M. (1993). The protein tyrosine kinase JAK1 complements defects in interferon-α/β and -γ signal transduction. *Nature* **366**, 129–135.

Müller, U., Steinhoff, U., Reis, L.F.L., Hemmi, S., Pavlovic, J., Zinkernagel, R.M. and Aguet, M. (1994). Functional role of type I and type II interferons in antiviral defense. *Science* **264**, 1918–1921.

Murray, H.W. (1981). Susceptibility of *Leishmania* to oxygen intermediates and killing by normal macrophages. *J. Exp. Med.* **153**, 1302–1315.

Murray, H.W., Spitalny, G.L. and Nathan, C.F. (1985). Activation of mouse peritoneal macrophages in vitro and in vivo by interferon-gamma. *J. Immunol.* **134**, 1619–1622.

Nathan, C.F. and Tsunawaki, S. (1986). Secretion of toxic oxygen products by macrophages: regulatory cytokines and their effects on the oxidase. In: *Biochemistry of Macrophages* (eds D. Evered, J. Nugent and M. O'Connor), Pitman, London, Ciba Foundation Symposium Vol. 118, pp. 211–230.

Nathan, C.F., Murray, H.W., Wiebe, M.E. and Rubin, B.Y. (1983). Identification of interferon-gamma as the lymphokine that activates human macrophage oxidative metabolism and antimicrobial activity. *J. Exp. Med.* **158**, 670–689.

Nathan, C.F., Horowitz, C.R., De La Harpe, J., Vadhan-Raj, S., Sherwin, S.A., Oetten, H.F. and Krown, S.E. (1985). Administration of recombinant interferon γ to cancer patients enhances monocyte secretion of hydrogen peroxide. *Proc. Natl Acad. Sci. U.S.A.* **82**, 8686–8690.

Naylor, S.L., Gray, P.W. and Lalley, P.A. (1984). Mouse immune interferon (IFN-gamma) gene is on chromosome 10. *Somat. Cell Mol. Genet.* **10**, 531–534.

Newport, M.J., Huxley, C.M., Huston, S., Hawrylowicz, C.M., Oostra, B.A., Williamson, R. and Levin, M. (1996). A mutation in the interferon-γ-receptor gene and susceptibility to mycobacterial infection. *N. Engl. J. Med.* **335**, 1941–1949.

Nilsen, T.W., Maroney, P.A. and Baglioni, C. (1981). Double-stranded RNA causes synthesis of 2′,5′-oligo(A) and degradation of messenger RNA in interferon-treated cells. *J. Biol. Chem.* **256**, 7806–7811.

Noma, T. and Dorf, M.E. (1985). Modulation of suppressor T cell induction with gamma-interferon. *J. Immunol.* **135**, 3655–3660.

Novick, D., Cohen, B. and Rubinstein, M. (1994). The human interferon alpha/beta receptor: characterization and molecular cloning. *Cell* **77**, 391–400.

Palombella, V.J. and Maniatis, T. (1992). Inducible processing of interferon regulatory factor-2. *Mol. Cell. Biol.* **12**, 3325–3336.

Philip, R. and Epstein, L.B. (1986). Tumour necrosis factor as immunomodulator and mediator of monocyte cytotoxicity induced by itself, gamma-interferon and interleukin-1. *Nature* **323**, 86–89.

Pine, R. (1992). Constitutive expression of an ISGF2/IRF1 transgene leads to interferon-independent activation of interferon-inducible genes and resistance to virus infection. *J. Virol.* **66**, 4470–4478.

Pine, R., Decker, T., Kessler, D.S., Levy, D.E. and Darnell, J.E., Jr. (1990). Purification and cloning of interferon-stimulated gene factor 2 (ISGF2): ISGF2 (IRF-1) can bind to the promoters of both beta interferon- and interferon-stimulated genes but is not a primary transcriptional activator of either. *Mol. Cell. Biol.* **10**, 2448–2457.

Proietti, E., Vanden Broecke, C., Di Marzio, P., Gessani, S., Gresser, I. and Tovey, M.G. (1992). Specific interferon genes are expressed in individual cells in the peritoneum and bone marrow of normal mice. *J. Interferon Res.* **12**, 27–34.

Quesada, J.R., Reuben, J., Manning, J.T., Hersh, E.M. and Gutterman, J.U. (1984). Alpha interferon for induction of remission in hairy-cell leukemia. *N. Engl. J. Med.* **310**, 15–18.

Reich, N.C. and Darnell, J.E., Jr. (1989). Differential binding of interferon-induced factors to an oligonucleotide that mediates transcriptional activation. *Nucleic Acids Res.* **17**, 3415–3424.

Reis, L.F.L., Lee, T.H. and Vilcek, J. (1989). Tumor necrosis factor acts synergistically with autocrine interferon-β and increases interferon-β mRNA levels in human fibroblasts. *J. Biol. Chem.* **264**, 16 351–16 354.

Rinderknecht, E., O'Connor, B.H. and Rodriguez, H. (1984). Natural human interferon-gamma. Complete amino acid sequence and determination of sites of glycosylation. *J. Biol. Chem.* **259**, 6790–6797.

Roberts, M.R., Cross, J.C. and Leaman, D.W. (1991). Unique features of the trophoblast interferons. *Pharmacol. Ther.* **51**, 329–345.

Rossi, G.B. (1985). Interferons and cell differentiation. *Interferon* **6**, 31–68.

Roy, S., Agy, M., Hovanessian, H.G., Sonenberg, N., and Katze, M.G. (1991). The integrity of the stem stucture of human immunodeficiency virus type 1 Tat-responsive sequence RNA is required for interaction with the interferon-induced 68,000-Mγ protein kinase. *J. Virol.* **65**, 632–640.

Rutherford, M.N., Kumar, A., Nissim, A., Chebath, J. and Williams B.R.G. (1991). The murine 2–5A synthetase locus: three distinct transcripts from two linked genes. *Nucleic Acids Res.* **19**, 1917–1924.

Ryals, J., Dierks, P., Ragg, H. and Weissmann, C. (1985). A 46-nucleotide promoter segment from an interferon alpha gene renders an unrelated promoter inducible by virus. *Cell* **41**, 497–507.

St Laurent, G., Yoshie, O., Floyd-Smith, G., Samanta, H., Sehgal, P.B. and Lengyel, P. (1983). Interferon action: two (2′–5′)(A)n synthetases specified by distinct mRNAs in Ehrlich ascites tumor cells treated with interferon. *Cell* **33**, 95–102.

Samid, D., Chang, E.H. and Friedman, R.M. (1984). Biochemical correlates of phenotypic reversion in interferon-treated mouse cells transformed by a human oncogene. *Biochem. Biophys. Res. Commun.* **119**, 21–28.

Samuel, C.E., Duncan, G.S., Knutson, G.S. and Hershey, J.W.B. (1984). Mechanisms of interferon action. *J. Biol. Chem.* **259**, 13 451–13 457.

Schindler, C., Fu, X.Y., Improta, T., Aebersold, R. and Darnell, J.E., Jr. (1992a). Proteins of transcription factor ISGF-3: One gene encodes the 91 and 84 kDa ISGF-3 proteins that are activated by interferon. *Proc. Natl Acad. Sci. U.S.A.* **89**, 7836–7839.

Schindler, C., Shuai, K., Prezioso, V.R. and Darnell, J.E., Jr. (1992b). Interferon-dependent tyrosine phosphorylation of a latent cytoplasmic transcription factor [see comments]. *Science* **257**, 809–813.

Seif, I. and De Maeyer-Guignard, J. (1986). Structure and expression of a new murine interferon-alpha gene: MuIFN-alpha9. *Gene* **43**, 111–121.

Seif, I., De Maeyer, E., Riviere, I. and De Maeyer-Guignard, J. (1991). Stable antiviral expression in BALB/c 3T3 cells carrying a beta interferon sequence behind a major histocompatibility complex promoter fragment. *J. Virol.* **65**, 664–671.

Sekellick, M.J. and Marcus, P.I. (1982). Interferon induction by viruses. *Virology* **117**, 280–285.

Shirayoshi, Y., Burke, P.A., Appella, E. and Ozato, K. (1988). Interferon-induced transcription of a major histocompatibility class I gene accompanies binding of inducible factors to the interferon consensus sequence. *Proc. Natl Acad. Sci. U.S.A.* **85**, 5884–5888.

Shuai, K., Schindler, C., Prezioso, V.R. and Darnell, J.E., Jr. (1992). Activation of transcription by IFN-γ: tyrosine phosphorylation of a 91-kD DNA binding protein. *Science* **258**, 1808–1812.

Shuai, K., Horvath, C.M., Tsai Huang, L.H., Qureshi, S.A., Cowburn, D. and Darnell, J.E., Jr. (1994). Interferon activation of the transcription factor Stat91 involves dimerization through SH2-phosphotyrosyl peptide interactions. *Cell* **76**, 821–828.

Silverman, R.H., Skehel, J.J., James, T.C., Wreschner, D. H. and Kerr, I.M. (1983). rRNA cleavage as an index of ppp(A2′p)nA activity in interferon-treated encephalomyocarditis virus-infected cells. *J. Virol.* **46**, 1051–1055.

Sims, S.H., Cha, Y., Romine, M.F., Gao, P.Q., Gottlieb, K. and Deisseroth, A.B. (1993). A novel interferon-inducible domain: structural and functional analysis of the human interferon regulatory factor 1 gene promoter. *Mol. Cell. Biol.* **13**, 690–702.

Soh, J., Donnely, R.J., Kotenko, S., Mariano, T.M., Cook, J.R., Wang, N., Emanuel, S., Schwartz, B., Miki, T. and Pestka, S. (1994). Identification and sequence of an accessory factor required for activation of the human interferon γ receptor. *Cell* **76**, 793–802.

Staeheli, P. (1990). The Mx locus. In: *Advances in Virus Research*, Academic Press, London, Vol. 38, pp. 147–200.

Svedersky, L.P., Benton, C.V., Berger, W.H., Rinderknecht, E., Harkins, R.N. and Palladino, M.A. (1984). Biological and antigenic similarities of murine interferon-gamma and macrophage-activating factor. *J. Exp. Med.* **159**, 812–827.

Taylor, M.W. and Feng, G. (1991). Relationship between interferon-γ, indoleamine 2,3-dioxygenase, and tryptophan catabolism. *FASEB J.* **5**, 2516–2522.

Taylor, P.M., Wraith, D.C. and Askonas, B.A. (1985). Control of immune interferon release by cytotoxic T-cell clones specific for influenza. *Immunology* **54**, 607–614.

Thomas, S.M., Garrity, L.F., Brandt, C.R., Schobert, C.S., Feng, G.S., Taylor, M.W., Carlin, J.M. and Byrne, G. (1993). IFN-γ-mediated antimicrobial response. *J. Immunol.* **150**, 5529–5534.

Tovey, M.G., Streuli, M., Gresser, I., Gugenheim, J., Blanchard, B., Guymarho, J., Vignaux, F. and Gigou, M. (1987). Interferon messenger RNA is produced constitutively in the organs of normal individuals. *Proc. Natl Acad. Sci. U.S.A.* **84**, 5038–5042.

Urban, J.L., Shepard, H.M., Rothstein, J.L., Sugarman, B.J. and Schreiber, H. (1986). Tumor necrosis factor: a potent effector molecule for tumor cell killing by activated macrophages. *Proc. Natl Acad. Sci. U.S.A.* **83**, 5233–5237.

Uzé, G., Lutfalla, G. and Gresser, I. (1990). Genetic transfer of a functional human interferon receptor into mouse cells: cloning and expression of its DNA. *Cell* **60**, 225–234.

Uzé, G., Lutfalla, G., Bandu, M.T., Proudhon, D. and Mogensen, K.E. (1992). Behavior of a cloned murine interferon receptor expressed in homospecific or heterospecific background. *Proc. Natl Acad. Sci. U.S.A.* **89**, 4774–4778.

Uzé, G., Lutfalla, G. and Mogensen, K.E. (1995). α and β interferons and their receptor and their friends and relations. *J. Interferon Cytokine Res.* **15**, 3–26.

Veals, S.A., Schindler, C., Leonard, D., Fu, X.Y., Aebersold, R., Darnell, J.E., Jr. and Levy, D.E. (1992). Subunit of an alpha interferon-responsive transcription factor is related to interferon regulatory factor and Myb families of DNA-binding proteins. *Mol. Cell. Biol.* **12**, 3315–3324.

Velazquez, L., Fellous, M., Stark, G.R. and Pellegrini, S. (1992). A protein tyrosine kinase in the interferon alpha/beta signaling pathway. *Cell* **70**, 313–322.

Verhaegen-Lewalle, M., Kuwata, T., Zhang, Z.X., DeClercq, E., Cantell, K. and Content, J. (1982). 2–5A synthetase activity induced by interferon alpha, beta, and gamma in human cell lines differing in their sensitivity to the anticellular and antiviral activities of these interferons. *Virology* **117**, 425–434.

Vieillard, V., Lauret, E., Rousseau, V. and De Maeyer, E. (1994). Blocking of retroviral infection at a step prior to reverse transcription in cells transformed to constitutively express interferon-beta. *Proc. Natl Acad. Sci. U.S.A.* **91**, 2689–2693.

Virelizier, J.L. and Gresser, I. (1978). Role of interferon in the pathogenesis of viral diseases of mice as demonstrated by the use of anti-interferon serum. *J. Immunol.* **120**, 1616–1619.

Wallach, D. (1983). Interferon-induced resistance to the killing by NK cells: a preferential effect of IFN-gamma. *Cell. Immunol.* **75**, 390–395.

Walter, M.R., Windsor, W.T., Nagabhushan, T.L., Lundell, D.J., Lunn, C.A., Zauodny, P.J. and Narula, S.K. (1995). Crystal structure of a complex between interferon-gamma and its soluble high-affinity receptor. *Nature* **376**, 230–235.

Wathelet, M., Moutschen, S., Cravador, A., DeWit, L., Defilippi, P., Huez, G. and Content, J. (1986). Full-length sequence and expression of the 42 kDa 2–5A synthetase induced by human interferon. *FEBS Lett.* **196**, 113–120.

Watling, D., Guschin, D., Muller, M., Silvennoinen, O., Witthuhn, B.A., Quelle, F.W., Rogers, N.C., Schindler, C., Stark, G.R., Ihle, J.N. *et al.* (1993). Complementation by the protein tyrosine kinase JAK2 of a mutant cell line defective in the interferon-gamma signal transduction pathway. *Nature* **366**, 166–170.

Weck, K.E., Dal Canto, A.J., Gould, G.D., O'Guin, A.K., Roth, K.A., Saffitz, J.E., Speck, S.H. and Virgin, H.W. (1997). Murine gamma-herpesvirus 68 causes severe large-vessel arteritis in mice lacking interferon-γ responsiveness: a new model for virus-induced vascular disease. *Nature Med.* **3**, 1346–1353.

Weissmann, C. and Weber, H. (1986). The interferon genes. *Progr. Nucleic Acid Res. Mol. Biol.* **33**, 251–300.

Williams, B.R., Saunders, M.E. and Willard, H.F. (1986). Interferon-regulated human 2–5A synthetase gene maps to chromosome 12. *Somat. Cell Mol. Genet.* **12**, 403–408.

Xanthoudakis, S., Cohen, L. and Hiscott, J. (1989). Multiple protein-DNA interactions within the human interferon-beta regulatory element. *J. Biol. Chem.* **15**, 1139–1145.

Yip, Y.K., Barrowclough, B.S., Urban, C. and Vilcek, J. (1982). Purification of two subspecies of human gamma (immune) interferon. *Proc. Natl Acad. Sci. U.S.A.* **79**, 1820–1824.

Zinkernagel, R.M. and Doherty, P.C. (1974). Restriction of *in vitro* T cell-mediated cytotoxicity in LCM within a syngeneic or semiallogeneic system. *Nature* **224**, 701–702.

Zhong, Z., Wen, Z. and Darnell, J. (1994). *Proc. Natl Acad. Sci. U.S.A.* **91**, 4806–4810.

Zinn, K. and Maniatis, T. (1986). Detection of factors that interact with the human beta-interferon regulatory region *in vivo* by DNAase I footprinting. *Cell* **45**, 611–618.

Chapter 19

Tumor Necrosis Factor

Minghuang Zhang and Kevin J. Tracey

Laboratory of Biomedical Science, North Shore University Hospital and The Picower Institute for Medical Research, Manhasset, New York, USA

INTRODUCTION

Tumor necrosis factor (TNF, TNF-α, cachectin), a primary mediator of immune regulation and the inflammatory response, has been implicated as a beneficial and injurious factor through its proinflammatory and cytotoxic effects. TNF research has moved at an astonishing rate resulting in voluminous information in the areas of molecular and cellular biology, as well as genetics. Here, we review some of the information in these areas, with emphasis on the mechanisms of TNF actions and TNF-related diseases. By necessity, not all of the important work in this field can be cited.

STRUCTURE

There are two forms of TNF, a 26-kDa transmembrane pro-TNF and a 17-kDa secreted mature TNF. Pro-TNF, the precursor of mature TNF, is a type II transmembrane protein with its *N*-terminus inside the cytoplasm and *C*-terminus outside the cells. Newly synthesized pro-TNF is first displayed on the plasma membrane and then cleaved proteolytically between alanine (−1) and valine (+1) in the extracellular domain to release the mature TNF (Kriegler *et al.*, 1988). The 17-kDa mature TNF is the active form mediating the biologic effects. The 26-kDa transmembrane pro-TNF, however, is also active and has been shown to mediate the cytotoxic effect of TNF through cell–cell contact (Kriegler *et al.*, 1988). It is possible that membrane TNF is more potent than mature TNF in activating TNF receptor 2 (TNFR2) (Grell *et al.*, 1995).

Bioactive secreted TNF is a trimer in solution as shown by analytic ultracentrifugation (Wingfield *et al.*, 1987), gel filtration chromatography and cross-linking (Smith and Baglioni, 1987). Crystallographic studies of TNF structure revealed TNF as a β-sheet protein that self-associates non-covalently into a symmetrically compact, bell-shaped trimer (Hakoshima and Tomita, 1988; Eck and Sprang, 1989; Jones *et al.*, 1989; Kindler *et al.*, 1989; Eck *et al.*, 1992). Each subunit consists almost entirely of antiparallel β-pleated sheets, organized into a 'jellyroll β sandwich'. The *C*-terminus of each subunit is embedded in the base of the trimer, and the *N*-terminus is relatively free from the base of

The Cytokine Handbook, 3rd ed.
ISBN 0-12-689662-3

the trimer. Thus the *N*-terminus does not participate in trimer interaction and is not critical for TNF biologic activity (Creasey *et al.*, 1987). Detailed structural analyses suggest that mutations that destabilize the trimeric association of monomers result in the loss of TNF biologic activity (Lin, 1992), suggesting that the trimeric structure of TNF is important for its biologic functions.

Several studies have been carried out to establish the relationship between TNF structure and its biologic activities. The results from mutation analysis of mature TNF indicate that each TNF trimer has three receptor interaction sites located in the intersubunit grooves near the base of the trimer (Van Ostade *et al.*, 1991; Zhang *et al.*, 1992). Two different mutants of TNF, L29S and R32W, exhibit decreased binding to TNFR1 (Loetscher *et al.*, 1993; Van Ostade *et al.*, 1993), and mutations of TNF at W32T86 and N143R145 render it TNFR1 or TNFR2 specific respectively (Loetscher *et al.*, 1993). These results indicate that it is theoretically possible to increase or decrease some particular TNF functions through genetic engineering. In fact, studies of TNF structure–function relationship have been mostly driven by trying to redesign a TNF molecule with maximum efficacy as an antitumor agent, with minimal side-effects. As TNFR2 was believed to be the major mediator of TNF toxicity, there has been widespread interest in identifying a TNFR1-specific TNF molecule. Unfortunately, TNFR1-specific mutants were found to exhibit similar toxicity *in vivo* (Van Zee *et al.*, 1994). Recently, one TNF mutein has been shown to have decreased toxicity in mice without any loss of its antitumor effects (Kuroda *et al.*, 1995).

Transmembrane pro-TNF is also biologically active, but less is known about its structure and structure–function relationship. Recently, Tang *et al.* (1996) found that human pro-TNF, like secreted mature TNF, also has a trimeric structure, which is assembled intracellularly before transport to the cell surface and is apparently required for its biologic activities. A structure–function model has been proposed in which the mature TNF domains of the pro-TNF are arranged as a compact bell-shaped trimer, very similar to the arrangement in the mature TNF crystal (Eck and Sprang, 1989; Jones *et al.*, 1989). The mobile *N*-terminus from each subunit is linked to the 20-residue linking sequence, which, in turn, is connected to the transmembrane domain. When effector cells are juxtaposed to the target cells, pro-TNF on the effector cells can interact with TNF receptors on the target cells, thereby inducing TNF responses via receptor aggregation. Thus, pro-TNF trimers may interact with the extracellular domains of TNF receptor, much as mature TNF does. The leader peptide sequence may be involved in facilitating trimer formation of pro-TNF, but it may not be required for trimer formation (Smith and Baglioni, 1987; Kriegler *et al.*, 1988).

Although TNF is a β-sheet protein, it can be a helical trimer at acidic pH (Hlodan and Pain, 1994; Narhi *et al.*, 1996). TNF is the only protein to date that has no α helix in its native structure but forms an α-helical structure in the molten globule (an intermediate structure produced by acid-induced unfolding of proteins). Under acidic conditions, TNF undergoes a conformational change which results in an increase in surface hydrophobicity and a simultaneous increase in cytolytic activity, membrane insertion and lipid vesicle binding (Yoshimura and Sone, 1987; Baldwin *et al.*, 1988; Chang and Wisnieski, 1990). The inserted proteins appear to be capable of functioning as an ion channel (Kagan *et al.*, 1992; Baldwin *et al.*, 1996). These data have led to a hypothesis that the increased cytolytic activity seen at low pH is due to an acid-induced conformational change which allows TNF to insert into lipid membranes and function as a Na^+

channel. This conformation may also allow TNF to penetrate directly into the target cells, where neutral pH in the cytosol restores the original structure and activity (Chang and Wisnieski, 1990; Kagan *et al.*, 1992). Importantly, the conformational change at low pH may be physiologically relevant, as a pH of about 3.6 has been reported in the space between macrophages and their substrates (Silver *et al.*, 1988).

BIOSYNTHESIS AND REGULATION

TNF has been shown to be produced by numerous immune cells including monocytic cells, natural killer (NK) cells, B cells, T cells, basophils, eosinophils, neutrophils, and non-immune cells including mast cells, Kupffer cells, astrocytes, granulosa cells, fibroblasts, osteoblasts, smooth muscle cells, spermatogenic cells, retinal pigment epithelial cells, keratinocytes, glia, neurons, and many kinds of tumor cells. TNF synthesis can be stimulated by a wide range of stimuli in different cell types. For example, in macrophages, TNF synthesis can be induced by biologic, chemical and physical stimuli including viruses, bacterial products, parasite products, tumor cells, cytokines (interleukin-1 (IL-1), IL-2, interferon-γ (IFN-γ), granulocyte–macrophage colony stimulating factor (GM-CSF), M-CSF, leukemia inhibitory factor (LIF) and TNF itself), complement, cGMP, X-ray radiation, ischemia and trauma. Among these stimuli, lipopolysaccharide (LPS) in monocytic cells, engagement of the T-cell receptor in T lymphocytes, cross-linking of surface immunoglobulin (sIg) in B lymphocytes, ultraviolet light in epithelial cells, and phorbol ester and viral infections in a variety of cell types have been studied most widely. Suppressors of TNF expression in macrophages are also highly diverse and include prostaglandin E_2 (PGE_2), transforming growth factor-β (TGF-β), IFN-α, IFN-β, IL-4, IL-6, IL-10, G-CSF, certain viral products such as adenoviral proteins, dexamethasone, glucocorticoids, cyclosporin A, spermine, pentoxifylline and cAMP.

The biosynthesis of TNF is tightly controlled at many different levels to ensure the silence of the TNF gene under normal circumstances. Therefore, TNF is produced in vanishingly small quantities in quiescent cells, but is one of the major secretory factors in activated cells (Beutler *et al.*, 1985). The TNF gene is one of the immediate early genes induced by a variety of stimuli, and TNF mRNA levels increase strikingly within 15–30 min with no requirement of *de novo* protein synthesis, suggesting that the factors necessary for the induction of TNF expression pre-exist in unstimulated cells. The cell-type-specific regulation of TNF synthesis is reflected not only by diverse distribution of TNF receptors, but also by the differential use of regulatory elements on TNF biosynthesis in different cells. Each one of the synthetic steps from sensing the presence of a stimulus to TNF secretion is controlled by different molecular events regulating the TNF gene, mRNA and protein.

Transcriptional Regulation

At the transcriptional level, besides a TATA box promoter located 20 base pairs (bp) upstream from the transcription start site, a number of regulatory sequences are also found upstream of the TNF gene (Fig. 1) including three nuclear factor (NF)-κB sites κ1, κ2 and κ3 (Goldfeld *et al.*, 1993); three nuclear factor of activated T cells (NFAT) binding sites for NFATp, NFAT-149, NFAT-117 and NFAT-76 (Tsai *et al.*, 1996a,b);

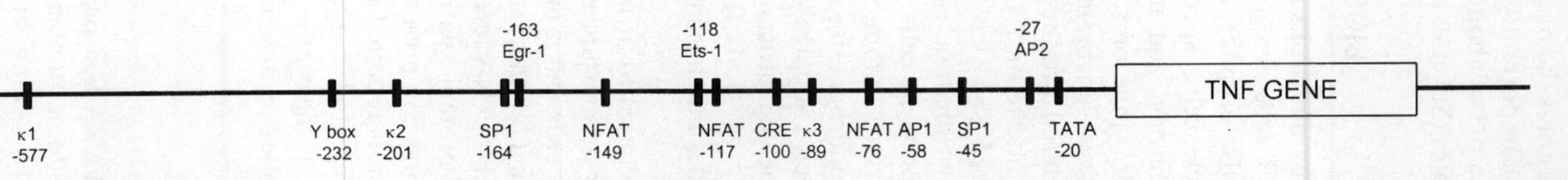

Fig. 1. Model of the TNF promoter. Putative cis-acting consensus sequences are denoted by boxes and numbers identify the position in relation to the transcription start site.

one 'Y' box; one cAMP-responsive element (CRE) for activation transcription factor-2 (ATF-2)/Jun (Leitman *et al.*, 1991; Tsai *et al.*, 1996a); two SV40 promotor-1 (SP-1) sites (Kramer *et al.*, 1994); one activating protein-1 (AP-1) and one AP-2 binding site for Fos/Jun (Economou *et al.*, 1989; Leitman *et al.*, 1992); one E26 transformation-specific (Ets) binding site (Kramer *et al.*, 1995); and an early growth responsive-1 (Egr-1) binding site (Kramer *et al.*, 1994). Although κ3 matches the binding consensus sequence for the transcription factor NF-κB, it is believed that κ3 mainly binds to NFATp instead of NF-κB. The CRE element in TNF promoter binds mainly to ATF-2/Jun instead of Fos/Jun. In addition, genetic studies discussed below have demonstrated that other regulatory elements such as −308 (Cavender and Edelbaum, 1988; Lattime *et al.*, 1988; Wong and Goeddel, 1988) may be important in regulating TNF transcription. Although TNF expression at the transcriptional level has also been shown to be negatively regulated (Rhoades *et al.*, 1992), the details are less well characterized.

The regulation of TNF synthesis is cell type specific, partially because of differential use of these regulatory elements on the TNF promoter. In calcium- or T-cell ligand-activated Ar-5 T lymphocytes, a functional interaction between NFATp on the κ3 site and ATF-2/Jun on the CRE element, but not AP-1 or AP-2, play a crucial role in the induction of TNF expression (Tsai *et al.*, 1996a,b). In calcium-activated A20 B lymphocytes, however, NFATp on the κ3 element is not required, but NFATp on NFAT-76 plays a more important role in the induction of TNF synthesis (Tsai *et al.*, 1996b). In phorbol myristate acetate (PMA)-activated T cells, the interaction between Ets and ATF/Jun is essential for both basal and induced TNF expression. TNF induction in LPS-activated macrophages occurs in association with activation of NF-κB transcription factor, but the role of NF-κB in TNF induction is still obscure. Goldfeld and colleagues (1990) have shown that the NF-κB sites are neither required nor sufficient for virus or LPS induction of TNF. However, TNF transcription inhibitors can inhibit the activation of NF-κB (see below).

As the factors participating in the regulation of TNF synthesis pre-exist in unstimulated cells, modification of the transcription factors able to bind to the regulatory elements on the TNF promoter, such as NFATp, NF-κB and AP-1, has been implicated in the mechanism of TNF induction. Protein kinases and phosphatases are also involved in the induction of TNF (Sung *et al.*, 1988; Chung *et al.*, 1992; Lee *et al.*, 1994), as well as proteases and phospholipase D (Balboa *et al.*, 1992). Some of these enzymes are able to modify the activity of TNF induction-related transcription factors such as NF-κB and AP-1. A pathway via mitogen-activated protein (MAP) kinase/extracellular signal-regulated kinase (ERK) kinase (MEKK) can activate a few TNF induction-associated transcription factors such NF-κB, c-Jun and p38 (Cohen *et al.*, 1996; S.Y. Lee *et al.*, 1996), and may play an important role in TNF regulation. Some of the transcriptional and translational control mechanisms in TNF synthesis are summarized in Fig. 2.

Translational Regulation

At the translational level, TNF synthesis is inhibited in quiescent cells but highly inducible upon stimulation. It is believed that the enforced inhibition is an important mechanism to protect the host from the harmful effects of TNF. In response to LPS stimulation, macrophages from C3H/HeJ mice or human monocytic cells pretreated with dexamethasone, spermine or CNI-1493 (a tetravalent guanyl-hydrazone) (Beutler

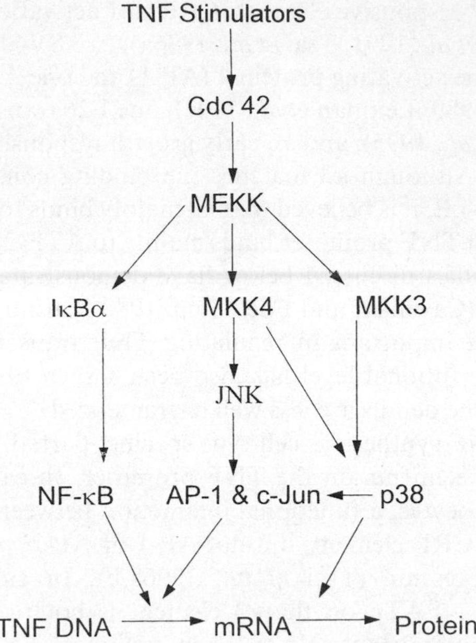

Fig. 2. Schematic illustration of signal transduction pathways regulating TNF transcription and translation in activated macrophages.

et al., 1986; Cohen *et al.*, 1996; Zhang *et al.*, 1997) fail to produce TNF protein, but TNF mRNA is induced dramatically. A key element for translational regulation of TNF has been identified in the 3′ untranslated region (UTR) of TNF mRNA, and is also found in the 3′ UTR of many other cytokines, growth factors and oncoproteins (Caput *et al.*, 1986). It is an AU-rich sequence, UUAUUUAU, and has been shown to decrease mRNA stability (Shaw and Kamen, 1986; Wilson and Treisman, 1988) or inhibit its translation. The mechanism of translational regulation by the AU-rich sequence has been suggested through physical interaction with the poly(A) tail, resulting in the failure of mRNA to form large polysomes (Grafi *et al.*, 1993). It has been recently shown that the minimal motif that confers mRNA instability is the UUAUUUAUU nonanucleotide (Lagnado *et al.*, 1994; Zubiaga *et al.*, 1995), but the presence of a single octanucleotide UUAUUUAU in the 3′ UTR is sufficient to inhibit translation significantly (Kruys *et al.*, 1989). The 3′ UTR of TNF has a TTATTTATTATTTATTTATTATTTATTTATTTATTTA sequence and, dependent on the cell system, TNF translation may be regulated either way through this sequence. Han and colleagues (1990a,b) demonstrated that the 3′ UTR of TNF caused a 600-fold decrease in translational efficiency of TNF mRNA. Although the stability of TNF mRNA can been modulated by IL-4, IFN-γ and amebic proteins (Seguin *et al.*, 1995; Suk and Erickson, 1996), there is no definite evidence to demonstrate the role of the AU-rich element in the stability of TNF mRNA. It has been shown that the AU-rich sequence has no effect on TNF mRNA stability in RAW 264.7 macrophages (Han *et al.*, 1991). Several factors have been identified that are capable of binding to the AU-rich sequence and may mediate the enforced inhibition of TNF translation (Malter, 1989; Bohjanen *et al.*, 1991; Vakalopoulou *et al.*, 1991; Zhang *et al.*, 1993). It has been suggested that p38 MAP kinase may be critical in the translational control of TNF synthesis (Lee *et al.*, 1994). The activation of this MAP kinase cascade enhances the translational efficiency of TNF mRNA (Fig. 2). However, the direct connection between the binding factors and the p38 protein kinase cascade or other pathways remains to be elucidated.

Post-translational Regulation

TNF is initially synthesized as a 233-amino-acid membrane-anchored prohormone which is processed proteolytically to yield the mature 157-amino-acid cytokine. Analysis of the cleavage site sequences in human TNF revealed significant homology with sites in collagen, α_1-inhibitory proteins, and α_1-macroglobulin known to be cleaved by matrix metalloproteinases (MMPs). The TNF-converting enzyme (TCE) has recently been identified as an MMP-like enzyme (Black *et al.*, 1997; Moss *et al.*, 1997). Metalloproteinase inhibitors block TNF processing (Gearing *et al.*, 1994). A serine protease, proteinase 3 (PR-3) has also been reported to be involved in TNF processing (Robache-Gallea *et al.*, 1995), but possibly by modulating the activity of MMP-like enzyme (Gearing *et al.*, 1995). Gearing and colleagues (1995) have found that TCE activity is expressed ubiquitously, even in HeLa cells that are believed not to produce TNF, and in insect cells where pro-TNF is expressed by means of a baculovirus system. Although the above result suggests that TCE is expressed constitutively, the expression and activity of TCE has also been reported to be regulated. Recent studies have demonstrated that LPS, TNF and other cytokines can modulate MMP levels (Panagakos *et al.*, 1996), and plasminogen and plasmin not only increase secretion of MMPs but also induce cleavage to their active forms (E. Lee *et al.*, 1996).

Application

Uncontrolled synthesis of TNF has been implicated in many human diseases. There has been a great deal of interest in developing therapeutic strategies to inhibit TNF synthesis. Several targets and mechanisms of actions have been identified, acting at different levels of TNF synthesis. Corticosteroids, the most studied class of drugs that suppress the production of TNF, inhibit the transcription of TNF and other cytokines. The major immunosuppressive mechanism of these drugs has been shown to be through stimulating IκBα production (Auphan *et al.*, 1995; Scheinman *et al.*, 1995) which in turn results in the inactivation of NF-κB. Corticosteroids have also been shown to inhibit gene transcription by preventing AP-1 transcription factor from binding to its target promoter (Marx, 1995). Pentoxifylline and other protein kinase C (PKC) inhibitors have also been reported to regulate TNF and other cytokine production through inhibiting PKC- or PKA-catalyzed activation of NF-κB (Biswas *et al.*, 1994). At the translational level, one class of cytokine synthesis inhibitors has recently been termed cytokine-suppressing anti-inflammatory drugs (CSAIDs) (Young *et al.*, 1994). These agents have no effect on transcription of TNF or other cytokines but rather exert their major suppression at the translational level by inhibiting the activity of MAP kinase p38 (Lee *et al.*, 1994). CNI-1493, a tetravalent guanylhydrazone which suppresses inflammation, inhibits TNF by preventing phosphorylation of p38 MAP kinase (Bianchi *et al.*, 1996). At the post-translational level, thalidomide has been found selectively to suppress the release of TNF but not of other cytokines (Sampaio *et al.*, 1991). IL-4 and IL-10 can both suppress the biosynthesis of matrix-destructive metalloproteinases (Lacraz *et al.*, 1992, 1995), and downregulate the proteolytic cleavage of pro-TNF to release mature TNF. IL-4 can also inhibit the activation of transcriptional factors NF-κB and AP-1. Like IL-4, IL-10 also inhibits TNF and other cytokines at multiple levels (Bogdan *et al.*, 1992) and can also suppress NF-κB.

RECEPTORS

The multiple activities of TNF are mediated through two distinct but structurally homologous TNF receptors, type I (p60 or p55) and type II (p80 or p75) with molecular masses of 60 and 80 kDa, respectively. Both are type I transmembrane glycoproteins and members of the TNF receptor superfamily characterized by the presence of multiple cystine-rich repeats of about 40 amino acids in the extracellular amino-terminal domain (Mallett and Barclay, 1991). TNFR1 and TNFR2 are present on virtually all cell types except for red blood cells (Hohmann *et al.*, 1989). TNFR1 is more ubiquitous, but TNFR2 is more abundant on endothelial cells and cells of hematopoietic lineage (Hohmann *et al.*, 1989; Brockhaus *et al.*, 1990; Porteu *et al.*, 1991). Both receptors can bind to TNF with high affinity, and the dissociation constants (K_d) are $2–5 \times 10^{-10}$ and $3–7 \times 10^{-11}$, respectively.

The exact roles of two TNF receptors in mediating the effects of TNF continue to be the focus of intensive study. It is generally believed that TNFR1 is responsible for the majority of biologic actions of TNF. Direct signaling through TNFR2 occurs less extensively and appears to be confined mainly to cells of the immune system. It is suggested by studies on fibroblasts from TNFR1-deficient mice that TNFR1 controls adhesion to leukocytes as well as intercellular adhesion molecule-1 (ICAM-1), vascular cell adhesion molecule-1 (VCAM-1), CD44 and major histocompatibility complex (MHC) class I upregulation, secretion of other cytokines, cell proliferation and NF-κB activation; however, TNFR2 stimulation in TNFR1-deficient fibroblasts has no effect on these functions (Mackay *et al.*, 1994). Studies on TNFR1-selective TNF mutants also suggest that TNFR1 is the sole mediator of TNF cytotoxicity and cytostasis, and a major mediator of neutrophil and endothelial activation, whereas TNFR2 by itself does not seem to be sufficient to stimulate these functions (Barbara *et al.*, 1994). The role of TNFR2 may be to potentiate the effects of TNFR1. A unique 'passing model' has been suggested to explain the role of TNFR2 in mediating TNF responses (Tartaglia *et al.*, 1993b). The rapid rate of dissociation of the TNF–TNFR2 complex may facilitate the interaction of TNF with TNFR1, suggesting a role of the TNFR2 in passing TNF to TNFR1, which has been postulated to be the main TNF signal transducer. In some cases, however, TNFR2 is also believed to mediate several TNF effects such as proliferation of T and B cells (Tartaglia *et al.*, 1991; Barbara *et al.*, 1994), induction of NF-κB (Hohmann *et al.*, 1990a) and cytotoxicity (Hohmann *et al.*, 1990a,b; Grell *et al.*, 1994). TNF receptor activities are also subject to regulation by a wide variety of agents such as PKC and PKA modulators, LPS, retinoid acid, glucocorticoids and cytokines including TNF, IL-2, IL-4, IL-6, IL-8, IFN-α, β, γ, GM-CSF and thyroid-stimulating hormone. The regulation of TNF receptor is cell type dependent; meanwhile, some agents can differentially regulate the expression of TNFR1 and TNFR2.

Soluble receptors for TNF (sTNFR, or TNF-BP for TNF binding protein) which bind and neutralize TNFR are derived from both TNFR1 and TNFR2 by means of proteolytic processing. The proteases responsible for this process have not yet been identified, but the metalloproteases implicated in processing TNF precursor have also been implicated in receptor processing: inhibitors of metalloproteinases can block the shedding of both the TNF receptors (Crowe *et al.*, 1995; Mullberg *et al.*, 1995). Soluble TNFRs retain the capacity to bind TNF and act as inhibitors of TNF in experimental systems (Engelmann *et al.*, 1989; Seckinger *et al.*, 1989; Kohno *et al.*, 1990). *In vivo* they may act to neutralize or modulate TNF activity (Aderka *et al.*, 1992; Mohler *et al.*, 1993). In some cases, however, low levels of sTNFRs may stabilize the activity of TNF by stabilizing the trimeric structure of TNF and providing a reservoir of TNF (Aderka *et al.*, 1992; Mohler *et al.*, 1993).

The proteolytic processing of TNFRs is a highly regulated event with important physiologic and pathologic significance. Both sTNFR1 and sTNFR2 are normally present in blood and urine (Seckinger *et al.*, 1989; Engelmann *et al.*, 1990b), and may be produced by monocytes and macrophages; promonocytic THP-1 cells and monocytes can release sTNFRs spontaneously *in vitro* (Gatanaga *et al.*, 1991; Leeuwenberg *et al.*, 1994a,b). Increased levels of sTNFRs are found in many human diseases including rheumatoid arthritis, systemic lupus erythematosus, malignancy, sepsis, after surgery and chronic infection (Aderka *et al.*, 1991; Brennan *et al.*, 1992; Chikanza *et al.*, 1993; Aderka, 1996). A number of stimuli have been identified that trigger the cleavage of the TNFRs, thereby causing an increase in the amount of sTNFRs in the circulation. These stimuli include TNF and other cytokines such as IL-1, PMA and LPS.

SIGNAL TRANSDUCTION

Understanding of TNF signaling pathways has been particularly challenging, because of the extremely wide variety of cell-specific TNF responses and the lack of similarities with other cell surface receptor signaling pathways. It is clear that signaling events of cellular responses to TNF are initiated by the interaction between TNF and its receptors. It is suggested by X-ray diffraction study (Banner *et al.*, 1993) that the primary function of TNF trimers is to mediate receptor aggregation, which activates receptors by inducing receptor trimerization. Receptor aggregation then activates receptor-associated effectors or facilitates recruitment of downstream factors to the receptor complex, which can then transduce signals via intracellular signaling molecules. Several downstream signaling events that are activated by TNF have been characterized, and signaling molecules that mediate the initial interaction with the ligand-occupied receptor have recently been identified.

TNFR1-Mediated Signaling Pathway

A majority of TNF activities can be mediated solely by TNFR1 (Espevik *et al.*, 1990; Wiegmann *et al.*, 1992; Tartaglia *et al.*, 1993a). These various activities and their coordinated induction are likely to be mediated by the heterogenicity of the intracellular functional domains of TNFR1 and of the adaptor proteins with which TNFR1 functional domains interact. Three functional domains, *C*-terminal death domain (Tartaglia *et al.*, 1993a) and adjacent N-SMase (neutral sphingomyelinase) and A-SMase (acidic sphingomyelinase) activating domains (NSD and ASD) (Schutze *et al.*, 1992; Wiegmann *et al.*, 1994), have been found in the intracellular region of TNFR1 to be responsible for transferring signals from TNF to intracellular adaptors. It appears that they act via protein–protein interaction. Three adaptors have been identified to link ligand-occupied TNFR1 to downstream signaling molecules, TRADD (TNFR1-associated death domain protein) binding to the death domain of TNFR1 via interaction between their death domains (Hsu *et al.*, 1995), FAN (factor associated with N-SMase activation) binding to the NSD domain (Adam-Klages *et al.*, 1996) and phosphatidylcholine-specific phospholipase c (PC-PLC) mediating the signal from ASD (Wiegmann *et al.*, 1994). Of note, there is an overlap between the death domain and ASD. It has also been reported that the death domain of TNFR1 is associated with the activation of PC-PLC (Machleidt *et al.*, 1996). RIP (receptor interacting protein), a serine–threonine kinase with a death domain, binds preferentially to TRADD by means of interaction between their death domains (Hsu *et al.*, 1996a) but can also bind weakly to TNFR1 (Stanger *et al.*, 1995). The proposed downstream pathways of TRADD, FAN and PC-PLC resulting in various cell activities are summarized briefly in Fig. 3. TRADD mediates both apoptosis and antiapoptosis pathways associated with interleukin-1β converting enzyme (ICE) and NF-κB respectively. FAN and PC-PLC can independently activate N-SMase and A-SMase (Wiegmann *et al.*, 1994). N-SMase mediates a number of cellular TNF responses including cell growth and inflammation through MAP kinase (ERK) and phospholipase A_2 (PLA2) respectively. A-SMase can induce activation of NF-κB, a factor involved in protection against apoptosis (Van Antwerp *et al.*, 1996; Wang *et al.*, 1996), but is also involved in induction of apoptosis (Pena *et al.*, 1997).

Death Domain-Mediated Pathways

TNFR1 death domain mediates both apoptotic and antiapoptotic pathways. TRADD plays an obligatory role in death domain-mediated TNF responses: overexpression of TRADD activates TNFR1-like signaling pathways, triggering both apoptosis, which is associated with ICE-activated protease cascade, and antiapoptosis, which is associated with activation of the transcription factor NF-κB (Hsu *et al.*, 1995). TRADD contains a death domain which alone, but not the rest of the portion of the protein, can trigger cell death, suggesting TRADD does not carry a death effector domain. This led to a view of TRADD's death domain as an 'adaptor' linking TNFR1 to other proteins. Once TRADD has bound to TNFR1 death domain in the presence of TNF, it causes recruitment of MORT1/FADD (Fas-associating protein with death domain) and RIP through the *C*-terminal death domain, and recruitment of TNF receptor-associated factor (TRAF2) through *N*-terminal TRAF-interacting domain (Hsu *et al.*, 1996b). Overexpression of FADD (Chinnaiyan *et al.*, 1995; Boldin *et al.*, 1996) or RIP (Stanger *et al.*, 1995) results in apoptosis; however, a dominant-negative mutant of FADD can inhibit only TNF-induced apoptosis but not TNF-mediated NF-κB activation (Hsu *et al.*, 1996b). In contrast, overexpression of TRAF2 activates NF-κB (Rothe *et al.*, 1995) but TRAF2 is not required for the induction of apoptosis (Hsu *et al.*, 1996b). Thus, these two TNFR1-TRADD signaling cascades appear to bifurcate at TRADD, thereby initiating two different pathways responsible for apoptosis induction and apoptosis protection respectively (Fig. 3). A key finding in the FADD apoptotic pathway is that overexpression of FADD's death domain alone, unlike TRADD, does not cause death, but the remainder of the protein minus the death domain induces apoptosis. This implies that FADD is the last step dependent on the death domain in this pathway. Recently, a

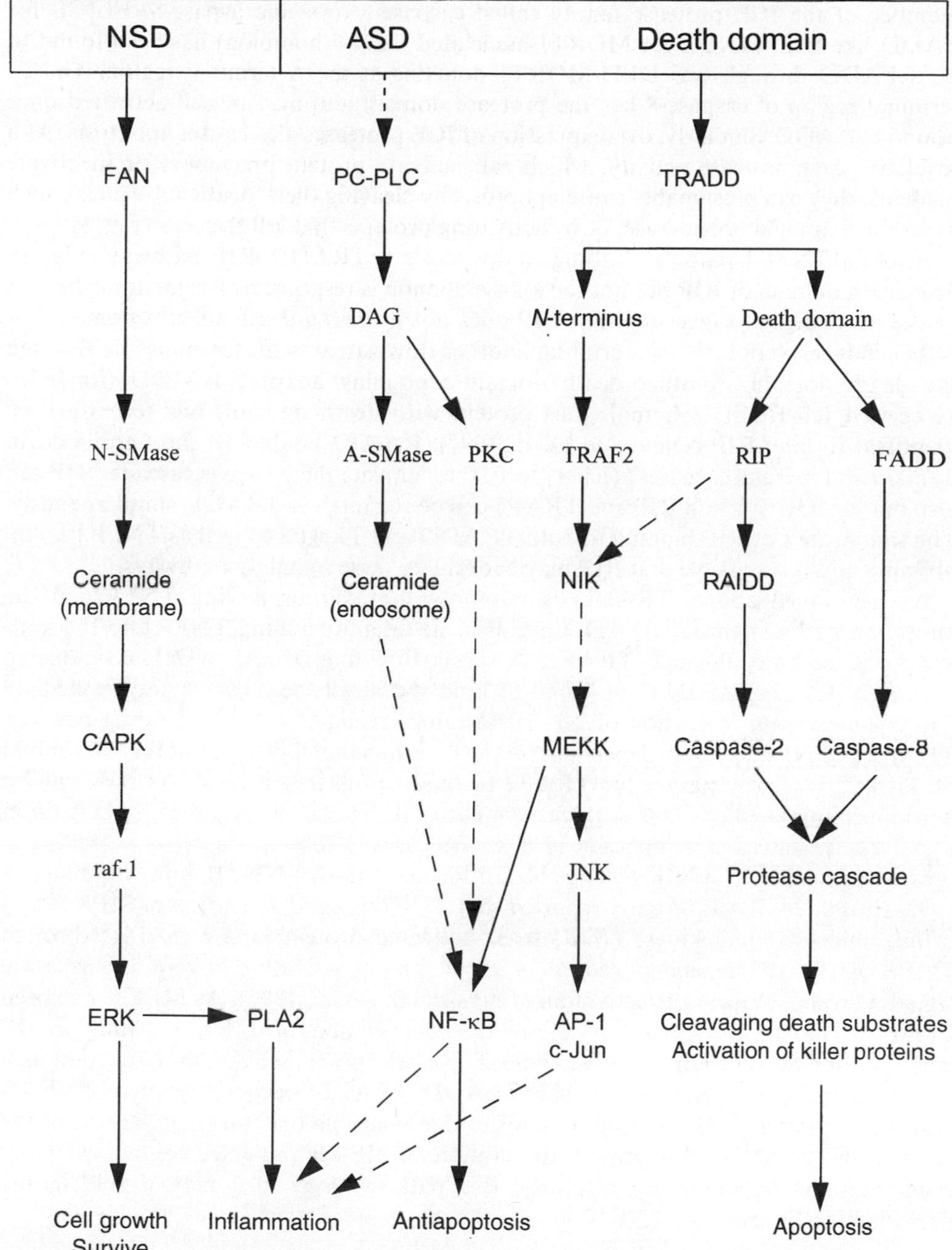

Fig. 3. TNF receptor 1-mediated TNF signaling pathways. TNF triggers different signal transduction pathways via three functional domains of TNFR1 intracellular region to mediate various activities.

member of the ICE protease family called caspase-8 (cysteine aspase-8, FLICE for FADD-like ICE; MACH for MORT1-associated CED-3 homolog) has been found to bind FADD through two DED/MORT1 domains at the *N*-terminal region. The *C*-terminal region of caspase-8 has the protease domain and may be self-activated once bound to FADD. Similarly, overexpression of ICE proteases also causes apoptosis. As a result of their protease activity, which can activate protein precursors or inactivate proteins, they can presumably cause apoptosis by cleaving their 'death substrates', such as lamin, actin and polymerase, or by activating proteins that kill the cells (Fig. 3).

Another TNFR1 pathway leading to apoptosis is TRADD–RIP pathway (Fig. 3). The death domain of RIP but not the kinase domain is responsible for the transduction of the death signal, suggesting that RIP does not possess a death effector domain but rather induces apoptosis by recruiting another downstream effector molecule through the death domain. Another death domain-containing adaptor RAIDD (for RIP-associated Ich-1/CED-3 homologous protein with death domain) has recently been identified to bind RIP (Duan and Dixit, 1997). RAIDD binds RIP through its death domain and recruits caspase-2 (Ich-1) to RIP to activate the protease cascade. RIP can also bind to TRAF2, and RIP and TRAF2 can be recruited to TRADD simultaneously. The significance of RIP binding to both TRADD and TRAF2 as well as TNFR1 is still obscure, and it is possible that RIP may coordinate these signaling pathways.

As mentioned above, TRADD is not only the adaptor linking TNFR1 to the apoptotic pathway through FADD but also the adaptor linking TNFR1 to the anti-apoptotic pathway through TRAF2. A serine–threonine kinase or NF-κB-inducing kinase (NIK) has recently been found to bind specifically to TRAF2 and to mediate TRAF2-dependent activation of NF-κB (Malinin *et al.*, 1997). The binding between TRAF2 and NIK is mainly dependent on the *C*-terminal of TRAF2, but the *N*-terminal of TRAF2 is also necessary for TRAF2 to bind to full-length NIK. As NIK can be autophosphorylated and has sequence similarity to several kinases that participate in MAP kinase cascades, it may act in a kinase cascade to activate NF-κB itself. The downstream cascade of NIK responsible for the activation of NF-κB, however, remains to be elucidated. It was recently reported that TNF-triggered activation of MEKK and c-Jun amino-terminal kinase (JNK)/stress-activated protein kinase (SAPK) through TNFR1 is TRAF2 dependent (Natoli *et al.*, 1997), but activation of JNK is a separate signaling response from the activation of NF-κB (Liu *et al.*, 1996). As MEKK has been shown to induce the phosphorylation and degradation of IκBα, resulting in the activation of NF-κB (Hirano *et al.*, 1996; Lee *et al.*, 1997), MEKK may function as a downstream factor of NIK and mediate TRADD–TRAF2-directed activation of NF-κB which can prevent TNF-induced apoptosis. It is plausible that this may be one of the mechanisms of TNF self-tolerance. In addition, as NF-κB and JNK are major factors regulating the inflammatory response, this pathway may also play a role in the proinflammatory effect of TNF.

NSD-Mediated Pathway

NSD (neutral sphingomyelin) mediates a number of TNF responses including inflammatory responses and cell proliferation through MAP kinase ERK and PLA2 (Fig. 3). A novel WD repeat ($\{X_{6-94}$-$[GH$-X_{23-41}-$WD]\}N_{4-8}$) protein FAN has recently been identified to couple TNFR1 to N-SMase through binding between the *C*-terminal WD repeats of FAN and a cytoplasmic nine-amino-acid binding motif (310–318) of TNFR1

(Adam-Klages *et al.*, 1996). The activation of N-SMase by FAN is strictly specific, as FAN does not have any effect on the activation of A-SMase (see ASD-mediated pathway below). This further supports the observation that N-SMase and A-SMase are activated independently by different cytoplasmic domains of TNFR1, and there is no apparent cross-talk between N-SMase and A-SMase in TNFR1 signaling (Wiegmann *et al.*, 1994). Once N-SMase has been activated by FAN, it in turn modulates the production of ceramide, a product of the sphingomyelin (SM) cycle. Ceramide can be generated by SM breakdown at various subcellular sites. It is interesting that ceramide action depends on the topology of its production (Wiegmann *et al.*, 1994). Activated N-SMase acts at the outer leaflet of the plasma membrane to trigger the degradation of sphingomyelin (SM) into ceramide, which in turn mediates further downstream events in NSD-mediated TNF response. Ceramide is also produced in endosomes by A-SMase where it mediates a distinct signaling pathway (see ASD-mediated pathway). Ceramide generated on membranes can further activate ceramide-activated protein kinase (CAPK) (Winston and Riches, 1995), which then phosphorylates cytoplasmic raf-1. Activated raf-1 can activate MAP kinase cascade and activate MAPK (ERK for extracellular signal-regulated kinase) followed by the activation of PLA2 through phosphorylation by MAPK (Lin *et al.*, 1993). PLA2 may be responsible for the generation of the arachidonic acid metabolites, leukotrienes and prostaglandins, which may contribute to TNF proinflammatory activities. Meanwhile, MAPK (ERK) also induces the expression of c-*myc* and other genes to regulate cell proliferation and differentiation. As inhibition of ERK has been reported to be critical for induction of apoptosis in some systems (Xia *et al.*, 1995), activation of ERK may prevent TNF-induced apoptosis.

ASD-Mediated Pathway

ASD is also suggested to play a role in TNF-induced apoptosis and NF-κB activation (Fig. 3), and it may be identical to the death domain. ASD mediates the activation of A-SMase via PC-PLC (Schutze *et al.*, 1992). The connection between ASD in TNFR1 and PC-PLC, however, is still obscure. Once PC-PCL is activated, 1,2-diacylglycerol (DAG) produced by PC-PLC activates A-SMase and PKC, and contributes to the activation of NF-κB. The functional coupling between PC-PLC and A-SMase is revealed by a selective inhibitor of PC-PLC, D609 (Schutze *et al.*, 1992). It has been shown that TNF-triggered A-SMase as well as NF-κB activation is blocked by this specific PC-PLC inhibitor, which effectively inhibits DAG production. Unlike N-SMase, A-SMase produces ceramide in the endosomal–lysosomal compartment (Merrill *et al.*, 1993; Spence, 1993) where endosomal vesicles are usually rapidly acidified, which conditions for A-SMase activation by DAG.

TNFR2-Mediated Signaling Pathway

Direct signaling through TNFR2 occurs less extensively and appears to be confined mainly to cells of the immune system. Although TNFR2 may function primarily to potentiate the effects of TNFR1, in some cases TNFR2 also mediates specific TNF effects including proliferation of T and B cells (Tartaglia *et al.*, 1991; Barbara *et al.*, 1994), induction of NF-κB, and cytotoxicity (Hohmann *et al.*, 1990a,b; Grell *et al.*, 1994). The signaling pathways mediated by TNFR2 are much less closely defined than those mediated by TNFR1. Recent evidence indicates that TNF, via interaction with

TNFR2, plays an important role in downregulating human and murine activated T cells (Lai *et al.*, 1995; Zheng *et al.*, 1995) by inducing apoptosis. The lack of intracellular death domain in TNFR2 indicates that TNFR2 may use a distinct signaling pathway to induce apoptosis. It has been suggested recently (Lin *et al.*, 1997) that TNFR2-induced apoptosis is associated with the downregulation of Bcl-xL, an antiapoptotic protein (Boise *et al.*, 1995; Boise and Thompson, 1997), but the precise mechanisms triggering TNFR2-induced apoptosis remain to be elucidated.

Whether TNFR2 alone can mediate the activation of NF-κB is controversial. Although TRAF2 also interacts with TNFR2 and TRAF2–TRAF1 (Rothe *et al.*, 1995), and TRAF2–TANK (Cheng and Baltimore, 1996) heterocomplexes are able to mediate the activation of NF-κB, the results of a recent study (Chainy *et al.*, 1996) indicate that TNFR2 does not seem to participate in the activation of NF-κB in many cell types, including immune cells.

BIOLOGIC EFFECTS

TNF was originally characterized as a protein inducing necrosis of methylcholanthrene induced (Meth A) sarcomas *in vivo* (Carswell *et al.*, 1975) and believed to be a growth modulatory agent that exhibits primarily antitumor activity. It subsequently became clear that TNF not only modulates tumor cell growth but also affects cellular growth, differentiation and metabolism in many cell types. It is produced during immune and host defense responses as a primary mediator of immune regulation and the inflammatory response. The biology of TNF is also characterized by its pathologic activities in many immune-mediated diseases. The net effects of TNF are influenced by a complex array of cell- and tissue-specific factors. The diverse role of TNF in mediating cellular responses is summarized in Table 1.

Central Nervous System

Early studies of TNF indicated that systemic administration of TNF produces fever and anorexia via hypothalamic centers that regulate body temperature and appetite (Plata-Salaman *et al.*, 1988; Tracey *et al.*, 1988; Plata-Salaman, 1991). The anorectic effect of TNF is attenuated by insulin and appears to be independent of serotonergic signals (Tracey and Cerami, 1990). Recently, the presence of TNF has been reported in various cells in the central nervous system (CNS) including microglia, astrocytes and neurons (Cheng *et al.*, 1994). Cells in CNS also express high-affinity TNF receptors (Smith and Baglioni, 1992). Furthermore, raised TNF levels have been detected in brain injury and certain CNS diseases including meningococcal meningitis (Leist *et al.*, 1988; Waage *et al.*, 1989b), human immunodeficiency virus infection (Grimaldi *et al.*, 1991), Alzheimer's disease (Fillit *et al.*, 1991) and multiple sclerosis (Hofman *et al.*, 1989; Sharief and Thompson, 1992). These findings strongly suggest that TNF is involved in various biologic processes within CNS.

The inflammatory effects of TNF are pivotal to the development of cerebral inflammation and edema during meningitis. High levels of TNF in cerebrospinal fluid during meningitis are predictive of poor outcome (Leist *et al.*, 1988; Mustafa *et al.*, 1989; Waage *et al.*, 1989a; Saukkonen *et al.*, 1990). The inflammatory effects of TNF have also been implicated in the development of the characteristic plaques in patients with multiple

Table 1. Some of the biologic effects of TNF.

Immune cells	Non-immune cells	*In vivo*
Monocytes–macrophages	*Vascular endothelial cells*	*Central nervous system*
Activation and autoinduction of TNF	Modulation of angiogenesis	Fever
Induction of IL-1, 6, 8, GM-CSF, MCSF, INF-γ, NGF, TGF-β, PDGF and PGE_2	Increasing permeability	Anorexia
Transmigration and chemotaxis	Enhanced expression of MHC I	Altered pituitary hormone secretion
Stimulation of metabolism	Procoagulant and antifibrinolytic	
Inhibition of differentiation	Increasing permeability of albumin and water	*Cardiovascular*
Suppression of proliferation	Suppression of proliferation	Shock
Internalization of complement-coated virus	Rearrangement of cytoskeleton	ARDS
	Induction of NO synthase, IL-1, IL-3 receptor, G-CSF, GM-CSF, ICAM-1, VCAM-1, P- and E-selectin, surface antigen, platelet-activating factor, prostacyclin	Capillary leakage syndrome
Polymorphonuclear leukocytes	Inhibition of integrin B30, thromomodulin, glutathione, protein S	
Priming of integrin response		*Gastrointestinal*
Release of granule components	*Fibroblasts*	Ischemia
Increasing phagocytic capacity	Induction of proliferation	Colitis
Enhanced production of superoxide	Induction of IL-1, 6, INF-β_2, leukemia inhibitory factor, metalloproteinases (MMTs)	Hepatic necrosis
Increased adherence to extracellular matrix	Suppression of respiratory activity	Inhibition of albumin expression
Suppression of chemotaxis to *N*-formyl-1-leucyl-1-phenylalanine	Inhibition of collagen synthesis, MMT inhibitor	Decreased catalase activity in liver
Suppression of cell surface expression of sialophlorin CD43		Suppression of HBV gene expression
	Adipocytes	
Lymphocytes	Enhanced release of free fatty acids and glycerol	*Metabolic*
Induction of T-cell colony formation	Suppression of lipoprotein lipase (LPL)	LPL suppression
Induction of superoxide in B cells		Net protein catabolism
Activation of cytotoxic T-cell invasiveness	*Endocrine system*	Net lipid catabolism
Induction of apoptosis in mature T cells	Stimulation of adrenocorticotrophic hormone and prolactin	Stress hormone release
	Inhibition of thyroid stimulating hormone, follicle stimulating hormone and growth hormone	Insulin resistance
	Enhancing IL-1 inhibition of steroidogenesis	
		Inflammatory
		Activation of cell cytotoxicity
		Enhanced NK cell function
		Mediation of IL-2 tumor toxicity
		Protective role in cutaneous leishmaniasis

sclerosis, and in ischemia in experimental models of stroke, in which application of TNF antibody can reduce the severity of brain damage (Saukkonen *et al.*, 1990; Meistrell *et al.*, 1997).

Although TNF may play a pivotal role in tissue injury, the accumulating evidence also indicates that TNF is very important in the repair and regeneration of damaged tissue in the CNS. It is generally believed that high levels of TNF are injurious, but low levels can be beneficial (Tracey *et al.*, 1989). TNF is able to stimulate the proliferation of astrocytes (Selmaj *et al.*, 1990) and neuronal progenitor cells (Bazan, 1991; Merrill, 1992; Mehler *et al.*, 1993), an effect that may be dependent on the activity of TNF as a stimulator to the production of nerve growth factor (NGF) (Gadient *et al.*, 1990), which is protective in tissue damage (Hefti *et al.*, 1989; Shigeno *et al.*, 1991; Mattson *et al.*, 1993). TNF can also prevent neuronal death following metabolic excitotoxic insults (Cheng *et al.*, 1994). Recently, a study using TNFR knockout (KO) mice (both TNFR1 and TNFR2 deleted) (Bruce *et al.*, 1996) has provided strong evidence that TNF can be protective in ischemic brains. The results from this study have shown that neurons from the TNFR KO mice exhibit reduced survival in culture compared with those from wild-type mice, and doses of kainic acid that cause minimal damage in wild-type mice produce extensive neuronal loss in TNFR KO mice. The damage in the cortex caused by middle cerebral artery (MCA) occlusion is significantly greater in TNFR KO mice than in normal mice. This contrasts with the overexpression of TNF during cerebral ischemia in normal rats subjected to focal infarction, where TNF significantly increases brain damage (Meistrell *et al.*, 1997).

Cardiovascular System

Normally, endothelial cells provide a blood vessel lining that reduces the coagulation of blood, but TNF causes endothelial cells to increase procoagulant activity by enhancing the expression of tissue factor and suppressing cofactor activity for the anticoagulant protein C (Bevilacqua *et al.*, 1986; Nawroth and Stern, 1986; Nawroth *et al.*, 1986; Bauer *et al.*, 1989). TNF can also induce rearrangement of actin filaments with structural damage of the endothelial cells and loss of tight junctions, which causes capillary leakage syndrome as plasma proteins and water leak into the tissues (Sato *et al.*, 1986; Stolpen *et al.*, 1986; Stephens *et al.*, 1988a,b; Brett *et al.*, 1989; Damle and Doyle, 1989; Kreil *et al.*, 1989). TNF exerts a negative inotropic effect on myocardial contractility. The exact mechanism for the negative inotropic effect of TNF is unknown. Nonetheless, there is increasing evidence that production of nitric oxide (NO) induced by TNF may be responsible. Both epithelial cells and cardiac myocytes express TNF receptors and are capable of synthesizing TNF under certain forms of stress (Giroir *et al.*, 1994; Kapadia *et al.*, 1995).

Raised levels of TNF have been identified in patients with advanced heart failure (Moriarty *et al.*, 1990; McMurray *et al.*, 1991; Dutka *et al.*, 1993; Katz *et al.*, 1994; Matsumori *et al.*, 1994). There is increasing evidence that TNF may play a primary pathophysiologic role in the development and pathogenesis of heart failure. Many of the clinical hallmarks of heart failure, including left ventricular dysfunction, cardiomyopathy and pulmonary edema, can be explained by the known biologic effects of TNF such as negative inotropic effects (Millar *et al.*, 1989; Natanson *et al.*, 1989; Suffredini *et al.*, 1989; Hegewisch *et al.*, 1990; Pagani *et al.*, 1992). Negative inotropic effects in isolated cardiac myocytes are reliably produced by TNF concentrations ranging from $0.7\text{–}1.4 \times 10^{-9}$ M (Yokoyama *et al.*, 1993; Kapadia *et al.*, 1995). These TNF concentrations have been reported in congestive heart failure (Levine *et al.*, 1990).

Direct injection of TNF produces hypotension, metabolic acidosis, hemaconcentration and death within minutes, mimicking the cardiovascular response in endotoxin-induced septic shock (Tracey *et al.*, 1986). This is caused by several TNF-mediated effects including a decrease in peripheral vascular resistance, falling cardiac output and loss of intravascular volume through capillary leakage (Tracey *et al.*, 1986, 1987b; Natanson *et al.*, 1989; Tracey and Lowry, 1990). Endothelium-derived NO induced by

TNF has been implicated in decreases of both peripheral vascular tone and cardiac function (Kilbourn *et al.*, 1990; Finkel *et al.*, 1992), whereas TNF-induced rearrangement of actin filaments and loss of tight junctions are responsible for the capillary leakage syndrome (Sato *et al.*, 1986; Stolpen *et al.*, 1986; Stephens *et al.*, 1988a,b; Brett *et al.*, 1989; Damle and Doyle, 1989; Kreil *et al.*, 1989).

TNF is extremely toxic to the lungs and is pivotal in the development of adult respiratory distress syndrome (ARDS), a syndrome characterized by pulmonary edema, hypoxia and a high mortality rate. ARDS develops because of TNF-induced activation of pulmonary endothelium, margination of leukocytes with degranulation of granulocytes, and capillary leakage, which precipitates the collection of edematous fluid in alveoli and prevents adequate perfusion and oxygeneration (Horvath *et al.*, 1988; Stephens *et al.*, 1988b; Zheng *et al.*, 1990).

Human Diseases

Although TNF has theoretically evolved to preserve cellular and biochemical homeostasis by participating in tissue remodeling and host defense responses, overproduction of TNF has also been shown to be very harmful, even lethal, to the host. Therefore, either insufficient or excessive production of TNF will result in abnormal conditions such as infection, tissue injury, shock and death. TNF has been directly implicated in the pathogenesis of many disease states, as listed in Table. 2.

Septic Shock

It is well established that TNF plays a critical role in septic shock. First, TNF is acutely overproduced systemically during overwhelming sepsis and affects its clinical outcome (Girardin *et al.*, 1988; Waage *et al.*, 1989a). Persistent increase seems to be associated with mortality (Girardin *et al.*, 1988; Calandra *et al.*, 1990) and multiple organ failure (van der Poll and Lowry, 1995). Second, administration of TNF itself causes shock and tissue injury indistinguishable from septic shock syndrome (Tracey *et al.*, 1986, 1987b). Meanwhile, TNFR1-deficient KO mice are resistant to endotoxic shock (Pfeffer *et al.*, 1993). Third, TNF is able to induce other mediators of sepsis such as IL-1, and infusion of TNF antibody attenuates these mediators (Tracey *et al.*, 1987a; Fong *et al.*, 1989). Finally, neutralization of TNF with antibodies prevents septic shock syndrome during lethal bacteremia (Tracey *et al.*, 1987a).

In sepsis, both TNF and soluble receptors of TNF (sTNFRs) are induced (Girardin *et al.*, 1992; Rogy *et al.*, 1994). As mentioned above, sTNFRs are inhibitors of TNF biologic activities but can also be a reservoir of TNF, and increased levels of sTNFRs are found in many TNF-associated diseases. In fact, increased levels of sTNFRs have been found to correlate with TNF levels and outcome in severe meningococcemia (Girardin *et al.*, 1992). The increase in sTNFR concentration is proportional to TNF increase when the TNF level is relatively low, but not when it is high. The imbalance between TNF and sTNFRs seems to be pathophysiologically important, as reflected by the observation that their ratio on admission appeared to be of predictive value for clinical outcome (Girardin *et al.*, 1992).

Several strategies have been proposed to prevent TNF toxicity in sepsis (Tracey and Cerami, 1994). One approach is directly to inhibit TNF activity with antibodies or native and chimeric sTNFRs (Tracey *et al.*, 1987a; Ashkenazi *et al.*, 1991; Peppel *et al.*, 1991).

Table 2. Some TNF-associated diseases.

Infectious diseases	Autoimmune diseases	Cancer and others
Bacterial	Glomerulonephritis	Hairy cell leukemia
Meningococcal disease	Guillain–Barré syndrome	T-cell leukemia
Leprosy	Inflammatory bowel disease	Chronic lymphocytic leukemia
Tuberculosis	Systemic lupus erythematosus	Malignant lymphoma
Septic shock syndrome	Insulin-dependent diabetes mellitus	Colorectal cancer
Nocardia brasiliensis	Rheumatoid arthritis	Lung cancer
Cryptogenic fibrosing alveolitis	Dermatitis	Cervical cancer
Pneumocystis carinii pneumonia	Celiac disease	
	Myasthenia gravis	Euthyroid sick syndrome
Viral	Systemic sclerosis	Hemorrhagic shock
AIDS		Disseminated intravascular coagulopathy
HIV-related lung disease		Anemia
Measles		Myocardial ischemia
Hepatitis		Obesity
Viral meningitis		Shock
Murine retrovirus infection		Stroke
		Rejection of transplanted organs
Parasitic		Pulmonary fibrosis
Cerebral malaria		Chronic osteomyelitis
Toxoplasma gondii		Graves' disease
Dysentery (*Shigella dysenteriae*)		Asthma
Children with pertussis		Injury
		Inflammatory hyperalgesia
		Down's syndrome
		Cachexia

Another is to suppress TNF synthesis using glucocorticoids, pentoxifylline, CNI-1493, spermine, thalidomide and cytokines such as IL-10 (Beutler *et al.*, 1986; Lilly *et al.*, 1989; Sampaio *et al.*, 1991; Giroir and Beutler, 1992; Cohen *et al.*, 1996; Zhang *et al.*, 1997). The third approach is to induce TNF tolerance with repetitive administration of low-dose TNF. Finally, TNF cytotoxicity may be inhibited by targeting its secondary mediators such as IL-1 (Arend *et al.*, 1990; Alexander *et al.*, 1991; McNamara *et al.*, 1993). Some of these therapies have been demonstrated to have salutory effects in animal models of sepsis, yet preliminary clinical trials have not been as encouraging (Fisher *et al.*, 1993; Natanson *et al.*, 1994; van der Poll and Lowry, 1995). Further experimental studies and clinical trials may provide clues to the possibility of fine-tuning TNF bioactivities in the clinical syndrome of sepsis. Recently, inhibitors of receptor-mediated endocytosis of TNF have been shown to decrease TNF-induced gene expression, thus providing yet another potential point of intervention (Bradley *et al.*, 1993). In addition, pharmacologic inhibitors of metalloproteinases have been developed to inhibit TNF processing, although clinical efficacy has not been addressed (McGeehan *et al.*, 1994).

Cancer

TNF was originally characterized as an antitumor protein inducing necrosis of Meth A sarcomas *in vivo* (Carswell *et al.*, 1975). TNF has been studied widely and considered as an antitumor drug; at present it is used clinically as an intratumoral agent and has some efficacy (Pfreundschuh *et al.*, 1989; Lienard *et al.*, 1992). The therapeutic application of TNF as a single and systemic antitumor agent is limited by the toxic side-effects at tumoricidal doses (Blick *et al.*, 1987; Spriggs *et al.*, 1988; Moritz *et al.*, 1989). Several strategies have been proposed to avoid the limitations imposed by toxicity, including: (1) high-dose adminstration of TNF intra-arterially directly into the tumor compartment (Ohkawa *et al.*, 1989); (2) identification of tumors sensitive to non-toxic levels of TNF; (3) administration of non-toxic doses of TNF in combination with other antitumor agents—synergistic effects have been identified (Winkelhake *et al.*, 1987; Zimmerman *et al.*, 1989) but enhanced side-effects have also been reported (Yang *et al.*, 1991; Schiller *et al.*, 1992); (4) redesign a TNF molecule with maximum efficacy as an antitumor agent but minimal side-effects as described in the section on TNF structure. In addition, chemical modification of TNF such as modification with polyethylene glycol (PEG) has been found selectively to increase its antitumor potency and effectively reduce its systemic toxic side-effects (Tsutsumi *et al.*, 1995).

A few mechanisms have been suggested for the tumoricidal effect of TNF, including direct cytotoxicity against tumor cells, activation of immune antitumor response and selective damage of tumor blood vessels (Nobuhara *et al.*, 1987; North and Havell, 1988). Direct cytotoxicity to tumor cells has been supported by most *in vitro* studies (Duerksen-Hughes *et al.*, 1992; the authors' unpublished data). TNF also activates other cytokines (Tracey and Cerami, 1994) and other cytotoxic factors such as NO (Estrada *et al.*, 1992), which in turn participate in immune antitumor responses and direct tumor cell killing. A recent study (Sato *et al.*, 1996) has demonstrated that it may be feasible to use TNF gene-transduced tumor cells as a vaccine owing to the effect of TNF in activation of the immune response. TNF selectively eradicates vascularized tumors but is much less effective in killing avascular implants (Yoshimura and Sone, 1987; Naredi *et al.*, 1993; Isaka *et al.*, 1996; Thomson, 1997). In contrast, TNF is produced by some tumors or tumor cells, and can act as an autocrine growth factor (Sugarman *et al.*, 1985; Buck *et al.*, 1990).

ASSOCIATED DISEASE SUSCEPTIBILITY

The TNF locus with two tandemly arranged and closely linked genes, TNF and lymphotoxin-α, lies in the class III region of the MHC on the short arm of human chromosome 6. A striking feature of the MHC is the high degree of polymorphism of genes in this region. At least five polymorphisms (Partanen and Koskimies, 1988; Ferencik *et al.*, 1992; Wilson *et al.*, 1992; D'Alfonso and Richiardi, 1994) and five microsatellites (Jongeneel *et al.*, 1991; Udalova *et al.*, 1993) have been described at the TNF locus. These polymorphisms have been linked to the variability of TNF production in different individuals (Molvig *et al.*, 1988; Jacob, 1992). As TNF production has been implicated as an important factor in immune regulation and the inflammatory response, and affects the outcome of human diseases, the polymorphism within the TNF locus is considered as a possible genetic basis for these diseases. Some studies have indicated that genetic variation of TNF and closely linked locuses is important in determining susceptibility to a significant number of human diseases, including autoimmune diseases such as diabetes, rheumatoid arthritis and multiple sclerosis; infectious diseases such as different parasitic, bacterial and viral infection; and cancer.

TNF responsiveness is controlled by variable genetic elements within the MHC region. A polymorphism located at −308 bp upstream of the TNF transcriptional start site, with two allelic forms referred to as TNF1 and TNF2, has been demonstrated to affect TNF production directly (Wilson *et al.*, 1992, 1993, 1997), and TNF2 is associated with higher constitutive and inducible transcriptional levels than the TNF1 allele (Wilson *et al.*, 1994, 1997). There is also a correlation of human leukocyte antigen (HLA)-DR2 with low TNF production (Bendtzen *et al.*, 1988; Molvig *et al.*, 1988), and of HLA-DR3 and HLA-DR4 with high TNF production (Jacob *et al.*, 1990; Abraham *et al.*, 1993). These polymorphisms have been shown to be associated with the susceptibility or severity of certain diseases. Thus, the TNF locus may contribute to the pathogenesis of diseases by influencing the TNF levels produced in those diseases.

Susceptibility to Infectious Diseases

MHC alleles affect outcome in several infectious diseases. TNF2 homozygotes associated with higher TNF levels are at increased risk of cerebral malaria (McGuire *et al.*, 1994). This is consistent with the phenotypic observation that Gambian children with cerebral malaria have higher TNF levels than children with mild malaria (Kwiatkowski *et al.*, 1990), and that TNF levels are highest in children who die or develop neurologic sequelae due to cerebral malaria. In contrast, several forms of malaria have been shown to be protected by HLA-B53 (Hill, 1992). The TNFB2 allele, also associated with higher levels of TNF production, is linked to lower survival in patients with multiple organ failure secondary to severe sepsis (Stuber *et al.*, 1996). Susceptibility to mucotaneous leishmaniasis, which is accompanied by high circulating TNF levels, is linked directly with the regulatory polymorphisms affecting TNF production (Cabrera *et al.*, 1995). In addition, HLA-DQ and HLA-DP alleles are associated with generalized and localized forms of onchocerciasis (Meyer *et al.*, 1994).

Susceptibility to Cancer

Alleles of HLA-A, -B and -DR loci are associated with resistance to lung cancer. Recently, investigators located one such element to the TNF locus and found that TNFB2 homozygosity is associated with disease resistance and a better prognosis in patients with lung cancer (Shimura *et al.*, 1994), in keeping with a previous study that demonstrated a strong association of HLA-A1 with a lower 1-year survival rate (Markman *et al.*, 1984). As TNFB2 is associated with higher levels of TNF production, these results are consistent with the original characterization of TNF as an antitumor factor.

It has been reported that germline allelic frequencies for the TNF locus are significantly different in patients with colorectal cancer compared with those in normal controls. Most strikingly, the α3 allele in patients with colorectal cancer has an allelic frequency of nearly 30% but is not detected in the normal control group (Campbell *et al.*, 1994). Other studies have shown that class II haplotypes such as DRB1*1501–DQB1*0602 are positively associated with the development of cervical cancer, but that DR13-positive haplotypes are negatively associated (Apple *et al.*, 1994).

Susceptibility to Autoimmune Diseases

The importance of TNF genetics in susceptibility to autoimmune diseases is suggested by a large number of studies. Celiac disease is an immune disease triggered by the cereal antigen gliadin, resulting in villus atrophy in the small intestine. Susceptibility to the development of celiac disease is strongly influenced by the genes in MHC, and associated with the HLA-A1–B8–DR3–DQ2 haplotype (Hall *et al.*, 1990; Bolsover *et al.*, 1991; Otley *et al.*, 1991; Ahmed *et al.*, 1993) and more strongly associated with DQA*0501/DQB*0201 heterodimer (Sollid *et al.*, 1989). Recently, Wilson and colleagues (1993) found that the TNF2 allele in the TNF locus is closely associated with HLA-A1–B8–DR3–DQ2, suggesting that the association of this haplotype with celiac disease and high TNF production may be related to polymorphism within the TNF locus. At the same time, susceptibility to celiac disease was demonstrated to be strongly related to TNF2 (Mansfield *et al.*, 1993), supporting this speculation. The role of TNF2 in the pathogenesis of CD is also supported by a recent study (Manus *et al.*, 1996). So far, the TNF2 allele is believed to be involved in a variety of MHC-linked diseases including systemic lupus erythematosus, dermatitis herpetiformis and insulin-dependent diabetes mellitus (IDDM), as well as parasitic infection. Another microsatellite polymorphism (TNFa2) close to the TNF gene has also been demonstrated to be independently associated with celiac disease, and could have functional significance as TNFa2 has also been correlated with high TNF production.

Two recent studies have located an RA element to the TNF locus and demonstrated that the TNFc1 allele (Mulcahy *et al.*, 1996) but not the TNFa allele (Brinkman *et al.*, 1996) appears to influence susceptibility to RA. Other studies have demonstrated that both TNFB1 (Bettinotti *et al.*, 1993) and TNF2 (Wilson *et al.*, 1993) are associated with systemic lupus erythematosus (SLE), but the association is not independent of HLA-DR3 (Welch *et al.*, 1988). In another study (D'Alfonso *et al.*, 1996), 123 patients with SLE and 199 matched controls were analyzed. Three TNF-238/A homozygous patients but no homozygous control were detected, suggesting that the −238/AA genotype is a marker of a particular clinical subtype. Because TNF has been implicated as a mediator of SLE, additional studies are now being undertaken to assess the contribution of the TNF locus in these diseases.

TNF has also been implicated in the pathogenesis of IDDM as a mediator in the immune-mediated destruction of pancreatic β cells (Mandrup-Poulsen *et al.*, 1987). Studies of TNF genetics in patients with IDDM show a weak association between TNF2 and IDDM (Pociot *et al.*, 1993; Cox *et al.*, 1994). However, another study demonstrated a higher frequency of the TNFa2 allele in DR3–DR4 heterozygotic patients with IDDM than in controls. As this marker is associated with higher TNF production independently of class II alleles, the TNF locus may play a direct role in susceptibility to IDDM (Pociot *et al.*, 1993a).

REFERENCES

Abraham, L.J., French, M.A. and Dawkins, R.L. (1993). Polymorphic MHC ancestral haplotypes affect the activity of tumour necrosis factor-α. *Clin. Exp. Immunol.* **92**, 14–18.

Adam-Klages, S., Adam, D., Wiegmann, K. *et al.* (1996). FAN, a novel WD-repeat protein, couples the p55 TNF-receptor to neutral sphingomyelinase. *Cell* **86**, 937–947.

Aderka, D. (1996). The potential biological and clinical significance of the soluble tumor necrosis factor receptors. *Cytokine Growth Factor Rev.* **7**, 231–240.

Aderka, D., Englemann, H., Hornik, V. *et al*. (1991). Increased serum levels of soluble receptors for tumor necrosis factor in cancer patients. *Cancer Res.* **51**, 5602–5607.

Aderka, D., Engelmann, H., Maor, Y. *et al*. (1992). Stabilization of the bioactivity of tumor necrosis factor by its soluble receptors. *J. Exp. Med.* **175**, 323–329.

Ahmed, A.R., Yunis, J.J., Marcus-Bagley, D. *et al*. (1993). Major histocompatibility complex susceptibility genes for dermatitis herpetiformis compared with those for gluten-sensitive enteropathy. *J. Exp. Med.* **178**, 2067–2075.

Alexander, H.R., Doherty, G.M., Buresh, C.M. *et al*. (1991). A recombinant human receptor antagonist to interleukin 1 improves survival after lethal endotoxemia in mice. *J. Exp. Med.* **173**, 1029–1032.

Apple, R.J., Erlich, H.A., Klitz, W. *et al*. (1994). HLA DR–DQ associations with cervical carcinoma show papillomavirus-type specificity. *Nature Genet.* **6**, 157–162.

Arend, W.P., Welgus, H.G., Thompson, R.C. *et al*. (1990). Biological properties of recombinant human monocyte-derived interleukin 1 receptor antagonist. *J. Clin. Invest.* **85**, 1694–1697.

Ashkenazi, A., Marsters, S.A., Capon, D.J. *et al*. (1991). Protection against endotoxic shock by a tumor necrosis factor receptor immunoadhesin. *Proc. Natl Acad. Sci. U.S.A.* **88**, 10 535–10 539.

Auphan, N., Di Donato, J.A., Rosette, C. *et al*. (1995). Immunosuppression by glucocorticoids: inhibition of NF-κB activity through induction of IκB synthesis. *Science* **270**, 286–290.

Balboa, M.A., Balsinde, J., Aramburu, J. *et al*. (1992). Phospholipase D activation in human natural killer cells through the KP43 and CD16 surface antigens takes place by different mechanisms. Involvement of the phospholipase D pathway in tumor necrosis factor α synthesis. *J. Exp. Med.* **176**, 9–17.

Baldwin, R.L., Chang, M.P., Bramhall, J. *et al*. (1988). Capacity of tumor necrosis factor to bind and penetrate membranes is pH-dependent. *J. Immunol.* **141**, 2352–2357.

Baldwin, R.L., Stolowitz, M.L., Hood, L. *et al*. (1996). Structural changes of tumor necrosis factor α associated with membrane insertion and channel formation. *Proc. Natl Acad. Sci. U.S.A.* **93**, 1021–1026.

Banner, D.W., D'Arcy, A., Janes, W. *et al*. (1993). Crystal structure of the soluble human 55 kD TNF receptor-human TNF β. *Cell* **73**, 431–445.

Barbara, J.A., Smith, W.B., Gamble, J.R. *et al*. (1994). Dissociation of TNF-α cytotoxic and proinflammatory activities by p55 receptor- and p75 receptor-selective TNF-α mutants. *EMBO J.* **13**, 843–850.

Bauer, K.A., ten Cate, H., Barzegar, S. *et al*. (1989). Tumor necrosis factor infusions have a procoagulant effect on the hemostatic mechanism of humans. *Blood* **74**, 165–172.

Bazan, J.F. (1991). Neuropoietic cytokines in the hematopoietic fold. *Neuron* **7**, 197–208.

Bendtzen, K., Morling, N., Fomsgaard, A. *et al*. (1988). Association between HLA-DR2 and production of tumour necrosis factor α and interleukin 1 by mononuclear cells activated by lipopolysaccharide. *Scand. J. Immunol.* **28**, 599–606.

Bettinotti, M.P., Hartung, K., Deicher, H. *et al*. (1993). Polymorphism of the tumor necrosis factor β gene in systemic lupus erythematosus: TNFB–MHC haplotypes. *Immunogenetics* **37**, 449–454.

Beutler, B., Greenwald, D., Hulmes, J.D. *et al*. (1985). Identity of tumour necrosis factor and the macrophage-secreted factor cachectin. *Nature* **316**, 552–554.

Beutler, B., Krochin, N., Milsark, I.W. *et al*. (1986). Control of cachectin (tumor necrosis factor) synthesis: mechanisms of endotoxin resistance. *Science* **232**, 977–980.

Bevilacqua, M.P., Pober, J.S., Majeau, G.R. *et al*. (1986). Recombinant tumor necrosis factor induces procoagulant activity in cultured human vascular endothelium: characterization and comparison with the actions of interleukin 1. *Proc. Natl Acad. Sci. U.S.A.* **83**, 4533–4537.

Bianchi, M., Bloom, O., Raabe, T. *et al*. (1996). Suppression of proinflammatory cytokines in monocytes by a tetravalent guanylhydrazone. *J. Exp. Med.* **183**, 927–936.

Biswas, D.K., Ahlers, C.M., Dezube, B.J. *et al*. (1994). Pentoxifylline and other protein kinase C inhibitors down-regulate HIV-LTR NF-κB induced gene expression. *Mol. Med.* **1**, 31–43.

Black, R.A., Rauch, C.T., Kozlosky, C.J. *et al*. (1997). A metalloproteinase disintegrin that releases tumour-necrosis factor-α from cells. *Nature* **385**, 729–733.

Blick, M., Sherwin, S.A., Rosenblum, M. *et al*. (1987). Phase I study of recombinant tumor necrosis factor in cancer patients. *Cancer Res.* **47**, 2986–2989.

Bogdan, C., Paik, J., Vodovotz, Y. *et al*. (1992). Contrasting mechanisms for suppression of macrophage cytokine release by transforming growth factor-β. *J. Biol. Chem.* **267**, 23 301–23 308.

Bohjanen, P.R., Petryniak, B., June, C.H. *et al*. (1991). An inducible cytoplasmic factor (AU-B) binds selectively to AUUUA multimers in the 3′ untranslated region of lymphokine mRNA. *Mol. Cell Biol.* **11**, 3288–3295.

Boise, L.H. and Thompson, C.B. (1997). Bcl-x(L) can inhibit apoptosis in cells that have undergone Fas-induced protease activation. *Proc. Natl Acad. Sci. U.S.A.* **94**, 3759–3764.

Boise, L.H., Minn, A.J., Noel, P.J. *et al*. (1995). CD28 costimulation can promote T cell survival by enhancing the expression of Bcl-xL. *Immunity* **3**, 87–98.

Boldin, M.P., Goncharov, T.M., Goltsev, Y.V. *et al*. (1996). Involvement of mach, a novel MORT1/FADD-interacting protease, in Fas/Apo-1- and TNF receptor-induced cell death. *Cell* **85**, 803–815.

Bolsover, W.J., Hall, M.A., Vaughan, R.W. *et al*. (1991). A family study confirms that the HLA-DP associations with celiac disease are the result of an extended HLA-DR3 haplotype. *Hum. Immunol.* **31**, 100–108.

Bradley, J.R., Johnson, D.R. and Pober, J.S. (1993). Four different classes of inhibitors of receptor-mediated endocytosis decrease tumor necrosis factor-induced gene expression in human endothelial cells. *J. Immunol.* **150**, 5544–5555.

Brennan, F.M., Maini, R.N. and Feldmann, M. (1992). TNF α—a pivotal role in rheumatoid arthritis? *Br. J. Rheumatol.* **31**, 293–298.

Brett, J., Gerlach, H., Nawroth, P. *et al*. (1989). Tumor necrosis factor/cachectin increases permeability of endothelial cell monolayers by a mechanism involving regulatory G proteins. *J. Exp. Med.* **169**, 1977–1991.

Brinkman, B.M., Huizinga, T.W., Breedveld, F.C. *et al*. (1996). Allele-specific quantification of TNFA transcripts in rheumatoid arthritis. *Hum. Genet.* **97**, 813–818.

Brockhaus, M., Schoenfeld, H.J., Schlaeger, E.J. *et al*. (1990). Identification of two types of tumor necrosis factor receptors on human cell lines by monoclonal antibodies. *Proc. Natl Acad. Sci. U.S.A.* **87**, 3127–3131.

Bruce, A.J., Boling, W., Kindy, M.S. *et al*. (1996). Altered neuronal and microglial responses to excitotoxic and ischemic brain injury in mice lacking TNF receptors. *Nature Med.* **2**, 788–794.

Buck, C., Digel, W., Schoniger, W. *et al*. (1990). Tumor necrosis factor-α, but not lymphotoxin, stimulates growth of tumor cells in hairy cell leukemia. *Leukemia* **4**, 431–434.

Cabrera, M., Shaw, M.A., Sharples, C. *et al*. (1995). Polymorphism in tumor necrosis factor genes associated with mucocutaneous leishmaniasis. *J. Exp. Med.* **182**, 1259–1264.

Calandra, T., Baumgartner, J.D., Grau, G.E. *et al*. (1990). Prognostic values of tumor necrosis factor/cachectin, interleukin-1, interferon-α and interferon-γ in the serum of patients with septic shock. Swiss–Dutch J5 immunoglobulin study group. *J. Infect. Dis.* **161**, 982–987.

Campbell, D.A., Field, M., McArdle, C.S. *et al*. (1994). Polymorphism at the tumour necrosis factor locus: a marker of genetic predisposition to colorectal cancer? *Lancet* **343**, 293–294 (letter).

Caput, D., Beutler, B., Hartog, K. *et al*. (1986). Identification of a common nucleotide sequence in the 3′-untranslated region of mRNA molecules specifying inflammatory mediators. *Proc. Natl Acad. Sci. U.S.A.* **83**, 1670–1674.

Carswell, E.A., Old, L.J., Kassel, R.L. *et al*. (1975). An endotoxin-induced serum factor that causes necrosis of tumors. *Proc. Natl Acad. Sci. U.S.A.* **72**, 3666–3670.

Cavender, D.E. and Edelbaum, D. (1988). Inhibition by IL-1 of endothelial cell activation induced by tumor necrosis factor or lymphotoxin. *J. Immunol.* **141**, 3111–3116.

Chainy, G.B., Singh, S., Raju, U. *et al*. (1996). Differential activation of the nuclear factor-κB by TNF muteins specific for the p60 and p80 TNF receptors. *J. Immunol.* **157**, 2410–2417.

Chang, M.P. and Wisnieski, B.J. (1990). Comparison of the intoxication pathways of tumor necrosis factor and diphtheria toxin. *Infect. Immun.* **58**, 2644–2650.

Cheng, B., Christakos, S. and Mattson, M.P. (1994). Tumor necrosis factors protect neurons against metabolic–excitotoxic insults and promote maintenance of calcium homeostasis. *Neuron* **12**, 139–153.

Cheng, G. and Baltimore, D. (1996). Tank, a co-inducer with TRAF2 of TNF- and CD 401-mediated NF-κB activation. *Genes Dev.* **10**, 963–973.

Chikanza, I.C., Roux-Lombard, P., Dayer, J.M. *et al*. (1993). Tumour necrosis factor soluble receptors behave as acute phase reactants following surgery in patients with rheumatoid arthritis, chronic osteomyelitis and osteoarthritis. *Clin. Exp. Immunol.* **92**, 19–22.

Chinnaiyan, A.M., O'Rourke, K., Tewari, M. *et al*. (1995). FADD, a novel death domain-containing protein, interacts with the death domain of Fas and initiates apoptosis. *Cell* **81**, 505–512.

Chung, I.Y., Kwon, J. and Benveniste, E.N. (1992). Role of protein kinase C activity in tumor necrosis factor-α gene expression. Involvement at the transcriptional level. *J. Immunol.* **149**, 3894–3902.

Cohen, P.S., Nakshatri, H., Dennis, J. *et al*. (1996). CNI-1493 inhibits monocyte/macrophage tumor necrosis factor by suppression of translation efficiency. *Proc. Natl Acad. Sci. U.S.A.* **93**, 3967–3971.

Cox, A., Gonzalez, A.M., Wilson, A.G. *et al.* (1994). Comparative analysis of the genetic associations of HLA-DR3 and tumour necrosis factor α with human IDDM. *Diabetologia* **37**, 500–503.

Creasey, A.A., Doyle, L.V., Reynolds, M.T. *et al.* (1987). Biological effects of recombinant human tumor necrosis factor and its novel muteins on tumor and normal cell lines. *Cancer Res.* **47**, 145–149.

Crowe, P.D., Walter, B.N., Mohler, K.M. *et al.* (1995). A metalloprotease inhibitor blocks shedding of the 80-kD TNF receptor and TNF processing in T lymphocytes. *J. Exp. Med.* **181**, 1205–1210.

D'Alfonso, S. and Richiardi, P.M. (1994). A polymorphic variation in a putative regulation box of the TNFA promoter region. *Immunogenetics* **39**, 150–154.

D'Alfonso, S., Colombo, G., Della Bella, S. *et al.* (1996). Association between polymorphisms in the TNF region and systemic lupus erythematosus in the Italian population. *Tissue Antigens* **47**, 551–555.

Damle, N.K. and Doyle, L.V. (1989). IL-2-activated human killer lymphocytes but not their secreted products mediate increase in albumin flux across cultured endothelial monolayers. Implications for vascular leak syndrome. *J. Immunol.* **142**, 2660–2669.

Duan, H. and Dixit, V.M. (1997). RAIDD is a new 'death' adaptor molecule. *Nature* **385**, 86–89.

Duerksen-Hughes, P.J., Day, D.B., Laster, S.M. *et al.* (1992). Both tumor necrosis factor and nitric oxide participate in lysis of simian virus 40-transformed cells by activated macrophages. *J. Immunol.* **149**, 2114–2122.

Dutka, D.P., Elborn, J.S., Delamere, F. *et al.* (1993). Tumour necrosis factor α in severe congestive cardiac failure. *Br. Heart J.* **70**, 141–143.

Eck, M.J. and Sprang, S.R. (1989). The structure of tumor necrosis factor-α at 2.6 Å resolution. Implications for receptor binding. *J. Biol. Chem.* **264**, 17 595–17 605.

Eck, M.J., Ultsch, M., Rinderknecht, E. *et al.* (1992). The structure of human lymphotoxin (tumor necrosis factor-β). *J. Biol. Chem.* **267**, 2119–2122.

Economou, J.S., Rhoades, K., Essner, R. *et al.* (1989). Genetic analysis of the human tumor necrosis factor α/cachectin promoter region in a macrophage cell line. *J. Exp. Med.* **170**, 321–326.

Engelmann, H., Aderka, D., Rubinstein, M. *et al* (1989). A tumor necrosis factor-binding protein purified to homogeneity from human urine protects cells from tumor necrosis factor toxicity. *J. Biol. Chem.* **264**, 11 974–11 980.

Engelmann, H., Holtmann, H., Brakebusch, C. *et al.* (1990a). Antibodies to a soluble form of a tumor necrosis factor (TNF) receptor have TNF-like activity. *J. Biol. Chem.* **265**, 14 497–14 504.

Engelmann, H., Novick, D. and Wallach, D. (1990b). Two tumor necrosis factor-binding proteins purified from human urine. Evidence for immunological cross-reactivity with cell surface tumor necrosis factor receptors. *J. Biol. Chem.* **265**, 1531–1536.

Espevik, T., Brockhaus, M., Loetscher, H. *et al.* (1990). Characterization of binding and biological effects of monoclonal antibodies against a human tumor necrosis factor receptor. *J. Exp. Med.* **171**, 415–426.

Estrada, C., Gomez, C., Martin, C. *et al.* (1992). Nitric oxide mediates tumor necrosis factor-α cytotoxicity in endothelial cells. *Biochem. Biophys. Res. Commun.* **186**, 475–482.

Ferencik, S., Lindemann, M., Horsthemke, B. *et al.* (1992). A new restriction fragment length polymorphism of the human TNF-B gene detected by ASPHI digest. *Eur. J. Immunogen.* **19**, 425–430.

Fillit, H., Ding, W.H., Buee, L. *et al.* (1991). Elevated circulating tumor necrosis factor levels in Alzheimer's disease. *Neurosci. Lett.* **129**, 318–320.

Finkel, M.S., Oddis, C.V., Jacob, T.D. *et al.* (1992). Negative inotropic effects of cytokines on the heart mediated by nitric oxide. *Science* **257**, 387–389.

Fisher, C.J., Jr., Opal, S.M., Dhainaut, J.F. *et al.* (1993). Influence of an anti-tumor necrosis factor monoclonal antibody on cytokine levels in patients with sepsis. The CB0006 sepsis syndrome study group. *Crit. Care Med.* **21**, 318–327.

Fong, Y., Tracey, K.J., Moldawer, L.L. *et al.* (1989). Antibodies to cachectin/tumor necrosis factor reduce interleukin 1β and interleukin 6 appearance during lethal bacteremia. *J. Exp. Med.* **170**, 1627–1633.

Gadient, R.A., Cron, K.C. and Otten, U. (1990). Interleukin-1β and tumor necrosis factor-α synergistically stimulate nerve growth factor (NGF) release from cultured rat astrocytes. *Neurosci. Lett.* **117**, 335–340.

Gatanaga, T., Hwang, C.D., Gatanaga, M. *et al.* (1991). The regulation of TNF receptor mRNA synthesis, membrane expression, and release by PMA- and LPS-stimulated human monocytic THP-1 cells *in vitro*. *Cell. Immunol.* **138**, 1–10.

Gearing, A.J., Beckett, P., Christodoulou, M. *et al.* (1994). Processing of tumour necrosis factor-α precursor by metalloproteinases. *Nature* **370**, 555–557.

Gearing, A.J., Beckett, P., Christodoulou, M. *et al.* (1995). Matrix metalloproteinases and processing of pro-TNF-α. *J. Leukoc. Biol.* **57**, 774–777.

Girardin, E., Grau, G.E., Dayer, J.M. *et al*. (1988). Tumor necrosis factor and interleukin-1 in the serum of children with severe infectious purpura. *N. Engl. J. Med.* **319**, 397–400.

Girardin, E., Roux-Lombard, P., Grau, G.E. *et al*. (1992). Imbalance between tumour necrosis factor-α and soluble TNF receptor concentrations in severe meningococcaemia. The J5 Study Group. *Immunology* **76**, 20–23.

Giroir, B.P. and Beutler, B. (1992). Effect of amrinone on tumor necrosis factor production in endotoxic shock. *Circ. Shock* **36**, 200–207.

Giroir, B.P., Horton, J.W., White, D.J. *et al*. (1994). Inhibition of tumor necrosis factor prevents myocardial dysfunction during burn shock. *Am. J. Physiol.* **267**, H118–124.

Goldfeld, A.E., Doyle, C. and Maniatis, T. (1990). Human tumor necrosis factor α gene regulation by virus and lipopolysaccharide. *Proc. Natl Acad. Sci. U.S.A.* **87**, 9769–9773.

Goldfeld, A.E., McCaffrey, P.G., Strominger, J.L. *et al*. (1993). Identification of a novel cyclosporin-sensitive element in the human tumor necrosis factor α gene promoter. *J. Exp. Med.* **178**, 1365–1379.

Grafi, G., Sela, I. and Galili, G. (1993). Translational regulation of human β interferon mRNA: association of the 3′ AU-rich sequence with the poly(A) tail reduces translation efficiency *in vitro*. *Mol. Cell Biol.* **13**, 3487–3493.

Grell, M., Zimmermann, G., Hulser, D. *et al*. (1994). TNF receptors TR60 and TR80 can mediate apoptosis via induction of distinct signal pathways. *J. Immunol.* **153**, 1963–1972.

Grell, M., Douni, E., Wajant, H. *et al*. (1995). The transmembrane form of tumor necrosis factor is the prime activating ligand of the 80 kDa tumor necrosis factor receptor. *Cell* **83**, 793–802.

Grimaldi, L.M., Martino, G.V., Franciotta, D.M. *et al*. (1991). Elevated α-tumor necrosis factor levels in spinal fluid from HIV-1-infected patients with central nervous system involvement. *Ann. Neurol.* **29**, 21–25.

Hakoshima, T. and Tomita, K. (1988). Crystallization and preliminary X-ray investigation reveals that tumor necrosis factor is a compact trimer furnished with 3-fold symmetry. *J. Mol. Biol.* **201**, 455–457.

Hall, R.P., Ward, F.E. and Wenstrup, R.J. (1990). An HLA class II region restriction fragment length polymorphism (RFLP) in patients with dermatitis herpetiformis: association with HLA-DP phenotype. *J. Invest. Dermatol.* **95**, 172–177.

Han, J., Brown, T. and Beutler, B. (1990a). Endotoxin-responsive sequences control cachectin/tumor necrosis factor biosynthesis at the translational level. *J. Exp. Med.* **171**, 465–475.

Han, J., Thompson, P. and Beutler, B. (1990b). Dexamethasone and pentoxifylline inhibit endotoxin-induced cachectin/tumor necrosis factor synthesis at separate points in the signaling pathway. *J. Exp. Med.* **172**, 391–394.

Han, J., Huez, G. and Beutler, B. (1991). Interactive effects of the tumor necrosis factor promoter and 3′-untranslated regions. *J. Immunol.* **146**, 1843–1848.

Hefti, F., Hartikka, J. and Knusel, B. (1989). Function of neurotrophic factors in the adult and aging brain and their possible use in the treatment of neurodegenerative diseases. *Neurobiol. Aging* **10**, 515–533.

Hegewisch, S., Weh, H.J. and Hossfeld, D.K. (1990). TNF-induced cardiomyopathy. *Lancet* **335**, 294–295 (letter).

Hill, A.V. (1992). Malaria resistance genes: a natural selection. *Trans. R. Soc. Trop. Med. Hyg.* **86**, 225–226, 232.

Hirano, M., Osada, S., Aoki, T. *et al*. (1996). MEK kinase is involved in tumor necrosis factor α-induced NF-κB activation and degradation of IκB-α. *J. Biol. Chem.* **271**, 13234–13238.

Hlodan, R. and Pain, R.H. (1994). Tumour necrosis factor is in equilibrium with a trimeric molten globule at low pH. *FEBS Lett.* **343**, 256–260.

Hofman, F.M., Hinton, D.R., Johnson, K. *et al*. (1989). Tumor necrosis factor identified in multiple sclerosis brain. *J. Exp. Med.* **170**, 607–612.

Hohmann, H.P., Remy, R., Brockhaus, M. *et al*. (1989). Two different cell types have different major receptors for human tumor necrosis factor (TNF α). *J. Biol. Chem.* **264**, 14927–14934.

Hohmann, H.P., Brockhaus, M., Baeuerle, P.A. *et al*. (1990a). Expression of the types A and B tumor necrosis factor (TNF) receptors is independently regulated, and both receptors mediate activation of the transcription factor NF-κB. TNF α is not needed for induction of a biological effect via TNF receptors. *J. Biol. Chem.* **265**, 22409–22417.

Hohmann, H.P., Remy, R., Poschl, B. *et al*. (1990b). Tumor necrosis factors-α and -β bind to the same two types of tumor necrosis factor receptors and maximally activate the transcription factor NF-κB at low receptor occupancy and within minutes after receptor binding. *J. Biol. Chem.* **265**, 15183–15188.

Honchel, R., McDonnell, S., Schaid, D.J. *et al*. (1996). Tumor necrosis factor-α allelic frequency and chromosome 6 allelic imbalance in patients with colorectal cancer. *Cancer Res.* **56**, 145–149.

Horvath, C.J., Ferro, T.J., Jesmok, G. *et al*. (1988). Recombinant tumor necrosis factor increases pulmonary vascular permeability independent of neutrophils. *Proc. Natl Acad. Sci. U.S.A.* **85**, 9219–9223.

Hsu, H., Xiong, J. and Goeddel, D.V. (1995). The TNF receptor 1-associated protein TRADD signals cell death and NF-κB activation. *Cell* **81**, 495–504.

Hsu, H., Huang, J., Shu, H.B. *et al*. (1996a). TNF-dependent recruitment of the protein kinase RIP to the TNF receptor-1 signaling complex. *Immunity* **4**, 387–396.

Hsu, H., Shu, H.B., Pan, M.G. *et al*. (1996b). TRADD–TRAF2 and TRADD–FADD interactions define two distinct TNF receptor 1 signal transduction pathways. *Cell* **84**, 299–308.

Isaka, T., Yoshimine, T., Maruno, M. *et al*. (1996). Morphological effects of tumor necrosis factor-α on the blood vessels in rat experimental brain tumors. *Neurol. Med. Chir. (Tokyo).* **36**, 423–427.

Jacob, C.O. (1992). Genetic variability in tumor necrosis factor production: relevance to predisposition to autoimmune disease. *Reg. Immunol.* **4**, 298–304.

Jacob, C.O., Fronek, Z., Lewis, G.D. *et al*. (1990). Heritable major histocompatibility complex class II-associated differences in production of tumor necrosis factor α: relevance to genetic predisposition to systemic lupus erythematosus. *Proc. Natl Acad. Sci. U.S.A.* **87**, 1233–1237.

Jones, E.Y., Stuart, D.I. and Walker, N.P. (1989). Structure of tumour necrosis factor. *Nature* **338**, 225–228.

Jongeneel, C.V., Briant, L., Udalova, I.A. *et al*. (1991). Extensive genetic polymorphism in the human tumor necrosis factor region and relation to extended HLA haplotypes. *Proc. Natl Acad. Sci. U.S.A.* **88**, 9717–9721.

Kagan, B.L., Baldwin, R.L., Munoz, D. *et al*. (1992). Formation of ion-permeable channels by tumor necrosis factor-α. *Science* **255**, 1427–1430.

Kapadia, S., Torre-Amione, G., Yokoyama, T. *et al*. (1995). Soluble TNF binding proteins modulate the negative inotropic properties of TNF-α *in vitro*. *Am. J. Physiol.* **268**, H517–525.

Katz, S.D., Rao, R., Berman, J.W. *et al*. (1994). Pathophysiological correlates of increased serum tumor necrosis factor in patients with congestive heart failure. Relation to nitric oxide-dependent vasodilation in the forearm circulation. *Circulation* **90**, 12–16.

Kilbourn, R.G., Gross, S.S., Jubran, A. *et al*. (1990). NG-methyl-L-arginine inhibits tumor necrosis factor-induced hypotension: implications for the involvement of nitric oxide. *Proc. Natl Acad. Sci. U.S.A.* **87**, 3629–3632.

Kindler, V., Sappino, A.P., Grau, G.E. *et al*. (1989). The inducing role of tumor necrosis factor in the development of bactericidal granulomas during BCG infection. *Cell* **56**, 731–740.

Kohno, T., Brewer, M.T., Baker, S.L. *et al*. (1990). A second tumor necrosis factor receptor gene product can shed a naturally occurring tumor necrosis factor inhibitor. *Proc. Natl Acad. Sci. U.S.A.* **87**, 8331–8335.

Kramer, B., Meichle, A., Hensel, G. *et al*. (1994). Characterization of a KROX-24/EGR-1-responsive element in the human tumor necrosis factor promoter. *Biochim. Biophys. Acta* **1219**, 413–421.

Kramer, B., Wiegmann, K. and Kronke, M. (1995). Regulation of the human TNF promoter by the transcription factor Ets. *J. Biol. Chem.* **270**, 6577–6583.

Kreil, E.A., Greene, E., Fitzgibbon, C. *et al*. (1989). Effects of recombinant human tumor necrosis factor α, lymphotoxin, and *Escherichia coli* lipopolysaccharide on hemodynamics, lung microvascular permeability, and eicosanoid synthesis in anesthetized sheep. *Circ. Res.* **65**, 502–514.

Kriegler, M., Perez, C., De Fay, K. *et al*. (1988). A novel form of TNF/cachectin is a cell surface cytotoxic transmembrane protein: ramifications for the complex physiology of TNF. *Cell* **53**, 45–53.

Kruys, V., Marinx, O., Shaw, G. *et al*. (1989). Translational blockade imposed by cytokine-derived UA-rich sequences. *Science* **245**, 852–855.

Kuroda, K., Miyata, K., Shikama, H. *et al*. (1995). Novel muteins of human tumor necrosis factor with potent antitumor activity and less lethal toxicity in mice. *Int. J. Cancer* **63**, 152–157.

Kwiatkowski, D., Hill, A.V., Sambou, I. *et al*. (1990). TNF concentration in fatal cerebral, non-fatal cerebral, and uncomplicated *Plasmodium falciparum* malaria. *Lancet* **336**, 1201–1204.

Lacraz, S., Nicod, L., Galve-de Rochemonteix, B. *et al*. (1992). Suppression of metalloproteinase biosynthesis in human alveolar macrophages by interleukin-4. *J. Clin. Invest.* **90**, 382–388.

Lacraz, S., Nicod, L.P., Chicheportiche, R. *et al*. (1995). IL-10 inhibits metalloproteinase and stimulates TIMP-1 production in human mononuclear phagocytes. *J. Clin. Invest.* **96**, 2304–2310.

Lagnado, C.A., Brown, C.Y. and Goodall, G.J. (1994). AUUUA is not sufficient to promote poly(A) shortening and degradation of an mRNA: the functional sequence within AU-rich elements may be UUAUUUA(U/A)(U/A). *Mol. Cell Biol.* **14**, 7984–7995.

Lai, N.S., Lan, J.L., Yu, C.L. *et al*. (1995). Role of tumor necrosis factor-α in the regulation of activated synovial

T cell growth: down-regulation of synovial T cells in rheumatoid arthritis patients. *Eur. J. Immunol.* **25**, 3243–3248.

Lattime, E.C., Stoppacciaro, A. and Stutman, O. (1988). Limiting dilution analysis of TNF producing cells in C3H/HEJ mice. *J. Immunol.* **141**, 3422–3428.

Lee, E., Vaughan, D.E., Parikh, S.H. *et al*. (1996). Regulation of matrix metalloproteinases and plasminogen activator inhibitor-1 synthesis by plasminogen in cultured human vascular smooth muscle cells. *Circ. Res.* **78**, 44–49.

Lee, F.S., Hagler, J., Chen, Z.J. *et al*. (1997). Activation of the IκB α kinase complex by MEKK1, a kinase of the JNK pathway. *Cell* **88**, 213–222.

Lee, J.C., Laydon, J.T., McDonnell, P.C. *et al*. (1994). A protein kinase involved in the regulation of inflammatory cytokine biosynthesis. *Nature* **372**, 739–746.

Leeuwenberg, J.F., Dentener, M.A. and Buurman, W.A. (1994a). Lipopolysaccharide LPS-mediated soluble TNF receptor release and TNF receptor expression by monocytes. Role of CD14, LPS binding protein, and bactericidal/permeability-increasing protein. *J. Immunol.* **152**, 5070–5076.

Leeuwenberg, J.F., Jeunhomme, T.M. and Buurman, W.A. (1994b). Slow release of soluble TNF receptors by monocytes *in vitro*. *J. Immunol.* **152**, 4036–4043.

Leist, T.P., Frei, K., Kam-Hansen, S. *et al*. (1988). Tumor necrosis factor α in cerebrospinal fluid during bacterial, but not viral, meningitis. Evaluation in murine model infections and in patients. *J. Exp. Med.* **167**, 1743–1748.

Leitman, D.C., Ribeiro, R.C., Mackow, E.R. *et al*. (1991). Identification of a tumor necrosis factor-responsive element in the tumor necrosis factor α gene. *J. Biol. Chem.* **266**, 9343–9346.

Leitman, D.C., Mackow, E.R., Williams, T. *et al*. (1992). The core promoter region of the tumor necrosis factor α gene confers phorbol ester responsiveness to upstream transcriptional activators. *Mol. Cell Biol.* **12**, 1352–1356.

Levine, B., Kalman, J., Mayer, L. *et al*. (1990). Elevated circulating levels of tumor necrosis factor in severe chronic heart failure. *N. Engl. J. Med.* **323**, 236–241.

Lienard, D., Ewalenko, P., Delmotte, J.J. *et al*. (1992). High-dose recombinant tumor necrosis factor α in combination with interferon γ and melphalan in isolation perfusion of the limbs for melanoma and sarcoma. *J. Clin. Oncol.* **10**, 52–60.

Lilly, C.M., Sandhu, J.S., Ishizaka, A. *et al*. (1989). Pentoxifylline prevents tumor necrosis factor-induced lung injury. *Am. Rev. Respir. Dis.* **139**, 1361–1368.

Lin, L.L., Wartmann, M., Lin, A.Y. *et al*. (1993). CPLA2 is phosphorylated and activated by MAP kinase. *Cell* **72**, 269–278.

Lin, L.S. (1992). In *Tumor Necrosis Factor: The Molecules and Their Emerging Role in Medince* (ed. B. Beutler), Raven Press, New York, pp. 33–48.

Lin, R.H., Hwang, Y.W., Yang, B.C. *et al*. (1997). TNF receptor-2-triggered apoptosis is associated with the down-regulation of Bcl-xL on activated T cells and can be prevented by CD28 costimulation. *J. Immunol.* **158**, 598–603.

Liu, Z.G., Hsu, H., Goeddel, D.V. *et al*. (1996). Dissection of TNF receptor 1 effector functions: JNK activation is not linked to apoptosis while NF-κB activation prevents cell death. *Cell* **87**, 565–576.

Loetscher, H., Stueber, D., Banner, D. *et al*. (1993). Human tumor necrosis factor α (TNF α) mutants with exclusive specificity for the 55-kDa or 75-kDa TNF receptors. *J. Biol. Chem.* **268**, 26 350–26 357.

Machleidt, T., Kramer, B., Adam, D. *et al*. (1996). Function of the p55 tumor necrosis factor receptor 'death domain' mediated by phosphatidylcholine-specific phospholipase C. *J. Exp. Med.* **184**, 725–733.

Mackay, F., Rothe, J., Bluethmann, H. *et al*. (1994). Differential responses of fibroblasts from wild-type and TNF-r55-deficient mice to mouse and human TNF-α activation. *J. Immunol.* **153**, 5274–5284.

Malinin, N.L., Boldin, M.P., Kovalenko, A.V. *et al*. (1997). MAP3K-related kinase involved in NF-κB induction by TNF, CD95 and IL-1. *Nature* **385**, 540–544.

Mallett, S. and Barclay, A.N. (1991). A new superfamily of cell surface proteins related to the nerve growth factor receptor. *Immunol. Today* **12**, 220–223.

Malter, J.S. (1989). Identification of an AUUUA-specific messenger RNA binding protein. *Science* **246**, 664–666.

Mandrup-Poulsen, T., Bendtzen, K., Dinarello, C.A. *et al*. (1987). Human tumor necrosis factor potentiates human interleukin 1-mediated rat pancreatic β-cell cytotoxicity. *J. Immunol.* **139**, 4077–4082.

Mansfield, J.C., Holden, H., Wilson, A.G. *et al.* (1993). Coeliac disease associates with a polymorphism in the promoter region of the TNFα gene and further defines the coeliac haplotype. *Gut* **34**, S20–S23 (abstract).

Manus, R.M., Wilson, A.G., Mansfield, J. *et al*. (1996). TNF2, a polymorphism of the tumour necrosis-α gene promoter, is a component of the celiac disease major histocompatibility complex haplotype. *Eur. J. Immunol.* **26**, 2113–2118.

Markman, M., Braine, H.G. and Abeloff, M.D. (1984). Histocompatibility antigens in small cell carcinoma of the lung. *Cancer* **54**, 2943–2945.

Marx, J. (1995). How the glucocorticoids suppress immunity. *Science* **270**, 232–233.

Matsumori, A., Shioi, T., Yamada, T. *et al*. (1994). Vesnarinone, a new inotropic agent, inhibits cytokine production by stimulated human blood from patients with heart failure. *Circulation* **89**, 955–958.

Mattson, M.P., Zhang, Y. and Bose, S. (1993). Growth factors prevent mitochondrial dysfunction, loss of calcium homeostasis, and cell injury, but not ATP depletion in hippocampal neurons deprived of glucose. *Exp. Neurol.* **121**, 1–13.

McGeehan, G.M., Becherer, J.D., Bast, R.C., Jr. *et al*. (1994). Regulation of tumour necrosis factor-α processing by a metalloproteinase inhibitor. *Nature* **370**, 558–561.

McGuire, W., Hill, A.V., Allsopp, C.E. *et al*. (1994). Variation in the TNF-α promoter region associated with susceptibility to cerebral malaria. *Nature* **371**, 508–510.

McMurray, J., Abdullah, I., Dargie, H.J. *et al*. (1991). Increased concentrations of tumour necrosis factor in 'cachectic' patients with severe chronic heart failure. *Br. Heart J.* **66**, 356–358.

McNamara, M.J., Norton, J.A., Nauta, R.J. *et al*. (1993). Interleukin-1 receptor antibody (IL-1RAB) protection and treatment against lethal endotoxemia in mice. *J. Surg. Res.* **54**, 316–321.

Mehler, M.F., Rozental, R., Dougherty, M. *et al*. (1993). Cytokine regulation of neuronal differentiation of hippocampal progenitor cells. *Nature* **362**, 62–65.

Meistrell, M., III, Botchkina, G.I., Wang, H. *et al*. (1997). TNF is a brain-damaging cytokine in cerebral ischemia. *Shock* **8**, 341–348.

Merrill, A.H., Jr., Hannun, Y.A. and Bell, R.M. (1993). Introduction: sphingolipids and their metabolites in cell regulation. *Adv. Lipid Res.* **25**, 1–24.

Merrill, J.E. (1992). Tumor necrosis factor α, interleukin 1 and related cytokines in brain development: normal and pathological. *Dev. Neurosci.* **14**, 1–10.

Meyer, C.G., Gallin, M., Erttmann, K.D. *et al*. (1994). HLA-D alleles associated with generalized disease, localized disease, and putative immunity in *Onchocerca volvulus* infection. *Proc. Natl Acad. Sci. U.S.A.* **91**, 7515–7519.

Millar, A.B., Foley, N.M., Singer, M. *et al*. (1989). Tumour necrosis factor in bronchopulmonary secretions of patients with adult respiratory distress syndrome. *Lancet* **ii**, 712–714.

Mohler, K.M., Torrance, D.S., Smith, C.A. *et al*. (1993). Soluble tumor necrosis factor (TNF) receptors are effective therapeutic agents in lethal endotoxemia and function simultaneously as both TNF carriers and TNF antagonists. *J. Immunol.* **151**, 1548–1561.

Molvig, J., Baek, L., Christensen, P. *et al*. (1988). Endotoxin-stimulated human monocyte secretion of interleukin 1, tumour necrosis factor α, and prostaglandin E2 shows stable interindividual differences. *Scand. J. Immunol.* **27**, 705–716.

Moriarty, T.M., Padrell, E., Carty, D.J. *et al*. (1990). G0 protein as signal transducer in the pertussis toxin-sensitive phosphatidylinositol pathway. *Nature* **343**, 79–82.

Moritz, T., Niederle, N., Baumann, J. *et al*. (1989). Phase I study of recombinant human tumor necrosis factor α in advanced malignant disease. *Cancer Immunol. Immunother.* **29**, 144–150.

Moss, M.L., Jin, S.L., Milla, M.E. *et al*. (1997). Cloning of a disintegrin metalloproteinase that processes precursor tumour-necrosis factor-α. *Nature* **385**, 733–736.

Mulcahy, B., Waldron-Lynch, F., McDermott, M.F. *et al*. (1996). Genetic variability in the tumor necrosis factor–lymphotoxin region influences susceptibility to rheumatoid arthritis. *Am. J. Hum. Genet.* **59**, 676–683.

Mullberg, J., Durie, F.H., Otten-Evans, C. *et al*. (1995). A metalloprotease inhibitor blocks shedding of the IL-6 receptor and the p60 TNF receptor. *J. Immunol.* **155**, 5198–5205.

Mustafa, M.M., Lebel, M.H., Ramilo, O. *et al*. (1989). Correlation of interleukin-1β and cachectin concentrations in cerebrospinal fluid and outcome from bacterial meningitis. *J. Pediatr.* **115**, 208–213.

Naredi, P.L., Lindner, P.G., Holmberg, S.B. *et al*. (1993). The effects of tumour necrosis factor α on the vascular bed and blood flow in an experimental rat hepatoma. *Int. J. Cancer* **54**, 645–649.

Narhi, L.O., Philo, J.S., Li, T. *et al*. (1996). Induction of α-helix in the β-sheet protein tumor necrosis factor-α: acid-induced denaturation. *Biochemistry* **35**, 11 454–11 460.

Natanson, C., Eichenholz, P.W., Danner, R.L. *et al*. (1989). Endotoxin and tumor necrosis factor challenges in dogs simulate the cardiovascular profile of human septic shock. *J. Exp. Med.* **169**, 823–832.

Natanson, C., Hoffman, W.D., Suffredini, A.F. *et al*. (1994). Selected treatment strategies for septic shock based on proposed mechanisms of pathogenesis. *Ann. Intern. Med.* **120**, 771–783.

Natoli, G., Costanzo, A., Ianni, A. *et al*. (1997). Activation of SAPK/JNK by TNF receptor 1 through a noncytotoxic TRAF2-dependent pathway. *Science* **275**, 200–203.

Nawroth, P.P. and Stern, D.M. (1986). Modulation of endothelial cell hemostatic properties by tumor necrosis factor. *J. Exp. Med.* **163**, 740–745.

Nawroth, P.P., Bank, I., Handley, D. *et al*. (1986). Tumor necrosis factor/cachectin interacts with endothelial cell receptors to induce release of interleukin 1. *J. Exp. Med.* **163**, 1363–1375.

Nobuhara, M., Kanamori, T., Ashida, Y. *et al*. (1987). The inhibition of neoplastic cell proliferation with human natural tumor necrosis factor. *Jpn. J. Cancer Res.* **78**, 193–201.

North, R.J. and Havell, E.A. (1988). The antitumor function of tumor necrosis factor (TNF). II. Analysis of the role of endogenous TNF in endotoxin-induced hemorrhagic necrosis and regression of an established sarcoma. *J. Exp. Med.* **167**, 1086–1099.

Ohkawa, S., Wright, K.C., Mahajan, H. *et al*. (1989). Hepatic arterial infusion of human recombinant tumor necrosis factor-α. An experimental study in dogs. *Cancer* **63**, 2096–2102.

Otley, C.C., Wenstrup, R.J. and Hall, R.P. (1991). DNA sequence analysis and restriction fragment length polymorphism (RFLP) typing of the HLA-DQW2 alleles associated with dermatitis herpetiformis. *J. Invest. Dermatol.* **97**, 318–322.

Pagani, F.D., Baker, L.S., Hsi, C. *et al*. (1992). Left ventricular systolic and diastolic dysfunction after infusion of tumor necrosis factor-α in conscious dogs. *J. Clin. Invest.* **90**, 389–398.

Panagakos, F.S., O'Boskey, J.F., Jr. and Rodriguez, E. (1996). Regulation of pulp cell matrix metalloproteinase production by cytokines and lipopolysaccharides. *J. Endod.* **22**, 358–361.

Partanen, J. and Koskimies, S. (1988). Low degree of DNA polymorphism in the HIA-linked lymphotoxin (tumour necrosis factor β) gene. *Scand. J. Immunol.* **28**, 313–316.

Pena, L.A., Fuks, Z. and Kolesnick, R. (1997). Stress-induced apoptosis and the sphingomyelin pathway. *Biochem. Pharmacol.* **53**, 615–621.

Penton, A., Selleck, S.B. and Hoffmann, F.M. (1997). Regulation of cell cycle synchronization by decapentaplegic during *Drosophila* eye development. *Science* **275**, 203–206.

Peppel, K., Crawford, D. and Beutler, B. (1991). A tumor necrosis factor (TNF) receptor-IGG heavy chain chimeric protein as a bivalent antagonist of TNF activity. *J. Exp. Med.* **174**, 1483–1489.

Pfeffer, K., Matsuyama, T., Kundig, T.M. *et al*. (1993). Mice deficient for the 55 kD tumor necrosis factor receptor are resistant to endotoxic shock, yet succumb to *L. monocytogenes* infection. *Cell* **73**, 457–467.

Pfreundschuh, M.G., Steinmetz, H.T., Tuschen, R. *et al*. (1989). Phase I study of intratumoral application of recombinant human tumor necrosis factor. *Eur. J. Cancer Clin. Oncol.* **25**, 379–388.

Plata-Salaman, C.R. (1991). Immunoregulators in the nervous system. *Neurosci. Biobehav. Rev.* **15**, 185–215.

Plata-Salaman, C.R., Oomura, Y. and Kai, Y. (1988). Tumor necrosis factor and interleukin-1β: suppression of food intake by direct action in the central nervous system. *Brain Res.* **448**, 106–114.

Pociot, F., Briant, L., Jongeneel, C.V. *et al*. (1993a). Association of tumor necrosis factor (TNF) and class II major histocompatibility complex alleles with the secretion of TNF-α and TNF-β by human mononuclear cells: a possible link to insulin-dependent diabetes mellitus. *Eur. J. Immunol.* **23**, 224–231.

Pociot, F., Wilson, A.G., Nerup, J. *et al*. (1993b). No independent association between a tumor necrosis factor-α promotor region polymorphism and insulin-dependent diabetes mellitus. *Eur. J. Immunol.* **23**, 3050–3053.

Porteu, F., Brockhaus, M., Wallach, D. *et al*. (1991). Human neutrophil elastase releases a ligand-binding fragment from the 75-kDa tumor necrosis factor (TNF) receptor. Comparison with the proteolytic activity responsible for shedding of TNF receptors from stimulated neutrophils. *J. Biol. Chem.* **266**, 18 846–18 853.

Preston, B.D. (1997). Reverse transcriptase fidelity and HIV-1 variation. *Science* **275**, 228–229 (letter); discussion 230–231.

Rhoades, K.L., Golub, S.H. and Economou, J.S. (1992). The regulation of the human tumor necrosis factor α promoter region in macrophage, T cell, and B cell lines. *J. Biol. Chem.* **267**, 22 102–22 107.

Robache-Gallea, S., Morand, V., Bruneau, J.M. *et al*. (1995). *In vitro* processing of human tumor necrosis factor-α. *J. Biol. Chem.* **270**, 23 688–23 692.

Rogy, M.A., Coyle, S.M., Oldenburg, H.S. *et al*. (1994). Persistently elevated soluble tumor necrosis factor receptor and interleukin-1 receptor antagonist levels in critically ill patients. *J. Am. Coll. Surg.* **178**, 132–138.

Rothe, M., Sarma, V., Dixit, V.M. *et al*. (1995). TRAF2-mediated activation of NF-κB by TNF receptor 2 and CD40. *Science* **269**, 1424–1427.

Sampaio, E.P., Sarno, E.N., Galilly, R. *et al*. (1991). Thalidomide selectively inhibits tumor necrosis factor α production by stimulated human monocytes. *J. Exp. Med.* **173**, 699–703.
Sato, N., Goto, T., Haranaka, K. *et al*. (1986). Actions of tumor necrosis factor on cultured vascular endothelial cells: morphologic modulation, growth inhibition, and cytotoxicity. *J. Natl Cancer Inst.* **76**, 1113–1121.
Sato, Y., Koshita, Y., Hirayama, M. *et al*. (1996). Augmented antitumor effects of killer cells induced by tumor necrosis factor gene-transduced autologous tumor cells from gastrointestinal cancer patients. *Hum. Gene Ther.* **7**, 1895–1905.
Saukkonen, K., Sande, S., Cioffe, C. *et al*. (1990). The role of cytokines in the generation of inflammation and tissue damage in experimental Gram-positive meningitis. *J. Exp. Med.* **171**, 439–448.
Scheinman, R.I., Cogswell, P.C., Lofquist, A.K. *et al*. (1995). Role of transcriptional activation of IκB α in mediation of immunosuppression by glucocorticoids. *Science* **270**, 283–286.
Schiller, J.H., Witt, P.L., Storer, B. *et al*. (1992). Clinical and biologic effects of combination therapy with γ-interferon and tumor necrosis factor. *Cancer* **69**, 562–571.
Schutze, S., Potthoff, K., Machleidt, T. *et al*. (1992). TNF activates NF-κB by phosphatidylcholine-specific phospholipase C-induced 'acidic' sphingomyelin breakdown. *Cell* **71**, 765–776.
Seckinger, P., Isaaz, S. and Dayer, J.M. (1989). Purification and biologic characterization of a specific tumor necrosis factor α inhibitor. *J. Biol. Chem.* **264**, 11 966–11 973.
Seguin, R., Keller, K. and Chadee, K. (1995). *Entamoeba histolytica* stimulates the unstable transcription of c-fos and tumour necrosis factor-α mRNA by protein kinase C signal transduction in macrophages. *Immunology* **86**, 49–57.
Selmaj, K.W., Farooq, M., Norton, W.T. *et al*. (1990). Proliferation of astrocytes *in vitro* in response to cytokines. A primary role for tumor necrosis factor. *J. Immunol.* **144**, 129–135.
Sharief, M.K. and Thompson, E.J. (1992). *In vivo* relationship of tumor necrosis factor-α to blood–brain barrier damage in patients with active multiple sclerosis. *J. Neuroimmunol.* **38**, 27–33.
Shaw, G. and Kamen, R. (1986). A conserved AU sequence from the 3′ untranslated region of GM-CSF mRNA mediates selective mRNA degradation. *Cell* **46**, 659–667.
Shigeno, T., Mima, T., Takakura, K. *et al*. (1991). Amelioration of delayed neuronal death in the hippocampus by nerve growth factor. *J. Neurosci.* **11**, 2914–2919.
Shimura, T., Hagihara, M., Takebe, K. *et al*. (1994). The study of tumor necrosis factor β gene polymorphism in lung cancer patients. *Cancer* **73**, 1184–1188.
Silver, I.A., Murrills, R.J. and Etherington, D.J. (1988). Microelectrode studies on the acid microenvironment beneath adherent macrophages and osteoclasts. *Exp. Cell Res.* **175**, 266–276.
Smith, R.A. and Baglioni, C. (1987). The active form of tumor necrosis factor is a trimer. *J. Biol. Chem.* **262**, 6951–6954.
Smith, R.A. and Baglioni, C. (1992). Characterization of TNF receptors. *Immunol. Ser.* **56**, 131–147.
Sollid, L.M., Markussen, G., Ek, J. *et al*. (1989). Evidence for a primary association of celiac disease to a particular HLA-DQ α/β heterodimer. *J. Exp. Med.* **169**, 345–350.
Spence, M.W. (1993). Sphingomyelinases. *Adv. Lipid Res.* **26**, 3–23.
Spriggs, D.R., Sherman, M.L., Michie, H. *et al*. (1988). Recombinant human tumor necrosis factor administered as a 24-hour intravenous infusion. A phase I and pharmacologic study. *J. Natl Cancer Inst.* **80**, 1039–1044.
Stanger, B.Z., Leder, P., Lee, T.H. *et al*. (1995). RIP: a novel protein containing a death domain that interacts with FAS/APO-1 (CD95) in yeast and causes cell death. *Cell* **81**, 513–523.
Stephens, K.E., Ishizaka, A., Larrick, J.W. *et al*. (1988a). Tumor necrosis factor causes increased pulmonary permeability and edema. Comparison to septic acute lung injury. *Am. Rev. Respir. Dis.* **137**, 1364–1370.
Stephens, K.E., Ishizaka, A., Wu, Z.H. *et al*. (1988b). Granulocyte depletion prevents tumor necrosis factor-mediated acute lung injury in guinea pigs. *Am. Rev. Respir. Dis.* **138**, 1300–1307.
Stolpen, A.H., Guinan, E.C., Fiers, W. *et al*. (1986). Recombinant tumor necrosis factor and immune interferon act singly and in combination to reorganize human vascular endothelial cell monolayers. *Am. J. Pathol.* **123**, 16–24.
Stuber, F., Petersen, M., Bokelmann, F. *et al*. (1996). A genomic polymorphism within the tumor necrosis factor locus influences plasma tumor necrosis factor-α concentrations and outcome of patients with severe sepsis. *Crit. Care Med.* **24**, 381–384.
Suffredini, A.F., Fromm, R.E., Parker, M.M. *et al*. (1989). The cardiovascular response of normal humans to the administration of endotoxin. *N. Engl. J. Med.* **321**, 280–287.

Sugarman, B.J., Aggarwal, B.B., Hass, P.E. *et al.* (1985). Recombinant human tumor necrosis factor-α: effects on proliferation of normal and transformed cells *in vitro*. *Science* **230**, 943–945.

Suk, K. and Erickson, K.L. (1996). Differential regulation of tumour necrosis factor-α mRNA degradation in macrophages by interleukin-4 and interferon-γ. *Immunology* **87**, 551–558.

Sung, S.S., Jung, L.K., Walters, J.A. *et al.* (1988). Production of tumor necrosis factor/cachectin by human B cell lines and tonsillar B cells. *J. Exp. Med.* **168**, 1539–1551.

Tang, P., Hung, M. and Klostergaard, J. (1996). Human pro-tumor necrosis factor is a homotrimer. *Biochemistry* **35**, 8216–8225.

Tartaglia, L.A., Weber, R.F., Figari, I.S. *et al.* (1991). The two different receptors for tumor necrosis factor mediate distinct cellular responses. *Proc. Natl Acad. Sci. U.S.A.* **88**, 9292–9296.

Tartaglia, L.A., Ayres, T.M., Wong, G.H. *et al.* (1993a). A novel domain within the 55 kD TNF receptor signals cell death. *Cell* **74**, 845–853.

Tartaglia, L.A., Pennica, D. and Goeddel, D.V. (1993b). Ligand passing: the 75-kDA tumor necrosis factor (TNF) receptor recruits TNF for signaling by the 55-kDa TNF receptor. *J. Biol. Chem.* **268**, 18542–18548.

Thomson, A.M. (1997). Neuroscience. More than just frequency detectors? *Science* **275**, 179–180.

Tracey, K.J. and Cerami, A. (1990). Metabolic responses to cachectin/TNF. A brief review. *Ann. N.Y. Acad. Sci.* **587**, 325–331.

Tracey, K.J. and Cerami, A. (1994). Tumor necrosis factor: a pleiotropic cytokine and therapeutic target. *Annu. Rev. Med.* **45**, 491–503.

Tracey, K.J. and Lowry, S.F. (1990). The role of cytokine mediators in septic shock. *Adv. Surg.* **23**, 21–56.

Tracey, K.J., Beutler, B., Lowry, S.F. *et al.* (1986). Shock and tissue injury induced by recombinant human cachectin. *Science* **234**, 470–474.

Tracey, K.J., Fong, Y., Hesse, D.G. *et al.* (1987a). Anti-cachectin/TNF monoclonal antibodies prevent septic shock during lethal bacteraemia. *Nature* **330**, 662–664.

Tracey, K.J., Lowry, S.F., Fahey, T.J., III *et al.* (1987b). Cachectin/tumor necrosis factor induces lethal shock and stress hormone responses in the dog. *Surg. Gynecol. Obstet.* **164**, 415–422.

Tracey, K.J., Wei, H., Manogue, K.R. *et al.* (1988). Cachectin/tumor necrosis factor induces cachexia, anemia, and inflammation. *J. Exp. Med.* **167**, 1211–1227.

Tracey, K.J., Vlassara, H. and Cerami, A. (1989). Cachectin/tumour necrosis factor. *Lancet* **i**, 1122–1126.

Tsai, E.Y., Jain, J., Pesavento, P.A. *et al.* (1996a). Tumor necrosis factor α gene regulation in activated T cells involves ATF-2/Jun and NFATP. *Mol. Cell Biol.* **16**, 459–467.

Tsai, E.Y., Yie, J., Thanos, D. *et al.* (1996b). Cell-type-specific regulation of the human tumor necrosis factor α gene in B cells and T cells by NFATP and ATF-2/Jun. *Mol. Cell Biol.* **16**, 5232–5244.

Tsutsumi, Y., Kihira, T., Tsunoda, S. *et al.* (1995). Molecular design of hybrid tumour necrosis factor α with polyethylene glycol increases its anti-tumour potency. *Br. J. Cancer* **71**, 963–968.

Udalova, I.A., Nedospasov, S.A., Webb, G.C. *et al.* (1993). Highly informative typing of the human TNF locus using six adjacent polymorphic markers. *Genomics* **16**, 180–186.

Vakalopoulou, E., Schaack, J. and Shenk, T. (1991). A 32-kilodalton protein binds to AU-rich domains in the 3′ untranslated regions of rapidly degraded mRNAs. *Mol. Cell Biol.* **11**, 3355–3364.

Van Antwerp, D.J., Martin, S.J., Kafri, T. *et al.* (1996). Suppression of TNF-α-induced apoptosis by NF-κB. *Science* **274**, 787–789.

van der Poll, T. and Lowry, S.F. (1995). Tumor necrosis factor in sepsis: mediator of multiple organ failure or essential part of host defense? *Shock* **3**, 1–12 (editorial).

Van Ostade, X., Tavernier, J., Prange, T. *et al.* (1991). Localization of the active site of human tumour necrosis factor (hTNF) by mutational analysis. *EMBO J.* **10**, 827–836.

Van Ostade, X., Vandenabeele, P., Everaerdt, B. *et al.* (1993). Human TNF mutants with selective activity on the p55 receptor. *Nature* **361**, 266–269.

Van Zee, K.J., Stackpole, S.A., Montegut, W.J. *et al.* (1994). A human tumor necrosis factor (TNF) α mutant that binds exclusively to the p55 TNF receptor produces toxicity in the baboon. *J. Exp. Med.* **179**, 1185–1191.

Waage, A., Brandtzaeg, P., Halstensen, A. *et al.* (1989a). The complex pattern of cytokines in serum from patients with meningococcal septic shock. Association between interleukin 6, interleukin 1, and fatal outcome. *J. Exp. Med.* **169**, 333–338.

Waage, A., Halstensen, A., Shalaby, R. *et al.* (1989b). Local production of tumor necrosis factor α, interleukin 1, and interleukin 6 in meningococcal meningitis. Relation to the inflammatory response. *J. Exp. Med.* **170**, 1859–1867.

Wang, C.Y., Mayo, M.W. and Baldwin, A.S., Jr. (1996). TNF- and cancer therapy-induced apoptosis: potentiation by inhibition of NF-κB. *Science* **274**, 784–787.

Welch, T.R., Beischel, L.S., Balakrishnan, K. *et al*. (1988). Major histocompatibility complex extended haplotypes in systemic lupus erythematosus. *Dis. Markers* **6**, 247–255.

Wiegmann, K., Schutze, S., Kampen, E. *et al*. (1992). Human 55-kDa receptor for tumor necrosis factor coupled to signal transduction cascades. *J. Biol. Chem.* **267**, 17 997–18 001.

Wiegmann, K., Schutze, S., Machleidt, T. *et al*. (1994). Functional dichotomy of neutral and acidic sphingomyelinases in tumor necrosis factor signaling. *Cell* **78**, 1005–1015.

Wilson, A.G., Di Giovine, F.S., Blakemore, A.I. *et al*. (1992). Single base polymorphism in the human tumour necrosis factor α (TNF α) gene detectable by NCOI restriction of PCR product. *Hum. Mol. Genet.* **1**, 353.

Wilson, A.G., de Vries, N., Pociot, F. *et al*. (1993). An allelic polymorphism within the human tumor necrosis factor α promoter region is strongly associated with HLA A1, B8, and DR3 alleles. *J. Exp. Med.* **177**, 557–560.

Wilson, A.G., Gordon, C., Di Giovine, F.S. *et al*. (1994). A genetic association between systemic lupus erythematosus and tumor necrosis factor α. *Eur. J. Immunol.* **24**, 191–195.

Wilson, A.G., Symons, J.A., McDowell, T.L. *et al*. (1997). Effects of a polymorphism in the human tumor necrosis factor α promoter on transcriptional activation. *Proc. Natl Acad. Sci. U.S.A.* **94**, 3195–3199.

Wilson, T. and Treisman, R. (1988). Removal of poly(A) and consequent degradation of c-fos mRNA facilitated by 3′ AU-rich sequences. *Nature* **336**, 396–399.

Wingfield, P., Pain, R.H. and Craig, S. (1987). Tumour necrosis factor is a compact trimer. *FEBS Lett.* **211**, 179–184.

Winkelhake, J.L., Stampfl, S. and Zimmerman, R.J. (1987). Synergistic effects of combination therapy with human recombinant interleukin-2 and tumor necrosis factor in murine tumor models. *Cancer Res.* **47**, 3948–3953.

Winston, B.W. and Riches, D.W. (1995). Activation of p42MAPK/ERK2 following engagement of tumor necrosis factor receptor CD120A (p55) in mouse macrophages. *J. Immunol.* **155**, 1525–1533.

Wong, G.H. and Goeddel, D.V. (1988). Induction of manganous superoxide dismutase by tumor necrosis factor: possible protective mechanism. *Science* **242**, 941–944.

Xia, Z., Dickens, M., Raingeaud, J. *et al*. (1995). Opposing effects of ERK and JNK-p38 MAP kinases on apoptosis. *Science* **270**, 1326–1331.

Yang, S.C., Fry, K.D., Grimm, E.A. *et al*. (1991). Phenotype and cytolytic activity of mouse tumor-bearer splenocytes and tumor-infiltrating lymphocytes from K-1735 melanoma metastases following anti-CD3, interleukin-2, and tumor necrosis factor-α combination immunotherapy. *J. Immunother.* **10**, 326–335.

Yokoyama, T., Vaca, L., Rossen, R.D. *et al*. (1993). Cellular basis for the negative inotropic effects of tumor necrosis factor-α in the adult mammalian heart. *J. Clin. Invest.* **92**, 2303–2312.

Yoshimura, T. and Sone, S. (1987). Different and synergistic actions of human tumor necrosis factor and interferon-γ in damage of liposome membranes. *J. Biol. Chem.* **262**, 4597–4601.

Young, P., McDonnell, P., Laydon, J. *et al*. (1994). A novel MAP kinase regulates the production of IL-1 and TNF in LPS activated human monocytes. *Cytokine* **6**, 564.

Zhang, M., Caragine, T., Wang, H. *et al*. (1997). Spermine inhibits proinflammatory cytokine synthesis in human mononuclear cells. *J. Exp. Med.* **185**, 1759–1768.

Zhang, W., Wagner, B.J., Ehrenman, K. *et al*. (1993). Purification, characterization, and cDNA cloning of an AU-rich element RNA-binding protein, AUF1. *Mol. Cell Biol.* **13**, 7652–7665.

Zhang, X.M., Weber, I. and Chen, M.J. (1992). Site-directed mutational analysis of human tumor necrosis factor-α receptor binding site and structure–functional relationship. *J. Biol. Chem.* **267**, 24 069–24 075.

Zheng, H., Crowley, J.J., Chan, J.C. *et al.* (1990). Attenuation of tumor necrosis factor-induced endothelial cell cytotoxicity and neutrophil chemiluminescence. *Am. Rev. Respir. Dis.* **142**, 1073–1078.

Zheng, L., Fisher, G., Miller, R.E. *et al.* (1995). Induction of apoptosis in mature T cells by tumour necrosis factor. *Nature* **377**, 348–351.

Zimmerman, R.J., Chan, A. and Leadon, S.A. (1989). Oxidative damage in murine tumor cells treated *in vitro* by recombinant human tumor necrosis factor. *Cancer Res.* **49**, 1644–1648.

Zubiaga, A.M., Belasco, J.G. and Greenberg, M.E. (1995). The nonamer UUAUUUAUU is the key AU-rich sequence motif that mediates mRNA degradation. *Mol. Cell Biol.* **15**, 2219–2230.

Chapter 20

Tumor Necrosis Factor-Related Ligands and Receptors

Carl F. Ware, Sybil Santee and Alison Glass

Division of Molecular Immunology, La Jolla Institute for Allergy and Immunology, San Diego, CA, USA

INTRODUCTION

The antitumor activity of tumor necrosis factor (TNF) sparked a close examination of the structure and function of this molecule. The collective result of this work has recast TNF as a central mediator of inflammatory and immune defenses. Furthermore, the multiple roles played by TNF in several pathogenic processes associated with cancer, infectious and autoimmune diseases strongly reinforce this concept (Beutler, 1992). TNF is now recognized as one of several related proteins that form a superfamily of secreted and membrane-bound ligands. Each member is paired with a specific cell surface receptor(s) that together form a corresponding family of receptors. These receptors generate the biochemical signals essential for eliciting cellular differentiation, growth or death in many tissues. Since 1995 several proteins have been discovered that mediate signal transduction by the TNF receptor (TNFR) family. Unique and unexpected aspects of the physiologic roles for the TNF-related cytokines have been revealed in recent studies using gene deletion techniques. This stands in contrast to the redundancy often observed in the response of cells in tissue culture elicited by TNF and its relatives (and in some cases unrelated cytokines). Here, we attempt to summarize some of the pertinent structural information as an aid to determining the functional roles of these molecules in physiology and pathogenesis.

THE TNF-RELATED CYTOKINES

The TNF superfamily currently consists of 13 molecularly defined ligand–receptor pairs (Bazan, 1990; Cosman *et al.*, 1990; Mallett and Barclay, 1991; Smith *et al.*, 1994), although several orphans have been recently discovered, one orphan ligand and five orphan receptors (Table 1), with more on the way. There are two groups of receptors: those that bind ligands related to TNF and whose functions appear to be linked primarily to the immune system, and a second group, currently with a single member, the low-affinity nerve growth factor receptor (neurotrophin receptor; NTR). NTR binds to several distinct neurotrophins, none of which shares significant structural homology

The Cytokine Handbook, 3rd ed.
ISBN 0–12–689662–3

Table 1. Members of the TNF-related cytokine-receptor superfamily.

Receptor	Ligand	Function
TNFR60 (type 1)	TNF/LT-α_3	Apoptosis, inflammation
TNFR80 (type 2)	TNF/LT-α_3	Apoptosis, proliferation
LTβR	LT-$\alpha_1\beta_2$	Apoptosis, lymph node development
Fas/CD95	Fas L	Apoptosis, immune privilege
CD40	CD40L	Cell survival, isotype switch
CD30	CD30L	Apoptosis, negative selection
CD27	CD27L	Costimulation
OX-40	OX-40L	Costimulation
4-1BB	4-1BBL	Costimulation
p75NTR	Neurotrophins, NGF	Cell survival
DR4 (TRAIL-R1)	TRAIL	Apoptosis
DR5 (TRAIL-R2)	TRAIL	Apoptosis
TRID/DcR1 (TRAIL-R3)	TRAIL	Decoy receptor for TRAIL
HVEM (ATAR/TR2)	?	Herpes virus entry, costimulation
TRAMP (DR3, WSL-1/LARD/APO3)	?	Apoptosis
CAR-1	?	Apoptosis, entry factor for avian leukosis virus (ALV)
GITR	?	Inhibits TCR-induced apoptosis
Osteoprotegerin	?	Regulation of bone mass
?	TWEAK	Apoptosis

CAR-1, cytopathic ALV receptor; CD, cluster of differentation; DR3, 4, 5, death receptor-3, -4, -5; HVEM, herpes virus entry mediator; LT, lymphotoxin; NTR, neurotropin (NGF) receptor; TNF, tumor necrosis factor; TRAIL, TNF-related apoptosis-inducing ligand; TRAMP, TNF receptor-associated membrane protein; TRID, TRAIL receptor without an intracellular domain; DcR1, decoy receptor 1; GITR, glucocorticoid-induced TNF receptor related gene.

with TNF-related ligands (for review see McDonald and Hendrickson, 1993). Both the ligand and receptor superfamilies are defined by the sequence and structural homologies of their respective extracellular domains.

A common functional feature among these receptors, including NTR, is the ability to regulate cell viability. Cell death by apoptosis (or programmed cell death) is a highly regulated process with specific participants and regulatory control points (Martin and Green, 1995; Jacobson *et al.*, 1997). Fas (Apo1/CD95) and the 55–60-kDa TNF receptor (TNFR60) share sequence homology in the cytoplasmic domain responsible for signaling apoptosis, termed the death domain (DD) (Nagata, 1997). Besides TNFR60 (Tartaglia *et al.*, 1993a) and Fas (Itoh and Nagata, 1993), two new receptor additions, death receptor-3 (DR3)/TNF receptor-associated membrane protein (TRAMP)/WSL-1/lymphocyte associated receptor of death (LARD) (Chinnaiyan *et al.*, 1996a; Kitson *et al.*, 1996; Bodmer *et al.*, 1997; Screaton *et al.*, 1997) and cytopathic avian leukosis virus receptor-1 (CAR-1) also contain a death domain region (Brojatsch *et al.*, 1996). Several other members, including the 75–80-kDa TNF receptor (type 2, CD120b) (Grell *et al.*, 1995), lymphotoxin-β receptor (LT-βR) (Browning *et al.*, 1996b) and CD30 (Smith *et al.*, 1993; Lee *et al.*, 1996) also induce cell death, but do not contain a DD, suggesting an alternate route to the apoptotic machinery. Most members of the TNF family enhance cell growth in tissue culture systems. For example, signals from CD27 (Sugita Kanji *et al.*, 1991; Goodwin *et al.*, 1993a; Sugita *et al.*, 1993), OX-40 (Paterson *et al.*, 1987) 4-1BB

(Goodwin *et al.*, 1993b; Pollok *et al.*, 1993) or TNFR80 (Tartaglia *et al.*, 1993b) enhance the proliferation of T lymphocytes activated via the T-cell antigen receptor.

However, in contrast to a true growth factor such as interleukin (IL)-2, neither TNF nor the other receptors sustain continued rounds of cell division in T cells and are thus classified as costimulatory factors for cellular proliferation induced through the antigen receptor. In this context, perhaps costimulation is a reflection of antiapoptotic signals (survival). Additional functional groups can be defined by interactions with signaling proteins that contain either a DD homology region, for example TNFR-associated death domain protein (TRADD), Fas-associated protein with death domain (FADD) and the serine kinase receptor interacting protein (RIP) (Cleveland and Ihle, 1995), or receptors that bind to the zinc RING finger proteins,TNFR-associated factor (TRAF) family, currently with six members. The ability of different receptors to bind the same sets of signaling proteins may account for some of the redundancy in cellular responses signaled by distinct ligand–receptor systems.

The current understanding of the molecular components in the signaling pathways does not account fully for the unique roles associated with these ligands and receptors *in vivo*. For example, several of the TNF-related systems play a significant role in developmental processes: genetic deletion of different components of the TNF/LT$\alpha\beta$ triad cause a failure of peripheral lymphoid organs to form (De Togni *et al.*, 1994; Banks *et al.*, 1995) and CD30 knockout mice have skewed negative selection during thymic development (Amakawa *et al.*, 1996).

STRUCTURAL FEATURES OF LIGANDS AND RECEPTORS

The TNF family of ligands are all type II transmembrane glycoproteins with the *C*-terminus displayed on the outside of the cell, a retained transmembrane region and cytoplasmic tails (Table 2). The genes encoding these proteins are scattered throughout the genome, although the genes for TNF, LT-α and LT-β form a cytokine locus in the major histocompatibility complex (MHC) (Spies *et al.*, 1986; Muller *et al.*, 1987; Browning *et al.*, 1993; Lawton *et al.*, 1995). The proteins are typically encoded by three or four exons (depending on the species). For human TNF, the fourth exon encodes the bulk of the extracellular receptor-binding domain. The TNF ligand family is defined by several discrete stretches of conserved sequences located primarily in the *C*-terminal receptor-binding region (Fig. 1). The sequence homology between members within the same species is modest (15–30% overall); however, substantial evidence indicates a high level of conservation in their three-dimensional shape.

The crystal structures of TNF (Jones *et al.*, 1989), LT-α (Eck *et al.*, 1992) and CD40 ligand (CD40L) (Karpusas *et al.*, 1995) provide useful models that guide structural predictions for other members of the family. The primary structure of TNF and LT-α is folded into an antiparallel β-sheet sandwich, with a 'Greek key' topology (Fig. 2). This subunit assembles into a compact trimer, with the end to face interaction about a threefold axis of symmetry (Fig. 2). This generates a molecule resembling a cone with a large flat base and narrow top. Both the *N*- and *C*-termini are located at the base and thus the top of the TNF molecule is thought to protrude outward from the cell surface. The internal β sheet is rich in hydrophobic residues which create a large surface area where subunits interact in the trimer. The discrete regions of homology that define this family are found primarily within the β-strand scaffold on the face of the internal β sheet

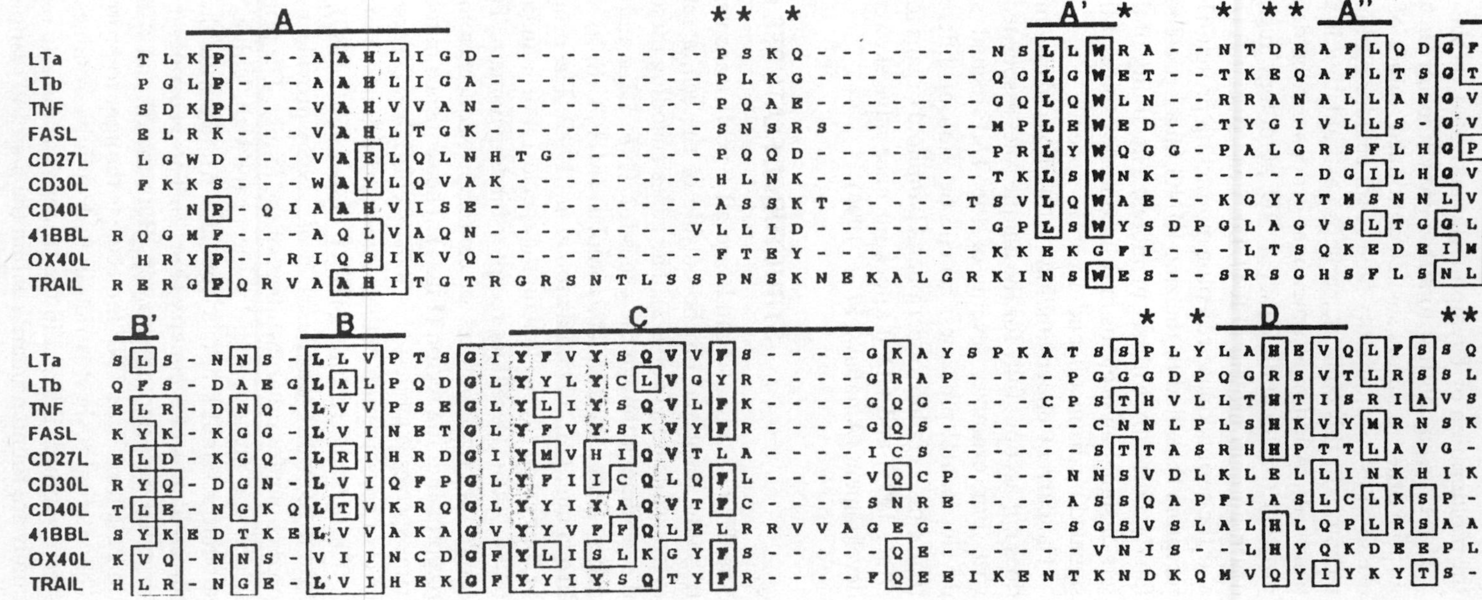

A
** *
A' *
* ** A"
LTa T L K P - - - A A H L I G D - - - - - - - - - - - - - P S K Q - - - - - - - - N S L L W R A - - N T D R A F L Q D G F
LTb P G L P - - - A A H L I G A - - - - - - - - - - - - - P L K G - - - - - - - - Q G L G W E T - - T K E Q A F L T S G T
TNF S D K P - - - V A H V V A N - - - - - - - - - - - - - P Q A E - - - - - - - - G Q L Q W L N - - R R A N A L L A N G V
FASL E L R K - - - V A H L T G K - - - - - - - - - - - - - S N S R S - - - - - - - M P L E W E D - - T Y G I V L L S - G V
CD27L L G W D - - - V A E L Q L N H T G - - - - - - - - - - P Q Q D - - - - - - - - P R L Y W Q G G - P A L G R S F L H G P
CD30L F K K S - - - W A Y L Q V A K - - - - - - - - - - - - H L N K - - - - - - - - T K L S W N K - - - - - - D G I L H G V
CD40L N P - Q I A A H V I S E - - - - - - - - - - - - A S S K T - - - - - - T S V L Q W A E - - K G Y Y T M S N N L V
41BBL R Q G M P - - - A Q L V A Q N - - - - - - - - - - V L L I D - - - - - - - G P L S W Y S D P G L A G V S L T G G L
OX40L H R Y P - - R I Q S I K V Q - - - - - - - - - - - F T E Y - - - - - - - K K E K G F I - - - L T S Q K E D E I M
TRAIL R E R G P Q R V A A H I T G T R G R S N T L S S P N S K N E K A L G R K I N S W E S - - S R S G H S F L S N L
B'
B
C
* * D
**
LTa S L S - N N S - L L V P T S G I Y F V Y S Q V V F S - - - - G K A Y S P K A T S S P L Y L A H E V Q L F S S Q
LTb Q F S - D A E G L A L P Q D G L Y Y L Y C L V G Y R - - - - G R A P - - - - P G G G D P Q G R S V T L R S S L
TNF E L R - D N Q - L V V P S E G L Y L I Y S Q V L F K - - - - G Q G - - - - C P S T H V L L T H T I S R I A V S
FASL K Y K - K G G - L V I N E T G L Y F V Y S K V Y F R - - - - G Q S - - - - - - C N N L P L S H K V Y M R N S K
CD27L E L D - K G Q - L R I H R D G I Y M V H I Q V T L A - - - I C S - - - - - - S T T A S R H H P T T L A V G -
CD30L R Y Q - D G N - L V I Q F P G L Y F I I C Q L Q F L - - - - V Q C P - - - - N N S V D L K L E L L I N K H I K
CD40L T L E - N G K Q L T V K R Q G L Y Y I Y A Q V T F C - - - S N R E - - - - A S S Q A P F I A S L C L K S P -
41BBL S Y K E D T K E L V V A K A G V Y Y V F F Q L E L R R V V A G E G - - - - - S G S V S L A L H L Q P L R S A A
OX40L K V Q - N N S - V I I N C D G F Y L I S L K G Y F S - - - - - Q E - - - - - - V N I S - - L H Y Q K D E E P L
TRAIL H L R - N G E - L V I H E K G F Y Y I Y S Q T Y F R - - - - F Q E E I K E N T K N D K Q M V Q Y I Y K Y T S -

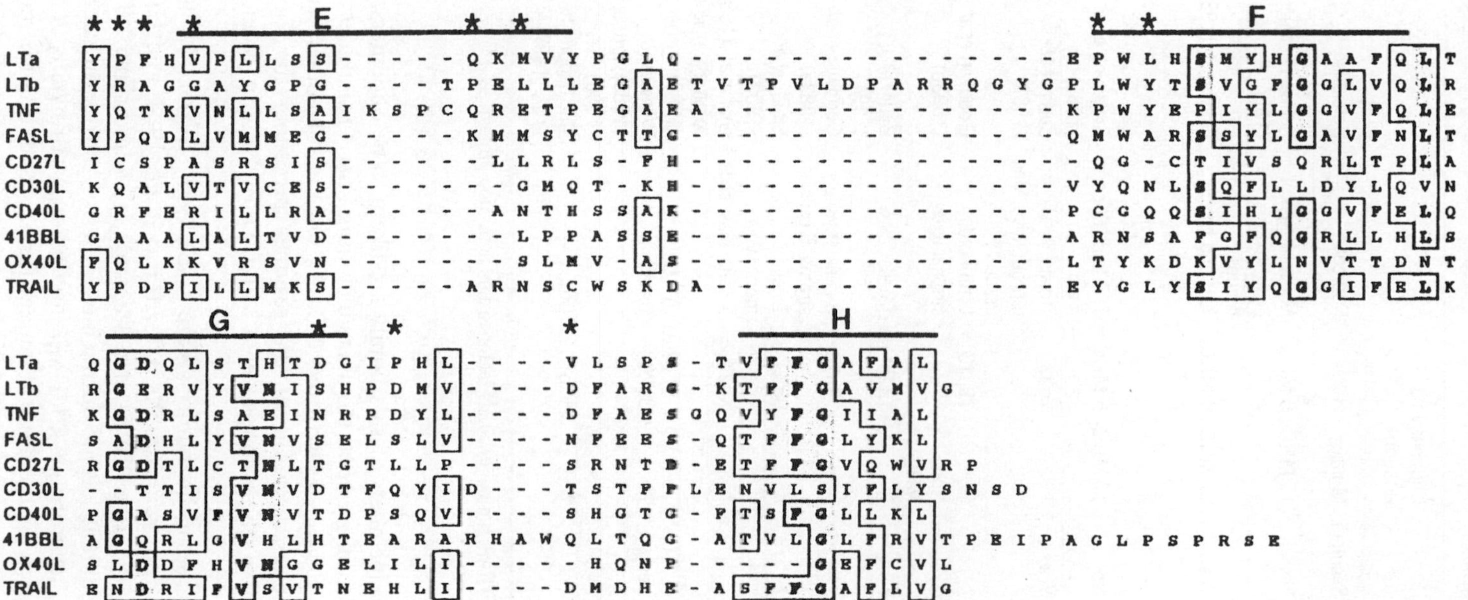

Fig. 1. Human TNF ligand superfamily. Alignment of TNF-related ligand receptor binding domain sequences were aligned with ClustalW (pam250 scoring matrix, MacVector). The position of β strands (A–H) in LTa (LT-α) are designated above the extracellular domain according to Eck *et al.* (1992). Shaded and boxed areas indicate identical and conserved residues respectively; * receptor contact residues with >20 Å^2 surface area. Genbank accession no.: LT-α, D12614; LT-β, L11015; TNF, M10988; FasL, U11821; CD27L, L08096; CD30L, L09753; CD40L, X67878; 4-1BBL, U03398; OX-40L, X79929; TRAIL, U37581.

Table 2. Properties of the TNF-related ligands.

Designation	Other names	Size (kDa)	Secreted	Chromosome location Mouse	Chromosome location Human	Sources	Reference
TNF	TNF-α, cachectin	17	Yes	17 MHC	6 MHC	Macrophages, T cells, many tissues	Pennica *et al.* (1984)
LT-α	TNF-β homotrimer; forms heterotrimers with LT-β, LT-$\alpha_1\beta_2$	25	Yes	17 MHC	6 MHC	T, B and NK cells	Gray *et al.* (1984)
LT-β	p33; heterotrimers with LT-α	33	No	17 MHC	6 MHC	T, B and NK cells	Browning *et al.* (1993)
FasL	Apo-1	45	Yes	1	1q23	T cells, reproductive, lens tissue	Suda *et al.* (1993)
4-1BBL		50	?	17 Non-MHC	19p13	Lymphoid, stromal lines	Goodwin *et al.* (1993b)
OX-40L		34	?	1	1q25	T cells	Baum *et al.* (1994)
CD27L	CD70	50	?	?	19p13	B cells, thymic stroma, T cells	Goodwin *et al.* (1993a)
CD30L		40	?	4	9q33	T cells, monocytes	Smith *et al.* (1993)
CD40L	gp39, T-BAM, TRAP	39	?	?	Xq26	T cells, mast cells	Graf *et al.* (1992), Armitage *et al.* (1992)
TRAIL	APO-2	calc. 32.5	Yes	?	3q26	Broad	Wiley *et al.* (1995), Pitti *et al.* (1996)
TWEAK		18	Yes	?	17p13	Broad	Chicheportiche *et al.* (1997)

T-BAM, T cell-B cell activation molecule; TRAP, tumour necrosis factor-related activation protein.

(Fig. 2). This suggests that other members of this family are likely to be oligomers. A trimer structure has been confirmed for several members including FasL (Tanaka *et al.*, 1995), LT-$\alpha\beta$ complex (Browning *et al.*, 1996a) and CD40L. On the solvent-exposed surface there is also significant conservation of the residues preceding the loops connecting the β strands. Two of these loops, A–A″ and D–E, contain residues important for receptor binding.

The surface form of LT is unique in this family (Ware *et al.*, 1995). Surface LT is a heteromeric complex assembled from two subunits, LT-α (Gray *et al.*, 1984) and LT-β (Browning *et al.*, 1993), whereas the other TNF-related ligands are homotrimers. The association between the two LT subunits occurs during biosynthesis (Androlewicz *et al.*, 1992) and generates two types of heterotrimers, LT-$\alpha_1\beta_2$ and LT-$\alpha_2\beta_1$ (Browning *et al.*, 1996a). The LT$\alpha_1\beta_2$ complex represents the most abundant form, accounting for more than 95% of the complex on the surface of activated T lymphocytes (Ware *et al.*, 1992).

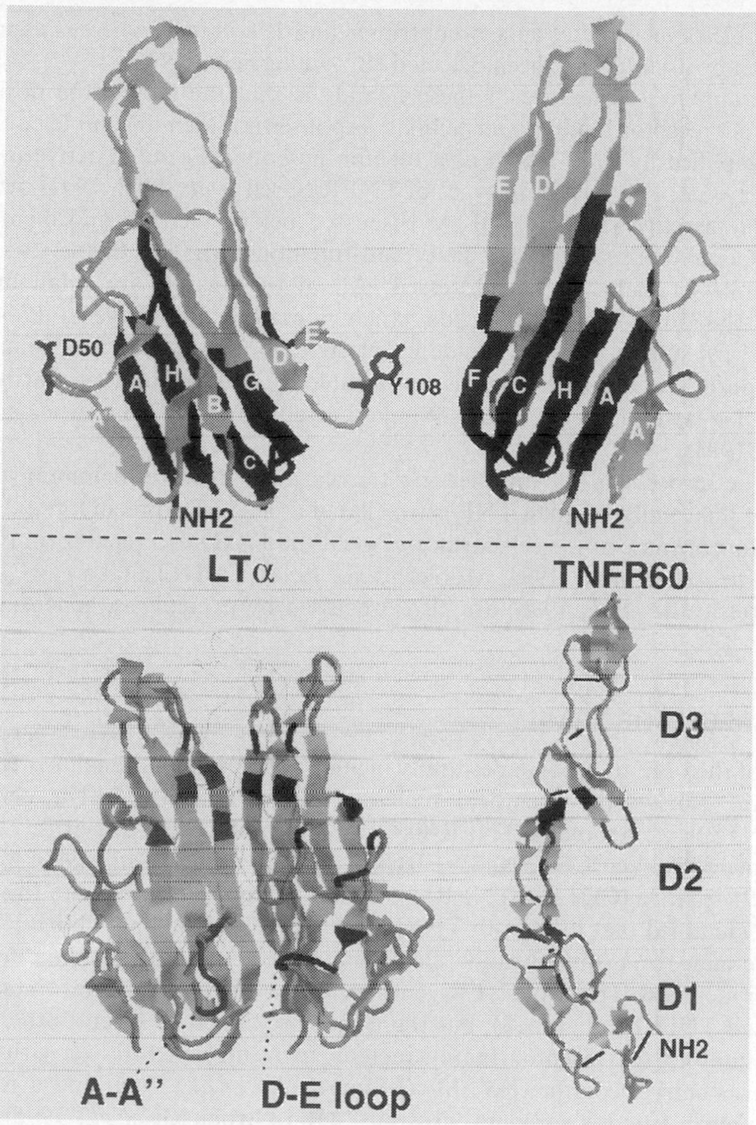

Fig. 2. Structural features of LT-α and TNFR60. Upper panels depict the β strands of a single LT-α subunit. The wide bands are the β strands and the smaller rope-like represent the connecting loops. The left upper panel shows the positions of D50 and Y108 in the solvent-exposed orientation; the two other subunits of the LT-α trimer are invisible. In the upper right panel the subunit has been rotated 180° revealing the interior β sheet, where the dark areas represent residues conserved among the superfamily (see Fig. 1). The lower left panel depicts a receptor binding site on LT-α composed of two subunits. The third LT-α subunit is in the background in outline. D50 is in the A-A″ loop and Y108 in the D–E loop. The dark bands identify the positions of residues that contact TNFR60. The lower right panel depicts the TNFR60 extracellular domain. The receptor orientation has been rotated counterclockwise approximately 90° (around the vertical axis) exposing the contact residues (dark areas); disulfide bonds are dashed lines; the demarcation of the cysteine-rich motifs (D1–D3) are indicated on the right side. In this orientation the ligand is attached to a T-cell membrane at the bottom, and the receptor is anchored to the target cell at the top. The views were generated from Banner *et al.* (1993) (available as 1TNR PDB, www.pdb.bnl.gov/cgi-bin/pdbmain) as visualized by RasMol (v2.6.1) available at www.umass.edu/microbio/rasmol/rasnew.htm.

LT-α assembles concurrently as a homotrimer and is secreted, whereas expression of LT-β protein by itself has not been detected (Browning *et al.*, 1995).

An underappreciated feature of these ligands is that the membrane position may confer distinct receptor binding and cellular responses not seen in soluble counterparts. This is exemplified by TNF which as a membrane-bound ligand is active in signaling death (Perez *et al.*, 1990) via both types of TNFR (Grell *et al.*, 1993, 1995), whereas the soluble form is active via TNFR60. As discussed below, secreted and membrane LT differ dramatically in receptor specificity and function—whether this is true for other family members remains to be addressed. All of the ligands are small proteins of approximately 250 to 300 amino acids which migrate in the range of 17–40 kDa in sodium dodecyl sulfate–polyacrylamide gel electrophoresis. Except for human TNF, a significant portion of the actual mass is contributed by glycosylation, primarily *N*-linked although LT-α is modified by *N*- and *O*-linked glycosylation (Androlewicz *et al.*, 1992; Chai *et al.*, 1993).

Proteolytic processing is another postranslational feature common to certain members of this family. Human TNF is translated as a 26-kDa precursor and processed by a metalloproteinase at the cell surface to yield the 17-kDa secreted form (Gearing *et al.*, 1994; McGeehan *et al.*, 1994; Mohler *et al.*, 1994). FasL (Tanaka *et al.*, 1995) and CD40L (Pietravalle *et al.*, 1996) are also processed by proteolysis, generating soluble forms.

THE TNF RECEPTOR FAMILY

Members of the TNF receptor superfamily include proteins of mammalian (Table 3) and viral origin. Noteworthy members include the low-affinity 75-kDa neurotrophin receptor (NTR) and the soluble TNFR-like proteins produced by poxvirus. Two distinct receptors of 55–60 kDa (CD120a; TNFR60 or RI) (Loetscher *et al.*, 1990; Schall *et al.*, 1990) and 75–80 kDa (CD120b; TNFR80 or RII) (Kohno *et al.*, 1990; Smith *et al.*, 1990) have been identified that bind both TNF and LT-α. TNFR60 and TNFR80, although binding the same two distinct ligands, share only 23% homology, less than that between some other members. The two TNFR genes reside on different chromosomes (TNFR80 on 1p36 and TNFR60 on 12p13), but there they are linked to several other members. New additions to the receptor family include two entry factors for viruses, herpes simplex virus entry mediator (HVEM) (Montgomery *et al.*, 1996) and a receptor for cytopathic avian leukosis sarcoma viruses, CAR-1 (Brojatsch *et al.*, 1996); the latter contains a DD. Other newly described DD-containing receptors are DR3 (also known as TRAMP, WSL-1, APO-3 and LARD) (Chinnaiyan *et al.*, 1996a; Kitson *et al.*, 1996; Bodmer *et al.*, 1997; Screaton *et al.*, 1997), DR4 (Pan *et al.*, 1997b), DR5. Three non-DD receptors that have been characterized recently are GITR (Nocentini *et al.*, 1997), osteoprotegerin (OPG) (Simonet *et al.*, 1997) and TRID/DcR1 (Pan *et al.*, 1997a; Sheridan *et al.*, 1997).

All the cellular receptors are type I membrane glycoproteins with the *N*-terminus located on the exterior of the cell. The defining feature of this family is a highly conserved cysteine-rich motif in the extracellular ligand-binding domain (Fig. 3). The family's signature motif is CXXCXXC. This cysteine-rich motif is repeated four times for most of the receptors; Fas and CD30 have three and six repeats respectively. The cytoplasmic domains show little sequence homology, yet several of the receptors bind common

Table 3. Properties of the TNF receptor family.

Receptor	Names	Size (kDa)	Ligands	Soluble forms	Chromosome Mouse	Chromosome Human	Tissue expression
CD120α	TNFR60, R1, A	55–60	TNF, LT-α, LT-$\alpha_2\beta_1$	Shed	6	12p13	Broad
CD120β	TNFR80, R2, B	75–80	TNF, LT-α, LT-$\alpha_2\beta_1$	Shed	4	1p36	Hematopoietic, broad
LT-βR	TNFrrp	61	LT-$\alpha_1\beta_2$	?	6	12p13	Broad
CD95	Fas, Apo1	43	FasL	Alternate splice	19	10	Lymphocytes, broad
4-1BB	ILA(hu)	33	4-1BBL	Alternate splice	4	1p36	T cells, broad
OX-40	ACT35(hu)	48	OX-40L	?	4	1p36	CD4$^+$ T cells
CD27	Tp55	50–55 dimer	CD70, CD27L	Shed	6	12p13	Resting T cells
CD30	Ki-1	120	CD30L	?	4	1p36	Hematopoietic, Hodgkin's lymphoma
CD40	Bp50, p50	43–47	CD40L, gp39, TBAM, TRAP	Shed	2	20q11–q13	B and T cells, carcinomas
NTR	p75, NGFR	75	NGF, neurotrophins	Shed	11	17q21–22	Nervous system
TRAMP	WSL-1, DR3, LARD, APO-3		?	Spliced		1p36.2	Broad
HVEM	ATAR, TR2	42	?	?	?	1p36	Lymphocytes, broad
CAR-1	?	?	?	?	?	?	
OPG		55 mono.	?	Secreted	?	8q23–24	Broad
GITR		?	?	?	?	?	T cells
DR4	TRAIL-R1	?	TRAIL	?	?	?	T cells, broad
DR5	TRAIL-R2	?	TRAIL	?	?	?	Broad
TRID	DcR1, TRAIL-R3	?	TRAIL	?	?	?	Broad, normal tissue

NGF, nerve growth factor; TNFrrp, TNF receptor-related protein.

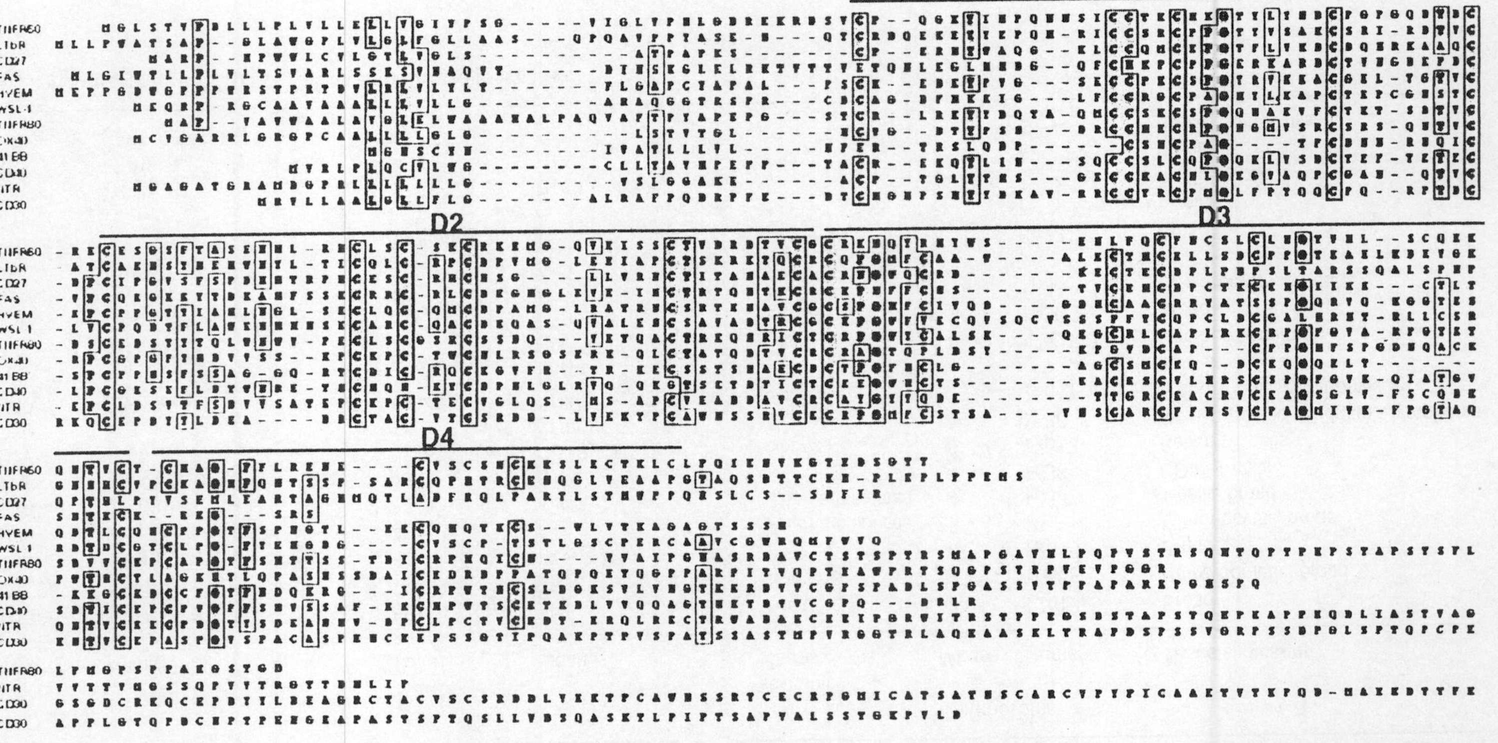

Fig. 3. TNF receptor superfamily. Alignment of the ligand binding domains of TNF receptor superfamily sequences of the extracellular domains were aligned with ClustalW (pam250 scoring matrix, MacVector) and maximized based on the TNFR60 structural model for positioning of cysteines. Shaded and boxed areas indicate identical and conserved residues respectively. D1–D4 represent the cysteine-rich motif repeats (CD30 contains six repeats). The connecting lines above D2 show the residues involved in disulfide bond formation. Genbank accession no.: NTR, M14764; OX-40, X75962; 4-1BB, U03397; CD40, X60592; LT-βR, L04270; TNFR80, M55994; TNFR60, M63121; CD27, M63928; Fas, M67454; CD30, M83554; WSL-1, Y09392; HVEM, U70321.

signaling proteins and initiate a similar spectrum of cellular responses. All of the receptors are glycoproteins, where 30–50% of the receptor mass may be due to glycosylation and/or phosphorylation and other post-translational modifications.

The extracellular portion of the receptor is the ligand-binding domain. Based on the crystal structure of the TNFR60 (Banner *et al.*, 1993), the cysteine-rich motif is a tightly knit domain formed around three disulfide bonds (Fig. 2). The cysteines bond in a pattern of 1–2, 3–5 and 4–6 (numbering from the *N*-terminus) in the second (D2) and third (D3) domains, which make contact with the ligand. Although the global architecture of the receptors is conserved, some variation in the CXXCXXC motif in D2 and D3 is found in 4-1BB, TNFR80 and CD40. The fourth motif, D4, exhibits a disulfide domain pattern of 1–2, 3–6, 4–5, differing from that of D2 and D3. Even more variation between members is seen in this membrane proximal domain. Substantial differences in the charged residues along the CXXCXXC sequence are also found among the different receptors. The three loops connected by the disulfide bonds also vary in size and residues.

THE LIGAND–RECEPTOR COMPLEX

The crystal structure of TNFR60 and LT-α homotrimer (Banner *et al.*, 1993) reveals the elongated receptor lays in the cleft formed between adjacent subunits with the receptor's *N*-terminus closest to the base of the ligand (Fig. 2). The D2 and D3 receptor domains contain the major contact regions although mutational analysis has indicated that D4 contributes minimally to binding (Marsters *et al.*, 1992). In this orientation, the receptor and ligand would be on opposing membranes. The loops connecting the D–E and the A–A″ β strands, located on opposite sides of the solvent-exposed surfacc, contain major receptor-binding residues for the LT-α–TNFR60 complex. The asymmetry of each subunit places the A–A″ loop of subunit 'a' next to the D–E loop of subunit 'c' to form a single receptor binding site. Thus, each ligand has three equivalent receptor binding sites (the exception is the LT-$\alpha\beta$ heterotrimers).

Mutational analysis of LT-α has demonstrated that certain residues in these loops (e.g. Asp-50 and Tyr-108) alter TNFR binding. In TNF, mutations in the equivalent loops distinguish between the contact residues required for binding TNFR60 and TNFR80. Saturation binding analysis of TNF to cell surface receptors shows multiple equilibria binding isotherms with high (K_d approximately 0.1–1 nM) and low (10–20 nM) affinity (Loetscher *et al.*, 1991; Pennica *et al.*, 1993). Aggregation or clustering of two or three receptors by the ligand appears to be a critical feature for the initiation of signal transduction, an event mimicked by antireceptor antibodies (Engelmann *et al.*, 1990). In the case of Fas, trimerization of receptors achieved with ligand, polyclonal or immobilized monoclonal antibodies is required for activating cell death (Dhein *et al.*, 1992).

The designation of ligand and receptor may be somewhat arbitrary in this superfamily given that all the ligands in their membrane-anchored state have cytoplasmic tails, notably long for FasL, CD30L and 4-1BBL. This is a conserved feature between species, suggesting a functional role for the cytoplasmic tail perhaps in metabolic regulation, cytoskeleton interactions or bidirectional signaling. Evidence for bidirectional signaling has been uncovered for CD40L (Yellin *et al.*, 1994), CD30L (Wiley *et al.*, 1996), CD27L (Bowman *et al.*, 1994), 4-1BBL (Pollok *et al.*, 1994) and OX-40L (Stuber *et al.*, 1995), and, interestingly, TNF is phosphorylated on its cytosolic *N*-terminus (Pocsik *et al.*,

1995). The responsible mechanisms involved in signaling via the ligand remain unclear at present.

SIGNAL TRANSDUCTION MOLECULES

A central theme in the initiation of signal transduction pathways by many cytokines is the ligand-dependent aggregation of receptors, for example the α-helical cytokines (e.g. growth hormone) induce dimerization of their receptors. Most TNF receptors are expressed at low levels (typically less than a few thousand per cell) (Ware *et al.*, 1991) and are probably at low density on the cell surface. Crystallographic evidence suggests that unligated TNFR60 may exist as a dimer in an inactive conformation (Rodseth *et al.*, 1994). The high-affinity binding of trimeric ligand is thought to cluster two or three receptors into a conformationally active complex. This results in the transmembrane aggregation of the cytoplasmic domains which, in turn, translocate (recruit) or activate receptor-associated signaling molecules. This concept has been confirmed over the past 2 years by the discovery of two distinct families of cytosolic signaling proteins, the DD and TRAF families, in addition to several other proteins, some with catalytic activity. Several studies indicate that the receptor signaling complex is formed in part by recruitment of different proteins to the ligand–receptor complex (Hsu *et al.*, 1996a; Muzio *et al.*, 1996; VanArsdale *et al.*, 1997).

The Death Domain Signaling Molecules

The DD family consists of receptors and several non-receptor proteins that contain a sequence homologous to the region responsible for apoptosis induced by Fas and TNFR60 (Table 4). The DD motif promotes self-association and heteroassociation with other DD proteins (Boldin *et al.*, 1995a). The approximately 100 amino acids of the Fas cytoplasmic domain consists of six antiparallel amphipathic α helices as revealed by nuclear magnetic resonance imaging NMR (Huang *et al.*, 1996). Sequence conservation would suggest that other DD-containing proteins will have a similar fold in this region. Five non-receptor DD proteins have been identified using the two-hybrid approach and include TRADD (Hsu *et al.*, 1995), FADD/mediator of receptor-induced toxicity (MORT-1) (Chinnaiyan *et al.*, 1995; Boldin *et al.*, 1995b) and the serine kinase RIP (Stanger *et al.*, 1995), MAP kinase-activating death domain protein (MADD) (Schievella *et al.*, 1997), and RAIDD (Duan and Dixit, 1997) (Table 4). TRADD is a 34-kDa protein that shows 25% identity with TNFR60 DD. TRADD association with TNFR60 requires an intact death domain in both molecules, which suggests this domain is a protein interaction motif. TRADD plays multiple roles for TNFR60 signaling, including acting as an adaptor for binding to TRAF2. TRADD overexpression induces apoptosis (Hsu *et al.*, 1995, 1996b; Rothe *et al.*, 1995b), which is dependent on FADD (Chinnaiyan *et al.*, 1996b) and nuclear factor (NF)-κB activation, an important transcription activator for genes involved in inflammation (Sha *et al.*, 1995). NF-κB also provides a protective effect against apoptosis (Beg and Baltimore, 1996; Van Antwerp *et al.*, 1996).

The Fas-associated protein, termed MORT-1 or FADD, associates specifically with Fas cytoplasmic death domain. FADD is a 23.3-kDa cytosolic protein, and within the death domain homology region it shares 28% identity with Fas. A point mutant of

Table 4. Some properties of the non-receptor death domain (DD) and death effector domain (DED) proteins.

Designation	Other names	No. of amino acids	Binding partners	Function
TRADD*		312	TNFR60, FADD, RIP, TRAF2	TNFR60 adaptor for FADD and TRAF2
MADD*		1588	TNFR60	Activates MAP kinase cascade and phospholipase A_2
RIP*		656	TNFR60, TRADD, FADD	Serine–threonine kinase
FADD*†	MORT-1	208	Fas, FLICE, TRADD, RIP	Couples Fas to FLICE
RAIDD*†	CRADD	200	RIP, ICH-1 (caspase 2)	Adaptor for RIP to caspases
Casper†	FLAME-1	480	Caspases 3 and 8, TRAF1 and 2, FADD	Regulates Fas/TNFR-mediated apoptotic pathways
Caspase 8†‡	FLICE, Mch5, Mach	479	FADD	Protease cleaves CPP32, LAP3, Tx, LAP6 and FLICE2
Caspase 10†‡	FLICE2, Mch4	479	FADD	Protease cleaves CPP32 and Mch3

These proteins contain *a death domain, †a death effector domain, ‡an interleukin 1β-converting enzyme (ICE)/CED-3 homology region.
CED-3, *C. elegans* gene designation; CPP, cysteine protease p32; FLICE, Fas-associated 'death' domain interleukin-1β-converting enzyme; LAP, latency-associated peptide; RAIDD, RIP-associated ICH-1/CED-3 homologous protein with a death domain.

FADD, analogous to the point mutant of Fas responsible for the *lpr*^{cg} (V238N) phenotype, fails to interact with Fas (Chinnaiyan *et al.*, 1995). However, this mutation probably alters the global protein structure as this residue is located within the interior of the DD (Huang *et al.*, 1996). FADD overexpression also induces cell death. Thus, TRADD and FADD mimic the responses of their respective receptors in self-association and cell death signaling. A mutant of FADD, but not TRADD, blocks both Fas- and TNFR60-induced apoptosis (Chinnaiyan *et al.*, 1995), indicating that FADD is the key link to the apoptosis pathway.

A serine–threonine protein kinase, termed RIP, is approximately 74 kDa with an *N*-terminal kinase domain and a *C*-terminal DD (Stanger *et al.*, 1995). RIP interacts with both Fas and TNFR60 through the DD and is recruited to TNFR60 by TRADD (Hsu *et al.*, 1996a). Overexpression of RIP cDNA, like that of TRADD and FADD, induces cell death and NF-κB activation. RIP also interacts with another DD protein, RIP-associated Ich/CED-3 homologous protein with death domain (RAIDD), which contains a *C*-terminal DD and an *N*-terminal domain homologous to ICE and CED-3 homologue (Ich/CED-3) caspases (Duan and Dixit, 1997).

Overexpression of any of the DD proteins induces apoptosis in certain cellular models. FADD- and TRADD-induced death, like that of TNFR60 and Fas, is blocked by the antiapoptotic protein, the product of *crmA*, a poxvirus gene (Tewari and Dixit, 1995). The product of *crmA* is an inhibitor of cysteine proteases (serpin) related to IL-1β convertase, a family of aspartic acid-specific proteases involved in apoptosis (caspases) (Martin and Green, 1995; Alnemri *et al.*, 1996).

The most recently described DD-containing signaling molecule is MADD, a 176-kDa protein with 14% identity to TRADD (Schievella *et al.*, 1997). MADD was isolated via its ability to directly bind TNFR60 through the interaction between the DD of the two proteins. Overexpression of MADD leads to activation of the mitogen-activated protein (MAP) kinase, extracellular signal-regulated kinase (ERK) and, to a lesser extent, c-jun *N*-terminal kinase (JNK), which suggests a possible link between TNFR60 and arachidonic acid release.

Included in this group of signaling molecules are those that contain a death effector domain (DED). FADD, unlike TRADD, is linked directly to the apoptotic machinery via an *N*-terminal domain termed DED. MACH1/FLICE, or caspase 8, indirectly binds to Fas, TNFR60 and DR3 via FADD (Boldin *et al.*, 1996; Chinnaiyan *et al.*, 1996a; Muzio *et al.*, 1996). Caspase 8 contains a distinct interaction domain for FADD, the DED, allowing it to link the receptor signaling complex to the apoptotic machinery. This work established proteolysis as a novel catalytic activity associated with these cell surface receptors (Muzio *et al.*, 1996). The evolutionary conservation of the apoptotic machinery is further underscored by several reports that the *Caenorhabditis elegans* CED-4 gene couples the death effector molecules to B-cell leukemia/lymphoma 2 (Bcl-2) in mammalian cells and to regulation of apoptosis and the integrity of mitochrondia and release of cytochrome c (Kluck *et al.*, 1997; Yang *et al.*, 1997).

Another protein with a similar structure to FADD has been cloned recently. RAIDD/caspase and RIP adaptor with death domain (CRADD) has a *C*-terminal DD and an *N*-terminal DED (Duan and Dixit, 1997). RAIDD interacts with RIP via its death domain and with caspase 2 via its DED. As expected, RAIDD induces apoptosis when overexpressed.

The newest member of the DED-containing family is caspase-eight-related protein (Casper)/FADD-like antiapoptotic molecule (FLAME-1) (Shu *et al.*, 1997; Srinivasula *et al.*, 1997). Casper/FLAME-1 interacts with FADD, caspases 2 and 8, and TRAFs 1 and 2. It is similar to caspase 8 but is believed not to have caspase activity as several conserved amino acids found in all caspases are absent in Casper/FLAME-1. Indeed, FLAME-1, when recruited to the Fas signaling complex, inhibits apoptosis and, therefore, may act as a competitive inhibitor (Srinivasula *et al.*, 1997).

A current model envisions the TNF receptor as a multisubunit complex. The signaling complex forms after TNF binding induces clusters of the receptor cytoplasmic domains that form neo (or high-affinity) binding sites which recruit key components, such as TRADD/FADD and RIP, from the cytosol, which assemble with RAIDD and/or FLICE into a membrane-localized enzymatic complex with distinct (kinase and protease) catalytic activities. The processes involved in forming the complex and substrates of the enzymes are currently an area of active research.

The TRAF Family

The TNFR-associated factors (TRAFs) are a distinct family of cytosolic RING finger proteins. At their *N*-terminus TRAFs contain multiple cysteine- and histidine-rich repeats predicted to form zinc ion coordination groups that fold into RING finger and/or zinc finger motifs. There are currently six members with distinct interaction patterns with the non-DD TNFR family members, CD40, LT-βR, TNFR80 and CD30. The zinc RING and finger motifs, in general, appear to be involved as effector domains as deletion causes loss of signaling function. The defining feature of this group is the *C*-terminal homology region containing the major receptor interaction 'TRAF' domain responsible for binding to several different TNFR-related molecules (Fig. 4). The TRAF domain is divided into *N*- and *C*-terminal regions. The TRAF-N region exhibits periodic spacing of hydrophobic residues with high α-helix potential (coiled coil motif/isoleucine zipper). The TRAF-C region is required for receptor binding; however, TRAF3 and 5 have an extended version of the TRAF-N-coiled coil which participates in binding to receptors. The TRAFs may interact indirectly with other receptors, since TRAF2 can associate with TNFR60 through a TRAF2 binding site on TRADD, a DD protein. Additionally, TRAF6 has been implicated as the activator of NF-κB by IL-1, and thus represents the first example of a TRAF molecule involved in a signaling pathway outside the TNFR family (Cao *et al*., 1996). The functions of the TRAF family are just emerging, with several TRAFs involved in gene regulation through NF-κB activation, apoptosis and developmental processes (Table 5).

TRAF1 and 2 were initially defined by their interaction with TNFR80 (Rothe *et al*., 1994). TRAF2 and TRAF1 form homo- and heterodimers with each other; TRAF2 binds directly to TNFR80, CD40 and CD30. TRAF1 lacks the zinc RING found in other TRAFs, and associates indirectly with TNFR via TRAF2; however, TRAF1 binds CD30 directly. TRAF1 expression is induced during T- and B-lymphocyte activation or transformation by Epstein–Barr virus (EBV) (Mosialos *et al*., 1995), whereas TRAF2 is expressed constitutively by most cells and tissues. The zinc RING and first three zinc finger motifs of TRAF2 are critical for activation of NF-κB (Takeuchi *et al*., 1996). TRAF2 is also implicated as an adaptor for the activation of the stress-activated protein kinase, SAPK (Liu *et al*., 1996; Natoli *et al*., 1997). Recent evidence has shown that TRAF2 interacts with a serine kinase, NF-κB inducing kinase (NIK), which is related to mitogen activated protein kinase kinase (MAPKK) kinase which leads to NF-κB activation (Malinin *et al*., 1997).

TRAF3 was initially identified by its binding interactions with CD40 (Hu *et al*., 1994; Cheng *et al*., 1995; Sato *et al*., 1995b) and to latent infection membrane protein 1

Table 5. Some characteristics of the human TRAF family

Designation	Other names	Chromosome location	No. of amino acids	Function	Binding partners
TRAF1	EBI-6	2	416	Forms heterocomplex with TRAF2	CD30, TRAF2
TRAF2	TRAP3	2	501	NF-κB activation, JNK activation	TNFR80, CD30, LT-βR, CD40, CD27, TRAF1, I-TRAF, IAP-1/2, NIK, TRADD, TRIP
TRAF3	CD40bp, LAP1, CAP1, CRAF1	17	568	Apoptosis	LT-βR, TNFR80, CD40, CD27, CD30
TRAF4	CART-1		470	Breast cancer	?
TRAF5		1q32	557	NF-κB	LT-βR, CD30, CD40
TRAF6			522	NF-κB	CD40, IRAK

I-TRAF, inhibitor of TRAF (TANK); IAP, inhibitor of apoptosis; NIK, MAPKK kinase related kinase; IRAK, IL-1β receptor associated kinase; JNK, c-jun *N*-terminal kinase; TRADD, TNFR-associated death domain protein; TRIP, TRAF-interacting protein; CART-1, C-rich domain associated with RING and TRAF domains.

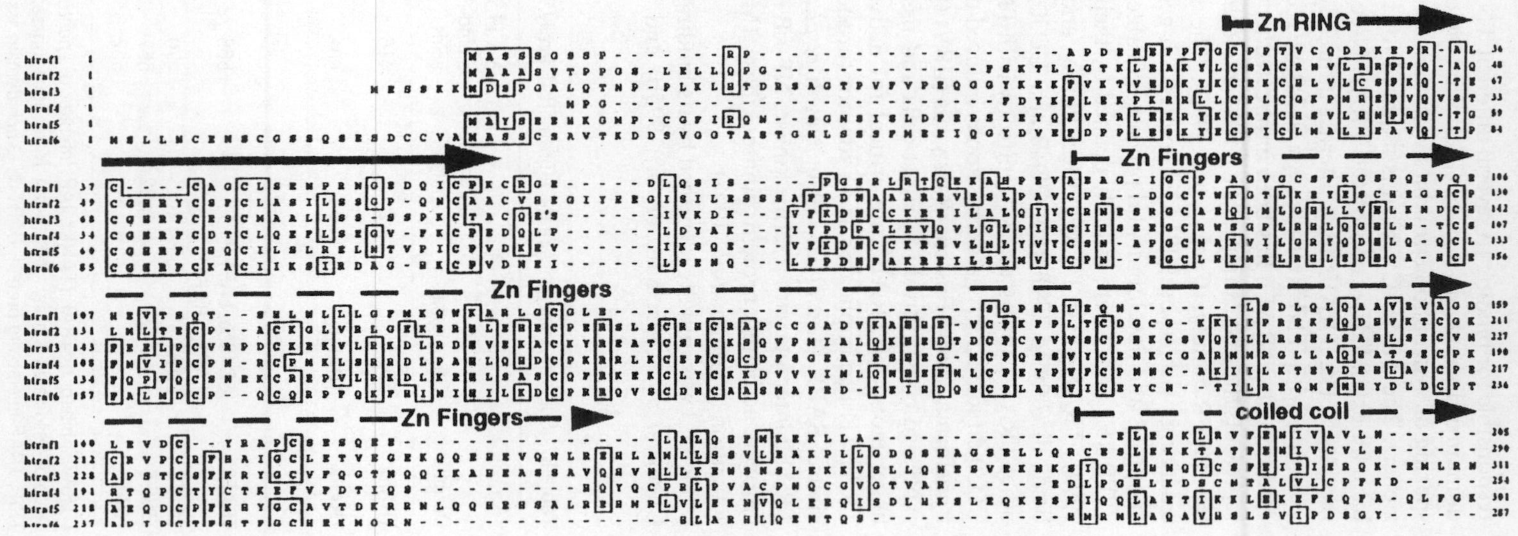
Zn RING
Zn Fingers
Zn Fingers
Zn Fingers
coiled coil
htraf1
htraf2
htraf3
htraf4
htraf5
htraf6

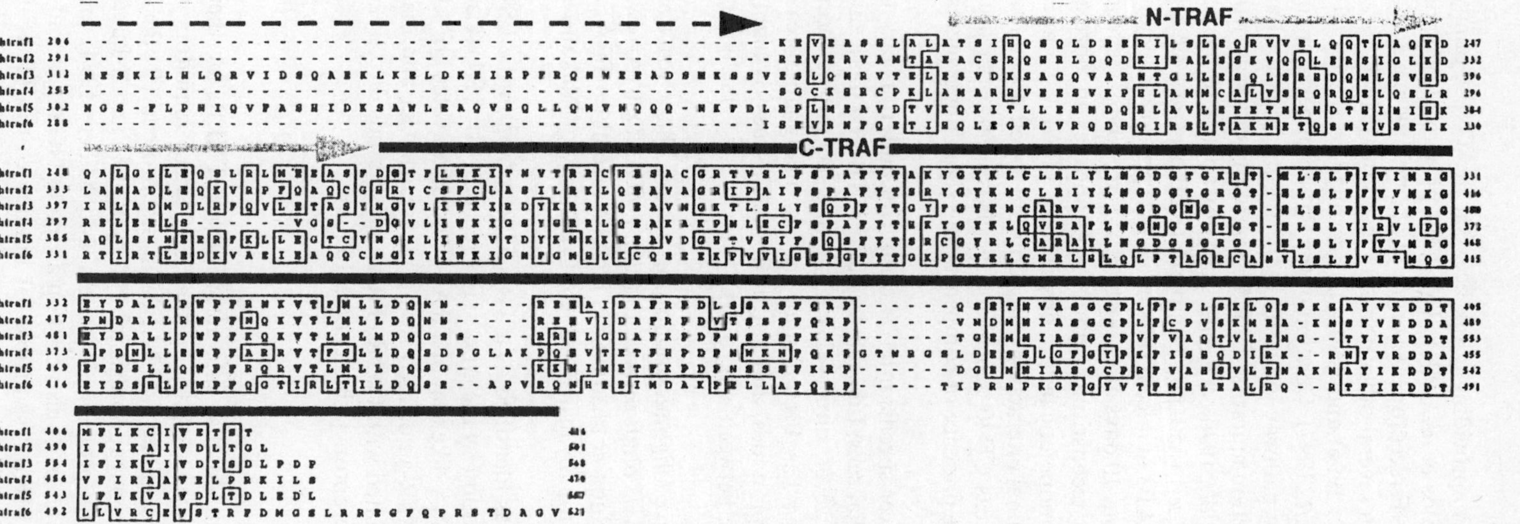

Fig. 4. Alignment of human TRAFs. TRAF family sequences were aligned with ClustalW (pam250 scoring matrix, MacVector). The functional domains are indicated with arrows. Genbank accession no. for human TRAF1, U19261; TRAF2, U12597; TRAF3, U19260; TRAF4, X80200; TRAF5, U69108; TRAF6, L81153.

(LMP-1), the dominant oncogene of EBV, providing a link between the TNFR family and a human transforming virus (Mosialos *et al.*, 1995). This viral protein also binds TRAF1/2 and thus mimicks a ligated TNF receptor (Devergne *et al.*, 1996; Kaye *et al.*, 1996). TRAF3 also interacts directly with several other non-DD receptors (Mosialos *et al.*, 1995; Gedrich *et al.*, 1996; Lee *et al.*, 1996) and is rapidly recruited to the LT-βR following ligand binding (VanArsdale *et al.*, 1997). Deletion of the RING finger domain of TRAF3 blocks the cell death signaling pathway induced via the LT-βR, but not Fas (VanArsdale *et al.*, 1997). *C*-terminal deletion mutants of TRAF3 partially block growth arrest mediated by CD40 (Eliopoulos *et al.*, 1996). TRAF3 does not appear to have a significant role in the activation of NF-κB (p50–p65 complex) (Rothe *et al.*, 1995b; VanArsdale *et al.*, 1997). Deletion of TRAF3 in mice results in a lethal phenotype where mice succumb to hemopoietic failure about 10 days after birth (Xu *et al.*, 1996). As fetal liver from TRAF3$^{-/-}$ mice is able to reconstitute hemopoietic lineages in normal irradiated recipients, this suggests the hemopoietic failure observed in TRAF3$^{-/-}$ mice is due to a deficiency in lymphocyte–stromal interactions. In addition, T-cell dependent responses to antigen are deficient, whereas CD40-dependent B-cell responses such as CD23 induction are intact, indicating the presence of compenstating signaling proteins, perhaps other TRAFs.

TRAF4 was identified as a breast cancer-specific mRNA transcript and, thus far, its expression is limited to malignant epithelial cells. Like TRAF2 and 3, TRAF4 has a zinc RING motif and a TRAF domain. However, direct TNFR binding has not yet been demonstrated. The finding that TRAF4 has been found only in malignant breast carcinomas and metastatic axillary lymph nodes, and not in normal tissues, is highly intriguing, and provides another link between the TRAF family and malignant processes.

TRAF5 was identified by degenerative oligonucleotide polymerase chain reaction amplification (Nakano *et al.*, 1996). TRAF5 binds directly to the LT-βR, CD30 (Aizawa *et al.*, 1997), and indirectly to CD40 (Ishida *et al.*, 1996b). TRAF5 resembles TRAF3 structurally, but resembles TRAF2 functionally in activating NF-κB activity (Nakano *et al.*, 1996).

The most recent member, TRAF6, was identified as a TRAF2 homolog in the EST database and is involved in NF-κB activation by the type 1 IL-1 receptor (IL-1R1) (Cao *et al.*, 1996). Transfection of TRAF6 cDNA, like that of TRAF2 and 5, activates NF-κB directly. TRAF6 interacts with the IL-1βR-associated serine kinase, IRAK, and may form a complex that is transiently associated with IL-1R. TRAF6 also binds directly to CD40, but not to several other related receptors (Ishida *et al.*, 1996a).

Other Receptor-Associated Proteins

Several other proteins have been identified that interact with the TNFRs or their signaling proteins. Two mammalian proteins related to the baculovirus inhibitors of apoptosis family (c-IAP 1,2) have been identified as TRAF2-binding proteins (Rothe *et al.*, 1995a). The IAP family includes neuronal apoptosis inhibitory protein and X-linked IAP (XIAP), in addition to insect virus protein p35 (Clem and Miller, 1994) and *Drosophila* proteins (Hay *et al.*, 1995). The IAPs also contain a zinc RING finger and a shared motif (baculovirus IAP repeat (BIR)) responsible for binding to the TRAF-N domain. IAP association with TNFR80 is indirect and requires the TRAF1 and 2 heterocomplex. XIAP was recently shown to be a direct inhibitor of caspases 3 and 7 (Deveraux *et al.*, 1997). The TNF receptor superfamily may link to multiple pathways, for example an apoptotic pathway (DD–DED interactions) and an antiapoptotic pathway (activation of NF-κB and of IAPs).

I-TRAF (also known as TRAF family member associated NF-κB activator (TANK)) was identified as a TRAF-binding protein by yeast two-hybrid analysis, and interacts with the TRAF-C domain of TRAF1, 2 and 3. I-TRAF shows both inhibitory or enhancing activity for TRAF2-dependent NF-κB activation (Cheng and Baltimore, 1996; Rothe *et al.*, 1996). The A20 zinc finger protein, known for its protective effect against TNF-mediated apoptosis (Opipari *et al.*, 1992), binds to the TRAF1/2 heterocomplex and also inhibits NF-κB activation (Song *et al.*, 1996).

Sensitivity to Fas-mediated apoptosis is increased when the *C*-terminal 15 residues of the receptor are deleted (Itoh and Nagata, 1993). A tyrosine phosphatase, known as fas-associated phosphatase 1 (FAP-1), binds to this region (Sato *et al.*, 1995a). FAP-1 (no DD) is a very large protein (2485 amino acids) containing multiple GLGF sequence repeats (PDZ domains). The GLGF repeat is thought to be important for targeting the protein to submembranous cytoskeleton or for enzyme regulation. FAP-1 also contains a membrane-binding site (not a transmembrane domain) similar to that found in cytoskeletal associated proteins, and also contains a *C*-terminal catalytic domain. Introduction of FAP-1 into tumor cells reduces their sensitivity to anti-Fas-induced apoptosis (Sato *et al.*, 1995a). Recently a ubiquitin-conjugating enzyme was shown to associate with DD of Fas (Wright *et al.*, 1996). UBc9 is a component of the ubiquitinating pathway that is important for cell cycle progression. In yeast mutants, human cDNA for Ubc9 restores growth, implicating a role for this enzyme in growth regulation.

In addition to the DD proteins interacting with TNFR60 cytoplasmic domain, factor associated with neutral sphingomyelinase activation (FAN), a protein responsible for activating neutral sphingomyelinases, enzymes that generate the lipid mediator ceramide, interacts with TNFR60 *N*-terminus of the DD (Adam-Klages *et al.*, 1996). Additionally, this same region of TNFR60 is involved in binding raf-1 kinase (Belka *et al.*, 1995).

It is clear that elucidation of signaling pathways for TNFR family members is at a formative stage; however, significant advances have been made that provide a glimpse at the events involved in signal transduction. Unlike some other cytokine and hormone receptors, TNFRs lack inherent enzymatic activity, yet the ligand–receptor complex contains multiple enzymes, including kinases, proteases and ubiquitin-conjugating catalytic activities, and their substrates, clearly indicating the TNF receptor is a multicomponent complex. How the receptor proximal components, their substrates and downstream effector pathways are linked to cellular responses will likely remain a focus for some time to come.

REGULATION AND EXPRESSION OF LIGANDS AND RECEPTORS

All of the TNF-related ligands are expressed by activated T cells, but several are also produced by other lymphoid and non-lymphoid tissue. TNF is a good example of a broadly expressed ligand. It is produced primarily by macrophages and T cells when activated by inflammatory or antigenic stimuli, but is also produced by mast cells and basophils. Ectopic expression in non-lymphoid tumors has also been observed. CD40L is restricted to production by $CD4^{+}$ T helper cells and basophils in response to stimulation via the high-affinity immunoglobulin E (IgE) Fc receptor. LT-$\alpha\beta$ expression is restricted to activated T and B lymphocytes, and to natural killer (NK) cells. CD30L is produced by macrophages, and CD27L by B cells and thymic stroma, as is 4-1BBL. FasL and OX-40L are expressed predominantly by T cells, yet FasL is detected in reproductive organs (Suda *et al.*, 1993; Bellgrau *et al.*, 1995), lens epithelium (Griffith *et al.*, 1995) and thyrocytes (Giordano *et al.*, 1997), as well as by melanoma cells (Hahne *et al.*, 1996). The non-lymphoid expression of FasL, with its dramatically potent apoptosis-inducing activity for T cells (Brunner *et al.*, 1995), is thought to provide an immunosuppressive function by destroying migrating inflammatory T cells, theoretically creating an immunologically privileged site within those organs.

Expression of TNF-related ligands is highly regulated with transcriptional, translational and post-translational control points. An example of one significant control point is production of TNF mRNA by T cells (Goldfeld *et al.*, 1991). TNF mRNA transcription is induced in response to engagement of the T cell with an antigen-presenting cell, and thus TNF expression is an antigen-specific response. TNF mRNA is short lived, giving rise to transient expression of the protein. Degradation of TNF and LT-α mRNA is controlled in part by AU-rich sequences in the 3′ untranslated region. Other ligands, such as FasL, LT-$\alpha\beta$, and CD40L, also exhibit inducible transient expression, remaining at the cell surface for only a few hours in tissue culture models. LT-α and LT-β are coexpressed during T-cell activation to form the LT-$\alpha\beta$ heterotrimers, yet LT-β mRNA can be detected in freshly isolated peripheral blood lymphocytes. A translation block may exist as LT-β protein is not detected on these cells. TNF mRNA is also found in a non-translated form in macrophages, and contributes to the ability of these cells to rapidly produce large amounts of TNF when stimulated by various activators, such as bacterial lipopolysaccharide (LPS). Other ligands in this family, including FasL and CD40L, are rapidly induced in T cells, but transiently expressed on the cell surface.

The membrane form of TNF is processed by a cell surface metalloproteinase (TNF-alpha converting enzyme (TACE)), which cleaves TNF on the extracellular side of its transmembrane domain, thus releasing a soluble product (Black *et al.*, 1997; Moss *et al.*, 1997). FasL also appears to be shed (Tanaka *et al.*, 1995), whereas LT-β is not cleaved during T-cell activation (Ware *et al.*, 1992; Browning *et al.*, 1995).

Cellular expression and tissue distribution of the different TNFR-related receptors varies substantially. TNF receptors are expressed at extremely low levels on most cells, typically less than a few thousand per cell. Several receptors are tissue restricted; for example, TNFR60 and LT-βR are expressed predominantly on epithelial cells, whereas TNFR80 is the most abundant receptor on T and B lymphocytes; macrophages appear to express all three. Many of the Chr 1p36-encoded receptors, for example TNFR80, 4-1BB and OX-40, are induced following T-cell activation. Freshly isolated T cells from peripheral blood have no binding sites for TNF; however, when activated via the T-cell receptor (TCR) or IL-2, the TNFR80 gene is transcribed resulting in peak surface expression of the protein in 48–72 h (Ware *et al.*, 1991; Gehr *et al.*, 1992; Crowe *et al.*, 1993). Activated T cells, when restimulated, rapidly (within 30 min) downregulate TNFR80 by two distinct mechanisms, one that involves shedding by a metalloproteinase, which releases a soluble extracellular receptor domain with ligand-binding activity, and a second mechanism involving internalization, which results in loss of surface expression independent of proteolysis (Crowe *et al.*, 1995). This metalloproteinase may be of the same class of enzyme that processes membrane TNF (Crowe *et al.*, 1995). Biosynthesis of TNFR80 is continuous throughout the downregulation process and surface expression of TNFR80 reappears by 12–24 h, suggesting that the shedding enzyme is regulated. As TNF is thought to function as an apoptosis signal for $CD8^+$ cytotoxic T cells (Zheng *et al.*, 1995), analogous to FasL-induced apoptosis of $CD4^+$ T cells, the downregulation of TNFR80 during TNF synthesis by $CD8^+$ T cells may exist to protect these cells from apoptosis.

Soluble receptors can have both stablizing and inhibitory effects on their ligands. For example, TNF is relatively unstable in solution, but is stabilized by binding a soluble monovalent form of TNFR (Aderka *et al.*, 1992), which is also shed during cellular activation (Crowe *et al.*, 1993). As concentrations of soluble TNFR approach saturation of the ligand, occupying two or three receptor binding sites, soluble TNFR inhibits binding to cellular receptors. This aspect of using soluble receptors has been exploited by creating fusion proteins with the Fc region of IgG. A dimeric receptor molecule is formed which effectively neutralizes TNF at a 1:1 molar stoichiometry (Crowe *et al.*, 1994). These types of antagonists are being tested in a variety of clinical situtations that may involve TNF-related molecules (Van Zee *et al.*, 1992; Eason *et al.*, 1995).

FUNCTIONAL ROLES IN THE IMMUNE RESPONSE

The TNF/LT-$\alpha\beta$ System

LT-deficient mice (either by gene disruption or soluble LT-βR–Ig fusion protein administration) and TNF- and TNFR60-deficient mice have sparked investigation of lymph node development and germinal center formation (Table 6). These model systems will allow the dissection of events that are required for primary and secondary phases of antibody responses, the role of the germinal center in class switching and affinity maturation. Additionally, the role of uncharacterized cell types such as metallophilic macrophages and follicular dendritic cells as well as regions in the lymph node and spleen such as the marginal zone will be elucidated further using these models.

TNF and LT-α are well recognized as pleuripotent mediators of inflammatory and immune defense mechanisms because of their ability to activate TNFR80 and TNFR60. These receptors trigger basic cellular responses including cell death and proliferation, and cellular differentiation required for innate inflammatory processes as exemplified by the role of TNF in disease states such as septic shock (for review see Beutler, 1992). The distinct receptor specificity of the LT-$\alpha1\beta2$ complex for the LT-βR hinted that new functions might be associated with this cytokine. The technique of producing mice deficient in specific genes has uncovered roles for LT in lymph node development and germinal center formation that were not indicated by *in vitro* systems. The production of LT-deficient mice has reawakened an interest in these molecules as regulators controlling fundamental processes central to the immune response (De Togni *et al.*, 1994; Banks *et al.*, 1995). LT-α-deficient mice have no detectable popliteal, inguinal, para-aortic, mesenteric, axillary or cervical lymph nodes. Both gross and histologic examination have failed to detect lymphoid aggregates or rudiments of lymph node stromal tissue. As mice deficient in TNF have normal lymph node structure (Pasparakis *et al.*, 1996), LT-α

Table 6. Genetic analysis of the TNF/TNFR superfamily.

	Phenotypes of TNF ligand knockouts						
	$TNF^{-/-}$	LT-$\alpha^{-/-}$	LT-$\beta^{-/-}$	LT-βRFc Tg	LT-βRFc Inj	$CD40L^{-/-}$	*gld* ($FasL^{-/-}$)
Lymph nodes	Yes	No	No, except mesenteric LN	Yes	No, except mesenteric LN	Yes	Yes
T or B cell splenic organization	No	No	No	No	No	Yes	NR
Germinal centers	No	No	No	No	NR	No	NR
Immunopathology	Similar to $TNFR60^{-/-}$	Reduced Ab/IgA response	Reduced Ab/IgA response	NR	NR	No Ab response to T-dependent Ag	Lymphoproliferative syndrome
References	Pasparakis *et al.* (1996)	De Togni *et al.* (1994), Banks *et al.* (1995)	Koni *et al.* (1997)	Ettinger *et al.* (1996)	Rennert *et al.* (1996)	Xu *et al.* (1994)	Hahne *et al.* (1995)

	Phenotype of TNF receptor knockouts				
	$TNFR60^{-/-}$	$TNFR80^{-/-}$	$CD40^{-/-}$	$CD30^{-/-}$	*lpr* ($Fas^{-/-}$)
Lymph nodes	Yes	Yes	Yes	Yes	Yes
T or B cell splenic organization	Slightly abnormal	Yes	Yes	NR	NR
Germinal center formation	No	Yes	No	NR	NR
Immunopathology	*L. monocytogenes* infection fatal; resistant to toxic effects of LPS	Milder phenotype of TNFR60	No antibody response to T-dependent Ag	Impaired negative selection	Lymphoproliferative syndrome
References	Neumann *et al.* (1996a)	Matsumoto *et al.* (1996b), Le Hir *et al.* (1996)	Kawabe *et al.* (1994)	Amakawa *et al.* (1996)	Watanabe-Fukunaga *et al.* (1992)

	Lymph node deficiencies in mice			
	Hox11	*BLR1*	*aly*	*Ikaros*
Mouse chromosome location	10q24	?	11	11
Protein product	Transcription factor	Putative chemokine receptor	Unknown	Transcription factor
Organ affected	Lack spleen	Lack inguinal LN; few or abnormal PP	Lack LN, PP; no splenic follicles or marginal zone	Lack LN, PP
References	Roberts *et al.* (1994)	Förster *et al.* (1996)	Miyawaki *et al.* (1994), Koike *et al.* (1996)	Georgopoulos *et al.* (1994)

LN, lymph nodes; PP, Peyer's patches; NR, not reported.

and TNF are not functionally redundant molecules. Interestingly, no defect in T lymphocyte number or function could be ascertained; proliferation, activation and cytotoxic activity were comparable to those of wild-type controls. There was, however, an increase in circulating B cells in the LT-α-deficient animals (De Togni *et al.*, 1994). Bone marrow cells obtained from LT-α-deficient mice were able to home to both spleen and lymph nodes of mice with severe combined immune deficiency or lethally irradiated wild-type mice, indicating no intrinsic defect in the hematopoietic compartment (Banks *et al.*, 1995; Mariathasan *et al.*, 1995). In contrast, normal bone marrow was unable to reconstitute lymph nodes in irradiated LT-deficient mice, indicating that lymph node organogenesis is fixed developmentally (Mariathasan *et al.*, 1995).

As mice deficient for the TNF receptors, TNFR60 and TNFR80, or TNF had no defects in lymph node organogenesis, it was speculated that an interaction between membrane-bound LT-αβ and LT-βR located on the stromal elements might play a role in lymph node genesis. To answer this question, three approaches were taken, two using an LT-βR–Fc fusion protein to neutralize membrane-bound LT-αβ complexes either expressed as a transgene (Ettinger *et al.*, 1996) or injected into pregnant mice for introduction into embryonic circulation (Rennert *et al.*, 1996), and the other by interruption of the LT-β locus (Alimzhanov *et al.*, 1997; Koni *et al.*, 1997). These inhibitors do not block soluble trimeric LT-α binding with TNFR60. LT-β-deficient mice and mice that were treated *in utero* with LT-βR–Fc fusion protein have no histologically detectable inguinal and popliteal lymph nodes, nor Peyer's patches. Temporal development of lymph nodes could be determined by administering the LT-βR–Fc fusion protein at different days of gestation; the full effect of LT-βR–Fc must be administered before day 12 of gestation. By this staging, Peyer's patches were the last in the sequence to develop. Interestingly, both mice possessed mesenteric lymph nodes, suggesting that LT-α may play a role in mesenteric lymph node development. As the LT-βR–Fc transgene was not expressed at neutralizing levels until day 3 after birth, these mice had normal lymph node development, but Peyer's patches were reduced or absent (Ettinger *et al.*, 1996). On closer inspection, TNFR60-deficient mice were found to lack Peyer's patches (Neumann *et al.*, 1996a), although this remains controversial, further suggesting that the LT-α homotrimer may play a role in the development of some lymph node structures. LT-α expressed as a transgene promotes the formation of lymph node-like structures (Kratz *et al.*, 1996), raising the possibility that LT-αβ may be a key cytokine system responsible for lymphoid follicles that form ectopically in tissues burdened with chronic immune and inflammatory reactions.

In addition to the LT system regulating lymph node genesis, several genes have been identified that also play a role in this process. Mice deficient for the early hematopoietic- and lymphocyte-restricted transcription factor, Ikaros, lack lymph nodes and Peyer's patches (Georgopoulos *et al.*, 1994). Mice that developed a spontaneous autosomal recessive mutation, termed *aly* (alymphoplasia), also lack lymph nodes (Miyawaki *et al.*, 1994). Interestingly, mice deficient for the putative chemokine receptor, Burkitt's lymphoma receptor 1 (BLR1), which is expressed on mature B cells and a subpopulation of T helper cells, lack inguinal lymph nodes and possess none or few phenotypically abnormal Peyer's patches, as well as showing a failure of B cells to migrate to splenic germinal centers (Förster *et al.*, 1996). This suggests that LT-αβ could mediate cellular migration during development via activation of certain chemokine genes as suggested by a recent report that LT-βR activation induces chemokine production (Degli-Esposti *et al.*, 1997a). Although lymph nodes are present in *Hox11* (a homeobox gene)-deficient mice, the spleen is absent (Roberts *et al.*, 1994). The LT-α-deficient mice, along with these others, will aid in determination of the signals involved in lymph node development.

In addition to the lymph node defect in the LT-α and LT-β deficient mice, abnormalities in splenic architecture were observed in these mice as well as in TNF–LT-α double-deficient mice (De Togni *et al.*, 1994; Banks *et al.*, 1995; Le Hir *et al.*, 1996; Matsumoto *et al.*, 1996b; Koni *et al.*, 1997). Mice rendered LT-$\alpha\beta$ deficient by expression of LT-βR–Fc transgene or LT-βR–Fc fusion protein administration also had disorganized splenic architecture (Ettinger *et al.*, 1996; Förster *et al.*, 1996). There is an absence of normal T cell–B cell segregation in the white pulp and loss of a well defined marginal zone. Both these mice and TNFR60-deficient mice also fail to generate germinal centers after immunization. LT-α- and LT-β-deficient mice, in contrast to TNFR60-deficient mice, are negative for MOMA-1 reactivity, a monoclonal antibody that detects metallophilic macrophages comprising the marginal zone (Matsumoto *et al.*, 1996b). Mucosal addressin cell adhesion molecule-1 (MAdCAM-1), another marker for the marginal zone, is absent in both TNFR60-deficient mice and those treated with LT-βR–Fc fusion protein (Neumann *et al.*, 1996b; Rennert *et al.*, 1996). Finally, the follicular dendritic cell network is absent in LT-α- and TNFR60-deficient mice (Fu *et al.*, 1997; Le Hir *et al.*, 1996; Matsumoto *et al.*, 1996a; Pasparakis *et al.*, 1996). These defects explain the deregulation of antibody responses to T-dependent antigens seen in LT-α-, LT-β-, TNF- and TNFR60-deficient mice (Banks *et al.*, 1995; Le Hir *et al.*, 1996; Pasparakis *et al.*, 1996) as well as to T-cell independent type 2 antigens observed in TNF–LT-α-deficient mice (Ryffel *et al.*, 1997).

In addition to the involvement of the LT-$\alpha\beta$ system in germinal center formation, several genes (CD40, CD40L, and CD28) have been shown to be involved in germinal center formation (Foy *et al.*, 1994; Kawabe *et al.*, 1994; Xu *et al.*, 1994; Han *et al.*, 1995; Ferguson *et al.*, 1996). Similar to the LT-$\alpha\beta$ system lacking a defined splenic marginal zone, *aly* mice also lack a splenic marginal zone (no antigen reactivity for marginal zone macrophages, fibroblastic dendritic reticular cells or marginal metallophils) (Koike *et al.*, 1996). Clearly, research involving lymph node organogenesis, structure and function will be aided by the generation of these mouse systems.

The CD40 System

CD40 ligand (CD40L) and its receptor, CD40, are also involved in the regulation of the immune system (for review see Banchereau *et al.*, 1994). CD40L is expressed on activated T cells (Graf *et al.*, 1992), human blood dendritic cells (Pinchuk *et al.*, 1996) and human vascular endothelial cells (ECs), smooth muscle cells (SMCs) and macrophages (Mach *et al.*, 1997). Strikingly, the genetic defect of activated T cells obtained from patients with X-linked hyper-IgM syndrome was mapped to Xq26 and confirmed to be due to mutations within the extracellular domain of CD40L (Aruffo *et al.*, 1993; Ramesh *et al.*, 1994). Aberrant expression of CD40L is also implicated in atherogenesis in that ECs, SMCs and macrophages at sites of human atherosclerotic lesions but not normal arterial tissues express CD40L and also coexpress CD40, which may contribute to secretion of proinflammatory cytokines via a CD40L–CD40 paracrine or autocrine pathway (Mach *et al.*, 1997). The receptor for CD40L, CD40, is expressed on B cells, follicular dendritic cells, epithelial cells, hematopoietic progenitor cells and some carcinomas (Banchereau *et al.*, 1994).

The outcome of CD40L-CD40 interaction has been predominantly characterized in B-cell activation where this interaction is required for a response to T-cell dependent antigens. However, recently, CD40 ligation of human keratinocytes was shown to inhibit proliferation and to induce differentiation (Péguet-Navarro *et al.*, 1997). Additionally, CD40L-deficient mice, CD40-deficient mice or mice treated with anti-CD40L antibodies show an absence of germinal center formation and impaired development of B-cell memory to T-cell dependent antigens (Foy *et al.*, 1994; Kawabe *et al.*, 1994; Xu *et al.*, 1994) as well as an enhanced susceptibility to *Leishmania* infection, indicating a role for CD40L–CD40 interaction in cell-mediated immunity as well (Campbell *et al.*, 1996; Kamanaka *et al.*, 1996; Soong *et al.*, 1996).

The Fas System

The Fas (CD95) system has emerged as an important component of the lytic activity of cytotoxic T lymphocytes and NK cells, as a mechanism controlling clonal expansion, and in the phenomenon of immune privilege. Fas contains a death domain that triggers a potent apoptotic response in many different cell types (Itoh *et al.*, 1991; Suda *et al.*, 1993; Tartaglia *et al.*, 1993a) and is expressed on a wide range of human and animal cells. Fas is upregulated on the surface of some cells after incubation with inflammatory cytokines (e.g. interferon-γ) (Yonehara *et al.*, 1989). All cells of T-cell lineage are able to express FasL transiently following activation with antigen or a combination of phorbol ester and ionomycin (Tanaka *et al.*, 1995). Non-lymphoid tissues in the eye, thyroid and testis, considered sites of immune privilege, can express FasL, as do some non-hematopoietic tumor cells (Hahne *et al.*, 1996; O'Connell *et al.*, 1996; Niehans *et al.*, 1997). $CD8^+$ T cells, whose primary function is lysis of virus-infected cells, use both the granule exocytosis pathway, which involves perforin and granzymes, and the Fas pathway to kill targets (Clark *et al.*, 1995). Activated T lymphocytes are particularly sensitive to Fas-induced apoptosis which has led to the idea that the Fas system is used as a self-regulatory mechanism to eliminate chronically activated T cells after an immune response (Brunner *et al.*, 1995). Additionally, the Fas pathway appears to be used by non-lymphoid tissue (e.g. lens epithelial, thyrocytes) to kill tissue-infiltrating leukocytes, preventing their accumulation in sensitive tissue. Such a process may create an immunologically privileged tissue (Cohen and Eisenberg, 1991). Expression of FasL by certain tumor cells, including melanoma, has raised the spector that FasL may function as an anti-immune surveillance mechanism. Thus strong selective pressures are placed on this cytokine system for both normal and aberrant situtations.

The importance of Fas in autoimmune disease is seen in mice that carry mutations in Fas or FasL genes. Mice with lymphoproliferative disease (*lpr*) or generalized lymphoproliferative disorder (*gld*) develop similar pathology with lymphadenopathy and splenomegaly owing to the accumulation of large numbers of non-malignant T cells (TCR-$\alpha\beta^+$, $CD3^+$, but $CD4^-CD8^-$ double negative), in addition to developing multiple autoantibodies. Interestingly, the Fas gene maps to the *lpr* locus (Watanabe-Fukunaga *et al.*, 1992). Analysis of the MRL *lpr*/*lpr* mouse showed that the liver and thymus lack Fas mRNA (Watanabe-Fukunaga *et al.*, 1992) owing to the insertion of a transposable element containing a polyadenylation signal that leads to termination after the second of the nine exons in *fas* (Adachi *et al.*, 1993). *lpr*^{cg} is a point mutation within the DD that probably disrupts the architecture of the Fas cytoplasmic domain, preventing efficient coupling to signaling proteins such as FADD (Chinnaiyan *et al.*, 1996b). Mice homozygous for *gld*, which arose independently from *lpr*, acquired a point mutation in the receptor binding domain of FasL (Hahne *et al.*, 1995). The phenotype of the *lpr* and *gld* mice suggests a role for the Fas system in the downregulation of immune responses via apopotic death. During an immune response both B and T cells become activated and clonally expanded. These activated effector cells need to be eliminated via peripheral clonal deletion, once the pathogen has been cleared. In mice lacking a functional Fas pathway, effector cells cannot be eliminated efficiently, and thus accumulate in the lymph nodes and spleen. Although these cells are double negative, they carry markers characteristic of mature lymphocytes, which differ from immature double-negative cells in the thymus, suggesting that these cells have downregulated CD4 and CD8 (Cohen and Eisenberg, 1991). Additionally, the formation of autoreactive antibodies shows that the Fas pathway is also involved in peripheral tolerance. However, positive and negative selection in the thymus for T-cell repertoire development is apparently normal, a feature that distinguishes Fas from CD30.

The CD30 System

CD30 ligand (CD30L) and its receptor, CD30, are involved in the regulation of the immune system. CD30L is expressed on T cells and monocytes–macrophages (Ellis *et al.*, 1993). CD30 was initially described as a marker for Hodgkin's lymphoma and Reed–Sternberg cells (Schwab *et al.*, 1982), but also is expressed on a subset of activated CD45RO^{+} T cells (Ellis *et al.*, 1993), CD8^{+} T-cell clones derived from human immunodeficiency virus (HIV)-positive individuals (Manetti *et al.*, 1994), EBV^{+} and human T-lymphotropic virus (HTLV)-I/II-transformed T cells, and B-cell lymphoma cell lines (Gruss *et al.*, 1994). CD30 can be cleaved proteolytically to produce a soluble form of the protein (Josimovic-Alasevic *et al.*, 1989). CD30L–CD30 interaction produces several outcomes depending on the cell type: proliferation of T cells after TCR triggering (Smith *et al.*, 1993), cell death in some lymphoma lines (Smith *et al.*, 1993), Fas-independent cell death in T-cell hybridomas (Lee *et al.*, 1996) and induction of HIV expression which is NF-κB dependent (Biswas *et al.*, 1995). Data from CD30-deficient mice indicate a role for CD30 in negative selection as evidenced by reduced susceptibility of thymocytes to CD3-mediated cell death and by an impairment of clonal deletion of H-Y TCR-transgenic and also $\gamma\delta$TCR-transgenic T cells in respective negative-selecting backgrounds (male H-2^{b} mice and H-2$^{b/b}$ mice) (Amakawa *et al.*, 1996). Thus far, CD30 is the second system recognized to provide a signal during development.

The 4-1BB System

4-1BB and its partner 4-1BB ligand represents one of the more distant systems in the TNF superfamily. 4-1BB ligand shares only 14–16% identity with other family members, although the tertiary structure seems to be conserved (Goodwin *et al.*, 1993b). 4-1BB ligand is found on activated macrophages and mature B cells, and can be induced with anti-IgM antibodies on naive B cells (Pollok *et al.*, 1994). Additionally, data obtained using a 4-1BB fusion protein suggest that 4-1BB ligand expressed on the surface of activated B cells may regulate B-cell proliferation in an antigen–MHC-independent fashion (Pollok *et al.*, 1994), an example of countersignaling. This suggests that the engagement of 4-1BB receptor–ligand may induce signals that simultaneously effect both T- and B-cell functions. 4-1BB has an extensive intracellular domain which includes a consensus binding site for $p56^{lck}$, a sarcoma (SRC) family tyrosine kinase (Kim *et al.*, 1993; Schwarz *et al.*, 1995). 4-1BB is expressed on activated T cells of both CD4 and CD8 lineage (Pollok *et al.*, 1993, 1994; Garni Wagner *et al.*, 1996) and functions as an accessory molecule during T-cell activation, enhancing CD3-induced T-cell proliferation (Pollok *et al.*, 1993, 1994). In mice, mRNA for 4-1BB appears within hours of stimulation, whereas protein does not appear on the surface for 2–3 days, but persists for 6 days (Pollok *et al.*, 1993, 1995). Murine 4-1BB expression is restricted to lymphoid tissue, but on human cells 4-1BB expression can also be induced in non-lymphoid cells such as epithelial and hepatoma cells after IL-1β stimulation (Schwarz *et al.*, 1995).

The OX-40 System

OX-40 was first described as an activation marker on $CD4^+$ blasts in the rat and appears to serve as a costimulator for T-cell activation (Paterson *et al.*, 1987). Mouse and human homologs of both the receptor and ligand have been identified (Calderhead *et al.*, 1993; Baum *et al.*, 1994). T cells express both OX-40 (Calderhead *et al.*, 1993) and OX-40 ligand (OX-40L) (Baum *et al.*, 1994), with higher levels of OX-40 on the $CD4^+$ subset (Baum *et al.*, 1994). The ligand is additionally found on activated B cells (Stuber *et al.*, 1995), although there is some controversy whether it is expressed on LPS-actviated B cells (Calderhead *et al.*, 1993; Baum *et al.*, 1994). Interaction of OX-40 on T cells with OX-40L on B cells serves as a costimulatory signal for a T-cell dependent B-cell response. The OX-40 signal results in B-cell proliferation and an increase in immunoglobulin production (Stuber and Strober, 1996). This research has suggested that activation via OX-40 leads to a distinct differentiation pathway resulting in high levels of immunoglobulin-producing cells, but is independent of class switching and memory B-cell differentiation (Stuber and Strober, 1996). OX-40L is also expressed on vascular endothelial cells. There is evidence that OX-40 on T cells can interact with OX-40L on vascular endothelial cells as an additional system mediating T-cell homing, independently but perhaps overlapping with E-selectin-mediated homing (Imura *et al.*, 1996). Mouse T cells costimulated with cells transfected with mouse OX-40L proliferate and secrete high levels of IL-2 and IL-4 (Baum *et al.*, 1994).

The CD27 System

CD27 and its ligand, CD27L (CD70), function as a costimulating system for T-cell activation and also in B-cell activation, NK cell effector activity and early thymic differentation. CD27L induces the proliferation of T cells costimulated with anti-CD3 and enhances the generation of cytolytic T cells (Goodwin *et al.*, 1993a). CD27 is a lymphocyte-specific TNFR family member (Camerini *et al.*, 1991; Loenen *et al.*, 1992) that is expressed on the majority of peripheral blood T cells with increased expression following activation via TCR–CD3. A soluble form is also shed during T-cell activation.

$CD4^+$ cells that have been stimulated persistently by antigen accumulate within the $CD45RA^- CD27^-$ subset (Hintzen *et al.*, 1993). Additionally, anti-CD27 antibodies can block IL-2 production (Sugita *et al.*, 1993). $CD4^+ CD45RA^+ CD45RO^-$ T cells (naive) were found to express CD27, whereas a small fraction of the $CD4^+ CD45RA^- CD45RO^+$ (memory) subset lacks the molecule. The $CD8^+ CD27^-$ subsets were found both in the $CD45RA^+$ and $CD45RA^-$ subpopulations. In addition, an increase of $CD27^-$ T cells has been observed under certain immunopathologic conditions and during aging (Hol *et al.*, 1993; Portegies *et al.*, 1993; van Oers *et al.*, 1993). The restricted upregulation of CD27 on $CD45RA^+$ cells after T-cell activation suggests a role for CD27–CD27L interaction during transition of $CD45RO^-$ to $CD45RA^-$ T cells. CD27L is expressed by B cells and is a marker for mature B cells that have recently been primed by antigen *in vivo* (Lens *et al.*, 1996). In addition, evidence using anti-CD27 suggests that CD27 may contribute to normal murine T-cell development by synergizing with the pre-TCR-mediated signal (Gravestein *et al.*, 1996). CD27 also appears to function in NK cell activity since CD27–CD70 interaction directly enhances NK cell activity in the presence of IL-2 or IL-12 by increasing effector and target conjugate formation (Yang *et al.*, 1996).

Recent Additions to the TNF Superfamily

HVEM/ATAR/TR2

An entry molecule for herpes simplex virus (HSV)-1 and 2, HVEM (Montgomery *et al.*, 1996), was identified using an expression cloning method to select for molecules that promote virus infection of Chinese hamster ovary (CHO) cells normally resistant to HSV infection. Although HSV infects a wide range of cell types, antibodies to HVEM block infection only of activated T cells, raising the possibility that this route of entry is an anti-immune response mechanism of the virus. HVEM was subsequently cloned as another TRAF-associated receptor (ATAR) (Hsu *et al.*, 1997) and TR2 (Kwon *et al.*, 1997) using expressed sequence tag (EST) database strategy and shown to bind TRAFs 2 and 5. HVEM is most closely related to Fas, with three cysteine-rich domains. The natural ligand of HVEM and its function in the immune system are under study.

CAR-1

A second virus entry factor, CAR-1 (Brojatsch *et al.*, 1996), was defined by transfer of infectivity of the subgroup B of the avian leukosis sarcoma virus (ALV). CAR-1 contains two cysteine-rich domains and a DD homology region. It functions as a specific entry factor for viral subgroups B and D and is unrelated to receptors–entry proteins for other retroviruses. A fusion protein made with the viral subunit surface B (SU-B) envelope protein and Fc of IgG induces apoptosis of cells expressing CAR-1.

DR3/WSL-1/TRAMP/APO-3/LARD

Another DD receptor was identified by two methods, as two-hybrid partners for cytoplasmic domain of TNFR60 (WSL-1) (Kitson *et al.*, 1996) and by database scanning for DD-containing proteins (and variously known as Apo-3, DR3, TRAMP or LARD) (Chinnaiyan *et al.*, 1996a; Bodmer *et al.*, 1997). The WSL-1 gene has three alternate spliced forms: a full-length receptor and two shorter forms that theoretically produce soluble molecules of only the ectodomain. These shorter forms contain an additional cysteine repeat, implicating a complicated binding scenario for WSL-1 with its ligand(s). This protein, like other DD molecules, induces apoptosis when overexpressed in human embryonic kidney-293 (HEK-293) cells which is inhibitable by crmA. DR3/WSL-1/TRAMP overexpression also induces NF-κB activity. This receptor does not bind to TRAIL (see below) or other known members of the ligand family, thus another apoptosis-inducing ligand remains to be uncovered.

TRAIL

A new member of the ligand family, TNF-related apoptosis-inducing ligand (TRAIL/Apo-2L) (Wiley *et al.*, 1995; Pitti *et al.*, 1996), was uncovered in the EST database with homology to Fas and TNF. TRAIL induces cell death characteristic of apoptosis, although independent of Fas. Cell death has been demonstrated in transformed cell lines of lymphoid and non-lymphoid origin (non-transformed lines are apparently insensitive); thus the signals required for TRAIL-induced death diverge somewhat from those for FasL. TRAIL mRNA is distributed widely in tissues, but is notably absent in liver, a pattern distinct from that of FasL. TRAIL does not appear to act as a costimulator for T-cell proliferation.

TRAIL Receptors (DR4, DR5, TRID/DcR1)

The newest members of the TNF receptor superfamily, death receptors 4 and 5 (DR4 and DR5) and decoy receptor 1 (DcR1)/TRAIL receptor without an intracellular domain (TRID) (Pan *et al.*, 1997a,b; Sheridan *et al.*, 1997) all bind TRAIL. Two of these, DR4 and DR5, are DD-containing receptors with 55% overall identity and 64% identity within the DD. This is in contrast to only approximately 29% identity with other DD receptors such as DR3 and TNFR1. Transcripts for DR4 and DR5 were detected in most human tissues and transformed cell lines. In contrast, transcripts for TRID/DcR1 were found in normal tissues but at much lower levels in transformed cell lines. TRID/DcR1 is a glycophospholipid cell surface protein and does not contain a DD, rather it appears to act as a decoy receptor for TRAIL. This might explain why normal tissue is resistant to the apoptotic effects of TRAIL. TRID/DcR1 is most related to DR4 and DR5 with 60% and 50% identity respectively.

GITR (Glucocorticoid-Induced Tumor Necrosis Factor Receptor)

Another new member of the TNF receptor family, GITR, was identified using the differential display technique comparing a dexamethasone-treated and an untreated murine T-cell hybridoma (Nocentini *et al.*, 1997). GITR has three extracellular cysteine repeats and an intracellular domain which shows strong similarity to that of 4-1BB and CD27 but no other members of the TNF receptor family. Thus GITR, CD27 and 4-1BB may represent a new subfamily. GITR was shown to be expressed in T cells, thymocytes, splenocytes and lymph nodes; however, expression in other cell types and tissues has not been ruled out. Constitutive expression of GITR in a $CD4^+$ T-cell hybridoma induced resistance to anti-CD3-induced apoptosis, but not apoptosis induced by Fas or ultraviolet light treatment. Speculation from these early results suggests GITR may be involved in lymphocyte protection against activation-induced death. No ligand for GITR has been identified.

Osteoprotegerin (OPG)

OPG is an unusual member of the TNF receptor superfamily in that it appears to occur only in a soluble form (Simonet *et al.*, 1997). It contains four cysteine-rich TNF receptor repeats but no hydrophobic transmembrane spanning region. Overexpression of OPG in mice results in ostropetrosis which correlates with a block in osteoclast differentiation. Osteoclasts resorb bone and are derived from the hematopoietic lineage (monocyte–macrophage). These studies indicate a possible clinical use of OPG for the treatment of osteopenic disorders.

These new members are probably not the last to be uncovered in the genome. The finding of several death and non-death domain-containing receptors speaks to the importance of the diversity of signaling processes that lead to cell death. Perhaps this diversity is necessary to stay ahead of the varied mechanisms employed by viruses to alter immunologically induced cell death.

VIRUS-DERIVED HOMOLOGS AND MODULATORS

Although our knowledge is limited, it seems that many different viruses disproportionately target the TNF superfamily (Table 7). Some of the viral genes are structural or functional homologs of TNF family members, and still others of unknown relationship. The poxvirus family, including the extinct virola (smallpox) contain two types of TNFR80 homolog (Smith *et al.*, 1991, 1996; Upton *et al.*, 1991). The T2 gene of Shope fibroma virus is an important virulence factor for this virus (Schreiber and McFadden, 1994). T2 contains an *N*-terminal domain with homology to TNFR80 extracellular domain and an unrelated *C*-terminal domain. The T2 protein is secreted and neutralizes TNF and LT-α, but blocks apoptosis of rabbit T cells (Macen *et al.*, 1996). A second poxvirus gene product binds TNF but not LT-α (Smith *et al.*, 1996). The poxvirus

Table 7. Virus homologs and modulators of the TNF superfamily.

Virus	Viral gene product	Effect	Reference
Adenovirus	E-1B	Inhibits Fas apoptosis	Hashimoto *et al.* (1991)
	E3-14.7K	Inhibits TNF apoptosis	Horton *et al.* (1991)
	E3-10.4K/14.5K	Downregulates surface Fas and inhibits apoptosis	Gooding *et al.* (1991)
Epstein–Barr virus	LMP-1	Sequesters TRAFs	Devergne *et al.* (1996)
Herpes simplex virus 1 and 2	gD	Entry protein via HVEM	Montgomery *et al.* (1996)
HSV-1	?	Downregulates Fas ligand	Sieg *et al.* (1996)
Equine herpes virus 2	E8	Inhibits Fas and TNF apoptosis	Bertin *et al.* (1997)
Poxvirus	T2	Binds TNF/LT-α; blocks T-cell apoptosis	Macen *et al.* (1996)
	A53R	Binds TNF, not LT-α	Smith *et al.* (1996)
	crmA	Blocks caspases in Fas/TNFR apoptosis	
	MC159	Inhibits Fas and TNF apoptosis	Bertin *et al.* (1997)
Hepatitis C virus	Core protein	Binds LT-βR	Matsumoto *et al.* (1997)
Avian leukosis virus	ALV-B Env SU	Entry factor via CAR-1 (tv-b^{s3}); induces apoptosis	Brojatsch *et al.* (1996)

genome is a veritable minigenomic database for gene products that inhibit cytokines, apoptosis and other immune defense mechanisms. That inactivation of the T2 gene in the rabbit poxvirus attenuates virulence speaks to the importance of T2 protein as an anti-immune defense mechanism. The EBV protein, LMP-1, encoded by the dominant oncogene, is a functional homolog of non-DD receptors. LMP-1, a transmembrane protein with six spanning regions, translocates TRAFs to the inner membrane, analogous to ligand–receptor complex (VanArsdale *et al.*, 1997). Apparently, LMP-1 may bind most of the available TRAF3, making little available for binding to receptors, thus effectively acting as an uncoupler of TRAF3-dependent signaling pathway(s) (Devergne *et al.*, 1996).

In addition to these structural or functional homologs of TNF superfamily members, these viruses also contain proteins that interfere with signal transduction. Adenovirus E3 region contains several genes that block TNF- and Fas-mediated apoptosis (Wold and Gooding, 1991; Gooding, 1992). The hepatitis C virus core protein, among other functions, binds to the cytoplasmic domain of LT-βR, implicating a possible mechanism of how this virus might inhibit immune responsiveness (Matsumoto *et al.*, 1997). Equine herpesvirus and molluscum contagiosum virus (poxvirus) contain proteins with similarity to the death effector domains of caspase 8 and FADD (Bertin *et al.*, 1997). These proteins block Fas and TNFR-mediated apoptosis. Clearly, the functional roles of these viral gene products as virulence factors remains to be established, yet the revelation that so many unrelated viruses target TNF superfamily members suggests their importance in virus–host interactions. One common pathogenetic feature among most of these viruses is their ability to persist in the presence of a functional immune system. Clearly, additional study in this area should reveal new insights into understanding pathogenicity.

ADDENDUM

New members of the TNF-related cytokine-receptor superfamily.

Receptor	Ligand	Function	References
HVEM	LIGHT	Herpes virus entry, costimulation of T cells	Mauri *et al.* (1998)
TACI	?	NFAT activation via CAML	von Bülow and Bram (1997)
TRANCE-R (RANK)	TRANCE (RANKL)	Costimulation of T cells, JNK activation	Wong *et al.* (1997), Anderson *et al.* (1997)
TRAIL-R4	TRAIL	Decoy receptor for TRAIL	Degli-Esposti *et al.* (1997)

LIGHT, Lymphotoxins, exhibits inducible expression, and competes with HSV glycoprotein D for HVEM, a receptor expressed by T lymphocytes; TACI, transmembrane activator and CAML-interactor; TRANCE, TNF-related activation-induced cytokine; RANK, Receptor activation of NF-κB; NFAT, nuclear factor of activated T cells; CAML, calcium-modulator and cyclophilin ligand.

REFERENCES

Adachi, M., Watanabe-Fukunaga, R. and Nagata, S. (1993). Aberrant transcription caused by the insertion of an early transposable element in an intron of the Fas antigen gene of lpr mice. *Proc. Natl Acad. Sci. U.S.A.* **90**, 1756–1760.

Adam-Klages, S., Adam, D., Wiegmann, K., Struve, S., Kolanus, W., Schneider-Mergener, J. and Krönke, M. (1996). FAN, a novel WD-repeat protein, couples the p55 TNF-receptor to neutral sphingomyelinase. *Cell* **86**, 937–947.

Aderka, D., Engelmann, H., Maor, Y., Brakebusch, C. and Wallach, D. (1992). Stabilization of the bioactivity of tumor necrosis factor by its soluble receptors. *J. Exp. Med.* **175**, 323–329.

Aizawa, S., Nakano, H., Ishida, T., Horie, R., Nagai, M., Ito, K., Yagita, H., Okumura, K., Inoue, J. and Watanabe, T. (1997). Tumor necrosis factor receptor-associated factor (TRAF) 5 and TRAF2 are involved in CD30-mediated NFκB activation. *J. Biol. Chem.* **272**, 2042–2045.

Alimzhanov, M.B., Kuprash, D.V., Kosco-Vilbois, M.H., Luz, A., Turetskaya, R.L., Tarakhovsky, A., Rajewsky, K., Nedospasov, S.A. and Pfeffer, K. (1997). Abnormal development of secondary lymphoid tissues in lymphotoxin β-deficient mice. *Proc. Natl Acad. Sci. U.S.A.* **94**, 9302–9307.

Alnemri, E.S., Livingston, D.J., Nicholson, D.W., Salvesen, G., Thornberry, N.A., Wong, W.W. and Yuan, J. (1996). Human ICE/CED-3 protease nomenclature. *Cell* **87**, 171.

Amakawa, R., Hakem, A., Kundig, T.M., Matsuyama, T., Simard, J.J., Timms, E., Wakeham, A., Mittruecker, H.W., Griesser, H., Takimoto, H., Schmits, R., Shahinian, A., Ohashi, P., Penninger, J.M. and Mak, T.W. (1996). Impaired negative selection of T cells in Hodgkin's disease antigen CD30-deficient mice. *Cell* **84**, 551–562.

Anderson, D.M., Maraskovsky, E., Billingsley, W.L., Dougall, W.C., Tometsko, M.E., Roux, E.R., Teepe, M.C., DuBose, R.F., Cosman, D. and Galibert, L. (1997). A homologue of the TNF receptor and its ligand enhance T-cell growth and dendritic-cell function. *Nature* **390**, 175–179.

Androlewicz, M.J., Browning, J.L. and Ware, C.F. (1992). Lymphotoxin is expressed as a heteromeric complex with a distinct 33-kDa glycoprotein on the surface of an activated human T cell hybridoma. *J. Biol. Chem.* **267**, 2542–2547.

Armitage, R., Fanslow, W., Strockbine, L., Sato, T., Clifford, K., Macduff, B., Anderson, D., Gimpel, S., Davis-Smith, T., Maliszewski, C., Clark, E., Smith, C., Grabstein, K., Cosman, D. and Spriggs, M. (1992). Molecular and biological characterization of a murine ligand for CD40. *Nature* **357**, 80–82.

Aruffo, A., Farrington, M., Hollenbaugh, D., Li, X., Milatovich, A., Nonoyama, S., Bajorath, J., Grosmaire, L.S., Stenkamp, R., Neubauer, M., Roberts, R.L., Noelle, R.J., Ledbetter, J.A., Francke, U. and Ochs, H.D. (1993). The CD40 ligand, gp39, is defective in activated T cells from patients with X-linked hyper-IgM syndrome. *Cell* **72**, 291–300.

Banchereau, J., Bazan, J.F., Blanchard, D., Briere, F., Galizzi, J.P., vanKooten, C., Liu, Y.J., Rousset, F. and Saeland, S. (1994). The CD40 antigen and its ligand. *Annu. Rev. Immunol.* **12**, 881–922.

Banks, T.A., Rouse, B.T., Kerley, M.K., Blair, P.J., Godfrey, V.L., Kuklin, N.A., Bouley, D.M., Thomas, J., Kanangat, S. and Mucenski, M.L. (1995). Lymphotoxin-α-deficient mice. Effects on secondary lymphoid organ development and humoral immune responsiveness. *J. Immunol.* **155**, 1685–1693.

Banner, D.W., D'Arcy, A., Janes, W., Gentz, R., Schoenfeld, H.J., Broger, C., Loetscher, H. and Lesslauer, W. (1993). Crystal structure of the soluble human 55 kD TNF receptor-human TNF beta complex: implications for TNF receptor activation. *Cell* **73**, 431–445.

Baum, P.R., Gayle, R.B., Ramsdell, F., Srinivasan, S., Sorensen, R.A., Watson, M.L., Seldin, M.F., Baker, E., Sutherland, G.R. and Clifford, K.N. (1994). Molecular characterization of murine and human OX40/OX40 ligand systems: identification of a human OX40 ligand as the HTLV-1-regulated protein gp34. *EMBO J.* **13**, 3992–4001.

Bazan, J.F. (1990). Structural design and molecular evolution of a cytokine receptor superfamily. *Proc. Natl Acad. Sci. U.S.A.* **87**, 6934–6938.

Beg, A.A. and Baltimore, D. (1996). An essential role for NF-κB in preventing TNF-α-induced cell death. *Science* **274**, 782–784.

Belka, C., Wiegmann, K., Adam, D., Holland, R., Neuloh, M., Herrmann, F., Kronke, M. and Brach, M.A. (1995). Tumor necrosis factor (TNF)-α activates c-raf-1 kinase via the p55 TNF receptor engaging neutral sphingomyelinase. *EMBO J.* **14**, 1156–1165.

Bellgrau, D., Gold, D., Selawry, H., Moore, J., Franzusoff, A. and Duke, R.C. (1995). A role for CD95 ligand in preventing graft rejection. *Nature* **377**, 630–632.

Bertin, J., Armstrong, R.C., Ottilie, S., Martin, D.A., Wang, Y., Banks, S., Wang, G.-H., Senkevich, T.G., Alnemri, E.S., Moss, B., Lenardo, M.J., Tomaselli, K.J. and Cohen, J.I. (1997). Death effector domain-containing herpesvirus and poxvirus proteins inhibit both Fas- and TNFR1-induced apoptosis. *Proc. Natl Acad. Sci. U.S.A.* **94**, 1172–1176.

Beutler, B. (1992). *Tumor Necrosis Factors. The Molecules and Their Emerging Role in Medicine*, Raven Press New York.

Biswas, P., Smith, C.A., Goletti, D., Hardy, E.C., Jackson, R.W. and Fauci, A.S. (1995). Cross-linking of CD30 induces HIV expression in chronically infected T cells. *Immunity* **2**, 587–596.

Black, R.A., Rauch, C.T., Kozlosky, C.J., Peschon, J.J., Slack, J.L., Wolfson, M.F., Castner, B.J., Stocking, K.L., Reddy, P., Srinivasan, S., Nelson, N., Boiani, N., Schooley, K.A., Gerhart, M., Davis, R., Fitzner, J.N., Johnson, R.S., Paxton, R.J., March, C.J. and Cerretti, D.P. (1997). A metalloproteinase disintegrin that releases tumour-necrosis factor-α from cells. *Nature* **385**, 729–733.

Bodmer, J.-L., Burns, K., Schneider, P., Hofmann, K., Steiner, V., Thome, M., Bornand, T., Hahne, M., Schröter, M., Becker, K., Wilson, A., French, L.E., Browning, J.L., MacDonald, H.R. and Tschopp, J. (1997). TRAMP, a novel apoptosis-mediating receptor with sequence homology to tumor necrosis factor receptor 1 and fas (apo-1/CD95). *Immunity* **6**, 79–88.

Boldin, M.P., Mett, I.L., Varfolomeev, E.E., Chumakov, I., Shemer-Avni, Y., Camonis, J.H. and Wallach, D. (1995a). Self-association of the 'death domains' of the p55 tumor necrosis factor (TNF) receptor and Fas/APO1 prompts signaling for TNF and Fas/APO1 effects. *J. Biol. Chem.* **270**, 387–391.

Boldin, M.P., Varfolomeev, E.E., Pancer, Z., Mett, I.L., Camonis, J.H. and Wallach, D. (1995b). A novel protein that interacts with the death domain of Fas/APO1 contains a sequence motif related to the death domain. *J. Biol. Chem.* **270**, 7795–7798.

Boldin, M.P., Goncharov, T.M., Goltsev, Y.V. and Wallach, D. (1996). Involvement of MACH, a novel MORT1/FADD-interacting protease, in Fas/APO-1- and TNF receptor-induced cell death. *Cell* **85**, 803–815.

Bowman, M.R., Crimmins, M.A.V., Yetz-Aldape, J., Kriz, R., Kelleher, K. and Herrmann, S. (1994). The cloning of CD70 and its identification as the ligand for CD27. *J. Immunol.* **152**, 1756–1761.

Brojatsch, J., Naughton, J., Rolls, M.M., Zingler, K. and Young, J.A. (1996). CAR1, a TNFR-related protein, is a cellular receptor for cytopathic avian leukosis-sarcoma viruses and mediates apoptosis. *Cell* **87**, 845–855.

Browning, J.L., Ngam-ek, A., Lawton, P., DeMarinis, J., Tizard, R., Chow, E.P., Hession, C., O'Brine-Greco, B., Foley, S.F. and Ware, C.F. (1993). Lymphotoxin β, a novel member of the TNF family that forms a heteromeric complex with lymphotoxin on the cell surface. *Cell* **72**, 847–856.

Browning, J.L., Dougas, I., Ngam-ek, A., Bourdon, P.R., Ehrenfels, B.N., Miatkowski, K., Zafari, M., Yampaglia, A.M., Lawton, P. and Meier, W. (1995). Characterization of surface lymphotoxin forms. Use of specific monoclonal antibodies and soluble receptors. *J. Immunol.* **154**, 33–46.

Browning, J.L., Miatkowski, K., Griffiths, D.A., Bourdon, P.R., Hession, C., Ambrose, C.M. and Meier, W. (1996a). Preparation and characterization of soluble recombinant heterotrimeric complexes of human lymphotoxins α and β. *J. Biol. Chem.* **271**, 8618–8626.

Browning, J.L., Miatkowski, K., Sizing, I., Griffiths, D.A., Zafari, M., Benjamin, C.D., Meier, W. and Mackay, F. (1996b). Signalling through the lymphotoxin-β receptor induces the death of some adenocarcinoma tumor lines. *J. Exp. Med.* **183**, 867–878.

Brunner, T., Mogil, R.J., LaFace, D., Yoo, N.J., Mahboubi, A., Echeverri, F., Martin, S.J., Force, W.R., Lynch, D.H. and Ware, C.F. (1995). Cell-autonomous Fas (CD95)/Fas–ligand interaction mediates activation-induced apoptosis in T-cell hybridomas. *Nature* **373**, 441–444.

Calderhead, D.M., Buhlmann, J.E., van den Eertwegh, A.J., Claassen, E., Noelle, R.J. and Fell, H.P. (1993). Cloning of mouse Ox40: a T cell activation marker that may mediate T–B cell interactions. *J. Immunol.* **151**, 5261–5271.

Camerini, D., Walz, G., Loenen, W.A., Borst, J. and Seed, B. (1991). The T cell activation antigen CD27 is a member of the nerve growth factor/tumor necrosis factor receptor gene family. *J. Immunol.* **147**, 3165–3169.

Campbell, K.A., Ovendale, P.J., Kennedy, M.K., Fanslow, W.C., Reed, S.G. and Maliszewski, C.R. (1996). CD40 ligand is required for protective cell-mediated immunity to *Leishmania major*. *Immunity* **4**, 283–289.

Cao, Z., Xiong, J., Takeuchi, M., Kurama, T. and Goeddel, D.V. (1996). TRAF6 is a signal transducer for interleukin-1. *Nature* **383**, 443–446.

Chai, H., Vasudevan, S.G., Porter, A.G., Chua, K.L., Oh, S. and Yap, M. (1993). Glycosylation and high-level secretion of human tumour necrosis factor-β in recombinant baculovirus-infected insect cells. *Biotechnol. Appl. Biochem.* **18**, 259–273.

Cheng, G. and Baltimore, D. (1996). TANK, a co-inducer with TRAF2 of TNF- and CD40L-mediated NF-κB activation. *Genes Dev.* **10**, 963–973.

Cheng, G., Cleary, A.M., Ye, Z.S., Hong, D.I., Lederman, S. and Baltimore, D. (1995). Involvement of CRAF1, a relative of TRAF, in CD40 signaling. *Science* **267**, 1494–1498.

Chicheportiche, Y., Bourdon, P.R., Scott, H., Hessio, C., Garcia, I. and Browning, J.L. (1997). TWEAK, a new secreted ligand in the tumor necrosis factor family that weakly induces apoptosis. *J. Biol. Chem.* **272**, 32401–32410.

Chinnaiyan, A.M., O'Rourke, K., Tewari, M. and Dixit, V.M. (1995). FADD, a novel death domain-containing protein, interacts with the death domain of Fas and initiates apoptosis. *Cell* **81**, 505–512.

Chinnaiyan, A.M., O'Rourke, K., Yu, G.L., Lyons, R.H., Garg, M., Duan, D.R., Xing, L., Gentz, R., Ni, J. and Dixit, V.M. (1996a). Signal transduction by DR3, a death domain-containing receptor related to TNFR-1 and CD95. *Science* **274**, 990–992.

Chinnaiyan, A.M., Tepper, C.G., Seldin, M.F., O'Rourke, K., Kischkel, F.C., Hellbardt, S., Krammer, P.H., Peter, M.E. and Dixit, V.M. (1996b). FADD/MORT1 is a common mediator of CD95 (Fas/APO-1) and tumor necrosis factor receptor-induced apoptosis. *J. Biol. Chem.* **271**, 4961–4965.

Clark, W.R., Walsh, C.M., Glass, A.A., Hayashi, F., Matloubian, M. and Ahmed, R. (1995). Molecular pathways of CTL-mediated cytotoxicity. *Immunol. Rev.* **146**, 33–44.

Clem, R.J. and Miller, L.K. (1994). Control of programmed cell death by the baculovirus genes *p35* and *iap*. *Mol. Cell Biol.* **14**, 5212–5222.

Cleveland, J.L. and Ihle, J.N. (1995). Contenders in FasL/TNF death signaling. *Cell* **81**, 479–482.

Cohen, P.L. and Eisenberg, R.A. (1991). *Lpr* and *gld*: single gene models of systemic autoimmunity and lymphoproliferative disease. *Annu. Rev. Immunol.* **9**, 243–269.

Cosman, D., Lyman, S.D., Idzerda, R.L., Beckmann, M., Park, L,S., Goodwin, R.G. and March, C.J. (1990). A new cytokine receptor superfamily. *Trends Biochem. Sci.* **15**, 265–270.

Crowe, P.D., VanArsdale, T.L., Goodwin, R.G. and Ware, C.F. (1993). Specific induction of 80-kDa tumor necrosis factor receptor shedding in T lymphocytes involves the cytoplasmic domain and phosphorylation. *J. Immunol.* **151**, 6882–6890.

Crowe, P.D., VanArsdale, T.L., Walter, B.N., Dahms, K.M. and Ware, C.F. (1994). Production of lymphotoxin (LT α) and a soluble dimeric form of its receptor using the baculovirus expression system. *J. Immunol. Methods* **168**, 79–89.

Crowe, P.D., Walter, B.N., Mohler, K.M., Otten-Evans, C., Black, R.A. and Ware, C.F. (1995). A metalloprotease inhibitor blocks shedding of the 80-kD TNF receptor and TNF processing in T lymphocytes. *J. Exp. Med.* **181**, 1205–1210.

De Togni, P., Goellner, J., Ruddle, N.H., Streeter, P.R., Fick, A., Mariathasan, S., Smith, S.C., Carlson, R., Shornick, L.P., Strauss-Schoenberger, J., Russell, J.H., Karr, R. and Chaplin, D.D. (1994). Abnormal development of peripheral lymphoid organs in mice deficient in lymphotoxin. *Science* **264**, 703–706.

Degli-Esposti, M.A., Davis-Smith, T., Din, W.S., Smolak, P., Goodwin, R.G. and Smith, C.A. (1997a). Activation of the lymphotoxin β receptor by cross-linking induces chemokine production and growth arrest in A375 melanoma cells. *J. Immunol.* **158**, 1756–1762.

Degli-Esposti, M.A., Dougall, W.C., Smolak, P.J., Waugh, J.Y., Smith, C.A. and Goodwin, R.G. (1997b). The novel receptor TRAIL-R4 induces NF-κB and protects against TRAIL-mediated apoptosis, yet retains an incomplete death domain. *Immunity* **7**, 813–820.

Devergne, O., Hatzivassiliou, E., Izumi, K.M., Kaye, K.M., Kleijnen, M.F., Kieff, E. and Mosialos, G. (1996). Association of TRAF1, TRAF2, and TRAF3 with an Epstein–Barr virus LMP1 domain important for B-lymphocyte transformation: role in NF-κB activation. *Mol. Cell. Biol.* **16**, 7098–7108.

Deveraux, Q.L., Takahashi, R., Salvesen, G.S. and Reed, J.C. (1997). X-linked IAP is a direct inhibitor of cell-death proteases. *Nature* **388**, 300–304.

Dhein, J., Daniel, P.T., Trauth, B.C., Oehm, A., Moller, P. and Krammer, P.H. (1992). Induction of apoptosis by monoclonal antibody anti-APO-1 class switch variants is dependent on cross-linking of APO-1 cell surface antigens. *J. Immunol.* **149**, 3166–3173.

Duan, H. and Dixit, V.M. (1997). RAIDD is a new 'death' adaptor molecule. *Nature* **385**, 86–89.

Eason, J.D., Wee, S.L., Kawai, T., Hong, H.Z., Powelson, J., Widmer, M. and Cosimi, A.B. (1995). Recombinant human dimeric tumor necrosis factor receptor (TNFR:Fc) as an immunosuppressive agent in renal allograft recipients. *Transplant. Proc.* **27**, 554.

Eck, M.J., Ultsch, M., Rinderknecht, E., de Vos, A.M. and Sprang, S.R. (1992). The structure of human lymphotoxin (tumor necrosis factor-β) at 1.9-A resolution. *J. Biol. Chem.* **267**, 2119–2122.

Eliopoulos, A.G., Dawson, C.W., Mosialos, G., Floettmann, J.E., Rowe, M., Armitage, R.J., Dawson, J., Zapata, J.M., Kerr, D.J., Wakelam, M.J., Reed, J.C., Kieff, E. and Young, L.S. (1996). CD40-induced growth inhibition in epithelial cells is mimicked by Epstein–Barr virus-encoded LMP1: involvement of TRAF3 as a common mediator. *Oncogene* **13**, 2243–2254.

Ellis, T.M., Simms, P.E., Slivnick, D.J., Jäck, H.-M. and Fisher, R.I. (1993). CD30 is a signal-transducing molecule that defines a subset of human activated CD45RO$^+$ T cells. *J. Immunol.* **151**, 2380–2389.

Engelmann, H., Holtmann, H., Brakebusch, C., Avni, Y.S., Sarov, I., Nophar, Y., Hadas, E., Leitner, O. and

Wallach, D. (1990). Antibodies to a soluble form of a tumor necrosis factor (TNF) receptor have TNF-like activity. *J. Biol. Chem.* **265**, 14497–14504.

Ettinger, R., Browning, J.L., Michie, S.A., van Ewijk, W. and McDevitt, H.O. (1996). Disrupted splenic architecture, but normal lymphnode development in mice expressing a soluble lymphotoxin-β receptor-IgG_1 chimeric fusion protein. *Proc. Natl Acad. Sci. U.S.A.* **93**, 13102–13107.

Ferguson, S.E., Han, S., Kelsoe, G. and Thompson, C.B. (1996). CD28 is required for germinal center formation. *J. Immunol.* **156**, 4576–4581.

Foy, T.M., Laman, J.D., Ledbetter, J.A., Aruffo, A., Claassen, E. and Noelle, R.J. (1994). gp39–CD40 interactions are essential for germinal center formation and the development of B cell memory. *J. Exp. Med.* **180**, 157–163.

Förster, R., Mattis, A.E., Kremmer, E., Wolf, E., Brem, G. and Lipp, M. (1996). A putative chemokine receptor, BLR1, directs B cell migration to defined lymphoid organs and specific anatomic compartments of the spleen. *Cell* **87**, 1037–1047.

Fu, Y.-X., Huang, G., Matsumoto, M., Molina, H. and Chaplin, D. (1997). Independent signals regulate development of primary and secondary follicle structure in spleen and mesenteric lymph node. *Proc. Natl Acad. Sci. U.S.A.* **94**, 5739–5743.

Garni Wagner, B.A., Lee, Z.H., Kim, Y.J., Wilde, C., Kang, C.Y. and Kwon, B.S. (1996). 4-1BB is expressed on CD45RAhiROhi transitional T cell in humans. *Cell. Immunol.* **169**, 91–98.

Gearing, A.J., Beckett, P., Christodoulou, M., Churchill, M., Clements, J., Davidson, A.H., Drummond, A.H., Galloway, W.A., Gilbert, R. and Gordon, J.L. (1994). Processing of tumour necrosis factor-α precursor by metalloproteinases. *Nature* **370**, 555–557.

Gedrich, R.W., Gilfillan, M.C., Duckett, C.S., Van Dongen, J.L. and Thompson, C.B. (1996). CD30 contains two binding sites with different specificities for members of the tumor necrosis factor receptor-associated factor family of signal transducing proteins. *J. Biol. Chem.* **271**, 12852–12858.

Gehr, G., Gentz, R., Brockhaus, M., Loetscher, H. and Lesslauer, W. (1992). Both tumor necrosis factor receptor types mediate proliferative signals in human mononuclear cell activation. *J. Immunol.* **149**, 911–917.

Georgopoulos, K., Bigby, M., Wang, J., Molnar, A., Wu, P., Winandy, S. and Sharpe, A. (1994). The Ikaros gene is required for the development of all lymphoid lineages. *Cell* **79**, 143–156.

Giordano, C., Stassi, G., De Maria, R., Todaro, M., Richiusa, P., Papoff, G., Ruberti, G., Bagnasco, M., Testi, R. and Galluzzo, A. (1997). Potential involvement of Fas and its ligand in the pathogenesis of Hashimoto's thyroiditis. *Science* **275**, 960–963.

Goldfeld, A.E., Strominger, J.L. and Doyle, C. (1991). Human tumor necrosis factor α gene regulation in phorbol ester stimulated T and B cell lines. *J. Exp. Med.* **174**, 73–81.

Gooding, L.R. (1992). Virus proteins that counteract host immune defenses. *Cell* **71**, 5–7.

Gooding, L., Ranheim, T., Tollefson, A., Aquino, L., Duerksen-Hughes, P., Horton, T. and Wold, W.S. (1991). The 10,400- and 14,500-dalton proteins encoded by region E3 of adenovirus function together to protect many but not all mouse cell lines against lysis by tumor necrosis factor. *J. Virol.* **65**, 4114–4123.

Goodwin, R.G., Alderson, M.R., Smith, C.A., Armitage, R.J., VandenBos, T., Jerzy, R., Tough, T.W., Schoenborn, M.A., Davis-Smith, T. and Hennen, K. (1993a). Molecular and biological characterization of a ligand for CD27 defines a new family of cytokines with homology to tumor necrosis factor. *Cell* **73**, 447–456.

Goodwin, R.G., Din, W.S., Davis-Smith, T. Anderson, D.M., Gimpel, S.D., Sato, T.A., Maliszewski, C.R., Brannan, C.I., Copeland, N.G. and Jenkins, N.A. (1993b). Molecular cloning of a ligand for the inducible T cell gene 4-1BB: a member of an emerging family of cytokines with homology to tumor necrosis factor. *Eur. J. Immunol.* **23**, 2631–2641.

Graf, D., Korthauer, U., Mages, H.W., Senger, G. and Kroczek, R.A. (1992). Cloning of TRAP, a ligand for CD40 on human T cells. *Eur. J. Immunol.* **22**, 3191–3194.

Gravestein, L.A., van Ewijk, W., Ossendorp, F. and Borst, J. (1996). CD27 cooperates with the pre-T cell receptor in the regulation of murine T cell development. *J. Exp. Med.* **184**, 675–685.

Gray, P., Aggarwal, B., Benton, C., Bringman, T., Henzel, W., Jarrett, J., Leung, D., Moffat, B., Ng, P., Svedersky, L., Palladino, M. and Nedwin, G. (1984). Cloning and expression of the cDNA for human lyphotoxin: a lymphokine with tumor necrosis activity. *Nature* **312**, 721–724.

Grell, M., Scheurich, P., Meager, A. and Pfizenmaier, K. (1993). TR60 and TR80 tumor necrosis factor (TNF)-receptors can independently mediate cytolysis. *Lymphokine Cytokine Res.* **12**, 143–148.

Grell, M., Douni, E., Wajant, H., Lohden, M., Clauss, M., Maxeiner, B., Georgopoulos, S., Lesslauer, W., Kollias, G. and Pfizenmaier, K. (1995). The transmembrane form of tumor necrosis factor is the prime activating ligand of the 80 kDa tumor necrosis factor receptor. *Cell* **83**, 793–802.

Griffith, T.S., Brunner, T., Fletcher, S.M., Green, D.R. and Ferguson, T.A. (1995). Fas ligand-induced apoptosis as a mechanism of immune privilege. *Science* **270**, 1189–1192.

Gruss, H.-J., DaSilva, N., Hu, Z.-B., Uphoff, C.C., Goodwin, R.G. and Drexler, H.G. (1994). Expression and regulation of CD30 ligand and CD30 in human leukemia–lymphoma cell lines. *Blood* **8**, 2083–2094.

Hahne, M., Peitsch, M.C., Irmler, M., Schroter, M., Lowin, B., Rousseau, M., Bron, C., Renno, T., French, L. and Tschopp, J. (1995). Characterization of the non-functional Fas ligand of gld mice. *Int. Immunol.* **7**, 1381–1386.

Hahne, M., Rimoldi, D., Schroter, M., Romero, P., Schreier, M., French, L.E., Schneider, P., Bornand, T., Fontana, A., Lienard, D., Cerottini, J. and Tschopp, J. (1996). Melanoma cell expression of Fas(Apo-1/CD95) ligand: implications for tumor immune escape. *Science* **274**, 1363–1366.

Han, S., Hathcock, K., Zheng, B., Kepler, T.B., Hodes, R. and Kelsoe, G. (1995). Cellular interaction in germinal centers: roles of CD40 ligand and B7–2 in established germinal centers. *J. Immunol.* **155**, 556–567.

Hashimoto, S., Ishii, A. and Yonehara, S. (1991). The *E1b* oncogene of adenovirus confers cellular resistance to cytotoxicity of tumor necrosis factor and monoclonal anti-Fas antibody. *Int. Immunol.* **3**, 343–351.

Hay, B.A., Wassarman, D.A. and Rubin, G.M. (1995). *Drosophila* homologs of baculovirus inhibitor of apoptosis proteins function to block cell death. *Cell* **83**, 1253–1262.

Hintzen, R.Q., de Jong, R., Lens, S.M., Brouwer, M., Baars, P. and van Lier, R.A. (1993). Regulation of CD27 expression on subsets of mature T-lymphocytes. *J. Immunol.* **151**, 2426–2435.

Hol, B.E., Hintzen, R.Q., van Lier, R.A., Alberts, C., Out, T.A. and Jansen, H.M. (1993). Soluble and cellular markers of T cell activation in patients with pulmonary sarcoidosis. *Am. Rev. Respir. Dis.* **148**, 643–649.

Horton, T.M., Ranheim, T.S., Aquino, L., Kusher, D.I., Saha, S.K., Ware, C.F., Wold, W.S.M. and Gooding, L.R. (1991). Adenovirus E3 14.7K protein functions in the absence of other adenovirus proteins to protect transfected cells from tumor necrosis factor cytolysis. *J. Virol.* **65**, 2629–2639.

Hsu, H., Xiong, J. and Goeddel, D.V. (1995). The TNF receptor 1-associated protein TRADD signals cell death and NF-κB activation. *Cell* **81**, 495–504.

Hsu, H., Huang, J., Shu, H.B., Baichwal, V. and Goeddel, D.V. (1996a). TNF-dependent recruitment of the protein kinase RIP to the TNF receptor-1 signaling complex. *Immunity* **4**, 387–396.

Hsu, H., Shu, H.B., Pan, M.G. and Goeddel, D.V. (1996b). TRADD–TRAF2 and TRADD–FADD interactions define two distinct TNF receptor 1 signal transduction pathways. *Cell* **84**, 299–308.

Hsu, H., Solovyev, I., Colobero, A., Elliott, R., Kelley, M. and Boyle, W.J. (1997). ATAR, a novel tumor necrosis factor receptor family member, signals through TRAF2 and TRAF5. *J. Biol. Chem.* **272**, 13471–13474.

Hu, H.M., O'Rourke, K., Boguski, M.S. and Dixit, V.M. (1994). A novel RING finger protein interacts with the cytoplasmic domain of CD40. *J. Biol. Chem.* **269**, 30069–30072.

Huang, B., Eberstadt, M., Olejniczak, E.T., Meadows, R.P. and Fesik, S.W. (1996). NMR structure and mutagenesis of the Fas (APO-1/CD95) death domain. *Nature* **384**, 638–641.

Imura, A., Hori, T., Imada, K., Ishikawa, T., Tanaka, Y., Maeda, M., Imamura, S. and Uchiyama, T. (1996). The human OX40/gp34 system directly mediates adhesion of activated T cells to vascular endothelial cells. *J. Exp. Med.* **183**, 2185–2195.

Ishida, T., Mizushima, S., Azuma, S., Kobayashi, N., Tojo, T., Suzuki, K., Aizawa, S., Watanabe, T., Mosialos, G., Kieff, E., Yamamoto, T. and Inoue, J. (1996a). Identification of TRAF6, a novel tumor necrosis factor receptor-associated factor protein that mediates signaling from an amino-terminal domain of the CD40 cytoplasmic region. *J. Biol. Chem.* **271**, 28745–28748.

Ishida, T., Tojo, T., Aoki, T., Kobayashi, N., Ohishi, T., Watanabe, T., Yamamoto, T. and Inoue, J.-I. (1996b). TRAF5, a novel tumor necrosis factor receptor-associated factor family protein, mediates CD40 signaling. *Proc. Natl Acad. Sci. U.S.A.* **93**, 9437–9442.

Itoh, N. and Nagata, S. (1993). A novel protein domain required for apoptosis. Mutational analysis of human Fas antigen. *J. Biol. Chem.* **268**, 10932–10937.

Itoh, N., Yonehara, S., Ishii, A., Yonehara, M., Mizushima, S., Sameshima, M., Hase, A., Seto, Y. and Nagata, S. (1991). The polypeptide encoded by the cDNA for human cell surface antigen Fas can mediate apoptosis. *Cell* **66**, 233–243.

Jacobson, M.D., Weil, M. and Raff, M.C. (1997). Programmed cell death in animal development. *Cell* **88**, 347–354.

Jones, E., Stuart, D. and Walker, N.P. (1989). Structure of tumor necrosis factor. *Nature* **338**, 225–228.

Josimovic-Alasevic, O., Durkop, H., Schwarting, R., Backe, E., Stein, H. and Diamantstein, T. (1989). Ki-1 (CD30) antigen is released by Ki-1-positive tumor cells *in vitro* and *in vivo*. I. Partial characterization of soluble Ki-1 antigen and detection of the antigen in cell culture supernatants and in serum by an enzyme-linked immunosorbent assay. *Eur. J. Immunol.* **19**, 157–162.

Kamanaka, M., Yu, P., Yasui, T., Yoshida, K., Kawabe, T., Horii, T., Kishimoto, T. and Kikutani, H. (1996). Protective role of CD40 in *Leishmania major* infection at two distinct phases of cell-mediated immunity. *Immunity* **4**, 275–281.

Karpusas, M., Hsu, Y.M., Wang, J.H., Thompson, J., Lederman, S., Chess, L. and Thomas, D. (1995). A crystal structure of an extracellular fragment of human CD40 ligand. *Structure* **3**, 1031–1039.

Kawabe, T., Naka, T., Yoshida, K., Tanaka, T., Fujiwara, H., Suematsu, S., Yoshida, N., Kishimoto, T. and Kikutani, H. (1994). The immune responses in CD40-deficient mice: impaired immunoglobulin class switching and germinal center formation. *Immunity* **1**, 167–178.

Kaye, K.M., Devergne, O., Harada, J.N., Izumi, K.M., Yalamanchili, R., Kieff, E. and Mosialos, G. (1996). Tumor necrosis factor receptor associated factor 2 is a mediator of NF-κB activation by latent infection membrane protein 1, the Epstein–Barr virus transforming protein. *Proc. Natl Acad. Sci. U.S.A.* **93**, 11 085–11 090.

Kim, Y.J., Pollok, K.E., Zhou, Z., Shaw, A., Bohlen, J.B., Fraser, M. and Kwon, B.S. (1993). Novel T cell antigen 4-1BB associates with the protein tyrosine kinase p56lck1. *J. Immunol.* **151**, 1255–1262.

Kitson, J., Raven, T., Jiang, Y.P., Goeddel, D.V., Giles, K.M., Pun, K.T., Grinham, C.J., Brown, R. and Farrow, S.N. (1996). A death-domain-containing receptor that mediates apoptosis. *Nature* **384**, 372–375.

Kluck, R.M., Bossy-Wetzel, E., Green, D.R. and Newmeyer, D. (1997). The release of cytochrome *c* from mitochondria: a primary site for Bcl-2 regulation of apoptosis. *Science* **275**, 1132–1136.

Kohno, T., Brewer, M.T., Baker, S.L., Schwartz, P.E., King, M.W., Hale, K.K., Squires, C.H., Thompson, R.C. and Vannice, J.L. (1990). A second tumor necrosis factor receptor gene product can shed a naturally occurring tumor necrosis factor inhibitor. *Proc. Natl Acad. Sci. U.S.A.* **87**, 8331–8335.

Koike, R., Nishimura, T., Yasumizu, R., Tanaka, H., Hataba, Y., Watanabe, T., Miyawaki, S. and Miyasaka, M. (1996). The splenic marginal zone is absent in alymphoplastic *aly* mutant mice. *Eur. J. Immunol.* **26**, 669–675.

Koni, P.A., Sacca, R., Lawton, P., Browning, J.L., Ruddle, N.H. and Flavell, R.A. (1997). Distinct roles in lymphoid organogenesis for lymphotoxins α and β revealed in lymphotoxin β-deficient mice. *Immunity* **6**, 491–500.

Kratz, A., Campos-Neto, A., Hanson, M.S. and Ruddle, N.H. (1996). Chronic inflammation caused by lymphotoxin is lymphoid neogenesis. *J. Exp. Med.* **183**, 1461–1472.

Kwon, B.S., Tan, K.B., Ni, J., Oh-OK-Kwi, Lee, Z.H., Kim, K.K., Kim, Y.-J., Wang, S., Gentz, R., Yu, G.-L., Harrop, J., Lyn, S.D., Silverman, C., Porter, T.G., Truneh, A. and Young, P.R. (1997). A newly identified member of the tumor necrosis factor receptor superfamily with a wide tissue distribution and involvement in lymphocyte activation. *J. Biol. Chem.* **272**, 14 272–14 276.

Lawton, P., Nelson, J., Tizard, R. and Browning, J.L. (1995). Characterization of the mouse lymphotoxin-β gene. *J. Immunol.* **154**, 239–246.

Le Hir, M., Bluethmann, H., Kosco-Vilbois, M.H., Müller, M., di Padova, F., Moore, M., Ryffel, B. and Eugster, H.-P. (1996). Differentiation of follicular dendritic cells and full antibody responses require tumor necrosis factor receptor-1 signaling. *J. Exp. Med.* **183**, 2367–2372.

Lee, S.Y., Park, C.G. and Choi, Y. (1996). T cell receptor-dependent cell death of T cell hybridomas mediated by the CD30 cytoplasmic domain in association with tumor necrosis factor receptor-associated factors. *J. Exp. Med.* **183**, 669–674.

Lens, S.M., de Jong, R., Hooibrink, B., Koopman, G., Pals, S.T., van Oers, M.H. and van Lier, R.A. (1996). Phenotype and function of human B cells expressing CD70 (CD27 ligand). *Eur. J. Immunol.* **26**, 2964–2971.

Liu, Z.G., Hsu, H.L., Goeddel, D.V. and Karin, M. (1996). Dissection of TNF receptor 1 effector functions: JNK activation is not linked to apoptosis while NF-κB activation prevents cell death. *Cell* **87**, 565–576.

Loenen, W.A., de Vries, E., Gravestein, L.A., Hintzen, R.Q., van Lier, R.A. and Borst, J. (1992). The CD27 membrane receptor, a lymphocyte-specific member of the nerve growth factor receptor family, gives rise to a soluble form by protein processing that does not involve receptor endocytosis. *Eur. J. Immunol.* **22**, 447–455.

Loetscher, H., Pan, Y., Lahm, H., Gentz, R., Brockhaus, M., Tabuchi, H. and Lesslauer, W. (1990). Molecular cloning and expression of the human 55 kD tumor necrosis factor receptor. *Cell* **61**, 351–359.

Loetscher, H., Gentz, R., Zulauf, M., Lustig, A., Tabuchi, H., Schlaeger, E.J., Brockhaus, M., Gallati, H., Manneberg, M. and Lesslauer, W. (1991). Recombinant 55-kDa tumor necrosis factor (TNF) receptor. Stoichiometry of binding to TNF α and TNF β and inhibition of TNF activity. *J. Biol. Chem.* **266**, 18 324–18 329.

Macen, J.L., Graham, K.A., Lee, S.F., Schreiber, M., Boshkov, L.K. and McFadden, G. (1996). Expression of the myxoma virus tumor necrosis factor receptor homologue and M11L genes is required to prevent virus-induced apoptosis in infected rabbit T lymphocytes. *Virology* **218**, 232–237.

Mach, F., Schönbeck, U., Sukhova, G.K., Bourcier, T., Bonnefoy, J.-Y., Pober, J.S. and Libby, P. (1997). Functional CD40 ligand is expressed on human vascular endothelial cells, smooth muscle cells, and macrophages: implications for CD40–CD40 ligand signaling in atherosclerosis. *Proc. Natl Acad. Sci. U.S.A.* **94**, 1931–1936.

Malinin, N.L., Boldin, M.P., Kovalenko, A.V. and Wallach, D. (1997). MAP3K-related kinase involved in NF-κB induction by TNF, CD95, and IL-1. *Nature* **385**, 540–544.

Mallett, S. and Barclay, A.N. (1991). A new superfamily of cell surface proteins related to the nerve growth factor receptor. *Immunol. Today* **12**, 220–223.

Manetti, R., Annunziato, F., Biagiotti, R., Giudizi, M.G., Piccinni, M.-P., Giannarini, L., Sampognaro, S., Parronchi, P., Vinante, F., Pizzolo, G., Maggi, E. and Romagnani, S. (1994). CD30 expression by $CD8^+$ T cells producing type 2 helper cytokines. Evidence for large numbers of $CD8^+$ $CD30^+$ T cell clones in human immunodeficiency virus infection. *J. Exp. Med.* **180**, 2407–2411.

Mariathasan, S., Matsumoto, M., Baranyay, F., Nahm, M.H., Kanagawa, O. and Chaplin, D.D. (1995). Absence of lymph nodes in lymphotoxin-α (LT-α)-deficient mice is due to abnormal organ development, not defective lymphocyte migration. *J. Inflamm.* **45**, 72–78.

Marsters, S.A., Frutkin, A.D., Simpson, N.J., Fendly, B.M. and Ashkenazi, A. (1992). Identification of cysteine-rich domains of the type 1 tumor necrosis factor receptor involved in ligand binding. *J. Biol. Chem.* **267**, 5747–5750.

Marsters, S.A., Sheridan, J.P., Donahue, C.J., Pitti, R.M., Gray, C.L., Goddard, A.D., Bauer, K.D. and Ashkenazi, A. (1996). Apo-3, a new member of the tumor necrosis factor receptor family, contains a death domain and activates apoptosis and NF-κB. *Curr. Biol.* **6**, 1669–1676.

Martin, S.J. and Green, D.R. (1995). Protease activation during apoptosis: death by a thousand cuts? *Cell* **82**, 349–352.

Matsumoto, M., Lo, S.F., Carruthers, C.J.L., Min, J., Mariathasan, S., Huang, G., Plas, D.R., Martin, S.M., Geha, R.S., Nahm, M.H. and Chaplin, D.D. (1996a). Affinity maturation without germinal centers in lymphotoxin-α-deficient mice. *Nature* **382**, 462–466.

Matsumoto, M., Mariathasan, S., Nahm, M.H., Baranyay, F., Preschon, J.J. and Chaplin, D.D. (1996b). Role of lymphotoxin and the type I TNF receptor in the formation of germinal centers. *Science* **271**, 1289–1291.

Matsumoto, M., Hsieh, T.-Y., Zhu, N., VanArsdale, T., Hwang, S.B., Jeng, K.-S., Gorbalenya, A.E., Lo, S.-Y., Ou, J.-H., Ware, C.F. and Lai, M.M.C. (1997). Hepatitis C virus core protein interacts with the cytoplasmic tail of lymphotoxin-β receptor. *J. Virol.* **71**, 1301–1309.

Mauri, D.N., Ebner, R., Montgomery, R.I., Kochel, K.D., Cheung, T.C., Yu, G.-L., Ruben, S., Murphy, M., Eisenbery, R.J., Cohen, G.H., Spear, P.G. and Ware, C.F. (1998). LIGHT, a new member of the TNF superfamily and lymphotoxin α are ligands for herpesvirus entry mediator. *Immunity* **8**, 21–30.

McDonald, N.Q. and Hendrickson, W.A. (1993). A structural superfamily of growth factors containing a cystine knot motif. *Cell* **73**, 421–424.

McGeehan, G.M., Becherer, J.D., Bast, R.C., Jr., Boyer, C.M., Champion, B., Connolly, K.M., Conway, J.G., Furdon, P., Karp, S. and Kidao, S. (1994). Regulation of tumour necrosis factor-α processing by a metalloproteinase inhibitor. *Nature* **370**, 558–561.

Miyawaki, S., Nakamura, Y., Suzuka, H., Koba, M., Yasumizu, R., Ikehara, S. and Shibata, Y. (1994). A new mutation, *aly*, that induces a generalized lack of lymph nodes accompanied by immunodeficiency in mice. *Eur. J. Immunol.* **24**, 429–434.

Mohler, K.M., Sleath, P.R., Fitzner, J.N., Cerretti, D.P., Alderson, M., Kerwar, S.S., Torrance, D.S., Otten-Evans, C., Greenstreet, T. and Weerawarna, K. (1994). Protection against a lethal dose of endotoxin by an inhibitor of tumour necrosis factor processing. *Nature* **370**, 218–220.

Montgomery, R.I., Warner, M.S., Lum, B. and Spear, P.G. (1996). Herpes simplex virus 1 entry into cells mediated by a novel member of the TNF/NGF receptor family. *Cell* **87**, 427–436.

Mosialos, G., Birkenbach, M., Yalamanchili, R., VanArsdale, T., Ware, C. and Kieff, E. (1995). The Epstein–Barr virus transforming protein LMP1 engages signaling proteins for the tumor necrosis factor receptor family. *Cell* **80**, 389–399.

Moss, M.L., Catherine Jin, S.-L., Milla, M.E., Burkhart, W., Carter, H.L., Chen, W.-J., Clay, W.C., Didsbury, J.R., Hassler, D., Hoffman, C.R., Kost, T.A., Lambert, M.H., Leesnitzer, M.A., McCauley, P., McGeehan, G., Mitchell, J., Moyer, M., Pahel, G., Rocque, W., Overton, L.K., Schoenen, F., Seaton, T., Su, J.-L., Warner, J., Willard, D. and Becherer, J.D. (1997). Cloning of a disintegrin metalloproteinase that processes precursor tumour-necrosis factor-α. *Nature* **385**, 733–736.

Muller, U., Jongeneel, C.V., Nedospasov, S.A., Lindahl, K.F. and Steinmetz, M. (1987). Tumour necrosis factor and lymphotoxin genes map close to H-2D in the mouse major histocompatibility complex. *Nature* **325**, 265–267.

Muzio, M., Chinnaiyan, A.M., Kischkel, F.C., O'Rourke, K., Shevchenko, A., Ni, J., Scaffidi, C., Bretz, J.D., Zhang, M., Gentz, R., Mann, M., Krammer, P.H., Peter, M.E. and Dixit, V.M. (1996). FLICE, a novel FADD-homologous ICE/CED-3-like protease, is recruited to the CD95 (Fas/APO-1) death-inducing signaling complex. *Cell* **85**, 817–827.

Nagata, S. (1997). Apoptosis by death factor. *Cell* **88**, 355–365.

Nakano, H., Oshima, H., Chung, W., Williams-Abbott, L., Ware, C., Yagita, H. and Okumura, K. (1996). TRAF5, an activator of NF-κB and putative signal transducer for the lymphotoxin-β receptor. *J. Biol. Chem.* **271**, 14 661–14 664.

Natoli, G., Costanzo, A., Ianni, A., Templeton, D.J., Woodgett, J.R., Balsano, C. and Levrero, M. (1997). Activation of SAPK/JNK by TNF receptor 1 through a noncytotoxic TRAF2-dependent pathway. *Science* **275**, 200–203.

Neumann, B., Luz, A., Pfeffer, K. and Holzmann, B. (1996a). Defective Peyer's patch organogenesis in mice lacking the 55-kD receptor for tumor necrosis factor. *J. Exp. Med.* **184**, 259–264.

Neumann, B., Machleidt, T., Lifka, A., Pfeffer, K., Vestweber, D., Mak, T.W., Holzmann, B. and Krönke, M. (1996b). Crucial role of 55-kilodalton TNF receptor in TNF-induced adhesion molecule expression and leukocyte organ infiltration. *J. Immunol.* **156**, 1587–1593.

Niehans, G.A., Brunner, T., Frizzelle, S.P., Liston, J.C., Salerno, C.T., Knapp, D.J., Green, D.R. and Kratzke, R. (1997). Human lung carcinomas express fas ligand. *Cancer Res.* **57**, 1007–1012.

Nocentini, G., Giunchi, L., Ronchetti, S., Krausz, L.T., Bartoli, A., Moraca, R., Migliorati, G. and Riccardi, C. (1997). A new member of the tumor necrosis factor/nerve growth factor receptor family inhibits T cell receptor-induced apoptosis. *Proc. Natl Acad. Sci. U.S.A.* **94**, 6216–6221.

O'Connell, J., O'Sullivan, G.C., Collins, J.K. and Shanahan, F. (1996). The Fas counter-attack: Fas-mediated T cell killing by colon cancer cells expressing fas ligand. *J. Exp. Med.* **184**, 1075–1082.

Opipari, A.W., Jr., Hu, H.M., Yabkowitz, R. and Dixit, V.M. (1992). The A20 zinc finger protein protects cells from tumor necrosis factor cytotoxicity. *J. Biol. Chem.* **267**, 12 424–12 427.

Pan, G., Ni, J., Wei, Y.-F., Yu, G.-L., Gentz, R. and Dixit, V.M. (1997a). An antagonist decoy receptor and a death domain-containing receptor for TRAIL. *Science* **277**, 815–817.

Pan, G., O'Rourke, K., Chinnaiyan, A.M., Gentz, R., Ebner, R., Ni, J. and Dixit, V.M. (1997b). The receptor for the cytotoxic ligand TRAIL. *Science* **276**, 111–113.

Pasparakis, M., Alexopoulou, L., Episkopou, V. and Kollias, G. (1996). Immune and inflammatory responses in TNFα-deficient mice: a critical requirement for TNFα in the formation of primary B cell follicles, follicular dendritic cell networks and germinal centers, and in the maturation of the humoral immune response. *J. Exp. Med.* **184**, 1397–1411.

Paterson, D.J., Jefferies, W.A., Green, J.R., Brandon, M.R., Corthesy, P., Puklavec, M. and Williams, A.F. (1987). Antigens of activated rat T lymphocytes including a molecule of 50,000 M_r detected only on CD4 positive T blasts. *Mol. Immunol.* **24**, 1281–1290.

Péguet-Navarro, J., Dalbiez-Gauthier, C., Moulon, C., Berthier, O., Réano, A., Gaucherand, M., Banchereau, J., Rousset, F. and Schmitt, D. (1997). CD40 ligation of human keratinocytes inhibits their proliferation and induces their differentiation. *J. Immunol.* **158**, 144–152.

Pennica, D., Nedwin, G.E., Hayflick, J.S., Seeburg, P.H., Derynck, R., Palladino, M.A., Kohr, W.J., Aggarwal, B.B. and Goeddel, D.V. (1984). Human tumour necrosis factor: precursor structure, expression and homology to lymphotoxin. *Nature* **312**, 724–729.

Pennica, D., Lam, V.T., Weber, R.F., Kohr, W.J., Basa, L.J., Spellman, M.W., Ashkenazi, A., Shire, S.J. and Goeddel, D.V. (1993). Biochemical characterization of the extracellular domain of the 75-kilodalton tumor necrosis factor receptor. *Biochemistry* **32**, 3131–3138.

Perez, C., Albert, I., DeFay, K., Zachariades, N., Gooding, L. and Kriegler, M. (1990). A nonsecretable cell surface mutant of tumor necrosis factor (TNF) kills by cell-to-cell contact. *Cell* **63**, 251–258.

Pietravalle, F., Lecoanet Henchoz, S., Blasey, H., Aubry, J.P., Elson, G., Edgerton, M.D., Bonnefoy, J.Y. and Gauchat, J.F. (1996). Human native soluble CD40L is a biologically active trimer, processed inside microsomes. *J. Biol. Chem.* **271**, 5965–5967.

Pinchuk, L.M., Klaus, S.J., Magaletti, D.M., Pinchuk, G.V., Norsen, J.P. and Clark, E.A. (1996). Functional CD40 ligand expressed by human blood dendritic cells is up-regulated by CD40 ligation. *J. Immunol.* **157**, 4363–4370.

Pitti, R.M., Marsters, S.A., Ruppert, S., Donahue, C.J., Moore, A. and Ashkenazi, A. (1996). Induction of apoptosis by Apo-2 ligand, a new member of the tumor necrosis factor cytokine family. *J. Biol. Chem.* **271**, 12687–12690.

Pocsik, E., Duda, E. and Wallach, D. (1995). Phosphorylation of the 26 kDa TNF precursor in monocytic cells and in transfected HeLa cells. *J. Inflamm.* **45**, 152–160.

Pollok, K.E., Kim, Y.J., Zhou, Z., Hurtado, J., Kim, K.K., Pickard, R.T. and Kwon, B.S. (1993). Inducible T cell antigen 4-1BB. Analysis of expression and function. *J. Immunol.* **150**, 771–781.

Pollok, K.E., Kim, Y.-J., Hurtado, J., Zhou, Z., Kim, K.K. and Kwon, B.S. (1994). 4-1BB T-cell antigen binds to mature B cells and macrophages and costimulates anti-μ-primed splenic B cells. *Eur. J. Immunol.* **24**, 367–374.

Pollok, K.E., Kim, S.H. and Kwon, B.S. (1995). Regulation of 4-1BB expression by cell–cell interactions and the cytokines, interleukin-2 and interleukin-4. *Eur. J. Immunol.* **25**, 488–494.

Portegies, P., Godfried, M.H., Hintzen, R.Q., Stam, J., van der Poll, T., Bakker, M., van Deventer, S.J., van Lier, R.A. and Goudsmit, J. (1993). Low levels of specific T cell activation marker CD27 accompanied by elevated levels of markers for non-specific immune activation in the cerebrospinal fluid of patients with AIDS dementia complex. *J. Neuroimmunol.* **48**, 241–247.

Ramesh, N., Fuleihan, R. and Geha, R. (1994). Molecular pathology of X-linked immunoglobulin deficiency with normal or elevated IgM (HIGMX-1). *Immunol. Rev.* **138**, 87–104.

Rennert, P.D., Browning, J.L., Mebius, R., Mackay, F. and Hochman, P.S. (1996). Surface lymphotoxin α/β complex is required for the development of peripheral lymphoid organs. *J. Exp. Med.* **184**, 1999–2006.

Roberts, C.W.M., Shutter, J.R. and Korsmeyer, S.J. (1994). *Hox11* controls the genesis of the spleen. *Nature* **368**, 747–749.

Rodseth, L.E., Brandhuber, B., Devine, T.Q., Eck, M.J., Hale, K., Naismith, J.H. and Sprang, S.R. (1994). Two crystal forms of the extracellular domain of type I tumor necrosis factor receptor. *J. Mol. Biol.* **239**, 332–335.

Rothe, M., Wong, S.C., Henzel, W.J. and Goeddel, D.V. (1994). A novel family of putative signal transducers associated with the cytoplasmic domain of the 75 kDa tumor necrosis factor receptor. *Cell* **78**, 681–692.

Rothe, M., Pan, M.G., Henzel, W.J., Ayres, T.M. and Goeddel, D.V. (1995a). The TNFR2–TRAF signaling complex contains two novel proteins related to baculoviral inhibitor of apoptosis proteins. *Cell* **83**, 1243–1252.

Rothe, M., Sarma, V., Dixit, V.M. and Goeddel, D.V. (1995b). TRAF2-mediated activation of NF-κB by TNF receptor 2 and CD40. *Science* **269**, 1424–1427.

Rothe, M., Xiong, J., Shu, H.-B., Williamson, K., Goddard, A. and Goeddel, D.V. (1996). I-TRAF is a novel TRAF-interacting protein that regulates TRAF-mediated signal transduction. *Proc. Natl Acad. Sci. U.S.A.* **93**, 8241–8246.

Ryffel, B., di Padova, F., Schreier, M.H., Le Hir, M., Eugster, H.-P. and Quesniaux, V.F.J. (1997). Lack of type 2 T cell-independent B cell responses and defect in isotype switching in TNF–lymphotoxin α-deficient mice. *J. Immunol.* **158**, 2126–2133.

Sato, T., Irie, S., Kitada, S. and Reed, J.C. (1995a). FAP-1: a protein tyrosine phosphatase that associates with Fas. *Science* **268**, 411–415.

Sato, T., Irie, S. and Reed, J.C. (1995b). A novel member of the TRAF family of putative signal transducing proteins binds to the cytosolic domain of CD40. *FEBS Lett.* **358**, 113–118.

Schall, T.J., Lewis, M., Koller, K.J., Lee, A., Rice, G.C., Wong, G.H., Gatanaga, T., Granger, G.A., Lentz, R. and Raab, H. (1990). Molecular cloning and expression of a receptor for human tumor necrosis factor. *Cell* **61**, 361–370.

Schievella, A.R., Chen, J.H., Graham, J.R. and Lin, L.-L. (1997). MADD, a novel death domain protein that interacts with the type 1 tumor necrosis factor receptor and activates mitogen-activated protein kinase. *J. Biol. Chem.* **272**, 12069–12075.

Schreiber, M. and McFadden, G. (1994). The myxoma virus TNF-receptor homologue (T2) inhibits tumor necrosis factor-α in a species-specific fashion. *Virology* **204**, 692–705.

Schwab, U., Stein, H., Gerdes, J., Lemke, H., Kirchner, H., Schaadt, M. and Diehl, V. (1982). Production of a monoclonal antibody specific for Hodgkin and Sternberg–Reed cells of Hodgkin's disease and a subset of normal lymphoid cells. *Nature* **299**, 65–67.

Schwarz, H., Valbracht, J., Tuckwell, J., von Kempis, J. and Lotz, M. (1995). ILA, the human 4-1BB homologue, is inducible in lymphoid and other cell lineages. *Blood* **85**, 1043–1052.

Screaton, G.R., Xu, X.-N., Olsen, A.L., Cowper, A.E., Tan, R., McMichael, A.J. and Bell, J.I. (1997). LARD: a new lymphoid-specific death domain containing receptor regulated by alternative pre-mRNA splicing. *Proc. Natl Acad. Sci. U.S.A.* **94**, 4615–4619.

Sha, W.C., Liou, H.C., Tuomanen, E.I. and Baltimore, D. (1995). Targeted disruption of the p50 subunit of NF-κB leads to multifocal defects in immune responses. *Cell* **80**, 321–330.

Sheridan, J.P., Marsters, S.A., Pitti, R.M., Gurney, A., Skubatch, M., Baldwin, D., Ramakrishnan, L., Gray, C.L., Baker, K., Wood, W.I., Goddard, A.D., Godowski, P. and Ashkenazi, A. (1997). Control of TRAIL-induced apoptosis by a family of signaling and decoy receptors. *Science* **277**, 818–821.

Shu, H.-B., Halpin, D.R. and Goeddel, D.V. (1997). Casper is a FADD- and caspase-related inducer of apoptosis. *Immunity* **6**, 751–763.

Sieg, S., Yildirim, Z., Smith, D., Kayagaki, N., Yagita, H., Huang, Y. and Kaplan, D. (1996). Herpes simplex virus type 2 inhibition of Fas ligand expression. *J. Virol.* **70**, 8747–8751.

Simonet, W.S., Lacey, D.L., Dunstan, C.R., Kelley, M., Cheng, M.-S., Lüthy, R., Nguyen, H.Q., Wooden, S., Bennett, L., Boone, T., Shimamato, G., DeRose, M., Elliott, R., Colombero, A., Tan, H.-L., Trail, G., Sullivan, J., Davy, E., Bucay, N., Renshaw-Gegg, L., Hughes, T.M., Hill, D., Pattison, W., Campbell, P., Sander, S., Van, G., Tarpley, J., Derby, P., Lee, R. and Boyle, W.J. (1997). Osteoprotegerin: a novel secreted protein involved in the regulation of bone density. *Cell* **89**, 309–319.

Smith, C., Davis, T., Anderson, D., Solam, L., Beckmann, M., Jerzy, R., Dower, S., Cosman, D. and Goodwin, R. (1990). A receptor for tumor necrosis factor defines an unusual family of cellular and viral proteins. *Science* **248**, 1019–1023.

Smith, C.A., Davis, T., Wignall, J.M., Din, W.S., Farrah, T., Upton, C., McFadden, G. and Goodwin, R.G. (1991). T2 open reading frame from the Shope fibroma virus encodes a soluble form of the TNF receptor. *Biochem. Biophys. Res. Commun.* **176**, 335–342.

Smith, C.A., Gruss, H.-J., Davis, T., Anderson, D., Farrah, T., Baker, E., Sutherland, G.R., Brannan, C.I., Copeland, N.G., Jenkins, N.A., Grabstein, K.H., Gliniak, B., McAlister, I.B., Fanslow, W., Alderson, M., Falk, B., Gimpel, S., Gillis, S., Din, W.S., Goodwin, R.G. and Armitage, R.J. (1993). CD30 antigen, a marker for Hodgkin's lymphoma, is a receptor whose ligand defines an emerging family of cytokines with homology to TNF. *Cell* **73**, 1349–1360.

Smith, C.A., Farrah, T. and Goodwin, R.G. (1994). The TNF receptor superfamily of cellular and viral proteins: activation, costimulation, and death. *Cell* **76**, 959–962.

Smith, C.A., Hu, F.-Q., Smith, T.D., Richards, C.L., Smolak, P., Goodwin, R.G. and Pickup, D.J. (1996). Cowpox virus genome encodes a second soluble homologue of cellular TNF receptors, distinct from CrmB, that binds TNF but not LTα. *Virology* **223**, 132–147.

Song, H.Y., Rothe, M. and Goeddel, D.V. (1996). The tumor necrosis factor-inducible zinc finger protein A20 interacts with TRAF1/TRAF2 and inhibits NF-κB activation. *Proc. Natl Acad. Sci. U.S.A.* **93**, 6721–6725.

Soong, L., Xu, J.-C., Grewal, I.S., Kima, P., Sun, J., Longley, B.J., Jr., Ruddle, N.H., McMahon-Pratt, D. and Flavell, R.A. (1996). Disruption of CD40–CD40 ligand interactions results in an enhanced susceptibility to *Leishmania amazonensis* infection. *Immunity* **4**, 263–273.

Spies, T., Morton, C.C., Nedospasov, S.A., Fiers, W., Pious, D. and Strominger, J.L. (1986). Genes for the tumor necrosis factors α and β are linked to the human major histocompatibility complex. *Proc. Natl Acad. Sci. U.S.A.* **83**, 8699–8702.

Stanger, B.Z., Leder, P., Lee, T.H., Kim, E. and Seed, B. (1995). RIP: a novel protein containing a death domain that interacts with Fas/APO-1 (CD95) in yeast and causes cell death. *Cell* **81**, 513–523.

Stuber, E. and Strober, W. (1996). The T cell–B cell interaction via OX40–OX40L is necessary for the T cell-dependent humoral immune response. *J. Exp. Med.* **183**, 979–989.

Stuber, E., Neurath, M., Calderhead, D., Fell, H.P. and Strober, W. (1995). Cross-linking of OX40 ligand, a member of the TNF/NGF cytokine family, induces proliferation and differentiation in murine splenic B cells. *Immunity* **2**, 507–521.

Suda, T., Takahashi, T., Golstein, P. and Nagata, S. (1993). Molecular cloning and expression of the Fas ligand, a novel member of the tumor necrosis factor family. *Cell* **75**, 1169–1178.

Sugita, K., Torimoto, Y., Nojima Y., Daley, J.F., Schlossman, S.F. and Morimoto, C. (1991). The 1A4 molecule (CD27) is involved in T cell activation. *J. Immunol.* **147**, 1477–1483.

Sugita, K., Tanaka, T., Doshen, J.M., Schlossman, S.F. and Morimoto, C. (1993). Direct demonstration of the CD27 molecule involved in the negative regulatory effect on T cell activation. *Cell. Immunol.* **152**, 279–285.

Srinivasula, S.M., Ahmad, M., Ottilie, S., Bullrich, F., Banks, S., Wang, Y., Fernandes-Alnemri, T., Croce, C.M., Litwack, G., Tomaselli, K.J., Armstrong, R.C. and Alnemri, E.S. (1997). FLAME-1, a novel FADD-like anti-apoptotic molecule that regulates fas/TNFR1-induced apoptosis. *J. Biol. Chem.* **272**, 18 542–18 545.

Takeuchi, M., Rothe, M. and Goeddel, D.V. (1996). Anatomy of TRAF2. Distinct domains for nuclear factor-κB activation and association with tumor necrosis factor signaling proteins. *J. Biol. Chem.* **271**, 19 935–19 942.

Tanaka, M., Suda, T., Takahashi, T. and Nagata, S. (1995). Expression of the functional soluble form of human fas ligand in activated lymphocytes. *EMBO J.* **14**, 1129–1135.

Tartaglia, L.A., Ayres, T.M., Wong, G.H. and Goeddel, D.V. (1993a). A novel domain within the 55 kD TNF receptor signals cell death. *Cell* **74**, 845–853.

Tartaglia, L.A., Goeddel, D.V., Reynolds, C., Figari, I.S., Weber, R.F., Fendly, B.M. and Palladino, M.A., Jr. (1993b). Stimulation of human T-cell proliferation by specific activation of the 75-kDa tumor necrosis factor receptor. *J. Immunol.* **151**, 4637–4641.

Tewari, M. and Dixit, V.M. (1995). Fas- and tumor necrosis factor-induced apoptosis is inhibited by the poxvirus crmA gene product. *J. Biol. Chem.* **270**, 3255–3260.

Upton, C., Macen, J., Schreiber, M. and McFadden, G. (1991). Myxoma virus expresses a secreted protein with homology to the tumor necrosis factor receptor gene family that contributes to viral virulence. *Virology* **184**, 370–382.

Van Antwerp, D.J., Martin, S.J., Kafri, T., Green, D.R. and Verma, I.M. (1996). Suppression of TNF-α-induced apoptosis by NF-κB. *Science* **274**, 787–789.

van Oers, M.H., Pals, S.T., Evers, L.M., van der Schoot, C.E., Koopman, G., Bonfrer, J.M., Hintzen, R.Q., von dem Borne, A.E. and van Lier, R.A. (1993). Expression and release of CD27 in human B-cell malignancies. *Blood* **82**, 3430–3436.

Van Zee, K.J., Kohno, T., Fischer, E., Rock, C.S., Moldawer, L.L. and Lowry, S.F. (1992). Tumor necrosis factor soluble receptors circulate during experimental and clinical inflammation and can protect against excessive tumor necrosis factor α *in vitro* and *in vivo*. *Proc. Natl Acad. Sci. U.S.A.* **89**, 4845–4849.

VanArsdale, T.L., VanArsdale, S.L., Force, W.R., Walter, B.N., Mosialos, G., Kieff, E., Reed, J.C. and Ware, C.F. (1997). Lymphotoxin-β receptor signaling complex: role of tumor necrosis factor receptor-associated factor 3 recruitment in cell death and activation of nuclear factor κB. *Proc. Natl Acad. Sci. U.S.A.* **94**, 2460–2465.

von Bülow, G.-U. and Bram, R.J. (1997). NF-AT activation induced by a CAML-interacting member of the tumor necrosis factor receptor superfamily. *Science* **278**, 138–141.

Ware, C.F., Crowe, P.D., VanArsdale, T.L., Andrews, J.L., Grayson, M.H., Jerzy, R., Smith, C.A. and Goodwin, R.G. (1991). Tumor necrosis factor (TNF) receptor expression in T lymphocytes. Differential regulation of the type I TNF receptor during activation of resting and effector T cells. *J. Immunol.* **147**, 4229–4238.

Ware, C.F., Crowe, P.D., Grayson, M.H., Androlewicz, M.J. and Browning, J.L. (1992). Expression of surface lymphotoxin and tumor necrosis factor on activated T, B, and natural killer cells. *J. Immunol.* **149**, 3881–3888.

Ware, C.F., VanArsdale, T.L., Crowe, P.D. and Browning, J.L. (1995). The ligands and receptors of the lymphotoxin system. In *Pathways for Cytolysis* (eds G.M. Griffiths and J. Tschopp), Springer, Basel, pp. 175–218.

Watanabe-Fukunaga, R., Brannan, C.I., Itoh, N., Yonehara, S., Copeland, N.G., Jenkins, N.A. and Nagata, S. (1992). The cDNA structure, expression, and chromosomal assignment of the mouse Fas antigen. *J. Immunol.* **148**, 1274–1279.

Wiley, S.R., Schooley, K., Smolak, P.J., Din, S.D., Huang, C., Nicholl, J.K., Sutherland, G.R., Smith, T.D., Rauch, C., Smith, C.A. and Goodwin, R.G. (1995). Identification and characterization of a new member of the TNF family that induces apoptosis. *Immunity* **3**, 673–682.

Wiley, S.R., Goodwin, R.G. and Smith, C.A. (1996). Reverse signaling via CD30 ligand. *J. Immunol.* **157**, 3635–3639.

Wold, W.S. and Gooding, L.R. (1991). Region E3 of adenovirus: a cassette of genes involved in host immunosurveillance and virus–cell interactions. *Virology* **184**, 1–8.

Wong, B.R., Rho, J., Arron, J., Robinson, E., Orlinick, J., Chao, M., Kalachikov, S., Cayani, E., Bartlett, F.S., III, Frankel, W.N., Lee, S.Y. and Choi, Y. (1997). TRANCE is a novel ligand of the tumor necrosis factor receptor family that activates c-Jun N-terminal kinase in T cells. *J. Biol. Chem.* **272**, 25 190–25 194.

Wright, D.A., Futcher, B., Ghosh, P. and Geha, R.S. (1996). Association of human fas (CD95) with a ubiquitin-conjugating enzyme (UBC-FAP). *J. Biol. Chem.* **271**, 31 037–31 043.

Xu, J., Foy, T.M., Laman, J.D., Elliott, E.A., Dunn, J.J., Waldschmidt, T.J., Elsemore, J., Noelle, R.J. and Flavell, R.A. (1994). Mice deficient for the CD40 ligand. *Immunity* **1**, 423–431.

Xu, Y., Cheng, G. and Baltimore, D. (1996). Targeted disruption of TRAF3 leads to postnatal lethality and defective T-dependent immune responses. *Immunity* **5**, 407–415.

Yang, F.C., Agematsu, K., Nakazawa, T., Mori, T., Ito, S., Kobata, T., Morimoto, C. and Komiyama, A. (1996). CD27/CD70 interaction directly induces natural killer cell killing activity. *Immunology* **88**, 289–293.

Yang, J., Liu, X., Bhalla, K., Kim, C.N., Ibrado, A.M., Cai, J., Peng, T.-I., Jones, D.P. and Wang, X. (1997a).

Prevention of apoptosis by Bcl-2: release of cytochrome *c* from mitochondria blocked. *Science* **275**, 1129–1132.

Yellin, M.J., Sippel, K., Inghirami, G., Covey, L.R., Lee, J.J., Sinning, J., Clark, E.A., Chess, L. and Lederman, S. (1994). CD40 molecules induce down-modulation and endocytosis of T cell surface T cell–B cell activating molecule/CD40-L. Potential role in regulating helper effector function. *J. Immunol.* **152**, 598–608.

Yonehara, S., Ishii, A. and Yonehara, M. (1989). A cell-killing monoclonal antibody (anti-Fas) to a cell surface antigen co-downregulated with the receptor of tumor necrosis factor. *J. Exp. Med.* **169**, 1747–1756.

Zheng, L., Fisher, G., Miller, R.E., Peschon, J., Lynch, D.H. and Lenardo, M.J. (1995). Induction of apoptosis in mature T cells by tumour necrosis factor. *Nature* **377**, 348–351.

Chapter 21

Transforming Growth Factor-β and its Receptors

Rik Derynck and Lisa Choy

Departments of Growth and Development, and Anatomy, University of California at San Francisco, San Francisco, CA, USA

INTRODUCTION

The initial identification of transforming growth factor-β (TGF-β), now more than 15 years ago, was based on its ability to reversibly induce phenotypic transformation of select fibroblast cell lines (Moses *et al.*, 1981; Roberts *et al.*, 1981). In some cell lines, this transforming activity was apparent only when both TGF-β and TGF-α were present (Anzano *et al.*, 1982, 1983) or when epidermal growth factor (EGF) was added to TGF-β (Roberts *et al.*, 1981). In spite of their names, however, TGF-α and TGF-β are unrelated structurally. TGF-α, which shares structural similarity with EGF and interacts with a common EGF–TGF-α receptor, will not be the subject of this chapter and has been reviewed elsewhere (Derynck, 1992). Rather, this review will only focus on TGF-β.

The initial detection of TGF-β activity in a transformation assay suggested a role for TGF-β in malignant transformation and tumor development. It is now well established that the activities of TGF-β are by no means restricted to tumor cells, but that TGF-β exerts a multiplicity of biologic activities on most cells, both normal or transformed, and regulates many cell physiologic processes. The biologic response to TGF-β is complex and depends on the cell type (and even on the individual cell line tested) and the physiologic conditions. Furthermore, TGF-β also plays an important role in the control of the immune response and wound healing, and in the development of various tissues and organs.

The recent accumulation of knowledge on the biology of TGF-β makes it a daunting task to review TGF-β in depth. For example, TGF-β exerts a large variety of biologic activities in various cell systems and only a few select examples will be discussed in this review. We will now primarily focus on some subject areas that are of great importance to understand the biology of TGF-β and have seen considerable progress in knowledge during the past few years. We will first outline our knowledge on the structure and post-translational processing of the different TGF-β isoforms, then elaborate on the structure and mechanism of signaling of the TGF-β receptors, next discuss the role of TGF-β in the control of cell proliferation and the immune response, and finally review our knowledge of the involvement of TGF-β in normal and tumor development.

The Cytokine Handbook, 3rd ed.
ISBN 0-12-689662-3

Since the original cDNA cloning of the first TGF-β, i.e. TGF-β_1 (Derynck *et al.*, 1985), extensive protein characterization and cDNA analysis have revealed a large group of structurally related secreted factors. Collectively, these factors are called the TGF-β superfamily, which contains not only the different TGF-β isoforms, but also the activins, bone morphogenetic proteins (BMPs) and several other secreted factors. TGF-β-related proteins are found in vertebrate species and the fruitfly *Drosophila*, and in the nematode *Caenorhabditis elegans*. They are thought to play important roles in cell differentiation and proliferation during development. For an overview of the TGF-β superfamily, which will not be discussed here, we refer to existing review articles (Massagué, 1990; Hoffmann, 1991; Hogan, 1996).

STRUCTURE OF TGF-β

It was originally assumed that there was only one type of TGF-β, i.e. the one now known as TGF-β_1, that could be purified from platelets (Assoian *et al.*, 1983). Protein purification and cDNA cloning approaches have subsequently revealed three TGF-β isoforms in mammalian cells, each encoded by their own gene. TGF-β_1 (Derynck *et al.*, 1985), TGF-β_2 (de Martin *et al.*, 1987) and TGF-β_3 (Derynck *et al.*, 1988; ten Dijke *et al.*, 1988) are made as larger precursors of 390 amino acids (TGF-β_1) or 412 amino acids (TGF-β_2 and TGF-β_3). Each precursor contains an *N*-terminal signal peptide, a long prosegment, also called 'latency-associated polypeptide' (LAP) and a 112-amino-acid *C*-terminal polypeptide that constitutes the mature TGF-β monomer. This monomer is cleaved from the remaining precursor segment following a tetrabasic peptide. The nature of this cleavage site suggests that the protease responsible for cleavage belongs to the KEX/furin-like family of proteases, which recognize multibasic sites in secreted proteins (Barr, 1991).

The active form of TGF-β is a hydrophobic, disulfide-linked dimer of the *C*-terminal segment of the pre-pro-TGF-β (Assoian *et al.*, 1983; Derynck *et al.*, 1985; Cheifetz *et al.*, 1987). The three TGF-β species, with a sequence identity of 70–80% (Derynck *et al.*, 1988; ten Dijke *et al.*, 1988), are thus generated as homodimers of two identical *C*-terminal polypeptides, although heterodimers between TGF-β_1 and TGF-β_2 have also been isolated (Cheifetz *et al.*, 1987). The biologic relevance of the heterodimeric species is unclear, especially as they presumably correspond to only minor species. The 112-amino-acid mature TGF-β polypeptide includes nine conserved cysteines, eight of which are paired. All cysteines form intrachain disulfide bonds, with the exception of one intermolecular cysteine bridge responsible for dimerization (Daopin *et al.*, 1992; Schlunegger and Grütter, 1992). Three-dimensional structure analysis has revealed an extended butterfly-like structure of the TGF-β_2 homodimer, rather than the compact globular structure frequently seen in highly stable and temperature- and protease-resistant growth factors (Daopin *et al.*, 1992; Schlunegger and Grütter, 1992). Considering the generally conserved sequence features and the number and relative positions of the cysteines among the different TGF-β superfamily members, the three TGF-β isoforms and all members of the TGF-β superfamily are likely to assume similar structural conformations, as is apparent from the structure of the TGF-β-related OP-1/BMP-7 (Griffith *et al.*, 1996). The sequence between the fifth and sixth cysteine, which has the most sequence divergence among the different TGF-β isoforms and the other members of the TGF-β superfamily, is exposed at the surface of this molecule and may play an important role in the determination of the specificity of the ligand–receptor recognition.

The prosegments of the three TGF-β precursors show a sequence identity of only 25–35% (Derynck *et al.*, 1988; ten Dijke *et al.*, 1988). Nevertheless, some structural elements are strongly conserved, suggesting critical roles either in the function or in the folding of this prosegment. Most noteworthy among these conserved features are three cysteine residues and two *N*-glycosylation sites (Derynck *et al.*, 1988). Two of these three cysteines are involved in intermolecular disulfide bridge formation (Miyazono *et al.*, 1988, 1991). The conserved *N*-linked carbohydrates are mannose-6-phosphorylated and mediate binding of the prosegment to the mannose-6-phosphate receptor (Purchio *et al.*, 1988; Kovacina *et al.*, 1989). Finally, a hydrophobic stretch of amino acids closely following the signal peptide is conserved as well and may be involved in the folding of the precursor segment (Lopez *et al.*, 1992).

Biochemical characterization has shown that TGF-β is normally secreted as a protein complex, consisting of the mature TGF-β homodimer and two prosegments which interact non-covalently with the mature dimer. The participation of mature TGF-β in this complex prevents it from interacting with the TGF-β receptors, thus rendering this complex inactive or 'latent' (Miyazono *et al.*, 1988; Wakefield *et al.*, 1988). Based on antibody recognition studies, it is likely that this interaction of TGF-β with the much larger prosegments leaves little of the mature TGF-β exposed at the surface of this complex. The TGF-β complex isolated from platelets has revealed the identity of yet another polypeptide that associates with this complex. This polypeptide, named latent TGF-β binding protein (LTBP), thus represents a fifth protein in this platelet-derived TGF-β complex (Miyazono *et al.*, 1988, 1991; Wakefield *et al.*, 1988). LTBP is a protein of 125–160 kDa and interacts through at least one disulfide bond with one of the pro-TGF-β segments in the latent complex (Miyazono *et al.*, 1991). Its amino acid sequence reveals 16 EGF-like repeats, the relevance of which is as yet unknown. In addition, it contains an Arg-Gly-Asp (RGD) sequence and an eight-amino-acid sequence identical to the proposed cell binding domain of the laminin B_2 chain (Kanzaki *et al.*, 1990). Northern hybridization suggests that this protein is made by a large variety of cell types, making it likely that many cells secrete the five-protein latent complex, rather than the smaller tetrameric complex, and that formation of this type of complex could be subject to regulation as well (Kanzaki *et al.*, 1990; Miyazono *et al.*, 1991; Olofsson *et al.*, 1992). Besides LTBP, which is now named LTBP-1, two related proteins, LTBP-2 (Moren *et al.*, 1994) and LTBP-3 (Yin *et al.*, 1995) have been identified as separate gene products. The differential biologic roles of these three LTBPs, including their possible roles as matrix proteins, remains to be determined.

The expression of the three TGF-β isoforms is differentially regulated. Thus, different inducers or exogenous agents have distinct effects on the secretion of the individual TGF-β species. The complexity of this regulation will not be discussed here, but has been illustrated by Roberts and Sporn (1992). As an example, treatment of cells with TGF-β_1 or TGF-β_2 results in differential changes in the synthesis of the different TGF-β isoforms (Bascom *et al.*, 1989). In addition, retinoic acid treatment results in a strong induction of TGF-β_2 but not of TGF-β_1 synthesis (Glick *et al.*, 1989). Finally, interaction of mammary epithelial cells with plastic as substrate induces transcription of TGF-β_1 but not of TGF-β_2 (Streuli *et al.*, 1993). This differential regulation of expression is largely due to a characteristic and independent pattern of transcriptional regulation of the three

TGF-β species, which is in turn regulated by distinct control elements in the promoters of the genes (Roberts *et al.*, 1991). In addition, all three TGF-β isoforms have long 5′ untranslated sequences (Derynck *et al.*, 1985; Arrick *et al.*, 1991; Kim *et al.*, 1992) which most likely play an isoform-specific role in the regulation of translation efficiency (Arrick *et al.*, 1991; Kim *et al.*, 1992; Romeo *et al.*, 1993).

THE ROLE OF ASSOCIATED PROTEINS IN THE ACTIVATION OF TGF-β

The synthesis of TGF-β as a latent complex raises questions regarding the normal mechanisms of activation of TGF-β and the role of the associated proteins. The association of the two prosegments with mature TGF-β may serve several functions: it may be required for secretion of TGF-β, maintenance of TGF-β in an inactive form, and targeting of the latent TGF-β. The role of the prosegment in secretion of TGF-β is suggested by two lines of evidence. TGF-β expression plasmids in which the sequence coding for mature TGF-β was deleted failed to produce secreted TGF-β, but synthesis of the prosegment from a cotransfected plasmid rescued the ability of the cells to secrete TGF-β, albeit at low efficiency (Gray and Mason, 1990). In addition, deletion of some conserved structural elements of the precursor segment, i.e. the hydrophobic sequence following the signal peptide cleavage site, and the two mannose-6-phosphorylated *N*-linked carbohydrates, prevents TGF-β secretion (Lopez *et al.*, 1992). Thus, the precursor segment may serve as a chaperone, required for secretion of TGF-β. Whether this chaperoning function merely facilitates the secretion process or also plays a role in the folding of mature TGF-β is unknown. Once secreted, the prosegments maintain mature and active TGF-β in the latent complex and prevent its interaction with the TGF-β receptors (Miyazono *et al.*, 1988; Wakefield *et al.*, 1988). Considering the widespread distribution of TGF-β receptors and the potent biologic effects of TGF-β, it is indeed strategically most important that TGF-β remains inactive until its required activation at the target site. Besides these functions of the latent complex, the presence of LTBP in the complex also allows a mechanism by which the large latent TGF-β complex can be localized preferentially into the extracellular matrix (Taipale *et al.*, 1994).

The release of active TGF-β from the latent complex probably occurs in a highly regulated manner. *In vitro*, latent TGF-β can be activated by treatment with heat and acid (Miyazono *et al.*, 1990). The activation of latent TGF-β under mildly acidic conditions (Lyons *et al.*, 1988) suggests that this mechanism may be of physiologic relevance. In addition, ionizing radiation can induce activation of latent TGF-β (Barcellos-Hoff and Dix, 1996) as apparent by the rapid increase of active TGF-β in irradiated breast tissue (Barcellos-Hoff *et al.*, 1994). However, normal physiologic activation of latent TGF-β is most likely due to the action of proteases, such as plasmin and cathepsins (Lyons *et al.*, 1988, 1990; for reviews see Harpel *et al.*, 1993; Munger *et al.*, 1997). Plasmin, presumably the major physiologic activator of latent TGF-β, may promote its activity in two ways: by degrading the prosegments and thereby allowing the release of the active TGF-β dimer (Lyons *et al.*, 1990) and by releasing latent TGF-β from its storage in the extracellular matrix (Taipale *et al.*, 1995). Induction of plasminogen activators, for example during angiogenesis and invasion, results in conversion of plasminogen into plasmin, which in turn allows activation of latent TGF-β at these sites.

The mechanism of physiologic activation of latent TGF-β has initially been studied mainly in cocultures of smooth muscle cells and endothelial cells, in which plasmin acts as the protease responsible for activation of TGF-β_1 (Antonelli-Orlidge *et al.*, 1989; Sato and Rifkin, 1989). Whereas the mechanism of plasmin activation needs to be further clarified in this system, the activation most likely occurs at the cell surface, where the prosegments of the latent TGF-β complexes are retained at least in part through interaction of their mannose-6-phosphorylated carbohydrates with the mannose-6-phosphate receptors. Accordingly, antibodies that interfere with the binding ability of this receptor block TGF-β activation (Dennis and Rifkin, 1991). In addition, antibodies to LTBP-1 and exogenous excess LTBP-1 block the activation of latent TGF-β, suggesting that LTBP plays a role, although as yet uncharacterized, in this activation (Flaumenhaft *et al.*, 1993a). Finally, the activation can also be inhibited by plasmin inhibitors such as antiplasmin (Flaumenhaft *et al.*, 1993b), as expected, and by inhibitors of type II transglutaminase, suggesting a requirement for the interaction of the latent complex with the matrix and a role of the latter enzyme in the activation mechanism (Kojima *et al.*, 1993).

Active-TGF-β is also released from other cell sources such as tumor cells, activated macrophages, osteoclasts and osteoblasts (Twardzik *et al.*, 1990; Flaumenhaft *et al.*, 1993b; Oursler *et al.*, 1993, 1994). Proteases are often involved in the activation of TGF-β by these cells as well, and plasmin is in many cases the effector protease, although the mechanisms of activation of latent TGF-β by these cells are poorly understood (Munger *et al.*, 1997). In addition to proteases, thrombospondin also has the ability to activate latent TGF-β by an as yet uncharacterized mechanism that does not involve proteolysis (Schultz-Cherry and Murphy-Ullrich, 1993; Schultz-Cherry *et al.*, 1994). Both thrombospondin and TGF-β are found in high concentrations in the α granules of platelets, and thrombospondin isolated from platelets is tightly associated with TGF-β (Murphy-Ullrich *et al.*, 1992). Finally, little is as yet known about how TGF-β_2 and TGF-β_3 are activated. Considering the similar nature of the latent complexes, it is likely that proteolytic mechanisms may be at work, but the extensive sequence differences in the prosegments suggest mechanistic differences, possibly even the involvement of different proteases.

Once released from the cells, latent and active TGF-β can interact with proteins in the extracellular milieu, especially with various components of the extracellular matrix or basement membrane. These interactions result in yet another type of control mechanism that governs the biologic availability and activity of TGF-β. One of the proteins with high affinity for TGF-β is α_2-macroglobulin, which is present at high levels in the circulation and interacts with various other growth factors and cytokines as well (for review, see James, 1990; LaMarre *et al.*, 1991a). Interaction of TGF-β with α_2-macroglobulin sequesters TGF-β into an inactive form that is unable to bind to the TGF-β receptors (O'Connor-McCourt and Wakefield, 1987; Huang *et al.*, 1988). Thus, α_2-macroglobulin may serve as an efficient scavenger that binds all free active TGF-β, especially as almost all TGF-β in plasma is sequestered in this type of complex (O'Connor-McCourt and Wakefield, 1987; Huang *et al.*, 1988). Following transportation to the liver, interaction of the complex with the α_2-macroglobulin receptor may result in the delivery of high concentrations of TGF-β to the liver, where it functions as an active molecule or, perhaps more likely, is degraded (LaMarre *et al.*, 1991b). Other soluble proteins that engage in a high-affinity interaction with TGF-β are decorin and biglycan, two secreted proteoglycans that are part of the extracellular matrix. Interaction of TGF-β with these proteins neutralizes the activity of TGF-β and, thus, could result in a physiologic inactivation of TGF-β in the extracellular matrix (Yamaguchi *et al.*, 1990). Finally, thrombospondin (Murphy-Ullrich *et al.*, 1992), fibronectin (Fava and McClure, 1987) and several collagens, including collagen type IV (Paralkar *et al.*, 1991) are also able to bind TGF-β with a high affinity. The interactions of these extracellular matrix components with TGF-β may result in sequestration of active TGF-β into the extracellular matrix, which thus could be considered as a reservoir for TGF-β (and many other growth and differentiation factors). The interactions of TGF-β with thrombospondin and collagen IV keep TGF-β in an active form (Murphy-Ullrich *et al.*, 1992).

TGF-β RECEPTORS AND RECEPTOR BINDING PROTEINS

Three major TGF-β receptors were initially identified by chemical crosslinking of ^{125}I-labeled TGF-β to cell surface proteins. These receptors, named type I, II and III on the basis of their decreasing electrophoretic mobility (Cheifetz *et al.*, 1986, 1987), are commonly observed on most cells in culture and are the best characterized receptors (Fig. 1). Whereas it was originally thought that the biologic activities of TGF-β were mediated through the type III receptors (Cheifetz *et al.*, 1987, 1988), later studies revealed that the type I and type II receptors mediate most if not all signaling activities of TGF-β (Laiho *et al.*, 1991a; Geiser *et al.*, 1992; Chen *et al.*, 1993). In addition to these three receptor types, other ^{125}I-TGF-β-crosslinked bands with different electrophoretic mobilities or properties have been reported (Massagué, 1992; Lin and Lodish, 1993). However, these receptors, named types IV–IX, have not been well characterized, and little or no evidence has been shown that they are involved in TGF-β signaling. The structural and functional analysis of the three receptor types has been reviewed extensively by Derynck and Feng (1997).

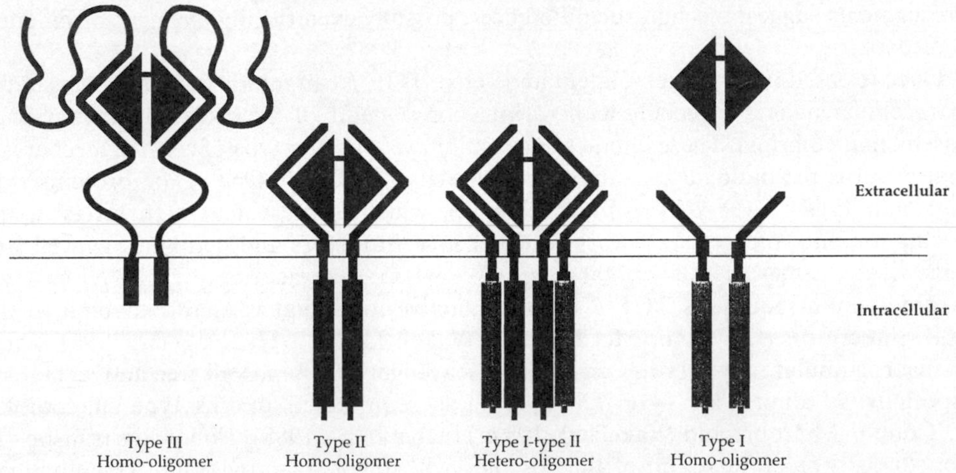

Fig. 1. Schematic model of the homomeric and heteromeric interactions of the type II receptor. The ligand binding to the type II but not the type I receptor is shown for TGF-β_1, although some TGF-β-related factors can bind to their type I receptors in the absence of type II receptors. Only the heteromeric receptor complex of type II and type I receptors is known to mediate the TGF-β response.

The predominant type III receptor, also called betaglycan, is present in a wide variety of cell types and lines, and is the most abundant TGF-β receptor. It is a cell surface proteoglycan containing covalently linked heparan sulfate and chondroitin sulfate (Segarini and Seyedin, 1988; Cheifetz and Massagué, 1989), but these side chains are not required for binding of TGF-β (Cheifetz and Massagué, 1989; López-Casillas *et al.*, 1994). This receptor binds all three TGF-β isoforms with high affinity ($3\text{–}30 \times 10^{-11}$), with TGF-$\beta_2$ being the highest (Cheifetz *et al.*, 1987, 1988; Segarini *et al.*, 1987; Segarini and Seyedin, 1988; López-Casillas *et al.*, 1994). Betaglycan is encoded as an 853-amino-acid protein with a short cytoplasmic domain without known signaling motifs, consistent with the current belief that this receptor does not directly mediate any signaling activities (López-Casillas *et al.*, 1991; Wang *et al.*, 1991). Both transmembrane and soluble forms of betaglycan can be detected (Andres *et al.*, 1989). Another type III receptor, endoglin, bears some homology to betaglycan, and binds TGF-β_1 and TGF-β_3 (Gougos and Letarte, 1990; Yamashita *et al.*, 1994a). This transmembrane glycoprotein has a more limited expression pattern than betaglycan and is expressed primarily by vascular endothelial cells and several hematopoietic cell types (Cheifetz *et al.*, 1992). Whereas the overall sequence identity is only about 25%, their intracellular and transmembrane domains are more highly conserved, although the biologic significance of this homology is unknown.

A likely role for the type III receptors is that of modulating ligand binding to the type II receptor. While the soluble form of betaglycan may competitively inhibit binding of TGF-β to cell surface receptors (López-Casillas *et al.*, 1994; Pepin *et al.*, 1994, 1995), expression of membrane-anchored betaglycan increases binding of TGF-β to the type II receptor (López-Casillas *et al.*, 1991; Wang *et al.*, 1991). The enhancement of ligand binding is particularly significant for the binding of the TGF-β_2 isoform, which has very low affinity for the type II receptor. When betaglycan is coexpressed with type II receptor, the antiproliferative response to TGF-β_2 increases tenfold (López-Casillas *et al.*, 1993). This situation is somewhat reminiscent of the heparan sulfate-enhanced presentation of basic fibroblast growth factor (bFGF) to the FGF receptor (Yayon *et al.*, 1991). Remarkably, betaglycan can also bind bFGF and could thus play a role in the presentation of this growth factor to its receptor as well (Andres *et al.*, 1991). Both betaglycan and endoglin have been found in association with the type II receptor in the presence of ligand (López-Casillas *et al.*, 1993; Moustakas *et al.*, 1993, Yamashita *et al.*, 1994a; Zhang *et al.*, 1996a). While the developmental consequences of targeted gene inactivation of the type III receptors have yet to be reported, a naturally occurring mutation in endoglin has been implicated as the cause of hereditary hemorrhagic telangiectasia type I in humans (McAllister *et al.*, 1994).

The type II TGF-β receptor is a transmembrane serine–threonine kinase encoded as a 567-amino-acid polypeptide (Lin *et al.*, 1992). Its glycosylated extracellular domain is rich in cysteines, and its intracellular domain contains all features and consensus sequences characteristic of serine–threonine kinases. Transfections of cells with a mutated receptor lacking the active kinase site indicate that the kinase activity is essential for signaling and subsequent biologic activities (Wrana *et al.*, 1992). Comparison of the type II receptor sequence with other serine–threonine kinase receptors (e.g. the activin type II receptors (Mathews and Vale, 1991; Attisano *et al.*, 1992), the BMP-2/4 type II receptor (Liu *et al.*, 1995; Nohno *et al.*, 1995; Rosenzweig *et al.*, 1995) and the *C. elegans* receptor daf-1 (Georgi *et al.*, 1990)) indicates striking sequence homologies in the extracellular and intracellular domains (Lin *et al.*, 1992). The extracellular domains are short and have an abundance of cysteines, including a characteristic cysteine cluster upstream from the transmembrane region, yet the overall sequence identity is limited, allowing for specificity in ligand binding and/or signaling by these different receptors.

The type II receptor independently binds TGF-β_1 and TGF-β_3 with high affinity, but requires either the type I or the type III receptor for binding of the TGF-β_2 isoform (Lin *et al.*, 1992, 1995; López-Casillas *et al.*, 1993; Lawler *et al.*, 1994; Rodriguez *et al.*, 1995). This lack of binding to TGF-β_2, together with the observation that the mouse type II receptor is not expressed at sites of TGF-β_2 expression during development, has raised the possibility that additional type II TGF-β receptors may exist (Lawler *et al.*, 1994).

Irrespective of ligand binding, the type II receptor exists as a constitutive homodimer (Chen and Derynck, 1994; Henis *et al.*, 1994), and is constitutively autophosphorylated (Wrana *et al.*, 1994). *In vivo*, autophosphorylation occurs primarily on five serine residues: Ser-223, Ser-226 and Ser-227 in the juxtamembrane domain, and Ser-549 and Ser-551 in the *C*-terminal tail beyond the kinase domain (Souchelnytskyi *et al.*, 1996); *in vitro*, autophosphorylation occurs primarily on threonine (Lin *et al.*, 1992). The recombinant receptor phosphorylates exogenous substrates and autophosphorylates not only on serine and threonine, but also on tyrosine. One of the autophosphorylated tyrosines is highly conserved among all serine–threonine kinase receptors and its replacement greatly impairs kinase activity; thus, tyrosine autophosphorylation may play an important regulatory role for the kinase (Lawler *et al.*, 1997). In addition, other kinases may exist which phosphorylate and potentially regulate the type II receptor kinase activity, because kinase-defective point mutants of the receptor expressed in mammalian cells are still phosphorylated, albeit to a significantly lesser extent than kinase-active receptor (Wrana *et al.*, 1994; Chen and Weinberg, 1995).

An important substrate for the type II receptor kinase is the type I receptor (Wrana *et al.*, 1994). The type I TGF-β receptors, called TβRI/R4/ALK-5 (Franzén *et al.*, 1993; ten Dijke *et al.*, 1994), Tsk7L/ActRI (Ebner *et al.*, 1993a,b) and TSR-1 (Attisano *et al.*, 1993), are also members of the transmembrane serine–threonine kinase receptor family. Their cytoplasmic domains do not have an extension beyond the kinase domain and are minimal in size. In addition, a highly conserved signature sequence among type I receptors, SGSGSGLP (GS domain), immediately precedes the kinase domain. The extracellular domain has a defined sequence similarity with the type II TGF-β receptor but is shorter and has only one *N*-glycosylation site. The type I receptors do not bind TGF-β independently; in fact, they require coexpression with the type II receptor for binding of TGF-β at the cell surface (Ebner *et al.*, 1993a,b). The required coexpression of the type II receptor for TGF-β binding to the type I receptor is consistent with observations in mutant cell lines that the type II receptor can rescue cell surface binding to the type I receptor (Wrana *et al.*, 1992). The type II receptor is often also important in determining the specificity of ligand bound to the type I receptor. Accordingly, Tsk7L/ActRI and TSR-1 bind activin or TGF-β, depending on which type II receptor is coexpressed (Attisano *et al.*, 1993; Ebner *et al.*, 1993a,b). To what extent each of these type I receptors actually functions as a TGF-β receptor remains to be characterized. It is known that the TβRI type I receptor binds TGF-β with a higher efficiency than the other type I receptors, is often the major type I receptor present, and mediates TGF-β responsiveness in Mv1Lu cells for several responses (Franzén *et al.*,

1993; Bassing *et al.*, 1994; ten Dijke *et al.*, 1994). However, Tsk7L is a functional TGF-β receptor in NMuMG cells which lack detectable TβRI (Miettinen *et al.*, 1994).

Similarly to the type II receptor, the type I receptors form ligand-independent homodimers (P.J. Miettinen and R. Derynck, unpublished results; R. Wells, Gilboa, H. Lodish and Y. Henis, unpublished results) and are constitutively phosphorylated by as yet unidentified cytoplasmic kinases (Wrana *et al.*, 1994). Coimmunoprecipitation analyses of ^{125}I-TGF-β-crosslinked cell surface receptors indicate that a fraction of the type II and type I receptors are associated with each other (Wrana *et al.*, 1992; Ebner *et al.*, 1993b; Franzén *et al.*, 1993; Tsuchida *et al.*, 1993; Bassing *et al.*, 1994). This heterotetrameric complex is the effector unit of ligand-induced signaling (Wrana *et al.*, 1992) and, based on both direct biochemical and functional complementation analyses, probably consists of two type II and two type I receptors (Chen and Derynck, 1994; Yamashita *et al.*, 1994b; Luo and Lodish, 1996; Weis-Garcia and Massagué, 1996). However, the type I and II receptors already associate with each other in a ligand-independent fashion (Ventura *et al.*, 1994; Chen and Weinberg, 1995; Chen *et al.*, 1995a; Liu *et al.*, 1995), even at endogenous levels of receptor expression (Wu *et al.*, 1996). Most likely, the type II and type I receptors have an inherent affinity for each other, and ligand binding stabilizes and strengthens the heterotetramer.

The kinase activity of the type II receptor is required for efficient interaction with type I receptor, whereas the kinase activity of the type I receptor is less important for this interaction (Chen *et al.*, 1995a). The type I receptor is phosphorylated by the type II receptor on serine and threonine following ligand binding (Wrana *et al.*, 1994). Five sites of ligand-induced phosphorylation have been identified in TβRI: Thr-186, Ser-187, Ser-189 and Ser-191 in the GS domain, and Ser-165 in the juxtamembrane region. With the exception of Ser-165, which is phosphorylated by the type II receptor kinase, it is unknown whether these sites are phosphorylated directly by the type II receptor kinase or by a subsequent autophosphorylation event (Souchelnytskyi *et al.*, 1996). Phosphorylation by the type II receptor is believed to initiate activation of the type I receptor and subsequent signaling by the receptor complex (Wrana *et al.*, 1994). The role of the type I receptor as effector of the TGF-β response is based on the observation that a specific point mutation in the GS domain of TβRI (T204D) confers increased kinase activity and constitutive activation of signaling (Wieser *et al.*, 1995). However, the signaling response of this mutant is lower than that of the ligand-bound heteromeric complex (Luo and Lodish, 1996). Furthermore, abolition of type II receptor signaling by overexpression of a dominant negative mutant of the type II receptor inhibited the antiproliferative effect of TGF-β, but not the induction of extracellular matrix proteins (Chen *et al.*, 1993). Thus, while the type I receptor plays an effector role downstream of the type II receptor, it should not be assumed that the type II receptor functions merely as the kinase that activates the type I receptor.

TGF-β RECEPTOR SIGNALING MEDIATORS

To identify the intracellular mediators of signal transduction by TGF-β receptors, much emphasis has been placed on the identification of the proteins that are associated with and are phosphorylated by the receptors. Several receptor-associated proteins have been identified using the yeast two-hybrid approach, utilizing the receptor cytoplasmic domains as bait. Using the type I receptor TβRI cytoplasmic domain as bait, FKBP12

(Wang *et al.*, 1994), the α chain of farnesyl transferase (Kawabata *et al.*, 1995), and apolipoprotein J (Reddy *et al.*, 1996) have been identified as interacting proteins; using the type II receptor as bait, a novel WD-repeat containing protein, TGF-β receptor interacting protein-1 (TRIP-1) (Chen *et al.*, 1995b), was identified.

FKBP12 associates with the TGF-β type I receptor, as well as other TGF-β superfamily type I receptors, but not with type II receptors, both *in vitro* and *in vivo*; this association is irrespective of the receptor's kinase activity. FKBP12 is an abundant 12-kDa protein that associates with two macrolides, FK506 and rapamycin, and is involved in their immunosuppressive activities. When complexed with either macrolide, FKBP12 inactivates the serine–threonine phosphatase calcineurin and the serine kinase FRAP/RAFT1, which is believed to affect regulation of translation (Brown and Schreiber, 1996). FKBP12 also has peptidyl-prolyl *cis–trans* isomerase activity, which may be important for protein folding (Kay, 1996). The association of FKBP12 with the receptor is believed to inhibit TGF-β signal transduction (Wang *et al.*, 1996a), but further studies will be required to elucidate better the mechanism and significance of the role of FKBP12 in TGF-β signaling. The α chain of farnesyl transferase (FT-α) is a subunit of both farnesyl transferase and geranylgeranyl transferase, and thus is involved in modification of p21ras and G proteins (Omer and Gibbs, 1994). Like FKBP12, FT-α interacts with other type I receptors besides TβRI (except with TSR-1) in yeast two-hybrid analyses, and not with type II receptors (Kawabata *et al.*, 1995; Ventura *et al.*, 1996; Wang *et al.*, 1996b). *In vivo*, FT-α can be found associated with TβRI when both proteins are transfected into cells, but not in the ligand-bound heteromeric receptor complex (Wang *et al.*, 1996b). No functional connection has yet been made between TGF-β responses, FT-α association, and protein farnesylation or prenylation levels (Ventura *et al.*, 1996); thus, the physiologic relevance of the FT-α type I receptor interaction in TGF-β signaling is unknown.

Apolipoprotein J, an abundant serum protein, and its ubiquitous intracellular form clusterin, have been found to interact with both the type I and type II TGF-β receptors, both *in vitro* and in coimmunoprecipitation analyses of transfected cells (Reddy *et al.*, 1996). Ligand treatment induces nuclear translocation of apolipoprotein J, and a possible role for this protein in TGF-β receptor trafficking has been suggested, but the physiologic relevance of this association has also not yet been demonstrated.

TRIP-1, a ubiquitously expressed protein with five WD repeat motifs but no other evident structural or enzymatic motifs, has been found to associate specifically with and to be phosphorylated by the TGF-β type II receptor, but not the type I receptor, both *in vitro* and in cotransfected cells (Chen *et al.*, 1995b). WD repeat motifs are found in a variety of proteins and are thought to be involved in protein–protein interactions (Neer *et al.*, 1995), suggesting that TRIP-1 may play a role as a docking protein associated with the type II receptor. TRIP-1 has subsequently also been identified as a subunit of eIF3 (Naranda *et al.*, 1997), which is required for initiation of translation. This function of TRIP-1 raises the possibility of a connection of TGF-β signaling to translational control, similarly to FKBP12; however, no data have yet been presented to address the actual role of TRIP-1 with respect to a function in the TGF-β signaling pathway.

Thus far, none of these receptor-associated proteins identified by direct interaction cloning methods has been shown to play a role in propagation of the TGF-β signal from the receptors. Genetic approaches have proven more fruitful in this regard. Genetic analyses of signaling by TGF-β-related factors in *Drosophila* and *C. elegans* have led to the identification of genes, called Mad (from *m*others *a*gainst *d*ecapentaplegic) and Sma respectively, involved in the signaling pathways for TGF-β-related factors (Raftery *et al.*, 1995; Savage *et al.*, 1996). Subsequent searches for vertebrate homologs for these genes have thus far uncovered several functional homologs, called Smads (Derynck and Feng, 1997). These genes do not contain any previously known structural or enzymatic motifs, but they do contain two domains in the *N*- and *C*-terminal thirds of the proteins that are highly conserved among the Smads, Mad and Sma genes. Smads have been shown to mimic biologic responses to TGF-β (Chen *et al.*, 1996; Lagna *et al.*, 1996; Zhang *et al.*, 1996b) and activin (Baker and Harland, 1996; Eppert *et al.*, 1996; Liu *et al.*, 1996; Zhang *et al.*, 1997) as well as BMP-2/4 (Graff *et al.*, 1996; Liu *et al.*, 1996; Newfeld *et al.*, 1996). Thus, Smad1 in cooperation with Smad4 induces a BMP-2/4 response, whereas Smad2 and 3, also in cooperation with Smad4, induce a TGF-β or activin-like response (Lagna *et al.*, 1996; Zhang *et al.*, 1996b, 1997). Furthermore, Smad2 and 3 are phosphorylated in response to TGF-β or activin, and can associate transiently with these receptor complexes, whereas Smad1 is phosphorylated following BMP-2/4 receptor activation (Macias-Silva *et al.*, 1996; Zhang *et al.*, 1996b; Kretzschmar *et al.*, 1997). Finally, the response to TGF-β is inhibited by overexpressing a dominant negative *C*-terminally truncated form of Smad3 or Smad4 (Zhang *et al.*, 1996b). These observations suggest that individual Smad types might be specific mediators for defined receptor complexes and that type I receptor activation results in phosphorylation and activation of these Smads (Fig. 2).

Following activation of the receptors, the phosphorylated Smads interact both physically and functionally with Smad4/DPC4 (deleted in pancreatic cancer 4). The functional interactions are illustrated by the synergy between Smad4 and the other Smads. Thus, Smad2 and Smad3 synergize with Smad4 to induce a strong ligand-independent TGF-β- or activin-like response, whereas Smad1 and Smad4 synergize to induce a BMP-2/4-like response (Lagna *et al.*, 1996; Zhang *et al.*, 1996b, 1997). The

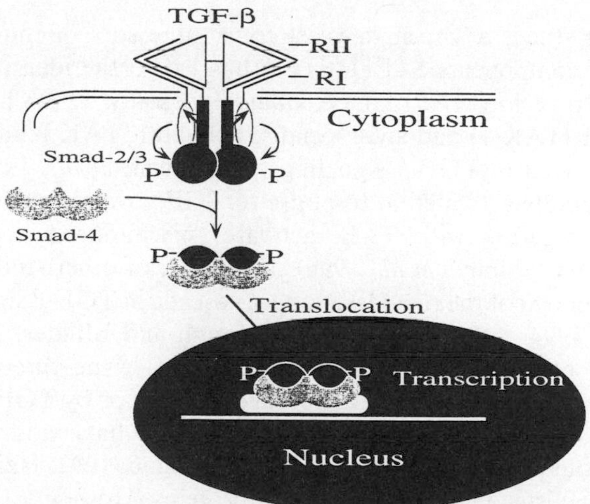

Fig. 2. Working model of TGF-β receptor activation and resulting Smad activity. TGF-β-related growth factors may possess similar signaling pathways, but utilize different Smads and Smad-interacting proteins to activate ligand-specific transcription.

required cooperativity is also apparent in experiments in which a *C*-terminally truncated Smad4 acts as a dominant negative inhibitor of not only the TGF-β response but also of the BMP-2/4 response (Lagna *et al.*, 1996; Zhang *et al.*, 1997). Finally, overexpression of a *C*-terminally truncated Smad4/DPC4 results in an inability of not only wild-type Smad4, but also wild-type Smad1, 2 and 3 to enter the nucleus (Zhang *et al.*, 1997).

Based on these observations, Smad4 is thought to function as a central and essential downstream mediator for the other Smads that are phosphorylated following receptor activation. A physical association of the phosphorylated Smads with Smad4/DPC4 is the basis for this functional cooperativity. Whereas individual Smads associate in a homomeric complex, they additionally engage in heteromeric interactions with Smad4 (Wu *et al.*, 1997).

Structural analysis suggests that the heteromeric complex consists of three phosphorylated Smad polypeptides such as Smad1, 2 or 3, and Smad4 proteins (Shi *et al.*, 1997). This heteromeric association is stabilized by the *C*-terminal Smad phosphorylation (Lagna *et al.*, 1996). Based on analyses of various mutants, both homomeric and heteromeric interactions are required for Smad activity (Wu *et al.*, 1997). Furthermore, deletion of the *C*-terminal sequence of Smads destabilizes the heteromeric and homomeric interactions, which results in their inactivity (Zhang *et al.*, 1996b) and allows these mutants to act as dominant negative mutants (Lagna *et al.*, 1996; Zhang *et al.*, 1996b). The heteromeric Smad complex has the ability to translocate into the nucleus, and increased nuclear accumulation of Smads is apparent following receptor activation (Macias-Silva *et al.*, 1996). Once in the nucleus, Smads probably function as transcription factors (Liu *et al.*, 1996; Wu *et al.*, 1997). Whereas it is as yet unclear how Smads activate (or inhibit) transcription, Smad2 has been shown to interact in an activin-dependent way with the winged helix protein FAST-1, which in turn binds to an activin-responsive promoter element of the Mix-1 gene (Chen *et al.*, 1996). The mechanisms of nuclear translocation and transcriptional activation by Smads are as yet largely unknown.

Another genetic study, a screen in yeast for suppressors of mutations in α-factor signaling pathway components STE11–STE7, has led to the identification of a novel mitogen-activated protein (MAP) kinase kinase kinase (MAPKKK), named TGF-β-associated kinase-1 (TAK-1) and an associated activator, TAK binding protein (TAB) which may be involved in TGF-β signaling (Yamaguchi *et al.*, 1995; Shibuya *et al.*, 1996). TAK-1 is rapidly activated in response to TGF-β (Yamaguchi *et al.*, 1995) and, when transfected together with TAB, activates plasminogen activator inhibitor-1 (PAI-1) transcription (Shibuya *et al.*, 1996). A number of other studies have suggested the possible involvement of the ras–MAP kinase cascade in TGF-β signaling in some cell types (Yan *et al.*, 1994; Atfi *et al.*, 1995; Hartsough and Mulder, 1995; Mucsi *et al.*, 1996), but not in others (Chatani *et al.*, 1995). The serine–threonine phosphatase, phosphatase 1, has also been reported to be rapidly activated by TGF-β in keratinocytes (Gruppuso *et al.*, 1991). Finally, a variety of reports have addressed the possible involvement of protein kinase C (Ohtsuki and Massagué, 1992; Halstead *et al.*, 1995), phospholipase C (Halstead *et al.*, 1995) and G proteins (Howe *et al.*, 1989; Kataoka *et al.*, 1993; Halstead *et al.*, 1995) in TGF-β signaling, but the data do not yet clearly support their direct involvement. More work is needed to determine the role of these known signal transduction proteins in the modulation or propagation of TGF-β signals.

THE MULTIFUNCTIONAL NATURE OF TGF-β

TGF-β exerts a large variety of biologic functions in most cells. The wide spectrum of target cells is due to the presence of cell surface TGF-β receptors in nearly all cells in culture. The nature of the biologic effects of TGF-β depends critically on several parameters including the cell type, culture conditions and cellular environment, the presence and nature of other growth factors, and the differentiation state, and, thus, on the general physiologic context. We will here only briefly outline the several classes of activities that have been extensively reviewed in detail by Roberts and Sporn (1990). In the next few sections, we will then focus on the role of TGF-β in some defined contexts.

Many TGF-β-induced effects can be grouped in several categories. One set of activities is related to the potent growth regulatory effect of TGF-β. TGF-β stimulates proliferation or acts as a potent antiproliferative factor depending on the cell type and cell line. We will further elaborate on this activity of TGF-β in a further section. An additional set of activities relates to the interaction of TGF-β with the extracellular matrix (Roberts and Sporn, 1990). In most cells of non-hematopoietic origin, TGF-β modifies the cellular interaction with the surrounding extracellular matrix and/or basement membrane in several different ways (although, as indicated before, the specific response will depend on the cell type and its physiologic environment).

First, TGF-β induces the synthesis and secretion of many proteins of the extracellular matrix, including several collagens, fibronectin, thrombospondin, tenascin, osteopontin, osteonectin and proteoglycans. The increased deposition of these proteins should result in an increased formation of extracellular matrix. Second, TGF-β increases the expression of many integrins. These are heterodimeric, transmembrane cell adhesion receptors, which, depending on the combination of the α and β polypeptide chains, interact with defined structural proteins in the extracellular matrix or basement membrane. These interactions mediate not only cell adhesion but also integrin-mediated intracellular signaling which affects gene expression and cellular differentiation (Damsky and Werb, 1992). The effect of TGF-β on integrin expression again strongly depends on the cell type and on the type of integrin, but frequently results in increased adhesiveness of the cells to the matrix. Third, TGF-β treatment also regulates the synthesis and activity of secreted proteases and protease inhibitors. The generally observed TGF-β-mediated decrease of protease secretion should inhibit degradation of extracellular matrix, whereas increased synthesis of protease inhibitors further accentuates the accumulation of extracellular matrix and/or basement membrane. Finally, TGF-β is also a potent chemoattractant for several cell types, especially monocytes (Wahl *et al.*, 1987) and fibroblasts (Postlethwaite *et al.*, 1987). This activity may be very important at sites of wound healing or other repair processes, where TGF-β is locally activated and may be a major effector of monocyte–macrophage and fibroblast influx.

EFFECTS OF TGF-β ON CELL PROLIFERATION

TGF-β induces a mitogenic or antiproliferative effect depending on the cell line and cell type (Massagué and Polyak, 1995). It promotes cell proliferation in culture in several cell types, predominantly of mesenchymal origin. Examples include fibroblasts (Leof *et al.*, 1986), osteoblasts (Centrella *et al.*, 1987), smooth muscle cells (Battegay *et al.*, 1990) and Schwann cells (Ridley *et al.*, 1989). In at least one fibroblast cell line, TGF-β induces an increased number of cell surface EGF receptors (Assoian *et al.*, 1984), suggesting that increased sensitivity to EGF or the related TGF-α in the medium could be the mechanistic basis for the TGF-β-induced increase in proliferation and the synergistic effect of TGF-β and EGF or TGF-α on anchorage independence (Roberts *et al.*, 1981; Anzano *et al.*, 1983).

The mitogenic effect of TGF-β on some cell types is indirect and may be due to stimulated production of platelet-derived growth factor (PDGF), which acts in an autocrine manner. TGF-β induces c-*sis*-mRNA which encodes the PDGF-B chain, and PDGF (or PDGF-like) protein in the AKR-2B fibroblastic cell line (Leof *et al.*, 1986), whereas, in human foreskin fibroblasts, TGF-β stimulates expression of the A-chain gene of PDGF (Soma and Grotendorst, 1989). In smooth muscle cells, fibroblasts and chondrocytes, TGF-β induces a bimodal response in proliferative behavior (Battegay *et al.*, 1990). Thus, treatment of these cells with a concentration of TGF-β that is ten times higher than that which stimulated DNA synthesis, nearly abolished the mitogenic response. At lower concentrations, TGF-β induced PDGF-A expression, whereas higher concentrations of TGF-β inhibited the expression of the PDGF receptor α subunit. In this manner, a PDGF-based autocrine growth loop stimulated by low concentrations of TGF-β was blocked at the receptor level by higher concentrations of TGF-β (Battegay *et al.*, 1990). Thus, at least in these cells, the effects of TGF-β on cell proliferation reflect the modulation by TGF-β of the autocrine growth regulation by PDGF; however, in other cells, other autocrine growth factors, such as connective tissue growth factor (Igarashi *et al.*, 1993), may be induced in response to TGF-β and contribute to the growth stimulatory response. In addition, the TGF-β-induced synthesis of several extracellular matrix proteins and their integrin receptors (Roberts and Sporn, 1993) is likely to result in altered cell matrix interactions which may affect the growth response (Morton and Barrack, 1995) and overcome direct TGF-β-induced effects on the cell cycle machinery (Herrera *et al.*, 1996).

Exogenous TGF-β administration *in vivo* has been shown to result in an increased cell density, fibrosis and angiogenesis (Roberts *et al.*, 1986), but there is as yet little evidence documenting increased cell proliferation in response to TGF-β *in vivo*. On the contrary, an increased cell mass in the chicken chorioallantoic membrane system following administration of TGF-β was observed concomitantly with decreased mitogenic activity in the newly formed tissue, suggesting that the increased cell mass was largely the result of cellular influx (Yang and Moses, 1990). Thus, hypercellular lesions at the site of TGF-β injection may be due to an influx of cells as a consequence of the chemotactic activity of TGF-β, rather than a TGF-β-induced increase in cell proliferation.

TGF-β is a potent inhibitor of cell proliferation of many cell types, which, in several cases, functions indirectly by inhibiting the proliferation induced by other growth factors. For example, TGF-β inhibits the FGF-stimulated proliferation of fibroblast cells without affecting the binding of the growth factor to its receptor, or many of the cellular events that occur within the first few hours after mitogen stimulation (Like and Massagué, 1986; Chambard and Pouysségur, 1988). However, in many cells, TGF-β inhibits cellular proliferation directly. The mechanism of TGF-β-induced growth arrest has been best studied in epithelial cells, most notably Mv1Lu cells and HaCaT keratinocytes. In epithelial cells, TGF-β arrests the cell cycle in late G1, when administered during early G1 (Laiho *et al.*, 1990a; Geng and Weinberg, 1993). Following

G1 to S transition, the cells become largely unresponsive to TGF-β and extracellular mitogenic factors. This growth arrest induced by TGF-β, like that induced by serum starvation, is accompanied by maintenance of the retinoblastoma protein, pRB, in its underphosphorylated form, which is the likely basis for growth inhibition at late G1 (Laiho *et al.*, 1990a). In actively dividing cells, pRB is dephosphorylated as cells exit mitosis, and is phosphorylated just before G1 to S transition. In its underphosphorylated form, pRB acts as a growth suppressor and prevents G1 to S transition, presumably via binding to E2F and several other transcriptional regulators. Phosphorylation of pRB results in release of E2F, and transcriptional activation of a set of genes required for entry into the S phase, cell cycle progression and DNA replication (Weinberg, 1995, 1996).

To gain insight into how TGF-β induces growth arrest and pRB phosphorylation, the effects of TGF-β on G1 cyclins and associated cyclin-dependent kinases (cdks) and cdk inhibitors have been studied. Among the many cyclins and cdks, cdk4 and cdk6 are the major catalytic partners of D-type cyclins, whereas cdk2 interacts with cyclin E. The cyclin D–cdk4 and cyclin E–cdk2 complexes are thought to phosphorylate pRB (Sherr, 1995; Weinberg, 1995). In the epithelial cells studied, TGF-β treatment did not affect the moderate and transient increases of cyclin D expression during G1 progression, but did inhibit the concomitant induction of cdk4 expression (Ewen *et al.*, 1993; Geng and Weinberg, 1993). In HaCaT cells, TGF-β treatment also nearly abolished the induction of cyclin E and its partner cdk2 that normally occurs later in G1 (Geng and Weinberg, 1993); however, this TGF-β effect did not occur in Mv1Lu cells (Ewen *et al.*, 1993; Koff *et al.*, 1993). Finally, TGF-β blocks induction of cyclin A expression, which normally occurs shortly before the late G1 stage of growth arrest induced by TGF-β (Geng and Weinberg, 1993). Constitutive overexpression of cdk4 but not of cdk2 in Mv1Lu cells results in phosphorylation of pRB in late G1 and entry into S phase even in the presence of TGF-β, thus effectively inducing resistance to the antiproliferative effect of TGF-β (Ewen *et al.*, 1993). Furthermore, TGF-β-induced downregulation of cdk4 inhibits cdk2 activation in Mv1Lu cells, event though the expression of cyclin E and cdk2 is not affected in these cells (Ewen *et al.*, 1993). These results emphasize the role of cdk4 in TGF-β-induced growth arrest in late G1 and underscore a functional link between cdk4 and cdk2 activation.

Several cdk inhibitors interact with cyclin–cdk complexes and inactivate their kinase activity. One class of inhibitors, which includes p21$^{\text{Cip1}}$, p27$^{\text{Kip1}}$ and p57$^{\text{Kip2}}$, inhibits the activity of cyclin D–cdk4 or cdk6 and cyclin E–cdk2 by binding directly to these complexes. These inhibitors also indirectly prevent cdk activation by Cak, the activating kinase complex. The second class of cdk inhibitors comprises p16$^{\text{Ink4}}$ and p15$^{\text{Ink4B}}$, and several other closely related cdk inhibitors, which prevent cdk activation by interfering with the interaction of cdk4 and cdk6 to cyclin D (for reviews see Hunter and Pines, 1994; Sherr, 1995; Sherr and Roberts, 1995). TGF-β-mediated growth arrest is thought to require the cooperative interactions of both types of cdk inhibitors. In Mv1Lu cells, TGF-β induces the synthesis of p15$^{\text{Ink4B}}$, which binds to and inhibits cdk4 and cdk6, and prevents their interaction with cyclin D (Hannon and Beach, 1994; Reynisdóttir *et al.*, 1995). As a result, p27$^{\text{Kip1}}$ is displaced from cyclin D–cdk4 and binds to the cyclin E–cdk2 complexes, inhibiting their activity. Thus, the increased action of p27$^{\text{Kip1}}$ against cdk2 complexes results not from increased synthesis, but from a displacement event caused by increased p15$^{\text{Ink4B}}$ synthesis and activity (Reynisdóttir *et al.*, 1995). The effects of TGF-β leading to cell cycle arrest in HaCaT cells are in many ways similar to those in Mv1Lu cells, but, in addition, TGF-β induces a rapid increase in p21 synthesis, which results in increased association of p21 with cdk2 and resulting inactivation of cdk2–cyclin E complexes (Datto *et al.*, 1995; Reynisdóttir *et al.*, 1995). In conclusion, the primary events that initiate the TGF-β-induced growth arrest may be associated with increased expression of $p15^{Ink4B}$ and $p21^{Cip1}$ (Reynisdóttir *et al.*, 1995).

TGF-β IN THE IMMUNE SYSTEM

There is considerable evidence for an immunomodulatory role of TGF-β and its function as a potent differentiation modulating and immunosuppressive agent (for a review see Letterio and Roberts, 1998). In culture, most cells of the immune system, such as lymphocytes, monocytes and dendritic cells, synthesize TGF-β that is almost exclusively TGF-β_1. The TGF-β released by these cells is mostly latent, suggesting that activation of the latent complex is required before TGF-β can exert its function. The target cells are not only the different leukoctye cell types, but also various other cell types in the immediate tissue environment, such as endothelial cells or tumor cells, indicative of autocrine and paracrine actions. Once activated, TGF-β exerts a variety of activities on proliferation, differentiation and function of various cell types of the immune system. TGF-β often inhibits cell proliferation as has been shown for T lymphocytes, B lymphocytes, thymocytes, large granular lymphocytes, natural killer (NK) cells, and lymphocyte-activated killer (LAK) cells (Kehrl *et al.*, 1986a,b; Ristow, 1986; Rook *et al.*, 1986; Kuppner *et al.*, 1988; Wahl *et al.*, 1988a,b; Ortaldo *et al.*, 1991). Furthermore, the response may depend strongly on the differentiation state of the cell and on the context of cytokines to which the cells are exposed. Accordingly, these different types of immune cells all have TGF-β receptors. How the expression of the receptors and the TGF-β responsiveness is regulated is poorly understood.

One of the cell populations which is strongly regulated by TGF-β are the T lymphocytes. Activated T cells secrete TGF-β and exogenous TGF-β inhibits interleukin-2-dependent proliferation (Kehrl *et al.*, 1986a). In addition, TGF-β also inhibits the secretion of a variety of cytokines and in this way inhibits the effector function of the activated cells. Thus, TGF-β inhibits the effects and/or the production of interferon-γ, tumor necrosis factors (TNF)-α and -β and interleukins-1, -2 and -3, and the expression of interleukin-2 receptor (Espevik *et al.*, 1987, 1988; Ohta *et al.*, 1987; Ranges *et al.*, 1987; for a review see Letterio and Roberts, 1998). The endogenous synthesis of TGF-β and the increased cytokine production in TGF-β_1-deficient mice (Shull *et al.*, 1992) strongly suggest that cytokine production is under autocrine control of TGF-β. The inhibition of cytokine production and the antiproliferative effect are presumably the major effector functions in the TGF-β-induced immunosuppression, especially as it relates to the cytolytic activities of the immune system. Accordingly, TGF-β has been shown to be a strong inhibitor of the generation and cytolytic activity of cytotoxic T cells, NK cells and LAK cells, and to suppress the natural and lymphokine-activated killing by large granular lymphocytes (Roberts and Sporn, 1990; Kehrl, 1991).

Further evidence for the strong immunosuppressive effect of TGF-β on lymphocytes comes from the observation that TGF-β downregulates interferon-γ-induced major histocompatibility complex (MHC) class II antigen expression by both lymphoid and non-lymphoid cells (Czarniecki *et al.*, 1988). In contrast to the inhibitory effects of

TGF-β on T lymphocytes, TGF-β also enhances the growth of immature lymphocytes (Cerwenka *et al.*, 1994) and inhibits T-cell apoptosis, thus allowing expansion of the effector cell population (Cerwenka *et al.*, 1996; Rich *et al.*, 1996). These effects may seem to contradict the previously mentioned activities, but they illustrate how the response to TGF-β depends on the differentiation of the responsive cell.

Finally, TGF-β may play an important regulatory role in differentiation of T cells. For example, TGF-β can induce expression of CD8 and can synergize with TNF-α to favor the development of cells expressing CD8. Conversely, TGF-β inhibits later stages of differentiation and thus inhibits the generation of $CD4^+ CD8^+$ cells (Suda and Zlotnik, 1992; Takahama *et al.*, 1994; Mossalayi *et al.*, 1995). Whereas TGF-β was initially shown strongly to promote the generation of T_{H1} cells, a subset of helper T cells with a characteristic cytokine secretion pattern (Swain *et al.*, 1991), TGF-β has recently been more clearly implicated as a positive regulator of differentiation of T_{H2} cells and a suppressor of differentiation of T_{H1} cells (Schmitt *et al.*, 1994; Strober *et al.*, 1997).

TGF-β also plays an important role in the function of B cells. The expression of TGF-β is regulated and is likely to depend on the differentiation state of these cells. For example, activation of B cells with *Staphylococcus aureus* Cowan mitogen strongly increases TGF-β_1 secretion without a major effect on mRNA levels (Kehrl *et al.*, 1986a), whereas lipopolysaccharide (LPS) stimulates the activation of TGF-β, which is required for LPS-induced immunoglobulin G (IgG) secretion by B cells (Snapper *et al.*, 1993). Furthermore, activation of the cells by binding of an anti-IgM antibody to cell surface IgM also induces the secretion of active TGF-β_1 (Warner *et al.*, 1992; Matthes *et al.*, 1993). The stimulation of TGF-β expression by B cells most likely plays an important autocrine role in B cell function, as TGF-β has a potent ability to inhibit cell proliferation and is able to induce apoptosis in B cells (Holder *et al.*, 1992; Gordon *et al.*, 1994). TGF-β also regulates B-cell differentiation and immunoglobulin expression. Indeed, TGF-β suppresses the expression by activated B lymphocytes of membrane immunoglobulin and decreases secretion of immunoglobulins of different classes (Kehrl *et al.*, 1986a, 1991; reviewed by Stavnezer, 1996). However, antibodies to TGF-β decrease the secretion of various immunoglobulin classes without affecting IgM secretion, suggesting that low levels of TGF-β may regulate and be required for immunoglobulin secretion (Snapper *et al.*, 1993). Furthermore, the presence of TGF-β promotes isotype switching of immunoglobulins, likely associated with differential transcriptional regulation (Coffman *et al.*, 1989; for a review see Stavnezer, 1996).

TGF-β also plays an important role in the physiology and functional regulation of monocytes and macrophages. As with cells of many other hematopoietic cell lineages, monocytes and macrophages secrete TGF-β_1 in a highly regulated fashion. In addition, the activation of the latent TGF-β complex is also highly regulated. For example, treatment of monocytes with LPS or macrophages with concanavalin A results in a drastically increased activation of TGF-β without changes in mRNA levels (Assoian *et al.*, 1987; Grotendorst *et al.*, 1989). Such stimulation of monocytes induces secretion of plasmin and cathepsin D, both of which are able to activate TGF-β (Nathan, 1987), and plasmin, urokinase, tissue type II transglutaminase and the mannose-6-phosphate receptor all have been shown to be involved in activation of latent TGF-β by macrophages (Nunes *et al.*, 1995). In addition, TGF-β_1 synthesis by monocytes is strongly induced by TGF-β itself (McCartney-Francis *et al.*, 1990). This autoinduction of TGF-β synthesis, which also occurs in many other cell types, may play a role in amplifying the autocrine response to TGF-β (Wahl *et al.*, 1989).

As with lymphocytes, TGF-β exerts a variety of effects on monocytes and macrophages, which depend on the differentiation state of the cells and the presence of other cytokines. Thus, TGF-β can induce or inhibit cell proliferation depending on the physiologic context (Celada and Maki, 1992; Fan *et al.*, 1992). Low concentrations of TGF-β induce a chemotactic response, which may be important for the recruitment of these cells at sites of inflammation, and activate these cells to secrete a variety of cytokines which may play a role in the inflammatory response and often associated angiogenesis (Wahl *et al.*, 1987; McCartney-Francis *et al.*, 1990), and to increase their expression of adhesion receptors (Bauvois *et al.*, 1992; Wahl *et al.*, 1993). The important role of TGF-β is further illustrated by the inflammatory response to subcutaneous injection of TGF-β, which elicits a sequence of inflammatory cell recruitment, fibroblast accumulation and vascular growth, similar to the normal inflammatory response to injury (Roberts *et al.*, 1986). It is thus conceivable that the release of TGF-β by activated degranulating platelets results in an influx of monocytes, which accompanies the initiation of the inflammation, and of fibroblasts, which may explain the appearance of fibroblasts at the site of injury (Postlethwaite *et al.*, 1987). Following recruitment of monocytes, the TGF-β at the site of inflammation then presumably activates monocytes to induce secretion of growth factors and inflammatory mediators, which are likely to amplify further the inflammatory response and to initiate the repair processes at the site of injury.

In contrast to these proinflammatory activities, TGF-β also inhibits a variety of activities of macrophages, which result in a suppression of macrophage function and are likely to contribute to a resolution of the inflammatory response. This drastic change in cell response may be causally related to decreased expression of cell surface TGF-β receptors as cells mature (McCartney-Francis and Wahl, 1994). Thus, TGF-β can inhibit the production of cytokines, including interferon-γ (Fargeas *et al.*, 1992) and increase the expression of the interleukin-1 receptor antagonist (Bauvois *et al.*, 1996) by macrophages. In addition, TGF-β is a potent inhibitor of the macrophage respiratory burst, as it suppresses hydrogen peroxide release by the activated macrophages. This inhibition can be reversed by activating agents such as interferon-γ and TNF-α (Tsunawaki *et al.*, 1988). The anti-inflammatory effect of TGF-β on macrophages is also illustrated by the suppression by TGF-β of TNF-α expression and Ia antigen expression in these cells (Tsunawaki *et al.*, 1988). At the site of inflammation, TGF-β also antagonizes the ability of TNF-α to increase the adhesiveness of the endothelial cells (Gamble and Vadas, 1988). Finally, TGF-β also strongly inhibits expression of nitric oxide synthetase and in this way is likely to downregulate the antimicrobial and tumoricidal response (Bogdan and Nathan, 1993; Vodovotz and Bogdan, 1994). The proinflammatory activity of TGF-β on monocytes and the subsequent deactivating activity on macrophages clearly illustrates the highly coordinated role of TGF-β in the complex response to injury and its modulatory role at the site of repair and inflammation.

The ability of TGF-β to suppress macrophage function also plays an important role in parasitic infection. The protozoan parasites *Trypanosoma cruzi*, which infects all nucleated cells, and *Leishmania*, which uses only macrophages as cellular hosts, both have the ability to induce secretion of active TGF-β, which in turn suppresses macrophage killing and allows intracellular proliferation of the parasite (Barral-Netto *et al.*, 1992; Gazzinelli *et al.*, 1992). Accordingly, exogenous administration of TGF-β in mice increases the infectivity of *Leishmania in vivo* (Barral-Netto *et al.*, 1992). In the case of *T. cruzi*, autocrine TGF-β signaling has been shown to play a role in parasite entry into epithelial cells, and cells lacking functional receptors (thus lacking TGF-β responsiveness) are resistant to trypanosome infection (Ming *et al.*, 1995). The ability of TGF-β to enhance the progression of leishmaniasis may further also depend on its ability to decrease cell-mediated immunity by suppressing the effects of interferon-γ or interleukin-4 on MHC class II antigen expression on antigen-presenting cells (Czarniecki *et al.*, 1988; Gollnick *et al.*, 1995). Furthermore, TGF-β also inhibits differentiation of cytotoxic $CD8^+$ T cells which are instrumental in eliminating infected cells (Barral *et al.*, 1995). Finally, the deactivating activity of TGF-β on macrophages is exploited by the intracellular bacterial pathogen *Mycobacterium avium*, which, similarly to protozoan parasites, increases the production of active TGF-β in macrophages and thereby suppresses their antibacterial activity (Champsi *et al.*, 1995).

Yet another leukocyte cell population whose differentiation and function are profoundly affected by TGF-β are the dendritic cells, which function as the primary antigen presenting cells in the activation of T-lymphocyte responses (Steinman, 1991). Among these, the Langerhans cells in the epidermis and follicular dendritic cells of the lymph nodes are regulated by TGF-β. Indeed, TGF-β is required for differentiation of dendritic cells and protects their viability (Strobl *et al.*, 1996; Riedl *et al.*, 1997). The requirement of TGF-β for their differentiation is clearly underscored by the absence of epidermal Langerhans cells in TGF-β_1-deficient mice (Borkowski *et al.*, 1996, 1997).

The potent activities of TGF-β on immune cells *in vitro* are also reflected in the various effects of TGF-β *in vivo*, such as in various animal model systems, which strongly suggest an important role for TGF-β in autoimmune disorders. Exogenous administration of TGF-β *in vivo* markedly depresses inflammatory and immunologic responses *in vivo*. For example, TGF-β_1 prolongs the time to rejection of rat islet transplants (Carel *et al.*, 1990, 1993), delays the onset of acute and chronic experimental allergic encephalomyelitis (Kuruvilla *et al.*, 1991; Racke *et al.*, 1992) and experimentally induced arthritis (Brandes *et al.*, 1991; Thorbecke *et al.*, 1992) and, furthermore, suppresses the generation of the cytotoxic T-cell response (Fontana *et al.*, 1989).

Endogenous TGF-β may also play a regulatory role in limiting immune function *in vivo* and in some pathologic dysfunctions of the immune system, as suggested by several correlative studies. Increased levels of TGF-β in the synovial fluid of joints of patients with rheumatoid arthritis may explain some of the arthritis-associated physiologic changes and contribute to the inflammation in this tissue context, and perhaps contribute to the functional changes in synovial lymphocytes (Wahl *et al.*, 1988b; Lotz *et al.*, 1990; Wilder *et al.*, 1990). The suppressed responsiveness of these cells in arthritic patients resembles the inhibited responsiveness in lymphocytes and macrophages from mice with experimental chronic arthritis (Wahl *et al.*, 1988b).

Another example of a pathologic role of active TGF-β in immune suppression associated with a specific disease is provided by adult T-cell leukemia (ATL). One of the clinical features of this disease, caused by the human T-lymphotropic virus-1 (HTLV-1), is a suppression of cellular and humoral immunity. ATL mononuclear cells secrete considerably higher levels of TGF-β_1 than their normal counterparts, which could be in part responsible for their inhibited immune function (Kim *et al.*, 1990). Excessive TGF-β production observed in peripheral blood mononuclear cells from patients with acquired immune deficiency syndrome (AIDS), and in human immunodeficiency virus (HIV)-infected monocytes may similarly play a causative role in impairment of the immune response in HIV-infected patients with AIDS (Kekow *et al.*, 1990, 1991; Wahl *et al.*, 1991).

Some new insights into the role of TGF-β in immune suppression and the mechanism of TGF-β presentation are also revealed through studies of MRL/lpr mice, which develop systemic autoimmune disease resembling human systemic lupus erythematosus, probably resulting in part from defective apoptosis (Steinberg, 1994). These mice have high levels of plasma TGF-β_1 bound to IgG antibodies (Rowley *et al.*, 1995; Caver *et al.*, 1996), which as a complex has a potent ability to suppress $CD8^+$ cytotoxic T-lymphocyte responses in the presence of macrophages (Stach and Rowley, 1993). Binding to IgG could restrict or target the activity of TGF-β to antigenic sites where it could play an important role in suppression of autoimmune disease (Stach and Rowley, 1993). Furthermore, the IgG-bound TGF-β complexes could increase the susceptibility to bacterial infection and defects in polymorphonuclear leukocyte function (Gresham *et al.*, 1991; Caver *et al.*, 1996). Finally, TGF-β bound to IgG results in a considerably higher (10–500-fold) activity than free TGF-β, suggesting that this recently discovered form of delivery of TGF-β results in an efficient targeting to the receptors (Stach and Rowley, 1993; Caver *et al.*, 1996).

Perhaps the best illustration of the endogenous role of TGF-β_1 in normal immune function and infiltration is provided by mice in which the TGF-β_1 gene is functionally inactivated by gene targeting. Shortly after birth, these mice develop a multifocal mixed inflammatory disease with rapid and massive infiltration of lymphocytes and neutrophils in many tissues, and subsequently die (Shull *et al.*, 1992; Kulkarni *et al.*, 1993). Thus, TGF-β_1 deficiency results in a severe pathology with dysfunction of the immune and inflammatory systems, which is accompanied by increased production of several cytokine mediators of inflammation, such as interferon-γ, TNF-α and macrophage inflammatory protein-1α (MIP-1α) (Shull *et al.*, 1992). These mice also show enhanced expression of MHC class I and II antigens, circulating antibodies to nuclear antigens, and pathogenic glomerular IgG deposits, thus resembling human autoimmune disease (Geiser *et al.*, 1993; Christ *et al.*, 1994; Dang *et al.*, 1995; for a review see Letterio and Roberts, 1998). The enhanced expression of MHC class II antigen in TGF-β_1 knockout mice is consistent with the ability of TGF-β to suppress its expression (Czarniecki *et al.*, 1988).

In summary, there is ample evidence that TGF-β is a major determinant of the immune and inflammatory response. Among the TGF-β isoforms, TGF-β_1 is probably the major effector molecule in most contexts, and TGF-β_2 and TGF-β_3 may exert their major normal effects outside the immune system.

TGF-β IN NORMAL MAMMALIAN DEVELOPMENT

To gain insight into the role of TGF-β in normal development, several approaches have been used. They include localization of expression of the three TGF-β isoforms and the receptors in the developing mouse, studies of the effects of TGF-β on cellular differentiation *in vitro*, and transgenic manipulation of the expression of TGF-β and its receptors in mice. A comprehensive coverage of the role of TGF-β in cell differentiation and development is not possible within the scope of this review; we will therefore only summarize the results on a few cell and tissue types.

Most studies on the role of TGF-β in normal embryonic development have so far focused on the localization of the expression of the three TGF-β species during embryonic development in the mouse. The resulting database on the temporal and spatial expression patterns (Heine *et al.*, 1987; Lehnert and Akhurst, 1988; Fitzpatrick *et al.*, 1990; Pelton *et al.*, 1990a,b,c; Unsicker *et al.*, 1991), while far from complex, suggests an important role of TGF-β in the development of mesenchymal tissues and skin. Some initial immunolocalization studies may not have accurately provided specificity for the individual TGF-β species, owing to the lack of isoform-specific antibodies. However, the specificity of *in situ* hybridization studies with isogenic cDNA probes and the use of some reliable isoform-specific antibodies have provided an accurate description of the expression patterns.

The three TGF-β isoforms have complex but well defined expression patterns during mouse development, which are often partially overlapping (Pelton *et al.*, 1990c). Most if not all organs express one or several TGF-β isoforms at defined stages during development, usually coinciding with active tissue differentiation and morphogenesis. Striking expression patterns of the different TGF-β species occur during development of many organs, such as the brain, heart, kidney, vascular system and developing hematopoietic system, as well as the skin and mesenchymally derived structures, as will be discussed below. Furthermore, the sites of TGF-β synthesis frequently coincide with the deposition of extracellular matrix components, such as collagens, fibronectin and glycosaminoglycans (Heine *et al.*, 1990). Thus, the localization and activities of TGF-β suggest that TGF-β may play an active role in various differentiation processes and morphogenetic events, which may be closely associated with the activities of TGF-β on extracellular matrix deposition and cell–matrix interactions. There is some evidence for differential localization of TGF-β mRNA *versus* protein in some organs (Pelton *et al.*, 1990c), but further substantiation of this conclusion is required. As an example, some epithelial cell populations synthesize TGF-β mRNA, whereas the protein itself can be found only in the basement membrane or the extracellular matrix of adjacent cell types (Heine *et al.*, 1987; Lehnert and Akhurst, 1988; Fitzpatrick *et al.*, 1990; Pelton *et al.*, 1990a,b,c). Differential localization of mRNA and protein may also occur in the brain, cartilage and bone.

High levels of TGF-β mRNA and protein synthesis are found in tissues of mesodermal origin. TGF-β synthesis is frequently associated with mesenchymal tissues and especially with cartilage and bone formation (Heine *et al.*, 1987; Lehnert and Akhurst, 1988; Fitzpatrick *et al.*, 1990; Pelton *et al.*, 1990a,b,c). In addition, TGF-β is synthesized in mesenchymal cells at sites of epithelial interactions (Heine *et al.*, 1987; Pelton *et al.*, 1990c). The pattern of TGF-β synthesis frequently depends on the stage of differentiation. Undifferentiated mesenchyme, an abundant source of TGF-β synthesis, makes all three TGF-β isoforms, but differentiation into cartilage and bone coincides with a drastic alteration of the expression pattern of the individual TGF-β isoforms. During mesenchymal condensation preceding cartilage formation, the TGF-β_1 synthesis is downregulated strikingly (Pelton *et al.*, 1990a,b,c). Furthermore, osteoblasts and chondrocytes synthesize all three TGF-β isoforms, whereas periosteal and perichondrial cells make predominantly TGF-β_1 and to a lesser extent TGF-β_3 (Pelton *et al.*, 1990a,b,c). Another example of differential expression of the three TGF-β isoforms is the developing tooth. Whereas the pulp cells and the mesodermally derived odontoblasts synthesize high levels of TGF-β_2, the ameloblasts from epithelial origin primarily make TGF-β_1 (Pelton *et al.*, 1990b,c; Heikinheimo *et al.*, 1993). Thus, the embryonic skeletal system is a major site for the activities of TGF-β.

Although the expression patterns of the receptors are less characterized, TGF-β receptors are also localized in developing mesenchymal tissues during development (Lawler *et al.*, 1994; Dewulf *et al.*, 1995). The type II receptor is expressed primarily in the undifferentiated mesenchyme in a pattern reminiscent of the TGF-β_1 expression pattern (Lawler *et al.*, 1994). This coincident expression of ligand and receptor confirms that TGF-β exerts its role in an autocrine and paracrine way during development. No type II receptor mRNA transcripts were detected in condensing mesenchyme that differentiates into cartilage. At these sites, the effects of TGF-β expression and responsiveness to TGF-β are likely to be functionally coordinated with the roles of BMPs in tissue differentiation and organ development (Hogan, 1996).

The epithelium and skin are additional major sites of TGF-β synthesis during development (Heine *et al.*, 1987; Lehnert and Akhurst, 1988; Fitzpatrick *et al.*, 1990; Pelton *et al.*, 1990a,b,c). Epithelium of the intestinal tract also synthesizes high TGF-β levels, whereas the developing epidermis makes all three different TGF-β types (Pelton *et al.*, 1990a,b,c). The expression pattern of the different TGF-β isoforms may again depend on the stage of development. Consistent with the expression of ligand, the developing skin also expresses high levels of type II TGF-β receptors, suggesting an autocrine role for TGF-β in skin development (Lawler *et al.*, 1994; Dewulf *et al.*, 1995). Furthermore, developing skin makes several BMPs, again suggesting the involvement of multiple TGF-β superfamily members in epidermal differentiation (Pelton *et al.*, 1990c).

The effects of TGF-β on mesenchymal cell differentiation have been evaluated primarily in cultured cells. Consistent with the normal mesenchymal expression of TGF-β during embryonic development, TGF-β has been shown to induce phenotypic alterations, and, whereas the results are sometimes conflicting, they nevertheless indicate the potent ability of TGF-β to influence cell differentiation. For example, TGF-β has been shown to inhibit myoblast maturation and concomitant myotube formation of several myoblast cell lines *in vitro*, the mechanism of which is unclear (Massagué *et al.*, 1986; Brennan *et al.*, 1991). However, under some culture conditions, TGF-β also stimulates myotube formation (Zentella and Massagué, 1992) and exposure of embryonic stem cells to TGF-β results in enhanced myogenic differentiation (Slager *et al.*, 1993). The stimulating effect of TGF-β on myoblast differentiation is also apparent from the inability of myoblasts that overexpress a cytoplasmically truncated type II receptor to differentiate into multinuclear myotubes. Thus, cells in which receptor signaling is inhibited still respond with growth arrest in response to culture conditions that normally induce myogenic differentiation; yet the myogenic differentiation program is blocked, presumably at the level of myogenic transcription factors. This observation suggests that TGF-β receptor signaling provides a competence to myoblasts to differentiate (Filvaroff *et al.*, 1994). The apparent contradictory stimulatory and inhibitory effects of TGF-β on myoblast differentiation could be explained by the interpretation that TGF-β stimulates early stage myoblasts, such as embryonic myoblasts, whereas TGF-β inhibits myoblast differentiation at a later stage, such as in fetal myoblasts (Cusella-de Angelis *et al.*, 1994).

In addition to muscle cell differentiation, mesenchymal cells also have the ability to differentiate into adipocytes, and this line of differentiation is also potently inhibited *in vitro* by TGF-β (Ignotz and Massagué, 1985). Mesenchymal cells can also differentiate into chondrocytes which give rise to cartilage, and osteoblasts which deposit bone matrix. Cell culture experiments suggest that TGF-β has the ability to guide undifferentiated mesenchymal cells into chondrocyte differentiation. In addition, TGF-β may induce osteoblastic differentiation under some conditions, but also inhibit maturation of established osteoblasts (Centrella *et al.*, 1994).

Besides the cell culture experiments implicating a role for TGF-β in mesenchymal differentiation, a role of TGF-β in mesenchymal differentiation has been documented by *in vivo* experiments. TGF-β is likely to play an important role in chondrocyte differentiation. For example, injection of TGF-β under the perichondrium stimulates the proliferation of these mesenchymal cells and induces their differentiation into chondrocytes. This cartilage is subsequently replaced with bone tissue (Joyce *et al.*, 1990). Consistent with a role of TGF-β in cartilage formation, overexpression of a truncated type II receptor in chondrocytes of transgenic mice results in defects in articular cartilage formation and organization, thus resembling human osteoarthritis (Serra *et al.*, 1997). The effect of increased expression of TGF-β ligand has also been studied in transgenic mice. Transgenic expression of the wild-type precursor of TGF-β did not induce any major effects, which is consistent with the knowledge that secreted latent TGF-β does not bind to the TGF-β receptors and requires proteolytic activation. However, replacement of two cysteines in the prosegment by serine allows expression of active TGF-β (Brunner *et al.*, 1989). Transgenic overexpression of such activated TGF-β_2 in osteoblasts has been shown to result in a skeletal phenotype. These mice show increased proliferation of preosteoblasts and increased density of osteocytes, and show a delayed ossification. Furthermore, they have an age-dependent progressive bone loss, reminiscent of high turnover osteoporosis, associated with an imbalance between increased bone deposition by osteoblasts and increased osteoclastic bone resorption (Erlebacher and Derynck, 1996). Taken together, these experiments strongly suggest that TGF-β plays an important role in mesenchymal differentiation, although this needs to be characterized further.

Whereas TGF-β is a potent inhibitor of epithelial cell and keratinocyte proliferation in culture (Massagué and Polyak, 1995), transgenic overexpression of TGF-β in mouse skin has revealed a more complex role of TGF-β in skin proliferation and differentiation. This complexity may be related to differential regulation of type I and type II receptors (Cui *et al.*, 1995) and to the differential expression of TGF-β_1 in the basal layer of the epidermis, compared with expression of TGF-β_2 in the suprabasal differentiating keratinocytes (Glick *et al.*, 1993). Consistent with the antiproliferative effect of TGF-β on keratinocytes, overexpression of active TGF-β from a promoter specific for suprabasal cells resulted in a hyperkeratotic epidermis with strongly inhibited proliferation and a reduced number of hair follicles, and resulted in early death (Sellheyer *et al.*, 1993). In addition, the increased epidermal proliferation in TGF-β_1 knockout mice (Glick *et al.*, 1993), and the similarly increased proliferation and the thickened and wrinkled hyperplastic phenotype of the skin in mice in which receptor signaling was inhibited by overexpressing a cytoplasmically truncated version of the type II TGF-β receptor (Wang *et al.*, 1997), are also consistent with a growth inhibitory role of TGF-β in skin. In contrast, suprabasal expression of TGF-β_1 from the keratin 10 promoter results in increased proliferation, suggesting a positive effect of TGF-β_1 on cell proliferation. However, induction of the type II but not the type I receptor by phorbol myristate treatment of the skin resulted in a TGF-β_1-induced growth inhibition, suggesting that the growth inhibitory effect of TGF-β on the epidermis correlates with the level of type II receptor (Cui *et al.*, 1995; Fowlis *et al.*, 1996).

Expression of TGF-β may play an important role in the branching morphogenesis of complex epithelial structures, such as in the developing lung and mammary gland (Robinson *et al.*, 1991), as will be discussed below. The effect of TGF-β in ductal morphogenesis has been studied in several transgenic mouse models. Overexpression of TGF-β from the mouse mammary tumor virus promoter in the mammary gland has been shown to inhibit ductal development (Pierce *et al.*, 1993), which is consistent with the ability of exogenous TGF-β to inhibit ductal elongation and to induce ductal involution (Robinson *et al.*, 1993). However, in these transgenic mice, alveolar outgrowths developed during pregnancy and lactation occurred (Pierce *et al.*, 1993). In contrast, expression of activated TGF-β_1 from the whey acidic protein promoter inhibited lobuloalveolar development and lactation (Jhappan *et al.*, 1993). A possible role of TGF-β as negative regulator of ductal morphogenesis is also apparent from the inhibitory effect of TGF-β_1 on branching morphogenesis of fetal lung explants *in vitro* (Serra *et al.*, 1994). In addition, transgenic overexpression of TGF-β_1 from the surfactant protein C promoter, which is activated during type II epithelial cell differentiation, also decreased ductal branching and saccular morphogenesis, as well as epithelial differentiation (Zhou *et al.*, 1996).

Finally, important information on the role of the different TGF-β isoforms in development has been obtained from the analysis of mice in which a gene for one of the TGF-β types was interrupted by gene targeting. These knockout mice have very different phenotypes, thus implying separate roles for TGF-β_1, β_2 and β_3 in development. However, the defects associated with each of these gene inactivations in these three types of mice resulted in lethality, thus often preventing extensive characterization of the developmental alterations and preventing the creation of mice that are defective in the synthesis of more than one type of TGF-β.

As discussed in the previous section, mice homozygous for the inactivated TGF-β_1 gene developed massive inflammation resulting in death within a few weeks after birth, a phenotype that is consistent with the important roles of TGF-β in the immune system (Shull *et al.*, 1992; Kulkarni *et al.*, 1993). However, only half of the TGF-$\beta_1^{-/-}$ mice were born, whereas the other half died *in utero* around embryonic day 10.5. The primary defects of the latter mice were restricted to embryonic tissues (i.e. the yolk sac vasculature and hematopoietic system). The defect in vasculature was related to impaired differentiation of endothelial cells which resulted in inadequate capillary tube formation, whereas defective hematopoiesis resulted in a reduced number of erythroid cells in the yolk sac (Dickson *et al.*, 1995). Thus, even though TGF-β_1 has been shown to be a potent growth inhibitor of endothelial (Heimark *et al.*, 1986; Muller *et al.*, 1987) and hematopoietic (Ohta *et al.*, 1987; Ottman and Pelus, 1988; Sing *et al.*, 1988) cells in culture, its primary role *in vivo* may be in regulating endothelial and hematopoietic differentiation, functions that are consistent with the high levels of endogenous TGF-β_1 expression in these cell types during early embryonic development (Wilcox and Derynck, 1988; Akhurst *et al.*, 1990). The early embryonic lethality associated with impaired vasculogenesis and hematopoiesis in the yolk sac was remarkably also the phenotype of mice in which the gene for the type II TGF-β receptor was functionally inactivated (Oshima *et al.*, 1996), further emphasizing the importance of TGF-β signaling in the differentiation of these two cell systems during early development.

Mice deficient in TGF-β_2 or TGF-β_3 expression are usually born alive but die shortly thereafter. In the case of the TGF-β_3 knockout mice, the mice suffer from a severely cleft palate, indicating that TGF-β_3 is required for fusion of the palatal shelves (Kaartinen *et al.*, 1995; Proetzel *et al.*, 1995), a function that is consistent with its localized expression in the medial edge epithelium (Pelton *et al.*, 1990b). In addition, TGF-β_3 null mice have impaired lung development, which presumably results from a generalized delay in maturation and is the cause of death shortly after birth. These data suggest an important role for TGF-β_3 in some epithelial–mesenchymal interactions (Kaartinen *et al.*, 1995). In contrast to the TGF-β_3 null mice, the phenotype of TGF-$\beta_2{}^{-/-}$ mice is much more complex and involves a large variety of organs and structures (Sanford *et al.*, 1997). These mice have also delayed lung development as well as a variety of congenital heart defects. Besides the impaired cardiovascular development, they also show a variety of defects in skeletal development, including cleft palate, various deformities of craniofacial bones generally associated with a reduction in bone size and ossification, and deformities and size reductions in a variety of bones of the axial and appendicular skeleton. Finally, various urogenital deformities and defects in inner ear and eye development arise from a deficiency in TGF-β_2 expression. Taken together, this plethora of abnormalities indicates an involvement of TGF-β_2 in various morphogenetic processes and strongly suggests a role of TGF-β_2 in inductive epithelial–mesenchymal interactions (Sanford *et al.*, 1997).

TGF-β IN TUMOR DEVELOPMENT

Because of the initial identification of TGF-β in a phenotypic transformation assay (Moses *et al.*, 1981; Roberts *et al.*, 1981), it was originally assumed that TGF-β might play a role in malignant transformation and tumor development. While this notion lost some enthusiasm as it became clear that TGF-β was made by nearly all cells in culture and that TGF-β inhibited the proliferation of many cell types in culture, further studies of TGF-β in tumorigenesis indicate that those initial assumptions were not spurious. Changes in TGF-β production and in TGF-β responsiveness are often apparent in tumor cells and these changes have been shown to confer advantages to tumor development *in vivo*. One set of changes relates to the inactivation of the TGF-β responsiveness, whereas overexpression of TGF-β is often apparent as well (reviewed in Arrick and Derynck, 1996; Markowitz and Roberts, 1996).

The development of resistance to TGF-β correlates with the progression of some tumors from a benign to a malignant phenotype (Manning *et al.*, 1991; Markowitz *et al.*, 1994). Components of the TGF-β signaling pathway have been identified as a target of functional inactivation in several tumor types. The type II receptor is inactivated in a high proportion of colonic (Markowitz *et al.*, 1995) and gastric (Myeroff *et al.*, 1995) tumors that exhibit microsatellite instability, wherein repetitive DNA sequences are prone to mutation. However, type II receptor mutations are rare in endometrial cancers with microsatellite instability, suggesting that this mutation is selected for in some tumor types, rather than a random occurrence resulting from genetic instability (Myeroff *et al.*, 1995). Furthermore, type II receptor mutations have also been found in gastric cancers that did not have microsatellite instability (Park *et al.*, 1994; Myeroff *et al.*, 1995), as well as in head and neck squamous carcinomas (Garrigue-Antar *et al.*, 1995), and in a T-cell lymphoma (Knaus *et al.*, 1996). Restoration of functional type II receptors to tumor cells in which this receptor was deleted has been shown to result in decreased tumorigenicity (Sun *et al.*, 1994; Markowitz *et al.*, 1995; Wang *et al.*, 1995), demonstrating directly that the type II receptor functions as a tumor suppressor. The type I receptor TβRI has also been found mutated in some cancers (Ciernik *et al.*, 1995; Kim *et al.*, 1996a), but this defect is found less frequently than type II receptor mutations.

Smads, which are effectors of the TGF-β response downstream from the receptors, have also been found mutated in tumors. The gene for Smad4 was originally cloned as *dpc4* for 'deleted in pancreatic cancer-4', a gene located at the 18q21 locus, which was deleted or inactivated in 50% of pancreatic carcinomas (Hahn *et al.*, 1996). Mutations in Smad4 have also been identified, albeit infrequently, in colorectal carcinomas (Riggins *et al.*, 1996; Thiagalingam *et al.*, 1996), breast and ovarian carcinomas (Schutte *et al.*, 1996), head and neck squamous carcinomas (Kim *et al.*, 1996b) and lung cancers (Nagatake *et al.*, 1996). Smad2, which also maps to the 18q21 locus, has been found mutated or deleted in some colorectal and lung cancers (Eppert *et al.*, 1996; Riggins *et al.*, 1996; Uchida *et al.*, 1996), although mutations in Smad2 are found less frequently than in Smad4. Taken together, inactivation of the TGF-β receptor signaling pathway by functional inactivation of the receptors or Smads is now considered as a step in the progression of several tumor types.

Mutations, or alterations in functions, of genes involved in the mechanism of TGF-β-mediated growth arrest frequently occur in tumors, resulting in resistance of TGF-β growth inhibition. Mutations or deletions in pRB occur in a wide variety of tumor types (Weinberg, 1995); this protein plays a central role in cell cycle regulation and is likely to be important for TGF-β-mediated growth arrest. Expression of some nuclear oncogenes, such as adenovirus E1A, the human papilloma virus E6–E7 proteins, the SV40 virus small t antigen, and polyoma middle T antigen, inactivates pRB as well as the growth inhibitory effect of TGF-β (Laiho *et al.*, 1990a, 1991b; Pietenpol *et al.*, 1990). The E1A protein also binds to and inactivates $p27^{Kip1}$, another cell cycle inhibitor that is regulated by TGF-β (Mai *et al.*, 1996). The p53 gene product may also be involved in growth inhibition by TGF-β in some cell types. SV40-immortalized human bronchial epithelial cells lose the growth inhibition response to TGF-β when transfected with a mutant p53 (Gerwin *et al.*,1992).

Although TGF-β receptor signaling or the ability of TGF-β to induce growth inhibition is often impaired in tumor cells, not all tumors are resistant to the growth inhibitory effects of TGF-β. Furthermore, many tumor cells show an enhanced TGF-β_1 production compared with that in normal cells, even though addition of TGF-β to these cells produces growth inhibition. This suggests that increased TGF-β production may provide an advantage to tumor development *in vivo* and this has been confirmed experimentally. For example, in one tumor cell line, overexpression of an antisense RNA that blocked TGF-β production resulted in increased tumor growth (Wu *et al.*, 1992). In addition, overexpression of TGF-β_1 in cells that are resistant to the effects of TGF-β on cell proliferation results in increased tumor formation (Arrick *et al.*, 1992), and this increased tumor growth is seen even with cells that are strongly growth inhibited by TGF-β in cell culture (Chang *et al.*, 1993). Thus, the antiproliferative effect of TGF-β

on tumor cell growth *in vitro* may be less important than its effect on tumor growth *in vivo*.

This increased tumor development *in vivo*, associated with increased TGF-β production by the tumor cells, may be due to several activities of TGF-β. Because of the potent effects of TGF-β on cell–matrix interactions, TGF-β may induce increased matrix formation and integrin synthesis not only in the tumor cells themselves, but also in the surrounding cells. TGF-β is chemotactic for fibroblasts, and tumor tissue is interspersed with large numbers of fibroblasts, which are responsive to TGF-β. In addition, the tumor cells may become more adhesive to the extracellular matrix. Increased extracellular matrix deposition and fibroblast infiltration and proliferation will result in extensive stroma formation, which favors tumor growth. Angiogenesis, required for tumor progression, may also be locally stimulated by TGF-β production (Fajardo *et al.*, 1996). In addition to local stromal effects, increased TGF-β synthesis and activation may result in a localized immunosuppressive environment; TGF-β suppresses T- and B-lymphocyte proliferation and function, and, perhaps most importantly in this context, inhibits the generation of cytotoxic T lymphocytes and NK cells (Torre-Amione *et al.*, 1990; Chang *et al.*, 1993). Thus, even though endogenous TGF-β synthesis might be predicted to inhibit tumor cell proliferation based on *in vitro* proliferation assays, increased tumor cell growth is likely to occur, as a consequence of changes in stroma deposition, increased extracellular matrix formation, increased angiogenesis and decreased immune surveillance, all of which act to create a more favorable environment for tumor growth (Arrick *et al.*, 1992; Steiner and Barrack, 1992; Chang *et al.*, 1993).

The relative importance of the different effects of TGF-β to tumor growth *in vivo* is hard to predict. TGF-β has been shown to act as both a tumor suppressor (Arteaga *et al.*, 1990; Wu *et al.*, 1992; Glick *et al.*, 1993, 1994; Pierce *et al.*, 1995; Cui *et al.*, 1996) and a tumor promoter (Sieweke *et al.*, 1990; Welch *et al.*, 1990; Steiner and Barrack, 1992; Arteaga *et al.*, 1993) at early or late stages of carcinogenesis *in vivo*. In skin, loss of TGF-β expression yields effects similar to inactivating receptor mutations, thus resulting in a high frequency of malignant conversion in chemically induced papillomas in transgenic mice (Glick *et al.*, 1993); in accordance, ras-transformed keratinocytes that lack TGF-β_1 expression show accelerated multifocal progression to squamous cell carcinoma, compared with wild-type cells (Glick *et al.*, 1994). Lack of TGF-β_1 expression in keratinocytes may greatly enhance genomic instability, which correlates with tumor progression (Cheng and Loeb, 1993). In a converse experiment, transgenic mice overexpressing TGF-β_1 in suprabasal keratinocytes were subjected to dimethylbenzathracene (DMBA)-induced chemical carcinogenesis (Cui *et al.*, 1996). These studies revealed that TGF-β_1 showed a biphasic action during chemical carcinogenesis, acting as a tumor suppressor at early stages but as an enhancer of the malignant phenotype at later stages: the transgenic mice were more resistant to induction of benign tumors than controls, whereas the malignant conversion rate was greatly increased (Cui *et al.*, 1996). Furthermore, increased TGF-β_1 expression resulted in a high incidence of spindle cell carcinoma formation (Cui *et al.*, 1996), consistent with the ability of TGF-β to induce an epithelial to mesenchymal conversion (Miettinen *et al.*, 1994). This fibroblastic phenotype is induced efficiently in mammary epithelial cells by cooperation of TGF-β with Ha-ras, thereby converting these cells into highly tumorigenic and invasive tumor cells *in vitro* and *in vivo*. This fibroblastic phenotype is maintained by an increased expression of

TGF-β_1 and resulting autocrine responsiveness, and is reverted to the epithelial phenotype in the presence of neutralizing antibodies (Oft *et al.*, 1996).

Based on the different effects of TGF-β on cell–matrix interaction, especially on extracellular matrix synthesis, cell adhesion, and secretion of proteases and protease inhibitors, TGF-β synthesis and responsiveness are likely to play a role in the invasive and metastatic behavior of tumor cells *in vivo*. As yet little is known about the importance of TGF-β in tumor metastasis. However, increased TGF-β_1 expression by tumor cells has been shown to confer increased metastasis, although this could have resulted from increased and accelerated tumor development (Steiner and Barrack, 1992). In addition, the TGF-β-induced epithelial to mesenchymal conversion of Ha-ras-expressing epithelial cells is also accompanied by increased invasiveness (Oft *et al.*, 1996), suggesting that increased TGF-β synthesis might result in increased metastasis. Clearly, the role of TGF-β expression in tumor cell metastasis remains to be further characterized and might be complex, depending on the tumor cell type.

The role of TGF-β in the different physiologic and pathologic processes *in vivo* is only now starting to be explored. Whereas a multiplicity of biologic activities of TGF-β has already been described, the coming years should bring us a better insight into the relevance of these activities *in vivo* and should define the role of TGF-β in the physiology of the cell, tissue and organism, and during normal development. Another area of progress will be in the further elucidation of how TGF-β induces its activities at the subcellular level, what signaling pathways mediate the biologic activities of TGF-β, and how modulation of the signaling pathways affects development and cell differentiation.

ACKNOWLEDGEMENTS

The authors thank Anita Roberts for providing helpful information. Work in the laboratory of the authors was supported by NIH grants AR41126, CA63101 and DE-10306.

REFERENCES

Akhurst, R.J., Lehnert, S.A., Faissner, A.J. and Duffie, E. (1990). The role of TGF-βs in mammalian development and neoplasia. *Development* **108**, 645–656.

Andres, J.L., Stanley, K., Cheifetz, S. and Massagué, J. (1989). Membrane-anchored and soluble forms of betaglycan, a polymorphic proteoglycan that binds transforming growth factor-β. *J. Cell Biol.* **109**, 3137–3145.

Andres, J.L., Rönnstrand, L., Cheifetz, S. and Massagué, J. (1991). Purification of the transforming growth factor-β (TGF-β) binding proteoglycan betaglycan. *J. Biol. Chem.* **266**, 23 282–23 287.

Antonelli-Orlidge, A., Saunders, K.B., Smith, S.R. and d'Amore, P.A. (1989). An activated form of transforming growth factor β is produced by cocultures of endothelial cells and pericytes. *Proc. Natl Acad. Sci. U.S.A.* **86**, 4544–4548.

Anzano, M.A., Roberts, A.B., Meyers, C.A., Komoriya, A., Lamb, L.C., Smith, J.M. and Sporn, M.B. (1982). Synergistic interaction of two classes of transforming growth factors from murine sarcoma cells. *Cancer Res.* **42**, 4776–4778.

Anzano, M.A., Roberts, A.B., Smith, J.M., Sporn, M.B. and DeLarco, J.E. (1983). Sarcoma growth factor from conditioned medium is composed of both type α and type β transforming growth factors. *Proc. Natl Acad. Sci. U.S.A.* **80**, 6264–6268.

Arrick, B.A. and Derynck, R. (1996). The biological role of transforming growth factor-β in cancer development. In *Molecular Endocrinology of Cancer* (ed. J. Waxman), Cambridge University Press, Cambridge, pp. 51–78.

Arrick, B.A., Lee, A., Grendell, R.L. and Derynck, R. (1991). Inhibition of translation of transforming growth factor-β3 mRNA by its 5′ untranslated region. *Mol. Cell. Biol.* **11**, 4306–4313.

Arrick, B.A., Lopez, A.R., Elfman, F., Ebner, R., Damsky, C.H. and Derynck, R. (1992). Altered metabolic and adhesive properties and increased tumorigenesis associated with increased expression of transforming growth factor β1. *J. Cell Biol.* **118**, 715–726.

Arteaga, C.L., Coffey, R.J., Jr., Dugger, T.C., McCutchen, C.M., Moses, H.L. and Lyons, R.M. (1990). Growth stimulation of human breast cancer cells with anti-transforming growth factor-β antibodies: evidence for negative autocrine regulation by transforming growth factor-β. *Cell Growth Differ.* **1**, 367–374.

Arteaga, C.L., Cart-Dugger, T., Moses, H.L., Hurd, S.D. and Pietenpol, J.A. (1993). Transforming growth factor β1 can induce estrogen-independent tumorigenicity of human breast cancer cells in athymic mice. *Cell Growth Differ.* **4**, 193–201.

Assoian, R.K., Komoriya, A., Meyers, C.A., Miller, D.M. and Sporn, M.B. (1983). Transforming growth factor-β in human platelets. Identification of a major storage site, purification, and characterization. *J. Biol. Chem.* **258**, 7155–7160.

Assoian, R.K., Frolik, C.A., Roberts, A.B., Miller, D.M. and Sporn, M.B. (1984). Transforming growth factor-β controls receptor levels for epidermal growth factor in NRK fibroblasts. *Cell* **36**, 35–41.

Assoian, R.K., Fleurdelys, B.E., Stevenson, H.C., Miller, D.J., Madtes, M.K., Raines, E.W., Ross, R. and Sporn, M.B. (1987). Expression and secretion of type β transforming growth factor by activated human macrophages. *Proc. Natl Acad. Sci. U.S.A.* **84**, 6020–6024.

Atfi, A., Lepage, K., Allard, P., Chapdelaine, A. and Chevalier, S. (1995). Activation of a serine/threonine kinase signaling pathway by transforming growth type β. *Proc. Natl Acad. Sci. U.S.A.* **92**, 12110–12114.

Attisano, L., Wrana, J.L., Cheifetz, S. and Massagué, J. (1992). Novel activin receptors: distinct genes and alternatives mRNA splicing generate a repertoire of serine/threonine kinase receptors. *Cell* **68**, 97–108.

Attisano, L., Cárcamo, J., Ventura, F., Weis, F.M.B., Massagué, J. and Wrana, J.L. (1993). Identification of human activin and TGF β type I receptors that form heteromeric kinase complexes with type II receptors. *Cell* **75**, 671–680.

Baker, J.C. and Harland, R.M. (1996). A novel mesoderm inducer, Madr 2, functions in the activin signal transduction pathway. *Genes Dev.* **10**, 1880–1889.

Barcellos-Hoff, M.H. and Dix, T.A. (1996). Redox-mediated activation of latent transforming growth factor-β1. *Mol. Endocrinol.* **10**, 1077–1083.

Barcellos-Hoff, M.H., Derynck, R., Tsang, M.L. and Weatherbee, J.A. (1994). Transforming growth factor-β activation in irradiated murine mammary gland. *J. Clin. Invest.* **93**, 892–899.

Barr, E. and Leiden, J.M. (1991). Systemic delivery of recombinant proteins by genetically modified myoblasts. *Science* **254**, 1507–1509.

Barral, A., Teixeira, M., Reis, P., Vinhas, V., Costa, J., Lessa, H., Bittencourt, A.L., Reed, S., Carvalho, E.M. and Barral-Netto, M. (1995). Transforming growth factor-β in human cutaneous leishmaniasis. *Am. J. Pathol.* **147**, 947–954.

Barral-Netto, M., Barral, A., Brownell, C.E., Skeiky, Y.A., Ellingsworth, L.R., Twardzik, D.R. and Reed, S.G. (1992). Transforming growth factor-β in leishmanial infection: a parasite escape mechanism. *Science* **257**, 545–548.

Bascom, C.C., Wolfshohl, J.R., Coffey, R.J., Madisen, L., Webb, N.R., Purchio, A.R., Derynck, R. and Moses, H.L. (1989). Complex regulation of transforming growth factor β1, β2 and β3 mRNA expression in mouse fibroblasts and keratinocytes by transforming growth factors β1 and β2. *Mol. Cell. Biol.* **9**, 5508–5515.

Bassing, C.H., Yingling, J.M., Howe, D.J., Wang, T., He, W.W., Gustafson, M.L., Shah, P., Donahoe, P.K. and Wang, X.-F. (1994). A transforming growth factor β type I receptor that signals to activate gene expression. *Science* **263**, 87–89.

Battegay, E.J., Raines, E.W., Seifert, R.A., Bowen-Pope, D.F. and Ross, R. (1990). TGF-β induces bimodal proliferation of connective tissue cells via complex control of autocrine PDGF loop. *Cell* **63**, 515–524.

Bauvois, B., Rouillard, D., Sanceau, J. and Wietzerbin, J. (1992). IFN-γ and transforming growth factor-β1 differentially regulate fibronectin and laminin receptors of human differentiating monocyte cells. *J. Immunol.* **148**, 3912–3919.

Bauvois, B., Van Weyenbergh, J., Rouillard, D. and Wietzerbin, J. (1996). TGF-β1-stimulated adhesion of human mononuclear phagocytes to fibronectin and laminin is abolished by IFN-gamma: dependence of alpha 5 beta 1 and beta 2 integrins. *Exp. Cell. Res.* **222**, 209–217.

Bogdan, C. and Nathan, C. (1993). Modulation of macrophage function by transforming growth factor β, interleukin-4, and interleukin-10. *Ann. N.Y. Acad. Sci.* **685**, 713–739.

Borkowski, T.A., Letterio, J.J., Farr, A.G. and Udey, M.C. (1996). A role for endogenous transforming growth factor β1 in Langerhans cell biology: the skin of transforming growth factor β1 null mice is devoid of epidermal Langerhans cells. *J. Exp. Med.* **184**, 2417–2422.

Borkowski, T.A., Letterio, J.J., Mackall, C.L., Saith, A., Wang, X.-J., Roop, D.R., Gress, R.E. and Udey, M.C. (1997). A role for TGF-β1 in Langerhans cell biology: further characterization of the epidermal Langerhans cell defect in TGF-β1 null mice. *J. Clin. Invest.* **100**, 575–581.

Brandes, M.E., Allen, J.B., Ogawa, Y. and Wahl, S.M. (1991). Transforming growth factor β1 suppresses acute and chronic arthritis in experimental animals. *J. Clin. Invest.* **87**, 1108–1113.

Brennan, T.J., Edmondson, D.G., Li, L. and Olson, E.N. (1991). Transforming growth factor β represses the actions of myogenin through a mechanism independent of DNA binding. *Proc. Natl Acad. Sci. U.S.A.* **88**, 3822–3826.

Brown, E.J. and Schreiber, S.L. (1996). A signaling pathway to translational control. *Cell* **86**, 517–520.

Brunner, A.M., Marquardt, H., Malacko, A.R., Lioubin, M.N. and Purchio, A.F. (1989). Site-directed mutagenesis of cysteine residues in the proregion of the transforming growth factor β1 precursor. Expression and characterization of mutant proteins. *J. Biol. Chem.* **264**, 13 660–13 664.

Carel, J.C., Schreiber, R.D., Falqui, L. and Lacy, P.E. (1990). Transforming growth factor β decreases the immunogenicity of rat islet xenografts (rat to mouse) and prevents rejection in association with treatment of the recipient with a monoclonal antibody to interferon γ. *Proc. Natl Acad. Sci. U.S.A.* **87**, 1591–1595.

Carel, J.C., Sheehan, K.C., Schreiber, R.D. and Lacy, P.E. (1993). Prevention of rejection of transforming growth factor beta-treated rat-to-mouse islet xenografts by monoclonal antibody to tumor necrosis factor. *Transplantation* **55**, 456–458.

Caver, T.E., O'Sullivan, F.X., Gold, L.I. and Gresham, H.D. (1996). Intracellular demonstration of active TGF-β1 in B cells and plasma cells of autoimmune mice. IgG bound TGF-β1 suppresses neutrophil function and host defense against *Staphylococcus aureus* infection. *J. Clin. Invest.* **98**, 2496–2506.

Celada, A. and Maki, R.A. (1992). Transforming growth factor-β enhances the M-CSF and GM-CSF-stimulated proliferation of macrophages. *J. Immunol.* **148**, 1102–1105.

Centrella, M., McCarthy, T.L. and Canalis, E. (1987). Transforming growth factor β is a bifunctional regulator of replication and collagen synthesis osteoblast-enriched cell cultures from fetal rat bone. *J. Biol. Chem.* **262**, 2869–2874.

Centrella, M., Horowitz, M.C., Wozney, J.M. and McCarthy, T.L. (1994). Transforming growth factor-β gene family members and bone. *Endocr. Rev.* **15**, 27–39.

Cerwenka, A., Bevec, D., Majdic, O., Knapp, W. and Holter, W. (1994). TGF-β1 is a potent inducer of human effector T cells. *J. Immunol.* **153**, 4367–4377.

Cerwenka, A., Kovar, H., Majdic, O. and Holter, W. (1996). Fas- and activation-induced apoptosis are reduced in human T cells preactivated in the presence of TGF-β1. *J. Immunol.* **156**, 459–464.

Chambard, J.-C. and Pouysségur, J. (1988). TGF-β inhibits growth factor-induced DNA synthesis in hamster fibroblasts without affecting the early mitogenic events. *J. Cell. Physiol.* **135**, 101–107.

Champsi, J., Young, L.S. and Bermudez, L.E. (1995). Production of TNF-α, IL-6 and TGF-β, and expression of receptors for TNF-α and IL-6, during murine *Mycobacterium avium* infection. *Immunology* **84**, 549–554.

Chang, H.-L., Gillett, N., Figari, I., Lopez, A.R., Palladino, M.A. and Derynck, R. (1993). Increased transforming growth factor β inhibits cell proliferation *in vitro*, yet increases tumorigenicity and tumor growth of Meth A sarcoma cells. *Cancer Res.* **53**, 4391–4398.

Chatani, Y., Tanimura, S., Miyoshi, N., Hattori, A., Sato, M. and Kohno, M. (1995). Cell type-specific modulation of cell growth by TGF-β1 does not correlate with mitogen-activated protein kinase activation. *J. Biol. Chem.* **270**, 30 686–30 692.

Cheifetz, S. and Massagué, J. (1989). Transforming growth factor-β (TGF-β) receptor proteoglycan. Cell surface expression and ligand binding in the absence of glycosaminoglycan chains. *J. Biol. Chem.* **264**, 12 025–12 028.

Cheifetz, S., Like, B. and Massagué, J. (1986). Cellular distribution of type I and type II receptors for transforming growth factor-β. *J. Biol. Chem.* **261**, 9972–9978.

Cheifetz, S., Weatherbee, J.A., Tsang, M.L.S., Andersen, J.K., Mole, J.E., Lucas, R. and Massagué, J. (1987). The transforming growth factor-β system, a complex pattern of cross-reactive ligands and receptors. *Cell* **48**, 409–415.

Cheifetz, S., Andres, J.L. and Massagué, J. (1988). The transforming growth factor-β receptor type III is a membrane proteoglycan. Domain structure of the receptor. *J. Biol. Chem.* **263**, 16 984–16 991.

Cheifetz, S., Bellón, T., Calés, C., Vera, S., Bernabeu, C., Massagué, J. and Letarte, M. (1992). Endoglin is a component of the transforming growth factor-β receptor system in human endothelial cells. *J. Biol. Chem.* **267**, 19027–19030.

Chen, F. and Weinberg, R.A. (1995). Biochemical evidence for the autophosphorylation and transphosphorylation of transforming growth factor β receptor kinases. *Proc. Natl Acad. Sci. U.S.A.* **92**, 1565–1569.

Chen, R.-H. and Derynck, R. (1994). Homomeric interactions between the type II TGF-β receptors. *J. Biol. Chem.* **269**, 22868–22874.

Chen, R.-H., Ebner, R. and Derynck, R. (1993). Inactivation of the type II receptor reveals two receptor pathways for the diverse TGF-β activities. *Science* **260** 1335–1338.

Chen, R.-H., Moses, H.L., Maruoka, E.M., Derynck, R. and Kawabata, M. (1995a). Phosphorylation-dependent interaction of the cytoplasmic domains of the type I and type II TGF-β receptors. *J. Biol. Chem.* **270**, 12235–12241.

Chen, R.-H., Miettinen, P.J., Maruoka, E.M., Choy, L. and Derynck, R. (1995b). A WD-domain protein that is associated with and phosphorylated by the type II TGF-β receptor. *Nature* **377**, 548–552.

Chen, Y., Lebrun, J.-J. and Vale, W. (1996). Regulation of transforming growth factor-β and activin-induced transcription by mammalian mad proteins. *Proc. Natl Acad. Sci. U.S.A.* **93**, 12992–12997.

Cheng, K.C. and Loeb, L.A. (1993). Genomic instability and tumor progression: mechanistic considerations. *Adv. Cancer Res.* **60**, 121–156.

Christ, M., McCartney-Francis, N.L., Kulkarni, A.B., Ward, J.M., Mizel, D.E., Mackall, C.L., Gress, R.E., Hines, K.L., Tian, H. and Karlsson, S. (1994). Immune dysregulation in TGF-β1 deficient mice. *J. Immunol.* **153**, 1936–1946.

Ciernik, I.F., Ciernik, B., Cockerell, C., Minna, J., Gazdar, A. and Carbone, D. (1995). Expression of transforming growth factor-β and transforming growth factor-β receptors on AIDS-associated Kaposi sarcoma. *Clin. Cancer Res.* **1**, 1119–1124.

Coffman, R.L., Lebman, D.A. and Schrader, B. (1989). Transforming growth factor β specifically enhances IgA production by lipopolysaccharide-stimulated murine B lymphocytes. *J. Exp. Med.* **170**, 1039–1044.

Cui, W., Fowlis, D.J., Cousins, F.M., Duffie, E., Bryson, S., Balmain, A. and Akhurst, R.J. (1995). Concerted action of TGF-β1 and its type II receptor in control of epidermal homeostasis in transgenic mice. *Genes Dev.* **9**, 945–955.

Cui, W., Fowlis, D.J., Bryson, S., Duffie, E., Ireland, H., Balmain, A. and Akhurst, R. (1996). TGF-β1 inhibits the formation of benign skin tumours, but enhances progression to invasive spindle carcinomas in transgenic mice. *Cell* **86**, 531–542.

Cusella-de Angelis, M.G., Molinari, S., Le Donne, A., Coletta, M., Vivarelli, E., Bouche, M., Molinaro, M., Ferrari, S. and Cossu, G. (1994). Differential response of embryonic and fetal myoblasts to TGF-β: a possible regulatory mechanism of skeletal muscle histogenesis. *Development* **120**, 925–933.

Czarniecki, C.W., Chiu, H.H., Wong, G.H.W., McCabe, S.M. and Palladino, M. (1988). Transforming growth factor-β1 modulates the expression of class II histocompatibility antigens on human cells. *J. Immunol.* **140**, 4217–4223.

Damsky, C.H. and Werb, Z. (1992). Signal transduction by integrin receptor for extracellular matrix: cooperative processing of extracellular information. *Curr. Opin. Cell Biol.* **4**, 772–781.

Dang, H., Geiser, A.G., Letterio, J.J., Nakabayashi, T., Kong, L., Fernandes, G. and Talal, N. (1995). SLE-like autoantibodies and Sjogren's syndrome-like lymphoproliferation in TGF-beta knockout mice. *J. Immunol.* **155**, 3205–3212.

Daopin, S., Piez, K.A., Ogawa, Y. and Davies, D.R. (1992). Crystal structure of transforming growth factor-β2: an unusual fold for the superfamily. *Science* **257**, 369–373.

Datto, M.B., Li, Y., Panus, J.F., Howe, D.J., Xiong, Y. and Wang, X.-F. (1995). Transforming growth factor β induces the cyclin-dependent kinase inhibitor p21 through a p53-independent mechanism. *Proc. Natl Acad. Sci. U.S.A.* **92**, 5545–5549.

de Martin, R., Haendler, B., Hofer-Warbinek, R., Gaugitsch, H., Wrann, M., Schlüsener, H., Seifert, J.M., Bodmer, S., Fontana, A. and Hofer, E. (1987). Complementary DNA for human glioblastoma-derived T cell suppressor factor, a novel member of the transforming growth factor-β gene family. *EMBO J.* **6**, 3673–3677.

Dennis, P.A. and Rifkin, D.B. (1991). Cellular activation of latent transforming growth factor β requires binding to the cation-independent mannose 6-phosphate/insulin-like growth factor type II receptor. *Proc. Natl Acad. Sci. U.S.A.* **88**, 580–584.

Derynck, R. (1992). The physiology of transforming growth factor-α. *Adv. Cancer Res.* **58**, 27–52.

Derynck, R. and Feng, X.-H. (1997). TGF-β receptor signaling. *BBA Reviews on Cancer* **1333**, F105–F150.

Derynck, R., Jarrett, J.A., Chen, E.Y., Eaton, D.H., Bell, J.R., Assoian, R.K., Roberts, A.B., Sporn, M.B. and

Goeddel, D.V. (1985). Human transforming growth factor-β cDNA sequence and expression in tumor cell lines. *Nature* **316**, 701–705.

Derynck, R., Lindquist, P.B., Lee, A., Wen, D., Tamm, J., Graycar, J.L., Rhee, L., Mason, A.J., Miller, D.A., Coffey, R.J., Moses, H.L. and Chen, E.Y. (1988). A new type of transforming growth factor-β, TGF-β3. *EMBO J.* **7**, 3737–3743.

Dewulf, N., Verschueren, K., Lonnoy, O., Moren, A., Grimsby, S., Vande Spiegle, K., Miyazono, K., Huylebroeck, D. and ten Dijke, P. (1995). Distinct spatial and temporal expression patterns of two type I receptors for bone morphogenetic proteins during mouse embryogenesis. *Endocrinology* **136**, 2652–2663.

Dickson, M.C., Martin, J.S., Cousins, F.M., Kulkarni, A.B., Karlsson, S. and Akhurst, R.J. (1995). Defective haematopoiesis and vasculogenesis in transforming growth factor-β1 knock out mice. *Development* **121**, 1845–1854.

Ebner, R., Chen, R.-H., Shum, L., Lawler, S., Zionchek, T.F., Lee, A., Lopez, A.R. and Derynck, R. (1993a). Cloning of a type I TGF-β receptor and its effect on TGF-β binding to the type II receptor. *Science* **260**, 1344–1348.

Ebner, R., Chen, R.-H., Lawler, S., Zioncheck, T. and Derynck, R. (1993b). The type II receptors for TGF-β or activin determine the ligand specificity of a single type I receptor. *Science* **262**, 900–902.

Eppert, K., Scherer, S.W., Ozcelik, H., Pirone, R., Hoodless, P., Kim, H., Tsui, L.-C., Bapat, B., Gallinger, S., Andrulis, I.L., Thomsen, G.H., Wrana, J.L. and Attisano, L. (1996). MADR2 maps to 18q21 and encodes a TGF-β-regulated MAD-related protein that is functionally mutated in colorectal carcinoma. *Cell* **86**, 543–552.

Erlebacher, A. and Derynck, R. (1996). Increased expression of TGF-β2 in osteoblasts results in an osteoporosis-like phenotype. *J. Cell Biol.* **132**, 195–210.

Espevik, T., Figari, I.S., Shalaby, R., Lackides, G.A., Lewis, G., Shepard, H.M. and Palladino, M.A. (1987). Inhibition of cytokine production by cyclosporin A and transforming growth factor β. *J. Exp. Med.* **166**, 571–576.

Espevik, T., Figari, I., Ranges, G.E. and Palladino, M.A. (1988). Transforming growth factor-β1 (TGF-β1) and recombinant human necrosis factor-α reciprocally regulate the generation of lymphokine-activated killer cell activity. Comparison between natural porcine platelet-derived TGF-β1 and TGF-β2, and recombinant human TGF-β1. *J. Immunol.* **140**, 2312–2316.

Ewen, M.E., Sluss, H.K., Whitehouse, L.L. and Livingston, D.M. (1993). TGF-β inhibition of Cdk4 synthesis is linked to cell cycle arrest. *Cell* **74**, 1009–1020.

Fajardo, L.F., Prionas, S.D., Kwan, H.H., Kowalski, J. and Allison, A.C. (1996). Transforming growth factor β1 induces angiogenesis *in vivo* with a threshold pattern. *Lab. Invest.* **74**, 600–608.

Fan, K., Ruan, Q., Sensenbrenner, L. and Chen, B. (1992). Transforming growth factor-β1 bifunctionally regulates murine macrophage proliferation. *Blood* **79**, 1679–1685.

Fargeas, C., Wu, C.Y., Nakajima, T., Cox, D., Nutman, T. and Delespesse, G. (1992). Differential effect of transforming growth factor β on the synthesis of Th1- and Th2-like lymphocytes by human T lymphocytes. *Eur. J. Immunol.* **22**, 2173–2176.

Fava, R. and McClure, D.B. (1987). Fibronectin-associated transforming growth factor. *J. Cell. Physiol.* **131**, 184–189.

Filvaroff, E.H., Ebner, R. and Derynck, R. (1994). Inhibition of myogenic differentiation in myoblasts expressing a truncated type II TGF-β receptor. *Development* **121**, 185–195.

Fitzpatrick, D.R., Denhez, F., Kondaiah, P. and Akhurst, R. (1990). Differential expression of TGF-β isoforms in murine palatogenesis. *Development* **109**, 585–595.

Flaumenhaft, R., Abe, M., Mignatti, P. and Rifkin, D.B. (1992). Basic fibroblast growth factor-induced activation of latent transforming growth factor β in endothelial cells: regulation of plasminogen activator activity. *J. Cell Biol.* **118**, 901–909.

Flaumenhaft, R., Abe, M., Sato, Y., Miyazono, K., Harpel, J.G., Heldin, C.-H. and Rifkin, D.B. (1993a). Role of the latent TGF-β binding protein in the activation of latent TGF-β by co-cultures of endothelial and smooth muscle cells. *J. Cell Biol.* **120**, 995–1002.

Flaumenhaft, R., Kojima, S., Abe, M. and Rifkin, D.B. (1993b). Activation of latent transforming growth factor β. *Adv. Pharmacol.* **24**, 51–76.

Fontana, A., Frei, K., Bodmer, S., Hofer, E., Schreier, M.H., Palladino, M.A. and Zinkernagel, R.M. (1989). Transforming growth factor-β inhibits the generation of cytotoxic T cells in virus-infected mice. *J. Immunol.* **143**, 3230–3234.

Fowlis, D.J., Cui, W., Johnson, S.A., Balmain, A. and Akhurst, R.J. (1996). Altered epidermal cell growth control *in vivo* by inducible expression of transforming growth factor β1 in the skin of transgenic mice. *Cell Growth Differ.* **7**, 679–687.

Franzén, P., ten Dijke, P., Ichijo, H., Yamashita, H., Schulz, P., Heldin, C.-H. and Miyazono, K. (1993). Cloning of a TGF-β type I receptor that forms a heteromeric complex with the TGF-β type II receptor. *Cell* **75**, 681–692.

Gamble, J.R. and Vadas, M.A. (1988). Endothelial adhesiveness for blood neutrophils is inhibited by transforming growth factor-β. *Science* **242**, 97–99.

Garrigue-Antar, L., Munoz-Antonia, T., Antonia, S.J., Gesmonde, J., Velluci, V.F. and Reiss, M. (1995). Missense mutations of the transforming growth factor-β type II receptor in human head and neck squamous carcinoma cells. *Cancer Res.* 3982–3987.

Gazzinelli, R.T., Oswald, I.P., Hieny, S., James, S.L. and Sher, A. (1992). The microbicidal activity of interferon-γ-treated macrophages against *Trypanosoma cruzi* involves an L-arginine-dependent, nitrogen oxide-mediated mechanism inhibitable by interleukin-10 and transforming growth factor-β. *Eur. J. Immunol.* **22**, 2501–2506.

Geiser, A.G., Burmester, J.K., Webbink, R., Robert, A.B. and Sporn, M.B. (1992). Inhibition of growth by transforming growth factor-β following fusion of two nonresponsive human carcinoma cell lines. *J. Biol. Chem.* **267**, 2588–2593.

Geiser, A.G., Busam, K.J., Kim, S.-J., Lafyatis, R., O'Reilly, M.A., Webbink, R., Roberts, A.B. and Sporn, M.B. (1993). Regulation of the transforming growth factor-β1 and -β3 promoters by transcription factor Sp1. *Proc. Natl Acad. Sci. U.S.A.* **90**, 9944–9948.

Geng, Y. and Weinberg, R.A. (1993). Transforming growth factor β effects on expression of G1 cyclins and cyclin-dependent protein kinases. *Proc. Natl Acad. Sci. U.S.A.* **90**, 10 315–10 319.

Georgi, L.L., Albert, P.S. and Riddle, D. (1990). daf.1, a *C. elegans* gene controlling dauer larva development, encodes a novel receptor kinase. *Cell* **61**, 635–645.

Gerwin, B.I., Spillare, E., Forrester, K., Lehman, T.A., Kispert, J., Welsh, J.A., Pfeifer, A.M.A., Lechner, J.F., Baker, S.J., Vogelstein, B. and Harris, C.C. (1992). Mutant p53 can induce tumorigenic conversion of human bronchial epithelial cells and reduce their responsiveness to a negative growth factor, transforming growth factor β1. *Proc. Natl Acad. Sci. U.S.A.* **89**, 2759–2763.

Glick, A.B., Flanders, K.C., Danielpour, D., Yuspa, S.H. and Sporn, M.B. (1989). Retinoic acid induces transforming growth factor-β2 in cultured keratinocytes and mouse epidermis. *Cell Regul.* **1**, 87–97.

Glick, A.B., Kulkarni, A.B., Tennenbaum, T., Hennings, H., Flanders, K.C., O'Reilly, M., Sporn M.B., Karlsson, S. and Yuspa, S.H. (1993). Loss of expression of transforming growth factor-β in skin and skin tumors is associated with hyperproliferation and high risk for malignant conversion. *Proc. Natl Acad. Sci. U.S.A.* **90**, 6076–6080.

Glick, A.B., Lee, M.M., Darwiche, N., Kulkarni, A.B., Karlsson, S. and Yuspa, S.H. (1994). Targeted deletions of the TGF-β1 gene causes rapid progression to squamous cell carcinoma. *Genes Dev.* **8**, 2429–2440.

Gollnick, S.O., Cheng, H.L., Grande, C.C., Thompson, D. and Tomasi, T.B. (1995). Effects of transforming growth factor-β on bone marrow macrophage Ia expression induced by cytokines. *J. Interferon Cytokine Res.* **15**, 485–491.

Gordon, J., Katira, A., Holder, M., MacDonald, I. and Pound, J. (1994). Central role of CD40 and its ligand in B lymphocyte responses to T-dependent antigens. *Cell Mol. Biol.* **40 (supplement 1)**, 1–13.

Gougos, A. and Letarte, M. (1990). Primary structure of endoglin, a KGD-containing glycoprotein of human endothelial cells. *J. Biol. Chem.* **265**, 8361–8364.

Graff, J.M., Bansal, A. and Melton, D.A. (1996). Xenopus Mad proteins transduce distinct subsets of signals for the TGF β superfamily. *Cell* **85**, 479–487.

Gray, A.M. and Mason, A.J. (1990). Requirement for activin A and transforming growth factor-β1 pro-regions in homodimer assembly. *Science* **247**, 1328–1330.

Gresham, H.D., Ray, C.J. and O'Sullivan, F.X. (1991). Defective neutrophil function in the autoimmune mouse string MRL/lpr. Potential role of transforming growth factor-β. *J. Immunol.* **146**, 3911–3921.

Griffith, D.L., Keck, P.C., Sampath, Rueger, D.C. and Carlson, W.D. (1996). Three-dimensional structure of recombinant human osteogenic protein 1: structural paradigm for the transforming growth factor β superfamily. *Proc. Natl Acad. Sci. U.S.A.* **93**, 878–883.

Grotendorst, G.R., Smale, G. and Pencev, D. (1989). Production of transforming growth factor β by human peripheral blood monocytes and neutrophils. *J. Cell. Physiol.* **140**, 396–402.

Gruppuso, P.A., Mikumo, R., Brautigan, D.L. and Braun, L. (1991). Growth arrest induced by transforming growth factor β1 is accompanied by protein phosphatase activation in human keratinocytes. *J. Biol. Chem.* **266**, 3444–3448.

Hahn, S.A., Schutte, M., Hoque, A.T., Moskaluk, C.A., da Costa, L.T., Rozenblum, E., Weinstein, C.L., Fischer, A., Yeo, C.J., Hruban, R.H. and Kern, S.E. (1996). DPC4, a candidate tumor suppressor gene at human chromosome 18q21.1. *Science* **271**, 350–353.

Halstead, J., Kemp, K. and Ignotz, R.A. (1995). Evidence for an involvement of phosphatidyl choline-phospholipase C and protein kinase C in transforming growth factor-β signaling. *J. Biol. Chem.* **270**, 13600–13603.

Hannon, G.J. and Beach, D. (1994). p15$^{ink4\beta}$ is a potential effector of TGF-β-induced cell cycle arrest. *Nature* **371**, 257–261.

Harpel, J.G., Metz, C.N., Kojima, S. and Rifkin, D.B. (1993). *Prog. Growth Factor Res.* **4**, 321–328.

Harper, J.W., Adam, G.R., Wei, N., Keyomarsi, K. and Elledge, S.J. (1993). The p21 CDK-interacting protein Cip1 is a potent inhibitor of G_1 cyclin-dependent kinases.

Hartsough, M.E. and Mulder, K.M. (1995). Transforming growth factor-β activation of p44^{mapk} in proliferating cultures of epithelial cells. *J. Biol. Chem.* **270**, 7117–7124.

Heikinheimo, K., Happonen, R.-P., Miettinen, P.J. and Ritvos, O. (1993). Transforming growth factor β2 in epithelial differentiation of developing teeth and odontogenic tumors. *J. Clin. Invest.* **91**, 1019–1027.

Heimark, R.L., Twardzik, D.R. and Schwartz, S.M. (1986). Inhibition of endothelial regeneration by type-β transforming growth factor from platelets. *Science* **233**, 1078–1080.

Heine, U.I., Munoz, E.F., Flanders, K.C., Ellingsworth, L.R., Lam, H.-Y.P., Thompson, N.L., Roberts, A.B. and Sporn, M.B. (1987). Role of transforming growth factor-β in the development of the mouse embryo. *J. Cell Biol.* **105**, 2861–2876.

Heine, U.I., Munoz, E.F., Flanders, K.C., Roberts, A.B. and Sporn, M.B. (1990). Colocalization of TGF-β1 and collagen I and III, fibronectin and glycosaminoglycans during lung branching morphogenesis. *Development* **109**, 29–36.

Henis, Y.I., Moustakas, A., Lin, H.Y. and Lodish, H.F. (1994). The type II and III transforming growth factor-β receptors form homo-oligomers. *J. Cell Biol.* **126**, 139–154.

Herrera, R.E., Mäkelä, T.P. and Weinberg, R.A. (1996). TGF-β-induced growth inhibition in primary fibroblasts requires retinoblastoma protein. *Mol. Biol. Cell* **7**, 1335–1342.

Hoffmann, F.M. (1991). Transforming growth factor-β-related genes in *Drosophila* and vertebrate development. *Curr. Opin. Cell Biol.* **3**, 947–952.

Hogan, B.M.L. (1996). Bone morphogenetic protein: multifunctional regulators of vertebrate development. *Genes Dev.* **10**, 1580–1594.

Holder, M.J., Knox, K. and Gordon, J. (1992). Factors modifying survival pathways of germinal center B cells. Glucocorticoids and transforming growth factor-beta, but not cyclosporin A or anti-CD19, block surface immunoglobulin-mediated rescue from apoptosis. *Eur. J. Immunol.* **22**, 2725–2728.

Howe, P.H., Bascom, C.C., Cunningham, M.R. and Loef, E.B. (1989). Regulation of transforming growth factor β1 action by multiple transducing pathways: evidence for both G protein-dependent and -independent signaling. *Cancer Res.* **49**, 6024–6031.

Huang, S.S., O'Grady, P. and Huang, J.S. (1988). Human transforming growth factor β: α2-macroglobulin complex is a latent form of transforming growth factor. *J. Biol. Chem.* **263**, 1535–1541.

Hunter, T. and Pines, J. (1994). Cyclins and cancer. II: Cyclin D and CDK inhibitors come of age. *Cell* **79**, 573–582.

Igarashi, A., Okochi, H., Bradham, D.M. and Grotendorst, G.R. (1993). Regulation of connective tissue growth factor gene expression in human skin fibroblasts and during wound repair. *Mol. Biol. Cell* **4**, 637–645.

Ignotz, R.A. and Massagué, J. (1985). Type β transforming growth factor controls the adipogenic differentiation of 3T3 fibroblasts. *Proc. Natl Acad. Sci. U.S.A.* **82**, 8530–8534.

James, K. (1990). Interactions between cytokines and α2-macroglobulin. *Immunol. Today* **11**, 163–166.

Jhappan, C., Geiser, A.G., Kordon, E.C., Bagheri, D., Hennighausen, L., Roberts, A.B., Smith, G.H. and Merlino, G. (1993). Targeting expression of a transforming growth factor β1 transgene to the pregnant mammary gland inhibits alveolar development and lactation. *EMBO J.* **12**, 1835–1846.

Joyce, M.E., Roberts, A.B., Sporn, M.B. and Bolander, M. (1990). Transforming growth factor-β and the initiation of chondrogenesis and osteogenesis in rat femur. *J. Cell Biol.* **110**, 2195–2207.

Kaartinen, V., Voncken, J.W., Shuler, C., Warburton, D., Bu, D., Heisterkamp, N. and Groffen, J. (1995). Abnormal lung development and cleft palate in mice lacking TGF-β3 indicates defects of epithelial–mesenchymal interaction. *Nature Genet.* **11**, 415–421.

Kanzaki, T., Olofsson, A., Morén, A., Wernstedt, C., Hellman, U., Miyazono, K., Claesson-Welsh, L. and Heldin, C.-H. (1990). TGF-β1 building protein: a component of the large latent complex of TGF-β1 with multiple repeat sequences. *Cell* **61**, 1051–1061.

Kataoka, R., Sherlock, J. and Lanier, S.M. (1993). Signaling events initiated by transforming growth factor-β1 that require Giα1. *J. Biol. Chem.* **268**, 19 851–19 857.

Kawabata, M., Imamura, T., Miyazono, K., Engel, M.E. and Moses, H.L. (1995). Cloning of a novel type II serine/threonine kinase receptor through interaction with the type I transforming growth factor-β receptor. *J. Biol. Chem.* **270**, 29 628–29 631.

Kay, J.E. (1996). Structure–function relationships in the FK506-binding protein (FKBP) family of peptidylprolyl cis-trans isomerases. *Biochem. J.* **314**, 361–385.

Kehrl, J.H. (1991). Transforming growth factor-β: an important mediator of immunoregulation. *Int. J. Cell Cloning* **9**, 438–450.

Kehrl, J.H., Roberts, A.B., Wakefield, L.M., Jakowlew, S., Sporn, M.B. and Fauci, A.S. (1986a). Transforming growth factor-β is an important immunomodulating protein for human B lymphocytes. *J. Immunol.* **137**, 3855–3860.

Kehrl, J.H., Wakefield, L.M., Roberts, A.B., Jakowlew, S., Alvarez-Mon, M., Derynck, R., Sporn, M.B. and Fauci, A.S. (1986b). The production of TGF-β by human T lymphocytes and its potential role in the regulation of T cell growth. *J. Exp. Med.* **163**, 1037–1050.

Kehrl, J.H., Thevenin, C., Rieckmann, P. and Fauci, A.S. (1991). Transforming growth factor-β suppresses human B lymphocyte Ig production by inhibiting synthesis and the switch from the membrane form to the secreted form of Ig mRNA. *J. Immunol.* **146**, 4016–4022.

Kekow, J., Wachsman, W., McCutchan, J.A., Cronin, M., Carson, D.A. and Lotz, M. (1990). Transforming growth factor β and noncytopathic mechanisms of immunodeficiency in human immunodeficiency virus infection. *Proc. Natl Acad. Sci. U.S.A.* **87**, 8321–8325.

Kekow, J., Wachsman, W., McCutchan, J.A., Gross, W.L., Zachariah, M., Carson, D.A. and Lotz, M. (1991). Transforming growth factor-β and suppression of humoral immune responses in HIV infection. *J. Clin. Invest.* **87**, 1010–1016.

Kim, I.-Y., Ahn, H.-J., Zelner, D.J., Shaw, J.W., Sensibar, J.A., Kim, J.H., Kato, M. and Lee, C. (1996a). Genetic change in transforming growth factor-β (TGF-β) receptor type I gene correlates with insensitivity to TGF-β1 in human prostrate cancer cells. *Cancer Res.* **56**, 44–48.

Kim, S.-J., Kehrl, J.H., Burton, J., Tendler, C.L., Jeang, K.T., Danielpour, D., Thevenin, C., Kim, K.Y., Sporn, M.B. and Roberts, A.B. (1990). Transactivation of the transforming growth factor β1 (TGF-β1) gene by human T lymphotropic virus type 1tax: a potential mechanism for the increased production of TGF-β1 in adult T cell leukemia. *J. Exp. Med.* **172**, 121–129.

Kim, S.-J., Park, K., Koeller, D., Kim, K.Y., Wakefield, L.M., Sporn, M.B. and Roberts. A.B. (1992). Post-transcriptional regulation of the human transforming growth factor-β1 gene. *J. Biol. Chem.* **267**, 13 702–13 707.

Kim, S.K., Fan, Y., Papadimitrakopulou, V., Clayman, G., Hittelman, W.N., Hong, W.K., Lotan, R. and Mao, L. (1996b). DPC4, a candidate tumor suppressor gene, is altered infrequently in head and neck squamous carcinoma. *Cancer Res.* **56**, 2519–2521.

Knaus, P.I., Lindemann, D., DeCoteau, J.F., Perlman, R., Yankelev, H., Hille, M., Kadin, M.E. and Lodish, H.F. (1996). A dominant inhibitory mutant of the type II transforming growth factor β receptor in the malignant progression of a cutaneous T-cell lymphoma. *Mol. Cell. Biol.* **16**, 3480–3489.

Koff, A., Ohtsuki, M., Polyak, K., Roberts, J.M. and Massagué, J. (1993). Negative regulation of G1 in mammalian cells: inhibition of cyclin E-dependent kinase by TGF-beta. *Science* **260**, 536–539.

Kojima, S. and Rifkin, D.B. (1993). Mechanism of retinoid-induced activation of latent transforming growth factor-β in bovine endothelial cells. *J. Cell. Physiol.* **155**, 323–332.

Kojima, S., Nara, K. and Rifkin, D.B. (1993). Requirement for transglutaminase in the activation of latent transforming growth factor-β bovine endothelial cells. *J. Cell Biol.* **12**, 439–448.

Kovacina, K.S., Steele, P.G., Purchio, A.F., Lioubin, M., Miyazono, K., Heldin, C.-H. and Roth, R.A. (1989). Interactions of recombinant and platelet transforming growth factor-β1 precursor with the insulin-like growth factor II/mannose 6-phosphate receptor. *Biochem. Biophys. Res. Commun.* **160**, 393–403.

Kretzschmar, M., Liu, F., Hata, A., Doody, J. and Massagué, J. (1997). The TGF-β family mediator Smad1 is phosphorylated directly and activated functionally by BMP receptor kinase. *Genes Dev.* **11**, 984–995.

Kulkarni, A.B., Huh, C.-G., Becker, D., Geiser, A., Light, M., Flanders, K.C., Roberts, A.B., Sporn, M.B., Ward, J.M. and Karlsson, S. (1993). Transforming growth factor-β1 null mutation in mice causes excessive inflammatory response and early death. *Proc. Natl Acad. Sci. U.S.A.* **90**, 770–774.

Kuppner, M.C., Hamou, M.F., Bodmer, S., Fontana, A. and De Tribolet, N. (1988). The glioblastoma-derived T-cell suppressor factor/transforming growth factor β2 inhibits the generation of lymphokine-activated killer (LAK) cells. *Int. J. Cancer* **42**, 562–567.

Kuruvilla, A.P., Shah, R., Hochwald, G.M., Liggitt, H.D., Palladino, M.A. and Thorbecke, G.J. (1991). Protective effect of transforming growth factor β1 on experimental autoimmune diseases in mice. *Proc. Natl Acad. Sci. U.S.A.* **88**, 2918–2921.

Lagna, G., Hata, A., Hemmati-Brivanlou, A. and Massagué, J. (1996). Partnership between DPC4 and SMAD proteins in TGF-β signaling pathways. *Nature* **383**, 832–836.

Laiho, M., DeCaprio, J.A., Ludlow, J.W., Livingston, D.M. and Massagué, J. (1990a). Growth inhibition by TGF β linked to suppression of retinoblastoma protein phosphorylation. *Cell* **62**, 175–185.

Laiho, M., Weis, F.M.B. and Massagué, J. (1990b). Concomitant loss of transforming growth factor (TGF)-β receptor types I and II in TGF-β-resistant cell mutants implicates both receptor types in signal transduction. *J. Biol. Chem.* **265**, 18518–18524.

Laiho, M., Weis, F.M.B., Boyd, F.T., Ignotz, R.A. and Massagué, J. (1991a). Responsiveness to transforming growth factor-beta (TGF-beta) restored by genetic complementation between cells defective in TGF-beta receptors I and II. *J. Biol. Chem.* **266**, 9108–9112.

Laiho, M., Rönnstrand, L., Heino, J., DeCaprio, J.A., Ludlow, J.W., Livingston, D.M. and Massagué, J. (1991b). Control of junB and extracellular matrix protein expression by transforming growth factor-beta 1 is independent of simian virus 40 T antigen-sensitive growth-sensitive growth-inhibiting events. *Mol. Cell. Biol.* **11**, 972–978.

LaMarre, J., Wollenberg, G.K., Gonias, S.L. and Hayes, M.A. (1991a). Cytokine binding and clearance properties of proteinase-activated α2-macroglobulins. *Lab. Invest.* **65**, 3–14.

LaMarre, J., Hayes, M.A., Wollenberg, G.K., Hussaini, I., Hall, S.W. and Gonias, S.L. (1991b). An α2-macroglobulin receptor-dependent mechanism for the plasma clearance of transforming growth factor-β1 in mice. *J. Clin. Invest.* **87**, 39–44.

Lawler, S., Candia, A.F., Ebner, R., Lopez, A.R., Moses, H.L., Wright, C.V.E. and Derynck, R. (1994). The murine type II TGF-β receptor has a coincident embryonic expression and binding preference for TGF-β1. *Development* **120**, 165–175.

Lawler, S., Feng, X.-H., Chen, R.-H., Maruoka, E.M., Turck, C.W., Griswold-Prenner, I. and Derynck, R. (1997). The type II transforming growth factor-β receptor autophosphorylates not only on serine and threonine but also on tyrosine residues. *J. Biol. Chem.* **272**, 14850–14859.

Lehnert, S.A. and Akhurst, E.J. (1988). Embryonic expression pattern of TGF-β type-I RNA suggests both paracrine and autocrine mechanisms of action. *Development* **104**, 263–273.

Letterio, J.J. and Roberts, A.B. (1998). Induction of c-sis mRNA and activity similar to platelet-derived growth factor by transforming growth factor β: a proposed model for indirect mitogenesis involving autocrine activity. *Annu. Rev. Immunol.* (in press).

Leof, E.B., Proper, J.A., Goustin, A.S., Shipley, G.D., DiCorleto, P.E. and Moses, H.L. (1986). Regulation of immune responses by TGF-β. *Proc. Natl Acad. Sci. U.S.A.* **83**, 2453–2457.

Like, B. and Massagué, J. (1986). The autoproliferative effect of type β transforming growth factor occurs at a level distal from receptors for growth activating factors. *J. Biol. Chem.* **261**, 13426–13429.

Lin, H.Y., Wang, X.-F., Ng-Eaton, E., Weinberg, R.A. and Lodish, H.F. (1992). Expression cloning of the TGF-β type II receptor, a functional transmembrane serine/threonine kinase. *Cell* **68**, 775–785.

Lin, H.Y. and Lodish, H.F. (1993). Receptors for the TGF-β superfamily: multiple polypeptides and serine/threonine kinases. *Trends Cell Biol.* **3**, 14–19.

Lin, H.Y., Moustakas, A., Knaus, P., Wells, R.G., Henism, Y.I. and Lodish, H.F. (1995). The soluble exoplasmic domain of the type II transforming growth factor (TGF)-β receptor. A heterogenous glycosylated protein with high affinity and selectivity for TGF-β ligands. *J. Biol. Chem.* **270**, 2747–2754.

Liu, F., Ventura, F., Doody, J. and Massagué, J. (1995). Human type II receptor for bone morphogenic proteins (BMPs): extension of the two-kinase receptor model to the BMPs. *Mol. Cell. Biol.* **15**, 3479–3486.

Liu, F., Hata, A., Baker, J.C., Doody, J., Cárcamo, J., Harland, R.M. and Massagué, J. (1996). A human Mad protein acting as a BMP-regulated transcriptional activator. *Nature* **381**, 620–623.

Lopez, A.R., Cook, J., Deininger, P.L. and Derynck, R. (1992). Dominant negative mutants of transforming growth factor-β1 inhibit the secretion of different transforming growth factor-β isoforms. *Mol. Cell. Biol.* **12**, 1674–1679.

López-Casillas, F., Cheifetz, S., Doody, J., Andres, J.L., Lane, W.S. and Massagué, J. (1991). Structure and expression of the membrane proteoglycan betaglycan, a component of the TGF-β receptor system. *Cell* **67**, 785–795.

López-Casillas, F., Wrana, J.L. and Massagué, J. (1993). Betaglycan presents ligand to the TGFβ signaling receptor. *Cell* **73**, 1435–1444.

López-Casillas, F., Payne, H.M., Andres, J.L. and Massagué, J. (1994). Betaglycan can act as a dual modulator of TGF-β access to signaling receptors: mapping of ligand binding and GAG attachment sites. *J. Cell Biol.* **124**, 557–568.

Lotz, M., Kekow, J. and Carson, D.A. (1990). Transforming growth factor-β and cellular immune responses in synovial fluids. *J. Immunol.* **144**, 4194–4198.

Luo, K. and Lodish, H.F. (1996). Signaling by chimeric erythropoietin-TGF-β receptors: homodimerization of the cytoplasmic domain of the type I TGF-β receptor and heterodimerization with the type II receptor are both required for intracellular signal transduction. *EMBO J.* **15**, 4485–4496.

Lyons, R.M., Keski-Oja, J. and Moses, H.L. (1988). Proteolytic activation of latent transforming growth factor-β from fibroblast-conditioned medium. *J. Cell Biol.* **106**, 1659–1665.

Lyons, R.M., Gentry, L.E., Purchio, A.F. and Moses, H.L. (1990). Mechanisms of activation of latent recombinant transforming growth factor β1 by plasmin. *J. Cell Biol.* **110**, 1361–1367.

Macias-Silva, M., Abdollah, S., Hoodless, P., Pirone, R., Attisano, L. and Wrana, J.L. (1996). MADR-2 is a substrate of the TGF-β receptor and its phosphorylation is required for nuclear accumulation and signaling. *Cell* **87**, 1215–1224.

Mai, A., Poon, R.Y.C., Howe, P., Toyoshima, H., Hunter, T. and Harter, M.L. (1996). Inactivation of $p27^{Kip1}$ by the viral E1A oncoprotein in TGF-β-treated cells. *Nature* **380**, 262–265.

Manning, A.M., Williams, A.C., Game, S.M. and Partskeva, C. (1991). Differential sensitivity of human colonic adenoma and carcinoma cells to transforming growth factor β (TGF-β): conversion of an adenoma cell line to a tumorigenic phenotype is accompanied by a reduced response to the inhibitory effects of TGF-β. *Oncogene* **6**, 1471–1476.

Markowitz, S.D. and Roberts, A.B. (1996). Tumor suppressor activity in the TGF-β pathway in human cancers. *Cytokine Growth Factor Rev.* **7**, 93–102.

Markowitz, S., Myeroff, L., Cooper, M.J., Traicoff, J., Kochera, M., Lutterbaugh, J., Swiriduk, M. and Willson, J.K. (1994). A benign cultured colon adenoma bears three genetically altered colon cancer oncogenes, but progresses to tumorigenicity and transforming growth factor-β independence without inactivating the p53 tumor suppressor gene. *J. Clin. Invest.* **93**, 1005–1013.

Markowitz, S., Wang, J., Myeroff, L., Parsons, R., Sun, L., Lutterbaugh, J., Fan, R.S., Zborowska, E., Kinzler, K.W., Vogelstein, B., Brattain, M.G. and Willson, J.K.V. (1995). Inactivation of the type II TGF-β receptor in colon cancer cells with microsatellite instability. *Science* **268**, 1336–1338.

Massagué, J. (1990). The transforming growth factor-β family. *Annu. Rev. Cell Biol.* **6**, 597–641.

Massagué, J. (1992). Receptors for the TGF-β family. *Cell* **69**, 1067–1070.

Massagué, J. and Polyak, K. (1995). Mammalian antiproliferative signals and their targets. *Curr. Opin. Genet. Dev.* **5**, 91–96.

Massagué, J., Cheifetz, S., Endo, T. and Nadal-Ginard, B. (1986). Type β transforming growth factor is an inhibitor of myogenic differentiation. *Proc. Natl Acad. Sci. U.S.A.* **83**, 8206–8210.

Mathews, L.S. and Vale, W.W. (1991). Expression cloning of an activin receptor, a predicted transmembrane serine kinase. *Cell* **65**, 973–982.

Matthes, T., Werner-Favre, C., Tang, H., Zhang, X., Kindler, V. and Zubler, R.H. (1993). Cytokine mRNA expression during an *in vitro* response of human B lymphocytes: kinetics of B cell tumor necrosis factor α, interleukin (IL) 6, IL-10, and transforming growth factor β1 mRNAs. *J. Exp. Med.* **178**, 521–528.

McAllister, K.A., Grogg, K.M., Johnson, D.W., Gallione, C.J., Baldwin, M.A., Jackson, C.E., Helmbold, E.A., Markel, D.S., McKinnon, W.C., Murrell, J., McKormick, M.K., Pericak-Vance, M.A., Heutink, P., Oostra, B.A., Haitjema, T., Westerman, C.J.J., Porteous, M., Guttmacher, A.E., Letarte, M. and Marchuk, D.A. (1994). Endoglin, a TGF-β binding protein of endothelial cells, is the gene for hereditary haemorrhagic telangiectasia type 1. *Nature Genet.* **8**, 345–351.

McCartney-Francis, N., Mizel, D., Wong, H., Wahl, L.M. and Wahl, S.M. (1990) TGF-β regulates production of growth factors and TGF-β by human peripheral blood monocytes. *Growth Factors* **4**, 27–32.

McCartney-Francis, N.L. and Wahl, S.M. (1994). Transforming growth factor-β: a matter of life and death. *J. Leukoc. Biol.* **55**, 401–409.

Miettinen, P.J., Ebner, R., Lopez, A.R. and Derynck, R. (1994). TGF-β-induced transdifferentiation of mammary epithelial cells to mesenchymal cells: involvement of type I receptors. *J. Cell Biol.* **127**, 2021–2036.

Ming, M., Ewen, M.E. and Pereira, M.E. (1995). Trypanosome invasion of mammalian cells requires activation of the TGF β signalling pathway. *Cell* **82**, 287–296.

Miyazono, K., Hellman, U., Wernstedt, C. and Heldin, C.H. (1988). Latent high molecular weight complex of transforming growth factor β1. Purification from human platelets and structural characterization. *J. Biol. Chem.* **263**, 6407–6415.

Miyazono, K., Yuki, K., Takaku, F., Wernstedt, C., Kanzaki, T., Olofsson, A., Hellman, U. and Heldin, C.-H. (1990). Latent forms of TGF-β: structure and biology. *Ann. N.Y. Acad. Sci.* **593**, 51–58.

Miyazono, K., Olofsson, A., Colosetti, P. and Heldin, C.-H. (1991). A role of the latent TGF-β1 binding protein in the assembly and secretion of TGF-β1. *EMBO J.* **10**, 1091–1101.

Moren, A., Olofsson, A., Stenman, G., Sahlin, P., Kanzaki, T., Claesson-Wesh, L., ten Dijke, P., Miyazon, K. and Heldin, C.-H. (1994). Identification and characterization of LTBP-2, a novel latent transforming growth factor-β-binding protein. *J. Biol. Chem.* **269**, 32469–32478.

Morton, D.M. and Barrack, E.R. (1995). Modulation of transforming growth factor β1 effects on prostate cancer cell proliferation by growth factors and extracellular matrix. *Cancer Res.* **55**, 2596–2602.

Moses, H.L., Branum, E.L., Proper, J.A. and Robinson, R.A. (1981). Transforming growth factor production by chemically transformed cells. *Cancer Res.* **41**, 2842–2848.

Moses, H.L., Yang, E.Y. and Pietenpol, J.A. (1990). TGF-β stimulation and inhibition of cell proliferation: new mechanistic insights. *Cell* **63**, 245–247.

Mossalayi, M.D., Mentz, F., Ouaaz, F., Dalloul, A.H., Blanc, C., Debre, F. and Ruscetti, F.W. (1995). Early human thymocyte proliferation is regulated by an externally controlled autocrine transforming growth factor-β1 mechanism. *Blood* **85**, 3594–3601.

Moustakas, A., Lin, H.Y., Henis, Y.I., Plamondon, J., O'Connor-McCourt, M.D. and Lodish, H.F. (1993). The transforming growth factor β receptors types I, II, and III form hetero-oligomeric complexes in the presence of ligand. *J. Biol. Chem.* **268**, 22215–22218.

Mucsi, I., Skorecki, K.L. and Goldberg, H.J. (1996). Extracellular signal-regulated kinase and the small GTP-binding protein, Rac, contribute to the effects of transforming growth factor-β1 on gene expression. *J. Biol. Chem.* **271**, 16567–16572.

Muller, G., Behrens, J., Nussbaumer, U., Bohlen, P. and Birchmeier, W. (1987). Inhibitory action of transforming growth factor β on endothelial cells. *Proc. Natl Acad. Sci. U.S.A.* **85**, 5600–5604.

Munger, J.S., Harpel, J.G., Gleizes, P.-E., Mazzieri, R., Nunes, I. and Rifkin, D.B. (1997). Latent transforming growth factor-β: structural features and mechanisms of activation. *Kidney Int.* **51**, 1376–1382.

Murphy-Ullrich, J.E., Schultz-Cherry, S. and Hook, M. (1992). Transforming growth factor-β complexes with thrombospondin. *Mol. Biol. Cell* **3**, 181–188.

Myeroff, L., Parsons, R., Kim, S.-J., Hedrick, L., Cho, K., Orth, K., Mathis, M., Kinzler, K.W., Lutterbaugh, J., Park, K., Barg, Y.-J., Lee, H., Park, J.-G., Lynch, H., Roberts, A., Vogelstein, B. and Markowitz, S.A. (1995). A TGF-β receptor type II gene mutation common in colon and gastric but rare in endometrial cancers with microsatellite instability. *Cancer Res.* **55**, 5545–5547.

Nagatake, M., Takagi, Y., Osada, H., Uchida, K., Mitsudomi, T., Saji, S., Shimokata, K., Takahashi, T. and Takahashi, T. (1996). Somatic *in vivo* alterations of the DPC4 gene at 18q21 in human lung cancers. *Cancer Res.* **56**, 2718–2720.

Naranda, T., Kainuma, M., MacMillan, S.E. and Hershey, J.W. (1997). The 39-kilodalton subunit of eukaryotic translation factor 3 is essential for the complex's integrity and for cell viability in *Saccharomyces cerevisiae*. *Mol. Cell. Biol.* **17**, 145–153.

Nathan, C.F. (1987). Secretory products of macrophages. *J. Clin. Invest.* **79**, 319–326.

Neer, E.J., Schmidt, C.J., Nambudripad, R. and Smith, T.F. (1994). The ancient regulatory-protein family of WD-repeat proteins. *Nature* **371**, 297–300.

Newfeld, S.J., Chartoff, E.H., Graff, J.M., Melton, D.A. and Gelbart, W.M. (1996). Mothers against dpp encodes a conserved cytoplasmic protein required in DPP/TGF-β responsive cells. *Development* **122**, 2099–2108.

Nohno, T., Ishikawa, T., Saito, T., Hosokawa, K., Noji, S., Wolsing, D.H. and Rosenbaum, J.S. (1995). Identification of a human type II receptor for bone morphogenetic protein-4 that differential heteromeric complexes with bone morphogenetic protein type I receptors. *J. Biol. Chem.* **270**, 22522–22526.

Nunes, I., Shapiro, R.L. and Rifkin, D.B. (1995). Characterization of latent TGF-β activation by murine peritoneal macrophages. *J. Immunol.* **155**, 1450–1459.

O'Connor-McCourt, M. and Wakefield, L.M. (1987). Latent transforming growth factor-β in serum. A specific complex with α2-macroglobulin. *J. Biol. Chem.* **262**, 14090–14099.

Oft, M., Peli, J., Rudaz, C., Schwarz, H., Beug, H. and Reichmann, E. (1996). TGF-β1 and Ha-Ras collaborate in modulating the phenotypic plasticity and invasiveness of epithelial tumor cells. *Genes Dev.* **10**, 2462–2477.

Ohta, M., Greenberger, J.S., Anklesaria, P., Bassols, A. and Massagué, J. (1987). Two forms of transforming growth factor-β distinguished by hematopoietic progenitor cells. *Nature* **329**, 539–541.

Ohtsuki, M. and Massagué, J. (1992). Evidence for the involvement of protein kinase activity in transforming growth factor-β signal transduction. *Mol. Cell. Biol.* **12**, 261–265.

Olofsson, A., Miyazono, K., Kanzaki, T., Colosatti, P., Engstrom, U. and Heldin, C.-H. (1992). Transforming growth factor-β1, -β2, and -β3 secreted by a human glioblastoma cell line. Identification of small and different forms of large latent complexes. *J. Biol. Chem.* **267**, 19482–19488.

Omer, C.A. and Gibbs, J.B. (1994). Protein prenylation in eukaryotic microorganisms: genetics biology and biochemistry. *Mol. Microbiol.* **11**, 219–225.

Ortaldo, J.R., Mason, A.T., O'Shea, J.J., Smith, M.J., Falk, L.A., Kennedy, A.C.S., Longo, D.L. and Ruscetti, F.W. (1991). Mechanistic studies of transforming growth factor-beta inhibition of IL-2-dependent activation of CD3-large granular lymphocyte functions. Regulation of IL-2Rβ (p75) signal transduction. *J. Immunol.* **146**, 3791–3798.

Oshima, M., Oshima, H. and Taketo, M.M. (1996). TGF-β receptor type II deficiency results in defects of yolk sac hematopoiesis and vasculogenesis. *Dev. Biol.* **179**, 297–302.

Ottman, O. and Pelus, L.M. (1988). Differential proliferative effects of transforming growth factor-β on human hematopoietic progenitor cells. *J. Immunol.* **140**, 2661–2665.

Oursler, M.J. (1994). Osteoclast synthesis and secretion and activation of latent transforming growth factor β. *J. Bone Miner. Res.* **9**, 443–452.

Oursler, M.J., Riggs, B.L. and Spelsberg, T.C. (1993). Glucocorticoid-induced activation of latent transforming growth factor-β by normal human osteoblast-like cells. *Endocrinology* **133**, 2187–2196.

Paralkar, M.V., Vukicevic, S. and Reddi, A.H. (1991). Transforming growth factor β type I binds to collagen IV of basement membrane matrix: implications for development. *Dev. Biol.* **143**, 303–308.

Park, J., Kim, S.-J., Bang, Y.-J., Park, J.-G., Kim, N.K., Roberts, A.B. and Sporn, M.B. (1994). Genetic changes in the transforming growth factor-β (TGF-β) type II receptor gene in human gastric cancer cells: correlation with sensitivity to growth inhibition by TGF-β. *Proc. Natl Acad. Sci. U.S.A.* **91**, 8772–8776.

Pelton, R.W., Dickinson, M.E., Moses, H.L. and Hogan, B.L.M. (1990a). *In situ* hybridization analysis of TGF-β3 RNA expression during mouse development: comparative studies with TGF-β1 and -β2. *Development* **110**, 609–620.

Pelton, R.W., Hogan, B.L.M., Miller, D.A. and Moses, H.L. (1990b). Differential expression of genes encoding TGFs β1, β2, and β3 during murine palate formation. *Dev. Biol.* **141**, 456–460.

Pelton, R.W., Saxena, B., Jones, M., Moses, H.L. and Gold, L.I. (1990c). Immunohistochemical localization of TGF β1, TGF β2, and TGF β3 in the mouse embryo: expression patterns suggest multiple roles during embryonic development. *J. Cell Biol.* **115**, 1091–1105.

Pepin, M.-C., Beauchemin, M., Plamondon, J. and O'Connor-McCourt, M.D. (1994). Mapping of the ligand binding domain of the transforming growth factor β receptor type III by deletion mutagenesis. *Proc. Natl Acad. Sci. U.S.A.* **91**, 6997–7001.

Pepin, M.-C., Beauchemin, M., Collins, C., Plamondon, J. and O'Connor-McCourt, M.D. (1995). Mutagenesis analysis of the membrane-proximal ligand binding site to the TGF-β receptor type III extracellular domain. *FEBS Lett.* **377**, 368–372.

Pierce, D.F., Jr., Johnson, M.C., Matsui, Y., Robinson, S.D., Gold, L., Purchio, A.F., Daneil, C.W., Hogan, B.L.M. and Moses. H.L. (1993). Inhibition of mammary duct development but not alveolar outgrowth during pregnancy in transgenic mice expressing active TGF-β1. *Genes Dev.* **7**, 2308–2317.

Pierce, D.F., Jr., Gorska, A.E., Chytil, A., Meise, K.S., Page, D.L., Coffey, R.J., Jr. and Moses, H.L. (1995). Mammary tumor suppression by transforming growth factor β1 transgene expression. *Proc. Natl Acad. Sci. U.S.A.* **92**, 4254–4258.

Pietenpol, J.A., Stein, R.W., Moran, E., Yaciuk, P., Schlegel, R., Lyons, R.M., Pittelkow, M.R., Münger, K., Howley, P.M. and Moses, H.L. (1990). TGF-β1 inhibition of c-myc transcription and growth in keratinocytes is abrogated by viral transforming proteins with pRB binding domains. *Cell* **61**, 777–785.

Postlethwaite, A.E., Keski-Oja, J., Moses, H.L. and Kang, A.H. (1987). Stimulation of the chemotactic migration of human fibroblasts by transforming growth factor β. *J. Exp. Med.* **165**, 251–256.

Proetzel, G., Pawlowski, S.A., Wiles, M.V., Yin, M., Boivin, G., Howles, P.N., Ding, J., Ferguson, M.W.J. and Doetschman, T. (1995). Transforming growth factor-β is required for secondary palate fusion. *Nature Genet.* **11**, 409–414.

Purchio, A.F., Cooper, J.A., Brunner, A.M., Lioubin, M.N., Gentry, L.E., Kovacina, K.S., Roth, R.A. and Marquardt, H. (1988). Identification of mannose-6-phosphate in two asparagine-linked sugar chains of recombinant TGF-β1 precursor. *J. Biol. Chem.* **263**, 14211–14215.

Racke, M.K., Cannella, B., Albert, P., Sporn, M.B., Raine, C.S. and McFarlin, D.E. (1992). Evidence of endogenous regulatory function of transforming growth factor-β1 in experimental allergic encephalomyelitis. *Int. Immunol.* **4**, 615–620.

Raftery, L.A., Twombly, V., Wharton, K. and Gelbart, W.M. (1995). Genetic screens to identify elements of the decapentaplegic signaling pathway in *Drosophila*. *Genetics* **139**, 241–254.

Ranges, G.E., Figari, I.S., Espevik, T. and Palladino, M.A. (1987). Inhibition of cytotoxic T cell development by transforming growth factor β and reversal by recombinant tumor necrosis factor α. *J. Exp. Med.* **166**, 991–998.

Reddy, K.B., Karode, M.C., Harmony, J.A.K. and Howe, P.H. (1996). Interaction of transforming growth factor β receptors with apolipoprotein J/clusterin. *Biochemistry* **35**, 309–314.

Reynisdóttir, I., Polyak, K., Iavarone, A. and Massagué, J. (1995). Kip/Cip and Ink4 cdk inhibitors cooperate to induce cell cycle arrest in response to TGF-β. *Genes Dev.* **9**, 1831–1845.

Rich, S., Van Nood, N. and Less, H.M. (1996). Role of $\alpha5\beta1$ integrin in TGF-β1 costimulated CD8+ T cell growth and apoptosis. *J. Immunol.* **157**, 2916–2923.

Ridley, A.J., Davis, J.B., Stroobant, P. and Land, H. (1989). Transforming growth factors-β1 and β2 are mitogens for rat Schwann cells. *J. Cell Biol.* **109**, 3419–3424.

Riedl, E., Strobl, H., Majdic, O. and Knapp, W. (1997). TGF-β1 promotes *in vitro* generation of dendritic cells by protecting progenitor cells from apoptosis. *J. Immunol.* **158**, 1591–1597.

Riggins, G.J., Thiagalingam, S.E.R., Weinstein, C.L., Kern, S.E., Hamilton, S.R., Willson, J.K.V., Markowitz, S.D., Kinzler, K.W. and Vogelstein, B. (1996). Mad-related genes in the human. *Nature Genet.* **13**, 347–349.

Ristow, H.J. (1986). BSC-1 growth inhibitor/type β transforming growth factor is a strong inhibitor of thymocyte proliferation. *Proc. Natl Acad. Sci. U.S.A.* **83**, 5531–5534.

Roberts, A.B. and Sporn, M.B. (1990). The transforming growth factor-βs. In *Peptide Growth Factors and their Receptors* (eds M.B. Sporn and A.B. Roberts), Springer, Heidelberg, pp. 421–472.

Roberts, A.B. and Sporn, M.B. (1992). Mechanistic interrelationships between two superfamilies: the steroid/retinoid receptors and transforming growth factor-β. *Cancer Surv.* **14**, 205–220.

Roberts, A.B. and Sporn, M.B. (1993). Physiological actions and clinical applications of transforming growth factor-β (TGF-β). *Growth Factors* **8**, 1–9.

Roberts, A.B., Anzano, M.A., Lamb, L.C., Smith, J.M. and Sporn, M.B. (1981). New class of transforming growth factors potentiated by epidermal growth factor: isolation from non-neoplastic tissues. *Proc. Natl Acad. Sci. U.S.A.* **78**, 5339–5343.

Roberts, A.B., Sporn, M. B., Assoian, R. K., Smith, J.M., Roche, N.S., Wakefield, L.M., Heine, U.I., Liotta, L.A., Falanga, V., Kehrl, J.H. and Fauci, A.S. (1986). Transforming growth factor β: rapid induction of fibrosis and angiogenesis *in vivo* and stimulation of collagen formation *in vitro*. *Proc. Natl Acad. Sci. U.S.A.* **83**, 4167–4171.

Roberts, A.B., Kim, S.-J., Noma, T., Glick, A.B., Lafyatis, R., Lechleider, R., Jakowlew, S.B., Geiser, A., O'Reilly, M.A., Danielpour, D. and Sporn, M.B. (1991). Multiple forms of TGF-β: distinct promoters and differential expression. *Ciba Found. Symp.* **157**, 7–15.

Robinson, S.D., Silberstein, G.B., Roberts, A.B., Flanders, K.C. and Daniel, C.W. (1991). Regulated expression and growth inhibitory effects of transforming growth factor-β isoforms in mouse mammary gland development. *Development* **113**, 867–878.

Robinson, S.D., Roberts, A.B. and Daniel, C.W. (1993). TGF β suppresses casein synthesis in mouse mammary explants and may play a role in controlling milk levels during pregnancy. *J. Cell Biol.* **120**, 245–251.

Rodriguez, C., Chen, F., Weinberg, R.A. and Lodish, H.F. (1995). Cooperative binding of transforming growth factor (TGF)-β2 to the types I and II TGF-β receptors. *J. Biol. Chem.* **270**, 15919–15922.

Romeo, D.S., Park, K., Roberts, A.B., Sporn, M.B. and Kim, S.-J. (1993). An element of the transforming growth factor-β1 5′-untranslated region represses translation and specifically binds a cytosolic factor. *Mol. Endocrinol.* **7**, 759–766.

Rook, A.H., Kehrl, J.H., Wakefield, L.M., Roberts, A.B., Sporn, M.B., Burlington, D.B., Lane, H.C. and Fauci, A.S. (1986). Effects of transforming growth factor β on the functions of natural killer cells: depressed cytolytic activity and blunting of interferon responsiveness. *J. Immunol.* **136**, 3916–3920.

Rosenzweig, B.L., Imamura, T., Okadome, T., Cox, G.N., Yamashita, H., ten Dijke, P., Heldin, C.-H. and Miyazono, K. (1995). Cloning and characterization of a human type II receptor for bone morphogenetic proteins. *Proc. Natl Acad. Sci. U.S.A.* **92**, 7632–7636.

Rowley, D.A., Becken, E.T. and Stach, R.M. (1995). Autoantibodies produced spontaneously by young lpr mice carry transforming growth factor β and suppress cytotoxic T lymphocyte responses. *J. Exp. Med.* **181**, 1875–1880.

Sanford, L.P., Ormsby, I., Gittenberger-de Groot, A.C., Sariola, H., Friedman, R., Boivin, G.P., Cardell, E.L. and Doetschman, T. (1997). TGFβ2 knockout mice have multiple developmental defects that are non-overlapping with other TGFβ knockout phenotypes. *Development* **124**, 2659–2670.

Sato, Y. and Rifkin, D.B. (1989). Inhibition of endothelial cell movement by pericytes and smooth muscle cells: activation of a latent transforming growth factor-β1-like molecule by plasmin during co-culture. *J. Cell Biol.* **109**, 309–315.

Savage, C., Das, P., Finelli, A.L., Townsend, S.R., Sun, C.Y., Baird, S.E. and Padgett, R.W. (1996). *Caenorhabditis elegans* genes sma-2, sma3, and sma-4 define a conserved family of transforming growth factor β pathway components. *Proc. Natl Acad. Sci. U.S.A.* **93**, 790–794.

Schlunegger, M.P. and Grütter, M.G. (1992). An unusual feature revealed by the crystal structure at 2.2 Å resolution of human transforming growth factor-β2. *Nature* **358**, 430–434.

Schmitt, E., Hoehn, P., Huels, C., Goedert, S., Palm, N., Rude, E. and Germann, T. (1994). T helper type 1 development of naive $CD4^+$ cells requires the coordinate action of interleukin-12 and interferon-γ and is inhibited by transforming growth factor β. *Eur. J. Immunol.* **24**, 793–798.

Schultz-Cherry, S. and Murphy-Ullrich, J.E. (1993). Thrombospondin causes activation of latent transforming growth factor-β secreted by endothelial cells by a novel mechanism. *J. Cell Biol.* **122**, 923–932.

Schultz-Cherry, S., Ribeiro, S., Gentry, L. and Murphy-Ullrich, J.E. (1994). Thrombospondin binds and activates the small and large forms of latent transforming growth factor-β in a chemically defined system. *J. Biol. Chem.* **269**, 26 775–26 782.

Schutte, M., Hruban, R.H., Hedrick, L., Cho, K., Nadasdy, G., Weinstein, C.L., Bova, G.S., Isaacs, W.B., Cairns, P., Nawroz, H., Sidransky, D., Casero, R.A., Meltzer, P.S., Hahn, S.A. and Kern, S.E. (1996). DPC4 gene in various tumor types. *Cancer Res.* **56**, 2527–2530.

Segarini, P.R. and Seyedin, S.M. (1988). The high molecular weight receptor to transforming growth factor-β contains glycosaminoglycan chains. *J. Biol. Chem.* **263**, 8366–8370.

Segarini, P.R., Roberts, A.B., Rosen, D.M. and Seyedin, S.M. (1987). Membrane binding characteristics of two forms of transforming growth factor-β. *J. Biol. Chem.* **262**, 14 655–14 662.

Sellheyer, K., Bickenbach, J.R., Rothnagel, J.A., Bundman, D., Longley, M.A., Krieg, T., Roche, N.S., Roberts, A.B. and Roop, D.R. (1993). Inhibition of skin development by overexpression of transforming growth factor β1 in the epidermis of transgenic mice. *Proc. Natl Acad. Sci. U.S.A.* 5237–5241.

Serra, R., Pelton, R.W. and Moses, H.L. (1994). TGF β1 inhibits branching morphogenesis and N-myc expression in lung bud organ cultures. *Development* **120**, 2153–2161.

Serra, R., Johnson, M., Filvaroff, E.H., LaBorde, J., Shehan, D., Derynck, R. and Moses, H.L. (1997). Expression of a cytoplasmically truncated TGF-β type II receptor in mouse skeletal tissue promotes terminal chondrocyte differentiation and an osteoarthritis-like phenotype. *J. Cell Biol.* (in press).

Sherr, C.J. (1995). D-type cyclins. *Trends Biochem. Sci.* **20**, 187–190.

Sherr, C.J. and Roberts, J.M. (1995). Inhibitors of mammalian G1 cyclin-dependent kinases. *Genes Dev.* **9**, 1149–1163.

Shi, Y., Hata, A., Lo, R.S., Massagué, J. and Pavletich, N.P. (1997). A structural basis for mutational inactivation of the tumour suppressor Smad4. *Nature* **388**, 87–93.

Shibuya, H., Yamaguchi, K., Shirakabe, K., Tonegawa, A., Gotoh, Y., Ueno, N., Irie, N., Nishida, E. and Matsumoto, K. (1996). TAB1: an activator of TAK1 MAPKKK in TGF-β signal transduction. *Science* **272**, 1179–1182.

Shull, M.M., Ormsby, I., Kier, A.B., Pawlowski, S., Diebold, R.J., Yin, M., Allen, R., Sidman, C., Proetzel, G., Calvin, D., Annunziata, N. and Doetschman, T. (1992). Targeted disruption of the mouse transforming growth factor-β1 gene results in multifocal inflammatory disease. *Nature* **359**, 693–699.

Sieweke, M.H., Thompson, N.L., Sporn, M.B. and Bissell, M.J. (1990). Mediation of wound related Rous sarcoma virus tumorigenesis by TGF-β. DNA sequence analysis by primed synthesis. *Science* **248**, 1656–1660.

Sing, G.K., Keller, J.R., Ellingsworth, L.R. and Ruscetti, F. (1988). Transforming growth factor β selectively inhibits normal and leukemic human bone marrow cell growth *in vitro*. *Blood* **72**, 1504–1511.

Slager, H.G., Van Inzen, W., Freund, E., van den Eijnden-van Raaij, A.J.M. and Mummery, C.L. (1993). Transforming growth factor-β in the early mouse embryo: implications for the regulation of muscle formation and implantation. *Dev. Genet.* **14**, 212–224.

Snapper, C.M., Waegell, W., Beernink, H. and Dash, J.R. (1993). Transforming growth factor-β1 is required for secretion of IgG of all subclasses by LPS-activated murine B cells *in vitro*. *J. Immunol.* **151**, 4625–4636.

Soma, Y. and Grotendorst, G.R. (1989). TGF-β stimulates primary human skin fibroblast DNA synthesis via an autocrine production of PDGF-related peptides. *J. Cell. Physiol.* **140**, 246–253.

Souchelnytskyi, S., ten Dijke, P., Miyazono, K. and Heldin, C.-H. (1996). Phosphorylation of Ser165 in TGF-β type I receptor modulates TGF-β1-induced cellular responses. *EMBO J.* **15**, 6231–6240.

Stach, R.M. and Rowley, D.A. (1993). A first or dominant immunization. II. Induced immunoglobulin carries transforming growth factor β and suppresses cytolytic T cell responses to unrelated alloantigens. *J. Exp. Med.* **178**, 841–852.

Stavnezer, J. (1996). Transforming growth factor-β. In *Cytokine Regulation of Humoral Immunity: Basic and Clinical Aspects* (ed. C.M. Snapper), John Wiley, pp. 289–324.

Steinberg, A.D. (1994). MRL-lpr/lpr disease: theories meet Fas. *Semin. Immunol.* **6**, 55–69.

Steiner, M.S. and Barrack, E.R. (1992). Transforming growth factor-β1 overproduction in prostate cancer: effects on growth *in vivo* and *in vitro*. *Mol. Endocrinol.* **6**, 15–25.

Steinman, R.M. (1991). The dendritic cell system and its role in immunogenicity. *Annu. Rev. Immunol.* **9**, 271–296.

Streuli, C.H., Schmidhauser, C., Kobrin, M., Bissell, M.J. and Derynck, R. (1993). Extracellular matrix regulates expression of the TGF-β1 gene. *J. Cell Biol.* **120**, 253–260.

Strober, W., Kelsall, B., Fuss, I., Marth, T., Ludviksson, B., Ehrhardt, R. and Neurath, M. (1997). Reciprocal IFN-γ and TGF-β responses regulate the occurrence of mucosal inflammation. *Immunol. Today* **18**, 61–64.

Strobl, H., Reidl, E., Scheinecker, C., Bello-Fernandez, C., Pickl, W.F., Rappersberger, K., Majdic, O. and Knapp, W. (1996). TGF-β1 promotes *in vitro* development of dendritic cells from $CD34^+$ progenitors. *J. Immunol.* **157**, 1499–1507.

Suda, T. and Zlotnik, A. (1992). *In vitro* induction of CD8 expression on thymic pre-T cells. II. Characterization of CD3-CD4-CD8 α^+ cells generated *in vitro* by culturing $CD25^+CD3^-CD4^-CD8^-$ thymocytes with T cell growth factor-β and tumor necrosis factor-α. *J. Immunol.* **149**, 71–76.

Sun, L., Wu, G., Willson, J.K.V., Zborowska, E., Yang, J., Rajkarunanayake, I., Wang, J., Gentry, L.E., Wang, X.F. and Brattain, M.G. (1994). Expression of transforming growth factor β type II receptor leads to reduced malignancy in human breast cancer MCF-7 cells. *J. Biol. Chem.* **269**, 26 449–26 455.

Swain, S.L., Huston, G., Tonkonogy, S. and Weinberg, A.D. (1991). Transforming growth factor-β and IL-4 cause helper T cell precursors to develop into distinct effector helper cells that differ in lymphokine secretion pattern and cell surface phenotype. *J. Immunol.* **147**, 2991–3000.

Taipale, J., Miyazono, K., Heldin, C.-H. and Keski-Oja, J. (1994). Latent transforming growth factor-β1 associates to fibroblast extracellular matrix via latent TGF-β binding protein. *J. Cell Biol.* **124**, 171–181.

Taipale, J., Lohi, J., Saarinen, J., Kovanen, P.T. and Keski-Oja, J. (1995). Human mast cell chymase and leukocyte elastase release latent transforming growth factor-β1 from the extracellular matrix of cultured human epithelial and endothelial cells. *J. Biol. Chem.* **270**, 4689–4696.

Takahama, Y., Letterio, J.J., Suzuki, H., Farr, A.G. and Singer, A. (1994). Early progression of thymocytes along the CD4/CD8 developmental pathway is regulated by a subset of thymic epithelial cells expressing transforming growth factor β. *J. Exp. Med.* **179**, 1495–1506.

ten Dijke, P., Hanson, P., Iwata, K.K., Pieler, C. and Foulkes, J.G. (1988). Identification of another member of the transforming growth factor type β gene family. *Proc. Natl Acad. Sci. U.S.A.* **82**, 4715–4719.

ten Dijke, P., Yamashita, H., Ichijo, H., Franzén, P., Laiho, M., Miyazono, K. and Heldin, C.-H. (1994). Characterization of type I receptors for transforming growth factor-β and activin. *Science* **264**, 101–104.

Thiagalingam, S., Lengauer, C., Leach, F.S., Schutte, M., Hahn, S.A., Overhauser, J., Willson, J.K.V., Markowitz, S., Hamilton, S.R., Kern, S.E., Kinzler, K.W. and Vogelstein, B. (1996). Evaluation of candidate tumour suppressor genes on chromosome 18 in colorectal cancers. *Nature Genet.* **13**, 343–346.

Thorbecke, G.J., Shah, R., Leu, C.H., Kuruvilla, A.P., Hardison, A.M. and Palladino, M.A. (1992). Involvement of endogenous tumor necrosis factor α and transforming growth factor β during induction of collagen type II arthritis in mice. *Proc. Natl Acad. Sci. U.S.A.* **89**, 7375–7379.

Torre-Amione, G., Beauchamp, R.D., Koeppen, H., Park, B.H., Schreiber, H., Moses, H.L. and Rowley, D.A. (1990). A highly immunogenic tumor transfected with a murine transforming growth factor type β1 cDNA escapes immune surveillance. *Proc. Natl Acad. Sci. U.S.A.* **87**, 1486–1490.

Tsuchida, K., Mathews, L.S. and Vale, W.W. (1993). Cloning and characterization of a transmembrane serine kinase that acts as an activin type I receptor. *Proc. Natl Acad. Sci. U.S.A.* **90**, 11 242–11 246.

Tsunawaki, S., Sporn, M.B., Ding, A. and Nathan, C. (1988). Deactivation of macrophages by transforming growth factor-beta. *Nature* **334**, 260–262.

Twardzik, D.R., Mikovits, J.A., Ranchalis, J.E., Purchio, A.F., Ellingsworth, L. and Ruscetti, F.W. (1990). γ-Interferon-induced activation of latent transforming growth factor-β by human monocytes. *Ann. N.Y. Acad. Sci.* **593**, 276–284.

Uchida, K., Nagatake M., Osada, H., Yatabe, Y., Kondo, M., Mitsudomi, T., Masuda, A. and Takahashi, T. (1996). Somatic *in vivo* alterations of the JV18-1 gene at 18q21 in human lung cancers. *Cancer Res.* **56**, 5583–5585.

Unsicker, K., Flanders, K.C., Cissel, D.S., Lafayatis, R. and Sporn, M.B. (1991). Transforming growth factor-β isoforms in the adult rat central and peripheral nervous system. *Neuroscience* **44**, 613–625.

Ventura, F., Doody, J., Liu, F., Wrana, J.J. and Massagué, J. (1994). Reconstitution and transphosphorylation of TGF-β receptor complexes. *EMBO J.* **13**, 5581–5589.

Ventura, F., Liu, F., Doody, J. and Massagué, J. (1996). Interaction of transforming growth factor-β receptor I with farnesyl-protein transferase-α in yeast and mammalian cells. *J. Biol. Chem.* **271** 13 931–13 934.

Vodovotz, Y. and Bogdan, C. (1994). Control of nitric oxide synthase expression by transforming growth-beta: implications for homeostasis. *Prog. Growth Factor Res.* **5**, 341–351.

Wahl, S.M., Hunt, D.A., Wakefield, L.M., McCartney-Francis, N., Wahl, L.M., Roberts, A.B. and Sporn, M.B. (1987). Transforming growth factor type β induces monocyte chemotaxis and growth factor production. *Proc. Natl Acad. Sci. U.S.A.* **84**, 5788–5792.

Wahl, S.M., Hunt, D.A., Wong, H.L., Dougherty, S., McCartney-Francis, N., Wahl, L.M., Ellingsworth, L., Schmidt, J.A., Hall, G. and Roberts, A.B. (1988a). Transforming growth factor-β is a potent immunosuppressive agent that inhibits IL-1-dependent lymphocyte proliferation. *J. Immunol.* **140**, 3026–3032.

Wahl, S.M., Hunt, D.A.., Bansal, G., McCartney-Francis, N., Ellingsworth, L. and Allen, J.B. (1988b). Bacterial cell wall-induced immunosuppression. Role of transforming growth factor β. *J. Exp. Med.* **168**, 1403–1417.

Wahl, S.M., McCartney-Francis, N.M. and Mergenhagen, S.E. (1989). Inflammatory and immunomodulatory roles of TGF-β. *Immunol. Today* **10**, 258–261.

Wahl, S.M., Allen, J.B., McCartney-Francis, N., Morganti-Kossmann, M.C., Kossmann, T., Ellingsworth, L., Mai, U.E., Mergenhagen, S.E. and Orenstein, J.M. (1991). Macrophage- and astrocyte-derived transforming growth factor β as a mediator of central nervous system dysfunction in acquired immune deficiency syndrome. *J. Exp. Med.* **173**, 981–991.

Wahl, S.M., Allen, J.B., Weeks, B.S., Wong, H.L. and Klotman, P.E. (1993). Transforming growth factor β enhances integrin expression and type IV collagenase secretion in human monocytes. *Proc. Natl Acad. Sci. U.S.A.* **90**, 4577–4581.

Wakefield, L.M., Smith, D.M., Flanders, K.C. and Sporn, M.B. (1988). Latent transforming growth factor-β from human platelets. A high molecular weight complex containing precursor sequences. *J. Biol. Chem.* **263**, 7646–7654.

Wang, J., Sun, L., Meyeroff, L., Wang, X.-F., Gentry, L.E., Yang, J., Liang, J., Zborowska, E., Markowitz, S., Willson, J.K.V. and Brattain, M.G. (1995). Demonstration that mutation of the type II TGF-β receptor inactivates its tumor suppressor activity in replication error-positive colon carcinoma cells. *J. Biol. Chem.* **270**, 22 044–22 049.

Wang, T., Donahoe, P.K. and Zervos, A.S. (1994). Specific interaction of type I receptors of the TGF-β family with the immunophilin FKBP-12. *Science* **265**, 674–676.

Wang, T., Li, B.-Y., Danielson, P.D., Shah, P.C., Rockwell, S., Lechleider, R.J., Martin, J., Manganaro, T. and Donahoe, P.K. (1996a). The immunophilin FKBP12 functions as a common inhibitor of the TGFβ family type I receptors. *Cell* **86**, 435–444.

Wang, T., Danielson, P.D., Li, B., Shah, P.C., Kim, S.D. and Donahoe, P.K. (1996b). The $p21^{Ras}$ farnesyltransferase α subunit in TGF-β and activin signaling. *Science* **271**, 1120–1122.

Wang, X.-F., Lin, H.Y., Ng-Eaton, E., Downward, J., Lodish, H.F. and Weinberg, R.A. (1991). Expression cloning and characterization of the TGF-β type III receptor. *Cell* **67**, 797–805.

Wang, X.-J., Greenhalgh, D.A., Bickenbach, J.R., Jiang, A., Bundman, D.S., Krieg, T., Derynck, R. and Roop, D.R. (1997). Expression of a dominant negative type II TGF-β receptor in the epidermis of transgenic mice blocks TGF-β-mediated growth inhibition. *Proc. Natl Acad. Sci. U.S.A.* **94**, 2386–2391.

Warner, G.L., Ludlow, J.W., Nelson, D.A., Gaur, A. and Scott, D.W. (1992). Anti-immunoglobulin treatment of murine B cell lymphomas induces active transforming growth factor β, but pRB hypophosphorylation is transforming growth factor β independent. *Cell Growth Differ.* **3**, 175–181.

Weinberg, R.A. (1995). The retinoblastoma protein and cell cycle control. *Cell* **81**, 323–330.

Weinberg, R.A. (1996). E2F and cell proliferation: a world turned upside down. *Cell* **85**, 457–459.

Weis-Garcia, F. and Massagué, J. (1996). Complementation between kinase-defective and activation-defective TGF-β receptors reveals a novel form of receptor cooperativity essential for signaling. *EMBO J.* **15**, 276–289.

Welch, D.R., Fabra, A. and Nakajima, M. (1990). Transforming growth factor β stimulates mammary adenocarcinoma cell invasion and metastatic potential. *Proc. Natl Acad. Sci. U.S.A.* **87**, 7678–7682.

Wieser, R., Wrana, J.L. and Massagué, J. (1995). GS domain mutations that constitutively activate TβRI, the downstream signaling component in the TGF-β receptor complex. *EMBO J.* **14**, 2199–2208.

Wilcox, J.N. and Derynck, R. (1988). Developmental expression of transforming growth factor-α and -β in mouse fetus. *Mol. Cell Biol.* **8**, 3415–3422.

Wilder, R., Lafyatis, R., Roberts, A.B., Case, J.P., Kumkumian, G.K., Sano, H., Sporn, M.B. and Reumers, E.F. (1990). Transforming growth factor-β in rheumatoid arthritis. *Ann. N.Y. Acad. Sci.* **593**, 197–207.

Wrana, J.L., Attisano, L., Carcamo, J., Zentella, A., Doody, J., Laiho, M., Wang, X.-F. and Massagué, J. (1992). TGF β signals through a heteromeric protein kinase receptor complex. *Cell* **71**, 1003–1014.

Wrana, J.L., Attisano, L., Wieser, R., Ventura, F. and Massagué, J. (1994). Mechanism of activation of the TGF-β receptor. *Nature* **370**, 341–347.

Wu, R.-Y., Zhang, Y., Feng, X.-H. and Derynck, R. (1997). Homomeric and heteromeric interactions are required for signaling activity and functional cooperativity of Smad-3 and -4. *Mol. Cell Biol.* **17**, 2521–2528.

Wu, S., Theodorescu, D., Kerbel, R.S., Willson, J.K.V., Mulder, K.M., Humphrey, L.E. and Brattain, M.G. (1992). TGF-β1 is an autocrine-negative growth regulator of human colon carcinoma FET cells *in vivo* as revealed by transfection of an antisense expression vector. *J. Cell Biol.* **116**, 187–196.

Wu, X., Robinson, C.E., Fong, H.W. and Gimble, J.M. (1996). Analysis of the native murine bone morphogenetic protein serine threonine kinase type I receptor (ALK-3). *J. Cell. Physiol.* **168**, 453–461.

Yamaguchi, K., Shirakabe, K., Shibuya, H., Irie, K., Oishi, I., Ueno, N., Taniguchi, T., Nishida, E. and Matsumoto, K. (1995). Identification of a member of the MAPKKK family as a potential mediator of TGF-β signal transduction. *Science* **270**, 2008–2011.

Yamaguchi, Y., Mann, D.M. and Ruoslahti, E. (1990). Negative regulation of transforming growth factor-β by the proteoglycan decorin. *Nature* **346**, 281–284.

Yamashita, H., Ichijo, H., Grimsby, S., Morén, A., ten Dijke, P. and Miyazono, K. (1994a). Endoglin forms a heteromeric complex with the signaling receptors for transforming growth factor-β. *J. Biol. Chem.* **269**, 1995–2001.

Yamashita, H., ten Dijke, P., Franzén, P., Miyazono, K. and Heldin, C.-H. (1994b). Formation of hetero-oligomeric complexes of type I and type II receptors for transforming growth factor-β. *J. Biol. Chem.* **269**, 20172–20178.

Yan, Z., Winawer, S. and Friedman, E. (1994). Two different signal transduction pathways can be activated by transforming growth factor β1 in epithelial cells. *J. Biol. Chem.* **269**, 13231–13237.

Yang, E.Y. and Moses, H.L. (1990). Transforming growth factor β1-induced changes in cell migration, proliferation, and angiogenesis in the chicken chorioallantoic membrane. *J. Cell Biol.* **111**, 731–741.

Yayon, A., Klagsbrun, M., Esko, J.D., Leder, P. and Ornitz, D. (1991). Cell surface, heparin-like molecules are required for binding of basic fibroblast growth factor to its high affinity receptor. *Cell* **64**, 841–848.

Yin, W., Smiley, E., Germiller, J., Mecham, R.P., Florer, J.B., Wenstrup, R.J. and Bonadio, J. (1995). Isolation of a novel latent transforming growth factor-β binding protein gene (LTBP-3). *J. Biol. Chem.* **270**, 10147–10160.

Zentella, A. and Massagué, J. (1992). Transforming growth factor β induces myoblast differentiation in the presence of mitogens. *Proc. Natl Acad. Sci. U.S.A.* **89**, 5176–5180.

Zhang, H., Shaw, A.R.E., Mak, A. and Letarte, M. (1996a). Endoglin is a component of the transforming growth factor (TGF)-β receptor complex of human pre-leukemic cells. *J. Immunol.* **156**, 565–573.

Zhang, Y., Feng, X.-H., Wu, R.-Y. and Derynck, R. (1996b). Receptor-associated Mad homologues synergize as effectors of the TGF-β response. *Nature* **382**, 168–172.

Zhang, Y., Musci, T. and Derynck, R. (1997). The tumor suppressor Smad4/DPC 4 as a central mediator of Smad function. *Curr. Biol.* **7**, 270–276.

Zhou, L., Dey, C.R., Wert, S.E. and Whitsett, J.A. (1996). Arrested lung morphogenesis in transgenic mice bearing an SP-C-TGF-β1 chimeric gene. *Dev. Biol.* **175**, 227–238.

Chapter 22

Granulocyte–Macrophage Colony Stimulating Factor

Valerie F.J. Quesniaux[1] and Thomas C. Jones[2]

[1]Preclinical Research, Novartis Pharma Ltd and [2]Clinical Research Consultants, Basel, Switzerland

INTRODUCTION

Granulocyte–macrophage colony stimulating factor (GM-CSF) was one of the first cytokines to be characterized and cloned. This molecule has elicited great interest (for example, there have been over 3000 publications on GM-CSF) and it keeps surprising us, as new data reveal novel aspects of its physiologic and pharmacologic activities. GM-CSF is already registered for clinical use in the areas of hematology, infectious diseases and oncology for prevention and reversal of neutropenia. However, the multifold actions of GM-CSF on different cell populations, such as dendritic cells or lung alveolar macrophages, have also suggested additional potential uses which would still need clinical validation.

In this review the essential features of the GM-CSF protein, current knowledge about the regulation of its expression and receptor-mediated signaling are summarized first. The main activities of GM-CSF identified experimentally *in vitro* on isolated cell populations and *in vivo* in animal models are then reviewed, and the present and potential clinical applications of GM-CSF are discussed.

PROTEIN STRUCTURE

The GM-CSF protein was first identified in mice (Burgess *et al.*, 1977) and subsequently cloned in several species, including mice (Gough *et al.*, 1984) and humans (Wong *et al.*, 1985). Murine and human GM-CSF are not cross-reactive in terms of biologic activity or receptor binding. They display only 56% amino acid sequence identity, being one of the less well conserved of the myeloid growth factors. Mature murine GM-CSF comprises 124-amino-acid residues (molecular mass 14 kDa), whereas human, gibbon and bovine GM-CSF have an insertion of three residues, between residues 27 and 28 of the murine sequence, within a region implicated in binding to the β chain of the receptor (for review see Rasko and Gough, 1994). The mature GM-CSF protein is preceded by a hydrophobic leader sequence of 25-amino-acid residues in length.

GM-CSF isolated from most murine and human sources is glycosylated, leading to an apparent mass of approximately 23 kDa. The polypeptide sequences contain two

The Cytokine Handbook, 3rd ed.
ISBN 0-12-689662-3

potential *N*-linked glycosylation sites at different locations. Glycosylation does not seem essential for activity, as recombinant GM-CSF proteins from various origins show similar activities: GM-CSF produced in Chinese hamster ovary (CHO) cells has an identical amino acid sequence to the native protein but is variably glycosylated on *O*-linked and *N*-linked sites, yielding a heterogeneous population of proteins with molecular weights of 18–30 kDa (Wong *et al.*, 1985); GM-CSF expressed in yeast can differ from the native protein (Leu for Arg-23), it is glycosylated on *N*-linked sites and yields three species of 15.5, 17 and 19.5 kDa (Cantrell *et al.*, 1985); GM-CSF expressed in *Escherichia coli* is not glycosylated and has a molecular weight of 15 kDa (Burgess *et al.*, 1987). In this chapter, no distinction between the source of GM-CSF will be made, since no difference in activity could be demonstrated, either *in vitro* or *in vivo* (Robison and Myers, 1993).

The crystal structure of human GM-CSF has been determined (Diederichs *et al.*, 1991). Similar to a number of other hematopoietic growth factors, GM-CSF comprises two pairs of antiparallel α helices. Residues responsible for interaction with the β chain of the receptor to form a high-affinity complex (see below) are located within the first α helix (residues 18–22).

The GM-CSF gene is located near other cytokine genes, namely M-CSF, interleukin (IL)-3, IL-4, IL-5, IL-9, platelet-derived growth factor (PDGF), both in humans (on long arm of chromosome 5) and in mice (on chromosome 11; see chapter 3).

RECEPTORS

The GM-CSF receptor is composed of two distinct chains: a ligand-specific α subunit which binds GM-CSF with low affinity (Gearing *et al.*, 1989) and a β chain (Hayashida *et al.*, 1990) which has no detectable binding to GM-CSF. The two subunits together generate a ligand-specific, high-affinity receptor (Hayashida *et al.*, 1990). The tyrosyl residue 421 was found to be essential for high-affinity binding and signaling of IL-3, GM-CSF and IL-5 (Woodcock *et al.*, 1996). While the α chain of the GM-CSF receptor is specific for GM-CSF, the β chain is also a component of the IL-3 and IL-5 receptors (for review see Bagley *et al.*, 1997). In mice, in addition to the originally shared β subunit (originally called AIC2B), a second β subunit (AIC2A) specific for IL-3 was found to associate with the IL-3Rα chain to generate a high-affinity receptor. The two murine β subunits, which probably arose from gene duplication, are highly homologous and no functional difference between them has been identified.

Both components of the GM-CSF receptor are related in structure and are members of the class I cytokine receptor superfamily (hematopoietin receptor family) characterized by conserved features in their extracellular domains such as conserved cysteine residues, 'WSXWS' motif and fibronectin type III-like domains (Bazan, 1990). The cytoplasmic domain of the β subunit is larger (approximately 430 amino acid residues) than the α subunit intracellular domain (54 amino acid residues), suggesting a preponderant role of the β subunit in signal transduction. The fact that GM-CSF, IL-3 and IL-5 use a common β subunit would thus explain some of their shared biologic activities.

GM-CSF receptors are found on hematopoietic cells and non-hematopoietic cells such as trophoblasts, endothelial cells, oligodendrocytes and on various malignant cells such as those associated with leukemia and renal carcinoma, but no receptor could be detected on T and B lymphocytes (Morrissey *et al.*, 1987).

The β subunit is encoded by a unique gene in the human genome (located at 22q12–q13; Shen *et al.*, 1992) whereas the murine genome has two closely related genes, *Aic2a* and *Aic2b*, which are closely linked to each other.

The gene encoding the α chain of the GM-CSF receptor is linked to the IL-3 receptor α chain gene in humans (Kremer *et al.*, 1993), but this is not the case in mice. The human GM-CSF receptor α-chain gene spans 44 kilobases (kb), has 13 exons, and is remarkably well conserved within the cytokine receptor superfamily (Nakagawa *et al.*, 1994). Its cell type-specific expression seems to involve the myeloid and B-cell transcription factor PU-1 and the CCAAT/enhancer binding protein C/EBP-α (Hohaus *et al.*, 1995). Like other genes encoding hematopoietin receptors, the GM-CSF receptor α gene is alternately spliced, and several mRNA isoforms with different translational efficiencies have been described (Chopra *et al.*, 1996). An α-chain isoform lacking the transmembrane region appears capable of transducing a proliferative signal in murine FDC-P1 cells, probably as a result of interaction with the cell-surface β chain in such cells (Crosier *et al.*, 1991).

EXPRESSION OF GM-CSF

GM-CSF can be produced by a number of different cell types including T lymphocytes, macrophages, endothelial cells, fibroblasts, stromal cells, leukemic and various solid organ tumor cells, and others (see Table 1) (Metcalf, 1984). It is usually not detected in blood, but may appear under certain pathologic conditions (such as asthma). GM-CSF production often requires stimulation of the producer cell, for example by antigens in T and B cells, or by inflammatory agents such as bacterial cell wall lipopolysaccharide (LPS), IL-1 and tumor necrosis factor (TNF) in macrophages, fibroblasts or endothelial cells (Table 1; reviewed in Rasko and Gough, 1994). GM-CSF release in response to cell stimulation can be the result of increased transcription of the GM-CSF gene, post-

Table 1. Cells capable of expressing GM-CSF.

Cell type	Inductive stimuli
T lymphocytes	Antigen, lectins, phorbol diester, IL-2
B lymphocytes	Antigen, LPS, IL-1
Macrophages	LPS, phagocytosis, adherence
Mesothelial cells	
Keratinocytes	
Osteoblasts	Parathyroid hormone, LPS
Uterine epithelial cells	
Synoviocytes	Rheumatoid arthritis
Mast cells	IgE
Fibroblasts	IL-1, TNF, LPS, oxidized lipoproteins
Stromal cells	Mitogens, TNF, IL-1
Endothelial cells (bone marrow or vascular)	IL-1, TNF, LPS, oxidized lipoproteins
Myeloblasts	IL-1
Leukemia cell lines	
Various solid tumors	

transcriptional stabilization of the mRNA (reviewed in Rasko and Gough, 1994), or secretion of stored protein (Brizzi *et al.*, 1995). The GM-CSF mRNA contains eight tandem copies of the 5′-AUUUA-3′ motif in the 3′ untranslated region, which are probably responsible for message instability and mediate the transient stabilization of the message after cellular stimulation. A number of sequences in proximal regions of the human GM-CSF gene have been implicated in its transcription, such as binding sites for transcription factors such as NG-GMa, NG-GMb, nuclear factor (NF)-κB/Rel, Elf-1, NF-AT/AP-1 (see Musso *et al.*, 1996). In addition, a distal regulatory element, located between the GM-CSF and IL-3 genes, has been identified as a strong cyclosporin A-sensitive enhancer (Cockerill *et al.*, 1993).

SIGNAL TRANSDUCTION

Like other members of the class I cytokine receptor superfamily, the GM-CSF receptor α and β subunits have no tyrosine kinase catalytic domains. The cytoplasmic part of the β subunit contains the conserved Box1 and Box2 motifs. Box1 was found to be critical for GM-CSF signaling, including interaction with the *N*-terminal portion and activation of Janus kinase JAK2, and phosphorylation of the β subunit (Watanabe *et al.*, 1996; reviewed in Itoh *et al.*, 1996). The *C*-terminal, membrane distal region of the β subunit was shown to be involved in activation of the Ras/Raf-1/mitogen-activated protein (MAP) kinase pathway. Recent evidence suggests that the α subunit can also mediate the signal cycle (Polotskaya *et al.*, 1993; Spielholz *et al.*, 1995).

Signaling events observed in response to GM-CSF, IL-3 and IL-5 include phosphorylation and/or activation of protein tyrosine kinases including JAK2, proteins involved in the Ras activation pathway such as Vav, Shc, Raf and MAP kinase, the Src family kinases Fyn and Lyn, Fps/Fes, phosphatidylinositol 3-kinase p85, protein kinase C, as well as changes in ion fluxes and inositol phosphate mobilization (reviewed in Mui *et al.*, 1995). The signaling pathways eventually reach the nucleus and result in the induction of specific genes, including the growth-related genes c-*fos*, c-*jun*, *egr-1* and c-*myc*, and genes important in cell cycle regulation such as various cyclins and cyclin-dependent kinases. GM-CSF-induced expression of the early growth response gene-1 (*egr-1*) is rapid and transient and involves binding to the cAMP response element (CRE) and phosphorylation of the CRE-binding protein, CREB (Sakamoto *et al.*, 1994).

JAK2 has been found to be associated with the shared β receptor subunit (Brizzi *et al.*, 1994). After GM-CSF or IL-3 binding, JAK2 is rapidly and transiently tyrosine phosphorylated and this is associated with an activation of its intrinsic kinase activity (Brizzi *et al.*, 1994). JAK kinases act upstream of signal transducers and activators of transcription (STATs), which exist in the cytoplasm in latent forms, dimerize upon phosphorylation on tyrosine residues, and translocate to the nucleus where they bind to DNA elements related to the γ-activated sequences found in the promoter of interferon-γ (IFN-γ)-inducible genes. STAT molecules implicated in GM-CSF signaling include STAT1, STAT3 (Brizzi *et al.*, 1996) and STAT5 isoforms (Azam *et al.*, 1995; Mui *et al.*, 1995) which are also activated in response to other cytokines, including IL-3 and IL-5. JAK2 activation seems to lead to the activation of c-*fos*, c-*jun* and c-*myc* genes (Watanabe *et al.*, 1996). JAK2 is also involved in GM-CSF-induced phosphorylation of the GM-CSF receptor β subunit, and of protein tyrosine phosphatase (PTP)-1D and Shc (Watanabe *et al.*, 1996).

Shc is a src homology domain-2 (SH2)-containing protein involved in Ras activation by tyrosine kinase receptors. GM-CSF, IL-3 and IL-5 induce tyrosine phosphorylation of Shc (Dorsch *et al.*, 1994). Mutational analysis suggested that GM-CSF stimulates at least two signaling pathways leading to the activation of Ras and ultimately to transcriptional activation of the c-*fos* gene, one mediated by Shc and the other by PTP-1D. When phosphorylated, Shc and PTP-1D bind to Grb2 and subsequently activate Ras, leading to Raf kinase activation and then to an increase in MAP kinase activity. MAP kinases appear to transduce a common signal elicited by many cytokines including GM-CSF, IL-3, IL-5 (Gomez-Cambronero *et al.*, 1992), c-kit ligand (Okuda *et al.*, 1992) and erythropoietin (Bittorf *et al.*, 1994), but not IL-4 (Welham *et al.*, 1992). In addition, the Src family protein tyrosine kinases Lyn and Fyn have been implicated in IL-3 signaling in certain cells (Torigoe *et al.*, 1992; see Mui *et al.*, 1995).

Because many of the intracellular signaling events induced by IL-3, GM-CSF and IL-5 are similar, including for instance the Src family kinases, JAK2/STAT5 and MAP kinase, it was suggested that the biologic differences between these cytokines may be due to differences in cellular sources or in cell type-restricted distribution of the respective α-chain receptors. In addition, the GM-CSF receptor α subunit can mediate specific signals. For instance, the short intracytoplasmic tail of the α chain was shown to be critical for cell growth, at the level of protein phosphorylation or entry into the cell cycle (Polotskaya *et al.*, 1993). Low-affinity GM-CSF receptor α chain expressed on melanoma cell lines at early stages of differentiation was also found to signal GM-CSF-stimulated glucose uptake in these cells, in a phosphorylation-independent manner (Spielholz *et al.*, 1995). Thus, some specificity of response to GM-CSF seems to be mediated through the GM-CSF receptor α chain.

IN VITRO BIOLOGIC ACTIVITIES

The activities of GM-CSF detected *in vitro* range from regulation of differentiation and proliferation of hematopoietic progenitors, to the regulation of survival and function of mature cells such as neutrophils, macrophages or dendritic cells, suggesting a prominent role of GM-CSF in hematopoiesis, antimicrobial defense and immune responses. The effects on various cell types and functions are summarized below (see Table 2).

Hematopoietic Progenitors

One of the first effects of GM-CSF documented was the regulation of differentiation and proliferation of granulocyte–macrophage progenitors *in vitro*.

GM-CSF stimulates the formation of a range of myeloid colony types in a dose-dependent manner in semisolid cultures *in vitro* (Metcalf *et al.*, 1986). Granulocyte and/or macrophage colony formation occurs at low concentrations whereas erythroid, multilineage and occasional megakaryocyte colonies occur at higher concentrations (Sieff *et al.*, 1985). Additive effects of GM-CSF have been demonstrated with other cytokines such as IL-3, G-CSF or KL (McNiece *et al.*, 1991).

Table 2. Main activities of GM-CSF identified experimentally *in vitro* or *in vivo*, and corresponding therapeutic indications for GM-CSF, either demonstrated or under clinical evaluation.

Main targets	*In vitro/ex vivo*	In animals	In humans
Hematologic progenitors	Increased differentiation Increased proliferation Adhesion molecule expression	Stimulation of hematopoiesis including increased peripheral blood count	Prevention or reversal of hematopoietic cell toxicity Peripheral blood progenitor mobilization
Neutrophils	Increased antimicrobial activity Increased chemotaxis Increased cytokine release	Control of infection	Control of infection
Monocytes–macrophages	Increased cytotoxicity Increased APC function Increased cytokine release	Control of infection	Control of infection
Dendritic cells	Increased production Increased differentiation Increased APC function	Increased antibody response Increased migration and proliferation of dendritic cells	Increased antibody response
Structural cells (endothelial cells, fibroblasts, keratinocytes)	Inhibition of chemotaxis Adhesion molecule expression Migration or proliferation of endothelial cells	Wound healing promotion	Wound healing promotion

Neutrophils

GM-CSF enhances neutrophil antimicrobial activity by increasing phagocytosis, oxidative metabolism, release of lysozyme and chemotactic factors, complement recruitment with enhanced opsonization, stimulating mechanisms for killing intracellular viruses, fungi, bacteria and protozoa, and by increased expression of surface antigens associated with Fc receptors and complement-mediated cell binding (FcγRI (CD64), CR-1 (CD35), CR-3 (CD11b)) involved in antibody-dependent cell-mediated cytotoxicity (ADCC) (reviewed in Ruef and Coleman, 1990; Jones, 1994). Some of these activities seem to be indirect (Liehl, 1991). They are relevant for the eradication of fungal, mycobacterial and protozoal pathogens. In a comparison between G-CSF and GM-CSF, equipotent antibacterial functions were found, whereas GM-CSF exhibited stronger polymorphonuclear neutrophil (PMN) chemotaxis, killing of *Candida albicans* blastospores, and surface expression of Fc and complement-mediated antigens (Bober *et al.*, 1995), suggestive of an interesting antifungal potential.

GM-CSF is able to induce neutrophil transmigration by increasing chemotaxis, enhancing adhesion to vascular endothelium. This effect is probably through increased surface expression of the β_2-integrin molecules CD11b/CD18, intercellular adhesion molecule-1 (ICAM-1) (CD54) and downregulation of L-selectin on neutrophils (Yong and Lynch, 1993).

GM-CSF induces the production of IL-1 and IL-1 receptor antagonist (Ra) by neutrophils *in vitro* (Lindemann *et al.*, 1988). Functional synergy with TNF has been demonstrated, in the enhanced complement-dependent phagocytosis of the fungal pathogen *Cryptococcus neoformans* for instance (Collins and Bancroft, 1992). GM-CSF was also shown to potentiate IL-8 priming of neutrophils (Yuo *et al.*, 1991).

Monocytes and Macrophages

GM-CSF has various effects on monocyte and macrophage functions (reviewed in Jones, 1996). GM-CSF promotes the differentiation of monocytes to large macrophage-like cells, increases the metabolism of mature cells and increases their function as antigen-presenting cells (APCs) by enhancing the expression of major histocompatibility complex (MHC) class II (Morrissey *et al.*, 1987; Fischer *et al.*, 1988) and CD1, a class of non-peptidic antigen-presenting molecules (Porcelli and Modlin, 1995). GM-CSF induces the production of TNF, IL-1α, IL-1β, IL-6, IL-8, IL-12, M-CSF and IL-1Ra, and the expression of IFNγR and of adhesion molecules such as CD11b and CD35 by monocytes (Cannistra *et al.*, 1988; Finbloom *et al.*, 1993; Danis *et al.*, 1995), whereas the induction of G-CSF seems more controversial (Cluitmans *et al.*, 1993).

GM-CSF promotes macrophage cytotoxicity and ADCC by upregulation of Fc receptors FcγRI and FcγRII (Grabstein *et al.*, 1986). Part of GM-CSF stimulation of monocyte cytotoxicity seems to be indirect through a TNF-dependent mechanism (Cannistra *et al.*, 1988).

GM-CSF enhances antimicrobial activity of monocytes–macrophages by increasing phagocytosis, pinocytosis, oxidative metabolism, numbers of Fc receptors and release of chemotactic factors (Ruef and Coleman, 1990). For instance, alveolar macrophages showed increased killing of *C. neoformans* after treatment with GM-CSF (Chen *et al.*, 1994). Also hyphae of *Aspergillus fumigatus* were shown to be damaged by monocytes stimulated with GM-CSF (Roilides *et al.*, 1996).

Dendritic Cells

Dendritic cells are the most efficient APCs, essential for the priming of both CD4$^+$ and CD8$^+$ T cells. They constitute only a small proportion of total cells in most tissues and are difficult to isolate and propagate. GM-CSF is the major stimulatory cytokine for the *in vitro* differentiation, viability and function of dendritic cells (Witmer-Pack *et al.*, 1987).

GM-CSF induces the production of dendritic cells, and specifically Langerhans-type cells, from monocytic cells, non-adherent MHC class II$^-$ murine peripheral blood (Inaba *et al.*, 1992), human CD34$^+$ progenitor cells (Caux *et al.*, 1992) or lymphocyte-depleted human tonsils (Clark *et al.*, 1992). Adding TNF to GM-CSF on cultures of CD34$^+$ progenitors results in a 10–20-fold increase of viable CD1a$^+$ dendritic cells (Caux *et al.*, 1992; Reid *et al.*, 1992). A further increased yield in bone marrow-derived immunostimulatory CD14$^-$HLA$^-$DR$^+$ dendritic cells is obtained in the presence of KL (Szabolcs *et al.*, 1995). Extensive phenotypic characterization revealed that CD34$^+$ hematopoietic progenitors from human cord blood differentiate into dendritic cells along two independent pathways in response to GM-CSF plus TNF (Caux *et al.*, 1996). Transforming growth factor (TGF-β_1) allows the development of dendritic cells from

CD34$^+$ under serum-free conditions, in the presence of GM-CSF, TNF and KL. The proportions and total yields of cells with typical dendritic cell morphology and CD1a expression and the allostimulatory capacity are higher in the presence of TGF-β_1 than in the presence of cord blood plasma (Strobl *et al.*, 1996).

GM-CSF plus IL-4 induces immature dendritic cells that are capable of capturing and processing antigen and express MHC class I and II, CD1, FcγRII, CD40, B7, CD44 and ICAM-1, but not CD14. Incubation of these cells with TNF or CD40L resulted in an increased surface expression of MHC class I and II, B7, ICAM-1 and CD44, characteristic of mature dendritic cells. This also resulted in an increased T-cell stimulatory capacity in mixed lymphocyte reaction together with a decreased capacity to present antigen (Sallusto and Lanzavecchia, 1994). IL-13 seems to be as effective as IL-4 in the generation of dendritic cells (Piemonti *et al.*, 1995).

The increased production and differentiation of dendritic cells induced by GM-CSF *in vitro* is also accompanied by increased function. In culture, GM-CSF augments the expression of MHC class II, CD1 and costimulatory molecules such as B7-1 (CD80) and B7-2 (CD86) (Larsen *et al.*, 1994) which are necessary for the interaction of APCs with T cells. Thus, GM-CSF could act as an adjuvant to promote immune responses.

Lymphokine-Activated Killer (LAK) Cells and Natural Killer (NK) Cells

Although GM-CSF receptors were not found on T cells, GM-CSF was shown to potentiate T-cell IL-2-driven proliferation (Santoli *et al.*, 1988) and IL-2-induced cytotoxicity (Masucci *et al.*, 1989). GM-CSF is synergistic with IL-2 in generating LAK cell activity in peripheral blood mononuclear cells (PBMCs) (Masucci *et al.*, 1989; Stewart-Akers *et al.*, 1993; Baxevanis *et al.*, 1995). In patients with cancer undergoing IL-2 immunotherapy *ex vivo*, generation of cytotoxic cells from PBMCs incubated with IL-2 plus GM-CSF was maximum after a 1-h pulse. GM-CSF increased the numbers of IL-2 receptor CD25$^+$ and CD8$^+$ cells and resulted in increased LAK cell activity (Baxevanis *et al.*, 1995). Enhanced activated killer cell function was also reported after GM-CSF administration in patients with acute myeloid leukemia (AML) undergoing bone marrow transplantation (Richard *et al.*, 1995).

Activity on Other Cell Types

GM-CSF induces proliferation and activation of eosinophils (reviewed in Gasson, 1991). GM-CSF, IL-3 and, to a lesser extent, IL-5 cause histamine release by basophilic granulocytes (Haak-Frendscho *et al.*, 1988).

GM-CSF also acts on structural cells such as fibroblasts, endothelial and smooth muscle cells. For instance, GM-CSF has been shown to induce functions in endothelial cells related to angiogenesis and inflammation such as migration, proliferation and expression of adhesion molecules (Bussolino *et al.*, 1989). This involves increased JAK2 association to GM-CSF receptor β chain and tyrosine phosphorylation (Soldi *et al.*, 1997).

Adhesion Molecule Expression

GM-CSF upregulates ICAM-1 and vascular cell adhesion molecule-1 (VCAM-1) expression on endothelial cells. It strongly upregulates ICAM-1 and enhances very late activation antigen-5 (VLA-5) expression involved in binding of activated T cells to synovial type B cells (Cicuttini *et al.*, 1994). GM-CSF augments the surface expression of ICAM-1 and CD18, but not that of ICAM-3, in monocytes and in tumor-associated macrophages in patients with ovarian cancer (Bernasconi *et al.*, 1995). GM-CSF regulates the functional adhesive state of VLA-4 expressed by eosinophilic granulocytes from a low- to a high-affinity state (Sung *et al.*, 1997).

Apoptosis

Mature hematopoietic cells are produced and eliminated continuously, mostly by rapid programming for cell death, by apoptosis. It was recognized early that GM-CSF prolongs the survival of neutrophils and eosinophils by inhibiting apoptosis (Brach *et al.*, 1992). This seems to involve the physical association with, and the early activation of, the Src family tyrosine kinase Lyn, and that of Syk, to the GM-CSF receptor (Wei *et al.*, 1996; Yousefi *et al.*, 1996). The modulation of apoptosis by GM-CSF seems to be twofold, as GM-CSF-enhanced apoptosis in the GM-CSF-dependent human leukemic cell line TF1 treated with inhibitors of protein or RNA synthesis such as cycloheximide or actinomycin D (Han *et al.*, 1995). Using the GM-CSF antagonistic mutant, E21R, it was recently shown that specific engagement of GM-CSF α-chain receptor promotes apoptosis whereas engagement of the common β chain is important for cell survival (Iversen *et al.*, 1996).

IN VIVO ACTIVITIES IN EXPERIMENTAL MODELS

A number of the pharmacologic activities of GM-CSF *in vivo* were foreseeable, based on its *in vitro* activities, such as stimulation of granulopoiesis, antimicrobial defense or acting as a vaccine adjuvant. However, some other effects of GM-CSF were not expected and were first unraveled by *in vivo* experiments. These include the peripheralization of hematopoietic progenitors, the effect on lung homeostasis, as well as the non-essential role of GM-CSF in steady-state hematopoiesis (see Table 2).

The *in vivo* effects of GM-CSF have been tested experimentally by several approaches: genetic models with overexpression of GM-CSF, deletion of GM-CSF or its receptor genes, and by pharmacologic administration of GM-CSF in healthy animals or in disease models.

Genetic Models

Transgenic Expression of GM-CSF

In transgenic mice constitutively expressing high GM-CSF levels, massive macrophage expansion and activation leads to blindness and to a fatal syndrome of tissue damage (Lang *et al.*, 1987). Constitutive overexpression of GM-CSF by bone marrow-derived cells was obtained in lethally irradiated mice reconstituted with transgenic hemato-poietic cells expressing GM-CSF. These mice showed a marked neutrophilia, and a fatal neutrophil and macrophage infiltration of spleen, lung, liver, peritoneal cavity, skeletal muscle and heart (Johnson *et al.*, 1989). Lung-directed expression of GM-CSF using adenoviral vectors which have a natural tropism for the respiratory epithelium led to accumulation of both eosinophils and macrophages in the lung with irreversible fibrotic reaction (Xing *et al.*, 1996).

Thus, the pathologic effects of GM-CSF overexpression, either systemic or local, are related to the functional activation of mature inflammatory effector cells rather than to hematopoietic progenitor cell derangement or leukemic transformation.

GM-CSF-Deficient Mice

Homozygous inactivation of the GM-CSF gene in mice did not affect fetal and postnatal development, nor steady-state hematopoiesis (Dranoff *et al.*, 1994). All animals developed an alveolar proteinosis-like pathology of the lung, namely a progressive accumulation of surfactant lipids and proteins in the alveolar space, probably due to defective function of alveolar macrophages. Extensive lymphoid hyperplasia associated with lung airways and blood vessels was also observed, in the absence of any detectable infective agent. These results demonstrated that GM-CSF is not essential for basal hematopoiesis and unraveled a unique role for GM-CSF in certain macrophage functions and in pulmonary homeostasis.

Expression of GM-CSF transgene in lung epithelium cells of GM-CSF-deficient mice corrected the alveolar proteinosis by restoring surfactant homeostasis (Huffman *et al.*, 1996). This study also demonstrated that GM-CSF regulates clearance and catabolism by alveolar macrophages, rather than the synthesis, of surfactant proteins and lipids.

In mice deficient for both GM-CSF and M-CSF, the alveolar proteinosis-like lung pathology of GM-CSF deficiency is combined with the osteopetrotic state of M-CSF deficiency (*op/op* spontaneous mutants). The accumulation of lipoproteinous alveolar material was more marked than in GM-CSF-deficient mice, and all mice died prematurely with pneumonia (Lieschke *et al.*, 1994). However, mice deficient for both GM-CSF and M-CSF had essentially normal numbers of circulating monocytes and lung macrophages, indicating that other cytokines can contribute to monocyte–macrophage production and function *in vivo*.

Receptor β-Chain-Deficient Mice

Mice deficient for the common *β*-chain receptor (*β*c, AIC2B) were unable to respond to either GM-CSF or IL-5, whereas IL-3 stimulation was normal, which is probably due to the additional IL-3-specific *β*-chain receptor (AIC2A) in mice (Nishinakamura *et al.*, 1995). The *β*c mutant mice showed a phenotype cumulative of GM-CSF and IL-5 pathway deficiency, namely lung pathology with lymphocytic infiltration and some alveolar proteinosis together with low eosinophil levels, either under steady state or after infection.

The influence of GM-CSF deficiency, both at the level of the cytokine and of the *β*-chain receptor, on *in vivo* dendritic cell numbers was minor, indicating that GM-CSF is not essential for *in vivo* dendritic cell development (Vremec *et al.*, 1997). However, GM-CSF overexpression led to an increase in dendritic cell levels, emphasizing the potential of GM-CSF for pharmacologic intervention.

Administration of GM-CSF in Healthy Animals

Rodents

In normal adult mice strains, GM-CSF injected intraperitoneally showed a twofold rise in peripheral neutrophil counts, a dramatic increase in peritoneal macrophages (up to 15-fold), neutrophils and eosinophils (up to 100-fold), and a several-fold increase in lung alveolar neutrophils (Metcalf *et al.*, 1987). This was accompanied by a 40% decrease in bone marrow total cellularity together with a 50% rise in spleen weight (Metcalf *et al.*, 1987). Some stimulation of megakaryopoiesis, however, not accompanied by an increase of peripheral blood platelet levels, was also reported after GM-CSF administration (Vannucchi *et al.*, 1990).

In the rat, injection of GM-CSF induced a transient neutropenia and monocytopenia, probably due to transient intravascular margination, followed by peripheral neutrophilia and monocytosis and a mild lymphopenia (Ulich *et al.*, 1990). There was some left-shifted myeloid hyperplasia in the bone marrow, suggestive of a myeloproliferative effect, but no generalized myeloid hyperplasis was seen after 7 days of daily treatment.

Non-Human Primates

In macaques the continuous intravenous or repeated subcutaneous administration of glycosylated or non-glycosylated human GM-CSF caused a marked (up to fivefold) leukocytosis and some reticulocytosis, usually within 1 day (Donahue *et al.*, 1986; Mayer *et al.*, 1987). The effect, manifested predominantly as a neutrophilia, eosinophilia and monocytosis, but also as lymphocytosis, could be sustained for as long as the infusion continued (up to 1 month) without apparent adverse effects; blood counts normalized on cessation of the treatment. There was no effect on platelet or erythrocyte counts, even after weeks of GM-CSF administration. As had been expected from *in vitro* experiments, the neutrophils were not only numerically increased, but they also manifested enhanced phagocytic and bacterial killing capacity (Mayer *et al.*, 1987). Some effects of GM-CSF on megakaryocytic maturation were also reported (Stahl *et al.*, 1991). Sequential administration of IL-3 followed by GM-CSF synergistically increased the effects of GM-CSF on leukocytosis (Donahue *et al.*, 1988), megakaryocytic maturation and platelet response (Stahl *et al.*, 1992).

Adverse effects in monkeys seen at high doses (roughly tenfold over pharmacologic doses) included facial erythema and edema, skin reaction, decreased motor activity and decreased body weight (Robison and Myers, 1993). Early death seen at high doses involved severe diffuse polyserositis with no associated detectable infection. Both the pharmacologic effects, bone marrow hyperplasia, raised leukocytes counts and PMN activation, as well as pathologic findings, were virtually identical with glycosylated CHO-derived and non-glycosylated *E. coli*-derived GM-CSF. Antibodies to GM-CSF, of both immunoglobulin M (IgM) and IgG class, appeared after 2 weeks of treatment in all experiments, indicating that human and non-human primate GM-CSF are not antigenically identical.

Effect of GM-CSF in Animal Disease Models

After Myelodepression in Mice or Monkeys

After bone marrow or peripheral blood stem cell transplantation. In sublethally irradiated rhesus monkeys receiving autologous bone marrow transplantation, a continuous infusion of GM-CSF resulted in a sustained and accelerated recovery of neutrophils and platelets (Monroy *et al.*, 1987; Nienhuis *et al.*, 1987). If the infusion was halted before recovery, neutrophil counts fell to untreated control values, which suggests that GM-CSF affects hematopoietic reconstitution by stimulating late progenitor cells in the graft, rather than early progenitors capable of repopulation (Clark and Kamen, 1987).

Extending the studies involving the direct radioprotective effects of cytokines in animals are murine models of allogeneic bone marrow transplantation using *ex vivo* donor graft incubation with combinations of cytokine (Muench and Moore, 1992). Engraftment of T cell-depleted donor marrow was effectively promoted by GM-CSF, although no effects were seen on post-transplantation mortality, marrow stem-cell capacity or the incidence of graft-*versus*-host disease (Blazar *et al.*, 1988).

After irradiation. When mice were infused or injected repeatedly with GM-CSF before the administration of sublethal doses of cytotoxic drugs or radiation, restoration of hematopoiesis was accelerated (Neta *et al.*, 1988). However, GM-CSF had no radioprotective effect when administered as a single intraperitoneal dose 3 h after lethal irradiation (Neta and Oppenheim, 1988). Nevertheless, GM-CSF was able to synergize with suboptimal doses of IL-1α and TNF to provide optimal radioprotection, perhaps indicating that the interaction between cytokines is important in the hematopoietic recovery following radiation damage (Neta and Oppenheim, 1988).

Infection Models

In vitro, GM-CSF has been shown to increase antimicrobial activity of monocytes or macrophages from different tissue sources. The *in vivo* studies of the response to fungal infections of GM-CSF-treated animals confirmed the potential role of GM-CSF in fungal sepsis. In a study in mice the use of GM-CSF led to significantly enhanced survival during 15 days associated with clearing of *C. albicans* from the liver and spleen, but not from the kidney (Liehl *et al.*, 1994). Mayer *et al.* (1991) studied GM-CSF in neutropenic mice and showed prolonged survival in comparison with controls after infection with *C. albicans*, as well as *Pseudomonas* and *Staphylococcus*. Hebert and O'Reilly (1996) showed that splenectomized mice treated with GM-CSF had increased alveolar macrophage bactericidal activity, and improved survival after challenge infection with pneumococci. Mandujano *et al.* (1995) gave GM-CSF to mice for 7 or 14 days beginning 4 weeks after lymphocyte depletion and infection with *Pneumocystis carinii*. Histologic examination of lung tissue showed a significant decrease in the intensity of infection, a significant increase in TNF secretion by alveolar macrophages, and a reduced inflammation score (which did not reach statistical significance) in the GM-CSF-treated animals compared with controls.

Cytokines, such as IFN-γ and GM-CSF, can inhibit the intracellular replication of bacteria or protozoa which rely on the intracellular microenvironment for their proliferation. Organisms such as *Mycobacterium, Salmonella, Listeria* and *Leishmania* utilize macrophages in tissue as a part of their lifecycle. Bermudez and colleagues showed that GM-CSF, TNF and IFN-γ significantly inhibited the intracellular growth of *Mycobacterium avium* complex (MAC) in macrophages from the intestine, whereas M-CSF did not (Hsu *et al.*, 1995).

In other studies (Bermudez *et al.*, 1994) the number of *M. avium* organisms in the liver and spleen of infected mice was determined after treatment of the animals for 14 days with the antibiotics amikacin or azithromycin, either alone or in combination with GM-CSF. Only the combinations of GM-CSF with either amikacin or azithromycin led to significant (50–100-fold) reductions in the number of tissue bacteria. Denis has shown similar mycobacteriostatic effects of GM-CSF against *M. tuberculosis* (Denis, 1991).

Wound Healing

Animal models of wound healing have been used to evaluate the role of GM-CSF by examining cytokine interactions and effects on cells such as fibroblasts (Vyalov *et al.*, 1993), effects on incisional wounds after radiation (Cajucom *et al.*, 1993) or in healthy tissue (Jyung *et al.*, 1994), in contaminated wounds (Robson *et al.*, 1994), and in wounds resulting from thermal injury (O'Reilly *et al.*, 1994; Molloy *et al.*, 1995). In the study by Vyalov *et al.* (1993) fibroblasts in a chamber implanted in a rodent underwent conversion to myofibroblasts after treatment with GM-CSF, but not with other cytokines. This was considered, however, to be an indirect effect because it occurred only in these tissue implants and not on isolated fibroblasts *in vitro*. It thus appears clear that GM-CSF not only has significant effects on cells in the cutaneous and subcutaneous tissue, such as fibroblasts, keratinocytes and inflammatory cells, but also affects features important in wound closure, such as wound tensile strength, infiltration of wounds by inflammatory cells (Cajucom *et al.*, 1993; Jyung *et al.*, 1994) and bacterial contamination of open wounds (Robson *et al.*, 1994). It has also been shown that among hematopoietic growth factors this feature is unique to GM-CSF, and is not shared by G-CSF for instance (Jyung *et al.*, 1994). After thermal injury in mice, survival was improved by treatment of the animals with GM-CSF on days 5–9 after the burn (Molloy *et al.*, 1995). This was correlated with improved T-cell function in the treated animals. In another study mice with burns that were contaminated with *Pseudomonas* had better survival after 1 week of systemic treatment with GM-CSF than did controls (O'Reilly *et al.*, 1994).

Vaccine Immunomodulation

GM-CSF has potent adjuvant effects, increasing both primary and secondary antibody responses in mice and primates (Morrissey *et al.*, 1987; Liehl *et al.*, 1994; reviewed in Tarr *et al.*, 1996a). Induction of tumor immunity by vaccination with irradiated tumor cells engineered to secrete various cytokine genes showed that GM-CSF is the most potent cytokine to stimulate strong, specific and long-lasting antitumor immunity (Dranoff *et al.*, 1993). Similarly, when the GM-CSF gene was introduced in human prostate cancer cells, the cells were shown to produce GM-CSF (Jaffee *et al.*, 1993; Mahvi *et al.*, 1996), and the concept of autologous tumor cell vaccination is currently being tested. This approach is also under evaluation with melanoma cells. In a mouse model, the direct fusion of GM-CSF to a B-cell lymphoma idiotype induced a strong anti-idiotypic immune response in the absence of other adjuvants, and protected the mice from challenge with tumor cells (Tao and Levy, 1993).

GM-CSF IN HOMEOSTASIS, DEFICIENCY OR EXCESS STATES IN HUMANS

The numerous activities of GM-CSF identified *in vitro* or *in vivo* in animals are suggestive of a prominent physiological role for this molecule. Current knowledge of the place of GM-CSF in human homeostasis and pathology is reviewed briefly.

GM-CSF in Homeostasis

It was a surprise when gene knockout experiments indicated that GM-CSF was not essential for steady-state hematopoiesis in mice (Dranoff *et al.*, 1994). Even double deficiency for both M-CSF and GM-CSF led to essentially normal monocytic function (Lieschke *et al.*, 1994). The role of cytokines in steady-state hematopoiesis, therefore, appears to be overlapping, redundant and incompletely defined. These experiments did, however, show the importance of GM-CSF in pulmonary homeostasis and helped to identify a potential use of GM-CSF in correcting a certain form of human alveolar proteinosis (Hallman and Merritt, 1996). The first patient with alveolar proteinosis was treated in 1996 in Australia (Seymour *et al.*, 1996). Although a genetic defect and GM-CSF deficiency were not proven in this patient, the similarity of the pathologic process to the disease in GM-CSF-deficient mice, and the response to GM-CSF treatment, makes it likely that this is indeed the first use of this cytokine in a patient with GM-CSF deficiency. Other considerations regarding the use of GM-CSF to correct potential deficiency states have been complicated by obvious redundancy and overlapping of effects within a complex cytokine network. Patients with the 5Q syndrome, who have a deletion of the region of chromosome 5 containing the GM-CSF gene, together to a series of other cytokine genes, and suffer from refractory anaemia (Huebner *et al.*, 1985), were considered potential candidates, but no special response to GM-CSF has been seen. The absence of specific diseases of homeostasis as a result of GM-CSF deficiency may indicate that the major role of GM-CSF is in emergency responses rather than in homeostasis.

Deficiencies in the Emergency Response

Potential deficiencies in the emergency cytokine response may exist in patients after cytotoxic drug or disease-induced damage to the hematopoietic system (Nemunaitis *et al.*, 1991; Gerhartz *et al.*, 1993; Gancer and Hoelzer, 1996), in subcutaneous tissue associated with some types of wound healing (Jones, 1994), or in antitumor immune surveillance (Nagler *et al.*, 1996). If GM-CSF is being used to address a deficiency in such conditions, pharmacologic doses of the drug are required. Debate will continue regarding whether such high doses are needed because the deficiency is in a micro-environment and tissue pharmacokinetics are inefficient, or whether a totally unphysiologic use of GM-CSF is employed to achieve transient, exogenous overstimulation of the cytokine cascade. Because these potential deficiency states are so common, clinical use of GM-CSF has been used frequently in these patients (see below).

Overexpression of GM-CSF

While there has been a search for transient deficiency states in which GM-CSF could be beneficial, diseases have also been identified in which an excess of GM-CSF may contribute to a continuing pathologic process. Increased levels of GM-CSF have been recorded in patients with asthma (Robinson *et al.*, 1993) and other pulmonary inflammatory diseases (Walker *et al.*, 1994). Excess local endogenous GM-CSF has been found in rheumatoid arthritis (Alvaro-Gracia *et al.*, 1989). Approaches aiming at inhibition of GM-CSF activity, through GM-CSF–toxin fusion protein or antagonistic peptides are under preclinical study. It has been particularly important to identify the conditions in which GM-CSF may contribute to the pathology, as the use of GM-CSF therapeutically could aggravate them. Aggravation of arthritis has been described in clinical trials of GM-CSF use in patients with cancer (de Vries *et al.*, 1991), and aggravation of other autoimmune diseases such as thyroiditis or idiopathic thrombocytopenic purpura has been reported (Lieschke *et al.*, 1989; Hoekman *et al.*, 1991). Expansion of myeloid leukemias that have GM-CSF receptors on the malignant clones (Vellanga *et al.*, 1987) is potentially associated with GM-CSF excess. Cases of autocrine leukaemia have been documented (Young and Griffin, 1986), a few with unusually large transcripts from rearranged GM-CSF loci (Cheng *et al.*, 1988). In general, as some leukemias remain factor dependent, it is clear that GM-CSF and other cytokines may be essential cofactors facilitating the *in vivo* expansion of the transformed myeloid clone. In addition, histocytosis (Risti *et al.*, 1994) may be an example of a disease in humans which resembles the condition observed in transgenic mice with the GM-CSF gene (Lang *et al.*, 1987).

CLINICAL APPLICATIONS OF GM-CSF

Based on the numerous potential activities of the molecule GM-CSF, studies have been conducted in humans to define the best role for the pharmacologic use of GM-CSF in clinical medicine (see Table 2). Studies relevant to its use in humans are described below.

Conditions in which GM-CSF is Most Commonly Used

Reversal of Neutropenia After Chemotherapy or Bone Marrow Transplantation

The two main effects of GM-CSF relevant to present clinical use are initiation of a proinflammatory cytokine response by activating monocytes and macrophages (Ruef and Coleman, 1990) and induction of the proliferation and release from the bone marrow of neutrophils, eosinophils and monocytes (Gasson, 1991). These effects allow restitution of a defective cytokine response particularly in patients in whom this response has been damaged by cytotoxic chemotherapy or disease. Two different formulations of GM-CSF (yeast and *E. coli* derived) have been registered worldwide for use in conjunction with cytotoxic chemotherapy for the treatment of cancer, or after bone marrow transplantation (BMT) to blunt and/or reverse as rapidly as possible the anticipated neutropenia. The main benefit derived from the use of GM-CSF has been the reduction of infectious complications of chemotherapy and diseases that cause bone marrow dysfunction (Nemunaitis *et al.*, 1991; Gerhartz *et al.*, 1993; Gancer and Hoelzer, 1996). Secondary benefits have included shorter periods of hospitalization, reduced costs of medical intensive care (Gulati and Bennett, 1992) and, in a few settings, such as bone marrow failure, improved survival (Nemunaitis *et al.*, 1990).

The common side-effects of GM-CSF in humans are those characteristic of pro-inflammatory cytokine stimulation, such as fever, myalgia and malaise. Infrequent side-effects have included rash, pruritus, arthralgia and cardiovascular events, as well as pulmonary events such as capillary leak syndrome and 'first-dose reaction'. These transient signs and symptoms are referred to as adverse events when a drug is given, but they are considered a part of the normal inflammatory response when an individual is exposed to an infectious agent. In a patient deficient in this response, a transient excess stimulation is appropriate. Careful selection of patients and gauging the best dose and duration of treatment to achieve this stimulation is a key task of the treating physician.

In practice, GM-CSF has been used in a more restricted way, by first identifying patients at special risk for profound or prolonged hematopoietic dysfunction after chemotherapy or BMT. The condition of bone marrow failure after intensive chemotherapy and BMT has been a focus of study (Gorin *et al.*, 1995; Weisdorf *et al.*, 1995; Dierdorf *et al.*, 1997). In view of the increasing number of complicating infections emerging with each week of prolonged neutropenia (Bodey *et al.*, 1966), it is clear that this approach has been life-saving in responding patients. Dexter and Heyworth (1994) and Wunder *et al.* (1991) have described a possible role for GM-CSF in restoring the microenvironment under certain conditions of cytokine imbalance. Caution is required because, as recorded in a recent report of four cases of engraftment failure, three of whom had received GM-CSF in the immediate post-transplant period, histiocytic infiltration of the marrow rather than normal hematopoiesis occurred (Rosenthal and Farhi, 1994). There is little doubt that in the patient with hematopoietic–cytokine cascade dysfunction after treatment for cancer, GM-CSF, used in the approriate dose and duration with careful patient monitoring, is beneficial and in some patients life-saving.

Neutropenia and Sepsis

The use of hematopoietic growth factors as an adjunct to management of the septic neutropenic patient has become almost routine even though only a few controlled studies have been done. The question of whether cytokine activation is dangerous in septic patients has been answered by these studies. In at least three studies of the use of GM-CSF in neutropenia and sepsis (Aviles *et al.*, 1994; Riikonen *et al.*, 1994; Mayordomo *et al.*, 1995), there has been no evidence of an increase in septic shock (such as release of TNF) (Tracey *et al.*, 1986), nor of increased complications such as adult respiratory distress syndrome or capillary leak syndrome. Similar results have been recorded during treatment with G-CSF (Maher *et al.*, 1994). On the other hand, the efficacy of cytokines in this condition has been considered marginal. For example, the mean duration of fever is not affected by the cytokine, primarily because the fever responds to the antibiotics used in a mean of 3 days and this cannot be shortened. The mean duration of neutropenia can be shortened by 2–3 days, but this seems rather an insignificant clinical benefit. Patients predicted to have prolonged neutropenia and septic complications are most likely to benefit. Riikonen *et al.* (1994) examined the proportion of patients with neutropenia lasting more than 10 days and found that the number requiring long hospitalization due to infection was reduced by 50% with the use of GM-CSF. Bodey *et al.* (1993) have presented the results of a small series in which

GM-CSF was added to amphotericin B in patients with fungal sepsis and neutropenia. A majority of patients had *Candida* sepsis, and several had been shown to be unresponsive to the antifungal drug alone. Other single patient reports of responses of an unusual and beneficial type have been made (Spielberger *et al.*, 1993; Schoepfer *et al.*, 1995). Although not proven in clinical trials, fungal sepsis is considered an indication for GM-CSF rather than G-CSF. Both can reverse the neutropenia, although GM-CSF, as described in the previous sections, also activates macrophages and neutrophils to increase fungicidal activity (Ruef and Coleman, 1990). The role of GM-CSF in the management of patients with microbial disease has recently been reviewed (Jones, 1997).

Idiosyncratic Drug-Induced Neutropenia and Neutropenia in Human Immunodeficiency Virus (HIV) Infection

The use of cytokines to reverse the hematopoietic toxicity of drugs other than those used in cancer chemotherapy has been well documented. The most dramatic responses have been in patients with idiosyncratic agranulocytosis, a setting in which a relatively intact bone marrow can show reversal of the neutropenia within a few days (Sprikkelman *et al.*, 1994). The same rapid effects have been seen when GM-CSF has been used to prevent the hematopoietic suppressive effects of the commonly used antiviral drugs, relevant particularly to the treatment of HIV infection or opportunistic infections in acquired immune deficiency syndrome, such as cytomegalovirus (CMV) retinitis (Levine *et al.*, 1991; Scadden, 1992). The major concern in the use of proinflammatory cytokines in HIV infection has been the evidence that activated macrophages support viral replication better than resting cells (Perno *et al.*, 1989). The best information at present is that GM-CSF can be used safely, particularly if the antiviral drug is continued, since GM-CSF also potentiates the activity of the antiviral drug (Perno *et al.*, 1989). One study showed a delay in the progression of CMV retinitis when GM-CSF was used with the antiviral drug, ganciclovir (Hardy *et al.*, 1994).

Peripheral Blood Progenitor Cell Mobilization

The use of peripheral blood as a source of cells to repopulate the bone marrow after chemical or radiation damage has become an increasingly important tool in the management of cancer (Gianni *et al.*, 1989). Four different major approaches have been used in evaluating the use of periperal blood progenitor cells (PBPCs): GM-CSF alone (Socinski *et al.*, 1988; Gianni *et al.*, 1989, 1990), G-CSF alone (Sheridan *et al.*, 1992), IL-3 plus GM-CSF or G-CSF (Brugger *et al.*, 1992), and mixtures of KL, IL-1, IL-3, and G-CSF or GM-CSF in *ex vivo* expansion of collected cells (Haylock *et al.*, 1992; Hocker *et al.*, 1993). At present the various combinations have reached the limit of an aplasia lasting 4–6 days, beginning 3–4 days after the ablative procedure and continuing until day 9–11 (Socinski *et al.*, 1988; Gianni *et al.*, 1989, 1990; Brugger *et al.*, 1992). After chemotherapy, thrombocytopenia can be a problem complicating the cell collection process, particularly if platelet transfusions need to be given at the same time. The leukocyte number must be monitored carefully as an overshoot to the 100 000 cells per mm^3 can occur and is potentially dangerous. The best cell collections after use of chemotherapy and cytokines occur in patients whose bone marrow has not been damaged previously by chemotherapy or radiation therapy (Campo *et al.*, 1993). In the future combinations of cytokines, both *in vivo* and *ex vivo*, as well as both before and after the autografting, will be used to achieve the proper balance of progenitor cells at each point (Siena *et al.*, 1993; Schneider *et al.*, 1994). It appears clear from several studies that a major advantage of the use of PBPCs, instead of or in addition to bone marrow aspirates, is an earlier platelet engraftment (Huan *et al.*, 1992). An additional advantage is that the increased number of cells harvested by PBPC collection allows reinfusion during several subsequent courses of intensive chemotherapy (Shea *et al.*, 1992; Patrone *et al.*, 1995).

Myelodysplastic Syndrome or Aplastic Anemia

Cytokines have been tested for use in patients with various serious, and often end-stage, illnesses involving bone marrow dysfunction. One of the conditions that is most important, although extremely heterogeneous with regard to the degree of bone marrow damage and therefore the likelihood of response to cytokines, is myelodysplastic syndrome (MDS). Several studies have examined the use of GM-CSF in this condition (Gancer and Hoelzer, 1996). One large study enrolled 116 patients with MDS who received GM-CSF and 108 who were observation controls (Schering-Plough International, personal communication). This study confirmed that GM-CSF is able to reverse neutropenia significantly in patients with MDS, and significantly fewer infections were recorded. The approach to use colony-stimulating factors in this way is in keeping with the recommendations of the recent guidelines published by the American Society of Oncology (1996).

The use of GM-CSF in aplastic anemia has also been tested, but with varied results. This may be due to the fact that aplastic anemia is a very heterogeneous condition (Camitta *et al.*, 1982). It is now recognized that cytokines are useful only for transient intervention, usually in conjunction with treatment to correct the autoimmune disease with antilymphocyte globulin (Gordon-Smith *et al.*, 1991), cyclosporin A (Stryckmans *et al.*, 1984), or BMT (Locasciulli *et al.*, 1990). Use of GM-CSF in combination with immunosuppressive drugs is a rational approach, in which the immunosuppressive drug is used to correct the basic autoimmune defect, and the cytokine is used transiently to correct the hematopoietic dysfunction associated with the damaged bone marrow.

A second way that GM-CSF is useful in patients with either myelodysplastic syndrome or aplastic anemia is in the management of a patient with neutropenia who then becomes febrile and requires antibiotics for an intercurrent infection (Stern and Jones, 1990). When used in this way, GM-CSF is given for 7–10 days, or for the duration of the infectious episode, to assist in overcoming the complicating illness. Another way to use GM-CSF has been suggested by studies of the effects of very low doses of GM-CSF (10–20 μg daily) on thrombopoiesis (Kurzrock *et al.*, 1992). Seven of 20 patients (35%) studied by Vesole *et al.* (1994) became platelet transfusion independent by the use of low-dose GM-CSF. Attempts to use GM-CSF for prolonged periods of many months or years has in general not been successful, usually because of adverse effects of the cytokine, development of antibodies to the cytokine, or progression of the underlying autoimmune disease.

Conditions Still Being Evaluated for Special Approaches to GM-CSF Use

Hematologic Malignancies

The use of GM-CSF in myeloid leukemia increases the risk of exacerbating the disease by cytokine stimulation. On the other hand, the patient has a high likelihood of death from the prolonged postchemotherapy aplasia and life-threatening infection. Since the beginning of the clinical use of GM-CSF, it has been recognized that malignant clones with receptors for GM-CSF could be expanded *in vitro* by cytokine stimulation (Gasson, 1991; Vellanga *et al.*, 1987). This observation actually led to a series of clinical trials attempting to induce rapid myeloid leukemia cell expansion just before administration of cytotoxic chemotherapy (Bettelheim *et al.*, 1991; Buchner *et al.*, 1993; Frenette *et al.*, 1995). Indeed, more cells were stimulated to initiate DNA synthesis, but the theoretic advantage of increased cell cytotoxicity did not result in an increased frequency or duration of remissions (Estey, 1995; Heil *et al.*, 1995). When the cytokine was used immediately after chemotherapy in patients with AML, a complicated picture emerged, in that some patients benefited by an earlier return of myeloid cell function while others had a relapse of the malignancy (Buchner *et al.*, 1993; Stone *et al.*, 1995). It became clear that the benefit of reversal of the aplasia could be realized only if those patients prone to relapse were identified and excluded from treatment. In a study by Rowe *et al.* (1996) this criteria was met by having a bone marrow biopsy done 9 days after the induction chemotherapy, and only when the histology of the biopsy showed that the malignant clone had disappeared was GM-CSF used to allow more rapid marrow function. To a large measure because of this study design, GM-CSF was observed to be very effective as adjunctive treatment in AML.

It is known from several studies that prolonged neutropenia, particularly in patients with leukemia, is a guarantee for serious and often fatal infection (Bodey *et al.*, 1966). In patients with bone marrow failure after BMT the mortality rate at 1–2 years is 80% (Nemunaitis *et al.*, 1990). It is also clear that GM-CSF can reverse this condition in approximately 60% of patients within 7–10 days (Dierdorf *et al.*, 1997). Based on these data, when a patient with AML has aplasia and infection or a high risk of infection, restoration of the myeloid cell number and function by use of GM-CSF is recommended. In this setting there is a high frequency of death due to infection, and this risk is more immediate and relevant than that of malignant clone expansion. After autologous BMT for AML, an increase in activated killer cell function has been shown in patients treated with GM-CSF (Richard *et al.*, 1995). This raises the potential for activity of the molecule in control of the malignant cell minimal residual disease state.

The use of GM-CSF as a differentiating agent in selected patients with AML has also been suggested (Bassan *et al.*, 1994). The cytokine is used to induce maturation of an immature malignant clone, thus stopping the uncontrolled cell replication. This method of using cytokines remains theoretically interesting, but a method of selecting patients for treatment remains problematic.

Patients with malignancies other than AML may have complications associated with cytokine stimulation. For example, some patients with lymphatic and monocytic malignancy, if these cells have receptors for GM-CSF (Ho *et al.*, 1990), have shown expansion of the malignant clone. In patients with multiple myeloma, beneficial effects of GM-CSF have been seen, but there is the potential for stimulating plasma cells by initiating IL-6 via the cytokine network (Barlogie *et al.*, 1990). Expansion of malignant clones in patients with preleukemic myelodysplastic syndrome has also been reported, although successful use of GM-CSF in these patients has been observed (Gancer *et al.*, 1989).

Cell Cycling of Hematopoietic Cells

Aglietta *et al*. (1993) first demonstrated that after 3 days of stimulation *in vivo* with GM-CSF the flux of cells into S phase doubled and the percentage of early granulopoietic cells in S phase increased from 26% to 41%. Three days after stopping the GM-CSF 17% of early granulopoietic cells were in S phase. Similar changes were seen in early erythroid cells and in the colony-forming units measured by CFU-GM assays. This meant that, on average, 10% more cells were temporarily protected from cell cycle-specific cytotoxic drugs compared with baseline steady-state conditions, and 24% more cells were protected compared with a stimulated marrow such as with cytokines, infection or other hematopoietic cell stimuli. This difference would not be helpful in protecting against intensive, prolonged, or non-cell-cycle-specific treatment, but it could be a useful method for allowing delivery of full chemotherapy on schedule. During the treatment of breast cancer or during use of standard therapy for lymphoma, delays are required because of hematopoietic toxicity in approximately 30% of treatment courses (Hovgaard and Nissen, 1992; Aglietta *et al*., 1993). The first evidence that this approach could be utilized successfully was shown in evaluating 60 courses of chemotherapy for breast cancer (Aglietta *et al*., 1993). A similar approach has been evaluated in patients with lymphoma (Tafuto *et al*., 1995).

One of the attractive features of the approach is that the duration of cytokine use is at a minimum; therefore, there is both reduced cost and fewer adverse events. Of particular note is that G-CSF does not have this progenitor cell stimulatory, followed by downregulator, effect. Indeed, one of the features of the effect of G-CSF, in contrast to GM-CSF, is that continued hematopoietic cell stimulation is recorded 7–10 days after the last dose (Lord *et al*., 1989). This observation has led to another approach in the management of more severe bone marrow toxicity by use of 4–5 days of G-CSF therapy early after the cytotoxic event, then switching to GM-CSF as the next cycle of chemotherapy is approaching, then stopping the cytokine 2–3 days before the next course (Lord *et al*., 1992; Donova *et al*., 1995). This approach utilizes the information on both the positive and negative features of cytokine effects on cell cycling to best advantage for patient care.

Wound Healing

There are many different ways to use cytokines on wounds. GM-CSF not only has significant effects on cells in the cutaneous and subcutaneous tissue, but it has also been shown to alter features important in wound closure, such as wound tensile strength, infiltration of wounds by inflammatory cells (Cajucom *et al*., 1993) and bacterial contamination of open wounds (Kucukcelebi *et al*., 1992). It is has also been shown that among hematopoietic growth factors this is a feature unique to GM-CSF, not shared by G-CSF for instance (Jyung *et al*., 1994). Wounds are complex because of their varied causes, and because the goals of treatment are different for different types of wounds. Four major categories of use of GM-CSF include: (1) use with damaged tissue in which the cause of the injury has been removed, (2) use in conditions in which the underlying cause of the wound continues, (3) use in special situations such as at donor site or recipient site associated with skin grafting, and (4) oral use for the topical treatment of mucositis.

In the first category it is clear that, because the insult to the tissue has been removed, the cytokine is used entirely to affect the rate at which the natural process of wound healing occurs. Examples of successful use of GM-CSF in this way include use in rapid healing after the extravasation of intravenous cytotoxic chemotherapy in patients with cancer (Shamseddin, 1994), in the healing of radiation wounds after head and neck surgery, and in the management of pressure or decubitus ulcers after the pressure has been removed. These wounds have a high frequency of successful results.

In the second category, the most important issue is that the patient with an ongoing underlying disease has had the process controlled sufficiently to allow the cytokine to be effective at all. Examples include use of GM-CSF in patients with diabetes with microvascular or neurogenic tissue damage, patients with venous stasis ulcers (Da Costa *et al.*, 1994; Arnold *et al.*, 1995; Jaschke *et al.*, 1996), those with vascular ulcers associated with hemoglobinopathies (Pieters *et al.*, 1995), and those with cutaneous malignancies such as Kaposi's sarcoma (Boente *et al.*, 1993). The endpoint for the assessment of groups of patients with these conditions is the proportion of patients that is able to achieve wound closure at all. Unfortunately, this percentage is more dependent on the condition of the patient at entry into the trial or treatment program than on the cytokine used. This has made comparison of different cytokines difficult and has led to confusion concerning their overall effectiveness. It is clear that, under the right conditions, GM-CSF can allow the healing of wounds which before its use were refractory to treatment (Da Costa *et al.*, 1994; Arnold *et al.*, 1995; Jaschke *et al.*, 1996). A major challenge in these patients is how to use the cytokine most efficiently. In some patients spraying GM-CSF topically has been the preferred method of application (Pieters *et al.*, 1995; Jaschke *et al.*, 1996); in others injection around the wound has been effective (Da Costa *et al.*, 1994; Arnold *et al.*, 1995).

In the third category, GM-CSF has been used primarily as an adjunct to skin grafting. There are many ways to use the cytokine in this way. One way has been to place the donor skin tissue in GM-CSF for several hours before grafting (Pojda and Struzyna, 1994). In other patients, GM-CSF has been used to achieve rapid wound healing by application to the skin-graft donor site, to the site of the applied graft after the graft has been placed, and in conjunction with the preparation of artificial skin grafts. In the fourth category, oral mucositis have been treated by use of GM-CSF mouthwash as topical treatment (reviewed in Masucci, 1997).

Vaccine Adjuvant/Immune Modulation

As described in previous sections, GM-CSF is a major stimulatory cytokine for the viability, differentiation and function of dendritic cells (Caux *et al.*, 1992; Inaba *et al.*, 1992); it has also been shown *in vitro* to augment the expression of the T-cell costimulatory molecules B7-1 and B7-2 (Larsen *et al.*, 1994) and of MHC class II molecules in APCs (Fischer *et al.*, 1988). Biopsies of GM-CSF injection sites in patients have demonstrated the accumulation of neutrophils and mononuclear cells, in agreement with the chemotactic and white blood cell-activating effects of GM-CSF (Gerhartz *et al.*, 1993). In macrophages GM-CSF also increases the secretion of cytokines, which modulate the immune responses (Ruef and Coleman, 1990; Gasson, 1991).

Anecdotal evidence for the immune-enhancing effects of GM-CSF in humans is derived from case reports of exacerbations of autoimmune disease in patients with cancer receiving GM-CSF (De Vries *et al.*, 1991; Hoekman *et al.*, 1991). Such reports have described the flare-up of rheumatoid arthritis (Nemunaitis *et al.*, 1990), worsening autoimmune hemolysis (Nathan and Besa, 1992), as well as increased levels of thyroid autoantibodies resulting in clinical hypothyroidism (Hoekman *et al.*, 1991).

In one recently published study, the effect of a single injection of GM-CSF on the antibody response to a single dose of a recombinant hepatitis B virus (HBV) vaccine, when given in short sequence, was investigated. GM-CSF was injected into the same site as the vaccine, in an attempt to stimulate resident APCs (i.e. Langerhans cells), to capture and present the locally available vaccine antigens in a more efficient manner (Tarr *et al.*, 1996b). One of 27 control subjects mounted a non-protective anti-HBVs response. In contrast, 5 of 27 and 6 of 27 subjects receiving GM-CSF by the subcutaneous and intramuscular route, respectively, mounted protective titers to a single dose of HBV vaccine. The results of this study confirm in human healthy volunteers that, under certain conditions, GM-CSF can have significant immune-augmenting activity, suggesting its clinical potential as a vaccine adjuvant. Approaches using GM-CSF–antigen fusion proteins, and tumor cells transduced with the GM-CSF gene, are now being tested in clinical trials.

Based on the data from animal studies, it may be expected that multiple injections of recombinant GM-CSF are also associated with a stronger immune-enhancing effect in humans. It might also be speculated that lower doses of GM-CSF would be associated with more pronounced immunostimulatory effects. GM-CSF is well tolerated when used as a vaccine adjuvant. The low incidence of adverse events is in agreement with clinical experience of low doses (5–30 μg/day) of GM-CSF in patients with myelodysplastic syndromes or aplastic anemia. Caution is required, however, as exacerbation of autoimmune conditions by GM-CSF remains possible at low doses, and induction of neutralizing antibodies to GM-CSF may occur. The development of antibodies has been reported after administration of GM-CSF in non-immunocompromised patients. These antibodies can significantly alter the pharmacokinetics of GM-CSF, induce allergic reactions, and/or neutralize its activities (Ragnhammar *et al.*, 1994; for review see Ragnhammar and Wadhwa, 1996).

Immune modulation by GM-CSF has been demonstrated by various *in vitro* and animal model studies. The use of these concepts in patients requires two steps, the first of which has been taken. That step is to confirm that the relevant correlate of immune response is changed during treatment of humans. In clinical trials cells from patients treated with GM-CSF have exhibited enhanced functions relevant to anti-infectious and antitumor effects. These functions include enhanced chemotaxis (Wang *et al.*, 1987), monocyte tumoricidal activity (Grabstein *et al.*, 1986, Grabstein and Alderson, 1993), activation of lymphocytes (Ho *et al.*, 1990), LAK cells (Baxevanis *et al.*, 1995), NK cell activity (Bendall *et al.*, 1995; Richard *et al.*, 1995), macrophage antibody-dependent cytotoxicity (Charak *et al.*, 1993) and other immunomodulatory functions (Aman *et al.*, 1996; San Miguel *et al.*, 1996). The best way to take maximum advantage of each of these functions needs further study. Use of GM-CSF in patients to control minimal residual disease is more likely than use as primary therapy for tumors, and the scheduling and dose of GM-CSF therapy will almost certainly be different from that presently recommended. Only one phase I/II study has suggested an effect of GM-CSF on survival of patients with melanoma when the drug was used as an immunomodulator (*Scrip*, 1997). This result is expected based on all the data available to date, but it needs confirmation and validation.

REFERENCES

Aglietta, M., Mongeglio, C., Pasquino, P., Carnino, F., Stern, A.C. and Garosto, F. (1993). Short term administration of GM-CSF decreases hematoietic toxicity of cytotoxic drugs. *Cancer* **72**, 2970–2973.

Alvaro-Gracia, H.G., Zvaifler, N.J. and Firestein, G.S. (1989). Cytokines in chronic inflammatory arthritis. IV. Granulocyte/macrophage colony stimulating factor-mediated induction of class II MHC antigen on human monocytes: a possible role in rheumatoid arthritis. *J. Exp. Med.* **170**, 865–875.

Aman, M.J., Stockdreher, K., Thews, A., Kienast, K, Aulitzky, W.E., Färber, L., Haus, U., Koci, B., Huber, C. and Peschel, C. (1996). Regulation of immunomodulatory functions by granulocyte–macrophage colony-stimulating factor and granulocyte colony-stimulating factor *in vivo*. *Ann. Hematol.* **73**, 231–238.

American Society of Oncology (1996). Update of recommendations for the use of hematopoietic colony-stimulating factors: evidence-based clinical practice guidelines. *J. Clin. Oncol.* **14**, 1957–1960.

Arnold, F., O'Brien, J. and Cherry, G. (1995). Granulocyte–macrophage colony-stimulating factor as an agent for wound healing. A study evaluating the use of local injections of a genetically engineered growth factor in the management of wounds with a poor healing prognosis. *Wound Care* **4**, 400–402.

Aviles, A., Rosas, A., Talavera, A. *et al*. (1994). Use of granulocyte–macrophage colony-stimulating factor in the treatment of infection and severe granulocytopenia. *Cancer Res. Therapy Control* **1**, 1–5.

Azam, M., Erdjument-Bromage, H., Kreider, B.L., Xia, M., Quelle, F., Basu, R., Saris, C., Tempst, P., Ihle, J.N. and Schindler, C. (1995). Interleukin-3 signals through multiple isoforms of Stat5. *EMBO J.* **14**, 1402–1411.

Bagley, C.J., Woodcock, J.M., Stomski, F.C. and Lopez, A.F. (1997). The structural and functional basis of cytokine receptor activation: lessons from the common β subunit of the granulocyte–macrophage colony-stimulating factor, interleukin-3 (IL-3), and IL-5 receptors. *Blood* **89**, 1471–1482.

Barlogie, B., Jagannath, S., Dixon, D.O., Cheson, B., Smallwood, L., Hendrikson, A., Purvis, J.D., Bonnen, E. and Alexanian. R. (1990). High-dose melphalan and granulocyte–macrophage colony-stimulating factor for refractory multiple myeloma. *Blood* **76**, 677–680.

Bassan, R., Rambaldi, A., Amaru, R., Motta, T. and Barbui, T. (1994). Unexpected remission of acute myeloid leukemia after GM-CSF. *Br. J. Haematol.* **87**, 835–838.

Baxevanis, C.N., Dedoussis, G.V.Z., Papadopoulos, N.G., Missitzis, I., Beroukas, C., Stathopoulos, G.P. and Papamichail, M. (1995). Enhanced human lymphokine-activated killer cell function after brief exposure to granulocyte–macrophage colony stimulating factor. *Cancer* **76**, 1253–1260.

Bazan, J.F. (1990). Structural design and molecular evolution of a cytokine receptor superfamily. *Proc. Natl Acad. Sci. U.S.A.* **87**, 6934–6938.

Bendall, L.J., Kortlepel, K. and Gottlieb, D.J. (1995). GM-CSF enhances IL-2-activated natural killer cell lysis of clonogenic AML cells by upregulating target cell expression of ICAM-1. *Leukemia* **9**, 677–684.

Bermudez, L.E., Martinelli, J., Petrovsky, M., Kolowski, P. and Young, L. (1994). Recombinant granulocyte–macrophage colony-stimulating factor enhances the effects of antibodies against *Mycobacterium avium* complex infection in the beige mouse model. *J. Infect. Dis.* **169**, 575–580.

Bernasconi, S., Matteucci, C., Sironi, M., Conni, M., Colotta, F., Mosca, M., Colombo, N., Bonazzi, C., Landoni, F., Corbetta, G., Mantovani, A. and Allavena, P. (1995). Effects of granulocyte–monocyte colony-stimulating factor (GM-CSF) on expression of adhesion molecules and production of cytokines in blood monocytes and ovarian cancer-associated macrophages. *Int. J. Cancer* **60**, 300–307.

Bettelheim, P., Valent, P., Andreff, M., Tafuri, A., Haimi, J., Gorischek, C., Muhm, M., Sillaber, C., Haas, O. and Vieder, L. (1991). Recombinant human granulocyte–macrophage colony-stimulating factor in combination with standard induction chemotherapy in *de novo* acute myeloid leukemia. *Blood* **77**, 700–711.

Bittorf, T., Jaster, R. and Brock, J. (1994). Rapid activation of the MAP kinase pathway in hematopoietic cells by erythropoietin, granulocyte–macrophage colony-stimulating factor and interleukin-3. *Cell Signal.* **6**, 305–311.

Blazar, B.R., Widmer, M.B., Soderling, C.C., Urdal, D.L., Gillis, S., Robison, L.L. and Vallera, D.A. (1988). Augmentation of donor bone marrow engraftment in histoincompatible murine recipients by granulocyte–macrophage colony stimulating factor. *Blood* **71**, 320–328.

Bober, L.A., Grace, M.J., Pugliese-Sivo, C., Rojas-Triana, A., Waters, T., Sullivan, L.M. and Narula, S.K. (1995). The effect of GM-CSF and G-CSF on human neutrophil function. *Immunopharmacology* **29**, 111–119.

Bodey, G.P., Buckley, M., Sathe, Y.S. and Freireich, E.J. (1966). Quantitative relationships between circulating leukocytes and infection in patients with acute leukemia. *Ann. Intern. Med.* **64**, 328.

Bodey, G.P., Anaissie, E., Gutterman, J. and Vadhan-Raj, S. (1993). Role of granulocyte–macrophage colony-stimulating factor as adjuvant therapy for fungal infection in cancer patients. *Clin. Infect. Dis.* **17**, 705–707.

Boente, P., Sampaio, C., Brandao, M.A., Moreira, E.D., Badaro, R. and Jones, T.C (1993). Local peri-lesional therapy with rhGM-CSF for Kaposi's sarcoma. *Lancet* **341**, 1154.

Brach, A., De Vos, S., Gruss, H. and Hermann, F. (1992). Prolongation of survival of human polymorphonuclear neutrophils by granulocyte–macrophage colony stimulating factor is caused by inhibition of programmed cell death. *Blood* **80**, 2920–2924.

Brizzi, M.F., Zini, M.G., Aronica, M.G., Blechman, J.M., Yarden, Y. and Pegoraro, L. (1994). Convergence of signaling by interleukin-3, granulocyte–macrophage colony-stimulating factor, and mast cell growth factor on JAK2 tyrosine kinase. *J. Biol. Chem.* **269**, 31 680–31 684.

Brizzi, M.F., Rossi, P.R., Rosso, A., Avanzi, G.C. and Pegoraro, L. (1995). Transcriptional and post-transcriptional regulatlon of granulocyte–macrophage colony-stimulating factor production in human growth factor dependent M-07e cells. *Br. J. Haematol.* **90**, 258–265.

Brizzi, M.F., Aronica, M.G., Rosso, A., Bagnara, G.P., Yarden, Y. and Pegoraro, L. (1996). Granulocyte–macrophage colony-stimulating factor stimulates JAK2 signaling pathway and rapidly activates p93^{fes}, STAT1 p91, and STAT3 p92 in polymorphonuclear leukocytes. *J. Biol. Chem.* **271**, 3562–3567.

Brugger, W., Bross, K., Frisch, J., Dern, P., Weber, B., Mertelsmann, R. and Kanz, L. (1992). Mobilization of peripheral blood progenitor cells by sequential administration of interleukin-3 and granulocyte–macrophage colony-stimulating factor following poly-chemotherapy with etoposide, ifosfamide and cisplatin. *Blood* **79**, 1193–1200.

Buchner, T., Hiddemann, W., Wormann, B., Rottmann, R., Maschmeyer, G., Ludwig, W.D., Zuhlsdorf, M., Buntkirchen, K., Sander, A. and Aswald, J. (1993). The role of GM-CSF in the treatment of acute myeloid leukemia. *Leukemia Lymphoma* **11**, 21–24.

Burgess, A.W., Camakaris, J. and Metcalf, D. (1977). Purification and properties of colony stimulating factor from mouse lung conditioned medium. *J. Biol. Chem.* **252**, 1998–2003.

Burgess, A.W., Begley, C.G., Johnson, G.R., Lopez, A.F., Williamson, D.J., Mermod, J.J., Simpson, R.J., Schmitz, A. and DeLamarter, J.F. (1987). Purification and properties of bacterially synthesized human granulocyte–macrophage colony stimulating factor. *Blood* **69**, 43.

Bussolini, F., Ming Wang, J., Defilippi, P., Turrini, F., Sanavio, F., Edgell, C.-J.S., Aglietta, M., Arese, P. and Mantovani, A. (1989). Granulocyte- and granulocyte–macrophage-colony stimulating factors induce human endothelial cells to migrate and proliferate. *Nature* **337**, 471–473.

Cajucom, C.C., Tia, R.C., Vega, G.P. and Avila, J.M. (1993). Molgramostim (GM-CSF) enhances wound healing after fractionated external beam radiation therapy in an experimental murine model. *Proceedings of the 9th Annual Convention of the Philippine Society of Oncologists*, December 1993, Manila.

Camitta, B.M., Storb, R. and Thomas, E.D. (1982). Aplastic anemia: pathogenesis, diagnosis, treatment, and prognosis. *N. Engl. J. Med.* **306**, 645–652 and 712–718.

Campos, L., Bastion, Y. and Roubi, N. (1993). Peripheral blood stem cells harvested after chemotherapy and GM-CSF for treatment intensification in patients with advanced lymphoproliferative diseases. *Leukemia* **7**, 1409–1415.

Cannistra, S.A. Vellenga, E. and Groshek, P. (1988). Human granulocyte–macrophage factor and interleukin-3 stimulate monocyte cytotoxicity through tumor necrosis factor-dependent mechanism. *Blood* **71**, 672–676.

Cantrell, M.A., Anderson, D. and Cerretti, D.P. (1985). Cloning, sequence and expression of a human granulocyte macrophage colony stimulating factor. *Proc. Natl Acad. Sci. U.S.A.* **82**, 6250.

Caux, C., Dezutter-Dambuyant, C., Schmitt, D. and Banchereau, J. (1992). GM-CSF and TNF-α cooperate in the generation of Langerhans cells. *Nature* **360**, 258–261.

Caux, C., Vanbervliet, B., Massacrier, C., Dezutter-Dambuyant, C., De Saint Vis, B., Jacquet, C., Yoneda, K., Imamura, S., Schmitt, D. and Banchereau, J. (1996). CD34$^+$ hematopoietic progenitors from human cord blood differentiate along two independent dendritic cell pathways in response to GM-CSF+TNFalpha. *J. Exp. Med.* **184**, 695–706.

Charak, B.S., Agah, R. and Mazumder, A. (1993). GM-CSF induced ADCC in bone marrow macrophages: application in bone marrow transplantation. *Blood* **81**, 3474–3479.

Chen, G.-H., Curtis, J.L., Mody, C.H., Christensen, P.J., Armstrong, L.R. and Toews, G.B. (1994). Effect of granulocyte–macrophage colony-stimulating factor on rat alveolar macrophage anticryptococcal activity *in vitro*. *J. Immunol.* **152**, 724–734.

Cheng, G.Y., Kelleher, C.A., Miyauchi, J., Wang, C., Wong, G., Clark, S.C., McCulloch, E.A. and Minden, M.D. (1988). Structure and expression of genes of GM-CSF and G-CSF in blast cells from patients with acute myeloblastic leukemia. *Blood* **71**, 204–208.

Chopra, R., Kendall, G., Gale, R.E., Thomas, N.S. and Linch, D.C. (1996). Expression of two alternatively spliced forms of the 5′ untranslated region of the GM-CSF receptor alpha chain mRNA. *Exp. Hematol.* **24**, 755–762.

Cicuttini, F.M., Martin, M. and Boyd, A.W. (1994). Cytokine induction of adhesion molecules on synovial type B cells. *J. Rheumatol.* **21**, 406–412.

Clark, E.A., Grabstein, K.H. and Shu, G.L. (1992). Cultured human follicular dendritic cells. Growth characteristics and interactions with B lymphocytes. *J. Immunol.* **148**, 3327–3335.

Clark, S.C. and Kammen, R. (1987). The human hematopoietic colony-stimulating factors. *Science* **236**, 1229–1237.

Cluitmans, F.H.M., Esendam, B.H.J., Landegent, J.E., Wollemze, R. and Falkenburg, J.H.F. (1993). Regulatory effects of T cell lymphokines on cytokine gene expression in monocytes. *Lymphokine Cytokine Res.* **12**, 457–464.

Cockerill, P.N., Shannon, M.F., Bert, A.G., Ryan, G.R. and Vadas, M.A. (1993). The granulocyte–macrophage colony-stimulating factor/interleukin 3 locus is regulated by an inducible cyclosporin A-sensitive enhancer. *Proc. Natl Acad. Sci. U.S.A.* **90**, 2466–2470.

Collins, H.L. and Bancroft, C.J. (1992). Cytokine enhancement of complement-dependent phagocytosis by macrophages: synergy of tumor necrosis factor-alpha and granulocyte–macrophage colony-stimulating factor for phagocytosis of *Cryptococcus neoformans*. *Eur. J. Immunol.* **22**, 1447–1454.

Crosier, K.E., Wong, G.G., Mathey, P.B., Nathan, D.G. and Sieff, C.A. (1991). A functional isoform of the human granulocyte/macrophage colony-stimulating factor receptor has an unusual cytoplasmic domain. *Proc. Natl Acad. Sci. U.S.A.* **88**, 7744–7748.

Da Costa, R.M., Aniceto, C., Jesus, F.M. and Mendes, M. (1994). Quick healing of leg ulcers after molgramostim. *Lancet* **344**, 481–482.

Danis, V.A., Mullington, M., Hyland, V.J. and Grenhan, D. (1995). Cytokine production by normal human monocytes: inter-subject variation and relationship to an IL-1 receptor antagonist (IL-IRa) gene polymorphism. *Clin. Exp. Immunol.* **99**, 303–310.

Danova, M., Rosti, V., Mazzini, G., De Renzis, M.R., Locatelli, F., Cazzola, M., Riccardi, A. and Ascari, E. (1995). Cell kinetics of CD34-positive hematopoietic cells following chemotherapy plus colony-stimulating factors in advanced breast cancer. *Int. J. Cancer* **63**, 646–651.

Denis, M. (1991). Involvement of cytokines in determining resistance and acquired immunity in murine tuberculosis. *J. Leukoc. Biol.* **50**, 495–501.

De Vries, E.G.E., Willemse, P.H.B., Biesma, B., Stern, A.C., Limburg, P.C. and Vellenga, E. (1991). Flare-up of rheumatoid arthritis during GM-CSF treatment after chemotherapy. *Lancet* **338**, 517–518.

Dexter, T.M. and Heyworth, C.M. (1994). Growth factors and molecular control of hematopoiesis. *Eur. J. Clin. Microbiol. Infect. Dis.* **2 (supplement)**, 2–8.

Diederichs, K., Boone, T. and Karplus, P.A. (1991). Novel fold and putative receptor binding site of granulocyte–macrophage colony-stimulating factor. *Science* **254**, 1779–1782.

Dierdorf, R., Kreuter, U. and Jones, T.C. (1997). Use of granulocyte–macrophage colony-stimulating factor in the treatment of prolonged hematopoietic dysfunction after chemotherapy alone or chemotherapy plus bone marrow transplantation. *Medical Oncology* (in press).

Donahue, R.E., Wang, E.A., Stone, D.K., Kamen, R., Wong, G.G., Sehgal, P.K., Nathan, D.G. and Clark, S.C. (1986). Stimulation of haematopoiesis in primates by continuous infusion of recombinant human GM-CSF. *Nature* **321**, 872–875.

Donahue, R.E., Seehra, J., Metzger, M., Lefebvre, D., Rock, B., Carbone, S., Nathan, D.G., Garnick, M., Sehgal, P.K., Laston, D., LaVallie, E., McCoy, J., Schendel, P.F., Norton, C., Turner, K., Yang, Y. and Clark, S.C. (1988). Human IL-3 and GM-CSF act synergistically in stimulating hematopoiesis in primates. *Science* **241**, 1820–1823.

Dorsch, M., Hock, H. and Diamantstein, T. (1994). Tyrosine phosphorylation of Shc is induced by IL-3, IL-5 and GM-CSF. *Biochem. Biophys. Res. Commun.* **200**, 562–568.

Dranoff, G., Jaffee, E., Lazenby, A., Golumbek, P., Levitsky, H., Brose, K., Jackson, V., Hamada, H., Pardoll, D. and Mulligan, R. (1993). Vaccination with irradiated tumor cells engineered to secrete murine granulocyte–macrophage colony-stimulating factor stimulates potent, specific and long lasting anti-tumor immunity. *Proc. Natl Acad. Sci. U.S.A.* **90**, 3539–3543.

Dranoff, G., Crawford, A.D., Sadelain, M., Ream, B., Rashid, A., Bronsin, R.T., Dickersin, G.R., Bachurski, C.J., Mark., E.L., Whitsett, J.A. and Mulligan, R.C. (1994). Involvement of granulocyte–macrophage colony-stimulating factor in pulmonary homeostasis. *Science* **264**, 713–716.

Estey, E.H. (1995). Granulocyte–macrophage colony-stimulating factor priming for acute myeloid leukemia. *Interferons Cytokines* **26**, 30–32.

Finbloom, D.S., Larner, A.C., Nakagawa, Y. and Hoover, D.L. (1993). Culture of human monocytes with granulocyte–macrophage colony-stimulating factor results in enhancement of IFN-γ receptor but suppression of IFN-γ-induced expression of the gene IP-10. *J. Immunol.* **150**, 2383–2390.

Fischer, H.G., Frosch, S., Reske, K. and Reske-Kunz, A.B. (1988). Granulocyte–macrophage colony-stimulating factor activates macrophages derived from bone marrow cultures to synthesis of MHC class II molecules and to augmented antigen presentation function. *J. Immunol.* **141**, 3882–3888.

Frenette, P.S., Desforges, J.F., Schenkein, D.P., Rabson, A., Slapack, C.A. and Miller, K.B. (1995). Granulocyte–macrophage colony-stimulating factor (GM-CSF) priming in the treatment of elderly patients with acute myelogenous leukemia. *Am. J. Hematol.* **49**, 48–55.

Ganser, A. and Hoelzer, D. (1996). Clinical use of hematopoietic growth factors in the myelodysplastic syndromes. *Semin. Hematol.* **33**, 186–195.

Ganser, A., Volkers, B., Greher, J., Ottmann, O.G., Walther, F., Becher, R., Bergmann, L., Schulz, G. and Hoelzer, D. (1989). Recombinant granulocyte–macrophage colony stimulating factor in patients with myelodysplastic syndromes—a phase I.II study. *Blood* **73**, 31–37.

Gasson, J.C. (1991). Molecular physiology of granulocyte–macrophage colony stimulating factor. *Blood* **77**, 1131–1145.

Gearing, D.P., King, J.A. and Gough, N.M. (1989). Expression cloning of a receptor for human granulocyte–macrophage colony-stimulating factor. *EMBO J.* **8**, 3667–3676.

Gerhartz, H.H., Engelhart, M., Meusers, P., Brittinger, G., Wilmanns, W., Schlimok, G., Mueller, P., Huhn, D., Musch, R., Siegert, W., Gerhatz, D., Hartlapp, J.H., Thiel, E., Huber, C., Peschl, C., Spann, W., Emmerich, B., Schadek, C., Westerhausen, M., Pees, H.W., Radtke, H., Engert, A., Terhardt, E., Schick, H., Binder, T., Fuchs, R., Hasford, J., Brandmeier, R., Stern, A., Jones, T.C., Ehrlich, F.J., Stein, H., Parwaresch, M., Tiemann, M. and Lennert, K. (1993). Randomized, double-blind, placebo-controlled, phase III study of recombinant human granulocyte–macrophage colony-stimulating factor (rhGM-CSF) as adjunct to induction treatment of high-grade malignant non-Hodgkin's lymphomas. *Blood* **82**, 2329–2339.

Gianni, A.M., Siena, S., Bergni, M. and Tarella, G. (1989). Granulocyte–macrophage colony-stimulating factor to harvest circulating hematopoietic stem cells for autotransplantation. *Lancet* **ii**, 580–584.

Gianni, A.M., Bregni, M., Siena, S., Orazi, A., Stern, A.C., Gandolo, L. and Bonadonna, G. (1990). Recombinant human granulocyte–macrophage colony-stimulating factor reduces hematopoietic toxicity and widens clinical application of high dose cyclophosphamide treatment in breast cancer and non-Hodgkin's lymphoma. *J. Clin. Oncol.* **8**, 768–778.

Gomez-Cambronero, J., Huang, C.-K., Gomez-Cambronero, T.M., Waterman, W.H., Becker, E.L. and Sha Afi, R.I. (1992). Granolocyte–macrophage colony-stimulating factor-induced protein tyrosine phosphorylation of microtubule-associated protein kinase in human neutrophils. *Proc. Natl Acad. Sci. U.S.A.* **89**, 7551–7555.

Gordon-Smith, E.C., Yandle, A., Milne, A., Speck, B., Marmont, A., Willenze, R. and Kolb, H. (1991). Randomized, placebo-controlled study of rhGM-CSF following ALG in the treatment of aplastic anemia. *Bone Marrow Transplant.* **7 (supplement 2)**, 78–80.

Gorin, N.C., Laporte, J.P. and Fouillard, L. (1995). Granulocyte–macrophage colony-stimulating factor as an adjunct to autologous bone marrow transplantation in hematology: considerations on increased safety on the transplant ward. In *Granulocyte–Macrophage Colony-Stimulating Factor (GM-CSF) Current Uses and Future Application*. Proceedings of an International Symposium, Lucerne, Switzerland, September 1993. Pennine Press, pp. 38–46.

Gough, N.M., Gough, J., Metcalf, D., Kelson, A., Grail, D., Nicola, N.A., Burgess, A.W. and Dunn, A.R. (1994). Molecular cloning of cDNA encoding a murine haematopoietic growth regulator, granulocyte–macrophage colony stimulating factor. *Nature* **309**, 763–767.

Grabstein, K.H. and Alderson, M.R. (1993). Regulation of macrophage tumoricidal activity by granulocyte–macrophage colony-stimulating factor. In *Hemopoietic Growth Factors and Mononuclear Phagocytes* (ed. R. Van Furth), Karger, Basel, pp. 140–147.

Grabstein, K.H., Urdal, D.L., Tushinski, R.J., Mochizuki, D.Y., Price, V.L., Cantrell, M.A., Gillis, S. and Conlon, P.J. (1986). Induction of macrophage tumoricidal activity by granulocyte–macrophage colony-stimulating factor. *Science* **232**, 506–508.

Gulati, S.C. and Bennett, C.L. (1992). Granulocyte–macrophage colony-stimulating factor (GM-CSF) as adjunct therapy in relapsed Hodgkin's disease. *Ann. Intern. Med.* **116**, 177–182.

Haak-Frendscho, M., Arai, N., Arai, K., Baeza, M.L., Finne, A. and Kaplan, A.P. (1988). Human recombinant granulocyte–macrophage colony-stimulating factor and interleukin 3 cause basophil histamine release. *Am. Soc. Clin. Invest.* **82**, 17–20.

Hallman, M. and Merritt, T.A. (1996). Lack of GM-CSF as a cause of pulmonary alveolar proteinosis. *J. Clin. Invest.* **97**, 589–590.

Han, J.H., Gileadi, C., Rajapsaka, R., Kosek, J. and Greenberg, P.L. (1995). Modulation of apoptosis in human myeloid leukemic cells by GM-CSF. *Exp. Hematol.* **23**, 265–272.

Hardy, D., Spector, S., Polesky, B., Crumpacker, C., van der Horst, C., Holland, G., Freeman, W., Heinemann, M.H., Sharuk, G. and Klystra, J. (1994). Combination of gancyclovir and granulocyte–macrophage colony-stimulating factor in the treatment of cytomegalovirus retinitis in AIDS patients. *Eur. J. Clin. Microbiol. Infect. Dis.* **13 (supplement 2)**, 34–40.

Hayashida, K., Kitamura, T., Gorman, D.M., Arai, K.I., Yokota, T. and Miyajima, A. (1990). Molecular cloning of a second subunit of the receptor for human granulocyte–macrophage colony-stimulating factor (GM-CSF): reconstitution of a high-affinity GM-CSF receptor. *Proc. Natl Acad. Sci. U.S.A.* **87**, 9655–9659.

Haylock, D.N., To, L.B., Dowse, T.L., Juttner, C.A. and Simmons, P.J. (1992). *Ex vivo* expansion and maturation of peripheral blood CD 34^+ cells into myeloid lineage. *Blood* **80**, 1405–1412.

Hebert, J.C. and O'Reilly, M. (1996). Granulocyte–macrophage colony-stimulating factor (GM-CSF) enhances pulmonary defenses against pneumococcal infections after splenectomy. *J. Trauma.* **41**, 663–666.

Heil, G., Chadid, L., Hoelzer, D., Seipelt, G., Mitrori, P., Huber, Ch., Kolbe, K., Mertelsmann, R., Lindemarin, A., Frisch, J., Nicolay, U., Gaus, W. and Heimpil, H. (1995). GM-CSF in a double-blind randomized, placebo controlled trial in therapy of adult patients with *de novo* acute myeloid leukemia. *Leukemia* **9**, 3–9.

Ho, A.D., Haas, R., Wulf, G., Knauf, W. and Ehrhardt, R. (1990). Activation of lymphocytes induced by recombinant granulocyte–macrophage colony-stimulating factor in patients with malignant lymphoma. *Blood* **75**, 203–212.

Hocker, P., Geissler, K., Kura, M., Wagner, A. and Gerhartl, K. (1993). Potentiation of GM-CSF or G-CSF induced mobilization of circulating progenitor cells by pre-treatment with IL-3 and harvest by apheresis. *Int. J. Artif. Organs* **16 (supplement 5)**, 25–29.

Hoekman K., Von Bloomberg-Van der Flier, B.M.E., Wagstaff, J., Drexhage, H.A. and Pinedo, H.M. (1991). Reversible thyroid dysfunction during treatment with GM-CSF. *Lancet* **338**, 541–542.

Hohaus, S., Petrovick, M.S., Voso, M.T., Sun, Z., Zhang, D. and Tenen, D.G. (1995). PU.I (Spi-I) and C/EBPα regulate expression of the granulocyte–macrophage colony-stimulating factor receptor α gene. *Mol. Cell. Biol.* **15**, 5830–5845.

Hovgaard, D. and Nissen, N.I. (1992). Effect of recombinant human granulocyte–macrophage colony-stimulating factor in patients with Hodgkin's disease. A phase I/II study. *J. Clin. Oncol.* **10**, 390–397.

Hsu, N., Young, L.S. and Bermudez, L.E. (1995). Response to stimulation with recombinant cytokines and synthesis of cytokines by murine intestinal macrophages infected with *Mycobacterium avium* complex. *Infect. Immun.* **63**, 528–533.

Huan, S., Hester, J., Spitzer, G., Yau, J.C., Dunphy, F.R., Wallerstein, R.O., Dicke, K., Spencer, V., LeMaistre, C.F. and Andersson, B.S. (1992). Influence of mobilized peripheral blood cells on the hematopoietic recovery by autologous marrow and recombinant human granulocyte–macrophage colony-stimulating factor after high-dose cyclophosphamide, etoposide, and cisplatin. *Blood* **79**, 3388–3393.

Huebner, K., Isobe, M., Croce, C.M., Golde, D.W., Kaufman, S.E. and Gasson, J.C. (1985). The human gene encoding GM-CSF is at 5q21-q32, the chromosome region deleted in the 5q-anomaly. *Science* **230**, 1282–1285.

Huffman, J.A., Hull, W.M., Dranoff, G., Mulligan, R.C. and Whitsett, J.A. (1996). Pulmonary epithelial cell expression of GM-CSF corrects the alveolar proteinosis in GM-CSF-deficient mice. *Am. Soc. Clin. Invest.* **97**, 649–655.

Inaba, K., Steinman, R.M., Pack, M.W., Aya, H., Inaba, M., Sudo, T., Wolpe, S. and Schuler, G. (1992). Identification of proliferating dendritic cell precursors in mouse blood. *J. Exp. Med.* **175**, 1157–1167.

Itoh, T., Muto, A., Watanabe, S., Miyajima, A., Yokota, T. and Arai, K. (1996). Granulocyte–macrophage colony-stimulating factor provokes RAS activation and transcription of c-*fos* through different modes of signaling. *J. Biol. Chem.* **271**, 7587–7592.

Iversen, P.O., To, L.B. and Lopez, A.F. (1996). Apoptosis of hemopoietic cells by the human granulocyte–macrophage colony-stimulating factor mutant E21R. *Proc. Natl Acad. Sci. U.S.A.* **93**, 2785–2789.

Jaffee, E.M., Dranoff, G., Cohen, L.K., Hauda, K.M., Clift, S., Marshall, F.F., Mulligan, R.C. and Pardoll, D.M. (1993). High efficiency gene transfer into primary human tumor explants without cell selection. *Cancer Res.* **53 (supplement 10)**, 2221–2226.

Jaschke, E., Zabernigg, A. and Gattringer, C. (1996). Low-dose recombinant human granulocyte–macrophage colony-stimulating factor in the local treatment of chronic wounds. *Proceedings of the 6th European Conference On Advances in Wound Management*, October 1996, Amsterdam, pp. 6–7.

Johnson, G.R., Gonda, T.J., Metcalf, D., Hariharan, I.K. and Cory, S. (1989). A lethal myeloproliferative syndrome in mice transplanted with bone marrow cells infected with a retrovirus expressing granulocyte–macrophage colony stimulating factor. *EMBO J.* **8**, 441–448.

Jones T.C. (1994). Future uses of granulocyte–macrophage colony-stimulating factor (GM-CSF). *Stem Cell* **12 (supplement 1)**, 229–240.

Jones, T.C. (1996). The effect of granulocyte–macrophage colony-stimulating factor (rGM-CSF) on macrophage function in microbial disease. *Med. Oncol.* **13**, 141–147.

Jyung, R.W., Wu, L, Pierce, G.F. and Mustoe, T.A. (1994). Granulocyte–macrophage colony-stimulating factor and granulocyte colony-stimulating factor: differential action on incisional wound healing. *Surgery* **115**, 325–334.

Kremer, E., Baker, E., D'Andrea, R.J., Slim, R., Phillips, H., Moretti, P.A.B., Lopez, A.F., Petit, C., Vadas, M.A. and Sutherland, G.R. (1993). A cytokine receptor gene cluster in the X–Y pseudoautosomal region? *Blood* **82**, 22–28.

Kucukcelebi, A., Carp, S.S., Hayward, P.G., Hui, P.S., Cowar, W.T., Cooper, D.M. and Robson, M.C. (1992). Granulocyte–macrophage colony-stimulating factor reverses the inhibition of wound contraction caused by bacterial contamination. *Wounds: A Compendium of Clinical Research and Practice* **4**, 241–247.

Kurzrock, R., Talpaz, M. and Gutterman, J.U. (1992). Very low doses of GM-CSF administration alone or with erythropoietin in aplastic anemia. *Am. J. Med.* **93**, 41–48.

Lang, R.A., Metcalf, D., Cuthbertson, R.A., Lyons, I., Stanley, E., Kelso, A., Kannourakis, G., Williamson, D.J., Klintworth, G.K., Gonda, T.J. and Dunn, A.R. (1987). Transgenic mice expressing a hemopoietic growth factor gene (GM-CSF) develop accumulations of macrophages, blindness, and a fatal syndrome of tissue damage. *Cell* **51**, 675–686.

Larsen, C.P., Ritchie, S.C., Hendrix, R., Linsly, P.S., Hathcock, K.S., Hodes, R.J., Lowry, R.P. and Pearson, T.C. (1994). Regulation of immunostimulatory function and costimulatory molecule (B7-1 and B7-2) expression on murine dendritic cells. *J. Immunol.* **152**, 5208–5219.

Levine, J.D., Allan, J.D, Tessitore, J.H., Falcone, N. and Galasso, F. (1991). Recombinant granulocyte–macrophage colony-stimulating factor ameliorates zidovuline-induced neutropenia in patients with AIDS/AIDS related complex. *Blood* **78**, 48–54.

Liehl, E. (1991). A preclinical perspective of GM-CSF. In *Breakthrough of Cytokine Therapy: An Overview of GM-CSF*, Part 2 (ed. J. Scarffe), International and Symposium Series Nb170. Royal Society of Medicine Services, London, pp. 17–25..

Liehl, E., Hildebrandt, J., Lam, C. and Mayer, P. (1994). Predication of the role of granulocyte–macrophage colony-stimulating factor in animals and man from *in vitro* results. *Eur. J. Clin. Microbiol. Infect. Dis.* **13 (supplement 2)**, 9–17.

Lieschke, G.J., Maher, D., Cebon, J., O'Conner, M., Green, M., Sheridan, W., Boyd, A., Rollings, M., Bonnen, E., Metcalf, D., Burgess, A.W., McGraith, K., Fox, R.M. and Morstyn, G. (1989). Effects of bacterially synthesized recombinant human granulocyte–macrophage colony-stimulating factor in patients with advanced malignancy. *Ann. Intern. Med.* **110**, 357–364.

Lieschke, G.J., Stanley, E., Grail, D., Hodgson, G., Sinickas, V., Gall, J.A.M., Sinclair, R.A. and Dunn, A.R. (1994). Mice lacking both macrophage- and granulocyte–macrophage colony-stimulating factor have macrophages and coexistent osteopetrosis and severe lung disease. *Blood* **84**, 27–35.

Lindemann, A., Riedel, D., Oster, W., Meuer, S.C., Blohn, D., Mertelsmann, R.M. and Hermann, F. (1988). Granulocyte–macrophage colony stimulating factor induces interleukin-1 production by human polymorphonuclear neutrophils. *J. Immunol.* **140**, 837–839.

Locasciulli, A., van't Veer, L., Bacigalupo, A., Hows, J., van Lint, M.T., Gluckman, E., Nissen, C., McCann, S., Vossen, J. and Schrezenmeier, A. (1990). Treatment with marrow transplantation or immunosuppression of childhood acquired severe aplastic anemia: a report of the EBMT-SAA working party. *Bone Marrow Transplant* **6**, 212–221.

Lord, B.I., Bronchud, M.H., Owens, S., Chang, J., Howell, A., Souza, L. and Dexter, T.M. (1989). The kinetics of human granulopoiesis following treatment with granulocyte colony stimulating factor *in vivo*. *Proc. Natl Acad. Sci. U.S.A.* **86**, 9499–9503.

Lord, B.I., Gurney, H., Chang, J., Thatcher, N., Crowther, D. and Dexter, T.M. (1992). Hematopoietic cell kinetics in humans treated with rGM-CSF. *Int. J. Cancer* **50**, 26–31.

Maher, D.W., Lieschke, G.J., Green, M., Bishop, J., Stuart-Harris, R., Wolf, M., Sheridan, W.P., Kefford, R.F., Cebon, J. and Olver, I. (1994). A double-blind placebo-controlled trial of filgrastim (r-metHuG-CSF) in addition to antibiotic therapy in patients with chemotherapy induced neutropenia complicated by fever. *Ann. Intern. Med.* **121**, 492–501.

Mahvi, D.M., Burkholder, J.K., Turner, J., Culp, J., Malter, J.S., Sondel, P.M. and Yang, N.S. (1996). Particle-mediated gene transfer of granulocyte–macrophage colony-stimulating factor cDNA to tumor cells: implications for a clinically relevant tumor vaccine. *Hum. Gene Ther.* **7**, 1535–1543.

Mandujano, J.F., D-Souza, N.B., Nelson, S., Summer, W.R., Beckerman, R.C. and Sellito, J.E. (1995). Granulocyte–macrophage colony-stimulating factor and *Pneumocystis carinii* pneumonia in mice. *Am. J. Respir. Crit. Care Med.* **151**, 1233–1238.

Masucci, G. (1996). New clinical applications of GM-CSF. *Med. Oncol.* **13**, 149–154.

Masucci, G., Wersaell, P., Ragnhammar, P. and Mellstedt, H. (1989). GM-CSF augments the cytotoxic capacity of lymphocytes and monocytes in ADCC. *Cancer Immunol. Immunother.* **29**, 288–292.

Mayer, P., Lam, C., Obenaus, H., Liehl, E. and Besemer, J. (1987). Recombinant human GM-CSF induces leucocytosis and activates peripheral blood polymorphonuclear neutrophils in nonhuman primates. *Blood* **70**, 206–213.

Mayer, P., Schütze, E., Lam, C., Kricek, F. and Liehl, E. (1991). Recombinant murine granulocyte–macrophage colony-stimulating factor augments neutrophil recovery and enhances resistance to infections in myelo-suppressed mice. *J. Infect. Dis.* **163**, 584–590.

Mayordomo, J.I., Zorina, T., Storkus, W.J., Zitvogel, L., Celluzi, C., Falo, L.D., Melief, C.J., Ildstad, S.T., Kast, W.M. and Deleo, A.B. (1995). Bone marrow derived dendritic cells pulsed with synthetic tumour peptides elicit protective and therapeutic antitumour immunity. *Nature Med.* **I**, 1297–1302.

McNiece, I.K., Langley, K.E. and Zsebo, K.M. (1991). Recombinant human stem cell factor synergizes with GM-CSF, G-CSF, IL-3 and erythropoietin to stimulate human progenitor cells of the myeloid and erythroid lineages. *Exp. Hematol.* **19**, 226–231.

Metcalf, D. (1984). *The Colony Stimulating Factors.* Elsevier, Amsterdam.

Metcalf, D., Begley, C.J., Johnson, G.R., Nicola, N.A., Vadas, M.A., Lopez, A.F., Williamson. D.J., Wong, G.G., Clark, S.C. and Wang, E.A. (1986). Biologic properties *in vitro* of a recombinant human granulocyte–macrophage colony-stimulating factor. *Blood* **67**, 37–45.

Metcalf, D., Begley, C.G., Williamson, D.J., Nice, E.C., DeLamarter, J., Mermod, J-J., Thatcher, D. and Schmidt, A. (1987). Hemopoietic responses in mice injected with purified recombinant murine GM-CSF. *Exp. Hematol.* **15**, 1–9.

Molloy, R.G., Holzheimer, R., Nestor, M. *et al*. (1995). Granulocyte–macrophage colony-stimulating factor modulates immune function and improves survival after experimental thermal injury. *Br. J. Surg.* **82**, 770–776.

Monroy, R.L., Skelly, R.R., MacVittie, T.J., Davis, T.A., Sauber, J.J., Clark, S.C. and Donahue, R.E. (1987). The effect of recombinant GM-CSF on the recovery of monkeys transplanted with autologous bone marrow. *Blood* **70**, 1696–1699.

Morrissey, P.J., Bressler, L., Park, L.S., Alpert, A. and Gillis, S. (1987). Granulocyte–macrophage colony-stimulating factor augments the primary antibody response by enhancing the function of antigen presenting cells. *J. Immunol.* **139**, 1113–1119.

Muench, M.O. and Moore, M.A. (1992). Accelerated recovery of peripheral blood cell counts in mice transplanted with *in vitro* cytokine-expanded hematopoietic progenitors. *Exp. Hematol.* **20**, 611–618.

Mui, A.L., Wakao, H., Harada, N., O'Farrell, A.-M. and Miyajima, A. (1995). Interleukin-3, granulocyte–macrophage colony-stimulating factor, and interleukin-5 transduce signals through two forms of STAT5. *J. Leukoc. Biol.* **57**, 799–803.

Musso, M., Ghiorzo, P., Fiorentini, P., Giuffrida, R., Ciotti, P., Carré, C., Ravazzolo, R. and Bianchi-Scarrà, G. (1996). An upstream positive regulatory element in human GM-CSF promoter is recognized by NF-κB/Rel family members. *Biochem. Biophys. Res. Commun.* **223**, 64–72.

Nagler, A., Shur, I., Barak, V. and Fabian, I. (1996). Granulocyte–macrophage colony-stimulating factor dependent monocyte-mediated cytotoxicity post-autologous bone marrow transplantation. *Leuk. Res.* **20**, 637–643.

Nakagawa, Y., Kosugi, H., Miyajima, A., Arai, K. and Yokota, T. (1994). Structure of the gene encoding the α subunit of the human granulocyte–macrophage colony stimulating factor receptor. *J. Biol. Chem.* **269**, 10905–10912.

Nathan, F.E. and Besa, E.C. (1992). GM-CSF and accelerated hemolysis. *N. Engl. J. Med.* **326**, 417.

Nemunaitis, J., Singer, J.W., Buckner, C.D., Hill, R., Storb, R., Thomas, E.D. and Appelbaum, F.R. (1990). Use of recombinant human granulocyte–macrophage colony-stimulating factor in graft failure after bone marrow transplantation. *Blood* **76**, 245–253.

Nemunaitis, J., Rabinowe, S.N., Singer, J.W., Bierman, P.J., Vose, J.M., Freedman, A.S., Onetto, N., Gillis, S., Oette, D., Gold, M., Buckner, D., Hansen, J.A., Ritz, J., Appelbaum, F.R., Armitage, J.O. and Nadler, L.M. (1991). Recombinant granulocyte–macrophage colony-stimulating factor after autologous bone marrow transplantation for lymphoid cancer. *N. Engl. J. Med.* **324**, 1773–1778.

Neta, R. and Oppenheim, J.J. (1988). Cytokines in therapy of radiation injury. *Blood* **72**, 1093–1095.

Neta, R., Oppenheim, J.J. and Douches, S.C. (1988). Interdependence of the radioprotective effects of human recombinant interleukin alpha, tumor necrosis factor alpha, granulocyte colony stimulating factor, and murine recombinant granulocyte–macrophage colony stimulating factor. *J. Immunol.* **140**, 108–110.

Nienhuis, A.W., Donahue, R.E., Karlsson, S., Clark, S.C., Agricola, B., Antinoff, N., Pierce, J.E., Turner, P., Anderson, W.F. and Nathan, D.G. (1987). Recombinant human granulocyte macrophage colony-stimulating factor (GM-CSF) shortens the period of neutropenia after autologous bone marrow transplantation in a primate model. *J. Clin. Invest.* **80**, 573–577.

Nishinakamura, R., Nakayama, N., Hirabayashi, Y., Inoue, T., Aud, D., McNeil, T., Azuma, S., Yoshida, S., Toyoda, Y., Arai, K., Miyajima, A. and Murray, R. (1995). Mice deficient for the IL-3/GM-CSF/IL-5 βc receptor exhibit lung pathology and impaired immune response, while β_{IL3} receptor-deficient mice are normal. *Immunity* **2**, 211–222.

Okuda, K., Sanghera, J.S., Pelech, S.L., Kanakura, Y., Hallek, M., Griffin, J.D. and Druker, B.J. (1992). Granulocyte–macrophage colony-stimulating factor, interleukin-3, and Steel factor induce rapid tyrosine phosphorylation of p42 and p44 MAP kinase. *Blood* **79**, 2880–2887.

O'Reilly, M., Silver, G.M., Gamelli, R.L., Davis, J.H. and Hebert, J.C. (1994). Dose dependency of granulocyte–macrophage colony-stimulating factor for improving survival following burn wound infection. *J. Trauma* **36**, 486–490.

Patrone, F., Ballestrero, A., Ferrando, F., Brema, F., Moraglio, L., Valbonesi, M., Basta, P., Ghio, R., Gobbi, M. and Sessarego, M. (1995). Four-step high-dose sequential chemotherapy with double hematopoietic progenitor-cell rescue for metastatic breast cancer. *J. Clin. Oncol.* **13**, 840–846.

Perno, C.F., Yarchoan, R., Cooney, D.A., Harman, N.R. and Webb, D.S.A. (1989). Replication of human immunodeficiency virus in monocytes; granulocyte–macrophage colony-stimulating factor potentiates viral replication yet enhances the anti-viral effect mediated by 3′azido-2′3′dideoxythymidine and other dideoxynucleoside congeners of thymidine. *J. Exp. Med.* **169**, 933–951.

Piemonti, L., Bernasconi, S., Luini, W., Trobonjaca, Z., Minty, A., Allavena, P. and Mantovani, A. (1995). IL-13 supports differentiation of dendritic cells from circulating precursors in concert with GM-CSF. *Eur. Cytokine Netw.* **6**, 245–252.

Pieters, R.C., Rojer, R.A., Saleh, A.W., Saleh, A.E. and Duits, A.J. (1995). Molgramostim to treat SS-sickle cell leg ulcers. *Lancet* **345**, 528.

Pojda, Z. and Struzyna, J. (1994). Treatment of non-healing ulcers with rhGM-CSF and skin grafts. *Lancet* **343**, 1100.

Polotskaya, A., Zhao, Y., Lilly, M.L. and Kraft, A.S. (1993). A critical role for the cytoplasmic domain of the granulocyte–macrophage colony-stimulating factor α receptor in mediating cell growth. *Cell Growth Differ.* **4**, 523–531.

Porcelli, S.A. and Modlin, R.L. (1995). CD1 and the expanding universe of T cell antigens. *Am. Assoc. Immunol.* **155**, 3709–3710.

Ragnhammar, P. and Wadhwa, M. (1996). Neutralising antibodies to granulocyte–macrophage colony stimulating factor (GM-CSF) in carcinoma patients following GM-CSF combination therapy. *Med. Oncol.* **13**, 161–166.

Ragnhammar, P., Friesen, H.-J., Frödin, J.-E., Lefvert, A.-K., Hassan, M., Oesterborg, A. and Mellstedt, H. (1994). Induction of anti-recombinant human granulocyte–macrophage colony-stimulating factor (*Escherichia coli*-derived) antibodies and clinical effects in nonimmunocompromised patients. *Blood* **84**, 4078–4087.

Rasko, J.E.J. and Gough, N.M. (1994). Granulocyte–macrophage colony stimulating factor. In *The Cytokine Handbook*, 2nd edn. (ed. A.W. Thomson), Academic Press, London, pp. 343–369.

Reid, C.D., Stackpoole, A., Meager, A. and Tikerpae, J. (1992). Interactions of tumor necrosis factor with granulocyte–macrophage colony-stimulating factor and other cytokines in the regulation of dendritic cell growth *in vitro* from early bipotent CD34+ progenitors in human bone marrow. *J. Immunol.* **149**, 2681–2688.

Richard, C., Baro, J., Bello-Fernandez, C., Hermida, G., Calavia, J., Olalla, I., Alsar, M.J., Loyola, I., Cuadrado, M.A. and Iriondo, A. (1995). Recombinant human granulocyte–macrophage colony-stimulating factor (rhGM-CSF) administration after autologous bone marrow transplantation for acute myeloblastic leukemia enhances killer cell function and may diminish leukemic relapse. *Bone Marrow Transplant.* **15**, 721–726.

Riikonen, P., Saarinen, U.M., Makipernaa, A., Hovi, L., Komulainen, A., Pihkala, J. and Jalanko, H. (1994). Recombinant human granulocyte–macrophage colony-stimulating factor in the treatment of fever and neutropenia: a double-blind placebo controlled study in children with malignancy. *Pediatr. Infect. Dis. J.* **13**, 197–200.

Risti, B., Flury, R.F. and Schaffner, A. (1994). Fatal hematophagic histiocytosis after granulocyte–macrophage colony-stimulating factor and chemotherapy for high-grade malignant lymphoma. *Clin. Invest.* **72**, 457–461.

Robinson, D.S., Durham, S.R. and Kay, A.B. (1993). Cytokines in asthma. *Thorax* **48**, 845–853.

Robison, R.L. and Myers, L.A. (1993). Preclinical safety assessment of recombinant human GM-CSF in rhesus monkeys. *Int. Rev. Exp. Pathol.* **34A**, 149–172.

Robson, M., Kucukcelebi, A., Carp S.S., Hayward, P.G., Hui, P.S., Cowan, W.T., Ko, F. and Cooper, D.M. (1994). Effects of granulocyte–macrophage colony-stimulating factor on wound contraction. *Eur. J. Clin. Microbiol. Infect. Dis.* **13 (supplement 2)**, S41–61.

Roilides, E., Blake, C., Holmes, A., Pizzo, A. and Walsh, T.J. (1996). Granulocyte–macrophage colony-stimulating factor and interferon-gamma prevent dexamethasone-induced immunosuppression of antifungal monocyte activity against *Aspergillus fumigatus* hyphae. *J. Med. Vet. Mycol.* **34**, 63–69.

Rosenthal, N.S. and Farhi, D.C. (1994). Failure to engraft after bone marrow transplantation: bone marrow morphologic findings. *Am. J. Clin. Pathol.* **102**, 821–824.

Rowe, J.M., Andersen, J.W., Mazza, J.J., Bennett, J.M., Paietta, E., Hayes, F.A., Oette, D., Cassileth, P.A., Stadtmauer, E.A. and Wiernik, P.H. (1996). A randomized placebo-controlled phase III study of granulocyte–macrophage colony-stimulating factor in adult patients (> 55 to 70 years of age) with acute myelogenous leukemia. *Blood* **86**, 457–462.

Ruef, C. and Coleman, D.L. (1990). Granulocyte–macrophage colony-stimulating factor: pleiotropic cytokine with potential clinical usefulness. *Rev. Infect. Dis.* **12**, 41–62.

Sakamoto, K.M., Fraser, J.F., Lee, H.J., Lehman, E. and Gasson, J.C. (1994). Granulocyte–macrophage colony-stimulating factor and interleukin-3 signaling pathways converge on the CREB-binding site in the human egr-1 promoter. *Mol. Cell. Biol.* **14**, 5975–5985.

Sallusto, F. and Lanzavecchia, A. (1994). Efficient presentation of soluble antigen by cultured human dendritic cells is maintained by granulocyte/macrophage colony-stimulating factor plus interleukin 4 and downregulated by tumor necrosis factor alpha. *J. Exp. Med.* **179**, 1109–1118.

San-Miguel, J.F., Hernandez, M.D., Gonzalez, M., Lopez-Berges, M.C., Caballero, M.D., Vazquez, L., Orfao, A., Nieto, M.J., Corral, M. and del Canizo, M.C. (1996). A randomized study comparing the effect of GM-CSF and G-CSF on immune reconstitution after autologous bone marrow transplantation. *Br. J. Haematol.* **94**, 140–147.

Santoli, D., Clark, S.C., Kreider, B.L., Maslin, P.A. and Rovera, G. (1988). Amplification of IL-2-driven T cell proliferation by recombinant human IL-3 and granulocyte–macrophage colony-stimulating factor. *J. Immunol.* **141**, 519–526.

Scadden, D.T. (1992). The use of GM-CSF in AIDS. *Infection* **20**, 103–106.

Schneider, J.G., Crown, J.P., Wasserheit, C., Kritz, A., Wong, G., Reich, L., Norton, L. and Moore, M.A. (1994). Factors affecting the mobilization of primitive and committed hematopoietic progenitors into the peripheral blood of cancer patients. *Bone Marrow Transplant.* **14**, 877–884.

Schoepfer, C., Carla, H., Bezou, M.J., Canbon, M., Girault, D., Demeocq, F. and Malpnech, G. (1995). Malassezia fungemia after bone marrow transplantation. *Arch. Pediatr.* **2**, 245–248.

Scrip (1997). Leukine increases survival in advanced melanoma patients. *Scrip* online database accession no. S00531014 19970321.

Seymour, J.F., Dunn, A.R., Vincent, J.M., Presneill, J.J. and Pain, M.C. (1996). Efficacy of granulocyte–macrophage colony stimulating factor in acquired alveolar proteinosis. *N. Engl. J. Med.* **335**, 1924–1925.

Shamseddin, A.I. (1994). Granulocyte–macrophage colony-stimulating factor (GM-CSF) for treatment of chemotherapy extravasation. *Proc. Am. Soc. Clin. Oncol.* **13**, 458, Abstract no. 1588.

Shea, T.C., Mason, J.R., Storniolo, A.M., Newton, B., Breslin, M., Mullen, M., Ward, D.M., Miller, L., Christian, M. and Taetle, R. (1992). Sequential cycles of high-dose carboplatin administered with recombinant human granulocyte–macrophage colony-stimulating factor and repeated infusions of autologous peripheral-blood progenitor cells: a novel and effective method for delivering multiple courses of dose-intensive therapy. *J. Clin. Oncol.* **10**, 464–473.

Shen, Y., Baker, E., Callen, D.F., Sutherland, G.R., Willson, T.A., Rakar, S. and Cough, N.M. (1992). Localization of the human GM-CSF receptor beta chain gene (CSF2RB) to chromosome 22q12.2→q13.1. *Cytogenet. Cell Genet.* **61**, 175–177.

Sheridan, W.P., Begley, C.G., Juttner, C.A., Szer, J., To, L.B., Maher, D., McGrath, K.M., Morstyn, G. and Fox, R.M. (1992). Effect of peripheral blood progenitor cells mobilized by filgrastin (G-CSF) on platelet recovery after high-dose chemotherapy. *Lancet* **339** (8794), 640–644.

Sieff, C.A., Emerson, S.G., Donahue, R.E., Nathan, D.G., Wang, E.A., Wong, G.G. and Clark, S.C. (1985). Human recombinant granulocyte–macrophage colony stimulating factor: a multilineage hematopoietin. *Science* **230**, 1171–1173.

Siena, S., Bregni, M., Bonsi, L., Strippoli, P., Peccatori, F., Magni, M., Di Nicola, M., Bagnara, G.P. and Massimo-Gianni, A. (1993). Clinical implications of the heterogeneity of hematopoietic progenitors elicited in peripheral blood by anticancer therapy with cyclophosphamide and cytokines. *Stem Cells* **11 (supplement 2)**, 72–75.

Socinski, M.A., Cannistra, S.A., Elias, A., Antman, K.H., Schipper, L. and Griffin, J.D. (1988). Granulocyte–macrophage colony-stimulating factor expands the circulating hematopoietic progenitor cell compartment in man. *Lancet* **i**, 1194–1198.

Soldi, R., Primo, L., Brizzi, M.F., Sanavio, F., Aglietta, M., Polentarutti, N., Pegoraro, L., Mantovani, A. and Bussolino, F. (1997). Activation of JAK2 in human vascular endothelial cells by granulocyte–macrophage colony-stimulating factor. *Blood* **89**, 863–872.

Spielberger, R., Falleroni, M.J., Coene, A.J. and Larson, R.A. (1993). Concomitant amphotericin B therapy, granulocyte transfusion and GM-CSF administration for disseminated infection with fusarium in a granulocytopenic patient. *Clin. Infect. Dis.* **16**, 528–530.

Spielholz, C., Haeney, M.L., Morrison, M.E., Houghthon, A.N., Vera, J.C. and Golde, D.W. (1995). Granulocyte–macrophage colony-stimulating factor signals for increased glucose uptake in human melanoma cells. *Blood* **85**, 973–980.

Sprikkelman, A., deWolf, J.T. and Vellenga, E. (1994). The application of hematopoietic growth factors in drug-induced agranulocytosis: a review of 70 cases. *Leukemia* **8**, 2031–2036.

Stahl, C.P., Winton, E.F., Monroe, M.C., Holman, R.C., Zelasky, M., Liehl, E., Myers, L.A., McClure, H., Anderson, D. and Evatt, B.L. (1991). Recombinant human granolocyte–macrophage colony-stimulating factor promotes megakaryocyte maturation in nonhuman primates. *Exp. Hematol.* **19**, 810–816.

Stahl, C.P., Winton, E.F., Monroe, M.C., Haff, E., Holman, R.C., Myers, L., Liehl, E. and Evatt, B.L. (1992). Differential effects of sequential, simultaneous, and single agent interleukin-3 and granulocyte–macrophage colony-stimulating factor on megakaryocyte maturation and platelet response in primates. *Blood* **80**, 2479–2485.

Stern, A.C. and Jones, T.C. (1990). Role of human recombinant GM-CSF in the prevention and treatment of leucopenia with special reference to infectious diseases. *Diagn. Microbiol. Infect. Dis.* **13**, 391–396.

Stewart-Akers, A.M., Cairns, J.S., Tweardy, D.J. and McCarthy, S.A. (1993). Effect of granulocyte–macrophage colony-stimulating factor on lymphokine-activated killer cell induction. *Blood* **81**, 2671–2678.

Stone, R.M., Berg, D.T., George, S.L., Dodge, R.K., Paciucci, P.A., Schulman, P., Lee, E.J., Moore, J.O., Powell, B.L. and Schiffer, C.A. (1995). Granulocyte–macrophage colony-stimulating factor after initial chemotherapy for elderly patients with primary acute myelogenous leukemia. *N. Engl. J. Med.* **332**, 1671–1677.

Strobl, H., Riedl, E., Scheinecker, C., Bello-Fernandez, C., Pickl, W.F., Rappersberger, K., Majdic, O. and Knapp, W. (1996). TGF-β promotes *in vitro* development of dendritic cells from CD34$^+$ hemopoietic progenitors. *J. Immunol.* **157**, 1499–1507.

Stryckmans, P.A., Dumont, J.P., Valu, T.H. and Debusscher, L. (1984). Cyclosporine in refractory severe aplastic anemia. *N. Engl. J. Med.* **310**, 656–657.

Sung, P.K., Yang, L., Elices, M., Jin, G., Sriramarao, P. and Broide, D.H. (1997). Granulocyte–macrophage colony-stimulating factor regulates the functional adhesive state of very late antigen-4 expressed by eosinophils. *J. Immunol.* **158**, 919–927.

Szabolcs, P., Moore, M.A.S. and Young, J.W. (1995). Expansion of immunostimulatory dendritic cells among myeloid progeny of human CD34$^+$ bone marrow precursors, cultured with c-kit ligand, granulocyte–macrophage colony-stimulating factor, and TNF-α. *J. Immunol.* **154**, 5851–5861.

Tafuto, S., Abate, G., D'Andrea, P., Silvestri, I., Marcolin, P., Volta, C., Monteverde A., Colombi, S., Andorno, S. and Aglietta, M. (1995). A comparison of two GM-CSF schedules to counteract the granulo-monocytopenia of carboplatin–etoposide chemotherapy. *Eur. J. Cancer* **31A**, 46–49.

Tao, M.H. and Levy, R. (1993). Idiotype/granulocyte–macrophage colony-stimulating factor fusion protein as a vaccine for B-cell lymphoma. *Nature* **362**, 755–758.

Tarr, P.E., Lin, R. and Jones, T.C. (1996a). Vaccine adjuvancy. In *Manual of GM-CSF* (ed. M. Marty), Blackwell Science, Oxford, pp. 219–232.

Tarr, P.E., Lin R., Mueller, E.A., Kovarik, J.M., Guillaume, M. and Jones, T.C. (1996b). Evaluation of tolerability and antibody response after recombinant human granulocyte–macrophage colony-stimulating factor (rhGM-CSF) and a single dose of recombinant hepatitis B vaccine. *Vaccine* **14**, 1199–1204.

Torigoe, T., O'Conner, R., Santoli, D. and Reed, J.C. (1992). Interleukin-3 regulates the activity of the Lyn protein-tyrosine kinase in myeloid-committed leukemic cell lines. *Blood* **80**, 617–624.

Tracey, K.J., Beutler, B., Lowry, SF., Merryweather, J., Wolpe, S., Milsork, I.W., Hariri, R.J., Fahey, T.J. III, Zentella, A., Albert, J.D., Shires, T. and Cerami, A. (1986). Shock and tissue injury induced by recombinant human cachetin. *Science* **234**, 470–474.

Ulich, T.R., del Castillo, J., McNiece, I., Watson, L., Yin, S. and Andresen, J. (1990). Hematologic effects of recombinant murine granulocyte–macrophage colony-stimulating factor on the peripheral blood and bone marrow. *Am. J. Pathol.* **137**, 369–376.

Vannucchi, A.M., Grossi, A., Rafanelli, D. and Rossi Ferrini, P. (1990). *In vivo* stimulation of megakaryocytopoiesis by recombinant murine granulocyte–macrophage colony-stimulating factor. *Blood* **76**, 1473–1480.

Vellanga, E., Young, D.C., Wagner, K., Wier, D., Ostapovicz, D. and Griffin, J.D. (1987). The effects of GM-CSF and G-CSF in promoting growth of clonogenic cells in acute myeloblastic leukemia. *Blood* **69**, 1771.

Vesole, D.H., Jagannath, S., Gleen, L.D. and Barlogie, B. (1994). Effect of low-dose granulocyte–macrophage colony-stimulating factor (LD-GM-CSF) on platelet transfusion-dependent thrombopenia. *Am. J. Hematol.* **47**, 203–207.

Vremec, D., Lieschke, G.J., Dunn, A.R., Robb, L., Metcalf, D. and Shorman, K. (1997). The influence of granulocyte/macrophage colony-stimulating factor on dendritic cell levels in mouse lymphoid organs. *Eur. J. Immunol.* **27**, 40–44.

Vyalov, S., Desmouliere, A. and Gabbiani, G. (1993). GM-CSF-induced granulation tissue formation: relationship between macrophage and myofibroblast accumulation. *Virchows Arch.* **63**, 231–239.

Walker, C., Bauer, W., Braun, R.K., Menz, G., Braun, P. Schwarz, F., Hansel, T.T. and Villiger, B. (1994). Activated T cells and cytokines in bronchoalveolar lavages from patients with various lung diseases associated with eosinophils. *Am. J. Respir. Crit. Care Med.* **150**, 1038–1048.

Wang, J.M., Colella, S., Alavena, P. and Mantovani, A. (1987). Chemotactic activity of human recombinant granulocyte–macrophage colony-stimulating factor. *Immunology* **60**, 439–444.

Watanabe, S., Itoh, T. and Arai, K. (1996). JAK2 is essential for activation of c-fos and c-myr promoters and cell proliferation through the human granulocyte–macrophage colony-stimulating factor receptor in BA/F3 cells. *J. Biol. Chem.* **271**, 12 681–12 686.

Wei, S., Liu, J.H., Epling-Burnette, P.K., Gamero, A.M., Ussery, D., Pearson, E.W., Elkabani, M.E., Diaz, J.I. and Djeu, J.Y. (1996). Critical role of Lyn kinase in inhibition of neutrophil apoptosis by granulocyte–macrophage colony-stimulating factor. *J. Immunol.* **157**, 5155–5162.

Weisdorf, D.J., Verfaillie, C.M., Davies, S.M., Filipovich, A.H., Wagner, J.E. Jr., Miller, J.S., Burroughs, J., Ramsay, N.K., Kersey, J.H. and McGlave, P.B. (1995). Hematopoietic growth factors for graft failure after bone marrow transplantation: a randomized trial of granulocyte–macrophage colony stimulating factor (GM-CSF) *versus* sequential GM-CSF plus G-CSF. *Blood* **85**, 3452–3456.

Welham, M.J., Duronio, V., Sanghera, J.S., Pelech, S.L. and Schrader, J.W. (1992). Multiple hemopoietic growth factors stimulate activation of mitogen-activated protein kinase family members. *J. Immunol.* **149**, 1683–1693.

Witmer-Pack, M.D., Olivier, W., Valinsky, J., Schuler, G. and Steinman, P.M. (1987). Granulocyte–macrophage colony stimulating factor is essential for the viability and function of cultured murine epidermal Langerhans cells. *J. Exp. Med.* **166**, 1484–1498.

Wong, G.C., Witek, J.S. and Temple, P.A. (1985). Human GM-CSF: molecular cloning of the complementary DNA and puring of the natural and recombinant proteins. *Science* **228**, 810–815.

Woodcock, J.M., Bagley, C.J., Zacharakis, B. and Lopez, A.F. (1996). A single tyrosine residue in the membrane-proximal domain of the granulocyte–macrophage colony-stimulating factor, interleukin (IL)-3, and IL-5 receptor common beta-chain is necessary and sufficient for high affinity binding and signaling by all three ligands. *J. Biol. Chem.* **271**, 25 999–26 006.

Wunder, E., Sovalat, H., Baerenzung, M., Herber, G., Schweitzer, C., Kim, A. and Henon, P.H. (1991). Monocytes participate in stimulation of GM-progenitors by rhGM-CSF and placental cell supernate. *Exp. Hematol.* **19**, 473.

Xing, Z., Braciak, T., Ohkawara, Y., Sallenave, J., Foley, R., Sime, P.J., Jordana, M., Graham, F.L. and Gauldie, J.

(1996). Gene transfer for cytokine functional studies in the lung: the multifunctional role of GM-CSF in pulmonary inflammation. *J. Leukoc. Biol.* **59**, 481–488.

Yong, K.L. and Linch, D.C. (1993). Granulocyte–macrophage-colony-stimulating factor differentially regulates neutrophil migration across IL-1-activated and nonactivated human endothelium. *J. Immunol.* **150**, 2449–2456.

Young, D.C. and Griffin, J.D. (1986). Autocrine secretion of GM-CSF in acute myeloblastic leukemia. *Blood* **68**, 1178–1181.

Yousefi, S., Hoessli, D.C., Blaser, K., Mills, G.B. and Simon, H.-U. (1996). Requirement of Lyn and Syk tyrosine kinases for the prevention of apoptosis by cytokines in human eosinophils. *J. Exp. Med.* **183**, 1407–1413.

Yuo, A., Kitagawa, S., Kasahara, T., Matsushima, K., Saito, M. and Takaku, F. (1991). Stimulation and priming of human neutrophils by interleukin-8: cooperation with tumor necrosis factor and colony-stimulating factors. *Blood* **78**, 2708–2714.

Chapter 23

Granulocyte Colony Stimulating Factor

Hiroshi Murakami[1] and Shigekazu Nagata[1,2]

[1]Osaka Bioscience Institute and [2]Department of Genetics, Osaka University Medical School, Suita, Osaka, Japan

INTRODUCTION

Neutrophilic granulocytes (neutrophils) play an important role in protecting mammals from bacterial infections. Neutrophils have a short half-life, and need to be constantly replenished from pluripotent stem cells in the bone marrow. Granulocyte colony-stimulating factor (G-CSF) is a 20 000 M_r glycoprotein that regulates the production of neutrophils and enhances their maturation. G-CSF is produced by activated macrophages, endothelial cells and fibroblasts, as well as by stroma cells in the bone marrow. Large quantities of G-CSF can now be produced in *Escherichia coli* and mammalian cells. Since 1991, when G-CSF was approved as a drug by the Food and Drug Administration in the USA, it has been widely used clinically to treat patients suffering from neutropenia, which is most frequently caused by treatment with anticancer drugs or immunosuppressive agents.

G-CSF stimulates cell proliferation and differentiation by binding to its receptor. The G-CSF receptor is a type I membrane protein belonging to the hemopoietic growth factor receptor family, and is specifically expressed in the neutrophilic progenitors and mature neutrophils. The binding of a single G-CSF induces the dimerization of the receptor, and the dimerized receptor transduces proliferation and differentiation signals in cells. The G-CSF receptor does not contain a kinase domain in its cytoplasmic region, but activates the Janus kinase (JAK) family of protein tyrosine kinases, which in turn induce the tyrosine phosphorylation of many cellular proteins, including the receptor itself and several members of the STAT (signal transducers and activators of transcription) transcription factor family. Here, we summarize the biochemical properties of G-CSF and the G-CSF receptor, the physiologic roles of G-CSF, and G-CSF-dependent signal transduction.

GRANULOCYTE COLONY-STIMULATING FACTOR AND ITS RECEPTOR

Biochemical Properties of G-CSF

G-CSF was initially identified as a factor that induced the terminal differentiation of murine myelomonocytic leukemia WEHI-3BD$^+$ cells into granulocytes and monocytes.

The Cytokine Handbook, 3rd ed.
ISBN 0-12-689662-3

Murine G-CSF was originally purified from medium conditioned with the lungs from mice previously injected with endotoxin (Nicola *et al.*, 1983). It is an acidic glycoprotein with an M_r of 24 000–25 000 (pI 4.5–5.8), and is relatively stable at extreme pH levels (pH 2–10), temperature (50% loss of the activity at 70°C for 30 min) and strong denaturation agents (6 M guanidine hydrochloride, 8 M urea, 0.1% sodium dodecyl sulfate). Human G-CSF was first purified from the conditioned medium of human malignant tumor cell lines that constitutively produce large amounts of G-CSF (Welte *et al.*, 1985; Nomura *et al.*, 1986). Human G-CSF is a glycoprotein with an apparent M_r of 19 000 and a pI of 5.5–6.1, depending on the degree of sialylation (Nomura *et al.*, 1986).

The amino acid sequence of the purified human G-CSF was used as a basis for isolating cDNA from human G-CSF-producing cell lines (Nagata *et al.*, 1986a,b; Souza *et al.*, 1986). Human G-CSF is composed of 204 amino acids including a 30-amino-acid signal sequence. The calculated molecular mass of the core protein is 18 671 Da. Human G-CSF is *O*-glycosylated at Thr-133. The structure of the sugar moiety is *N*-acetylneuramic acid α(2–6)[galactose, β(1–3)] *N*-acetylgalactosamine (Souza *et al.*, 1986; Oheda *et al.*, 1988). The sugar moiety is not essential for biologic activity, as the recombinant G-CSF produced in *E. coli* is as active as native glycosylated G-CSF. However, it contributes to the stability of the molecule by preventing its aggregation (Oheda *et al.*, 1990). Human G-CSF contains five cysteine residues, four of which are connected by disulfide bonds (Cys-36–Cys-42 and Cys-64–Cys-74). They are essential for the proper folding of the molecule, and thus for its biologic activity (Lu *et al.*, 1989). Mouse, rat, bovine and dog G-CSF cDNAs have also been isolated, based on their homology with human G-CSF (Tsuchiya *et al.*, 1986; Lovejoy *et al.*, 1993; Han *et al.*, 1996). The G-CSFs from different species are more than 70% identical at the amino acid sequence level. A large quantity of recombinant human, bovine and canine G-CSF is now produced in *E. coli* and mammalian cells, and their activities fully crossreact.

G-CSF shares significant sequence homology with other cytokines such as interleukin-6 (IL-6), leukemia inhibitory factor and oncostatin M (Kishimoto *et al.*, 1994). The X-ray diffraction analysis of human, canine and bovine G-CSF, and nuclear magnetic resonance (NMR) analysis of human G-CSF have indicated that it has a four-α-helical bundle structure, as found for other cytokines such as interferon (IFN), IL-2, growth hormone and granulocyte–macrophage colony stimulating factor (GM-CSF) (Hill *et al.*, 1993; Lovejoy *et al.*, 1993; Zink *et al.*, 1994). The extensive mutational analysis of human G-CSF has shown that the *N*-terminal portion as well as the *C*-terminal portion are important for its binding to G-CSF receptor (Reidhaar-Olson *et al.*, 1996).

Genetic Properties of G-CSF and its Expression

A single G-CSF gene is localized on human chromosome 17q21–22 (Kanda *et al.*, 1987) and on mouse chromosome 11 (Buchberg *et al.*, 1988). The human and mouse G-CSF genes are about 2.5 kilobases (kb) in length and split by four introns (Nagata *et al.*, 1986b; Tsuchiya *et al.*, 1987).

Peritoneal macrophages or macrophage cell lines can be induced *in vitro* to produce G-CSF in response to tumor necrosis factor-α (TNF-α), IL-1 and bacterial endotoxin such as lipopolysaccharide (LPS) (Metcalf and Nicola, 1985; Lu *et al.*, 1988; Vellenga *et al.*, 1988; Nishizawa and Nagata, 1990). Accordingly, calvariae from the *op/op* mice, which are deficient in producing M-CSF for macrophage development, do not produce

G-CSF upon LPS stimulation (Felik *et al.*, 1995). Fibroblasts and endothelial cells produce G-CSF following stimulation by IL-1 or TNF-α (Broudy *et al.*, 1987; Koeffler *et al.*, 1987; Seelentag *et al.*, 1987; Kaushansky *et al.*, 1988). These results suggest that during the inflammatory process endotoxins stimulate macrophages to produce not only G-CSF, but also several monokines, including IL-1 and TNF-α, which then induce the release of G-CSF from fibroblasts and endothelial cells. Thus, bacterial infection causes an accumulation of G-CSF in the serum (Kawakami *et al.*, 1990), which then induces the granulopoiesis that accompanies inflammation. On the other hand, various tumor cells produce large quantities of G-CSF constitutively. These tumor cells include squamous carcinoma, bladder carcinoma, melanoma, glioblastoma, hepatoma, chronic B-cell leukemia, and human T-lymphotropic virus-1-transformed endothelial cells (Tweardy *et al.*, 1987; Nagata, 1990; Takashima *et al.*, 1996; Wang *et al.*, 1996).

G-CSF gene expression is transcriptionally and post-transcriptionally regulated (Koeffler *et al.*, 1987; Ernst *et al.*, 1989; Nishizawa and Nagata, 1990). G-CSF mRNA carries several AUUUA sequences in its 3′ non-coding region (Shaw and Kamen, 1986), through which IL-10 stimulation destabilizes G-CSF mRNA (Brown *et al.*, 1996). The transcriptional activation of the G-CSF gene is mediated by a promoter sequence located at the 5′ flanking region of the G-CSF gene. Three regulatory elements (G-CSF promoter elements GPE1, GPE2, and GPE3) in the region about 300 nucleotides upstream of the ATG initiation codon are responsible for the inducible expression of the G-CSF gene in macrophages and its constitutive expression in carcinoma cells (Nishizawa and Nagata, 1990; Nishizawa *et al.*, 1990). GPE1 is composed of a nuclear factor (NF)-κB (PuGAGPuTTCCACPu) and an NF–IL-6 binding site (TT/GNNGNAAT/G). GPE2 is the octamer transcription factor (OTF) binding site (ATTTGCAT), and GPE3 is a unique element for the G-CSF gene (Asano and Nagata, 1992). During the activation of macrophages by endotoxins, all three GPEs and their cognate transcription factors seem to be activated in a concerted fashion (Asano *et al.*, 1991). The involvement of NF–IL-6 in the expression of the G-CSF gene was recently demonstrated in cells lacking NF–IL-6: fibroblasts and macrophages from NF–IL-6-null mice (Tanaka *et al.*, 1995) or cells treated with antisense NF–IL-6 oligonucleotides (Kiehntopf *et al.*, 1995) could not produce G-CSF upon activation with LPS or TNF-α.

Biochemical Properties of the G-CSF Receptor

Expression of the G-CSF receptor is restricted to neutrophilic progenitors, mature neutrophils and various myeloid leukemia cells, all of which respond to G-CSF (Nicola and Peterson, 1986; Park *et al.*, 1989; Fukunaga *et al.*, 1990a). This agrees with the fact that G-CSF specifically regulates the production of neutrophilic granulocytes. The number of receptors is relatively low, around 300–1000 molecules per cell, and G-CSF binds to the receptor with a dissociation constant (K_d) of about 100 pM. This value is much higher than the concentration (10 pM) required for the half-maximal biologic response, indicating that biologic responses to G-CSF occur at low levels of receptor occupancy. In addition to myeloid cells, the G-CSF receptor is expressed abundantly in the placenta. As the placenta also expresses G-CSF, a role for G-CSF–G-CSF receptor in the development of the placenta has been suggested (McCracken *et al.*, 1996). However, as described below, mice deficient in G-CSF or the G-CSF receptor develop normally, suggesting that G-CSF is not required for development of the placenta.

The murine G-CSF receptor cDNA has been isolated using expression cloning (Fukunaga *et al.*, 1990b). It consists of 837 amino acids, including a 25-amino-acid signal sequence at the *N*-terminus. The polypeptide contains a single transmembrane region of 24 amino acids, which divides the molecule into extracellular and cytoplasmic regions of 601 and 187 amino acids respectively. The extracellular region appears to be heavily glycosylated, which may explain the difference between the M_r predicted from the amino acid sequence (90 814) and that measured for the native protein (100 000–130 000). The extracellular region of the G-CSF receptor has a composite structure. The *N*-terminal domain of 100 amino acids has an immunoglobulin (Ig)-like sequence. The following 200-amino-acid domain (cytokine receptor homology (CRH) domain) is related to the domain of the extracellular region of various hemopoietic growth factor receptors, as described below. This region contains four conserved cysteine residues in its *N*-terminal half, and a consensus Trp–Ser–X–Trp–Ser element in its *C*-terminal portion. Between the CRH and the transmembrane domains, the G-CSF receptor contains three fibronectin type III domains (FNIII). The cytoplasmic region of the G-CSF receptor carries no motif for enzymatic activity such as tyrosine kinase. However, two stretches of amino acids with limited sequence homology to other receptors can be found in the membrane-proximal region, and they are designated as Box1 and Box2 (Fukunaga *et al.*, 1991). The cytoplasmic region of the G-CSF receptor also contains four tyrosine residues in the membrane-distal region, which are phosphorylated upon G-CSF stimulation. The human G-CSF receptor is a protein consisting of 813 amino acids with a calculated M_r of 89 743. (Fukunaga *et al.*, 1990c; Larsen *et al.*, 1990). The overall structure of the human G-CSF receptor is similar to that of the mouse, with an amino acid sequence similarity of 62.5%.

The G-CSF receptor is a member of the hemopoietic growth factor receptor family, which includes the receptors for interleukins and colony stimulating factors, thrombopoietin, growth hormone/prolactin, and leptin (Kishimoto *et al.*, 1994). A domain of about 200 amino acids (CRH domain) in the extracellular region is conserved on the amino acid sequence level among the members of this family. The tertiary structure of the CRH domains of the growth hormone and G-CSF receptors, as revealed by X-ray and NMR analyses, indicates that they are remarkably similar (de Vos *et al.*, 1992; Yamasaki *et al.*, 1997). The receptors for G-CSF, LIF and leptin, and gp130 (the signal transducer of the IL-6 receptor) have a similar overall structure, and constitute a cytokine receptor subfamily (Nagata and Fukunaga, 1993).

Genetic Properties of the G-CSF Receptor and its Expression

The human genome carries a single G-CSF receptor gene per haploid genome, which is located on chromosome lp35–34.3 (Inazawa *et al.*, 1991). Mouse genome carries a functional gene located on chromosome 4, and a non-functional pseudogene (Ito *et al.*, 1994). Both the human gene and the mouse functional gene are composed of 17 exons (Seto *et al.*, 1992; Ito *et al.*, 1994), with exons 3–17 coding for the G-CSF receptor protein. The subdomains of the receptor are coded for by a set of exons, and the organization is similar to that of the chromosomal genes for other hemopoietic growth factor receptors. Two regions in the promoter seem to be important for G-CSF receptor gene expression. One region, located at −49 from the transcription initiation site, contains a GCAAT sequence that specifically binds the C/EBP α transcription factor

(Smith *et al.*, 1996). Recently, mice lacking the C/EBP α gene were generated. Neutrophils in the C/EBP α-null mice cannot express the G-CSF receptor, which confirms an important role for C/EBP α in G-CSF receptor expression (Zhang *et al.*, 1997). The other *cis*-regulatory elements are in the 5′ untranslated region (at +36 and +43), where there are two binding sites for the ets family member PU.1. Mutation of these sites reduces the promoter activity by 75% (Smith *et al.*, 1996). However, mice lacking PU.1 can express the G-CSF receptor (Olson *et al.*, 1995), suggesting that this element is not essential for the expression of the G-CSF receptor gene.

PHYSIOLOGIC ROLES OF G-CSF AND ITS RECEPTOR

Biologic Activities of G-CSF and its Clinical Use

As per its definition, G-CSF specifically stimulates the neutrophilic granulocyte colony formation of bone marrow cells in semisolid cultures. It also supports the survival of the mature neutrophils and enhances their functional capacity (Kitagawa *et al.*, 1987; Yuo *et al.*, 1989, 1990; Williams *et al.*, 1990). Several mouse and human myeloid leukemia cell lines respond to G-CSF. For example, mouse NFS-60 cells proliferate in response to G-CSF without differentiation (Weinstein *et al.*, 1986). On the other hand, G-CSF causes the growth arrest of mouse myeloid leukemia WEHI-3BD$^+$ (Nicola *et al.*, 1983), 32Dcl3 (Valtieri *et al.*, 1987) and L-G cells (Lee *et al.*, 1991), and induces their morphologic differentiation into neutrophils. This differentiation is accompanied by the downregulation of c-*myc* and c-*myb* gene expression and the upregulation of the c-*fos* gene, as well as the expression of various neutrophil-specific genes such as myeloperoxidase, lactoferrin and chloroacetate esterase.

The specific role of G-CSF in the development of neutrophils has also been demonstrated *in vivo* (Cohen *et al.*, 1987; Tsuchiya *et al.*, 1987; Welte *et al.*, 1987). When G-CSF was administered to mice, hamsters and monkeys, the number of neutrophils in the blood increased immediately after injection, and reached five to ten times the basal level within 24 h. These effects were specific to neutrophilic granulocytes, and no significant change was observed in the number of other cell types such as erythrocytes, monocytes, lymphocytes and platelets. The injected G-CSF was cleared in a biexponential manner with a distribution half-life of 30 min and an elimination half-life of 3.8 h (Cohen *et al.*, 1987). When the administration of G-CSF ceased, the number of neutrophils in the blood returned to the basal level within 48 h. It is now possible to administer G-CSF rectally using a rectal dosage vehicle (Watanabe *et al.*, 1996), by infecting bone marrow cells with a G-CSF expression vector (Chang *et al.*, 1989) or by implanting fibroblasts or myoblasts transfected with a similar expression vector (Tani *et al.*, 1989; Bonham *et al.*, 1996; Lejnieks *et al.*, 1996). Under these conditions, animals show sustained neutrophilia and are protected from lethal bacterial infection.

After extensive clinical trials, the Food and Drug Administration approved the use of G-CSF as a drug in 1991. G-CSF is now widely used clinically in patients with granulopenia from various causes (Ganser and Karthaus, 1996; Welte *et al.*, 1996). G-CSF is administered to patients with cancer receiving chemotherapy or radiotherapy with or without bone marrow transplantation, and to patients receiving immunosuppressive agents after organ transplantation. In both types of patient, G-CSF accelerates the recovery of neutrophilic granulocytes and diminishes the risk of severe bacterial and fungal infections after chemotherapy. G-CSF has almost no adverse side-effects; at most, there may be slight bone pain, which is tolerated well by patients. G-CSF is also being used successfully to treat patients with congenital cyclic neutropenia (Hammond *et al.*, 1989). In addition to being used to directly increase the neutrophil counts in patients, G-CSF is also used to mobilize precursor cells from the bone marrow to the periphery: the administration of G-CSF mobilizes hemopoietic stem cells from the bone marrow to the periphery (Varas *et al.*, 1996). Thus, the peripheral blood, instead of bone marrow cells, can be used for the transfusion of patients who receive chemotherapy or radiotherapy (Winter *et al.*, 1996).

The Function of the G-CSF Receptor

As discussed above, G-CSF stimulates the proliferation and differentiation of the progenitor cells of neutrophilic granulocytes. The cloned G-CSF receptor can mediate these signals in a reconstituted system. G-CSF receptor cDNA has been expressed in IL-3-dependent cell lines such as the myeloid precursor cells L-GM (Dong *et al.*, 1993; Yoshikawa *et al.*, 1995) as well as FDC-P1, the pro-B-cell line BAF-B03 and the IL-2-dependent T-cell line CTLL2 (Fukunaga *et al.*, 1991). These cell lines normally do not express the G-CSF receptor, but the transformants acquire the ability to bind G-CSF with high affinity, indicating that a single polypeptide coded by the cloned G-CSF receptor cDNA is sufficient to bind G-CSF with high affinity. FDC-P1 and BAF-B03 transformants proliferate in response to G-CSF, but CTLL2 transformants do not, indicating that the cloned G-CSF receptor can transduce the growth signal in myeloid and pro-B cells, but not in T cells. In addition to the growth promoting activity, G-CSF can induce the terminal differentiation of neutrophilic granulocytes. When FDC-P1 and L-GM1 cell transformants expressing the G-CSF receptor are shifted from culture medium containing IL-3 to medium containing G-CSF, the cells start to express neutrophil-specific genes such as myeloperoxidase and leukocyte elastase (Fukunaga *et al.*, 1993; Yoshikawa *et al.*, 1995). Furthermore, the L-GM1 but not FDC-P1 cell transformants stop growing on G-CSF stimulation, and undergo morphologic differentiation with lobulated and segmented nuclei. This differentiation-inducing activity is specific for G-CSF, and GM-CSF and IL-3 work as inhibitors of the G-CSF receptor-mediated differentiation signal. Moreover, this differentiation signal is observed specifically in myeloid precursor cell lines, but not in the pro-B-cell line, although pro-B cells can transduce the G-CSF-mediated proliferation signal.

Mice Lacking G-CSF or its Receptor

G-CSF- or G-CSF receptor-deficient mice have been generated by the targeted disruption of the respective genes in embryonic stem cells (Lieschke *et al.*, 1994; F. Liu *et al.*, 1996). These mice are viable, fertile and apparently healthy, but have chronic neutropenia: their peripheral blood neutrophil levels are 20–30% of those of wild-type mice, indicating that G-CSF is required for the maintenance of neutrophil production during 'steady-state' granulopoiesis. In addition, G-CSF plays an important role in the 'emergency' granulopoiesis that occurs during infections, as the infection of the G-CSF-deficient mice with *Listeria monocytogenes* results in diminished and delayed granulopoiesis. The neutrophilic progenitor cells in the bone marrow are also decreased in G-CSF- or G-CSF receptor-deficient mice, confirming that G-CSF functions not only in the terminal differentiation of neutrophils, but also in the initial stage of neutrophil development. Although all these results indicate that G-CSF is a major regulator of neutrophil development, there is also a G-CSF-independent mechanism(s) for granulopoiesis, as the G-CSF- or G-CSF receptor-null mice still have some apparently normal neutrophils in their circulation.

SIGNAL TRANSDUCTION THROUGH THE G-CSF RECEPTOR

Dimerization of the G-CSF Receptor

Most hemopoietic growth factor receptors such as the IL-2, IL-3, IL-5, IL-6 and GM-CSF receptors are composed of two or three subunits, and function as a heterodimer or heterotrimer to bind their respective ligands with high affinity (Nicola and Metcalf, 1991; Miyajima *et al.*, 1992; Kishimoto *et al.*, 1994). The receptors for growth hormone, prolactin and G-CSF, however, seem to function as a homodimer. As described above, G-CSF binds to the receptor at two sites (Reidhaar-Olson *et al.*, 1996). Mutational analysis of the receptor has indicated that the CRH domain of the extracellular region of the G-CSF receptor is its major G-CSF binding site, although the Ig-like domain at the *N*-terminal portion of the receptor also contributes to the binding of G-CSF (Fukunaga *et al.*, 1991). Recently, the CRH- and Ig-like domains were produced in *E. coli*, and their ability to interact with G-CSF was demonstrated (Hiraoka *et al.*, 1994). Furthermore, when the CRH- and Ig-like domains were incubated with G-CSF, they formed a 1:1:1 complex, indicating that a single G-CSF molecule interacts with two G-CSF receptor molecules (Hiraoka *et al.*, 1994). This is similar to the mechanism found for the growth hormone system (de Vos *et al.*, 1992).

The dimeric or oligomeric form, but not the monomeric form, of the G-CSF receptor transduces the signal. This was demonstrated using chimeric receptors composed of portions from the growth hormone and G-CSF receptors (Ishizaka-Ikeda *et al.*, 1993). In these chimeric receptors, the extracellular region of the G-CSF receptor was replaced with that of the growth hormone receptor by exon swapping. The chimeric receptors could transduce the proliferation signal in FDC-P1 cells upon growth hormone stimulation (Fuh *et al.*, 1992; Ishizaka-Ikeda *et al.*, 1993). However, the growth response occurred only at growth hormone concentrations from 10 pM to 100 nM. At higher concentrations, growth hormone could not stimulate the proliferation of the transformants expressing the chimeric receptor. The growth hormone induces dimerization of the receptor at its optimal concentration, but prevents receptor dimerization at higher concentrations (Fuh *et al.*, 1992). Therefore, the above results were interpreted to mean that the dimerized cytoplasmic region of the G-CSF receptor in the chimeric construct transduces the signal, but that its monomeric form cannot. A chimeric receptor composed of the extracellular region from the Fas receptor and the cytoplasmic region from the G-CSF receptor was also expressed in FDC-P1 cells (Takahashi *et al.*, 1996). When the transformants were incubated with Fas ligand or anti-Fas antibodies, the cells proliferated. Because Fas ligand and anti-Fas antibodies cause trimerization and oligomerization of the receptor, these results indicate that trimers or oligomers of the G-CSF receptor still can transduce the signal.

Distinct Domains of the G-CSF Receptor for the Growth and Differentiation Signals

As mentioned above, the G-CSF receptor can transduce the signals for both proliferation and differentiation. To examine which regions of the G-CSF receptor are responsible for the transduction of different signals, various deletion and point mutations were introduced into the cytoplasmic region of the receptor. These studies indicated that a 76-amino-acid stretch proximal to the transmembrane domain, containing the Box1 and Box2 motifs, is essential for transducing the growth signal (Fukunaga *et al.*, 1993; Ziegler *et al.*, 1993). On the other hand, both the *N*- and *C*-terminal domains of the cytoplasmic region are indispensable for transducing the differentiation signal (Dong *et al.*, 1993; Fukunaga *et al.*, 1993), although the *N*-terminal domain containing the Box1 and Box2 motifs can be replaced by the analogous domain from the growth hormone receptor (Fukunaga *et al.*, 1993). These results indicate that different domains of the G-CSF receptor, and thus different signal transducing systems, are utilized for the G-CSF-induced proliferation and differentiation of neutrophilic progenitor cells. Two models have been proposed for the cytokine-induced differentiation of hemopoietic cells. In the stochastic model (Till *et al.*, 1964; Korn *et al.*, 1973; Nakahata *et al.*, 1982), cytokines provide proliferative and survival signals to the differentiating hematopoietic cells, but they do not provide specific lineage commitment signals. In the instructive model (Curry and Trentin, 1987), cytokines transmit specific signals to multipotent hemopoietic cells, thereby directing lineage commitment. The fact that a specific region of the G-CSF receptor mediates the differentiation signal supports the instructive model for the G-CSF-induced terminal differentiation of progenitor cells into neutrophils.

The *C*-terminal domain of the cytoplasmic region of the mouse G-CSF receptor contains four tyrosine residues (Tyr-703, Tyr-728, Tyr-743 and Tyr-763), which are phosphorylated upon stimulation by G-CSF. The effects of mutating each tyrosine residue on the G-CSF-dependent differentiation signal have been examined. The mutation of either Tyr-703 or Tyr-728 eliminated the G-CSF-dependent growth suppression during differentiation, induction of neutrophilic nuclear lobulation and neutrophil-specific myeloperoxidase gene induction (Yoshikawa *et al.*, 1995). On the other hand, mutation of the most distal tyrosine residue abolished the G-CSF-induced tyrosine phosphorylation of Shc protein (de Koning *et al.*, 1996b) (H.M. and S.N., unpublished observation). Therefore, it appears that the different tyrosine-containing regions of the G-CSF receptor transduce distinct signals. This is schematically shown in Fig. 1.

Molecules Involved in G-CSF Signaling

Upon the binding of G-CSF to its receptor, various cellular proteins including the receptor itself are tyrosine phosphorylated (Pan *et al.*, 1993). As the G-CSF receptor does not have an apparent protein kinase domain, G-CSF stimulation seems to activate cellular tyrosine kinases. So far, various protein kinases have been reported to be activated upon G-CSF stimulation. Among them, involvement of JAK family kinases in

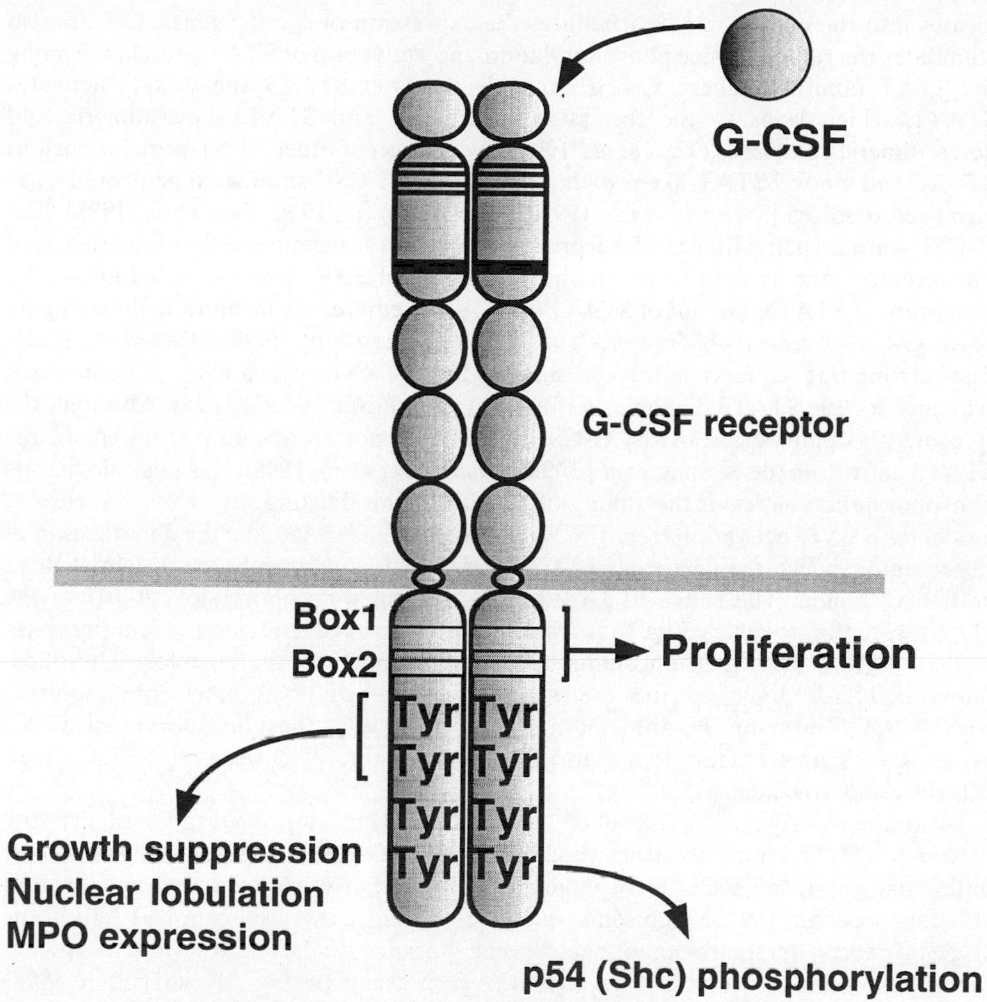

Fig. 1. Distinct signals from each tyrosine residue of G-CSF receptor. MPO, myeloperoxidase.

the signal transduction from the G-CSF receptor has been demonstrated convincingly. That is, JAK1 is associated constitutively with the G-CSF receptor and becomes activated by the binding of G-CSF to the receptor (Nicholson *et al.*, 1994). G-CSF also activates JAK2 and TYK2 (Shimoda *et al.*, 1994; Tian *et al.*, 1996; Avalos *et al.*, 1997). The membrane-proximal region containing the Box1 and Box2 motifs is required for the activation of the JAK kinases (Nicholson *et al.*, 1995). In the interferon-γ (IFN-γ) receptor system (Briscoe *et al.*, 1996), the activation of JAK1 and JAK2 is interdependent. Whether all JAK kinases (JAK1, JAK2 and TYK2) are required for the signal transduction through the G-CSF receptor, or whether they are redundant, remains to be established.

In most of the cytokine receptor systems, activation of JAK kinase leads to the tyrosine phosphorylation of STAT proteins. The phosphorylated STAT then translocates into the nucleus, where it induces the expression of specific genes. G-CSF also stimulates the rapid tyrosine phosphorylation and activation of STAT proteins. Among the STAT family members, G-CSF strongly activates STAT3 and weakly activates STAT1, which leads to the formation of STAT1 and STAT3 homodimeric and heterodimeric complexes (Tian *et al.*, 1994). Activation of other STAT proteins such as STAT5 and a novel STAT-like protein, STAT G, in G-CSF-stimulated neutrophils has also been reported (Tweardy *et al.*, 1995; Nicholson *et al.*, 1996; Tian *et al.*, 1996). The G-CSF-induced activation of STAT proteins requires the membrane-proximal region of the receptor, and is dependent on the activation of JAK kinases. In addition, the activation of STAT3, but not of STAT1 or STAT5, requires the membrane-distal region of the G-CSF receptor which carries Tyr-703 (de Koning *et al.*, 1996a; Tian *et al.*, 1996). The surrounding sequence of this tyrosine residue is YXXQ, which fits to the consensus sequence for the STAT3 docking site found in gp130 (Stahl *et al.*, 1995). Although the tyrosine phosphorylation of the G-CSF receptor is not an absolute requirement for STAT3 activation (de Koning *et al.*, 1996a; Nicholson *et al.*, 1996), it is possible that its phosphorylation increases the affinity of STAT3 to this docking site. Thus, the current model for STAT3 activation by G-CSF is as follows: G-CSF induces the dimerization of the receptor, which then activates JAK kinases through its membrane-proximal Box1 and Box2 region. The activated JAK kinases tyrosine-phosphorylate Tyr-703 of the G-CSF receptor, to which STAT3 is recruited through its SH2 region (de Koning *et al.*, 1996a). JAK kinases then phosphorylate STAT3 bound to the receptor. The phosphorylated STAT3 releases from the receptor by forming a homodimer or heterodimer with STAT1, probably because the association between the phosphorylated STAT proteins is stronger than that with G-CSF receptor. The released STAT3 then translocates to the nucleus.

What kinds of signals are transduced by the STAT3 activation? Mutation of Tyr-703, to which STAT3 binds, abolishes the ability of the G-CSF receptor to transduce the differentiation signals such as morphologic change and myeloperoxidase gene induction (Yoshikawa *et al.*, 1995; Nicholson *et al.*, 1996). Thus, an involvement of STAT3 in the G-CSF-induced differentiation of neutrophilic granulocytes has been proposed. In fact, the overexpression of dominant-negative STAT3 prevents the G-CSF-induced morphologic changes of mouse myeloid L-GM cells (Shimozaki *et al.*, 1997), indicating that STAT3 activation is essential for the G-CSF-induced differentiation of myeloid cells.

In addition to the JAK family kinases, an src-related protein tyrosine kinase, Lyn, and a non-src-related Syk tyrosine kinase have been reported to be associated with the G-CSF receptor and activated upon G-CSF stimulation (Corey *et al.*, 1994). However, Lyn-deficient mice do not show an abnormality in neutrophil development, and irradiated mice reconstituted with Syk-deficient fetal liver show no gross perturbations in G-CSF responsiveness, suggesting no requirement of Lyn and Syk for G-CSF signal (Turner *et al.*, 1995), although the possibility that these kinases are redundant cannot be ruled out. Several other tyrosine kinases such as Tec, a cytoplasmic src-related protein kinase, and $p72^{sak}$ tyrosine kinase, are tyrosine phosphorylated and activated specifically by G-CSF (Matsuda *et al.*, 1995; Miyazato *et al.*, 1996). However, their physiologic roles in G-CSF-induced signal transduction are unknown.

The ras/mitogen-activated protein (MAP) kinase pathway, which is rapidly being elucidated, is another signaling cascade activated by G-CSF. In the pro-B-cell line BAF-B03, G-CSF activates ras and MAP kinase (Bashey *et al.*, 1994; Nicholson *et al.*,

1995). This activation seems to be mediated by various molecules such as the Shc, Grb2 and Syp adaptors, and the vav guanine nucleotide exchanger, as these molecules are phosphorylated by G-CSF in various cells responding to G-CSF (de Koning *et al.*, 1996b). The activation of ras requires the membrane-proximal region of the receptor (Barge *et al.*, 1996), as well as the membrane-distal region, specifically the domain containing Tyr-763 (de Koning *et al.*, 1996b). It is likely that the JAK family kinases, activated through the membrane-proximal region of the receptor, tyrosine phosphorylate Tyr-763 on the receptor, to which adaptor molecules are recruited to activate the ras/MAP kinase pathway. So far, it is not clear what kinds of signals are transduced by the ras/MAP kinase. As the mutation of Tyr-763 does not affect the ability of the G-CSF receptor to transduce the proliferation or differentiation signals (Yoshikawa *et al.*, 1995), it is unlikely that this pathway is involved in the transduction of these signals.

In addition to tyrosine kinases and ras/MAP kinase, G-CSF activates c-*rel*, a proto-oncogene belonging to the NF-κB family (Druker *et al.*, 1994; Avalos *et al.*, 1995). Although the Box1 motif in the membrane-proximal region of the receptor is required for NF-κB activation, it is also not clear what roles NF-κB plays in the G-CSF-mediated signal. One possibility is that NF-κB activation by G-CSF induces the antiapoptotic signal, as found in the IL-1 and TNF systems (Z.-G. Liu *et al.*, 1996).

Mutations in the Gene for the G-CSF Receptor in Patients with Severe Congenital Neutropenia

Severe congenital neutropenia (Kostmann's syndrome) is characterized by profound neutropenia and a maturation arrest of marrow progenitor cells at the promyelocyte–myelocyte stage. Bone marrow cells from such patients frequently display a reduced responsiveness to G-CSF. Somatic point mutations in one allele of the G-CSF receptor gene have been identified in some patients with severe congenital neutropenia (Dong *et al.*, 1994, 1995, 1997). These mutations result in the truncation of the *C*-terminal domain of the receptor, which is required for the G-CSF-dependent neutrophilic differentiation signaling (Dong *et al.*, 1993; Fukunaga *et al.*, 1993). To date, five severe cases of congenital neutropenia have been identified with G-CSF receptor mutations affecting the maturation signaling function of the receptor (Dong *et al.*, 1997). Four of these patients have developed acute myeloblastic leukemia (AML). The mutations in the G-CSF receptor gene occur either congenitally or sporadically in AML cells, and some of the mutants are dominant-negative. These mutations in the G-CSF receptor gene arrest the maturation to neutrophils and cause an accumulation of the neutrophilic progenitor cells; a secondary mutation in some oncogene or tumor suppressor gene in those accumulated cells seems to lead to leukemic transformation.

CONCLUSION AND PERSPECTIVES

More than 15 years have elapsed since G-CSF was identified as a factor that stimulates the proliferation and differentiation of neutrophilic progenitors. The extensive biochemical and biologic characterization of this factor, as well as its extensive clinical use have demonstrated that G-CSF is one of the most valuable cytokines for use in treating patients. This is because G-CSF has a rather strict target cell specificity, limited to regulating the production of neutrophils. In fact, G-CSF has proved to be very beneficial in treating patients suffering from neutropenia. Furthermore, it has been possible to use the specific effect of G-CSF on neutrophils to develop an ideal model system for examining the molecular mechanisms underlying cell growth and differentiation. Recently, many molecules that seem to be involved in the G-CSF-induced signaling cascade have been identified (Avalos, 1996). Although it is not yet clear how the activation of these molecules leads to the growth and differentiation of neutrophils, more elaborate biochemical analyses and the establishment of mouse lines deficient in each participating molecule should allow the entire signaling cascade to be elucidated in the near future.

ACKNOWLEDGEMENTS

The authors thank Dr O. Hayaishi, Director of Osaka Bioscience Institute, for support and encouragement. The work in their laboratories was supported in part by Grants-in-Aid from the Ministry of Education, Science and Culture of Japan. The authors also thank Ms H. Fujiwara for secretarial assistance.

REFERENCES

Asano, M. and Nagata, S. (1992). Constitutive and inducible factors bind to regulatory element 3 in the promoter of the gene encoding mouse granulocyte colony-stimulating factor. *Gene* **121**, 371–375.

Asano, M., Nishizawa, M. and Nagata, S. (1991). Three individual regulatory elements of the promoter positively activate the transcription of the murine gene encoding granulocyte colony-stimulating factor. *Gene* **107**, 241–246.

Avalos, B.R. (1996). Molecular analysis of the granulocyte colony-stimulating factor receptor. *Blood* **88**, 761–777.

Avalos, B.R., Hunter, M.G., Parker, J.M., Ceselski, S.K., Druker, B.J., Corey, S.J. and Mehta, V.B. (1995). Point mutations in the conserved box 1 region inactivate the human granulocyte colony-stimulating factor receptor for growth signal transduction and tyrosine phosphorylation of p75c-rel. *Blood* **85**, 3117–3126.

Avalos, B.R., Parker, J.M., Ware, D.A., Hunter, M.G., Sibert, K.A. and Druker, B.J. (1997). Dissociation of the Jak kinase pathway from G-CSF receptor signaling in neutrophils. *Exp. Hematol.* **25**, 160–168.

Barge, R.M., de Koning, J.P., Pouwels, K., Dong, F., Lowenberg, B. and Touw, I.P. (1996). Tryptophan 650 of human granulocyte colony-stimulating factor (G-CSF) receptor, implicated in the activation of JAK2, is also required for G–CSF-mediated activation of signaling complexes of the p21ras route. *Blood* **87**, 2148–2153.

Bashey, A., Healy, L. and Marshall, C.J. (1994). Proliferative but not nonproliferative responses to granulocyte colony-stimulating factor are associated with rapid activation of the p21ras/MAP kinase signalling pathway. *Blood* **83**, 949–957.

Bonham, L., Palmer, T. and Miller, A.D. (1996). Prolonged expression of therapeutic levels of human granulocyte colony-stimulating factor in rats following gene transfer to skeletal muscle. *Hum. Gene Ther.* **7**, 1432–1439.

Briscoe, J., Rogers, N.C., Witthuhn, B.A., Watling, D., Harpur, A.G., Wilks, A.F., Stark, G.R., Ihle, J.N. and Kerr, I.M. (1996). Kinase-negative mutants of JAK1 can sustain interferon-gamma-inducible gene expression but not an antiviral state. *EMBO J.* **15**, 799–809.

Broudy, V.C., Kaushansky, K., Harlan, J.M. and Adamson, J.W. (1987). Interleukin 1 stimulates human endothelial cells to produce granulocyte–macrophage colony-stimulating factor and granulocyte colony-stimulating factor. *J. Immunol.* **139**, 464–468.

Brown, C.Y., Lagnado, C.A., Vadas, M.A. and Goodall, G.J. (1996). Differential regulation of the stability of cytokine mRNAs in lipopolysaccharide-activated blood monocytes in response to interleukin-10. *J. Biol. Chem.* **271**, 20108–20112.

Buchberg, A.M., Bedigian, H.G., Taylor, B.A., Brownell, E., Ihle, J.N., Nagata, S., Jenkins, N.A. and Copeland,

N.G. (1988). Localization of Evi-2 to chromosome 11: linkage to other proto-oncogene and growth factor loci using interspecific backcross mice. *Oncogene Res.* **2**, 149–165.

Chang, J.M., Metcalf, D., Gonda, T.J. and Johnson, G.R. (1989). Long-term exposure to retrovirally expressed granulocyte-colony-stimulating factor induces a nonneoplastic granulocytic and progenitor cell hyperplasia without tissue damage in mice. *J. Clin. Invest.* **84**, 1488–1496.

Cohen, A.M., Zsebo, K.M., Inoue, H., Hines, D., Boone, T.C., Chazin, V.R., Tsai, L., Ritch, T. and Souza, L.M. (1987). *In vivo* stimulation of granulopoiesis by recombinant human granulocyte colony-stimulating factor. *Proc. Natl Acad. Sci. U.S.A.* **84**, 2484–2488.

Corey, S.J., Burkhardt, A.L., Bolen, J.B., Geahlen, R.L., Tkatch, L.S. and Tweardy, D.J. (1994). Granulocyte colony-stimulating factor receptor signaling involves the formation of a three-component complex with Lyn and Syk protein-tyrosine kinases. *Proc. Natl Acad. Sci. U.S.A.* **91**, 4683–4687.

Curry, J.L. and Trentin, J.J. (1987). Haemopoietic spleen colony-stimulating factors. *Science* **236**, 1229–1237.

de Koning, J.P., Dong, F., Smith, L., Schelen, A.M., Barge, R.M., van der Plas, D.C., Hoefsloot, L.H., Lowenberg, B. and Touw, I.P. (1996a). The membrane-distal cytoplasmic region of human granulocyte colony-stimulating factor receptor is required for STAT3 but not STAT1 homodimer formation. *Blood* **87**, 1335–1342.

de Koning, J.P., Schelen, A.M., Dong, F., van Buitenen, C., Burgering, B.M., Bos, J.L., Lowenberg, B. and Touw, I.P. (1996b). Specific involvement of tyrosine 764 of human granulocyte colony-stimulating factor receptor in signal transduction mediated by p145/Shc/GRB2 or p90/GRB2 complexes. *Blood* **87**, 132–140.

de Vos, A.M., Ultsch, M. and Kossiakoff, A.A. (1992). Human growth hormone and extracellular domain of its receptor: crystal structure of the complex. *Science* **255**, 306–312.

Dong, F., van Buitenen, C., Pouwels, K., Hoefsloot, L.H., Lowenberg, B. and Touw, I.P. (1993). Distinct cytoplasmic regions of the human granulocyte colony-stimulating factor receptor involved in induction of proliferation and maturation. *Mol. Cell. Biol.* **13**, 7774–7781.

Dong, F., Hoefsloot, L.H., Schelen, A.M., Broeders, C.A., Meijer, Y., Veerman, A.J., Touw, I.P. and Lowenberg, B. (1994). Identification of a nonsense mutation in the granulocyte-colony-stimulating factor receptor in severe congenital neutropenia. *Proc. Natl Acad. Sci. U.S.A.* **91**, 4480–4484.

Dong, F., Brynes, R.K., Tidow, N., Welte, K., Lowenberg, B. and Touw, I.P. (1995). Mutations in the gene for the granulocyte colony-stimulating-factor receptor in patients with acute myeloid leukemia preceded by severe congenital neutropenia. *N. Engl. J. Med.* **333**, 487–493.

Dong, F., Dale, D.C., Bonilla, M.A., Freedman, M., Fasth, A., Neijens, H.J., Palmblad, J., Briars, G.L., Carlsson, G., Veerman, A.J., Welte, K., Lowenberg, B. and Touw, I.P. (1997). Mutations in the granulocyte colony-stimulating factor receptor gene in patients with severe congenital neutropenia. *Leukemia* **11**, 120–125.

Druker, B.J., Neumann, M., Okuda, K., Franza, B.J. and Griffin, J.D. (1994). rel is rapidly tyrosine-phosphorylated following granulocyte-colony stimulating factor treatment of human neutrophils. *J. Biol. Chem.* **269**, 5387–5390.

Ernst, T.J., Ritchie, A.R., Demetri, G.D. and Griffin, J.D. (1989). Regulation of granulocyte- and monocyte-colony stimulating factor mRNA levels in human blood monocytes is mediated primarily at a post-transcriptional level. *J. Biol. Chem.* **264**, 5700–5703.

Felik, R., Genolet, C.L., Lowik, C., Cecchini, M.G. and Hofstetter, W. (1995). Cytokine production by calvariae of osteopetrotic mice. *Bone* **17**, 5–9.

Fuh, G., Cunningham, B.C., Fukunaga, R., Nagata, S., Goeddel, D.V. and Wells, J.A. (1992). Rational design of potent antagonists to the human growth hormone receptor. *Science* **256**, 1677–1680.

Fukunaga, R., Ishizaka-Ikeda, E. and Nagata, S. (1990a). Purification and characterization of the receptor for murine granulocyte colony-stimulating factor. *J. Biol. Chem.* **265**, 14 008–14 015.

Fukunaga, R., Ishizaka-Ikeda, E., Seto, Y. and Nagata, S. (1990b). Expression cloning of a receptor for murine granulocyte colony-stimulating factor. *Cell* **61**, 341–350.

Fukunaga, R., Seto, Y., Mizushima, S. and Nagata, S. (1990c). Three different mRNAs encoding human granulocyte colony-stimulating factor receptor. *Proc. Natl Acad. Sci. U.S.A.* **87**, 8702–8706.

Fukunaga, R., Ishizaka-Ikeda, E., Pan, C.X., Seto, Y. and Nagata, S. (1991). Functional domains of the granulocyte colony-stimulating factor receptor. *EMBO J.* **10**, 2855–2865.

Fukunaga, R., Ishizaka-Ikeda, E. and Nagata, S. (1993). Growth and differentiation signals mediated by different regions in the cytoplasmic domain of granulocyte colony-stimulating factor receptor. *Cell* **74**, 1079–1087.

Ganser, A. and Karthaus, M. (1996). Clinical use of hematopoietic growth factors. *Curr. Opin. Oncol.* **8**, 265–269.

Hammond, W., Price, T.H., Souza, L.M. and Dale, D.C. (1989). Treatment of cyclic neutropenia with granulocyte colony-stimulating factor. *N. Engl. J. Med.* **320**, 1306–1311.

Han, S.W., Ramesh, N. and Osborne, W.R. (1996). Cloning and expression of the cDNA encoding rat granulocyte colony-stimulating factor. *Gene* **175**, 101–104.

Hill, C.P., Osslund, T.D. and Eisenberg, D. (1993). The structure of granulocyte-colony-stimulating factor and its relationship to other growth factors. *Proc. Natl Acad. Sci. U.S.A.* **90**, 5167–5171.

Hiraoka, O., Anaguchi, H., Yamasaki, K., Fukunaga, R., Nagata, S. and Ota, Y. (1994). Ligand binding domain of granulocyte colony-stimulating factor receptor. *J. Biol. Chem.* **269**, 22412–22419.

Inazawa, J., Fukunaga, R., Seto, Y., Nakagawa, H., Misawa, S., Abe, T. and Nagata, S. (1991). Assignment of the human granulocyte colony-stimulating factor receptor gene (*CSF3R*) to chromosome 1 at region p35–p34.3. *Genomics* **10**, 1075–1078.

Ishizaka-Ikeda, E., Fukunaga, R., Wood, W.I., Goeddel, D.V. and Nagata, S. (1993). Signal transduction mediated by growth hormone receptor and its chimeric molecules with the granulocyte colony-stimulating factor receptor. *Proc. Natl Acad. Sci. U.S.A.* **90**, 123–127.

Ito, Y., Seto, Y., Brannan, C.I., Copeland, N.G., Jenkins, N.A., Fukunaga, R. and Nagata, S. (1994). Structural analysis of the functional gene and pseudogene encoding the murine granulocyte colony-stimulating-factor receptor. *Eur. J. Biochem.* **220**, 881–891.

Kanda, N., Fukushige, S., Murotsu, T., Yoshida, M.C., Tsuchiya, M., Asano, S., Kaziro, Y. and Nagata, S. (1987). Human gene coding for granulocyte-colony stimulating factor is assigned to the q21–q22 region of chromosome 17. *Somat. Cell Mol. Genet.* **13**, 679–684.

Kaushansky, K., Lin, N. and Adamson, J.W. (1988). Interleukin 1 stimulates fibroblasts to synthesize granulocyte–macrophage and granulocyte colony-stimulating factors. Mechanism for the hematopoietic response to inflammation. *J. Clin. Invest.* **81**, 92–97.

Kawakami, M., Tsutsumi, H., Kumakawa, T., Abe, H., Hirai, M., Kurosawa, S., Mori, M. and Fukushima, M. (1990). Levels of serum granulocyte colony-stimulating factor in patients with infections. *Blood* **76**, 1962–1964.

Kiehntopf, M., Herrmann, F. and Brach, M.A. (1995). Functional NF-IL6/CCAAT enhancer-binding protein is required for tumor necrosis factor alpha-inducible expression of the granulocyte colony-stimulating factor (CSF), but not the granulocyte/macrophage CSF or interleukin 6 gene in human fibroblasts. *J. Exp. Med.* **181**, 793–798.

Kishimoto, T., Taga, T. and Akira, S. (1994). Cytokine signal transduction. *Cell* **76**, 253–262.

Kitagawa, S., Yuo, A., Souza, L.M., Saito, M., Miura, Y. and Takaku, F. (1987). Recombinant human granulocyte colony-stimulating factor enhances superoxide release in human granulocytes stimulated by the chemotactic peptide. *Biochem. Biophys. Res. Commun.* **144**, 1143–1146.

Koeffler, H.P., Gasson, J., Ranyard, J., Souza, L., Shepard, M. and Munker, R. (1987). Recombinant human TNF alpha stimulates production of granulocyte colony-stimulating factor. *Blood* **70**, 55–59.

Korn, A.P., Henkelman, R.M., Ottensmeyer, F.P. and Till, J.E. (1973). Investigations of a stochastic model of haemopoiesis. *Exp. Hematol.* **1**, 362–375.

Larsen, A., Davis, T., Curtis, B.M., Gimpel, S., Sims, J.E., Cosman, D., Park, L., Sorensen, E., March, C.J. and Smith, C.A. (1990). Expression cloning of a human granulocyte colony-stimulating factor receptor: a structural mosaic of hematopoietin receptor, immunoglobulin, and fibronectin domains. *J. Exp. Med.* **172**, 1559–1570.

Lee, K.H., Kinashi, T., Tohyama, K., Tashiro, K., Funato, N., Hama, K. and Honjo, T. (1991). Different stromal cell lines support lineage-selective differentiation of the multipotential bone marrow stem cell clone LyD9. *J. Exp. Med.* **173**, 1257–1266.

Lejnieks, D.V., Han, S.W., Ramesh, N., Lau, S. and Osborne, W.R. (1996). Granulocyte colony-stimulating factor expression from transduced vascular smooth muscle cells provides sustained neutrophil increases in rats. *Hum. Gene Ther.* **7**, 1431–1436.

Lieschke, G.J., Grail, D., Hodgson, G., Metcalf, D., Stanley, E., Cheers, C., Fowler, K.J., Basu, S., Zhan, Y.F. and Dunn, A.R. (1994). Mice lacking granulocyte colony-stimulating factor have chronic neutropenia, granulocyte and macrophage progenitor cell deficiency, and impaired neutrophil mobilization. *Blood* **84**, 1737–1746.

Liu, F., Wu, H.Y., Wesselschmidt, R., Kornaga, T. and Link, D.C. (1996). Impaired production and increased apoptosis of neutrophils in granulocyte colony-stimulating factor receptor-deficient mice. *Immunity* **5**, 491–501.

Liu, Z.-G., Hsu, H., Goeddel, D. and Karin, M. (1996). Dissection of TNF receptor 1 effector functions: JNK activation is not linked to apoptosis while NF-κB activation prevents cell death. *Cell* **87**, 565–576.

Lovejoy, B., Cascio, D. and Eisenberg, D. (1993). Crystal structure of canine and bovine granulocyte-colony stimulating factor (G-CSF). *J. Mol. Biol.* **234**, 640–653.

Lu, H.S., Boone, T.C., Souza, L.M. and Lai, P.H. (1989). Disulfide and secondary structures of recombinant human granulocyte colony stimulating factor. *Arch. Biochem. Biophys.* **268**, 81–92.

Lu, L., Walker, D., Graham, C.D., Waheed, A., Shadduck, R.K. and Broxmeyer, H.E. (1988). Enhancement of release from MHC class II antigen-positive monocytes of hematopoietic colony stimulating factors CSF-1 and G-CSF by recombinant human tumor necrosis factor-alpha: synergism with recombinant human interferon-gamma. *Blood* **72**, 34–41.

Matsuda, T., Takahashi, T.M., Fukada, T., Okuyama, Y., Fujitani, Y., Tsukada, S., Mano, H., Hirai, H., Witte, O.N. and Hirano, T. (1995). Association and activation of Btk and Tec tyrosine kinases by gp130, a signal transducer of the interleukin-6 family of cytokines. *Blood* **85**, 627–633.

McCracken, S., Layton, J.E., Shorter, S.C., Starkey, P.M., Barlow, D.H. and Mardon, H.J. (1996). Expression of granulocyte-colony stimulating factor and its receptor is regulated during the development of the human placenta. *J. Endocrinol.* **149**, 249–258.

Metcalf, D. and Nicola, N.A. (1985). Synthesis by mouse peritoneal cells of G-CSF, the differentiation inducer for myeloid leukemia cells: stimulation by endotoxin, M-CSF and multi-CSF. *Leuk. Res.* **9**, 35–50.

Miyajima, A., Kitamura, T., Harada, N., Yokota, T. and Arai, K. (1992). Cytokine receptors and signal transduction. *Annu. Rev. Immunol.* **10**, 295–331.

Miyazato, A., Yamashita, Y., Hatake, K., Miura, Y., Ozawa, K. and Mano, H. (1996). Tec protein tyrosine kinase is involved in the signaling mechanism of granulocyte colony-stimulating factor receptor. *Cell Growth Differ.* **7**, 1135–1139.

Nagata, S. (1990). Granulocyte colony-stimulating factor. In *Handbook of Experimental Pharmacology* (eds M.B. Sporn and A.B. Roberts), Springer, Berlin, pp. 699–722.

Nagata, S. and Fukunaga, R. (1993). Granulocyte colony-stimulating factor receptor and its related receptors. *Growth Factors* **8**, 99–107.

Nagata, S., Tsuchiya, M., Asano, S., Kaziro, Y., Yamazaki, T., Yamamoto, O., Hirata, Y., Kubota, N., Oheda, M., Nomura, H. and Ono, M. (1986a). Molecular cloning and expression of cDNA for human granulocyte colony-stimulating factor. *Nature* **319**, 415–418.

Nagata, S., Tsuchiya, M., Asano, S., Yamamoto, O., Hirata, Y., Kubota, N., Oheda, M., Nomura, H. and Yamazaki, T. (1986b). The chromosomal gene structure and two mRNAs for human granulocyte colony-stimulating factor. *EMBO J.* **5**, 575–581.

Nakahata, T., Gross, A.J. and Ogawa, M. (1982). A stochastic model of self-renewal and commitment to differentiation of the primitive hemopoietic stem cells in culture. *J. Cell. Physiol.* **113**, 455–458.

Nicholson, S.E., Oates, A.C., Harpur, A.G., Ziemiecki, A., Wilks, A.F. and Layton, J.E. (1994). Tyrosine kinase JAK1 is associated with the granulocyte-colony-stimulating factor receptor and both become tyrosine-phosphorylated after receptor activation. *Proc. Natl Acad. Sci. U.S.A.* **91**, 2985–2988.

Nicholson, S.E., Novak, U., Ziegler, S.F. and Layton, J.E. (1995). District regions of the granulocyte colony-stimulating factor receptor are required for tyrosine phosphorylation of the signaling molecules JAK2, Stat3, and p42, p44MAPK. *Blood* **86**, 3698–3704.

Nicholson, S.E., Starr, R., Novak, U., Hilton, D.J. and Layton, J.E. (1996). Tyrosine residues in the granulocyte colony-stimulating factor (G-CSF) receptor mediate G-CSF-induced differentiation of murine myeloid leukemic (M1) cells. *J. Biol. Chem.* **271**, 26 947–26 953.

Nicola, N.A. and Metcalf, D. (1991). Subunit promiscuity among hemopoietic growth factor receptors. *Cell* **67**, 1–4.

Nicola, N.A. and Peterson, L. (1986). Identification of distinct receptors for two hemopoietic growth factors (granulocyte colony-stimulating factor and multipotential colony-stimulating factor) by chemical cross-linking. *J. Biol. Chem.* **261**, 12 384–12 389.

Nicola, N.A., Metcalf, D., Matsumoto, M. and Johnson, G.R. (1983). Purification of a factor inducing differentiation in murine myelomonocytic leukemia cells. Identification as granulocyte colony-stimulating factor. *J. Biol. Chem.* **258**, 9017–9023.

Nishizawa, M. and Nagata, S. (1990). Regulatory elements responsible for inducible expression of the granulocyte colony-stimulating factor gene in macrophages. *Mol. Cell. Biol.* **10**, 2002–2011.

Nishizawa, M., Tsuchiya, M., Watanabe-Fukunaga, R. and Nagata, S. (1990). Multiple elements in the promoter of granulocyte colony-stimulating factor gene regulate its constitutive expression in human carcinoma cells. *J. Biol. Chem.* **265**, 5897–5902.

Nomura, H., Imazeki, I., Oheda, M., Kubota, N., Tamura, M., Ono, M., Ueyama, Y. and Asano, S. (1986). Purification and characterization of human granulocyte colony-stimulating factor (G-CSF). *EMBO J.* **5**, 871–876.

Oheda, M., Hase, S., Ono, M. and Ikenaka, T. (1988). Structures of the sugar chains of recombinant human granulocyte-colony-stimulating factor produced by Chinese hamster ovary cells. *J. Biochem. (Tokyo)* **103**, 544–546.

Oheda, M., Hasegawa, M., Hattori, K., Kuboniwa, H., Kojima, T., Orita, T., Tomonou, K., Yamazaki, T. and Ochi, N. (1990). *O*-linked sugar chain of human granulocyte colony-stimulating factor protects it against polymerization and denaturation allowing it to retain its biological activity. *J. Biol. Chem.* **265**, 11 432–11 435.

Olson, M.C., Scott, E.W., Hack, A.A., Su, G.H., Tenen, D.G., Singh, H. and Simon, M.C. (1995). PU.1 is not essential for early myeloid gene expression but is required for terminal myeloid differentiation. *Immunity* **3**, 703–714.

Pan, C.X., Fukunaga, R., Yonehara, S. and Nagata, S. (1993). Unidirectional cross-phosphorylation between the granulocyte colony-stimulating factor and interleukin 3 receptors. *J. Biol. Chem.* **268**, 25 818–25 823.

Park, L.S., Waldron, P.E., Friend, D., Sassenfeld, H.M., Price, V., Anderson, D., Cosman, D., Andrews, R.G., Bernstein, I.D. and Urdal, D.L. (1989). Interleukin-3, GM-CSF, and G-CSF receptor expression on cell lines and primary leukemia cells: receptor heterogeneity and relationship to growth factor responsiveness. *Blood* **74**, 56–65.

Reidhaar-Olson, J.F., De Souza-Hart, J.A. and Selick, H.E. (1996). Identification of residues critical to the activity of human granulocytes colony-stimulating factor. *Biochemistry* **35**, 9034–9041.

Seelentag, W.K., Mermod, J.J., Montesano, R. and Vassalli, P. (1987). Additive effects of interleukin 1 and tumour necrosis factor-alpha on the accumulation of the three granulocyte and macrophage colony-stimulating factor mRNAs in human endothelial cells. *EMBO J.* **6**, 2261–2265.

Seto, Y., Fukunaga, R. and Nagata, S. (1992). Chromosomal gene organization of the human granulocyte colony-stimulating factor receptor. *J. Immunol.* **148**, 259–266.

Shaw, G. and Kamen, R. (1986). A conserved AU sequence from the 3′ untranslated region of GM-CSF mRNA mediates selective mRNA degradation. *Cell* **46**, 659–667.

Shimoda, K., Iwasaki, H., Okamura, S., Ohno, Y., Kubota, A., Arima, F., Otsuka, T. and Niho, Y. (1994). G-CSF induces tyrosine phosphorylation of the JAK2 protein in the human myeloid G-CSF responsive and proliferative cells, but not in mature neutrophils. *Biochem. Biophys. Res. Commun.* **203**, 922–928.

Shimozaki, K., Nakajima, K., Hirano, T. and Nagata, S. (1997). Involvement of STAT3 in the granulocyte colony stimulating factor-induced differentiation of myeloid cells. *J. Biol. Chem.* **272**, 25184–25189.

Smith, L.T., Hohaus, S., Gonzalez, D.A., Dziennis, S.E. and Tenen, D.G. (1996). PU.1 (Spi-1) and C/EBP alpha regulate the granulocyte colony-stimulating factor receptor promoter in myeloid cells. *Blood* **88**, 1234–1247.

Souza, L.M., Boone, T.C., Gabrilove, J., Lai, P.H., Zsebo, K.M., Murdock, D.C., Chazin, V.R., Bruszewski, J., Lu, H., Chen, K.K., Barendt, J., Platzer, E., Moore, M.A.S., Mertelsmann, R. and Welte, K. (1986). Recombinant human granulocyte colony-stimulating factor: effects on normal and leukemic myeloid cells. *Science* **232**, 61–65.

Stahl, N., Farruggella, T.J., Boulton, T.G., Zhong, Z., Darnell, J.J. and Yancopoulos, G.D. (1995). Choice of STATs and other substrates specified by modular tyrosine-based motifs in cytokine receptors. *Science* **267**, 1349–1353.

Takahashi, T., Tanaka, M., Ogasawara, J., Suda, T., Murakami, H. and Nagata, S. (1996). Swapping between Fas and granulocyte colony-stimulating factor receptor. *J. Biol. Chem.* **271**, 17 555–17 560.

Takashima, H., Eguchi, K., Kawakami, A., Kawabe, Y., Migata, K., Sakai, M., Origuchi, T. and Nagataki, S. (1996). Cytokine production by endothelial cells infected with human T cell lymphotropic virus type 1. *Ann. Rheum. Dis.* **55**, 632–637.

Tanaka, T., Akira, S., Yoshida, K., Umemoto, M., Yoneda, Y., Shirafuji, N., Fujiwara, H., Suematsu, S., Yoshida, N. and Kishimoto, T. (1995). Targeted disruption of the NF-IL6 gene discloses its essential role in bacteria killing and tumor cytotoxicity by macrophages. *Cell* **80**, 353–361.

Tani, K., Ozawa, K., Ogura, H., Takahashi, T., Okano, A., Watari, K., Matsudaira, T., Tajika, K., Karasuyama, H., Nagata, S., Asano, S. and Takaku, F. (1989). Implantation of fibroblasts transfected with human granulocyte colony-stimulating factor cDNA into mice as a model of cytokine-supplement gene therapy. *Blood* **74**, 1274–1280.

Tian, S.S., Lamb, P., Seidel, H.M., Stein, R.B. and Rosen, J. (1994). Rapid activation of the STAT3 transcription factor by granulocyte colony-stimulating factor. *Blood* **84**, 1760–1764.

Tian, S.S., Tapley, P., Sincich, C., Stein, R.B., Rosen, J. and Lamb, P. (1996). Multiple signaling pathways induced by granulocyte colony-stimulating factor involving activation of JAKs, STAT5, and/or STAT3 are required for regulation of three distinct classes of immediate early genes. *Blood* **88**, 4435–4444.

Till, J.E., McCulloch, E.A. and Siminovitch, L. (1964). A stochastic model of stem cell proliferation based on the growth of spleen colony forming cells. *Proc. Natl Acad. Sci. U.S.A.* **51**, 29–36.

Tsuchiya, M., Asano, S., Kaziro, Y. and Nagata, S. (1986). Isolation and characterization of the cDNA for murine granulocyte colony-stimulating factor. *Proc. Natl Acad. Sci. U.S.A.* **83**, 7633–7637.

Tsuchiya, M., Nomura, H., Asano, S., Kaziro, Y. and Nagata, S. (1987). Characterization of recombinant human granulocyte-colony-stimulating factor produced in mouse cells. *EMBO J.* **6**, 611–616.

Turner, M., Mee, P.J., Costello, P.S., Williams, O., Price, A.A., Duddy, L.P., Furlong, M.T., Geahlen, R.L. and Tybulewicz, V.L. (1995). Perinatal lethality and blocked B-cell development in mice lacking the tyrosine kinase Syk. *Nature* **378**, 298–302.

Tweardy, D.J., Caracciolo, D., Valtieri, M. and Rovera, G. (1987). Tumor-derived growth factors that support proliferation and differentiation of normal and leukemic hemopoietic cells. *Ann. N.Y. Acad. Sci.* **511**, 30–38.

Tweardy, D.J., Wright, T.M., Ziegler, S.F., Baumann, H., Chakraborty, A., White, S.M., Dyer, K.F. and Rubin, K.A. (1995). Granulocyte colony-stimulating factor rapidly activates a distinct STAT-like protein in normal myeloid cells. *Blood* **86**, 4409–4416.

Valtieri, M., Tweardy, D.J., Caracciolo, D., Johnson, K., Mavilio, F., Altmann, S., Santoli, D. and Rovera, G. (1987). Cytokine-dependent granulocytic differentiation. Regulation of proliferative and differentiative responses in a murine progenitor cell line. *J. Immunol.* **138**, 3829–3835.

Varas, F., Bernad, A. and Bueren, J.A. (1996). Granulocyte colony-stimulating factor mobilizes into peripheral blood the complete clonal repertoire of hematopoietic precursors residing in the bone marrow of mice. *Blood* **88**, 2495–2501.

Vellenga, E., Rambaldi, A., Ernst, T.J., Ostapovicz, D. and Griffin, J.D. (1988). Independent regulation of M-CSF and G-CSF gene expression in human monocytes. *Blood* **71**, 1529–1532.

Wang, S.Y., Chen, L.Y., Tsai, T.F., Su, T.S., Choo, K.B. and Ho, C.K. (1996). Constitutive production of colony-stimulating factors by human hepatoma cell lines: possible correlation with cell differentiation. *Exp. Hematol.* **24**, 437–444.

Watanabe, Y., Kiriyama, M., Ito, R., Kikuchi, R., Mizufune, Y., Nomura, H., Miyazaki, M. and Matsumoto, M. (1996). Pharmacodynamics and pharmacokinetics of recombinant human granulocyte colony-stimulating factor (rhG-CSF) after administration of rectal dosage vehicle. *Biol. Pharm. Bull.* **19**, 1059–1063.

Weinstein, Y., Ihle, J.N., Lavu, S. and Reddy, E.P. (1986). Truncation of the c-*myb* gene by a retroviral integration in an interleukin 3-dependent myeloid leukemia cell line. *Proc. Natl Acad. Sci. U.S.A.* **83**, 5010–5014.

Welte, K., Platzer, E., Lu, L., Gabrilove, J.L., Levi, E., Mertelsmann, R. and Moore, M.A. (1985). Purification and biochemical characterization of human pluripotent hematopoietic colony-stimulating factor. *Proc. Natl Acad. Sci. U.S.A.* **82**, 1526–1530.

Welte, K., Bonilla, M.A., Gillio, A.P., Boone, T.C., Potter, G.K., Gabrilove, J.L., Moore, M.A., O'Reilly, R.J. and Souza, L.M. (1987). Recombinant human granulocyte colony-stimulating factor. Effects on hematopoiesis in normal and cyclophosphamide-treated primates. *J. Exp. Med.* **165**, 941–948.

Welte, K., Gabrilove, J., Bronchud, M.H., Platzer, E. and Morstyn, G. (1996). Filgrastim (r-metHuG-CSF): the first 10 years. *Blood* **88**, 1907–1929.

Williams, G.T., Smith, C.A., Spooncer, E., Dexter, T.M. and Taylor, D.R. (1990). Haemopoietic colony stimulating factors promote cell survival by suppressing apoptosis. *Nature* **343**, 76–79.

Winter, J.N., Lazarus, H.M., Rademaker, A., Villa, M., Mangan, C., Tallman, M., Jahnke, L., Gordon, L., Newman, S., Bryrd, K., Cooper, B.W., Horvath, N., Crum, E., Stadtmauer, E.A., Conklin, E., Bauman, A., Martin, J., Goolsby, C., Gerson, S.L., Bender, J. and O'Gorman, M. (1996). Phase I/II study of combined granulocyte colony-stimulating factor and granulocyte–macrophage colony-stimulating factor administration for the mobilization of hematopoietic progenitor cells. *J. Clin. Oncol.* **14**, 277–286.

Yamasaki, K., Naito, S., Anaguchi, H., Ohkubo, T. and Ota, Y. (1997). Solution structure of an extracellular domain containing the WSxWS motif of the granulocyte colony-stimulating factor receptor and its ligand interaction. *Nature Struct. Biol.* **4**, 498–504.

Yoshikawa, A., Murakami, H. and Nagata, S. (1995). Distinct signal transduction through the tyrosine-containing domains of the granulocyte colony-stimulating factor receptor. *EMBO J.* **14**, 5288–5296.

Yuo, A., Kitagawa, S., Ohsaka, A., Ohta, M., Miyazono, K., Okabe, T., Urabe, A., Saito, M. and Takaku, F. (1989). Recombinant human granulocyte colony-stimulating factor as an activator of human granulocytes: potentiation of responses triggered by receptor-mediated agonists and stimulation of C3bi receptor expression and adherence. *Blood* **74**, 2144–2149.

Yuo, A., Kitagawa, S., Ohsaka, A., Saito, M. and Takaku, F. (1990). Stimulation and priming of human neutrophils by granulocyte colony-stimulating factor and granulocyte–macrophage colony-stimulating factor: qualitative and quantitative differences. *Biochem. Biophys. Res. Commun.* **171**, 491–497.

Zhang, D.E., Zhang, P., Wang, N.D., Hetherington, C.J., Darlington, G.J. and Tenen, D.G. (1997). Absence of granulocyte colony-stimulating factor signaling and neutrophil development in CCAAT enhancer binding protein alpha-deficient mice. *Proc. Natl Acad. Sci. U.S.A.* **94**, 569–574.

Ziegler, S.F., Bird, T.A., Morella, K.K., Mosley, B., Gearing, D.P. and Baumann, H. (1993). Distinct regions of the human granulocyte-colony-stimulating factor receptor cytoplasmic domain are required for proliferation and gene induction. *Mol. Cell. Biol.* **13**, 2384–2390.

Zink, T., Ross, A., Luers, K., Cieslar, C., Rudolph, R. and Holak, T.A. (1994). Structure and dynamics of the human granulocyte colony-stimulating factor determined by NMR spectroscopy. Loop mobility in a four-helix-bundle protein. *Biochemistry* **33**, 8453–8463.

Chapter 24

The Biochemistry of Colony Stimulating Factor-1 Action

John A. Hamilton
Inflammation Research Centre, Department of Medicine, University of Melbourne, Royal Melbourne Hospital, Parkville, Australia

INTRODUCTION

The chief emphasis in this review is to attempt to summarize the most recent information on the signal transduction cascades that are set in train upon the interaction of colony stimulating factor-1 (CSF-1) (or macrophage colony stimulating factor (M-CSF)) with its receptor. Where possible the relevance to cellular function will be mentioned. For further information on CSF-1 and the CSF-1 receptor (CSF-1R) biology, structure, etc., as well as on earlier signal transduction studies not covered, the reader is referred to previous reviews (Stanley, 1985, 1994; Lee, 1991; Sherr, 1991; Vairo and Hamilton, 1991). I shall begin by detailing many of the biochemical responses that CSF-1 elicits and then refer to the receptor biochemistry that has been connected to these responses. It will be realized that there are many disparities in the literature and these will be highlighted so that the reader is at least aware of the problems; where possible, some reasons for these differences will be put forward.

BIOCHEMICAL RESPONSES TO CSF-1

Introduction

CSF-1 is a homodimeric glycoprotein that binds to high-affinity receptors expressed mainly on cells of the monocyte–macrophage lineage (Sherr, 1991). Homozygous disruption of CSF-1 coding sequences in osteopetrotic (*op*/*op*) mice severely impairs production of certain macrophage populations, underlining the importance of CSF-1 for their development (Wiktor-Jedrzejczak *et al.*, 1990; Yoshida *et al.*, 1990). Early studies have shown that CSF-1 exerts pleiotropic effects (e.g. survival, proliferation, differentiation and activation on mononuclear phagocytes) (reviewed by Stanley, 1985, 1994; Vairo and Hamilton, 1991). A number of morphological, biological and biochemical changes are evident soon after the addition of CSF-1. For macrophages to enter S phase, CSF-1 must be present throughout the G1 phase of the cell cycle until they pass a late G1

The Cytokine Handbook, 3rd ed.
ISBN 0-12-689662-3

restriction point, after which they become committed to DNA synthesis (Tushinski and Stanley, 1985; Matsushime *et al.*, 1991). The rate-limiting steps that control this transition and lead to DNA synthesis are regulated, at least in part, by G1 cyclins and their associated catalytic subunits, the cyclin-dependent kinases (cdks) (Sherr, 1993). A critical substrate of the cyclin D-dependent kinases is likely to be the retinoblastoma protein whose phosphorylation late in G1 reverses its growth suppressive function (Nevins, 1992). Addition of CSF-1 to quiescent macrophages results in synthesis of cyclin D1 (Matsushime *et al.* 1991; Cocks *et al.*, 1992; Vadiveloo *et al.*, 1996) and cdk4 (Matsushime *et al.*, 1992; Vadiveloo *et al.*, 1996), as well as phosphorylation of retinoblastoma protein (Vadiveloo *et al.*, 1996). Many studies are trying to link such delayed changes with early events triggered after interaction of CSF-1 with its receptor.

The CSF-1 receptor (CSF-1R) is encoded by the c-*fms* proto-oncogene, the cellular counterpart of the v-*fms* oncogene encoded by the genomes of both Susan McDonough and Hardy-Zuckerman's five strains of feline sarcoma virus, and consists of an extracellular ligand-binding domain joined through a single membrane-spanning helix to a cytoplasmic protein tyrosine kinase (PTK) domain (Coussens *et al.*, 1986; Sherr, 1988). The CSF-1R is more closely related structurally to the α and β receptors for platelet-derived growth factor (PDGF), the c-*kit* proto-oncogene product (the receptor for stem cell factor) and flt3/flk2 than to other PTK receptor subfamilies. The extracellular domains of these receptors consist of the five immunoglobulin-like loops (Wang *et al.*, 1993) and their PTK domains are each interrupted by kinase insert (KI) sequences of variable lengths (Sherr, 1991) (see below). Many of the downstream signalling cascades that have been claimed to be modulated following c-Fms activation will now be discussed.

The Role of Ras

Ras Activation

The *ras* genes encode small GTP-binding proteins whose functions are important for both cell growth and differentiation. Ras proteins have low instrinsic GTPase activity that can be stimulated by a GTPase-activating protein (GAP) by converting active p21 Ras (GTP bound) into the inactive form (GDP bound). CSF-1 activates Ras in NIH3T3 cells expressing human c-Fms (Gibbs *et al.*, 1990) and in BAC1.2F5 cells (Xu *et al.*, 1993, 1995; Büscher *et al.*, 1995). Transgenic mice expressing dominant suppressors of Ras signalling display altered macrophage differentiation, whereas the mature peritoneal macrophages exhibited changed responses to CSF-1, including increased dependence on CSF-1 for survival, changes in the morphological responses and suppression of the increased mRNA expression of a CSF-1-responsive gene (Jin *et al.*, 1995).

It has been found that CSF-1 rapidly and reproducibly induces tyrosine phosphorylation of GAP in both NIH3T3 fibroblasts and in the murine myeloid cell line 32D, when transfected with human c-*fms* (Heidaran *et al.*, 1992). These findings are fairly consistent with those obtained in the NIH3T3 cells transfected with the murine c-*fms* (Reedijk *et al.*, 1990), although there are disagreements about the role of the KI domain in the two studies. In contrast to these results showing tyrosine phosphorylation of GAP, no such phosphorylation was observed in Rat-2 cells expressing murine c-Fms or in BAC1.2F5 cells (Reedijk *et al.*, 1990).

Ras and Cell Proliferation

The growth of fibroblasts transformed by the constitutively active product of the v-*fms* oncogene can be suppressed by microinjection of antibodies to $p21^{Ras}$ (Smith *et al.*, 1986) or by overexpression of the catalytic domain of GAP (Bortner *et al.*, 1991). Similarly, the enforced expression of the GAP catalytic domain in NIH3T3 cells expressing human c-Fms inhibits the activation of genes controlled by 'Ras-responsive elements' and suppresses CSF-1-induced mitogenesis (Bortner *et al.*, 1991). Using NIH3T3 cells transfected with human c-*fms* and expression of an Ets–LacZ fusion protein, it was concluded that the Ets family of transcription factors plays a central role in integrating both c-Fms and Ras-induced mitogenic signals and in modulating the c-*myc* response to CSF-1 stimulation (Langer *et al.*, 1992). The working model developed from these experiments was that CSF-1 stimulation of c-Fms results in a flow of information from a cytoplasmic effector that interacts with the receptor in the region surrounding tyrosine 809 (Y809) through a GTP exchange factor to $p21^{Ras}$, through Ets transcription factors, and ultimately through c-Myc (Fig. 1). However, it was concluded that the proposed Ras–Ets–Myc pathway was probably not sufficient but rather necessary for the proliferative response to CSF-1 (Langer *et al.*, 1992). More recent data with NIH3T3 cells containing the human c-Fms Y809F mutant indicated that the defective mitogenic signalling and concomitant c-Myc expression could be rescued by D-type cyclins, leading to the proposal that c-Myc and cyclin D1 must either

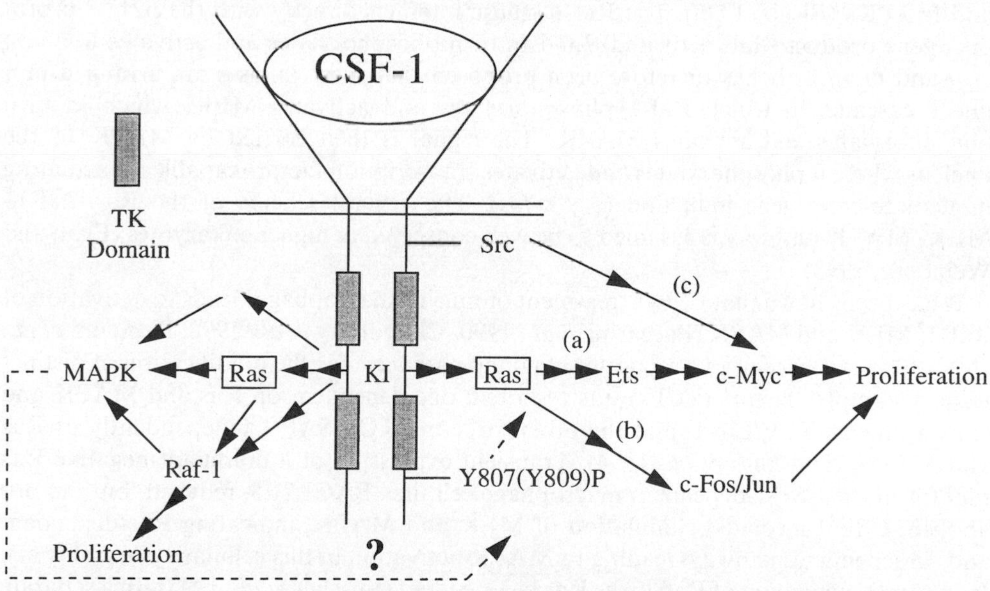

Fig. 1. The role(s) of Ras in CSF-1R signalling? There is debate about whether c-Myc or c-Fos expression lies downstream of Ras activation (pathways (a) and (b), respectively) (Langer *et al.*, 1992; Barone and Courtneidge, 1995); whether or not MAPK (Erk?) activity is involved (Hipskind *et al.*, 1994; Treisman, 1995) is unknown. Alternatively, c-Myc expression has been proposed to lie downstream of Src activation (pathway (c)) (Barone and Courtneidge, 1995). The classical Ras–Raf-1–MEK–MAPK (Erk) pathway may not be of major significance as there is evidence for Ras independence for MAPK and Raf-1 activation (Büscher *et al.*, 1995; Xu *et al.*, 1995); Raf-1 may not be the MEK kinase in CSF-1-activated macrophages.

act in parallel cross-concomitant pathways or, alternatively, that Myc function may be required both upstream and downstream of cyclin D1 for cells to proliferate continuously (Roussel *et al.*, 1995).

Again using NIH3T3 cells expressing human c-Fms and microinjection techniques, Barone and Courtneidge (1995) appear to have drawn different conclusions in that they suggested that, in response to CSF-1, Src family kinases activate a pathway (most probably Ras independent) which culminates in the transcription of c-*myc* and which is required for DNA synthesis (Fig. 1); they also concluded that an important downstream component of the Ras pathway is the induction of c-Fos expression and not that of c-Myc (Fig. 1). Neither the Src pathway nor the Ras pathway alone is sufficient to induce DNA synthesis. In the same cells, CSF-1 activated three Src family kinases, Src, Fyn and Yes, as well as their concomitant association with the ligand-activated c-Fms (Courtneidge *et al.*, 1993); the Y809F mutation reduced both their binding and enzymatic activation. It should be noted from the discussion later that the binding site for the Src family kinases on human c-Fms has been mapped to Y561 and surrounding amino acids (Alonso *et al.*, 1995).

Ras and MAP Kinase

A variety of mitogens lead to the formation of membrane-associated signalling complexes that deliver an essential signal via Ras to a protein kinase network involving the Raf-1 kinase, the mitogen-activated protein kinase (MAPK) activator, viz. MEK, and MAPKs (Blenis, 1993). The Raf-1 kinase interacts directly with the $p21^{Ras}$ proto-oncogene product while activated Raf-1 in turn phosphorylates and activates MEK *in vivo* and *in vitro*. It has therefore been proposed that these kinases are arranged in a linear cascade, in which Raf-1 phosphorylates and activates MEK, which in turn phosphorylates and activates MAPK. The signal is then carried by MAPK to the nucleus where it phosphorylates and activates transcription factors capable of mediating immediate-early gene induction (e.g. c-*fos*). The general scheme of the Ras–Raf-1–MEK–MAPK pathway is assumed to be well conserved in higher eukaryotes (Egan and Weinberg, 1993).

It has been shown that CSF-1 treatment of murine macrophages leads to activation of Raf-1, MEK and MAPK (Baccarini *et al.*, 1990; Choudhury *et al.*, 1990; Reimann *et al.*, 1994). The existence of two separate signalling pathways for the modification of distinct ternary complex factors (TCFs) was reported: one dependent on Ras and MAPK and converging on TCF/Elk-1, and the other targeting TCF/SAP-1 independently of Ras and MAPK (Hipskind *et al.*, 1994). Transient expression of a dominant-negative Ras mutant in the CSF-1-dependent macrophage cell line, BAC1.2F5, reduced, but did not abolish, CSF-1-mediated stimulation of MEK and MAPK, indicating Ras-dependent and -independent pathways leading to MAPK activation in this cellular system (Fig. 1). In contrast, activation of Raf-1 was Ras independent (Büscher *et al.*, 1995). In a separate study with BAC1.2F5 cells treated with CSF-1, such a Ras-independent mechanism for Raf-1 activation was claimed (Xu *et al.*, 1995). More recent findings in CSF-1-treated murine macrophages indicate that the protein tyrosine phosphatase SHP-1 is involved in Ras-mediated activation of the MAPK pathway but not of Raf-1 kinase (Krautwald *et al.*, 1996). All of the above data are consistent with the hypothesis that Raf-1 may not be the CSF-1-induced MEK kinase in macrophages (Fig. 1). It is likely that MAPK (Erk) activation by CSF-1 will be involved in c-Fos and/or c-Myc expression (Hipskind *et al.*, 1994; Treisman, 1995).

As regards cell proliferation, oncogenically active Raf-1 protein (v-Raf) conferred CSF-1-independent growth on the BAC1.2F5 cells and led to immediate early gene expression without MAPK activation (Büscher *et al.*, 1993). This v-Raf expression suppressed CSF-1-stimulated MAPK activity, which appeared to be due to a feedback mechanism whereby v-Raf induced a MAPK-specific phosphatase, MKP-1 (Krautwald *et al.*, 1995). The implication is that MAPK activation may not be necessary for CSF-1-induced proliferation, a conclusion that is consistent with our studies of Erk-1 activation in cycling and poorly cycling murine macrophages (Jaworowski *et al.*, 1996).

In murine macrophages evidence was obtained for two kinetically distinct pathways of Erk-1 activation by CSF-1 with peak activation at 5 and 15 min and which could be distinguished from the action of lipopolysaccharide (LPS), which activated Erk-1 with a single peak corresponding to the later peak of activation by CSF-1 (Jaworowski *et al.*, 1996). For the activation of the other MAPK family members, p38 and JNK-1, CSF-1, unlike LPS, was a poor stimulator (A. Jaworowski *et al.*, unpublished results). As the p38 MAPK is implicated in the control of cytokine translation in LPS-stimulated monocytes–macrophages (Lee *et al.*, 1994), the weak stimulation by CSF-1 is consistent with the observation that CSF-1 is a poor inducer of cytokine secretion in monocyte–macrophages (Strassmann *et al.*, 1991; Hamilton, 1993; Hamilton *et al.*, 1993); however, others have claimed that CSF-1 stimulates p38 activity in FD-*Mycobacterium avium* complex (MAC) II cells (Foltz *et al.*, 1997).

The discussion outlined above in this section illustrates that the Ras–Raf-1–MEK–MAPK pathway may not be that significant in CSF-1-mediated signalling in macrophages, including the signalling relevant for the proliferative response.

Phospholipase and Protein Kinase C Activation

In CSF-1-treated murine bone marrow-derived macrophages (BMMs) and human monocytes, no evidence for phospholipase C-mediated hydrolysis of phosphoinsositides (i.e. phospholipase Cγ activity) could be found (Whetton *et al.*, 1986; Hamilton *et al.*, 1989; Imamura *et al.*, 1990), in contrast to what has been found in response to PDGF; it has been proposed that v-Fms could be coupled to a guanine nucleotide-dependent polyphoinositide-specific phospholipase C generating diacylglycerol and inositol trisphosphate (Jackowski *et al.*, 1986). In NIH3T3 cells transfected with human c-*fms*, and in BAC1.2F5 cells, no tyrosine phosphorylation of phospholipase Cγ could be detected (Downing *et al.*, 1989).

Several reports have shown that CSF-1 stimulates diacylglycerol formation in a variety of cell systems (Imamura *et al.*, 1990; Choudhury *et al.*, 1991a; Veis and Hamilton, 1991; Xu *et al.*, 1993, 1995); however, further experiments with murine BMMs in our laboratory could not confirm this (Jaworowski *et al.*, 1994). In BAC1.2F5 cells the stimulation of diglyceride production via phospholipase C hydrolysis of phosphatidylcholine was proposed to be an early event in the mitogenic action of CSF-1 (Xu *et al.*, 1993). Cell extracts from CSF-1-treated receptor-bearing cells did not show evidence for activation of a phosphatidylcholine-specific phospholipase C (Choudhury *et al.*, 1991b). We could find no evidence for phospholipase D or protein kinase C (PKC) activation in murine BMMs in response to CSF-1 (Jaworowski *et al.*,

1994); in human monocytes, however, PKC is activated by CSF-1 (Imamura *et al.*, 1990). When PKC activity (particularly PKCδ activity) is monitored in BAC1.2F5 cells, CSF-1 has been claimed to be able to activate it in one report (Xu *et al.*, 1995) but not in others (Büscher *et al.*, 1993, 1995; Xu *et al.*, 1993; Reimann *et al.*, 1994).

In human monocytes, CSF-1 triggered the phosphorylation and activation of cytosolic phospholipase A_2 that appears to correlate with arachidonic acid release and prostaglandin E_2 (PGE_2) formation (Nakamura *et al.*, 1992); however, we could not confirm PGE_2 formation in this system (Hamilton *et al.*, 1993). In murine macrophages phosphorylation of cytosolic phospholipase A_2 by CSF-1 was not accompanied by arachidonic acid release and PGE_2 formation (Jackowski *et al.*, 1990; Puri *et al.*, 1991; Strassmann *et al.*, 1991; Xu *et al.*, 1994).

S6 Kinase

Phosphorylation of the S6 protein of the 40*S* ribosomal subunit is a highly conserved response of animal cells to treatment with a wide range of stimuli, including growth factors. $p70^{S6K}$ is the physiological S6 kinase activity in mammalian cells. Rapamycin, which can prevent activation of $p70^{S6K}$ in many cells via its ultimate target mTOR/RAFT/FRAP, inhibits CSF-1-stimulated DNA synthesis in BAC1.2F5 cells (Kato *et al.*, 1994) and cell proliferation in murine BMMs (Cooper *et al.*, 1994). However, for CSF-1-stimulated BMM DNA synthesis, we found that rapamycin had a relatively incomplete inhibitory effect (Hamilton *et al.*, in press). In this same study data were obtained that were consistent with phosphatidylinositol-3 (PI-3) kinase lying upstream of $p70^{S6K}$ activity.

G Proteins

G proteins are known signal transducers for many hormones, neurotransmitters and chemotactic substances, but not for cytokines. Pertussis toxin can inhibit the activity of some G proteins by ADP ribosylation of the active α subunit of the heterotrimeric G-protein complex. CSF-1 increased GTP binding to and raised GTPase activity in human monocyte membrane preparations and a variety of responses of these cells to CSF-1 are pertussis toxin sensitive (Imamura and Kufe, 1988; Imamura *et al.*, 1990). A similar conclusion was reached based on studies of signal transduction in hamster fibroblasts expressing human c-Fms (Hartmann *et al.*, 1990) and in studies with murine BMMs but not with murine macrophage cell lines (Xie *et al.*, 1993). In some cases the actions of CSF-1 on murine haemopoietic cells, including macrophages, appear to be less sensitive to the effects of pertussis toxin (He *et al.*, 1988; Hamilton *et al.*, 1991). Inhibition of G_{i2} function inhibits CSF-1-mediated signal transduction in BAC1.2F5 cells, leading to the conclusion that the metabolic pathways regulated by G_{i2} proteins and c-Fms converge on a common effector necessary for the regulation of macrophage survival and proliferation (Corre and Hermouet, 1995).

Na^+/H^+ Antiport Activity

We have shown that CSF-1 induces a rapid increase in Na^+/H^+ exchange activity in murine BMMs, the accompanying Na^+ influx subsequently increasing Na^+/H^+-

ATPase (Na^+ pump) activity (Vairo and Hamilton, 1988; Vairo *et al.*, 1990a); persistent stimulation of this 'Na^+ cycle' occurred for most of the BMM cycle. Inhibition of Na^+/H^+ exchange activity with dimethylamiloride completely abolished CSF-1-induced BMM proliferation even when added up to 8 h after the growth factor, suggesting that Na^+/H^+ antiport activity late in the G1 phase of the BMM cell cycle is required for S-phase entry (Vairo *et al.*, 1990a). It was subsequently found that dimethylamiloride and other amiloride analogues suppressed the induction of mRNA expression for M1 and M2 subunits of ribonucleotide reductase but not for proliferating cell nuclear antigen or cyclin D1, indicating selective suppression of 'cell cycle' genes by Na^+/H^+ antiport inhibitors in this system (Vairo *et al.*, 1992a).

Tyrosine Phosphorylation

Several proteins are tyrosine phosphorylated after c-Fms activation and many are beginning to be identified (Sengupta *et al.*, 1988; Yeung *et al.*, 1992; Lioubin *et al.*, 1994). Several of these are now discussed.

PI-3 Kinase Activation and Function

There are reports in various cell populations that CSF-1 can activate PI-3 kinase activity including in antiphosphotyrosine immunoprecipitates (Varticovski *et al.*, 1989; Reedijk *et al.*, 1990, 1992; Shurtleff *et al.*, 1990; Choudhury *et al.*, 1991c; van der Geer and Hunter, 1991; Yusoff *et al.*, 1994). We have found that the PI-3 kinase p85α subunit is tyrosine phosphorylated on CSF-1 stimulation in primary cultures of murine macrophages (Kanagasundaram *et al.*, 1996), agreeing with a recent report on human monocytes (Saleem *et al.*, 1995). These findings seem to differ from observations made in fibroblasts engineered to express human c-Fms where there was little if any tyrosine phosphorylation of PI-3 kinase following receptor activation (Shurtleff *et al.*, 1990). A stable association of PI-3 kinase with c-Fms, c-Cbl and several other tyrosine phosphorylated proteins has recently been reported for CSF-1-stimulated murine macrophages (Kanagasundaram *et al.*, 1996). These last observations are interesting given the view that 'the tyrosine phosphorylated CSF-1R does not appear to physically associate with many other intracellular signalling molecules' (Stanley, 1994). The evidence for this notion comes from the observations that activated c-Fms does not associate stably with phospholipase Cγ (Downing *et al.*, 1989), Raf-1 (Baccarini *et al.*, 1990), ras GTPase activating protein (GAP) (Reedijk *et al.*, 1990), SHP-1 (Yeung *et al.*, 1992) or other proteins that can be labelled with ^{35}S-methionine or ^{32}P-phosphate (Stanley, 1994). In the same study, where the stable association of activated c-Fms with several tyrosine-phosphorylated proteins was observed (Kanagasundaram *et al.*, 1996), we found that the only detectable ^{35}S-labelled protein found in anti-c-Fms immunoprecipitates whose levels increased consistently upon CSF-1 stimulation appeared to be the PI-3 kinase p85α subunit.

As regards the involvement of PI-3 kinase in the proliferative response, early reports using either the murine c-Fms Y721F mutant in Rat-2 fibroblasts (Reedijk *et al.*, 1990; van der Geer and Hunter, 1993) or the KI insert mutant of human c-Fms in NIH3T3 cells (Shurtleff *et al.*, 1990; Choudhury *et al.*, 1991c) suggested a relationship based on concordant reduction in kinase activity and cell proliferation (Y721 in the KI domain is believed to be the binding site for the PI-3 kinase p85 regulatory subunit (Reedijk *et al.*,

1992) (see below)). However, data with both human c-Fms Y809F in NIH3T3 cells (Roussel *et al.*, 1990) and with murine c-Fms Y807F in Rat-2 fibroblasts (van der Geer and Hunter, 1991) show that the reduction in the proliferative response (see later) was not paralleled by a change in PI-3 kinase activity. More recent reports using microinjection of antibodies to the p110α PI-3 kinase catalytic subunit into NIH3T3 cells expressing human c-Fms (Roche *et al.*, 1994) or using the PI-3 kinase inhibitor, wortmannin, with CSF-1-treated BMMs (Hamilton *et al.*, in press) failed to find a role for PI-3 kinase in the control of DNA synthesis; in the latter study, and by the same approach, it was concluded that PI-3 kinase activity does not have a role in the control of c-*fos* mRNA expression and activation of Erk-1 activity, in contrast to other growth factor cellular responses.

PI-3 kinase has been shown to play a role in the endocytic process for growth factor receptors, such as PDGF-R (Joly *et al.*, 1995). However, wortmannin does not significantly affect c-Fms degradation in CSF-1-treated murine macrophages, suggesting a lack of requirement for PI-3 kinase activity in this case (V. Kanagasundaram *et al.*, unpublished results).

c-Cbl and Ubiquitination

Among the cellular proteins that are tyrosine phosphorylated in response to CSF-1 in murine macrophages is the proto-oncogene product c-Cbl (Tanaka *et al.*, 1995; Kanagasundaram *et al.*, 1996; Wang *et al.*, 1996). Upon tyrosine phosphorylation, c-Cbl associates with tyrosine-phosphorylated Shc and an unknown phosphorylated protein, pp80, and is transiently ubiquitinated and translocated to the plasma membrane (Wang *et al.*, 1996). Under these conditions, c-Fms is also ubiquitinated (Mori *et al.*, 1995; Wang *et al.*, 1996). As noted earlier, a stable association of c-Cbl with c-Fms, PI-3 kinase and several other tyrosine phosphorylated proteins has been observed in CSF-1-stimulated murine macrophages (Kanagasundaram *et al.*, 1996).

Shc and Associated Proteins

For the positive regulation of Ras activity in many cells, the guanine nucleotide exchange factors, for example Sos and other Ras guanine nucleotide-releasing factors, act by increasing the rate of exchange of GDP for GTP. Two adaptor molecules, Grb2 and Shc, seem to be involved in coupling Ras regulators, such as Sos, to PTK receptors. After CSF-1 stimulation of murine myeloid cell lines expressing the murine c-Fms, $p52^{Shc}$ and $p46^{Shc}$ are tyrosine phosphorylated and associated with Grb2 and c-Fms (Lioubin *et al.*, 1994; Welham *et al.*, 1994), although for some reason this association was not noted in CSF-1-treated BAC1.2F5 cells (Xu *et al.*, 1995). The conclusion from one of these publications (Lioubin *et al.*, 1994) was that CSF-1 stimulation of myeloid cells could activate Ras through Sos1 by Grb2 binding to Fms, Shc or p150 (subsequently identified as $p150^{Ship}$; see below) and that Fms signal transduction in myeloid cells differs from that in fibroblasts. With CSF-1-treated human monocytes, Kufe's laboratory has evidence for at least three Grb2 pools with independent association of Grb2 with c-Fms and Shc as well as separate Grb2 complexes with pp125 focal adhesion kinase (FAK) and dynamin, and with PI-3 kinase (Kharbanda *et al.*, 1995; Saleem *et al.*, 1995). The last complex was proposed as providing a potential mechanism for CSF-1-induced interaction of PI-3 kinase and Ras (Saleem *et al.*, 1995). If this were so, and if Ras is controlling Fos/Jun expression as discussed above (Barone and Courtneidge, 1995), then it would be predicted that PI-3 kinase inhibition would give rise to a concomitant inhibition of Fos expression; however, wortmannin failed to suppress CSF-1-stimulated c-*fos* mRNA expression in murine macrophages (Hamilton *et al.*, in press).

Very recently it has been shown (Chen *et al.*, 1996) that Grb2 becomes associated with the tyrosine-phosphorylated phosphatase SHP-1 and other tyrosine-phosphorylated proteins in CSF-1-treated murine macrophages (see below), the purpose of which may be to recruit to SHP-1 substrates that bind to the Grb2 Src homology region 3 (SH3) domains as a potential new function for Grb2 (Chen *et al.*, 1996). The relationships of the various Grb2-containing complexes listed above in CSF-1-treated cells are unclear; as mentioned, Y697 in murine c-Fms would appear to be the Grb2 association site based on experiments with Rat-2 fibroblasts.

p150^{Ship}

As referred to above, in CSF-1-treated FDC-P1 myeloid cells expressing c-Fms (but not in the analogous fibroblasts), there is induction of the tyrosine phosphorylation and association of a 150-kDa protein with the phosphotyrosine-binding (PTB) domain of Shc; this novel protein has been termed p150^{Ship} and has inositol polyphosphate-5-phosphatase activity (Lioubin *et al.*, 1994, 1996). Mutations in c-Fms that abolish its interactions with Grb2/Sos (Y697F) and PI-3 kinase (Y721F) (see below) do not abrogate p150^{Ship} phosphorylation and its association with Grb2 and Shc. Likewise the Y706F and Y807F mutations do not affect these parameters but the triple mutation, Y697–706–807F, does. Data have been presented for a role for p150^{Ship} as a negative regulator of cell growth (Lioubin *et al.*, 1996).

SHP-1

SHP-1 (PTP1C, HCP, SHPTP1) is an SH2 domain-containing tyrosine phosphatase expressed predominantly in haemopoietic cells. It has been reported that SHP-1 becomes tyrosine phosphorylated following CSF-1 stimulation of BAC1.2F5 cells, although for some reason it was a major substrate in one study but was weakly phosphorylated in the other (Yeung *et al.*, 1992; Yi and Ihle, 1993); in contrast to c-Kit, association between SHP-1 and c-Fms was not observed (Yi and Ihle, 1993). Upon CSF-1 stimulation of murine macrophages, SHP-1 has been found to be associated with an unidentified 130-kDa tyrosine phosphoprotein (Chen *et al.*, 1996).

Genetic evidence suggests that SHP-1 is a negative regulator of haemopoietic cell signal transduction. Homozygous mutations in the SHP-1 gene give rise to motheaten (*me*/*me*) and motheaten viable (*me^v*/*me^v*) mice (Shultz *et al.*, 1993). Because of an early frameshift mutation, *me*/*me* mice are essentially SHP-1 null; the *me^v*/*me^v* mutation results in SHP-1 species with markedly decreased phosphatase activity. Motheaten mice display multiple defects in haemopoietic cells, which result in immunodeficiency, systemic autoimmunity and premature death. One study claimed that splenic macrophages from *me*/*me* mice proliferate in the absence of CSF-1 (McCoy *et al.*, 1984), while it was found recently that their BMMs retain an absolute requirement for CSF-1 for both viability and proliferation and are hyper-responsive to CSF-1 (Chen *et al.*, 1996). BMMs from *me*/*me* mice were used to show that Grb2 is associated, via its SH3 domain, with several tyrosine-phosphorylated proteins, leading to the concept (see above) that Grb2 may recruit substrates for SHP-1 which may therefore be a critical negative regulator of CSF-1 signalling *in vivo* (Chen *et al.*, 1996). By using macrophages expressing dominant-negative Ras and the me^v/me^v mouse mutant, it has recently been shown that SHP-1 is activated by CSF-1 in a Ras-dependent manner and that its activity is necessary for the Ras-dependent activation of the MAPK pathway but not that of Raf-1 kinase (Krautwald *et al.*, 1996).

The serine–threonine protein phosphatase 2A (PP2A) is rapidly activated in CSF-1-treated BAC1.2F5 cells and can be associated with Raf-1 (M. Baccarini, personal communication).

JAK–STAT Pathways

It has been shown that CSF-1 causes the tyrosine phosphorylation of TYK2, a protein kinase belonging to the family of Janus kinase (JAK) kinases, in murine BMMs as well as in NIH3T3 cells expressing human c-Fms (Novak *et al.*, 1995); in the latter cell type, but not in the former, JAK1 was also phosphorylated. In addition, CSF-1 activated the signal transducers and activators of transcription, STAT1, STAT3, STAT5A and STAT5B (Novak *et al.*, 1995, 1996).

Vav

The proto-oncogene product, Vav, has been shown to associate with receptors by SH2 domains and to be tyrosine phosphorylated (Margolis *et al.*, 1992). However, CSF-1 treatment of BAC1.2F5 cells did not lead to increased tyrosine phosphorylation of the constitutively expressed protein (Margolis *et al.*, 1992; Xu *et al.*, 1995).

Effects of cAMP and other Inhibitors of CSF-1-stimulated Cell Cycle Progression

A number of agents have been found to suppress the proliferative response of macrophages to CSF-1 and used as tools to understand CSF-1 action (Vairo and Hamilton, 1988; Vairo *et al.*, 1990a,b, 1991, 1992a,b, 1995; Cocks *et al.*, 1992; Hamilton *et al.*, 1992; Rock *et al.*, 1992; Kato *et al.*, 1994; Vadiveloo *et al.*, 1996). The studies have indicated that different antiproliferative agents (8BrcAMP, LPS, interferon-γ (IFN-γ), dimethylamiloride) arrest CSF-1-treated murine macrophages at distinct checkpoints in the G1 phase of the cell cycle and that their effects on cell cycle machinery are linked to the position at which they arrest cells in G1 (Hamilton *et al.*, 1992; Vadiveloo *et al.*, 1996).

We have evidence that increased cAMP activity, which results in increased protein kinase A activity, reduces cyclin D1 and cdk4 expression in CSF-1-treated murine macrophages (Cocks *et al.*, 1992; Vadiveloo *et al.*, 1996), in contrast to a study where no such reduction occurred but induction of $p27^{Kip1}$ was given as the explanation for the cAMP-induced G1 phase arrest (Kato *et al.*, 1994). 8BrcAMP also resulted in down-modulation of c-*myc* mRNA expression. Overexpression of c-Myc was not able to abrogate the antiproliferative effect of 8BrcAMP; however, it did reverse the proliferation inhibition caused by the addition of IFN-γ, indicating that, while both agents shared an ability to lower c-Myc expression, they also acted via distinct mechanisms to bring about growth arrest (Vairo *et al.*, 1995).

It was also observed that endogenous type I interferon in murine macrophage cultures could suppress drastically the degree of proliferation found in cultures of CSF-1-treated murine macrophages, thereby masking the proliferative response (Hamilton *et al.*, 1996); the varying interferon levels might explain the variability in the proliferative response in such cultures.

We have found that tumour necrosis factor-α (TNF-α) suppresses CSF-1-induced murine macrophage proliferation (Vairo *et al.*, 1991), whereas others have reported that it, in fact, can enhance this proliferation (Branch *et al.*, 1989; Shieh *et al.*, 1989).

CSF-1 RECEPTOR

c-Fms Dimerization, Internalization and Activating Mutations

Dimerization

Upon addition of CSF-1 to the murine macrophage cell line BAC1.2F5, it has been reported that cell surface c-Fms undergoes the following sequence of changes: non-covalent dimerization → autophosphorylation and signalling → covalent dimerization via disulphide bonds → modification (homodimer to heterodimer) → tyrosine dephosphorylation and internalization (Li and Stanley, 1991). It was proposed that ligand-induced non-covalent dimerization activated c-Fms kinase activity, whereas the covalent dimerization led to the subsequent changes. In contrast, for the myeloid FDC-P1 cells engineered to express murine c-Fms, dimers could be detected only using cross-linkers (Carlberg and Rohrschneider, 1994).

Internalization

In the absence of CSF-1, c-Fms has little or no tyrosine kinase activity and remains, apparently unaltered, on the cell surface for several hours (Rettenmier *et al.*, 1987; Carlberg *et al.*, 1991). Internalization of CSF-1-bound receptors occurs rapidly; this internalization is via clathrin-coated pits and vesicles (Manger *et al.*, 1984) and targeted to lysosomes, where both the receptor and ligand are degraded (Guilbert and Stanley, 1986). A mutant c-Fms protein lacking tyrosine kinase activity was rapidly internalized after CSF-1 binding but not degraded (Carlberg *et al.*, 1991). Another mutant c-Fms molecule that lacked most of the KI region was similarly internalized after CSF-1 binding and was also not degraded. These data indicate that the signal for internalization is separate from that directing degradation of the receptor. It was also claimed (Carlberg *et al.*, 1991) that CSF-1-induced degradation of c-Fms may require tyrosine phosophorylation of a protein other than that of c-Fms itself and that the kinase insert region may be necessary for recognition of this substrate.

An amino acid segment near the cytoplasmic juxtamembrane region surrounding tyrosine-569 in murine c-Fms would appear to be important for internalization because its mutation to alanine eliminates ligand-induced rapid endocytosis of receptor molecules (Myles *et al.*, 1994). The mutant Fms Y569A also lacks tyrosine kinase activity. Thus tyrosine-569 within the juxtamembrane region of murine c-Fms is essential for kinase activity and CSF-1-dependent internalization. However, receptor tyrosine kinase activity was shown not to be essential for its endocytosis as the kinase-inactive receptor c-Fms K614A underwent endocytosis, albeit at a reduced rate.

Activating Mutations

The transforming mutations in v-*fms*, which are missing in c-*fms*, lead to activation even in the absence of CSF-1. These transforming or activating mutations were identified by assaying the ability of c-*fms* and v-*fms* chimaeric genes to induce foci formation of Rat-2 or NIH3T3 fibroblasts (Roussel *et al.*, 1988; Woolford *et al.*, 1988). In this way, it was shown that *C*-terminal deletion in v-Fms had a transforming effect. In addition, v-Fms has transforming point mutations in the extracellular domain. Many additional single point mutations, which alter the encoded extracellular domain, can also activate c-Fms (Wetters *et al.*, 1992). Mutations at amino acids 301 and 374 in the fourth immunoglobulin-like loop of the extracellular domain of murine c-Fms are necessary, but not sufficient, for receptor dimerization in the absence of CSF-1 (Carlberg and Rohrschneider, 1994). Only receptors with a truncated *C*-terminus as well as the extracellular domain mutations, dimerized efficiently in the absence of CSF-1, suggesting that the *C*-terminus of c-Fms also regulated receptor oligomerization. Receptors with mutations in both regions underwent ligand-independent aggregation, transphosphorylation, and phosphorylation of cellular proteins, followed by rapid internalization and degradation. These results suggest that CSF-1 binding to c-Fms initiates activation by inducing conformational changes in both the cytoplasmic *C*-terminal domain and the fourth immunoglobulin-like loop of the extracellular domain, leading to the formation of stable receptor dimers (Carlberg and Rohrschneider, 1994).

There are data to indicate that the cell type in which a mutated c-Fms is expressed can also determine the biological outcome of the mutation (see also below). In the myeloid FDC-P1 cells, c-Fms D802V is a very potent 'transforming' mutation (Glover *et al.*, 1995). However, this mutation in c-Fms could not induce Rat-2 fibroblast foci, even in the presence of CSF-1; in this respect c-Fms D802V has the characteristics of an inactivating mutation in Rat-2 fibroblasts (Glover *et al.*, 1995), despite the fact that it was selected as an activating mutation in FDC-P1 cells.

c-Fms Tyrosine Phosphorylation

Upon CSF-1 binding and dimerization the tyrosine kinase domain of c-Fms is activated, resulting in transphosphorylation of specific tyrosine residues within the cytoplasmic domain (Ohtsuka *et al.*, 1990) with the potential to create high-affinity binding sites for Src homology region 2 (SH2) domains, PTB domains, etc., contained within cytoplasmic signalling proteins. Several tyrosine 'autophosphorylation' sites have been identified to date in the Fms protein, either directly by tryptic phosphopeptide analysis or inferred by other approaches. For the murine receptor these are: Y697, Y706, Y721 (each located in the KI region), Y807 and Y973 in the tyrosine kinase domain, and Y559 in the juxtamembrane region (Tapley *et al.*, 1990; van der Geer and Hunter, 1990; Reedijk *et al.*, 1992; Joos *et al.*, 1996) (Fig. 2); the equivalent tyrosines in human Fms are at positions 699, 708, 723, 809, 969 and 561, respectively (Sherr, 1991). However, it should be noted that it is unlikely that all of the tyrosine phosphorylation sites in activated c-Fms have been discovered (van der Geer and Hunter, 1990; Downing *et al.*, 1991); this consideration is given further credibility by recent studies with v-Fms where an additional 'major autophosphorylation site' has been claimed (Joos *et al.*, 1996) (see below). One of the difficulties in assigning all of the sites is that the c-Fms tyrosine

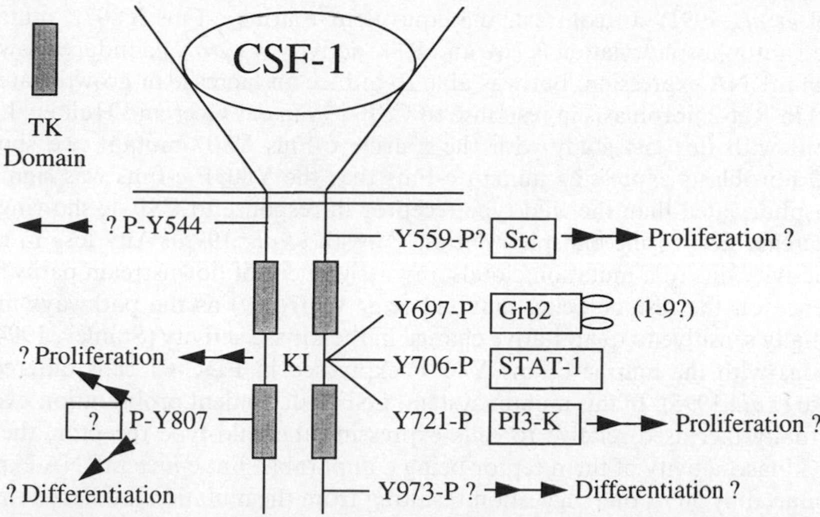

Fig. 2. Cytoplasmic regions of c-Fms and downstream pathways? The activated murine c-Fms protein is depicted with the TK and KI domains shown. It would appear that phosphorylated Y559, 697, 706 and 721 are the binding sites for the Src family, Grb2, STAT1 and PI-3 K p85α subunit respectively. A role for Y807/809 in the CSF-1 proliferative response has been proposed (Roussel *et al.*, 1990), although it may be important for differentiation rather than proliferation (Bourette *et al.*, 1995); for v-Fms there is association with tyrosine-phosphorylated p120 Ras GAP (Trouliaris *et al.*, 1995), but there are no reports for any protein to be associated with Y807/809 in c-Fms. The roles of the KI domain and PI-3 kinase (PI3-K) in the proliferative response are debated. There is the suggestion that Y973 is phosphorylated and involved in differentiation signalling (Roussel *et al.*, 1987; Borzillo *et al.*, 1990; Joos *et al.*, 1996). Numerous complexes (nine or more) containing Grb2 have been described; their relationships are unclear. The equivalent tyrosine to Y544 in v-Fms has been shown to be a 'major autophosphorylation site' (Joos *et al.*, 1996) but not reported in c-Fms.

phosphorylation resulting from CSF-1 action is transient; also, it is still possible that some of the assigned sites may not actually result from an autophosphorylation reaction but be due to an indirect effect of another tyrosine kinase (e.g. TYK2). It would seem that further studies should be undertaken.

The roles of the identified tyrosine 'autophosphorylation' sites in cellular function will be discussed in turn, as will the binding of specific proteins.

Y809 (Human) and Y807 (Murine)

Ligand-activated human c-Fms can induce continuous proliferation when ectopically expressed in murine NIH3T3 fibroblasts, replacing their requirement for PDGF and insulin-like growth factor-1 (Roussel and Sherr, 1989). In contrast, a receptor mutant containing a phenylalanine for tyrosine substitution at codon 809 (here designated c-Fms Y809F) was mitogenically inactive (Roussel *et al.*, 1990) and ligand-stimulated cells remained arrested in early G1 phase (Roussel *et al.*, 1994). Although the cells expressing c-Fms Y809F exhibited normal receptor tyrosine kinase (RTK) activity *in vitro* and *in vivo*, and unimpaired c-*fos* and c-*jun* expression (Roussel *et al.*, 1990), c-*myc* mRNA was poorly induced and its levels were unsustained. Enforced expression of c-Myc in these cells overcame the G1 arrest and restored the proliferative response, suggesting that c-Myc is required for CSF-1-induced mitogenesis in this cellular system

(Roussel *et al.*, 1991). In contrast, the equivalent murine c-Fms Y807F mutant had decreased autophosphorylation *in vivo* and PTK activity *in vitro*, and induced lower c-*fos* and c-*jun* mRNA expression, but was able to induce an increase in growth rate (albeit reduced) in Rat-2 fibroblasts in response to CSF-1 (van der Geer and Hunter, 1991). In agreement with this last study with the murine c-Fms Y807 mutant, we showed in NIH3T3 fibroblasts expressing human c-Fms that the Y809F c-Fms was significantly less phosphorylated than the wild-type receptor in response to CSF-1, showing only a small increase in tyrosine phosphorylation (Novak *et al.*, 1995). Any loss in receptor kinase activity due to a mutation clouds any assignment of downstream pathways to a dependence on the residue being mutated (e.g. Y807/809) as the pathways might be differentially sensitive to quantitative change in the kinase activity (Stanley, 1994).

Findings with the murine c-Fms Y807F expressed in FDC-P1 cells differed again (Bourette *et al.*, 1995). In this mutant system, CSF-1-dependent proliferation over a few days actually increased relative to cells expressing the wild-type receptor, the *in vivo* tyrosine kinase activity of the receptor being comparable, but c-*myc* mRNA expression being reduced by 50%; one suggestion resulting from the mutation studies in the FDC-P1 cells was that c-Myc induction is not absolutely required for growth (Bourette *et al.*, 1995). Interestingly in the FDC-P1 cells the mutation in the Y807 site totally abrogated CSF-1-induced differentiation (instead of proliferation) indicating, in these cells at least, that Y807 is required for initiation of a differentiation signal and not a proliferation one. The different conclusions regarding the effects of the Y807F (or Y809F) mutation on the proliferative response to CSF-1 are depicted in Fig. 2.

Tyrosines in the c-Fms KI Domain

In NIH3T3 cells genetically engineered to express human c-Fms mutants, deletion of 67 amino acids from the KI domain, including Y699, Y708 and Y723, resulted in a partially impaired ligand-dependent proliferative response with an intact *in vivo* and *in vitro* tyrosine kinase activity (Shurtleff *et al.*, 1990; Choudhury *et al.*, 1991c); however, in Rat-2 fibroblasts transfected with mutated murine c-*fms*, CSF-1-dependent growth signalling was abrogated in cells containing the Y697F mutant, the Y721F mutant or the mutant with Y697, Y706 and Y721 all changed to phenylalanine (Reedijk *et al.*, 1990; van der Geer and Hunter, 1993); for the last mutant there was a reduction in autophosphorylation activity (van der Geer and Hunter, 1993). Again, in contrast to the above studies with transfected fibroblasts, for FDC-P1 cells transfected with the receptor with Y697, Y706 and Y721 all mutated to phenylalanine, CSF-1-dependent proliferation was slightly increased with *in vivo* tyrosine kinase activity intact (Bourette *et al.*, 1995). This uncertainty over the role of the tyrosine phosphorylation sites in the KI domain in the proliferative response to CSF-1 in the various cell systems is indicated in Fig. 2.

Y973

It has been implied that Y973 in the *C*-terminal region of murine c-Fms (Y969 in human c-Fms) is tyrosine phosphorylated (Joos *et al.*, 1996) and has been implicated in the lineage switching of pre-B cells to macrophages (i.e. differentiation) (Borzillo *et al.*, 1990) and in the negative regulation of growth in CSF-1-treated NIH3T3 cells expressing human c-Fms (Roussel *et al.*, 1987) (Fig. 2).

Substrate Binding to Tyrosine Phosphorylated c-Fms

Following tyrosine phosphorylation after activation, RTKs acquire the ability to bind to proteins containing SH2 domains, PTB domains, etc. The specificity of the interactions is determined both by these SH2 domains, for example, and by the sequence surrounding the phosphorylated tyrosine (Reedijk *et al.*, 1992). The evidence for the binding of SH2 domain-containing proteins to activated c-Fms is now listed and the conclusions are summarized schematically in Fig. 2.

Grb2. In Rat-2 fibroblasts transfected with murine c-*fms*, the adaptor protein Grb2 associates with the receptor upon CSF-1 treatment and Y697 had been identified as its binding site (van der Geer and Hunter, 1993). As discussed earlier, at least nine (different?) Grb2-containing complexes have been claimed (Lioubin *et al.*, 1994; Kharbanda *et al.*, 1995; Saleem *et al.*, 1995; Chen *et al.*, 1996).

PI-3 kinase. As mentioned above, there is evidence in various cell populations that CSF-1 can activate PI-3 kinase activity (Varticovski *et al.*, 1989; Reedijk *et al.*, 1990, 1992; Shurtleff *et al.*, 1990; Choudhury *et al.*, 1991c; van der Geer and Hunter, 1991; Yusoff *et al.*, 1994); it has been reported that Y721 regulates this activity in Rat-2 fibroblasts containing mutant murine c-Fms and is the binding site for the p85 regulatory subunit, most likely through SH2 domains (Reedijk *et al.*, 1992). In CSF-1-treated murine macrophages, the p85 subunit of PI-3 kinase is the major [^{35}S]methionine-labelled protein associated with the receptor (Kanagasundaram *et al.*, 1996).

STAT1. In a study with FDC-P1 cells transfected with c-*fms* mutants, it has been reported that Y706 is required for efficient activation by CSF-1 of the transcription factor STAT1, but not of STAT3 (Novak *et al.*, 1997); also, addition of phosphopeptides spanning Y708 of human c-Fms (Y706 in murine c-Fms) to electrophoretic mobility shift assays led to competition of STAT1-containing complexes with the DNA probe, suggesting that this tyrosine is the binding site for STAT1. The Y807F mutation in murine c-Fms expressed in FDC-P1 cells led to a block in the activation of STAT1, STAT3, STAT5A and STAT5B in response to CSF-1, but Y807 does not seem to be a STAT binding site.

Src. It has been proposed that Src family members may interact with the phosphorylated Y561 residue in the juxtamembrane region of activated human Fms (Y559 in murine Fms) (Alonso *et al.*, 1995). However, functional studies have not been performed to confirm this interaction.

v-Fms Tyrosine Phosphorylation

The v-*fms* oncogene encodes a modified PTK which differs in only seven amino acid positions and in the *C*-terminal sequence from c-Fms (Sherr, 1988). The entire cytoplasmic domain of v-Fms contains 18 tyrosine residues that are conserved in c-Fms. All c-Fms proteins, however, contain an additional tyrosine residue at the *C*-terminus. Y-543 has been shown to be a 'major autophosphorylation site' in v-Fms (Joos *et al.*, 1996). Strangely there has been no claim for autophosphorylation of the equivalent tyrosine in c-Fms (Y544 in the mouse, Y546 in the human) after activation

(Fig. 2). Interestingly, in the same study Y558 (equivalent to Y559 in murine c-Fms and to Y561 in human c-Fms) was not considered to be a major autophosphorylation site, even though it has been inferred for c-Fms (Alonso *et al.*, 1995).

No association of any protein with Y807 has been reported for c-Fms (Fig. 2), but for v-Fms there is association with p120 Ras GAP, which is tyrosine phosphorylated (Trouliaris *et al.*, 1995). In v-Fms, it has been published that Y543 associates with a tyrosine-phosphorylated p55 protein (Joos *et al.*, 1996).

It would be of interest to see whether further studies with activated c-Fms can confirm the observation made for v-Fms.

CONCLUDING REMARKS

From the above discussion it can be seen that there are serious disagreements regarding key issues in CSF-1R signalling, for example the role of Y807 and the degree of kinase activity retained upon its mutation, pathways governing c-Myc expression, the degree of significance of the 'classical' Ras→Raf-1→MEK→MAP kinase (Erk) pathway in CSF-1-treated macrophages, the role of the KI region in the proliferative response, the nature of the Grb2/Shc-containing complexes (e.g. whether $p150^{Ship}$ is present), the effects of the c-Fms D802V, the degree of tyrosine phosphorylation of PI-3 kinase, etc.

For certain types of experiment, particularly those involving mutational analysis, it has been necessary to insert c-Fms into cell types in which the receptor is not usually found. An important caveat of this type of study is that the ability of the receptor to induce a response in such cells (e.g. proliferation) may involve the recruitment of second messenger systems that are not normally engaged in cells in which the receptor is normally expressed (Sherr, 1991; Stanley, 1994); perhaps differences in the actual contents of signal transducers in various cell types may be a significant contributing factor. It could be argued that non-monocyte/macrophage myeloid cells, such as FDC-P1 cells, although not ideal, may more closely reflect the normal situation than fibroblast-type cells. Findings from these heterologous systems will have to be checked where possible to determine their relevance for CSF-1-dependent monocyte/macrophage biology.

While the discordant findings obtained from different cell types, which were outlined above, could result from the presence of c-Fms in vastly different cell types, there are other inconsistent findings using the same or similar cellular systems to analyse the CSF-1 response which are more difficult to explain: for example, the relationship of Ras activity to subsequent c-*fos* or c-*myc* expression, the role of c-Myc and PI-3 kinase in proliferation, the degree of tyrosine phosphorylation of SHP-1 and GAP, whether PKC and phospholipase C are activated, whether PGE_2 is produced, the differing actions of 8BrcAMP on cyclin D1 expression, etc. Obviously further experimentation is needed before these issues are clarified.

One other conclusion that can be made from the above review of the literature is that the same or very similar molecules are activated by different mechanisms and function in different pathways depending on the cellular background and on the stimulus used. One example of this, which is referred to above, is the activation of the MAPK pathway where, in macrophages, it could be that the paradigm Ras–Raf-1–MEK–MAPK pathway may not be that significant (Büscher *et al.*, 1995; Krautwald *et al.*, 1996) even though each of the components appears able to be activated. In other words, the pathways linking 'classical' signal transducers may not be able to be generalized for each cellular system, again reinforcing the need to analyse each biological system independently.

c-Fms is a RTK and more specifically is included, along with PDGF-R, c-kit, and flt3/flt2, in the RTK family containing a KI domain. In spite of the similarities in the biochemical responses that this structural homology can lead to, there are examples where each receptor can interact with a different set of signalling proteins which might reflect the specific biological responses to receptor activation *in vivo* (Yi and Ihle, 1993). In other words, each of these RTK family members needs to be studied in its own right, again preferably in its normal milieu, and caution needs to be exercised before extrapolating conclusions from observations made on one of these receptors to another. Examples where c-Fms activation can lead to different responses from other RTKs, including those with the KI domain, are the lack of activation of phospholipase Cγ, lack of association of c-Fms with SHP-1, lack of suppression of DNA synthesis by microinjection of blocking antibody to PI-3 kinase, lack of involvement of PI-3 kinase in receptor degradation, lack of tyrosine phosphorylation of Vav, etc.

There is strong evidence that CSF-1 is involved in the survival, proliferation and differentiation of at least some monocyte–macrophage populations (Cecchini *et al.*, 1994; Stanley, 1994); there is also evidence that CSF-1 may function as a macrophage activator, for example as part of host defence or in inflammation (Hamilton *et al.*, 1993). Also, when monocytes/macrophages are treated *in vitro* with CSF-1 there is increased cell spreading, membrane ruffling and receptor turnover (Boocock *et al.*, 1989). Many of the *in vitro* biochemical responses to CSF-1 outlined in this review may be relevant to these acute membrane-associated events. If the CSF-1 signal transduction studies of the type presented here are to have any significance, they will have to explain ultimately the *in vivo* role of CSF-1 in the development and function of cells of the monocyte/macrophage lineage and of the other cell types in which c-Fms is expressed, for example trophoblasts (Stanley, 1994). Considering some of the issues raised above, it would seem a reasonable approach, where possible, to study CSF-1R-directed signal transduction cascades in the monocyte/macrophage lineage, i.e. in cells in which CSF-1R is normally expressed. However, many of these studies have involved cells from different tissues and species. Many have also been with the CSF-1-dependent murine macrophage cell line, BAC1.2F5. Perhaps not surprisingly, some differences in the properties of various BAC1.2F5 lines have been reported (e.g. LPS sensitivity), as well as some differences from primary macrophage cultures. Although a powerful tool, the observations with this line(s) need to be confirmed at some stage, if possible in primary cells.

It seems that the validity of the following observations for the effects of CSF-1 on monocyte/macrophage cultures is probably established, and they are presented as a conservative and partial listing: c-Fms dimerizes and autophosphorylates, leading to the tyrosine phosphorylation of the p85 subunit of PI-3 kinase, c-Cbl, SHP-1 and TYK2 but not of Vav; the activities of PI-3 kinase, SHP-1, STAT proteins, Raf-1, MEK, Erks, S6 kinase and the Na^+ cycle, but not of phospholipase Cγ, are enhanced; Grb2 and PI-3 kinase can associate with many proteins, as can Shc; Ras-dependent and -independent pathways exist for Erk activation, and Raf-1 may not be involved in Erk activation; PI-3 kinase is not involved in proliferation; G proteins are involved; c-*fos*, c-*myc* and cyclin D1 mRNA expression are enhanced.

In conclusion, if the signal transduction studies of the type presented in this review are to have any significance, they will have to explain ultimately the *in vivo* role of CSF-1 in the development and function of cells of the monocyte/macrophage lineage and of the other cell types in which c-Fms is expressed, for example trophoblasts (Stanley, 1994). As examples of recent approaches, data from c-*src* and c-*fos* knockout mice (Soriano *et al.*, 1991; Grigoriadis *et al.*, 1994) suggest that the products of these genes are at least dispensable for macrophage development.

ACKNOWLEDGEMENTS

The author thanks Melissa Anderson for typing the manuscript and U. Novak, A. Jaworowski, V. Kanagasundaram and P. Vadiveloo for helpful discussions. M. Baccarini is also thanked for providing unpublished data. The work is supported by a Program Grant from the National Health and Medical Research Council of Australia from which the author also receives a Senior Principal Research Fellowship.

REFERENCES

Alonso, G., Koegl, M., Mazurenko, N. and Courtneidge, S.A. (1995). Sequence requirements for binding of Src family tyrosine kinases to activated growth factor receptors. *J. Biol. Chem.* **270**, 9840–9848.

Baccarini, M., Sabatini, D.M., App, H., Rapp, U.R. and Stanley, E.R. (1990). Colony stimulating factor-1 (CSF-1) stimulates temperature dependent phosphorylation and activation of the RAF-1 proto-oncogene product. *EMBO J.* **9**, 3649–3657.

Barone, M.V. and Courtneidge, S.A. (1995). Myc but not Fos rescue of PDGF signalling block caused by kinase-inactive Src. *Nature* **378**, 509–512.

Blenis, J. (1993). Signal transduction via the MAP kinases: proceed at your own RSK. *Proc. Natl Acad. Sci. U.S.A.* **90**, 5889–5892.

Boocock, C.A., Jones, G.E., Stanley, E.R. and Pollard, J.W. (1989). Colony-stimulating factor-1 induced rapid behavioural responses in the mouse macrophage cell line, BAC1.2F5. *J. Cell Sci.* **93**, 447–456.

Bortner, D.M., Ulivi, M., Roussel, M.F. and Ostrowski, M.C. (1991). The carboxy-terminal catalytic domain of the GTPase-activating protein inhibits nuclear signal transduction and morphological transformation mediated by the CSF-1 receptor. *Genes Dev.* **5**, 1777–1785.

Borzillo, G.V., Ashmun, R.A. and Sherr, C.J. (1990). Macrophage lineage switching of murine early pre-B lymphoid cells expressing transduced *fms* genes. *Mol. Cell. Biol.* **10**, 2703–2714.

Bourette R.P., Myles G.M., Carlberg K., Chen A.R. and Rohrschneider L.R. (1995). Uncoupling of the proliferation and differentiation signals mediated by the murine macrophage colony-stimulating factor receptor expressed in myeloid FDC-P1 cells. *Cell Growth Differ.* **6**, 631–645.

Branch, D.R., Turner, A.R. and Guilbert, L.J. (1989). Synergistic stimulation of macrophage proliferation by the monokines tumor necrosis factor-alpha and colony-stimulating factor-1. *Blood* **73**, 307–311.

Büscher, D., Sbarba, P.D., Hipskind, R.A., Rapp, U.R., Stanley, E.R. and Baccarini, M. (1993). v-raf confers CSF-1 independent growth to a macrophage cell line and leads to immediate early gene expression without MAP-kinase activation. *Oncogene* **8**, 3323–3332.

Büscher, D., Hipskind, R.A., Krautwald, S., Reimann, T. and Baccarini, M. (1995). Ras-dependent and -independent pathways target the mitogen-activated protein kinase network in macrophages. *Mol. Cell. Biol.* **15**, 466–475.

Carlberg, K. and Rohrschneider, L. (1994). The effect of activating mutations on dimerization, tyrosine phosphorylation and internalization of the macrophage colony stimulating factor receptor. *Mol. Biol. Cell.* **5**, 81–95.

Carlberg, K., Tapley, P., Haystead, C. and Rohrschneider, L. (1991). The role of kinase activity and the kinase insert region in ligand-induced internalization and degradation of the c-fms protein. *EMBO J.* **10**, 877–831.

Cecchini, M.G., Dominguez, M.G., Mocci, S., Wetterwald, A., Felix, R., Fleisch, H., Chisholm, O., Hofstetter, W., Pollard, J.W. and Stanley, E.R. (1994). Role of colony stimulating factor-1 in the establishment and regulation of tissue macrophages during postnatal development of the mouse. *Development* **120**, 1357–1372.

Chen, H.E., Chang, S., Trub, T. and Neel, B.G. (1996). Regulation of colony-stimulating factor-1 receptor signaling by the SH2 domain-containing tyrosine phosphatase SHPTP1. *Mol. Cell. Biol.* **16**, 3685–3697.

Choudhury, G.G., Sylvia, V.L., Pfeifer, A., Wang, L.-M., Smith, E.A. and Sakaguchi, A.Y. (1990). Human colony stimulating factor-1 receptor activates the C-raf-1 proto-oncogene kinase. *Biochem. Biophys. Res. Commun.* **172**, 154–159.

Choudhury, G.G., Sylvia, V.L., Wang, L.-M., Pierce, J. and Sakaguchi, A.Y. (1991a). The kinase insert domain of colony stimulating factor-1 receptor is dispensable for CSF-1 induced phosphatidylcholine hydrolysis. *FEBS Lett.* **282**, 351–354.

Choudhury, G.G., Sylvia, V.L. and Sakaguchi, A.Y. (1991b). Activation of a phosphatidylcholine-specific phospholipase C by colony stimulating factor 1 receptor requires tyrosine phosphorylation and a guanine nucleotide-binding protein. *J. Biol. Chem.* **266**, 23147–23151.

Choudhury, G.G., Wang, L.-M., Pierce, J., Harvey, S.A. and Sakaguchi, A.Y. (1991c). A mutational analysis of phosphatidylinositol-3-kinase activation by human colony-stimulating factor-1 receptor. *J. Biol. Chem.* **266**, 8068–8072.

Cocks, B.G., Vairo, G., Bodrug, S.E. and Hamilton, J.A. (1992). Suppression of growth factor-induced CYL1 cyclin gene expression by antiproliferative agents. *J. Biol. Chem.* **267**, 12307–12310.

Cooper, M.H., Gregory, S.H., Starzl, T.E. and Wing, E.J. (1994). Rapamycin but not FK506 inhibits the proliferation of mononuclear phagocytes induced by colony-stimulating factors. *Transplantation* **57**, 433–439.

Corre, I. and Hermouet, S. (1995). Regulation of colony-stimulating factor 1-induced proliferation by heterotrimeric G_{i2} proteins. *Blood* **86**, 1776–1783.

Courtneidge, S.A., Dhand, R., Pilat, D., Twamley, G.M., Waterfield, M.D. and Roussel, M.F. (1993). Activation of Src family kinases by colony stimulating factor-1, and their association with its receptor. *EMBO J.* **12**, 943–950.

Coussens, L., Van Beveren, C., Smith, D., Chen, E., Mitchell, R.L., Isacke, C.M., Verma, I.M. and Ullrich, A. (1986). Structural alteration of viral homologue of receptor proto-oncogene *fms* at carboxyl terminus. *Nature* **320**, 277–280.

Downing, J.R., Margolis, B.L., Zilberstein, A., Ashmun, R.A., Ullrich, A., Sherr, C.J. and Schlessinger, J. (1989). Phospholipase C-γ, a substrate for PDGF receptor kinase, is not phosphorylated on tyrosine during the mitogenic response to CSF-1. *EMBO J.* **8**, 3345–3350.

Downing, J.R., Shurtleff, S.A. and Sherr, C.J. (1991). Peptide antisera to human colony-stimulating factor 1 receptor detect ligand-induced conformational changes and a binding site for phosphatidylinositol 3-kinase. *Mol. Cell. Biol.* **11**, 2489–2495.

Egan, S.E. and Weinberg, R.A. (1993). The pathway to signal achievement. *Nature* **365**, 781–783.

Foltz, I.N., Lee, J.C., Young, P.R. and Schrader, J.W. (1997). Hemopoietic growth factors with the exception of interleukin-4 activate the p38 mitogen-activated protein kinase pathway. *J. Biol. Chem.* **272**, 3296–3301.

Gibbs, J.B., Marshall, M.S., Scolnick, E.M., Dixon, R.A.F. and Vogel, U.S. (1990). Modulation of guanine nucleotides bound to Ras in NIH3T3 cells by oncogenes, growth factors, and the GTPase activating protein (GAP). *J. Biol. Chem.* **265**, 20437–20442.

Glover, H.R., Baker, D.A., Celetti, A. and Dibb, N.J. (1995). Selection of activating mutations of c-*fms* in FDC-P1 cells. *Oncogene* **11**, 1347–1356.

Grigoriadis, A.E., Wang, Z.-Q., Cecchini, M.G., Hofstetter, W., Felix, R., Fleisch, H.A. and Wagner, E.F. (1994). c-Fos: a key regulator of osteoclast–macrophage lineage determination and bone remodeling. *Science* **266**, 443–448.

Guilbert, L.J. and Stanley, E.R. (1986). The interaction of ^{125}I-colony-stimulating factor-1 with bone marrow-derived macrophages. *J. Biol. Chem.* **261**, 4024–4032.

Hamilton, J.A. (1993). Colony stimulating factors, cytokines and monocyte–macrophages—some controversies. *Immunol. Today* **14**, 18–24.

Hamilton, J.A., Veis, N., Bordun, A.M., Vairo, G., Gonda, T.J. and Phillips, W.A. (1989). Activation and proliferation signals in murine macrophages: relationships among c-*fos* and c-*myc* expression, phosphoinositide hydrolysis, superoxide formation, and DNA synthesis. *J. Cell. Physiol.* **141**, 618–626.

Hamilton, J.A., Vairo, G., Knight, K.R. and Cocks, B.J. (1991). Activation and proliferation signals in murine macrophages. Biochemical signals controlling the regulation of macrophage urokinase-type plasminogen activator activity by colony stimulating factors and other agents. *Blood* **77**, 616–627.

Hamilton, J.A., Vairo, G. and Cocks, B.G. (1992). Inhibition of S-phase progression in macrophages is linked to G1/S-phase suppression of DNA synthesis genes. *J. Immunol.* **148**, 4028–4035.

Hamilton, J.A., Whitty, G.A., Stanton, H. and Meager, A. (1993). Effects of macrophage-colony stimulating factor (M-CSF or CSF-1) on human monocytes. Induction of expression of urokinase-type plasminogen activator, but not of secreted prostaglandin E_2, interleukin-6, interleukin-1 or tumor necrosis factor-α. *J. Leukoc. Biol.* **53**, 707–714.

Hamilton, J.A. Whitty, G.A., Kola, I. and Hertzog, P.J. (1996). Endogenous interferon α/β suppresses CSF-1 stimulated macrophage DNA synthesis and mediates inhibitory effects of lipopolysaccharide and tumor necrosis factor α. *J. Immunol.* **156**, 2553–2557.

Hamilton, J.A., Byrne, R., Whitty, G., Vadiveloo, P.K., Marmy, N., Pearson, R.B., Christy, E. and Jaworowski, A. Effects of wortmannin and rapamycin on CSF-1 mediated responses in macrophages (in press).

Hartmann, T., Seuwen, K., Roussel, M.F., Sherr, C.J. and Pouyssegur, J. (1990). Functional expression of the human receptor for colony-stimulating factor-1 (CSF-1) in hamster fibroblasts: CSF-1 stimulates Na^+/H^+ exchange and DNA synthesis in the absence of phosphoinositide breakdown. *Growth Factors* **2**, 289–300.

He, Y., Hewlett, E., Temeles, D. and Quesenberry, P. (1988). Inhibition of interleukin 3 and colony-stimulating factor 1-stimulated marrow cell proliferation by pertussis toxin. *Blood* **71**, 1187–1195.

Heidaran, M.A., Molloy, C.J., Pangelinan, M., Choudhury, G.G., Wang, L.-M., Fleming, T.P., Sakaguchi, A.Y. and Pierce, J.H. (1992). Activation of the colony-stimulating factor 1 receptor leads to the rapid tyrosine phosphorylation of GTPase-activating protein and activation of cellular $p21^{ras}$. *Oncogene* **7**, 147–152.

Hipskind, R.A., Büscher, D., Nordheim, A. and Baccarini, M. (1994). Ras/MAP kinase-dependent and -independent signaling pathways target distinct ternary complex factors. *Genes Dev.* **8**, 1803–1816.

Imamura, K. and Kufe, D. (1988). Colony-stimulating factor 1-induced Na^+ influx into human monocytes involves activation of a pertussis toxin-sensitive GTP-binding protein. *J. Biol. Chem.* **263**, 14 093–14 098.

Imamura, K., Dianoux, A., Nakamura, T. and Kufe, D. (1990). Colony-stimulating factor 1 activates protein kinase C in human monocytes. *EMBO J.* **9**, 2423–2429.

Jackowski, S., Rettenmier, C.W., Sherr, C.J. and Rock, C.O. (1986). A guanine nucleotide-dependent phosphatidylinositol 4,5-diphosphate phospholipase C in cells transformed by the v-*fms* and v-*fes* oncogenes. *J. Biol. Chem.* **261**, 4978–4985.

Jackowski, S., Rettenmier, C.W. and Rock, C.O. (1990). Prostaglandin E_2 inhibition of growth in a colony-stimulating factor 1-dependent macrophage cell line. *J. Biol. Chem.* **265**, 6611–6616.

Jaworowski, A., Argyriou, S., Yusoff, P. and Hamilton, J.A. (1994). Phospholipase D is activated by phorbol ester but not CSF-1 in murine bone marrow derived macrophages. *Biochem. Biophys. Res. Commun.* **201**, 733–739.

Jaworowski, A., Christy, E., Yusoff, P., Byrne, R. and Hamilton, J.A. (1996). Differences in the kinetics of activation of protein kinases and ERK-1 in CSF- and LPS-stimulated macrophages. *Biochem. J.* **320**, 1011–1016.

Jin, D.I., Jameson, S.B., Reddy, M.A., Schenkman, D. and Ostrowski, M.C. (1995). Alterations in differentiation and behavior of monocytic phagocytes in transgenic mice that express dominant suppressors of ras signaling. *Mol. Cell. Biol.* **15**, 693–703.

Joly, M., Kazlauskas, A. and Corvera, S. (1995). Phosphatidylinositol 3-kinase activity is required at a postendocytic step in platelet-derived growth factor receptor trafficking. *J. Biol. Chem.* **270**, 13 225–13 230.

Joos, H., Trouliaris, S., Helftenbein, G., Niemann, H. and Tamura, T. (1996). Tyrosine phosphorylation of the juxtamembrane domain of the v-Fms oncogene product is required for its association with a 55-kDa protein. *J. Biol. Chem.* **271**, 24 476–24 481.

Kanagasundaram, V., Jaworowski, A. and Hamilton, J.A. (1996). Association between phosphatidylinositol-3 kinase, Cbl and other tyrosine phosphorylated proteins in CSF-1-stimulated macrophages. *Biochem. J.* **320**, 69–77.

Kato, J.-Y., Matsuoka, M., Polyak, K., Massagué, J. and Sherr, C.J. (1994). Cyclic AMP-induced G1 phase arrest mediated by an inhibitor (p27Kip1) of cyclin-dependent kinase 4 activation. *Cell* **79**, 487–496.

Kharbanda, S., Saleem, A., Yuan, Z., Emoto, Y., Prasad, K.V.S. and Kufe, D. (1995). Stimulation of human monocytes with macrophage colony-stimulating factor induces a Grb2-mediated association of the focal adhesion kinase $pp125^{FAK}$ and dynamin. *Proc. Natl Acad. Sci. U.S.A.* **92**, 6132–6136.

Krautwald, S., Büscher, D., Dent, P., Ruthenberg, K. and Baccarini, M. (1995). Suppression of growth factor-mediated MAP kinase activaton by v-*raf* in macrophages: a putative role for the MKP-1 phosphatase. *Oncogene* **10**, 1187–1192.

Krautwald, S., Büscher, D., Kummer, V., Buder, S. and Baccarini, M. (1996). Involvement of the protein tyrosine phosphatase SHP-1 in ras-mediated activation of the mitogen-activated protein kinase pathway. *Mol. Cell. Biol.* **16**, 5955–5963.

Langer, S.J., Bortner, D.M., Roussel, M.F., Sherr, C.J. and Ostrowski, M.C. (1992). Mitogenic signaling by colony-stimulating factor 1 and ras is suppressed by the ets-2 DNA-binding domain and restored by myc overexpression. *Mol. Cell. Biol.* **12**, 5355–5362.

Lee, A.W.-M. (1991). Signal transduction by the colony-stimulating factor-1 receptor; comparison to other receptor tyrosine kinases. *Curr. Top. Cell. Regul.* **32**, 73–181.

Lee, J.C., Laydon, J.T., McDonnell, P.C., Gallagher, T.F., Kumar, S., Green, D., McNulty, D., Blumenthal, M.J., Heys, J.R., Landvatter, S.W., Strickler, J.E., McLaughlin, M.M., Siemens, I.R., Fisher, S.M., Livi, G.P., White, J.R., Adams, J.L. and Young, P.R. (1994). A protein kinase involved in the regulation of inflammatory cytokine biosynthesis. *Nature* **372**, 739–746.

Li, W. and Stanley, E.R. (1991). Role of dimerization and modification of the CSF-1 receptor in its activation and internalization during the CSF-1 response. *EMBO J.* **10**, 277–288.

Lioubin, M.N., Myles, G.M., Carlberg, K., Bowtell, D. and Rohrschneider, L.R. (1994). Shc, Grb2, Sos1, and a 150-kilodalton tyrosine-phosphorylated protein form complexes with Fms in hematopoietic cells. *Mol. Cell. Biol.* **14**, 5682–5691.

Lioubin, M.N., Algate, P.A., Tsai, S., Carlberg, K., Aebersold, R. and Rohrschneider, L.R. (1996). p150^{Ship}, a signal transduction molecule with inositol polyphosphate-5-phosphatase activity. *Genes Dev.* **10**, 1084–1095.

Manger, R., Najita, L., Nichols, E.J., Hakomori, S. and Rohrschneider, L. (1984). Cell surface expression of the McDonough strain of feline sarcoma virus fms gene product (gp140fms). *Cell* **34**, 327–337.

Margolis, B., Hu, P., Katzav, S., Li, W., Oliver, J.M., Ullrich, A., Weiss, A. and Schlessinger, J. (1992). Tyrosine phosphorylation of *vav* proto-oncogene product containing SH2 domain and transcription factor motifs. *Nature* **356**, 71–74.

Matsushime, H., Roussel, M.F., Ashmun, R.A. and Sherr, C.J. (1991). Colony-stimulating factor 1 regulates novel cyclins during the G1 phase of the cell cycle. *Cell* **65**, 701–713.

Matsushime, H., Ewen, M.E., Strom, D.K., Kato, J.-Y., Hanks, S.K., Roussel, M.F. and Sherr, C.J. (1992). Identification and properties of an atypical catalytic subunit (p34$^{psk.j3}$/cdk4) for mammalian D types G1 cyclins. *Cell* **71**, 323–334.

McCoy, K.L., Nielson, K. and Clagett, J. (1984). Spontaneous production of colony-stimulating activity by splenic Mac-1 antigen-positive cells from autoimmune motheaten mice. *J. Immunol.* **132**, 272–276.

Mori, S., Claesson-Welsh, L., Okuyama, Y. and Saito, Y. (1995). Ligand-induced polyubiquitination of receptor tyrosine kinases. *Biochem. Biophys. Res. Commun.* **213**, 32–39.

Myles, G.M., Brandt, C.S., Carlberg, K. and Rohrschneider, L.R. (1994). Tyrosine 569 in the c-Fms juxtamembrane domain is essential for kinase activity and macrophage colony-stimulating factor-dependent internalization. *Mol. Cell. Biol.* **14**, 4843–4854.

Nakamura, T., Lin, L.-L., Kharbanda, S., Knopf, J. and Kufe, D. (1992). Macrophage colony stimulating factor activates phosphatidylcholine hydrolysis by cytoplasmic phospholipase A2. *EMBO J.* **11**, 4917–4922.

Nevins, J.R. (1992). E2F: a link between the Rb tumor suppressor protein and viral oncoproteins. *Science* **258**, 424–429.

Novak, U., Harpur, A.G., Paradiso, L., Kanagasundaram, V., Jaworowski, A., Wilks, A.F. and Hamilton, J.A. (1995). CSF-1 induced STAT1 and STAT3 activation is accompanied by phosphorylation of Tyk2 in macrophages and Tyk2 and JAK1 in fibroblasts. *Blood* **86**, 2948–2956.

Novak, U., Mui, A., Miyajima, A. and Paradiso, L. (1996). Formation of STAT5 containing DNA binding complexes in response to colony stimulating factor-1 and platelet-derived growth factor. *J. Biol. Chem.* **271**, 26 947–26 953.

Novak, U., Paradiso, L., Nice, E. and Hamilton, J.A. (1997). Requirement for Y706 of the murine (or Y708 of the human) CSF-1 receptor for STAT1 activation in response to CSF-1. *Oncogene* **13**, 2607–2613.

Ohtsuka, M., Roussel, M.F., Sherr, C.J. and Downing, J.R. (1990). Ligand-induced phosphorylation of the colony-stimulating factor 1 receptor can occur through an intermolecular reaction that triggers receptor down modulation. *Mol. Cell. Biol.* **10**, 1664–1671.

Puri, J., Pierce, J.H. and Hoffmann, T. (1991). Transduction of a signal for arachidonic acid metabolism by CSF-1 receptor and its ligand: separate regulation of phospholipase A_2 and cyclooxygenase by CSF-1 receptor/CSF-1. *Prostaglandins Leukot. Essent. Fatty Acids* **45**, 43–48.

Reedijk, M., Liu, X. and Pawson, T. (1990). Interactions of phosphatidylinositol kinase, GTPase-activating protein (GAP), and GAP-associated proteins with the colony-stimulating factor 1 receptor. *Mol. Cell. Biol.* **10**, 5601–5608.

Reedijk, M., Liu, X., van der Geer, P., Letwin, K., Waterfield, M.D., Hunter, T. and Pawson, T. (1992). Tyr721 regulates specific binding of the CSF-1 receptor kinase insert to PI3-kinase SH2 domains: a model for SH2-mediated receptor–target interactions. *EMBO J.* **11**, 1365–1372.

Reimann, T., Büscher, D., Hipskind, R.A., Krautwald, S., Lohmann-Matthes, M.-L. and Baccarini, M. (1994). Lipopolysaccharide induces activation of the Raf-1/MAP kinase pathway. *J. Immunol.* **153**, 5740–5749.

Rettenmier, C.W., Roussel, M.F., Asmun, R.A., Ralph, P., Price, V. and Sherr, C.J. (1987). Synthesis of membrane-bound colony stimulating factor-1 (CSF-1) and downregulation of CSF-1 receptors in NIH3T3 cells transformed by contransfection of the human CSF-1 and c-*fms* (CSF-1 receptor) genes. *Mol. Cell. Biol.* **7**, 2378–2387.

Roche, S., Koegl, M. and Courtneidge, S.A. (1994). The phosphatidylinositol 3-kinase α is required for DNA synthesis induced by some, but not all, growth factors. *Proc. Natl Acad. Sci. U.S.A.* **91**, 9185–9189.

Rock, C.O., Cleveland, J.L. and Jackowski, S. (1992). Macrophage growth arrest by cyclic AMP defines a distinct checkpoint in the mid-G1 stage of the cell cycle and overrides constitutive c-*myc* expression. *Mol. Cell. Biol.* **12**, 2351–2358.

Roussel, M.F. and Sherr, C.J. (1989). Mouse NIH 3T3 cells expressing human colony-stimulating factor 1 (CSF-1) receptors overgrow in serum-free medium containing human CSF-1 as their only growth factor. *Proc. Natl Acad. Sci. U.S.A.* **86**, 7924–7927.

Roussel, M.F., Dull, T.J., Rettenmier, C.W., Ralph, P., Ullrich, A. and Sherr, C.J. (1987). Transforming potential of the c-*fms* proto-oncogene (CSF-1 receptor). *Nature* **325**, 549–552.

Roussel, M.F., Downing, J.R., Rettenmier, C.W. and Sherr, C.J. (1988). A point mutation in the extracellular domain of the human CSF-1 receptor (c-*fms* proto-oncogene product) activates its transforming potential. *Cell* **55**, 979–988.

Roussel, M.F., Shurtleff, S.A., Downing, J.R. and Sherr, C.J. (1990). A point mutation at tyrosine-809 in the human colony-stimulating factor 1 receptor impairs mitogenesis without abrogating tyrosine kinase activity, association with phosphatidylinositol 3-kinase, or induction of c-*fos* and *junB* genes. *Proc. Natl Acad. Sci. U.S.A.* **87**, 6738–6742.

Roussel, M.F., Cleveland, J.L., Shurtleff, S.A. and Sherr, C.J. (1991). Myc rescue of a mutant CSF-1 receptor impaired in mitogenic signalling. *Nature* **353**, 361–363.

Roussel, M.F., Davis, J.N., Cleveland, J.L., Ghysdael, J. and Hiebert, S.W. (1994). Dual control of *myc* expression through a single DNA binding site targeted by ets family proteins and E2F-1. *Oncogene* **9**, 405–415.

Roussel, M.F., Theodoras, A.M., Pagano, M. and Sherr, C.J. (1995). Rescue of defective mitogenic signalling by D-type cyclins. *Proc. Natl Acad. Sci. U.S.A.* **92**, 6837–6841.

Saleem, A., Kharbanda, S., Yuan, Z.-M. and Kufe, D. (1995). Monocyte colony-stimulating factor stimulates binding of phosphatidylinositol 3-kinase to Grb2-Sos complexes in human monocytes. *J. Biol. Chem.* **270**, 10380–10383.

Sengupta, A., Liu, W.K., Yeung, Y.G., Yeung, D.C., Frackelton, A.R., Jr. and Stanley, E.R. (1988). Identification and subcellular localization of proteins that are rapidly phosphorylated in tyrosine in response to colony-stimulating factor 1. *Proc. Natl Acad. Sci. U.S.A.* **85**, 8062–8066.

Sherr, C.J. (1988). The *fms* oncogene. *Biochim. Biophys. Acta* **948**, 225–243.

Sherr, C.J. (1991). Mitogenic response to colony-stimulating factor 1. *Trends Genet.* **7**, 398–402.

Sherr, C.J. (1993). Mammalian G1 cyclins. *Cell* **73**, 1059–1065.

Shieh, J.H., Peterson, R.H., Warren, D.J. and Moore, M.A. (1989). Modulation of colony-stimulating factor-1 receptors on macrophages by tumor necrosis factor. *J. Immunol.* **143**, 2534–2539.

Shultz, L.D., Schweitzer, P.A., Rajan, T.V., Yi, T., Ihle, J.N., Matthews, R.J., Thomas, M.L. and Beier, D.R. (1993). Mutations at the murine montheaten locus are within the hematopoietic cell protein-tyrosine phosphatase (*Hcph*) gene. *Cell* **73**, 1445–1454.

Shurtleff, S.A., Downing, J.R., Rock, C.O., Hawkins, S.A., Roussel, M.F. and Sherr, C.J. (1990). Structural features of the colony-stimulating factor 1 receptor that affect its association with phosphatidylinositol 3-kinase. *EMBO J.* **9**, 2415–2421.

Soriano, P., Montgomery, C., Geske, R. and Bradley, A. (1991). Targeted disruption of the c-*src* proto-oncogene leads to osteoporosis in mice. *Cell* **64**, 693–702.

Smith, M.R., De Gudicibus, S.J. and Stacey, D.W. (1986). Requirement for c-ras proteins during viral oncogene transformation. *Nature* **320**, 540–543.

Stanley, E.R. (1985). The macrophage colony-stimulating factor, CSF-1. *Methods Enzymol.* **116**, 564–587.

Stanley E.R. (1994). Colony stimulating factor 1 (macrophage colony stimulating factor). In *The Cytokine Handbook*, 2nd edn. (ed. A.W. Thomson), Academic Press, San Diego, pp. 387–418.

Strassmann, G., Bertolini, D.R., Kerby, S.B. and Fong, M. (1991). Regulation of murine mononuclear phagocyte inflammatory products by macrophage colony-stimulating factor. Lack of IL-1 and prostaglandin E_2 production and generation of a specific IL-1 inhibitor. *J. Immunol.* **147**, 1279–1285.

Tanaka, S., Neff, L., Baron, R. and Levy, J.B. (1995). Tyrosine phosphorylation and translocation of the c-Cbl protein after activation of tyrosine kinase signaling pathways. *J. Biol. Chem.* **270**, 14 347–14 351.

Tapley, P., Kazlauskas, A., Cooper, J.A. and Rohrschneider, L.R. (1990). Macrophage colony-stimulating factor-induced tyrosine phosphorylation of c-fms proteins expressed in FDC-P1 and BALB/c 3T3 cells. *Mol. Cell. Biol.* **10**, 2528–2538.

Treisman, R. (1995). Journey to the surface of the cell: Fos regulation and the SRE. *EMBO J.* **14**, 4905–4913.

Trouliaris, S., Smola, U., Chang, J.-H., Parsons, S.J., Niemann, H. and Tamura, T. (1995). Tyrosine 807 of the v-Fms oncogene product controls cell morphology and association with p120RasGAP. *J. Virol.* **69**, 6010–6020.

Tushinski, R.J. and Stanley, E.R. (1985). The regulation of mononuclear phagocyte entry into S-phase by the colony stimulating factor CSF-1. *J. Cell. Physiol.* **122**, 221–228.

Vadiveloo, P.K., Vairo, G., Novak, U., Royston, A.K., Whitty, G., Filonzi, E.L., Cragoe, E.J., Jr. and Hamilton, J.A. (1996). Differential regulation of cell cycle machinery by various antiproliferative agents is linked to macrophage arrest at distinct G1 checkpoints. *Oncogene* **13**, 599–608.

Vairo, G. and Hamilton, J.A. (1988). Activation and proliferation signals in murine macrophages: stimulation of Na^+,K^+-ATPase activity by hemopoietic growth factors and other agents. *J. Cell. Physiol.* **134**, 13–24.

Vairo, G. and Hamilton, J.A. (1991). Signaling through CSF receptors (a review). *Immunol. Today* **12**, 362–369.

Vairo, G., Argyriou, S., Bordun, A.-M., Gonda, T.J., Cragoe, E.J., Jr. and Hamilton, J.A. (1990a). Na^+/H^+ exchange involvement in colony-stimulating factor-1-stimulated macrophage proliferation. Evidence for a requirement during late G1 of the cell cycle but not for early growth factor responses. *J. Biol. Chem.* **265**, 16 929–16 939.

Vairo, G., Argyriou, S., Bordun, A.-M., Whitty, G. and Hamilton, J.A. (1990b). Inhibition of the signaling pathways for macrophage proliferation by cyclic AMP. Lack of effect on early responses to colony stimulating factor-1. *J. Biol. Chem.* **265**, 2692–2701.

Vairo, G., Argyriou, S., Knight, K.R. and Hamilton, J.A. (1991). Inhibition of colony-stimulating factor-stimulated macrophage proliferation by tumor necrosis factor α, interferon γ, and lipopolysaccharide is not due to a general loss of responsiveness to growth factor. *J. Immunol.* **146**, 3469–3477.

Vairo, G., Cocks, B.G., Cragoe, E.J., Jr. and Hamilton, J.A. (1992a). Selective suppression of growth factor-induced cell cycle gene expression by Na^+/H^+ antiport inhibitors. *J. Biol. Chem.* **267**, 19 043–19 046.

Vairo, G., Royston, A.K. and Hamilton, J.A. (1992b). Biochemical events accompanying macrophage activation and the inhibition of colony-stimulating factor-1 induced macrophage proliferation by tumor necrosis factor-α, interferon-γ and lipopolysaccharide. *J. Cell. Physiol.* **151**, 630–641.

Vairo, G., Vadiveloo, P.K., Royston, A.K., Filonzi, E.L., Rockman, S.P., Rock, C.O., Jackowski, S. and Hamilton, J.A. (1995). Deregulated c-*myc* expression overrides interferon γ-induced macrophage growth arrest. *Oncogene* **10**, 1969–1976.

Van der Geer, P. and Hunter, T. (1990). Identification of tyrosine 706 in the kinase insert as the major colony-stimulating factor 1 (CSF-1)-stimulated autophosphorylation site in the CSF-1 receptor in a murine macrophage cell line. *Mol. Cell. Biol.* **10**, 2991–3002.

Van der Geer, P. and Hunter, T. (1991). Tyrosine 706 and 807 phosphorylation site mutants in the murine colony-stimulating factor-1 receptor are unaffected in their ability to bind or phosphorylate phosphatidylinositol-3 kinase but show differential defects in their ability to induce early response gene transcription. *Mol. Cell. Biol.* **11**, 4698–4709.

Van der Geer, P. and Hunter, T. (1993). Mutation of Tyr697, a GRB2-binding site, and Tyr721, a P1 3-kinase binding site, abrogates signal transduction by the murine CSF-1 receptor expressed in Rat-2 fibroblasts. *EMBO J.* **12**, 5161–5172.

Varticovski, L., Druker, B., Morrison, D., Cantley, L. and Roberts, T. (1989). The colony stimulating factor-1 receptor associates with and activates phosphatidylinositol-3 kinase. *Nature* **342**, 699–702.

Veis, N. and Hamilton, J.A. (1991). Colony stimulating factor-1 stimulates diacylglycerol generation in murine bone marrow-derived macrophages, but not in resident peritoneal macrophages. *J. Cell. Physiol.* **147**, 298–305.

Wang, Z., Myles, G.M., Brandt, C.S., Lioubin, M.N. and Rohrschneider, L. (1993). Identification of the ligand-binding regions in the macrophage colony-stimulating factor receptor extracellular domain. *Mol. Cell. Biol.* **13**, 5348–5359.

Wang, Y., Yeung, Y.-G., Langdon, W.Y. and Stanley, E.R. (1996). c-Cbl is transiently tyrosine-phosphorylated, ubiquitinated, and membrane-targeted following CSF-1 stimulation of macrophages. *J. Biol. Chem.* **271**, 17–20.

Welham, M.J., Duronio, V., Leslie, K.B., Bowtell, D. and Schrader, J.W. (1994). Multiple hemopoietins, with the exception of interleukin-4, induce modification of Shc and mSos1, but not their translocation. *J. Biol. Chem.* **269**, 21 165–21 176.

Wetters, T. van D., Hawkins, S.A., Roussel, M.F. and Sherr, C.J. (1992). Random mutagenesis of CSF-1 receptor (FMS) reveals multiple sites for activating mutations within the extracellular domain. *EMBO J.* **11**, 551–557.

Whetton, A.D., Monk, P.N., Consalvey, S.D. and Downes, C.P. (1986). The haemopoietic growth factors interleukin 3 and colony stimulating factor-1 stimulate proliferation but do not induce inositol lipid breakdown in murine bone-marrow-derived macrophages. *EMBO J.* **5**, 3281–3286.

Wiktor-Jedrzejczak, W., Bartocci, A., Ferrante, A.W., Jr., Ahmed-Ansari, A., Sell, K.W., Pollard, J.W. and Stanley, E.R. (1991). Total absence of colony stimulating factor-1 in the macrophage deficient osteopetrotic (*op/op*) mouse. *Proc. Natl Acad. Sci. U.S.A.* **87**, 4828–4832.

Woolford, J., McAuliffe, A. and Rohrschneider, L.R. (1988). Activation of the feline c-*fms* proto-oncogene: multiple alterations are required to generate a fully transformed phenotype. *Cell* **55**, 965–977.

Xie, Y., von Gavel, S., Cassady, A.I., Stacey, K.J., Dunn, T.L. and Hume, D.A. (1993). The resistance of macrophage-like tumour cell lines to growth inhibition by lipopolysaccharide and pertussis toxin. *Br. J. Haematol.* **84**, 392–401.

Xu, J., Kim, S., Chen, M., Rockow, S., Yi, S.E., Wagner, A.J., Hay, N., Weichselbaum, R.R. and Li, W. (1995). Blockage of the early events of mitogenic signaling by interferon-γ in macrophages in response to colony-stimulating factor-1. *Blood* **86**, 2774–2788.

Xu, X.-X., Tessner, T.G., Rock, C.O. and Jackowski, S. (1993). Phosphatidylcholine hydrolysis and c-*myc* expression are in collaborating mitogenic pathways activated by colony-stimulating factor 1. *Mol. Cell. Biol.* **13**, 1522–1533.

Xu, X.-X., Rock, C.O., Qiu, Z-H., Leslie, C.C. and Jackowski, S. (1994). Regulation of cytosolic phospholipase A_2 phosphorylation and eicosanoid production by colony-stimulating factor 1. *J. Biol. Chem.* **269**, 31 693–31 700.

Yeung, Y.-G., Berg, K.L., Pixley, F.J., Angeletti, R.H. and Stanley, E.R. (1992). Protein tyrosine phosphatase-1C is rapidly phosphorylated in tyrosine in macrophages in response to colony stimulating factor-1. *J. Biol. Chem.* **267**, 23 447–23 450.

Yi, T. and Ihle, J.N. (1993). Association of hematopoietic cell phosphatase with c-Kit after stimulation with c-Kit ligand. *Mol. Cell. Biol.* **13**, 3350–3358.

Yoshida, H., Hayashi, S.-I., Kunisada, I., Ogawa, M., Nishikawa, S., Okamaru, H., Sudo, T., Schultz, L.D. and Nishikawa, S.-I. (1990). The murine mutation 'osteoporosis' is a mutation in the coding region of the macrophage colony stimulating factor (*Csf m*) gene. *Nature* **345**, 442–444.

Yusoff, P., Hamilton, J.A., Nolan, R.D. and Phillips, W.A. (1994). Haematopoietic colony stimulating factors CSF-1 and GM-CSF increase phosphatidylinositol 3-kinase activity in murine bone marrow-derived macrophages. *Growth Factors* **10**, 181–192.

Chapter 25

Stem Cell Factor

Graham Molineux[1] and Ian K. McNiece[2]

[1]Amgen Inc., Thousand Oaks, CA, [2]University Hospital, Denver, CO, USA

INTRODUCTION

For several decades, hematologists and developmental biologists have been intrigued by the similarities in defective hematopoiesis, reduced fertility and reduced skin pigmentation that are exhibited by mice carrying mutations at the W locus on chromosome 5 and on the Sl locus on chromosome 10 (Sarvella and Russell, 1956; Bernstein *et al.*, 1990; Copeland *et al.*, 1990). Among the phenotypic similarities of these animals are hematopoietic defects manifest as a macrocytic anemia and a total lack of mast cells (Kitamura and Go, 1979). The infertility in these mutant mice arises as a consequence of defective migration and/or survival of primordial germ cells in midterm fetal development (Dolci *et al.*, 1991; Godin *et al.*, 1991), and the absence of coat pigmentation is due to a deficiency in melanocyte distribution during embryogenesis (Mayer and Green, 1968).

The close relationship between these two independent phenotypes was further defined by early experiments which showed that the hematopoietic deficiency in heterozygous viable mutant mice of W/W^v genotype could be cured if the animals were transplanted with bone marrow cells from mice carrying mutations at the Sl locus (Sl/Sl^d) (Bernstein *et al.*, 1968). Thus the hematopoietic stem cells in Sl/Sl^d mice appeared to be functionally normal and, despite the normally defective stem cells in W/W^v mice, their bone marrow environment was fully capable of sustaining hematopoiesis. It appeared, therefore, that a product of the Sl locus, which was normally expressed in W/W^v tissues, could sustain stem cells from Sl/Sl^d mice but could not support W/W^v stem cells. These data suggested that defective expression of an Sl gene product (ligand) was responsible for the Sl/Sl^d or Steel phenotype, whereas W/W^v mutant mouse produced normal amounts of this product but was itself incapable of responding to the material due to a defective receptor (c-kit).

Evidence to support this hypothesis appeared in 1990 with the simultaneous publication by several groups of details of the cloning of a hematopoietic cell growth factor. Thus, *kit* ligand (KL) (Huang *et al.*, 1990; Nocka *et al.* 1990) described a protein that bound to a previously identified tyrosine kinase-associated receptor known as c-*kit*; mast cell growth factor (MGF) (Williams *et al.*, 1990) identified a molecule that could stimulate the *in vitro* growth of mast cells; and stem cell factor (SCF) (Zsebo *et al.*, 1990b) described a hemopoietin that influenced the growth of primitive hematopoietic

The Cytokine Handbook, 3rd ed.
ISBN 0-12-689662-3

progenitor cells. It was clear, however, that KL, MGF and SCF were identical products of the same gene of the mouse Sl locus and that c-*kit*, the receptor for the ligand, was located at the W locus.

Purification

SCF was originally isolated from an activity present in medium conditioned by buffalo rat liver cells (BRL-3A) (Martin *et al.*, 1990; Zsebo *et al.*, 1990b). Using a series of purification steps including anion exchange chromatography, gel filtration, immobilized lectin chromatography, cation exchange chromatography and reverse phase chromatography (Zsebo *et al.*, 1990b), SCF was purified and found to run over a 28–35 000 molecular weight range on sodium dodecyl sulfate–polyacrylamide gel electrophoresis. Amino acid sequencing of the purified rat material led to the isolation of partial and full-length cDNAs and genomic clones encoding these sequences in both rat and human tissues (Martin *et al.*, 1990). Based on the primary structure of the protein purified from BRL-3A cells, various truncated forms were expressed in *Escherichia coli* and COS-1 (mammalian) cells and found to possess the same biologic activities as the BRL-3A product (Martin *et al.*, 1990).

BIOCHEMICAL PROPERTIES OF SCF

Full-length SCF protein comprises 248 amino acids, and is cleaved in the extracellular domain to form the soluble form of SCF consisting of 165 amino acids (Anderson *et al.*, 1990; Bernstein *et al.*, 1991). A shorter form of 220 amino acids results in a membrane-bound form of SCF (Anderson *et al.*, 1990; Bernstein *et al.*, 1991). Soluble native SCF is secreted as a glycoprotein monomer. The core protein is 18.4 kDa, but is glycosylated with both *N*-linked and *O*-linked carbohydrate residues to increase the apparent molecular weight to 30–35 kDa (Lu *et al.*, 1991). Glycosylation is not necessary for biologic activity as recombinant *E. coli*-derived material is active *in vitro* (Zsebo *et al.*, 1990b). The recombinant soluble form of SCF exists possibly as a dimer linked by two intramolecular disulfide bonds (Cys-4–Cys-89 and Cys-43–Cys-138; see Fig. 1).

Genetic Properties of SCF and its Expression

Murine and human SCF map to chromosomes 10 and 12 respectively (Copeland *et al.*, 1990; Zsebo *et al.*, 1990a). This gene yields predominantly a single mRNA species of 6.5 kb. Several cDNA clones resulting from alternate mRNA splicing events have been reported from both human and murine sources. In both humans and mice cDNAs encoding SCF forms with a 28-amino-acid deletion exist, and an additional 16-amino-acid deletion form has also been cloned from mice (Lyman and Williams, 1992).

CIRCULATING LEVELS OF SCF

Many tissues are known to secrete SCF, including bone marrow stromal cells (Andrews *et al.*, 1992; Heinrich *et al.*, 1992), fibroblasts (Broudy *et al.*, 1992a; Buzby *et al.*, 1992; Heinrich *et al.*, 1992) and endothelial cells. In studies of serum levels in 257 normal individuals, the circulating level has been measured by enzyme-linked immunosorbent

Human	Glu	Gly	Ile	Cys	Arg	Asn	Arg	Val	Thr	Asn	Asn	Val	Lys	Asp	Val	Thr	Lys	Leu	Val	Ala
Rat	Gln	Gly	Ile	Cys	Arg	Asn	Pro	Val	Thr	Asp	Asn	Val	Lys	Asp	Ile	Thr	Lys	Leu	Val	Ala
																				40
Human	Asn	Leu	Pro	Lys	Asp	Tyr	Met	Ile	Thr	Leu	Lys	Tyr	Val	Pro	Gly	Met	Asp	Val	Leu	Pro
Rat	Asn	Leu	Pro	Asn	Asp	Tyr	Met	Ile	Thr	Leu	Asn	Tyr	Val	Ala	Gly	Met	Asp	Val	Leu	Pro
																				60
Human	Ser	His	Cys	Trp	Ile	Ser	Glu	Met	Val	Val	Gln	Leu	Ser	Asp	Ser	Leu	Thr	Asp	Leu	Leu
Rat	Ser	His	Cys	Trp	Leu	Arg	Asp	Met	Val	Thr	His	Leu	Ser	Val	Ser	Leu	Thr	Thr	Leu	Leu
																				80
Human	Asp	Lys	Phe	Ser	Asn	Ile	Ser	Glu	Gly	Leu	Ser	Asn	Tyr	Ser	Ile	Ile	Asp	Lys	Leu	Val
Rat	Asp	Lys	Phe	Ser	Asn	Ile	Ser	Glu	Gly	Leu	Ser	Asn	Tyr	Ser	Ile	Ile	Asp	Lys	Leu	Gly
																				100
Human	Asn	Ile	Val	Asp	Asp	Leu	Val	Glu	Cys	Val	Lys	Glu	Asn	Ser	Ser	Lys	Asp	Leu	Lys	Lys
Rat	Lys	Ile	Val	Asp	Asp	Leu	Val	Ala	Cys	Met	Glu	Glu	Asn	Ala	Pro	Lys	Asn	Val	Lys	Glu
																				120
Human	Ser	Phe	Lys	Ser	Pro	Glu	Pro	Arg	Leu	Phe	Thr	Pro	Glu	Glu	Phe	Phe	Arg	Ile	Phe	Asn
Rat	Ser	Leu	Lys	Lys	Pro	Glu	Thr	Arg	Asn	Phe	Thr	Pro	Glu	Glu	Phe	Phe	Ser	Ile	Phe	Asn
																				140
Human	Arg	Ser	Ile	Asp	Ala	Phe	Lys	Asp	Phe	Val	Val	Ala	Ser	Glu	Thr	Ser	Asp	Cys	Val	Val
Rat	Arg	Ser	Ile	Asp	Ala	Phe	Lys	Asp	Phe	Met	Val	Ala	Ser	Asp	Thr	Ser	Asp	Cys	Val	Leu
																				160
Human	Ser	Ser	Thr	Leu	Ser	Pro	Glu	Lys	Asp	Ser	Arg	Val	Ser	Val	Thr	Lys	Pro	Phe	Met	Leu
Rat	Ser	Ser	Thr	Leu	Gly	Pro	Glu	Lys	Asp	Ser	Arg	Val	Ser	Val	Thr	Lys	Pro	Phe	Met	Leu
					165															
Human	Pro	Pro	Val	Ala	Ala															
Rat	Pro	Pro	Val	Ala	Ala															

Fig. 1. Amino acid sequence of soluble human and rat stem cell factor (SCF).

assay at around 3.3 + 1.1 ng/ml (Langley, 1993). Patients with aplastic anemia tend to have serum SCF levels in the lower part of the normal distribution, although a wide variation in levels from one patient to another makes it difficult to state with certainty that the SCF level in patients with aplastic anemia is significantly lower than that in normals (Wodnar Filipowicz *et al.*, 1993). Patients with myelodysplastic syndrome have significantly reduced levels of circulating SCF (Bowen *et al.*, 1993), although no correlation was found between FAB classification and SCF levels. The significance of the relatively high concentrations of SCF in the blood of both normal and diseased individuals is not immediately apparent. It is interesting, however, that the concentration of SCF required for half-maximum effect in *in vitro* colony-forming assays is in the range of 2–9 ng/ml (110–480 pmol/l) (Langley *et al.*, 1993), the circulating level is 3.3 ng/ml (180 pmol/l) and the K_d for the interaction of SCF with c-*kit* is in the range of 50–100 pmol/l (Abkowitz *et al.*, 1992; Broudy *et al.*, 1992b). This would indicate a significant occupancy of receptors *in vivo* and suggests a role for SCF in normal homeostasis.

IN VITRO ACTIONS OF SCF

The initial biologic characterization of SCF included exhaustive study of its abilities as a hematopoietic colony stimulating factor (CSF). A detailed analysis of these data is beyond the scope of this review; however, a substantial amount of work can be summarized relatively briefly.

As a single agent, SCF is a modest CSF. In semisolid cultures of murine bone marrow cells, SCF induces the formation of limited numbers of small colonies containing predominantly neutrophils. In longer-term liquid culture of the same target cells, mast cells emerge as the dominant population. SCF has relatively little effect on human bone marrow cells (Bernstein *et al.*, 1991; McNiece *et al.*, 1991a; Ulich *et al.*, 1991a,b; Williams *et al.*, 1992).

SCF is, however, a potent costimulatory or synergistic factor when added in the presence of other growth factors such as erythropoietin, thrombopoietin, granulocyte–macrophage colony stimulating factor (GM-CSF), G-CSF, interleukin-3 (IL-3), IL-7 and M-CSF (Bernstein *et al.*, 1991; McNiece *et al.*, 1991a,b; Namen *et al.*, 1991; Ulich *et al.*, 1991a,b; Williams *et al.*, 1992; Hunt *et al.*, 1994). The effect of SCF is to increase the size and perhaps number of colonies obtained without influencing the lineage differentiation of the progenitors.

It is not immediately apparent how the *in vitro* synergistic actions of SCF might be exploited to clinical advantage. To date the most promising applications of SCF to manipulation of human tissues *in vitro* are in the area of cell processing before transplantation. Data are accumulating to indicate that SCF is a ubiquitous addition to the cytokine cocktails used to expand the progenitor content of hematopoietic grafts (McNiece *et al.*, 1991a; Williams *et al.*, 1992; Shieh *et al.*, 1994). Flt3 ligand, a molecule with many biologic features in common with SCF, may also show some potential in this area (McKenna *et al.*, 1997). A second area, possibly also linked to altering the proliferative status of hematopoietic cells, is the manipulation of transduction frequency in retroviral vector-mediated gene transfer (Walsh *et al.*, 1995; Bernad *et al.*, 1997). It would appear from preliminary data that, by combining *ex vivo* 'expansion' technologies with gene transfer methods, significant graft engineering may be possible, resulting in a graft tailored to the specific indication, whether that indication is the correction of a genetic disease or support of multicycle chemotherapy in patients with cancer. SCF would appear to offer a significant supporting role in all settings of the manipulation of hematopoietic cells that proceed in the absence of hematopoietic stroma.

IN VIVO ACTIONS OF SCF

The actions of administered SCF have been studied in various species including rodents, dogs, non-human primates and, most recently, humans. The first, and most obvious, application of SCF was in attempting to treat the disorders of Sl/Sl^d mice. Thus, Zsebo *et al.* (1990a) treated Sl/Sl^d mice with recombinant rat SCF (rrSCF) and partially corrected the macrocytic anemia and mast cell deficiency typical of these mice. This validated the contention that SCF was the product of the Sl locus and that injecting the product of this gene may go some way to correcting the phenotype of the mutant mouse. It did not, however, lead to suggestions for the clinical application of the material, a human equivalent of the mouse Sl/Sl^d defect not having been reported.

When administered to normal rodents, SCF which had been chemically modified by the addition of polyethylene glycol (PEG–rrSCF) induced a modest (in comparison to G-CSF or GM-CSF) but significant leukocytosis (Molineux *et al.*, 1991; Ulich *et al.*, 1991a,b). Baboons also showed a significant lymphocytosis in response to administered PEG–rhSCF and increases in the numbers of circulating platelets, basophils, mast cells, eosinophils and monocytes were documented in different species (Andrews *et al.*, 1991; Tsai *et al.*, 1991; Wershil *et al.*, 1992; Galli *et al.*, 1993; Dale *et al.*, 1997). Most of these observations were made in a single or in a few species, the neutrophilia seen first in rodents being the only peripheral change that was consistent between all the species studied. Baboons also showed altered erythrocyte parameters in response to SCF: increases in the number of reticulocytes and a transient erythrocytosis, leading under protracted treatment (more than 2 weeks) to iron deficiency anemia (Andrews *et al.*, 1991). These effects in other species were either too modest for detection or insignificant. Despite the increased number of megakaryocytes seen in the marrow of dogs, baboons and mice given PEG–SCF, only Sl/Sl^d mice showed a significant thrombocytosis. Indeed, dogs treated with rcSCF had significant thrombocytopenia. Megakaryocytes and their precursors are known to have significant numbers of SCF receptors (Avraham *et al.*, 1992; Briddell *et al.*, 1991, 1993; Tanaka *et al.*, 1992; Yan *et al.*, 1994).

It remains a possibility that the true potential of SCF in manipulating the platelet lineage may lie in combined treatment with the newly available mpl ligands (also called thrombopoietins) (Bartley *et al.*, 1994; de Sauvage *et al.*, 1994; Kaushansky *et al.*, 1994; Lok *et al.*, 1994; Wending *et al.*, 1994) which appear to represent the long-sought lineage restricted regulator of thrombopoiesis.

Marrow cellularity is largely unchanged in rodents treated with SCF, although the spleen has been reported to increase in size (Molineux *et al.*, 1991; Ulich *et al.*, 1991a,b). Dogs and baboons show increases in marrow cellularity and particularly in megakaryocyte number (Rosen *et al.*, 1990; Andrews *et al.*, 1991, 1992a,b) There is a temporary preponderance of more primitive marrow populations in rodents and non-human primates associated with the beginning of SCF treatment. This shift, which in primates is associated with increased $CD34^+$ cell numbers, corrects with continued treatment. Rodents and primates also show increased marrow mast cell numbers in association with SCF treatment (Ulich *et al.*, 1991a,b). These changes in recognizable marrow cells of all treated species are accompanied by alterations in the number of progenitor cells. Typically, rodents show increased progenitor numbers in the spleen and blood, and perhaps reduced numbers in the marrow, when treated with SCF. Primates and dogs, on the other hand, show increases in marrow progenitor content which parallel the increased cellularity in these species (Andrews *et al.*, 1991; Schuening *et al.*, 1993). Blood progenitor numbers are increased under the influence of SCF (Andrews *et al.*, 1992a,b).

In keeping with the documented *in vitro* activities of SCF the most marked effects of SCF result from its coadministration with other cytokines. The largest, and to date the most useful, effect of SCF is in a two-cytokine combination with simultaneous G-CSF (Molineux *et al.*, 1991; Ulich *et al.*, 1991a). G-CSF administered to many different species induces not only the neutrophilia that was perhaps expected from the *in vitro* activities for which the material was named but also the relocation of numerous progenitor cells from bone marrow to the blood.

The mobilization of transplantable progenitor populations under the influence of G-CSF is employed widely in experimental and clinical protocols as the best current method of obtaining large numbers of cells for hematologic rescue after myeloablative or myelosuppressive protocols (Welte *et al.*, 1996). Interestingly, by coadministering high-dose SCF with G-CSF to mice, 1.5-fold higher numbers of leukocytes (compared with G-CSF alone) were accompanied by fivefold greater numbers of GM-CFC, twofold more high proliferative potential CFC (HPP-CFC) and more than fivefold more cells with long-term repopulating potential (de Revel *et al.*, 1994).

However, initial data that emerged from early clinical trials of SCF showed that high-dose treatment with SCF was unlikely to be possible in patients. It was then shown (Briddell *et al.*, 1993; de Revel *et al.*, 1994) that lower-dose treatment with SCF was a feasible option. Using 25 μg/kg, a dose of SCF that itself does not mobilize significant numbers of peripheral blood progenitor cells (PBPCs) as a single agent, a potent interaction with G-CSF was seen and up to threefold greater numbers of PBPCs were collected compared with G-CSF alone (Briddell *et al.*, 1993).

CLINICAL APPLICATION

Despite the success of autologous bone marrow transplantation for re-establishing hematopoiesis after myeloablative chemotherapy, widespread application of the technique is hampered by several limitations. These limitations include the rate of hematopoietic recovery, the associated liability to infection, and the need for a general anesthetic for marrow harvest. By using PBPC transplants, these limitations can be overcome to some extent. PBPCs are more easily collected and offer superior rates of short-term recovery with no apparent loss in long-term reconstituting potential when compared with conventionally harvested bone marrow grafts. As an inadvertent side-effect, PBPCs may also have a reduced tumor cell load compared with a marrow harvest from a patient in which infiltration of the bone marrow is a significant feature of the disease.

In the normal situation few PBPCs can be found circulating in the blood. Consequently, large volumes of blood need to be processed to obtain sufficient cells for transplantation. This limitation can be overcome by 'mobilizing' transplantable progenitor cells from their normal sites to the peripheral blood. There exist several methods for mobilizing PBPCs. Among the longer established of these methods is the collection of PBPCs during the rebound from exposure to cancer chemotherapy, particularly high-dose cyclophosphamide. This is obviously applicable only to patients who may be receiving chemotherapy as part of their treatment and is not without risks—occasional deaths have been reported. An alternative strategy is to collect PBPCs after treatment with G-CSF, either as a single agent or combined with chemotherapy. In either situation G-CSF mobilizes considerable numbers of PBPCs—enough, in some studies, to allow hematopoietic reconstitution from the product of a single leukapheresis.

Preclinical studies, some of which are discussed above, indicated that SCF may be a powerful single agent or synergistic factor for the mobilization of PBPCs in combination with G-CSF. However, exploration of this possibility for the benefit of patients required the satisfactory completion of preliminary clinical trials.

Clinical Trials of SCF

Pharmacokinetics

Detailed pharmacokinetic data were obtained from a phase I study in patients with cancer prior to myeloablative chemotherapy. These patients were injected intravenously with 5, 10, 25 or 50 μg rhSCF per kg body weight per day for 14 days. Using immunoassay, the baseline SCF concentration in the blood was found to be 1 μg/ml in this study. The time to maximum serum SCF concentration (T_{max}) was 12–17 h on the first day of the study, but had fallen to 2–11 h by day 14. The maximum serum concentration (C_{max}) of SCF increased about twofold during the study, although equal trough concentrations before and 24 h after the day 14 dose indicated that a steady-state concentration had been reached (Young *et al.*, 1993). Further data were obtained from patients treated with a single dose of 25 μg rhSCF per kg body weight. From measurement made in the 72 h following injection, a serum half-life of 35 h was calculated for rhSCF.

These data indicate that the pharmacokinetic model for SCF is linear and that slow absorption from repeated injections leads to a cumulative twofold increase in serum concentration at steady state.

Phase I Studies

Patients suffering advanced lung cancer were enrolled in a phase I study of SCF before being treated with chemotherapy (Crawford *et al.*, 1993). For 14 days they received rhSCF and were observed for a further 14 days. Later, patients were treated with carboplatin–etoposide chemotherapy and received a further 14 subcutaneous injections of rhSCF at 5, 10, 25 or 50 μg/kg daily. Some patients also received combination antihistamine treatment, with or without ephedrine, in an attempt to control adverse events. From 35 patients enrolled in the study, ten were removed prematurely: seven because of adverse events and three because of progression of the underlying disease.

Similarly, 23 patients were withdrawn from an enrollment of 26 with advanced breast carcinoma (21 received SCF, five received G-CSF) (Demetri *et al.*, 1993). Ten of the premature withdrawals resulted from adverse reactions, four were due to disease progression, seven were withdrawn as a result of administrative or investigator decisions and two withdrew for other reasons.

The most frequently documented adverse effects were skin reactions at the site of injection including edema, erythema, pruritus, skin hyperpigmentation and urticaria (Crawford *et al.*, 1993; Demetri *et al.*, 1993). The local raised pruritic wheal and surrounding erythema reaction, which occurred within 90–120 min at the injection site, did not intensify during the treatment period. The rest of the symptoms were manifestations of a systemic anaphylactoid reaction to the injected SCF. These symptoms are consistent with reports documenting mast cell hyperplasia and mediator release (Costa *et al.*, 1996) in response to SCF.

Hematologic data indicated only a modest dose-related increase in leukocyte and platelet counts in patients after chemotherapy. No observable changes occurred in the bone marrow, but patients receiving the highest (50 μg/kg) dose showed increases in circulating PBPCs as assessed by GM-CFC and erythroid burst-forming unit (BFU-E) assays. Blood chemistry showed no overall trends and serum samples showed no evidence of antibodies to SCF in any patient.

Phase I–II Studies

A population consisting of patients at high risk of breast carcinoma was used to evaluate rhSCF in phase I–II trial (McNiece *et al.*, 1993). This population represented appropriate candidates for high-dose chemotherapy with cellular support. PBPCs were harvested by apheresis after mobilization with increasing doses of SCF in combination with a standard dose (10 μg/kg daily) of G-CSF. The comparison group received G-CSF alone. Enrollment into the group receiving low-dose rhSCF (5 μg/kg per day) with no G-CSF was suspended after the first five patients failed to achieve stable engraftment when mobilized PBPCs were transplanted following high-dose chemotherapy. These patients also failed to show significant leukocytosis, which was noted in all groups receiving G-CSF either with or without rhSCF. The leukocytosis in G-CSF- and G-CSF–SCF-treated recipients comprised mainly neutrophils. Mononuclear cell numbers also increased more with the G-CSF–SCF combination than in either G-CSF- or SCF-treated cohorts. Erythroid parameters were unchanged, but platelet numbers were predictably affected by leukapheresis. The number of $CD34^+$ cells in the blood was increased by G-CSF (to 3.16×10^4 per ml on day 5), but was increased a further two- to threefold in G-CSF/SCF combination recipients. $CD34^+$ count in the blood of SCF-alone recipients never exceeded 0.3×10^4 per ml, tenfold lower than that in the G-CSF group.

Leukapheresis was performed on three occasions in this study: on day 5–7, day 8–10 or day 11–13, after being dosed with the G-CSF–SCF combinations for 7, 10 or 13 days. The G-CSF-alone cohort received G-CSF for 7 days and was apheresed on day 5–7. For each leukapheresis product, mononuclear cells (MNCs), GM-CFC and BFU-E progenitors, and CD34 counts were measured. A total yield figure was obtained by adding the products from each of the rounds of apheresis for each of the above parameters. Median MNC yield per kilogram increased in all SCF groups where patients received more than 10 μg/kg per day in combination with G-CSF. Median CD34 yield was 3.2×10^6 per kg for the G-CSF-alone cohort and the G-CSF–SCF cohorts yielded at least 7.2×10^6 per kg, approximately twice that seen with G-CSF alone. Increasing the dose of SCF to more than 10 μg/kg daily did not result in greater returns, suggesting that the dose response for mobilization of PBPCs by PEG–rhSCF reaches a plateau at a concentration of around 10 μg per kg per day.

As part of the treatment protocol in this study, PBPCs mobilized by G-CSF alone,

SCF alone or G-CSF–SCF were transplanted back to the donors, who in the interim had been treated with high-dose chemotherapy (McNiece *et al.*, 1993; Glaspy *et al.*, 1994a,b). Patients who received PBPCs mobilized with SCF alone, despite receiving the target dose of 4×10^8 MNCs per kg, showed significantly delayed recovery in re-establishing leukocyte and platelet levels. Indeed, all patients in this treatment group were given a backup transplant of bone marrow cells which had been frozen for just this eventuality. In the recipients of G-CSF–SCF-mobilized PBPC, recovery of leukocytes and platelets was excellent. Absolute neutrophil counts (ANC) reached 0.5×10^9/l at day 10 in patients who received a G-CSF-mobilized PBPC product, and this compared with 10.5, 10, 9 and 8.5 days in patients in whom PBPCs were mobilized with the same dose of G-CSF combined with 5, 10, 15 or 20 μg/kg rhSCF respectively. Similar values were obtained for return to an ANC of 1×10^9/l, with a comparable reduction in the period with incrementally increasing doses of rhSCF. There was also a suggestion of improved platelet recovery in the groups that received PBPCs mobilized with the higher doses of SCF. Recipients of G-CSF-mobilized PBPC took 13 days to attain 20×10^9 and 16 days to attain 50×10^9 platelets/μl, whereas recipients of G-CSF–SCF recovered platelets to 20 and 50×10^9/l by days 10.5 and 12 respectively with the highest dose of SCF (20 μg), intermediate-dose PBPC products sustaining recovery at intermediate timepoints.

Phase III Studies

Results from a phase III trial of rhSCF in breast cancer have been announced, although, to date, detailed data have not been published. These studies extend the data obtained in the phase I–II trial outlined above. The target of the phase III study was to determine whether the inclusion of rhSCF with G-CSF could reduce the number of aphereses required to obtain sufficient $CD34^+$ cells for successful transplantation. It was shown that the number of leukaphereses required to collect 5×10^6 $CD34^+$ cells per kg was reduced by an average of two leukapheresis cycles when rhSCF was included in the mobilization protocol. Blood cell recovery in transplanted recipients was very rapid. Studies presented recently (Tricot *et al.*, 1996) indicate that in heavily pretreated patients with multiple myeloma the inclusion of rhSCF with rhG-CSF allowed target numbers of $CD34^+$ cells to be obtained in fewer aphereses. This in itself is a significant finding, but the studies also support the contention that by including SCF with G-CSF in the mobilization protocol it should be possible to collect, on average, more PBPCs per donor should the normal number of aphereses be performed. In the future it may be possible to exploit these greater numbers of PBPCs in various settings:

(1) to facilitate postmobilization cell selection (which inevitably loses many cells) for either increasing the concentration of progenitor cells or reducing the incidence of tumor cells or alloreactive lymphoid populations;
(2) splitting an apheresis product to support accelerated hematologic recovery over successive rounds of a multicycle chemotherapy regimen to allow greater dose intensity;
(3) transplants in the allogeneic setting (which appear to require more cells);
(4) as a tissue source for *ex vivo* expansion studies;
(5) as a target tissue for gene therapy protocols that have low infectivity rates;
(6) for use in poor mobilization groups such as heavily pretreated donors or patients with diseases such as multiple myeloma.

CONCLUSIONS

SCF is a pleiotropic cytokine with effects in cellular systems as diverse as primordial germ cells, melanocyte precursors and hematopoietic stem cells. In hematopoietic systems, SCF typically exerts its influence in concert with other regulators, where the effects are often synergistic between the two factors. The mechanism of this synergy is not yet known. However, preclinical studies which suggested that SCF might be useful in combination with G-CSF for mobilizing transplantable cells from the bone marrow to the blood (PBPCs) have been confirmed in humans. Development of SCF as a therapeutic agent has occurred quickly and the results of phase III studies are expected soon.

REFERENCES

Abkowitz, J.L., Broudy, V.C., Bennett, L.G., Zsebo, K.M. and Martin, F.H. (1992). Absence of abnormalities of c-kit or its ligand in two patients with Diamond–Blackfan anemia. *Blood* **79**, 25–28.

Anderson, D.M., Lyman, S.D., Baird, A., Wignall, J.M., Eisenman, J., Rauch, C., March, C.J., Boswell, H.S., Gimpel, S.D., Cosman, D. and Williams, D.E. (1990). Molecular cloning of mast cell growth factor, a hematopoietin that is active in both membrane bound and soluble forms. *Cell* **63**, 235–243.

Andrews, D.F., Montgomery, R.B., Moran, D.J., Leung, D., Harris, W.E., Bursten, S.L., Bianco, J.A. and Singer, J.W. (1992). Tumor necrosis factor-alpha (TNFalpha) suppression of c-kit ligand (KL) is mediated by phospholipid (PL) intermediates in marrow stromal cells (MSC). *Blood* **80**, 365A (abstract).

Andrews, R.G., Knitter, G.H., Bartelmez, S.H., Langley, K.E., Farrar, D., Hendren, R.W., Appelbaum, F.R., Bernstein, I.D. and Zsebo, K.M. (1991). Recombinant human stem cell factor, a c-kit ligand, stimulates hematopoiesis in primates. *Blood* **78**, 1975–1980.

Andrews, R.G., Bartelmez, S.H., Knitter, G.H., Myerson, D., Bernstein, I.D., Appelbaum, F.R. and Zsebo, K.M. (1992a). A c-kit ligand, recombinant human stem cell factor, mediates reversible expansion of multiple CD34+ colony-forming cell types in blood and marrow of baboons. *Blood* **80**, 920–927.

Andrews, R.G., Bensinger, W.I., Knitter, G.H., Bartelmez, S.H., Longin, K., Bernstein, I.D., Appelbaum, F.R. and Zsebo, K.M. (1992b). The ligand for c-kit, stem cell factor, stimulates the circulation of cells that engraft lethally irradiated baboons. *Blood* **80**, 2715–2720.

Avraham, H., Vannier, E., Cowley, S., Jiang, S.X., Chi, S., Dinarello, C.A., Zsebo, K.M. and Groopman, J.E. (1992). Effects of the stem cell factor, c-kit ligand, on human megakaryocytic cells. *Blood* **79**, 365–371.

Bartley, T.D., Bogenberger, J., Hunt, P., Li, Y.-S., Lu, H.S., Martin, F., Chang, M.-S., Samal, B., Nichol, J.L., Swift, S., Johnson, M.J., Hsu, R.-Y., Parker, V.P., Suggs, S., Skrine, J.D., Merewether, L.A., Clogston, C., Hsu, E., Hokom, M.M., Hornkohl, A., Choi, E., Pangelinan, M., Sun, Y., Mar, V., McNinch, J., Simonet, L., Jacobsen, F., Xie, C., Shutter, J., Chute, H., Basu, R., Selander, L., Trollinger, D., Sieu, L., Padilla, D., Trail, G., Elliot, G., Izumi, R., Covey, T., Crouse, J., Garcia, A., Xu, W., del Castillo, J., Biron, J., Cole, S., Hu, M.C.-T., Pacifici, R., Ponting, I., Saris, C., Wen, D., Yung, Y.P., Lin, H. and Bosselman, R.A. (1994). Identification and cloning of a megakaryocyte growth and development factor that is a ligand for the cytokine receptor mpl. *Cell* **77**, 1117.

Bernad, A., Varas, F., Gallego, J.M., Almendral, J.M. and Bueren, J.A. (1997). Ex vivo expansion and selection of retrovirally transduced bone marrow: an efficient methodology for gene-transfer to murine lymphohaemopoietic stem cells. *Br. J. Haematol.* **87** (1), 6–17.

Bernstein, A., Chabot, B., Dubreuil, P., Reith, A., Nocka, K., Majumder, S. and Besmer, P. (1990). The mouse W/c-kit locus. *CIBA Found. Symp.* **148**, 158–166.

Bernstein, I.D., Andrews, R.G. and Zsebo, K.M. (1991). Recombinant human stem cell factor enhances the formation of colonies by CD34+ and CD34+lin− cells, and the generation of colony-forming cell progeny from CD34+lin− cells cultured with interleukin-3, granulocyte colony-stimulating factor, or granulocyte-macrophage colony-stimulating factor. *Blood* **77**, 2316–2321.

Bernstein, S.E., Russell, E.S. and Keighley, G.H. (1968). Two hereditary mouse anemias (Sl/Sld and W/Wv) deficient in response to erythropoietin. *Ann. N.Y. Acad. Sci.* **149**, 475–485.

Bowen, D., Yancik, S., Bennett, L., Culligan, D. and Resser, K. (1993). Serum stem cell factor concentration in patients with myelodysplastic syndromes. *Br. J. Haematol.* **85**, 63–66.

Briddell, R.A., Bruno, E., Cooper, R.J., Brandt, J.E. and Hoffman, R. (1991). Effect of c-kit ligand on in vitro human megakaryopoiesis. *Blood* **78**, 2854–2859.

Briddell, R.A., Hartley, C.A., Smith, K.A. and McNiece, I.K. (1993). Recombinant rat stem cell factor synergizes with recombinant human granulocyte colony-stimulating factor in vivo in mice to mobilize peripheral blood progenitor cells that have enhanced repopulating potential. *Blood* **82**, 1720–1723.

Broudy, V.C., Kovach, N., Lin, N., Jacobsen, F.W. and Bennett, L.G. (1992a). Human umbilical vein endothelial cells (HUVECs) display high affinity c-kit receptors and produce stem cell factor (SCF). *Blood* **80**, 362A (abstract).

Broudy, V.C., Smith, F.O., Lin, N., Zsebo, K.M., Egrie, J. and Bernstein, I.D. (1992b). Blasts from patients with acute myelogenous leukemia express functional receptors for stem cell factor. *Blood* **80**, 60–67.

Buzby, J.S., Knoppel, E., Yancik, S., Bhullar, A. and Cairo, M.S. (1992). Differential expression of stem cell factor, kit receptor, ICAM-1 and ELAM-1 in mesenchymal cells from newborns compared to adults. *Blood* **80**, 1571A (abstract).

Copeland, N.G., Gilbert, D.J., Cho, B.C., Donovan, P.J., Jenkins, N.A., Cosman, D., Anderson, D., Lyman, S.D. and Williams, D.E. (1990). Mast cell growth factor maps near the Steel locus on mouse chromosome 10 and is deleted in a number of Steel alleles. *Cell* **63**, 158–183.

Costa, J.J., Demetri, G.D., Harrist, T.J., Dvorak, A.M., Hayes, D.F., Marica, E.A., Menchaca, D.M., Gringeri, A.J., Schwartz, L.B. and Galli, S.J. (1996). Recombinant human stem cell factor (kit ligand) promotes human mast cell and melanocyte hyperplasia and functional activation in vivo. *J. Exp. Med.* **183**, 2681–2686.

Crawford, J., Lau, D., Erwin, R., Rich, W., McGuire, B. and Mayers, F. (1993). A phase 1 trial of recombinant methionyl human stem cell factor (SCF) in patients with advanced non-small cell lung carcinoma (NSCLC). *Proc. ASCO* **12**, 135 (abstract).

Dale, D.C., Hammond, W.P. and Zsebo, K.M. (1997). Stem cell factor therapy for cyclic hematopoiesis in grey collie dogs. *Blood* **78 (supplement)**, 375A.

de Revel, T., Appelbaum, F.R., Storb, R., Schuening, F., Nash, R., Deeg, J., McNiece, I., Andrews, R. and Graham, T. (1994). Effects of granulocyte colony-stimulating factor and stem cell factor, alone and in combination, on the mobilization of peripheral blood cells that engraft lethally irradiated dogs. *Blood* **83**, 3795–3799.

de Sauvage, F.J., Hass, P.E., Spencer, S.D., Malloy, B.E., Gurney, A.L., Spencer, S.A., Darbonne, W.C., Henzel, W.J., Wong, S.C., Kuang, W.-J., Oles, K.J., Kultgren, B., Solberg, L.A., Goeddel, D.V. and Eaton, D.L. (1994). Stimulation of megakaryopoiesis and thrombopoiesis by the c-Mpl ligand. *Nature* **369**, 533.

Demetri, G., Costa, J., Hayes, D., Sledge, G., Galli, S., Hoffman, R., Merica, E., Rich, W., Harkins, B., McGuire, B. and Gordon, M. (1993). A phase 1 trial of recombinant methionyl human stem cell factor (SCF) in patients with advanced breast carcinoma pre- and post-chemotherapy (chemo) with cyclophosphamide (C) and doxorubicin (A). *Proc. ASCO* **12**, 142 (abstract).

Dolci, S., Williams, D.E., Ernst, M.K., Resnick, J.L., Brannan, C.I., Lock, L.F., Lyman, S.D., Boswell, H.S. and Donovan, P.J. (1991). Requirement for mast cells growth factor for primordial germ cell survival in culture. *Nature* **352**, 809–811.

Galli, S.J., Lemura, A., Garlick, D.S., Gamba Vitalo, C., Zsebo, K.M. and Andrews, R.G. (1993). Reversible expansion of primate mast cell populations in vivo by stem cell factor. *J. Clin. Invest.* **91**, 148–152.

Glaspy, J., McNiece, I., LeMaistre, F., Menchaca, D., Briddell, R., Lill, M., Jones, R., Tami, J., Morstyn, G., Brown, S. and Shpall, E.J. (1994a). Effects of stem cell factor (rhSCF) and Filgrastim (rhG-CSF) on mobilization of peripheral blood progenitor cells (PBPC) and on hematological recovery post-transplant: early results from a phase I/11 study. *Proc. ASCO* **13**, 68 (abstract).

Glaspy, J., McNiece, I.K., LeMaistre F., Menchaca, D., Briddell, R., Lill, M., Jones, R., Tami, J., Morstyn, G., Brown, S. and Shpall, E.J. (1994b). Effects of stem cell factor (rhSCF) and Filgrastim (rhG-CSF) on mobilization of peripheral blood progenitor cells (PBPC) and on hematological recovery post-transplant: preliminary phase I/II study results. *Br. J. Haematol.* **87 (supplement 1)**, 156 (abstract).

Godin, I., Deed, R., Cooke, J., Zsebo, K., Dexter, M. and Wylie, C.C. (1991). Effects of the steel gene product on mouse primordial germ cells in culture. *Nature* **352**, 807–809.

Heinrich, M.C., Dooley, D.C., Freed, A.C., Band, L., Keeble, W.K., Oppernlander, B., Spurgin, P. and Bagby, G.C. (1992). TGF-β1 represses steel factor (SF) gene expression in long term human bone marrow culture (LTBMC) adherent cells. *Blood* **80**, 369A (abstract).

Huang, E., Nocka, K., Beier, D.R., Chu, T.Y., Buck, J., Lahm, H.W., Wellner, D., Leder, P. and Besmer, P. (1990).

The hematopoietic growth factor KL is encoded by the Sl locus and is the ligand of the c-kit receptor, the gene product of the W locus. *Cell* **63**, 225–233.

Hunt, P., Bartley, T., Li, Y.-S., Bogenberger, J., Lu, H., Samal, B., Martin, F., Chang, M.S., Parker, V. and Bosselman, B. (1994). Purification and cloning of a megakaryocytic growth and development factor: A novel cytokine found in aplastic plasma. *Exp. Hematol.* **22**, 838 (abstract).

Kaushansky, K., Lok, S., Holly, S., Broudy, V.C., Lin, N., Bailey, M.C., Forstrom, J.W., Buddle, M.M., Oort, P.J., Hagen, F.S., Roth, G.J., Papayannopoulou, T. and Foster, D.C. (1994). Promotion of megakaryocyte progenitor expansion and differentiation by the c-Mpl ligand thrombopoietin. *Nature* **368**, 568.

Kitamura, Y. and Go, S. (1979). Decreased production of mast cells in Sl/Sld anemic mice. *Blood* **53**, 492–497.

Langley, K.E., Bennett, L.G., Wypych, J., Yancik, S.A., Liu, X.D., Westcott, K.R., Chang, D.G., Smith, K.A. and Zsebo, K.M. (1993). Soluble stem cell factor in human serum. *Blood* **81**, 656–660.

Lok, S., Kaushansky, K., Holly, R.D., Kuilper, J.L., Lofton-Day, C.E., Oort, P.J., Grant, F.J., Helpel, M.D., Burkhead, S.K., Kramer, J.M., Bell, L.A., Sprecher, C.A., Blumberg, H., Ching, A.F.T., Mathewes, S.L., Bailey, M.C., Forstrom, J.W., Buddle, M.M., Osborn, S.G., Evans, S.J., Sheppard, P.O., Presnell, S.R., O'Hara, P.J., Hagen, F.S., Roth, G.J. and Foster, D.C. (1994). Cloning of murine thrombopoietin cDNA and stimulation of platelet production in vivo. *Nature* **369**, 565.

Lu, H.S., Clogston, C.L., Wypych, J., Fausset, P.R., Lauren, S., Mendiaz, E.A., Zsebo, K.M. and Langley, K.E. (1991). Amino acid sequence and post-translational modification of stem cell factor isolated from buffalo rat liver cell-conditioned medium. *J. Biol. Chem.* **266**, 8102–8107.

Lyman, S.D. and Williams, D.E. (1992). Biological activities and potential therapeutic uses of steel factor. A new growth factor active on multiple hematopoietic lineages. *Am. J. Pediatr. Hematol. Oncol.* **14**, 1–7.

Martin, F.H., Suggs, S.V., Langley, K.E., Lu, H.S., Ting, J., Okino, K.H., Morris, C.F., McNiece, I.K., Jacobsen, F.W. and Mendiaz, E.A. (1990). Primary structure and functional expression of rat and human stem cell factor DNAs. *Cell* **63**, 203–211.

Mayer, T.C. and Green, M.C. (1968). An experimental analysis of the pigment defect caused by mutations at the W and Sl loci in mice. *Dev. Biol.* **18**, 62–75.

McKenna, H.J., de Vries, P., Brasel, K., Lyman, S.D. and Williams, D.E. (1997). Effect of flt3 ligand on the ex vivo expansion of human CD34+ hematopoietic progenitor cells. *Blood* **86**(9), 3413–3420.

McNiece, I., Glaspy, J., LeMaistre, F., Briddell, R., Menchaca, D. and Shpall, E.J. (1993). Effects of recombinant methionyl human stem cell factor (rhSCF) and filgrastim (rhG-CSF) on mobilization of peripheral blood progenitor cells: preliminary laboratory results from phase I/II study. *Blood* **82 (supplement)**, 84 (abstract).

McNiece, I.K., Langley, K.E. and Zsebo, K.M. (1991a). Recombinant human stem cell factor synergises with GM-CSF, G-CSF, IL-3 and epo to stimulate human progenitor cells of the myeloid and erythroid lineages. *Exp. Hematol.* **19**, 226–231.

McNiece, I.K., Langley, K.E. and Zsebo, K.M. (1991b). The role of recombinant stem cell factor in early B cell development: synergistic interaction with IL-7. *J. Immunol.* **146**, 3785–3790.

Molineux, G., Migdalska, A., Szmitkowski, M., Zsebo, K. and Dexter, T.M. (1991). The effects on hematopoiesis of recombinant stem cell factor (ligand for c-kit) administered in vivo to mice either alone or in combination with granulocyte colony-stimulating factor. *Blood* **78**, 961–966.

Namen, A.E., Widmer, M.B., Voice, R., Christensen, S., Braddy, S., Lyman, S.D. and Williams, D.E. (1991). A ligand for the c-kit proto-oncogene (MGF) stimulates lymphoid progenitor cells in vitro. *Exp. Hematol.* **19**, 497 (abstract).

Nocka, K., Buck, J., Levi, E. and Besmer, P. (1990). Candidate ligand for the c-kit transmembrane kinase receptor: KL, a fibroblast derived growth factor stimulates mast cells and erythroid progenitors. *EMBO J.* **9**, 3287–3294.

Rosen, B., Catchatourian, R., Egrie, J., Gould, S.A., Greenberg, R.A., Langley, K.E., Sehgal, L.R. and Zsebo, K.M. (1990). The in vivo effects of recombinant human stem cell factor (rhSCF) on hematopoiesis in nonhuman primates. *Blood* **76 (supplement)**, 163a.

Sarvella, P.A. and Russell, L.B. (1956). Steel, a new dominant gene in the house mouse. *J. Hered.* **47**, 123–128.

Schuening, F.G., Appelbaum, F.R., Deeg, H.G., Sullivan-Pepe, M., Graham, T.C., Hackman, R. and Zsebo, K.M. (1993). Effects of recombinant canine stem cell factor, a c-kit ligand, and recombinant granulocyte colony-stimulating factor on hematopoietic recovery after otherwise lethal total body irradiation. *Blood* **81**, 20–26.

Shieh, J.-H., Chen, Y.-F., Briddell, R., Stoney, G. and McNiece, I.K. (1994). High purity of blast cells in CD34 selected populations are essential for optimal ex vivo expansion of human GM-CFC. *Exp. Hematol.* **22**, 756 (abstract).

Tanaka, R., Koike, K., Imai, T., Shiobara, M., Kubo, T., Amano, Y., Komiyama, A. and Nakahata, T. (1992). Stem cell factor enhances proliferation, but not maturation of murine megakaryocytic progenitors in serum free culture. *Blood* **79**, 365–371.

Tricot, G., Jagannath, S., Desikan, K.R., Siegel, D., Munshi, N., Olson, E., Wyres, M., Parker, W. and Barlogie, B. (1996). Superior mobilization of peripheral blood progenitor cells (PBPC) with r-metHuSCF (SCF) and r-metHU (G-CSF) (Filgrastim) in heavily pretreated multiple myeloma (MM) patients. *Blood* **88 (supplement)**, 388a (abstract).

Tsai, M., Takeishi, T., Thompson, H., Langley, K.E., Zsebo, K.M., Metcalfe, D.D., Geissler, E.N. and Galli, S.J. (1991). Induction of mast cell proliferation, maturation, and heparin synthesis by the rat c-kit ligand, stem cell factor. *Proc. Natl Acad. Sci. U.S.A.* **88**, 6382–6386.

Ulich, T.R., del Castillo, J., McNiece, I.K., Yi, E.S., Alzona, C.P., Yin, S.M. and Zsebo, K.M. (1991a). Stem cell factor in combination with granulocyte colony-stimulating factor (CSF) or granulocyte-macrophage CSF synergistically increases granulopoiesis in vivo. *Blood* **78**, 1954–1962.

Ulich, T.R., del Castillo, J., Yi, E.S., Yin, S.M., McNiece, I., Yung, Y.P. and Zsebo, K.M. (1991b). Hematological effects of stem cell factor in vivo and in vitro in rodents. *Blood* **78**, 645–650.

Walsh, C.E., Mann, M.M., Emmons, R.V., Wang, S. and Liu, J.M. (1995). Transduction of CD34-enriched human peripheral and umbilical cord blood progenitors using a retroviral vector with the Fanconi anemia group C gene. *J. Invest. Med.* **43**, 379–385.

Welte, K., Gabrilove, J., Bronchud, M.H., Platzer, E. and Morstyn, G. (1996). Filgrastim (r-metHuG-CSF)—the first 10 years. *Blood* **88**, 1907–1929.

Wendling, F., Marakovsky, E., Debill, N., Florindo, C., Teepe, M., Titeux, M., Mathia, N., Breton-Gorius, J., Cosman, D. and Vainchenker, W. (1994). c-Mpl ligand is a humoral regulator of megakaryopoiesis. *Nature* **369**, 571.

Wershil, B.K., Tsai, M., Geissler, E.N., Zsebo, K.M. and Galli, S.J. (1992). The rat c-kit ligand, stem cell factor, induces c-kit receptor-dependent mouse mast cell activation in vivo. Evidence that signaling through the c-kit receptor can induce expression of cellular function. *J. Exp. Med.* **175**, 245–255.

Williams, D.E., Eisenman, J., Baird, A., Rauch, C., Van Ness, K., March, C.J., Park, L.S., Martin, U., Mochizuki, D.Y. and Boswell, H.S. (1990). Identification of a ligand for the c-kit proto-oncogene. *Cell* **63**, 167–174.

Williams, N., Bertoncello, I., Kavnoudias, H., Zsebo, K. and McNiece, I. (1992). Recombinant rat stem cell factor stimulates the amplification and differentiation of fractionated mouse stem cell populations. *Blood* **63**, 58–64.

Wodnar Filipowicz, A., Yancik, S., Moser, Y., dalle Carbonare, V., Gratwohl, A., Tichelli, A., Speck, B. and Nissen, C. (1993). Levels of soluble stem cell factor in serum of patients with aplastic anemia. *Blood* **81**, 3259–3264.

Yan, X.Q., Briddell, R., Hartley, C., Stoney, G., Samal, B. and McNiece, I.K. (1994). Mobilization of long-term hematopoietic reconstitution cells in mice by the combination of SCF plus G-CSF. *Blood* **84**, 795–799.

Young, J.D., Crawford, J., Gordon, M., Hsieh, A., Yyres, M., Mayres, F., Fare, J. and Demetri, G. (1993). Pharmacokinetics (pk) of recombinant methionyl stem cell factor (SCF) in patients (pts) with lung and breast cancer in phase I trials. *Proc. Am. Assoc. Cancer Res.* **34**, 217 (abstract).

Zsebo, K.M., Williams, D.A., Geissler, E.N., Broudy, V.C., Martin, F.H., Atkins, H.L., Hsu, R.Y., Birkett, N.C., Okino, K.H., Murdock, D.C., *et al*. (1990a). Stem cell factor is encoded at the Sl locus of the mouse and is the ligand for the c-kit tyrosine kinase receptor. *Cell* **63**, 213–224.

Zsebo, K.M., Wypych, J., McNiece, I.K., Lu, H.S., Smith, K.A., Karkare, S.B., Sachdev, R.K., Yuschenkoff, V.N., Birkett, N.C., Williams, L.R., *et al*. (1990b). Identification, purification, and biological characterization of hematopoietic stem cell factor from buffalo rat liver-conditioned medium. *Cell* **63**, 195–201.

Chapter 26

Flt3 Ligand

Stewart D. Lyman, Eugene Maraskovsky and Hilary J. McKenna
Immunex Corporation, Seattle, WA, USA

INTRODUCTION

The survival, growth and differentiation of hematopoietic cells is regulated primarily by proteins that interact with specific receptors on the surface of these cells. Among these proteins are the interleukins and the colony stimulating factors (CSFs), which bind to different families of receptors. In 1991 the flt3 (*f*ms-*l*ike *t*yrosine kinase 3) tyrosine kinase receptor was isolated independently by two groups using different strategies (Matthews *et al.*, 1991; Rosnet *et al.*, 1991a,b). One group used low-stringency hybridization with a DNA probe from the kinase domain of the CSF-1 receptor (c-*fms*) to isolate a partial cDNA clone of a related DNA sequence (Rosnet *et al.*, 1991b). The partial clone was then used to isolate a full-length receptor clone (Rosnet *et al.*, 1991a) that was named flt3. A second group used a polymerase chain reaction (PCR)-based strategy (again taking advantage of the highly conserved sequences in the kinase domain of tyrosine kinase receptors) to isolate a novel receptor fragment from purified mouse fetal liver stem cells (Matthews *et al.*, 1991). This fragment was used to isolate a full-length receptor clone that was given the name flk-2 (*f*etal *l*iver *k*inase 2).

A comparison of the published mouse flt3 and flk-2 receptor sequences (Lyman *et al.*, 1993b) has shown that these sequences differ by only two amino acids in their extracellular domain. However, a large number of differences at the amino acid level have been noted in a region near their C-termini. The sequence of the mouse flt3 receptor has been confirmed independently by several groups (Lyman *et al.*, 1993b; Rossner *et al.*, 1994; Zeigler *et al.*, 1994), and the human flt3 receptor sequence is homologous to the mouse flt3, but not the mouse flk-2, sequence (Rosnet *et al.*, 1993a; Small *et al.*, 1994). No independent confirmation of the sequence identified as flk-2 has been reported. While this issue of the different flt3 and flk-2 sequences has never been fully resolved, the group that originally cloned flk-2 has reported that 'the validity of the sequence reported as flk2 is still unclear' (Dosil *et al.*, 1993). Therefore, we refer to the receptor as flt3, and to its ligand as flt3 ligand (Flt3-L). The flt3 receptor has also been referred to as Stk-1 (*St*em cell *k*inase-1) (Small *et al.*, 1994), but this name is not used widely, perhaps because it has been designated previously to denote a gene regulating stem cell kinetics (Van Zant *et al.*, 1983) as well as a different receptor tyrosine kinase of the met/sea/ron family (Iwama *et al.*, 1994).

The Cytokine Handbook, 3rd ed.
ISBN 0-12-689662-3

STRUCTURE OF THE FLT3 RECEPTOR

The flt3 receptor belongs to the same subfamily of tyrosine kinase receptors as c-*kit*, c-*fms* and the platelet-derived growth factor (PDGF) A and B receptors. Each of the receptors in this family has five immunoglobulin-like domains in its extracellular region, as well as an 'insert region' of 75–100 amino acids in the middle of its cytoplasmic kinase domains. The flt3 receptor has nine potential sites for *N*-linked glycosylation in its extracellular domain (Matthews *et al.*, 1991; Rosnet *et al.*, 1991a, 1993a; Small *et al.*, 1994) and is glycosylated at one or more of these sites (Lyman *et al.*, 1993b). Immunoprecipitation of the flt3 receptor shows two major protein bands of 143 and 158 kDa (Lyman *et al.*, 1993b). Pulse chase analysis reveals that the 158-kDa protein arises from the 143-kDa protein. This size change is likely to reflect glycosylational processing of the protein from one containing high mannose carbohydrates to one containing complex carbohydrates. Consistent with this interpretation is the finding that only the 158-kDa species is found on the cell surface (Lyman *et al.*, 1993b).

An isoform of the mouse flt3 receptor has been identified recently that is missing the fifth of the five immunoglobulin-like regions (the one nearest the transmembrane region) in the extracellular domain as a result of the skipping of two exons during transcription (Lavagna *et al.*, 1995). This isoform is present at much lower levels than the wild-type receptor. The receptor isoform lacking the fifth immunoglobulin-like domain is functional in that it binds ligand and is phosphorylated as a result of this binding. However, the physiologic significance of this isoform, if any, is presently unknown.

A partial clone of the downstream part of the human flt3 receptor genomic locus has been isolated and compared to the genes encoding the tyrosine kinase receptors c-*fms* and c-*kit* (Agnès *et al.*, 1994). All three receptors share structural similarities, including the number of exons, size of exons and exon–intron boundaries. These data support the idea that the flt3, c-*fms* and c-*kit* receptors, all of which are involved in hematopoiesis, are ancestrally related and are derived as a result of gene duplication and divergence.

CLONING OF THE LIGAND FOR THE FLT3 RECEPTOR

Cloning of Flt3-L was achieved by two groups using different strategies. Lyman and coworkers (1993a) used a soluble form of the flt3 receptor to screen a variety of cell lines, looking for one that was capable of binding the soluble receptor. A mouse T-cell line was identified which specifically bound the soluble flt3 receptor; the ligand was then cloned from a cDNA expression library made from mRNA isolated from these cells. An alternative approach employed by Hannum and coworkers (1994) was to purify the Flt3-L from medium conditioned by a mouse thymic stromal cell line. Purification of the protein was facilitated by coupling the extracellular domain of the flt3 receptor to a matrix to generate an affinity column. *N*-terminal sequencing of the purified protein generated a short amino acid sequence. This information was used to construct degenerate oligonucleotide primers to amplify by PCR a portion of the Flt3-L gene, which eventually led to the isolation of a full-length mouse cDNA clone.

The mouse Flt3-L cDNA was then used in hybridization experiments to isolate cDNAs encoding the human gene (Hannum *et al.*, 1994; Lyman *et al.*, 1994b). The mouse and human Flt3-L proteins are 72% identical at the amino acid level. Both human and mouse Flt3-L cDNAs encode a transmembrane protein that is biologically

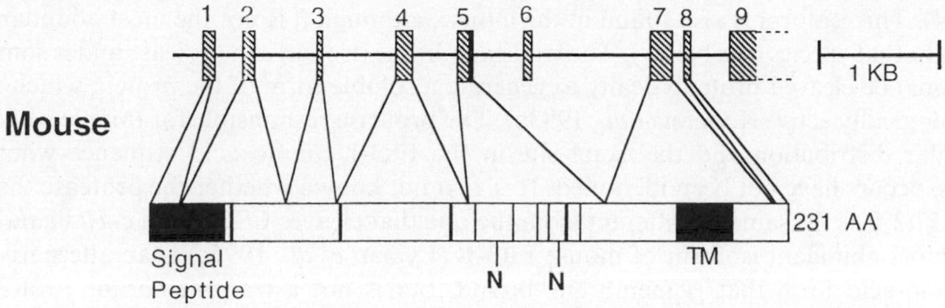

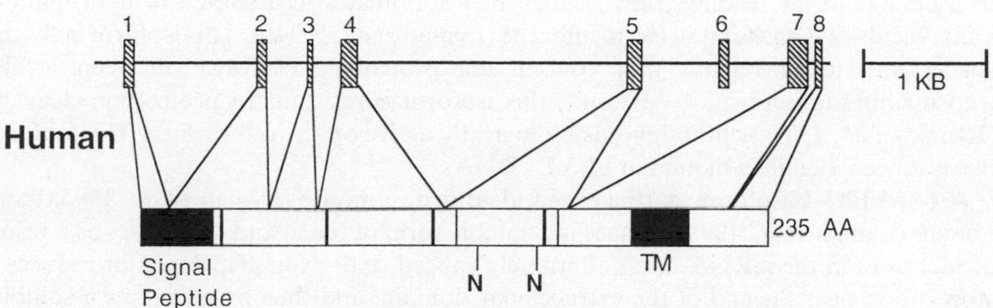

Fig. 1. Structure of the mouse and human flt3 ligand genomic loci and the encoded proteins. Exons and introns are drawn to the same scale for both the mouse and human loci. The human Flt3-L locus is larger than the mouse locus as a result of having larger introns (Lyman *et al.*, 1995c). The sequence information encoding the transmembrane form of the flt3 ligand is contained within exons 1–5 plus 7 and 8; an alternatively spliced exon 6 is used to generate a soluble form of the protein (Lyman *et al.*, 1995c). The vertical black lines in the protein structures represent cysteine residues; those that are conserved between Flt3-L, CSF-1 and c-*kit* ligand are in bold. The symbol N is used to denote the position of consensus sites for the addition of asparagine-linked carbohydrates; TM denotes the transmembrane region.

active on the cell surface and can also give rise to a soluble, biologically active form of the ligand; the structure of the proteins is shown in Fig. 1. Flt3-L is structurally related to CSF-1 and c-*kit* ligand (see below).

To date there appears to be no species specificity with regard to Flt3-L binding or biologic activity. Both the mouse and human ligand are fully active on cells bearing either the mouse or human receptors (Lyman *et al.*, 1994a), and the human Flt3-L protein is active on mouse, rat, cat, rabbit, primate and human cells (unpublished data). This is in contrast to either CSF-1 or c-*kit* ligand, which show species-specific limitations in their activities (reviewed in Lyman *et al.*, 1994a).

MULTIPLE ISOFORMS OF FLT3 LIGAND

Multiple isoforms of both the mouse and human Flt3-L have been identified by analysis of multiple cDNA clones and PCR (Lyman *et al.*, 1993a, 1994b, 1995a; Hannum *et al.*, 1994). The biologic significance of these multiple isoforms is presently unknown. The predominant isoform of human Flt3-L is the transmembrane protein (described above) which is biologically active on the cell surface (Lyman *et al.*, 1993a, 1994b; Hannum *et al.*, 1994). This isoform is also found in the mouse, although it is not the most abundant isoform in that species (see below). The transmembrane protein can (at least under some conditions) be cleaved proteolytically to generate a soluble form of the protein which is also biologically active (Lyman *et al.*, 1993a). The protease responsible for this cleavage, its cellular distribution and the exact site in the Flt3-L amino acid sequence where cleavage occurs have not been identified. It is also not known whether the protease that cleaves Flt3-L is the same or different from the one that cleaves CSF-1 and c-*kit* ligand.

The most abundant isoform of mouse Flt3-L (Lyman *et al.*, 1995c) is an alternative 220-amino-acid form that is membrane bound, but is not a transmembrane protein (Hannum *et al.*, 1994; Lyman *et al.*, 1995a; McClanahan *et al.*, 1996). This form arises as a result of a failure to splice an intron out of the mRNA (Lyman *et al.*, 1995a). This leads to a change in the reading frame, which then terminates in a stretch of hydrophobic amino acids that anchor it in the membrane (Lyman *et al.*, 1994b). This isoform lacks the spacer and tether regions that contain the proteolytic cleavage site seen in the transmembrane isoform. As a result, this isoform is resistant to proteolytic cleavage (Lyman *et al.*, 1995a), although it is biologically active on the cell surface. This isoform has not been identified in human Flt3-L cDNAs.

A third Flt3-L isoform that has been identified in mouse (Lyman *et al.*, 1995a) and human (Lyman *et al.*, 1995c) tissues is a soluble form of the ligand that arises as a result of inclusion in the mRNA of an alternately spliced sixth exon. This exon introduces a stop codon near the end of the extracellular domain, and thereby generates a soluble, biologically active protein. This isoform appears to be relatively rare compared with other isoforms (Lyman *et al.*, 1995c). A variant of this isoform (seen only in the mouse) is one in which exon 6 is present, but 16 base pairs of the preceding exon have been deleted (Lyman *et al.*, 1995a). The resulting protein is a transmembrane protein that is not biologically active, most likely because it is missing the last of the four conserved cysteine residues mentioned above. Another method of generating soluble Flt3-L in the human is to splice out the transmembrane domain (Lyman *et al.*, 1994b), but the relative abundance of this isoform has not been quantitated.

It is interesting to note the difference between c-*kit* ligand and Flt3-L in regard to their alternatively spliced sixth exons. The amino acid sequences of exon 6 in mouse and human c-*kit* ligand are nearly identical, whereas amino acids in exon 6 of mouse and human Flt3-L have virtually no homology to each other (Lyman *et al.*, 1995c). There is no homology between the sixth exon of c-*kit* ligand and the sixth exon of either mouse or human Flt3-L. In the case of c-*kit* ligand, the sixth exon is normally part of the transmembrane protein and contains the proteolytic cleavage site. In the case of Flt3-L, the sixth exon is not normally a part of the transmembrane protein, but inclusion of the sixth exon results in the generation of a soluble protein due to a shift in the reading frame. Thus, two different uses have been made evolutionarily of the sixth exon, each of which may contribute to the generation of a soluble protein.

COMPARISON OF THE STRUCTURES OF FLT3-L, C-KIT LIGAND AND CSF-1

Flt3-L, CSF-1 and c-*kit* ligand are all transmembrane proteins that are similar in size and structure, but are not very homologous at the amino acid level (Lyman *et al.*, 1993a, 1994b; Hannum *et al.*, 1994). Despite this lack of primary sequence conservation, it seems likely that these three proteins are ancestrally related since there are four cysteine residues that are completely conserved in position between all three proteins. The Flt3-L and CSF-1 proteins share an additional two conserved cysteine residues that are not found in c-*kit* ligand. All of these cysteine residues are believed to play a critical role in the formation of intramolecular disulfide bonds that serve to stabilize the three-dimensional structure of the protein. c-*kit* ligand has two intramolecular bonds (Lu *et al.*, 1991), whereas CSF-1 has three (Pandit *et al.*, 1992); it is likely that Flt3-L also has three intramolecular bonds. All three of these proteins undergo alternate splicing of a sixth exon (reviewed in Lyman *et al.*, 1995a), and mouse and human Flt3-L genomic clones have a similar exon–intron structure to c-*kit* ligand and CSF-1 (Lyman *et al.*, 1995c).

CHROMOSOMAL LOCATION OF FLT3 RECEPTOR AND LIGAND GENES

The mouse flt3 receptor gene has been mapped to region G of chromosome 5 (Rosnet *et al.*, 1991b). It is located near the mouse flt1 receptor (Rosnet *et al.*, 1993b) but is separated from the clustered c-*kit*, PDGFRA and flk-1/kinase insert domain-containing receptor (KDR) receptors that have been localized to region D–E of this chromosome. The human flt3 receptor genomic locus is about 100 kb in size and maps to chromosome 13q12 (Rosnet *et al.*, 1991b), again near the flt1 receptor locus (Shibuya *et al.*, 1989). The human flt3 and flt1 genes are linked physically (Rosnet *et al.*, 1993b) in a head to tail fashion, and are separated by about 150 kb (Imbert *et al.*, 1994). This region of chromosome 13q12 is deleted in some patients with myeloproliferative disease. However, examination of several leukemic samples with (del)13 have shown no evidence of deletion of the flt3 receptor (Birg *et al.*, 1992). The flt1 receptor has a similar overall structure compared with the flt3 receptor, except that it has seven immunoglobulin-like domains in its extracellular region (Shibuya *et al.*, 1989), compared with five in flt3.

The Flt3-L gene has been mapped to the proximal portion of mouse chromosome 7 and to human chromosome 19q13.3 (Lyman *et al.*, 1995a). There are no genetic diseases associated with these loci in either mice or humans that have a phenotype suggestive of a defect in Flt3-L. Interestingly, it has been reported that trisomy 19 is strongly associated with myeloid malignancy (Johansson *et al.*, 1994). Whether overexpression of Flt3-L is in some way responsible for the increased incidence of leukemia remains to be determined.

EXPRESSION OF FLT3 LIGAND IN MOUSE AND HUMAN TISSUES

Flt3-L is widely expressed in both mouse and human tissues (Hannum *et al.*, 1994; Lyman *et al.*, 1994b, 1995a). Highest levels of Flt3-L mRNA on human tissue Northern blots have been reported in peripheral blood mononuclear cells, but the ligand is also expressed in heart, placenta, lung, spleen, thymus, prostate, testis, ovary, intestine, liver, skeletal muscle, kidney and pancreas. Expression of Flt3-L in total brain was reported to be low by one group (Hannum *et al.*, 1994) and absent by another (Lyman *et al.*, 1994b). This apparent lack of expression in human brain may be an artefact as it was not specified which regions of the brain were removed and used to isolate the mRNA. Flt3-L would be expected to be found in brain because flt3 receptor has been shown to be expressed widely in that tissue (Ito *et al.*, 1993). Expression of Flt3-L on cell lines mirrors that seen in tissues: it is widely expressed on a great number of mouse and human cell lines (Brasel *et al.*, 1995a; Meierhoff *et al.*, 1995). This includes myeloid, monocytic, megakaryocytic, erythroid and B-cell lines. Flt3-L has also been shown to be widely expressed on mouse stromal cell lines (Wineman *et al.*, 1996), as are a number of other hematopoietic growth factors.

FLT3 RECEPTOR EXPRESSION IN MOUSE AND HUMAN TISSUES

Several early reports showed apparent widespread expression of the mouse flt3 receptor in various hematopoietic (bone marrow, spleen, thymus, lymph node, fetal liver) as well as non-hematopoietic (brain, testis, kidney, skin, ovary, placenta, liver) tissues (Rosnet *et al.*, 1991a; Maroc *et al.*, 1993; deLapeyriere *et al.*, 1995). However, these results were not consistent in that expression (determined by Northern blot analysis) seen in tissues such as liver, testis and ovary was not observed by *in situ* hybridization or Western blotting. Other investigators reported that flt3 receptor expression was not widespread, and was in fact restricted to mouse brain, bone marrow and a very small subpopulation of early hematopoietic progenitor cells from fetal liver and thymus (Matthews *et al.*, 1991). The more restricted expression of flt3 receptor in mice was confirmed by a third group which showed expression in bone marrow, brain, thymus, fetal liver and spleen, but not kidney, adult liver or testis (Rossner *et al.*, 1994), or adult thymus or spleen (Rasko *et al.*, 1995). It is possible that a relatively minor subset of cells that are flt3 receptor positive within the given tissues could give the impression of widespread receptor expression, when in fact it now seems rather restricted. Expression of flt3 receptor in human tissues has been reported in liver, spleen, thymus, bone marrow, and weakly in placenta (Rosnet *et al.*, 1993a).

Most mature mouse hematopoietic cells have been shown to be negative for flt3 receptor expression. These include promyelocytes, myelocytes, promonocytes, metamyelocytes, polymorphic mononuclear cells, eosinophils and nucleated red blood cells (Rasko *et al.*, 1995). Mouse cells reported to express flt3 receptor mRNA are peritoneal macrophages, B cells, T lymphocytes from thymus, spleen and peripheral blood (Rosnet *et al.*, 1991a), and monocytes (Rasko *et al.*, 1995). Within the B cell lineage, flt3 receptor expression is seen in pre-pro-B cells and pro B-cells, but not in pre-B, immature B or mature B cells (Wasserman *et al.*, 1995). Flt3 receptor expression has been reported in the most immature population of mouse thymocytes (defined as $CD4^-$ $CD8^-$ Thy-1^{lo}–IL-$2R^-$), but not in the more mature $CD4^-$ $CD8^+$, $CD4^+$ $CD8^-$ or $CD4^+$ $CD8^+$ cells (Matthews *et al.*, 1991). In the stem and progenitor cell compartment, flt3 receptor has been found in Lin^- Sca-1^+ $AA4.1^+$ cells (Matthews *et al.*, 1991), Lin^- $AA4.1^+$ cells (Rasko *et al.*, 1995) and WGA^+15-1.1^- Rh123 bright and dull cells (Visser *et al.*, 1993).

In humans, no expression of the flt3 receptor has been seen on mature lymphohematopoietic cells fractionated from human peripheral blood (Small *et al.*, 1994) or on human B cells, T cells, monocytes or granulocytes (Birg *et al.*, 1992). However, a later report shows monocytes and granulocytes to be weakly positive for receptor expression (Rosnet *et al.*, 1993a). The different results in these studies likely reflect either differential sensitivity of measuring techniques or contamination of cells with a minor flt3 receptor positive cell population. In the stem and progenitor cells compartment, flt3 receptor has been found in the $CD34^+$ fraction of human bone marrow cells (Small *et al.*, 1994; Rosnet *et al.*, 1996), but not in $CD34^-$ cells (Small *et al.*, 1994). Flt3 receptor has been detected on the surface of stem and progenitor cells by using either antibodies or biotinylated flt3 ligand (Zeigler *et al.*, 1994; McKenna *et al.*, 1995).

EFFECTS OF FLT3-L ON THE DEVELOPMENT OF HEMATOPOIETIC CELLS

Before the cloning of Flt-3-L, is was reported that the flt3 receptor was expressed on hematopoietic stem and progenitor cells, and it was hypothesized that Flt3-L may have a role in hematopoietic cell development (Matthews *et al.*, 1991; Rosnet *et al.*, 1993a). Indeed, this hypothesis was confirmed once Flt3-L was cloned, expressed, and its biologic activities determined (Lyman *et al.*, 1993a, 1994b; Hannum *et al.*, 1994). Flt3-L alone weakly stimulated proliferation or colony formation by hematopoietic progenitor cells (HPCs) *in vitro*, but when added with other hematopoietic growth factors synergistic effects were noted. These observations suggested Flt3-L was a growth factor with a role in hematopoietic cell development, and that it shared some biologic similarities with c-*kit* ligand.

Efforts to characterize the biologic effects of Flt3-L using *in vitro* assays have demonstrated that Flt3-L affects the growth of primitive multipotent HPCs as well as myeloid and lymphoid committed progenitors. The effects of Flt3-L on mouse (Table 1) and human (Table 2) myelopoiesis have been summarized. Flt3-L alone has weak proliferative activity on HPCs, but is able to support the survival of HPCs, which retain their responsiveness to cytokines that induce growth and differentiation (Gabbianelli *et al.*, 1995; Rasko *et al.*, 1995; Jacobsen *et al.*, 1996; Veiby *et al.*, 1996). Flt3-L synergizes with a wide range of CSFs and interleukins (Hudak *et al.*, 1995; Jacobsen *et al.*, 1995; Muench *et al.*, 1995; Rasko *et al.*, 1995), as well as with soluble thrombopoietin (TPO) and granulocyte colony stimulating factor (G-CSF) receptors (Ku *et al.*, 1996) to promote colony growth of both committed and primitive progenitor cells. Both transforming growth factor-β (TGF-β) and tumor necrosis factor-α (TNF-α) inhibit the effects of Flt3-L (Ohishi *et al.*, 1996; Veiby *et al.*, 1996). *Ex vivo* expansion of progenitors is not supported by Flt3-L alone, although progenitor cell survival is observed (McKenna *et al.*, 1995). Flt3-L synergizes with cytokine combinations to augment the *ex vivo* expansion of colony forming units (CFUs); these include G-CSF or interleukin-11 (IL-11) (Jacobsen *et al.*, 1995), IL-3, IL-6, G-CSF or c-*kit* ligand (Hudak *et al.*, 1995), IL-1 + IL-3 + IL-6 + erythropoietin (EPO) (McKenna *et al.*, 1995), IL-3 + granulocyte–macrophage colony stimulating factor (GM-CSF) + EPO (Koller *et al.*, 1996), and IL-3 + IL-6 + G-CSF (Shapiro *et al.*, 1996). However, the most primitive human progenitors (long-term culture-initiating cells), derived from bone marrow, are induced to expand by Flt3-L alone (Gabbianelli *et al.*, 1995; Petzer *et al.*, 1996), and a 30-fold expansion of these cells is achieved with a cocktail containing Flt3-L + IL-3 + c-*kit* ligand (Petzer *et al.*, 1996). The implication that Flt3-L has effects on the most primitive HPCs is supported by data obtained by Mackarehtschian *et al.* (1995) from mice (generated by homologous recombination) that were deficient in flt3 receptor expression. These animals were viable and healthy, but had a defect in primitive stem cells, as measured in long-term competitive repopulation assays. When bone marrow cells from the flt3 receptor-deficient mice were transferred to irradiated recipients, their potential to reconstitute the myeloid and lymphoid lineages was reduced compared with wild-type bone marrow cells.

Table 1. Effects of Flt3-L on murine myelopoiesis.

Cells	Assay	Results	Reference
mu BM $Lin^- Sca\text{-}1^+$	Single cell liquid cultures	Flt3-L alone had weak activity, but synergized with IL-3, IL-6, IL-11, IL-12, KL, G-CSF, M-CSF or GM-CSF to promote growth. Flt3-L + G-CSF or Flt3-L + IL-11 expanded progenitors 40-fold after 2 weeks *in vitro*.	Jacobsen *et al.* (1995)
mu BM $Lin^- Thy\text{-}1^{low} Sca\text{-}1^+$	CFU and liquid cultures	Flt3-L + IL-3, IL-6, G-CSF or KL induced an expansion of CFUs, but only Flt3-L + IL-6 expanded CFU-S_{12}. No effect of Flt3-L on erythroid, megakaryocyte, eosinophil or mast cell progenitors was noted.	Hudak *et al.* (1995)
mu BM $Lin^- Sca\text{-}1^+$ c-*kit*$^+$	CFU and liquid cultures	Flt3-L synergized with GM-CSF, G-CSF, M-CSF, IL-3 and IL-6 to promote colony growth. Flt3-L alone supported the survival of primitive cells.	Rasko *et al.* (1995)
mu BM Post 5-FU Lin^-	Serum-free CFU	Flt3-L shortened the cell cycle time and increased the rate of growth of IL-3-dependent blast cell colonies. TGF-β abrogated the effects of Flt3-L	Ohishi *et al.* (1996)
mu BM $Lin^- Sca\text{-}1^+$	Single cell liquid cultures	Flt3-L synergized with G-CSF, KL, IL-3, IL-6, IL-7, IL-11 or IL-12 to promote cell growth. TGF-β and TNF-α inhibited both myeloid and B-lymphoid colony formation in the presence of Flt3-L cytokine. From single cell studies it was concluded that Flt3-L had a direct effect on precursors.	Jacobsen *et al.* (1996)
mu BM $Lin^- Sca\text{-}1^+$	Single cell liquid cultures	Flt3-L alone promoted the survival of B-cell and myeloid cell precursors. Flt3-L suppressed apoptosis of $Lin^- Sca\text{-}1^+$ cells, and TGF-β or TNF-α counteracted this.	Veiby *et al.* (1996)
mu BM Post 5-FU $Lin^- Sca\text{-}1^+$ c-*kit*$^+$	CFU	Flt3-L synergized with soluble TPO receptor, and weakly with soluble G-CSF receptor, to promote growth of colonies comprised of undifferentiated cells.	Ku *et al.* (1996)
mu BM $Lin^- Sca\text{-}1^+$	Liquid cultures	Flt3-L synergized with TPO to increase the number and size of clones. No increase in megakaryocyte production was noted.	Ramsfjell *et al.* (1997)
mu d14 FL $AA4.1^+$ $Sca\text{-}1^+ CD43^+$	CFU	Flt3-L alone had no activity but synergized with IL-6, IL-11, IL-12 and G-CSF to promote colony growth.	Fujimoto *et al.* (1996)

mu, Murine; BM, bone marrow; CFU, colony forming unit; IL, interleukin; 5-FU, 5-fluorouracil; KL, c-*kit* ligand; FL, fetal liver; CFU-S_{12}, colony forming unit-spleen (day 12).

Table 2. Effects of Flt3-L on human myelopoiesis.

Cells	Assay	Results	Reference
hu fetal liver Lin^{-} $CD34^{+}$ CD38	CFU, serum-free liquid cultures	Flt3-L alone had weak activity but synergized with GM-CSF, IL-3 and KL to promote HPP and LPP colony growth.	Muench *et al.* (1995)
hu BM and PB	CFU, liquid cultures, LTCIC	Flt3-L augmented CFU-GM colony growth when added to IL-3 + GM-CSF + EPO, but not BFU-E nor CFU-GEMM growth. Flt3-L enhanced CFU-blast and CFU-HPP growth. Flt3-L supported survival of LTCIC *in vitro*.	Gabbianelli *et al.* (1995)
hu BM and CB $CD34^{+}$ and MNCs	CFU	Flt3-L synergized with GM-CSF or IL-3 ± KL to increase CFU-GM growth, and with CSF-1 to increase CFU-M growth. Flt3-L synergized with EPO ± IL-3 + KL to increase CFU-GEMM, BFU-E and CFU-HPP growth.	Broxmeyer *et al.* (1995)
hu BM $CD34^{+}$	CFU, liquid cultures	Flt3-L alone had weak activity but synergized with Pixy321 + EPO to enhance CFU-GM, CFU-GEMM and CFU-HPP growth, but not BFU-E growth. Flt3-L alone supported survival of CFU *ex vivo*, and synergized with IL-1 + IL-3 + IL-6 + EPO to promote *ex vivo* expansion of the progenitors.	McKenna *et al.* (1995)
hu BM $CD34^{+}$	CFU	Flt3-L synergized with IL-3, G-CSF, GM-CSF, CSF-1, IL-6 or KL to augment CFU growth. Flt3-L did not affect BFU-E growth in the presence of serum, but in its absence some effect was noted.	Rusten *et al.* (1996)
hu BM $CD34^{+}$	CFU	Flt3-L alone had no effect on megakaryocyte development, but augmented the effects of IL-3 on CFU-MK and BFU-MK growth, and GM-CSF and KL on CFU-MK growth.	Piacibello *et al.* (1996)
hu BM and CB $CD34^{+}$ $CD38^{-}$	Dexter-type cultures, CFU	Flt3-L alone augmented total cell and CFU production by primitive progenitors in the presence of irradiated BM stroma. Flt3-L synergized with cytokine cocktails to increase this effect.	Shah *et al.* (1996)
hu BM $CD34^{+}$ $CD38^{-}$	LTCIC	A panel of ten cytokines was examined to determine whether LTCIC could be expanded *ex vivo*. Flt3-L + IL-3 + KL were necessary. A 30-fold expansion of LTCIC in 10 days was noted. As single factors, only Flt3-L and TPO induced LTCIC expansion.	Petzer *et al.* (1996)
hu BM $CD34^{+}$	CFU, *ex vivo* expansion	Flt3-L alone had weak activity but synergized with IL-3, IL-6, G-CSF, GM-CSF or KL to augment CFU growth.	Brashem-Stein *et al.* (1996)
hu BM MNCs	CFU, *ex vivo* expansion	Flt3-L augmented *ex vivo* expansion of total cells, CFU-GM and LTCIC by IL-3 + GM-CSF + EPO.	Koller *et al.* (1996)
hu BM, CB, FL and PBPCs	*Ex vivo* expansion	Flt3-L synergized with IL-3 + IL-6 + G-CSF to increase the numbers of CFU generated *ex vivo*.	Shapiro *et al.* (1996)

mu, murine; hu human; BM, bone marrow; CB, cord blood; CFU, colony forming unit; BFU, burst forming unit; LTCIC, long-term culture initiating cell; IL, interleukin; LPP, low proliferative potential; HPP, high proliferative potential; MK, megakaryocyte; KL, c-*kit* ligand; MNC, mononuclear cell; FL, fetal liver; PB, peripheral blood.

Although there are a number of reports demonstrating that Flt3-L stimulates myeloid committed precursors, as well as multipotent precursors, there is little evidence that Flt3-L has a significant effect on erythroid development (Broxmeyer *et al.*, 1995; Gabbianelli *et al.*, 1995; Hirayama *et al.*, 1995; Hudak *et al.*, 1995; McKenna *et al.*, 1995). Similarly, Flt3-L does not appear to play a role in either eosinophil or mast cell development (Lyman *et al.*, 1994a; Hudak *et al.*, 1995; Hjertson *et al.*, 1996). Flt3-L shares a number of biologic characteristics with c-*kit* ligand, including the ability to synergize with a wide range of growth factors to stimulate hematopoietic development, while having only weak stimulatory activity as a single agent. However, the absence of an effect of Flt3-L on erythropoiesis and mast cell production, two lineages that c-*kit* ligand has potent effects on (Galli *et al.*, 1994), indicates that Flt3-L and c-*kit* ligand do not have entirely overlapping biologic functions. This is also reflected in differential regulation of Flt3-L and c-*kit* ligand serum levels in healthy individuals and in those with hematopoietic disorders (see below).

The effects of Flt3-L on T and B lymphoid progenitor cells has also been examined. One group originally purified soluble Flt3-L from mouse thymic stromal cells, suggesting that Flt3-L may have a role in T-cell development (Hannum *et al.*, 1994). Flt3-L alone, or in combination with IL-7, stimulates the proliferation of 14-day-old mouse fetal thymocytes (Hannum *et al.*, 1994). More recently, Freedman and coworkers (1996) have described a model of human T-cell development *in vitro*. When human bone marrow-derived $CD34^+$ cells are cultured on a primary irradiated fetal–thymic stroma layer, development of T-cell receptor-positive cells is observed, provided IL-12 is added to the culture. The addition of Flt3-L and IL-12 to the culture augments cell production without altering the phenotype of the cells produced.

Flt3 receptor expression has been studied in highly purified subsets of B-cell precursors derived from mouse bone marrow. Flt3 receptor is expressed in the primitive pre-pro-B cells, and as B-cell development progresses through the pre-B cell stage, flt3 receptor expression is progressively lost (Wasserman *et al.*, 1995). Hunte and coworkers (1996) have reported that Flt3-L, in combination with IL-7 or c-*kit* ligand, stimulates proliferation of the most primitive $B220^{low}$, $CD43^+$, $CD24^-$ bone marrow-derived B-cell precursors, but no effect of Flt3-L is noted on the more differentiated pro- or pre-B-cell progenitors. Uncommitted progenitors from mouse yolk sac or fetal liver develop into B-lineage cells in response to Flt3-L + IL-7 + IL-11 and Flt3-L + IL-7 stimulated expansion of pro-B cells (Ray *et al.*, 1996). Purified progenitors (Lin^- $Sca\text{-}1^+$) from mouse bone marrow develop into B-lineage cells in the presence of Flt3-L + IL-7, although development proceeds only to the pro-B stage (Veiby *et al.*, 1996). Flt3-L effects on human B lymphopoiesis from fetal liver-derived cells have also been reported. Flt3-L synergizes with IL-3 + IL-7 to promote pro-B-cell growth and differentiation into pre-B cells (Namikawa *et al.*, 1996). The hypothesis that Flt3-L influences the development of immature B-cell precursors is supported by studies on flt3 receptor-deficient mice, which have reduced numbers of both pro-B and pre-B cells (Mackarehtschian *et al.*, 1995).

Collectively, these results demonstrate that Flt3-L has an important role in early events in B-cell lymphopoiesis and, although less well documented, Flt3-L is also implicated in T-cell development. To date there have been no published reports on Flt3-L effects *in vitro* on natural killer (NK) cell development. However, Flt3-L does appear to increase NK cell numbers in peripheral blood *in vivo* (Winton *et al.*, 1995).

HEMATOPOIETIC EFFECTS OF FLT3-L *IN VIVO*

The administration of Flt3-L to mice demonstrates the potent ability of this protein to expand HPCs *in vivo* (Brasel *et al.*, 1996a). In these studies, hematologic parameters were followed over 2 weeks of Flt3-L administration. Over the first 5 days, Flt3-L causes a significant expansion of the number of progenitors in the bone marrow, including an increase in colony forming unit-granulocyte macrophage (CFU-GM) and the more primitive colony forming unit-granulocyte erythrocyte macrophage megakaryocyte (CFU-GEMM) progenitors (approximately five-fold). Flt3-L also induces expansion of clonal HPCs in the spleen (approximately 100-fold) and expands the number of immature B cells in the bone marrow and spleen (Brasel *et al.*, 1996a). Upon cessation of Flt3-L treatment, all measurable hematopoietic parameters return to normal levels, indicating that the effects of Flt3-L *in vivo* are transient.

Daily administration of Flt3-L to mice results in an increase in the peripheral blood cell counts and a 200- to 500-fold increase in the number of HPCs in the circulation. A number of cytokines cause mobilization of hematopoietic stem and progenitor cells from the bone marrow to the circulation when administered to animals and humans. Mobilized progenitor cells can be harvested via apheresis, and have been transplanted successfully in both autologous and allogeneic settings (Juttner *et al.*, 1988; Kessinger and Armitage, 1991). Some cytokines, such as IL-8 (Laterveer *et al.*, 1995) and macrophage inflammatory protein-1α (MIP-1α) (Lord *et al.*, 1995), induce an increase in circulating progenitor cells 30 min after injection. Other factors, including GM-CSF (Lane *et al.*, 1995), and G-CSF (Roberts and Metcalf, 1994) act more slowly, taking 2–4 days to induce mobilization. A study of the kinetics of mobilization with Flt3-L shows that the number of progenitor cells in the circulation peaks after 10 days of growth factor administration, indicating that Flt3-L is a relatively slow-acting mobilizing agent (Brasel *et al.*, 1996a).

Transplantation experiments demonstrate that the hematopoietic precursors present in Flt3-L-mobilized blood include primitive HPCs. A 200-fold increase in the number of day 12 colony forming unit-spleen (CFU-S_{12}) in the blood was noted (Brasel *et al.*, 1996a). Transfer of mobilized mononuclear cells to lethally irradiated mice leads to survival of the animals and both short-term (1 month) and long-term (6 months) multilineage hematopoietic reconstitution, demonstrating that primitive hematopoietic stem cells are mobilized (Brasel *et al.*, 1996b).

FLT3-L SERUM LEVELS IN HEMATOPOIETIC DISORDERS

The results from both *in vitro* and *in vivo* experimentation with Flt3-L support the idea that Flt3-L is a critical regulator of early events in the development of hematopoietic cells. Further support for this hypothesis has come from studies on the levels of Flt3-L in the blood of healthy individuals compared with those in patients with a variety of hematopoietic disorders. Serum levels of Flt3-L in normal healthy individuals average less than 100 pg/ml, which is the limit of detection of this protein in enzyme-linked immunosorbent assay (Lyman *et al.*, 1995b). This is in contrast to the structurally related proteins c-*kit* ligand and CSF-1, which are present in serum at levels in the range of 2–6 ng/ml (Langley *et al.*, 1993) and 2–4 ng/ml (Shadle *et al.*, 1989) respectively. Flt3-L blood levels are not raised in a variety of anemias that predominantly affect only the erythroid lineage, such as acquired pure red cell aplasia, congenital pure red cell aplasia (Diamond–Blackfan anemia) or α-thalassemia (Lyman *et al.*, 1995b). Flt3-L serum levels are also normal in patients with rheumatoid arthritis, systemic lupus erythematosus, acute myeloblastic leukemia (AMI), acute lymphoblastic leukemia (ALL) or human immunodeficiency virus (S.D. Lyman *et al.*, unpublished observations).

In contrast to these findings (and to what has been observed with c-*kit* ligand), serum levels of Flt3-L were greatly raised in patients with hematopoietic disorders that specifically affect the stem cell compartment, i.e. pancytopenias. Thus, the majority of patients with Fanconi anemia or acquired aplastic anemia have greatly increased levels of Flt3-L (up to 10 ng/ml) (Lyman *et al.*, 1995b). Flt3-L serum levels returned to normal in a patient with Fanconi anemia following a cord blood transplant that cured the pancytopenia (Lyman *et al.*, 1995b). Similarly, successful treatment of patients with aplastic anemia with either a bone marrow or peripheral blood transplant, or with immunosuppressive treatment (antilymphocyte globulin + cyclosporin A), results in Flt3-L serum concentrations returning to normal (Wodnar-Filipowicz *et al.*, 1996). These studies were extended to patients undergoing chemotherapy and radiation treatment for various leukemias, and who consequently had marrow aplasia. The level of Flt3-L in their serum appears to be inversely correlated to the peripheral blood white cell count.

It has been hypothesized that in situations where HPC numbers are severely reduced (Fanconi anemia, acquired aplastic anemic or chemotherapy–irradiation), Flt3-L levels become raised in an attempt to stimulate the production of hematopoietic stem cells *in vivo* (Wodnar-Filipowicz *et al.*, 1996). One interpretation of these data is that the loss of functional stem/progenitor cells leads to a loss of a negative regulator of Flt3-L production. The levels of Flt3-L in the blood of patients with hematopoietic disorders (with a suspected stem cell component) may be increased to a level that is physiologically active. In summary, these data suggest that restoration of stem cells in these patients is associated with a return of Flt3-L serum levels to those measured in normal healthy individuals, and that Flt3-L serum levels may be a surrogate marker for stem cell content of bone marrow.

EXPRESSION OF THE FLT3 RECEPTOR IN LEUKEMIAS

Within the hematopoietic system flt3 receptor expression appears to be restricted to immature precursor cells (Matthews *et al.*, 1991; Small *et al.*, 1994; Zeigler *et al.*, 1994; Rasko *et al.*, 1995; McKenna *et al.*, 1996a). Like the related receptors c-*kit* and c-*fms*, flt3 receptor expression has also been described on leukemic blast cells. However, unlike c-*kit* and c-*fms*, whose expression appears to be restricted to myeloid leukemic cells (Birg *et al.*, 1994), flt3 receptor expression has been detected on both myeloid and lymphoid primary leukemic blasts (Birg *et al.*, 1992; Carow *et al.*, 1996; McKenna *et al.*, 1996b; Stacchini *et al.*, 1996). In addition, a variety of myeloid and lymphoid leukemic cell lines express flt3 receptor (Da Silva *et al.*, 1994; Brasel *et al.*, 1995a; Meierhoff *et al.*, 1995; Turner *et al.*, 1995). Flt3 receptor expression is particularly prevalent on pre-B and monocytic–myeloid leukemic cells. It is presently unclear what role, if any, flt3 receptor expression on leukemic blasts has in the pathogenesis of leukemia. Recently, in a screen of 80 patients with AML or ALL, leukemic cells from five of the patients with AML were found to have internal tandem duplications within the flt3 receptor gene (Nakao *et al.*, 1996). These duplications resulted in the production of larger flt3 receptor proteins (with amino acids added in the juxtamembrane region) because the duplications were in frame. The flt3 receptor alterations were not detected in samples of cells taken from these five patients when in remission, suggesting that the alterations were tumor specific. The clinical significance of this observation remains to be determined.

There have been numerous reports of various cytokine receptors expressed on leukemic blast cells, but the *in vitro* response of these cells to the respective ligands has not been consistent. Expression of a cytokine receptor has not been predictive of a proliferative response, and it has been difficult to determine whether specific cytokine–cytokine receptor interactions are important in the pathogenesis of leukemia (Zola *et al.*, 1994). The published reports of flt3 receptor expression on leukemic cells, and the response of those cells to Flt3-L, are consistent with these observations (Drexler, 1996). Flt3-L alone has been reported to stimulate the proliferation of some leukemic cells, including AML, B-ALL and T-ALL cells (Piacibello *et al.*, 1995; Eder *et al.*, 1996; McKenna *et al.*, 1996b). Synergistic proliferative effects between Flt3-L and other cytokines have also been noted, as has been discussed for normal hematopoiesis (Tables 1 and 2). More recently, it has been suggested that Flt3-L can prevent apoptosis in AML blast cells (Lisovsky *et al.*, 1996). The prevention of apoptosis, which is augmented with the addition of G-CSF or GM-CSF, is associated with decreased Bax expression in the surviving cells.

Flt3-L expression has been detected on the surface of leukemic cell lines (Brasel *et al.*, 1995a; Meierhoff *et al.*, 1995), and some examples of concomitant flt3 receptor and ligand expression have been described in these reports. However, there is as yet no evidence of an autocrine loop in these cells, or that the cells are transformed as a result of this concomitant expression. To date there have been no reports of Flt3-L expression on the surface of leukemic blast cells.

FLT3-L IS A KEY REGULATOR OF DENDRITIC CELLS

Dendritic cells (DCs) are rare hematopoietic-derived cells that form a cellular network involved in immune surveillance, antigen capture and presentation (Steinman, 1991). DCs are predominantly found in the T cell-dependent areas of lymphoid tissue (Metlay *et al.*, 1990; Steinman, 1991), although a subset of DCs has also been identified in the B cell-dependent germinal centers (Grouard *et al.*, 1996). DCs acquire antigens in peripheral tissues, migrate to lymphatic organs and present these antigens as processed peptides to T or B cells in the context of major histocompatibility complex (MHC) antigens (Metlay *et al.*, 1990; Steinman, 1991; Austyn, 1996; Grouard *et al.*, 1996). This can initiate a cascade of events that results in an antigen-specific immune response. In this way DCs are key regulators of immune responses. In addition to their capacity for processing and presenting foreign or altered self-antigens to induce immunity (Steinman, 1991), certain DC subsets are believed to negatively regulate T-cell immunity (Matzinger and Guerder, 1989; Inaba *et al.*, 1991; Mazda *et al.*, 1991; Süss and Shortman, 1996). Furthermore, there is emerging evidence that DCs may regulate the cytokine repertoire of T cells during the induction of immunity (Macatonia *et al.*, 1993).

Although the lineage derivation of DCs continues to be clarified, it appears that they can be derived from either a myeloid- or a lymphoid-committed precursor. The relationship of DCs to the myeloid lineage is based mainly on *in vitro* experimentation showing that DCs can be generated from bone marrow or cord blood progenitors, as well as peripheral blood mononuclear cells, using a combination of GM-CSF (Inaba *et al.*, 1992a,b, 1993) and other cytokines such as TNF-α (Caux *et al.*, 1992; Romani *et al.*, 1994; Sallusto and Lanzavecchia, 1994; Szabolcs *et al.*, 1995; Young *et al.*, 1995), IL-4 (Romani *et al.*, 1994; Sallusto and Lanzavecchia, 1994), and c-*kit* ligand (Siena *et al.*, 1995; Szabolcs *et al.*, 1995; Young *et al.*, 1995).

In contrast, adoptive transfer of highly purified mouse thymic precursor cells has shown that certain DC subsets can arise from the most immature T-cell precursors, which can also generate NK and B cells, but not cells of the myeloid lineage (Ardavin *et al.*, 1993; Galy *et al.*, 1995; Wu *et al.*, 1996; reviewed by Caux and Banchereau, 1997). In contrast to myeloid precursor-derived DCs, DCs derived from the putative lymphoid-committed precursor have been shown to express high levels of CD8α homodimer (Vremec *et al.*, 1992; Ardavin *et al.*, 1993; Wu *et al.*, 1996) and do not require GM-CSF for *in vitro* development (Galy *et al.*, 1995; Saunders *et al.*, 1996).

Interestingly, although DCs can be generated *in vitro* using GM-CSF, the increased level of GM-CSF in GM-CSF-expressing transgenic mice does not increase the number of DCs in lymphoid tissue, suggesting that other growth factors are important for DC generation *in vivo* (Metcalf *et al.*, 1996).

FLT3-L STIMULATES DC DEVELOPMENT IN MICE

Mice were injected with Flt3-L to determine whether this growth factor induces the generation of DCs *in vivo*. As reported previously, daily subcutaneous injections of recombinant human Flt3-L (produced by expressing the extracellular domain of the protein in Chinese hamster ovary cells) (Brasel *et al.*, 1996a) for 10 days resulted in a dramatic increase in the numbers of hematopoietic progenitors (Brasel *et al.*, 1995b, 1996a). In addition, a significant increase in class II MHC$^+$ Thy-1$^-$ CD8α^+ cells was also observed in the peripheral blood of Flt3-L-treated mice, suggesting that Flt3-L increases the number of DCs in these mice (Vremec *et al.*, 1992; Ardavin *et al.*, 1993; Wu *et al.*, 1996).

DCs are very rare in dissociated spleen cell suspensions and require enzymatic digestion to be released from the splenic stroma, but even then they constitute a small proportion of the suspension (Crowley *et al.*, 1989). However, unlike spleen cell suspensions from control mice, which contained few phenotypically identifiable DCs, 20% of spleen cells from Flt3-L-treated animals coexpressed class II MHC and the DC marker, CD11c and DEC205 (Maraskovsky *et al.*, 1996). Further phenotypic analysis revealed that these class II$^+$, CD11c$^+$ cells were CD3$^-$, CD19$^-$, Gr-1$^-$, NK1.1$^-$, Ter119$^-$, CD80$^-$ and CD86$^+$, and could be separated into three subpopulations using CD11b expression (CD11b$^-$, CD11b^{dull} and CD11b^{bright}) (Maraskovsky *et al.*, 1996). The absolute number of class II$^+$, CD11c$^+$ cells was increased by 17-fold in the spleen, four-fold in the lymph nodes and six-fold in the peripheral blood by day 9. These cells were detected as early as day 5 in the spleen and by day 7 in lymph nodes and peripheral blood (Maraskovsky *et al.*, 1996).

Lymphocyte-depleted spleen cells from Flt3-L-treated mice were examined for the expression of CD11b and CD11c (Maraskovsky *et al.*, 1996). Five distinct spleen populations were identified and designated A, B, C, D and E (Fig. 2). Only populations

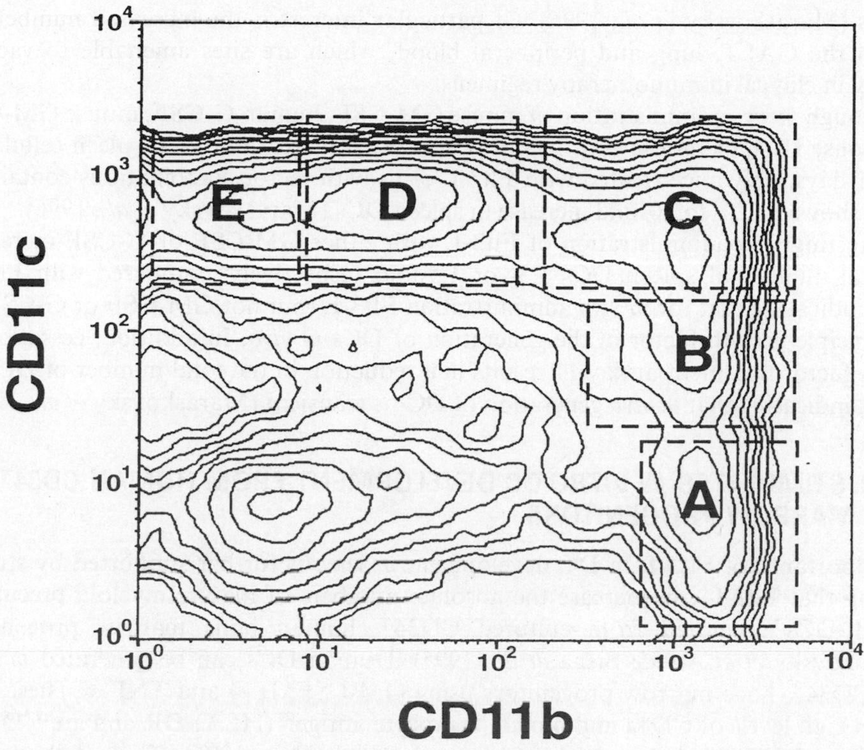

Fig. 2. Flow cytometric analysis of spleen cells isolated from mice treated with Flt3-L. The distribution of CD11c and CD11b on depleted spleen cells from Flt3-L-treated mice showing the gates used for cell sorting. C57BL/6 mice (five per group) were injected once daily, subcutaneously, with mouse serum albumin (MSA) (1 μg) plus human Flt3-L derived from Chinese hamster ovary (CHO) cells (10 μg) for 9 consecutive days.

C, D and E expressed the highest levels of class II MHC, CD11c and CD86, and were morphologically indistinguishable from the rare DCs that can be freshly isolated from the spleens of untreated mice (control DCs) (Maraskovsky *et al.*, 1996). Populations A and B were composed of immature myeloid cells and myeloblasts respectively. Cells within populations A, B and C were CD8α^-, whereas populations D, E and control DCs could be separated into CD8α^- and CD8α^+ subsets (Maraskovsky *et al.*, 1996). This indicated that Flt3-L was able to expand both myeloid-derived and CD8α^+ lymphoid precursor-derived DCs. Examination of these cells for *in vitro* antigen-presenting capacity revealed that cells from populations C, D and E were as efficient as control DCs at stimulating the proliferation of allogeneic T cells in a mixed lymphocyte reaction or of syngeneic keyhole limpet hemocyanin (KLH)-specific T cells in a KLH presentation assay (Maraskovsky *et al.*, 1996). In contrast, cells from populations A and B were 30-fold less efficient as antigen-presenting cells compared with control DCs. Furthermore, Flt3-L-generated spleen DCs were as efficient as control DCs at generating antigen-specific T cells *in vivo* after being pulsed with antigen *ex vivo* (Maraskovsky *et al.*, 1996). Raised numbers of DCs were detected in both lymphoid and non-lymphoid tissue in Flt3-L-treated mice, including the bone marrow, gastro-associated lymphoid tissue (GALT), liver, lymph nodes, lung, peripheral blood, peritoneal cavity, spleen and thymus (Maraskovsky *et al.*, 1996). Of particular interest is the increased numbers of DCs in the GALT, lung and peripheral blood, which are sites amenable to vaccine delivery in clinical immunotherapy regimens.

Although *in vivo* administration of mouse GM-CSF, human G-CSF, mouse GM-CSF plus mouse IL-4, or mouse c-*kit* ligand results in an increase in total spleen cellularity over 11 days, only mice treated with Flt3-L or growth factor combinations containing Flt3-L showed a 20- to 30-fold increase in spleen DCs (Maraskovsky *et al.*, 1996).

Interestingly, coadministration of Flt3-L with either GM-CSF or G-CSF increased the total number of spleen DCs 1.3- or 1.5-fold respectively, compared with Flt3-L alone, indicating that for *in vivo* administration Flt3-L (but not GM-CSF or G-CSF) is the principle growth factor in the generation of DCs *in vivo*. In addition, cessation of growth factor treatment at day 11 results in a reduction in the total number of DCs by day 17, indicating that *in vivo* generation of DCs is transient (Maraskovsky *et al.*, 1996).

FLT3-L STIMULATES *IN VITRO* DC DEVELOPMENT FROM HUMAN CD34$^+$ BONE MARROW PROGENITORS

The importance of Flt3-L in DC development *in vitro* is further supported by studies showing that Flt3-L can increase the absolute numbers of mature myeloid precursor-derived DCs generated from cultured CD34$^+$ human bone marrow progenitors (Maraskovsky *et al.*, 1995; Siena *et al.*, 1995). Human DCs can be generated *in vitro* from CD34$^+$ bone marrow progenitors using GM-CSF, IL-4 and TNF-α. These cells express high levels of CD1a and human leukocyte antigen (HLA)-DR and are CD13$^+$, CD86$^+$ and CD80$^+$ (reviewed by Caux and Banchereau, 1996). These cells do not express CD3, CD14 or CD19. As shown in Table 3, the addition of Flt3-L to GM-CSF, IL-4 and TNF-α containing cultures enhances the total expression of CD34$^+$ bone marrow progenitors by eight-fold. However, no significant change has been observed in the percentage of cells coexpressing CD1a, HLA-DR and CD86. As a result, the enhanced expansion of the starting progenitor pool translates into an eight-fold increase in the absolute numbers of DCs (Table 3). Interestingly, sorted CD1a$^+$, HLA-DR$^+$, CD86$^+$ DCs from Flt3-L-containing cultures are as efficient at stimulating the proliferation of alloreactive T cells, or processing and presenting tetanus toxoid (TTX) to TTX-specific autologous T cells, as DCs generated in the absence of Flt3-L (Maraskovsky *et al.*, 1995; Siena *et al.*, 1995). This suggests that the addition of Flt3-L enhances the expansion of DCs, but has little effect on their acquisition of functional capacity.

Table 3. Analysis of proliferation and generation of dendritic cells from *in vitro* cultured CD34$^+$ bone marrow progenitors.

Growth factor conditions	Total number of cultured cells ($\times 10^{-6}$)	Percentage of CD1a$^+$ HLA-DR$^+$ CD86$^+$ cells	Total number of CD1a$^+$ HLA-DR$^+$ CD86$^+$ cells ($\times 10^{-6}$)
GM-CSF + IL-4 + TNF-α	6.0 ± 0.8	48.3 ± 12.3	2.9 ± 0.9
+ Flt3-L	48.5 ± 1.6	46.2 ± 10.2	22.4 ± 1.6

Values are the mean ± SD of five experiments. Cultures were initiated with 1 × 10^6 CD34$^+$ cells.

DC DEVELOPMENT IN FLT3-L-DEFICIENT MICE

The importance of Flt3-L for DC generation from progenitors is further highlighted by examination of mice rendered genetically deficient in Flt3-L. Mice that lack Flt3-L as a result of homologous recombination show a significant reduction in the numbers of primitive bone marrow progenitors as well as myeloid cells, B cells and NK cells (McKenna *et al.*, 1996a). In addition, deficient DC development and reduced numbers of mature DCs in the peripheral lymphoid tissues were observed in these animals. Both myeloid precursor-derived and lymphoid precursor-derived DCs are reduced, although a significantly greater reduction in CD8α^+ lymphoid precursor-derived DCs is observed (McKenna *et al.*, 1996a). The residual spleen DCs isolated from Flt3-L-deficient mice are as efficient at stimulating the proliferation of alloreactive T cells *in vitro* as DCs isolated from the spleen of wild-type animals.

MECHANISM OF FLT3-L-MEDIATED DC DEVELOPMENT AND IMMUNE CONSEQUENCE

The exact mechanism by which Flt3-L influences DC generation *in vivo* and *in vitro* is not understood completely. However, it is unlikely that Flt3-L treatment simply results in the mobilization of existing mature DCs from other sites into the lymphoid tissue. First, flt3 receptor is not detected on mature DCs, as assessed by flow cytometry (E. Maraskovsky, unpublished observation), and mature DCs do not proliferate when cultured in Flt3-L alone, as assessed either by microphysiometric analysis of cytoplasmic pH changes or by thymidine incorporation of proliferating cells. Second, increased numbers of DCs have been detected in multiple organs and tissues in Flt3-L-treated mice, indicating that there is a generalized expansion of DCs throughout these animals. Third, the effects of Flt3-L on the expansion of DCs from bone marrow progenitors, its described effects on primitive progenitors *in vitro*, and the potent effects of Flt3-L on hematopoiesis *in vivo* suggest that Flt3-L targets and expands the primitive progenitor cell pool in the bone marrow and other hematopoietic organs. Other signals may potentiate and induce DC maturation *in vivo*. In this way Flt3-L may facilitate the terminal development of a primitive, Flt3-L-sensitive, rapidly expanding progenitor population into functionally mature DCs.

Does the expansion of distinct DC subpopulations *in vivo* result in a more robust or sensitive immune response to antigenic challenge? Recent studies have shown that Flt3-L-treated mice respond more efficiently to a tumor challenge *in vivo* (Lynch *et al.*, 1997). This appears to be associated with the induction of a T cell-specific antitumor response, which can mediate the rejection of even established tumors. Similar results have been obtained with breast cancer cells transduced with a cDNA expressing Flt3-L. Injection of these cells into mice leads not only to their rejection, but to the subsequent rejection of freshly transplanted, untransduced, breast cancer cells (Chen *et al.*, 1995).

POTENTIAL USES OF FLT3-L IN THE CLINIC

Although still under intensive study, current knowledge of Flt3-L suggests that it has the potential to be used in a number of different areas in clinical medicine, as indicated below. There appears to be no overt toxicity associated with the administration of Flt3-L to animals *in vivo*, and as a result of this the clinical potential of this molecule is promising.

Ex vivo Stem Cell Expansion

Ex vivo expansion of stem cells is an area receiving intense clinical study at present, and Flt3-L would be an excellent candidate for this setting. Incubation of Thy-1^{lo} Sca-1^{+} mouse bone marrow cells in liquid culture with Flt3-L + IL-3 or Flt3-L + c-*kit* ligand generates large numbers of mature myeloid cells (Jacobsen *et al.*, 1995). Culture of these same cells in the presence of Flt3-L + G-CSF or IL-11 results in the generation of cells with an immature blast cell phenotype, and clonogenic progenitors expand over 40-fold after 14 days in culture (Jacobsen *et al.*, 1995). *Ex vivo* expansion of progenitors is also noted with the combination of Flt3-L with IL-3, c-*kit* ligand, IL-6 or G-CSF (Hudak *et al.*, 1995). Flt3-L is more potent than c-*kit* ligand in combination with IL-6 or G-CSF at generating CFUs in liquid culture. The combination of Flt3-L + IL-6 is the most potent at generating CFU-S_{12} cells *in vitro*, demonstrating that primitive cells could be expanded with Flt3-L + IL-6 and still retain functional potential (Hudak *et al.*, 1995).

Flt3-L as a single factor is able to maintain human bone marrow-derived CFU-GM and high proliferative potential (HPP) *ex vivo* for 3–4 weeks, and when combined with IL-1 + IL-3 + IL-6 + EPO results in a similar level of CFU-GM expansion as seen when c-*kit* ligand is added to this combination of factors (McKenna *et al.*, 1995). Flt3-L acts to stimulate the expansion of long-term culture initiating cells (Gabbianelli *et al.*, 1995; Petzer *et al.*, 1996). A very promising combination of factors for the *ex vivo* expansion of stem–progenitor cells from cord blood is the combination of Flt3-L and TPO, which allows for continuous expansion of these cells for a 5-month period (Piacibello *et al.*, 1997).

Gene Therapy

Flt3-L may be of some use in the setting of gene therapy. Transduction of stem cells with retroviruses requires that the cells be cycling actively; therefore Flt3-L may be used to stimulate the proliferation of candidate stem cell populations to facilitate this process. Several papers have now been published showing that Flt3-L does increase the capacity of retroviruses to transduce hematopoietic stem cells (Elwood *et al.*, 1996; Dao *et al.*, 1997).

Stem Cell Mobilization

Flt3-L has potent effects on the mobilization of bone marrow stem cells to the peripheral blood for transplantation. Data in both mice (Brasel *et al.*, 1996a) and primates (Winton *et al.*, 1995, 1996) indicate that Flt3-L is capable of increasing the number of colony forming cells (CFU-GM, CFU-GEMM and BFU-E) in peripheral blood. Furthermore, transplantation studies in the mouse have demonstrated that primitive hematopoietic stem cells capable of long-term multilineage repopulation are mobilized by Flt3-L (Brasel *et al.*, 1996b).

Organ Transplantation

Long-term survivors of organ transplants, particularly liver transplants, have recently been found to exhibit multilineage donor cell microchimerism in a variety of tissues (Starzl *et al.*, 1992). This implies that hematopoietic stem cells in the liver have contributed to hematopoiesis in the organ recipient. Examination of the donor-derived cells found in host tissue reveals that DCs are commonly present. It has been hypothesized that donor-derived DCs are involved in the long-term tolerance to the allogeneic organ in the recipients (Starzl *et al.*, 1993; Lu *et al.*, 1995b; Thomson *et al.*, 1995). Immature DCs in particular (classified by their low level of MHC class II antigens, and low or absent levels of CD80 and CD86) are postulated to be involved in the process of tolerance (Lu *et al.*, 1995a). This putative role of immature DCs in tolerance induction raises the possibility that Flt3-L-generated DCs may be used to augment tolerance in organ transplantation settings.

Immunotherapy

By virtue of their highly developed capacity to initiate or regulate immunity, the use of DCs as cellular vectors for either antitumor and infectious disease vaccines or as inducers of transplantation tolerance are promising immunotherapy strategies (Mayordomo *et al.*, 1995; Thomson *et al.*, 1995; Young and Inaba, 1996). However, the clinical feasibility of using DCs as immunotherapy vectors continues to be limited by the extremely small numbers of DCs that can be procured from peripheral blood of normal individuals, and the uncertainty as to which DC subsets are the most appropriate for the induction or regulation of immunity *in vivo*. The majority of current immunotherapy regimens rely heavily upon the generation of DCs *in vitro* (from either bone marrow progenitors or peripheral blood mononuclear cells) using GM-CSF, which is known preferentially to bias the outgrowth of myeloid precursor-derived DCs (Inaba *et al.*, 1991, 1992a,b; Caux *et al.*, 1992; Romani *et al.*, 1994; Sallusto and Lanzavecchia, 1994; Szabolcs *et al.*, 1995; Young *et al.*, 1995; Caux and Banchereau, 1997). The functional stability, trafficking properties and the qualitative nature of the T- or B-cell immune responses that *in vitro*-derived myeloid DCs generate is not well understood. The identification of ontogenically distinct DC subpopulations may lead to the identification of differing functions (i.e. stimulation or regulation of T-cell function and control of cytokine repertoire). Therefore, the simultaneous generation of both lymphoid precursor- and myeloid precursor-derived DC subpopulations *in vivo* using Flt3-L may provide a more physiologically relevant representation of DC subpopulations to mediate the most appropriate immune response for any given antigen. Preliminary results in mice indicate that Flt3-L treatment significantly enhances the generation of antigen-specific responses *in vivo* to limiting doses of protein antigen in the absence of chemical adjuvants (C.R. Maliszewski, personal communications). One can envisage using Flt3-L as a vaccine adjuvant whereby DC subsets would be expanded *in vivo* by Flt3-L treatment, and then antigen-based vaccines would be injected systemically. Alternatively, larger numbers of circulating DCs from Flt3-L-treated individuals could be isolated for *ex vivo* manipulation (such as vaccine or tolerogen exposure) followed by reinfusion of these DCs into the individual.

ACKNOWLEDGEMENTS

The authors acknowledge the contributions of many colleagues at Immunex who contributed to this body of work, including Ken Brasel, Bali Pulendran, Dave Lynch, Doug Williams, Charlie Maliszewski, Mark Teepe and Eileen Roux. They also thank Anne Bannister for excellent editorial work on the manuscript.

REFERENCES

Agnès, F., Shamoon, B., Dina, C., Rosnet, O., Birnbaum, D. and Galibert, F. (1994). Genomic structure of the downstream part of the human *FLT3* gene: exon/intron structure conservation among genes encoding receptor tyrosine kinases (RTK) of subclass III. *Gene* **145**, 283–288.

Ardavin, C., Wu, L., Li, C.-L. and Shortman, K. (1993). Thymic dendritic cells and T cells develop simultaneously in the thymus from a common precursor population. *Nature* **362**, 761–763.

Austyn, J.M. (1996). New insights into the mobilization and phagocytic activity of dendritic cells.. *J. Exp. Med.* **183**, 1287–1292.

Birg, F., Courcoul, M., Rosnet, O., Bardin, F., Pébusque, M.-J., Marchetto, S., Tabilio, A., Mannoni, P. and Birnbaum, D. (1992). Expression of the *FMS/KIT*-like gene *FLT3* in human acute leukemias of the myeloid and lymphoid lineages. *Blood* **80**, 2584–2593.

Birg, F., Rosnet, O., Carbuccia, N. and Birnbaum, D. (1994). The expression of FMS, KIT and FLT3 in hematopoietic malignancies. *Leuk. Lymphoma* **13**, 223–227.

Brasel, K., Escobar, S., Anderberg, R., de Vries, P., Gruss, H.-J. and Lyman, S.D. (1995a). Expression of the flt3 receptor and its ligand on hematopoietic cells. *Leukemia* **9**, 1212–1218.

Brasel, K., McKenna, H.J., Charrier, K., Morrissey, P., Williams, D.E. and Lyman, S.D. (1995b). Mobilization of peripheral blood progenitor cells with Flt3 ligand. *Blood* **86**, 463A.

Brasel, K., McKenna, H.J., Morrissey, P.J., Charrier, K., Morris, A.E., Lee, C.C., Williams, D.E. and Lyman, S.D. (1996a). Hematologic effects of flt3 ligand *in vivo* in mice. *Blood* **88**, 2004–2012.

Brasel, K., McKenna, H.J., Williams, D.E. and Lyman, S.D. (1996b). Flt3 ligand mobilized hematopoietic progenitor cells are capable of reconstituting multiple lineages in lethally irradiated recipient mice. *Blood* **88 (Supplement 1)**, 601A.

Brashem-Stein, C., Flowers, D.A. and Bernstein, I.D. (1996). Regulation of colony forming cell generation by flt-3 ligand. *Br. J. Haematol.* **94**, 17–22.

Broxmeyer, H.E., Lu, L., Cooper, S., Ruggieri, L., Li, Z.H. and Lyman, S.D. (1995). Flt3 ligand stimulates/costimulates the growth of myeloid stem/progenitor cells. *Exp. Hematol.* **23**, 1121–1129.

Carow, C.E., Levenstein, M., Kaufmann, S.H., Chen, J., Amin, S., Rockwell, P., Witte, L., Borowitz, M.J., Civin, C.I. and Small, D. (1996). Expression of the hematopoietic growth factor receptor FLT3 (STK-1/Flk2) in human leukemias. *Blood* **87**, 1089–1096.

Caux, C. and Banchereau, J. (1996). *In vitro* regulation of dendritic cell development and function. In *Blood Cell Biochemistry, Vol. 7: Hemopoietic Growth Factors and Their Receptors* (eds A. Whetton and J. Gordon), Plenum Press, London, p. 263.

Caux, C., Dezutter-Dambuyant, C., Schmitt, D. and Banchereau, J. (1992). GM-CSF and TNF-α cooperate in the generation of dendritic Langerhans cells. *Nature* **360**, 258–261.

Chen, K., Braun, S.E., Wiebke, E., Lyman, S.D., Chiang, Y., Broxmeyer, H.E. and Cornetta, K. (1995). Retroviral-mediated gene transfer of the Flt3 ligand into murine breast cancer cells prevents tumor growth *in vivo*. *Blood* **86 (supplement 1)**, 244A.

Crowley, M., Inaba, K., Witmer-Pack, M. and Steinman, R.M. (1989). The cell surface of mouse dendritic cells: FACS analyses of dendritic cells from different tissues including thymus. *Cell. Immunol.* **118**, 108–125.

Da Silva, N., Hu, Z.B., Ma, W., Rosnet, O., Birnbaum, D. and Drexler, H.G. (1994). Expression of the FLT3 gene in human leukemia–lymphoma cell lines. *Leukemia* **8**, 885–888.

Dao, M.A., Hannum, C.H., Kohn, D.B. and Nolta, J.A. (1997). FLT3 ligand preserves the ability of human $CD34^+$ progenitors to sustain long-term hematopoiesis in immune-deficient mice after *ex vivo* retroviral-mediated transduction. *Blood* **89**, 446–456.

deLapeyriere, O., Naquet, P., Planche, J., Marchetto, S., Rottapel, R., Gambarelli, D., Rosnet, O. and Birnbaum, D. (1995). Expression of Flt3 tyrosine kinase receptor gene in mouse hematopoietic and nervous tissues. *Differentiation* **58**, 351–359.

Dosil, M., Wang, S. and Lemischka, I.R. (1993). Mitogenic signalling and substrate specificity of the Flk2/Flt3 receptor tyrosine kinase in fibroblasts and interleukin 3-dependent hematopoietic cells. *Mol. Cell. Biol.* **13**, 6572–6585.

Drexler, H.G. (1996). Expression of FLT3 receptor and response to FLT3 ligand by leukemic cells. *Leukemia* **10**, 588–599.

Eder, M., Hemmati, P., Kalina, U., Ottmann, O.G., Hoelzer, D., Lyman, S.D. and Ganser, A. (1996). Effects of Flt3 ligand and interleukin-7 on *in vitro* growth of acute lymphoblastic leukemia cells. *Exp. Hematol.* **24**, 371–377.

Elwood, N.J., Zogos, H., Willson, T. and Begley, C.G. (1996). Retroviral transduction of human progenitor cells: use of granulocyte colony-stimulating factor plus stem cell factor to mobilize progenitor cells *in vivo* and stimulation by Flt3/Flk-2 ligand *in vitro. Blood* **88**, 4452–4462.

Freedman, A.R., Zhu, H., Levine, J.D., Kalams, S. and Scadden, D.T. (1996). Generation of human T lymphocytes from bone marrow CD34$^+$ cells *in vitro. Nature Med.* **2**, 46–51.

Fujimoto, K., Lyman, S.D., Hirayama, F. and Ogawa, M. (1996). Isolation and characterization of primitive hematopoietic progenitors of murine fetal liver. *Exp. Hematol.* **24**, 285–290.

Gabbianelli, M., Pelosi, E., Montesoro, E., Valtieri, M., Luchetti, L., Samoggia, P., Vitelli, L., Barberi, T., Testa, U. and Lyman, S. (1995). Multi-level effects of flt3 ligand on human hematopoiesis: expansion of putative stem cells and proliferation of granulomonocytic progenitors/monocytic precursors. *Blood* **86**, 1661–1670.

Galli, S.J., Zsebo, K.M. and Geissler, E.N. (1994). The kit ligand, stem cell factor. *Adv. Immunol.* **55**, 1–96.

Galy, A., Travis, M., Cen, D. and Chen, B. (1995). Human T, B, natural killer, and dendritic cells arise from a common bone marrow progenitor cell subset. *Immunity* **3**, 459–473.

Grouard, G., Durand, I., Filgueira, L., Banchereau, J. and Liu, Y.-J. (1996). Dendritic cells capable of stimulating T cells in germinal centres. *Nature* **384**, 364–367.

Hannum, C., Culpepper, J., Campbell, D., McClanahan, T., Zurawski, S., Bazan, J.F., Kastelein, R., Hudak, S., Wagner, J., Mattson, J., Luh, J., Duda, G., Martina, N., Peterson, D., Menon, S., Shanafelt, A., Muench, M., Kelner, G., Namikawa, R., Rennick, D., Roncarolo, M.-G., Zlotnick, A., Rosnet, O., Dubreuil, P., Birnbaum, D. and Lee, F. (1994). Ligand for FLT3/FLK2 receptor tyrosine kinase regulates growth of haematopoietic stem cells and is encoded by variant RNAs. *Nature* **368**, 643–648.

Hirayama, F., Lyman, S.D., Clark, S.C. and Ogawa, M. (1995). The flt3 ligand supports proliferation of lymphohematopoietic progenitors and early B-lymphoid progenitors. *Blood* **85**, 1762–1768.

Hjertson, M., Sundström, C., Lyman, S.D., Nilsson, K. and Nilsson, G. (1996). Stem cell factor, but not flt3 ligand, induces differentiation and activation of human mast cells. *Exp. Hematol.* **24**, 748–754.

Hudak, S., Hunte, B., Culpepper, J., Menon, S., Hannum, C., Thompson-Snipes, L. and Rennick, D. (1995). FLT3/FLK2 ligand promotes the growth of murine stem cells and the expansion of colony-forming cells and spleen colony-forming units. *Blood* **85**, 2747–2755.

Hunte, B.E., Hudak, S., Campbell, D., Xu, Y. and Rennick, D. (1996). *flk2/flt3* ligand is a potent cofactor for the growth of primitive B cell progenitors. *J. Immunol.* **156**, 489–496.

Imbert, A., Rosnet, O., Marchetto, S., Ollendorff, V., Birnbaum, D. and Pebusque, M.J. (1994). Characterization of a yeast artificial chromosome from human chromosome band 13q12 containing the FLT1 and FLT3 receptor-type tyrosine kinase genes. Cytogenet. *Cell Genet.* **67**, 175–177.

Inaba, K., Inaba, M., Romani, N., Aya, H., Deguchi, M., Ikehara, S., Muramatsu, S. and Steinman, R.M. (1992a). Generation of large numbers of dendritic cells from mouse bone marrow cultures supplemented with granulocyte/macrophage colony-stimulating factor. *J. Exp. Med.* **176**, 1693–1702.

Inaba, K., Steinman, R.M., Pack, M.W., Aya, H., Inaba, M., Sudo, T., Wolpe, S. and Schuler, G. (1992b). Identification of proliferating dendritic cell precursors in mouse blood. *J. Exp. Med.* **175**, 1157–1167.

Inaba, K., Inaba, M., Deguchi, M., Hagi, K., Yasumizu, R., Ikehara, S., Muramatsu, S. and Steinman, R.M. (1993). Granulocytes, macrophages, and dendritic cells arise from a common major histocompatibility complex class II-negative progenitor in mouse bone marrow. *Proc. Natl Acad. Sci. U.S.A.* **90**, 3038–3042.

Inaba, M., Inaba, K., Hosono, M., Kumamoto, T., Ishida, T., Muramatsu, S., Masuda, T. and Ikehara, S. (1991). Distinct mechanisms of neonatal tolerance induced by dendritic cells and thymic B cells. *J. Exp. Med.* **173**, 549–559.

Ito, A., Hirota, S., Kitamura, Y. and Nomura, S. (1993). Developmental expression of flt3 mRNA in the mouse brain. *J. Mol. Neurosci.* **4**, 235–243.

Iwama, A., Okano, K., Sudo, T., Matsuda, Y. and Suda, T. (1994). Molecular cloning of a novel receptor tyrosine kinase gene, *STK*, derived from enriched hematopoietic stem cells. *Blood* **83**, 3160–3169.

Jacobsen, S.E.W., Okkenhaug, C., Myklebust, J., Veiby, O.P. and Lyman, S.D. (1995). The FLT3 ligand potently and directly stimulates the growth and expansion of primitive murine bone marrow progenitor cells *in vitro*: synergistic interactions with interleukin (IL) 11, IL-12, and other hematopoietic growth factors. *J. Exp. Med.* **181**, 1357–1363.

Jacobsen, S.E.W., Veiby, O.P., Myklebust, J., Okkenhaug, C. and Lyman, S.D. (1996). Ability of flt3 ligand to stimulate the *in vitro* growth of primitive murine hematopoietic progenitors is potently and directly inhibited by transforming growth factor-β and tumor necrosis factor-α. *Blood* **87**, 5016–5026.

Johansson, B., Billstrom, R., Mauritzson, N. and Mitelman, F. (1994). Trisomy 19 as the sole chromosomal anomaly in hematologic neoplasms. *Cancer Genet. Cytogenet.* **74**, 62–65.

Juttner, C.A., To, L.B., Ho, J.Q., Bardy, P.G., Dyson, P.G., Haylock, D.N. and Kimber, R.J. (1988). Early lymphohemopoietic recovery after autografting using peripheral blood stem cells in acute non-lymphoblastic leukemia. *Transplant. Proc.* **20**, 40–42.

Kessinger, A. and Armitage, J.O. (1991). The evolving role of autologous peripheral stem cell transplantation following high-dose therapy for malignancies. *Blood* **77**, 211–213.

Koller, M.R., Oxender, M., Brott, D.A. and Palsson, B.Ø. (1996). *flt-3* ligand is more potent than c-*kit* ligand for the synergistic stimulation of *ex vivo* hematopoietic cell expansion. *J. Hematother.* **5**, 449–459.

Ku, H., Hirayama, F., Kato, T., Miyazaki, H., Aritomi, M., Ota, Y., D'Andrea, A.D., Lyman, S.D. and Ogawa, M. (1996). Soluble thromobopoietin receptor (Mpl) and granulocyte colony-stimulating factor receptor directly stimulate proliferation of primitive hematopoietic progenitors of mice in synergy with steel factor or the ligand for Flt3/Flk2. *Blood* **88**, 4124–4131.

Lane, T.A., Law, P., Maruyama, M., Young, D., Burgess, J., Mullen, M., Mealiffe, M., Terstappen, L.W.M.M., Hardwick, A., Moubayed, M., Oldham, F., Corringham, R.E.T. and Ho, A.D. (1995). Harvesting and enrichment of hematopoietic progenitor cells mobilized into the peripheral blood of normal donors by granulocyte–macrophage colony-stimulating factor (GM-CSF) or G-CSF: potential role in allogeneic marrow transplantation. *Blood* **85**, 275–282.

Langley, K.E., Bennett, L.G., Wypych, J., Yancik, S.A., Liu, X.-D., Westcott, K.R., Chang, D.G., Smith, K.A. and Zsebo, K.M. (1993). Soluble stem cell factor in human serum. *Blood* **81**, 656–660.

Laterveer, L., Lindley, I.J., Hamilton, M.S., Willemze, R. and Fibbe, W.E. (1995). Interleukin-8 induces rapid mobilization of hematopoietic stem cells with radioprotective capacity and long-term myelolymphoid repopulating ability. *Blood* **85**, 2269–2275.

Lavagna, C., Marchetto, S., Birnbaum, D. and Rosnet, O. (1995). Identification and characterization of a functional murine FLT3 isoform produced by exon skipping. *J. Biol. Chem.* **270**, 3165–3171.

Lisovsky, M., Estrov, Z., Zhang, X., Consoli, U., Sanchez-Williams, G., Snell, V., Munker, R., Goodacre, A., Savchenko, V. and Andreeff, M. (1996). Flt3 ligand stimulates proliferation and inhibits apoptosis of acute myeloid leukemia cells: regulation of Bcl-2 and Bax. *Blood* **88**, 3987–3997.

Lord, B.I., Woolford, L.B., Wood, L.M., Czaplewski, L.G., McCourt, M., Hunter, M.G. and Edwards, R.M. (1995). Mobilization of early hematopoietic progenitor cells with BB-10010: a genetically engineered variant of human macrophage inflammatory protein-1 alpha. *Blood* **85**, 3412–3415.

Lu, H.S., Clogston, C.L., Wypych, J., Fausset, P.R., Lauren, S., Mendiaz, E.A., Zsebo, K.M. and Langley, K.E. (1991). Amino acid sequence and post-translational modification of stem cell factor isolated from buffalo rat liver cell-conditioned medium. *J. Biol. Chem.* **266**, 8102–8107.

Lu, H.S., Chang, W.C., Mendiaz, E.A., Mann, M.B., Langley, K.E. and Hsu, Y.R. (1995a). Spontaneous dissociation–association of monomers of the human-stem-cell-factor dimer. *Biochem J.* **305**, 563–568.

Lu, L., Rudert, W.A., Qian, S., McCaslin, D., Fu, F., Rao, A.S., Trucco, M., Fung, J.J., Starzl, T.E. and Thomson, A.W. (1995b). Growth of donor-derived dendritic cells from the bone marrow of murine liver allograft recipients in response to granulocyte/macrophage colony-stimulating factor. *J. Exp. Med.* **182**, 379–387.

Lyman, S.D., James, L., Vanden Bos, T., de Vries, P., Brasel, K., Gliniak, B., Hollingsworth, L.T., Picha, K.S., McKenna, H.J., Splett, R.R., Fletcher, F.F., Maraskovsky, E., Farrah, T., Foxworthe, D., Williams, D.E. and Beckmann, M.P. (1993a). Molecular cloning of a ligand for the flt3/flk-2 tyrosine kinase receptor: a proliferative factor for primitive hematopoietic cells. *Cell* **75**, 1157–1167.

Lyman, S.D., James, L., Zappone, J., Sleath, P.R., Beckmann, M.P. and Bird, T. (1993b). Characterization of the protein encoded by the flt3 (flk2) receptor-like tyrosine kinase gene. *Oncogene* **8**, 815–822.

Lyman, S.D., Brasel, K., Rousseau, A.M. and Williams, D.E. (1994a). The flt3 ligand: a hematopoietic stem cell factor whose activities are distinct from steel factor. *Stem Cells* **12**, 99–107.

Lyman, S.D., James, L., Johnson, L., Brasel, K., de Vries, P., Escobar, S.S., Downey, H., Splett, R.R., Beckmann, M.P. and McKenna, H.J. (1994b). Cloning of the human homologue of the murine flt3 ligand: a growth factor for early hematopoietic progenitor cells. *Blood* **83**, 2795–2801.

Lyman, S.D., James, L., Escobar, S., Downey, H., de Vries, P., Brasel, K., Stocking, K., Beckmann, M.P., Copeland, N.G., Cleveland, L.S., Jenkins, N.A., Belmont, J.W. and Davison, B.L. (1995a). Identification of soluble and membrane-bound isoforms of the murine flt3 ligand generated by alternative splicing of mRNAs. *Oncogene* **10**, 149–157.

Lyman, S.D., Seaberg, M., Hanna, R., Zappone, J., Brasel, K., Abkowitz, J.L., Prchal, J.T., Schultz, J.C. and Shahidi, N.T. (1995b). Plasma/serum levels of flt3 ligand are low in normal individuals and highly elevated in patients with Fanconi anemia and acquired aplastic anemia. *Blood* **86**, 4091–4096.

Lyman, S.D., Stocking, K., Davison, B., Fletcher, F., Johnson, L. and Escobar, S. (1995c). Structural analysis of human and murine flt3 ligand genomic loci. *Oncogene* **11**, 1165–1172.

Lynch, D.H., Andreasen, A., Maraskovsky, E., Whitmore, J., Miller, R.E. and Schuh, J.C.L. (1997). Flt3 ligand induces tumor regression and anti-tumor immune responses *in vivo. Nature Med.* **3**, 625.

Macatonia, S.E., Hsieh, C.-S., Murphy, K.M. and O'Garra, A. (1993). Dendritic cells and macrophages are required for Th1 development of $CD4^+$ T cells from $\alpha\beta$ TCR transgenic mice: IL-12 substitution for macrophages to stimulate IFN-γ production is IFN-γ-dependent. *Int. Immunol.* **5**, 1119–1128.

Mackarehtschian, K., Hardin, J.D., Moore, K.A., Boast, S., Goff, S.P. and Lemischka, I.R. (1995). Targeted disruption of the *flk2/flt3* gene leads to deficiencies in primitive hematopoietic progenitors. *Immunity* **3**, 147–161.

Maraskovsky, E., Roux, E., Tepee, M., McKenna, H.J., Brasel, K., Lyman, S.D. and Williams, D.E. (1995). The effect of Flt3 ligand and/or c-kit ligand on the generation of dendritic cells from human $CD34^+$ bone marrow. *Blood* **86**, 420a.

Maraskovsky, E., Brasel, K., Teepe, M., Roux, E.R., Lyman, S.D., Shortman, K. and McKenna, H.J. (1996). Dramatic increase in the numbers of functionally mature dendritic cells in Flt3 ligand-treated mice: multiple dendritic cell subpopulations identified. *J. Exp. Med.* **184**, 1953–1962.

Maroc, N., Rottapel, R., Rosnet, O., Marchetto, S., Lavezzi, C., Mannoni, P., Birnbaum, D. and Dubreuil, P. (1993). Biochemical characterization and analysis of the transforming potential of the FLT3/FLK2 receptor tyrosine kinase. *Oncogene* **8**, 909–918.

Matthews, W., Jordan, C.T., Wiegand, G.W., Pardoll, D. and Lemischka, I.R. (1991). A receptor tyrosine kinase specific to hematopoietic stem and progenitor cell-enriched populations. *Cell* **65**, 1143–1152.

Matzinger, P. and Guerder, S. (1989). Does T-cell tolerance require a dedicated antigen-presenting cell? *Nature* **338**, 74–76.

Mayordomo, J.I., Zorina, T., Storkus, W.J., Zitvogel, L., Celluzzi, C., Falo, L.D., Melief, C.J., Ildstad, S.T., Kast, W.M., Deleo, A.B. and Lotze, M.T. (1995). Bone marrow-derived dendritic cells pulsed with synthetic tumour peptides elicit protective and therapeutic antitumour immunity. *Nature Med.* **1**, 1297–1302.

Mazda, O., Watanabe, Y., Gyotoku, J.-I. and Katsura, Y. (1991). Requirement of dendritic cells and B cells in the clonal deletion of Mls-reactive T cells in the thymus. *J. Exp. Med.* **173**, 539–547.

McClanahan, T., Culpepper, J., Campbell, D., Wagner, J., Franz-Bacon, K., Mattson, J., Tsai, S., Luh, J., Guimaraes, M.J., Mattei, M.-G., Rosnet, O., Birnbaum, D. and Hannum, C.H. (1996). Biochemical and genetic characterization of multiple splice variants of the Flt3 ligand. *Blood* **88**, 3371–3382.

McKenna, H.J., de Vries, P., Brasel, K., Lyman, S.D. and Williams, D.E. (1995). Effect of flt3 ligand on the *ex vivo* expansion of human $CD34^+$ hematopoietic progenitor cells. *Blood* **86**, 3413–3420.

McKenna, H.J., Miller, R.E., Brasel, K.E., Maraskovsky, E., Maliszewski, C., Pulendran, B., Lynch, D., Teepe, M., Roux, E.R., Smith, J., Williams, D.E., Lyman, S.D., Peschon, J.J. and Stocking, K. (1996a). Targeted disruption of the flt3 ligand gene in mice affects multiple hematopoietic lineages, including natural killer cells, B lymphocytes, and dendritic cells. *Blood* **88 (supplement 1)**, 474A.

McKenna, H.J., Smith, F.O., Brasel, K., Hirschstein, D., Bernstein, I.D., Williams, D.E. and Lyman, S.D. (1996b). Effects of flt3 ligand on acute myeloid and lymphocytic leukemic blast cells from children. *Exp. Hematol.* **24**, 378–385.

Meierhoff, G., Dehmel, U., Gruss, H.-J., Rosnet, O., Birnbaum, D., Quentmeier, H., Dirks, W. and Drexler, H.G. (1995). Expression of flt3 receptor and flt3-ligand in human leukemia–lymphoma cell lines. *Leukemia* **9**, 1368–1372.

Metcalf, D., Shortman, K., Vremec, D., Mifsud, S. and Di Rago, L. (1996). Effects of excess GM-CSF levels on hematopoiesis and leukemia development in GM-CSF/*max* 41 double transgenic mice. *Leukemia* **10**, 713–719.

Metlay, J.P., Witmer-Pack, M.D., Agger, R., Crowley, M.T., Lawless, D. and Steinman, R.M. (1990). The distinct leukocyte integrins of mouse spleen dendritic cells as identified with new hamster monoclonal antibodies. *J. Exp. Med.* **171**, 1753–1771.

Muench, M.O., Roncarolo, M.G., Menon, S., Xu, Y., Kastelein, R., Zurawski, S., Hannum, C.H., Culpepper, J., Lee, F. and Namikawa, R. (1995). FLK-2/FLT-3 ligand regulates the growth of early myeloid progenitors isolated from human fetal liver. *Blood* **85**, 963–972.

Nakao, M., Yokota, S., Iwai, T., Kaneko, H., Horiike, S., Kashima, K., Sonoda, Y., Fujimoto, T. and Misawa, S. (1996). Internal tandem duplication of the flt3 gene found in acute myeloid leukemia. *Leukemia* **10**, 1911–1918.

Namikawa, R., Muench, M.O., de Vries, J.E. and Roncarolo, M.-G. (1996). The FLK2/FLT3 ligand synergizes with interleukin-7 in promoting stromal-cell-independent expansion and differentiation of human fetal pro-B cells *in vitro*. *Blood* **87**, 1881–1890.

Ohishi, K., Katayama, N., Itoh, R., Mahmud, N., Miwa, H., Kita, K., Minami, N., Shirakawa, S., Lyman, S.D. and Shiku, H. (1996). Accelerated cell-cycling of hematopoietic progenitors by the *flt3* ligand that is modulated by transforming growth factor-β. *Blood* **87**, 1718–1727.

Pandit, J., Bohm, A., Jancarik, J., Halenbeck, R., Koths, K. and Kim, S.-H. (1992). Three-dimensional structure of dimeric human recombinant macrophage colony-stimulating factor. *Science* **258**, 1358–1362.

Petzer, A.L., Zandstra, P.W., Piret, J.M. and Eaves, C.J. (1996). Differential cytokine effects on primitive ($CD34^{+}CD38^{-}$) human hematopoietic cells: novel responses to Flt3-ligand and thrombopoietin. *J. Exp. Med.* **183**, 2551–2558.

Piacibello, W., Fubini, L., Sanavio, F., Brizzi, M.F., Severino, A., Garetto, L., Stacchini, A., Pegoraro, L. and Aglietta, M. (1995). Effects of human FLT3 ligand on myeloid leukemia cell growth: heterogeneity in response and synergy with other hematopoietic growth factors. *Blood* **86**, 4105–4114.

Piacibello, W., Garetto, L., Sanavio, F., Severino, A., Fubini, L., Stacchini, A., Dragonetti, G. and Aglietta, M. (1996). The effects of human FLT3 ligand on *in vitro* human megakaryocytopoiesis. *Exp. Hematol.* **24**, 340–346.

Piacibello, W., Sanavio, F., Garetto, L., Severino, A., Bergandi, D., Ferrario, J., Fagioli, F., Berger, M. and Aglietta, M. (1997). Extensive amplification and self-renewal of human primitive hematopoietic stem cells from cord blood. *Blood* **89**, 2644–2653.

Ramsfjell, V., Borge, O.J., Cardier, J., Murphy, M.J., Jr., Lyman, S.D. and Jacobsen, S.E.W. (1997). Thrombopoietin, but not erythropoietin, directly stimulates multilineage growth of primitive murine bone marrow progenitor cells in synergy with early acting cytokines: distinct interactions with the ligands for c-*kit* and FLT3. *Blood* **88**, 4481–4492.

Rasko, J.E.J., Metcalf, D., Rossner, M.T., Begley, C.G. and Nicola, N.A. (1995). The flt3/flk-2 ligand: receptor distribution and action on murine haemopoietic cell survival and proliferation. *Leukemia* **9**, 2058–2066.

Ray, R.J., Paige, C.J., Furlonger, C., Lyman, S.D. and Rottapel, R. (1996). Flt3 ligand supports the differentiation of early B cell progenitors in the presence of interleukin-11 and interleukin-7. *Eur. J. Immunol.* **26**, 1504–1510.

Roberts, A.W. and Metcalf, D. (1994). Granulocyte colony-stimulating factor induces selective elevations of progenitor cells in the peripheral blood of mice. *Exp. Hematol.* **22**, 1156–1163.

Romani, N., Gruner, S., Brang, D., Kämpgen, E., Lenz, A., Trockenbacher, B., Konwalinka, G., Fritsch, P.O., Steinman, R.M. and Schuler, G. (1994). Proliferating dendritic cell progenitors in human blood. *J. Exp. Med.* **180**, 83–93.

Rosnet, O., Marchetto, S., deLapeyriere, O. and Birnbaum, D. (1991a). Murine *Flt3*, a gene encoding a novel tyrosine kinase receptor of the PDGFR/CSF1R family. *Oncogene* **6**, 1641–1650.

Rosnet, O., Mattei, M.G., Marchetto, S. and Birnbaum, D. (1991b). Isolation and chromosomal localization of a novel FMS-like tyrosine kinase gene. *Genomics* **9**, 380–385.

Rosnet, O., Schiff, C., Pébusque, M.-J., Marchetto, S., Tonnelle, C., Toiron, Y., Birg, F. and Birnbaum, D. (1993a). Human *FLT3/FLK2* gene: cDNA cloning and expression in hematopoietic cells. *Blood* **82**, 1110–1119.

Rosnet, O., Stephenson, D., Mattei, M.-G., Marchetto, S., Shibuya, M., Chapman, V.M. and Birnbaum, D. (1993b). Close physical linkage of the *FLT1* and *FLT3* genes on chromosome 13 in man and chromosome 5 in mouse. *Oncogene* **8**, 173–179.

Rosnet, O., Bühring, H.-J., Marchetto, S., Rappold, I., Lavagna, C., Sainty, D., Arnoulet, C., Chabannon, C., Kanz, L., Hannum, C. and Birnbaum, D. (1996). Human FLT3/FLK2 receptor tyrosine kinase is expressed at the surface of normal and malignant hematoietic cells. *Leukemia* **10**, 238–248.

Rossner, M.T., McArthur, G.A., Allen, J.D. and Metcalf, D. (1994). Fms-like tyrosine kinase 3 catalytic domain can transduce a proliferative signal in FDC-P1 cells that is qualitatively similar to the signal delivered by c-Fms. *Cell Growth Differ.* **5**, 549–555.

Rusten, L.S., Lyman, S.D., Veiby, O.P. and Jacobsen, S.E.W. (1996). The FLT3 ligand is a direct and potent stimulator of the growth of primitive and committed human CD34$^+$ bone marrow progenitor cells *in vitro*. *Blood* **87**, 1317–1325.

Sallusto, F. and Lanzavecchia, A. (1994). Efficient presentation of soluble antigen by cultured human dendritic cells is maintained by granulocyte/macrophage colony-stimulating factor plus interleukin 4 and downregulated by tumor necrosis factor α. *J. Exp. Med.* **179**, 1109–1118.

Saunders, D., Lucas, K., Ismaili, J., Wu, J., Maraskovsky, E., Dunn, A. and Shortman, K. (1996). Dendritic cell development in culture from thymic precursor cells in the absence of granulocyte/macrophage colony-stimulating factor. *J. Exp. Med.* **184**, 2185–2196.

Shadle, P.J., Allen, J.I., Geier, M.D. and Koths, K. (1989). Detection of endogenous macrophage colony-stimulating factor (M-CSF) in human blood. *Exp. Hematol.* **17**, 154–159.

Shah, A.J., Smogorzewska, E.M., Hannum, C. and Crooks, G.M. (1996). Flt3 ligand induces proliferation of quiescent human bone marrow CD34$^+$CD38$^-$ cells and maintains progenitor cells *in vitro*. *Blood* **87**, 3563–3570.

Shapiro, F., Pytowski, B., Rafii, S., Witte, L., Hicklin, D.J., Yao, T.J. and Moore, M.A.S. (1996). The effects of Flk-2/flt3 ligand as compared with c-kit ligand on short-term and long-term proliferation of CD34$^+$ hematopoietic progenitors elicited from human fetal liver, umbilical cord blood, bone marrow, and mobilized peripheral blood. *J. Hematother.* **5**, 655–662.

Shibuya, M., Matsushime, H., Yamane, A., Ikeda, T., Yoshida, M.C. and Tojo, A. (1989). Isolation and characterization of new mammalian kinase genes by cross hybridization with a tyrosine kinase probe. *Int. Symp. Princess Takamatsu Cancer Res. Fund* **20**, 103–110.

Siena, S., Di Nicola, M., Bregni, M., Mortarini, R., Anichini, A., Lombardi, L., Ravagnani, F., Parmiani, G. and Gianni, A.M. (1995). Massive *ex vivo* generation of functional dendritic cells from mobilized CD34$^+$ blood progenitors for anticancer therapy. *Exp. Hematol.* **23**, 1463–1471.

Small, D., Levenstein, M., Kim, E., Carow, C., Amin, S., Rockwell, P., Witte, L., Burrow, C., Ratajczak, M.Z., Gewirtz, A.M. and Civin, C.I. (1994). STK-1, the human homolog of Flk-2/Flt-3, is selectively expressed in CD34$^+$ human bone marrow cells and is involved in the proliferation of early progenitor/stem cells. *Proc. Natl Acad. Sci. U.S.A.* **91**, 459–463.

Stacchini, A., Fubini, L., Severino, A., Sanavio, F., Aglietta, M. and Piacibello, W. (1996). Expression of type III receptor tyrosine kinases FLT3 and KIT and responses to their ligands by acute myeloid leukemia blasts. *Leukemia* **10**, 1584–1591.

Starzl, T.E., Demetris, A.J., Murase, N., Ildstad, S., Ricordi, C. and Trucco, M. (1992). Cell migration, chimerism, and graft acceptance. *Lancet* **339**, 1579–1582.

Starzl, T.E., Demetris, A.J., Trucco, M., Murase, N., Ricordi, C., Ildstad, S., Ramos, H., Todo, S., Tzakis, A., Fung, J.J., Nalesnik, M., Rudert, W.A. and Kocova, M. (1993). Cell migration and chimerism after whole-organ transplantation: the basis of graft acceptance. *Hepatology* **17**, 1127–1152.

Steinman, R.M. (1991). The dendritic cell system and its role in immunogenicity. *Annu. Rev. Immunol.* **9**, 271–296.

Süss, G. and Shortman, K. (1996). A subclass of dendritic cells kills CD4 T cells via Fas/Fas-ligand-induced apoptosis. *J. Exp. Med.* **183**, 1789–1796.

Szabolcs, P., Moore, M.A.S. and Young, J.W. (1995). Expansion of immunostimulatory dendritic cells among the myeloid progeny of human CD34$^+$ bone marrow precursors cultured with c-*kit* ligand, granulocyte–macrophage colony-stimulating factor, and TNF-α. *J. Immunol.* **154**, 5851–5861.

Thomson, A.W., Lu, L., Murase, N., Demetris, A.J., Rao, A.S. and Starzl, T.E. (1995). Microchimerism, dendritic cell progenitors and transplantation tolerance. *Stem Cells* **13**, 622–639.

Turner, A.M., Bennett, L.G., Lin, N.L., Wypych, J., Bartley, T.D., Hunt, R.W., Atkins, H.L., Langley, K.E., Parker, V. and Martin, F. (1995). Identification and characterization of a soluble c-kit receptor produced by human hematopoietic cell lines. *Blood* **85**, 2052–2058.

Van Zant, G., Eldridge, P.W., Behringer, R.R. and Dewey, M.J. (1983). Genetic control of hematopoietic kinetics revealed by analyses of allophenic mice and stem cell suicide. *Cell* **35**, 639–645.

Veiby, O.P., Lyman, S.D. and Jacobsen, S.E.W. (1996). Combined signaling through interleukin-7 receptors and flt3 but not c-*kit* potently and selectively promotes B-cell commitment and differentiation from uncommitted murine bone marrow progenitor cells. *Blood* **88**, 1256–1265.

Visser, J.W., Rozemuller, H., de Jong, M.O. and Belyavsky, A. (1993). The expression of cytokine receptors by purified hemopoietic stem cells. *Stem Cells* **11**, 49–55.

Vremec, D., Zorbas, M., Scollay, R., Saunders, D.J., Ardavin, C.F., Wu, L. and Shortman, K. (1992). The surface phenotype of dendritic cells purified from mouse thymus and spleen: investigation of the CD8 expression by a subpopulation of dendritic cells. *J. Exp. Med.* **176**, 47–58.

Wasserman, R., Li, Y.-S. and Hardy, R.R. (1995). Differential expression of the Blk and Ret tyrosine kinases during B lineage development is dependent on Ig rearrangement. *J. Immunol.* **155**, 644–651.

Wineman, J., Moore, K., Lemischka, I. and Müller-Sieburg, C. (1996). Functional heterogeneity of the hematopoietic microenvironment: rare stromal elements maintain long-term repopulating stem cells. *Blood* **87**, 4082–4090.

Winton, E.F., Bucur, S.Z., Bray, R.A., Toba, K., Williams, D.E., McClure, H.M. and Lyman, S.D. (1995). The hematopoietic effects of recombinant human (rh) Flt3 ligand administered to non-human primates. *Blood* **86 (supplement 1)**, 424A.

Winton, E.F., Bucur, S.Z., Bond, L.D., Hegwood, A.J., Hillyer, C.D., Holland, H.K., Williams, D.E., McClure, H.M., Troutt, A.B. and Lyman, S.D. (1996). Recombinant human (rh) Flt3 ligand plus rhGM-CSF or rhG-CSF causes a marked $CD34^+$ cell mobilization to blood in rhesus monkeys. *Blood* **88 (supplement 1)**, 642A.

Wodnar-Filipowicz, A., Lyman, S.D., Gratwohl, A., Tichelli, A., Speck, B. and Nissen, C. (1996). Flt3 ligand level reflects hematopoietic progenitor cell function in multilineage bone marrow failure. *Blood* **88**, 4493–4499.

Wu, L., Li, C.-L. and Shortman, K. (1996). Thymic dendritic cell precursors: relationship to the T lymphocyte lineage and phenotype of the dendritic cell progeny. *J. Exp. Med.* **184**, 903–911.

Young, J.W. and Inaba, K. (1996). Dendritic cells as adjuvants for class I major histocompatibility complex-restricted antitumor immunity. *J. Exp. Med.* **183**, 7–11.

Young, J.W., Szabolcs, P. and Moore, M.A.S. (1995). Identification of dendritic cell colony-forming units among normal human $CD34^+$ bone marrow progenitors that are expanded by c-*kit*-ligand and yield pure dendritic cell colonies in the presence of granulocyte/macrophage colony-stimulating factor and tumor necrosis factor α. *J. Exp. Med.* **182**, 1111–1120.

Zeigler, F.C., Bennett, B.D., Jordan, C.T., Spencer, S.D., Baumhueter, S., Carroll, K.J., Hooley, J., Bauer, K. and Matthews, W. (1994). Cellular and molecular characterization of the role of the FLK-2/FLT-3 receptor tyrosine kinase in hematopoietic stem cells. *Blood* **84**, 2422–2430.

Zola, H., Siderius, N., Flego, L., Beckman, I. and Seshadri, R. (1994). Cytokine receptor expression in leukaemic cells. *Leuk. Res.* **18**, 347–355.

Chapter 27

The Expanding Universe of C, CX_3C and CC Chemokines

Kevin B. Bacon[1], David R. Greaves[2], Daniel J. Dairaghi[3] and Thomas J. Schall[3]

[1]Neurocrine Bioscience, La Jolla, CA, USA, [2]Sir William Dunn School of Pathology, University of Oxford, Oxford, UK and [3]Molecular Medicine Research Institute, Mountain View, CA, USA

INTRODUCTION

Many cosmologists profess that the universe is both finite and boundless. This paradox, they explain, is more apparent than real, and they resolve it with a simple mental model. They maintain that the universe which we perceive is like the thin membrane of an expanding soap bubble. It started as a tiny bubble at the Big Bang, and has been getting bigger ever since. Our perceptions, in both time and space, are limited to the surface of the soap bubble. Thus, from our point of view, the universe seems to be expanding, with everything in it moving away from everything else in it at roughly the same rate. Moreover, try as we might, we can find no boundary. Lastly, there are 'alternate' Universes, those 'meta-dimensions' within the bubble, and the unknown—that is to say unknowable—regions without.

Cosmology may also provide the ideal metaphor for the biology of the chemokine system in the present era. Since the appearance of the last version of this chapter in the second edition of *The Cytokine Handbook*, the chemokine field has undergone its own big bang, expanding into practically every aspect of physiology and medicine. The first part of the expansion has been sheerly numeric. The number of independent chemokine gene products approaches 60, and will doubtless go higher. The number of receptors, of which only three or four were known in 1994, now approaches two dozen. Thus the bioinformaticians, like the cosmologists, are busily at work, plying their numbers and expanding the surface of the chemokine universe in a strictly numeric sense. At the same time, the bubble of our understanding of the range of chemokine function has been distorted to the point of bursting. Historically, the predominant paradigm has been one of chemokines regulating leukocyte trafficking and its attendant pathologic inflammatory correlates. But other dimensions of chemokine function, only vaguely hinted at previously, are beginning to be revealed. For example, the explosion of chemokines into the area of infectious disease, most notably the understanding that chemokine receptors are at the center of human immunodeficiency virus (HIV) pathogenesis, sent shock waves through the field, which are still being felt.

The Cytokine Handbook, 3rd ed.
ISBN 0-12-689662-3

Because expansion and complexity have become hallmarks of this field, a detailed treatment of all of chemokine biology, or even its history, goes far beyond what can be offered in this one chapter. The objective of this section, then, is to provide an update on certain key molecules discovered since the last edition of this *Handbook*, and on emerging themes concerning the roles of the chemokines in leukocyte trafficking and beyond. We focus here on the C, CC and CX_3C chemokines (and almost exclusively those of the human system), leaving the examination of the CXC molecules to Chapter 10. These two chapters, used in conjunction with the information contained in the second edition of *The Cytokine Handbook*, should provide a reasonable glimpse of chemokine biology today.

NEW CHEMOKINES AND CHEMOKINE SUBFAMILIES

The rapid expansion in the number of known chemokine-encoding sequences has been aided in part by random sequencing and the accessibility of public databases and the World Wide Web, allowing investigators to piece together chemokines and their receptors from expressed sequence tags representing short fragments of DNA. The repertoire of family members seems to increase every day, and careful scrutiny of all the available databases suggests that the number of distinct chemokine gene products currently approaches 60 in the human system. A firm number is difficult to come by because a number of putatively unique chemokines are certainly duplicated through confusing nomenclature (sometimes reflecting irresponsible attempts to claim priority for a clone's 'discovery') and separately deposited accession numbers; however, it is clear that the superfamily consists of an extensive array of proteins of four major subclasses: C, CC, CXC and CX_3C. A partial roster of the CC chemokines is given in Table 1. As will be discussed later, the classic appreciation of chemokines as solely involved in inflammation and leukocyte trafficking will have to be revised in order to accommodate such a variety of protein factors, as it has been determined that many chemokines can fulfill extremely important functions in such areas as homeostasis and development.

Two of the more significant developments in the past 3 years have been the discovery of two additional subfamilies within the chemokine superfamily. While these subfamilies consist of a sole member each, the biologic functions thus far attributable to these molecules suggest that they may play vital roles in both physiologic leukocyte trafficking and lymphocyte activation.

Fractalkine: New Structure, New Paradigm

Molecular regulation of the interface between the endothelium and the leukocyte cell surface determines the balance between health and disease. Nature has achieved this regulation through the orchestrated actions of the adhesions molecules (mainly the selectins and integrins and their counter-receptors) and the promigratory chemokines (see reviews by Springer, 1994; Butcher and Picker, 1996). The recent elucidation of a new type of chemokine, fractalkine, and its cell surface receptor reveals that the traditional distinction between adhesion molecules and chemokines may be less than absolute.

Fractalkine (Bazan *et al.*, 1997) is structurally unique among chemokines in two ways:

Table 1. CC chemokine family members.

Chromosome 17	
MCP-1	Gen X14768
MCP-2	NCBI 1924937
MCP-3	EMBL X72308
MCP-4	Gen U46767
muMCP-5	Gen U66670
Eotaxin	NCBI 1552240
RANTES	Gen M21121
MIP-1α	Swissprot P10147
MIP-1β	Swissprot P13236
I-309	Gen M57506
Chromosome 2	
LARC/MIP-3α/Exodus	Gen U77035
Chromosome 9	
ELC/MIP-3β	NCBI 2189952
Chromosome 16	
TARC	Gen D43767
MDC	Gen U83171
Fractalkine (CX_3C)	Gen U91835
Other	
CKβ-8/MPIF-1	Gen U85767
CKβ-6/MPIF-2/Eotaxin-2	Gen U85768
HCC-1	EMBL Z49270
muMIP-1γ	Gen U49513
CC1	EMBL Z70292
CC2	EMBL Z70293
CC3	EMBL Z70293
STCP-1	Gen U83239
Exodus 2	Gen U88320
Exodus 3	Gen U88321

Genbank, EMBL and NCBI files can be accessed from http://www.ncbi.nlm.nih.gov/ using the accession numbers given in this table. Swissprot entries can be accessed from http://expasy.hcuge.ch/sprot/.

it possesses three intervening amino acids between the first two cysteine residues of the conserved chemokine motif, and it exists as a membrane-bound molecule with the chemokine domain attached to a long mucin-like stalk. Fractalkine is made by activated endothelial cells, and can be cleaved by unknown mechanisms to be released as a soluble molecule (Fig. 1). Fractalkine's unique molecular structure seems to impart intrinsic adhesion and chemokine activities. While acting as a soluble chemoattractant for lymphoid cells and monocytes, this molecule also functions as a solid-phase adhesion molecule. These properties are interesting, particularly in light of early studies by Rot (1993) demonstrating that the chemokine interleukin-8 (IL-8) stimulated neutrophils to migrate by a mechanism known as haptotaxis, where the soluble chemoattractant was immobilized by glycosaminoglycan (GAG) moieties, thus establishing a solid gradient of

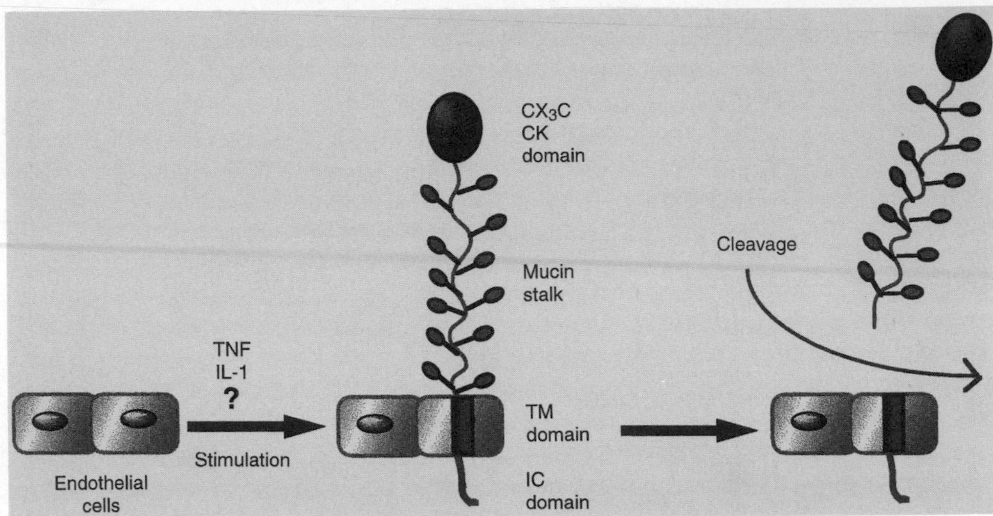

Fig. 1. Fractalkine structure and processing. Upon stimulation, endothelial cells express fractalkine, as a 'CX_3C' chemokine sitting atop a mucin stalk. This chemokine–mucin structure is attached to the cell by a single membrane span and a small intracellular (IC) domain. Cleavage of the mucin stalk presumably allows additional regulatory control of fractalkine expression by removing its solid-phase expression gradient.

attractant factor. The chemokine entity of fractalkine, expressed on its cleavable stalk, thus may fulfill the roles of chemokine and GAG in eliciting and facilitating leukocyte migration. Elucidation of the fractalkine receptor CX_3CR1, as discussed below, supports the view that the receptor–ligand pair act directly as adhesion molecules, distinct from the action of selectins or integrins. Thus, fractalkine and its receptor may suggest a new mechanism of leukocyte adhesion and extravasation at the leukocyte–endothelial cell interface.

Lymphotactin, the Lone C Chemokine

In early 1994, Kelner *et al.* cloned lymphotactin (Ltn; independently characterized by Yoshida *et al.* (1995) as SCM-1α/SCM-1β) from a pro-T-cell cDNA library. Lymphotactin exhibited sequence similarities to known chemokines such as macrophage inflammatory protein-1α (MIP-1α) and (MIP-1β). Further biologic analysis revealed potent chemoattractant activity induced by this molecule, the first example of chemoattractant selectivity for one class of leukocyte, the lymphoid cell. Lymphotactin was originally shown to be produced mainly by $CD8^+$ lymphocytes, thymocytes and later natural killer (NK) cells (Hedrick *et al.*, 1997). One study has, however, reported the expression of mRNA for lymphotactin in mast cells following cross-linking of FcεRI (Runsaeng *et al.*, 1997). Further studies revealed significant induction of $\alpha\beta$ T-cell migration in response to lymphotactin expression by resident $\gamma\delta$ dendritic epidermal T cells in skin (Biosmenu *et al.*, 1996). To date, this chemokine remains a lymphoid-specific chemoattractant and has been shown in a murine tumor model to attract NK and CD4 T cells to tumor sites, preventing subsequent growth of the tumor (Dilloo *et al.*, 1996).

New Members of the CC Class

In the previous edition of this book, all CC chemokine genes known at that time had been localized to a gene cluster on human chromosome 17. Since then, new CC chemokines encoded in this cluster have been identified, and additional CC chemokine gene loci have been identified on chromosomes 2, 9 and 16. A brief overview of a few key additions to the CC roster follows.

Eotaxin

Eotaxin came to prominence in early 1994 after Jose *et al.* cloned a novel eosinophil-specific chemoattractant from allergen-challenged lungs of the guinea-pig. This was followed by cloning of the human and murine genes (Garcia-Zepeda *et al.*, 1996; Gonzalo *et al.*, 1996). The great attraction of eotaxin initially was that it was purported to be an 'eosinophil-specific' chemokine. While this is almost certainly not true (for example, effects on thymocytes and T cells have now been noted), the effects of eotaxin on eosinophils are profound, and doubtless of great importance in human pathophysiology (Kitaura *et al.*, 1996; Ponath *et al.*, 1996). An examination of some of eotaxin's effects *in vivo* is given below.

Monocyte Chemotactic Proteins

The previous edition of this chapter described monocyte chemotactic protein-1 (MCP-1), 2 and 3. Subsequent work has revealed an expansion of what might be called the 'MCP cluster' of CC chemokines. MCP-4 was originally described following random sequencing of a human fetal cDNA library (Uguccioni *et al.*, 1996). Although multiple variants were identified, the mature secreted MCP-4 was shown to have approximately 60% sequence identity with MCP-1–3, and eotaxin. In a manner similar to MCP-3, MCP-4 is a chemoattractant for monocytes and T lymphocytes, and exhibits equipotent activity with eotaxin and MCP-3 on eosinophils, suggesting that this chemokine is a functional hybrid of MCP-3 and eotaxin. MCP-4 has also, more recently, been shown to be present in a number of bronchial epithelial cell lines, and to induce histamine release from IL-3-primed basophils (Stellato *et al.*, 1997). MCP-4 mRNA expression was regulated by inflammatory cytokines and inhibited by glucocorticoid pretreatment *in vitro*. Analysis of receptor usage has demonstrated the potential for MCP-4 binding to CCR2b and CCR3, completely desensitizing MCP-3-induced eosinophil calcium mobilization, and ^{125}I-eotaxin binding to eosinophils and CCR3-transfected L1.2 murine lymphoma cells (Garcia-Zepeda *et al.*, 1996; Stellato *et al.*, 1997).

MCP-5, a novel murine CC chemokine, was cloned from a cDNA library generated from the inflamed lungs of animals sensitized with ovalbumin (Jia *et al.*, 1995), as well as from screening of a mouse genomic library using the MCP-4 gene as a probe (Sarafi *et al.*, 1997). Reports of biologic activity are conflicting, with the first group claiming chemoattractant capacity for monocytes, T lymphocytes and eosinophils *in vitro* and *in vivo*, whereas the latter study shows apparent exclusivity for monocytes. Both studies apparently agree on the constitutive expression of MCP-5 in the lymph nodes and thymus, whereas additional information is provided for expression in mast cells by Jia *et al.* (1995). Using different strains of mice, Jia and coworkers also demonstrated apparent T-cell dependency for expression of MCP-5 mRNA by alveolar macrophages and smooth muscle cells. Both groups have shown the upregulation of mRNA for MCP-5 in the lungs of mice under antigen challenge (ovalbumin), and Sarafi *et al.* (1997) additionally reported results using mice infected with *Nippostrongylus brasiliensis*.

HCC Cluster

Another new cluster of CC chemokines is emerging, the HCC cluster. Molecules in this cluster are characterized by the possession of six cysteines rather than the canonical four in the chemokine motif. HCC-1 was first isolated from plasma of patients with chronic renal failure and subsequently cloned from human bone marrow, although it appears to be expressed constitutively in spleen, liver, heart and gut tissue (Schulze-Knappe *et al.*, 1996). HCC-1 shares 46% sequence identity with MIP-1α and β. While there is some weak activity on monocytes, the only other biologic data reported is an enhancement of CD34$^+$ myeloid progenitor proliferation. Other '6-C' chemokines are contained within this cluster (for example, the human homolog of MIP-1γ/CCF-18 which also has at least two other designations, thankfully elusive at present); in fact it seems likely that the cluster contains HCC-1 through HCC-5. Details of this cluster are just beginning to emerge.

MIP-3α/LARC/Exodus

Another product of random sequencing efforts, this multiply identified chemokine is widely diverged from the rest of the family, with its closest relatives being MIP-1α and RANTES (Regulated on Activation, Normal T-cell Expressed and Secreted) (26% and 28% identity respectively at the amino acid level (Hieshima *et al.*, 1997; Hromas *et al.*, 1997; Rossi *et al.*, 1997). Clones encoding this chemokine have been isolated from cDNA libraries generated from isolated pancreatic islet cells, fetal liver and other fetal tissues, and mRNA has been shown to be expressed preferentially in lymphoid tissue, with significant regulation of expression in HUVEC by TNF-α also observed. This molecule, under the guise of Exodus, was shown to stimulate the migration of peripheral blood lymphocytes (PBLs) and to be an inhibitor of proliferation of a cytokine-dependent myeloid cell line. Also, Heishima *et al.* (1997) demonstrated that the gene for this chemokine is located on human chromosome 2q33–q37, and reported significant activity of the protein for HUT78 T-cell line chemotaxis.

MIP-3β/ELC

More unfortunate nomenclature since MIP-3β is not related at all to MIP-3α (itself wholly misnamed as it has no discernible connection to other 'macrophage inflammatory proteins'), this molecule was another orphan love child of bioinformatics. The most notable features of MIP-3β/ELC (Epstein–Barr virus-induced gene 1 ligand chemokine) are its chromosomal location (chromosome 9) and the fact that it is a ligand for CCR7, formerly the orphan clone EBI1, leading to the sensible designation of ELC (EBI1 ligand chemokine).

TARC

TARC (thymus and activation-regulated chemokine) is another novel CC chemokine discovered from a cDNA library of phytohemagglutinin (PHA)-stimulated peripheral blood mononuclear cells (PBMCs) and shown to be expressed constitutively in thymus and transiently in PHA-stimulated PBMCs (Imai *et al.*, 1996). More recently, it has been shown to bind CCR-4 and the Duffy antigen (Imai *et al.*, 1997). While TARC also binds to peripheral blood T cells and Jurkat cells, and chemotaxis of the HUT78 T-cell line has been demonstrated, the exact function of this chemokine remains to be determined.

Myeloid progenitor inhibitory factors

MPIF-1/Ckβ-8 (myeloid progenitor inhibitory factor) and MPIF-2/Ckβ-6/eotaxin-2 are recent products of a random cDNA sequencing effort from aortic endothelium and monocytes respectively (Patel *et al.*, 1997). MPIF-1 is chemotactic for resting T cells, monocytes and neutrophils, while MPIF-2 is apparently chemotactically active only on resting T cells, although it does activate eosinophil calcium mobilization through CCR-3, consistent with the characterization of Ckβ-6 as eotaxin-2 (Forssmann *et al.*, 1997). As the names suggest, MPIF-1 was found to inhibit myeloid progenitors that give rise to granulocytes and monocytes. In contrast, MPIF-2 potently inhibits the colony formation of multipotential hematopoietic progenitors in *in vitro* culture.

Macrophage-derived chemokine

Macrophage-derived chemokine (MDC) was isolated by random sequencing of cDNA clones from monocytes–macrophages (Godiska *et al.*, 1997), and at most shows less than 35% identity with other CC chemokines. It is expressed constitutively in thymus, lung and spleen, and the gene has been localized to chromosome 16, not far from the CX_3C chemokine fractalkine gene. Biologic activity (chemotaxis) has thus far been demonstrated for monocyte-derived dendritic cells and IL-2-activated NK cells.

CC Chemokine Targeted Gene Ablations

A small number of CC chemokine 'knockout' mice have been generated. Two have been published, the ablation of the MIP-1α and eotaxin genes. The MIP-1$\alpha^{-/-}$ mouse exhibits no obvious phenotype in the basal state, but on infection with coxsackie virus wild-type mice succumb to lethal consequences of cardiopulmonary inflammatory events. The knockout mice, however, are unaffected. Eotaxin$^{-/-}$ mice exhibit a reduction of about one-third in the numbers of eosinophils in the periphery and a similar diminution of eosinophils in the lung during pulmonary challenge (Rothenberg *et al.*, 1997). From this, the authors conclude that eotaxin is essential for recruitment of eosinophils to the lungs, but the interpretation that it is needed merely for eosinophil maturation or egress from the bone marrow seems equally plausible.

CC and CX_3C Chemokine Receptors

Since publication of the previous chapter in 1994, the cloning of chemokine receptors has constituted one of the major advances in the field, not only from the point of view of assigning many of the specific receptors of the chemokines, but also from the contributions made to understanding the biology and mechanisms of inflammatory and infectious diseases. The receptors CXCR1 (IL-8RA), CXCR2 (IL-8RB), CCR1, CCR2, US28 and Duffy have been considered in detail in other reviews, and will not be discussed here. In the past 3 years, the cloning efforts of numerous groups have now identified receptors CCR3–CCR9, which bind to an overlapping spectrum of CC chemokine ligands, depicted in Fig. 2, and the fractalkine receptor, the first CX_3C chemokine receptor. Brief biographic sketches of each of these receptors are given below.

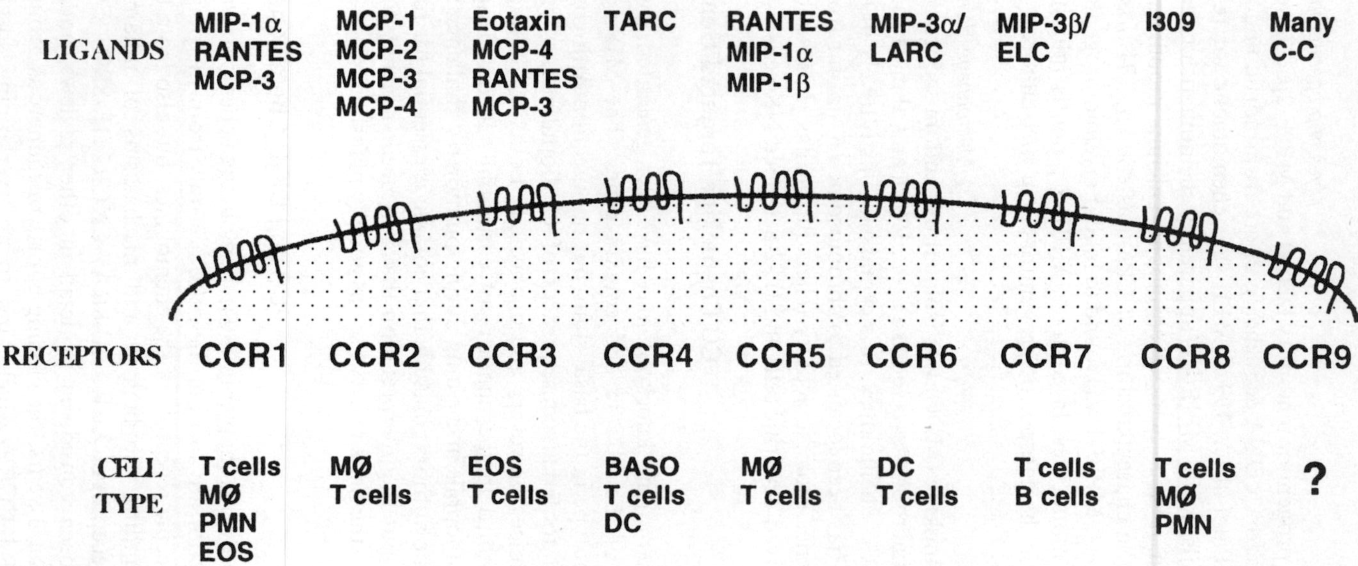

Fig. 2. The chemokine–receptor pairs. Shown are the nine CC chemokine receptors (designated CCR1–CCR9) that are all seven-transmembrane GCRs. Above each receptor are the cognate ligand(s), roughly in rank order of potency or binding affinity. The cell types known to express the specific receptors are indicated below, again in an order corresponding roughly to the relative expression levels. PMN, polymorphonuclear neutrophil; DC, dendritic cell.

CCR3

The existence of this receptor was originally indicated in Southern blots performed for studies of CCR-1 (Neote *et al.*, 1993) and definitively identified by a number of groups, who postulated it to be an eotaxin-specific receptor (Daugherty *et al.*, 1996; Kitaura *et al.*, 1996; Ponath *et al.*, 1996). Now it is known to bind many CC chemokines including eotaxin, MCP-2, MCP-3 and MCP-4, and others. This receptor is highly expressed on eosinophils and basophils, as well as some glial cells (He *et al.*, 1997). Originally publicized as the targetable receptor in inflammatory diseases of the lung, owing to its specificity of expression and the actions of the above-mentioned chemokines, the story has become somewhat less clear. For example, the exact function of the CCR3-binding chemokines on glial cells is unclear. In addition, we have demonstrated clear expression of CCR3 as well as the actions of eotaxin on human thymocytes (manuscript submitted), and we and others have recognized eotaxin actions through CCR3 on T-cell clones (Ugiccioni *et al.*, 1996). In fact, in addition to being highly expressed on eosinophils, it is likely that CCR3 also represents a T helper 2 T_{H2} T-cell marker.

CCR4

Originally cloned from a basophilic cell line, CCR4 was shown to induce chloride currents in transfected *Xenopus* oocytes in response to RANTES, MCP-1 and MIP-1α (Power *et al.*, 1995). These findings have been contentious, however, as subsequent equilibrium binding done on mammlian cells transfected with CCR4 show binding only to the novel chemokine TARC (Imai *et al.*, 1997).

CCR5

CCR5 has been shown to bind MIP-1β, MIP-1α and RANTES. Biologically, the significance of this receptor was highlighted by the discovery in early 1996 that this receptor is the primary coreceptor for M-tropic strains of HIV, as discussed below.

CCR6

Reported as an orphan chemokine receptor clone by several groups, CCR6 was ultimately characterized in a systematic analysis of chemokine receptors expressed by human dendritic cells (Greaves *et al.*, 1997) and PBLs (Baba *et al.*, 1997). CCR6 shows 42% amino acid identity with the CXC chemokine receptors CXCR2 and 39% identity with the CC chemokine receptor CCR4. Unlike the genes for CCR1–CCR5, which are closely linked on chromosome 3p21, the CCR6 gene maps to 6q26–27. Whereas CCR1–CCR5 seem to interact with multiple CC chemokine ligands, CCR6 thus far appears to be MIP-3α/LARC specific. The role of CCR6 in dendritic cell migration and function has yet to be determined but it is of great interest that CCR6 is expressed only by $CD34^+$ cord blood-derived dendritic cells and not by monocyte-derived dendritic cells (Greaves *et al.*, 1997).

CCR7

Recent studies typifying the 'reverse biology' of EST fishing exercises have led to the identification of this novel receptor and its CC chemokine ligand (Yoshida *et al.*, 1997). An orphan seven-transmembrane receptor, termed EBI1 (Epstein–Barr virus-induced gene 1), expressed in lymphoid tissues and immortalized B and T cells was shown to bind ELC (EBI1 ligand chemokine) after systematic analysis of equilibrium binding to known and orphan chemokine receptors, using a novel chemokine cloned entirely from EST fragments. The biologic significance of this receptor–ligand pair remains elusive, however, with the only analyses to date showing chemotaxis and calcium flux analyses on the T-cell line HUT78.

CCR8

An orphan receptor cDNA variously known as ter-1, CKR-L2, or CY6 has now been shown to bind the CC chemokine I-309 (Tiffany *et al.*, 1997). This clone is also notable in being a cofactor for HIV infection.

CCR9

It is likely that the human homolog of the murine receptor identified by Nibbs *et al.* (1996) constititutes CCR9 (K.A. Graham *et al.*, personal communication).

CX_3CR1

The first CX_3C receptor has been identified as the specific molecule that binds the new chemokine fractalkine (Imai *et al.*, 1997). Notable in its high expression on NK cells, T cells and monocytes, CX_3CR1 works in concert with fractalkine to form an adhesion molecule conjugate pair. This conjugate induces directly the adhesion of cells bearing the receptor (leukocytes in the native state) to cells exhibiting the cell surface fractalkine (to include the endothelium *in vivo*). The fractalkine–CX_3CR-mediated adhesion events are independent of the action of selectins and integrins (Imai *et al.*, 1997).

CC Chemokine Receptor Knockouts

The ability to ablate specific murine genes by homologous recombination in ES cells is a powerful approach to elucidate gene function in an experimental animal system. In the interpretation of results with chemokine receptor knockout mice one should always be mindful that the chemokines of mice and humans may not play identical roles and that genes exhibiting a high degree of sequence homology between mouse and humans may not be true functional homologs or paralogs. This is well illustrated by the first reported chemokine receptor gene knockout, the ablation of the murine IL-8RB (CXCR2) gene (Cacalano *et al.*, 1994). $CXCR2^{-/-}$ mice exhibit marked lymphadenopathy characterized by increased numbers of B cells in lymph nodes and increased numbers of neutrophils in spleen. $CXCR2^{-/-}$ mice are deficient in neutrophil recruitment in response to a inflammatory stimuli and show defects in myeloid progenitor cell production. Equally fascinating has been the ablation of an orphan chemokine receptor, BLR1, whose absence results in severe disruption of normal lymphoid organ architecture with an almost complete absence of Peyer's patches and defects in B-cell homing in $BLR1^{-/-}$ mice (Forster *et al.*, 1996).

To date, only one CC chemokine receptor knockout has been reported in the literature. CCR1 is expressed on multiple hematopoietic cell types (monocytes, neutrophils, lymphocytes and eosinophils) and binds multiple CC chemokines (MIP-1α, RANTES and MCP-3). $CCR1^{-/-}$ mice have a normal distribution of leukocytes in adult animals but $CCR1^{-/-}$ neutrophils are defective in chemotaxis and Ca^{2+} flux in response to MIP-1α, while $CCR1^{-/-}$ macrophages are able to respond to MIP-1α, presumably via CCR4 and CCR5 receptors. $CCR1^{-/-}$ mice are defective in neutrophil recruitment *in vivo* and are more susceptible to an experimental fungal infection than wild-type animals (Gao *et al.*, 1997). Much more interesting are the data of Gerard and colleagues, showing that in pancreas injury model, the $CCR1^{-/-}$ animals develop distal disease, manifesting adult respiratory distress syndrome-like symptoms of neutrophil damage in the lungs.

At the time of writing, details of no other chemokine receptor knockout animals have been published, but the generation of chemokine gene knockout animals and the detailed analysis of their defects in leukocyte recruitment and function will provide an insight into the non-redundant functions that can be ascribed to individual CC chemokine receptors.

HIV AND THE CHEMOKINE SYSTEM

By now the dramatic developments of late 1995 and 1996, revealing the central role of the chemokine system in the pathogenesis of HIV infection, have been eloquently and elegantly described. The picture that first developed was simple yet stunning: one form of HIV-1, so-called M-tropic isolates, used the CC chemokine receptor CCR5 in conjunction with CD4 to bind and infect a target cell. Another form of HIV, the 'T-cell line-tropic' isolates, used another chemokine receptor, the CXC chemokine receptor 4 (CXCR4), sometimes called 'fusin'. The predominance of M-tropic forms as the clinically relevant transmission agent was supported by observation that $CCR5^{-/-}$ individuals were, with some rare exceptions, resistant to HIV infection.

As with everything involving chemokines, expansion and complexity quickly became the order of the day, and this simple model has been made much more complex of late. For example, it has been shown with steady regularity that increasing numbers of different chemokine receptors appear to support infection of various types of viruses. The relevance of this greater complexity to actual pathophysiology in humans is a topic of intense debate. One view is given in synopsis in Fig. 3. Here we depict the 'evolution' of viral forms *in vivo* as the infection progresses over the course of some years. Initially M-tropic HIV predominates, exploiting CCR5 as its major infection cofactor. However, under some circumstances, CCR3 may also play a role in the spread of M-tropic virus. Nevertheless, over the course of many years, it is known that, eventually, T-tropic forms of HIV predominate in the patient; their emergence is roughly the inverse of the decline in the number of $CD4^+$ T cells and the attendant diminution of immune function. The view espoused in Fig. 3 posits a number of intermediate dual tropic forms, each using a different, or combination of different, chemokine receptors as infective cofactors. Thus, the virus might always stay one step ahead of host defense, a canny moving target until such time as host defense is destroyed.

The emergence of T-tropic forms late in the disease may reflect some more efficient infective and replicative capacity. Indeed, it has been shown that in some cases T-tropic HIV does not exhibit an absolute requirement for CD4. But the paradox is that, even though T-tropic HIV predominates late in acquired immune deficiency syndrome (AIDS), viral transmission from these individuals to a new host still seems to be a function of CCR5, hence dependent on M-tropic virus. There is a subtle but important suggestion that may resolve this paradox. The data seem to fit a model in which the T-tropic form can be efficiently cleared, or at least held in check, by a healthy immune response. The stealthy, less prevalent, M-tropic form may be more difficult to stop. As

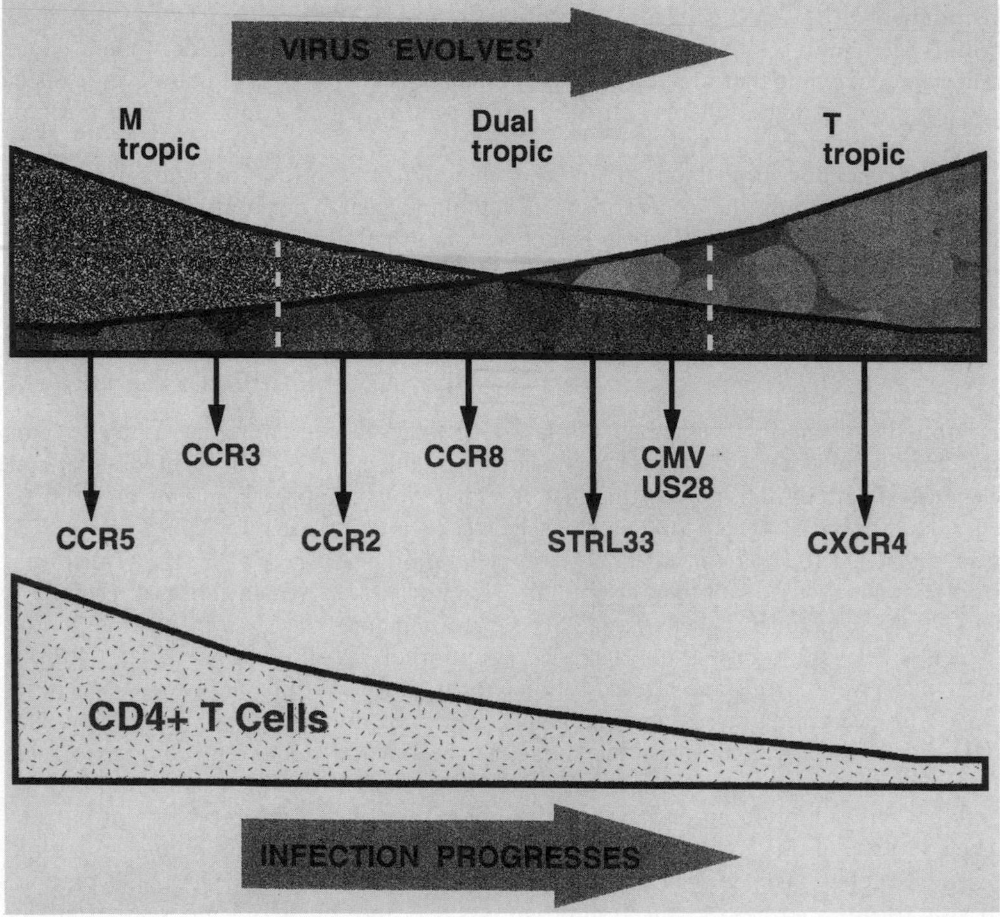

Fig. 3. Chemokines receptors as cofactors in the progression of HIV infection. The time course of HIV infection, marked by a drop in circulating $CD4^+$ T cells, displays a progression in the usage of chemokine receptors as infective cofactors. The initial infection by macrophage tropic HIV strains progresses almost exclusively through CCR5, whereas in the final stages the virus is composed predominantly of T-cell tropic variants employing CXCR4. The evolution of the HIV 'quasispecies' during disease progression goes through stages, and at times dual tropic virions can be isolated. It is during this transition that HIV variants make use of multiple chemokine receptors as cofactors, as indicated.

this form slowly plunders the immune system over the course of years, perhaps evolving into dual tropic forms to increase the range of cofactor usage and hence infectivity, more efficient viruses emerge directly proportional with decay of the deterrent power of the immune system.

OTHER VIRUSES AND CHEMOKINES

Viruses other than HIV also seem to be fond of the chemokine system. The herpesvirus family is perhaps the best example. So far, chemokine and chemokine receptor orthologs have been identified in members of the β and γ herpesvirus subfamilies. The cytomegalovirus (CMV) open reading frames (ORFs) US28, US27 and UL33 encode seven-transmembrane receptors and have been the most studied, with US28 being identified as a high-affinity receptor for a broad range of CC chemokines. US28 ORF has the added distinction of also serving as a cofactor for certain HIV subtypes (Pleskoff *et al.*, 1997). Murine CMV, which does not encode a CC chemokine receptor like US28, does encode a CC chemokine ortholog, m131 (MacDonald *et al.*, 1997). Human herpesvirus-6 (HHV-6) encodes a gene most similar to CMV UL33, as well as a possible CC chemokine in U83 (Gompels *et al.*, 1995). One of the most compelling stories surrounds HHV-8, also known as Kaposi's sarcoma-associated herpesvirus (KSHV). HHV-8 has been shown to encode three potential CC chemokines (vMIP-I, vMIP-II, and BCK), as well as a possible CXC chemokine receptor in ORF74 (Moore *et al.*, 1996; Guo *et al.*, 1997). As HHV-8 is implicated in causing Kaposi's sarcoma in patients with AIDS, it is interesting that vMIP-I is able to block HIV infection, presumably through CCR-5 (Moore *et al.*, 1996). The family of herpesviruses, then, has shown a strong inclination to make use of the host trafficking system. How they do this and what the actual mechanisms are remain to be elucidated.

Not to be outdone, the poxviruses have developed their own strategies based on the chemokine system. Similar to herpesviruses, the poxvirus causing molluscum contagiosum encodes a chemokine ortholog (Senkevich *et al.*, 1996). Moreover, myxoma virus and vaccinia virus appear to control the chemokine-mediated influx of leukocytes involved in a viral infection by producing virally encoded chemokine binding and inhibitory proteins. These are 35-kDa proteins, bearing no resemblance to known cellular proteins, which can bind both CC and CXC chemokines with high affinity (Graham *et al.*, 1997). Thus, it is becoming clear that many classes of viruses are exploiting various aspects of chemokine biology to further their own propagation.

SIGNAL TRANSDUCTION

Classic signaling by heterotrimeric G protein-linked seven-transmembrane receptors via immediate intracellular calcium release, phospholipase Cβ hydrolysis and protein kinase C activation has, in many cases, been demonstrated for chemokines and their receptors. Recently, however, it has become more apparent that this is just the tip of the iceberg. In addition, the dogma that exists concerning seven-transmembrane receptor coupling to pertussis toxin-sensitive G proteins has been questioned (Wu *et al.*, 1993; Arai and Charo, 1996; Kuang *et al.*, 1996). The observation that CKRs may link to multiple, PTX-sensitive and PTX-insensitive G proteins suggests a potential mechanism for specificity of chemokine actions amongst different cell types.

In addition, it has been demonstrated that RANTES, IL-8 and MCP-1, amongst other chemokines, can induce tyrosine phosphorylation, mitogen-activated protein kinase (MAPK) and MEK activation, and phosphatidylinositol-3 kinase (PI-3K) activation (Gaudry *et al.*, 1992; Thompson *et al.*, 1994; Bacon *et al.*, 1995, 1996; Turner *et al.*, 1995; Dubois *et al.*, 1996). More recently, Knall *et al.* (1996, 1997) demonstrated that IL-8-induced MAPK activation was dependent on PI-3K, but that IL-8-induced chemotaxis was dependent solely on the PI-3K activation, adding further complexity to the issue of chemokine-induced signaling of migration. Other functional consequences of this activity, aside from migration and adhesion have included T-cell activation and proliferation (Bacon *et al.*, 1995, 1996; Taub *et al.*, 1996; Szabo *et al.*, 1997), most likely via the activation of T cell-specific kinases, such as ZAP-70, as well as the MAPK cascade.

THE EXPANDING SCOPE OF CHEMOKINE ACTIVITIES

For many years, ideas about chemokine function have focused mainly on their promigratory properties in the constitutive trafficking of white blood cells, and attendant inflammatory pathologies when this process becomes dysregulated. But it is clear that, while leukocyte trafficking is a substantial part of chemokine biology, the scope of chemokine function is likely to be much more expansive than that. Fig. 4 is a representation of the growing constellation of the potential roles of chemokines, many of which are outside the realm of leukocyte trafficking. Below we focus on several of the most notable and emerging trends.

Chemokine Action in Animal Models of Pulmonary Inflammation

Animal models to test the roles of chemokines (mostly in inflammation) have become more refined and efficient, and have been used to good purpose to investigate the *in vivo* properties of newly discovered chemokines. Some of the most interesting models concern lung function. A great many publications have appeared attempting to link the expression of chemokines with the induction of lung inflammation in pulmonary disorders such as asthma, and the reader is referred to an earlier review for relevant references (Bacon and Schall, 1996). In contrast to early studies investigating the effects of IL-8, MIP-1α or RANTES, emphasis has shifted to the more recently cloned and potent eosinophil-activating chemokines, eotaxin, MCP-3, MCP-4 and MCP-5. While

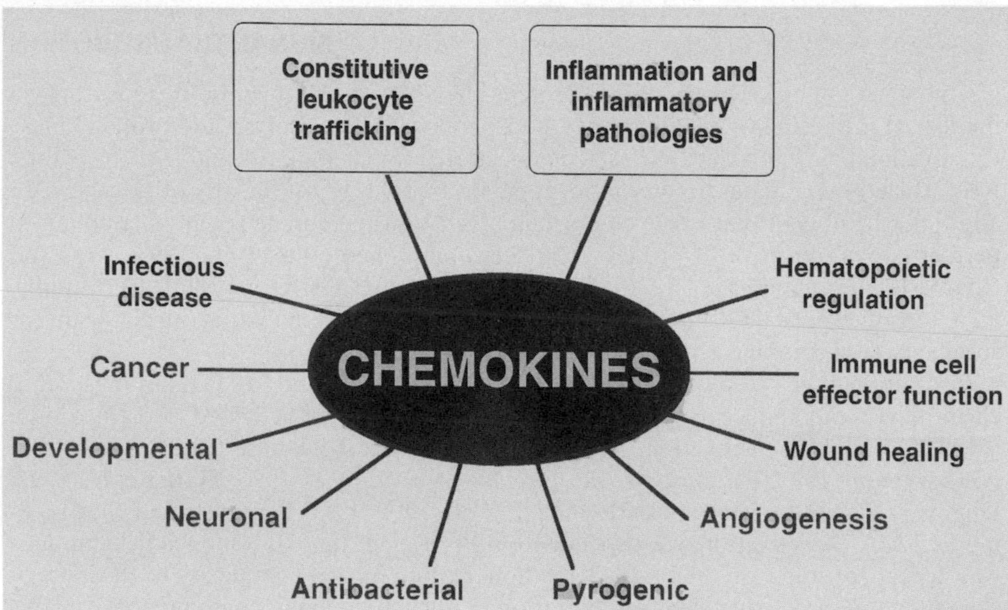

Fig. 4. The big bang of chemokine biology. Representation of the many roles for chemokines hinted at in published reports, with emphasis on leukocyte trafficking and inflammation.

the *in vitro* effects of these chemokines is undisputed, little hard evidence exists to elevate any one of these molecules to prime position, and the only proof of concept for chemokine involvement remains studies where specific neutralizing monoclonal antibodies have been shown to ameliorate the inflammation.

Eotaxin came to prominence in early 1994 after Jose *et al.* cloned a novel eosinophil-specific chemoattractant from allergen-challenged lungs of the guinea-pig. Followed closely by cloning of the human and murine genes, eotaxin mRNA expression has been shown to closely parallel early eosinophilia in models of allergic asthma, and human inflammatory bowel disease (ulcerative colitis and Crohn's disease) (Garcia-Zepeda *et al.*, 1996; Gonzalo *et al.*, 1996), while bioactivity has revealed highly specific activities on eosinophils (Kitaura *et al.*, 1996; Ponath *et al.*, 1996). Thus, despite the expression of RANTES mRNA and protein in airway epithelium, the ability of RANTES and MIP-1α to induce immunoglobulin E production in B cells (Kimata *et al.*, 1996), and the prominent role for RANTES and MIP-1α in schistosome egg antigen-induced allergic inflammation (Lukacs *et al.*, 1996, 1997), and the eosinophil, monocyte and lymphocyte chemoattractant ability of these two chemokines (Bacon and Schall, 1996; Braciak *et al.*, 1996), there has been a somewhat premature bias towards the therapeutic potential of antagonists of eotaxin and the CCR3 receptor.

In addition, Stafford *et al.* (1997) have provided further evidence against the exclusivity of eotaxin in an ovalbumin-challenged mouse model. This study demonstrated high levels of MCP-1, MIP-1α and MCP-3, with significant eosinophil and T-cell infiltration. Pretreatment of mice with polyclonal anti-MCP-3 resulted in reduced eosinophilia, suggesting a prominent role for this chemokine. It is unclear, however, whether antibodies to MIP-1α or MCP-1 would reduce the eosinophilia and T-cell infiltration. In addition, because MCP-3 is a T-cell chemoattractant, it would be interesting to know whether or not the pAb to MCP-3 reduced the T-cell infiltrate. Despite the ability of MCP-4, RANTES and MCP-3 to bind CCR-3, MIP-1α uses predominantly CCR1 and CCR5.

As only two studies exist to show antibody-induced reduction in leukocyte infiltration in murine models of allergic lung inflammation, and MIP-1α is one of the activating chemokines, it is possible that the effects of eotaxin, while being important for tissue and blood eosinophilia (Mould *et al.*, 1997), are not exclusive. Evidence supporting such a theory comes from studies analyzing targeted disruption of the eotaxin gene (Rothenberg *et al.*, 1997). In two models of tissue eosinophilia, ovalbumin-challenged mice and antigen-induced corneal eosinophilia using the parasitic helminth *Onchocerca volvulus* (generating stromal keratitis), eotaxin was shown to have a contributory role to the early-onset eosinophilia in both cases. While the knockout mice clearly indicate a role for eotaxin in the disease state, as noted by the authors, there was still significant eosinophilia in the absence of eotaxin. This partial reduction in eosinophil infiltration suggests that the gross pathology in allergic disease characterized by early-onset tissue eosinophilia may indeed be the product of numerous chemoattractant factors.

Beyond Trafficking: Novel Bioactivities for Old Clones

Using rat vascular smooth muscle, TCA3/I309 along with MIP-1α and MCP-1 stimulates pertussis toxin-dependent chemotaxis and adhesion to collagen (Luo *et al.*, 1996). Of major significance was that TCA3/I309 alone induced proliferation of smooth muscle cells, potentially through an high-affinity (K_d 3 nM) interaction. It is now known that CCR-8 may represent this receptor, although the binding competition by MIP-1α, MCP-1 and RANTES may be unique to the smooth muscle cells.

In addition to its chemoattractant activities for monocytes and microglia (see below), TCA3/I309 has also been shown to prevent dexamethasone-induced apoptosis of a human thymic lymphoma cell line (Van Snick *et al.*, 1996). This activity, which was exclusive to TCA3/I309 among chemokines, was similar to that of IL-9, the most potent antiapoptotic factor for these cells.

Both CXC and CC chemokines have been shown to be inhibitors of megakaryocytopoiesis (Gerwitz *et al.*, 1995). Specifically, NAP-2, IL-8, MIP-1α, MIP-1β and C10 inhibited megakaryocyte colony formation at equivalent doses, and were all more potent than PF4, which was originally shown to inhibit colony formation.

Chemokines in Neurobiology

One of the most exciting expansions in chemokine biology concerns their activities with respect to neuronal biology. A limited number of publications has appeared, detailing the production of chemokines by neuronal cell types; however, the scope of biologic systems has been limited thus far. One of the most commonly investigated systems involves the activity of chemokines in murine models of experimental encephalomyelitis. Astrocytes and microglia have been assayed for chemokine message and migratory capacity in response to chemokines released by mononuclear cells. Using lipopolysaccharide (LPS) as a primary activator, astrocytes and microglia have been shown to produce MIP-1α, MIP-1β and RANTES (Sun *et al.*, 1997), MCP-1, IP-10 and KC (Glabinski *et al.*, 1997). In addition, MBP-reactive T cells have been shown to induce MIP-1α, RANTES, inducible protein-10 (IP-10) and MCP-1 in a major histocompatibility complex II-restricted interaction with the glial cells (Sun *et al.*, 1997). Further studies have identified specific mRNA for RANTES, MIP-1α, MIP-1β, TCA3/I309, IP-10, MCP-1, KC and MCP-3 in spinal cord (Godiska *et al.*, 1995). The apparent significance of chemokine production relates to the attraction of mononuclear cells during development and relapse of the chronic inflammatory reactions in this murine model. Significant proof of this concept was provided in a study investigating the actions of monoclonal antibodies against the β chemokines MIP-1α and MCP-1 (Karpus *et al.*, 1995). Administration of monoclonal antibodies against MIP-1α, but not MCP-1, prevented the development of acute and chronic relapsing demyelinating disease in a murine model, reportedly due to the inhibition of mononuclear cell infiltration.

In similar fashion, targeted gene expression of the neutrophil attractant KC/N51 to oligodendrocytes in mice resulted in significant disease, characterized by neutrophil influx, microglial activation and disruption of the blood–brain barrier (Tani *et al.*, 1996). These studies therefore attribute the leukocyte migration induction capacity to chemokines, which clearly can be expressed by cells of the nervous system.

Further studies have actually looked at the activation of the astrocytes and microglia by the chemokines, either in autocrine fashion or produced by infiltrating mononuclear cells. LPS, IL-1β, or tumor necrosis factor-α induced MCP-1 expression by astrocytes, which stimulated microglia chemotaxis. In addition TCA3/I309 produced by encephalitogenic T lymphocytes also caused microglia chemotaxis (Hayashi *et al.*, 1995). Another study reported the production of MIP-1α, MIP-1β and MCP-1 by both stimulated astrocytes and microglia; however, only the microglia were chemotactically responsive to the chemokines (Peterson *et al.*, 1997).

Although specific function was not addressed, Meda *et al.* (1996) have demonstrated the production of MCP-1 by β-amyloid (25–35) peptide and shown its production by monocytes and microglia, suggesting a possible mechanism for mononuclear cell recruitment into amyloid plaques in aging and dementia. Cerebrospinal fluid samples from patients with bacterial meningitis contained IL-8, Gro-α, MCP-1, MIP-1α and MIP-1β, suggesting a possible mechanism for the presence of neutrophils and monocytes in this fluid (Spanaus *et al.*, 1997). While chemoattractant activity for leukocytes was inhibited by specific monoclonal antibodies to the chemokines, there was little statistical correlation between the actual leukocyte counts and chemokine concentrations, suggesting the activity of additional factors.

In searching for functions of chemokines beyond the recruitment of leukocytes in the central nervous system (CNS), a number of studies have been published demonstrating significant biologic activities pertinent to the neuronal cells themselves. Of particular interest was the early report showing that IL-8 was capable of enhancing the neuronal survival of hippocampal cultures, increasing the numbers of astrocytes as microglia (Araujo and Cotman, 1993). In addition, IL-8 has been shown to be produced by malignant human astrocytes present in astrocytomas (Kasahara *et al.*, 1991; Van Meir *et al.*, 1992), suggesting a prominent role for this chemokine in CNS tumor biology.

Bolin *et al.* (1997) have recently demonstrated that RANTES, but not MIP-1α or MCP-1/JE, can induce migration and differentiation of dorsal root ganglia neurons, the progenitors of sensory neurons. The migrated neurons expressed neuropeptide Y and calcitonin gene-related peptide, phenotypes classically associated with pain and temperature sensation in the periphery. Interestingly, RANTES was also shown to be expressed at early developmental stages in the mouse, in the DRG anlagen, spinal cord and skin, providing some evidence for a connection between DRG development and sensory innervation of the skin.

More recently, Horuk *et al.* (1997) have demonstrated the expression of CXCR2 in numerous regions of the brain and spinal cord in normal subjects. In addition, CXCR2 was expressed around amyloid plaques in tissue derived from patients with Alzheimer's disease, suggesting some correlation between neurodegenerative disorders and IL-8 or Gro-α bioactivity. In addition to CXCR2, expression of the Duffy antigen receptor for chemokines was found exclusively in Purkinje cells of the cerebellum. While the biologic significance of the Duffy receptor has yet to be determined, this highly specific pattern of expression in the brain does suggest that this receptor may fulfill an important role in CNS physiology.

Subsequently, Hesselgesser *et al.* (1997) demonstrated the presence of receptors CXCR4, CCR1 and CCR5 on a neuronal cell line, and the activation of neuronal cell migration by chemokines known to bind these receptors. Importantly, the authors were able to demonstrate CXCR4 association of the viral envelope protein of HIV strain IIIB, and thus the potential for infection of neuronal cells, significantly, in the absence of CD4.

SUMMARY AND PERSPECTIVES

The chemokine universe we have historically inhabited is that slippery and unstable space at the surface of the expanding soap bubble. In the present era much of this expansion is being driven by increased information, which we sometimes mistakenly take for enhanced understanding. Just as cosmologists yearn for ever more powerful telescopes, believing that more precise measurements will give them a privileged view of the metaphysical, the chemoinformatician believes that appreciation of the numeric expansion of the superfamily constitutes the whole science. Thus the quest for information can sometimes obscure the drive for understanding, with the body of accumulated chemoinformatics becoming an electronic and printed 'litter-ature' strewn with discarded EST fragments, each claimed by many as their own unique gems.

But biology, like cosmology, is a bit like having déja vu and amnesia at the same time. We feel certain that we have forgotten this all before. For cosmologists, that feeling might come from the knowledge that in 1783 Imammuel Kant provided us with epistemic prophylaxis for the kind of twisted reasoning to which we are now sometimes subjected. Kant proved that rather than being external independent entities, space and time were pure forms of intuition: a kind of epistemic hardwiring through which the universe took shape, and through which it was made real—not in and of itself—but as a consequence of the perceiving process. This is not as strange as it might at first sound. These principles anticipated, indeed are validated, by later ideas. Think of Einstein's 'special relatives': the identical twins who turn out to be of dramatically different ages when one takes a space trip at near light speed, while the other remains earthbound. Kant's edict was clear: the language of physics could not be used in the realm of metaphysics, any more than the language of physics could explain the metaphysical. For him, the issue is not one of universes in multidimensions, but of much dementia. Thus such questions as 'what happened before the Big Bang', while making grammatical sense, may actually be no more meaningful than such statements as 'green is wetter than the inside of a memory', or the nomenclatural implication that MIP-3α must somehow be related to MIP-3β, or the finding of 'perma-press' setting on an iron.

So we must turn now from the question of 'How many chemokine elements are there?', to the more fascinating 'Why are there so many chemokine elements?' The enabling endeavor of rapid and informative assessment of the actual biologic properties of chemokines, and their workings as a system, requires more than counting; rather it requires an integrated and integrative process. So the next challenge for chemokines is both technological and intellectual: how can the enormous amount of information generated from genomics efforts be coupled to a process that will result in the enhanced understanding of how the elements revealed by that information act in health and disease?

Some of this is already happening, revealing glimpses of a new model of the chemokine universe. This model is increasingly less suggestive of the soap bubble, and perhaps more reasonably viewed as a cluster of semi-independent galaxies, each comprising its own collection of biologic processes (inflammation, development, wound healing), but each exerting gravitational influences on the other, and each originating from universal themes. For example, if chemokines direct the migration of motile cells (e.g. leukocytes) in a mature organism, then why not exploit the same ability to move around various cells during embryogenesis and development? Or to provide architectural supervision during tissue remodeling or repair? Or to orchestrate the dynamic environment of the bone marrow during hematopoiesis? The biology and properties of this system are more than a little complex but, given the time and the tools, a unified field theory explaining much of the chemokine universe may not be an unreasonable idea.

ACKNOWLEDGEMENTS

The authors thank John Mak for assistance with the figures.

REFERENCES

Aiuti, A., Webb, I.J., Bluel, C., Springer, T. and Gutterrez-Ramos, J.-C. (1997). The chemokine SDF-1 is a chemoattractant for human CD34$^+$ hematopoietic progenitor cells and provides a new mechanism to explain the mobilization of CD34$^+$ progenitors to peripheral blood. *J. Exp. Med.* **185**, 111–120.

Araujo, D.M. and Cotman, C.W. (1993). Trophic effects of interleukin-4, -7 and -8 on hippocampal neuronal cultures: potential involvement of glial-derived factors. *Brain Res.* **600**, 49–55.

Arenberg, D.A., Kunkel, S.L., Polverini, P.J., Morris, S.B., Burdick, M.D., Glass, M.C., Taub, D.D., Lannettoni, M.D., Whyte, M.D. and Strieter, R.M. (1996). Interferon-γ-inducible protein 10 (IP-10) is an angiostatic factor that inhibits human non-small cell lung cancer (NSCLC) tumorigenesis and spontaneous metastases. *J. Exp. Med.* **184**, 981–992.

Baba, M., Imai, T., Nishimura, M., Kakizaki, M., Takagi, S., Hieshima, K., Nomiyama, H. and Yoshie, O. (1997). Identification of CCR6, the specific receptor for a novel lymphocyte-directed CC chemokine LARC. *J. Biol. Chem.* **272**, 14893–14898.

Bacon, K.B. and Schall T.J. (1996). Chemokines as mediators of allergic inflammation. *Int. Arch. Allergy Immunol.* **109**, 97–109.

Bacon, K.B., Premack, B., Gardner, P. and Schall, T.J. (1995). Activation of dual T cell signaling pathways by the chemokine RANTES. *Science* **269**, 1727–1730.

Bacon, K.B., Szabo, M.C., Yssel, H., Bolen, J.B. and Schall, T.J. (1996). RANTES induces tyrosine kinase activity of stably complexed p125FAK and ZAP-70 in human T cells. *J. Exp. Med.* **184**, 873–882.

Bazan, J.F., Bacon, K.B., Hardiman, G., Wang, W., Soo, K., Rossi, D., Greaves, D.R., Zlotnik, A. and Schall, T.J. (1997). Fractalkine: a new class of membrane-bound chemokine with a CX_3C motif. *Nature* **385**, 640–644.

Biosmenu, R., Feng, L., Xia, Y.Y., Chang, J.C.C. and Havran, W.L. (1996). Chemokine expression by intraepithelial gd cells. *J. Immunol.* **157**, 985–992.

Bluel, C.C., Farzan, M., Choe, H., Parolin, C., Clark-Lewis, I., Sodroski, J. and Springer, T.A. (1996a). The lymphocyte chemoattractant SDF-1 is a ligand for LESTR/fusin and blocks HIV-1 entry. *Nature* **382**, 829–833.

Bluel, C.C., Fuhlbrigge, R.C., Casanovas, J.M., Aiuti, A. and Springer, T.A. (1996b). A highly efficacious lymphocyte chemoattractant, stromal cell-derived factor (SDF-1). *J. Exp. Med.* **184**, 1101–1109.

Bluel, C.C., Wu, L., Hoxie, J.A., Springer, T.A. and Mackay, C.R. (1997). The HIV coreceptors CXCR4 and CCR5 are differentially expressed and regulated on human T lymphocytes. *Proc. Natl Acad. Sci. U.S.A.* **94**, 1925–1930.

Braciak, T.A., Bacon, K., Xing, Z., Torry, D.J., Graham, F.L., Schall, T.J., Richards, C.D., Croitoru, K. and Gauldie, J. (1996). Overexpression of RANTES using a recombinant adenovirus vector induces the tissue-directed recruitment of monocytes to the lung. *J. Immunol.* **157**, 5076–5084.

Butcher, E.C. and Picker, L.J. (1996). Lymphocyte homing and homeostasis. *Science* **272**, 60–66.

Cacalano, G., Lee, J., Kikly, K., Ryan, A.M., Pitts-Meek, S., Hultgren, B., Wood, W.I. and Moore, M.W. (1994). Neutrophil and B cell expansion in mice that lack the murine IL-8 receptor homolog. *Science* **265**, 682–684.

Daugherty, B.L., Siciliano, S.J., DeMartino, J.A., Malkowitz, L., Sirotina, A. and Springer, M.S. (1996). Cloning, expression and characterization of the human eosinophil eotaxin receptor. *J. Exp. Med.* **183**, 2349–2354.

Deichmann, M., Kronenwett, R. and Haas, R. (1997). Expression of the human immunodeficiency virus type-1 coreceptors CXCR-4 (fusin, LESTR) and CKR-5 in CD34$^+$ hematopoietic progenitor cells. *Blood* **89**, 3522–3528.

Dilloo, D., Bacon, K., Holden, W., Zhong, W., Burdach, S., Zlotnik, A. and Brenner, M. (1996). Combined chemokine and cytokine gene transfer enhances antitumor immunity. *Nature Med.* **2**, 1090–1095.

Dubois *et al.* (1996). Early signal transduction by the receptor to the chemokine monocyte chemotactic protein-1 in a murine T cell hybrid. *J. Immunol.* **156**, 1356–1361.

Forssmann, U., Uguccioni, M., Loetscher, P., Dahinden, C.A., Langen, H., Thelen, M. and Baggiolini, M. (1997). Eotaxin-2, a novel CC chemokine that is selective for the chemokine receptor CCR3, and acts like eotaxin on human eosinophil and basophil leukocytes. *J. Exp. Med.* **185**, 2171–2176.

Forster, R., Mattis, A.E., Kremmer, E., Wolf, E., Brem, G. and Lipp, M. (1996). A putative chemokine receptor, BLR1, directs B cell migration to defined lymphoid organs and specific anatomic compartments of the spleen. *Cell* **87**, 1037–1047.

Gao, J.L., Wynn, T.A., Chang, Y., Lee, E.J., Broxmeyer, H.E., Cooper, S., Tiffany, H.L., Westphal, H., Kwon-Chung, J. and Murphy, P.M. (1997). Impaired host defense, hematopoiesis, granulomatous inflammation and type 1–type 2 cytokine balance in mice lacking CC chemokine receptor 1. *J. Exp. Med.* **185**, 1959–1968.

Garcia-Zepeda, E.A., Rothenberg, M.E., Ownbey, R.T., Celestin, J., Leder, P. and Luster, A.D. (1996a). Human eotaxin is a specific chemoattractant for eosinophil cells and provides a new mechanism to explain tissue eosinophilia. *Nature Med.* **2**, 449–456.

Garcia-Zepeda, E.A., Combadiere, C., Othenberg, M.E., Sarafi, M.N., Lavigne, F., Hamid, Q., Murphy, P.M. and Luster, A.D. (1996b). Human monocyte chemoattractant protein (MCP)-4 is a novel CC chemokine with activities on monocytes, eosinophils and basophils induced in allergic and non-allergic inflammation that signals through the CC chemokine receptors (CCR)-2 and -3. *J. Immunol.* **157**, 5613–5626.

Gaudry, M. *et al.* (1992). Evidence for the involvement of tyrosine kinases in the locomotory responses of human neutrophils. *J. Leukoc. Biol.* **51**, 103–108.

Gerwitz, A.M., Zhang, J., Ratajczak, J., Ratajczak, M., Park, K.S., Li, C., Yan, Z. and Poncz, M. (1995). Chemokine regulation of human megakaryocytopoiesis. *Blood* **86**, 2559–2567.

Glabinski, A.R., Tani, M., Strieter, R.M., Tuohy, V.K. and Ransohoff, R.M. (1997). Synchronous synthesis of a- and b-chemokines by cells of diverse lineage in the central nervous system of mice with relapses of chronic experimental autoimmune encephalomyelitis. *Am. J. Pathol.* **150**, 617–630.

Godiska, R., Chantry, D., Deitsch, G.N. and Gray, P.W. (1995). Chemokine expression in murine experimental allergic encephalomyelitis. *J. Neuroimmunol.* **58**, 167–176.

Godiska, R., Chantry, D., Raport, C.J., Sozzani, S., Allavena, P., Leviten, D., Mantovani, A. and Gray, P.W. (1997). Human macrophage-derived chemokine (MDC), a novel chemoattractant for monocytes, monocyte-derived dendritic cells and natural killer cells. *J. Exp. Med.* **185**, 1595–1604.

Gompels, U.A., Nicholas, J., Lawrence, G., Jones, M., Thonson, B.J., Martin, M.E.D., Efstathiou, S., Craxton, M. and Macaulay, H.A. (1995). The DNA sequence of human herpesvirus-6: structure, coding content, and genome evolution. *Virology* **209**, 29–51.

Gonzalo, J.-A., Jia, G.-Q., Aguirre, V., Friend, D., Coyle, A.J., Jemkins, N.A., Lin, G.-S., Katz, H., Lichtman, A., Copeland, N., Kopf, M. and Guttierrez-Ramos, J.-C. (1996). Mouse eotaxin expression parallels eosinophil accumulation during lung allergic inflammation but it is not restricted to a Th2-type resposne. *Immunity* **4**, 1–14.

Graham, K.A., Lalani, A.S., Macen, J.L., Ness, T.L., Barry, M., Liu, L.-Y., Lucas, A., Clark-Lewis, I., Moyer, R.W. and McFadden, G. (1997). The T1/35kDa family of poxvirus-secreted proteins bind chemokines and modulate leukocyte influx into virus-infected tissues. *Virology* **229**, 12–24.

Guo, H.G., Browning, P., Nicholas, J., Hayward, G.S., Tschachler, E., Jiang, Y.W., Sadowska, M., Raffeld, M., Colombini, S., Gallo, R.C. and Reitz, M.S., Jr. (1997). Characterization of a chemokine receptor-related gene in human herpesvirus 8 and its expression in Kaposi's sarcoma. *Virology* **228**, 371–378.

Hayashi, M., Luo, Y., Laning, J., Strieter, R.M. and Dorf, M.E. (1995). Production and function of monocyte chemoattractant protein-1 and other b-chemokines in murine glial cells. *J. Neuroimmunol.* **60**, 143–150.

He, J., Chen, Y., Farzan, M., Choe, H., Ohagen, A., Gartner, S., Busciglio, J., Yang, X., Hofmann, W., Newman, W., Mackay, C.R., Sodroski, J. and Gabuzda, D. (1997). CCR3 and CCR5 are co-receptors for HIV-1 infection of microglia. *Nature* **385**, 645–649.

Hedrick, J.A., Saylor, V., Figueroa, D., Mizoue, L., Xu, Y., Menon, S., Abrams, J., Handel, T. and Zlotnik, A. (1997). Lymphotactin is produced by NK cells and attracts both NK cells and T cells *in vivo*. *J. Immunol.* **158**, 1533–1540.

Hesselgesser, J., Halks-Miller, M., DelVecchio, V., Peiper, S.C., Hoxie, J., Kolson, D.L., Taub, D. and Horuk, R. (1997). CD4-independent association between HIV-1 gp120 and CXCR4: functional chemokine receptors are expressed in human neurons. *Curr. Biol.* **7**, 112–121.

Hieshima, K., Imai, T., Opdenakker, G., Van Damme, J., Kusuda, J., Tei, H., Sakaki, Y., Takatsuki, K., Miura, R., Yoshie, O. and Nomiyama, H. (1997). Molecular cloning of a novel human CC chemokine liver and activation-regulated chemokine (LARC) expressed in liver. *J. Biol. Chem.* **272**, 5846–5853.

Horuk, R., Martin, A.W., Wang, Z., Schweitzer, L., Gerassimides, A., Guo, H., Lu, Z., Hesselgesser, J., Perez,

H.D., Kim, J., Parker, J., Hadley, T.J. and Peiper, S.C. (1997). Expression of chemokine receptors by subsets of neurons in the central nervous system. *J. Immunol.* **158**, 2882–2890.

Hromas, R., Gray, P.W., Chantry, D., Godiska, R., Krathwohl, M., Fife, K., Bell, G.I., Takeda, J., Aronica, S., Gordon, M., Cooper, S., Broxmeyer, H.E. and Klemsz, M.J. (1997). Cloning and characterization of exodus, a novel β-chemokine. *Blood* **89**, 3315–3322.

Imai, T., Yoshida, T., Baba, M., Nishimura, M., Kakizaki, M. and Yoshie, O. (1996). Molecular cloning of a novel T cell-directed CC chemokine expressed in thymus by signal sequence trap using Epstein–Barr virus vector. *J. Biol. Chem.* **271**, 21 514–21 521.

Imai, T., Baba, M., Nishimura, M., Kakizaki, M., Takagi, S. and Yoshie, O. (1997). The T cell-directed CC chemokine TARC is a highly specific biological for CC chemokine receptor 4. *J. Biol. Chem.* **272**, 15 036–15 042.

Jia, G.-Q., Gonzalo, J.A., Lloyd, C., Kremer, L., Lu, L., Martinez-A., C., Wershil, B.K. and Gutierrez-Ramos, J.C. (1996). Distinct expression and function of the novel mouse chemokine monocyte chemotactic protein-5 in lung allergic inflammation. *J. Exp. Med.* **184**, 1939–1951.

Karpus, W.J., Lukacs, N.W., McRae, B.L., Strieter, R.M., Kunkel, S.L. and Miller, S.D. (1995). An important role for the chemokine macrophage inflammatory protein-1a in the pathogenesis of the T cell-mediated autoimmune disease, experimental autoimmune encephalomyelitis. *J. Immunol.* **155**, 5003–5010.

Kasahara, T., Mukaida, N., Yamashita, K., Yagisawa, H., Akahoshi, T. and Matsushima, K. (1991). IL-1 and TNF induction of IL-8 and monocyte chemotactic and activating factor (MCAF) mRNA expression in a human astrocytoma cell line. *Immunology* **74**, 60.

Kimata, H., Yoshida, A., Ishioka, C., Fujimoto, M., Lindley, I. and Furusho, K. (1996). RANTES and macrophage inflammatory protein 1a selectively enhance immunoglobulin (IgE) and IgG_4 production by human B cells. *J. Exp. Med.* **183**, 2397–2402.

Kitaura, M., Nakajima, T., Imai, T., Harada, T., Combadiere, C., Tiffany, H.L., Murphy, P.M. and Yoshie, O. (1996). Molecular cloning of human eotaxin, an eosinophil-selective CC chemokine, and identification of a specific eosinophil eotaxin receptor, CC chemokine receptor 3. *J. Biol. Chem.* **271**, 7725–7730.

Knall, C. *et al.* (1996). Interleukin-8 regulation of the Ras/Raf/mitogen-activated protein kinase pathway in human neutrophils. *J. Biol. Chem.* **271**, 2832–2838.

Knall, C., Worthen, G.S. and Johnson, G.L. (1997). Interleukin-8-stimulated phosphoinositol-3-kinase activity regulates the migration of human neutrophils independent of extracellular signal-regulated kinase and p38 mitogen-activated protein kinases. *Proc. Natl Acad. Sci. U.S.A.* **94**, 3052–3057.

Kuang, Y. *et al.* (1996). Selective G-protein coupling by CC chemokine receptors. *J. Biol. Chem.* **271**, 3975–3978.

Loetscher, M., Gerber, B., Loetscher, P., Jones, S.A., Piali, L., Clark-Lewis, I., Baggiolini, M. and Moser, B. (1996). Chemokine receptor specific for IP-10 and Mig: structure, function, and expression in activated T-lymphocytes. *J. Exp. Med.* **184**, 963–969.

Lukacs, N.W., Standiford, T.J., Chensue, S.W., Kunkel, R.G., Strieter, R.M. and Kunkel, S.L. (1996). CC chemokine-induced eosinophil chemotaxis during allergic airway inflammation. *J. Leukoc. Biol.* **60**, 573–578.

Lukacs, N.W., Strieter, R.M., Warmington, K., Lincoln, P., Chensue, S.W. and Kunkel, S.L. (1997). Differential recruitment of leukocyte populations and alteration of airway hyperreactivity by CC family chemokines in allergic airway inflammation. *J. Immunol.* **158**, 4398–4404.

Luo, Y., D'Amore, P.A. and Dorf, M.E. (1996). β-chemokine TCA3 binds to and activates rat vascular smooth muscle. *J. Immunol.* **157**, 2143–2148.

MacDonald, M.R., Li, X.-Y. and Virgin, H.W. (1997). Late expression of a β chemokine homolog by murine cytomegalovirus. *J. Virol.* **71**, 1671–1678.

Margulies, B., Browne, H. and Gibson, W. (1996). Identification of the human cytomegalovirus G protein-coupled receptor homologue encoded by UL33 in infected cells and enveloped virus particles. *Virology* **225**, 111–125.

Meda, L., Bernasconi, L., Bonaiuto, C., Sozzani, S., Zhou, D., Otvos, L., Mantovani, A., Rossi, F. and Cassatella, M.A. (1996). β-amyloid (25–35) peptide and IFN-γ synergistically induce production of the chemotactic cytokine MCP-1/JE in monocytes and microglial cells. *J. Immunol.* **157**, 1213–1218.

Moore, J.P. (1997). Coreceptors: implications for HIV pathogenesis and therapy. *Science* **276**, 51–52.

Moore, J.P. and Koup, R.A. (1996). Chemoattractants attract HIV researcher. *J. Exp. Med.* **184**, 311–313.

Moore, P.S., Boshoff, C., Weiss, R.A. and Chang, Y. (1996). Molecular mimicry of human cytokine and cytokine response pathway genes by KSHV. *Science* **274**, 1739–1744.

Mould, A.W., Matthaei, K.I., Young, I.G. and Foster, P.S. (1997). Relationship between interleukin-5 and eotaxin in regulating blood and tissue eosinophilia in mice. *J. Clin. Invest.* **99**, 1064–1071.

Nagasawa, T., Hirota, S., Tachibana, K., Takakura, N., Nishikawa, S., Kitamura, Y., Yoshida, N., Kikutani, H. and Kishimoto, T. (1996). Defects of B-cell lymphopoiesis and bone marrow myelopoiesis in mice lacking the CXC chemokine PBSF/SDF-1. *Nature* **382**, 635–638.

Nibbs, R.J.B., Wylie, S.M., Pragnell, I.B. and Graham, G.J. (1997). Cloning and characterization of a novel murine β chemokine receptor D6. *J. Biol. Chem.* **272**, 12495–12504.

Oberlin, E., Amara, A., Bachelerie, F., Bessia, C., Virelizier, J.-L., Arenzana-Seisdedos, F., Schwartz, O., Heard, J.-M., Clark-Lewis, I., Legler, D.F., Loetscher, M., Baggiolini, M. and Moser, B. (1996). The CXC chemokine SDF-1 is the ligand for LESTR/fusin and prevents infection by T-cell-line-adapted HIV-1. *Nature* **382**, 833–835.

Patel, V.P., Kreider, B.L., Li, Y., Li, H., Leung, K., Salcedo, T., Nardelli, B., Pippalla, V., Gentz, S., Thotakura, R., Parmelee, D., Gentz, R. and Garotta, G. (1997). Molecular and functional characterization of two novel human C-C chemokines as inhibitors of two distinct classes of myeloid progenitors. *J. Exp. Med.* **185**, 1163–1172.

Peterson, P.K., Hu, S., Salak-Johnson, J., Molitor, T.W. and Chao, C.C. (1997). Differential production of and migratory response to β chemokines by human microglia and astrocytes. *J. Infect. Dis.* **175**, 478–481.

Pleskoff, O., Treboute, C., Brelot, A., Heveker, N., Seman, M. and Alizon, M. (1997). Identification of a chemokine receptor encoded by human cytomegalovirus as a cofactor for HIV-1 entry. *Science* **276**, 1874–1878.

Ponath, P.D., Qin, S., Post, T.W., Wang, Wu, L., Gerard, N.P., Newman, W., Gerard, C. and Mackay, C.R. (1996). Molecular cloning and characterization of a human eotaxin receptor expressed selectively on eosinophils. *J. Exp. Med.* **183**, 2437–2448.

Power, C.A., Meyer, A., Nemeth, K., Bacon, K.B., Hoogewerf, A.E., Proudfoot, A.E. and Wells, T.N. (1995). Molecular cloning and functional expression of a novel CC chemokine receptor cDNA from a human basophilic cell line. *J. Biol. Chem.* **270**, 19495–19500.

Raport, C.J., Gosling, J., Schweickart, V.L., Gray, P.W. and Charo, I.F. (1996). Molecular cloning and functional characterization of a novel human CC chemokine receptor (CCR5) for RANTES, MIP-1beta, and MIP-1alpha. *J. Biol. Chem.* **271**, 17161–17166.

Rossi, D.L., Vicari, A.P., Franz-Bacon, K., McClanahan, T.K. and Zlotnik, A. (1997). Identification through bioinformatics of two macrophage proinflammatory human chemokines, MIP-3a and MIP-3b. *J. Immunol.* **158**, 1033–1036.

Rot, A. (1993). Neutrophil attractant/activation protein-1: (interleukin-8) induced *in vitro* neutrophil migration by haptotactic mechanism. *Eur. J. Immunol.* **23**, 303–306.

Rothenberg, M.E., Maclean, J.A., Pearlman, E., Luster, A.D. and Leder, P. (1997). Targeted disruption of the chemokine eotaxin partially reduces antigen-induced tissue eosinophilia. *J. Exp. Med.* **185**, 785–790.

Runsaeng, V., Vliagoftis, H., Oh, C.K. and Metcalfe, D.D. (1997). Lymphotactin gene expression in mast cells following Fcε receptor 1 aggregation. *J. Immunol.* **158**, 1353–1360.

Samson, M., Labbe, O., Mollereau, C., Vassart, G. and Parmentier, M. (1996). Molecular cloning and functional expression of a new human CC-chemokine receptor gene. *Biochemistry* **35**, 3362–3367.

Sarafi, M.N., Garcia-Zepeda, E.A., MacLean, J.A., Charo, I.F. and Luster, A.D. (1997). Murine monocyte chemoattractant protein (MCP)-5: a novel CC chemokine that is a structural and functional homolog of human MCP-1. *J. Exp. Med.* **185**, 99–109.

Schall, T.J. and Bacon, K.B. (1994). Chemokines, leukocyte trafficking, and inflammation. *Curr. Opin. Immunol.* **6**, 865–873.

Schulz-Knappe, P., Magert, H.-J., Dewald, B., Meyer, M., Cetin, Y., Kubbies, M., Tomeczkowski, J., Kirchhoff, K., Raida, M., Adermann, K., Kist, A., Reinecke, M., Sillard, R., Pardigol, A., Uguccioni, M., Baggiolini, M. and Forssmann, W.-G. (1996). HCC-1, a novel chemokine from human plasma. *J. Exp. Med.* **183**, 295–299.

Senkevich, T.G., Bugert, J.J., Sisler, J.R., Koonin, E.V., Darai, G. and Moss. (1996). Genome sequence of a human tumorigenic poxvirus: prediction of specific host response-evasion genes. *Science* **273**, 813–816.

Spanaus, K.-S., Nadal, D., Pfister, H.-W., Seebach, J., Widmer, U., Frei, K., Gloor, S. and Fontana, A. (1997). CXC and CC chemokines are expressed in the cerebrospinal fluid in bacterial meningitis and mediate chemotactic activity on peripheral blood-derived polymorphonuclear and mononuclear cells *in vitro*. *J. Immunol.* **158**, 1956–1964.

Springer, T.A. (1994). Traffic signals for lymphocyte recirculation and leukocyte emigration: the multistep paradigm. *Cell* **76**, 301–314.

Stafford, S., Li, H., Forsyth, P.A., Ryan, M., Bravo, R. and Alam, R. (1997). Monocyte chemotactic protein-3 (MCP-3)/fibroblast-induced cytokine (FIC) in eosinophilic inflammation of the airways and the inhibitory effects of an anti-MCP-3/FIC antibody. *J. Immunol.* **158**, 4953–4960.

Stellato, C., Collins, P., Ponath, P.D., Soler, D., Newman, W., La Rosa, G., Li, H., White, J., Schweibert, L.M., Bickel, C., Liu, M., Bochner, B.S., Williams, T. and Schleimer, R.P. (1997). Production of the novel C-C chemokine MCP-4 by airway cells and comparison of its biological activity with other C-C chemokines. *J. Clin. Invest.* **99**, 926–936.

Strieter, R.M., Polverini, P.J., Kunkel, S.L., Arenberg, D.A., Burdick, M.D., Kasper, J., Dzuiba, J., Van Damme, J., Walz, A., Marriott, D., Chan, S.-Y., Roczniak, S. and Shanafelt, A.B. (1995). The functional role of the ELR motif in CXC chemokine-mediated angiogenesis. *J. Biol. Chem.* **270**, 27 348–27 357.

Sun, D., Hu, X., Liu, X., Whitaker, J.N. and Walker, W.S. (1997). Expression of chemokine genes in rat glial cells: the effect of myelin basic protein-reactive encephalitogenic T cells. *J. Neurosci. Res.* **48**, 192–200.

Szabo, M.C., Butcher, E.C., McIntyre, B.W., Schall, T.J. and Bacon, K.B. (1997). RANTES stimulation of T lymphocyte adhesion and activation: role for LFA-1 and ICAM-3. *Eur. J. Immunol.* **27**, 1061–1068.

Tani, M., Fuentes-M.E., Peterson, J.W., Trapp, B.D., Durham, S.K., Loy, J.K., Bravo, R., Ransohoff, R.M. and Lira, S. (1996). Neutrophil infiltration, glial reaction and neurological disease in transgenic mice expressing the chemokine N51/KC in oligodendrocytes. *J. Clin. Invest.* **98**, 529–539.

Thompson, H.L., Marshall, C.J. and Saklatvala, J. (1994). Characterization of two different forms of mitogen-activated protein kinase-kinase induced in polymorphonuclear leukocytes following *N*-formylmethionyl-leucyl-phenyl-alanine or granulocyte–macrophage colony-stimulating factor. *J. Biol. Chem.* **269**, 9486–9492.

Tiffany, H.L., Lautens, L.L., Gao, J.-L., Pease, J., Locati, M., Combadiere, C., Modi, W., Bonner, T.I. and Murphy, P.M. (1997). Identification of CCR8: a human monocyte and thymus receptor for the CC chemokine I-309. *J. Exp. Med.* **186**, 165–170.

Turner, L., Ward, S.G. and Westwick, J. (1995). RANTES-activated human T lymhocytes. A role for phosphoinositide 3-kinase. *J. Immunol.* **155**, 2437–2444.

Uguccioni, M., Loetscher, P., Forssmann, U., Dewald, B., Li, H., Lima, S.H., Li, Y., Kreider, B., Garotta, G., Thelen, M. and Baggiolini, M. (1996). Monocyte chemotactic protein 4 (MCP-4), a novel structural and functional analogue of MCP-3 and eotaxin. *J. Exp. Med.* **183**, 2379–2384.

Van Meir, E., Ceska, M., Effenberger, F., Walz, A., Grouzmann, E., Desbaillets, I., Frei, K., Fontana, A. and de Tribolet, N. (1992). Interleukin-8 is produced in neoplastic and infectious disease of the human central nervous system. *Cancer Res.* **52**, 4297.

Van Snick, J., Houssiau, F., Proost, P., Van Damme, J. and Renauld, J. C. (1996). I-309/T cell activating gene-3 chemokine protects murine T cell lymphomas against dexamethasone-induced apoptosis. *J. Immunol.* **157**, 2570–2576.

Wu, D., La Rosa, G.L. and Simon, M.I. (1993). G-protein-coupled signal transduction pathways for interleukin-8. *Science* **261**, 101–104.

Yoshida, T., Imai, T., Takagi, S., Nishimura, M., Ishikawa, I., Yaoi, T. and Yoshie, O. (1996). Structure and expression of two highly related genes encoding SCM-1/human lymphotactin. *FEBS Lett.* **395**, 82–88.

Yoshida, T., Imai, T., Kakizaki, M., Nishimura, M. and Yoshie, O. (1995). Molecular cloning of a novel C or gamma type chemokine, SCM-1. *FEBS Lett.* **360**, 155–159.

Yoshida, R., Imai, T., Hieshima, K., Kusuda, J., Baba, M., Kitaura, M., Nishimura, M., Kakizaki, M., Nomiyama, H. and Yoshie, O. (1997). Molecular cloning of a novel CC chemokine EBI1-ligand chemokine that is a specific functional ligand for EBI1, CCR7. *J. Biol. Chem.* **272**, 13 803–13 809.

Chapter 28

Cytokine and Anti-Cytokine Therapy

Catherine Haworth[1], Ravindir Nath Maini[2] and Marc Feldmann[2]
[1]Department of Haematology, Leicester Royal Infirmary, Leicester and
[2]Kennedy Institute of Rheumatology, London, UK

INTRODUCTION

At the time that cytokines became available for clinical use it was anticipated that there would be a wide variety of diseases in which the aspects of cell biology that are controlled by these factors are disturbed and hence would provide opportunities for cytokine therapy. There was always caution that the difficulties in realizing this potential would mean that therapeutic applications would be slow to materialize and, although there are no examples of the widespread use of cytokines in the management of common clinical problems, there has been a steady but significant increase in use in a wide variety of conditions. This can be expected to increase as a result of better understanding of the pathophysiology of diseases which 10 years ago would have seemed unlikely candidates for therapeutic options that involved targeting cytokines, for example stroke (Rothwell and Relton, 1993), and the extension to the clinic of newer cytokines such as interleukin-12 (IL-12) and IL-11.

Activation at the site of inflammation of both resident cells (e.g. endothelial cells, macrophages and fibroblasts) and migratory cells (e.g. T cells, B cells, neutrophils and monocytes) leads to cytokine production. The final result under optimal circumstances is the limitation of the cause of the inflammation and the repair of any damaged tissue, which is mediated by the action of cytokines on target cells. Activation of target cells results in the production of a cascade of cytokines acting on neighbouring cells and including positive and negative feedback loops. This results in induction of a 'repertoire' of cytokines; however, the cytokine profile may vary depending on the method of initiating the inflammatory response. Within the inflammatory response cytokines do not act in isolation. Their effect depends on:

1. the other cytokines they induce. Although some cytokines (e.g. tumour necrosis factor (TNF) and IL-1) are more powerful inducers of cytokines than others, most if not all cytokines are inducers of other cytokines under some circumstances, and there are probably no truly 'end-stage' cytokines;
2. coexisting cytokines, that may be synergistic or inhibitory;
3. levels of cytokine inhibitors, such as soluble receptors or IL-1 receptor antagonist (IL-1ra);

The Cytokine Handbook, 3rd ed.
ISBN 0-12-689662-3

4. the target cells present, whether or not they express receptors, their physiological state (i.e. active or quiescent) (Gerlach *et al.*, 1989), possibly the orientation of the target cells to the cytokine (Poo *et al.*, 1988) and the mode of activation of the target cells (e.g. interferon-γ inhibits mouse B-cell proliferation induced by soluble but not bound anti-immunoglobulin antibodies (Mond *et al.*, 1985);
5. examples of interspecies variation, and in laboratory animals interstrain variation, suggest that in outbred species individual variation may be important;
6. interaction with non-cytokine mediators (e.g. neuropeptides and eicosanoids), which may also induce and be induced by cytokines.

The multiple biological effects of many cytokines have been a major problem in their widespread introduction to therapy. Thus the haemopoietic growth factors with their relative specificity are the factors that have seen the greatest application. Other cytokines, especially those involved in the acute inflammatory response (e.g. the proinflammatory molecules, TNF and IL-1, the interferons, IL-4 and IL-6) have a wider variety of effects in a number of cell systems. Finer targeting of effects has always been a major goal. The greater understanding of inflammation, growth and repair which has come about because of newer strategies (e.g. transgenic and knockout mice) and the concept of immune deviation have resulted in strategies with potential for greater specificity.

IL-1

IL-1 is a proinflammatory molecule, the major source of which is activated macrophages. Its wide spectrum of activity is summarized in Table 1. It acts on target cells both directly and via the induction of further cytokines (e.g. IL-6, TNF, granulocyte–

Table 1. Immunotherapy with IL-1.

Property	Therapeutic potential	Pathological potential
Induction of haemopoietic growth factors Neutrophil adhesion Neutrophil and monocyte/macrophage activation Induction of IL-6	Protection from infections	
Endogenous pyrogen		
Bone and cartilage lysis		Malignant bone destruction Rheumatoid arthritis
Radioprotection Induction of MnSOD	Radioprotection	
T-cell activation Activation of NK cells	May enhance IL-2 therapy	
Stimulation of fibroblast proliferation		Implicated in fibrosis (?)
Stimulation of division of haemopoietic precursors		(Autocrine) leukaemic cell growth factor
Basophil degranulation		Histamine release

macrophage colony stimulating factor (GM-CSF), M-CSF, G-CSF and also by the induction of further IL-1 (Seelentag *et al.*, 1987; Howells *et al.*, 1988; Sironi *et al.*, 1989). In addition it synergises with TNF in a wide variety of effects. IL-1 has a short half-life which localizes the effects of the molecule adjacent to the site of production, and the use of IL-1 in systemic therapy is limited by its systemic toxicity.

The properties of IL-1 that suggested therapeutic potential were radioprotection, enhancement of haemopoiesis and stimulation of T cells and natural killer (NK) cells. As a non-specific radioprotectant, one therapeutic application considered was protection from massive whole-body irradiation in normal individuals. This strategy has been superseded by the introduction of haemopoietic growth factors with greater activity on stem cells (e.g. stem cell factor (SCF)), by novel approaches to protect cells from induction of apoptosis (e.g. the haemopoietic stem cell inhibitors macrophage inflammatory protein-1α (MIP-1α) and transforming growth factor-β (TGF-β) (which may also protect skin and colon crypt stem cells), and by transfer of genetic resistance (e.g. by the transfection into stem cells of genes that allow protection from specific cytotoxic agents) (Jelinek *et al.*, 1996).

IL-1 is itself implicated in many pathologies and thus it is appropriate to downregulate the molecule. This may be achieved via the soluble receptor (Mullarkey *et al.*, 1993), by the natural inhibitor IL-1ra (Arend *et al.*, 1989) or by antibodies (Catalano, 1993). Thus inhibitors of IL-1 are likely to be of therapeutic potential, for example in downregulating the immune response. Initial studies led to clinical trials in patients with overwhelming sepsis and RA. More recently IL-1 blockade has been shown to be effective in experimental models of stroke (Rothwell and Relton, 1993).

IL-2

IL-2 acts on cells expressing high-affinity IL-2 receptors. The heterogeneity of the IL-2 receptor brought about by the combination of α, β, γ components results in appropriate targeting of the cytokine. Because of its role as a major activator of T and NK cell function, IL-2 has received much attention with regard to its therapeutic potential in conditions where enhancement of T and NK cell responses may be advantageous. These include malignancy, immunodeficiency and chronic infection. This cytokine has one of the longest track records in clinical evaluation and its progress may be used as an example of the 'rites of passage' of cytokine therapy.

The use of IL-2 in trials on refractory malignancy stems from the observations that prolonged stimulation of peripheral lymphocytes by high doses of IL-2 results in the production of a population of cytotoxic cells in which cell killing is restricted neither by major histocompatibility complex (MHC) specificity, as is the case with conventional cytotoxic T cells, or to conventional NK cell targets (Burns *et al.*, 1984). These lymphokine-activated killer (LAK) cells are heterogeneous phenotypically ($CD3^+ CD16^-$ and $CD3^- CD16^+$ cells) and functionally, but amongst them are population(s) that will kill tumour targets. IL-2 induces the target cell population to produce a variety of cytokines including TNF and interferon-γ which may be the agents of the tumour cell killing (Chong *et al.*, 1989). Various therapeutic regimens have been studied, such as administration of LAK cells alone, cytokine alone, the two together in combination and intrasplenic cytokine. It is possible that the differing routes of therapy may generate differing populations of effector cells. The precise mechanism of the IL-2 response remains ill understood (Parmiani, 1990). The main responses have been seen on tumours that have been recognized as susceptible to immunomodulation—melanomas and renal carcinomas.

In pilot studies of IL-2 and LAK cells in a variety of tumours, but mainly melanomas, renal carcinomas and colorectal tumours, 8 of 106 patients had a complete remission, the average duration of which was 10 months. A further 25 patients showed some improvement. In the same study, when IL-2 was used alone in 49 patients, there was one complete remission and a further six patients showed some evidence of response (Rosenberg *et al.*, 1987). These results were updated (Rosenberg *et al.*, 1994), and 15 of 283 patients were found to have complete remission of 7–91 months.

The IL-2–LAK cell therapy schedule as originally studied comprised 5 days of 8-hourly pulses of IL-2, followed by 5 days of leucophoresis to generate LAK cells *in vitro*, and immediately after the last leucophoresis the daily infusion of the *in vitro* generated LAK cells together with further IL-2. The therapy was toxic: 'capillary leak syndrome', cardiovascular damage, renal toxicity and cholestatic jaundice were major problems. Continuous infusion of IL-2 intravenously appears to be less toxic than the original 8-hourly pulses, but it is accepted that the toxicity, low response rate and cost of the supportive care of current regimens means that IL-2 has not yet become standard therapy. However, therapeutic strategies have continued to evolve and currently low-dose low-toxicity regimens are being studied, for example in the improvement of immune responses in patients with acquired immune deficiency syndrome (AIDS).

The critical role of IL-2 has also made it a target for downregulation of the immune response in allograft rejection and autoimmune disorders. The possible mechanisms of downregulating IL-2 effects include antibodies to the IL-2 receptor, and IL-2 conjugated to a toxic module (e.g. IL-2 diphtheria toxin (Bacha *et al.*, 1988) and *Pseudomonas* enterotoxin (IL-2 PE40)). By deleting activated T cells the latter has been used to improve the survival of cardiac allografts in mouse (Lorberboum-Galski *et al.*, 1989) and to improve collagen arthritis in the rat, but results of IL-2 diphtheria toxin in patients with RA have been disappointing (Moreland *et al.*, 1995). Human adult T-cell leukaemia (ATLL) is characterized by the expression of high-affinity IL-2 receptors and in the early stage of the disease the cells are dependent on IL-2 for growth. It is therefore possible that downregulation of IL-2 may have a role in the management of this disease; preliminary studies with anti-CD25 antibodies have shown a response in some patients but other regimens are still the primary management for this condition.

IL-4

As can be seen from Table 2, IL-4 has some but not all of the properties of IL-2 on T cells but not on B cells or macrohages. It is an autocrine growth factor for T cells (Kupper *et al.*, 1987) and appears to play a critical role during thymic ontogeny. IL-4 has been investigated for therapeutic activity in similar situations to IL-2. In the mouse IL-4 is an adjuvant to IL-2 in the induction of LAK cells. Tumours carrying a high expression vector with IL-4 when transplanted into congenic mice are less lethal than tumours not expressing IL-4. Local delivery of IL-4 at the site of experimentally implanted tumours (e.g. by implantation with high IL-4-expressing fibroblasts) results in reduced tumorigenicity and animals that completely reject the initial challenge can reject subsequent challenges. This phenomenon is enhanced by the coadministration of IL-2.

Table 2. Immunotherapy with IL-4.

Property	Therapeutic potential	Pathological potential
Stimulates T_{H2} cells		Allergy
Generates cytotoxic T cells	Generation of tumour-infiltrating lymphocytes	
Costimulates haemopoietic progenitor cells		? Growth factor for leukaemic cells
Stimulates B-cell IgE production		Allergy
Inhibits cytokine production and superoxide production in monocytes		Immunosuppression
Induces B-cell IL-6 production		
Mast-cell growth factor (with IL-3)		? Systemic mastocytosis

IL-4 has been used as a B-cell differentiation agent in lymphoid malignancy. As an inhibitor of the production of proinflammatory cytokines from macrophages it has been used as an anti-inflammatory agent, and it is effective in animal models of arthritis. Clinical trials of IL-4 in patients with RA have been initiated.

The increasing knowledge of the role IL-4 plays in 'immune deviation' adds to its therapeutic potential. It may be used in an attempt to skew the immune response in favour of $CD4^+$ T helper 2 (T_{H2}) cells, either to enhance a T_{H2} response or to suppress a T_{H1} response when this is pathologically relevant. However, initial studies of experimental allergic encephalomyelitis (EAE) in mice, normally considered a T_{H1} disease, have given surprising results, with worsening of the condition if IL-4 was given alone late in the disease process. IL-4 given repeatedly before onset was able to alter the T_{H1}:T_{H2} ratio and was beneficial (Racke *et al.*, 1994).

The profile of the immune response from T_{H1} and T_{H2} cells is now recognized to be markedly dependent on the cytokine milieu, with IL-12 influencing development along the T_{H1} profile and IL-4 influencing development along the T_{H2} profile. Excess IL-4 may be lethal in some infections (e.g. leishmaniasis) (Lehn *et al.*, 1989), which may be due to its critical role in downregulating interferon-γ (Sadick *et al.*, 1990). Because of its role in stimulating immunoglobulin E (IgE) production (Morris *et al.*, 1993), IL-4 is important in the response to parasitic infections, as has been demonstrated in the mouse against *Nippostrongylus brasiliensis*.

IL-6

The properties of IL-6 are summarized in Table 3. Many cells are capable of both producing and responding to IL-6 and therefore it is capable of being an autocrine regulator of growth and/or differentiation in many systems. Within the immune system it has been shown to be an (autocrine) activator of peripheral T and NK cells, which in part is mediated via IL-2 (Garman *et al.*, 1987). In thymic ontogeny IL-6 may be important alongside IL-2, IL-4 and IL-7 in thymic development. It also stimulates function, but not growth in normal activated B cells, inducing IgG secretion. In addition it is an important mediator of the acute phase response, inducing the liver to produce

Table 3. Immunotherapy with IL-6.

Property	Therapeutic potential	Pathological potential
Synergizes with other haemopoietic growth factors for differentiation	Induces differentiation of some leukaemic cells	
Endogenous pyrogen Stimulates acute-phase response Induces IgG production in activated B cells	Septic shock	? Hyperimmunoglobulinaemia in HIV Associated with autoimmunity Growth factor for myeloma
Induces IL-2R in T cells Induces cytotoxic T cells	Antitumour in combination with IL-2	

acute-phase proteins such as C-reactive protein (CRP) and inhibiting the production of albumin (Morrone *et al.*, 1988). Increased production of IL-6 secondary to some tumours (e.g. cardiac myxomas and cervical carcinomas) has been implicated in causing autoimmune phenomena in these diseases (e.g. rheumatoid factor and anti-nuclear factor production) (Hirano *et al.*, 1988). High levels have also been observed in other autoimmune conditions, such as the synovial fluid of RA. As an important mediator of the acute inflammatory response IL-6 has been sought in other conditions and demonstrated, for example, in high levels in psoriatic skin (Grossman *et al.*, 1989).

Because IL-6 has been implicated in many cell–cell interactions, it is natural that it should be considered either as the pathological agent or a therapeutic modality for many disease states. It has been considered as a potential immune stimulant in many situations. More recently experimental studies have suggested that some malignant cell lines transfected with the IL-6 genes together with the cytosine deaminase gene can be used as an autologous tumour vaccine to enhance the survival of animals with metastases from wild-type tumour (Mullen *et al.*, 1996). In clinical trials recombinant human IL-6 given daily with relatively minor toxic complications had little impact on metastatic renal cell carcinoma; two of 40 assessable patients had a partial remission (Stouthard *et al.*, 1996). A further phase I–II study also failed to demonstrate any antitumour effect in advanced malignant disease at doses of IL-6 given to the limits of acceptable toxicity (Weber *et al.*, 1994).

However, IL-6 continues to be assessed for its role in stimulating haemopoiesis. Following bone marrow transplantation, IL-6 enhances the haemopoietic constitution, particularly at low cell numbers, but enhances graft-*versus*-host disease (GVHD). If this phenomenon can be harnessed to enhance the graft *versus* leukaemia effect without increasing GVHD to clinically unacceptable levels then this may be of therapeutic significance. The thrombopoietic effect of IL-6 was the focus of much clinical study particularly before the discovery of thrombopoietin (TPO, megakaryocyte colony stimulating factor) and is still studied in this clinical setting (D'Hondt *et al.*, 1995; Lazarus *et al.*, 1995; Nagler *et al.*, 1995).

The role of IL-6 in downregulating the immune response has received attention in animal studies as, when given prophylactically, it reduces toxicity in models of septic shock (Barton *et al.*, 1996), neonatal B streptococcal disease (Mancuso *et al.*, 1994) and multiple sclerosis (Rodriguez *et al.*, 1994). However, this role in prophylaxis will not have any therapeutic application unless the downstream events are better understood and open to manipulation.

IL-6 therapy is limited by pyrexia, fall in serum albumin levels and influenza-like symptoms; plasma cell proliferation has been documented occasionally (Csaki *et al.*, 1996), but autoimmune phenomena have not been seen at the doses limited by other toxicities. Neutralizing IL-6 activity in multiple myeloma, where it had been implicated as a growth factor, has been disappointing and has not proceeded to clinical use.

IL-10

IL-10 a cytokine with chiefly immunosuppressive properties, both anti-immune and anti-inflammatory. From the pathology of the IL-10 knockout mice it has been shown to be a major endogenous regulator of the immune response in gut and skin, and is also important in rheumatoid joints. Exogenous IL-10 is effective in animal models of autoimmune diseases including arthritis and therefore has been evaluated in clinical trials. The most extensive of these has been in Crohn's disease. Repeated (daily) injections given for 1 month led to a high frequency and degree of response (Van Deventer *et al.*, 1997). IL-10 has also successfully completed a phase 1 study in RA (Maini *et al.*, unpublished observations).

IL-11

IL-11 is a member of the 'IL-6 family' as it uses gp130 for signal transduction. Its best known effects are on the generation of platelets, but it also has anti-inflammatory activities and direct effects on the gastrointestinal tract.

Daily injection of IL-11 increases the platelet count and megakaryocyte size and ploidy, and after combination chemotherapy and radiotherapy led to significantly increased survival and rapid recovery of small intestinal mucosa (Du *et al.*, 1994). These studies have led to clinical trials of IL-11 in patients receiving chemotherapy, including those who have thrombocytopenia. These have shown that IL-11 can prevent thrombocytopenia if given before and during chemotherapy, and can be used to reduce the need for platelet transfusion after chemotherapy (Gordon *et al.*, 1996; Tepler *et al.*, 1996). Results of this type have led to the product being licensed for this application.

The effectiveness of IL-11 in models of inflammatory bowel disease, for example in B27 transgenic rats (Keith *et al.*, 1994), has led to clinical trials in Crohn's disease. Phase I has been completed successfully.

IL-12

IL-12 is a heterodimeric cytokine produced by phagocytic cells and professional antigen-presenting cells, B cells and mast cells in response to infection. It has an important role in immunoregulatory function, being responsible for the induction of interferon-γ in T and NK cells, and inducing T_{H1} cells and consequently inhibiting the induction of T_{H2} cells. This strategic role has led to the relatively rapid progress from discovery (1989) to phase I–II clinical trials (1994). Potential therapeutic applications have been based around its enhancement of protection against mycobacterial (Kobayashi *et al.*, 1995), protozoal (Urban *et al.*, 1996) and fungal (Kawakami *et al.*, 1996) infections, its antiangiogenesis properties, and its ability to block T_{H2} responses, which has been studied in animal models of asthma (Gavett *et al.*, 1995). The antitumour effects of IL-12 have been studied extensively. In part this may be due to the antiangiogenic effect of interferon-γ, but is also likely to be due to other immunomodulatory effects. When induced by IL-12-transfected cells the antitumour response occurs distant to the site of IL-12 production, occurs to wild-type as well as IL-12-transfected tumour cells, may persist to protect in rechallenge experiments and may synergize with IL-2. Hence there is considerable interest in its role in anticancer therapy.

Paradoxically, in a mouse model of GVHD, IL-12 is protective. The mechanism is complicated and not fully understood, being associated with numerous variations in subpopulations of donor T cells post-transplant (Sykes *et al.*, 1995). The observation that an anti-GVHD effect may be obtained without the loss of anti-graft *versus* leukaemia effect will undoubtedly lead to further study in this field.

Toxicity has been a significant problem, including hepatic and renal dysfunction, and unexplained cardiotoxicity occasionally resulting in death. Dose modification including dose reduction and predosing ameliorate these problems to some extent, and progress has continued despite initial setbacks (Brunda and Gately, 1995).

Anti-IL-12 has been effective in various animal models of chronic inflammation, for example diabetes and colitis (Neurath *et al.*, 1995; Trembleau *et al.*, 1995). IL-12 knockout mice develop a much milder form of arthritis following collagen injection (McIntyre *et al.*, 1996).

TNF-α

TNF is produced by a wide variety of cells (e.g. T and B lymphocytes and macrophages), and has been demonstrated to play important roles in many physiological and pathological states. Although lymphotoxin (TNF-β) has similar properties, TNF-α is the only one of the two that has been available in large amounts for experimental and therapeutic assessment. Therefore, most of this section is devoted to the potential use of TNF-α.

TNF is one of the major proinflammatory cytokines, having major direct and indirect effects via further cytokine induction, especially of IL-1 and IL-6, in many systems (Dinarello *et al.*, 1986). It thus plays a pivotal role in many pathologies and similarly its blockade is of potential benefit in many disorders.

TNF has many effects on cell growth and differentiation. Its currently accepted terminology is based on its property for inducing haemorrhagic tumour necrosis in mouse sarcomas *in vivo*. The tumour necrosis is now realized to be due to altered endothelial cell function and not to directly toxic effects. The preliminary observation of tumour necrosis led to numerous experimental and clinical studies of the effect of TNF on tumours, with a view to developing it as a therapeutic agent. TNF has growth inhibiting effects and growth stimulatory effects on a variety of cell types. In normal cells, TNF is directly growth inhibitory on haemopoietic precursors (Broxmeyer *et al.*, 1986) but is stimulatory on a variety of lymphoid cells (Ranges *et al.*, 1988). In addition it induces a wide variety of target cells including fibroblasts, endothelial cells and macrophages to produce further cytokines which may have properties that oppose its direct effect. For example, it induces the synthesis of many of the haemopoietic growth factors that are stimulatory for those target cells on which it is directly inhibitory (Koeffler *et al.*, 1987).

The picture is made more complicated as TNF may stimulate growth in malignant cells but be inhibitory in the normal counterpart. For example, myeloid leukaemic cells may be stimulated by TNF whereas the normal counterparts are growth inhibited (Hoang *et al.*, 1989). TNF resistance of tumours emerges rapidly both *in vivo* and *in vitro*. In addition the effect of TNF on some target cells may vary depending on the current state of the targets. It has been shown to have a greater effect on endothelium that is growing actively than on confluent endothelium (Gerlach *et al.*, 1989). With these difficulties in predicting cell growth in response to TNF, it is not surprising that initial trials of TNF in cancer therapy have had disappointing results. In one study (Selby *et al.*, 1987) 3 of 18 patients had a short-lived response. Toxicity problems included rigors and hypotension.

To overcome some of the systemic toxicity associated with TNF therapy, attempts have been made to attain high levels of TNF locally, for example by limb perfusion (Eggimann *et al.*, 1995), isolated lung perfusion (Pass *et al.*, 1996), endoscopically for gastric carcinoma and intravesically for bladder tumours (Karcz *et al.*, 1994; Glazier *et al.*, 1995). Limb perfusion has resulted in high systemic levels as a result of vascular leak. Responses to TNF perfusion alone have been disappointing but combination with other biological response modifiers (e.g. interferons and chemotherapy) has been more successful (Lejeune *et al.*, 1994).

TNF plays a pivotal role in the acute inflammatory response both by inducing further cytokines and by direct action on cells, such as stimulation of human leukocyte antigen (HLA) class II expression, neutrophil and macrophage activation, and killing virally infected cells. Inhibition of its production leads to inadequate resolution of the acute response, which may be fatal. In mice, differential ability to produce TNF in different strains of mice has led to pathology that may be ameliorated by TNF therapy (e.g. cutaneous leishmaniasis in BALB/c mice).

TGF-β FAMILY

The TGF-β superfamily of genes includes hormones involved in the pituitary–gonadal axis, activin and inhibin, and bone morphogenic proteins. In spite of the name the TGF-βs (1, 2, 3) are one of the few well-defined growth inhibitory cytokines. TGF-β has some growth stimulatory properties (see Table 4), for example for fibroblasts possibly via induction of platelet-derived growth factor (PDGF) or PDGF receptor. Interest in TGF-β was initially focused on diseases where TGF-β-induced fibrosis was important (e.g. pulmonary fibrosis). Recent developments have also been concerned with using the growth inhibitory properties of TGF-β in the protection of clonogenic cells from cytotoxic damage. Stem cells that are inhibited by TGF-β include bone marrow, skin and gastrointestinal tract crypt cells, and therapy is easiest where the TGF-β can be targeted locally to cells (e.g. mucosa) (Sonis *et al.*, 1994) and may lead to further strategies to protect some stem cells populations from chemotherapy.

The bone morphogenic proteins are members of the TGF-β superfamily involved in the embryonic induction of chondroblasts and osteoblasts. They can induce new bone formation *in vivo* and *in vitro*, and as such have been studied in a variety of experimental

Table 4. Immunotherapy with TGF-β.

Property	Therapeutic potential*	Pathological potential
Inhibits activation of T cells	Inhibits graft rejection	
Stimulates fibroblast proliferation and collagen deposition		Kidney pulmonary and retinal fibrosis, etc.
Inhibits cell division in some target cells	Cancer	Loss of responsiveness may accompany malignant change
Induces bone resorption		
Induces integrins		

*All theoretical therapeutic applications are limited by toxicity.

pathological situations including bone defects in a variety of animals and more recently in cases of bone non-union in humans (Kirker-Head, 1995; Wozney, 1995; Riley *et al.*, 1996). They can be applied locally on a variety of carriers. Whether they will have any role in systemic bone disease (e.g. osteoporosis) has yet to be determined (Prestwood *et al.*, 1995).

Activin and inhibin have indirect effects on erythropoiesis. In some studies activin appeared to induce higher levels of haemoglobin F and so may have potential benefits in disorders of defective β-chain production (e.g. sickle cell anaemia or thalassaemia). However, clinical studies have not supported the initial *in vitro* studies and this is no longer considered a potential therapeutic option.

INTERFERONS

It is now over 40 years since interferon was first recognized as an antiviral agent, but it is only during the past 10 years that the interferons have become established therapeutic agents with an ever increasing role. The type 1 interferons, acting through their common receptor, have growth inhibitory effects on a wide variety of target cells, including normal haemopoietic cells, and numerous malignant cell lines; in addition they are associated with the induction of higher levels of MHC class I antigens on some target cells. Interferon-γ acts through its own specific receptor. It has weaker antiviral properties than the type 1 interferons. Its growth inhibitory properties are less than those of class 1 interferons on many target cells, with the exception of haemopoietic precursors. It differs in several properties, including the ability to induce MHC class II antigens on some target cells and the fact that it is a major activator of macrophage function (Boraschi *et al.*, 1984; Groenewegen *et al.*, 1986).

The antiviral role of the interferons is only one of the many properties that have been exploited for therapeutic use; others include growth inhibitory properties which have been studied in malignant disease, macrophage activation in chronic intracytoplasmic infections, and enzyme induction as an adjuvant to anticancer chemotherapy. In spite of the long period of study of the interferons both in the laboratory and the clinic, new observations continue to be made that confound previously accepted beliefs: the difference in activity of the type 1 interferons (α and β), for example, suggests that even though they bind to a common receptor downstream events are significantly different and can result in different clinical outcomes, for example in relapsing or remitting multiple sclerosis.

The anticancer effects of the interferons may be mediated by one or more of the above-known properties of the interferons. The clinically responsive tumours have not been those predicted from animal studies. Interferon-α is now recognized as effective therapy for hairy cell leukaemia (overall response rate 80%, with 10–15% complete remissions) (Quesada *et al.*, 1984). The recent availability of effective cytotoxic drugs (e.g. 2-chloro-2-deoxyadenosine), combined with significant toxicity with systemic influenza-like symptoms and myelosuppression (Saven and Piro, 1993), means that this role is under review. Since 1993 the role of interferon-α in Philadelphia chromosome positive (Ph^+) chronic myeloid leukaemia (CML) has been consolidated. Several studies have demonstrated improved survival and reduction in the incidence of blast transformation in patients treated with interferon-α. In some studies this has been confined to patients showing a significant reduction in the incidence of Ph^+ cells, higher in the interferon-treated than in the chemotherapy-treated group (Ohnishi *et al.*, 1995), whereas in others the improvement has been more general (Allan *et al.*, 1995). It is interesting to note that in patients treated with interferon-α, blast transformation to acute disease is associated with a higher incidence of lymphoid blast crises, suggesting that the interferon therapy may inhibit myeloid cells more than lymphoid cells. Interferon has been widely studied in multiple myeloma following initial case reports that it induced remission as a single agent. However, there is still no evidence that it has a role to play either in induction as a single agent or in combination with other agents (Ahre *et al.*, 1984) in the maintenance of remission following conventional chemotherapy (Cooper *et al.*, 1993); although it prolonged the duration of response, there was no evidence for increased survival (Belch *et al.*, 1988; Peest *et al.*, 1995). However, meta-analysis supports a significant impact on prolongation of plateau phase (Joshua *et al.*, 1997). It also seems to improve survival following high-dose chemotherapy with stem cell rescue (Powles *et al.*, 1995). In solid tumours, interferon has been evaluated and some positive responses noted in tumours found to be sensitive to immunomodulation (e.g. renal carcinoma and melanoma) (Kirkwood, 1991).

Interferon-γ has received most attention for its role in phagocytic cell activation. It is now the treatment of choice in a rare group of genetically variable inherited disorders of neutrophil function associated with defective intracellular bacterial killing—chronic granulomatous disease (Ezekowitz, 1991a,b). Interferon-γ therapy is associated with induction of oxidative enzymes within the neutrophils. A dose of 0.01–0.05 mg/m^2 subcutaneously has been shown to be effective in the management of all the genetic variants of this disease, reconstituting defective bactericidal activity in monocytes and neutrophils, although within each group some patients have failed to respond. Similarly it has been tested in osteopetrosis where failure of bone remodelling leads to growth abnormalities, nerve entrapment syndromes including deafness and blindness, and bone marrow dysfunction. By enhancing activity of osteoclasts, interferon-γ may have some role in the management of this condition.

The antiviral properties of the type 1 interferons have been studied widely and the indications for their use are becoming more clarified. They have no role in acute viral infections but their effectiveness in the management of persistent viral infections is becoming established. The hepatitis viruses are the most widely studied because of their high incidence worldwide and their mortality rate in relation to the development of cirrhosis and hepatoma. In chronic hepatitis B infections a 12-week course of lymphoblastoid interferon thrice weekly has shown response in 48% of patients, as assessed by conversion to core antibody positivity (Scully *et al.*, 1987). This has been associated with improvement of liver function and histology. Patients who acquired hepatitis B neonatally do not appear to respond to interferon therapy. Following from the therapeutic efficacy of interferon in hepatitis B management, its use in hepatitis C has been widely studied, and again therapy has resulted in temporary clearance of hepatitis C virus in approximately 50% of patients and apparently permanent cure in about half of the responders. Patients least likely to respond fall into predictable groups including hepatitis subtype (type 1b, where a non-structural protein is associated with interferon non-responsiveness (Enomoto *et al.*, 1996), patients with high hepatic iron concentrations and those who are immunologically compromised including patients with haemophilia (Olynyk *et al.*, 1995; Hanley *et al.*, 1996). However, it is probably the initial response to treatment in terms of clearance of virus that is the best predictor of outcome (Booth *et al.*, 1995).

Treatment of condylomata acuminata by local injection three times per week led to a higher rate of complete clearance of the lesion and smaller average lesion size than in untreated controls (Eron *et al.*, 1986). Laryngeal papillomas have shown an initial high response to intravenous interferon-α, but this is only short term and the papillomas progress even though treatment is continued (Healy *et al.*, 1988). Kaposi's sarcoma, which is suspected to be virally induced, may occur sporadically or be associated with human immunodeficiency virus (HIV) infection; in the latter group six of 20 patients have been reported to respond to interferon (Groopman *et al.*, 1984).

One of the major questions being assessed currently is the role of interferon-β in the management of relapsing–remitting multiple sclerosis. This form of the disease accounts for approximately 40% of cases and results have continued to accumulate since the IFN-β Multiple Sclerosis Study Group first published the results of a 3-year study in 1993. In this study patients receiving 8×10^6 units thrice weekly had an improved relapse rate compared with placebo-treated controls. This was paralleled by a reduced burden of disease detected by magnetic resonance imaging. However, progression of disability has not been improved and significant side-effects, particularly depression including suicide attempts, have been important. Immunogenicity of interferon-β may be important, with 45% of patients developing significant antibody response in one series. It will become clear in the short term what role, if any, interferon-β plays in the management of this condition. These studies have also highlighted the differences between the interferons in this condition, where interferon-γ has caused deterioration, interferon-α has no effect but interferon-β has produced some changes indicating improvement.

Because a great deal of information has accumulated on interferons as single agents, it has been possible to study them in combination with many agents. Studying the properties of the interferons, however, shows that there are many theoretical interactions between interferons and other cytokines that may enhance therapeutic effectiveness. In the previous edition of this book, the synergism of interferon-γ and TNF was highlighted as a potentially useful combination. A phase I trial studied 200 mg/m^2 of interferon-γ in combination with variable doses of TNF-α, both as 24-h infusions overlapping by 12 h. The maximum tolerated dose of TNF was 205 mg/m^2 with indomethacin cover. Of 36 patients with a variety of tumours only two of six patients with sarcoma showed evidence of response (Demetri *et al.*, 1989). The therapeutic effects of intratumoral injections of interferons α and γ were potentiated by TNF in a dose-dependent manner in human breast cancer xenografts into nude mice, where complete tumour regression occurred in the injected lesion. However, there was no widespread effect as shown by lack of response in (non-injected) contralateral tumours.

Interferon–drug therapy combinations have also been studied. This has mainly been around two interferon properties. First, combinations of interferon-α and antiviral drugs have been studied for combined antiviral effect, as for example in ATLL where zivuridine and interferon-α in combination have shown encouraging results (Gill *et al.*, 1995). The second synergizing property that has been studied in the clinic has been the ability of interferon-α to modulate the antitumour effect of 5-fluorouracil (5-FU) via the induction of thymidine phosphorylase. In one study of advanced colorectal carcinoma there was no obvious advantage over 5-FU alone (Ferguson *et al.*, 1995). *In vivo* experiments in nude mice (Schwartz *et al.*, 1995) have suggested that the synergism is optimal when the thymidine phosphorylase is not rate limited by cosubstrate availability and that a non-cytotoxic pyrimidine analogue can enhance the effect of the interferon–5-FU combination.

ANTI-CYTOKINE THERAPY

There have been impediments to the rapid development of anti-cytokine therapy. With the major involvement of cytokines in host protective pathways, especially immunity and inflammation, anti-cytokine therapy could be detrimental. Furthermore, the considerable overlaps in cytokine function, often termed ‘redundancy’, has made it difficult to define the optimal cytokines to block for any given indication. Hence there have been numerous publications suggesting that cytokines are poor targets for therapy. The early results of anti-TNF-α antibody in sepsis appeared to substantiate this view.

Nevertheless, anti-TNF-α therapy has been doing well since 1992 in RA, and since 1994 in Crohn’s disease. These results have been reproducible with four different biological agents, and have changed the perception of anti-cytokine therapy, and of the use of biological agents. Relatively little is yet known about blockade of IL-1, IL-6 or other cytokines in humans.

Rationale for Anti-TNF-α Therapy in Rheumatoid Arthritis

As cytokines were cloned in the 1980s, there was an opportunity to document cytokine gene expression in accessible human autoimmune–inflammatory disease sites. The reasons for doing so have been amply reviewed (Feldmann *et al.*, 1996, 1997). The rheumatoid joint provided a good opportunity to document cytokine expression at the height of the disease process, not possible in less accessible diseases, such as insulin-dependent diabetes or Graves’ disease.

The results were interesting: virtually all cytokines assessable are present and produced locally, as judged by mRNA expression. Furthermore the degree of expression was not dependent on disease duration or the type or intensity of therapy, even with drugs able to inhibit cytokine synthesis such as corticosteroids. These results suggested that, in a chronic disease site, cytokine production was persistent and not transient as in normal immune or inflammatory states induced by mitogens or antigens. This hypothesis was tested by culturing dissociated RA synovial joint cell mixtures in the absence of extrinsic stimulation. Prolonged and high-level cytokine synthesis was detected, and these cultures provided a system for attempting to unravel the abnormally upregulated cytokine production (Feldmann *et al.*, 1996).

At the time (late 1980s) there was considerable belief that IL-1 was of major importance in arthritis, owing to its proinflammatory activities and capacity to induce the destruction of cartilage and bone in model systems. Hence with our new 'model system' of cultured human synovial joint cell mixtures obtained from the rheumatoid synovium, we investigated the regulation of IL-1. As it is bioactive IL-1 that can induce biological activity, this was measured by bioassay. As a tool for inhibiting signals, antibodies were used. In view of the reports from Dinarello's and Cerami's groups that TNF-α was a powerful inducer of IL-1 synthesis, antibodies to TNF were used: anti-TNF-α had a dramatic effect on IL-1 production within 3 days, whereas anti-TNF-β (also known as anti-lymphotoxin-α (anti-LT-α) was ineffective (Brennan *et al.*, 1989). This suggested that the proinflammatory cytokines were linked together and led to studies of what other proinflammatory cytokines were also regulated by TNF-α. These included GM-CSF (Haworth *et al.*, 1991), and more recently IL-6 and IL-8 (Butler *et al.*, 1995).

While these *in vitro* studies provided our first understanding of the 'cytokine cascade', and suggested that TNF-α was a therapeutic target, it was necessary to perform some 'reality testing' before the performance of clinical trials of anti-TNF therapy. Two approaches were used. The first was *in situ* analysis, using tissues frozen within moments of biopsy. This excludes the possibility that the cytokine expression detected was induced *in vitro*, during tissue processing. Abundant TNF and TNF receptors are expressed *in vivo* (Chu *et al.*, 1991). The second approach to verify that TNF-α was a therapeutic target was to analyse the effects of anti-TNF-α antibody on a mouse model of arthritis, collagen type II-induced arthritis. This ameliorated arthritis, even if used after disease onset, provided biweekly doses of 10 mg/kg of hamster anti-immune TNF-α TN3.19.2 were used (Williams *et al.*, 1992).

Clinical Trials

Based on the above rationale, clinical trials of anti-TNF-α in RA were initiated in 1992, using the cA2 chimaeric (mouse Fv, human IgG_1, K) high-affinity neutralizing antibody. Based on previous work on human volunteers and patients with sepsis, it was possible to use relatively high doses of anti-TNF-α (10 mg/kg, twice) in the first clinical trial, which had dramatic results and, despite being an open trial and hence subject to potential bias, appeared to substantiate that TNF-α was a good therapeutic target. This was because, as well as a marked and rapid onset of clinical benefit, there was a rapid reduction in the cytokine-dependent acute-phase protein, CRP (Elliott *et al.*, 1993).

The duration of clinical benefit after 'pulse therapy' with 20 mg/kg over 2 weeks was variable, ranging from 8 to 26 weeks before all the long-standing patients with active RA relapsed. To evaluate whether these results were clinically important, it was necessary to ascertain whether clinical benefit could be reinduced on multiple occasions. After relapse, eight patients from this trial were reinjected with 10 mg/kg up to three times and it was found that on each occasion the relapse was treatable. The importance of this result is that it indicates that, if TNF is blocked, other proinflammatory cytokines do not emerge to replace its function. Thus the mechanism of the disease does not alter (Elliott *et al.*, 1994a).

The most rigourous clinical trial, necessary to prove efficacy, is the randomized blinded placebo-controlled trial. This was performed, with a high and low dose of cA2. Both doses were significantly better than placebo, with 79% and 48% meeting the preset efficacy response criteria (20% Paulus response) at 4 weeks, compared with 12% of the placebo group (Elliott *et al.*, 1994b).

These results have led other groups that had produced anti-TNF-α agents for use in sepsis to initiate trials in RA. These have now been reported. A humanized antibody CDP571, developed by Celltech, has been shown to be effective, also administered by intravenous infusion (Rankin *et al.*, 1995). A p75 TNF receptor IgG Fc fusion protein developed by Immunex has been shown to be effective if injected biweekly subcutaneously for up to 3 months (Moreland *et al.*, 1997). A p55 TNF receptor IgG Fc fusion protein is being developed by Roche. A variety of protocols for its use was recently reported, with varying degrees of efficacy (Cutolo *et al.*, 1996).

The mechanism of anti-TNF-α therapy has been investigated, and is discussed in detail elsewhere. Important comments include downregulation of the proinflammatory cascade *in vivo*, and diminution of leukocyte trafficking to joints (Paleolog *et al.*, 1996). In view of the success of anti-TNF-α antibody therapy in RA it has been tried in other putatively TNF-α-dependent diseases. Multiple trials in Crohn's disease have also been reported to be successful. Clinical effects of anti-TNF-α have recently been reviewed in detail (Feldmann *et al.*, 1997).

Anti-TNF-α antibodies were first used in sepsis, and the results so far of TNF blockade are not conclusive. In some instances there is a trend towards benefit, but in others there is no clear benefit, and in one example there was even increased mortality. Some groups are persevering, and the results of a 1900-patient phase III trial with murine anti-TNF-α should clarify whether there are prospects for TNF-α in sepsis.

Anti-IL-1 Therapy

IL-1 has two active forms, IL-1α and IL-1β, the latter being more abundant. There have been no reports of the clinical use of anti-IL-1 antibodies, probably because two antibodies may need to be used for clinical benefit. However, the IL-1ra, which blocks the binding of both IL-1β and α to the stimulatory type I IL-1 receptor, has been used in sepsis and RA.

The results in sepsis, while initially encouraging, did not pan out in larger trials. More recently IL-1ra has been used in RA. The results so far obtained have been difficult to interpret. The anti-inflammatory effect of multiple regimens has not been dramatic (Bresnihan *et al.*, 1996). However, a large trial of IL-1ra in patients with relatively early RA over 6 months again revealed a minimal anti-inflammatory effect but evaluation of joint destruction by radiography revealed an improvement that was just statistically significant. These results are encouraging, but need to be confirmed (Watt and Cobby, 1996).

Anti-IL-6 Therapy

There have been reports of anti-IL-6 therapy in RA. A murine neutralizing anti-human IL-6 antibody has been used in a small number of patients (Wendling *et al.*, 1993). Short-term improvement was noted. Anti-IL-6 receptor antibody has been reported by T. Kishimoto (personal communication). A dose of 50 μg per week yielded incremental benefit over several weeks. Onset of benefit was slower than with anti-TNF-α but after about 4 weeks the same degree of benefit was noted.

FUTURE DEVELOPMENTS

Previously we have suggested that the more appropriate application of cytokines to therapy would require the better prediction of patient responses to individual cytokines, and the search for maximum tolerated doses. Phase I–II studies have been accomplished and the agents tested as single agents in the major disease categories where therapy is likely to be of benefit. It can been seen that there have been few changes to conventional first-line therapy over the past 10 years because of cytokine availability. However, the lessons from interferons and IL-2 suggest that the appropriate use of cytokines in therapy evolves over many years with refinement of indications and dosage regimens.

What recent advances are going to enable this?

Extension of Studies to Combination Therapy

With greater knowledge of the effects of single cytokines it will be possible to study the effects of combinations of cytokines:

1. with other cytokines: Cytokine combinations could be chosen in which the desired properties were shared but the undesired effects (therapeutic side-effects) were not. In addition cytokines may be used sequentially to maximize the effect (e.g. by induction of receptor molecules or adhesion molecules);
2. with chemotherapy;
3. with inhibition of unwanted coinduced cytokines (e.g. by soluble receptors, inhibitors or monoclonal antibodies).

The logistics of evaluating these combinations are formidable but, now that initial studies have been assessed, combination therapy is more feasible.

Drug Delivery

Because of the multisystemic effects of most cytokines, it was predicted that physical localization may prove to be the most effective therapy where the situation allowed. It has already been stated that topical therapy has been tested for buccal mucosa (TGF-β), bladder and stomach (TNF), bone lesions (TGF-β, bone morphogenic protein (BMP)), and localized limb, liver and lung perfusion (TNF and IL-2, both as single agents and in combination with systemic chemotherapy).

Cytokine therapy to enhance the immune response to malignancy has been one of the main therapeutic objectives. The nature of the immune response in malignancy and an understanding of its apparent failure have led to optimism for enhancing the immunogenicity of tumour cells. In animal experiments tumour cells transfected with cytokine genes (IL-2, IL-4, IL-5, IL-6, IL-7, IL-10, IL-12, GM-CSF, TNF, interferon-γ) have led to tumour rejection associated with a variety of inflammatory cell infiltrates according to the nature of the cytokine, implying that a variety of immune response can bring about this endpoint. Similarly non-tumour cells (e.g. fibroblasts) stably transfected with cytokine genes can also induce tumour regression. These models have little relevance to the clinical situation where primary tumours are resectable and the clinical problem is one of management of metastatic disease. Animal models, whereby transfected tumour cells (IL-2, IL-4, GM-CSF, interferon-γ) can induce regression of established tumour at distant sites, are more appropriate models for established metastases, and models that are associated with long-term immmunological memory are more appropriate to the clinical situation (IL-2, IL-12).

Some clinical studies have already been started, for example with IL-2-transfected melanoma cells in patients with advanced metastatic disease, but the development of effective immunity is outweighed by tumour progression in this situation. Reducing the time to effective therapy may be achieved by reducing or eliminating the *in vitro* culture period, for instance by *in vivo* transfection of genes targeted specifically to tumour cells via antibodies or receptors, as may the introduction of therapy when tumour burden is low (e.g. as adjuvant therapy for minimal residual disease).

Immune Deviation

It is now accepted that T-cell populations can be classified into three groups on the basis of the profile of cytokine production. T_{H1} cells producing predominantly IL-2 and interferon-γ, T_{H2} producing predominantly IL-4, IL-5 and IL-10, and T_{H0} cells which have a wider cytokine profile not restricted along the above lines. It has long been recognized that certain pathological states are the result of either overproduction of type 1 (autoimmune diseases such as RA, allergic encephalitis and contact hypersensitivity) or type 2 cytokines (atopic disease such as asthma and antibody-related autoimmune diseases) or underproduction (e.g. failure of type 1 response in established intracellular infections such as leishmaniasis or type 2 in extracellular parasitic infections). It is now recognized that the cytokine milieu at the time of antigenic challenge is a contributory factor to the subsequent T-cell differentiation and that IL-12 and IL-4 are pivotal in the type 1 and type 2 differentiation routes respectively. This has led to modification of the immune response in experimental situations (e.g. IL-4 administration at the time passive transfer of autoimmune T cells in experiments, as in EAE (Racke *et al.*, 1994).

There is still considerable debate as to whether established T-cell responses can be altered on a sufficiently long-term basis for this strategy to be effective therapeutically in the management of some of the above diseases. The results of cytokine and cytokine–antigen combinations study aimed at this objective are eagerly awaited.

Apoptosis

Over the past 5 years the importance of apoptosis has become established and the interrelationships of cytokines, oncoproteins, antioncoproteins and the components of the signalling system(s) involved in the induction and suppression of apoptosis are becoming understood. At the biochemical level it is possible to distinguish between growth promoting and cell survival properties of some cytokines (e.g. IL-3 and GM-CSF). The complexity of the balance between proapoptotic and antiapoptotic signals is now appreciated and with fuller understanding of the biochemistry it may be feasible in the not too distant future to use cytokines more effectively, possibly in conjunction with modulators of the downstream pathways to augment cell death or survival.

Reduction of Incidence of Anti-Cytokine Antibodies

Anti-cytokine antibodies are observed in long-term therapy, as seen in the interferon-β study in multiple sclerosis (MS). It is less certain whether these are clinically relevant, but recombinant proteins with lower antigenicity may be more useful.

AIDS

As one of the major health problems of the decade, it is perhaps important to address separately the question of the use of cytokine in the management of AIDS. Previously we have considered that the evidence pointed away from the likely success of cytokine therapy in AIDS because of the likelihood that potentially beneficial cytokines would also stimulate viral replication. The current strategy for management includes a multi-disciplinary approach aimed at reducing viral load, correcting T-cell deficiency (now known to be T_{H1} deficiency) and effective management of infections. Cytokines may have a role in the management of the T-cell deficiency. Low-dose non-toxic schedules of IL-2 therapy (Bernstein *et al.*, 1995), which do not increase the viral load but which enhance delayed-type hypersensitivity response and increase $CD4^+$ T cells, have been studied. The important role of IL-12 in T_{H1} cell production means that this cytokine too is being assessed as a potentially important therapy. Recently the recognition that members of the CC family of chemokine receptors act as coreceptors for some groups of HIV (Cocchi *et al.*, 1995) will allow for the study of strategies for blocking these receptors, possibly using natural or modified ligands (e.g. MIP-1α, MIP-1β or RANTES). The relatively recently recognized molecule controversially identified as IL-16 binds to CD4 and may be similarly exploited to block viral entry into cells.

CONCLUDING REMARKS

It is now more than 15 years since cytokines became available. There have been some conditions where they have become standard therapy, but outside the fields of the haemopoietic growth factors these are rare—the role of interferon-γ in chronic granulocytes disease, for example. In other conditions their role alongside conventional therapy is established (e.g. interferon-α in hairy cell leukaemia, and CML). The levels of efficacy of interferons in inflammatory states, for example interferon-α in viral hepatitis and to a lesser extent interferon-β in relapsing–remitting multiple sclerosis, are documented, but in these common diseases the questions of where to find the financial resources necessary to implement this therapy are probably more difficult to answer than the preceding scientific and clinical questions.

In common malignancies, phase I–II studies have been completed for most of the cytokines likely to have any activity: TNF, IL-2, IL-4, IL-6, IL-10 and the relatively recently available IL-12. For all but the latter, which is still relatively untested (but with encouraging preliminary results), a role as a single agent in therapy of common cancers has been excluded. IL-12 may be found to be such an agent, but adjuvant therapy as part of a strategy for minimal residual disease is a more likely role, and the occasional success of IL-2-activated donor lymphocytes in the management of leukaemia relapse following allogeneic bone marrow transplantation (Slavin *et al.*, 1996) and as vaccine adjuvants in animal models may support this optimism (McLaughlin *et al.*, 1996).

The effectiveness of anti-TNF in RA has encouraged further the use of cytokine-neutralizing therapies. The same antibody has been effective in Crohn's disease, but the use of anti-TNF agents in septic shock has not been effective. The methods of neutralizing cytokines have also to be studied seriously, with chimeras of TNF-R p75 and IgG_1 Fc worsening septic shock possibly by retaining TNF in the circulation. Antibodies to other pivotal cytokines (e.g. IL-12) may also have a therapeutic potential and have been shown to be effective when administered at critical time points in animal models, for example of inflammatory bowel disease (Neurath *et al.*, 1995).

ACKNOWLEDGEMENTS

The assistance of P. Wells and A. Wilcox with the manuscript, and of K. Bull with references, is gratefully acknowledged.

REFERENCES

Ahre, A., Bjorkholm, M., Mellstedt, H., Brenning, G., Engstedt, L., Gahrton, G., Gyllenhammar, H., Holm, G., Johansson, B. and Jarnmark, M. (1984). Human leukocyte interferon and intermittent high-dose melphalan–prednisone administration in the treatment of multiple myeloma: a randomized clinical trial from the Myeloma Group of Central Sweden. *Cancer Treat. Rep.* **68**, 1331–1338.

Allan, N.C., Richards, S.M. and Shepard, P.C. (1995). UK Medical Research Council randomised, multicentre trial of interferon-α n1 for chronic myeloid leukaemia: improved survival irrespective of cytogenetic response. The UK Medical Research Council's Working Parties for Therapeutic Trials in Adult Leukaemia. *Lancet* **345**, 1392–1397.

Arend, W.P., Joslin, F.G., Thompson, R.C. and Hannum, C.H. (1989). An IL-1 inhibitor from human monocytes. Production and characterization of biologic properties. *J. Immunol.* **143**, 1851–1858.

Bacha, P., Williams, D.P., Waters, C., Williams, J.M., Murphy, J.R. and Strom, T.B. (1988). Interleukin 2 receptor-targeted cytotoxicity. Interleukin 2 receptor-mediated action of a diphtheria toxin-related interleukin 2 fusion protein. *J. Exp. Med.* **167**, 612–622.

Barton, B.E., Shortall, J. and Jackson, J.V. (1996). Interleukins 6 and 11 protect mice from mortality in a staphylococcal enterotoxin-induced toxic shock model. *Infect. Immun.* **64**, 714–718.

Belch, A., Shelley, W., Bergsagel, D., Wilson, K., Klimo, P., White, D. and Willan, A. (1988). A randomized trial of maintenance *versus* no maintenance melphalan and prednisone in responding multiple myeloma patients. *Br. J. Cancer* **57**, 94–99.

Bernstein, Z.P., Porter, M.M., Gould, M., Lipman, B., Bluman, E.M., Stewart, C.C., Hewitt, R.G., Fyfe, G., Poiesz, B. and Caligiuri, M.A. (1995). Prolonged administration of low-dose interleukin-2 in human immunodeficiency virus-associated malignancy results in selective expansion of innate immune effectors without significant clinical toxicity. *Blood* **86**, 3287–3294.

Booth, J.C., Foster, G.R., Kumar, U., Galassini, R., Goldin, R.D., Brown, J.L. and Thomas, H.C. (1995). Chronic hepatitis C virus infections: predictive value of genotype and level of viraemia on disease progression and response to interferon α. *Gut* **36**, 427–432.

Boraschi, D., Censini, S. and Tagliabue, A. (1984). Interferon-γ reduces macrophage-suppressive activity by inhibiting prostaglandin E_2 release and inducing interleukin 1 production. *J. Immunol.* **133**, 764–768.

Brennan, F.M., Chantry, D., Jackson, A., Maini, R. and Feldmann, M. (1989). Inhibitory effect of TNF α antibodies on synovial cell interleukin-1 production in rheumatoid arthritis. *Lancet* **ii**, 244–247.

Bresnihan, B., Lookbaugh, J., Witt, K. and Musikic, P. (1996). Treatment with recombinant human interleukin-1 receptor antagonist (rhIL-1ra) in rheumatoid arthritis (RA): results of a randomized double-blind, placebo-controlled multicenter trial. *Arthritis Rheum.* **39**, S282.

Broxmeyer, H.E., Williams, D.E., Lu, L., Cooper, S., Anderson, S.L., Beyer, G.S., Hoffman, R. and Rubin, B.Y. (1986). The suppressive influences of human tumor necrosis factors on bone marrow hematopoietic progenitor cells from normal donors and patients with leukemia: synergism of tumor necrosis factor and interferon-gamma. *J. Immunol.* **136**, 4487–4495.

Brunda, M.J. and Gately, M.K. (1995). Interleukin-12: potential role in cancer therapy. *Important Adv. Oncol.* 3–18.

Burns, G.F., Triglia, T. and Werkmeister, J.A. (1984). *In vitro* generation of human activated lymphocyte killer cells: separate precursors and modes of generation of NK-like cells and 'anomalous' killer cells. *J. Immunol.* **133**, 1656–1663.

Butler, D.M., Maini, R.M., Feldmann, M. and Brennan, F.B. (1995). Modulation of proinflammatory cytokine release in rheumatoid synovial membrane cell cultures. Comparison of monoclonal anti-TNFα antibody with the IL-1 receptor antagonist. *Eur. Cytokine Netw.* **6**, 225–230.

Catalano, M.A. (1993). Clinical use of human recombinant IL-1 receptor antagonist. *J. Cell Biochem. Suppl.* **17B**, 55.

Chong, A.S., Scuderi, P., Grimes, W.J. and Hersh, E.M. (1989). Tumor targets stimulate IL-2 activated killer cells to produce interferon-γ and tumor necrosis factor. *J. Immunol.* **142**, 2133–2139.

Chu, C.Q., Field, M., Feldmann, M. and Maini, R.N. (1991). Localization of tumor necrosis factor α in synovial tissues and at the cartilage–pannus junction in patients with rheumatoid arthritis. *Arthritis Rheum.* **34**, 1125–1132.

Cocchi, F., De Vico, A.L., Garzino-Demo, A., Arya, S.K., Gallo, R.C. and Lusso, P. (1995). Identification of RANTES, MIP-1α and MIP-1β as the major HIV-suppressive factors produced by CD8$^+$ T cells. *Science* **270**, 1811–1815.

Cooper, M.R., Dear, K., McIntyre, O.R., Ozer, H., Ellerton, J., Canellos, G., Bernhardt, B., Duggan, D., Faragher, D. and Schiffer, C. (1993). A randomized clinical trial comparing melphalan/prednisone with or without interferon α-2b in newly diagnosed patients with multiple myeloma: a Cancer and Leukemia Group B study. *J. Clin. Oncol.* **11**, 155–160.

Csaki, C., Ferencz, T., Sipos, G., Kopper, L., Schuler, D. and Borsi, J.D. (1996). Diffuse plasmacytosis in a child with brainstem glioma following multiagent chemotherapy and intensive growth factor support. *Med. Pediatr. Oncol.* **26**, 367–371.

Cutolo, M., Kirkham, B., Bologna, C., Sany, J., Scott, D., Brooks, P., Førre, O., Jain, R., Kvient, T., Markenson, J., Seibold, J., Sturrock, R., Veys, E., Edwards, J., Zaug, M., Durwell, L., Bisschops, C., St. Clair, P., Milnarik, P., Baudink, M. and Van der Auwera, P. (1996). Loading/maintenance doses approach to neutralization of TNF by lenercept in patients with rheumatoid arthritis treated for 3 months, results of a double-blind placebo-controlled phase II trial. *Arthritis Rheum.* **39 (supplement S243)**; Abstract 1294.

Demetri, G.D., Spriggs, D.R., Sherman, M.L., Arthur, K.A., Imamura, K. and Kufe, D.W. (1989). A phase I trial of recombinant human tumor necrosis factor and interferon-γ: effects of combination cytokine administration *in vivo*. *J. Clin. Oncol.* **7**, 1545–1553.

D'Hondt, V., Humblet, Y., Guillaume, T., Baatout, S., Chatelain, C., Berliere, M., Longueville, J., Feyens, A.M., de Greve, J. and Van Oosterom, A. (1995). Thrombopoietic effects and toxicity of interleukin-6 in patients with ovarian cancer before and after chemotherapy: a multicentric placebo-controlled, randomized phase Ib study. *Blood* **85**, 2347–2353.

Dinarello, C.A., Cannon, J.G., Wolff, S.M., Bernheim, H.A., Beutler, B., Cerami, A., Figari, I.S., Palladino, M.A. and O'Connor, J.V. (1986). Tumor necrosis factor (cachectin) is an endogenous pyrogen and induces production of interleukin 1. *J. Exp. Med.* **163**, 1433–1450.

Du, X.X., Doerschuk, C.M., Orazi, A. and Williams, D.A. (1994). A bone marrow stromal-derived growth factor, interleukin-11, stimulates recovery of small intestinal mucosal cells after cytoablative therapy. *Blood* **83**, 33–37.

Eggimann, P., Chiolero, R., Chassot, P.G., Lienard, D., Gerain, J. and Lejeune, F. (1995). Systemic and hemodynamic effects of recombinant tumor necrosis factor α in isolation perfusion of the limbs. *Chest* **107**, 1074–1082.

Elliott, M.J., Maini, R.N., Feldmann, M., Long-Fox, A., Charles, P., Katsikis, P., Brennan, F.M., Walker, J., Bijl, H., Ghrayeb, J. and Woody, J. (1993). Treatment of rheumatoid arthritis with chimeric monoclonal antibodies to TNFα. *Arthritis Rheum.* **36**, 1681–1690.

Elliott, M.J., Maini, R.N., Feldmann, M., Kalden, J.R., Antoni, C., Smolen, J.S., Leeb, B., Breedveld, F.C., *et al.* Macfarlane, J.D., Bijl, H. and Woody, J.N. (1994a). Randomised double blind comparison of a chimaeric monoclonal antibody to tumour necrosis factor α (cA2) *versus* placebo in rheumatoid arthritis. *Lancet* **344**, 1105–1110.

Elliott, M.J., Maini, R.N., Feldmann, M., Long-Fox, A., Charles, P., Bijl, H. and Woody, J.N. (1994b). Repeated therapy with monoclonal antibody to tumour necrosis factor α (cA2) in patients with rheumatoid arthritis. *Lancet* **344**, 1125–1127.

Enomoto, N., Sakuma, I., Asahina, Y., Kurosaki, M., Murakami, T., Yamamoto, C., Ogura, Y., Izumi, N., Marumo, F. and Sato, C. (1996). Mutations in the nonstructural protein 5A gene and response to interferon in patients with chronic hepatitis C virus 1b infection. *N. Engl. J. Med.* **334**, 77–81.

Eron, L.J., Judson, F., Tucker, S., Prawer, S., Mills, J., Murphy, K., Hickey, M., Rogers, M., Flannigan, S. and Hien, N. (1986). Interferon therapy for condylomata acuminata. *N. Engl. J. Med.* **315**, 1059–1064.

Ezekowitz, R.A.B. (1991a). A controlled trial of interferon-γ to prevent infection in chronic granulomatous disease. *N. Engl. J. Med.* **324**, 509–516.

Ezekowitz, R.A.B. (1991b). Interferon γ for chronic granulomatous disease. *N. Engl. J. Med.* **325**, 1516–1517 (letter).

Feldmann, M., Brennan, F.M. and Maini, R.N. (1996). Role of cytokines in rheumatoid arthritis. *Annu. Rev. Immunol.* **14**, 397–440.

Feldmann, M., Elliott, M.J., Woody, J.N. and Maini, R.N. (1997). Anti-tumor necrosis factor α therapy of rheumatoid arthritis. *Adv. Immunol.* **64**, 283–550.

Ferguson, J.E., Hulse, P., Lorigan, P., Jayson, G. and Scarffe, J.H. (1995). Continuous infusion of 5-fluorouracil with α2b interferon for advanced colorectal carcinoma. *Br. J. Cancer* **72**, 193–197.

Garman, R.D., Jacobs, K.A., Clark, S.C. and Raulet, D.H. (1987). B-cell-stimulatory factor 2 (β2 interferon) functions as a second signal for interleukin 2 production by mature murine T cells. *Proc. Natl Acad. Sci. U.S.A.* **84**, 7629–7633.

Gavett, S.H., O'Hearn, D.J., Li, X., Huang, S.K., Finkelman, F.D. and Wills-Karp, M. (1995). Interleukin 12 inhibits antigen-induced airway hyperresponsiveness, inflammation and Th2 cytokine expression in mice. *J. Exp. Med.* **182**, 1527–1536.

Gerlach, H., Lieberman, H., Bach, R., Godman, G., Brett, J. and Stern, D. (1989). Enhanced responsiveness of endothelium in the growing/motile state to tumor necrosis factor/cachectin. *J. Exp. Med.* **170**, 913–931. (Erratum appears in *J. Exp. Med.* 1989, **170**, 1793.)

Gill, P.S., Harrington, W., Jr., Kaplan, M.H., Ribeiro, R.C., Bennett, J.M., Liebman, H.A., Bernstein-Singer, M., Espina, B.M., Cabral, L. and Allen, S. (1995). Treatment of adult T-cell leukemia-lymphoma with a combination of interferon α and zidovudine. *N. Engl. J. Med.* **332**, 1744–1748.

Glazier, D.B., Bahnson, R.R., McLeod, D.G., von Roemeling, R.W., Messing, E.M. and Ernstoff, M.S. (1995). Intravesical recombinant tumor necrosis factor in the treatment of superficial bladder cancer: an Eastern Cooperative Oncology Group study. *J. Urol.* **154**, 66–68.

Gordon, M.S., McCaskill-Stevens, W.J., Battiato, L.A., Loewy, J., Loesch, D., Breeden, E., Hoffman, R., Beach, K.J., Kuca, B., Kaye, J. and Sledge, G.W.J. (1996). A phase I trial of recombinant human interleukin-11 (neumega rhIL-11 growth factor) in women with breast cancer receiving chemotherapy. *Blood* **87**, 3615–3624.

Groenewegen, G., de-Ley, M., Jeunhomme, G.M. and Buurman, W.A. (1986). Supernatants of human leukocytes contain mediator, different from interferon γ, which induces expression of MHC class II antigens. *J. Exp. Med.* **164**, 131–143.

Groopman, J.E., Gottlieb, M.S., Goodman, J., Mitsuyasu, R.T., Conant, M.A., Prince, H., Fahey, J.L., Derezin, M., Weinstein, W.M. and Casavante, C. (1984). Recombinant α-2 interferon therapy for Kaposi's sarcoma associated with the acquired immunodeficiency syndrome. *Ann. Intern. Med.* **100**, 671–676.

Grossman, R.M., Krueger, J., Yourish, D., Granelli-Piperno, A., Murphy, D.P., May, L.T., Kupper, T.S., Sehgal, P.B. and Gottlieb, A.B. (1989). Interleukin 6 is expressed in high levels in psoriatic skin and stimulates proliferation of cultured human keratinocytes. *Proc. Natl Acad. Sci. U.S.A.* **86**, 6367–6371.

Hanley, J.P., Jarvis, L.M., Andrew, J., Dennis, R., Hayes, P.C., Piris, J., Lee, R., Simmonds, P. and Ludlam, C.A. (1996). Interferon treatment for chronic hepatitis C infection in hemophiliacs—influence of virus load, genotype and liver pathology on response. *Blood* **87**, 1704–1709.

Haworth, C., Brennan, F.M., Chantry, D., Turner, M., Maini, R.N. and Feldmann, M. (1991). Expression of granulocyte–macrophage colony-stimulating factor in rheumatoid arthritis: regulation by tumor necrosis factor-α. *Eur. J. Immunol.* **21**, 2575–2579.

Healy, G.B., Gelber, R.D., Trowbridge, A.L., Grundfast, K.M., Ruben, R.J. and Price, K.N. (1988). Treatment of recurrent respiratory papillomatosis with human leukocyte interferon. Results of a multicenter randomized clinical trial. *N. Engl. J. Med.* **319**, 401–407.

Hirano, T., Matsuda, T., Turner, M., Miyasaka, N., Buchan, G., Tang, B., Sato, K., Shimizu, M., Maini, R., Feldmann, M. and Kishimoto, T. (1988). Excessive production of interleukin 6/B cell stimulatory factor-2 in rheumatoid arthritis. *Eur. J. Immunol.* **18**, 1797–1801.

Hoang, T., Levy, B., Onetto, N., Haman, A. and Rodriguez-Cimadevilla, J.C. (1989). Tumor necrosis factor α stimulates the growth of the clonogenic cells of acute myeloblastic leukemia in synergy with granulocyte/macrophage colony-stimulating factor. *J. Exp. Med.* **170**, 15–26.

Howells, G.L., Chantry, D. and Feldmann, M. (1988). Interleukin 1 (IL-1) and tumour necrosis factor synergise in the induction of IL-1 synthesis by human vascular endothelial cells. *Immunol. Lett.* **19**, 169–173.

Jelinek, J., Fairbairn, L.J., Dexter, T.M., Rafferty, J.A., Stocking, C., Ostertag, W. and Margison, G.P. (1996). Long-term protection of hematopoiesis against the cytotoxic effects of multiple doses of nitrosourea by retrovirus-mediated expression of human *O*6-alkylguanine-DNA-alkyltransferase. *Blood* **87**, 1957–1961.

Joshua, D.E., MacCallum, S. and Gibson, J. (1997). Role of alpha interferon in multiple myeloma. *Blood Rev.* **11**, 191–200.

Karcz, D., Popiela, T., Szczepanik, A.M., Czupryna, A., Szymanska, B., Siedlar, M. and Zembala, M. (1994). Preoperative endoscopic intratumor application of tumor necrosis factor α in patients with locally advanced resectable gastric cancer. *Endoscopy* **26**, 369–370 (letter).

Kawakami, K., Tohyama, M., Xie, Q. and Saito, A. (1996). IL-12 protects mice against pulmonary and disseminated infection caused by *Cryptococcus neoformans. Clin. Exp. Immunol.* **104**, 208–214.

Keith, J.C., Jr., Albert, L., Sonis, S.T., Pfeiffer, C.J. and Schaub, R.G. (1994). IL-11, a pleiotropic cytokine: exciting new effects of IL-11 on gastrointestinal mucosal biology. *Stem Cells* **1**, 79–89.

Kirker-Head, C.A. (1995). Recombinant bone morphogenetic proteins: novel substances for enhancing bone healing. *Vet. Surg.* **24**, 408–419.

Kirkwood, J.M. (1991). Studies of interferon in the therapy of melanoma. *Sem. Oncol.* **18 (supplement 7)**, 83–89.

Kobayashi, K., Kasama, T., Yamazaki, J., Hosaka, M., Katsura, T., Mochizuki, T., Soejima, K. and Nakamura, R.M. (1995). Protection of mice from *Mycobacterium avium* infection by recombinant interleukin-12. *Antimicrob. Agents Chemother.* **39**, 1369–1371.

Koeffler, H.P., Gasson, J., Ranyard, J., Souza, L., Shepard, M. and Munker, R. (1987). Recombinant human TNF α stimulates production of granulocyte colony-stimulating factor. *Blood* **70**, 55–59.

Kupper, T., Horowitz, M., Lee, F., Robb, R. and Flood, P.M. (1987). Autocrine growth of T cells independent of interleukin 2: identification of interleukin 4 (IL 4, BSF-1) as an autocrine growth factor for a cloned antigen-specific helper T cell. *J. Immunol.* **138**, 4280–4287.

Lazarus, H.M., Winton, E.F., Williams, S.F., Grinblatt, D., Campion, M., Cooper, B.W., Gunn, H., Manfreda, S. and Isaacs, R.E. (1995). Phase I multicenter trial of interleukin 6 therapy after autologous bone marrow transplantation in advanced breast cancer. *Bone Marrow Transplant* **15**, 935–942.

Lehn, M., Weiser, W.Y., Engelhorn, S., Gillis, S. and Remold, H.G. (1989). IL-4 inhibits H_2O_2 production and antileishmanial capacity of human cultured monocytes mediated by IFN-γ. *J. Immunol.* **143**, 3020–3024.

Lejeune, F., Lienard, D., Eggermont, A., Schraffordt Koops, H., Rosenkaimer, F., Gerain, J., Klaase, J., Kroon, B., Vanderveken, J. and Schmitz, P. (1994). Rationale for using TNF α and chemotherapy in regional therapy of melanoma. *J. Cell Biochem.* **56**, 52–61.

Lorberboum-Galski, H., Barrett, L.V., Kirkman, R.L., Ogata, M., Willingham, M.C., FitzGerald, D.J. and Pastan, I. (1989). Cardiac allograft survival in mice treated with IL-2-PE40. *Proc. Natl Acad. Sci. U.S.A.* **86**, 1008–1012.

Mancuso, G., Tomasello, F., Migliardo, M., Delfino, D., Cochran, J., Cook, J.A. and Teti, G. (1994). Beneficial effects of interleukin-6 in neonatal mouse models of group B streptococcal disease. *Infect. Immun.* **62**, 4997–5002.

McIntyre, K.W., Shuster, D.J., Gillooly, K.M., Warrier, R.R., Connaughton, S.E., Hall, L.B., Arp, L.H., Gately, M.K. and Magram, J. (1996). Reduced incidence and severity of collagen-induced arthritis in interleukin-12-deficient mice. *Eur. J. Immunol.* **26**, 2933–2938.

McLaughlin, J.P., Schlom, J., Kantor, J.A. and Greiner, J.W. (1996). Improved immunotherapy of a recombinant carcinoembryonic antigen vaccinia vaccine when given in combination with interleukin-2. *Cancer Res.* **56**, 2361–2367.

Mond, J.J., Finkelman, F.D., Sarma, C., Ohara, J. and Serrate, S. (1985). Recombinant interferon-γ inhibits the B cell proliferative response stimulated by soluble but not by Sepharose-bound anti-immunoglobulin antibody. *J. Immunol.* **135**, 2513–2517.

Moreland, L.W., Sewell, K.L., Trenthan, D.E., Bucy, R.P., Sullivan, W.F., Schrohenloher, R.E., Shmerling, R.H., Parker, K.C., Swartz, W.G., Woodworth, T.G. and Koopman, W.J. (1995). Interleukin-2 diphtheria fusion protein in refractory rheumatoid arthritis: a double blind placebo-controlled trial with open label extension. *Arthritis Rheum.* **38**, 1177–1186.

Moreland, L.W., Baumgartner, S.W., Schiff, M.H., Tindall, E.A., Fleischmann, R.M., Weaver, A.L., Ettlinger, R.E., Cohen, S., Koopman, W.J., Maher, K., Widmer, M.B. and Blosch, C.M. (1997). Treatment of RA with a recombinant human tumor necrosis factor receptor (p75)–Fc fusion protein. *N. Engl. J. Med.* **337**, 141–147.

Morris, L., Troutt, A.B., McLeod, K.S., Kelso, A., Handman, E. and Aebischer, T. (1993). Interleukin-4 but not γ interferon production correlates with the severity of murine cutaneous leishmaniasis. *Infect. Immun.* **61**, 3459–3465.

Morrone, G., Ciliberto, G., Oliviero, S., Arcone, R., Dente, L., Content, J. and Cortese, R. (1988). Recombinant interleukin 6 regulates the transcriptional activation of a set of human acute phase genes. *J. Biol. Chem.* **263**, 12 554–12 558.

Mullarkey, M.F., Rubin, A.S., Roux, E.R., Hanna, R.K. and Jacobs, C.A. (1993). Modification of allergic late-phase response by soluble human IL-1 receptor (RHU IL-1R). *J. Cell. Biochem. Suppl.* **17B**, 55.

Mullen, C.A., Petropoulos, D. and Lowe, R.M. (1996). Treatment of microscopic pulmonary metastases with recombinant autologous tumor vaccine expressing interleukin 6 and *Escherichia coli* cytosine deaminase suicide genes. *Cancer Res.* **56**, 1361–1366.

Nagler, A., Deutsch, V.R., Varadi, G., Pick, M., Eldor, A. and Slavin, S. (1995). Recombinant human interleukin-6 accelerates *in-vitro* megakaryocytopoiesis and platelet recovery post autologous peripheral blood stem cell transplantation. *Leuk. Lymphoma* **19**, 343–349.

Neurath, M.F., Fuss, I., Kelsall, B.L., Stuber, E. and Strober, W. (1995). Antibodies to interleukin 12 abrogate established experimental colitis in mice. *J. Exp. Med.* **182**, 1281–1290.

Ohnishi, K., Ohno, R., Tomonaga, M., Kamada, N., Onozawa, K., Kuramoto, A., Dohy, H., Mizoguchi, H., Miyawaki, S. and Tsubaki, K. (1995). A randomized trial comparing interferon-α with busulfan for newly diagnosed chronic myelogenous leukemia in chronic phase. *Blood* **86**, 906–916.

Olynyk, J.K., Reddy, K.R., Di Bisceglie, A.M., Jeffers, L.J., Parker, T.I., Radick, J.L., Schiff, E.R. and Bacon, B.R. (1995). Hepatic iron concentration as a predictor of response to interferon α therapy in chronic hepatitis C. *Gastroenterology* **108**, 1104–1109.

Paleolog, E.M., Hunt, M., Elliott, M.J., Feldmann, M., Maini, R.N. and Woody, J.N. (1996). Deactivation of vascular endothelium by monoclonal anti-tumor necrosis factor α antibody in rheumatoid arthritis. *Arthritis Rheum.* **39**, 1082–1091.

Parmiani, G. (1990). An explanation of the variable clinical response to interleukin 2 and LAK cells. *Immunol. Today* **11**, 113–115.

Pass, H.I., Mew, D.J., Kranda, K.C., Temeck, B.K., Donington, J.S. and Rosenberg, S.A. (1996). Isolated lung perfusion with tumor necrosis factor for pulmonary metastases. *Ann. Thorac. Surg.* **61**, 1609–1617.

Peest, D., Deicher, H., Coldewey, R., Leo, R., Bartl, R., Bartels, H., Braun, H.J., Fett, W., Fischer, J.T. and Gobel, B. (1995). A comparison of polychemotherapy and melphalan/prednisone for primary remission induction, and interferon-α for maintenance treatment, in multiple myeloma. A prospective trial of the German Myeloma Treatment Group. *Eur. J. Cancer* **2**, 146–151.

Poo, W.J., Conrad, L. and Janeway, C., Jr. (1988). Receptor-directed focusing of lymphokine release by helper T cells. *Nature* **332**, 378–380.

Powles, R., Raje, N., Cunningham, D., Malpas, J., Milan, S., Horton, C., Mehta, J., Singhal, S., Viner, C. and Treleaven, J. (1995). Maintenance therapy for remission in myeloma with Intron-A following high-dose melpholan and either autologous bone marrow transplantation or peripheral stem cell rescue. *Stem Cells* **13**, 114–117.

Prestwood, K.M., Pilbeam, C.C. and Raisz, L.G. (1995). Treatment of osteoporosls. *Annu. Rev. Med.* **46**, 249–256.

Quesada, J.R., Reuben, J., Manning, J.T., Hersh, E.M. and Gutterman, J.U. (1984). Alpha interferon for induction of remission in hairy-cell leukemia. *N. Engl. J. Med.* **310**, 15–18.

Racke, M.K., Bonomo, A., Scott, D.E., Cannella, B., Levine, A., Raine, C.S., Shevach, E.M. and Rocken, M. (1994). Cytokine-induced immune deviation as a therapy for inflammatory autoimmune disease. *J. Exp. Med.* **180**, 1961–1966.

Ranges, G.E., Zlotnik, A., Espevik, T., Dinarello, C.A., Cerami, A. and Palladino, M., Jr. (1988). Tumor necrosis factor α/cachectin is a growth factor for thymocytes. Synergistic interactions with other cytokines. *J. Exp. Med.* **167**, 1472–1478.

Rankin, E.C., Choy, E.H., Kassimos, D., Kingsley, G.H., Sopwith, A.M., Isenberg, D.A. and Panayi, G.S. (1995). The therapeutic effects of an engineered human anti-tumour necrosis factor α antibody (CDP571) in rheumatoid arthritis. *Br. J. Rheumatol.* **34**, 334–342.

Riley, E.H., Lane, J.M., Urist, M.R., Lyons, K.M. and Lieberman, J.R. (1996). Bone morphogenetic protein-2: biology and applications. *Clin. Orthop.* **324**, 39–46.

Rodriguez, M., Pavelko, K.D., McKinney, C.W. and Leibowitz, J.L. (1994). Recombinant human IL-6 suppresses demyelination in a viral model of multiple sclerosis. *J. Immunol.* **153**, 3811–3821.

Rosenberg, S.A., Lotze, M.T., Muul, L.M., Chang, A.E., Avis, F.P., Leitman, S., Linehan, W.M., Robertson, C.N., Lee, R.E. and Rubin, J.T. (1987). A progress report on the treatment of 157 patients with advanced cancer using lymphokine-activated killer cells and interleukin-2 or high-dose interleukin-2 alone. *N. Engl. J. Med.* **316**, 889–897.

Rosenberg, S.A., Yang, J.C., Topalian, S.L., Schwartzentruber, D.J., Weber, J.S., Parkinson, D.R., Seipp, C.A., Einhorn, J.H. and White, D.E. (1994). Treatment of 283 consecutive patients with metastatic melanoma or renal cell cancer using high-dose bolus interleukin 2. *JAMA* **271**, 907–913.

Rothwell, N.J. and Relton, J.K. (1993). Involvement of cytokines in acute neurodegeneration in the CNS. *Neurosci. Biobehav. Rev.* **17**, 217–227.

Sadick, M.D., Heinzel, F.P., Holaday, B.J., Pu, R.T., Dawkins, R.S. and Locksley, R.M. (1990). Cure of murine leishmaniasis with anti-interleukin 4 monoclonal antibody. Evidence for a T cell-dependent, interferon γ-independent mechanism. *J. Exp. Med.* **171**, 115–127.

Saven, A. and Piro, L.D. (1993). 2-Chlorodeoxyadenosine in the treatment of hairy cell leukemia and chronic lymphocytic leukemia. *Leuk. Lymphoma.* **2**, 109–114.

Schwartz, E.L., Baptiste, N., Megati, S., Wadler, S. and Otter, A. (1995). 5-Ethoxy-2′-deoxyuridine, a novel substrate for thymidine phosphorylase, potentiates the antitumor activity of 5-fluorouracil when used in combination with interferon, an inducer of thymidine phosphorylase expression. *Cancer Res.* **55**, 3543–3550.

Scully, L.J., Shein, R., Karayiannis, P., McDonald, J.A. and Thomas, H.C. (1987). Lymphoblastoid interferon therapy of chronic HBV infection. A comparison of 12 *vs.* 24 weeks of thrice weekly treatment. *J. Hepatol.* **5**, 51–58.

Seelentag, W.K., Mermod, J.J., Montesano, R. and Vassalli, P. (1987). Additive effects of interleukin 1 and tumour necrosis factor-α on the accumulation of the three granulocyte and macrophage colony-stimulating factor mRNAs in human endothelial cells. *EMBO J.* **6**, 2261–2265.

Selby, P., Hobbs, S., Viner, C., Jackson, E., Jones, A., Newell, D., Calvert, A.H., McElwain, T., Fearon, K. and Humphreys, J. (1987). Tumour necrosis factor in man: clinical and biological observations. *Br. J. Cancer* **56**, 803–808.

Sironi, M., Breviario, F., Proserpio, P., Biondi, A., Vecchi, A., Van-Damme, J., Dejana, E. and Mantovani, A. (1989). IL-1 stimulates IL-6 production in endothelial cells. *J. Immunol.* **142**, 549–553.

Slavin, S., Naparstek, E., Nagler, A., Ackerstein, A., Samuel, S., Kapelushnik, J., Brautbar, C. and Or, R. (1996). Allogeneic cell therapy with donor peripheral blood cells and recombinant human interleukin-2 to treat leukemia relapse after allogeneic bone marrow transplantation. *Blood* **87**, 2195–2204.

Sonis, S.T., Lindquist, L., Van Vugt, A., Stewart, A.A., Stam, K., Qu, G.Y., Iwata, K.K. and Haley, J.D. (1994). Prevention of chemotherapy-induced ulcerative mucositis by transforming growth factor β3. *Cancer Res.* **54**, 1135–1138.

Stouthard, J.M., Goey, H., de Vries, E.G., de Mulder, P.H., Groenewegen, A., Pronk, L., Stoter, G., Sauerwein, H.P., Bakker, P.J. and Veenhof, C.H. (1996). Recombinant human interleukin 6 in metastatic renal cell cancer: a phase II trial. *Br. J. Cancer* **73**, 789–793.

Sykes, M., Szot, G.L., Nguyen, P.L. and Pearson, D.A. (1995). Interleukin-12 inhibits murine graft-*versus*-host disease. *Blood* **86**, 2429–2438.

Tepler, I., Elias, L., Smith, J.W., II, Hussein, M., Rosen, G., Chang, A.Y., Moore, J.O., Gordon, M.S., Kuca, B., Beach, K.J., Loewy, J.W., Garnick, M.B. and Kaye, J.A. (1996). A randomized placebo-controlled trial of recombinant human interleukin-11 in cancer patients with severe thrombocytopenia due to chemotherapy. *Blood* **87**, 3607–3614.

Trembleau, S., Penna, G., Bosi, E., Mortara, A., Gately, M.K. and Adorini, L. (1995). Interleukin 12 administration induces T helper type 1 cells and accelerates autoimmune diabetes in NOD mice. *J. Exp. Med.* **181**, 817–821.

Urban, J.F., Jr., Fayer, R., Chen, S.J., Gause, W.C., Gately, M.K. and Finkelman, F.D. (1996). IL-12 protects immunocompetent and immunodeficient neonatal mice against infection with *Cryptosporidium parvum*. *J. Immunol.* **156**, 263–268.

Van Deventer, S.J.H., Elson, C.O. and Fedorak, R.N. (1997). Multiple doses of intravenous interleukin-10 in steroid-refractory Crohn's disease. *Gastroenterology* **113**, 383–389.

Watt, O. and Cobby, M. (1996). Recombinant human IL-1 receptor antagonist (rhIL-1ra) reduces the rate of joint erosion in rheumatoid arthritis (RA). *Arthritis Rheum.* **39**, S576.

Weber, J., Gunn, H., Yang, J., Parkinson, D., Topalian, S., Schwartzentruber, D., Ettinghausen, S., Levitt, D. and Rosenberg, S.A. (1994). A phase I trial of intravenous interleukin-6 in patients with advanced cancer. *J. Immunother. Emphasis. Tumor Immunol.* **15**, 292–302.

Wendling, D., Racadot, E. and Wijdenes, J. (1993). Treatment of severe rheumatoid arthritis by anti-interleukin 6 monoclonal antibody. *J. Rheumatol.* **20**, 259–262.

Williams, R.O., Feldmann, M. and Maini, R.N. (1992). Anti-tumor necrosis factor ameliorates joint disease in murine collagen-induced arthritis. *Proc. Natl Acad. Sci. U.S.A.* **89**, 9784–9788.

Wozney, J.M. (1995). The potential role of bone morphogenetic proteins in periodontal reconstruction. *J. Periodontol.* **66**, 506–510.

Chapter 29

Clinical Applications of Hematopoietic Growth Factors

Robert S. Negrin

Bone Marrow Transplant Program, Department of Medicine (Hematology), Stanford University Hospital, Stanford, CA, USA

INTRODUCTION

The hematopoietic growth factors (HGFs) represent a novel class of drugs with potent activity in stimulating the production and activity of hematopoietic cells. These agents are increasingly being used in the treatment of patients with hematologic and neoplastic disease. Currently, two members of this family of drugs have been approved for clinical use, namely granulocyte–macrophage colony stimulating factor (GM-CSF) (Leukomax™, Leukine™) and granulocyte colony stimulating factor (G-CSF) (Neupogen™). The optimal roles for these potent myeloid growth factors are still being defined. In this chapter, the potential uses and toxicities of HGFs will be discussed, focusing on the two HGFs that have been released for use and for which the most clinical data is available. In addition, several novel HGFs will be discussed including interleukin-3 (IL-3), stem cell factor, flt ligand and thrombopoietin, which are entering clinical trials.

G-CSF and GM-CSF are polypeptide growth factors that have been cloned and expressed by recombinant technology. *In vitro*, the effects of G-CSF are limited to neutrophil development, whereas GM-CSF has broader activity including stimulation of neutrophils, monocytes and eosinophils. Both growth factors function by binding to specific receptors found on hematopoietic progenitor cells and mature neutrophils. In addition to acting on immature hematopoietic progenitor cells to increase the number of mature neutrophils, G-CSF also stimulates bacterial activity and chemotaxis. Interestingly, GM-CSF also increases bacterial activity but inhibits chemotaxis of mature neutrophils.

CLINICAL USES OF THE HGFs FOLLOWING CYTOTOXIC CHEMOTHERAPY

The potential uses of HGFs are listed in Table 1. The most common use of G-CSF and GM-CSF is to accelerate myeloid recovery following high-dose cytotoxic chemotherapy. There are numerous phase I–II clinical trials which demonstrate that treatment of patients with G-CSF shortens the neutrophil nadir following high-dose chemotherapy as compared with historical controls (Bronchud *et al.*, 1987; Antman *et al.*, 1988; Gabrilove

The Cytokine Handbook, 3rd ed.
ISBN 0–12–689662–3

Table 1. Clinical uses of the colony stimulating factors.

	GM-CSF	G-CSF
Acquired neutropenias		
Chemotherapy induced	+	Approved
Autologous BMT	Approved	Approved
Allogeneic BMT	Approved	+
Graft failure post BMT	Approved	NA
Aplastic anemia	+/−	+/−
Myelodysplastic syndromes	+	+
Acute myelogenous leukemia	Approved	+
AIDS	+	+
Congenital neutropenias		
Cyclic neutropenia	−	+
Kostmann's syndrome	−	Approved
Mobilization of stem cells		
Pre-autologous BMT	Approved	Approved

NA, not tested; +, efficacy demonstrated in phase II–III trials; −, efficacy not demonstrated in phase II trials; approved indication for use in the USA.

et al., 1988; Glaspy *et al.*, 1988; Morstyn *et al.*, 1988; Steward *et al.*, 1990). Randomized placebo-controlled trials have confirmed these results. Crawford *et al.* (1991) performed a randomized double-blind placebo-controlled trial in patients with small cell lung carcinoma who received multiple cycles of high-dose chemotherapy with or without G-CSF. Patients who developed fever and neutropenia during any cycle were allowed to cross over to treatment with open-label G-CSF. In this study, there was a significant decrease in incidence of fever and neutropenia (57% *versus* 28%), median duration of neutropenia, days of antibiotic use, incidence of confirmed infections and days in hospital in favor of the group of patients treated with G-CSF (Fig. 1). There was no impact on overall survival as neutropenic deaths were relatively uncommon. Another prospective randomized clinical trial was performed in patients with non-Hodgkin's lymphoma who were treated with vincristine, adriamycin, prednisone, etoposide, cyclophosphamide and bleomycin (VAPEC-B) chemotherapy. In this study of 80 patients, those patients who were randomized to receive G-CSF had a lower incidence of neutropenia than the control group (37% *versus* 85%, $P = 0.00001$), less fever and fewer treatment delays. Again, no differences in overall survival were noted (Pettengell *et al.*, 1992).

The question as to whether HGFs will help reduce infectious complications in patients who develop neutropenia has been addressed. In one study, 218 patients with malignancy who developed febrile neutropenia following chemotherapy were randomized to receive G-CSF or placebo. Although G-CSF reduced the duration of febrile neutropenia by 1 day, which reached statistical significance, the clinical utility was marginal (Maher *et al.*, 1994). Thus, the major use of HGFs is in the attempt to prevent rather than treat febrile neutropenia.

Upon approval of G-CSF and GM-CSF, the relatively broad potential clinical situations where these drugs could be used has led to considerable debate as to what is the appropriate use of these potent yet expensive drugs. This led to the development of

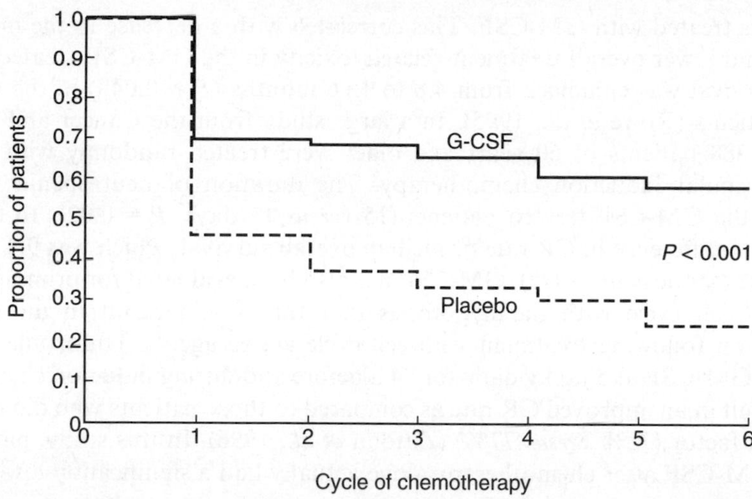

Fig. 1. Kaplan–Meier curve for the proportion of patients with lung cancer who did not develop fever and neutropenia following administration of cytotoxic chemotherapy who were treated with either G-CSF or placebo. From Crawford *et al.* (1991) with permission.

clinical practice guidelines to attempt to define the clinical settings for the most appropriate use following chemotherapy. These guidelines recommend that HGFs be used to reduce the likelihood of febrile neutropenia in clinical situations where the expected incidence is greater than 40%, in patients who have already developed febrile neutropenia in a previous cycle of that regimen, and to maintain dose intensity in treatment cycles where it is considered that dose reduction is not appropriate (Ozer, 1994).

HGFs have also been used to hasten neutrophil recovery following induction chemotherapy for patients with acute leukemia. This course of treatment is cautioned by the observation that leukemic cells do express receptors for a variety of HGF receptors including G-CSF and GM-CSF (Begley *et al.*, 1987; Cheng *et al.*, 1988). In the first reported trial of 108 patients with relapsed or refractory acute leukemia, patients who were randomized to treatment with G-CSF had accelerated recovery of neutrophils and less frequent infections without any effect on relapse rate (Ohno *et al.*, 1990). In 173 elderly patients with acute myeloblastic leukemia (AML) randomly assigned to received glycosylated human G-CSF, more rapid recovery of neutrophils was observed with a higher incidence of complete remission although this did not result in an overall survival benefit (Dombret *et al.*, 1995). G-CSF has also resulted in more rapid recovery of neutrophils in both adults and children undergoing induction chemotherapy for treatment of acute lymphoblastic leukemia. In these studies, there was a reduction in infections and a tighter adherence to planned chemotherapy schedules (Ottmann *et al.*, 1995; Welte *et al.*, 1996).

GM-CSF has also been studied extensively in leukemic patients undergoing induction chemotherapy, with positive results. In a study performed by the Eastern Cooperative Oncology Group (ECOG), 124 elderly patients were treated with standardized chemotherapy including daunorubicin and cytarabine followed by randomization to either GM-CSF or placebo. Median times to neutrophil recovery were significantly shorter in the patients treated with GM-CSF. This correlated with a decrease in the incidence of infection and lower overall treatment-related toxicity in the GM-CSF-treated arm. The median survival was enhanced from 4.8 to 10.6 months ($P = 0.048$) in the GM-CSF-treated patients (Rowe *et al.*, 1995). In a large study from the Cancer and Leukemia Group B, 388 patients of 60 years and older were treated randomly with GM-CSF following similar induction chemotherapy. The duration of neutropenia was again shorter in the GM-CSF-treated patients (15 *versus* 17 days, $P = 0.02$). In this study, there was no difference in CR rate or median overall survival, which was 9.4 months in both groups (Stone *et al.*, 1995). GM-CSF has also been evaluated for priming leukemic blasts into cell cycle with the hypothesis that this would result in more effective cytoreduction following treatment with cell cycle active agents. Unfortunately, treatment with GM-CSF at 5 μg/kg daily for 24 h before and during inducton chemotherapy did not result in an improved CR rate as compared to those patients who did not receive the growth factor (72% *versus* 77%) (Zittoun *et al.*, 1996). In this study, patients who received GM-CSF after chemotherapy alone actually had a significantly lower CR rate of 48%.

Taken together, these studies indicate that both G-CSF and GM-CSF are probably safe to administer following induction chemotherapy and result in more rapid acceleration of neutrophil recovery.

AUTOLOGOUS BONE MARROW OR PERIPHERAL BLOOD CELL TRANSPLANTATION

Autologous bone marrow transplantation (ABMT) is being increasingly utilized. Randomized studies have shown improved efficacy for transplanted patients in both non-Hodgkin's lymphoma and multiple myeloma (Philip *et al.*, 1995; Attal *et al.*, 1996). HGFs have been utilized in several settings in transplant patients including following transplantation to accelerate engraftment, to treat patients with graft failure and to mobilize hematopoietic stem cells.

GM-CSF has been studied following ABMT in several phase I–II trials (Brandt *et al.*, 1988; Nemunaitis *et al.*, 1988). In these studies, accelerated myeloid engraftment was observed compared with that in historic controls. As a result, several randomized placebo-controlled trials have been performed in patients following ABMT. In one collaborative multicenter study, 128 patients with lymphoid neoplasia were randomized to receive either GM-CSF or placebo. More rapid recovery of neutrophils, fewer infections, less antibiotic use and decreased length of stay in the hospital was found in the GM-CSF-treated group (Nemunaitis *et al.*, 1991b). Long-term follow-up of this study demonstrated no difference in overall survival for both treatment arms (Rabinowe *et al.*, 1993). In a second study performed at Stanford University, 69 consecutive patients with non-Hodgkin's lymphoma or Hodgkin's disease were treated with either GM-CSF or placebo. More rapid neutrophil recovery was observed in the GM-CSF-treated group (Advani *et al.*, 1992). In addition, the incidence of bacterial infection was significantly decreased in patients treated with GM-CSF. A third randomized study of 24 patients with Hodgkin's disease demonstrated similar reductions in neutropenia in the GM-CSF-treated group compared with patients treated with placebo. In addition, hospital stay and in-hospital costs were reduced in GM-CSF-treated patients (Gulati and Bennett, 1992). Two additional randomized studies observed similar results in patients undergoing ABMT for the treatment of acute lymphoblastic leukemia and non-Hodgkin's lymphoma (Gorin *et al.*, 1992; Link *et al.*, 1992). In all of these studies, there is no evidence that GM-CSF results in accelerated engraftment of any lineage other than myeloid cells and no overall survival advantage has been observed.

Similar studies have been performed utilizing G-CSF following ABMT. Phase I–II studies demonstrated safety and initial efficacy compared with that in historic control patients (Sheridan *et al.*, 1989; Taylor *et al.*, 1989). Randomized placebo-controlled clinical trials have also been performed. In one study of 315 patients who underwent randomization, the patients treated with G-CSF had more rapid recovery of neutrophils following transplantation (16 *versus* 27 days, $P < 0.001$), fewer days of infection, antibiotic treatment and hospitalization. As transplant-related mortality is rare, this did not translate into overall survival advantage at 1 year of follow-up (Gisselbrecht *et al.*, 1994). Similar data were reported in a smaller trial of patients undergoing autologous transplantation for high-risk lymphoid malignancy (Stahel *et al.*, 1994).

Recently, a major clinical advance has been the observation that hematopoietic stem cells circulate in the peripheral blood and these cells can be 'mobilized' by treating patients with HGFs and/or chemotherapy.

These mobilized peripheral blood progenitor cells (PBPCs) can then be collected by apheresis (Gianni *et al.*, 1989, 1990; Siena *et al.*, 1989, 1991; Molineux *et al.*, 1990). Reinfusion of the mobilized stem cells following myeloablative chemotherapy has resulted in more rapid myeloid and platelet engraftment (Gianni *et al.*, 1989, 1990; Sheridan *et al.*, 1992).

The stem cell content of the apheresis products can be assessed by measuring the absolute number of $CD34^+$ cells in the graft. Rapid recovery has been demonstrated when between 2 and 5×10^6 $CD34^+$ cells/kg are reinfused (Bensinger *et al.*, 1994; Negrin *et al.*, 1995; Tricot *et al.*, 1995; Weaver *et al.*, 1995). Infusion of greater numbers of $CD34^+$ cells does not result in more rapid recovery, indicating that a threshold value exists (Fig. 2). This has resulted in dramatic reductions in transplant-related mortality to approximately 1% for patients who receive this threshold value of hematopoietic stem cells (K.S. Goldstein, unpublished results). A randomized clinical trial has been performed comparing bone marrow with mobilized peripheral blood as the source of stem cells for 58 patients undergoing autologous transplantation for lymphoma. The patients receiving mobilized peripheral blood had more rapid recovery of neutrophils and platelets, required fewer transfusions and had shorter hospitalization than patients who received bone marrow (Schmitz *et al.*, 1996).

Several studies have addressed the question as to whether HGFs are necessary following mobilized peripheral blood transplants. These studies have demonstrated that post-transplant G-CSF hastens recovery and reduces antibiotic requirements in this setting as well (Spitzer *et al.*, 1994; Klumpp *et al.*, 1995). A dose of 5 μg/kg daily has been demonstrated to be adequate (Bolwell *et al.*, 1997). In one study, the administration of G-CSF was delayed until day 6 following the transplant, and recovery was no different from that in patients having treatment on day 1 (Faucher *et al.*, 1996).

GM-CSF has also been used to treat graft failure following bone marrow transplantation. Nemunaitis *et al.* (1990) compared results in patients who failed to engraft, defined by an absolute neutrophil count (ANC) of less than 100×10^6 liters by day +28 following bone marrow transplantation, who were treated with GM-CSF compared with an untreated historic control group. In this study, patients treated with GM-CSF

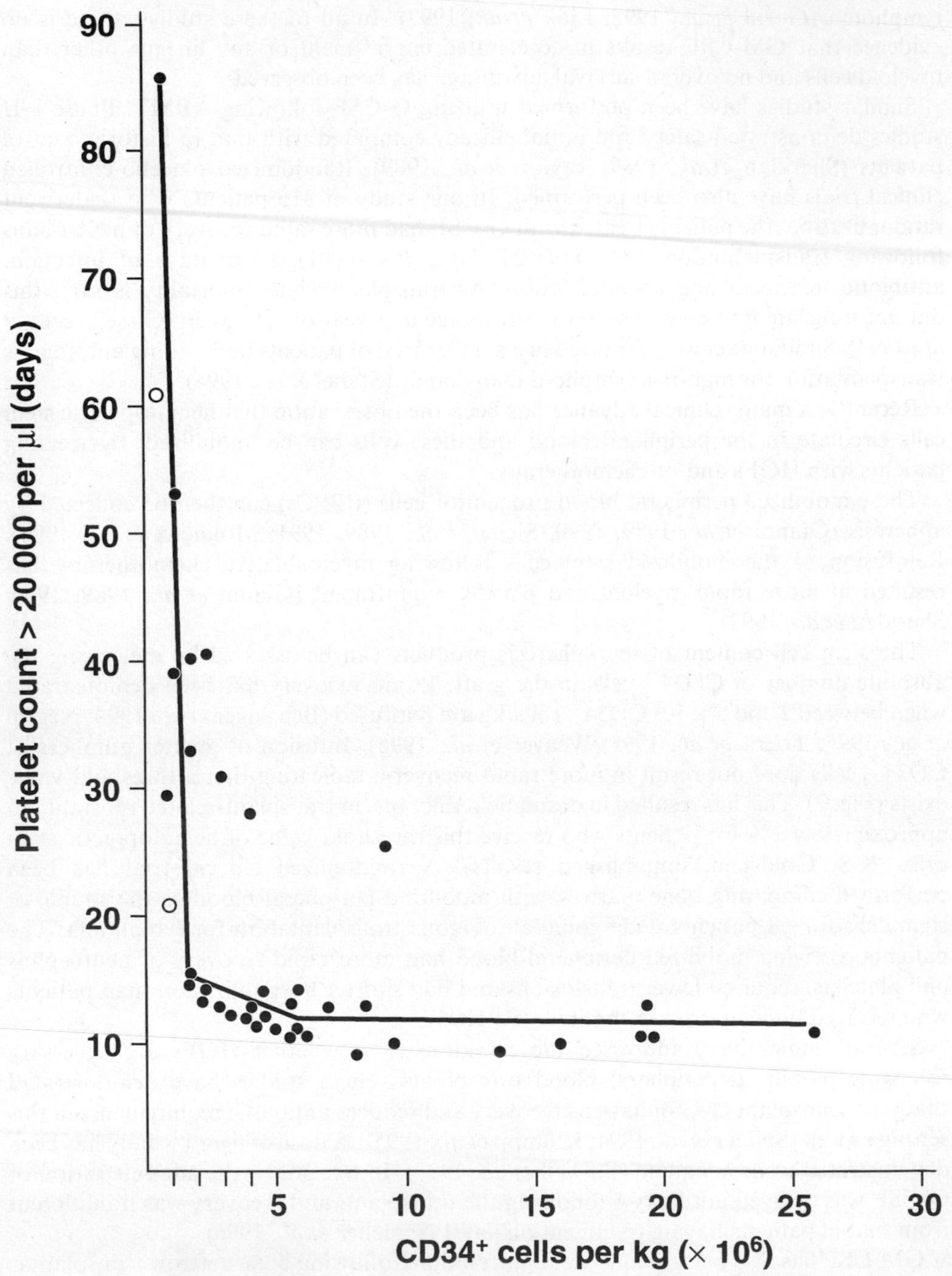

Fig. 2. Days required for platelet recovery following autologous transplantation with grafts containing variable numbers of $CD34^+$ cells per kg of recipient weight. ●, PBPCs; ○, bone marrow plus PBPCs.

had improved survival without an increase in graft-*versus*-host disease (GVHD) in patients who had received allogeneic bone marrow transplants. GM-CSF was found to be superior to the sequential use of GM-CSF and G-CSF (Weisdorf *et al.*, 1995).

ALLOGENEIC BONE MARROW TRANSPLANTATION

There has been less use of hematopoietic colony stimulating factors in recipients of allogeneic bone marrow transplantation (BMT) because of the concern that this treatment may increase the incidence or severity of GVHD and graft rejection. A phase I–II trial has been performed in patients with advanced hematologic malignancies undergoing BMT who received escalating doses of GM-CSF (30–500 μg/m^2 daily) with significantly faster recovery of neutrophils in patients who received GM-CSF than in historic controls (Nemunaitis *et al.*, 1991a). The severity of GVHD was unchanged based on historic controls. A similar phase I–II trial has been performed using G-CSF which demonstrated significantly faster myeloid engraftment without detectable differences in GVHD and relapse again compared with historic controls (Masaoka *et al.*, 1989).

Two randomized controlled clinical trials using GM-CSF in allogeneic BMT have been reported. In both studies, GM-CSF was used at 8 μg per kg per day by continuous intravenous infusion for 14 days, beginning on the day of transplant. In one study, myeloid engraftment was accelerated in the GM-CSF-treated group (although not statistically significantly different from findings in the group of patients treated with placebo). There was no increase in the incidence or severity of GVHD. There were no significant differences in survival between the groups and none of the 13 patients with myeloid leukemias treated with GM-CSF have relapsed (Powles *et al.*, 1990). DeWitte *et al.* (1992) in a multicenter trial treated 57 patients who received T cell-depleted bone marrow grafts and cyclosporin for GVHD prophylaxis. They observed a statistically significant acceleration in myeloid recovery. There was no increase in incidence or severity of GVHD or relapse. A trend to better disease-free survival was observed in patients who were treated with GM-CSF.

Owing to the success of mobilized peripheral blood in the autologous transplant setting, this has also been applied to patients undergoing allogeneic transplantation. The major concern has been that the high numbers of T cells in mobilized peripheral blood grafts, which are 10–100 times greater than those of a bone marrow graft, may increase the risk of GVHD. Several studies have demonstrated that sufficient CD34$^+$ cells can be collected from normal donors treated with G-CSF and that this appears to be well tolerated (Matsunaga *et al.*, 1993; Tjonnfjord *et al.*, 1994; Lane *et al.*, 1995). Several pilot studies have been performed demonstrating the feasibility of G-CSF-mobilized PBPCs for allogeneic transplantation. In these studies, rapid hematologic engraftment of neutrophils has been observed (Bensinger *et al.*, 1995; Korbling *et al.*, 1995; Schmitz *et al.*, 1995). Interestingly, the incidence and severity of acute GVHD has been approximately that observed with bone marrow as the source of the graft. The relative rates of chronic GVHD remain to be determined; however, in a retrospective analysis, the incidence of chronic GVHD also appeared to be the same irrespective of the stem cell source (Bensinger *et al.*, 1996). Randomized clinical trials are presently underway.

MYELODYSPLASTIC SYNDROMES

HGFs have also been utilized in an effort to treat patients with chronic neutropenic disorders such as the myelodysplastic syndromes. These disorders are characterized by cellular dysfunction and maturation defects resulting in ineffective hematopoiesis, ultimately culminating in infection, bleeding or transformation to acute leukemia (Greenberg, 1983). A number of phase I–II trials using HGFs have been reported for the treatment of patients with myelodysplastic syndrome (Vadhan-Raj *et al.*, 1987; Antin *et al.*, 1988; Ganser *et al.*, 1989; Herrmann *et al.*, 1989; Negrin *et al.*, 1989, 1990; Thompson *et al.*, 1989). In all these studies, improvements in neutrophil counts were demonstrated during the duration of growth factor treatment. Some studies show a decrease in infection rate compared with that found in historical experience. A randomized clinical trial has been performed with G-CSF and has been reported in abstract form only. In this study of 102 patients with advanced myelodysplastic syndrome (refractory anemia with excess blasts (RAEB) and refractory anemia with excess blasts in transformation (RAEB-T)), virtually all the patients randomized to treatment with G-CSF had an improvement in neutrophil counts. Importantly, this did not result in acceleration of disease but also did not result in improvement in overall survival (Greenberg *et al.*, 1993). Combination therapy with G-CSF and erythropoietin (EPO) have also been performed. EPO alone results in a 10–20% erythroid response rate (Stein *et al.*, 1991). *In vitro*, G-CSF and EPO are synergistic for erythroid colony formation both in normal individuals and patients with myelodysplastic syndrome (Souza *et al.*, 1986; Greenberg *et al.*, 1991). Owing to this synergistic effect, combined therapy with G-CSF and EPO has been pursued. In two independent studies, an erythroid response rate of approximately 40% has been observed (Hellstrom-Lindberg *et al.*, 1993; Negrin *et al.*, 1993). These erythroid responses have been durable in some patients and many require both G-CSF and EPO (Negrin *et al.*, 1996). Responding patients generally had lower serum EPO levels before treatment and had a lower red blood cell transfusion requirement before treatment (Hellstrom-Lindberg *et al.*, 1995; Negrin *et al.*, 1996).

APLASTIC ANEMIA

HGFs have also been used in the treatment of patients with aplastic anemia. In the five reported phase I–II trials, 41 patients have been treated with variable doses of GM-CSF (Antin *et al.*, 1988; Nissen *et al.*, 1988; Vadhan-Raj *et al.*, 1988; Champlin *et al.*, 1989; Guinan *et al.*, 1990). Thirty-three patients have had an increase in white blood cell counts consisting primarily of neutrophils and eosinophils. No clear dose response was evident, nor was it clear that GM-CSF caused a decrease in infections. G-CSF has been used in patients with aplastic anemia and short-term improvements in neutrophil counts have been found (Kojima *et al.*, 1991; Sonoda *et al.*, 1992). There were no effects on red blood cell and platelet counts with either G-CSF or GM-CSF. Randomized controlled clinical trials are necessary to determine whether GM-CSF or G-CSF treatment impacts on morbidity and mortality in patients with aplastic anemia. No such trials have been performed.

ACQUIRED IMMUNE DEFICIENCY SYNDROME

In patients with the acquired immune deficiency syndrome (AIDS) or AIDS-related complex (ARC) treated with zidovudine (AZT), the dose-limiting toxicity in the majority of patients is neutropenia. HGFs have been beneficial in limiting AZT-induced neutropenia and allowing continued treatment (Levine *et al.*, 1991; Van der Wouw *et al.*, 1991). In patients who received GM-CSF with AZT, there was no increase in viral replication as measured by P24 antigen levels (Levine *et al.*, 1991). *In vitro* tests have suggested that GM-CSF in combination with AZT may enhance the effect of AZT by increasing the concentration of AZT within monocytes and macrophages (Kitano *et al.*, 1991).

CYCLIC AND CONGENITAL NEUTROPENIAS

Cyclic neutropenia and severe congenital neutropenia are rare diseases in which the primary disorder is impairment in neutrophil production. Patients typically have neutropenia and suffer from frequent episodes of fever and infection, and often die at a young age. Cyclic neutropenia is characterized by regular 21-day cyclic fluctuation in all hematopoietic cells, the most prominent clinical manifestation being neutropenia. During the periods of neutropenia, mucosal ulcers, fevers and malaise are common and patients may encounter life-threatening infections.

Patients with cyclic neutropenia treated with G-CSF have demonstrated persistent cycling but a decrease in cycle length, decreased days of severe neutropenia, as well as decrease in mucosal inflammation, fever and infections (Hammond *et al.*, 1989). Patients with severe congenital neutropenia have also been treated with G-CSF, resulting in a dose-dependent increase in ANC (Welte *et al.*, 1990). Interestingly, GM-CSF has not been effective in increasing the neutrophil count in these patients (Welte *et al.*, 1990). A randomized clinical trial has been performed where patients with cyclic neutropenia either started G-CSF immediately or were observed for 4 months followed by G-CSF treatment. In this study, 123 patients were treated. The patients had a 50% reduction in the incidence and duration of infections and a 70% reduction in antibiotic usage while being treated with G-CSF (Dale *et al.*, 1993). Asymptomatic splenic enlargement was commonly observed. Similar effects with G-CSF have been demonstrated in patients with congenital agranulocytosis (Bonilla *et al.*, 1989).

NOVEL HEMATOPOIETIC GROWTH FACTORS

Although G-CSF and GM-CSF are the only two HGFs currently approved for clinical use, a number of other growth factors are in clinical trials. These HGFs have the potential of stimulating other populations of precursor cells for potential clinical applications. These novel HGFs are summarized below.

Interleukin-3

IL-3 has been shown to stimulate myelopoiesis, erythropoiesis and thrombopoiesis, suggesting an earlier site of action than G-CSF or GM-CSF (Leary *et al.*, 1987; Saeland *et al.*, 1988; Sonoda *et al.*, 1988). Phase I–II trials using IL-3 for patients with normal hematopoiesis (Ganser *et al.*, 1990b; Lindemann *et al.*, 1991), aplastic anemia (Ganser *et al.*, 1990c; Nimer *et al.*, 1994) and myelodysplastic syndromes (Ganser *et al.*, 1990a) have demonstrated a consistent rise in neutrophil count which is slower than that seen with G-CSF or GM-CSF. Theoretically, this is due to the effect of IL-3 on earlier progenitor cells in the hematopoietic system. In patients with advanced malignancy and normal hematopoiesis, and those with bone marrow failure secondary to chemotherapy or tumor infiltration, the majority of patients had an increase in reticulocyte and platelet count in response to IL-3. This rise in platelets and reticulocytes was not seen in patients with aplastic anemia or myelodysplastic syndromes. Toxicity secondary to IL-3 appears to be greater than that seen with either GM-CSF or G-CSF, with virtually all patients developing fevers and approximately one-third of patients developing a unique toxicity consisting of headaches with associated nuchal rigidity. Because of toxicity and marginal clinical activity, development of IL-3 has been slow.

Stem Cell Factor

Human stem cell factor (SCF) or c-kit ligand has been utilized in the autologous transplant setting to enhance the mobilizing effects of G-CSF. *In vitro*, SCF alone stimulates little colony growth but in combination with later-acting growth factors dramatically increases proliferation of colony forming cells (Bernstein *et al.*, 1991). In primates increases in the levels of erythrocytes, neutrophils, monocytes and lymphocytes have been demonstrated (Andrews *et al.*, 1991). In primates, SCF plus G-CSF was more effective than G-CSF alone in mobilizing hematopoietic progenitor cells which were capable of engrafting lethally irradiated primates and dogs (Andrews *et al.*, 1995). In humans, phase I–II studies have demonstrated a similar effect both following chemotherapy or in combination with G-CSF (Glaspy *et al.*, 1994; Weaver *et al.*, 1996). A randomized phase III trial comparing G-CSF alone with G-CSF plus SCF is underway. SCF is complicated by the fact that this HGF stimulates mast cells, which can result in significant allergic toxicities (Costa *et al.*, 1996). A regimen of prophylactic medication can alleviate this problem in the majority of patients, however.

Flt3 Ligand

The flt3 ligand (flt3L) is a novel HGF that binds to the tyrosine kinase receptor flt3 and stimulates primitive hematopoietic stem cells (Hannum *et al.*, 1994; Lyman *et al.*, 1994). Like SCF, the flt3L synergizes with a number of other HGFs, greatly augmenting colony growth *in vitro* (Jacobsen *et al.*, 1995). Unlike SCF, flt3L does not activate mast cells and the expectation is that this HGF will have less potential toxicity *in vivo*. Phase I–II trials in humans are underway.

Thrombopoietin

A major limitation of several of the HGFs is that they do not stimulate megakaryocyte growth and platelet production. For many years, factors that effect platelet production have been sought. Recently several groups identified a new cytokine that binds to the mpl receptor, which has profound activity on megakaryocyte growth and development (Bartley *et al.*, 1994; de Sauvage *et al.*, 1994; Lok *et al.*, 1994; Wendling *et al.*, 1994). This

HGF has been termed thrombopoietin (TPO) mpl ligand, and megakaryocyte growth and differentiation factor (MGDF), and these terms will be utilized interchangeably. Knockout mice deficient in the receptor c-mpl are profoundly thrombocytopenic, with an 85% decrease in the number of megakaryocytes and platelets (Gurney *et al.*, 1994). Recombinant MGDF stimulates thrombopoiesis in mice and non-human primates which are either normal or pretreated with cytotoxic chemotherapy (Farese *et al.*, 1995; Hokom *et al.*, 1995; Ulich *et al.*, 1995; Harker *et al.*, 1996). Early clinical trials have been performed in humans. In one study of patients with advanced malignancy, pegylated recombinant human MGDF stimulated platelet production, which peaked 16 days after administration of the cytokine (O'Malley *et al.*, 1996). No apparent toxicity was noted and the platelets were fully functional. A recent study has documented the profound biologic activity of MGDF in patients with lung cancer who have been treated with chemotherapy. In this study of 38 patients, some of which were randomized to MGDF whereas others received placebo, the treated patients had a significantly higher platelet nadir following chemotherapy (Fanucchi *et al.*, 1997). The challenge will be to identify the clinical setting in which MGDF is clinically efficacious, especially as the effects are relatively delayed.

Fusion and Synthetic Cytokines

With the advent of molecular cloning techniques, it has become possible to engineer recombinant cytokines with desired clinical effects. The first HGF in this class of agents is PIXY321 which is a fusion protein combining the activities of IL-3 and GM-CSF (Vadhan-Raj *et al.*, 1995). Unfortunately, this molecule has not proven superior to GM-CSF alone in randomized studies (O'Shaughnessy *et al.*, 1996).

The synthetic cytokine SC55494 (Synthokine) which is an IL-3 receptor ligand has been produced and is in clinical trials. This cytokine has impressive activity in a radiation-induced bone marrow aplasia model in rhesus monkeys, both alone and in combination with G-CSF (Farese *et al.*, 1996; MacVittie *et al.*, 1996). This novel cytokine is currently in clinical trials in humans.

TOXICITIES OF THE CSFs

Human colony stimulating factors have generally been well tolerated in clinical use. Toxicities associated with G-CSF and GM-CSF are outlined in Table 2.

An interesting syndrome has been observed in some patients following the first but not subsequent doses of GM-CSF (Lieschke *et al.*, 1989b). This reaction is characterized by flushing, tachycardia, hypotension, dyspnea, nausea, vomiting and musculoskeletal pain. This reaction occurs typically within the first hour of GM-CSF administration and is more commonly seen with intravenous infusion. Treatment with oxygen, intravenous fluids and acetaminophen relieves symptomology.

GM-CSF shows clear dose-related toxicity with a capillary leak syndrome with pericardial and pleural effusions, ascites, weight gain and peripheral edema seen at dosages greater than 20 μg/kg daily (Antman *et al.*, 1988; Brandt *et al.*, 1988; Steward *et al.*, 1989). This syndrome is not observed at dosage levels of GM-CSF that are generally used clinically. The mechanism of this syndrome is thought to be secondary to IL-1 and

Table 2. Toxicities of CSFs.

	GM-CSF	G-CSF
Non-dose related		
Fever	Common	Rare
Bone pain	Common (10%)	Common (10%)
Myalgia	Common	Rare
Catheter thrombosis	Rare	Not reported
Splenomegaly	Rare	Rare
Dose related (> 32 µg/kg daily)		
Effusion		
Pericardial	+	–
Pleural	+	–
Ascites	+	–
Pulmonary emboli	+	–
Edema	+	–
Weight gain	+	–
Laboratory		
Increased LDH	+	+
Increased LAP	+	+
Increased uric acid	+	+
Increased ALP	+	+
Increased eosinophils	+	–

tumor necrosis factor (TNF) production from activated monocytes. G-CSF has not been shown to have any dose-limiting side-effects and does not produce this syndrome.

GM-CSF use has been associated with fever and bone pain. GM-CSF bone pain typically occurs at the time of initial infusion and is usually relieved with acetaminophen. Fevers are also a common side-effect following GM-CSF, occurring in up to 50% of patients usually 4–6 h following GM-CSF infusion. In the five randomized control trials comparing GM-CSF with placebo (Powles *et al.*, 1990; Nemunaitis *et al.*, 1991b; Advani *et al.*, 1992; Dewitte *et al.*, 1992; Gulati and Bennett, 1992), these were the only symptoms that were significantly more common in the GM-CSF group. Patients treated with subcutaneous injections of GM-CSF may experience erythematous lesions at injection sites (Lieschke *et al.*, 1989a). Case reports have appeared where administration of GM-CSF to patients with rheumatoid arthritis, hemolysis and idiopathic thrombocytopenic purpura have resulted in flares of the condition (Lieschke *et al.*, 1989a; De Vries *et al.*, 1991; Nathan and Besa, 1992).

In clinical trials using G-CSF, the only side-effect regularly observed is bone pain. This generally occurs 1–2 days before rebound of the neutrophil count rather than with infusion of growth factor, as seen with GM-CSF. In randomized trials, mild to moderate bone pain was reported in approximately 10% of patients receiving G-CSF (Crawford *et al.*, 1991). This side-effect is generally controllable with acetaminophen. There is one reported case of Sweets' syndrome occurring after G-CSF in a patient with hairy cell leukemia (Glaspy *et al.*, 1988). In addition, flare of a patient's underlying psoriasis was reported in one patient with myelodysplastic syndrome treated with G-CSF (Negrin *et al.*, 1989). In patients with congenital neutropenia, splenomegaly may develop concomitant with an improved blood count (Hammond *et al.*, 1989; Welte *et al.*, 1990).

Both G-CSF and GM-CSF may result in mild laboratory abnormalities such as increases in lactate dehydrogenase (LDH), leukocyte alkaline phosphatase (LAP), alkaline phosphatase (ALP) and uric acid are common and decrease upon discontinuation of the drug. GM-CSF treatment is invariably associated with eosinophilia which is not seen with G-CSF. A transient decrease in leukocyte count has been observed immediately after administration of G-CSF and GM-CSF (Lindemann *et al.*, 1989; Lieschke *et al.*, 1989a).

THEORETIC CONCERNS WITH THE CLINICAL USE OF CSFs

Although toxicities of the colony stimulating factors at dosages used clinically are minimal, a number of theoretic concerns remain (Table 3). The major theoretic concern is the possibility that these growth factors could stimulate clonogenic tumor cells. This concern is raised most in treatment of myelodysplastic syndromes and acute myelogenous leukemia.

Although concern over acceleration of disease is most prominent in the myeloid leukemias, several other neoplasms have been demonstrated to have receptors for G-CSF and GM-CSF (Baldwin *et al.*, 1989, 1991; Berdel *et al.*, 1989; Salmon and Lui, 1989).

The use of hematopoietic growth factors in patients with myeloid leukemia raises special concerns. A number of *in vitro* studies have demonstrated the presence of receptors for G-CSF and GM-CSF on leukemic blasts and that growth factors *in vitro* may stimulate proliferation of these cells (Griffin *et al.*, 1986; Hoang *et al.*, 1986; Begley *et al.*, 1987; Vellenga *et al.*, 1987; Motoji *et al.*, 1989). *In vivo*, several studies involving AML have been reported, none of which has reported an increased incidence of disease progression in patients treated with HGFs.

At present, there is no evidence that hematopoietic colony stimulating factors increase leukemic growth *in vivo* when given in combination with chemotherapy. It seems prudent to limit the use of colony stimulating factors in patients with myeloid leukemia to individuals at high risk of infection and those with few blasts present following chemotherapy.

A major concern in the treatment of myelodysplastic syndrome with CSFs is the potential for acceleration of disease to acute leukemia. However, in one randomized trial, there was no increase in leukemia progression in patients treated with G-CSF (Greenberg *et al.*, 1993).

Table 3. Theoretic toxicities of the colony stimulating factors.

Activation of disease
Myeloid leukemia
Myelodysplastic syndromes
Solid tumors
HIV
Graft-*versus*-host disease
Bone marrow exhaustion
Biased stem cell commitment

Biased Stem Cell Commitment and Bone Marrow Exhaustion

Two other theoretic toxicities, namely biased stem cell commitment (also referred to as stem cell steal) and bone marrow exhaustion, are of concern. Biased stem cell commitment refers to the possibility that colony stimulating factors may drive progenitor cells along one pathway of differentiation and thereby limit differentiation into other lineages. Randomized control trials in a variety of settings have not shown this effect. Bone marrow exhaustion is another theoretic concern. With repeated stimulation of stem cells by the use of colony stimulating factors, it is theoretically possible that the pool of stem cells may become exhausted. One trial with long-term maintenance therapy of patients with myelodysplastic syndrome has not demonstrated such a phenomenon (Negrin *et al.*, 1990). These potential concerns are heightened by the use of CSFs active on more immature progenitor cells or when combinations of HGFs are utilized.

SUMMARY

In summary, the HGFs are currently being used in a variety of clinical situations. These potent agents have clear biologic efficacy which has translated into clinical benefit in a variety of settings. Toxicities at doses used clinically are mild, although theoretic concerns remain. New and promising factors are currently under development. Future efforts may involve combinations of growth factors to minimize toxicity and maximize effectiveness.

REFERENCES

Advani, R., Chao, N.J., Horning, S.J. *et al.* (1992). Granulocyte–macrophage colony-stimulating factor (GM-CSF) as an adjunct to autologous hemopoietic stem cell transplantation for lymphoma. *Ann. Intern. Med.* **116**, 183–189.

Andrews, R., Briddell, R.A., Knitter, G. *et al.* (1991). Recombinant human stem cell factor, a c-kit ligand stimulates hematopoiesis in primates. *Blood* **78**, 1975–1980.

Andrews, R.G., Briddell, R.A., Knitter, G.H., Rowley, S.D., Appelbaum, F.R. and McNiece, I.K. (1995). Rapid engraftment by peripheral blood progenitor cell mobilized by recombinant human stem cell factor and recombinant human granulocyte colony-stimulating factor in nonhuman primates. *Blood* **85**, 15–20.

Antin, J.H., Smith, B.R., Holmes, W. and Rosenthal, D. (1988). Phase I/II study of recombinant human granulocyte–macrophage colony-stimulating factor in aplastic anemia and myelodysplastic syndrome. *Blood* **72**, 705–713.

Antman, K.S., Griffin, J.D., Elias, A. *et al.* (1988). Effect of recombinant human granulocyte–macrophage colony stimulating factor on chemotherapy induced myelosuppression. *N. Engl. J. Med.* **319**, 593–598.

Attal, M., Harousseau, J.-L., Stoppa, A.-M. *et al.* (1996). A prospective, randomized trial of autologous bone marrow transplantation and chemotherapy in multiple myeloma. *N. Engl. J. Med.* **335**, 91–97.

Baldwin, G.C., Gasson, J.C., Kaufman, S.E. *et al.* (1989). Nonhematopoietic tumor cells express functional CSF receptors. *Blood* **83**, 1033–1037.

Baldwin, G., Golde, D., Widhopf, G., Economou, J. and Gasson, J.C. (1991). Identification and characterization of a low affinity granulocyte macrophage colony stimulating factor receptor on primary and cultured human melanoma cells. *Blood* **78**, 609–615.

Bartley, T.D., Bogenberger, J., Hunt, P. *et al.* (1994). Identification and cloning of a megakaryocyte growth and development factor that is a ligand for the cytokine receptor Mpl. *Cell* **77**, 1117–1124.

Begley, C.G., Metcalf, D. and Nicola, N.A. (1987). Primary human myeloid leukemia cells: comparative responsiveness to proliferative stimulation by GM-CSF or G-CSF and membrane expression of CSF receptors. *Leukemia* **1**, 1–8.

Bensinger, W.I., Longin, K.G., Appelbaum, F. *et al.* (1994). Peripheral blood stem cells (PBSCs) collected after recombinant granulocyte colony stimulating factor (rhG-CSF): an analysis of factors correlating with the tempo of engraftment after transplantation. *Br. J. Haematol.* **87**, 825–831.

Bensinger, W.I., Weaver, C.H., Appelbaum, F.R. *et al.* (1995). Transplantation of allogeneic peripheral blood stem cells mobilized by recombinant human granulocyte colony-stimulating factor. *Blood* **85**, 1655–1658.

Bensinger, W.I., Clift, R., Martin, P. *et al.* (1996). Allogeneic peripheral blood stem cell transplantation in patients with advanced hematologic malignancies: a retrospective comparison with marrow transplantation. *Blood* **88**, 2794–2800.

Berdel, W.E., Danhauser-Riedl, S., Steinhauser, G. and Winton, E.F. (1989). Various human hematopoietic growth factors (interleukin-2, GM-CSF, G-CSF) stimulate clonal growth of nonhematopoietic tumor cells. *Blood* **73**, 80–83.

Bernstein, I.D., Andrews, R.G. and Zsebo, K.M. (1991). Recombinant human stem cell factor enhances the formation of colonies by $CD34^+$ and $CD34^+lin^-$ cells and the generation of colony-forming cell progeny from $CD34^+lin^-$ cells cultured with IL-3, G-CSF or GM-CSF. *Blood* **77**, 2316–2321.

Bolwell, B., Goormastic, M., Dannley, R. *et al.* (1997). G-CSF postautologous progenitor cell transplantation: a randomized study of 5, 10 and 16 μg/kg/day. *Bone Marrow Transplant.* **19**, 215–219.

Bonilla, M.A., Gillio, A.P., Ruggeiro, M. *et al.* (1989). Effects of recombinant human granulocyte colony-stimulating factor on neutropenia in patients with congenital agranuylocytosis. *N. Engl. J. Med.* **320**, 1574–1580.

Brandt, S.J., Peters, W.P., Atwater, S.K. *et al.* (1988). Effect of recombinant human granulocyte–macrophage colony-stimulating factor on hematopoietic reconstitution after high-dose chemotherapy and autologous bone marrow transplantation. *N. Engl. J. Med.* **318**, 869–876.

Bronchud, M.H., Scarffe, J.H., Thatcher, N. *et al.* (1987). Phase I/II study of recombinant human granulocyte colony-stimulating factor in patients receiving intensive chemotherapy for small cell lung cancer. *Br. J. Cancer* **36**, 809–813.

Champlin, R.E., Nimer, S.D., Ireland, P., Oette, D.J. and Golde, D.W. (1989). Treatment of refractory aplastic anemia with recombinant human granulocyte–macrophage colony-stimulating factor. *Blood* **73**, 694–699.

Cheng, G.Y.M., Kelleher, C.A., Miyauchi, J. *et al.* (1988). Structure and expression of genes with GM-CSF and G-CSF in blast cells from patients with acute myeloblastic leukemia. *Blood* **71**, 204–208.

Costa, J.J., Demetri, G.D., Harrist, T.J. *et al.* (1996). Recombinant human stem cell factor (kit ligand) promotes human mast cell and melanocyte hyperplasia and functional activation *in vivo*. *J. Exp. Med.* **183**, 2681–2686.

Crawford, J., Ozer, H., Stoller, R. *et al.* (1991). Reduction by granulocyte colony-stimulating factor of fever and neutropenia induced by chemotherapy in patients with small-cell lung cancer. *N. Engl. J. Med.* **315**, 164–170.

Dale, D.C., Bonilla, M.A., Davis, M.W. *et al.* (1993). A randomized controlled phase III trial of recombinant human granulocyte colony-stimulating factor (filgrastim) for treatment of severe chronic neutropenia. *Blood* **81**, 2496–2502.

De Sauvage, F.J., Hass, P.E., Spencer, S.D. *et al.* (1994). Stimulation of megakaryocytopoiesis and thrombopoiesis by the c-Mpl ligand. *Nature* **369**, 533–536.

De Vries, E.G.E., Willemse, P.H.B., Biemsa, B., Stern, A.C., Limburg, P.C. and Vellenga, E. (1991). Flare-up of rheumatoid arthritis during GM-CSF treatment after chemotherapy. *Lancet* **338**, 517–518.

DeWitte, T., Gratwohl, A., Van Der Lely, N. *et al.* (1992). Recombinant human granulocyte–macrophage colony-stimulating factor accelerates neutrophil and monocyte recovery after allogeneic T-cell depleted bone marrow transplantation. *Blood* **79**, 1359–1365.

Dombret H., Chastang, C., Fenaux, P. *et al.* (1995). A controlled study of recombinant human granulocyte colony-stimulating factor in elderly patients after treatment for acute myelogenous leukemia. *N. Engl. J. Med.* **332**, 1678–1683.

Fanucchi, M., Glaspy, J., Crawford, J. *et al.* (1997). Effects of polyethylene glycol-conjugated recombinant human megakaryocyte growth and development factor on platelet counts after chemotherapy for lung cancer. *N. Engl. J. Med.* **336**, 404–409.

Farese, A.M., Hunt, P., Boone, T. and MacVittie, T.J. (1995). Recombinant human megakaryocyte growth and development factor stimulates thrombocytopoiesis in normal nonhuman primates. *Blood* **86**, 54–59.

Farese, A.M., Herodin, F., McKearn, J.P., Baum, C., Burton, E. and MacVittie, T.J. (1996). Acceleration of hematopoietic reconstitution with a synthetic cytokine (SC-55494) after radiation-induced bone marrow aplasia. *Blood* **87**, 581–591.

Faucher, C., Le Corroller, A.G., Chabannon, C. *et al.* (1996). Administration of G-CSF can be delayed after transplantation of autologous G-CSF-primed blood stem cells: a randomized study. *Bone Marrow Transplant.* **17**, 533–536.

Gabrilove, J.L., Jakubowski, A., Scher, H. *et al.* (1988). Effect of granulocyte colony stimulating factor on neutropenia and associated morbidity due to chemotherapy for transitional cell carcinoma of the urothelium. *N. Engl. J. Med.* **318**, 1414–1422.

Ganser, A., Volkers, B., Greher, J. *et al.* (1989). Recombinant human granulocyte–macrophage colony-stimulating factor in patients with myelodysplastic syndromes—a phase I/II trial. *Blood* **76**, 31–37.

Ganser, A., Seipelt, G., Lindemann, A. *et al.* (1990a). Effect of recombinant human interleukin-3 in patients with myelodysplastic syndromes. *Blood* **3**, 455–462.

Ganser, A., Lindemann, A., Seipelt, G. *et al.* (1990b). Effects of recombinant human interleukin-3 in patients with normal hematopoiesis and in patients with bone marrow failure. *Blood* **76**, 666–676.

Ganser, A., Lindemann, A., Seipelt, G. *et al.* (1990c). Effects of recombinant human interleukin-3 in aplastic anemia. *Blood* **76**, 1287–1292.

Gianni, A.M., Siena, S., Bregni, M. *et al.* (1989). Granulocyte–macrophage colony-stimulating factor to harvest circulating hematopoietic stem cells for autotransplantation. *Lancet* **ii**, 580–584.

Gianni, A.M., Tarella, C., Siena, S. *et al.* (1990). Durable and complete engraftment of rhGM-CSF exposed peripheral blood progenitor cells. *Bone Marrow Transplant.* **6**, 143–145.

Gisselbrecht, C., Prentice, H.G., Bacigalupo, A. *et al.* (1994). Placebo-controlled phase III trial of lenograstim in bone-marrow transplantation. *Lancet* **343**, 696–700.

Glaspy, J.A., Baldwin, G.C., Robertson, P.A. *et al.* (1988). Therapy for neutropenia in hairy cell leukemia with recombinant human granulocyte colony-stimulating factor. *Ann. Intern. Med.* **110**, 789–795.

Glaspy, J., McNiece, I., LeMaistre, F. *et al.* (1994). Effects of stem cell factor (rhSCF) and filgrastim (rhG-CSF) on mobilization of peripheral blood progenitor cells (PBPC) and on hematological recovery posttransplant: early results from a phase I/II study. *J. Clin. Oncol.* **13**, 76A.

Gorin, N.C., Coiffier, B., Hayat, M. *et al.* (1992). Recombinant human granulocyte–macrophage colony-stimulating factor after high-dose chemotherapy and autologous bone marrow transplantation with unpurged and purged marrow in non-Hodgkin's lymphoma: a double-blind placebo-controlled trial. *Blood* **80**, 1149–1157.

Greenberg, P.L. (1983). The smoldering myeloid leukemia states: clinical and biologic features. *Blood* **61**, 1035–1044.

Greenberg, P.L., Negrin, R.S. and Ginzton, N.L. (1991). G-CSF synergizes with erythropoietin for enhancing erythroid colony-formation in myelodysplastic syndromes. *Blood* **78 (supplement 1)**, 38A.

Greenberg, P., Taylor, K., Larson, R. *et al.* (1993). Phase III randomized multicenter trial of G-CSF *vs.* observation for myelodysplastic syndromes (MDS). *Blood* **82 (supplement 1)**, 196A.

Griffin, J.D., Young, D., Herrmann, F., Wiper, D., Wagner, K. and Sabbath, K.D. (1986). Effects of recombinant GM-CSF on proliferation of clonogenic cells in acute myeloblastic leukemia. *Blood* **67**, 1448–1453.

Guinan, E., Sieff, C., Oette, D. and Nathan, D.G. (1990). A phase I–II trial of recombinant granulocyte–macrophage colony-stimulating factor for children with aplastic anemia. *Blood* **76**, 1077–1082.

Gulati, S.C. and Bennett, CL. (1992). Granulocyte–macrophage colony-stimulating factor (GM-CSF) as adjunct therapy in relapsed Hodgkin disease. *Ann. Intern. Med.* **116**, 177–182.

Gurney, A.L., Carber-Moore, K., de Sauvage, F. and Moore, M.W. (1994). Thrombocytopenia in c-mpl-deficient mice. *Science* **264**, 1445–1447.

Hammond, W.P., Price, T.H., Souza, L.M. and Dale, D.C. (1989). Treatment of cyclic neutropenia with granulocyte colony-stimulating factor. *N. Engl. J. Med.* **320**, 1306–1311.

Hannum, C., Culpepper, J., Campbell, D. *et al.* (1994). Ligand for FLT3/FLK2 receptor tyrosine kinase regulates growth of haematopoietic stem cells and is encoded by variant RNAs. *Nature* **368**, 643–646.

Harker, L.A., Marzec, U.M., Hunt, P. *et al.* (1996). Dose–response effects of pegylated human megakaryocyte growth and development factor on platelet production and function in nonhuman primates. *Blood* **88**, 511–521.

Hellstrom-Lindberg, E., Birgegard, G., Carlsson, M. *et al.* (1993). A combination of G-CSF and erythropoietin may synergistically improve the anemia in patients with MDS. *Leukemia Lymphoma* **11**, 221–228.

Hellstrom-Lindberg, E., Negrin, R., Stein, R. *et al.* (1995). Efficacy of G-CSF and EPO on the anemia in patients with myelodysplastic syndromes. *Blood* **86 (supplement 1)**, 338A.

Herrmann, F., Lindemann, A., Klein, H., Lubbert, M., Schulz, G. and Mertelsmann, R. (1989). Effect of recombinant granulocyte–macrophage colony-stimulating factor in patients with myelodysplastic syndrome with excess blasts. *Leukemia* **3**, 335–338.

Hoang, T., Nara, N., Wong, G., Clark, S., Minden, S.G. and McCulloch, E.A. (1986). Effects of recombinant GM-CSF on the blast cells of acute myeloblastic leukemia. *Blood* **68**, 313–316.

Hokom, M.M., Lacey, D., Kinstler, O.B. *et al.* (1995). Pegylated megakaryocyte growth and development factor abrogates the lethal thrombocytopenia associated with carboplatin and irradiation in mycelex troches. *Blood* **86**, 4486–4492.

Jacobsen, S.E.W., Okkenhaug, C., Myklebust, J., Veiby, O.P. and Lyman, S.D. (1995). The FLT3 ligand potently and directly stimulates the growth and expansion of primitive murine bone marrow progenitor cells *in vitro*: synergistic interactions with interleukin (IL) 11, IL-12, and other hematopoietic growth factors. *J. Exp. Med.* **181**, 1357–1363.

Kitano, K., Abboud, C., Ryan, D., Quan, S.G., Baldwin, G.C. and Golde, D.W. (1991). Macrophage-active colony stimulating factors enhance human immunodeficiency virus type 1 infection in bone marrow stem cells. *Blood* **77**, 1699–1705.

Klumpp, T.R., Mangan, K.F., Goldberg, S.L., Pearlman, E.S. and Macdonald, J.S. (1995). Granulocyte colony-stimulating factor accelerates neutrophil engraftment following peripheral-blood stem-cell transplantation: a prospective, randomized trial. *J. Clin. Oncol.* **13**, 1323–1327.

Kojima, S., Fukuda, M., Miyajima, Y., Matsayama, T. and Horibi, K. (1991). Treatment of aplastic anemia in children with recombinant human granulocyte colony stimulating factor. *Blood* **77**, 937–941.

Korbling, M., Przepiorka, D., Huh, Y.O. *et al.* (1995). Allogeneic blood stem cell transplantation for refractory leukemia and lymphoma: potential advantage of blood over marrow allografts. *Blood* **85**, 1659–1665.

Lane, T.A., Law, P., Maryuama, M. *et al.* (1995). Harvesting and enrichment of hematopoietic progenitor cells mobilized into the peripheral blood of normal donors by granulocyte–macrophage colony-stimulating factor (GM-CSF) or G-CSF: potential role in allogeneic marrow transplantation. *Blood* **85**, 275–282.

Leary, A.G., Yang, Y.-C., Clark, S.C., Gasson, J.C., Golde, D.W. and Ogawa, M. (1987). Recombinant gibbon interleukin-3 supports formation of human multilineage colonies and blast cell colonies in culture: comparison with recombinant human granulocyte–macrophage colony-stimulating factor. *Blood* **70**, 1343–1348.

Levine, J., Allan, J., Tessitore, J. *et al.* (1991). Recombinant human granulocyte–macrophage colony stimulating factor ameliorates zidovudine induced neutropenia in patients with acquired immunodeficiency syndrome (AIDS), AIDS related complex. *Blood* **78**, 3148–3154.

Lieschke, G., Maher, D., Cebon, J. *et al.* (1989a). Effects of bacterially synthesized recombinant human granulocyte macrophage colony stimulating factor in patients with advanced malignancy. *Ann. Intern. Med.* **110**, 357–364.

Lieschke, G., Cebon, J. and Morstyn, G. (1989b). Characterization of the clinical effects after the first dose of bacterially synthesized recombinant human granulocyte–macrophage colony stimulating factor. *Blood* **74**, 2634–2643.

Lindemann, A., Herrmann, F., Oster, W. *et al.* (1989). Hematologic effects of recombinant human granulocyte colony stimulating factor in patients with malignancy. *Blood* **74**, 2644–2651.

Lindemann, A., Ganser, F., Herrmann, F. *et al.* (1991). Biologic effects of recombinant human interleukin-3 *in vivo*. *J. Clin. Oncol.* **9**, 2120–2127.

Link, H., Boogaerts, M.A., Carella, A.M. *et al.* (1992). A controlled trial of recombinant human granulocyte–macrophage colony-stimulating factor after total body irradiation, high-dose chemotherapy, and autologous bone marrow transplantation for acute lymphoblastic leukemia or malignant lymphoma. *Blood* **80**, 2188–2195.

Lok, S., Kaushansky, K., Holly, R.D. *et al.* (1994). Cloning and expression of murine thrombopoietin cDNA and stimulation of platelet production *in vivo*. *Nature* **369**, 565–569.

Lyman, S.D., James, L., Johnson, L. *et al.* (1994). Cloning of the human homologue of the murine flt3 ligand: a growth factor for early hematopoietic progenitor cells. *Blood* **83**, 2795–2801.

MacVittie, T.J., Farese, A.M., Herodin, F., Grab, L.B., Baum, C.M. and McKearn, J.P. (1996). Combination therapy for radiation-induced bone marrow aplasia in nonhuman primates using synthokine SC-55494 and recombinant human granulocyte colony-stimulating factor. *Blood* **87**, 4129–4135.

Maher, D.W., Lieschke, G.J., Green, M. *et al.* (1994). Filgrastim in patients with chemotherapy-induced febrile neutropenia. *Ann. Intern. Med.* **121**, 492–501.

Masaoka, T., Takaku, F., Kato, S. *et al.* (1989). Recombinant human granulocyte colony stimulating factor in allogeneic bone marrow transplant. *Exp. Haematol.* **17**, 1047–1050.

Matsunaga, T., Sakamaki, S., Kohgo, Y., Ohi, S., Hirayama, Y. and Niitsu, Y. (1993). Recombinant human granulocyte colony-stimulating factor can mobilize sufficient amounts of peripheral blood stem cells in healthy volunteers for allogeneic transplantation. *Bone Marrow Transplant.* **11**, 103–108.

Molineux, G., Pojda, Z., Hampson, J., Lord, B.I. and Dexter, T.M. (1990). Transplantation potential of peripheral blood stem cells induced by granulocyte colony stimulating factor. *Blood* **76**, 2153–2158.

Morstyn, G., Souza, L.M., Keech, J. *et al.* (1988). Effect of granulocyte colony stimulating factor on neutropenia induced by cytotoxic chemotherapy. *Lancet* **i**, 667–672.

Motoji, T., Takanashi, M., Fuchinoue, M., Masuda, M., Oshimi, K. and Mizoguchi, H. (1989). Effect of recombinant GM-CSF and recombinant G-CSF on colony formation of blast progenitors in acute myeloblastic leukemia. *Exp. Haematol.* **17**, 56–60.

Nathan, F.E. and Besa, E.C. (1992). GM-CSF and accelerated hemolysis. *N. Engl. J. Med.* **326**, 417. (Erratum *N. Engl. J. Med.* 1992, **326**, 904.)

Negrin, R.S., Haeuber, D.H., Nagler, A. *et al.* (1989). Treatment of myelodysplastic syndromes with recombinant human granulocyte-colony-stimulating factor. A phase I/II trial. *Ann. Intern. Med.* **110**, 976–985.

Negrin, R.S., Haeuber, D.H., Nagler, A. *et al.* (1990). Maintenance treatment of patients with myelodysplastic syndromes using recombinant human granulocyte colony-stimulating factor. *Blood* **76**, 36–43.

Negrin, R.S., Stein, R., Vardiman, J. *et al.* (1993). Treatment of the anemia of myelodysplastic syndromes using recombinant human granulocyte colony-stimulating factor in combination with erythropoietin. *Blood* **82**, 737–743.

Negrin, R.S., Kusnierz-Glaz, C.R., Still, B.J. *et al.* (1995). Transplantation of enriched and purged peripheral blood progenitor cells from a single apheresis product in patients with non-Hodgkin's lymphoma. *Blood* **85**, 3334–3341.

Negrin, R.S., Stein, R., Doherty, K.G. *et al.* (1996). Maintenance treatment of the anemia of myelodysplastic syndromes with recombinant human granulocyte colony-stimulating factor and erythropoietin: evidence for *in vivo* synergy. *Blood* **87**, 4076–4081.

Nemunaitis, J., Singer, J.W., Buckner, C.D. *et al.* (1988). Use of recombinant human granulocyte–macrophage colony-stimulating factor in autologous marrow transplantation for lymphoid malignancies. *Blood* **72**, 834–836.

Nemunaitis, J., Singer, J.W., Buckner, C.D. *et al.* (1990). Use of recombinant human granulocyte–macrophage colony-stimulating factor in graft failure after bone marrow transplantation. *Blood* **76**, 245–253.

Nemunaitis, J., Buckner, C.D., Appelbaum, F.R. *et al.* (1991a). Phase I/II trial of recombinant human granulocyte–macrophage colony stimulating factor following allogeneic bone marrow transplantation. *Blood* **77**, 2065–2071.

Nemunaitis, J., Rabinowe, S.N., Singer, J.W. *et al.* (1991b). Recombinant granulocyte–macrophage colony-stimulating factor after autologous bone marrow transplantation for lymphoid cancer. *N. Engl. J. Med.* **324**, 1773–1778.

Nimer, S.D., Paquette, R.L., Ireland, P., Resta, D., Young, D. and Golde, D.W. (1994). A phase I/II study of interleukin-3 in patients with aplastic anemia and myelodysplasia. *Exp. Hematol.* **22**, 875–880.

Nissen, C., Tichelli, A., Gratwohl, A. *et al.* (1988). Failure of recombinant human granulocyte–macrophage colony-stimulating factor therapy in aplastic anemia patients with very severe neutropenia. *Blood* **72**, 2045–2047.

Ohno, R., Tomonaga, M., Kobayashi, T. *et al.* (1990). Effect of granulocyte colony-stimulating factor after intensive induction therapy in relapsed or refractory acute leukemia. *N. Engl. J. Med.* **323**, 871–877.

O'Malley, C.J., Rasko, J.E.J., Basser, R.L. *et al.* (1996). Administration of pegylated recombinant human megakaryocyte growth and development factor to humans stimulates the production of functional platelets that show no evidence of *in vivo* activation. *Blood* **88**, 3288–3298.

O'Shaughnessy, J.A., Tolcher, A., Riseberg, D. *et al.* (1996). Prospective, randomized trial of 5-fluorouracil, leucovorin, doxorubicin, and cyclophosphamide chemotherapy in combination with the interleukin-3/G granulocyte–macrophage colony-stimulating factor (GM-CSF) fusion protein (PIXY321) *versus* GM-CSF in patients with advanced breast cancer. *Blood* **87**, 2205–2211.

Ottmann, O.G., Hoelzer, D., Gracien, E. *et al.* (1995). Concomitant granulocyte colony-stimulating factor and induction chemoradiotherapy in adult acute lymphoblastic leukemia: a randomized phase III trial. *Blood* **86**, 444–450.

Ozer, H., Miller, L.L., Anderson, J.R. *et al.* (1994). American Society of Clinical Oncology recommendations for the use of hematopoietic colony-stimulating factors: evidence-based clinical practice guidelines. *J. Clin. Oncol.* **12**, 2471–2508.

Pettengell, R., Gurney, H., Radford, J.A. *et al.* (1992). Granulocyte colony-stimulating factor to prevent dose-limiting neutropenia in non-Hodgkin's lymphoma: a randomized controlled trial. *Blood* **71**, 204–208.

Philip T., Guglielmi, C., Hagenbeek, A. *et al.* (1995). Autologous bone marrow transplantation as compared with salvage chemotherapy in relapses of chemotherapy-sensitive non-Hodgkin's lymphoma. *N. Engl. J. Med.* **333**, 1540–1545.

Powles, R., Smith, C., Milan, S. *et al.* (1990). Human recombinant GM-CSF in allogeneic bone marrow transplantation for leukaemia: double-blind, placebo-controlled trial. *Lancet* **ii**, 1417–1420.

Rabinowe, S.N., Neuberg, D., Bierman, P.J. *et al.* (1993). Long-term follow-up of a phase III study of recombinant human granulocyte–macrophage colony-stimulating factor after autologous bone marrow transplantation for lymphoid malignancies. *Blood* **81**, 1903–1908.

Rowe, J.M., Andersen, J.W., Mazza, J.J. *et al.* (1995). A randomized placebo-controlled phase III study of granulocyte–macrophage colony-stimulating factor in adult patients (>55 to 70 years of age) with acute myelogenous leukemia: a study of the Eastern Cooperative Oncology Group. *Blood* **86**, 457–462.

Saeland, S., Caux, C., Favre, C. *et al.* (1988). Effects of recombinant human interleukin 3 on CD34-enriched normal hematopoietic progenitors and on myeloblastic leukemia cells. *Blood* **72**, 1580–1588.

Salmon, S.E. and Liu, R. (1989). Effects of granulocyte–macrophage colony-stimulating factor on *in vitro* growth of human solid tumors. *J. Clin. Oncol.* **7**, 1346–1350.

Schmitz, N., Dreger, P., Suttorp, M. *et al.* (1995). Primary transplantation of allogeneic peripheral blood progenitor cells mobilized by filgrastim (granulocyte colony-stimulating factor). *Blood* **85**, 1666–1672.

Schmitz, N., Linch, D.C., Dreger, P. *et al.* (1996). Randomized trial of filgrastim-mobilized peripheral blood progenitor cell transplantation *versus* autologous bone-marrow transplantation in lymphoma patients. *Lancet* **347**, 353–357.

Sheridan, W.P., Morstyn, G., Wolf, M. *et al.* (1989). Granulocyte colony-stimulating factor and neutrophil recover after high-dose chemotherapy and autologous bone marrow transplantation. *Lancet* **ii**, 891–895.

Sheridan, W.P., Begley, C.G., Juttner, C.A. *et al.* (1992). Effect of peripheral blood progenitor cells mobilized by filgrastim (G-CSF) on platelet recovery after high dose chemotherapy. *Lancet* **339**, 640–644.

Siena, S., Bregni, M., Brando, M., Ravagnani, F., Bonadonna, G. and Gianni, A.M. (1989). Circulation of $CD34^+$ hematopoietic stem cells in the peripheral blood of high-dose cyclophosphamide-treated patients: enhancement by intravenous recombinant human granulocyte macrophage colony-stimulating factor. *Blood* **74**, 1905–1914.

Siena, S., Bregni, M., Brando, B. *et al.* (1991). Flow cytometry for clinical estimation of circulating hematopoietic progenitors for autologous transplantation in cancer patients. *Blood* **77**, 400–409.

Sonoda, Y., Yang, Y.-C., Wong, G.G., Clark, S.C. and Ogawa, M. (1988). Analysis in serum free culture of targets of recombinant human hematopoietic growth factors: interleukin 3 and granulocyte/macrophage-colony-stimulating factor are specific for developmental stages. *Proc. Natl Acad. Sci. U.S.A.* **85**, 4360–4364.

Sonoda, Y., Yashige, H., Fuuii, H. *et al.* (1992). Bilineage response in refractory aplastic anemia patients following long-term administration of recombinant human granulocyte colony-stimulating factor. *Eur. J. Haematol.* **48**, 41–48.

Souza, L., Boone, T., Gabrilove, J. *et al.* (1986). Recombinant human granulocyte colony-stimulating factor: effects on normal and leukemic myeloid cells. *Science* **232**, 61–65.

Spitzer, G., Adkins, D.R., Spencer, V. *et al.* (1994). Randomized study of growth factors post-peripheral-blood stem-cell transplant: neutrophil recovery is improved with modest clinical benefit. *J. Clin. Oncol.* **12**, 661–670.

Stahel, R.A., Jost, L.M., Cerny, T. *et al.* (1994). Randomized study of recombinant human granulocyte colony-stimulating factor after high-dose chemotherapy and autologous bone marrow transplantation for high-risk lymphoid malignancies. *J. Clin. Oncol.* **12**, 1931–1938.

Stein, R.S., Abels, R.I. and Krantz, S.B. (1991). Pharmacologic doses of recombinant human erythropoietin in the treatment of myelodysplastic syndromes. *Blood* **232**, 61–65.

Steward, W.P., Scarffe, J.H., Austin, R. *et al.* (1989). Recombinant human granulocyte macrophage colony stimulating factor given as daily short infusions—a phase I dose toxicity study. *Br. J. Cancer* **59**, 142–145.

Steward, W.P., Scarffe, J.H., Dirix, L.Y. *et al.* (1990). Granulocyte–macrophage colony stimulating factor (GM-CSF) after high-dose melphalan in patients with advanced colon carcinoma. *Br. J. Cancer* **61**, 749–754.

Stone, R.M., Berg, D.T., George, S.L. *et al.* (1995). Granulocyte–macrophage colony-stimulating factor after initial chemotherapy for elderly patients with primary acute myelogenous leukemia. *N. Engl. J. Med.* **332**, 1671–1677.

Taylor, K., Jagannath, S., Spitzer, G. *et al.* (1989). Recombinant human granulocyte colony stimulating factor hastens granulocyte recovery after high dose chemotherapy and autologous bone marrow transplantation in Hodgkin's disease. *J. Clin. Oncol.* **7**, 1791–1799.

Thompson, J.A., Lee, D.J., Kidd, P. *et al.* (1989). Subcutaneous granulocyte–macrophage colony-stimulating factor in patients with myelodysplastic syndrome: toxicity, pharmacokinetics and hematologic effects. *J. Clin. Oncol.* **7**, 629–637.

Tjonnfjord, G.E., Steen, R., Evensen, S.A., Thorsby, E. and Egeland, T. (1994). Characterization of $CD34^+$ peripheral blood cells from healthy adults mobilized by recombinant human granulocyte colony-stimulating factor. *Blood* **84**, 2795–2801.

Tricot, G., Jagannath, S., Vesole, D. *et al.* (1995). Peripheral blood stem cell transplants for multiple myeloma: identification of favorable variables for rapid engraftment in 225 patients. *Blood* **85**, 589–596.

Ulich, T.R., del Castillo, J., Yin, S. *et al.* (1995). Megakaryocyte growth and development factor ameliorates carboplatin-induced thrombocytopenia in mice. *Blood* **86**, 971–976.

Vadhan-Raj, S., Keating, K., LeMaistre, A. *et al.* (1987). Effects of recombinant human granulocyte–macrophage colony stimulating factor in patients with myelodysplastic syndromes. *N. Engl. J. Med.* **317**, 1545–1552.

Vadhan-Raj, S., Buescher, S., Broxmeyer, H.E. *et al.* (1988). Stimulation of myelopoiesis In patients with aplastic anemia by recombinant human granulocyte–macrophage colony-stimulating factor. *N. Engl. J. Med.* **319**, 1628–1634.

Vadhan-Raj, S., Broxmeyer, H.E., Andreeff, M. *et al.* (1995). *In vivo* biologic effects of PIXY321, a synthetic hybrid protein of recombinant human granulocyte–macrophage colony-stimulating factor and interleukin-3 in cancer patients with normal hematopoiesis: a phase I study. *Blood* **86**, 2098–2105.

Van Der Wouw, P., Van Leeuwen, R., Van Oers, R. and Lange, J.M.A. (1991). Effects of recombinancy human granulocyte colony stimulating factor on leucopenia in zidovudine-treated patients with AIDS and AIDS related complex, a phase I/II study. *Br. J. Haematol.* **78**, 319–324.

Vellenga, E., Young, D.C., Wagner, K., Wiper, D., Ostapovicz, D. and Griffin, J.D. (1987). The effect of GM-CSF and G-CSF in promoting growth of clonogenic cells in acute myeloblastic leukemia. *Blood* **69**, 1771–1776.

Weaver, C.H., Hazelton, B., Birch, R. *et al.* (1995). An analysis of engraftment kinetics as a function of the CD34 content of peripheral blood progenitor cell collections in 692 patients after the administration of myeloablative chemotherapy. *Blood* **86**, 3961–3969.

Weaver, A., Ryder, D., Crowther, D., Dexter, T.M. and Testa, N.G. (1996). Increased numbers of long-term culture-initiating cells in the apheresis products of patients randomized to receive increasing doses of stem cell factor administered in combination with chemotherapy and a standard dose of granulocyte colony-stimulating factor. *Blood* **88**, 3323–3328.

Weisdorf, D.J., Verfaillie, C.M., Davies, S.M. *et al.* (1995). Hematopoietic growth factors for graft failure after bone marrow transplantation: a randomized trial of granulocyte–macrophage colony-stimulating factor (GM-CSF) *versus* sequential GM-CSF plus granulocyte-CSF. *Blood* **85**, 3452–3456.

Welte, K., Zeidler, C., Reiter, A. *et al.* (1990). Differential effects of granulocyte–macrophage colony-stimulating factor and granulocyte colony-stimulating factor in children with severe congenital neutropenia. *Blood* **75**, 1056–1063.

Welte, KG., Reiter, A., Mempel, K.G., Pfetsch, M., Schwab, G. and Schrappe, M. (1996). A randomized phase-III study of the efficacy of granulocyte colony-stimulating factor in children with high-risk acute lymphoblastic leukemia. *Blood* **87**, 3143–3150.

Wendling, F., Maraskovsky, E., Debill, N. *et al.* (1994). C-Mpl ligand is a humoral regulator of megakaryocytopoiesis. *Nature* **369**, 571–573.

Zittoun, R., Suciu, S., Mandelli, F. *et al.* (1996). Granulocyte–macrophage colony-stimulating factor associated with induction treatment of acute myelogenous leukemia: a randomized trial by the European Organization for Research and Treatment of Cancer Leukemia Cooperative Group. *J. Clin. Oncol.* **14**, 2150–2159.

Chapter 30

Cytokine Gene Therapy

Ian D. Davis and Michael T. Lotze
Department of Surgical Oncology, University of Pittsburgh, Pittsburgh, PA, USA

INTRODUCTION

Background

The use of immunostimulation for cancer now dates back over 100 years (Coley, 1893). In the modern era of cancer immunotherapy, recombinant cytokines such as interleukin-2 (IL-2) or interferon-α (IFN-α) have been administered systematically to patients with cancer to replicate the effects of the non-specific immunostimulants. The underlying rationale for using cytokines in patients with cancer is to improve the immune response of the patient to the autologous tumor. This may be achieved in several ways (Fig. 1). Whether it is administered systematically or delivered locally, the cytokine may act to enhance local antigen presentation by the tumor by means of inducing expression of human leukocyte antigen (HLA) class I or II antigens or by increasing expression of the antigen. High levels of local expression of cytokines may also act to enhance antigen uptake and processing by tissue antigen-presenting cells (APCs) such as Langerhans' cells in the skin, with subsequent migration of these cells to secondary lymphoid tissue and presentation of the antigen to resident T and B cells. Depending on the local cytokine milieu, the immune response to the antigen may be skewed in favor of a T_{H1} or T_{H2} response, or towards a predominantly B-cell response. Local expression of cytokines may act to increase the permeability of endothelial barriers in the vicinity of the tumor cell. When tumor-specific or non-specific effector cells migrate into the tumor mass, their function may be enhanced by high local levels of particular cytokines. Ultimately these effects come down to one of two aims: either the tumor cell must become a better APC (and thus also be visible to the effector arm of the immune system), or the body's normal APCs must become better at presenting tumor antigens. The latter is the favored current interpretation of available data from murine models.

Despite promising *in vitro* data, however, studies using cytokines in humans with malignancy have generally given limited results (10–30% response rates in only a few malignancies, e.g. renal cell cancer and melanoma) when they are used as single-agent anticancer drugs. The cost of using recombinant proteins is often prohibitive. In most cases, either antitumor activity has been mininal or the side-effects of treatment have been unacceptable. There are several possible reasons for this. First, a minimal

The Cytokine Handbook, 3rd ed.
ISBN 0-12-689662-3

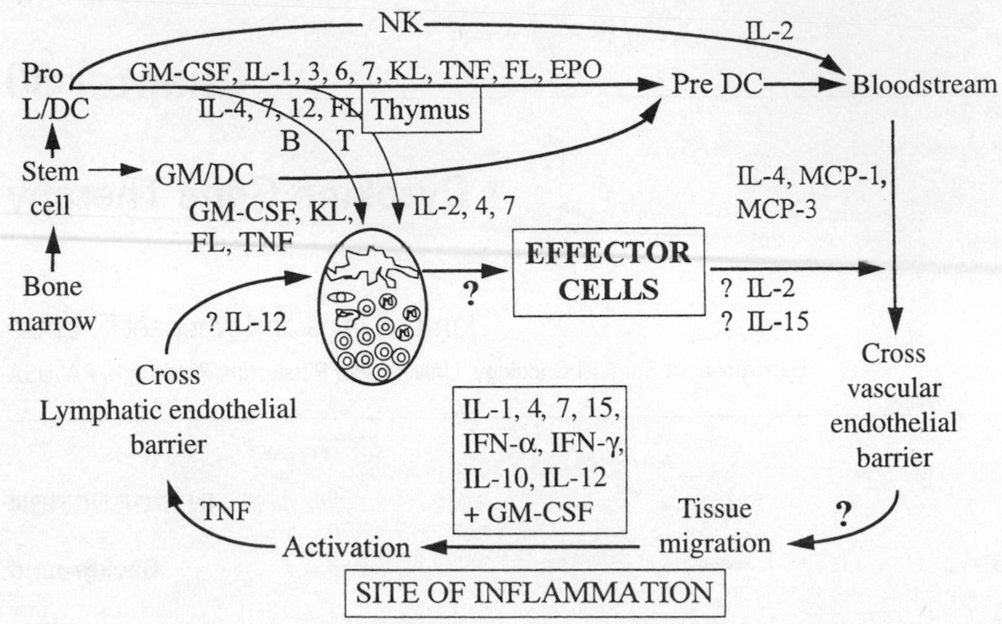

Fig. 1. Dendritic cell (DC) maturation: role of cytokines. DCs occupy a pre-eminent position in initiating the immune response. Cytokines act at many different points in the maturation pathway of DCs to enhance differentiation, migration, antigen uptake and processing, and antigen presentation, and also act on effector cells and endothelium. DCs are thought to be derived from bone marrow precursors and develop a DC phenotype either in the bone marrow or thymus. DC progenitors and 'immature' DCs circulate in the peripheral blood and may reside in tissues. Both 'myeloid' and 'lymphoid' DCs are now recognized. Upon stimulation by inflammatory cytokines, DCs migrate into the site of inflammation, phagocytose and process antigen, and then migrate to secondary lymphoid tissue where processed antigen is presented to T and B cells. Expanded antigen-specific effector cells then circulate and enter the site of inflammation. Pro L/DC, lymphoid/DC progenitor; KL, kit ligand (stem cell factor); FL, Flt-3 ligand; EPO, erythropoietin; MCP, monocyte chemoattractant protein.

therapeutic dose may exist for a particular agent, and toxicity of the cytokine may preclude giving a high enough dose for an effect to be seen. Second, cytokines differ from hormones in that they are most often secreted locally and usually provide local paracrine effects rather than systemic effects (Kupfer and Singer, 1989). For this reason, treating patients using systemically administered cytokines may not be the most appropriate method of optimizing local antigen presentation and effector cell function at the sites where they are required. Third, the tumor may be inherently unresponsive or the host may be unable to mount a meaningful antitumor response; and fourth, the cytokine simply may not be an active antitumor agent. Animal studies using engrafted tumors in which cytokines show no effect are not always helpful in distinguishing between these possibilities. Gene transfer techniques are able to provide high local levels of cytokine without systemic toxicity and may allow some of these problems to be overcome. As most gene therapy studies have been performed in the context of malignant disease, this review concentrates primarily on cancer; however, mention is made where appropriate of cytokine gene transfer studies for non-malignant disease.

Gene Transfer Vectors

Gene transfer techniques have been used for over a decade to deliver high-level expression of cytokines and other gene products within the tumor microenvironment. Early studies used plasmids containing recombinant cytokine genes, and these were introduced into tumor cells by physical methods such as electroporation (Potter, 1988; Bubeník *et al.*, 1990), calcium phosphate DNA coprecipitation (Graham and van der Eb, 1994), lipofection (Felgner *et al.*, 1987), DEAE–dextran (Lopata *et al.*, 1984), protoplast fusion (Schaffner, 1980) or receptor-mediated transfection (Cotten *et al.*, 1993). These methods were hampered by low efficiency of transfection and often low levels of gene expression (Heslop, 1994). There was also concern that the insertion of multiple copies of the plasmids into the genome might increase the risk of activation of cellular proto-oncogenes, causing so-called insertional mutagenesis. Because of this, replication-defective retroviral vectors were developed that allowed high-efficiency single-copy gene transfer into the target cells with a concomitant lower risk of insertional mutagenesis (Donahue *et al.*, 1992). Such vectors are first introduced into a packaging cell line which provides the missing viral protein products and introns to allow the production of the defective viruses that are capable of only one round of infection (Cornetta *et al.*, 1991). The risk of helper virus contamination, or of recombinant events allowing reconstitution of replication-competent retrovirus, is minimal and such events are easily detectable in quality assurance programs (Anderson, 1994).

Higher levels of production of protein may now be achieved as a result of the strong promoter activity of the retroviral long terminal repeat (LTR) (Danos and Mulligan, 1988; Ohashi *et al.*, 1992; Cepko, 1993; Dranoff *et al.*, 1993), although there is evidence for some cytokines that continuous low levels of expression are more effective than sustained high level expression (Orange *et al.*, 1994). For efficient production of transfected cells, a selectable marker is included in the vector construct. This is usually a gene for G418 resistance (neomycin phosphotransferase (NPT; Neo)). However, this strategy necessitates the expression of two protein products, so retroviral vectors were constructed that would drive expression of one gene from an internal promoter and the other gene from the LTR (Hawley *et al.*, 1991; Cepko, 1993). In practice the gene driven from the internal promoter would often be relatively underexpressed (Cepko, 1993). Researchers were therefore faced with the problem of choosing between good selection with inefficient expression of the gene of interest, or good expression of the cytokine gene but difficulty in selecting transfected cells. More recently, retroviral-based polycistronic vectors containing one or more internal ribosomal entry sites have been developed. These vectors allow for two or three gene products to be expressed at high levels (Tahara *et al.*, 1994), and have allowed the study of coordinate high-level local production of more than one cytokine, or the study of heterodimeric cytokines such as IL-12 (Tahara *et al.*, 1994).

Other modified viral vectors such as herpesviruses, vaccinia viruses or adenoviruses have also been developed and are being used in preclinical and clinical trials. The use of these vectors is also encumbered by complex regulatory issues, but under certain circumstances they may be more appropriate for *in vivo* gene transfer (Nilaver *et al.*, 1995; Richards *et al.*, 1995; Lattime *et al.*, 1996).

Other non-infectious transfection methods are now also enjoying popularity. The injection of 'naked' DNA has been shown to lead to expression of the gene of interest *in vivo* (Wolff *et al.*, 1990; Pardoll and Beckerleg, 1995). DNA can also be delivered as a calcium phosphate precipitate, as a complex with various liposome formulations, or combined with ligands for particular cell surface receptors as a means to improve tissue specificity (for a review see Cooper, 1996). Efficient transfection of dividing or non-dividing cells can also be obtained using so-called 'bioballistic' or 'gene gun' approaches, in which the cDNA of interest is coated on to gold beads and then shot into the cell or tissue of interest using gas under pressure (Sun *et al.*, 1995). This technique gives transfection efficiencies that are at least comparable to those of other methods and may be higher (Sun *et al.*, 1995). RNA can also be coated on these beads and used to transfect cells both *in vivo* and *in vitro* (Qui *et al.*, 1996). Human clinical trials using this technology are in progress.

New vectors may also include inducible or repressible expression cassettes. These systems are effective both *in vitro* and *in vivo* and may allow control over the expression of the transgene at different time points (Gossen and Bujard, 1992; Dhawan *et al.*, 1995). This in turn may allow expression of a cytokine at different anatomic sites when the gene is expressed by cells capable of migration.

Target Cells

For the most part, gene transfer studies of cytokines have involved expression of genes in tumor cell lines. This has theoretical advantages in that expression of the cytokine will be occurring in the same microenvironment as expression of tumor antigens, which should facilitate a heightened immune response (Rosenthal *et al.*, 1994). In studies examining the effects of cytokine gene transfer *in vitro*, or the biologic properties of transfected cell lines which are subsequently inoculated into animals, these methods have generally worked well. Practical problems arise, however, when planning anticancer studies in humans.

It is often difficult to generate tumor cell lines from primary tumors (Jaffee *et al.*, 1993), and because successful *in vitro* gene transfer usually requires a proliferating cell line, many patients may not be able to be treated using current methods. In addition, some tumor-associated antigen epitopes are expressed only in association with certain major histocompatibility complex (MHC) phenotypes (Berzofsky and Berkower, 1993; Germain, 1993). Therefore, depending on the HLA profile of the population, only a minority of patients may be suitable for treatment based on fully characterized tumor antigens.

Ethical difficulties arise when contemplating injecting malignant cells back into patients, even those who already have an extensive tumor load, particularly if these cells have been infected with a retrovirus and have not been irradiated (Gansbacher *et al.*, 1992). These concerns have largely been ameliorated by the demonstrated safety of the techniques being used, and the meticulous scrutiny under which such trials have been placed by bodies such as the US Recombinant DNA Advisory Committee (Anderson, 1994; Ostrove, 1994). As a result, *in vivo* human gene transfer studies are now becoming more common (Epstein, 1996; Ross *et al.*, 1996). More novel means of gene transfer, such as intra-arterial administration in order to improve targeting of the vector to a particular tissue (Hurford *et al.*, 1995), are likely to be investigated.

More recently, interest has turned towards transfection of other cell types, such as tumor-infiltrating lymphocytes (TILs) (Rosenberg *et al.*, 1990; Itoh *et al.*, 1995) or autologous fibroblasts (Zitvogel *et al.*, 1995; Elder *et al.*, 1996). One interesting approach has been to take advantage of the fact that retroviruses integrate only into dividing cells, and therefore should be safe to use where host cells are not dividing. This observation has been put to use in the setting of glioma, where the malignant cells are dividing while surrounding normal brain cells are not in cell cycle. Several groups (Martuza *et al.*, 1991; Culver *et al.*, 1992; Boviatsis *et al.*, 1994) have used this system successfully in experimental brain tumors.

Culver *et al.* (1992) have transfected the gene for herpes simplex virus thymidine kinase (HSV-TK) into murine fibroblasts and injected these modified cells into established tumors. The retroviral vector was incorporated into nearby proliferating glioma cells, rendering them sensitive to ganciclovir. Subsequent treatment of the animals with ganciclovir, which produces a toxic product when metabolized by HSV-TK, led to complete regression of the tumor (Culver *et al.*, 1992). Surrounding non-dividing neurons were not infected by the retrovirus. Interestingly, nearby non-transduced glioma cells were also affected indicating that a ‘bystander effect’ of toxicity on neighboring cells exists in this situation. This effect has been shown to be mediated at least in part by the transfer of toxic metabolites from transfected to non-transfected cells (Bi *et al.*, 1993). Such an effect is important in this type of study, as it is not possible to ensure that every malignant cell has been transduced with the gene of interest. Human studies with this system are ongoing (Oldfield *et al.*, 1993).

Antigen-Presenting Cells

Gene transfer technology has led to a major shift in the understanding of tumor immunology in recent years. Initial studies were based on the premise that expression of cytokines by transfected tumor cells, or by admixed cells such as fibroblasts, enhanced the ability of the tumor cells to act as APCs. In a recent study, Huang *et al.* (1994) demonstrated that MHC class I-restricted tumor antigens were presented to the immune system exclusively by bone marrow-derived APCs and not directly by tumor cells. It now seems likely that many of the effects of cytokines expressed by transfected tumor cell vaccines may be mediated through effects on host APCs and effector cells. However, some cytokines such as IFN-γ or IFN-α may still act directly on tumor cells to enhance MHC expression and thereby render them more easily recognizable by MHC-restricted lymphocytes.

As dendritic cells (DCs) occupy such a pivotal position in the initiation of the immune response (Fig. 1), it seems logical that these cells might be a useful target for cytokine gene transfer. Under appropriate conditions it might be possible to use these cells to provide enhanced expression of the cytokine of interest in the environment of antigen presentation within secondary lymphoid tissue. We have used murine (m) IL-12-gene-transduced murine bone marrow-derived DCs to treat 8-day established MCA205 tumors in syngeneic C57BL/6 mice (Zitvogel *et al.*, 1996). DCs transfected with the mIL-12 gene expressed 1–2 ng per 10^6 cells per 48 h. When these DCs were pulsed with MCA205 MHC class I peptides, transduced DCs were able to induce a longer period of tumor stasis than non-transduced DCs (Zitvogel *et al.*, 1996). We are currently investigating the effects of other genes introduced into DCs, including the effects of an inducible–repressible promotor system which may allow gene expression to be induced after the DC has entered secondary lymphoid tissue (Boczowski *et al.*, 1996; Henderson *et al.*, 1996).

EFFICACY AND TOXICITY OF CYTOKINES USED IN *IN VIVO* GENE TRANSFER STUDIES

The cDNAs for virtually every described cytokine have been introduced into tumor cells and studied for their efficacy *in vivo* as anticancer agents. With a few notable exceptions, all of these molecules mediate antitumor activity to varying degrees, and the results of these studies have given insights into the mechanisms underlying the host response to tumors. In some cases, the effects seen were not predicted by the previously known actions of the cytokine. For example, in the studies of Tepper *et al.* (1989, 1992), the antitumor effects of IL-4 were shown to be mediated by eosinophils and macrophages, and it was shown that IL-5 played a part in the response. In contrast, studies using IL-5 gene transfer have not demonstrated any antitumor efficacy (Blankenstein *et al.*, 1990; Dranoff *et al.*, 1993; Krüger-Krasagakes *et al.*, 1993). This suggests that the effects of IL-4 are mediated through other cytokines and that, although IL-5 is one of those cytokines, local expression of IL-5 alone cannot elicit the same host response.

Similarly, macrophages appear to be one of the mediators of non-T-cell-dependent responses to tumors, and gene transfer experiments using granulocyte–macrophage colony stimulating factor (GM-CSF) (Dranoff *et al.*, 1993) and IFN-γ (Watanabe *et al.*, 1989; Gansbacher *et al.*, 1990a; Esumi *et al.*, 1991; Porgador *et al.*, 1991; Restifo *et al.*, 1992; Dranoff *et al.*, 1993; Lollini *et al.*, 1993) demonstrated antitumor activity. Interestingly, M-CSF was ineffective in inducing tumor rejection although it had potent chemotactic activity (Dorsch *et al.*, 1993). These tumor-infiltrating macrophages could not be activated by administering systemic IFN-γ or lipopolysaccharide (LPS) (Dorsch *et al.*, 1993). The reasons for these disparate effects are unclear, but it may be that activation of tumor-infiltrating macrophages requires an additional signal, such as a tumor-specific antibody which could induce macrophage-mediated antibody-dependent cell-mediated cytotoxicity (ADCC) (Munn and Cheung, 1989; Bock *et al.*, 1991).

Transforming growth factor-β_1 (TGF-β_1) has complex actions on cell lines *in vitro* and its actions *in vivo* are sometimes difficult to predict. For example, TGF-β_1 has antiproliferative effects on certain cell types (Roberts and Sporn, 1990; Arrick and Derynck, 1993), but also has immunosuppressive effects *in vivo* (Wahl *et al.*, 1989; Kehrl, 1991; Ruscetti and Palladino, 1991) which might counteract its latent antitumor activity. In addition, TGF-β_1 is often produced by tumor cells (Knabbe *et al.*, 1987; Liu *et al.*, 1988; Jennings *et al.*, 1991) and some tumors may even require it as an autocrine growth signal (Chang *et al.*, 1993).

Three studies have used TGF-β_1 gene transfer into different tumor types, all of which were strongly immunogenic (Torre-Amione *et al.*, 1990; Arrick *et al.*, 1992; Chang *et al.*, 1993). In two of these studies (Arrick *et al.*, 1992; Chang *et al.*, 1993), TGF-β_1 expression led to tumor growth earlier than controls and to larger tumors that grew more rapidly. In the third study (Torre-Amione *et al.*, 1990), transfected cells did not grow in immune-competent mice but did in irradiated mice, and tumors that grew had lost the cytotoxic T lymphocyte (CTL) target antigen. Another study using immune competent mice (Chang *et al.*, 1993) showed a protumor effect and suggested that TGF-β_1 had autocrine activity, despite evidence that the *in vitro* growth rate was reduced. In a rat model using 9L glioma cells injected subcutaneously, a TGF-β_1 antisense expression vector was shown to lead to tumor rejection (Fakhrai *et al.*, 1996).

Another group (Wu *et al.*, 1992) transfected a human (h) TGF-β_1 antisense expression vector into a human colonic carcinoma cell line. This cell line expresses mRNA only for hTGF-β_1 and not for hTGF-β_2 or hTGF-β_3 (Wu *et al.*, 1992), expresses all three types of hTGF-β_1 receptor (Hoosein *et al.*, 1989) and is inhibited by exogenous hTGF-β_1 *in vitro* (Hoosein *et al.*, 1989). Tumor take of transfected cells was enhanced in this study and tumors grew more rapidly (Wu *et al.*, 1992), implying a negative autocrine regulatory role for hTGF-β_1 in this situation. The reason for this apparently opposite effect is not clear, but it may be due to differences in effects on surrounding cells (Roberts and Sporn, 1990; Arrick and Derynck, 1993) with alterations in expression of growth factors such as platelet-derived growth factor (PDGF) (Leof *et al.*, 1986), or growth factor receptors such as PDGF receptor (Gronwald *et al.*, 1989).

TGF-β_1 has also been used in gene transfer studies for non-malignant conditions, with variable effects. In the setting of the kidney, TGF-β_1 overexpression in the glomerulus after *in vivo* gene transfer leads to reduced mitogenic activity and reduced metalloproteinase expression in response to IL-1 (Kitamura *et al.*, 1995). In contrast, when the TGF-β_1 gene is introduced into arterial walls *in vivo*, it leads to intimal and medial hyperplasia (Nabel *et al.*, 1993). In cardiac allograft models, TGF-β_1 expression leads to prolongation of graft survival (Qin *et al.*, 1994, 1995). In a liver transplant model, adenovirus-mediated transfer of the TGF-β_1 gene led to decreased level of production of TNF-α and IFN-γ within the allograft (Drazan *et al.*, 1996), although long-term graft rejection was not assessed in this paper as the animals were killed by the fifth day after transplantation. These studies exemplify the difficulties of *in vivo* cytokine studies, where complex interactions may take place and simple autocrine loops may be modified by interactions with host cells.

Some cytokines, although able to produce an antitumor effect, also caused toxicity similar to that seen in studies using the systemically administered cytokine. TNF-α in particular caused dose-dependent toxicity, notably cachexia (Oliff *et al.*, 1987; Teng *et al.*, 1991; Dranoff *et al.*, 1993), which in some cases led to death of the animal. Other studies did not observe this (Asher *et al.*, 1991; Blankenstein *et al.*, 1991), although in both of these cases TNF-α was produced at low levels by the transfected cells (14–40 pg/ml and 10 ng per 10^6 cells per day, respectively). In the studies of Dranoff *et al.* (1993), toxicity was seen with mIL-6 (hepatosplenomegaly and death), TNF-α (wasting, shivering and death) and GM-CSF (fatal leukocytosis, hepatosplenomegaly and pulmonary hemorrhage). mIL-5 led to a marked eosinophilia and splenomegaly. These cytokines were expressed at the highest amounts *in vitro* in their system, aside from mIL-2-transfected cells which were rejected completely and probably did not produce large amounts of cytokine *in vivo* for sustained periods.

Cytokine gene transfer has also provided insights into non-malignant pathophysiology. For example, when the mGM-CSF gene was introduced into rat pulmonary epithelium using an adenoviral vector, early infiltration by eosinophils and macrophages was observed, with irreversible fibrosis ensuing later (Xing *et al.*, 1996). This response was in contrast to that observed with other proinflammatory cytokines such as IL-5, IL-6, macrophage inflammatory protein-2 (MIP-2) or RANTES (Xing *et al.*, 1996), and confirmed that GM-CSF has many other functions in the lung than simply acting as a hematopoietic growth factor. In the same fashion, overexpression of cytokines in transgenic animals, or knockout of cytokine genes by homologous recombination, provides information about the function of the cytokine in a variety of tissues although the redundancy of many systems may cloud interpretation of the results.

IMMUNOTHERAPY OF ESTABLISHED TUMORS

Injecting a bolus of malignant cells into a host is an artificial situation which does not reflect the complex interrelationships that occur in an established tumor between the malignant cells and the supporting stroma and vasculature. Other infiltrating cells may have protumor or antitumor effects (Mantovani, 1994). Indeed, it has been demonstrated that, if cancer cells are injected into a host as part of a tumor fragment, these tumors will often grow, whereas more than ten-fold higher numbers of suspended pure cancer cells fail to form tumors (Singh *et al.*, 1992). In patients with cancer, tumors are established and have evaded first-line immune surveillance. When contemplating the clinical use of cytokine gene transfer, then, it is important to ascertain whether or not cytokines expressed by transfected cells can influence the host response to an established tumor by means of effects on nearby cells.

Initial cytokine gene transfer studies concentrated on the effects of tumor-secreted cytokines on establishment of tumors. More recently, greater emphasis has been placed on the more biologically relevant model of established cancer. In the minority of studies in which this has been investigated, delayed or slowed growth of established primary or metastatic tumors has been observed (Zitvogel *et al.*, 1995; Coveney *et al.*, 1996; Irvine *et al.*, 1996); most studies found no effect on established primary or metastatic tumors. Some studies used neutralizing antibodies to block the effects of the cytokine and could demonstrate tumor growth which subsequently slowed again on removal of the antibody (Tepper *et al.*, 1989, 1992; Hock *et al.*, 1991; Aoki *et al.*, 1992). One group in two reports showed tumor growth in irradiated mice inoculated with a G-CSF-transfected adenocarcinoma, with tumor regression seen after neutrophil recovery (Colombo *et al.*, 1991; Stoppacciaro *et al.*, 1993). In a few studies some cures of established tumors were seen (Golumbek *et al.*, 1991; Cavallo *et al.*, 1992; Chen *et al.*, 1992; Ferrantini *et al.*, 1993; Vieweg *et al.*, 1994), although this tumor rejection was not always tumor specific (Cavallo *et al.*, 1992), suggesting an upregulation of non-specific antitumor effector mechanisms. One study showed that the addition of systemic IL-2 enhanced the antitumor effect of mIL-4-transfected tumor or fibroblast cell lines against established tumors (Pippin *et al.*, 1994). Another more recent report, in which β-galactosidase was used as a model tumor antigen, showed that the addition of systemic adjuvant IL-2, IL-6, IL-7 or IL-12 enhanced the antitumor effect of the vaccine, with IL-12 being the most effective adjuvant (Irvine *et al.*, 1996). The potent effects on established tumor growth of IL-12 secreted either by the tumor or by admixed transfected syngeneic fibroblasts have also been demonstrated by our group (Tahara *et al.*, 1994; Zitvogel *et al.*, 1995). These more biologically relevant studies indicate that tumor vaccines generated using cytokine gene transfer may yet be useful in the treatment of established human malignancy.

THE MIXED TUMOR TRANSPLANTATION ASSAY

Owing to the difficulties in obtaining transfection of every malignant cell within a tumor *in vivo*, it is necessary to demonstrate that transfection of only some of the cells within the tumor may still lead to useful clinical effects. The mixed tumor transplantation assay (MTTA), a term coined by Tepper *et al.* (1989), allows the detection of paracrine or 'bystander' antitumor effects. Briefly, transfected cells are mixed with control parental or mock-transfected cells or with cells transfected with an irrelevant gene (such as vector alone), and then injected into the host. If the antitumor effect is directed only against the cells producing the cytokine of interest, these cells will be selected against and tumors consisting only of control cells will grow at the site of injection. However, if high local concentrations of the gene product also induce an effective antitumor response against nearby control cells, then tumor take and/or tumor growth rates of both cell types at that site will be reduced or abrogated.

In certain circumstances the MTTA may also allow the comparison of local *versus* systemic antitumor effects. For example, Tepper *et al.* (1989) demonstrated abrogation of growth when non-producing cells were mixed with mIL-4-producing cells, but no effect on growth of non-producing cells on the contralateral side of the same animal was observed. This may be due either to the fact that insufficient levels of systemic mIL-4 were achieved, or that the effect of mIL-4 is truly local.

Most published studies in which the MTTA was used showed either rejection or delay of growth of admixed tumor cells. However, five reports, involving hIL-2, mIFN-γ, mGM-CSF, hTNF-α and mIL-10, provided exceptions to this rule.

When expressed by non-immunogenic murine fibrosarcoma cells, hTNF-α does not lead to tumor rejection in immunocompetent mice, and has no effect on admixed non-transfected cells in the MTTA (Karp *et al.*, 1992). In another report on hTNF-α from the same group, using a related weakly immunogenic fibrosarcoma, the MTTA showed initial growth followed by regression (Asher *et al.*, 1991). Another group examined the effects of mIFN-γ in the murine CMS-5 fibrosarcoma cell line (Gansbacher *et al.*, 1990a). They found that although cells expressing mIFN-γ did not grow in immunocompetent mice, these cells did not inhibit tumor growth in the MTTA (Gansbacher *et al.*, 1990a). In another study from the same group, hIL-2, mIFN-γ and mGM-CSF, alone and in pairs, were transfected into a variety of cell lines including the murine CMS-5 fibrosarcoma (Rosenthal *et al.*, 1994). In this study, cells expressing only a single cytokine did not lead to rejection in the MTTA, even when cells secreting IL-2 were mixed with cells secreting IFN-γ (Rosenthal *et al.*, 1994). Clones expressing both IL-2 and IFN-γ in the same cell were rejected, however (Rosenthal *et al.*, 1994). The reason for the discrepancy in results from the same group (no growth of mIFN-γ-transfected cells in one report, growth in the other) (Gansbacher *et al.*, 1990a; Rosenthal *et al.*, 1994), and from other groups studying the same cytokines, is that different numbers of cells were used, and tumors may be able to grow if a large enough inoculum of cells is injected.

A Japanese study (Watanabe *et al.*, 1989) examined the effects of transfecting mIFN-γ into a neuroblastoma cell line and also found that rejection was not enhanced in the MTTA. This study, like that of Gansbacher *et al.* (1990a) also demonstrated that mIFN-γ-transfected cells are rejected by immunocompetent mice (Watanabe *et al.*, 1989; Gansbacher *et al.*, 1990a), indicating that the antitumor effect is directed only against cells expressing mIFN-γ and not nearby non-transfected cells.

Finally, a xenograft study in which mIL-10 was transfected into Chinese hamster ovary (CHO) cells and inoculated into immunodeficient mice showed that although mIL-10-transfected tumors did not grow *in vivo*, these cells did not inhibit the growth of admixed J558L plasmacytoma cells (Richter *et al.*, 1993).

Overall, these results indicate that in general local antitumor effects appear to be mediated by local expression of the cytokine and are not always restricted to the cell secreting the cytokine, but can also affect nearby malignant cells which do not necessarily have to be of the same type. A notable exception was in the study of Tsai *et al.* (1993) which compared hIL-2 transfection into murine mammary tumor cells with hIL-2 transfection into syngeneic murine mammary-derived fibroblasts. They found that when IL-2 was expressed by tumor cells, rejection was seen, but that mixing parental tumor with IL-2-secreting fibroblasts did not lead to rejection of the tumor by mice. This implies that coordinated expression of some cytokines by tumor cells will be required for optimal antitumor effect. The study by Rosenthal *et al.* (1994), discussed above, is in accord with this theory.

The MTTA is an important technique for evaluating the efficacy of gene transfer into tumors; however, other controls are also critical. Only a small minority of published studies had included control tumors (i.e. cells not expressing the transfected gene) injected into the same mouse as the transfected tumors, and in two of these studies (Bubeník *et al.*, 1990; Yu *et al.*, 1993) control cells were injected into the same site as the transfected cells rather than at a distant site. Antitumor effects on control tumor cells implanted in a distant site in the same animal have now been demonstrated, however (Zitvogel *et al.*, 1995; Coveney *et al.*, 1996; Irvine *et al.*, 1996), although results have not been uniform. For example, in one study several cytokines were transfected into a murine mammary carcinoma cell line, and mIL-2, mIL-7, mIL-10 and mIFN-γ led to cure of the contralateral control tumor in some mice only (Allione *et al.*, 1994). Other studies found that contralateral control tumors were not inhibited, indicating a local rather than systemic effect on tumor rejection. Interestingly, in one study (Mullen *et al.*, 1992), the presence of an established distant parental tumor reduced the growth of mIL-6-transduced fibrosarcoma cells. These results serve to emphasize that without comparable and well-designed controls, results may be difficult to interpret. The study by Bubeník *et al.* (1990) was somewhat unusual, in that transfected cells were injected around established parental or unrelated tumors, and so did not properly represent a MTTA. Another study, although employing mixed inocula of cells, did not strictly perform a MTTA. Tahara *et al.* (1994) mixed tumorigenic cells with non-tumorigenic cytokine producer cells that were of a different lineage (fibroblasts). This method tests the hypothesis of paracrine antitumor effects of the studied cytokine, but clearly does not address the question of mixing transfected and non-transfected cells of the same type.

In some cases, late outgrowth of the transfected tumor was noted (Colombo *et al.*, 1991; Golumbek *et al.*, 1991; Cignetti *et al.*, 1994). In one case it was demonstrated that the growing tumor cells lacked the transfected gene, indicating that negative selection had occurred in the MTTA and that parental cells had survived rejection (Colombo *et al.*, 1991). A study of hIL-2 showed that outgrowing tumors had lost expression of the cytokine (Cignetti *et al.*, 1994). In another study TGF-β_1-expressing cells were rejected in immune competent mice but grew in irradiated mice; however, growing cells had lost a CTL target antigen suggesting selection in favor of a less immunogenic subclone (Torre-Amione *et al.*, 1990). Such *in vivo* immunoselection is now well described in both murine (Dudley and Roopenian, 1996) and human (Jäger *et al.*, 1996; Maeurer *et al.*, 1996) tumors.

EFFECTOR MECHANISMS

Background

A major difficulty in interpreting the results of the published studies is that in many cases the cells found histologically at the site of tumor rejection are assumed to be the cells effecting the rejection. This is not always the case, however, and in only a minority of published studies were effector mechanisms formally elucidated. For example, although in many cases macrophages were seen at the site of tumor rejection, and were shown in some cases to be important, this is not always the case, as evidenced by the lack of responsiveness to M-CSF (Dorsch *et al.*, 1993) in which no rejection was seen despite a dense macrophage infiltration. In other cases, the time of biopsy may have had an influence. Colombo *et al.* (1992) distinguish between an induction phase and an effector phase of antitumor activity, and note that different cell types are important in each phase. The induction phase may be T cell-dependent or -independent (Colombo *et al.*, 1992), and the importance of T cells may be seen in studies comparing athymic nude nice with T cell-replete mice. Clearly, if rejection can occur in athymic nude mice, then a T cell-independent mechanism must be playing a role. This does not exclude T cell-dependent pathways of tumor rejection, however, which may also be active at the same time in an immunologically intact host.

Only studies that have used hosts with specific immune defects can be interpreted as defining the effector cells relevant to that particular experimental situation. These immune defects may be naturally occurring such as athymic nude mice (Flanagan, 1966), severe combined immune deficiency (SCID) mice (Bosma *et al.*, 1983) which lack T- and B-cell function, or beige mice which lack natural killer (NK) cells (Roder, 1979), or induced defects such as in studies where specific T-cell subsets, granulocytes, NK cells or other cells have been depleted by genetic means or by monoclonal antibodies. These studies are outlined in Table 1, which lists studies demonstrating effector cells in a given experimental situation and excludes those simply documenting cellular infiltrates. It must be borne in mind that the demonstration of a role for a particular cell type does not imply that that cell type is unique in its action: other pathways and mechanisms may also be active concurrently.

Non-Specific Effectors

In some cases, the cells seen infiltrating sites of tumor rejection may represent non-specific reactions to local inflammation. Colombo *et al.* (1992) have analyzed their interesting data involving G-CSF-transfected tumors (Colombo *et al.*, 1991; Stoppacciaro *et al.*, 1993). They concluded that cell-to-cell crosstalk occurs between diverse cell types in both directions, and this may involve cells whose effects had not been thought to be interconnected previously. Neutrophils activated by G-CSF gene transfer into a colonic tumor model can prevent tumor take but are unable to cause regression of an established tumor (Colombo *et al.*, 1991; Stoppacciaro *et al.*, 1993). Functional T-cell interactions appear to be required for rejection of established tumors (Colombo *et al.*, 1991; Stoppacciaro *et al.*, 1993). Neutrophils, eosinophils, macrophages and NK cells all play important roles in tumor immunity. As can be seen in Table 1, all of these cells have been shown to be critical in studies of gene transfer.

Table 1. Effector cells associated with tumor rejection.

Effector cell	Cytokine	Tumor cell type	Reference
CD4$^+$ T cells	mIL-3	Lung carcinoma	Pulaski *et al.* (1993)
	mIL-4	Fibrosarcoma	Marincola *et al.* (1994)
	mIL-7	Plasmacytoma	Hock *et al.* (1991)
		Mammary carcinoma	Hock *et al.* (1993a)
	mIFN-γ	Fibrosarcoma	Marincola *et al.* (1994)
	hTNF-α	Fibrosarcoma	Asher *et al.* (1991)
	mTNF-α	Fibrosarcoma	Marincola *et al.* (1994)
	IL-12	Melanoma	Tahara *et al.* (1994)
CD8$^+$ T cells	hIL-2	Fibrosarcoma	Karp *et al.* (1993)
	mIL-2	Mammary carcinoma	Fearon *et al.* (1990)
		Lung carcinoma	Pulaski *et al.* (1993)
		Plasmacytoma	Hock *et al.* (1993a)
	mIL-3	Lung carcinoma	Pulaski *et al.* (1993)
	mIL-4	Fibrosarcoma	Marincola *et al.* (1994)
	mIL-7	Ependymoblastoma	Aoki *et al.* (1992)
		Plasmacytoma	Hock *et al.* (1993a)
	mIL-12	Reticulum cell sarcoma	Brunda *et al.* (1993)
		Melanoma	Tahara *et al.* (1994)
	mIFN-γ	Fibrosarcoma	Marincola *et al.* (1994)
	hTNF-α	Fibrosarcoma	Asher *et al.* (1991)
	mTNF-α	Plasmacytoma	Marincola *et al.* (1994)
			Hock *et al.* (1993a)
	hG-CSF	Colonic carcinoma	Colombo *et al.* (1991)
			Stoppacciaro *et al.* (1993)
Macrophages	hIL-2*	Human renal carcinoma	Gastl *et al.* (1992)
			Belldegrun *et al.* (1993)
	mIL-4*	Plasmacytoma	Tepper *et al.* (1989)
		Mammary carcinoma	Tepper *et al.* (1992)
		Melanoma	
	mIL-7	Plasmacytoma	Hock *et al.* (1991)
		Mammary carcinoma	
	mIL-12	Melanoma	Tahara *et al.* (1994)
	mTNF-α	Plasmacytoma	Blankenstein *et al.* (1991)
	mIFN-γ	Mammary carcinoma	Lollini *et al.* (1993)
	mGM-CSF*	Various	Dranoff *et al.* (1993)
	hMCP-1*	CHO ovary	Rollins and Sunday (1991)
	mMCP-1*		
NK cells	mIL-2	Lung carcinoma	Ohe *et al.* (1993)
		Plasmacytoma	Hock *et al.* (1993a)
	hIL-2	Fibrosarcoma	Karp *et al.* (1993)
		Various	Rosenthal *et al.* (1994)
	mIL-4	CHO ovary	Platzer *et al.* (1992)
		Lung carcinoma	Ohe *et al.* (1993)
		Plasmacytoma	Hock *et al.* (1993a)
	mIL-7	Plasmacytoma	Hock *et al.* (1993b)
			Hock *et al.* (1993a)
	mIFN-γ	CHO ovary	Platzer *et al.* (1992)
		Plasmacytoma	Hock *et al.* (1993a)
		Various	Rosenthal *et al.* (1994)
	mTNF-α	Plasmacytoma	Blankenstein *et al.* (1991)
			Hock *et al.* (1993a)

Table 1. (*continued*)

Effector cell	Cytokine	Tumor cell type	Reference
Eosinophils	mIL-4	Plasmacytoma	Tepper *et al.* (1992)
		Melanoma	
	mIL-7	Fibrosarcoma	McBride *et al.* (1992)
	hMCP-1	CHO ovary	Rollins and Sunday (1991)
	mMCP-1		
Neutrophils	hG-CSF	Colonic carcinoma	Colombo *et al.* (1991)
			Stoppacciaro *et al.* (1993)
	?mTNF-α	Plasmacytoma	Blankenstein *et al.* (1991)

*Denotes likely, not proven. IL, interleukin; TNF, tumour necrosis factor; IFN, interferon; GM-CSF, granulocyte–macrophage colony-stimulating factor; MCP, monoctye chemoattractant protein; G-CSF, granulocyte colony-stimulating factor.

Specific Effectors

Because many cytokines show different activities depending on whether the host is T cell-deplete or immunologically intact, it is clear that T cells play an important role in tumor rejection. The presence or absence of CTLs that recognize tumor cells specifically has been examined. In some cases, CTLs specific for the tumor of interest have been isolated from the spleens of the animals (Watanabe *et al.*, 1989; Gansbacher *et al.*, 1990a,b; Ley *et al.*, 1991; Aoki *et al.*, 1992; McBride *et al.*, 1992; Porgador *et al.*, 1992; Qin *et al.*, 1993), or in some cases from TILs (Restifo *et al.*, 1992; Karp *et al.*, 1993). In other cases CTLs could be isolated but were not tumor-specific (Esumi *et al.*, 1991; McBride *et al.*, 1992; Pulaski *et al.*, 1993). Hock *et al.* (1993a) have pointed out that a T-cell infiltrate into a tumor does not necessarily imply a role for these cells in tumor rejection; indeed, $CD8^+$ CTLs are not required early in the process of tumor rejection (Hock *et al.*, 1993a). This is in contrast to the work of others (Fearon *et al.*, 1990; Gansbacher *et al.*, 1990b) who suggested that CTLs were required early in the process. This discrepancy may be due to the presence of NK cells in the CTL population, the different tumor cell lines being studied, the time of tumor sampling, or the level of cytokine expression.

There is now a substantial body of evidence indicating that a variety of signaling defects may occur in the T cells of patients with cancer (Zier *et al.*, 1996). When animals with such defects are given tumor vaccines engineered to express IL-2, these defects may be reversed (Salvadori *et al.*, 1994). The use of parenteral cytokines may also be of benefit in this situation (Irvine *et al.*, 1996). Indeed, some patients with melanoma with demonstrated defects in IL-2 signaling associated with reduced expression of the IL-2R ζ chain have been shown to normalize expression of the ζ chain after IL-2 therapy (Rabinowich *et al.*, 1996).

$CD4^+$ T cells have in general not been found to be critical as tumor rejection occurred even when $CD4^+$ T cells had been depleted using monoclonal antibodies. Exceptions include IL-3, IL-4, IL-7, IFN-γ and TNF-α as outlined in Table 1. In these cases tumor rejection was impaired when $CD4^+$ T cells were depleted, indicating that at least under those particular experimental conditions helper T cells were required for optimum tumor rejection.

$CD8^+$ CTLs are generated in response to antigenic peptides presented in association with MHC class I molecules. Classically, this response was thought to require help from $CD4^+$ T cells; however, some studies have provided evidence that under certain circumstances such T cells may not be required (Fearon *et al.*, 1990). As indicated above, tumor-specific splenic CTL production could often be induced by a variety of cytokines. Whether these cells are direct mediators of tumor rejection, however, is not always clear. Despite impressive *in vitro* activity (Rosenberg, 1992), even tumor-specific CTLs derived from tumor explants (TILs) are by definition unable to induce tumor rejection *per se* (if they were so capable then the tumor nodule from which they were derived would not exist). Studies using TILs to treat patients with cancer have in some cases shown impressive initial response rates although responses were usually of short duration (Rosenberg *et al.*, 1988; Topalian *et al.*, 1988). *In vivo* studies examining the homing of reinfused tumor-specific TILs back to tumors have been disappointing (Rosenberg *et al.*, 1990; Ribeiro and Brenner, 1992) in that, although reinfused TILs could be detected around tumors, they were not present in high numbers and this distribution could have been random. In some instances TILs derived from various tumor types probably represent NK cells and as such are non-specific effector cells (Blankenstein *et al.*, 1991; Platzer *et al.*, 1992; Karp *et al.*, 1993; Pulaski *et al.*, 1993). This phenomenon of ineffective TIL function probably reflects inefficiencies in antigen presentation; however, there is also evidence for global T-cell dysfunction in patients with cancer.

Another important question relates to the viability of the cellular vaccine. Dranoff *et al.* (1993) studied the effects of mIL-2, mIL-4, mIL-5, mIL-6, mGM-CSF, mIFN-γ, murine intercellular adhesion molecule-1 (mICAM-1), mCD2, mIL-1RA and hTNF-α gene transfer on the immunogenicity of murine B16 melanoma cells, a cell line that had previously been widely studied and found to be of low immunogenicity (Tanaka *et al.*, 1988; Bottazzi *et al.*, 1992; Sun *et al.*, 1992; Tahara *et al.*, 1994). They found that, when using viable transfected cells, only modest protection was seen with only some cytokines (IL-2, IL-4, IL-6, IFN-γ) and that parental wild-type cells were not rejected. However, when irradiated transfected cells were injected, they found that GM-CSF was the most potent protective cytokine, that IL-4 and IL-6 were less powerful, and that the other cytokines were not protective. IL-2 was effective only in combination with GM-CSF. An important point from this study was the finding that other tumor types (CT-26, CMS-5, RENCA and WP-4), all of which had previously been thought to be of low immunogenicity, were also potently immunogenic when irradiated. This suggests that damaged cells are much more effective at eliciting an immunologic response, and that the mechanism of this damage is to some extent less relevant. It is possible that intact wild-type tumors *in vivo*, although bearing tumor-specific antigens towards which an immune response may be mounted under appropriate conditions (Ostrand-Rosenberg *et al.*, 1990; Chen *et al.*, 1992; Baskar *et al.*, 1993), do not ordinarily precipitate an effective antitumor response despite the presence of TILs. The B16 melanoma line used in the study of Dranoff *et al.* (1993) was not immunogenic when irradiated, and hence is a valid control for the transfection studies. These investigators were able to document that specific immune effector cells ($CD4^+$ and $CD8^+$ T cells) were required for an immune response and that non-specific effectors such as NK cells did not influence tumor rejection. On the other hand, for some cytokines it may be more important to use viable non-irradiated cells for the initial vaccination. This would then allow the growth of the tumor with a corresponding increase in the number of cytokine-producing cells and an increase in the amount of cytokine produced locally for a longer period of time, which may be critical for the priming of the immune response. In the case of cytokines such as IFN-α which may cause direct inhibitory effects on the growth of the tumor (Ferrantini *et al.*, 1993; Kaido *et al.*, 1995), while enhancing host response, it is possible that unless the inoculated cells have time to proliferate then they may be eliminated too rapidly and therefore be unable to prime an effective immune response. In clinical practice, however, injecting viable autologous tumor cells is unlikely to meet regulatory approval unless they also express an appropriate suicide gene such as *HSV-TK*.

These results highlight three points. First, the type of control used in the experiment is very important. In some studies (Chen *et al.*, 1992; Porgador *et al.*, 1993; Tahara *et al.*, 1994) parental cells were irradiated and transfected cells were not, although in other studies (Esumi *et al.*, 1991; Ley *et al.*, 1991; Chen *et al.*, 1992; Porgador *et al.*, 1992; Cavallo *et al.*, 1993; Connor *et al.*, 1993; Yu *et al.*, 1993) the conditions were more comparable for the transfected and control cells. Second, the definition of immunogenicity is imprecise. Some investigators have documented the immunogenicity of the studied lines in detail (Dranoff *et al.*, 1993; Karp *et al.*, 1993), whereas others presume that immunogenicity is low because tumors grow *in vivo*. This is more likely to be a function of the inherent aggressiveness of that particular cell line than a true reflection of immunogenicity. Finally, although the 'suicide genes' such as *HSV-TK* provide a useful approach, such a strategy used in isolation is unlikely to be successful: live non-immunogenic HSV-TK^{+} tumor vaccines are not always completely eliminated *in vivo* when the host is treated with ganciclovir (Golumbek *et al.*, 1992).

LONG-ACTING TUMOR IMMUNITY

An immediate antitumor effect with the rejection of a bolus of injected cells may not necessarily provide protection against a later challenge of a lethal dose of cells. Certain cytokines may induce a state of long-lasting and specific antitumor immune memory (Bosco *et al.*, 1990; Zitvogel *et al.*, 1995). For the purposes of this review, long-acting tumor immunity (LATI) is defined as the ability of a host that has been immunized against a specific tumor type to reject a subsequent challenge of tumor cells of the same type (Bosco *et al.*, 1990; Zitvogel *et al.*, 1995). The initial vaccination may be with non-viable irradiated cells, or with cytokine-transfected cells that are subsequently rejected. The duration of LATI may vary from only a few weeks to an indefinite period, but the second challenge must occur after the initial period of rejection.

A distinction needs to be made between studies looking at immediate tumor rejection and those in which LATI could be demonstrated. In the latter, specific memory T cells would need to be generated and so it would be expected that LATI would not be seen in athymic nude mice. In most studies in which it was examined, LATI was detectable, although generally rechallenge took place only a few weeks after immunization. This is an important point since LATI can be lost with increasing time (Fearon *et al.*, 1990), or is inducible only with multiple immunizations (Plaksin *et al.*, 1988; Esumi *et al.*, 1991). In some cases LATI was induced using parental cells alone (Russell *et al.*, 1991; Hock *et al.*, 1993b), indicating that although that particular cell line was thought to be weakly immunogenic, a significant antitumor response could be raised against parental cells. In other cases LATI could be detected, for example against a mammary adenocarcinoma (Cavallo *et al.*, 1993), but was non-specific in that other cell types (fibrosarcoma) were also inhibited (Cavallo *et al.*, 1993). Some studies found that LATI could not be generated with transfected hIL-2 (Karp *et al.*, 1993) or mIL-4 (Tepper *et al.*, 1992; Dranoff *et al.*, 1993), although other studies with those cytokine genes did detect LATI (Gansbacher *et al.*, 1990b; Golumbek *et al.*, 1991; Russell *et al.*, 1991; Connor *et al.*, 1993; Porgador *et al.*, 1993; Pippin *et al.*, 1994; Tahara *et al.*, 1994). Non-specific NK cells were found to be important in the study of hIL-2 where LATI was not detected (Karp *et al.*, 1993). Perhaps the NK response pre-empts the need for a LATI response.

One group (Allione *et al.*, 1994) compared the effectiveness of cytokine-transfected mammary tumor cells in inducing systemic immunity against a later challenge of untransfected parental tumor cells. They found that, for some cytokines (mIL-2, mIL-4, mIL-7, mIL-10), replicating cytokine-transfected tumor cells were more effective than non-replicating transfected cells. They also found that a conventional adjuvant (*Corynebacterium parvum*), when mixed with untransfected cells, was also more effective in inducing systemic immunity when replicating cells were used. This phenomenon may be due to the growth of the cells, leading to a greater load of tumor antigens and greater expression of the cytokine than would occur with non-proliferating cells.

FACTORS AFFECTING INTERPRETATION OF CYTOKINE GENE TRANSFER STUDIES

There are many factors that may cloud the interpretation of the tumor responses in cytokine gene transfer experiments in cancer. Some of these factors are outlined in Table 2.

Some tumors are more likely than others to become established once injected into a host (Hock *et al.*, 1991; Russell *et al.*, 1991). This phenomenon of 'tumor take' or the 'threshold effect' influences whether or not an expressed cytokine shows efficacy. If a tumor has a low take, it will be difficult to determine whether lack of tumor growth is due to the expressed cytokine or not. Most of the tumor types that are used in gene transfer models are aggressively growing tumors which grow in mice at a relatively low inoculum of cells; however, some required higher cell numbers to induce tumors in a high proportion of hosts. For example, the HSNLV rat sarcoma line has a very high tumor

Table 2. Important factors to consider in the interpretation of gene transfer studies.

'Threshold' effect (number of cells required for tumor take)
Immunogenicity of tumor
Other tumor factors
Cell dose *versus* gene dose
Loss of gene expression
Autocrine effects on the tumor
Host immune status
Differences in *in vitro* growth
Species specificity
Exogenous antigens as agents provoking an immune response
Disparate effects of cytokines *in vivo*
Effects of stromal cells and infiltrating cells
Vector
Potency of vaccine adjuvant

take (Russell *et al.*, 1991), with the TD_{50} (cell dose at which 50% of animals develop tumors) being only 100 cells. In contrast, even at a dose of 2×10^6 cells, only 13 of 15 mice injected with parental J558L cells grew tumors (Hock *et al.*, 1991). Similarly, if fewer than 2×10^6 human FET colonic carcinoma cells were injected into nude mice, tumors did not form within 6 weeks (Wu *et al.*, 1992). An immunogenic fibrosarcoma would not grow in C3H mice and only 30–40% of the inoculated tumors grew in irradiated mice (Torre-Amione *et al.*, 1990), although the inoculum dose was not specified in this study.

Other tumor factors may also come into play. For example, some melanoma cell lines have been shown to secrete IL-10 in response to tumor-specific CTLs (Chen *et al.*, 1994), which may lead to downregulation of a T_{H1}-type T-cell response. Of greater concern is the recent report that many melanoma cell lines express CD95 (Fas ligand) (Hahne *et al.*, 1996). Indeed, in hepatocellular carcinoma, one mechanism of immune resistance by tumor cells may be downregulation of the Fas receptor and upregulation of Fas ligand (Strand *et al.*, 1996). Activated T cells coming into contact with these cells are likely to be induced to enter apoptosis, effectively producing an immunologically privileged zone around the $CD95^+$ tumor. Treatment with chemotherapeutic drugs may be one means of inducing Fas ligand expression (Friesen *et al.*, 1996). Successful therapeutic tumor vaccines will have to employ strategies to overcome this problem.

It is important to distinguish between cell number and cytokine dose. There is probably an interrelationship between the two, whereby if a high enough number of cells is given then a tumor may become established regardless of cytokine production. The reason for this is not clear, but it is possible that factors such as the timing of infiltration or lack of accessibility of the effector cells to the tumor cells may be relevant. In many studies, antitumor effects depended on the dose of cytokine given, which in turn correlated in most cases with the number of cells inoculated. If a high enough cell dose were given this could, in some situations, override the antitumor effects of the cytokine, such as with mIL-7 (Aoki *et al.*, 1992; Hock *et al.*, 1993b).

When a high cell inoculum is given, it is conceivable that the rejection process could take longer, allowing time for spontaneous tumorigenic mutants to arise. This was seen in some studies where there was late outgrowth of tumors that lacked the transfected DNA or expression of the gene product (Colombo *et al.*, 1991; Cavallo *et al.*, 1992; Cignetti *et al.*, 1994), or of tumor-specific antigens (Torre-Amione *et al.*, 1990). Also, when a particular gene product provides an autocrine proliferative signal to a cell, this stimulation may override any stimulation of host antitumor effects especially if the host is immunocompromised, as with hIL-2 or mIL-2 and T-cell lines in nude mice (Yamada *et al.*, 1987; Karasuyama *et al.*, 1989), mIL-3 mutants and a myeloblastic cell line (Dunbar *et al.*, 1989), or IL-5 on a B-cell line (Blankenstein *et al.*, 1990). This 'swamping' of host antitumor effects by cytokine-induced autocrine growth may also be seen with immunologically intact hosts, such as with hTGF-β_1 (Chang *et al.*, 1993) or mGM-CSF (Lang *et al.*, 1985).

In experiments using cytokine-transfected tumors to immunize hosts, *in vitro* growth characteristics of cytokine-transfected and control cells need to be assessed to ensure that any differences of growth observed *in vivo* are not due to direct effects of the cytokine on the cells. If this is not done then an alteration in growth rate observed *in vivo* may be erroneously attributed to an immunologic effect of the cytokine. Of the published studies that reported *in vitro* growth characteristics of cytokine-transfected tumor cells, most showed no difference in growth rates or morphology, and therefore this could not have contributed to the lack of tumorigenicity in those transfected cells. However, hTNF-α gene transfer was associated with a reduction of *in vitro* growth rate in one study (Karp *et al.*, 1993), although in similar studies with hTNF-α, no difference in *in vitro* growth rates was found by others (Asher *et al.*, 1991; Blankenstein *et al.*, 1991; Teng *et al.*, 1991; Qin *et al.*, 1993). Similarly, hIL-6 expression in Lewis lung carcinoma cells led to a reduction in *in vitro* growth rate (Porgador *et al.*, 1992). Transfected IL-6 increased the *in vitro* adherence of B16 murine melanoma cells (Sun *et al.*, 1992). This was associated with upregulation of fibronectin and vitronectin receptors (Sun *et al.*, 1992). Human or murine TGF-β_1 expression was associated with a reduction in growth rate or an alteration in morphology in various cell lines (fibrosarcomas (Torre-Amione *et al.*, 1990; Chang *et al.*, 1993) or human embryonic kidney (Wu *et al.*, 1992)). It is possible that these effects could have influenced the results of the *in vivo* studies, although with hTGF-β_1 the effects that were observed *in vitro* and *in vivo* were in the opposite direction. Isolated reports of reduced *in vitro* growth rates with hIFN-α-transfected human renal carcinoma cells (Belldegrun *et al.*, 1993), hIL-2-transfected murine fibrosarcomas (Rosenthal *et al.*, 1994), mIFN-α_1-transfected murine B16 melanoma cells (Kaido *et al.*, 1995), or increased growth with hIL-2-transfected murine fibroblasts (Tsai *et al.*, 1993), have also appeared, but these observations remain to be confirmed. If these observations were consistently reported then it would be more difficult to be certain that the growth of transfected tumor cells *in vivo* is inhibited by host immunologic responses induced by the cytokine.

In most of the published studies to date, the issue of species specificity has not arisen. Either the cytokine being expressed was of the same species as the host, or was reactive with both the tumor and host cells. In the cases where a xenografted tumor was being studied, the transfected cytokine gene was of the same species as the host (Platzer *et al.*, 1992; Yu *et al.*, 1993), with the exceptions of one study examining hIL-2 transfected into a rat sarcoma in a mouse host (Russell *et al.*, 1991), another report studying hTNF-α in CHO cells in a nude mouse host (Oliff *et al.*, 1987), and two other reports involving hTGF-β_1 gene transfer into human kidney or colonic cell lines in nude mouse hosts (Arrick *et al.*, 1992; Wu *et al.*, 1992). In each of these cases, however, the cytokine is not species specific in its action (Oliff *et al.*, 1987; Segarini, 1990; Arrick *et al.*, 1992). To date, there have been three reports using human cytokines transfected into human tumors and then implanted in mice. One is the study by Belldegrun *et al.* (1993), in which human IFN-α was transfected into a human renal cell line and implanted into mice. IFN-α, like IFN-γ, is species specific (Langer and Pestka, 1988), so the antitumor effects seen in this experiment must be due to direct effects of the hIFN-α on the tumor cells, and in fact a reduction in the *in vitro* growth rate was also seen. In two other studies hIFN-γ was used to transfect human renal or melanoma cell lines (Gansbacher *et al.*, 1992; Gastl *et al.*, 1992). In both cases no antitumor activity was seen although one study demonstrated alterations in tumor surface antigens with hIFN-γ (Gastl *et al.*, 1992). In other reported cases the human cytokine is cross-reactive with murine receptors and is active on murine cells. It is possible, however, that subtle species differences may exist and that other effects may come into play when using cytokines of a different species to the host, particularly in immune-competent hosts (Davis *et al.*, 1995). For example, no study using immunologically intact hosts has tested for the development of anticytokine antibodies, although these have been well documented to develop and to limit the efficacy of cytokine treatment in human clinical trials (Steis *et al.*, 1988, 1991). For these reasons, to date it has not been possible to understand the relative contributions of the actions of cytokines upon the host cells and upon the tumor cells in mediating rejection of tumors. The use of human and murine IL-4 and their species-restricted activities has allowed a dissection of the roles of the host and of the tumor cells in the antitumor response (I.D. Davis *et al.*, unpublished results).

Some cytokines such as M-CSF have disparate effects under different circumstances and these may be emphasized by the experimental conditions. The presence of other factors such as tumor-specific antibody may determine whether or not the cytokine is effective (Munn and Cheung, 1989; Bock *et al.*, 1991). It is possible that cytokines such as M-CSF, which appear to be inactive in these studies, may yet have activity in combination with other cytokines or therapeutic manipulations.

Many gene transfer studies have used virus-derived vectors to transfect tumor cells, which are useful for giving high-level expression of cytokine. However, it is possible that the viral vectors may also contribute antigens to the cell which may influence the host response to the tumour. For example, mB7 gene transfer into a melanoma cell line led to an immune response directed against the human papilloma virus-derived E7 antigen with which these cells had been transfected previously, and no antitumor effect was seen against E7-negative cells (Chen *et al.*, 1992). SV40 large T antigen, adenovirus early region genes E1A and E1B, papilloma viruses, hepatitis B virus, Epstein–Barr virus (EBV) and some RNA viruses can also encode proteins or peptides that may be recognized by CTLs and induce an immune response against infected cells (Schreiber and Paul, 1993). If non-infected control cells are used then it may be difficult to interpret the results of the experiment. Thus it is essential to use control cells that have been infected with the viral vector. Similarly, if control cells are irradiated and cytokine-transfected cells are not, difficulties in interpretation may arise (Dranoff *et al.*, 1993).

To the surprise of many investigators in the field, a recent report (Sato *et al.*, 1996) indicated that plasmid vectors may induce non-specific tumor responses. This effect is due to the presence within the vector of short immunostimulatory DNA sequences (ISSs) containing a CpG dinucleotide in a specific base context. These sequences are found within the ampicillin resistance gene contained in many plasmid vectors. When monocytes were transfected with plasmids containing ISSs, large amounts of INF-α, INF-β and IL-12 were produced, whereas these cytokines were not produced when plasmids lacking ISSs were used. Although ISSs resulted in enhanced CTL and T_{H1} responses when used in intradermal vaccines, the induction of expression of interferons may in fact downregulate expression of the gene of interest (Sato *et al.*, 1996). Many newer plasmid vectors have been designed without ampicillin resistance genes, but the effects of ISSs still need to be considered in vector design, particularly when DNA is being administered intradermally.

Lastly, a sobering caveat has been made by Hock *et al.* (1993b) who analyzed the results of their studies of transfer of murine IL-2, IL-4, IL-7, TNF-α and IFN-γ genes into J558L plasmacytoma cells. They found that, although these cytokines were effective in inducing LATI, the efficacy was similar to using *C. parvum* as a vaccine adjuvant, and that no synergy could be detected when combining pairs of cytokines. However, other reports have found significant differences between the efficacy of cytokine-transfected tumor cells and the effects of conventional immune adjuvants (Allione *et al.*, 1994). Numerous aspects of the experimental designs of these studies may account for these differences. On the other hand, conventional cancer treatments may be no more effective in eradicating tumors. One study has compared the effects of an immunotherapy strategy using IL-2-transfected tumor cells with the use of cisplatin chemotherapy for experimental murine bladder carcinoma (Connor *et al.*, 1993). This study was designed to mimic closely the biology of *de novo* bladder cancer in that tumor cells were implanted directly into the bladder wall. Mice treated with irradiated IL-2-secreting tumor cells intraperitoneally had a superior survival to those treated with a single dose of cisplatin alone (Connor *et al.*, 1993), and treatment with cisplatin in addition to IL-2-secreting tumor cells did not further enhance the efficacy of the cellular immunotherapy (Connor *et al.*, 1993). These results suggest that immunotherapy for cancer may be a valid alternative or complementary modality to conventional anticancer treatment.

STUDIES USING COMBINATIONS OF CYTOKINES AND OTHER GENES

It may be possible to improve responsiveness by combining cytokines other than those used by Hock *et al.* (1993b), or by using other costimulatory signal molecules (such as B7.1 or B7.2 (CD80/CD86)) together with cytokine secretion in order to improve tumor antigen presentation. Some such experiments have now been reported. In a tumor establishment model, Cayeux *et al.* (1996) reported that J558L plasmacytoma tumors transfected with the genes for both mIL-4 and B7.1 were rejected by 100% of injected BALB/c mice, compared with 73–82% of mice injected with single-transfectant J558L cells or mice treated with *C. parvum* as a conventional adjuvant (Hock *et al.*, 1993b). Rechallenge of cured mice showed that mice receiving tumor cells bearing both genes had improved tumor-specific antitumor memory, although if the cells in the initial vaccine were irradiated the effect was lost. If tumor cells are unable to act as effective APCs *in vivo*, the mechanism of action of B7.1 expressed in the periphery in this fashion is unclear. It is likely to be a result of the activation of naive T cells and of memory T and NK cells (Van de Velde *et al.*, 1993; Kuiper *et al.*, 1994; Denfeld *et al.*, 1995; Wu *et al.*, 1995). Along similar lines, our group has demonstrated that the combination of B7.1 and IL-12 allows the development of an effective antitumor response and the induction of specific antitumor immunologic memory (Rao *et al.*, 1996; Zitvogel *et al.*, 1996).

Combination cytokine gene transfer experiments may also be aimed at enhancing the maturation and migration of tissue APCs. For example, *in vitro* culture of human peripheral blood-derived mononuclear cells in IL-4 and GM-CSF gives rise to cells with the morphologic and functional characteristics of dendritic cells (Sallusto and Lanzavecchia, 1994). When transfected B16 melanoma cells were injected intracerebrally into syngeneic C57BL/6 mice, mice receiving cells producing GM-CSF survived significantly longer than those receiving non-transfected cells (Wakimoto *et al.*, 1996). When cells expressing both GM-CSF and IL-4 were used, a further significant improvement in survival was seen (Wakimoto *et al.*, 1996). This effect could be abrogated by depleting $CD4^+$ or $CD8^+$ T cells or asialo $GM1^+$ cells, suggesting that the effects were mediated through cells very proximal in the immune response, such as tissue APCs.

HUMAN GENE TRANSFER STUDIES

Human gene transfer clinical trials began 7 years ago with the administration of genetically marked TILs in combination with systemic cyclophosphamide and IL-2

(Rosenberg *et al.*, 1990). The first therapeutic cytokine gene transfer study commenced soon afterwards and involved treating patients with advanced cancer with autologous TILs transfected with the cDNA for TNF-α (Anonymous, 1990). At the time of writing, the US Recombinant DNA Advisory Committee had approved 149 studies of which 40 involved the transfer of cytokine genes. Details are incomplete for gene transfer studies outside the US; however, at least 13 of 55 known studies involve cytokine gene transfer. The vast majority of these studies were for patients with malignancy and involved gene transfer into autologous or allogeneic tumor cells, TILs, peripheral blood lymphocytes or fibroblasts (usually autologous but which in at least one study (Culver *et al.*, 1994) are xenografts). Most studies involve gene transduction *ex vivo*, although an increasing number are looking towards transduction of cells *in vivo*. At least three US studies involve the use of cytokine genes for the treatment of non-malignant conditions, such as IL-1RA for rheumatoid arthritis (Evans *et al.*, 1996) and vascular endothelial growth factor for arterial disease (two studies).

To date, most cytokine gene transfer studies involving humans have been phase I or II studies, and meaningful clinical responses have been sparse. Some responses have been reported, however, including one patient with melanoma who received a vaccine consisting of irradiated autologous tumor cells that had been transfected with the IFN-γ gene (Ross *et al.*, 1996). This patient has remained free of tumor for at least 7 months. In our own experience in a study of patients with metastatic melanoma treated with irradiated autologous tumor cells mixed with irradiated hIL-4-transfected autologous fibroblasts, high local levels of IL-4 production were obtained (Lotze and Rubin, 1994; Suminami *et al.*, 1995; Elder *et al.*, 1996), and one patient of 18 experienced regression of tumor. A similar study using IL-12-transfected autologous skin fibroblasts injected directly into tumor masses *in situ* is ongoing, with several patients having had meaningful clinical responses including one patient who has had a surgical complete remission sustained for more than 16 months following resection of the residual nodule.

An *in vivo* gene transfer study using an IL-2 plasmid in a liposome formulation in patients with advanced cancer was recently reported in abstract form (Stopeck *et al.*, 1996). Despite documentation of *in vivo* gene transfer, only one of 24 patients demonstrated a partial response to treatment.

A randomized double-blind phase I study using a retroviral vector to introduce the hGM-CSF gene into autologous renal cell carcinoma cells has recently been reported (Jaffee *et al.*, 1996). Seventeen patients underwent nephrectomy and were vaccinated with autologous GM-CSF-transfected or mock-transfected irradiated tumor cells. No dose-limiting toxicity was observed. Patients receiving GM-CSF-transduced vaccines experienced greater delayed-type hypersensitivity (DTH) responses to both autologous tumor and normal kidney cells, although no clinically apparent autoimmune reactions were seen. A partial clinical response was seen in one patient who experienced the greatest DTH response.

In another interesting study, patients with stage IV metastatic melanoma received vaccinations with IL-2-transduced irradiated allogeneic melanoma cells (Arienti *et al.*, 1996). Twelve patients were treated at two cell dose levels. Three clinical mixed responses were observed. In most patients, an increase in MHC-unrestricted cytotoxicity was seen in peripheral blood lymphocytes, although two patients developed CTLs against either the tyrosinase 368–376 or gp100 280–288 peptides. In addition, in one patient in whom an autologous melanoma cell line was available, tumor-specific $CD4^+$ T cells were induced after vaccination. Although clinical and laboratory responses were seen in only a minority of patients, this study raises hope that allogeneic cytokine gene-transduced vaccines may be of use in this patient population, making this type of treatment feasible for the majority of patients in whom an autologous tumor cell line is not available. As host APCs are critical in initiating the immune response (Huang *et al.*, 1994), perhaps such a strategy may be combined with the use of autologous dendritic cells in future vaccine designs.

Conclusions

With very few exceptions, cytokines are unlikely to be used as single drugs given at pharmacologic doses. Initial studies using gene transfer to provide high local levels of cytokines have shown promising results in some experimental systems. However, there is still a large gulf between inducing the rejection of genetically modified tumor cells in a mouse host, and the treatment of metastatic cancer in humans. Our developing understanding of the fundamental mechanisms of tumor immunology as demonstrated by cytokine gene transfer techniques provides hope for the future of tumor immunotherapy.

ACKNOWLEDGEMENTS

The authors thank Elaine Elder, Theresa Whiteside, Paul Robbins, Hideaki Tahara, Walter Storkus and Joseph Glorioso for their invaluable contributions. This work was supported by NIH grant PO1 CA68067-01, and is derived in part from I.D.D.'s PhD thesis supported by the Australian National Health and Medical Research Council.

REFERENCES

Allione, A., Consalvo, M., Nanni, P., Lollini, P.L., Cavallo, F., Giovarelli, M., Forni, M., Gulino, A., Colombo, M.P., Dellabona, P., Hock, H., Blankenstein, T., Rosenthal, F.M., Gansbacher, B., Bosco, M.C., Musso, T., Gusella, L. and Forni, G. (1994). Immunizing and curative potential of replicating and nonreplicating murine mammary adenocarcinoma cells engineered with interleukin (IL)-2, IL-4, IL-7, IL-10, tumor necrosis factor α, granulocyte–macrophage colony-stimulating factor, and γ-interferon gene or admixed with conventional adjuvants. *Cancer Res.* **54**, 6022–6026.

Anderson, W.F. (1994). Was it stupid or are we poor educators? *Hum. Gene Ther.* **5**, 791–792.

Anonymous (1990). TNF/TIL human gene therapy clinical protocol. *Hum. Gene Ther.* **1**, 441–480.

Aoki, T., Tashiro, K., Miyatake, S., Kinashi, T., Nakano, T., Oda, Y., Kikuchi, H. and Honjo, T. (1992). Expression of murine interleukin 7 in a murine glioma cell line results in reduced tumorigenicity *in vivo. Proc. Natl Acad. Sci. U.S.A.* **89**, 3850–3854.

Arienti, F., Sulé-Suso, J., Belli, F., Mascheroni, L., Rivoltini, L., Melani, C., Maio, M., Cascinelli, N., Colombo, M.P. and Parmiani, G. (1996). Limited antitumor T cell response in melanoma patients vaccinated with interleukin-2 gene-transduced allogeneic melanoma cells. *Hum. Gene Ther.* **7**, 1955–1963.

Arrick, B.A. and Derynck, R. (1993). Growth regulation by transforming growth factor-β. In *Oncogenes and Tumor Suppressor Genes in Human Malignancies* (eds C. Benz and E. Liu), Kluwer Academic Publishers, Norwell, MA, pp. 255–264.

Arrick, B.A., Lopez, A.R., Elfman, F., Ebner, R., Damsky, C.H. and Derynck, R. (1992). Altered metabolic and adhesive properties and increased tumorigenesis associated with increased expression of transforming growth factor β_1. *J. Cell Biol.* **118**, 715–726.

Asher, A.L., Mulé, J.J., Kasid, A., Restifo, N.P., Salo, J.C., Reichert, C.M., Jaffe, G., Fendly, B., Kriegler, M. and

Rosenberg, S.A. (1991). Murine tumor cells transduced with the gene for tumor necrosis factor-α. Evidence for paracrine immune effects of tumor necrosis factor against tumors. *J. Immunol.* **146**, 3227–3234.

Baskar, S., Ostrand-Rosenberg, S., Nabavi, N., Nadler, L.M., Freeman, G.J. and Glimcher, L.H. (1993). Constitutive expression of B7 restores immunogenicity of tumor cells expressing truncated major histocompatibility complex class II molecules. *Proc. Natl Acad. Sci. U.S.A.* **90**, 5687–5690.

Belldegrun, A., Tso, C.-L., Sakata, T., Duckett, T., Brunda, M.J., Barsky, S.H., Chai, J., Kaboo, R., Lavey, R.S., McBride, W.H. and deKernion, J.B. (1993). Human renal carcinoma line transfected with interleukin-2 and/or interferon α gene(s): implications for live cancer vaccines. *J. Natl Cancer Inst.* **85**, 207–216.

Berzofsky, J.A. and Berkower, I.J. (1993). Immunogenicity and antigen structure. In *Fundamental Immunology* (ed. W.E. Paul), Raven Press, New York, pp. 235–282.

Bi, W.L., Parysek, L.M., Warnick, R. and Stambrook, P.J. (1993). *In vitro* evidence that metabolic cooperation is responsible for the bystander effect observed with HSV-tk retroviral gene therapy. *Hum. Gene Ther.* **4**, 725–731.

Blankenstein, T., Li, W., Überla, K., Qin, Z., Tominaga, A., Takatsu, K., Yamaguchi, N. and Diamantstein, T. (1990). Retroviral interleukin 5 gene transfer into interleukin 5-dependent growing cell lines results in autocrine growth and tumorigenicity. *Eur. J. Immunol.* **20**, 2699–2705.

Blankenstein, T., Qin, Z., Überla, K., Müller, W., Rosen, H., Vok, H.-D. and Diamanstein, T. (1991). Tumor suppression after tumor cell-targeted tumor necrosis factor α gene transfer. *J. Exp. Med.* **173**, 1047–1052.

Bock, S.N., Cameron, R.B., Kragel, P., Mulé, J.J. and Rosenberg, S.A. (1991). Biological and antitumor effects of recombinant human macrophage colony-stimulating factor in mice. *Cancer Res.* **51**, 2649–2654.

Boczowski, D., Nair, S.K., Snyder, D. and Gilboa, E. (1996). Dendritic cells pulsed with RNA are potent antigen-presenting cells *in vitro* and *in vivo*. *J. Exp. Med.* **184**, 465–472.

Bosco, M., Giovarelli, M., Forni, M., Modesti, A., Scarpa, S., Masuelli, L. and Forni, G. (1990). Low doses of IL-4 injected perilymphatically in tumor-bearing mice inhibit the growth of poorly and apparently non-immunogenic tumors and induce a tumor-specific immune memory. *J. Immunol.* **145**, 3136–3143.

Bosma, G.C., Custer, R.P. and Bosma, M.J. (1983). A severe combined immune deficiency mutation in the mouse. *Nature* **301**, 527–530.

Bottazzi, B., Walter, S., Govoni, D., Colotta, F. and Mantovani, A. (1992). Monocyte chemotactic cytokine gene transfer modulates macrophage infiltration, growth, and susceptibility to IL-2 therapy of a murine melanoma. *J. Immunol.* **148**, 1280–1285.

Boviatsis, E.J., Park, J.S., Sena-Esteves, M., Kramm, C.M., Chase, M., Efird, J.T., Wei, M.X., Breakefield, X.O. and Chiocca, E.A. (1994). Long-term survival of rats harboring brain neoplasms treated with ganciclovir and a herpes simplex virus vector that retains an intact thymidine kinase gene. *Cancer Res.* **54**, 5745–5751.

Brunda, M.J., Luistro, L., Warrier, R.R., Wright, R.B., Hubbard, B.R., Murphy, M., Wolf, S.F. and Gately, M.K. (1993). Antitumor and antimetastatic activity of interleukin 12 against murine tumors. *J. Exp. Med.* **178**, 1223–1230.

Bubeník, J., Simová, J. and Jandlová, T. (1990). Immunotherapy of cancer using local administration of lymphoid cells transformed by IL-2 cDNA and constitutively producing IL-2. *Immunol. Lett.* **23**, 287–292.

Cavallo, F., Giovarelli, M., Gulino, A., Vacca, A., Stoppacciaro, A., Modesti, A. and Forni, G. (1992). Role of neutrophils and $CD4^+$ T lymphocytes in the primary and memory response to nonimmunogenic murine mammary adenocarcinoma made immunogenic by IL-2 gene. *J. Immunol.* **149**, 3627–3635.

Cavallo, F., Di Pierro, F., Giovarelli, M., Gulino, A., Vacca, A., Stoppacciaro, A., Forni, M., Modesti, A. and Forni, G. (1993). Protective and curative potential of vaccination with interleukin-2-gene-transfected cells from a spontaneous mouse mammary adenocarcinoma. *Cancer Res.* **53**, 5067–5070.

Cayeux, S., Beck, C., Dörken, B. and Blankenstein, T. (1996). Coexpression of interleukin-4 and B7.1 in murine tumor cells leads to improved tumor rejection and vaccine effect compared to single gene transfectants and a classical adjuvant. *Hum. Gene Ther.* **7**, 525–529.

Cepko, C. (1993). Preparation of a specific retrovirus producer cell line. In *Current Protocols in Molecular Biology* (eds F.M. Ausubel, R. Brent, R.E. Kingston, D.D. Moore, J.G. Seidman, J.A. Smith and K. Struhl), John Wiley and Sons, Canada, pp. 9.11.1–9.11.12.

Chang, H.-L., Gillett, N., Figari, I., Lopez, A.R., Palladino, M.A. and Derynck, R. (1993). Increased transforming growth factor β expression inhibits cell proliferation *in vitro*, yet increases tumorigenicity and tumor growth of meth A sarcoma cells. *Cancer Res.* **53**, 4391–4398.

Chen, L., Ashe, S., Brady, W.A., Hellström, I., Hellström, K.E., Ledbetter, J.A., McGowan, P. and Linsley, P.S. (1992). Costimulation of antitumor immunity by the B7 counterreceptor for the T lymphocyte molecules CD28 and CTLA-4. *Cell* **71**, 1093–1102.

Chen, Q., Daniel, V., Maher, D.W. and Hersey, P. (1994). Production of IL-10 by melanoma cells: examination of its role in immunosuppression mediated by melanoma. *Int. J. Cancer* **56**, 755–760.

Cignetti, A., Guarini, A., Carbone, A., Forni, M., Cronin, K., Forni, G., Gansbacher, B. and Foa, R. (1994). Transduction of the IL2 gene into human acute leukemia cells: induction of tumor rejection without modifying cell proliferation and IL2 receptor expression. *J. Natl Cancer Inst.* **86**, 785–791.

Coley, W.B. (1893). The treatment of malignant tumors by repeated inoculations of erysipelas: with a report of ten original cases. *Amer. J. Med. Sci.* **105**, 487–511.

Colombo, M.P., Ferrari, G., Stoppacciaro, A., Parenza, M., Rodolfo, M., Mavilio, F. and Parmiani, G. (1991). Granulocyte colony-stimulating factor gene transfer suppresses tumorigenicity of a murine adenocarcinoma *in vivo*. *J. Exp. Med.* **173**, 889–897.

Colombo, M.P., Modesti, A., Parmiani, G. and Forni, G. (1992). Local cytokine availability elicits tumor rejection and systemic immunity through granulocyte–T-lymphocyte cross-talk. *Cancer Res.* **52**, 4853–4857.

Connor, J., Bannerji, R., Saito, S., Heston, W., Fair, W. and Gilboa, E. (1993). Regression of bladder tumors in mice treated with interleukin 2 gene-modified tumor cells. *J. Exp. Med.* **177**, 1127–1134.

Cooper, M.J. (1996). Noninfectious gene transfer and expression systems for cancer gene therapy. *Semin. Oncol.* **23**, 172–187.

Cornetta, K., Morgan, R.A. and Anderson, W.F. (1991). Safety issues related to retrovirus-mediated gene transfer in humans. *Hum. Gene Ther.* **2**, 5–14.

Cotten, M., Wagner, E. and Birnstiel, M.L. (1993). Receptor-mediated transport of DNA into eukaryotic cells. *Methods in Enzymology* **217**, 618–644.

Coveney, E., Clary, B., Iacobucci, M., Philip, R. and Lyerly, K. (1996). Active immunotherapy with transiently transfected cytokine-secreting tumor cells inhibits breast cancer metastases in tumor-bearing animals. *Surgery* **120**, 265–272.

Culver, K.W., Ram, Z., Wallbridge, S., Ishii, H., Oldfield, E.H. and Blaese, R.M. (1992). *In vivo* gene transfer with retroviral vector–producer cells for treatment of experimental brain tumors. *Science* **256**, 1550–1552.

Culver, K.W., Van Gilder, J., Link, C.J., Carlstrom, T., Buroker, T., Yuh, W., Koch, K., Schabold, K., Doornbas, S. and Wetjen, B. (1994). Gene therapy for the treatment of malignant brain tumors with *in vivo* tumor transduction with the herpes simplex thymidine kinase gene/ganciclovir system. *Hum. Gene Ther.* **5**, 343–379.

Danos, O. and Mulligan, R.C. (1988). Safe and efficient generation of recombinant retroviruses with amphotropic and ecotropic host ranges. *Proc. Natl Acad. Sci. U.S.A.* **85**, 6460–6464.

Davis, I.D., Treutlein, H., Friedrich, K. and Burgess, A.W. (1995). A potent human interleukin-4 antagonist stimulates the proliferation of murine cells expressing the human interleukin-4 binding chain. *Growth Factors* **12**, 69–83.

Denfeld, R.W., Dietrich, A., Wuttig, C., Tanczos, E., Weiss, J.M., Vanscheidt, W., Schopf, E. and Simon, J.C. (1995). *In situ* expression of B7 and CD28 receptor families in human malignant melanoma: relevance for T-cell-mediated anti-tumor immunity. *Int. J. Cancer* **62**, 259–265.

Dhawan, J., Rando, T.A., Elson, S.L., Bujard, H. and Blau, H.M. (1995). Tetracycline-regulated gene expression following direct gene transfer into mouse skeletal muscle. *Somat. Cell Mol. Genet.* **21**, 233–240.

Donahue, R.E., Kessler, S.W., Bodine, D., McDonagh, K., Dunbar, C., Goodman, S., Agricola, B., Byrne, E., Raffeld, M., Moen, R., Bacher, J., Zsebo, K.M. and Nienhuis, A.W. (1992). Helper virus induced T cell lymphoma in nonhuman primates after retroviral mediated gene transfer. *J. Exp. Med.* **176**, 1125–1135.

Dorsch, M., Hock, H., Kunzendorf, U., Diamantstein, T. and Blankenstein, T. (1993). Macrophage colony-stimulating factor gene transfer into tumor cells induces macrophage infiltration but not tumor suppression. *Eur. J. Immunol.* **23**, 186–190.

Dranoff, G., Jaffee, E., Lazenby, A., Golumbek, P., Levitsky, H., Brose, K., Jackson, V., Hamada, H., Pardoll, D. and Mulligan, R.C. (1993). Vaccination with irradiated tumor cells engineered to secrete murine granulocyte–macrophage colony-stimulating factor stimulates potent, specific, and long-lasting anti-tumor immunity. *Proc. Natl Acad. Sci. U.S.A.* **90**, 3539–3543.

Drazan, K.E., Olthoff, K.M., Wu, L., Shen, X.-D., Gelman, A. and Shaked, A. (1996). Adenovirus-mediated gene transfer in the transplant setting. *Transplantation* **62**, 1080–1084.

Dudley, M.E. and Roopenian, D.C. (1996). Loss of a unique tumor antigen by cytotoxic T lymphocyte immunoselection from a 3-methylcholanthrene-induced mouse sarcoma reveals secondary unique and shared antigens. *J. Exp. Med.* **184**, 441–447.

Dunbar, C.E., Browder, T.M., Abrams, J.S. and Nienhuis, A.W. (1989). COOH-terminal-modified interleukin-3 is retained intracellularly and stimulates autocrine growth. *Science* **245**, 1493–1496.

Elder, E.M., Lotze, M.T. and Whiteside, T.L. (1996). Successful culture and selection of cytokine gene-modified human dermal fibroblasts for the biologic therapy of patients with cancer. *Hum. Gene Ther.* **7**, 479–487.

Epstein, S. (1996). Addendum to the points to consider in human somatic cell and gene therapy (1991). *Hum. Gene Ther.* **7**, 1181–1190.

Esumi, N., Hunt, B., Itaya, T. and Frost, P. (1991). Reduced tumorigenicity of murine tumor cells secreting γ-interferon is due to nonspecific host responses and is unrelated to class I major histocompatibility complex expression. *Cancer Res.* **51**, 1185–1189.

Evans, C.H., Robbins, P.D., Ghivizzani, S.C., Herndon, J.H. and Kang, R. (1996). Clinical trial to assess the safety, feasibility, and efficacy of transferring a potentially anti-arthritic cytokine gene to human joints with rheumatoid arthritis. *Hum. Gene Ther.* **7**, 1261–1280.

Fakhrai, H., Dorigo, O., Shawler, D.L., Lin, H., Mercola, D., Black, K.L., Royston, I. and Sobol, R.E. (1996). Eradication of established intracranial rat gliomas by transforming growth factor β antisense gene therapy. *Proc. Natl Acad. Sci. U.S.A.* **93**, 2909–2914.

Fearon, E.R., Pardoll, D.M., Itaya, T., Golumbek, P., Levitsky, H., Simons, J.W., Karusuyama, H., Vogelstein, B. and Frost, P. (1990). Interleukin-2 production by tumor cells bypasses T helper function in the generation of an antitumor response. *Cell* **60**, 397–403.

Felgner, P.L., Gadek, T.R., Holm, M., Roman, R., Chan, H.W., Wen, M., Northrop, J.P., Ringold, G.M. and Danielson, M. (1987). Lipofectin: a highly efficient, lipid-mediated DNA/transfection procedure. *Proc. Natl Acad. Sci. U.S.A.* **84**, 7413–7417.

Ferrantini, M., Proietti, E., Santodonato, L., Gabriele, L., Peretti, M., Plavec, I., Meyer, F., Kaido, T., Gresser, I. and Belardelli, F. (1993). α_1-Interferon gene transfer into metastatic Friend leukemia cells abrogated tumorigenicity in immunocompetent mice: antitumor therapy by means of interferon-producing cells. *Cancer Res.* **53**, 1107–1112.

Flanagan, S.P. (1966). 'Nude', a new hairless gene with pleiotropic effects in the mouse. *Genet. Res.* **8**, 295–309.

Friesen, C., Herr, I., Krammer, P.H. and Debatin, K.-M. (1996). Involvement of the CD95 (APO-1/Fas) receptor/ligand system in drug-induced apoptosis in leukemia cells. *Nature Med.* **2**, 574–577.

Gansbacher, B., Bannerji, R., Daniels, B., Zier, K., Cronin, K. and Gilboa, E. (1990a). Retroviral vector-mediated γ-interferon gene transfer into tumor cells generates potent and long lasting antitumor immunity. *Cancer Res.* **50**, 7820–7825.

Gansbacher, B., Zier, K., Daniels, B., Cronin, K., Bannerji, R. and Gilboa, E. (1990b). Interleukin 2 gene transfer into tumor cells abrogates tumorigenicity and induces protective immunity. *J. Exp. Med.* **172**, 1217–1224.

Gansbacher, B., Zier, K., Cronin, K., Hantzopoulos, P.A., Bouchard, B., Houghton, A., Gilboa, E. and Golde, D. (1992). Retroviral gene transfer induced constitutive expression of interleukin-2 or interferon-γ in irradiated human melanoma cells. *Blood* **80**, 2817–2825.

Gastl, G., Finstad, C.L., Guarini, A., Bosl, G., Gilboa, E., Bander, N.H. and Gansbacher, B. (1992). Retroviral vector-mediated lymphokine gene transfer into human renal cancer cells. *Cancer Res.* **52**, 6229–6236.

Germain, R.N. (1993). Antigen processing and presentation. In *Fundamental Immunology* (ed. W.E. Paul), Raven Press, New York, pp. 629–676.

Golumbek, P.T., Lazenby, A.J., Levitsky, H.I., Jaffee, L.M., Karasuyama, H., Baker, M. and Pardoll, D. (1991). Treatment of established renal cancer by tumor cells engineered to secrete interleukin-4. *Science* **254**, 713–716.

Golumbek, P.T., Hamzeh, F.M., Jaffee, E.M., Levitsky, H., Lietman, P.S. and Pardoll, D.M. (1992). Herpes simplex-1 thymidine kinase gene is unable to completely eliminate live, nonimmunogenic tumor cell vaccines. *J. Immunother.* **12**, 224–230.

Gossen, M. and Bujard, H. (1992). Tight control of gene expression in mammalian cells by tetracycline-responsive promoters. *Proc. Natl Acad. Sci. U.S.A.* **89**, 5547–5551.

Graham, F.L. and van der Eb, A.J. (1994). Cloning of a T cell growth factor that interacts with the β chain of the interleukin-2 receptor. *Virology* **52**, 456.

Gronwald, R.G.K., Seifert, R.A. and Bowen-Pope, D.F. (1989). Differential regulation of expression of two platelet-derived growth factor receptor subunits by transforming growth factor-β. *J. Biol. Chem.* **264**, 8120–8125.

Hahne, M., Rimoldi, D., Schroter, M., Romero, P., Schreier, M., French, L.E., Schneider, P., Bornand, T., Fontana, A., Lienard, D., Cerottini, J.C. and Tschopp, J. (1996). Melanoma cell expression of Fas (Apo-1/Cd95) ligand—implications for tumor immune escape. *Science* **274**, 1363–1366.

Hawley, T.S., Lach, B., Burns, B.F., May, L.T., Sehgal, P.B. and Hawley, R.G. (1991). Expression of retrovirally transduced IL-1α in IL-6-dependent B cells: a murine model of aggressive multiple myeloma. *Growth Factors* **5**, 327–338.

Henderson, R.A., Nimgaonkar, M.T., Watkins, S.C., Robbins, P.D., Ball, E.D. and Finn, O.J. (1996). Human dendritic cells genetically engineered to express high levels of the human epithelial tumor antigen mucin (MUC-1). *Cancer Res.* **56**, 3763–3770.

Heslop, H.E. (1994). Cytokine gene transfer in the therapy of malignancy. *Baillieres Clin. Haematol.* **7**, 135–151.

Hock, H., Dorsch, M., Diamantstein, T. and Blankenstein, T. (1991). Interleukin 7 induces $CD4^+$ T cell-dependent tumor rejection. *J. Exp. Med.* **174**, 1291–1298.

Hock, H., Dorsch, M., Kunzendorf, U., Qin, Z., Diamantstein, T. and Blankenstein, T. (1993a). Mechanisms of rejection induced by tumor cell-targeted gene transfer of interleukin 2, interleukin 4, interleukin 7, tumor necrosis factor, or interferon γ. *Proc. Natl Acad. Sci. U.S.A.* **90**, 2774–2778.

Hock, H., Dorsch, M., Kunzendorf, U., Überla, K., Qin, Z., Diamantstein, T. and Blankenstein, T. (1993b). Vaccinations with tumor cells genetically engineered to produce different cytokines: effectivity not superior to a classical adjuvant. *Cancer Res.* **53**, 714–716.

Hoosein, N.M., McKnight, M.K., Levine, A.E., Mulder, K.M., Childress, K.E., Brattain, D.E. and Brattain, M.G. (1989). Differential sensitivity of subclasses of human colon carcinoma cell lines to the growth inhibitory effects of transforming growth factor-β_1. *Exp. Cell Res.* **181**, 442–453.

Huang, A.Y.C., Golumbek, P., Ahmadzadeh, M., Jaffe, E., Pardoll, D. and Levitsky, H. (1994). Role of bone marrow-derived cells in presenting MHC class I-restricted tumor antigens. *Science* **264**, 961–965.

Hurford, R.K., Jr., Dranoff, G., Mulligan, R.C. and Tepper, R.I. (1995). *Nature Genet.* **10**, 430–435.

Irvine, K.R., Rao, J.B., Rosenberg, S.A. and Restifo, N.P. (1996). Cytokine enhancement of DNA immunization leads to effective treatment of established pulmonary metastases. *J. Immunol.* **156**, 238–245.

Itoh, Y., Koshita, Y., Takahashi, M., Watanabe, N., Kohgo, Y. and Niitsu, Y. (1995). Characterization of tumor-necrosis-factor-gene-transduced tumor-infiltrating lymphocytes from ascitic fluid of cancer patients: analysis of cytolytic activity, growth rate, adhesion molecule expression and cytokine production. *Cancer Immunol. Immunother.* **40**, 95–102.

Jaffee, E.M., Dranoff, G., Cohen, L.K., Hauda, K.M., Clift, S., Marshall, F.F., Mulligan, R.C. and Pardoll, D.M. (1993). High efficiency gene transfer into primary human tumor explants without cell selection. *Cancer Res.* **53**, 2221–2226.

Jaffee, E.M., Marshall, F., Weber, C., Pardoll, D.M., Levitsky, H., Nelson, W., Carducci, M., Mulligan, R. and Simons, J. (1996). Bioactivity of a human GM-CSF tumor vaccine for the treatment of metastatic renal cell cancer. In *Proceedings of the American Society of Clinical Oncology*, p. 237 (abstract 585).

Jäger, E., Ringhoffer, M., Karbach, J., Arand, M., Oesch, F. and Knuth, A. (1996). Inverse relationship of melanocyte differentiation antigen expression in melanoma tissues and $CD8^+$ cytotoxic-T-cell responses: evidence for immunoselection of antigen-loss variants *in vivo*. *Int. J. Cancer* **66**, 470–476.

Jennings, M.T., Maciunas, R.J., Carver, R., Bascom, C.C., Juneau, P., Misulis, K. and Moses, H.L. (1991). TGF-β_1 and TGF-β_2 are potent growth regulators for low-grade and malignant gliomas *in vitro*: evidence in support of an autocrine hypothesis. *Int. J. Cancer* **49**, 129–139.

Kaido, T., Bandu, M.-T., Maury, C., Ferrantini, M., Belardelli, F. and Gresser, I. (1995). IFN-α_1 gene transfection completely abolishes the tumorigenicity of murine B16 melanoma cells in allogeneic DBA/2 mice and decreases their tumorigenicity in syngeneic C57Bl/6 mice. *Int. J. Cancer* **60**, 221–229.

Karasuyama, H., Tohyama, N. and Tada, T. (1989). Autocrine growth and tumorigenicity of interleukin 2-dependent helper T cells transfected with IL-2 gene. *J. Exp. Med.* **169**, 13–25.

Karp, S.E., Hwu, P., Farbe, A., Restifo, N.P., Kriegler, M., Mulé, J.J. and Rosenberg, S.A. (1992). *In vivo* activity of tumor necrosis factor (TNF) mutants. Secretory but not membrane-bound TNF mediates the regression of retrovirally transduced murine tumor. *J. Immunol.* **149**, 2076–2081.

Karp, S.E., Farber, A., Salo, J.C., Hwu, P., Jaffe, G., Asher, A.L., Shiloni, E., Restifo, N.P., Mulé, J.J. and Rosenberg, S.A. (1993). Cytokine secretion by genetically modified nonimmunogenic murine fibrosarcoma. Tumor inhibition by IL-2 but not tumor necrosis factor. *J. Immunol.* **150**, 896–908.

Kehrl, J.H. (1991). Transforming growth factor-β: an important mediator of immunoregulation. *Int. J. Cell Cloning* **9**, 438–450.

Kitamura, M., Burton, S., English, J., Kawachi, H. and Fine, L.G. (1995). Transfer of a mutated gene encoding active transforming growth factor-β_1 suppresses mitogenesis and IL-1 response in the glomerulus. *Kidney Int.* **48**, 1747–1757.

Knabbe, C., Lippman, M.E., Wakefield, L.M., Flanders, K.C., Kasid, A., Derynck, R. and Dickson, R.B. (1987).

Evidence that transforming growth factor-β is a hormonally regulated negative growth factor in human breast cancer cells. *Cell* **48**, 417–428.

Krüger-Krasagakes, S., Li, W., Richter, G., Diamantstein, T. and Blankenstein, T. (1993). Eosinophils infiltrating interleukin-5 gene-transfected tumors do not suppress tumor growth. *Eur. J. Immunol.* **23**, 992–995.

Kuiper, H., Brouwer, M., de Boer, M., Parren, P. and van Lier, R.A. (1994). Differences in responsiveness to CD3 stimulation between naive and memory $CD4^+$ T cells cannot be overcome by CD28 costimulation. *Eur. J. Immunol.* **24**, 1956–1960.

Kupfer, A. and Singer, S.J. (1989). Cell biology of cytotoxic and helper T cell functions: immunofluorescence microscopic studies of single cells and cell couples. *Annu. Rev. Immunol.* **7**, 309–337.

Lang, R.A., Metcalf, D., Gough, N.M., Dunn, A.R. and Gonda, T.J. (1985). Expression of a hemopoietic growth factor cDNA in a factor-dependent cell line results in autonomous growth and tumorigenicity. *Cell* **43**, 531–542.

Langer, J.A. and Pestka, S. (1988). Interferon receptors. *Immunol. Today* **9**, 393–400.

Lattime, E.C., Lee, S.S., Eisenlohr, L.C. and Mastrangelo, M.J. (1996). *In situ* cytokine gene transfection using vaccinia virus vectors. *Semin. Oncol.* **23**, 88–100.

Leof, E.B., Proper, J.A., Goustin, A.S., Shipley, G.D., DiCorleto, P.E. and Moses, H.L. (1986). Induction of c-sis mRNA and activity similar to platelet-derived growth factor by transforming growth factor-β: a proposed model for indirect mitogenesis involving autocrine activity. *Proc. Natl Acad. Sci. U.S.A.* **83**, 2453–2457.

Ley, V., Langlade-Demoyen, P., Kourilsky, P. and Larsson-Sciard, E.-L. (1991). Interleukin 2-dependent activation of tumor-specific cytotoxic T lymphocytes *in vivo*. *Eur. J. Immunol.* **21**, 851–854.

Liu, C., Tsao, M.S. and Grisham, J.W. (1988). Transforming growth factors produced by normal and neoplastically transformed rat liver epithelial cells in culture. *Cancer Res.* **48**, 850–855.

Lollini, P.L., Bosco, M.C., Cavallo, F., De Giovanni, C., Giovarelli, M., Landuzzi, L., Musiani, P., Modesti, A., Nicoletti, G., Palmieri, G., Santoni, A., Young, H.A., Forni, G. and Nanni, P. (1993). Inhibition of tumor growth and enhancement of metastasis after transfection of the γ-interferon gene. *Int. J. Cancer* **55**, 320–329.

Lopata, M.A., Cleveland, D.W. and Sollner-Webb, B. (1984). High-level expression of a chloramphenicol acetyltransferase gene by DEAE-dextran-mediated DNA transfection coupled with a dimethylsulfoxide or glycerol shock treatment. *Nucleic Acids Res.* **12**, 5707.

Lotze, M.T. and Rubin, J.T. (1994). Gene therapy of cancer: a pilot study of IL-4-gene-modified fibroblasts admixed with autologous tumor to elicit an immune response. *Hum. Gene Ther.* **5**, 41–55.

Maeurer, M.J., Gollin, S.M., Storkus, W.J., Swaney, W., Karbach, J., Martin, D., Castelli, C., Salter, R., Knuth, A. and Lotze, M.T. (1996). Tumor escape from immune recognition—loss of HLA-A2 melanoma cell surface expression is associated with a complex rearrangement of the short arm of chromosome 6(1). *Clin. Cancer Res.* **2**, 641–652.

Mantovani, A. (1994). Tumor-associated macrophages in neoplastic progression: a paradigm for the *in vivo* function of chemokines. *Lab. Invest.* **71**, 5–16.

Marincola, F.M., Ettinghausen, S., Cohen, P.A., Cheshire, L.B., Restifo, N.P., Mulé, J.J. and Rosenberg, S.A. (1994). Treatment of established lung metastases with tumor-infiltrating lymphocytes derived from a poorly immunogenic tumor engineered to secrete human TNF-α. *J. Immunol.* **152**, 3500–3513.

Martuza, R.L., Malick, A., Markert, J.M., Ruffner, K.L. and Coen, D.M. (1991). Experimental therapy of human glioma by means of a genetically engineered virus mutant. *Science* **252**, 854–856.

McBride, W.H., Thacker, J.D., Comora, S., Economou, J.S., Kelley, D., Hogge, D., Dubinett, S.M. and Dougherty, G.J. (1992). Genetic modification of a murine fibrosarcoma to produce interleukin 7 stimulates host cell infiltration and tumor immunity. *Cancer Res.* **52**, 3931–3937.

Mullen, C.A., Coale, M.M., Levy, A.T., Stetler-Stevenson, W.G., Liotta, L.A., Brandt, S. and Blaese, R.M. (1992). Fibrosarcoma cells transduced with the IL-6 gene exhibited reduced tumorigenicity, increased immunogenicity, and decreased metastatic potential. *Cancer Res.* **52**, 6020–6024.

Munn, D.H. and Cheung, N.-K.V. (1989). Antibody-dependent antitumor cytotoxicity by human monocytes cultured with recombinant macrophage colony-stimulating factor. *J. Exp. Med.* **170**, 511–526.

Nabel, E.G., Shum, L., Pompili, V.J., Yang, Z.Y., San, H., Shu, H.B., Liptay, S., Gold, L., Gordon, D., Derynck, R. and Nabel, G.J. (1993). Direct transfer of transforming growth factor β_1 gene into arteries stimulates fibrocellular hyperplasia. *Proc. Natl Acad. Sci. U.S.A.* **90**, 10759–10763.

Nilaver, G., Muldoon, L.L., Kroll, R.A., Pagel, M.A., Breakefield, X.O., Davidson, B.L. and Neuwelt, E.A. (1995). Delivery of herpesvirus and adenovirus to nude rat intracerebral tumors after osmotic blood–brain barrier disruption. *Proc. Natl Acad. Sci. U.S.A.* **92**, 9829–9833.

Ohashi, T., Boggs, S., Robbins, P., Bahnson, A., Patrene, K., Wei, F.-S., Wei, J.-F., Li, J., Lucht, L., Fei, Y., Clark,

S., Kimak, M., He, H., Mowery-Rushton, P. and Barranger, J.A. (1992). Efficient transfer and sustained high expression of the human glucocerebrosidase gene in mice and their functional macrophages following transplantation of bone marrow transduced by a retroviral vector. *Proc. Natl Acad. Sci. U.S.A.* **89**, 11 332–11 336.

Ohe, Y., Podack, E.R., Olsen, K.J., Miyahara, Y., Ohira, R., Miura, K., Nishio, K. and Saijo, N. (1993). Combination effect of vaccination with IL2 and IL4 cDNA transfected cells on the induction of a therapeutic immune response against Lewis lung carcinoma cells. *Int. J. Cancer* **53**, 432–437.

Oldfield, E.H., Ram, Z., Culver, K.W., Blaese, R.M., DeVroom, H.L. and Anderson, W.F. (1993). Gene therapy for the treatment of brain tumors using intratumoral transduction with the thymidine kinase gene and intravenous ganciclovir. *Hum. Gene Ther.* **4**, 36–69.

Oliff, A., Defeo-Jones, D., Boyer, M., Martinez, D., Kiefer, D., Vuocolo, G., Wolfe, A. and Socher, S.H. (1987). Tumors secreting human TNF/cachectin induce cachexia in mice. *Cell* **50**, 555–563.

Orange, J.S., Wolf, S.F. and Biron, C.A. (1994). Effects of IL-12 on the response and susceptibility to experimental viral infections. *J. Immunol.* **152**, 1253–1264.

Ostrand-Rosenberg, S., Thakur, A. and Clements, V. (1990). Rejection of mouse sarcoma cells after transfection of MHC class II genes. *J. Immunol.* **144**, 4068–4071.

Ostrove, J.M. (1994). Safety testing programs for gene therapy viral vectors. *Cancer Gene Ther.* **1**, 125–131.

Pardoll, D.M. and Beckerleg, A.M. (1995). Exposing the immunology of naked DNA vaccines. *Immunity* **3**, 165–169.

Pippin, B.A., Rosenstein, M., Jacob, W.F., Chiang, Y. and Lotze, M.T. (1994). Local IL-4 delivery enhances immune reactivity to murine tumors: gene therapy in combination with IL-2. *Cancer Gene Ther.* **1**, 35–42.

Plaksin, D., Gelber, C., Feldman, M. and Eisenbach, L. (1988). Reversal of the metastatic phenotype in Lewis lung carcinoma cells after transfection with syngeneic H-2Kb gene. *Proc. Natl Acad. Sci. U.S.A.* **85**, 4463–4467.

Platzer, C., Richter, G., Überla, K., Hock, H., Diamantstein, T. and Blankenstein, T. (1992). Interleukin-4-mediated tumor suppression in nude mice involves interferon-γ. *Eur. J. Immunol.* **22**, 1729–1733.

Porgador, A., Brenner, B., Vadai, E., Feldman, M. and Eisenbach, L. (1991). Immunization by γ-IFN-treated B16-F10.9 melanoma cells protects against metastatic spread of the parental tumor. *Int. J. Cancer Suppl.* **6**, 54–60.

Porgador, A., Tzehoval, E., Katz, A., Vadai, E., Revel, M., Feldman, M. and Eisenbach, L. (1992). Interleukin 6 gene transfection into Lewis lung carcinoma tumor cells suppresses the malignant phenotype and confers immunotherapeutic competence against parental metastatic cells. *Cancer Res.* **52**, 3679–3686.

Porgador, A., Gansbache, B., Bannerji, R., Tzehoval, E., Gilboa, E., Feldman, M. and Eisenbach, L. (1993). Anti-metastatic vaccination of tumor-bearing mice with IL-2-gene-inserted tumor cells. *Int. J. Cancer* **53**, 471–477.

Potter, H. (1988). Electroporation in biology: methods, applications, and instrumentation. *Anal. Biochem.* **174**, 361–373.

Pulaski, B.A., McAdam, A.J., Hutter, E.K., Biggar, S., Lord, E.M. and Frelinger, J.G. (1993). Interleukin 3 enhances development of tumor-reactive cytotoxic cells by a CD4-dependent mechanism. *Cancer Res.* **53**, 2112–2117.

Qin, L., Chavin, K.D., Ding, Y., Woodward, J.E., Favaro, J.P., Lin, J. and Bromberg, J.S. (1994). Gene transfer for transplantation. Prolongation of allograft survival with transforming growth factor-β_1. *Ann. Surg.* **220**, 508–518.

Qin, L., Chavin, K.D., Ding, Y., Favaro, J.P., Woodward, J.E., Lin, J., Tahara, H., Robbins, P., Shaked, A., Ho, D.Y., Sapolsky, R.M., Lotze, M.T. and Bromberg, J.S. (1995). Multiple vectors effectively achieve gene transfer in a murine cardiac transplantation model. Immunosuppression with TGF-β_1 or vIL-10. *Transplantation* **59**, 809–816.

Qin, Z., Krüger-Krasagakes, S., Kunzendorf, U., Hock, H., Diamantstein, T. and Blankenstein, T. (1993). Expression of tumor necrosis factor by different tumor cell lines results either in tumor suppression or augmented metastasis. *J. Exp. Med.* **178**, 355–360.

Qiu, P., Ziegelhoffer, P., Sun, J. and Yang, N.S. (1996). Gene gun delivery of mRNA *in situ* results in efficient transgene expression and genetic immunization. *Gene Ther.* **3**, 262–268.

Rabinowich, H., Banks, M., Reichert, T.E., Logan, T.F., Kirkwood, J.M. and Whiteside, T.L. (1996). Expression and activity of signaling molecules in T lymphocytes obtained from patients with metastatic melanoma before and after interleukin 2 therapy. *Clin. Cancer Res.* **2**, 1263–1274.

Rao, J.B., Chamberlain, R.S., Bronte, V., Carroll, M.W., Irvine, K.R., Moss, B., Rosenberg, S.A. and Restifo, N.P.

(1996). IL-12 is an effective adjuvant to recombinant vaccinia virus-based tumor vaccines: enhancement by simultaneous B7-1 expression. *J. Immunol.* **156**, 3357–3365.

Restifo, N.P., Spiess, P.J., Karp, S.E., Mulé, J.J. and Rosenberg, S.A. (1992). A nonimmunogenic sarcoma transduced with the cDNA for interferon γ elicits $CD8^+$ T cells against the wild-type tumor: correlation with antigen presentation capability. *J. Exp. Med.* **175**, 1423–1431.

Ribeiro, R.C. and Brenner, M.K. (1992). Interleukin-2 and solid tumours. In *Interleukin-2* (eds J. Waxman and F. Balkwill), Blackwell Scientific Publications, Oxford, pp. 145–161.

Richards, C.D., Braciak, T., Xing, Z., Graham, F. and Gauldie, J. (1995). Adenovirus vectors for cytokine gene expression. *Ann. N.Y. Acad. Sci.* **762**, 282–292.

Richter, G., Krüger-Krasagakes, S., Hein, G., Hüls, C., Schmitt, C., Diamantstein, T. and Blankenstein, T. (1993). Interleukin 10 transfected into Chinese hamster ovary cells prevents tumor growth and macrophage infiltration. *Cancer Res.* **53**, 4134–4137.

Roberts, A.B. and Sporn, M.B. (1990). The transforming growth factor-βs. In *Peptide Growth Factors and Their Receptors. Handbook of Experimental Pharmacology* (eds M.B. Sporn and A.B. Roberts), Springer, Heidelberg, pp. 419–472.

Roder, J.C. (1979). The beige mutation in the mouse. I. A stem cell predetermined impairment in natural killer cell function. *J. Immunol.* **123**, 2168–2173.

Rollins, B.J. and Sunday, M.E. (1991). Suppression of tumor formation *in vivo* by expression of the JE gene in malignant cells. *Mol. Cell Biol.* **11**, 3125–3131.

Rosenberg, S.A. (1992). The immunotherapy and gene therapy of cancer. *J. Clin. Oncol.* **10**, 180–199.

Rosenberg, S.A., Packard, B.S., Aebersold, P.M., Solomon, D., Topalian, S.L., Toy, S.T., Simon, P., Lotze, M.T., Yang, J.C., Seipp, C.A., Simpson, C., Carter, C., Bock, S., Schwartzentruber, D., Wei, J.P. and White, D.E. (1988). Use of tumor-infiltrating lymphocytes and interleukin-2 in the immunotherapy of patients with metastatic melanoma. *N. Engl. J. Med.* **319**, 1676–1680.

Rosenberg, S.A., Aebersold, P., Cornetta, K., Kasid, A., Morgan, R.A., Moen, R., Karson, E.M., Lotze, M.T., Yang, J.C., Topalian, S.L., Merino, M.J., Culver, K., Miller, A.D. and Blaese, R.M. (1990). Gene transfer into humans—immunotherapy of patients with advanced melanoma, using tumor-infiltrating lymphocytes modified by retroviral gene transduction. *N. Engl. J. Med.* **323**, 570–578.

Rosenthal, F.M., Cronin, C., Bannerji, R., Golde, D.W. and Gansbacher, B. (1994). Augmentation of antitumor immunity by tumor cells transduced with a retroviral vector carrying the interleukin-2 and interferon-γ cDNAs. *Blood* **83**, 1289–1298.

Ross, G., Erickson, R., Knorr, D., Motulsky, A.G., Parkman, R., Samulski, J., Straus, S.E. and Smith, B.R. (1996). Gene therapy in the United States: a five-year status report. *Hum. Gene Ther.* **7**, 1781–1790.

Ruscetti, F.W. and Palladino, M.A. (1991). Transforming growth factor-β and the immune system. *Prog. Growth Factors Res.* **3**, 159–175.

Russell, S.J., Eccles, S.A., Flemming, C.L., Johnson, C.A. and Collins, M.K. (1991). Decreased tumorigenicity of a transplantable rat sarcoma following transfer and expression of an IL-2 cDNA. *Int. J. Cancer* **47**, 244–251.

Sallusto, F. and Lanzavecchia, A. (1994). Efficient presentation of soluble antigen by cultured human dendritic cells is maintained by granulocyte/macrophage colony-stimulating factor plus interleukin 4 and down-regulated by tumor necrosis factor α. *J. Exp. Med.* **179**, 1109–1118.

Salvadori, S., Gansbacher, B. and Zier, K. (1994). Functional defects are associated with abnormal signal transduction in T cells of mice inoculated with parental but not IL-2 secreting tumor cells. *Cancer Gene Ther.* **1**, 165–170.

Sato, Y., Roman, M., Tighe, H., Lee, D., Corr, M., Nguyen, M.-D., Silverman, G.J., Lotz, M., Carson, D.A. and Raz, E. (1996). Immunostimulatory DNA sequences necessary for effective intradermal gene immunization. *Science* **273**, 352–354.

Schaffner, W. (1980). Direct transfer of cloned genes from bacteria to mammalian cells. *Proc. Natl Acad. Sci. U.S.A.* **77**, 2163–2167.

Schreiber, H.I.F.I. and Paul W.E. (eds) (1993). Tumor immunology. In *Fundamental Immunology*, Raven Press, New York, pp. 1143–1178.

Segarini, P.R. (1990). Cell type specificity of TGF-β binding. In *Transforming Growth Factor-βs. Chemistry, Biology, and Therapeutics* (eds K.A. Piez and M.B. Sporn), New York Academy of Sciences, New York, pp. 73–90.

Singh, S., Ross, S.R., Acena, M., Rowley, D.A. and Schreiber, H. (1992). Stroma is critical for preventing or permitting immunological destruction of antigenic cancer cells. *J. Exp. Med.* **175**, 139–146.

Steis, R.G., Smith, J., II, Urba, W., Clark, J.W., Itri, L.M., Evans, L.M., Schoenberger, C. and Longo, D.L. (1988).

Resistance to recombinant interferon α_{2a} in hairy cell leukemia associated with neutralizing anti-interferon antibodies. *N. Engl. J. Med.* **318**, 1409–1413.

Steis, R.G., Smith, J.W., II, Urba, W.J., Venzon, D.J., Longo, D.L., Barney, R., Evans, L.M., Itri, L.M. and Ewel, C.H. (1991). Loss of interferon antibodies during prolonged continuous interferon-2a therapy in hairy cell leukemia. *Blood* **77**, 792–798.

Stopeck, A., Hersh, E., Warneke, J., Unger, E., Rinehart, J., Schreiber, A. and Stahl, S. (1996). Results of a phase I study of direct gene transfer of interleukin-2 (IL-2) formulated with cationic lipid vector, Leuvectin, inpatients with metastatic solid tumors. In *Proceedings of the American Society of Clinical Oncology*, Philadelphia, PA, p. 234 (abstract 577).

Stoppacciaro, A., Melani, C., Parenza, M., Mastracchio, A., Bassi, C., Baroni, C., Parmiani, G. and Colombo, M.P. (1993). Regression of an established tumor genetically modified to release granulocyte colony-stimulating factor requires granulocyte-T cell cooperation and T cell-produced interferon γ. *J. Exp. Med.* **178**, 151–161.

Strand, S., Hofmann, W.J., Hug, H., Müller, M., Otto, G., Strand, D., Mariani, S.M., Stremmel, W., Krammer, P.K. and Galle, P.R. (1996). Lymphocyte apoptosis induced by CD95 (APO-1/Fas) ligand-expressing tumor cells—a mechanism of immune evasion? *Nature Med.* **2**, 1361–1366.

Suminami, Y., Elder, E.M., Lotze, M.T. and Whiteside, T.L. (1995). *In situ* interleukin-4 gene expression in cancer patients treated with genetically modified tumor vaccine. *J. Immunother.* **17**, 238–248.

Sun, W.H., Kreisle, R.A., Phillips, A.W. and Ershler, W.B. (1992). *In vivo* and *in vitro* characteristics of interleukin 6-transfected B16 melanoma cells. *Cancer Res.* **52**, 5412–5415.

Sun, W.H., Burkholder, J.K., Sun, J., Culp, J., Turner, J., Lu, X.G., Pugh, T.D., Ershler, W.B. and Yang, N.S. (1995). *In vivo* cytokine gene transfer by gene gun reduces tumor growth in mice. *Proc. Natl Acad. Sci. U.S.A.* **92**, 2889–2893.

Tahara, H., Zeh, H.J., III, Storkus, W.J., Pappo, I., Watkins, S.C., Gubler, U., Wolf, S.F., Robbins, P.D. and Lotze, M.T. (1994). Fibroblasts genetically engineered to secrete murine interleukin 12 can suppress tumor growth and induce antitumor immunity to a murine melanoma *in vivo. Cancer Res.* **54**, 182–189.

Tanaka, K., Gorelik, E., Watanabe, M., Hozumi, N. and Jay, G. (1988). Rejection of B16 melanoma induced by expression of a transfected major histocompatibility complex class I gene. *Mol. Cell Biol.* **8**, 1857–1861.

Teng, M.N., Park, B.H., Koeppen, H.K., Tracey, K.J., Fendly, B.M. and Schreiber, H. (1991). Long-term inhibition of tumor growth by tumor necrosis factor in the absence of cachexia or T-cell immunity. *Proc. Natl Acad. Sci. U.S.A.* **88**, 3535–3539.

Tepper, R.I., Pattengale, P.K. and Leder, P. (1989). Murine interleukin-4 displays potent anti-tumor activity *in vivo. Cell* **57**, 503–512.

Tepper, R.I., Coffman, R.L. and Leder, P. (1992). An eosinophil-dependent mechanism for the antitumor effect of interleukin-4. *Science* **257**, 548–551.

Topalian, S.L., Solomon, D., Avis, F.P., Chang, A.E., Freerksen, D.L., Linehan, W.M., Lotze, M.T., Robertson, C.N., Seipp, C.A., Simon, P., Simpson, C.G. and Rosenberg, S.A. (1988). Immunotherapy of patients with advanced cancer using tumor-infiltrating lymphocytes and recombinant interleukin-2: a pilot study. *J. Clin. Oncol.* **6**, 839–853.

Torre-Amione, G., Beauchamp, R.D., Koeppen, H., Park, B.H., Schreiber, H., Moses, H.L. and Rowley, D.A. (1990). A highly immunogenic tumor transfected with a murine transforming growth factor type β_1 cDNA escapes immune surveillance. *Proc. Natl Acad. Sci. U.S.A.* **87**, 1486–1490.

Tsai, S.-C.J., Gansbacher, B., Tait, L., Miller, F.R. and Heppner, G.H. (1993). Induction of antitumor immunity by interleukin-2 gene-transduced mouse mammary tumor cells *versus* transduced mammary stromal fibroblasts. *J. Natl Cancer Inst.* **85**, 546–553.

Van de Velde, H., Lorre, K., Bakkus, M., Thielemans, K., Ceuppens, J.L. and de Boer, M. (1993). CD45RO$^+$ memory T cells but not CD45RA$^+$ naive T cells can be efficiently activated by remote co-stimulation with B7. *Int. Immunol.* **5**, 1483–1487.

Vieweg, J., Rosenthal, F.M., Bannerji, R., Heston, W.D.W., Fair, W.R., Gansbacher, B. and Gilboa, E. (1994). Immunotherapy of prostate cancer in the Dunning rat model: use of cytokine gene modified tumor vaccines. *Cancer Res.* **54**, 1760–1765.

Wahl, S.M., McCartney-Francis, N. and Mergenhagen, S.E. (1989). Inflammatory and immunoregulatory roles of TGF-β. *Immunol. Today* **10**, 258–261.

Wakimoto, H., Abe, J., Tsunoda, R., Aoyagi, M., Hirakawa, K. and Hamada, H. (1996). Intensified antitumor immunity by a cancer vaccine that produces granulocyte–macrophage colony-stimulating factor plus interleukin 4. *Cancer Res.* **56**, 1828–1833.

Watanabe, Y., Kuribayashi, K., Miyatake, S., Nishihara, K., Nakayama, E.-I., Taniyama, T. and Sakata, T.-A. (1989). Exogenous expression of mouse interferon γ cDNA in mouse neuroblastoma C1300 cells results in reduced tumorigenicity by augmented anti-tumor immunity. *Proc. Natl Acad. Sci. U.S.A.* **86**, 9456–9460.

Wolff, J.A., Malone, R.W., Williams, P., Chong, W., Acsadi, G., Jani, A. and Felgner, P.L. (1990). Direct gene transfer into mouse muscle *in vivo. Science* **247**, 1465–1468.

Wu, S., Theodorescu, D., Kerbel, R.S., Willson, J.K.V., Mulder, K.M., Humphrey, L.E. and Brattain, M.G. (1992). TGF-β_1 is an autocrine-negative growth regulator of human colon carcinoma FET cells *in vivo* as revealed by transfection of an antisense expression vector. *J. Cell Biol.* **116**, 187–196.

Wu, Y., Xu, J., Shinde, S., Grewal, I., Henderson, T., Flavell, R.A. and Liu, Y. (1995). Rapid induction of a novel costimulatory activity on B cells by CD40 ligand. *Curr. Biol.* **5**, 1303–1311.

Xing, Z., Braciak, T., Ohkawara, Y., Sallenave, J.M., Foley, R., Sime, P.J., Jordana, M., Graham, F.L. and Gauldie, J. (1996). Gene transfer for cytokine functional studies in the lung: the multifunctional role of GM-CSF in pulmonary inflammation. *J. Leukoc. Biol.* **59**, 481–488.

Yamada, G., Kitamura, Y., Sonoda, H., Harada, H., Taki, S., Mulligan, R.C., Osawa, H., Diamantstein, T., Yokoyama, S. and Taniguchi, T. (1987). Retroviral expression of the human IL-2 gene in a murine T cell line results in cell growth autonomy and tumorigenicity. *Eur. Mol. Biol. Assoc. J.* **6**, 2705–2709.

Yu, J.S., Wei, M.X., Chiocca, E.A., Martuza, R.L. and Tepper, R. (1993). Treatment of glioma by engineered interleukin 4-secreting cells. *Cancer Res.* **53**, 3125–3128.

Zier, K., Gansbacher, B. and Salvadori, S. (1996). Preventing abnormalities in signal transduction of T cells in cancer: the promise of cytokine gene therapy. *Immunol. Today* **17**, 39–45.

Zitvogel, L., Tahara, H., Robbins, P.D., Storkus, W.J., Clarke, M.R., Nalesnik, M.A. and Lotze, M.T. (1995). Cancer immunotherapy of established tumors with IL-12. Effective delivery by genetically engineered fibroblasts. *J. Immunol.* **155**, 1393–1403.

Zitvogel, L., Robbins, P.D., Storkus, W.J., Clarke, M.R., Maeurer, M.J., Campbell, R.L., Davis, C.G., Tahara, H., Schreibe, R.D. and Lotze, M.T. (1996a). Interleukin-12 and B7.1 co-stimulation cooperate in the induction of effective antitumor immunity and therapy of established tumors. *Eur. J. Immunol.* **26**, 1335–1341.

Zitvogel, L., Couderc, B., Mayordomo, J.I., Robbins, P.D., Lotze, M.T. and Storkus, W.J. (1996b). IL-12 engineered dendritic cells serve as effective tumor vaccine adjuvants *in vivo.* In *Annals of the New York Academy of Sciences. Interleukin 12. Cellular and Molecular Immunology of an Important Regulatory Cytokine* (eds M.T. Lotze, G. Trinchieri, M. Gately and S. Wolf), New York Academy of Sciences, New York, pp. 284–293.

Chapter 31

Assays for Cytokines

Meenu Wadhwa and Robin Thorpe
Division of Immunobiology, National Institute for Biological Standards and Control, Potters Bar, UK

Introduction

Accurate and sensitive methods for the measurement and detection of cytokines are an obvious prerequisite for understanding the biology of cytokines, to determine the clinical potential of these molecules and their role in the pathogenesis of numerous diseases. Various types of assays are now available for estimation of the amount of cytokines present in a wide range of biological fluids. The biological activities elicited by cytokines usually become the basis for biological detection and assay of the molecules (e.g. formation of granulocyte colonies in the case of granulocyte colony stimulating factor or protection from cell killing by cytocidal viruses in the case of interferons). However, none of the bioassays fulfils the criteria of an ideal assay system in terms of specificity, sensitivity, simplicity, speed and reliability, and this has stimulated the development of immunoassays or alternative immunochemical or biochemical techniques for measuring and identifying cytokines.

BIOLOGICAL ASSAYS FOR CYTOKINES

Numerous bioassay formats and types based on use of either *in vivo* whole animal assays or *in vitro* cell-based assays have been developed for quantitation of cytokines.

Whole Animal-Based Assays

In vivo assays can provide invaluable information about the biological activity of cytokines as well as a useful assessment of their biological potency. Common examples are the pyrogen assay for interleukin-1 (IL-1) (Dinarello *et al.*, 1988) and the post-hypoxic polycythaemic mouse assay for erythropoietin (Cotes and Bangham, 1961). However, such assays can be expensive, involve large numbers of animals, and are often imprecise and tedious to perform. Therefore, these assays are now seldom used and have been largely superseded by *in vitro* cell-based bioassays which require either primary cultures of cells (Tables 1 and 2) or continuous cell lines (Table 3).

The Cytokine Handbook, 3rd ed.
ISBN 0-12-689662-3

Table 1. Whole animal or primary cell culture-based assays for cytokines.

Cytokine	Target system	Biological assay	Reference
	Thymocytes	Mitogenesis assay	Gery *et al*. (1972)
	Fibroblasts	Induction of cytokines (e.g. IL-6)	Van Snick *et al*. (1986)
IL-1	Skeletal plants	Bone resorption	Gowen *et al*. (1983)
	Synovial cells	Release of collagenase or prostaglandin E_2	Mizel *et al*. (1981)
	Hepatocytes	Induction of acute-phase proteins	Dinarello (1984)
	In vivo: mouse, rat	Induction of acute-phase proteins	Dinarello (1984)
	In vivo: mouse, rabbit	Pyrogen assay	Dinarello *et al*. (1988)
IL-2	Mitogen-activated T cells	Proliferation	
IL-3	Progenitor cells of bone marrow	Stimulation of erythroid, granulocyte–macrophage colony formation—colony assay	Metcalf (1984)
IL-4	B-cells costimulated with either SAC, anti-Ig or anti-CD40 monoclonal antibodies or phorbol esters	Proliferation assay	Callard *et al*. (1987)
	Tonsillar B cells	Enhancement of activation antigens (e.g. CD23 or surface IgM expression)	Callard *et al*. (1987)
IL-5	Bone marrow	Formation of eosinophil colonies in a colony assay or by estimation of eosinophil peroxidase	O'Garra and Sanderson (1987)
IL-6	B cells costimulated with anti-CD40 monoclonal antibodies	Enhancement of IgG secretion	Callard *et al*. (1987)
	Hepatocytes	Induction of acute-phase proteins	May *et al*. (1988)
IL-7	Precursor B cells	Proliferation	Namen *et al*. (1988)
	Mitogen-stimulated T cells	Proliferation	Watanabe *et al*. (1992)
IL-9	Long-term cultured PBLs (activated)	Proliferation	Houssiau *et al*. (1993)
	Bone marrow	Formation of erythroid colonies in the presence of erythropoietin	Donahue *et al*. (1990)
	Mast cells, mast cell lines	Proliferation	Hultner *et al*. (1990), Williams *et al*. (1990)
IL-10	Activated PBMCs, mitogen/antigen-activated T_{H1} clones	Inhibition of IFN-γ production	Fiorentino *et al*. (1989)
	Mast cells	Enhances IL-3/IL-4-induced proliferation	Thomson-Snipes *et al*. (1991)
	Macrophages	Inhibition of cytokine production following LPS activation	Fiorentino *et al*. (1991)

(continued)

Table 1. *Continued.*

Cytokine	Target system	Biological assay	Reference
IL-11	Bone marrow	Stimulation of megakaryocyte colony formation in synergy with IL-3	Paul *et al*. (1990)
	B cells	Proliferation	Paul *et al*. (1990)
IL-12	Phytohaemagglutinin-activated lymphoblasts	Proliferation	Stern *et al*. (1990)
	NK cells	Stimulates IFN-γ production	Kobayashi *et al*. (1989)
	NK cells	Augmentation of NK cell activity	Kobayashi *et al*. (1989)
IL-13	Monocytes	Upregulation of MHC class II expression	McKenzie *et al*. (1993)
		Inhibits production of inflammatory cytokines in response to LPS/IFN-γ	Minty *et al*. (1993)
	B cells costimulated with anti-Ig and anti-CD40 antibodies	Proliferation	Defrance *et al*. (1994)
IL-14	B cells activated with SAC	Proliferation	Ambrus *et al*. (1993)
IL-15	Activated T lymphocytes	Proliferation	Giri *et al*. (1994), Grabstein *et al*. (1994)
	B cells	Proliferation	Matthews *et al*. (1995)
	NK cells	Activation	Carson *et al*. (1994)
IL-16	T cells	Chemotaxis assay	Lim *et al*. (1996)
IL-17	T cells	Proliferation	Yao *et al*. (1995)
	Fibroblasts	Secretion of cytokines (IL-6, G-CSF)	Yao *et al*. (1995), Fossiez *et al*. (1996)
IL-18	PBMCs	Induction of IFN-γ production	Ushio *et al*. (1996)
		Augmentation of NK cell activity	Micallef *et al*. (1996), Ushio *et al*. (1996)
TNF-α/β	Endothelial cells	Induction of HLA class I expression, increased expression of adhesion molecules	Vilcek and Lee (1991), Meager (1996)
	Hepatocytes	Upregulates expression of acute-phase proteins	Perlmutter *et al*. (1986)
	Fibroblasts	Proliferation	Vilcek *et al*. (1986)
	Synovial cells	Production of collagenase and prostaglandin E_2	Dayer *et al*. (1985)
Erythropoietin	*In vivo*: Post-hypoxic polycythaemic mouse	Stimulation of erythropoiesis	Cotes and Bangham (1961)
	Bone marrow	Formation of erythroid colonies	Kennedy *et al*. (1980)
	Bone marrow	Stimulates haemoglobin synthesis	Clarke *et al*. (1979)
G-CSF	Bone marrow	Formation of granulocyte colonies	Nomura *et al*. (1986)

(*continued*)

Table 1. *Continued.*

Cytokine	Target system	Biological assay	Reference
G-CSF	Bone marrow	Proliferation	Nomura *et al*. (1986), Wadhwa *et al*. (1990)
	Neutrophils	Enhancement of superoxide production	Yuo *et al*. (1989)
GM-CSF	Bone marrow	Formation of granulocyte–macrophage colonies	Metcalf *et al*. (1986)
	Monocytes, granulocytes	Increased expression of adhesion molecules	Arnaout *et al*. (1986), Grabstein *et al*. (1986)
	Neutrophils	Respiratory burst activity	Khwaja *et al*. (1992)
M-CSF	Bone marrow	Macrophage colony formation	Stanley *et al*. (1978, 1983)
	Monocytes	Production of cytokines (e.g. IFN, TNF)	Warren and Ralph (1986)
Stem cell factor	Bone marrow, cord blood	Synergizes with CSFs and cytokines in multilineage colony formation	Zsebo *et al*. (1990), Pietsch *et al*. (1992)
	Mast cells	Proliferation	Tsai *et al*. (1991)
LIF	Bone marrow	Synergizes with IL-3 in megakaryocyte colony formation	Metcalf *et al*. (1991)
Thrombo-poietin	Bone marrow	Formation of megakaryocyte colonies	Kaushansky *et al*. (1994), Broudy *et al*. (1995), Gurney *et al*. (1995)
Flt3 ligand	Bone marrow, fetal liver	Synergizes with CSFs in multilineage colony formation	Lyman *et al*. (1993), Gabbianelli *et al*. (1995), Muench *et al*. (1995)

LIF, leukaemia inhibitory factor; LPS, lipopolysaccharide; PBL, peripheral blood lymphocyte; PBMC, peripheral blood mononuclear cell; SAC, *Staphylococcus aureus* Cowan type 1.

Primary Cell Culture-Based Assays

These assays require primary cultures of cells, for example cells of the haemopoietic or immune system which can be obtained from bone marrow, blood or other tissues, organs and glands (e.g. liver, spleen, thymus) from human or other animal sources. In these assays, inconsistencies due to donor variation can be problematical. However, in many cases, and particularly in early research involving newly identified cytokines, these assays are very useful and are often the assays of choice for showing biological effects of particular cytokines. The bone marrow colony assay, for example, for assessment of colony stimulating activity has been shown to be very useful for investigating the effects of the cytokine or combination of cytokines on haemopoiesis *in vitro*. In certain instances, purified cells are used for assays such as B- or T-lymphocyte proliferation, neutrophil chemotaxis and natural killer (NK) cell stimulatory activity, and can provide a quantitative estimate of cytokine levels. A drawback with these assays, however, is that the separation of purified cell preparation is not straightforward and contamination of the cell preparation can often influence the results from the bioassays.

Table 2. Primary cell culture-based assays for chemokines.

Chemokine	Target cell types in chemotaxis assay	Other assays
IL-8	Neutrophils, basophils, T cells	Degranulation of neutrophils; respiratory burst activity and lysosomal enzyme release from neutrophils; induces Ca^{2+} fluxes in monocytes; releases histamine from basophils
PF-4	Monocytes, neutrophils, fibroblasts	Inhibition of angiogenesis
β-Thromboglobulin	Neutrophils, fibroblasts	
RANTES	Monocytes, eosinophils, basophils, memory T cells	Induces histamine release from basophils
MIP-1α	Monocytes, B cells, cytotoxic T cells, $CD4^+$ T cells, basophils, eosinophils	Induces histamine release from basophils and mast cells; stem cell inhibition; induces Ca^{2+} fluxes in monocytes
MIP-1β	Monocytes, $CD4^+$ T cells (preferentially the 'naive' T cells)	Lysosomal enzyme release and degranulation of neutrophils
MCP-1, 2 and 3	Monocytes, basophils	Induces histamine release from basophils; augments superoxide production and lysosomal enzyme release from monocytes
GRO-α, β and γ	Neutrophils	Induces Ca^{2+} fluxes in monocytes; induces respiratory burst activity and lysosomal enzyme release from neutrophils
ENA-78	Neutrophils	Induces Ca^{2+} fluxes in neutrophils
I-309	Monocytes	Induces Ca^{2+} fluxes in monocytes
IP-10	Monocytes, T cells	
NAP-2	Neutrophils	Induces Ca^{2+} fluxes in neutrophils, elastase release from neutrophils

GRO, growth-regulated oncogene; IP, interferon-induced protein; MCP, monocyte chemotactic protein; MIP, macrophage inflammatory protein; NAP, neutrophil activating protein; PF, platelet factor; RANTES, Regulated upon Activation, Normal T-cell Expressed and Secreted; ENA, epithelial derived neutrophil activating 78 amino acid peptide.

CELL LINE-BASED ASSAYS

Extensive research has enabled the development of bioassays using continuously growing cell lines and/or transfected lines derived either from laboratory-induced leukaemias or from leukaemic patients. Generally, cell line-based assays are economical, easy to use and analyse, and usually produce reliable, reproducible and accurate estimates of biologically active cytokine. Although cell line-based assays have distinct advantages over other biological assays for cytokines, they are still prone to problems that need to be addressed before reliable estimates of cytokine levels can be derived. The major problem associated with such cell lines is that they are not specific for a single cytokine. For example, the TF1 erythroleukaemia cell line proliferates in response to a range of cytokines and is also susceptible to substances such as transforming growth

Table 3. Cell line-based assays for cytokines.

Cytokine	Cell line	Origin	Biological readout	Range (per ml)	Cytokine responsiveness	References
IL-1	D10S	Murine T helper cell	Proliferation	0.01–10 IU	IL-2	Orencole and Dinarello (1989)
	A375	Human melanoma	Inhibition of proliferation		TNF-α, TNF-β, TGF-β_1, TGF-β_2, IL-6, LIF	Nakai *et al*. (1988)
	NOB-1	Murine thymoma	Production of IL-2	0.01–10 IU	TNF	Gearing *et al*. (1987)
IL-2	CTLL-2	Murine cytotoxic T cell	Proliferation	0.01–10 IU	IL-15, TGF-β_1, TGF-β_2	Gillis *et al*. (1978)
	KIT-225	Human chronic T lymphocytic leukaemia	Proliferation	0.1–20 IU	IL-12, IL-7, IL-15	Hori *et al*. (1987)
IL-3	M-07e	Human megakaryoblastic leukaemia	Proliferation	0.05–20 IU	GM-CSF, SCF, IL-9, IL-15, TPO, TNF-α, TNF-β, IFN-α, IFN-β, TGF-β_1, TGF-β_2	Avanzi *et al*. (1990)
	TF-1	Human erythroleukaemia	Proliferation	0.02–10 IU	IL-4, IL-5, IL-6, IL-13, IL-15, GM-CSF, NGF, SCF, LIF, oncostatin M, CNTF, EPO, TNF-α, TNF-β, IFN-α, IFN-β, TGF-β_1, TGF-β_2	Kitamura *et al*. (1989)
IL-4	CT.h4S	Murine cytotoxic T cell transfected with hIL-4 receptor	Proliferation	0.1–25 IU	IL-2, TNF-α, TNF-β, IFN-α, IFN-β, TGF-β_1, TGF-β_2	Hu-Li *et al*. (1989)
	CCL-185	Human lung tumour	Inhibition of proliferation	0.05–10 IU	IL-13, IL-1, TNF-α, TNF-β	Page *et al*. (1996a)
	RAMOS	Human B-cell lymphoma	Augmentation of CD23 expression	1–20 IU	TNF-α, TNF-β	Siegel and Mostowski (1990)
IL-5	TF-1	Human erythroleukaemia	Proliferation	0.1–50 U	IL-4, IL-5, IL-6, IL-13, IL-15, GM-CSF, NGF, SCF, LIF, oncostatin M, CNTF, EPO, TNF-α, TNF-β, IFN-α, IFN-β, TGF-β_1, TGF-β_2	Kitamura *et al*. (1989)
IL-6	B9	Murine hybridoma	Proliferation	0.05–5 IU	IL-13, IL-11	Aarden *et al*. (1987)
	CESS	Human B cell	Production of IgG	20–5000 IU		Callard *et al*. (1987)
	7TD1	Murine hybridoma	Proliferation	0.1–15 IU		Van Snick *et al*. (1986)
IL-7	2bx	Murine pre-B cell	Proliferation	10–5000 U		Park *et al*. (1990)
	KIT-225	Human chronic T lymphocytic leukaemia	Proliferation	5–1000 U	IL-2, IL-12, IL-15	Authors' unpublished observations

(*continued*)

Table 3. *Continued.*

Cytokine	Cell line	Origin	Biological readout	Range (per ml)	Cytokine responsiveness	References
IL-9	M-07e	Human megakaryoblastic leukaemia	Proliferation	0.5–100 U	GM-CSF, SCF, IL-9, IL-15, TPO, TNF-α, TNF-β, IFN-α, IFN-β, TGF-β_1, TGF-β_2	Yang *et al*. (1989)IL-10
MC-9	Murine mast cell line	Proliferation	1–100 U	IL-5	Thompson-Snipes *et al*. (1991)	
	Ba8.1c1	Murine pro-B cell transfected with hIL-10 receptor	Proliferation	0.5–500 U		Liu *et al*. (1994)
IL-11	B9–11	Murine hybridoma	Proliferation	0.2–50 U	IL-6	Lu *et al*. (1994)
	T10	Murine plasmacytoma	Proliferation	0.5–50 U	IL-6	Nordan and Potter (1986)
IL-12	KIT-225	Human chronic T lymphocytic leukaemia	Proliferation	0.01–20 U	IL-2, IL-7, IL-15	Hori *et al*. (1987)
IL-13	B9.1.3	Murine hybridoma	Proliferation	0.3–20 U	IL-6, IL-11	Bouteiller *et al*. (1995)
	TF-1	Human erythroleukaemia	Proliferation	0.5–100 U	IL-4, IL-5, IL-6, IL-13, IL-15, GM-CSF, NGF, SCF, LIF, oncostatin M, CNTF, EPO, TNF-α, TNF-β, IFN-α, IFN-β, TGF-β_1, TGF-β_2	
IL-15	CTLL-2	Murine cytotoxic T cell	Proliferation	0.2–10 U	IL-15, TGF-β_1, TGF-β_2	Giri *et al*. (1994)
	KIT-225	Human chronic T lymphocytic leukaemia	Proliferation	1–1000 U	IL-2, IL-7, IL-12	
GM-CSF	M-07e	Human megakaryoblastic leukaemia	Proliferation	0.1–10 IU	GM-CSF, SCF, IL-9, IL-15, TPO, TNF-α, TNF-β, IFN-α, IFN-β, TGF-β_1, TGF-β_2	Avanzi *et al*. (1990)
	TF-1	Human erythroleukaemia	Proliferation	0.01–5 IU	IL-4, IL-5, IL-6, IL-13, IL-15, GM-CSF, NGF, SCF, LIF, oncostatin M, CNTF, EPO, TNF-α, TNF-β, IFN-α, IFN-β, TGF-β_1, TGF-β_2	Avanzi *et al*. (1990)
G-CSF	GNFS-60	Murine myeloid leukaemia	Proliferation	0.5–100 IU	IL-6, TGF-β_1, TGF-β_2, M-CSF, oncostatin M, LIF, IL-13	Weinstein *et al*. (1986)
	WEHI 3BD$^+$	Murine myelomonocytic leukaemia	Differentiation	10–100 IU	GM-CSF	Nicola *et al*. (1983)
M-CSF	MNFS-60	Murine myeloid leukaemia	Proliferation	1.5–150 IU	IL-6, TGF-β_1, TGF-β_2, G-CSF, oncostatin M, LIF, IL-13	Nakoinz *et al*. (1989)

(*continued*)

Table 3. *Continued.*

Cytokine	Cell line	Origin	Biological readout	Range (per ml)	Cytokine responsiveness	References
M-CSF	BAC1.2F5	SV40-transformed murine macrophage line	Proliferation	10–100 IU	GM-CSF	Morgan *et al*. (1987)
SCF	M-07e	Human megakaryoblastic leukaemia	Proiferation	1–100 U	GM-CSF, SCF, IL-9, IL-15, TPO, TNF-α, TNF-β, IFN-α, IFN-β, TGF-β_1, TGF-β_2	Hendrie *et al*. (1991)
	TF-1	Human erythroleukaemia	Proliferation	1–100 U	IL-4, IL-5, IL-6, IL-13, IL-15, GM-CSF, NGF, SCF, LIF, oncostatin M, CNTF, EPO, TNF-α, TNF-β, IFN-α, IFN-β, TGF-β_1, TGF-β_2	Thorpe *et al*. (1992)
LIF	DA-1a	Murine myeloid leukaemia	Proliferation	0.01–100 U	G-CSF	Moreau *et al*. (1986)
	TF-1	Human erythroleukaemia	Proliferation	0.05–50 U	IL-4, IL-5, IL-6, IL-13, IL-15, GM-CSF, NGF, SCF, LIF, oncostatin M, CNTF, EPO, TNF-α, TNF-β, IFN-α, IFN-β, TGF-β_1, TGF-β_2	
TPO	M-07e	Human megakaryoblastic leukaemia	Proliferation	20 pg to 20 ng	GM-CSF, SCF, IL-9, IL-15, TPO, TNF-α, TNF-β, IFN-α, IFN-β, TGF-β_1, TGF-β_2	Page *et al*. (1996b)
	32D/Mpl$^+$	Murine myeloid leukaemia transfected with human mpl receptor	Proliferation	20 pg to 2 ng	G-CSF	Bartley *et al*. (1994)
Oncostatin M	A375	Human melanoma	Inhibition of proliferation	0.5–100 U	IL-1, IL-6, LIF, TNF-α, TNF-β, TGF-β_1, TGF-β_2	Nakai *et al*. (1988)
	TF-1	Human erythroleukaemia	Proliferation	0.5–500 U	IL-4, IL-5, IL-6, IL-13, IL-15, GM-CSF, NGF, SCF, LIF, oncostatin M, CNTF, EPO, TNF-α, TNF-β, IFN-α, IFN-β, TGF-β_1, TGF-β_2	
TGF-β	MuLv1	Mink lung fibroblasts	Proliferation	10 pg to 1 ng	TNF-α, TNF-β	Meager (1991a)
	TF-1	Human erythroleukaemia	Proliferation	500 fg to 1 ng	IL-4, IL-5, IL-6, IL-13, IL-15, GM-CSF, NGF, SCF, LIF, oncostatin M, CNTF, EPO, TNF-α, TNF-β, IFN-α, IFN-β, TGF-β_1, TGF-β_2	Randall *et al*. (1993)

(*continued*)

Table 3. *Continued.*

Cytokine	Cell line	Origin	Biological readout	Range (per ml)	Cytokine responsiveness	References
TNF-α	KYM-1D4	Human rhabdo-myosarcoma	Cytotoxicity	0.2–4 IU	No other cytokine	Meager (1991b)
	L-M	Murine fibroblasts	Cytotoxicity	0.2–4 IU		Kramer and Carver (1986)
TNF-α	WEHI164	Human fibrosarcoma	Cytotoxicity	0.04–2 IU		Espevik and Nissen-Meyer (1986)
TNF-β	KYM-1D4	Human rhabdo-myosarcoma	Cytotoxicity	4–75 IU		Meager (1991b)
	WEHI164	Human fibrosarcoma	Cytotoxicity	0.075–40 IU		Espevik and Nissen-Meyer (1986)
	L-M	Murine fibroblasts	Cytotoxicity	0.10–30 IU		Kramer and Carver (1986)
CNTF	BAF-LRgpCR	Murine CNTF receptor transfected cell line	Proliferation	1 pg to 1 ng	LIF, oncostatin M	Gearing *et al*. (1994)
RANTES	THP-1	Human acute monocytic leukaemia	Chemotaxis	0.02–10 U	MIP-1a, MCP-1, 2, 3	Tsuchiya *et al*. (1980), Wang *et al*. (1993a)
MIP-1a	THP-1	Human acute monocytic leukaemia	Chemotaxis	0.2–10 U	RANTES, MCP-1, 2, 3	Wang *et al*. (1993b)
MCP-1, 2, 3	THP-1	Human acute monocytic leukaemia	Chemotaxis	0.02–100 U	RANTES, MIP-1a	Wang *et al*. (1993b)
GRO	Hs294T	Human melanoma	Proliferation	100 pg to 100 ng	IL-8	Horuk *et al*. (1993)
	U937	Human monocytic leukaemia	Ca^{2+} mobilization	10–1000 ng	IL-8	Horuk *et al*. (1993)
IFN-α	2D9 + ECV	Human glioblastoma, encephalo-mycarditis virus	Antiviral	0.1 pg to 2 pg	IFN-α, IFN-β	Däubener *et al*. (1994)
	TF-1	Human erythroleukaemia	Inhibition of proliferation	0.5 pg to 10 ng	IL-4, IL-5, IL-6, IL-13, IL-15, GM-CSF, NGF, SCF, LIF, oncostatin M, CNTF, EPO, TNF-α, TNF-β, IFN-α, IFN-β, TGF-β_1, TGF-β_2	Mire-Sluis *et al*. (1996c)
	Daudi	Human B lymphoblastoid cell line	Inhibition of proliferation	1–100 pg		Prummer *et al*. (1994)
IFN-β	2D9 + EMCV	Human glioblastoma cell line + encephalo-myocarditis virus	Antiviral	0.3–6 pg	IFN-α, IFN-β	Däubener *et al*. (1994)
	TF-1	Human erythroleukaemia	Inhibition of proliferation	0.5 pg to 10 ng	IFN-α, IFN-β	Mire-Sluis *et al*. (1996c)

(*continued*)

Table 3. *Continued.*

Cytokine	Cell line	Origin	Biological readout	Range (per ml)	Cytokine responsiveness	References
IFN-β	Daudi	Human B lymphoblastoid cell line	Inhibition of proliferation	10 pg to 1 ng		Prummer *et al*. (1994)
IFN-γ	2D9 + EMCV	Human glioblastoma cell line + encephalo-myocarditis virus	Antiviral	20–200 pg	IFN-α, IFN-β	Däubener *et al*. (1994)
	COLO 205	Human colorectal carcinoma	MHC class II expression	20–2000 pg		Gibson and Kramer (1989)
EPO	UT-7/EPO	Megakaryoblastic leukaemia	Proliferation	200 pg to 10 ng	TGF-β_1, TGF-β_2	Komatsu *et al*. (1993)
	TF-1	Human erythroleukaemia	Proliferation		IL-4, IL-5, IL-6, IL-13, IL-15, GM-CSF, NGF, SCF, LIF, oncostatin M, CNTF, EPO, TNF-α, TNF-β, IFN-α, IFN-β, TGF-β_1, TGF-β_2	Kitamura *et al*. (1989)

CNTF, ciliary neurotrophic factor; EPO, erythropoietin; NGF, nerve growth factor; SCF, stem cell factor; TPO, thrombopoietin; EMCV, encephalomycarditis virus.

factor-β (TGF-β) that inhibit the proliferative response to stimulatory cytokines (Randall *et al.*, 1993; Wadhwa *et al.*, 1995). Interpretation of results obtained for cytokine content of biological fluids using this cell line is therefore difficult as the measured proliferation could be due to one or more cytokines present in variable proportions which are possibly masked by the effects of inhibitory factors or cytokines (Table 3). In certain instances, bioassays may underestimate the cytokine activity in biological samples owing to the presence of inhibitors such as IL-1 receptor antagonist, soluble cytokine receptors or other antagonistic molecules. It is also possible that bioassays may overestimate cytokine content due to synergistic interactions between individual cytokines. As an example, the synergistic effect of IL-4 and TNF-α on proliferation of the M-07e cell line is illustrated in Fig. 1.

One approach to identifying which cytokine(s) are causing effects on multispecific cell lines is specifically to neutralize the effects of single cytokines using antibodies, and the present authors and others have used this approach successfully to dissect the response of such cell lines to individual cytokines (Fig. 2).

In some cases, it is possible to increase specificity by using cell lines from a different species from the source of the cytokine. The mouse CTLL-2 cell line proliferates in response to both murine and human IL-2 and murine IL-4, but is unresponsive to human sequence IL-4. It can therefore be used specifically to measure human IL-2 in biological samples with little interference from other molecules. The murine NFS-60 line responds to murine granulocyte colony stimulating factor (G-CSF), granulocyte–macrophage colony stimulating factor (GM-CSF) and IL-3, but can be used to discriminate human G-CSF from the human sequence IL-3 and GM-CSF, as the latter cytokines do not affect mouse cells.

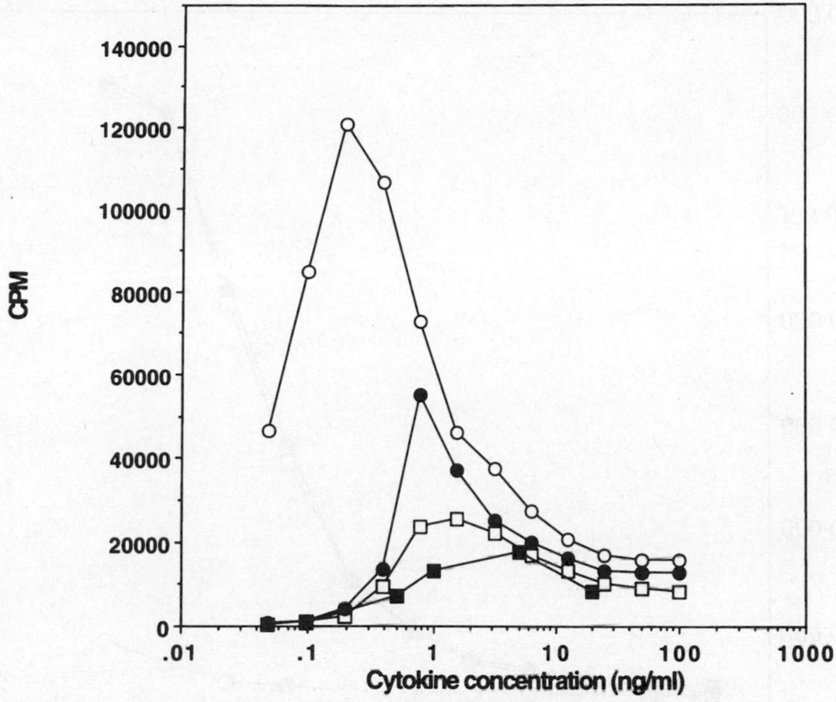

Fig. 1. Synergistic effect of TNF and IL-4 on proliferation of M-07e cells. Dose–response curve of TNF-α alone (□), IL-4 alone (■), TNF-α in combination with IL-4 at 5 ng/ml (○) and IL-4 at 0.1 ng/ml (●) were obtained as described by Wadhwa *et al*. (1996).

The recent development of receptor-transfected cell lines has produced several more specific cell lines, but even these can still respond to other cytokines (Hu-Li *et al*., 1989). It is difficult to find a parent line that does not respond in one way or another to more than one cytokine.

TYPES OF BIOASSAYS

There are several different approaches available for cytokine bioassay.

Proliferation and Antiproliferation Assays

These assays provide the most versatile format for measurement of the majority of cytokines. In such assays, one simply estimates the levels of cytokines by their ability to stimulate or inhibit cellular proliferation. Several methods can be used for measuring the proliferation of cells. Direct counting of cell numbers is probably the simplest method; however, it is tedious, often inaccurate and not practical for a large number of samples. Measurement of tritiated thymidine incorporation into DNA is the most commonly used method for estimating the increase or decrease in proliferative activity of cells. This method is easily automated and can be used routinely for handling large numbers of samples.

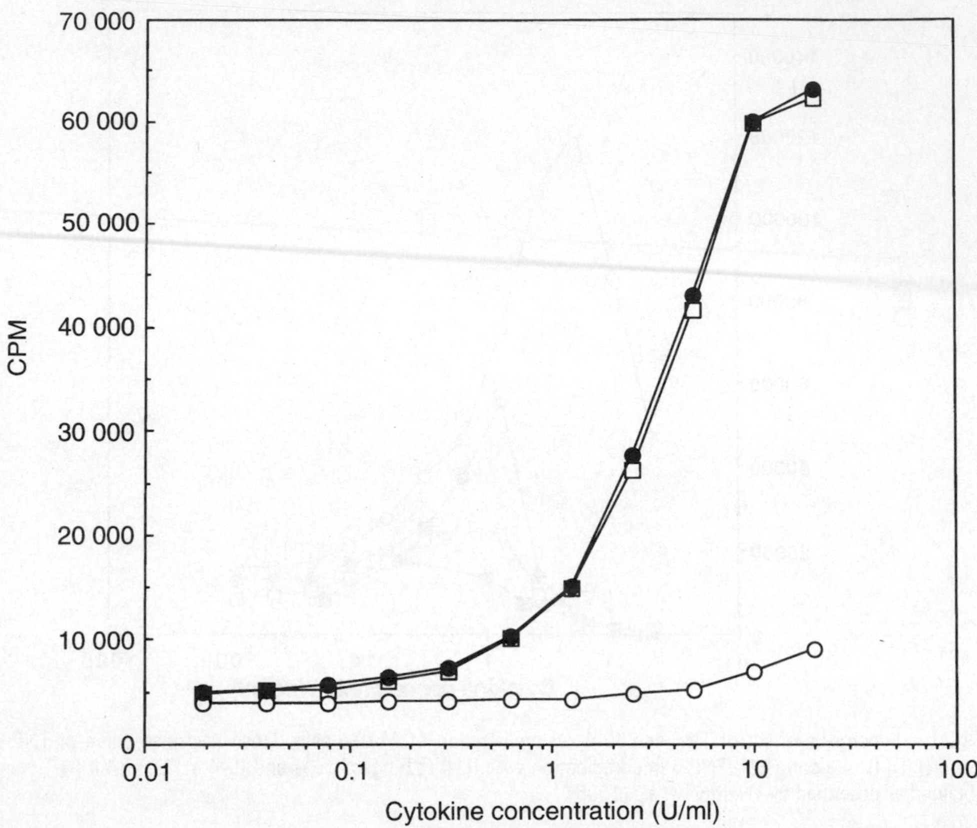

Fig. 2. Neutralization of IL-15-induced proliferation of the KIT-225 cell line by an IL-15-specific polyclonal antiserum. Proliferation of cells in response to IL-15 (□) alone or in the presence of a combination of IL-15 and anti-IL-15 polyclonal antiserum (○) (1:1000 dilution), or IL-15 and an irrelevant polyclonal antiserum (●) (1:1000 dilution).

Use of alternative non-radioactive procedures that measure cell metabolism as an index of proliferation have now become very popular. Several studies report the use of the redox-sensitive formazan (3-(4,5-dimethylthiazol-2-yl)-2,5-diphenyl tetrazolium bromide) (MTT), which forms purple–dark blue crystals on exposure to the dehydrogenase activity in metabolizing cells and requires solublization with detergent before estimation of the absorbance of the resulting solution by an enzyme-linked immunosorbent assay (ELISA) reader (Mosmann, 1983). Recently, tetrazolium salts such as XTT (Roehm *et al.*, 1991), MTS (Buttke *et al.*, 1993) and WST-1 that produce soluble formazan products (this eliminates the need for a solublization step) have become available and shown to be useful for measuring proliferation of lymphokine-dependent cell lines (Buttke *et al.*, 1993).

Other newly developed formats for assessment of proliferation include detection of metabolic products such as adenosine 5′-triphosphate (ATP) or cyclic adenosine monophosphate (cAMP) using bioluminescence (Crouch *et al.*, 1993). As for tritiated thymidine-based systems, automated systems have been devised to deal with these non-radioactive procedures. It should be noted that a relatively lower stimulation index is often found with non-radioactive procedures compared with methods using tritiated thymidine incorporation.

Induction of Cell Surface Molecules

Alterations in expression of specific cell surface molecules can be used to bioassay some cytokines. Such assays have been based on upregulation of major histocompatibility complex (MHC) class I by interferon-γ (IFN-γ) using various cell types (Gibson and Kramer, 1989), or CD23 on B cells or cell lines by IL-4 (Siegel and Mostowski, 1990), or expression of cellular adhesion molecules (e.g. intercellular adhesion molecule-1 (ICAM-1)) on glioblastoma or astrocytoma cell lines with cytokines such as IL-1α/β, tumour necrosis factor-α/β (TNF-α/β) and IFN-γ (Meager, 1996). Such biological responses can be quantified immunochemically by microscopy, flow cytometry or more conveniently using the Pandex format (Custer and Lotze, 1990) or in ELISAs or immunoradiometric assays (IRMAs) (Rousset *et al.*, 1988).

Induction of Secretion of 'Secondary' Molecules

The ability of some cytokines to enhance or inhibit secretion of some substances from appropriate cell types can form the basis of a bioassay (Tables 1 and 3). Common examples are measurement of IL-6 by its ability to enhance immunoglobulin G (IgG) secretion by some B or plasmacytoma cell lines (Callard *et al.*, 1987), determination of IL-1 by its induction of IL-2 secretion from T cells (Gearing *et al.*, 1987), assay of IL-6 by synthesis and secretion of acute-phase proteins such as α_1-antichymotrypsin from hepatoma cell lines or hepatocytes (May *et al.*, 1988). Induction of Mx protein in fibroblast cell lines can be used as a bioassay for interferons. The levels of secreted proteins can be measured using specific bioassays or immunoassays.

Cytokines such as GM-CSF and IL-8 can be assayed by their ability to induce neutrophil degranulation, which is measured by release of granule constituents such as β-glucuronidase, elastase and myeloperoxidase (Schroder *et al.*, 1987; Peveri *et al.*, 1988). Less commonly used 'secondary' procedures are assay of IL-1 and TNF by measurement of prostaglandin release from cells of the synovium (Mizel *et al.*, 1981) or carcinoma cell lines (Last-Barney *et al.*, 1988).

Antiviral Assays

The potent antiviral activity of some cytokines (e.g. interferons) constitutes the basic principle for these assays (Lewis, 1995a). Procedures involve incubating susceptible cells with interferon before addition of a cytopathic virus and estimating the increase in cell survival due to a reduction in cytopathic effect of the virus. Generally, various combinations of target cells such as primary human fibroblasts or cell lines (e.g. 2D9, Hep 2C, WISH, A549) and challenge viruses (encephalomycarditis virus, vesicular stomatitis virus, Semliki Forest virus) are available for antiviral assays, with some combinations providing accurate, sensitive and reliable results. The number of viable cells at the end of the assay can be determined by a variety of methods such as tritiated thymidine incorporation, tetrazolium salt reduction or vital staining, for example with naphthol blue–black or neutral red.

Bone Marrow Colony Formation Assays

Assessment of bone marrow colony formation remains the definitive assay for the CSFs. The technique involves culture of bone marrow cells with dilutions of CSFs for 7–14 days followed by counting numbers of colonies obtained (Metcalf *et al.*, 1979; Metcalf, 1984). The type of colony produced depends on the factor. IL-3 and GM-CSF stimulate the production of mixed colonies of different cell lineages, whereas with cytokines such as G-CSF and M-CSF colonies of only one cell type predominate (but are not exclusive). The procedure is time consuming, tedious, requires considerable skill to perform consistently and is easily influenced by contaminating factors which may enhance or inhibit colony formation (Suda *et al.*, 1988; Munker *et al.*, 1986). The biological readout is subjective (size and type of colony) and very variable between operators. Attempts at refining the basic procedures to produce more efficient methods capable of at least semiautomation have not been particularly successful.

Cytotoxic Assays

Some cytokines, notably TNF-α and TNF-β, can be assayed by their potent cytotoxic effect on susceptible cell lines (Table 3). Such assays involve culture of these cells (usually as an adherent monolayer) with dilutions of cytokine and assessment of cytotoxic activity by uptake of a dye (e.g. naphthol blue–black) by the residual viable cells at the end of the incubation period. The amount of dye associated with the cells is proportional to the number of viable cells and can be quantitated using an ELISA reader; this is inversely proportional to cytotoxic activity (Meager *et al.*, 1989; Meager, 1991b). For partially adherent and non-adherent cells, however, alternative methods employing uptake of tritiated thymidine or reduction of tetrazolium salts can be used for assessing residual cells, and is inversely related to cytotoxic activity.

Inhibition of Cytokine Secretion and Activity

A few cytokines inhibit secretion of other cytokines sufficiently for this to be used for their assay. Examples are IL-10 inhibition of IFN-γ secretion (Fiorentino *et al.*, 1989) and TGF-β-mediated inhibition of the proliferative effects of IL-5 (Randall *et al.*, 1993).

Chemotaxis Assays

For chemokines, chemotaxis is often the only suitable biological assay. Chemotaxis involves quantifying the levels of chemokines via their ability to attract cells through membranes (Boyden, 1962). The number of cells migrating through the membrane serves as a measure of the potency of the chemotactic cytokine present in the sample. There are several commercially available chambers that can be used for chemotaxis assays (Van Damme and Conings, 1995). A chamber that incorporates a standard 96-well microtitre plate is the most suitable for quantitative biological assays for cytokines

(Junger *et al.*, 1993). Again, cell number can be measured by any of the methods described above.

Assessment of Respiratory Burst Induction

Some cytokines induce a respiratory burst in neutrophils and this can be used for their measurement (e.g. IL-8 and TNF). GM-CSF primes neutrophils for a potent respiratory burst which can be triggered by other molecules (e.g fmet–leu–phe). The effect can be measured in neutrophils by spectrophotometric estimation of the reduction of cytochrome C (by change in absorbance at 550 nm) caused by release of superoxide anions (Doyle *et al.*, 1995).

Measurement of Ionic Flux

Measurement of a cytokine mediated Ca^{2+} flux using appropriate fluorescent calcium indicator dyes (quin2, fura2, indo3) can be used for assay of some cytokines, in particular for members of the chemokine family.

Reporter Gene-based Assays

Reporter gene assays for cytokines have been developed. For this, appropriate cells are transfected with a construct consisting of the promoter of a gene known to be induced by the cytokine fused to a rcporter gene system. Such an assay has been described for TGF-β using a truncated plasminogen activator inhibitor-I promoter fused to the gene for firefly luciferase (Abe *et al.*, 1994). Addition of TGF-β to the cells results in a dose-dependent increase in secretion of luciferase into the cell supernatant. This enzymatic activity can be quantitated using either tritium-labelled substrates or chromogenic procedure (Abe *et al.*, 1994; Lewis, 1995a,b). These assays may require minimum specialized equipment and can be adapted to 96-well microtitre plate format, which facilitates multiple assays.

Kinase Receptor Activation Assay (KIRA)

A recent development for the measurement of cytokines is the KIRA procedure. In this assay, interaction of the cytokine with its receptor triggers phosphorylation of a specific protein, which can be detected with an appropriate and sensitive immunoassay (Sadick *et al.*, 1996).

IMMUNOASSAYS

Immunoassays provide a convenient method for cytokine measurement and are usually easier and quicker to perform than bioassays. They can be used to distinguish cytokines that have similar biological activities which may be difficult using bioassays (e.g. IL-1α and IL-β; TNF-α and TNF-β).

Most types of immunoassay have been devised for various cytokines including radioimmunoassays (RIAs), IRMAs and ELISAs. Conventional competitive binding RIAs are not usually sufficiently sensitive to measure physiological or even pathological levels of cytokines; they also often suffer from problematical 'matrix' effects when used for assaying cytokines in complex biological fluids, which necessitates some kind of sample extraction before assay (Cannon *et al.*, 1988). Probably the most useful immunoassays are immunometric assays based on a 'two-site' principle: one antibody is used to capture cytokine antigen(s) and another (suitably labelled) antibody is used to detect bound antigen. This format can be particularly specific if at least one of the antibody pair is a carefully characterized monoclonal antibody; such assays can be used to measure cytokines in complex biological fluids without pretreatment of samples (Thorpe *et al.*, 1988). These assays can be IRMAs or ELISAs using radiolabelled or enzyme-labelled detector antibodies respectively. A problem that sometimes occurs with two-site assays is that immobilization of 'capture' antibody or capture of antigen causes steric changes that severely compromise subsequent binding events essential for the assay. One possible solution to this is to modify the procedure to allow antibody recognition to occur in liquid phase followed by capture of antibody–antigen (cytokine) complexes with a solid-phase anti-immunoglobulin reagent (Wadhwa *et al.*, 1990). Some ELISA methods have been coupled to enzyme amplification cascades which are claimed to increase sensitivity.

A main problem with immunoassays for cytokines is that they can detect biologically inactive or partially active cytokine molecules. This problem can be particularly important when analysing complex biological samples and may result in a significant and variable lack of correlation between the biological activity of a particular cytokine (as detected by bioassay) and the quantity of cytokine estimated using immunoassays. The loss of biological activity can be due to denaturation by proteases or other factors or possibly by complexing with inhibitors such as soluble receptors. Incorrect folding or refolding of rDNA derived molecules can also influence biological activity.

A further problem with immunoassays is that they may exhibit differential detection of different subforms of a particular cytokine because of the differing recognition of various cytokine subspecies by the antibodies used (Bird *et al.*, 1991). Factors such as variable glycosylation, amidation and primary structure variations (both intentionally introduced in rDNA-derived materials and naturally occurring) as well as differences in secondary, tertiary or even quaternary structure can influence the interaction of cytokine molecules with antibodies, and therefore results obtained using immunoassays (Fig. 3). These effects can be 'all or none', i.e. a particular antibody either reacts well or will not bind a particular subspecies or may be graded in effect between different cytokine forms. The result of this phenomenon is that a particular immunoassay can fail to detect a particular form of a cytokine whereas other forms are detected in amounts that variably reflect the total or partial biological activity of cytokine preparations. This results in an overall assessment of cytokine content which can be unrelated to biological activity and will vary according to the precise subspecies composition of different preparations.

In addition, immunoassays can be influenced by 'matrix effects' when used for estimating the cytokine content of some biological fluids (e.g. serum and plasma). As with all immunoassays, the specificity, sensitivity and other characteristics relating to the quality of the procedure are determined mainly by the particular antibodies used. Use of different immunoassay formats invoving different antibodies and standards can result in

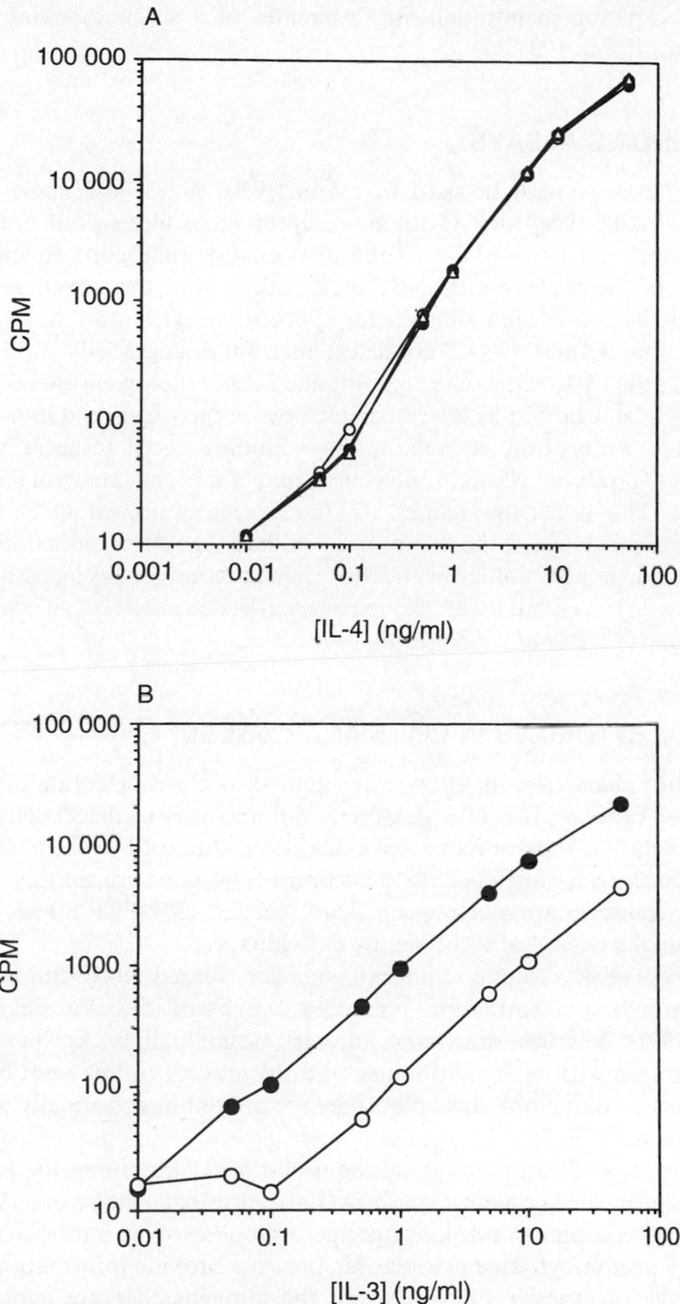

Fig. 3. Immunoradiometric assays (IRMAs) which show essentially identical (A) or preferential (B) reactivity with different molecular forms of analyte. (A) An IRMA for human IL-4 which detects equally three different subspecies of the cytokine, which differ in primary sequence and/or glycosylation. (B) An IRMA for human IL-3 which reacts differently to subspecies that differ slightly in primary structure. Note that considerably different (incorrect) estimates of analyte content for the two preparations would be obtained if the assay were standardized using the 'incorrect' species.

a considerable variation in immunoassay estimates of a single cytokine preparation (Mire-Sluis *et al.*, 1996a, 1997).

RECEPTOR BINDING ASSAYS

Receptor binding assays can be used to assess levels of cytokine molecules able to interact with cytokine receptors. Cytokine receptor molecules (often produced using rDNA technology) can be used in a solid-phase assay analogous to immunometric assays in place of the capture antibody; these assays have the advantages that such receptors usually have very high affinities for appropriate cytokines; they are with some exceptions very specific; and they do not detect aberrant or degraded cytokine molecules that have lost the ability to react with receptor and induce biological effects. Such assays can in some cases be regarded as intermediate between bioassays and immunochemical assays. However, results obtained using receptor binding assays (especially those using isolated or rDNA-produced receptor molecules) may not correleate with data generated using bioassays. This is because some cytokine molecules appear to be able to bind receptor, but cannot 'trigger' the receptor to initiate signal transduction mechanism essential for production of biological effect. Receptor-based assays using other formats (e.g. competitive displacement assays using receptor-bearing cells) have been devised for cytokine detection (Park *et al.*, 1986, 1987).

DETECTION OF CYTOKINES IN CELLS AND TISSUES

Specific procedures have been developed or adapted to allow detection of cytokines in cells and tissues (Table 4). It is now possible simultaneously to detect cellular cytokine mRNA and also the presence of translated cytokine product and/or the phenotype of the producer cell. Such techniques include detection of cytokine transcripts in individual cells by reverse transcriptase–polymerase chain reaction (RT-PCR) (Fig. 4) and those that demonstrate the release of cytokines by individual cells.

Immunohistochemistry and immunocytochemistry based procedures can provide useful semiquantitative information regarding the tissue distribution of cytokines (Sander *et al.*, 1991). A serious drawback, however, is that high 'background' artefactual staining is often seen with many antibodies, and the procedures may not be sufficiently sensitive to detect cytokine proteins unless they are present in abnormally high amounts (Lewis and Campbell, 1995).

Flowcytometry (cytofluorography) can be useful for determining the proportion of cells in a population that contain cytokines (Labalette-Houache *et al.*, 1991). Double labelling methods combining cytokine-specific antibodies with antibodies that detect surface markers or non-cytokine cytoplasmic proteins provide information concerning the cell types that synthesize cytokines and the antigens that are coproduced with cytokines following specific stimulii (Lewis and Campbell, 1995). The wide range of immunostaining methods now available offers the facility of not only identifying producer cells and visualizing production of more than one cytokine within an individual cell but also investigating receptor expression (Pusztai *et al.*, 1994) and, in some instances, even distinguishing the subcellular organelles involved in cytokine synthesis.

Table 4. Some other techniques for the detection of cytokines.

Parameter assessed	Technique	Advantages	Disadvantages
Production or expression of cytokine mRNA transcripts	*In situ* hybridization Isotopic Non-isotopic	Can be used in combination with other methods for identification of producer cell type(s). Detection does not rely on presence of threshold numbers of producer cells. Non-isotopic method is rapid	Non-quantitative. Isotopic method may be hazardous
	Polymerase chain reaction	Semiquantitative. Highly sensitive	Relatively difficult and cumbersome. Contamination with other RNA or DNA molecules may be problematical
Detection of cytokine release by single cells	Reverse haemolytic plaque assay	Semiquantitative. Sensitive. Rapid. Can be adapted to detect release of more than one cytokine per cell	May detect only immunoreactive and not necessarily bioactive protein. Presence of some antibodies may cause abnormal patterns of cytokine secretion
	Cell blotting	Quantitative. Sensitive. Rapid. Can detect release of more than one cytokine per cell	Same as above
	ELISPOT	Enumeration of producer cells. Highly sensitive, simple, reproducible and rapid	May detect only immunoreactive and not bioactive material. Properly characterized antibodies crucial to avoid artefactual results
Production of intracellular cytokine	Immuno-cytochemistry	Rapid. Identification of producer cell type(s)	Non-quantitative. Non-specific staining may produce artefactual results; cells responsive to the stimulatory cytokine may be labelled as producer cells. May detect only immunoreactive and not necessarily bioactive cytokine molecules
	Immuno-fluorescence	Same as above. Simultaneous production of more than one cytokine can be visualized	Same as above

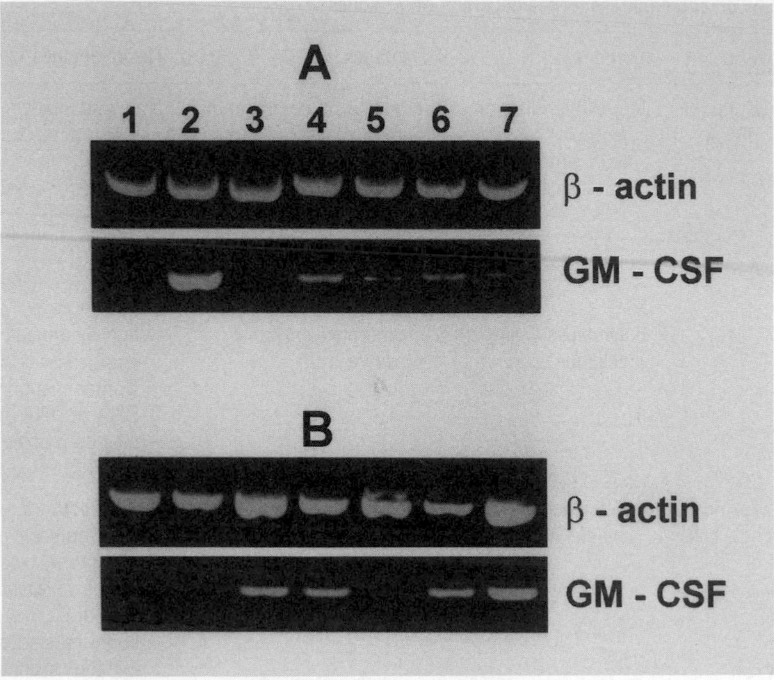

Fig. 4. Expression of GM-CSF mRNA in M-07e cells following cytokine stimulation for 2 h (A) and 18 h (B). Lane 1: unstimulated cells; lane 2: treatment with SCF (20 ng/ml); lane 3: rhIL-4 (20 ng/ml); lane 4: rhTNF-α (2 ng/ml); lane 5: rhTNF-α (1 ng/ml); lane 6: rhTNF-α (5 ng/ml) and rhIL-4 (20 ng/ml); lane 7: rhTNF-α (1 ng/ml) and rhIL-4 (20 ng/ml). A β-actin primer was used as a control in all experiments.

Secretion of cytokines by individual cells can be assessed using the enzyme-linked immunospot (ELISPOT) procedure (Lewis, 1991) or reverse haemolytic plaque assays (RHPA) (Lewis *et al.*, 1991). In the ELISPOT procedure, single cells are overlaid on immobilized cytokine-specific antibody, usually in tissue culture plates. After a suitable period, the cells are washed away and secreted cytokine molecules 'captured' by the antibody are detected by routine immunocytochemical methods using an enzyme-labelled cytokine-specific antibody followed by an appropriate substrate. This procedure has been used to detect and enumerate human lymphocytes secreting IFN-γ (Czerkinsky *et al.*, 1988) and human cells from kidney allografts secreting IFN-γ, IL-6 and IL-10 (Merville *et al.*, 1993). For RHPA, a similar strategy as the ELISPOT is adopted, except protein A-treated red cells coated with appropriate cytokine-specific antibodies are used to detect secreted cytokine molecules from individual cells. In the ELISPOT technique, cytokine-secreting cells are identified by spots of chromogenic substrate, whereas plaques of haemolysis are produced by cells in the RHPA procedure. The diameter of plaques in RHPA is usually approximately proportional to the amount of cytokine produced by secreting cells. This approach has been used to quantify the release of various cytokines by human blood and tumour-infiltrating leukocytes (Lewis *et al.*, 1991).

As an alternative to assaying cytokine protein in cells and tissues, levels of cytokine mRNA can be assessed and this used as an indication of the level of localized cytokine synthesis. It should be emphasized, however, that production of cytokine mRNA and protein may not be correlated (Wadhwa *et al.*, 1996). In addition, a particular cytokine associated with a given cell may not have been synthesized by that cell, but may represent an internalized product of another cell type. Various *in situ* hybridization procedures involving the use of appropriately radiolabelled or enzyme-labelled oligonucleotide probes can be used for detection of cytokine mRNA (Zheng *et al.*, 1991; Naylor *et al.*, 1995).

SAMPLE PREPARATION

The preparation and collection of samples intended for cytokine measurement is important if artefacts are to be avoided. Cytokine levels are often measured in culture supernatants or biological fluids such as serum or plasma. The presence of cytotoxic or cytostatic agents, cell culture additives and mitogenic materials in supernatants can influence many bioassay procedures. Serum or plasma can cause matrix effects in many assays, and it is important to validate methods for appropriate use with particular biological fluids.

If blood levels of cytokines are to be assessed, it is essential to ensure that post-collection processes, such as coagulation, do not result in cell activation, causing release of cytokines and possible misinterpretation of results. In most cases, it is best to avoid clotting and to use plasma for assay; however, several anticoagulants can interfere with bioassays (e.g. ethylene diamine tetra-acetic acid (EDTA) and citrate (by chelation of divalent cations such as Ca^{2+})). A low amount of preservative-free heparin (2 units/ml), in the authors' experience, is the best anticoagulant for most cytokine assays. Plasma or serum should be separated from blood cells as soon as possible, and this should be stored at −20°C or lower.

A frequent problem with tissue fluids is the presence of soluble receptors, cytokine autoantibodies or other binding proteins which may mask activity, and so sample extraction may be necessary before assessment of cytokine activity (Cannon *et al.*, 1988). Some biological samples (e.g. plasma) may require special treatment before detection of certain cytokines by bioassay, for example heat treatment (56°C for 30 min) for IL-6 and acid activation (0.12 M hydrochloric acid for 30 min) for TGF-β.

STANDARDIZATION AND STATISTICAL EVALUATION OF CYTOKINE ASSAYS

As with all assays for biological materials, it is essential that cytokine assays are standardized carefully and correctly. It is usually difficult to calibrate bioassays in 'real' mass units, and calibration in such terms as '50% maximum of proliferation' units can vary considerably between laboratories, assay systems and even experiments carried out in the same laboratories by the same methods at different times. The best approach for most systems is to use an arbitrary unitage that is related directly to a standard preparation; a range of such preparations is available from the National Institute for Biological Standards and Control (Table 5).

Standardization of immunoassays for cytokines is made problematical by their possible microheterogeneity in structure (see above). In many cases, it is virtually impossible to standardize such methods in a way that allows a valid comparison between different assays and possibly different preparations of a cytokine.

Table 5. Current list of available cytokine standards and their status.

Preparation	Product code	Status*	Depository†
Interleukin 1α rDNA	86/632	IS	NIBSC
Interleukin 1β rDNA	86/680	IS	NIBSC
Interleukin 2 cell line derived	86/504	IS	NIBSC
Interleukin 2 rDNA	86/564		NIBSC
Interleukin 3 rDNA	91/510	IS	NIBSC
Interleukin 4 rDNA	88/656	IS	NIBSC
Interleukin 5 rDNA	90/586	RR	NIBSC
Interleukin 6 rDNA	89/548	IS	NIBSC
Interleukin 7 rDNA	90/530	RR	NIBSC
Interleukin 8 rDNA	89/520	RR	NIBSC
Interleukin 9 rDNA	91/678	RR	NIBSC
Interleukin 10 rDNA	93/722	RR	NIBSC
Interleukin 11 rDNA	92/788	RR	NIBSC
Interleukin 12 rDNA	95/544	RR	NIBSC
Interleukin 13 rDNA	94/622	RR	NIBSC
Interleukin 15 rDNA	95/554	RR	NIBSC
M-CSF rDNA	89/512	IS	NIBSC
G-CSF rDNA	88/502	IS	NIBSC
GM-CSF rDNA	88/646	IS	NIBSC
Leukaemia inhibitory factor rDNA	93/562	RR	NIBSC
Oncostatin M	93/564	RR	NIBSC
Stem cell factor/MGF rDNA	91/682	RR	NIBSC
Flt3 ligand rDNA	96/532	RR	NIBSC
Bone morphogenetic protein-2	93/574	RR	NIBSC
RANTES rDNA	92/520	IR	NIBSC
GRO-α rDNA	92/722	IR	NIBSC
MCP-1 rDNA	92/794	IR	NIBSC
MCP-2 rDNA	96/594	IR	NIBSC
MIP-1α rDNA	92/518	IR	NIBSC
MIP-1β rDNA	96/588	IR	NIBSC
IFN-α leukocyte	69/19	IS	NIBSC
IFN-β fibroblast	Gb23-902-531	IS	NIAID
IFN-γ rDNA	Gxg01-902-535	IS	NIAID
TGF-β_1 rDNA	89/514	IR	NIBSC
TGF-β_1 (Nat bovine)	89/516		NIBSC
TGF-β_2 rDNA	90/696	IR	NIBSC
TNF-α rDNA	87/650	IS	NIBSC
TNF-β rDNA	87/640	RR	NIBSC
Interleukin-1 soluble receptor type 1 rDNA	96/616	IR	NIBSC

*IS, International standard; RR, WHO Reference Reagent; IR, Interim Reference Reagent. †NIBSC, National Institute for Biological Standards and Control, Blanche Lane, South Mimms, Potters Bar, Hertfordshire EN6 3QG, UK; NIAID, National Institute for Allergy and Infectious Diseases, Solar Building, 6003 Executive Drive, Maryland, USA. MGF, mast cell growth factor.

It is essential that bioassay data are analysed correctly and evaluated statistically. This is particularly important when a valid biological potency is to be ascribed to a preparation and if samples are to be considered to have significantly different cytokine activities. Analysis of assays by the parallel line approach is used widely; for this, the unknown samples are titrated and compared with a standard curve of known unitage.

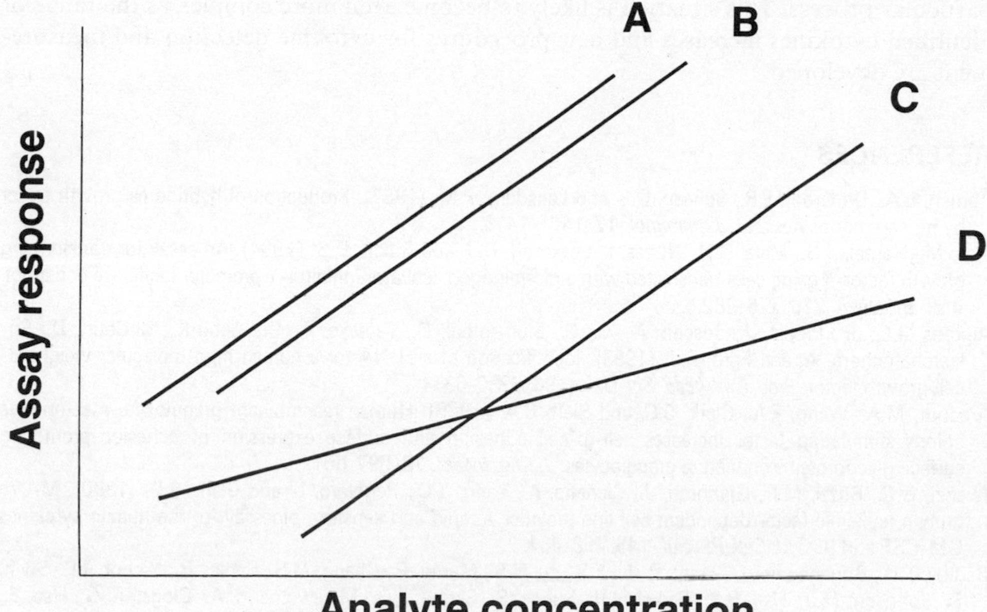

Fig. 5. Comparison of the linear portions of dose–response curves obtained by bioassays. Preparations A, B and C produce parallel lines, showing that the assay is responding similarly to the stimulatory activity in the preparations. The difference in displacement of the curves along the analyte concentration axis is due to different potencies of the three preparations. Preparation D is not parallel to the other curves, which clearly indicates that the bioassay responds to this material differently. It is therefore not valid to compare the potency of preparation D with preparations A, B and C.

The parallel portions of these curves are then used to measure the displacement of unknowns from the standard, which is proportional to the biologically active cytokine content of the samples (Fig. 5). These curves should be parallel if the molecule responsible for the activity in samples and standards is the same. Alternatively, an approximate estimate of activity can be made by taking two or three points from the titration curve and reading these from the standard curve.

For deriving a potency for cytokine products intended for therapeutic use, a particularly careful assay design and a thorough statistical evaluation are essential (Mire-Sluis *et al.*, 1996b).

CONCLUSIONS

In several instances, cytokine measurements have contributed significantly to the understanding of cytokine biology and biochemistry, and to the assessment of cytokine involvement in pathology and physiological processes. However, in a few cases, inappropriate assay choice or design has led to confusion or even erroneous conclusions in these important areas. Only if well characterized, validated methods are used, and if results are analysed carefully, can unambiguous information be obtained. Often it is necessary to use more than one assay system to confirm cytokine involvement in a particular process. This situation is likely to become even more complex as the range of identified cytokines increases and new procedures for cytokine detection and measurement are developed.

REFERENCES

Aarden, L.A., De Groof, E.R., Schaap, O.L. and Lansdorp, P.M. (1987). Production of hybridoma growth factor by human monocytes. *Eur. J. Immunol*. **17**, 1411–1416.

Abe, M., Harpel, J.G., Metz, D.N., Nunes, I., Loskutoff, D.J. and Rifkin, D.B. (1994). An assay for transforming growth factor-β using cells transfected with a plasminogen activator inhibitor-1 promoter luciferase construct. *Anal. Biochem*. **216**, 276–282.

Ambrus, J.L., Jr., Pippin, J., Joseph, A., Xu, C., Blumenthal, D., Tamayo, A., Claypool, K., McCourt, D., Srikiatchatochorn, A. and Ford, R.J. (1993). Identification of a cDNA for a human high-molecular-weight B-cell growth factor. *Proc. Natl Acad. Sci. U.S.A*. **90**, 6330–6334.

Arnaout, M.A., Wang, E.A., Clark, S.C. and Sieff, C.A. (1986). Human recombinant granulocyte–macrophage colony-stimulating factor increases cell-to-cell adhesion and surface expression of adhesion-promoting surface glycoproteins on mature granulocytes. *J. Clin. Invest*. **78**, 597–601.

Avanzi, G.C., Brizzi, M.F., Giannotti, J., Ciarietta, A., Yang, Y.C., Pegoraro, L. and Clark, S.C. (1990). M-07e human leukemic factor-dependent cell line provides a rapid and sensitive bioassay for the human cytokines GM-CSF and IL-3. *J. Cell. Physiol*. **145**, 458–464.

Bartley, T.D., Bogenberger, J., Hunt, P., Li, Y.S., Lu, H.S., Martin, F., Chang, M.S., Samai, B., Nichol, J.L., Swift, S., Johnson, M.J., Hsu, R.Y., Parker, V.P., Suggs, S., Skrine, J.D., Merewether, L.A., Clogston, C., Hsu, E., Hokom, M.M., Hornkohl, A., Choi, E., Pangelinan, M., Sun, Y., Mar, V., McNinch, J., Simonet, L., Jacobsen, F., Xie, C., Shutter, J., Chute, H., Basu, R., Selander, L., Trollinger, D., Sieu, L., Padilla, D., Trail, G., Elliott, G., Izumi, R., Covey, T., Crouse, J., Garcia, A., Xu, W., Del Castillo, J., Biron, J., Cole, S., Hu, M.C.-T., Pacifici, R., Ponting, I., Saris, C., Wen, D., Yung, Y.P., Lin, H. and Bosselman, R.A. (1994). Identification and cloning of a megakaryocyte growth and development factor that is a ligand for the cytokine receptor Mpl. *Cell* **77**, 1117–1124.

Bird, C., Wadhwa, M. and Thorpe, R. (1991). Development of immunoassays for human IL-3 and IL-4 some of which discriminate between different recombinant DNA derived molecules. *Cytokine* **3**, 562–567.

Bouteiller, C.L., Astruc, R., Minty, A., Ferrara, P. and Lupker, J.H. (1995). Isolation of an IL-13-dependent subclone of the B9 cell line useful for the estimation of human IL-13 bioactivity. *J. Immunol. Methods* **181**, 29.

Boyden, S. (1962). The chemotactic effect of mixtures of antibody and antigen of polymorphonuclear leukocytes. *J. Exp. Med*. **115**, 453.

Broudy, V.C., Lin, N.L. and Kaushansky, K. (1995). Thrombopoietin (c-mpl ligand) acts synergistically with erythropoietin, stem cell factor, and interleukin-11 to enhance murine megakaryocyte colony growth and increases megakaryocyte ploidy *in vitro*. *Blood* **85**, 1719–1726.

Buttke, T.M., McCubrey, J.A. and Owen, T.C. (1993). Use of an aqueous soluble tetrazolium/formazan assay to measure viability and proliferation of lymphokine-dependent cell lines. *J. Immunol. Methods* **157**, 233.

Callard, R.E., Shields, J.G. and Smith, S.H. (1987). Assays for human B cell growth and differentiation factors. In *Lymphokines and Interferons. A Practical Approach* (eds M.J. Clemens, A.G. Morris and A.J.H. Gearing), IRL Press, Oxford, pp. 345–364.

Cannon, J.G., Van Der Meer, J.W.M., Kwiatkowski, D., Endres, S., Lonneman, G., Burke, J.F. and Dinarello, C.A. (1988). Interleukin-1b in human plasma: optimisation of blood collection, plasma extraction and radioimmunoassay methods. *Lymphokine Res*. **7**, 457–467.

Carson, W.E., Giri, J.G., Lindemann, M.J., Linett, M.L., Ahdieh, M., Paxton, R., Anderson, D., Eisenmann, J., Grabstein, K. and Caligiuri, M.A. (1994). Interleukin (IL) 15 is a novel cytokine that activates human natural killer cells via components of the IL-2 receptor. *J. Exp. Med*. **180**, 1395–1403.

Clarke, B.J., Nathan, D.G., Alter, B.P., Forget, B.G., Hillman, D.G. and Housman, D. (1979). Hemoglobin synthesis in human BFU-E and CFU-E-derived erythroid colonies. *Blood* **54**, 805–817.

Cotes, P.M. and Bangham, D.R. (1961). Bioassay of erythropoietin in mice made polycythaemic by exposure to air at a reduced pressure. *Nature* **191**, 1065–1067.

Crouch, S.P., Kozlowski, R., Slater, K.J. and Fletcher, J. (1993). The use of ATP bioluminescence as a measure of cell proliferation and cytotoxicity. *J. Immunol. Methods* **160**, 81.

Custer, M.C. and Lotze, M.T. (1990). A biologic assay for IL-4. Rapid fluorescence assay for IL-4 detection in supernatants and serum. *J. Immunol. Methods* **128**, 109–117.

Czerkinsky, C., Andersson, G., Ekre, H.P., Nilsson, L.A., Klareskog, L. and Ouchterlony, O. (1988). Reverse ELISPOT assay for clonal analysis of cytokine production. I. Enumeration of γ-interferon-secreting cells. *J. Immunol. Methods* **110**, 29–36.

Däubener, W., Wanagat, N., Pilz, K., Seghrouchni, S., Fischer, H.G. and Hadding, U. (1994). A new, simple, bioassay for human IFN-γ. *J. Immunol. Methods* **168**, 39–47.

Dayer, J.M., Beutler, B. and Cerami, A. (1985). Cachectin/tumor necrosis factor stimulates collagenase and prostaglandin E_2 production by human synovial cells and dermal fibroblasts. *J. Exp. Med*. **162**, 2163–2168.

Defrance, T., Carayon, P., Billian, G., Guillemot, J.C., Minty, A., Caput, D. and Ferrara, P. (1994). Interleukin 13 is a B cell stimulating factor. *J. Exp. Med*. **179**, 135–143.

Dinarello, C.A. (1984). Interleukin-1. *Rev. Infect. Dis*. **6**, 51–95.

Dinarello, C.A., Cannon, J.G. and Wolff, S.M. (1988). New concepts on the pathogenesis of fever. *Rev. Infect. Dis*. **10**, 168–189.

Donahue, R.E., Yang, Y.C. and Clark, S.C. (1990). Human P40 T-cell growth factor (interleukin-9) supports erythroid colony formation. *Blood* **75**, 2271–2275.

Doyle, A., Stein, M., Keshav, S. and Gordon, S. (1995). Assays for macrophage activation by cytokines. In *Cytokines. A Practical Approach*, 2nd edn. (ed. F.R. Balkwill), IRL Press, Oxford, pp. 269–278.

Espevik, T. and Nissen-Meyer, J. (1986). A highly sensitive cell line, WEHI 164 clone 13, for measuring cytotoxic factor/tumour necrosis factor from human monocytes. *J. Immunol. Methods* **95**, 99–106.

Fiorentino, D.F., Bond, M.W. and Mosmann, T.R. (1989). Two types of mouse T helper cell. IV. Th2 clones secrete a factor that inhibits cytokine production by Th1 clones. *J. Exp. Med*. **170**, 2081–2095.

Fiorentino, D.F., Zlotnik, A., Mosmann, T.R., Howard, M. and O'Garra, A. (1991). IL-10 inhibits cytokine production by activated macrophages. *J. Immunol*. **147**, 3815–3822.

Fossiez, F., Djossou, O., Chomarat, P., Flores Romo, L., Ait Yahia, S., Maat, C., Pin, J.J., Garrone, P., Garcia, E., Saeland, S., Blanchard, D., Gaillard, C., Das Mahapatra, B., Rouvier, E., Golstein, P., Banchereau, J. and Lebecque, S. (1996). T cell interleukin-17 induces stromal cells to produce proinflammatory and hematopoietic cytokines. *J. Exp. Med*. **183**, 2593–2603.

Gabbianelli, M., Pelosi, E., Montesoro, E., Valtieri, M., Luchetti, L., Samoggia, P., Vitelli, L., Barberi, T., Testa, U. and Lyman, S. (1995). Multi-level effects of flt3 ligand on human hematopoiesis: expansion of putative stem cells and proliferation of granulomonocytic progenitors/monocytic precursors. *Blood* **86**, 1661–1670.

Gearing, A.J.H., Bird, C.R., Bristow, A., Poole, S. and Thorpe, R. (1987). A simple sensitive bioassay for interleukin-1 which is unresponsive to 10^3 U/ml of interleukin 2. *J. Immunol. Methods* **99**, 7–11.

Gearing, D.P., Ziegler, S.F., Comeau, M.R., Friend, D., Thoma, B., Cosman, D., Park, L. and Mosley, B. (1994). Proliferative responses and binding properties of hematopoietic cells transfected with low-affinity receptors for leukemia inhibitory factor, oncostatin M, and ciliary neurotrophic factor. *Proc. Natl Acad. Sci. U.S.A*. **91**, 1119–1123.

Gery, L., Gershon, R.K. and Waksmann, B.H. (1972). Potentiation of the T lymphocyte response to mitogens. I. The responding cell. *J. Exp. Med*. **136**, 128–142.

Gibson, U.E. and Kramer, S.M. (1989). Enzyme-linked bio-immunoassay for IFN-γ by HLA-DR induction. *J. Immunol. Methods* **125**, 105–113.

Gillis, S., Ferm, M.M., Ou, W.E. and Smith, K.A. (1978). T cell growth factor: parameters of production and a quantitative microassay for activity. *J. Immunol*. **120**, 2027.

Giri, J.G., Ahdieh, M., Eisenman, J., Shanebeck, K., Grabstein, K., Kumaki, S., Namen, A., Park, L.S., Cosman, D. and Anderson, D. (1994). Utilization of the β and γ chains of the IL-2 receptor by the novel cytokine IL-15. *EMBO J*. **13**, 2822–2830.

Gowen, M., Wood, D.D., Ihrie, E.J., McGuire, M.K. and Russell, R.G. (1983). An interleukin 1 like factor stimulates bone resorption *in vitro*. *Nature* **306**, 378–380.

Grabstein, K.H., Urdal, D.L., Tushinski, R.J., Mochizuki, D.Y., Price, V.L., Cantrell, M.A., Gillis, S. and Conlon, P.J. (1986). Induction of macrophage tumoricidal activity by granulocyte–macrophage colony-stimulating factor. *Science* **232**, 506–508.

Grabstein, K.H., Eisenman, J., Shanebeck, K., Rauch, C., Srinivasan, S., Fung, V., Beers, C., Richardson, J., Schoenborn, M.A. and Ahdieh, M. (1994). Cloning of a T cell growth factor that interacts with the β chain of the interleukin-2 receptor. *Science* **264**, 965–968.

Gurney, A.L., Kuang, W.J., Xie, M.H., Malloy, B.E., Eaton, D.L. and De Sauvage, F.J. (1995). Genomic structure, chromosomal localization, and conserved alternative splice forms of thrombopoietin. *Blood* **85**, 981–988.

Hendrie, P.C., Miyazawa, K., Yang, Y.-C. and Langefeld, C.D. (1991). Mast cell growth factor (c-*kit* ligand) enhances cytokine stimulation of proliferation of the human factor-dependent cell line, MO7e. *Exp. Hematol*. **19**, 1031–1037.

Hori, T., Uchiyama, T., Tsuda, M., Umadome, H., Ohno, H., Fukuhara, S., Kitamura, K. and Uchino, H. (1987). Establishment of an interleukin-2-dependent human T cell line from a patient with T cell chronic lymphocytic leukaemia who is not infected with human T cell leukaemia/lymphoma virus. *Blood* **70**, 1069–1078.

Horuk, R., Yansura, D.G., Reilly, D., Spencer, S., Bourell, J., Henzel, W., Rice, G. and Unemori, E. (1993). Purification, receptor binding analysis, and biological characterization of human melanoma growth stimulating activity (MGSA). Evidence for a novel MGSA receptor. *J. Biol. Chem*. **268**, 541–546.

Houssiau, F.A., Renauld, J.C., Stevens, M., Lehmann, F., Lethe, B., Coulie, P.G. and Van-Snick, J. (1993). Human T cell lines and clones respond to IL-9. *J. Immunol*. **150**, 2634–2640.

Hu-Li, J., Ohara, J., Watson, C., Tsang, W. and Paul, W.E. (1989). Derivation of a T cell line that is highly responsive to IL-4 and IL-2 (CT.4R) and of an IL-2 hyporesponsive mutant of that line (CT.4S). *J. Immunol*. **142**, 800.

Hultner, L., Druez, C., Moeller, J., Uyttenhove, C., Schmitt, E., Rude, E., Dormer, P., Van Snick, J. (1990). Mast cell growth-enhancing activity (MEA) is structurally related and functionally identical to the novel mouse T cell growth factor P40/TCGFIII (interleukin 9). *Eur. J. Immunol*. **20**, 1413–1416.

Junger, W.G., Cardoza, T.A., Liu, F.C., Hoyt, D.B. and Goodwin, R. (1993). Improved rapid photometric assay for quantitative measurement of PMN migration. *J. Immunol. Methods* **160**, 73.

Kaushansky, K., Lok, S., Holly, R.D., Broudy, V.C., Lin, N., Bailey, M.C., Forstrom, J.W., Buddle, M.M., Oort, P.J. and Hagen, F.S. (1994). Promotion of megakaryocyte progenitor expansion and differentiation by the c-Mpl ligand thrombopoietin. *Nature* **369**, 568–571.

Kennedy, W.L., Alpen, E.L. and Garcia, J.F. (1980). Regulation of red blood cell production by erythropoietin: normal mouse marrow *in vitro*. *Exp. Hematol*. **8**, 1114–1122.

Khwaja, A., Carver, J.E. and Linch, D.C. (1992). Interactions of granulocyte–macrophage colony-stimulating factor (CSF), granulocyte CSF, and tumor necrosis factor α in the priming of the neutrophil respiratory burst. *Blood* **79**, 745–753.

Kitamura, T., Tange, T., Terasawa, T., Chiba, S., Kuwaki, T., Miyagawa, K., Piao, Y.-F., Miyazono, K., Urabe, A. and Takaku, F. (1989). Establishment and characterisation of a unique human cell line that proliferates dependently on GM-CSF, IL-3, or erythropoietin. *J. Cell. Physiol*. **140**, 323–334.

Kobayashi, M., Fitz, L., Ryan, M., Hewick, R.M., Clark, S.C., Chan, S., Loudon, R., Sherman, F., Perussia, B. and Trinchieri, G. (1989). Identification and purification of natural killer cell stimulatory factor (NKSF), a cytokine with multiple biologic effects on human lymphocytes. *J. Exp. Med*. **170**, 827–845.

Komatsu, N., Yamamoto, M., Fujita, H., Miwa, A., Hatake, K., Endo, T., Okano, H., Katsube, T., Fukumaki, Y., Sassa, S. and Miura, Y. (1993). Establishment and characterization of an erythropoietin-dependent subline, UT-7/epo, derived from human leukaemia cell line, UT-7. *Blood* **82**, 456–464.

Kramer, S.M. and Carver, M.E. (1986). Serum-free *in vitro* bioassay for the detection of tumor necrosis factor. *J. Immunol. Methods* **93**, 201–206.

Labalette-Houache, M., Torpier, G., Capron, A. and Dessaint, J.P. (1991). Improved permeabilisation procedure for flow cytometric detection of internal antigens. Analysis of interleukin-2 production. *J. Immunol. Methods* **138**, 143–153.

Last-Barney, K., Homon, C.A., Faanes, R.B. and Merluzzi, V.J. (1988). Synergistic and overlapping activities of tumor necrosis factor-α and IL-1. *J. Immunol*. **141**, 527–530.

Lewis, C.E. (1991). Cytokine production by individual cells. In *Cytokines. A Practical Approach* (ed. F.R. Balkwill), IRL Press, Oxford, pp. 279–297.

Lewis, C.E. and Campbell, A. (1995). Visualising the production of cytokines and their receptors by single human cells. In *Cytokines. A Practical Approach*, 2nd edn. (ed. F.R. Balkwill), IRL Press, Oxford, pp. 339–356.

Lewis, C.E., McCracken, D., Ling, R., Richards, P.S., McCarthy, S.P. and McGee, J.O'D. (1991). Cytokine release by single, immunophenotyped human cells: use of the reverse hemolytic plaque assay. *Immunol. Rev*. **119**, 23–39.

Lewis, J.A. (1995a). Antiviral activity of cytokines. In *Cytokines. A Practical Approach*, 2nd edn. (ed. F.R. Balkwill), IRL Press, Oxford, pp. 129–141.

Lewis, J.A. (1995b). A sensitive biological assay for interferons. *J. Immunol. Methods* **185**, 9–17.

Lim, K.G., Wan, H.C., Bozza, P.T., Resnick, M.B., Wong, D.T., Cruikshank, W.W., Kornfeld, H., Center, D.M. and Weller, P.F. (1996). Human eosinophils elaborate the lymphocyte chemoattractants IL-16 (lymphocyte chemoattractant factor) and RANTES. *J. Immunol*. **156**, 2566–2570.

Liu, Y., Wei, S.H., Ho, A.S., De Waal Malefyt, R. and Moore, K.W. (1994). Expression cloning and characterization of a human IL-10 receptor. *J. Immunol*. **152**, 1821–1829.

Lu, Z.-Y., Zhang, X.-G., Gu, Z.-J., Yasukawa, K., Amiot, M., Etrillard, M., Bataille, R. and Klein, B. (1994). A highly sensitive quantitative bioassay for human interleukin-11. *J. Immunol. Methods* **173**, 19.

Lyman, S.D., James, L., Vanden-Bos, T., de-Vries, P., Brasel, K., Gliniak, B., Hollingsworth, L.T., Picha, K.S., McKenna, H.J. and Splett, R.R. (1993). Molecular cloning of a ligand for the flt3/flk-2 tyrosine kinase receptor: a proliferative factor for primitive hematopoietic cells. *Cell* **75**, 1157–1167.

Matthews, D.J., Clark, P.A., Herbert, J., Morgan, G., Armitage, R.J., Kinnon, C., Minty, A., Grabstein, K.H., Caput, D. and Ferrara, P. (1995). Function of the interleukin-2 (IL-2) receptor γ-chain in biologic responses of X-linked severe combined immunodeficient B cells to IL-2, IL-4, IL-13, and IL-15. *Blood* **85**, 38–42.

May, L.T., Ghrayeb, J., Santhanam, U., Tatter, S.B., Sthoeger, Z., Helfgott, D.C., Chiorazzi, N., Grieninger, G. and Sehgal, P.B. (1988). Synthesis and secretion of multiple forms of β_2-interferon/B-cell differentiation factor 2/ hepatocyte-stimulating factor by human fibroblasts and monocytes. *J. Biol. Chem*. **263**, 7760–7766.

McKenzie, A.N., Culpepper, J.A., De Waal Malefyt, R., Briere, F., Punnonen, J., Aversa, G., Sato, A., Dang, W., Cocks, B.G. and Menon, S. (1993). Interleukin 13, a T-cell-derived cytokine that regulates human monocyte and B-cell function. *Proc. Natl Acad. Sci. U.S.A*. **90**, 3735–3739.

Meager, A. (1991a). A cytotoxicity assay for tumour necrosis factor using a human rhabdomyosarcoma cell line. *J. Immunol. Methods* **144**, 141.

Meager, A. (1991b). Assays for transforming growth factor-β. *J. Immunol. Methods* **141**, 1.

Meager, A. (1996). Bioimmunoassays for proinflammatory cytokines involving cytokine-induced cellular adhesion molecule expression in human glioblastoma cell lines. *J. Immunol. Methods* **190**, 235–244.

Meager, A., Leung, H. and Woolley, J. (1989). Assay for TNF and related cytokines: a review. *J. Immunol. Methods* **116**, 1.

Merville, P., Pouteil-Noble, C., Wijdenes, J., Potaux, L., Touraine, J.L. and Banchereau, J. (1993). Detection of single cells secreting IFN-γ, IL-6, and IL-10 in irreversibly rejected human kidney allografts, and their modulation by IL-2 and IL-4. *Transplantation* **55**, 639–646.

Metcalf, D. (1984). *The Haemopoietic Colony Stimulating Factors*. Elsevier/North-Holland Biomedical Press, Amsterdam.

Metcalf, D., Johnson, G.R. and Mandel, T.E. (1979). Colony formation in agar by multipotential hemopoietic cells. *J. Cell. Physiol*. **98**, 401–420.

Metcalf, D., Begley, C.G., Johnson, G.R., Nicola, N.A., Vadas, M.A., Lopez, A.F., Williamson, D.J., Wong, G.G., Clark, S.C. and Wang, E.A. (1986). Biologic properties *in vitro* of a recombinant human granulocyte–macrophage colony-stimulating factor. *Blood* **67**, 37–45.

Metcalf, D., Hilton, D. and Nicola, N.A. (1991). Leukemia inhibitory factor can potentiate murine megakaryocyte production *in vitro*. *Blood* **77**, 2150–2153.

Micallef, M.J., Ohtsuki, T., Kohno, K., Tanabe, F., Ushio, S., Namba, M., Tanimoto, T., Torigoe, K., Fujii, M., Ikeda, M., Fukuda, S. and Kurimoto, M. (1996). Interferon-γ-inducing factor enhances T helper 1 cytokine production by stimulated human T cells: synergism with interleukin-12 for interferon-γ production. *Eur. J. Immunol*. **26**, 1647–1651.

Minty, A., Chalon, P., Derocq, J.M., Dumont, X., Guillemot, J.C., Kaghad, M., Labit, C., Leplatois, P., Liauzun, P. and Miloux, B. (1993). Interleukin-13 is a new human lymphokine regulating inflammatory and immune responses. *Nature* **362**, 248–250.

Mire-Sluis, A.R., Gaines-Das, R. and Thorpe, R. (1996a). Implications for the assay and biological properties of interleukin-4: results of a WHO international collaborative study. *J. Immunol. Methods* **194**, 13–25.

Mire-Sluis, A.R., Gaines-Das, R.E., Gerrard, T., Padilla, A. and Thorpe, R. (1996b). Biological assays: their role in the development and quality control of biological medicinal products. *Biologicals* **24**, 351–361.

Mire-Sluis, A.R., Page, L.A., Meager, A., Igaki, J., Lee, J., Lyons, S. and Thorpe, R. (1996c). An anti-cytokine bioactivity assay for interferons-α, -β and -ω. *J. Immunol. Methods* **195**, 55–61.

Mire-Sluis, A.R., Gaines-Das, R. and Thorpe, R. (1997). Implications for the assay and biological properties of interleukin-8: results of a WHO international collaborative study. *J. Immunol. Methods* **200**, 1–16.

Mizel, S.B., Dayer, J.M., Krane, S.M. and Mergenhagen, S.E. (1981). Stimulation of rheumatoid synovial cell collagenase and prostaglandin production by partially purified lymphocyte-activating factor (interleukin 1). *Proc. Natl Acad. Sci. U.S.A*. **78**, 2474–2477.

Moreau, J.F., Bonneville, M.,. Peyrat, M.A., Jacques, Y. and Soulillou, J.P. (1986). Capacity of alloreactive human T clones to produce factor(s) inducing proliferation of the IL-3-dependent DA-1 murine cell line. *Ann. Inst. Pasteur/Immunol*. **137C**, 25–37.

Morgan, C., Pollard, J.W. and Stanley, E.R. (1987). Isolation and characterization of a cloned growth factor dependent macrophage cell line, BAC1.2F5. *J. Cell. Physiol.* **130**, 420–427.

Mosmann, T. (1983). Rapid colorimetric assay for cellular growth and survival: application to proliferation and cytotoxicity assays. *J. Immunol. Methods* **65**, 55.

Muench, M.O., Roncarolo, M.G., Menon, S., Xu, Y., Kastelein, R., Zurawski, S., Hannum, C.H., Culpepper, J., Lee, F. and Namikawa, R. (1995). FLK-2/FLT-3 ligand regulates the growth of early myeloid progenitors isolated from human fetal liver. *Blood* **85**, 963–972.

Munker, R., Gasson, J., Ogawa, M. and Koeffler, H.P. (1986). Recombinant human TNF induces production of granulocyte monocyte colony stimulating factor. *Nature* **323**, 79–82.

Nakai, S., Mizuno, K., Kaneta, M. and Hirai, Y. (1988). A simple, sensitive bioassay for the detection of interleukin-1 using human melanoma A375 cell line. *Biochem. Biophys. Res. Commun.* **154**, 1189–1196.

Nakoinz, I., Lee, M.-T., Weaver, J.F. and Ralph, P. (1989). An M-CSF responsive subline of an IL-3-dependent myeloid cell line. In *7th International Congress of Immunology: Berlin (West)*, Fischer, Stuttgart, p. 227.

Namen, A.E., Lupton, S., Hjerrild, K., Wignall, J., Mochizuki, D.Y., Schmierer, A., Mosley, B., March, C.J., Urdal, D. and Gillis, S. (1988). Stimulation of B-cell progenitors by cloned murine interleukin-7. *Nature* **333**, 571–573.

Naylor, M.S., Relf, M. and Balkwill, F.R. (1995). Northern analysis, ribonuclease protection, and in situ analysis of cytokine messenger RNA. In *Cytokines. A Practical Approach*, 2nd edn. (ed. F.R. Balkwill), IRL Press, Oxford, pp. 35–56.

Nicola, N.A., Metcalf, D., Matsumoto, M. and Johnson, G.R. (1983). Purification of a factor inducing differentiation in murine myelomonocytic leukemia cells. Identification as granulocyte colony-stimulating factor. *J. Biol. Chem.* **258**, 9017–9023.

Nomura, H., Imazeki, I., Oheda, M., Kubota, N., Tamura-M., Ono, M., Ueyama, Y. and Asano, S. (1986). Purification and characterization of human granulocyte colony-stimulating factor (G-CSF). *EMBO J.* **5**, 871–876.

Nordan, R.P. and Potter, M. (1986). A macrophage-derived factor required by plasmacytomas for survival and proliferation *in vitro*. *Science* **233**, 566–569.

O'Garra, A. and Sanderson, C.J. (1987). Eosinophil differentiation factor and its associated B cell growth factor cytokines. In *Lymphokines and Interferons. A Practical Approach* (eds M. Clemens, A. Morris and A. Gearing), IRL Press, Oxford, pp. 323–344.

Orencole, S.F. and Dinarello, C.A. (1989). Characterization of a subclone (D10S) of the D10.G4.1 helper T-cell line which proliferates to attomolar concentrations of interleukin-1 in the absence of mitogens. *Cytokine* **1**, 14–20.

Page, L.A., Topp, M.S., Thorpe, R. and Mire-Sluis, A.R. (1996a). An antiproliferative bioassay for interleukin-4. *J. Immunol. Methods* **189**, 129–135.

Page, L.A., Thorpe, R. and Mire-Sluis, A.R. (1996b). A sensitive human cell line based bioassay for megakaryocyte growth and development factor or thrombopoietin. *Cytokine* **8**, 66–69.

Park, L.S., Friend, D., Gillis, S. and Urdal, D.L. (1986). Characterisation of the cell surface receptor for human granulocyte/macrophage colony stimulating factor. *J. Exp. Med.* **164**, 251–262.

Park, L.S., Friend, D., Sassenfeld, H.M. and Urdal, D.L. (1987). Chacterisation of the human B cell stimulatory factor 1 receptor. *J. Exp. Med.* **166**, 476–488.

Park, L.S., Friend, D.J., Schmierer, A.E., Dower, S.K. and Namen, A.E. (1990). Murine interleukin-7 (IL-7) receptor: characterization on an IL-7-dependent cell line. *J. Exp. Med.* **171**, 1073–1079.

Paul, S.R., Bennett, F., Calvetti, J.A., Kelleher, K., Wood, C.R., O'Hara, R.M., Jr., Leary, A.C., Sibley, B., Clark, S.C. and Williams, D.A. (1990). Molecular cloning of a cDNA encoding interleukin 11, a stromal cell-derived lymphopoietic and hematopoietic cytokine. *Proc. Natl Acad. Sci. U.S.A.* **87**, 7512–7516.

Perlmutter, D.H., Dinarello, C.A., Punsal, P.I. and Colten, H.R. (1986). Cachectin/tumor necrosis factor regulates hepatic acute-phase gene expression. *J. Clin. Invest.* **78**, 1349–1354.

Peveri, P., Walz, A., Dewald, B. and Baggiolini, M. (1988). A novel neutrophil activating factor produced by human mononuclear phagocytes. *J. Exp. Med.* **167**, 1547–1559.

Pietsch, T., Kyas, U., Steffens, U., Yakisan, E., Hadam, M.R., Ludwig, W.D., Zsebo, K. and Welte, K. (1992). Effects of human stem cell factor (c-kit ligand) on proliferation of myeloid leukemia cells: heterogeneity in response and synergy with other hematopoietic growth factors. *Blood* **80**, 1199–1206.

Prummer, O., Streichan, U., Heimpel, H. and Porzsolt, F. (1994). Sensitive antiproliferative neutralization assay for the detection of neutralizing IFN-α and IFN-β antibodies. *J. Immunol. Methods* **171**, 45–53.

Pusztai, L., Clover, L.M., Cooper, K., Starkey, P.M., Lewis, C.E. and McGee, J.O. (1994). Expression of tumour necrosis factor α and its receptors in carcinoma of the breast. *Br. J. Cancer* **70**, 289–292.

Randall, L.A., Wadhwa, M., Thorpe, R. and Mire-Sluis, A.R. (1993). A novel sensitive bioassay for transforming growth factor-β. *J. Immunol. Methods* **164**, 61.

Roehm, N.W., Rodgers, G.H., Hatfield, S.M. and Glasebrook, A.L. (1991). An improved colorimetric assay for cell proliferation and viability utilizing the tetrazolium salt XTT. *J. Immunol. Methods* **142**, 257.

Rousset, F., De Waal Malefyt, R., Slierendregt, B., Aubry, J.P., Bonnefoy, J.Y., DeFrance T., Banchereau, J. and deVries, J.E. (1988). Regulation of Fc receptor for IgE (CD23) and class II MHC antigen expression on Burkitt's lymphoma cell lines by human IL4 and IFN. *J. Immunol.* **140**, 2625–2632.

Sadick, M.D., Sliwkowski, M.X., Nuijens, A., Bald, L., Chiang, N., Lofgren, J.A. and Wong, W.L.T. (1996). Analysis of heregulin-induced ErbB2 phosphorylation with a high-throughput kinase receptor activation enzyme-linked immunosorbant assay. *Anal. Biochem.* **235**, 207–214.

Sander, B., Andersson, J. and Andersson, U. (1991). Assessment of cytokines by immunofluorescence and the paraformaldehyde–saponin procedure. *Immunol. Rev.* **119**, 65–93.

Schroder, J.M., Mroweitz, U., Morita, E. and Christophers, E. (1987). Purification and partial biochemical characterisation of a human monocyte-derived, neutrophil-activating peptide that lacks interleukin 1 activity. *J. Immunol.* **139**, 3474–3483.

Siegal, J.P. and Mostowski, H.S. (1990). A bioassay for the measurement of human IL-4. *J. Immunol. Methods* **132**, 287–295.

Stanley, E.R., Chen, D.M. and Lin, H.S. (1978). Induction of macrophage production and proliferation by a purified colony stimulating factor. *Nature* **274**, 168–170.

Stanley, E.R., Guilbert, L.J., Tushinski, R.J. and Bartelmez, S.H. (1983). CSF-1—a mononuclear phagocyte lineage-specific hemopoietic growth factor. *J. Cell. Biochem.* **21**, 151–159.

Stern, A.S., Podlaski, F.J., Hulmes, J.D., Pan, Y.C., Quinn, P.M., Wolitzky, A.G., Familletti, P.C., Stremlo, D.L., Truitt, T. and Chizzonite, R. (1990). Purification to homogeneity and partial characterization of cytotoxic lymphocyte maturation factor from human B-lymphoblastoid cells. *Proc. Natl Acad. Sci. U.S.A.* **87**, 6808–6812.

Suda, T., Yamaguchi, Y., Suda, J., Miura, Y., Okano, A. and Akiyama, Y. (1988). Effect of interleukin-6 (IL-6) on the differentiation and proliferation of murine and human hemopoietic progenitors. *Exp. Haematol.* **16**, 891–895.

Thompson-Snipes, L., Dhar, V., Bond, M.W., Mosmann, T.R., Moore, K.W. and Rennick, D.M. (1991). Interleukin 10: a novel stimulatory factor for mast cells and their progenitors. *J. Exp. Med.* **173**, 507–510.

Thorpe, R., Wadhwa, M., Gearing, A.J.H., Mahon, B. and Poole, S. (1988). Sensitive and specific immunoradiometric assays for human interleukin-1α. *Lymphokine Res.* **7**, 119–127.

Thorpe, R., Wadhwa, M., Bird, C.R. and Mire-Sluis, A.R. (1992). Detection and measurement of cytokines. *Blood Rev.* **6**, 133.

Tsai, M., Takeishi, T., Thompson, H., Langley, K.E., Zsebo, K.M., Metcalfe, D.D., Geissler, E.N. and Galli, S.J. (1991). Induction of mast cell proliferation, maturation, and heparin synthesis by the rat c-kit ligand, stem cell factor. *Proc. Natl Acad. Sci. U.S.A.* **88**, 6382–6386.

Tsuchiya, S., Yamabe, M., Yamaguchi, Y., Kobayashi, Y., Konno, T. and Tada, K. (1980). Establishment and characterization of a human acute monocytic leukemia cell line (THP-1). *Int. J. Cancer* **26**, 171–174.

Ushio, S., Namba, M., Okura, T., Hattori, K., Nukada, Y., Akita, K., Tanabe, F., Konishi, K., Micallef, M., Fujii, M., Torigoe, K., Tanimoto, T., Fukuda, S., Ikeda, M., Okamura, H. and Kurimoto, M. (1996). Cloning of the cDNA for human IFN-γ-inducing factor, expression in *Escherichia coli*, and studies on the biologic activities of the protein. *J. Immunol.* **156**, 4274–4279.

Van Damme, J. and Conings, R. (1995). Assays for chemotaxis. In *Cytokines. A Practical Approach*, 2nd edn. (ed. F.R. Balkwill), IRL Press, Oxford, pp. 215–224.

Van Snick, J., Cayphas, S., Vink, A., Uyttenhove, C., Coulie, P.G., Rubira, M.R. and Simpson, R.J. (1986). Purification and NH_2-terminal amino acid sequence of a T-cell-derived lymphokine with growth factor activity for B-cell hybridomas. *Proc. Natl Acad. Sci. U.S.A.* **83**, 9679–9683.

Vilcek, J. and Lee, T.H. (1991). Tumor necrosis factor. New insights into the molecular mechanisms of its multiple actions. *J. Biol. Chem.* **266**, 7313–7316.

Vilcek, J., Palombella, V.J., Henriksen-DeStefano, D., Swenson, C., Feinman, R., Hirai, M. and Tsujimoto, M. (1986). Fibroblast growth enhancing activity of tumor necrosis factor and its relationship to other polypeptide growth factors. *J. Exp. Med.* **163**, 632–643.

Wadhwa, M., Thorpe, R., Bird, C.R. and Gearing, A.J.H. (1990). Production of polyclonal and monoclonal antibodies to human granulocyte colony-stimulating factor (G-CSF) and development of immunoassays. *J. Immunol. Methods* **128**, 211–217.

Wadhwa, M., Bird, C., Page, L., Mire-Sluis, A.R. and Thorpe, R. (1995). Quantitative biological assays for individual cytokines. In *Cytokines. A Practical Approach*, 2nd edn. (ed. F.R. Balkwill), IRL Press, Oxford, pp. 357–392.

Wadhwa, M., Dilger, P., Meager, A., Walker, B., Gaines-Das, R. and Thorpe, R. (1996). IL-4 and TNF α mediated proliferation of the human megakaryocytic line M-07e is regulated by induced autocrine production of GM-CSF. *Cytokine* **8**, 900–909.

Wang, J.M., McVicar, D.W., Oppenheim, J.J. and Kelvin, D.J. (1993a). Identification of RANTES receptors on human monocytic cells: competition for binding and desensitization by homologous chemotactic cytokines. *J. Exp. Med*. **177**, 699–705.

Wang, J.M., Sherry, B., Fivash, M.J., Kelvin, D.J. and Oppenheim, J.J. (1993b). Human recombinant macrophage inflammatory protein-1α and -β and monocyte chemotactic and activating factor utilize common and unique receptors on human monocytes. *J. Immunol*. **150**, 3022–3029.

Warren, M.K. and Ralph, P. (1986). Macrophage growth factor CSF-1 stimulates human monocyte production of interferon, tumor necrosis factor, and colony stimulating activity. *J. Immunol*. **137**, 2281–2285.

Watanabe, Y., Mazda, O., Aiba, Y., Iwai, K., Gyotoku, J., Ideyama, S., Miyazaki, J. and Katsura, Y. (1992). A murine thymic stromal cell line which may support the differentiation of $CD4^-8^-$ thymocytes into $CD4^+8^-$ $\alpha\beta$ T cell receptor positive T cells. *Cell. Immunol*. **142**, 385–397.

Weinstein, Y., Ihle, J.N., Lavu, S. and Reddy, E.P. (1986). Truncation of the c-myb gene by a retroviral integration in an interleukin-3-dependent myeloid leukemia cell line. *Proc. Natl Acad. Sci. U.S.A*. **83**, 5010–5017.

Williams, D.E., Morrissey, P.J., Mochizuki, D.Y., De-Vries, P., Anderson, D., Cosman, D., Boswell, H.S., Cooper, S., Grabstein, K.H. and Broxmeyer, H.E. (1990). T-cell growth factor p40 promotes the proliferation of myeloid cell lines and enhances erythroid burst formation by normal murine bone marrow cells *in vitro*. *Blood* **76**, 906–911.

Yang, Y.C., Ricciardi, S., Ciarletta, A., Calvetti, J., Kelleher, K. and Clark, S.C. (1989). Expression cloning of cDNA encoding a novel human hematopoietic growth factor: human homologue of murine T-cell growth factor p40. *Blood* **74**, 1880–1884.

Yao, Z., Painter, S.L., Fanslow, W.C., Ulrich, D., Macduff, B.M., Spriggs, M.K. and Armitage, R.J. (1995). Human IL-17: a novel cytokine derived from T cells. *J. Immunol*. **155**, 5483–5486.

Yuo, A., Kitagawa, S., Ohsaka, A., Ohta, M., Miyazono, K., Okabe, T., Urabe, A., Saito, M. and Takaku, F. (1989). Recombinant human granulocyte colony-stimulating factor as an activator of human granulocytes: potentiation of responses triggered by receptor-mediated agonists and stimulation of C3bi receptor expression and adherence. *Blood* **74**, 2144–2149.

Zheng, R.Q.H., Abney, E., Chu, C.Q., Field, M., Grubeck-Loebenstein, B., Maini, R.N. and Feldmann, M. (1991). Detection of interleukin-6 and interleukin-1 production in human thyroid epithelial cells by non-radioactive *in situ* hybridization and immunohistochemical methods. *Clin. Exp. Immunol*. **83**, 314–319.

Zsebo, K.M., Wypych, J., McNiece, I.K., Lu, H.S., Smith, K.A., Karkare, S.B., Sachdev, R.K., Yuschenkoff, V.N., Birkett, N.C. and Williams, L.R. (1990). Identification, purification, and biological characterization of hematopoietic stem cell factor from buffalo rat liver-conditioned medium. *Cell* **63**, 195–201.

Zurawski, S.M., Vega, F.J., Huyghe, B. and Zurawski, G. (1993). Receptors for interleukin-13 and interleukin-4 are complex and share a novel component that functions in signal transduction. *EMBO J.* **12**, 2663.

Chapter 32

Cytokines and their Receptors as Potential Therapeutic Targets

R. Geoffrey P. Pugh-Humphreys[1] and Angus W. Thomson[2]

[1]Cell and Immunobiology Unit, Department of Zoology, University of Aberdeen, Aberdeen, UK, and [2]Departments of Surgery and Molecular Genetics and Biochemistry, University of Pittsburgh, Pittsburgh, PA, USA

GENERAL FEATURES OF CYTOKINES

The previous chapters in this book have been concerned with biochemical and biological considerations of individual cytokines. This chapter not only addresses the nature of the interactions of cytokines with their receptors and the pathophysiological roles of cytokines, but also considers cytokines and their receptors as therapeutic targets in the treatment of various pathological conditions.

Cytokines normally exert their diverse biological activities within the context of a cytokine network to maintain homeostatic mechanisms. Cytokine interactions play pivotal roles in many normal and pathological events, including the generation of immune responses, inflammatory reactions, remodelling of tissues, angiogenesis and neoplastic transformation of cells.

An imbalance in the production and/or action of cytokines within the network provides the basis for generating immune deficiency states as well as pathological processes including inflammatory disorders such as rheumatoid arthritis, fibrotic disorders, growth factor-related diseases and neoplasia, all of which can be ameliorated by intervening in the production of cytokines using drugs and in the interaction of cytokines with their respective receptors through the use of antibodies, antagonists and other blocking factors. We commence this review with a consideration of the operation of cytokines within the cytokine network and its role in health and disease, and then review research work that has been concerned with cytokines and their receptors as potential therapeutic targets in the treatment of a variety of pathological states.

THE CYTOKINE NETWORK AND ITS ROLE IN HEALTH AND DISEASE

Cytokines are a diverse group of small, 6–30 kDa, pleiotropic cell regulatory (glyco)-protein molecules involved in cell signalling within a dynamic cellular communication network (Arai *et al.*, 1990) (Table 1). Studies of their molecular architecture, coupled

The Cytokine Handbook, 3rd ed.
ISBN 0-12-689662-3

Table 1. General properties of cytokines.

Cytokines are proteins or glycoproteins that function as signalling molecules within a dynamic cellular communication, or cytokine, network.
Cytokines are normally produced for short periods of time, and at extremely low concentrations, being active at nanomolar or picomolar concentrations during the effector phases of both natural and acquired immune reactions, as well as during developmental processes.
Most cytokines are secreted as soluble molecules, but some can be expressed in a membrane-anchored form, and yet others bind to molecules within tissue extracellular matrices.
Cytokines are pleiotropic mediators, being produced by a wide variety of cell types and often exerting effects on many different types of target cells.
Cytokines can exert effects upon the producer cell (autocrine activity), neighbouring target cells (juxtacrine and paracrine activity) or, in some instances, on distant target cells (endocrine activity).
Cytokines mediate their actions by binding to their specific cognate receptors on target cells.
Within the cytokine network, cytokines can act in concert with each other (synergize) or antagonize each other.
Cytokines can regulate the development, state of activation, differentiation and effector functions of a wide variety of target cells.

with knowledge of cytokine gene organization, have enabled molecular biologists to categorize cytokines into six major superfamilies, namely the haemopoietins (four α-helical bundles), the epidermal growth factor (EGF) superfamily (β sheet and β trefoil), the tumour necrosis factor (TNF) superfamily ('jelly roll' motif), the cysteine knot superfamily, and the chemokine superfamily (triple-stranded antiparallel β sheet) (Callard and Gearing, 1994) (Table 2).

Some of these superfamilies can be subdivided; for instance, the haemopoietins contain a core of four α helices forming a bundle with unique 'up-up-down-down' topology (Wlodawer *et al.*, 1993). They can be divided into two divergent subgroups: the 'short-chain' cytokines 105 to 140 amino acids in length with two short antiparallel β strands and helices of approximately 15 amino acids in length, and the 'long-chain'

Table 2. Structural superfamilies of cytokines.

Superfamily	Members
The haemopoietins (four α-helical bundles)	IL-2, IL-3, IL-4, IL-5, IL-6, IL-7, IL-9, IL-10, IL-13, G-CSF, GM-CSF, M-CSF, CNTF, OSM, LIF, IFN-α, IFN-β, IFN-γ
Epidermal growth factor superfamily	
β sheet	EGF, TGF-α
β trefoil	FGF-1, FGF-2
	IL-1α, IL-1β, IL-1ra
TNF superfamily (jellyroll motif)	TNF-α, TNF-β
Cysteine knot superfamily	NGF, TGF-β_1, TGF-β_2, TGF-β_3, PDGF, VEGF
Chemokine superfamily (triple-stranded, antiparallel β sheet)	IL-8, eotaxin, fractalkine, lymphotactin,neurotactin, MIP-1α, MIP-1β, MIP-2, PF4, I309/TCA-3, MCP-1, MCP-2, MCP-3, γIP-10, RANTES, GRO-α, GRO-β, GRO-γ, etc.

cytokines with helices some 25 amino acids in length, but lacking the β chains found within the short-chain cytokines (Sprang and Bazan, 1993). One rapidly growing structural superfamily includes members of four families of cysteine-rich dimeric polypeptide growth factors (nerve growth factor (NGF), transforming growth factor-β (TGF-β), platelet-derived growth factor (PDGF) and HCG) which show little or no sequence similarity, but whose structural subunits display similarities, including a characteristic cluster of three disulphide bridges called the 'cysteine knot' that provide a framework for two parallel twisted β hairpins (Murzin and Clothia, 1992; McDonald and Hendrickson, 1993).

Much of the folded structure of cytokines serves as a conserved scaffold providing a unique topographic surface that is 'read' by a receptor through specific ligand–receptor pairing, during which the cytokines must present a distinct symmetry-related surface to their cognate receptors (Sprang and Bazan, 1993). Each cytokine molecule contains distinct functional domains or epitopes that induce different biological responses; for instance, the four helix-bundle cytokines are all composed of a common structural framework within which there are receptor subunit binding epitopes (Chaiken and Williams, 1996) and in the case of the neuregulins, the EGF-like domain is the functional domain for activating the erbB family of receptors (Chang *et al.*, 1997). However, not all epitopes on a cytokine molecule are necessarily involved in receptor binding, and signal transduction domains are quite distinct (Billiau, 1996). Nevertheless, unravelling the details of cytokine–receptor interactions is pivotal in the design of specific cytokine antagonists (Debets and Savelkoul, 1994), and through advances in engineered protein construction, coupled with high resolution structural analyses of both cytokines and cytokine–receptor interactions, considerable progress has been made in the design of *de novo* mimetics (including cytokine hybrids, structure-excerpted scaffolds and contact residue topology mimetics) which can function as antagonists (Chaiken and Williams, 1996).

The precursors of the majority of secreted cytokines contain a typical hydrophobic signal peptide sequence and are secreted by the classical secretion pathway involving the endoplasmic reticulum (ER) and Golgi complexes. However, some cytokines such as fibroblast growth factor-1 (FGF-1), FGF-2 and the FGF homologues do not have a typically glycosylated signal sequence as seen in the other secreted FGFs and which allows them to cross biomembranes in the intracellular secretory pathway, and consequently they are retained inside the cell (Smallwood *et al.*, 1996). For these cytokines to be released extracellularly, the producer cell must either lyse or an alternative mechanism for protein secretion must operate (Kuchler, 1993; Wiedmann *et al.*, 1994), possibly involving an energy-dependent, non-ER–Golgi pathway involving heat shock proteins as molecular chaperones (Bikfalvi *et al.*, 1997).

Some cytokines (e.g. interleukin-1α (IL-1α), EGF, stem cell factor (SCF), SLF, TNF-α, TGF-α and macrophage colony stimulating factor (M-CSF)) exist in a membrane-anchored form and may mediate juxtacrine signalling between adjacent cells (Flanagan and Leder, 1990; Huang *et al.*, 1990; Perez *et al.*, 1990; Dinarello, 1991; Massague and Pandiella, 1993). All members of the TNF family, with the exception of TNF-β, are type II transmembrane proteins anchored into the lipid bilayer by an amino-terminal hydrophobic anchor sequence; the membrane anchored form of TNF-α is involved in intercellular communication (Aversa *et al.*, 1993; Macchia *et al.*, 1993) as well as exerting a cytotoxic activity on target cells (Perez *et al.*, 1990). Proteolytic cleavage and release ('shedding') of the extracellular domains of cell surface proteins by adamalysins (ADAMs) is a common event in mammalian cells (Rose-John and Heinrich, 1994; Wolfsberg and White, 1996). The extracellular domain of the membrane-anchored 26-kDa TNF-α can be cleaved to generate a 17-kDa secreted form of TNF-α (Perez *et al.*, 1990) by a membrane-bound disintegrin metalloproteinase of the ADAM family called TNF-α converting enzyme (TACE) (Black *et al.*, 1997; Moss *et al.*, 1997). The release of TNF-α from cells can be inhibited by serine protease inhibitors (Suffys *et al.*, 1988) and broad-spectrum hydroxamic acid-based inhibitors of metalloproteinases (Gearing *et al.*, 1994; McGeehan *et al.*, 1994; Mohler *et al.*, 1994). Since TNF-α contributes to the pathogenesis of many inflammatory diseases (Vassali, 1992), protease inhibitors that target TACE may prove effective in the treatment of these diseases.

Although monomeric forms of cytokines do exist, most cytokines are active as dimers or trimers. For instance, interferon-γ (IFN-γ) is a homodimer of N-glycosylated glycoproteins. Natural IFN-γ is heterogeneous in size and charge due to enzymatic trimming of the *C*-terminus and variation in the degree of glycosylation. While apparently unimportant for bioactivity, the molecular heterogeneity of IFN-γ may well influence the dynamics of its tissue distribution (Billiau, 1996). Different isoforms of particular cytokines may arise as splice variants of gene transcripts. For instance, the neuregulins are a family of multipotent EGF-like polypeptides that arise from splice variants of a single gene (Carraway and Burden, 1995) and the cytokine vascular endothelial growth factor (VEGF) has at least four isoforms ($VEGF_{121}$, $VEGF_{165}$, $VEGF_{189}$ and $VEGF_{206}$) which result from alternative splicing of the VEGF gene transcript (Ferrera *et al.*, 1991). Multiple RNA species ranging in size from 1.5 to 4.4 kb are produced by alternative splicing of the M-CSF gene transcript generating different forms of M-CSF called M-CSFα, M-CSFβ and M-CSFγ containing 256, 554 and 438 amino acids respectively (Cerretti *et al.*, 1988). After translation the M-CSF polypeptides are glycosylated, form dimers and are inserted into biomembranes as membrane-anchored M-CSF (mM-CSF) within the ER–Golgi system. The mM-CSF is then cleaved proteolytically within secretory vesicles to release soluble (mature) M-CSF (Kawasaki *et al.*, 1985; Rettenmeir *et al.*, 1987). The TGF-β family of multifunctional dimeric polypeptides, includes five structurally and functionally related isoforms called TGF-β_1, TGF-β_2, TGF-β_3, TGF-β_4 and TGF-β_5 which influence cell proliferation, differentiation and extracellular matrix production (Miyazono and Heldin, 1991; Johnson and Williams, 1993).

Many cytokines (e.g. interleukins 1–13, granulocyte–macrophage colony stimulating factor (GM-CSF), G-CSF, M-CSF, EGF, hepatocyte growth factor (HGF), scatter factor (SF), HGF–SF, NGF, leukaemia inhibitory factor (LIF), oncostatin M (OSM), TNF-β and TGF-β) are synthesized initially as inactive precursors which are cleaved subsequently to form the active proteins. All the TGF-β isoforms are secreted as inactive, latent complexes. TGF-β_1 is secreted as a large precursor protein containing 390 amino acids and consists of a latency-associated 40-kDa *N*-terminal peptide, an active 25-kDa homodimer and a 125–160-kDa binding protein linked by disulphide bonds to the latency-associated peptide (Miyazono and Heldin, 1991; Rifkin *et al.*, 1993). The active 25-kDa homodimer of TGF-β is released from the inactive precursor through proteolytic cleavage by proteases such as plasmin (Lyons *et al.*, 1990): IL-1β and IL-18 (interferon-γ-inducing factor (IGIF)) are synthesized as biologically inactive precursors (pro-IL-1β and pro-IGIF respectively), and are processed at Asp–X cleavage sites by interleukin-1β converting enzyme (ICE or caspase-I) (Fraser and Evans, 1996; Ghayur *et al.*, 1997; Gu *et al.*, 1997). HGF–SF is a plasminogen-related growth factor that is translated as a 90-kDa single-chain precursor and is activated proteolytically by an arginine–valine cleavage either by urokinase (u-Pa) (Naka *et al.*, 1992; Naldini *et al.*, 1992, 1995) or by a serine protease called HGF activator (Shimomura *et al.*, 1995) to produce a heterodimer consisting of a 60-kDa α subunit and a 30-kDa β subunit linked by a single disulphide bond. HGF–SF lacks protease activity because of amino acid replacements at the catalytic triad of the β subunits (Nakamura *et al.*, 1989).

The chemokines are inflammatory mediators that stimulate the directional migration of leukocytes (Alam, 1997). The chemokine superfamily contains polypeptides some 70 to 80 amino acids in length that share substantial sequence similarity, with four cysteine (C) residues at near identical positions. The chemokine families categorized to date are the CXC (α), CC (β), CX_3C (δ) and C (γ) families, based on the spacings of the *N*-terminal cysteine residues (Mackay, 1997; Wells and Peitsch, 1997) (Table 3). The α chemokines have the cysteines separated by one amino acid residue (C-X-C), whereas in the β chemokines, the cysteines are adjacent (C-C) (Clark-Lewis *et al.*, 1995). The C-X-C family of chemokines is further subdivided into two subfamilies: those with a characteristic glutamic acid–leucine–arginine (ELR) sequence immediately preceding the first cysteine residue near the *N*-terminus, and the second subfamily of C-X-C chemokines lacking the ELR sequence. Members of the α family are chemotactic for neutrophils and lymphocytes, whereas the β family chemokines are chemotactic for monocytes, lymphocytes and eosinophils (Teran and Davies, 1996). Eotaxin is an eosinophil-selective β chemokine which is apparently unique among the chemokines in binding specifically to only one receptor, the CCR-3 receptor (Jose *et al.*, 1996). The only C, or γ, chemokine identified to date is lymphotactin which has only one pair of cysteines (Kelner *et al.*, 1994; Kennedy *et al.*, 1995). The CX_3C or δ chemokines include neurotactin, a type I membrane-anchored protein which is chemotactic for neutrophils and is upregulated in the microglia of mice with experimental autoimmune encephalomyelitis (Pan *et al.*, 1997), and fractalkine, which is a membrane-bound molecule with a chemokine domain perched on a mucin-like stalk and which engages in juxtacrine signalling (Premack and Schall, 1996).

Table 3. The chemokines.

α (C-X-C)		β (C-C)	γ (C)	δ (C-X_3-C)
With ELR sequence	Without ELR sequence			
IL-8	IP-10/murine CRG	MIP-1α	Lymphotactin	Neurotactin
GRO-α, β, γ	Mig	MIP-1β		Fractalkine
Murine KC	PBSF/SDF-1	RANTES		
Murine MIP-2	PF-4	MCP-1, 2, 3, 4		
ENA-78		Murine JE/MARC		
GCP-2		Eotaxin		
PBP/CTAPIII/ β-TG/NAP-2		I-309/TCA-3		
		HCC-1		

CYTOKINE RECEPTORS

Receptors for cytokines have been purified and cloned, and have been grouped into seven superfamilies on the basis of structural analyses and identification of amino acid sequence homologies clustered into domains or repeats. They are the haemopoietin receptor superfamily (also known as the cytokine receptor superfamily type I); the immunoglobulin receptor superfamily; the protein kinase receptor superfamily; the TNF–NGF factor receptor superfamily; the interferon receptor superfamily (also known as the cytokine receptor superfamily type II); the G protein-coupled seven-transmembrane spanning receptor superfamily; and the complement control protein superfamily (Table 4).

The receptor kinases are classified according to their substrate specificity, with one family phosphorylating tyrosine residues (the receptor tyrosine kinases), while the other family phosphorylates serine and threonine residues (the receptor serine–threonine kinases). The receptor tyrosine kinases (RTKs) are a family of more than 50 different, single-pass transmembrane proteins with an intrinsic protein tyrosine kinase (PTK) domain in their cytoplasmic portions (Kavanaugh and Williams, 1996). The RTK family can be subdivided into several subclasses including the epidermal growth factor (EGFR) subclass; the insulin receptor (IR) subclass; the platelet derived growth factor receptor

Table 4. The superfamilies of cytokine receptors.

Cytokine receptor superfamily type I
IL-2Rβ, IL-2Rα, IL-3Rα, IL-3Rβ, IL-4R, IL-5Rα, IL-5Rβ, IL-6R, gp130, IL-9R, IL-12R, G-CSFR, GM-CSFR, CNTFR, LIFR, EpoR, PRLR and GHR

Immunoglobulin receptor superfamily
IL-1R, IL-6R, FGFR, PDGFR, M-CSFR and SCFR (c-kit)

Cytokine receptor superfamily type II
The type I IFN-αR, IFN-βR and IFN-ωR; the type II IFN-γR; IL-10R

Protein kinase receptor superfamily

(a) Protein tyrosine kinase receptor family

Subclass I	EGFR
Subclass II	Insulin R, IGF-1R
Subclass III	PDGFR, CSF-1R, SCFR
Subclass IV	VEGFRs
Subclass V	HGFR/SFR
Subclass VI	NTR
Subclass VII	FGFRs

(b) Protein serine–threonine kinase receptor family
Activin and inhibin receptors; types I and II TGF-βRs

Tumour necrosis factor/nerve growth factor receptor superfamily
TNF-R1 (p55), TNF-R2 (p75), NGFR

G protein-coupled seven-transmembrane spanning receptor superfamily
PAFR, CCRs and CXCRs

Complement control protein superfamily
IL- 2Rα (p55)

(PDGFR)–macrophage colony stimulating factor-1 receptor (M-CSFR) family–c-kit protein (Steel ligand) receptor subclass; the vascular endothelial cell growth factor receptor (VEGFR) subclass; the hepatocyte growth factor receptor (HGFR) receptor subclass; the neurotrophin receptor (NTR) subclass; and the fibroblast growth factor receptor (FGFR) subclass. The receptors in a given family share common structural features distinct from those of the other families (Fantl *et al.*, 1993).

Ligation of the RTKs, such as EGFRs, results in receptor dimerization and activation of the intrinsic PTK resulting in trans- or autophosphorylation of the receptor, followed by the recruitment into multiunit complexes, and activation, of intracellular signal transducers containing amino acid motifs called Src homology (SH) domains or phosphotyrosine binding (PTB) domains which recognize phosphotyrosine residues (Fantl *et al.*, 1993; Geet *et al.*, 1994; Heldin, 1995). EGF binding to its cognate receptor also triggers EGFR downregulation by recruitment of EGFRs into clathrin-coated vesicles, followed by endocytosis of the EGF–EGFR complexes and their delivery to the lysosome system for degradation (Sorkin and Waters, 1993; Benmerak *et al.*, 1995; Wong and Moran, 1996). Endocytotic trafficking of ligated EGFRs also plays a critical role in controlling specific signalling pathways (Vieira *et al.*, 1996).

The receptors for the activin/inhibin/TGF-β cytokines are serine–threonine kinases (RSTKs) (Altisano *et al.*, 1993; ten-Dijke *et al.*, 1994; Massague, 1996). The kinase domain of the RSTKs shares structural features in common with the RTKs, and the RSTKs may associate to form functional multichain complexes as seen with the type I and type II TGF-β receptors (Wrana *et al.*, 1992; Ebner *et al.*, 1993). The type III TGF-β receptor, which is not an RSTK but rather a transmembrane proteoglycan, also interacts with the type II TGF-β receptor (Wang *et al.*, 1991; Lopez-Casillas *et al.*, 1993). Members of the TGF-β superfamily transduce their signals through two serine–threonine kinase receptors, the 53-kDa type I (TβR-I) and the 75-kDa type II (TβR-II) receptors. The third receptor is the >200-kDa type III receptor called betaglycan, which is a proteoglycan whose glycosaminoglycan chains are attached to a 100-kDa core protein (Wang *et al.*, 1991). Instead of betaglycan some cells express endoglin, a 170-kDa disulphide-linked dimeric protein related to betaglycan (Cheifetz *et al.*, 1992). Both betaglycan and endoglin facilitate the receptor binding of TGF-βs by presenting these molecules to their type I and type II signalling receptors (Lopez-Casillas *et al.*, 1993). Whereas the type II receptors bind the TGF-β ligands independently, the type I receptors bind the ligands only when the type II receptors are present, resulting in the assembly of signal transducing heteroligomeric complexes of type I and type II receptors (Wrana *et al.*, 1994; Miyazono, 1996). Signalling by the receptors is mediated by SMAD proteins (Wrana and Pawson, 1997) which, upon phosphorylation by activated receptors, form complexes and then translocate into the nucleus where they associate with DNA binding proteins and activate gene transcription (Chen *et al.*, 1996; Massague *et al.*, 1997). SMAD2 and SMAD4 are suppressor proteins which, when stimulated by TGF-β, form a complex to inhibit cell growth (Lagna *et al.*, 1996). Missense mutations in these SMAD proteins inhibit SMAD2–SMAD4 interactions and inhibit TGF-β signalling (Hata *et al.*, 1997).

The FGFR genes encode related glycoproteins with a common structure consisting of a signal peptide with three immunoglobulin (Ig) domains containing characteristic cysteine residues, a single transmembrane segment and an intracellular split kinase domain. Four major FGF receptors (FGFR-1 (flg), FGFR-2 (bek), FGFR-3 and

FGFR-4) have been identified (Basilico and Moscatelli, 1992; Johnson and Williams, 1993; Green *et al.*, 1996) and spliced variants exist that differ in the composition of the ligand binding domains and intracellular portions (McKeehan *et al.*, 1993). For instance, the third Ig loop in the ligand binding domain is encoded by a common 5′ exon (IIIa) which is spliced to either of the two 3′ exons (IIIb or IIIc); the choice of the 3′ exon determines the ligand binding specificity of the FGFR (Jaye *et al.*, 1992; Partanen *et al.*, 1993). The existence of several FGFR isoforms, particularly of FGFR-1 and FGFR-2, generates receptor diversity and is important for ligand specificity (Ornitz *et al.*, 1996). A fifth FGFR has been described which is structurally distinct from the other receptors (Burrus *et al.*, 1992); it is a 150-kDa integral membrane glycoprotein called CFR, which is concentrated within the membranes of the Golgi complex and appears to be involved in intracellular FGF trafficking and regulation of cell responses to FGF (Zuber *et al.*, 1997). Another FGFR isoform is the E-selectin ligand (ESL-1) (Steegmaier *et al.*, 1994). FGFs bind to glycopolymers in tissue matrices and body fluids, and these molecules can influence the bioavailability and activities of the FGFs (Bikfalvi *et al.*, 1997). For instance, heparan sulphate proteoglycans bind FGFs avidly, assist in the presentation of FGFs to their receptors and hence are important non-signalling molecules required for FGF signalling (Rapraeger *et al.*, 1994; Bikfalvi *et al.*, 1997).

The TNF receptors, TNF-R1 (p55) and TNF-R2 (p75), belong to the TNF–NGF receptor superfamily which contains more than 12 members, all of which are type I transmembrane proteins with cysteine-rich repeats in their extracellular domains (Mallet and Barclay, 1991; Vandenabeele *et al.*, 1995; Aggarwal and Natarajan, 1996). Although approximately the same size, the cytoplasmic domains of TNF-R1 and TNF-R2 show little sequence homology and may be linked to different signal transduction pathways (Tartaglia *et al.*, 1991). TNF-R1 and TNF-R2 can recognize two or more different ligands, and receptor activation involves receptor clustering induced by binding of their respective oligomeric ligands (Tewari and Dixit, 1996). The membrane-anchored form of TNF-α binds preferentially to TNF-R2 (Grell *et al.*, 1995) and TNF-R2 binds soluble TNF-α and then presents it to TNF-R1 (Tartaglia *et al.*, 1993). The cytoplasmic portions of the Fas–Apo-1 (CD95) receptor and the TNF-R1 (CD120a) contain a 'death domain' consisting of a 90-amino-acid protein–protein interaction sequence (Wallach, 1997). TNFR death domain homologues, and associated factors (TRAFs) involved in downstream signalling, include those adaptor proteins associated with either TNF-R1 (i.e. TNFR-associated death domain protein (TRADD), Fas-associating protein with death domain/MORT1, TRAF2, I-TRAF, receptor interacting protein (RIP), FAN, TRAP-1, FLICE/MACH and sentrin) or TNF-R2 (TRAF1, TRAF2, cIAP-1 and cIAP-2) (Darnay and Aggarwal, 1997). These adaptor proteins mediate the multifunctional effects of TNF-α on target cells. TNF-α modulates cell growth and induces gene activation through activation of signalling intermediates including protein kinases, protein phosphatases, reactive oxygen intermediates, phospholipases, proteases (such as the caspases or ICE/CED3 proteases), sphingomyelinases and transcription factors (Wallach, 1997). The deleterious effects of TNF-α appear to be mediated by TNF-R1, whereas the role of TNF-R2 requires elucidation (Lucas *et al.*, 1997).

Most functional high-affinity receptors of the cytokine receptor superfamily type I are multichain complexes comprising high (10–100 pM) and lower (1–10 nM) affinity receptor subunits, with the lower affinity subunits behaving as affinity converters which are shared by more than one receptor. The ligand binding subunits associate with signal transducing subunits (Miyajima *et al.*, 1992; Taga and Kishimoto, 1992; Ihle *et al.*, 1994). Members of the cytokine receptor superfamily type I often utilize common receptor subunits with signal transducing activity (e.g. the common β chain (βc) of GM-CSFR, IL-3R, IL-5R; gp130 of IL-6R, LIFR, OSMR, IL-11R and CNTFR; the common γ chain (γc) of IL-2R, IL-4R, IL-7R, IL-9R and IL-15R; the IL-4α chain, which is an important ligand binding and signal transducing component of both IL-4R and IL-13R). This event provides a structural basis for the shared, overlapping, activities displayed by many cytokines and accounts for their functional redundancy (Kishimoto *et al.*, 1994; Callard *et al.*, 1996). Although IL-2 and IL-15 display no sequence homologies, they have related bioactivities because their tripartite receptors share two common chains, IL-2Rβ and γc (DiSanto, 1997). The unique third chains, IL-2Rα and IL-15Rα, provide specific binding sites for IL-2 and IL-15 respectively, and aid the formation of high-affinity receptor complexes with IL-2Rβ and γc which signal through the same signal transduction pathways. Although both IL-2 and IL-15 exhibit redundant roles in lymphoid development, IL-2 is critical for thymus-dependent T-cell homeostasis, whereas IL-15 promotes the extrathymic development of both T cells and natural killer (NK) cells. DiSanto (1997) has postulated that this specificity may be achieved by a combination of compartmentalized cytokine production and differential expression of the IL-2Rα and IL-15Rα chains.

The receptors of the cytokine receptor superfamilies lack intrinsic PTK domains and they compensate for this by recruiting and activating cytoplasmic PTKs in response to ligand binding (Kishimoto *et al.*, 1994). Ligand binding mediates not only receptor component dimerization, but also their association with Janus kinases (JAKs) as well as homotypic or heterotypic aggregation and activation of the JAKs (Ziemiecki *et al.*, 1994; Ihle, 1995, 1996). The JAKs couple ligand binding to tyrosine phosphorylation of signalling molecules recruited to the receptor complex, as will be discussed below. The IL-2R consists of three subunits, α (55 kDa; known as CD25, or Tac antigen), β (75 kDa) and γc (64 kDa), and tyrosine kinases such as those of the Src family (lck, fyn or lyn), JAK1 and Syk are associated with the IL-2Rβ chain (Taniguchi, 1995; Theze *et al.*, 1996). The γc chain of the IL-2R is also a member of the cytokine receptor superfamily type I, and dysfunction through mutation of the γc chain is associated with profound defects in T-cell development such as that observed in X-linked severe combined immunodeficiency (XSCID) (Noguchi *et al.*, 1993). JAK3 is physically associated with the γc chain, and JAK3–γc interactions are important in IL-2, IL-4, IL-7, IL-9 and IL-15-dependent responses (Johnston *et al.*, 1996).

The IL-1 family of cytokines consists of two agonists, IL-1α and IL-1β (both of which are important inflammatory mediators) and one antagonist, IL-1ra (Dinarello, 1994). These three naturally occurring ligands bind to the type I IL-1R, which contains three immunoglobulin domains; domains 1 and 2 are tightly linked, but domain 3 is completely separate and connected by a flexible linker. The IL-1β agonist contains two receptor binding domains, one of which binds to the first two immunoglobulin domains of IL-1R, while the other binds exclusively to the third immunoglobulin domain (Vigers *et al.*, 1997). A 'receptor trigger site' on IL-1β is thought to induce movement of domain 3, which allows the IL-1R to wrap around IL-1β (Schreuder *et al.*, 1997).

Some transmembrane receptors signal through heterotrimeric GTP-dependent regulatory G proteins which contain three subunits: the 39–46-kDa α subunits (which bind guanine nucleotides), the 35–36-kDa β subunits and the 8-kDa γ subunits. G proteins dissociate into α subunits and $\beta\gamma$ subunit complexes, both of which can act individually as regulators of downstream signalling molecule activity (Nerr, 1995; Raymond, 1995; Wess, 1997). The $\beta\gamma$ complexes regulate the activities of effector enzymes (such as phospholipase C-β (PLC-β) isoforms, adenylyl cyclase isozymes, phospholipase A_2), ion channels and G protein receptor kinases (Raymond, 1995; Sternweiss, 1996) eventually leading to altered physiological responses (Wess, 1997).

The chemokine receptors belong to the family of single-chain, seven-helix transmembrane spanning, G protein-coupled receptors which bind the α chemokines (α receptors), β chemokines (β receptors) or both ($\alpha\beta$ receptors) (Teran and Davies, 1996). There are four α-chemokine receptors (CXCR1–4) and five β-chemokine receptors (CCR1–5), and these receptors exhibit overlapping ligand specificities (Schall and Bacon, 1994; Power and Wells, 1996). Signal transduction by the chemokine receptors involves G proteins and cell-specific catalytic enzymes (Bokoch, 1995). Following G-protein 'activation' the G_α subunit activates PLC-β_1 and PLC-β_2, while the G $\beta\gamma$ subunit preferentially activates PLC-β_2. PLC activation in turn results in the formation of inositol trisphosphate (Ins(1,4,5)P_3) and diacylglycerol (DAG), which mobilize intracellular calcium (iCa^{2+}) and activate protein kinase C (PKC) respectively. The expression level of the G_α subunit may determine the degree and type of signal transduced (Neer, 1994). Following G-protein activation, multiple signalling pathways may be activated, involving the MAPK cascade and both the serine–threonine and tyrosine kinases (Howard *et al.*, 1996) (see below).

In addition to their membrane-anchored forms, many cytokine receptors exist also as soluble cytokine receptors (sCRs), lacking the residues that anchor the transmembrane forms, but which still retain high-affinity binding (Fernandez-Botran *et al.*, 1996). The sCRs are generated either by proteolytic cleavage by metalloproteinases of the transmembrane receptors ('receptor shedding') (e.g. sIL-2Rα, sTNF-R1 and sTNF-R2) or by the *de novo* synthesis from alternatively spliced receptor gene mRNA transcripts specific for the sCR and different from those encoding the transmembrane receptors (e.g. sIL-4R, sIL-5R, sIL-6R, sIL-7R, sIFNαR, sIFNβR and sGM-CSFR) (Rose-John and Heinrich, 1994). In the latter case, the sCRs are truncated membrane receptors lacking both the cytoplasmic and transmembrane-anchoring domains.

The activation of cells, such as lymphocytes and macrophages, is frequently accompanied by upregulated surface expression of cytokine receptors and enhanced release of sCRs (Fernandez-Botran *et al.*, 1996) and in certain diseases, such as cancer, the levels of sCR released by tumour cells may provide a prognostic indicator of tumour progression (Viac *et al.*, 1996). The sCRs play important roles as regulators of cytokine activity and can function as competitive inhibitors by sequestering cytokines away from their respective transmembrane receptors (Fernandez-Botran *et al.*, 1996). However, although sCRs can function as natural antagonists for their respective cytokines, in some instances the sCR (e.g. sIL-6R) can couple with its ligand (e.g. IL-6) and subsequently complex with a membrane-anchored signal transducing unit (e.g. gp130) to generate a signal (Hibi *et al.*, 1990). sCRs may also function as 'carrier proteins' and increase the stability of cytokines through reduced activity decay by protecting them from tissue fluid proteases (Aderka *et al.*, 1992).

CYTOKINE SIGNALLING

Cells respond to signalling molecules such as cytokines and other environmental stimuli through programmes of altered gene expression mediated by signal transduction pathways between ligated cell surface receptors and transcriptional control elements within genes (Karin, 1992; Horseman and Yu-Lee, 1994; Woodgett, 1996). There are three general pathways for information flow from the cell surface receptors to the nucleus. The first is the mitogen-activated peptide (MAP) kinase pathway, in which an activated kinase phosphorylates a cytoplasmic transcription factor which then binds to a *cis* regulatory element within a gene. The second pathway involves activation, through phosphorylation by a tyrosine kinase, of a latent transcription factor which then translocates to the nucleus. The third pathway involves phosphorylation by a cytoplasmic kinase of an inhibitory protein subunit within a latent transcription factor complex; the complex then dissociates and the functional transcription factor subunit translocates to the nucleus (Ihle, 1996; Woodgett, 1996).

Receptor-mediated signal transduction proceeds by activation of components of intracellular signalling pathways involving cytoplasmic kinase cascades (Robbins *et al.*, 1994; Pawson, 1995; Ihle, 1996; Sells and Chernoff, 1997) (Fig. 1). Every cell contains families of protein kinases, and may express several structurally related, yet genetically distinct, kinases of each family. There are several discrete families of cytoplasmic kinases including cAMP-dependent protein kinases (PKA); cGMP-dependent kinases; protein kinase C (PKC); protein tyrosine kinases (PTK) including the novel family of Janus kinases (JAK); cyclin-dependent kinases (cdk); calcium–calmodulin-dependent protein kinases (CaMK); MAP kinases (MAPK) including the subgroup of extracellular signal regulated kinases (ERK); MAPK kinases (MAPKK or MEK); the stress-activated protein kinases (SAPK) or c-jun terminal kinases (JNK); the p21-activated protein kinases (PAK); and oncogene products with kinase activity. The activity of the protein kinases is highly regulated by chaperone proteins, such as cdc 37 and the heat-shock protein Hsp90 (Hunter and Poon, 1997).

Numerous cytoplasmic proteins involved in signal transduction contain sequences some 50 to 100 amino acids in length, called src homology 2 (SH2), src homology 3 (SH3) and pleckstrin homology (PH) domains. The SH2 and SH3 domains recognize short peptide motifs bearing phosphotyrosine (pTyr) and proline residues respectively, whereas the PH domain may associate with lipids or bind specific proteins and thus promote the association of signalling proteins with the plasma membrane (Pawson, 1995). Stimulation of the growth factor (RTK) receptors causes tyrosine phosphorylation of the receptors themselves and several cytoplasmic proteins including MAP kinase and the oncoproteins ras, Shc, vav and raf by activation of RTK and non-receptor type tyrosine kinases (Kishimoto *et al.*, 1994). Tyrosine phosphorylation of these target proteins generates the binding sites for the SH2 or phosphotyrosine binding (PTB) domains to form signal transduction complexes involving the adaptor proteins (Pawson, 1995; Rosen *et al.*, 1995). Phosphorylation of the membrane-distal regions of cytokine receptors creates docking sites for the p85 subunit of phosphatidylinositol 3′-kinase (PI-3K). Activation of PI-3K results from phosphorylation of the p85 subunit which contains two SH2 domains and functions as an adaptor molecule that targets the catalytic 110-kDa subunit to the activated receptor complex (Ihle, 1996). One of the major targets for phosphorylation by PI-3K is the protein serine–threonine kinase Akt

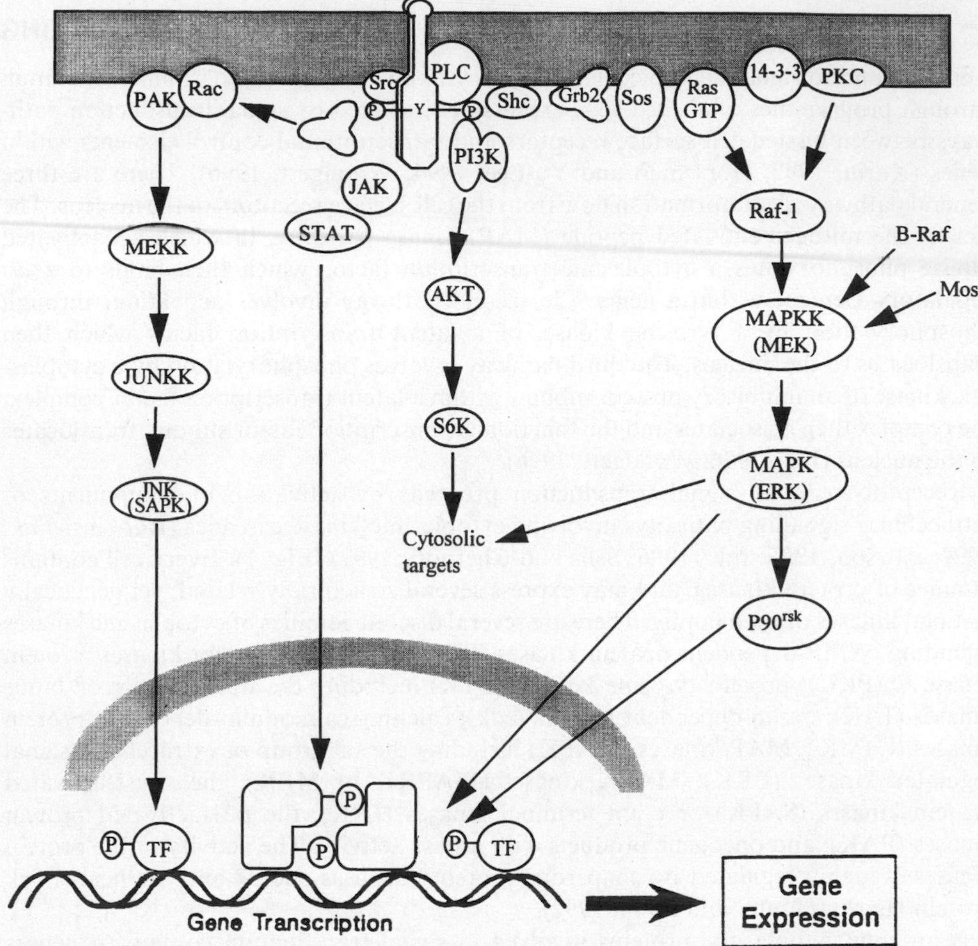

Fig. 1. Diagrammatic overview of signal transduction from the cell surface to the nucleus via the JNK, JAK–STAT, Ras, MAP kinase and ERK signalling pathways. Receptor tyrosine kinases (Y) cause Ras activation by binding to the adaptor protein Grb2 and the Ras-GTP–GDP exchange protein Sos. Ras binds to, and activates, protein kinase Raf-1 which activates MEK; MEK in turn activates ERKs. Ras also causes partial activation of the JNK signal transduction pathway which is activated by proinflammatory cytokines; these cytokines also cause activation of the p38 group of MAPKs (ERKs). The JNK and ERK phosphorylate transcription factors (TFs), which bind to regulatory elements within the promoter regions of genes to initiate gene expression. Although lacking catalytic domains, members of the cytokine receptor superfamilies use the JAK–STAT signalling pathway, in which Janus kinases (JAK) mediate gene expression by phosphorylating TFs called STATs; the STATs dimerize, bind to regulatory elements within genes and activate gene expression. See text for further details.

(protein kinase B, PKB) which participates in growth factor maintenance of cell survival (Hemmings, 1997a,b).

The adaptor protein Shc, which contains the SH2 and PTB domains, binds to pTyr on RTKs (Blaikie *et al.*, 1994) and becomes tyrosine phosphorylated; it then interacts with the SH2 domains of Grb2 which leads to the activation of the ras pathway via the guanine nucleotide exchange factor SOS ('son of sevenless') which binds constitutively to the SH3 domains of Grb2 (Lowenstein *et al.*, 1992; Buday and Downward, 1993; Chardin *et al.*, 1993; Cutler *et al.*, 1993) (Fig. 1). The ras protein is a member of the GTPase superfamily (Leevers, 1996) and is a central component of RTK-mediated mitogenic signal transducing pathways (Marshall, 1995). Activation of the ras signalling pathway is essential for cells to leave G0 and pass through the G1/S transition point in the cell cycle (Pronk and Bos, 1994; Laird and Shalloway, 1997). The activity of ras is regulated by GTPase activating proteins (GAPs) and guanine nucleotide exchange factors (GNEFs) such as SOS; the adaptor protein Grb2 binds to SOS, and the Grb2–Sos complex activates ras, which initiates a phosphorylation cascade which activates the Raf/MEK/MAP kinase signalling pathway (Pawson, 1995; Huang and Erickson, 1996). The MAP kinases then phosphorylate a variety of effector molecules including serine–threonine kinases and transcription factors; in effect, ras behaves as an amplifier switch, converting signals from tyrosine kinases into serine kinase activation (Ihle, 1996). The mechanisms by which G protein-coupled mitogenic receptors activate the MAP kinase signalling pathway are poorly understood, although it has been suggested that tyrosine phosphorylation of the tyrosine kinase Pyk2 allows it to bind to, and activate, the kinase src; the activated src then activates the map kinase signalling pathway via Grb2 and SOS (Dikic *et al.*, 1996).

Vav is a 95-kDa oncoprotein which contains SH2 and SH3 domains, and is heavily and rapidly phosphorylated on its tyrosine residues in response to the ligation of a variety of cell surface receptors; the large number of signalling pathways that activate vav in this way underline its importance as a signalling molecule, particularly since vav interacts with a large number of intracellular PTKs (including lck, fyn, syk, ZAP-70, JAKs and TEC) in the formation of signalling complexes (Collins *et al.*, 1997). Raf-1 is a 72–76-kDa oncoprotein with serine–threonine kinase activity and belongs to the Raf gene family; there is evidence that when ras proteins are in their active GTP binding state they bind Raf-1, and that this interaction mediates MEK activation (Robbins *et al.*, 1994; Pawson, 1995; Huang and Erickson, 1996).

The JAKs are involved in the signalling mechanisms of cytokine receptors that lack intrinsic kinase activity (Heldin, 1995; Ihle, 1995). In mammalian cells there are four members of the JAK family (JAK1, JAK2, JAK3 and TYK2) and they contain two kinase homology domains (JH1 and JH2) (Ziemiecki *et al.*, 1994). The JAKs associate with cytokine receptors in one of three ways (Ihle, 1995). The single-chain receptors (e.g. G-CSFR) associate with JAKs by their conserved membrane proximal domains, and the association may be constitutive or be enhanced by ligand binding. With the two-chain receptors the common chain associates with one or more JAKs. For instance, JAK2 associates with the membrane proximal domain of the βc chain in IL-3R , IL-5R and GM-CSFR, whereas in the receptors for IL-6, CNTF, OSM, LIF and IL-11, the shared gp130 chain (or the related protein LIFR-β) associates with JAK1, JAK2 or TYK2; aggregation of gp130 can bring together, and activate, the JAKs. The third association involves the two-chain receptors where both chains are required for signalling (e.g. the IFN family of receptors) and the JAKs associate with both signalling chains.

The family of transcription factors called STATs ('signal transducers and activators of transcription') consists of six members which participate in numerous cytokine signalling pathways (Finbloom and Larner, 1995; Ihle, 1995; Ihle and Kerr, 1995). Members of the STAT family contain SH2 and SH3 domains, and activation of STATs requires phosphorylation of specific tyrosines. The recruitment of specific STATs to receptor complexes, and subsequent STAT activation, is dependent on STAT SH2 domain recognition of receptor phosphotyrosine docking sites (Pawson, 1995; Stahl *et al.*, 1995). STATs normally exist in the cytoplasm in latent form and upon phosphorylation of their tyrosine residues by JAKs they form homo- or heterodimers through SH2-phosphotyrosyl peptide interactions (Pawson, 1995) and then translocate to the nucleus where they bind to the promoter regions of genes (Darnell *et al.*, 1994; Ihle, 1995; Ihle and Kerr, 1995). STATs recognize the γ activation sequence (GAS) motif and regulate the expression of a variety of genes that confer specific functional properties on cells (Schindler and Darnell, 1995). JAK–STAT signalling pathways are activated by IL-2–7, IL-9–13, IL-15, IFN-α/β, IFN-γ, LIF, OSM, G-CSF, GM-CSF, PDGF and EGF, and the existence of multiple STATs involved in various cytokine signalling pathways may explain the pleiotropic effects of many cytokines (Ihle, 1996). Constitutive activation of the JAK–STAT signalling pathway has been observed in cells transformed by human T-lymphotropic virus-1 (HTLV-1) (Migone *et al.*, 1995), v-ab1 (Danial *et al.*, 1995) and v-src (Yu *et al.*, 1995), and the dysregulation of STAT-responsive genes may play a role in the pathogenesis of certain types of lymphoid malignancies, such as B-cell lymphomas (Dent *et al.*, 1997). Thus, interference with the JAK–STAT signalling pathway may be useful in the treatment of certain malignancies.

Although the key events in cytokine receptor signal transduction are well documented (Ihle, 1995), the mechanism by which cytokine signal transduction is 'switched off' has been less well explored. However, recent reports reveal that a family of cytokine-inducible, SH2 domain-containing proteins called 'suppressors of cytokine signalling' (SOCS), which inhibit JAK activity and suppress signal transduction processes in a classical negative feedback loop, can control the intensity and duration of the cellular responses to cytokines and other external stimuli (Starr *et al.*, 1997). Other inducible SOCS include the JAK-binding protein, JAB (Endo *et al.*, 1997) and the STAT-induced STAT inhibitor, SSI-1 (Naka *et al.*, 1997).

The stress-activated protein kinases (SAPKs or JNKs) are stimulated by inflammatory cytokines (Minden *et al.*, 1994; Pombo *et al.*, 1994; Sluss *et al.*, 1994; Westwick *et al.*, 1995) and are also believed to be part of an osmosensing signal transduction pathway involved in cellular responses to osmotic stress (Galcheva-Gargova *et al.*, 1994). Certain cytokine-suppressive anti-inflammatory drugs (CSAIDs) target the osmosensing signal transduction pathway and bind to CSAID binding proteins (CSBPs) which are involved in cytokine production, leading to inhibition in the production of proinflammatory cytokines such as IL-1 and TNF-α (Lee *et al.*, 1994).

The MAP kinases (MAPKs), including p42–p44 MAPKs, p38 MAPK and p46–p54 JNKs, are a large family of serine–threonine-specific protein kinases which channel extracellular signals into specific cellular responses (Neiman, 1993). Each MAPK is specifically activated through phosphorylation by upstream kinases (MKK1 and MKK2 for p42–p44 MAPKs; MKK3 for p38 MAPK; MKK4 for JNK) (Sanchez *et al.*, 1994; Derijard *et al.*, 1995). The p42–p44 MAPKs mediate cell proliferation in response to growth factors (Cowley *et al.*, 1994; Mansour *et al.*, 1994) whereas p38 MAPK and JNKs mediate signals in response to cytokines and environmental stress (Freshney *et al.*, 1994; Han *et al.*, 1994; Lee *et al.*, 1994). The MAP kinase pathway, involving the oncoproteins ras and Raf-1, as well as MAPKKs and ribosomal S6 kinase (RSK), transduces mitogenic signals from the RTKs to the nucleus during growth factor stimulation of cells; the MAPKs translocate to the nucleus during mitogen stimulation and, by phosphorylating the appropriate transcription factors, they stimulate the expression of a spectrum of cell growth-specific genes (Huang and Erickson, 1996).

The cell division cycle contains a series of transition points, and progression through the cell cycle is regulated by a family of serine–threonine complexes composed of cyclins and cyclin-dependent kinases (cdks) (Sherr, 1995). The retinoblastoma tumour suppressor protein (Rb) is a regulator of G1 exit (Weinberg, 1995) and is an essential mediator that links ras-dependent mitogenic signalling to cell cycle regulation (Peeper *et al.*, 1997). Cyclin–cdk complexes phosphorylate Rb, and the phosphorylated Rb releases the transcription factor E2F which induces the expression of genes required for both G1 to S transition and DNA synthesis (Miyazono, 1996). Through inhibitory actions on cyclin–cdk complexes, TGF-β prevents cell cycle progression at late G1 through the increased production of the cdk 4/6 by increased tyrosine phosphorylation through repressed expression of the cdk tyrosine phosphatase, cdc25A (Iavarone and Massague, 1997).

Signal transduction pathways regulate cell metabolism, proliferation and differentiation, and dysregulation of these pathways can lead to the development of diseases resulting from aberrant signal transduction including the proliferative diseases (psoriasis, atherosclerosis), inflammatory diseases (rheumatoid arthritis) and malignant diseases (Saltiel, 1994, 1995). Dysregulation of cell growth occurs when structural alterations occur in the regulatory enzymes and signal transducing proteins through genetic mutations, and many of the targets for antiproliferative therapeutic intervention have emerged from studies of signal transduction pathways (Saltiel, 1994, 1995).

Many intracellular signalling pathways are regulated by reversible phosphorylation and dephosphorylation of substrates mediated by the concerted actions of protein kinases and protein phosphatases respectively; the protein phosphatases include phosphotyrosine phosphatases (PTPs) which occur as both receptor-like transmembrane proteins, with intracellular segments containing one or two phosphatase domains, and as cytoplasmic phosphatases (Tonks, 1996). PTPs modulate signalling processes through dephosphorylation of phosphotyrosines in both cytoplasmic signalling molecules and within multiprotein RTK-associated signalling complexes (Tonks, 1996). Members of the PTP family, such as SHP-2, bind to signal regulatory proteins (SIRPs) (Stein-Gerlach *et al.*, 1995) and, as negative or positive regulators of receptor-mediated signalling involving both RTKs and receptors of cytokine receptor superfamilies I and II (Yi *et al.*, 1993; Saltiel, 1994; Ihle, 1996; Matozaki and Kasuga, 1996; Streuli, 1996; Kharitonenkov *et al.*, 1997), can regulate signals defining different physiological processes (Zolnierowicz and Hemmings, 1996). The protein serine–threonine phosphatases (PSTPs) exist as the cytoplasmic holoenzymes PP1, PP2A, PP2B and PP2C, each containing catalytic subunits which associate with non-catalytic regulatory subunits (Zolnierowicz and Hemmings, 1996). PSTPs may act as cell growth inhibitors, as enhanced cell growth following transformation of cells with oncogenic DNA viruses involves the deregulation of PSTP function (Zolnierowicz and Hemmings, 1996).

CYTOKINES AND GENE EXPRESSION

Most genes contain promoters or enhancers within their regulatory elements which contain binding sites for multiple transcription factors including the B Zip proteins (containing basic (B) and leucine zipper (Zip) domains), the rel family of oncoproteins, and the c-myc oncoprotein (Karin and Smeal, 1992; Kato and Dang, 1992; Rushlow and Warrior, 1992; Crabtree and Clipstone, 1994; Kishimoto *et al.*, 1994; Woodgett, 1996). Because of their importance in modulating the expression of cytokine genes, transcription factors are targets for therapeutic intervention in the treatment of a range of immune and inflammatory disorders (Peterson and Baichwal, 1993).

The transcription factor, nuclear factor (NF)-κB, consists of homodimers or heterodimers of proteins belonging to the rel family (Verma *et al.*, 1995; Baldwin, 1996). In mammalian cells, the rel proteins include the c-rel, p105 (p50), p100 (p52), p65 and rel B (Baldwin, 1996). NF-κB/rel transcription factors are present in most cell types as inactive cytosolic complexes of NF-κB and the inhibitory proteins called IκBs. Signals from mitogens, lipopolysaccharide and viruses/viral proteins target the NF-κB–IκB complexes leading to phosphorylation, ubiquitation and degradation of the IκB proteins, thereby releasing NF-κB which translocates to the nucleus and activates gene transcription (Verma *et al.*, 1995; Baldwin, 1996). Protein kinase A (PKA)-mediated phosphorylation of the NF-κB p65 subunit is involved in the constitutive and inducible activation of NF-κB (Verma *et al.*, 1995; Zhong *et al.*, 1997). Reactive oxygen intermediates (ROIs) are also involved in signal transduction pathways leading to NF-κB activation (Schreck *et al.*, 1991).

Many cytokines, including IL-1, TNF-α, LT-α and LT-β, signal through their respective receptors to activate NF-κB, and activated NF-κB is critical for the inducible expression of the genes for IL-1, IL-2, IL-6, IL-8, TNF-α, TNF-β, IFN-β, GM-CSF and G-CSF (Kopp and Ghosh, 1995; Baldwin, 1996). Thus, NF-κB plays a key role in immune and inflammatory responses (Baeuerle and Baltimore, 1996) and provides a target in the therapy of chronic inflammatory diseases such as asthma (Barnes and Adcock, 1997). The TNFR-associated factors (TRAFs) are a family of signal transducing proteins (Rothe *et al.*, 1994) which are involved in the activation of NF-κB by members of the TNFR superfamily (Hsu *et al.*, 1996; Takeuchi *et al.*, 1996) and IL-1R (Cao *et al.*, 1996). TNF-α activates NF-κB exclusively through TNFR1 via interaction with TRAF2 (Rothe *et al.*, 1995), and the signal from TRAF2 results in phosphorylation of IκB causing the release of NF-κB which translocates to the nucleus to activate genes carrying NF-κB response elements (Nagata, 1997).

Many cells undergo 'programmed cell death', or apoptosis, following interaction of selected ligands with cell surface receptors, such as the Fas ligand (FasL) with the Fas/APO-1 receptor (Thornberry *et al.*, 1992; Enari *et al.*, 1995) and TFN-α with TNFR-1 (Smith *et al.*, 1994; Hsu *et al.*, 1996). In both instances, apoptosis is mediated by signal cascades culminating in the activation of ICE-related proteases (IRPs). TNFR-1 contains a *C*-terminal TNFR-associated death domain (TRADD) which interacts with TRAFs, and TNF-α activates NF-κB exclusively through the interaction of TRAF2 with TRNF-1 (Rothe *et al.*, 1995). The signal from TRAF2 results in the phosphorylation of IκB, and the active NF-κB translocates to the nucleus and binds to NF-κB response elements within various 'suicide' genes (Nagata, 1997). TNFR-1 activation also induces apoptosis via activation of the JNK/SAPK cascade (Cuvillier *et al.*, 1996). Interestingly, however, other reports (Beg and Baltimore, 1996; Wang *et al.*, 1996) indicate that NF-κB may have a role in cell survival and prevent apoptosis. In this context, NF-κB could be an attractive target for therapeutic intervention against cellular proliferative diseases such as arthritis and cancer.

The transcription factor, nuclear factor-IL6 (NF-IL6), belongs to the basic leucine zipper family of transcription factors, and binds to the IL-1-responsive element (IRE) in the human IL-6 gene; NF-IL6 is an inducible transcriptional activator and is activated by phosphorylation of its tyrosines by the ras-dependent MAP kinase cascade. In addition to the NF-IL6 binding site, the promoter region of the IL-6 gene also contains an NF-κB binding site, and both the NF-IL6 and NF-κB binding sites are essential for IL-6 gene expression (Matsusaka *et al.*, 1993). NF-IL6 and NF-κB are also essential for IL-8 gene induction by IL-1, TNF-α and phorbol myristate acetate (PMA) (Mukaida *et al.*, 1990), and the cooperative interactions between NF-IL6 and NF-κB result in synergistic activation of many genes (Kishimoto *et al.*, 1994).

cAMP response element binding protein (CREB) is a bZip transcription factor for genes that are responsive to cAMP and mitogens (Arias *et al.*, 1994; Kwok *et al.*, 1994; Sassone and Corsi, 1995). CREB binding protein (CBP) and its homologue p300 are proteins with intrinsic histone acetyl transferase (HAT) activity (Ogryzko *et al.*, 1996) and are essential for signal-dependent activation of transcription through association with regulated transcription factors such as CREB (Arias *et al.*, 1994; Kwok *et al.*, 1994), activator protein-1 (AP-1) (Arias *et al.*, 1994) and STATs (Horvai *et al.*, 1997). CBP and p300 are recruited to signal-dependent promoters and stimulate gene expression both through their association with RNA polymerase II complexes (Pol II) and their HAT activities (Bannister and Kouzarides, 1996; Ogryzko *et al.*, 1996). There is a critical STAT interaction domain within the first 100 amino acids of CBP (Horvai *et al.*, 1997), and peptides corresponding to the *N*-terminal regions of CBP can markedly inhibit IFN-γ-dependent gene activation (Torchia *et al.*, 1997).

The AP-1 transcription factor consists of either homodimers of the jun oncoprotein, or heterodimers of jun and fos oncoproteins which bind to the palindromic TRE sequence TGA (C/G) TCA (Woodgett, 1996). The Jun gene family members include c-jun, jun B and jun D, while the Fos gene family members include c-fos, fos B, Fra-1 and Fra-2. The leucine zipper domains mediate dimerization of nuclear factors such as jun, fos and CREB through a heptad repeat of leucine residues (Landschultz *et al.*, 1988; Hai and Curran, 1991).

The phospholipases A_2, C and D degrade membrane phospholipids during the generation of the derivatives arachidonate, lysophospholipids, inositol (1,4,5) triphosphate (Ins(1,4,5)P_3), DAG and phosphatidic acid, some of which function as second messengers. Ins(1,4,5)P_3 releases Ca^{2+} from intracellular stores and promotes capacitative Ca^{2+} entry across the plasma membrane to provide the intracellular Ca^{2+} signals required to initiate gene transcription (Clapham, 1995, 1997). Ca^{2+} activates the phosphatase calcineurin leading to dephosphorylation of the transcription factor NF-ATc which then translocates to the nucleus, couples with NF-ATn and promotes gene transcription (Crabtree and Clipstone, 1994). DAG not only induces activation of members of the Bzip superfamily of signal-regulated transcription factors (Schutze *et al.*, 1992), but also activates the Ca^{2+} and phospholipid-dependent protein serine–threonine kinase C (PKC); there are 11 isotypes of PKC, each of which catalyses differential phosphorylation of intracellular proteins. Activation of PKC results in phosphorylation of the nuclear Bzip transcription factor AP-1 that binds to the *cis* promoter elements of a number of genes. PKC-ζ has been implicated in the pathway of IκB phosphorylation, and hence NF-κB activation (Diazmeco *et al.*, 1994).

The myc family of proto-oncogenes belongs to the basic helix–loop–helix leucine zipper (bHLHZ) class of transcription factors; myc oncoproteins dimerize with the bHLHZ protein Max and form transcription complexes which influence cell growth and proliferation through direct activation of genes involved in DNA and RNA synthesis as well as cell cycle progression (Grandori and Eisenman, 1997). Activating mutations of myc oncoproteins lead to autonomous proliferation and are involved in the development of human cancer (Kato and Dang, 1992).

THE ACTIONS OF CYTOKINES WITHIN TISSUES AND BODY FLUIDS

In the complex milieu of tissues, the actions of cytokines will be affected by macromolecules present within the microenvironment, which consists of a solid phase (the extracellular matrix or ECM) and a plasma-derived tissue fluid phase. Certain cytokines engage in the process of 'crinopexy' and become bound to, and stabilized within, the ECM; for instance, latent TGF-β and its binding protein (LTBP-1) bind to ECM microfibrils (Taipale *et al.*, 1996). Proteoglycans, including syndecans, perlecans and glypicans (David, 1993; Iozzo, 1994), can bind specific growth factors and cytokines, thereby creating local reservoirs within tissue matrices by sequestering these potent signalling molecules within specific tissue locations (Witt and Lander, 1994). Proteoglycans and glycosaminoglycans not only present matrix-bound cytokines to cells in biologically active forms (Gilat *et al.*, 1996; Jackson, 1997) but, as an integral component of cell surface glycocalyces, can also facilitate cytokine–receptor interactions and thus assist cytokine-mediated signal transduction processes within target cells (Raghow, 1994; Friesel and Maciag, 1995). For instance, binding of chemokines to proteoglycans and glycosaminoglycans both on cell surfaces and within extracellular matrices is an important event in the maintenance of chemokine gradients required for leukocyte activation, transendothelial diapedesis and extravasation into tissues (Tanaka *et al.*, 1993; Schall and Bacon, 1994).

The ECM is a complex and dynamic reservoir of cytokines, and active cytokines can be released enzymatically from extracellular matrices by proteolysis of the proteoglycan core proteins or by partial degradation of the glycosaminoglycan (usually heparan sulphate) component (Ruoslahti and Yamaguchi, 1991). The matrix metalloproteinases (MMPs) are a family of enzymes, including collagenases, gelatinases and stromelysins which can release matrix-bound cytokines by proteolytic degradation of extracellular matrices (Matrisian, 1992) and whose expression is normally tightly regulated at the transcriptional level by cytokines (Ries and Petrides, 1995). The liberation of cytokines, such as FGF-2 and TGF-β, from the ECM plays an important role in angiogenesis and the repair of wounded tissue (Vlodavsky *et al.*, 1991; Falcone *et al.*, 1993).

The plasma-derived tissue fluid contains a number of cytokine binding proteins which can modulate the activities of cytokines *in vivo*, and these include soluble cytokine receptors (Fernandez-Botran *et al.*, 1996), IgG (Edgington, 1993), autoantibodies (Bendtzen *et al.*, 1990), proteins such as α_2-macroglobulin (James, 1990) and naturally occurring cytokine antagonists. In addition to soluble type II IL-1 receptors (Colotta *et al.*, 1993; Symons *et al.*, 1995), naturally occurring antagonists for IL-1 exist (Arend, 1993; Dinarello, 1994). The cDNA for the receptor antagonist IL-1ra has been cloned (Eisenberg *et al.*, 1990; Dinarello, 1994) and the protein shows some amino acid homology with IL-1α and IL-1β (Dinarello, 1994); it binds to IL-IR1 and IL-IR2 without initiating signal transduction (Dripps *et al.*, 1991) and can block the actions of both IL-1α and IL-1β while not appearing to compromise host immune responsiveness (Dinarello and Thomson, 1991; Arend, 1993). The increased production of IL-1ra and its increased titre within the circulation appears to be a natural host response in certain clinical conditions, such as endotoxaemia, sepsis and rheumatoid arthritis, and reduces the severity of these diseases (Dinarello and Thomson, 1991; Arend, 1993; Dinarello, 1994).

CYTOKINES IN INFLAMMATION AND IMMUNITY

The multifunctionality of cytokines is amply illustrated in the molecular and cellular events associated with inflammation, angiogenesis and wound repair, and the host immune response. The inflammatory response is a host defence reaction which mobilizes leukocytes to tissues traumatized either by invasion of potentially pathogenic micro-organisms (viruses, bacteria, fungi) and parasites, or as a consequence of hypersensitivity reactions or invasion by malignant neoplasms, or alternatively to tissues damaged by chemicals or physical trauma (Wong and Wahl, 1990; Whaley and Burt, 1992; Fresno *et al.*, 1997). Within developing inflammatory lesions, blood plasma, containing fibrinogen and fibronectin, extravasates and clots extravascularly to form a hydrated fibrin–fibronectin gel. This gel serves as a provisional matrix which is subsequently invaded by inflammatory leukocytes (polymorphonuclear neutrophil (PMN) leukocytes, monocytes and lymphocytes), fibroblasts and capillary endothelial cells; these events are regulated by cytokines released by both platelets and inflammatory leukocytes (Nathan, 1990; Wong and Wahl, 1990; Baumann and Gauldie, 1994) (Fig. 2).

Infection and injury are associated with increases in the numbers of circulating PMN leukocytes and monocytes in response to CSFs generated within inflammatory lesions (Wong and Wahl, 1990). Both the PMN leucocytes (Lloyd and Oppenheim, 1992) and the monocytes–macrophages (Nathan, 1990; Baumann and Gauldie, 1994) release cytokines that coordinate and regulate inflammatory and immune responses within inflammatory lesions (Nathan, 1990; Wong and Wahl, 1990; Adams and Hamilton, 1992; Fresno *et al.*, 1997). PMN leucocytes and monocytes are recruited into lesions where they debride the lesion of infectious agents by their phagocytic activities, and where monocyte-derived macrophages participate in tissue repair and remodelling by releasing cytokines that induce fibroblast chemotaxis, activation and proliferation (e.g. IL-1α/β, TGF-β) (Nathan, 1990; Adams and Hamilton, 1992) and which are mediators of general inflammation. These subsequently stimulate the production of IL-1α/β, IL-6, IL-8, M-CSF, IFN-α and IFN-β (Wong and Wahl, 1990). These cytokines are produced early in the response to injury and infection, and not only express expert pro-inflammatory activities (Wong and Wahl, 1990; van Deuren *et al.*, 1992), but if their production becomes dysregulated, can also be involved in the pathogenesis of inflammatory diseases (Baggiolini *et al.*, 1995; Strieter *et al.*, 1995; Dinarello, 1996). IL-1α/β, IL-6 and TNF-α induce the production of haemopoietic colony stimulating factors by fibroblasts, endothelial cells and bone marrow stromal cells (Wong and Wahl, 1990). Induction of IL-6 expression is part of a general response within tissues traumatized by inflammation, and results in the increased production within the liver of many acute-phase reactant proteins (APRPs) (e.g. C-reactive protein, α_1-antitrypsin, haptoglobin,

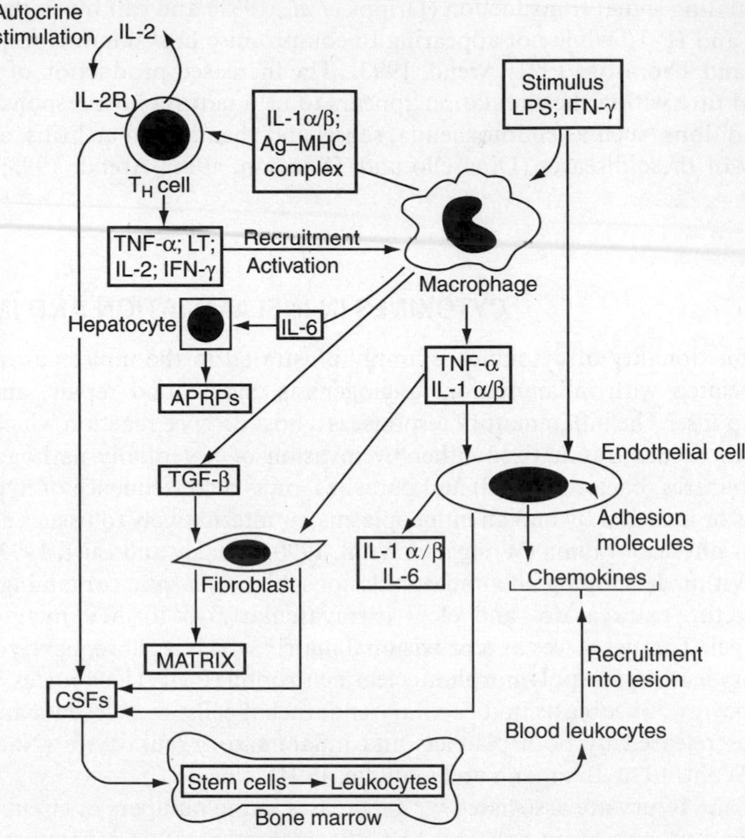

Fig. 2. Diagrammatic representation of the interactions between endothelial cells, fibroblasts, T-helper lymphocytes and cells of the monocyte–macrophage lineage in the generation of cytokines, especially colony stimulating factors (CSFs), and acute-phase reactant proteins (APRPs) during the host response to an inflammatory stimulus. The cytokines involved in these interactions, particularly the effects of cytokines (IL-1α/β and TNF-α) on vascular endothelial cells which promote transendothelial leukocyte migration into an inflammatory lesion, as well as the effects of CSFs in promoting leukopoiesis within the bone marrow during host inflammatory reactions, are indicated.

serum amyloid A protein and fibrogen) as well as fibronectin (Beutler, 1990; Hagiwara *et al.*, 1990; Akira *et al.*, 1993).

Cytokines are involved in the response to acute septic shock and can contribute to the cascade of events leading to death (Bone, 1991). Acute sepsis (endotoxaemia) provokes a local inflammatory response with coordinated and sequential activation of a series of pro-inflammatory cytokines (van Deveter *et al.*, 1990) that impinge on the hypothalmic–pituitary–adrenal (HPA) axis (Chrousos, 1995). As well as the production of pro-inflammatory factors that potentiate endotoxaemia (Ray and Melmed, 1997), the pituitary response to toxic shock involves the production of inflammation suppressing agents which activate the HPA axis with consequent increased adrenal glucocorticoid production (Chrousos, 1995). Glucocorticoids have long been considered the most effective anti-inflammatory agents through their suppressive action on AP-1 and NF-κB activity involved in the production of pro-inflammatory cytokines (Auphan *et al.*, 1995; Mori *et al.*, 1997).

Both macrophages and endothelial cells produce cytokines that orchestrate many facets of inflammatory reactions. Endothelial cells are active participants in the pathogenesis of infectious diseases and are both targets and sources of inflammatory cytokines (Pober and Cotran, 1990; Mantovani *et al.*, 1997). Leukocyte traffic into inflammatory lesions involves a complex relationship between the generation of chemokines (and other cytokines) and the upregulated expression of endothelial leukocyte adhesion molecules (selectins, integrins and immunoglobulin-related molecules) which regulate leukocyte–endothelial cell interactions and the transendothelial migration of leukocytes into inflammatory lesions (Pober and Cotran, 1990; Albelda *et al.*, 1994; Mantovani *et al.*, 1997). The differential trafficking and selective recruitment of T_{H1} and T_{H2} cells into inflammatory lesions is mediated by P- and E-selectins expressed on endothelial cells (Austrup *et al.*, 1997).

The host immune response within an inflammatory lesion involves interactions between T and B cells, macrophages and other effector cells, and is determined, in part, by the pattern of cytokines produced by different subsets of T helper cells (Romagnani, 1995, 1996; Fresno *et al.*, 1997). In addition to their function as generators of inflammatory mediators and enzymes, mast cells function in immune responses as both antigen-presenting cells (APCs) and as important sources of cytokines, which are not only pro-inflammatory but also immunoregulatory, and can influence the development of T- and B-cell responses (Mecheri and David, 1997).

The type of antigenic specific immune response that follows antigen administration depends upon the preferential activation of $CD4^+$ T helper (T_H) cells which are functionally heterogeneous and consist of precursor cells (T_{HP}), naive cells (T_{H0}), T_{H1} and T_{H2} cells (Romagnani, 1994, 1995, 1996, 1997; Mosmann and Sad, 1996). The basic effector T_{H1} and T_{H2} cells are both derived from a common pool of precursor cells (T_{HP}) which differentiate into naive T_{H0} cells. The differentiation of T_{H0} cells into T_{H1} or T_{H2} cells is a multistep process in which the naive cells pass through an intermediate stage (Fig. 3); the final differentiation pathway followed is determined by the environment in which the T_H cells react to antigen stimulation, and the most important cytokines that influence T_{H1} and T_{H2} differentiation are IL-12 and IL-4 respectively (Romagnani, 1994, 1995, 1996, 1997; Trinchieri, 1995; Abbas *et al.*, 1996). In the absence of polarizing signals the T_{H0} cells represent a heterogeneous cell population, but under the influence of strong microenvironmental signals provided by APCs, the cells within the T_{H0} population differentiate into subsets of polarized T_{H1} or T_{H2} cells able to secrete defined patterns of cytokines and to exhibit distinct functional properties (del Prete *et al.*, 1994; Mosmann and Sad, 1996). T_{H1} cells activate pro-inflammatory effector mechanisms, whereas T_{H2} cells induce humoral and allergic responses and downregulate local inflammation (Mosmann and Sad, 1996). Murine T_{H1} cells produce IL-2, TNF-β (LT) and IFN-γ, whereas T_{H2} cells produce IL-4, IL-5, IL-6, IL-9, IL-10 and IL-13 (Seder and Paul, 1994; Romagnani, 1995, 1996, 1997) (Fig. 4). Human T_{H1} and T_{H2} cells exhibit similar cytokine profiles to their murine counterparts, although compared with murine T_H cells the synthesis of IL-2, IL-6, IL-10 and IL-13 does not appear to be so tightly restricted to a single subset of T_H cells.

The differentiation pathways of the T_H cells are selectively regulated by cytokines produced by APCs including dendritic cells (DCs), macrophages and B cells (Mosmann

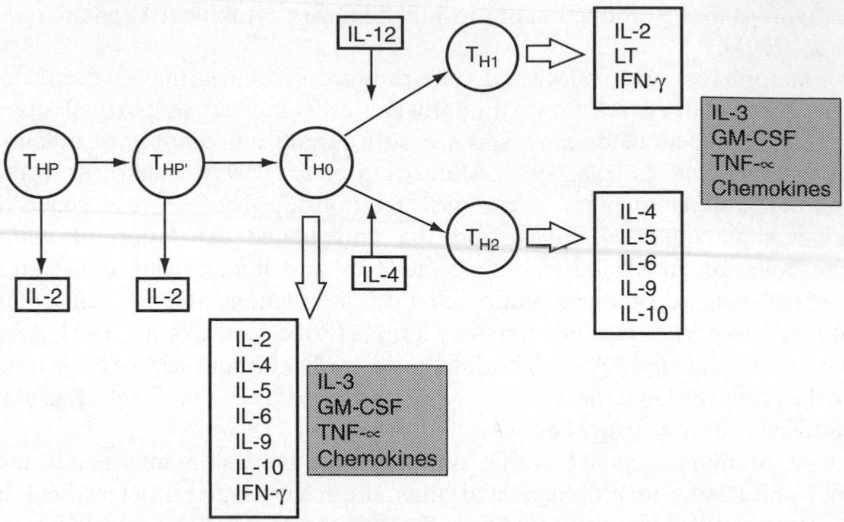

Fig. 3. Schematic representation of the differentiation of T_{H1} and T_{H2} cells from naive precursors (T_{HP} and T_{HP}') indicating the important roles of IL-12 and IL-4 in T_{H1} and T_{H2} cell development, respectively. The cytokines produced by the T_H cells at the different stages in their development are indicated, and those produced in common by both the T_{H0} as well as the T_{H1} and T_{H2} cells are shown in shaded boxes.

and Sad, 1996). Whereas the differentiation of T_{H0} cells into T_{H2} cells is dependent on IL-4 signalling through STAT6 (Kopf *et al.*, 1993; Shimoda *et al.*, 1996; Takeda *et al.*, 1996), T_{H1} is an IL-12-dependent event involving the interaction of the CD40 ligand on T cells with CD40 on DCs which results in the production of IL-12 (Adorini and Sinigaglia, 1997). IL-12 induces phosphorylation of STAT4 in developing T_{H1} cells but not in T_{H2} cells (Bacon *et al.*, 1995; Jacobson *et al.*, 1995; Szabo *et al.*, 1995), whereas IL-4 activates STAT6 in T_{H2} cells (Hou *et al.*, 1994; Schindler *et al.*, 1994). T_H cell subset development is dependent on activation of these STAT proteins, and STAT4 or STAT6 deficient mice are unable to generate T_{H1} and T_{H2} responses (Kaplan *et al.*, 1996; Thierfelder *et al.*, 1996). DCs are derived from $CD34^+$ precursors (colony forming unit (CFU)-DC) in the bone marrow which, under the influence of appropriate stimulation and specific cytokines, can differentiate into professional APCs (Steinman, 1991); these cells can influence the intensity and direction of immune responses and play a critical role in transplantation immunity (Rao *et al.*, 1996). Using the C-C chemokine DC-CK1, DCs preferentially attract naive T_H cells (Adema *et al.*, 1997) and then initiate immune responses by presenting antigens to the T_H cells (Marland *et al.*, 1997). The two ligands B7.1 (CD80) and B7.2 (CD86) are expressed on APCs, and bind CD28 and cytotoxic T lymphocyte-associated antigen-4 (CTLA-4) on T cells. Whereas CD28 is expressed constitutively on T cells, CTLA-4 is induced following T cell activation. These costimulatory molecules play very important roles in the differentiation and activation of T cells and can influence the profile of cytokines produced by T_H cells (Gause *et al.*, 1997); DC progenitors deficient in B7.1 or B7.2 can suppress T cell-mediated responses (Lu *et al.*, 1995; Fu *et al.*, 1996).

Whereas inflammatory immune ('delayed-type hypersensitivity' (DTH)) responses are mediated by T_{H1} cells, T_{H2} cells mediate humoral immune responses (Romagnani, 1995,

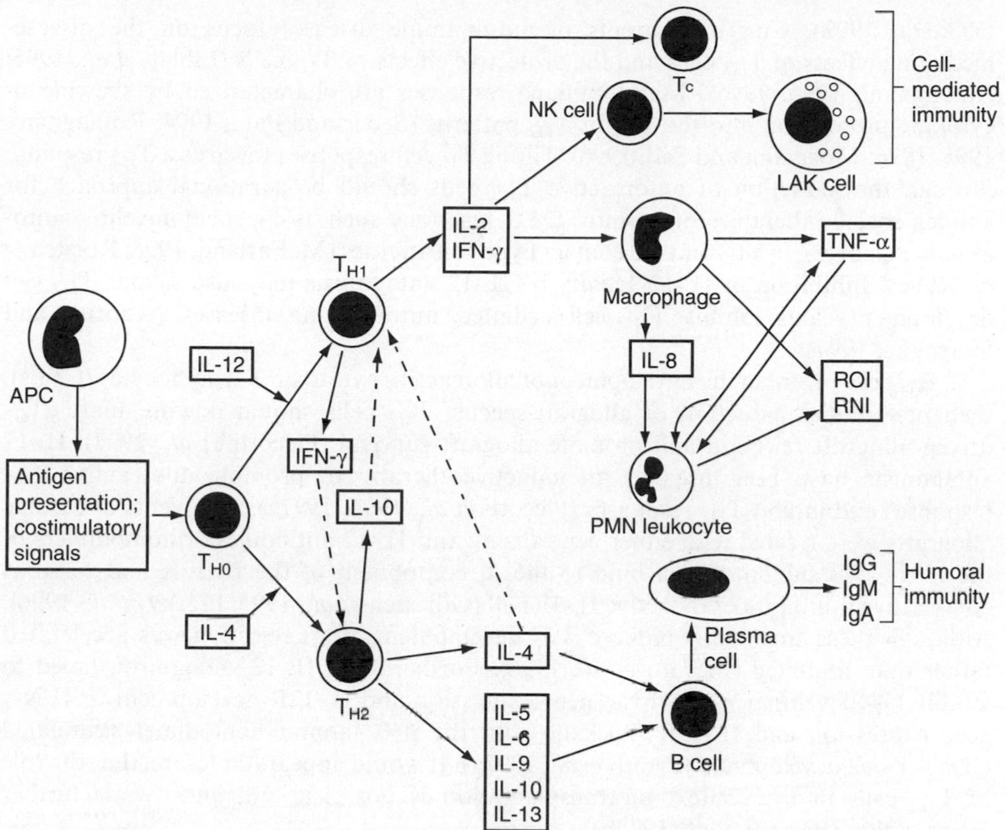

Fig. 4. Schematic representation of the roles of T_{H1} and T_{H2} cells in orchestrating events in cell-mediated and humoral immune responses, respectively, through the release of cytokines. Through the release of IL-2 and IFN-γ, T_{H1} cells activate Tc cells and NK cells to become effector cells, whereas macrophages and PMN leukocytes are activated to release TNF-α, ROI and RNI, which participate in the killing of neoplastic cells, microorganisms and parasites. Through the release of cytokines (IL-4, IL-5, IL-6, IL-10), T_{H2} cells promote the differentiation of B cells into antibody-secreting plasma cells. Both T_{H1} and T_{H2} cells regulate each other's functions through the suppressive effects (- - -) of IFN-γ and IL-4/IL-10 respectively.

1996; Mosmann and Sad, 1996). Both T_{H1} and T_{H2} cells appear able to regulate each other's functions (Mosmann and Sad, 1996). The fact that IL-4 and IL-10 produced by the T_{H2} cells inhibit the activities of T_{H1} cells, whereas IFN-γ produced by T_{H1} cells suppresses the activities of T_{H2} cells, provides an explanation for the inverse relationship observed between cell-mediated and humoral immune responses (Bretscher *et al.*, 1992). However, T_{H1} and T_{H2} cells represent opposite ends of the T_H cell spectrum and it is an oversimplification to state that these cells have totally distinct patterns of cytokine secretion since both T_{H1} and T_{H2} cells produce IL-3, GM-CSF and TNF-α (Mosmann and Sad, 1996). Through the production of IL-3, IL-4, IL-5 and GM-CSF, T_{H2} cells are involved in the pathogenesis of allergic inflammatory reactions (Ricci *et al.*, 1993). The action of IL-10 as a powerful inhibitor of the secretion of the proinflammatory cytokines IL-1α/β, IL-6, IL-8 and TNF-α, as well as being a mediator of IL-1ra expression, makes IL-10 an important regulator of inflammatory and immune responses (Howard and

O'Garra, 1992). Current concepts of autoimmune diseases focus on the disease-mediating effects of T_{H1} cells and the protective effects of T_{H2} cells (Liblau *et al.*, 1995; Romagnani, 1995, 1996). Many immune responses are characterized by skewing of cytokine production into the T_{H1} or T_{H2} patterns (Seder and Paul, 1994; Romagnani, 1995, 1996; Mosmann and Sad, 1996). Tilting T_H cell responses towards a T_{H2} response through the induction of autoreactive T_{H2} cells should be a rational approach for antigen-specific therapies of harmful DTH reactions such as encephalomyelitis, autoimmune diabetes, uveitis and the contact hypersensitivities (McFarland, 1996; Rocken *et al.*, 1996). Inhibition of IL-12 activity by IL-12 antagonists may also favour T_{H2} cell development, and inhibit T_{H1} cell-mediated autoimmune diseases (Adorini and Sinigaglia, 1997).

If T_{H1} cells promote the development of alloreactive cytotoxic T lymphocytes (CTLs), then preferential induction of allograft-specific T_{H2} cells should downregulate T_{H1}-driven allograft rejection and promote allograft survival (Piccotti *et al.*, 1997). IL-12 antagonists have been used as an inductive therapy to promote alloreactive T_{H2} responses and inhibit T_{H1} responses (Piccotti *et al.*, 1996, 1997a). Recipients of cardiac allografts were treated with either neutralizing anti-IL-12 antibodies or homodimers of the IL-12 p40 subunit which bind to the β_1 component of the IL-12R and serve as competitive inhibitors of bioactive IL-12 p70 (Gillessen *et al.*, 1995; Presky *et al.*, 1996). Although these treatments induced T_{H2} development, graft rejection was accelerated rather than inhibited (Piccotti *et al.*, 1996). Furthermore, IL-12 antagonism failed to inhibit T_{H1} development or IFN-γ gene expression and IL-12R development or IFN-γ gene expression and IL-12R blockade with the p40 subunit homodimer stimulated $CD8^+$ T-cell development (Picotti *et al.*, 1997a). It would appear, therefore, that the role of T_{H2} cells in the context of transplantation is not clear-cut, and awaits further investigation (Piccotti *et al.*, 1997b).

Macrophages play an important role in the afferent and efferent arms of host immune responses through their actions as antigen processing, presenting and cytokine secreting cells (Steinman, 1991; Stein and Keshav, 1992), as well as their capacity to act as effector cells in host defence against infections with viruses (Gendelman and Morahan, 1992), bacteria (Speert, 1992) and parasites (Sadick, 1992), and against malignancy (Nathan, 1990). The effector functions of macrophages are modulated by cytokines produced by T_H cells (Nathan, 1990; Adams and Hamilton, 1992), and macrophages activated by cytokines such as IFN-γ play a major role in host defence against infection and cancer through the production of ROI (e.g. superoxide anions, OH^- and H_2O_2), reactive nitrogen intermediates (RNIs) (e.g. nitric oxide, nitrite and nitrate) and TNF-α (Nathan, 1990, 1992; Adams and Hamilton, 1992). ROIs play an important role in the bactericidal activities of macrophages (Speert, 1992). Nitric oxide (NO) is a highly reactive product of activated macrophages and is derived by the oxidative cleavage of guanidino nitrogens from L-arginine leaving L-citrulline as a coproduct of L-arginine metabolism (Nathan, 1992). The reaction is catalysed by the enzyme NO synthase, of which there are two forms, the constitutive form and the inducible form. The cytokines TNF-α, IL-1 and IFN-γ upregulate the expression of inducible NO synthase (iNOS) which is a Ca^{2+} independent enzyme requiring tetrahydrobiopterin as a cofactor. Induction of iNOS by cytokines contributes to the enhanced cytotoxicity of macrophages towards tumour cells, bacteria, fungi and protozoa, and also provides a defensive mechanism against viral infection (Liew *et al.*, 1990; Nathan, 1992; Sadick, 1992).

The local paracrine effects of macrophage-derived TNF-α are often beneficial to the host (Cerami, 1992) and include actions that directly influence the host's immune response, such as inducing effector cell activation, cytokine induction and receptor modulation, and degranulation of cytotoxic effector cells (Cerami, 1992). TNF-α also affords a protective effect against neoplastic cells, parasites and facultative intracellular microorganisms (Liew *et al.*, 1990; Nathan, 1990; Pesanti, 1991; Sadick, 1992). A number of cytokines including IFN-β, IFN-γ, TNF-α and TGF-β exert an antiviral role (Ramsay *et al.*, 1993). IL-2 can induce an antiviral activity through the chemotactic recruitment and activation of NK cells. NK cells not only kill virus-infected cells, but also secrete IFN-γ and TNF-α. Following class I major histocompatibility complex (MHC)-restricted recognition of virus-infected cells, $CD8^+$ CTLs also release antiviral cytokines such as IFN-γ and TNF-α within virus-infected tissues, thereby mediating an antiviral effect (Ramsay *et al.*, 1993). In instances where there is overproduction of pro-inflammatory cytokines (e.g. severe infection leading to clinical disease), a systemic cytokine release syndrome occurs, in which there are deleterious effects of TNF-α on the lungs, liver and brain by induction of microvascular pathology through endothelial cell apoptosis (Lucas *et al.*, 1997). Pro-inflammatory cytokines also contribute to the cascade of events leading to multiple organ failure and fatality in cases of sepsis (Bone, 1991) and acute respiratory distress syndrome (Yokoi *et al.*, 1997).

Following neutralization of the inflammatory agent by PMN leucocytes and macrophages within an inflammatory lesion, the lesion resolves. The fibrin–fibronectin gel is degraded and replaced first by loose, vascularized connective tissue (called granulation tissue) and later by dense, collagenous, hypocellular and hypovascular scar tissue (Whaley and Burt, 1992). The formation of granulation tissue, which involves fibroblast proliferation (fibroplasia), the biosynthesis of matrix molecules by fibroblasts and endothelial cells, as well as angiogenesis, is regulated by cytokines released by platelets and inflammatory leukocytes (Wong and Wahl, 1990; Kovacs, 1991). If the inflammatory agent persists, the inflammatory lesion becomes chronic, does not heal, and the granulation tissue not only persists, but may hypertrophy.

TGF-β plays a dual role during the course of inflammatory reactions, being pro-inflammatory at first and immunosuppressive later (Wahl, 1992). Dysregulated production of TGF-β is usually associated with fibrotic and proliferative inflammatory diseases (Kilkarni and Karlsson, 1993; Sanderson *et al.*, 1995). For instance, a central pathological feature of glomerulonephritis is the excessive accumulation of ECM within the glomeruli due to overexpression of TGF-β (Okuda *et al.*, 1990). Administration of antibodies to TGF-β to glomerulonephritic rats suppresses glomerular matrix production and accumulation (Border *et al.*, 1990). Decorin has been shown to protect against scarring in experimental glomerulonephritis in rats (Border *et al.*, 1992) and may prove useful in the clinical treatment of diseases associated with overexpression of TGF-β in organs such as kidney, lung and liver.

CYTOKINES AND THE PATHOGENESIS OF DISEASES

Cytokines play a pivotal role in maintaining homeostatic mechanisms required for the well-being of the host, but any imbalance in the production and action of cytokines and/or their receptors and cellular response elements will disturb homeostatic processes and have pathological consequences (Duff, 1989; Meager, 1990). Thus, cytokines can be

Table 5. Diseases related to angiogenesis and excessive FGF activity.

Angiogenesis-related diseases	Diseases associated with excessive FGF activity
Diabetic retinopathy	Atherosclerosis
Neovascular glaucoma	Rheumatoid arthritis
Trachoma	Cirrhosis
Retrolental fibroplasia	Psoriasis
Psoriasis	Scleroderma
Pyogenic granuloma	Sarcoidosis
Burn granulations	Idiopathic pulmonary fibrosis
Hypertrophic scars	Tumour development
Delayed wound healing	
Atherosclerotic plaques	
Haemangioma	
Angiofibroma	
Tumour growth	
Arthritis	

viewed as double-edged swords, benefiting the host when their production and actions are regulated, but posing a threat to the host when their production and actions are unregulated (Baggiolini *et al.*, 1995; Strieter *et al.*, 1995; Dinarello, 1996). Such imbalances can have profound influences on acute and chronic inflammation and can contribute to the pathogenesis of of diseases such as those related to angiogenesis and excessive FGF activity (Table 5), as well as to the pathogenesis of autoimmune diseases such as rheumatoid arthritis, and the development of malignant tumours.

CYTOKINES AND RHEUMATOID ARTHRITIS

Within synovial joints the lining tissue, or synovium, consists of loose connective tissue with occasional blood vessels and scattered fibroblasts and macrophages; in rheumatoid arthritis (RA) the synovium becomes inflamed and hyperplastic, and the unrestrained proliferation of fibroblasts and angiogenesis leads to the formation of granulation tissue, whose degradative enzymes contribute to profound polyarticular destruction involving cartilage and bone within the joint (Harris, 1990; Ivashkiv, 1996). Histologically the rheumatoid synovium is characterized by synovial layer hyperplasia resulting primarily from accumulation of type A and type B synoviocytes, T cells, B cells, plasma cells, mast cells and neutrophils (Cush and Lipsky, 1991; Yanni *et al.*, 1992; Firestein, 1996), and interactions between these cells serve to perpetuate synovitis through complex cytokine networks involving autocrine and paracrine signalling.

The rheumatoid synovium and synovial fluid are enriched in dendritic cells with potent APC function, and the disease process is initiated and perpetuated by the responses of autoreactive $CD4^+$ T cells to peptides presented by the DCs (Thomas and Lipsky, 1996a,b; Kinne *et al.*, 1997). The rheumatoid synovium is the site of extensive production of the pro-inflammatory cytokines, IFN-α, -β, -γ, interleukins 1–3 and 6–9, IL-12, IL-15, GM-CSF, M-CSF, macrophage inflammatory protein-1α (MIP-1α), MIP-1β, OSM, PDGF, FGF-2, LIF, macrophage inhibitory factor (MIF), LT, TNF-α, VEGF, RANTES and growth-regulated oncogene-α (GRO-α), as well as the immunoregulatory cytokines IL-4, IL-10, IL-11, IL-13 and TGF-β (Firestein, 1996; Ivashkiv, 1996; Feldmann *et al.*, 1997). In RA, there is an excess of pro-inflammatory cytokines; this is not a static equilibrium, but rather a dynamic one with fluctuating levels of pro- and anti-inflammatory cytokines accounting for the flares and remissions in disease activity. It is the activated T cells that drive the production of pro-inflammatory cytokines by synoviocytes within the inflamed synovium (Ivashkiv, 1996; Kinne *et al.*, 1997) (Fig. 5). Specific regulation of STAT activity occurs during inflammatory responses, and STAT3 appears to be activated in synoviocytes during the arthritogenic process through the actions of IL-6 (Wand *et al.*, 1995).

Defining the cytokine networks that contribute to the perpetuation of RA has revealed the abundance of macrophage and fibroblast synoviocyte-derived cytokines which are sufficient to recruit inflammatory leukocytes and lymphocytes, and which then perpetuate the inflammatory process. The synovial macrophages are responsible for much of the IL-1, TNF-α, GM-CSF and chemokine production, whereas the fibroblasts produce IL-6, FGFs and VEGFs (Firestein, 1996). The fibroblasts also produce natural antiarthritogenic inhibitors, such as sTNF-R and IL-1ra, which serve to limit the severity of the disease (Firestein, 1996). Measurements of levels of soluble cytokine receptors, such as sTNF-R, in serum and synovial fluid are useful in monitoring disease progression within patients with rheumatic disease (Cope *et al.*, 1992). GM-CSF upregulates not only the expression of the relevant costimulatory molecules, such as the B7 family of molecules, critical for the effective presentation of antigens by APCs, but also the expression of MHC class II molecules, and there is an association between RA and certain human leukocyte antigen (HLA)-DR alleles (Winchester, 1992). TNF-α accelerates collagen-induced arthritis through a combination of stimulated production of prostaglandin E_2 (PGE_2) and matrix degrading proteinases, fibroblast proliferation and the production of the cytokines IL-6, IL-8, LIF, TGF-β and MCP-1 (Ito *et al.*, 1992; Williams *et al.*, 1994).

The synovial milieu contains the cytokines IL-1, IL-6, TNF-α, GM-CSF, IL-8 and IL-10 (Firestein, 1996; Ivashkiv, 1996) as well as free radicals, prostaglandins, autoantibodies and proteolytic enzymes (Zvaifler and Firestein, 1994). The hyperplastic synovium forms pannus tissue which invades the articular cartilage and bone through the actions of proteolytic enzymes produced not only by the type A and type B synoviocytes (Allard *et al.*, 1990; McCachren *et al.*, 1990) but also by mast cells and neutrophils (Zvaifler and Firestein, 1994). The proteolytic enzymes include matrix metalloproteinases (collagenases, stromelysins and gelatinases), the serine proteinases (plasminogen activator, proteinase-3, elastase and cathepsin G) and the cysteine proteinases (papain, cathepsins B, H, L and S), and there is increasing evidence that the chronicity of RA is related to an imbalance in the activities of both calpains (calcium-activated papain-like proteases) and their natural inhibitor calpastatin (Croall and DeMartino, 1991) whose production is cytokine driven (Menard and El-Amine, 1996).

A distinct hierarchy exists in the effects of the pro-inflammatory cytokines involved in the pathogenesis of RA, and both IL-1 and TNF-α are of pivotal importance (Ivashkiv, 1996). The actions of the pro-inflammatory cytokines are inhibited by IL-4 (Miossec *et al.*, 1992) and by IL-10, and the exogenous production of IL-10 within RA joints exerts a potent anti-inflammatory effect upon synovial fluid mononuclear cells (Isomaki *et al.*, 1996). Consequently targeting the pro-inflammatory cytokines would appear to be a rational strategy in the therapy of the arthritides. Experiments with animal models of

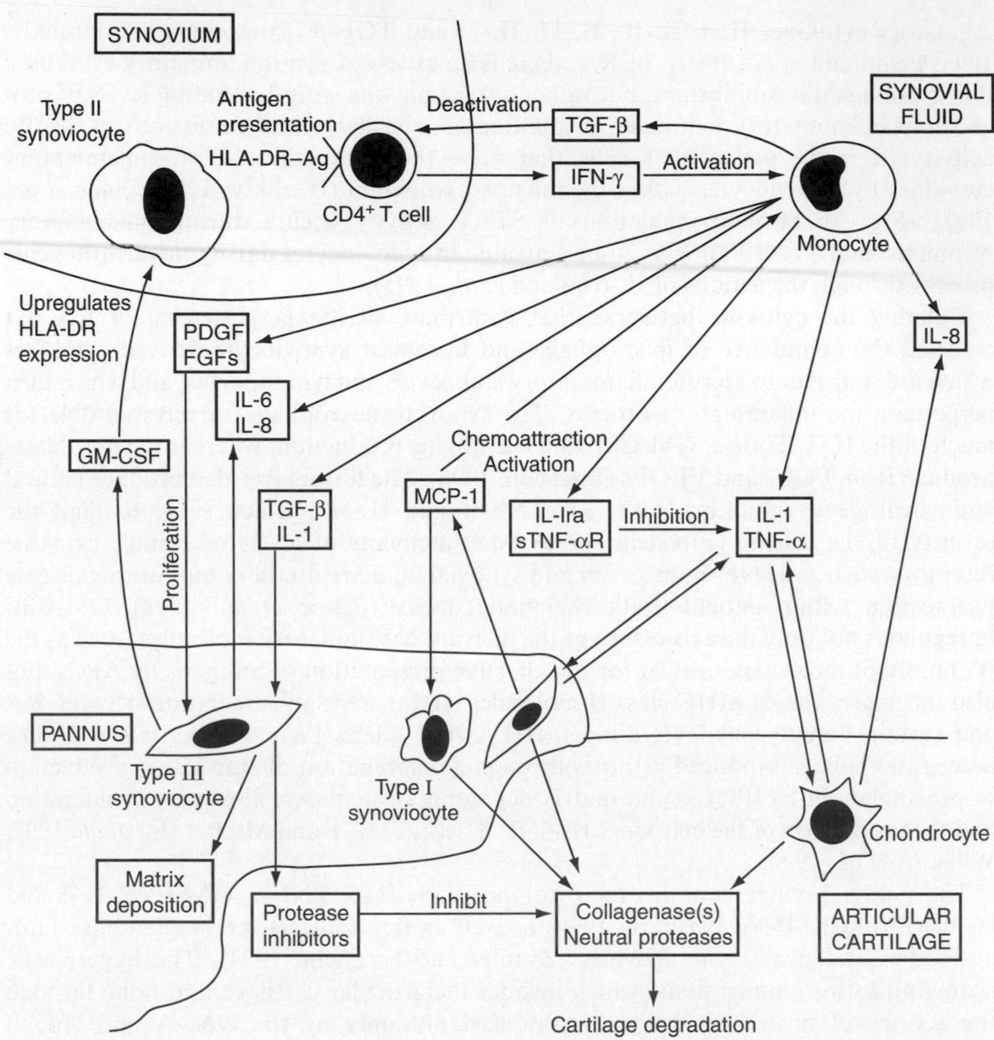

Fig. 5. Diagrammatic representation of key cellular interactions involving chondrocytes, CD4⁺ T cells and synoviocytes (types I, II, III) involved in facets of the pathogenesis of rheumatoid arthritis within articulated joints, which are orchestrated by inflammatory monocytes and mediated by monocyte-derived cytokines (IL-1, IL-6, IL-8, TNF-α, PDGF, FGFs and TGF-β). The proinflammatory effects of cytokines such as IL-1 and TNF-α are partially inhibited by the endogenous inhibitors IL-1Ra and sTNF-αR. The continued influx of inflammatory monocytes into the inflamed joint in response to the chemotactic protein MCP-1, generated by type I synoviocytes, amplifies the pathogenic events within the rheumatoid joint. Articular cartilage erosion results from the stimulation by IL-1 and TNF-α of the production of proteases (collagenase(s) and neutral proteases) by type I synoviocytes and chondrocytes. The actions of the proteases are regulated by the production of protease inhibitors. Pannus tissue formation results from stimulation of proliferation of type III synoviocytes by PDGF and FGFs, and from the TGF-β- and IL-1-mediated deposition of extracellular matrix by the type III synoviocytes. The production of GM-CSF by type III synoviocytes induces the upregulation of HLA-DR molecule expression on type II synoviocytes, thereby enhancing their ability to present autoantigens to CD4⁺ T cells within lymphoid follicles in the inflamed synovium of arthritic joints.

RA, as well as clinical trials, indicate that interventions aimed at inhibiting the actions of pro-inflammatory cytokines may have considerable therapeutic potential (Ivashkiv, 1996; Feldmann *et al.*, 1997). Several experimental therapeutic approaches to the treatment of the arthritides have targeted the pro-inflammatory cytokines through the use of anticytokine antibodies, soluble cytokine receptors, natural cytokine inhibitors (e.g. IL- 1ra) and anticytokine drugs, such as tenidap (Brennan *et al.*, 1989; Piquet *et al.*, 1992; Arend, 1993; Arend and Dayer, 1995; Chikanza *et al.*, 1995; Joosten *et al.*, 1996; Kingsley *et al.*, 1996; Matsukawa *et al.*, 1997). TNF-α is considered to be of central importance in RA pathogenesis and should be a therapeutic target (Feldmann *et al.*, 1996). Antibodies to TNF-α not only ameliorate synovitis in experimental collagen-induced arthritis (Williams *et al.*, 1992), especially when combined with $CD4^+$ T_H cell depletion and IL-1 receptor blockage (Williams *et al.*, 1994), but also prevent the onset of autoimmune arthritis (Arend and Dayer, 1995) and have been used successfully in the clinical treatment of patients with RA (Elliott *et al.*, 1993, 1994). Thus anti-TNF-α antibody therapy, possibly combined with other approaches for inhibiting TNF-α signalling, should be effective therapeutically and of widespread benefit in the treatment of RA (Feldmann *et al.*, 1997).

It has been suggested that an unbalanced expression of T_{H1} or T_{H2} cytokines can be associated with the pathogenesis of inflammatory diseases (Romagnani, 1995, 1996), and, although it is tempting to correlate the pattern of synoviocyte cytokine expression into the T_{H1}–T_{H2} paradigm, the complexity of the arthritogenic process precludes such a simplification (Miossec *et al.*, 1996; Kinne *et al.*, 1997), especially as studies of experimental collagen-induced arthritis indicate that cytokine regulation of the inflammatory process may change through the course of the disease (Boissier *et al.*, 1995; Kinne *et al.*, 1997).

CYTOKINES AND CANCER

Malignant transformation of cells is a complex multistep process involving deletion of tumour suppressor genes and activation of dominant oncogenes, and arises from oncoprotein-induced changes in signalling pathways which generate deregulated cell growth (Frame, 1997). Twenty oncogenes have been described in human cancers and have been implicated in cancer pathogenesis; these oncogenes deregulate cell proliferation and induce virocrine transformation of cells (Drummond-Barbosa and DiMaio, 1997) by encoding oncoproteins with autocrine growth factor activity (e.g. sis, hst, int-2 and fms), modified receptors with protein tyrosine kinase activity (e.g. erbB1, erbB2, met and trk), and downstream components of intracellular signalling pathways such as non-receptor protein tyrosine kinases (e.g. myc, fos, myb, jun, ets-I, hox-II and lyl). The oncoproteins myc and fos are rapidly induced following activation of RTKs (van der Geer *et al.*, 1994) and alterations in the expression of these nuclear oncoproteins in tumorigenesis and tumour progression are well documented (Garte, 1993; Gee *et al.*, 1995).

The human EGFR (c-erbB1) is a transmembrane receptor expressed characteristically on epithelial cells and malignant carcinoma cells, where it is an important prognostic indicator; tumour cells that overexpress EGFR frequently coexpress the ligand, leading to growth stimulation (Davies and Chamberlin, 1996). Autocrine growth stimulation, by an autocrine loop involving growth promoting oncoproteins, is a critical step in carcinogenesis, and diverts transformed cells into a mode of continuous proliferation (Roberts and Sporn, 1990). Cell proliferation is also enhanced in cells expressing members of the src family of protein tyrosine kinase oncoproteins (including src, yes, fyn, lck, hck, fgr, blk and trk) which are required for growth factor-stimulated progression from G0 to S and from G2 to M (Shalloway and Taylor, 1997); the src oncoprotein is a non-receptor tyrosine kinase which perturbs normal cell cycle control and induces transit of the G1 checkpoint by increasing immediate–early response gene activities associated with the AP-1 transcription factor production and the MAP kinase cascade (Wyke *et al.*, 1996). Receptor mutations that confer growth factor independence can result from the transformation of cells with oncogenic leukaemia viruses, and the mutated receptors, which include truncated receptors with constitutively active PTK domains, or activating mutations, allow ligand-independent receptor oligomerization (Ihle, 1996). A mutation involving duplication of 37 amino acids in the extracellular domain has been identified in the βc chains of IL-3R, IL-5R and GM-CSFR which mediates haemopoietic growth factor-independent growth of myeloid cells (D'Andrea *et al.*, 1994). Neuregulins (NRGs) are multipotent EGF-like polypeptide factors that signal through members of the erbB subfamily of tyrosine kinase receptors to regulate cell growth and differentiation in many tissues (Carraway and Burden, 1995; Marchionni, 1995), and deregulation of these signalling pathways has been implicated in the pathogenesis of a variety of cancers (Hynes and Stern, 1994). Acquisition of TGF-β_1 resistance is an important progression factor in human renal cell carcinoma (Ramp *et al.*, 1997) and deregulation of TGF-β signalling through missense mutations in SMAD proteins generates tumorigenic transformation of cells, in which there is a loss of cell growth control by TGF-β (Hata *et al.*, 1997; Shi *et al.*, 1997).

The growth of solid tumours is dependent on their ability to recruit a connective tissue stroma (Dvorak, 1986; van den Hooff, 1988; Liotta *et al.*, 1991; Sieweke and Bissell, 1994; Hanahan and Folkman, 1996). The stromal compartment may contribute more than 90% of the mass of certain tumours and consists of endothelial cells, mast cells, inflammatory leukocytes and a heterogeneous population of fibroblasts. Stromal reactions within tumours represent complex events involving orchestrated interactions between tumour cells and stromal cells (van den Hooff, 1988; Hsu *et al.*, 1993; Leek *et al.*, 1994; Dickson and Lippman, 1995; Lewis *et al.*, 1995; Ronnov-Jessen *et al.*, 1996). The loss of normal cell and tissue homeostasis, which characterizes tumour tissue, is reflected in alterations in the amounts and composition of ECM molecules which, in turn, are related to changes in the ability of cells to receive and integrate ECM-derived signals (Ronnov-Jessen *et al.*, 1996).

In tumours the vascular endothelial cells respond to cytokines generated within the tumour mass by expressing chemokines and adhesion molecules which play key roles in promoting leukocyte traffic into tumours; the inflammatory leukocyte content of a tumour depends on the kinds of chemokines expressed within the tumour, with the C-C chemokines MCP-1, MIP-1α and RANTES promoting infiltrations of lymphocytes and monocytes into the tumour (Negus *et al.*, 1995, 1997; Zhang *et al.*, 1997). Monocyte–macrophage infiltrates can aid or inhibit tumour growth according to the state of activation of these cells (Mantovani, 1994). Tumour cells produce cytokines that are trophic for stromal cells and which can alter the phenotype of the macrophages to one that augments the stromal response (and hence tumour growth) through the production of stromagenic cytokines (Pugh-Humphreys, 1992; Leek *et al.*, 1994; Mantovani, 1994;

Polverini and Nickoloff, 1995). The stromal events associated with tumours are analogous to those observed in granulation tissue formation observed during wound repair (Dvorak, 1986; Whalen, 1990). Both the tumour stroma and granulation tissue contain specialized fibroblasts called 'reactive stromal fibroblasts' which express the surface-bound serine protease called 'fibroblast activation protein' (FAP) and display an altered pattern of gene expression not seen in resting fibroblasts within normal connective tissues (Mathew *et al.*, 1995; Niedermeyer *et al.*, 1997). These reactive stromal fibroblasts produce growth factors (Cullen *et al.*, 1991), distinct extracellular matrix proteins and ECM degrading enzymes (Liotta *et al.*, 1991) which facilitate tumour invasion into host tissues. In addition, tumour invasion can be promoted by activation of cancer cell plasminogen-dependent proteolysis through the paracrine actions of IL-1 and TNF-α generated by tumour-associated macrophages (Tran-Thang *et al.*, 1996).

Malignant tumours of the immune system are classified as Hodgkin's disease (HD) or non-Hodgkin's lymphoma (NHL) (Harris *et al.*, 1994). Hodgkin's lymphoma is characterized by the presence of mononuclear Hodgkin's (H) cells and bi- or multi-nucleated Reed–Sternberg (RS) cells; the malignant H–RS cells make up less than 1% of the total tumour mass, the remainder being a non-neoplastic, reactive stroma (Hsu *et al.*, 1985) within which infiltrations of T cells, eosinophils, neutrophils, plasma cells, macrophages, endothelial cells and fibroblasts are evident (Haluska *et al.*, 1994; Kadin, 1994). Within the lymphoma cells, the usually tightly regulated production of cytokines may be dysregulated and certain cytokines may be produced constitutively (Hsu *et al.*, 1993). The H–RS cells express IL-1, IL-4, IL-5, IL-6, IL-9, TNF-α, G-CSF, M-CSF and TGF-β, and these cytokines are involved in cascades of paracrine interactions between H–RS cells and reactive stromal cells (Hsu *et al.*, 1993). Reactive stromal cells encourage the infiltration of inflammatory leukocytes, particularly neutrophils, through the production of IL-8 (Foss *et al.*, 1996). Many of the clinical symptoms and pathological features of HD can be linked to an unbalanced production of cytokines, particularly IL-1, IL-2 and IL-6 (Blay *et al.*, 1994), through interactions of H–RS cells with the reactive stromal cells (Gruss *et al.*, 1994, 1997). The reactive T cells in the stroma express cytokines of the T_{H2} pattern, stimulating an ineffective humoral immune response in which the H–RS cells escape a CTL response (Gruss *et al.*, 1997). The overexpression of cytokines by the H–RS cells and surrounding reactive stromal cells can be related to constitutive activation of NF-κB (Gruss *et al.*, 1992, 1996) and strategies to facilitate an effective CTL response would involve reversing the ineffective T_{H2} response and blocking the constitutive activation of NF-κB (Gruss *et al.*, 1997). IL-12 is a key cytokine involved in establishing the T_{H1} *versus* T_{H2} dominance by promoting T_{H1} development and inhibiting T_{H2} development (Trinchieri, 1995). IL-12 induces anti-tumour effects *in vivo* (Brunda *et al.*, 1993; Zou *et al.*, 1993; Nastala *et al.*, 1994) by promoting the maturation of T_{H1} cells (Trinchieri, 1994) and enhancing the activities and secretion of IFN-γ by CTLs and NK cells (Brunda, 1994). The synergistic interactions of the T_{H1} cytokines, IL-2 and IFN-γ, promote the generation of tumour-reactive cytotoxic cells (McAdam *et al.*, 1996a,b).

The progressive local growth, malignancy and metastatic potential of solid tumours is angiogenesis dependent (Folkman, 1992, 1995) and the turnover of the tumour vasculature is determined by a balance between stimulatory and inhibitory molecules generated within the tumour mass (Klagsbrun and Folkman, 1990; Leek *et al.*, 1994;

Polverini and Nickoloff, 1995). Among the many angiogenic inhibitors within tumours can be listed thrombospondin and angiostatin (O'Reilly *et al.*, 1994; Folkman, 1995) and the C-X-C chemokines GRO-β and IP-10 (Angiolillo *et al.*, 1995; Cao *et al.*, 1995). A number of angiogenic factors have been purified, sequenced and cloned; within tumours, angiogenesis is stimulated by members of the FGF family (FGF-1, FGF-2, int-2 and Hst-1) (Friesel and Macaig, 1995; Ronnov-Jessen *et al.*, 1996), VEGFs (Kim *et al.*, 1993; Klagsbrun and Soker, 1993; Millauer *et al.*, 1994; Dvorak *et al.*, 1995) and subfamilies of the C-X-C chemokines (Strieter *et al.*, 1995). Midkine and pleiotrophin are secreted heparin-binding neurokines (Bohlen and Kovesdi, 1991) which are expressed in a variety of human tumours (Garver *et al.*, 1994; Riegel and Wellstein, 1994; Nakagawara *et al.*, 1995; O'Brien, 1996) and play a direct role as angiogenic factors (Chaudhuri *et al.*, 1997). HGF–SF is a 90-kDa single-chain glycoprotein, the overexpression of which within tumours enhances tumour cell motility, invasion and angiogenesis (Bellusci *et al.*, 1994; Rosen and Goldberg, 1995; Madeddu *et al.*, 1996; Lamszus *et al.*, 1997; Laterra *et al.*, 1997). When activated, oncogenes can influence the angiogenic phenotype of the cells within which they are expressed, either by encoding for secreted angiogenic factors, such as VEGFs, FGFs and HGF–SF, or by stimulating the production of angiogenic factors; oncogenes with the latter capacity include v-*erbA*, v-*abl*, v-*fos*, v-*mos*, c-*myb*, H-*ras*, K-*ras*, N-*ras*, *ros*, *sis*, *src* and v-*yes* (Bouck *et al.*, 1996).

A role for FGF–FGFR interactions has been invoked in the development of several types of cancer (Saxena and Ali, 1992; Yamanaka *et al.*, 1993; Leung *et al.*, 1994; Ueki *et al.*, 1995). Transfection with FGF-9 cDNA stimulates the proliferation of a number of cell types (Naruo *et al.*, 1993; Matsumoto-Yoshitomi *et al.*, 1997) and the transformation of solid tumours following transplantation of these transfected cells is inhibited by the neutralizing anti-FGF9 monoclonal antibody (mAb) Iso-59 (Matsumoto-Yoshitomi *et al.*, 1997). An immunoneutralizing mAb against FGF-2 suppresses the growth of tumours transplanted within nude mice by an antiangiogenic effect (Hori *et al.*, 1991).

Overexpression of FGFRs has been detected in a number of human cancers (Leung *et al.*, 1994; Yamaguchi *et al.*, 1994; Luqmani *et al.*, 1995; Penault-Llorca *et al.*, 1995; Myoken *et al.*, 1996) and the overexpression of FGFRs, or activating mutations of FGFRs, has been implicated in various angiogenic pathologies, including tumour angiogenesis (Klagsbrun and D'Amore, 1991). The expression of FGF-2 is enhanced within malignant tumours (Takahashi *et al.*, 1989) and the endogenous production of FGFs by cancer cells promotes their invasive potential (Jouanneau *et al.*, 1991; Jayson *et al.*, 1994; Galzie *et al.*, 1997). Tumour cells secreting FGF-2 stimulate stromal cells to produce MMP-2 which promotes tumour cell invasion and metastasis (Miyake *et al.*, 1996), and neutralizing antibodies to FGF-2 have been shown to reduce the constitutive invasiveness of colorectal carcinoma cells (Galzie *et al.*, 1997). Variant forms of FGFR, especially FGFR2, may contribute to tumorigenesis (Kobrin *et al.*, 1993; Yan *et al.*, 1993; Itoh *et al.*, 1994). The FGFR2 gene is amplified in breast tumours (Adnane *et al.*, 1991) and expression of the IIIb or IIIc mRNA spliced variant forms of FGFR2 (called BEK and K-SAM respectively) has been found in breast tumours (Luqmani *et al.*, 1996) where they exert a profound influence on the ligand-binding specificity of FGFR2 from one (BEK) binding FGF-7 to one (K-SAM) binding FGF-2. Tumour necrosis is often caused by ischaemia due to vascular shutdown, and this phenomenon is being exploited in the development of novel therapeutic strategies for tumours that target the tumour vasculature (Mannel *et al.*, 1996). Strategies to inhibit FGF–FGFR interactions should exert beneficial therapeutic effects on angiogenesis in some types of cancer (Morrison, 1991) and represent helpful adjuvant therapies in treating breast cancer (Lappiu, 1995; Wellstein and Czubayko, 1996; Coll-Fresno *et al.*, 1997; Luqmani *et al.*, 1997).

Current gene therapy for cancer involves strategies to enhance specific host antitumour immune responses through the use of vaccines containing cells transfected with either cytokine genes or metabolic suicide genes (Tepper and Mule, 1994; Dranoff and Mulligan, 1995). The suicide genes code for enzymes that can activate otherwise nontoxic prodrugs and confer lethal sensitivity to the appropriate antimetabolites (Mullen, 1994). A variety of tumour cells have been transfected with genes for the cytokines IL-1, IL-2, IL-3, IL-4, IL-6, IL-7, IL-12, TNF-α, IFN-α, IFN-γ, GM-CSF, JE and IP-10, and then used as a vaccine in the hope that the engineered cells secrete cytokines which will profoundly influence the host's antitumour responses, and thus prove to be therapeutically useful (Tepper and Mule, 1994; Dranoff and Mulligan, 1995). Whereas the parental tumour cells grew progressively when transplanted into hosts, the genetically engineered cells were rejected (Russel, 1990; Pardoll, 1993; Colombo and Forni, 1994; Rosenthal *et al.*, 1994; Tepper and Mule, 1994) and proved to be an effective vaccine which augmented antitumour immunity not only to subsequent transplants of parental tumour cells (Connor *et al.*, 1993; Allione *et al.*, 1994; Saito *et al.*, 1994; Vieweg *et al.*, 1994), but also to existing tumours, probably as a consequence of cytokine-driven alterations in the cytokine network operating within the tumour environment (Dranoff and Mulligan, 1995).

The effector cells involved in the rejection of the genetically modified tumour cells varied with the cytokine gene(s) introduced, which evoked distinct reactions involving dendritic cells, macrophages, eosinophils, neutrophils, NK cells, $CD4^+$ T cells and $CD8^+$ T cells (Dranoff and Mulligan, 1995; Hanada *et al.*, 1996). In the case of IL-2-secreting tumour cells, tumour cell rejection was achieved through an enhanced antigen-specific T-cell response (Gansbacher *et al.*, 1990), whereas with GM-CSF-secreting tumour cells, the GM-CSF provoked an inflammatory response at the site of transplantation which promoted the maturation and activation of DCs, which in turn attracted $CD4^+$ T cells to the site of transplantation (Dranoff *et al.*, 1993; Levitzky *et al.*, 1994; Wakimoto *et al.*, 1996). Considerable attention is currently focused on DCs which are of critical importance for the effectiveness of $CD4^+$ T cell-mediated specific antitumour immune responses; the DCs can be isolated from tumour patients, activated with GM-CSF and pulsed with tumour-specific antigens *in vitro*, and then transplanted back into the patient to induce protective immunity (Grabbe *et al.*, 1995). Another strategy for use with tumour cells lacking the expression of MHC and/or costimulatory B7 molecules is to use MHC and/or B7-transfected cells as a vaccine; the rationale of this form of immunotherapy is that T cells require both an antigen-specific signal, delivered through the T-cell receptor, and a costimulatory signal delivered by B7 molecules to become fully activated (L. Chen *et al.*, 1992, 1993; Baskar *et al.*, 1993a,b, 1995; Townsend and Allison, 1993; Grabbe *et al.*, 1995; Hayakawa *et al.*, 1997).

CYTOKINES AND THEIR RECEPTORS AS THERAPEUTIC TARGETS

The involvement of cytokines in the pathogenesis of disease has prompted the development of strategies for targeting cytokines and their receptors as therapeutic manoeuvres for inhibiting cytokine action in the treatment of cytokine-mediated disease processes.

The actions of cytokines can be inhibited at several levels (Fig. 6) through the use of: (1) signal transduction inhibitors (i.e. inhibitors of second messenger production); (2) drugs that function as transcription and translation inhibitors; (3) agents that prevent cytokine action by binding soluble cytokines (e.g. mAbs, soluble cytokine receptors, cytokine binding proteins); and (4) agents that interfere with the binding of cytokines to their receptors (e.g. mAbs or antagonists). These strategies are proving effective in the treatment of diseases like arthritis and cancer (Davies and Chamberlin, 1996; Feldmann *et al.*, 1997; Moreland *et al.*, 1997).

INTERFERING WITH THE ACTIONS OF CYTOKINES BY THE USE OF TRANSDUCTION INHIBITORS

Signal transduction inhibitors are low-molecular-weight, non-protein drugs that interfere with the generation of intracellular signals following cytokine–receptor binding. Both PKC and PTK are targets for inhibitor development as they are involved in the signal transduction mechanisms associated with cytokine receptors.

One approach to the treatment of cell growth disorders involves selective inhibition of PTK activity (Krontiris, 1995; Levitzki and Gazit, 1995). Tyrophostins are small hydrophobic molecules which resemble tyrosine and inhibit tyrosine kinases either by direct competition for substrate, or through ATP binding (Levitzki and Gazit, 1995). The tyrophostin erbstatin can traverse cell membranes easily and can block both PTK activity and intracellular Ca^{2+} mobilization (Levitzki and Gazit, 1995). Tyrophostins of the quinoxaline class (e.g. compounds AG1295 and AG1296) selectively inhibit the RTK of the PDGFR at an IC_{50} of less than 1 μM (Kovalenko *et al.*, 1994) and can inhibit tumour cell growth (Krystal *et al.*, 1997). The tyrophostin AG17 induces apoptosis in human OCI-Ly8 immunoblastic lymphoma cells by targeting cdk2 (Palumbo *et al.*, 1997). The JAK2 PTK inhibitor, AG-490, is a tyrophostin which selectively blocks leukaemic cell growth *in vitro* and *in vivo* by inducing apoptosis without any deleterious effect on normal haemopoiesis (Meyden *et al.*, 1996). The cell growth inhibitory indolinones are a novel class of PTK inhibitors which inhibit the kinase activity of FGFR-1 and PDGFR by occupying the ATP binding pocket of the receptor kinases (Mohammadi *et al.*, 1997), and a novel series of quinazoline inhibitors has been developed which prevent the autophosphorylation of the EGF RTK, and hence downstream EGFR signalling, with consequent inhibition of the growth of EGF–TGF-α-dependent cancer cells (Wakeling *et al.*, 1996; Jones *et al.*, 1997). The PTK inhibitors, herbimycin A and genistein, inhibit the proliferation of human breast cancer cells, MCF-7 and SKBR3, through inhibitory effects on the proteins Shc and Grb2 which mediate signal transduction from activated growth factor receptors through the ras signalling pathways (Clark *et al.*, 1996). The glycosylated indole carbazole alkaloid K252 compounds are structurally similar to staurosporine and exert inhibitory effects on cell growth through selective inhibition of RTKs (Knusel and Hefti, 1992; Tapley *et al.*, 1992; Chin *et al.*, 1997). PKC inhibitors such as chelerythrine and hypericin have been used clincally in the treatment of diseases including periodontal inflammatory disease (chelerythrine; Cerna, 1989), and as antiviral agents (hypericin; Takahashi *et al.*, 1989; Carpenter and Kraus, 1991). Both PKC and PTK inhibitors have also proved useful in the clinical treatment of osteosarcomas, systemic lupus erythematosus and cerebral malaria (Mire-Sluis, 1993).

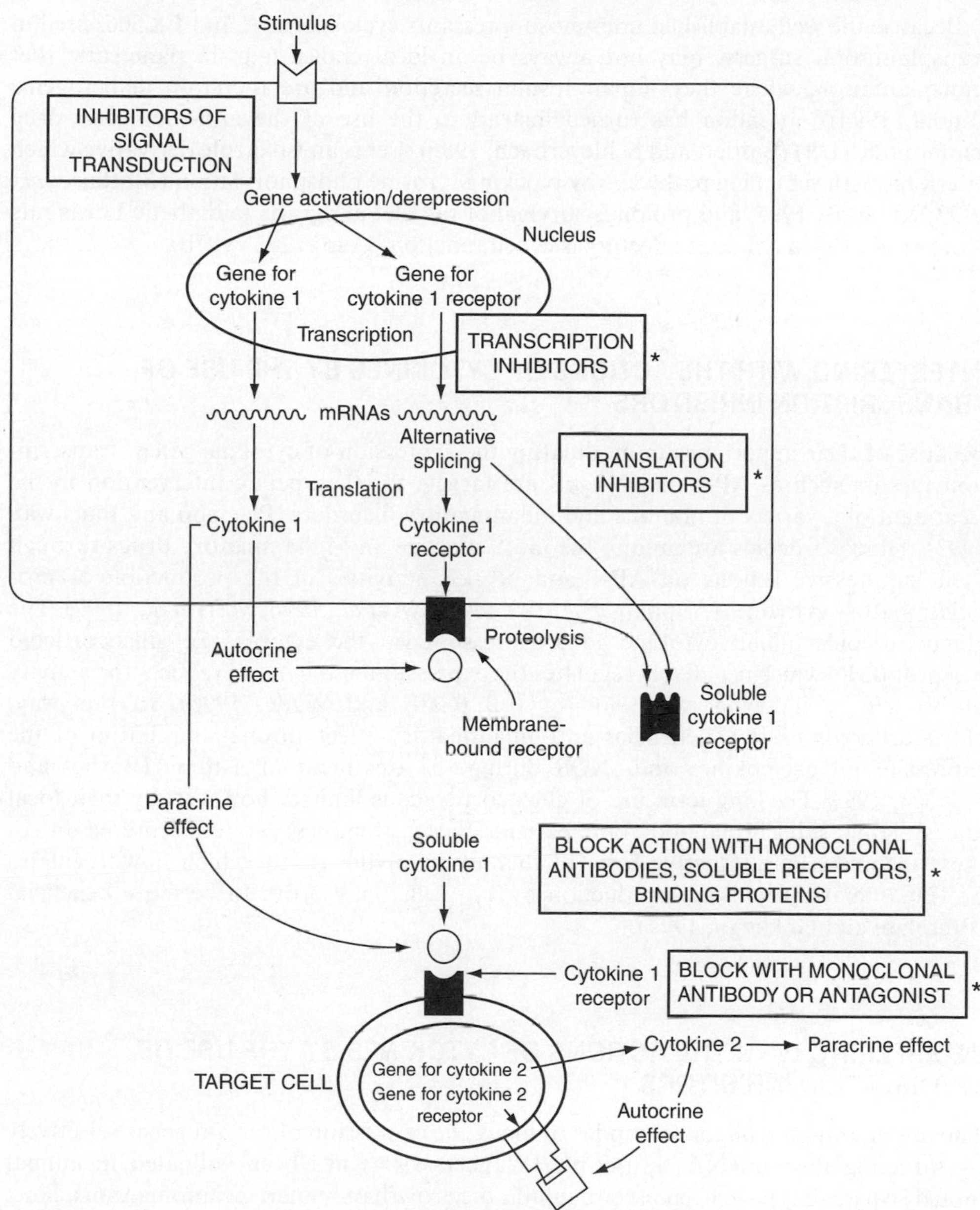

Fig. 6. Diagrammatic representation of the actions of cytokines on cells within the cytokine network. The activation of cytokine and cytokine receptor genes with cytokine-secreting cells, and the sites of interference with gene expression using inhibitors of second-messenger production, as well as transcription and translation inhibitors, are indicated. Cytokine-secreting cells not only release cytokines which can exert autocrine or paracrine actions, but they can also synthesize cytokine receptors, some of which are membrane bound but can be cleaved extracellularly to be released as soluble cytokine receptors, whereas others are released directly as soluble cytokine receptors. The interactions of cytokines with cells can be blocked using antibodies directed against the cytokines and/or against their receptors. Cytokines can induce the release of other cytokines within target cells through activation of the relevant cytokine genes.

Because the well established immunosuppressants cyclosporin A and FK506, used in transplantation surgery, may not always be an ideal choice (e.g. in pancreatic islet transplantation, where they impair insulin secretion and are therefore diabetogenic (Jindal, 1994)), attention has turned instead to the use of the anti-rheumatic drug leflunomide (Lef) (Bartlett and Schleyerbach, 1985). Lef is an isoxazole derivative which interferes with signalling pathways by blocking tyrosine phosphorylation (Mattar *et al.*, 1993; Xu *et al.*, 1995) and prolongs survival of rat islet allografts in diabetic Lewis rats (Guo *et al.*, 1997a) without affecting islet cell function (Guo *et al.*, 1997b).

INTERFERING WITH THE ACTIONS OF CYTOKINES BY THE USE OF TRANSCRIPTION INHIBITORS

Because of their importance in modulating the expression of cytokine genes, transcription factors such as AP-1 and NF-κB are targets for therapeutic intervention in the treatment of a variety of immune and inflammatory disorders (Peterson and Baichwal, 1993). Glucocorticoids are among the most effective anti-inflammatory drugs through their suppressive actions on AP-1 and NF-κB activities in the production of pro-inflammatory cytokines (Auphan *et al.*, 1995; Almawi *et al.*, 1996; Mori *et al.*, 1997). The glucocorticoids inhibit cytokine gene expression via the cytoplasmic glucocorticoid receptor (GR) which negatively regulates the expression of AP-1 and reduces the activity of NF-κB by inducing expression of IκB (Cato and Wade, 1996). In this way, glucocorticoids exert a beneficial anti-inflammatory effect through inhibition of the expression of chemokines and iNOS during the treatment of asthma (Barnes and Adcock, 1997).The long-term use of glucocorticoids is limited, however, by their local and systemic side-effects, and some patients (e.g. asthmatics), are corticoresistant. In these instances an alternative form of therapy involving IL-10, which downregulates pro-inflammatory cytokine production by T_{H1} cells, may prove to be more beneficial (Pretolani and Goldman, 1997).

INTERFERING WITH THE ACTIONS OF CYTOKINES BY THE USE OF TRANSLATION INHIBITORS

The use of antisense oligonucleotides to block the expression of certain genes selectively by targeting their mRNA transcripts (Wagner, 1994) has been validated in animal models where they have been used to inhibit the growth of human melanoma transplants in nude mice (Leonetti, 1993) and immunodeficient mice (Jansen, 1995), to inhibit IL-8-induced tumour angiogenesis (Strieter *et al.*, 1995) and to block chemokine production (Howard *et al.*, 1996). Animal studies have revealed that the antisense oligonucleotides called phosphorothioate (PS) oligonucleotides are well tolerated, having an LD_{50} in excess of 500 mg/kg body weight, and the first generation of PS oligonucleotides have reached clinical trials in leukaemia and patients infected with human immunodeficiency virus (HIV) (Agrawal, 1996). Other drugs that affect mRNA stability will also affect cytokine production; for instance, thalidomide inhibits TNF-α synthesis by enhancing TNF-α mRNA degradation (Moreira *et al.*, 1993).

IMMUNOSUPPRESSIVE DRUGS THAT INTERFERE WITH CYTOKINE PRODUCTION OR THE RESPONSES OF LYMPHOCYTES TO GROWTH FACTORS

The function of APCs is to process complex antigens and then display complexes of peptide antigens (Ag) and major histocompatibility complex (MHC) molecules on their surfaces which T cells can monitor for foreign peptides using their T-cell receptors (TCRs) (Cresswell, 1994). As few as 50 to 100 specific peptide-MHC complexes per APC can trigger up to 18,000 TCRs per T cell, and sustained signalling requires continuous TCR engagements (Harding and Unanue, 1990; Cantrell, 1996; Valitutti and Lanzavecchia, 1997). Adhesion between the T cell and APC is mediated by integrins, and this intracellular adhesion creates opportunities for TCRs to engage Ag–MHC complexes within the contact zone (Paul and Seder, 1994; Shaw and Austin, 1997). Engagement of the TCR by the Ag–MHC complexes alone does not directly trigger T-cell activation, and two signals are required; the ligated TCR provides the first signal and the B7–CD28/CTLA-4 costimulatory pathway provides the second signal (Gause *et al.*, 1997). As a consequence of the high constitutive expression of MHC and costimulatory molecules, as well as the production of the appropriate bioactive cytokines, DCs are the most potent stimulatory APCs for naive T_H cells (Steinman, 1991; Austyn, 1993), especially $CD4^+$, $CD11c^+$, $CD3^-$ DCs found within germinal centres of lymphoid organs (Grouard *et al.*, 1997). Following T_H cell activation, the T_H cells generate a number of immunoregulatory cytokines and the inducible expression of these cytokine genes requires members of the NF-AT (nuclear factor of activated T cells) transcription factor family (Rooney *et al.*, 1995). NF-AT is a multisubunit transcription complex that contains a cytoplasmic phosphoprotein component (NF-ATc) and an inducible nuclear component (NF-ATn) composed of proteins of the AP-1 family (Flanagan *et al.*, 1991; Jain *et al.*, 1992). NF-ATc is the direct substrate of the serine–threonine phosphatase calcineurin (CN), and dephosphorylation of NF-ATc is required for its translocation to the nucleus, where it complexes with NF-ATn to form the transcription factor, NF-AT, which binds to NF-AT-responsive elements to promote gene transcription (Crabtree and Clipstone, 1994).

The immunosuppressants cyclosporin A (CsA), tacrolimus (formerly called FK-506) and sirolimus (formerly called rapamycin) exert their effects by inhibiting the signalling pathways responsible for the inducible expression of immunoregulatory cytokines in T_H cells by binding to cytoplasmic immunophilins (Thomson *et al.*, 1995, 1997; Thomson, 1997). There are two ubiquitous families of immunophilins called cyclophilins (Cyp), which bind the undecapeptide fungal metabolite cyclosporin A, and the FK binding proteins (FKBP) which bind to the macrolides tacrolimus and sirolimus (Braun *et al.*, 1995). There are two cyclophilins, CypA and CypB, whereas there are four major isoforms of mammalian FKBPs (FKBP12 (cytosolic), FKBP (ER lumen), FKBP25 (nuclear) and FKBP52), each designated according to their molecular mass. The immunophilins exhibit *cis–trans*, peptidyl–prolyl isomerase activity, and are involved in protein folding (Galat, 1993; Fruman *et al.*, 1994). When CsA and tacrolimus form complexes with their respective immunophilins, the complexes inhibit signal transduction pathways by interacting specifically with CN; the inhibitory interactions of CsA–Cyp and FK506–FKBP complexes with CN provide key steps in the cascade of intracellular events leading to immunosuppression (Braun *et al.*, 1995). The immunosuppressive properties of CsA and tacrolimus correlate with both the extent of inhibition of CN phosphatase activity (Schreiber and Crabtree, 1992) and the consequential failure to activate cytokine genes regulated by NF-AT (Ho *et al.*, 1996). Despite its structural similarity with tacrolimus, the chemically related macrolide sirolimus binds FKBP even more strongly than tacrolimus, but appears to exert its immunosuppressive effects directly (Schreiber, 1991) by blocking IL-2-mediated elimination of the cyclin-dependent kinase inhibitor $p27^{kipl}$ (Nourse *et al.*, 1994). The clinical applications of CsA, tacrolimus and sirolimus have been reviewed elsewhere (Thomson, 1997; Thomson *et al.*, 1997).

AGENTS THAT PREVENT CYTOKINE ACTION BY BINDING SOLUBLE CYTOKINES, MONOCLONAL ANTIBODIES AND FUSION PROTEINS

The application of antibodies in research and in the clinic is widespread and many are being used in clinical trials for conditions as diverse as cancer, transplant rejection, viral infections, autoimmunity and sepsis (Chester and Hawkins, 1995). Numerous references have been made throughout this chapter to the use of neutralizing antibodies to inhibit the actions of various cytokines. In the clinical context mAbs directed against specific cytokines have been used in the treatment of a number of diseases (Mire-Sluis, 1993) (Table 6). However, since murine monoclonal antibodies are immunogenic in humans, and induce the formation of anti-idiotypic antibodies (LoBuglio *et al.*, 1989), their therapeutic value in the clinic has been disappointing and consequently attempts have been made to engineer the antibodies to make them more human-like to improve their clinical performance (Chester and Hawkins, 1995).

The most complete humanization of a murine antibody has been achieved by grafting the CDRs from a murine antibody on to a human immunoglobulin framework to create a functional chimaeric antibody. The chimaeric anti-TNF-α antibody, cA2, is a chimaerized mAb to human TNF-α and was constructed by grafting the variable regions of a murine anti-human TNF-α mAb(A2) on to a human IgG_1 with κ light chains. The benefits of cA2 treatment of patients with RA are not only inhibition of arthritogenic TNF-α activity itself, but also reduced production of other proinflammatory cytokines induced by TNF-α (Feldmann *et al.*, 1997).

To date, chimaeric antibodies have met with limited success because they retain immunogenicity in humans; consequently attention is now focused on phage–antibody display technology to select human mAbs and construct derivatives with immunotherapeutic applications (de Kruif *et al.*, 1996). Among the constructs are single chain Fv

Table 6. Use of monoclonal antibodies directed against cytokines in the treatment of diseases.

Disease or disorder	Pathogenic cytokine	Therapeutic intervention
Myeloma	IL-6	Anti-IL-6 mAb
Breast carcinoma	TGF-β	Anti-TGF-β mAb
Fibrosis	TGF-β	Anti-TGF-β mAb
Lymphoid leukaemia	TNF-α	Anti-TNF-α mAb
Rheumatoid arthritis	TNF-α	Anti-TNF-α mAb
Multiple sclerosis	TNF-α	Anti-TNF-α mAb
Septic shock	TNF-α	Anti-TNF-α mAb

(scFv) molecules with heavy and light chain variable regions tethered by a flexible 15-amino-acid 'linker' and which retain full monovalent antibody binding activity (Chester and Hawkins, 1995; de Kruif *et al.*, 1996). The next stage in development will be the generation of bivalent scFv molecules, and to extend the use of antibody engineering technology to generate scFv–cytokine fusions. Success has already been achieved in the generation of scFv–IL-12 products which enhance antitumour immune responses (Savaage *et al.*, 1993; Sabzevari *et al.*, 1994).

SOLUBLE CYTOKINE RECEPTORS AND IMMUNOADHESINS

The generation of soluble cytokine receptors appears to be a mechanism for regulating cytokine activity *in vivo* (Fernadez-Botran *et al.*, 1996). The circulating levels of certain soluble cytokine receptors (e.g. sIL-2R and sIL-6R), are raised in clinical patients with particular disease states (Tables 7 and 8), and there is reason to believe that soluble cytokine receptors may have clinical significance as cytokine inhibitors (Fernandez-

Table 7. Diseases associated with raised plasma levels of soluble IL-2 receptors.

Haematological malignancies
Adult T-cell leukaemia
T-cell chronic lymphocytic leukaemia
T-cell acute lymphoblastic leukaemia
Hairy cell leukaemia
Malignant lymphoma
Multiple myeloma
Non-Hodgkin's lymphoma
Hodgkin's lymphoma
Solid tumours
Oesophageal, gastric and colonic carcinomas
Pancreatic carcinoma
Breast cancer
Lung cancer (primary and small cell carcinomas)
Nasopharyngeal carcinoma
Ovarian cancer
Melanoma
Autoimmune and inflammatory diseases
Autoimmune chronic hepatitis
Myasthenia gravis
Graves' disease
Scleroderma
Rheumatoid arthritis
Systemic lupus erythematosus
Multiple and systemic sclerosis
Type I diabetes
Glomerulonephritis
Inflammatory bowel disease
Other diseases
Asthma
Cystic fibrosis
Sarcoidosis

Table 8. Diseases associated with a raised level of circulating soluble IL-6 receptors.

Monoclonal gammopathy (benign, and both early and established multiple myeloma)
Rheumatoid arthritis
Interstitial lung diseases (interstitial pneumonia, sarcoidosis)
Inflammatory bowel disease (ulcerative colitis, Crohn's disease)
Cerebral malaria

Botran *et al.*, 1996). Soluble receptors that bind IL-1, IL-4, TNF-α and IFN-γ have been used in the treatment of a number of clinical and experimentally induced pathologies (Table 9).

Although soluble cytokine receptors can inhibit the binding of cytokines to their respective membrane-anchored receptors, there are instances where the soluble receptors can enhance the action of the respective cytokine. For instance, sIL-6R enhances the action of IL-6 when the IL-6/sIL-6R complex associates with the plasma membrane-anchored gp130 signal transducer; under these circumstances the IL-6/sIL-6R complex can mimic the actions of OSM, LIF or CNTF (Akira *et al.*, 1993).

Immunoadhesins are molecules that combine 'immune' and 'adhesion' functions, and are generated by fusion of the hinge and Fc regions of an immunoglobulin molecule with the ligand binding domain of a receptor or an adhesion molecule (Chamow and Ashkenazi, 1996). Immunoadhesins based on cytokine receptors have been generated and used as research reagents (Table 10). From the therapeutic viewpoint, immunoadhesins hold promise as antiviral agents, modulators of inflammation and immunity, and as anticancer agents (Chamow and Ashkenazi, 1996).

A potent TNF-α antagonist is the TNF receptor–IgG fusion protein construct called IgG–TNFR immunoadhesin, constructed by fusion of the extracellular portion of TNF-R1 with the constant domains of human IgG heavy chains (Agosti *et al.*, 1994). The immunoadhesin has been shown to be effective in protection of mice against

Table 9. Soluble cytokine receptors in the treatment of clinical and experimentally induced pathologies.

Soluble cytokine receptor	Pathology	Reference
sIL-IR1	Autoimmune encephalomyelitis	Jacobs *et al.* (1991)
	LPS-induced bronchoalveolar inflammation	Ulich *et al.* (1994)
	IL-1α-induced ocular inflammation	Rosenbaum and Boney (1991)
	Allergic reactions	Mullarkey *et al.* (1994)
sIL-4R	Murine cutaneous leishmaniasis	Gessner *et al.* (1994)
	Candidiasis	Pucetti *et al.* (1994)
sTNF-R	Septic shock	Ashkenazi *et al.* (1991)
		Lesslauer *et al.* (1991)
	IL-2-induced pulmonary vascular leak in mice	Dubinett *et al.* (1994)
	Collagen-induced arthritis in mice	Piquet *et al.* (1992)
sIFN-γR	Graft *versus* host disease	Ozmen *et al.* (1993)
	Glomerulonephritis in NZB/W autoimmune mice	Ozmen *et al.* (1995)

Table 10. Immunoadhesions based on cytokine receptors.

Immunoadhesin cytokine receptor	Ligands	Reference
TNFR-1	TNF-α, LT	Ashkenazi *et al.* (1991) Lesslauer *et al.* (1991) Butler (1994) Crowe *et al.* (1994)
TNFR-2	TNF-α, LT	Howard *et al.* (1993) Wooley *et al.* (1993)
Fas/Apo-1	Fas ligand	Suda *et al.* (1993) Hanabuchi *et al.* (1994) Nagata and Goldstein (1995)
IFN-γRα chain	IFN-γ	Kurschner *et al.* (1992) Haak-Frendscho *et al.* (1993) Marsters *et al.* (1995)
IL-1R	IL-1	Pitti *et al.* (1994)
IL-4R	IL-4	Morrison and Leder (1992)

endotoxic shock (Ashkenazi *et al.*, 1991) and has been used to treat patients with sepsis (Wherry *et al.*, 1993; van der Poll and Lowry, 1995). A fusion protein consisting of the ligand binding domains of two p75 sTNFR joined to the Fc portion of human IgG_1 (sTNFR : Fc) has also been used to block TNF-α function (Moreland *et al.*, 1994).

AGENTS THAT TARGET CYTOKINE RECEPTORS

Antibodies to TNF Receptors and Other Cytokine Receptors

Although TNF-α plays an important role as a regulator of inflammatory responses, its overproduction induces the pathogenesis of autoimmune, inflammatory and non-inflammatory diseases (Vassalli, 1992) and blocking TNF receptors with antibodies should be an effective strategy for inhibiting the pathological effects of TNF-α. Antibodies to TNF receptors can be agonistic as well as antagonistic (Engelmann *et al.*, 1990). Antagonistic antibodies to TNF-R1 completely block several TNF-α activities (Espevik *et al.*, 1990; Shalaby *et al.*, 1990; Naume *et al.*, 1991), whereas monoclonals directed against TNF-R2 only partially antagonize TNF-α activities (Hohmann *et al.*, 1990; Shalaby *et al.*, 1990; Naume *et al.*, 1991). These observations suggest that TNF-R1 is the biologically relevant receptor, and that the binding of TNF-α to TNF-R2 is not sufficient to initiate TNF-α responses within target cells (Thoma *et al.*, 1990). Polyclonal antibodies directed against TNF-R1 and TNF-R2 have been described which behave as specific receptor agonists, and each induces a subset of TNF-α activities (Tartaglia *et al.*, 1991).

Immunosuppression has been achieved through the use of antibodies directed against IL-2R (Kelley *et al.*, 1988; Kupiec-Weglinski *et al.*, 1989), and in a number of instances the immunosuppressive effects of the anti-IL-2R antibodies have been enhanced by the simultaneous administration of cyclosporin A (Kupiec-Weglinski *et al.*, 1989, 1991a,b; Hancock *et al.*, 1990; Tananka *et al.*, 1990; Ueda *et al.*, 1990; Higuchi *et al.*, 1991; Steinbruchel *et al.*, 1991, 1992). Monoclonal antibodies to IL-2R prolong allograft survival across MHC barriers within various animal species (Strom *et al.*, 1988; Thedrez *et al.*, 1989; Okada *et al.*, 1992; Steinbruchel *et al.*, 1992) and suppress both graft-*versus*-host (GVH) diseases (Herve *et al.*, 1990, 1992; Blaise *et al.*, 1991) and certain autoimmune reactions (Braley-Mullen *et al.*, 1991). Treatment of mice with the anti-IL-2R mAb, AMT-13, not only prolonged the survival of cardiac allografts, but also caused an increase in the serum levels of sIL-2R (Burkhardt *et al.*, 1989). The 20-fold increase in the titre of sIL-2R in the serum of AMT-13-treated mice is believed to play a role in the immunosuppressive action of AMT-13. Activated T cells and macrophages play an important role in the induction of streptotocin-induced diabetes in C57BL/6J mice, and the anti-IL-2R mAb M7/20, which recognizes the α-subunit of IL-2R, inhibits autoimmune insulitis in non-obese diabetic (NOD) mice and lupus nephritis within NZB × NZWF$_1$ murine hybrids (Kelley *et al.*, 1988). Following the administration of the first or second dose of the murine anti-IL-2R mAb called OKT3, unwanted systemic effects occur, such as chills, tremors, fever, nausea and vomiting, and neurotoxicity (Horneff *et al.*, 1991), which arise owing to an imbalance between the levels of TNF-α and its soluble receptors (Herbelin *et al.*, 1995). Soluble TNFR–Fc treatment partially antagonizes the adverse effects of OKT treatment (Wee *et al.*, 1997).

The formation of neutralizing anti-idiotypic antibodies is a limiting factor in the use of murine mAbs for immunosuppressive therapy in humans; consequently chimaeric mouse–human anti-IL-2R mAbs have been constructed which are not only less immunogenic than the murine mAbs, but which are more effective immunological reagents that inhibit human T-cell proliferation (Rose *et al.*, 1992) and elicit antibody-dependent cell-mediated cytotoxicity (ADCC) activity not observed with the original murine monoclonals (Waldmann *et al.*, 1990). The hyperchimaeric anti-IL-2Rα (anti-Tac) mAb is more than 95% human and has minimal immunogenicity in humans (Hale *et al.*, 1988; LoBuglio *et al.*, 1989), providing a versatile reagent with clinical applications in the treatment of lymphoid malignancies. Anti-Tac is a murine IgG$_{2\alpha}$ mAb that recognizes the IL-2Rα chain (Uchiyama *et al.*, 1981) which is overexpressed in various lymphoid malignancies (Kay *et al.*, 1988; Ambrosetti *et al.*, 1989; Motoi *et al.*, 1989; Richards *et al.*, 1990) and provides a target for immunotherapy (Waldmann *et al.*, 1995). Although circulating levels of sIL-2Rα can interfere with targeting the IL-2Rα chain with anti-Tac (Waldmann *et al.*, 1995), a preinjection of humanized anti-Tac IgG antibody improves the targeting of anti-Tac to the IL-2Rα$^+$ tumour cells (Kobayashi *et al.*, 1997). Humanized anti-Tac mAbs armed with β-emitting strontium-90 or α-emitting bismuth-212 have been used clinically in radiotherapy of T-cell leukaemias (Pastan *et al.*, 1992).

Despite its beneficial effects in physiological defence mechanisms (Kishimoto *et al.*, 1995), the overproduction of IL-6 can lead to several pathological conditions including autoimmune disorders, defects in bone metabolism such as osteoporosis and B-cell malignancies (Hirano *et al.*, 1990; Kishimoto *et al.*, 1995). IL-6 is a major growth factor for myeloma cells *in vitro* (Klein *et al.*, 1989; Hilbert *et al.*, 1995) and is involved in the pathogenesis of multiple myelomas (Klein *et al.*, 1995) and B-cell lymphomas (Emilie *et al.*, 1997). Although antibodies to IL-6 are therapeutically effective and exert an antitumour effect in patients with multiple myeloma (Klein *et al.*, 1991; Bataille *et al.*, 1995), they are often unable to neutralize the large amounts of IL-6 produced in the majority of patients and reach only a limited clinical efficiency because of the formation of IL-6–antibody complexes (Bataille *et al.*, 1995). IL-6 variants, which behave as IL-6R antagonists and prevent IL-6 signalling by inhibiting dimerization of the IL-6R with gp130 (Savino *et al.*, 1994), have been shown to inhibit the proliferation of IL-6-dependent myeloma cells and act as potent inducers of myeloma cell apoptosis (Demartis *et al.*, 1996).

CYTOKINE ANTAGONISTS AS THERAPEUTIC REAGENTS

Alternative strategies for blocking cytokine action involve the use of proteins that bind specific receptors, but which lack the agonist activity of cytokines. As a pro-inflammatory cytokine, IL-1 is involved in the effector phase of inflammatory responses and induces a number of pathophysiological effects during the pathogenesis of various diseases, and consequently strategies for inhibiting IL-1 activities provide a focal point for the therapy of clinical diseases (Dinarello, 1991, 1994; Dinarello and Wolff, 1993).

A naturally occurring antagonist for IL-1, called secreted IL-1 receptor antagonist (sIL-1ra), has been described which is a β-trefoil member of the EGF superfamily (Arend, 1993; Dinarello, 1994). sIL-1ra shows 26% amino acid homology with IL-1β and 19% homology with IL-1α (Dinarello, 1994). sIL-1ra binds to the type I IL-1R with an affinity comparable to that of IL-1α or IL-1β (K_d approximately 200 pM), thus competing with both IL-1α and IL-1β for binding to the type I IL-1R; sIL-1ra binds to the type II IL-1R with considerably lower affinity than IL-1β. The binding of sIL-1ra to the type I and type II IL-1Rs does not initiate signal transduction (Dripps *et al.*, 1991) and secreted IL-1ra can block the pro-inflammatory effects of IL-1 (Dinarello and Thomson, 1991; Fischer *et al.*, 1992; Arend, 1993). sIL-1ra is generated in the course of a number of diseases (Fischer *et al.*, 1992) and circulating levels of IL-1 and sIL-1ra vary in parallel (Hirsch *et al.*, 1996).

IL-1 plays a major role in the pathogenesis of chronic synovitis, as well as in damage to articular cartilage and bone within rheumatoid joints (Arend, 1993), and the balance between the amounts of bioactive IL-1 and sIL-1ra may be a critical determinant in the severity and clinical course of RA; patients with high concentrations of sIL-1ra, but low levels of IL-1β, in the synovial fluid have rapid resolution in attacks of RA, whereas, in the reverse situation, patients have long intervals to recovery from attacks of RA (Miller *et al.*, 1993). However, although the production of sIL-1ra by neutrophils, monocytes, macrophages and fibroblasts is upregulated by IL-4, IL-6, IL-13, IFN-γ and GM-CSF in patients with active RA, the rate of production does not attain levels sufficient to neutralize all of the bioactive IL-1 generated within the RA joint (Firestein *et al.*, 1992; Feldmann *et al.*, 1997), since IL-1 antagonism by sIL-1ra requires more than a tenfold molar excess of the antagonist (Dinarello and Wolff, 1993). Nevertheless, clinical trials have shown that administration of exogenous sIL-1ra can provide beneficial effects in the treatment of RA, where it not only inhibits the influx of inflammatory leukocytes into the synovium and joint cavity, but also blocks the IL-1-mediated degradation of proteoglycans within the articular cartilage, and therefore beneficially influences the clinical course of the disease (Arend, 1993). In addition to the treatment of RA, sIL-1ra has found clinical application to the treatment of sepsis, cachexia, ankylosing spondylitis, type I diabetes, multiple sclerosis, asthma, chronic myelogenous leukaemia, and GVH disease during bone marrow transplantation (Antin *et al.*, 1994; Dinarello, 1994; Hirsch *et al.*, 1996).

Cytokine Peptides and Muteins as Cytokine Antagonists

Unravelling the molecular details of the structures of, and interactions between, cytokines and their receptors has been pivotal in the design of specific cytokine peptides and muteins that can function as cytokine antagonists (Chaiken and Williams, 1996).

IL-1 peptides have been generated which act as IL-1 antagonists. The *C*-terminal portion of mature IL-1β retains some biological activity and an IL-1β subpeptide, consisting of residues 237–269, antagonizes the effects of IL-1β on T cells (Palaszynski, 1987), whereas a 21-amino-acid subpeptide (residues 165–186) of human IL-1β exhibits fibroblast stimulatory properties (Herzbeck *et al.*, 1989). A nonapeptide of IL-1β consisting of residues 163–171 activates T cells, stimulates glycosaminoglycan (GAG) synthesis, and recruits antitumour reactivity *in vivo*, but lacks the proinflammatory and pyrogenic activities of IL-1. A pentapeptide, consisting of amino acids 165–169, is more potent than the nonapeptide. Various muteins of IL-1α/β have been generated by amino acid substitutions within the intact molecules. A single point mutation of aspartic acid 151 to tyrosine in mature IL-1β generates a mutein that functions as a receptor antagonist for both IL-1α and IL-1β. *N*-terminal mutations have yielded IL-1β muteins with altered biological properties and receptor-binding charcteristics (Huang *et al.*, 1987). A single point mutation of arginine 127 to glycine in human IL-1β results in a 100-fold loss of bioactivity in the IL-1β mutein on T cells without diminishing receptor–ligand binding, and this mutein could function as an IL-1 receptor antagonist.

PDGF acts as an autocrine or paracrine modulator of neovascularization and can act synergistically with other cytokines (e.g. FGF-2, TGF-β and insulin-like growth factor-1 (IGF-1)) in the process of neovascularization (Satoh *et al.*, 1992). Coexpression of the PDGF-β chain and the PGDF-R has been demonstrated in hyperproliferating microvascular endothelial cells in malignant gliomas (Hermansson *et al.*, 1988). The basic *C*-terminal sequence 211 amino acids within PDGF-A is involved in the binding of PDGF-A to cell surface heparan sulphate proteoglycans (LaRochelle *et al.*, 1990a,b; Ostman *et al.*, 1991) and a peptide identical to this *C*-terminal region is able to compete with, and displaces, attached forms of PDGF and other growth factors (Khachigian *et al.*, 1992; Raines and Ross, 1992) and could have clinical application in the control of angiogenesis-related diseases.

The haemopoietin GM-CSF has a core structure consisting of a left-handed, antiparallel four α-helix bundle with up-up-down-down connectivity (Diederichs *et al.*, 1991); common β sheet elements connecting helices A and B, and helices C and D, are also found in GM-CSF. Separate binding epitopes for recognition of the α and βc subunits of the GM-CSFR have been mapped within GM-CSF; the epitope for GM-CSFRα binding is centred in the C helix of GM-CSF, and a synthetic peptide, based on the amino acid sequence of the C helix, binds the receptor and displays antagonistic activities (Chaiken and Williams, 1996). The A helix of GM-CSF binds to the high-affinity GM-CSFRα–βc complex but not to GM-CSFRα alone, and it has been proposed, therefore, that the A helix in GM-CSF contains the binding site for the βc chain of GM-CSFR (Shanafelt *et al.*, 1991); mutants of Glu 21 in the A helix function as

GM-CSF antagonists (von Feldt *et al.*, 1994). From knowledge of the diverse geometrics of the contact residues for receptor recognition within the four α-helical bundle cytokines (Clackson and Wells, 1995; Morton *et al.*, 1995), immunomolecular techniques, involving recombinant antibody mimicry of the GM-CSF molecule, have been adapted to design a peptide with a reverse-turn geometry which behaves as a GM-CSF mimetic (Monfardi *et al.*, 1995). Similar molecular approaches for the development of other cytokine mimetics should provide useful cytokine antagonists for use in the clinic.

Chemokines, such as IL-8, play a pivotal role in many aspects of inflammation as well as in the pathogenesis of neoplastic lesions and chronic inflammatory diseases such as arthritis (Baggiolini *et al.*, 1994; Strieter *et al.*, 1995). Among the many therapeutic strategies that have been developed to neutralize the harmful effects of chemokines such as IL-8 are the use of analogues and peptides that can interfere with IL-8–receptor interactions (Howard *et al.*, 1996) including several chimaeric and deletion mutants (Clark-Lewis *et al.*, 1994; Suzuki *et al.*, 1994) and a synthetic hexapeptide with acetylated *N* and *C* termini (Hayashi *et al.*, 1995). The Glu–Leu–Arg (ELR) motif is found in all chemokines except PF4, Mig and IP-10, and is necessary for high-affinity binding of IL-8 and related α chemokines to IL-8Rβ (Ben-Baruch *et al.*, 1995). Several IL-8 mutants and chimaeras that lack the ELR motif do not bind IL-8Rβ and also inhibit the binding of other chemokines (Clark-Lewis *et al.*, 1994; Suzuki *et al.*, 1994).

Other uses for chemokine antagonists are currently being investigated in acquired immune deficiency syndrome (AIDS) research. The T-cell tropic retroviruses use the orphan chemokine receptor LESTR/fusin for infective entry (Choe *et al.*, 1996; Doranz *et al.*, 1996; Feng *et al.*, 1996), whereas macrophage and dendritic cell trophic HIV uses CCR-5 and CCR-3 (Alkhatib *et al.*, 1996; Choe *et al.*, 1996; Deng *et al.*, 1996; Doranz *et al.*, 1996; Dragic *et al.*, 1996). The capacity of different HIV strains to use different chemokine receptors as coreceptors for entry, influences their tropism and cytopathicity (Clapham, 1997; Clapham and Weiss, 1997). Retroviral infection of DCs renders them poor stimulators of T cells (Gabrilovich *et al.*, 1996) and the impaired function of the DCs may underlie the numerous immunodeficiencies identified within HIV-infected patients (Helbert *et al.*, 1993) and possibly be responsible for abnormalities in the T_{H1} to T_{H2} switch which is regarded as a pathogenic mechanism in HIV disease progression (Clerici and Shearer, 1993). Since HIV infection can be suppressed by chemokines and is dependent on chemokine receptors as a cell entry cofactor (Premack and Schall, 1996), pharmaceutical companies are currently exploiting the HIV–chemokine receptor nexus to develop therapies for AIDS (Balter, 1996a,b; Cohen, 1997).

The lymphocyte chemoattractant SDF-1 (stromal cell-derived factor) is a CXC chemokine (Tashiro *et al.*, 1993; Nagasawa *et al.*, 1994) which blocks HIV infection of $CD4^+$ T cells (Bleul *et al.*, 1996; Oberlin *et al.*, 1996). Although the β chemokines can inhibit HIV infection of target cells through blockade of CCR-5 (Cocchi *et al.*, 1995; Alkhatib *et al.*, 1996; Deng *et al.*, 1996; Dragic *et al.*, 1996; Paxton, 1996), the undesirable inflammation promoting activities of these chemokines would render them unsuitable as chemotherapeutic agents, and attention has focused instead on the development of chemokine receptor antagonists, such as variant chemokines which lack the chemotactic and leukocyte activating properties of natural chemokines, and which can function as suitable candidates for treatment of HIV-infected individuals (Arenzana-Seisdedos *et al.*, 1996; Simmons *et al.*, 1997), as well as for a variety of chemokine-driven diseases (Howard *et al.*, 1996).

Immunotoxins as Novel Therapeutic Reagents

Cytokine receptors are suitable targets for directed cytotoxin therapy, and a variety of immunotoxins have been generated which target cytokine receptors (Frankel *et al.*, 1996). *Pseudomonas* exotoxin (PE) is a single-chain toxin secreted by *Pseudomonas aeruginosa* and must be endocytosed in order to kill a cell; a 40-kDa protein (PE40) derived from PE has been used to generate an immunotoxin by coupling it to scFv molecules (Pastan *et al.*, 1992). Several single-chain immunotoxins have been generated from these single-chain antibodies, which have a high affinity for their target antigens and have PE40 fused to their carboxyl ends (Pastan *et al.*, 1992). For instance, the single-chain antibodies scFv (FRP5) (anti-erbB2) and scFv(225) (anti-EGFR) both coupled to PE (Wels *et al.*, 1992; Schmidt *et al.*, 1996) and the scFv constructed from mAb 55.1 and PE are immunotoxins that are cytotoxic to cells bearing EGFR family receptors and a mucin-like antigen respectively (Frankel *et al.*, 1996). A recombinant bispecific single-chain toxin containing a scFv antibody domain specific for erbB2, as well as TGF-α linked to PE, has been shown both to be cytotoxic *in vitro* for human tumour cells expressing erbB2 and to inhibit the growth of human A431 erbB2$^+$ epidermoid tumour cell xenografts in nude mice (Schmidt and Wels, 1996). Other agents that have been used in the production of immunotoxins include ricin, shaga toxin, diphtheria toxin and the ribonuclease, α-sarcin (Frankel *et al.*, 1996).

Targeting angiogenesis is proving to be an effective antitumour strategy, and the tumour vascular endothelial cells are especially vulnerable targets for immunotoxins (Bicknell and Harris, 1992; Burrows and Thorpe, 1993, 1994). Toxins can be directed to target cells by coupling them to cytokines (Ramakrishnan *et al.*, 1992). Chimaeric cytotoxins have been made by fusing PE40 to cytokines such as IL-2, IL-6, VEGF and TGF-α (Pastan *et al.*, 1992). TGF-α–PE40 targets epidermoid carcinoma cells and adenocarcinoma cells, and proliferating smooth muscle cells, whereas IL-6–PE40 targets myeloma and hepatoma cells. IL-2–PE40 targets adult T-cell leukaemia cells and lymphoma cells, and has been used in the treatment of inflammatory diseases such as autoimmune arthritis, uveitis and encephalitis (Pastan *et al.*, 1992). An endothelial cell-specific cytotoxic conjugate, consisting of VEGF coupled to truncated diphtheria toxin (VEGF165–DT385), has been used to inhibit FGF-2-induced endothelial cell proliferation *in vitro* and angiogenesis *in vivo* (Ramakrishnan *et al.*, 1996).

Suramin as a Cytokine Antagonist and Anticancer Agent

Suramin is a polysulphonated naphthylurea that interferes with cell proliferation-inducing activities of various cytokines (including IGFs, EGF, PDGF, IL-2 and FGFs) by uncoupling G proteins from receptors and by inhibiting the binding of the cytokines to their respective receptors (Berns *et al.*, 1990; Kopp and Pfeiffer, 1990; Mills *et al.*, 1990; Pollak and Richard, 1990; Vignon *et al.*, 1992; Foekens *et al.*, 1993). Suramin blocks PKC activity (Hensey *et al.*, 1989) and can block the expression of mRNA for FGFRs (Kasayama *et al.*, 1993). Androgen stimulation of SC-3 murine mammary carcinoma cells induces them to secrete a heparan-binding autocrine growth factor which binds to FGF receptors and, by inhibiting the androgen-inducible autocrine loop, suramin exerts an antiproliferative effect (Kasayama *et al.*, 1993). However, intracellular effects of suramin cannot be excluded since suramin is internalized and can sequester within the Golgi–lysosome system, where it causes lysosomal perturbations, and can accumulate within both mitochondria and the nucleus (Baghdiguinan *et al.*, 1996). Suramin can adversely affect mitochondrial functions (Calcaterra *et al.*, 1988; Rago *et al.*, 1991, 1994) and affects RNA and DNA metabolism through inhibitory effects on RNA polymerase (Hawking, 1987) and DNA topoisomerase II (Bojanowski *et al.*, 1992). The antineoplastic effects of suramin may also reflect its ability to inhibit the reduced folate carrier system responsible for folate transport into cancer cells (Rideout *et al.*, 1995). Thus suramin exerts its effects by several mechanisms (Stein, 1993).

Suramin has gained renewed interest as an anti-tumour agent (Myers *et al.*, 1992) as it inhibits the proliferation of a variety of malignant cells *in vitro* (Freuhauf *et al.*, 1990; Olivier *et al.*, 1990; Vignon *et al.*, 1992; Foekens *et al.*, 1993). Other clinical applications of suramin centre on its ability to interfere with growth factor activity; for instance, FGFs are involved in angiogenesis through stimulation of endothelial cell proliferation and motility (Klagsbrun and Folkman, 1990) and suramin may have application in the treatment of pathological conditions that are angiogenesis dependent (Pesenti *et al.*, 1992). Suramin inhibits the proliferation of fibroblasts in response to monocyte-derived growth factors (Kelley and Hildebran, 1987) and may have clinical application in the treatment of fibrotic disorders that develop within several organ systems through cytokine-driven proliferation of fibroblasts and excessive deposition of extracellular matrix (Kovacs, 1991) (see Table 5).

REFERENCES

Aaronson, S.A. (1991). *Science* **254**, 1146–1153.

Abbas, A.K., Murphy, K.M. and Sher, A. (1996). *Nature* **383**, 787–793.

Adams, D.O. and Hamilton, T.A. (1992). In *The Macrophage* (eds C.E. Lewis and J.O'D. McGee), IRL Press, Oxford, pp. 77–114.

Adema, G.J., Hartgers, F., Verstraten, R., de Vries, E., Marland, G., Menon, S., Foster, J., Xu, Y., Nooyen, P., McClanachan, T., Bacon, K.B. and Figdor, C.G. (1997). *Nature* **387**, 713–717.

Aderka, D., Engelmann, H., Maor, Y., Brakebusch, C. and Wallach, D. (1992). *J. Exp. Med.* **175**, 323–329.

Adorini, L. and Sinigaglia, F. (1997). *Immunol. Today* **18**, 209–211.

Aggarwal, B.B. and Natarajan, K. (1996). *Eur. Cytokine Netw.* **7**, 93–124.

Agosti, J.M., Fisher, C.J., Opal, S.M., Lowry, S.F., Balk, R.A. and Sadoff, J.C. (1994). *Proceedings of the 34th Ann. ICAAC* **34**, 65.

Agrawal, S. (1996). *TIBTech.* **14**, 376–387.

Akira, S., Taga, T. and Kishimoto, T. (1993). *Adv. Immunol.* **54**, 1–78.

Alam, R. (1997). *J. Allergy Clin. Immunol.* **99**, 273–277.

Albelda, S.M., Smith, C.W. and Ward, P.A. (1994). *FASEB J.* **8**, 504–512.

Alkhatib, G., Combadiere, C., Broder, C.C., Feng, Y., Kennedy, P.E., Murphy, P.M. and Berger, E.A. (1996). *Science* **272**, 1955–1958.

Allard, S.A., Bayliss, M.T. and Maini, R.N. (1990). *Arthritis Rheum.* **33**, 1170–1179.

Allione, A., Consalvo, M., Nanni, P., Lollini, P.L., Cavallo, F., Giovarelli, M., Forni, M., Crulino, A., Colombo, M.P., Dellabona, P., Hock, H., Blankenstein, T., Rosenthal., F.M., Gansbacher, B., Bosco, M.C., Musso, T., Gusella, L. and Forni, G. (1994). *Cancer Res.* **54**, 6022–6026.

Almawi, W.Y., Beyhum, H.N., Rahme, A.A. and Rieder, M.J. (1996). *J. Leukocyte Biol.* **60**, 563–572.

Altisano, L., Cacamo, J., Ventura, F., Weiss, F.M.B., Massague, J. and Wrana, J.L. (1993). *Cell* **75**, 671–680.

Ambrosetti, A., Semenzato, G., Prior, M., Chilosi, M., Vinante, F., Vincenzi, C., Zanotti, R., Trentin, L., Portuese, A., Menestrina, F., Perona, G., Agostini, C. and Todeschini, G. (1989). *Br. J. Haematol.* **73**, 181–186.

Angiolillo, A.I., Sgadari, C., Taub, D.D., Liao, F., Farber, J.M., Maheshwari, S., Kleinman, H.K., Reaman, G.H. and Tosato, G. (1995). *J. Exp. Med.* **182**, 155–162.

Ankrapp, D.P. and Bevan, D.R. (1993). *Cancer Res.* **53**, 3399–3404.

Antin, J.H., Weinstein, H.J., Guinan, E.C., McCarthy, P., Bierer, B.E., Gilliland, D.G., Parsons, S.K., Ballen, K.K., Rimm, I.J., Falzarano, G., Bloedow, D.C., Abate, I., Lebsack, M., Burakoff, S.J. and Ferrara, J.L.M. (1994). *Blood* **84**, 1342–1348.

Arai, K., Lee, F., Hiyajima, A., Miyatake, S., Arai, N. and Yokota, T. (1990). *Annu. Rev. Biochem.* **59**, 783–836.

Arend, W.P. (1993). *Adv. Immunol.* **54**, 167–227.

Arend, W.P. and Dayer, J.-M. (1995). *Arthritis Rheum.* **38**, 151–160.

Arenzana-Siededos, F., Virelizier, J.-L., Rousset, D., Clark-Lewis, I., Loetscher, P., Moser, B. and Baggiolini, M. (1996). *Nature* **383**, 400.

Arias, J., Alberts, A.S., Brindle, P., Claret, F.X., Smeal, T., Karin, M., Fermisco, J. and Montminy, M. (1994). *Nature* **370**, 226–229.

Ashkenazi, A., Marsters, S.S.A., Capon, A.L., Chamow, S.M., Figari, I.S., Pennica, D., Goeddel, D.V., Palladino, M.A. and Smith, D.H. (1991). *Proc. Natl Acad. Sci. U.S.A.* **88**, 10535–10539.

Auphan, N., Didonato, J.A., Rosette, C., Helmberg, A. and Karin, M. (1995). *Science* **270**, 286–289.

Austrup, F., Vestweber, D., Borges, E., Lohning, M., Brauer, R., Herz, U., Renz, H., Hallman, R., Scheffold, A., Radbruch, A. and Hamann, A. (1997). *Nature* **385**, 81–83.

Austyn, J.M. (1993). In *Immunology of Heart and Lung Transplantation* (eds M.L. Rose and M.H. Jacoub), Little Brown, Boston, MA, pp. 22–41.

Aversa, G., Punnonen, J. and Devries, J.E. (1993). *J. Exp. Med.* **177**, 1575–1585.

Bacon, C.M., Petricoin, E.F., Ortaldo, J.R., Rees, R.C., Larner, A.C., Johnston, J.A. and O'Shea, J.-J. (1995). *Proc. Natl Acad. Sci. U.S.A.* **92**, 7307–7311.

Baeuerle, P.A. and Baltimore, D. (1996). *Cell* **87**, 13–20.

Baeuerle, P.A. and Henkel, T. (1994). *Ann. Rev. Immunol.* **12**, 141–179.

Baggiolini, M.P. and Dahinden, C.A. (1994). *Immunol. Today* **15**, 127–133.

Baggiolini, M.P., Dewald, B. and Moser, B. (1994). *Adv. Immunol.* **55**, 97–197.

Baggiolini, M.P., Loetscher, P. and Moser, B. (1995). *Int. J. Immunopharmacol.* **17**, 103–108.

Baghdiguian, S., Boudier, J.L., Boudier, J.A. and Fantini, J. (1996). *Eur. J. Cancer* **32A**, 525–532.

Bagley, C.J., Woodcock, J.M., Stromski, F.C. and Lopez, A.F. (1997). *Blood* **89**, 1471–1482.

Baldwin, A.S. (1996). *Ann. Rev. Immunol.* **14**, 649–681.

Balter, M. (1996a). *Science* **274**, 1464–1465.

Balter, M. (1996b). *Science* **274**, 1988–1989.

Bannister, A.J. and Kouzarides, T. (1996). *Nature* **384**, 641–643.

Barnes, P.J. and Adcock, I.M. (1997). *Trends Pharm Sci.* **18**, 46–50.

Bartlett, R. R. and Schleyerbach, R. (1985). *Int. J. Immunopharmacol.* **7**, 7–16.

Basilico, C. and Moscatelli, D. (1992). *Adv. Cancer Res.* **59**, 115–165.

Baskar, S., Nabavi, N., Glimcher, L. and Ostrand-Rosenberg, S. (1993a). *J. Immunother.* **14**, 209–215.

Baskar, S., Ostrand-Rosenberg, S., Nabavi, N., Nadler, L.M., Freeman, G.J. and Glimcher, L. (1993b). *Proc. Natl Acad. Sci. U.S.A.* **90**, 5687–5690.

Baskar, S., Glimcher, L., Nabavi, N., Jones, R.T. and Ostrand-Rosenberg, S. (1995). *J. Exp. Med.* **181**, 610–629.

Bataille, R., Barlogie, B., Lu, Z., Rossi, J.-F., Lavabre-Bertand, T., Beck, T., Wijdenes, F., Brochier, J. and Klein, B. (1995). *Blood* **86**, 685–691.

Baumann, H. and Gauldie, J. (1994). *Immunol. Today* **15**, 74–80.

Bazan, J.F. (1993). *Curr. Biol.* **3**, 603–606.

Beck, L. and D'Amore, P.A. (1997). *FASEB J.* **11**, 365–373.

Bedard, P.-A. and Golds, E.E. (1993). *J. Cell. Physiol.* **154**, 433–441.

Beg, A.A. and Baltimore, D. (1996). *Science* **274**, 782–784.

Bellusci, S., Moens, G., Gaudino, G., Comoglio, P.M., Nakamura, T., Thiery, J.P. and Jouannaeu, J. (1994). *Oncogene* **9**, 1091–1099.

Ben-Baruch, A., Michiel, D.F. and Oppenheimer, J.J. (1995). *J. Biol. Chem.* **270**, 11 703–11 706.

Bendtzen, K., Svenson, M., Jonsson, V. and Hippe, E. (1990). *Immunol. Today* **11**, 167–169.

Benmerak, A., Gagnon, J., Begue, B., Megarbane, B., Dautry-Varsat and Cerf-Bensussan, N. (1995). *J. Cell Biol.* **131**, 1831–1838.

Benveniste, E.N. and Benos, D.J. (1995). *FASEB J.* **9**, 1577–1584.

Bergmann, U., Funatomi, H., Yokoyama, M., Berger, H.G. and Kore, M. (1995). *Cancer Res.* **55**, 2007–2011.

Berns, E.M.J.J., Schuurmans, A.L.G., Bolt, J., Lamb, D.J., Foekens, J.A. and Mulder, E. (1990). *Eur. J. Cancer* **26**, 470–474.

Besedovsky, H.O. and Del Rey, A. (1996). *Endocr. Rev.* **17**, 64–102.

Betsholtz, C., Westermark, B., Ek, B. and Heldin, C.H. (1984). *Cell* **39**, 447–457.
Beutler, B. (1990). In *Peptide Growth Factors and Their Receptors 2* (eds M.B. Sporn and A.B. Roberts), Springer, London, pp. 39–70.
Bicknell, R. and Harris, A.L. (1992). *Semin. Cancer Biol.* **3**, 399–407.
Bikfalvi, A., Klein, S., Pintucci, G. and Rifkin, D.B. (1997). *Endocr. Rev.* **18**, 26–45.
Billiau, A. (1996). *Adv. Immunol.* **62**, 61–130.
Bishop, J.M. (1987). *Science* **235**, 305–311.
Black, R.A., Rauch, C.T., Kozlosky, C.J., Peschon, J.J., Slack, J.L., Wolfson, M.F., Castner, B.J., Stocking, K.L., Reddy, P., Srinivasan, S., Nelson, N., Boiani, N., Schooley, K.A., Gerhardt, M., Davis, R., Fitzner, J.N., Johnson, R.S., Paxton, R.J., March, C.J. and Cerretti, D.P. (1997). *Nature* **385**, 729–733.
Blaikie, P., Immanuel, D., Wu, J., Li, N.X., Yajnik, V. and Margolis, B. (1994). *J. Biol. Chem.* **269**, 32 031–32 034.
Blaise, D., Olive, D., Hirn, M., Viens, P., Lafage, M. and Attal, M. (1991). *Bone Marrow Transpl.* **8**, 105–111.
Blay, J.-Y., Farcet, J.-P., Lavaud, A., Radoux, D. and Chouaib, S. (1994). *Eur. J. Cancer* **30A**, 321–324.
Bleul, C.C., Farzan, M., Choe, H., Parolin, C., Clark-Lewis, I., Sodroski, J. and Springer, T.A. (1996). *Nature* **382**, 829–833.
Bohlen, P. and Kovesdi, I. (1991). *Prog. Growth Factor Res.* **3**, 143–157.
Boissier, M.C., Chiocchia, G., Bessis, N., Hajnal, J., Garotta, G., Nicoletti, F. and Fournier, C. (1995). *Eur. J. Immunol.* **25**, 1184–1190.
Bojanowski, K., Lelievre, S., Markovits, J., Couprie, J., Jacqueminsablon, A. and Larsen, A.K. (1992). *Proc. Natl Acad. Sci. U.S.A.* **89**, 3025–3029.
Bokoch, G.M. (1995). *Blood* **86**, 1649–1660.
Bone, R.C. (1991). *Ann. Intern. Med.* **115**, 457–469.
Bonner, J.C. and Brody, A.R. (1995). *Am. J. Physiol.* **268**, L869–L878.
Border, W.A., Okuda, S., Languino, L.R., Sporn, M.B. and Ruoslahti, E. (1990). *Nature* **346**, 371–374.
Border, W.A., Noble, N.A., Yamamoto, T., Harper, J.R., Yamaguchi, Y., Pierschbacher, M.D. and Ruoslahti, E. (1992). *Nature* **360**, 361–364.
Bottaro, D.P., Rubin, J.S., Faletto, D.L., Chan, A.M.-L., Kmiecik, T.E., van de Woude, G.F. and Aaronson, S.A. (1991). *Science* **251**, 802–804.
Bouck, N., Stellmach, V. and Hsu, S.C. (1996). *Adv. Cancer Res.* **69**, 135–174.
Bradley-Mullen, H., Sharp, G.C., Bickel, J.T. and Kyriakos, M. (1991). *J. Exp. Med.* **173**, 899–912.
Braun, W., Kallen, J., Mikol, V., Walkinshaw, M.D. and Wuthrich, K. (1995). *FASEB J.* **9**, 63–72.
Brennan, M., Chantry, D., Jackson, A., Maini, R. and Feldmann, M. (1989). *Lancet* **ii**, 244–247.
Bretscher, P.A., Wei, G., Menon, J.N. and Bielefold-Ohmann, H. (1992). *Science* **257**, 539–542.
Brunda, M. (1994). *J. Leukocyte Biol.* **55**, 280–288.
Brunda, M.J., Luistro, L., Warrier, R.R., Wright, R.B., Hubbard, B.R., Murphy, M., Wolf, S.F. and Gately, M.K. (1993). *J. Exp. Med.* **178**, 1223–1230.
Brunet, A. and Pouyssegur, J. (1996). *Science* **272**, 1650–1655.
Buday, L. and Downward, J. (1993). *Cell* **73**, 611–620.
Burkhardt, K., Mandel, T.E., Diamantstein, T. and Loughnan, M.S. (1989). *Immunology* **66**, 183–189.
Burrows, F.J. and Thorpe, P.E. (1993). *Proc. Natl Acad. Sci. U.S.A.* **90**, 8996–9000.
Burrows, F.J. and Thorpe, P.E. (1994). *Pharmacol. Ther.* **64**, 155–174.
Burrus, L.W., Zuber, M.E., Lueddecke, B.A. and Olwin, B.B. (1992). *Mol. Cell. Biol.* **12**, 5600–5609.
Butler, D.M. (1994). *Cytokine* **6**, 616–623.
Calcaterra, N.B., Vicario, L.R. and Roveri, O.A. (1988). *Biochem. Pharmacol.* **37**, 2521–2527.
Callard, R. and Gearing, A. (1994). *The Cytokine Facts Book.* Academic Press, London.
Callard, R.E., Mathews, D.J. and Hibbert, L. (1996). *Immunol. Today* **17**, 108–110.
Cantley, L.C., Auger, K.R., Carpenter, C., Duckworth, B., Graziani, A., Kapeller, R. and Soltoff, S. (1991). *Cell* **64**, 281–302.
Cantrell, D. (1996). *Annu. Rev. Immunol.* **14**, 259–274.
Cao, Y., Chen, C., Weatherbee, J.A., Tsang, M. and Folkman, J. (1995). *J. Exp. Med.* **182**, 2069–2077.
Cao. Z., Xiong, J., Takeuchi, M., Kurama, T. and Goeddel, D.V. (1996). *Nature* **383**, 443–446.
Carpenter, S. and Kraus, G.A. (1991). *Photochem. Photobiol.* **53**, 169–174.
Carraway, K.L. and Burden, S.J. (1995). *Curr. Opin. Neurobiol.* **5**, 606–612.
Cato, A.C.B. and Wade, E. (1996). *Bioessays* **18**, 371–378.
Cerami, A. (1992). *Clin. Immunol. Immunopathol.* **62**, S3–S10.

Cerna, H. (1989). *Acta Univ. Palacki. Olomuc. Fac. Med.* **123**, 293–302.
Cerretti, D.P., Wignall, J., Anderson, D., Tushinski, R.J., Gallis, B.M., Stya, M., Gillis, S., Urdal, D.L. and Cosman, D. (1988). *Mol. Immunol.* **25**, 761–770.
Chaiken, I.M. and Williams, W.V. (1996). *TIBTech.* **14**, 369–375.
Chamow, S.M. and Ashkenazi, A. (1996). *TIBTech.* **14**, 52–60.
Chang, H., Riese, D.J., Gilbert, W., Stern, D.F. and McMahan, U.J. (1997). *Nature* **387**, 509–512.
Chardin, P., Camonis, J.H., Gale, N.W., Vanaelst, L., Schlessinger, J., Wigler, M.H. and Barsagi, D. (1993). *Science* **260**, 1338–1343.
Chen, E., Keystone, E.C. and Fish, E.N. (1993). *Arthritis Rheum.* **36**, 901–910.
Chen, L., Ashe, S., Brady, W.A., Hellstrom, I., Hellstrom, K.E., Lebbetter, J.A., McGowan, P. and Linsley, P.S. (1992). *Cell* **71**, 1093–1102.
Chen, L., Linsley, P.S. and Hellstrom, K.E. (1993). *Immunol. Today* **14**, 483–486.
Chen, R.-H., Sarnecki, C. and Blenis, J. (1992). *Mol. Cell. Biol.* **12**, 915–927.
Chen, S.-H., Kosai, K.-I., Xu, B., Pham-Nguyen, K., Contant, C., Finegold, M.J. and Woo, S.L.C. (1996). *Cancer Res.* **56**, 3758–3762.
Chen, X., Rubock, M.J. and Whitman, M. (1996). *Nature* **383**, 691–696.
Chen, Z., Fisher, R.J., Riggs, C.W., Rhim, J.S. and Lautenberger, J.A. (1997). *Cancer Res.* **57**, 2013–2019.
Chester, K.A. and Hawkins, R.E. (1995). *TIBTech.* **13**, 294–300.
Chiefetz, S., Bellon, T., Cales, C., Vera, S., Bernabeu, C., Nassague, J. and Letarte, M. (1992). *J. Biol. Chem.* **267**, 19 027–19 030.
Chikanza, I.C., Roux-Lombard, P., Dayer, J.-M. and Panayi, G.S. (1995). *Arthritis Rheum.* **38**, 642–648.
Chin, L.S., Murray, S.F., Zitnay, K.M. and Rami, B. (1997). *Clin. Cancer Res.* **3**, 771–776.
Choe, H., Farzan, M., Sun, Y., Sullivan, N., Rollins, B., Ponath, P.D., Wu, L.J., Mackay, C.R., Larosa, G., Newman, W., Gerard, N., Gerard, C. and Sodroski, J. (1996). *Cell* **85**, 1135–1148.
Choudhuri, R., Zhang, H.-T., Donnini, S., Ziche, M. and Bicknell, R. (1997). *Cancer Res.* **57**, 1814–1819.
Chrousos, G.P. (1995). *N. Engl. J. Med.* **332**, 1351–1362.
Clackson, T. and Wells, J.A. (1995). *Science* **267**, 383–386.
Claffrey, K.P., Brown, L.F., del Aguila, L.F., Tognazzi, K., Yeo, K.-T., Manseau, E.J. and Dvorak, H.F. (1996). *Cancer Res.* **56**, 172–181.
Clapham, P.R. (1995). *Cell* **80**, 259–268.
Clapham, P.R. (1997). *Trends Cell Biol.* **7**, 264–268.
Clapham, P.R. and Weiss, R.A. (1997). *Nature* **388**, 230–231.
Clark, J.W., Santos-Moore, A., Stevenson, L.E. and Frackleton, A.R. (1996). *Int. J. Cancer* **65**, 186–191.
Clarke, P.R. (1994). *Curr. Biol.* **4**, 647–650.
Clark-Lewis, I., Dewald, B., Loetscher, M., Moser, B. and Baggiolini, M. (1994). *J. Biol. Chem.* **269**, 16 075–16 081.
Clark-Lewis, I., Kim, K.S., Rajarathnam, K., Gong, J.H., Dewald, B., Moser, B., Baggiolini, M. and Sykes, B.D. (1995). *J. Leukocyte Biol.* **57**, 703–711.
Clerici, M. and Shearer, G.M. (1993). *Immunol. Today* **14**, 107–111.
Clore, G.M. and Gronenborn, A.M. (1995). *FASEB J.* **9**, 57–62.
Cocchi, F., Devico, A.L., Garzinodemo, A., Arya, S.K., Gallo, R.C. and Lusso, P. (1995). *Science* **270**, 1811–1815.
Cohen, J. (1997). *Science* **275**, 1261–1264.
Coll-Fresno, P.M., Batoz, M., Tarquin, S., Birnbaum, D. and Coulier, F. (1997). *Oncogene* **14**, 243–247.
Collins, T.L., Deckert, M. and Altman, A. (1997). *Immunol. Today* **18**, 221–225.
Colombo, M.P. and Forni, G. (1994). *Immunol. Today* **15**, 48–51.
Colotta, F., Re, F., Muzio, M., Bertini, R., Polentarutti, N., Sironi, M., Giri, J.G., Dower, S.K., Sims, J.E. and Mantovani, A. (1993). *Science* **261**, 472–475.
Connor, J., Bannerji, R., Saito, S., Heston, W., Fair, W. and Gilboa, E. (1993). *J. Exp. Med.* **177**, 1127–1134.
Cooper, R.N. (1992). *J. Cell Biochem.* **16G**, 131–136.
Cope, A.P., Aderka, D., Doherty, M., Engelmann, H., Gibbons, D., Jones, A.C., Brennan, F.M., Maini, R., Wallach, D. and Feldmann, M. (1992). *Arthritis Rheum.* **35**, 1160–1169.
Cowley, S., Paterson, H., Kemp, P. and Marshall, C.J. (1994). *Cell* **77**, 841–852.
Crabtree, G.R. and Clipstone, N.A. (1994). *Ann. Rev. Biochem.* **63**, 1045–1083.
Cresswell, P. (1994). *Ann. Rev. Immunol.* **12**, 259–293.
Croall, D.E. and DeMartino, G.N. (1991). *Physiol. Rev.* **71**, 813–847.

Croft, M., Carter, L., Swain, S.L. and Dutton, R.W. (1995). *J. Exp. Med.* **180**, 1715–1728.
Crowe, P.D., Vanarsdale, T.L., Walter, B.N., Dahms, K.M. and Ware, C.F. (1994). *J. Immunol. Methods* **168**, 79–89.
Cullen, K.J., Smith, H.S., Hill, S., Rosen, N. and Lippman, M.E. (1991). *Cancer Res.* **51**, 4978–4985.
Cush, J.J. and Lipsky, P.E. (1991). *Clin. Orthop.* **265**, 9–22.
Culter, R.L., Lui, L., Damen, J.E. and Krystal, G. (1993). *J. Biol. Chem.* **268**, 1463–1469.
Cuvillier, O., Pirianov, G., Kleuser, B., Vanek, P.G., Coso, O.A., Gutkind, J.S. and Spiegel, S. (1996). *Nature* **381**, 800–803.
Czubayko, F., Riegel, A.T. and Wellstein, A. (1994). *J. Biol. Chem.* **269**, 21 358–21 363.
D'Andrea, R., Rayner, J., Moretti, P., Lopez, A., Goodall, G.J., Gonda, T.J. and Vadas, M. (1994). *Blood* **83**, 2802–2808.
Danial, N.N., Pernis, A. and Rothman, P.B. (1995). *Science* **269**, 1875–1877.
Darnay, B.G. and Aggarwal, B.B. (1997). *J. Leukocyte Biol.* **61**, 559–566.
Darnell, J.J., Kerr, I.M. and Stark, G.R. (1994). *Science* **264**, 1415–1421.
David, G. (1993). *FASEB J.* **1**, 1023–1030.
Davies, D.E. and Chamberlin, S.G. (1996). *Biochem. Pharmacol.* **51**, 1101–1110.
Davies, R.J. (1994). *TIBS* **19**, 470–473.
Dayal, A.K. and Kammer, G.M. (1996). *Arthritis Rheum.* **39**, 23–33.
Debets, R. and Savelkoul, H.F.J. (1994). *Immunol. Today* **15**, 455–458.
de Kruif, J., van der Vuurst de Vries, A.-R., Cilenti, L., Boel, E., van Ewijk, W. and Logtenberg, T. (1996). *Immunol. Today* **17**, 453–455.
del Prete, G., Maggi, E. and Rogagnani, S. (1994). *Lab. Invest.* **70**, 299–306.
Demartis, A., Bernassola, F., Savino, R., Melino, G. and Gilberto, G. (1996). *Cancer Res.* **56**, 4213–4218.
Deng, H.K., Liu, R., Ellmmeier, W., Choe, S., Unutmaz, D., Burkhart, M., Dimarzio, P., Marmon, S., Sulton, R.E., Hill, C.M., Davis, C.B., Pieper, S.C., Schall, T.J., Littman, D.R. and Landau, N.R. (1996). *Nature* **381**, 661–666.
Dent, A.L., Shaffer, A.L., Yu, X., Allman, D. and Staudt, L.M. (1997). *Science* **276**, 589–592.
de Paulis, A., Ciccarelll, A., Marino, I., de Crescenzo, G., Marino, D. and Marone, G. (1997). *Arthritis Rheum.* **40**, 469–478.
Derijard, B., Hibi, M., Wu, I.H., Barrett, T., Su, B., Deng, T.L., Karin, M. and Davies, R.J. (1994). *Cell* **76**, 1025–1037.
Derijard, B., Raingeaud, J., Barrett, T., Wu, I.H., Han, J.H., Ulevitch, R.J. and Davies, R.J. (1995). *Science* **267**, 682–685.
DeVita, V.T., Hellman, S. and Rosenberg, S.A. (1995). *Biological Therapy of Cancer.* J.B. Lippincott, Philadelphia, PA.
Diazmeco, M.T., Dominguez, I., Sang, L., Dent, P., Lazano, J., Municio, M.M., Berra, E., Hay, R.T., Sturgill, T.W. and Moscat, J. (1994). *EMBO J.* **13**, 2842–2848.
Dickson, R.B. and Lippman, M.E. (1995). *Endocr. Rev.* **16**, 559–589.
Diederichs, K., Boone, T. and Karplus, P.A. (1991). *Science* **254**, 1779–1782.
Dikic, I., Tokaiwa, G., Lev, S., Courtneidge, S.A. and Schlessinger, J. (1996). *Nature* **383**, 547–550.
Dinarello, C.A. (1991). *Blood* **77**, 1627–1652.
Dinarello, C.A. (1994). *FASEB J.* **8**, 1314–1325.
Dinarello, C.A. (1996). *Blood* **87**, 2095–2147.
Dinarello, C.A. and Thomson, R.C. (1991). *Immunol. Today* **12**, 404–410.
Dinarello, C.A. and Wolff, S.M. (1993). *N. Engl. J. Med.* **328**, 106–112.
DiSanto, J.P. (1997). *Curr. Biol.* **7**, R424–R426.
Dobrowolski, S., Harter, M. and Stacey, D.W. (1994). *Mol. Cell. Biol.* **14**, 5441–5449.
Doranz, B.J., Rucker, J., Yi, Y.J., Smyth, R.J., Samson, M., Peiper, S.C., Parmentier, M., Collman, R.G. and Doms, R.W. (1996). *Cell* **85**, 1149–1158.
Dragic, T., Litwin, V., Allaway, G.P., Martin, S.R., Huang, Y.X., Nagashima, K.A., Cayanan, C., Maddon, P.J., Koup, R.A., Moore, J.P. and Paxton, W.A. (1996). *Nature* **381**, 667–673.
Dranoff, G. and Mulligan, R.C. (1995). *Adv. Immunol.* **58**, 417–454.
Dranoff, G., Jaffee, E., Lazenby, A., Golumbek, P., Levitsky, H., Broze, K., Jackson, V., Hamada, H., Pardoll, D. and Mulligan, R.C. (1993). *Proc. Natl Acad. Sci. U.S.A.* **90**, 3539–3543.
Dripps, D.J., Brandhuber, B.J., Thomson, R.C. and Eisenberg, S.P. (1991). *J. Biol. Chem.* **266**, 10 331–10 336.
Drummond-Barbosa, D. and DiMaio, D. (1997). *Biochim. Biophys. Acta* **1332**, M1–M17.

Duff, G.W. (1989). In *Peptide Regulatory Factors* (ed. A.R. Green), Edward Arnold, London, pp. 112–120.
Dvorak, H.F. (1986). *N. Engl. J. Med.* **315**, 1650–1659.
Dvorak, H.F., Brown, L.F., Detmar, M. and Dvorak, A.M. (1995). *Am. J. Pathol.* **146**, 1029–1039.
Ebner, R., Chen, R.-H., Shum, L., Lawler, S., Zioncheck, T.F., Lee, A., Lopez, A.R. and Derynck, R. (1993). *Science* **260**, 1344–1348.
Edgington, S.M. (1993). *Biotechnology* **11**, 676–681.
Edwards, D.R. (1994). *Trends Pharmacol. Sci.* **15**, 239–244.
Eisenberg, S.P., Evans, R.J., Arend, W.P., Verderber, E., Brewer, M.T., Hannum, C.H. and Thompson, R.C. (1990). *Nature* **343**, 341–346.
Eisenman, R.N. and Cooper, J.A. (1995). *Nature* **378**, 438–439.
Elliott, M.J., Maini, R.N., Feldmann, M., Long-Fox, A., Charles, P., Katsikis, P., Brennan, F.M., Walker, J., Bijl, H., Ghrayeb, J. and Woody, R.C. (1993). *Arthritis Rheum.* **36**, 1681–1690.
Elliott, M.J., Maini, R.N., Feldmann, M., Kalden, J.R., Antoni, C., Smolen, J.S., Leeb, B., Breedveld, F.C., MacFarlane, J.D., Bijl, H. and Woody, R.C. (1994). *Lancet* **344**, 1105–1110.
Emilie, D., Zou, W., Fior, R., Llorente, L., Durandy, A., Crevon, M.-C., Maillot, M.-C., Durand-Gasselini, I., Raphael, M., Peuchmaur, M. and Galamaud, P. (1997). *Companion Methods Enzymol.* **11**, 133–142.
Enari, M., Hug, H. and Nagata, S. (1995). *Nature* **375**, 78–81.
Endo, T.A., Masuhara, M., Yokouchi, M., Susuki, R., Sakamoto, H., Mitsui, K., Matsumoto, A., Tanimura, S., Ohtsubo, M., Misawa, H., Miyazaki, T., Leonor, N., Taniguchi, T., Fujita, T., Kanakura, Y., Komiya, S. and Yoshimura, A. (1997). *Nature* **387**, 921–924.
Engelmann, H., Novick, D. and Wallach, D. (1990). *J. Biol. Chem.* **265**, 1531–1536.
Espevik, T.M., Brockhaus, M., Loetscher, H., Nonstad, U. and Shalaby, R. (1990). *J. Exp. Med.* **171**, 415–426.
Falcone, D.J., McCaffery, T.A., Haimovitz-Friedman, A. and Garcia, M. (1993). *J. Cell. Physiol.* **155**, 595–605.
Fantl, W.J., Johnson, D.E. and Williams, L.T. (1993). *Annu. Rev. Biochem.* **62**, 453–481.
Feldmann, M., Brennan, F.M. and Maini, R.N. (1996). *Ann. Rev. Immunol.* **14**, 397–440.
Feldmann, M., Elliott, M.J., Woody, J.N. and Maini, R.N. (1997). *Adv. Immunol.* **64**, 283–350.
Feng, Y., Broder, C.C., Kennedy, P.E. and Berger, E.A. (1996). *Science* **272**, 872–877.
Fernandez-Botran, R., Chilton, P.M. and Ma, Y. (1996). *Adv. Immunol.* **63**, 269–336.
Ferrera, N., Houck, K.A., Jakeman, L.B., Winer, J. and Leung, D.W. (1991). *J. Cell. Biochem.* **47**, 211–218.
Fidler, I.J. and Ellis, L.M. (1994). *Cell* **79**, 185–188.
Finbloom, D.S. and Larner, A.C. (1995). *Cell. Signal.* **1**, 739–745.
Firestein, G.S. (1996). *Arthritis Rheum.* **39**, 1781–1790.
Firestein, G.S., Berger, A.E., Tracey, D.E., Chosay, J.G., Chapman, D.L., Paine, M.M., Yu, C. and Zvaifler, N.J. (1992). *J. Immunol.* **149**, 1054–1062.
Fischer, E., Zanzee, K.J., Marano, M.A., Rock, C.S., Kenney, J.S., Poutsiaka, D.D., Dinarello, C.A., Lowry, S.F. and Moldawer, L.L. (1992). *Blood* **79**, 2916–2200.
Fisher, T.C., Milner, A.E., Gregory, C.D., Jackman, A.L., Aherne, G.W., Hartley, J.A., Dive, C. and Hickman, J.A. (1993). *Cancer Res.* **53**, 3321–3326.
Flanagan, J.G. and Leder, P. (1990). *Cell* **63**, 185–194.
Flanagan, W.M., Corthesy, B., Bram, R.J. and Crabtree, G.R. (1991). *Nature* **352**, 803–807.
Foekens, J.A., Sieuwerts, A.M., Stuurman-Smeets, E.M.J., Peters, H.A. and Klijn, J.G.M. (1993). *Br. J. Cancer* **67**, 232–236.
Folkman, J. (1992). *Semin. Cancer Biol.* **3**, 65–71.
Folkman, J. (1995). *Mol. Med.* **1**, 120–122.
Folkman, J. (1996). *N. Engl. J. Med.* **334**, 921–927.
Folkman, J. and Shing, Y.J. (1992). *J. Biol. Chem.* **267**, 10931–10936.
Foss, H.-D., Herbst, H., Gottstein, S., Demel, G., Araujo, I. and Stein, H. (1996). *Am. J. Pathol.* **148**, 1229–1236.
Frame, M.C. (1997). In *Encyclopaedia of Cancer* (ed. J.R. Bertino), Academic Press, San Diego, CA, pp. 1172–1191.
Frankel, A.E., Fitzgerald, D., Siegall, C. and Press, O.W. (1996). *Cancer Res.* **56**, 926–932.
Fraser, A. and Evans, G. (1996). *Cell* **85**, 781–784.
Freshney, N.W., Rawlingson, L., Guesdon, F., Jones, E., Cowley, S., Hsuan, J. and Saklatvala (1994). *Cell* **78**, 1039–1049.
Fresno, M., Kopf, M. and Rivas, L. (1997). *Immunol. Today* **18**, 56–59.
Friesel, R.E. and Macaig, T. (1995). *FASEB J.* **9**, 919–925.
Fruehauf, J.P., Myers, C.E. and Sinha, B.K. (1990). *J. Natl Cancer Inst.* **82**, 1206–1209.

Fruman, D.A., Burakoff, S.F. and Bierer, B.E. (1994). *FASEB. J.* **8**, 391–400.
Fu, F., Li, Y., Qian, S., Lu, F., Chambers, F., Starzl, T.E., Fung, J.J. and Thomson, A.W. (1996). *Transplantation* **62**, 659–665.
Fu, X.-Y. (1992). *Cell* **70**, 323–335.
Gabbianelli, M., Pelsoi, E., Montesoro, E., Valtieri, M., Luchetti, L., Samoggia, P., Vitelli, L., Barberi, T., Testa, U., Lyman, S. and Peschle, C. (1995). *Blood* **86**, 1661–1670.
Gabriele, L., Kaido, T., Woodrow, D., Moss, J., Ferrantini, M., Proletti, E., Santodonato, L., Rozera, C., Maury, C., Belardelli, F. and Gresser, I. (1995). *Am. J. Pathol.* **147**, 445–460.
Gabrilovich, D.I., Patterson, S., Timofeer, A.V., Harvey, J.J. and Knight, S.C. (1996). *Clin. Immunol. Immunopathol.* **80**, 139–146.
Galat, A. (1993). *Eur. J. Biochem.* **216**, 689–707.
Galcheva-Gargova, Z., Derijard, B., Wu, I.-H. and Davies, R.J. (1994). *Science* **265**, 806–808.
Galzie, Z., Fernig, D.G., Smith, J.A., Poston, G.J. and Kinsella, A.R. (1997). *Int. J. Cancer* **71**, 390–395.
Ganem, D. (1995). *Curr. Biol.* **5**, 469–471.
Gansbacher, B., Zier, K., Daniels, B., Cronin, K., Bannerji, R. and Gilboa, T. (1990). *J. Exp. Med.* **172**, 1217–1224.
Garte, S.J. (1993). *Crit. Rev. Oncog.* **4**, 435–449.
Garver, R.I., Radford, D.M., Doris-Kellen, H., Wick, M.R. and Miller, P.G. (1994). *Cancer* **74**, 1584–1590.
Gause, W.C., Halvorson, M.J., Lu, P., Greenwald, R., Linsley, P., Urban, J.F. and Finkelman, F.D. (1997). *Immunol. Today* **18**, 115–120.
Gearing, A.J.H., Beckett, P., Christodoulou, M., Churchill, M., Clements, J., Davidson, A.H., Drummond, A.H., Galloway, W.A., Gilbert, R., Gordon, J.L., Leber, T.M., Mangan, M., Miller, K., Nayee, P., Owen, K., Patel, S., Thomas, W., Wells, G., Woog, L.M. and Woolley, K. (1994). *Nature* **370**, 555–557.
Gee, J.M., Ellis, I.O., Robertson, J.F., Willsher, P., McClelland, R.A., Hewitt, K.N., Blamey, R.W. and Nicholson, R.I. (1995). *Int. J. Cancer* **64**, 269–273.
Gendelman, H.E. and Morahan, P.S. (1992). In *The Macrophage* (eds C.E. Lewis and J.O'D. McGee), IRL Press, Oxford, pp. 157–195.
Ghayur, T., Banerjee, S., Hugunin, M., Butler, D., Herzog, L., Carter, A., Quintai, L., Sekut, L., Talanian, R., Paskind, M., Wong, W., Kamen, R., Tracey, D. and Allen, H. (1997). *Nature* **386**, 619–623.
Gilat, D., Cahalou, L., Hershkoviz, R. and Lider, O. (1996). *Immunol. Today* **17**, 16–20.
Gillessen, S., Carvajal, D., ling, P., Podlaski, F.J., Stremlo, D.L., Familletti, P.C., Gubler, U., Presky, D.H., Stern, A.S. and Gately, M.K. (1995). *Eur. J. Immunol.* **25**, 200–206.
Girolomoni, G. and Ricciardi-Castagnoli, P. (1997). *Immunol. Today* **18**, 102–104.
Gonda, T.J. and D'Andrea, R.J. (1997). *Blood* **89**, 355–369.
Gooding, L.R. (1992). *Cell* **71**, 5–7.
Gotis-Graham, I. and McNeil, H.P. (1997). *Arthritis Rheum.* **40**, 479–489.
Grabbe, S., Biessert, S., Schwartz, T. and Granstein, R.D. (1995). *Immunol. Today* **16**, 117–121.
Gracia-Zepeda, E.A., Rothenberg, M., Ownbey, R.T., Celestin, J., Leder, P. and Luster, A.D. (1996). *Nature Med.* **2**, 449–456.
Grandori, C. and Eisenman, R.N. (1997). *TIBS* **22**, 177–181.
Green, P., Walsh, F. and Doherty, P. (1996). *Bioessays* **18**, 639–646.
Grell, M., Douni, E., Wajant, H., Lohden, M., Clauss, M., Maxeiner, B., Georgopoulos, S., Lesslauer, W., Kollias, G., Pfizenmaier, K. and Scheurich, P. (1995). *Cell* **83**, 793–802.
Grouard, G., Durand, I., Filegueira, L., Banchereau, J. and Liu, Y.-J. (1997). *Nature* **384**, 364–367.
Gruss, H.-J., Brach, M.A., Drexler, H.G., Bonifer, R., Mertelmann, R.H. and Herrmann, F. (1992). *Cancer Res.* **52**, 3353–3360.
Gruss, H.-J., Herrmann, F. and Drexler, H.G. (1994). *Crit. Rev. Oncog.* **5**, 23–88.
Gruss, H.-J., Pinto, A., Duyster, J., Poppema, S. and Herrmann, F. (1997). *Immunol. Today* **18**, 156–163.
Gruss, H.-J., Ulrich, D., Dower, S.K., Herrmann, F. and Brach, M.A. (1996). *Blood* **87**, 2443–2449.
Gu, Y., Kuida, K., Tsutsui, H., Ku, G., Hsiao, K., Fleming, M. A., Hayashi, N., Higashino, K., Okamura, H., Nakanishi, K., Kurimoto, M., Tanimoto, T., Flavell, R.A., Sato, V., Harding, M. W., Livingston, D.J. and Su, M.S.S. (1997). *Science* **275**, 205–209.
Guo, Z., Chong, A.S.F., Shen, J., Foster, P., Sankary, H.N., McChesney, L., Mital, D., Jensik, S.C. and Williams, J.N. (1997a). *Transplantation* **63**, 711–716.
Guo, Z., Chong, A.S.F., Shen, J., Foster, P., Sankary, H.N., McChesney, L., Mital, D., Jensik, S.C., Gebel, H. and Williams, J.N. (1997b). *Transplantation* **63**, 716–721.

Haak-Frendscho, M., Marsters, S.A., Chamow, S.M., Peers, D.H., Simpson, N.J. and Ashkenazi, A. (1993). *Immunology* **79**, 594–599.

Hai, T.Y. and Curran, T. (1991). *Proc. Natl Acad. Sci. U.S.A.* **88**, 3720–3724.

Hale, G., Dyer, M.J., Clark, M.R., Phillips, J.M., Riechmann, L. and Waldmann, H. (1988). *Lancet* **ii**, 1394–1399.

Haluska, F.G., Brufsky, A.M. and Canellos, G.P. (1994). *Blood* **84**, 1005–1019.

Han, J., Brown, T. and Beutler, B. (1990). *J. Exp. Med.* **171**, 465–475.

Han, J., Lee, J.D., Bibbs, L. and Ulevitch, R.J. (1994). *Science* **265**, 808–811.

Han, J., Jiang, Y., Li, Z., Kravchenko, V.V. and Ulevitch, R.J. (1997). *Nature* **386**, 296–299.

Hanada, K., Tsunoda, R. and Hamada, H. (1996). *J. Leukocyte Biol.* **60**, 181–190.

Hanahan, D. and Folkman, J. (1996). *Cell* **86**, 353–364.

Hanbuchi, S., Koyanagi, M., Kawasaki, A., Shinohara, N., Matsuzawa, A., Nishimura, Y., Kobayashi, Y., Yonehara, S., Yagita, H. and Okumura, K. (1994). *Proc. Natl Acad. Sci. U.S.A.* **91**, 4930–4934.

Hancock, W.W., DiStefano, R., Braun, P., Schweizer, R.T., Tilney, N.L. and Kupiec-Weglinski, J.W. (1990). *Transplantation* **49**, 416–421.

Hagiwara, T., Suzuki, H., Kono, I., Kashiwagi, H., Akiyama, Y. and Onozaki, K. (1990). *Am. J. Pathol.* **136**, 39–47.

Harada, A., Sekido, N., Akahoshi, T., Wada, T., Mukaida, N. and Matsushima, K. (1994). *J. Leukocyte Biol.* **56**, 559–564.

Haran, E.F., Maretzek, A.F., Goldberg, I., Horowitz, A. and Degani, H. (1994). *Cancer Res.* **54**, 5511–5514.

Harbour, J., Lai, S., Whang-Peng, J., Gadzar, A., Minna, J. and Kaye, F. (1988). *Science* **241**, 353–357.

Harding, C.V. and Unanue, E.R. (1990). *Nature* **346**, 574–576.

Harris, E.D. (1990). *N. Engl. J. Med.* **322**, 1277–1289.

Harris, N.L., Jaffe, E.S., Stein, H., Banks, P.M., Chan, J.K.C., Cleary, M.L., Delsol, G., Dewolfpeeters, C., Falini, F., Gatter, K.C., Grogan, T.M., Isaacson, P.G., Knowles, D.M., Mason, D.Y., Mullerhermelink, H.K., Pileri, S.A., Piris, M.A., Ralfkiaer, E. and Warnke, R.A. (1994). *Blood.* **84**, 1361–1392.

Hata, A., Lo, R.S., Wotoon, D., Lagna, G. and Massague, J. (1997). *Nature* **388**, 82–87.

Hawking, F. (1987). *Adv. Pharmacol. Chemother.* **15**, 289–301.

Hayakawa, M., Kawaguchi, S., Ishii, S., Marakami, M. and Uede, T. (1997). *Int. J. Cancer* **71**, 1091–1102.

Hayashi, S., Kurdowska, A., Miller, E.J., Allbright, M.E., Girten, B.E. and Cohen, A.B. (1995). *J. Immunol.* **154**, 814–824.

Helbert, M.R., Lagestehr, J. and Mitchison, N.A. (1993). *Immunol. Today* **14**, 340–344.

Heldin, C.H. (1995). *Cell* **80**, 213–223.

Hemmings, B.A. (1997a). *Science* **275**, 628–630.

Hemmings, B.A. (1997b). *Science* **275**, 1899.

Hensey, C.F., Boscoboinik, D. and Azzi, A. (1989). *FEBS Lett.* **258**, 156–158.

Herbelin, A., Chatenoud, L., Roux-Lombard, P., Degroote, D., Legendre, C., Dayer, J.M., Descampslatcha, B., Kreis, H. and Bach, J.F. (1995). *Transplantation* **59**, 1470–1475.

Hermansson, M., Nister, M., Betsholtz, C., Heldin, C.-H., Westermark, B. and Funa, K. (1988). *Proc. Natl Acad. Sci. U.S.A.* **85**, 7748–7752.

Herve, P., Flesch, M., Tiberghien, P., Wijdenes, J., Racadot, E., Bordigoni, P., Plouvier, E., Stephen, J.L., Bordeau, H., Holler, E., Loioure, B., Roche, C., Vilmer, E., Bemeocq, F., Kuentz, M. and Cahn, J.Y. (1992). *Blood* **79**, 3362–3368.

Herzbeck, H., Blum, B., Ronspeck, W., Frenzel, B., Brandt, E., Ulmer, A.J. and Flad, H.-D. (1989). *Scand. J. Immunol.* **30**, 549–562.

Hibi, M., Murakami, M., Saito, M., Hirano, T., Taga, T. and Kishimoto, T. (1990). *Cell* **63**, 1149–1157.

Higuchi, M., Diamantstein, T., Osawa, H. and Caspi, R.R. (1991). *J. Autoimmun.* **4**, 113–124.

Hilbert, D.M., Kopf, M., Mock, B.A., Kohler, G. and Rodikoff, S. (1995). *J. Exp. Med.* **182**, 243–248.

Hill, C.S. and Treisman, R. (1995). *Cell* **80**, 199–211.

Hirano, T., Akira, S., Taga, T. and Kishimoto, T. (1990). *Immunol. Today* **11**, 443–449.

Hirsch, E., Irikura, V.M., Paul, S.M. and Hirsch, D. (1996). *Proc. Natl Acad. Sci. U.S.A.* **93**, 11 008–11 013.

Ho, S., Clipstone, N., Timmermann, I., Northop, J., Graef, I., Fiorentino, D., Nourse, J. and Crabtree, G.R. (1996). *Clin. Immunol. Immunopathol.* **80**, 540–545.

Hohmann, H.-P., Brockhaus, M., Baeuerle, P.A., Remy, R., Kolbeck, R. and van Loon, A.P.G.M. (1990). *J. Biol. Chem.* **265**, 22 409–22 417.

Hori, A., Sasada, R., Matsutani, E., Naito, K., Sakura, Y., Fujita, T. and Kozai, Y. (1991). *Cancer Res.* **51**, 6180–6184.

Horneff, G., Krause, A., Emmich, F., Kalden, J.R. and Burmester, G.R. (1991). *Cytokine* **3**, 266–267.

Horseman, N.D. and Yu-Lee, L.-Y. (1994). *Endocr. Rev.* **15**, 627–649.

Horvai, A.E., Xu, L., Korsus, E., Brard, G., Kalafus, D., Mullen, T.M., Rose, D.W., Rosenfield, M.G. and Glass, C.K. (1997). *Proc. Natl Acad. Sci. U.S.A.* **94**, 1074–1079.

Hou, J., Schindler, U., Henzel, W.J., Ho, T.C., Brasseur, M. and McKnight, S.L. (1994). *Science* **265**, 1701–1706.

Howard, M. and O'Garra, A. (1992). *Immunol. Today* **13**, 198–200.

Howard, O.M.Z., Clouse, K.A., Smith, C., Goodwin, R.G. and Farrar, W.L. (1993). *Proc. Natl Acad. Sci. U.S.A.* **90**, 2335–2339.

Howard, O.M.Z., Ben-Baruch, A. and Oppenheim, J.J. (1996). *TIBTech.* **14**, 46–51.

Hsu, H.L., Shu, H., Pan, M.G. and Goeddel, D.V. (1996). *Cell* **84**, 299–308.

Hsu, S.-M., Yang, K. and Jaffe, E.S. (1985). *Am. J. Pathol.* **118**, 209–217.

Hsu, S.-M., Waldron, J.W., Hsu, P.-L. and Hough, A.J. (1993). *Hum. Pathol.* **24**, 1040–1057.

Huang, E., Nocka, K., Beier, D.R., Chu, T.Y., Buck, J., Lahm, H.W., Wellner, D., Leder, P. and Besmer, P. (1990). *Cell* **63**, 225–233.

Huang, J.J., Newton, R.C., Horuk, R., Mathew, J.B., Corington, M., Pezzella, K. and Lin, Y. (1987). *FEBS Lett.* **223**, 294–298.

Huang, W. and Erickson, R.L. (1996). In *Signal Transduction* (eds C.-H. Heldin and M. Purton), Chapman and Hall, London, pp. 159–172.

Hunter, T. and Poon, R.Y.C. (1997). *Trends Cell Biol.* **7**, 157–161.

Huston, J.S. (1993). *Int. Rev. Immunol.* **10**, 195–217.

Hynes, N.E. and Stern, D.F. (1994). *Biochim. Biophys. Acta* **1198**, 165–184.

Iavarone, A. and Massague, J. (1997). *Nature* **387**, 417–422.

Ihle, J.N. (1995). *Adv, Immunol.* **60**, 1–35.

Ihle, J.N. (1996). *Adv. Cancer Res.* **68**, 23–65.

Ihle, J.N. and Kerr, I.M. (1995). *Trends Genet.* **11**, 69–74.

Ihle, J.N., Witthuhn, B.A., Quelle, F.W., Yamamoto, K., Thierfelder, W.E., Kreider, B. and Silvennoinen, O. (1994). *TIBS* **19**, 222–227.

Iozzo, R.V. (1994). *Matrix Biol.* **14**, 203–208.

Isner, J.M. (1996). *Clin. Immunol. Immunopathol.* **80**, 582–591.

Isomaki, P., Luukkainen, R., Saario, R., Toivanen, P. and Punnonen, J. (1996). *Arthritis Rheum.* **39**, 386–395.

Ito, A., Itoh, Y., Sasaguri, Y., Morimatsu, M. and Mori, Y. (1992). *Arthritis Rheum.* **35**, 1197–1201.

Itoh, H., Haltori, Y., Sakamoto, H., Ishii, H., Kishi, T., Sasaki, H., Yoshida, T., Koono, M., Sugimura, T. and Terada, M. (1994). *Cancer Res.* **54**, 3237–3241.

Ivashkiv, L.B. (1996). *Adv. Immunol.* **63**, 337–376.

Jackson, D.G. (1997). *Biochem. Soc. Trans.* **25**, 220–224.

Jacobson, N.G., Szabo, S.J., Weder-Nordt, R.M., Zhong, Z., Schreiber, R.D., Darnell, J.E. and Murphy, K.M. (1995). *J. Exp. Med.* **181**, 1755–1762.

Jadus, M.R., Irwin, M.C.N., Irwin, M.R., Horansky, R.D., Sekhon, S., Pepper, K.A., Kohn, D.B. and Wepsic, H.T. (1996). *Blood* **87**, 5232–5241.

Jain, J., McCaffery, P.G., Valgeaarcher, V.E. and Rao, A. (1992). *Nature* **356**, 801–804.

James, K. (1990). *Immunol. Today* **11**, 163–166.

Jansen, B. (1995). *Antisense Res. Dev.* **5**, 271–277.

Jaye, M., Schlessinger, J. and Dionne, C.A. (1992). *Biochim. Biophys. Acta* **1135**, 185–199.

Jayson, G.C., Evans, G.S., Pemberton, P.W., Lobley, R.W. and Allen, T. (1994). *Cancer Res.* **54**, 5718–5723.

Jindal, R.M., Popescu, I., Schwartz, M.E., Emre, S., Boccagni, P. and Miller, C.M. (1994). *Transplantation* **58**, 370–372.

Johnson, D. and Williams, L. (1993). *Adv. Cancer Res.* **60**, 1–41.

Johnston, J.A., Bacon, C.M., Riedy, M.C. and O'Shea, J.J. (1996). *J. Leukocyte Biol.* **60**, 441–452.

Jones, H.E., Dutkowski, C.M., Barrow, D., Harper, M.E., Wakeling, A.E. and Nicholson, R.I. (1997). *Int. J. Cancer* **71**, 1010–1018.

Joosten, L.A.B., Helsen, M.M.A., van de Loo, F.A.J. and van den Berg, W.B. (1996). *Arthritis Rheum.* **39**, 797–809.

Jose, P., Griffiths-Johnson, D.A., Collins, P.D., Walsh, D.T., Moqbel, R., Totty, N.F., Truong, O., Hsuan, J.J. and Williams, T.J. (1994). *J. Exp. Med.* **179**, 881–887.

Jouanneau, J., Gavrilovic, J., Caruelle, D., Jaye, M., Moens, G., Caruelle, J.P. and Thiery, J.P. (1991). *Proc. Natl Acad. Sci. U.S.A.* **88**, 2893–2897.

Kadin, M.E. (1994). *Curr. Opin. Oncol.* **6**, 456–463.
Kaplan, D. (1996). *Immunol. Today* **17**, 303–304.
Kaplan, M.H., Schindler, U., Smiley, S.T. and Grusby, M.J. (1996). *Immunity* **4**, 313–319.
Karin, M. (1992). *FASEB J.* **6**, 2581–2590.
Karin, M. and Smeal, T. (1992). *TIBS* **17**, 418–421.
Kasayama, S., Saito, H., Kouhra, H., Sumitani, S. and Sato, B. (1993). *J. Cell. Physiol.* **154**, 254–261.
Kato, G.J. and Dang, C. V. (1992). *FASEB J.* **6**, 3064–3072.
Kato, K., Tamura, N., Okumura, K. and Yagita, H. (1993). *Eur. J. Immunol.* **23**, 1412–1415.
Kavanaugh, W.M. and Williams, L.T. (1996). In *Signal Transduction* (eds C.-H. Heldin and M. Purton), Chapman and Hall, London, pp. 3–18.
Kawasaki, E.S., Ladner, M.B., Wang, A.M., van Arsdell, J., Warren, M.K., Coyne, M.Y., Schweichert, V.L., Lee, M.T., Wilson, K.J., Boosman, A., Stanley, E.R., Ralph, P. and Mark, D.F. (1985). *Science* **230**, 291–296.
Kay, N.E., Burton, J., Wagner, D. and Nelson, D.L. (1988). *Blood* **72**, 447–450.
Kelley, J. and Hildenbran, J.N. (1987). *Am. Rev. Respir. Dis.* **135**, A65.
Kelley, V.E., Gaulton, G.N., Haltori, M., Ikegami, H., Eisenbarth, G. and Strom, T.B. (1988). *J. Immunol.* **140**, 59–61.
Kelly, J.L., Sanchez, A., Brown, G.S., Chesterman, C.N. and Sleigh, M.J. (1993). *J. Cell Biol.* **121**, 1153–1163.
Kelner, G.S., Kennedy, J., Bacon, K.B., Kleyensteuber, S., Largaespada, D.A., Jenkins, N.A., Copeland, N.G., Bazan, J.F., Moore, K.W., Schall, T.J. and Zlotnik, A. (1994). *Science* **266**, 1395–1399.
Kennedy, J., Kelner, G.S., Kleyensteuber, S., Schall, T.J., Weiss, M.C., Yssel, H., Schneider, P.V., Cocks, B.G., Bacon, K.B. and Zlotnik, A. (1995). *J. Immunol.* **155**, 203–209.
Khachigian, L.M., Owensby, D.A. and Chesterman, C.N. (1992). *J. Biol. Chem.* **267**, 1660–1666.
Kharitonenkov, A., Chen, Z., Sures, I., Wang, H., Schilling, J. and Ullrich, A. (1997). *Nature* **386**, 181–186.
Kilkarni, A.B. and Karlsson, S. (1993). *Am. J. Pathol.* **143**, 3–9.
Kim, K.J., Li, B., Winer, J., Armanini, M., Gillett, N., Phillips, H.S. and Ferrara, N. (1993). *Nature* **362**, 841–844.
Kingsley, G. and Sieper, J. (1993). *Immunol. Today* **14**, 387–391.
Kingsley, G., Lanchbury, J. and Panayi, G. (1996). *Immunol. Today* **17**, 9–12.
Kinne, R.W., Palombo-Kinne, E. and Emmrich, F. (1997). *Biochim. Biophys. Acta* **1360**, 109–141.
Kishimoto, T., Taga, T. and Akira, S. (1994). *Cell* **76**, 253–262.
Kishimoto, T., Akira, S., Narazaki, M. and Taga, T. (1995). *Blood* **86**, 1243–1254.
Klagsbrun, M. and D'Amore, P.A. (1991). *Ann. Rev. Physiol.* **53**, 217–239.
Klagsbrun, M. and Folkman, J. (1990). In *Peptide Growth Factors and Their Receptors 2* (eds M.B. Sporn and A.B. Roberts), Springer, London, pp. 549–586.
Klagsbrun, M. and Soker, S. (1993). *Curr. Biol.* **3**, 699–702.
Klein, B., Zhang, X.G., Jourdan, M., Content, J., Houssiau, F., Aarden, L., Piechaczyk, M. and Bataille, R. (1989). *Blood* **73**, 517–526.
Klein, B., Wijdenes, J., Zhang, X.-G., Jourdan, M., Boiron, J.-M., Brochier, J., Liautard, J., Merlin, M., Clement, C., Mourel-Fournier, B., Zhao-Liang, L., Mannoni, P., Sany, J. and Bataille, R. (1991). *Blood* **78**, 1198–1204.
Klein, B., Zhang, X.G., Lu, Z.Y. and Bataille, R. (1995). *Blood* **85**, 863–872.
Knusel, B. and Hefti, F. (1992). *J. Neurochem.* **59**, 1987–1996.
Kobayashi, H., Yoo, T.M., Drumm, D., Kim, M.-K., Sun, B.-F., Le, N., Webber, K.O., Pastan, I., Waldmann, T.A., Paik, C.H. and Carrasquillo, J.A. (1997). *Cancer Res.* **57**, 1955–1961.
Kobrin, M., Yamanaka, M., Friess, M., Lopez, M. and Korc, M. (1993). *Cancer Res.* **53**, 4741–4744.
Kondo, M., Takeshita, T., Ishii, N., Nakamura, M., Watanabe, S., Arai, K. and Sugamura, K. (1993). *Science* **262**, 1874–1877.
Kondo, M., Takeshita, T., Higuchi, M., Nakamura, M., Sudo, T., Nishikawa, S. and Sugamura, K. (1994). *Science* **263**, 1453–1454.
Kopf, M., Legros, G., Bachmann, M., Lamers, M.C., Bluethmann, H. and Kohler, C. (1993). *Nature* **362**, 245–248.
Kopp, E.B. and Ghosh, S. (1995). *Adv. Immunol.* **58**, 1–27.
Kopp, R. and Pfeiffer, A. (1990). *Cancer Res.* **50**, 6490–6496.
Kovacs, E.J. (1991). *Immunol. Today* **12**, 17–23.
Kovalenko, M., Gazit, A., Bohmer, A., Rorsman, C., Ronnstrand, L., Heldin, C.-H., Waltengerger, J., Bohmer, F. and Levitski, A. (1994). *Cancer Res.* **54**, 6106–6114.
Krontiris, T.G. (1995). *N. Engl. J. Med.* **333**, 303–306.

Krystal, G.W., Carbon, P. and Litz, J. (1997). *Cancer Res.* **57**, 2203–2208.
Kuchler, K. (1993). *Trends Cell Biol.* **3**, 421–426.
Kupiec-Weglinski, J.W., van der Meide, P., Stunkel, K.G., Tanaka, K., Diamanstein, T. and Tilney, N.L. (1989). *Transplant. Proc.* **21**, 992–993.
Kupiec-Weglinski, J.W., Sablinski, T., Hancock, W.W., DiStefano, R., Mariani, G., Mix, C.T. and Tilney, G.L. (1991a). *Transplantation* **51**, 300–305.
Kupiec-Weglinski, J.W., Sablinski, T., Hancock, W.W., Mix, C.T. and Tilney, G.L. (1991b). *Transplant. Proc.* **23**, 285–286.
Kurschner, C., Ozmen, L., Garotta, G. and Dembic, Z. (1992). *J. Immunol.* **149**, 4096–4100.
Kwok, R.P.S., Lundblad, J.R., Chrivia, J.C., Richards, J.P., Bachinger, H.P., Brennan, R.G., Roberts, S.G.E., Green, M.R. and Goodman, R.H. (1994). *Nature* **370**, 223–226.
Kyrakis, J.M., Banerjee, P., Nikolakaki, E., Dai, T.A., Rubie, E.A., Ahmad, M.F., Avruch, J. and Woodgett, J.R. (1994). *Nature* **369**, 156–160.
Lagna, G., Hata, A., Hemmati-Brivanlou, A. and Massague, J. (1996). *Nature* **383**, 832–836.
Laird, A.D. and Shalloway, D. (1997). *Cell. Signal.* **9**, 249–255.
Lamszus, K., Joseph, A., Jin, L., Yao, Y., Chowdhury, S., Fuchs, A., Polverini, P.J., Goldberg, I.D. and Rosen, E.M. (1996). *Am. J. Pathol.* **149**, 805–819.
Lamszus, K., Jin, L., Fuchs, A., Shi, E., Chowdhury, S., Yao, Y., Polverini, P.J., Laterra, J., Goldberg, I.D. and Rosen, E.M. (1997). *Lab. Invest.* **76**, 339–353.
Landschultz, W.H., Johnson, P.F. and McKnight, S.L. (1988). *Science* **240**, 1759–1764.
Lappi, D. (1995). *Cancer Biol.* **6**, 279–288.
LaRochelle, W.J., Giese, N., May-Siroff, M., Robbins, K.C. and Aaronson, S.A. (1990a). *Science* **248**, 1541–1544.
LaRochelle, W.J., Giese, N., May-Siroff, M., Robbins, K.C. and Aaronson, S.A. (1990b). *J. Cell Sci. Suppl.* **13**, 31–42.
Latchman, D.S. (1993). *Int. J. Exp. Pathol.* **74**, 417–422.
Laterra, J., Nam, M., Rosen, E., Rao, J.S., Lamszus, K., Goldberg, I.D. and Johnston, P. (1997). *Lab. Invest.* **76**, 565–577.
Lee, J.C., Laydon, J.T., McDonnell, P.C., Gallagher, T.F., Kumar, S., Green, D., McNulty, D., Blumenthal, M.J., Heys, J.R., Landvatter, S.W., Strickler, J.E., McLaughlin, M.M., Siemens, I.R., Fisher, S.M., Livi, G.P., White, J.R., Adams, J.L. and Young, P.R. (1994). *Nature* **372**, 739–746.
Leek, R.D., Harris, A.L. and Lewis, C.E. (1994). *J. Leukocyte Biol.* **56**, 423–435.
Leevers, S.J. (1996). In *Signal Transduction* (eds C.-H. Heldin and M. Purton), Chapman and Hall, London, pp. 143–158.
Leone, G., DeGregori, J., Sears, R., Jakoi, L. and Nevins, J.R. (1997). *Nature* **387**, 422–426.
Leonetti, C. (1993). *J. Natl Cancer Inst.* **88**, 419–429.
Lesslauer, W., Tabuchi, H., Gentz, R., Brockhaus, M., Schlaeger, E.J., Grau, G., Piguet, P.F., Pointaire, P., Vassalli, P. and Loetscher, H. (1991). *Eur. J. Immunol.* **21**, 2883–2886.
Leung, H., Gullick, W. and Lemoine, N. (1994a). *Int. J. Cancer* **59**, 669–675.
Leung, H.Y., Hughes, C.M., Kloppel, G., Williamson, R.C.N. and Lemoine, N.R. (1994b). *Int. J. Oncol.* **4**, 1219–1223.
Levitzki, A. and Gazit, A. (1995). *Science* **267**, 1782–1788.
Levitzky, H.I., Lazenby, A., Hayashi, R.J. and Pardoll, D.M. (1994). *J. Exp. Med.* **179**, 1215–1224.
Lewis, C.E., Leek, R., Harris, A.L. and McGee, J.O'D. (1995). *J. Leukocyte Biol.* **57**, 747–751.
Liblau, R.S., Singer, S.M. and McDevitt, H. (1995). *Immunol. Today* **16**, 34–39.
Lichtman, A.H. and Abbas, A.K. (1997). *Curr. Biol.* **7**, R242–R244.
Liew, F.Y., Millot, S., Parkinson, C., Palmer, R.M.J. and Moncada, S. (1990). *J. Immunol.* **144**, 4794–4797.
Liotta, L.A., Steeg, P.S. and Stetler-Stevenson, W.G. (1991). *Cell* **64**, 327–336.
Lipsky, P.E., Davies, L.S., Cush, J.-J.T. and Oppenheimer-Marks, N. (1989). *Springer Semin. Immunopathol.* **11**, 123–162.
Liu, C. and Tsao, M.S. (1993). *Am. J. Pathol.* **142**, 1155–1162.
Liu, J., Albers, M.W., Wandless, T.J., Luan, S., Alberg, D.G., Belshaw, P.J., Cohen, P., Mackintosh, C., Klee, C.B. and Schrieber, S.L. (1992). *Biochemistry* **31**, 3896–3901.
Lloyd, A.R. and Oppenheim, J.J. (1992). *Immunol. Today* **13**, 169–172.
LoBuglio, A.F., Wheeler, R.H., Trang, J., Haynes, A., Rogers, K., Harvey, E.B., Sun, L., Ghrayeb, J. and Khazaeli, M.B. (1989). *Proc. Natl Acad. Sci. U.S.A.* **86**, 4220–4224.

Lopez-Casillas, F., Wrana, J.L. and Massague, J. (1993). *Cell* **73**, 1435–1444.
Lowenberg, B. and Touw, I.P. (1993). *Blood* **81**, 281–292.
Lowenstein, E.J., Daly, R.J., Batzer, A.G., Li, W., Margolis, B., Lammers, R., Ullrich, A., Skolnik, E.Y., Barsagi, D. and Schlessinger, J. (1992). *Cell* **70**, 431–442.
Lu, L., McCaslin, D., Starzl, T.E. and Thomson, A.W. (1995). *Transplantation* **60**, 1539–1545.
Lucas, R., Lou, J., Morel, D.R., Ricou, B., Suter, P.M. and Grau, G.E. (1997). *J. Leukocyte Biol.* **61**, 551–558.
Luqmani, Y.A., Mortimer, C., Yiangou, C., Johnston, C.L., Bansal, G.S., Sinnet, D., Law, M. and Coombes, R.C. (1995). *Int. J. Cancer* **64**, 274–279.
Luqmani, Y.A., Bansal, G., Mortimer, C., Buluwela, L. and Coombes, R.C. (1996). *Eur. J. Cancer* **32A**, 578–524.
Luqmani, Y.A., Chandler, S., Coope, R., Gommi, J., Lappi, D. and Coombes, R.C. (1997). *Oncol. Rep.* **4**, 425–428.
Lyons, R.M., Gentry, L.E., Purchio, A.F. and Moses, H.L. (1990). *J. Cell. Biol.* **110**, 1361–1367.
Macchia, D., Almerigogna, F., Parronchi, P., Ravina, A., Maggi, E. and Romagnani, S. (1993). *Nature* **363**, 464–466.
Mackay, C.R. (1997). *Curr. Biol.* **7**, R384–R386.
Mackintosh, C. and Mackintosh, R.W. (1994). *TIBS J.* **19**, 444–448.
Madeddu, R., Casadio, C., Pennacchietti, S., Nicotra, M.R., Prat, M., Maggi, G., Arena, N., Natali, P.G., Comoglio, P.M. and Di Renzo, M.F. (1996). *Br. J. Cancer* **74**, 1862–1868.
Mallett, S. and Barclay, A.N. (1991). *Immunol. Today* **12**, 220–223.
Mannel, D., Murray, C., Risau, W. and Clauss, M. (1996). *Immunol. Today* **17**, 254–256.
Mansour, S.J., Matten, W.T., Hermann, A.S., Candia, J.M., Rong, S., Fukasawa, K., Vandewoude, G.F. and Ahn, N.G. (1994). *Science* **265**, 966–970.
Mantovani, A. (1994). *Lab. Invest.* **71**, 5–16.
Mantovani, A., Bussolino, F. and Introna, M. (1997). *Immunol. Today* **18**, 231–240.
Marchionni, M. (1995). *Nature* **378**, 334–355.
Marland, G., Bakker, A.B.H., Adema, G.-J. and Fidgor, C.G. (1997). *Stem Cells* **14**, 501–507.
Marshall, C.J. (1995). *Cell* **80**, 179–185.
Marsters, S.A., Pennica, D., Bach, E., Schreiber, R.D. and Ashkenazi, A. (1995). *Proc. Natl Acad. Sci. U.S.A.* **92**, 5401–5405.
Massague, J. (1996). *Cell* **85**, 947–950.
Massague, J. and Pandiella, A. (1993). *Ann. Rev. Biochem.* **62**, 515–541.
Massague, J., Hata. A. and Liu, F. (1997). *Trends Cell Biol.* **7**, 187–192.
Mathew, S., Scanlan, M.J., Mohan-Raj, B.K., Murty, V.V.V.S., Garin-Chesa, P., Old, L.J., Rettig, W.J. and Chaganti, R.S.K. (1995). *Genomics* **25**, 335–337.
Matozaki, T. and Kasuga, M. (1996). *Cell. Signal.* **8**, 13–19.
Matrisian, L.M. (1992). *Bioessays* **14**, 455–463.
Matsuda, S., Gotch, Y. and Nishida, E. (1994). *J. Leukocyte Biol.* **56**, 548–553.
Matsukawa, A., Yoshimura, T., Miyamoto, K., Ohkawara, S. and Yoshinaga, M. (1997). *Lab. Invest.* **76**, 629–638.
Matsumoto-Yoshitomi, S., Habashita, J., Nomura, C., Kuroshima, K.-I. and Kurokawa, T. (1997). *Int. J. Cancer* **71**, 442–450.
Matsusaka, T., Fujikawa, K., Nishio, Y., Mukaida, N., Matsushima, K., Kishimoto, T. and Akira, S. (1993). *Proc. Natl Acad. Sci. U.S.A.* **90**, 10193–10197.
Mattar, T., Kochhar, K., Barlett, R., Bremer, E.G. and Finnegan, A. (1993). *FEBS Lett.* **334**, 61–66.
McAdam, A.J., Pulaski, B.A., Harkins, S.S., Hutter, E.K., Lord, E.M. and Frelinger, J.G. (1995a). *Int. J. Cancer* **61**, 628–634.
McAdam, A.J., Pulaski, B.A., Storozynsky, E., Yeh, K.-Y., Sickel, J.Z., Frelinger, J.G. and Lord, E.M. (1995b). *Cell. Immunol.* **165**, 183–192.
McCachren, S.S., Haynes, B.F. and Niedel, J.E. (1990). *J. Clin. Immunol.* **10**, 19–27.
McCaffery, P.C., Perrino, B.A., Sorderling, T.R. and Rao, A. (1993). *J. Biol. Chem.* **268**, 3747–3752.
McDonald, N.Q. and Hendrickson, W.A. (1993). *Cell* **73**, 421–424.
McFadden, G. (1995). In *Human Cytokines: Their Role in Health and Disease* (eds B.B. Aggarwal and R.K. Puri), Blackwell Press, London, pp. 403–422.
McFadden, G., Graham, K., Ellison, K., Barry, M., Macen, J., Schrieber, M., Mossman, K., Nash, P., Lalani, A. and Everett, H. (1995). *J. Leukocyte Biol.* **57**, 731–738.
McFarland, H.F. (1996). *Science* **274**, 2037–2038.

McGeehan, G.M., Becherer, J.D., Bast, R.C., Boyer, C.M., Champion, B., Connolly, K.M., Conway, J.G., Furdon, P., Karp, S., Kidao, S., McElroy, A.B., Nichols, J., Pryzwansky, K.M., Schoenon, F., Sekut, L., Truesdale, A., Verghese, M., Warner, J. and Ways, J.P. (1994). *Nature* **370**, 558–561.

McKeehan, W.L., Hou, J., Adams, P., Wang, F., Yan, G.C. and Kan, M. (1993). *Adv. Exp. Med. Biol.* **330**, 203–213.

McKeon, F. (1991). *Cell* **66**, 823–826.

Meager, A. (1990). In *Cytokines*. Open University Press, Milton Keynes.

Mecheri, S and David, B. (1997). *Immunol. Today* **18**, 212–215.

Menard, H.-A. and El-Amine, M. (1996). *Immunol. Today* **17**, 545–547.

Meyden, N., Grunberger, T., Dadi, H., Shahar, M., Arpaia, E., Lapidot, Z., Leeder, J.S., Freedman, M., Cohen, A., Gazit, A., Levitzki, A. and Roifman, C.M. (1996). *Nature* **379**, 645–648.

Migone, T.S., Lin, J.X., Cereseto, A., Mulloy, J.C., O'Shea, J.J., Franchini, G. and Leonard, W.J. (1995). *Science* **269**, 79–81.

Millauer, B., Shawver, L.K., Plate, K.H., Risau, W. and Ullrich, A. (1994). *Nature* **367**, 576–579.

Miller, I.C., Lynch, E.A., Isa, S., Logan, J.W., Dinarello, C.A. and Steere, A.C. (1993). *Lancet* **341**, 146–148.

Mills, G.B., Zhang, N., May, C., Hill, M. and Chung, A. (1990). *Cancer Res.* **50**, 3036–3042.

Minden, A., Lin, A., McMahon, M., Langecarter, C., Derijard, B., Davis, R.J., Johnson, G.L. and Karin, M. (1994). *Science* **266**, 1719–1723.

Miossec, P., Briolay, J., Dechanet, J., Wijdenes, J., Martinez-Valdez, H. and Banchereau, J. (1992). *Arthritis Rheum.* **35**, 874–881.

Miossec, P., Chomarat, P. and Dechanet, J. (1996). *Immunol. Today* **17**, 170–173.

Mire-Sluis, A. (1993). *TIBTech.* **11**, 74–77.

Miyajima, A., Hara, T. and Kitamura, T. (1992). *TIBS* **17**, 378–382.

Miyake, H., Hara, I., Yoshimura, K., Eto, H., Arakawa, S., Wada, S., Chihara, K. and Kamidono, S. (1996). *Cancer Res.* **56**, 2440–2445.

Miyazono, K. (1996). In *Signal Transduction* (eds C.-H. Heldin and M. Purton), Chapman and Hall, London, pp. 65–78.

Miyazono, K. and Heldin, C.-H. (1991). Clinical applications of TGF-β. *Ciba Found. Symp.* **157**, 81–92.

Modjtahedi, H., Eccles, S., Box, G., Styles, J. and Dean, C. (1993). *Br. J. Cancer* **67**, 254–261.

Mohammadi, M., McMahon, G., Sun, L., Tang, C., Hirth, P., Yen, B.K., Hubbard, S.R. and Schlessinger, J. (1997). *Science* **276**, 955–960.

Mohler, K.M., Sleath, P.R., Fitzner, J.N., Cerretti, D.P., Alderson, M., Kerwar, S.S., Torrance, D.S., Ottenevans, C., Greenstreet, T., Weerawarna, K., Kronheim, S.R., Petersen, M., Gerhart, M., Kozlosky, C.J., March, C.J. and Black, R.A. (1994). *Nature* **370**, 218–220.

Monfardi, C., Kieber-Emmons, T., von Feldt, J.M., O'Malley, B., Rosenbaum, H., Godillot, A.P., Kaushansky, K., Brown, C.B., Voet, D., McCallus, D.E., Weiner, D.B. and Williams, W.S. (1995). *J. Biol. Chem.* **270**, 6628–6638.

Moreira, A.L., Sampaio, E.P., Zmuidzinas, A., Frindt, P., Smith, K.A. and Kaplan, G. (1993). *J. Exp. Med.* **177**, 1675–1680.

Moreland, L.W., Margolies, G.R., Heck, L.W., Saway, P.A., Jacobs, C., Beck, C., Blash, C. and Koopman, W.J. (1994). *Arthritis Rheum.* **37**, S295.

Moreland, L.W., Heck, L.W. and Koopman, W.J. (1997). *Arthritis Rheum.* **40**, 397–409.

Mori, A., Kaminuma, O., Suko, M., Inoue, S., Ohmura, T., Hashino, A., Asakura, Y., Miyazawa, K., Yokota, T., Okumura, Y., Ito, K. and Okudaira, H. (1997). *Blood* **89**, 2891–2900.

Morisson, B.W. and Leder, P. (1992). *J. Biol. Chem.* **267**, 11 957–11 963.

Morisson, R.S. (1991). *J. Biol. Chem.* **266**, 728–734.

Morton, T., Li, J., Cook, R. and Chaiken, I.M. (1995). *Proc. Natl Acad. Sci. U.S.A.* **92**, 10 879–10 884.

Moser, B. (1997). *Trends Microbiol.* **5**, 88–90.

Mosmann, T.R. and Sad, S. (1996). *Immunol. Today* **17**, 138–146.

Moss, M.L., Jin, S.L.Cl., Milla, M.E., Burkhart, W., Carter, H.L., Chen, W.J., Clay, W.C., Didsbury, J.R., Hassler, D., Hoffman, C.R., Kost, T.A., Lambert, M.H., Leesnitzer, M.A., McCauley, P., McGeehan, G., Mitchell, J., Moyer, M., Pahel, G., Rocque, W., Overton, L.K., Schoenen, F., Seaton, T., Su, J.L., Warner, J., Willard, D. and Becherer, J.D. (1997). *Nature* **385**, 733–736.

Motoi, T., Uchiyama, T., Hori, T., Itoh, K., Uchino, H. and Ueda, R. (1989). *Blood* **74**, 1052–1057.

Mukaida, N., Mahe, Y. and Matsushima, K. (1990). *J. Biol. Chem.* **265**, 21 128–21 133.

Mulherin, D., Fitzgerald, O. and Bresnihan, B. (1996). *Arthritis Rheum.* **39**, 115–124.

Mullen, C.A. (1994). *Pharmacol. Ther.* **63**, 199–207.
Mullen, C.A., Petropoulos, D. and Lowe, R.M. (1996). *Cancer Res.* **56**, 1361–1366.
Muller, M., Briscoe, J., Laxton, C., Guschin, D., Ziemiecki, A., Silvennoinen, O., Harpur, A.G., Barbieri, G., Witthuhn, B.A., Schindler, C., Pellegrini, S., Wilks, A.F., Ihle, J.N., Stark, G.R. and Kerr, I.M. (1993). *Nature* **366**, 129–135.
Murakami, M., Hibi, M., Nakagawa, N., Nakagawa, T., Yasakawa, K., Yamanishi, K., Taga, T. and Kishimoto, T. (1993). *Science* **260**, 1808–1810.
Murzin, A.G. and Clothia, C. (1992). *Curr. Opin. Struct. Biol.* **2**, 895–903.
Musiani, P., Allione, A., Modica, A., Lollini, P.L., Giovarelli, M., Cavallo, F., Belardelli, F., Forni, G. and Modesti, A. (1996). *Lab. Invest.* **74**, 146–157.
Musiani, P., Modesti, A., Giovarelli, M., Cavallo, F., Lollini, P.L. and Forni, G. (1997). *Immunol. Today* **18**, 32–36.
Myers, C., Cooper, M., Stein, C., Lacorra, R., Walther, M.M., Weiss, G., Choyke, P., Dawson, N., Steinberg, S., Uhrich, M.M., Cassidy, J., Kohler, D.R., Trepel, J. and Linehan, W.M. (1992). *J. Clin. Oncol.* **10**, 881–889.
Myoken, Y., Myoken, Y., Okamoto, T., Kan, M., McKeehan, W.L., Sato, J.D. and Takada, K. (1996). *Int. J. Cancer* **65**, 650–657.
Nagasawa, T., Kikutani, H. and Kishimoto, T. (1994). *Proc. Natl Acad. Sci. U.S.A.* **91**, 2305–2309.
Nagata, S. (1997). *Cell* **88**, 355–365.
Nagata, S. and Goldstein, P. (1995). *Science* **267**, 1449–1456.
Naka, D., Ishii, T., Yoshiyama, Y., Miyazawa, K., Hara, H., Hishida, T. and Kitamura, N. (1992). *J. Biol. Chem.* **267**, 20114–20119.
Naka, T., Narazaki, M., Hirata, M., Matsumoto, T., Minamoto, S., Aono, A., Nishimoto, N., Kajita, T., Taga, T., Yoshizaki, K., Akira, S. and Kishimoto, T. (1997). *Nature* **387**, 924–929.
Nakagawara, A., Milbrandt, J., Muramatsu, T., Deuel, T.F., Zhao, H., Canan, A. and Brodeur, G.M. (1995). *Cancer Res.* **55**, 1792–1797.
Nakamura, T., Nishizawa, T., Hagiya, M., Seki, T., Shimonishi, M., Sugimara, A. and Shimizu, S. (1989). *Nature* **342**, 440–443.
Naldini, L., Tamagnone, L., Vigna, E., Sachs, M., Hartmann, G., Birchneier, W., Daikuhara, Y., Tsubouchi, H., Blavi, F. and Comoglio, P.M. (1992). *EMBO J.* **11**, 4825–4833.
Naldini, L., Vigna, E., Bardelli, A., Follenzi, A., Galimi, F. and Comoglio, P.M. (1995). *J. Biol. Chem.* **270**, 603–611.
Naruo, K., Seko, C., Kuroshima, K., Matsutani, E., Sasada, R., Kondo, T. and Kurokawa, T. (1993). *J. Biol. Chem.* **268**, 2857–2864.
Nash, R.A., Pineiro, L.A., Storb, R., Deeg, H.J., Fitzsimmons, W.E., Furlong, T., Hansen, J.A., Godey, T., Maher, R.M., Martin, P., McSweeney, P.A., Sullivan, K.M., Anasetti, C. and Fay, J.W. (1996). *Blood* **88**, 3634–3641.
Nastala, C.L. Edington, H.D., McKinney, T.G., Tahar, H., Nalesnik, M.A., Brunda, M.J., Gately, M.K., Wolf, S.F., Schreiber, R.D., Storkus, W.J. and Lotze, M.T. (1994). *J. Immunol.* **153**, 1697–1706.
Nathan, C.F. (1990). In *Peptide Growth Factors and Their Receptors* (eds M.B. Sporn and A.B. Roberts), Springer, London, pp. 427–462.
Nathan, C. (1992). *FASEB J.* **6**, 3051–3055.
Nathan, C. and Sporn, M. (1991). *J. Cell Biol.* **113**, 981–987.
Naume, B., Shalaby, R., Lesslauer, W. and Espevik, T. (1991). *J. Immunol.* **146**, 3045–3048.
Neer, E.J. (1994). *Protein Sci.* **3**, 3–14.
Neer, E.J. (1995). *Cell* **80**, 249–257.
Negus, R.P.M., Stamp, G.W.H., Relf, M.G., Burke, F., Malik, S.T., Bernasconi, S., Allavena, P., Sozzani, S., Mantovani, A. and Balkwill, F.R. (1995). *J. Clin. Invest.* **95**, 2391–2396.
Negus, R.P.M., Stamp, G.W.H., Hadley, J. and Balkwill, F.R. (1997). *Am. J. Pathol.* **150**, 1723–1734.
Neiman, A.M. (1993). *Trends Genet.* **9**, 390–394.
Nickerson, P., Steiger, J., Zheng, X.X., Steele, A.W., Steurer, W., Roy-Chaudhury, P. and Strom, T.B. (1997). *Transplantation* **63**, 489–494.
Niedermeyer, J., Scanlan, M.J., Garin-Chesa, P., Daiber, C., Fiebig, H.H., Odl, L.J., Rettig, W.J. and Schnapp, A. (1997). *Int. J. Cancer* **71**, 383–389.
Noguchi, M., Nakamura, Y., Russell, S.M., Ziegler, S.F., Tsang, M., Cao, X. and Leonard, W.J. (1993a). *Science* **262**, 1877–1880.
Noguchi, M., Yi, H., Rosenblatt, H.M., Filipovich, A.H., Adelstein, S., Modi, W.S., McBride, O.W. and Leonard, W.J. (1993b). *Cell* **73**, 147–157.

Nourse, J., Firpo, E., Flanagan, W.M., Coats, S., Polak, K., Lee, M.H., Massague, J., Crabtree, G.R. and Roberts, J.M. (1994). *Nature* **372**, 570–573.
Oberlin, E., Amara, A., Bachelerie, F., Bessia, C., Virelizier, J.-L., Arenzana-Seisdedos, F., Schwantz, O., Heard, J.-M., Clack-Lewis, I., Legler, D.F., Loetscher, M., Baggiolini, M. and Moser, B. (1996). *Nature* **382**, 833–835.
O'Brien, T., Cranston, D., Fuggle, S., Bicknell, R. and Harris, A.L. (1996). *Cancer Res.* **56**, 2515–2518.
Ogryzko, V.V., Shiltz, R.L., Russanova, V., Howard, B.H. and Nakatani, Y. (1996). *Cell* **87**, 953–960.
Okada, M., Senod, Y. and Teramoto, S. (1992). *Acta Med. Okayama* **46**, 37–44.
Okuda, S., Languino, L.R., Ruoslahti, E. and Border, A.W. (1990). *J. Clin. Invest.* **86**, 453–462.
Olivier, S., Formento, P., Fischel, J.L., Etienne, M.C. and Milano, G. (1990). *Eur. J. Cancer* **26**, 867–871.
Olivero, M., Rizzio, M., Madeddu, R., Casadio, C., Pennacchietti, S., Nicotra, M.R., Prat, M., Maggi, G., Arena, N., Natalia, P.G., Comoglio, P.M. and DiRenzo, M.F. (1996). *Br. J. Cancer* **74**, 1862–1868.
Opdenakker, G. and van Damme, J. (1994). *Immunol. Today* **15**, 103–107.
O'Reilly, M.S., Holmgren, L., Shing, Y., Chen, C., Rosenthal, R.A., Moses, M., Lane, W.S., Cao, Y., Sage, E.H. and Folkman, J.A. (1994). *Cell* **79**, 315–328.
Ornitz, D., Xu, J., Colvin, J., McEwen, D., MacArthur, G., Coulier, F., Gao and Goldfarb, G.M. (1996). *J. Biol. Chem.* **271**, 15 292–15 297.
Ostman, A., Andersson, M., Betsholtz, C., Westermark, B. and Heldin, C.-H. (1991). *Cell Regul.* **2**, 503–512.
Packham, G. and Cleveland, J. (1995). *Biochem. Biophys. Acta* **1242**, 11–28.
Palaszynski, E.W. (1987). *Biochem. Biophys. Res. Commun.* **147**, 204–209.
Palumbo, G.A., Yarom, N., Gazit, A., Sandalon, Z., Baniyash, M., Kleinberger-Doron, N., Levitzki, A. and Ben-Yehuda, D. (1997). *Cancer Res.* **57**, 2434–2439.
Pan, Y., Lloyd, C., Zhou, H., Dolich, S., Deeds, J., Gonzalo, J.-A., Vath, J., Gosselin, M., Ma, J., Dussault, B., Woolf, E., Alperin, G., Culpepper, J., Gutierrez-Ramos, J.C. and Gearing, D. (1997). *Nature* **387**, 611–617.
Pankewycz, O.G., Guan, J.-X. and Benedict, J.F. (1995). *Endocr. Rev.* **16**, 164–176.
Pardoll, D.M. (1993). *Immunol. Today* **14**, 310–316.
Partanen, J., Vainikka, S. and Alitalo, K. (1993). *Philos. Trans. R. Soc. London [Biol.]* **340**, 297–303.
Pastan, I., Chaudhary, V and Fitzgerald, D.J. (1992). *Annu. Rev. Biochem.* **61**, 331–354.
Pattersen, H. (1992). *Eur. J. Cancer* **28**, 258–263.
Paul, W.E. and Seder, R.A. (1994). *Cell* **76**, 241–251.
Pawson, T. (1995). *Nature* **373**, 573–580.
Paxton, W.A. (1996). *Nature Med.* **2**, 412–416.
Peeper, D.S., Upton, T.M., Ladha, M.H., Neuman, E., Zalvide, J., Bernards, R., DeCaprio, J.A. and Ewen, M.E. (1997). *Nature* **386**, 521.
Pellegrini, S. and Schindler, C. (1993). *TIBS* **18**, 338–342.
Penault-Llorca, F., Bertucci, F., Adelaide, J., Parc, P., Coulier, F., Jacquemier, J., Birnbaum, D. and DeLapeyriere, O. (1995). *Int. J. Cancer* **61**, 170–176.
Perez, C., Albert, I., Defay, K., Zacharides, N., Gooding, L. and Kriegler, M. (1990). *Cell* **63**, 251–258.
Perosio, P.M. and Brooks, J.J. (1989). *Lab. Invest.* **60**, 245–253.
Pesenti, E.L. (1991). *J. Infect. Dis.* **163**, 611–616.
Pesenti, E., Sola, F., Mongelli, N., Grandi, M. and Spreafico, F. (1992). *Br. J. Cancer* **66**, 367–372.
Peterson, M.G. and Baichuwal, V.R. (1993). *TIBTech.* **11**, 11–18.
Pfeffer, L.M., Mullersman, J.E., Pfeffer, S.R., Murti, A., Shi, W. and Shang, C.H. (1997). *Science* **276**, 1418–1420.
Piccotti, J.R., Chan, S.Y., Goodman, R.E., Magram, J., Eichwald, E.J. and Bishop, D.K. (1996). *J. Immunol.* **157**, 1951–1957.
Piccotti, J.R., Chan, S.Y., Li, K., Eichwald, E.J. and Bishop, D.K. (1997a). *J. Immunol.* **158**, 643.
Piccotti, J.R., Chan, S.Y., van Buskirk, A.M., Eichwald, E.J. and Bishop, D.K. (1997b). *Transplantation* **63**, 619–624.
Piquet, P.F., Grau, G.E., Vesin, C., Loetscher, H., Gentz, R. and Lesslauer, W. (1992). *Immunology* **77**, 510–514.
Pitti, R.M., Marsters, S.A., Haakfrendscho, M., Osaka, G.C., Mordenti, J., Chamow, S.M. and Ashkenazi, A. (1994). *Mol. Immunol.* **31**, 1345–1351.
Pober, J.S. and Cotran, R.S. (1990). *Physiol. Rev.* **70**, 427–451.
Pollak, M. and Richard, M. (1990). *J. Natl Cancer Inst.* **82**, 1349–1352.
Polverini, P.J. and Nickoloff, B.J. (1995). *Adv. Cancer Res.* **66**, 235–253.
Pombo, C.M., Bonventre, J.V., Avruch, J., Woodgett, J.R., Kyriakis, J.M. and Force, T. (1994). *J. Biol. Chem.* **269**, 26 546–26 551.

Power, C.A. and Wells, T.N.C. (1996). *Trends Pharmacol. Sci.* **17**, 209–213.
Powis, G. (1991). *Trends Pharmacol.* **12**, 188–194.
Premack, B.P. and Schall, T.J. (1996). *Nature Med.* **2**, 1174–1178.
Presky, D.H., Yang, H., Minetti, L.J., Chua, A.O., Navabi, N., Wu, C.Y., Gately, M.K. and Gubler, U. (1996). *Proc. Natl Acad. Sci. U.S.A.* **93**, 14 002–14 007.
Pretolani, M. and Goldman, M. (1997). *Immunol. Today* **18**, 277–280.
Pronk, G.J. and Bos, J.L. (1944). *Biochim. Biophys. Acta* **1198**, 131–147.
Pugh-Humphreys, R.G.P. (1992). *FEMS Microbiol. Immunol.* **105**, 289–308.
Raghow, R. (1994). *FASEB J.* **8**, 823–831.
Rago, R., Mitchen, J., Cheng, A.L., Oberley, T. and Wilding, G. (1991). *Cancer Res.* **51**, 6629–6635.
Rago, R.P., Miles, J.M., Sufit, R.L., Spriggs, D.R. and Wilding, G. (1994). *Cancer* **73**, 1954–1959.
Raines, E.W. and Ross, R. (1992). *J. Cell Biol.* **116**, 533–543.
Ramakrishnan, S., Fryxell, D., Mohanraj, D., Olson, M. and Li, B.-Y. (1992). *Ann. Rev. Pharmacol. Toxicol.* **32**, 579–621.
Ramakrishnan, S., Olson, M., Bautch, V.L. and Mohanraj, D. (1996). *Cancer Res.* **56**, 1324–1330.
Ramp, U., Jaquet, K., Reinecke, P., Nitsch, T., Gabbert, H.E. and Gerharz, C.-D. (1997). *Lab. Invest.* **76**, 739–749.
Ramsay, A.J., Ruby, J. and Ramshaw, I.A. (1993). *Immunol. Today* **14**, 155–157.
Ramshaw, I.A., Ruby, J. and Ramsay, A. (1992). *TIBTech.* **10**, 424–426.
Rao, A. (1991). *Crit. Rev. Immunol.* **10**, 495–519.
Rao, A. (1994). *Immunol. Today* **15**, 274–281.
Rao, A.S., Starzl, T.E., Demetris, A.J., Trucco, M., Thomson, A., Qian, S., Murase, N. and Fung, J.J. (1996). *Clin. Immunol. Immunopathol.* **80**, S46–S51.
Rapraeger, A.C., Guimonds, S., Krufka, A. and Olwin, B.B. (1994). *Methods Enzymol.* **245**, 219–240.
Ray, D. and Melmed, S. (1997). *Endocr. Rev.* **18**, 206–228.
Raymond, J.R. (1995). *Am. J. Physiol.* **269**, F141–F158.
Reddy, G.P.V. (1994). *J. Cell Biochem.* **54**, 379–385.
Rettenmier, C.W., Roussel, M.F., Ashman, R.A., Ralph, P., Price, K. and Sherr, C.J. (1987). *Mol. Cell Biol.* **7**, 2378–2387.
Ricci, M., Rossi, O., Bertoni, M. and Matucci, A. (1993). *Clin. Exp. Allergy* **23**, 360–369.
Richards, J., Mick, R., Latta, J.M., Daly, K., Ratain, J., Vardiman, J.W. and Colomb, H.M. (1990). *Blood* **70**, 1941–1945.
Rideout, D.C., Bustamente, A., Patel, R. and Henderson, G.B. (1995). *Int. J. Cancer* **61**, 840–847.
Riegel, A.T. and Wellstein, A. (1994). *Breast Cancer Res. Treat.* **31**, 309–314.
Ries, C. and Petrides, P.E. (1995). *Biol. Chem. Hoppe Seyler* **376**, 345–355.
Rifkin, D.B., Kojima, S., Abe, M. and Harpel, J.G. (1993). *Thromb. Haemost.* **70**, 177–179.
Robbins, D.J., Zhen, E., Cheng, M., Xu, S., Ebert, D. and Cobb, M.H. (1994). *Adv. Cancer Res.* **63**, 93–116.
Roberts, A.B. and Sporn, M.B. (eds) (1990). *Peptide Growth Factors and Their Receptors 1.* Springer, London.
Rocken, M., Racke, M. and Shevach, E.M. (1996). *Immunol. Today* **17**, 225–231.
Romagnani, S., (1994). *Ann. Rev. Immunol.* **12**, 227–257.
Romagnani, S. (1995). *J. Clin. Immunol.* **15**, 121–129.
Romagnani, S. (1996). *Clin. Immunol. Immunopathol.* **80**, 225–235.
Romagnani, S. (1997). *Immunol. Today* **18**, 263–266.
Ronnov-Jessen, L., Peterson, O. W. and Bissel, M.J. (1996). *Physiol. Rev.* **76**, 69–125.
Rooney, J.W., Hoey, T. and Glimcher, L.H. (1995). *Immunity* **2**, 473–483.
Rorth, P., Nerlov, C., Blasi. F. and Johnsen, M. (1990). *Nucleic Acids* **18**, 5009–5017.
Rose, B., Gillespie, A., Wunderlich, D., Kelley, K., Dzuiba, J., Shedd, D., Cahill, K. and Zerler, B. (1992). *Immunology* **76**, 452–459.
Rose-John, S. and Heinrich, P.C. (1994). *Biochem. J.* **300**, 281–290.
Rosen, E.M. and Goldberg, I.D. (1995). *Adv. Cancer Res.* **67**, 257–279.
Rosen, M.K., Yamazaki, T., Gish, G.D., Kay, C.M., Pawson, T. and Kay, L.E. (1995). *Nature* **374**, 477–479.
Rosenthal, F., Zier, K.S. and Gansbacher, B. (1994). *Curr. Opin. Oncol.* **6**, 611–615.
Rothe, M., Wong, S.C., Henzel, W.J. and Goeddel, D.V. (1994). *Cell* **78**, 681–692.
Rothe, M., Sarma, V., Dixit, V.W. and Goeddel, D.V. (1995). *Science* **269**, 1424–1427.
Ruoslahti, E. and Yamaguchi, Y. (1991). *Cell* **64**, 867–869.
Rushlow, C and Warrior, R. (1992). *Bioessays* **14**, 89–95.

Russell, S.J. (1990). *Immunol. Today* **11**, 196–200.

Russell, S.M., Keegan, A.D., Harada, N., Nakamura, Y., Noguchi, M., Leland, P., Friedmann, M.C., Miyajima, A., Puri, R.X., Paul, W.E. and Leonard, W.J. (1993). *Science* **262**, 1880–1883.

Sabzevari, H., Gillies, S.D., Mueller, B.M., Pancock, J.D. and Reisfeld, R.A. (1994). *Proc. Natl Acad. Sci. U.S.A.* **91**, 9626–9630.

Sadick, M.D. (1992). In *The Macrophage* (eds C.E. Lewis and J.O'D. McGee), IRL Press, Oxford, pp. 265–284.

Saito, S., Bannerji, R., Gansbacher, B., Rosenthal, F.M., Romanenko, P., Heston, W.D.W., Fair, W.R. and Gilbao, E. (1994). *Cancer Res.* **54**, 3516–3520.

Saltiel, A.R. (1994). *FASEB J.* **8**, 1034–1040.

Saltiel, A.R. (1995). *Sci. Am. Sci. Med.* **2**, 58–67.

Sanchez, I., Hughes, R.T., Mayer, B.J., Yee, K., Woodgett, J.R., Avruch, J., Kyriakis, J.M. and Zon, L.I. (1994). *Nature* **372**, 794–798.

Sanderson, N., Factor, V., Nagy, P., Kopp, J., Kondaiah, P., Wakefield, L., Roberts, A.B., Sporn, M.B. and Thorgeirsson, S.S. (1995). *Proc. Natl Acad. Sci. U.S.A.* **92**, 2572–2576.

Sassone-Corsi, P. (1995). *Annu. Rev. Cell Dev. Biol.* **11**, 355–377.

Satoh, T., Uehara, Y. and Kaziro, Y. (1992). *J. Cell Biol.* **267**, 2537-2541.

Savaage, P., So, A., Spooner, R.A. and Epenetos, A.A. (1993). *Br. J. Cancer* **67**, 304–310.

Savino, R., Ciapponi, L., Lahm, A., Demartis, A., Cabibbo, A., Toniatti, C., Delmastro, P., Altamura, S. and Ciliberto, G. (1994a). *EMBO J.* **13**, 5863–5870.

Savino, R., Lahm, A., Salvatti, A.L., Ciapponi, L., Sporeno, E., Altamura, S., Paonessa, G., Toniatti, C. and Ciliberto, G. (1994b). *EMBO J.* **13**, 1357–1367.

Saxena, A. and Ali, I.U. (1992). *Oncogene* **7**, 243–247.

Schall, T. (1997). *Immunol. Today* **18**, 147.

Schall, T.J. and Bacon, K.B. (1994). *Curr. Opin. Immunol.* **6**, 865–873.

Schindler, C. and Darnell, J.E. (1995). *Annu. Rev. Biochem.* **64**, 621–651.

Schindler, C., Shuai, K., Prezioso, V.R. and Darnell, J.E.J. (1992). *Science* **257**, 809–813.

Schindler, C., Kashleva, H., Pernis, A., Pine, R. and Rothman, P. (1994). *EMBO J.* **13**, 1350–1356.

Schmidt, M. and Wels, W. (1996). *Br. J. Cancer* **74**, 853–862.

Schmidt, M., Hynes, N.E., Groner, B. and Wels, W. (1996). *Int. J. Cancer.* **65**, 538–546.

Schreck, R., Rieber, P. and Bauerle, P.A. (1991). *EMBO. J.* **10**, 2247–2258.

Schreiber, S.L. (1991). *Science* **251**, 283–287.

Schreiber, S.L. and Crabtree, G.R. (1992). *Immunol. Today* **13**, 136–142.

Schreuder, H., Tardif, C., Trump-Kallmeyer, S., Soffientini, A., Sarubbi, E., Akeson, A., Bowlin, T., Yanofsky, S. and Barrett, R.W. (1997). *Nature* **386**, 194–200.

Schutze, S., Potthoff, K., Machleidt, T., Berkovic, D., Wiegmann and Kronke, M. (1992). *Cell* **71**, 765–776.

Seder, R.A. and Paul, W. (1994). *Ann. Rev. Immunol.* **12**, 635–673.

Sells, M.A. and Chernoff, J. (1997). *Trends Cell Biol.* **7**, 162–167.

Semler, J., Wachtel, H. and Endres, S. (1993). *Int. J. Immunopharmacol.* **15**, 409–413.

Shalaby, M.R., Sundan, A., Loetscher, H., Brockhaus, M., Lesslauer, W. and Esperik, T.M. (1990). *J. Exp. Med.* **172**, 1517–1520.

Shalloway, D. and Taylor, S.J. (1997). *Trends Cell Biol.* **7**, 215–217.

Shanafelt, A.B., Miyajima, A., Kitamura, T. and Kastelein, R.A. (1991). *EMBO J.* **10**, 4105–4112.

Shaw, A.S. and Austin, M.L. (1997). *Immunity* **6**, 361–369.

Sherr, C.J. (1995). *Cell* **79**, 551–555.

Shi, Y., Hata, A., Lo, R.S., Massague, J. and Pavletich, N.P. (1997). *Nature* **388**, 87–93.

Shimoda, K., Vandeursen, J., Sangster, M.Y., Sarawar, S.R., Carson, R.T., Tripp, R.A., Chu, C., Quelle, F.W., Nosaka,T., Vignali, D.A.A., Doherty, P.C., Grosveld, G., Paul, W.E. and Ihle, J.N. (1996). *Nature* **380**, 630–633.

Shimomura, T., Miyazawa, K., Komiyama, Y., Hikaora, N., Naka, D., Morimoto, Y. and Kitamura, N. (1995). *Eur. J. Biochem.* **229**, 257–261.

Shuai, K., Schindler, C., Prezioso, V.R. and Darnell, J.E. (1992). *Science* **258**, 1808–1812.

Sieper, J. and Kingsley, G. (1996). *Immunol. Today* **17**, 160–163.

Sieweke, M.H. and Bissell, M.J. (1994). *Crit. Rev. Oncog.* **5**, 297–311.

Simmons, G., Clapham, P.R., Picard, L., Offord, R.E., Rosenkilde, M.M., Schwartz, T.W., Buser, R., Wells, T.N.C. and Proudfoot, A.E.J. (1997). *Science* **276**, 276–279.

Sluss, H.K., Barrett, T., Derijard, B. and Davis, R.J. (1994). *Mol. Cell. Biol.* **14**, 8376–8384.

Smallwood, P.M., Munoz-Sanjuan, I., Tong, P., Macke, J.P., Hendry, S.H.C., Gilbert, D.J., Copeland, N.G., Jenkins, N.A. and Nathans, J. (1996). *Proc. Natl Acad. Sci. U.S.A.* **93**, 9850–9857.

Smith, C.A., Farrah, T. and Goodwin, R.G. (1994). *Cell* **76**, 959–962.

Sorkin, A. and Waters, C.M. (1993). *Bioessays* **15**, 375–382.

Speert, D.P. (1992). In *The Macrophage* (eds C.E. Lewis and J.O'D. McGee), IRL Press, Oxford, pp. 215–263.

Sprang, S.R. and Bazan, J.F. (1993). *Curr. Opin. Struct. Biol.* **3**, 815–827.

Sprenger, H., Rosler, A., Tonn, P., Braune, H.J., Huffman, G. and Gemsa, D. (1996). *Clin. Immunol. Immunopathol.* **80**, 155–161.

Springer, T.A. (1994). *Cell* **76**, 301–314.

Stahl, N., Boulton, T.G., Farrugella, T., Ip, N.Y., Davis, S., Witthuhn, B.A., Quelle, F.W., Silvenoinen, O., Barbieri, G., Pelligrini, S., Ihle, J.N. and Yancopoulos, G.D. (1994). *Science* **263**, 92–95.

Stahl, N., Farruggella, T.J., Boulton, T.G., Zhong, Z., Darnell, J.E. and Yancopoulos, G.D. (1995). *Science* **267**, 1349–1353.

Starr, R., Willson, T.A., Viney, E.M., Murray, L.J.L., Rayner, J.R., Jenkins, B.J., Gonde, T.J., Alexander, W.S., Metcalf, D., Nicola, N.A. and Hilton, D.J. (1997). *Nature* **387**, 917–921.

Steegmaier, M., Levinovitz, A., Isenmann, S., Borges, E., Lenter, M., Kocher, P., Kleuser, B. and Vestweber, D. (1994). *Nature* **373**, 615–620.

Stein, C.A. (1993). *Cancer Res.* **53**, 2239–2248.

Steinbruchel, D.A., Hoch, C., Kristensen, T. and Kemp, E. (1991). *Scand. J. Immunol.* **34**, 627–633.

Steinbruchel, D.A., Larsen, S., Kristensen, T., Starklint, H., Koch, C. and Kemp, E. (1992). *Acta Path. Microbiol. Immunol. Scand.* **100**, 682–694.

Stein-Gerlach, M., Kharitonenkov, A., Vogel, W., Ali, S. and Ullrich, A. (1995). *J. Biol. Chem.* **270**, 24 635–24 637.

Steinman, R.M. (1991). *Annu. Rev. Immunol.* **9**, 271–296.

Sternweiss, P.C. (1996). In *Signal Transduction* (eds C.-H. Heldin and M. Purton), Chapman and Hall, London, pp. 287–301.

Stetler-Stevenson, M., Mansoor, A., Lim, M., Fukushima, P., Kehrl, J., Marti, G., Ptaszynski, K., Wang, J. and Stetler-Stevenson, W.G. (1997). *Blood* **89**, 1708–1715.

Streuli, M. (1996). *Curr. Opin. Cell Biol.* **8**, 182–188.

Strieter, R.M., Polverini, P.J., Arenberg, D.A., Walz, A., Opdenakker, G., Vandamme, J. and Kunkel, S.L. (1995). *J. Leukocyte Biol.* **57**, 752–762.

Strom, T.B., Kelley, V.E., Murphy, J.R., Osawa, H., Tilney, N.L. and Shapiro, M.E. (1988). In *Interleukin 2* (ed K.A. Smith), Academic Press, New York, pp. 223–236.

Su, B., Jacinto, E., Hibi, M., Kallunki, T., Karin, M. and Ben-Neriah, Y. (1994). *Cell* **77**, 727–736.

Suda, T., Takahashi, T., Golstein, P. and Nagata, S. (1995). *Cell* **75**, 1169–1178.

Suffys, P., Beyaert, R., van Roy, F. and Fiers, W. (1988). *Eur. J. Biochem.* **178**, 257–265.

Sun, H. and Tonks, N.K. (1994). *TIBS* **19**, 480–485.

Suzuki, A., Takahashi, T., Okuno, Y., Nakamura, K., Tashiro, H., Futumoto, M., Konaka, Y. and Imura, H. (1991). *Int. J. Cancer* **48**, 428–433.

Suzuki, H., Prado, G.N., Wilkinson, N. and Navarro, J. (1994). *J. Biol. Chem.* **269**, 18 263–18 266.

Symons, J.A., Young, P.R. and Duff, G.W. (1995). *Proc. Natl Acad. Sci. U.S.A.* **92**, 1714–1718.

Szabo, S.J., Gold, J.S., Murphy, T.L. and Murphy, K.M. (1993). *Mol. Cell. Biol.* **13**, 4793–4805.

Szabo, S.J., Jacobson, N.G., Dighe, A.S., Gubler, U. and Murphy, K.M. (1995). *Immunity* **2**, 665–675.

Taga, T. and Kishimoto, T. (1992). *FASEB J.* **6**, 3387–3396.

Taipale, J., Saharinen, J., Hedman, K. and Keski-Oja, J. (1996). *J. Histochem. Cytochem.* **44**, 875–889.

Tak, P.P., Taylor, P.C., Breedveld, F.C., Smeets, T.J.M., Daha, M.R., Kliun, P.M., Meinders, A.E. and Maini, R.N. (1996). *Arthritis Rheum.* **39**, 1077–1081.

Tak, P.P., Smeets, T.J.M., Daha, M.R., Kluin, P.M., Miejers, K.A.E., Brand, R., Meinders, A.E. and Breedvelt, F.C. (1997). *Arthritis Rheum.* **40**, 217–225.

Takahashi, I., Nakanishi, S., Kobayashi, E., Nakano, H., Suzuki, K. and Tamaoki, T. (1989). *Biochem. Biophys. Rev. Commun.* **165**, 1207–1212.

Takeda, K., Tanaka, T., Shi, W., Matsumoto, M., Minami, M., Kashiwamura, S., Nakanishi, K., Yoshida, N., Kishimoto, T. and Akira, S. (1996). *Nature* **380**, 627–630.

Takeshita, T., Arita, T., Asao, H., Tanaka, N., Higuchi, M., Kuroda, H., Kaneko, K., Munakata, H., Endo, Y., Fujita, T. and Sugamura, K. (1996). *Biochem. Biophys. Res. Commun.* **225**, 1035–1039.

Takeshita, T., Arita, T., Higuchi, M., Asao, H., Endo, K., Kuroda, H., Tanaka, N., Murata, K., Ishii, N. and Sugamura, K. (1997). *Immunity* **6**, 449–457.

Takeuchi, M., Rothe, M. and Goeddel, D.V. (1996). *J. Biol. Chem.* **271**, 19935–19942.
Tanaka, K., Tilney, N.L., Stunkel, K.G., Hancock, W.W., Diamantstein, T. and Kupiec-Weglinski, J.W. (1990). *Proc. Natl Acad. Sci. U.S.A.* **87**, 7375–7379.
Tanaka, Y., Adams, D.H. and Shaw, S. (1993). *Immunol. Today* **14**, 111–115.
Taniguchi, T. (1995). *Science* **268**, 251–255.
Tapley, P., Lamballe, F. and Barbacid, M. (1992). *Oncogene* **7**, 371–381.
Tartaglia, L.A., Weber, R.F., Figari, I.S., Reynolds, C., Palladino, M.A. and Goeddel, D.V. (1991). *Proc. Natl Acad. Sci. U.S.A.* **88**, 9292–9296.
Tartaglia, L.A., Pennica, D. and Goeddel, D.V. (1993). *J. Biol. Chem.* **268**, 18542–18548.
Tashiro, K., Tada, H., Heilker, R., Shirozu, M., Nakano, T. and Honjo, T. (1993). *Science* **261**, 600–603.
Tepper, R.L. and Mule, J.J. (1994). *Hum. Gene Ther.* **5**, 153–164.
Teran, L.M. and Davies, D.E. (1996). *Clin. Exp. Allergy* **26**, 1005–1019.
Tewari, M. and Dixit, V.M. (1996). In *Signal Transduction* (eds C.-H. Heldin and M. Purton), Chapman and Hall, London, pp. 79–92.
Thedrez, P., Paineau, J., Jacques, Y., Chatal, J.F., Pelegrin, A., Bouchaud, G. and Soulillou, J.P. (1989). *Transplantation* **48**, 367–371.
Theze, J., Alzari, P.M. and Bertoglio, J. (1996). *Immunol. Today* **17**, 480–486.
Thierfelder, W.E., van Deursen, J.M., Yamamoto, K., Tripp, R.A., Sarawar, S.R., Carson, R.R., Sangster, M.Y., Vignali, D.A.A., Deherly, P.C., Grosveld, G.C. and Ihle, J.N. (1996). *Nature* **382**, 171–174.
Thoma, B., Grell, M., Pfizenmaier, K. and Sheurich, P. (1990). *J. Exp. Med.* **172**, 1019–1023.
Thomas, R. and Lipsky, P.E. (1996a). *Immunol. Today* **17**, 559–564.
Thomas, R. and Lipsky, P.E. (1996b). *Arthritis Rheum.* **39**, 183–190.
Thomson, A.W. (1997). In *Renal Transplantation* (ed. R. Shapiro), Appleton and Lange, New York, pp. 163–209.
Thomson, A.W., Bonham, C.A. and Zeevi, A. (1995). *Ther. Drug Monit.* **17**, 584–591.
Thomson, A.W., Ilstad, S.T. and Simmons, R.L. (1997). In *Textbook of Surgery: The Biological Basis of Modern Surgical Practice*, 15th edn (ed. D.C. Sabiston), W.B. Saunders, Philadelphia, PA (in press).
Thornberry, N.A., Bull, H.G., Calaycay, J.R., Chapman, K.T., Howard, A.D., Kostura, M.J., Miller, D.K., Molineaux, S.M., Weidner, J.R., Aunins, J., Elliston, K.O., Ayala, J.M., Casano, F.J., Chin, J., Ding, G.J.F., Egger, L.A., Gaffney, E.P., Yamin, T.T., Lee, T.D., Shively, J.E., Maccross, M., Mumford, R.A., Schmidt, J.A. and Tocci, M.J. (1992). *Nature* **356**, 768–774.
Tonks, N.K. (1996). In *Signal Transduction* (eds C.-H. Heldin and M. Purton), Chapman and Hall, London, pp. 253–269.
Torchia, J., Rose, D.W., Inostroza, J., Kamel, Y., Westin, S., Glass, C.K. and Rosenfeld, M.G. (1997). *Nature* **387**, 677–684.
Townsend, S.E. and Allison, J.P. (1993). *Science* **259**, 368–370.
Tran-Thang, C., Kruithof, E.K.O., Lahm, K., Schuster, W.-A., Tada, M. and Sordat, B. (1996). *Br. J. Cancer* **74**, 846–852.
Trinchieri, G. (1993). *Immunol. Today* **14**, 335–338.
Trinchieri, G. (1994). *Blood* **84**, 4008–4027.
Trinchieri, G. (1995). *Annu. Rev. Immunol.* **13**, 251–278.
Uchiyama, T., Broder, S. and Waldmann, T.A. (1981). *J. Immunol.* **126**, 1393–1397.
Ueda, H., Hancock, W.W., Cheung, Y.C., Diamantstein, T., Tilney, N.L. and Kupiec-Weglinski, J.W. (1990). *Transplantation* **50**, 545–560.
Ueki, T., Koji, T., Tamiya, S., Nakane, P. and Tsuneyoshi, M. (1995). *J. Pathol.* **177**, 353–361.
Valitutti, S. and Lanzavecchia, A. (1997). *Immunol. Today* **18**, 299–305.
van Antwerp, D.J., Martin, S.J., Kafri, T., Green, D.R. and Verma, I.M. (1996). *Science* **274**, 786–789.
Vandenabeele, P., Declercq, W., Beyaert, R. and Fiers, W. (1995). *Trends Cell. Biol.* **5**, 392–399.
van den Hooff, A. 1988. *Adv. Cancer Res.* **50**, 159–196.
van der Geer, P., Hunter, T. and Linberg, R.A. (1994). *Annu. Rev. Cell Biol.* **10**, 251–337.
van der Poll, T. and Lowry, S.F. (1995). *Shock* **3**, 1–12.
van Deuren, M., Doffer, A.S.M. and van der Meer, J.W.M. (1992). *J. Pathol.* **168**, 349–356.
van Deveter, S.J.H., Buller, H.R., ten Cate, J.W., Aarden, L.A., Hack, C.E. and Sturk, A. (1990). *Blood* **76**, 2500–2526.
Vassalli, P. (1992). *Annu. Rev. Immunol.* **10**, 411–452.
Velaquez, L., Fellous, M., Stark, G.R. and Pelligrini, S. (1992). *Cell* **70**, 313–322.

Verma, I.M., Stevenson, J.K., Schwartz, E.M., van Antwerp, D. and Miyamoto, S. (1995). *Genes Dev.* **9**, 2723–2735.

Viac, J., Vincent, C., Palacio, S., Schmitt, D. and Claudy, A. (1996). *Eur. J. Cancer* **32A**, 447–449.

Vieira, A.V., Lamaze, C. and Schmid, S.L. (1996). *Science* **274**, 2086–2089.

Vieweg, J., Rosenthal, F.M., Bannerji, R., Heston, W.D.W., Fair, W.R., Gansbacher, B. and Gilboa, E. (1994). *Cancer Res.* **54**, 1760–1765.

Vigers, G.P.A., Anderson, L.J., Caffes, P. and Brandhuber, B.J. (1997). *Nature* **386**, 190–194.

Vignon, F., Prebois, C. and Rochefort, H. (1992). *J. Natl Cancer Inst.* **84**, 38–42.

Viola, A.A. and Lanzavecchia, A. (1996). *Science* **273**, 104–106.

Vlodavsky, I., Fuks, Z., Ishai-Michaeli, R., Bashkin, P., Levi, E. and Komer, G. (1991). *J. Cell Biochem.* **45**, 167–176.

von Feldt, J.M., Monfardini, C., Kieber-Emmons, T., Voet, D., Weiner, D.B. and Williams, W.V. (1994). *Immunol. Res.* **13**, 96–109.

Wagner, R.M. (1994). *Nature* **372**, 333–335.

Wahl, S.M. (1992). *J. Clin. Immunol.* **12**, 61–74.

Wakeling, A.E., Barker, A.J., Davies, D.H., Brown, D.S., Green, L.R., Cartildge, S.A. and Woodburn, J.R. (1996). *Breast Cancer Res. Treat.* **38**, 67–73.

Wakimoto, H., Abe, J., Tsunoda, R., Aoyagi, M., Hirakawa, W. and Hamada, H. (1996). *Cancer Res.* **56**, 1828–1833.

Waldmann, T. A., Grant, A., Tendler, C., Greenberg, S., Goldman, C. and Bamford, R. (1990). *J. Clin. Immunol.* **10 (supplement 6)**, 195–285.

Waldmann, T.A., White, J.D., Carrasquillo, J.A., Reynolds, J.C., Paik, C.H., Gansow, O.A., Brechbiel, M.W., Jaffe, E.S., Fleisher, T.A., Goldman, C.K., Top, L.E., Bamford, R., Zaknoen, S., Roessler, E., Kastensportes, C., England, R., Litou, H., Johnson, J.A., Jacksonwhite, T., Manns, A., Hanchard, B., Junghans, R.P. and Nelson, D.L. (1995). *Blood* **86**, 4063–4075.

Wallach, D. (1997). *TIBS* **22**, 107–109.

Wallis, R.S. and Ellner, J.S. (1994). *J. Leukocyte Biol.* **55**, 676–681.

Wang, C.-Y., Mayo, M.W. and Baldwin, A.S. (1996). *Science* **274**, 784–787.

Wang, F., Sengupta, T.K., Zhong. Z. and Ivashkiv, L.B. (1995). *J. Exp. Med.* **182**, 1825–1831.

Wang, X.F., Lin, H.Y., Ngeaton, E., Downward, J., Lodish, H.F. and Weinberg, R.A. (1991). *Cell* **67**, 797–805.

Watanabe, M., McCormick, K.L., Volker, K., Ortaldo, J.R., Wigginton, J.M., Brunda, M.J., Wiltrout, R.H. and Fogler, W.E. (1997). *Am. J. Pathol.* **150**, 1869–1880.

Watling, D., Guschin, D., Muller, M., Silvennoinen, O., Witthuhn, B.A., Quelle, F.W., Rogers, N.C., Schindler, C., Stark, G.R., Ihle, J.N. and Kerr, I.M. (1993). *Nature* **366**, 166–170.

Wee, S., Pascual, M., Eason, J.D., Schoenfeld, D.A., Phelan, J., Boskovic, S., Blosch, C., Mohler, K. and Cosimi, A.B. (1997). *Transplantation* **63**, 570–577.

Weinberg, R.A. (1995). *Cell* **81**, 323–330.

Wells, T.N.C. and Peitsch, M.C. (1997). *J. Leukocyte Biol.* **61**, 545–550.

Wellstein, A. and Czubayko, F. (1996). *Breast Cancer Res. Treat.* **38**, 109–119.

Wels, W., Harwerth, I.M., Mueller, M., Groner, B. and Hynes, N.E. (1992). *Cancer Res.* **52**, 6310–6317.

Wen, Z.L., Zhong, Z. and Darnell, J.E. (1995). *Cell* **82**, 241–250.

Wera, S. and Hemmings, B.A. (1995). *Biochem. J.* **311**, 17–29.

Wernert, N., Raes, M.B., Lasalle, P., Dehouck, M.P., Gosselin, B., Vandenbunder, B. and Stehelin, D. (1992). *Am. J. Pathol.* **140**, 119–127.

Wernert, N., Gilles, F., Fafeur, V., Boulai, F., Raes, M.B., Pyke, C., Dupressoir, T., Seitz, G., Vanderbunder, B. and Stehelin, D. (1994). *Cancer Res.* **54**, 5683–5688.

Wess, J. (1997). *FASEB J.* **11**, 346–354.

Westwick, J.K., Bielawska, A.E., Dbaibo, G., Hannun, Y.A. and Brenner, D.A. (1995). *J. Biol. Chem.* **270**, 22 689–22 692.

Whalen, G.F. (1990). *Lancet* **336**, 1489–1492.

Whaley, K. and Burt, A.D. (1992). In *Muir's Textbook of Pathology*, 13th edn (eds R.N.M. MacSween and K. Whaley), Edward Arnold, London, pp. 112–165.

Wherry, J.C., Pennington, J.E. and Wenzel, R.P. (1993). *Crit. Care Med.* **21**, S436–S440.

Wick, Hu, Y., Schwarz, S. and Kroemer, G. (1993). *Endocr. Rev.* **14**, 539–563.

Wiedmann, B., Sakai, H., Davis, T.A. and Wiedmann, M. (1994). *Nature* **370**, 434–440.

Williams, R.O., Feldmann, M. and Maini, R.N. (1992). *Proc. Natl Acad. Sci. U.S.A.* **89**, 9784–9788.

Williams, R.O., Mason, L.J., Feldmann, M. and Maini, R.N. (1994). *Proc. Natl Acad. Sci. U.S.A.* **91**, 2762–2766.
Winchester, R. (1992). *Ann. Intern. Med.* **117**, 869–871.
Witt, D.P. and Lander, A.D. (1994). *Curr. Biol.* **4**, 394–400.
Wlodawer, A., Pavlovsky, A. and Gustchina, A. (1993). *Protein Sci.* **2**, 1375–1384.
Wolfsberg, T.G. and White, J.M. (1996). *Dev. Biol.* **180**, 389–401.
Wong, H. and Wahl, S.M. (1990). In *Peptide Growth Factors and Their Receptors 2* (eds M.B. Sporn and A.B. Roberts), Springer, London, pp. 509–548.
Wong, Z. and Moran, M.F. (1996). *Science* **272**, 1935–1939.
Woodgett, J.R. (1996). In *Signal Transduction* (eds C.-H. Heldin and M. Purton), Chapman and Hall, London, pp. 321–333.
Wooley, P.H., Dutcher, J., Widmer, M.B. and Gillis, S. (1993). *J. Immunol.* **151**, 6602–6607.
Wrana, J. and Pawson, T. (1997). *Nature* **388**, 28–29.
Wrana, J.L. , Altisano, L., Wieser, R., Ventura, F. and Massague, J. (1994). *Nature* **370**, 341–347.
Wyke, A.W., Lang, A. and Frame, M.C. (1996). *Cell. Signal.* **8**, 131–139.
Xu, X., Williams, J.W., Bremer, E.G., Finnegan, A. and Chong, A.S. (1995). *J. Biol. Chem.* **270**, 12 398–12 403.
Yamaguchi, F., Saya, H., Bruner, J.M. and Morrison, R.S. (1994). *Proc. Natl Acad. Sci. U.S.A.* **91**, 484–488.
Yamanaka, Y., Friess, H., Buchler, M., Beger, H., Uchida, E., Onda, M., Kobrin, M. and Korc, M. (1993). *Cancer Res.* **53**, 5289–5296.
Yan, G., Fukabori, Y., McBride, G., Nikolaropolou, S. and McKeehan, W. (1993). *Mol. Cell. Biol.* **13**, 4513–4522.
Yanni, G., Whelan, A., Feighery, C. and Bresnihan, B. (1992). *Semin. Arthritis Rheum.* **21**, 393–399.
Yi, T., Mui, A.L., Krystal, G. and Ihle, J.N. (1993). *Mol. Cell. Biol.* **13**, 7577–7586.
Yokoi, K., Mukaida, N., Harada, A., Watanabe, Y. and Matsushima, K. (1997). *Lab. Invest.* **76**, 375–384.
Yoneda, T., Nakai, M., Moriyama, Y., Scott, L., Ida, N., Kunitomo, T. and Mundy, G.R. (1993). *Cancer Res.* **53**, 737–740.
Yoshiji, H., Gomez, D.E., Shibuya, M. and Thorgeirsson, U.P. (1996). *Cancer Res.* **56**, 2013–2016.
Yu, C.L., Meyer, D.J., Campbell, G.S., Larner, A.C., Carter-Su, C., Schwartz, J. and Jove, R. (1995). *Science* **269**, 81–83.
Yu, D., Wang, S.S., Dulski, K.M., Tsai, C.M., Nicholson, G.L. and Hung, M.C. (1994). *Cancer Res.* **54**, 3260–3266.
Zhang, L., Khayat, A., Cheng, H. and Graves, D.T. (1997). *Lab. Invest.* **76**, 579–590.
Zhong, H., SuYang, H., Erdjument-Bromage, H., Tempst, P. and Ghosh, S. (1997). *Cell* **89**, 413–424.
Ziemiecki, A., Harpur, A.G. and Wilks, A.F. (1994). *Trends Cell Biol.* **4**, 207–211.
Zier, K.S. and Gansbacher, B. (1996). *Eur. J. Cancer* **32A**, 1408–1412.
Zinck, R., Cahill, M.A., Kracht, M., Sachsemaier, C., Hipskins, R.A. and Nordheim, A. (1995). *Mol. Cell. Biol.* **15**, 4930–4938.
Zolnierowicz, S. and Hemmings, B.A. (1996). In *Signal Transduction* (eds C.-H. Heldin and M. Purton), Chapman and Hall, London, pp. 271–286.
Zou, J.P., Yamamoto, N., Fujii, T., Takenaka, H., Kabayashi, M., Herrman, S.H., Brunda, M.J., Luistro, L., Warrier, R.R., Wright, R.B., Hubbard, B.R., Murphy, M., Wolf, S.F. and Gately, M.K. (1993). *J. Exp. Med.* **178**, 1223–1230.
Zuber, M.E., Zhou, Z., Burrus, L.W. and Olwin, B.B. (1997). *J. Cell. Physiol.* **170**, 217–227.
Zvaifler, N.J. and Firestein, G.S. (1994). *Arthritis Rheum.* **37**, 783–789.
Zvaifler, N.J., Tsai, V., Alsalameh, S., von Kempis, J., Firestein, G.S. and Lotz, M. (1997). *Am. J. Pathol.* **150**, 1125–1138.

Chapter 33

The Phylogeny of Cytokines

Christopher J. Secombes
Department of Zoology, University of Aberdeen, Aberdeen, UK

INTRODUCTION

The largest dichotomy in animal defences is seen in early vertebrates (Klein, 1989; Andersson and Matsunaga, 1996; Litman, 1996). Only jawed vertebrates possess anticipatory immune responses, characterized by the presence of lymphocytes with antigen-specific receptors (immunoglobulins and T-cell receptors) that can undergo clonal proliferation in response to specific antigens. In addition, all jawed vertebrates possess both class I and class II major histocompatibility complex (MHC) genes (Horton and Ratcliffe, 1996; Stet *et al.*, 1996), necessary for antigen presentation to T cells. Lymphocyte heterogeneity appears to exist throughout the jawed vertebrates, with conclusively demonstrated T- and B-cell subpopulations being present in amphibians and higher vertebrates (Horton and Ratcliffe, 1996). In fish it has been more difficult to confirm the thymic-dependence of T cells, although cells functionally equivalent to T and B cells are clearly present (Graham and Secombes, 1990a). Vertebrates also possess a variety of non-specific cellular and humoral defences based on phagocytes, natural killer (NK) cells and a large array of soluble molecules (Secombes, 1996; Yano, 1996). Invertebrates similarly possess non-specific cellular and soluble defences, many of which are the forerunners to those seen in vertebrates (Soderhall *et al.*, 1996).

That cytokines are required to initiate and regulate immune responses is well established in mammals. Intuitively, this seems a likely scenario in all animals with complex cellular defences, whether or not such defences are anticipatory in nature. This does not imply that all mammalian cytokines will be present in all animals. Indeed, those released from T cells that act primarily on lymphocytes may prove to be unique to the jawed vertebrates. On the other hand, cytokines released from cells of the non-specific defences or that are important in their regulation may well be universal, or have a functionally equivalent analogue in 'lower' animals. In this chapter what is known about non-mammalian cytokines is reviewed under the following categories: interferons, interleukins, migration inhibition factors, tumor necrosis factors and growth factors.

INTERFERONS

Interferons (IFNs) are substances able to inhibit virus replication and in mammals three classes exist: α (leukocyte), β (fibroblast) and γ (immune), based on their biological and

The Cytokine Handbook, 3rd ed.
ISBN 0-12-689662-3

biochemical properties and gene sequences. IFN-α and IFN-β have many similarities and are grouped together as type I interferons. They are induced in cells infected with virus, are acid stable, and probably any cell type can produce them. In contrast, IFN-γ is unrelated to type I IFNs, is induced following antigen or mitogen (concanavalin-A, phytohaemagglutinin (PHA)) stimulation of T lymphocytes and has a wide spectrum of biological activities, including macrophage activation and upregulation of class II MHC antigens. It is therefore referred to as type II or immune IFN. Although recent studies report antiviral activity in invertebrates (Beckage, 1996; Washburn *et al.*, 1996) and even the presence of functional IFN consensus response elements (Georgel *et al.*, 1995), to date IFN activity appears to be unique to vertebrates. The possible divergence of IFN-α from IFN-β at the level of the reptiles has also been indicated.

Type I Interferon

Fish

It is well established that fish fibroblast or epithelial cell lines (Gravel and Malsberger, 1965; DeSena and Rio, 1975) and isolated leukocytes (Rogel-Gaillard *et al.*, 1993; Snegaroff, 1993; Congleton and Sun, 1996) can secrete IFN in response to virus infection and poly I:C, particularly when confluent before infection in the former case. IFN activity can be detected in the supernatants 24–72 h following stimulation, as assessed by their ability to reduce viral cytopathic effects (CPE) in cell lines pretreated with the supernatants for 24 h before addition of a challenge virus. This IFN activity is acid (pH 2) stable, relatively temperature resistant but destroyed by trypsin treatment (Table 1), all properties typical of a type I IFN. In addition, this activity is species specific with respect to the cell line being protected but non-specific with regard to the challenge virus, i.e. it can provide protection against a virus unrelated to that which induced it.

Similarly, *in vivo* studies have shown that IFN activities with biological properties identical to those described above can be detected in the serum of fish following natural or experimental viral infection (Dorson *et al.*, 1975). In the latter case IFN activity is detectable within 1 day postinfection, begins to decrease by 4 days and is undetectable by day 14. The IFN is protective *in vivo,* as demonstrated by an increased survival of rainbow trout (*Oncorhynchus mykiss*) alevins bath challenged with viral haemorrhagic septicaemia (VHS) virus following an injection of IFN-containing serum, compared

Table 1. Physicochemical characterization and species specificity of purified type 1 interferon from rainbow trout fibroblasts.

Treatment	Type of cells protected	IFN activity (units/ml)	Reduction in IFN activity (%)
None	Trout	8831	0
None	Goldfish	449	95
pH 2	Trout	9120	−3
25°C	Trout	8710	1
56°C	Trout	5623	36
65°C	Trout	398	96
Trypsin	Trout	1023	88

Adapted from DeSena and Rio (1975) with permission.

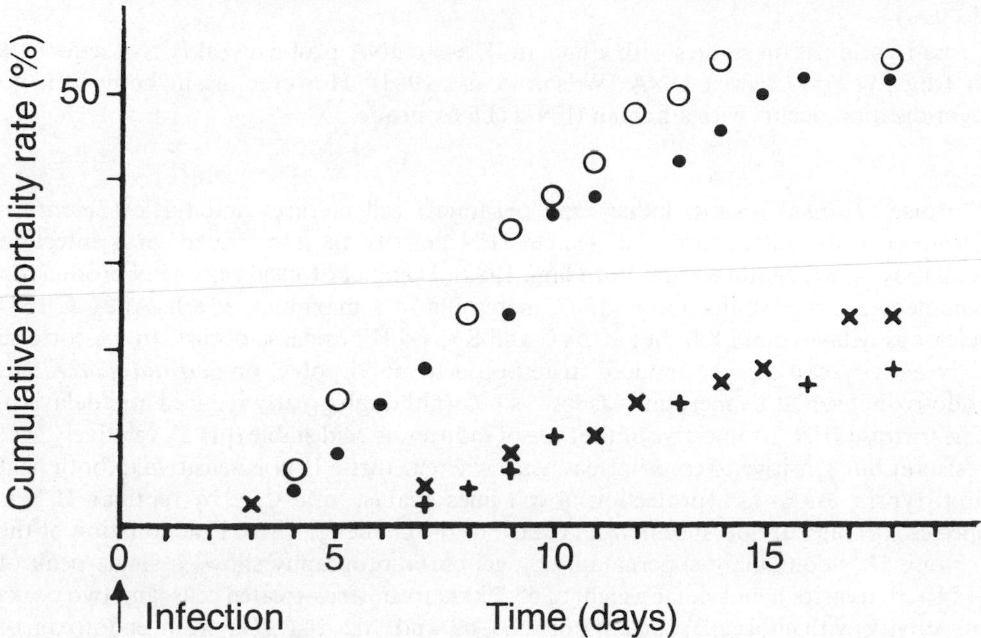

Fig. 1. Protection of young rainbow trout against VHS virus by transfer of IFN. Two groups of 50 fish were injected intraperitoneally with 20–30 IFN units contained in either serum or its low-molecular-weight fraction. Two control groups received either the high-molecular-weight serum fraction of the IFN-containing serum or whole normal trout serum. All groups were then infected immediately by immersion in VHS virus and the mortality was noted. ○, Antibody fraction; ●, control; ×, total serum; +, IFN fraction. From DeKinkelin *et al.* (1982), *Dev. Comp. Immunol. (Suppl.)* **2**, 167–174, with permission.

with fish injected with control serum (Fig. 1) (DeKinkelin *et al.*, 1982). The IFN activity is short lived following passive transfer, being undetectable 6–8 h postinjection. Protection is also seen against infectious pancreatic necrosis (IPN) virus *in vivo*, demonstrating that the *in vivo* antiviral activity is also non-specific in nature.

Physicochemical analysis of these *in vitro* and *in vivo* induced IFNs has shown large differences. Serum IFN has a molecular weight of 26 kDa, a pI of 5.3 and a sedimentation coefficient of 2.3 (Dorson *et al.*, 1975), whereas fibroblast IFN has a molecular weight of 94 kDa and a pI of 7.1 (DeSena and Rio, 1975).

Cross-hybridization studies with a human IFN-β cDNA probe show that IFN-β sequences are present in bony fish (Wilson *et al.*, 1983; Tengelsen *et al.*, 1991) but studies with a human IFN-α cDNA probe failed to reveal hybridization (Wilson *et al.*, 1983). A recent study has reported an IFN sequence from the Japanese flatfish (*Paralichthys olivaceus*), obtained by screening an expression library for IFN activity (Tamai *et al.*, 1993). The cloned product encoded a 138-amino-acid polypeptide with a molecular weight of 16 kDa, including a glycosylation site and a signal peptide. However, the amino acid sequence homology with other known mammalian and bird IFNs is typically less than 20%, and the recombinant molecule (produced in BHK cells) did not show species specificity, questioning the validity of calling this molecule an IFN.

Amphibia

Cross-hybridization studies with a human IFN-β cDNA probe reveal IFN-β sequences in frog (*Xenopus laevis*) DNA (Wilson *et al.*, 1983). However, as in bony fish, no hybridization occurs with a human IFN-α cDNA probe.

Reptiles

Tortoise (*Testudo graeca*) kidney and peritoneal cell cultures and turtle (*Terrapene carolina*) heart cell cultures can secrete IFN activity in response to virus infection (Galabov, 1981; Mathews and Vorndam, 1982). Using peritoneal cells, IFN production commences 2 h postinfection at 37°C, increasing to a maximum at 8 h. At 26°C IFN release is delayed until 8 h, but at 20°C and 8°C no IFN release occurs. In the tortoise IFN activity can also be induced in response to polyI:polyC or *Serratia marcescens* endotoxin, even at temperatures as low as 8°C (although greatly reduced and delayed). The tortoise IFN, irrespective of the type of inducer, is acid stable (pH 2), relatively heat resistant but sensitive to trypsin treatment, whereas turtle IFN is sensitive to both heat and trypsin. As in fish, protection of cell lines against viral CPE by reptilian IFN is species specific but non-specific with regard to the challenge virus. Fractionation of the tortoise IFN-containing supernatants by gel chromotography shows a single peak of IFN activity with a molecular weight of 58.5 kDa from virus-treated cells, and two peaks of activity with molecular weights of 43.5–62 and 102–144 kDa from endotoxin or polyI:polyC-treated cells.

Cross-hybridization of a human IFN-α cDNA probe with lizard (*Lacerta viridis*) DNA reveals faint but consistently labelled DNA fragments (Wilson *et al.*, 1983). Thus, different classes of type I IFN are likely to be present in amniotes. Human IFN, however, has no antiviral effects on tortoise kidney cells.

Birds

Hens' eggs were the first known source of IFN, and IFN activity is readily demonstrated in allantoic fluid from eggs inoculated with virus for 48 h (Moehring and Stinebring, 1981). IFN is also produced in adult birds following intravenous inoculation with virus. In addition, virally induced IFN is produced by chicken embryo fibroblast and macrophage cultures between 1 and 4 days postinfection. This IFN activity is pH stable, relatively heat resistant and trypsin sensitive (Lampson *et al.*, 1965) as in other vertebrate classes, indicating that it is a type I IFN. The genes for type I IFN have recently been sequenced in several bird species. In chickens, at least ten IFN genes are present in the genome (Sick *et al.*, 1996). Three intron-less IFN genes have been sequenced and each found to consist of a 31-amino-acid signal peptide and a mature protein of 162 amino acids, with four potential glycosylation sites and an estimated molecular weight of 19 kDa (Fig. 2) (Sekellick *et al.*, 1994; Sick *et al.*, 1996). These genes have 24% amino acid homology to IFN-α, 20% homology to IFN-β and 3% homology to IFN-γ, and have been termed IFN-1. An additional intron-less chicken (Ch) IFN gene has been sequenced, that exists as a single gene with 57% homology to chicken IFN1 (Fig. 2), termed IFN-2. It contains a 27-amino-acid signal peptide and a 176-amino-acid mature protein, with an estimated MW of 20 kDa. Similarly, in ducks an intron-less IFN gene has been sequenced, and contains a 30-amino-acid signal peptide and a 161-amino-acid mature protein (Fig. 2) (Schultz *et al.*, 1995a), with 50% homology to chicken IFN-1 and 61% homology to chicken IFN-2. Lastly, turkey IFN has been sequenced

```
Ch IFN-1   MAVPASPGHPRGYGILLLTLLLKALATTASACNHLRPQDATFSHDSLQLL
Ch IFN-2   --MTANHQSPGMHSILLLLLLPALTTT---FCNHLRHQDANFSWKSLQLL
Duck IFN   MPGPSAPPPPAIYSALALLLLLTPPAN-AFSCSPLRLHDSAFAWDSLQLL
Tk IFN     MAVPASPQHPRGYGILLLTLLMKALAA-AAACNHLRPQDATFSRDSLQLL
                    *     * * **          *  ** .*. *.  *****
               .                    .

Ch IFN-1   RDMAPTLPQLCPQHNASCSFNDTILDTSNTRQADKTTHDILQHLFKILSS
Ch IFN-2   QNTAPPPPQPCPQQDVTFPFPETLLKSKDKKQAAITTLRILQHLFNMLSS
Duck IFN   RNMAPSPTQPCPQQHAPCSFPDTLLDTNDTQQAAHTALHLLQHLFDTLSS
Tk IFN     RDMAPSPPQPCPQHNAPCSFNDTVLDTNNTQQADKTTHNILQHLFKILSG
              **   * ***.     * .*.* .   .**  *.  .*****  **
           .

Ch IFN-1   PSTPAHWNDSQRQSLLNRIHRYTQHLEQCLDSSDTRSRTRWPRNLHLTIK
Ch IFN-2   PHTPKHWIDRTRHSLLNQIQHYIHHLEQCFVNQGTRSQRRGPRNAHLSIN
Duck IFN   PSTPAHWLHTARHDLLNQLQHHIHHLERCFPADAARLHRRGPRNLHLSIN
Tk IFN     PTTPAHWIDSQRQSLLNQIQRYAQHLEQCLADSHTRSRTRWPHNPHLTIN
           * ** **    *. ***....  .***.*     .* . * *.* **.*

Ch IFN-1   KHFSCLHTFLQDNDYSACAWEHVRLQARAWFLHIHNLTGNTRT-------
Ch IFN-2   KYFRSIHNFLQHNNYSACTWDHVRLQARDCFRHVDTLIQWMKSRAPLTAS
Duck IFN   KYFGCIQHFLQNHTYSPCAWDHVRLEAHACFQRIHRLTRTMR--------
Tk IFN     KHFSCLHAFLHDNDYSACAWDHVRLRARAWLLHIHDLVRNTRT-------
           * * ... **. . ** *.*.**** *.    ..  *    .

Ch IFN-1   -------
Ch IFN-2   SKRLNTQ
Duck IFN   -------
Tk IFN     -------
```

Fig. 2. Alignments of the predicted amino acid sequences of bird type I interferons. Asterisks indicate identity. Ch, chicken; Tk, turkey.

and also has a 30-amino-acid signal peptide and a 162-amino-acid mature protein (Fig. 2) (Suresh *et al.*, 1995).

Chicken recombinant IFN-1 (rIFN-1) has potent antiviral activity whether produced in *Escherichia coli* or COS cells (Schultz *et al.*, 1995b), with approximate activity of 10^8 IU per mg protein. The COS cell-produced IFN-1 is glycosylated, giving a glycoprotein of 23–28 kDa. Chicken rIFN-1 also has high Mx promoter-inducing activity but lacks macrophage activating factor (MAF) activity (Schultz *et al.*, 1995c). Chicken rIFN-2 has an antiviral activity about 20–100-fold lower than ChIFN1 (approximately 10^6 IU per mg), which is not inhibited by an antiserum able to neutralize chicken rIFN-1 (Sick *et al.*, 1996). Similarly, *E. coli*-produced duck rIFN is biologically active, being antiviral and able to induce Mx gene expression in duck embryo cells (Schultz *et al.*, 1995a).

Interferons induce a broad array of genes involved in antiviral activity and immune responses, and several are well characterized in birds. For example, the Mx protein is IFN induced in birds (Bernasconi *et al.*, 1995; Schultz *et al.*, 1995a), although its relationship to antiviral activity is unclear, and 2′,5′-oligoadenylate synthetase activity is increased by IFN (Fulton *et al.*, 1995). Such IFN-inducible genes possess IFN-stimulated response elements (ISREs) in their 5′ flanking regions, and several IFN regulatory factors are known to bind to these ISREs. This IFN-inducible transcription system is very conserved in birds (Jungwirth *et al.*, 1995), despite the IFN sequence divergence with mammals.

Type II Interferon

Fish

In rainbow trout IFN activity is present in 48 supernatants from concanavalin-A-pulsed head kidney leukocytes (Graham and Secombes, 1990b). The highest degree of protection of trout epithelial (RTH) cells against the CPE of IPN virus is seen using relatively low virus titres, 5×10^3 $TCID_{50}$ per well and lower. IFN activity in these crude preparations is significantly decreased following exposure to acid conditions (pH 2) or relatively high temperatures (60°C). However, some antiviral activity does remain in these supernatants, possibly because of the presence of a small amount of type I IFN. Fractionation of the supernatants by high-performance liquid chromatography (HPLC) size-exclusion chromatography reveals IFN activity in two protein peaks with molecular weights of 19 and 32 kDa, with most activity being in the former peak. Exposure of the 19-kDa peak to acid conditions (pH 2), 60°C for 1 h and trypsin removes the antiviral activity completely. The acid and temperature sensitivity, combined with the mode of induction, suggest that this is a type II IFN. In addition, the IFN activity coelutes with MAF activity in this species, which is also acid, temperature and trypsin sensitive but relatively freeze–thaw resistant. MAF activities recorded to date include increases in macrophage spreading, phagocytosis, respiratory burst activity, nitric oxide (NO) production and bactericidal activity (Graham and Secombes, 1988; Neumann *et al.*, 1995; Hardie *et al.*, 1996; Yin *et al.*, 1997), and a decrease in 5′ nucleotidase activity (Hepkema and Secombes, 1994). Interestingly, there is a biphasic effect of MAF on goldfish macrophage respiratory burst activity and NO production, with the former peaking before the latter (Fig. 3) (Neumann and Belosevic, 1996). MAF-containing supernatants can synergize with lipopolysaccharide (LPS) (Neumann *et al.*, 1995) or

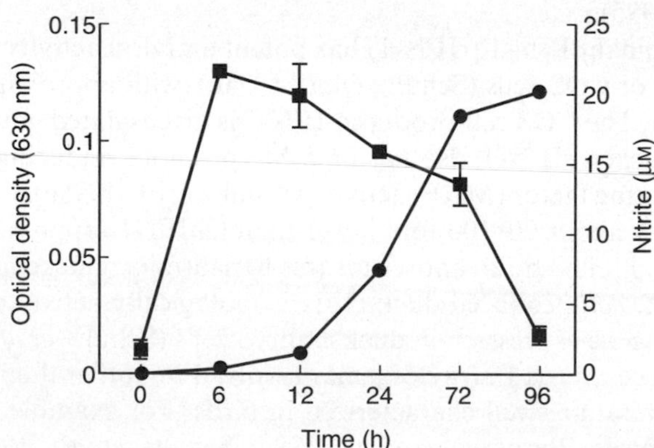

Fig. 3. Biphasic production of oxygen (OD 630 nm) and nitrogen radicals (μM nitrite) by a goldfish macrophage cell line stimulated with MAF (diluted 1:4) and LPS (5 μg/ml). ■, Respiratory burst; ●, nitric oxide. Reprinted from Neumann and Belosevic (1996), *Dev. Comp. Immunol.* **20**, 427–439, with kind permission of Elsevier Science-NL, Sara Burgerhartstraat 25, 1055 KV Amsterdam, The Netherlands.

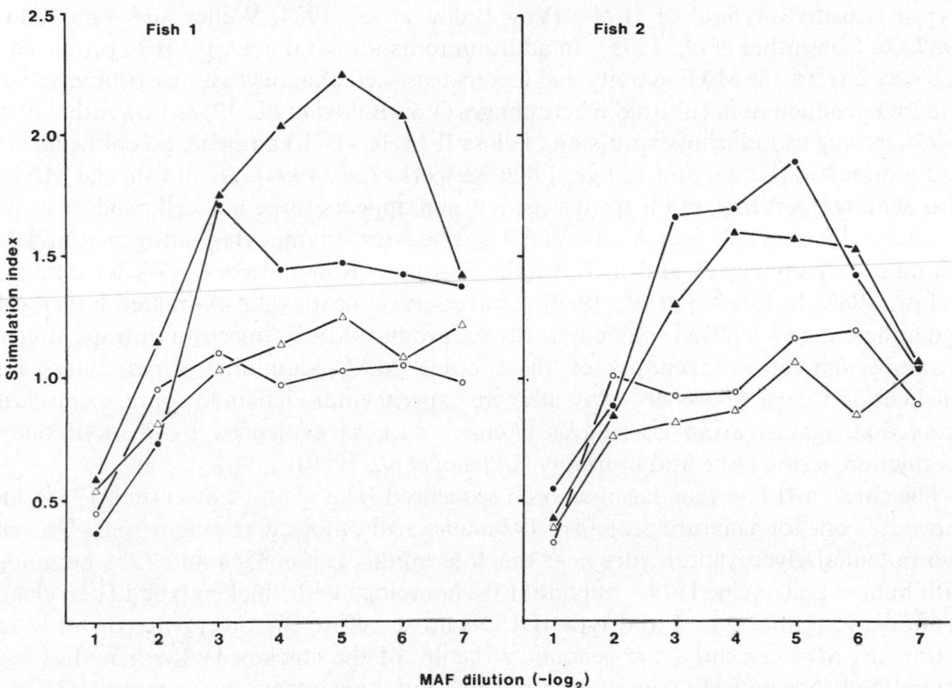

Fig. 4. MAF production by unfractionated rainbow trout blood leukocytes (●), macrophages (○), sIg⁻ lymphocytes plus macrophages (▲) and sIg⁺ lymphocytes plus macrophages (△). MAF activity was assessed by the respiratory burst activity of target macrophages following a 48-hour incubation with the MAF-containing supernatant. The results were expressed as a stimulation index by dividing the value from macrophages incubated with test supernatants with that from macrophages incubated with control supernatants obtained from leukocytes not stimulated with concanavalin-A/PMA. From Graham and Secombes (1990a), *Dev. Comp. Immunol.* **14**, 59–68, with permission.

human TNF-α (Hardie *et al.*, 1994a) to enhance further respiratory burst activity and NO production, and such combinations can overcome the inhibitory effects of prostaglandin E_2 (PGE_2) on respiratory burst activity (Novoa *et al.*, 1996). MAF has been shown to be a T-cell product in fish, using lymphocytes fractionated with anti-trout immunoglobulin (Ig) monoclonal antibody (mAb) by panning (Fig. 4), although they require the presence of accessory cells (which themselves do not secrete MAF) (Graham and Secombes, 1990a). Furthermore, MAF production is temperature sensitive (Hardie *et al.*, 1994b), as with other T-cell functions in fish, and can be primed by vaccination against bacterial pathogens (Francis and Ellis, 1994; Marsden *et al.*, 1994). Whether these two activities (MAF and IFN) are really due to the same molecular species awaits further study. Mammalian IFN-γ does not have an effect on fish cells, in that it does not increase the respiratory burst activity of cultured rainbow trout macrophages (L. Nagelkerke, personal communication).

Birds

In chickens it is well known that concanavalin-A-stimulated splenocytes and thymocytes, and cloned T-cell lines, secrete IFN that is acid (pH 2), temperature (56°C) and trypsin sensitive, typical of IFN-γ (Von Bulow *et al.*, 1984; Weiler and Von Bulow, 1987a,b; Lowenthal *et al.*, 1995). In addition to its antiviral activity, IFN produced in this way has potent MAF activity and can increase cell phagocytosis, cytostatic activity and NO production in cultured macrophages (Von Bulow *et al.*, 1984; Lowenthal *et al.*, 1995), as well as inducing expression of class II MHC (B–L) antigens on epithelial cells and monocytes (Kogut and Lange, 1989; Kaspers *et al.*, 1994). As in fish, the MAF is also acid, temperature and trypsin sensitive, and appears to be a T-cell product as it is produced by concanavalin-A-stimulated splenocytes, thymocytes, antigen-stimulated immune T lymphocytes and a T lymphoblastoid cell line, JMV-1 (Weiler and Von Bulow, 1987a,b; Lillehoj *et al.*, 1989). Conversely, concanavalin-A-treated bursal cells and adherent cell-depleted splenocytes do not produce MAF. Injection (intraperitoneal, intramuscular or intravenous) of these crude MAF-containing supernatants into chickens 1–5 days before or 1 day after an experimental challenge confers significant protection against avian coccidiosis (*Eimeria* sp.), as evidenced by reduced oocyst production, lesion score and mortality (Lillehoj *et al.*, 1989).

The chicken IFN-γ gene has now been sequenced (Digby and Lowenthal, 1995), and shown to code for a mature protein of 145 amino acids, molecular weight 16.8 kDa, with two potential glycosylation sites near the *N*-terminus. It has 35% and 32% homology with human and equine IFN-γ but only 15% homology with chicken type I IFN, clearly demonstrating that type I and type II IFNs arose before the divergence of birds and mammals. More recently, the genomic structure of the chicken IFN-γ gene has been determined and shown to consist of four exons and three introns as in mammals (Kaiser *et al.*, 1996), with only intron 3 being dramatically different (shorter) in size. Chicken rIFN-γ is able to induce NO production and increase class II MHC expression in macrophages, is heat and pH sensitive, but is a far less potent antiviral agent relative to recombinant chicken type I IFN (Digby and Lowenthal, 1995; Weining *et al.*, 1996).

INTERLEUKINS

The term interleukin was introduced over a decade ago to describe cytokines able to act on leukocytes in a specific manner. The number of interleukins discovered has been increasing steadily, with possibly more receiving interleukin status in the near future. Except in mammals few cytokines with interleukin activity have been described, probably due to the lack of long-term lymphocyte cell lines, especially B-cell lines which are commonly used to detect the activity of several of the interleukins. Those best described appear to correlate functionally with interleukin-1 (IL-1) and IL-2, and both of these activities have been detected in fish, amphibia and birds. IL-1 activity in particular may be quite conserved phylogenetically, with several reports of invertebrates producing an IL-1-like protein. The biological activity and sources of these molecules are now considered in more detail.

Interleukin-1

Stimulation of phagocytes with LPS has been the main approach taken to generate supernatants with IL-1-like activity. A thymocyte costimulation assay which examines thymocyte proliferation in the presence of submitogenic concentrations of T-cell mitogens (concanavalin-A, PHA), has commonly been used to test for IL-1 activity.

Invertebrates

Both Deuterostome* and Protostome invertebrates have been shown to produce molecules with IL-1-like activity. Within Deuterostomes, tunicates of the class Ascidiacea (Beck *et al.*, 1989) and the starfish *Asterias forbesi* (Beck and Habicht, 1986) have had this activity detected by means of a murine thymocyte costimulation assay (where a two- to three-fold increase in proliferation is seen) and a fibroblast proliferation assay. In addition, the tunicate factor can increase vascular permeability when injected intradermally into rabbits. The function of these factors in Deuterostomes is not known.

IL-1-like activity can be detected in starfish freeze–thaw extracts of coelomocytes (the principal phagocytes in this species) and coelomic fluid, or from tunicate tissue homogenates and blood. However, in the starfish samples activity is detectable only following fractionation by gel chromatography. The presence of activity in the blood and coelomic fluid suggests that the IL-1-like factor is a secreted protein, probably from the invertebrate phagocytes. The molecular weights of the secreted form are approximately 30 and 20 kDa for the starfish and tunicate respectively, with pI values of 4.8, 5.9 and 7.5, or 5.5 and 7.0. The intracellular activity from the starfish coelomocytes is more heterogeneous and can be detected at molecular weights of approximately 10, 30 and 70 kDa (Beck and Habicht, 1986). Both starfish and tunicate IL-1-like activities can be neutralized significantly using polyclonal antiserum to human IL-1. Lastly, anti-IL-1R mAbs bind a significant number of axial organ cells in the sea star *Asterias rubens*, by fluorescence activated cell sorter (FACS) analysis, interpreted as showing the presence of IL-1R (Legac *et al.*, 1996).

In Protostomes, stimulation of haemocytes of the mussel *Mytilus edulis* with LPS induces cell spreading *in vitro*, and this can be inhibited with anti-IL-1α (Hughes *et al.*, 1991). Human rIL-1α induces similar effects in *M. edulis* and other molluscan haemocytes (Ottaviani *et al.*, 1995a) and immunoreactive IL-1 can be detected in concentrated haemolymph (Hughes *et al.*, 1990). More recently, haemolymph from the snail *Biomphalaria glabrata* has been shown to possess IL-1 bioactivity, using the murine D10.G4.1 cell line, and to possess factors that antigenically cross-react with anti-human IL-1β sera (Granath *et al.*, 1994). Strain differences in haemolymph levels were apparent after exposure of the snails to *Schistosoma mansoni*, with resistant snails maintaining significantly higher IL-1 levels. In addition, human rIL-1 can prime haemocytes from resistant snails for superoxide production but has no effect on haemocytes from susceptible strains (Granath *et al.*, 1994; Connors *et al.*, 1995). Such data support the notion that IL-1 is a functionally conserved molecule, vital for host resistance, and possibly interacting with the invertebrate neuroendocrine system as in vertebrates (Ottaviani *et al.*, 1995b).

Fish

The requirement for macrophages or monocytes in fish (catfish and salmonid) immune responses has been known for many years. *In vitro* studies have shown that proliferation in response to T-cell mitogens or allogeneic leukocytes (mixed leukocyte reaction (MLR)), MAF production and secretion of antihapten antibodies are all accessory cell dependent (Graham and Secombes, 1990a; Clem *et al.*, 1991). Indeed, supernatants from

*The Deuterostomes diverged at an early stage in evolution from the Protostome invertebrates, and are thought to be on the evolutionary line that gave rise to the vertebrates.

LPS-stimulated monocytes or spontaneous monocyte-like cell lines of channel catfish (*Ictalurus punctatus*) allow macrophage-depleted lymphocytes to undergo concanavalin-A-induced mitogenesis and *in vitro* antibody responses to both TD and TI antigens, without the need for intact accessory cells (Clem *et al.*, 1991). Similarly, supernatants containing murine or human IL-1 (but not rIL-1) can substitute for intact accessory cells in fish, and in the latter case this effect can be inhibited using antibodies against human IL-1 (Hamby *et al.*, 1986; Clem *et al.*, 1991). Murine IL-1 also has more general physiological effects in fish, suggestive of an immunoendocrine axis, as in mammals (Balm *et al.*, 1995). Thus, fish cells appear to possess a receptor(s) that can recognize and respond to IL-1. Antihuman IL-1α and IL-1β also inhibit the fish response to fish supernatants. Similarly, fish supernatants, in concert with PHA, can induce proliferation of murine thymocytes and IL-2 production by a murine T lymphoma (LBRM33-1AS).

An IL-1-like factor has also been found in supernatants from a carp (*Cyprinus carpio*) epithelial cell line (Sigel *et al.*, 1986), from carp macrophages and neutrophils (Verburg van Kemenade *et al.*, 1995) and a carp macrophage cell line (Weyts *et al.*, 1997). This cytokine has been shown to be costimulatory in a murine thymocyte proliferation assay, to stimulate proliferation of a murine IL-1-dependent cell line (D10(N4)M), to induce IL-2 production from murine T cells and to cause an increase in thromboxane production from rabbit corneal epithelial cells and chicken chondrocytes.

Fractionation of the catfish monocyte-derived IL-1-like activity by gel filtration has shown two forms of the molecule to be present in catfish, with molecular weights of approximately 15 and 70 kDa (Clem *et al.*, 1991; Ellsaesser and Clem, 1994). The high-molecular-weight form is active on catfish cells but not mouse T cells, whereas the low-molecular-weight form has the reverse activity. Similar studies with culture supernatants from a mouse monocyte cell line (P388D1) have also found IL-1 activity for catfish and mouse T cells in either a high- or low-molecular-weight fraction respectively; both fractions react with antimouse IL-1α. Western blot analysis of the catfish material with anti-human IL-1 has shown that both forms possess α and β determinants. Furthermore, cytoplasmic mRNA from catfish monocytes hybridizes with a murine IL-1α cDNA probe, and the hybridizing species has a mobility similar to that observed with P388D1 cell mRNA. While these findings suggest that a polymeric IL-1-like molecule is required for stimulation of cells from some fish species, in other species this may not be the case, as in carp where the 15-kDa molecular species is the predominant form (Verburg van Kemenade *et al.*, 1995).

An IL-1β gene has recently been isolated in rainbow trout (Secombes, 1997; Zou *et al.*, unpublished data). The full length gene is 1343 bp and translates into a 260-amino-acid precursor, with three glycosylation sites at positions 6, 12 and 142. A *C*-terminal mature peptide of 166 amino acids is also identifiable, comparable with mammalian IL-1s (Table 2), and has an estimated size of 19 kDa. The amino acid homology of the mature peptide with known mammalian forms of IL-1β ranges from 49% to 55%. The trout IL-1β gene is inducible, as evidenced by the increased expression following stimulation of leukocytes with LPS or PHA (Fig. 5).

Amphibia

In *X. laevis*, 24-h supernatants from LPS-stimulated adherent peritoneal cells (containing 50–80% macrophages) are active in a thymocyte costimulation assay (Watkins *et al.*, 1987). Both adult and larval thymocyte proliferation is only marginal in response to the

Table 2. Characteristics of mature IL-1 proteins.

	Species	No. of amino acids	*N*-terminal residue	Mol. wt. (kDa)	Percentage homology to trout IL-1
IL-1α	Cow	150	Gln-119	17.2	47.3
	Human	159	Ser-113	17.5	44.1
	Mouse	156	Ser-115	18.0	45.0
	Pig	152	Gln-119	17.4	47.0
	Rabbit	155	Ser-113	17.8	45.8
	Rat	156	Ser-115	18.0	44.4
	Sheep	150	Gln-119	17.2	46.7
IL-1β	Cow	153	Ala-114	17.7	54.7
	Human	153	Ala-117	17.5	55.3
	Mouse	151	Val-119	17.4	53.6
	Pig	152	Ala-115	17.6	52.6
	Rabbit	152	Ala-117	17.4	49.0
	Rat	152	Val-117	17.3	53.6
	Sheep	153	Ala-114	17.7	54.1
	Trout	166	Ala-95	18.7	

All mammalian IL-1β sequences are cut by interleukin 1β-converting enzyme (ICE) after aspartic acid, to release the mature protein. This does not appear to be the case with trout IL-1β.

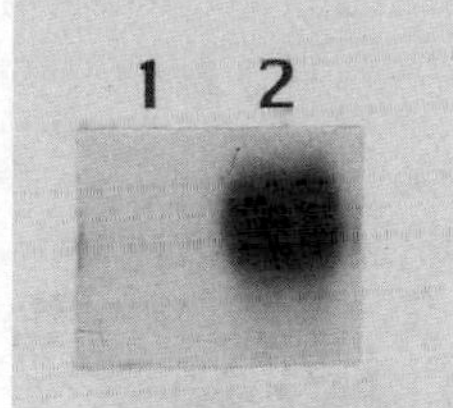

Fig. 5. Northern blot of rainbow trout head kidney leukocyte RNA hybridized with a trout-specific IL-1β probe after stimulation of the cells *in vitro* with PHA (at 5 μg/ml) for 0 h (lane 1) or 4 h (lane 2). Note induction of a strong band of approximately 1300 bp after 4 h stimulation.

supernatants or low concentrations of PHA alone, or to a combination of LPS and PHA, but a marked costimulatory effect is seen when the supernatants plus PHA are used (Table 3). Culture with LPS for 48 h before harvesting the supernatants results in lower activity. These supernatants do not stimulate the proliferation of thymic lymphoblasts or support their long-term growth, which suggests that they do not contain T-cell growth factors (cf. IL-2). Furthermore, as LPS is not a T-cell mitogen in amphibia, as in mammals, it seems likely that this costimulatory cytokine is a macrophage product.

Human IL-1-containing supernatants do not costimulate *Xenopus* thymocytes, and *Xenopus* supernatants do not stimulate the mouse D.10.G4.1 cell line, suggesting a lack of cross-reactivity *in vitro*. However, human rIL-1 does appear to have effects *in vivo*. During metamorphosis from tadpole to adult, high levels of endogenous corticosteroids are thought to impair T-cell functions, including an antigen-dependent thymus suppressor function. Injection of human rIL-1 can restore this suppressor function in a dose-dependent manner, possibly by overcoming a lesion in the antigen-presenting cells

Table 3. Thymocyte costimulatory activity of 24-h supernatants from LPS-stimulated macrophages (SN) in two *Xenopus* clones.

	Mean (s.e.m.) c.p.m. when assayed on thymocytes	
Medium supplementation	Adults	Larvae
*Experiment 1**		
Nothing	480(131)	1886(1012)
PHA	1149(141)	1564(853)
SN	4650(413)	4331(1222)
SN + PHA	24 605(1932)	8214(1663)
Experiment 1†		
Nothing	400(12)	913(153)
PHA	860(217)	1699(358)
SN	3437(85)	2855(309)
Sn + PHA	10 820(293)	7424(810)

*Clone LG-6 stage 60–61 larvae; outbred 10 month old. †Clone LG-17 stage 56–57 larvae; 7-month-old adults. From Cohen *et al.* (1987).

(Highet and Ruben, 1987). Human rIL-1 also has effects *in vivo* in the common American newt (*Notophthalmus viridescens*), where it can substitute for carrier priming in antihapten responses (Ruben *et al.*, 1988).

Reptiles

Nothing is known about the production of IL-1 by reptile cells. However, the lizard *Dipsosaurus dorsalis* does appear to be able to respond to rabbit IL-1, as evidenced by induction of fever following an intracardiac injection (Bernheim and Kluger, 1977). The lizard can also produce an endogenous pyrogen in response to stimulation with thioglycollate or killed bacteria, but the relationship of this pyrogen to IL-1 is unclear.

Birds

In chickens 72-h supernatants from LPS-pulsed (4-h) adherent spleen cells are costimulatory for chicken thymocytes stimulated with concanavalin-A or PHA, enhancing the mitogen response approximately twofold (Hayara *et al.*, 1982). Similarly, macrophages from infected (with *Eimeria*) chickens release a factor with IL-1 activity (Byrnes *et al.*, 1993a). These factors also show costimulatory effects on the concanavalin-A response of mouse thymocytes, and mouse IL-1 is costimulatory in the chicken response. Mammalian IL-1 is also able to induce PGE_2 and cAMP formation from chicken ileal mucosa *in vitro* (Chang *et al.*, 1987), and to stimulate increases in short-circuit current (a measure of anion secretion). Since the latter is blocked using a specific inhibitor of cyclo-oxygenase it appears that IL-1 induces the formation of inflammatory mediators such as PGE_2 by activating arachidonic acid metabolism.

The chicken IL-1 type I receptor has been cloned and shown to code for a 557-amino-acid mature protein with 61% and 64% homology to murine and human IL-1 respectively (Guida *et al.*, 1992). Interestingly, it is the intracellular domain that shows the highest homology (76–79%) to mammalian IL-1R, and to the Toll membrane receptor protein of *Drosophila melanogaster*.

Interleukin-2

Stimulation of leukocytes/lymphocytes with a T-cell mitogen has been the main approach adopted to generate supernatants containing IL-2-like activity in the 'lower' vertebrates. Owing to the essential role of IL-2 in T-cell growth, a blast-cell assay is normally employed to test for IL-2 activity.

Invertebrates

Human rIL-2 has been shown to increase phagocytosis and the induction of molecules cross-reactive with an antiserum to rat cerebellar nitric oxide synthetase (NOS) in molluscan haemocytes (Ottaviani *et al.*, 1995a). In addition, IL-2R mAbs binds to axial organ* cells in the sea star *A. rubens*, as determined by FACS analysis (Legac *et al.*, 1996).

Fish

In carp, supernatants from leukocyte cultures stimulated by mitogen (PHA) or alloantigen (MLR) induce the proliferation of purified PHA-activated lymphoblasts (T-cell blasts?), analogous to the effect of IL-2 (Fig. 6) (Caspi and Avtalion, 1984; Grondel and Harmsen, 1984). Only cells that are already activated undergo proliferation in response to such supernatants; freshly harvested (virgin) leukocytes are not stimulated. Furthermore, adsorption of the supernatants with blast cells significantly reduces the activity of this growth factor, whereas adsorption with virgin leukocytes has little effect. Such data suggest that receptors for this factor are induced upon activation of the virgin cells. Supernatants from two-way MLR cultures show a lower and more variable growth-promoting effect than do those from mitogen-stimulated cultures (Caspi and Avtalion, 1984). As mitogens have a polyclonal stimulatory effect upon lymphocytes this may induce the secretion of more lymphokine, explaining the higher activity in such supernatants. The variability of the MLR supernatants may be due to variability of the histocompatibility antigens on the lymphocytes used for allogeneic stimulation, as outbred animals are used. Pooling supernatants from at least four donors in several combinations has been found to give more consistent results, and significant enhancement of the growth-promoting activity can be obtained by having PMA present in the MLR cultures.

Mouse IL-2 has no effect on carp lymphoblasts (Grondel and Harmsen, 1984), although supernatants containing human, rat and monkey IL-2 have been reported to stimulate their proliferation (Caspi and Avtalion, 1984). Human IL-2 has also been reported to increase specific cellular cytotoxicity by carp lymphocytes, but fails to increase binding to, or killing of, target cells by rainbow trout NK cells. Whether the positive effect of mammalian IL-2-containing supernatants is due solely to the activity of IL-2 is unclear, especially as more purified preparations are often less active. The use of defined reagents is clearly warranted.

Amphibia

In *X. laevis*, 24–48-h supernatants from splenic leukocytes stimulated with PHA, concanavalin-A and neuraminidase plus galactose (NAGO) are costimulatory in a

*A tissue involved in the defence system of this organism and considered to be an ancestral lymphoid organ.

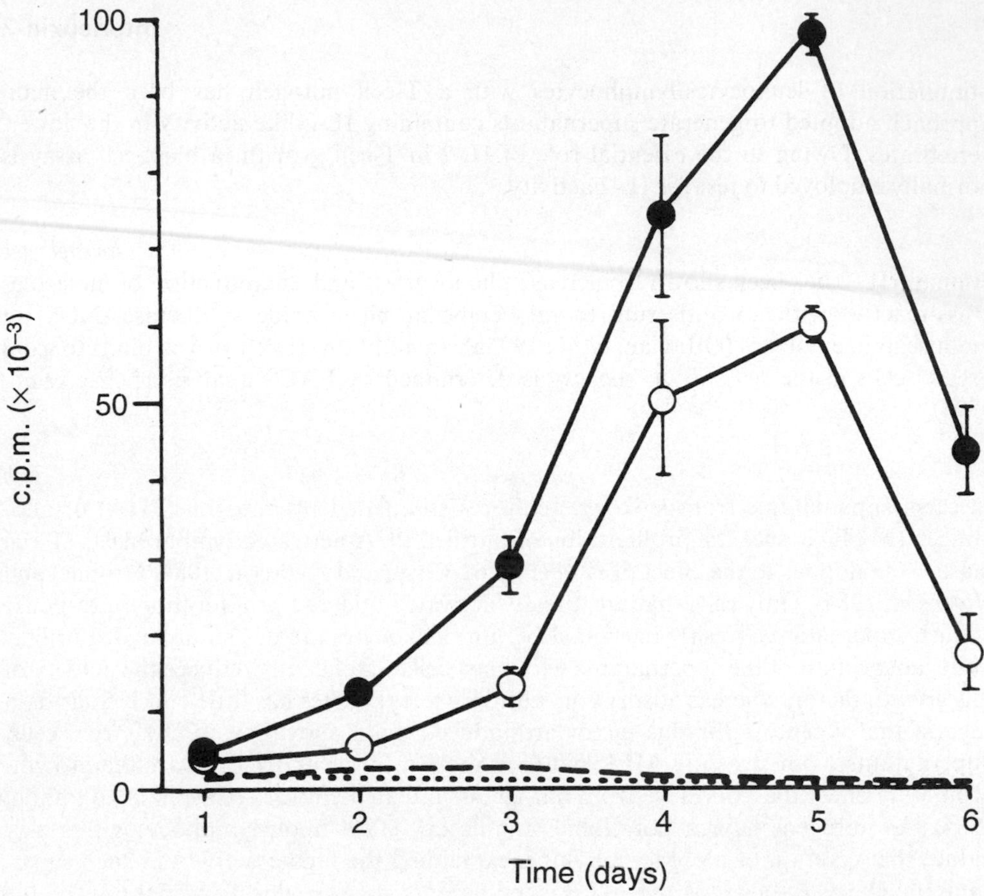

Fig. 6. Effect of growth factor-containing supernatants on the kinetics of [^{3}H]thymidine uptake into carp lymphoblasts. Blast cells (2.5×10^4 per well) were incubated with growth factor-containing supernatants from PHA-activated head kidney leucocytes (●), control supernatants from non-mitogen treated cells (— — —), control supernatants supplemented with PHA (○) or fresh medium (- - - -). Both types of control culture incorporated only low levels of [^{3}H]thymidine. PHA by itself was mitogenic for blasts, but significantly more incorporation was seen with the growth factor-containing supernatants. From Grondel and Harmsen (1984), *Immunology* **52**, 477–482, with permission.

thymocyte proliferation assay (using adult or larval cell supernatants), induce the proliferation of PHA-induced lymphoblasts and splenocytes from coccobacilloid infected animals (Haynes *et al.*, 1992), and support the growth of alloreactive T-cell lines (Watkins and Cohen, 1987). Splenocytes from infected animals also show constitutive production of this factor. The supernatants do not induce proliferation of unstimulated splenocytes, and adsorption with splenocytes does not affect the IL-2-like activity. However, adsorption with blasts does reduce this activity, the decrease being proportional to the number of blasts used. Therefore, as in fish, it appears that only lymphoblasts can bind to or react with this growth factor, suggesting the presence of an inducible receptor on frog T cells. Similarly, in the Axolotl (*Ambystoma americanum*) splenic blasts but not resting splenocytes can be costimulated with homologous growth factor-containing supernatants (Koniski and Cohen, 1994).

Purification of the *Xenopus* growth factor by diethylaminoethyl (DEAE) ion-exchange chromatography and sodium dodecyl sulphate–polyacrylamide gel electrophoresis (SDS-PAGE) have shown that it has a molecular weight of 16 kDa, and lectin affinity chromatography indicates it has a three-dimensional configuration of carbohydrates similar to that of human IL-2 (Watkins and Cohen, 1985; Haynes and Cohen, 1993a). SDS-PAGE analysis of [^{35}S]methionine labelled Axolotl growth factor containing supernatants reveals a labelled band of between 14 and 21 kDa (Koniski and Cohen, 1994).

Evidence to suggest that T cells are involved in producing, as well as responding to, this amphibian growth factor comes from several sources. First, T-cell mitogens induce secretion of this factor. Second, supernatants (48–120 h) from MLR cultures of MHC-distinct *Xenopus* strains and from alloreactive T-cell lines stimulated with PHA or irradiated stimulator cells also have growth-promoting activity (Watkins and Cohen, 1987). Finally, thymectomy studies have shown that early thymectomy impairs production of this factor by PHA stimulation. Thymocytes, however, apparently do not secrete this factor (Cohen *et al.*, 1987). Whether this is due to a complete inability of these cells to produce this factor or to an inefficient activation of the cells by low IL-1 secretion from thymic macrophages is unclear.

The *Xenopus* IL-2-like factor has no effect on human T-cell blasts or on a murine IL-2-dependent cell line, HT-2 (Watkins and Cohen, 1987). Cross-reactivity of mammalian IL-2 and anti-IL-2 receptor (anti-Tac) antibody with the *Xenopus* cells, however, is controversial. Some authors have elegantly shown by flow cytometry and immunogold electron microscopy that *Xenopus* splenocytes and PHA-stimulated thymocytes (when fixed with paraformaldehyde) bind both human rIL-2 and murine anti-Tac antibody (Fig. 7), and that rIL-2 will compete with the anti-Tac antibody for the binding sites (Ruben *et al.*, 1990a,b). PHA activation of splenocytes increases the binding of anti-Tac and is inhibitable with cycloheximide, suggesting that this treatment induces the synthesis of new receptors. Immunoprecipitation of labelled PHA-stimulated thymocytes with the anti-Tac antibody reveals a molecule of approximately 55 kDa (Ruben *et al.*, 1990b), similar to the human IL-2R Tac protein. Despite these findings, human rIL-2- and human or mouse IL-2-rich supernatants fail to stimulate proliferation of *Xenopus* blasts or T-cell mitogenesis *in vitro* (Watkins and Cohen, 1985). Nevertheless, factors such as PHA or phorbol dibutyrate (PDB), which do increase mitogenesis, increase the binding of anti-Tac (Ruben *et al.*, 1990b), and factors such as dexamethasone or cyclosporin A, which inhibit PHA- or PDB-induced mitogenesis, decrease binding of anti-Tac. *In vivo* it has been shown that injection of rIL-2 can (1) substitute for carrier priming of helper function in antihapten responses, (2) increase (12-fold) the antibody response to TI-type 2 antigens, and (3) break hapten-specific tolerance (although not in thymectomized frogs) (Ruben, 1986; Ruben *et al.*, 1987). In addition, it can restore T-cell functions that are inhibited by high levels of endogenous corticosteroids during metamorphosis, as seen with human rIL-1, and can break tolerance to adult cells (when accompanied by antigenic costimulation) resulting in mortality (Ruben *et al.*, 1996).

In contrast, other authors have failed to immunoprecipitate molecules in the growth factor-containing supernatants with anti-human IL-2 antibodies, to stain *Xenopus* splenic PHA blasts with human anti-Tac, and to immunoprecipitate cell surface

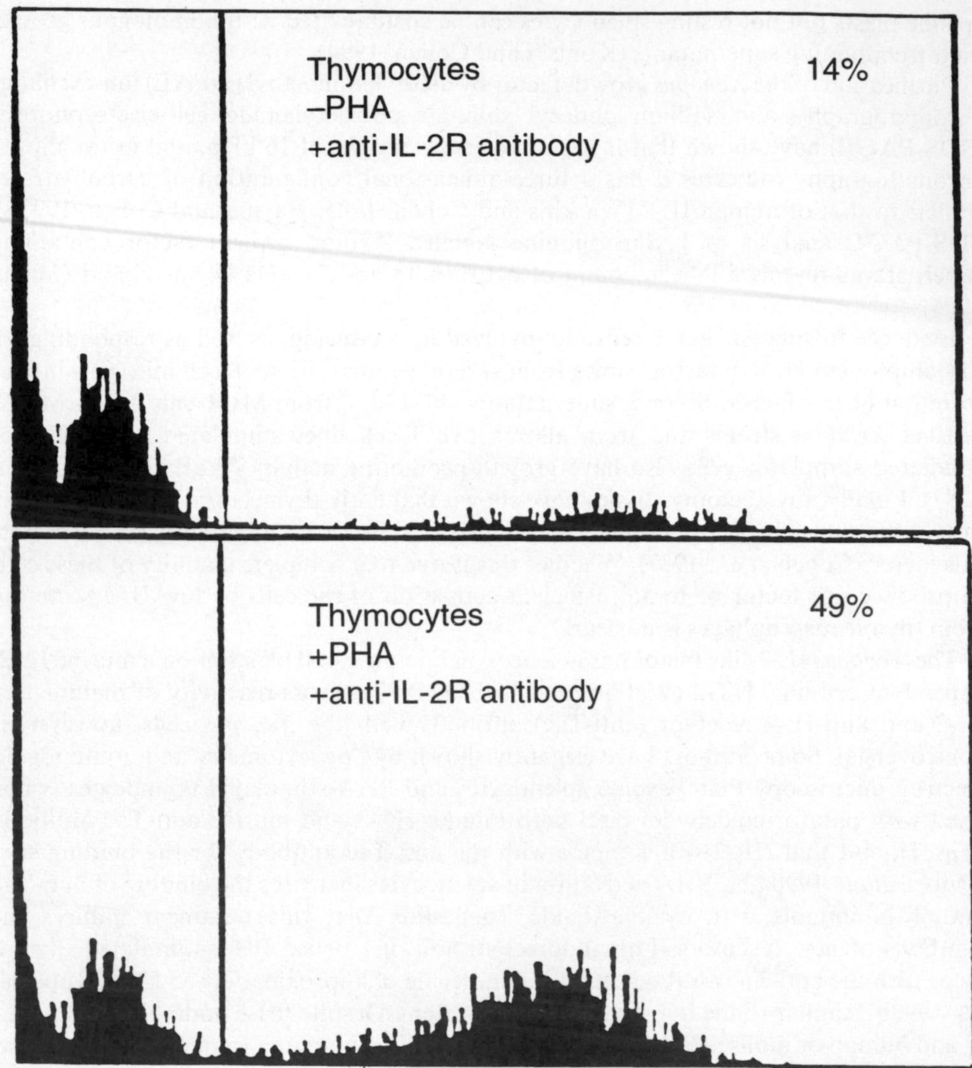

Fig. 7. Effect of PHA stimulation on the reactivity of *Xenopus* thymocytes with antihuman IL-2R antibody, analysed by flow cytometry. The histograms depict the relative cell number on the ordinate and log fluorescence intensity on the abscissa. The percentage of cells binding the antibody is shown in the upper right corner of each histogram. From Ruben *et al.* (1990a), in *Defense Molecules* (eds J.J. Marchalonis and C. Reinisch), Alan R. Liss, New York, pp. 133–147. Reprinted by permission of John Wiley & Sons, Inc. All rights reserved.

molecules with anti-Tac antibody (Haynes *et al.*, 1990). These authors suggest that binding occurs only following membrane damage and does not reflect ligand binding on the cell surface. Until these controversies are resolved, whether IL-2 and its receptor are conserved phylogenetically must remain an enigma.

Reptiles

In the snake *Spalerosophis diadema*, supernatants from concanavalin-A-stimulated splenocytes enhance the concanavalin-A-induced proliferation of splenic lymphoblasts

(El Ridi *et al.*, 1986). Fractionation of these supernatants reveals activity in two peaks, with molecular weights of 15 and 39–42 kDa. Only the low-molecular-weight protein is detectable by SDS-PAGE, which suggests that the high-molecular-weight molecule is possibly a polymer of the former. The pI values of these molecules lie in the ranges 5.5–5.8 and 6.4–6.6.

Birds

As in other vertebrate classes, supernatants from mitogen- or antigen-stimulated chicken splenocytes and peripheral blood leukocytes (PBLs) contain IL-2-like activity (Schauenstein *et al.*, 1982; Schnetzler *et al.*, 1983). Such supernatants can induce the proliferation of concanavalin-A-induced lymphoblasts, are costimulatory in a thymocyte proliferation assay and can maintain long-term cultured T cells. However, they do not induce proliferation of freshly isolated (unstimulated) splenocytes. Cyclosporin A (CsA) does not block this factor-mediated blast stimulation, suggesting that it acts specifically on activated T cells that are unresponsive to the effects of CsA (Schnetzler *et al.*, 1983). Indeed, both cultured T cells and concanavalin-A-induced blasts die within 1–2 days in the absence of this growth factor, demonstrating that their growth is totally factor dependent. In addition, adsorption studies show that cultured T cells and concanavalin-A-induced blasts can remove virtually all of this IL-2-like activity; thymocytes can remove about 45% but virgin (unstimulated) PBLs and bursal (B) cells remove insignificant amounts. Thus, again, an inducible growth factor receptor appears to be present on T cells. The use of T-cell mitogens to generate these growth factor-containing supernatants, coupled with the fact that T cell-enriched (84–88%) PBLs (Schnetzler *et al.*, 1983) and PBLs from agammaglobulinaemic birds can also be used for this purpose (Vainio *et al.*, 1986), suggests that this factor is a T-cell product.

IL-2-like activity is highest in 24-h supernatants, and concanavalin-A is slightly better at inducing its release than PHA. Supplementation of diets with low levels (0.1 ppm) of triiodothyronine (T_3) (Chandratilleke *et al.*, 1994) and treatment of lymphocytes *in vitro* with concanavalin-A-coated chicken erythrocytes (Kromer *et al.*, 1984) markedly enhances secretion of this IL-2-like molecule. Activity of the supernatants is stable at $-20°C$ but is lost after 5 weeks at 4°C. The half-life at 40°C is approximately 10 ± 2 h, significantly shorter than for mammalian IL-2. Fractionation of the supernatants by gel chromatography reveals two major peaks of activity, with molecular weights of 9–13 and 19.5–21.5 kDa, the former having a pI of 5.9. Using SDS-PAGE a single polypeptide of 13 kDa is found, indicating that the larger form may be a dimer of the 13-kDa molecule. The chicken IL-2-like factor is species restricted, as it cannot induce proliferation of mouse or rat lymphoblasts, and mouse, rat and human IL-2 cannot induce proliferation of chicken lymphoblasts.

The IL-2-like factor appears to have a crucial role in T-cell activation in chickens, as there is a close correlation between proliferative responses of lymphocytes in culture and detectable growth factor activity of the supernatants (Fig. 8) (Schauenstein and Kromer, 1987). Growth factor activity can be detected in cultures where the cells are not proliferating due to irradiation or cytostatic treatment, but proliferation cannot occur without the IL-2-like factor. This has also been shown for genetically determined high and low responsiveness to concanavalin-A of certain chicken strains, which parallels differences in the production and responsiveness to the growth factor. Similarly, during ontogeny of this species, the acquisition of proliferative responses to concanavalin-A is

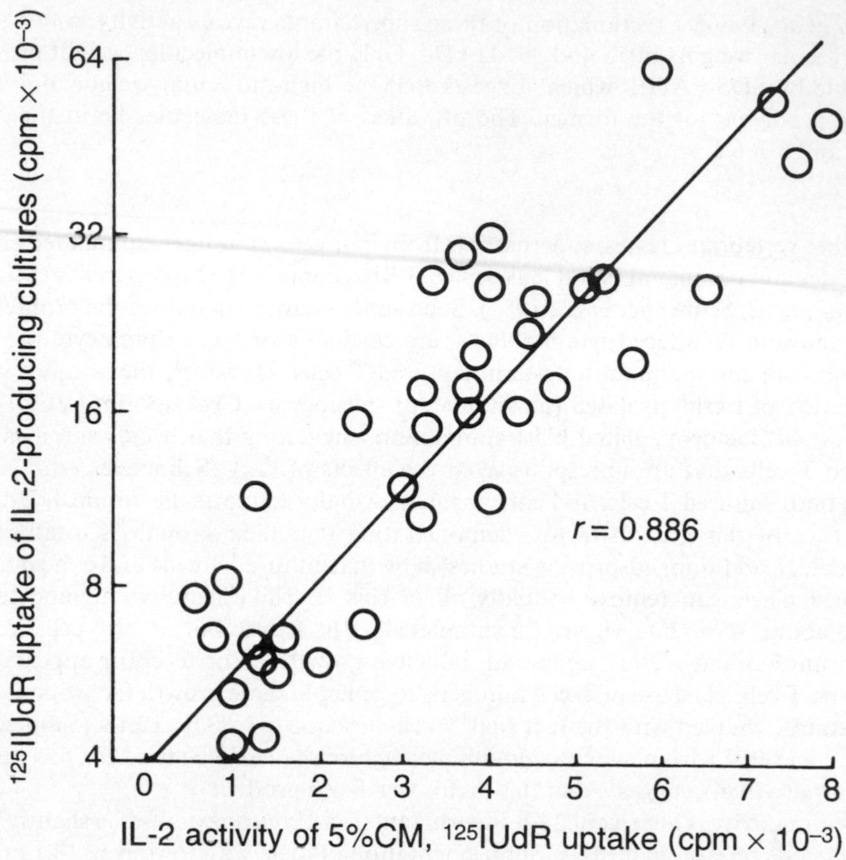

Fig. 8. Correlation of the proliferative response to concanavalin-A (ordinate) and the production of IL-2 (abscissa) of individual chicken splenocyte cultures. From Kromer *et al.* (1984), *J. Immunol. Methods* **73**, 273–281, with permission.

closely linked with the capacity to produce this IL-2-like factor. Thus the production or secretion of this growth factor largely determines the magnitude of mitogen- or antigen-driven chicken T-cell responses.

Recently a mAb has been raised against concanavalin-A-activated lymphoblasts that recognizes a 'chicken-activated T-lymphocyte antigen' (CATLA) (Schauenstein *et al.*, 1988; Lee and Tempelis, 1991). The CATLA molecule has been shown to have a molecular weight of 45–50 kDa by SDS-PAGE following immunoprecipitation of radiolabelled lymphocyte surface components. The number of splenocytes and PBLs expressing CATLA ranges from 1% of freshly isolated cells to 15–30% of 3-h concanavalin-A-stimulated cells and more than 90% of 72-h concanavalin-A-stimulated cells. The presence of the anti-CATLA antibody markedly inhibits, in a competitive manner, the proliferative response of T lymphoblasts to the IL-2-like molecule. Similarly, pretreatment of lymphoblasts with this antibody reduces their ability to absorb the IL-2-like activity from conditioned medium. Such data suggest strongly that CATLA is the chicken equivalent of an IL-2 receptor or an associated structure, and that the anti-CATLA antibody is equivalent to an anti-Tac antibody. More recently, a second mAb has been raised in the same manner to a 110-kDa receptor on T-cell blasts, thought to represent the IL-2Rγ subunit (Lee and Tempelis, 1992).

Pathological disturbances of T-cell regulation are known in chickens, particularly in the Obese strain (OS) of chickens which exhibit Hashimoto-like spontaneous autoimmune thyroiditis (SAT) (Schauenstein *et al.*, 1985; Kroemer and Wick, 1989). Animals suffering from SAT have increased T-cell proliferation, IL-2 production and IL-2 receptor expression following mitogen stimulation. The hypersecretion of IL-2-like molecules is thought to be an accelerating factor modulating SAT, and not a secondary event of autoimmunity. Morphological studies have shown that lymphocytes infiltrating the thyroid, which initiate the disease process, contain a high percentage of IL-2R-positive T cells and, if isolated, these thyroid-infiltrating lymphocytes can be shown to secrete large amounts of IL-2 spontaneously. This T-cell defect appears to be encoded by a single recessive gene (Kroemer *et al.*, 1990). T-cell dysfunction has also been seen in avian scleroderma, where raised IL-2 activity is seen in serum but blood leukocytes show impaired IL-2 production and IL-2R expression (Gruschwitz *et al.*, 1991).

Interleukin-3 and Colony Stimulating Factors (CSFs)

No reports of IL-3-like activity have been described outside of the mammals. Nevertheless, biological cross-reactivity of mammalian IL-3 has been noted in some species and factors with CSF activity have been described.

Invertebrates

In the starfish *A. forbesi*, 48-h supernatants from mixed (xenogeneic) cultures of non-adherent axial organ cells are mitogenic for the entire axial organ cell population but not for adherent and non-adherent cells cultured separately (Luquet *et al.*, 1989). The released molecules are probably glycoproteins with a molecular weight in excess of 10 kDa. Whether this due to a CSF-like molecule is not known.

Fish

Anti-IL-3 sera have been used to detect antigenically cross-reactive molecules in the sera of virus-challenged trout by enzyme-linked immunosorbent assay (ELISA) (Ahne, 1993). Serum CSF activity is present in rainbow trout following injection with LPS (Kodama *et al.*, 1994), and in carp following injection with Freund's complete adjuvant (Moritomo *et al.*, 1994). In addition, carp macrophage supernatants have CSF activity (Yoshikawa *et al.*, 1994) and a recently developed goldfish cell line requires autologous conditioned medium for optimal proliferation (Wang *et al.*, 1995). Although no CSF sequences are yet known in fish, the macrophage colony stimulating factor (M-CSF) receptor (CSF1R) has been cloned in the puffer fish (*Fugu rubripes*). The gene consists of 21 exons and has 39% amino acid homology with human CSF1R, with the kinase domain having much higher homology (63%) (How *et al.*, 1996). The gene is linked tandemly in a head-to-tail array with the platelet-derived growth factor receptor-β (PDGFRβ) gene (see Transforming Growth Factors).

Birds

A single report exists of the cross-reactivity of mammalian IL-3 on chicken cells where, as with IL-1, IL-3 induces PGE_2 and cAMP formation and increases short-circuit current from isolated ileal mucosa (Chang *et al.*, 1987). However, factors with CSF activity are well known. For example, serum CSF activity is detectable during and immediately after coccidial infection in chickens, using a soft agarose bone marrow assay (Byrnes *et al.*, 1993b), and stromal cell lines secrete factors that induce the proliferation and differentiation of precursor cells in embryonic and haematopoietic tissues (Obranovich and Boyd, 1996; Siatskas *et al.*, 1996). Stem cell factor (SCF) has been sequenced in chickens and quail (Zhou *et al.*, 1993; Petitte and Kulik, 1996), with both the long (soluble) and short (membrane bound) isoforms being present in the latter case. Chicken and quail SCF have 98% amino acid homology but only 53% homology with mammalian forms of SCF. Chicken myelomonocytic growth factor (MGF), necessary for the survival and growth of normal and transformed avian myeloid precursor cells, has also been cloned (Leutz *et al.*, 1989; Sterneck *et al.*, 1992). It consists of 201 amino acids, with a 23-amino-acid signal peptide, and has significant homology to both mammalian G-CSF (52–56%) and IL-6 (40%). The MGF gene has five exons and four introns, as with both G-CSF and IL-6. Expression of constructs containing the MGF promoter is specific to myelomonocytic cells, and can be activated by kinases (Sterneck *et al.*, 1992).

Interleukin-6

Invertebrates

Concentrated coelomic fluids from the starfish *A. forbesi* have been found to contain a factor that supports the growth of a murine IL-6-dependent B-cell hybridoma (B9 cells), and which is recognized in Western blots with a polyclonal antihuman IL-6 serum (Beck and Habicht, 1996). This factor has a molecular weight of 30 kDa and a pI of 5.5. Isolated coelomocytes can release this factor after stimulation with LPS, with a peak release after 12 h. Furthermore, IL-6R mAbs have been shown to bind to axial organ cells in the sea star *A. rubens* by FACS analysis (Legac *et al.*, 1996).

Fish

Anti-IL-6 sera have been used to detect antigenically cross-reactive molecules in the sera of virus-challenged carp and trout (Ahne, 1993) and in supernatants from concanavalin-A-stimulated trout blood leukocytes (Ahne, 1994a), by ELISA.

Birds

IL-6 bioactivity has been described in chicken ascitic fluid, using the B9 hybridoma proliferation assay (Rath *et al.*, 1995). In addition, human rIL-6 has been shown to increase fibronectin production from chicken hepatocytes, as can supernatants from LPS-stimulated chicken fibroblasts (Kaiser, 1996). Both reactions are inhibitable with anti-IL-6 mAb. Chicken MGF, necessary for the survival and growth of normal and transformed avian myeloid precursor cells, has been cloned and found to have significant amino acid homology to both mammalian IL-6 (40%) and G-CSF (52–56%) (Leutz *et al.*, 1989; Sterneck *et al.*, 1992). This molecule is discussed further under 'Interleukin-3 and Colony Stimulating Factors'. A further member of the IL-6 family of cytokines has been described in birds; growth promoting activity (GPA), equivalent to ciliary neurotrophic factor (CNTF) (Koshlukova *et al.*, 1996). Cultured chick ciliary ganglion neurons possess receptors capable of binding both GPA and CNTF but GPA is three to five times more potent than human CNTF in promoting chick neuronal survival (Koshlukova *et al.*, 1996).

Interleukin-8 and Chemokines

IL-8 belongs to a family of low-molecular-weight cytokines that are potent chemoattractants for leukocytes, termed chemokines. The migration of leukocytes has been studied in two main ways. In the agarose method, cells migrate from a well cut in agarose gel towards wells containing the various stimuli. In the Boyden chamber method, cells migrate from one compartment, through a micropore filter, towards a blind compartment containing the stimuli. Whether an increased migration is chemotactic (directed) or chemokinetic (random) can be determined by the Boyden chamber method, which permits migration to occur with increasing concentration of the stimulus in the presence or absence of a gradient (chequerboard assay). Since this has been rarely carried out with non-mammalian cytokines, the term chemotactic factor (CTF) is used throughout this section and does not imply that a distinction between these activities has been demonstrated.

Fish

That fish can produce cytokines able to influence leukocyte migration has been demonstrated in carp (Howell, 1987). CTF-containing supernatants can be generated by incubation of sensitized kidney cells from carp immunized for 1–2 weeks, with specific antigen for 24–48 h at 28°C. Alternatively, the mitogen PHA can be used to generate such supernatants in the same manner. In a Boyden-type assay both types of supernatants induce significant migration of kidney cells from unsensitized animals, but require rather long migration times (16–22 h).

Amphibia

In amphibia, supernatants (48–72 h) from mitogen- and specific antigen-stimulated splenocytes of the frog *Rana temporaria* are able to increase the migration of peritoneal cells in both the agarose and Boyden-type assays (Gearing and Rimmer, 1985a). Predominantly polymorphonuclear leukocytes migrate in response to these supernatants, possibly due to a faster migration rate. Fractionation of the supernatants reveals that the CTF activity has a molecular weight range of 16–27 kDa.

Birds

Chicken supernatants from splenocytes, thymocytes and blood leukocytes incubated with concanavalin-A (5–10 μg/ml) for 48 h significantly increase the migration of monocytes (Altman and Kirchner, 1974). Similarly, lymphocytes from chick embryos stimulated with mitogen *in vivo* are able to produce a factor that enhances eosinophil migration *in vitro*, using either embryonic or adult eosinophils (Moriya and Ichikawa, 1981). The production of this chicken monocyte CTF correlates with a delayed skin reaction (wattle swelling), suggesting that it is a good *in vitro* correlate of delayed hypersensitivity in this species (Altman and Kirchner, 1974). Physicochemical analysis of this cytokine has shown that it has a molecular weight of 12.5 kDa (Fig. 9), and is relatively heat and freeze–thaw stable. No cross-reactivity is apparent for chicken CTF with human, guinea-pig and rabbit monocytes (Altman and Kirchner, 1974), although

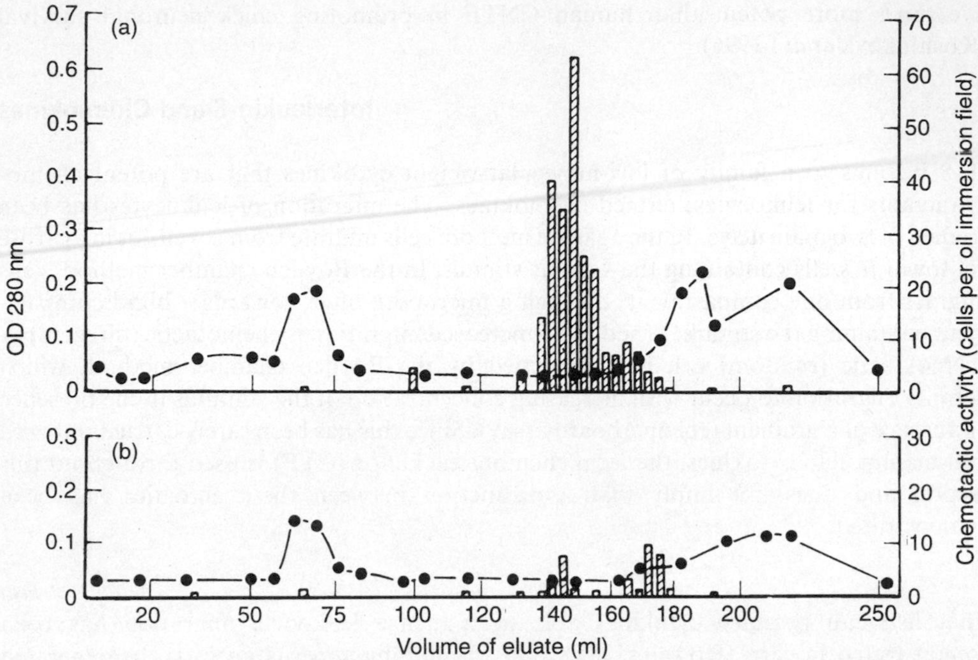

Fig. 9. Fractionation of chemotactic activity in 48-h supernatants from (a) concanavalin-A-stimulated chicken splenocytes or (b) non-mitogen-treated cells reconstituted with concanavalin-A, by gel chromatography. Fractions were tested for absorbance (●) and monocyte chemotactic activity (▨). The highest column in graph (a) has a molecular weight of 12.5 kDa. From Altman and Kirchner (1974), *Immunology* **26**, 393–405, with permission.

human rIL-8 has been shown to induce inducible NOS (iNOS) expression in chicken osteoclasts (Sunyer *et al.*, 1996). The concanavalin-A-induced monocyte CTF is not produced by bursal cells but is produced by spleen cells from agammaglobulinaemic chickens (Altman and Kirchner, 1972), suggesting that it is a T-cell product. The eosinophil CTF, on the other hand, appears to be produced by thymocytes or splenocytes in response to concanavalin-A and by bursal cells in response to LPS.

A gene with homology to IL-8 (51%) and *gro* α (45%) has been sequenced in chickens recently, termed 9E3/CEF4 (Barker *et al.*, 1993; Martins-Green *et al.*, 1996). 9E3/CEF4 is synthesized as a 9-kDa precursor molecule that is cleaved by plasmin to release a 6–7-kDa peptide. The molecule is synthesized and secreted in less than 10 min following stimulation of chick embryo fibroblasts with tetradecanoylphorbol acetate (TPA). The smaller peptide binds to extracellular matrix molecules and may be involved in wound healing and tumour suppression, situations in which it is overexpressed.

Other Interleukin Activities

Invertebrates

Isolated sea star factor (SSF) from coelomocyte lysates of the starfish *A. forbesi* inhibits the lymphokine-induced expression of class II MHC antigens on murine macrophages *in vitro* (Donnelly *et al.*, 1985). Intraperitoneal injection of SSF also reduces class II MHC antigen expression on peritoneal exudate macrophages in *Listeria*-infected mice. The presence of indomethacin reverses this effect, suggesting that SSF affects arachidonic acid metabolism in macrophages, which in turn inhibits class II MHC antigen expression.

Fish

In fish, *in vitro* stimulation of immunoglobulin production by PHA has been shown and may be a result of interleukins secreted from T cells which subsequently influence B cells. Similarly, phorbol myristate acetate (PMA)/calcium-ionophore-induced T-like cell lines of channel catfish secrete factors that can enhance the mitogenic responses of B (and T) lymphocytes, but non-specifically suppress *in vitro* antibody responses (Clem *et al.*, 1991). These putative interleukins do not function in mouse bioassays for B-cell growth factors (IL-4), and rIL-4 does not exhibit functional activity with fish cells. Lastly, hybridization with a monkey erythropoietin (EPO) cDNA probe occurs in Northern blots, using rainbow trout kidney, spleen and liver RNA, and antigenic cross-reactivity can be detected in concentrated (four-fold) trout serum using a human EPO radioimmunoassay (Shiels and Wickramasinghe, 1995).

Amphibians

In *Xenopus*, supernatants from PHA-stimulated splenocytes support the growth of homologous B cells (Cohen and Haynes, 1991). In addition, a striking costimulation of these supernatants with LPS and an anti-*Xenopus* IgM mAb is seen.

Birds

Chicken supernatants from concanavalin-A-stimulated splenocytes induce the formation of multinucleated giant cells in macrophage cultures (Weiler and Von Bulow, 1987b). This macrophage fusion factor (MFF) activity is optimal when added to 5–7-day-old cultures. MFF activity in mammals has been ascribed to IL-4 (McInnes and Rennick, 1988). In addition, human rIL-10 has been shown to increase the production of NO and iNOS mRNA expression in chicken osteoclasts (Sunyer *et al.*, 1996).

MIGRATION INHIBITORY FACTORS

Migration inhibitory factor (MIF) activity was one of the earliest cytokine bioactivities discovered, and has been studied in many animal groups because of the relatively simple biological assay used, the capillary tube technique. This involves centrifuging a capillary tube containing leukocytes in order to pellet the cells, cutting the tube at the cell–medium interface and then allowing the cells to migrate out into a culture dish containing medium for 18–24 h. The medium may contain preformed MIFs (indirect assay) or, if *in vivo* sensitized leukocytes are used for migration, specific antigen to generate MIFs *in situ* (direct assay). In either case the degree of migration inhibited is noted, usually by planimetry. In mammals, MIF has been sequenced and is a peptide of 114–115 amino acids, with a molecular weight of 12.5 kDa.

Invertebrates

Lysates from starfish *A. forbesi* coelomocytes (phagocytes) induce prompt aggregation of circulating coelomocytes when injected into the arm of intact starfish (Prendergast and Suzuki, 1970). The cells become fixed to the walls of the respiratory papulae for 20–40 min, after which they slowly disperse. When tested *in vitro* for MIF activity, significant inhibition of guinea-pig macrophage migration is seen. Exposure of SSF to 56°C for 30 min completely destroys its MIF activity. Biochemical characterization of SSF has shown it to have a two-polypeptide chain structure with a molecular weight of 38 kDa and a pI of 8.2 (Prendergast and Liu, 1976).

Fish

MIF activity has been demonstrated in a large range of fish species using T-cell mitogens and a variety of antigens, including bacterial and parasite antigens, alloantigens and even autoantigens. Most studies have employed direct assays to detect MIF release from sensitized blood, head kidney or peritoneal leukocytes. Only few studies have used MIF-containing supernatants to confirm the presence of soluble mediators. For example, in carp, supernatants from blood leukocytes stimulated with 10 μg/ml of concanavalin-A for 24–48 h induce significant inhibition of head kidney leukocyte migration (Secombes, 1981). The migratory cells have been identified as macrophages by their morphology and characteristic adherence to the migration culture plates when the plates are inverted.

The ability to release MIF appears to vary inversely with the ability to secrete specific antibody. This occurs over a range of timings after immunization and antigen doses, with MIF production predominating early in the immune response or in response to low antigen doses (Jayaraman *et al.*, 1979). On the other hand, MIF production correlates very closely with allograft rejection in the garpike (*Lepisosteus platyrhincus*) (McKinney *et al.*, 1981) and with resistance against furunculosis in vaccinated brown trout (*Salmo trutta*) (Smith *et al.*, 1980). Thus its detection is a good *in vitro* indicator of an ongoing cell-mediated immune response.

Supernatants from eel (*Anguilla japonica*) spleen, kidney or blood leukocytes incubated with specific antigen for 72 h can induce migration inhibition of guinea-pig peritoneal cells, suggesting a significant degree of conservation of MIF cytokines (Song *et al.*, 1989).

Amphibia

Both urodele (Axolotls) and anuran (frogs and toads) amphibians have been shown to produce MIF (Tahan and Jurd, 1979; Gearing and Rimmer, 1985b). MIF activity has been detected by direct and indirect assays, using antigen- and mitogen-stimulated splenocyte supernatants in the latter. In the axolotl *A. mexicanum,* splenocytes sensitized to *Mycobacterium tuberculosis* release MIF in direct assays following exposure to the purified protein derivative of tuberculin (PPD) (Tahan and Jurd, 1979). Sensitization can be transferred to unsensitized axolotls by injection of splenocytes (but not serum) from immunized donors, which suggests that this is a cell-mediated phenomenon. In *R. temporaria,* antigen-specific MIF release is seen 1–3 weeks postinjection in direct assays and is dependent on the antigen concentration in the medium (Gearing and Rimmer, 1985b). In indirect assays MIF is detectable in 72-h supernatants from antigen-stimulated (primed) splenocytes and in 48-h supernatants from concanavalin-A-stimulated control cells. In both assays its activity is blocked by α-L-fucose (but not by β-D-galactose) which is thought to compete with MIF for a cell surface receptor.

Gel chromatography of the frog concanavalin-A supernatants reveals MIF activity in the 27–50-kDa molecular weight range (Gearing and Rimmer, 1985b). Frog MIF activity does not appear to be species specific, and a supernatant inducing 39% migration inhibition of *Rana* peritoneal cells will induce significant inhibition of rat (25%), *Xenopus* (15%) and carp (11%) leukocyte migration. Finally, injection of a MIF-rich mammalian preparation into *R. esculenta* causes a marked reduction in the number of peritoneal macrophages, similar to that seen following injection of specific antigen into sensitized *Rana*. This macrophage disappearance reaction suggests that *Rana* cells can also react to mammalian MIF.

Reptiles

In the lizard *Calotes versicolor*, antigen-specific MIF release from sensitized splenocytes is detectable by direct assay 1–5 weeks after immunization or sensitization to skin allografts (Jayaraman and Muthukkaruppan, 1977, 1978). An inverse relationship exists between the migration inhibition response and humoral antibody response (Jayaraman and Muthukkaruppan, 1978); for example, low antigen doses result in a high degree of migration inhibition but low numbers of antibody-secreting cells, whereas high doses have the reverse effect. Addition of 5% sensitized splenocytes to unsensitized cells is sufficient to effect significant migration inhibition in the presence of antigen. Thymocytes from sensitized animals can also bring about this effect (although a higher cell ratio is required), suggesting active generation of MIF-producing cells in the thymus. Thymus-derived cells have also been implicated as the MIF-producing cells in cell separation experiments. Only nylon-wool non-adherent splenocytes from immunized animals, which are sensitive to antithymocyte serum, are effective in inhibiting the migration of control splenocytes in the presence of specific antigen (Manickasundari *et al.*, 1984).

Birds

MIF production by sensitized avian leukocytes incubated with specific antigen can be demonstrated by both direct and indirect assays some 2–3 weeks after immunization (Vlaovic *et al.*, 1975; Trifonov *et al.*, 1977). Both splenocytes and blood leukocytes have been used in direct assays, and a positive correlation with wattle tests (a measure of chicken delayed-type hypersensitivity) has been found (Vlaovic *et al.*, 1975). Using indirect assays, chicken splenocytes, thymocytes, bursal cells and ileocaecal tonsil lymphocytes have all been shown to secrete MIF, as evidenced by migration inhibition of target splenocytes, thymocytes, thrombocytes, bursacytes or peritoneal cells (Trifonov *et al.*, 1977; Stinson *et al.*, 1979; Martin and Glick, 1983). Turkey splenocytes and ileocaecal tonsil lymphocytes also secrete MIF, and in both species splenocyte migration is inhibited to a greater extent than that of peritoneal cells. The ability of B (bursal) cells to secrete a MIF is rather intriguing. Bursal MIF does not appear to be identical to thymocyte-derived MIF, in that it has a wider target specificity (Martin and Glick, 1983). However, as with mammalian MIF, both cytokines have a molecular weight range of 10–50 kDa, are sensitive to chymotrypsin and neuraminidase treatment, but are relatively stable to heat and pH. In contrast to other vertebrate groups, avian MIF is species specific. Turkey MIF has no effect on the migration of chicken or goose peritoneal and spleen cells, and chicken MIF has no effect on turkey or goose cells (Trifonov *et al.*, 1977).

TUMOR NECROSIS FACTORS

Tumor necrosis factor (TNF) is the principal mediator of the host response to Gram-negative bacteria and is typically released from LPS-stimulated macrophages and monocytes (Abbas *et al.*, 1991). It can exert a paracrine and autocrine effect on leukocytes during inflammatory reactions, and can act as an endogenous pyrogen. Many of its effects are enhanced by IFN-γ, possibly through its ability to increase TNF receptor expression.

Invertebrates

Human rTNF induces cell spreading, and increases phagocytosis and migratory activity of molluscan haemocytes (Hughes *et al.*, 1990; Ottaviani *et al.*, 1995a). Immunoreactive TNF can also be detected in concentrated haemolymph by ELISA (Hughes *et al.*, 1990; Ouwe-Missi-Oukem-Boyer *et al.*, 1994). In *Biomphalaria glabrata* a band of 53 kDa is labelled in Western blots with anti-TNF-α serum, and detectable levels of TNF (by ELISA) are decreased significantly following infection of the snails with *Schistosoma mansoni* (Ouwe-Missi-Oukem-Boyer *et al.*, 1994). That mussel (*Mytilus edulis*) haemocytes may themselves produce TNF has been suggested from two observations; first, that stimulation *in vitro* with LPS induces haemocyte spreading and is inhibitable with anti-TNF and, second, a reduction in haemocyte number in response to injection of LPS into the posterior adductor muscle of this species is also inhibitable with anti-TNF (Hughes *et al.*, 1991).

Fish

Human rTNF-α synergizes with rainbow trout MAF to increase macrophage respiratory burst activity (Fig. 10) and with mitogenic stimuli to heighten proliferative responses of

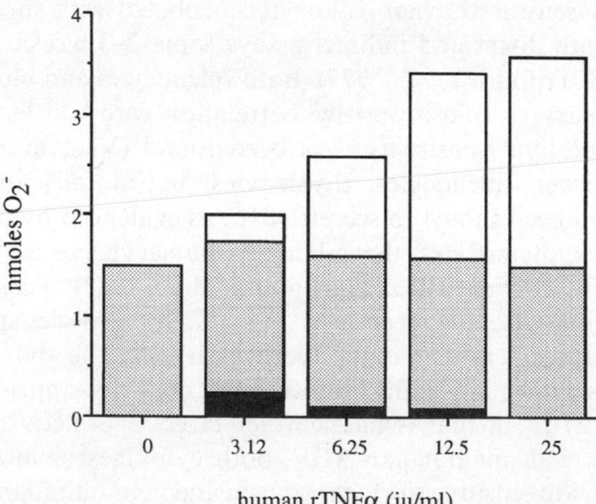

Fig. 10. Respiratory burst activity (superoxide anion production) of rainbow trout head kidney macrophages treated with human rTNF and 1:2 diluted MAF individually and together to show their synergistic effect. Values represent the mean for four fish. ■, rTNF alone; ▨, MAF alone; □, synergy between MAF and rTNF. From Hardie *et al.* (1994a), *Vet. Immunol. Immunopathol.* **40**, 73–84, with permission.

trout head kidney leukocytes (Hardie *et al.*, 1994a). In addition, human rTNF-α induces a dose-dependent increase in trout neutrophil migration (Jang *et al.*, 1995a). These effects are inhibited with anti-TNF serum, soluble TNFR1 and anti-TNFR-1 mAb but not by treatment with polymixin B (Hardie *et al.*, 1994a; Jang *et al.*, 1995a). The anti-TNFR-1 mAbs are also able to inhibit the MAF activity of supernatants from stimulated trout macrophages (Jang *et al.*, 1995b), suggesting the presence of an endogenous TNF in trout. Similarly, the stimulatory effect of M-glucan on Atlantic salmon (*Salmo salar*) macrophages is inhibited in the presence of anti-TNF-α sera (Robertsen *et al.*, 1994). Lastly, anti-TNF-α sera have been used to detect antigenically cross-reactive molecules in the sera of virus-challenged carp and trout by ELISA (Ahne, 1993).

Birds

TNF release from chicken macrophages can be detected after infection with *Eimeria* species or Marek's disease, as demonstrated using mammalian cellular cytotoxicity assays (Qureshi *et al.*, 1990; Byrnes *et al.*, 1993a; Zhang *et al.*, 1995). Injection of chickens with such TNF-like factors enhances weight loss due to *Eimeria* infection, and this is partially reversible by treatment with anti-human rTNF-α sera (Zhang *et al.*, 1995). Human rTNF-α has been shown to cross-react with chicken cells (Kaiser, 1996; Sunyer *et al.*, 1996), and antimurine TNF-α immunoprecipitates a 50-kDa protein in chickens, as in mammals (Gendron *et al.*, 1991). A partial gene sequence of a chicken TNF leukotriene (LT) homologue is in the European Molecular Biology Laboratory (EMBL) database (accession no. M80573) but details of how it was obtained are lacking.

GROWTH FACTORS

Transforming Growth Factor-β

TGF-βs belong to a superfamily of structurally related proteins, including activins and inhibins, antimüllerian hormone and bone morphogenetic proteins (BMPs) (Burt and Law, 1994). The conserved cysteines give a characteristic 'cysteine knot' crystal structure, as seen in other cytokines such as PDGF, nerve growth factor and brain-derived neurotrophic factor. Within the immune system, TGF-βs have both pro- and anti-inflammatory activities, and are particularly well known for their ability to inhibit T- and B-cell proliferation, macrophage activation and NK cell activity. They exist as three isoforms in mammals, termed TGF-β_1 to TGF-β_3.

Invertebrates

Whilst TGF-β has not been found in invertebrates to date, members of the superfamily are present, such as decapentaplegic protein, which participates in dorsal–ventral speciation in *Drosophila*, and *Drosophila* 60A protein, which participates in development of the embryonic gut (Burt and Law, 1994).

Fish

It has been known for several years that mammalian TGF-β_1 can cross-react to fish leukocytes, and is able to inhibit activated trout macrophages (Jang *et al.*, 1994).

Furthermore, the addition of anti-TGF-β_1 to supernatants from mitogen-stimulated leukocytes increases their ability to stimulate trout macrophages to release factors with MAF activity (Jang *et al.*, 1995b). Subsequent to this work, two forms of TGF-β have been sequenced in fish: a molecule in trout with similarities to TGF-β_1 (Hardie *et al.*, 1998: accession no. X993030) and a molecule in carp with similarities to TGF-β_2 (Sumathy *et al.*, 1997: accession no. U66874). The trout TGF-β has been fully cloned and has 62–66% homology to the known mammalian isoforms. As with these forms, it is produced as a precursor molecule with a signal peptide, that is cleaved to give a 112-amino-acid mature peptide with 78% amino acid homology to TGF-β_1 and TGF-β_5 (Table 4). It is expressed in both lymphoid (kidney, gill, spleen) and nervous (brain) tissue but is absent from the liver. The carp molecule is incomplete (104 amino acids of the mature peptide are sequenced, and 256 amino acids of the precursor), and currently has 97.1% amino acid homology with human TGF-β_2 and 69.2% homology with the trout TGF-β (Table 4).

In addition to TGF-β, activins (Ge *et al.*, 1993) and BMPs (Martinez-Barbera *et al.*, 1997) have been cloned in fish and are members of the TGF-β superfamily.

Whilst the TGF-βRs have not been sequenced in fish, a receptor for another member of the cysteine knot cytokine family, PDGFRβ, has been cloned in the puffer fish (How *et al.*, 1996). The gene consists of 21 exons and has 45% amino acid homology with human PDGFRβ, with the kinase domain having much higher homology as with CSF1R (How *et al.*, 1996). The gene is linked tandemly in a head-to-tail array with the CSF1R (see Interleukin-3 and Colony Stimulating Factors).

Amphibians

In *Xenopus*, human rTGF-β_1 inhibits T-cell growth factor-induced proliferation of splenic blasts (Haynes and Cohen, 1993b) and has mesoderm-inducing activity on amphibian ectoderm (Kimelman and Kirschner, 1987). In addition, conditioned medium from a cell line derived from the tadpole stage of *Xenopus* has mesoderm inducing activity, suggesting the presence of endogenous TGF-β (Smith *et al.*, 1988). Two forms of TGF-β have now been sequenced in *Xenopus*, TGF-β_5 (equivalent to TGF-β_1) and TGF-β_2 (Burt and Law, 1994). The mature peptide of *Xenopus* TGF-β_5 has

Table 4. TGF-β mature peptide amino acid homologies.

	hTGF-β_1	hTGF-β_2	hTGF-β_3	chTGF-β_4	chTGF-β_2	chTGF-β_3	xTGF-β_5	xTGF-β_2	tTGF-β	caTGF-β
hTGF-β_1	100	71.4	76.8	**80.4**	72.3	76.8	**75.0**	71.4	**77.7**	82.7
hTGF-β_2		100	79.5	63.4	**99.1**	79.5	66.1	**95.5**	70.5	**97.1**
hTGF-β_3			100	69.6	79.5	**99.1**	68.8	76.8	74.1	91.4
chTGF-β_4				100	63.4	69.6	68.8	76.8	75.9	78.9
chTGF-β_2					100	79.5	66.1	94.6	70.5	**97.1**
chTGF-β_3						100	68.8	75.8	74.1	91.4
xTGF-β_5							100	64.3	**77.7**	82.7
xTGF-β_2								100	68.8	94.2
tTGF-β									100	69.2
caTGF-β										100

Values in bold represent the highest homologies with other forms. h, Human; ch, chicken; x, *Xenopus*; t, trout; ca, carp.

approximately 75% amino acid homology with mammalian TGF-β_1, whereas *Xenopus* TGF-β_2 is 95% homologous to mammalian TGF-β_2 (Table 4). *Xenopus* rTGF-β_5 also inhibits T-cell growth factor-induced proliferation of splenic blasts and, using an anti-TGF-β_5, it can be shown that *Xenopus* leukocytes release TGF-β_5 in response to mitogen stimulation (Haynes and Cohen, 1993b).

BMPs and activins have also been cloned in *Xenopus* (Thomsen *et al.*, 1990; Dale *et al.*, 1992; Clements *et al.*, 1995).

Birds

Three forms of TGF-β have been cloned from chickens: TGF-β_4 (equivalent to TGF-β_1) (Burt and Jakowlew, 1992), TGF-β_2 (Burt and Paton, 1991) and TGF-β_3 (Burt *et al.*, 1995). The mature peptide of cTGF-β_4 is unusual in having 114 amino acids instead of 112, and has approximately 80% amino acid homology with mammalian TGF-β_1 (Table 4). Chicken TGF-β_2 shows 96–99% homology with mammalian TGF-β_2, whilst chicken TGF-β_3 shows 97–99% with mammalian TGF-β_3. All of the chicken TGF-βs have a signal peptide and thus the potential to be secreted. The precursor molecule (minus the signal peptide) is 372, 392 and 389 amino acids for chicken TGF-β_4, TGF-β_2 and TGF-β_3 respectively. The gene organization of chicken TGF-β_2 and TGF-β_3 is known, and consists of seven exons and six introns, as in the equivalent mammalian genes. Three different forms of chicken TGF-β_2 are known to be expressed in chicken embryos, of 3900, 4300 and 8000 bp, owing to differently sized 3′ untranslated regions.

Other members of the TGF-β superfamily have also been cloned in chickens, as with the BMPs (Houston *et al.*, 1994) and inhibins–activins (Chen and Johnson, 1996).

Fibroblast Growth Factor

The fibroblast growth factors (FGFs) are a group of heparin-binding single-chain polypeptides that play a pivotal role in development, cell growth, tissue repair and transformation. Currently nine members of the FGF family exist in mammals, with FGF-1, FGF-2 and FGF-9 lacking a signal peptide. Four FGF receptors of the tyrosine kinase class are known, with two or three extracellular immunoglobulin-like domains present, dependent upon the particular splice variant. Whilst the different FGFs bind to the four FGFRs with different affinities, there is no strict specificity between FGF and the FGFR.

Invertebrates

In *Drosophila melanogaster* two FGFRs have been sequenced, *Drosophila* FGF receptor homolog (DFR) 1 and DFR2, which are distantly related to vertebrate FGFRs (Shishido *et al.*, 1993). They have two and five extracellular immunoglobulin-like domains respectively, and a highly conserved intracellular kinase domain (60% amino acid homology to vertebrate FGFR kinase domains). The two genes show 79% homology and DFR2 is virtually identical to a previously described DFGF-R (Glazer and Shilo, 1991). They are expressed at all stages of *Drosophila* development and appear to be particularly important for development of the tracheal system (Klambt *et al.*, 1992). A member of the FGFR family has also been cloned in a sea urchin (McCoon *et al.*, 1996) and in the nematode *Caenorhabditis elegans* (DeVore *et al.*, 1995). The sea urchin molecule has 47–51% homology with mammalian FGFR and, interestingly, when expressed in COS cells does not bind FGF-2 or FGF-7.

Fish

FGF-2 has been cloned and sequenced in rainbow trout (Hata *et al.*, 1997). It contains 155 amino acids (approximately 17.3 kDa) and has more than 70% amino acid homology with mammalian FGF-2. Trout rFGF-2 binds tightly to heparin–Sepharose and promotes the proliferation of fibroblasts. FGF-3 has been cloned in zebrafish and the rFGF-3 shown to be mitogenic for murine BALB/MK cells known to express FGFR (Kiefer *et al.*, 1996). An FGF has also been isolated biochemically from the swimbladder of red seabream (*Pagrus major*) by heparin affinity (Suzuki *et al.*, 1994). It has a molecular weight of 22.5 kDa and an isoelectric point of 9.4. It promotes the growth of fibroblasts and mesoderm induction (in a *Xenopus* assay), and is heat and acid labile. Antisera to the red seabream FGF have shown that FGF is also produced in the pharynx of developing fish (flounder), where it stimulates cartilage formation (Suzuki and Kurokawa, 1996).

Four FGFR genes have been sequenced in the Medaka (*Oryzias latipes*), corresponding to human FGFR-1, FGFR-2, FGFR-3 and FGFR-4 (Emori *et al.*, 1992). The four FGFRs are highly homologous to each other (76–87%) and to known mammalian FGFRs (approximately 80% homology). Each is a single copy gene that appears to be under different transcriptional control. FGFR-4 has also been sequenced in zebrafish embryos, and uniquely contains four immunoglobulin domains in its *N*-terminal region (Thisse *et al.*, 1995).

Amphibians

It has been known for many years that heparin binding growth factors are present in *Xenopus* embryos, and have mesoderm-inducing activity inhibitable with anti-FGF serum (Slack and Isaacs, 1989). Five members of the FGF family have been cloned in *Xenopus* to date: FGF-2, FGF-3, FGF-8 (partial sequence), FGF-9 and an embryonic (e) FGF closely related to FGF-4 and FGF-6 (Isaacs *et al.*, 1992; Song and Slack, 1996; Christen and Slack, 1997). *Xenopus* FGF-2 and FGF-3 have 84% and 71% amino acid homology with human FGF-2 and FGF-3 respectively, with Xenopus FGF-2 lacking a signal peptide as in mammals. *Xenopus* FGF-9 shows 93% amino acid homology with human FGF-9 and also lacks a signal peptide, but appears to be secreted as it is glycosylated in an *in vitro* translation system (possibly using its central hydrophobic domain to cross the endoplasmic reticulum membrane). *Xenopus* eFGF has a signal peptide and shows 57–58% amino acid homology with FGF-4 and 61% homology with FGF-6. All four FGFs have mesoderm-inducing activity in isolated animal caps, with FGF-2, FGF-9 and eFGF being expressed maternally (i.e. by persistent maternal transcripts), whereas FGF-3 is not expressed until the early gastrula stage. FGF-9 and eFGF can also influence the anteroposterior axis in *Xenopus* embryos. FGF-1 has also been cloned in the newt *Notophthalmus viridescens*, and has amino acid homology of 79–83% with mammalian and avian FGF-1 (Patrie *et al.*, 1997). Lastly, a partial Axolotl FGF-8 sequence is in the databank (Han, 1997).

Four FGFR genes has also been cloned in amphibians (Musci *et al.*, 1990; Friesel and Brown, 1992; Shi *et al.*, 1992, 1993; Poulin and Chin, 1994). In newts the amino acid homologies to the respective human receptors are 85% for FGFR-1, 73–78% for

FGFR-3, 75% for FGFR-3 and 66% for FGFR-4. The variation in FGFR-2 arises from the numerous splice variants of this gene (Shi *et al.*, 1994). Interestingly, the exon IIIb-containing receptors appear to play a role in development of epithelial tissues, whereas the exon IIIc-containing receptors appear to be more important for neural tissue formation. It is clear that there is differential expression and regulation of FGFR during amphibian development (Launay *et al.*, 1994), and that they have distinct roles in amphibian limb regeneration (Poulin *et al.*, 1993).

Birds

Five FGFs have been cloned in chickens to date: FGF-1, FGF-2, FGF-3, FGF-4 and FGF-8 (Borja *et al.*, 1993; Niswander *et al.*, 1994; Han, 1995; Mahmood *et al.*, 1995; Vogel *et al.*, 1996), with amino acid homologies of 90%, 89%, 67%, 81% and 86% with mammalian FGF-1, FGF-2, FGF-3, FGF-4 and FGF-8 respectively. FGF-1 is highly expressed in developing eyes, suggesting a role in eye development, and both FGF-1 and FGF-2 have a role in skin appendage induction, as seen with developing chicken feather buds (Widelitz *et al.*, 1996). FGFs play a role in limb skeletal development, and FGF-2 and FGF-4 will support virtually complete limb outgrowth of limb buds from which the apical ectodermal ridge has been removed surgically (Szebenyi *et al.*, 1995). In quail, expression of FGF-2 in neurons of the neural tube and crest, and later in the spinal cord and dorsal root ganglia, is seen, suggesting a role in neural development. Similarly, FGF-3 is expressed within the developing cranial neural tube and in pharyngeal endoderm and ectoderm.

As in the other vertebrate classes, four FGFRs have been cloned in birds: FGFR-1 (Cek1), FGFR-2 (Cek3), FGFR-3 (Cek2) and FGFR-like embryonic kinase (FREK) (Pasquale, 1990; Marcelle *et al.*, 1994). Cek1, Cek3 and Cek2 show 93%, 96% and 97% amino acid homology with their respective human FGFR. FREK is most closely related to the newt FGFR-4 (80% amino acid homology) and human FGFR-4 (72% homology), but is thought to be distinct from FGF-4 as it is expressed strongly in myotomes early in development, followed by expression in all embryonic skeletal muscle and in satellite cells of adult muscle. Whether this form also exists in mammals has still to be determined. Splice variants of these FGFRs appear to be under tissue- or area-specific regulation (Sato *et al.*, 1992), as in other vertebrates.

CONCLUSIONS

The cloning of cytokines in non-mammalian vertebrates is slowly increasing and significant advances have been made in recent years. Representatives of many of the cytokine groups are now known in birds, amphibians or fish, including IFN, IL-1, SCF, MGF, 9E3/CEF4, FGF and TGF-β. Based on known mammalian crystal structures (Sprang and Bazan, 1993), some of these cytokines will be expected to belong to the same structural family, as with IFN, SCF and MGF (four antiparallel helical bundle) and IL-1 and FGF (β trefoil), and it will be interesting to model these sequences to gain an insight into this aspect of cytokine evolution. A few cytokine receptor sequences are also now known and show remarkable conservation. Despite these advances in vertebrates, there is still a complete lack of cytokine sequence information for invertebrates, and this is particularly disappointing at the present time. In species for which no cytokine sequences currently exist, there is still a tendency to rely on biological cross-reactivity to demonstrate the presence of a particular cytokine activity. This is a major concern when conditioned media are used in mammalian assays, potentially containing numerous uncharacterized molecules, many of which are probably not cytokines. Similarly, with antigenic cross-reactivity using antisera raised to mammalian cytokines it is clear that problems exist with false positives, as seen in invertebrate (Hahn *et al.*, 1996) and lower vertebrate (Ahne, 1994b) studies. Nevertheless, homologous assay systems have convincingly demonstrated cytokine activities in most animal groups, and the sequences obtained above have confirmed the molecular identity of the molecules responsible for the biological activities measured in these species. This is both reassuring and inspiring, and should be a major impetus to learn more about the origins of cytokines.

ACKNOWLEDGEMENTS

The author thanks Mr J. Zou and Mr G. Daniels for help in preparing Figures 2 and 5, and Tables 2 and 4.

NOTE ADDED IN PROOF

Since this chapter was written, a paper has been published on the isolation of a chicken IL-2 gene by expession cloning (Sundick and Gill-Dixon, 1997). The gene shows homology to both mammalian IL-2 (44%) and IL-15 (46%) and is expressed only by activated T cells. Genetic distance analysis reveals that it is closer to IL-2 than IL-15.

REFERENCES

Abbas, A.K., Lichtman, A.H. and Pober, J.S. (1991). *Cellular and Molecular Immunology.* W.B. Saunders, Philadelphia.

Ahne, W. (1993). Presence of interleukins (IL-1, IL-3, IL-6) and the tumour necrosis factor (TNF α) in fish sera. *Bulletin of the European Association of Fish Pathologists* **13**, 106–107.

Ahne, W. (1994a). Lectin (ConA) induced interleukin (IL-1α, IL-2, IL-6) production *in vitro* by leukocytes of rainbow trout (*Oncorhynchus mykiss*). *Bulletin of the European Association of Fish Pathologists* **14**, 33–35.

Ahne, W. (1994b). Significant ConA-interference in quantification of interleukin 2 (IL-2) by sandwich type enzyme immunoassays (ELISA). *Bulletin of the European Association of Fish Pathologists* **14**, 76.

Altman, L.C. and Kirchner, H. (1972). The production of a monocyte chemotactic factor by agammaglobulinemic chicken spleen cells. *J. Immunol.* **109**, 1149–1151.

Altman, L.C. and Kirchner, H. (1974). Mononuclear leucocyte chemotaxis in chickens. Definition of a phylogenetically specific cytokine. *Immunology* **26**, 393–405.

Andersson, E. and Matsunaga, T. (1996). Jaw, adaptive immunity and phylogeny of vertebrate antibody V_H gene family. *Res. Immunol.* **147**, 233–240.

Balm, P.H.M., van Lieshout, E., Lokate, J. and Bonga, S.E.W. (1995). Bacterial lipopolysaccharide (LPS) and interleukin-1 (IL-1) exert multiple physiological effects in the tilapia *Oreochromis mossambicus* (Teleostei). *J. Comp. Physiol.* **165B**, 85–92.

Barker, K.A., Hampe, A., Stoeckle, M.Y. and Hanafusa, H. (1993). Transformation-associated cytokine 9E3/CEF4 is chemotactic for chicken peripheral blood mononuclear cells. *J. Virol.* **67**, 3528–3533.

Beck, G. and Habicht, G.S. (1986). Isolation and characterization of a primitive IL-1-like protein from an invertebrate, *Asterias forbesi. Proc. Natl Acad. Sci. U.S.A.* **83**, 7429–7433.

Beck, G. and Habicht, G.S. (1996). Characterization of an IL-6-like molecule from an echinoderm (*Asterias forbesi*). *Cytokine* **8**, 507–512.

Beck, G., Vasta, G.R., Marchalonis, J.J. and Habicht, G.S. (1989). Characterization of interleukin-1 activity in tunicates. *Comp. Biochem. Physiol.* **92B**, 93–98.

Beckage, N.E. (1996). Interactions of viruses with invertebrate cells. In *New Directions in Invertebrate Immunology* (eds K. Soderhall, S. Iwanaga and G.R. Vasta), SOS Publications, Fair Haven, USA, pp. 375–399.

Bernasconi, D., Schultz, U. and Staeheli, P. (1995). The interferon-induced Mx protein of chickens lacks antiviral activity. *J. Interferon Cytokine Res.* **15**, 47–53.

Bernheim, H.A. and Kluger, M.J. (1977). Endogenous pyrogen-like substance produced by reptiles. *J. Physiol.* **267**, 659–666.

Borja, A.Z.M., Zeller, R. and Meijers, C. (1993). Expression of alternatively spliced bFGF coding exons and antisense mRNAs during chicken embryogenesis. *Dev. Biol.* **157**, 110–118.

Burt, D.W. and Jakowlew, S.B. (1992). A new interpretation of a chicken transforming growth factor-β_4 complementary DNA. *Mol. Endocrinol.* **6**, 989–992.

Burt, D.W. and Law, A.S. (1994). Evolution of the transforming growth factor-β superfamily. *Prog. Cytokine Factor Res.* **5**, 99–118.

Burt, D.W. and Paton, I.R. (1991). Molecular cloning and primary structure of the chicken transforming growth factor-β_2 gene. *DNA Cell Biol.* **10**, 723–734.

Burt, D.W., Dey, B.R., Paton, I.R., Morrice, D.R. and Law, A.S. (1995). The chicken transforming growth factor-β_3 gene: genomic structure, transcriptional analysis, and chromosomal location. *DNA Cell Biol.* **14**, 111–123.

Byrnes, S., Eaton, R. and Kogut, M. (1993a). *In vitro* interleukin-1 and tumor necrosis factor-α production by macrophages from chickens infected with either *Eimeria maxima* or *Eimeria tenella*. *Int. J. Parasitol.* **23**, 639–645.

Byrnes, S., Emerson, K. and Kogut, M. (1993b). Dynamics of cytokine production during coccidial infections in chickens: colony-stimulating factors and interferon. *FEMS Immunol. Medical Microbiol.* **6**, 45–52.

Caspi, R.R. and Avtalion, R.R. (1984). Evidence for the existence of an IL-2 like lymphocyte growth promoting factor in bony fish, *Cyprinus carpio*. *Dev. Comp. Immunol.* **8**, 51–60.

Chandratilleke, D., Scanes, C.G. and Marsh, J.A. (1994). Effect of triiodothyronine and *in vitro* growth hormone on avian interleukin-2. *Dev. Comp. Immunol.* **18**, 353–362.

Chang, E.B., Wang, N.S. and Mayer, L.F. (1987). Lymphokines stimulate intestinal secretion: possible role as initiators of arachidonic acid (AA) metabolism in IBD. *Gastroenterology* **92**, 1342.

Chen, C.C. and Johnson, P.A. (1996). Molecular cloning of inhibin activin β (A) subunit complementary deoxyribonucleic acid and expression of inhibin activin α subunits and β (A) subunits in the domestic hen. *Biol. Reprod.* **54**, 429–435.

Christen, B. and Slack, J.M.W. (1997). Expression of FGF-8 reveals a cryptic apical ectoderm ridge in the *Xenopus* limb bud. Accession no. Y10312.

Clem, L.W., Miller, M.W. and Bly, J.E. (1991). Evolution of lymphocyte subpopulations, their interactions and temperature sensitivities. In *Phylogenesis of Immune Functions* (eds G.W. Warr and N. Cohen), CRC Press, Boca Raton, FL, pp. 191–213.

Clements, J.H., Fettes, P., Knochel, S., Lef, J. and Knochel, W. (1995). Bone morphogenetic protein-2 in the early development of *Xenopus laevis*. *Mech. Dev.* **52**, 357–370.

Cohen, N. and Haynes, L. (1991). The phylogenetic conservation of cytokines. In *Phylogenesis of Immune Functions* (eds G.W. Warr and N. Cohen), CRC Press, Boca Raton, FL, pp. 241–268.

Cohen, N., Watkins, D. and Parsons, S.V. (1987). Interleukins and T-cell ontogeny in *Xenopus laevis*. In *Developmental and Comparative Immunology* (eds E.L. Cooper, C. Langlet and J. Bierne), Alan R. Liss, New York, pp. 53–68.

Congleton, J. and Sun, B.L. (1996). Interferon-like activity produced by anterior kidney leukocytes of rainbow trout stimulated *in vitro* by infectious hematopoietic necrosis virus or Poly I-C. *Dis. Aquat. Organisms* **25**, 185–195.

Connors, V., DeBuron, I. and Granath, W. (1995). *Schistosoma mansoni*: interleukin-1 increases phagocytosis and superoxide production by hemocytes and decreases output of cercariae in schistosome-susceptible *Biomphalaria glabrata*. *Exp. Parasitol.* **80**, 139–148.

Dale, L., Howes, G., Price, B.M.J. and Smith, J.C. (1992). Bone morphogenetic protein 4: a ventralizing factor in early *Xenopus* development. *Development* **115**, 573–585.

DeKinkelin, P., Dorson, M. and Hattenberger-Baudouy, A.M. (1982). Interferon synthesis in trout and carp after viral infection. *Dev. Comp. Immunol. (Suppl.)* **2**, 167–174.

DeSena, J. and Rio, G.J. (1975). Partial purification and characterization of RTG-2 fish cell interferon. *Infect. Immunol.* **11**, 815–822.

DeVore, D.L., Horvitz, H.R. and Stern, M.J. (1995). An FGF receptor signalling pathway is required for the normal cell migrations of the sex myoblasts in *C. elegans* hermaphrodites. *Cell* **83**, 611–620.

Digby, M.R. and Lowenthal, J.W. (1995). Cloning and expression of the chicken interferon-γ gene. *J. Interferon Cytokine Res.* **15**, 939–945.

Donnelly, J.J., Vogel, S.N. and Prendergast, R.A. (1985). Down-regulation of Ia expression on macrophages by sea star factor. *Cell. Immunol.* **90**, 408–415.

Dorson, M., Barde, A. and DeKinkelin, P. (1975). Egtved virus-induced rainbow trout serum interferon: some physicochemical properties. *Ann. Microbiol. Inst. Pasteur* **126B**, 485–489.

Ellsaesser, C.F. and Clem, L.W. (1994). Functionally distinct high and low molecular weight species of channel catfish and mouse IL-1. *Cytokine* **6**, 10–20.

El Ridi, R., Wahby, A.F. and Saad, A.-H. (1986). Characterization of snake interleukin 2. *Dev. Comp. Immunol.* **10**, 128.

Emori, Y., Yasuoka, A. and Saigo, K. (1992). Identification of four FGF receptor genes in Medaka fish (*Oryzias latipes*). *FEBS Lett.* **314**, 176–178.

Francis, C.H. and Ellis, A.E. (1994). Production of a lymphokine (macrophage activating factor) by salmon (*Salmo salar*) leucocytes stimulated with outer membrane protein antigens of *Aeromonas salmonicida. Fish Shellfish Immunol.* **4**, 489–499.

Friesel, R. and Brown, S.A.N. (1992). Spatially restricted expression of fibroblast growth factor receptor-2 during *Xenopus* development. *Development* **116**, 1051–1058.

Fulton, R.W., Morton, R.J., Burge, L.J., Short, E.C. and Payton, M.E. (1995). Action of quail and chicken interferons on a quail cell line, QT35. *J. Interferon Cytokine Res.* **15**, 297–300.

Galabov, A.S. (1981). Induction and characterization of tortoise interferon. *Methods Enzymol.* **78A**, 196–208.

Ge, W., Gallin, J., Strobeck, C. and Peter, R.E. (1993). Cloning and sequencing of goldfish activin subunit genes: strong structural conservation during vertebrate evolution. *Biochem. Biophys. Res. Commun.* **193**, 711–717.

Gearing, A. and Rimmer, J.J. (1985a). Amphibian lymphokines. I. Leucocyte chemotactic factors produced by amphibian spleen cells following antigenic and mitogenic challenge *in vitro. Dev. Comp. Immunol.* **9**, 281–290.

Gearing, A. and Rimmer, J.J. (1985b). Amphibian lymphokines. II. Migration inhibition factor produced by antigenic and mitogenic stimulation of amphibian leucocytes. *Dev. Comp. Immunol.* **9**, 291–300.

Gendron, R.L., Nestel, F.P., Lapp, W.S. and Baines, M.G. (1991). Expression of tumor necrosis factor α in the developing nervous system. *Int. J. Neurosci.* **60**, 129–136.

Georgel, P., Kappler, C., Langley, E., Gross, I., Nocolas, E., Reichhart, J. and Hoffmann, J. (1995). *Drosophila* immunity. A sequence homologous to mammalian interferon consensus response element enhances the activity of the diptericin promoter. *Nucleic Acids Res.* **23**, 1140–1145.

Glazer, L. and Shilo, B.Z. (1991). The *Drosophila* FGF-R homolog is expressed in the embryonic tracheal stem and appears to be required for directed tracheal cell extension. *Genes Dev.* **5**, 697–705.

Graham, S. and Secombes, C.J. (1988). The production of a macrophage-activating factor from rainbow trout *Salmo gairdneri* leucocytes. *Immunology* **65**, 293–297.

Graham, S. and Secombes, C.J. (1990a). Cellular requirements for lymphokine secretion by rainbow trout *Salmo gairdneri* leucocytes. *Dev. Comp. Immunol.* **14**, 59–68.

Graham, S. and Secombes, C.J. (1990b). Do fish lymphocytes secrete interferon-γ? *J. Fish Biol.* **36**, 563–573.

Granath, W.O., Jr., Connors, V.A. and Tarleton, R.L. (1994). Interleukin 1 activity in haemolymph from strains of the snail *Biomphalaria glabrata* varying in susceptibility to the human blood fluke, *Schistosoma mansoni*: presence, differential expression, and biological function. *Cytokine* **6**, 21–27.

Gravel, M. and Malsberger, R.G. (1965). A permanent cell line from the fathead minnow (*Pimephales promelas*). *Ann. N.Y. Acad. Sci.* **126**, 555–565.

Grondel, J.L. and Harmsen, E.G.M. (1984). Phylogeny of interleukins: growth factors produced by leucocytes of the cyprinid fish, *Cyprinus carpio* L. *Immunology* **52**, 477–482.

Gruschwitz, M.S., Moormann, S., Kromer, G., Sgonc, R., Dietrich, H., Boeck, G., Gershwin, M.E., Boyd, R. and Wick, G. (1991). Phenotypic analysis of skin infiltrates in comparison with peripheral blood lymphocytes, spleen cells and thymocytes in early avian scleroderma. *J. Autoimmun.* **4**, 577–593.

Guida, S., Heguy, A. and Melli, M. (1992). The chicken IL-1 receptor: differential evolution of the cytoplasmic and extracellular domains. *Gene* **111**, 239–243.

Hahn, U.K., Fryer, S.E. and Bayne, C.J. (1996). An invertebrate (mulluscan) plasma protein that binds to vertebrate immunoglobulins and its potential for yielding false-positives in antibody-based detection systems. *Dev. Comp. Immunol.* **20**, 39–50.

Hamby, B.A., Huggins, E.M., Jr., Lachman, L.B., Dinarello, C.A. and Sigel, M.M. (1986). Fish lymphocytes respond to human IL-1. *Lymphokine Res.* **5**, 157–162.

Han, J.K. (1995). Molecular-cloning of a chicken-embryo fibroblast-growth-factor-1 cDNA and its expression during early embryogenesis. *Mol. Cells* **5**, 579–585.

Han, J.K. (1997). *Ambystoma mexicanum* mRNA for fibroblast growth factor 8, partial sequence. Accession no. Y11093.

Hardie, L.J., Chappell, L.H. and Secombes, C.J. (1994a). Human tumor necrosis factor α influences rainbow trout *Oncorhynchus mykiss* leucocyte responses. *Vet. Immunol. Immunopathol.* **40**, 73–84.

Hardie, L.J., Fletcher, T.C. and Secombes, C.J. (1994b). Effect of temperature on macrophage activation and the production of macrophage activating factor by rainbow trout (*Oncorhynchus mykiss*) leucocytes. *Dev. Comp. Immunol.* **18**, 57–66.

Hardie, L.J., Ellis, A.E. and Secombes, C.J. (1996). *In vitro* activation of rainbow trout macrophages stimulates inhibition of *Renibacterium salmoninarum* growth concomitant with augmented generation of respiratory burst products. *Dis. Aquat. Organisms* **25**, 175–183.

Hardie, L.J., Laing, K.J., Daniels, G.D., Grabowski, P.S., Cunningham, C. and Secombes, C.J. (1998). Isolation of the first piscine transforming growth factor β gene: analysis reveals tissue specific expression and a potential regulatory sequence in rainbow trout (*Oncorhynchus mykiss*). *Cytokine* (in press).

Hata, J., Takeo, J., Segawa, C. and Yamashita, S. (1997). A cDNA encoding fish fibroblast growth factor-2, which lacks alternative translation initiation. *J. Biol. Chem.* **272**, 7285–7289.

Hayara, Y., Schauenstein, K. and Globerson, A. (1982). Avian lymohokines. II. IL-1 activity in supernatants of stimulated adherent splenocytes of chickens. *Dev. Comp. Immunol.* **6**, 785–789.

Haynes, L. and Cohen, N. (1993a). Further characterization of an interleukin-2-like cytokine produced by *Xenopus laevis* T lymphocytes. *Dev. Immunol.* **3**, 231–238.

Haynes, L. and Cohen, N. (1993b). Transforming growth factor β (TGFβ) is produced by and influences the proliferative responses of *Xenopus laevis* lymphocytes. *Dev. Immunol.* **3**, 223–230.

Haynes, L., Moynihan, J.A. and Cohen, N. (1990). A monoclonal antibody against the human IL-2 receptor binds to paraformaldehyde-fixed but not viable frog (*Xenopus*) splenocytes. *Immunol. Lett.* **26**, 227–232.

Haynes, L., Harding, F.A., Koniski, A.D. and Cohen, N. (1992). Immune system activation associated with a naturally occurring infection in *Xenopus laevis. Dev. Comp. Immunol.* **16**, 453–462.

Hepkema, F.W. and Secombes, C.J. (1994). 5′ Nucleotidase activity of rainbow trout *Oncorhynchus mykiss* macrophages: correlation with respiratory burst activity. *Fish Shellfish Immunol.* **4**, 301–309.

Highet, A.B. and Ruben, L.N. (1987). Corticosteroid regulation of IL-1 production may be responsible for deficient immune suppressor function during metamorphosis of *Xenopus laevis*, the South African clawed toad. *Immunopharmacology* **13**, 149–155.

Horton, J. and Ratcliffe, N. (1996). Evolution of immunity. In *Immunology,* 4th edn (eds I. Roitt, J. Brostoff and D. Male), Mosby, Barcelona, pp. 15.1–15.22.

Houston, B., Thorp, B.H. and Burt, D.W. (1994). Molecular cloning and expression of bone morphogenetic protein-7 in the chick epiphyseal growth plate. *J. Mol. Endocrinol.* **13**, 289–301.

How, G.-F., Venkatesh, B. and Brenner, S. (1996). Conserved linkage between the puffer fish (*Fugu rubripes*) and human genes for platelet-derived growth factor receptor and macrophage colony-stimulating factor receptor. *Genome Res.* **6**, 1185–1191.

Howell, C.J.St.G. (1987). A chemokinetic factor in the carp *Cyprinus carpio. Dev. Comp. Immunol.* **11**, 139–146.

Hughes, T.K., Jr., Smith, E.M., Chin, R., Cadet, P. Sinisterra, J., Leung, M.K., Shipp, M.A., Scharrer, B. and Stefano, G.B. (1990). Interaction of immunoactive monokines (interleukin 1 and tumor necrosis factor) in the bivalve mollusc *Mytilus edulis. Proc. Natl Acad. Sci. U.S.A.* **87**, 4426–4429.

Hughes, T.K., Jr., Smith, E.M., Barnett, J.A., Charles, R. and Stefano, G.B. (1991). LPS stimulated invertebrate hemocytes: a role for immunoreactive TNF and IL-1. *Dev. Comp. Immunol.* **15**, 117–122.

Isaacs, H.V., Tannahill, D. and Slack, J.M.W. (1992). Expression of a novel FGF in *Xenopus* embryo—a new candidate inducing factor for mesoderm formation and anteroposterior specification. *Development* **114**, 711–720.

Jang, S.I., Hardie, L.J. and Secombes, C.J. (1994). The effects of transforming growth factor β_1 on rainbow trout *Oncorhynchus mykiss* macrophage respiratory burst activity. *Dev. Comp. Immunol.* **18**, 315–323.

Jang, S.I., Mulero, V., Hardie, L.J. and Secombes, C.J. (1995a). Inhibition of rainbow trout phagocyte responsiveness to human tumor necrosis factor α (hTNFα) with monoclonal antibodies to the hTNFα 55 kDa receptor. *Fish Shellfish Immunol.* **5**, 61–69.

Jang, S.I., Hardie, L.J. and Secombes, C.J. (1995b). Elevation of rainbow trout *Oncorhynchus mykiss* macrophage respiratory burst activity with macrophage-derived supernatants. *J. Leukocyte Biol.* **57**, 943–947.

Jayaraman, S. and Muthukkaruppan, V.R. (1977). *In vitro* correlate of transplantation immunity: spleen cell migration inhibition in the lizard, *Calotes versicolor. Dev. Comp. Immunol.* **1**, 133–144.

Jayaraman, S. and Muthukkaruppan, V.R. (1978). Detection of cell-mediated immunity to sheep erythrocytes by the capillary migration inhibition technique in the lizard, *Calotes versicolor. Immunology* **34**, 231–240.

Jayaraman, S., Mohan, R. and Muthukkaruppan, V.R. (1979). Relationship between migration inhibition and plaque-forming cell responses to sheep erythrocytes in the teleost, *Tilapia mossambica. Dev. Comp. Immunol.* **3**, 67–75.

Jungwirth, C., Rebbert, M., Ozato, K., Degen, H.J., Schultz, U. and Dawid, I.B. (1995). Chicken interferon consensus sequence-binding protein (ICSBP) and interferon regulatory factor (IRF) 1 genes reveal evolutionary conservation in the IRF gene family. *Proc. Natl Acad. Sci. U.S.A.* **92**, 3105–3109.

Kaiser, P. (1996). Avian cytokines. In *Poultry Immunology* (eds T.F. Davidson, T.R. Morris and L.N. Payne), Carfax Publishing Company, Abingdon, pp. 83–114.

Kaiser, P., Rothwell, L., Wain, H., Davidson, F. and Kaufman, J. (1996). Structure of the chicken interferon-γ gene, and comparison to mammalian homologues. *Immunology* **89 (supplement 1)**, OZ268.

Kaspers, B., Lillehoj, H.S., Jenkins, M.C. and Pharr, G.T. (1994). Chicken interferon-mediated induction of major histocompatibility complex class II antigens on peripheral blood monocytes. *Vet. Immunol. Immunopathol.* **44**, 71–84.

Kiefer, P., Mathieu, M., Mason, I. and Dickson, C. (1996). Secretion and mitogenic activity of zebrafish FGF3 reveal intermediate properties relative to mouse and *Xenopus* homologues. *Oncogene* **12**, 1503–1511.

Kimelman, D. and Kirschner, M. (1987). Synergistic induction of mesoderm by FGF and TGF-β and the identification of an mRNA coding for FGF in the early *Xenopus* embryo. *Cell* **51**, 869–877.

Klambt, C., Glazer, L. and Shilo, B.Z. (1992). Breathless, a *Drosophila* FGF receptor homolog, is essential for migration of tracheal and specific midline glial-cells. *Genes Develop.* **6**, 1668–1678.

Klein, J. (1989). Are invertebrates capable of anticipatory immune responses? *Scand. J. Immunol.* **29**, 499–505.

Kodama, H., Mukamoto, M., Baba, T. and Mule, D.M. (1994). Macrophage-colony stimulating activity in rainbow trout (*Oncorhynchus mykiss*) serum. *Modulators of Fish Immune Responses* **1**, 59–66.

Kogut, M.H. and Lange, C. (1989). Recombinant interferon-γ inhibits cell invasion by *Eimeria tenella. J. Interferon Res.* **9**, 67–77.

Koniski, A. and Cohen, N. (1994). Mitogen-activated Axolotl (*Ambystoma mexicanum*) splenocytes produce a cytokine that promotes growth of homologous lymphoblasts. *Dev. Comp. Immunol.* **18**, 239–250.

Koshlukova, S., Finn, T.P., Nishi, R. and Halvorsen, S.W. (1996). Identification of functional receptors for ciliary neurotrophil factor on chick ciliary ganglion neurons. *Neuroscience* **72**, 821–832.

Kroemer, G. and Wick, G. (1989). The role of interleukin-2 in autoimmunity. *Immunol. Today* **10**, 246–251.

Kroemer, G., Gastinel, L.N., Neu, N., Auffray, C. and Wick, G. (1990). How many genes code for organ-specific autoimmunity? *Autoimmunity* **6**, 215–233.

Kromer, G., Schauenstein, K. and Wick, G. (1984). Avian lymphokines: an improved method for chicken IL-2 production and assay. A Con A–erythrocyte complex induces higher T cell proliferation and IL-2 production than does free mitogen. *J. Immunol. Methods* **73**, 273–281.

Lampson, G.P., Tytell, A.A., Nemes, M.M. and Hilleman, M.R. (1965). Purification and characterization of chick embryo interferon. *Proc. Soc. Exp. Biol. Med.* **118**, 441–448.

Launay, C., Fromentoux, V., Thery, C., Shi, D.L. and Boucaut, J.C. (1994). Comparative analysis of the tissue distribution of 3 fibroblast growth-factor messenger-RNAs during amphibian morphogenesis. *Differentiation* **58**, 101–111.

Lee, T.-H. and Tempelis, C.H. (1991). Characterization of a receptor on activated chicken T cells. *Dev. Comp. Immunol.* **15**, 329–339.

Lee, T.-H. and Tempelis, C.H. (1992). Possible 110 kDa receptor for interleukin 2 in the chicken. *Dev. Comp. Immunol.* **16**, 463–472.

Legac, E., Vaugier, G.-L., Bousquet, F., Bajelan, M. and LeClerc, M. (1996). Primitive cytokines and cytokine receptors in invertebrates: the sea star *Asterias rubens* as a model of study. *Scand. J. Immunol.* **44**, 375–380.

Leutz, A., Damm, K., Sterneck, E., Kowenz, E., Ness, S., Frank, R., Gausepohl, H., Pan, Y.C.E., Smart, J., Hayman, M. and Graf, T. (1989). Molecular cloning of the chicken myelomonocytic growth factor (cMGF) reveals relationship to interleukin 6 and granulocyte colony stimulating factor. *EMBO J.* **8**, 175–181.

Lillehoj, H.S., Kang, S.Y., Keller, L. and Sevoian, M. (1989). *Eimeria tenella* and *Eimeria acervulina*: lymphokines secreted by an avian T cell lymphoma or by sporozoite-stimulated immune T lymphocytes protect chickens against avian coccidiosis. *Exp. Parasitol.* **69**, 54–64.

Litman, G.W. (1996). Sharks and the origins of vertebrate immunity. *Sci. Am.* **275**, 47–51.

Lowenthal, J.W., Digby, M.R. and York, J.J. (1995). Production of interferon-γ by chicken T cells. *J. Interferon Cytokine Res.* **15**, 933–938.

Luquet, G., Lemaitre, J. and LeClerc, M. (1989). Evidence for the production of an interleukin-like protein by *Asterias rubens* (axial organ cells) in mixed leukocyte culutures. *Lymphokine Res.* **8**, 451–458.

Mahmood, R., Kiefer, P., Guthrie, S., Dickson, C. and Mason, I. (1995). Multiple roles for FGF-3 during cranial neural development in the chicken. *Development* **121**, 1399–1410.

Manickasundari, M., Selvaraj, P. and Pitchappen, R.N. (1984). Studies on T-cells of the lizard, *Calotes versicolor*: adherent and non-adherent populations of the spleen. *Dev. Comp. Immunol.* **8**, 367–374.

Marcelle, C., Eichmann, A., Halevy, O., Breant, C. and LeDouarin, N.M. (1994). Distinct developmental expression of a new avian fibroblast growth factor receptor. *Development* **120**, 683–694.

Marsden, M.J., Cox, D. and Secombes, C.J. (1994). Antigen-induced release of macrophage activating factor from rainbow trout *Oncorhynchus mykiss* leucocytes. *Vet. Immunol. Immunopathol.* **42**, 199–208.

Martin, D.E. and Glick, B. (1983). Physiochemical characterization of lymphocyte inhibitory factor (LyIF) isolated from avian B and T cells. *Cell. Immunol.* **79**, 383–388.

Martinez-Barbera, J.P., Toresson, H., Rocha, S.D. and Krauss, S. (1997). Cloning and expression of three members of the zebrafish Bmp family: Bmp2a, Bmp2b and Bmp4. *Gene* **198**, 53–59.

Martins-Green, M., Stoeckle, M., Hampe, A., Wimberly, S. and Hanafusa, H. (1996). The 9E3/CEF4 cytokine: kinetics of secretion, processing by plasmin, and interaction with extracellular matrix. *Cytokine* **8**, 448–459.

Mathews, J.H. and Vorndam, A.V. (1982). Interferon-mediated persistent infection of Saint Louis encephalitis virus in a reptilian cell line. *J. Gen. Virol.* **61**, 177–186.

McCoon, P.E., Angerer, R.C. and Angerer, L.M. (1996). SpFGFR, a new member of the fibroblast growth factor receptor family, is developmentally regulated during early sea urchin development. *J. Biol. Chem.* **271**, 20119–20125.

McInnes, A. and Rennick, D.M. (1988). Interleukin 4 induces cultured monocytes/macrophages to form giant multinucleated cells. *J. Exp. Med.* **167**, 598–611.

McKinney, E.C., McLeod, T.F. and Sigel, M.M. (1981). Allograft rejection in a holostean fish, *Lepisosteus platyrhinus*. *Dev. Comp. Immunol.* **5**, 65–74.

Moehring, J.M. and Stinebring, W.R. (1981). General procedures for the induction and production of avian interferons. *Methods Enzymol.* **78A**, 189–192.

Moritomo, T., Noda, H., Yokota, R., Cho, M. and Watanabe, T. (1994). Kidney colony forming cells and serum colony stimulating factors of carp. *Dev. Comp. Immunol.* **18**, IX–X.

Moriya, O. and Ichikawa, Y. (1981). Production of an eosinophil migration factor in lymphocytes of mitogen-treated chick embryos. *Microbiol. Immunol.* **25**, 417–421.

Musci, T.J., Amaya, E. and Kirschner, M.W. (1990). Regulation of the fibroblast growth factor receptor in early *Xenopus* embryos. *Proc. Natl Acad. Sci. U.S.A.* **87**, 8365–8369.

Neumann, N.F. and Belosevic, M. (1996). Deactivation of primed respiratory burst responses of goldfish macrophages by leukocyte-derived macrophage activating factor(s). *Dev. Comp. Immunol.* **20**, 427–439.

Neumann, N.F., Fagan, D. and Belosevic, M. (1995). Macrophage activating factor(s) secreted by mitogen stimulated goldfish kidney leukocytes synergize with bacterial lipopolysaccharide to induce nitric oxide production in teleost macrophages. *Dev. Comp. Immunol.* **19**, 473–482.

Niswander, L., Jeffrey, S., Martin, G.R. and Tickle, C. (1994). A positive feedback loop coordinates growth and patterning in the vertebrate limb. *Nature* **371**, 609–612.

Novoa, B., Figueras, A., Ashton, I. and Secombes, C.J. (1996). *In vitro* studies on the regulation of rainbow trout (*Oncorhynchus mykiss*) macrophage respiratory burst activity. *Dev. Comp. Immunol.* **20**, 207–216.

Obranovich, T.D. and Boyd, R.L. (1996). A bursal stromal derived cytokine induces proliferation of MHC class II bearing cells. *Dev. Comp. Immunol.* **20**, 61–75.

Ottaviani, E., Franchini, A., Cassanelli, S. and Genedani, S. (1995a). Cytokines and invertebrate immune responses. *Biol. Cell* **85**, 87–91.

Ottaviani, E., Caselgrandi, E. and Franceschi, C. (1995b). Cytokines and evolution: *in vitro* effects of IL-1α, IL-1β, TNF-α and TNF-β on an ancestral type of stress response. *Biochem. Biophys. Res. Commun.* **207**, 288–292.

Ouwe-Missi-Oukem-Boyer, O., Porchet, E., Capron, A. and Dissous, C. (1994). Characterization of immunoreactive TNFα molecules in the gastropod *Biomphalaria glabrata*. *Dev. Comp. Immunol.* **18**, 211–218.

Pasquale, E.B. (1990). A distinctive family of embryonic protein-tyrosine receptors. *Proc. Natl Acad. Sci. U.S.A.* **87**, 5812–5816.

Patrie, K.M., Botelho, M.J., Ray, S.K., Mehta, V.B. and Chiu, I.M. (1997). Amphibian FGF-1 is structurally and functionally similar to but antigenically distinguishable from its mammalian counterpart. *Growth Factors* **14**, 39–57.

Petitte, J.N. and Kulik, M.J. (1996). Cloning and characterization of cDNAs encoding two forms of avian stem cell factor. *Biochim. Biophys. Acta* **1307**, 149–151.

Poulin, M.L. and Chin, I.-M. (1994). Nucleotide sequences of two newt (*Notophthalmus viridescens*) fibroblast growth factor receptor-2 variants. *Biochim. Biophys. Acta* **1220**, 209–211.

Poulin, M.L., Patrie, K.M., Botelho, M.J., Tassava, R.A. and Chiu, I.-M. (1993). Heterogeneity in the expression of fibroblast growth factor receptors during limb regeneration in newts (*Notophthalmus viridescens*). *Development* **119**, 353–361.

Prendergast, R.A. and Liu, S.H. (1976). Isolation and characterization of sea star factor. *Scand. J. Immunol.* **5**, 873–880.

Prendergast, R.A. and Suzuki, M. (1970). Invertebrate protein stimulating mediators of delayed hypersensitivity. *Nature* **227**, 277–279.

Qureshi, M.A., Miller, A., Lillehoj, H.S. and Ficken, M.D. (1990). Establishment and characterization of a chicken mononuclear cell line. *Vet. Immunol. Immunopathol.* **26**, 237–250.

Rath, N.C., Huff, W.E., Bayyari, G.R. and Balog, J.M. (1995). Identification of transforming growth-factor β and interleukin-6 in chicken ascites-fluid. *Avian Dis.* **39**, 382–389.

Robertsen, B., Engstad, R.E. and Jorgensen, J.B. (1994). β-Glucans as immunostimulants in fish. *Modulators of Fish Immune Responses* **1**, 83–99.

Rogel-Gaillard, C., Chilmonczyk, S. and de Kinkelin, P. (1993). *In vitro* induction of interferon-like activity from rainbow trout leucocytes stimulated by Egtved virus. *Fish Shellfish Immunol.* **3**, 383–394.

Ruben, L.N. (1986). Recombinant DNA-produced IL-2 injected will substitute for carrier priming of helper function in the South African clawed toad, *Xenopus laevis. Immunol. Lett.* **13**, 227–230.

Ruben, L.N., Clothier, R,H., Mirchandani, M., Wood, P. and Balls, M. (1987). Signals provided *in vivo* by human rIL-2 and Con A can switch hapten-specific tolerance from unresponsiveness to responsiveness in the South African clawed toad. *Immunology* **61**, 235–241.

Ruben, L.N., Beadling, C., Langeberg, L., Shiigi, S. and Selden, N. (1988). The substitution of carrier priming of helper function in the common American newt, *Notophthalmus viridescens* by lectins and human lymphokines. *Thymus* **11**, 77–87.

Ruben, L.N., Langeberg, L., Lee, R., Clothier, R.H., Malley, A., Holenstein,C. and Shiigi, S. (1990a). An example of the conservation of IL-2 and its receptor. In *Defense Molecules* (eds J.J. Marchalonis and C. Reinisch), Alan R. Liss, New York, pp. 133–147.

Ruben, L.N., Langeberg, L., Malley, A., Clothier, R.H., Beadling, C., Lee, R. and Shiigi, S. (1990b). A monoclonal mouse anti-human IL-2 receptor antibody (anti-Tac) will recognize molecules on the surface of *Xenopus laevis* immunocytes which specifically bind rIL-2 and are only slightly larger than the human Tac protein. *Immunol. Lett.* **24**, 117–126.

Ruben, L.N., DeLeon, R.T., Johnson, R.O., Bowman, S. and Clothier, R.E. (1996). Interleukin-2-induced mortality during the metamorphosis of *Xenopus laevis. Immunol. Lett.* **51**, 157–161.

Sato, M., Kitazawa, T., Katsumata, A., Mukamoto, M., Okada, T. and Takeya, T. (1992). Tissue-specific expression of 2 isoforms of chicken fibroblast growth-factor receptor, bek and Cek3. Cell *Growth Differ.* **3**, 355–361.

Schauenstein, K. and Kromer, G. (1987). Avian lymphokines. In *Avian Immunology: Basis and Practice*, Vol. 1 (eds A. Toivanen and P. Toivanen), CRC Press, Boca Raton, FL, pp. 213–227.

Schauenstein, K., Globerson, A. and Wick, G. (1982). Avian lymphokines. I. Thymic cell growth factor in supernatants of mitogen stimulated chicken spleen cells. *Dev. Comp. Immunol.* **6**, 533–540.

Schauenstein, K., Kromer, G., Sundick, R.S. and Wick, G. (1985). Enhanced response to Con A and production of TCGF by lymphocytes of obese strain (OS) chickens with spontaneous autoimmune thyroiditis. *J. Immunol.* **134**, 872–879.

Schauenstein, K., Kromer, G., Hala, K., Bock, G. and Wick, G. (1988). Chicken-activated T-lymphocyte-antigen (CATLA) recognized by monoclonal antibody INN-CH 16 represents the IL-2 receptor. *Dev. Comp. Immunol.* **12**, 823–831.

Schnetzler, M., Oommen, A., Nowak, J.S. and Franklin, R.M. (1983). Characterization of chicken T cell growth factor. *Eur. J. Immunol.* **13**, 560–566.

Schultz, U., Kock, J., Schlicht, H.-J. and Staeheli, P. (1995a). Recombinant duck interferon: a new reagent for studying the mode of interferon action against hepatitis B virus. *Virology* **212**, 641–649.

Schultz, U., Rinderle, C., Sekellick, M.J., Marcus, P.I. and Staeheli, P. (1995b). Recombinant chicken interferon from *Escherichia coli* and transfected COS cells is biologically active. *Eur. J. Biochem.* **229**, 73–76.

Schultz, U., Kaspers, B., Rinderle, C., Sekellick, M.J., Marcus, P.I. and Staeheli, P. (1995c). Recombinant chicken interferon: a potent antiviral agent that lacks intrinsic macrophage activating factor activity. *Eur. J. Immunol.* **25**, 847–851.

Secombes, C.J. (1981). Comparative studies on the structure and function of teleost lymphoid organs. PhD thesis, University of Hull, UK.

Secombes, C.J. (1996). The nonspecific immune system: cellular defenses. In *The Fish Immune System. Organism, Pathogen and Environment* (eds G. Iwama and T. Nakanishi), Academic Press, San Diego, CA, pp. 63–103.

Secombes, C.J. (1997). Rainbow trout cytokine genes. *Dev. Comp. Immunol.* **21**, 138.

Sekellick, M.J., Ferrandino, A.F., Hopkins, D.A. and Marcus, P.I. (1994). Chicken interferon gene: cloning, expression, and analysis. *J. Interferon Res.* **14**, 71–79.

Shi, D.-L., Feige, J.-J., Riou, J.-F., DeSimone, D.W. and Boucaut, J.-C. (1992). Differential expression and regulation of two distinct fibroblast growth factor receptors during early development of the urodele amphibian *Pleurodeles waltl. Development* **116**, 261–273.

Shi, D.-L., Fromentoux, V., Launay, C., Umbhauer, M. and Boucaut, J.-C. (1993). Isolation and developmental expression of the amphibian homolog of the fibroblast growth factor receptor 3. *J. Cell Sci.* **107**, 417–425.

Shi, D.-L., Launay, C., Fromentoux, V., Feige, J.-J. and Boucaut, J.-C. (1994). Expression of fibroblast growth-factor receptor-2 splice variants is developmentally and tissue-specifically regulated in the amphibian embryo. *Dev. Biol.* **164**, 173–182.

Shiels, A. and Wickramasinghe, S.N. (1995). Expression of an erythropoietin-like gene in the trout. *Br. J. Haematol.* **90**, 219–221.

Shishido, E., Higashijima, S., Emori, Y. and Saigo, K. (1993). 2 FGF receptor homologs of *Drosophila*—one is expressed in mesodermal primordium in early embryos. *Development* **117**, 751–761.

Siatskas, C., McWaters, P.G., Digby, M., Lowenthal, J.W. and Boyd, R.L. (1996). *In vitro* characterization of a novel avian haemopoietic growth factor derived from stromal cells. *Dev. Comp. Immunol.* **20**, 139–156.

Sick, C., Schultz, U. and Staeheli, P. (1996). A family of genes coding for two serologically distinct chicken interferons. *J. Biol. Chem.* **271**, 7635–7639.

Sigel, M.M., Hamby, B.A. and Huggins, E.M., Jr. (1986). Phylogenetic studies on lymphokines. Fish lymphocytes respond to human IL-1 and epithelial cells produce an IL-1-like factor. *Vet. Immunol. Immunopathol.* **12**, 47–58.

Slack, J.M.W. and Isaacs, H.V. (1989). Presence of basic fibroblast growth factor in the early *Xenopus* embryo. *Development* **105**, 147–153.

Smith, J.C., Yaqoob, M. and Symes, K. (1988). Purification, partial characterization and biological effects of the XTC mesoderm-inducing factor. *Development* **103**, 591–600.

Smith, P.D., McCarthy, D.H. and Paterson, W.D. (1980). Further studies on furunculosis vaccination. In *Fish Diseases*, 3rd COPRAQ Session (ed. W. Ahne), Springer, Berlin, pp. 113–119.

Snegaroff, J. (1993). Induction of interferon synthesis in rainbow trout leucocytes by various homeotherm viruses. *Fish Shellfish Immunol.* **3**, 191–198.

Soderhall, K., Iwanaga, S. and Vasta, G.R. (1996). *New Directions in Invertebrate Immunology.* SOS Publications, Fair Haven.

Song, J. and Slack, J.M.W. (1996). XFGF-9: A new fibroblast growth factor from *Xenopus* embryos. *Develop. Dynamics* **206**, 427–436.

Song, Y.-L., Lin, T. and Kou, G.-H. (1989). Cell-mediated immunity of the eel, *Anguilla japonica* (Temminck and Schlegel), as measured by the migration inhibition test. *J. Fish Diseases* **12**, 117–123.

Sprang, S.R. and Bazan, J.F. (1993). Cytokine structural taxonomy and mechanisms of receptor engagement. *Curr. Opin. Struct. Biol.* **3**, 815–827.

Sterneck, E., Blattner, C., Graf, T. and Leutz, A. (1992). Structure of the chicken myelomonocytic growth factor gene and specific activation of its promoter in avian myelomonocytic cells by protein kinases. *Mol. Cell. Biol.* **12**, 1728–1735.

Stet, R.J.M., Dixon, B., van Erp, S.H.M., van Lierop, M.-J.C., Rodrigues, P.N.S. and Egberts, E. (1996). Inference of structure and function of fish major histocompatibility complex (MHC) molecules from expressed genes. *Fish Shellfish Immunol.* **6**, 305–318.

Stinson, R.S., Mashaly, M.M. and Glick, B. (1979). Thrombocyte migration and the release of thrombocyte inhibitory factor (ThrIF) by T and B cells in the chicken. *Immunology* **36**, 769–774.

Sumathy, K., Desai, K.V. and Kondaiah, P. (1997). Isolation of transforming growth factor-β2c DNA from a fish, *Cyprinus carpio*, by RT-PCR. *Gene* **191**, 103–107.

Sundick, R.S. and Gill-Dixon, C. (1997). A cloned chicken lymphokine homologous to both mammalian IL-2 and IL-5. *J. Immunol.* **159**, 720–725.

Sunyer, T., Rothe, L., Jiang, X., Osdoby, P. and Collin-Osdoby, P. (1996). Proinflammatory agents, IL-8 and IL-10, upregulate inducible nitric oxide synthase expression and nitric oxide production in avian osteoclast-like cells. *J. Cell. Biochem.* **60**, 469–483.

Suresh, M., Karaca, M., Foster, D. and Sharma, J.M. (1995). Molecular and functional characterization of turkey interferon. *J. Virol.* **69**, 8159–8163.

Suzuki, T and Kurokawa, T. (1996). Functional analyses of FGF during pharyngeal cartilage development in flounder (*Paralichthys olivaceus*) embryo. *Zool. Sci.* **13**, 883–891.

Suzuki, T., Kurokawa, T. and Asashima, M. (1994). Identification of a heparin-binding, mesoderm-inducing peptide in the swim-bladder of the red seabream, *Pagrus major*—a probable fish fibroblast growth-factor. *Fish Physiol. Biochem.* **13**, 343–352.

Szebenyi, G., Savage, M.P., Olwin, B.B. and Fallon, J.F. (1995). Changes in the expression of fibroblast growth factor receptors mark distinct changes of chondrogenesis *in vitro* and during chick limb skeletal patterning. *Develop. Dynamics* **204**, 446–456.

Tahan, A.M. and Jurd, R.D. (1979). Delayed hypersensitivity in *Ambystoma mexicanum. Dev. Comp. Immunol.* **3**, 299–306.

Tamai, T., Shirahata, S., Noguchi, T., Sato, N., Kimura, S. and Murakami, H. (1993). Cloning and expression of flatfish (*Paralichthys loivaceus*) interferon cDNA. *Biochim. Biophys. Acta* **1174**, 182–186.

Tengelsen, L.A., Trobridge, G.D. and Leong, J.C. (1991). Characterization of an inducible interferon-like antiviral activity in salmonids. *Proceedings of the Second International Symposium on Viruses of Lower Vertebrates*, 29–31 July, Oregon State University, Corrollis, USA, pp. 219–226.

Thisse, B., Thisse, C. and Weston, J.A. (1995). Novel FGF receptor (Z-FGFR4) is dynamically expressed in mesoderm and neurectoderm during early zebrafish embryogenesis. *Develop. Dynamics* **203**, 377–391.

Thomsen, G., Woolf, T., Whitman, M., Sokol, S., Vaughan, J., Vale, W. and Melton, D. (1990). Activins are expressed early in *Xenopus* embryogenesis and can induce axial mesoderm and anterior structure. *Cell* **63**, 495–501.

Trifonov, S., Sotirov, N. and Filchev, A. (1977). *In vitro* migration of peritoneal and spleen cells and its inhibition in some avian species. *Cell. Immunol.* **32**, 361–369.

Vainio, O., Ratcliffe, M.J.H. and Leanderson, T. (1986). Chicken T cell growth factor—use in the generation of a long term cultured T cell line and biochemical characterization. *Scand. J. Immunol.* **23**, 135–142.

Verburg van Kemenade, B.M.L., Weyts, F.A.A., Debets, R. and Flik, G. (1995). Carp macrophages and neutrophilic granulocytes secrete an interleukin-1-like factor. *Dev. Comp. Immunol.* **19**, 59–70.

Vlaovic, M.S., Buening, G.M. and Loan, R.W. (1975). Capillary tube leukocyte migration inhibition as a correlate of cell-mediated immunity in the chicken. *Cell. Immunol.* **17**, 335–341.

Vogel, A., Rodriguez, C. and Izpisua-Belmonte, J.C. (1996). Involvement of FGF-8 in initiation, outgrowth and patterning in the vertebrate limb. *Development* **122**, 1737–1750.

Von Bulow, V., Weiler, H. and Klasen, A. (1984). Activating effects of interferons, lymphokines and viruses on cultured chicken macrophages. *Avian Pathol.* **13**, 621–637.

Wang, R., Neumann, N.F., Shen, Q. and Belosevic, M. (1995). Establishment and characterisation of a macrophage cell line from the goldfish. *Fish Shellfish Immunol.* **5**, 329–346.

Washburn, J.O., Kirkpatrick, B.A. and Volkman, L.E. (1996). Insect protection against viruses. *Nature* **383**, 767.

Watkins, D. and Cohen, N. (1985). The phylogeny of interleukin-2. *Dev. Comp. Immunol.* **9**, 819–824.

Watkins, D. and Cohen, N. (1987). Mitogen-activated *Xenopus laevis* lymphocytes produce a T-cell growth factor. *Immunology* **62**, 119–125.

Watkins, D., Parsons, S.C. and Cohen, N. (1987). A factor with interleukin-1-like activity is produced by peritoneal cells from the frog, *Xenopus laevis. Immunology* **62**, 669–673.

Weiler, H. and Von Bulow, V. (1987a). Development of optimal conditions for lymphokine production by chicken lymphocytes. *Vet. Immunol. Immunopathol.* **14**, 257–267.

Weiler, H. and Von Bulow, V. (1987b). Detection of different macrophage-activating factor and interferon activities in supernatants of chicken lymphocyte cultures. *Avian Pathol.* **16**, 439–452.

Weining, K.C., Schultz, U., Munster, U., Kaspers, B. and Staeheli, P. (1996). Biological properties of recombinant chicken interferon-γ. *Eur. J. Immunol.* **26**, 2440–2447.

Weyts, F.A.A., Rombout, J.H.W.M., Flik, G. and Verburg van Kemenade, B.M.L. (1997). A common carp (*Cyprinus carpio* L.) leucocyte cell line shares morphological and functional characteristics with macrophages. *Fish Shellfish Immunol.* **7**, 123–133.

Widelitz, R.B., Jiang, T.X., Noveen, A., Chen, C.W.J. and Chuong, C.M. (1996). FGF induces new feather buds from developing avian skin. *J. Invest. Dermatol.* **107**, 797–803.

Wilson, V., Jeffreys, A.J., Barrie, P.A., Boseley, P.G., Slocombe, P.M., Easton, A. and Burke, D.C. (1983). A comparison of vertebrate interferon gene families detected by hybridization with human interferon DNA. *J. Mol. Biol.* **166**, 457–475.

Yano, T. (1996). The nonspecific immune system: humoral defense. In *The Fish Immune System. Organism, Pathogen and Environment* (eds G. Iwama and T. Nakanishi), Academic Press, San Diego, CA, pp. 105–157.

Yin, Z., Lam, T.J. and Sin, Y.M. (1997). Cytokine-mediated antimicrobial immune response of catfish, *Clarias gariepinus*, as a defence against *Aeromonas hydrophila. Fish Shellfish Immunol.* **7**, 93–104.

Yoshikawa, N., Morimoto, T. and Watanabe, T. (1994). Granulocyte/macrophage growth stimulation by a supernatant from adherent cells of carp pronephros. *Dev. Comp. Immunol.* **18**, XVIII.

Zhang, S.P., Lillehoj, H.S. and Ruff, M.D. (1995). *In vivo* role of tumor necrosis-like factor in *Eimeria tenella* infection. *Avian Dis.* **39**, 859–866.

Zhou, J.H., Ohtaki, M. and Sakurai, M. (1993). Sequence of a cDNA encoding chicken stem cell factor. *Gene* **127**, 269–270.

Index

Numbers in italic refer to illustrations or tables. Numbers in bold indicate main discussion.